A DICTIONARY OF ENTOMOLOGY

Compiled by

Gordon Gordh
Honorary Professor,
The University of Queensland,
St Lucia, Queensland, Australia

with assistance by

David Headrick
Crop Science Department,
California Polytechnic State University,
San Luis Obispo, USA

CABI *Publishing*

CABI Publishing is a division of CAB International

CABI Publishing
CAB International
Wallingford
Oxon OX10 8DE
UK

Tel: +44 (0)1491 832111
Fax: +44 (0)1491 833508
Email: cabi@cabi.org
Web site: www.cabi-publishing.org

CABI Publishing
875 Massachusetts Avenue
7th Floor
Cambridge, MA 02139
USA

Tel: +1 617 395 4056
Fax: +1 617 354 6875
Email: cabi-nao@cabi.org

A catalogue record for this book is available from the British Library, London, UK

Library of Congress Cataloging-in-Publication Data
Gordh, Gordon
A dictionary of entomology / compiled by Gordon Gordh ; with the assistance of David Headrick.
p. cm.
Includes bibliographical references (p.).
ISBN 0-85199-291-9 (alk. paper)
1. Entomology—Dictionaries. I. Headrick, David. II. Title.
QL462.3 .G67 2000
595.7′03—dc21

00-044427

Hardback edition
ISBN-13: 978-0-85199-291-4
ISBN-10: 0-85199-291-9

Paperback edition
ISBN-13: 978-0-85199-655-4
ISBN-10: 0-85199-655-8

First published 2000
Reprinted 2001, 2002

First paperback edition 2002
Reprinted 2005

Printed and bound in the UK by Biddles Ltd, King's Lynn

Contents

Terms are the tools of the teacher;
and only an inferior hand persists in toiling
with a clumsy instrument when a
better one lies within one's reach.

(Richard Owen 1866, I: xiii)

Foreword

The title of this book implies an ambitious and in a sense, perhaps pretentious work. A dictionary serves many useful roles, but should always strive for accuracy and completeness. Entomological dictionaries and glossaries have been published in several languages, but only a few have been comprehensive in breadth of terms, nuance of meaning or number of definitions. Accuracy has been inconsistent, and no doubt that remains the case.

The foundation of this Dictionary began more than 30 years ago when I was a graduate student. Jose Torre-Bueno's *Glossary of Entomology* was my primary source of entomological definitions. The text was published in 1937 and reprinted with a modest supplement in 1962 by George Tulloch. Torre-Bueno was not a professional entomologist or lexicographer. His selection of terms was shaped by his specific interests and limited readings in entomology. His text was incomplete and stylistically obsolete, which prompted work on expanding the scope of his *Glossary* and correcting some of the deficiencies. The present Dictionary began more than 25 years ago, with progress slow and sporadic. During the interim, an update of Torre-Bueno was produced (Stephen Nichols and Randall Schuh, 1989). Their work (*The Torre-Bueno Glossary of Entomology*) added considerably to the volume of terms, but differs in many fundamental features from the present Dictionary.

Content and organization

Entries include the terms from Torre-Bueno's Glossary for historical completeness. Other entries have been encountered in the primary entomological literature at least twice. The breadth of the literature includes all major subdisciplines of entomology (behaviour, biochemistry, ecology, morphology, physiology and taxonomy). We have attempted to include the recent proliferation of terms in new areas of science that have an impact on entomology (notably biotechnology and molecular biology). The depth of our literature review reaches the origins of entomological references with Greeks and Romans. We hope that readers consulting this Dictionary will accept our brief explanations of ancient and medieval writers and their philosophies. Each classical entry has been prompted by a question from a student of a statement in the literature that has required the research explained under each statement. We have not included references from ancient texts in Chinese, but this should be considered in a future edition of this Dictionary.

Terms are arranged alphabetically. In instances of variable spellings, we have identified the lesser-used spelling at the end of the Definition as 'Alt.' We have made no attempt to determine which spelling is orthographically, grammatically or etymological correct; we have relied upon common or prevailing usage to determine 'correctness'. In some instances involving Latin and Greek terms, the word is multiply listed under variant spellings.

The part of speech for each term is identified. Most terms are nouns or adjectives. We have included some verbs, adverbs, prepositions and conjunctions that are regarded as essential, particularly Latin words which are frequently encountered and which may be important in understanding older taxonomic literature. We feel this inclusion is important, particularly for readers whose native language is not English. Similarly, the plural form of nouns is provided. This may seem trivial to native English speakers, but we feel it is important for non-native English speakers and especially important for some words derived from Latin or Greek because the formation of plurals in these languages is not always straightforward (*cf. Sensillum, Soma*). Sadly, scholarship in classical languages is not well understood by modern entomologists (including the present writers) and a source of this information must be provided. Synonyms and antonyms are given where appropriate.

Orthography. All scientific names are italicized and generic names are capitalized. This is in keeping with accepted practice and the code of zoological nomenclature. We have capitalized the Latin and Greek terms of anatomy. This practice has the advantage of instantly identifying the word as a Greek or Latin noun that deals with a part of the insect's body. We have not capitalized the adjectival forms of these words.

Etymology. We believe the etymology of entomological words is important. Tracing the etymology of words is a complicated task that requires specialized linguistic training and a comprehensive library of dictionaries and seminal publications. We provide etymologies for words including the compound words synthesized from Latin or Greek. However, we are not lexicographers, our source books are sometimes unreliable, and the etymology of some words remains obscure. We feel that including our etymological conclusions will serve as a point of reference for future editions of this Dictionary and classical scholars to correct our mistakes.

Definitions. Some words have more than one meaning, and some meanings are context-dependent. Often, meanings are radically different and depend upon the subdiscipline of Entomology in which the word is used. Different definitions are enumerated with reference to subdiscipline (Taxonomy, Medical Entomology, Behaviour) when appropriate. We have tried to identify the context in which words are used, but undoubtedly this Dictionary has omissions and errors of fact. The second edition will provide many new definitions, some definitions will be refined and errors will be corrected.

Names of entomologists. We have included the names of many entomologists, because entomologists have shaped and defined our discipline, and their accomplishments are integral to the language we use to discuss entomology. The lives, work and accomplishments of entomologists must be recognized somewhere. Perhaps a dictionary that purports to collect the words that embody entomology is an appropriate place to list them. Many local, national or ethnic biographies have been tabulated, but we have not had access to many of these. Our list of entomologists is not complete and many names do not appear here. We have included the names of prominent North American entomologists, based on our reading of the entomological literature. The names of many entomologists from other countries are included from lists published elsewhere (e.g. Gilbert, 1977; Hasegawa, 1967). Some names have been included although they do not appear on lists. The roles or accomplishments of these entomologists are briefly recorded or a reference to a biography or obituary is provided. Lists have not been developed for entomologists from many countries. The next edition of this Dictionary should include the names of entomologists from other countries, and ideally should include entomologists from all countries. This task will require international cooperation.

Names of insects. Names are important as a reference for existing knowledge and communicating new knowledge. More than 1,000 families and 1.5 million species of insects have been described and information collected about them. All of these insects have scientific names and many have common names. Confusion begins when names are not used consistently or correctly. Some families, genera and species of insects have more than one common name; some families, genera and species have synonymous scientific names. Some common names may apply to more than one species of insect, and one species of insect may have more than one common name. Many insects have common names in languages other than English. We have included the common names of insects that occur in North America and Australia. These names have been taken from lists of common names. In many cases, common names remain to be given to insects that only have scientific names. In some instances we must resolve the ambiguities surrounding common names within and between countries.

Names of families. We include a diagnosis for family-level names with a list of their synonyms. We recognize that the validity of family-level names is debated, and taxonomists often hotly contest the validity and scope of some family names. Undoubtedly, some scientific names and taxon names given in this Dictionary will be challenged by various taxonomic experts. However, as a first approximation, the list of family-level names is relatively complete and robust. The diagnoses of these names are variable, some being long and detailed while others are short and vague. This variation is a consequence of our lack of familiarity with some groups and a paucity of knowledge about other groups. We hope that the list will improve and entries will be refined in a second edition of this Dictionary.

Conceptual paths. Definitions often may be ambiguous without benefit of peripheral information. In an attempt to enhance the appreciation of words and explain their contextual implications, we have directed the reader to other words following many definitions. We call these prompts 'conceptual paths', and refer to other relevant words by 'See', 'Cf.' and 'Rel.'. Terms following 'See' refer to a connection in which the defined term is at a lower level in a conceptual hierarchy (e.g. DDT... See Organochlorine Insecticide; Protocerebrum... See Brain). By reading the entry under the additional word, the reader will appreciate the definition of the first word in a broader context. Terms following 'Cf.' refer to a connection in which the defined term is at a coordinate conceptual level (e.g. Pharate... Cf. Exarate; Mandibular Gland... Cf. Labial Gland). Terms following 'Rel.' refer to a connection in which the defined term is at a subordinate conceptual level or related to the term in a more abstract way (e.g. Photophase... Rel. Circadian Rhythm). One term may embrace all three levels of organization (e.g. Maceration... See Histology. Cf. Clearing. Rel. Fixation). We hope that this system of conceptual paths serves as a useful feature of this Dictionary.

Acknowledgements

The effort in this Dictionary has been confined mostly to persistent plodding over a protracted period of time. Few people have contributed directly to this Dictionary, and thus few people can be criticized for omission or errors of fact. The Dictionary started as a collection of hand-written entries on index cards that were transferred to a computer and has moved from mainframe to Unix to Macintosh to PC. Some errors of transcription may remain in the printed text.

Gordon Gordh
Mission, Texas

AALBORG, NIELS NIKKELSEN (1562–1645) (Henriksen 1921, Entomol. Medd. 15: 25.)

AALL, NICOLAI BENJAMIN (1805–1888) (Natvig 1944, Norsk. Entomol. Tidsskr. 7: 45–47.)

AARON, SAMUEL FRANCIS (1862–1947) (Calvert 1947, Entomol. News 58: 137–140; Weiss 1947, Entomol. News 58: 235–236.)

AATACK® A registered biopesticide derived from *Bacillus thuringiensis* var. *kurstaki*. See *Bacillus thuringiensis*.

AAZOMAT® See Benzoximate.

ABAMECTIN® Trade name for a commercially available chemical compound introduced in 1985 as an agricultural pesticide and antiparasitic drug for humans and livestock. Abamectin is a member of the Avermectin family of compounds. See Avermectin; Ivermectin.

A-BAND In muscle cells under magnification, the A-Band forms the anisotropic (doubly refracting) band of a Sarcomere. The A-Band appears dark and is composed of Actin. Syn. A-disc. See Muscle Cell.

ABANDON Noun. (Latin, *ad* = to; *bandon* = authority, power.) To relinquish control to natural impulses. To leave territory, nest or habitat. A term that is often used to characterize the nests of social insects (particularly bees or wasps) that have been vacated.

ABAPICAL Adj. (Latin, *ab* = from; *apex* = summit.) Anatomy: Descriptive of a movement or feature positioned away from the apex of an elongate structure. See Orientation. Cf. Adapical.

ABAT® See Temephos.

ABATE® Trade name for a non-systemic organophosphate insecticide marketed by American Cyanamid (1964.) See Temephos. Syn. Abat, Abathion, Difenthos, Ecopro, Lypor, Nimitex, Swebate, Temeguard, Tiempo.

ABATHION® See Temephos.

ABAXIAL Adj. (Latin, *ab* = from; *axis* = axel.) Anatomy: Concerning the surface of a structure that is turned away from the structure's primary axis. See Orientation. Cf. Adaxial.

ABBOT, JOHN (1751–1840) (Hagen 1863, Stettin. Entomol. Ztg. 24: 369–378; Scudder 1888, Canad. Entomol. 20: 150–154; Kirby 1888, Canad. Entomol. 20: 230–232.)

ABBOTT, WALTER SYDNEY (1879–1943) (Siegler *et al.* 1943, Proc. Entomol. Soc. Wash. 45: 92.)

ABBREVIATE Adj. (Latin, *abbreviatus; ad* = to; *brevis* = short.) 1. Descriptive of structure that has been shortened or is disproportionately short. 2. Pertaining to structure that appears shorter than 'normal' as compared with other individuals in a population or coordinate Taxa within a taxonomic group (Species, Genus, Family). Cf. Brachypterous; Micropterous. Trans. Verb, Abbreviate.

ABBREVIATED FASCIA A band of colour, a texture or distinctive feature that traverses less than half the length of the structure on which it occurs (*e.g.* a wing fascia).

ABBREVIATED WIREWORM *Hypolithus abbreviatus* (Say) [Coleoptera: Elateridae].

ABDOMEN Noun. (Latin, *abdomen* = belly. Pl., Abdomens.) Anatomy: The third or posterior-most Tagma of the insect body. The abdominal Groundplan is composed of 11 segments, with a trend of reduction in number of segments in higher Orders of insects. In many insects, the abdomen is less strongly sclerotized than the head or Thorax. Each abdominal segment consists of a sclerotized Tergum, Sternum and sometimes a Pleurite. Terga are separated from each other and from the adjacent Sterna or Pleura by membrane. Spiracles occur in the pleural area. Modifications of this groundplan include the fusion of Terga, or Terga with Sterna, to form continuous dorsal or ventral shields or a conical tube. Apterygotes (*e.g.* Thysanura) possess two plural sclerites; the ventral-most is called the Coxopodite and contains appendages (Styli). Some insects bear a sclerite (Laterotergite) in the pleural area. Ventral sclerites are sometimes called 'Laterosternites.' Spiracles are often situated in the definitive Tergum, Sternum, Laterotergite or Laterosternite. The Abdomen is more conspicuously segmented than the head or Thorax. Abdominal segmentation is evident and serially uniform. Typically abdominal Terga show secondary segmentation with the posterior part of a segment overlapping the anterior part of the segment behind it. Functionally, this overlap prevents damage or injury to the animal while it moves through the environment, particularly in confined spaces. Secondary segmentation with associated membrane permits rapid change in size (volume) because the Abdomen must accommodate change in size associated with respiration, digestion and production of gametes. Superficially, the Abdomen is the least specialized body Tagma, but notable exceptions include scale insects (Coccidae, Diaspididae). Most Pterygota lack abdominal appendages except Cerci, and external genitalia and pregenital appendages in adult Apterygota and larval Pterygota. Pregenital segments in a male are 1–8; pregenital segments in a female are 1–7. Among Apterygota, male genitalia in Collembola are positioned between segments 5–6, and between segments 11 and Periproct in Protura. Genital segments of Pterygota include 9 in males and 8–9 in females. Postgenital segments of Pterygota are 10–11 in females and 9–10 in males. The adult's Abdomen is primarily responsible for digestion, defence, respiration (predominantly), excretion and reproduction. These activities are conducted in different areas of the Abdomen. In most insects, the anterior area is predominantly involved with digestion and an accumulation of waste in some insects. Gas exchange during respiration occurs generally over the entire Abdomen, with one pair of spiracles per segment. The posterior abdominal area is predominantly involved with reproduction and excretion. Etymology of the word 'Abdomen' is uncertain. Abdomen is a Latin word, but its precise original meaning in Entomology is not clear. In Latin, 'abdomen' is perhaps derived from adipomen (adips) that means the soft fat of animals. Alternatively abdomen could be derived from 'abdo' that means to conceal or cover. (Problems in entomological word usage are common. The word 'head' originates in old English; the word 'thorax' originates in Greek. Thus, the three principal words representing the three regions of the insect body are derived from three languages.) Alt. Abdomin. See Tagma. Cf. Head; Thorax. Rel. Metasoma.

ABDOMERE Noun. (Latin, *abdomen* = belly; Greek, *meros* = part. Pl., Abdomeres.) 1. Morphology: Strictly, a part of an abdominal segment, and not an entire segment. 2. Taxonomy and other non-morphological disciplines: An entire abdominal segment or sclerite. See Abdomen; Segmentation. Cf. Antennomere; Tarsomere.

ABDOMINAL Adj. (Latin, *abdominalis* = belly.) Pertaining to the Abdomen or third Tagma of the insect body.

ABDOMINAL APPENDAGE We presume that the hypothetical ancestor of the Insecta was a myriapod with one pair of appendages for each body segment. Head appendages for modern insects include antennae, mandibles, first and second maxillae. Thorax appendages include legs, and wings are regarded as secondary origin. Most Apterygota show paired abdominal appendages. In most Pterygota, embryological appendages are formed and lost before eclosion. Embryonic appendages apparently represent ancestral conditions that are not expressed in postembryonic stages of modern insects. Abdominal appendages include Styli, Gonostyli, Cerci, Cornicles. AA of Entognatha, Apterygota and Paleoptera include: Colophore, Retinaculum, Furcula, eversible vesicles and Appendix Dorsalis. See Abdomen. Cf. Cercus; Colophore; Cornicle; Gonostylus; Retinaculum. Rel. Pleuropod.

ABDOMINAL FEET See Proleg.

ABDOMINAL GANGLION (Pl., Ganglia.) Enlargements of the Ventral Nerve Cord that contain Interneurons and connect with Efferent Neurons that innervate the Abdomen. Groundplan condition: One ganglion per segment and positioned between the Alimentary Canal and large ventral muscles. AG number constant within Species but variable among Orders (1–8 ganglia per Abdomen) with tendency toward larger size and consolidation or reduction in number among the higher Orders. Each ganglion typically sends a pair of principal nerves to nearby muscles of each abdominal segment. Abdominal Ganglia typically are smaller than Thoracic Ganglia. The posterior-most Abdominal Ganglion is usually largest because it represents the fusion of ganglia of the last four abdominal segments. Usually ganglia innervate the muscles of the segment in which a ganglion is located. Interganglionic connectives are paired and extend between ganglia. See Gangion. Cf. Thoracic Ganglion; Ventral Nerve Cord. Rel. Central Nervous System.

ABDOMINAL GLANDS The insect Abdomen contains many glands and many names have been proposed for these glands. Many abdominal glands originate as modified epidermal cells beneath abdominal sclerites or intersegmental membrane (purported integumental glands); some abdominal glands are associated with the reproductive system. Tergal Glands occur in Orthoptera and Blattaria; Pygidial Glands occur in Homoptera and Hymenoptera; Sternal Glands occur in Hymenoptera. See Accessory Gland; Gland. Rel. Integument.

ABDOMINAL GROOVE Lepidoptera: A concave lobe along the inner margin of the hindwing; AG envelops part of the Abdomen in some butterflies.

ABDOMINAL MUSCLE Snodgrass (1931) proposed a system for naming muscles. Accordingly, a Dorsal Muscle Group includes longitudinal muscles responsible for retraction and protraction. External Dorsal Muscles are relatively short and rarely extend the length of a segment. These muscles often occur in the posterior half of a segment, and sometimes become oblique or transverse in orientation. Origin of External Dorsals may shift to a position posterior of the segment in which they act antagonistically with Internal Muscles, and become responsible for protraction of the segment. Internal Dorsal Muscles are relatively long and sometimes extend the length of a segment. The IDMs sometimes shift to an oblique orientation, but remain responsible for retraction because they shorten the distance between adjacent segments. Dorsal muscles attach from antecostal ridge to antecostal ridge in adults and to intersegmental folds in larvae. Segments are pulled forward then contract, and therefore muscles have their origin on the anterior fold (ridge) and their insertion on the posterior fold (ridge). A Transverse Muscle Group includes Transverse Dorsal Muscles that form the Dorsoventral Diaphragm and Transverse Ventral Muscles that form the Ventral Diaphragm. The Dorsal Diaphragm consists of muscles that originate along the Tergum's anterolateral margin and insert on the heart's ventral wall. Some insects develop Dorsal Diaphragm Muscles along the entire lateral tergal margin. Alternatively, musculature may form an anterior and posterior group in each segment. In Hymenoptera and some Orthoptera, the Ventral Diaphragm forms a continuous sheet of tissue extending the length of the abdomen and separating the Ventral Nerve Cord from the Abdomen's main cavity. In Orthoptera (crickets) ventral muscles form distinct bands originating in the Sternum's anterolateral region and crossing the anterior part of each Sternum. The Ventral Muscle Group includes Longitudinal Ventral Muscles that mirror development of the Dorsal Muscle Group. Longitudinal Muscles are subdivided into External and Internal Longitudinal Muscles. Ventral Internal Muscles are long and act as retractors; Ventral External Muscles are shorter and act as protractors. Ventral Muscles insert on intersegmental folds or sternal Antecostae. Lateral Muscles also are called 'dorsoventrals' or 'tergosternals.' These muscles are more variable in development compared with dorsal and ventral groups. Lateral muscles are also subdivided into External Laterals and Internal Laterals; internals are often absent or reduced. Lateral Muscles may be intersegmental and extend obliquely between segments or intrasegmentally. Most Lateral Muscles compress the Abdomen during expiration because they are moderately long and attach to the Sternum's lateral margin and extend some distance along the Tergum. Other lateral muscles are relatively short and extend between the Tergum's ventral margin and Sternum's lateral margin. When these muscles contract, they expand the distance between Tergum and Sternum, thereby serving to dilate the Abdomen.

ABDOMINAL PORE Openings on the surface of sclerites or within intersegmental membrane of the Abdomen. AP often is microscopic and evident only under high LM magnification or SEM. Pores presumably secrete chemicals of diverse nature and function. See Exocrine Gland; Pheromone.

ABDOMINAL POUCH Female Parnassida: A saclike ventral cavity formed by material secreted during copulation.

ABDOMINAL PUMPING A mechanism involved in Active Ventilation through which the Abdomen is rhythmically contracted and expanded with dorsoventral movement of the Terga and Sterna. In Coleoptera and Heteroptera the abdominal Terga are lowered; in most other insects the abdominal Sterna are raised. Active Ventilation consists of an Expiration Phase and an Inspiration Phase. Expiration is an active process and is the first phase of ventilation; expiration is characterized by the Abdomen initially in an expanded condition. During contraction of dorsoventral (Tergo-Sternal) muscles, the Tergum is lowered or the Sternum is raised. This action expels gas from the Tracheal System. The Inspiration Phase may or may not involve muscles. Inspiration may actively involve muscles in some insects such as grasshoppers; in other insects the Abdomen passively returns to its resting condition through the elastic recoil of Sterna or Terga to their natural shape, or through the action of cuticular proteins such as Resilin. See Active Ventilation. Cf. Autoventilation; Auxiliary Respiratory Mechanisms; Spiracular Synchronization. Rel. Respiratory Muscles; Tracheal Ventilation.

ABDOMINAL REGION See Abdomen.

ABDOMINAL SCENT GLAND A type of Exocrine Gland on the Abdomen of all nymphal and some adult Heteroptera. ASG composed of Secretory Cells that produce substances that may be ejected with aid of stretch muscles attached to A Scent Gland Sac; substance is passed through an aperture (Ostiole) between Terga 4–5 and 5–6. ASG typically produce defensive secretions. See Gland; Scent Gland System. Cf. Metathoracic Scent Gland. Rel. Exocrine Gland.

ABDOMINAL SEGMENTATION Variation in abdominal segment number is considerable. Adult Protura display 12 segments; adult Collembola display six segments; Diplura display 10 segments. Archaeognatha and Thysanura display 11 segments. The fossil zorapteran *Zorotypus palaeus* Poinar from Dominican Amber (lower Miocene – upper Eocene) displays a 10-segmented abdomen. Sucking lice have 9 abdominal segments; biting lice have 8–10 abdominal segments. Diptera (Muscidae) have nine segments with segments 2–5 visible and segments 6–9 telescoped. Most Hymenoptera display seven exposed segments with the remainder concealed. In a few instances, the groundplan may be obscured by profound reorganization of the abdomen. In adult Holometabola, nymphal and adult Paurometabola, the abdominal segmentation is secondary with each segment called a Uromere. The Abdomen is clearly differentiated from the Thorax in most insects. A notable exception is in the apocritious Hymenoptera where the first abdominal segment has become incorporated into the thoracic region and separated from the remainder of the abdomen by a constriction. In these insects the first abdominal segment is called the Propodeum. The constriction (second segment) is called the Petiole and the remaining segments of the Abdomen are called the Gaster. Because the middle region of the body is no longer a primitive Tagma, the term Mesosoma is sometimes used. The term 'Metasoma' is applied to remaining segments of the Abdomen. Anamorphosis (addition of segments to the body after eclosion) is noted among some primitive (?ancestral) hexapods (*e.g.* Protura) that emerge from the egg with eight abdominal segments and a terminal telson. Subsequently, three segments are added (one per moult) between the Telson and last abdominal segment. Most insects undergo Epimorphosis in which the definitive number of segments is present at hatching. Postembryonic developmental conditions also obscure the groundplan of abdominal segmentation. The larval stage of Holometabola is radically different from the adult, and the abdomen is completely reorganized between the two stages. However, nymphs of Paurometabola resemble the adult. Thus we cannot follow continuous development of the Abdomen in Holometabola, or trace the course of evolution between Paurometabola and Holometabola.

ABDUCENT See Adducent.

ABDUCTION Noun. (Latin, *abductio*, *abducere* = to lead away; English, *-tion* = result of an action. Pl., Abductions.) The action of moving away from the median or primary axis a structure. Drawing away from the body; retracting or removing from a point near the body. See Orientation. Cf. Adduction.

ABDUCTOR Noun. (Latin, *abductus* = led away; *or* = an activity. Pl., Abductors.) 1. Something that abducts or takes away. 2. Any one of several muscles that move the distal part of an appendage away from the midline of the body. See Muscle. Cf. Adductor; Rotator.

ABDUCTOR MANDIBULA A muscle that spreads Mandibles apart or moves the apical portion laterad. Cf. Adductor Mandibula.

ABDUCTOR MUSCLE Any muscle that induces abduction. A muscle that moves a structure, organ or appendage away from the median longitudinal line of the body or away from the primary axis of a body. Cf. Adductor Muscle.

ABDUCTOR OF COXA The second primary coxal muscle that moves the Coxa. See Coxal Muscle. Alt. Coxal Abductor.

ABEILLE DE PERRIN, ELEAZAR (1843–1910) (Caillel 1911, Ann. Soc. Entomol. Fr. 80: 492–502.)

ABEL, L (–1940) (Anon. 1941, Zeit. Wien. Entomol. Ver. 26: 32.)

ABERRANT Adj. (Latin, *aberrans, aberrare* = to stray.) 1. Pertaining to an anatomical feature or behavioural characteristic of organisms that deviates from the typical condition or expectation. 2. Taxonomy: Descriptive of a specimen that is not typical or does not conform to the concept of the type. Syn. Abnormal. Cf. Exceptional; Unique. Rel. Species Concept.

ABERRATION Noun. (Latin, *aberratio;* English, *-tion* = result of an action. Pl., Aberrations.) Any body, form (shape) or structure that deviates from the normal or typical condition. The term is often applied subjectively in a negative context.

ABG-6185® See *Bacillus sphaericus.*

ABIENCE Noun. (Latin, *abire* = to depart. Pl., Abiences.) An avoidance reaction to a stimulus. Avoiding reaction. Cf. Adience.

ABIENT Adj. (Latin, *abire* = to depart.) Concerning the avoidance of a source of stimulation. Cf. Adient. Rel. Taxis.

ABILGAARD, PETER CHRISTIAN (1740–1801) (Henriksen 1923, Entomol. Meddr. 15: 103–105.)

ABIOGENESIS Noun. (Greek, *a* = without; *bios* = life; *genesis* = creation. Pl., Abiogeneses.) The concept of Spontaneous Generation that asserts that life develops from lifeless matter. A concept understood by Aristotle and which prevailed as a popular concept through the 19th Century. Alt. Archebiosis; Archigenesis; Autogenesis; Heterogenesis. See Genesis; Spontaneous Generation. Cf. Biogenesis. Rel. Evolution.

ABIOLOGY Noun. (Greek, *a* = not; *bios* = life; *logos* = discourse. Pl., Abiologies.) The study of nonliving things. Cf. Biology.

ABIOSIS Noun. (Greek, *a* = not; *biosis* = living. Pl., Abioses.) The apparent suspension of life. See Biosis. Cf. Anhydrobiosis; Antibiosis; Archebiosis; Calobiosis; Cleptobiosis; Hamabiosis; Kleptobiosis; Lestobiosis; Parabiosis; Phylacobiosis; Plesiobiosis; Synclerobiosis; Trophobiosis; Xenobiosis.

ABIOTIC Adj. (Greek, *a* = not; *biotic* = life; *-ic* = of the nature of.) Pertaining to or descriptive of an absence of life. See Biosis. Cf. Biotic; Ecobiotic. Rel. Ecology.

ABJUGAL Adj. (Latin, *ab* = from; *jugum* = yoke; *-alis* = having the character of.) The furrow (Sulcus) separating the Aspidosoma (Prodorsum) and Podosoma.

ABLATION Noun. (Latin, *ablatio;* English, *-tion* = result of an action. Pl., Ablations.) Remove from the body. Trans. Verb, Ablate.

ABLE® A registered biopesticide derived from *Bacillus thuringiensis* var. *aizawai* strain GC-91. See *Bacillus thuringiensis.*

ABNEURAL Adj. (Latin, *ab* = from; *neuron* = nerve; *-alis* = pertaining to.) Descriptive of structure positioned away from the central nervous system.

ABNORMAL Adj. (Latin, *abnormis* = irregular; *-alis* = having the character of.) Deviating from the 'typical' condition; atypical. A term applied to structure, biological process or behaviour of an individual organism, population or Species and used to describe a condition notably different from the condition observed in other, related organisms. Not normal. See Normal. Cf. Aberrant.

ABNORMITY Noun. (Latin, *abnormis* = irregular; English, *-ity* = suffix forming abstract nouns. Pl., Abnormities.) Abnormal. A monstrosity compared with a typical condition.

ABORAD Adv. (Latin, *ab* = away; *os* = mouth.) Moving away from the mouth.

ABORAL Adj. (Latin, *ab* = away; *os* = mouth.) Pertaining to structure or action away from or opposite the mouth. Cf. Adoral.

ABORTED Adj. (Latin, *abortus* = premature birth.) Pertaining to life terminated in early stages of development.

ABORTION Noun. (Latin, *abortio* = to miscarry; English, *-tion* = result of an action. Pl., Abortions.) The condition of arrested development of structure, organ or organism such that development is imperfect, malformed or prematurely terminated. Verb, Abort.

ABRAHAM, RUDOLF (1905–1942) (Sachtleben 1942, Arb. Physiol. Angew. Entomol. Berl. 9: 55.)

ABRAMS, GEORGE, JENVEY (1902–1965) (Johansson & Johansson 1967, J. Econ. Entomol. 60: 896.)

ABRANCHIAL Adj. (Greek, *a* = without; *brangchia* = gills.) Not brachial; descriptive of organisms, or a stage of development, that lack gills. Alt. Abranchiate.

ABRANCHIATE Adj. (Greek, *a* = without; *brangchia* = gills; *-atus* = adjectival suffix.) Pertaining to organisms that lack gills. A term sometimes applied to naiads or aquatic immatures. Alt. Abranchious.

ABRASION Noun. (Latin, *abrasio* = wearing off; English, *-tion* = result of an action. Pl., Abrasions.) The surface of a structure that has been abraded, scared, scraped, scratched or worn. Trans. Verb, Abrade, Abraded, Abrading.

ABRASIVE Noun. (Latin, *abradere* = scrape away. Pl., Abrasives.) A natural or synthetic substance that causes a wearing away (abrading) of another substance or surface through the action of friction or rubbing. Natural abrasives include Alumina, Silica (diatomaceous earth), Silicates (pumice) and Carbon (diamond). Some pesticides rely on abrasives to penetrate the Integument or cause desiccation. Alt. Rasping.

ABRUPT Adj. (Latin, *abruptus, abrumpere* = to break off.) 1. A sudden or sharp transition. 2. Characterized by or causing a sharp break in structure or sudden end in action. 3. Some thing, act or process without gradation. Cf. Gradual. Rel. Blunt; Truncate.

ABRUPTLY ACUMINATE Structure with a broad apex that tapers to a sharp point. See Acuminate.

ABSCISSA Noun. (Latin, *abscindere* = to cut off. Pl., Abscissae, Abscissas.) Wings: A portion of a wing vein that is delimited by the intersection of another vein. A segregated or distinct part (section) of a vein (*e.g.,* Abscissa of the Radius).

ABSCISSION Noun. (Latin, *abscissio, abscindere* = to cut off; English, *-tion* = result of an action. Pl., Abscissions.) The process of terminating or cutting off. The separation of parts. Alt. Excision; Removal. Arcane Abscissus.

ABSCISSION SCAR The place on a plant where a kernel of grain had been attached.

ABSCOND Trans., Intrans. Verb. (Latin, *abscondere* = to hide.) 1. To hide or conceal. 2. Social Insects: The mass migration of highly eusocial bees from an active colony to a new nest site. Arcane Absconditus. Adv. Abscondedly.

ABSCONDING See Budding; Fission; Swarming. Cf. Nuptial Flight.

ABSOLUTE Adj. (Latin, *absolutus*; from *ab* + *solvere* = to loose.) Free from contamination; existing in and of itself.

ABSOLUTE ALCOHOL See Alcohol.

ABSOLUTE ASYMMETRY A body that has no primary axis, *e.g.,* sponges.

ABSORB Trans. Verb. (Latin, *absorbere*; from *ab* + *sorbere* = to drink in.) To cause to disappear; to cause to lose identity.

ABSORPTION Noun. (Latin, *absorptio, absorbere* = to suck in; English, *-tion* = result of an action. Pl., Absorptions.) 1. Assimilation, incorporation or engulfment of a lesser body or substance by a larger (different) body or substance. 2. The passage of nutrients through living cells. 3. The acceptance and containment of light that is not then reflected or transmitted. Cf. Adsorption.

ABSTRACT Noun. (Latin, *abstractus* = dragged away. Pl., Abstracts.) 1. A section or part of a scientific paper (usually following the Title and preceding the Introduction) published in a journal, in which essential details of the paper are provided. 2. The Summary of a speech delivered at a scientific meeting. See Publication; Summary.

ABT, VELI KURT (1892–1920) (Anon. 1921, Notulae Entomol. 1: 26.)

ABYSSAL Adj. (Greek, *abyssos* = unfathomable; Latin, *-alis* = belonging to.) Pertaining to aquatic organisms that live at great depths in total darkness beyond the continental shelf. See Habitat. Cf. Benthos; Pelagic.

ABYSSAL ZONE The biogeographic marine realm characterized by open sea at a deep level with a perpetual lack of light and an absence of photosynthetic plants. AZ is inhabited by organisms capable of withstanding extremely high pressure. Animals in the Abyss are typically carnivorous, blind or light-generating. See Habitat. Cf. Littoral Zone; Neritic Zone; Pelagic Zone; Photic Zone.

AC 303 630® See Pyrrol.

ACACIA PSYLLID *Acizzia uncatoides* (Ferris & Klyver) [Hemiptera: Psyllidae]: A native of Australia that became a pest in California during the late 1980s. Control of AP has been achieved with imported coccinellid beetles that act as predators of the pest.

ACACIA SPOTTING BUG *Eucerocoris tumidiceps* Horvath [Hemiptera: Miridae].

ACADREX® See Amitraz.

ACALYPTERAE See Acalyptrata.

ACALYPTERATAE Acalypterate muscoid Diptera that are characterized by small or linear Squamae (Curran). The Acalyptrata of authors.

ACALYPTRATA Muscoid Diptera whose Alulae are absent or rudimentary. Acalypteratae *sensu* Curran; Acalypterae *sensu* Imms.

ACANACEOUS Adj. (Greek, *akanos* = thistle; Latin, *-aceus* = of or pertaining to.) Thorny; prickly. Pertaining to structure with bristles.

ACANTHA Noun. (Greek, *akantha* = thorn. Pl., Acanthae.) Single-celled cuticular projections that lack a socket and sense cell. Acanthae are always sclerotized to maintain their shape, but their size and shape can be highly variable. Acanthae are external or internal and have been identified in most insect Orders. Richards & Richards (1979) recognize two types. 'Type A' forms a process nearly as great as the cross-sectional area of the cell producing it. One example is called a Ctenidium. 'Type B' forms a relatively small process. These may form so-called 'denticles' of the Proventriculus. See Projection; Protrusion. Cf. Ctenidium; Denticle; Spine. Rel. Sensillum; Seta.

ACANTHACEOUS Adj. (Greek, *akantha* = thorn; Latin, *-aceus* = of or pertaining to.) Pertaining to structure that bears thorns or spines.

ACANTHOCEPHALA Noun. (Greek, *akantha* = spine; *kephale* = head. Acanthocephalae.) Spiny-Headed Worms; Thorny-Headed Worms. A Phylum of parasitic worms with taxonomic affinities not clear and whose members demonstrate a complicated life cycle. Adults live in the intestinal tract of vertebrate predators; immatures develop within arthropods as intermediate hosts. (Conway & Crompton 1982, Biol. Rev. 57: 85–115.)

ACANTHOCERIDAE See Ceratocanthidae.

ACANTHOCNEMIDAE Plural Noun. A small Family of polyphagous Coleoptera assigned to the Superfamily Cleroidea. Adult <6.0 mm long; body elongate, flattened, black, covered with erect, stiff Setae; Antenna with 11 segments including club of three segments; tarsal formula 5-5-5; Abdomen with five Ventrites. Adults taken at lights and campfire ashes. Larva unknown.

ACANTHODION Noun. (Greek, *acanthodes* = thorny. Pl., Acanthodia.) Acarina: A tarsal Seta.

ACANTHOID Adj. (Greek, *akantha* = spine; *eidos* = shape.) Spine-like; descriptive of a conical cuticular process or projection that resembles a spine or thorn. See Acanthus. Cf. Aculeiform; Penicilliform; Spiniform; Spiciform; Styliform; Trichoid. Rel. Eidos; Form; Shape.

ACANTHOPARIA Noun. (Greek, *akantha* = spine; Latin, *paries* = wall. Pl., Acanthopariae.) Scarab larva: A spinose marginal area of the Paria; lateral, paired regions of the Epipharnyx (Boving).

ACANTHOPTEROCTETIDAE Davis 1978. Plural Noun. A small Family of primitive moths assigned to Superfamily Eriocranioidea. Apparently related to Eriocraniidae and include a Genus and five Species that occur in western USA. Adult small, slender-winged, Ocelli absent, Mandible vestigial; vom Rath's Organ absent; Maxillary Palpus with five segments; Labial Palpus with two segments. Epiphysis absent; wing scales hollow; wing-coupling Jugate; forewing-metathoracic locking device present; vaginal sclerite absent. Larva mines leaves; pupation within cocoon in debris beneath host plant. Pupa with hypertrophied Mandibles.

ACANTHOSOMATIDAE Signoret 1863. Plural Noun. A numerically small Family of Hemiptera assigned to Superfamily Pentatomoidea. Body moderately sclerotized; Tarsi with two segments; Metapleuron concealing the second abdominal spiracle; male with Sternum VIII large and exposed.

ACANTHUS Noun. (Greek, *akantha* = spine, thorn. Pl., Acanthae.) 1. A thorn or spine. 2. A single-celled cuticular projection from the body of an insect. Variable in size and shape, but characterized by not bearing a socket (Tormogen) or sense cells. (Richards & Richards 1979, J. Ins. Morph. Embryo. 8: 143–157.) See Projection. Cf. Ctenidium; Spine.

ACARABEN® See Chlorobenzilate.

ACARAC® See Amitraz.

ACARAPIS Noun. A Genus of tarsonemid mites that includes several Species that are parasitic on honey bees; one Species (Tracheal Mite) transmits Isle of Wight Disease among European Honey Bees. Cf. Tracheal Mite; *Tropilaelaps clareae*; Varroa Mite.

ACARAPIS DORSALIS A widespread external parasite that lives within a 'V' shaped Sulcus on the Mesoscutellum of the European Honey Bee, *Apis mellifera.* Anatomically, *Ad* is very similar to other Species of *Acarapis* that live on honey bees. Life cycle of *Ad* ca 11 days; mites feed on bee Haemolymph and die within 3 days if on a dead bee. Cf. Varroa mite; Tracheal Mite.

ACARAPIS EXTERNUS A widespread tarsonemid mite that lives as an external parasite on the cervical region of adult *Apis cerana* and *A. mellifera*. The mite feeds on Haemolymph of bees. Anatomically very similar to other Species of *Acarapis* that live on honey bees. Eggs of *Ae* laid on body of the host; eclosion within 4 days; become adults within 3–4 days. Cf. Varroa mite; Tracheal Mite.

ACARAPIS VAGANS A tarsonemid mite that lives as an external parasite on the axillary region of the hindwing and over the body of drones of the honeybee *Apis mellifera.* When population densities are high, the mites may be found on the Propodeum and first gastral segment. Anatomically *Av* is very similar to other Species of *Acarapis* that live on honey bees; *Av* feeds on Haemolymph of adult bees. Cf. Varroa Mite; Tracheal Mite.

ACARAPIS WOODI See Tracheal Mite.

ACARI Plural Noun. (Greek, *akares* = tiny, minute; Latin, *acarus* = mite. Sl., Acarus.) The Acarina.

ACARIASIS Noun. (Greek, *akares* = tiny, minute; Latin, *acarus* = mite; *sis* = a condition or state. Pl., Acariases.) The phenomenon or condition of being infested by mites. Cf. Enteric Acariasis. Rel. Delusory Cleptoparasitosis; Delusory Parasitoisis.

ACARICIDE Noun. (Latin, *acarus* = mite; *-cide* > *caedere* = to kill. Pl., Acaricides.) Chemical compounds used to kill mites and ticks. Some common compounds available commercially include: Aldicarb, Amitraz, Azinphos-Ethyl, Azinphos-Methyl, Chlorobenzilate, Cryolite, Diazinon, Dichlorvos, Dicrotophos, Dimethoate, Disulfoton, EPN, Fenamiphos, Fenthiocarb, Fenitrothion, Fenpropathrin, Formothion, Pyrrol, Fenproximate, Fenthion, Formetanate, Malathion, Mecarbam, Mephosfolan, Methamidophos, Methiocarb, Methidathion, Mevinphos, Monocrotophos, Naled, Omethoate, Oxydemeton-Methyl, Parathion, Phenthoate, Phorate, Phosalone, Phosmet, Phosphamidon, Profenofos, Promecarb, Prothoate, Pyraclofos, Pyridaben, Tetrachlorvinphos, Tetradifon, Thiometon, Triazophos, Trichlorfon, Vamidothion. See Pesticide. Cf. Biopesticide; Filaricide; Fungicide; Herbicide; Insecticide; Nematicide. Rel. Insect Growth Regulator.

ACARIFLOR® See Hexythiazox.

ACARIFORM Adj. (Greek, *acaris* = mite; Latin, *forma* = shape.) Mite-shaped; pertaining to a structure or an organism shaped as a mite. See Acaroid. Rel. Eidos; Form; Shape.

ACARIN® See Dicofol.

ACARINA Plural Noun. (Greek, *akares* = tiny, minute; Latin, *acarus* = mite. Sl., Acarus.) A taxonomic category that includes mites and ticks. See Acari.

ACARINARIUM Noun. (Latin, *Acarina* = mite; *arium* = a place for. Pl., Acarinaria.) Cuticular depressions on the body of solitary aculeate Hymenoptera that harbour non-parasitic mites. Cf. Mycangia. Rel. Symbiosis.

ACARISTOP® See Clofentezine.

ACARMATE® See Benzoximate.

ACAROCECIDIUM Noun. (Greek, *akares* = minute; *kekis* = gall. Latin, *-ium* = diminutive > Greek, *-idion*. Pl., Acarocecidia.) A plant gall induced by eriophyid mites.

ACARODOMATIUM Noun. (Greek, *akares* = minute; *domation* = small house. Pl., Acarodomatia.) A term proposed by Lundstrom (1887, Nova Acta Reg. Soc. Sci. Uppsal. 2: 1–88) for modifications of plant structure that accommodate habitation by Acarina. See Domatium.

ACAROID Adj. (Greek, *akares* = minute; *eidos* = form.) Acarine-like; descriptive of an organism that resembles mites or ticks in form, behaviour or biology. See Acarina. Cf. Acariform; Arachniform. Rel. Eidos; Form; Shape.

ACAROL® See Bromopropylate.

ACAROLOGY Noun. (Greek, *akares* = minute; Latin, *acarus* = mite; Greek, *logos* = discourse. Pl., Acarologies.) A subdiscipline of Zoology that involves study of Acarina (mites and ticks).

ACAROPHILY Noun. (Latin, *acarus* = mite; Greek, *philein* = to love. Pl., Acarophilies.) A symbiotic relationship between mites and plants. Alt. Acarophilous.

ACARTOPHTHALMIDAE Plural Noun. A small and rarely collected Family of schizophorous Diptera assigned to the Superfamily Opomyzoidea. Species occur in the Holarctic; two Species known in North America. Adults have been taken on rotting fungi and carrion.

ACARYL® See Propargite.

ACAUDAL Adj. (Greek, *a* = not; Latin, *cauda* = tail; *-alis* = pertaining to.) Descriptive of organisms that lack a tail. Alt. Acaudate. See Ecaudate.

ACCEPTABLE DAILY INTAKE The level of a chemical residue on edible agricultural products that does not cause appreciable health risks with daily exposure over a lifetime. Acronym ADI.

ACCEPTOR Noun. (Latin, *accipere* = to accept; *or* = a condition. Pl., Acceptors.) A substance that receives and unites with another substance.

ACCESSORY APPENDAGE 1. Any secondarily-derived appendage. 2. Odonata: Genital appendages on the venter of the second abdominal segment (Garman).

ACCESSORY BURROW Tunnels near the nesting burrow of ground-nesting sphecid wasps that are not provisioned or completely developed. AB regarded as an adaptive device that can deter predation or parasitism by other insects or animals.

ACCESSORY CARINA Orthoptera: Lateral Carinae on the face that are sometimes diagnostically important or taxonomically useful.

ACCESSORY CELL 1. A wing cell not commonly present in a group of insects; AC in some Orders occurs in definite positions on the wing. 2. Lepidoptera: Usually a small cell at the apex of the Subcosta that gives rise to veins 7–10. The first Radius *sensu* Comstock. The Areole. A small cell occurring in wings of some Lepidoptera, between R2+3 and R4+5, and closed by Sectorial Crossvein (Klots).

ACCESSORY CIRCULATORY ORGAN Sac-like, frequently membranous, delicate structures associated with the circulatory system of insects. ACOs appear in different anatomical forms: Small ovoid Ampullae, small fan-shaped muscles, membranous tubes derived from connective tissue, or thin, pulsating muscle filaments. ACO is usually independent of the Dorsal Vessel and ACO contractions are independent of the Heart. An ACO typically is positioned at base of an appendage (antenna, wing, leg). Syn. Accessory Heart; Antennal Heart; Cercal Heart; Leg Heart; Pulsatile Organ; Wing Heart. See Dorsal Vessel; Circulatory System. Rel. Hydrostatic Skeleton.

ACCESSORY CLAWS Acari: Paired ventrolateral, subapical 'clawlets' of pretarsal claws (Goff *et al.* 1982. J. Med. Entomol. 19: 222). See Claw; Pretarsus. Rel. Leg.

ACCESSORY DOUBLET The paired microtubules of a spermatozoon's axoneme. Number of pairs of ADs in Spermatozoa typically 9 but can become extraordinarily large (ca. 70–90 doublets in fly *Sciara coprophila* and *Rhynchosciara*). AD has not been found in some cecidomyiid flies. Central Tubules and ADs typically extend length of Flagellum. In most insects accessory doublets are parallel and straight; in *Psocus* sp. and cat flea, *Ctenocephalides felis* (Bouché), ADs spiral around Central Tubules and form a helical configuration. Syn. Accessory Tubule. See Axoneme; Microtubule. Cf. Central Tubule; Peripheral Singlets. Rel. Sperm Flagellum.

ACCESSORY GENITALIA Odonata: See Accessory Appendages.

ACCESSORY GLAND 1. Any secondarily-derived gland of a glandular system (*e.g.* a gland associated with the Salivary Gland of Hemiptera). See Salivary Gland. 2. Insect reproductive system: Glands derived from Ectoderm. 3. Female reproductive system: AGs secrete venom, pheromones, lubricants, adhesives and other biologically relevant chemicals. AG called by numerous names. See Gland. Cf. Alkaline Gland; Acid Gland; Colleterial Gland; Dufour's Gland; Milk Gland. Rel. Male Accessory Gland; Female Accessory Gland.

ACCESSORY GONOPORE See Collatoria.

ACCESSORY HEART See Accessory Circulatory Organ.

ACCESSORY LOBES Ventral lobes of the Protocerebrum. See Brain. Cf. Protocerebrum.

ACCESSORY NUCLEI Diptera, Hymenoptera, Mallophaga: Small nuclei that develop from the primary nucleus of an Oocyte.

ACCESSORY PROCESS Immature insects: Any cuticular eversion from the apex of the first or second Antennal segment.

ACCESSORY PRONG See Palpal Claw.

ACCESSORY PULSATORY ORGAN See Accessory Circulatory Organ.

ACCESSORY SAC Any glandular structure of the female reproductive system that contains a secretory product.

ACCESSORY SUBCOSTAL VEIN Odonata (Perlidae): A wing vein diverging from the Subcosta and branching toward the wing's apex. See Vein; Wing Venation.

ACCESSORY TUBULE See Accessory Doublet; Axoneme. Rel. Microtubule; Sperm Flagellum.

ACCESSORY TYMPANAL MEMBRANE Lepidoptera: A secondary membrane (Fenestra Media) of the Tympanum; anatomically, a membranous part of the Postnotum (Richards) in Geometridae. Syn. Accessory Tympanum.

ACCESSORY VEINS 1. Secondarily developed longitudinal veins in wings of insects (Comstock & Kellogg). 2. Hymenoptera (Symphyta): The third Anal Vein or posterior-most vein in the anal region of the forewing. See Vein; Wing Venation.

ACCLIMATION Noun. (Latin, *ad* = to; Greek, *klima* = climate; English, *-tion* = result of an action. Pl., Acclimations.) Habituation of an organism to different environmental conditions. Syn. Acclimatization.

ACCLINATE Adj. (Latin, *acclinatus*; *-atus* = adjectival suffix.) See Acclivous.

ACCLIVITOUS Adj. (Latin, *acclivitas* = a slope; from *ad* = to + *kli* = lean; *-osus* = possessing the qualities of.) See Acclivous.

ACCLIVITY Noun. (Latin, *acclivitas*; English, *-ity* = suffix forming abstract nouns. Pl., Acclivities.) A sloped surface rising from the point of observation. Ant. Declivity.

ACCLIVOUS Adj. (Latin, *acclivis* = ascending; *-osus* = full of.) Descriptive of a surface that slopes upward. Ant. Declivous.

ACCOMMODATION Noun. (Latin, *accommodatio, ad* = to; *commodus* = fitting; English, *-tion* = result of an action. Pl., Accommodations.) A physiological, behavioural or social phenomenon in which physiological state, behavior or function is modifiied as the result of continuous exposure to stimuli.

ACCOMMODATIVE Adj. (Latin, *accommodatio, ad* = to; *commodus* = fitting.) Inclined toward accommodation.

ACCRESCENCE Noun. (Latin, *accrescentia, accrescere* = to increase. Pl., Accrescenses.) Continuous growth.

ACCRESCENT Adj. (Latin, *accrescere* = to increase.) 1. Growing continuously. 2. As an anatomical descriptor: Increasing gradually in thickness toward the apex. Ant. Tapering.

ACCRETION Noun. (Latin, *accretio, accrescere* = to increase; English, *-tion* = result of an action. Pl., Accretions.) 1. Growth or enlargement through the addition of external parts. 2. Adding new matter to the external surface of structure. *e.g.* Sand grains accumulating in a caddisfly larval case. Alt. Accumulation; Growth.

ACUTE POISONING Illness or death resulting from a single exposure of a toxicant. See Chronic Poisoning.

ACELLULAR Adj. (Latin, *a* = without; *cellula* = a small room.) Descriptive of structure not composed of cells.

ACENTRIC Adj. (Greek, *a* = without; *kentron* = center; *-ic* = characterized by.) Pertaining to chromosomes that lack a Centromere. Cf. Metacentric; Telocentric.

ACEPHALIA Noun. (Greek, *a* = without; *kephale* = head. Pl., Acephali.) A headless organism.

ACEPHALOUS Adj. (Greek, *a* = without; *kephale* = head; Latin, *-osus* = possessing the qualities of.) Describing a headless condition. A term applied to higher Diptera and apocritious Hymenoptera whose larvae lack a well defined, sclerotized head.

ACEPHATE An organic-phosphate (thiophosphoric acid) compound {O,S-dimethyl acetylphosphoramidothioate} used as a contact and systemic insecticide against aphids, thrips, armyworms, bagworms, loopers, cutworms, leafminers, caterpillars and grasshoppers. Compound applied to ornamentals, celery, cotton, cranberries, lettuce, peppers, soybeans, peanuts and vegetables. Phytotoxic to American elm, sugar maple, cottonwood, some varieties of chrysanthemums. Toxic to bees. Trade names include: Aimthane®, Acesul®, Asataf®, Asatal®, Kitron® Ortan®, Orthene®, Ortran®, Ortril®, Payload®, Tornado®, Racet®, Vital®. See Organophosphorus Insecticide.

ACERATA Plural Noun. (Greek, *a* = without; *keras* = horn.) 1. A Class of organisms in older classifications that embraced current concepts of the Arachnida and Merostomata. 2. Arthropods that lack true Antennae.

ACERATE Adj. (Latin, *acerosus*, *acer* = sharp; *-atus* = adjectival suffix.) Needle-shaped; pointed; descriptive of a slender structure with a sharp point. Alt. Acerous; Acerose. See Outline Shape. Cf. Setiform. Rel. Eidos; Form; Shape.

ACERBI, GUISEPPE (JOSEPH) (1773–1846) (Rose 1850, *New General Biographical Dictionary* 1: 71.)

ACERENTOMIDAE Plural Noun. A relatively large Family of Protura assigned to Suborder Acerentomoidea. Body and mouthparts slender; Maxillary Lacinea slender and apically pointed; middle and hind leg claws boat-shaped; Operculum of gland on Abdomen segment VIII with pectinate ornamentation; tracheal system absent. Acerentomids typically inhabit soil and leaf litter where they probably feed on mycorrhizae. Some classifications may include the Acerellidae and Berberentomide within the Acerentomidae.

ACEROUS Adj. (Greek, *a* = without; *keras* = horn; Latin, *-osus* = full of.) 1. Pertaining to insects without Antennae. 2. Hornless; without tentacles.

ACERVATE Adj. (Latin, *acervare* = to amass; *-atus* = adjectival suffix.) Clustered; pertaining to collections or mounds of compact, discrete clusters. Cf. Acinose; Adeniform; Botryoidal. Trans. Verb. (Latin, *acervare* = to amass.) To heap.

ACERVATION Noun. (Latin, *acervatio*; English, *-tion* = result of an action. Pl., Acervations.) An accumulation of collections; heaps.

ACERVULUS Noun. (Latin, diminutive *acervus* = heap. Pl., Acervuli.) A small heap or cluster.

ACESUL® See Acephate.

ACETABULAR Adj. (Latin, *acetabulum* = vinegar cup.) Pertaining to the Acetabulum.

ACETABULAR CAP Hemiptera: The coxal cavity. See Acetabulum.

ACETABULIFORM Adj. (Latin, *acetabulum* = vinegar cup; *forma* = shape.) Cup-shaped; descriptive of structure shaped as a shallow saucer, depression or socket with sides curved inward. See Acetabulum. Cf. Ampulliform; Calathiform; Patelliform; Vasiform. Rel. Form 2; Shape 2; Outline Shape.

ACETABULUM Noun. (Latin, *Acetabulum* = vinegar cup. Pl., Acetabula.) 1. A concave surface or cavity in the Integument or surface of a structure into which the protuberance of an appendage is fitted, thereby creating a ball-and-socket type of articulation (*e.g.* coxal cavity, Mandible). 2. Diptera: A cup-shaped cavity in which maggot mouthparts are held at repose. 3. Hymenoptera: Transverse Carinae located on the Mesothorax of some aculeate wasps. See Articulation; Segmentation. Cf. Cotyla; Diarthrosis; Enarthrosis; Socket. Rel. Condyle; Ginglymus.

ACETATE Noun. (Latin, *Acetum* = vinegar; *-atus* = adjectival suffix. Pl., Acetates.) An ester or salt of acetic acid.

ACETIC ACID A colourless liquid (CH_3COOH) with a pungent odour that resembles vinegar and acidic taste. AA is part of an histological fixative when diluted with water or alcohol. Alt. Glacial Acetic Acid.

ACETOGENINS Plural Noun. (Latin, *acteum* = vinegar; Greek suffix = *gen.*) A complex of alkaloid biopesticides isolated from bark of the pawpaw (papaya) tree *Asimina triloba* Dunal and seeds of *Annona squamosa* Linnaeus [Annonaceae]. Insecticidal compounds called Acetogenins include: Annonin I, Annonin IV, Annonacin, Asimisin, Bullacticin, Goniothalamicin, Neoannonin, Sylvaticin and Uvaricin. The mode-of-action of Acetogenins is cytotoxic and may involve electron transport inhibition. See Biopesticide; Botanical Insecticide; Natural Insecticide; Insecticide; Pesticides. Cf. Alkaloids; Chromenes; Cyclopepsipeptides; Domoic Acid; Furanocumarins; Mammeins; Polyacetylenes; Terthienyl.

ACETONE Noun. (Formed from chemical nomenclature. Pl., Acetones.) Volatile liquid ketone (CH_3COCH_3) naturally occurring in crude wood alcohol. A solvent for many organic compounds.

ACETOUS Adj. (Latin, *acetum* = vinegar; *-osus* = possessing the qualities of.) Pertaining to vinegar; something acidic or sour tasting.

ACETYLCHOLINE Noun. (Latin, *acetum* = vinegar; *hyte* = substance; Greek, *chole* = bile; *ine* = chemical suffix.) A neurotransmittor that perpetuates an action potential from one neuron to another by its release from a presynaptic neuron, flooding the synaptic junction, and reacting with receptors on the post-synaptic neuron, causing a change in the ionic potential of the neuronal membrane which starts anew a wave of depolarization.

ACETYLCHOLINESTERASE Noun. (Latin, *acetum* = vinegar; *hyte* = substance; Greek, *chole* = bile; *aether* = ether; *saeure* = acid; *ase* = chemical suffix for an enzyme.) An enzyme present in synaptic junctions that breaks down acetylcholine, a neurotransmitter, into acetic acid and choline, thus freeing the post-synaptic receptors for future interaction with acetylcholine.

ACHAETOUS Adj. (Greek, *a* = without; *chaete* = hairs; Latin, *-osus* = possessing the qualities of.) An arcane term that pertains to a structure that lacks Setae, spines or bristles.

ACHARIUS, ERIK (1757–1819) (Anon. 1819, Svenska Vetensk Akad. Handl. 40: 299–305; Anon. 1820, Meisners' Naturw. Anz. 3: 65–66; Rose 1850, *New General Biographical Dictionary* 1: 72–73.)

ACHATINE Adj. (Greek, *achates*.) Resembling the lines of an agate in more or less concentric circles. Alt. Achatinus.

ACHEIRIA Noun. (Greek, *a* = without; *cheir* = hand.) 1. Without hands. 2. Without the ability to perceive the sense of touch upon the skin.

ACHEIROUS Adj. (Greek, *a* = without; *cheir* = hand;

Latin, *-osus* = possessing the qualities of.) Pertaining to an inability to perceive the sense of touch.

ACHELATE Adj. (Greek, *a* = without; *chele* = claw; *-atus* = adjectival suffix.) Without claws; pertaining to clawless appendages.

ACHEMON SPHINX *Eumorpha achemon* (Drury) [Lepidoptera: Sphingidae].

ACHENE Noun. (Greek, *a* = without; *chainen* = to gape. Pl., Achenes.) Naked seed. Typically, a seed that is not fused to the fruit wall.

ACHILIDAE Plural Noun. A cosmopolitan Family of auchenorrhynchous Hemiptera assigned to Superfamily Fulgoroidea and including about 375 Species. Antenna positioned beneath compound eye; Pedicel enlarged with conspicuous Sensilla; apical segment of Rostrum elongate; wings typically held flat and overlapping, rarely tectiform; anal area of hindwing not reticulate; second Tarsomeres of hindleg with apical spines. Biology not well studied; all Species are phytophagous but not host-plant specific; a few Species are associated with fungi; nymphs are associated with wood, sometimes under bark or in crevices.

ACHOR Noun. (Greek, *acchor* = dandruff; *or* = a condition. Pl., Achores.) A pustule.

ACHREIOPTERA Noun. (Etymology obscure.) An Order-level name proposed for the Family Platypsyllidae (Coleoptera).

ACHROMA Noun. (Greek, *a* = without; *chroma* = colour. Pl., Achromata.) A condition characterized by a lack of pigmentation.

ACHROMASIA Noun. (New Latin, *a* = without; *chroma* = colour; *sia* = condition.) A physical/chemical condition of biological material in which a structure or substance loses its characteristic staining properties. See Achroma.

ACHROMATE Noun. (Greek, *a* = without; *chroma* = colour; *-atus* = adjectival suffix. Pl., Achromates.) 1. Unable to perceive colour. 2. Colour blind.

ACHROMATIC Adj. (Greek, *achromatos, a* = without; *chroma* = colour; *-ic* = of the nature of.) 1. Descriptive of tissue that is not stained by conventional histological procedures. 2. Colourless; describing structure that is free of colour. 3. Refracted light that does not separate into constituent colours. Adv. Achromatically. See Coloration. Cf. Monochromatic; Polychromatic.

ACHROMATIC CONDENSER A lens system used as an optical condenser.

ACHROMATIC LENS A lens system composed of two glass elements: Usually a biconvex crown-glass lens that is fitted into a plano-concave flint-glass lens. The chromatic aberration created by one lens is corrected by the other.

ACHROMATICITY Noun. (Greek, *a* = without; *chroma* = colour; English, *-ity* = suffix forming abstract nouns. Pl., Achromaticities.) The quality of being achromatic.

ACHROMATIN Noun. (Greek, *a* = without; *chroma* = colour. Pl., Achromatins.) Material within a cell's nucleus that is not coloured by basic stains.

ACHROMATIZE Trans. Verb. (Greek, *a* = without; *chroma* = colour.) To render colourless.

ACHROMATO- Greek prefix indicating colourless properties of the noun stem. *e.g.* Achromatophore. Alt. Achromat-.

ACHROMIC Adj. (Greek, *a* = without; *chroma* = colour; *-ic* = of the nature of.) Colourless; paleness due to a lack of pigmentation.

ACHROMOTRICHIA Noun. (Greek, *a* = without; *chroma* = colour; *thrix* = hair. Pl., Achromotrichiae.) Setae without colour.

ACHROMOUS Adj. (Greek, *a* = without; *chroma* = colour; Latin, *-osus* = possessing the qualities of.) See Achromic.

ACIA Noun. (Etymology Obscure. Pl., Aciae.) A thin cuticular sclerite of the Mandible (MacGillivray).

ACICULA Noun. (Latin diminutive of *acus, acicula* = small needle. Pl., Aciculae.) A small needle-like Seta; a needle-like projection from the surface of body. Characteristically small in relation to the body or surrounding structures.

ACICULAR Adj. (Latin, *acicula* = small needle.) Needle-shaped; structure shaped into a long, slender point. Syn. Aciculate. Cf. Acuminate; Trichoid. Adv. Acicularly.

ACICULATE Adj. (Latin, *acicula* = small needle.) 1. With aciculae. 2. Any surface that appears superficially scratched, scored or etched. Exceedingly fine or narrow cuticular etching, such as expressed in some Integument sculpturing. Alt. Aciculatus. See Sculpture Pattern. Cf. Acuductate; Alveolate; Asperites; Baculate; Clavate; Corrugated; Echinate; Favose; Gemmate; Psilate; Punctate; Reticulate; Rugulate; Scabrate; Scobinate; Shagreened; Smooth; Striate; Verrucate; Wrinkle.

ACICULUM Noun. (Latin, *acicula* = small pin for headress. Pl., Aciculae.) A needle-like spine or bristle on the surface of an animal or plant. Cf. Seta; Sensillum.

ACID Noun. (Latin, *acidus* = sour. Pl., Acids.) A chemical compound sour to the taste and turning blue litmus a red colour when in solution. Containing the H^+ radical; a salt of hydrogen.

ACID ALCOHOL Typically, 70% alcohol to which is added 0.1–1.0% hydrochloric acid (HCl). In histological protocols, acid alcohol is often used for destaining.

ACID GLAND Aculeate Hymenoptera: One of a pair of elongate Accessory Glands that opens via a duct into the poison sac. Acid glands secrete the acid constituent of venom that is used in defence and the immobilization of prey. See Accessory Gland. Cf. Alkaline Gland.

ACID SCENT A pungent, sour smell.

ACIDIC Adj. (Latin, *acidus* = sour; *-ic* = characterized by.) 1. Descriptive of subtances with the properties of an acid. 2. Pertaining to histological stains that typically affect the cytoplasm and cytoplasmic content of cellular preparations. Cf. Basic.

ACIDOPHILIC Adj. (Latin, *acidus* = sour; Greek, *philein* = to love; *-ic* = characterized by.) 1. Descriptive of plants that grow in an acidic medium. 2. Histology: Pertaining to tissue that easily or rapidly stains with acidic compounds. Alt. Aciduric.

ACIDOPHILOUS Adj. (Latin, *acidus* = sour; Greek, *philein* = to love; *-osus* = possessing the qualities of.) Microbiology: Descriptive of bacteria and related organisms that survive and reproduce in acidic media. See Habitat. Cf. Halophilous; Hydrophilous; Hypogenous; Nitrophilous. Rel. Ecology.

ACIDOTHECA Noun. (Latin, *acidus* = sour; Greek, *theke* = case. Pl., Acidothecae.) The pupal sheath covering the Ovipositor.

ACIES (Latin, *acies* = edge.) The extreme termination of a margin.

ACIFON® See Azinphos-Methyl.

ACIFON-E® See Azinphos-Ethyl.

ACIFORM Adj. (Latin, *acus* = needle; *forma* = shape.) Needle-shaped. Syn. Acicular. Rel. Form 2; Shape 2; Outline Shape.

ACINACICATE Adj. See Acinaciform.

ACINACIFORM Adj. (Latin, *acinaces* = short sword; *forma* = shape.) Scimitar-shaped; Sickle-shaped. Descriptive of structure that is relatively long, flattened, curved and apically tapered. Syn. Acinacicate. Alt. Sword-like; Dagger-like. See Falcate. Rel. Form 2; Shape 2; Outline Shape.

ACINAR SALIVARY GLAND A type of Salivary Gland regarded as more common in primitive insects. ASG contain three cell types: Duct Cells, Peripheral Cells and Central Cells. DC usually comprise epithelium of gland; PC and CC usually comprise Acinus (secretory part) of gland. DCs contain a cuticular lining. Secretory cells in duct of cockroach *Nauphoeta* positioned near Acinus; secretory cells have not been found in duct of Orthoptera. Non-secretory DC display a highly convoluted apical plasma membrane; various organelles occur near this membrane. Basal plasma membrane is convoluted and holds mitochondria. CC contain large amounts of secretory material, but organelles found in this cell type vary among Species and gland activity. See Gland. Cf. Salivary Gland.

ACINIC Adj. (Latin, *acinus* = berry; *-ic* = consisting of.) Pertaining to an Acinus.

ACINIFORM Adj. (Latin, *acinus* = berry; *forma* = shape.) A cluster of connected spheres; Descriptive of a structure resembling a cluster (bunch) of grapes. Cf. Acinose; Botryoidal. Rel. Form 2; Shape 2; Outline Shape.

ACINOSE Adj. (Latin, *acinosus, acinus* = berry; *-osus* = full of.) 1. Pertaining to Acini. 2. Descriptive of structure superficially set with Acini. 3. Descriptive of structure shaped as a cluster (bunch) of grapes. Alt. Acinous. Cf. Acervate; Aciniform; Adeniform; Botryoidal.

ACINUS Noun. (Latin, *acinus* = berry. Pl., Acini.) 1. A small sac filled with secretory cells. Often part of a system of sacs that connect via ducts to a larger lumen. 2. Granulations resembling those on a blackberry. 3. The terminal secreting tubes of glands.

ACKERMAN, ARTHUR JOHN (1889–1937) (Porter 1937, J. Econ. Entomol. 30: 385–386.)

ACLASTIC Adj. (Greek, *a* = without; *klastos* = break.) Pertaining to substance incapable of refracting light.

ACLERDIDAE Plural Noun. A small Family of sternorrhynchous Hemiptera assigned to Superfamily Coccoidea.

ACLINAL Adj. (Greek, *a* = without; *klinein* = slope; Latin, *-alis* = characterized by.) Descriptive of structure or surface that lacks an inclination (horizontal). Cf. Clinal.

ACME Noun. (Greek, *akme* = prime. Pl., Acmies.) 1. The summit or highest point. 2. In the development of phyletic lines: The geological period during which the largest number of Species are extant. 3. The highest phase of development in a lineage. Alt. Apex; Peak; Summit; Zenith. Cf. Epacme; Paracme.

ACMIC Adj. (Greek, *akme* = prime.) Pertaining to the Acme or top of a structure.

ACOELOMATA Plural Noun. (Greek, *a* = without; *koiloma* = hollow.) Taxonomic division of Metazoa that includes members that lack a Coelom (body cavity). Sponges and Coelenterates.

ACOELOMATE Adj. (Greek, *a* = without; *koiloma* = hollow; *-atus* = adjectival suffix.) Descriptive of a body that lacks a Coelom. Alt. Acoelous.

ACOLOUS Adj. (Greek, *a* = without; *kolon* = the colon.) Descriptive of a body without limbs or appendages.

ACONDYLOSE Adj. (Greek, *a* = without; *kondylos* = knuckle; *-osus* = possessing the qualities of.) 1. Jointless. 2. Botany: Structure without nodes. Alt. Acondylous.

ACONIC Adj. (Greek, *a* = without; *konos* = cone; *-ic* = characterized by.) Without a cone; pertaining to an Aconic Eye.

ACONIC EYE A morphological type of compound eye in which the Ommatidium lacks a crystalline or liquid-filled cone. Instead, a group of elongate, transparent cells occur. Aconic condition (Grenacher 1879) is characterized by Semper Cells that do not form a crystalline cone. Instead, proximal part of lens projects conically inward toward the Rhabdom. Alternatively, the space between lens and Rhabdom may be filled with glycogen-rich cytoplasm and nuclei of Semper Cells. This condition may be a primitive feature of insect vision because it occurs in primitive groups such as Apterygota. Aconic eyes also occur in ancient groups such as Hemiptera, Diptera and Coleoptera (Polyphaga, cucujiforms and universal in beetles less than 1 mm long). See Compound Eye. Cf. Eucone; Pseudocone. Rel. Ommatidium.

ACORIT® See Hexythiazox.

ACOSTA, JOSE D' (1539–1600) (Dusmet y Alonso

1919, Boln. Soc. Entomol. Espan. 2: 76–77; Hagen 1948, *Green World of the Naturalist*, pp. 56–73.)

ACOUSTIC NERVE A nerve that connects the auditory pits or other organs of hearing with special ganglia.

ACQUA, CAMILLO (1863–1936) (Lombardi 1936, Riv. Biol. (Firenze) 21: 150–158.)

ACQUIT® See Alphacypermethrin.

ACRIDIDAE Plural Noun. Grasshoppers. The largest Family of chaeliferous Orthoptera and nominant Family of Acridoidea; perhaps including 17 Subfamilies; difficult to characterize. Acridids are a cosmopolitan Family with many Species of economic importance, including locust pests. See Acridoidea. Cf. Locust.

ACRIDOIDEA Noun. Grasshoppers and Locusts. A Superfamily of caeliferous Orthoptera including about 8,000 Species within 1,500 Genera. Adult body usually more than 12 mm long; Antenna not longer than body and without Sensilla; wings present or absent; hind Basitarsus lacking spines, each Pretarsus with an Arolium; Cerci short, unsegmented; Paraprocts without Cercus-like processes; Ovipositor present. Species are exclusively phytophagous and typically diurnal; mating, moulting and migration may occur nocturnally; stridulation by various mechanisms and best developed in male; some Species stridulate with Mandibles. Family-level classification fluctuates but common Families include: Acrididae, Lathiceridae, Lentulidae, Ommexechidae, Pamphagidae, Pamphagodidae, Pauliniidae, Pyrgomorphidae, Romaleidae and Tristiridae. See Orthoptera.

ACRIDOLOGY Noun. (Greek, *akridion* = locust; *logos* = discourse. Pl., Acridologies.) The study of locusts and grasshoppers. See Orthoptera. Cf. Crickets.

ACRIDOPHAGUS Adj. (Greek, *akridion* = locust; *phagein* = to eat.) Pertaining to organisms that are predaceous upon grasshoppers.

ACRIDOXENIDAE Plural Noun. Dead-leaf Crickets. A monogeneric Family of ensiferous Orthoptera that includes two Species in West Africa. Body large; head large and round; Antenna longer than body. Prosternum spined; legs with lamellate, spined Tibiae and Femora. Species live on spiny plants; biology poorly known. Common name (Dead-leaf Crickets) is derived from the Tegmina and hindwings which resemble brown, withered and chewed leaves.

ACRINATHRIN A synthetic-pyrethroid compound with use as a contact and stomach poison and acaracide. Compound is used on vegetables, fruits and ornamentals in some countries but is not registered in USA. Acrinathrin shows minor phytotoxicity. Trade names include: Ardent® and Rufast®. See Synthetic Pyrethroids.

ACROBAT ANTS Members of the Genus *Crematogaster* [Hymenoptera; Formicidae]. Workers ca 3–5 mm long; shiny brown or black; Pedicel composed of two nodes; AAs assume a characteristic pose by positioning the Gaster over head or Thorax when alarmed. A few Species are regarded as urban pests; colonies of some Species become established inside tunnels of termites, carpenter ants or wood-boring beetles; some Species that occur in wooden door frames and window sills are affected by fungal decay; other Species nest in decaying trees. AAs are omnivorous with workers foraging day or night; some Species tend aphids for honeydew. Bites inflicted by workers may be painful.

ACROBE® A registered biopesticide derived from *Bacillus thuringiensis* var. *israelensis*. See *Bacillus thuringiensis.*

ACROCENTRIC Adj. (Greek, *akros* = tip; *kentron* = center.) Pertaining to chromosomes with an apical centromere. Cf. Acentric; Telocentric; Metacentric.

ACROCERATIDAE Plural Noun. See Acroceridae.

ACROCERIDAE Plural Noun. Bladder Flies; Small-Headed Flies. A cosmopolitan Family of orthorrhaphous Diptera including ca 500 Species and 50 Genera assigned to Superfamily Tabanoidea. Acrocerids resemble Bombyliidae but pulvilliform Empodia separate these Families. Adult small to moderate-sized with small head, humped Thorax and globose Abdomen; without bristles; compound eyes large (holoptic); Antenna with three segments; Calypter large; simple unbranched vein R2; pulvilliform Empodia. Larval stage develops as an internal parasite of spiders. Eggs frequently deposited in high numbers (ca 5,000 per female); eggs are dropped by female during flight or deposited on dead twigs, grass stems or tree trunks; eggs are always deposited away from intended hosts. First-instar larva (Planidium) seeks its host and burrows inside the body through intersegmental membrane at base of leg. Larva then moves to book lungs where it becomes attached. Larva remains in diapause several months to a few years; dispause broken when spider moults to adult stage. Then, the acrocerid larva moults, feeds voraciously and usually kills spider within short time. Larva exits posterior-end-first from spider while still feeding internally for several hours. Mature larva frees itself from host and pupates in spider's premoulting web; adult flies emerge ca 7–10 days later. Syn. Acroceratidae; Cyrtidae; Henopidae; Oncodidae.

ACROGENESIS Noun. (Greek, *akro* = tip; *genes* = producing. Pl., Acrogeneses.) The phenomenon of adding additional sements to the body or appendage at the anterior or posterior end. Acrogenesis is manifest in several groups of arthropods, including trilobites, polychaetes, some chilopods, all diplopods, pauropods, symphylans and proturans. The dipluran *Campodea* (an entognathan) adds antennal segments via acrogenous growth with each postembryonic moult until the definitive number of antennal segments is achieved. Syn. Telogenesis.

ACROGENOUS Adj. (Greek, *akros* = tip; *genes* = producing; *-osus* = with the qualities of.) 1. Pertaining to Acrogenesis. 2. Growth at the apex (distal point) of a structure. Adv. Acrogenously.

ACROLEPIIDAE Plural Noun. Acrolepiid Moths. A Family of ditrysian Lepidoptera assigned to Superfamily Yponomentoidea with three North American spp. Similar to Plutellidae; larvae skeletonize leaves of *Smilax* or bore stems of lilies.

ACROLOPHIDAE Plural Noun. A Family of ditrysian Lepidoptera assigned to Superfamily Tineoidea, and consisting of about 275 Species. Larvae develop and feed within silken tubes among plant debris and roots. Acrolophidae are synonymized with Tineidae in some classifications.

ACRON Noun. (Greek, *akron* = top. Pl., Acrons.) 1. The anteriormost, preoral segment of an insect's body. 2. The Prostomium. Arthropod embryo: The head region anterior of the tritocerebral somite (Snodgrass). 3. The first segment of an ant (Jardine).

ACRONYM Noun. (Greek, *akron* = top; *onoma* = name. Pl., Acronyms.) A combination of letters that forms a word or scientific epithet. Acronym is based on the initial letter(s) of sequential words in a compound term. *e.g.* PAUP: Phylogenetic Analysis Using Parsimony; DDT: Dichloro-diphenyl-trichloroethane. Cf. Eponym; Patronym; Synonym.

ACROPARIA Noun. (Greek, *acro* = tip; *paria* = paria Pl., Acropariae.) Scarabaeoid beetle larvae: The anterior part of the Paria that is usually long and bears bristles (Boving).

ACROSBRISTLE Noun. (Greek, *acro* = tip; Middle English, *brostle* = bristle. Pl., Acrosbristles.) Diptera: Two rows of bristles on the middle of the dorsum. Specifically, Acrosbristles are minute, peculiar bristles on the dorsocentral region of Dolichopodidae.

ACROSBRISTLE HAIRS Diptera: Setae positioned between the dorsocentral bristles (Curran). Rel. Sensillum.

ACROSBRISTLE SETULAE Diptera: Very short Setae between the dorsocentral bristles (Curran).

ACROSOME Noun. (Greek, *akros* = tip; *soma* = body. Pl., Acrosomes.) An organelle at the apex of the spermatozoon head, and regarded as (at least partially) derived from Golgi Apparatus. Acrosome is highly variable in size, shape and structure; it attaches to the egg and dissolves egg membrane. When Acrosome is absent, there appears to be no substitute for it. Acrosome occurs in primitive Apterygota including Diplura, Collembola and Thysanura, but absent from many Protura; Acrosome occurs in most pterygote insects but it has not been identified in Isoptera, some Coccoidea and some insects with a Micropyle. Many Neuroptera (including chrysopid Genera *Chrysoperla* and *Eumicromus*) lack an Acrosome. Among Diptera, Acrosome is absent from most Species of gall midges. From widespread occurrence of Acrosome and its omission in several groups, we infer that an Acrosome is a primitive feature of insectan Spermatozoa and absence of an Acrosome from spermatozoon head is a derived or specialized feature. Alt. Apical Body; Head-Cap. See Sperm Anatomy. Rel. Spermatozoa.

ACROSOME REACTION A process involving vesiculation of the Plasma and Outer Membranes that permits leakage of enzymes from the spermatozoon that in turn facilitates spermatozoon penetration of the eggshell. Other changes may also be involved. Acrosomal region forms a large disk-shaped swelling (the leaf) that encloses a cup-like structure that covers a dense axial rod. Cell membrane forms a 'conspicuous bag around the Acrosome.' Acrosome within female is compact, homogeneous and the cell membrane is near the Acrosome. See Spermatozoa. Cf. Sperm Capacitation.

ACROSTERNITE Noun. (Greek, *akros* = tip; *sternon* = chest. *-ites* = constituent. Pl., Acrosternites.) The narrow, marginal flange anterior of the Antecosta of a definitive sternal sclerite. Acrosternites are characteristic of the abdominal Sterna of insects but apparently are absent from thoracic Sterna (Snodgrass). Cf. Acrotergite.

ACROSTICHAL BRISTLES Diptera: One or more longitudinal rows of small, thick Setae on the central part of the Mesonotum.

ACROTERGAL Adj. (Greek, *akros* = tip; Latin, *tergum* = back; *-alis* = having the character of.) Pertaining to or descriptive of an Acrotergite. See Acrotergite.

ACROTERGITE Noun. (Greek, *akros* = tip; Latin, *tergum* = back; *-ites* = constituent. Pl., Acrotergites.) The anterior, precostal part of the tergal sclerite of a secondary segment. An Acrotergite is usually in the form of a narrow flange; sometimes it is greatly enlarged but usually reduced or obliterated (Snodgrass). See Pretergum; Tergum. Cf. Acrosternite. Rel. Secondary Segmentation.

ACROTROPHIC Adj. (Greek, *akros* = tip; *trophe* = nourishment.) Pertaining to meroistic-type ovarioles that maintain trophic cells within the apical chamber. Syn. Telotrophic Ovariole. See Ovariole. Cf. Polytrophic Ovariole. Rel. Panoistic Ovary.

ACRYDIAN Adj. Pertaining to the qualities, features or characteristics of a grasshopper; grasshopper-like.

ACTA Plural Noun. (Latin, *actum*, *agere* = to do.) Recorded proceedings; transactions. A term applied to some journal titles.

ACTALETIDAE Borner 1902. Plural Noun. A small Family of arthropleonous Collembola assigned to Superfamily Entomobryoidea.

ACTELLIC® See Pirimiphos-Methyl.

ACTELLIFOG® See Pirimiphos-Methyl.

ACTIN Noun. (Greek, *aktis* = ray. Pl., Actins.) A myofibrillar protein with molecular weight about 43,000 and which combines with myosin to form actinomyosin. Cf. Arthrin; Myosin. Rel. Muscle.

-ACTINAL A Latin adjectival suffix meaning 'rayed'.

ACTINE Noun. (Greek, *aktis* = ray. Pl., Actines.) A star-shaped spicule or structure.

ACTINIC RAY Blue, green, ultraviolet light rays demonstrating photochemical effects.

ACTINIFORM Adj. (Greek, *aktis* = ray; Latin, *forma* = shape.) Star-shaped; pertaining to a radiating or star-shaped form; resembling a sea anemone. See Actine. Cf. Actinoid; Actinomorphic; Maniform; Rotuliform. Rel. Form 2; Shape 2; Outline Shape.

ACTINOID Adj. (Greek, *aktis* = ray; *eidos* = shape.) Ray-like; star-like. Descriptive of an organism with radial symmetry. Cf. Actiniform; Actinomorphic. Rel. Eidos; Form; Shape; Symmetry.

ACTINOMERE Noun. (Greek, *aktis* = ray; *meros* = part. Pl., Actinomeres.) A radial arm or segment of a radially symmetrical body.

ACTINOMERIC Adj. (Greek, *aktis* = ray; *meros* = part; *-ic* = of the nature.) Pertaining to an Actinomere.

ACTINOMORPHIC Adj. (Greek, *aktis* = ray; *moprhe* = form; *-ic* = of the nature of.) Radially symmetrical; descriptive of structure or organisms that have similar parts arranged and radiating away from a central axis (*e.g.* starfish). Syn. Actinomorphous. See Actine. Cf. Actinophorous; Zygomorphic. Rel. Symmetry; Form 2; Shape 2; Outline Shape.

ACTINOMORPHY Noun. (Greek, *aktis* = ray; *moprhe* = form. Pl., Actinomorphies.) The state or act of being actinomorphic. See Symmetry. Cf. Zygomorphy.

ACTINOMYSIN D Noun. (Greek, *aktis* = ray; *mykes* = fungus.) An antibiotic produced by *Streptomyces chrysomallus* that prevents the formation of mRNA.

ACTINOPHOROUS Adj. (Greek, *aktis* = ray; *phora* = producing; Latin, *-osus* = possessing the qualities of.) Pertaining to structure with radially-oriented spines or projections. Syn. Actinomorphic. See Actine. Cf. Actinoid. Rel. Form 2; Shape 2; Outline Shape.

ACTINOPILIN Noun. The axial, anisotropic matter of the Actinotrichid Setae.

ACTION LEVEL Economic Entomology: The level or relative abundance of a pest population in an agricultural crop (commodity) at which some control measure should be administered to maintain profitability of the agricultural operation. See Economic Injury Level; Economic Threshold.

ACTIVE CONSTITUENT See Active Ingredient.

ACTIVE INGREDIENT The biologically active component of a pesticide that is responsible for killing or repelling organisms. Syn. Active Constituent. Cf. Diluent; Inert Ingredient; Solvent; Surfactant; Synergist. Rel. Acaracide; Fungicide; Herbicide; Insecticide; Nematicide; Pesticide.

ACTIVE VENTILATION A form of Tracheal Ventilation in which compression of any region of the body elevates internal pressure that subsequently acts on any compressible part of the Respiratory System. Active Ventilation consists of an Expiration Phase and an Inspiration Phase. During the Expiration Phase, compression forces gas (carbon dioxide) from the Tracheal System and out of the body via the open Spiracle. During the Inspiration Phase, gas (oxygen) is drawn into the Respiratory System through abdominal expansion. Abdomen movement is typically through the action of Dorsoventral muscles and may be augmented by telescoping (extension and retraction) of the Abdomen. Patterns of ventilation vary among insects. AV is characteristic of large-bodied and powerful-flying insects, but is not restricted to them. AV involves four mechanistic processes: Abdominal Pumping, Autoventilation, Auxiliary Respiratory Mechanisms and Spiracular Synchronization. See Tracheal Ventilation. Cf. Passive Suction Ventilation. Rel. Respiratory Muscles; Respiratory System; Abdominal Pumping; Autoventilation; Auxiliary Respiratory Mechanisms; Spiracular Synchronization.

ACTIVOL® See Gibberellic Acid.

ACUATE Adj. (Latin, *acus* = needle; *-atus* = adjectival suffix.) Pointed; needle-shaped. Cf. Aciculate. Trans. Verb. To sharpen; to quicken.

ACUDUCTATE Adj. (Latin, *acuate* = needle shaped; *-atus* = adjectival suffix.) Descriptive of a surface with fine scratches as if made with a needle. Alt. Acuducted; Acuductus. See Sculpture Pattern. Cf. Alveolate; Asperites; Baculate; Clavate; Corrugated; Echinate; Favose; Gemmate; Psilate; Punctate; Reticulate; Rugulate; Scabrate; Scobinate; Shagreened; Smooth; Striate; Verrucate; Wrinkle.

ACUITY Noun. (Latin, *acutus* = sharp, pointed; English, *-ity* = suffix forming abstract nouns. Pl., Acuities.) A sharpness of sense or quickness of action (*e.g.* visual acuity).

ACULAGNATHIDAE Plural Noun.

ACULEA Noun. (Latin, *aculeus* = sting, prickle. Pl., Aculeae.) 1. Adult insects: Microtrichia or immobile projections on the wing of some Lepidoptera (particularly Zeugloptera and Dachnonypha). A minute, needle-like spine (spicule) on Lepidoptera wings (*sensu* Tillyard). 2. Immature insects: Minute spines, spicules or cuticular protuberances. Cf. Acanthae. Rel. Sharp.

ACULEATA Latreille 1802. The taxonomic subdivision of Hymenoptera (usually considered a Suborder) that includes bees, wasps and ants. Classification of these insects is based upon the ability of females to sting. See: Hymenoptera. Cf. Apoidea; Chrysidoidea; Formicoidea; Pompiloidea; Sphecoidea; Vespoidea. Rel. Apocrita; Parasitica; Symphyta.

ACULEATE Adj. (Latin, *aculeus* = sting, prickle;

-*atus* = adjectival suffix.) 1. A member of the Aculeata. 2. Bearing a Sting. 3. A structure invested with spines.

ACULEATE-SERRATE Adj. (Latin, *aculeus* = sting, *serra* = saw.) Bearing numerous short points inclined toward one end, as the teeth of a saw. Cf. Biserrate; Dentate-Serrate; Multiserrate; Subserrate; Uniserrate.

ACULEIFORM Adj. (Latin, *aculeus* = sting, prickle; *forma* = shape.) Needle-shaped; descriptive of structure shaped as a short spine. Alt. Aculiform. See Aculeus. Cf. Acanthiform; Penicilliform; Setiform; Spiciform; Spiniform. Rel. Form 2; Shape 2; Outline Shape.

ACULEOLATE Adj. (Latin, *aculeolus* = small needle; -*atus* = adjectival suffix.) Spinose; descriptive of structure invested with small or minute spines. See Aculeus. Cf. Aculeiform.

ACULEUS Noun. (Latin, *aculeus* = sting, prickle. Pl., Aculei.) 1. Hymenoptera: The Sting. A sclerotized, tapered, hollow cylindrical part of the female reproductive system in aculeates. Aculeus is concealed at apex of Abdomen and everted by muscular action. The sting delivers venom in a manner analogous to a hypodermic needle, but is not used for the passage of eggs during oviposition. Structure is incorrectly called a 'stinger'. Cf. Gorgeret. 2. Diptera: A slender, sclerotized, often curved and apically pointed projection in the reproductive system of male Tipulidae. 3. Lepidoptera: Minute, hair-like projections on the wings and other structures beneath the scales of primitive moths. Syn. Microtrichia.

ACULIFORM Adj. (Latin, *aculeus* = sting, prickle; *forma* = shape.) Shaped as a short spine. Syn. Aculeiform.

ACUMINATE Adj. (Latin, *acumen* = point; -*atus* = adjectival suffix.) Pointed; descriptive to structure that tapers to a point. Alt. Acuminated; Acuminatus. See Acanthus. Cf. Acicular; Aculeiform; Cacuminous. Verb. Transitive: To sharpen; to point. Intransitive: To terminate in an acute point.

ACUMINATE EGG Characteristically long, narrow and generally adapted for extrusion from the long Ovipositor of parasitic Hymenoptera that attack insects that form galls or live in galleries or tunnels. This form of egg has been described for some Ichneumonoidea and Chalcidoidea. See Egg.

ACUMINATE SCALE *Kilifia acuminata* (Signoret) [Hemiptera: Coccidae].

ACUPUNCTATE Adj. (Latin, *acutus* = sharp; *punctus* = pointed.) With fine superficial punctures as if made with a needle. Alt. Acupunctatus.

ACUS Noun. (Latin, *acus* = a needle. Pl., Acusus.) Needle. See Aculeus.

ACUTATE Adj. (Latin, *acutus* = sharp; -*atus* = adjectival suffix.) Somewhat pointed.

ACUTE Adj. (Latin, *acutus* = sharpened.) 1. Pertaining to structure that is sharp or pointed apicad. 2. Structure whose margin is angled <90°. Cf. Aciculate; Aciform; Aculeiform.

ACUTE ANGULATE Forming an acute angle.

ACUTE BEE-PARALYSIS VIRUS A viral disease of adult honey bees first noted by Bailey *et al.* (1963). ABP Virus can be transmitted from infected adult workers to larvae. Action resembles Chronic Bee-Paralysis but kills bees more rapidly at lower temperatures (ca 30°C). ABPV is vectored by Varroa Mites. See Honey Bee Viral Disease. Cf. Chronic Bee-Paralysis Virus. Rel. Varroa Mite.

ACUTE-ANGLED FUNGUS BEETLE *Cryptophagus acutangulus* Gyllenhal [Coleoptera: Cryptophagidae].

ACUTENACULUM Noun. (Latin, *acutus* = sharpened; *tenax* = holding. Pl., Acutenacula.) Any structure that holds a needle or needle-like process.

ACUTILINGUAE Plural Noun. Hymenoptera: In older classifications, bees with a short, sharp tongue.

ACUTILINGUAL Adj. (Latin, *acutus* = sharpened; *lingua* = tongue; Latin, -*alis* = having the character of.) Pertaining to a sharp-pointed mouth structure (*e.g.* some bees).

ACZEL, MARTIN LADISLAU (1906–1958) (Hardy 1959, Proc. Entomol. Soc. Wash. 61: 139–140; Milhalyi 1959, Folia Entomol. Hung. 12: 13–16.)

ADACTYL Adj. (Greek, *a* = without; *daktylos* = finger.) An appendage, Ambulacrum or Apotele with Ungues absent or reduced.

ADACYLOUS Adj. (Greek, *a* = without; *daktylos* = finger; Latin, -*osus* = possessing the qualities of.) Pertaining to organisms that lack appendages such as claws.

ADAMETZ, LOTTE (1879–1966) (Zapfe 1966, 69: 11–13.)

ADAMS, CHARLES CHRISTOPHER (1873–1955) American academic. (Calvert 1956, Entomol. News 67: 169–171.)

ADAMS, CHARLES FREDERICK (1877–1950) (Cattell 1949, American Men of Science 1: 8; Anon. 1950, Jefferson City, Missouri Sunday News and Tribune 22 Jan 1950, p. 1.)

ADAMS, FREDERICK CHARLSTROM (–1920) (Austen 1920, Entomol. Mon. Mag. 56: 256.)

ADAMS, HERBERT JORDAN (1838–1912) (Anon. 1912, Entomol. Mon. Mag. 48: 243.)

ADAMSON, ALASTAIR MARTIN (1901–1945) (Anon. 1946, Trop. Agric. Trin. 23: 3–4.)

ADANAL SEGMENT Acarology: The anal segment of the Actinotrichida.

ADANALE Noun. The fourth Axillary Sclerite of the wing (Crampton).

ADANSON, MICHAEL (1727–1806) French botanist and early explorer of Senegal. Developed natural classification of plants and worked in numerical taxonomy. (Fée 1832, Mém. Soc. Sci. Agric. Lille 1: 155–158; Rose 1850, *New General Biographical Dictionary* 1: 97–98.)

ADAPICAL Adj. (Latin, *ad* = to; *apex* = summit.)

Anatomy: Descriptive of a movement or feature positioned toward the apex of an elongate structure or appendage. See Orientation. Cf. Abapical.

ADAPTATION Noun. (Latin, *adaptare* = to fit to; English, *-tion* = result of an action. Pl., Adaptations.) 1. The process by which an organism or a lineage of organisms becomes suited to the environment (climate, habitat). 2. Tissue Culture: Metabolic changes that occur when cells are placed in a new culture medium. Changes may involve induction or repression of specific enzymes. Cf. Coadaptation. Rel. Evolution.

ADAPTIVE Adj. (Latin, *adaptare* = to fit to.) The capacity of fitting to different conditions. Regarded as a positive attribute.

ADAPTIVE OCELLI 1. The grouped simple-eyes or Ocelli of most larvae (Comstock). See Stemmata. 2. Ocellae, Ocellalae or Ocellanae of MacGillivray.

ADAPTIVE RADIATION The rapid evolutionary expansion of plant or animal Taxa within lineages (Genera, Families, Classes, Orders) as manifest by the fossil record displayed within stratigraphic zones during relatively short intervals of geological time. *e.g.* Families of corals during the Jurassic Period. See Geological Time Scale. Cf. Mass Extinction.

ADAXIAL Adj. (Latin, *ad* = to; *axis* = axle; *-alis* = appropriate to.) Located on the side or directed toward the axis. See Orientation. Cf. Abaxial.

ADDENDUM Noun. (Latin, *addere* = to add. Pl., Addenda.) The supplement to a book, monograph or comprehensive published treatment of a subject.

ADDIS, CLARENCE JOHN (1921–1946) (Chandler 1946, J. Parasit. 32: 585–586.)

ADDLED BROOD Honey Bees: A condition descriptive of bee eggs that fail to hatch, larvae that fail to pupate or pupae that fail to develop without infection or apparent cause.

ADDORSAL Adj. (Latin, *ad* = toward; *dorsum* = back; *-alis* = pertaining to.) Descriptive of structure, colour, texture or substance that is adjacent to the midline of the dorsum.

ADDORSAL LINE Lepidoptera larvae: An imaginary longitudinal line, parallel to the primary axis and dorsal of the subdorsal line.

ADDORSED Adj. (Latin, *ad* = toward; *dorsum* = back.) Pertaining to placement of structures back-to-back.

ADDUCE Trans. Verb. (Latin, *adducere* = to lead or bring to.) To advance; to lead; to bring forward. Adduced.

ADDUCENT Adj. (Latin, *adducere* = to lead or bring to.) Pertaining to adduction. See Abducent.

ADDUCIBLE Adj. (Latin, *adducere* = to lead or bring to.) Something that can be adduced. Adduceable.

ADDUCTION Noun. (Latin, *ad* = to; *ducere* = to lead; English, *-tion* = result of an action. Pl., Adductions.) The movement of a structure toward or beyond the imaginary midline or primary axis of a body. The movement of 1 of 2 similar structures toward the other. Cf. Abduction.

ADDUCTOR Noun. (Latin, *adducio* = to bring forward; *or* = an activity. Pl., Adductors.) 1. Something that adducts or brings toward the median line or major axis of a body or structure. 2. Any one of several muscles that move the distal part of an appendage toward the body. Cf. Abductor; Rotator.

ADDUCTOR COXAL MUSCLE Muscles inserted on the Coxa that draw the leg toward the body. Cf. Abductor Muscle.

ADDUCTOR MANDIBULAE Muscles that draw Mandibles together or bring them into apposition. Syn. Mandibular Adductor. Cf. Abductor Mandibulae.

ADDUCTOR MUSCLE Any muscle that induces adduction. A muscle that moves a structure, organ or appendage toward the body or brings body parts into apposition. See Abductor Muscle.

ADECTICOUS Adj. (Greek, *a* = without; *dekticos* = biting; *-osus* = with the qualities of.) Insects in the pupal stage that lack functional Mandibles capable of cutting through the pupal cocoon. Cf. Decticous.

ADECTICOUS PUPA A form of holometabolous development in which Mandibles in the pupal stage are immobile. See Pupa. Cf. Decticous Pupa; Obtect Pupa. Rel. Metamorphosis.

ADELARTHROSOMATA Plural Noun. (Greek, *adel* = concealed, *arthro* = joint, *soma* = body.) The solpugids, harvestmen and book scorpions in some taxonomic classifications.

ADELGIDAE Plural Noun. Adelgids. A numerically small, widespread Family of sternorrhynchous Hemiptera assigned to Superfamily Aphidoidea. Polymorphic generations; four nymphal instars; Antenna of Apterae with three segments, Alatae with five segments, Sexuales with four segments; wings tectiform; hindwing with an oblique vein; wax glands in all stages; Malpighian Tubules absent; Ovipositor present; Cauda not evident. Heteroecy (alternation of generations) present; sexual and parthenogenetic females oviparous. Host plants: Conifers with primary hosts *(Picea)* and secondary hosts (*Abies*, *Pinus*, *etc.*); galls formed on primary hosts; high incidence of host plant Species specificity. Complex life-history requires five generations on two host plants during two years. First instar fundatrices overwinter on *Picea*; second generation nymphal Gallicollae induce galls and as adults migrate to secondary host; third generation overwinter as first instar on secondary host; some fourth generation become alate Sexuparae that fly to primary host and produce Sexulaes; Fundatrices hatch from eggs of these Sexuales. Sister-group of Phylloxeridae. Cf Aphididae.

ADELIDAE Wocke 1871. Plural Noun. A small Family of heteroneurous Lepidoptera assigned to Superfamily Incurvarioidea. Adult small bodied; wingspan 12–15 mm; head with elevated, hair-like scales; compound eye typically large, with Ommatidia of different sizes; male nearly holoptic; Ocelli and Chaetosemata absent; Antenna longest of Lepidoptera (longer than forewing, some five times forewing length); Proboscis long, scales basad; Maxillary Palpus typically long with 4–5 segments; Labial Palpus with three segments; foreleg with Epiphysis; tibial spur formula 0–2–4; forewing Chorda well-developed; apex of Ovipositor slender and nearly pointed. Larva long and cylindrical; head typically prognathous; six Stemmata on each side of head; thoracic legs rather broad; Prolegs on abdominal legs 3–6; anal Prolegs absent; Crochets often multiserial in later instars. Pupa with appendages exarate or weakly obtect; Maxillary Palpus present; Antenna long, surrounding Abdomen. Family predominantly temperate in distribution. Adult nocturnal; female penetrates leaf tissue or flower bud with Ovipositor; first instar larva mines leaf or feeds inside developing flower bud; later instar larvae form portable case and inhabit flowers and leaf litter.

ADELOCEROUS Adj. (Greek, *adel* = concealed; *keras* = horn; Latin, *-osus* = possessing the qualities of.) Pertaining to concealed Antennae. Alt. Adeloceratous. See Cryptocerata.

ADELOMORPHIC Adj. (Greek, *adel* = concealed; *morphe* = form; *-ic* = characterized by.) Pertaining to an indefinite form. Cf. Pleomorphic. Rel. Shape.

ADELPHO- Greek prefix meaning 'brother.'

ADELPHOGAMY Noun. (Greek, *adelphos* = brother; *gamos* = marriage. Pl., Adelphogamies.) A form of polyandry in which sibling males have a common female partner or partners. Males that mate with their sisters. See Polyandry; Sib Mating.

ADELPHOPARASITE Noun. (Greek, *adelphos* = brother; *parasitos* = parasite; *-ites* = inhabitant. Pl., Adelphoparasites.) A larval male that develops as a hyperparasite of a conspecific female larva. Syn. Adelphoparasitoid; Autoparasite. See Hyperparasite. Cf. Alloparasitoid.

ADELPHOPARASITIC Adj. (Greek, *adelphos* = brother; *parasitos* = parasite; *-ic* = of the nature of.) Pertaining to an adelphoparasite. Syn. Autoparasitic. See Hyperparasitic.

ADELPHOPARASITISM Noun. (Greek, *adelphos* = brother; *parasitos* = parasite; English, *-ism* = condition. Pl., Adelphoparasitisms.) A form of hyperparasitism in which the larval stage of a male develops as a hyperparasite of a conspecific female larva. Adelphoparasitism is a phenomenon expressed in some parasitic Hymenoptera (Aphelinidae: *Coccophagoides*). Syn. Autoparasitism. See Parasitism. Cf. Hyperparasitism.

ADELPHOUS Adj. (Greek, *adelphous* = brother; *-osus* = possessing the qualities of.) Descriptive of similar structures that are bound together in bundles. Cf. Aggregate; Conglomerate.

ADEMOSYNIDAE Plural Noun. A Family of fossil archostematan Coleoptera found in Mesozoic deposits.

ADENIA Noun. (Greek, *aden* = gland.) An enlarged gland.

ADENIFORM Adj. (Greek, *aden* = gland; Latin, *forma* = shape.) Gland-shaped; resembling a gland; descriptive of structure that has the three-dimensional shape (form) of a gland. Adeniform glands are, by implication, pendulous or globular. See Gland. Cf. Acervate; Acinose; Adenoid; Botryoidal; Napiform; Ovaliform; Lageniform; Pyriform; Scrotiform. Rel. Form 2; Shape 2; Outline Shape.

ADENO- Greek prefix forming a word descriptive of glands. Alt. Aden-.

ADENOID Adj. (Greek, *aden* = gland; *eidos* = shape.) Gland-like; structure that resembles a gland in shape, substance or texture. See Gland. Cf. Acervate; Acinose; Adeniform. Rel. Eidos; Form; Shape.

ADENOSE Adj. (Greek, *aden* = gland; *-osus* = full of.) Gland-like; Glandular in shape or appearance. Alt. Adenous. See Form. Shape.

ADENOTAXY Noun. (Greek, *aden* = gland; *taxis* = arrangement. Pl., Adenotaxies.) Acarology: Arrangement of the orifices of the glands of the integument.

ADENOTROPHIC VIVIPARITY (Greek, *aden* = gland; *trophe* = nourishment; *vivus* = living; *parere* = to beget.) A method of viviparous reproduction restricted to pupiparous Diptera (Hippoboscidae, Nycteribiidae, Streblidae and *Glossina*). The egg hatches inside the mother and the larva feeds on uterine secretions produced by the mother. Pupation occurs shortly after larviposition. See Viviparity. Cf. Haemocoelic Viviparity; Pseudoplacental Viviparity. Rel. Paedogenesis.

ADENSAMER, THEODOR (1867–1900) (Anon. 1900, Ver. Zool. -Bot. Ges. Wien 50: 579.)

ADEPHAGA Noun. (Greek, *adephagos* = gluttonous, voracious.) (Triassic Period-Recent.) A moderately large, biologically specialized Suborder of Coleoptera containing Carabidae, Cicindelidae, Gyrinidae and Dytiscidae. Adult hind Coxae fused to Metasternum; first abdominal Sternum divided into two sternites; six abdominal Ventrites; pygidial defence glands; polytrophic ovarioles; testes coiled, tubular. Larva with fluid-feeding mouthparts; Labrum fused; legs with 1–2 claws and five segments; Urogomphi usually present. Predominantly predaceous as larvae and adults. Traditionally divided into terrestrial (Geadephaga) and aquatic (Hydradephaga) forms; argument persists regarding the primitive condition. Cf. Polyphaga.

ADEPHAGAN Adj. or Noun. (Greek, *adephagos* = gluttonous, voracious. Pl., Adephagans.) Pertaining to the Adephaga in form, habit or biology. An Adephaga. Alt. Adephagid.

ADEPHAGOUS Adj. (Greek, *adephagos* = gluttonous, voracious; Latin, *-osus* = possessing the qualities of.) Belonging to the Adephaga. See Hydradephagous.

ADERIDAE Plural Noun. A small Family of polyphagous Coleoptera assigned to Superfamily Tenebrionoidea. Adult resembling Anthicidae with narrow Pronotum and lacking lateral Carinae; body elongate, convex, somewhat flattened; head constricted behind eyes; fore Coxae internally open; Ventrites I and II connate or fused. Larva narrow, elongate, parallel-sided. Adults taken on vegetation; larvae found under bark, in rotten wood or leaf litter. Syn. Euglenidae.

ADERMATA Noun. (Greek, *a* = without; *derma* = skin. Pl., Adermatae.) Pupae in which wings and other adult structures are visible through the pupal integument, after casting off the larval integument.

ADFRONTAL AREA Lepidoptera larvae: An oblique sclerite lateral of the Frons that extends from the base of the Antenna to the Epicranial Suture. A median Epicranial Suture is formed by the union of oblique sclerites from each side of the head; an Epicranial Sulcus (notch) is formed when the sclerites do not fuse. Syn. Adfrontal Sclerite.

ADFRONTAL SETA Lepidoptera larvae: A pair of Setae found on adfrontal areas.

ADFRONTAL SUTURE Lepidoptera larvae: The suture formed by the medial fusion of Adfrontal Sclerites. Rel. Suture.

ADHESION Noun. (Latin, *adhaesio, ad* = to; *haerere* = stick; *-ion* = state, result of. Pl., Adhesions.) 1. The act, state or condition of adhering. 2. The condition of substances sticking together when in contact without growing together. Cf. Cohesion.

ADHESIVE CELL Glandular cells used for the purpose of adhesion.

ADHESIVE SETA Setae adapted to climb on highly inclined or smooth surfaces or to adhere to objects. Adhesive Setae on the Maxillae of malachiid beetles capture smooth pollen grains. In some Species, Adhesive Setae are restricted to one sex and may operate in courtship or copulation. Male fleas possess Adhesive Setae on their Antennae, but females of the same Species do not. Male Colorado Potato Beetle has scoop-like Setae on his Tarsomeres that females do not have on their Tarsomeres. These may be Sensilla Spatulatae. The anatomy of AS varies in response to different requirements and different ways of achieving adhesion. AS show at least three shapes: Spatulate Setae (an enlarged or capitate apex); Recurved Setae (curved along apical third); Branched Setae (tines or furcations along the shaft or apex.) Basic components of the AS include a proximal shaft and distal sclerite. Microscopically, the sclerite articulates with the shaft. Molecular cohesion probably contributes significantly to adhesion in many insects. Adhesive Setae are small and many can be placed in a small area. The number of points of contact of Adhesive Setae with the substrata determine the force of adhesion: The greater the number of adhesive Setae per unit area, the stronger the force holding the appendage. AS may adhere to one another or the substrate. This creates a potential problem that has been solved in different ways. Rigid Setae may not come in contact. Alternatively, antimatting devices occur on the dorsal surface of shaft or sclerite. The antimatting devices (setules or nodules) reduce the area of contact between adjacent Setae, thereby preventing unwanted adhesion. Other surface features may prevent adhesion. Adhesive Setae must display an element of flexibility and rigidity. Flexibility of the Seta increases the surface area that contacts the substrata; rigidity enables efficient detachment from the substrata. The two attributes are at opposite ends of a continuum, but their physical expression can be manifest at different sites on one Seta. Some anatomical features of a Seta may reflect adaptations to promote localized flexibility or rigidity. For instance, Setae are hollow. Biomechanically, strength is provided by the cylindrical, hollow Seta. Setae that are flattened or oval in cross-section lose the strength provided in a cylindrical design. To compensate for this loss, a central depression in the distal sclerite may provide strengthening. Strength to resist buckling or compression of the distal sclerite may be provided by a thickened rim. Additional strength to resist buckling or compression is provided to the sclerite or shaft through longitudinal ribs or flutes. Flexibility is expressed through bending, which is usually directional; most Adhesive Setae bend only in one direction. Flattening of structure is a method of promoting flexibility; flattening is typically dorsoventrally oriented and promotes dorsoventral flexibility but not necessarily lateral flexibility. Flexibility is also promoted by reducing Cuticle thickness that permits bending. Transverse cuticular ridges or wrinkles also creates alternatively strong and weak areas that promote bending. See Sensilla; Seta. Cf. Defence Seta; Prey-Capture Seta.

ADI See Acceptable Daily Intake.

ADIABATIC Adj. (Greek, *adiabatos, a* = not; *diabatos* = passable; *-ic* = of the nature of.) Descriptive of an isothermal process or processes in which heat is not lost or gained. Adv. Adiabatically.

ADIANTIFORM Adj. (Greek, *adiantos* = a plant, maidenhair; Latin, *forma* = shape.) Fern-shaped; descriptive of structure resembling the shape or form of a maidenhair fern, *Adiantum*. Cf. Bipectinate; Filiciform. Rel. Form 2; Shape 2; Outline Shape.

ADIAPHANOUS Adj. (Greek, *a* = without, *diaphanos* = to see through; Latin, *-osus* = possessing the qualities of.) Displaying a surface transparency; opaque; not transmitting light (Kirby & Spence). Alt. Adiaphanus.

ADIENCE Noun. (Latin, *adire* = to approach.) Advancing toward a stimulus. Cf. Abience.

ADIENT Adj. (Latin, *adire* = to approach.) Approaching a source of stimulation. Cf. Abient.

ADIHETEROTHRIPIDAE Plural Noun. A numerically small Family of Thysanoptera assigned to the Terebrantia.

ADION® See Permethrin.

ADIPO- (Latin, *adeps* = fat.) A combining form (usually a prefix) that pertains to fat. Alt., Adip-.

ADIPOHAEMOCYTE Noun. (Latin, *adeps* = fat; Greek, *haima* = blood; *kytos* = hollow vessel; container. Pl., Adipohaemocytes.) A basic anatomical form of insect Haemocyte. Adipohaemocyte cells are variable in size and spherical-oval in shape; nucleus is small relative to other Haemocyte, round to slightly elongate and eccentrically positioned; Cytoplasm contains fat droplets, Polyribosomes, Mitochondria and Golgi Bodies. Histochemical analysis shows granules contain PAS-positive substance. Syn. Adipocyte; Adipoleucocyte; Spheroidocyte. See Haemocyte. Cf. Coagulocyte; Granulocyte; Oenocyte; Plasmatocyte; Prohaemocyte; Spherulocyte. Rel. Circulatory System.

ADIPOLEUCOCYTE Noun. (Latin, *adeps* = fat; Greek, *leukos* = white; *kytos* = hollow vessel; container. Pl., Adipoleucocytes.) A form of Haemocyte in which the cytoplasm has become charged with small, oily fat droplets. An adipose cell found in some Heteroptera (Wardle). Cf. Haematocyte. Rel. Circulatory System.

ADIPOSE TISSUE The kind of animal tissue that stores fat. See Fat Body.

ADISCOTA Noun. (Latin, *a* = without; *discus* = disk; *-ota* = noun suffix. Pl., Adiscotae.) Insects that develop into adults without forming imaginal discs. See Discota.

ADITUS Noun (Latin, *aditus* = entrance. Pl., Aditus; Adituses.) 1. Anatomical structure that forms the access to another anatomical part. Cf. Aperture; Mouth; Oriface. 2. A horizontal opening to a burrow that permits ventillation or drainage. Cf. Tunnel. 3. A passage or opening to a burrow or mine. Alt. Entrance.

ADJACENT Adj. (Latin, *adiaceo, adjecere* = to lie by the side of.) Pertaining to structures that maintain their relative position near, close or juxtaposed. Alt. Abutting; Amplectant; Bordering; Contiguous; Touching. Ant. Distant. Rel. Subcontiguous. Adv. Adjacently.

ADJOIN Verb. (Latin, *adjungere*.) Transitive: To join physically; Intransitive: Contiguous or adjacent.

ADJUST® See Chlormequat.

ADJUSTOR NEURON See Interneurons.

ADJUVANT Noun. (Latin, *adjuvans* = to aid; *-antem* = an agent of something. Pl., Adjuvants.) 1. A substance injected with antigens that increases the production and longevity of antibodies. 2. Any ingredient that improves the solubility or handling properties of a pesticide formulation. Adjuvants are sometimes called 'stickers' or 'spreaders'. See Surfactant. Rel. Formulation.

ADKIN, BENAIAH WHITLEY (1865–1948) (Williams 1949, Proc. Roy. Entomol. Soc. London (C) 13: 60.)

ADKIN, ROBERT (1849–1935) (Sheldon 1935, Entomologist 68: 145–147.)

ADLER, SAUL (1895–1966) (Shortt 1967, Biogr. Mem. Fellows Roy. Soc. 13: 1–34.)

ADLERZ, ALEXANDRE (1835–1910) (Perrier 1910, Bull. Mus. Nat. Hist. Nat. Paris 16: 117–118.)

ADMAXILLARY Adj. (Latin, *ad* = to, *maxilla* = jawbone.) Descriptive of structure that is adjacent or connected to the Maxilla.

ADMEDIAL Adj. (Latin, *ad* = to; *medius* = middle; *-alis* = pertaining to, appropriate to.) Pertaining to structure or colour near an imaginary median line. Alt. Admedian.

ADMEDIAL LINES Parallel, longitudinal impressions on the Mesothorax of some aculeate Hymenoptera (wasps). Cf. Notaulix; Parapsidal Suture; Sulcus.

ADMINICULUM Noun. (Latin, *adminiculum* = support. Pl., Adminicula.) 1. A small tooth-like process or Acantha on the dorsal abdominal surface of subterranean pupae. An Adminiculum is used in support or movement for positioning. 2. The elevated or indented lines found in some pupae.

ADMIRAL, JACOB Dutch naturalist and author of a folio on butterflies published in 1740.

ADMIRAL® See Fonofos; Pyriproxyfen.

ADMIRE® See Imidacloprid.

ADNASCENCE Noun. (Latin, *adnascens* from *ad* = to; *nasci* = to be born, to grow to. Pl., Adnascences.) Something adnascent.

ADNASCENT Adj. (Latin, *adnascens* from *ad* = to; *nasci* = to be born, to grow to.) Descriptive of something growing on another structure, object or body.

ADNATE Adj. (Latin, *adnatus, ad* = to; *nasci* = born; *-atus* = adjectival suffix.) 1. Descriptive of structures adjoining. Cf. Juxtaposed. 2. Dissimilar structures adhering or growing together. 3. Structures closely or intimately connected. Ex. Adnate Setae.

ADNATION Noun. (Latin, *ad* = to; *nasci* = born; English, *-tion* = result of an action. Pl., Adnations.) The state of being adnate.

ADNEURAL Adj. (Latin, *ad* = to; Greek, *neuron* = nerve; Latin, *-alis* = pertaining to, appropriate to.) Descriptive of something adjacent to a nerve (*e.g.* an adneural tumour).

ADOLAPIN Noun. A polypeptide toxin derived from the venom of the European honey bee, *Apis mellifera* L., is a strong anti-inflammatory substance and inhibits cyclooxygenase, thus providing analgesic activity. Cf. Apamin.

ADOLPH, CARL EDUARD (1828?–1895) (Papavero 1975, *Essays on the History of Neotropical Dipterology* 2: 322–323. São Paulo.)

ADORAL Adj. (Latin, *ad* = to; *os* = mouth; *-alis* = belonging to, pertaining to.) Near the mouth (*e.g.* adoral setae). Cf. Aboral.

ADOSCULATION Noun. (Latin, *adosculatus* = to kiss; English, *-tion* = result of an action. Pl.,

Adosculations.) The state of being inseminated by external contact and without intromission.

ADPRESS Trans. Verb. (Latin, *ad* = to; *premere* = to press.) To press close; to press against; to lay flat.

ADRECTAL Adj. (Latin, *at* = to; *rectus* = rectum; *-alis* = belonging to, pertaining to.) Descriptive of structure that is adjacent to the Rectum. See Rectum. Rel. Alimentary System.

ADRIAANSE, ALOYS (1908?–1946) (Wricke 1946, Entomol. Ber. Amst. 12: 33.)

ADSORB Trans. Verb. (Latin, *ad* = to; *sorbere* = to suck in.) To condense and hold by adsorption.

ADSORBATE Noun. (Latin, *ad* = to; *sorbere* = to suck in; *-atus* = adjectival suffix. Pl., Adsorbates.) A substance that has been adsorbed. See Adsorption.

ADSORPTION Noun. (Latin, *ad* = to; *sorbere* = to suck in; English, *-tion* = result of an action. Pl., Adsorptions.) Thin-layer molecular adhesion of gas, liquid, or dissolved substance to the surface of a solid body contacting the molecular layer. Cf. Absorption. Adv. Adsorptively.

ADSPERSE Adj. (Latin, *ad* = to; *aspersus* = a sprinkling.) Descriptive of structure marked with closely crowded small spots. Alt. Adspersus. Cf. Punctate.

ADSTERNAL Adj. (Latin, *ad* = to; Greek, *sternon* = chest; Latin, *-alis* = belonging to, pertaining to, having the character of.) Descriptive of structure or colour that is positioned near or adjacent to the Sternum.

ADTERMINAL Adj. (Latin, *ad* = to; *terminus* = end; *-alis* = belonging to, pertaining.) Toward the ends of a muscle or its point of insertion. Cf. Muscle Origin.

ADULT Adj. (Latin, *adultus* = grown; *adolescere* = to grow up.) Descriptive of an organism that has attained full size, stature, strength and sexual maturity, *e.g.* adult insect. Noun. (Latin, *adultus* = grown. Pl., Adults.) 1. An individual (plant or animal) that has achieved physical maturity in terms of size, strength and sex. 2. Insects: The imaginal stage of development. Alt. Imago. Cf. Naiad; Nymph; Larva. Rel. Metamorphosis.

ADULTOID Noun. (Latin, *adultus* = grown; Greek, *eidos* = form. Pl., Adultoids.) An immature individual resembling an adult in behaviour or biology. See Neoteny 2, 3; Paedogenesis. Rel. Eidos; Form; Shape.

ADUMBRANT Adj. (Latin, *adumbro* = to shade in; *-antem* = adjectival suffix.) Producing a faint shadow. Noun. (Latin, *adumbro* = to shade in; *-antem* = an agent of something. Pl., Adumbrants.) A faint shadow.

ADUMBRATE Trans. Verb. (Latin, *ad* = to; *umbrare* = to shade; *-atus* = adjectival suffix.) To produce a faint shadow.

ADUMBRATION Noun. (Latin, *adumbrari* = to shade; English, *-tion* = result of an action. Pl., Adumbrations.) Shade; overshadowing.

ADUMBRATIVE Adj. (Latin, *adumbro* = to shade in.) Tending to adumbrate. Adv. Adumbratively.

ADUNCATE Adj. (Latin, *aduncus* = hooked; *-atus* = adjectival suffix.) Hook-like. Pertaining to a structure curved or bent inward. See Hook. Cf. Ankistroid; Ankyroid; Hamate; Uncinate.

ADUNCITY Noun. (Latin, *aduncitas* = bending, curvature; English, *-ity* = suffix forming abstract nouns. Pl., Aduncities.) A structure that is bent inward.

ADVANCING COLORATION Any colour (*e.g.* yellow) or combination of colours that appears nearer the observer than other colours viewed in the same plane. See Coloration. Cf. Alluring Coloration; Anticryptic Coloration; Apetetic Coloration; Aposematic Coloration; Combination Coloration; Cryptic Coloration; Diffraction Colour; Directive Coloration; Disruptive Coloration; Epigamic Coloration; Episematic Coloration; Procryptic Coloration; Protective Coloration; Pseudepisematic Coloration; Pseudoaposematic Coloration; Seasonal Coloration; Sematic Coloration. Rel. Crypsis; Mimicry.

ADVANTAGE® See Carbosulfan.

ADVECTITIOUS Adj. (Latin, *advecticius* = something brought from a distance; foreign; *-osus* = possessing the qualities of.) Pertaining to passive movement from one place to another. See Dispersal; Migration. Rel. Phoresy.

ADVEHENT Adj. (Latin, *adveho* = to carry; to convey to a place.) Afferent; carrying to an organ, appendage or structure.

ADVENE Intrans. Verb. (Latin, *advenire* = to come to.) To come; to reach.

ADVENIENT Adj. (Latin, *advenire* = to come to.) Pertaining to something influenced by or resulting from external stimuli or causes. Alt. Advential.

ADVENTITIA Noun. (Latin, *adventicius* = extraneous, coming from without, from outside sources. Pl., Adventitiae.) 1. The external investment or outer covering of an organ. 2. The elastic connective tissue surrounding the insect Heart (Folsom & Wardle).

ADVENTITIOUS Adj. (Latin, *adventicius* = extraneous; Latin, *-osus* = possessing the qualities of.) 1. Pertaining to an accidental occurrence. 2. Extraordinary or without apparent reason. 3. Structure found in an unusual place. Alt. Accidental; Casual; Fortuitious; Unplanned.

ADVENTITIOUS VEIN 1. Any of several markings on the wing that appear as nebulous or tubular veins. AVs occur in areas such that they cannot be homologized with conventional venation in existing systems of nomenclature. 2. Secondary wing veins that are not Accessory Veins. Intercalary Veins *sensu* Comstock. See Vein; Wing Venation. Rel. Wing.

ADVENTIVE Adj. (Latin, *adventitius* = extraneous, coming from without.) 1. Plants, animals or microorganisms not originating or not naturally occurring in a geographically defined area in which they are found. For example, the housefly is adventive to many regions of the world. Syn. Introduced. Cf. Endemic; Exotic; Non-Indigenous. 2. Populations of organisms that occur in a specific location, but did not originate there and arrived from elsewhere (= exotic, non-indigenous, endemic.) Two types of adventive organisms: Immigrants and Introduced. Immigrant Organisms arrive without the deliberate, purposeful aid of humans (including actively dispersing organisms); organisms that arrive as stowaways on plants or other commodities moved by commerce. Introduced organisms are brought to a location by the conscious choice of humans (including food crops, ornamental or forage crops, pets and domestic animals and biological control agents) (VanDriesche & Bellows 1996). Noun. (Latin, *adventus* = an arrival. Pl., Adventives.) 1. An immigrant Species. 2. An organism in a new habitat, but not necessarily established.

ADVENTRAL LINE Lepidoptera larva: An imaginary line extending along the venter between the midline and base of the legs.

ADVENTRAL TUBERCLE Lepidoptera larva: Cuticular protuberance on each abdominal segment along the medial base of the leg, and correspondingly set on the apodal segments, constant, number VIII of the abdominal series (Dyar).

ADYE, JAMES MORTIMER (1860?–1895) (Anon. 1895, Entomol. Mon. Mag. 31: 122.)

AEDEAGAL APODEME A sclerotized rod-like structure, presumably of ectodermal origin, in the male Aedeagus. AA supports or manipulates the Aedeagus during exsertion, intromission and copulation.

AEDEAGAL BLADDER An epithelial layer of cells in association with the Ejaculatory Duct of male ants. The Aedeagal Bladder contains a milky substance in mature, sterile and male ants; the function of the substance is unknown.

AEDEAGAL ROTATION The rotation along the central axis by male genitalia in some insects. Rotation is 180° in most brachycerous Diptera and strophandrous Symphyta; Rotation may be 360° in Syrphidae.

AEDEAGUS Noun. (Greek, *aidoia* = genitals; New Latin. Pl., Aedeagi; Aedeaguses.) 1. The male's intromittent organ of copulation. The Aedeagus is typically long, cylindrical, tapered and sclerotized. When not deployed, the Aedeagus is retracted within the Phallobase and surrounded by a membranous sheath (Phallotheca). The distal aperture of the Aedeagus is called the Phallotreme. 2. A term restricted to a muscular intromittent organ found in other groups of animals than insects (Klots). Alt. Aedaegus; Aedoeagus. See Phallus. Cf. Phallobase; Penis.

AEDES Noun. (New Latin.) A numerically large, cosmopolitan Genus of mosquitoes including several Species that are involved in the transmission of Yellow Fever and Dengue to humans. Most *Aedes* Species are considered annoying but do not transmit disease. Cf. *Anopheles*; *Culex*.

AEDOEAGUS See Aedeagus.

AEGERIIDAE Plural Noun. (Latin, *aegeria* = a nymph.) Clearwing Moths. See Sesiidae.

AELUROPODOUS Adj. (Greek, *ailouros* = cat; *pous* = foot; Latin, *-osus* = possessing the qualities of.) Pertaining to animals with retractile claws.

AENEOUS Adj. (Latin, *aeneus* = brazen, made of copper or brass; *-osus* = full of.) Descriptive of structure that is bright brassy or golden green in colour. Term used frequently in taxonomic descriptions to indicate a lustrous appearance. See Bronze. Cf. Verdigris. Rel. Coloration.

AENESCENT Adj. (Latin, *aeneus* = made of copper or brass.) Descriptive of structure in which the colour becomes or appears brassy. See Bronze. Cf. Aeneous. Rel. Coloration.

AENICTOPECHEIDAE Plural Noun. A small Family of heteropterous Hemiptera assigned to Superfamily Enicocephaloidea. Aenictopecheids are predominantly Gondwanan in distribution with an Holarctic element. Adult body small to medium sized and reduviid-like in appearance; Pronotum not divided by transverse Sulci into regions; forewing with Costal fracture; Ovipositor usually present. Species are probably predaceous upon other insects within leaf litter and decaying wood.

AEOLOTHRIPIDAE Plural Noun. (Latin, *Aeolus* = god of the winds.) Banded-winged Thrips; Banded Thrips. A cosmopolitan, primitive Family of Thysanoptera assigned to the Suborder Terebrantia. Aeolothripids are predominantly Holarctic in distribution. Antenna and Palpi elongate; forewing large, conspicuously veined, often banded with marginal fringe often reduced; female with saw-like Ovipositor, apex of Abdomen conical. Many Species are facultatively predaceous.

AEOLOTROPIC Adj. (Greek, *aiolos* = quick moving, changeful of hue; *trepein* = to turn; *-ic* = characterized by.) Substance showing different physical properties with respect to light, heat and compressibility. Cf. Isotropic.

AEOLOTROPISM Noun. (Greek, *aiolos* = changeful; English, *-ism* = condition. Pl., Aeolotropisms.) The condition of being aeolotropic. See Tropism. Cf. Anemotropism; Chemotropism; Electrotropism; Galvanotropism; Geotropism; Heliotropism; Hydrotropism; Phototropism; Rheotropism; Stereotropism; Thermotropism; Thigmotropism; Tonotropism. Rel. Taxis.

AEPOPHILIDAE Plural Noun. A small Family of heteropterous Hemiptera assigned to Superfamily Saldoidea.

AEQUALE Plural Noun. (Latin, *aequalis* = equal.) Equal. (Archaic.) Alt. Aequata.

AEQUILATE Adj. (Latin, *aequalis* = equal; *-atus* = adjectival suffix.) Of equal breadth throughout. (Archaic) Alt. Aequilatus.

AEQUIPALPIA Plural Noun. (New Latin, *aequalis* = equal; *palpus* = palm of the hand.) A Suborder of Trichoptera whose members have an equal number of palpal segments.

AEQUOR Noun. (Latin, *aequo* = calm; *or* = a conditon. Pl., Aequors.) A flat or planar surface. See Shape.

AERAGE Noun. (Latin, *aer* = air; *age* = action. Pl., Aerages.) Ventilation.

AERATE Trans. Verb. (Latin, *aer* = air; *-atus* = adjectival suffix.) To supply air or common gas; to oxygenate.

AERATION Noun. (Latin, *aer* = air; English, *-tion* = result of an action. Pl., Aerations.) The process of aerating.

AERATOR Noun. (Latin, *aer* = air; *-ate* = to become; *or* = one that does something. Pl., Aerators.) An apparatus used to bleach grain and kill grain pests, including insects.

AERENCHYMA Noun. (Greek, *aer* = air; *engchyma* = infusion. Pl., Aerenchymae.) Parenchyma tissue with large intercellular spaces used to store air. Aerenchymae are used as a secondary respiratory tissue by many aquatic plants. See Tissue. Cf. Parenchyma. Rel. Aeropyle.

AERIAL Adj. (Latin, *aer* = air; *-alis* = belonging to, pertaining to, appropriate to.) 1. Pertaining to flying forms of insects. The term is used in comparison with non-winged or otherwise flightless forms. 2. Plant roots used for growth above ground. Adv. Aerially. Noun Aerialness.

AERIAL ROOT Roots of epiphytic plants that do not anchor in soil and often serve in photosynthesis.

AERIAL YELLOWJACKET *Dolichovespula arenaria* (Fabricius) [Hymenoptera: Vespidae]: A common wasp in North America and an occasional urban pest. Colonies typically begin in mid-spring, peak in mid-summer and conclude in autumn. Typically, AY nests are aerial in trees or ornamental vegetation; nest may form 5–6 combs and contain ca 6,000 workers. AY is predatory on insects but may feed on carrion. See Yellowjacket. Cf. Baldfaced Hornet; Common Yellowjacket; German Yellowjacket; Western Yellowjacket.

AERIDUCT Noun. (Latin, *aer* = air; *ducere* = to lead. Pl., Aeriducts.) A respiratory spiracle; specifically, tracheal gill-like structures and tail-like extensions of rat-tailed maggots and some aquatic Heteroptera. Alt. Aeriductus.

AERO- Greek combining form meaning 'air.'

AEROBE Noun. (Greek, *aer* = air; *bios* = life. Pl., Aerobes.) An organism requiring oxygen to sustain life. See Respiration 1. Cf. Anaerobe.

AEROBIC Adj. (Greek, *aer* = air; *bios* = life; *-ic* = characterized by.) Pertaining to organisms living only in the presence of atmospheric air. Term is used in conjunction with microbial life. See Respiration. Cf. Anaerobic.

AERODONETICS Singular Noun. (Greek, *aer* = air; *donekos* = a reed shaken by the wind; *etics* = the study of.) The study of gliding and soaring flight. See Passive Flight.

AERODYNAMIC Adj. (Greek, *aer* = air; *dynamikos* = powerful; *-ic* = of the nature of.) Pertaining to the force of air in motion. Alt. Aerodynamical. Rel. Flight. Noun. (Greek, *aer* = air; *dynamikos* = powerful.) Study of gas motion and the influence of gas motion on bodies moving through gasses or forces of gasses moving past bodies.

AEROPHILE Noun. (Greek, *aer* = air; *philos* = to love. Pl., Aerophiles.) An organism that seeks, uses or associates with wind or air currents. Cf. Aerophobe. Rel. Orientation.

AEROPHILIC Adj. (Greek, *aer* = air; *philos* = to love; *-ic* = of the nature of.) Describing an attraction to or frequent association with air currents. Cf. Aeorphobic. Rel. Orientation.

AEROPHOBE Noun. (Greek, *aer* = air; *phobos* = fearing. Pl., Aerophobes.) An organism that avoids association with wind or air currents. Cf. Aerophile. Rel. Orientation.

AEROPHOBIC Noun. (Greek, *aer* = air; *phobia* = fear; *-ic* = characterized by. Pl., Aerophobics.) A condition or state in which an organism displays an aversion to air currents. Cf. Aerophilic. Rel. Orientation.

AEROPHORE Noun. (Greek, *aer* = air; *phorein* = to carry. Pl., Aerophores.) 1. Lepidoptera larva: Long Setae that may facilitate dispersal via wind. Cf. Anemochore. Alt. Aerophor. 2. A type of Seta with an associated vesicle that contains liquid. Cf. Urticating Hair. Rel. Defence Gland.

AEROPHYTE Noun. (Greek, *aer* = air; *phyton* = plant. Pl., Aerophytes.) An Epiphyte. See Habitat. Cf. Halophyte; Hydrophyte; Psammophyte.

AEROPLANKTON Noun. (Greek, *aer* = air; *plangktos* = wandering. Pl., Aeroplanktons.) Organisms (virus, bacteria, minute insects) and non-living material (dust, silk, spores) that drift passively through the air. Alt. Aerial Plankton. See Plankton. Cf. Aneomoplankton.

AEROPYLE (Greek, *aer* = air; *pylos* = pillar. Pl., Aeropyles.) A modification of eggshell structure that facilitates gas exchange (respiration). Some insects (*e.g.* stick insect *Bacillus rossiusi*) apparently lack an Aeropyle. Most insects have at least one aperture through which gas is exchanged. Aeropyles are exceedingly small and may be overlooked, even when an eggshell is ornate and examined with Scanning Electron Microscope. At least two anatomical forms of Aeropyles occur on insect eggshell. One type forms an Aeroscopic Plate on the egg surface. This occurs in some parasitic Hymenoptera (Encyrtidae) whose egg is immersed in host fluids. Aeroscopic Plate consists of a series of chorionic pillars projecting from egg Pedicel. The pillars are parallel sided and each apically expands to form a head or plate. Air is contained along the surface of the egg between the pillars. Another type of Aeropyle consists of two parts: An aperture in the eggshell and a complex subsurface porous meshwork of Endochorion that holds gas beneath the Exochorion and above the Vitelline Membrane. The latter type of Aeropyle forms the respiratory system of many terrestrial insect eggs. A layer of gas in eggshell Chorion provides oxygen for the developing embryo. The arrangement of apertures of Aeropyle type 2 differs among insects. The apertures may be sparsely scattered over the egg surface, positioned on elevations above the surface or arranged in distinct bands or rings. Aeropyle apertures of the pentatomids *Euschistus* spp. are stalked; when the egg is covered with water (morning dew, rain), respiration can be affected. In such eggs, the stalked Aeropyle functions as a diver's snorkel to permit the exchange of free gas between embryo and atmosphere when separated by a surface of water. In some insects the meshwork fills with gas as the egg dries after being laid (grasshoppers). In other insects, gas replaces liquid in the meshwork while the egg is inside the Median Oviduct. See Eggshell. Cf. Capitulum; Hydropyle; Micropyle; Operculum. Rel. Plastron; Respiration.

AEROSCEPSIS Noun. (New Latin from Greek, *aer* = air; *skepsasthai* = to explore; *-sis* = a condition or state. Pl., Aeroscepses.) 1. The ability of some animals to perceive the quality or contents of air with the aid of specialized sensory receptors. 2. Insects: An hypothetical sense located in the Antenna. Cf. Dielectric Waveguide Theory.

AEROSCEPSY Noun. (New Latin from Greek, *aer* = air; *skepsasthai* = to explore. Pl., Aeroscepsies.) Perception of sound or odour through the medium of the air.

AEROSCOPIC PLATE An anatomical form of Aeropyle that occurs on the surface of some insect eggs (*e.g.* Encyrtidae). See Aeropyle. Rel. Gas Exchange; Respiration.

AEROSE Adj. (Latin, *aerose, aeris* = copper; brass; *-osus* = full of.) Pertaining to colour or quality of brass; brassy. See Coloration. Cf. Brass.

AEROSOL Noun. (Greek, *aer* = air; solution. Pl., Aerosols.) A type of insecticide formulation in which the active ingredient is dissolved in an oil solvent (or liquid carbon dioxide) and propellant. The insectcide is applied (distributed) as a gas-propelled spray into closed or confined spaces. See Fumigant. Cf. Dust; Emulsifiable Concentrate; Microencapsulated Concentrate; Oil Concentrate; Suspension Concentrate; Wettable Powder. Rel. Adjuvant.

AEROSTAT Noun. (Latin, *aer* = air; Latin, *stare* = stand. Pl., Aerostats.) 1. Any air sac within the body of an insect. 2. Diptera: A large air sac at the base of the Abdomen. See Ventilation.

AEROSTATIC Adj. (Greek, *aer* = air; *stare* = stand; *-ic* = characterized by.) Pertaining to aerostatics. Containing air spaces. An insect's ability to maintain balance in air. Alt. Aerostatical. Cf. Hydrostatic. Noun. (Greek, *aer* = air; *statikos* = causing to stand.) The study of equilibrium between gaseous fluids and the solids within them.

AEROTACTIC See Aerotaxis.

AEROTAXIS Noun. (Greek, *aer* = air; *taxis* = arrangement. Pl., Aerotaxes.) 1. The action of air on aerobic and anaerobic microorganisms. See Taxis. Cf. Anemotaxis; Chemotaxis; Geotaxis; Menotaxis; Osmotaxis; Phototaxis; Rheotaxis; Rotaxis; Scototaxis; Strophotaxis; Telotaxis; Thermotaxis; Thigmotaxis; Tonotaxis; Tropotaxis. 2. The arrangement of microorganism toward or away from air. Alt. Aerotactic.

AERUGINOUS Adj. (Latin, *aeruginosus* = covered with verdigris or copper rust; *-osus* = full of.) Pertaining to the colour of rusted copper; verdigris. Alt. Aeruginose; Aeruginus; Eruginous. See Coloration. Cf. Verdigris.

AERUGO Noun. (Latin, *aeruginis* = copper rust. Pl., Aerugoes.) The rust of metal, particularly copper or brass. Cf. Verdigris.

AES Noun. (Latin, *aes* = copper.) 1. Copper; bronze. 2. Things made of copper or brass. See Coloration; Copper.

AESCHYNOMENOUS Adj. (Greek, *aischynein* = to disfigure, or shame; Latin, *-osus* = full of.) Pertaining to the sensitivity displayed by the leaves of some plants.

AESHNIDAE Plural Noun. Darners. A moderate-sized, cosmopolitan Family of anisopterous Odonata assigned to Superfamily Aeshnoidea. Adult large to very large; compound eyes in contact medially; wings long and narrow; Pterostigma narrow and elongate; median basal space sometimes crossed; forewing and hindwing triangles elongate and similar in shape; hindwing Anal Vein forms a conspicuous Anal Loop; Abdomen elongate and slender; female with well-developed Ovipositor. Adults are powerful fliers and can migrate. Naiad elongate, head flattened; Antenna slender, typically 7-segmented; legs short, robust; Tarsi with three segments; Abdomen cylindrical.

AESHNOIDEA A Superfamily of Odonata. Adults variable in size. Male with lateral projections (Auricles) on segment II. Antenodal Crossveins not in line in Costal and Subcostal Spaces. Triangles similar in all wings. Included Families: Aeshnidae, Cordulegastridae, Gomphidae and Petaluridae.

AESTHETIC PEST 1. A pest whose presence is culturally objectionable to humans. 2. A pest that is objectionable to humans due to psychological reasons.

AESTIVAL Adj. (Latin, *aestivus* = of summer; *-alis* = appropriate to, having the character of.) Descriptive of activities occurring during the summer.

AESTIVATION Noun. (Latin, *aestivus* = of summer; English, *-tion* = result of an action. Pl., Aestivations.) 1. Summer Dormancy. 2. A physiological condition or expression of Dormancy. 3. A physical state in which an organism is metabolically inactive or physically dormant during summer or during periods of continued high temperatures in temperate areas or during a dry season in tropical areas (Folsom & Wardle). Alt. Estivation. See Dormancy. Cf. Athermopause;

Hibernation; Overwinter. Rel. Diapause; Oligopause; Quiescence.

AETALIONIDAE Plural Noun. A small Family of auchenorrhynchous Hemiptera assigned to Superfamily Cicadelloidea.

AETIOLOGY The cause(s) or manner of causation of a disease or condition.

-AETUS New Latin noun termination corresponding to *aeto-* in the formation of some zoological names.

AFFERENT Adj. (Latin, *afferre* = to bring.) 1. Physiology: Pertaining to an electrical impulse transmitted from the periphery inward. 2. An impulse toward the center or Central Nervous System. Cf. Efferent.

AFFERENT NERVE See Afferent Neuron.

AFFERENT NEURON A nerve cell that connects a Sensory Neuron with a Motor Neuron and transmits an electrical impulse (stimulus) from a sense organ at the periphery toward a nerve center. Syn. Afferent Nerve; Afferent Neurone. See Nervous System; Nerve Cell. Cf. Efferent Neuron; Interneuron; Motor Neuron. Rel. Central Nervous System.

AFFINITY Noun. (Latin, *affinis* = related, adjacent; English, *-ity* = suffix forming abstract nouns. Pl., Affinities.) 1. Taxonomy: Kinship determined by similarity in structure or development. Typically inferred in higher classification of organisms (members of a Genus, Tribe, Family, *etc.*) with genetical relationship also inferred. Cf. Consanguinity. 2. The relationship between Species depending upon resemblance of structural plan and suggesting a common origin.

AFFLICT Trans. Verb (Latin, *affligere* = to cast down.) To inflict injury, hurt or pain.

AFFRICATE Tran. Verb. (Latin, *affricatus; -atus* = suffix.) To rub.

AFICIDA® See Primicarb.

AFILENE® See Butocarboxim.

AFLAGELLAR Adj. (Greek, *a* = without; Latin, *flagellum* = whip.) 1. An Antenna lacking a Flagellum. 2. A Spermatozoon lacking a tail. Cf. Flagellum.

AFOS® See Mecarbam.

AFOSSA Noun. (Greek, *a* = without; Latin, *fossa* = ditch. Pl., Afossae.) The Antennal fossa or place of attachment between Antenna and head (MacGillivray). Cf. Torulus.

AFRICAN ARMYWORM *Spodoptera exempta* (Walker) [Lepidoptera: Noctuidae]: A major pest of cereals and other graminaceous crops in Africa. See Armyworms. Cf. Beet Armyworm; Fall Armyworm.

AFRICAN BALL-ROLLING DUNG BEETLE *Kheper nigroaeneus* (Boheman) [Coleoptera: Scarabaeidae]: A large bodied, diurnal, ball-rolling dung beetle in warm to hot summer-rainfall areas of southern Africa.

AFRICAN BEE See Africanized Honey Bee.

AFRICAN BOLLWORM *Helicoverpa armigera* (Hübner) [Lepidoptera: Noctuidae]: The most significant pest of cotton in Australia, China, and India. This Species has been reported under numerous other common names, including American Bollworm; Corn Earworm; Cotton Bollworm; Tobacco Budworm; Tomato Grub. See Noctuidae. Cf. Cotton Bollworm.

AFRICAN HORSE-SICKNESS An arboviral disease endemic to Africa, but now widespread in the Middle East and Spain. AHS is caused by an arbovirus assigned to Reoviridae. AHS is a highly virulent disease of horses, mules and donkeys, but zebras are resistant. AHS virus affects the vascular endothelium. Vectors implicated in transmission of AHS include *Culicoides* and ticks. Acronym AHS. See Reoviridae. Cf. Bluetongue Disease; Colorado Tick Fever; Epizootic Haemorrhagic Disease.

AFRICAN MOLE-CRICKET *Gryllotalpa africana* Palisot de Beauvois [Orthoptera: Gryllotalpidae]: A minor pest of lawns that occasionally enters homes and human habitations in some areas. AMC presumably is endemic to Africa and occurs in southern Asia, islands of Pacific and Japan; AMC repeatedly is intercepted by USA quarantine. Records of AMC in Australia refer to *G. monanka* Otte & Alexander, a Species presumably endemic to Australia. Adult 35–40 mm long, body cylindrical, brown to dark brown in colour; Ocelli small and circular in outline shape; Antenna about as long as head and Prothorax combined; Pronotum very large with Mesonotum and Metanotum very small; fore leg fossorial; fore Tibia with four dactyls (digging spurs); hind Femur longer than Pronotum; base of fore Femur with a short, bladeless process; apex of hind Tibia with seven spines and dorsomedial margin with five spines; forewing very short and hardened; hindwing large and folded at rest with apical part rolled and projecting to apex of Abdomen; Cerci multisegmented; less than half as long as Abdomen. Males stridulate to attract females; mate when one week old; female lays ca 100 eggs in loose clusters of 15–75 eggs in egg chambers ca 4 cm diam and 3–30 cm deep in soil. Eggs oval, ca 3 mm long initially white, then yellow, finally brown; eclosion ca 2–4 weeks. Nymphal stage 6–12 months; 10 nymphal instars. AMCs typically burrow into the ground and feed on grass roots, but prefer insects, earthworms and spiders; cannibalistic; development clearly fastest on carnivorous diet. See Gryllotalpidae; Mole Crickets. Cf. Dark-night Mole Cricket; European Mole Cricket.

AFRICAN SLEEPING SICKNESS See Human Trypanosomiasis.

AFRICAN SWINE FEVER An arboviral disease of uncertain taxonomic placement. Viron an icosahedron ca 200 nm diam and covered with lipoprotein coat. ASF highly virulent and highly contagious in domestic pigs with nearly 100% mortality; disease produced no clinical signs in wild porcines. ASF is enzootic in Africa and established in Spain and Portugal; ASF spread elsewhere through commerce but has been eradicated. Disease is naturally vectored by the argasid tick *Ornithodoros moubata* and clinically is transmitted by other Species of *Ornithodoros.* See Arbovirus.

AFRICAN-VIOLET MEALYBUG *Rhizoecus dianthi* Green [Hemiptera: Pseudococcidae].

AFRICANIZED HONEY BEE *Apis mellifera scutellata* (Lepetelier) (Hymenoptera: Apidae). AHB is similar in appearance and size to European Honey Bee, but workers are much more defensive of their hives. Counter to popular notion, AHB are not behaviourally more aggressive than European Honey Bees. Alt. African Bee; Killer Bee. See Apidae; European Honey Bee; Eastern Honey Bee. (See *Killer Bees: The Africanized Honey Bee in the Americas*. Mark Winston 1992, 162 pp. Harvard Univ. Press, Cambridge, Mass.)

AFROTROPICAL REGION Adj. A major zoogeographic area of the world delimited to the north by the Sahara Desert and extending south to the Cape of Good Hope. Syn. Ethiopian Region.

AFTERBODY Noun. (Greek, *aptoero* = further off; English = *body*. Pl., Afterbodies.) Coleoptera: The body posterior of the Pronotum (Tillyard). Alt. After-body.

AFTERFORM Noun. (Greek, *aptoero* = further off; Latin, *forma* = form. Pl., Afterforms.) A secondary or subsequent shape assumed by an organ or organism. See Form 2; Shape 2; Outline Shape.

AFTERNOSE Noun. (Greek, *aptoero* = further off; -*osus* = full of. Pl., Afternoses.) A triangular sclerite below the Antenna and above the Clypeus. See Postclypeus.

AFUNCTION Noun. (Latin, *a* = without; *funcion* from *fungi* = to perform; English, *-tion* = result of an action. Pl., Afunctions.) 1. A loss of function by a structure, process or mechanism in an organism during its lifetime. See Disfunctional. Rel. Senescence. 2. A loss of function by a structure in a lineage during evolutionary time. Rel. Atrophy.

AFUNCTIONAL Adj. (Latin, *a* = without; *fungi* = to perform; *-alis* = belonging to, pertaining to, appropriate to, having the character of.) Pertaining to structure that has apparently lost its function. See Function. Cf. Disfunctional.

AGA SOLUTION Alcohol-Glycerin-Acetic Acid. A preservative consisting of ethanol (ethyl alcohol) 8 parts; water 5 parts; Glycerin 1 part; Glacial Acetic Acid 1 part.

AGAMIC Adj. (Greek, *a* = without; *gamos* = marriage; *-ic* = characterized by.) Pertaining to parthenogenetic development. Asexually reproducing. Alt. Agamous. Adv. Agamically. Rel. Parthenogenesis.

AGAMOGENESIS Noun. (Greek, *a* = without; *gamos* = marriage; *genesis* = descent. Pl., Agamogeneses.) Any form of reproduction without fertilization by a male. See Gamogenesis; Parthenogenesis.

AGAMOGENETIC Adj. (Greek, *a* = without; *gamos* = marriage; *genesis* = descent; *-ic* = characterized by.) Asexual; produced asexually.

AGAMOGONY Noun. (Greek, *a* = without; *gamos* = marriage; *gonos* = generation. Pl., Agamogonies.) Any form of reproduction without the sexual process.

AGAMOSPECIES Plural and Singular Noun. (Greek, *a* = without; *gamos* = marriage; Latin, *species* = particular kind.) A Species that reproduces non-sexually; a parthenogenetic Species. (Cain 1953, Evolution 7: 76–83.) See Species. Cf. Morphospecies. Rel. Arrhenotoky; Heterogony; Thelytoky.

AGANAIDAE Plural Noun. A small Family (ca 100 Species) of ditrysian Lepidoptera assigned to Superfamily Noctuoidea that occurs within Ethiopian and Indoaustralian Realms; Aganaids are placed within Hypsidae of some classifications. Adult medium to large sized; wingspan 25–75 mm; head with short, lamellar scales; Ocelli present; Chaetosemata absent; Antenna bipectinate or ciliate in male; filiform or bipectinate in females; Proboscis well developed; Maxillary Palpus with one minute segment; Labial Palpus large, upturned. Thorax with smooth scales; Microtymbals absent from metathoracic Episternum; metathoracic Tympanal Organs present; fore Tibia with Epiphysis; tibial spur formula 0-2-4. Egg hemispherical or subglobular; Chorion with microscopic sculpture; eggs laid individually or in clusters. Mature larva with head hypognathous but lacking secondary Setae; Prolegs on abdominal segments 3–6, 10. Pupa well sclerotized; Maxillary Palpus absent; Cremaster absent. Larvae and adults of some Species aposematically coloured; larvae feed upon Apocynaceae and Moraceae. Adults of most Species are nocturnal with wings held flat over Abdomen at repose.

AGAONIDAE Plural Noun. Fig Wasps. A moderate sized, pantropical Family (ca 40 Genera, 360 Species) of apocritous Hymenoptera assigned to Superfamily Chalcidoidea. Agaonids are difficult to characterized, but typically display a small body and profound sexual dimorphism. Female dark coloured, head prognathous; Mandibles often with parallel grooves or ridges but lack apical teeth. Corbiculae occur on legs; Ovipositor elongate and exserted. Male typically pale coloured, without metallic coloration, eyeless; apterous; Gaster elongate or tubular. All Species are phytophagous within the receptacles of *Ficus* spp. Individual female wasps penetrate a fig's ostiole and pollinate flowers of figs within which wasp progeny develop. Agaonids are sometimes regarded as Subfamily of Torymidae.

AGAPETIDAE Plural Noun. (New Latin, from Greek, *agapetos* = loveable.) See Satyridae.

AGAR Noun. (Malay, *agar-agar* = a seaweed. Pl., Agars.) Any of several non-nutritional gelatinous culturing media that utilize agar-agar as a solidifying agent. Specifically, a sulfuric-acid ester of a complex galactose polysaccharide extracted from several Species of *Gelidium* and other algae. Agar is relatively expensive in artificial diets for mass-reared insects and substitutes such as algin or carrageenan may be used as gelling agent. See Algin; Artificial Diet; Carrageenan.

AGARIC GNAT Fungus gnats. See Mycetophilidae.

AGARIC Adj. (Latin, *agaricum* = fungus; *-ic* = of the nature of.) Fungus-like. Pertaining to fungi of the Family Agaricaceae, particularly the Genera *Agaricus* and *Fomes*.

AGARICIFORM Adj. (Latin, *agaricum* = fungus; *forma* = shape.) Mushroom-shaped; descriptive of structure with the form of an Agaric. Specifically, an umbrella-like cap attached to an elongate, cylindrical stem. See Agaricoid. Cf. Adeniform; Pileiform; Pyriform. Rel. Form 2; Shape 2; Outline Shape.

AGARICOID Adj. (Latin, *agaricum* = fungus; Greek, *eidos* = form.) 1. Mushroom-like; descriptive of structure that is mushroom-like in function or three-dimensional form. 2. Pertaining to a structure or substance that has the properties of an Agaric. See Agaric. Cf. Agariciform; Fungoid; Rel. Eidos; Form; Shape.

AGARISTIDAE Plural Noun. Forster Moths. A small Family of noctuoid moths that is sometimes placed as a Subfamily of Noctuidae. Agaristids are predominantly diurnal. Adult body brightly coloured; Ocelli present; Antenna dilated apicad; tympanal hood small or absent. Larvae typically are aposematically coloured, often with hump on posterior end of body. Pupa heavily sclerotized and concealed within cocoon.

AGASSIZ, ALEXANDRE (1835–1910) American academic (Harvard University), naturalist and businessman. Director of the Museum of Comparative Zoology (1874–1885) with appointment upon death of his father, J. L. R. Agassiz. (Gardiner 1910, Proc. Linn. Soc. London 1910: 83–86.)

AGASSIZ, JEAN LOUIS RODOLPHE (1807–1873) Swiss born; educated at Laussane, Zurich, Heidelberg and Munich; worked with Cuvier 1830–1832; Professor of Natural History at Neuchatel (1832–1846) where he published *Recherches sur les Poissons Fossiles* (5 vols, 1833–1842) and *Nomenclator Zoologicus Index* (1842–1846). Agassiz emigrated to America during 1846 and was appointed Chair of Natural History in Lawrence Scientific School (Harvard University, 1848), naturalist and geologist. Agassize became a founding member of the Museum of Comparative Zoology at Harvard. He was a prolific writer on many topics but probably best remembered among biologists for his opposition to Darwinian Evoution (Evolution and Permanence of Type, 1873). (Anon. 1874, Amer. Nat. 8: 62–63; Bliss 1874, Pop. Sci. Mon. 4: 608–618; Mareon 1972, *Life, Letters and Works of L. Agassiz.* 318 pp.)

AGASTOPARASITISM Noun. (Greek, *a* = not; *gastro* = stomach; *parasitos*; *-ism* = condition. Pl., Agastoparasitisms.) Parasitism between closely related Species (Ronquist 1994, Evolution 48: 241–266.) See Parasitism.

AGATHIPHAGIDAE Plural Noun. Kauri Moths. A numerically small, primitive Family of monotrysian Lepidoptera assigned to Suborder Aglossata; sometimes placed in Superfamily Agathiphagoidea. Species occur in Fiji, Solomon Islands, New Caledonia and Australia. Adult superficially resembles caddisflies; body small; head elongate, dorsoventrally compressed with hair-like scales; Ocelli and Chaetosemata absent; Antenna filiform; Mandibles large, dicondylic with well developed musculature; Maxilla with Lacinea, Galea prominent; Maxillary Palpus with five segments, folded; Labial Palpus with three segments; Paraglossa absent. Epiphysis present, tibial spur formula 1-4-4; wing venation homoneurous, Sc forked to form Sc1 and Sc2; M4 present. Ovipositor extensible, nonpiercing. Larva apodous, stout, body wider than head; head hypognathous; Stemmata vestigial; Antenna of one segment; Spinneret absent. Pupa decticous, exarate; Mandibles large, asymmetrical; pharate adult uses Mandibles to cut exit hole. Adult nocturnal; wings held tectiform over Abdomen at repose; larva hollows and feeds on seeds of *Agathis* (Araucariaceae); pupation within excavated seed; pupa may remain in cone for three years.

AGATHO- Greek combining form meaning 'good.' Alt. Agath-.

AGE DISTRIBUTION Ecology: 1. The distribution of age classes within a population at a particular time. 2. The proportion of individuals of an age class present in a population at any given time. A population exhibiting exponential growth is characterized as having a stable age distribution; a population that is not changing in size is characterized as having a stationary age distribution.

AGE POLYETHISM Social Insects. A division of labour within a colony involving one individual whose forms of specialization or social responsibilities change with age. See Polytheism. Cf. Caste Polytheism. Rel. Social.

AGELENIDAE Plural Noun. Platform Spiders; Funnel-Web Spiders. Ecribellate spiders with three claws on each leg, Spinnerets close-set with tracheal spiracle adjacent, tarsal Trichobothria linearly arranged.

AGENNESIS Noun. (Greek, *a* = not; *gennesis* = engendering. Pl., Agenneses.) 1. Failure to develop. 2. Sterility.

AGERATUM BLUE (Greek, *ageraton* = a plant.) A colour of low saturation and high brilliance.

AGEUSIA Noun. (Greek, *a* = not; *geusia* = taste.

Pl., Ageusiae.) An inability to taste. Alt. Ageustia

AGGER Noun. (Latin, *aggerare* = to heap up. Pl., Aggers.) An elevation formed above a flat surface from material moved to a site for purposes such as a road, dam, bulwark, mound or wall.

AGGEROSE Adj. (Latin, *aggerare* = to heap up; *-osus* = full of.) Pertaining to material in heaps.

AGGLOMERATE Adj. (Latin, *ad* = to; *glomerare* = to form into a ball; *-atus* = adjectival suffix.) Clustered; pertaining to material collected into a ball, mass or heap. Alt. Agglomeratic. Noun. (Latin, *ad* = to; *glomerare* = to form into a ball; *-atus* = suffix. Pl., Agglomerates.) A ball, mass or heap. Syn. Conglomerate; Conglomeration. Verb, Agglomerate. (-ated, ating).

AGGLOMERATION Noun. (Latin, *agglomerare*; English, *-tion* = result of an action. Pl., Agglomerations.) The act of agglomerating.

AGGLOMERATIVE Adj. (Latin, *agglomeratus* = to heap up, join.) Tending to agglomerate.

AGGLUTINANT Adj. (Latin, *agglutinare* = to glue to; *-antem* = adjectival suffix.) 1. Uniting as glue; tending to cause adhesion. Noun. (Latin, *ad* = to; *glutinare* = to glue; *-antem* = an agent of something. Pl., Agglutinants.) Substance that causes particles to adhere.

AGGLUTINATE Adj. (Latin, *agglutinare* = to glue to a thing; *-atus* = adjectival suffix.) Particles bound, stuck or glued together and thereby fused into a mass. Trans. Verb. Agglutinate. (-nated; -nating).

AGGLUTINATION Noun. (Latin, *ad* = to; *glutinare* = to glue; English, *-tion* = result of an action. Pl., Agglutinations.) 1. The act or process of agglutinating. 2. Object composed of agglutinated parts. 3. A process involving Spermatozoa adhering to one another. Typically, head-to-head contact but other types of agglutination are known. The process can be reversed in some vertebrates, so aggultination probably does not involve gross structural changes in Spermatozoa, but perhaps changes in their membranes. Cf. Sperm Activation.

AGGLUTINATIVE Adj. (Latin, *ad* = to; *glutinare* = to glue.) Adhesive.

AGGREGATE Adj. (Latin, *ad* = to; *gregare* = to herd; *-atus* = adjectival suffix.) 1. Descriptive of structure formed by a collection of related objects and shaped into a mass (*e.g.* aggregate gland). 2. Clustered. 3. Descriptive of small to microscopic colonial organisms that are densely clustered. Adv. Aggregately. Verb, Aggregate (-gated; gating). Noun, Aggregateness. See Aggregation. Cf. Adelphous; Conglomerate.

AGGREGATION Noun. (Latin, *ad* = to; *gregare* = to collect; English, *-tion* = result of an action. Pl., Aggregations.) 1. A mass or composite of smaller objects bound together. 2. Eusocial Insects: A group of individuals of one Species gathered in one place or small area but the individuals of which do not cooperatively construct nests or cooperatively rear offspring. See Communal; Eusocial; Highly Social; Parasocial; Primitively Social Quasisocial; Subsocial. Cf. Isolation; Colony; Social.

AGGREGATION PHEROMONE Chemical compounds synthesized and released by members of a Species and which attract members of the same sex or same Species to the source of the pheromone. Aggregation pheromones have been reported in several Orders and many Species of insects. APs may be responsible for forming Leks, attracting individuals to feeding sites or colony defence.

AGGREGATIVE Adj. (Latin, *ad* = to; *gregare* = to collect.) Collective.

AGGRESSION Noun. (Latin, *aggredi* = to attack; English, *-tion* = result of an action. Pl., Aggressions.) 1. Unprovoked attack. 2. Social Insects: A physical act, threat or action by an individual, group of individuals or colony against other individuals, groups or colonies. Rel. War.

AGGRESSIVE Adj. (Latin, *aggressio* = to attack.) Characterized by aggression. Adv. Aggressively. Noun Aggressiveness.

AGGRESSIVE MIMICRY A type of mimicry in which predaceous Species mimic their prey to facilitate the capture of prey. See Mimicry. Rel. Alluring Coloration.

AGGRESSOR Noun. (Latin, *aggressio* = to attack; *or* = one who does something. Pl., Aggressors.) An individual who initiates aggression.

AGILE Adj. (Latin, *agere* = to move, to act.) Ability to move quickly with dexterity.

AGILITY Noun. (Latin, *agere* = to move, to act; English, *-ity* = suffix forming abstract nouns. Pl., Agilities.) The act or state of being agile. Alt. Dexterity.

AGITATION Noun. (Latin, *agitare* = to put into motion; English, *-tion* = result of an action. Pl., Agitations.) 1. An act of causing disturbance. 2. The state of being agitated. 3. A phenomenon of collective excitement, disturbance or aggressive action within a population. Alt. Disturbance; Excitement. Verb. Agitate, -tated, -tating. Adv. Agitatedly. Adj. Agitational.

AGITATOR Noun. (Latin, *agitare* = to put into motion; *or* = one who does something. Pl., Agitators.) An individual who causes disturbance or disruption in the actions or behaviour of another individual or group of individuals. Alt. Activists.

AGLOSSATA Noun. (Greek, *a* = without; *glossa* = tongue. Pl., Aglossatae.) A Suborder of primitive Lepidoptera including the Agathiphagidae.

AGLOSSATE Adj. (Greek, *a* = without; *glossa* = tongue; *-atus* = adjectival suffix.) Tongueless; descriptive of groups of Lepidtopera or bees that lack a 'tongue.'

AGLYCYDERIDAE Plural Noun. A small Family of polyphagous Coleoptera assigned to Superfamily Curculionoidea.

AGNATHOUS Adj. (Greek, *a* = without; *gnathos* = jaw; Latin, *-osus* = possessing the qualities of.) Pertaining to organisms that lack Mandibles (*e.g.* Neuropteroids that lack mouth structures.) See

Mouthparts. Cf. -gnathous. Rel. Mandible.

AGNOMICAL Adj. (Greek, *a* = without; *gnome* = thought, intelligence; *-alis* = appropriate to, having the character of.) Not expressing purpose or design.

AGOMPHIOUS Adj. (Greek, *a* = without; *gomphios* = molar; Latin, *-osus* = possessing the qualities of.) Toothless.

AGONISTIC BEHAVIOUR Exhibition of combative displays between conspecific or non-conspecific individuals. These displays may be exhibited under the context of reproduction, territoriality, defence, or retainment of a required resource and may involve non-contact displays of the body or its parts or actual contact between individuals.

AGONOXENIDAE Plural Noun. Nut Borers; Fruit Borers. A very small Family of ditrysian Lepidoptera assigned to Superfamily Gelechioidea and apparently related to Elachistidae. Agonoxenids include Species in Indonesia, tropical Australia, New Guinea and islands of Pacific. Adult small, pale coloured with reddish-brown longitudinal marks on forewing; wingspan 9–15 mm; head with smooth scales; Ocelli and Chaetosemata absent; Antenna filiform, nearly as long as forewing, Scape lacking Pecten; Proboscis small, densely scaled basad; Maxillary Palpus minute, of one segment; Labial Palpus long, recurved; Epiphysis present; forewing lanceolate, without Pterostigma; female with Frenulum of two bristles. Larva forms weak web and feeds beneath it; host plant include Species of Arecaceae with larva feeding on fronds; pupation occurs within a double-walled silk cocoon. Pupa flattened with Antenna projecting beyond apex of wings. Adult runs rapidly; at rest the Antenna projects forward and wings are compact against Abdomen.

AGRARIAN Adj. (Latin, *agrarius* from *ager* = field.) 1. Of or pertaining to Agriculture. 2. Descriptive of crops grown in fields or in lands organized for the purpose of crop production. 3. Botany: Descriptive of plants that grow wild in fields; campestral. Cf. Agronomic.

AGREE® A biopesticide used to control lepidopterous pests attacking sugarbeets, herbs, spices, peanuts and a variety of vegetables. Active ingredient *Bacillus thuringiensis* var. *aizawai* strain GC-91. See *Bacillus thuringiensis*. Cf. Javelin®; Teknar®. Rel. Biopesticide.

AGREN, HUGO (1878–1956) (Lindroth 1957, Opusc. Entomol. 22: 49–50.)

AGRICERE Noun. (Latin, *ager* = field; *cere* = articles made of wax. Pl., Agriceres.) Particles of soil enveloped in wax or resin.

AGRICOLA, GEORGIUS (1494–1555) German physician, naturalist, geologist, scientist and academic. True name Georg Bauer. (Eiselt 1836, *Geschichte, Systematik und Literatur der Insektenkunde.* 255 pp., Leipzig.)

AGRICOLOUS Adj. (Latin, *agricola* = a tiller of the fields, farmer; *colere* = living in or on.) Descriptive of or pertaining to agricultural habitats. See Agriculture; Habitat. Cf. Caespiticolous; Deserticolous; Ericeticolous; Gallicolous; Lignicolous; Madicolous; Nemoricolous; Pratinicolous; Silvicolous. Rel. Ecology.

AGRICON-G® See Gossyplure.

AGRICULTURAL ANT *Pogonmyrmex barbatus.* [Hymenoptera: Formicidae].

AGRICULTURAL ENTOMOLOGY The subdiscipline of Entomology concerned with the study of insects affecting horticultural plants, timber and crops. See Entomology. Cf. Applied Entomology; Economic Entomology. Rel. Agronomy.

AGRIMEK® See Avermectin.

AGRIO- Greek combining form meaning 'wild.'

AGRIONIDAE Plural Noun. A Family of Odonata.

AGRIOPOCORINAE Noun. A Subfamily of Coreidae.

AGRIOTYPIDAE Plural Noun. A monotypic Family of parasitic Hymenoptera sometimes placed in the Ichneumonidae. Adult with an aculeate-like Gaster. Larva acts as an external parasite of prepupae and pupae of caddisflies in Family Goeridae. Female wasp walks underwater and holds a plastron for respiration that enables her to remain submerged for half an hour. Larva remains inside caddis case and produces a long silken filament that holds a plastron for respiration. See Ichneumonoidea.

AGRITOL® A registered biopesticide derived from *Bacillus thuringiensis* var. kurstaki. See *Bacillus thuringiensis.*

AGROECOSYSTEM Noun. (Latin, *ager* = a field; Greek, *oikos* = dwelling; + system. Pl., Agroecosystems.) An ecosystem created by humans that includes cultivated plants and domesticated animals associated with organized agricultural and whose products and efforts are for human benefit. See Ecosystem. Cf. Urban Ecosystem. Rel. Habitat.

AGROMYZIDAE Plural Noun. Leafminer Flies. A cosmopolitan Family of economically important acalypterate Diptera. Body small. Larvae form and feed within stem or leaf mines or galls. Syn. Phytomyzidae.

AGRONOMY Noun. (Greek, *agronomos* = rural. Pl., Agronomies.) The subdiscipline of agriculture that concerns the theory and practise of crop production and soil management. See Agriculture. Cf. Horticulture.

AGRONOMIST Noun. (Greek, *agronomos* = overseer of land. Pl., Agronomists.) A person concerned with Agronomy. Cf. Botanist; Entomologist.

AGROTHRIN® See Cypermethrin.

AGYRTIDAE Plural Noun. A small Family of polyphagous Coleoptera assigned to Superfamily Staphylinoidea.

AHENEUS Adj. (Greek, *ahenius* = look back.) Bright brassy or golden-green.

AHLBERG, OLOF (1893–1967) (Lindroth 1967,

Opusc. Entomol. 32: 182–183.)

AHLWARDT, KARL (1873–1915) (Schumacher 1915, Deut. Entomol. Zeitsch. 1915: 222–224.)

AHMAD, IMTIAZ (1941–) Pakistani academic and taxonomic specialist of Hemiptera.

AHRENS, AUGUST (1779–1841) (Germar 1842, Stettin. Entomol. Ztg. 3: 45–48.)

AHS See African Horse Sickness.

AHUATLE Noun. (Nahuatl, *ahuautli.*) Ephydrid eggs used as food by indians in Mexico.

AI Acronym for 'active ingredient,' pertaining to the chemical in pesticides responsible for killing or repelling organisms. See Active Ingredient. Cf. Inert Ingredient; Synergist. Rel. Pesticide.

AICHHORN, SIGMUND JOHANN NEPOMUK (1814–1892) (Rumpf 1892, Mitt. naturw. Ver. Steierm. 29: 246–261.)

AIGNER-ABAFI, LUDWIG VON (1840–1909) (Anon. 1909, Entomol. Rdsch. 26: 86; Csiki 1910, Rovart. Lap. 17: 34–37.)

AIKEN, ARTHUR (1773–1854) (Lisney 1960, *Bibliography of British Lepidoptera 1608–1799*, 315 pp. London.)

AIKEN, JOHN (1747–1822) (Linsey 1960, *Bibliography of British Lepidoptera 1608– 1799.* 315 pp, London.)

AIKINS, JOHN SAM (1929–1966) (Kennedy 1968, Proc. Roy. Entomol. Soc. Lond. (C) 32: 58.)

AILANTHUS WEBWORM *Atteva punctella* (Cramer) [Lepidoptera: Yponomeutidae].

AILE Adj. (Latin, *ala* = wing.) Winged; pertaining to wings.

AILERON Noun. (French diminutive, *aile.* Pl., Ailerons.) A scale-like sclerite that covers the forewing base in some insects. Diptera: Alula and Squama. See Tegula.

AIM® See Chlorflurazuron.

AIMOCRON® See Monocrotophos.

AIMSAN® See Phenthoate.

AIMTHANE® See Acephate.

AINO VIRUS An arbovirus assigned to the Bunyaviridae and which is known to occur in Australia and Japan. Aino Virus affects cattle, sheep and goats. See Arbovirus; Bunyaviridae. Cf. Akabane Virus.

AINSLEE, GEORGE GOODING (1886–1939) (Osborn 1952, A *Brief History of Entomology.* 305 pp., Columbus, Ohio.)

AINSLIE, CHARLES NICHOLAS (1856–1939) (Mickel 1940, Ann. Entomol. Soc. Amer. 33: 215–216.)

AIR SACS Pouch-like expansions of tracheal tubes in some winged insects (*e.g.* bumblebees). Air sacs vary in size, number and position among Species. AS typically lack Taenidia and are capable of inflation; inflation is supposed to lessen specific gravity, aids in flight, and probably serves in thermoregulation. Syn. Air Bladder; Air Vesicles.

AIR TUBE See Respiratory Siphon.

AIRFOIL Noun. (Latin, *aer* = air; *folium* = leaf. Pl., Airfoils.) A flat or curved surface (wings, in the case of insects) that obtains an aerodynamic reaction to the air through which the surface moves. Cf. Drag; Lift; Thrust.

AITKEN, EDWARD HAMILTON (1851–1909) (Anon. 1909, J. Bombay Nat. Hist. Soc. 19: 540–543.)

AIZENBERG, E E (1899–1967) (Narzikulov 1968, Izv. Akad. Nauk Tadzhik SSR 4: 99–100.)

AKABANE VIRUS An arbovirus of the Bunyaviridae known to occur in Australia, Korea, Africa, Israel and Japan. AV causes calving losses in cattle. AV vectored in Australia by biting midge *Culicoides brevitarsus;* female midge feeds on cow, then oviposits in dung; eggs hatch and larvae complete development in dung. See Arbovirus; Bunyaviridae. Cf. Aino Virus.

AKASHI, HIROSHI (1881–1946) (Hasegawa 1967, Kontyû 35 (suppl.): 3–4.)

AKERMAN, CONRAD (1878?–1946) (Anon. 1975, Samab 11: 301.)

AKHURST, JOHN (1816–1902) (Dos Passos 1951, J. N. Y. Entomol. Soc. 59: 134.)

AKIN Adj. (Middle English, *kin* = kind.) Genetically related; consanguineous.

AKNAMA, MOTOSHI (1853–) (Hasegawa 1967, Kontyû 35 (supl.) : 4.)

AKOLUTHIC Adj. (Greek, *akolouthia* = sequence; *-ic* = characterized by.) Pertaining to non-phyletic and non-phenetic biological classifications in which the interpretation of sequential morphological resemblance is independent of the fossil record. (Inglis 1966, 15: 224.)

AKRE, ROGER D (1937–1994) American academic (Washington State University) and Economic Entomologist specializing in yellowjackets, spiders, carpenter ants and syrphid fly Genus *Microdon.* Author of 130 papers including 11 book chapters; editor of Melanderia, the journal of the Washington State Entomological Society.

ALA Noun. (Latin, *ala* = wing. Pl., Alae.) Literally, the wing. In older literature developed from more classically trained workers, the wing blade that contains veins.

ALACARDO Noun. (Latin, *ala* = wing; *cardo* = hinge; Pl., Alacardoes.) Distal sclerite of the Cardo, along the lateral margin of the Subcardo. The Eucardo *sensu* MacGillivray.

ALACERCUS Noun. (Latin, *ala* = wing; Greek, *kerkos* = tail. Pl., Alacerci.) 1. The Caudal Filament or Telofilum. 2. The 'Median Cercus' when three Cerci are present (MacGillivray). See Appendix Dorsalis.

ALACOXASUTURE Noun. (Latin, *ala* = wing; *coxa* = hip; *sutura* = seam. Pl., Alacoxasutures.) A Suture between the Coxa Genuina and Meron of the insect leg (MacGillivray). See Coxa.

ALAFORAMEN Noun. (Latin, *ala* = wing; *foramen* = opening. Pl., Alaforamens.) The aperture through which the Alimentary Canal passes from the head to the Thorax (MacGillivray). Syn. Occipital Foramen; Foramen Magnum.

ALAKALI BEE *Nomia melanderi* Cockerell [Hymenoptera: Halictidae]: A monovoltine Species endemic to western USA that nests in alkali soils. Adults active during July-August; female constructs nest of 15–20 cells; females construct nests in aggregations. AB overwinters as a Prepupa. In some areas, AB is an important pollinator of alfalfa grown for seed.

ALANYCARB A carbamate compound {S-methyl-N-[[N-methyl-N-[N-benzyl-N-(2-ethoxycarbonylethyl)aminothio]carbamoyl] thioacetimidate} used as a contact insecticide and stomach poison against aphids and leaf chewing insects, particularly Lepidoptera. Alanycarb is applied to fruit trees, citrus, cotton, corn, grapes, peanuts, potatoes, soybeans, sugar beets, tea, tobacco and vegetables in some countries. Compound is not registered for use in USA. Alanycarb is toxic to fishes. Trade names include Onic®, Orion® and Rumbline®. See Carbamate Insecticide.

ALAR Adj. (Latin, *alaris, ala* = wing.) Pertaining to wings; wing-like; axillary. Alt. Alary.

ALAR® See Daminozide.

ALAR APPENDAGE See Alulet.

ALAR FRENUM Diptera: A small ligament dividing the Supra-Alar Cavity into anterior and posterior parts (Comstock). Hymenoptera: A small ligament crossing the Supra-Alar Groove toward the wing base (Smith).

ALARALIAE Plural Only. Prealaraliae and Postalaraliae collectively (MacGillivray).

ALARIA Noun. (Latin, *alarius* = wing. Pl., Alariae.) 1. Projections of the notal margin against which the wing articulates. The Notal Wing Processes (MacGillivray). 2. A Genus of olive-brown seaweed assigned to Family Laminariaceae.

ALARIMA Noun. (Etymology obscure. Pl., Alarimae.) A fissure between the Paraglossae (MacGillivray).

ALARM Noun. (Middle English, *alarme* = alarm. Pl., Alarms.) 1. A warning signal (typically vibration or chemical) produced by individuals to increase a sense of attention. 2. A sound or information used to warn of danger. Alt. Warning. Rel. Pheromone.

ALARM PHEROMONE A chemical substance, exchanged among members of a Species, that induces a behavioral state of alertness or alarm. The glandular cells that produce and secrete alarm pheromones are found on various parts of the insect body (*e.g.* Cornicles of aphids). Syn. Alarm Substance. See Cornicle. Cf. Aggregation Pheromone; Sex Pheromone; Trail Pheromone. Rel. Chemical Communication.

ALARM SUBSTANCE See Alarm Pheromone.

ALARMED Adj. (Middle English, *alarme*.) Individuals of a population that are disturbed, aroused or excited and prepared for agonistic encounters. Adv. Alarmedly.

ALARY Adj. (Latin, *ala* = wing.) Wing-like; pertaining to wings.

ALARY MUSCLE Dorsal Transverse Muscles that originate on the lateral wall of thoracic and abdominal Terga and insert medially on the Heart. Primitive number of AM is 12 in orthopteroids; trend toward reduction in number of AMs in more highly evolved or powerful flying insects. In many insects AMs may consist of modified connective tissue rather than muscle cells. AMs are paired muscles that actuate and support the insect heart, but their role in heartbeat modulation is debated. See Muscle. Cf. Dorsal Vessel. Rel. Circulation; Haemocoel; Haemolymph.

ALARY POLYMORPHISM Two or more forms of wings in a Species. AP is a condition independent of sexual dimorphism. See Polymorphism.

ALATAE Plural Noun. (Latin, *alatus* = winged.) The winged form of insects (*e.g.* aphids and ants) that have winged and wingless forms.

ALATATE Noun. (Latin, *alatus* = winged; *-atus* = adjectival suffix. Pl., Alatates; Alatatae.) Lateral wing-like expansions on the Tibiae of some insects, or on some insect eggs.

ALATE Adj. (Latin, *alatus* = winged; *-atus* = adjectival suffix.) Pertaining to wings; describing a winged condition. Alt. Alated. See Wing. Cf. Aptera. Noun. (Latin, *alatus* = winged; *-atus* = adjectival suffix. Pl., Alates.) 1. Winged insects. 2. The winged reproductive castes of termites and ants, specifically the winged female and male when workers are wingless. 3. In non-social insects, a winged form when the apterous condition also occurs. See Wing. Cf. Aptera. Rel. Dealate.

ALATION Noun. (Latin, *alatus* = winged; English, *-tion* = result of an action. Pl., Alations.) The state or condition of bearing wings.

ALATUS Noun. (Latin, *alatus* = winged.) An archaic term for Wingedness.

ALAVERTEX Noun. (Latin, *ala* = wing; + *vertex*. Pl., Alavertexes; Alavertices.) 1. Part of the Vertex along its ventral aspect. 2. The Occiput of taxonomists *teste* MacGillivray.

ALBARDA, E (1859–1936) (Knoch 1936, Ent. Z. Frankf. a. M. 50: 393–395.)

ALBARDA, J HERMANN (–1899) (Anon. Tijdschr. Entomol. 42: 38–39.)

ALBEDO Noun. (Latin, *albus* = white. Pl., Albedos; Albedoes.) Any white tissue, structure or substance.

ALBERS, GUSTAV FRIDOLIN (1822–1893) (Barthe 1894, Miscnea Entomol. 2: 107.)

ALBERS, THEODOR (1893–1960) (de Lattin 1961, Verh. Ver. Naturw. Heimatforsch. 35: i-vi.)

ALBERTIS ENRICO A D' (1846–1932?) (Vinciguerra 1932, Ann. Mus. Civ. Stor. Nat. Genova 56: 5–10.)

ALBERTIS, LUIGI MARIA D' (1841–1901) (Vinciguerra 1901, Boll. Soc. Geogr. Ital. (4) 2: 849–855.)

ALBERTUS, MAGNUS (1200–1280) Also known as 'Albert the Great.' German scholastic, philosopher, theologian and teacher of Thomas Aquinas. Albertus studied at Padua and became a Dominican monk, later the Rector of the school in

Cologne (1249) and subsequently the Bishop of Ratisbon (1260). Albertus is probably best known as a proponent of Scholasticism and author of *De Animalibus* (1255–1270), a collection of 26 books (including 19 books by Aristotle and 7 by Albertus). Albertus' work in biology included a comentary on Aristotle's work; he advocated careful observation and experimentation. (Henriksen 1921, 15: 8–9.)

ALBESCENT Adj. (Latin, *albescere* = to grow white.) Becoming white. Albescence, noun.

ALBICANT Adj. (Latin, *ablicare* = to be white; *-antem* = adjectival suffix.) Becoming white.

ALBICATION Noun. (Latin, *ablicare* = to be white; English, *-tion* = result of an action. Pl., Albications.) The act, process or phenomenon of substance becoming white.

ALBIDUS Adj. (Latin, *ablicare* = to be white.) White with a dusky tinge.

ALBIFICATION Noun. (Medieval Latin, *albificatio;* English, *-tion* = result of an action. Pl., Albifications.) The act or process of making something white.

ALBIFIY Transitive Verb. (Medieval Latin, *albificare.*) To whiten. -fied; - fying.

ALBIN, ELEAZAR (1713–1759) English painter and author of *Insectorum Angliae Naturalis Historia* (1720). (Weiss 1926, Sci. Mon. 23: 558–560.)

ALBINISM Noun. (Spanish, *albino* = white, from Latin, *albus* = white; English, *-ism* = condition. Pl., Albinisms.) The state or condition of an organism being white through an absence of colour imparted by pigmentation. Albinism is considered a genetical defect.

ALBINISTIC Adj. (Spanish, *albino* = white, from Latin, *albus* = white; *-ic* = characterized by.) Tending to whiteness or fading to white in normally dark forms. Affected with albinism.

ALBINO Noun. (Spanish, *albino;* Latin, *albo* = white. Pl., Albinoes.) An animal that lacks pigment in its body and thus appears startlingly white when compared with other animals of the same Species or population. Albinism is considered a 'normal' condition of long-term dark-adapted Species (*e.g.* cave-inhabiting troglobites), such as some crickets and ground beetles; albinism is considered a genetical defect for light-adapted Species.

ALBINUS, BERNARD SIEGFRIED (1697–1770) (Nordenskiöld 1935, *History of Biology.* 629 pp. (258–259), London.)

ALBRECHT, JOHANN (1695–1774) (Rose 1850, *New General Biographical Dictionary* 1: 240.)

ALBUGINEA Noun. (Latin, *albugineus* = white covering. Pl., Albugineae.) A white fibrous tissue covering certain organs.

ALBUGINEOUS Adj. (Latin, *albugineus* = white covering; *-osus* = possessing the qualities of.) Descriptive of structure that resembles the white of an egg in texture or colour.

ALBUMEN Noun. (Latin, *albumen* = white of an egg. Pl., Albumens.) 1. Egg Albumen; the white of an egg that is used in histology to adhere microtome sections to microscope slides. See Albumin. 2. Parenchymous tissue of plants.

ALBUMIN Noun. (Latin, *albumen* = white of an egg. Pl., Albumins.) Structurally complex, sulfur-rich proteins that form a constituent of blood serum, muscles, milk and the white of egg. Albumin is soluble in distilled water and weak salt solutions; albumin precipitates in concentrated salt solutions and mineral acids. In earlier literature, Albumin is confused with Albumen.

ALBUMINIOUS Adj. (Latin, *albumen* = white of an egg; *-osus* = possessing the qualities of.) Possessing the properties, characteristics or features of albumin. Albuminousness, noun.

ALBUMINOID Adj. (Latin, *albumen* = white of an egg; Greek, *eidos* = form.) 1. Albumin-like; descriptive of structure resembling albumin in form or substance. 2. Proteins such as elastin, keratin and collagen. See Albumen. Rel. Eidos; Form; Shape.

ALBUS Noun. (Latin, *albus* = white.) Pure white or chalk white.

ALCHEMY Noun. (Latin, *alchimia,* Greek, *chymos* = plant juice. Pl., Alchemies.) 1. Medieval chemical science and speculative philosophy obsessed with the transmutation of base metals to gold and the discovery of a universal cure for disease to infinitely prolong life. 2. The act or process of transforming.

ALCHITRAN Noun. (Arabic, *alquitran.*) 1. Resin produced by fir trees. 2. Oil of Cedar. 3. Tar or bitumen. Alt. Alkitran.

ALCOCK, ALFRED WILLIAM (1859–1933) English soldier (Lt. Col.) and Head of London School of Tropical Medicine. (Anon. 1933, Nature 131: 573–575.)

ALCOHOL Noun. (Arabic, *alkuhl.* Pl., Alcohols.) Any of several hydroxides of organic radicals. Chemical nature makes alcohols important in industry and science. Alcohol is commonly used as a preservative for entomological museum specimens or as a component of fixatives for histology.

ALCOHOL ACID A compound that is an alcohol and an acid.

ALCOHOL, BUTYL A colourless liquid, C_4H_9OH, used as an histological dehydrant that works rapidly and does not remove stains. Undesirable features of BA include its high toxicity and imperfect solubilty with water.

ALCOHOL, ETHYL A clear liquid, C_2H_5OH, obtained from potatoes and some grains by fractional distillation. ETOH is characterized by miscibility with water in any proportion, and is used to harden and preserve animal tissues. In histological fixatives, ethyl alcohol is used to dehydrate tissues for microscopic preparations, and as a solvent for fats and oils.

ALCOHOL, METHYL A colourless liquid, CH_3OH, that is produced by the destructive distillation of wood. MeOH is ill-smelling when impure and is used as a substitute for ethyl alcohol in a limited way. MeOH hardens tissues and is less desirable as a substitute for ETOH in histological work. Syn. Wood Alcohol.

ALCOHOL, POLYVINYL An ingredient in some water-soluble mountants. Acronym PVA.

ALCOHOL, TERTIARY BUTYL A dehydration agent used in histological studies of plant material. Acronym TBA.

ALDEHOL Noun. A liquid used for denaturing alcohol.

ALDEHYDE Noun. A colourless, volatile liquid, CH_3CHO, obtained from alcohol by oxidation.

ALDER BARK-BEETLE *Alniphagus aspericollis* (LeConte) [Coleoptera: Scolytidae].

ALDER FLEA-BEETLE *Macrohaltica ambiens* (LeConte) [Coleoptera: Chrysomelidae].

ALDER FLIES See Megaloptera.

ALDER SPITTLEBUG *Clastoptera obtusa* (Say) [Hemiptera: Cercopidae].

ALDICARB A carbamate compound {2-methyl-2-(methylthio)propionaldehyde O-(methylcarbamoyl) oxime} used as an Acaricide, Nematicide and contact insecticide; systemic activity only upward in plant. Aldicarb is applied as an in-furrow treatment for beans, citrus, coffee, cotton, sorghum, soybeans, sugarcane and sugar beets. Aldicarb is phytotoxic to plant foliage. Effective against mites, plant-sucking insects (aphids, scale insects, leafhoppers, psyllids), thrips and soil nematodes. Aldicarb is highly toxic to mammals and is responsible for human deaths through inappropriate use. Trade names include Sanacarb® and Temik®. See Carbamate Insecticide.

ALDRICH, JOHN MERTON (1866–1934) (Melander 1934, Psyche 41: 133–149.)

ALDRIN A synthetic, cyclodiene insecticide {1R, 4S, 4aS, 5S, 8R, 8aR) 1, 2, 3, 4, 10, 10-hexachloro-1, 4, 4a, 5, 8, 8a-hexahydro-1, 4:5, 8-dimethanonaphthalene} used as a seed treatment and foliar applicant. Aldrin is stable in soil and used as a soil insecticide to prevent and control subterranean termites. Aldrin has been used in Africa for control of desert locust. Effective against ants, armyworms, chinch bugs, crickets, grasshoppers, Jap Beetles, snails, slugs, sowbugs, spittlebugs, leaf miners and thrips. Compound enters insect as a contact or oral poison that affects the nervous system. Aldrin is applied to bananas, corn, grains, potatoes, sorghum, sugar beets, sugarcane, tobacco and orchards in some countries. Aldrin is toxic to fishes; plants treated with Aldrin are not for livestock consumption; compound not for use in glasshouses. Aldrin was removed from registration in USA and many developed countries. See Cyclodiene Insecticide.

ALDROVANDI (ALDROVANDUS), ULYSSES (1522–1605) Italian aristocrat, physician, academic and naturalist. Aldrovandi travelled widely and collected plant and animal specimens. He established the Botanical Garden at Bologna where his collection served as the foundation of the Bolonga Museum. Aldrovandi observed that bed bugs were less common in the homes of rich people because they had resources necessary to keep their homes clean. Aldrovandi was a contemporary of Conrad Gesner and the author of *De Animalibus Insectis libri septem* (1602) which was the first work specifically dealing with insects and the first published dichotomous key. (Wilson 1835, A Treatise on Insects, p. 5.)

ALECITHAL Adj. (Greek, *a* = without; *lekithos* = yolk of an egg; Latin, *-alis* = belonging to, pertaining to.) Descriptive of eggs without yolk or with a small amount of yolk uniformly distributed. See Egg. Cf. Centrolecithal; Telolecithal.

ALEMMAL Adj. (Greek, *a* = without; *eilema* = a covering; Latin, *-alis* = belonging to, appropriate to.) Pertaining to nerves cells or fibres that lack a Neurilemma. See Nerve.

ALEPPO GALL Galls produced on *Quercus infectoria* in southeastern Europe by the wasp *Cynips tinctoria*. AG is high in tannic acid and used for making permanent writing ink.

ALERT® See Pyrrol.

ALEURITIC ACID An organic acid contained in lac.

ALEURO- Greek combining form derived from the word 'aleuron', meaning 'flour.'

ALEXANDER, CHARLES PAUL (1889–1981) American academic administrator and taxonomic specialist on the Diptera. CPA described about 11,000 Species of Tipulidae and published more than 1,000 scientific papers. (Wheeler 1985, J. New York Entomol. Soc. 93: 1141–1164; Byers 1982, J. Kans. Entomol. Soc. 55: 409–417.)

ALEXANDER, CLAIR CAHILL (1911–1959) (Ferguson 1959, J. Econ. Entomol. 52: 1033.)

ALEXANDER, GORDON (–1974) American academic (University of Colorado) and naturalist who published some work on Orthoptera of arctic-alpine zones. (Liebermann 1974, Revita Entomol. Soc. Argent. 34: 248.)

ALEXANDRE, AUGUSTE (1811–1886) (Burgeois 1886, Bull. Soc. Entomol. Fr. 6: cxcvii.)

ALEXIIDAE Plural Noun. A small Family of polyphagous Coleoptera assigned to Superfamily Cucujoidea.

ALEYRODIDAE Plural Noun. (Greek, *aleuron* = flour.) Whiteflies. A large, cosmopolitan Family of sternorrhynchous Hemiptera assigned to Superfamily Aleyrodoidea. Family includes about 1,200 Species. Adults are psyllid-like, small, fragile and often covered with wax; compound eyes transversely divided into dorsal and ventral regions of different facet size and shape; two Ocelli; Antenna with seven segments; Rostrum with four segments. Forewing and hindwing similar with reduced venation but without coupling mechanism and lying flat over Abdomen at repose; Tarsi with two segments; two thoracic and two abdominal spiracles; abdominal venter with paired, segmental wax plates and vasiform orifice. Parthenogenesis common in Aleyrodidae. Egg pedicellate, hygroscopic, inserted at least partially into plant tissue or stomata. Nymphs coccoid-like; first instar with Antennae and legs; subsequent instars lack appendages. All Species phytophagous with most attacking angiosperms; many Species of aleyrodids are regarded as serious economic pests, but only a few Species vector diseases; a few Species are tended by ants. Cf. Coccidae; Diaspididae; Eriococcidae; Pseudococcidae.

ALEYRODIFORM Adj. (*Aleurodes*, *forma* = shape.) Aleyrodid-shaped; descriptive of insects resembling *Aleyrodes* in form. See Aleyrodidae. Rel. Form 2; Shape 2; Outline Shape.

ALEYRODOIDEA A Superfamily of sternorrhynchous Hemiptera including Aleyrodidae.

ALFACRON® See Azamethiphos.

ALFALFA BLOTCH LEAFMINER *Agromyza frontella* (Rondani) [Diptera: Agromyzidae].

ALFALFA CATERPILLAR 1. *Colias eurytheme* Boisduval [Lepidoptera: Pieridae]: A Species widespread in North America but a multivoltine, primary pest of alfalfa in southwestern USA. AC also feeds on clover and other legumes. Female oviposits 200–500 eggs on underside of leaves. Neonate larva brown, then becomes green; larval stage completed ca 12–15 days. Pupation occurs on alfalfa stalk; AC overwinters as pupa except in extreme southern part of range where generations are continuous. 2. Larvae of the butterfly *Colias* spp. that feed on alfalfa, *Medicago sativa*. See Pieridae.

ALFALFA GALL-MIDGE *Asphondylia websteri* Felt [Diptera: Cecidomyiidae].

ALFALFA HOPPER *Stictocephala festina* [Hemiptera: Cicadellidae]: A pest that causes damage to *Medicago sativa*.

ALFALFA LEAFCUTTING-BEE *Megachile rotundata* (Fabricius) [Hymenoptera: Megachilidae]: An important pollinator of alfalfa in Canada and the USA. Alt. Alfalfa Leafcutter Bee. See Megachilidae.

ALFALFA LEAFTIER *Dichomeris acuminata* (Standinger) [Lepidoptera: Gelechiidae].

ALFALFA LOOPER *Autographa californica* (Speyer) [Lepidoptera: Noctuidae]: A pest of *Medicago sativa*.

ALFALFA PLANT-BUG *Adelphocoris lineolatus* (Goeze) [Heteroptera: Miridae].

ALFALFA-SEED CHALCID *Bruchophagus roddi* (Gussakovsky) [Hymenoptera: Eurytomidae]: A pest of alfalfa in North America. Eggs laid in seeds; larva overwinters in seed. ASC multivoltine and completes 2–3 generations per year. See Eurytomidae. Cf. Clover-Seed Chalcid.

ALFALFA SNOUT-BEETLE *Otiorhynchus ligustici* (Linnaeus) [Coleoptera: Curculionidae].

ALFALFA WEBWORM *Loxostege cerealis* (Zeller) [Lepidoptera: Pyralidae]: In North America, a pest of many succulent plants to include alfalfa and sugarbeets. See Pyralidae. Cf. Garden Webworm.

ALFALFA WEEVIL *Hypera postica* (Gyllenhal) [Coleoptera Curculionidae]: A significant pest of *Medicago sativa* in USA and also attacks clovers. AW probably endemic to southern Europe; first detected in Utah during 1904. AW overwinters as an adult; adult becomes active during spring, feeds and mates. Eggs lemon-yellow, oval, laid in clusters (to 40/cluster) within stems excavated by female; during spring female can lay 800 eggs. Neonate Larva white, then becomes green with white dorsomedian stripe, and legless but displays ridges on venter that function as legs. Larva appears during April, feeds on interior of stalk ca 3–4 days, then skeletonizes leaves; 3–4 instars; pupation within spherical net-like cocoon on plant or fallen leaf. Adult feeds on leaves and buds. See Curculionidae. Cf. Clover-Leaf Weevil; Lesser Clover-Leaf Weevil.

ALFIERI, ANASTASE (1893–1971) A government economic Entomologist who worked in Egypt on control of pink bollworm. Alfieri served as General Secretary and Curator of the Royal Entomological Society of Egypt. (Butler 1972, Proc. Roy. Entomol. Soc. London (C) 36: 61.)

ALFKEN, JOHANN DIEDRICH (1862–1945) (Amsel. 1949, Beitr. Naturk. Niedersachs. 8: 26–28.)

ALFREY, J I (1898–1959) (Uvarov 1960, Proc. Roy. Entomol. Soc. London (C) 24: 52.)

ALGESIA Noun. (Greek, *algesis* = sense of pain. Pl., Algesiae.) Sensitive to pain.

ALGESIS Noun. (Greek, *algesis* = sense of pain; *sis* = a condition or state. Pl., Algeses.) The sensation of pain. Adj. Algesic.

ALGESTHESIS Noun. (Greek, *algein* = to suffer; *aisthesis* = perception. Pl., Algesthesis.) 1. The perception of pain. 2. Pain.

ALGETIC Adj. (Greek, *algesis* = sense of pain; *-ic* = of the nature of.) Pertaining to pain; pain-inducing.

ALGICOLOUS Adj. (Latin, *alga* = seaweed; *colere* = to inhabit.) Pertaining to organisms that live in or around algae. See Habitat. Cf. Arenicolous; Fimicolous; Fungicolous; Lignicolous; Madicolous; Muscicolous; Paludicolous; Pratinicolous; Ripicolous; Silvicolous. Rel. Ecology.

ALGIN Noun. (Etymology obscure. Pl., Algins.) A polymer of mannuronic and guluronic acids extracted from kelp and other seaweeds. Algin is used as an inexpensive substitute for agar as a gelling agent in artificial diets. Algin forms gel at room temperature and does not affect heat-sensitive components of diets. Algin also prevents segregation of heavy dietary constituents that often settle during cooling process. Cf. Carrageenan.

ALGIO- A Greek combining form used to signify pain or pertain to pain.

ALGIVOROUS Adj. (Latin, *alga* = seaweed; *vorare* = to devour; *-osus* = full of.) Pertaining to organisms that feed on algae.

ALGOLOGICAL Adj. (Latin, *alga* = seaweed; Greek, *logos* = discourse; Latin, *-alis* = having the character of.) Descriptive of structure that bears the characteristics or properties of seaweeds or algae.

ALGOLOGIST Noun. (Latin, *alga* = seaweed; *logos* = discourse. Pl., Algologists.) A person who studies algae.

ALGOLOGY Noun. (Latin, *alga* = seaweed; *logos* = discourse. Pl., Algologies.) The study of algae.

ALGOPHAGOUS Adj. (Latin, *alga* = seaweed; Greek, *phagein* = devour; Latin, *-osus* = full of.) Feeding on algae.

ALGOUS Adj. (Latin, *algosis; -osus* = with the qualities of.) Pertaining to algae.

-ALIA (Greek, *halia*) A Greek suffix used to identify zoogeographical areas. *e.g.* Australia.

ALIEN Adj. (Latin, *alienus*.) Pertaining to foreign status. Noun. (Latin, *alienus* = belonging to another. Pl., Aliens.) An entity (organism, form, composition) that is not indigenous or native to an area, region or place. Foreign.

ALIENICOLA Noun. (Latin, *alienus* = another; *cola* = inhabitant. Pl., Alienicolae.) 1. Literally, a foreign inhabitant. 2. Multivoltine, migratory Aphididae: Parthenogenetic, viviparous females that typically develop on secondary host plant Species. Morphologically different from other forms in the complex life cycle, and capable of producing several generations that are winged or wingless. Cf. Fundatrigenia; Fundartrix; Migrantes; Sexupare; Sexuales.

ALIFER Noun. (Latin, *ala* = wing; *ferre* = to bear. Pl., Alifers.) The plural fulcrum of the wing (Crampton). The Plural Wing Process.

ALIFERA Noun. (Latin, *ala* = wing; *ferre* = to bear. Pl., Aliferae.) Projections of the Pleuron against which the Pteralia articulate (MacGillivray). See Notal Wing Process.

ALIFEROUS Adj. (Latin, *ala* = wing; *ferre* = to bear; *-osus* = full of.) Pertaining to wings; wing bearing.

ALIFORM Adj. (Latin, *ala* = wing; *forma* = shape.) 1. Wing-shaped; with wing-like extensions. A term descriptive of structure that is broad, flat, narrow at the base and oblately rounded at the apex. See Wing. Rel. Form 2; Shape 2; Outline Shape.

ALIGEROUS Adj. (Latin, *ala* = wing; *gerere* = to carry; *-osus* = full of.) Winged; Wing bearing.

ALIGN® An emulsifiable concentrate containing 3.0% azadirachtin developed by AgriDyne Technologies and approved on non-food crops in the USA. See Azadirachtin.

ALIMENT Noun. (Latin, *alimentum*, from *alere* = to nourish. Pl., Aliments.) Food; nourishment. Alt. Diet; Food.

ALIMENTAL Adj. (Latin, *alimentarius* = sustenance; *-alis* = pertaining to, having the character of.) Providing food or nourishment. Adv. Alimentally.

ALIMENTARY Adj. (Latin, *alimentarius* = sustenance.) Pertaining to the function of nutrition.

ALIMENTARY SYSTEM An internal tube that extends from the mouth to the Anus and open on both ends. Morphologically, the AS is subdivided into regions and areas based on structure and function. The following comments are descriptive of the generalized adult pterygote insect: The anterior region (foregut or Stomodaeum) is located primarily in the head and is responsible for ingestion, mastication, limited digestion and filtering. The middle region (midgut or Mesenteron) is located primarily in the Thorax and is responsible for digestion. The posterior region (hindgut or Proctodaeum) is located primarily in the Abdomen and is responsible for processing waste products, resorption of metabolites and excretion. Syn. Alimentary Canal. See Digestion. Cf. Circulatory System; Nervous System; Reproductive System; Respiratory System. Rel. Mouth; Buccal Cavity; Pharynx; Oesophagus; Crop; Proventriculus; Cardiac Valve; Gastric Caeca; Peritrophic Membrane; Pyloric Valve; Anterior Intestine; Ilium; Malpighian Tubules; Colon; Posterior Intestine; Rectum; Anus.

ALIMENTARY WATER Water ingested with food or water used as a source of food.

ALIMENTATION Noun. (Latin, *alimentum* = nourishment; English, *-tion* = result of an action. Pl., Alimentations.) The act or process of feeding or ingesting nourishment. Alt. Provision; Subsistance.

ALIMENTATIVE Adj. (Latin, *alimentum* = nourishment.) Descriptive of a thing that is nutritive. Alimentatively, adv.; Alimentativeness, noun.

ALIMENTIC Adj. (Latin, *alimentum* = nourishment; *-ic* = characterized by.) Alimentary.

ALIMENTIVE Adj. (Latin, *alimentum* = nourishment.) Pertaining to food.

ALIN, WASSILIJ NIKOLAJEWITSCH (1905–) (Rohlfien 1975, Beitr. Entomol. 25: 266.)

ALINOTUM Noun. (Latin, *ala* = wing; Greek, *noton* = back. Pl., Alinota.) The wing-bearing sclerite on the dorsum of the Mesothorax or Metathorax of winged insects. See Thorax. Cf. Alitrunk. Rel. Wing.

ALIPHATIC Adj. (Greek, *aleiphar* = oil, unguent, *aleiphein* = to smear; *-ic* = characterized by.) Open-chained fatty acids and unsaturated compounds containing fat. An organic chemical compound in which the carbon atoms are linked in chains rather than rings. A deriviative of organophosphate pesticides. See Heterocyclic; Phenyl.

ALITRUNK Noun. (Latin, *ala* = wing; *truncus* = trunk. Pl., Alitrunks.) 1. The portion of the Thorax to which the wings are attached. 2. In some aculeate Hymenoptera (such as ants) the Alitrunk includes the first abdominal segment (Propodeum). The combined Thorax and first abdominal segment. Syn. Mesosoma. See Thorax.

ALIVE Adj. (Middle English, *alive* from Old English, *on live*.) Descriptive or organisms that display the characteristics of life, to include growth, respiration and reproduction. Not dead. Alt. Living; Vital. See Life. Cf. Dead.

ALIZARIN Noun. (French, *alizari*. Pl., Alizarins.) A transparent dye, $C_{14}H_6O_2$–$(OH)_2$, orange-red as crystal or reddish-orange as a powder. Originally produced from madder, but more recently manufactured synthetically from anthracene. Alt. Alizarine.

ALK Noun. (Arabic, *ilk* = resin. Pl., Alks.) Resin of Chian turpentine.

ALKAHEST Noun. (Medieval Latin, *alchahest* = becomes. Pl., Alkahests.) A mythical universal solvent of alchemists. Adj. Alkahestic; Alkahestical.

ALKALESCENCE Noun. (Arabic, *al-qili* = ashes of saltwort; *-escent* = slightly. Pl., Alkalescences.) The quality of being alkaline. Alt. Alkalescency.

ALKALESCENT Adj. (Arabic, *al-qili* = ash; *-escent* = slightly.) Descriptive of substances that are slightly alkaline.

ALKALI Noun. (Arabic, *al-qili* = ash of saltwort plant. Pl., Alkalies; Alkalis.) Any substance with chemically basic properties or containing the OH-radical and turning red litmus blue when in solution. Alt. Alkaine; Caustic.

ALKALI BEE *Nomia melanderi* Cockerell [Hymenoptera: Halictidae].

ALKALINE Adj. (Arabic, *al-qili* = ash of saltwort plant.) Pertaining to alkali properties or pertaining to alkali metals.

ALKALINE ALCOHOL 70% ethyl alcohol with a few drops of 0.1% solution of soda bicarbonate ($NaHCO_3$).

ALKALINE CELL A battery whose cell contains an alkaline electrolyte.

ALKALINE EARTHS Oxides of barium, calcium and strontium.

ALKALINE GLAND Aculeate Hymenoptera: An apparently unpaired gland associated with the Median Oviduct. The Alkaline Gland discharges an alkaline secretion into the Sting. AG secretions may serve as a lubricant or react with the acid to make venom used in defence or immobilization of prey. Syn. Dufour's Gland. See Gland. Cf. Accessory Gland; Acid Gland. Rel. Exocrine Gland.

ALKALOID Noun. (Arabic, *al-qili* = ash of saltwort plant; Greek, *eidos* = form. Pl., Alkaloids.) 1. Pertaining to alkaloids. Alt. Alkaloidal. 2. Naturally produced organic compounds that incorporate a nitrogen atom in a heterocyclic ring. Alkaloids are base substances that display alkaline properties and which naturally occur in plants and animals. Notable plant alkaloids include nicotine, strychnine, morphine and cocaine. See Insecticides.

ALKANES Noun. Saturated hydrocarbons of geologically ancient origin. See Paraffin.

ALKRON® Parathion.

ALKSANKIN, YAKOV VASIL'EVICH (–1968) (Savzsarg 1969, Zashch. Rast. 1969: 17–19.)

ALLAN, PHILIP BERTRAM MURRAY (1884–1973) (1974. Entomol. Rec. J. Var. 86: 119–120.)

ALLAN, WILLIAM (1820–1915) (Hamilton 1916, Proc. Linn. Soc. N.S.W. 41: 5.)

ALLANTOIC ACID Waste product of nitrogen metabolism that occurs in the excrement of Lepidoptera and Hymenoptera.

ALLARD, ERNEST (1820–1900) (Girard 1900, Bull. Soc. Entomol. Fr. 1900: 42.)

ALLARD, GESTON (1838?–1918) (Edwards 1919, Proc. Soc. London Entomol. Nat. Hist. Soc. 1918–1919: 37.)

ALLECULIDAE Plural Noun. See Tenebrionidae.

ALLEGHANIAN FAUNAL AREA Part of the North American transition zone comprising most of New England, southeastern Ontario, New York, Pennsylvania, Michigan, Wisconsin, Minnesota, eastern N. Dakota, northeastern S. Dakota, and the Alleghanies from Pennsylvania to Georgia.

ALLEGHENY MOUND-ANT *Formica exsectoides* Forel [Hymenoptera: Formicidae]: An ant Species widespread throughout eastern North America. Workers ca 2–5 mm long; head and Mesosoma red; Metasoma and legs black. Nests display large conical mounds (<2 m diam, 1 m high). Young queens must have workers to establish colony, or colonies develop by fission. Food includes honeydew and insects. See Formicidae. Cf. Silky Ant.

ALLEGHENY SPRUCE-BEETLE *Dendroctonus punctatus* LeConte [Coleoptera: Scolytidae].

ALLELE Noun. (Greek, *allelon* = one another. Pl., Alleles.) A gene that occupies a locus on a chromosome. Alt. Allelomorph. Cf. Gene.

ALLELOCHEMIC See Allelochemical.

ALLELOCHEMICAL Noun. (Greek, *allelon* = one another; *-alis* = belonging to, having the character of. Pl., Allelochemicals.) Non-nutritional chemicals used by one Species to affect growth, health, behaviour or population biology of another Species (Cf. Pheromone). Allelochemicals potentially are present in all plants and in most tissues. Common allelochemicals include alkaloids, benzoxazinones, cyanogenic compounds, ethylene, flavonoids and polyacetylenes. Entomologists divide allelochemicals into four categories: Allomones, Apneumones, Kairmones and Synomones. Syn. Allelochemic; Xenomone *sensu* Chernin 1970, BioScience 20: 845. Whittaker 1970, *Chemical Ecology*, pp 43–70, E. & B. Semeone, eds.) See Allomones; Apneumone; Kairomones; Synomones. Rel. Exocrine Gland.

ALLELOMORPH Noun. (Greek, *allelon* = one another; *morphe* = form. Pl., Allelomorphs.) Alternative, contrasting genetic conditions expressing Mendelian unit characters (*e.g.* smooth or wrinkled seed coats of the garden pea).

ALLELOMORPHIC Adj. (Greek, *allelon* = one another; *morphe* = form; *-ic* = consisting of.) Pertaining to an allelomorph.

ALLELOPATHIC Adj. (Greek, *allelon* = one another;

pathos = suffering; *-ic* = characterized by.) The influence or effect of one living plant upon another living plant, manifest by the reduction in emergence, growth or performance. Allelopathic influences are caused by allelochemicals.

ALLELOTYPE Noun. (Greek, *allelon* = one another; *typos* = pattern. Pl., Allelotypes.) The frequency with which alleles occur in a population. See Type. Cf. Genotype.

ALLEN, ANSON (1829–1884) (Fernald 1881, Canad. Entomol. 16: 43–44.)

ALLEN, JAMES EDWARD ROTHWELL (1866–1916) (Burrows 1917, Entomol. Rec. J. Var. 29: 67–68.)

ALLEN, WILLARD ROSS (1913–1973) (Westal 1973, J. Econ. Entomol. 66: 1242.)

ALLEN'S RULE A generalization concerning environmental conditions and morphological characteristics. AR observes that appendages of a Species are relatively short in the cooler parts of that Species' geographical range. Cf. Bergmann's Rule.

ALLEON-DULAC, JEAN LOUIS (1723–1768) (Rose 1850, *New General Biographical Dictionary*. 1: 348.)

ALLERON® Parathion.

ALLETHRIN The first synthesized pyrethroid insecticide. A synthetic-pyrethroid compound used as contact insecticide and fumigant for control of flies, mosquitoes, midges, fleas, lice, cockroaches, wasps and other urban pests. Allethrin is toxic to bees and fishes. Trade names include: Bio-Allethrin®, Alleviate® and Esbiothrion®. See Synthetic Pyrethroid.

ALLETHRIN® See Allethrin.

ALLEVIATE® See Allethrin.

ALLIACEOUS Adj. (Latin, *allium* = garlic; *-aceus* = of or pertaining to.) Pertaining to the odour of garlic or onions.

ALLICIENT Adj. (Latin, *allicere* = to draw gently to.) Pertaining to something that attracts. Cf. Repellant.

ALLIGATE Transitive Verb. (Latin, *alligare* = to bind, from *ad* = to, *ligare* = to tie up; *-atus* = suffix.) To unite; to tie. *e.g.* Fastened or suspended by a thread, as the Chrysalis of *Papilio*.

ALLIGATION Noun. (Latin, *alligatio*; English, *-tion* = result of an action. Pl., Alligations.) The act of attaching. The state of being attached.

ALLIGATOR Adj. (Spanish, *el lagarto; or* = a condition.) Pertaining to or descriptive of the surface features of the alligator. Intransitive Verb. To develop intersecting cracks or ridges. Alligatored, adj.

ALLIGATOR WEED *Alternanthera philoxeroides* (Martius) Griseback [Amaranthaceae]: An aquatic perennial herb to one meter long and generally regarded as one of world's most serious aquatic weeds. Endemic to South America and introduced into southeastern USA (1894), Indonesia (1875), New Zealand (1906), Burma (1932) and Australia (ca 1940s). AW infests still or slow-moving water and wetlands in many tropical and subtropical areas of the world. AW grows as a rooted, emergent, aggressive competitor in water to 0.5m deep; stems hollow and may form mats on surface of water; may grow in moist terrestrial habitats. Reproduction via seeds or vegetatively. AW does not set seed in Australia or USA; a single stem node can initiate a new plant. AW controlled with Alligator-Weed Flea Beetle, *Agasicles hygrophila* Selman & Vogt, and pyralid *Vogita malloi* Pastrana; both Species endemic to South America and moved to USA and Australia. Cf. Salvinia; Water Hyacinth; Water Lettuce.

ALLIGATOR-WEED FLEA BEETLE *Agasicles hygrophila* Selman & Vogt [Coleoptera: Chrysomelidae]: A beetle endemic to South America and imported into USA for biocontrol of Alligator Weed. AWFB is the most effective control agent of AW; feeding damage by larva and adult reduces plant vigour, growth and competitive ability.

ALLIGATOR-WEED MOTH *Vogtia malloi* Pastrana [Lepidoptera: Pyralidae]: A moth endemic to South America and imported into USA for biocontrol of Alligator Weed. Larval feeding damage sometimes is impressive.

ALLIOGENESIS See Alloiogenesis.

ALLIONI, CARLO (1725–1805) Allioni's collection of insects originally was placed in the Zoological Museum at the University of Torino. The collection apparently was destroyed by fire. (Rose 1850, *New General Bibliographical Dictionary* 1: 352.)

ALLIS, THOMAS HENRY (1817–1870) (Anon. 1870, Entomol. Mon. Mag. 7: 90–91.)

ALLOCHIRAL Adj. (Greek, *allos* = other; *cheir* = hand; Latin, *-alis* = belonging to, having the character of.) Symmetrically similar parts that are reversed in position and arrangement (*e.g.* left and right human hands). Adv. Allochirally.

ALLOCHROIC Adj. (Greek, *allos* = other; *chros* = colour; *-ic* = characterized by.) Pertaining to colour change or colour variation.

ALLOCHROMATIC Adj. (Greek, *allos* = other; *chromatos* = colour; *-ic* = of the nature of.) Pertaining to the variability in colour displayed by a substance not through inheritance.

ALLOCHRONIC Adj. (Greek, *allos* = other; *chronos* = time; *-ic* = of the nature of.) Pertaining to things separated by time or season. Cf. Allopatric; Synchronic.

ALLOCHTHONOUS Adj. (Greek, *allos* = other; *chthon* = the ground; Latin, *-osus* = full of.) Not autochthonous. Pertaining to a Species with origin in one area but which occurs in other areas. Alt. Allocthonous.

ALLOCORYNIDAE Plural Noun. A small Family of polyphagous Coleoptera assigned to Superfamily Curculionoidea.

ALLOCRYPTIC Adj. (Greek, *allos* = other; *kryptos* = hidden; *-ic* = characterized by.) Pertaining to a kind of concealment (Crypsis) in which the shape, size, colour or texture of an animal resembles familiar objects in the environment. Objects are typically inanimate (rocks) or inedible (bark) to a predator. See Coloration; Crypsis. Cf. Allosematic; Aposomatic. Rel. Mimicry.

ALLOGENEITY Noun. (Greek, *allos* = other; *genos* = descent; English, *-ity* = suffix forming abstract nouns. Pl., Allogeneities.) Varying in nature or kind.

ALLOGENEOUS Adj. (Greek, *allos* = other; *genos* = descent; Latin, *-osus* = possessing the qualities of.) 1. Pertaining to floras persisting from an older environmental period. 2. Pertaining to relationship not based on heritable traits.

ALLOIOGENESIS Noun. (Greek, *alloios* = different; *genesis* = descent. Pl., Alloiogeneses.) Cyclical parthenogenesis or alternation of generations. A genetical condition in which entire generations are composed of females and other generations of the same Species display males and females (*e.g.* Cynipidae). A term more often applied to flatworms than insects. See Parthenogenesis. Cf. Heterogony.

ALLOISOMERISM Noun. (Greek, *allos* = other; *iso* = equal; *meros* = part; English, *-ism* = condition. Pl., Alloisomerisms.) In chemistry, isomerism that cannot be explained by understood structural formulas.

ALLOKINESIS Noun. (Greek, *allos* = other; *kinesis* = movement. Pl., Allokineses.) Reflexive movements or seemingly purposeless passive movement. Adj. Allokinetic. See Kinesis. Cf. Blastokinesis; Diakinesis; Klinokinesis; Ookinesis; Orthokinesis; Thigmokinesis. Rel. Taxis; Tropism.

ALLOMETRIC Adj. (Greek, *allos* = other; *metron* = measure; *-ic* = characterized by.) 1. Pertaining to allometry. 2. Descriptive of the disproportional change in size of structure during development of an individual. 3. Relating to change measured among members of a population, representatives of Species or a collection of coordinate-level taxa. See Allometry. Cf. Isometric. Rel. Morphometric.

ALLOMETRIC GROWTH EQUATION The power function advanced by Snell (1892) which states: $Y = b(X)\alpha$ where Y is a dependent variable; b is a constant or value of Y when $X = 0$; X is an independent variable; and α is the slope of a line generated by data points on a rectilinear plot from the study. When $\alpha = 1$, the condition is isometric and geometrical similarity is maintained and the smaller organism is a scale model of the larger. When $\alpha > 1$, the condition is termed 'positive allometry' and a disproportionate increase in Y (the dependent variable) exists for the comparative value of X (the independent variable.) When $\alpha < 1$, the condition is termed 'negative allometry' and there is a disproportionate decrease in the value of Y for a corresponding value of X. An example of negative allometry can be seen when comparing the body size with eye size for mammals including shrews, mice, rabbits, elks, rhinos, elephants and whales. In this example, the size of the eye becomes disproportionately small with an increase in body size. The AGE is used in morphometric studies to quantify deviations from isometric growth of structures, appendages, organs and bodies. See Allometry. Rel. Morphology.

ALLOMETRY Noun. (Greek, *allos* = other; *metrios*, *metron* = measure. Pl., Allometries.) 1. A subdiscipline of morphometrics that involves the study of changes in form and relative size of body parts during the processes of growth and development. 2. The study of changes in size and shape. See Morphometry. Cf. Isometry. Rel. Morphology.

ALLOMONE Noun. (Greek, *allos* = other; *hormaein* = to excite. Pl., Allomones.) Chemical substance (produced or acquired by an organism) which, when contacting an individual of another Species in the natural context, evokes in the receiver a behavioural or physiological reaction adaptively favourable to the emitter. (Brown 1968, Amer. Nat. 102: 188–191.) See Semiochemical. Cf. Hormone; Kairomone; Pheromone. Syn. Alloiohormone *sensu* Beth 1932, Naturwissenschaften 20: 177–183.

ALLOMORPH Noun. (Greek, *allos* = other; *morphe* = form. Pl., Allomorphs.) Any of two or more forms of a compound or crystal. Allomorph is a term more commonly used in Mineralogy than Entomology.

ALLOMORPHIC Adj. (Greek, *allos* = other; *morphe* = form.) Pertaining to more than one form. See Morphometric. Cf. Dimorphic; Monomorphic; Polymorphic.

ALLONYM Noun. (Greek, *allos* = other; *onyma* = name. Pl., Allonyms.) An assumed name used by the author of a scientific work or taxonomic description instead of his/her real name. Syn. Pseudonym. Rel. Nomenclature; Taxonomy.

ALLONYMOUS Adj. (Greek, *allos* = other; *onyma*; Latin, *-osus* = possessing the qualities of.) Descriptive of a publication (article, paper or book) that was published by the author under a false name. See Publication. Cf. Pseudonym.

ALLOPARASITOID Noun. (Greek, *allos* = other; *parasitos* = one who eats at the table of another, from *para* = beside, *sitos* = food ; *eidos* = form. Pl., Allparasitoids.) A parasitoid that attacks and feeds upon members of an unrelated taxon. See Parasite; Parasitoid. Cf. Adelphoparasite; Hyperparasite.

ALLOPATRIC Adj. (Greek, *allos* = other; *patra* = fatherland; *-ic* = characterized by.) Pertaining to things or organisms separated in space or geographical distribution. See Zoogeography. Cf. Allotopic; Parapatric; Sympatric. Rel. Allochronic; Distribution.

ALLOPATRY Noun. (Greek, *allos* = other; *patra* = fatherland. Pl., Allopatries.) The act or condition of populations of related organisms being separate in space or geographical distribution. See Biogeography. Cf. Parapatry; Sympatry. Rel. Distribution.

ALLOPELAGIC Adj. (Greek, *allos* = other; *pelagos* = sea; *-ic* = characterized by.) Descriptive of marine organisms that are distributed at the surface of water and at great depth with their spatial distribution not influenced by temperature. See Biogeography. Rel. Zone.

ALLOPLASM Noun. (Greek, *allos* = other; *plasma* = mold. Pl., Alloplasms.) Specialized living substances that are differentiated and less active than protoplasm. Alloplasm occurs within Cilia, Flagella and some other fibrils.

ALLOPLASMIC Adj. (Greek, *allos* = other; *plasma* = moulded; form; *-ic* = of the nature of.) Pertaining to Alloplasm. Alt. Alloplasmatic.

ALLOPLASTIC Noun. (Greek, *allos* = other; *plastos* = formed; *-ic* = consisting of. Pl., Alloplastics.) A shape or form determined by external environmental conditions. Cf. Autoplastic.

ALLOSEMATIC Adj. (Greek, *allos* = other; *sema* = sign; *-ic* = characterized by.) Pertaining to coloration that is protective because it resembles the coloration of dangerous animals or inedible plants.

ALLOSOME Noun. (Greek, *allos* = other; *soma* = body. Pl., Allosomes.) See Heterochromosome. Rel. Chromosome.

ALLOTOPIC Adj. (Greek, *allos* = other; *topos* = place; *-ic* = characterized by.) Pertaining to organisms that occupy different microhabitats. See Biogeography. Cf. Allopatric; Syntopic.

ALLOTRIO- A Greek combining form meaning 'strange' (in the negative sense), perverted or abnormal. Alt. Allotri-.

ALLOTROPHIC Adj. (Greek, *allos* = other; *trophe* = nourishment; *-ic* = characterized by.) 1. Descriptive of food that has a nutritive value different from a typical (normative) condition. 2. Pertaining to a condition in which an organism is dependent upon unrelated organism for nutrition. Syn. Heterotrophic. Ant. Autotrophic. Cf. Saprophytic.

ALLOTROPHY Noun. (Greek, *allos* = other; *trophe* = nourishment. Pl., Allotrophies.) Populations of a herbivorous Species that feed on plants other than the host plant preferred by the herbivorous Species. (Jolivet 1953, Mem. Inst. Roy. Sci. Nat. Belg. Ser. 2, 50: 1–88.)

ALLOTROPIC Adj. (Greek, *allos* = other; *trophe* = nourishment; *-ic* = characterized by.) Pertaining to allotropy. Allotropical, allotropically, adv.

ALLOTROPY Noun. (Greek, *allos* = other; *tropos* = manner. Pl., Allotropies.) Mineralogy and Chemistry: An element, compound or crystal that exists in more than one form.

ALLOTYPE Noun. (Greek, *allos* = other; *typos* = type. Pl., Allotypes.) Zoological Taxonomy: In the description of a new Species, a designated member of the type-series (a Paratype) selected by the original describer as typical of the sex opposite of the Holotype. See Type. Cf. Holotype; Lectotype; Neotype; Paratype; Syntype. Rel. Taxonomy

ALLOTYPICAL Adj. (Greek, *allos* = other; *typos* = type; Latin, *-alis* = belonging to.) Pertaining to an Allotype. See Allotype.

ALLOXYSTIDAE Plural Noun. A numerically small, widespread Family of parasitic Hymenoptera assigned to the Cynipoidea. Arguably synonymous with Charipidae. Adult body less than 2.0 mm long, smooth, shiny without sculpture; female Antenna with 13 segments; male with 14 segments; Parapsidal Grooves seldom present; Thorax smooth, seldom with sculpture or ornamentation; forewing Areola absent; some Species apterous; Gaster compressed with Tergum II large and pubescent. Larvae develop as hyperparasites on Hymenoptera attacking aphids, particularly Aphidiidae, and some records of Aphelinidae; relationship influenced by host plants and host Species of aphids; host specificity not conclusively demonstrated. See Charipidae.

ALLOZYME Noun. (Greek, *allos* = other; *zyme* = leaven. Pl., Allozymes.) Alternative enzyme forms encoded by different alleles at the same gene locus.

ALLRED, DONALD MERVIN (1923–1996) American academic (Brigham Young University) and specialist in Siphonaptera. (Lewis 1996, Flea News, no 53: 625–626.)

ALLURE Noun. (Middle English, *aluren*, from Old French, *aleurrer* = to lure. Pl., Allures.) Act of alluring. The power of attracting; fascination. Alt. Allurement; Attraction. Transitive, Intransitive Verb. (Middle English, *aluren*. Allured, Alluring.) To attract or entice with an offer of something real, desirable or beneficial. Cf. Repel.

ALLURER Noun. (Middle English, *aluren* = to lure. Pl., Allurers.) An individual, substance or object capable and responsible for attracting an individual to a source of the attraction.

ALLURING COLORATION Any colour, colour pattern or combination of colours that are attractive to some Species of insects (prey) and displayed (used) by predaceous Species as part of their feeding strategy. Prey are attracted to predator with alluring coloration. See Coloration. Cf. Advancing Coloration; Anticryptic Coloration; Apetetic Coloration; Aposematic Coloration; Coloration; Combination Coloration; Directive Coloration; Disruptive Coloration; Epigamic Coloration; Episematic Coloration; Procryptic Coloration; Protective Coloration; Pseudepisematic Coloration; Pseudoaposematic Coloration; Seasonal Coloration; Sematic Coloration. Rel. Aggressive Mimicry; Cryptic Coloration.

ALLURING Adj. (Middle English, *aluren*.) Pertaining to the capacity to attract, entice or tempt. Alluringly, Adv.

ALLURING GLANDS Glandular structures (frequently found on males) that diffuse odours attractive to the opposite sex. (Obsolete). See Sex

Pheromone. Rel. Androconium.

ALLUVIAL Adj. (Latin, *alluvion*, from *alluere* = to wash upon; *-alis* = appropriate to, having the character of.) Pertaining to Alluvium. Noun. (Latin, *alluvius*. Pl., Alluvials.) Alluvial soil.

ALLUVION Noun. (Latin, *alluvio*, from *alluere* = to wash against. Pl., Alluvions.) 1. The flow of water against the bank of a flowing river, stream or creek; the wash of water on the shore of a pond, lake, sea, or ocean. 2. An inundation or flood.

ALLUVIOUS Adj. (Latin, *alluere* = to wash upon; *-osus* = full of.) Alluvial.

ALLUVIUM Noun. (Latin, *alluere* = to wash upon; *-ium* = diminutive > Greek, *-idion*. Pl., Alluvia; Alluviums.) Soil, sand, gravel or other material deposited by flowing water, typically at the junction of mountainous areas and flat terrain.

ALLUX Noun. (Latin, *hallex* or *allex* = the great toe.) Toe Ball. The penultimate Tarsomeres in Rhynchophora. Alt. Hallux.

ALLYN, ARTHUR C (1913–1985) (Miller & Miller 1985, Bull. Allyn Mus. 97: 1–6.)

ALMEIDA, ROMUALDE FERREIRA D' (1891–1969) (Mieike 1974, J. Lepid. Soc. 28: 293–296.)

ALMOND MOTH *Cadra cautella* (Walker) [Lepidoptera: Pyralidae]: A cosmopolitan pest of stored products including grains, nuts and seeds, and especially dried fruits; AM is common in warm areas, not common in arid areas and invades heated facilities in cold-temperate areas during winter. AM life history and habits resemble Mediterranean Flour Moth. Female lays ca 200–300 globular, white eggs during lifetime; eggs laid in produce and eclose within three days. Larvae are pale grey with dark head capsule and spots, body with numerous Setae; Larva webs food extensively as it feeds; mature larva 12–15 mm; five larval instars; larval period 22 days; in cooler climates larva overwinters inside silk cocoon. Pupation occurs within a cocoon; if infestation of product is high, then pupation occurs on walls of the storage facility; pupal period ca seven days. Life cycle 1–3 months. Syn. *Ephestia cautella* (Walker); Dried Currant Moth; Tropical Warehouse Moth. Cf. Rasin Moth; Tobacco Moth.

ALOE MITE *Eriophyes aloinis* Keifer [Acari: Eriophyidae].

ALOI, ANTONIO (1845–1900) (Sommier 1900, Boll. Soc. Bot. Ital. 1900: 160.)

AIP A stored-product fumigant (Aluminium Phosphide + $2H_2O$ AlO(OH) + PH_3) used to control stored grain beetles, moths and rodents. Slow acting, but the most effective penetrant available. Trade names include: Alutal®, Delecia®, Detia®, Celphos®, Excel®, Fumitoxin®, Gastion®, Gastoxin®, Phosfinon®, Phostek®, Phostoxin®, Quick-Phos®, Synfume®, Talunex®. See Fumigant.

ALPHA TAXONOMY The technical process of describing and naming of Taxa that are new to science and identifying organisms known to science. See Taxonomy. Cf. Beta Taxonomy, Gamma Taxonomy.

ALPHA-CHLOROPHYLL A form of chlorophyll that produces colours in insects (Wardle). See Betachorophyll.

ALPHA CYPERMETHRIN Noun. A synthetic-pyrethroid compound used as a contact and stomach poison against many diverse insects. AC is used on numerous agricultural crops and for public-health pests in some regions of the world; the compound is not registered in USA. AC displays minor phytotoxicity, and is toxic to bees and fishes. Trade names include: Acquit®, Alpha-Guard®, Bala®, Bestox®, Bonsul®, Concord®, Dominex®, Efitax®, Fastac®, Fendona®, Littac®, Pominex®, Renegade®. See Synthetic Pyrethroids.

ALPHA-FEMALE Formicidae: A normal female when it coexists with an aberrant beta-Female (Imms).

ALPHA-GUARD® See Alpha cypermethrin.

ALPHATE® See Monocrotophos.

ALPHERAKY, SERGHYI NIKOLAEVICH (1850–1902) (Oberthür 1915, Etud. Lepid. Comp. 10: 5.)

ALPINE CASE-MOTH *Lomera caespitosae* (Oke) [Lepidoptera: Psychidae].

ALPINE GRASS-GRUB *Oncopera alpina* Tindale [Lepidoptera: Hepialidae].

ALPINE HAIRY-CICADA *Tettigarcta crinita* Distant [Hemiptera: Tettigarctidae].

ALPINE THERMOCOLOUR-GRASSHOPPER *Kosciuscola tristis* Sjöstedt [Orthoptera: Arctiidae].

ALPINE ZONE A biogeographical concept corresponding to the Arctic Zone of mountains. Arctic-Alpine Zone.

ALPINI, PROSPERO (1553–1617) (Rose 1850, *New General Biographical Dictionary* 1: 361–362.)

ALSTED (ALSTEDIUS), JOHANN HEINRICH (1588–1638) (Rose 1850, *New General Biographical Dictionary* 1: 364.)

ALTACERATUBAE Noun. Some coccids: The Ceratubae modified into large, broad cylinders, with usually oblique mouths and located at or near the margin of the Pygidium (MacGillivray).

ALTERANT Adj. (Medieval Latin, *alterans*; *-antem* = adjectival suffix.) Something with the capacity to alter.

ALTERNATE Adj. (Latin, *alternare* = to do by turns; *-atus* = adjectival suffix.) 1. Occurring in predictable succession at a given place and/or during a specified interval of time. 2. Members of a series that precede, follow or replace other members of the series. Alternately, adv. Noun. (Latin, *alternare* = to do by turns; *-atus* = suffix. Pl., Alternates.) Alternative entities. A substitute. Alt. Alternateness. Verb. (Latin, *alternatus*. Alternated; Alternating.) 1. Transitive: To succeed or to replace in position or time. 2. Intransitive: To happen, to occur, or to vary in reciprocity; followed by with.

ALTERNATION Noun. (Latin, *alternare* = to do by turns; English, *-tion* = result of an action.) Act of alternating or state of being an alternate.

ALTERNATION OF GENERATIONS A form of reproduction found in many groups of invertebrates, including insects. AoG is characterized by inseminated females of a population producing males and females, and subsequent females of this population parthenogenetically producing only females. Fatherless females of the population subsequently produce males and females that reproduce sexually. The phenomenon occurs in many Cynipidae (Hymenoptera), and some Hemiptera. In general, sexual and asexual generations alternate cyclically and regularly. See Heterogeny.

ALTICINAE Flea Beetles. A cosmopolitan, numerically large Subfamily (ca 55 Genera, 8,000 Species) of Chrysomelidae (Coleoptera), characterized by small body size and ability to jump with the aid of a 'metafemoral spring' mechanism. All Species are phytophagous; some Species are pests, and some Species are potentially useful in the control of weeds. Cf. Flea Beetles.

ALTODEL® See Kinoprene.

ALTOSID® See Methoprene.

ALTRICAL Adj. (Latin, *altrix* = nurse; *-alis* = appropriate to, having the character of.) Descriptive of young (immature stages) that require parental care after eclosion from the egg. Cf. Presocial. 2. Relating to the Altrices (Bird Division).

ALTRUISM Noun. (French, *altruisme,* Italian, *altrui* = of or to others; from Latin, *alter* = other; English, *-ism* = condition.) 1. An ethical concept, philosophy or practice of placing the needs or concerns of other individuals of the same Species before the needs of self. 2. Behavioural motivation based on selflessness in which an individual will sacrifice itself for the welfare of the group (colony). Ant., Individualism; Egoism. Rel. Elete.

ALTRUIST Noun. (French, *altruisme.* Pl., Altruists.) An individual with altruism.

ALTRUISTIC BEHAVIOUR Behaviour on the part of one organism that benefits an unrelated organism and at the same time works to the detriment of the organism performing the behaviour.

ALTRUISTIC Adj. (French, *altruisme; -ic* = of the nature of.) Pertaining to altruism. Not selfish. Altruistically, adv.

ALTUM, BERNARD (1827–1900) Author of *Forest Entomology* published in Berlin (1881) (Anon. 1900, Leopoldina 36: 45–46.)

ALTUS An archane Latin term meaning 'high' or 'elevated.'

ALUCITIDAE Plural Noun. Many-Plume Moths. A small Family of ditrysian Lepidoptera assigned to Superfamily Alucitoidea. Alucitids include ca 50 Species that occur in most regions; Family is closely related to Tineodidae. Adult small; wingspan 8–28 mm; head with smooth scales and lateral tufts; Antenna filiform, ca half length of forewing, Scape without Pecten; Ocelli sometimes absent; Chaetosemata absent; Proboscis without scales; Maxillary Palpus small (3–5 segments) or absent; Labial Palpus moderately long; fore Tibia with Epiphysis; tibial spur formula 0-2-4; hind Tibia with long, piliform scales along dorsal margin; wings broad, plumed, spread flat at rest; wings of most Species divided into several regions with each region resembling a plume; Pterostigma absent in forewing. Eggs are flattened, cylindrical with lower ends rounded and laid individually on host plant. Larva is stout with head subprognathous; Prolegs with crochets forming circle. Larva tunnels within flowers, fruits and shoots; some Species form galls. Pupa is well sclerotized; Epicranial Suture present; Maxillary Palpus absent; Cremaster absent but hooked Setae on dorsum of A10; pupation occurs inside cocoon on ground or within gall. Cf. Pterophoridae.

ALULA Noun. (Latin, *ala* = little wing. Pl., Alulae.) 1. A small membranous lobe attached to the base of the forewing in Diptera and some Coleoptera (Tillyard). 2. Some Diptera and Coleoptera: The expanded axillary membrane of the wing. Diptera: Alulet, Squama, Calypter (Comstock). 3. Diptera: A pair of membranous scales above the Halteres. Syn. Aileron; Scale; Axillary Lobe; Calyptra; Cuilleron; Lobulus; Squamula; Tegula.

ALULAR Adj. (Latin, *ala* = little wing.) Pertaining to the Alula.

ALULET Noun. (Latin, *ala* = little wing. Pl., Alulets.) Diptera: A posteriobasal lobe of the wing. Syn. Alar Appendage; Posterior Lobe.

ALUM Noun. (Latin, *alumen* = alum.) Potassium aluminium sulphate, $KAl(SO_4)_2.12H_2O$.

ALUM CARMINE An histological stain composed of carmine, alum, ammonia and water.

ALUM COCHINEAL An histological stain composed of alum, cochineal and water.

ALUM HAEMATOXYLIN An histological stain composed of haematoxylin, alum, alcohol and water.

ALUMINIFORM Adj. (Latin, *alumen* = alum; *forma* = shape.) Descriptive of material that is alum-like in chemical properties or histological characteristics. See Form 2; Shape 2; Outline Shape.

ALUMINIUM Noun. (Latin, *alumen, aluminis.*) A lightweight, very malleable, ductile, rust-resistant metal that comprises about 8% of the earth's crust. Aluminium occurs naturally as mineral (bauxite, cryolite, corundum, *etc.*) or silicate (felspar, mica, kaolin). First isolated by Oersted in 1824, and named 'Alumium' by Davey. Davey changed the name to aluminum and others modified it to aluminium, to conform by analogy with sodium and potassium.

ALUMINIUM CHLORIDE $AlCl_3$ A white, deliquescent crystalline substance obtained by heating aluminium in the presence of hydrogen chloride. Used as a catalyst in organic synthesis and cracking oils.

ALUMINIUM HYDROXIDE $Al(OH)_3$ A gelatinous, white precipitate formed by treating aluminium salts with alkalis. AH is used in histology as a mordant, as a filter and in making lake pigments. AH acts as a weak base and weak acid.

ALUMINIUM SULPHATE $Al_2(SO_4)_3$ A colourless salt produced by treating bauxite or kaolin with sulphuric acid. AS is used as a mordant, in the purification of water, in dyeing and sizing paper. Syn. Alum; Concentrated Alum.

ALURE Noun. (Old French, *aleure*.) A passageway or gallery.

ALUTA Noun. (Latin, *aluta* = leather.) A pliable, alum-dressed leather.

ALUTACEOUS Adj. (Latin, *aluta* = leather; *-aceus* = of or pertaining to.) Taxonomic descriptor pertaining to the surface properties of Aluta. Rather pale leather brown, covered with minute cracks, like the human skin. See Sculpture; Texture. Cf. Coriaceous. Alt. Alutaceus.

ALUTAL® See AlP.

ALVAREZ, RICARDO ZARIQUIEY (1897–1965) (Espanol 1965, Eos Madr. 41: 7–8.)

ALVEARY Noun. (Latin, *alvearium* = a beehive. Pl., Alvearies.) A beehive. See Apiary.

ALVEOLA Noun. (New Latin, from *alveolus* = a small cavity. Pl., Alveolae.) A small pit or depression.

ALVEOLAR Adj. (Latin, *alveolus* = a small cavity.) Descriptive of structure that is pitted. Pertaining to a surface with numerous small pits or depressions. See Alveolate. Cf. Punctate.

ALVEOLATE Adj. (Latin, *alveolus* = a small cavity; *-atus* = adjectival suffix.) 1. Furnished with cells or alveoli. 2. Descriptive of a surface that is deeply pitted. 3. Descriptive of surface sculpture, usually the insect's integument, that is pitted like a honeycomb. Alveolate implies a surface with circular depressions of similar diameter and depth. Alt. Alveolated; Alveolatus. See Sculpture Pattern. Cf. Baculate; Clavate; Echinate; Favose; Gemmate; Psilate; Punctate; Reticulate; Rugulate; Scabrate; Shagreened; Smooth; Striate; Verrucate.

ALVEOLATION Noun. (Latin, *alveolus* = a small cavity; English, *-tion* = result of an action.) A formation of cells or alveoli. The condition of displaying cells or alveoli.

ALVEOLE Noun. (Latin, *alveolus* = a small cavity. Pl., Alveoli.) An Alveolus. See Alveola.

ALVEOLIFORM Adj. (Latin, *alveolus* = a small cavity; *forma* = shape.) Alveolus-shaped; descriptive of structure shaped as an alveolus; structure that is pitted or honeycomb-like in appearance. Alt. Alveolariform. See Alveola. Cf. Alveolate; Favose; Punctate. Rel. Form 2; Shape 2; Outline Shape.

ALYDIDAE Amyot & Serville 1843. Plural Noun. Broad Headed Bugs. A numerically small, widespread Family of Coreoidea (Hemiptera). Body elongate, narrow; head more than half as wide as base of Pronotum; Bucullae very short; legs long, slender; Metaplural Scent Glands with Peritremes. Nymphs of some Species resemble ants in habitus and habits; adults of some Species are wasp mimics. All Species are phytophagous and occur on flowers and foliage; some Species are economic pests; a few records suggest carrion feeding habits.

ALZONA, CARLO (1881–1961) (Guiglia 1961, Ann. Mus. Civ. Stor. Nat. Giacomo Doria 72: 353–357, bibliography.)

AMACRINE CELLS Local Interneurons (nerve cells) of the antennal lobe (Deutocerebrum) of the Brain that display arborizations but lack a conspicuous Axon. See Brain; Deutocerebrum. Cf. Centrifugal Interneurons; Local Interneuron; Projection Interneuron.

AMADELPHOUS Adj. (Greek, *ama* = at the same time, *adelphous* = brother, sibling; Latin, *-osus* = full of.) Gregarious.

AMALGAMATION Noun. (Latin, *amalgama* from Arabic *al-jama-ha* = assembly. Pl., Amalgamations.) 1. The act or process of amalgamating. 2. Uniting races, varieties, strains or biotypes. Cf. Coadunation; Composite; Fusion.

AMANSHAUSERS, HERMAN (–1959) (Schüller 1959, Zeit. Wien. Entomol. Ges. 42: 144.)

AMASTIGOTE Noun. (Greek, *a* = without; *mastig* = whip. Pl., Amastigotes.) An anatomical form manifest at a specific phase in the complex life cycle of some parasitic Protozoa *(Leishmania; Trypanosoma)*. The Amastigote lacks a free Flagellum but the flagellar base and Kinetoplast are present anterior of the nucleus. See Leishmania; Trypanosoma. Cf. Epimastigote; Opisthomastigote; Promastigote; Trypomastigote.

AMATIDAE Plural Noun. See Arctiidae.

AMAUROBIIDAE Plural Noun. White-eyed spiders. Widely distributed. A Family of Araneae in the Suborder Labidognatha. Form irregular webs in concealed places on the ground (rocks, debris). The eyes appear light coloured.

AMAZON ANT Ants of the Genus *Polyergus* that remove immature stages from the nests of other ant Species and take them to the *Polyergus* nest. Ultimately, the immatures are used as slaves in *Polyergus* nests.

AMBER Adj. (Arabic, *anbar* = ambergris.) Consisting of amber. Pertaining to the qualities of amber, particularly in colour. Noun. (Arabic, *anbar* = ambergris. Pl., Ambers.) A fossilized resin of coniferous trees in alluvial soils, on seashores and associated with beds of lignite. Amber is physically characterized as hard, somewhat brittle, transparent, pale yellowish to brown in colour and 2.0–3.0 on the Moh's Scale. Amber resembles plastic (both are polymerized hydrocarbons, hard and take a high polish); amber is distinguished from plastic in that amber floats in a saturated salt solution but plastic does not. Amber will smell resinous when penetrated with a hot needle, but plastic will smell acrid. Amber became abundant during the Late Cretaceous

Period from plants of the Araucariaceae, Taxodiaceae, Cupressaceae and Pinaceae; *Liquidambar* (Hamamelidaceae) and related Genera are found in tropics now. Oldest amber is Cretaceous Period. Most commonly preserved insects included by pine resin, but Amber can occur from angiosperms. Distribution of amber is worldwide, but notable deposits occur in Northern Hemisphere: Baltic Amber (Eocene/Oligocene), Burmese Amber (Eocene, ?Miocene), Canadian Amber (Cedar Lake, Edmonton, Medicine Hat: Upper Cretaceous Period), Chiapas Amber (Oligocene/Miocene), Dominican Amber (Oligocene-Miocene), Lebanese Amber (Lower Cretaceous Period), Philippine Amber (Miocene/Pliocene), Russian Amber (Taimyrian, Yantardakh: Upper Cretaceous Period), Sicilian Amber (Miocene). Amber is an important material for the preservation of insect fossils. See Fossil; Resin. Cf. Chert; Copal.

AMBERIFEROUS Adj. (Arabic, *anbar* = amber; *-iferous* = bearing.) Amber producing.

AMBEROUS Adj. (Arabic, *anbar* = amber; Latin, *-osus* = possessing the qualities of.) Amber-like in colour, texture or appearance.

AMBIDEXTROUS Adj. (Latin, *ambi* = both; *dexter* = right hand; Latin, *-osus* = possessing the qualities of.) Descriptive of humans with the ability to use either hand with equal facility. (However... the word 'dexter' pertains to the right hand or the right side. Thus, formation of 'ambidextrous' is curious because of the implied bias for the right hand.)

AMBIENT Adj. (Latin, *ambiens* from *ambire* = to go around.) Surrounding; Encompassing entirely. Noun. (Latin, *ambiens* from *ambire* = to go around. Pl., Ambients.) An encompassing sphere, such as the atmosphere.

AMBIENT TEMPERATURE The temperature surrounding an object. The temperature of a habitat or place of habitation for an animal or plant.

AMBIENT VEIN 1. Hemiptera: A vein partially encircling the wing near its margin (Tillyard). 2. The vein-like structure stiffening the margin of the wing (Comstock). See Vein; Wing Venation.

AMBIGUITY Noun. (Latin, *ambiguitas* = to wander; English, *-ity* = suffix forming abstract nouns.) Having the quality of being ambiguous.

AMBIGUOUS Adj. (Latin, *ambiguus,* from *ambigere* = to wander about; *-osus* = possessing the qualities of.) A work, condition or concept with two or more possible meanings or outcomes.

AMBIPHARYNX Noun. (Latin, *ambi-* = around; Greek, *pharyngx* = gullet.) A membrane along the mesial margin of the proximal end of each Mandible, forming the lateral aspect of the Prepharynx (MacGillivray).

AMBLYCERA Noun. (Greek, *amblys* = dull; *keras* = horn.) A large Suborder of Phthiraptera (Mallophaga in part) including the Families Boopiidae, Gyropidae, Laemobothriidae, Menoponidae, Ricinidae and Trimenoponidae. See Mallophaga. Cf. Ischnocera.

AMBROCIDE® See Lindane.

AMBROSI, FRANCESCO (1821–1891) (Saccardo 1898, Boll. Soc. Ven. Trent. Sci. Nat. 6: 117–119.)

AMBROSIA Noun. (Greek, *ambrosia* = food of the gods.) 1. Bee-bread. 2. Fungal food cultures of some wood-boring scolytid beetles.

AMBROSIA BEETLES Wood inhabiting Ipidae and all Platypodidae. AB tunnel in sapwood and heartwood of unseasoned timber, construct characteristic galleries but do not feed upon the wood; AB adults carry symbiotic fungi (Ambrosia) that are cultivated within galleries; high moisture content is necessary for fungal development. AB larvae develop inside small cells adjacent to galleries; adult females maintain galleries and give fungi to larvae as food. Female AB remains with offspring until they pupate. Neonate adults carry fungal conidia to new site and thereby spread disease. Damage to timber is caused by holes and galleries constructed by adults and the stain of wood is caused by fungi; larvae do not damage wood. See Scolytidae. Cf. Bark Beetles; Engravers.

AMBULACRA Plural Noun. (Latin, *ambulare* = to walk.) 1. The walking legs of insects. 2. The tube-feet of echinoderms.

AMBULACRAL Adj. (Latin, *ambulare* = to walk; *-alis* = pertaining to, having the character of.) Pertaining to walking.

AMBULACRAL SETA Setae, typically in a ventrolateral position, used to support the body or facilitate locomotion. Syn. Ambulatorial Seta. Cf. Ambulacral Wart.

AMBULACRAL WART Blunt, short, cuticular protuberances positioned on the ventral (sometimes dorsal) surface of the Abdomen. AWs are used to facilitate locomotion in Lepidoptera larvae. Cf. Ambulacral Seta.

AMBULATE Adj. (Latin, *ambulare* = to walk.) To move backward and forward; to walk.

AMBULATORIA Noun. (Latin, *ambulare* = to walk.) Orthoptera in which legs are adapted for walking only (*e.g.* phasmids).

AMBULATORIAL Adj. (Latin, *ambulare* = to walk; *-alis* = appropriate to, having the character of.) Adapted for walking or moving on a substrate or surface. Alt. Ambulatory.

AMBULATORIAL SETA Specialized Setae or bristles on the ventral segments of the Abdomen of some Coleoptera. Syn. Ambulacral Seta.

AMBUSH® See Permethrin.

AMBUSH BUGS See Phymatidae; Reduviidae.

AMDRO® See Hydramethylnon.

AMELETOPSIDAE Plural Noun. A small Family of schistonotous Ephemeroptera assigned to Superfamily Baetoidea. Family currently with three Species and restricted to Australia. Adult wing red or purple; forewing vein MA forked; Cubital intercalaries not forked; Basitarsus partially fused to Tibia and half as long as Tibia;

terminal filament longer than Cercus. Naiads are nocturnal and inhabit stony streams; they are predaceous upon midges and other Ephemeroptera.

AMERICAN ANTEATERS See Myrmecophagidae.

AMERICAN ASPEN-BEETLE *Gonioctena americana* (Schaeffer) [Coleoptera: Chrysomelidae].

AMERICAN BLACK FLOUR-BEETLE *Tribolium audax* Halstead [Coleoptera: Tenebrionidae].

AMERICAN BOLLWORM *Helicoverpa armigera* (Hübner) [Lepidoptera: Noctuidae]. See Bollworm.

AMERICAN BROOD DISEASE See American Foulbrood Disease. Cf. European Brood Disease.

AMERICAN BUMBLE BEE *Bombus pennsylvanicus* (DeGeer) [Hymenoptera: Apidae]: A polylectic Species widespread in grasslands of North America. ABB queens appear during May and nest in the ground or in rodent burrows. Colonies typically are small and workers are aggressive. See Bumble Bee. Cf. Black-Faced Bumble Bee; Yellow Bumble Bee; Yellow-Faced Bumble Bee.

AMERICAN COCKROACH *Periplaneta americana* (Linnaeus) [Blattodea: Blattidae]: A cosmopolitan, urban pest that is probably endemic to Africa and transported to the New World with slave trade and moved elsewhere via commerce. In tropical-subtropical environments, AC tends to live outdoors; in temperate environments AC tends to live indoors. The Species prefers warm, humid, dark habitats and is common in wall voids, sewers, grease traps and garbage dumps. Adult ca 40 mm long, resembles Australian Cockroach but lacks pale stripe along anterobasal margin of forewing; wings macropterous and adult flies during warm weather. Adult lives ca one year; female lays 10–50 Oothecae that are dropped or glued to substrate; each Ootheca contains 12–15 eggs. Nymphs undergo 7–10 moults. AC life cycle requires 6–12 months, depending upon temperature and nutrition. Nymphs and adults eat starch, sugar, leather, parchment and fermented food. See Blattoidea. Cf. Australian Cockroach; Smokybrown Cockroach.

AMERICAN COCKROACHS See Blattidae.

AMERICAN DAGGER-MOTH *Acronicta americana* (Harris) [Lepidoptera: Noctuidae].

AMERICAN DOG-TICK *Dermacentor variabilis* (Say) [Acari: Ixodidae].

AMERICAN FOULBROOD DISEASE A widespread bacterial disease affecting honey bee larvae. The disease is caused by *Paenibacillus* larvae and is nearly always fatal. Larval queens are more susceptible than larval workers to infection. AFB only affects larvae and propupae in capped cells; infected larvae slump to lower side of the cells, become brown-coloured, putrefy and emanate an objectionable odour. The bacterium is rod-shaped, gram-positive, motile with peritrichous flagella; endospores are oval-shaped, heat-tolerant, and resist chemical disinfectants and desiccation for many years. Bee larvae become infected when they ingest spore-contaminated food. Neonate larvae are infected by very few spores; mature larvae require millions of spores to become infected. Spores germinate and become rod-shaped in the larva's intestine; when the larva begins pupation, the bacterial rods pass through the gut epithelium and invade the haemolymph. One infected bee larva can produce 2.5 billion bacterial spores. AFB disease typically is treated by oxytetracycline hydrochloride. See European Honey Bee. Cf. European Foulbrood.

AMERICAN GRASSHOPPER *Schistocera americana* (Drury) [Orthoptera: Acrididae].

AMERICAN HORNET-MOTH *Sesia tibialis* (Harris) [Lepidoptera: Sesiidae].

AMERICAN HOUSE-DUST MITE *Dermatophagoides farinae* Hughes [Acari: Pteronyssidae].

AMERICAN MOUSE FLEA *Stenoponia americana* [Siphonaptera: Hystrichopsyllidae]: A parasite of insectivores and rodents in North America. Adult 4–5 mm long; genal comb with 13 teeth on each side; pronotal comb with 25–26 teeth on each side. See Hystrichopsyllidae. Cf. Chipmunk Flea.

AMERICAN MUCOCUTANEOUS LEISHMANIASIS A form of Cutaneous Leishmaniasis endemic to tropical America. AML affects the mucosa, destroys tissue and bone, and causes horrific disfigurement. Syn. Espundia; Uta. See Leishmaniasis.

AMERICAN PALM-CIXIID *Myndus crudus* Van Duzee [Hemiptera: Cixiidae].

AMERICAN PLUM-BORER. *Euzophera semifuneralis* (Walker) [Lepidoptera: Pyralidae].

AMERICAN SOLDIER FLY *Hermetia illucens* (Linnaeus) [Diptera: Stratiomyidae]: A widespread synanthropic fly that can become a nuisance in some situations. Adult feeds on flowers; female deposits eggs on decaying vegetables, fruit, animal matter or excrement. Larvae feed ca 2 weeks; mature larva ca 20 mm. Enteric Pseudomyiasis may result when ASF eggs are ingested by humans and the larval stage develops within the intestine. Syn. Black Soldier Fly.

AMERICAN SPIDER-BEETLE *Mezium americanum* (Laporte) [Coleoptera: Ptinidae].

AMES, SARAH (–1902) (Anon.1902, Entomol. Rec. J. Var. 14: 139.)

AMETABOLA Noun. (Greek, *a* = without; *metabole* = change.) Insects that develop without Metamorphosis. Specifically, the Apterygota that express a primitive form of development (lack Metamorphosis) and the adult closely resembles the immature stages. See Metamorphosis. Cf. Hemimetabola; Holometabola; Paurometabola.

AMETABOLIC Adj. (Greek, *a* = without; *metabole* = change; *-ic* = characterized by.) Pertaining to insect development that lacks apparent Metamorphosis. Alt. Ametabolous.

AMETABOLION Noun. (Greek, *a* = without; *metabole* = change. Pl., Ametabolions.) An insect that has no distinct Metamorphosis.

AMETABOLOUS Adj. (Greek, *a* = without; *metabole* = change). Pertaining to a form of development that lacks Metamorphosis. Specifically, Ametaboly is characteristic of Apterygota in which the adult closely resembles immature stages. See Development; Metamorphosis. Cf. Hemimetabolous; Holometabolous; Paurometabolous.

AMETABOLY Noun. (Greek, *a* = without; *metabole* = change.) The condition of being Ametabolous. Ametaboly is the groundplan condition among hexapods. More highly evolved hexapods (pterygote insects) have developed patterns of Metamorphosis (probably based on the apterygote groundplan) that are biologically adaptive and anatomically distinct. See Metamorphosis. Cf. Hemimetaboly; Holometaboly; Paurometaboly. Rel. Groundplan.

AMETHYSTINE Adj. (Middle English, *ametist* > Latin, *amethysius* > Greek, *amethystos* not drunken.) 1. Something amethyst-like in colour, bright blue with a reddish admixture. 2. Something made of amethyst. Alt. Amethystinus.

AMETROPODIDAE Plural Noun. A small Family of schistonotous Ephemeroptera assigned to Superfamily Baetoidea or Heptagenioidea, depending upon classification. Adult wing venation resembles Heptageniidae but hind Tarsus with four segments and first segment of male Foretarsus slightly longer than second segment; abdominal Sternum IX without a median notch along its posterior margin. In areas where ametropodids occur, the naiad is commonly taken in streams.

AMIDOCID® See Isofenphos.

AMINATRIX® See Dichlorvos.

AMINO ACID A nitrogen-containing component of proteins; AAs arise from decomposing proteins through exposure to digestive enzymes, acids and alkalis.

AMINOFURACARB A carbamate compound {Ethyl N-[2,3-dihydro-2,2-dimethyl-benzofuran-7-yloxycarbonyl (methyl) aminothio]-*N*-isopropyl-B-alaninate} used as a nematicide, contact insecticide and stomach poison for control of nematodes, aphids, scale insects, leafhoppers, thrips, cutworms, cabbage worms, wireworms and related insects. Aminofuracarb is applied to fruit trees, citrus, corn, rice, potatoes, sorghum, soybeans, sugarbeets, tobacco, vegetables, and ornamentals in some countries. Aminofuracarb is not registered for use in USA because it is toxic to fishes. Syn. Benfuracarb. Trade names include: Furacon® and Oncol®. See Carbamate Insecticide.

AMISEGINAE Noun. (Eytmology obscure.) A predominantly tropical Subfamily of Chrysididae (aculeate Hymenoptera). Adult Frons without medial Sulcus ventrad of median Ocellus; Pronotum usually lacking transverse Sulcus; females frequently brachypterous or apterous with ant-like habitus; gastral venter convex, combined length of Terga I-II much greater than combined length of Terga III-IV, Ovipositor slender and needle-like. Females apparently chew a hole in Chorion of phasmatid eggs and oviposit into eggs; larval stage develops as parasites on embryonic phasmatids.

AMITOSIS Noun. (Greek, *a* = without; *mitos* = thread. Pl., Amitoses.) Cell division by cleavage without change in structure of nucleus. See Mitosis.

AMITOTIC Adj. (Greek, *a* = without; *mitos* = thread; *-ic* = characterized by.) Not mitotic; without mitosis.

AMITRAZ An amidene compound {(N'-2,4-dimethylphenyl)-*N*-[[(2,4-dimethylphenyl)imino]methyl]-N-methylmethanimidamide} used as an acaricide and insecticide. Target pests of Amitraz include lice, sheep keds, mange mites and ticks on livestock, plant-sucking insects (aphids, psyllids, leaf hoppers, whiteflies), bollworms, loopers and perforators. Amitraz is effective as a vapour against ticks and causes mouthparts to be withdrawn from livestock; some ovicidal activity has been observed. Amitraz is registered on cotton, pears and cattle in USA; the compound is applied to other tree crops, ornamentals, vegetables and livestock in other countries. Amitraz is phytotoxic to young peppers and pears at high temperatures and toxic to fishes; Amitraz is not toxic to bees and is used to control apiary pests. Compound is compatible with other insecticides. Trade names include: Acadrex®, Acarac®, Azadieno®, Bumetran®, Danicut®, Ectodex®, Edrizar®, Garial®, Mitac®, Metex®, Ovasyn®, Ovidrex®, Romitraz®, Taktic®, Topline®, Triatix®, Triazid®, Vapcozin®.

AMMAN, JOSEF (1873–1940) (Sachtleben 1941, Arb. Morph. Taxon. Entomol. Ber. 8: 77.)

AMMITZBØLL, IVAR FREDERIK CHRISTIAN (1847–1934) (Henrikson 1935, Entomol. Meddr. 19: 181–183.)

AMMO® See Cypermethrin.

AMMOCHAETA Noun. (Greek, *ammos* = sand; *chaite* = hair; Pl., Ammochaetae.) Deserticolous or xeric adapted ants: Specialized Setae that are elongated and directed forward on the lower surface of the head (Gular Ammochaetae) or in a tuft on the Mentum (Mental Ammochaetae). These Setae are used to remove dust and sand from the foreleg Strigil (Wheeler). See Chaeta. Cf. Psammophore.

AMMONIA Noun. (Greek, *ammoniakon* = resinous gum.) An alkaline, colourless gaseous compound of nitrogen and hydrogen formed by decompositon of protein and urea.

AMMONIUM CARBONATE An unstable salt of ammonia and carbon that decomposes readily, liberating ammonia.

AMMOPHILE Noun. (Greek, *ammos* = sand; *philein* = to love. Pl., Ammophiles.) An organism that lives within or frequently inhabits sandy habitats. *e.g.* Sand Wasps. See Habitat. Cf. Psammophilous.

AMMOPHILOUS Adj. (Greek, *ammos* = sand; *philein* = to love; *-osus* = full of.) Sand-loving; a term used to describe organisms that frequently visit sand habitats, organisms that inhabit sandy areas. See Habitat. Cf. Arenicolous; Deserticolous; Eremophilous; Ericeticolous; Lapidicolous; Psammophilous; Rupicolous; Saxicolous. Rel. Ecology.

AMNEMUS WEEVIL *Amnemus quadrituberculatus* (Boheman) [Coleoptera: Curculionidae]: In Australia, a pest of leguminous pasture plants especially white clover.

AMNION Noun. (Greek, *amnion* = fetal membrane. Pl., Amnions.) 1. The inner 'envelope' or membrane that covers the germ band and ultimately the entire embryo during embryonic development. 2. An extraembryonic membrane between the embryo and serosal membrane of some Species. See Embryo.

AMNION CAVITY See Amniotic Cavity.

AMNION FOLD An extension of the Amnion that closes the mouth of the amnion cavity in the embryo (Smith). See Amniotic Folds.

AMNIOS Noun. (Greek, *amnion* = fetal membrane.) The first cast 'skin' of the larva when a moult occurs almost immediately after emergence from the egg. The Amnios is probably embryonic membrane and not larval integument.

AMNIOTIC Adj. (Greek, *amnion* = fetal membrane; *-ic* = characterized by.) Descriptive of or pertaining to the Amnion.

AMNIOTIC CAVITY In the developing egg: A space or cavity formed by the amniotic folds and which contains the germ band (Imms).

AMNIOTIC FOLDS Folds arising from the edge of the germ band in the Ovum; the folds usually meet medially (Imms).

AMNIOTIC PORE A permanent opening in the amniotic cavity during the development of certain insects (Imms).

AMOEBA Noun. (Greek, *amoibe* = a change; alternation. Pl., Amoebae.) A Genus of microscopic, aquatic, single-celled Rhizopods, the Species of which propel themselves by means of Pseudopodia.

AMOEBA DISEASE 1. *Malpighamoeba mellificae* Prell, an amoeba that infests Malpighian Tubules of adult honey bees. 2. *Melemba locustae* (King & Taylor), an amoeba that infests Malpighian Tubules of acridid Orthoptera.

AMOEBIFORM Adj. (Greek, *amoibe* = a change; alternation. Latin, *forma* = appearance.) 1. Amoeba-shaped; descriptive of structure without rigid form. 2. Structure with the appearance, properties or behavior of an amoeba. Syn. Amoeboid. Alt. Amebiform. See Amoeba. Cf. Amoeboid; Proteiform. Rel. Form 2; Shape 2; Outline Shape.

AMOEBOCYTE Noun. (Greek, *amoibe* = a change, alternation; *kytos* = hollow vessel; container. Pl., Amoebocytes.) A kind of blood cell mentioned in some classifications of Insect Haemocytes. Amoebocytes typically are nucleated, vary in shape and display amoeba-like characteristics when active. Syn. Plasmatocyte. Cf. Leucocyte. Rel. Circulatory System.

AMOEBOID Adj. (Greek, *amoibe* = a change; alternation; *eidos* = form.) Amoeba-like in outline shape or movements; descriptive of something that moves and changes shape as it moves. See Amoeba. Cf. Amoebiform; Malacoid. Rel. Eidos; Form; Shape.

AMORBUS BUGS *Amorbus* spp. [Hemiptera: Coreidae].

AMOREAUX, PIERRE JOSEPH (1741–1824) (Rose 1850, *New General Biographical Dictionary* 1: 405.)

AMORETTI, CARLO (1741–1816) (Rose 1850, *New General Biographical Dictionary* 1: 405.)

AMORPHA Noun. (Greek, *a* = not; *morphe* = form. Pl., Amorphae.) 1. Holometabolous insects in which the pupa bears no resemblance to the Imago (adult). See Metamorphosis. 2. A small genus of Fabaceae, including false indigo and lead plant.

AMORPHOSCELIDAE Plural Noun. A small Family of Mantodea including 11 Genera and a few Species that occur around the Mediterranean area and within the Ethiopian and Indoaustralian Realms. Adult small bodied, cryptically coloured, and occur on ground and tree trunks; a few Species mimic ants. Head usually with ocellar tubercles. Fore Femur spinose and Tibia with apical claw; proximal outer spine of Femur subequal to other spines. Males usually macropterous; female usually brachypterous or apterous; hindwing with one vein in fold between main blade and anal fold.

AMORPHOUS Adj. (Greek, *amorphos* > *a* = not; *morphe* = shape, form.) Descriptive of something without apparent organization or definite shape. Alt. Formless; Shapeless; Unstructured. Rel. Decomposed.

AMOS, D W (–1949) (Manson-Bahr 1950, Nature 166: 422.)

AMPHIBICORIZAE A Suborder of semiaquatic (shore-inhabiting) Hemiptera whose members have conspicuous Antennae and three pairs of Trichobothria on the head.

AMPHIBIOTIC Adj. (Greek, *amphi* = both; *biotikos* = living; *-ic* = of the nature of.) Insects whose active immature stage (naiad) is aquatic and whose adult stage is terrestrial. Cf. Amphibious.

AMPHIBIOTICA Noun. (Greek, *amphi* = both; *biotikos* = living. Pl., Amphibioticae.) An obsolete taxonomical or ecological arrangement of insects that includes Species with aquatic nymphs (naiads) or larvae and winged (aerial) adults.

AMPHIBIOUS Adj. (Greek, *amphi* = both; *bios* = life; Latin, *-osus* = possessing the qualities of.) Pertaining to organisms that can live in aquatic or terrestrial environments, irrespective of life stage.

AMPHIENTOMETAE Plural Noun. A small Family of troctomorph Psocoptera assigned to Superfamily Amphientometae. Body and wings with scales; Antenna with fewer than 20 segments and secondary annulations.

AMPHIMIXIS Noun. (Greek, *amphi* = both; *mixis* = mingling. Pl., Amphimixes.) 1. The mingling of the germ plasm of two individuals. 2. The fusion of the egg pronucleus and sperm pronucleus to form the zygote nucleus. Cf. Automixis.

AMPHIODONT Noun. (Greek, *amphi* = both; *odous* = tooth. Pl., Amphidonts.) An anatomical form of male lucanid beetle with Mandibles of medium size and intermediate form between the teleodont and priodont forms. See Mandible. Cf. Mesodont.

AMPHIPNEUSTIC Adj. (Greek, *amphi* = both sides; *pneusis* = breathing; *-ic* = of the nature of.) Pertaining to larval respiration.

AMPHIPNEUSTIC RESPIRATION A type of larval respiratory system in which only anterior thoracic and posterior 1–2 abdominal spiracles are functional. See Respiration. Cf. Apneustic; Oligopneustic; Polypneustic.

AMPHIPSOCIDAE Plural Noun. A small Family of psocomorph Psocoptera assigned to Superfamily Caecilietae. Antenna with 13 segments and lacking secondary annulations; wing often brachypterous, setose with venation reduced; Pterostigma long and narrow; Areola Postica and M apically fused; Tarsi with two segments. Species colonial under webbing they construct on leaves.

AMPHIPTERYGIDAE Plural Noun. A small Family of zygopterous Odonata assigned to Superfamily Calyopterygoidea. Adult medium to large size, body robust; wings narrow and petiolate with two basal Antenodal Crossveins projecting across Costal and Subcostal spaces; Arculus halfway between base and Nodus. Naiad with large, conspicuous eyes; Antenna with six segments; inhabits moving water.

AMPHIS LIONS See Neuroptera.

AMPHITOKY See Deuterotoky.

AMPHIZOIDAE Plural Noun. Trout-stream Beetles. A minute Family of adephagous Coleoptera assigned to Superfamily Caraboidea. Amphidoids display a disjunct distribution: One Species in Tibet and four Species in North America. All Species are predaceous; adults and immatures feed on Plecoptera larvae in backwaters of fast-moving streams.

AMPHORA Noun. (Greek, *amphiphoreus* = jar with two handles. Pl., Amphorae.) An ancient Greek jar with characteristic design: A large cylindrical body with constricted neck and two handles that originate near the region of constriction and insert near the mouth (aperture) of the jar. The prototypical design for the Roman Ampulla. The term is used to characterize structural shape among insects. See Ampulla. Rel. Form, Shape.

AMPLE Adj. (Latin, *amplus*.) Descriptive of structure that is broad or large; sufficient in size.

AMPLECTANT Adj. (Latin, *amplecti* = to embrace; *-antem* = adjectival suffix.) Clasping. Pertaining to structure embracing or closely appressed to another body. Alt. Adjacent; Juxtaposed;

AMPLECTED Adj. (Latin, *amplexus* = embrace.) Descriptive of the head set into a concavity on the Prothorax (*e.g.* beetles of the genus *Hister*).

AMPLEXIFORM COUPLING Lepidoptera: A type of wing-locking mechanism that lacks a Frenulum. The hindwing humeral lobe is enlarged and projects beneath the forewing.

AMPLIATE Adj. (Latin, *amplio* = to make large; *-atus* = adjectival suffix.) Descriptive of a structure that is moderately dilated. Alt. Ampliatus.

AMPLIFICATE Adj. (Latin, *amplificatio* = an enlarging; *-atus* = adjectival suffix.) Dilated; enlarged. Alt. Amplificatus.

AMPLIFYING HOST Medical Entomology: A vertebrate host that is infected with pathogens at a level sufficiently high for a blood-feeding arthropod vector to be infected. Cf. Dead-End Host; Resistant Host; Silent Host; Susceptible Host.

AMPLIXICATE Adj. (Latin, *amplus* = large; *-atus* = adjectival suffix.) Dilated. Alt. Amplicated; Amplixicatus.

AMPULICIDAE Plural Noun. A small Family of sphecoid Hymenoptera; sometimes placed as a Subfamily within the Sphecidae. Adult ampulicid body typically less than 10 mm long, black and red coloured. North American Species nests are concealed under bark, leaves, stones or in twigs; nest are provisioned with cockroaches. Female wasp paralyzes cockroach, leads cockroach to nesting site, then constructs nest; one cockroach per nest is used as provision. Included Genera: *Ampulex, Aphelotoma, Austrotoma, Dolichurus, Paradolichurus* and *Trirogma*.

AMPULLA Noun. (Latin, *ampulla* = flask. Pl., Ampullae.) 1. A glass flask consisting of a globular body and a constricted neck with two handles. See Amphora. 2. Any membranous vesicle, blister or blister-like surface structure. 3. Orthoptera: An extensile sac between the head and Prothorax used by the embryo to escape from the Ootheca, and later, in moulting. 4. Heteroptera: A blister-like enlargement at the middle of the anterior margin of the Prothorax. 5. Genitalia of male Lepidoptera: A process (often finger-like) arising from the medial surface of the Harpe, near the base and extending more-or-less dorsad. See Transitilla (Klots).

AMPULLACEOUS Adj. (Latin, *ampulla* = flask; *-aceus* = of or pertaining to.) Ampulla-like; flask-shaped. See Ampulliform.

AMPULLACEOUS SENSILLUM See Ampulliform Sensilla.

AMPULLIFORM Adj. (Latin, *ampulla* = flask; *forma* = shape.) Ampulla-shaped; bladder-shaped. Descriptive of a globular structure with a constricted neck and single aperture. Syn. Ampulaceous. See Ampulla. Cf. Acetabuliform; Adeniform; Guttiform; Lageniform; Napiform; Vasiform. Rel. Form 2; Shape 2; Outline Shape.

AMPULLIFORM SENSILLUM A sense organ in which the sense cone is in a more-or-less flask-shaped cavity with an axial rod. AS is thought to be auditory. Syn. Sensillum Ampullaceum; Ampullaceous Sensilla. See Sensillum.

AMSTEIN, JOHANN RUDOLPH (1777–1862) (Stein 1862, Jber. Naturf. Ges. Graubünden 7: 178–187.)

AMYDRIIDAE See Tineidae.

AMYGDALA Noun. (Greek, *amygdale* = almond. Pl., Amygdalae.) 1. A tonsil of the Pharynx. 2. A rounded prominence on the lower part of a hemisphere of the Cerebellum.

AMYGDALACEOUS Adj. (Latin, *amygdala* = almond; *aceus* = of or pertaining to.) Almond-like.

AMYGDALIFORM Adj. (Latin, *amygdala* = almond; *forma* = shape.) Almond-shaped; See Almond. Cf. Lentiform; Napiform. Rel. Form 2; Shape 2; Outline Shape.

AMYL ALCOHOL $C_5H_{11}OH$ Colourless liquid with a cough-inducing odour. A source of amyl compounds, including amyl nitrate.

AMYLASE Noun. (Greek, *amylon* = starch, fine meal. Pl., Amylases.) A digestive enzyme that converts starch into maltose or malt sugar, a constituent of saliva.

AMYLOLITIC Adj. (Greek, *amylon* = starch, fine meal; *lysis* = to break down; *-ic* = characterized by.) Descriptive of or pertaining to the enzyme amylase that has the action or effect of amylase and aids in digesting starch.

AMYOT, CHARLES JEAN BAPTISTE (1799–1866) French taxonomic Entomologist and early worker on insects of India. (Signoret 1866, Ann. Soc. Entomol. Fr. (4) 6: 603–606, bibliography.)

ANABOLIC Adj. (Greek, *ana* = up; *bole* = throw; *-ic* = of the nature of.) Pertaining to the constructive metabolic change of food material to animal tissue. See Metabolism. Cf. Catabolic.

ANABOLISM Noun. (Greek, *ana* = up; *bole* = throw; English, *-ism* = condition. Pl., Anabolisms.) A productive metabolic process in which biochemical reactions within the animal body produce proteins, fats and carbohydrates from food materials. See Metabolism. Cf. Catabolism.

ANACERORES Noun. Coccids: Cerores located in the Rectum; Structures also known as rectal wax pores, rectal Spinnerets of honey dew glands (MacGillivray).

ANACHARITIDAE Plural Noun. A small Family of parasitic Hymenoptera often placed as Subfamily of Figitidae. Adult body slender, medium-sized; wedge-shaped head distinctly wider than Thorax; Gena not margined (except *Paraegilies* Kieffer); basal Tergum longest but shorter than half Gaster. Larvae develop as parasites of Neuroptera (Hemerobiidae).

ANAEROBE Noun. (Greek, *an* = without; *aer* = air; *bios* = life. Pl., Anaerobes.) An organism (some bacteria) that respires without air or oxygen. See Respiration. Cf. Aerobe.

ANAEROBIC Adj. (Greek, *an* = without; *aer* = air; *bios* = life; *-ic* = characterized by.) Pertaining to organisms able to live without air containing levels of oxygen necessary to sustain most forms of life. Alt. Anaerobiosis. See Respiration. Cf. Aerobic.

ANAGENESIS Noun. (Greek, *ana* = again; *genesis* = birth. Pl., Anageneses.) 1. Progressive evolution: The evolutionary process in organisms seen through natural or sexual selection. A term applied to a structure or organ that leads to improvement in terms of its function. Rensch (1959) notes essential features of Anagenesis: Increased complexity of structure; 'rationalization of structures and functions' (oligomerization); special complexity and rationaliziation of the nervous system; increased plasticity of structures and functions; modification permitting improvement; enhanced independence from the environment. See Evolution. Cf. Cladogenesis; Katagenesis. Rel. Transformation Series. 2. Anatony: Regeneration of tissue.

ANAJAPYGIDAE Plural Noun. A small and rare Family of Diplura.

ANAL Adj. (Latin, *anus* = anus; *-alis* = characterize by.) Descirpive of structure or process associated with the Anus. In the direction of, pertaining or attached to the Anus or to the last abdominal segment. Cf. Oral.

ANAL ANGLE 1. Hindwing: The angle nearest the apex of the Abdomen when the wings are expanded. 2. The angle between the inner and outer margins of any wing. 3. Hind angle of the forewings.

ANAL APPENDAGES Odonata: Movable appendages at the apex of the Abdomen (Garman). The external genitalia (Smith).

ANAL AREA The posterior or anal part of a wing that is supported by the Anal Veins. The axillary area.

ANAL CELLS Spaces between the Anal Veins (Comstock). Diptera: Space in the wing membrane nearest the body and enclosed by fifth and sixth veins. AC are sometimes called the third Basal Cell (Coquillet) or first Anal Cell (Comstock).

ANAL CERARI Coccoids: The last pair of Cerari, positioned on the anal lobes (MacGillivray).

ANAL CLEFT Coccoids: A deep incision in females that extends from the anal orifice.

ANAL COMB 1. Lepidoptera larvae: A mesial, sclerotized ventral projection on the anal sclerite and is used to eject frass. 2. Siphonaptera larvae: A few rows of Setae on the tenth abdominal segment. 3. Coleoptera larvae: Cerci-like appendages. Syn. Anal Fork; Anal Catapult.

ANAL CROSSVEIN 1. Diptera: The crossvein closing the Anal Cell apically. 2. The CU2 of Comstock-Needham *teste* Curran. See Vein; Wing Venation.

ANAL DISC Simuliidae larvae: The sucker-like area of attachment at the apex of the Abdomen. Cf. Anal Foot.

ANAL FAN A fan-like extension of the anal area of the insect hindwing (Tillyard).

ANAL FIELD Orthoptera: The area on the Tegmina corresponding to the anal area of the hindwing.

ANAL FILAMENTS See Caudal Seta.

ANAL FIMBRIA Hymenoptera: A setal fringe surrounding the Anus.

ANAL FOLD See Plica Vannalis.

ANAL FOOT Chironomidae: The posterior end of the larval body that is modified to serve as a holdfast. Cf. Anal Disc.

ANAL FORCEPS Coleoptera: A strongly chitinized structure of the genitalia that projects from the Abdomen (Tillyard).

ANAL FORK See Anal Comb.

ANAL FURROW 1. The Vannal Fold. See Plica Vannalis. 2. The suture-like groove in the wing membrane, usually in the Cubito-Anal folds. 3. Heteroptera: The Anterior part of the Cubitus (Comstock).

ANAL GLANDS Ectodermal glands of the Alimentary Canal, opening into it near the posterior extremity or the Anus, secreting either a lubricant, a silk-gum, or some other non-fecal material. See Exocrine Gland.

ANAL HOOKS Lepidoptera pupae: Hooked or clubbed Setae at the apex of the Abdomen that attach the pupa to the cocoon or silk pad.

ANAL HORNS Collembola: Small processes on the posterior-most abdominal segment. Sphingidae larvae: A spine-like mesial process on the eighth abdominal segment.

ANAL LEGS Lepidoptera larva: The appendages (or legs) of the tenth abdominal segment. See Leg. Cf. Proleg.

ANAL LOBE 1. Hymenoptera: The posterior lobe of the wings (Comstock). 2. Diptera: The basal part of the wing behind the Anal Vein (Curran). 3. Coccidae: A pair of small, triangular, hinged processes forming a valve that covers the anal orifice. 4. Immatures: Any protrusion of the Integument near the Anus. See Lobe.

ANAL LOOP Odonata: Wing area including a few to several cells between the branches of the Anal Vein, or between Cubitus and first Anal Vein (Garman). AL is also present in Trichoptera.

ANAL MEMBRANE Odonata: An opaque membrane in the hindwing of some dragonflies beginning at the articulation of the wing and extending along the hind margin. The membranule *sensu* Garman.

ANAL NERVURE The nerve of the wing separating the Cubitus from the anal area (Jardine).

ANAL OPENING The anal orifice; Anus. The posterior opening of the Alimentary Canal (Tillyard).

ANAL OPERCULUM 1. Lepidoptera larva: The dorsal arch of the tenth abdominal segment. 2. A supra-anal sclerite. See Operculum.

ANAL ORGANS Collembola: Two modified Setae arising from a tubercle ventro-cephalad of the Anus and usually curving caudo-dorsad.

ANAL ORIFICE See Anus.

ANAL PAPILLAE Culicidae (anopheline larvae): Four soft, white protuberances on abdominal segment IX. Collembola: See Anal Tubercle.

ANAL PLATE 1. Lepidoptera larva: The shield-like covering of the dorsum of the last segment. 2. Embryonic larva: Tergum XI. Coccoids (Lecaniinae): A pair of triangular or semicircular sclerites at the cephalic end of the caudal cleft (Comstock). 3. Operculum *sensu* MacGillivray; valves of the Operculum *sensu* Green. See Sclerite.

ANAL PROLEG Immature Holometabola: Prolegs located on the last abdominal segment (*e.g.* Trichoptera.) See Proleg.

ANAL REGION An area of the wing separated from the Remigium by an anal (vannal) fold. In insects that flex their wings, this fold permits the Remigium to cover the anal region. The anal area is variable in size and shape. AR is often triangular and best expressed in the posterior wing. AR is reduced in powerful flying insects but relatively large and flexible in slow-flying insects. Syn. Vannal Region. See Wing. Cf. Axillary Region; Jugal Region; Remigium.

ANAL RING Coccidae: An elevated ring-like structure surrounding the Anus (MacGillivray); the Genital Ring or Genito-Anal Ring.

ANAL SCALE Cynipidae: A process of the Ovipositor positioned lateral and ventral of the lateral scale (Smith).

ANAL SETAE Coccids: One or more prominent Setae of the Anal Lobes (Comstock).

ANAL SHIELD See Anal Sclerite.

ANAL SIPHON The anal breathing tube of culicid larvae.

ANAL SLIT Immatures: The Anus or narrow anal opening.

ANAL STRUT Siphonaptera larvae: Elongate cuticular appendages directed caudad and ventrad from the tenth abdominal segment.

ANAL STYLE Hemiptera: A slender process on or within the terminal segment of the Abdomen.

ANAL TRIANGLE Wings of some Anisoptera: A well-marked triangular area bounded anteriorly by vein A' and distally by vein A3 (Comstock).

ANAL TUBERCLE 1. Collembola: The tubercle bearing the anal organs; Anal Papilla. 2. Coccoidea: A pair of prominent, rounded or conical processes, one on each side of the Anus.

ANAL VALVE See Podical Plates.

ANAL VEINS 1. Longitudinal, unbranched veins that extend from the base of the insect wing to the outer margin below the Cubitus. 2. The first anal; Vena Dividens *sensu* Smith. 3. All veins between the Cubitus and Jugal region, according to the Comstock–Needham System. 4. Odonata:

Longitudinal veins 6–9 behind the Cubitus and commonly short or abbreviated (Garman). See Vein; Wing Venation.

ANAL VESICLE Parasitic Hymenoptera: A membranous sac at the posterior end of an early instar endoparasitic larva. AV formed by an evagination of the Rectum.

ANALIS Noun. (Latin, *anus* = anus.) A projection formed by the Anal and Cubital Veins (MacGillivray).

ANALOGOUS Adj. (Latin, from Greek, *analogos* = ratio, proportionate; Latin, *-osus* = possessing the qualities of.) Pertaining to similarity in function, but differing in origin and structure. Classical examples include the wings of birds and insects. See Homologous; Paralogous.

ANALOGUE Noun. (Greek, *analogos* = ratio, proportionate. Pl., Analogues.) 'A part or organ in one animal which has the same function as another part or organ in a different animal' (Owen 1843). See Analogy. Cf. Homologue. Syn. Analog.

ANALOGY Noun. (Greek, *analogia* = proportion. Pl., Analogies.) 1. 'A part or organ in one animal which has the same function as another part or organ in a different animal.' (Owen (1843). 2. Resemblance in function between two or more structures among organisms, but not through common descent or construction. (See Boyden 1947, Homology and Analogy: A critical review of the meanings and implications of these concepts in biology. Amer. Midl. Nat. 37: 648–669.) See Evolution. Cf. Homology; Paralogy. Rel. Phylogeny.

ANAL-RING HAIRS Coccids: A variable number of long, stout Setae on the Anal Ring (Comstock). Syn. Ring Setae; Ring Spines.

ANAMORPH Noun. (Greek, *ana* = throughout; *morphe* = form. Pl., Anamorphs.) The sexual phase in the life cycle of fungi that produces asexual spores such as conidia.

ANAMORPHOSIS Noun. (Greek, *ana* = throughout; *morphosis* = shaping. Pl., Anamorphoses.) 1. The evolution of one 'form' to another 'form' through a series of gradual changes. See Anagenesis. 2. An increase in the number of segments during moulting in an arthropod after emergence from the egg. The addition of body segments occurs at a zone of growth anterior of the terminal segment of the body. The phenomenon is expressed in Protura and regarded as a primitive condition in metamorphosis. See Metamorphosis. Cf. Acrogenesis; Epimorphosis. Rel. Development; Growth.

ANAPHYLAXIS Noun. (Greek, *ana* = upon; *phylax* = guard. Pl., Anaphylaxes.) A human medical condition of hypersensitization induced by foreign proteins obtained by bite, invasion or sting. Aculeate Hymenoptera often implicated in Anaphylactic Shock.

ANAPHYTE Noun. (Greek, *ana* = up; *phyton* = plant. Pl., Anaphytes.) An internode.

ANAPLASMOSIS Noun. (Greek, *ana* = up; *plasma* = formed, molded; *sis* = a condition or state. Pl., Anaplasmoses.) Infectious anaemia: An often fatal disease of cattle caused by the bacterium *Anaplasma marginale*. Ticks serve as biological vectors and tabanid flies serve as mechanical vectors for the pathogen.

ANAPLEURITE Noun. (Greek, *ana* = up; *pleura* = side; *-ites* = constituent. Pl., Anapleurites.) The sclerotized area above the Coxa (supracoxal area) in a generalized thoracic Pleuron. The Eupleuron *sensu* Snodgrass.

ANAPLURAL Adj. (Greek, *ana* = up; *pleura* = side.) Descriptive of or pertaining to an Anapleurite.

ANAPTERYGOTA Noun. (Greek, *ana* = backward; *pterygota*. Pl., Anapterygotes; Anapterygotae.) Wingless adult insects that have evolved from winged ancestors (*e.g.* fleas, lice).

ANASTOMOSE Adj. (Greek, *anastomosis* = an opening; coming together; *-osus* = full of.) Running into each other (*e.g.* like converging lines or veins of an insect's wing).

ANASTOMOSIS Noun. (Greek, *anastomosis* = coming together; an outlet. Pl., Anastomoses.) 1. The intimate (complex) connection of similar structures such as wing veins, blood vessels or nerves. In Entomology, the term is usually applied to wing veins, often to markings; sometimes equivalent to the term Stigma in the broad sense. 2. Plecoptera: Transverse Cord (Comstock).

ANATOMIST Noun. (Greek, *ana* = up; *tome* = cutting. Pl., Anatomists.) A student of anatomy. Correct venereal term: A *corps* of anatomists. See Anatomy. Cf. Morphologist; Histologist; Pathologist.

ANATOMY Noun. (Greek, *ana* = up; *tome* = cutting. Pl., Anatomies.) The discipline of Biology concerned with naming the parts and describing the structure of organisms based upon gross observation, dissection and microscopical examination. See Morphology. Cf. Comparative Anatomy; Histology.

ANATREPSIS Noun. (Greek, *anatrepein* = to turn over; *-sis* = a condition or state, Pl., Anatrepses.) During blastokinesis of the insect embryo, the passage of the embryo from the ventral to the dorsal aspect of the Ovum (Wheeler). Cf. Katatrepsis. Rel. Blastokinesis.

ANAUTOGENOUS Adj. (Greek, *a* = without; *auto* = self; *genes* = producing; Latin, *-osus* = full of.) Pertaining to adult insects (including parasitic Diptera and Hymenoptera) that require a meal or ingested food before eggs can mature or oviposition can occur. See Development. Cf. Autogenous.

ANAXIAL Adj. (Greek, *an* = without; Latin, *axis* = axis; *-alis* = having the character of.) Descriptive of structure that lacks an apparent axis. Asymmetrical. See Orientation.

ANAXIMANDER (611–547 BCE) Greek mathematician, astronomer and philosopher from Miletus. Student of Thales and credited with the invention of geographical maps.

ANAXIMENES of MILETUS (585–528 BCE) Greek philosopher, student of Anaximander and advocate of air as the elemental substance of which all physical objects are composed.

ANAXYELIDAE Martynov 1925. Plural Noun. Anaxyelid Wasps. A fossil Family of symphytous Hymenoptera assigned to Superfamily Siricoidea.

ANCEPS Adj. (Greek, *an* = without; *kephale* = head.) Two-edged; ensiform (Say).

ANCESTOR Noun. (Middle English, *ancestre* > Latin, *antecessor* = one who goes before; *or* = one who does something. Pl., Ancestors.) 1. An organism from whom another organism is descended. 2. An early form of organism from which other forms evolve. Cf. Descendant.

ANCESTRAL Adj. (Latin, *antecedere* = to go before; *-alis* = pertaining to, appropriate to.) Pertaining to a primitive stage of development; inherited from an earlier form or ancestor. See Classification.

ANCESTRAL INGROUP NODE Cladistics: A branching point on a cladogram that represents the ancestor of a Taxon targeted for phylogenetic analysis. The ancestral ingroup node is established once polarities of the transformation series have been established. The ancestral group is comprised of plesiomorphic or primitive characters. (See *The Compleat Cladist. A Primer of Phylogenetic Procedures*. E. O. Wiley *et al.* 1991, University of Kansas Special Publication 19: 158.)

ANCEY, CÉSAR MARIE FÉLIX (1860–1906) (Fisher 1908, J. Conch. London 55: 404–412.)

ANCHOR PROCESS A fleshy process at the anterior end of some dipterous larvae; the breastbone. Syn. Sternal Spatula.

ANCILOTTO, ENRICO (1902–1971) (Anon. 1972, Boll. Soc. Ent. Ital. 104: 21.)

ANCIPITAL Adj. (Latin, *anceps* = double.) Descriptive of structure that is flattened and with two opposite edges or angles.

ANCISTROPSYLLIDAE Plural Noun. A small and rare Family of Siphonaptera assigned to Superfamily Ceratophylloidea.

ANCYROID See Ankyroid.

ANDALIN® See Flucycloxuron.

ANDEAN PLAGUE A form of Plague caused by the bacterium *Yersinia pestis* and vectored by fleas. Disease is localized in Andes Mountains of South America. An inflammatory form of Plague that produces malaise, vomiting and sometimes death. See Plague.

ANDERSEN, JOHANNES (1885–1961) (Anon. 1961, Ent. Meddr 31: 1–2.)

ANDERSON, EDWIN J (–1975) (Anon. 1975, Bee World 56: 54.)

ANDERSON, JAMES (1739–1808) (Rose 1850, *New General Biographical Dictionary* 1: 430–431.)

ANDERSON, JOHN (1873–1939) (Anon. 1939, Amer. Bee J. 79: 340.)

ANDERSON, WILLIAM ESCO (1892–1952) (Anon. 1952, J. Econ. Entomol. 45: 560.)

ANDERSSON, CARL JOHAN (1827–1867) (MacGillivray 1947, Entomol. Ber. Amst. 12: 173–175.)

ANDERSSON, SNICKARE ERIC (1890–1939) (Anon. 1940, Bitidningen 2(2): 42.)

ANDRADE, NUNO FREIRE DE (1924?–1958) (R. F. D'A. 1960, Mem. Estud. Mus. Zool. Univ. Coimbra 262: (2–4), bibliography.)

ANDRE, ALBERT ARGOT-VALLON (1859–1936) (Anon. 1937, Arb. Morph. Taxon. Entomol. Berl. 4: 160.)

ANDRE, ERNST (1838–1914) (Alland 1914, Bull. Entomo. Soc. Fr. 1914: 221.)

ANDRE, FLOYD (1909–1972) (Brindley 1972, J. Econ. Entomol. 65: 938.)

ANDRE, JAQUES ERNST EDMOND (1844–1891) (Godman 1891, Proc. Entomol. Soc. London 1891: l-li.)

ANDRE, MARC (1900–1966) (Anon.1966, Entomol. Berl. Amst. 26: 193–194.)

ANDREAS, KARL (1867–1932) (Gaul 1932, Entomol. Zeit. Frankf. A. M. 46: 133.)

ANDREINI, ALFREDO (1870–1943) (DeBeaux 1945, Ann. Mus. Civ. Stor. Nat. Giacomo Doria 62: = 1–2.)

ANDREINI, ALFREDO (1891–1948) (Anon. 1948, Mem. Soc. Entomol. Ital. 27: 52–63, bibliography.)

ANDRENID BEES See Andrenidae.

ANDRENIDAE Latreille 1802. Plural Noun. Andrenid Bees. A Family of ground-nesting, solitary, aculeate Hymenoptera assigned to Superfamily Apoidea (bees). Adult small to moderate sized; two Subantennal Sutures beneath antennal socket; Facial Foveae present; Labrum wider than long, subtriangular; Flabellum absent; Galea short basad of Palpus and usually short apical of Palpus; marginal cell normal, Stigma large and elongate; basitibial sclerite present in female; Scopa present on hind Tibia, absent from abdominal Sterna; middle or posterior part of second Recurrent Vein not arcuate outward; Pygidium present in female, rarely present in male; Arolia well developed in male and female. Andrenids nest in soils as Halictidae and often in large aggregations; their diet typically is oligolectic. Subfamilies include Andreninae and Panurginae. See Apoidea.

ANDRENINAE The nominant Subfamily of Andrenidae, consisting of about five Genera; forewing typically with three Submarginal Cells and Marginal Cell apically pointed.

ANDRES, ADOLF (1894–1921) (Gassner 1931, Koleopt. Rdsch. 17: 255–256.)

ANDREWES, FREDERICK (1859–1932) (E.B.P. 1932, Obit. Not. Fell. Roy. Soc. London 1: 37–44.)

ANDREWES, HERBERT EDWARD (1863–1950) (Britton 1950, Entomol. Mon. Mag. 87: 64.)

ANDREWS, E A (–1972) (Lees 1974, Proc. Roy.

Entomol. Soc. London (C) 38: 58.)

ANDREWS, H L (–1946) (Williams 1949, Proc. Roy. Entomol. Soc. London (C) 13: 66.)

ANDREWS, HENRY WILLIAM (1876–1955) (E.A.C. 1955, Entomologist 88: 166–167.)

ANDREWS, WILLIAM VALENTINE (1811–1878) (Anon. 1878, Canad. Entomol. 10: 240.)

ANDROCHROMATYPIC Adj. (Greek, *aner* = man; *chromo* = colour; *typos* = type; *-ic* = characterized by.) Pertaining to male-mimicking females in a sexually dimorphic population. (Hilton 1987, Entomol. News 98 (5): 221–223.) Syn. Andromorphic; Homochrome; Homeochromatic; Isochromatic; Isomorphous. Cf. Gynochromatypic. Rel. Sexual Dimorphism.

ANDROCONIUM Noun. (Greek, *aner* = male; *konia* = dust; Latin, *-ium* = diminutive > Greek, *-idion*. Pl., Androconia.) Lepidoptera: So-called 'scent scales' which are anatomically different from other scales and localized on some male butterflies. Syn. Androconia Scales. Cf. Coremata; Hair Pencil; Stobbe's Gland. Rel. Pheromone Gland.

ANDROERGATOGYNOMORPH Noun. (Greek, *aner* = male; *ergates* = worker; *gyne* = female; *morphe* = form. Pl., Androergatogynomorphs.) A genetical condition in which an ant displays anatomical features of a male and worker. (Berndt & Kremer 1983, Ins. Soc. 30: 461–465.) See Gynandromorph. Rel. Caste.

ANDROGYNE Noun. (Greek, *aner* = male; *gyne* = female. Pl., Androgynes.) An hermaphrodite or individual with sexual characteristics or anatomical features of male and female. See Gynandromorph.

ANDROGYNOUS See Gynandromorph.

ANDROID Adj. (Greek, *aner* = male; *eidos* = form.) Male-like in form or habits. Cf. Arrhenoid; Gynecoid. Rel. Eidos; Form; Shape.

ANDROMEDA LACE-BUG *Stephanitis takeyai* (Drake & Maa) [Hemiptera: Tingidae]: A native of Asia and adventive in North America where it is a pest of numerous ornamental plants.

ANDROPOLYMORPHISM (Greek, *aner* = man; *polys* = many; *morphe* = form; *-ism* = condition. Pl., Andropolymorphisms.) Mites: A condition manifest in which a male has a pair of legs that are disproportionately large when compared to other legs on the body. See Polymorphism.

ANDROTYPE Noun. (Greek, *aner* = male; *typos* = pattern. Pl., Androtypes.) A male specimen in the type-series of a formally described Species. See Type. Cf. Allotype; Cotype; Gynetype; Holotype; Lectotype; Neotype; Paratype; Syntype. Rel. Nomenclature.

ANELCANIDAE Carpenter 1958. Plural Noun. A replacement name for Parelcanidae Carpenter 1966, a monogeneric Family of fossil Orthoptera found in the Permian beds of Kansas, USA.

ANELECTROUS Adj. (Greek, *an* = without; *elytron* = sheath; Latin, *-osus* = possessing the qualities of.) Descriptive of beetles that lack Elytra. See Elytron.

ANELLUS Noun. (Latin, *anellus* = little ring. Pl., Anelli.) 1. Hymenoptera: A short, annular (ring-like) segment distad of the Pedicel in the Antenna. Some Species display several Anelli. 2. A sclerotization of the inner wall of the Phallocrypt or Phallotheca that often forms a ring or tube surrounding the base of the Aedeagus (Snodgrass). 3. Lepidoptera: A chitinized structure supporting and often surrounding the terminal part of the Aedeagus. The Anellus may articulate with the bases of the Harpes and its ventral part, and thereby forms a median sclerite (the Juxta) below the Aedeagus (Klots).

ANEMOCHOROUS Adj. (Greek, *anemos* = air; *chorein* = to disperse; Latin, *-osus* = possessing the qualities of.) Pertaining to wind-mediated dispersal, such as seeds and spores. Wind-blown dispersal. Cf. Dispersal; Migration. Rel. Pollination.

ANEMOCHORY Noun. (Greek, *anemos* = air; *chorein* = to disperse. Pl., Anemochories.) The process or phenomenon involving dispersal of organisms, seeds or pollen by the wind. Wind-blown dispersal. Rel. Dispersal.

ANEMOPHILY Noun. (Greek, *anemos* = wind; *philein* = to love. Pl., Anemophilies.) The phenomenon or process of wind pollination. Alt. Anemophilous. See Pollination.

ANEMOPLANKTON Noun. (Greek, *anemos* = wind; *plangktos* = wandering. Pl., Anemoplanktons.) Small organisms, pollen and spores that move with the wind, by implication, as part of the life cycle or developmental process. Alt. Aerial Plankton. See Plankton. Cf. Aeroplankton. Rel. Dispersal.

ANEMOTAXIS Noun. (Greek, *anemos* = air; *taxis* = arrangement. Pl., Anemotaxes.) Orientation, tactic reaction or movement of a free-living organism in response to air currents or wind. See Orientation; Taxis. Cf. Aerotaxis; Chemotaxis; Geotaxis; Menotaxis; Osmotaxis; Phototaxis; Rheotaxis; Rotaxis; Scototaxis; Strophotaxis; Telotaxis; Thermotaxis; Thigmotaxis; Tonotaxis; Tropotaxis.

ANEMOTROPISM Noun. (Greek, *anemos* = air; *trope* = turn; English, *-ism* = state. Pl., Anemotropisms.) 1. Orientation of the body or a turning reaction in response to currents of air or wind movement. 2. Curvature of a plant in response to wind. See Tropism. Cf. Aeolotropism; Chemotropism; Electrotropism; Galvanotropism; Geotropism; Heliotropism; Hydrotropism; Phototropism; Rheotropism; Stereotropism; Thermotropism; Thigmotropism; Tonotropism. Rel. Taxis.

ANENCEPHALY Noun. (Greek, *an* = without; *engkephalon* = brain. Pl., Anencephalies.) The pathological condition in which an organism has no brain. See Brain.

ANEPIMERON Noun. (Greek, *ana* = up; *epi* = upon; *meros* = upper thigh. Pl., Anepimera.) The upper sclerotized part of an Epimeron above a distinct suture (Comstock). See Pleuron. Cf.

Anepisternum.

ANEPISTERNITE Noun. (Greek, *ana* = up; *epi* = upon; *-ites* = constituent. Pl., Anepisternites.) A portion (sclerotized subdivision) of an Anepisternum. The term is often used to indicate the entire Anepisternum. See Pleuron. Cf. Anepimeron; Mesopleuron.

ANEPISTERNUM Noun. (Greek, *ana* = up; *epi* = upon; *sternon* = chest. Pl., Anepisterna.) The upper part of the Episternum when it is divided by a suture or cleft into two parts. Syn. Supreapisternum *sensu* Garman. See Sternum. Cf. Infraepisternum; Katepisternum; Pleuron.

ANER Noun. (Greek, *aner* = male. Pl., Aners.) 1. General: An uncommon term for a male insect. 2. Hymenoptera: A male ant.

ANEURONIC Adj. (Greek, *a* = without; *neuron* = nerve; *-ic* = characterized by.) Pertaining to an anatomical condition in which structure is not innervated (without nerve endings).

ANEUROSE Adj. (Greek, *a* = without; *neuron* = nerve; *-osus* = full of.) Pertaining to a wing without veins except near the Costa.

ANGAS, GEORGE FRENCH (1822–1886) (J. D. H. 1887, Proc. Linn. Soc. Lond. 1886–1887: 33–34; Anon. 1927, Australian Encyclopaedia 1: 58; Musgrave 1932, *A Bibliography of Australian Entomology 1775–1930,* 380 pp. (5), Sydney.)

ANGEL RIESGO ORDONEZ, DON (–1969) (N. R. 1969, Boln. Serv. Plagas Forest. 12(23): 73.)

ANGELL, JOHN WATSON (1885–1946) (Bromley 1946, J. N. Y. Ent. Soc. 54: 290.)

ANGIOSPERM Noun. (Greek, *anggeion* = vessel; *spema* = seed. Pl., Angiosperms.) A member of the Angiospermae. Cf. Gymnosperm.

ANGIOSPERMAE Plural Noun. Flowering Plants. The larger of the two major divisions of the seed-bearing plants; Angiospermae include members whose reproductive organs are contained within flowers. Angiosperm seeds develop in a closed Ovary composed of carpels. Cf. Gymnospermae.

ANGIOSPOROUS Adj. (Greek, *anggeion* = vessel; *sporos* = seed; Latin, *-osus* = possessing the qualities of.) Descriptive of plants whose spores are contained within a spore capsule (Theca).

ANGLE OF TEGMINA 'The longitudinal ridge formed along the internomedian by the sudden flexure from the horizontal to the vertical portion when closed' (*sic* in Smith).

ANGLEMAN, JOHN BARTON (1860–1926) (Engelhardt 1926, Bull. Brooklyn Ent. Soc. 21: 181.)

ANGORA GOAT BITING-LOUSE *Bovicola crassipes* (Rudnow) [Mallophaga: Trichodectidae].

ANGORA GOAT LOUSE *Damalinia limbata* (Gervais) [Mallophaga: Trichodectidae]. (Australia).

ANGOUMOIS GRAIN-MOTH *Sitotroga cerealella* (Oliver) [Lepidoptera: Gelechiidae]: A cosmopolitan, multivoltine pest associated with stored grains; AGM prefers corn and wheat over other cereals; only intact grain is attacked. AGM was named after the Province of Angoumois in France where this insect was first reported a pest ca 1736. Adult pale brown, 5–7 mm long (wingspan 10–15 mm); Labial Palpus long; wing fringe long; wings narrow and apically pointed. Adults are strong fliers but short lived. Female lays 80–200 eggs during 5–10 days; eggs white, deposited on wheat head, exposed tops of corn in field and kernels of grain in storage; eggs laid in cracks, grooves or holes made by other insects. Solitary larva on most grains; 2–3 larvae per kernel of corn. Larva spins small web, bores into kernel, feeds, cuts a circular flap in kernel surface; overwinters as larva within stored kernel or field litter; pupation occurs within cocoon inside kernel. Infestations of AGM increase temperature of stored grain; larva increases moisture content of grain. See Gelechiidae. Cf. Clothes Moth; Mediterranean Flour Moth.

ANGRAECUM SCALE *Conchaspis angraeci* Cockerell [Hemiptera: Conchaspidae]: A native of tropical America and pest of several woody plants and orchids.

ÅNGSTROM Noun. (Pl., Angstroms.) A unit of measurement named in honour of the Swedish physicist A. J. Ångström. 1 Ångstrom = 1 nm. Symbol: Å, ÅU.

ANGULAR AREA Hymenoptera: The posterior of the three areas on the Metanotum between the lateral and pleural Carinae; the third pleural area *teste* Torre Bueno.

ANGULAR PUPA See Pupae Angulares.

ANGULAR-WINGED KATYDID *Microcentrum retinerve* (Burmeister) [Orthoptera: Tettigoniidae]. Cf Broad-Winged Katydid; Citrus Katydid; Crested Katydid; Forktailed Bush-Katydid; Inland Katydid; Mottled Katydid; Philippine Katydid; Spotted Katydid.

ANGULATE Adj. (Latin, *angulus* = angle; *-atus* = adjectival suffix.) Structure forming an angle; when two margins or lines meet in an angle.

ANGULATE LEAFHOPPER *Acinopterus angulatus* Lawson [Hemiptera: Cicadellidae].

ANGULOSE Adj. (Latin, *angulus* = corner, angle; *-osus* = full of.) Angled; having angles.

ANGULOSO-UNDULATE Extending in more-or-less zig-zags or having alternating acute sinuses.

ANGULUS Adj. (Latin, *angulus* = corner, angle.) An angle; forming an angle; angulate.

ANGUS, JAMES (–1902?) (Moffat 1903, Rep. ent. Soc. Ont. 34: 103–108.)

ANGUSTATE Adj. (Latin, *angustus* = narrow; *-atus* = adjectival suffix.) Disproportionately narrow. Alt. Angustatus.

ANGUSTATE ANTENNA An Antenna with middle and apical segments thinner than basal segments. See Antenna. Cf. Clavate.

ANGUSTIFOLIATE Adj. (Latin, *angustus* = narrow; *folium* = leaf; *-atus* = adjectival suffix.) Descriptive of plants whose leaves are narrow. Cf. Latifoliate.

ANGUSTIROSTRATE Adj. (Latin, *angustus* = narrow; *rostrum* = beak; *-atus* = adjectival suffix.) Descriptive of insects with a narrow beak. Alt. Angustirostral.

ANHOLOCYCLIC Adj. (Greek, *an* = not; *holo* = complete; *kyklos* = circle.) Pertaining to aphid Species that have eliminated the sexual phase from the life cycle. A phenomenon induced by genetic or environmental change. Net effect: Host plants entirely of summer herbaceous type and reproduction always parthenogenetic.

ANHYDROBIOSIS Noun. (Greek, *anydros* = waterless: *an* = without; *hydro* = water; *biosis* = way of living. Pl., Anhydrobioses.) Dormancy in insects induced by low humidity or by desiccation. Alt. Anhydrobiose. See Biosis. Cf. Abiosis; Antibiosis; Archebiosis; Calobiosis; Cleptobiosis; Hamabiosis; Kleptobiosis; Lestobiosis; Parabiosis; Phylacobiosis; Plesiobiosis; Synclerobiosis; Trophobiosis; Xenobiosis.

ANHYDROPIC Adj. (Greek, *an* = without; *hydro* = water; *-ic* = characterized by.) Descriptive of insect eggs that contain abundant amounts of yolk (Flanders, 1942, Ann. Ent. Soc. Amer. 35: 241–256.) Syn. Lecithal. Cf. Hydropic.

ANISEMBIIDAE Plural Noun. A small Family of Embiidina (Embioptera) characterized by a Mandible without teeth, male Tergum X divided by median longitudinal membrane and left Cercus with blunt spines on basal segment.

ANISO- Greek prefix meaning 'unequal.'

ANISOLABIDIDAE Plural Noun. A small Family of forficuline Dermaptera assigned to Superfamily Anisolabidodea. Body size variable with one Species 55 mm long; adult typically apterous; Pygidium sometimes concealed. Syn. Carcinophoridae.

ANISOLABIDOIDEA A Superfamily of Dermaptera that includes two Families, Anisolabididae and Labiduridae.

ANISOMORPHA Noun. (Greek, *anisos* = unequal; *morphe* = form. Pl., Anisomorphae.) Taxonomic groups in which the Metamorphosis is a varying character, or not all Species in a taxonomic group have identical metamorphic development.

ANISOMORPHIC Adj. (Greek, *anisos* = unequal; *morphe* = form; *-ic* = characterized by.) Pertaining to differences in shape, size or structure. See Heteromorphic. Cf. Isomorphic.

ANISOPLEURAL Adj. (Greek, *anisos* = unequal; *pleuron* = side.) Pertaining to organisms that are bilaterally asymmetical. See Symmetry.

ANISOPODIDAE Plural Noun. Window Flies. A small, widespread Family of nematocerous Diptera assigned to Bibionomorpha and sometimes regarded as similar to hypothetical ancestral Brachycera. Adult body stout, Ocelli present; Mesonotum without 'V'-shaped transverse suture, with discal cell, 2A not extending to wing margin. Syn. Phryneidae; Rhyphidae; Sylvicolidae.

ANISOPTERA Dragonflies. Suborder of Odonata typically robust as adults that vary considerably in size. Compound eyes typically in contact but may be separated in some Families. Wings not petiolate; hindwing more broad and triangular than forewing; Pterostigma generally elongate; Discoidal Cell typically triangular. Adults rest with wings horizontal and perpendicular to body. Naiad Abdomen flattened and gills enclosed in Rectum. Young naiad with three abdominal processes that taper and form a 'pyramid'. Cerci formed in older naiads. Cf. Zygoptera. Anisozygoptera.

ANISOPTEROUS Adj. (Greek, *anisos* = unequal; *pteron* = wing; Latin, *-osus* = possessing the qualities of.) 1. Descriptive of or pertaining to Anisoptera (dragonflies *sensu stricto*). 2. Pertaining to insects with wings of unequal size.

ANISOTOMIDAE Plural Noun. See Leiodidae.

ANISOTROPIC Adj. (Greek, *anisos* = unequal; *trope* = turn; *-ic* = characterized by.) Descriptive of structure with optical properties that are doubly refractive.

ANISOZYGOPTERA A Suborder (included Family: Epiophlebiidae) that is transitional between dragonflies and damselflies. Adult body and naiadal characters resemble the Anisoptera (dragonflies) and wings are suggestive of the Zygoptera (damselflies).

ANKER, LUGWIG (1822–1881) (Aigner 1901, Rovart Lap. 8: 197–203.)

ANKISTROID Adj. (Greek, *agkistrion* = fish hook; *eidos* = form, shape.) Hook-like; descriptive of structure that is shaped as a fish hook and barbed. Alt. Ancistroid. See Hook. Cf. Aduncate; Hamate; Uncinate. Rel. Eidos; Form; Shape.

ANKYLOSE Adj. (Greek, *ankylosis* = stiffening of joints.) Descriptive of structure that has grown together at a joint. Alt. Anchylosed. Rel. Fused.

ANKYLOSIS Noun. (Greek, *ankylosis* = stiffening of joints. Pl., Ankyloses.) 1. Union, fusion or welding together of hard parts to form one structure. 2. Stiffening or growing together of a joint. Alt. Anchylosis.

ANKYROID Adj. (Greek, *agkyra* = anchor; *eidos* = shape.) Anchor-like; Hook-like. Alt. Ancyroid. See Ankistroid. Rel. Eidos; Form; Shape.

ANLAGE Noun. (German, *anlage* = predisposition. Pl., Anlagen.) Embryology: The initial group of cells that will ultimately develop into an organ, appendage or structure. Syn. Primordium.

ANNAND, PERCY NICOL (1898–1950) (Morrison 1950, Ann. Ent. Soc. Am. 43: 460; Gordon & White 1951, J. Econ. Entomol. 44: 269–270, portrait; Osborn 1952, *A Brief History of Entomology.* 305 pp., Columbus, Ohio.)

ANNANDALE, THOMAS NELSON (1876–1924) (Calvert 1924, Ent. News 35: 264; Anon. 1925, J. Bombay Nat. Hist. Soc. 30: 213–214; S. K. 1925, Rec. Indian Mus. 27: 1–28, bibliography.)

ANNECKE, DAVID P South African government Entomologist (Department of Agricultural and Technical Services, Pretoria) and taxonomic specialist in parasitic Hymenoptera, particularly the chalcidoid Families Encyrtidae and Aphelinidae.

ANNECTENT Adj. (Latin, *annectere* = to bring together.) 1. Anatomy: Pertaining to a physical connection of anatomical parts. 2. Taxonomy: Pertaining to linkage or intermediate Taxa between related Species, Genera or other higher taxonomic categories.

ANNELID Noun. (Latin, *annelus* = ring; Greek, *eidos* = form. Pl., Annelids.) Any of the segmented worms. Alt. Worm. See Annelida. Rel. Arthropod.

ANNELIDA Noun. (Latin, *annelus* = ring. Pl., Annelidae.) The Class of animals containing the segmented worms. Member Taxa are characterized by a relatively thin and flexible cuticle, annulated body, preoral Prostomium and postanal Pygidium. Rel. Arthropoda.

ANNELLA Noun. (Latin, *annelus* = ring. Pl., Anellae.) The Fundita associated with the Anal Vein (MacGillivray).

ANNELLUS Noun. (Latin, *annelus* = ring. Pl., Anelli; Pl., Annelluses.) Hymenoptera: Small ring-like segments (Antennomeres) between the Pedicel and Funicle of the Antenna. Alt. Annelet. Syn. Ring Segment. See Antenna.

ANNONACIN See Acetogenins.

ANNONE, JEAN JACQUES d' (1728–1804) (Rose 1850, *New General Biographical Dictionary* 1: 483.)

ANNONIN See Acetogenins.

ANNUAL COLONY Social Hymenoptera: A colony that is active only one season per year.

ANNULAR Adj. (Latin, *annulus* = ring.) Pertaining to structures that are ring-shaped or ring-like. Syn. Annuliform. See Ring Segment.

ANNULAR LAMINA The sternal sclerite or shield of ants (segment IX), that lies anteriad of the genitalia.

ANNULAR SPIRACLE Immatures: An unmodified circular or oval spiracle with one opening.

ANNULAR ZONE See Peripheral Pad.

ANNULATA Noun. (Latin, *annulus* = ring. Pl., Annulatae.) Animals with bilateral symmetry and metameric segmentation, including Annelida, Arachnida, Crustacea, Insecta and Myriapoda. Alt. Annulosa.

ANNULATE Adj. (Latin, *annulus* = ring; *-atus* = adjectival suffix.) 1. Descriptive of structure (such as an appendage) that is ringed or ring-like, but not demonstrating primary or secondary segmentation. See Segmentation. Rel. Mere. 2. Pertaining to antennal segments that are ring-like in appearance. 3. A narrow circle (ring) of a colour different from the adjacent region. Cf. Striped. 4. A ring-shaped spot (Kirby & Spence). Alt. Annulatus; Annulated. See Annular. Cf. Armillate.

ANNULET Noun. (Latin, *annulus* = ring. Pl., Annulets.) Immatures: 1. Any small, secondary ring that subdivides a segment or appendage. Cf. Annellus. 2. A partial dorsal subdivision of a body segment formed by transverse inflections in the integument.

ANNULIFORM Adj. (Latin, *annulus* = ring; *forma* = shape.) Ring-shaped; descriptive of structure in the form of rings or ring-like segments. Syn. Annular. See Annulate. Rel. Form 2; Shape 2; Outline Shape.

ANNULUS Noun. (Latin, *annulus* = ring. Pl., Annuli.) 1. A ring of membrane, sclerite or pigment surrounding a joint, segment, spot or mark. 2. A ring-like marking, or a ring of hard cuticle. 3. The sclerotized ring on the head into which the basal segment (Scape) of the Antenna is inserted. The Antennal sclerite. Alt. Antennalis.

ANOBIIDAE Plural Noun. Death-Watch Beetles; Drugstore Beetles. A Family of polyphagous Coleoptera assigned to the Bostrichoidea. Adult <9 mm long, elongate, clyindrical to oval and globose; body typically setose or with scales; head deflexed, typically concealed from above; Antenna filiform, serrate or pectinate; tarsal formula 5-5-5; Abdomen with five Ventrites. Larva scarabaeiform, lightly sclerotized; head hypognathous; 0–1 Stemmata; Antenna minute, with 1–2 segments; legs typically with five segments, rarely absent; Urogomphi absent. Larvae bore into wood or bark of trees; a few Species are significant pests of furniture and stored products. Syn. Ectrephidae.

ANOCELLATE Adj. (Greek, *an* = without; *ocellus* = little eye; *-atus* = adjectival suffix.) Pertaining to insects that lack Ocelli. See Ocellus. Cf. Anophthalmus. Rel. Ommatidium.

ANOMALADENSAE Noun. Coccoids: Thickenings of the Pygidium that appear to be proximal prolongations of the lobes (MacGillivray).

ANOMALOPSYCHIDAE Plural Noun. A small Family of Trichoptera assigned to Superfamily Limnephiloidea.

ANOMALOUS Adj. (Greek, *anomalos* = irregular, uneven; Latin, *-osus* = possessing the qualities of.) Unusual; departing widely from the usual type. Ant. Typical. Alt. Abnormal; Atypical; Unusual.

ANOMOPTERELLIDAE Rasnitsyn 1975. Plural Noun. A Family of parasitic Hymenoptera assigned to the Evanioidea.

ANOMOSETIDAE Turner 1922. Plural Noun. A small Family of glossatous Lepidoptera endemic to Australia and assigned to Superfamily Hepialoidea; in some classifications anomosetids are placed with African Prototheoridae. Adult small; head with lamellate scales; Ocelli and Chaetosemata absent; Mandible small, non-functional; Antennal segments long with pointed hair-like scales; Proboscis small; Maxillary Palpus minute; Labial Palpus with three segments and longer than head; Epiphysis absent; tibial spur formula 0-2-4. The Australian Species from eastern coastal rainforest; biology poorly known; immature stages unknown.

ANOPHELES A Genus of mosquitoes [Diptera: Culicidae] consisting of ca 420 described Species. About 40 Species are regarded as primary vectors of Malaria; about 30 Species are regarded as secondary vectors of Malaria. See Malaria. Cf. *Aedes*; *Culex*.

ANOPHTHALMIC Adj. (Greek, *an* = without; *ophthalmos* = eye; *-ic* = of the nature of.) Pertaining to eyeless organisms. Alt. Blind.

ANOPHTHALMUS Noun. (Greek, *an* = without; *ophthalmos* = eye. Pl., Anophthalmuses.) An organism without eyes. Insects lacking compound eyes in the adult stage. See Compound Eye. Cf. Anocellate.

ANOPLEURE Noun. (Greek, *an* = without; *pleuron* = side. Pl., Anopleures.) The Epimeron *sensu* MacGillivray. See Pleuron.

ANOPLODERMATIDAE Plural Noun. A small Family of polyphagous Coleoptera assigned to Superfamily Chrysomeloidea.

ANOPLURA Leach 1815. (Recent. No Fossil Record.) (Greek, *an* = without; *hoplon* = weapon; *oura* = tail.) Sucking Lice. A cosmopolitan Order of about 250 Species of lice. Adult body small, dorsoventrally compressed; compound eye reduced to a facet or absent; mouthparts consisting of three, sclerotized piercing stylets; stylets concealed when not in use; Maxillary and Labial Palpi reduced or absent; thoracic segments fused; wingless; Tarsi 1-segmented with large claw opposable to thumb-like process on apex of Tibia; Abdomen 9-segmented; Ovipositor, Cerci absent. Metamorphosis simple; nymphs resemble adults; all stages found on host. Eggs are operculate and cemented individually to hair shaft near skin of host. Anoplura complete three nymphal instars that feed by sucking blood from the capillaries within the skin of their hosts; they are parasites of mammals only. Anoplura do not attack bats, marsupials or carnivores (except dogs). Anopluran Species attack humans: Crab louse = *Phthirus pubis* (Linnaeus), head louse = *Pediculus humanus capitis* De Geer, body louse = *P. humanus humanus* Linnaeus. Body louse is an important vector of Typhus and other diseases. Three Families occur in the USA: Pediculidae, Echinophthridae (on marine mammals), and Haematopinidae (on horses, cattle, sheep, hogs, and other animals). Mallophaga and Siphunculata are included in the Order (*sensu* Imms). See Insecta. Cf. Ischnocera; Mallophaga; Phthiraptera. Rel. Parasite.

ANOPTHALMIC Adj. (Greek, *an* = without; *ophthalmos* = eyes; *-ic* = characterized by.) Pertaining to eyeless organisms.

ANOVA See Analysis of Variance.

ANSLIJN, NICOLAAS (1777–1838) (MacGillivray 1938, Ent. Ber., Amst. 10: 87–88.)

ANSORGE, CARL (1873–1964) (Weidner 1967, Abh. Verh. Naturw. Ver. Hamburg Suppl. 9: 266.)

ANSULA Noun. (Latin, *ansa* = handle. Pl., Ansulae.) Small projections on the eggshell of ceratopogonid flies and other Diptera. The Ansula probably functions as a plastron or holdfast structure.

ANT BEETLES See Anthicidae.

ANT BIRDS Tropical birds (Formicariidae) that follow raids of army ants to feed upon prey disturbed by the ants.

ANT BUTTERFLIES Butterflies (Nymphalidae, Ithomininae) that follow raids of army ants to feed upon the excrement of antbirds.

ANT CRICKETS See Myrmecophilidae.

ANT GARDEN Epiphytic plants inhabited by ants with mutual benefit derived from the association. Rel. Symbiosis.

ANT LIONS See Neuroptera.

ANT LOVING BEETLES See Pselaphidae.

ANT Noun. (Middle English, *amte*. Pl., Ants.) See Formicidae. Correctly: A colony of ants. Cf. Bee, Wasp.

ANTACAVA Noun. A deep socket into which an Antenna articulates (MacGillivray). Syn. Antennal Socket. Cf. Torulus.

ANTACOILA Noun. See Antennifer (MacGillivray).

ANTACORIUM Noun. (Greek, *anti* = against; Latin, *corium* = leather. Pl., Antacoria.) 1. The Corium or segmental membrane of each Antennal segment. 2. The narrow ring of membrane connecting an Antenna with the head (MacGillivray). See Corium.

ANTARCTOPERLARIA Noun. A Suborder of Plecoptera including Austroperlidae, Diamphipnoidae, Eustheniidae and Gripopterygidae.

ANTAROLIUM Noun. (Greek, *anti* = against; Latin, *areola* = small space; *-ium* = diminutive. Pl., Antarolia.) The large Pulvillus (MacGillivray). See Arolium.

ANTARTIS Noun. (Pl., Antartes.) The basal 'joint' of an Antenna (MacGillivray). Cf. Radicle.

ANT-BLOOD BEETLES See Dermestidae.

ANTEALAR Adj. (Latin, *ante* = before; *ala* = wing.) Pertaining to structure positioned anterior of the forewing (Tillyard). Alt. Ante-alar. Rel. Post-alar; Supra-alar.

ANTEALAR SINUS Odonata: A grooved area extending transversely immediately anterior of the base of each forewing.

ANTEAPICAL Adj. (Latin, *ante* = before; *apex* = summit.) Pertaining to structure proximad of the apex; preceding the apex. Alt. Ante-apical.

ANTECLYPEUS Noun. (Latin, *ante* = before; *clypeus* = shield. Pl., Anteclypeuses.) 1. The anterior portion of the Clypeus attached to the Labrum. 2. Odonata: Lower of two divisions of the Clypeus; the lower half of the Clypeus whenever any line of demarcation appears between the Clypeus and Labrum. Syn. Anterior Clypeus; Infraclypeus; Rhinarium; Second Clypeus. 3. Larvae: An anterior, usually transverse, more sclerotized portion of the Clypeus that serves as an articulation for the Labrum. Syn. Anticlypeus. See Clypeus.

ANTECOSTA Noun. (Latin, *ante* = before; *costa* = rib. Pl., Antecostae.) 1. The anterior submarginal or marginal ridge on the inner surface of a tergal or sternal sclerite corresponding to the primary intersegmental fold, and to which the longitudinal muscles typically are attached (Snodgrass). 2. Pretergite *sensu* Needham. See Costa.

ANTECOSTAL SUTURE 1. The groove that extends through the base of a Phragma and delimits the base of the Antecosta. 2. The external groove of the Antecosta (Snodgrass). See Antecosta.

ANTECOXAL PIECE 1. A sclerite between the single or divided Trochantin and the Episternum, or between the Trochantin and the Precoxal Bridge. 2. Mandible: The paired lateral sclerites of the Clypeus, one on each side (Smith).

ANTECUBITAL CROSSVEINS Odonata: Antenodal Crossveins (Comstock). See Crossvein; Vein. Rel. Wing Venation.

ANTECUBITAL Adj. (Latin, *ante* = before; *cubitus* = elbow; *-alis* = belonging to, pertaining to.) See Antenodal Crossvein; Antenodal Spaces.

ANTEFUNGIVORIDAE Plural Noun. A small Family of fossil Diptera.

ANTEFURCA Noun. (Latin, *ante* = before; *furca* = fork. Pl., Antefurcae.) An internal forked process from the Prosternum and to which muscles are attached. Cf. Furca.

ANTEHUMERAL STRIPE Odonata: A discoloured stripe medial of the Humeral Suture.

ANTEHUMERAL Adj. (Latin, *ante* = before; *humerus* = shoulder; *-alis* = belonging to, pertaining to.) Pertaining or relating to the space immediately anterior of the wing base. See Humerus. Cf. Posthumeral.

ANTEMEDIAL LINE A transverse line anterior of the midline which bisects a structure.

ANTEMEDIAN Adj. (Latin, *ante* = before; *medius* = middle.) Pertaining to bristles or conspicuous large Setae positioned anterior of the middle of a segment or sclerite.

ANTENNA Noun. (Latin, *antenna* = sailyard, from Greek, *ana* = up; *teinein* = to stretch. Pl., Antennae.) Paired, segmented, flexible, sensory appendages located on the head of most arthropods. Antennae are missing from Chelicerata; two pairs occur in Crustacea; a pair occur in Insecta. Among hexapods, Antennae are missing from Protura but are present in Diplura and Collembola. Insect Antenna is the anteriormost appendage of postembryonic head. Antenna is innervated by deutocerebral lobe of Brain and usually attached to head in facial region. Antennal segment number varies considerably (3–70+); Antenna is subdivided into three parts: Scape (attached to head), Pedicel (second segment) and Flagellum (all distal segments). Scape is attached to head via a membranous socket (Torulus) and articulates with head via a small sclerotic process (Antennifer). Antenna is sometimes divided into two types: segmented and annulated. Segmented Antenna occurs in non-insect groups, Symphyla, Collembola and Diplura; each Antennal segment has intrinsic musculature. Annulated Antenna occurs in Pterygota and Thysanura; intrinsic Antennal musculature only occurs in Scape and Pedicel. Flagellar segments of pterygotes are sometimes called Flagellomeres. Antenna functions in olfaction, contact chemoreception, courtship and other activities. Antenna form is variable. Flagellomere size and shape are serially monotonous or undifferentiated in most lower Orders, highly variable among groups of Holometabola and frequently a source of profound sexual dimorphism. For explantions of Antennal shapes see: Aristate, Capitate, Clavate (Clubbed), Filiform, Flabellate, Geniculate (Elbowed), Lamellate, Moniliform, Pectinate, Ramose, Setaceous. Antenna is moved in part by extrinsic musculature that usually originates on Anterior Arms or Dorsal Rami of the Tentorium and insert on base of antennal Scape. Colloquial: Feelers; Horns.

ANTENNA CLEANER Hymenoptera: A grooming device intended to clean the Antenna. The AC consists of a basomedial fringe of Setae, 'hairs' or spines on the medial surface of the anterior Basitarsus; the Setae form a comb-like structure through which the Antenna may be drawn. The Antenna is held in place against the comb by an enlarged, often curved or forked tibial spur that is at the tibial apex. Similar structures are on the fore Tibia of other insects. Rel. Calcar; Strigil.

ANTENNAL APPENDAGE Mallophaga: A process that projects from antennal segments of the male.

ANTENNAL CIRCULATORY ORGAN Dermaptera: Paired circulatory organs in the head. ACO consists of a pulsatile Ampulla connected to an antennal blood vessel. Ampulla attached to frontal Cuticle medial of Antenna base, forming a thin-walled sac with a valved Ostium on the ventral surface. Pumping movements are effected by a precerebral, Frontopharyngeal Muscle that causes systolic compression of the Ampulla upon contraction. Elasticity of the Ampulla and suspensory connective tissue passively effects diastole. A valve flap at the base of the vessel near Ampulla prevents Haemolymph backflow during diastole. See Antennal Heart.

ANTENNAL CLUB A variable number of segments of the antennal Flagellum usually identified by a change in shape or form from preceding segments. The AC is always apical, is sometimes arbitrarily delimited by segment number (*e.g.* segments 8–11, 9–12) and always includes the terminal segment. See Antenna. Rel. Funicle; Clavus.

ANTENNAL COMB Grooming structure found in many insects such as Hymenoptera. See Antenna Cleaner. Rel. Calcar; Strigil.

ANTENNAL FORMULA Any code or convention by which the relative lengths of antennal segments is communicated, typically in taxonomic description or taxonomic key. Coccidae: AF is made by enumerating the antennal segments in the order of their length, beginning with the longest and bracketing together those of the same length. Other insects: The arrangement of the antennal segments (joints) in point of length, or in a proportion in serial order, beginning with the first. See Antenna.

ANTENNAL FOSSA A Sulcus, depression, groove or cavity in which Antennae are located or concealed. Syn. Antennal Grooves; Antennary Fossa (Diptera); Facial Depression (Comstock). See Antenna.

ANTENNAL FOVEA 1. Diptera: A longitudinal groove or oblique convergent grooves positioned in middle of the Face for reception of the Antenna. AF is bounded laterad by facial ridges. 2. Hymenoptera: Depressed areas surrounding antennal sockets, frequently connected with antennal furrows and lateral Foveae. See Antenna.

ANTENNAL FOVEOLAE Orthoptera: Pits or depressions between the frontal Costa and the lateral Carinae into which the Antennae are inserted. See Antenna. Cf. Torulus.

ANTENNAL FURROWS Hymenoptera: Furrows (grooves, sulci) extending from the Tentorial Pits along the lateral margin of the antennal sockets across the cephalic aspect of the head toward the lateral Ocelli, thence across the dorsal aspect and ending just below the ridge separating the dorsal and caudal margins of the head, most frequently complete, but at times with certain sections obsolete. See Vertical Furrows, Lateral Foveae, Ocellar Furrows, Interocellar Furrows. Rel. Antenna.

ANTENNAL GROOVE 1. Hymenoptera: The curved portion of an antennal furrow (groove, Sulcus) extending on each side of the head between tentorial pits and frontal crest around lateral margin of antennal sockets (toruli). See Scrobal Impression. 2. Siphonaptera: A Sulcus behind the compound eye that divides head into two regions. See Antennal Fossa. 3. Diptera: See Antennal Fovea.

ANTENNAL HEART A pulsatile organ specialized for Haemolymph circulation within the Antenna of many insects, including Orthoptera, Phasmatoidea and Dermaptera. AH first reported by Pawlova (1895) for cockroaches and grasshoppers. AH typically consists of a short, bulbous ampullary sac connected to long, distal nonmusculated antennal vessel that is provided with 1–2 valvular Ostia. Insects with lamellate Antennae in which branches are spread when active, AHs may be responsible for spreading branches. AHs are not musculated in Amyocerata. Innervation of AH differs among insect groups, including: Suboesophageal Ganglion in *Rhodnius* (Pinet 1964), Frontal Ganglion in *Periplaneta* (Pawlowa 1895), antennal nerve branch in *Cionus* (Donges 1954) and *Thermobia* (Chaudonneret 1950). AH is not innervated in the mosquito *Culex* (Clements 1956). Blood enters AH during diastolic phase; diastole of ampulla is in response to contraction of a well-developed transverse muscle. Systolic phase is caused by contraction of ampulla wall in response to elasticity of the wall. See Antenna; Antennal Circulatory Organ.

ANTENNAL LOBES See Deutocerebrum.

ANTENNAL MECHANOSENSORY AND MOTOR CENTER See Deutocerebrum. Syn. Dorsal Lobe.

ANTENNAL MUSCULATURE Flexibility of Antenna enables animal to explore surrounding environment. Capacity to explore environment is provided by musculature. Antennal musculature is divided into two types. Extrinsic musculature includes up to 12 muscles of which at least three are attached to Scape: A Flexor, an Extensor, and a Levator. Intrinsic musculature of Apterygota includes Scape, Pedicel and flagellar segments; Intrinsic musculature of Pterygota includes Scape and Pedicel. Intrinsic musculature of Antenna reflects broad evolutionary trends involving arthropod groups. Segmented Antenna with intrinsic musculature occurs in many non-insectan groups including Symphyla, Collembola, and Diplura. Dipluran *Campodea* adds antennal segments through acrogenous growth with each postembryonic moult until definitive number of antennal segments is attained. Phenomenon also occurs in Zoraptera: Nymphal Antenna has eight segments; adult Antenna has nine segments. Annulated Antenna lacks intrinsic musculature in Flagellum; Annulated Antenna occurs in Pterygota and Thysanura. Intrinsic antennal musculature is restricted to the Scape and Pedicel. Flagellar segments of pterygotes are sometimes called Flagellomeres because they lack intrinsic musculature. See Antenna.

ANTENNAL ORGANS Collembola: Sensory structures on the distal segment of the Antenna. See Antenna.

ANTENNAL PAPILLA Immatures: A short, unsegmented appendage on the Antennal orbit. Similar to a one-segmented Antenna but of secondary origin. See Antenna.

ANTENNAL PROCESS Diptera: The frontal protuberance upon which the Antenna is inserted. See Antenna.

ANTENNAL SCLERITE A sclerotized ring into which the basal joint of each Antenna is inserted. See Antenna. Cf. Annulus; Antennalis.

ANTENNAL SEGMENT The second (deutocerebral) segment of the head of the generalized (hypothetical) insect. See Brain. Cf. Protocerebrum; Tritocerebrum.

ANTENNAL SENSILLA Antenna employs olfaction and contact chemoreception to mediate courtship and other activities. Insects show numerous different types of receptive organs and sensillar types. See Antenna; Sensilla. Cf. Bohm's Organ; Janet's Organ; Johnson's Organ; Multiporous Plate Sensilla.

ANTENNAL SOCKET The membranous area of the cranial wall, reinforced by a marginal ridge, in which the Antenna is set (Snodgrass). See Antenna. Cf. Torulus.

ANTENNAL SUTURE An external inflection forming an internal ridge to strengthen the rim of the Antennal socket (Snodgrass). See Antenna; Suture.

ANTENNAPEDIA Noun. (Latin, *antenna* = sailyard; *pes* = foot.) An error in development that results in a leg appearing where an Antenna should appear.

ANTENNARIUM Noun. (Latin, *antenna* = sailyard; *arium* = place of a thing. Pl., Antennaria.) 1. An Antennal suture or ring at the base of the Antenna. 2. An annular sclerite forming the edge of the Antacoria. 3. The Antennale *sensu* MacGillivray.

ANTENNARY FOSSA See Antennal Fossa.

ANTENNARY FURROW Mallophaga: A groove on ventral surface of the head adapted for the reception of the Antenna. See Antenna. Cf. Scrobe.

ANTENNARY NERVES Antennal nerves. Rel. Deutocerebrum.

ANTENNATE Adj. (Latin, *antenna* = sailyard; *-atus* = adjectival suffix.) Descriptive of an organism bearing Antennae. Alt. Antennatus.

ANTENNATION Noun. (Latin, *antenna* = sailyard; English, *-tion* = result of an action. Pl., Antennations.) The act by which insects touch a substrate or other organisms with the Antennae, presumably to explore, examine or communicate. See Antenna. Cf. Palpation; Tarsation.

ANTENNIFER Noun. (Latin, *antenna* = sailyard; *ferre* = to carry. Pl., Antennifers.) A pivot-like process on the rim of the antennal socket. The Antennifer forms a special support and articular point for the base of the Scape and enables the Antenna to move in all directions (Snodgrass).

ANTENNIFEROUS Adj. (Latin, *antenna* = sailyard; *fere* = to carry; Latin, *-osus* = possessing the qualities of.) Bearing Antennae.

ANTENNIFORM Adj. (Latin, *antenna* = sailyard; *forma* = shape.) Antenna-shaped; descriptive of Antenna-like structures or those with the appearance of an Antenna. See Antenna. Cf. Antennate; Palpiform. Rel. Form 2; Shape 2; Outline Shape.

ANTENNO-OCULAR POUCHES Muscid larvae: A pair of lateral pouches that contain the histoblastic rudiments of the Antennae and compound eyes. The pouches are formed by invagination of the dorsal wall of the head on which the Antennae and eyes form (Snodgrass).

ANTENNULAR REGION See Deutocerebral Region.

ANTENNULE Noun. (Latin, *antenna* = sail yard. Pl., Antennules.) 1. A small Antenna or Antenna-like process. 2. Crustacea: The first Antenna.

ANTENODAL Adj. (Latin, *ante* = before; *nodus* = knob; *-alis* = pertaining to, appropriate to.) Before or preceding a Node, specifically, the Node in the dragonfly wing. A crossvein positioned between C, Sc and R, between the base and Nodus in Odonata (Tillyard).

ANTENODAL CELLS Agrionidae (Odonata): The cells included between the short sector (M4, Comstock) and the upper sector of the triangle (Cu1, Comstock), and between the quadrilateral (or quadrangle) and the vein descending from the Nodus.

ANTENODAL COSTAL SPACES Odonata: The cells between the Costa and Subcosta, from the base to the Nodus. Syn. Antecubitals. Alt. Antinodal.

ANTENODAL CROSSVEINS Odonata: Veins extending between the Costa and Subcosta, and between the Subcosta and Media, from the base to the Nodus, forming the Antenodal or Antecubital Cells. Syn. Ante-cubitals. See Vein; Wing Venation.

ANTEOCULAR Adj. (Latin, *ante* = before; *oculus* = eye.) Before the eye. Collembola: A peculiar structure of undefined function anterior of the eyes; prostemmatic.

ANTEPECTUS Noun. (Latin, *ante* = before; *pectus* = breast. Pl., Antepectuses.) The ventral surface of the Prothorax. Cf. Prepectus.

ANTEPENULTIMATE Noun. (Latin, *ante* = before; *paene* = almost, *ultimus* = last. Pl., Antepenultimates.) 1. The second-to-the-last segment preceding the apical segment in any serially homodynamous appendage or body Tagma. The third segment from the end. Cf. Penultimate. Rel. Apical; Distal. 2. The next-to-the-last moult.

ANTEPISTERNUM Noun. (Latin, *ante* = before; *epi* = upon; *sternum* = breast bone. Pl., Antepisterna.) The Episternum of the Mesothorax and Episternum of the Metathorax (MacGillivray).

ANTEPRONOTUM Noun. (Latin, *ante* = before; *pro* = before; *notos* = back. Pl., Antepronota.) The anterior region of the Pronotum (Crampton).

ANTERIAD Adv. (Latin, *anterior* = former; *ad* = toward.) Toward the anterior end or front of a body. See Orientation. Cf. Apicad; Basad; Caudad; Centrad; Cephalad; Craniad; Dextrad; Dextrocaudad; Dextrocephalad; Distad; Dorsad; Ectad; Entad; Laterad; Mediad; Mesad; Neurad; Orad; Proximad; Rostrad; Sinistrad; Sinistrocaudad; Sinistrocephalad; Ventrad.

ANTERIOR Adj. (Latin, *anterior* = former; *or* = a condition.) 1. In front; before. 2. A term of position pertaining to a structure or colour located in front of the midline. 3. Toward the front or cephalic end of the insect. 4. Pertaining to the surface of an appendage that is visible from the frontal aspect. See Orientation. Cf. Medial; Posterior; Lateral. Ant. Posterior. Adv. Anteriad.

ANTERIOR ACROSTICHALS Diptera: See Preacrostichals.

ANTERIOR ANGLE The angle of the Thorax near the head.

ANTERIOR APOPHYSIS Female Lepidoptera: One of a pair of slender, chitinized rods that extend cephalad within the Abdomen from the abdominal segment IX and serve for muscle attachment

(Klots).

ANTERIOR ARCULUS In wing veins, that part of the Arculus that is a section of the Media (Comstock).

ANTERIOR BRANCH OF THIRD VEIN Diptera: Radius 4 of Comstock. See Vein; Wing Venation.

ANTERIOR CLYPEUS See Anteclypeus.

ANTERIOR CORNU See Precornua (MacGillivray).

ANTERIOR CROP Hemiptera: The anterior part of the Filter Chamber. Cf. Crop.

ANTERIOR CROSSVEIN Diptera: The short crossvein connecting the third and fourth longitudinal veins on the basal half of the wing (r-m) (Curran). See Vein; Wing Venation.

ANTERIOR DORSOCENTRAL BRISTLES Diptera: A row of bristles on each side anterior of the Transverse Suture (Comstock).

ANTERIOR FIELD In the Tegmina of Orthoptera. Cf. Costal Field.

ANTERIOR INTERCALARY VEIN An old and uncommon term of the 'Media 2' in Diptera. In the ground plan to the dipteran wing; the anterior branch of the Media. See Vein; Wing Venation.

ANTERIOR INTESTINE 1. Part of the Hindgut (Proctodaeum) between the Ventribulus and Rectum, or between the Pylorus and Rectum when a proctodaeal pyloric region is distinct. 2. The Small Intestine *sensu* Snodgrass. See Alimentary System.

ANTERIOR LABRAL MUSCLES One of two pairs of long muscles that move the Labrum. ALM is inserted on the anterior margin of the labral base (Snodgrass). See Labrum. Cf. Posterior Labral Muscles.

ANTERIOR LAMINA Odonata: The anterior sternal border of abdominal segment II, modified to form the anterior margin of the genital pocket.

ANTERIOR MESENTERON RUDIMENT Insect embryo: The anterior group of cells of the ventral Endoderm remnant that regenerates the Mesenteron (Snodgrass).

ANTERIOR NOTAL WING-PROCESS The anterior lobe of the lateral margin of the Alinotum supporting the first Axillary Sclerite. Syn. Suralare; Procondilo *sensu* Berlese. See Wing Articulations. Cf. Posterior Notal Wing Process.

ANTERIOR ORBIT See Facial Orbit (MacGillivray).

ANTERIOR PALPUS See Labial Palpi.

ANTERIOR PHARYNX The precerebral part of the Pharynx, in insects showing a differentiated posterior Pharynx behind the Brain (Snodgrass). See Pharnyx. Rel. Foregut.

ANTERIOR PHRAGMA One of a pair of internal, sclerotized lobes that originate on the Antecosta and project into the mesothoracic cavity. AP serves as a surface for muscle attachment (Needham). See Phragma. Rel. Antecosta.

ANTERIOR SQUAMA Antisquama. See Squama.

ANTERIOR STIGMATAL TUBERCLE Lepidoptera larvae: A prominence on the thoracic and abdominal segments. See Stigmata; Tubercle.

ANTERIOR TENTORIAL ARMS 1. The paired, cuticular invaginations of the head that originate from the Anterior Tentorial Pits along the Subgenal (Epistomal) Suture. ATPs form part of the Tentorium. 2. The apodemes on each side of the junction of the Clypeus with the Frons, or on the antecoxal parts of the Mandible. Syn. Anterior Arms of the Tentorium (Imms). See Tentorium. Cf. Dorsal Rami. Rel. Endoskeleton.

ANTERIOR TENTORIAL PITS The internally-directed depressions in the Subgenal (Epistomal) Sutures. ATPs are sometimes seen as elongate depressions in the Subgenal Suture. See Tentorium. Cf. Posterior Tentorial Pits. Rel. Endoskeleton.

ANTERIOR TRAPEZOIDAL TUBERCLE Lepidoptera larva: A protuberance on the thoracic and abdominal segments. See Tubercle.

ANTERIOR TUBEROSITY An arcane term used to describe the anteriormost of two prominent elevated, shoulder-like areas at the base of the wing in some insects (Comstock).

ANTERIOR WING Forewing. The front wings or paired wings that originate on the insect's Mesothorax. AWs are modified into structures of various forms in many groups of insects. See Wings. Cf. Elytron; Hemelytron; Tegmina. Rel. Flight; Posterior Wing.

ANTERIOVENTRAL SETA Siphonaptera larvae: A small Seta anterior of the ventrolateral Setae on abdominal segment X.

ANTERO- Latin prefix, before, or to the front.

ANTERODORSAL Adj. (Latin, *anterior* = former; *dorsum* = back; *-alis* = pertaining to, appropriate to.) 1. Toward the anterior and dorsal surface. 2. Diptera: Descriptive of any coarse or large leg Seta (bristle) at the junction of the anterior and the dorsal faces of a leg segment.

ANTEROVENTRAL Adj. (Latin, *anterior* = former; *venter* = belly; *-alis* = pertaining to, appropriate to.) 1. Descriptive of or pertaining to structure, colour or form located toward the front and ventral surfaces of a body or appendage. 2. Diptera: Any 'coarse' or large leg Seta (bristle) at the junction of the anterior and the ventral surfaces of a leg segment.

ANTHELIDAE Turner 1920 Plural Noun. A Family including ca 150 Species of ditrysian Lepidoptera assigned to Superfamily Bombycoidea; Anthelids are known only from Australia and New Guinea. Adult moderate to large sized; wingspan 20–160 mm; Ocelli and Chaetosemata absent; Antenna bipectinate in male, variable in female; Proboscis typically absent; Maxillary Palpus vestigial; Labial Palpus porrect; Thorax with piliform scales; forewing triangular in outline shape or falcate in some Species; Epiphysis present in male, reduced or absent in female; tibial spurs short, formula typically 0–2–2, occasionally 0–2–4. Egg slightly wider at one end with Micropyle; Chorion not sculptured; eggs laid individually or in small clusters. Larval head hypognathous, well sclerotized with body typically verrucate. Pupa concealed in double-walled cocoon composed of silk and larval Setae. Adults nocturnal with males attracted to lights.

ANTHELMINTIC Adj. (Greek, *anth* > *anti* = against; *helminis, inthos* = worm, tapeworm.) Pertaining to an expelling or destroying of intestinal worms. Alt. Antihelmintic.

ANTHESIS Noun. (Greek, *antensis* = bloom; *anthe* = flower; *sis* = a condition or state. Pl., Antheses.) Full Bloom. The period during which flower parts are open (expanded) and receptive to pollination. Alt. Florescence.

ANTHICIDAE Plural Noun. Ant Beetles; Ant-like Flower Beetles. A cosmopolitan Family of polyphagous Coleoptera assigned to Superfamily Tenebrionoidea and which includes more than 1,800 nominal Species. Anthicid taxonomic placement is controversial: Superfamily Cucujoidea, related to Pedilidae; also placed in Tenebrionoidea and Mordelloidea. Adult 1–13 mm long; elongate, slender, dark coloured sometimes with yellow or red spots, setose; head deflexed and constricted behind eyes giving head a 'neck-like' appearance; Antenna with 11 segments; Pronotum without lateral Carinae; tarsal formula 5-5-4; Abdomen with five free abdominal Sterna; Coxae separated posteriad. Larva elongate, parallel sided, weakly sclerotized; Antenna with three segments; 1–5 Stemmata; legs with five segments, Pretarsi each with two claws; Urogomphi present or absent. Biology poorly studied. Most Species are phytophagous and found on flowers, in vegetable detritus, in stored products and associated with dried fruits. A few predacious Species attack the egg cases of sialids or pupae of *Archips argyrospila* Walker. Adults apparently chew a hole in egg case of prey and deposit their eggs inside egg case. Mature larvae of predator drops to ground and pupates within soil. Adults also have been recovered feeding on eggs of sialids. Syn. Cononotidae; Pedilidae. See Colydiidae; Nosodendridae.

ANTHIO® See Formothion.

ANTHOBIAN Adj. (Greek, *antho* = flower; *bios* = life.) Descriptive of organisms that feed on flowers. A term applied to some lamellicorn Coleoptera in which the Labium extends beyond the Mentum.

ANTHOCORIDAE Fieber 1837. Plural Noun. Anthocorids; Flower Bugs; Minute Pirate Bugs. A numerically large, cosmopolitan Family of heteropterous Hemiptera assigned to Superfamily Cimicoidea; anthocorids are most abundant in tropical regions. Adult body small (2–5 mm long) and typically flattened; Labium with three apparent segments; compound eyes, Ocelli present. Fully-developed wings are typically macropterous and forewing with Cuneus; Hemelytral membrane veinless. Haemocoelic insemination is common among anthocorids. Most Species occur on plants or inhabit leaf litter; some Taxa occupy bird nests or inhabit mammal burrows. Most Species are predaceous upon insect eggs or Haemolymph of soft-bodied prey; a few Species feed facultatively upon mammal blood; a few Species feed upon plant tissue or pollen. Predatory Species are regarded as important natural control agents of some pest insects; a few Species have been tested in applied biological control of agricultural pests. Rel. Biological Control; Predator.

ANTHOMYIDAE See Anthomyiidae.

ANTHOMYIIDAE Plural Noun. Anthomyiid Flies. A large, structurally diverse, cosmopolitan Family of muscoid Diptera containing about 1,500 Species. This Family often is confused with Muscidae; however, most anthomyiids are smaller than the housefly and recognized by wing vein M1 not curved forward beyond middle and the lower Calypter is not longer than upper Calypter. Taxonomically, anthomyiids are fairly stable with about 50 Genera distributed among at least three Subfamilies; Genera *Hylema* and *Pegomya* contain greatest number of Species. Separation of Genera and Species is based mainly on chaetotaxy and characters of male genitalia. Faunas of anthomyiids that occur in Oriental, Australian and Neotropical regions lack modern revisionary studies. Anthomyiids are best known for their phytophagous habits. Adults of predaceous anthomyiid Species primarily capture adults of other insects, usually smaller flies. A similar habit is expressed in Asilidae. Anthomyiid mouthparts are well-adapted for piercing and sucking. Larvae of entomophagous Species are parasitic or predaceous upon other insects. Predaceous larvae typically inhabit dung or are aquatic; some predaceous larvae feed upon small Lepidoptera larvae. Parasitic Species attack nymphs and adults of Acrididae and devour egg capsules. Syn. Anthomyidae. Rel. Asilidae; Muscidae.

ANTHOMYZIDAE Plural Noun. A moderate-sized Family of cyclorrhaphous Diptera assigned to Superfamily Asteioidea. Anthomyzids are cosmopolitan in distribution but are most abundant in Holarctic. Adults are small-sized with face weakly sclerotized; Vibrissae are present and Postvertical Bristles ususally convergent; second antennal segment lacks a dorsal slit; third antennal segment with coarse Setae on apex and directed ventrad; Tibiae lack a preapical dorsal bristle; Prosternum broad, and lacks a Precoxal Bridge. Biology of anthomyzids is poorly known; adult and larva often are taken in grass and marsh-like habitats; larvae of some Species live in grass stems. See Asteioidea.

ANTHON® See Trichlorfon.

ANTHOPHILA Noun. (Greek, *anthos* = flower; *philos* = to love. Pl., Anthophilae.) Hymenoptera: Bee Species in which the hind leg Basitarsus is dilated and pubescent. Rel. Leg.

ANTHOPHILOUS Adj. (Greek, *anthos* = flower; *philein* = to love; Latin, *-osus* = possessing the qualities of.) Flower-loving. Descriptive of insects attracted to flowers, or insects feeding on flowers. Most adult bees and butterflies are anthophilous; some adult beetles and flies are specialized for herbivory on flowers or feeding on flower parts. See Feeding Strategies. Cf. Phytophilous. Rel. Ecology.

ANTHOPHORE Noun. (Greek, *anthophoros* = flower-baering; *anthos* = flower; *pherein* = to carry. Pl., Anthophores.) An elongated area (internode) of the floral receptacle between the Calyx and Corolla. See Flower; Perianth.

ANTHOPHORIDAE Plural Noun. Carpenter Bees; Cuckoo Bees; Digger Bees; Long-tongued Bees (in part). A large, cosmopolitan Family of aculeate Hymenoptera assigned to Superfamily Apoidea (bees). Adult body size, pubescence, key anatomical features and behaviour are highly variable; adult Antenna with one Subantennal Suture; Facial Foveae absent; Labrum typically wider than long and Flabellum typically present; Labial Palpus segments 1–2 elongate and flattened; forewing with 1–3 Submarginal Cells and Marginal Cell apically closed; Pterostigma typically reduced; Corbicula absent in parasitic forms or reduced in size and restricted to hind Tibia and Basitarsus. Some Species of anthophorids are cleptoparasitic on ground-nesting bees; some Species are social parasites of other bees. Some taxonomists assert that Anthophoridae are paraphyletic assemblage and constitute the stem from which Apidae and Megachilidae evolved. Subfamilies include Anthophorinae, Nomadinae and Xylocopinae. See Apoidea; Bees.

ANTHOPHORINAE The nominant Subfamily of Anthophoridae (Hymenoptera) that is geographically widespread and numerically large, consisting of about 75 Genera. Clypeus conspicuous; forewing marginal cell apically rounded and separated from margin of wing. Some Species are cleptoparasitic upon other bees. See Anthophoridae.

ANTHRACINE Adj. (Greek, *anthrax* = coal; *ene* = having the properties of.) Coal black; descriptive of a colour that is a deep shiny black with a bluish tinge. Alt. Anthracinus. See Colour.

ANTHRACNOSE Noun. (Greek, *anthrakos* = carbuncle; *nosos* = disease. Pl., Anthracnoses.) A fungal disease of plants that displays characteristic depressed spots or lesions on leaves, stems or fruit.

ANTHRAX Noun. (Greek, a burning coal. Pl., Anthraces.) The bacterial pathogen (*Bacillus anthracis* Cohn) that affects humans and nearly all domestic animals. In particular, a splenic fever of sheep and cattle is caused by the pathogen in blood and tissue. In humans, the pathogen enters the body through insect bites, inhalation or the ingestion of contaminated food or water; incubation period of disease is 3–6 days. Tabanid flies may act as mechanical vectors of Anthrax, but evidence of their capacity as vectors is not conclusive.

ANTHRIBIDAE Plural Noun. Fungus Weevils. A cosmopolitan Family of polyphagous Coleoptera with ca 2,400 nominal Species and assigned to Superfamily Curculionoidea. Adult anthribids are <30 mm long, convex; body frequently dull coloured (some tropical Species brightly coloured) with Setae or scales that form a pattern; Rostrum flat and Labrum free; Antenna not geniculate; third tarsal segment strongly bilobed. Larval body is somewhat 'C' shaped and the Antenna is minute; Frons wider than Clypeus; legs reduced without claws. Larvae feed within dead wood, on fungi and seeds; the adult stage occupies the same habitat as larva. A few Species of anthribids are significant economic pests of stored products. See Curculionoidea.

ANTHROPOBIOCENOSE Noun. (Greek, *anthropos* = man; *bios* = life; *koinos* = in common. Pl., Anthropobiocenoses.) Urban Entomology: A habitat or community created by human activity. See Habitat. Rel. Ecosystem.

ANTHROPOCENTRIC Adj. (Greek, *anthropos* = man; Medieval Latin, *centricus* = centering upon.) Regarding humans or human objectives as the central focus of a system or situation. Related term: Ethnocentrism.

ANTHROPOMETRY Noun. (Greek, *anthropos* = man; *metros* = measure. Pl., Anthropometries.) The study of proportional measurements in humans. Cf. Biometry. Rel. Allometry; Morphometrics.

ANTHROPONOSIS Noun. (Greek, *anthropos* = man; *nosos* = disease: *-sis* = a condition or state. Pl., Anthroponoses.) Medical Entomology: Any pathogen-induced disease that is exclusively associated with humans as the vertebrate host. Examples include Epidemic Typhus and Bancroftian Filariasis. Cf. Zoonosis. Rel. Medical Entomology.

ANTHROPOPHAGOUS Adj. (Greek, *anthropos* = man; *phagein* = to devour; Latin, *-osus* = possessing the qualities of.) Medical Entomology: Pertaining to Species of organisms, especially insects and ticks, that bite or feed upon humans. Syn. Anthropophily. See Feeding Strategy. Cf. Zoophily.

ANTHROPOPHILY Adj. (Greek, *anthropos* = man; *philein* = to love.) See Anthropophagous. Alt. Anthropophilous. Cf. Ornithophily; Zoophily.

ANTHROPOZOONOSIS See Zoonosis.

ANTIBACTERIAL PEPTIDES Insects exhibit strong resistance to bacterial infection following injury due to the rapid synthesis of cationic peptides. These antibacterial peptides are regulated at the transcriptional level and the genes that produce them otherwise remain silent until induced by bacterial infection. Cecropin genes produce antibacterial peptides and are so named as they were originally isolated and identified from the pupal stage of the moth *Hyalophora cercropia*. See Cercropins; Myotropic Peptides.

ANTIBIOSIS Noun. (Greek, *anti* = against; *biosis* = life; Pl., Antibioses.) 1. An association between two or more organisms that is injurious or counterproductive to one of them. 2. Plant resistance: plant characteristics that affect insects in a negative manner. See Biosis. Cf. Abiosis; Anhydrobiosis; Archebiosis; Calobiosis; Cleptobiosis; Hamabiosis; Kleptobiosis; Lestobiosis; Parabiosis; Phylacobiosis; Plesiobiosis; Synclerobiosis; Trophobiosis; Xenobiosis. Rel. Antixenosis; Non-preference.

ANTIBIOTIC Noun. (Greek, *anti* = against; *bios* = life; *-ic* = consisting of. Pl., Antibiotics.) Specific chemical substances used to eliminate microbial infections from the bodies of animals. Antibiotics are typically produced by microorganisms; semi-synthetic antibiotics are chemically-modified natural antibiotics. Rel. Immune Response.

ANTIBODY Noun. (Greek, *anti* = against; Anglo Saxon, *bodig* = body. Pl., Antibodies.) A glycoprotein released by the immune system (a Lymphocyte) in response to a foreign compound (antigen) in the body, and characterized by a specific reactivity to a specific antigen. See Immune Response. Cf. Antigen. Rel. Cellular Defence Mechanism.

ANTICLYPEUS See Anteclypeus.

ANTICOAGULANT Noun. (Greek, *anti* = against; Latin, present part. of *coagulans* = to curdle; *-antem* = an agent of something. Pl., Anticoagulants.) Any substance that prevents the clotting or coagulation of blood.

ANTICOAGULATORY Adj. (Greek, *anti* = against; Latin, *coagulare* = to curdle.) Pertaining to a chemical that acts to prevent coagulation. See Coagulation.

ANTICOAGULIN Noun. (Greek, *anti* = against; *coagulare* = to curdle; *in* = an agent of substance. Pl., Anticoagulins.) 1. An animal secretion that acts to prevent the coagulation of blood. 2. A substance secreted by some blood-sucking arthopods that prevents the blood of their prey or hosts from coagulating or inhibits the action of Coagulin. Cf. Coagulin.

ANTICRYPTIC COLORATION Any combination of colours on the body that expose, reveal or draw attention to an organism. See Coloration. Cf. Advancing Coloration; Alluring Coloration; Aposematic Coloration; Coloration; Apetetic Coloration; Combination Coloration; Cryptic Coloration; Directive Coloration; Disruptive Coloration; Epigamic Coloration; Episematic Coloration; Procryptic Coloration; Protective Coloration; Pseudepisematic Coloration; Pseudoaposematic Coloration; Seasonal Coloration. Rel. Crypsis; Mimicry.

ANTICUS Adj. (Greek, *anticus* = forward.) Frontal; belonging to or directed toward the front.

ANTIGA, Y SUNYER PEDRO (1854–1904) (Anon. 1904, Butll. Inst. Catal. Hist. Nat. 4: 81; Morice 1904, Entomol. mon. Mag. 40: 239; Bofill 1905, Butll. Inst. Catal. Hist. Nat. 5: 22–24; Dusmet y Alonso 1919, Boln. Soc. Ent. Esp. 2: 174–175.)

ANTIGEN Noun. (Greek, *anti* = against; *genes* = born. Pl., Antigens.) A chemical compound (toxin, enzyme) or substance introduced into a body (organism) and which is recognized as 'foreign' or not naturally of that body. Recognition of the foreign substance by the immune system results in the production (release) of antibodies to chemically or physically attack the foreign substance. See Immunity. Cf. Antibody. Rel. Cellular Defence Mechanism.

ANTIGENY Noun. (Greek, *anti* = against; *genos* = birth. Pl., Antigenies.) 1. Opposition or antagonism of the sexes; a concept that embraces all forms of sexual diversity. 2. Sexual Dimorphism.

ANTILIA Noun. (Etymology obscure.) 1. Lepidoptera: The spiral Proboscis or Haustellum. 2. Diptera: The dilated part of the Postpharynx; the Oesophageal Pump (MacGillivray). 3. A former land mass that included Central America and the West Indies.

ANTIMICROBIAL PEPTIDES Inducible peptides that attack microorganisms other than bacteria. See Antibacterial Peptides.

ANTIMILACE® See Metaldehyde.

ANTIPERISTALSIS Noun. (Greek, *anti* = against; *peristaltikos* = clasping and compressing; *-sis* = a condition or state. Pl., Antiperistalses.) The reversed muscular contractions (peristalisis) of the gut. Cf. Peristalsis.

ANTIPODOECIIDAE Plural Noun. A monotypic Family of Trichoptera assigned to Superfamily Limnephiloidea and known only from Australia. Adult lacks Ocelli; female Maxillary Palpus with five segments; male Maxillary Palpus with three segments and held curved vertically in front of face; Antenna as long as forewing; wing slender, apically pointed, span ca 10 mm; wing Discoidal and Median Cells absent; tibial spur formula 2–2–4; middle leg with row of short, dark spines. Larval Labrum conspicuous; mesonotal and metanotal sclerites obscure; legs stout; abdominal gills and lateral fringe absent. Larval case tubular, slightly curved and composed of small sand grains.

ANTIPYGIDIAL BRISTLES Siphonaptera: Large bristles positioned at the dorsal angle of abdominal segment VII (Comstock). Rel. Pygidium.

ANTISENSE GENES Genes that contain the mirror image of a normal nucleotide base sequence and which prevent the expression of natural genes. See Gene. Rel. Biotechnology.

ANTISEPTIC Adj. (Greek, *anti* = against; *sepsis* = disease.) Preventing infection; inhibiting the action of microorganisms. Cf. Aseptic.

ANTISQUAMA Noun. (Greek, *anti* = against; Latin, *squama* = flake. Pl., Antisquamae.) Diptera: The upper of the two Alulae which moves with the wings; Antitegula. See Squama.

ANTITEGULA Noun. (Greek, *anti* = against; *tegula* = tile. Pl., Antitegulae.) See Antisquama.

ANTIXENOSIS Noun. (Greek, *anti* = against; *xenos* = stranger; *-sis* = a condition or state. Pl., Antixenoses.) Plant resistance: Plant characteristics that drive parasitic or predaceous insects away from a particular host. Syn. Non-preference. See Antibiosis. Rel. Behavioural Resistance; Host Plant Resistance.

ANTLERED LARVAE Lepdioptera: Neonate larvae of some *Heterocampa* that display a pair of large antler-like (horn-like) cuticular processes on the first thoracic segment, and other 'horns' on the abdominal segments (Comstock).

ANTLIATA Noun. (Latin, *antlia* = a pump; Greek, *antlia* = bilge water from the hold of a ship. Pl., Antliatae.) Insects with sucking mouthparts. The term originally was applied to Lepidoptera and Diptera, but subsequently more specifically to Diptera.

ANT-LIKE FLOWER BEETLES See Anthicidae.

ANTLION Noun. (English, ant + lion. Pl., Antlions.) The larval stage of Neuroptera in the Myrmeleontidae, particularly *Myrmelion* and related Genera. So-called because the larva constructs a conical pit within sand or sandy soil and waits at the bottom of the pit for ants and other insects. Alt. Ant lion. See Myrmeleontidae. Cf. Aphid Lion; Aphid Midge; Aphid Wolves.

ANT-LOVING BEETLES See Pselaphidae.

ANT-LOVING CRICKETS See Myrmecophilidae.

ANTOINE, MAURICE M (1887–1962) (Anon. 1962, Bull. Soc. Ent. Fr.)

ANTRAM, CHARLES BEAUMONT (1864–1955) (A. B. A. 1955, Entomologist's Rec. J. Var. 67: 160.)

ANTHROPOMORPHISM Noun. (Greek, *anthropos* = man; *morphe* = shape; *-ism* = condition. Pl., Anthropomorphisms.) The attribution of human motivations, characteristics or behaviours to non-humans, inanimate objects or natural phenomena. *e.g.* The bee was angry. The rock was stubborn and refused to move. The rain was gentle.

ANTRORSE Adj. (Latin, *antrorsis* > *antero* + *versus* = turned.) Directed toward the front. Directed forward. Alt. Antrorsum.

ANTS See Formicidae; Formicoidea.

ANUS Noun. (Latin, *anus* = anus. Pl., Anuses.) 1. General: The opening at the posterior end of the digestive tract, through which excrement is passed. See Alimentary System; Hindgut. Cf. Orificium. 2. Coccidae: A more-or-less circular opening on the dorsal surface of Pygidium, varying in position relative to circumgenital gland orifices. Syn. Anal Orifice. 3. Acarina: Structure marking a non-functional opening of the hindgut and positioned ventrally along midline on posterior 1/3 of Idiosoma. Syn. Uropore. (Goff *et al.* 1982, J. Med. Entomol. 19: 222.)

AO-GVS® See Summer Fruit-Tortrix Granulosis Virus.

AOKI, AKIRA (1917–1951) (Hasegawa 1967, Kontyû 35 (Suppl.) 3).

AORTA Noun. (Greek, *aorte* = great artery. Pl., Aortas.) The anterior, non-chambered, narrow part of an insect Heart (Dorsal Vessel) that is confined to Thorax and head. See Circulatory System. Cf. Dorsal Vessel; Heart. Rel. Haemolymph.

AORTAL CHAMBER A thoracic enlargement of the Aorta (Packard).

AORTIC VALVE The closing mechanism of an insect's Dorsal Vessel. The AV is positioned at the junction between the Aorta and Heart. See Valve.

APACHE® Cadusafos.

APACHYIDAE Plural Noun. A small Family of forficuline Dermaptera assigned to Superfamily Apachyoidea. Adult body is large and strongly flattened as an adaptation for living under bark; Pygidium directed downward. See Dermaptera.

APAMIN Noun. A small, basic, polypeptide toxin derived from the venom of the European honey bee, *Apis mellifera* L., that blocks a class of Ca^{2+}-activated K^{+} channels, crosses the blood–brain barrier, and possesses a convulsant activity. The peptide is an 18-amino acid chain with two disulphide bridges. Honey bee venom contains at least 18 such active substances. See Adolapin; Melittin.

APATELODIDAE Plural Noun. A small Family of ditrysian Lepidoptera assigned to Superfamily Bombycoidea. See Lepidoptera.

APERDEX® See NAA.

APERTUM Noun. (Latin, *apertum* = open. Pl., Apertia.) A cell opening basally in the hindwing of Coleoptera (Tillyard).

APERTURE Noun. (Latin, *apertura* from *aperire* = to uncover; to open. Pl., Apertures.) Any opening in a wall, surface or tube. Syn. Hole; Opening; Pore. Cf. Oriface. Rel. Duct.

APETETIC COLORATION Combinations of colour or colour patterns that resemble some part of the environment or the appearance of another Species (Folsom & Wardle). See Coloration. Cf. Advancing Coloration; Alluring Coloration; Anticryptic Coloration; Aposematic Coloration; Combination Coloration; Cryptic Coloration; Directive Coloration; Disruptive Coloration; Epigamic Coloration; Episematic Coloration; Procryptic Coloration; Protective Coloration; Pseudepisematic Coloration; Pseudoaposematic Coloration; Seasonal Coloration; Sematic Coloration. Rel. Crypsis; Mimicry.

APEX Noun. (Latin, *apex* = summit. Pl., Apices.) 1. The part of an appendage, joint or segment farthest from the base of the part. 2. The part of a wing furthest from the base. 3. The highest topographcal point on a surface. Syn. Peak; Summit; Tip; Top. See Orientation. Cf. Base. Adv. Apicad.

APEX® See Methoprene.

APEX ABDOMINIS 1. An archane term for the apical part of the Abdomen. 2. The Cremaster.

APEX OF THE WING The angle of the insect wing between the Costal margin and the outer margin (Comstock).

APFELBECK, VICTOR (1859–1934) (Apfelbeck 1923, Glasn. Zamalj. Mus. Bosni Herceg. 35: 1–

47, bibliography only 1889–1922; Horn 1934, Arb. physiol. angew. Ent. Berl. 1: 306–307.)

APHANIPTERA Noun. (Greek, *aphanes* = unapparent; *-optera* = wing.) Indistinctly winged. See Siphonaptera.

APHELEN, HANS VON (1719–1779) (Henriksen 1921, Ent. Meddr 15: 26.)

APHELINIDAE Plural Noun. A moderately large (ca 45 Genera, 1,000 nominal Species), cosmopolitan Family of parasitic Hymenoptera assigned to Superfamily Chalcidoidea. Aphenlinids are typically very small (0.5–1.5 mm long), fragile with body variable in coloration and very rarely metallic; Antenna with 5–8 segments; Parapsidal Sutures present; Mesopleuron rather large but not strongly convex; marcropterous forms with long Marginal Vein, sessile Stigmal Vein and Postmarginal Vein absent; middle tibial spur rather long but not robust; legs with 4–5 Tarsomeres. Anatomy of larvae is variable. Aphelinids are biologically diverse, acting as primary and secondary ectoparasites, endoparasites with males of some Species adelphoparasitic; aphelinids predominantly attack sternorrhynchous Hemiptera. Several Species are important in biological control of scale insects and whiteflies (90 of 216 programmes *teste* Greathead 1986). Included Subfamilies: Aphelininae, Aphytinae, Azotinae, Calesinae, Coccophaginae, Eriaphytinae, Eriaporinae, Phycinae and Prospaltellinae. See Chalcidoidea. Rel. Heteronomy.

APHELININAE Howard 1896. The nominant Subfamily of Aphelinidae whose morphological features include: Antenna with six segments and club fused; male Antennal Scape with rounded Sensilla; Pronotum and Prepectus of two sclerites; forewing with Linea Calva; apical Tergum medially divided; female Sternum VII posteromedially produced. Biology: Internal Parasites of Aphididae. See Aphelinidae.

APHICIDE® See Thiometon.

APHID See Aphididae.

APHID LION The larval stage of Chrysopidae. See Chrysopidae. Cf. Antlion; Aphid Midge; Aphid Wolves.

APHID MIDGE *Aphidolestes aphidimyza* (Rondani) [Diptera: Cecidomyiidae]: A predaceous fly used in biological control of aphids that occur in glasshouse culture.

APHID WOLVES See Coniopterygidae.

APHIDEINE Noun. See Aphidilutein.

APHIDICOLOUS Adj. (Greek, *apheides* = lavish; *colere* = living or growing on.) Pertaining to organisms associated with aphids. A term often applied to ants. See Habitat. Cf. Agricolous; Caespiticolous; Gallicolous; Myrmecophilous; Silvicolous. Rel. Ecology.

APHIDIDAE Plural Noun. Aphids; Plant Lice. A numerically large, cosmopolitan Family of sternorrhynchous Hemiptera assigned to Superfamily Aphidoidea. Polymorphic; Antenna with 4–6 segments; apterous and macropterous populations with wings usually tectiform, rarely held flat; Siphunculus (Cornicle) on posterior margin of fifth abdominal segment emits defensive wax secretions or alarm pheromones; Cauda modified to discharge honeydew; Ovipositor absent. Aphids display a complex biology including four nymphal instars and distinct generations (Ovipare, Fundatrices, Vivipare, Gynopare). All aphids are phytophagous and predominantly phloem feeders; some Species form galls. Many Species are regarded as significant agricultural or horticultural pests; some Species vector plant diseases. Heteroecy (alternation of hosts) has been reported in some Species. Most Species are parthenogenetic with viviparous females; sexual females are oviparous. Generalized aphid life-cycle for temperate Species: Winter: Aphids pass winter as egg on perennial host plant (tree). Early spring: Eggs hatch and several generations of viviparous, apterous, parthenogenetic females feed upon spring leaves. Late spring: Leaves mature and become unsuitable for aphids. Spring migrants: Next aphid generation viviparous, winged, parthenogenic females. Summer: Annual plants growing and colonized by winged spring migrants. Viviparous, apterous, parthenogenic females produced on annual plants for several generations. Late fall: Annual plants deteriorate as hosts. Fall migrants: Generation includes male and oviparous, winged female that migrate to overwintering perennial host plant, mate and lay overwintering eggs. Viviparous reproduction shortens generation time and maximizes population growth; oviparous reproduction during last generation provides a resistant stage (egg) to pass winter. Parthenogenic generations result in accelerated population growth; sexual reproduction occurs once a year for genetic recombination. See Aphidoidea. Cf. Bean Aphid; Corn Root Aphid; Green Peach Aphid; Melon Aphid. Rel. Adelgidae.

APHIDIIDAE Haliday 1833. Plural Noun. A numerically small (ca 300 Species) geographically cosmopolitan Family of apocritous Hymenoptera assigned to the Ichneumonoidea. All Species purported to parasitize aphids; some Species used in biological control of aphids. Aphidiidae are sometimes considered a Subfamily within the Braconidae. See Braconidae.

APHIDIVOROUS Adj. (Greek, *apheides* = unsparing; *vorare* = to devour Latin, *-osus* = possessing the qualities of.) Pertaining to insects that feed on aphids. Alt. Aphidophagous. See Feeding Strategy. Rel. Coccidiphagous.

APHIDOIDEA A Superfamily of sternorrhynchous Hemiptera including about 4,700 nominal Species included within the Families Adelgidae, Aphididae and Phylloxeridae. Superfamily is cosmopolitan in distribution but most speciose in Holarctic. Polymorphic generations; Antenna with six or fewer segments; beak (Rostrum) with 4–5 segments; compound eye modified with ocular tubercle (Triommatidion); three Ocelli in Alatae only, other morphs are Anocellate; forewing and hindwing dissimilar in size and venation; Tarsi with two segments; Malpighian Tubules absent. Aphidoids are phloem-feeding insects with complex life-cycles including four nymphal instars, alternation of parthenogenetic and sexual generations, polymorphism and generations shifting between unrelated Taxa of host plants. Some aphidoids are regarded as economically important agricultural or horticultural pests through sap feeding, vectoring plant disease and disease associated with excessive excretion (honeydew). Fossil record of aphidoids extends into Triassic Period when early Taxa apparently fed on conifers; rapid diversification of aphidoids during late Cretaceous Period was associated with shifting to angiosperms. See Sternorrhyncha.

APHIDOPHAGOUS Adj. (Medieval Latin, *apheides* = unsparing; Greek, *phagein* = to devour; Latin, *-osus* = possessing the qualities of.) Pertaining to or descriptive of insects that feed upon aphids, typically as predators (*e.g.* some Coccinellidae, Chrysopidae). Alt. Aphidivorous. See Feeding Strategy. Rel. Aphid Lion; Aphid Midge; Aphid Wolves.

APHIS The acronym for **A**nimal **P**lant **H**ealth **I**nspection **S**ervice. The branch of the US Department of Agriculture responsible for port inspection for the detection and eradication of unwanted organisms from the USA. Cf. AQIS; ARS; USDA.

APHISTAR® See Triazamate.

APHOX® See Primicarb.

APHRODISIAC Adj. (Greek, *aphrodisios* = pertaining to Aphrodite = born of foam). Behaviour: Chemical compounds secreted by members of one sex and intended to facilitate copulation after attraction of the opposite sex. (Sreng 1984, J. Morph. 182: 279–294.) See Pheromone. Cf. Sex Pheromone.

APHROPHORIDAE Plural Noun. A cosmopolitan Family of auchenorrhynchous Hemiptera assigned to Superfamily Cercopoidea, and sometimes regarded as Subfamily of Cercopidae. When recognized as distinct, aphrophorids constitute the largest Family of Cercopoidea. Adult Pronotum flat with posterior margin 'W' shaped; forewing appendix reduced or absent; hindwing with R branched; two cells between M and Costal margin. Nymphs are enclosed within spittle. See Cercopidae.

APHYLIDAE Plural Noun. See Pentatomidae.

APHYTINAE Yasnosh 1976. A cosmopolitan Subfamily of Aphelinidae. Morphological features include: Antenna with 3–6 segments including fused club; Pronotum entire or divided longitudinally along medial surface; Prepectus divided into two sclerites; Linea Calva usually present, sometimes poorly developed. Host spectrum and biology of aphytines is diverse: Most Species are internal parasites of Diaspididae and Aleyrodidae; someSspecies are egg parasites of Orthoptera and Hemiptera; a few Species are hyperparasites of Coccidae and Psyllidae. See Aphelinidae.

APICAD Adv. (Latin, *apex* = summit; *ad* = toward.) Toward the apex of an appendage, structure or body. See Orientation. Cf. Anteriad; Basad; Caudad; Centrad; Cephalad; Craniad; Dextrad; Dextrocaudad; Dextrocephalad; Distad; Dorsad; Ectad; Entad; Laterad; Mediad; Mesad; Neurad; Orad; Proximad; Rostrad; Sinistrad; Sinistrocaudad; Sinistrocephalad; Ventrad.

APICAL Adj. (Latin, *apex* = summit, New Latin, *apicalis*; *-alis* = belonging to, pertaining to.) 1. At or near the apex. 2. Pertaining to the apex or distalmost part of a structure. See Orientation. Cf. Basal; Distal.

APICAL ANGLE The angle of the wing at its apex (Wardle).

APICAL AREA Apical cells of the wing in some Hemiptera. See Petiolar Area.

APICAL CELL 1. A relatively large, 'trophic cell' of the upper end of the testicular tube in some insects *sensu* Snodgrass. Syn. Versonian Cell. 2. Trichoptera: The series of cells along the outer margin of the wing from Pterostigma to Arculus (Smith). 3. Diptera: The first posterior cell of the wing; the space between the third and fourth longitudinal veins beyond the anterior crossvein (R3) (Curran). 4. Hymenoptera: A cell near or at the apex of a wing *teste* Norton; Comstock: The Medial Cell; Outer Apical Cell; second Medial; Inner Apical Cell.

APICAL CHAMBER The Germarium of the Acrotrophic-type egg tube *teste* Snodgrass. See Ovariole. Cf. Germarium.

APICAL MARGIN The outer margin of the insect wing (Wardle). See Wing. Cf. Distal. Rel. Marginal Fringe.

APICAL PLATE An external sensory structure of the primitive annelid-arthropod nervous system or group of nerve cells at the anterior pole of the body (Snodgrass). See Nervous System.

APICAL SCUTELLARS Diptera: The apical pair of marginal bristles on the Scutellum. A term loosely applied and often meaning the 'sub-apical scutellars' when the true apicals are absent (Curran).

APICAL SECTOR Neuroptera: A longitudinal vein in the apical part of the wing. See Wing.

APICAL SETA Acarina: The distal Seta on the palpal Tarsus. (Goff *et al.* 1982, J. Med. Entomol. 19: 222.)

APICAL SPURS Diptera: Short, rather stout bristles that vary in number and are often present on the ventral surface of the Tibia (Curran).

APICAL TRANSVERSE CARINA Hymenoptera: The Carina that crosses the Metanotum behind the middle and separates the median from the posterior cells or areas.

APICALLY Adj. (Latin, *apex* = summit.) Toward or directed toward the apex of a structure. See Orientation.

APICOMPLEXA Noun. A Class of parasitic Protozoa, the members of which lack Cilia or Flagella (except some male gametes). Apicomplexa is the most important Class of Protozoa from the viewpoint of medical and veterinary Entomology. The Class includes Suborder Haemosporonia whose members are characterized by merogony (asexual reproduction) in vertebrate red blood cells, and gametogony (sexual reproduction) and sporogony (sporozoite formation) in blood sucking flies. See Plasmodiidae.

APICULATE Adj. (Latin, *apex* = tip; *-atus* = adjectival suffix.) Covered with erect, fleshy, short points. Alt. Apiculatus.

APICULTURE Noun. (English, from Latin, *apis* = bee; *cultura*, care. Pl., Apicultures.) The care and management of honey bees. See Agriculture. Cf. Agronomy; Horticulture.

APICULUS Noun. (Latin, diminutive of Apex.) 1. Any short, small tip or point. 2. Botany: An erect, fleshy, short point. A structure terminating in a point.

APIDAE Latreille 1802. Plural Noun. Bumble Bees; Carpenter Bees; Honey Bees; Stingless Bees. A comparatively small, cosmopolitan Family of aculeate Hymenoptera assigned to Superfamily Apoidea (bees). Adult head with one Subantennal Suture; Facial Foveae absent; Labrum longer than wide; Flabellum present; Maxilla with short preceding Palpus and long following Palpus; Labial Palpus with first two segments elongate and flattened; Basitibial Plate absent; forewing Marginal Cell open at apex; hind Tibia with Corbicula on lateral surface for collection of pollen; Scopa not present on abdominal Sterna; Pygidium absent. Family includes honey bee (*Apis mellifera*), stingless bees and all highly-social bees. Oldest fossil bee: Worker *Trigona prisca* from Cretaceous Period deposits of New Jersey (Michener & Grimaldi 1988, PNAS 85 (17): 6424–6426.) Subfamilies include: Apinae, Bombinae and Meliponinae. See Apoidea.

APIMYIASIS Noun. (Greek, *apis* = bee; *myia* = fly; *-iasis* = suffix for names of diseases. Pl., Myiases.) The infestation of adult bees with the larval stage of Diptera. Diptera commonly associated with honey bees include *Borophaga incrassata, Braula coeca, Melaloncha ronnai, Sarcophaga tricuspis* and *Senotainia tricuspis.* See Myiasis. Rel. Braulidae.

APINAE Honey Bees. The monogeneric, nominant Subfamily of Apidae. See Honey Bee.

APIOCERATIDAE See Apioceridae.

APIOCERIDAE Plural Noun. A Family of orthorrhaphous Diptera assigned to Superfamily Asiloidea, and which contains ca. 150 Species. Nearly all described Species of apiocerids occur in North America or Australia. The Family is widely distributed in arid and semiarid regions but is unknown in Palearctic. Apiocerids superficially resemble Asilidae with body medium to large-sized, fast flying and elusive. Antennal Flagellum ovoid, much shorter than head length; wing rather short, venation with R4, R5 and M curved anteriorly before apex of wing; tarsal Empodium setiform. Adult Apioceridae do not feed or take nectar. Eggs deposited into sandy soil where all immature stages develop. Larvae probably are predaceous; first-instar larva verimiform.

APIONIDAE See Brentidae; Curculionidae.

APIS IRIDESCENT-VIRUS An iridoviral disease of *Apis cerana*. Iridoviruses are poorly documented in Hymenoptera. AIV is characterized by crystalline masses in tissues that appear blue-violet or green when illuminated with bright white light; the virus can be detected within individual infected-cells by their distinctive coloration. Heavily infested colonies become inactive during summer, with flightless individuals forming clusters or walking upon the ground. The disease seems restricted to *A. cerana* in the Himalayan region. The virus can multiply in *A. mellifera.*

APISTAN® See Fluvalinate.

APIVOROUS Adj. (Latin, *apis* = bee; *vovare* = to devour; Latin, *-osus* = possessing the qualities of.) Bee-eating. Descriptive of organisms that feed on bees. See Feeding Strategy.

APM-ROPES® A synthetic pheromone {(Z)-II, hexadecenal} used to disrupt sex pheromones of Artichoke Plume Moth on artichokes. See Synthetic Pheromones.

APNEUMONE Noun. (Greek, *a* = without; *pnein* = to breathe, from *pneuma* = breath. Pl., Apneumones.) A substance emitted by non-living material that evokes a behavioural or physiological reaction adaptively favourable to a receiving organism, but detrimental to an organism of another Species that may be found on or in the nonliving material. (Nordlund & Lewis 1976, J. Chem. Ecol. 2: 211–220.)

APNEUSTIC Adj. (Greek, *apneustos* = breathless; *-ic* = characterized by.) Pertaining to a type of Tracheal System in which insects lack functional Spiracles. 2. Pertaining to insects that respire through the Integument or through Tracheal Gills but lack specific external breathing organs (Spiracles). 3. Pertaining to insects in which the Tracheal System is usually absent or rudimentary. See Tracheal System; Spiracle. Cf. Hemipneustic; Holopneustic; Oligopneustic; Polypneustic; Propneustic. Rel. Respiratory System.

APNOEA Noun. The resting position of the Abdomen during the first phase of Active Ventilation. See Respiration.

APOCRITA Noun. Ants; Bees; Wasps. A numerically large, holophyletic, cosmopolitan Suborder of Hymenoptera, including ants, bees and wasps. Morphologically characterized by adult without closed Anal Cells in the wings and first abdominal segment (Propodeum) functionally incorporated with Thorax and separated from remainder of Abdomen by a constriction formed through reduction of second abdominal segment (Petiole). Larva of second and subsequent instars are hymenopteriform with head capsule and Antenna present or absent, Mandibles usually are present, body apodous, midgut and hindgut are not connected and excretion is restricted to prepupa or last larval instar. Syn. Clistogastra; Petiolata. See Aculeta; Parasitica. Cf. Symphyta.

APODAL Adj. (Greek, *a* = without; *pous* = feet; Latin, *-alis* = pertaining to, having the character of.) Pertaining to organisms that lack feet. A condition typical of Diptera larvae. Larvae with single, simple tubercles instead of ambulatory feet. Alt. Apodous. Cf. Pedate.

APODE Noun. (Greek, *a* = without; *pous* = feet.) An organism that has no feet or legs.

APODEMA Noun. (Pl., Apodemata.) 1. Coccidae: A conspicuous transverse band crossing the Thorax anterior of the Scutellum in males. 2. Coccids: A thin, vertical, cuticular sclerite extending into the body cavity from the inner surface of each Episternum *sensu* MacGillivray.

APODEMAL Adj. (Greek, *apo* = away; *demas* = body; Latin, *-alis* = having the character of.) Descriptive of structure with the character of an Apodeme.

APODEME Noun. (Greek, *apo* = away; *demas* = body. Pl., Apodemes.) 1. Any rigid process that forms the insect endoskeleton (Imms). 2. A ridge or inflection of the Cuticle that provides surface for the attachment of muscles (Tillyard). 3. Any invagination of the body wall. 4. Any inward pleural projection of the Thorax. 5. Acarina: Any chitinous ridge or process extending internally from the body wall, appendage or claw; Apodemes provide attachment sites for muscles. Apodemes of leg segments form internal annulations. (Goff *et al.* 1982, J. Med. Entomol. 19: 222.) See Apophysis; Endoskeleton; Furca. Cf. Fovea; Suture; Sulcus.

APODOUS Adj. (Greek, *a* = without; *pous* = feet; Latin, *-osus* = possessing the qualities of.) Without feet; legless. See Apodal.

APOID Adj. (Latin, *apis* = a bee; Greek, *eidos* = shape.) Bee-like; resembling a bee larva (*e.g.* white, cylindrical, headless and legless). Rel. Eidos; Form; Shape.

APOIDEA The Superfamily of Hymenoptera called 'bees.' A cosmopolitan Taxon consisting of 11 Families and ca 450 Genera; ca 70 Family-group names and 2,700 Genus-group names have been proposed; ca 20,000 Species of bees have been estimated. Diagnostic anatomical features are highly variable; bees display anatomy and habitus of sphecoid wasps except bees possess branched Setae on body; hind Basitarsus of female frequently wider than other Tarsomeres; metasomal Tergum VIII divided into four Hemitergites. Larva blind, without Antenna or legs (apodous); Maxilla typically with one papilla; Galea absent. Female adult constructs nest as site for oviposition. Nests typically in soil; some nests in wood or pithy stems of plants; a few nests constructed of plant resin or mud. Nest consists of one to several cells in which egg is laid and provisions for larva are provided. Nests are provisioned with pollen (source of protein) and nectar (source of carbohydrate); only females provision nests. Most bees nest individually; some bees nest communally; some bees are social; a few bees are cleptoparastic. Sociality in Apoidea is highly developed with several gradations and evolution of castes. Bees represent the most important group of plant pollinators; pollen collection by bees is termed polylectic or oligolectic. Apoids apparently arose and flourished with proliferation of angiosperms (Cretaceous Period ca 140 million years before present); fossil record of bees originates about 85 million years before present with *Trigona*. Family-level classification of bees is controversial with 1–11 Families recognized; Families here include: Andrenidae, Anthophoridae, Apidae, Ctenoplectridae, Colletidae, Halictidae, Megachilidae, Melittidae, Oxaeidae, Pararhophitidae and Stenotritidae. See Social Insects. Cf. Sphecoidea; Vespoidea.

APOLAR Adj. (Greek, *a* = without; Latin, *polus* = pole.) Structure without differentiated poles; without apparent radiating processes. A term applied to cells. Cf. Polarity.

APOLLO® See Clofentezine.

APOLLO BUTTERFLIES See Papilionidae.

APOLO® See Clofentezine.

APOLYSIS Noun. (Greek, *apo* = away: *lysis* = loosening. Pl., Apolyses.) 1. The physical separation of the Epidermis from the old Cuticle before the formation of the new cuticle. 2. The intermoult period during which the epidermal cells of the Integument secrete moulting fluid and new cuticle. See Moulting. Cf. Ecdysis. Rel. Integument.

APOMIXIS Noun. (Greek, *apo* = away from; *mixis* = mixing. Pl., Apomixes.) A cytological form of thelytokous parthenogenesis in which meiosis is absent. See Parthenogenesis; Thelytoky. Cf. Amphimixis; Automixis.

APOMORPHIC Adj. (Greek, *apo* = away from; *morph* = form; *-ic* = of the nature of.) A relatively derived character-state or condition within a morphological transformation series of a homologous character. See: Autapomorphic; Synapomorphic. Cf. Pleisiomorphy.

APOMORPHOUS Adj. (Greek, apo = away from; *morph* = form; Latin, *-osus* = possessing the qualities of.) Pertaining to a derived character state. Cf. Plesiomorphous.

APOMORPHY Noun. (Greek, *apo* = away from; *morph* = form. Pl., Apomorphies.) Cladistic classification: A derived character state. Apomorphies are divided into two types: Autapomorphy and Synapomorphy. See Classification. Cf. Autapomorphy; Synapomorphy.

APOPHYSIS Noun. (Greek, *apo* = away; *phyein* = to grow; *-sis* = a condition or state. Pl., Apophyses.) 1. Any tubercular or elongate process that projects internally or externally from the body wall of an insect. 2. An evagination of the Cuticle and forming an integral part of it (Comstock). 3. A ventral thoracic projection of the insect endoskeleton (Folsom & Wardle). 4. The lower (distal) of the two joints of the Trochanter in the Ditrocha. The Trochanterellus. 5. The dorsolateral metathoracic spines in Hymenoptera. Apophysis also used synonymously with Entothorax (Smith); Coxa (Say); a ventral apodeme of the Thorax, Furca, Furcella *sensu* MacGillivray. Cf. Apodeme; Fovea.

APOPHYSTEGAL PLATES Orthoptera: Flattened blade-like or plate-like sclerites covering the Gonapophyses.

APOPROGONIDAE Plural Noun. A small Family of ditrysian Lepidoptera of uncertain placement. See Lepidoptera.

APOPTOSIS (Greek, *apo* = away from; *-sis* = a condition or state.) Cell death caused by intracellular enzymes. Syn. Autolysis.

APOSEMATIC COLORATION Warning coloration; Combination of colours or colour patterns displayed by animals and plants to indicate unpalatability to potential predators or herbivores. Aposematic coloration is abundantly documented by Insecta. See Coloration. Cf. Advancing Coloration; Alluring Coloration; Anticryptic Coloration; Apetetic Coloration; Combination Coloration; Cryptic Coloration; Directive Coloration; Disruptive Coloration; Epigamic Coloration; Episematic Coloration; Procryptic Coloration; Protective Coloration; Pseudepisematic Coloration; Pseudoaposematic Coloration; Seasonal Coloration; Sematic Coloration. Rel. Crypsis; Mimicry.

APOTELE Noun. (Greek, *apotelein* = to complete. Pl., Apoteles.) Acarology: The terminus of an appendage. The Apotele of the Chelicera is constituted by the movable jaws; the Apotele of the legs by the Ungues.

APOTOME Noun. (Greek, *apo* = away; *tome* = cutting. Pl., Apotomes.) Immatures: A sclerotized part of the head capsule that is separated at ecdysis; a functional rather than homolous subdivision. 2. Japygidae: A short anterior subdivision of each abdominal Sternum separated by a membranous fold from the rest of the sclerite (Snodgrass).

APOTYPE Noun. (Greek, *apo* = away; *typos* = pattern. Pl., Apotypes.) See Type. Cf. Hypotye; Plesiotype. Rel. Taxonomy.

APOTYPIC Adj. (Greek, *apo* = away; *typos* = pattern; *-ic* = of the nature of.) Characterized as an Apotype.

APOZYGIDAE Mason 1978. Plural Noun. A numerically small Family of apocritous Hymenoptera assigned to Superfamily Ichneumonoidea. Apozygidae occur in Chile. Adult with paired spiracles on gastral segments 1–8. Biology of the Family remains unknown. See Ichneumonoidea.

APPA® See Phosmet.

APPARENT RESISTANCE Plant resistance: Temporary plant resistance characteristics that develop in response to favorable changes in environmental conditions for plant growth; induced plant vigour leads to induced insect resistance.

APPEASEMENT SUBSTANCE Social Insects: A secretion provided by a social parasite to a host that reduces aggression by the host and facilitates acceptance by the host colony. See Kairomone. Rel. Trichome.

APPEL, HERMAN (1892–1966) (Kãtter 1969, Proc. Int. Union Study Soc. Inns. Congr. 6: 275–279.)

APPENDAGE Noun. (Latin, *appendere; ad* = to; *pendere* = to hang; *-age* = collection. Pl., Appendages.) 1. Any part, portion, piece or organ attached by a joint to the body or other primary unit of structural organization (*e.g.* leg, Antenna, wing). 2. A subordinate part or organ.

APPENDAGE VEIN Diptera: A short vein at the angle of a bend (Curran). See Vein; Wing Venation.

APPENDICIAL Adj. (Latin, *appendere; -alis* = having the character of.) Something supplementary; relating to appendices.

APPENDICLE Noun. (Latin, *ad* = to; *pendere* = to hang. Pl., Appendicles.) 1. A small appendix. 2. In some insects, a small sclerite at the apex of the Labrum. Alt. Appendicula. See Labrum.

APPENDICULAR Adj. (Latin, *appendere; ad* = to; *pendere* = to hang.) Descriptive of or pertaining to an appendage.

APPENDICULATE Adj. (Latin, *ad* = to; *pendere* = to hang; *-atus* = adjectival suffix.) Pertaining to structure that bears appendages.

APPENDICULATE CELL Hymenoptera: A wing cell on the Costa distad of the second Radius.

APPENDIGEROUS Adj. (Latin, *ad* = to; *pendere* = to hang; *gerere* = to carry; Latin, *-osus* = possessing the qualities of.) Bearing appendages.

APPENDIX Noun. (Latin, *ad* = to; *pendere* = to hang. Pl., Appendices; Appendixes.) 1. A supplementary or additional part that has been added to or attached to another structure. 2. Heteroptera: The Cuneus. Cf. Appendage.

APPENDIX DORSALIS A median, segmented, slender filament arising from the distal end of the abdominal Epiproct (Tergum XI) of primitive groups of living insects such as Thysanura and Ephemeroptera, and some fossil Paleodictyoptera. The AD appears annulated and similar in shape to the Cerci that are lateral of the AD; functional significance obscure. AD is attached to the supra-anal sclerite according to Tillyard. Syn. Caudal Style; Media Cercus; Filum Terminale. Rel. Cercus.

APPEX® See Tetrachlorvinphos.

APPLAUD® See Buprofezin.

APPLE-AND-THORN SKELETONIZER *Choreutis pariana* (Clerck) [Lepidoptera: Choreutidae].

APPLE APHID 1. *Aphis pomi* DeGeer [Hemiptera:

Aphididae]: A widespread pest of apples in North America. Eggs first green then black; during autumn, eggs typically laid on water sprouts and eclosion during spring when buds swell; nymphs feed on terminal twigs and leaves, causing stunting and leaf curl. Winged generation migrates to other apple trees. During autumn, winged male and oviparous female mate to produce overwintering eggs. AA is multivoltine with as many as 19 generations per year. Cf. Apple-Grain Aphid. Alt. Green Apple Aphid. 2. *Aphis citricola* van der Goot [Hemiptera: Aphididae] (Australia).

APPLE-BARK BORER *Synanthedon pyri* (Harris) [Lepidoptera: Sesiidae]: A pest of apple, hawthorne, mountain ash and pear in North America. Larva loosens bark and feeds on cambium. See Sesiidae. Cf. Peach-Tree Borer.

APPLE BARKMINER *Marmara elotella* (Busck) [Lepidoptera: Gracillariidae].

APPLE BLOTCH LEAFMINER *Phyllonorycter crataegella* (Clemens) [Lepidoptera: Gracillariidae].

APPLE CLEARWING-MOTH *Synanthedon myopaeformis* Borkausen [Lepidoptera: Sesiidae].

APPLE CURCULIO *Tachypterellus quadrigibbus* (Say) [Coleoptera: Curculionidae]: A monovoltine pest of apple, hawthorn, pear and quince in eastern USA. Eggs laid within punctures on fruit; larvae feed and pupate within fruit. Adults emerge during mid-July-September and feed on ripening fruit until overwintering. See Curculionidae. Cf. Plum Curculio; Quince Curculio.

APPLE DIMPLING-BUG *Campylomma liebknechti* (Girault) [Hemiptera: Miridae]: An important secondary pest of cotton in Australia. See Yellow Mirid.

APPLE FLEA-WEEVIL *Rhynchaenus pallicornis* (Say) [Coleoptera: Curculionidae]: A pest of apple and ornamental trees in northeastern USA. Adults overwinter in trash and become active during budding in spring. Adults feed on new leaves ca 10–15 days.

APPLE FRUITMINER *Marmara pomonella* Busck [Lepidoptera: Gracillariidae].

APPLE-FRUIT MOTH *Argyresthia conjugella* Zeller [Lepidoptera: Argyresthiidae].

APPLE-GRAIN APHID *Rhopalosiphum fitchii* (Sanderson) [Hemiptera: Aphididae]: An early season pest of apples and pear in North America. Cf. Apple Aphid.

APPLE-GRASS APHID *Rhopalosiphum insertum* (Walker) [Hemiptera: Aphididae]: A minor pest of apples, pears, grains and grasses in North America. Adult yellow-green with darker bands on Abdomen; appendages yellow with apices darker. Cf. Apple Aphid.

APPLE LEAFHOPPER *Empoasca maligna* (Walsh) [Hemiptera: Cicadellidae]: A Species widespread in North America where it is a pest of apple, gooseberry, potato, sugarbeet, beans, grasses and roses. 2. *Typholcyba froggatti* Baker [Hemiptera: Cicadellidae] (Australia).

APPLE-LEAF MINER *Phyllonorycter ringoniella* [Lepidoptera: Gracillariidae].

APPLE-LEAF SEWER *Ancylis nubeculana* Clemens [Lepidoptera: Tortricidae].

APPLE-LEAF SKELETONIZER *Psorosina hammondi* (Riley) [Lepidoptera: Pyralidae]: A bivoltine pest of apples, plum and quince in eastern North America. ALS overwinters as pupa; pupation occurs in webs on leaves on ground or orchard trash. Eggs are laid on leaves during June-July. Larva skeletonizes upper surface of leaves, predominantly affecting leaves at ends of branches and tops of trees; feeding is completed within 20–25 days.

APPLE-LEAF TRUMPET MINER *Tischeria malifoliella* Clemens [Lepidoptera: Tischeriidae]: A multivoltine, common pest of apples in eastern USA. ALTM larva constructs trumpet-shaped mines on the upper surface of leaves; Species overwinters as mature larva within mine and pupation occurs during spring. See Tischeriidae.

APPLE LOOPER *Chloroclystis laticostata* (Walker) [Lepidoptera: Geometridae].

APPLE MAGGOT *Rhagoletis pomonella* (Walsh) [Diptera: Tephritidae]: A Species endemic to North America and reported as a serious pest of apples for more than 100 years; AM also attacks haws, blueberries, cherries and European plums. AM serves as a vector for Bacterial Rot. Ancestral form of AM apparently was a pest of *Crateagus* spp. that shifted host plant preference to apple. AM overwinters as pupa within soil; some AMs may pass two years in pupal stage; adults emerge over summer. Female undergoes 15–20 day preovipositional period, then oviposits eggs individually in fruit via punctures made with Ovipositor; eclosion ca 5–10 days. Neonate larva feeds in green fruit; larva completes development in fallen fruit. See Tephritidae. Cf. Cherry Fruit Fly; Mediterranean Fruit Fly; Melon Fly; Mexican Fruit Fly; Oriental Fruit Fly; Papaya Fruit Fly; Queensland Fruit Fly.

APPLE MEALYBUG *Phenacoccus aceris* (Signoret) [Hemiptera: Pseudococcidae].

APPLE MUSSEL-SCALE *Lepidosaphes ulmi* (Linnaeus) [Hemiptera: Diaspididae] (Australia).

APPLE REDBUG *Lygidea mendax* Reuter [Hemiptera: Miridae]: A pest of apple and pear in eastern North America. Overwinters in egg stage within bark of host plant. Neonate nymphs feed on foliage and older nymphs feed on fruit. Feeding induces dwarf fruit with pitted surface and hardened area around feeding punctures.

APPLE ROOT-WEEVILS *Perperus* spp. [Coleoptera: Curculionidae].

APPLE RUST-MITE *Aculus schlechtendali* (Nalepa) [Acari: Eriophyidae].

APPLE SAWFLY *Hoplocampa testudinea* (Klug) [Hymenoptera: Tenthredinidae].

APPLE SEED-CHALCID *Torymus varians* (Walker) [Hymenoptera: Torymidae].

APPLE SUCKER *Cacopsylla mali* (Schmidberger) [Hemiptera: Psyllidae].

APPLE THRIPS See Plague Thrips.

APPLE TORTRIX *Archips fuscocupreana* Walsingham [Lepidoptera: Tortricidae].

APPLE TWIG-BEETLE *Hypothenemus obscurus* (Fabricius) [Coleoptera: Scolytidae].

APPLE TWIG-BORER *Amphicerus bicaudatus* (Say) [Coleoptera: Bostrichidae]: A pest of deciduous fruit trees in eastern USA. Larvae occur on diseased or dying trees; adults bore into twigs of live apple but do not oviposit.

APPLE WEEVIL *Otiorhynchus cribicollis* Gyllenhal [Coleoptera: Curculionidae]: A monovoltine pest of apple and other pome fruits, citrus and ornamental plants in Australia. Adult ca 9 mm long, dark brown and shiny. Larvae feed on roots without apparent damage to plant. Adults feed on foliage during night and reside in soil during day.

APPLIED ENTOMOLOGY The subdiscipline of Entomology concerned with the study of insects (anatomy, biology, behaviour) with the expectation that the knowledge gained can be utilized by humans and applied to solve problems that involve insects. See Entomology. Cf. Agricultural Entomology; Economic Entomology.

APPOSED Adj. (Latin, *appositus* = united; placed near.) Descriptive of surfaces contiguous, juxtaposed or against each other. Alt. Amplectant. Ant. Distant. Cf. Disjunct.

APPOSITION EYE A form of compound eye typical of insects active in bright light. The eye restricts passage of light through pigment cells of one Ommatidium to another Ommatidium. Characteristic of diurnal insects. See Day-Eye. Cf. Superposition Eye.

APPOSITION IMAGE In insect vision, an image developed in eyes adapted entirely for seeing by day, by apposed points of light falling side-by-side and not overlapping. Alt. Apposed image. See Compound Eye. Cf. Superposition Image.

APPRESSED Adj. Adv. (Latin, *ad* = to; *pressare* = to press) Closely applied. Pertaining to Setae parallel or nearly parallel to the body surface.

APRICOT-SCALE See European Fruit Lecanium.

APPROVED COMMON NAME 1. Pesticides: One of three names given to a pesticide in the United States; see Chemical name and Trade name. 2. Insects: The non-scientific name of an insect. Common names of pesticides and insects are approved by the Entomological Society of America. There is only one approved common name given to a pesticide or insect, although an insect may have many different non-scientific names associated with it depending on geographic distribution.

APPROXIMATE Adj. (Latin, *approximatus* = approached; *-atus* = adjectival suffix.) Descriptive of structures which are near or adjacent; juxtaposed. Applied to any body parts that are close together. Alt. Amplectant.

APPUN, KARL FERDINAND (1820–1972) (Papavero 1975, *Essays on the History of Neotropical Dipterology*. 1: 291, São Paulo.)

APRICOT RUSSETING MITE *Phyllocoptes abaenus* Keifer [Acari: Eriophyidae].

APROCARB® See Proproxur.

APSTEIN, CARL HEINRICH (1862–1950) (Anon. 1930, Reichshandgel. Dt. Ges. 1: 31–32.)

APTERA Plural Noun. (Greek, *a* = without; *pteron* = wing.) Wingless Insects: An ordinal term formerly employed for fleas, lice and other wingless forms. Presently not a taxonomic category but considered to include the more primitive Orders of insects and their closest relatives (Thysanura, Entognatha, Collembola, Diplura.)

APTERODICERA Noun. Wingless hexapods with two Antennae.

APTEROPANORPIDAE Plural Noun. A small Family of Mecoptera known from Australia. Adult superficially resemble Boreidae; apterous; Mesothorax and Metathorax sclerites small and fused; Tarsi each with two claws. Adult found in mossy habitat, and can be taken on snow.

APTEROUS Adj. (Greek, *a* = without; *pteron* = wing; Latin, *-osus* = possessing the qualities of.) Pertaining to insects without wings; a wingless condition. Alt. Apterus.

APTERY Noun. (Greek, *a* = without; *pteron* = wing. Pl., Apteries.) The state of being totally wingless. Examples include the parasitic Mallophaga, Anoplura, Siphonaptera and free-living Zoroaptera. Contemporary members of these Orders lack wings, the ancestors of these Orders came from at least three distinct lineages and were winged. Cf. Brachyptery; Macroptery; Microptery.

APTERYGOGENEA Noun. (Greek, *a* = without; *pteron* = wing; *genesis* = descent. Pl., Apterygogeneae.) Insects that are wingless in all stages and hence presumed to be derived from ancestors that were not winged. See Pterygogenea.

APTERYGOTA Noun. (Greek, *a* = without; *pterygion* = little wing. Pl., Apterygotae.) 1. Formerly a Subclass of Insecta including Diplura, Protura, Collembola, Microcoryphia and Thysanura (Imms, Ross). 2. A Subclass of Insecta including Archaeognatha and Thysanura (Mackerras). 3. A concept abandoned by Borror *et al.* in favour of Class Hexapoda. When enacted, a small to moderate-sized, primitively apterous group with slight or no Metamorphosis (Imms); Apterygogenea. Cf. Pterygota.

APTERYGOTOUS Adj. (Greek, *a* = without; *pterygion* = little wing; Latin, *-osus* = possessing the qualities of.) Pertaining to the Apterygota or wingless adult insects.

APTERYGYNY Noun. (Greek, *a* = without; *pteron* = wing; *gyne* = woman. Pl., Apterygynies.) A form of sexual dimorphism in Species for which the male is winged and the female is wingless. Apterygyny is expressed in Mutillidae (Hymenoptera).

APYRENE SPERM A type of Spermatozoa produced by Lepidoptera during spermatogenesis. Apyrene Spermatozoa were first noted by Meves (1903), and are the product of meiosis in which chromatin is not normally developed. Apyrene Spermatozoa migrate to the Spermatheca, but they do not fertilize the eggs. Anatomically, Apyrene Spermatozoa are smaller than normal Spermatozoa and lack a nucleus. Apyrene Spermatozoa bundles disassociate before they reach the duplex region of the male reproductive system. Function of Apyrene Sperm not completely understood, but they may facilitate transfer of Eupyrene Sperm to female's Copulatory Bursa. See Spermatozoa. Cf. Eupyrene Sperm; Hyperpyrene Sperm; Oligopyrene Sperm.

APYRENE Adj. (Greek, *a* = without; *pyren* = fruit stone.) 1. Pertaining to Spermatozoa that lack a nucleus. 2. Seedless cultivated fruits.

AQIS The acronym for Australian Quarantine Inspection Service. The branch of the Australian government responsible for port inspection for the detection and eradication of unwanted organisms from Australia. Cf. APHIS.

AQMATRINE® See Fenvalerate.

AQUATIC Adj. (Latin, *aquaticus* = water; *-ic* = of the nature of.) Pertaining to organisms that live in fresh or salt water; salt-water forms are usually described as marine. See Habitat.

AQUATILIA Noun. Cryptoceratous Heteroptera of aquatic habit.

AQUINAS, THOMAS (1227–1274) Student of Albertus Magnus and Scholastic theologian whose philosophy was the culmination of philosophic efforts of the Christian schools of the Middle Ages. Wrote several works influential in the Middle Ages but little that directly involved insects.

ARACHNIDA Noun. (Greek, *arachne* = spider.) The Division or Class of air-breathing Arthropoda that includes the Orders Scorpiones (scorpions), Araneae (spiders), Acari (mites, ticks), Palpigradi and Pseudoscorpiones.

ARACHNIFORM Adj. (Greek, *arachne* = spider; Latin, *forma* = shape.) Spider-shaped; stellate; descriptive of a large globular body and relatively long, thin legs radiating from the body. Syn. Araneiform. See Arachnida; Cf. Acaroid; Acariform. Rel. Form 2; Shape 2; Outline Shape.

ARACHNOIDEOUS Adj. (Greek, *arachne* = spider; Latin, *-osus* = possessing the qualities of.) Cobweb-like; resembling cobweb. Alt. Arachnoideus.

ARACONOL-F® See Cyhexatin.

ARADIDAE Spinola 1837. Plural Noun. Flat Bugs; Bark Bugs. A numerically moderate-sized, cosmopolitan Family of heteropterous Hemiptera assigned to Superfamily Aradoidea; most abundant in tropical forests. Body dark coloured, dorsoventrally flattened and often with cryptic coloration or ornamentation; Clypeus enlarged; mouth Stylets long and coiled at repose within clypeal region; Ocelli absent; typically macropterous, sometimes apterous. Most Species are gregarious, associated with dead, decaying wood under high humidity; some Species found under bark or fallen tree limbs or in leaf litter. Species are mycetophagous feeding upon polypore fungus mycelia or fruiting bodies.

ARADOIDEA Noun. A Superfamily of heteropterous Hemiptera including Aradidae and Termitaphididae. Body strongly flattened; elongate Stylets; gut without Gastric Caecae, Ileum present; ejaculatory bulb with 2–3 layers. Mycetophagous.

ARAKAWA, YASUI (1893–1933) (Hasegawa 1967, Kontyû 35 (Suppl.): 4–5, bibliography. [In Japanese].)

ARAKI, HARUTUGU (1914–1945) (Hasegawa 1967, Kontyû 35 (Suppl.) : bibliography. [In Japanese].)

ARALDITE Noun. (*-ites* = substance. Pl., Araldites.) An epoxy resin developed during the 1950s as an embedding medium used in microtomy for electron microscopy. (Glauert & Glauert 1958, J. Biophys. Biochem. Cytol. 4: 291.) Cf. Epon 812.

ARALIACEAE A predominantly tropical Family of trees and shrubs presently consisting of ca 60 Genera and 1,150 Species. Gondwanaland distribution and closely related to Apiaceae. Familiar Genera include *Dizygotheca* and *Schefflera*.

ARANEAE Noun. Spiders. An Order of arthropods assigned to the Class Arachnida. The Order includes ca 40,000 Species; cosmopolitan in distribution, predaceous and predominantly terrestrial. Body of two Tagma (Prosoma = head + Thorax; Opisthosoma = Abdomen), lacking antenna; possessing several simple eyes, Chelicerae, eight legs, Pedicel, unsegmented Abdomen, Book Lungs and Spinnerets at the apex of the Abdomen; male Palpus modified into a sperm-containing device for insemination. Many Species construct webs of silk from Spinnerets; webs used as retreats, hibernacula or snares for prey; some Species do not construct webs. Spiders predaceous primarily upon insects, but also attack other spiders, small vertebrates; some Species known to take pollen and the eggs of insects. Habitats include all terrestrial space and intertidal zone; some Species live on surface of water or adapted for living in nests below water.

ARANEIDAE Plural Noun. Orb-weaving Spiders. A numerically large Family of spiders. Species typically active at night build large, planar webs of orb-shape; some Species do not construct orb webs. Syn. Argiopidae. Rel. Bolas Spider; Dome Web Spider.

ARANEIFORM Adj. (Latin, *aranea* = spider; *forma* = shape.) Spider-shaped; resembling a spider. Syn. Arachniform. See Araneae. Rel. Form 2; Shape 2; Outline Shape.

ARAUCARIA APHID *Neophyllaphis araucariae* Takahashi [Hemiptera: Aphididae].

ARBOREAL Adj. (Latin, *arbor* = tree; *-alis* = belonging to, appropriate to.) Descriptive of organisms living in, on, or among trees. Cf. Terrestrial.

ARBORESCENT (Latin, *arbor* = tree; *-escent*.) Adj. Branching like the twigs of a tree.

ARBORIO DI GATTINARA, FERDINANDO (1807–1869) (Targioni Tozzetti 1869, Bull. Soc. Ent. Ital. 1: 83.)

ARBORIZATIONS Adj. The fine branching terminal fibres of axons or collaterals (Snodgrass).

ARBORVITAE LEAFMINER *Argyresthia thuiella* (Packard) [Lepidoptera: Argyresthiidae]: A monovoltine pest of arborvitae in eastern USA and Canada. ALM eggs green and deposited between leaf scales; larva overwinters within leaf mines.

ARBORVITAE SCALE See Fletcher's Scale.

ARBORVITAE WEEVIL *Phyllobius intrusus* Kono [Coleoptera: Curculionidae].

ARBOVIRUS Noun. (Acronym, ar = arthropod; bo = bourne; virus. Pl., Arboviruses.) Any virus that is biologically carried by insects or acari and transmitted to humans or other vertebrates. Term excludes mechanical vectors of virus. About 350 Species of arboviruses recognized and assigned to five Families: Bunyaviridae, Flaviviridae, Reoviridae, Rhabdoviridae and Togaviridae. Most Arboviruses assigned to Bunyaviridae and a few Arboviruses are taxonomically unplaced. See Bunyaviridae, Flaviviridae, Reoviridae, Rhabdoviridae and Togaviridae. Cf. African Swine Fever; Myxomatosis; Rickettsia.

ARBUTHNOT, KENNETH DERWIN (1902–1962) (Chada 1962, J. Econ. Ent. 55: 1026.)

ARCANGELI, GIOVANNI (1840–1921) (Mattei 1906, Naturalista Sicil. 19: 14–22; Bottini 1922, Memorie Soc. Tosc. Sci. Nat. 34: 1–vii.)

ARCAYA, MANUEL DIAZ DE (1841–1916) (Lopez de Zuazo 1917, Boln. Soc. Aragon. Cienc. Nat. 16: 96–113.)

ARCHAEOCYNIPIDAE Rasnitsyn & Kovalev 1988. A small Family of apocritious Hymenoptera assigned to the Cynipoidea.

ARCHAEOGNATHA (Greek, *archaeo* = old; *gnatha* = jaw.) (Triassic Period - Recent.) Bristletails. A primitive Order including about 450 Species of apterygotes once assigned to the Thysanura, and including two modern Families: Machilidae (primitive) and Meinertellidae (derived). Earliest fossil record from USSR (Triassomachilidae). Body moderate sized and laterally compressed; head hypognathous; compound eye well developed, medially contiguous; Ocelli present; Mandible monocondylic. Thorax arched; Coxae often with Styli; Trochanter with two segments; Tarsi with three segments. Adults lack wings and their ancestors were wingless. Abdomen with 11 segments, Styli on some segments; Appendix Dorsalis longer than Cerci; female with long, slender Ovipositor; Ovaries each with several panoistic Ovarioles; Malpighian Tubules and tracheal system well developed. Bristletails occur under bark, stones and in leaf litter where they feed on algae, moss and other plant material. Bristletails are typically nocturnal and capable of jumping by rapid flexion of Abdomen. Syn. Microcoryphia. Cf. Thysanura.

ARCHAIC Adj. (Greek, *archaios* = ancient; *-ic* = of the nature of.) Ancient, generalized; pertaining to organisms no longer dominant as a life form. Ant. Advanced.

ARCHARIUS, ERIK (1757–1819) (Anon. 1819, Svenska Vetensk Akad. Handl. 40: 299–305.)

ARCHEAN EON (4.6–2.5 BYBP.) The first or earliest interval in the earth's Geological Time Scale. AE comprises ca 45% of the earth's history. AE interval apparently devoid of life, characterized by significant physical change in the earth's surface but with no unequivocal biostratigraphic indicators to separate Archean and Proterozoic Eons. Direct evidence of cellular life forms indicated by Stromatolites appear ca 3.4–3.5 BYBP in Pilbara Shield (Australia). See Geological Time Scale. Cf. Precambrian; Proterozoic Eon.

ARCHEBIOSIS See Abiogenesis.

ARCHED NERVES Two nerves arising from the two upper basal parts of the Cruua Cerebri which connect the Central Nervous System and Peripheral Nervous System. See Nervous System.

ARCHEDICTYON Noun. (Greek, *archein* = primitive; *dictyon* = net. Pl., Archedictyons.) The primitive (original) network of veins that characterize the wings of many of the most ancient fossil insects (Needham).

ARCHENCEPHALON Noun. (Greek, *archein* = primitive; *cephalon* = head.) See Archicerebrum.

ARCHENTERIC Adj. (Greek, *archein* = primitive; *enteron* = intestine; *-ic* = characterized by.) Descriptive of or pertaining to the Archenteron.

ARCHENTERON Noun. (Greek, *archein* = primitive; *enteron* = intestine. Pl., Archenterons.) 1. The Gastrocoele or primitive stomach. 2. A simple food-pocket.

ARCHEOCRYPTIDAE Plural Noun. A small Family of polyphagous Coleoptera assigned to Superfamily Tenebrionoidea. Adult dark-coloured, somewhat ovate, slightly flattened and finely pubescent; Frontoclypeal Suture complete; Tibiae spinose. Larva elongate, parallel-sided, slightly flattened. Archeocryptids inhabit leaf litter or fungi.

ARCHER, FRANCIS (1839–1892) (Capper 1892, J. Entomol. mon. Mag. 28: 112; Reade 1892, Naturalist, Hull 1892: 113–116.)

ARCHESCYTINIDAE Plural Noun. The oldest geologically datable Family of Hemiptera, represented in the Lower Permian.

ARCHETYPE Noun. (Greek, *arche* = beginning; *typos* = pattern. Pl., Archetypes.) 1. The first or 'primitive,' ancestral type of any lineage. 2. The original (usually hypothetical) type or form from which an entire group of existing forms is supposed to have developed (Tillyard). See Groundplan. Rel. Plesiomorphic.

ARCHICEPHALON Noun. (Greek, *archi* = first; *kephale* = head. Pl., Archicephalons.) 1. The primitive annelid-arthropod head. 2. The Prostomium *sensu* Snodgrass.

ARCHICEREBRUM Noun. (Greek, *archi* = first; Latin, *cerebrum* = brain. Pl., Archicerebrata.) 1. The ganglionic nerve mass of the Prostomium in Annelida; the primitive supreastomodaeal nerve mass of the Prostomium. 2. The Archencephalon *sensu* Snodgrass.

ARCHIDERMAPTERA A Suborder that includes the fossil *Protodiplatys fortis* Martynov (1925), the oldest known dermapteran described from Jurassic Period beds in Turkestan. Cerci segmented and Tarsi apparently with five segments. See Dermaptera.

ARCHIGENESIS See Abiogenesis.

ARCHIMYMENIDAE Enderlein 1914, Plural Noun. A Family-group name of aculeate Hymenoptera. See Aculeata. Cf. Plumariidae.

ARCHIPSOCIDAE Plural Noun. A small Family of psocomorphous Psocoptera assigned to Superfamily Homilopsocidea. Wings setose, often brachypterous with reduced venation; tarsal formula 2-2-2. Species are colonial under silken sheets of web; colonies may be large with their web covering trees. Some Species are viviparous. See Homilopsocidea

ARCHIPTERA Noun. (Greek, *archi* = first; *pteron* = wing. Pl., Archipterae.) Neuroptera with incomplete Metamorphosis. The so-called Pseudoneuroptera.

ARCHIZELMIRIDAE Rohdendorf 1962 Plural Noun. A monotypic Genus of fossil Diptera known from Upper Jurassic Period deposits in Kazakhstan. The Family appears near the Pleciofungivoridae (Bibionomorpha). See Diptera.

ARCHOSTEMATA Noun. A primitive Suborder of Coleoptera with fossil record in Triassic Period. Archostemata lack cervical sclerites, possess an external prothoracic Pleuron, a hindwing apical margin rolled, Oblongum Cell present, major transverse fold crosses MP and adult hind Coxa movable with Trochantin visible. Larvae bore into wood with a large mandibular Mola and sclerotized Ligula. Included Families: Cupedidae, Micromalthidae, Ommatidae, Tetraphaleridae. See Coleoptera.

ARCTIC REALM See Holarctic Realm.

ARCTIC SULPHUR *Colias nastes* Boisduval 1832 [Lepidoptera: Pieridae].

ARCTIC ZONE Part of the Boreal Region characterized by Tundra (above the limits of tree growth) in latitude or altitude and having mean summer maxima not exceeding 10°C. Cf. Alpine; Boreal. Rel. Biogeography.

ARCTIIDAE Plural Noun. Tiger Moths. A Family of ditrysian Lepidoptera assigned to Superfamily Noctuoidea. Arctiids are cosmopolitan in distribution, best represented in tropics and include ca 6,000 Species. Adult small to medium-sized, often aposematically coloured; wingspan 10–85 mm; head with short, lamellar scales; Ocelli present or absent; Chaetosemata absent; Proboscis small or absent; Maxillary Palpus minute, with one segment; Labial Palpus typically short; Antenna sexually dimorphic, female filiform, male often bipectinate or ciliate. Prothoracic defensive glands sometimes present; fore tibial Epiphysis present; tibial spur formula 0-2-4 or 0-2-2; Metathorax with tympanal organs that produce sound disruptive to bat echolocation; countertympanal hood sometimes present on Abdomen. Egg subglobular, Micropyle centrally positioned on top of egg and Chorion with reticulate sculpture. Eggs laid upright individually or in small clusters with a covering of scales. Larval head hypognathous; usually with dense vestiture of secondary Setae from Verrucae; abdominal Prolegs on segments 3–6, 10. Pupa well sclerotized, glabrous; Maxillary Palpus absent; Cremaster typically absent; cocoon poorly developed with little silk and many larval Setae. Larvae feed externally upon herbaceous foliage, lichen or dead leaves; a few Species of arctiids are economically significant. Adult typically are nocturnal; a few Species are diurnal or crepuscular.

ARCTOPERLARIA Noun. A Suborder of Plecoptera, characterized by a lobe on Sternum IX of male. Families included: Capniidae, Chloroperlidae, Leuctridae, Nemouridae, Notonemouridae, Peltoperlidae, Perlidae, Perlodidae, Pteronarcyidae, Scopuridae and Taeniopterygidae. All but Notonemouridae are found in Northern Hemisphere. See Plecoptera.

ARCTOPSYCHIDAE Plural Noun. A small Family of Trichoptera assigned to Superfamily Hydropsychoidea. See Hydropsychoidea.

ARCUATE Adj. (Latin, *arcuatus* = curved.) Pertaining to structure that is arched or bow-like. Alt. Arcuatus.

ARCUATE VEIN See Vena Arcuata (Snodgrass). See Vein; Wing Venation.

ARCUATO-EMARGINATE Descriptive of a shape with a bow-like or curved excision.

ARCULUS Noun. (Latin, *arculus* = little bow, arch.) 1. Odonata: A small crossvein between the Radius and Cubitus near base, leaving an elongate triangle between them. 2. Trichoptera: A point (often hyaline) on the forewing where the Cubitus (or Post Cubitus) meets the margin. 3. Hemiptera: A crossveinlet nearly reaching the wing's posterior margin at same point as in Trichoptera. 4. In other orders, the term is applied to a crossvein in similar position, apparently giving rise to the Median.

ARCUS Noun. (Latin, *arcus* = bow. Pl., Arcus.) 1. A bow. 2. Part of a circle, but less than one half.

ARDENT® See Acrinathrin.

AREA MEDIASTINAL, SCAPULARIS and ULNARIS 1, The areas anterior of the Mediastinal. 2. Orthoptera: The Scapular and Ulnar Veins *teste* Smith.

AREA POROSA Acarology: A distinctive area of the

Integument or appendage traversed by pore canals and regarded as sensory, glandular or respiratory in function.

AREATE Adj. Furnished with areas.

ARECHAVALETA, JOSEPH (–1912) (Anon. 1912, Dt. Ent. Z. 1912: 608.)

ARENA® See Tecnazene.

ARENACEOUS Adj. (Latin, *arenaceus* > *arena* = sand; *-aceus* = of or pertaining to.) 1. Sandy or sand-like. 2. Containing sand. 3. Descriptive of plants growing in sandy areas. See Habitat.

ARENBER, PIERRE D' (–1958) (Anon. 1958, Bull. Soc. Ent. Fr. 63: 113.)

ARENICOLOUS Adj. (Latin, *arenosus* > *arena* = sand; a sandy place; *colere* = living in or on.) 1. Descriptive of organisms that inhabit sandy areas. 2. Pertaining to organisms that burrow in sand. See Habitat. Cf. Agricolous; Caespiticolous; Deserticolous; Geophilous; Lapidicolous; Paludicolous; Psammophilous; Saxicolous; Xerophilous.

ARENOSE Adj. (Latin, *arenosus* > *arena* = sand; a sandy place; *osus* = full of.) Superficially sandy or gritty. Descriptive of a surface with the texture of sand or grit.

ARENS, L Y (1890–1967) (Stre'nekov 1969, Ent. Obozr. 48: 223–226; transl. Ent. Rev. Wash. 48: 130–131.)

AREOCEL Noun. (Latin, *areola* = small space. Pl., Areocels.) Lepidoptera: The closed cell formed by fusion of the Areole and Basal Cell (Tillyard).

AREOLA Noun. (Latin, *areola* = small space. Pl., Areolae.) 1. General: Wing cells or spaces in membrane between the veins. 2. Hemiptera: A small cell on the wing of some Species. 3. Hymenoptera: Three Median Areas on the Metanotum; second Median Area; upper Median Area. Ichneumonidae: The median area on the surface of the Propodeum and enclosed by Carinae that form a pentagonal, hexagonal or polygonal shape.

AREOLA POSTICA Psocoptera: A cell in the wing formed by the fork of CuA. The AP is a characteristic feature of the forewing of Psocoptera.

AREOLATE Adj. (Latin, *areola* = small space.) 1. A wing with closed cells. 2. Descriptive of structure with a network or fissures or cracks. 3. Pit-like sculptural patterns on the integument.

AREOLE Noun. (Latin, *areola* = small space. Pl., Areoles.) Lepidoptera: The closed Radial Cell of the forewing between veins R3 and R4. See Accessory Cell; Cell; Cellule (Smith).

AREOLET Noun. (Latin, *areola* = small space. Pl., Areolets.) 1. General: A small wing cell. A small space between veins of net-veined insects. 2. Hymenoptera: A small cell in the centre of the forewing, bounded by the Intercubital Veins. Ichneumonidae: The first Radial Sector Cell.

ARGASIDAE Plural Noun. Soft Ticks. A Family of acarines including ca five Genera and 170 Species. See Ixodida. Cf. Ixodidae; Nuttalliellidae.

ARGENTATE Adj. (Latin, *argenteus* = silver; *-atus* = adjectival suffix.) Shining; silvery white.

ARGENTEOUS Adj. (Latin, *argenteus* = silver; *-osus* = possessing the qualities of.) Silver-like in colour; silvery. Cf. Argentate.

ARGENTINE ANT *Linepithema humile* (Mayr) [Hymenoptera: Formicidae]: A widespread urban pest in many parts of Europe, USA, South Africa and Australia. AA is endemic to South America and adventive to New Orleans, USA during 1890 and more recently to Australia. AA is world's most significant ant pest. Workers aggressive; colonies opportunistic and form large nests in soil, wood, houses, under stones. AA displaces native ants, kills nesting birds and is invasive to homes where it will forage throughout house. Workers 2.2–2.8 mm long, monomorphic, stingless with weak bite; head, Thorax, Antenna and legs pale brown; Gaster dark brown; live ca 1 year. Workers predaceous, carnivorous, granivorous, but prefer honeydew and tend or protect aphids and scale insects. Workers slow but efficient foragers along pheromone trails during day and night. Several queens per colony; mating presumably in nest because nuptial flights have not been observed. Males are produced throughout year and are taken at lights; queens are not taken at lights. Colony formation by sociotomy; egg-to-adult cycle ca 75–80 days. Syn. *Iridomyrmex humilis* (Mayr). See Formicidae. Cf. Black Carpenter-Ant; Fire Ant; Larger Yellow-Ant; Little Black Ant; Odorous House Ant; Pavement Ant; Pharaoh Ant; Thief Ant.

ARGENTINE STEM-WEEVIL *Listronotus bonariensis* (Kuschel) [Coleoptera: Curculionidae].

ARGIDAE Konow 1890 (Newman 1834.) Plural Noun. Argid Sawflies. A large cosmopolitan Family (ca 800 Species) of symphytous Hymenoptera assigned to Superfamily Tenthredinoidea; argids are best represented in tropics and poorly represented in Australia. Adult 5–12 mm long; Antenna with three segments, apical segment undifferentiated, clavate or bifid; Hypostomal Bridge absent; Pronotum short with posterior margin curved; forewing without 2R-Rs vein; fore Tibia with two unmodified spurs; middle Tibia sometimes without subapical spur; Ovipositor short, not conspicuously projecting beyond apex of Abdomen; male genitalia strophandrous. Larva with Stemmata; Antenna with 1–4 segments; thoracic legs present, claws present or absent. Larvae phytophagous on various angiosperms.

ARGILLACEOUS Adj. (Latin, *argilla* = clay; *-aceus* = of or pertaining to.) Clay-like. Structure displaying the texture, appearance or colour of clay. Alt. Argillaceus.

ARGOD, ANDRÉ ALBERT (1859–1936) (Barthe1936, Miscnea Ent. 37: 123–124; Horne 1937, Arb. Morph. Taxon. Ent. Berl. 4: 160.)

ARGOOD, ROBERT (1897–1915) (Berland 1920, Ann. Soc. Ent. Fr. 89: 418–419.)

ARGUS TORTOISE-BEETLE *Chelymorpha cassidea* (Fabricius) [Coleoptera: Chrysomelidae].

ARGYREIDAE See Lycaenidae.

ARGYRESTHIIDAE Plural Noun. Argyresthiid Moths. A small, predominantly Holarctic, Family of ditrysian Lepdioptera assigned to Superfamily Yponomeutoidea and placed within Yponomeutidae in older classifications. Adult small; wingspan 6–15 mm; Vertex with rough lamellar scales and Frons with appressed lamellar scales; Ocelli and Chaetosemata absent. Antenna usually filiform, sometimes weakly serrate, shorter than forewing, with Pecten on Scape. Proboscis without scales; Maxillary Palpus of one small segment; Labial Palpus long, curved upward or porrect; fore Tibia with Epiphysis; tibial formula 0-2-4; forewing lanceolate with Pterostigma; hindwing lanceolate with reduced venation. Egg oval in outline shape, flattened on ventral surface with Chorion pitted. Adult of some Species rests with head near substrate and apex of Abdomen elevated. Larva typically feeds upon trees, boring into fruits and twigs and mining leaves. Pupa without Cremaster, with apical hooks; pupation often in larval gallery.

ARID Adj. (Latin, *aridus* = dry.) Dry; an area in which desert conditions prevail. Term pertains to any region in which rainfall cannot produce ordinary farm crops without irrigation. Cf. Humid.

ARID TRANSITION AREA A North American faunal division that comprises the western part of the Dakotas, northern Montana east of the Rockies, southern Assiniboia, small areas in southern Manitoba and Alberta, the higher parts of the Great Basin and the plateau region generally, the eastern base of Cascade Sierras and local areas in Oregon and California.

ARIOTOX® See Metaldehyde.

ARISTA Noun. (Latin, *arista* = awn. Pl., Aristae.) Higher Diptera: A specialized bristle or hair-like process on the Antenna. The Arista is usually dorsal, long and flexible but rarely apical in position. See Antenna. Cf. Stylus.

ARISTATE Adj. (Latin, *aristatus* from *arista* = awn; *-atus* = adjectival suffix.) 1. Displaying or bearing an Arista or Setae in Diptera. 2. A type of Antenna that bears an Arista.

ARISTIFORM Adj. (Latin, *arista* = beard of grain; *forma* = *shape*.) Arista-shaped; descriptive of structure with the appearance of an Arista. See Arista. Cf. Penniform; Plumiliform; Plumose; Scopiform. Rel. Form 2; Shape 2; Outline Shape.

ARISTOTLE (384–322 BCE) A disciple of Plato, tutor of Alexander and member of the Teleological School. Aristotle was regarded as preeminent among ancient scientists. He was founder of general Entomology and Comparative Anatomy; first writer to describe in detail the anatomy of many Species; he attempted to develop a classification based on his anatomical observations. Aristotle recognized a unity of anatomical plan (groundplan) among and within major groups and had a sense of homology, but he did not distinguish between what modern workers regard as Homology and Analogy. Aristotle was a prolific writer whose ideas in biology were influential for more than 1,500 years. His most notable publications included *Historia Animalium* (the earliest known biological text), *De Generatione Animalium* (developmental biology), *De Partibus Animalium* (comparative anatomy and physiology of animals), *De Motu Animalium* (dealing with metaphysics and psychology), *De Incessu Animalium* (dealing with locomotion). Aristotle's classification of insects was based first upon wings and then upon mouthparts. (Eiselt 1836, *Geschichte, Systematik und Literatur der Insektenkunde*, 255 pp. (3–10), Leipzig. Sachse 1847, Allgem. Dt. Naturh. Zig. 2: 444–448; Rose 1853, *New General Biographical Dictionary* 2: 142–153; Conrad 1940, Sber. Ges. Naturf. Freunde Berl. 1940: 147–152.)

ARIXINIIDAE Plural Noun. A small Family of Dermaptera assigned to Suborder Arixeniina. Included Taxa are robust bodied, conspicuously setose, compound eyes nearly absent, apterous and viviparous; Forceps typically rod-like. Arixiniids are associated with bats in the Philippines, Malaysia and Indonesia. Species often are regarded as parasitic, but evidence has not been provided to substantiate this assertion. See Dermaptera.

ARIZONA PINE BEETLE See Southern Pine Beetle.

ARKANSAS BEE-VIRUS See Honey Bee Viral Disease. Cf. Kashmir Bee-Virus.

ARKHANGELSKY, PETRA (1893–1960) (Yakhontov 1961, Ent. Obozr. 40: 710–712, bibliography.)

ARMADILLO BUGS See Pillbugs.

ARMATURE Noun. (Latin, *armatura* = armour. Pl., Armatures.) 1. Any spinous or sclerotized process on the body, leg or wing that presumably functions in defence. 2. The sclerotized portion of genitalic structures.

ARMATUS Adj. (Latin, *armatura* > *armare* = to arm.) An archane term used to describe structure that displays spines, claws or other chitinous processes.

ARMBRUSTER, LUDWIG (1886–1973) (Anon. 1974, Bee Wld 55: 30–31.)

ARMILLATE Adj. (Latin, *armilla* = armlet; *-atus* = adjectival suffix.) Descriptive of structure with a fringe, ring or annulus of elevated or differentiated tissue. Alt. Armillatus. Cf. Annulate.

ARMISTEAD, WILSON (1819–1868) (Muller 1868, Zoologist (2) 3: 1196–1208; Newman 1868, Entomologist 4: 1.)

ARMITAGE, EDWARD (1817–1896) (Anon. 1896, Proc. Ent. Soc. Lond. 1896: xcii–xciii; Anon. 1896, Entomol. mon. Mag. 32: 164.)

ARMITAGE, H M Chief, Bureau of Entomology and

Plant Quarantine, California Department of Agriculture.

ARMOR® See Cyromazine.

ARMOURED SCALE INSECTS See Diaspididae.

ARMOURED SCALES See Diaspididae. Cf. Soft Scales. Rel. Mealybugs; Whiteflies.

ARMY ANT Any of many Species of tropical ants in the Subfamilies Ponerinae, Ecitoninae, Leptanillinae, Myrmicinae and Dorylinae. The numbers of individuals per colony among the better known Species range from 10,000 to 20 million. AAs nest in underground bivouac sites, sometimes excavating large amounts of soil, and among leaf litter and fallen logs; one Species is arboreal. Their diets consist mostly of arthropods, but include small vertebrates and some vegetable matter. AAs display two basic kinds of swarming patterns: Columnar and swarming. Colony cycles include frequent emigration from one per month to five during one day. Syn. Foraging Ant; Legionary Ant; Soldier Ant; Visiting Ant.

ARMY CUTWORM *Euxoa auxiliaris* (Grote) [Lepidoptera: Noctuidae]: A monovoltine, surface-feeding pest of wheat in the western USA. AC eggs are laid individually on soil. Larvae are pale green-grey to brown with pale stripes and white spots on dorsum; Integument finely granulate. See Cutworms; Noctuidae. Cf. Armyworms.

ARMYTAGE, EDWARD OSCAR (–1946) (Carpenter 1947, Proc. R. Ent. Soc. Lond. (C) 11: 60.)

ARMYWORM *Pseudaletia unipuncta* (Haworth) [Lepidoptera: Noctuidae]. See Armyworms.

ARMYWORMS Plural Noun. (Sl., Armyworm.) A general term for caterpillars that are smooth bodied, longitudinally striped and locally abundant. When present in large numbers, AWs move or migrate collectively into pastures and crops, consuming all plant material along a broad front. Armyworm food includes cereals, graminaceous forage crops and grasses; armyworms consume stems, leaves and seed heads. In some instances, term 'armyworm' is restricted to larval Nocutidae. See Common Armyworm; Lawn Armyworm; Noctuidae; Southern Armyworm. Cf. Cutworms.

ARNASON, ARNI PALL (1903–1964) (Peterson 1965, J. Econ. Ent. 58: 182.)

ARNOLD GEORGE (1881–1962) (Varley 1963, Proc. R. Ent. Soc. Lond. (C) 27: 50.)

ARNOLD, CHARLES (–1883) (Anon. 1883, Rep. Ent. Soc. Ont. 14: 81; Anon. 1883, Can. Ent. 15: 177–178.)

ARNOLD, EUGEN (1866–1939) (Kell 1940, Mitt. Münch. Ent. Ges. 30: 347–439; Sachtleben 1940, Arb. Morph. Taxon. Ent. Berl. 7: 170.)

ARNOLD, WALTER (1892–1938) (Heikertinger 1939, Koleopt. Rdsch. 25: 78.)

ARO, JOHN EMIL (1874–1928) (Forsius 1928, Notul. Ent. 8: 63–64.)

AROLANNA Noun. (Latin, *areola* = small space. Pl., Arolannae.) Hymenoptera: The Arolium *sensu* MacGillivray. See Pretarsus. Cf. Arolium. Rel Leg.

AROLELLA Noun. (Latin, *areola* = small space. Pl., Arolellae.) Hemiptera: The Arolium *sensu* MacGillivray. See Pretarsus. Cf. Arolium. Rel. Leg.

AROLIUM Noun. (Greek, *arole* = protection; Latin, *areola* = small space; Latin, *-ium* = diminutive > Greek, *-idion*. Pl., Arolia.) 1. A cushion-like, unpaired, apical 'pad' positioned at the apex of the Pretarsus of many insects. The Arolium arises between the bases of the tarsal claws (Ungues) and is sometimes membranous and capable of inflation. The Arolium is present in Phasmatoidea, most Orthoptera and some primitive Isoptera. The Arolium is absent from Grylloblattoidea, probably absent from Heteroptera (replaced with the Empodium) and from most Diptera. Some Lepidoptera possess an Arolium, Empodium and Pulvillus. Syn. Arolanna. See Pretarsus. Cf. Arolella; Empodium; Euplantula; Onychium; Palmula; Paronychium; Plantulla; Pseudonychium; Pulvillus. Rel. Leg.

AROLOIDEA Noun. (Latin, *areola* = small space. Pl., Aroloideae.) Hemiptera: The distended bag-like Coria (MacGillivray). See Corium.

AROTEX-EXTRA® See Chlormequat.

ARPPE, TOIMI AURO AKSELI (1917–1941) (Kangas 1944, Suomen Hyönt. Aikak 10: 156.)

ARRESTANT Noun. (Middle English, *arrest; -antem* = an agent of something. Pl., Arrestants.) A chemical that causes insects to aggregate in contact with it (Dethier *et al.* 1960, 53: 135.) See Semiochemical. Cf. Attractant; Repellant.

ARRHENOID Adj. (Greek, *arrhen* = male; *eidos* = form.) Male-like; descriptive of an organism exhibiting male anatomical features or behavioural characteristics in an otherwise genetically female animal. See Parthenogenesis. Cf. Android; Androgyne; Gynecoid; Gynandromorph. Rel. Eidos; Form; Shape.

ARRHENOPHANIDAE See Psychidae.

ARRHENOTOKOUS Adj. (Greek, *arrhen* = male; Latin, *-osus* = possessing the qualities of.) 1. Descriptive of a form of parthenogenetic reproduction in which male offspring are produced as internal or external hyperparasitoids of conspecific, congeneric or confamilial female parasitoids. See Parthenogenesis. Cf. Oligotokous; Thelytokous. Rel. Reproduction.

ARRHENOTOKY Noun. (Greek, *arrhen* = male; *tokos* = birth. Pl., Arrhenotokies.) 1. A form of parthenogenetic reproduction in which male progeny arise from unfertilized eggs. Arrhenotoky is common in Hymenoptera and Acarina. 2. A form of 'heteronomous ontogeny' *sensu* Walter (1983). See Parthenogenesis. Cf. Deuterotoky; Heterogony; Thelyotoky.

ARRIBALZAGA, E L See Lynch-Arribalzaga, E.

ARRIVO® See Cypermethrin.

ARROW, GILBERT JOHN (1873–1948) (Horn 1939,

Arb. Morph. Taxon. Ent. Berl. 6: 69–76; Bacchus 1974, Bull. Br. Mus. Nat. Hist. (Ent.) 31: 26–44, partial bibliography.)

ARROWHEAD SCALE *Unaspis yanonensis* (Kuwana) [Hemiptera: Diaspididae].

ARS An acronym for Agricultural Research Service, the agency within the US Department of Agriculture that is responsible for agricultural reseach. Cf. APHIS; USDA.

ARSENIOUS TRIOXIDE An inorganic compound that functions as a stomach poison and commonly is used to control termites. Symptoms of human poisoning include gastrointestinal upset, vomiting, diarrhoea, low fever, salivation and pallor. First aid treatment: avoid or remove contact with material and seek medical advice. AT is suspected of causing skin cancer and is known to attack cells of the intestine. Syn. White Arsenic. Alt. Arsenic Trioxide. See Inorganic Insecticide.

ARTABAN® See Benzoximate.

ARTATENDON Noun. (Pl., Artatendons.) The tendon operating the Articularis (MacGillivray).

ARTEFACT Noun. (Latin, *ars* = art; *factus* = made. Pl., Artifacts.) A structure due to the method of preparation and not formed as part of an organism. Alt. Artifact.

ARTEMATOPIDAE Plural Noun. A small Family of polyphagous Coleoptera assigned to Superfamily Elateroidea. See Coleoptera; Elateroidea.

ARTENKREIS Noun. (German, *art* = species; *kreis* = circle.) A complex of related Species that replace one another geographically. See Species.

ARTERIAL Adj. (Latin, *arteria* = windpipe, artery; *-alis* = pertaining to, having the character of.) Pertaining to the system of tubular vessels that move blood from the Heart and around the body. Cf. Veinal. Rel. Circulation; Haemolymph.

ARTERIAL WORM See Elaeophorosis.

ARTERIOLE Noun. (Latin, *arteriola* = small artery. Pl., Arterioles.) A small artery. See Artery. Rel. Circulation.

ARTERY Noun. (Latin, *arteria* = windpipe, artery > Greek, *arteria*. Pl., Arteries.) A vessel that conducts blood away from the Heart in vertebrates. Arteries have thicker, more elastic (muscular) walls that veins. The distal most arteries are thin-walled and form capillaries that connect with veins. Cf. Vein. Rel. Circulation.

ARTHRIN Noun. (Greek, *arthron* = joint. Pl., Arthrins.) A myofibrillar protein in the flight muscles of some insects. Arthrin is related to Actin, but with a molecular weight of about 55,000. See Muscle. Cf. Actin.

ARTHRIUM Noun. (Greek, *arthron* = joint; Latin, *-ium* = diminutive > Greek, *-idion*. Pl., Arthria.) Coleoptera: The minute, concealed tarsal joint in Pseudotetramera and Trimera.

ARTHRODERM Noun. (Greek, *arthron* = joint; *derma* = skin. Pl., Arthroderms.) The body-wall, outer 'skin' or cuticular covering of articulated invertebrates (arthropods). See Integument. Cf. Scleroderm.

ARTHRODIA Noun. (Greek, *arthrodes* = articulated. Pl., Arthrodiae.) A gliding joint in which the parts glide upon one another without axial motion.

ARTHRODIAL Adj. (Greek, *arthrodes* = articulated; Latin, *-alis* = having the character of.) Pertaining to a gliding joint that permits motion in any direction during the process of articulation.

ARTHROMERE Noun. (Greek, *arthron* = joint; *meros* = part. Pl., Arthromeres.) 1. A ring-like joint or segment of the insect body. 2. A somite. See Mere. Cf. Antennomere; Podomere.

ARTHROPLEURON Noun. (Greek, *arthron* = joint; *pleura* = rib, side. Pl., Arthropleurons.) 1. The limb-bearing part of the insect body. 2. The part of an Arthromere that is positioned between the Tergum and Sternum. Alt. Arthropleure.

ARTHROPOD Noun. (Greek, *arthron* = joint; *pous* = foot. Pl., Arthropods.) 1. A member of the Arthropoda. 2. An organism with metameric segmentation, jointed appendages and an exoskeleton. See Appendage. Cf. Integument. Rel. Segmentation.

ARTHROPODA Plural Noun. (Greek, *arthron* = joint; *pous* = foot.) A Phylum of metamerically segmented metazoan organisms with jointed legs and a structurally and chemically complex, external skeleton composed of numerous compounds including chitin. Rel. Insecta.

ARTHROPODIASIS Noun. (Greek, *arthron* = joint; *pous* = foot; *-iasis* = suffix for names of diseases. Pl., Arthropodiases.) A term used to describe the direct effect of arthropods on vertebrates, including: 1. Local and systemic reactions to bites and stings; 2. allergenic effects from contact with arthropods; 3. psychological conditions; 4. loss of body weight; and 5. extreme avoidance of arthropods. Cf. Acariasis; Entomophobia.

ARTHROPODIN Noun. (Greek, *arthron* = joint; *pous* = foot. Pl., Arthopodins.) A category of water-soluble proteins found in the Integument of arthropods and which is associated with chitin. See Integument. Cf. Resilin; Sclerotin.

ARTHROPODS See Arthropoda.

ARTICHOKE APHID *Capitophorus elaeagni* (del Guercio) [Hemiptera: Aphididae].

ARTICHOKE PLUME-MOTH *Platyptilia carduidactyla* (Riley) [Lepidoptera: Pterophoridae]: A significant pest of artichokes in California.

ARTICLE Noun. (Latin, *articulus* = division, part, diminutive of *artus* = joint. Pl., Articles.) 1. A sclerotized portion of any articulated segment of an arthropod's body. 2. A 'joint' or jointed body part. See Joint; Segment. 3. A scientific report of variable length and form.

ARTICULAR AREA The basal area of the insect wing. See Ala.

ARTICULAR CORIUM The membrane of the insect leg joint (*e.g.* Coxal Corium). See Leg Articulation.

ARTICULAR MEMBRANE 1. The non-sclerotized

flexible area of a joint. 2. The ring of thin membrane uniting a Seta at its base with the wall of the Trichopore (Comstock). Syn. Articular Corium.

ARTICULAR PAN A cup-like or dish-like depression forming the socket into which an articulation is fitted. Cf. Acetabulum; Cotyla.

ARTICULAR SCLERITE A sclerite occupying an intermediate position between the body and its appendage (Comstock). See Sclerite.

ARTICULARIS Noun. (Latin, *articularis* = joint.) A term used by early workers for the Pretarsus or claw-bearing portion of the insect leg. The Posttarsus *sensu* MacGillivray. See Pretarsus. Rel. Leg.

ARTICULATA Plural Noun. (Latin, *articulus* = joint.) A group of animals whose bodies are constructed of annulations (rings) and segments or articulations. The Articulata are broadly viewed as including the worms, crustaceans and insects. See Arthropoda.

ARTICULATE Adj. (Latin, *articulus* = joint; *-atus* = adjectival suffix.) 1. Jointed or segmented. See Joint; Segment. 2. Descriptive of a person who is eloquent or well spoken. Verb. (Latin, *articulare* = jointed, uttered distinctly. Articulated, Articulating, Articulates.) To connect or unite by means of a joint.

ARTICULATE FASCIA 1. A fascia composed of contiguous spots. 2. The *fascia articulata* of Kirby & Spence. See Fascia.

ARTICULATED APEX See Clasp Filament.

ARTICULATION Noun. (Latin, *articulus* = joint. English, *-tion* = result of an action. Pl., Articulations.) 1. A moveable point or place where two parts or segments are joined. 2. An individual joint or segment. Rel. Acetabulum; Condyle.

ARTICULATORY EPIDEME The partly chitinized membrane by which the wings are attached to the Thorax. See Wing Base. Cf. Axillary Sclerites; Humeral Sclerites. Rel. Wing.

ARTIFACT Noun. (Latin, *artis* = art; *factus* = made. Pl., Artifacts.) See Artefact.

ARTIFICIAL CLASSIFICATION A classification of organisms that is based upon taxonomic characters (anatomical, behavioral, biological) which are convenient to see, score, interpret or use but the characters are not considered in relation to their phylogenetic significance. See Classification. Cf. Natural Classification. Rel. Cladistics; Phylogeny; Taxonomy.

ARTIFICIAL DIET A combination of nutritional ingredients provided to an insect instead of its natural food. AD may be dry, moist or fluid. Rel. Chemically Defined Diet; Crude Diet; Semisynthetic Diet; Synthetic Diet.

ARTIFICIAL SELECTION The process of altering plants and animals through controlled breeding programs managed by humans. Selection made by animal and plant breeders (*e.g.* deliberate selection made in strains of *Drosophila* for genetic study). See Evolution; Selection. Cf. Natural Selection; Orthogenetic selection. Rel. Genetics.

ARTIS Noun. The point of articulation between the base of an appendage and the body to which the appendage is attached (MacGillivray).

ARTUS Noun. (Latin, *artus* = joint.) A joint, limb or locomotor appendage.

ARVALL, HARRY (1886–1966) (Lindroth 1967, Opusc. Ent. 32: 2.)

ARVEST® See Ethephon.

ARYLMATE® See Ethiofencarb.

ASANA® See Esfenvalerate.

ASATAF® See Acephate.

ASATAL® See Acephate.

ASCALAPHIDAE Plural Noun. Owlflies. A moderate sized, widespread Family of Neuroptera assigned to Myrmeleontoidea. Adult medium to large bodied; Antenna elongate, clubbed; compound eyes divided by longitudinal groove; Pterostigma present; wings typically hyaline. Eggs are large and laid in clusters on grass, twigs and stems. Repagula are deposited by some Species. Larvae are predaceous and exposed on bark and ground. See Neuroptera.

ASCANUS, PETRUS (1723–1803) (Anon. 1832, Mem. Soc. Sci. Agric. Lille 1 (1831): 161–169; Henriksen 1921, Ent. Meddr 15: 35–37.)

ASCEND® See Avermectin.

ASCENDING FRONTAL BRISTLES Diptera: The uppermost of 1–4 Frontal Bristles (Comstock). See Frontal Bristles.

ASCETASPORIN® A registered biopesticide derived from *Bacillus thuringiensis* var. *kurstaki*. See *Bacillus thuringiensis*.

ASCHIZA Noun. A taxonomically heterogenous Series of cyclorrhaphous Diptera including the Superfamilies Lonchopteroidea, Phoroidea and Sryphoidea. Members lack a Ptilinal Fissure but display male genitalia which are rotated (circumverted) along their primary axis. See Cyclorrhapha.

ASCOSPORE Noun. (Latin *asco* = bag; *sporos* = seed. Pl., Ascospores.) A spore formed during the sexual phase in the life cycle of some fungi. Ascospores carry a structure called an Ascus.

ASEPTIC Adj. (Latin, *a* = not; New Latin, *septic* from Greek, *sepsis* = putrefication; *-ic* = of the nature of.) 1. Pertaining to physical condition or an environment that is free of pathogenic microorganisms. 2. Preventing infection. Cf. Antiseptic; Septic. Rel. Disease; Pathogen.

ASEXUAL Adj. (Greek, *a* = without; Latin, *sexus* = sex; *-alis* = pertaining to, having the character of.) 1. Without sex; producing eggs or young by cell-budding. 2. Pertaining to organisms whose reproductive organs are incompletely developed. See Parthenogenetic. Cf. Sexual.

ASEXUAL REPRODUCTON 1. Vegetative reproduction. 2. Reproduction that does not involve the exchange of genetic material between two parents or union of gametes. AR is common among the bacteria, fungi and parthenogenetic insects. See Reproduction. Cf. Parthenogenesis; Sexual Reproduction.

ASH BORER *Podosesia syringae* (Harris) [Lepidoptera: Sesiidae]: A pest of ash and mountain ash in North America. AB sometimes is regarded as a Subspecies of Lilac Borer. See Sesiidae. Cf. Lilac Borer.

ASH-GRAY LEAF BUGS See Piesmatidae.

ASHGREY BLISTER-BEETLE *Epicauta fabricii* (LeConte) [Coleoptera: Meloidae].

ASH GREYLEAF BUGS See Piesmatidae.

ASH PLANT BUG *Tropidosteptes amoenus* Reuter [Heteroptera: Miridae].

ASH WHITEFLY *Siphonius phillyreae* [Hemiptera: Aleyrodidae]: An old world, polyphagous pest of trees and shrubs.

ASHBY, EDWARD BERNARD (1876–1936) (Turner 1936, Entomologist's Rec. J. Var. 48: 12.)

ASHBY, EDWIN (1861–1941) (Winckworth 1941, Proc. Linn. Soc. Lond. 153: 287–288.)

ASHBY, SIDNEY ROBERT (1864–1944) (Jarvis 1944, Entomol. mon. Mag. 80: 264; Jarvis 1945, Proc. Trans. S. Lond. Ent. Nat. Hist. Soc. 1944–45: xxv-xxvii.)

ASHDOWN, W J (1855–1919) (Anon. 1920, Entomol. mon. Mag. 56: 17.)

ASHLOCK, PETER D (1929–1989) American academic (University of Kansas), taxonomic specialist of Lygaeidae and advocate of Cladism. (Polhemus 1989, Pan-Pac. Entomol. 65: 310–311.)

ASHMEAD, WILLIAM HARRIS (19 September 1855, Philadelphia - 18 October 1908, Washington, D.C.) American taxonomic Entomologist who, early in his career, established 'The Florida Dispatch' in Jacksonville, Florida which dealt with agricultural matters. Ashmead was appointed field Entomologist by C.V. Riley of USDA during 1887, became Assistant Entomologist, Division of Insects USDA during 1889 and studied in Berlin during 1890–1891. Ashmead was appointed Assistant Curator of Insects US National Museum (1895). He was a taxonomic specialist working predominantly in Hymenoptera, particularly within the Parasitica. He was the author of insect names and published more than 260 papers. (Crawford 1909, Proc. Entomol. Soc. Wash. 10 (3–4): 125–156.)

ASHTON, THOMAS BEVERIDGE (1826–1895) (Knaus 1896, Ent. News 7: 96; Knaus 1928, Kans. Ent. Soc. 1: 19.)

ASIAN COCKROACH *Blattella asahinai* Mizukubo [Blattodea: Blattellidae]: Described from sugarcane field on Okinawa and presumably endemic to Asia; AC was first detected in North America ca 1985. AC is anatomically very similar to German Cockroach with some reproductive compatability between Species; distinguished on the basis of cuticular carbon profiles, habitat preference, activity patterns and flight ability. AC prefers leaf litter in shaded habitat and often is taken in wooded areas near urban development. Species is highly active at sunset for about one hour, and active until 90 minutes before sunrise; daylight flights are less than a few metres and typically downward or toward nearby plants; night time flights can be hundreds of metres without landing. Adults are attracted to lights and accumulate on windows, television screens and lighting; fly onto persons wearing light-coloured clothing. Food includes dead insects, fungi, fruit, pollen, honeydew and vegetables. See Blattoidea. Cf. German Cockroach.

ASIAN CORN-BORER *Ostrinia furnacalis* (Guenèe) [Lepidoptera: Pyralidae]: A significant pest of maize in the orient.

ASIAN GYPSY MOTH A Subspecies or biotype of the Gypsy Moth. AGM is morphologically very similar to GM but differs in biological and behavioral characteristics. See Gypsy Moth.

ASIAN HONEY BEE See Indian Honey Bee.

ASIAN TIGER-MOSQUITO *Aedes albopictus* (Skuse) [Diptera: Culicidae]: A Species endemic to southeast Asia and spread widely in recent years. ATM was discovered in USA during 1985 and now is established in many states. Recent movement throughout world has been attributed to distribution of used tyres. ATM is a significant vector of Dengue Fever and Dengue Haemorrhagic Fever, and may also vector other arboviruses. See Culicidae. Cf. Yellow Fever Mosquito.

ASIATIC GARDEN-BEETLE *Maladera castanea* (Arrow) [Coleoptera: Scarabaeidae]: A polyphagous pest of more than 100 plants including vegetables, tree crops and ornamental plants. Endemic in the Orient and first reported in USA during 1922. AGB overwinters as an early instar larva; pupates during spring and adults appear during June-July. Adults chew holes in leaves; larvae feed on roots. See Scarabaeidae.

ASIATIC OAK-WEEVIL *Cyrtepistomus castaneus* (Roelofs) [Coleoptera: Curculionidae].

ASIATIC RICE-BORER *Chilo suppressalis* (Walker) [Lepidoptera: Pyralidae]. See Rice Stem Borer.

ASIATIC ROSE-SCALE *Aulacaspis rosarum* Borchsenius [Hemiptera: Diaspididae].

ASILIDAE Leach (in Samuelle 1819.) Plural Noun. Robber Flies. A cosmopolitan Family of orthorrhaphous Diptera with ca 5,000 Species in 400 Genera assigned to Superfamily Asiloidea. Adult size highly variable (3–50 mm long); habitus bristly; head short and freely moveable; Antennae diverse in structure, usually first Flagellomere subequal to head length, 1–2 additional Flagellomeres of varying lengths and thickness; Thorax stout; legs relatively stout; Empodium usually setiform; Abdomen long and tapering caudad. All Asilidae are predaceous as adults and rarely are selective in prey capture; representatives of many Families and several Orders are consumed by a Species. Factors in prey capture are: Size, flight speed and agility, local abundance and toughness of prey integument. Oviposition behavior is diverse: Groups utilize different methods and oviposition sites; oviposit into soil with Ovipositor and acanthophorite spines to assist digging; oviposit on vegetation with Ovipositor lengthened and laterally flattened to assist in placing eggs within crevices; groups drop eggs while in flight or perching. Larvae are predaceous and many Species in various Subfamilies may feed on scarabaeid larvae that live in soil. Larval Laphriinae live within dead trees, woody shrubs and prey on larvae of wood-burrowing insects, mainly beetles. Five contemporary Subfamilies are recognized by most taxonomic authorities: Asilinae, Dasypogoninae, Laphryiinae, Leptogastrinae and Megapodinae. Asilid fossil record begins during Eocene Epoch.

ASILOIDEA Noun. A Superfamily of orthorrhaphous Diptera including Apioceridae (Apioceratidae), Asilidae, Bombyliidae, Mydaidae (Mydasidae, Mydidae), Scenopinidae (Omphalidae) and Therevidae. See Diptera.

ASIMISIN See Acetogenins.

ASIOPSOCIDAE Plural Noun. A small Family of New World psocomorph Psocoptera assigned to Superfamily Caecilietae. Adult head short and wide; Mandible elongate, posteriorly concave with Stipes and Galea filling concavity; Labrum broad; Antenna with 13 segments; tarsal claws without preapical denticle; abdominal Sterna lacking transverse vesicles. Species inhabit twigs and branches of shrubs and small trees. See Psocoptera.

ASLAM, NAZIR AHMAD (1918–1971) (Butler 1972, Proc. Ent. Soc. Lond. (C) 36: 61.)

ASPARAGUS APHID *Brachycorynella asparagi* (Mordvilko) [Hemiptera: Aphididae].

ASPARAGUS BEETLE *Crioceris asparagi* (Linnaeus) [Coleoptera: Chrysomelidae]: A widespread, multivoltine pest of asparagus in North America and Europe. AB apparently is host plant specific and overwinters as an adult in sheltering habitats then becomes active during April-May. Eggs are black, laid on end in rows on asparagus with eclosion ca 7 days. Larvae feed on asparagus tips and excrete a black fluid that stains plants. Pupation occurs in soil and adults chew out buds from tips causing scaring and discoloration. AB completes 2–5 generations per year, depending upon climate. See Chrysomelidae. Cf. Spotted Asparagus Beetle.

ASPARAGUS MINER *Ophiomyia simplex* (Loew) [Diptera: Agromyzidae]: A widespread, bivoltine pest of asparagus in North America. AM overwinters within a puparium in larval mines. Adults appear during May. Eggs are inserted into epidermis and egg stage requires 2–3 weeks. Larvae mine stems of host plant near base and may girdle plant to cause death. Syn. *Melanagromyza simplex* (Loew). See Agromyzidae.

ASPARAGUS SPIDER-MITE *Schizotetranychus asparagi* (Oudemans) [Acari: Tetranychidae].

ASPECT Noun. (Latin, *aspicere* = to look forward. Pl., Aspects.) 1. The physical appearance of a body, structure, organ or organism. 2. The direction in which a fixed surface faces (northern aspect). 3. An observer's view of a mobile surface. The insect body (appendage, structure) has several surfaces that are called 'aspects.' Thus, cephalic aspect refers to the vertical surface at the anterior part of the body. Dorsal aspect refers to the upper portion (top) of the body (generally interpreted in terms of an animal's natural posture). Ventral aspect refers to the lower part of the body. Lateral aspect refers to either side of the body. Caudal aspect refers to the vertical surface at the posterior part of the body. Mesal aspect refers to structure positioned near an imaginary midline along the primary axis of a bilaterally symmetrical body. In addition, several terms refer to direction. Apical refers to structure positioned most distant from the body. Proximal (basal) refers to structure positioned nearest the midline of a body or the point of connection of an appendage to a body. Dextral refers to the right side of the body or appendages found on the right side of the body. Sinistral refers to the left side of the body or appendages found on the left side of the body. See Axis; Orientation. Rel. Order; Organization.

ASPECTION Noun. (Latin, *aspectio, aspicere* = to look forward; English, *-tion* = result of an action. Pl., Aspections.) A viewing.

ASPEN BLOTCH-MINER *Phyllonorycter tremuloidiella* (Braun) [Lepidoptera: Gracillariidae]: A pest of aspen and poplar in western North America. Larvae form irregularly shaped tunnels that periodically form blotches and cause premature leaf drop. ABM pupates within its mine. See Gracillariidae.

ASPEN LEAF-BEETLE *Chrysomela crotchi* Brown [Coleoptera: Chrysomelidae].

ASPER Adj. (Latin, *asper* = rough.) Pertaining to a rough and uneven surface. Alt. Asperous.

ASPERATE Adj. (Latin, *asper* = rough; *-atus* = adjectival suffix.) Roughened. Alt. Asperatus. Cf. Smooth. Rel. Sculpture; Texture.

ASPERITES Plural Noun. (Latin, *asper* = rough.) Small spine-like or peg-like structures on Integument and sometimes arranged in geometrical patterns. See Microspine; Microtricha; Spinula; Spinule. See Surface Sculpture. Adj. (Latin, *asper* = rough.) Pertaining to surface roughenings or dot-like elevations. Cf. Shagreened.

ASPERSUS Noun. (Latin, *asper* = rough.) Rugged; structure with distinct elevated dots or bumps.

ASPHYXIATION Noun. (Greek, *asphyyxia,* from *a* = not; *sphyzein* = to throb. English, *-tion* = result of an action. Pl., Asphyxiations.) The act or process that causes asphyxia; smothering.

ASPID- (Greek, *aspidos* = shield.) A prefix used in anatomical terms which refers to a shield or shield-like appearance. Cf. Scutellum.

ASPIDOSOMA Noun. (Greek, *aspidos* = shield; *soma* = body. Pl., Aspidosomata.) Acarology: A dorsal division of the body that is bordered by the Abjugal Furrow. The Stethosoma without the Podosoma. The dorsal region of the anterior Idiosoma in acariform mites. The name is most commonly applied to boxmites [Euphthiracaroidea; Phthiracaroidea].

ASPIDOTHORACIDAE Plural Noun. A Family of Megasecoptera with tubular projections on the Abdomen. See Megasecoptera.

ASPIRATOR Noun. (Latin, *aspirare* = to breathe toward; *or* = one who does something. Pl., Aspirators.) A hand-held device for collecting small terrestrial arthropods. Aspirator designs are variable, but typically constructed of a test tube or glass vial that is closed at one end and sealed with a cork or rubber stopper projecting into the neck at the other end. Two small glass or metal tubes (of equal diameter) project through the stopper and penetrate the atrium of the larger vial. One small tube is covered with muslin or screen on the end inside the vial; the opposite end of the tube is placed in the mount and lung power is used to generate an air current that draws air and the insect into the tube via the aperture at the end of the other small tube. The device is popular for collecting small insects, particularly Microhymenoptera, some Hemiptera and small Diptera. Syn. Pooter. Cf. Slurp Gun.

ASPITAN® See Tau-Fluvalinate.

ASSASSIN BUG *Pristhesancus plagipennis* Walker [Hemiptera: Reduviidae]: In Australia, arguably the largest Species of Reduviidae and regarded as a beneficial predator (especially of Lepidoptera larvae) but also a significant predator of honey bees. Eggs are reddish brown, cylindrical, operculate, ca 2.5 mm long and laid in clutches of 30–40; incubation requires 15–45 days. Life cycle requires 85–100 days; adults live 90–330 days. See Reduviidae.

ASSASSIN FLIES See Asilidae.

ASSATEAGUE INSECT TRAP A device for collecting mosquitoes and biting flies. Mosq. News 34: 196–199. See Trap.

ASSEMBLY Noun. (Middle English, *assemble*. Pl., Assemblies.) Social Insects: Gathering of colony members for communal activity. 2. Gathering of males near a virgin female.

ASSEMBLY ZONE An area of the Integument distinct from the cuticle. The Assembly Zone is positioned above the epidermal cells and beneath the cuticle. Term was first applied by Delbecque *et al.* (1978). Little is known about AZ structure or function. The region has been identified in the dipteran *Drosophila* and lepidopteran *Manduca*. AZ is fundamental to the process of Cuticle formation; antigenically distinct from the Cuticle and does not contain lamellae in the process of formation. Probably a stable component of the Integument and a permeable matrix through which the lamellar elements move during Cuticle formation. See Cuticle; Integument. Cf. Basement Membrane; Cement Layer; Cuticulin Layer; Dermal Glands; Epidermal Cells; Endocuticle; Epidermis; Exocuticle; Pore Canals; Wax Layer. Rel. Skeleton; Skin.

ASSOCIATION Noun. (Latin, *associatus* > *associare* = to join, unite. Pl., Associations.) 1. Ecology: A major or fundamental unit in ecological community organization. 2. Groups of strata that are uniform over a large area (*teste* Shelford). 3. Any organization of people brought together through a common interest (*e.g.* American Association for the Advancement of Science). Cf. Society.

ASSOCIATION NEURON A neuron positioned within the central nervous system that mediates between sensory and motor neurons, or between other association neurons. An internuncial neuron *sensu* Snodgrass. See Neuron.

ASSOCIATIVE LEARNING The mechanism by which a stimulus produces direct effects determined by its nature and the effects of different stimuli that earlier attacked the organisms at the same time as the given stimulus. See Learning. Cf. Habituation; Imprinting.

ASSORTIVE MATING A mating system in which conspecific sexes of a reproducing population tend to mate based on similarity of some measurable parameter such as overall body size (*e.g.* small males tend to mate with small females and large males mate with large females).

ASSURGENT Adj. (Latin, *assurgere* = to rise up.) Curved downward at the base, then curved upward to a vertical position.

ASTAUROV, BOROS LIVOVICH (1904–1974) (Ghilyarov 1975, Ent. Obozr. 54: 233–236.)

ASTEGASIMOUS Adj. (Greek, *a* = without; *stegos* = roof > *stegein* = to cover; *-osus* = possessing the qualities of.) Acarology: The condition of the rostral Tectum absent or reduced such that the Gnathosoma is exposed.

ASTEIIDAE Plural Noun. A small-sized, widespread Family of schizophorous Diptera forming the nominant Family in Superfamily Asteioidea. Adult body <2.0 mm long; Propleuron without vertical Carina; wing vein reduced, Costa not broken; R2+3 ending in Costa near R1. Biology poorly known; larvae of some Species in Europe are associated with fungi. See Asteioidea; Schizophora.

ASTEIOIDEA Noun. A Superfamily of schizophorous Diptera including Anthomyzidae, Asteiidae (Astiidae), Aulacigastridae, Neurochaetidae, Periscelididae and Teratomyzidae. Body small to minute; lower part of face concave and visible in ventral aspect; Vibrissae typically present; Tibiae lacking preapical bristles; female Cerci medially separated. See Diptera.

ASTER LEAFHOPPER *Macrosteles quadrilineatus* Forbes [Hemiptera: Cicadellidae].

ASTER LEAFMINER *Calycomyza humeralis* (Roser) [Diptera: Agromyzidae].

ASTER YELLOWS A viral disease transmitted by leafhoppers to asters and other ornamental plants. Rel. Disease.

ASTERIFORM Adj. (Latin, *aster* = star; *forma* = shape.) Star-shaped; descriptive of a relatively large central body with radiating projections. Syn. Asteroid. See Aster. Rel. Form 2; Shape 2; Outline Shape.

ASTEROID Adj. (Greek, *asteroeides* = star-like.) See Asteriform.

ASTEROLECANIIDAE Plural Noun. Pit Scales. A numerically small, widespread Family of sternorrhynchous Hemiptera assigned to Superfamily Coccoidea. Female with 8-shaped pores; tubular ducts without slender filament but with inner, short, bent base; legs absent or evanescent; abdominal spiracles absent. Some Species induce galls on plants in Australia. See Coccoidea; Sternorrhyncha.

ASTHENIA Adj. (Greek, *astheneia* = weakness.) Loss of strength. Debility. Rel. Disease.

ASTHENOBIOSE Adj. (Greek, *asthenes* = without strength; *bios* = life; *-osus* = full of. Pl., Asthenobioses.) Dormancy in certain insects that is induced by autointoxication from non-eliminated uraemic poisons. Alt. Asthenobiosis.

ASTIIDAE See Asteiidae.

ASTRO® See Permethrin.

ASTRONOMICAL TWILIGHT See Twilight.

ASUNTOL® See Coumaphos.

ASYMMETRIC Adj. (Greek, *a* = not; *symmetros* > *syn* = together + *metron* = measure; *-ic* = of the nature of.) Pertaining to a structure whose sides are not mirror images; not symmetrical. Alt. Asymmetrical. See Asymmetry; Symmetry.

ASYMMETRY Noun. (Greek, *a* = not; *symmetria*, from *syn* = together; *metron* = measure. Pl., Asymmetries.) 1. A state of unlikeness in lateral development. 2. An absence of symmetry in form or in development. 3. Asymmetry is the antithesis of symmetry and exists in several forms, including behavioral asymmetry, biological asymmetry and anatomical asymmetry. In anatomical asymmetry structures have no plane of symmetry. An asymmetrical object can never be made to coincide with its reflection in a mirror. For instance, a symmetrical image is formed by the letter 'O'; an asymmetrical image is formed by the letter 'F'. In essence, an asymmetric object and its image are reversed. The object and its image form an entantiomorphic pair (*e.g.* shoes, pairs of gloves, left and right handed scissors). Allochiral refers to a condition in which symmetrically similar parts are reversed in position and arrangement. An example would be the left and right human hands. Ambidexterous is a behavioral term and refers to the ability to use either hand with equal facility. See Symmetry. Cf. Absolute Asymmetry; Behavioral Asymmetry; Biological Asymmetry; Developmental Asymmetry; Morphological Asymmetry; Pattern Asymmetry; Skeletal Asymmetry. Rel. Organization.

ASYNCHRONOUS GENERATIONS Population dynamics: A circumstance in which adult females of a population lay eggs more-or-less continuously and all or most developmental stages of the Species are present within that population at any time. Cf. Synchronous Generations.

ATABRON® See Chlorflurazuron.

ATAVISM Noun. (Latin, *atavus* = ancestor; English, *-ism* = condition. Pl., Atavisms.) 1. The spontaneous occurrence of an ancestral characteristic (feature) in an organism and the feature was not expressed in immediate progenators. 2. Reversion to ancestral characteristics.

ATAVISTIC Adj. (Latin, *atavus* = ancestor; *-ic* = characterized by.) Descriptive of or pertaining to Atavism; of the nature of Atavism.

ATELURIDAE See Nicoletiidae.

ATERRIMUS Adj. (Latin, *ater* = black.) Deep black in coloration.

ATGARD® See Dichlorvos.

ATHERICEROUS See Aristate.

ATHERICIDAE Plural Noun. A small Family of orthorrhaphous Diptera assigned to Superfamily Tabanoidea. Athericidae are cosmopolitan in distribution with ca 50 Species in seven Genera and most Species found in the Orient. Adults nondescript, medium-sized 7–8 mm long and fuscous to black almost always with fasciated Abdomen. Athericidae are separated from Rhagionidae and other tabaniform Diptera by a well developed Postscutellum or Subscutellum; scale-like elevation behind posterior spiracle; antennal Flagellum with slender, non-annulated Arista; marginal cell closed apically; Tibiae without erect bristles. Adults occur in moist habitats; larvae occur in streams. Larvae are partially predaceous on other aquatic insects; adults of some *Suragina* suck blood of mammals and other Species suck blood from owls; *Athrichops* suck blood of frogs. Adult female oviposits in clusters on undersides of leaves of plants overhanging water; female dies and often becomes stuck to egg mass. Initial egg mass acts as an attractant for other females who deposit their eggs in and on attracting egg mass. Largest egg mass cover ca 50 sq. ft. and at least 0.5' in depth. Eggs hatch in about six days and young larvae drop into water. Pupation occurs in soil along stream banks. See Tabanoidea.

ATHERMOBIOSE Adj. (Greek, *a* = not; *therme* = heat; *bios* = life; *-osus* = full of.) Dormancy in insects that is induced by cold or by relatively low temperatures in relation to the organism. Alt. Athermobiosis.

ATHERMOPAUSE Noun. (Greek, *a* = not; *therme* = heat; *bios* = life; *pausi* = ending. Pl., Athermopauses.) A physiological condition or expression of Dormancy that is influenced by one or more non-thermal factors (Mansingh 1971, Canad. Entomol. 103: 1002). See Dormancy. Cf. Aestivation; Hibernation. Rel. Diapause; Oligopause; Quiescence.

ATKIN, THOMAS (1813–1879) (Carrington 1880, Entomologist 13: 24.)

ATKINSON, DAVID JACKSON (1897–1952) (Riley 1953, Proc. R. Ent. Soc. Lond. (C) 17: 71.)

ATKINSON, EDWIN FELIX THOMAS (1840–1890) (Anon. 1890, Entomol. mon. Mag. 26: 329; Walshingham 1890, Proc. Ent. Soc. Lond. 1890: lix.)

ATKINSON, H S (1897–1963) (Wigglesworth 1964, Proc. R. Ent. Soc. Lond. (C) 28: 57.)

ATKINSON, NORMAN JEFCOATE (1902–1933) (Hungerford 1933, Ann. Ent. Soc. Am. 26: 187–188.)

ATKINSON, WILLIAM STEPHEN (–1876) (Westwood 1876, Proc. Ent. Soc. Lond. 1876: xliv.)

ATMORE, EDWARD A (1855–1930) (Turner 1930, Entomologist's Rec. J. Var. 42: 160; Jordan 1931, Proc. Ent. Soc. Lond. 1931: 130.)

ATMOSPHERE Noun. (Greek, *atmos* = vapour; Latin, *sphaera* = sphere. Pl., Atmospheres.) 1. The medium (often a gas) which surrounds an object. 2. Arcane: The exterior circle of an ocellate spot.

ATOM Noun. (Latin, *atomus* > Greek, *atomos* = indivisible, > *a* = not; *tom* = to cut. Pl., Atoms.) 1. A small, indivisible unit generally regarded as a discrete, independent member of a group. 2. A minute particle. Alt. Atomus.

ATOMARIUS Adj. (Latin, *atomus* > Greek, *atomos* = indivisible, > *a* = not; *tom* = to cut.) An archane term used to characterize minute dots, spots or points.

ATOX® See Rotenone.

ATRACHEATE Adj. (Latin, *trachia* > Greek, *a* = without; *tracheia* = windpipe; *-atus* = adjectival suffix.) Descriptive of an organism that lacks Tracheae or a tracheal system. See Trachea. Rel. Respiratory System.

ATRACHELIA Noun. (Greek, *a* = without; *trachelos* = neck.) Coleoptera in which there is no visible constriction between head and Prothorax, including the Rhynchophora and some Heteromera.

ATRIAL ORIFICE The external opening of the spiracular Atrium; Porta Atrii (Snodgrass).

ATRIATE Adj. (Latin, *atrium* = chamber; *-atus* = adjectival suffix.) Pertaining to structure with an Atrium. Alt. Atrial.

ATRIPLECTIDIDAE Plural Noun. A small Family of Trichoptera assigned to Superfamily Limnephiloidea. Known only from one monotypic Genus on Seychelles Islands and Australia. Adult medium-size with wing span ca 25 mm; Ocelli absent; Maxillary Palpus with five segments; Antenna longer than forewing; forewing narrow; Discoidal Cell present; Hamuli on anterior margin of hindwing; tibial spur formula 2-4-4. Larval head long and narrow; Antenna long; Pronotum slender with two pairs of dorsal sclerites; hind leg long; abdominal gills filiform. Larva occurs on bottom of lakes and slow-flowing rivers; case tubular, slightly curved and formed of sand grains. See Limnephiloidea.

ATRIUM Noun. (Latin, *atrium* = chamber; Latin, *-ium* = diminutive > Greek, *-idion*. Pl., Atria.) 1. Any chamber at the entrance of a body opening. 2. The preoral cavity in muscoid larvae. 3. A chamber immediately below the respiratory spiracle. Cf. Felt Chamber; Stigmatic Chamber.

ATROBAN® See Permethrin.

ATROCERULEOUS Adj. (Latin, *ater* = black; *caeruleus* = blue; *-osus* = possessing the qualities of.) A deep blue-black. Alt. Atroceruleus; Atrocoeruleus.

ATROPHY Noun. (Greek, *a* = without; *trophe* = food > *trephein* = to nourish. Pl., Atrophies.) 1. Tissue, muscles, appendages, parts or organs that seem diminished in size or lack growth, implicitly from insufficient nourishment, exercise, disuse or age. 2. A structure that is reduced in size; withered. 3. The process of progressive decline. See Development. Cf. Hypertrophy.

ATROPURPUREUS Adj. (Latin, *ater* = black; *purpureus* = purple.) Dark purplish; nearly black.

ATROVELUTINUS Adj. (Latin, *ater* = black, *vellus* = cut wool fleece.) Velvety black.

ATROVIRENS Adj. (Latin, *ater* = black; *virideus* = green.) Dark green, approaching blackish.

ATTA Noun. (Hindi, *ata*.) 1. Unsorted wheat flour or meal. 2. A Genus of leafcutting ants that occurs in the New World. See Formicidae.

ATTACHMENT ORGAN An elaborate holdfast in some oestrid and gasterophillid flies that parasitize vertebrate hosts. Flies 'glue' their eggs to hair shaft of their host. Egg is maintained on hair via a complex structure that consists of a basal, short, flexible pedicel attached to egg's body and a distal 'clasper'. The claspers are elongate, paired and medially grooved; medial grooves are spread apart during oviposition and embrace the hair after the egg is released by female fly. Medial grooves contain an adhesive substance secreted by follicle cells; the adhesive dries and effectively bonds the egg to the hair shaft. Cf. Pedicel 5. Attachment Organs are important for eggs of some aquatic insects. Mayflies: AO on lateral surfaces of egg and come in many shapes, adhesive, 'sucker like discs or plates', or threads with apical knobs or adhesive projections. Cf. Polar Cap.

ATTELABIDAE Plural Noun. Leaf-rolling Weevils. A small Family of polyphagous Coleoptera assigned to Superfamily Curculionoidea and closely related to Rhynchitidae. Adult 2–7 mm long, robust, narrow anteriad; coloration black, red or a combination of black and red; Rostrum short with Gular Sutures fused; Mandibles exodont; Labrum concealed; Maxillary Palpus well developed; Labial Palpus reduced. Prothorax without lateral Carina and narrower than base of Elytra. Female oviposits single egg near tip of leaf, then cuts and rolls leaf to form ball in which egg is concealed. Larvae are short, wide and curved; most anatomical features are variable. Larvae feed on leaves and pupation occurs in leaf roll or in ground. Biology of attelabids is diverse: Some Species are leaf-rollers, some Species are leafminers and some Species feed on flower buds or fruits. Syn. Rhynchitidae. See Curculionoidea.

ATTENUATE Adj. (Latin, *attenuare* = to thin; *-atus* = adjectival suffix.) 1. Descriptive of structure that gradually tapers apicad into a long, slender point. 2. Descriptive of structure that is drawn out or slender. Alt. Attenuated; Attenuatus.

ATTINGENT Adj. (Latin, *attingere* = to touch.) Touching. Cf. Contiguous.

ATTRACT 'N' KILL PBW® See Gossyplure.

ATTRACTANT Noun. (Latin, *attrahere* = to draw to; *-antem* = an agent of something. Pl., Attractants.) A chemical that causes an insect to make oriented movements toward its source. (Dethier *et al.* 1960, 53: 135.) See Semiochemical. Cf. Arrestant; Repellant.

ATTRIBUTE Noun. (Latin, *attribuere* = to assign. Pl., Attributes.) 1. A quality, trait, characteristic or feature of an organism. Attributes may be biochemical, behavioural, ultrastructural or anatomical. 2. Systematic theory: The demonstration of a specific character state in an organism. Cf. Character State.

ATTWOOD, R W (–1941) (Wakely 1942, Proc. Trans. S. Lond. Ent. Nat. Hist. Soc. 2: 73.)

-ATUS A Latin suffix that denotes possession of a quality or structure.

ATYPICAL Adj. (Greek, *a* = without; *typos* = pattern; Latin, *-alis* = pertaining to, having the character of.) 1. Pertaining to structure, behaviour or colour that does not conform to the usual condition. 2. Anatomical features not of the usual or expected form. Alt. Atypic. Syn. Anomalous; Aberrant; Abnormal. Cf. Typical.

ATZE, CHRISTIAN GOTLIEB (–1826) (Rose 1853, *New General Biographical Dictionary* 2: 313.)

AUBE, CHARLES NICOLAS (1802–1869) Author of *Pselaphiorum Monographia cum Synonymia extricata* (1835). (See Marseul 1882, Abeille 20: 1–5, bibliography.)

AUBERTIN, DAPHNE (–1970) (Hinton 1971, Proc. R. Ent. Soc. Lond. (C) 35: 52.)

AUBRIET, CLAUDE (1651–1743) (Rose 1853, *New General Biographical Dictionary* 2: 320–321; Larousse 1865, *Grand Dictonnaire Universale Du XIX Siècle* 1: 917.)

AUBYN, JOHN G ST (1889?–1975) (Gunn 1976, Proc. R. Ent. Soc. Lond. (C) 40: 52.)

AUCHENORRHYNCHA Noun. (Greek, *aucheno* = throat; *rhynchus* = beak.) A numerically large, economically important Suborder of Hemiptera (division of Homoptera in some classifications). Auchenorrhyncha are characterized by a Rostrum (beak) that appears to arise from the lower part of the head. See Hemiptera. Cf. Sternorrhyncha.

AUCHENORRHYNCHUS Adj. (Greek, *aucheno* = throat; *rhynchus* = beak.) Hemiptera: Pertaining to Taxa in which the Rostrum (beak) originates on the posteroventral portion of the head near the fore Coxa. Cf. Sternorrhynchous.

AUCTORUM A Latin term meaning 'of authors' and used in taxonomy to refer to the sense in which a scientific name is applied. Frequently, 'of authors' draws attention to the misapplication of scientific names or the application of names considered incorrect or idiosyncratic. See Systematic Name. Rel. Nomenclature.

AUDCENT, HENRI L F (1875–1951) (Anon. 1951, Entomologist's Rec. J. Var. 63: 29; R. W. L. 1951, Entomol. mon. Mag. 87: 159.)

AUDEOUD, GEORGES (1874–1943) (Olivier 1943, Verh. Schweiz. Naturf. Ges. 123: 293–295; J. R. 1945, Mitt. Schweiz. Ent. Ges. 19: 209–216, bibliography.)

AUDINET-SERVILLE, JEAN GUILLAUME (1775–1858) Author of *Nouvelle Classification de la Famille des Longicornes* (In: Ann. Soc. Entomol. Fr., 1833–1834.) Also contributed to *Faune Francaise and Encyclopedie Methodique.* (See Anon.1889, Abeille 26: 232–239, bibliography.)

AUDITORY Adj. (Latin, *audire* = to hear.) Pertaining to the sense of hearing.

AUDITORY ORGAN 1. Any structure that functions as an ear. 2. Orthoptera: Specialized structures that consist of a tympanic membrane on the fore Tibia or base of the Abdomen. 3. Lepidoptera: A tympanic membrane that is stretched across a sclerotized (cuticular) frame and covers a tympanal air sac. The Tympanum vibrates in response to vibrations in the air. A chordotonal organ attached to the Tympanum receives the vibrations and transduces them into electrical impulses that are communicated to the afferent nervous system. AOs typically occur on the Metathorax or Abdomen of many Species in several Families. AOs are best known in Noctuoidea, Geometroidea and Pyraloidea; they also occur in some Species of Cossoidea, Drepanoidea and Tineoidea. AOs may occur at the wing base of some Nymphalidae and Hedylidae, or mouthparts of adult Sphingidae. See Ear; Tympanal Organ. Rel. Acoustical Communication; Stridulation.

AUDITORY PEG See Scolopale.

AUDITORY SENSE The sense of hearing. The perception of vibrations through membranes. See Acoustical Communication.

AUDOUIN, JEAN VICTOR (1797–1841) French physician, Professor of Entomology (Musee d'Histoire Naturelle, Paris), librarian, naturalist at Jardin du Roi and early worker in economic Entomology. Audouin published on forest insects, vine insects, diseases of the the silkworm and insects that attack wooden structures (*Histoire des insectes nuisible a la vigne,* 1840–1842). Audouin was the author of numerous entomological publications including *Recherches Anatomiques sur le Thorax des Animaux articules, et celui des insectes hexapodes en* *particulier* (Annales des Science, 1: 97–416), *Lettres sur la generation des Insectes* (Annales des Science., 2: 281), *Recherches Anatomiques pour servir a l'historire naturelle des Cantharides* (Annales des Science, 9: 31.) He was coauthor of *Resume d'Entomologie, ou d'Histoire Naturelle des Animaux Articules* (2 vols, 1829). (Westwood 1842, Arcana Entomol. 1: 155–159.)

AUDY, J RALPH (1914–1974) (Traub 1975, J. Med. Ent. 11: 765–768.)

AUGER BEETLE *Xylodeleis obispa* Germar [Coleoptera: Bostrychidae].

AUGMENTATIVE BIOLOGICAL CONTROL A strategy for pest management involving the mass propagation and organized release of beneficial natural enemies for the regulation of a pest population. See Biological Control. Cf. Classical Biological Control.

AUGMENTATIVE RELEASE The organized release of beneficial natural enemies to aid or supplement resident natural enemies for the control of a pest population. AR typically is viewed as a long-term strategy with beneficial organisms reproducing in field or crop. See Augmentative Biological Control. Cf. Innoculative Release; Inundative Release.

AUGUSTIN, CARL WILHELM (Weidner, Abh. Verh. Naturw. Ver. Hamburg Suppl. 9: 174–177.)

AULACIDAE Schuckard 1841. Plural Noun. A numerically small, cosmopolitan Family of apocritious Hymenoptera assigned to Superfamily Evanioidea and sometimes placed within Gasteruptiidae. About 150 Species of aulacids have been described. Adult medium sized (<25 mm long) with antennal Torulus near Clypeus; Propleuron elongate and forming a neck; forewing venation extensive with second Recurrent Vein and several Intercubitals; Propodeum pyramid-shaped with gastral Petiole attached at apex; Ovipositor exceptionally long and held in place during oviposition by opposible grooves formed on medial surfaces of the hind Coxae. Species are solitary endoparasites of wood boring Coleoptera and Hymenoptera. Female aulacids oviposit into the egg of a host but development is arrested until the host has completed larval development. Aulacid larva then consumes mature host larva, emerges from host's integument, spins a cocoon and pupates. See Evanioidea.

AULACIGASTRIDAE Plural Noun. A numerically small and geographically widespread Family of schizophorus Diptera assigned to Superfamily Asteioidea. Adult body small to minute; face convex with one reclinate and one proclinate Fronto-orbital Bristle; Antenna correct with second segment displaying a dorsal slit and third segment short; Wing with Costa broken at end of Sc; CuP complete with CuA+1A projecting beyond apex. Biology of aulacigasterids is poorly known; larvae of some Species in New Guinea are associated with *Eucalyptus*. Adults of some Species occur at sap flux of tree wounds where larvae feed on sap; some Species occur in grasses. Syn. Aulacigasteriidae. See Asteioidea.

AULAX Noun. (Etylomology obscure.) A groove that extends the length of the dorsolateral surface of the Ovipositor valves (Gonophysis, first Valvulae) on abdominal segment VIII. The groove (Aulax) engages an elongate 'tongue' (Rachis) along the ventrolateral surface of Ovipositor valves on abdominal segment IX (second Valvulae.) Together, the Aulax and Rhachis form a 'tongue-in-groove' that permits a sliding action between the Valvulae, but which holds them together during the act of oviposition. The collective holdfast-sliding mechanism is called an Olistheter. See Ovipositor. Cf. Olistheter; Rachis.

AUMONT, LOUISE CAROLINE D' (1827–1853) (Mulsant 1855, Opusc. Ent. 6: 184–196, portrait; Mulsant 1855, Ann. Soc. Linn. Lyon 2: 284–296.)

AURANTIACEOUS Adj. (Latin, *aurantium* = specific epithet of orange; *-aceous* = resembling.) 1. The colour of an orange. 2. Pertaining to the sour orange. 3. Pertaining to the Rutaceae. Alt. Aurantius.

AURATE Adj. (Latin, *auris* = ears; *ate* = having the quality of; *-atus* = adjectival suffix.) With ears or ear-like expansions. Alt. Auratus.

AURATUS Adj. (Latin, *aureus* = golden.) Golden yellow.

AURELIA Noun. (Latin, *aureolus* = golden.) The pupa or Chrysalis of Lepidoptera.

AURELIAN Adj. A lepidopterist.

AUREOLAT Adj. (Latin.) Descriptive of structure with a diffuse coloured ring.

AUREOLE Noun. (Latin, *aureola* > *aureolus*, dim. *aureus*, = golden. Pl., Aureoles.) A ring of colour that is usually diffuse outwardly like a halo.

AUREOUS Adj. (Latin, *aureus* = golden; *-osus* = possessing the qualities of.) Gold coloured; descriptive of structure that is golden yellow.

AURICHALCEOUS Adj. (Latin, *aurichalcum*, from Greek, *oros* = mountain; *chalkos* = brass or copper; Latin, *-aceus* = of or pertaining to.) Brassy; the metallic yellow of brass. Correctly, Orichalceous; Orichalceus. Alt. Aurichalceus.

AURICLE Noun. (Latin, *auricula* = small ear. Pl., Auricles.) 1. Apoidea, Andrenidae: A short, membranous process laterally on the Ligula (Smith). Apidae: A structure in the anterior end of the first tarsal segment of the hind leg that pushes up the pollen mass into the Corbicula (pollen basket) of the hind Tibia. 2. A chamber of the insect heart. 3. Anisoptera: Either one of a pair of lateral appendages of the second segment of the Tergum in the males that have angulated hindwings, of smaller size in the females (Imms).

AURICULAR Adj. (Latin, *auricula* = small ear.) 1. Descriptive of or pertaining to the Auricle. 2. The space or cavity surrounding the Dorsal Vessel.

AURICULAR OPENINGS The lateral openings into the Heart by means of which blood is admitted into the dorsal vessel. Syn. Ostia. See Heart.

AURICULAR VALVE A slit-like valve of the insect Heart at each Ostium and which prevents the return flow of blood into the dorsal sinus.

AURICULATE Adj. (Latin, *auricula* = small ear; *-atus* = adjectival suffix.) 1. Ear-like in shape or form. Descriptive of organisms with an ear-like appendage. 2. Descriptive of Antennae in which the basal segment is distended into a concave, plate-like sclerite that envelops other antennal segments. Cf. Auriform.

AURICULATE ANTENNA An Antenna in which one of the basal segments (joints) is dilated into an ear-like shield (cap) partly covering the other segments (Westwood).

AURICULO-VENTRICULAR Adj. The outer valves of the Heart between the auricular space and the chamber. See Heart.

AURIFORM Adj. (Latin, *aura* = ear; *forma* = shape.) Ear-shaped; descriptive of structures that resemble the external shape of a human ear. See Ear. Cf. Auriculate. Rel. Form 2; Shape 2; Outline Shape.

AURITUS Adj. (Latin, *aura* = ear.) Descriptive of structures that display ear-like spots or appendages.

AURIVILLIUS PER OLAF CHRISTOPHER (1853–1928) Swedish economic Entomologist and taxonomist working at Royal Museum of Natural History (Stockholm) specializing in Coleoptera. (Bryk 1923, Ent. Tijdskr. 44: 1–55, bibliography; Tullgren 1928, Ent. Tijdskr. 49: 171–178, bibliography.)

AURORAL SPOT Applied to the bright orange-coloured spot at the apical area of Anthocharis.

AUROREOUS Adj. (Latin, *aurora* = dawn; *-osus* = possessing the qualities of.) An arcane term used to describe the colour red as it appears in the Aurora Borealis. Alt. Auroreus.

AUSSERER, ANTON (1843–1889) (Ausserer, Bonnet 1945, *Bibliographia Araneorum* 1: 47–48.)

AUSSERER, CAROL (1844–1920) ([Menestrina] 1920, Studi Trent Sci. Nat. 1: 355.)

AUSTEN, ERNEST EDWARD (1867–1938) English military (Major) and early worker on Tsetse Fly. (Musgrave 1932, *Bibliography of Australian Entomology 1775–1930*. 380 pp. (8, partial bibliography), Sydney; Blair 1938, Entomol. mon. Mag. 74: 42–43.)

AUSTRAL Adj. (Latin, *australis* = southern; *-alis* = belonging to.) Biogeography: Pertaining to the faunal area that covers the USA and Mexico except boreal mountains and tropical lowlands. Austral Region divided into Transition, Upper, Lower and Gulf Strip. See Boreal; Tropical.

AUSTRALASIAN REGION A biogeographical region that includes islands south and east of Wallace's Line, to include New Guinea, Australia and New Zealand. See Biogeography.

AUSTRALEMBIIDAE Plural Noun. A Family of Embioptera known from eastern Australia and Tasmania which is, in terms of Species, numerically the largest Family in Australia. Both sexes apterous; male with articulated lobe on right lateral margin of Sternum VII; male left Cercus of one segment; male with large Maxillary Palpus and elongate Submentum. Species inhabit leaf litter but their biology is poorly known. See Embioptera.

AUSTRALIAN ADMIRAL *Vanessa itea* (Fabricius) [Lepidoptera: Nymphalidae].

AUSTRALIAN ATLAS-MOTH See Hercules Moth.

AUSTRALIAN CARPET-BEETLE *Anthrenocerus australis* (Hope) [Coleoptera: Dermestidae]: A widespread, minor pest of wool, carpet, jute, fabrics, animal hides, stored meat and milk products. ACB is endemic to Australasia and adventive in Europe. Adult 2–3 mm long, oval in outline shape and black with pale markings. Larva is reddish-brown and elongate-oval in shape; body covered with bristles with a few long bristles trailing from apex of Abdomen. ACB pupates within last larval exuvia; neonate adult is quiescent within exuvia ca 2–4 days. Adults mate several times and live ca 40–50 days; female lays 25–35 eggs. See Dermestidae. Cf. Black Carpet Beetle; Common Carpet Beetle; Furniture Carpet Beetle; Varied Carpet Beetle.

AUSTRALIAN CITRUS WHITEFLY *Orchamoplatus citri* (Takahashi) [Hemiptera: Aleyrodidae]: An endemic pest of *Citrus* in Australia. Adult ca 2.5 mm long, white, with body covered in wax; wings slightly separated medially when held tentiform in repose; adults swarm when distrurbed. ACW eggs ovoid, yellow and deposited in circular arrangement on underside of *Citrus* leaf. Crawlers flattened; three 'larval' instars; pupa often attached near midrib of leaf and mistaken for juvenile Soft Brown Scale. ACW prefers underside of young leaves inside tree canopy; Species is multivoltine with up to six generations per year, depending upon climate. ACWs produce honeydew that serves as substrate for sooty mould and other fungi. See Aleyrodidae.

AUSTRALIAN COCKROACH *Periplaneta australasiae* (Fabricius) [Blattodea: Blattidae]: A cosmopolitan urban pest that is endemic to Africa, transported to America via slave ships and moved elsewhere through commerce. AC is most common in tropical and subtropical environments. Adult ca 35 mm long and resembles the American Cockroach but body is more dark brown, yellow markings on Pronotum are clearly defined and the anterobasal margin of forewing displays a yellow stripe. Adult lives 4–8 months; female oviparous and lays 12–20 Oothecae. An Ootheca contains 16–24 eggs and is dropped or glued to substrate. Nymphs undergo 10–12 moults; life cycle 6–12 months. Nymphs and adults prefer food of plant origin. AC inhabits greenhouses, factories and building voids. See Blattodea. Cf. American Cockroach; Smokybrown Cockroach.

AUSTRALIAN FAIRY-MARTIN ARGASID *Argas lagenoplastis* Froggatt [Acari: Argasidae].

AUSTRALIAN FERN-WEEVIL *Syagrius fulvitarsis* Pascoe [Coleoptera: Curculionidae].

AUSTRALIAN GOAT-MOTH *Culama caliginosa* (Walker) [Lepidoptera: Cossidae].

AUSTRALIAN GRASS-LEAFHOPPER *Nesoclutha pallida* (Evans) [Hemiptera: Cicadellidae].

AUSTRALIAN KING-CRICKET *Australostoma australasiae* (Gray) [Orthoptera: Stenopelmatidae].

AUSTRALIAN MALARIA-MOSQUITO *Anopheles farauti* Laveran [Diptera: Culicidae]: A pest of humans, other mammals and birds. AMM is an important vector of Malaria and Filariasis in Pacific. Resembles Common Australian Anopheline in anatomy and biology. Adult female speckled grey with spotted wings; Proboscis all dark; Palp as long as Proboscis; apical half of Palp with four white and three black bands. See Culicidae. Cf. Common Australian Anopheline.

AUSTRALIAN MANTID *Tenodera australasiae* (Leach) [Mantodea: Mantidae]. See Mantidae.

AUSTRALIAN MUSHROOM PIGMY-MITE *Brennandania lambi* Krczal [Acari: Pygmephoridae].

AUSTRALIAN PAINTED-LADY *Vanessa kershawi* (McCoy) [Lepidoptera: Nymphalidae].

AUSTRALIAN PARALYSIS-TICK *Ixodes holocyclus* Neumann [Acari: Ixodidae].

AUSTRALIAN PINE-BORER *Chrysobothris tranquebarica* (Gmelin) [Coleoptera: Buprestidae].

AUSTRALIAN PLAGUE-LOCUST *Chortoicetes terminifera* (Walker) [Orthoptera: Acrididae]: The most injurious Australian locust and a major pest of cereal crops (wheat, barley, oats, millet) in Australia. APL spreads during outbreak seasons of abundant rainfall and grass growth. Adult 25–44 mm long (females larger than males), elongate and green to brown; hindwings transparent with apex black; hind leg with scarlet (red) coloration. APL Antenna short, Pronotum large and hind legs adapted for jumping. Adult female forms a tunnel in bare, compact ground with the apex of her Abdomen. APL Ovipositor is spined for laying eggs in soil; egg pod contains 30–50 eggs banana-shaped eggs in frothy accessory gland secretion that also serves as plug for hole. Adults die after oviposition and the egg stage requires 18–90 days depending on temperature. Eclosion ca 2–3 weeks following oviposition; hopper stage 5–6 weeks; wings require several days before adequately hardened for sustained flight. Swarms are strongly migratory and flights of 500 miles have been reported. See Acrididae. Cf. Spur-Throated Locust.

AUSTRALIAN PRIVET-HAWKMOTH *Psilogramma menephron menephron* (Cramer) [Lepidoptera: Sphingidae].

AUSTRALIAN RAT-FLEA *Xenopsylla vexabilis* (Jordan) [Siphonaptera: Pulicidae].

AUSTRALIAN REALM The faunal region including Australia, New Zealand, the eastern Malay Islands and Polynesia. See Realm. Cf. Indo-Australian Realm. Rel. Biogeography.

AUSTRALIAN SHEEP BLOW-FLY See Sheep Blow-Fly.

AUSTRALIAN SPIDER-BEETLE 1. *Ptinus ocelus* Brown [Coleoptera: Ptinidae]. 2. *Ptinus tectus* Boieldieu [Coleoptera: Ptinidae]: A cosmopolitan pest of wheat, milled grains, spices; ASB is not common in tropical areas. Female lays ca 100 eggs during life; eggs ca 0.5 mm long, sticky, covered with debris; incubation 5–7 days. Larva white, curved with small legs and surface with golden Setae. ASP undergoes three larval instars with development ca 40 days; mature larvae leave food source to pupate, and often bore through packaging material; produce pupal chamber and spin silk cocoon inside chamber; pupal stage requires ca 20–30 days. Adult remains inside pupal chamber 7–20 days. Adult globular, 3–4 mm long, reddish brown; Elytra with longitudinal rows of pits. See Ptinidae. Cf. White-Marked Spider Beetle.

AUSTRALIAN WHEAT-WEEVIL *Rhyzopertha dominica* (Fabricius) [Coleoptera: Bostrichidae]. Syn. Lesser Grain Borer.

AUSTRALIAN WHIRLIGIG-MITE *Walzia australica* Womersley [Acari: Anystidae].

AUSTRIONIIDAE Kozlov 1975. Plural Noun. A small Family of apocritous Hymenoptera assigned to Superfamily Proctotrupoidea. Trupochalcididae sometimes included in Austrioniidae. See Proctorupoidea.

AUSTROPERIDAE Plural Noun. A small Family of Plecoptera assigned to Suborder Antarctoperlaria. Species inhabit southern South America, Australia and New Zealand (Arctogaeic-Neotropic). Adult medium sized; Pronotum rectangular with acute corners; hindwing with seven veins. Naiad with 3–5 simple anal gills or Cerci and anal sclerites with gills; Pronotum rectangular with acute corners; Sternum X absent. Naiads occupy streams under stones or gravel. Syn. Penturoperlidae. See Plecoptera.

AUSTRORIPARIAN FAUNAL AREA Part of the lower Austral Zone of the USA that covers the greater part of the South Atlantic and Gulf States. AFA begins near the mouth of Chesapeake Bay, covers half or more of Virginia, North and South Carolina, Georgia, Florida, Alabama, all of Mississippi and Louisiana, east Texas, more than half of Arkansas and parts of Oklahoma, Southeastern Kansas, Southern Missouri, Southern Illinois, Southwestern corner of Indiana and the bottom lands of Kentucky and Tennessee.

AUSTROSERPHIDAE Plural Noun. Very small, primitive Family of extant Proctotrupoidea that is endemic to Australia. Adult body ca 6.0 mm long; Antenna geniculate with Scape pointed and short; forewing with three basal cells, large Pterostigma. M, Cu and 1A veins in basal portion of wing. See Proctotrupoidea.

AUTAPOMORPHIC Adj. (Greek, *autos* = self; *apo*

= away; *morphe* = form; *-ic* = of the nature of.) Pertaining to a unique, derived character. Alt. Autapomorphous. See Apomorphy. Cf. Synapomorphic.

AUTAPOMORPHY Noun. (Greek, *autos* = self; *apo* = away; *morphe* = form. Pl., Autapomorphies.) Taxonomy: A unique, derived character, which, in cladistic classification, is derived from an ancestral condition and unique to a taxon. See Apomorphy. Cf. Synapomorphy.

AUTECOLOGY Noun. (Greek, *autos* = self; *oikos* = home; *logos* = discourse. Pl., Autecologies.) 1. The ecology of individuals. 2. The study of the influence of physical and chemical stimuli of an environment upon the individual animal or plant (Folsom & Wardle). Alt. Auto-Ecology. See Ecology. Cf. Synecology.

AUTHORITY CITATION The practice of citing the name of the author of a scientific name or name combination (*e.g. Apis mellifera* Linnaeus).

AUTOCHTHONOUS Noun. (Greek, *autochthon* = sprung from the land itself > *autos* = self; *chthon* = ground, earth; Latin, *-osus* = possessing the qualities of.) Native or aboriginal. A term used for Species that are considered to have arisen as a part of the native or aboriginal fauna or flora, as contrasted with Species that have immigrated from outside regions (Tillyard). See Endemic. Cf. Migration.

AUTOGENESIS See Abiogenesis.

AUTOGENOUS Adj. (Greek, *autos* = self; *genes* = producing; Latin, *-osus* = possessing the qualities of.) 1. Self-generated or produced without external aid or influence; endogenous. 2. Pertaining to something that originates within one individual. 3. A condition descriptive of some parasitic Diptera and Hymenoptera that do not require a blood meal before oviposition. 4. Insects that do not require food before oviposition. See Feeding Strategy. Cf. Anautogenous.

AUTOGENY Noun. 1. Pertaining to some parasitic Diptera and Hymenoptera that do not require a blood meal before oviposition. 2. Insects that do not require food before oviposition. See Autogenous.

AUTOMIXIS Noun. (Greek, *autos* = self; *mixis* = mixing. Pl., Apomixes.) A cytological form of thelytokous parthenogenesis in which meiosis is present. See Parthenogenesis; Thelytoky. Cf. Amphimixis; Apomixis.

AUTONOMIC GANGLIA Ganglia of the Central Nervous System that control respiration and circulation in an insect. See Ganglion; Nervous System.

AUTONOMIC NERVES Nerves that control the respiratory and circulatory functions of an insect. See Nerve.

AUTONOMIC NERVOUS SYSTEM The Sympathetic Nervous System. ANS innervates the alimentary, circulatory and respiratory organs (Wardle). See Nervous System. Cf. Central Nervous System; Peripheral Nervous System.

AUTOPARASITISM Noun. (Greek, *autos* = self; *parasitos* = parasite; English, *-ism* = condition. Pl., Autoparasitisms.) 1. Formicidae: The phenomenon in which Queens return to the parent nest after being inseminated in contrast to queens founding new nests after being inseminated. Colonies composed of autoparasites frequently undergo fission with queens absconding with workers. 2. A phenomenon in which members of a Species parasitize other members of the same Species. Syn. Adelphoparasitism. See Parasitism.

AUTOPARASITOID Noun. (Greek, *autos* = self; *parasitos* = parasite; *eidos* = form. Pl., Autoparasitoids.) A Species in which individuals develop parasitically upon or within individuals of the same Species. Syn. Autoparasitoid. See Parasitoid.

AUTOPHAGOCYTOSIS Noun. (Greek, *autos* = self; *phagein* = to eat; *kytos* = hollow vessel; container; *sis* = a condition or state. Pl., Autophagocytoses.) The absorption of contractile muscular tissue by cells originating from the muscular fibre rather than Leucocytes (Henneguy). Cf. Phagocytosis.

AUTOPHYTE Noun. (Greek, *autos* = self; *phyte* = plant. Pl., Autophytes.) A plant that synthesizes its food. Cf. Heterophyte; Parasite; Saprophyte.

AUTOPLASTIC Adj. (Greek, *autos* = self; *plastos* = molded; *-ic* = characterized by.) Pertaining to grafts on another place of the same inivdidual. Cf. Heteroplastic; Homoplastic.

AUTOPSY Noun. (Greek, *autopsia* = seeing with one's own eyes. Pl., Autopsies.) A critical analysis of the cause of death of an individual. Cf. Biopsy; Necropsy. Rel. Pathology.

AUTORADIOGRAPHY Noun. (Greek, *autos* = self; Latin, *radius* = ray; Greek, *graphein* = to write. Pl., Autoradiographies.) A technique in which a photographic plate is exposed to the disintegration of radioisotopes. The technique is used for the determination of the primary structure of macromolecules such as DNA.

AUTOSYNTHESIS Noun. (Greek, *autos* = self; *synthesis* = putting together. Pl., Autosyntheses.) The condition in which an Oocyte is capable of synthesizing yolk. Autosynthesis is regarded as a primitive mechanism of supplying yolk to ovarian follicles. Cf. Heterosynthesis. Rel. Nutrition.

AUTOTHYSIS Noun. (Greek, *autos* = self; *lysis* = to loosen; *-sis* = a condition. Pl., Autothyses.) Self rupture. A defensive mechanism of some social insects (termites and ants) in which individuals rupture the body and release a sticky glandular secretion. The individual involved in autothysis dies in defence of nestmates by this action. (Maschwitz & Maschwitz 1974, Oecologia 14: 289–294.) See Reflex Bleeding. Cf. Abdominal Dehiscence; Autotomy; Defaecation.

AUTOTOMIZE Noun. (Greek, *autos* = self; *tome* = cutting. Pl., Autotomizes.) To perform the act of self-mutilation (*e.g.* spiders that cast off legs). Cf. Autothysis.

AUTOTOMY Noun. (Greek, *autos* = self; *tome* = cutting. Pl., Autotomies.) 1. Self-mutilation. 2. The ability of an insect to detach its legs or other appendages for self-defence or to evade predation. For example, the Pronotum of membracid Genera *Oeda* and *Anchistrous* is enlarged and pressure on the inflated portion causes a rupture, thereby permitting the animal to escape. Alt. Autospasy.

AUTOTROPH Noun. (Greek, *autos* = self; *trophein* = feed. Pl., Autotrophs.) An organism that uses CO_2 or carbonates and inorganic nitrogen compounds for metabolic synthesis (*e.g.* Green Plants). Cf. Heterotroph. Rel. Nutrition.

AUTOTYPE Noun. (Greek, *autos* = self; *typos* = pattern. Pl., Autotypes.) Any specimen identified by the describer as an example of 'his Species' and compared with the type or co-type (Smith); Heautotype (Banks & Caudell). See Type. Cf. Cotype; Genoholotype; Genolectotype; Genosyntype; Genotype; Holotype; Lectotype; Neotype; Paralectotype; Paratype; Syntype; Topotype. Rel. Nomenclature; Taxonomy.

AUTOVENTILATION Noun. (Greek, *autos* = self; Latin, *ventilus* = breeze. Pl., Autoventilations.) A mechanism involved in Active Ventilation by which gas is moved through or within the Respiratory System as a consequence of body movement (locomotion). Autoventilation is most apparent through action of flight muscles upon the thoracic tracheal system. See Active Ventilation. Cf. Abdominal Pumping; Auxiliary Respiratory Mechanisms; Spiracular Synchronization. Rel. Respiratory Muscles; Tracheal Ventilation.

AUTUMN GUM-MOTH *Mnesampela privata* (Guenée) [Lepidoptera: Geometridae].

AUTUMN LEAF-ROLLER *Syndemis musculana*. [Lepidoptera: Tortricidae].

AUXER, SAMUEL (1835–1909) (Anon. 1909, Ent. News 20: 96.)

AUXILIA Noun. (Latin, *auxiliarius* = help; Pl., Auxiliae.) 1. Small sclerites beneath the bases of the pretarsal claws, bearing the Pulvilli when the latter are present (Snodgrass). 2. Hymenoptera: The Basipulvilli *sensu* MacGillivray. See Prearasus. Cf. Arolium. Rel. Leg.

AUXILIARY Adj. (Latin, *auxilium* = assistance.) Additional or supplementing.

AUXILIARY RESPIRATORY MECHANISMS A mechanism involved in Active Ventilation through which dorsoventral abdominal movement is augmented by longitudinal action (telescoping) of the Abdomen. ARM further increases pressure in the Abdomen to facilitate expiration of gas (carbon dioxide) from the Respiratory System and thereby permit a greater amount of gas exchange during a respiratory cycle. ARM involves Tergal and Sternal Longitudinal Muscles that shorten to retract the Abdomen during dorsoventral contraction (Expiration Phase) in ventilation cycle. ARM occurs in some grasshoppers and flies, but not all insects. Other ARMs seen in 'Neck Ventilation,' 'Prothoracic Ventilation' and 'Metathoracic Pumping' of some Acrididae and Coleoptera. See Active Ventilation. Cf. Abdominal Pumping; Autoventilation; Spiracular Synchronization. Rel. Respiratory Muscles; Tracheal Ventilation.

AUXILIARY VEIN 1. Diptera: The Subcosta *sensu* Comstock. 2. The Subcostal Vein or vein positioned between the Costal and the first vein. 3. The Mediastinal *sensu* Curran. See Vein; Wing Venation.

AUXILIARY WORKER Formicidae: The slaves of slave-making ants.

AVEBURY, LORD See Lubbock, John.

AVERILL, ALICE W (–1946) (Patch 1946, J. Econ. Ent. 6: 1–4.)

AVERMECTIN Noun. A 16-membered macrocyclic lactone developed from fermentation involving the soil-dwelling actinomycete *Streptomyces avermitilis* (Culture MA-4680.) Regarded as a broad-spectrum, non-phytotoxic, slow-acting natural pesticide with high potency against insects, mites and nematodes. Toxic to bees and fish. Mode-of-action through release and/or binding of GABA at inhibitory synapses in neuromuscular junctions; antihelminthic activity causes an increase in membrane permeability to chloride ions in the target organism. Chronic exposure causes ataxia, tremors and seizures. Avermectins have been used worldwide for control of parasitic organisms in livestock, companion animals and humans; avermectins also used against insect and mite pests of agriculture. Characteristics include: non mobile, degrade well, do not accumulate in food chains. Marketed under many trade names including: Abamectin®, Agrimek®, Ascend®, Avert®, Avid®, Dynamex®, Vertimec®, Zephyr® and Zimectrin®. See Botanical Insecticide; Natural Insecticide; Insecticide. Cf. Abamectin; Ivermectin.

AVERT® See Avermectin.

AVIAN Adj. (Latin, *aves* = birds.) Descriptive of or pertaining to birds.

AVICADE® See Cypermethrin.

AVID® See Avermectin.

AVIDOV, ZVI (1896–) (Harpaz 1971, Israel J. Ent. 6: 1–4.)

AVINOFF, ANDRÉ (1884–1949) (Cattell 1944, American Men of Science 1: 61; Hellman 1948, New Yorker Mag. Aug. 21, pp. 32–44.)

AVOCADO BARK-BEETLE *Paleticus* sp. [Coleoptera: Curculionidae].

AVOCADO BROWN-MITE *Oligonychus punicae* (Hirst) [Acari: Tetranychidae].

AVOCADO LEAF-ROLLER *Homona spargotis* Meyrick [Lepidoptera: Tortricidae].

AVOCADO RED-MITE *Oligonychus yothersi* (McGregor) [Acari: Tetranychidae].

AVOCADO THRIPS *Scirtothrips perseae* Mound [Thysanoptera: Thripidae]: A recent introduction into California, presumably from Mexico. Immature stages and adults feed on new foliage and cause significant leaf drop when population numbers are high; fruit feeding causes direct damage and economic losses. Currently, AT is not known to occur outside California and Mexico.

AVOCADO WHITEFLY *Trialeurodes floridensis* (Quaintance) [Hemiptera: Aleyrodidae].

AWARD® See Fenoxycarb.

AXENIC Adj. (Greek, *a* = without; *xenos* = stranger; *-ic* = characterized by.) Developing or reared under microbe-free conditions.

AXIAL Adj. Descriptive of or pertaining to an axis.

AXIAL FILAMENT See Axoneme.

AXIAL POSITION See Head Orientation.

AXIAL PRONG See Palpal Claw.

AXIIDAE Plural Noun. A small Family of ditrysian Lepidoptera assigned to Superfamily Axioidea or Geometroidea in some classifications. See Axioidea.

AXILLA Noun. (Latin, *axilla* = armpit. Pl., Axillae.) 1. The convex vertical or subvertical area of the Scutum between the Medacoria and Caudacoria (MacGillivray). 2. Hymenoptera: Two small, subtriangular sclerites at the lateral basal angles of the Mesoscutellum. See Axillula.

AXILLARIES See Axillary Sclerites.

AXILLARIS Noun. (Latin, *axilla* = armpit.) The second and third Anal Veins of Comstock *teste* Enderlein.

AXILLARY Adj. (Latin, *axilla* = armpit.) Placed in the angle of origin of two structures; arising from the angle of ramification.

AXILLARY AREA See Anal Area.

AXILLARY CALLI See Calli Axillares.

AXILLARY CELL Diptera: The second Anal Cell *sensu* Comstock; the area behind the Anal Vein *sensu* Curran.

AXILLARY CORDS The produced posterior angles of the Notum that form the posterior margins of the basal membranes of the wings (Comstock).

AXILLARY EXCISION Diptera: An axillary incision or notch in the inner margin of the insect wing near its base (Comstock).

AXILLARY FURROW See Plica Jugalis.

AXILLARY INCISION Diptera: An incision on the inner margin of the wing, near the base, which separates the Alula from the main part.

AXILLARY LOBE Diptera: The sclerite covering the base of the wing. See Alula; Axillary Cell.

AXILLARY MUSCLES The only muscles attached directly on the wing bases (except Odonata) that arise on the Pleuron and are inserted on the first and third Axillary Sclerites (Snodgrass).

AXILLARY PLATE Odonata: The posterior sclerite of the wing base which supports the Subcostal, Radial, Medial, Cubital and Vannal Veins (Snodgrass).

AXILLARY REGION The region of the wing base containing the Axillary Sclerites (Snodgrass). AR is typically triangular and best developed in insects that flex their wings. See Wing. Cf. Anal Region; Jugal Region; Remigium.

AXILLARY SCLERITES Sclerites of the axillary region in wing-flexing insects. Flexion is achieved via muscles attached to a series of sclerites in the basal region of the wing. The First Axillary Sclerite serves as a hinge for the wing. The anterior portion of the First Axillary Sclerite is supported by the Anterior Notal Process. The proximal portion of the FAS usually forms a finger-like process whose apex is associated with the Subcostal Vein. The lateral surface of the FAS articulates with the proximal part of the Second Axillary Sclerite; the distal portion articulates with the Third Axillary Sclerite. Supplemental points of articulation include the median and postmedian processes. The SAS is variable in shape; its anterior portion is connected to the proximal portion of the Radial Vein. The SAS articulates with the Proximal Median Plate and Third Axillary Sclerite. The SAS is pivotal and rests upon the Pleural Wing Process. The TAS is variable in shape and typically includes three processes. The anterior process articulates with the SAS; the proximal process is connected to the Posterior Notal Process by a ligament; the distal process articulates with the base of the Anal Vein(s). The TAS provides attachment for the muscle that flexes the wing. Some authors believe that the axillary sclerites are subdivisions of the basal portion of the wing. Under this scheme, the First Axillary Sclerite is a proximal separation of the Subcostal Vein; the Second Axillary Sclerite is a proximal separation of the Radial Vein; the Third Axillary Sclerite is a portion of the Anal Vein. The Proximal Median Plate attaches to the Median and Cubital Veins, and is probably a subdivision of them. The Distal Median Plate is somewhat more difficult to homologize in winged insects because it is presumably secondary in origin and not found in Ephemeroptera that hold their wings vertically over their thoracic dorsum or some insects that do not flex their wings (Odonata). See Wing Articulation. Rel. Median Plate.

AXILLARY VEIN 1. Ephemeridae: 1–2 longitudinal veins toward the inner margin from the Anal Vein. 2. Orthoptera: A group of 10–20 radiate veins that occupy the anal field (Smith). 3. Diptera: The second Anal Vein *sensu* Curran. See Vein; Wing Venation.

AXILLULA Noun. Latin, *axilla* = armpit. Pl., Axillulae.) Hymenoptera: A small sclerite lateral of the Mesoscutellum and posterior of the Axilla. See Axilla.

AXILLULAR Adj. (Latin, *axilla* = armpit.) Pertaining to the Axillula.

AXIMIIDAE Plural Noun. A small Family of nematocerous Diptera assigned to Division Bibionomorpha. See Bibionomorpha; Nematocera.

AXIS Noun. (Latin, *axis* = axle. Pl., Axes.) 1. Any process at the base of a structure upon which it turns. 2. An imaginary central line of a structure, around which the structure is constructed or around which it turns or rotates. Axes through the body are determined by orientation or shape.

Often, details of internal anatomy must be reported. For insects that are bilaterally symmetrical, relative positions are defined by sections through the body. The sagittal plane is vertical, longitudinal and passes through the imaginary midline; several parasagittal planes pass parallel to the midline in the same plane as the sagittal section. If the insect body is cut in infinitely thin sections, then only one sagittal section and an infinite number of parasagittal sections exist. A transverse plane (cross section) is any one of several mutually parallel planes that pass through the body at a right angle to the long axis and at a right angle to the sagittal plane. In theory, one section bisects the insect into an anterior and posterior half. In reality, morphologists do not consider this relevant and view many cross sections. A frontal plane (horizontal plane) is any one of several mutually parallel planes that pass through the body at a right angle to the transverse plane and sagittal plane. See Aspect; Organization; Orientation. Cf. Primary Axis; Secondary Axis; Tertiary Axis. Rel. Plane; Section.

AXIS-CYLINDER Part of a nerve-fibre composed of Fibrillae and covered with a membrane sheath (Folsom & Wardle).

AXON Noun. (Greek, *axon* = axle. Pl., Axons.) The principal process or nerve fibre of a neurone (neurite) (Snodgrass). Alt. Axone.

AXONAL Adj. (Greek, *axon* = axle; Latin, *-alis* = pertaining to, having the character of.) Pertaining to an axon. See Axon; Nerve.

AXONEME Noun. (Greek, *axon* = axle; *nema* = thread. Pl., Axonemes.) 1. Animals: An anatomically and functionally complex organelle that is the basic ciliary unit of somatic cells. 2. Insects: The propulsive engine within the tail of an insect spermatozoon that causes movement of the Flagellum (tail) and thereby locomotion. Ultrastructure of Axoneme reveals a series of microtubules that extend the length of the sperm tail. Sychronized contraction of microtubules cause undulatory movement (bending) of the Flagellum that translates into spermatozoon motility. Cross-section through sperm tail shows two large Central Tubules surrounded by a series of smaller microtubules (Accessory Doublets, Accessory Tubules); a Central Sheath surrounds Central Tubules. The CTs ADs are found in most insect Spermatozoa. Structure of ADs is similar to CTs; CTs ca 200–300 Å in diameter and ADs are considerably smaller. A ring of Peripheral Singlets may surround the accessory doublets. PS may occupy a position similar to the singlets in mammalian Spermatozoa; PS of insects are hollow but PS in mammals are dense or solid. See Microtubule; Sperm Anatomy; Sperm Flagellum. Rel. Accessory Doublets; Central Tubules; Peripheral Singlets.

AXOR® See Lufenuron.

AXYMYIIDAE Plural Noun. A small Family of nematocerous Diptera assigned to Division Bibionomorpha. See Bibionomorpha; Nematocera.

AYYAR, P N KRISHNA See KRISHNA AYYAR.

AYYAR, T V RAMAKRISHNA (See RAMAKRISHNA AYYAR.) Government Entomologist stationed in Coimbatore, India.

AYZENBERG, EVGENIYA ERSEYEVICH (1899–1967) (Shaposhnikov 1968, Ent. Obozr. 47: 947–949 (trans. Ent. Rev. Wash. 47: 580–581.))

AZADIENO® See Amitraz.

AZADIRACHTIN Noun. (New Latin, from generic name *Azadirachta*. Pl., Azadirachtins.) A complex of several triterpenoids derived from the seeds, leaves, wood and bark of the Indian Neem tree (*Azadirachta indica* A. Jussieu) of India, Philippine Neem Tree *Azadirachta excelsa* (Jack) Jacobs, and chinaberry tree (*Melia azedarach* Linnaeus) of Asia [Meliaceae]. Principal insecticidal compound, Azadirachtin A, is derived from seeds and leaves of *A. indica*. Azadirachtin is used as an oviposition deterrent for some insects, feeding-deterrent for chewing insects and growth inhibitor or Internal Growth Regulator. IGR affects ecdysteroid and juvenile hormone titres and thus interferes with neuroendocrine control of metamorphosis. Identity and mode-of-action of other triterpinoid compounds from neem not clearly understood. Formulated and sold under name 'Margosan' for use in greenhouse, nurseries and crops in some countries; target pests include Lepidoptera larvae. Azadirachtin is unstable in UV light, at high temperature and under high humidity. Trade names include: Align®, Azatin®, Bioneem®, Bionim®, Margosan-O®, Meen®, Neem Benefit®, NeemAzal-T/S@, Neemisis®, Neemix®, Trilogy®, Turplex®. See Neem; Neem Oil; Botanical Insecticide; Natural Insecticide; Insecticide; Pesticides.

AZALEA BARK-SCALE *Eriococcus azaleae* Comstock [Hemiptera: Eriococcidae]: A pest of rhododendron, azalea and Japanese adromeda in North America. Adult elongate-oval and white; honeydew produced by ABS provides a substrate for sooty mould. See Eriococcidae.

AZALEA CATERPILLAR *Datana major* Grote & Robinson [Lepidoptera: Notodontidae].

AZALEA LACE-BUG *Stephanitis pyrioides* (Scott) [Hemiptera: Tingidae]: A pest of azalea in North America that causes discoloration of upper leaf surface. ALB overwinters in the egg stage.

AZALEA LEAFMINER *Caloptilia azaleella* (Brants) [Lepidoptera: Gracillariidae]: A multivoltine pest of azalea in North America. Larva mines leaves, then folds edge and feeds on surface; ALM overwinters as a larva. See Gracillariidae. Cf. Lilac Leafminer.

AZALEA PLANT-BUG *Rhinocapsus vanduzeei* Uhler [Heteroptera: Miridae].

AZALEA WHITEFLY *Pealius azaleae* (Baker & Moles) [Hemiptera: Aleyrodidae].

AZALEA WHITE-MITE *Eotetranychus clitus* Prichard & Baker [Acari: Tetranychidae].

AZAMETHIPHOS An organic-phosphate (thiophosphoric acid) compound {s-(6-chloro-2-oxooxazolo(4,5-6)pyridin-3(2H)-yl)methyl)O,O-dimethyl phosphorothioate} used as a contact insecticide and stomach poison against public-health pests in some countries. Compound applied as a paint to barns and other buildings; also applied as surface spray for control of stored-grain pests, cockroaches, spiders and flies. Azamethiphos is not registered for use in USA, in part because it is toxic to bees. Trade names include Alfacron® and Snip®. See Organophosphorus Insecticide.

AZARA, FELIZ D' (1746–1821) (Rose 1853, *New General Biogeographical Dictionary* 2: 416; Larousse 1865, *Grand Dictionnaire Universal du XIX siècle.* 1: 1106.)

AZATIN® A broad-spectrum biopesticide that can be applied to cole crops, bulb vegetables, citrus, curcurbits and other vegetables, tropical vegetables, stone fruits and other crops. Active ingredient 3% Azadirachtin. See Azadirachtin.

AZEVEDO MARQUES, LUIS ANGUSTO See MARQUES, A. L. A.)

AZIINOS® See Azinphos-Ethyl.

AZINPHOS-ETHYL An organic-phosphate (dithiophosphate) compound {O,O-dimethyl-S-((4-oxo-3H-1,2,3-benzotriazin-3-yl)-methyl)-dithiophosphate} used as a contact insecticide, stomach poison and acaricide against spider mites, aphids, thrips, plant-sucking insects, leaf-chewing insects and bollworms. Compound applied to fruit, cotton, corn, citrus, grapes, onions, potatoes, rice, tomatoes, vegetables and ornamentals. The compound is not registered for use in USA in part because it is toxic to bees and fishes. Trade names include: Acifon-E®, Aziinos®, Bionex®, Cotnion-Ethyl®, Gusathnion A®, Triaotion®. See Organophosphorus Insecticide.

AZINPHOS-METHYL An organic-phosphate (dithiophosphate) compound {O,O-dimethyl S-(4-oxo-1,2,3-benzotriazin-3 (4H)-ylmethyl) phosphorodithioate} used as a contact insecticide, stomach poison and acaricide against mites, aphids, thrips, plant-sucking insects, leaf-chewing insects, armyworms and bollworms. Compound applied to fruit trees, alfalfa, cotton, cucumbers, cherries, grapes, onions, peppers, potatoes, rice, tomatoes, strawberries, vegetables, ornamentals. Compound phytotoxic to hawthorn and American linden, as well as toxic to bees, wildlife and fishes. Trade names include: Acifon®, Azitox®, Carfene®, Cotnion-Methyl®, Gothnion®, Gusathion®, Guthion®, Valefos®. See Organophosphorus Insecticide.

AZITOX® See Azinphos-Methyl.

AZOCYCLOTIN A synthetic, heterocyclic tin compound {1-tricyclohexylstannyl)-1H-1,2,4-triazole} used as a contact Acaricide. Compound is applied to vegetables, cotton, grapes, fruit and citrus in some countries. Azocyclotin is not registered for use in USA. Toxic to fishes but does not affect bees; long-term effectiveness on treated plants. Trade names include Clearmite®, Clermait® and Peropal® See Insecticide.

AZODRIN® See Monocrotophos.

AZOFENE® See Phosalone.

AZOTINAE Nikolskaya 1966. A Subfamily of Aphelinidae. Adult Antenna with seven segments, Flagellum frequently bicolorous, short third Flagellomere and unsegmented club; Pronotum entire, Prepectus of two segments; forewing Radial Vein elongate with long marginal fringe; gastral Terga IX, X medially divided; apical Sternum trapeziform, nearly projecting to apex. Larvae develop as egg parasites of Lepidoptera and Cicadellidae or developed as hyperparasites of Diaspididae and Aleyrodidae.

AZTEC® See Triazamate.

AZUKI BEAN-WEEVIL *Callosobruchus chinensis* Linnaeus [Coleoptera: Bruchidae].

AZURE Adj. (Arabic, *lazaward* = lapis lazuli, blue). Clear sky-blue. Alt. Azureus. Cf. Caeruleus.

AZYGOUS Adj. (Greek, *a* = without; *zygoun* = to yoke; Latin, *-osus* = possessing the qualities of.) Unpaired; without a partner. A term sometimes applied to an unpaired oviduct, specifically the enlarged portion of the Vagina at the junction of the Lateral Oviducts.

B_{12} A vitamin required by vertebrates for metabolism of propionate.

BABCOCK, AMORY LELAND (1826–1903) (Morse 1903, Psyche 10: 187.)

BABESIASIS Noun. (Patronym of V. Babes, Rumanian pathologist (1854–1926) *Babesia*; Greek suffix *-iasis* = disease. Pl., Babesiases.) A medical condition characterized by the presence of the parasitic protozoan Genus *Babesia* in a vertebrate host and typified by a piroplasm. See Babesiosis. Cf. Theileriasis.

BABESIIDAE Plural Noun. A Family of parasitic Protozoa assigned to the Subclass Piroplasmasina and including the Genus *Babesia*. Babesiidae lack pigment, exhibit a piroplasm, undergo merogony within erythrocytes or lymphocytes of vertebrate host, and gametogony and sporogony within ticks. Babesiidae include about 100 *Babesia* Species that have been associated with carnivores, rodents and ruminants. See Babesiosis.

BABESIOSIS Noun. (Babesia; Greek suffix *-iasis* = disease. Pl., Babesioses.) A term that includes several clinical diseases, primarily of domesticated mammals, caused by infection with *Babesia*. Babesiosis is vectored by ixodid ticks; a tick becomes infected while feeding upon diseased host, through trans-oviarian transmission or trans-stadial transmission. *Babesia* replicates by schizogony in vertebrate, gamogony within tick intestine, and sporogony in tick salivary glands. See Bovine Babesiosis; Canine Babesiosis; Equine Babesiosis. Cf. Theileriosis.

BABINGTON, CHARLES CARDALE (1808–1895) (J.B.B. 1896, Proc. Roy. Soc. Lond. 59: viii–x.)

BACCATE Adj. (Latin, *bacca* = berry; *-atus* = adjectival suffix.) Berry-like. Term applied to bladder-like ovaries from the surface of which the short ovarian tubules arise. Alt. Bacciform; Baccatus.

BACCHUS, ARTHUR DOUGLAS (–1912) (Anon. 1924, Entomol. Mon. Mag. 60: 157–158.)

BACCIFORM Adj. (Latin, *bacca* = berry; *forma* = shape.) Berry-shaped. Alt. Baccate. See Berry. Cf. Morula. Rel. Form 2; Shape 2; Outline Shape.

BACH, MICHAEL (1808–1878) (Osborn 1952, *A Brief History of Entomology*. 305 pp. (176), Columbus, Ohio.)

BACHINGER, I (1856–1939) (Heikertinger 1939, Koleopt. Rdsch. 25: 206.)

BACHMETEV, PORTRIG IVANOVICH (1860–1913) (Kuhnt 1914, Dt. Entomol. Z. 1914: 92.)

BACILEX® A registered biopesticide derived from *Bacillus thuringiensis* var. *kurstaki*. See *Bacillus thuringiensis*.

BACILLUS CHITINIVOROUS The soil microorganism that attacks and breaks up the chitinous skeleton of dead insects in the soil (Folsom & Wardle).

BACILLUS POPILLIAE A living bacterium that is used as a biopesticide in the control of scarab larvae, particularly the Japanese Beetle. BP is applied to grass, turf and pasture; it is regarded as non-phytotoxic and not toxic to warm-blooded animals. BP was first commercially marketed by Fairfax Biological Labs (1948) and subsequently by other companies. Trade names include: Doom®, Grub Attack®, Grub Killer®, Japidemic® and Milky Spore®. See Biopesticide. Cf. *Bacillus sphaericus; Bacillus thuringiensis*.

BACILLUS SPHAERICUS A bacterium used as an experimental biopesticide in the control of mosquito larvae. BS was developed by Abbott Labs and Novo Labs ca 1986, but is not available in USA. Trade names include: ABG-6185®, Farimos®, Spherimos® and Vectrolex® See Biopesticide. Cf. *Bacillus popilliae; Bacillus thuringiensis*.

BACILLUS THURINGINESIS Berliner. A bacterium used as a biopesticide in the control of numerous chewing-insect pests, particularly Lepidoptera larvae with alkaline pH in gut. Arguably the most popular naturally-derived product used for insect-pest control. BT is applied to virtually all tree, field and vegetable crops. Photosensitive with life limited to a few days at most; only effective when ingested by insect; acts as stomach poison; most effective against early larval instars. First marketed commercially by Nutrilite Products (1961) and subsequently by numerous chemical companies throughout the world. BT genes responsible for endotoxin incorporated into cotton genome to develop transgenic cotton. BT marketed as various strains or varieties under many trade names including: Aatack®, Able®, Acrobe®, Agree®, Agritol®, Ascetasporin®, Bacilex®, Bactimos®, Bactir®, Bactis-G®, Bactis-P®, Bactis-S®, Bactospeine®, Bactucide®, Baturad®, Bennam IV®, Berman I®, Berman III®, Bernam I®, Bernam II®, Biobit®, Biocot®, Biodart®, Biolar®, Biotrol®, Bitoxibacillin®, BMC®, BTB®, BTV®, Bug-Time®, Certan®, Collapse®, Condor®, Crystaline®, Cutlass®, Delfin®, Delta BT®, Design®, Dibeta®, Dipel®, Ecotech®, Ecotech Extra®, Ecotech Pro®, Florbac®, Foil®, Foray®, Full-Bac®, Futura®, Gnatrol®, Gomelin®, Gut-Buster®, Javelin®, Larvatrol®, Larvo-BT®, Laser®, Lepid®, Lepidocide®, Leptox®, Magnam®, M-Peril®, M-Trak®, Moskitocid®, MVP®, Novabac-3®, Novodor®, Nubilacid®, Resecticid®, Skeetocid®, Sperimos®, Sporeine®, Stan-Guard®, Steward®, Teknar®, Thuricide®, Thuringiensin®, Tribactur®, Turex®, Vault®, Vectobac®, Vectocide®, Victory®, Worm Buster® and Xentari®. See Biopesticide. Cf. *Bacillus popillae; Bacillus sphaericus*.

BACK OFF-1® See *Metarhizium anisopliae* ESF 1.

BACK, ERNEST ADNA (1880–1959) American economic Entomologist (USDA); 1913–1918 served as inspector for Mediterranean Fruit Fly in Hawaii; most of career devoted to stored products and their pests. (Mallis 1971, *American Entomologists*. 549 pp., New Brunswick.)

BACKMUND, FRITZ (1901–1975) (Anon. 1975,

Prakt. Schädlingsk. 27: 74.)

BACKSWIMMERS See Notonectidae.

BACLASH® See Cycloprothrin.

BACON BEETLE See Larder Beetle.

BACON FLIES See Piophilidae.

BACOT, ARTHUR WILLIAM (1886–1922) (Page 1923, Entomol. Rec. J. Var. 34: 99–100.)

BACTERIAL ROT A bacterial disease of apples caused by *Pseudomonas melophthora* and transmitted by the Apple Maggot. Adult flies transmit the bacterium while ovipositing. See Apple Maggot.

BACTERIAL SOFT ROT A bacterial disease of vegetables caused by *Erwinia carotovora* and transmitted by maggots that innoculate seeds and roots.

BACTERICIDAL Adj. (Greek, *bakterion* = little rod; Latin, *caedere* = to kill; *-alis* = having the character of.) Bacteria-killing. Alt. Bacteriocidal.

BACTERIOD Adj. (Greek, *bakterion* = rod; *eidos* = form.) Bacteria-like. Descriptive of organisms that appear or act as bacteria. See Bacterium. Cf. Viroid.

BACTERIOLOGY Noun. (Greek, *bakterion* = small rod; *logos* = discourse. Pl., Bacteriologies.) The study of Bacteria. Bacteriology was regarded as a subdiscipline of Botany in the past, but contemporary science regards it as a distinct science. See Bacteria. Cf. Virology. Rel. Entomology; Parasitology.

BACTERIOLYSIS Noun. (Greek, *bakterion* = small rod; *lysis* = to loosen. Pl., Bacteriolyses.) The process or phenomenon by which bacteria are killed through dissolution. See Lysis. Cf. Histolysis; Hydrolysis.

BACTERIOPHAGE Noun. (Greek, *bakterion* = small rod; *phagein* = to devour. Pl., Bacteriorphages.) A virus that multiplies within bacteria. Syn. Phage. Rel. Biotechnology.

BACTERIOSIS Adj. (Greek, *bakterion* = small rod; *-sis* = a condition or state.) Any morbid or disease-condition caused by bacteria.

BACTERIOSTATIC Adj. (Greek, *bakterion* = small rod; *statikos* = causing to stand; *-ic* = characterized by.) Pertaining to the inhibition of bacterial development.

BACTERIUM Noun. (Greek, *bakterion* = small rod; Latin, *-ium* = diminutive > Greek, *-idion*. Pl., Bacteria.) Microscopic, single-celled prokaryotic organisms that lack chlorophyll, display a cell wall and are capable of rapid reproduction via fission, asexual spores and sexual process. Bacteria occur in other organisms, water and soil. Some kinds of bacteria occur as a natural component of the insect body; other bacteria are pathogenic of insects and may be used in biological control. Cf. Fungus; Rickettsiae; Virus. Rel. Symbiosis; Pathogen.

BACTEROID Noun. (Greek, *bakterion* = small rod; *eidos* = form. Pl., Bacteroids.) 1. Bacterium-like in form or action. 2. An irregularly-shaped form of certain bacteria, especially those formed in root cells during nodule formation. Alt. Bacterioid. See Bacterium. Cf. Parasitoid; Viroid. Rel. Eidos; Form; Shape.

BACTIMOS® A registered biopesticide derived from *Bacillus thuringiensis* var. *israelensis*. See *Bacillus thuringiensis*.

BACTIR® A registered biopesticide derived from *Bacillus thuringiensis* var. *kurstaki*. See *Bacillus thuringiensis*.

BACTIS-G® A registered biopesticide derived from *Bacillus thuringiensis* var. *israelensis*. See *Bacillus thuringiensis*.

BACTIS-P® A registered biopesticide derived from *Bacillus thuringiensis* var. *israelensis*. See *Bacillus thuringiensis*.

BACTIS-S® A registered biopesticide derived from *Bacillus thuringiensis* var. *israelensis*. See *Bacillus thuringiensis*.

BACTOSPEINE® A registered biopesticide derived from *Bacillus thuringiensis* var. *kurstaki*. See *Bacillus thuringiensis*.

BACTUCIDE® A registered biopesticide derived from *Bacillus thuringiensis* var. *kurstaki*. See *Bacillus thuringiensis*.

BACULATE Adj. (Latin, *baculum* = rod; *-atus* = adjectival suffix.) Descriptive of surface sculpture, usually the insect's Integument, that displays rod-like or post-like projections. See Sculpture Pattern. Cf. Alveolate; Carinate; Clavate; Echinate; Favose; Gemmate; Psilate; Punctate; Reticulate; Rugulate; Scabrate; Shagreened; Smooth; Striate; Verrucate.

BACULIFORM Adj. (Latin, *baculum* = rod; *forma* = shape.) Rod-shaped; descriptive of structure that is elongate and shaped as a rod (staff). See Rod. Cf. Palpiform. Rel. Form 2; Shape 2; Outline Shape.

BACULOVIRIDAE Plural Noun. A Family of pathogenic viruses whose members are typified by a rod-shaped envelope (nucleocapsid) and relatively large, circular, double-stranded DNA. Baculoviridae are divided into two Genera (Nuclearpolyhedrosis viruses and Granuloviruses) or three groups (nuclear polyhedrosis virus, granulosis virus and nonoccluded-type virus) depending upon classification. Baculoviruses are infectious only to arthropods and most attack Lepidoptera. Baculovirids typically product two virion phenotypes (Budded Virion and Occlusion Derived Virion) at different times and different locations within the infected cell. See Virus. Cf. Cytoplasmic Polyhedrosis Virus; Insect Poxvirus; Parvoviridae.

BACULOVIRUS Noun. (Latin, *baculum* = rod; *virus* = poisonous liquid. Pl., Baculoviruses.) A member of the Baculoviridae. See Virus. Rel. Pathogen.

BADEN, JOHAN ANDRE FERDINAND (1828–1914) (Weidner 1967, Abh. Erh. Naturw. Ver. Hamburg Suppl. 9: 185.)

BADGE HUNTSMAN SPIDERS *Olios* spp. [Aranaea: Heteropodidae].

BADIUS Adj. (Latin, *badius* = chestnut coloured.) Bay-coloured; the colour of a bay horse.

BAENOMERE Noun. (Greek, *baino* = to walk, step, go; *meros* = part. Pl., Baenomeres.) An arcane term for a leg-bearing (thoracic) segment.

BAENOPOD Noun. (Greek, *baino* = to walk, step, go; *pous* = foot. Pl., Baenopoda; Baenopods.) An obsolete term for a thoracic leg.

BAENOSOME Noun. (Greek, *baino* = to walk, step, go; *soma* = body. Pl., Baenosomes.) An obsolete term for the Thorax.

BAER, CARL ERNST VON (1792–1876) Russian Zoologist and founder of modern science of Embryology. Among his numerous accomplishments, Baer proposed the Germ-Layer Theory. (Oppenheimer 1940, Quart. Rev. Biol. 15: 2–4, bibliogr.)

BAER, GUSTAVE ADOLPHE (1839–1918) (Anon. 1918, Entomol. News. 29: 280.)

BAER, WILLIAM (1867–1934) (Zwolfer 1935, Z. Angew. Entomol. 22: 516.)

BAETIDAE Plural Noun. A large, widespread Family of baetoid Ephemeroptera whose adults are small bodied with reduced wing venation and hindwings small, narrow or absent. Baetids display sexual dimorphism in body size and coloration. Male compound eye large, divided with upper part sometimes turbinate; females with unmodified compound eyes. Male genitalia reduced; Penis lobes reduced to internal membranes. Nymphs small, slender and active swimmers; Abdomen with three caudal filaments. See Ephemeroptera.

BAETISCIDAE Plural Noun. A small Family of pannotous Ephemeroptera assigned to Superfamily Caenoidea. Adult moderate sized; front wing lacks Cubital intercalarys; Abdomen with two caudal filaments (Cerci). Naiads occur in cool, flowing streams. See Ephemeroptera.

BAETOIDEA A Superfamily of schizonotous Ephemeroptera including Baetidae, Siphlonuridae, Heptageniidae, Metretopodidae and Oligoneuridae. See Ephemeroptera.

BAG MOTHS See Psychidae.

BAGDAD BOIL See Oriental Sore. Syn. Delhi Boil.

BAGNALL, RICHARD SIDDOWAY (1889–1962) (Mound 1968, Bull. Br. Mus. Nat. Hist. (Entomol.) Suppl. 11: 5–6.)

BAGPIPES CICADA *Lembeja paradoxa* (Karsch) [Hemiptera: Tibicinidae].

BAG-SHELTER MOTH *Ochrogaster lunifer* Herrich-Schaeffer [Lepidoptera: Notodontidae].

BAGWELL PUREFOY, EDWARD (1868–1960) (Uvarov 1961, Proc. Roy. Entomol. Soc. Lond. (C) 25: 49.)

BAGWORM *Thyridopteryx ephemeraeformis* (Haworth) [Lepidoptera: Psychidae]: A monovolitne, polyphagous pest of deciduous and evergreen trees that occurs from Canada to Argentina. Adults are dimorphic with male body black and wings pale; female wingless and remains within their bag. Male is attracted to female in bag and then mates; female oviposits in bag, emerges and dies. Bagworm overwinters as an egg within female's bag. Neonate larva emerges from mother's bag and constructs new bag of silk that it carries and enlarges throughout larval development; larva projects head from base of bag and feeds on host plant foliage; bits of foliage are woven into the bag by the developing larva; physical extraction of larva from bag is impossible without killing larva. Pupation occurs within bag. See Psychidae.

BAGWORM MOTHS See Psychidae.

BAHAMAN SWALLOWTAIL *Papilio andraemon bonhotei* Sharpe [Lepidoptera: Papilionidae].

BAIKIE, WILLIAM BALFOUR (1825–1864) (Pascoe 1866, Proc. Entomol. Soc. Lond. (3) 2: 71.)

BAILEY, FREDERICK MARSHAM (1882–1967) (Anon. 1967, The Times, London 19.4.1967.)

BAILEY, JAMES HAROLD (1870–1909) (Tomlin 1909, Entomol. Mon. Mag. 45: 260–261.)

BAILEY, JAMES SPENCER (1830–1883) (Anon. 1884, Bull. Brooklyn Entomol. Soc. 6: 48.)

BAILEY, JOHN W (1895–1967) (Anon. 1968, Bull. Entomol. Soc. Amer. 14 (1): [i].)

BAILLY, M H G (–1958) (Anon. 1958, Bull. Soc. Entomol. Fr. 63: 57.)

BAINBRIDGE, WILLIAM GEORGE (1867–1935) (Anon. 1935, Arb. Morph. Taxon. Entomol. Berl. 2: 308.)

BAINBRIGGE, THOMAS FLETCHER. See FLETCHER, T. B.

BAIRD, SPENCER FULLERTON (1823–1887) American naturalist and founder of United States National Museum. (Dos Passos 1951, J. N. Y. Entomol. Soc. 59: 136.)

BAIRD, WILLIAM (1803–1872) (Anon. 1872, Proc. Roy. Soc. Lond. 1871–1872: xxiii–xxiv.)

BAIRSTOW, SAMUEL DENTON (–1889) (Anon. 1901, Entomol. Jb. 10: 243.)

BAISAS, FRANCISCO E (1896–1973) (Knight & Pugh 1974, Mosquito Syst. 6: 74–77, bibliogr.)

BAISSIDAE Rasnitsyn 1975. A Family of parasitic Hymenoptera assigned to the Evanioidea; sometimes placed within the Gasteruptiidae. See Evalioidea.

BAISSODIDAE Rasnitsyn 1975. A Family of aculeate Hymenoptera assigned to the Superfamily Sphecoidea. See Aculeata; Sphecoidea.

BAIT Noun. (Middle English, from Old Norse, *beita* = food. Pl., Baits.) 1. A pesticide formulation that combines an edible or attractive substance with a pesticide. Designated as '**B**' on USA Environmental Protection Agency labels and Material Safety Data Sheet documentation. 2. A lure intended to attract specific organisms.

BAJARI, ELIZABETH (1912–1963) (Moczár 1964, Ann. Hist.-Nat. Mus. Natn. Hung. 56: 5–7, bibliogr.)

BAKER, ALBERT WESLEY (JACK) (1891–1974) (Anon. 1975, Proc. Entomol. Soc. Ont. 105: 2.)

BAKER, ALEX D (1894–1974) (Anon. 1975, Bull. Ent. Soc. Can. 7: 68)

BAKER, ARTHUR C (1875–1959) (Anon. 1959, Bull Entomol. Soc. Amer. 5: 154.)

BAKER, CHARLES FULLER (1872–1927) (Born Lansing, Michigan; died Manila, PI.) American academic; educated Michigan Agricultural College (1892); worked with Gillette in Colorado and Cook at Pomona College; served institutions in Cuba and Brazil; appointed Professor of Agronomy, University of Philippines 1912 and subsequently appointed Dean. Baker was a general collector with prodigious efforts and made extensive collections of Western Pacific insects. (Mallis 1971, *American Entomologists*. 549 pp., New Brunswick.)

BAKER, GEORGE (1830–1913) (Sheldon 1913, Entomologist 46: 120.)

BAKER, WILLIAM A (1899–1973) (Anon. 1973, Bull. Entomol. Soc. Amer. 19 (3): [i].)

BAKERS' ITCH A hypersensitivity in humans to stored products caused by the mite *Acarus siro* Linnaeus. Syn. Grocers' Itch.

BAKEWELL, ROBERT (1810–1867) (Musgrave 1932, *A Bibliogr. of Australian Entomology 1775–1930,* 380 pp. (10), Sydney.)

BAKKENDORF, O (1893–1972) (Tuxen 1972, Entomol. Meddr 40: 75–80, bibliogr.)

BALA® See Alphacypermethrin.

BALANCE OF NATURE A fluctuating biotic equilibrium in which interdependent organisms bear quantitative relations each to all the others, as modified by variable elements, such as weather, food, physiography, vegetation, *etc.* (Folsom & Wardle).

BALANCERS See Halteres.

BALASSOGLO, VLADIMIR ALEXANDROVICH (1841–1900) (Jacobson 1901, Horae Soc. Entomol. Ross. 35: li–lvi.)

BALBIANI, EDOUARD GÉRARD (1823–1899) (Anon. 1901, Entomol. Jb. 10: 238.)

BALBIANI RINGS Diptera: The largest puffs observed on polytene chromosomes. Named in honor of the 19th century microscopist who discovered polytene chromosomes.

BALD Adj. (Middle English, *balde*.) Structure without Setae or other surface vestiture, especially in spots. Syn. Hairless; Smooth. Cf Calvescens; Glabrous.

BALDASSERONI, VINCENZO (1884–1963) (Zocchi 1964, Monti Boschi 2: 76)

BALD-CYPRESS CONEWORM *Dioryctria pygmaeella* Ragonot [Lepidoptera: Pyralidae].

BALDFACED HORNET *Dolichovespula maculata* (Linnaeus) [Hymenoptera: Vespidae]: Probably the most common vespid wasp in North America and not a true hornet. Queen establishes her colony during late spring; workers are ca 15–16 mm long and construct paper nest exposed and suspended from branch of tree or overhang and covered by thick, multi-layered envelope of macerated fibre. The nest is large with 3–5 combs and 500–600 workers. Workers typically act as aerial predators of other insects, may take sweets or scavenge for meat. Alt. Bald-Face Hornet. See Hornet. Cf. Aerial Yellowjacket; Yellowjacket.

BALDI, EDGARDO (1899–1951) (Brunelli 1951, Boll. Pesca Piscic. Idrobiol. 6: 214–217, bibliogr.)

BALDINGER, ERNST GOTTFRIED (1738–1804) (Larousse 1865, *Grand dictionnaire universale du XIX siècle* 2: 98.)

BALDUF, WALTER VALENTINE (1889–1969) (Anon. 1970, Ohio J. Sci. 70: 239.)

BALDUS, KARL (1898–1927) (Spek 1927, Zool. Anz. 73: 48.)

BALFOUR, ALICE B (–1936) (Imms 1937, Proc. Roy. Entomol. Soc. Lond. (C) 1: 54–55.)

BALFOUR, ANDREW (1873–1931) English Peer of Wellcome Research Laboratories and London School of Tropical Medicine. (P.H.M.B. 1931, Br. Med. J. 1: 245–246.)

BALFOUR-BROWNE, WILLIAM ALEXANDER FRANCIS (1874–1967) (Waterston 1968, Yrb. Roy. Soc. Edinb (Sess. 1966–67): 8–10.)

BALIANI, ARMANDO (1874–1945) (Guiglia *et al.* 1969, Memorie Soc. Entomol. Ital. 48: 668–669, bibliogr.)

BALKWILL, JOHN A (–1908) (Jarvis 1908, Rep. Entomol. Soc. Ont. 39: 34.)

BALL GALL-FLY *Eurosta solidaginis* (Fitch) [Diptera: Tephritidae]: A fly that oviposits on several Species of goldenrod in North America.

BALL, ELMER DARWIN (1870–1943) (Osborn 1937, *Fragments of Entomological History*. 394 pp. (198, portrait), Columbus, Ohio.)

BALL, ROBERT (1802–1857) (Patterson 1859, Proc. Dublin Univ. Zool. Bot. Ass. 1: 7–48, bibliogr.)

BALL Noun. (Middle English, *bal,* Old High German, *balla* = ball. Pl., Balls.) 1. A spherical or rounded object. 2. A small seed pod.

BALL-AND-SOCKET Morphology: A form of articulation in which an hemispherical surface fits into a cup-like depression. Ball-and-socket articulation permits rotation or movement in several planes. Cf. Acetabulum; Condyle; Cotyla.

BALLARD, JULIA PERKINS (–1894) (W.H.E. 1894, Can. Entomol. 26: 234.)

BALLENSTEDT, JOHANN GEORG JUSTUS (–1837) (Rose 1850, *New General Biographical Dictionary* 3: 84.)

BALLION, ERNST ERNESTOVICH (1816–1901) (Anon. 1902, Entomol. Jb. 12: 252.)

BALLOU, CHARLES H (1890–1961) (Anon. 1961, Bull. Entomol. Soc. Amer. (4): [i].)

BALLOU, HENRY ARTHUR (–1937) American economic Entomologist who worked for Imperial Department of Agriculture, West Indies, Barbados (1903–1937); Professor of Entomology, Imperial College of Tropical Agriculture (1922–1937). Ballou studied insects affecting cotton, sugarcane and cocoa; he wrote an important early article on Entomology of West Indies (West Indian Bull. 1911, 11 (4): 282–317; Anon. 1938, Mass. Sta. Coll. Fernald Club N. Y. 7: 26.)

BALSAM Noun. (Latin, *balsamum* = balsam. Pl.,

Balsams.) Any of several substances found in many Species of plants. Balsam exuded from wounds to protect the plant from insects and fungi. See Amber; Copal; Gum Arabaic; Gum Damar; Lac; Resin; Rosin.

BALSAM FIR SAWFLY *Neodiprion abietis* (Harris) (Hymenoptera: Diprionidae].

BALSAM FIR SAWYER *Monochamus marmorator* Kirby [Coleoptera: Cerambycidae].

BALSAM GALL MIDGE *Paradiplosis tumifex* Gagné [Diptera: Cecidomyiidae].

BALSAM SHOOT-BORING SAWFLY *Pleroneura brunneicornis* Rohwer [Hymenoptera: Xyelidae].

BALSAM TWIG-APHID *Mindarus abietinus* Koch [Hemiptera: Aphididae].

BALSAM WOOLLY-ADELGID *Adelges piceae* (Ratzeburg) [Hemiptera: Adelgidae]: A serious pest of fir trees in North America and Europe. Introduced from Europe to North America about 1900. Salivary secretions injected into cambium layer cause hyperplasia and hormonal imbalance for infested trees. Sistentes are wingless with long stylets, a darkly sclerotized Cuticle with four instars and two generations in Canada. Overwintering generation is called the 'hiemosistens'; summer generation is called 'aestivosistens.' BWA eggs hatch; crawlers settle, feed, and then enter first instar diapause which is obligatory for hiemosistens and facultative for aesitvosistens. First instar is called 'neosistens' in both generations. Aestivosistens adult lays ca 50 eggs during September which gives rise to hiemosistens generation. In early spring feeding begins, moulting occurs and adult hiemosistens lays about 200 eggs that give rise to aestivosistens. Syn. Balsam Woolly Aphid. See Adeligidae.

BALSBAUGH, E U (1933–1991) American academic (University of North Dakota) and taxonomic specialist on Chrysomelidae (Carlson & Schultz 1991, Proc. Entomol. Soc. Wash. 93: 996.)

BALTHASAR, VLADIMIR (1897–1967) (Heyrovsky 1967, Acta Entomol. Bohemoslovaca 64: 402–404, bibliogr.)

BALTIC AMBER Amber taken from the Baltic region of Europe, typically containing fossil deposits of Lower Oligocene age. Approximately 2,500 Species and 250 Families have been described from BA; More than 170 Species of Hymenoptera have been described from Baltic Amber. See Fossil. Cf. Burmese Amber; Canadian Amber; Chiapas Amber; Dominican Amber; Lebanese Amber; Taimyrian Amber.

BALY, JOSEPH SUGAR (1816–1890) (Musgrave 1932, *A Bibliogr. of Australian Entomology 1775–1930,* 380 pp. (11–12), Sydney.)

BAMBOO APHID *Melanaphis bambusae* (Fullaway) [Hemiptera: Aphididae].

BAMBOO BORER 1. *Chlorophorus annularis* (Fabricius) [Coleoptera: Cerambycidae]. See Bamboo Longicorn. 2. *Dinoderus minutus* (Fabricius) [Coleoptera: Bostrichidae].

BAMBOO LONGICORN *Chlorophorus annularis* (Fabricius) [Coleoptera: Cerambycidae]: A pest of bamboo and sugarcane in southeast Asia and the USA. Adult ca 14 mm long.

BAMBOO MEALYBUG *Chaetococcus bambusae* (Maskell) [Hemiptera: Pseudococcidae].

BAMBOO POWDERPOST-BEETLE *Dinoderus minutus* (Fabricius) [Coleoptera: Bostrichidae]: A significant tropicopolitan pest of bamboo and bamboo products. Larvae and adults cause damage. Egg stage 6 days; larval stage 21–76 days; pupal stage 4–5 days; adult stage 31–85 days. See Bostrichidae.

BAMBOO SPIDER-MITE *Schizotetranychus celarius* (Banks) [Acari: Tetranychidae].

BANANA APHID *Pentalonia nigronervosa* Coquerel [Hemiptera: Aphididae]: An important pest of *Musa* spp. in the Pacific and southeast Asia. BA causes crop loss through direct damage or as a vector of banana bunchy top virus. See Aphididae.

BANANA FLOWER-THRIPS *Thrips hawaiiensis* (Morgan) [Thysanoptera: Thripidae].

BANANA FRUIT-CATERPILLAR *Tiracola plagiata* (Walker) (Lepidoptera: Noctuidae]: A minor pest of *Citrus* in Australia; BFC also feeds upon banana, corn, passion fruit, pawpaw and pumpkin. BFC is endemic to southeast Asia. Adult ca 35 mm long, wingspan ca 60 mm; wings grey-brown with dark V-shaped mark along margin. Caterpillar to 60 mm long, pale brown with interrupted, narrow yellow-white band on side of body and two pairs of black markings on dorsum near posterior end. Infestations usually patchy and most frequent during late summer when larvae chew holes in young fruit; six larval instars. Pupation occurs in soil or leaf litter. BFC life cycle requires about three months; the Species is multivoltine with 3–4 generations per year. See Noctuidae.

BANANA FRUIT-FLY *Dacus musae* (Tryon) [Diptera: Tephritidae].

BANANA MITE *Phyllocoptruta musae* Keifer [Acari: Eriophyidae].

BANANA ROOT-BORER *Cosmopolites sordidus* (Germar) [Coleoptera: Curculionidae].

BANANA RUST-THRIPS *Chaetanoaphothrips signipennis* (Bagnall) [Thysanoptera: Thripidae].

BANANA SCAB-MOTH *Nacoleia octasema* (Meyrick) [Lepidoptera: Pyralidae]: A widespread pest of *Musa* and *Pandanus* that is endemic to the South Pacific and perhaps the Orient. BSM is characterized by several biotypes. Eggs are laid in clusters (to 30) on leaves or bracts; larvae feed primarily on inflorescence. See Pyralidae.

BANANA SILVERING-THRIPS *Hercinothrips bicinctus* (Bagnall) [Thysanoptera: Thripidae].

BANANA SKIPPER *Erionota thrax* (Linnaeus) [Lepidoptera: Hesperiidae]: A Species endemic to southeast Asia whose larva causes damage to bananas through feeding upon and rolling leaves. See Hesperiidae.

BANANA SPIDER-MITE *Tetranychus lambi* Pritchard & Baker [Acari: Tetranychidae]: Adult 1–2 mm long; body oval and reddish with several pairs of spots on either side of the body. Eggs less than 0.1 mm in diameter, transparent and reddish. Females lay their eggs on underside of leaves. Egg stage requires 3–7 days depending on temperature. First instar (larva) 0.1–0.2 mm long, pink and has six legs. Two nymphal stages follow. Nymphal mites are pale green or brown and have eight legs. Mites feed by inserting mouthparts into leaf tissue; feeding near veins produces yellow patches which progress to reddening of tissue and leaf death. Syn. Strawberry Spider-Mite. See Tetranychidae.

BANANA SPOTTING-BUG *Amblypelta lutescens lutescens* (Distant) [Hemiptera: Coreidae].

BANANA STALK-FLY *Teleostylinus bivittatus* Cresson [Diptera: Neriidae].

BANANA WEEVIL *Cosmopolites sorididus* (Germar) [Coleoptera: Curculionidae]: A pest originating in southeast Asia and which is widespread in tropical Africa and America. BW larvae bore into banana rhizomes, weaken older rhizomes and kill younger rhizomes; plants are affected by secondary rot organisms; sucker growth and production is reduced. Syn. Banana Weevil Borer. See Curculionidae.

BANCOL® See Bensultap.

BANCROFT, EDWARD (1744–1821) (Rose 1850, *New General Biographical Dictionary* 3: 105–106.)

BANCROFT, JOSEPH (1836–1894) Founder of Medical Entomology (1877) as discoverer of microfilarial worms from person survived in gut of mosquito. (Mackerras & Marks 1974, Changing patterns in Entomology. A symposium. 76 pp. (7), Sydney.)

BANCROFT, THOMAS LANE (1860–1933) Son of Joseph Bancroft, naturalist, M.D. and Medical Entomologist. In 1898 TLB demonstrated that filarial larvae complete development of life cycle to infective stage in *Culex fatigans.* He also believed that *Aedes aegypti* was implicated in transmission of Dengue Fever. TLB collected many insects, particularly mosquitoes and fruit flies for British Museum (Mackerras & Marks 1974, Changing patterns in Entomology. A symposium. 76 pp. (7–8), Sydney.)

BANCROFTIAN FILARIASIS A human disease caused by the onchocercid filarial worm *Wuchereria bancrofti.* BF has been reported in the neotropics (northeast Brazil, Guianas), Africa, and Indo-Pacific region. Adult worms occupy lymphatic vessels and nodes; female worms release microfilariae into host's circulatory system. Microfilariae are ingested along with the blood meal taken by female mosquitoes. Inside mosquito midgut, microfilariae shed their sheaths, penetrate midgut epithelium, enter haemocoel and migrate to flight muscles. Within flight muscles, worms shorten, undergo two moults and develop into infective larvae. These larvae enter the haemocoel, migrate to the mosquito's head, enter the Labium and subsequently are transmitted to the vertebrate host during feeding. Principal urban vector of BF is *Culex quinquefasciatus* and principal rural vectors include *Anopheles* spp. and *Aedes* spp. Disease develops slowly, often over decades, and if untreated can lead to Elephantiasis. See Brugian Filariasis.

BAND Noun. (Middle English, *band* from Old English, *bindan* = to bind; Pl., Bands.) 1. Something that constricts while permitting a limited amount of flexibility or movement. 2. A transverse marking, (usually coloration) wider than a line. 3. A stripe, streak or elongated mark on an animal.

BANDED ALDER BORER *Rosalia funebris* Motschulsky [Coleoptera: Cerambycidae].

BANDED ANTLION *Glenoleon pulchellus* Rambur [Neuroptera: Myrmeleontidae].

BANDED ASH CLEARWING *Podosesia aureocincta* Purrington & Nielsen [Lepidoptera: Sesiidae].

BANDED CATERPILLAR PARASITE *Ichneumon promissorius* Erichson [Hymenoptera; Ichneumonidae]: An Australian Species. Adult wingspan 12–15 mm; body elongate, black with cream markings; head black with Antenna longer than head and Thorax. Fore and middle legs pale; hind leg banded black and pale. Abdomen black with dorsum of first segment red; Pedicle produces a 'waist.' Pale cream coloration occurs on posterior portion of each abdominal segment. Female Ovipositor is short but eggs are deposited within the host. Larva lives within host feeding on organs. Pupation occurs within host. BCP use pupal stage of lepidopterans as their host. See Ichneumonidae.

BANDED CUCUMBER BEETLE *Diabrotica balteata* LeConte [Coleoptera: Chrysomelidae].

BANDED GREENHOUSE THRIPS *Hercinothrips femoralis* (O. M. Reuter) [Thysanoptera: Thripidae]: A pest of bean, sugarbeet, celery, cucumber, cotton and banana that is widespread in North America. Syn. Sugarbeet Thrips. See Thirpidae.

BANDED HICKORY BORER *Knulliana cincta* (Drury) [Coleoptera: Cerambycidae].

BANDED LICHEN MOTH *Eutane terminalis* Walker [Lepidoptera: Arctiidae].

BANDED SUNFLOWER MOTH *Cochylis hospes* Walsingham [Lepidoptera: Cochylidae]: A pest of cultivated sunflower.

BANDED THRIPS See Aeolothripidae.

BANDED-WINGED THRIPS See Aeolothripidae.

BANDED-WINGED WHITEFLY *Trialeurodes abutilonea* (Haldeman) [Hemiptera: Aleyrodidae]: In North America, a pest of cotton, beans and crucifers. Egg ca 1 mm long, yellowish; deposited on underside of leaf and eclosion within 5–6 days. Neonate nymph sheds legs after first moult; nymph ca 2 mm long, oval, pale green with white, sinuous wax filaments radiating from body margin. Nymphal and pseudopupal stage ca 10–12 days. Feeding by BWW causes premature defoliation; accumulation of honeydew can cause sooty mould on leaves and lint. Adult ca 1.6 mm long, white with three narrow, brown bands across wings. See Aleyrodidae. Cf. Greenhouse Whitefly.

BANDED WOOLLYBEAR *Pyrrharctia isabella* (J. E. Smith) [Lepidoptera: Arctiidae].

BANDICOOT TICK *Haemaphysalis humerosa* Warbuton & Nuttall [Acari: Ixodidae].

BANFIELD, EDMUND JAMES (1852–1923) (Musgrave 1932, *A Bibliogr. of Australian Entomology 1775–1930,* 380 pp. (12), Sydney.)

BANG-HAAS, ANDREAS (1846–1925) (Tuxen 1968, Entomol. Meddr. 36: 18.)

BANKES, EUSTACE RALPH (1861–1929) (Jordan 1931, Proc. Entomol. Soc. Lond. 5: 130–131.)

BANKS, CHARLES JOHN (1915–1975) (Gunn 1976, Proc. Roy. Entomol. Soc. Lond. (C) 40: 51.)

BANKS GRASS-MITE *Oligonychus pratensis* (Banks) [Acari: Tetranychidae].

BANKS, JOSEPH (1743–1820) English aristocrat, natural historian, traveller and collector. JB was educated at Eton and Oxford. He travelled with James Cook (1768–1771) as principal naturalist on first trip around the world. JB was active in founding of Kew Gardens; President of Royal Gardens; President of Royal Society (1778–1820). He collected many insect Species described by Fabricius; specimens collected by Banks are deposited in British Museum. (Carter 1974, Bull. Br. Mus. Nat. Hist. (hist. ser.) 4: 4 pls.)

BANKS, NATHAN (1868–1953) American academic (Harvard University), Curator of Insects (Museum of Comparative Zoology) and government taxonomist specializing in arthropods. (Essig 1931, *History of Entomology,* 1029 pp. (549–552); Musgrave 1932, *A Bibliogr. of Australian Entomology 1775–1930,* 380 pp. (14), Sydney.)

BANKSIA JEWEL BEETLE *Cyria imperialis* (Fabricius) [Coleoptera: Buprestidae].

BANKSIA LONGICORN *Paroplites australis* (Erichson) [Coleoptera: Buprestidae].

BANKSIA MOTH *Danima banksiae* (Lewin) [Lepidoptera: Notodontidae].

BANNIGER, MAX (–1964) (Anon. 1964, Mitt. Dt. Ent. Entomol. Ges. 23: 62.)

BANNISTER, JOHN (1650–1692) (Ewan & Ewan 1970, *John Bannister and the Natural History of Virginia.* 485 pp. Chicago.)

BANTI, ADOLF (–1934) (Anon. 1934, Boll. Soc. Entomol. Ital. 66: 182.)

BANYAN APHID *Thoracaphis fici* (Takahashi) [Hemiptera: Aphididae].

BAR, CONSTANT (1817–1884) (Oberthür 1916, Etudes de lepidopterologie comparee. 9. (Portraits de lépidoptéristes): [8])

BAR Noun. (Middle English, *barre.* Pl., Bars.) 1. A short, straight band of uniform width; may refer to a colour. 2. A sclerite that is long relative to its width and thickness; bar has rigidity and may be used as a lever or for support.

BARAN, GABRIEL DE See Migeot de Baran.

BARATTINI, LUIS P (1903–1965) (San Guinetti 1968, Revta. Soc. Urug. Entomol. 7: 104–106.)

BARB Noun. (Latin, *barba* = beard. Middle English, *barbe.*) 1. Any sharp projection whose point is oblique, crosswise or directed backward from the tip. 2. A bristle ending in a hook. 3. A spine armed with teeth pointing obliquely rearward. 4. Acarina: See Setule; Uncus. Cf. Acantha; Barbule; Spicule.

BARBADOS LEG See Elephantiasis.

BARBAROCHTHONIDAE Plural Noun. A small Family of Trichoptera assigned to Superfamily Limnephiloidea. See Trichoptera.

BARBASCO® See Rotenone.

BARBATE Adj. (Latin, *barbatus*, from *barba* = beard; *-atus* = adjectival suffix.) 1. Pertaining to structure which is covered with Setae and give the appearance of a beard. 2. Antenna with tufts or fascicles of Seta or short bristles on sides of each segment. 3. Abdomen with flat tufts at the sides or apex.

BARBED Adj. (Latin, *barba* = beard.) Hooked; descriptive of structure with barbs or hooks. See Barb. Cf. Hamate; Spinose; Uncinate.

BARBELLION, W N P See Cummings, B.F.

BARBER BROWN-LACEWING *Sympherobius barberi* (Banks) [Neuroptera: Hemerobiidae].

BARBER, GEORGE WARE (1890–1948) (Mallis 1971, *American Entomologists.* 549 pp., New Brunswick.)

BARBER, HARRY GARDNER (1871–1960) A school teacher in New York and subsequently curator in the U. S. National Museum. Specialist in taxonomy of Hemiptera, especially Lygaeidae. (Mallis 1971, *American Entomologists.* 549 pp., New Brunswick.)

BARBER, HERBERT SPENCER (1882–1950) (Osborn 1952, *A Brief History of Entomology.* 305 pp., Columbus, Ohio.)

BARBER, MARSHALL A (–1953) (Anon. 1953, Revta. Ecuat. Entomol. Parasit. 1 (4): 128–130.)

BARBER-POLE CATERPILLAR *Mimoschinia rufofascialis* (Stephens) [Lepidoptera: Pyralidae].

BARBERRY APHID *Liosomaphis berberidis* (Kaltenbach) [Hemiptera: Aphididae].

BARBER'S FLUID An insect preservative consisting of ethanol (ethyl alcohol) 53 parts; Water 49 parts; Ethyl acetate (acetic ether) 19 parts; Benzene (benzol) 7 parts. BF also used to relax specimens that have become stiff, brittle and easily broken.

BARBEY, AUGUSTE (1872–1948) (Anon. 1948, Bull. Soc. Entomol. Fr. 53: 113.)

BARBIELLINI, AMADEU AMIDEI CONDE (1877–1955) (Carrera 1956, Revta. Bras. Entomol. 4: 213–214, bibliogr.)

BARBOUR, THOMAS (–1946) (Gardiner 1946, J.

S. Nature 157: 220.)

BARBULA Noun. (Latin, *barbula* diminutive of *barba* = beard. Pl., Barbulae.) Scarabaeoid larvae: Tufts or patches of Setae or short bristles along the sides of the Abdomen near anal region (Boving). Alt. Barbule.

BARBULE Noun. (Latin, *barbula* diminutive of *barba* = beard. Pl., Barbules.) A small barb, beard or filiform appendage. Alt. Barbula. See Barb. Cf. Spine.

BARBUT, JAMES (fl 1781) (Lisney 1960, *Bibliogr. of British Lepidoptera 1608–1799,* 315 pp. London.)

BARCA, EMIL (–1959) (Knaben 1961, Norsk Entomol. Tidsskr. 11: 285–286.)

BARCLAY, RICHARD D (–1938) (Anon. 1939, Amer. Bee J. 79: 36–37.)

BARDEE *Bardistus cibarius* Newman [Coleoptera: Cerambycidae].

BARDOHYMENIDAE Zalessky 1937. Plural Noun. Paleozoic (Lower Permian) Family of Megasecoptera with tubular projections from Thorax and Abdomen. Composed of Genera including *Actinohymen* Carpenter, *Alexahymen* Kukalova-Peck, *Bardohymen* Zalessky, *Calohymen* Carpenter, *Sylvohymen* Martynov. Apparently related to Protohymenidae (Carpenter 1962, 69: 37–41). See Megasecoptera.

BARE Adj. (Anglo Saxon, *baer* = naked.) Naked; descriptive of structure without natural covering. Exposed; uncovered; nude. See Bald.

BARE, ORLANDO S (1880–1958) (Hill 1958, J. Econ. Entomol. 51: 565.)

BARFOED, ERIK CHRISTIAN (1847–1937) (Henriksen 1937, Entomol. Meddr. 15: 543.)

BARGAGLI, PIETRO (1844–1918) (Guiglia 1969, *et al.* Memorie Soc. Entomol. Ital. 48: 669–670, 787.)

BARGEN, AUGUST VON (1889–1961) (Weidner 1967, Abh. Verh. naturw. Ver. Hamburg Suppl. 9: 292.)

BARICELLI (BARICELLO), GIULIO CESARE (–1800?) (Rose 1853, *New General Biographical Dictionary* 3: 176.)

BARK Noun. (Danish, *bark*; Old Norse, *borkr*; Old English, *beorcan* = bark. Pl., Barks.) 1. The tough, external, dead, cellular covering of a perennial woody stem or root. All tissues outside the cambium layer of a vascular plant. 2. A colour that is reddish-yellow in hue, of low brilliance and saturation; Mocha; dark olive-brown colour. 3. The rind or husk of seeds, fruits or nuts.

BARK BEETLES See Curculionidae; Scolytidae.

BARK BUGS See Aradidae.

BARK GNAWING BEETLES See Trogossitidae.

BARK LICE See Psocoptera.

BARKER, HENRY WILLIAM (1860–1909) (Hall 1909, Entomologist 42: 228.)

BARLEY GRUB *Persectania eqingii* (Westwood) [Lepidoptera: Noctuidae].

BARLEY JOINTWORM *Tetramesa hordei* (Harris) [Hymenoptera: Eurytomidae].

BARLEYQUAT-B® See Chlormequat.

BARLOW, JOHN (1872–1944) (Weiss 1946, Ann. Entomol. Soc. Amer. 39: 3–4.)

BARNACLE SCALE *Ceroplastes cirripediformis* Comstock [Hemiptera: Coccidae].

BARNARD, FRANCIS GEORGE ALLMAN (1857–1932) (E.E.P. 1932, Victorian Nat. 49: 69–73, bibliogr.)

BARNARD, GEORGE (1830–1894) (Musgrave 1932, *A Bibliogr. of Australian Entomology 1775–1930,* 380 pp. (14), Sydney.)

BARNARD, WILLIAM STEBBINS (1849–1887) (Wilder 1887, Amer. Nat. 21: 1136–1137.)

BARNES, DWIGHT FLETCHER (1890–1957) (Simmons 1958, J. Econ. Ent. 51: 117)

BARNES, HORACE FRANCIS (1902–1960) (Uvarov 1961, Proc. Roy. Entomol. Soc. Lond. (C) 25: 50.)

BARNES, MAURICE DRUMMOND (1912–1945) (H[incks] & Stubbs 1945, Naturalist, Hull No. 815: 151–152, portrait; Carpenter 1946, Proc. R. Ent. Soc. Lond. (C) 10: 53.)

BARNES, WILLIAM (1860–1930) (Mallis 1971, *American Entomologists*. 549 pp. New Brunswick.)

BARNEVILLE, HENRI BRISOUT DE (–1887) (Dimmock 1888, Psyche 5: 35.)

BARNS, THOMAS ALEXANDER (–1930) (Talbot 1930, Bull. Hill Mus. Witley 4: 145–153.)

BARON, OSCAR THEODOR (1847–1926) (Brown 1965, Journ. Lepid. Soc. 19: 35–46.)

BARON® See Phosdiphen.

BARRAL® See Chlorpyrifos.

BARRAUD, PHILIP JAMES (1879–1948) (Williams 1949, Proc. Roy. Entomol. Soc. Lond. (C) 13: 66.)

BARRERE, PIERRE (1690–1755) (Larousse 1865, *Grand dictionnaire universale du XIX siècle.* 2: 261.)

BARRETT, CHARLES GOLDING (1836–1904) (Anon. 1906, Entomol. Jb. 15: 192.)

BARRETT, J PLATT (1838–1916) (Porritt 1907, Entomol. mon. Mag. 53: 69–70.)

BARRETT, LUCAS (1837–1862) (Anon. 1863, Proc. Linn. Soc. London. 1863: xxxi–xxxiv.)

BARRICADE® See Cypermethrin.

BARRICIDE® See Cypermethrin.

BARROTS, THEODORE (1857–1920) (Viets 1955, Milben 1: 32–33)

BARROW, REGINALD LONSADU (1879–1967) (Kennedy 1968, Proc. Roy. Entomol. Soc. London. (C) 32: 58.)

BARROWS, WALTER BRADFORD (1855–1923) (W.A.H. 1925, J. Econ. Entomol. 18: 563–564.)

BARTELS, CARL (–1901) (Anon. 1903, Entomol. Jb. 12: 252.)

BARTHE, EUGENE (1862–1945) (Favarel 1947, Miscnea Entomol. 44: 113–118.)

BARTHE, RENÉ (1894–1957) (Rivalier 1957, Bull. Soc. Entomol. Fr. 62: 105, 161–162.)

BARTLETT, BLAIR R (–1977) American economic Entomologist at University of California Experiment Station and specializing in Biological Control of homopterous pests. Early proponent of Integrated Pest Management concept.

BARTON, STEPHEN (–1898) (Hudd 1899, Entomol. Mon. Mag. 35: 16.)

BARTONELLOSIS A disease of humans in South America caused by the protozoan *Bartonella bacilliformis*. Bartonellosis may be a form of Carrion's Disease or Oroya Fever. In Peru, disease is transmitted by the sandfly *Lutzomyia verrucarum* (Townsend).

BARTOS, EMANUEL (–1962) (Sramek & Husek 1962, Cas. Csl. Spol. Entomol. 59: 207–208.)

BARTRAM, WILLIAM (1739–1823) Official Botanist for George III in the Floridas and founder of the botanical garden near Philadelphia. (Osborn 1952, *A Brief History of Entomology*. 305 pp. (178), Columbus, Ohio.)

BARTSCH, AMBROS (1828–1904) (Anon. 1906, Entomol. Jb. 15: 194.)

BASAD Adv. (Latin, *basis* = base.) In the direction of or toward the base of a structure. See Orientation. Cf. Anteriad; Apicad; Caudad; Centrad; Cephalad; Craniad; Dextrad; Dextrocaudad; Dextrocephalad; Distad; Dorsad; Ectad; Entad; Laterad; Mediad; Mesad; Neurad; Orad; Proximad; Rostrad; Sinistrad; Sinistrocaudad; Sinistrocephalad; Ventrad.

BASAL Adj. (Latin, *basis* = base; *-alis* = pertaining to, appropriate to.) 1. At or near the base. 2. Forming the base. 3. Pertaining to the base or point of attachment nearest the main body. See Proximal. Cf. Apical.

BASAL ANAL AREA Dragon-fly: An outlined area at the base of the wing which is bounded by Cu+A, the anal crossing and the secondary Anal Vein A' (Comstock).

BASAL ANAL CELL Plecoptera: A wing cell near wing base closed by a cross-vein which extends from first Anal Vein to second Anal Vein (Comstock).

BASAL APODEME An apodeme of the Phallobase (Snodgrass).

BASAL AREA 1. Wings: A space nearest the place where a wing attaches to Metanotum. 2. Hymenoptera: Anterior of the three median cells or areas; the first median area.

BASAL BRIDGE Hymenoptera: Dorsal bridge of Penis valves (Tuxen).

BASAL CELL 1. Odonata: An elongate cell between Radius and Cubitus, just before the Arculus. 2. Trichoptera: 1–3 cells enclosed by branches that form the Postcostal or Anal Vein. 3. Diptera: 1st (Williston), Radial 2 (Comstock), 2nd (Williston), Media (Comstock).

BASAL FOLD See Plica Basalis.

BASAL LINE Lepidoptera: A transverse line extending half way across the primaries near the base.

BASAL LOBE Culicidae genitalia: See Claspette.

BASAL METABOLISM The slow rate of metabolism of the resting animal.

BASAL PIGMENT CELL A type of Pigment Cell that surrounds the base of the Rhabdom. Cf. Primary Pigment Cell; Secondary Pigment Cell.

BASAL PLATES Sclerites of the Phallobase (Snodgrass).

BASAL POSTCOSTAL VEIN Agrioninae: One of the cubito-anal crossveins. See Vein; Wing Venation.

BASAL RING 1. Diptera (Nematocera): Fusion of abdominal Tergum IX with Sternum IX; Culicidae larvae: Supporting base of Setae (Harbach & Knight). 2. Hymenoptera: See Gonobase.

BASAL SEGMENT OF CLASP See Side Piece.

BASAL SPACE Lepidoptera: Area on the forewing between the base and transverse anterior line.

BASAL STREAK Noctuidae: Extends from the base of the wing, through the submedian interspace to the transverse anterior line.

BASAL SUTURE Termitidae: A line of weakness along which the fracture and consequent shedding of the wings takes place.

BASAL TRANSVERSE CARINA Hymenoptera: A Carina that crosses anterior of the middle of the Metanotum and separates the anterior and median areas. See Carina.

BASAL VEIN 1. Diptera (Chironomidae): See Brachiolum. 2. Hymenoptera: A short vein connecting M + CuA and Rs + M in the forewing. See Vein; Wing Venation.

BASALAR MUSCLES Muscles attached to the Basalar Sclerites or basalar lobe of the Episternum; usually two BMs on each side, but sometimes three or one only (Snodgrass).

BASALAR PLATE Miridae (Capsidae): The sclerite at the base of the forewing below the posterior edge of the Pronotum and behind the upper margin of the Propleura.

BASALAR SCLERITES Small pleural sclerites located at the base of the wings (Crampton). Syn. Basalare.

BASALARE Noun. (New Latin, *basis* = base; *ala* = wing. Pl., Basalares.) A thoracic sclerite (sometimes double) near the base of the wing and anterior of the Pleural Wing Process. Basalare serves as a place of insertion for the anterior pleural muscle of the wing. Sometimes, Basalare is represented by an undetached or partially detached lobe of the Episternum anterior of the Pleural Wing Process. Syn. Preparapteron (Snodgrass). Cf. Subalare.

BASALIS Noun. The principal mandibular sclerite (when sclerites are distinguishable) to which all other parts are joined. Basalis corresponds to the Maxillary Stipes. See Mandible; Maxilla.

BASAMID® See Dazomet.

BASANALE Noun. (Greek, Latin, *basis* = base, step. Pl., Basanales.) The third Axillary Sclerite at the base of the Anal Veins (Crampton).

BASANTENNA See Antacoria.

BASARCUS Noun. (Greek, Latin, *basis* = base, step; *arcus* = arc. Pl., Basarcuses.) The basal arch in cockroaches (MacGillivray).

BASCOMBE, VICTOR HASTINGS DARE (1904–

1967) (Kennedy 1968, Proc. Roy. Entomol. Soc. Lond (C) 32: 58.)

BASE Noun. (Latin, *basis* = base, from Greek, *bainein* = to go, step. Pl., Bases.) 1. Part of any appendage or structure that is nearest the body. 2. Part of the Thorax nearest the Abdomen; part of the Abdomen nearest the Thorax. 3. The bottom part or foundation upon which anything stands or is structurally supported. Cf. Apex.

BASEMENT MEMBRANE 1. Innermost layer of Integument and secreted by Haemocytes. BM forms a continuous sheath of connective tissue that separates internal organs from remainder of Integument. Transmission electron microscopy shows BM forms straight or sinuous parallel fibres about 0.5 Å thick. Actual thickness may vary considerably, depending upon part of body and Species under study. Chemically, BM is composed of mucopolysaccharides and collagen. Physically, BM is inelastic, white and birefringent. Functionally, BM separates Haemocoel from Integument, provides support for Integument, anchors epidermal cells in position and serves as a permeability barrier with physical and electrical charge selectivity. 2. The membrane covering the inner surface of insect eye, and continuous with that of the surrounding Epidermis. Syn. Membrana Fenestrata (Snodgrass). See Integument. Cf. Cement Layer; Cuticulin Layer; Dermal Glands; Epidermal Cells; Epidermis; Endocuticle; Exocuticle; Pore Canals; Wax Layer. Rel. Skeleton; Skin.

BASENDITE Noun. (Latin, *basis* = base; Greek, *endon* = within; *-ites* = constituent. Pl., Basendites.) An Endite of the basis of an appendage. See Endite.

BASIC Adj. (Greek, *basis* = base; *-ic* = characterized by.) 1. Descriptive of substance with the properties of a base. Pertaining to histological stains that typically affect the nucleus and nuclear content of cellular preparations. Cf. Acidic.2. Pertaining to things, acts or processes that are fundamental or essential.

BASICARDO Noun. (Latin, *basis* = base; *cardo* = hinge. Pl., Basicardos.) The basal region of the Cardo (Crampton). See Mouthparts.

BASICONIC SENSILLUM (Pl., Basiconic Sensilla.) A sense organ whose external part forms an immobile cone or peg which projects above the surface of the Integument. Thin walled Basiconic Sensilla have one neuron; thick-walled Basiconic Sensilla have several neurons. Basiconic Sensilla may be hair-like, longitudinally ridged and saddle-shaped in cross-section. A bifurcate or forked Sensillum serves as an apparent variant of the Basiconic Sensillum in Lepidoptera, Thrips, the Tarsi of female *Simulium venustum*, beetles including *Monochamus* sp. and scorpion flies. In Lepidoptera, BS are regarded as general olfactory receptors, not specific pheromone receptors. Coleoptera possess BS on the Antenna and mouthparts; BS on the alfalfa weevil are multiporous olfactory receptors; Some BS are uniporous contact chemoreceptors. Lepidoptera larval Antennae typically bear Basiconic and Styloconic Sensilla which are used for temperature and chemoreception sensitivity. Syn. Sensillum Basiconicum. See Sensillum.

BASICOSTA Noun. (Latin, *basis* = base; *costa* = rib. Pl., Basicostae.) 1. Diptera: The bare second distinct 'scale' at the base of the wing in muscoids following the basal 'scale,' (the setose epaulet *sensu* Curran). 2. The proximal submarginal ridge of the inner wall of a leg segment (Snodgrass).

BASICOSTAL SUTURE The external groove on a leg segment that forms the Basicosta (Snodgrass). The proximal end of the Coxa is surrounded by the Basicostal Suture. When the suture completely surrounds the Coxa, it forms a narrow basal zone (Basicoxite) and a larger, more distal zone (Disticoxite). See Suture. Rel. Leg.

BASICOXITE Noun. (Latin, *basis* = base; *coxa* = hip; *-ites* = constituent. Pl., Basicoxites.) The usually narrow basal rim of the Coxa proximal of the Basicostal Suture and its internal ridge. The Coxomarginale *sensu* Snodgrass.

BASIFEMUR Noun. (Latin, *basis* = base; *femur* = thigh. Pl., Basifemora; Basifemurs.) Acarina: The leg segment between the Trochanter and Telofemur (Goff *et al.* 1982, J. Med. Entomol. 19: 222.) See Femur. Rel. Leg.

BASIGALEA Noun. (Greek, *basis* = base; *galea* = helmet. Pl., Basigaleae.) The basal portion of the Galea (Crampton). See Galea. Rel. Mouthparts.

BASILABIUM Noun. (Greek, *basis* = base; *labium* = lip. Pl., Basilabia.) The basal sclerite of the Labium. See Labium. Rel. Mouthparts.

BASILAR Adj. (Latin, *basis* = base; *ala* = wing.) Descriptive of or pertaining to the structure near the base, such as a process or style.

BASILAR CROSSVEIN Odonata: The vein crossing the basilar space. See Vein; Wing Venation.

BASILAR MEMBRANE A thin, fenestrate membrane separating the rods and cones of the insect eye from the optic tract (Packard).

BASILAR SPACE Odonata: A cell at the base of the wing bounded by the Radius, Cubitus, Arculus and base of the wing; median space *sensu* Garman.

BASILARE Noun. (Latin, *basis* = base; *ala* = wing. Pl., Basilares.) The Jugulum *sensu* Straus.

BASILEMMA Noun. (Greek, *basis* = base; *lemma* = skin. Pl., Basilemmae.) The Basement Membrane of the Integument. See Basement Membrane.

BASIMANDIBULA Noun. (New Latin, *basilaris* = situated at the base; *mandibula* = Mandible. Pl., Basimandibulae.) 1. A small sclerite at the base of the Mandible. 2. The Trochantin in the groundplan condition explaining the hypothetical evolution of the Mandible as an appendage (MacGillivray).

BASIPARAMERE Noun. (Greek, *basis* = base; *para* = beside; *meros* = part. Pl., Basiparameres.) Hymenoptera: See Gonocoxite.

BASIPHARYNX Noun. (Greek, *basis* = base; *pharyngx* = throat. Pl., Basipharnices.) 1. A slender, chitinized tube formed by the greatly reduced Epigusta and Subgusta (MacGillivray). 2. Diptera: The proximal portion of the Pharnyx that connects with the Oesophagus. Cf. Postpharynx

BASIPODITE Noun. (Greek, *basis* = base; *pous* = foot; *-ites* = constituent. Pl., Basipodites.) 1. The segment following the Coxopodite in the insect Maxilla (Imms). 2. The second of the two basal sclerites of a segmented appendage in Arthropoda, when the Protopodite is divided (Tillyard). 3. The first Trochanter (Snodgrass). See Appendage. Cf. Telopodite.

BASIPROBOSCIS Noun. (Greek, *basis* = base; *proboskis* = trunk. Pl., Basiproboscides.) 1. Diptera: The basal part of the flexed Proboscis in muscid flies. 2. The region proximal of the constriction of the Proboscis (MacGillivray).

BASIPULVILLUS Noun. (Greek, *basis* = base; *pulvillus* = small cushion. Pl., Basipulvilli.) 1. Diptera: A small, paired, lateral sclerite at the base of Pulvillus (Holway 1935). 2. Heteroptera: Basipulvillus forms an extension called Distipulvillus (Goel & Schaefer 1970, AESA 63: 308.) See Pretarsus. Cf. Pulvillus.

BASIS Noun. (Greek, *basis* = base. Pl., Bases.) 1. The foundation or base of a structure or concept. 2. The entire lower part of the Theca from mouth to Labella (Jardine). See Base.

BASIS CAPITULI Ticks: The basal part of the Capitulum (Matheson).

BASIS CAPUTULUM See Gnathobase.

BASISTERNUM Noun. (Greek, *basis* = base; *sternon* = chest. Pl., Basisterna.) The principal sclerite of the Sternum. Basisternum is positioned anterior of the Sternal Apophyses or Sternacostal Suture and laternally connected with pleural region of Precoxal Bridge. See Sternum. Cf. Eusternum.

BASITARSUS Noun. (Greek, *basis* = base; *tarsos* = sole of foot. Pl., Basitarsi.) The proximal or basal segment of the Tarsus. See Tarsus. Cf. Pretarsus.

BASITIBIAL PLATE Hymenoptera (Apoidea): A small, flat, elevated, basidorsal portion of the middle Tibia.

BASIVALVULA Noun. (Greek, *basis* = base; *valva* = fold. Pl., Basivalvulae.) Insect Genitalia: Small sclerites positioned at the bases of the first Valvulae; often confused with the first Valvifers (Snodgrass). See Genitalia.

BASIVOLSELLA Noun. (Greek, *basis* = base; Latin, *volsella* = a pair of tweezers. Pl., Basivolsellae.) Hymenoptera genitalia: The primary sclerite of the Volsella.

BASIVOLSELLAR APODEME Hymenoptera: An apodeme projecting from the apex of the volsellar strut (Tuxen).

BASKING Verb. (Middle English, *basken* = bathe.) A thermoregulatory process in which an insect changes the orientation of its body or posture in relation to incident solar radiation in order to increase internal body temperature. See Thermoregulation. Cf. Evaporative Cooling. Rel. Circulatory System.

BASOLON® See Dazomet.

BASONYM Noun. (Latin, *basis* = foundation; Greek, *onyma* = name. Pl., Basonyms.) A generic name to which a suffix has been added. Rel. Taxonomy; Nomenclature.

BASSA® See Fenobucarb.

BASSETT, HOMER FRANKLIN (1826–1902) American teacher, librarian and businessman. First taxonomic specialist in Cynipidae. (Essig 1931, *History of Entomology*. 1029 pp., New York; Beard 1974, In 25th Anniversary Memoirs Cornell Entomol. Soc. p.13.)

BASSI, AGOSTINO (1773–1856) (Osborn 1952, *A Brief History of Entomology*. 305 pp. (178, portrait, pl. 1), Columbus, Ohio.)

BASSI, CARLO (1807–1856) (Conci 1967, Atti Soc. Ital. Sci. Nat. 106: 37–38, 54.)

BASSINGER, A J American Entomologist (University of California 1923–1951) working in biological control of *Citrus* pests.

BASSWOOD LACE BUG *Gargaphia tiliae* (Walsh) [Heteroptera: Tingidae].

BASSWOOD LEAFMINER *Baliosus nervosus* (Panzer) [Coleoptera: Chrysomelidae].

BASSWOOD LEAFROLLER *Pantographa limata* Grote & Robinson [Lepidoptera: Pyralidae].

BASTELBERGER, MAX JOSEPH (1851–1916) (Anon. 1916, Mitt. Münch. Entomol. Ges. 7: 3–7, bibliogr.)

BASTER, JOB (1711–1775) (Swainson 1840, *Taxidermy: With Biogr. Of Zoologists*. 392 pp. (122), London. Volume 12 of *Cabinet Cyclopaedia*. Edited by D. Lardner.)

BASTIANELLI, GIUSEPPE (1862–1959) Italian Medical Entomologist worked on human Malaria and *Anopheles*. (Raffaele 1959, Riv. Parassit. 20: 233–228.)

BASUDIN® See Diazinon.

BAT BUGS See Cimicidae; Polyctenidae.

BAT FLEA *Myodopsylla insignis* (Rothschild) [Siphonaptera: Ischnopsyllidae]: The most common and widespread bat flea in North America. Adult with 4–5 bristles on head near base of Antenna; Postantennal Bristles in four rows; Prontoal Ctenidium with 18–21 spines on each side. See Ischnopsillidae.

BAT FLIES See Nycteribiidae; Streblidae.

BAT TICK *Ixodes simplex* Neumann [Acarina: Ixodidae].

BATES, FREDERICK (1829–1903) (Musgrave 1932, *A Bibliogr. of Australian Entomology 1775–1930*, 380 pp., (15–16, bibliogr.), Sydney.)

BATES, GEORGE LATIMER (–1940) (Riley 1940, Entomologist 73: 95–96.)

BATES, HENRY WALTER (1825–1892) English

natural historian and prolific collector of insects. Worked for 11 years in Amazon basin (1848–1861), including two years with Alfred Wallace. HWB added knowledge to our understanding of Mimicry and described the phenomenon of Batesian Mimicry. Published *The Naturalist on the River Amazons* (2 vols, 1863) and *A Naturalist in Nicaragua*. (Osborn 1952, *A Brief History of Entomology*. 305 pp. (178), Columbus, Ohio.)

BATESIAN MIMICRY A type of mimicry in which an edible or innocuous Species (the mimic) obtains protection from predators by resembling the appearance of an inedible or obnoxious Species (the model). The phenomenon was described by Henry Bates and subsequently named in his honor. See Mimicry. Cf. Müllerian Mimicry. Rel. Crypsis.

BATESON, WILLIAM (1861–1926) (J. B. F. 1927, Proc. Roy. Soc. Lond. (B) 101: i–iv.)

BATH, WOLFGANG (1882–1932) (Freise 1932, Insektenbörse 49: 153.)

BATHMIS See Pterostigma.

BATHURST BURR SEE FLY *Euaresta bullans* Wiedemann [Diptera: Tephritidae].

BATI, LEONTINA (1885–1952) (Sykora 1953, Roc. Csl. Spol. Entomol. 50: 240–241.)

BATRACHEDRIDAE Plural Noun. A small, widespread Family of ditrysian Lepidoptera assigned to Superfamily Gelechioidea. Batrachedridae includes a diverse assemblage of Genera, some of which suggest placement among Coleophoridae, Oecophoridae, Momphidae and Cosmopterygidae. Adult small; wingspan 8–18 mm; head with smooth scales; Ocelli and Chaetosemata absent; Antenna filiform, ca 0.75 times as long as forewing; Proboscis covered with overlapping scales; Maxillary Palpus small, folded; Labial Palpus recurved with smooth scales; fore Tibia with Epiphysis; tarsal formula 0-2-4; wings unusually narrow with long marginal fringe. Pupa with Maxillary and Labial Palps exposed; Abdomen with long Setae on dorsal surface. Larval feeding habits are diverse and many Species feed on various plant tissues including flowers, stems, leaves; some Species are predaceous on scale insects. See Gelechioidea.

BATTLE® See Lambda Cyhalothrin.

BATTLEY, ARTHUR UNWIN (1866–1905) (Anon. 1906, Entomol. Jb. 15: 193.)

BATTY JAMES (1831–1893) (Hall 1893, Entomol. Mon. Mag. 29: 287–288.)

BATUMEN Noun. (Latin, *battuo* = to fence. Pl., Batumens.) The layer of material (mud, cerumen) surrounding the nest cavity of stingless bees.

BATUMEN PLATES Thick batumen subdividing part of a larger cavity from the space used as a nest cavity by stingless bees.

BATURAD® A registered biopesticide derived from *Bacillus thuringiensis* var. *kurstaki*. See *Bacillus thuringiensis*.

BAU, ARMINIUS (1861–1932) (Alfken 1932, Mitt. Entomol. Ver. Bremen 19: 14.)

BAUDI DI SELVE, FLAMINIO (1821–1901) (Anon. 1903, Entomol. Jb. 12: 252.)

BAUDYS, EDUARD (1886–1968) (Caca 1968, Marcellia 35: 221–224.)

BAUDYS, ZEMREL (1886–1968) (Skuhravá 1968, Acta Entomol. Bohemosl. 65: 399.)

BAUER, GEORG LUDWIG (1809–1882) (Anon. 1883, Abh. Ber. Ver. Naturk. Cassel 29–30: 7–8.)

BAUER, HEINRICH (–1915) (Hubenthal 1916, Entomol. Bl. Biol. Syst. Käfer 12: 138.)

BAUER, VIKTOR (–1927) (Anon. 1927, Zool. Anz. 74: 144.)

BAUM, JIRI (1900–1943) (Novak 1946, Cas. Csl. Spol. Entomol. 10: 35–38, bibliogr.)

BAUMAN, HARRY H (1911–1961) (Anon. 1962, Bull. Entomol. Soc. Amer. 8: 47.)

BAUMEISTER, JOHANNES (–1891) (Anon. 1891, Insektenbörse 8 (18), unpaginated.)

BAUMER, JOHANNES PAULLUS (1725–1771) (Rose 1853, *New General Biographical Dictionary* 3: 379.)

BAUMHOFER, LYNN G (1895–1942.) (Osborn 1946, *Fragments of Entomological History*, Part II. 232 pp. (59), Columbus, Ohio.)

BAUPLAN Noun. (German, *bauen* = to build; *plan* = plan. Pl., Bauplans.) A generalized, abstract plan of biological organization that is based on absolute relationships of body part form and structure, Bauplan is used as a conceptual framework for hypothesis formulation and testing. Syn: Morphotype; Basic Plan. See General Homology. Cf. Groundplan.

BAUR, GEORGE HERMAN CARL LUDWIG (1859–1898) (Wheeler 1899, Amer. Nat. 33: 15–30, bibliogr.)

BAVE Noun. (French, drivel.) 1. The double strand of silk used by the silkworm to construct its cocoon. 2. Fluid silk as it is spun by caterpillars. See Silk. Cf. Brin; Gres; Grege. Rel. Fibroin; Sericin.

BAWEJA, K D (1896–1962) (Pradhan 1962, Indian J. Entomol. 24: 293.)

BAY MAT7484® See Tebupirimfos.

BAYARD, ANDRÉ (–1967) (Anon. 1967, Bull. Soc. Entomol. Fr. 72: 222.)

BAYCID® See Fenthion.

BAYER, EMIL (1875–1947) (Anon. 1947, Vest. Csl. Zool. Spol. 11: 89, bibliogr.)

BAYER, LORENZ (1847–1932) (Pfaff 1933, Entomol. Z., Frankf. a M. 46: 109–110.)

BAYFORD, EDWIN GOLDTHORP (1865–1958) (Uvarov 1960, Proc. Roy. Entomol. Soc. Lond. (C) 24: 52.)

BAYGON® See Proproxur.

BAYLIS, ERNEST (1877–1930) Lutz 1930, Entomol. News 41: 285–286.)

BAYMIX® See Coumaphos.

BAYNE, ARTHUR F (1869–1947) (Anon. 1948, Revta. Soc. Entomol. Argent. 14: 124–128.)

BAYNES, EDWARD STUART AUGUSTUS (–1972) (Butler 1973, Proc. Roy. Entomol. Soc. Lond. (C)

37: 55.)

BAY-NTN-33893® See Imidacloprid.

BAY-O-CIDE® See Proproxur.

BAYRUSIL® See Quinalphos.

BAYTEX® See Fenthion.

BAYTHROID® See Cyfluthrin.

BAZILLE, LOUIS (1828–1886) (Riley 1887, Scient. Amer. 56: 64.)

BEACHGRASS SCALE *Eriococcus carolinae* Williams [Hemiptera: Eriococcidae].

BEAD Noun. (Middle English, *bede* = prayer bead. Pl., Beads.) Small, round objects of stone, glass, wood or other material that are drilled or pierced for threading with string, wire or silk to form a necklace. The catenation of a series of beads gives a characteristic appearance that may be used to characterized some anatomical structures (Antenna, Ovariole). Rel. Moniliform.

BEAD GLAND See Pearl Body.

BEADED LACEWINGS See Berothidae.

BEAK Noun. (Middle English, *beeke*. Pl., Beaks.) 1. Any notable prolongation of the front of the head. 2. The snout in Rhynchophora. 3. The jointed structure or Rostrum covering the Maxillae in the mouth of any piercing-sucking insect. See Mouthparts. Cf. Rostrum.

BEAM® See Carbosulfan.

BEAMER, RAYMOND HILL (1889–1957) American academic (University of Kansas) and taxonomic specialist on the Cicadellidae. (Mallis 1971, *American Entomologists*. 549 pp., New Brunswick.)

BEAN APHID *Aphis fabae* Scopoli [Hemiptera: Aphididae]: In North America, a widespread pest of beans, beet, chard. In northern states, BA overwinters as egg on *Euonymus*; in southern states, BA apparently produces only females and prefers dock. Syn. Black Bean Aphid. See Aphididae. Cf. Bean Root Aphid.

BEAN BRUCHID *Acanthoscelides obtectus* (Say) [Coleoptera: Bruchidae]: A cosmopolitan, serious pest of all legumes particularly beans and cowpeas; peas and lentils in storage. BB is fungivorous and capable of developing on numerous fungi associated with stored grains. BB is endemic to South and Central America and became established in Europe and Africa. Adult 2–3 mm long, mottled brown; eyes emarginate; Antenna serrate; Elytra with pale areas, not covering apical segments of Abdomen; hind Femur with a large spine and a few smaller spines on ventral surface. Life history of BB is similar to pea weevil. Female lays 40–60 white eggs during lifetime on produce or bean pods in field; incubation requires 3–9 days. Larvae are white, curved and legless; feed inside beans; four larval instars complete development over 12–150 days. Pupation occurs within bean behind a 'window' that facilitates adult emergence; pupation requires 8–25 days. Life cycle one to several months; BB is multivoltine. Syn. Bean Weevil. See Bruchidae. Cf. Pea Weevil.

BEAN BUG *Riptortus clavatus* Thunberg [Hemiptera: Coreidae].

BEAN BUTTERFLY *Lampides boeticus* (Linnaeus) [Lepidoptera: Lycaenidae].

BEAN CAPSID *Pycnoderes quadrimaculatus* Guérin-Méneville [Hemiptera: Miridae].

BEAN FLY *Ophiomyia phaseoli* (Tryon) [Diptera: Agromyzidae]: A sporadic and sometimes significant summer pest of French beans and peas. BF is endemic to Australia. Eggs are laid in leaf tissue near leaf stalk; larvae tunnel into petioles and lower part of stems; pupation occurs within mines. Life cycle requires about three weeks during summer. Feeding damage causes loss of vigor, plant collapse and reduced yield. See Agromyzidae.

BEAN LADY BEETLE See Mexican Bean Beetle.

BEAN LEAFROLLER *Urbanus proteus* (Linnaeus) [Lepidoptera: Hesperiidae].

BEAN MAGGOT See Seed Corn Maggot.

BEAN MOSAIC A viral disease of beans which is transmitted by several Species of aphids.

BEAN PODBORER *Maruca testulalis* (Geyer) [Lepidoptera: Pyralidae]: A minor and sporadic pest of French beans in Australia. Larvae damage flowers, flower buds or bore into pods to consume seeds. Eggs are cream-coloured, oval, scale-like and laid near flower buds, BP larvae frequently web pods, flowers or leaves. Pupation typically occurs inside pod, sometimes within webbing or within ground debris. See Pyralidae.

BEAN ROOT-APHID *Smynthurodes betae* Westwood [Hemiptera: Aphididae]: Adult 1.6–2.9 mm long; body globular, yellowish-white and dusted with wax; head, legs and Antenna pale brown. BRA has complex life-cycle. Wingless adults form galls around leaves of host plant; winged forms migrate from galls to roots. Aphids can overwinter in ant nests where they are tended by ants. Ants construct small chambers to allow aphid movement around roots; chambers are covered with white, waxy dust from aphids. In Australia, BRA rarely is found on cotton, but may attack roots at seedling stage. See Aphididae. Cf. Bean Aphid.

BEAN SPIDER-MITE *Tetranychus ludeni* Zacher [Acari: Tetranychidae]: Adult 0.37–0.46 mm long; body oval in female and slender to diamond shaped in male. Adults are red and females are larger than males. During summer, actively feeding nymphs have two dark spots on the Abdomen, but adults are uniformly red. The overwintering adult usually is red. Eggs are less than 0.1 mm in diameter, transparent and reddish. Females lay eggs on the underside of leaves. Egg stage requires 3–7 days depending on temperature. The first instar larva is 0.1–0.2 mm long, pink and has six legs, followed by two nymphal stages. Nymphal mites are pale green or brown with eight legs. Mites feed by inserting mouthparts into leaf tissue. Feeding by BSM near veins produces yellow patches which progress to reddening of the tissue and leaf death. See Tetranychidae. Cf. Strawberry Spider-Mite.

BEAN THRIPS 1. *Caliothrips fasciatus* (Pergande). 2. *Megalurothrips usitatus* (Bagnall) [Thysanoptera: Thripidae].

BEAN WEEVIL See Bean Bruchid.

BEAN WEEVILS See Bruchidae.

BEAN-BLOSSOM THRIPS *Megalurothrips usitatis* (Bagnall) [Thysanoptera: Thripidae]: A major pest of French beans in Australia. Adult ca 1.5 mm long; eggs laid in plant tissue; nymph pale-yellow; pupation in soil. BBT causes twisting and curling of pods that is induced by adults or nymphs feed on flowers when pods are being formed. See Thripidae.

BEAN-FLOWER CATERPILLAR *Jamides phaseli* (Mathew) [Lepidoptera: Lycaenidae].

BEAN-LEAF BEETLE *Cerotoma trifurcata* (Forster) [Coleoptera: Chrysomelidae]: A pest of legumes in eastern North America, soybeans and cotton. BLB eggs are lemon-shaped, orange in colour and deposited in small clusters on host plant. Larvae are whitish with head and apex of Abdomen dark coloured; segmentation conspicuous. Pupation occurs within cell in soil; pupa soft-bodied and white. Adult generally reddish-yellow, ca 5–6.5 mm long; head black, Elytra with 3–4 dark spots along medial margin and dark strip along lateral margin. Adult female oviposits ca 40 eggs during 30 days in soil near plants; eclosion occurs within 7–21 days. Larvae feed on roots and nodules, and girdle plants just beneath surface of soil while preparing their pupal chambers. Adults feed on underside of leaves and form circular holes in them; pod-feeding causes lesions that increase seed vulnerability to pathogens, reduction in seed weight and reduction in seed quality. See Chrysomelidae.

BEAN-LEAF SKELETONIZER *Autoplusia egena* (Guenée) [Lepidoptera: Noctuidae].

BEAN-ROOT APHID *Smynthurodes betae* Westwood [Hemiptera: Pemphididae].

BEAN-SEED MAGGOT *Delia florilega* (Zetterstedt) [Diptera: Anthomyiidae].

BEAN-STALK WEEVIL *Sternechus paludatus* (Casey) [Coleoptera: Curculionidae].

BEARBERRY APHID *Tamalia (Phyllaphis) coweni* (Cockerell) [Hemiptera: Aphididae].

BEARD Noun. (Latin, *barba*. Pl., Beards.) 1. Hair on the chin, lips or on the face of an individual. 2. Diptera: See Mystax. 3. Any group of Setae or processes which resemble a beard.

BEARDED Adj. (Anglo Saxon, *beard* = beard.) Invested with Setae such that a beard-like appearance is presented. See Barbated.

BEARDSLEY LEAFHOPPER *Balclutha saltuella* (Kirschbaum) [Hemiptera: Cicadellidae].

BEARE, THOMAS HUDSON (1859–1940) (Donisthorpe 1940, Entomol. Mon. Mag. 76: 187.)

BEATING SHEET A device for collecting sedentary insects in an arboreal habitat. BS is constructed of durable cloth, such as canvas, cut into a one-metre square. A reinforced pocket is sewn into each corner. Two wooden dowels insert their ends into adjacent pockets, intersect one another in the centre of the sheet, and insert their opposite ends in the pouch along a diagonal of the cloth. The sheet is held beneath the branch or shrub which is struck sharply with a stick. Insects fall onto the sheet and are collected with moistened brush, forceps or aspirator. An effective tool for collecting small insects or when the weather is cool and insects are inactive. Cf. Ground Cloth. Rel. Trap.

BEAULNE, JOSEPH ISIDORE (1886–1971) (Maltais 1972, Ann. Soc. Entomol. Quebec 17: 170–173, bibliogr.)

BEAUMONT, ALFRED (1832–1905) (Anon. 1906, Entomol. Jb. 15: 192.)

BEAUMONT, E. See ELIEDE BEAUMONT.

BEAUMONT, JACQUES DE (1901–1985) (See Besuchet 1986, Bull. Soc. Vaud. Sc. Nat. 78: 81–89.)

BEAUPERTHUY, LOUIS DANIEL French physician in West Indies who first suggested (1854) that mosquitoes were linked to Malaria and Yellow Fever. (Sanabria & Beauperthuy 1969, Beauperthuy Ensayo Biografico. 171 pp. São Paulo.)

BEAURIEU, GASPARD GUILLARD DE (1728–1795) (Rose 1853, *New General Biographical Dictionary* 3: 453.)

BEAUVERIA BASSIANA A naturally-occurring fungal pathogen whose spores attack a wide range of insects and mites. BB is not registered in USA but is regarded as non-phytotoxic and not toxic to warm-blooded animals. BB is relatively slow-acting (ca 7–14 days). BB spores attach to an insect, secrete an enzyme that dissolves Cuticle and enables pathogen to penetrate the Integument. Insect dies through desiccation. See Biopesticide. Cf. *Beauveria brongniartii; Nosema locustae; Verticillium lecanii.*

BEAUVERIA BRONGNIARTII A fungal pathogen whose spores are used as a biopesticide in the control of scarab beetle larvae in some countries. BB is not registered in USA. Regarded as non-phytotoxic and not toxic to warm-blooded animals; relatively slow-acting (ca 7–14 days). BB spores atttach to an insect, germinate and penetrate the Integument. Trade names include: Betel® and Engerling Spilz®. See Biopesticide. Cf. *Metarhizium anisopliae* ESF 1®; *Nosema locustae*; *Verticillium lecanii.*

BEAUVOIS, PALISOT DE See PALISOT DE BEAUVOIS.

BECCARI, NELLO (1883–1957) (Levi 1958, Atti Accad. naz. Lincei 13: 102–113.)

BECCARI, ODOARDO (1843–1920) (Anon. 1921, Entomol. News 32: 160.)

BECCARIAN BODIES Bodies produced on young leaves or stipules of plants such as euphorbes, and which provide food for ants. (Rickson 1980, Amer. J. Bot. 67: 285–292.) See Pearl Bodies.

Cf. Beltian Bodies; Müllerian Bodies.

BECHER, EDUARD (1856–1886) (Steindachner 1897, Ann. Naturh. Mus. Wien 2: 7–9, biogr.)

BECHSTEIN, JOHANN MATTHIAS (1757–1810) (Anon. 1858, *Accentuated List of British Lepidoptera*. 118 pp. (xi), Oxford & Cambridge Entomological Societies, London.)

BECHYNE, JAN KAREL (1920–1973) (Anon. 1975, Studia Entomol. 18: 622–630, bibliogr. only.)

BECK, D ELDON (1906–1967) (Tanner 1967, Great Basin Nat. 27: 230–239.)

BECK, HENRICH HENRICHSEN (1799–1863) (Henriksen 1923, Entomol. Meddr 15: 145–146.)

BECK, RICHARD (1827–1866) (Newman 1866, Entomologist 3: 168.)

BECKER, ALEXANDER (1818–1901) (Wiren 1901, Rev. Russe Entomol. 1: 130–133.)

BECKER, ERNEST GEORGIYEVICH (1874–1962) (Makhotin & Pravdin 1963, Entomol. Obozr. 42: 226–233, bibliogr.; trans. Entomol. Rev., Wash. 42: 123–126.)

BECKER, JOHANN JOSEPH MARIA (1788–1859) (Heyden 1860, Stettin. Entomol. Ztg 21: 37.)

BECKER, LEON (1826–1909) (Bonnet 1945, Bibliog. Araneorum 1: 37.)

BECKER, THEODOR (1840–1928) (Collin 1929, Proc. Entomol. Soc. Lond. 3: 103.)

BECKERS, GUSTAV (1847–1895) (Anon. 1896, Entomol. Jb. 6: 243.)

BECKHAM, CLIFFORD MYRON (1815–1917) (Tippins 1971, J. Econ. Entomol. 64: 1343.)

BECKWITH, CHARLES S (1891–1944) (Osborn 1946, *Fragments of Entomological History*. Pt. II. 232 pp. (60), Columbus, Ohio.)

BED BUG *Cimex lectularius* Linnaeus [Hemiptera: Cimicidae]: A widespread, obligate, blood-sucking parasite of humans that is gregarious in human habitations. Adult to 6 mm long, oval in outline shape and reddish-brown in colour; Antenna with four segments, apical three segments long and slender. Compound eyes are small and widely separated at lateral margin of head; Ocelli absent. Labium with three segments, projecting postero-ventrally to fore Coxa. Prothorax recessed medially to surround posterior and lateral margin of head; apterous; tarsal formula 3-3-3. Abdomen with 11 segments; female with notch (paragential sinus) along right posterior margin of fifth Sternum, notch opens into Ectospermalege. Male with corresponding left Paramere (right Paramere absent) and engages in 'traumatic insemination.' BB is nocturnal and rests away from hosts during daylight; bug feeds at night via several linearly arranged, closely spaced wounds; the parasite rapidly engorges and detaches from host. BB Life cycle typically requires 6–8 weeks, ca three weeks at 30°C. BB lives ca six months and can live one year without feeding. BB infestations are associated with poor sanitation or housekeeping; high-population infestations create a distinctive, sickly sweet odour associated with bed bug excrement. Female BB lays 200–300 eggs; eggs are white, operculate, 1 mm long and 0.5 mm wide. Female BB prefers rough surfaces as oviposition site; eggs are held in place with transparent, glue-like accessory gland secretion. Eggs are placed individually in cracks and crevices near the host's sleeping area; eclosion from eggs occurs within seven days (ca 4–5 days @ 32°C); eggs do not hatch at temperature extremes (<13°C, >37°C). BB undergoes five nymphal instars and each requires a blood meal; first instar can feed within 24 hours following eclosion. BB bites usually are not felt by the host; saliva (anticoagulant) is injected at feeding sites, but rarely cause allergic reaction. Feeding requires 5-10 minutes. BB does not display site preference on host's body. BB can feed upon mice, rats, rabbits and chickens. See Cimicidae. Cf. Tropical Bed Bug. Rel. Parasitism.

BEDEGUAR Noun. (Persian, through French, *bedeguar*. Pl., Bedeguars.) A moss-like, rounded gall induced on Rose by the cynipid wasp *Diplolepis rosae* (Hymenoptera: Cynipidae). Bedeguar is regarded as useful in treatment of sleeplessness, scurvy and diarrhoea. Syn. Pincushion Gall; Sweetbriar Gall. Cf. Pea Gall.

BEDEL, ERNEST MARIE LOUIS (1849–1922) (Bedel 1933, Rev. Biol. 8: 325–330, translation by W. M. Wheeler from Livre Cent. Entomol. Soc. Fr.)

BEDELL, GEORGE (1805–1877) (Douglas 1877, Entomol. Mon. Mag. 14: 22.)

BEDFORD, GERALD AUGUSTUS HAROLD (1891–1938) South African Entomologist. (Sachtleben 1939, Arb. Physiol. Angew. Entomol. Berl. 6: 209.)

BEDWELL, ERNEST CHARLES (1875–1945) (Carpenter 1946, Proc. Roy. Entomol. Soc. Lond. (C) 10: 53.)

BEE Noun. (Greek, *beine* > Anglo Saxon, *beo* = bee. Pl., Bees.) Any member of the Hymenoptera Superfamily Apoidea. Correctly: A swarm of bees. See Apoidea; Hymenoptera. Cf. Ant; Wasp. Rel. Social Insect.

BEE FLIES See Bombyliidae.

BEE HAWK MOTH *Cephonodes kingii* (Macleay) [Lepidoptera: Sphingidae].

BEE KILLER *Pristhesancus papuensis* Stål [Hemiptera: Reduviidae].

BEE LICE See Braulidae.

BEE LOUSE *Braula coeca* Nitzsch [Diptera: Braulidae].

BEE MILK Brood food secreted by *Apis* nurse bees and mixed with Crop contents.

BEE MITES See Scutacaridae.

BEE MOTHS Members of the Galleriinae (Pyralidae).

BEE VIRUS-X A viral disease of adult honey bees that occurs primarily in Europe. BVX replicates when ingested but not when injected into bees. BVX shortens the life of infected individuals and is associated with the protozoan *Mapighamoeba* *mellificae,* but is not dependent upon it. BVX is distantly related to Bee Virus-Y. See Honey Bee Viral Disease. Cf. Bee Virus-Y.

BEE VIRUS-Y A viral disease of honey bees that occurs during early summer. BVY displays no symptoms and may be restricted to the gut epithelium. BVY occurs in England, North America and Australia. BVY is associated with *Nosema apis* and may enhance the pathogenicity of *Nosema*. See Honey Bee Viral Disease. Cf. Bee Virus-X.

BEE WOLF *Philanthus triangulum* [Hymenoptera: Sphecidae]: A black-and-yellow wasp, endemic to Europe. BW stings honey bees in gular region (throat), malaxates the victim to feed on honey from its Crop. Honey bee victim is then stored as provender for BW larva. See Sphecidae. Cf. Burrowing Wasp. Rel. Parasitism; Predation.

BEECH APHID *Phyllaphis fagi* (Linnaeus) [Hemiptera: Aphididae].

BEECH BLIGHT APHID *Fagiphagus imbricator* (Fitch) [Hemiptera: Aphididae].

BEECH SCALE *Cryptococcus fagisuga* Lindinger [Hemiptera: Eriococcidae].

BEESONIIDAE Plural Noun. A small Family of sternorrhynchous Hemiptera assigned to Superfamily Coccoidea. See Coccoidea.

BEESWAX Fatty acids, esters and hydrocarbons manufactured by workers and used in the construction of bee comb. Wax precursors may be developed in oenocytes and fat body, and subsequently incorporated into the wax gland. The beeswax is secreted through wax mirrors on the gastral Sterna. The glandular epithelium is positioned dorsad of the wax mirror. See Gland; Wax. Rel. Wax Mirror.

BEET ARMYWORM *Spodoptera exigua* (Hübner) [Lepidoptera: Noctuidae]: A pest of cotton in parts of USA. BA egg is pale green to white, 0.3–0.4 mm dia., spherical or dome-shaped with rib-like ridges; eggs are typically laid during night in masses on underside of cotton leaves and covered with whitish scales from female's Abdomen. Female lays 500–600 eggs during lifetime; eclosion occurs within ca 3–4 days. BA larvae usually are pale to dark olive-green with pale and dark longitudinal bands on dorsum and black spot on pleural area of second thoracic segment. Early-instar larvae feed gregariously on leaves to cause skeletonizing damage (called a 'hit'); third instar larvae disperse. Late-season larvae feed on leaves, terminals, squares, blooms and young bolls; 6 larval instars complete feeding within 17–21 days. Pupa is brown with two spines along posterior; pupation occurs below surface of soil and requires 7–10 days. Adult 12–18 mm long with wingspan ca 30 mm. See Armyworms; Noctuidae. Cf. African Armyworm; Fall Armyworm.

BEET ARMYWORM NPV A naturally-occurring, host-specific multicapsid Nuclear Polyhedrosis Virus used as a biopesticide against larva of Beet Armyworm *(Spodoptera exigua)*. The NPV is effective only when ingested by the host; NPV attacks the stomach and kills the host larva within 5–7 days of ingestion. BANPV is registered for application on vegetable crops. Trade names include: Se NPV®, Instar® and Spod-X® See Biopesticide; Nuclear Polyhedrosis Virus. Cf. Douglas-Fir Tussock Moth NPV; European-Pine Sawfly NPV; Heliothis NPV; Mamestra brassicae NPV; Redheaded Pine Sawfly NPV; Spodoptera littoralis NPV.

BEET BUGS See Piesmatidae.

BEET LEAF-BEETLE *Erynephala puncticollis* (Say) [Coleoptera: Chrysomelidae].

BEET LEAFHOPPER *Circulifer tenellus* (Baker) [Hemiptera: Cicadellidae]: A significant vector of Curly Top Virus and other diseases of vegetable crops in the western USA. BLH is strongly migratory and usually univoltine. BLH is presumed native to Afghanistan and neighbouring states of Russia; the pest was discovered and described during 1890s in USA. See Cicadellidae. Cf. Grape Leafhopper.

BEET LEAFMINER 1. *Pegomya betae* Curtis [Diptera: Anthomyiidae]. 2. *Liriomyza chenopodii* (Watt) [Diptera: Agromyzidae].

BEET WEBWORM 1. *Loxostege sticticalis* (Linnaeus), 2. *Spoladea recurvalis* (Fabricius) [Lepidoptera: Pyralidae]: A pest in North America that attacks field crops including sugarbeet and alfalfa. Female lays eggs end-to-end in single rows on underside of leaves. Larvae skeletonize leaves and form sheltering tubes by tying leaves together. BWW overwinters as a pupa or late larva within a silk-lined cell in soil. See Pyralidae.

BEETLE Noun. (Anglo Saxon, *bitula* from *bitan* = to bite. Pl., Beetles.) See Coleoptera.

BEETLE CRICKETS See Cachoplistidae.

BEETLE FLIES See Celyphidae.

BEETLE WASPS See Sphecidae.

BEF See Bovine Ephemeral Fever.

BEFFA, GIUSEPPE DELLA (1885–1969) (Osella 1970, Memorie Soc. Entomol. Ital. 49: 195–203, bibliogr.)

BEGG, J (1916–1965) (Wressel 1965, Proc. Entomol. Soc. Ont. 96: 127–128.)

BEGUINOT, AUGUSTO (1875–1940) (De Beaux. 1943, Annali Mus. Civ. Stor. Nat. Giacomo Doria 61: 1–2.)

BEHAVIOUR Noun. (Middle English, *behaven*; *-or* = a condition. Pl., Behaviours.) 1. An organism's muscular or glandular response to stimulation, especially responses that can be observed. 2. The way that an organism responds to stimulation. 3. The unique or predictable reaction of a 'thing' to specific conditions. Rel. Social Insects.

BEHAVIOURAL ASYMMETRY A widespread phenomenon involving patterns of movement and best explained through examples. 1. Flagella that turn in left-handed spirals will not entangle when they rotate clockwise (when viewed from the body). Among Protozoa, some Species rotate clockwise and other Species rotate counterclockwise. 2. Species without eyes or statocysts rotate in one direction; Species with sensory orientation equipment can rotate in either direction. 3. Animals such as crabs walk sideways: Legs on both body sides are morphologically identical, but 'behave' differently. 4. Flight by insects represents another example of behavioural asymmetry. An unequal distribution of indirect flight muscles in mosquitoes causes them to fly in an inappropriate way. The problem is corrected by complex sensory reflexes. 5. Milkweed bugs will walk in circles if the motor output to the hindlegs and optomotor reactions do not compensate for the problem. See Asymmetry; Symmetry. Cf. Absolute Asymmetry; Biological Asymmetry; Developmental Asymmetry; Morphological Asymmetry; Pattern Asymmetry; Skeletal Asymmetry.

BEHAVIOURAL RESISTANCE The ability of an insect population to change its behaviour to avoid insecticides or other detrimental circumstances. Rel. Pest Management.

BEHN, WILHELM FRIEDRICH GEORG (1808–1878) (Henriksen 1926, Entomol. Meddr. 15: 220.)

BEHNINGIIDAE Plural Noun. A small Family of Ephemeroptera, known in North America only on the basis of a few nymphs of one Species.

BEHR, HANS HERMAN (1818–1904) (Born in Colthen, Duchy of Anhalt, Germany; died in San Francisco, CA. German–American physician, scientist and Lepidopterist. Behr made extensive collections that were destroyed in San Francisco during the earthquake of 1906. (Essig 1931, *History of Entomology*. 1029 pp. (553–536), New York; Mallis 1971, *American Entomologists*. 549 pp., New Brunswick.)

BEHRENS, JAMES (1824–1898) Born in Lübeck, Germany; died in San Jose, CA. German–American collector. (Essig 1931, *A History of Entomology*, 1029 pp., New York.)

BEI-BIENKO, GREGORII YAKOVLEVICH (1903–1971) Soviet Russian taxonomic specialist on Diptera. (Zajanckauskas & Skirevicius 1973, Acta Entomol. Lithuanica 2: 217–218.)

BEIJERINCK, MARTINUS WILHELM (1851–1931) (Trotter 1932, Marcellia: 27: 123–124.)

BEIJERINCK, W (1892–1960) (Anon. 1960, Entomol. Ber., Amst. 20: 81–83.)

BEINLUNG, THEODOR (1825–1900) (Kletke 1900, Zeit. Entomol. 25: 26–27.)

BEKKER, ERNEST GEORGIEV (1874–1962) (Makhotin & Pravdin 1963, Entomol. Obozr. 42: 226–233, bibliogr.; transl. Rev. Entomol., Wash. 42: 123–126.)

BELFRAGE, GUSTAF WILHELM (1834–1882) Swedish professional collector who emigrated to USA and collected in Texas (1867–1882.) His collections were sold to European Museums and served as the basis for descriptions of new Species by Packard, Cresson and taxonomic entomologists of the period. (Nowell 1975, Entomol. News 86: 88–90.)

BELIDAE Plural Noun. A small Family of polyphagous Coleoptera assigned to Superfamily Curculionoidea. Adult <22 mm long, elongate with parallel sides and decumbent Setae; Antenna straight, few Species with weakly developed club; Labrum concealed; Pronotum without lateral Carina; Pygidium concealed. Larva curved ventrally, setose and weakly sclerotized; Prothorax declivous. Larvae are wood borers. See Curculionoidea; Curculionidae.

BELING, THEODOR (1816–1898) (Alexander 1920, Mem. Cornell Univ. Agric. Exp. Stn. 38: 693.)

BELIZIN, VLADIMIR IVANOVICH (1905–1970) (Kovalev & Tobias 1971, Entomol. Obozr. 50: 469–471, bibliogr.)

BELL BIRD PSYLLID *Glycaspis baileyi* Moore [Hemiptera: Psyllidae].

BELL ORGAN See Campaniform Sensillum.

BELL, ALWIN S (–1877) (Hewitson 1877, Entomol. Mon. Mag. 14: 141.)

BELL, ERNEST LAYTON (1876–1964) (Brown & Heineman 1972, *Jamaica and its Butterflies*. 478 pp. (22–23), London.)

BELL, THOMAS (1792–1880) (Anon. 1880, Zool. Anz. 3: 168.)

BELL, THOMAS REID DAVYS (1863–1948) (Williams 1949, Proc. Roy. Entomol. Soc. Lond. (C) 13: 66–67.)

BELL, WILLIAM J (1943–1998) American academic (University of Kansas) and administrator; specialist on cockroach biology and physiology; author of ca 100 publications on physiology and behaviour; author of two books and editor of four books on insect biology, orientation behaviour and chemical ecology. Bell developed a specialized locomotion–compensation device (servosphere). Editor, Journal of the Kansas Entomological Society (1982–1984), Environmental Entomology (1984–1987); member of Editorial Board Journal of Insect Physiology; initiated Journal of Insect Behaviour and served as Editor (1988–1998).

BELLA MOTH *Utetheisa bella* (Linnaeus) [Lepidoptera: Arctiidae].

BELLANI, ANGELO (1776–1852) (Veladini 1856, G. R. Istit. Lombardo 9: 485–489.)

BELLARDI, LUIGI (1818–1889) (Papavero 1975, *Essays on the History of Neotropical Dipterology*. 2: 349–351, São Paulo.)

BELLEVOYE, ADOLPHE NICOLAS (1830–1908) (Delamaison 1908, Bull. Soc. Hist. Nat. Rheims 17: 115–116.)

BELLIER DE LA CHAVIGNERIE, JEAN BAPTISTE EUGENE (1819–1888) (Oberthür 1916, Etudes de lepidopterologie comparee 11: (Portraits de Lépidoptèristes 3me sér.) : [2].)

BELLING, HERMAN (1851–1939) (Sachtleben 1939, Arb. Morph. Taxon. Entomol. Berl. 6: 349.)

BELLIO, GIUSEPPE (1901–1949) (Melis 1949

Redia 34: xviii.)

BELL-MARLEY, H W (1872–1945) (Whicher 1949, Entomol. Mon. Mag. 85: 49.)

BELLONCI, GIUSEPPE (–1888) (Anon 1891, Psyche 5: 156.)

BELLOTTI, CRISTOFORO (1823–1919) (De Marchi 1919, Atti Soc. Ital. Sci. Nat. 58: 365–370, bibliogr.)

BELLY Noun. (Middle English, *bely* > Old English, *baelg* = bag, skin. Pl., Bellies.) 1. Venacular term for the ventral surface of the Abdomen. The venter. 2. The general term for the Abdomen, abdominal cavity or abdominal organs. See Abdomen.

BELMARK® See Fenvalerate.

BELON, MAURICE JOSEPH PAUL (1839–1912) (Musgrave 1932, *A Bibliogr. of Australian Entomology 1775–1930*. 380 pp. (17–18), Sydney.)

BELONOID Adj. (Greek, *belone* = needle; *eidos* = form.) Needle-like; pertaining to structure shaped as a needle. Alt. Aciform. See Needle. Rel. Eidos; Form; Shape.

BELOSTOMATIDAE Plural Noun. Giant Water Bugs; Electric Light Bugs; Toe Biters. A numerically small, cosmopolitan Family of heteropterous Hemiptera assigned to Superfamily Nepoidea. Adult body very large (50–100 mm long), flattened, elongate-oval in outline shape, yellow to dull-brown in coloration. Fore legs are raptorial and adapted for prey capture; hind legs flattened and fringed with long Setae that are adpated for swimming. Apex of Abdomen with two short respiratory filaments, but not a conspicuous respiratory siphon. Adults and nymphs inhabit ponds, lakes and streams where they are predaceous upon other insects, snails and vertebrates including fish and frogs. See Nepoidea.

BELT, THOMAS (1832–1878) (Osborn 1952, *A Brief History Of Entomology 1775–1930*. 305 pp. (179), Columbus, Ohio.)

BELTED SKIMMERS See Macromiidae.

BELTIAN BODIES Characteristic structures at the apex of Pinnules or Rachises of *Acacia* which are consumed by ants (*Pseudomyrmex*). See Pearl Bodies. Cf. Beccarian Bodies; Müllerian Bodies.

BEMIS, WILLARD G (–1950) (Sasscer 1950, J. Econ. Entomol. 43: 405.)

BEND OF FOURTH VEIN Diptera: The curve of the fourth vein beyond the posterior crossvein in muscoids (Curran). See Vein; Wing Venation.

BENDERITTER, EUGENE (1869–1940) (Anon. 1940, Bull. Soc. Entomol. Fr. 45: 30.)

BENDEX® See Fenbutatin-Oxide.

BENDIOCARB A carbamate compound {2,2-dimethyl-1,3-benzodioxol-4-yl methylcarbamate} used as a contact insecticide and stomach poison for control of household pests (ants, cockroaches, fleas, flies, silverfish, spiders, termites, wasps). Bendiocarb is applied to non-bearing fruit trees and ornamentals for control of agricultural pests. Phytotoxic to coleus and cardoon. Trade names include: Dycarb®, Ficam®, Garvox®, Multamata®, Seedox®, Seedoxin®, Turcam®. See Carbamate Insecticide.

BENE, GIOVANNI DI (1867–1938) (Anon. 1939, Apicultore Ital. 6: 293–294.)

BENECKE, (–1886) (Anon. 1887, Schr. Phys.-Okon. Ges. Konigsb. 27: 17–18.)

BENEDEN, PIERRE JOSEPH (1809–1894) (Anon. 1895, Entomol. Jb. 4: 247–248.)

BENEDICENTI, ALBERICO (1866–1961) (Bovet 1963, Atti Accad. Naz. Lincei Sci. Fis. Mat. Nat. (8) 34: 320–327.)

BENEFICIAL ORGANISM Biological Control: An organism (pathogen, predator, parasite) that helps reduce a pest population. Syn. Natural Enemies. See Biological Control. Cf. Pathogen; Predator; Parasite.

BENELUZ® See Thiofanox.

BENESAN See Lindane.

BENFURACARB See Aminofuracarb.

BENGTSSON, SIMON FREDERICK (1860–1939) (Brekke 1941, Norsk Entomol. Tidsskr. 6: 47–48.)

BENICK, LUDWIG (1874–1951) (Sokolowsi 1954, Verh. Ver. Naturw. Heimatforsch. 31: xv–xvi, biogr.)

BENIGN TERTIAN MALARIA A moderately serious form of human Malaria caused by the protozoan *Plasmodium vivax*. BTM occurs in cooler areas and is more persistent than other forms of the disease. *Plasmodium vivax* merozoite stage attacks young red blood cells (reticulocytes); hypnozoites may persist in liver for several years. See Malaria. Cf. Malignant Tertian Malaria.

BENJAMIN, FORSTER HENDRIKSON (1895–1936) (Mallis 1971, *American Entomologists*. 549 pp. (334–335, portrait), New Brunswick.)

BENNAM IV® A registered biopesticide derived from *Bacillus thuringiensis* var. *israelensis*. See *Bacillus thuringiensis*.

BENNEFIELD, B L (1937–1962) (Anon. 1962, Bull. Entomol. Soc. Amer. 8: 177.)

BENNETT, GEORGE (1804–1893) (Musgrave 1932, *A Bibliogr. of Australian Entomology 1775–1930*. 380 pp. (18), Sydney.)

BENNETT, STELMON EMERSON (1924–1974) (Southards *et al.* 1974, J. Econ. Entomol. 67: 567.)

BENNETT, WILLIAM HENRY (1862–1931) (Donisthorpe 1931, Entomol. Rec. J. Var. 54: 92.)

BENNINGSEN, H E RUDOLPH VON (–1912) (Morice 1912, Proc. Entomol. Soc. Lond. 1912: clxix.)

BENOIST, RAYMOND (–1970) (Anon. 1970, Bull. Soc. Entomol. Fr. 75: 6–7.)

BENSA, PAOLA (–1964) (Invrea 1964, Memorie Soc. Entomol. Ital. 43: 145–146.)

BENSON, ROBERT BERNARD (1904–1967) (Quinlan 1974, Bull. Brit. Mus. Nat. Hist. (Entomol.) 30: 220.)

BENSULTAP A sulphonate compound {S,S-[2-(dimethylamino) trimethylene]bis (benzenethiosulphonate)} used as a contact insecticide and stomach poison against thrips, Lepidoptera and Coleoptera larvae and aphids. Bensultap is used on vegetables, fruit trees and field crops in some countries, but it is not registered for use in USA. Compound is unusual in acting as synaptic blocking agent and does not act as a cholinesterase inhibitor. Trade names include: Bancol®, Malice®, Ruban®, Victenon®.

BENTHIC Adj. (Greek, *benthos* = depth of the sea; *-ic* = characterized by.) Descriptive of or pertaining to the sea-bottom; by extension, the bottom of any permanent body of water, such as lakes or ponds. Pertaining to organisms near the sea-bottom. Syn. Benthonic. Cf. Estuarine; Pelagic; Littoral; Planktonic.

BENTHOS Noun. (Greek, *benthos* = depth of the sea.) 1. The bottom of a body of water. Cf. Abyssal. 2. The flora and fauna at the bottom of the sea. Cf. Plankton. Rel. Ecology.

BENTINCK, MARGARET CAVENDISH Duchess of Portland. (Griffin 1942, Proc. Roy. Entomol. Soc. Lond. (A) 17: 3.)

BENTIVOGLIO, TITO (1868–1945) (Bertolani 1945, Atti Soc. Nat. Mat. 76: 164–170 bibliogr.)

BENTLEY, GORDAN MANSIR (1875–1954) (Bickley 1957, J. Econ. Entomol. 50: 115.)

BENTON, FRANK (1852–1919) (Anon. 1936, Proc. Entomol. Soc. Wash. 38: 105–106.)

BENTWING GHOST MOTH *Zelotypia stacyi* Scott [Lepidoptera: Hepialidae].

BENTWING SWIFT MOTH See Bentwing Ghost Moth.

BENZ, FRITZ (1907–1975) (Anon. 1976, Bull. Soc. Entomol. Mulhouse 1976 (Apr–June) Suppl.: 1.)

BENZENE A colourless, oily liquid produced from coal. Characterized as miscible with alcohol, volatile, inflammable; often used in clearing tissues in histology. Alt. Benzol.

BENZILAN® See Chlorobenzilate.

BENZOEPIN® See Endosulfan.

BENZOFOS® See Phosalone.

BENZOXIMATE An hydroximate compound {Ethyl O-benzoyl 3-chloro-2,6-dimetoxy-benzohydroximate} used as a contact acaricide in many countries; also, it is effective as a residual compound on all life stages. Benzoximate is applied to fruit trees and grapes in some countries. The compound is toxic to fishes but regarded as non-toxic to beneficial insects. Compound is not registered for use in USA. Trade names include: Aazomat®; Acarmate®; Artaban®; Citrazon®.

BENZYL BENZOATE A repellent for use against ticks and chiggers.

BENZYLISOQUINOLINE ALKALOIDS (See Miller & Feeney 1983, Oecologia 58: 332–339.) Cf. Hydroxycoumarin.

BEOSIT® See Endosulfan.

BERAEIDAE Plural Noun. A small Family of Trichoptera assigned to Superfamily Limnephiloidea. See Limnephiloidea.

BERCE, JEAN ETIENNE (1802–1879) (McLachlan 1880, Entomol. Mon. Mag. 16: 236.)

BERCIO, HANS (–1938) (Heikertinger 1939, Koleopt. Rdsch. 25: 78.)

BERD, CLAUDE (1813–1878) (Nordenskiold 1935, *History of Biology*. 629 pp. (377–380), London.)

BERDEN, SVEN (–1954) (Lindroth 1954, Opusc. Entomol. 19: 238.)

BERENGER, ADOLFO DI (1815–1895) (Micheletti 1895, Boll. Soc. Bot. Ital. 1895: 132–137.)

BERESHCHKOV, ROSTISLAVA PETROCICHA (1881–1961) (Kovalenok & Ioghanzen 1962, Entomol. Obozr. 41: 699–704, bibliogr.; transl. Entomol. Rev., Wash. 41: 434–436.)

BEREZINA, VALENTINYA MIKHAILOVNYA (1899–1960) (Stark 1961, Entomol. Obozr. 40: 454–459; transl. Entomol. Rev., Wash. 40: 238–240.)

BERG, CARL HEINRICH EDMUND VON (1800–1874) (Mobius 1943, Dt. Entomol. Z. Iris 57: 3.)

BERG, CLIFFORD (–1987) American academic (Cornell University) and specialist on the dipterous Family Sciomyzidae.

BERG, FREDERICO GUILLERMO CARLOS (1843–1902) (Osborn 1952, *A Brief History of Entomology*. 305 pp. (179), Columbus, Ohio.)

BERG, JENS FREDERIK (1807–1847) (Natvig 1944, Norsk Entomol. Tidsskr. 7: 19–21.)

BERGE, KARL FREDRICH (1811–1883) (Rebel 1925, Entomol. Z., Frankf. A. M. 37: 22–23.)

BERGENSTAMM, JULIUS EDLER VON (1837–1896) (Brauer 1896, Ann. Naturh. Mus. Wien. 11: 55.)

BERGER, EDWARD WILLIAM (1869–1944) (Osborn 1952, *A Brief History of Entomology*. 305 pp. (179), Columbus, Ohio.)

BERGMANN, ARNO (–1960) (Alberti 1961, Mitt. Dt. Entomol. Ges. 20: 33–35.)

BERGMANN'S RULE A general statement concerning ecological conditions and morphological characteristics which observes that body size of geographic races, Subspecies (or closely related Species) is affected by climate. Organisms with 'Forms' living in cold climates are larger-bodied than related organisms whose 'forms' live in warm climates. Cf. Allen's Rule. Rel. Allometry; Morphometrics.

BERGROTH, ERNST EVALD (1857–1925) Finnish physician and amateur Hemipterist. (China 1926, Entomol. Mon. Mag. 62: 63–64; Lindberg 1928, Mem. Soc. Fauna Flora Fenn. 4: 292–317; Torre-Bueno 1948, Bull. Brooklyn Entomol. Soc. 43: 152–153.)

BERKELEY, M S (–1949) (Wigglesworth 1950, Proc. Roy. Entomol. Soc. Lond. (C) 14: 64.)

BERKENHOUT, JOHN English physician, naturalist and author of *Synopsis of the Natural History of Great Britain and Ireland* (2 vols, 1795.) (Lisney 1960, *Bibliogr. of British Lepidoptera 1608–1799*. 315 pp. (192–196, bibliogr.), London.)

BERLEPSCH, AUGUSTUS VON 1815–1877) (Vogel 1878, Bienenvater Wien 10: 20–25, 103–107, 121–122.)

BERLESE, ANTONIO (1863–1927) Italian Academic; educated at Lycée Foscarini in Venice and Faculty of Natural Sciences at University of Padua; Professor of General and Agricultural Zoology at Agricultural College in Portici and Director of the Agricultural Entomology Station. Berlese was a prolific writer and scientific illustrator who focused his attention on Economic Entomology, taxonomy and biology of insects and mites; he studied life histories of many insects and mites. Berlese established Acarology as a research discipline (with his teacher Giovanni Canestrini) and was a pioneer in biological control. Major achievements include *Gli Insetti;* founded *Redia.* AB did not accept Darwinian evolution but developed his own ideas on polymorphism and change. (Paoli 1928, Mem. Soc. Entomol Ital. 6: 55–84, bibliogr.)

BERLESE, AUGUSTO NAPOLEONE (1864–1903) Botanist and phytopathologist; brother of Antonio Berlese and co-founder of *Rivista di Patologia Vegetale.* (Saccardo 1903, Malpighi 17: 117–128, biogr.)

BERLESE FUNNEL A separation and extraction device that relies on positive geotaxis in response to decreasing humidity to extract and collect small arthropods from leaf litter, duff, and similar material. Basic design involves a funnel with screen material whose size corresponds to the largest diameter of the funnel. The screen is held in place along the shoulder of the funnel and supports vegetative material providing refuge for arthropods. As the sample dries, arthropods move downward through the neck of the funnel and into a collecting container filled with alcohol or other suitable preservative. The original Berlese Funnel used a hot water bath to produce a heat and humidity gradient in the sample; modern methods use an incandescent light bulb above the funnel. The latter device is more properly referred to as a Tullgren Funnel or modified Tullgren Apparatus. See Trap. Cf. Tullgren Funnel Malaise Trap.

BERLESEATE Noun. (Berlese; *-atus* = suffix. Pl., Berleseates.) The arthropod specimens extracted from a Berlese or Tullgren Funnel.

BERMUDAGRASS MITE *Eriophyes cynodoniensis* Sayed [Acari: Eriophyidae].

BERNAM I® A registered biopesticide derived from *Bacillus thuringiensis* var. *kurstaki.* See *Bacillus thuringiensis.*

BERNAM II® A registered biopesticide derived from *Bacillus thuringiensis* var. *kurstaki.* See *Bacillus thuringiensis.*

BERNAM III® A registered biopesticide derived from *Bacillus thuringiensis* var. *kurstaki.* See *Bacillus thuringiensis.*

BERNAN BT II® A biopesticide derived from *Bacillus thuringiensis* var. *morrisoni.* See *Bacillus thuringiensis.*

BERNET-KEMPERS, KAREL JAN WILLEM (1867–1945) (MacGillivray 1945, Entomol. Ber., Amst. 11: 258–260, bibliogr.)

BERNHAUER, MAX (1866–1946) (Jarvis 1946, Entomol. Mon. Mag 82: 160.)

BERNUTH, EMIL VON (1807–) (Ratzeburg 1874, Forstwissensch. Schriftsteller-Lexicon 1: 43–45, bibliogr.)

BEROTHIDAE Plural Noun. Beaded Lacewings. A Family of mantispoid Neuroptera. Adult body small to moderate sized; Antenna shorter than forewing with Scape enlarged and lacking a club; forewing without Nygmata; wings conspicuously setose or with scale-like Setae on veins; hindwing with a few crossveins between R1 and Rs; forelegs not raptorial or rarely so. Berothids are poorly known biologically; some Species are predaceous on termites and some Species are predaceous under bark. See Neuroptera.

BERRILL, GEORGE BATES (1886–1971) (Denmark 1971, Fla. Entomol. 54: 314.)

BERRO AGUILERA, JESUS MARIA DEL (1888–1932) (Canizo 1933, Reseñ. Cient. Roy. Soc. Esp. Hist. Nat. 7: 148.)

BERRY Noun. (Middle English, berye. Pl., Berries.) A fleshy or pulpy fruit with more than one seed and which typically does not split (*e.g.* strawberry, blackberry).

BERTERO, CARLO GIUSEPPE (1789–1831) (Papavero 1975, *Essays on the History of Neotropical Dipterology.* 2: 348–349, São Paulo.)

BERTHA ARMYWORM *Mamestra configurata* Walker [Lepidoptera: Noctuidae]. See Armyworms.

BERTHELOT, SABIN (1794–1880) (Anon. 1881, Leopoldina 17: 45–46.)

BERTHOUMIEU, G VICTOR (–1916) (Bonnet 1945, *Bibliographia Araneorum* 1.)

BERTOLINI, STEFANO DI (1832–1905) (Porta 1905, Riv. Coleott. Ital. 3: 105.)

BERTOLONI, GUISEPPE (1804–1878) (Sforza 1911, G. Storico Lunigiana 3: 128–144, bibliogr.)

BERTRAM, BERT (1895–1928) (Musgrave 1932, *A Bibliogr. of Australian Entomology 1775–1930.* 380 pp. (20), Sydney.)

BERWIG, WIHELM (1899–1973) (Schimitschek 1974, Anz. Schädlingsk. Pflanzenn-Umwelt 47: 78.)

BERYTIDAE Fieber 1851. Plural Noun. Stilt Bugs. A numerically small Family of heteropterous Hemiptera assigned to Superfamily Piesmatoidea or Lygaeoidea. Adult and nymph delicate bodied with elongate, slender, thread-like legs and Antennae. Berytids often are confused with Hydrometridae. However, Berytidae never occur on surface of water and do not have raptorial front legs. Ocelli usually present; Femora and antennal Scape apically enlarged. Eggs adhere to plants via accessory gland secretion or project from plant surface via stalk. Species are phytophagous on plant fluids, Haemolymph of soft-bodied insects or Ooplasm of insect eggs. Syn. Berytinidae; Neididae. See Piesmatoidea.

BERYTINIDAE See Berytidae.

BESCKE, CARL HENRICH (–1851) (Papavero 1971, *Essays on the History of Neotropical Dipterology*. 1: 87–89. São Paulo.)

BESEKE (BESECKE), JOHANN MELCHIOR GOTTLIEB (1746–1802) (Rose 1850, *New General Biographical Dictionary* 4: 183.)

BESET Transitive Verb. (Middle English, *besetten*.) Densely or thickly set with cuticular outgrowths such as tubercles or Setae. Surrounded or ornamented with projections or Setae.

BESS BEETLES See Passalidae.

BESSEMER, HENRY DOUGLAS (1895–1968) (Kennedy 1969, Proc. Roy. Entomol. Soc. Lond. (C) 33: 53.)

BESSER, WILIBALD SWIBERT JOSEPH GOTTLIEB VON (1784–1842) (Trautvetter 1843, Bull. Soc. Imp. Nat. Mosc. 16: 341–360, bibliogr.)

BESSEY, CHARLES EDWIN (1845–1915) (Osborn 1937, *Fragments of Entomological History*. 394 pp. (148–150), Columbus, Ohio.)

BEST, DUDLEY (1843–1928) (Musgrave 1932, *A Bibliogr. of Australian Entomology 1775–1930*. 380 pp. (12), Sydney.)

BESTER, W (–1929) (Auel 1929, Entomol. Z., Frankf. A. M. 43: 29.)

BEST-GARDINER, C (–1965) (Pearson 1966, Proc. Roy. Entomol. Soc. Lond. (C) 30: 62.)

BESTOX® See Alphacypermethrin.

BETA TAXONOMY The arrangement of Species into higher levels of taxonomic categories. See Taxonomy. Cf Alpha Taxonomy; Gamma Taxonomy. Rel. Classification.

BETA-CHLOROPHYLL A form of chlorophyll that produces colour in insects (Wardle). The chlorophyll of all plants is identical, and consists of a mixture of two related compounds: a-chlorophyll and b-chlorophyll. About one molecule of 'b' exists relative to three molecules of 'a.'

BETARAL® See NAA.

BETEL® See *Beauveria brongniartii*.

BETHUNE, CHARLES JAMES STEWART (1838–1932) Canadian cleric and pioneer in economic Entomology. Bethune was Headmaster of Trinity College School (at Port Hope), a founding member Ontario Entomological Society and the American Association of Economic Entomologists. First paper published 1867 (Cutworms Destroying Spring Wheat.) (Mallis 1971, *American Entomologists* 549 pp. (110–113), New Brunswick.)

BETHUNE-BAKER, GEORGE THOMAS (1857–1944) (Wainwright 1945, Entomol. Mon. Mag. 81: 48.)

BETHYLIDAE Haliday 1839. Bethylid Wasps. A cosmopolitan Family of aculeate Hymenoptera assigned to Chrysidoidea and including ca 100 nominal Genera and 2,000 Species. Important diagnostic features of bethylids include body medium sized (1–15 mm), usually black or dark coloured, very rarely with metallic reflections; head subprognathous and body somewhat dorsoventrally compressed; Antenna with 12–13 segments in both sexes, funicular segments usually moniliform; Pronotum extending to Tegula; Parapsidal Sutures and Notauli often present or incipient; Femora often incrassate; wings typically macropterous, but many brachypterous or apterous Species; wing venation reduced; forewing very rarely with closed Submarginal Cell; Stigma typically large or conspicuous; Discoidal Cell usually absent, at most one present; hindwing lacks closed cell but possess Anal Lobe; Propodeum typically large and with Carinae. Fossil record of bethylids begins in Cretaceous Period and the Family is regarded as among the most primitive aculeate Hymenoptera. Host is paralysed with venom from female; eggs are deposited on body of host. Larvae develop as primary, gregarious external parasites of Coleoptera and Lepidoptera larvae. A cocoon is spun by the parasite larva after its host has been exsanguinated. Included Subfamilies: Bethylinae, Epyrinae, Pristocerinae, Galodoxinae, Protopristocerinae, Mesitinae. See Chrysidoidea.

BETHYLOIDEA Haliday 1839. See Chrysioidea.

BETHYLONYMIDAE Rasnitsyn 1975. A Family of aculeate Hymenoptera known from fossil record only.

BETHYLONYMOIDEA Rasnitsyn 1975. A Superfamily of aculeate Hymenoptera.

BETSY BEETLES See Passalidae.

BETTAQUAT-B® See Chlormequat.

BETTONI, EUGENIO (1845–1898) (Pavesi 1898, Rc. Ist. lomb. Sci. Lett. (2) 31: 1285–1299, bibliogr.)

BEURET-MADEUX, LEON (1876–1951) (Vogt 1951, Mitt. Entomol. Ges. Basel 1: 45.)

BEURET-STADELMANN, HENRY (1901–1961) (Anon. 1961, Mitt. Entomol. Ges. Basel 11: 67–77, bibliogr.)

BEUTENMÜLLER, WILLIAM (1864–1919) American curator of insects (American Museum of Natural History) and taxonomic specialist in Lepidoptera. (Weiss 1943, J. N. Y. Entomol. Soc. 51: 286.)

BEUTHIN, HEINRICH (1838–) (Weidner 1967, Abh. Verh. naturw. Ver. Hamburg Suppl. 9: 184–185.)

BEY, PIOT (1857–1953) (Brumpt 1953, Bull. Soc. Path. Exot. 28: 49–50.)

BEYER, GEORGE E (1861–1926) (Viosca 1926, Science 64: 151.)

BEYER, GUSTAV (1840–1924) (Leng 1924, J. N. Y. Entomol. Soc. 32: 165–166.)

BEZZI, MARIO (1868–1927) (Musgrave 1932, *A Bibliogr. of Australian Entomology 1775–1930*. 390 pp. (21–22), Sydney.)

BHATIA, HIRA LAL (1901–1941) (Anon. 1941, Indian J. Entomol. 3: 350–351.)

BHC See Lindane.

BIALAR Alt. Two-winged. Alt. Bialate; Bialatus.

BIALINITZKY-BIRULA, ALEXEI ANDREEVITSCH (1864–1937) (Anon. 1938, Festschrift E. Strand. 4: 660–661.)

BIANCHI, LEO V (–1936) (Anon. 1936, Arb. Morph. Taxon. Entomol. Berl. 3: 151.)

BIANCONI, GIOVANNI GIUSEPPE (1809–1879) (Anon. 1878, Bull. Com. Geol. Ital. 9: 548–549.)

BIARBINEX® See Heptachlor.

BIARCUATE Adj. (Latin, *bis* = twice; *arcus* = bow; -*atus* = adjectival suffix.) Twice curved. Alt. Biarcuatus.

BIAREOLATE Adj. (Latin, *bis* = twice; *areola* = small area; -*atus* = adjectival suffix.) Descriptive of structure with two cells or Areoles. See Bilocular. Alt. Biareolatus.

BIARTICULATE Adj. (Latin, *bis* = twice; *articulus* = joint; -*atus* = adjectival suffix.) Pertaining to structure or appendage with two joints. Alt. Biarticulatus

BIBIONIDAE Plural Noun. Harlequin Flies; Love Bugs; March Flies. A cosmopolitan Family of nematocerous Diptera containing ca. 1,000 extant and fossil Species; most extant Species are tropical. Bibionidae form the most speciose Family of fossil Diptera with more than 340 described Species; oldest Species taken in Upper Jurassic Period deposits of Dan River Formation. Adult body small to moderately large, mainly dark-coloured and rather setose. Head distinctively elongate. Compound eyes are holoptic in males and widely separate in females; Ocelli positioned on a well developed prominence. Scutum is conspicuously raised and dome-like. Abdomen long and slender with apex upturned in male. Adults commonly are observed in mating swarms. Adults of some Species feed upon pollen and nectar. Larvae live and feed in decaying vegetation. See Bibionomorpha; Diptera.

BIBIONOMORPHA A Division of nematocerous Diptera including Anisopodidae (Phryneidae, Rhyphidae, Silvicolidae), Aximiidae, Bibionidae, Blephariceridae, (Blepharoceridae, Blepharoceratidae), Cecidomyiidae, (Cecidomyidae, Itonididae), Deuterophlebiidae, Hyperoscelididae, (Corynoscelidae), Mycetophilididae, (Bolitophilidae, Ceroplatidae, Diadociidae, Ditomyiidae, Fungivoridae, Macroceratidae, (Sciophilidae), Nymphomyiidae, Pacyneuridae, Perissommatidae, Scatopsidae, Sciaridae. See Nematocera.

BIBRA, ERNST VON (1806–1878) (Anon. 1878, Amer. J. Sci. (3) 16: 164.)

BICARINATE Adj. (Latin, *bis* = twice; *carina* = keel; -*atus* = adjectival suffix.) Pertaining to Integument with two Carinae or elevated cuticular ridges. See Carinate. Alt. Bicarinatus. See Carina. Rel. Sculpture.

BICAUDATE Adj. (Latin, *bis* = twice; -*atus* = adjectival suffix.) Descriptive of insects with two 'tails' or anal processes. Alt. Bicaudatus. See Cauda.

BICHLORIDE OF MERCURY $HgCl_2$. A chemical compound of mercury and chlorine in the form of crystals or a white powder. Compound is highly poisonous, soluble in water, glycerine and alcohol. It is used as a preservative of tissues in aqueous solution.

BICKHARDT, HEINRICH (1873–1920) (Weber 1921, Entomol. Bl. Biol. Syst. Käfer 17: 1–5, bibliogr.)

BICOLOUR Adj. (Latin, *bis* = twice; *colour*, perhaps *celare* = to conceal.) Pertaining to structure with two colours that contrast to some extent. Alt. Bicolourate; Bicolouratus; Bicoloured; Bicolourous.

BICONVEX Adj. (Latin, *bis* = twice; *convexus* = arched, vaulted). Double convex, *i.e.,* lenticular or lens-shaped. See Lens. Cf. Lenticular.

BICORNUTE Adj. (Latin, *bis* = two; *cornua* = horn.) Pertaining to structure with two 'horns' or projections. A term often used to describe cephalic processes. Alt. Bicornutus. See Cornua.

BICUSPIDATE Adj. (Latin, *bis* = twice; *cuspis* = point; -*atus* = adjectival suffix.) Two-pointed; pertaining to structure with two cusps. Alt. Bicuspidatus. See Cusp.

BIDACTYLATE Adj. (Latin, *bis* = twice; Greek, *daktylos* = finger; -*atus* = adjectival suffix.) With two fingers or finger-like processes. Alt. Bidactylus. See Dactyl.

BIDDIES See Cordulesgastridae.

BIDENS BORER *Epiblema otiosana* (Clemens) [Lepidoptera: Tortricidae].

BIDENTATE Adj. (Latin, *bis* = twice; *dens* = tooth; -*atus* = adjectival suffix.) Pertaining to structure with two teeth or tooth-like projections. Alt. Bidentatus. See Dens. Rel. Tooth.

BIDENTICULATE Adj. (Latin, *bis* = twice; *denticule* = small tooth; -*atus* = adjectival suffix.) Pertaining to structure with two small teeth. Alt. Bidenticulatus. See Denticle.

BIDERON® See Prothiofos.

BIDRIN® See Dicrotophos.

BIEBERDORFF, G A (1898–1961) (Anon. 1961, Bull. Entomol. Soc. Amer. 7(4): [i].)

BIEBL, AMANDUS (–1949) (Anon. 1949, Anz. Schädlingsk. 22: 78.)

BIEBUYCK, ANDRÉ (–1945) (Collart 1945, Bull. Ann. Soc. Entomol. Belg. 81: 53.)

BIEDERMANN, ROBERT (–1956) (Anon. 1956, Bull. Soc. Entomol. Fr. 61: 7.)

BIEGELEBEN, FRANCESCO (1881–1942) (Bononi 1942, Studi trent. Sci. Nat. 23: 157–158.)

BIELZ, EDUARD ALBERT (1827–1898) (Csiki 1900, Rovart. Lap. 7: 1–4, bibliogr.)

BIEMARGINATE Adj. (Latin, *bis* = twice; *ex* = out; *marginare* = to delimit; -*atus* = adjectival suffix.) Pertaining to the margin of a stucture which has two excisions or notches. Alt. Biemarginatus. See Emarginate. Rel. Margin.

BIENER, CHRISTIAN GOTTLOB (1748–1828) (Rose 1854, *New General Biographical Dictionary* 4: 232.)

BIFARIOUS Adj. (Latin, *bis* = twice; *fariam* = in rows; -*osus* = full of.) Pointing in opposite direction. Alt. Bifarius.

BIFASCIATE Adj. (Latin, *bis* = twice; *fascia* = band; -*atus* = adjectival suffix.) Descriptive of structure with two bands of sheet-like membranous connective tissue. Alt. Bifasciatus. See Fascia.

BIFENTHRIN A synthetic-pyrethroid compound used as a contact insecticide, stomach poison or acaricide on ornamental plants and cotton in some countries. Phytotoxic on several ornamentals. Toxic to bees and fishes. Trade names include: Biflex®, Brigade®, Brigata®, Brookade®, Capture®, Flee®, Talstar®. See Synthetic Pyrethroids.

BIFID Adj. (Latin, *bis* = twice; *findere* = to split.) Pertaining to structure that is cleft or divided into two parts or lobes; forked. Alt. Bifidus. See Biparted.

BIFLABELLATE Adj. (Latin, *bis* = twice; *flabellum* = fan; *-atus* = adjectival suffix.) Twice-flabellate; a term applied to Antennae in which both sides of the funicular segments are fan-like or display flabellate processes. Alt. Biflabellatus. See Flabellum.

BIFLEX® See Bifenthrin.

BIFOLLICULAR Adj. (Latin, *bis* = two; *follis* = bag, sac.) Consisting of two follicles.

BIFOROUS Adj. (Latin, *biforous* = with two openings.) Pertaining to structure with two openings. A term usually applied to spiracles. Alt. Biforate.

BIFOROUS SPIRACLES Spiracles with two pouches in the Atrium. BS are seen in some Coleoptera larvae and hypothetically originally opened separately to the exterior (Snodgrass). See Spiracle. Rel Respiration.

BIFURCATE Adj. (Latin, *bis* = twice; *furca* = fork; *-atus* = adjectival suffix.) Pertaining to structure that is divided partly, or forked into two arms which emanate from a basomedial stem. Alt. Bifurcatus; Bifurcous. See Furca.

BIFURCATION Noun. (Latin, *bis* = twice; *furca* = fork; English, *-tion* = result of an action. Pl., Bifurcations.) 1. A division into two arms. 2. The point at which a forking occurs. See Furca.

BIG HEADED FLIES See Pipunculidae.

BIG-EYED BUG 1. *Geocoris lubra* (Kirkaldy) [Hemiptera: Lygaeidae]: A beneficial insect found in western North America. Adult and nymph are predaceous upon aphids and leafhoppers in field crops (alfalfa, cotton, sugarbeet). 2. *Geocoris lubra* Kirkaldy [Hemiptera: Lygaeidae]: An Australian Species. Adult 2.5–3.5 mm long, generally black with dark brown markings on the Pronotum, head and wing margins. The head, underside of body and legs are covered in golden hairs. The distinctive large eyes are dark brown approaching black or a plain mixture of black and red. Eyes are positioned on large lateral lobes of the head. The mouthparts extend to the middle Coxa. Eggs are white, cylindrical and laid singly on terminals and under leaves. Nymphs resemble the adult but lack wings. Adults and nymphs are predatory on soft-bodied insects and mites. Adults and nymphs are predatory during the day and found over the cotton plant. 3. *Germalus* sp. [Hemiptera: Lygaeidae]: An Australian Species. Adult 3.0–4.0 mm long, generally brown and green; head, underside of body and legs covered with golden Setae; compound eyes are large, dark brown approaching black or a plain mixture of black and red; eyes positioned on large lateral lobes of the head; mouthparts extend to middle Coxa. Eggs white, cylindrical and laid singly on terminals and under leaves. Nymphs resemble adult but lack wings. Adult and nymphs are diurnal predators on soft-bodied insects and mites. See Lygaeidae.

BIG-EYED CLICK BEETLE See Eyed Click Beetle.

BIG-EYED FLIES See Pipunculidae.

BIG-HEADED ANT *Pheidole megacephala* (Fabricius) [Hymenoptera: Formicidae].

BIG-HEADED FLIES See Pipunculidae.

BIG-HEADED GRASSHOPPER *Aulocara elliotti* (Thomas) [Orthoptera: Acrididae]: In North America, a common, economically important pest of rangeland grasses. BHG overwinters in egg-pods within two cm from top of soil. See Acrididae.

BIGEYES See Priacanthidae.

BIGNELL, GEORGE CARTER (1826–1910) (Morley 1910, Entomologist 43: 128.)

BIGNONIACEAE A pantropical Family of flowering trees and vines with more than 800 described Species. Nearly 650 Species are Neotropical.

BIGOT, JACQUES MARIE FRANGILE (1818–1893) (Papavero 1971, *Essays on the History of Neotropical Dipterology*. 1: 196. São Paulo.)

BIGUTTATE Adj. (Latin, *bis* = twice: *gutta* = drop; *-atus* = adjectival suffix.) Pertaining to structure with two drop-like spots. Alt. Biguttatus.

BIJUGATE Adj. (Latin, *bis* = twice; *jugare* = to join.) Descriptive of branches with two pairs of leaflets. See Jugum.

BILAMELLAR Adj. (Latin, *bis* = twice; *lamina* = plate.) Pertaining to structure which is divided into two laminae or sclerites. Alt. Bilamellate; Bilamellatus. See Lamella.

BILATERAL Adj. (Latin, *bis* = twice; *latus* = side; *-alis* = pertaining to.) Descriptive of structure with two equal or symmetrical sides. See Symmetry.

BILATERAL GYNANDROMORPH An individual that expresses a type of gynandromorphism in which one side of an individual shows female features and the other side shows male features. BGs appear widespread and are the most common type of Gynandromorph in most Orders of insects. In orthopteran *Valanga irregularis*, BG probably develops from a binucleate egg as a result of double fertilization. BG also known in *Melanoplus differentialis* and *Chorthippus montanus*. See Gynandromorphism. Cf. Sexual Mosaic; Transverse Gynandromorph. Rel. Development; Metamorphosis; Mutant.

BILATERAL SYMMETRY Anatomy: Similarity of shape with one side a mirror image of the other. BS is apparent when an imaginary line is drawn through the primary axis of a body: The left and right halves of the organism display BS symmetry. BS is the most common form of symmetry in biological systems. See Asymmetry; Symmetry. Cf. Pentameral Symmetry; Radial Symmetry; Spherical Symmetry; Zonal Symmetry.

BILBERG (BILLBERG), JOHANN (1650–1717) (Rose 1854, *New General Biographical Dictionary* 4: 239.)

BILEK, ALOIS (1909–1974) (Jurzitza 1975, Odonatoiogica 4: 31–33, bibliogr.)

BILIARY FEVER See Equine Babesiosis.

BILIARY VESSELS See Malpighian Tubules.

BILIMEK, DOMINIK (Papavero 1975, *Essays on the History of Neotropical Dipterology*. 2: 291–292. São Paulo.)

BILINEATE Adj. (Latin, *bis* = twice; *linea* = line; *-atus* = adjectival suffix.) Descriptive of structure marked with two lines. Alt. Bilineatus.

BILLBUG *Sphenophorus brunnipennis* (Germar) [Coleoptera: Curculionidae]. Any of several Species of the curculionid beetle Genus *Sphenophorus* that occur primarily in the grasslands of North America. Billbugs are significant agricultural pests in that larvae feed on fibrous roots of grains and cultivated grasses and stems of small grains. Adults chew distinctive-shaped holes in foliage and chew holes in stems to feed and oviposit. Eggs are reniform and deposited in a small hole in stem chewed by female; eclosion requires 4–15 days. Adults display thanatotic behaviour when disturbed; typically univoltine and overwinter as adults. See Curculionidae.

BILLIARD, GEORGES (–1960) (Anon. 1960, Bull. Soc. Entomol. Fr. 66: 6.)

BILLINGS, BRODDISH (1819–1871) (Anon. 1872, Can. Entomol. 4: 70–73.)

BILLINGS, ELKANAH (1820–1876) (Whiteaves 1878, Can. Entomol. 8: 251–261.)

BILLIOUD, GABRIEL (1806–1893) (Anon. 1894, Ann. Soc. Agric. Lyon (7) 1: 521–523.)

BILLUPS, THOMAS RICHARD (1841–1919) (R. A. 1920, Entomol. Mon. Mag. 56: 66.)

BILOBATE Adj. (Latin, *bis* = twice; *-atus* = adjectival suffix.) Pertaining to structure which is divided into two lobes. Alt. Bilobatus; Bilobed. See Lobe. Cf. Lobate.

BILOBRAN® See Monocrotophos.

BILOCULAR Adj. (Latin, *bis* = twice; *oculus* = eye.) Pertaining to structure with two cells or compartments. Cf. Biareolate.

BILSING, SHERMAN WEAVER (1885–1954) (Mallis 1971, *American Entomologists*. 549 pp. (429–430), New Brunswick.)

BIMACULATE Adj. (Latin, *bis* = twice; *macula* = spot; *-atus* = adjectival suffix.) With two spots or maculae. Alt. Bimaculatus. See Maculate.

BINARY Adj. (Latin, *binarius* = pair.) Zoological nomenclature: Two similar or related things or concepts; binominal.

BINARY NAME A name consisting of two parts, namely, a generic name and a specific name. See Binomen. Cf. Uninomial.

BINARY NOMENCLATURE A system of designating Species in Zoology with a binary name.

BINATE Adj. (Latin, *bis* = twice; *-atus* = adjectival suffix.) Consisting of a single pair; in pairs.

BINDER, ADOLF (1876–1935) (Müller 1935, Ost. EntVer. 20: 30–32.)

BINGHAM, CHARLES THOMAS (1848–1908) (Musgrave 1932, *A Bibliogr. of Australian Entomology 1775–1930*, 380 pp., (23, bibliogr.), Sydney.)

BINOMEN Noun. (Latin, *bis* = twice; *nomen* = name. Pl., Binomens.) The Latin or latinized scientific name of an organism. The binomen is comprised of two parts, the generic name and the specific epithet. See Nomenclature. Cf. Common Name; Scientific Name.

BINOMINAL Adj. (Latin, *bis* = twice; *nomen* = name.) 1. Descriptive of or pertaining to two names; consisting of two names. 2. Something with two terms, but not necessarily names.

BINOMIAL NOMENCLATURE A double-term name method which is the ordinarily accepted system of zoological nomenclature, consisting of the name of a Genus and the name of a Species coupled together.

BINOT, JEAN (1867–1909) (Horn 1910, Dt. Entomol. Z. 1910: 212.)

BINOTATE Adj. (Latin, *bi* = two; *nota* = mark; *-atus* = adjectival suffix.) Descriptive of structure with two rounded spots of colour, size, shape or texture that are distinctive from the surrounding area. Alt. Binotatus.

BINUS Adj. Paired; doubled.

BIO 1020® See *Metarhizium anisopliae.*

BIOBIT® A registered biopesticide derived from *Bacillus thuringiensis* var. *kurstaki*. See *Bacillus thuringiensis*.

BIOBLAST® See *Metarhizium anisopliae* ESF 1.

BIOCHEMICAL RESISTANCE The ability of an insect population to develop enzymes that detoxify an insecticide or other injurious material before it can reach the anatomical site of action.

BIOCHEMISTRY Biological chemistry. The chemistry of living organisms and of their functions, secretions and parts. The branch of chemistry that is concerned with the formation, constitution and reactions of the chemical components of the living organism.

BIOCLIMATIC LAW See Hopkins Bioclimatic Law.

BIOCOENOSIS Noun. (Greek, *bios* = life; *koinos* = general, common; *sis* = a condition or state. Pl., Biocoenoses.) 'A community of living beings where the sum of the Species and individuals, being mutually limited and selected under the average external conditions of life, have by means of transmission, continued in possession of a certain territory' (K. Mobius). Alt . Biocenose; Biocoenose.

BIOCONTROL-1® See Red-headed Pine Sawfly NPV.

BIOCOT® A registered biopesticide derived from

Bacillus thuringiensis var. *kurstaki*. See *Bacillus thuringiensis*.

BIODART® A registered biopesticide derived from *Bacillus thuringiensis* var. *kurstaki*. See *Bacillus thuringiensis*.

BIODETERIORATION Noun. (Greek, *bios* = life; Latin, *deteriorare* = to make worse. Pl., Biodeteriorations.) The process of decay or disintegration involving the action of microorganisms and higher organisms such as insects.

BIODIVERSITY Noun. (Greek, *bios* = life; Latin, *diversus*. Pl., Biodiversities.) 1. The condition of being different biologically. 2. '...the variety of the world's organisms, including their genetic diversity and the assemblages they form. A blanket term for the natural biological wealth that undergirds human life and well being. The breadth of the concept reflects the interrelatedness of genes, Species and ecosystems.' (Reid & Miller 1989, *Keeping Opinions Alive: The scientific basis for conserving biodiversity*. World Research Institute, Washington.) 3. The variety of living organisms considered at all levels, from genetics through Species, to higher taxonomic levels, and including the variety of habitats and ecosystems (Meffe & Carroll *Conservation Biology*.) 4. United States Office of Technology Assessment (OTA): '...the variety and variability among living organisms and the ecological complexes in which they occur.' In *Technologies to Maintain Biological Diversity*. See Diversity.

BIOECOLOGY Noun. (Greek, *bios* = life; *oikos* = dwelling. Pl., Bioecologies.) 1. The sociology of organisms. 2. The study of living organisms from the perspective of where they live and what they do. See Ecology. Rel. Habitat.

BIOFLEA® See Entomogenous Nematodes.

BIOGAS Noun. (Greek *bios* = life; Latin, *chaos*. Pl., Biogasses.) Methane produced by the fermentation of animal excrement. Biogas is a flammable mixture (ca 70% methane) that is produced by biomethanation involving symbiotic microorganisms under anaerobic conditions. Biogas is a significant small-scale source of energy in many developing countries. Rel. Biotechnology.

BIOGENESIS Noun. (Greek *bios* = life; *genesis* = descent. Pl., Biogeneses.) The production of life from antecedent life. See Genesis. Cf. Abiogenesis. Rel. Evolution.

BIOGENETIC LAW See Haeckel's Law.

BIOGEOGRAPHICAL Adj. (Greek, *bios* = life; *geographia* > *gaia* = earth, *graphein* = to write; Latin, *-alis* = pertaining to.) Pertaining to biogeography.

BIOGEOGRAPHIC REGION Any of several parts of the earth's geography that are identified as distinctive due to their native (endemic) plants and animals. See Biogeography. Cf. Phytogeography; Zoogeography. Rel. Ecology.

BIOGEOGRAPHY Noun. (Greek, *bios* = life; *geographia* > *gaia* = earth, *graphein* = to write. Pl., Biogeographies.) The study of the distribution of organisms in space through time (Wiley 1981). Component elements of Biogeography include Zoogeography (which refers to the distribution of animals) and Phytogeography (which refers to the distribution of plants). Endemism, centre of origin, dispersal, and range are all concepts related to Biogeography. See Distribution. Cf. Phytogeography; Zoogeography. Rel. Continental Drift.

BIOLAR® A registered biopesticide derived from *Bacillus thuringiensis* var. *israelensis*. See *Bacillus thuringiensis*. Rel. Biopesticide.

BIOLEACHING Noun. (Greek, *bios* = life; Anglo Saxon, *leccan* = to moisten > Old Norse, *leka* = trickle. Pl., Bioleachings.) The process by which microbes (bacteria, yeasts, fungi, algae, protozoa) are used to solubilize and extract commercially important metals from ores. Rel. Biotechnology.

BIOLISTICS Plural Noun. A technique that is used to introduce DNA into plant cells. Biolistics involves precipitating DNA onto microscopic particles. The DNA-treated particles are then accelerated to penetrate the plant cells and deposit the DNA. Rel. Biotechnology.

BIOLLEY, PAUL (1862–1909) (Rehn 1908, Entomol. News 19: 394–395.)

BIOLOGICAL ASYMMETRY A type of asymmetry involving placement in part as influenced by substrate and hierarchical complexity of standard parts. For instance, the sperm Flagellum consists of a basal portion, tubules, subtubules, arms, connecting fibres, giant molecules, peptides and amino acids. The functional position of standard parts is influenced by 'burden.' The degree of burden is genetically specified by the number of subsequent decisions that depend upon a preliminary decision. Standard Parts are hierarchical in nature, which means that a defect will cause problems at a higher level and not at a lower level. This is the basis of biological asymmetry. See Asymmetry; Symmetry. Absolute Asymmetry; Behavioural Asymmetry; Developmental Asymmetry; Morphological Asymmetry; Pattern Asymmetry; Skeletal Asymmetry. Rel. Organization.

BIOLOGICAL CLOCK The general term used to describe behavioural, biological and physiological phenomena that are periodic in nature and predictable. See Circadian Rhythm; Diel Periodicity. Rel. Endogenous Ryhthm; Exogenous Rhythm; Photoperiodism.

BIOLOGICAL CONTROL 1. The control of pests (pathogens, mites, insects, vertebrates, weeds) by employing natural enemies including predators, parasites and pathogens. See Classical Biological Control, Augmentative Biological Control. Cf. Chemical Control, Cultural Control, Integrated Pest Management, Microbial Control, Natural Control, Regulatory Control. 2. The use of natural enemies such as parasitoids, predators, pathogens, antagonist, or competitor populations to suppress a pest population making it less abundant and thus less damaging (VanDriesche & Bellows 1996.)

BIOLOGICAL SPECIES A population with a common heredity (Kinsey). See Species.

BIOLOGICAL SPECIES CONCEPT A concept that develops the notion that Species are interbreeding natural populations reproductively isolated from other similar, natural populations. (Mayr 1963, *Animal Species and Evolution*.) Cf. Species Concepts. Rel. Evolution.

BIOLOGICAL VECTOR An arthropod that serves an essential role in the passage of a pathogen from one vertebrate host to another host. Biological vectors may have one of three transmission relationships with the pathogen: Propagative, Cyclopropagative or Cyclodevelopmental. Arthropods that serve as biological vectors are intermediate hosts for the pathogens and for a period of time are not capable of transmitting the pathogen to another host. When the pathogen has completed part of its life cycle or multiplied within the biological vector, the pathogen can be effectively transmitted. Eliminating the biological vector of a pathogen will eliminate transmission of the pathogen. *e.g.* Mosquitoes serve as biological vectors of Malaria to vertebrates. See Cyclopropagative Transmission; Cyclodevelopmental Transmission; Propagative Transmission. Cf. Mechanical Vector; Passive Transmission.

BIOLOGY Noun. (Greek, *bios* = life; *logos* = discourse. Pl., Biologies.) The study of life to include all of its theoretical and practical aspects. Biology is broadly divided into Botany and Zoology, with each branch including numerous subdisciplines. Cf. Botany; Zoology. Rel. Anatomy; Behaviour; Biochemistry; Ecology; Genetics; Morphology; Physiology;

BIOLUMINESCENCE Noun. (Greek, *bios* = life; Latin, *lumen* = light; *escent* = beginning. Pl., Bioluminescences.) The phenomenon of living organisms producing visible light. In one process, light is produced by an oxidative reaction within specialized cells or tissues in which luciferin is broken down by luciferase. Energy from the reaction is released in the form of a heatless light. The phenomenon exists in many groups of plants and animals including algae, arthropods, bacteria, fungi, fish, mollusks and plankton. Insect bioluminescence is known in Collembola, Diptera and Coleoptera. Fireflies (Lampyridae) of the Genera *Photuris* and *Photinus* are best-known luminescent insects. Luminescence that is excited by ultraviolet has been studied in some insects.

BIOMAGNIFICATION Noun. (Greek, *bios* = life; Latin, *magnificatus* = magnificent, splendid. Pl., Biomagnifications.) The increase in concentration of a substance (especially a pesticide) in animal tissues as related to the animal's higher position in the food chain. Rel. Ecology.

BIOMASS Plural Noun. (Greek, *bios* = life; *maza* = lump. Pl., Biomasses.) The collective organic matter that results from the photosynthetic conversion of solar energy. Annual plant biomass production from photosynthesis on land is about 120 billion tons and oceanic production is about 50 billion tons. Agriculture accounts for about 6% of the photosynthetic productivity.

BIOME Noun. (Greek, *bios* = life. Pl., Biomes.) A major ecological community that consists of climax plants and animals in a large area or region (*e.g.* grassland biome, desert biome). See Ecosystem. Rel. Habitat.

BIOMECHANICS Plural Noun. (Greek, *bios* = life; Latin, *mechanicus* > Greek, *mechanikos* > *mechane* = machine.) The study of the mechanics of operation or motion of structure during its function. See Morphology. Cf. Ergonomics.

BIOMETER Noun. (Greek, *bios* = life; *metron* = to measure. Pl., Biometers.) An organism that may be used as an indicator of the conditions or suitability of climate to the establishment or continued existence of life under a range of climatic conditions (Shelford).

BIOMETRICIAN Noun. Greek, *bios* = life; *metron* = measure. Pl., Biometricians.) 1. A student of biometrics or biometry. 2. A person who uses statistical information or methods to study organisms. See Biometry.

BIOMETRY Noun. (Greek, *bios* = life; *metron* = measure. Pl., Biometries.) The application of statistical methods to the study of biological facts or phenomena and their variation. Cf. Allometry; Morphometrics.

BIOMORPHOTICA Noun. Neuroptera: Species with a Pupa that can move.

BION Noun. (Greek, *bion* = living. Pl., Bions.) The physiological individual characterized by definiteness and independence of function. An organism.

BIONEEM® See Azadirachtin.

BIONEX® See Azinphos-Ethyl.

BIONICS Plural Noun. (Greek, *bion* = living.) The application of engineering principles to the analysis or study of biological systems.

BIONIM® A neem insecticide derived by ethanolic extraction from neem seed. Developed by Gabrol Produkter for use in Sweden against aphids, thrips, whiteflies and mites on ornamental plants. See Azadirachtin.

BIONOMIC Adj. (Greek, *bios* = life; *nomos* = law; *-ic* = of the nature of.) Pertaining to bionomics. Adv. Bionomically.

BIONOMICS Noun. (Greek, *bios* = life; *nomos* = law.) A branch of biology that involves the relationship of animals to their environment. Ecology in the broad sense. Alt. Bionomy. Cf. Ergonomics.

BIONT Noun. (Greek, *bion* = living.) See Bion.

BIOPATH FLY CONTROL® See *Metarhizium anisopliae* ESF 1.

BIOPATH ROACH CONTROL® See *Metarhizium anisopliae* ESF 1.

BIOPESTICIDE Noun. (Greek, *bios* = living; *pestis* = plague; Latin, *-cide* > *caedere* = to kill. Pl., Biopesticides.) 1. Secondary plant compounds which have chemical properties that kill insects. See Botanical Insecticide. 2. Organisms (Viruses, Bacteria, Fungi, Nematodes) that develop as pathogens of arthropods, weeds and vertebrates. See *Bacillus thuringiensis*; Entomogenous Nematodes; Entomogenous Fungi; Fungicide; Granulosis Virus; Nuclear Polyhedrosis Virus. Cf. Acaricide; Filaricide; Fungicide; Herbicide; Insecticide; Nematicide; Pesticide. Rel. Microbial Pesticide; Insect Pathogen.

BIOPHARMACEUTICAL Noun. (Greek, *bios* = living; *pharmakeutikos*. Pl., Biopharmaceuticals.) Any pharmaceutical drug derived from biological sources, including recombinant protein drugs, recombinant vaccines and monoclonal antibodies. Cf. Synthetic Pharmaceutical. Rel. Biotechnology.

BIOPHORE Noun. (Greek, *bios* = life; *pherein* = to carry. Pl., Biophores.) An hypothetical, supposed ultimate, constituent of germ plasm or hereditary substance.

BIOPSY Noun. (Greek, *bios* = living; *-opsis* = sight, appearance, view. Pl., Biopsies.) The removal (excision) of tissue from an organism for the purpose of diagnostic examination. Cf. Autopsy; Necropsy. Rel. Pathology.

BIORDINAL CROCHETS Lepidoptera larva: Proleg Crochets arranged in a uniserial circle of two lengths, alternating.

BIOREACTOR Noun. (Greek, *bios* = life; Latin, *re* = backward; *act; or* = one who does something. Pl., Bioreactors.) A vessel (fermenter) that can support large-scale production of cell cultures. Bioreactors are used in industrial production of pathogens such as *Bacillus thuringiensis*.

BIOS Noun. (Greek, *bios* = life.) Any form of organic life.

BIOSAFE® See Entomogenous Nematodes.

BIOSAFE-N® See Entomogenous Nematodes.

BIOSENSOR Noun. (Greek, *bios* = life; *or* = one who does something. Pl., Biosensors.) An electronic device that uses biological molecules to detect specific chemical compounds.

BIOSIS Noun. (Greek, *biosis* = way of life > *bios* = life. Pl., Bioses.) 1. A mode of living. 2. Vitality. Cf. Abiosis; Anhydrobiosis; Antibiosis; Archebiosis; Calobiosis; Cleptobiosis; Hamabiosis; Kleptobiosis; Lestobiosis; Parabiosis; Phylacobiosis; Plesiobiosis; Synclerobiosis; Trophobiosis; Xenobiosis.

BIOSPECIES See Biological Species.

BIOSPELEOLOGY Noun. (Greek, *bios* = life; *spelaion* = cave; *logos* = discourse. Biospeleologies.) The study of cave-dwelling organisms.

BIOSPHERE Noun. (Greek, *bios* = life; *sphaira* = globe. Pl., Biospheres.) The portion of the world that contains living organisms. Rel. Ecology.

BIOSTASIS Noun. (Greek, *bios* = life; *stasis* = standing. Pl., Biostases.) The capacity of an organism to withstand environmental change without being altered by the changing environment. Cf. Selection.

BIOSYNTHESIS Noun. (Greek, *bios* = life; *synthesis* > *syntithenai* = to put together. Pl., Biosyntheses.) The formation of organic compounds by living organisms. Cf. Autolysis.

BIOSYSTEMATICS Plural Noun. (Greek, *bios* = life; *systema* = whole made of several parts.) The development of taxonomic classifications that are based on similarities in anatomical characters and biological features. See Classification; Systematics. Cf. Taxonomy; Cladistics.

BIOTA Noun. (Greek, *bios* = life. Pl., Biotas.) The fauna and flora of a given habitat, area or zoogeographical region. See Ecology. Cf. Fauna; Flora. Rel. Biome.

BIOTECHNOLOGY Noun. (Greek, *bios* = life + technology. Pl., Biotechnologies.) 1. The use of living organisms and their constituents for agriculture, food and industrial process. 2. The application of scientific and engineering principles to develop or process biological organisms or their products to provide goods or services. 3. The integration of biochemistry, microbiology and engineering to develop or refine industrial processes that involve microorganisms, cultured cells or cell products.

BIOTHERAPY Noun. (Greek, *bios* = life; *therapia* = treatment. Pl., Biotherapies.) The medicinal use of live organisms (*e.g* maggots, leeches) in the treatment of human disorders. Cf. Medical Debridement Therapy.

BIOTHRIN® See Permethrin.

BIOTIC POTENTIAL 1. The maximum rate of population increase based on maximum natality and minimum mortality. 2. '...the power an organism has to reproduce and survive in its environment' (Chapman).

BIOTIC Adj. (Greek, *biotikos* = pertaining to life; *-ic* = characterized by.) Descriptive of or pertaining to the biota or the dynamical properties of living organisms. See Biota. Cf. Abiotic; Ecobiotic.

BIOTICALLY Adv. (Greek, *biotikos* = pertaining to life.) In a biotic manner or way.

BIOTOPE Noun. (Greek, *bios* = life; *topos* = place. Pl., Biotopes.) The habitat, environment or ecological setting (ecotope) in which an organism (population, biotype, Species, ecospecies) resides. Cf. Biome; Ecosystem. Rel. Ecology.

BIOTROL® A registered biopesticide derived from *Bacillus thuringiensis* var. *kurstaki*. See *Bacillus thuringiensis*.

BIOTYPE Noun. (Greek, *bios* = life; *typos* = pattern. Pl., Biotypes.) Entomology: A genetically cohesive population of insects that demonstrates biological and phenological differences from morphologically identical forms. The terms 'race,' 'strain' and 'variety' have been used interchangeably with the term biotype. However, each term has a subtly different meaning that is not applicable to Entomology, has not been defined for Entomology and is well established in other fields of biology (Anthropology, Botany, Microbiology). A biotype has no standing in Zoological Nomenclature. See Species. Cf. Race; Strain; Variety. Rel. Biological Control; Taxonomy.

BIOVECTOR® A biopesticide used for control of various insects. Material is marketed under the name Biovector® 25 for control of weevils, girdlers and borers in cranberry and mint; marketed under the name Biovector® 335 for control of Citrus Root Weevil, Sugarcane Rootstalk Borer and Blue-Green Weevil. Active ingredient the nematode *Steinernema riobravis*. See Entomogenous Nematodes. Cf. Millenium® Rel. Biopesticide.

BIPARTITE Adj. (Latin, *bis* = twice; *partitus* = divided; *-ites* = constituent.) Pertaining to structures, protuberances or appendages that are divided into two parts. See Bifid. Alt. Bipartitus.

BIPECTINATE Adj. (Latin, *bis* = twice; *pecten* = comb; *-atus* = adjectival suffix.) Two-combed; descriptive of structure with comb-like teeth or processes on each side. Alt. Bipectinatus. See Pectinate. Cf. Filiciform.

BIPENNATE Adj. (Latin, *bis* = twice; *penna* = feather.) Descriptive of muscle whose tendon of insertion extends along one side of a bone (vertebrate) or apodeme/sclerite (invertebrate). See Muscle. Cf. Pennate. Rel. Musculature.

BIPHYLIDAE Plural Noun. A small Family of polyphagous Coleoptera assigned to Superfamily Cucujoidea. Adult oblong to elongate, somewhat flattened, densely setose; 1.5–8.5 mm long; Antenna with 11 segments; Ommatidia coarse; lateral margin of Pronotum well developed; tarsal formula 5-5-5 or 4-4-4; Abdomen with five Ventrites. Larva elongate, cylindrical, dorsally sclerotized; head prognathous; six Stemmata on each size of head; Antenna with three segments; legs with five segments, Pretarsi each with one claw; Urogomphi present or absent. Occur beneath bark; feed upon fungal fruiting bodies. See Cucujoidea.

BIPOLAR CELL A nerve cell with two neurons proceeding from the cell body. See Neuron; Nerve Cell; Nervous System. Cf. Multipolar Cell; Unipolar Cell. Rel. Muscle.

BIPOLAR Adj. (Latin, *bis* = twice; *polus* = pole.) Pertaining to cells or structures with two poles, one at each end of an axis. See: Symmetry. Cf. Unipolar; Multipolar. Rel. Organization.

BIPUPILLATE Adj. (Latin, *bis* = twice; *pupilla* = dim. of pupa; *-atus* = adjectival suffix.) Pertaining to two pupils; ocellate spots in insects which have two pupils or central spots, sometimes of different colours. Alt. Bipupillatus.

BIRADIAL Adj. (Latin, *bis* = twice; *radius* = ray; *-alis* = having the character of.) Pertaining to bodies, structures or appendages which are radially and bilaterally symmetrical. See Radial; Symmetry.

BIRADIATE Adj. (Latin, *bis* = twice; *radius* = ray; *-atus* = adjectival suffix.) Descriptive of structure consisting of, or with two rays or spokes. Alt. Biradiatus. See Radiate. Cf. Neuroid.

BIRAMOUS Adj. (Latin, *bis* = twice; *ramus* = branch; *-osus* = possessing the qualities of.) Descriptive of elongate appendages or structures which are composed of two-branches. Alt. Biramose; Biramosus. See Ramose. Cf. Uniramous.

BIRCH APHID *Euceraphis punctipennis* (Zetterstedt), *Calaphis flava* Mordvilko [Hemiptera: Aphididae].

BIRCH BARK BEETLE *Dryocoetes betulae* Hopkins [Coleoptera: Scolytidae].

BIRCH CASEBEARER *Coleophora serratella* (Linnaeus) [Lepidoptera: Coleophoridae]. Syn. Cigar Casebearer.

BIRCH LEAFMINER *Fenusa pusilla* (Lepeletier) [Hymenoptera: Tenthredinidae]: Endemic to Europe and adventive to North America during 1923. A bivoltine pest of birch; eggs are laid on young leaves.

BIRCH SAWFLY *Arge pectoralis* (Leach) [Hymenoptera: Argidae].

BIRCH SKELETONIZER *Bucculatrix canadensisella* Chambers [Lepidoptera: Lyonetiidae]: A pest of alder, birch and oak in northeastern USA and eastern Canada. Eggs are laid individually on leaves and young larvae form serpentine leaf mines; older larvae feed externally on underside of leaves. BS overwinters as a pupa on ground within a white, ribbed cocoon.

BIRCH TUBEMAKER *Acrobasis betulella* Hulst [Lepidoptera: Pyralidae].

BIRCHALL, EDWIN (1819–1884) (Dunning 1884, Proc. Entomol. Soc. Lond. 1884: xlii.)

BIRCKELL, JOHN (fl 1750) (Weiss 1936, *Pioneer Century of American Entomology,* 320 pp. (18–21), New Brunswick.)

BIRD BUGS See Cimicidae.

BIRD CHERRY-OAT APHID *Rhopalosiphum padi* (Linnaeus) [Hemiptera: Aphididae].

BIRD LICE See Mallophaga.

BIRD TICK *Haemaphysalis chordeilis* (Packard) [Acari: Ixodidae].

BIRD TICKS See Hippoboscidae.

BIRD, HENRY (1887–1961) (Anon. 1961, Bull. Entomol. Soc. Amer. 7: (4): [i].)

BIRD, JOHN FRANCIS (1874–1958) (Richards 1959, Proc. Roy. Entomol. Soc. Lond. (C) 23: 73.)

BIRD, RALPH DURHAM (1901–1972) (Smith 1972, Bull. Can. Entomol. 4: 36–37)

BIRD-DROPPING SPIDER *Celaenia kingbergi* Thorell [Araneae: Araneidae].

BIRD-OF-PARADISE FLIES *Callipappus* spp. [Hemiptera: Margarodidae].

BIREFRINGENCE Noun. (Latin, *bis* = twice; *refringere* = to break up. Pl., Birefringence.) The optical property of double refraction. Refringence is high or low, according as the difference between the refractive indices is large or small.

BIRKMANN, G (1855–1944) (Anon. 1944, Science

99: 463.)

BIRLANE® See Chlorfenvinphos.

BIRO, LAJOS (1856–1931) (Szekessy 1956, Ann. hist.-nat. Mus. Natn. Hung. 7: 7–14.)

BIRULA, ALEKSEII ANDREEVICH (1804–1937) (Anon. 1938, Festschr. E. Strand. 4: 660–661.)

BISCHOF, CARL GUSTAV CHRISTOPH (1792–1870) (D. F. 1871, Geol. Mag. 8: 45–47.)

BISCHOF, EDWIN A (1866–1923) (Leng 1924, Entomol. News 35: 114.)

BISCHOFF, HANS (1890–1960) (Anon. 1961, Entomol. News 62: 132.)

BISCHOFF-EHINGER, ANDREAS (1812–1875) (Rüttmeyer, Verh. naturf. Ges. Basel 6: 549–554.)

BISCUIT BEETLE See Drugstore Beetle.

BISERIATELY Adv. (Latin, *bis* = twice; *serra* = saw.) Arranged in double rows or series. See Serrate.

BISERRATE Adj. (Latin, *bis* = twice; *serra* = saw; *-atus* = adjectival suffix.) 1. Descriptive of structure with two rows of tooth-like projections along a margin. 2. Double saw-toothed; descriptive of an Antenna with a saw tooth on both sides of each segment. Alt. Biserratus. See Serrate. Cf. Aculeate-Serrate; Dentate-Serrate; Multiserrate; Subserrate; Uniserrate.

BISETOSE Adj. (Latin, *bis* = twice; *seta* = bristle; *-osus* = full of.) Descriptive of structure with two bristle-like or setaceous appendages. Alt. Bisetosus; Bisetous. Cf. Setose.

BISEXUAL Adj. (Latin, *bis* = twice; *sexus* = sex; *-alis* = having the character of.) Pertaining to an individual possessing sexual organs of male and female. See Sex. Cf. Asexual; Hermaphrodite.

BISHOPP, FRED CORRY (1884–1970) (Smith 1971, J. Econ. Entomol. 64: 1341–1342.)

BISINUATE Adj. (Latin, *bis* = twice; *sinuosus* > *sinus* = bent curve; *-atus* = adjectival suffix.) Displaying two sinuations or incisions. Alt. Bisinuatus. See Sinuous.

BISSETT, G A (–1944) (Cockayne 1945, Proc. Roy. Entomol. Soc. Lond. (C) 9: 48.)

BITING LICE See Mallophaga.

BITING MIDGES Members of the Ceratopogonidae, particularly the Genera *Austroconops, Culicoides, Lasiohelea, Leptoconops* and *Styloconops.* See Ceratopogonidae.

BITOU SEED FLY *Mesoclanis polana* Munro [Diptera: Tephritidae]: A Species introduced into eastern Australia from South Africa for the control of the invasive weed *Chrysanthemoidea monilifera* (Linnaeus) [Asteraceae]. Larvae feed within the flowerhead and destroy developing seeds.

BITOXIBACILLIN® A registered biopesticide derived from *Bacillus thuringiensis* var. *kurstaki*. See *Bacillus thuringiensis*.

BITTACIDAE Plural Noun. Hanging flies; Hanging Scorpionflies. A numerically small, geographically widespread Family of Mecoptera whose Species resemble craneflies. Adult body moderate-sized with large compound eyes; Ocelli large and on a protuberance; Rostrum long and slender; legs each with one tarsal claw; male Basistyles enlarged and fused ventrally; male Dististyles small. Larva with two dorsal rows of branched, fleshy processes. Adults rest suspended from vegetation by forelegs and middle legs; hind Tarsi raptorial. Adults are predaceous and often carry prey in flight. See Mecoptera.

BITUBERCULAR Adj. (Latin, *bis* = twice; *tuberculum* = small swelling.) Bearing two distinct tubercles. Alt. Bituberculate; Bituberculatus. See Tubercle.

BIUNCINATE Adj. (Latin, *bis* = twice; *uncus* = hook; *-atus* = adjectival suffix.) Descriptive of structure with hooks. Alt. Biuncinatus.

BIVALVE Adj. (Latin, *bis* = twice; *valva* = fold.) With two longitudinal stripes or vittae. Alt. Bivittatus. See Valve.

BIVOLTINE Adj. (Latin, *bis* = twice; Italian, *volta* = time.) Animals with two generations during a year or during a season. Alt. Digoneutism. See Voltanism. Cf. Multivoltine; Univoltine.

BIVOUAC Noun. (German, *beiwacht* > *bei* near; *wachen* = to watch. Pl., Bivouacs.) 1. Social Insects: The mass of army ant workers which surrounds the queen and brood. 2. An encampment for a short period of time under deprived conditions.

BIZARRE LOOPER *Eucyclodes pieroides* (Walker) [Lepidoptera: Geometridae].

BJERKANDER, CLAUDIUS (1735–1795) (Anon. 1858, *Accentuated list of British Lepidoptera*. 118 pp. (xi), Cambridge & Oxford Entomological Societies, London.)

BLABERIDAE Plural Noun. Giant Cockroaches. Numerically, the second largest cockroach Family with 155 Genera and 1,020 Species. Female ovoviviparous or viviparous; body usually broad, moderate to large sized; Antenna often less than half body length; Tegmina and wings reduced or absent; legs short, stout; Femora and Tarsi sometimes smooth; Cerci often short, not segmented. See Blaberoidea.

BLABEROIDEA A Superfamily or Suborder of cockroaches including Blaberidae, Blatellidae, Nocticolidae and Polyphagidae. See Blattoidea.

BLACHIER, CHARLES THÉODORE (1859–1915) (Bethune-Baker 1916, Entomol. Rec. J. Var. 28: 23–24.)

BLACK ARMY-CUTWORM *Actebia fennica* (Tauscher) [Lepidoptera: Noctuidae]. See Cutworms.

BLACK AUSTRALIAN-ANOPHELINE *Anopheles bancroftii* Giles [Diptera: Culicidae].

BLACK BAMBOO-BORER *Bostrychopsis parallela* [Coleoptera: Bostrichidae]: A pest of bamboo in southeast Asia.

BLACK BEAN-APHID *Aphis fabae* Scopoli [Hemiptera: Aphididae]: A common pest of beans and sugar beets in Europe. See Bean Aphid.

BLACK BEETLE *Metanastes vulgivagus* (Olliff) [Coleoptera: Scarabaeidae].

BLACK BLISTER-BEETLE *Epicauta pennsylvanica*

(DeGeer) [Coleoptera: Meloidae].

BLACK BLOWFLY *Phormia regina* (Meigen) [Diptera: Calliphoridae]: A significant urban pest in North America. BB has the general biology and habits of blow flies; larvae typically feed in carrion, but have been associated with human myiasis. Adult BB coloration is variable from greenish-black to blue green; adults are capable of moving ca 50 km from the place of their origin. See Calliphoridae. Cf. Blowfly.

BLACK BORERS 1. *Heterobostrychus* spp. [Coleoptera: Bostrichidae]: A Genus of beetles whose Species bore into wood, structural timbers and stored foods in southeast Asia and Africa. 2. *Apate* spp. [Coleoptera: Bostrichidae]: A Genus whose Species bore into stored products in Africa and tropical South America.

BLACK BUG *Corimelaena pulicaria* (Gemar) [Heteroptera: Thyreocoridae].

BLACK CARPENTER-ANT *Camponotus pennsylvanicus* (DeGeer) [Hymenoptera: Formicidae]: A Species widespread and endemic to eastern North America and a sporadic pest of wooden structures, particularly old and abandoned buildings, tree stumps and fallen logs. Workers 6–15 mm; body dark brown to black; Gaster with vestiture of grey or yellow Setae; Petiole with one node. BCAs construct their nests in wood but the colony does not necessarily originate in wood; one colony may contain several thousand workers. Workers are polymorphic, bite but do not sting and emit strong odour of formic acid when crushed. Food includes insects, honeydew and juice of fruit. BCA do not feed upon wood but cause structural damage through the colonial tunnels constructed in wood. See Formicidae. Cf. Brown Carpenter Ant; Florida Carpenter Ant; Red Carpenter Ant.

BLACK CARPET-BEETLE *Attagenus unicolor* (Brahm) [Coleoptera: Dermestidae]: A cosmopolitan larval pest of wool, hair, silk, feathers, fur, leather, book bindings and cereal products; BCB is an occasional serious pest in flour mills and often is associated with birds nesting in attics. Adult 3–5 mm long; body shiny black to dark brown with brown legs; elongate-oval in outline shape. Larva to 7 mm long; reddish brown with erect bristles with tuft of long bristles from tip of Abdomen. Egg stage requires 10 days at 24°C. BCB undergoes 7–17 larval instars, depending upon diet; larval stage requires 258–639 days. Pupal stage lasts 8–14 days. Adult stage occupies 270–650 days depending upon diet, mating and humidity; a female may lay 100 eggs; BCB feeds on pollen. Syn. *Attagenus megatoma.* See Dermestidae. Cf. Australian Carpet Beetle; Common Carpet Beetle; Furniture Carpet Beetle; Varied Carpet Beetle.

BLACK CARRION-FLY *Australophyra rostrata* (Robineau-Desvoidy) [Diptera: Muscidae]: A minor pest around poultry houses in Australia. Adult shiny black and about the size of a House Fly. See Muscidae.

BLACK CITRUS-APHID *Toxoptera aurantii* (Boyer de Fonscolombe). *Toxoptera citricida* (Kirkaldy) [Hemiptera: Aphididae]: A widespread minor pest of *Citrus,* camellia, holly and coffee in the tropics and subtropics; restricted to California in North America; not north of Mediterranean. Adult black, ca 2 mm long; winged adult of *T. citricida* with third antennal segment black but same segment of *T. aurantii* clear with apex black. BCA on *Citrus* foliage most of year; abundant in spring and autumn. Small colonies overwinter on young shoot growth near trunk of trees; usually in spring, move to new shoots, young leaves and blossom. Leaves become distorted; sooty mould fungus may develop on honeydew deposited on leaves and twigs by aphids. Shoot growth is deformed by aphids curling very young leaves and stems. BCA is most efficient vector of tristeza. BCA is multivoltine with up to 30 generations per year; one generation can be completed in a week. See Aphididae.

BLACK COCKROACH-WASP *Dolichurus stantoni* (Ashmead) [Hymenoptera: Sphecidae].

BLACK COWPEA-MOTH *Cydia ptychora* (Meyrick) [Lepidoptera: Tortricidae].

BLACK CUTWORM *Agrotis ipsilon* (Hufnagel) [Lepidoptera: Noctuidae]: A sporadic pest of cotton and several Species of vegetable plants throughout the world. BCW larva consumes leaves, cuts leaves from plant at petiole, or cuts seedling at base of plant. Adult BC ca 25 mm long with wingspan 35–55 mm; Forewing pale purplish-brown to grey black; Hindwing greyish to pale brown with brown-black veins. Egg dome-shaped, ca 0.5–0.6 mm dia, with radially developed ridges in Chorion; egg changes colour with development (white to reddish then grey-black.) Eggs deposited individually or in clusters 2–3 rows deep on plant or in soil; eclosion ca 3–7 days in warm weather to several weeks in cold climates. Neonate and early-instar larvae are pale yellow-green and mature larvae grey-brown or mottled with faint longitudinal stripes. Larvae reside in soil during day and forage on plants during night. BCW with six larval instars and typically 3–4 generations per year. See Noctuidae; Cutworms.

BLACK DEATH The second pandemic of Bubonic Plague (14–17 centuries); purported to have killed ca 30% of the human population in Europe. BD attributed to the aetiological agent *Yersinia pestis mediaevalis* and vectored by Oriental Rat Flea. BD probably originated in Central Asia (Alma Ata) and spread east to China, south to India and west to Europe. Syn. Bubonic Plague. See Oriental Rat Flea; Plague.

BLACK DESERT-GRASSHOPPER. *Taeniopoda eques* (Burmeister) [Orthoptera: Acrididae]

BLACK DISEASE See Kala-azar. Syn. Dumdum Fever; Tropical Splenomagaly.

BLACK DUNG-BEETLE *Copris incertus prociduus* (Say) [Coleoptera: Scarabaeidae].

BLACK EARWIG *Chelisoches morio* (Fabricius) [Dermaptera: Chelisochidae].

BLACK EARWIGS See Chelisochidae.

BLACK FIELD-CRICKET *Teleogryllus commodus* (Walker) [Orthoptera: Gryllidae]: A Species endemic and common in eastern Australia. BFC is an occasional pest of pasture, field crops, horticultural crops and vegetables. Adults and nymphs feed on leaves and stems of seedling plants, flowers and developing fruits. Adult dark brown to black, 20–25 mm long; Antenna filiform, longer than body; hindleg large and adapted for jumping; forewing small and leathery; hindwing long, fanlike and folded beneath forewing with apical portion filled and projecting rearward; Cerci multisegmented, tapering apically and about as long as hind Tibia and Tarsus; Ovipositor ca 1.5 times as long as Cercus. BFC is attracted to lights at night. Female lays ca 1,000 eggs during lifetime. Eggs are banana-shaped, white to pale yellow, 2.5–3 mm long; deposited singly 1–4 cm deep in soil; incubation requires 1–3 weeks through most of year; Species overwinters in egg stage. Nymph completes 8–12 instars, typically 9–10. BFC is multivoltine with 2–3 generations per year. See Gryllidae.

BLACK FIELD-EARWIG *Nala lividipes* (Dufour) [Dermaptera: Labiduridae]: A minor pest of crops in Queesland, Australia.

BLACK FLOWER-THRIPS *Haplothrips gowdeyi* (Franklin) [Thysanoptera: Phlaeothripidae].

BLACK FLY. See Simuliidae.

BLACK FUNGUS-BEETLE *Alphitobius laevigatus* (Fabricius) [Coleoptera: Tenebionidae]: In Australia, a pest of stored products.

BLACK FUNGUS-GNATS See Sciaridae.

BLACK GARBAGE-FLY See Bronze Dump Fly.

BLACK GRAIN-STEM SAWFLY *Trachelus tabidus* (Fabricius) [Hymenoptera: Cephidae]: A pest of grasses which is endemic in Europe and adventive to North America. See Cephidae. Cf. European Wheat Stem Sawfly; Wheat Stem Sawfly.

BLACK HILLS BEETLE See Mountain Pine Beetle.

BLACK HORSE-FLY *Tabanus atratus* Fabricius [Diptera: Tabanidae]: Pest of mammals that is widespread in North America. Adult ca 20–26 mm long; black; Thorax with whitish pubescence; Abdomen with bluish-white tinge; wing not spotted, black to brown; Ocelli absent; hind Tibia without apical spur. Eggs are elongate, laid vertically in masses on leaves or stems of aquatic vegetation or trees overhanging water. Larvae are predaceous in mud; body pointed at anterior and posterior ends, whitish with black bands. See Tabanidae. Cf. Striped Horse-Fly. Rel. Deer Fly.

BLACK HOUSE-ANT *Ochetellus glaber* (Mayr) [Hymenoptera: Formicidae]: A minor urban pest which appears endemic to Australia, probably Japan, Philippines, and Indonesia; adventive to USA and New Zealand. Common in Australian *Citrus* and an important and frequent pest which feeds on honeydew and protects homopterous pests. BHA nests in crevices and cavities such as rockeries, paving and in brickwork; workers invade homes and creates problems in kitchens. Adult worker 2.5–3.0 mm long, black. Syn. *Iridomyrmex glaber* (Mayr). See Formicidae.

BLACK HOUSEHOLD-ANT *Technomyrmex albipes* (F. Smith) [Hymenoptera: Formicidae].

BLACK HOUSE-SPIDER *Badumna insignis* (L. Koch) [Araneae: Desidae].

BLACK HUNTER-THRIPS *Leptothrips mali* (Fitch) [Thysanoptera: Phlaeothripidae]: A predaceous Thysanoptera assigned to the Phlaeothripidae and common in North America. See Phlaeothripidae.

BLACK IMPORTED FIRE-ANT *Solenopsis richteri* Forel [Hymenoptera: Formicidae].

BLACK LADY-BEETLE *Rhyzobius ventralis* (Erichson) [Coleoptera: Coccinellidae].

BLACK LADYBIRD *Rhyzobius forestieri* (Mulsant) [Coleoptera: Coccinellidae].

BLACK LARDER-BEETLE *Dermestes ater* DeGeer [Coleoptera: Dermestidae]: A cosmopolitan pest of bones, carcasses, wool and wood. Larvae are pests. Egg stage 3–4 days; larval stage 42–63 days; pupal stage 56–63 days; adult stage 60–90 days. See Dermestidae.

BLACK LEAF-BEETLE *Rhyparida morosa* Jacoby [Coleoptera: Chrysomelidae].

BLACK MAIZE-BEETLE *Heteronychus arator* (Fabricius) [Coleoptera: Scarabaeidae].

BLACK PARLATORIA-SCALE *Parlatoria pergandii* Comstock [Hemiptera: Diaspididae].

BLACK PEACH-APHID *Brachycaudus persicae* (Passerini) [Hemiptera: Aphididae]: A widespread pest of peach, apricot and almond in North America. Wingless adult forms live on roots throughout year; during late winter or early spring some individuals move to new growth above ground and establish colonies on fresh twigs and young shoots. Winged forms develop, migrate to other trees, feed and reproduce. Subsequent generation returns to peach during autumn, produce males and oviparious females which mate and lay overwintering eggs. See Aphididae.

BLACK PECAN-APHID *Melanocallis caryaefoliae* (Davis) [Hemiptera: Aphididae]: A pest of pecans in North America. BPA feeds on upper and lower sides of leaves and causes premature leaf drop. BPA prefers the inner, shaded part of tree. See Aphididae.

BLACK PINE BARK BEETLE *Hylastes ater* (Paykull) [Coleoptera: Curculionidae].

BLACK PINE-LEAF SCALE *Nuculaspis californica* (Coleman) [Hemiptera: Diaspididae].

BLACK PLAGUE See Bubonic Plague.

BLACK PLAGUE-THRIPS *Haplothrips froggatti* Hood [Thysanoptera: Phlaeothripidae].

BLACK POTTER-WASP *Delta pyriformis philippinensis* (Bequaert) [Hymenoptera: Vespidae].

BLACK PRINCE *Psaltoda plaga* (Walker) [Hemiptera: Cicadellidae].

BLACK QUEEN-CELL VIRUS A viral disease associated with honey bee queen-cells, the walls of which become dark brown or black. Contaminated queen cells contain dead propupae or pupae that contain large numbers of virus particles; infected pupae are pale yellow with a tough Integument. The disease does not readily infect or multiply when ingested by drones, worker larvae or young adult workers. See Honey Bee Viral Disease. Cf. Sacbrood. Rel. Pathogen.

BLACK SAGE *Cordia macrostachys* (Jacquin) Roemer & Schultes. An introduced weed pest of sugarcane on Mauritius and controlled with the chrysomelid *Metrogaleruca obscura* DeGeer imported from Trinidad.

BLACK SALT-MARSH MOSQUITO *Aedes taeniorhynchus* Wiedemann [Diptera: Culicidae]. See Culicidae. Cf. Salt-Marsh Mosquito.

BLACK SCALE *Saissetia oleae* (Oliver) [Hemiptera: Coccidae]: A cosmopolitan, polyphagous softscale in orchard crops. BS host plants include almond, apple, apricot, citrus, grape, oleander, olive, pear, walnut and numerous ornamental trees. BS is probably endemic to Africa and is now a cosmopolitan pest of many commercially important trees. Adult female black or dark brown, dome-shaped with an 'H'-shaped ridge on dorsum of cover. Males are rare, with pupal stage and two-wings. Female oviposits ca 2,000 eggs beneath cover; eclosion within 1–20 days; crawlers prefer to settle on midrib of leaves or young shoots. BS undergoes two nymphal instars and shifts to permanent location on a host plant after the second moult. Life cycle is complete within 4–8 months; BS completes 2–4 generations per year depending upon host plant and climate. Nymphs and adult females produce copious honeydew which serves as substrate for sooty mould and other fungi. BS is controlled in many regions with cocinellid beetles and parasitic Hymenoptera. See Coccidae. Cf. Brown Soft Scale; Hemispherical Scale.

BLACK SCAVENGER FLIES See Sepsidae.

BLACK SLUG CUP-MOTH *Doratifera casta* Scott [Lepidoptera: Limacodidae]: A bivoltine, minor pest of fruit and native ornamental trees in eastern Australia. Female oviposits on leaves in rows of 8–12 eggs per row; eggs are bright yellow, flattened and covered with short, dark Setae from female. Mature caterpillar is large, black with conspicuous yellow fleshy protuberances from the sides and dorsum of each segment, and four stellate tubercles with defensive spines near head. Syn. Chinese-Junk Caterpillar. See Limacodidae.

BLACK SOLDIER-FLY See American Soldier Fly.

BLACK SPIDER-BEETLE *Mezium americanum* (Laporte) [Coleoptera: Anobiidae]: A cosmopolitan pest of stored foods. Adult 1.5–3.5 mm long; head and Thorax golden; Abdomen shiny black; head, Pronotum and base of Elytra densely setose; Pronotum with median longitudinal Sulcus deepening and expanding posteriad; basal collar of Elytra interrupted on each side. Abdomen with five visible Sterna; hind Trochanter less than 0.33 as long as Femur; Elytra with basal tomentose collar interrupted. BSB feeds on dried animal products. In Australia, BSB feeds on beans, dried fruits and grains. See Anobiidae.

BLACK STINK-BUG *Coptosoma xanthogramma* (White) [Hemiptera: Plataspidae].

BLACK STRAWBERRY-BEETLE *Clivina tasmaniensis* Sloane [Coleoptera: Carabidae].

BLACK SUNFLOWER-STEM WEEVIL *Apion (Fallapion) occidentale* Fall [Coleoptera: Curculionidae].

BLACK SWALLOWTAIL *Papilio polyxenes asterius* Stoll [Lepidoptera: Papilionidae]: A common butterfly throughout North America which feeds upon plants in carrot Family and acts as an occasional pest of garden and field crops. BS overwinters as a pupa and completes 2–4 generations per year. Syn. American Swallowtail; Celeryworm; Carrotworm; Parsleyworm. See Papilionidae.

BLACK SWARMING LEAF-BEETLE *Rhyparida discopunctulata* Blackburn [Coleoptera: Chrysomelidae].

BLACK THREAD-SCALE *Ischnaspis longirostris* (Signoret) [Hemiptera: Diaspididae].

BLACK TURFGRASS ATAENIUS *Ataenius spretulus* (Haldeman) [Coleoptera: Scarabaeidae].

BLACK TURPENTINE BEETLE *Dendroctonus terebrans* (Oliver) [Coleoptera: Scolytidae].

BLACK TWIG BORER *Xylosandrus compactus* (Eichoff) [Coleoptera: Scolytidae].

BLACK VINE-WEEVIL *Otiorhnychus sulcatus* (Fabricius) [Coleoptera: Curculionidae]: A widespread beetle which is an important pest of grapes and other horticultural crops. Adults feed on stems and berry pedicels; larvae feed on roots. See Curculionidae.

BLACK WISHBONE SPIDER *Aname diversicolor* (Hogg) [Araneae: Nemistrinidae].

BLACK WITCH *Ascalapha odorata* (Linnaeus) [Lepidoptera: Noctuidae].

BLACK, JAMES EBENEZER (1865–1924) (B. D. J. 1926, Proc. Linn. Soc. Lond. 1925–6: 74–75.)

BLACK-BARREL MANTID *Mantis octospilota* Westwood [Mantodea: Mantidae]. See Mantidae.

BLACK-BELLIED CLERID *Enoclerus lecontei* (Wolcott) [Coleoptera: Cleridae].

BLACKBERRY LEAFHOPPER *Dikrella californica* (Lawson) [Hemiptera: Cicadellidae].

BLACKBERRY SKELETONIZER *Schreckensteinia festaliella* (Hübner) [Lepidoptera: Heliodinidae].

BLACKBURN BUTTERFLY *Vaga blackburni* (Tuely) [Lepidoptera: Lycaenidae].

BLACKBURN DAMSEL BUG *Nabis blackburni* White [Hemiptera: Nabidae].

BLACKBURN DRAGONFLY *Nesogonia blackburni* (McLachlan) [Odonata: Libellulidae].

BLACKBURN, CHARLES V (1857–1944) (Anon. 1944, Psyche 51: 7.)

BLACKBURN, JOHN BICKETON (1845–1881) (Anon. 1882, Zool. Jb. 1881: 5.)

BLACKBURN, NORRIS DWIGHT (1902–1962) (Frost 1962, J. Econ. Entomol. 55: 1–24.)

BLACKBURN, THOMAS (1844–1912) (Anon. 1949, Proc. Roy. Soc. Qd. 60: 71–72.)

BLACKBURNE, ANNA (–1793) (Anon. 1794, Gentleman's Mag. 64: 180.)

BLACKBUTT LEAFMINER *Acrocercops laciniella* (Meyrick) [Lepidoptera: Gracillariidae].

BLACK CHERRY APHID *Myzus cerasi* (Fabricius) [Hemiptera: Aphididae]: A pest of sweet and sour cherry, peppergrass and crucifers in western North America. BCA overwinters as an egg among buds; eclosion occurs during bud development; nymphs feed in curled leaves and cover leaves with honeydew. Winged adults migrate to other host plants during midsummer. Later generations return to cherry during autumn, produce wingless males and oviparous females which lay overwintering eggs. See Aphididae.

BLACK-CHERRY FRUIT FLY *Rhagoletis fausta* (Osten Sacken) [Diptera: Tephritidae]: A pest of sour cherries in North America. See Tephritidae. Cf. Cherry Fruit Fly; Western Cherry Fruit Fly.

BLACK-ELM BARK WEEVIL *Magdalis barbita* (Say) [Coleoptera: Curculionidae.) A pest of weakened elm, hickory, oak and walnut in eastern North America. Larvae tunnel in bark and cambium; adults feed on leaves. Cf. Red Elm Bark Beetle.

BLACK-FACED BUMBLE BEE *Bombus californicus* Smith [Hymenoptera: Apidae]: A Species widespread in western North America. Adult predominantly black; anterior aspect of Thorax and Dorsum of fourth gastral segment yellow; body with moderate vestiture of long, coarse Setae. Queens ca 15–22 mm long, typically appear during June and nest on ground. See Bumble Bees. Cf. Yellow-Faced Bumble Bee.

BLACK-FACED LEAFHOPPER *Graminella nigrifrons* (Forbes) [Hemiptera: Cicadellidae].

BLACK-HEADED ASH SAWFLY *Tethida barda* (Say) [Hymenoptera: Tenthridindae].

BLACK-HEADED BUDWORM *Acleris variana* Fernald [Lepidoptera: Tortricidae].

BLACK-HEADED CATERPILLAR *Opisina arenosella* Walker (= *Nephantis serinopa* Meyrick) [Lepidoptera: Cryptophasidae = Xyloryctidae]: A severe pest of coconut palm and palmyra palm in Burma, India and Sri Lanka. A female can lay 350 eggs, typically depositing them on lower surface of leaflets and leaf tips. Gregarious larvae feed on leaflets and spin silken tunnels under shelter of which they live and feed. See Xyloryctidae.

BLACK-HEADED FIREWORM *Rhopobota naevana* (Hübner) [Lepidoptera: Tortricidae].

BLACK-HEADED PASTURE COCKCHAFER 1. *Aphodius tasmaniae* Hope [Coleoptera: Scarabaeidae]: Endemic to Australia; larvae pasture pests in southern Australia. 2. *Aphodius pseudotasmaniae* Given [Coleoptera: Scarabaeidae].

BLACK-HEADED PINE SAWFLY *Neodiprion excitans* Rohwer [Hymenoptera: Diprionidae].

BLACK-HORNED PINE BORER *Callidium antennatum hesperum* Casey [Coleoptera: Cerambycidae].

BLACK-HORNED TREE CRICKET *Ocanthus nigricornis* Walker [Orthoptera: Gryllidae].

BLACKJACKET *Vespula consobrina* (Saussure) [Hymenoptera: Vespidae].

BLACKLEG Noun. (Middle English, *blak* Anglo Saxon, *blaec*; Middle English, *legge*. Pl., Blacklegs.) A fungal disease of cabbage that is caused by *Phoma lingam* and vectored by the Cabbage Maggot. See Cabbage Maggot.

BLACKLEGGED TICK *Ixodes scapularis* (Say) [Acari: Ixodidae].

BLACKLEGGED TORTOISE BEETLE *Jonthonota nigripes* (Oliver) [Coleoptera: Chrysomelidae].

BLACKMAN, MAULSBY WILLETT (1876–1943) (Mallis 1971, *American Entomologists*. 549 pp. (430–431), New Brunswick.)

BLACKMAN, VERNON HERBERT (1872–1967) (Porter 1968, Biog. Mem. Fellows Roy. Soc. 14: 37–60, bibliogr.)

BLACKMARGINED APHID. *Monellia caryella* (Fitch) [Hemiptera: Aphididae.]

BLACKMORE, ERNEST HENRY (1878–1929) (Hatch 1949, *A Century of Entomology in the Pacific Northwest*. 43 pp. (17, portrait), Seattle.)

BLACKMORE, TROVEY (1835–1876) (Westwood 1876, Proc. Entomol. Soc. Lond. 1876: xliii.)

BLACKSOIL ITCH MITE *Eutrombicula sarcina* (Womersley) [Acarina: Trombiculidae].

BLACKSOIL SCARAB *Othnonius batesi* Olliff [Coleoptera: Scarabaeidae].

BLACK-STRIPED MOSQUITO See Southern Saltmarsh Mosquito.

BLACKWALL, JOHN (1789–1881) (Anon. 1882, Proc. Linn. Soc. Lond. 1880–2: 17.)

BLACK-WALNUT CURCULIO *Conotrachelus retentus* (Say) [Coleoptera: Curculionidae].

BLACKWELL, ELIZABETH (fl1737) (Lisney 1960, *Bibliogr. of British Lepidoptera 1608–1799*, 315 pp. (114–120, bibliogr.), London.)

BLACKWELL, JOHN (1790–1880) (Bonnet 1945, *Bibliographia Araneorum* 1: 32–33.)

BLACK-WIDOW SPIDER *Latrodectus mactans* (Fabricius) [Araneae: Theridiidae]: A highly venomous spider widespread in warm regions of the world. BWS is regarded as an urban pest in North America, but more significant as a pest in agricultural areas; true impact of BWS difficult to assess but phobia to this spider is widespread and notable. Common names in different parts of world include: Malmignatte (southern Europe), knoppie spider, button spider (South Africa). See Latrodectism. Cf. Brown-Widow Spider; Redback

Spider.

BLACK-WINGED DAMSELFLY See Broad Winged Damselfly.

BLADAN® Parathion.

BLADDER CICADA *Cytosoma schmeltzi* Distant [Hemiptera: Cicadidae]: An endemic Species of Australia which attacks many native plants and citrus. Adult 30–40 mm long, green with leaf-like green wings and enlarged Abdomen. Females oviposit into bark of twigs and branches; upon eclosion, nymphs fall to ground, burrow into soil and feed on roots of trees. BC generation requires 2–3 years. Males produce sound to attract mates. See Cicadidae.

BLADDER FLIES See Acroceridae.

BLADE Noun. (Anglo Saxon, *blaed* = leaf; Pl., Blades.) **1** Any thin, flat structure resembling a leaf, sword or knife. 2. The Lacinia of many insects such as bees. 3. The surface of the insect wing.

BLADE® See Oxamyl.

BLAIR, KENNETH GLOYNE (1882–1952) (Hawkins 1954, Proc. Trans. Soc. Lond. Entomol. Nat. Hist. Soc. 1952–53: xliii–xliv.)

BLAISDELL, FRANK ELLSWORTH (1862–1946) (Mallis 1971, *American Entomologists.* 549 pp., New Brunswick.)

BLAKE, CHARLES ALFRED (1834–1903) (Anon. 1903, Entomol. News 14: 213–215.)

BLANC, EDWARD (–1923) (Roubaud 1923, Bull. Soc. Entomol. Fr. 1923: 48–49.)

BLANC, M (1857–1944) (Mars 1969, Bull. Mus. Hist. Nat. Marseille 29: 49–50.)

BLANCHARD, CHARLES EMILE (1819–1900) French author of *Metamorphoses des Insectes.* (Papavero 1971, *Essays on the History of Neotropical Dipterology.* 1: 193. São Paulo.)

BLANCHARD, EVARADO EELS (1899?–1971) Argentinian government Entomologist who was educated at Maine State Agricultural College. (Cortes 1973, Revta. Chil. Entomol. 7: 261–262.)

BLANCHARD, FREDERICK (1843–1912) (Sherman 1913, J. N. Y. Entomol. Soc. 21: 69–71.)

BLANCHARD, RAPHAEL ANATOLE EMILE (1857–1919) French parasitologist who wrote major work on mosquitoes. (Howard 1930, Smithson. Misc. Collns. 84: 134, 485.)

BLAND, JAMES H B (1833–1911) (Osborn 1937, *Fragments of Entomological History.* 394 pp. (23), Columbus, Ohio.)

BLAND, THOMAS (1809–1885) (Dimmock 1888, Psyche 5: 35.)

BLASTEM Noun. (Greek, *blastos* = stem. Pl., Blastems.) A nucleated protoplasmic layer preceding the Blastoderm.

BLASTICOTOMIDAE Thomson 1871. Plural Noun. A small (ca 10 Species) Palearctic Family of Symphyta (Hymenoptera) assigned to Tenthredinoidea. Adult body less than 10 mm long; Antenna short, with four segments (segment three largest; segment four minute); Hypostomal Bridge absent; Pronotum short with posterior margin curved; forewing M1 cell large, Rs and M veins strongly curved; fore Tibia with two apical spurs, medial spur bifid; middle Tibia without subapical spur; Abdomen carinate laterally; Ovipositor strongly projecting beyond apex of Abdomen; male genitalia orthandrous. Larvae bore within stems of Filicales. See Tenthredinoidea.

BLASTOBASIDAE Meyrick 1894. Plural Noun. A cosmopolitan Family of ditrysian Lepidoptera assigned to Superfamily Gelechioidea and including ca 300 Species. Blastobasids are closely related to Oecophoridae and best represented in North America. Adult body small, drab coloured with few diagnostic wing patterns; wingspan 8–30 mm; head with smooth scales; Ocelli and Chaetosemata absent; Antenna shorter than forewing, Scape usually with Pecten; Proboscis well developed with scales on base; Maxillary Palpus with four segments, folded over base of Proboscis; Labial Palpus recurved; fore Tibia with Epiphysis; tibial spur formula 0-2-4; forewing lanceolate with Pterostigma and subradial Retinaculum in female. Larvae scavenge dead or dying plant material; some feed in flower or seed heads on fallen conifer seeds or fruits. See Gelechioidea.

BLASTOCOELE Noun. (Greek, *blastos* = bud; *koilos* = hollow. Pl., Blastocoels.) The segmentation cavity of the embryonic blastula which contains yolk cells and is surrounded by Blastoderm.

BLASTOCYTE Noun. (Greek, *blastos* = bud; *kytos* = hollow vessel; container. Pl., Blastocytes.) An undifferentiated embryonic cell.

BLASTODACNIDAE Plural Noun. A small Family of ditrysian Lepidoptera assigned to Superfamily Gelechioidea.

BLASTODERM Noun. (Greek, *blastos* = bud; *derma* = skin. Pl., Blastoderms.) A continuous cellular layer surrounding the blastocoel which contains egg yolk. The germinal membrane from which organs of the embryo are formed.

BLASTODERMIC CELLS Cells which form Blastoderm.

BLASTOGENESIS Noun. (Greek, *blastos* = bud; *genesis* = descent.) Social Insects: The origin of caste traits from variation in ovarian environment of the egg or nongenetic contents of the egg. Cf. Trophogenesis.

BLASTOGENIC Adj. (Greek, *blastos* = bud; *genesis* = descent; *-ic* = of the nature of.) Relating to or inherent in the germ or blast.

BLASTOKINESIS Noun. (Greek, *blastos* = bud; *kinein* = to move; *-sis* = a condition or state. Pl., Blastokineses.) 1. Embryology: All displacements, rotations or revolutions of the embryo within the egg (Johannsen & Butt 1941). 2. Movements of the embryo that are Species characteristic and predictable in terms of time and direction of orientation in the eggshell. Movements are an apparent adaptation to utilize yolk and protect the embryo from desiccation. See Embryology. Cf. Anatrepsis; Katatrepsis. Rel. Allokinesis; Diakinesis; Klinokinesis; Ookinesis; Orthokinesis; Thigmokinesis.

BLASTOMERE Noun. (Greek, *blastos* = bud; *meros* = part. Pl., Blastomeres.) Cleavage cells or cells produced by division of the egg or its nucleus which form the Blastoderm (Snodgrass).

BLASTOPORE Noun. (Greek, *blastos* = bud; *poros* = passage. Pl., Blastopores.) The mouth of the gastrulation cavity in embryonic development.

BLASTULA Noun. (Greek, *blastos* = bud; Latin suffix *-ula* = place. Pl., Blastulae; Blastulas.) An early stage of embryonic development in which the Blastoderm represents the only layer of cells (Snodgrass).

BLATCH, WILLIAM GABRIEL (1840–1900) (Verrall 1900, Proc. Entomol. Soc. Lond. 1900: xliii.)

BLATCHLEY, WILLIS STANLEY (1859–1940) American professional geologist (State of Indiana), amateur Entomologist specializing in Hemiptera and traditional field naturalist who prepared comprehensive manuals for Coleoptera, Orthoptera and Heteroptera. Blatchley is best remembered for his publication 'Heteroptera of Eastern North America.' (Osborn 1946, *Fragments of Entomological History*. Pt. II. 232 pp. (61), Columbus, Ohio.)

BLATEX® See Hydramethylnon.

BLATTARIA Burmeister 1829. (Upper Carboniferous-Recent.) (Latin, *blatta* = an insect that shuns the light.) An ordinal name for cockroaches. A cosmopolitan Order consisting of about 460 Genera and 3,800 nominal Species. Adult dorsoventrally compressed with head concealed by Pronotum when viewed from dorsal aspect; omnivorous with mouthparts adapted for chewing; Antenna filiform usually with more than 30 segments. Cursorial legs with hindlegs similar in shape and size to middle legs; many Species are extremely rapid runners. Some Species lack wings; when wings are present the forewings are modified into moderately sclerotized Tegmina. Abdomen with 10 apparent segments; Cerci usually contain many segments; male usually with a pair of Styli on Sternum IX, occasionally absent. A few Species are parthenogenetic. Female with panoistic Ovary; manufactures an Ootheca (egg case) that is rather hard in Species that deposit it on a substrate. Some Species are ovoviviparous with a membranous Oothecae containing eggs within a brood pouch; a few Species are viviparous with an Ootheca incompletely developed. Metamorphosis is simple; nymphs and adults occupy the same habitats. Nymphs resemble the adult in shape but can differ considerably in coloration. Domestic cockroaches are nocturnal and usually avoid light; some outdoor Species are attracted to light at night. During day, most Species remain concealed under loose bark of trees, beneath rotting logs, or in similar, fairly moist habitats. Included Families: Blaberidae, Blattidae, Blattellidae, Cryptoceridae, Nocticolidae, Polyphagidae.

BLATTELLIDAE Plural Noun. Wood Cockroaches. Largest Family of cockroaches with 210 Genera, 1,750 Species. Adult small-bodied; Antenna usually longer than half body length; legs long, spinose, slender; Cerci long and tapered; male Styli usually present; female oviparous. See Blattaria.

BLATTIDAE Plural Noun. Oriental Cockroaches; American Cockroaches. A cosmopolitan Family of about 45 Genera and 525 Species including some of economic importance. Adult large-bodied; fore Femur usually spinose; male Tergal Glands absent or concealed on Tergum 1. See Blattaria.

BLATTODEA Brunner 1882 (Latin, *blatta* = an insect that shuns the light.) A superordinal name applied to the Orders including mantids and cockroaches in some classifications; the Order of cockroaches in other classifications. Syn: Cursoria Westwood 1839; Oothecaria Verhoeff 1903; Blattiformia Werner 1906.) See Blattaria. Cf. Blaberoidea.

BLATTOIDEA Noun. (Latin, *blatta* = an insect that shuns the light.) A Suborder or Superfamily of cockroaches. Included Families Blattidae and Cryptocercidae. Cf. Blaberoidea.

BLAU, CARL (1839–1901) (Anon. 1901, Insektenbörse 18: 90.)

BLEDOWSKI, RYSZARD (1886–1932) Polish academic (Free University of Poland.) (M. K. 1932, Polski Pismo Entomol. 11: 17–21.)

BLENKARN, STANLEY ARTHUR (1882–1927) (Anon. 1928, Entomol. Mon. Mag. 64: 171.)

BLEPHARICERIDAE Plural Noun. Net-Winged Midges. A small Family of nematocerous Diptera assigned to Superfamily Culicoidea. Adult resembles Tipulidae or Culicidae; small, slender-bodied, long-legged, resembling mosquito; wing with pseudovenation and held at right angle to body. Adults occur near fast-flowing water where they rest suspended from overhanging vegetation, rocks and bridges. Larva and pupa aquatic; larvae use ventral suckers to cling to rocks in fast-flowing water. Syn. Blepharoceratidae; Blepharoceridae.

BLEPHAROCERATIDAE Plural Noun. See Blephariceridae.

BLEPHAROCERIDAE Plural Noun. See Blephariceridae.

BLESZINSKI, STANISLAW (1927–1967) (Roesler 1971, Entomol. Nachr. Wien 18: 33–36, bibliogr.)

BLEX® See Pirimiphos-Methyl.

BLIGHT Noun. (Anglo Saxon, *blaecan* = to grow pale. Pl., Blights.) A plant disease that causes rapid death of leaves, flowers and stems. Blight is induced by insects or fungi. See Pathogen.

BLIND Noun. (Middle English, *blind*.) Eyeless; without eyes.

BLIND OCELLUS An ocellate spot without a central spot.

BLINDING FILARIAL DISEASE See Onchocerciasis.

BLISS, ARTHUR (1858–1890) (Anon. 1890, Entomologist 23: 104.)

BLISSINAE A Subfamily of Lygaeidae. (Slater 1979, Bull. Amer. Mus. Nat. Hist. 165: 1–180.)

BLISTER BEETLES See Meloidae.

BLISTER CONEWORM *Dioryctria clarioralis* (Walker) [Lepidoptera: Pyralidae].

BLOCHMANN'S CORPUSCLES Insect egg: Minute greenish bodies that are independent organisms capable of cultivation in artificial media (Imms).

BLOCKMANN, FRIEDRICH (1858–1931) (Anon. 1933, Entomol. Jarhrb. 42: 180.)

BLOESCH, CHARLES (1819–1908) (Bloesch 1909, Mitt. Aargau. Naturf. Ges. 11: 99–113, bibliogr.)

BLOMEFIELD, LEONARD (1880–1893) (Winwood 1894, Proc. Bath Nat. Hist. Antiq. Fld. Club 8: 35–55.)

BLOOD Noun. (Middle English, *blod* > Anglo Saxon, *blōd* = blood. Pl., Bloods.) The fluid (plasma) which circulates in the body and contains essential elements for defence, nutrition, waste and sometimes respiration. 2. Insects: The body-cavity liquid and its contents. See Haemolymph. Rel. Circulation.

BLOOD CELLS Cellular elements of the blood. See Haemocyte; Leucocytes. Rel. Haemolymph.

BLOOD CORPUSCLES See Blood Cells.

BLOOD GILLS Aquatic larvae: Hollow, non-tracheated, usually tubular filamentous or digitiform respiratory evaginations of the body wall or Proctodaeum within which Haemolymph circulates.

BLOOD PLASMA The fluid part of the blood. See Haemolymph. Cf. Serum.

BLOOD-SUCKING CONENOSE *Triatoma sanguisuga* (LeConte) [Hemiptera: Reduviidae]: A blood-sucking parasite of vertebrates widespread in the New World. Adults mate, find host and do not fly. BSC life cycle requires about three years and eight nymphal instars Eggs are pearly with batches of eggs laid following ingestion of a blood meal; eggs are individually placed. Wood rats are a favoured host. BSC transmits Chagas' Disease to small mammals and man. Syn. Big Bed Bug; Mexican Bed Bug. See Reduviidae; Chagas Disease.

BLOOD WORM See Elaeophorosis.

BLOOD WORMS Larval stage of Genus *Chironomus*. BW are so-named due to the Haemoglobin which occurs in Haemolymph.

BLOOM Noun. (Middle English, *blome* > Anglo Saxon, *blowan* = to bloom. Pl., Blooms.) 1. A fine violet dusting similar to that on plums; pruinescence. 2. Sawflies: The Abdomen and the Thorax. (Obsolete). 3. A blossom or the stage of bud development for a flowering plant. Syn. Flower; Blossom; Bud.

BLOOMFIELD, EDWIN NEWSON (1827–1914) (E. A. B. 1914, Entomol. Mon. Mag. 50: 157.)

BLOT, FREDERIC (1795–1884) (Eudes-Longchamps 1884, Mem. Soc. Linn. Normandie 7: 70–84, bibliogr.)

BLOTCH Noun. (Modern English > Old French, *bloche* = a clod. Pl., Blotches.) 1. A large or irregular-shaped spot. 2. Any disease of plants that causes large, irregular spots on leaves or fruit. Syn. Mark; Blemish; Spot; Imperfection.

BLOTIC® See Propetamphos.

BLOW FLIES See Calliphoridae. Typically members of *Calliphora, Chrysomya* and *Lucilia*. Adults 8–12 mm long, usually with metallic coloration. BF are active in warm, sunny weather and can travel several km from their development site; BF are attracted indoors through windows. Female oviposits batches of ca 100–200 eggs upon meat, excrement or garbage and eclosion occurs within 24 hours. Larvae feed 3–10 days and pupation occurs away from food in drier habitat. BF life cycle requires 2–5 weeks. Sudden large populations sometimes are due to oviposition in dead rodents and birds in houses. See Calliphoridae. Cf. Black Blowfly; Eastern Goldenhaired Blowfly; Lesser Brown-Blowfly; Sheep Blowfly. Rel. Bottle Flies.

BLOWFLY STRIKE See Sheep Blowfly.

BLUE ALFALFA-APHID *Acyrthosiphon kondoi* Shinji [Hemiptera: Aphididae].

BLUE BOTTLE FLY *Calliphora vomitaria* (Linnaeus) [Diptera: Calliphoridae]: A widespread Species that is attracted to flowers, excrement and decaying fruit. BBF life cycle requires ca 15–20 days. BBF was the first insect to be reared successfully from egg to adult on an artificial diet (Bogdanow 1908, Arch. Anat. Physiol. Abt. Suppl. 173–200.) See Calliphoridae. Cf. Green Bottle Flies. Rel. Blow Flies.

BLUE BUTTERFLIES See Lycaenidae.

BLUE CACTUS-BORER *Melitara dentata* (Grote) [Lepidoptera: Pyralidae].

BLUE HORNTAIL *Sirex cyaneus* Fabricius [Hymenoptera: Siricidae]: A pest of pine, fir and spruce in North America. Adult metallic blue-black. See Siricidae.

BLUE MUD DAUBER *Chalybion californicum* (Saussure) [Hymenoptera: Sphecidae]: A metallic blue-black wasp, ca 12–15 mm long, that is endemic to North America and adventive to the Caribbean and Hawaii. BMD does not construct nest but uses nests of other mud daubers, notably *Sceliphron caementarium*. BMD uses water to dissolve partitions in nest, removes spiders and eggs of resident and deposits its own prey and eggs. Prey include Black Widow Spiders. See Sphecidae.

BLUE SOLDIER-FLY *Exaireta spiniqera* (Wiedemann) [Diptera: Stratiomyidae] (USA). Syn. Garden Soldier Fly.

BLUE STAIN A fungal disease of Norway pine caused by *Ceratostomella ips* and vectored by bark beetles, *Ips* spp. See Scolytidae.

BLUE-BANDED BEE *Amegilla pulchra* (Smith) [Hymenoptera: Anthophoridae]: A bee native and widespread along coastal Australia which becomes a minor pest in some urban situations. Adult 10–12 mm long; head and Mesosoma with golden Setae; Metasoma banded with black and blue. BBB primitively eusocial with solitary construction and maintenance of nests but females nest gregariously in soil along creek banks and in wall cavities by tunnelling through mortar between bricks. Large colonies can develop over several years if not controlled. Males sometimes cluster. See Anthophoridae.

BLUEBERRY BUD-MITE *Acalitus vaccinii* (Keifer) [Acari: Eriophyidae].

BLUEBERRY CASE-BEETLE *Neochlamisus cribripennis* (LeConte) [Coleoptera: Chrysomelidae].

BLUEBERRY FLEA-BEETLE *Altica sylvia* Malloch [Coleoptera: Chrysomelidae].

BLUEBERRY LEAFTIER *Croesia curvalana* (Kearfoot) [Lepidoptera: Tortricidae].

BLUEBERRY MAGGOT *Rhagoletis mendax* Curran [Diptera: Tephritidae]: A pest of blueberries, and closely related Species, in eastern North America. See Tephritidae.

BLUEBERRY THRIPS *Frankliniella vaccinii* Morgan [Thysanoptera: Thripidae].

BLUEBERRY TIP-MIDGE *Prodiplosis vaccinii* (Felt) [Diptera: Cecidomyiidae].

BLUEBOTTLES See Calliphoridae.

BLUECHER, FRIEDRICH (–1894) (Kraatz 1894, Dt. Entomol. Z. 1894: 7–8.)

BLUEGRASS BILLBUG *Sphenophorus parvulus* Gyllenhal [Coleoptera: Curculionidae].

BLUEGRASS WEBWORM *Parapediasia teterrella* (Zincken) [Lepidoptera: Pyralidae].

BLUE-GUM PSYLLID *Ctenarytaina eucalypti* (Maskell) [Hemiptera: Psyllidae]: An oligophagous Species which feeds and develops on juvenile foliage of many Species of *Eucalyptus*. BGP is native to Australia and accidentally introduced into other *Eucalyptus* growing regions including New Zealand, Sri Lanka and California where it is a pest. See Psyllidae.

BLUES See Lycaenidae.

BLUETONGUE DISEASE A widespread arthropod-borne viral disease of ruminants caused by an *Orbivirus* (Reoviridae). BtD is endemic to Africa and now is widespread between latitudes 40° N and 35°S. About 25 serotypes of bluetongue virus (BTV) have been isolated, which differ in their pathogenicity for hosts and infectivity for vectors. Cattle are important reservoirs for BTV; sheep are more susceptible to disease and death from BTV; goats are not affected clinically and causes little clinical effect in cattle. BTD is manifest through fever, swelling of mucous membranes, enteritis and lameness. Most important vectors of BTV are ceratopogonid flies of Genus *Culicoides*. Acronym BTD. See Reoviridae. Cf. African Horse Sickness; Colorado Tick Fever; Epizootic Haemorrhagic Disease.

BLUNCK, JOHANN CHRISTIAN (HANS) (1885–1958) (Weidner 1958, Verh. Dt. Zool. Ges. (Zool. Anz. Suppl.) 22: 427–429.)

BLUNT Adj. (Old Norse, *blunda* = to doze.) 1. Not sharp; obtuse at the edge or apex. 2. Not perceptive; insensitive. See Dull. Cf. Acute; Sharp.

BLUNT-NOSED CRANBERRY LEAFHOPPER *Scleroracus viccinii* (Van Duzee) [Hemiptera: Cicadellidae].

BLUTEL, JEAN PIERRE ESPRIT (1782–1858) (Buquet 1858, Ann. Soc. Fr. (3) 6: 905–911.)

BLÜTHGREN, P (1880–1967) (Konigsman 1968, Mitt. Dt. Entomol. Ges. 27: 2–3.)

BMC® A registered biopesticide derived from *Bacillus thuringiensis* var. *israelensis*. See *Bacillus thuringiensis*.

B-NINE® See Daminozide.

BOAS, JOHN ERIK VESTI (1885–1935) Danish Zoologist and Economic Entomologist. Boas was a Docent of Zoology at the Royal Veterinary and Agricultural College and author of 'Danish Forest Zoology' (1896) and several papers dealing with agricultural and nursery pests. Boas recommended measures for control of warble fly. (Howard 1930, Smiths. Misc. Collns. 84: 278–279; Henriksen 1935, Ent. Meddr 19: 186–190.)

BOBRETZKII, NIKOLAY VASELEVICH (1843–1887) (Bogdanov 1888, Izv. imp. obshch. Lyub. Estest. Antrop. Etnogr. imp. Mosc. Univ. 57: [1–2], bibliogr.)

BOCK, GEORGE, W (1856–1940) (Meiners 1941, Entomol. News 52: 119.)

BÖCKMAN, FRIEDRICH (Weidner 1967, Abh. Verh. naturw. Ver. Hamburg. Suppl. 9: 130.)

BODE, OTTO (1860–1904) (Ziegler 1905, Berl. Entomol. Z. 50: 183–185.)

BODEMEYER, AUGUST RUDOLF EDWARD VON (1854–1918) (Bodemeyer 1928, Ent. Z., Frankf. a. M. 42: 197–198.)

BODEMEYER, BODO VON (1883–1929) (Meissner 1930, Entomol. Z., Frankf. A. M. 43: 255.)

BODENHEIMER, FREDERICK SIMON (1897–1959) (Anon. 1961, Bull. Res. Council Israel (B) 10: iii–iv)

BODY Noun. (Middle English, *bodi;* Anglo Saxon, *bodig*. Pl., Bodies.) 1. The physical structure of an organism or something which resembles an organism. 2. Collectively, the structure and substance which comprise the cells, tissues, organs and organ systems of an organism. 3. Insects: The head, Thorax and Abdomen (Kirby & Spence 1828). See Integument; Skeleton. Cf. Corpus.

BODY CAVITY The definitive hollow portion of the body and appendages that is contained by the Integument; not stricly equivalent in all animals (Snodgrass). Cf. Haemocoel.

BODY LOUSE *Pediculus humanus humanus* Linnaeus [Phthiraptera: Pediculidae]: A cosmopolitan blood-sucking, obligatory parasite of humans. Adult 2.0–3.6 mm long; dorso-ventrally flattened; head small, narrower than Thorax; haustellate mouthparts, without evident Mandibles; compound eyes nearly absent, not capable of forming an image; apterous; all legs same size and shape; Tarsi each with one claw; Abdomen length greater than basal width; lateral lobes of Abdomen segments 3–8 with sclerotized paratergal plates. Eggs are operculate, yellowish and cemented to clothing; eclosion occurs within a week at temperature range 24–37°C; eclosion is affected by humidity with optimal rate at 75% RH. Three nymphal instars complete development within 9 days when continuously on host; complete development within 16–19 days when removed from host during night. Adults copulate throughout life. Nymph and adult respond to warmth, odour and are positively thigmotactic; their preferred temperature range is 29–30°C. Fecundity is 80–200 eggs, depending upon conditions. BL causes irritation and scratching that can lead to secondary infection. BL is a vector of Epidemic Typhus (*Rickettsia prowazecki* Da Rocha-Lima) and Relapsing Fever (*Borrelia recurrentis*). Alt. Cootie; Seam Squirrel. See Pediculidae. Cf. Head Louse. Rel. Epidemic Typhus.

BODY OF CHELICERA Acarology: The principle segment of the Chelicera.

BODY OF THE TENTORIUM A median sclerite of the Tentorium.

BODY WALL The Integument of the insect body which is formed from embryonic Ectoderm and consists of Epidermis, Cuticle and Basement Membrane. See Body; Integument. Rel. Skeleton.

BOECK, CHRISTIAN PETER BIANCO (1789–1877) (Natvig 1944, Norsk Entomol. Tidsskr. 7: 15.)

BOESIGER, ERNEST (1914–1975) (Babeyrie 1976, Ann. Zool-ecol. Anim. 8: 130–131.)

BOETTGER, OSKAR (1844–1910) (Haas 1900, Entomol. Bl. Biol. Syst. Käfer 6: 267–268.)

BOG Noun. (Celtic, *bog* = soft. Pl., Bogs.) An ecological habitat characterized by wet, very acidic peat. Syn. Marsh; Marshland; Swamp. See Habitat. Cf. Fen; Marsh; Swale.

BOGANIIDAE Plural Noun. A small Family of polyphagous Coleoptera assigned to Superfamily Cucujoidea and whose members occur in southern Africa and Australia. Adult <4 mm long, oblong; Antennal with 11 segments, club absent or weakly developed; Labrum membranous and concealed; Mandible large with dorsal process and setose cavity. Larva elongate, narrowing posteriad with thoracic and abdominal tergites sclerotized; Urogomphi short or absent; Pygopods on abdominal Tergum X. Adult and larva feed on cycad pollen and *Eucalyptus* flowers. See Cucujoidea.

BOGDANOV, ANATOLII PETROVICH (1834–1896) (Anon. 1896, Zool. Anz. 19: 176.)

BOGONG MOTH *Agrotis infusa* [Lepidoptera: Noctuidae]: An Australian moth whose adult stage aggregates in montane caves and is collected by aboriginal men and used as food by aboriginal peoples. Adult moth wingspan ca 38 mm; forewing black to rusty brown with a row of three paler spots; hindwing pale brown to beige. Females lay eggs on leaves or on the soil surface near the host plant. Eggs are spherical with a ribbed appearance, ca 0.45 mm high. Egg stage requires 3–6 days. Larvae 38–50 mm long, light grey to black. See Noctuidae.

BOHATSCH, OTTO (1843–1912) (Rebel 1912, Verh. Zool.-bot. Ges. Wien 62: (204)–(209), bibliogr.)

BOHEMAN, CARL HEINRICH (1796–1868) Swedish coleopterist specializing in Chrysomelidae and Rhynchophora; Boheman's collection is maintained in Stockholm. Boheman named the cotton bollweevil *Anthonomus grandis* Boheman. (Musgrave 1932, *A Bibliogr. of Australian Entomology 1775–1930*. 380 pp. (27, bibliogr.), Sydney.)

BOHM'S ORGAN Lepidoptera: A mechanoreceptive organ on the Antenna. (Bohm 1911, Arb. Zool. Inst. Univ. Wien 19: 219–246.) See Antenna. Cf. Janet's Organ; Johnston's Organ; Multiporous Plate Sensilla.

BOHOLDOYIDAE Kovalev 1985 Plural Noun. A small Family of fossil Diptera known from two Genera in Jurassic Period and Cretaceous Period deposits of Siberia, and apparently closely related to Eopleciidae.

BOIE, FRIEDRICH (1789–1870) (Weidner 1967, Abh. Verh. naturw. Ver. Hamburg Suppl. 9: 112.)

BOIELDIEU, ANATOLE AUGUSTE (1824–1886) (Bourgeois 1886, Bull. Soc. Entomol. Fr. (6) 6: lxxxix.)

BOILEAU, HENRI (1866–1924) (Didier 1928, Etudes des coleopteres lucanides du globe 1: 1–32, bibliogr.)

BOISDUVAL SCALE *Diaspis boisduvalii* Signoret [Hemiptera: Diaspididae].

BOISDUVAL, JEAN BAPTISTE ALPHONSE DECHAUFFOUR DE (1799–1879) French lepidopterist who described several insects common in North America. Boisduval was a founding member of Entomological Society of France (1832) and founder and president of Society for Agricultural Entomology (1868). He was an early worker in economic Entomology: and established the journal 'Insectologie Agricole' (1867). Boisduval's monographic *Entomologie Horticole* (1867, 650 pp, 426 woodcuts) consisted of a series of essays on insects injurious to agricuture. (Dos Passos 1951, J. N. Y. Entomol. Soc. 59: 138.)

BÖKEL, JOHANNES (1535–) (Weidner 1967, Arb. Verh. Naturw. Ver. Hamburg Suppl. 9: 68–70.)

BOLD, THOMAS JOHN (1816–1874) (Wright 1889, Trans. Nat. Hist. Soc. Northumb. 8: 33–46, bibliogr.)

BOLD'REVA, VASILIYA FEDOROVICHA (1883–1957) (Uvarov 1957, Nature 179: 758.)

BOLDT, RUDOLF (1874–1952) (Lempke 1953, Entomol. Ber., Amst. 14: 301–302.)

BOLITOPHILIDAE Plural Noun. A small Family of

Diptera related to Mycetophilidae. The larval stage of Species occur in fungi. See Diptera.

BOLIVAR Y URRUTIA, IGNACIO (1850–1944) (Cabrere 1946, Revta. Mus. La Plata 1944: 225–228.)

BOLL, ERNEST FRIEDRICH AUGUST (1817–1868) (Boll 1869, Arch. Ver. Freunde Naturg. Mecklenb. 22: 1–34.)

BOLL, JACOB (1828–1900) (Geiser 1929, *Naturalists of the Frontier.* 341 pp. (11–37), Dallas, Texas.)

BOLL WEEVIL *Anthonomus grandis grandis* Boheman [Coleoptera: Curculionidae]: A Species endemic to Mexico that entered USA during 1892. BW is a persistent, multivoltine pest of cotton in the southeastern USA and parts of Brazil; BW is arguably the most significant pest of cotton in USA. BW overwinters as an adult near cotton fields or in cotton trash and causes damage by feeding on bolls, leaves and fruting structures, and by oviposition. Adult ca 8–10 mm long, greyish-brown with sparse vestiture of golden Setae and prolonged snout; Elytra with longitudinal stripes. Female chews hole in square or boll, oviposits and refills the site. Egg is smooth, white and ca 0.8 mm long. Eggs are laid in groups of 1–3 and eclosion occurs within three days. Larval head pale-brown, body white and becoming brown with maturity; three larva instars feed inside a square and pupate within a fruit. BW life cycle requires 15–25 days. See Curculionidae.

BOLLES LEE, ARTHUR (1849–1927) (Anon. 1927, Nature 119: 432.)

BOLLMAN, CHARLES HARVEY (1868–1898) (Riley 1893, Bull. U. S. Natn. Mus. 46: 7, bibliogr.)

BOLLMAN, WILHELM (1887–1913) (Weidner 1967, Abh. Verh. Naturw. Ver. Hamburg Suppl. 9: 263.)

BOLLWORM *Helicoverpa zea* (Boddie) [Lepidoptera: Noctuidae]. Cf. Cotton Bollworm; Egyptian Bollworm; Pink Bollworm; Spiny Bollworm; Spotted Bollworm.

BOLSTAR® See Sulprofos.

BOLSTER, PERCY GARDNER (1865–1932) (Hungerford 1933, Ann. Entomol. Soc. Amer. 26: 188.)

BOLTAGE® See Pyraclofos.

BOLTER, ANDREW (1820–1900) (Rapp 1945, Entomol. News 56: 209.)

BOLTON, JAMES (fl 1794) (Lisney 1960, *Bibliogr. of British Lepidoptera 1608–1799.* 315 pp. (278–283, bibliogr.), London.)

BOLTSHAUSER, HEINRICH (1859–1899) (Wegelin 1900, Mitt. Thurgau. Naturf. Ges. 14: 156–159.)

BOLWIG'S ORGAN The remnant of a Stemmatum in cyclorrhaphous Diptera larvae. BO consists of an internal, pigmented cup-like structure positioned in the Tentorium of maggots. See Stemma.

BOMATE-C® See Codelure.

BOMBARDIER BEETLE 1. Any member of carabid Genus *Brachinus* which discharges a malodorous gas when disturbed. 2. *Brachinus americanus* [Coleoptera: Carabidae]: A widespread Species in western USA. Eggs are laid individually in mud cells on plants and stones; larvae are parasitic on pupae of whirligig beetles and other aquatic beetles. See Carabidae.

BOMBIFRONS. Adj. (Latin, *bombus* = a deep, hollow sound; *frons* = forehead.) Displaying a blister-like swelling on the front of the head.

BOMBINAE The Bumble Bees. A Subfamily of Apidae (Apoidea) consisting of about 10 Genera. See Bumble Bees.

BOMBOUS Adj. (Latin, *bombus* = buzzing; a hollow, deep sound; *-osus* = full of.) Blister-like; spherically enlarged or dilated. Descriptive of a convex, rounded surface.

BOMBYCIC ACID The acid constituent of the fluid that some moths use to dissolve gum that binds the silk threads of the cocoon when the adult emerges (Packard).

BOMBYCIDAE Plural Noun. Silkworm Moths. A Family of ditrysian Lepidoptera assigned to Superfamily Bombycoidea. Adult medium-sized; wingspan 20–60 mm; head small with piliform scales; Ocelli and Chaetosemata absent; Antenna bipectinate in male and female; Proboscis and Maxillary Palpus absent; Labial Palpus very short; Thorax with long piliform scales; Epiphysis variable; tibial spurs absent or formula 0-2-2; wings broad, forewing often falcate. Egg flattened with Micropyle at apex; Chorion smooth; eggs are laid individually or in small groups. Pupa within cocoon of dense silk. Family includes commercial silkworm, *Bombyx mori* (Linnaeus). See Bombycoidea; Silkworm.

BOMBYCINE Adj. (Latin, *bombycinus* = silken.) Of silk.

BOMBYCINOUS Adj. (Latin, *bombycinus* = silken; *-osus* = full of.) A very pale yellow, like fresh spun silk.

BOMBYCOIDEA Noun. A Superfamily of ditrysian Lepidoptera including Anthelidae, Bombycidae, Brahmaeidae, Carthaeidae, Cercophanidae, Endromidae, Eupterotidae, Lacosomidae, Lasiocampidae, Lemoniidae, Oxytenidae, Ratardidae and Saturnidae. The Superfamily is regarded as a complex assemblage including some of the most highly evolved Lepidoptera. See Lepidoptera.

BOMBYLIIDAE Plural Noun. Bee Flies. A cosmopolitan Family of 6,000 Species of brachycerous Diptera assigned to Superfamily Asiloidea, related to Scenopinidae and Therevidae with some microbombyliids showing affinity to Empididae. Bombyliids are best represented in temperate regions. Adults typically are stout-bodied, usually black and with setose/scaly Integument; wings usually patterned. Bombyllids are separated from therevids and empidids by an absence of a small crossvein (apex of second Basal Cell not touching two Posterior Cells); Scenopinidae share this characteristic. Bee flies also are distinguished by a Costa that continues around apex of wing and vein M ending beyond apex of wing. Bombyliids are predaceous or parasitic as larvae; parasitic Species are primary or secondary; hosts attacked by primary parasitic Species include Coleoptera, Lepidoptera, Hymenoptera, Neuroptera and Diptera; hyperparasitic Species attack Tiphiidae, Ichneumonidae and Tachinidae. Predaceous Species feed on egg pods of Acridoidea. Females of most Species with perivaginal 'sand pouch'; Acanthophorites of Tergum IX possess strong, apically hooked spines which pick up fine sand and deposit then in the sand pouch. Eggs pass through the sand pouch and become coated with sand; this adaptation apparently prevents desiccation. Eggs are laid near host or directly on host. Females hover above substrate near hosts and flick eggs into holes, burrows, cracks in soil or bark of bushes. Females attacking grasshopper egg pods place eggs in soil near the grasshopper's oviposition site. First instar larva is minute, planidial-type and searches for hosts. With contact of host, bombyliid larvae moult and feed or remain quiescent until host reaches last larval instar when feeding begins. Pupae are obtect and armed with cephalic horns, caudally projecting spines and Setae on abdominal dorsum. Horns, spines and Setae are utilized by pupa in making its way to surface of substrate before emergence of adult. See Asiloidea.

BOND, FREDERICK (1811–1889) (Balding 1890, Entomologist 23: 97–98.)

BONDAR, GREGORIO (1881–1959) (Reis 1959, Studia Entomol. 2: 475.)

BONDY, FLOYD F (1894–1950) (Anon.1950, J. Econ. Entomol. 43: 574.)

BONELLI, FRANCIS Italian academic and Director of the Cabinet of Natural History at Turin. Author of *Observations Entomologiques* (published in Mem. l'Acad. Sci.) and *Descrizione de sei nouvi insetti lepidopteri della Sardegna.*

BONELLI, FRANCO ANDREA (1784–1830) (Musgrave 1932, *A Bibliogr. of Australian Entomology 1775–1930.* 380 pp. (28), Sydney.)

BONHAG, PHILIP FREDERIC (1923–1959) (Linsley & Smith 1960, Ann. Entomol. Soc. Amer. 53: 282.)

BONHOURE, ALFONSE (1864–1909) (Perrier 1909, Bull. Mus. Natn. Hist. Nat. Paris 15: 53.)

BONNAIRE, ACHILLE (1824–1907) (Anon. 1908, Entomol. Jb. 17: 202.)

BONNEFOIS, ALOYS (1845–1895) (Giard 1895, Bull. Soc. Entomol. Fr. 1895: ccclxxiv.)

BONNET, AMÉDÉE (–1944) (Stempffer 1944, Bull. Soc. Entomol. Fr. 49: 2.)

BONNET, CHARLES (1720–1793) French-Swiss philosopher and naturalist who published *Traite d'Insectologie* (9 vols, 1779), *Contemplation de la Nature* (2 parts, 1764–1765) and *Oeuvres d'histoire naturelle et de philosophie* (8 parts, 1779–1783) Bonnet provided observations on caterpillar respiration and experimentally demonstrated parthenogenesis. (Wheeler 1931, *Demons of the Dust.* 378 pp. (20–27).

BONNIER, JULES (1859–1908) (Keilin 1923, Parasitology 15: 110–111.)

BONOMI, LINO (1893–1964) (Conci 1965, Natura, Milano 56: 209–211.)

BONPLAND, AIMÉ JACQUES ALEXANDRE GOUJAND (1773–1858) (Breyer 1942, Revta. Soc. Entomol. Argent. 11: 380–381.)

BONSUL® See Alphacypermethrin.

BONVOULOIR, HENRI ACHARD DE (1839–1914) (Anon. 1915, Entomol. News 26: 191–192.)

BONY, GASTON DE (–1935) (Fage 1935, Bull. Soc. Entomol. Fr. 40: 97.)

BOOK LUNG Arachnida. Paired respiratory organs, located in the Abdomen and consisting of saclike invaginations that contain a series of sheetlike leaves. Book lung openings are located at lateral ends of the Epigastric Furrow on the anterio-ventral aspect of the Abdomen. Some spider Species have two pairs of BL; Caponiidae lack BL. See Respiration.

BOOK LOUSE 1. *Liposcelis corrodens* Heymons [Psocoptera: Liposcelidae]: A cosmopolitan, multivoltine pest of damp or mouldy books, paper, new plaster and cereal products. Eggs are white, oval-shaped and incubate for 6–9 days. Species overwinters in egg stage and apparently is parthenogenetic. 2. *Liposcelis bostrychophila* (Badonnel) [Psocoptera: Liposcelidae]: A global pest of stored products, food factories and processing plants; Species is apparently endemic to tropical Africa and adventive elsewhere; parthenogenetic in Great Britain. Species is often confused or misidentified in the literature with *L. divinatoria.* In Australia, a pest of insect collections and stored material which has become mouldy. In Great Britain, Species attacks the eggs of other insects in culture. Adult 1–2 mm long; white or greyish in colour; apterous and soft bodied; life cycle requires 1–4 months. Female lays ca 110 eggs during lifetime at rate of 2–3 per day; eggs are ovoid and ca 0.3 times as large as female; dust and particulate matter adhere to surface of egg; oviposition occurs between 18–35°C. Four nymphal instars; development 9–30 days; generation time ca 9–56 days. 3. *Lepinotus inquilinus* Heyden [Psocoptera: Trogiidae] (Australia). 4. *Trogium pulsatorium* (Linnaeus) [Psocoptera: Trogiidae] (Australia).

BOOPIIDAE Plural Noun. A small Family of amblycerous Phthiraptera including ca 50 Species, nearly all found in Australia and New Guinea where they act as external parasites on marsupials. Mandibulate mouthparts; Antenna typically capitate, with four segments, concealed in grooves; Maxillary Palpus with 2–3 segments; Tarsi with paired claws. Family includes the Dog Louse. See Phthiraptera.

BOORBORIDAE Plural Noun. See Sphaeroceridae.

BORA-CARE® See Inorganic Borate.

BORBORIDAE Plural Noun. See Sphaeroceridae.

BORCHERS, FRIEDRICH (1880–1957) (Gerneck 1958, Anz. Schädlingsk. 31: 11.)

BORCHGRAVE, (BORCHGRAVE D'ALTONA) LÉONE DE (1824–1873) (Anon. 1873. C. R. Soc. Entomol. Belg. 16: ix–x.)

BORCHMAN, FRITZ HEINRICH CHRISTIAN (1870–1944) (Weidner 1967, Abh. Verh. Naturw. Ver. Hamburg Suppl. 9: 226–229.)

BORDAGE, EDMOND (–1924) (Picard 1924, Bull. Soc. Entomol. Fr. 1924: 33.)

BOREAL Adj. (Latin, *boreas* = north wind.) From or belonging to the north. A faunal region that extends from the polar sea southward to near the northern boundary of the USA and farther south occupies a narrow strip along Pacific Coast and higher parts of the Sierra-Cascade, Rocky and Allegheny Mountain ranges; divided into Arctic, Hudsonian and Canadian. Cf. Austral; Tropical.

BOREIDAE Plural Noun. Snow Scorpionflies. A small Family of Mecoptera. Adults common in winter. See Mecoptera.

BORELLI, ALFREDO (1858–1943) (Anon. 1946, Entomol. News 57: 99.)

BOREMAN, THOMAS (1730–1743) (Lisney 1960, *Bibliogr. of British Lepidoptera 1608–1799*. 315 pp. (86–113, bibliogr.), London.)

BORER Noun. (Old Norse, *bora* = hole made by boring. Pl., Borers.) 1. Any adult or larval insect that burrows or makes channels in woody or other plant tissue. 2. The insect Ovipositor when it is adapted for boring.

BORG, KARL ANDERS HJALMER (1859–1910) (Anon. 1911, Entomol. Tidsskr. 37: 109.)

BORGMEIER, THOMAS (1892–1975) (Reichart 1975, Revta. Bras. Entomol. 19: 37.)

BORIC ACID An inorganic compound (H_3BO_3) that functions as a slow-acting, contact insecticide in buildings, ships, sewers and food-handling facilities; placement of BA is critical in food-handling facilities. BA used as a dust to control urban insect pests such as cockroaches, termites, silverfish, ants and termites. Pest absorbs material through Integument or ingests it when grooming or feeding at bait; death occurs 2–10 days after contact. BA is non-flammable, odourless, non-staining, non-corrosive and long-lasting when dry. Trade-names include: Borid®, Drax®, Roach Kil®, Roach-Prufe®. See Inorganic Insecticide.

BORID® See Boric Acid.

BORIDAE Plural Noun. A small Family of polyphagous Coleoptera assigned to Superfamily Tenebrionoidea. See Tenebrionoidea.

BORING BRISTLE (Colloquial.) The appendicular, tubular, often elongate Ovipositor of some Hymenoptera. Cf. Borer.

BORKHAUSEN, MORITZ BALTHASAR (1732–1807) (Anon. 1858, *Accentuated list of British Lepidoptera*. 118 pp. Oxford & Cambridge Entomological Societies, London.)

BORKHSENIUS, NIKOLAY SERGEEVICH (1906–1965) Soviet specialist on scale insects, particularly the Diaspididae. (Kryzhanovskiy 1965, Entomol. Obozr. 44: 951–957, bibliogr. Translation: Entomol. Rev., Wash. 44: 551–555.)

BORMANS, AUGUSTE DE (M. B. 1901, Entomol. Rec. J. Var. 13: 85–88.)

BORN, PAUL (1859–1928) (Bodemeyer 1928, Entomol. Anz. 8: 117–118, 129–132, 137–140, 149–150, 157–160.)

BORNEBUSCH, CARL HEINRICH HARTMANN (1886–1951) (Tuxen 1952, Entomol. Meddr 26: 245–246, bibliogr.)

BORNEMANN, GUSTAV (–1920) (Holze 1921, Int. Entomol. Z. 14: 185–186.)

BÖRNER, KARL J B (1880–1953) (Buxton 1954, Proc. Roy. Entomol. Soc. Lond. (C) 18: 80.)

BORRA, OTTAVIO (–1954) (Anon. 1954, Boll. Soc. Entomol. Ital. 84: 97.)

BORREL, AMÉDÉE (1867–1936) (Magron 1936, La Presse Med. 44: 1697–1698.)

BORRI, CELSO (1887–1940) (Pardi 1941, Atti Soc. Toscana Sci. nat. 50: 6–10, bibliogr.)

BORRIES, HERMAN (1860–1896) (Henriksen 1918, Entomol. Meddr 12: 24–25.)

BORRMANN, FRIEDRICH (1877–1960) (Anon. 1961, Mitt. Dt. Entomol. Ges. 20: 20.)

BORROR, DONALD JOYCE (1907–1988) American academic (Ohio State University) with diverse interests. Coauthor of *An Introduction to the Study of Insects* and *A Field Guide to the Insects*. (White 1989, PESW 91: 304–306, photograph.)

BORTHWICK, THOMAS (1860–1924) (Musgrave 1932, *A Bibliogr. of Australian Entomology 1775–1930*. 380 pp. (29), Sydney.)

BOS, JAN RITZEMA (1850–1928) (Howard 1928, J. Econ. Entomol. 21: 636–637.)

BOSC D'ANTIC, LOUIS AUGUSTE GUILLAUME (1759–1828) Frenchman who lived in colonial America during the Revolutionary War and who collected many insects, especially in Carolina. His collections were sent to the Museum National d'Historie Naturelle, Paris where the types were lost. (Papavero 1971, *Essays on the History of Neotropical Dipterology* 1: 20. São Paulo.)

BOSCH, CARL (1874–1940) (Heikertinger 1940, Entomol. Z., Frankf. A. M. 54: 88.)

BOSELLI, FRANCESCO (1867–1964) (Russo 1965, Boll. Lab. Entomol. Agr. Portici 23: 304–307, bibliogr.)

BOSSI, FEDRIGOTTI FILIPPO (1838–1907) (Gerosa 1908, Atti Accad. Agiati (3) 14: xix–xxvii.)

BOSTOCK, EDWIN DILLON (–1953) (Buxton 1954, Proc. Roy. Entomol. Soc. Lond. (C) 18: 78.)

BOSTRICHIDAE See Bostrychidae.

BOSTRYCHIDAE Plural Noun. False Powder-post Beetles; Auger Beetles. A cosmopolitan Family of polyphagous Coleoptera assigned to Bostrychoidea. Adult <20 mm long with body dark coloured and cylindrical; head prognathous or strongly deflexed; Antenna short with 9–11 segments and loosely clavate; eyes circular in outline, strongly protuberant; Prothorax forming hood concealing head from above; Elytra broadly beveled posteriorly; tarsal formula 5-5-5; Abdomen with five Ventrites. Larva scarabaeiform and weakly sclerotized; head small, strongly retracted; Stemmata absent; Antenna with three segments; Thorax swollen; legs with five segments; Urogomphi usually absent. Adults bore into moribund trees, freshly felled trees, wood and grain; moisture is required for oviposition. Larvae cannot produce cellulase, thus restricted to moist sapwood containing starch which they reduce to powder. Syn. Bostrichidae. See Bostrychoidea.

BOSTRYCHIFORMIA A paraphyletic series of polyphagous Coleoptera including Bostrichoidea and Derodontoidea.

BOSTRYCHOIDEA A Superfamily of polyphagous Coleoptera including the Families Anobiidae, Bostrychidae, Dermestidae, Jacobsoniidae (Sarothriidae) and Nosodendridae. See Coleoptera.

BOSWELL, JOHN THOMAS (1822–1888) (Dimmock 1889, Psyche 5: 156.)

BOT FLIES *Gasterophilus* [Oestridae]: A Genus of parasitic Diptera apparently endemic to Asia and Africa with three Species adventive to Neotropical Realm and other parts of world where horses are bred. Adult bee-like with body densely setose but with few long bristles; mouthparts reduced and not adapted for feeding; Maxillary Palp with unusual sensory receptors apparently adapted for host finding or host selection; Abdomen curved ventrally. Adult is short-lived (ca few days), most active during sunny afternoons and does not fly on cloudy, windy or rainy days; BF vibrates its wings to increase body temperature for flight. BF male adopts aerial territory and hovers for long periods of time; female is embraced in air and the couple fall to ground to copulate for 3–4 minutes. Larva typically found attached to various parts of the digestive system of Equidae; feeds on tissue exudates but does not take blood. See Oestridae; Gasterophilidae; Hypodermatidae. Cf. Horse Bot-Fly; Nose Bot-Fly; Throat Bot-Fly.

BOTANICAL INSECTICIDE A category of insecticide whose members typically act as contact nerve-poisons or affect Metamorphosis. BI are plant-derived compounds from specific parts of relatively few plant Species. BI characteristics include: broad-spectrum, non-residual, fast acting and low mammalian toxicity. BIs are typically expensive. BIs often formulated with Piperonyl Butoxide which acts as a synergist to increase effectiveness of active ingredient. Examples include: Azadirachtin, Chromenes, Domoic Acid; Furanocumarins, Mammeins, Nicotine, Polyacetylenes, Pyrethrins, Rotenone, Ryania, Sabadilla, Terthienyl. See Biopesticide; Insecticide; Natural Insecticide. Cf. Carbamate; Inorganic Insecticide; Insect Growth Regulators; Organochlorine; Organophosphorus; Pyrethrin; Synthetic Pyrethroid.

BOTANIST Noun. (Greek, *botanikos*. Pl., Botanists.) A person who studies plants. See Botany. Cf. Agronomist; Entomologist.

BOTANY Noun. (Greek, *botanikos* > *botane* = herb > *boskein* = to feed.) The branch of Biology that involves the study of plants. See Biology. Cf. Zoology. Rel. Anatomy; Behaviour; Biochemistry; Ecology; Genetics; Morphology; Physiology.

BOTHMER, ULRICH VON (–1975) (Wyniger 1975, Mitt. Entomol. Ges. Basel 25: 76–77.)

BOTHRIDERIDAE Plural Noun. A small Family of polyphagous Coleoptera assigned to Superfamily Cucujoidea. Adult 1.5–13 mm long, oblong to narrowly elongate, subcylindrical to flattened; Antenna short with 10–11 segments, including compact club of 1–2 segments. Tibiae apically expanded and spinose; tarsal formula 4-4-4, rarely 3-3-3. Abdomen with five Ventrites; male with ring-like Aedeagus. Larval head prognathous; 0–6 Stemmata present on each side of head; legs with five segments, Pretarsi each with 0–1 claw; Urogomphi present or absent. Larval feeding habits are diverse: Some Species eat fungal fruiting bodies; other Species are predaceous upon immature wood-boring insects; some Species are parasitic. See Cucujoidea.

BOTHRIOTRICHIA Noun. (Greek, *bothros* = trench, pit, trough; *tricho* = hair. Pl., Bothriotrichiae.) Collembola: Long slender Setae that arise from Integumental depressions or pits.

BOTRYOIDAL Adj. (Greek, *botrys* = cluster of grapes.) Pertaining to structures clustered as a bunch of grapes. Cf. Acinose.

BÖTTCHER, AUGUST (–1900) (Anon. 1900, Insektenbörse 17: 385.)

BOTTLE FLIES Members of Calliphoridae. Adults are large-bodied and metallic coloured. Larvae scavenge dead animals, excrement and garbage. BF life history and habits resemble Blow Flies. See Calliphoridae. Cf. Blue Bottle Fly; Green Bottle Flies.

BOUBAS See Yaws.

BOUCHARD, PETER (–1865) (Pascoe 1865, Entomol. Mon. Mag. 2: 167.)

BOUCHÉ, PETER FRIEDRICH (1784–1856) (Anon. 1858, *Accentuated list of British Lepidoptera*. 118 pp. (xii–xiii), Oxford & Cambridge Entomological Societies, London.)

BOUCLIER Noun. (Etymology obscure.) The Pronotum.

BOUCOMONT ANTOINE (1868–1936) (Anon. 1937, Koleopt. Rdsch. 23: 56.)

BOUDIER, EMILE (1828–1920) (Simon 1920. Ann. Soc. Entomol. Fr. 89: 376.)

BOUIN'S FLUID A general purpose, widely used histological fixative involving various combinations of picric acid, formaldehyde and acetic acid.

BOULANGER, LEO WILFRED (1924–1970) (Dean 1971, J. Econ. Entomol. 64: 563.)

BOULAY, FRANCIS HOUSSEMAYNE DE (1837–

1914) (Musgrave 1932, *A Bibliogr. of Australian Entomology 1775–1930*. 380 pp. (72), Sydney.)

BOULAY, R (–1975) (Péricart 1976, Bull. Soc. Entomol. Fr. 80: 298.)

BOULEY, HENRI MARIE (1814–1885) (Anon. 1885, Bull. Soc. Acclim. Paris (4) 2: i–xxvii, 699–701.)

BOULLET, EUGENE (–1923) (Rabard 1923, Bull. Soc. Entomol. Fr. 1923: 48–49.)

BOULT, JAMES WILLIAM (1847–1924) (Robinson 1906, Trans. Hull Scient. Fld. Nat. Club 3: 236–246.)

BOUND PUPA See Pupa Contigua.

BOUND WATER In any chemical compound, such as protoplasm or water molecules, so incorporated in the compound that when they are driven off by evaporation, the chemical structure of the remaining compound is changed.

BOUNDARY LAYER 1. Entomology: The stratum of air extending from ground level upward through increasing wind speeds to a height where wind speed and insect flight speed are equal. 2. Botany: The layer of air around a leaf surface in which temperature and humidity are affected by leaf morphology and plant physiology. The boundary layer often serves as a microenvironment allowing for certain insect habitation.

BOURGEOIS, JULES (1846–1911) (Anon. 1912, Entomol. News 23: 48.)

BOURNE, ARTHUR ISRAEL (1886–1961) (Lilly & Shaw 1962, J. Econ. Entomol. 55: 1025–1026.)

BOURSIN, CHARLES (1901–1971) (De Bos 1972, Mitt. Entomol. Ges. Basel 22: 101–102; Dujardin 1972, Entomops 25: 1–2.)

BOUTON Noun. (French, *bouton*. Pl., Boutons.) 1. A button. 2. The terminal lappet-like process at the apex of the Ligula in bees. See Labellum.

BOUTONNEUSE FEVER A rickettsial disease in Europe caused by *Rickettsia conori*. Disease typically is vectored by ticks with dogs serving as a reservoir host; apparently, fleas can also vector the disease.

BOUVIER, EUGENE LOUIS (1856–1944) (Queney 1957, Bull. Mens. Soc. Linn. Lyon 26: 199–201.)

BOVAID® See Fenvalerate.

BOVIEN, PROSPER (1894–1963) (Tuxen 1968, Entomol. Meddr 36: 102.)

BOVINE BABESIOSIS An important form of the disease Babesiosis expressed in cattle primarily by *Babesia bovis* and secondarily by *B. bigemina, B. divergens, B. jakimovi, B. major, B. oculans* and *B. ovata*. BB agent is transmitted by ixodid ticks, most notably *Boophilus annulatus;* other vectors include *Boophilus* spp. in tropical and subtropical areas and *Rhipicephalus bursa* in Europe. BB is transmitted transovarially with vertebrate host erythrocyte cytoplasm ingested and Haemoglobin digested. BB incubation period 2–3 weeks, followed by fever, anaemia and blood in urine; often, the disease is fatal and death can occur within 24 hours of symptom appearance. BB is called Redwater in some areas. See Babesiosis. Cf. Canine Babesiosis; Equine Babesiosis.

BOVINE EPHEMERAL FEVER A disease of cattle and water buffalo in Africa and Asia, and epizootic in Australia. BEF is caused by arboviruses assigned to the Rhabdoviridae. Transmission of BEF to humans may occur through the vectors *Culicoides* and culicine mosquitoes. Acronym BEF.

BOVINE ONCHOCERCAS A widespread filarial disease of cattle caused by any one of several Species of *Onchocerca*. BO generally is regarded as a relatively minor disease with microfilariae occuring in skin, blood and subcutaneous lymph; adults are encapsulated in skin. Microfilariae are ingested by blackflies *(Culicoides, Lasiohelea, Simulium)* with the blood meal. Cf. Equine Onchocercas.

BÖVING, ADAM GIEDE (1869–1957) (Anon. 1957, Bull. Entomol. Soc. Amer. 3 (2): 4; Mallis 1971, *American Entomologists*. 549 pp. (431–433), New Brunswick.)

BOWATER, WILLIAM (1880–1973) (Noble 1974, Proc. Birm. Nat. Hist. Soc. 22: 230–231.)

BOWER, HAROLD M (1888–1963) (Powell 1964, J. Lepid. Soc. 18: 192–194.)

BOWERBAND, JAMES SCOTT (1797–1877) (Tyler 1878, J. Roy. Microscop. Soc. 1: 28–30.)

BOWES, ANTHONY JOHN LEE (1913–1942) (de Worms 1943, Entomol. Rec. J. Var. 55: 42.)

BOWKER, JAMES HENRY (–1900) (R. T. 1901, Proc. Linn. Soc. Lond. 113: 40–42.)

BOWLES, EDWARD AUGUSTUS (1865–1954) (Allen 1973, *E. A. Bowles and His Garden*. 264 pp., bibliogr. London.)

BOWLES, GEORGE JOHN (1837–1888?) (Goding 1889, Rep. Entomol. Soc. Ont. 20: 20–21.)

BOWLES, GUILLERMO (–1780) (Dusmet y Alonso 1919, Boln. Soc. Entomol. Esp. 2: 81–82.)

BOWMAN, JOHN RIDGEWAY (1910–1962) (Wallace 1963, Entomol. News: 74: 142.)

BOWRING, JOHN CHARLES (1821–1893) (Anon. 1894, Proc. Linn. Soc. Lond. 1893–4: 29–30.)

BOX, HAROLD EDMUND (1897?–1972?) (Gunn 1976, Proc. Roy. Entomol. Soc. Lond (C) 40: 51.)

BOXELDER APHID *Periphyllus negundinis* (Thomas) [Hemiptera: Aphididae].

BOXELDER BUG *Boisea trivittata* (Say) [Hemiptera: Coreidae]: A pest of boxelder *(Acer negundo)* throughout North America and an occasional pest on ornamental plants, ash, maple, fruit trees and in homes during autumn and winter. BEB completes 1–2 generations per year; female adult overwinters. Body 11–14 mm long, black with conspicuous red markings along lateral and medial margins of Pronotum and lateral and apical corial margins. Eggs are laid in small clutches preferentially on seed pods of female trees or bark or leaves; incubation requires ca 10 days. BEB undergoes five nymphal instars; adults feed on foliage. Syn. *Leptocoris trivittatus* (Say).

BOXELDER LEAF-ROLLER *Caloptilia negundella*

(Chambers) [Lepidoptera: Gracillariidae].

BOXELDER PSYLLID *Cacopsylla negundinis* Mally [Hemiptera: Psyllidae].

BOXELDER TWIG-BORER *Proteoteras willingana* (Kearfott) [Lepidoptera: Tortricidae].

BOXWOOD LEAFMINER *Monarthropalpus flavus* (Schrank) [Diptera: Cecidomyiidae].

BOXWOOD PSYLLID *Cacospylla buxi* (Linnaeus) [Hemiptera: Psyllidae].

BOY, HUGO CARLOS (–1937) (Wacherpfermig 1938, Entomol. Rdsch. 55: 435.)

BOYCE, ALFRED M (1901–1997) American political Entomologist and Director of Citrus Experiment Station (University of California); Ph.D. Cornell University; published minor work in pesticide evaluation and fruitfly control.

BOYCE, RUBERT WILLIAM (1864–1911) English Peer and Medical Entomologist who worked on Malaria and mosquitoes. (Howard 1930, Smithson. Misc. Coll. 84: 482–483.)

BOYD, ARNOLD WHITWORTH (1885–1960) (Uvarov 1961, Proc. Roy. Entomol. Soc. Lond. (C) 25: 49.)

BOYD, THOMAS (1829–1913) (Morice 1913, Proc. Entomol. Soc. Lond. 1913: clxiii–clxiv.)

BOYD, WILLIAM CHRISTOPHER (–1906) (Anon. 1908, Entomol. Jb. 17: 98.)

BOYDYLLA, GEORG (–1911) (Soldanski 1912, Dt. Entomol. Z. 1912: 100–101.)

BPMC® See Fenobucarb.

BRABANT, GEORGES CÉLESTIN EDOUARD (1849–1912) (Anon. 1913, Entomologist 46: 24.)

BRABEC, RUDOLF (1924–1942) (Weidner 1967, Abh. Verh. naturw. Ver. Hamburg Suppl. 9: 263.)

BRACE® See Isazofos.

BRACHELYTRON Noun. (Greek, *brachys* = short; *elytron* = cover, sheath. Pl., Brachelytra.) Elytra or 'wing covers' that are disportionately short and do not completely cover the hindwings at repose. See Elytron.

BRACHELYTROUS Adj. (Greek, *brachys* = short; *elytron* = cover, sheath; Latin, *-osus* = possessing the qualities of.) Pertaining to Coleoptera with short Elytra. See Elytron. Cf. Brachypterous.

BRACHIA Noun. (Greek, *brachion* = arm-like process. Pl., Brachiae.) 1. Orthoptera: The finger-like structures surrounding the Aedeagus. 2. In general: A clasper; a Paramere. 3. The Aedeagus *sensu* MacGillivray. See Brachium.

BRACHIAL Adj. (Greek, *brachion* = arm-like process; Latin, *-alis* = characterized by.) Descriptive of structure or function resembling an arm; arm-like.

BRACHIAL CELLS Hymenoptera, 1st (Norton), Costal and Subcostal (Comstock), 2nd (Norton), Medial (Comstock), 3rd (Norton), Cubital (Comstock), 4th (Norton), 2nd Anal (Comstock).

BRACHIAL NERVES The nerves or veins of the forewing.

BRACHIAL NERVURES See Brachial Veins.

BRACHIAL VEINS Hymenoptera: BVs originate at the base of the forewing, extend parallel to the inner edge toward anal angle. BVs are often connected with Cubital Cells via Recurrent Veins. See Vein; Wing Venation.

BRACHIUM Noun. (Greek, *brachion* = arm-like process; Latin, *-ium* = diminutive > Greek, *-idion*. Pl., Brachia.) 1. Arm; a raptorial foreleg; sometimes the fore Tibia. 2. Heteroptera: A wing vein on the Corium that lies nearest the Claval Suture and is continued on the membrane to bound the Areoles (Cubitus).

BRACHODIDAE Plural Noun. A Family including less than 100 Species of ditrysian Lepidoptera assigned to Superfamily Sesioidea; widespread distribution but not reported from Nearctic. Adult small; wingspan 10–40 mm; head with loose assemblage of scales; Ocelli large; Chaetosemata absent; Antenna considerably shorter than forewing and typically filiform in female but of diverse form in male. Proboscis without scales; Maxillary Palpus minute, with 1–3 segments; Labial Palpus short and slightly recurved; fore Tibia with Epiphysis; tibial spur formula 0-2-4; forewing triangular and lacking Pterostigma; hindwing ovate, larger than forewing and orange in some Species. Larval head hypognathous; prothoracic shield weakly developed. Pupal head with projection; most abdominal segments each with two transverse rows of spines. Adults of many Species are diurnal; biology of larvae is poorly known.

BRACHYCENTRIDAE Plural Noun. A small Family of Trichoptera assigned to Superfamily Limnephiloidea. See Trichoptera.

BRACHYCERA The largest Suborder of Diptera, apparently derived from bibionid-like forms in Nematocera. Adult typically stout-bodied; Antenna short, usually with less than seven segments and apical style or Arista; wing vein CuA directed toward or meeting 1A; discal cell often present; Brachycera are classified into Divisions the Orthorrhapha and Cyclorrhapha. Included Brachycera Families: Acartophthalmidae, Acroceridae (Acroceratidae, Cyrtidae, Henopidae, Oncodidae), Agromyzidae (Phytomyzidae), Anthomyiidae (Anthomyidae, Scopeumatidae), Anthomyzidae, Apioceridae (Apioceratidae), Asilidae, Asteiidae (Astiidae), Athericidae, Aulacigastridae, Bombyliidae, Boorboridae, Braulidae, Calliphoridae, Camillidae, Campichoetidae, Canacidae, Carnidae, Celyphidae, Chloropidae (Oscinidae, Siphonellopsidae, Titaniidae), Chyromyidae (Chyromyiidae), Clusiidae, (Clusiodidae, Heteroneuridae), Coelopidae (Phycodromidae), Conopidae, Cryptochaetidae, Curtonotidae (Cyrtonotidae), Cuterebridae, Cypselidae, Cypselosomatidae, Diastatidae, Diopsidae, Dolichopodidae (Dolichopidae), Drosophilidae, Dryomyzidae, Empididae (Empidae), Ephydridae (Hydrellidae, Notiophilidae), Fergusonidae, Gasterophilidae (Gastrophilidae), Helcomyzidae, Hilarimorphidae, Hippoboscidae, Lauxaniidae (Sapromyzidae), Lonchaeidae,

Lonchopteridae (Musidoridae), Megamerinidae, Micropezidae (Tylidae, Calobatidae), Milichiidae (Phyllomyzidae), Muscidae (Glossinidae, Fanniidae), Mydidae (Mydasidae, Mydaidae), Nemestrinidae, Neriidae, Nycteribiidae, Odiniidae (Odinidae), Oestridae (Hypodermatidae.) Opomyzidae, Otitidae (Ortalidae, Ortalididae, Ulidiidae, Doryceridae, Cephaliidae, Ceroxydidae), Pallopteridae, Pantophthalmidae, Pelecorhynchidae, Periscelidae (Periscelididae), Phoridae, Piophilidae (Neottiophilidae), Pipunculidae (Dorilaidae, Dorylaidae), Platypezidae (Clythiidae), Pseudopomyzidae, Psilidae, Pyrgotidae, Rhagionidae (Leptidae), Rhinotoridae, Ropalomeridae (Rhopalomeridae), Scenopinidae (Omphralidae), Sciadoceridae, Sciomyzidae (Tetanoceridae, Tetanoceratidae), Sepsidae, Sphaeroceridae (Sphaeroceratidae), Stratiomyidae (Stratiomyiidae, Chiromyzidae), Streblidae, Syrphidae, Tabanidae, Tachinidae, Tanypezidae, Tethinidae, Tetratomyzidae, Thyreophoridae, Trixoscelidae (Trichoscelidae), Vermileonidae, Xylomyidae (Xylomyiidae), Xylophagidae (Erinnidae, Coenomyiidae, Rachiceridae.) See Cyclorrhapha; Orthorrhapha. Cf. Nematocera.

BRACHYCEROUS Adj. (Greek, *brachys* = short; *keros* = horn.) Pertaining to Diptera with short Antennae.

BRACHYOSTOMATA Noun. (Greek, *brachys* = short; *stomatos* = mouth.) Brachycerous Diptera with a short Proboscis.

BRACHYPSECTRIDAE Plural Noun. A numerically small, infrequently collected Family of polyphagous Coleoptera assigned to Superfamily Elateroidea. The Family is monogeneric with a disjunct distribution (USA, India, Malaysia and Australia). Brachypsectrids were originally placed in the Rhipiceridae (LeConte 1874), subsequently placed in Dascyllidae (LeConte & Horn 1883) and finally in their own Family. Adults somewhat resemble flattened elaterids. Adult < 7 mm long, oblong with posterior angle of Pronotum acute and carinate; female Antenna short and somewhat enlarged apicad; male Antennal club weakly pectinate; Mandibles rather small or reduced in size. Elytra are weakly striate; tarsal formula 5-5-5. Larvae are oval-shaped and dorsoventrally compressed; head prognathous with a pair of Stemmata forming eyes; Mandibles falcate; Thorax and abdominal segments each with elaborate process; thoracic spiracles biforous; abdominal spiracles small and annular. Biology is poorly known; larvae occur under bark or in crevices, and are predaceous with preoral digestion. North American adults and larvae occur under bark where they feed on ants. Larvae are extremely slow moving and feed on *Solenopsis xyloni* McL. See Elateroidea.

BRACHYPTERISM Noun. (Greek, *brachys* = short; *pteron* = wing; English, *-ism* = condition. Pl., Brachypterisms.) In insects, the condition of short wings.

BRACHYPTEROUS Adj. (Greek, *brachys* = short; *pteron* = wing; Latin, *-osus* = possessing the qualities of.) Pertaining to insects with unusually short wings. Term often is associated with (but not restricted to) insects that possess limited or feeble powers of flight. Syn. Micropterous; Subapterous. See Wing. Cf. Apterous; Macropterous. Rel. Brachyelytrous.

BRACHYPTERY Noun. (Greek, *brachys* = short; *pteron* = wing. Pl., Brachypteries.) The genetical condition in which the wing is disproportionately (allometrically) small in relation to the body size. See Wing. Cf. Aptery; Micropterous.

BRACHYSTOMELLIDAE Plural Noun. A widespread Family of arthropleonous Collembola assigned to Superfamily Poduroidea. Adult body elongate with most segments visibly separated by membrane; Mandible absent; Maxilla apically toothed; first thoracic segment setose. Biology of Species is poorly known; specimens taken in humid habitats and possibly feed on fungal spores. See Collembola.

BRACHYTRACHEA Noun. Acarology: Elongate respiratory structure.

BRACKEN, C W (1868–1950) (Wigglesworth 1951, Proc. Roy. Entomol. Soc. Lond. (C) 15: 75.)

BRACONIDAE Nees 1814. Plural Noun. Braconid Wasps. A large, cosmopolitan Family of apocritious Hymenoptera assigned to Superfamily Ichneumonoidea. Presently about 10,000 Species are recognized with 40,000 Species in total; about 2,000 Species known from North America. Braconid classifications include 4–31 Subfamilies. Braconids are distinguished from Ichneumonidae by only one Recurrent Vein in forewing (some exceptions exist). Range of body size is considerable: Smallest braconid (*Aspilot insignis* Stelfox & Graham) is about 0.8 mm long; largest braconids are in Genera *Trachybracon* and *Archibracon,* ca 25 mm long, and act as ectoparasites. Typically, large body size is characteristic of ectoparasitic 'generalized' braconids; small body size characteristic of endoparasitic, specialized groups (*e.g.* Alysiinae, Euphorinae, Microgasterinae). Large body size is typical of tropical regions; small body size is typical of temperate regions. Cf. Aphidiidae.

BRACT Noun. (Latin, *bractea* = thin metal plate. Pl., Bracts.) Botany: A reduced or modified leaf near a flower or inflorescence.

BRACTEA Noun. (Latin, *bractea* = thin metal plate. Pl., Bracteae.) Some coccids: Projections of the lateral portions of preabdomen segments (MacGillivray).

BRADFORD, GEORGE DEXTER (1873–1894) (Anon. 1895, Entomol. News 6: 64.)

BRADLEY, JAMES CHESTER (1884–1975) American academic (Cornell University) and taxonomic specialist on aculeate Hymenoptera. (Gunn 1976, Proc. Roy. Entomol. Soc. Lond. (C) 40: 51.)

BRADLEY, JOHN W (–1918) (Burgess 1918, J. Econ. Entomol. 11: 390.)

BRADLEY, RICHARD (fl 1721–1732) (Lisney 1960, *Bibliogr. of British Lepidoptera 1608–1799*, 315 pp. (83–85, bibliogr.), London.)

BRADSHAW, BENJAMIN (1848–1883) (Trimen 1887, Trans. So. Afr. Phil. Soc. 4: xviii–xix.)

BRADY, L S (1867–1921) (G. T. P. 1921, Naturalist, Hull No. 772: 221.)

BRADYNOBAENIDAE Saussure 1892. Plural Noun. A small Family of aculeate Hymenoptera assigned to the Scolioidea or Vespoidea, depending upon classification.

BRADYPORIDAE Plural Noun. (Greek, *bradys* = slow.) A small Family of ensiferous Orthoptera with 28 Genera and about 200 Species. Family is best represented in the Palaearctic and Ethiopian. Adult body bulky with head round and small, and eyes widely separated; Antenna inserted below ventral margin of compound eye and shorter than the body; Species are micropterous or apterous. Defensive haemorrhage is common in Hetrodinae. Fluid originating in the Trochanteral-femoral membrane can be ejected 40–50 cm. Most Species are ground-dwelling and phytophagous; under some circumstances, bradyporids are cannibalistic. See Orthoptera.

BRAGA, JOSÉ MARIA (1892–1972) (Anon. 1972, Anais Fac. Ciene., Porto 55: 1–6, bibliogr.)

BRAGDON, KENNETH EDWARD (1886–1974) (Denmark 1974, Fla. Entomol. 57: 224.)

BRAHMAEIDAE Plural Noun. A small Family of ditrysian Lepidoptera assigned to Superfamily Bombycoidea.

BRAIN Noun. (Anglo Saxon, *braegen*. Pl., Brains.) Insects: The Ganglion of the Nervous System which lies in the head above the Oesophagus. Brain is a composite structure, but the precise number of primitive, segmental ganglia remains unknown. Presumably the Brain is formed from the first three (anteriormost) Ganglia in the Head. Brain originates essentially as the Ventral Nerve Cord (from Epidermal Neuroblasts). Daughter cells of the Neuroblasts form rudiments that are continuous with the Ventral Nerve Cord. Subsequently, daughter cells form three Brain regions: Anterior Lobe (Protocerebrum), middle lobe (Deutocerebrum), and posterior lobe (Tritocerebrum). During embryo development, Brain lobes fuse medially except Tritocerebrum which connects by Commissures beneath Pharynx. See Nervous System; Nerve Cell. Cf. Deutocerebrum; Protocerebrum; Tritocerebrum. Rel. Central Nervous System. Supra-oesophageal Ganglion.

BRAIN APPENDAGES Muscid larvae: Two sacs, one applied to each side of the Brain (Comstock).

BRAKELEY, JONAH TURNER (1847–1915) (Dow 1915, Bull. Brooklyn Entomol. Soc. 10: 84–86.)

BRAKMAN, PIETER JACOBUS (1910–1968) (Poot 1968, Entomol. Ber., Amst. 28: 202–204.)

BRAMBLE LEAFHOPPER *Ribautiana tenerrima* (Herrich–Schäffer) [Hemiptera: Cicadellidae].

BRAMSON, CONSTANTIN LUDWIG (1842–1909) (Anon. 1910, Entomol. Rdsch. 17: 22, 45.)

BRAN BUG See Confused Flour Beetle.

BRANCH Noun. (Latin, *branca* = paw. Pl., Branches.) 1. Something which extends from or enters into a main body or source. 2. A division or fork of a body, *e.g.* branches of tracheoles; branches of a nerve. Cf. Dendrite; Furca. Rel. Anastomosis. 3. A stem growing from the trunk or limb of a tree.

BRANCHEAE Respiratory tubes in arachnids. See Book Lung.

BRANCHIA Noun. (Greek, *branchium* = a gill. Pl., Branchiae.) A gill, air-tubes or gill-like processes of aquatic larvae (Smith).

BRANCHIAL BASKET In rectal gills of Anisoptera: A barrel-like chamber formed by the expanded anterior two-thirds of the Rectum and containing the rectal gills.

BRANCHIAL Adj. (Greek, *branchium* = a gill; *-alis* = characterized by.) Descriptive of structure that resembles gills (Branchiae).

BRANCHIATE Adj. (Greek, *branchium* = a gill; *-atus* = adjectival suffix.) Supplied with gills (Branchiae).

BRANCHIOPNEUSTIC Adj. (Greek, *branchium* = a gill; *pneumono* = lungs; *-ic* = characterized by.) A system of respiration in larvae in which spiracles are functionally supplanted by gills.

BRANCSIK, KARL (1842–1915) (Hetschko 1932, Wien. Entomol. Ztg. 49: 51–55, bibliogr.)

BRAND Noun. (Anglo Saxon, *beornan* = to burn. Pl., Brands.) 1. Pamphilinae (Lepidoptera): A conspicuous patch crossing the forewing disc and appearing as a scorched streak. 2. A complicated organ formed by tubular scales, Androconia and other scales (Comstock).

BRANDES, GUSTAV (1862–1941) (Sachtleben 1942, Arb. Physiol. Angew. Entomol. Berl. 9: 55.)

BRANDSTATTER, ENGEBERT (–1955) (Anon. 1955, Zeits. Wien. Entomol. Ges. 40: 208.)

BRANDT, A VAN DEN (1829–1909) (Willemse 1960, Naturhist. Maandbl. 49: 286–195.)

BRANDT, ALEXANDR FEDOROVIC (1844–) (Bogdanov 1889, Izv. Imp. Obshch. Lyub. Estest. AnTrop. Etnogr. Imp. Mosc. Univ. 55: [13–16], bibliogr.)

BRANDT, EDUARD KARLOVICH (1839–1891) (Portschinsky 1893, Horae Soc. Entomol. Ross. 27: i–iv, bibliogr.)

BRANDT, JOHANN FRIEDRICH (1802–1879) (Geinitz 1880, Leopoldina 16: 20–21.)

BRANDZA, MARCEL (1868–1934) (Houard 1935, Marcellia 28: xliv.)

BRANHAM, HENRY G (–1927) (Anon. 1927, Fla. Entomol. 11: 12.)

BRANNON, DAVID HOMER (1903–1969) (Telford 1969, J. Econ. Entomol. 62: 1253.)

BRANNON, LLOYD WILLIAM (1900–1948) (Howard 1948, J. Econ. Entomol. 41: 527.)

BRASAVOLA DE MASSA, ALBERTO (1886–1956)

(Tamanini 1957, Natura alpina 1957: 50–51.)

BRASS Noun. (Middle English, *bras;* Anglo Saxon, *braes;* Latin, *ferrum* = iron. Pl., Brasses.) 1. Historically, an alloy composed of copper and a base metal (tin or zinc). 2. Presently, an alloy of two parts copper, one part zinc and traces of other metals. Brass is typically yellow in colour, harder than copper, malleable and with other properties variable. Cf. Bronze. 3. A colour that is reddish-yellow in hue, medium saturation and high brilliance; more reddish than Canary Yellow. See Colour.

BRAUER, AUGUST (1863–1917) (Vanhöffen 1918, Mitt. zool. Mus. Berl. 9: 1–2, bibliogr.)

BRAUER, FRIEDRICH MORITZ (1832–1904) (Papavero 1975, *Essays on the History of Neotropical Dipterology* 2: 319–322. São Paulo.)

BRAULA COECA Nitzsch. A widespread Species of braulid fly; adult less than 2 mm long, flattened; eyes vestigial; apterous; tarsal claws pectinate. Species is associated with honey bees; typically phoretic on worker bees; take nectar and pollen from bee. Eggs are laid on inner side of cap of a cell filled with honey; larval development occurs in honey cell, not among brood cells. Larva tunnels between cells and pupation occurs within tunnels/cells. Adults are associated with adult workers, queen and sometimes drones. Bc is regarded as an inquiline.

BRAULIDAE Plural Noun. A small Family of brachycerous Diptera assigned to Superfamily Brauloidea. Species are associated with honey bees.

BRAUN, CARL FRIEDRICH WILHELM (1800–1864) (Ratzeburg 1874, Forstwissenschaft. Schriftsteller-Lexicon 1: 80.)

BRAUN, KARL (1870–1935) (Speyer 1935, Anz. Schädlingsk. 11: 144.)

BRAUNS, HANS HEINRICH JUSTUS CARL ERNST (1857–1929) (Anon. 1929, Nature 123: 499–500.)

BRAUSE, HEINZ (–1964) (Anon. 1832, Ann. Soc. Entomol. Fr. 1: 114.)

BRAZILIAN LEAFHOPPER *Protalebrella brasiliensis* (Baker) [Hemiptera: Cicadellidae].

BREAST Noun. (Middle English, *brest*; Pl., Breasts.) 1. The ventral surface of the Thorax. 2. The Sternum. See Sternum. Cf. Pectus.

BREAST-BONE Cecidomyiid larvae: A horny, more-or-less elongate process of the ventrum behind the mouth opening. BB is supposed to represent the Labium.

BREATHING PORE See Spiracle.

BREBISSON, JEAN BAPTISTE GILLES (–1832) (Anon. 1832, Ann. Soc. Entomol. Fr. 1: 114.)

BREDDIN, GUSTAV (1864–1909) (Horn 1910, Dt. Entomol. Z. 1910: 212.)

BREHM, ALFRED EDMUND (1829–1884) (Viets 1955, Die Milben 1: 52.)

BREHM, HERMAN, H (1869–1924) (Rummel 1925, Bull. Brooklyn Entomol. Soc. 20: 96.)

BREINDLE, VCLAC (1890–1948) (Komarek 1949, Cas. csl. Spol Entomol. 13: 3–7, bibliogr.)

BREIT, JULIUS (1847–1903) (Anon. 1904, Entomol. Jb. 13: 241.)

BRELLIN® See Gibberellic Acid.

BREMER, OTTO (–1873) (Anon. 1873, Horae Soc. Entomol. Ross. 10: xxi.)

BRENCHLEY, WINIFRED (–1953) (Buxton 1954, Proc. Roy. Entomol. Soc. Lond. (C) 18: 79.)

BRENSKE, ERNST (1845–1904) (Auel 1929, Entomol. Z., Frankf. a. M. 43: 113.)

BRENTHIDAE See Brentidae.

BRENTIDAE Plural Noun. Brentid Beetles; Straight-snouted Weevils. A moderate-sized Family of polyphagous Coleoptera assigned to Superfamily Curculionoidea, and which is predominantly tropical in distribution. Adult <40 mm long and elongate; Rostrum typically long and projecting straight forward; snout of female longer than snout of male. Antenna typically long, straight with short Scape; Labrum concealed; Prothorax without lateral Carinae; Ventrites 1 and 2 fused, longer than Ventrites 3–5; Pygidium concealed. Larvae phytophagous on numerous plant Species where they bore into wood and sometimes are regarded as pests.

BRETHES, JUAN (1871–1928) Argentinian taxonomic specialist in parasitic Hymenoptera. (Howard 1930, Smiths. Misc. Coll. 84: 354, 422.)

BRETSCHE, KONRAD (1858–1943) (Hotz 1943, Vjschr. naturf. Ges. Zürich 88: 219–222, bibliogr.)

BRETSCHNEIDER, RICHARD (1876–1959) (Anon. 1961, Z. wien. Entomol. Ges. 46: 29–30, bibliogr.)

BREVACERATUBAE Coccids: Ceratubae which do not open at the margin of the Pygidium (MacGillivray).

BREVIATE Adj. (Latin, *breviare* = to shorten; *-atus* = adjectival suffix.) Shortened. Pertaining to Antennae about as long as the head.

BREVIORATE Adj. (Latin, *breviare* = to shorten; *-atus* = adjectival suffix.) Pertaining to Antennae that are longer than the head but shorter than the body.

BREVIS A Latin term meaning short.

BREVISSIMATE ANTENNA An Antenna shorter than the head.

BREYER, ADOLFO (1889–1936) (Imms 1937, Proc. Roy. Entomol. Soc. Lond. (C) 1: 55.)

BREYER, ALBERT (1812–1876) (Katter 1877, Entomol. Kal. 2: 73.)

BREYER, ALBERTO (1890–1963) (Wigglesworth 1964, Proc. Roy. Entomol. Soc. Lond (C) 28: 57.)

BRIAN, ALESSANDRO (1873–1969) (Guiglia 1969, Annali Mus. civ. Stor. nat. Giacomo Doria 77: 751–770, bibliogr.)

BRICK, CARL (Weidner 1967, Abh. Verh. naturw. ver. Hamburg Suppl. 9: 304–306.)

BRIDWELL, JOHN C (1878–1957) American Hymenopterist who served in several positions. (Anon. 1957, Bull. Entomol. Soc. Amer. 3 (3): 51.)

BRIDGE Noun. (Middle English, *brigge*. Pl., Bridges.) Odonata: A secondary longitudinal vein connecting the Radial Sector (Comstock) and M1+2, with the Media apparently forming a continuous part of the Radial Sector. The proximal portion of the Subnodal Sector *sensu* de Selys and Hagen.

BRIDGE CROSSVEINS Odonata: Crossveins, one or more in number, extending between M1+2, and the bridge (in de Selys, between the principal and the Subnodal Sectors) proximal of the Oblique Vein. See Vein; Wing Venation.

BRIDGES, CALVIN BLACKMAN (1889–1938) (Osborn 1946, *Fragments of Entomological History*. Pt. II. 232 pp. (62–63), Columbus, Ohio.)

BRIDGMAN, JOHN BROOKS (1836–1899) (Morley 1900, Entomol. Mon. Mag. 36: 15–16.)

BRIFUR® See Carbofuran.

BRIGADE® See Bifenthrin.

BRIGANT, [BRIGATI] VINCENZIO (1766–1836) (Costa 1847, Trans. Entomol. Soc. Lond. 4: xviii, Appendix.)

BRIGATA® See Bifenthrin.

BRIGGS, CHARLES ADOLPHUS (1849–1916) (Lucas 1917, Entomologist 50: 23–24.)

BRIGHT, PERCY MAY (1865–1941) (Russel 1946, Proc. S. Lond. Entomol. Nat. Hist. Soc. 1946: 1–4.)

BRIGHTWEN, ELIZABETH (–1906) (Merrifield 1906, Proc. Entomol. Soc. Lond. 1906: xlviii.)

BRILLIANCE Noun. (French, *brilliant* = shine, sparkle.) 1. One of three attributes of colour; brilliance comprises the greys and ranges from black (zero brilliance) to white (total brilliance). 2. Radiant brightness. See Colour. Cf. Hue; Saturation.

BRILL-ZINNSER DISEASE See Epidemic Typhus.

BRIMLEY, CLEMENT SAMUEL (1863–1946) (Metcalf 1947, J. Econ. Entomol. 40: 141.)

BRIMSTONE BUTTERFLIES See Pieridae.

BRIN Noun. (French. Pl., Brins.) 1. Silk-worms: The fluid silk thread from each salivary gland. 2. One of two filaments issuing from silk glands in the fluid phase which combines with the other filament to form the bave. See Bave. Cf. Gres; Grege. Rel. Fibroin; Serecin.

BRINCKLE, WILLIAM D (1798–1862) (Weiss 1936, *Pioneer Century of American Entomology,* 320 pp. (272), New Brunswick.)

BRINDLEY, HAROLD HULME (1865–1944) (Gardiner 1944, Nature 153: 309–310.)

BRINDLEY, MAUD DORIA HAVILAND (1891–1941) (Blair 1942, Proc. Roy. Entomol. Soc. Lond. (C) 6: 40.)

BRINDLEY'S GLAND A sac-like Scent Gland which occurs in many adult Reduviidae; not found in Tribocephalinae and members of other Subfamilies. BG is associated with Metathoracic Scent Gland and Carayon's Gland, but is independent of them. BG originates in Metathorax, projects into abdominal Haemocoel and displays an aperture on the metathoracic Episternum. (See Barrett *et al.* 1979, Canad. J. Zool. 57: 1109–1119.) A similar gland occurs in Thaumastellidae, Tingidae and Pachynomidae. See Exocrine Gland. Cf. Carayon's Gland; Metathoracic Scent Gland.

BRINE FLIES See Ephydridae.

BRINK, ROBERT (–1923) (Anon. 1926, Verh. naturhist. Ver. preuss. Rheinl. 83: 209.)

BRIOSI, GIOVANNI (1846–1919) Italian academic (Professor of Botany, University of Pavia) and founder of the Chemical Research Station for Viticulture in Palermo (1873). Briosi published on mites of importance on grapes. (Monti 1922, Rc. Ist. lomb. Sci. lett. 55: 489–498.)

BRIPOXUR® See Proproxur.

BRISCHKE, CARL GUSTAV ALEXANDER (1814–1897) (van Rossum 1903, Entomol. Ber., Amst. 1: 93.)

BRISOUT DE BARNVILLE, CHARLES N F (1822–1893) (Bonvouloir 1894, Ann. Soc. Entomol. Fr. 63: 439–448, bibliogr.)

BRISTLE Noun. (Anglo Saxon, *byrst*. Pl., Bristles.) 1. Setae that appear stiff, based on an impression conveyed by size and shape. Bristles are often short and blunt. 2. Setae that are significantly larger in length or diameter than surrounding Setae or nearby Setae. Cf. Acantha; Microseta; Seta.

BRISTLETAILS See Archaeognatha; Thysanura.

BRISTLY CUTWORM *Lacinipolia renigera* (Stephens) [Lepidoptera: Noctuidae]. See Cutworms.

BRISTLY ROSESLUG *Cladius difformis* (Panzer) [Hymenoptera: Tenthredinidae].

BRITTEN, HARRY (1870–1954) (Parker 1973, Brit. Arachnol. Soc. Newsl. No. 8: 4–5.)

BRITTON, G F J M (–1940) (D. M. 1940, Trans. Proc. Roy. Soc. N. Z. 70: xxxvii–xxxviii.)

BRITTON, WILTON EVERETT (1868–1939) (Mallis 1971, *American Entomologists*. 549 pp. (433–434), New Brunswick.)

BROAD HEADED BUGS See Alydidae.

BROAD MITE *Poyphagotarsonemus latus* (Banks) [Acari: Tarsonemidae]: BM attacks young growth of many dicotyledonous crops, ornamentals and weed plants. In Australia, a major and sporadic pest of lemons and Hickson and Ellendale mandarins. Damaged fruit have a silvery-green appearance. Eggs are oval, translucent, ornamented with white tubercles and laid on underside of leaves. Life cycle (egg, two nymphal stages adult) requires 6–9 days.

BROAD-BEAN WEEVIL *Bruchus rufimanus* Boheman [Coleoptera: Bruchidae].

BROAD-HORNED FLOUR BEETLE *Gnathocerus cornutus* (Fabricius) [Coleoptera: Tenebrionidae]: A cosmopolitan, secondary pest of stored cereal and animal products. In Australia, a minor pest of flour and other processed grain products; adult and larva also are predaceous on other stored product pests. Adult 4.0–4.5 mm long and pale reddish brown; Elytra smooth; male with mandibular processes resembling horns. Mature larva ca 8 mm long, white-yellow with brown bands near posterior end of body. Life cycle requires 8–10 weeks. Egg incubation is 5–8 days and high humidity facilitates eclosion. Adult lives 4–8 months; female lays to 360 eggs. See Tenebrionidae. Cf. Long-headed Flour Beetle; Slender-Horned Flour Beetle.

BROAD-NECKED ROOT BORER *Prionus laticollis* (Drury) [Coleoptera: Cerambycidae].

BROAD-NOSED GRAIN WEEVIL *Caulophilus oryzae* (Gyllenhal) [Coleoptera: Curculionidae]: A pest of stored maize in Central America and the southern USA and ginger in the West Indies. See Curculionidae.

BROAD-SHOULDERED WATER STRIDERS See Veliidae.

BROAD-SPECTRUM PESTICIDE Chemicals that affect pests and benefical organisms. See Pesticide.

BROAD-WINGED DAMSELFLIES See Calopterygidae.

BROAD-WINGED DAMSELFLY *Agrion maculatum* [Agrionidae: Odonata]. AKA Black Winged Damselfly.

BROAD-WINGED KATYDID *Microcentrum rhombifolium* (Saussure) [Orthoptera: Tettigoniidae]: A minor pest of citrus, willow and cottonwood in southern USA west to California. Eggs are flat, oval, overlapping in 1–2 rows and glued to twigs or leaves; BWK overwinters in the egg stage. See Tettigoniidae. Cf. Angular-Winged Katydid; Citrus Katydid; Crested Katydid; Forktailed Bush-Katydid; Inland Katydid; Mottled Katydid; Philippine Katydid; Spotted Katydid.

BROCHER, FRANK (1866–1936) (1936. Pictet Mitt. schweiz. Entomol. Ges. 16: 749–761, bibliogr.)

BROCHUS Noun. (Latin, *brochus* = with projecting teeth. Pl., Brochi.) Hymenoptera: Dorsal Serrula at the apex of Gonapophysis IX *teste* Tuxen.

BROCKES, BARTHOLD HEINRICH (1680–1747) (Weidner 1967, Abh. Verh. Naturw. Ver. Hamburg Suppl. 9: 23–26)

BRODAN® See Chlorpyrifos.

BRODIE, PETER BELLINGER (1815–1857) (Trimen 1897, Proc, Entomol. Soc. Lond. 1897: 72.)

BRODIE, WILLIAM (1831–1909) (Anon. 1910, Can. Entomol. 42: 47–48.)

BRODIIDAE Plural Noun. A Family of Megasecoptera with tubular projections on the Abdomen.

BROERSE, J (1875–1950) (Corporaal 1940, Entomol. Ber., Amst. 10: 213–214.)

BROGI, SIGISMONDO (–1889) (Anon. 1900, Entomol. Jb. 9: 296.)

BROKEN Adj. (Middle English, *brocen* > *brecan* = to break.) Pertaining to structure interrupted in continuity. Often, the term is applied to a line of Setae, wing venation, sclerite or band of coloration. Syn. Fractured; Ruptured; Discontinuous.

BROKENBACKED BUG *Taylorilygus pallidulus* (Blanchard) [Hemiptera: Miridae]: Adult 5 mm long; brown with a greenish tint; brown markings around base of legs and Abdomen; forewing with a dark blotch in centre near the membranous part of wing; second antennal segment has a dark mark; Abdomen curved so that apex almost touches leaf surface, hence common name 'brokenbacked.' Adult female lays eggs in soft plant tissue; eggs are sac-shaped and 0.75 mm long; egg stage requires ca eight days. Nymphs 1–5 mm long and resemble adults but lack wings. Nymphs not known to cause damage, but do occur in large numbers on cotton in Australia.

BRÖLEMANN, HENRI (1860–1933) (J. C. 1935, Mitt. Schweiz. Entomol. Ges. 16: 611.)

BROMATIUM Noun. (Greek, *broma* = food; Latin, *-ium* = diminutive > Greek, *-idion*. Pl., Bromatia.) The hyphal swellings of fungi which are cultivated by ants and used for food.

BROMBACHER, ERNST (1879–1935) (Von der Goltz 1935, Int. Entomol. Z. 28: 549–550.)

BROME-GRASS SEED MIDGE *Contarinia bromicola* (Marikovskij & Agafonova) [Diptera: Cecidomyiidae].

BROMLEY, STANLEY WILLLARD (1899–1954) (Weiss 1955, J. N. Y. Entomol. Soc. 63: 1–7, bibliogr.)

BROMOFLOR® See Ethephon.

BROMO-FUME® See Ethylene Dibromide.

BROMOMETHANE® See Methyl Bromide.

BROMOPROPYLATE A chlorinated hyrocarbon {Isopropyl 4,4'-dibromobenilate} used as a residual, contact acaricide in some countries. Not registered in USA. Trade names include Acarol®, Folbex® and Neoron®. See Organochlorine Insecticide.

BRONCHIAE Plural Noun. (Greek, *bronchos* = wind pipe.) See Tracheole.

BRONGNIART, ALEXANDRE (1770–1847) (Dumas 1877, Mem. Acad. Sci Inst. Fr. 39: xxxvii–xcix, bibliogr.)

BRONGNIART, CHARLES JULES EDME (1859–1899) (Wilderman 1900, Jb. naturw. 15: 476.)

BRONGNIARTIELLIDAE Plural Noun. A Mesozoic Family of Neuroptera. (See Martynova 1962.)

BRONN, HEINRICH GEORG (1800–1862) (Anon. 1962, Amer. J. Sci. (2) 34: 304.)

BRONZE APPLE-TREE WEEVIL *Magdalis aenescens* LeConte [Coleoptera: Curculionidae].

BRONZE BIRCH-BORER *Agrilus anxius* Gory [Coleoptera: Buprestidae]: A pest endemic to Nearctic which attacks aspen, birch and poplar in boreal North America. BBB first causes browning of ends of branches in crown of tree, then tree death. Adult ca 8–12 mm long, greenish-black with bronze tinge. Eggs are laid in small clutches within bark fissures. Larvae cut zig-zag tunnels through cambium, leaving frass and causing dieback. BBB overwinters as mature larva within a cell constructed at end of tunnel; pupation occurs during spring. See Buprestridae. Cf. Bronze Poplar Borer; Pacific Flathead Borer; Roundheaded Apple-Tree Borer.

BRONZE DUMP-FLY *Ophyra aenescens* [Diptera: Muscidae]: A New World and Oceanic pest of carrion and excrement. Syn. Black Garbage Fly. See Muscidae.

BRONZE LEAF-BEETLE *Diachus auratus* (Fabricius) [Coleoptera: Chrysomelidae].

BRONZE ORANGE BUG *Musgraveia sulciventris* (Stål) [Hemiptera: Tessaratomidae]: A univoltine pest of *Citrus* which is endemic to Australia and important in coastal *Citrus* orchids where it damages growing shoots, fruit and flower pedicels. Adults and nymphs cause wilt and fruit drop during spring and early summer. Adult ca 25 mm long, shield-like and body bronze-coloured with orange-coloured Antenna. Eggs are deposited in clusters of 10–20 on underside of leaves; female can produce a few hundred eggs per lifetime. Neonate egg is pale green, spherical and ca 3 mm diameter; eclosion occurs within 8–14 days. First nymphal instar is gregarious near eggshells and does not feed; second instar is thin, flat and overwinters on underside of leaves. BOB undergoes five nymphal instars; instars 1–3 pale green; instars 4–5 orange with black spot at apex of Scutellum, apex of Abdomen and multicoloured Antenna. See Tessaratomidae.

BRONZE POPLAR-BORER *Agrilus liragus* Barter & Brown [Coleoptera: Buprestidae]: A pest of poplars in North America. BPB resembles Bronze Birch Borer in anatomy and biology; larva girdles twigs of poplar. BPB prefers to attack trees in weakend condition but does not attack dead trees. See Buprestidae. Cf. Bronze Birch Borer; Pacific Flathead Borer; Roundheaded Apple-Tree Borer; Sinuate Pear Tree Borer.

BRONZED CUTWORM *Nephelodes minians* Guenée [Lepidoptera: Noctuidae]: A monovoltine pest of grain, corn and grasses in northern states of USA. Larvae are granulated, dark bronze with five pale longitudinal lines. See Noctuidae; Cutworms.

BROOD Noun. (Middle English, *brood*. Pl., Broods.) 1. Social Insects: Collectively, all immature members of a colony (eggs, larvae, nymphs, pupae), 2. General: A cohort of individuals that eclose at one time from eggs laid by a female and which normally mature concurrently. Hymenopterists studying social behaviour sometimes refer to first brood, second brood, *etc.* 3. Several cohorts of offspring produced by a parent population at different times in different places.

BROOD CANNIBALISM Social Insects: The behavioural phenomenon in which workers consume immatures of the same colony.

BROOD CHAMBER Stylops females: The space between the venter and the puparium, into which the triungulinds escape (Comstock).

BROOD FOOD Social Hymenoptera: Glandular secretions from the heads of workers which are fed to larvae in the colony. See Royal Jelly.

BROOKADE® See Bifenthrin.

BROOKS, ALBERT EUGENE (1876–1955) (Hall 1956, J. Proc. Roy. Entomol. Soc. Lond. (C) 20: 74.)

BROOKS, CECIL JOSLYN (1875–1953) (Remington 1955, Lepid. News 9: 22.)

BROOKS, GEORGE SHIRLEY (1872–1947) (Wallis 1948, Lepid. News 2:1.)

BROOKS, JOHN GEORGE (1910–1975) (Gunn 1976, Proc. Roy. Entomol. Soc. Lond. (C) 40: 51.)

BROOKS, RICHARD (fl 1750–1763) (Lisney 1960, *Bibliogr. of British Lepidoptera 1608–1799.* 315 pp. (180–191, bibliogr.), London.)

BROOKS, WILLIAM (–1916) (Anon. 1916, Entomol. Rec. J. Var. 28: 117.)

BROOME, RODERICK R (1900–1966) (Pearson 1967, Proc. Roy. Entomol. Soc. Lond. (C) 31: 62.)

BROSCHK, WILHELM (1896–1974) (Gruber 1974, Prakt. Schädlingsbekampf 26: 33.)

BROSSE Noun. 1. A brush of Setae. 2. Apoidea: The Scopa.

BROUN, THOMAS (1838–1919) (Howard 1930, Smiths. Misc. Coll. 84.)

BROWN-BANDED COCKROACH *Supella longipalpa* (Fabricius) [Blattodea: Blattellidae]: A cosmopolitan urban pest of African origin. Adult 10–15 mm long, pale brown with pale bands across Thorax and Abdomen; female wings somewhat shortened; male wings extend past apex of Abdomen. BBC adults are active and fly when disturbed. Adult lives 3–6 months and female lays 6–12 Oothecae during lifetime. An Ootheca contains 10–16 eggs and is glued to substrate. BBC nymphs undergo 6–8 moults and the life cycle requires six months. BBC is a domicilary pest that is spread through house, hospital, restaurant, office and storage facility. Nymphs and adults feed on starch, sugar, leather, parchment and fermented foods. See Blattodea. Cf. German Cockroach.

BROWN BUTTERFLIES See Nymphalidae.

BROWN CARPENTER-ANT *Camponotus castaneus* [Hymenoptera: Formicidae]: An ant Species common in southeastern USA. See Formicidae. Cf. Black Carpenter-Ant; Florida Carpenter-Ant; Red Carpenter-Ant.

BROWN CHICKEN-LOUSE *Goniodes dissimilis* Denny [Mallophaga: Philopteridae]: A pest of poultry to include chickens. See Philopteridae. Cf. Chicken Body-Louse; Fluff Louse; Shaft Louse; Wing Louse.

BROWN CITRUS-APHID *Toxoptera citricida* (Kirkaldy) [Hemiptera: Aphididae]: A pest of *Citrus* that is endemic to Asia, adventive to South America during 1970s, USA during 1990s and probably global where *Citrus* is grown. BCA feeds on new-flush growth of *Citrus* and is an important vector of Citrus Tristeza Virus. See Aphididae.

BROWN COCKROACH *Periplaneta brunnea* Burmeister [Blattodea: Blattidae].

BROWN COTTON-LEAFWORM *Acontia dacia* Druce [Lepidoptera: Noctuidae].

BROWN CUTWORM See Pink Cutworm.

BROWN DOG-TICK *Rhipicephalus sanguineus* (Latreille) [Acari: Ixodidae].

BROWN-DOTTED CLOTHES MOTH *Niditinea fuscella* (Linnaeus) [Lepidoptera: Tineidae]: An Holarctic pest of wollen products.

BROWN DUNG-BEETLE *Onthophagus gazella* Fabricius [Coleoptera: Scarabaeidae].

BROWN EAR-TICK *Rhipicephalus appendiculatus* [Acari: Ixodidae]: The primary vector of a lethal form of Theileriosis that attacks cattle and water buffalo in eastern tropical Africa.

BROWN, EDWIN (1819–1876) (Musgrave 1932, *A Bibliogr. of Australian Entomology 1775–1930*, 380 pp. (32), Sydney.)

BROWN, ERIC SEPTIMUS (1912–1972) (Betts 1974, Entomol. Mon. Mag. 103: 65–71, bibliogr.)

BROWN FLOUR-MITE *Gohieria fusca* (Oudemans) [Acari: Glycyphagidae].

BROWN, FRANÇOIS ROBERT FENWICK (1837–1915) (Oberthür 1916, Etudes de lépidopterologie comparée 11 (Portraits de Lépidoptèristes 3 Sér.)

BROWN-HEADED ASH SAWFLY *Tomostethus multicinctus* (Rohwer) [Hymenoptera: Tenthridinidae].

BROWN-HEADED JACK-PINE SAWFLY *Neodiprion dubiosus* Schedl [Hymenoptera: Diprionidae].

BROWN-HOODED COCKROACHES See Cryptocercidae.

BROWN HOUSE-ANT 1. *Doleromyrma darwiniana* (Forel) [Hymenoptera: Formicidae]: Endemic to Australia and minor urban pest in New South Wales. Worker 2.0–3.0 mm long with brown body; gastral Pedicel consisting of one node. BHA is odorous when crushed. Aka Darwin's Ant. Syn. *Iridomyrmex darwinianus*. 2. *Pheidole megacephala* (Fabricius) [Formicidae] (South Africa). See Formicidae.

BROWN HOUSE-MOSQUITO *Culex quinquefasciatus* Say [Diptera: Culicidae]: Closely related to Northern House Mosquito *(Culex pipiens)* in North America and regarded as a Subspecies of *C. pipiens* in some classifications. BHM occurs below 39°N in New World and widely distributed in Old World. BHM becoming more common in Urban Africa and Asia; females invade homes and bite at night. BHM serves as vector of human Filariasis *(Wucheria bancrofti)*, Ross River Virus, Encephalitis, Dog Heartworm and Fowl Pox. Adult brown; Thorax with white scales but not on postspiracular area; Tibia apically pale; Tarsi not banded; Abdomen apically truncate. Female lays eggs in 'rafts' on surface of water; incubation ca 3 days, Larvae develop in still, polluted water (septic tanks, liquid manure, drainage systems); larval development requires 7–10 days; pupation in water. Syn. *Culex fatigans*. See Culicidae.

BROWN HOUSE-MOTH *Hofmannophila pseudospretella* (Stainton) [Lepidoptera: Oecophoridae]: A cosmopolitan, univoltine pest of mills, storage areas and houses. Larvae are exceptionally polyphagous feeding on cereals, pulses, seeds, dried fruit, wool, fur, feathers, corks, book bindings, house carpets and underlay. Adult female ca 14 mm long (wingspan 20–25 mm); male 8 mm long (wingspan 17–19 mm); bronze coloured; male flies while female tends to run. Eggs are laid in produce and resist desiccation; eclosion highly varible (8–110 days). Larvae are white with brown head capsule; mature larva ca 20 mm long; diapause is common. Pupation occurs inside a tough cocoon which includes debris. See Oecophoridae.

BROWN, JAMES MEIKLE (1875–1951) (H[incks] 1951, Naturalist, Hull 1951: 147–148.)

BROWN, JAY ELMER (–1903) (Anon. 1940, Entomol. News 15: 56.)

BROWN, JOHN HUGH (–1968) (Kennedy 1969, Proc. Roy. Entomol. Soc. Lond. (C) 33: 53.)

BROWN LACEWING *Micromus tasmaniae* Walker [Neuroptera: Hemerobiidae]: An Australian Species. Adult 2.5–3.5 mm long; wingspan 5–8 mm; body elongate and brown; Thorax cream-coloured with brown stripe in center. Head triangular; Antenna as long as forewing and pale brown; eyes large. Wings transparent with lace-like venation, may be spotted with dark brown on veins; terminal segments of Abdomen pale coloured. Eggs are cream coloured, oval and laid singly on underside of leaves. Larvae are elongate but tapered distally, appear humped and smooth with brown and white markings; mouthparts sickle-shaped, large and at apex of head. Pupation within leaf litter. Adults and larvae predaceous; feed on aphids, soft-bodied insects and eggs. Larvae on foliage seeking prey. See Hemerobiidae. Cf. Green Lacewing.

BROWN-LEGGED GRAIN MITE *Aleuroglyphus ovatus* (Troupeau) [Acari: Acaridae].

BROWN LYCTID BEETLE *Lyctus brunneus* [Coleoptera: Lyctidae]: A widespread, significant pest of recently seasoned hardwoods. Eggs are long, slender and deposited in sapwood vessels; eggs are not deposited in wood containing <3% starch in xylem parenchyma cells; eclosion within ca 21 days; first instar larva tunnels in outer layers of sapwood; larvae lack cellulase and cannot digest cell walls; food consists of cell contents (sugars, starches, some protein); last instar larva tunnels near surface of wood and constructs pupal chamber; pupation within 2–4 weeks; life cycle 1–3 years. See Lyctidae.

BROWN MIRID *Creontiades pacificus* (Stål) [Hemiptera: Miridae]: Adult 3–7 mm long, yellow-green to green with dark pigments; body elongate-oval; head and Pronotum triangular; second antennal segment longest. Adult stage up to 28 days. Adult female lays eggs singly within plant tissue with an oval egg cap projecting above leaf or petiole surface. Egg stage requires 7–10 days. Nymphs resemble the adult but lack wing buds. Nymphs 1–6 mm long, green to yellow-green with long Antenna. BMs feed on flower buds as they approach full size and cause shedding; plant tissue is pierced by stylets and pectinase is released into plant. Pectinase destroys plant cells in feeding zone; occurs in many field crops such as lucerne, safflower, sunflowers and potatoes. See Miridae. Cf. Green Mirid; Yellow Mirid

BROWN MITE *Bryobia rubrioculus* (Scheuten) [Acari: Tetranychidae].

BROWN PINEAPPLE-SCALE *Melanaspis bromiliae* (Leonardi) [Hemiptera: Diaspididae].

BROWN PLANTHOPPER *Nilaparvata lugens* (Stål) [Hemiptera: Delphacidae]: A major pest of rice in tropical Asia and some temperate areas.

BROWN, ROBERT (1773–1858) (Edwards 1976, J. Soc. Bibl. Nat. Hist. 7: 385–407.)

BROWN-RECLUSE SPIDER *Loxosceles reclusa* Gertsch & Mulaik [Araneae: Loxoscelidae].

BROWN ROT A fungal disease of cherry, peach and plum that is caused by *Sclerotinia fructicola* and vectored by the Plum Curculio.

BROWN SALTMARSH-MOSQUITO *Aedes cantator* (Coquillett) [Diptera: Culicidae].

BROWN SCAVENGER BEETLES See Lathridiidae.

BROWN SHIELD BUG *Dictyotus caenosus* (Westwood) [Hemiptera: Pentatomidae]: In Australia, a minor pest of cotton, but does not feed on cotton; primary hosts include grasses, beans and lucerne. Adults yellowish-brown or yellowish with brown punctations; body 7–11 mm long; apical segments of Antennae darkened. Head rounded and triangular-shaped. Mouthparts extend to the middle Coxa. Body segments visible beneath wings and are darkened at the margins. No information is available on the egg stage. Nymphs are similar to adults but lack wings. See Pentatomidae. Cf. Glossy Shield Bug; Predatory Shield Bug.

BROWN SMUDGE BUG *Deraeocoris signatus* (Distant) [Hemiptera: Miridae]: An Australian Species common in cotton. Adult 2–6 mm long, red to maroon-coloured and elongate-oval; head and pronotum triangular; second antennal segment longest. Adult stage to 28 days. Adult female lays eggs singly within plant tissue with oval egg cap showing above leaf or petiole surface. Egg stage requires 7–10 days. Nymphs are similar to adult but lack wing buds. Nymphs 1–6 mm long and red to maroon-coloured with long Antennae. Adult and nymphal BSB are predatory on *Helicoverpa* eggs, aphids, mites and the apple dimpling bug in cotton and cause no damage to cotton. See Miridae.

BROWN SOFT-SCALE *Coccus hesperidum* Linnaeus [Hemiptera: Coccidae]: A cosmopolitan minor pest of ornamentals and *Citrus*. Eggs complete development within females who produce crawlers. BSS resembles Black Scale in habitus and janots. More than 25 Species of parasitic Hymenoptera have been recovered from BSS. See Coccidae.

BROWN SPECKLED-LEAFHOPPER *Paraphlepsius irroratus* (Say) [Hemiptera: Cicadellidae].

BROWN SPIDER-BEETLE *Ptinus clavipes* Panzer [Coleoptera: Ptinidae].

BROWN STINK-BUG 1. *Euschistus servus* (Say) [Hemiptera: Pentatomidae]: A pest of several plant Species in the USA. BSB overwinters up to nine months as adult; female deposits ca 600 eggs during lifetime; barrel-shaped eggs are deposited in clusters of 30–70 on host plant foliage and eclosion occurs within 12 days. Nymphs form aggregations and undergo five instars; mature nymph ca 12 mm long, yellow or pale brown with darker brown spots that form a median longitudinal line on Abdomen. Adult ca 16 mm long, brown with yellow or pale-green venter. See Pentatomidae. Cf. Green Stink Bug; Southern Green Stink Bug. 2. Japan: *Hylamorpha mista* [Hemiptera: Pentatomidae]: A pest of apple, pear, cherry and peach trees in Japan. BSB forms large aggregations as first and second instar nymphs during growing season and as adults during winter. See Pentatomidae.

BROWN-TAIL MOTH *Euproctis chrysorrhoea* (Linnaeus) [Lepidoptera: Lymantriidae]: A monovoltine pest of apple, cherry, pear and deciduous ornamental trees. BTM is endemic to Palaearctic and introduced into North America ca 1897 on nursery stock. BTM overwinters gregariously as early-instar larva within colonial web on leaves and twigs; larvae resume activity during spring and feed on new foliage; early-instar larvae return to web at night while later instars remain on foliage. Larval body is covered with urticating Setae that cause discomfort, illness and severe reaction in some people. Mature larva dark brown or black with red tubercles on dorsum of posterior abdominal segments. Adults appear during July. Eggs are laid in masses on underside of host plant leaves; eclosion occurs during August–September. Syn. *Nygmia phaeorrhoea* (Donovan). See Lymantriidae. Cf. Gypsy Moth.

BROWN, THOMAS NESMITH (1851–1921) (Miller 1930, Entomol. News 41: 29–30.)

BROWN WHEAT-MITE *Petrobia latens* (Müller) [Acari: Tetranychidae]: Adult 0.2–0.5 mm long; body brown and spherical. First pair of legs distinctly longer than other legs and with two long Setae on apex. Adult stage ca 20 days. Females lay 300 long-term and 100 short-term eggs. Long-term and short-term eggs are laid singly in leaf litter or soil. Long-term eggs are white with a radial pattern and are viable for several seasons. Short-term eggs are red and eclosion occurs ca 7 days after deposition. Nymphs resemble adults. Nymphal stage requires ca 7–10 days. Mites insert mouthparts to feed on leaf tissue. Feeding produces mottling of leaves. See Tetranychidae.

BROWN WIDOW SPIDER *Latrodectus geometricus* (Fabricius) [Araneae: Theridiidae]: A spider presumably endemic to Africa and now widespread in tropical and subtropical areas. BWS is a minor pest in Australia. BWS resembles the Black Widow in shape and size; body and legs are brown to black; hourglass mark on venter of Abdomen yellow-coloured. Egg sac brown, pea-sized with numerous small points projecting from surface. BWS bite is considerably less dangerous than Black Widow or Redback. See Therididae. Cf. Black-Widow Spider; Redback Spider.

BROWN, WILLIAM L JR (–1997) American academic (Cornell University) and specialist on ant systematics.

BROWNE, GEORGE B (1851–1920) (A. L. R. 1920, Entomologist 54: 24.)

BROWNE, M G M (–1944) (Cockayne 1945, Proc. Roy. Entomol. Soc. Lond. (C) 9: 48.)

BROWNE, W A F BALFOUR See BALFOUR-BROWNE.

BRUAND D'UZELLE, CHARLES THÉOPHILE (1806–1861) (Milliere 1861, Ann. Soc. Entomol. Fr. (4) 1: 651–656, bibliogr.)

BRUCE, ADAM TODD (1860–1887) (WIlderman 1888, Jb. Naturw. 3: 540.)

BRUCE, DAVID (1833–1903) Pioneer in Medical Entomology who discovered the role of Tsetse Fly in transmission of Nagana. (Browne 1966, J. N. Y. Entomol. Soc. 74: 126–133.)

BRUCE SPANWORM *Operophtera bruceata* (Hulst) [Lepidoptera: Geometridae]: A minor pest of apple, aspen, beech, maple, *Prunus* spp. and willow, in parts of Canada and the northeastern USA.

BRUCE, WESLEY GORDON (1892–1966) (Knipling 1967, J. Econ. Entomol. 60: 310–311.)

BRUCH, CARLOS (1869–1943) Chilean Entomologist. (Lizer y Trelles 1947. Curso Entomol. 1: 41–42.)

BRUCHIDAE Plural Noun. Seed Beetles. A Family of Coleoptera assigned to Superfamily Chrysomeloidea. Adult typically less than 6 mm long, dull-coloured with head forming a short snout; Elytra do not cover apex of Abdomen. Adults oviposit on seeds; larvae feed inside seeds and typically pupate in seeds. Some Species are pests of agriculture; other Species are pests of stored products. See Chrysomeloidea.

BRÜEL, LUDWIG (1871–1949) (Spek 1951, J. Naturw. Rdsch. 4: 278–279.)

BRUES, CHARLES THOMAS (1870–1955) (American academic. Mallis 1971, *American Entomologists*. 549 pp. (434–435), New Brunswick.)

BRÜGGEMANN, FRIEDRICH (1850–1879) (Buchenaau 1880, Abh. Naturw. Ver. Bremen 6: 310–328, bibliogr.)

BRUGGER-MEIER, ADOLF (1913–1970) (Stocklin 1970, Mitt. Entomol. Ges. Basel 20: 47–48.)

BRUGIAN FILARIASIS A human disease caused by the onchocercid filarial worms *Brugia malayi* and *B. timori.* Disease occurs in Malaysia, Indonesia and Philippines. BF resembles Bancroftian Filariasis but is less common, less severe and vectored by *Anopheles* spp. and *Aedes* spp. See Bancroftian Filariasis.

BRUGNATELLI, GASPARE (1795–1852) (Veladini 1901, G. R. Istit. Lombardo 9: 489–490.)

BRUGNATELLI, LUIGI VALENTINO (1761–1818) (Pavesi 1856, Memorie Soc. Ital. Sci. Nat. 6: 1–68, bibliogr.)

BRULLE, GASPARD AUGUSTE (1809–1873) (Marchant 1887, Abeille (Les Entomol. et leurs écrits) 24: 175–177, bibliogr.)

BRUMPT, EMILE (1877–1951) (Gaillard 1952, Ann. Parasit. hum. comp. 27: 5–46, bibliogr.)

BRUN, RUDOLF (1885–1969) (Kutter 1970, Mitt. schweiz. Entomol. Ges. 43: 68–71.)

BRUNER, LAWRENCE (1856–1937) (Mallis 1971, *American Entomologists*. 549 pp. (191–195), New Brunswick.)

BRUNER, STEPHEN COLE (1891–1951) (de Zayas 1951, Boln. Hist. nat. Soc. Filipe Poey 2: 111–122, bibliogr.)

BRUNETTI, ENRICO ADELEMO (1862–1927) (Smart 1945, J. Soc. Biblio. nat. Hist. 2: 35–38, bibliogr. only.)

BRUNICKI, JULIAN (1864–1924) (Schille 1924, Polskie Pismo Entomol. 3: 1–2.)

BRUNN, K O MAX VON (1858–1942) (Weidner 1967, Abh. Verh. naturw. Ver. Hamburg Suppl. 9: 213–216.)

BRUNNER VON WATTENWYL, CARL (1823–1914) (Rehn 1915, Entomol. News 26: 285–288.)

BRUNNER, LAWRENCE American academic (University of Nebraska) with interests in Orthoptera. Brunner served as a USDA Special Agent and made observations on migratory locust. He was the father-in-law and an influential teacher of Harry Scott Smith.

BRUNNEUS Adj. (Latin, *brunus.*) Dark brown coloration.

BRUNNICH, MARTIN THOMAS Danish academic, naturalist and author of *Entomolgia, Sistens Insectorum Tabulas Systematicas* (1764).

BRUNNICH, MARTIN THRANE (1737–1827) (Henriksen 1921, Entomol. Meddr 15: 41–47.)

BRUNOL® See Permethrin.

BRUSH FOOTED BUTTERFLIES See Nymphalidae.

BRUSH ORGANS See Hair Pencil.

BRUSHES Plural Noun. (Middle English, *brusch.*) 1. Culicidae larvae: Tufts of stout Setae positioned on each side of the head. 2. Fin-like processes of the anal end or respiratory siphon.

BRUSH-FOOTED BUTTERFLIES See Nymphalidae. Butterflies with fore Tibia short and clothed with long Setae.

BRUSTIA Noun. (Etymology obscure. Pl., Brustiae.) Bands or areas of Spinulae or Setae on Mandibles (MacGillivray).

BRUYANT, CHARLES (1869–1916) (Berland 1920, Ann. Soc. Entomol. Fr. 1920. 89: 420–421, bibliography.)

BRYANT, ELIZABETH BANGS (1875–1953) (Deichman 1958, Psyche 65: 3–10, bibliogr.)

BRYANT, OWEN (1882–1958) (Anon. 1959, Bull. Entomol. Soc. Amer. 5: 51.)

BRYK, FELIX (1882–1957) Swedish Entomologist. (Paclt 1957, Entomologist 90: 192.)

BRYOPHYTE Noun. (Greek, *bryo* = moss; *phyton* = plant. Pl., Bryophytes.) A geologically ancient group of non-vascular plants that are characterized by rhyzoids (not true roots), multicellular archegonia and antheridia, and by a clearly defined alternation-of-generations that involves the sporophyte lacking chlorophyll and being attached to the gametophyte. Extant bryophyte forms include the mosses, liverworts and horworts.

BRYSON, HARRY RAY (1892–1956) (Smith 1957, J. Kans. Entomol. Soc. 33: 97.)

BT See *Bacillus thuringinesis* Berliner, a bacterium that is used in the control of some insects, particularly Lepidoptera larvae.

BTB® A registered biopesticide derived from *Bacillus thuringiensis* var. *kurstaki*. See *Bacillus thuringiensis*.

BTD See Bluetongue Disease.

BTV® A registered biopesticide derived from *Bacillus thuringiensis* var. *kurstaki*. See *Bacillus thuringiensis*.

BUBACEK, OTTO (1872–1934) (Reisser 1934, Z. ost. Ent Ver. 19: 57–62.)

BUBONIC PLAGUE The most common clinical form of Plague. BP is caused by the bacteria *Yersinia pestis* and vectored by many Species of fleas, including *Xenopsylla* spp. and *Pulex irritans*. BP incubation period requires 2–8 days. Symptoms include headache, fever and painful swellings (buboes) of axillary lymph nodes; BP is often fatal to humans. AKA Black Death; Black Plague. See Human Flea; Oriental Rat Flea; Plague.

BUCCA Noun. (Latin, *bucca* = cheek. Pl., Buccae.) The mouth in insects. Diptera: Part of head-wall below the transverse impression and eye to the edge of the mouth opening, up to the vibrissal ridge and back on the Gena to the caudal margin of the head (Comstock).

BUCCAL Adj. (Latin, *bucca* = cheek; *-alis* = belonging to, pertaining to.) 1. Descriptive of structure associated with the mouth cavity. 2. Pertaining to the cheeks (Gena).

BUCCAL APPENDAGES The mouthparts excluding the Labrum. See Trophi.

BUCCAL CAVITY The region of the foregut immediately inside the mouth (preoral cavity) The Buccal Cavity provides points of insertion for dilator muscles that originate on the Clypeus. Cf. Proboscidial Fossa. Rel. Cavity.

BUCCAL FISSURE The mouth slit or opening; the opening on both sides of the Mentum. Rel. Fissure.

BUCCAL FUNNEL Mallophaga: The tubular foregut in the region of the head. The BF extends into the Pharynx. Cf. Sac Tube. Rel. Funnel.

BUCCAL TEETH Lice: Minute denticles (acanthae) that occur in the forgut and which may be responsible for filtering. See Tooth.

BUCCATE Adj. (Latin, *bucca* = cheek; *-atus* = adjectival suffix.) Distended; especially pertaining to the Gena. Alt. Buccatus.

BUCCOPHARYNGEAL Adj. (Latin, *bucca* = cheek; Greek, *pharynx* = throat; *-alis* = pertaining to.) Descriptive of or pertaining to the Bucca (mouth) and Pharynx collectively. See Mouthparts.

BUCCOPHARYNGEAL ARMATURE See Pharyngeal Skeleton.

BUCCULA Noun. (Latin, *buccula* = little cheek. Pl., Bucculae.) A small, distended area consisting of elevated sclerites or ridges on the ventral part the head and side of the Rostrum. An anatomical feature used in the classification and taxonomy of Heteroptera. See Mouthparts.

BUCCULATRICIDAE Mosher 1916 Plural Noun. A cosmopolitan Family of ditrysian Lepidoptera assigned to Superfamily Tineoidea. Adult small with wingspan 5–15 mm; head with erect hair-like scales on Vertex; Ocelli absent; Chaetosemata typically absent; Antenna filiform, Scape expanded; dense Pecten forming eye-cap. Proboscis short and unscaled; Maxillary Palpus small, of one segment; Labial Palpus very small, pendulous, of three segments. Fore Tibia with Epiphysis; tibial spur formula 0-2-4; forewing lanceolate with apex pointed; hindwing narrowly lanceolate. Egg typically ovoid, flattened with reticulate sculpture; typically eggs are laid on leaf. First instar larva mines leaves; later instar forms galls in leaves or bark, or shelters in leaves. Pupa with ribbed cocoon, dorsal spines on most of abdominal Terga. Some Species are agricultural pests. See Tineoidea

BUCHECKER, HEINRICH (1829–1894) (Anon. 1898, Ber. Friedländer U. Sohn 40: 1690–1691.)

BUCHOLZ, FRIEDRICH (1911–1967) (Roesler 1968, Entomol. Z., Frankf. A. M. 78: 29–32.)

BUCHOLZ, OTTO (1874–1958) (Muller 1959, J. Lepid. Soc. 13: 27–29.)

BUCHOLZ, REINHOLD WILHELM (1837–1876) (Kraatz 1876, Entomol. Mbl. 1: 79.)

BUCK, FREDERICK (–1974) (Hillaby 1975, N. Sci. 66 (944): 91–92.)

BUCK MOTH *Hemileuca maia* (Drury) [Lepidoptera: Saturniidae].

BUCK, PIO (1883–1972) (Anon. 1973, Revta. bras. Entomol. 17: vi–vii, bibliogr.)

BUCKELL, EDWARD RONALD (1889–1962) (Spencer 1963, Proc. Entomol. Soc. Br. Columb. 60: 55–56.)

BUCKEYE Noun. (Old English, *buc* = stag; *eye*. Pl., Buckeyes.) Any of several shrubs or trees that resemble the horse chestnut (*Aesculus glabrata*).

BUCKEYE BUTTERFLIES *Junonia* and *Precis* spp. [Lepidoptera: Nymphalidae]: Butterflies noted for their colourful eyespots. *Junonia coenia* (Hübner) is the most common Species in the USA; its larvae feed on *Plantago*.

BUCKHOUT, WILLIAM ARMSTRONG (1846–1912) (Anon. 1912, Entomol. News 24: 48.)

BUCKLAND, WILLIAM (1784–1856) (Buckland 1858, Buckland's Geol. Mineral. 1: xix–lxxxiii,

bibliogr.)

BUCKLE, CLAUDE W (–1904) (Anon. 1904, Irish Nat. 13: 156.)

BUCKLER, SAMUEL BOTSFORD (1809–1883) (Weiss 1936, *Pioneer Century of American Entomology*. 320 pp. (224–225), New Brunswick.)

BUCKLER, WILLIAM (1814–1884) (Russel 1944, Entomol. Rec. J. Var. 56: 38.)

BÜCKLING, HERMAN (1885–1928) (Ochs 1928, Entomol. Bl. Biol. Syst. Käfer 24: 97–98.)

BUCKMASTER, C J (–1925) (Anon. 1925, Entomologist 58: 176.)

BUCKNILL, EDWIN G (–1970) (Butler 1972, Proc. Roy. Entomol. Soc. Lond. (C) 36: 61.)

BUCKTHORN APHID *Aphis nasturtii* Kaltenbach [Hemiptera: Aphididae].

BUCKTON, GEORGE BOWDLER (1818–1905) (Musgrave 1932, *A Bibliogr. of Australian Entomology 1775–1930*. 380 pp. (34), Sydney.)

BUD Noun. (Middle English, *budde* = a bud. Pl., Buds.) 1. Agamic viviparous aphids: An embryo that develops into a new individual. See Aphididae. 2. Botany: An undeveloped shoot or stem. A flower or leaf that has not fully opened. 3. Anatomy: A primordial stage of development for an organ or structure (*e.g.* wing bud).

BUD-AND-ROSETTE GALL Plant galls characterized by enlargement of buds into many shapes. BG typically develop during spring. See Gall. Cf. Covering Gall; Filz Gall; Mark Gall; Pit Gall; Pouch Gall; Roll-and-Fold Gall.

BUD MOTHS Any of several Species of tortricid (olethreutid) moths whose larvae consume the opening buds of fruit trees. See Eye-Spotted Bud Moth.

BUD ROT A fungal disease of carnations that is caused by *Sporotrichum poae* and vectored by the mite *Pediculopsis graminum*.

BUDDEBERG, CARL DIETRICH (1840–1909) (Roi 1912, Sber. naturh. Ver. preuss. Rheinl. Westf. 1911: 179–180, bibliogr.)

BUDDE-LUND, GUSTAV (1846–1911) (Tuxen 1968, Entomol. Meddr 36: 18, portrait only.)

BUDDENBROCK, WOLFGANG VON (–1959) (Grob 1959, Entomol. Z., Frankf. A. M. 69: 181–184.)

BUDDING Noun. (Middle English, *budde* > Middle Low German, *budecke* = swollen; Anglo Saxon, *-ing* = descended from.) 1. Social Insects: Colony multiplication by the departure of a queen and a small group of workers from a larger colony. Cf. Absconding; Fission; Swarming. 2. Agamic reproduction by aphids. See Aphididae.

BUDGET Noun. (Middle English, *bougette,* Latin, *bulga* = leather bag. Pl., Budgets.) 1. A pouch or bag including its contents. 2. A plan, schedule or record of the expenditure of energy, time or resources (*e.g.* energy budget).

BUENO, J R DE LA TORRE- See TORRE-BUENO.

BUFFA, PIETRO (1871–1941) Italian Thysanopterist and Acarologist. (Anon. 1941, Studi Trent. Sci. nat. 22: 213.)

BUFFALO BEETLE See Carpet Beetle.

BUFFALO FLIES See Simuliidae; Tabanidae.

BUFFALO FLY *Haematobia irritans exigua* (De Meijere) [Diptera: Muscidae]: A small-bodied biting muscoid fly that develops as an obligate parasite of cattle and buffalo. BF is endemic to the Orient and was introduced into Australia during 1838 where it is currently considered a significant pest of cattle and horses in northern Australia; BF also attacks dogs and humans. Adult serrates skin of host, inflicts a painful bite and takes blood. Female BF lays eggs laid in fresh cattle or buffalo dung. See Muscidae. Cf. Horn Fly.

BUFFALO GNATS See Simuliidae.

BUFFALO-GRASS WEBWORM *Surattha indentella* Kearfott [Lepidoptera: Pyralidae].

BUFFALO TREEHOPPER *Stictocephala bisonia* Kopp & Yonke [Hemiptera: Membracidae]: A univoltine pest that is widespread in North America. BTh eggs are laid in crescent-shaped slits in bark of trees, with 6–12 eggs deposited per slit. BTh overwinters in egg stage. Neonate nymphs drop to ground where they feed on alfalfa and grasses. BTh displays five nymphal instars and is univoltine.

BUFFON, GEORGE LOUIS LE CLERC (1707–1788) (Comte de Buffon.) French naturalist and Superintendent of Jardin des Plantes (1739). Buffon was among the most noted biologists of his time. He published a multivolume treatise entitled *Natural History* (1749–1788) in 36 volumes, which summarized existing information on biology and biogeography. (Greenwood 1886, *Eminent Naturalists*. 202 pp. (140–159), London.)

BUG Noun. (Etymology Uncertain, Middle English, *bugge,* confused form of Anglo Saxon *budda* = beetle. Pl., Bugs.) 1. The term 'bug' is used loosely to refer to all insects. In the strict sense, only Species of Hemiptera are correctly called 'bugs'. See Hemiptera. 2. A bed bug. See Cimicidae. 3. A microscopic organism that transmits disease.

BUG PIN® See Butoxycarboxim.

BUG-BUSTER® See Pyrethrin.

BUGHDANOV, GEORGIYA BUGHDANOVICH (1886–1963) (Borodenok *et al.* 1964, Entomol. Obozr. 43: 942–944, bibliogr. (transl. Entomol. Rev. Wash. 43: 479–480.))

BUGNION, CHARLES JUSTE JEAN MARIE (1811–1897) (Bugnion 1897, Ann. Soc. Entomol. Fr. 66: 317–318.)

BUGNION, FRÉDÉRIC EDOUARD (1845–1939) (Anon. 1958, Mitt. schweiz. Entomol. Ges. 31: 119)

BUG-TIME® A registered biopesticide derived from *Bacillus thuringiensis* var. *kurstaki*. See *Bacillus thuringiensis*.

BUHK, FERDINAND (1880–1953) (Weidner 1967, Abh. Verh. naturw. Ver. Hamburg Suppl. 9: 244–245.)

BUHR, HERBERT (1902–1968) (Haase 1968, Entomol. Ber., Berlin 1968: 93–95.)

BULB Noun. (Latin, *bulbus* = bulb, onion > Greek, *bolbos*. Pl., Bulbs.) Botany: A corm, tuber or other fleshy plant structure that resembles a bulb. 2. General: Any pear-shaped protuberance or structure that resembles a bulb. Cf. Corm. Rel. Shape.

BULB MITE *Rhizoglyphus echinopus* (Fumouze & Robin) [Acari: Acaridae]: A pest of bulb-plants and cereals; BM is endemic to Europe and adventive to North America. Host plants include amaryllus, crocus, gladiolus, hyacinth, iris, narcissus and onion. Female BM lays 50–100 eggs during 30–60 day life. Eggs are deposited behind bud scales. Larva and nymphs bore into bud scales; BM has an hypopal stage that is dispersive.

BULB OF STING Aculeate Hymenoptera: The basal portion of the fused second Valvulae. See Aculeus; Sting.

BULB SCALE-MITE *Stenotarsonemus laticeps* (Halbert) [Acari: Tarsonemidae].

BULBAPEX See Solenidion.

BULBOUS Adj. (Latin, *bulbus* = bulb; *-osus* = full of.) Bulb-like. A term descriptive of structure that is swollen and resembles a bulb. See Bulb.

BULBUS ARTERIOSUS A swelling of the insect Aorta at its junction with the Heart (Imms).

BULBUS EJACULATORIUS Male Lepidoptera: Termination of the Ejaculatory Duct at the base of the Aedeagus (Imms).

BULBUS Noun. (Latin, *bulbus* = bulb.) The base of the Antennal Scape which fits into the Torulus; frequently globular and appearing as distinct segment. See Radicle.

BULL, GEORGE VERNON (1873–1959) (Scott 1960, Entomologist 93: 72.)

BULLA Noun. (Latin, *bulla* = bubble. Pl., Bullae.) 1. A blister or blister-like structure. 2. Ephemeridae: A stigma-like enlarged part of the Costal area of the wing near the apex. 3. Weak spots on some of the wing veins where they are crossed by furrows (Comstock). 4. Diaspidinae: A minute knob or nipple in the truncate inner end of the Ceratubae (MacGillivray). 5. Coleoptera: A shield-like sclerite closing the tracheal aperture in lamellicorn beetles. 6. Hymenoptera; Ichneumonidae: A translucent section of wing vein marking area folds or flexion lines in the wing membrane. Ants: The thin, convex, roof-like sclerite over the Metaplural Gland.

BULLACTICIN See Acetogenins.

BULLATE Adj. (Latin, *bullatus*, *bulla* = bubble; *-atus* = adjectival suffix.) Blistered; blister-like. A term sometimes used in Botany to describe the surface of a leaf. Arcane, Bullatus.

BULLDOG ANTS *Myrmecia* spp. [Hymenoptera: Formicidae]: Ants endemic to Australia and New Caledonia; one Species was accidentally introduced into New Zealand. BA is not an urban pest but its sting is powerful and can cause death in humans. Syn: Sergeant Ants; Jumper Ants. See Formicidae.

BULLER, ARTHUR HENRY REGINALD (1874–1944) (Brooks 1945, Obit. Not. Fell. Roy. Soc. Lond. 5: 51–59, bibliogr.)

BULLULE Noun. (Latin, *bullatus* = blistered. Pl., Bullules.) Small blisters or bullae.

BUMBLE BEE WAX-MOTH *Aphomia sociella* Linnaeus [Lepidoptera: Galleriidae].

BUMBLE BEES Members of the Subfamily Bombinae (Hymenoptera: Apidae). BBs are relatively large bodied, robust, black-and-yellow and important pollinators of some kinds of clover because of their elongate mouthparts. Most BBs nest in ground, colonies are annual, and only fertilized queens overwinter. BB genus *Psithyrus* is parasitic on other bumble bees. See Apidae. Cf. American Bumble Bee; Black-Faced Bumble Bee; Yellow Bumble Bee; Yellow-Faced Bumble Bee. Rel. Carpenter Bee; Honey Bee; Stingless Bee.

BUMBLE FLOWER-BEETLE *Euphoria inda* (Linnaeus) [Coleoptera: Scarabaeidae].

BUMELIA FRUIT-FLY *Anastrepha pallens* Coquillett [Diptera: Tephritidae].

BUMETRAN® See Amitraz.

BUNGE, HERMANN (1870–1929) (Weidner 1967, Abh. Verh. naturw. Ver. Hamburg Suppl. 9: 293.)

BUNKER, ROBERT (1821–1892) (Anon. 1892, Entomol. News 3: 104.)

BUNNETT, E J (1865–1949) (T. R. E. 1950, Proc. Trans. Soc. Lond. Entomol. nat. Hist. Soc. 1949–50: xli.)

BUNYAVIRIDAE Plural Noun. A Family of viruses that contains about 250 arboviruses that are placed in the Genera *Bunyavirus, Phlebovirus* and *Nairovirus*. Virons ca 80–100 nm diameter, spherical with lipid envelope and glycoprotein projections. Bunyavirid virons are vectored by mosquitoes, sand flies and ticks. Viruses produce several diseases in humans and livestock, including Rift Valley Fever, Akabane, Sandfly Fever and Epidemic Haemorrhagic Fever. See Arbovirus. Cf. Flaviviridae; Reoviridae; Rhabdoviridae; Tongaviridae.

BUPRESTIDAE Plural Noun. Flatheaded Wood Borers; Flatheaded Borers; Jewel Beetles; Metallic Wood Borers. A large, cosmopolitan Family of polyphagous Coleoptera assigned to Buprestoidea; many Species are regarded as significant economic pests of trees and shrubs. Adult body to 65 mm long, compact, elongate, heavily sclerotized and with contrasting coloration patterns that are often metallic. Head somewhat globular, deflexed and deeply embedded in Prothorax; Antenna short, serrate with 11 segments; compound eye reniform. Prothorax broadly attached to Mesothorax; Mesosternum with large depression for reception of Prosternal Process; Metasternum with transverse suture; legs short, tarsal formula 5-5-5, segments 1–4 with membranous lobe on ventral surface; Empodium absent; hindwing fully developed, functional; Abdomen with five Ventrites and first two Ventrites fused. Larva soft-bodied, pale; head small; Prothorax enlarged and flattened; Antenna short, with two segments; legs absent; Urogomphi absent. Adults are active in sunlight and taken in nectar-bearing flowers. Larvae feed in living plant tissue as borers in wood, roots of trees/shrubs or form galls/mines on leaves. See Bronze Birch Borer; Pacific Flathead Borer; Roundheaded Apple-Tree Borer; Sinuate Pear Tree Borer. Rel. Cerambycidae; Scolytidae.

BUPRESTOIDEA A Superfamily of polyphagous Coleoptera including the Buprestidae.

BUPROFEZIN A thiadiazine compound {2-tert-butylimino-3-isopropyl-5-phenyl-3,4,5,6-tetrahydro-2H-1,3,5-thiadiazin-4-one} used as a contact insecticide, stomach poison or Insect Growth Regulator for control of plant-sucking insects (aphids, leafhoppers, planthoppers, scale insects, whiteflies), mites and beetles. Regarded as slow-acting and persistent; most effective against Hemiptera but no apparent effect on Lepidoptera, Hymenoptera and Diptera; compound is not effective on adult insects or eggs. Compound is applied to fruit trees, cotton, rice, ornamentals, potatoes and vegetables in some countries. Some phytotoxicity to Chinese cabbage and turnips has been observed; low toxicity to mammals and fishes. Buprofezin is not available in USA or Europe. Trade names include Applaud® and Nichino®

BUQUET, JEAN BAPTISTE LUCIEN (1807–1899) (Anon. 1890, Entomol. Mon. Mag. 26: 53.)

BURCHELL, WILLIAM JOHN (1782–1863) (Papavero 1975, *Essays on the History of Neotropical Dipterology* 2: 221–229. São Paulo.)

BURDETTE, ROBERT CARLTON (1898–1935) (Headlee 1935, J. Econ. Entomol. 28: 252–253.)

BURDOCK BORER *Papaipema cataphracta* (Grote) [Lepidoptera: Noctuidae].

BUREAU, LOUIS (–1937) (Anon. 1937, Bull. Soc. Entomol. Fr. 42: 49.)

BURES, IVAN (1865–) (Maran 1966, Acta Entomol. Bohemoslovaca 63: 170–172.)

BURGEON, LOUIS (1844–1947) (Schouteden 1951, Revue Zool. Bot. afr. 44 (Suppl.) (30–32).

BURGER, OWEN FRANCIS (1885–1928) (Weber 1928, Mon. Bull. Sta. Pl., Bd. Florida 12: 158–162.)

BURGESS, ALBERT FRANKLIN (1873–1953) (Mallis 1971, *American Entomologists*. 549 pp. (435–437), New Brunswick.)

BURGESS EDWARD (1848–1891) (Osborn 1937, *Fragments of Entomological History*. 394 pp. (207), Columbus, Ohio.)

BURGESS SHALE A Cambrian deposit of rock that Is found in British Columbia, Canada and notable for its preservation of detail in soft-bodied fossils. The importance of this site was revealed by C. D. Walcott during fieldwork in 1909. Walcott developed the descriptions of many Taxa of enigmatic invertebrates based on the collections taken from Burgess Shale. Rel. Paleozoic Era.

BURKART, HERMAN JOSEPH (1798–1874) (V. D. 1874, Verh. naturhist. Ver. preuss. Rheinl. (4)1: 112–121.)

BURKE, HARRY E (1878–1963) (Mallis 1971, *American Entomologists*. 549 pp. (437–438, portrait), New Brunswick.)

BURLESON, THOMAS WILLIAM (–1944) Director of Honeybee Research Institute. (Root 1945, Gleanings in Bee Culture 73: 55.)

BURMEISTER, CARL HERMANN CONRAD (1807–1892) Born in Stralsund, Germany and died in Buenos Aires, Argentina. German academic (University of Berlin, University of Halle) and later director of National Museum in Buenos Aires. Burmeister was a prolific writer on all aspects of natural history. He was author of *Handbuch der Entomologie* and specialist in anatomy and systematics of Scarabaeidae. He conducted an extensive study of migratory locust in Argentina that was published as *Reise durch die Plata Staaten* (1861, 2 vols, Halle). (Papavero 1975, *History of Neotropical Dipterology,* 2: 292–293. São Paulo.)

BURMEISTER MANTID *Orthodera burmeisteri* Wood-Mason [Mantodea: Mantidae]. See Mantidae.

BURMESE AMBER See Burmite.

BURMITE Noun. (Burma; *-ites* = mineral. Pl., Burmites.) Burmese amber. A fossilized resin which differs from Baltic Amber in being flourescent and lacking succinic acid. Burmite resembles Sicilian amber in that both are often red. Burmite is apparently of Miocene age. (Helm 1893, Records of the Geological Survey of India 26 (2): 61.) See Amber; Fossil. Cf. Baltic Amber; Canadian Amber; Chiapas Amber; Dominican Amber; Lebanese Amber; Taimyrian Amber.

BURNET MOTHS See Zygaenidae.

BURNETT, WALDO IRVIN (1828–1854) American physician in Boston who wrote about cotton worm and boll weevil. (Weiss 1936, *Pioneer Century of American Entomology*. 320 pp. (178–181), New Brunswick.)

BURNEY, HENRY (1813–1893) (Elwes 1893, Proc. Entomol. Soc. Lond. 1893: lvi.)

BURNS, ALEXANDER NOBLE (1899–1994) Australian taxonomic Entomologist and curator (National Museum of Victoria) whose primary research interests involved Lepidoptera and Cicadidae. Burns was founding member of Entomological Society of Queensland (1923).

BURNSIDE, CARLTON E (1895–1949) (Sturtevant 1950, J. Econ. Entomol. 43: 113.)

BURR, MALCOLM (1878–1954) (Anon. 1956, Entomol. Rec. J. Var. 68: 300.)

BURRAS, ALFRED E (1871–1903) (Wigglesworth 1964, Proc. Roy. Entomol. Soc. Lond. (C) 28: 57.)

BURROW Noun. (Middle English, *borow* = a hole for a shelter. Pl., Burrows.) 1. A hollow place in the earth dug by an animal for shelter, habitation or refuge. 2. A passage beneath the skin that is made by the movement of an internal parasite.

Syn. Tunnel.

BURROWER BUGS See Cydnidae.

BURROWING BUGS See Cydnidae.

BURROWING COCKROACHES Members of the blaberid Genera *Macropanesthia* and *Geoscapheus* in Australia. See Blaberidae.

BURROWING WASP *Philanthus ventilabris* Fabricius [Hymenoptera: Sphecidae]: A wasp ca 13 mm long, with body black-and-yellow and legs yellow; wings with yellowish tinge. The adult female provisions her nest with halicitid bees. See Sphecidae. Cf. Bee Wolf.

BURROWING WATER BEETLES See Noteridae.

BURROWS, CHARLES RICHARD NELSON (1851–1936) (Imms 1937, Proc. Roy. Entomol. Soc. Lond. (C) 1: 55.)

BURSA Noun. (Greek, *bursa* = hyde; Latin, *bursa* = pouch. Pl., Bursae.) 1. Any pouch-like, bladder-like or sac-like structure of the body. Cf. Sac. 2. Trichoptera: A wing pouch in male caddis flies in connection with a stalked Hair Pencil. 3. Diptera: A dorsal invagination of the Genital Chamber. 4. Hymenoptera: The proximal portion of the Endophallus. Syn. Bag; Pouch; Sac; Vessicle.

BURSA COPULATRIX The copulatory pouch of the female in some insect Orders and which is a modification of the Vagina. Alt. Copulatory Bursa. See Genital Chamber; Reproductive System. Cf. Vagina.

BURSICON Noun. (Latin, diminutive of *bursa* = purse. Pl., Bursicons.) A hormone secreted by cells of the nervous system into an insect's Haemolymph after moulting. Bursicon plays a role in hardening and darkening of the new cuticle.

BURSIFORM Adj. (Latin, *bursa* = pouch; *forma* = shape.) Purse-shaped; subspherical. Descriptive of structure that has a single aperture and which is elongate and closed at the bottom. See Bursa 1. Cf. Saccate; Scrotiform. Rel. Form 2; Shape 2; Outline Shape.

BURTON, RICHARD FRANCIS LINGEN (1864–1922) (M[elville] 1922, Entomologist 55: 144.)

BURY, HERBERT (1871–1944) (Jackson 1940, NWest Nat. 15: 71–72.)

BURYING BEETLES See Silphidae.

BUSAN 1020® See Metham-Sodium.

BÜSCH, JOHANN GEORG (1728–1800) (Weidner 1967, Abh. Verh. naturw. Ver. Hamburg Suppl 9: 75–78.)

BUSCK, AUGUST (1870–1944) (Gates-Clarke 1974, J. Lepid. Soc. 28: 183, 185–186.)

BUSE, ULRICH (1890–1972) (Rudwick 1974, Entomol. Berichte, Berlin 1973: 121–124)

BUSH Noun. (Middle English, *busch*. Pl., Bushes.) 1. A densely branched shrub, generally of low height (< 3 metres), with one or several main stems (trunks). Cf. Herb; Shrub; Tree. Rel. Plant. 2. Vernacular: The arid interior region of Australia.

BUSH CRICKETS See Eneopteridae; Saltatoria; Tettigonioidea.

BUSH FLY *Musca vetustissima* Walker [Diptera: Muscidae]: Endemic to Australia and adjacent islands of Papua New Guinea, and the most widely encountered fly in Australia during summer. Adult transmits eye infections and enteric diseases to humans. Adult 5–6 mm long and resembles housefly in habitus but displays dark stripes on Thorax that form two 'Y'-shaped marks. BF is attracted to saliva, sweat and tears of large mammals. BF annoys humans and livestock for extended periods, particularly in inland areas. The female requires protein for egg development. Female and male are not sexually mature at emergence and the female is not receptive to copulation for 3–4 days following emergence. Mating is not well documented in field. In lab, copulation requires 1.3 hours during which male passes accessory-gland secretions that inhibit subsequent mating and stimulates oviposition. BF female undergoes 4–5 ovarian cycles during her lifetime. She lays clusters of 5–50 eggs in mammal excrement (dog, dingo, kangaroo, cow, horse, sheep, human), typically in surface fissures or interface between soil and excrement. The ovipositing female attracts numerous other gravid females to site. Eggs are creamy white, elongate and slightly curved; eclosion occurs within 24 hours. Larvae develop in excrement, preferring liquid and soft-solid parts; larvae feed along the surface, periodically taking air at surface; as excrement dries, larva burrows into pile. BF undergoes three larval instars with stage completed ca 3–4 days. Pupation occurs in soil adjacent to excrement. Life cycle 2–10 weeks; adults live ca 60–90 days, or longer under optimal conditions. BF does not overwinter in cooler parts of Australia and migrates yearly to south. Cf. Egyptian House Fly; House Fly; Lesser House Fly.

BUSH TICK *Haemaphysalis longicornis* [Acari: Ixodidae]: A Species found in Australia.

BUSHELL, HUGH SIDNEY (1893–1974) (Lees 1975, Proc. Roy. Entomol. Soc. Lond. (C) 39: 55.)

BUSSART, JAMES EVERETT (1903–1965) (Still 1968, *et al.* J. Econ. Entomol. 61: 588)

BUTACIDE® Trade-name for Piperonyl Butoxide that is registered to mix with synthetic-pyrethroids or other insecticides and which acts as a synergist. See Pyrethrin.

BUTAMIN® See Tetramethrin.

BUTCHER, FRED DUNAWAY (1897–1969) (Reed 1970, J. Econ. Entomol. 63: 345)

BUTLER, ARTHUR GARDINER (1844–1925) (Musgrave 1932, *A Bibliogr. of Australian Entomology 1775–1930*. 380 pp. (36–37), Sydney.)

BUTLER, EDWARD ALBERT (1845–1925) (China 1926, Entomol. Mon. Mag. 62: 24.)

BUTLER, GEORGE American government economic Entomologist specializing in insects associated with cotton.

BUTLER, HENRY ALBERT (1892–1974) (Morris 1975, Bull. Entomol. Soc. Can. 7: 10–11.)

BUTOCARBOXIM A carbamate compound {3-

methylthio-O-[(methylamino) carbomyl]oxime-2-butanone} that is used as a systemic/contact insecticide and stomach poison against plant-sucking insects (aphids, scale insects, mealybugs, whiteflies), thrips and mites. Butocarboxim is applied to numerous fruit tree crops, citrus, field crops and ornamental plants. Butocarboxim is not registered for use in the USA. Trade names include Afilene® and Darwin 755®. See Carbamate Insecticide.

BUTOFLIN® See Deltamethrin.

BUTOSS® See Deltamethrin.

BUTOX® See Deltamethrin.

BUTOXYCARBOXIM A carbamate compound {3-(methylsulphonyl)-O-[(methylamino) carbomyl]-oxine-2-butanone} that is used as a systemic/contact insecticide and stomach poison against plant-sucking insects (aphids, scale insects, mealybugs, whiteflies), thrips and mites. Butoxycarboxim is applied as a soil treatment for potted ornamental plants in some countries, but is not registered for use in the USA. Trade names include Bug Pin® and Plant Pin®. See Carbamate Insecticide.

BUTTE Noun. (French, *butte* = knoll. Pl., Buttes.) A small, isolated, steep-sided flat or round-topped hill. Cf. Mesa.

BUTTERFIELD, ROSSE (1875–1935) (Anon. 1939, Entomol. Mon. Mag. 75: 19.)

BUTTERFLY Noun. (Anglo Saxon, *buterflege*. Pl., Butterflies.) A common name applied to any member of the Lepidoptera with clubbed Antennae. See Lepidoptera. Cf. Moth.

BUTTERFLY MOTHS See Castniidae.

BUTTERNUT CURCULIO *Conotrachelus juglandis* LeConte [Coleoptera: Curculionidae].

BÜTTNER, FRIEDRICH OTTO (1824–1880) (Anon. 1939, Stettin. Entomol. Ztg. 100: 39–40.)

BUTTNER, KURT (1881–1967) (Anon. 1967, Mitt. dt. Entomol. Ges. 26: 38.)

BUXTON, E C (–1878) (Carrington 1879, Entomologist 12: 63–64.)

BUXTON, PATRICK ALFRED (1892–1955) (Anon. 1956, Parasitology 47: 1–15.)

BUYSSON, HENRI DU (1856–1927) (Anon. 1928, Entomol. News 39: 296.)

BUZZARD, CHARLES NORMAN (1873–1961) (Brangham 1961, Entomol. Mon. Mag. 97: 142.)

BYARS, LOREN FREELAND (1908–1956) (Gregg 1956, Proc. Entomol. Soc. Wash. 58: 6.)

BYBP Acronym meaning 'Billions of Years Before Present.' See Geological Time Scale. Cf. MYBP.

BYERS, GEORGE American academic (University of Kansas) with taxonomic interest in Mecoptera and Tipulidae.

BYGRAN® See Tecnazene.

BYNUM, ELI KENNIE (1890–1969) (Ingram 1970, J. Econ. Entomol. 63: 1035.)

BYRD, ELON E (1905–1974) (Anon. 1974, J. Ga. Entomol. Soc. 9: 115.)

BYRRHIDAE Plural Noun. Pill Beetles. A small Family of polyphagous Coleoptera assigned to Superfamily Byrrhoidea. Body 1.5–6.0 mm long and rotund; head and legs retractile with head deflexed; Antenna short, clavate with 11 segments; tarsal formula 5-5-5, rarely 4-4-4; Abdomen with five Ventrites. Larva scarabaeiform; head hypognathous with several Stemmata; Antenna with three segments; Prothorax large; legs with five segments and each with two tarsal claws. Adult and larva inhabit moss, litter or soil near roots. See Byrrhoidea.

BYRRHOIDEA A Superfamily of polyphagous Coleoptera of variable composition, but including at least Byrrhidae, Callirhipidae, Chelonariidae, Elmidae, Heteroceridae, Limnichidae, Psephenidae and Ptilodactylidae. See Coleoptera.

BYTHELL, W J S (1872–1950) (Wigglesworth 1951, Proc. Roy. Entomol. Soc. Lond. (C) 15: 75.)

BYTINSKI-SALZ, HANAN (HANS) (1903–1939) German born and educated; moved to Palestine 1939. Agricultural Entomologist in Israel, general collector and academic with broad interests. (Anon., Israel J. Entomol. 4: 207–215; Kugler 1986, Israel J. Entomol. 20: 95–97.)

BYTURIDAE Plural Noun. Fruit-worm Beetles. A small Family of polyphagous Coleoptera assigned to Superfamily Cucujoidea.

CABBAGE APHID *Brevicoryne brassicae* (Linnaeus) [Hemiptera: Aphididae]: In North America and the UK, a widespread pest of cruciferous plants. Adults distinguished from other aphids by powdery-wax covering. In northern parts of distribution, males and oviparous females are produced during autumn; overwintering eggs are laid in crop residue. In southern parts of distribution, as many as 30 generations of winged and wingless females are produced each year. See Aphididae.

CABBAGE BORER *Hellula undalis* (Fabricius) [Lepidoptera: Pyralidae].

CABBAGE BUTTERFLY *Pieris rapae* (Linnaeus) [Lepidoptera: Pieridae]. See Imported Cabbageworm.

CABBAGE CLUSTER CATERPILLAR *Crocidolomia binotalis* [Lepidoptera: Pyralidae]: A pest of crucifers in the Old World tropics.

CABBAGE CURCULIO *Ceutorhynchus rapae* Gyllenhal [Coleoptera: Curculionidae]: A widespread, multivoltine pest in North America and Europe. CC eggs are grey, oval and deposited in stems; larvae feed within stems, petioles and on edge of seedling leaves; pupation occurs within an earthen cell. CC overwinters as an adult. Host plants include cabbage, cauliflower, horseradish, mustard and turnip. See Curculionidae. Cf. Cabbage Seedpod Weevil.

CABBAGE LOOPER *Trichoplusia ni* (Hübner) [Lepidoptera: Noctuidae]: An important multivoltine pest of many crops in North America. Adult ca 25 mm long with wingspan 30–40 mm; forewing dark greyish-brown with silver spot and U-shape near middle. Eggs are white and become grey before eclosion; ca 0.6 mm, somewhat flattened with ridges radiating from apex. Eggs are laid individually on terminals and lower surface of leaves; eclosion occurs within 2–4 days. Larva with two pairs of abdominal Prolegs (excluding Anal Prolegs) and walks with looping motion; neonate larva white; subsequent instars become green with longitudinal white lines; head pale green; mature larva ca 35 mm long. Larval feeding causes large holes in leaves or ragged edges. CL forms fine membrane on underside of leaf; pupation can transpire within 6–7 days or CL can overwinter as brown Pupa within a thin cocoon. See Noctuidae.

CABBAGE MAGGOT Larval stage of *Delia radicum* (Linnaeus) [Diptera: Anthomyiidae]. See Cabbage Root-Fly. Cf. Onion Maggot. Syn. Radish Maggot in some parts of North America.

CABBAGE MOTH See Diamondback Moth.

CABBAGE ROOT-FLY *Delia radicum* (Linnaeus) [Diptera: Anthomyiidae]: An Holarctic pest of seedling stands of brussel sprouts, cabbage and other cole crops; CRF is endemic to Europe and probably adventive to North America. Adult 8–9 mm long and resembles house fly but smaller; grey with black stripes on Thorax; black bristles over body; medial surface of hind Femur with a tuft of short bristles. Post-emergent seedlings are most vulnerable for ca one month. Female lays white, finely reticulate eggs in cracks in the soil near base of seedlings. Larvae burrow down to roots and begin feeding. Damage to root system can cause wilting, chlorosis, stunting and plant death. Feeding damage also allows penetration of fungal pathogens. Pupation occurs within soil. CRF completes 2–3 generations per year. Alt. Cabbage Maggot. *Erioischia brassicae; Hylemya brassicae.* See Anthomyiidae. Cf. Seed-Corn Maggot.

CABBAGE SEED-POD WEEVIL *Ceutorhynchus assimilis* (Paykull) [Coleoptera: Curculionidae]: A serious pest of cabbage, turnip and rutabaga seed crops in northwestern USA. Eggs are inserted in seed pods; larva feeds inside pod. See Curculionidae. Cf. Cabbage Curculio.

CABBAGE SEED-STALK CURCULIO *Ceutorhynchus quadridens* (Panzer) [Coleoptera: Curculionidae].

CABBAGE WEBWORM 1. *Hellula rogatalis* (Hulst) [Lepidoptera: Pyralidae]. 2. *Hellula phidilealis* (Walker) in Africa, Central and South America. Cf. Old World Cabbage Webworm.

CABBAGE WHITE-BUTTERFLY *Pieris brassicae* (Linnaeus) [Lepidoptera: Pieridae.]

CABBAGE WHITEFLY *Aleyrodes proletella* (Linnaeus) [Hemiptera: Aleyrodidae]

CABINET BEETLE *Trogoderma sternale* Jayne, *T. variabile* Ballion, *T. inclusum* LeConte, *T. ornatum* (Say) [Coleoptera: Dermestidae]: Pests of museum collections and cereal products. See Dermestidae.

CABLE MITE See Delusory Dermatitis.

CACAO WEEVIL See Coffee Bean Weevil.

CACHOPLISTIDAE Plural Noun. Beetle Crickets. A small Family (three Species) known from India and Australia. Adult Frons transverse, Pronotum cuboid with lateral keels, hind legs with slender Femora and serrated, spineless Tibia; male stridulatory organs well developed; female Tegmina Elytra-like. Biology of Family unknown. See Coleoptera.

CACTUS FLY *Odontoloxozus longicornis* (Coquillett) [Diptera: Neriidae]: A Species known from southwestern USA, Mexico and Central America. Adult 8–10 mm long and brown with face and ventral aspect of head yellow; antennal Arista with white pubescence; Scutellum and Abdomen with broad, brown, medial longitudinal stripe; legs yellowish; wings hyaline. Larvae feed on roots of papaya and other plants. See Neriidae.

CACTUS MOTH *Cactoblastis cactorum* (Bergroth) [Lepidoptera: Pyralidae]: A moth endemic to Argentina whose larval stage feeds on *Opuntia* spp. CM was imported into Australia during 1925 and was responsible for biological control of *O. stricta* and *O. inermis* DeCandolle directly through damage of vital plant parts and indirectly through opening tissues to infection by bacterial soft rots. Syn. Cactoblastis (Australia). See Pyralidae. Cf.

Common Prickly Pear; Spiny Prickly Pear.

CACTUS SCALE *Diaspis echinocacti* (Bouché) [Hemiptera: Diaspididae]: A minor pest of *Opuntia* spp. in North America and some areas where cactus has been introduced. See Diaspididae.

CACUMINOUS Adj. (Latin, *cacuminis* = the limit; *-osus* = possessing the quality of.) Pertaining to an elongate structure that is pointed at the apex. See Shape; Acanthus. Cf. Acicular; Aculeiform; Acuminate.

CADDISFLIES See Trichoptera.

CADELLE *Tenebroides mauritanicus* (Linnaeus) [Coleoptera: Trogositidae]: A cosmopolitan pest of grains in mills, warehouses, ships and stores. Species also occurs under bark or within the galleries of wood-boring insects. Adult 5–12 mm long, dark brown or black and flattened; antennal club loosely defined with three segments; head and Prothorax incised from remainer of body; Elytra with conspicuous striae. Female may deposit ca 1,300 eggs during lifetime; eggs are laid in groups of 10–60 within concealed situations near food supply and eclosion occurs within 7–15 days. Larvae are campodeiform, soft-bodied, white with black head and spots on thoracic segments. Larvae are predaceous or omnivorous and prefers to consume germ of grain; last larval instar may bore into soft wood to form pupal chamber. Larval development requires 70–90 days under favourable conditions or longer under cool temperatures; larvae can survive 3.5 years. Adults and larvae bore into wood when grains are not present; Cadelle overwinters as an adult or larva. See Trogositidae. Rel. Dark Mealworm.

CADUCOUS Adj. (Latin, *caducous* = falling; deciduous; *-osus* = possessing the qualities of.) Pertaining to any structure or appendage that is easily detached or shed. Typically, the term is descriptive and applied to plants more readily than insects. Alt. Caducus. Cf. Fugacious; Marcescent.

CADUSAFOS An organic-phosphate (dithio-phosphate) compound {O-ethyl S,S-di-sec-butyl phosphorodithioate} used as a soil insecticide and nematicide against nematodes, rootworms, wireworms, cutworms and beetle grubs in some countries. Compound is not registered for use in USA. Trade names include: Edufos®, Apache®, Rugby®, Taredan®. See Organophosphorus Insecticide.

CAECAL Adj. (Latin, *caecus* = blind; *-alis* = belonging to.) Descriptive of structure, function or process involving tubular structures that end 'blind' or lack an outlet.

CAECAL TUBE A sac or blind tube-like structure surrounding the chylific ventricle at its junction with the Crop. CT secretes a digestive fluid. Syn: Caecal Pouch. See Alimentary System; Foregut.

CAECILIIDAE Plural Noun. A numerically large, cosmopolitan Family of Psocoptera assigned to the Suborder Psocomorpha and Superfamily Caecilietae. Head short and wide; Antenna with 13 segments and lacking secondary annulations; Areola Postica present, free abdominal Sternum with 2–3 inflatable vessicles. Species often are found on fresh foliage and occur on conifers and broad-leafed trees. Syn. Polypsocidae. See Psocoptera.

CAECUM Noun. (Latin, *caecus* = blind. Pl., Caeca.) A blind sac or tube-like structure that serves as one of the caecal tubes or pouches (*e.g.* intestinal caecum). See Coecum.

CAELATE Adj. (Latin, *caelatura* = to sculpt in relief; *-atus* = adjectival suffix.) Pertaining to an engraved surface; structure with superficial plane elevations of varying forms. See Sculpture Pattern.

CAELIFERA Noun. A Suborder of Orthoptera in which the hind Femur is enlarged and adapted for jumping. Caelifera include short-horned grasshoppers and pigmy mole crickets. Included Taxa display short Antennae, short Cerci and a short Ovipositor; Tarsi with three or fewer segments. See Orthoptera. Cf. Ensifera.

CAENIDAE Plural Noun. A Family of mayflies assigned to Ephemeroidea. Sexual dimorphism is minimal or absent and body size very small; lateral Ocellus more than half size of compound eye; Thorax exceptionally large; Abdomen small; forewing developed but hindwing absent; Tarsi with five segments; Cerci and Appendix Dorsalis present. Naiads burrow. See Ephemeroptera.

CAENOGENESIS Noun. (Greek, *kainos* = new; *genesis* = descent. Pl., Caenogeneses) The developmental processes within an organism which are not actively involved with phylogenetics. Cf. Paedogenesis.

CAENOGENIC Adj. (Greek, *kainos* = recent; *genesis* = descent; *-ic* = characterized by.) Descriptive of structure or organisms of recent origin. See Evolution. Cf. Paedogenic. Rel. Phylogeny.

CAENOZOIC ERA See Cenozoic Era.

CAERULESCENT Adj. (Latin, *caeruleus* = dark blue; *escens* = beginning, slightly.) With a tinge of sky-blue.

CAERULEUS Adj. (Latin, *caeruleus* = dark blue.) Sky-blue in colour. Alt. Coeruleous; Coeruleus. Cf. Azure.

CAESAR, LAWSON (1870–1952) (Baker 1952, J. Econ. Ent. 45: 1113–1114.)

CAESEOUS (Latin, *caesius* = pale grey; *-osus* = possessing the qualities of.) A greyish or dirty blue; a very pale blue with a little black (Kirby & Spence). Alt. Caesious; Caesius. Cf. Azure; Caeruleus.

CAESPITICOLOUS Adj. (Latin, *caespes* = turf; *colere* = living in or on.) Descriptive of organisms that frequently visit or inhabit meadows, grassy habitats, pastures or lawns. See Habitat. Cf. Agricolous; Caespiticolous; Dendrophilous; Eremophilous; Ericeticolous; Lignicolous; Paludicolous; Silvicolous. Rel. Ecology.

CAESPITOSE Adj. (Latin, *caespes* = turf; *-osus* = full of.) Descriptive of or pertaining to turf. Tufted; matted together; pestose. Obsolete: Cespitosus. Alt. Cespitose.

CAFFREY, DONALD JOHN (1886–1960) (Reed & Vance 1963, Proc. ent. Soc. Wash. 65: 169–171.)

CAFLISCH, J LUCIUS (1847–1900) (Anon. 1900, Jber. naturf. Ges. Graubünders 43: xxxix–xlii.)

CAGLE, LEROY R (1893–1964) (Woodside 1964, J. Econ. Ent. 57: 1017.)

CAILLE, JEAN DE LA (–1720) (Larousse 1866, Grand dictionnaire universale du XIX Siècle 3: 79.)

CAIN, BRIGHTON CLARK (–1951) (Usinger 1952, Pan-Pacif. Ent. 28: 125.)

CAJANDER, AIMO KAARLO (1879–1943) (Palmgren 1945, Arsb/Vuosik. Soc. Sci. fenn. 23: 3, 26, bibliogr.)

CALABAR SWELLINGS Swellings of the human wrist and ankle joints that are indicative of Loiasis. Characteristically CS are hot, painful and last for a few days before disappearing or may periodically occur throughout the course of the disease. See Loiasis.

CALAMISTRAL Adj. (Latin, *calamistrum* = curling iron; *-alis* = characterized by.) Descriptive of structure that resembles the Calamistrum or functions similar to those perfomed by the Calamistrum.

CALAMISTRUM Noun. (Latin, *calamistrum* = curling iron. Pl., Calamistra.) A comb-like arrangement of curved Setae on the Metatarsus of the fourth pair of legs in cribellate spiders; sometimes reduced or absent in males. Calamistrum used to 'card' silk produced by Spinnerets. See Cribellum.

CALAMOCERATIDAE Plural Noun. A small Family of Trichoptera containing about 100 Species and assigned to Superfamily Limnephiloidea; cosmopolitan distribution but most abundant in tropics. Body moderate sized; Ocelli absent; Antenna filiform, longer than forewing; Maxillary Palpus long, with six segments and with dense vestiture of errect Setae; Mesoscutum with two bands of setigerous punctures; Scutellum lacking Setal warts; tibial spur formula 2-4-3. Larvae construct compressed, tubular cases and inhabit slow-moving rivers, lakes and swamps. Larvae feed upon decaying vegetation. See Trichoptera.

CALATHIFORM Adj. (Latin, *calathus* = a basket; *forma* = shape.) Basket-shaped. Cf. Acetabuliform. Rel. Form 2; Shape 2; Outline Shape.

CALBERLA, HEINRICH WILHELM (1839–1916) (Heller 1919, Dt. ent. Z., Iris 31: 1–4, bibliogr.)

CALCANEA Noun. (Latin, *calcaneum* = heel. Pl., Calcaneae.) 1. A concealed sclerite on the ventral side of the proximal end of the Articularis (MacGillivray). See Articularis; Unguitractor.

CALCAR Noun. (Latin, *calcar* = spur. Pl., Calcaria; Calcars.) Hymenoptera and Orthoptera: A moveable spur or spine-like process at apex of the fore Tibia. Calcar is used for grooming the Antenna and head. Alt. Calcarium. See Tibia. Cf. Epiphysis; Strigil. Rel. Leg.

CALCEIFORM Adj. (Latin, *calceolus* = small shoe; *forma* = shape.) Shoe-shaped; descriptive of structure oblong in shape with a somewhat constricted middle. Syn. Calceoliform. Cf. Panduriform; Soleaform; Unguliform. Rel. Form 2; Shape 2; Outline Shape.

CALCEOLIFORM See Calceiform.

CALCIATI, CESARE (1885–) (Conci 1963, Atti Soc. ital. Sci. nat.102: 327.)

CALCINO Noun. (Italian, from Latin, *calx, calcis* = lime. Pl., Calcinos.) A disease of silkworms and some other caterpillars that is caused by the parasitic fungi *Beauvaria bassiana* and *B. tenella*. Disease first repoted by Agostino Bassi (1834). After death, the insect body dries and becomes covered with a white efflorescence giving it a chalky appearance. Syn. Muscardine. Cf. Chalk Brood. Rel. Pathogen.

CALCIUM Noun. (Latin, *calx, calcis* = lime.) A soft, white metal not found in a free state but common in carbonates (lime, limestone), sulphates (gypsum) and phosphates. Calcium is a vital element in bone (calcium phosphate). See Bone; Skeleton.

CALCIUM CHLORIDE A white salt of calcium and chlorine. CC is used to absorb water in desiccators.

CALCOSPHERITES Plural Noun. (Latin, *calx* = lime; *sphaera* = globe.) 1. Diptera: Certain bodies (presumably calcareous) that are found in the fat body cells of some Species. 2. Laminated granules of calcium carbonate in Malpighian Tubules of some insects.

CALDAN® See Cartap.

CALDER, EDWIN EDDY (1853–1929) (Engelhardt 1929, Bull. Brooklyn Ent. Soc. 24: 115.)

CALEDONIA SEED BUG *Nysius caledoniae* Distant [Hemiptera: Lygaeidae].

CALESINAE Mercet 1929. A Subfamily of Aphelinidae important in biological control. Morphological features include: Female Antenna with six segments, basal segments short and club not segmented; male Antenna with five segment and long Setae. Labial Palpi rudimentary; Pronotum short, deeply incised; forewing narrow, sparsely setose, long marginal fringe; tarsal formula 4-4-4. Biology: Parasites of Aleyrodidae with *Cales* as the most important Genus.

CALIBRE® See Hexythiazox.

CALICO SCALE *Eulecanium cerasorum* (Cockerell) [Hemiptera: Coccidae]: A pest of walnut, pear and other stone fruit in California.

CALIFORNIA DOG-FACE *Colias eurydice* Boisduval [Lepidoptera: Pieridae].

CALIFORNIA FIVE-SPINED IPS *Ips paraconfusus* Lanier [Coleoptera: Scolytidae]: A multivoltine pest of several pine Species throughout western USA and Mexico. Adult 3–4 mm long, black dorsally and brown ventrally with appendages pale brown; Frons with coarse granulations in female large median tubercle in male; Pronotum with dense punctures; Elytra with coarse striate punctures. Egg galleries composed of 3–5 straight tunnels radiating from central chamber. CFSI overwinters as adult under bark of dead tree and completes 2–5 broods per year. See Scolytidae. Cf. Monterey-Pine Engraver; Pine Engraver; Six-Spined Ips.

CALIFORNIA FLATHEADED-BORER *Melanophila californica* Van Dyke [Coleoptera: Buprestidae].

CALIFORNIA GROUND-SQUIRREL FLEA *Diamanus montanus* [Siphonaptera: Ceratophyllidae]: A pest of ground squirrels and rarely other rodents in western North America. A moderately efficient vector of Sylvatic Plague. See Ceratophyllidae.

CALIFORNIA HARVESTER-ANT *Pogonomyrmex californicus* (Buckley) [Hymenoptera: Formicidae]: A Species endemic to southwestern USA. Workers 4–6 mm long, reddish brown; workers sting and bite vigorously; male red and black. CHA constuct nests in sand with fan-shaped crater on one side of entrance. Feeds on seed but does not cut vegetation. See Formicidae. Cf. Florida Harvester-Ant; Red Harvester Ant; Harvester-Ant; Western Harvester-Ant.

CALIFORNIA OAKWORM *Phryganidia californica* Packard [Lepidoptera: Dioptidae].

CALIFORNIA PEAR-SAWFLY *Pristophora abbreviata* (Hartig) [Hymenoptera: Tenthredinidae]: A monovoltine pest of pear in the USA. Eggs are white and inserted in leaf tissue. Larvae are green, slimy and feed on foliage. Pupation occurs within soil inside a thin, brown cocoon. See Tenthredinidae.

CALIFORNIA PRIONUS *Prionus californicus* (Motschulsky) [Coleoptera: Cerambycidae].

CALIFORNIA RED SCALE *Aonidiella aurantii* (Maskell) [Hemiptera: Diaspididae]: A pest of *Citrus* which is endemic to China but now globally distributed with *Citrus*; CRS develops on many ornamental plants, including roses. Adult female scale-cover is reddish-brown, circular, flattened and ca 2 mm in diameter. Male cover is elongate and somewhat pale; male undergoes prepupal and pupal stage with winged adult. Female moults from second instar nymph directly into wingless, sedentary adult beneath scale cover of earlier stages. CRS resembles Yellow Scale but reddish coloured, occupies outer canopy and attacks fruit. Adult female is viviparous and produces ca 100–150 crawlers during 6–8 weeks; crawlers disperse, insert mouthparts into host plant and produce a white, flocculent, waxy 'whitecap'. CRS is multivoltine with up to six generations per year depending upon climate. See Diaspididae. Cf. Glovers Scale; Purple Scale; San Jose Scale; Yellow Scale.

CALIFORNIA SALT-MARSH MOSQUITO *Aedes squamiger* (Coquillett) [Diptera: Culicidae]: A relatively large-bodied pest that inhabits tide pools along the western coast of North America; CSMM swarms may invade communities. See Culicidae. Cf. Black Salt-Marsh Mosquito; Salt-Marsh Mosquito.

CALIFORNIA TORTOISE-SHELL *Nymphalis californica* (Boisduval) [Lepidoptera: Nymphalidae]: A butterfly endemic to western USA whose larval host plants include *Ceanothus* spp.

CALIMYRNA FIG A female, long-styled *Ficus caricae* plant. CF is endemic to Mediterranean region and pollinated by the fig wasp *Blastophaga psenes* (Linnaeus). Wasp larva develops within syconium of inedible caprifig and pupates within the caprifig. The adult wasp collects pollen, flies to Calimyrna fig, enters the syconium and pollinates the fig. See Caprifig.

CALIPERS Plural Noun. (French, *caliber*.) Dermaptera: Anal Forceps. Alt. Calliper.

CALKINS, LAVERE A (1906–1962) (Anon. 1962, Bull. ent. Soc. Am. 8: 87.)

CALLES Noun. (Latin, *callum* = hard skinned.) 1. In certain bodies, presumably calcareous, found in the fat-body cells of some insects (Henneguy). 2. Coccids: A curved, transverse band of thickenings in the cephalic region of the dorsal aspects of the Pygidium; the Calli of Leonardi (MacGillivray).

CALLI AXILLARY Odonata: Thickenings at bases of wings; distinguished as anterior at base of Costa, and posterior at base of Radius + Medius and Cubitus. Syn. Axillary Calli.

CALLIDULIDAE Plural Noun. A Family of ditrysian Lepidoptera including ca 100 Species assigned to Superfamily Calliduloidea. Species occur in tropics through Africa, southeast Asia and Australia. Adult small to medium-sized; Chaetosemata well developed; Ocelli present but reduced; Antenna filiform; Proboscis well developed; wings typically dark brown with orange band or patch; tibial spur formula 0-2-4. Eggs are laid individually on leaflet margins. Larvae feed on ferns. Adults are diurnal. See Calliduloidea.

CALLIDULOIDEA A Superfamily of ditrysian Lepidoptera including Callidulidae and Pterothysanidae. Antenna not pectinate; base of Proboscis not scaled; Maxillary Palpus reduced; Labial Palpus upturned; Epiphysis present; male lacking Retinaculum; Tympanal Organs absent. See Lepidoptera.

CALLIMOMIDAE See Torymidae.

CALLING ORGAN Male Orthoptera: The stridulatory apparatus. See Stridulation.

CALLIPHORA Robineau-Desvoidy 1830. Diptera: Calliphoridae. A Genus of blow fly.

CALLIPHORID Adj. (Calliphora.) 1. Any fly of the Family Calliphoridae, especially the Genus *Calliphora*. 2. Pertaining to biological or anatomical features of Calliphoridae. See Calliphoridae.

CALLIPHORIDAE Plural Noun. Blow Flies. A cosmopolitan Family of ca 1,000 Species of brachycerous Diptera assigned to Superfamily Muscoidea and which is predominately tropical in distribution. Calliphords are closely related to Muscidae, Tachinidae and Sarcophagidae; calliphorids are separated from Tachinidae by absence of Postscutellum; separated from Sarcophagidae by presence of two Notopleural and two Sternopleural Bristles; separated from Muscidae by presence of Hypopleural Bristles. Calliphorid adult body moderate sized and often metallic blue or green; second antennal segment with longitudinal seam; Arista plumose; transverse Mesonotal Suture complete; Postalar Calli distinct; bristles on Hypopleura and Pteropleura; Sternopleura with 2–3 bristles; wing apical cell narrowed but not closed; Squamae well developed. Calliphorids cause Myiasis in man and animals; they are best known for saprophytic habits, feeding on carrion, dung and other decaying animal matter. Several Species exhibit parasitic or predaceous habits. Members of Rhiniinae are predaceous upon egg pods of grasshoppers; Species in other Subfamilies capture and feed on winged termites; others feed on immature ants. Parasitic Species develop in earthworms; one Species is parasitic on snails. Insect hosts are unknown for parasitic calliphorids. Females oviposit or larviposit near host and larvae search for a host. Females usually deposit large numbers of eggs or larvae; females of some Species may contain 500 mature eggs in ovaries. See Muscoidea.

CALLIRHIPIDAE Plural Noun. A small Family of Coleoptera assigned to Superfamily Dryopoidea in some classifications. See Coleoptera.

CALLONI, SALVIO (1851–1931) (Jaggli 1940, Boll. Soc. ticin. Sci. nat. 34: 19–111, bibliogr.)

CALLOSE Noun. (Latin, *callum* = hardened, thick skin; *ose* = possessing the qualities of. Pl., Calloses.) 1. An amorphous polysaccharide that yields glucose upon hydrolysis. 2. Furnished with Callin.

CALLOSITY Noun. (Latin, *callum* = hard skin. Pl., Callosities.) 1. A callus or thick, swollen lump, harder than surrounding structure. 2. A rather flattened elevation not necessarily harder than the surrounding tissue. See Callus.

CALLOW Noun. (Middle English, *claewe* > Anglo Saxon, *calu* = bald. Pl., Callows.) The condition immediately following moulting when the Integument has not attained the final coloration as when hardened. Callow is a term sometimes applied to a newly emerged worker ant. See Metamorphosis. Cf. Teneral. Rel. Pharate.

CALLOW WORKER A newly emerged (neonate) worker ant whose Integument is soft and not completely pigmented.

CALLUS Noun. (Latin, *callum* = hard skin. Pl., Calli.) 1. General: A hard lump or mound-like, rounded swelling of the integument. 2. A swelling at the base of the wing articulating with the Thorax by means of an Axillary (Tillyard). 3. Heteroptera: The thickened or raised spots on the Thorax, especially of Pentatomidae. 4. An unorganized plant-cell mass with *in vitro* capacity of cell division and growth.

CALMAN, WILLIAM THOMAS (1871–1952) (Gordon 1953, Proc. Linn. Soc. Lond. 165: 83–87, bibliogr.)

CALMBACH, VIKTOR (–1941) (Sachtleben 1941, Arb. morph. taxon Ent. Berl. 8: 77.)

CALOBATIDAE. See Micropezidae.

CALOBIOSIS Noun. (Greek, *kalos* = beautiful; *bios* = life; *-sis* = a condition or state. Pl., Calobioses.) A form of symbiosis between social insects in which one Species (often only the female) lives within the nest of and at the expense of another Species. The association may be for a short period (Temporary Calobiosis) or as a perpetual relationship (Permanent Calobiosis) (Smith). See Biosis; Symbiosis; Parasitisim. Cf. Abiosis; Anhydrobiosis; Antibiosis; Archebiosis; Cleptobiosis; Hamabiosis; Kleptobiosis; Lestobiosis; Parabiosis; Phylacobiosis; Plesiobiosis; Synclerobiosis; Trophobiosis; Xenobiosis.

CALOCIDAE Plural Noun. A small Family of Trichoptera assigned to Superfamily Limnephiloidea, known only from Australia and New Zealand. Adult moderate to large-sized; Ocelli absent; Antenna about as long as forewing; Pronotum typically with two pairs of setal warts; Mesoscutum without setal warts; Scutellum with a pair of setal warts; tibial spur formula 2-2-4. Larvae form their cases of sand or plant debris and inhabit cool streams; one Australian Species is terrestial. See Limnephiloidea; Trichoptera.

CALOPHYIDAE Plural Noun. A small Family of Psylloidea (Hemiptera). Vena Spuria absent; forewing with M and CuA with common origin; Mercanthus present, apically pointed; hind Basitarsus spineless or with one spine. See Psylloidea.

CALOPSOCIDAE Plural Noun. A Family of Psocoptera assigned to the Suborder Psocomorpha. Antenna with 13 segments but without secondary annulations; forewing venation reticulate distally. See Psocoptera.

CALOPTERYGIDAE Plural Noun. Broad-winged Damselflies. A Family of Odonata assigned to Superfamily Calopterygoidea. Adult is typically large, metallic coloured with slender Abdomen and short Thorax; wings usually broad and not petiolate and densely veined with numerous Antenodal Crossveins and supplementary veins in distal part of wing; Arculus near wing base. Larva with small head; Antenna with seven segments and basal segment as long as other segments combined; dorsal gill short and lamellate, lateral gills long and triangular in cross-section; mask long and incised. See Calopterygoidea; Odonata.

CALOPTERYGOIDEA A Superfamily of Odonata.

Included Families are: Amphipterygidae, Calopterygidae, Chlorocyphidae, Epallagidae, Heliocharitidae and Polyphoridae. See Odonata.

CALTROPS SPINES The branched and otherwise specialized irritating spines in limacodid larvae. Cf. Urticaria.

CALVA Noun. (Latin, *calvaria* = a bare skull. Pl., Calvas; Calvae.) The upper part of the head. See Epicranium.

CALVALLI-SFORZA GENETIC DISTANCE INDEX (Cavalli-Sforza & Edwards 1967, Evolution 21: 550–570; Nei *et al.* 1983, J. Mol. Evol. 19: 153–170.)

CALVERLEY, STEPHEN H (Dos Passos 1951, J. N.Y. Entomol. Soc. 59: 139.)

CALVERT, PHILIP POWELL (1871–1961) (Mallis 1971, *American Entomologists*. 549 pp. (178–180), New Brunswick.)

CALVERT, ROBERT (–1891) (South 1891, Entomologist 24: 104.)

CALVESCENS Adj. (Latin, *calvescens* = becoming bald.) Pertaining to baldness; becoming bald; growing bald.

CALVUS Adj. (Latin, *calvus* = bald; hairless; smooth.) 1. Bald; glabrous. 2. Pertaining to structure that lacks a vestiture of Setae.

CALWER, C G (–1874) (Anon. 1874, NachBl. dt. malak. Ges. 6: 14.)

CALX Noun. (Latin, *calx* = limestone. Pl., Calyces.) 1. Heel. 2. The cup-shaped or head-like portion of a pedunculate body. 3. The distal end of the Tibia. 4. The curving basal portion of the first Tarsomere. 5. The enlarged apical portion of the Lateral Oviduct which connects with the Pedicel of the Ovariole. See Lateral Oviduct.

CALYCIFORM CELL Larval Lepidoptera, Trichoptera and Ephemeroptera, and possibly midgut of Thysanura: A digestive cell of the stomach having a large Ampulla or swelling in its mesial part, opening by a narrow neck through a small aperture on the inner surface. CC interspersed between epithelium cells of the midgut. Also called 'Goblet Cells' with name derived from the deep cavity called the 'Goblet Chamber' that is formed by an invagination of the apical border of the cell. CC chamber confines the nucleus to the basal region of the cell. Cup is invested with Microvilli that contain elongate mitochondria. CC contains mitochondria, the number of which depends on the region of midgut. The cytoplasmic side of membrane that holds mitochondria is studded with small particles. Apparently anterior and middle part absorb water and posterior region secretes water. This result is a countercurrent flux of fluid that recovers enzymes from undigested food. CC of insects apparently take excess potassium from the Haemolymph and move calcium ions from adjacent Columnar Cells into the midgut lumen. In the moth Genus *Tineola*, Goblet Cells accumulate metal ions and dye. Syn Goblet Cell. See Midgut.

CALYCIFORM Adj. (Greek, *kalyx* = cup; Latin, *forma* = shape.) Calyx-shaped; goblet-shaped. See Calyx. Cf. Campaniform; Napiform; Poculiform. Rel. Form 2; Shape 2; Outline Shape.

CALYCOID Adj. (Greek, *kalyx* = cup; *eidos* = form.) Calyx-like in outline shape, form or appearance. See Calyx. Cf. Calyciform. Rel. Form 2; Shape 2; Outline Shape.

CALYCULATE Adj. (Latin, *calyculus* = little calyx.) Cup-like; descriptive of Antennae with cup-shaped segments arranged such that one fits into the adjacent segment. Alt. Calyciform. See Antenna.

CALYCULUS Noun. (Latin, *calyculus* = little calyx.) A cup-shaped or bud-shaped structure.

CALYPSO® See Deltamethrin.

CALYPTER Noun. (Greek, *kalypter* = covering, Pl., Calypters.) Diptera: The Alula or Squama when it covers the Haltere. See Alula.

CALYPTERAE Plural Noun. Calypterate (Curran).

CALYPTRA Noun. (Greek, *kalyptra* = covering; *kalyptos* = hidden. Pl., Calyptrae.) A hood or cap. See Hood.

CALYPTRATA Noun. (Greek, *kalyptra* = covering. Pl., Calyptratae.) Flies that have Alulae or membranous scales above the Halteres.

CALYPTRON Noun. (Greek, *kalyptra* = covering. Pl., Calyptrons.) See Calypter.

CALYX Noun. (Greek, *kalyx* = cup or calyx of a flower; Sanskrit, *kalika* = a bud; Pl., Calyces; Calyxes.) 1. A cup into which certain structures are set. 2. Insect Brain: The cap or crown of the mushroom bodies of the Protocerebrum. See Protocerebrum. 3. Reproductive System: In primitive pterygote insects (*e.g.* Orthoptera) the Pedicel of each Ovariole that may serially open into the Calyx. The Calyx of some parasitic Hymenoptera is the site of microorganism aggregation and replication (a polydnavirus); The Calyx secretes an egg adhesive in chrysomeline beetles that lack Colleterial Glands. 4. The outer group of leaves surrounding a flower. See Eddcalyx. The external-most part of the flower; the Calyx is part of the Perianth and typically green. See Flower; Perianth. Cf. Corolla.

CALYX GLAND Hymenoptera: Paired glands, one of which occurs at the distal end of each Lateral Oviduct. See Accessory Gland. Cf. Acid Gland; Dufour's Gland.

CALYX GLOMERULI The Calyx or cup of the glomerules.

CAMBERWELL BEAUTY See Mourningcloak Butterfly.

CAMBIUM Noun. (Latin, *cambio* = to exchange; Latin, *-ium* = diminutive > Greek, *-idion*. Pl., Cambia; Cambiums.) A thin layer of actively dividing cells between the xylem and phloem systems in most vascular plants. Cambium is responsible for growth (increasing diameter) of plants. Cf. Xylem; Phloem.

CAMBRIAN PERIOD (*Cambria* = Roman name for Wales, where CP first investigated.) The first Period (570–510 MYBP) of the Palaeozoic Era.

The interface between CP and Precambrian is difficult to define, and now is based on vote of paleostratigraphers at a meeting in Bristol, England. Traditionally, CP was defined on basis of trilobites being present; now, CP is viewed as a chronostratigraphic concept - age and content of deposits are not involved. Early Cambrian is well studied in North America and Scandanavia. All main divisions of marine invertebrates occur in Cambrian: Protozoa (Radiolaria), Porifera (sponges), Coelenterata (jelly-fish), Echinodermata (crinoids and star fish), Brachiopoda, Bryozoa, Mollusca, Arthropoda (trilobites), Annelida and Graptozoa. Most soft-bodied animals left no fossil records, but numerous Taxa of unusual forms were deposited in the Burgess Shale of British Columbia. Late Cambrian experienced mass extinction (but not elimination) of trilobites. See Geological Time Scale; Palaeozoic Era. Cf. Devonian Period; Lower Carboniferous Period; Upper Carboniferous Period; Ordovician Period; Permian Period; Silurian Period. Rel. Fossil.

CAMBRIDGE, OCTAVIUS PICKARD (1835–1917) (Anon. 1917, Physis 3: 313.)

CAMEL CRICKETS See Rhaphidophoridae.

CAMELLIA SCALE *Lepidosaphes camelliae* Hoke [Hemiptera: Diaspididae].

CAMELNECK FLIES See Raphidoidea.

CAMERA Noun. (Greek, *kamara* = vaulted chamber. Pl., Cameras; Camerae.) 1. Hymenoptera: The curved, narrow sclerite supporting paired lobes of the Arolium. Term also applied in other orders where the sclerite exists. 2. A curved band of Cuticle serving as a support for the proximal end of a Pulvillus. 3. A chamber used for the humidification of dried insect specimens.

CAMERA LUCIDA An image-deflecting device attached to the eyepiece of a microscope which projects a microscopic image to a mirror and then onto a piece of paper for drawing.

CAMERANO, LORENZO (1856–1917) (Cognetti de Martis 1927, Boll. Mus. Zool. Anat. comp. R. Univ. Genova 7: 1–15.)

CAMERON, JAMES WILLIAM MACBAIN (–1975) (Anon. 1975, Proc. ent. Soc. Ont. 105: 2–3.)

CAMERON, MALCOLM (1873–1954) (Barros Machado 1959, De Publiçoes cult. Diam. Angola 48: 111–112.)

CAMERON, PETER (1847–1912) (Musgrave 1932, *A Bibliography of Australian Entomology 1775–1930*. 380 pp. (38), Sydney.)

CAMEROSTOME Noun. (Latin, *camera* = chamber; Greek, *stoma* = mouth. Pl., Camerostomes.) Ticks: An emargination at the anterior end of the body in which the specialized head (false head, Capitulum) is set.

CAMILLIDAE Plural Noun. A small Family of brachycerous Diptera assigned to Superfamily Ephydroidea related to Drosophilidae. Adult metallic coloured, Sternopleural Bristles absent, with Anal Cell open apically. See Ephydroidea.

CAMIN, JOSEPH American academic (University of Kansas), acarologist and taxonomic theorist.

CAMINACULE Noun. Any member of a group of hypothetical organisms used to test putative phylogenetic relationships. (Camin & Sokal 1965, Evolution 19: 311–326.)

CAMIN-SOKAL PARSIMONY (Camin & Sokal, 1965, Evolution 19: 311–326. See Parsimony.

CAMPANIFORM Adj. (Latin, *campana* = bell; *forma* = shape.) Bell-shaped. Cf. Calyciform. Rel. Form 2; Shape 2; Outline Shape.

CAMPANIFORM ORGAN See Campaniform Sensillum.

CAMPANIFORM SENSILLUM A sense organ that lacks an external process and whose cuticular part typically displays the form of a bell or hollow cone and receives the distal process of the sense cell. CS are distributed over the insect body and appendages, and probably function in stretch reception. CS are dome shaped with two layers: A thin outer exocuticle and relatively thick inner Endocuticle. CS always with one neuron, but frequently several Sensilla are clumped or aggregated. The dendrite penetrates the Endocuticle. CS display a definite orientation with respect to the major axis of an appendage. The Sensilla respond to mechanical stress in the Cuticle. Syn. Bell Organ; Cupola Organ; Dome Organ; Sense Dome; Sensillum Campaniformium; Umbrella Organ. See Sensilla; Proprioceptors.

CAMPANULATE Adj. (Latin, diminutive of *campana* = bell; *-atus* = adjectival suffix.) Bell-shaped. Pertaining to structure more-or-less ventricose at the base and a little recurved at the margin. Alt. Campanulatus.

CAMPBELL, CHARLES (–1873) (Anon. 1874, Entomol. mon. Mag.)

CAMPBELL, ROBERT WALKER (1872–1912) (Anon. 1912, Dt. ent. Z. 1912: 608.)

CAMPBELL SHELFORD, ROBERT WALTER (–1912) English Entomologist and curator (Sarawak Museum, Hope Museum 1905–1912) interested in Blattidae.

CAMPBELL-TAYLOR, J E (1873–1959) (Wigglesworth Proc. R. ent. Soc. Lond. (C) 15: 75.)

CAMPESTRAL Adj. (Latin, *campestris* = concerning a field or plain; *-alis* = appropriate to, belonging to.) Descriptive of organisms that inhabit open fields. See Habitat.

CAMPHOR Noun. (French, *camphre* from Latin, *camphora; -or* = one who does something. Pl., Camphors.) A fragrant, hard crystalline compound obtained from the bark and wood of the camphor tree and which has medicinal uses.

CAMPHOR SCALE *Pseudaonidia duplex* (Cockerell) [Hemiptera: Diaspididae].

CAMPHOR THRIPS *Liothrips floridensis* (Watson) [Thysanoptera: Phlaeothripidae].

CAMPICHOETIDAE Plural Noun. A small Family of brachycerous Diptera assigned to Superfamily Ephydroidea.

CAMPION, HERBERT (1870–1924) (Musgrave 1932, *A Bibliography of Australian Entomology 1775–1930*. 380 pp. (39–40, bibliogr.), Sydney.)

CAMPODEIDAE Plural Noun. A common Family of relatively large bodied Diplura. Species lack compound eyes and display a moderately large (4–7 mm long), scale-less body; Appendix Dorsalis absent; Cercus multisegmented and as long as Antenna, abdominal segments 2–7 with Styli but Palpi absent. Campodeids occur in damp habitats under stones, bark and logs or in rotting wood and debris. See Diplura.

CAMPODEIFORM Adj. (Greek, *kampe* = caterpillar; *eidos* = form; Latin, *forma* = shape.) *Campodea*-shaped. A term applied to elongate larval forms that are characterized by a prognathous head, long thoracic legs and multisegmented Urogomphi. Campodeiform larvae are often predaceous. Alt. Leptiform. See Campodeidae. Rel. Form 2; Shape 2; Outline Shape.

CAMPOS, FRANCISCO (1878–1943) (Anon. 1943, J. Econ. Ent. 36: 247.)

CAMPOS, LUCIANO ELIOT (1927–1989) Chilean academic born in São Paulo, Brazil and a student of Hymenoptera.

CAMPOSTAN® See Ethephon.

CAMPUS Noun. (Latin, *campus* = field. Pl., Campuses; Campi.) Scarabaeoid larvae: The bare or almost bare ventral region of the abdominal segment X or fused IX and X anterior of an entire or anteriorly split edges or anterior of the paired Tegilla (a.v.); Palidia with Septula between, sometimes found extended into the Campus medianly (Boving).

CAMUS, GUILIO (1847–1917) (De Toni 1918, Memorie R. Acad. Sci. Lett. Modena (3) 13: 219–229, bibliogr.)

CANACEIDAE Plural Noun. See Canacidae.

CANACIDAE Hendel 1913. Plural Noun. Beach Flies. A cosmopolitan Family (ca 100 Species in 11 Genera) of schizophorous Diptera assigned to Superfamily Chloropoidea and closely related to Tethinidae. Adult body <3 mm long; head lacks convergent Postvertical Bristles; face typically with bulbous Carina. Most canacids are found near intertidal marine habitats; a few Species occupy freshwater habitats. The larvae of *Canace* feed on algae and respire via a Plastron. Syn. Canaceidae. See Chloropoidea.

CANADA BALSAM A sap-like aromatic oleoresin or combination of resin acids, esters and terpenes that is exuded from wounds on the North American fir tree *Abies balsamea*. CB is used as a mounting medium for histological preparations on microscope slides because it hardens to form a transparent material of good refractive index (ca 1.5). However, CB darkens over long periods of time (ca 100 years) and may become black. CB is soluble in organic solvents such as Benzene and Toluene. Syn. Canada Turpentine. See Balsam. Cf. Gum Damar; Resin.

CANADA TURPENTINE See Canada Balsam.

CANADIAN AMBER Fossilized resin that is Cretaceous Period in age and contains some insect inclusions. See Amber; Fossil. Cf. Baltic Amber; Burmese Amber; Chiapas Amber; Dominican Amber; Lebanese Amber; Taimyrian Amber.

CANADIAN PONDWEED *Elodea canadensis* Michaux [Hydrocharitaceae]: A totally submerged, perennial aquatic herb that is endemic to North America. CP roots in mud and inhabits still or slow-moving water. CP apparently invaded New Zealand and Tasmania during 1960s; date of establishment on mainland Australia not known.

CANADIAN ZONE Part of the Boreal Region comprising southern part of transcontinental coniferous forests of Canada, northern parts of Maine, New Hampshire and Michigan, and a strip along the Pacific Coast extending south to Cape Mendocino and the greater part of the high mountains of the USA and Mexico. In the east, CZ covers Green, Adirondack and Catskill Mountains and the higher mountains of Pennsylvania, West Virginia, Virginia, western North Carolina and eastern Tennessee. CZ in the Rockies extends continuously from British Columbia to western Wyoming, and in the Cascades from British Columbia to southern Oregon with a narrow interruption along the Columbia River.

CANAL OF FECUNDATION Coleoptera: A passage or canal which leads from the Spermatheca or its duct and opens into the Vagina near the point of union of the Lateral Oviducts (Imms).

CANALICULAR Adj. (Latin, *canaliculus* = little channel.) Channelled. A term applied to structure that is longitudinally grooved, with a deeper concave line in the middle. Alt. Canaliculate; Canaliculatus.

CANALICULUS Noun. (Latin, *canaliculus* = little channel. Pl., Canaliculi.) A minute canal.

CANCELLATE Adj. (Latin, *cancellosus* = latticed; *-atus* = adjectival suffix.) Cross-barred; latticed. A term applied to structure with longitudinal lines decussate by transverse lines. Alt. Cancellatus.

CANDEZE, ERNEST CHARLES AUGUSTE (1827–1898) Belgian physician, student of Lacordaire, founding member of Belgian Entomological Society and amateur Coleopterist noted for his work on Elateridae. (Musgrave 1932, *A Bibliography of Australian Entomology 1775–1930*. 380 pp., (39–40, bibliogr.), Sydney.)

CANDEZE, LEON (1863–1926) (Anon. 1926, Bull. Ann. Soc. ent. Belg. 66: 325–326.)

CANDIDA, GIULIO (–1784) (Salfi 1963, Annuar. Ist. Mus. Zool. Univ. Napoli 15: 6–7.)

CANDOLLE, AUGUSTIN PYRAMUS de (1778–1841) Swiss botanist trained in Paris who became Chair of Botany at Montpellier 1808 and returned to Geneva 1815. Candolle published many influential botantical works *(History of Succulent Plants; Essay on the Medicinal Properties of Plants; Introduction to a Natural System* *of the Vegetable Kingdom)* and defined basic notions of endemism and biogeographic regions. Candolle also distinguished between ecological and historical biogeography.

CANE Noun. (Latin, *canna* > Greek, *kaneon* = wicker basket. Pl., Canes.) 1. Sugarcane. 2. Grapevine: Mature woody shoot from leaf-fall through the second year. 3. A slender rod or glass tube.

CANE, HARRY (1860–1935) (Foxlee 1947, Proc. ent. Soc. Br. Columb. 43: 45–46.)

CANEGRUB Noun. (Latin, *canna* > Greek, *kaneon* = wicker basket; Middle English, *grubben*. Pl., Canegrubs.) Larvae of the scarab beetle Genera *Antitrogus, Dermolepida, Lepidiota* and *Rhopaea* in eastern Australia. Canegrubs destroy roots of sugarcane, thereby depriving plants of soil moisture, soil nutrients and mechanical support. Cane is damaged when grubs reach third instar. Species have 1–2 year life-cycle; overwinter deep in soil. Cf. Caudata Canegrub; Consobrina Canegrub; French's Canegrub; Froggatt's Canegrub; Grey-Back Canegrub; Grisea Canegrub; Nambour Canegrub; Negatoria Canegrub; Noxia Canegrub; Picticollis Canegrub; Rhopaea Canegrub; Sororia Canegrub; Southern One-Year Canegrub; Squamulata Canegrub.

CANESCENT Adj. (Latin, *canescens* = becoming grey.) Hoary; descriptive of structure with more white than gray.

CANESTRINI, GIOVANNI (1835–1900) Italian academic (University of Modena, Chair of Zoology, Anatomy and Comparative Physiology, University of Padova) with extensive and broad scientific interests. Canestrini established the Trento-Venetian Society of Natural Sciences and founded the modern study of Acarology in Italy with Berlese. Canestrini began the first mite collection in Italy and worked in spider systematics and biology. (Bonnet 1945, *Bibliographia Araneorum* 1: 41.)

CANESTRINI, RICARDO (1858–1892) Brother of Giovanni Canestrini, working in Entomology, Acarology and Bacteriology. (Castelli 1892, Boll. Soc. Ven.-Trent. Sci. Nat. 5: 47–54.)

CANINE BABESIOSIS A widespread form of the disease Babesiosis as expressed in dogs primarily by *Babesia canis* and *B. gibsoni*. CB agent is transmitted by ixodid ticks, including *Haemaphysalis, Rhipicephalus* and *Dermacentor*. CB is trans-ovarially transmitted in ticks. The disease is often fatal. See Babesiosis. Cf. Bovine Babesiosis; Equine Babesiosis.

CANINE EHRLICHIOSIS A cosmopolitan, typically mild, disease of dogs. Aetiological agent rickettsia-like *Ehrlichia canis* assigned to Tribe Ehrlichieae of Rickettsiaceae. CE can cause haemorrhagic condition (tropical canine pancytopenia) which kills dogs. Vector red dog-tick *(Rhipicephalus sanguineus)* which can transmit CE trans-stadially but not trans-ovarially. See Rickettsiaceae. Cf. Heartwater; Potomac Horse Fever.

CANINE TEETH Sharp, conical teeth of Mandibles in predatory Species. Syn. Dentes Caninae.

CANINES Plural Noun. (Latin, *canis* = dog; *caninus* = pertaining to a dog.) Mayflies: Two heavily chitinized spines arising from the Mandibles which are adapted for holding food (Needham).

CANKER Noun. (Latin, *cancer* = crab, cancer. Pl., Cankers.) 1. Botany: A pathogen-induced, dead, often sunken or cracked area on a stem, twig, limb or trunk that is surrounded by healthy plant tissue. 2. General: A necrotic tissue or a sore that increases in size and does not heal.

CANKER WORMS See Geometridae.

CANNELLA Noun. (French, *canneler* = to flute. Pl., Cannellae.) Coccids: The furrow extending from a spiracle to the lateral margin of the body (MacGillivray).

CANNIBAL Noun. (New Latin, *canibalis* > Spanish, *canibales*, from *caribes* = brave and daring men. Pl., Cannibals.) An animal which feeds upon members of the same Species.

CANNIBALISM Noun. (New Latin, *canibalis* > Spanish, *canibales*, from *caribes* = brave and daring men; *-ism* = state. Pl., Cannibalisms.) A behavioural phenomenon in which an insect feeds upon members of the same Species. The extent of cannibalism is not clear, but many insects feed cannibalistically when under conditions of starvation. Apparently, other insects are cannibalistic as a part of their lifestyle. In terms of diet and matrix, cannibalism probably differs little, if any, from conventional predation. See Feeding Strategies. Cf. Parasitism; Predation.

CANNULA Noun. (Latin, diminutive of *canna* = reed, tube.) A tubule or small tube that is inserted into a body to serve as a drain for fluid or blood.

CANNULAR Adj. (Latin, *canna* = reed, tube.) In the form of a cannula (tubular).

CANOGARD® See Dichlorvos.

CANOLA Noun. Canadian cultivars of rapeseed (*Brassica campestris* Linnaeus, *B. napus* L.) with low levels of glucosinolate and erucic acid.

CANT, ARTHUR (1863–1924) (Riley 1925, Entomologist 58: 72.)

CANTHARIASIS Noun. (Greek, *kantharos* = a scarab beetle; *-iasis* = suffix for names of diseases. Pl., Canthariases.) The invasion of man and animals by beetle larvae. Cf. Enteric Myiasis; Myiasis.

CANTHARID Adj. (Greek, *kantharos* = a scarab beetle.) 1. A blister-beetle assigned to the Cantharidae. 2. Descriptive of or pertaining to the Cantharidae.

CANTHARIDAE Plural Noun. Soldier Beetles. A cosmopolitan Family of polyphagous Coleoptera with ca 3,500 nominal Species. Adult small to moderate sized, soft bodied and pubescent with head exposed, Epipleura narrow or absent with margin of Metasternum keel-like; middle Coxae contiguous; tarsal formula 5-5-5 with claws simple; Elytra thin and pliant; Abdomen without luminous organs. Cantharids are frequently taken at flowers feeding on nectar and pollen; adults are predaceous and feed on soft-bodied insects (aphids and mealybugs). Larvae are campodeiform, somewhat flattened, densely clothed with Setae and live in concealed situations. Larvae presumably are predaceous on insect eggs and small arthropods. Eggs are deposited in soil or other concealed situations. At eclosion, an embryonic 'Prelarva' appears which does not feed, but digests yolk in gut. See Coleoptera.

CANTHARIDIN Noun. (Greek, *kantharis* = beetle-fly; Latin, *cantharis* = Spanish fly. Pl., Cantharidins.) An irritating substance produced by blister beetles (Cantharidae and Meloidae). Cantharidin occurs mainly in the Elytra, produces blisters on the human skin and is used medicinally in keratolytic preparations.

CANTHUS Noun. (Greek, *kanthos* = corner of eye. Pl., Canthi.) The chitinous process more-or-less completely dividing the eyes of some insects into an upper and lower half (*e.g.* some adult Ephemeroptera).

CANTHYLOSCELIDAE Plural Noun. A small Family of nematocerous Diptera assigned to Division Bibionomorpha. See Diptera.

CANUS Adj. (Latin, *canutus* = frosty white; gray-haired.) The colour of gray hair.

CAP Noun. (Middle English, *cappe* > Anglo Saxon, *caeppe* > Latin, *cappa*. Pl., Caps). 1. The top or uppermost part of a structure. 2. The Tormogen (socket-forming) cell of a Seta (Snodgrass). 3. A covering for the head.

CAP SUBSTANCE Diptera: A mucoid secretion from the anterior end of the ovarian follicle. The cap substance is removed before fertilization. Removal of the CS is probably accomplished by cuticular spines positioned within the female's Median Oviduct. See Egg.

CAPACITATION See Sperm Capacitation.

CAPELOUTO, RUEBEN (1920–1980) A student of Ephemeroptera. (See Peters 1982, Fla. Ent. 65: 200.)

CAPEX® See Summer Fruit-Tortrix Granulosis Virus.

CAPFOS® See Fonofos.

CAPILLACEOUS Adj. (Latin, *capillus* = hair, *-aceus* = of or pertaining to.) Capilla or hair-like. Alt. Capillaceus.

CAPILLARITY Noun. (Latin, *capillus* = hair; English, *-ity* = suffix forming abstract nouns. Pl., Capillarities.) Capillary action. The movement of liquids within minute tubules by surface tension.

CAPILLARY Noun. (Latin, *capillus* = hair. Pl., Capillaries.) 1. A long, slender hair-like tube. 2. Antennae with long, slender, loosely articulated Flagellomeres. Alt. Capillaris.

CAPILLATE Adj. (Latin, *capillus* = hair; *-atus* = adjectival suffix.) Clothed with long slender Setae; Coryphatus. Alt. Capillatus.

CAPILLI Head Setae.

CAPILLITIUM Noun. (Latin, *capillus* = hair; Latin, *-ium* = diminutive > Greek, *-idion*. Pl., Capillitia; Pl., Capillitiums.) 1. The hood-like collar in some noctuid moths, *e.g., Cucullia*. See Cucullus. 2. A collection or network of strands formed of waste material in sporangia of slime mould and fruiting bodies of some gaseromycetes.

CAPILURE® See Trimedlure.

CAPIOMONT, GUILLAUME (1812–1871) (Newman 1872, Entomologist 6: 32.)

CAPITATE Adj. (Latin, *caput* = head; *-atus* = adjectival suffix.) 1. Descriptive of elongate structure with a terminal knob-like or head-like enlargement. 2. Reference to a type of clavate Antenna in which the Flagellum is relatively slender and the club is abruptly enlarged to form a spherical mass. Alt. Capitatus. See Capitulum.

CAPITATE ANTENNA An Antenna in which the Capitulum is abrupt and strongly marked. See Antenna.

CAPITATE SETA Setae with knobbed apices (apically enlarged or dilated) that appear hollow. CS occur on the body and appendages of some Psyllidae. Syn. Capitate Hairs. See Seta. Cf. Spatulate Setae.

CAPITULAR SADDLE Acarology: An axial convexity of the Cervix that separates the cheliceral grooves.

CAPITULAR STERNUM See Gnathobase.

CAPITULUM Noun. (Latin, *capitulum* = small head. Pl., Capitula.) 1. Any small 'headed' structure. 2. The enlarged apex of an Antenna. 3. Diptera: The small knob at the apex of the Haltere. The Labella or lapping part of the mouthparts of some flies. 4. Ticks: The false head. Acarina: See Gnathosoma. 5. A small cap-like structure on the Operculum of phasmids which can be easily detached. When eggs are treated with NaOH, the Capitulum swells while the Operculum remains unaffected. Despite its widespread occurrence in Phasmida, large taxonomic groups of phasmids lack a Capitulum: It is not found in Phyllidae (leaf insects); among Phasmatidae (stick insects), Capitulum is not found among several Subfamilies. Early workers attributed a respiratory function to the Capitulum, a notion rejected by Hinton (1981). Eggs of phasmid *Bacillus coccyx* Westwood are attractive to ants. The Capitulum is a protuberance that worker ants of *Acantholepis capensis* Mayr and *Pheidole megacephala* Fabricius use to carry the phasmid egg into their nests without damage. Worker subsequently removes the Capitulum and consumes it. Capitulum-free eggs hatch within ant nests and first instar nymphs apparently are not harmed. See Egg. Cf. Operculum.

CAPNIIDAE Plural Noun. Winter Stoneflies; Snowflies. A Family of Plecoptera with 16 Genera and about 250 nominal Species predominantly found in mountainous regions. Adults with most reduced wing venation in the Plecoptera and typically slender Cerci. Adults emerge during winter. Larvae are benthic and taken only before adult emergence; believed hyporheic. All Species apparently are herbivorous. See Plecoptera.

CAPORIACCO, LUDOVICO DE (1900–1951) (Salfi 1954, Atti Accad. Naz. ital. Ent. Rc. 2: 24–26.)

CAPPER, SAMUEL JAMES (1825–1912) (Turner 1912, Entomologist's Rec. J. Var. 24: 52.)

CAPRICORN BEETLE See Cerambycidae.

CAPRIFICATION Noun. (Latin, *caprificus* = wild fig; English, *-tion* = result of an action. Pl., Caprifications.) The process by which Smyrna figs are fertilized by *Blastophaga psenes* through the medium of wild, inedible or 'caprifigs'.

CAPRIFIER Noun. (Latin, *caprificus* = wild fig. Pl., Caprifiers.) A female agaonid wasp (Hymenoptera: Chalcidoidea) which fertilizes figs by carrying pollen from the male flowers to the female flowers in another receptacle.

CAPRIFIG Noun. (Latin, *caprificus* = wild fig. Pl., Caprifigs.) A form of male, pollen-producing, short-styled inedible fig (*Ficus caricae*). Caprifig pollen is carried to the female, edible long-styled Calimyrna fig by the fig wasp *Blastophaga psenes*. A few caprifig trees (3–5) can supply wasps to pollinate 100 Calimyrna fig trees.

CAPRONNIER, JEAN BAPTISTE (1814–1891) (Thiene 1911, Allg. Lex-bild. Künstler. 5: 558–559.)

CAPSID Noun. (Latin, *capsa* = box; Pl., Capsids.) 1. The protein coat surrounding the nucelic acid of a virus. 2. The common name of capsid bugs. See Miridae.

CAPSIDAE See Miridae.

CAPSULAR Adj. (Latin, *capsula* = little box.) Structure in the form of a capsule or small cup-like container; of or pertaining to a capsule. See Capsule.

CAPSULE Noun. (Latin, *capsula* = little box. Pl., Capsules.) 1. Any small, closed vessel or structure enveloping another structure. 2. A sac-like membrane that envelopes an organ. 3. A closed vessel containing spores, seeds or fruits.

CAPSULE BORER *Antigastra catalaunalis* (Dupon).

CAPTURE® See Bifenthrin.

CAPUT Noun. (Latin, *caput* = head. Pl., Capita.) 1. A knob-like protuberance. 2. The head and its appendages.

CAPUT MORTUUM A Latin phrase meaning 'worthless remains.'

CAPYLUS Noun. (Etymology obscure. Pl., Capyla.) A hump on the dorsal surface of the segments of many larvae.

CARABIDAE Plural Noun. Ground Beetles. A numerically large (ca 20,000 spp.), cosmopolitan Family of adephagous Coleoptera. Adult 1–60 mm long; Antenna typically filiform with 11 segments; fore Coxa globose and Trochantin concealed; Metasternum with distinct transverse suture; tarsal formula 5-5-5. Abdomen with six Ventrites, three connate basal Ventrites. Larva campodeiform with legs well developed; Antenna and Mandible well developed; Urogomphi present and sometimes segmented. Adults are cursorial, active, predatory. In terms of their Ecology, Carabids are divided into geophiles, hydrophiles and arboricoles.

CARABIDOID Adj. (Greek, *karabos, karabis* = a horned beetle, *eidos* = resembling.) Descriptive of the second instar of a hypermetamorphic meloid larva, when it resembles a larval carabid.

CARABOID Adj. (Greek, *karabos, karabis* = a horned beetle, *eidos* = resembling.) Carabus-like; resembling a carabid beetle. See Carabidae. Rel. Eidos; Form; Shape.

CARADJA, ARISTIDE (1861–1955) (Popescu-Gorgi 1970, Revta. Muzeelor 1970, 299–303.)

CARAGANA APHID *Acyrthosiphon caraganae* (Cholodkovsky) [Hemiptera: Aphididae].

CARAGANA BLISTER-BEETLE *Epicauta subglabra* (Fall) [Coleoptera: Meloidae].

CARAGANA PLANT-BUG *Lopidea dakota* Knight [Hemiptera: Miridae].

CARAMBOLA FRUIT-BORER *Eucosma notanthes* Meyrick [Lepidoptera: Tortricidae]: A significant multivoltine pest of carambola in China.

CARAPACE Noun. (Spanish, *carapacho* = covering. Pl., Carapaces.) The hard covering of the Thorax in Crustacea including the shell of crabs.

CARAWAY WEBWORM *Depressaria nervosa* (Haworth) [Lepidoptera: Oecophoridae].

CARAYON'S GLAND Paired, sac-like Scent Glands that occur in adult Elasmodemiinae, Holoptilinae and Phymatinae. CG is associated with the Metathoracic Scent Gland but is independent of it. CGs project into the abdominal Haemocoel and display an aperture on ventral membrane between the Thorax and Abdomen. (See Staddon 1979, Advances in Insect Physiology, 14: 377.) Syn. Ventral Gland. See Exocrine Gland. Cf. Brindley's Gland; Metathoracic Scent Gland.

CARBAMATE INSECTICIDE A group of synthetic insecticides derived from carbamic acid (NH_2COOH). Members act as contact and oral poisons which inhibit cholinesterase activity and display moderate to high mammalian toxicity. Carbamates are susceptible to oxidative and hydrolytic detoxification. They vary in persistence from a few weeks to a few months. Some carbamates are used as surface sprays or baits for control of urban pests. Examples include: Aldicarb, Bendiocarb, Carbaryl, Methomyl, Propoxur. Symptoms of carbamate poisoning include: Headache, salivation, sweating, confusion, constricted pupils and abdominal pain (if swallowed). Treatments include: If ingested, then induce vomiting; if contacted, then remove contaminated clothing, wash with soap and water; consult medical advice. See Insecticide; Cf. Botanical Insecticide; Inorganic Insecticide; Insect Growth Regulator; Organochlorine; Organophosphorus; Synthetic Pyrethroid.

CARBAMULT® See Promecarb.

CARBARYL A carbamate compound {1-naphthyl N-methylcarbamate} used as a contact insecticide and stomach poison. Carbaryl is applied to numerous fruit tree crops, grapes, vegetable crops, field crops, cotton, forest trees, peanuts, sorghum, tobacco, pets and livestock. The compound is phytotoxic to some varieties of apples, pears and watermelons. Carbaryl is effective against plant-sucking insects (aphids, scale insects, leafhoppers, psyllids, whitelfies), plant chewing insects (armyworms, Jap beetle, grasshoppers) thrips, and fleas. Trade names include: Carpolin®, Crunch®, Denapon®, Hexavin®, Karbaspray®, Microcarb®, Murvin®, Patrin®, Ravion®, Scattercarb®, Septene®, Sevin®, Tercyl®, Thinsec®, Tricarnam®, Twister®. See Carbamate Insecticide.

CARBAX® See Dicofol.

CARBICRON® See Dicrotophos; Monocrotophos.

CARBODAN® See Carbofuran.

CARBOFURAN A carbamate compound {2,3-dihydro-2,2-dimethyl-7-benzofuranyl methyl-carbamate} that is used as a Nematicide, contact/systemic insecticide and stomach poison. Carbofuran is applied to vegetable crops, field crops, peanuts, sorghum, sugar beets, tobacco and potatoes. The compound is effective against aphids, scale insects, plant chewing insects, thrips, rootworms, and wireworms, but it is toxic to bees, birds, fishes and wildlife. Trade names include: Brifur®, Carbodan®, Carbo-sect®, Carbosip®, Curaterr®, Fruacarb®, Furadan®, Futura®, Kenofuran®, Nex®, Rampart®, Sunfuran®, Throttle®, Yaltox®. See Carbamate Insecticide.

CARBOHYDRASE Noun. (Latin, *carbo* = coal; Greek, *hydro* = water. Pl., Carbohydrases.) A digestive enzyme that splits or breaks carbohydrates in food.

CARBOHYDRATE Noun. (Latin, *carbo* = coal; Greek, *hydro* = water; Chemistry *-ate* = salt formed from acid. Pl., Carbohydrates.) Any member of a class of compounds including starches and sugars. Carbohydrates are prominent constituents of plants and animals and contain carbon, hydrogen and oxygen in the proportion to form water. Carbohydrates are the most available source of energy for the living cell.

CARBOLIC ACID When pure, needle-shaped crystals that form a crystalline, colourless, white or light pink mass. CA is extremely hygroscopic and caustic, soluble in water, alcohol, glycerin, ether, chloroform and in oils, such as xylol. CA is used in dehydration in microscopy and histology.

CARBOL-XYLENE See Carbol-Xylol.

CARBOL-XYLOL A mixture of three parts xylol and one part melted crystal carbolic acid. The mixture is used to clear whole mounts of insects, chitinous parts (*e.g.,* genitalia) and other preparations.

CARBON BISULPHIDE CS_2 is an offensive smelling, colourless liquid compound used as a solvent of greases and oils. In entomological museums, CB has been used to clean greasy insects, as a disinfectant and insecticide in collections. CB acts as a rapid acting fumigant with good penetrating characteristics for application against pests of the grain handling industry. CB is explosively inflammable when vaporized and poisonous to humans. Syn. Carbon Disulphide. See Insecticide; Fumigant.

CARBON DIOXIDE See CO_2.

CARBON TET® See Carbon Tetrachloride.

CARBON TETRACHLORIDE An aromatic, non-flammable, non-explosive, colourless liquid (CCl_4) used as a multipurpose compound during the first half of the 20th century. In Entomology, CT is used as a solvent to clean grease and dirt from insects, to disinfect collections and as a killing agent. CT also has been used as an internal medicine against hook-worm. As a killing agent, CT made specimens brittle and difficult to pin. CT is carcinogenic and a liver toxin. CT is used as a fumigant or mixed with other fumigants because it is a good solvent and suppresses flammability. Compound is applied to stored barley, corn, oats, rice, rye and sorghum in some countries, but it is not registered for use in USA as a fumigant. Trade names include Carbon Tet®, Tetrachloromethane® and Perchloromethane®.

CARBONARIUS Adj. (Latin, *carbonarius* = of or relating to charcoal.) Coal black.

CARBONATE Noun. (French, *carbone* = carbon; *-ate*. Pl., Carbonates.) A salt of carbonic acid.

CARBONIFEROUS Adj. (Latin, *carbo* = coal; *ferre* = to carry; *-osus* = full of.) The Palaeozoic Period during which most of the world's coal deposits were formed. See Palaeozoic Period. Cf. Lower Carboniferous Period; Upper Carboniferous Period.

CARBOPHOS® Malathion.

CARBOSECT® See Carbofuran.

CARBOSIP® See Carbofuran.

CARBOSULFAN A carbamate compound {2,3-dihydro-2,2-dimethyl-7-benzofuranyl [(dibutylamino)thio] methylcarbamate} used as a contact/systemic insecticide and stomach poison. Carbosulfan is applied to apples, *Citrus*, alfalfa, corn, cotton, sorghum, soybeans, sugarbeets, vegetables and potatoes in some countries. Compound is used for control of mites, nematodes, plant-sucking insects (aphids, scale insects, leafhoppers, whiteflies), plant chewing insects, thrips, rootworms, and wireworms. Carbosulfan is not registered for use in USA because it is toxic to wildlife. Trade names include: Advantage®, Beam®, Gazelle®, Marshall®, Posse®, Suscon®. See Carbamate Insecticide. See Insecticide.

CARCEL, PIERRE 1800–1831) (Anon. 1832, Ann. Soc. ent. Fr. 1: 114.)

CARCINOGEN Noun. (Greek, *karkinos* = cancer; *gennaein* = to produce. Pl., Carcinogens.) A cancer causing substance.

CARCINOPHORIDAE See Anisolabididae.

CARDEN, GEORGE (–1894) (H. G. 1894, Entomologist 27: 168.)

CARDENOLIDES Cardiac glucosides that occur in some plants, such as milkweed (*Asclepias* spp.), and cause emetic response in vertebrates. Cardenolides confer protection from vertebrate predation because the predators learn to avoid insects that may have fed upon plants containing cardenolides.

CARDER BEE A bee so named because of its habit of smoothing or carding certain mosses for use in nest building. See *Bombus*.

CARDIA Noun. (Greek, *kardia* = heart; upper orifice of the stomach. Pl., Cardiae; Cardias.) The anterior end of the Mesenteron which frequently is mistermed the Proventriculus (Snodgrass). The 'gizzard.' The Heart *sensu* Smith. See Heart.

CARDIAC SINUS The haemocoelic channel in the embryo dorsal of the yolk or Alimentary Canal. Part of the Cardiac Sinus becomes the lumen of the dorsal blood vessels (Snodgrass). See Sinus.

CARDIAC VALVE 1. The Oesophageal Valve at the junction of foregut and midgut (Imms). 2. The Stomodaeal Valve *sensu* Snodgrass. Alt. Cardiac Valvule. See Valve.

CARDINAAL, H (–1971) (Nieuwenhuis 1971, Ent. Ber., Amst. 31: 204.)

CARDINAL Adj. (Latin, *cardo* = hinge.) 1. Any 'thing' of fundamental or basic importance (*e.g.* cardinal principle). 2. Central or basic to a construction, framework, organization, plan or process. 3. Descriptive of or pertaining to the Cardo.

CARDINAL BEETLES See Pyrochroidae.

CARDINAL CELL Odonata: See Cardinal Triangle.

CARDINALES Plural Noun. Sclerotized rods joining Labium to Head; the Cardines.

CARDINOSTERNAL Adj. (Latin, *cardo* = hinge; *sternum* = breast bone.) Descriptive of or pertaining to Cardo and Sternum of the labial segment (Snodgrass).

CARDINOSTIPITAL Adj. (Latin, *cardo* = hinge; *stipes* = stalk.) Descriptive of or pertaining to Cardo and Stipes collectively.

CARDIOBLAST Noun. (Greek, *kardia* = heart; *blastos* = bud. Pl., Cardioblasts.) A row of cells in the embryo which develops into the body cavity. Cf. Myoblast.

CARDO Noun. (Latin, *cardo* = hinge. Pl., Cardines.) 1. Immatures: The basal (proximal) segment of the Maxilla between head and Stipes; the so-called 'hinge' of the Maxilla. 2. Adult: The basal segment of the Maxilla which articulates with the Cranium via one Condyle and broadly attached to the Stipes. The Cardo is variable in size and shape. 3. Hymenoptera: A basal ring in the Genitalia. 4. The proximal sclerite of the Protopodite. See Maxilla.

CARDOSUBMENTAL Adj. (Latin, *cardo* = hinge; *sub* = under; *mentum* = chin.) Pertaining to the Cardo and Submentum.

CARETTE, EDUARDO A (–1946) (Anon. 1947, Revta. Soc. ent. argent. 13: 344.)

CARFENE® See Azinphos-Methyl.

CARIBBEAN BLACK-SCALE *Saissetia neglecta* DeLotto [Hemiptera: Coccidae].

CARIBBEAN FRUIT-FLY *Anastrepha suspensa* (Loew) [Diptera: Tephritidae].

CARIBBEAN POD-BORER *Fundella pellucens* Zeller [Lepidoptera: Pyralidae].

CARIBOU WARBLE-FLY *Oedemagena tarandi* [Diptera: Cuterebridae]: A Holarctic pest of reindeer. Adult yellow in colour and bee-like in appearance. CWF biology and life history resemble Cattle Grub. A high incidence of mortality is seen in young animals with resistance apparently developing in older animals. Syn. Reindeer Warble-Fly.

CARINA Noun. (Latin, *carina* = keel., Pl., Carinae.) An elevation (ridge) of the Cuticle (Integument). Carinae are sometimes tall, acute or otherwise distinctive and taxonomically important. Sometimes, Carinae may be precipitous and called 'inflections'; relatively broad elevations may be called 'crests.' Syn. Keel. Cf. Culmen; Ridge.

CARINAL Adj. (Latin, *carina* = keel.) Keel-like; pertaining to a Carina.

CARINATE Adj. (Latin, *carina* = keel.) 1. Descriptive of surface sculpture (usually the insect's Integument) that displays keel-like elevations or ridges. 2. Pertaining to any structure bearing one to several longitudinal, narrow, raised ridges. Alt. Carinatus. See Sculpture Pattern. Cf. Alveolate; Baculate; Biarinate; Clavate; Echinate; Favose; Gemmate; Psilate; Punctate; Reticulate; Rugulate; Scabrate; Shagreened; Smooth; Striate; Verrucate.

CARINULA Noun. (Latin, diminutive of *carina* = small keel. Pl., Carinulae.) 1. A small Carina or keel-like ridge. 2. The longitudinal elevation on the middle of the snout in Rhynchophora.

CARINULATE Adj. (Latin, diminutive of *carina* = small keel; *-ate* = adjectival suffix.) Descriptive of structure superficially with small and numerous Carinae. Arcane: Carinulatus.

CARIOSE Adj. (Latin, *cariosus* from *caries* = decay.) Corroded; appearing as if worm-eaten. Alt. Cariosus; Carious. Rel. Decay.

CARL, JEAN (1877–1944) (Anon. 1945, Ent. News 56: 279–280.)

CARLET, GASTON (1845–1892) (Kilian 1892, Naturaliste 14: 152–153.)

CARLIER, STUART EDMOND WACE (1899–1962) (Wigglesworth 1964, Proc. R. ent. Soc. Lond. (C) 28: 57.)

CARMINATE Adj. (Latin, *carminium* = vivid red.) Descriptive of colour that is tinged with carmine. Alt. Carminated.

CARMINE Noun. (French, *kermes* from Arabic, *qirmiz* = crimson. Pl., Carmines.) 1. A red coloured, insoluble compound formed by treating an extract of cochineal with a solution of alum. 2. A stain used in histology.

CARMINE SPIDER-MITE *Tetranychus cinnabarinus*

(Boisduval) [Acari: Tetranychidae].

CARNATION MAGGOT *Delia brunnescens* (Zetterstedt) [Diptera: Anthomyiidae].

CARNATION TIP-MAGGOT *Delia echinata* (Séguy) [Diptera: Anthomyiidae].

CARNEOSE Adj. (Latin, *caro* = flesh; *-osus* = full of.) 1. Flesh-like in substance. 2. Flesh coloured, or white tinged with red (Kirby & Spence). Alt. Carneous; Carneus.

CARNIDAE Plural Noun. A small Family of brachycerous Diptera assigned to Superfamily Opomyzoidea and placed in Milichiidae in some classifications. Most Species found in Holarctic. Wing with CuA +1A more distinct that 2A; female Cerci fused. See Opomyzoidea.

CARNIDAE See Milichiidae.

CARNIVORE Noun. (Latin, *caro, carnis* = flesh; *vorare* = to devour. Pl., Carnivores.) A flesh-eating organism such as an insect (*e.g.* mantid) that feeds on other living insects. Carnivores are secondary or tertiary-level trophic consumers. See Feeding Strategies. Cf. Herbivore; Omnivore; Phytophage; Saprovore. Rel. Ecology.

CARNIVOROUS Adj. (Latin, *caro, carnis* = flesh; *vorare* = to devour; *-osus* = possessing the qualities of.) Descriptive of organisms (plants or animals) that feed upon the bodies of animals. See Carnivore; Feeding Strategies. Cf. Herbivorous; Phytophagous; Saprophagous.

CARNIVORY Noun. (Latin, *carnivorous* from *carnis* = flesh + *vorare* = to devour.) 1. The act or process of being carnivorous. 2. A type of feeding strategy that requires animal tissue as a source of nutrition. See Feeding Strategies; Digestion; Extra-oral Digestion. Cf. Herbivory; Phytophagy; Saprophagy.

CARNOSE Adj. (Latin, *carnosus* = fleshy; *-osus* = full of.) Descriptive of a substance which is soft or fleshy. Alt. Carneous.

CAROB MOTH *Spectrobates ceratoniae* (Zeller) [Lepidoptera: Pyralidae]: A moth whose larva is a polyphagous agricultural pest on fruit trees; CM is a serious pest of stored produce including dried fruits, nuts, seeds and beans. CM larvae feed inside produce. Apparently, CM is endemic to Mediterranean region and established in New World and South Africa; it commonly infests produce shipped to temperate areas. Adult CM resembles Mediterranean Flour Moth; wingspan 20–28 mm and hindwings whitish with dark veins. Syn. *Ectomyelois ceratoniae* (Zeller); Locust-Bean Moth. See Pyralidae.

CAROLDSFELD-KRAUSE, A G (1905–1969) (Jacobs 1969, Entomol. mon. Mag. 81: 150–151.)

CAROLINA CONIFER-APHID *Cinara atlantica* (Wilson) [Hemiptera: Aphididae].

CAROLINA GRASSHOPPER *Dissosteira carolina* (Linnaeus) [Orthoptera: Acrididae]: A minor pest of grasses, grains and other cultivated crops in North America. CG pods contain 20–70 eggs; males stridulate. See Acrididae.

CAROLINA MANTID *Stagmomantis carolina* (Johannson) [Mantodea: Mantidae]: A Species regarded as beneficial, endemic to North America and found in southern USA west to Arizona. See Mantidae.

CAROLINIAN FAUNAL AREA The Upper Austral Zone comprising the larger part of the Middle States (except the mountains), southeast South Dakota, eastern Nebraska, Kansas and part of Oklahoma, nearly all of Iowa, Missouri, Illinois, Indiana, Ohio, Maryland and Delaware, more than half of West Virginia, Kentucky, Tennessee and New Jersey and large areas in Alabama, Georgia, the Carolinas, Virginia, Pennsylvania, New York, Michigan and South Ontario: extends along Atlantic Coast from near the mouth of the Chesapeake Bay to southern Connecticut and sends narrow arms up the valleys of the Hudson and Connecticut, a narrow arm follows the east shore of Lake Michigan to Grand Traverse Bay.

CARON, JAN ROBERT (1919–1973) (Oorschot 1974, Ent. Ber., Amst. 34: 41–42.)

CAROTIN Noun. (Latin, *carota* = carrot. Pl., Carotins.) An organic pigment that produces the yellows and reds of certain insects (Wardle). See Pigmentary Colours. Cf. Xanthin.

CAROTIN ALBUMIN A substance in insect Haemolymph (blood) that produces the pink, purple and green colours in certain insects. See Albumin.

CAROTINOID Adj. (Latin, *carota* = carrot; Greek, *eidos* = form.) A carotin-like compound. Carotenoid pigments are soluble in fats and contain no nitrogen. Basic types of carotenoids include carotenes and xanthophils. Carotenoids are important in chloroplasts and the colouring of plant parts. Carotenoids originate within plants and are not synthesized by insects or other animals. Animals ingest carotenoids and deposit them unchanged inside the body or add side groups to enhance visual intensity. Carotenoids produce yellow, red and orange colour in insects; carotenoids are responsible for yellow coloration in pierid butterflies, some wasps and orange coloration in ladybird beetles. In combination with a blue pigment (mesobilverdin), carotenoids can produce greens called insectoverdin. The functions of carotenoids in insect metabolism are poorly known; quantities of carotenoids are responsible for production of the visual pigment retinene. See Pigmentary Colour. Cf. Flavinoids; Melanins; Ommochromes; Pterines (Pteridines); Quinones. Rel. Structural Colour.

CARPENTER ANTS Members of the Genus *Camponotus* [Hymenoptera: Formicidae]: Widespread, predominantly tropical with ca 1,000 Species. CAs inhabit decaying and decayed parts of trees, stumps. CAs nest in wood but do not feed on it; as urban pests, they nest in decaying timber and subfloors of homes. Most Species are predaceous or feed on aphid honeydew. CAs may protect aphids in agricultural situations or become pests of Honey Bee colonies. See Formicidae.

CARPENTER BEE *Xylocopa virginica* (Linnaeus) [Hymenoptera: Xylocopidae (Anthophoridae)]: A monovoltine, polylectic Species widespread and endemic to eastern USA. Adult 20–25 mm long, black and resembles Bumble Bees. Female Antenna with 12 segments and hind Tibia with one spur. Male Antenna with 13 segments and hind Tibia with two spurs; Female lacks Corbiculae but hind Tibia with long brush of Setae that holds pollen. CBs nest in solid wood excavated and partitioned into several cells with cemented wood chips; each cell is provisioned with pollen and nectar. CB overwinters in prepupal stage. See Anthophoridae. Cf. Little Carpenter Bee; Mountain Carpenter Bee. Rel. Bumble Bee.

CARPENTER BEES Members of Subfamily Xylocopinae. Adults lack protuberant Clypeus and triangular sclerite at apex of Metasoma. CBs nest in wood and plant stems, and sometimes are regarded as a nuisance or urban pest. See Anthophoridae; Xylocopinae. Cf. Bumble Bee.

CARPENTER MOTHS See Cossidae.

CARPENTER WORM *Prionoxystus robiniae* (Peck) [Lepidoptera: Cossidae]: A pest of living hardwood and shade trees that resides throughout the USA and southern Canada. Eggs are deposited in bark fissures. Larvae bore into trunks of tree, and construct mines that undermine strength and stain wood; sawdust typically is exuded from holes in bark. CW overwinters within tunnels and pupation occurs within cells excavated from mines. CW requires three years to complete one generation. See Cossidae. Cf. Leopard Moth.

CARPENTER WORMS See Cossidae.

CARPENTER, FRANK MORTON (1902–1993) American academic (Harvard University): Fisher Professor of Natural History and Alexander Agassiz Professor of Zoology. Carpenter was preeminent among palaeoentomologists. He was coauthor of *Classification of Insects* (1954) and Honorary Fellow of the Royal Entomological Society. (Furth 1994, Psyche 101: 126–144.)

CARPENTER, GEOFFREY DOUGLAS HALE (1882–1953) English academic and Hope Professor of Zoology (Oxford University). Carpenter served in Colonial Medical Service (1910–1930) as M.D. and became interested in bionomics of tsetse fly. His other interests included Lepidoptera and mimicry. (Remington 1954, Lepid. News 8: 31–43, bibliogr.)

CARPENTER, GEORGE HERBERT (1865–1939) (Anon. 1939, Anon. Indian J. Ent. 1: 128.)

CARPENTER-WORM MOTHS See Cossidae.

CARPENTIER, CHARLES (–1935) (Anon. 1935, Miscnea ent. 36: 92.)

CARPENTIER, FRITZ (1873–1968) (Bovey 1969, Mitt. schweiz. ent. Ges. 42: 237–238.)

CARPENTIER, LEON (1838–1914) (Anon. 1915, Miscnea ent. 22: 66–67.)

CARPET BEETLE Common name for one of several Species of Coleoptera which feed on carpet. Syn. Buffalo Beetle. See Australian Carpet-Beetle; Black Carpet-Beetle; Common Carpet-Beetle; Furniture Carpet-Beetle; Varied Carpet-Beetle.

CARPET BEETLES See Dermestidae.

CARPET MOTH *Trichophaga tapetzella* (Linnaeus) [Lepidoptera: Tineidae]: Cosmopolitan pest of carpets and storage rooms.

CARPET MOTHS See Geometridae.

CARPOCAPSA® See Hexaflumuron.

CARPOLIN® See Carbaryl.

CARPOPHAGOUS Adj. (Greek, *karpos* = fruit; *phagein* = to eat; Latin, *-osus* = possessing the qualities of.) Fruit-eating. Descriptive of animals which feed on fruit or obtain nutrients from the ingestion of fruit or fruiting bodies of plants. See Feeding Strategy. Cf. Cannibalistic; Carnivorous; Frugivorous; Parasitic; Predaceous; Phytophagous; Saprophagous; Xylophagous. Rel. Ecology.

CARPOPHILOUS Adj. (Greek, *karpos* = fruit; *philein* = to love; Latin, *-osus* = full of.) Fruit-loving. Descriptive of animals which are often taken or found in association with fruit.

CARPOPODITE Noun. (Latin, *carpus* = wrist; *pous* = foot; *-ites* = constituent. Pl., Carpopodites.) 1. The fifth segment of a generalized appendage. 2. The Tibia (Snodgrass). See Appendage. Cf. Coxopodite; Meropodite; Telopodite.

CARPOSINIDAE Plural Noun. A small Family (ca 200 Species) of ditrysian Lepidoptera sometimes placed in its own Superfamily (Carposinoidea) or Copromorphoidea. Family is cosmopolitan in distribution but best represented in Australia, New Guinea, New Zealand and Hawaii. A few Species are of minor economic importance as borers in fruits, flower buds, bark and galls. Adult small with wingspan 10–40 mm; Frons scales are smooth, Ocelli nearly always absent and Chaetosemata absent; Antenna ca 0.75 times as long as forewing; Proboscis without scales; Maxillary Palpus minute, with one segment; Labial Palpus large; fore Tibia with Epiphysis; tibial spur formula 0-2-4; forewing with raised tufts of scales; hindwing without vein M2. Larvae form galls or tunnel in fruit or bark. Pupa are sclerotized and occur in larval habitat or within cocoon in soil. Adults are nocturnal and attracted to lights. See Lepidoptera.

CARPOVIRUSINE® See Codling-Moth Granulosis Virus.

CARPUS Noun. (Greek, *karpos* = wrist. Pl., Carpi.) 1. Odonata: The Pterostigma; the extremity of the forewing Radius and Cubitus, or the place on the wing at which they are transversely folded. 2. The fifth segment from the base of the generalized crustacean appendage.

CARR, JOHN WESLEY (1862–1939) (Leivers 1939, Entomologist 72: 248.)

CARRAGEENAN Noun. (From Carrageen, a town

near Waterford, Ireland, Pl., Carrageenans.) A polysaccharide extracted from seaweeds (*Chondrus crispus*) and which consists of galactose, dextrose and levulose. A cold-water form of Carrageenan produces a viscous solution; a hot-water form produces a gel that is used as substitute for agar in artificial diets. Cf. Agar; Algin.

CARRENO, EDUARDO (–1842) (Dusmet y Alonso 1919, Boln. Soc. ent. Esp. 2: 83.)

CARRET, A (–1907) (Daniel 1908, Münch. Koleopt. Z. 3: 399.)

CARRIERE, JUSTUS (–1893) (Wilderman 1894, Jber. Naturw. 9: 501.)

CARRIKER, MELBOURNE A (1879–1965) (Emerson 1967, Bull. U. S. natn Mus. 248: 1–150, bibliogr.)

CARRINGTON, JOHN THOMAS (1846–1908) (Briggs 1908, Entomologist 41: 73–74.)

CARRION BEETLES See Silphidae.

CARRION'S DISEASE See Bartonellosis.

CARROLL, JAMES (1854–1907) American Medical Entomologist and member of the team (Ross, Carrol, Lazear) that worked on transmission of Yellow Fever by *Anopheles*. (Howard 1930, Smithson. misc. Collns. 84: 133, 174, 466, 481–482, 496.)

CARROT BEETLE *Ligyrus gibbosus* (DeGeer) [Coleoptera: Scarabaeidae]: A widespread, monovoltine pest in USA that attacks carrots, celery, cotton, corn, cabbage, parsnips, potatoes and beets. CB overwinters as an adult within soil; during spring, female oviposits nocturnally in soil. Eggs increase in size before eclosion. Larvae feed on roots of host plant. See Scarabaeidae.

CARROT FLIES See Psilidae.

CARROT RUST-FLY *Psila rosae* (Fabricius) [Diptera: Psilidae]: A Palaearctic pest of carrot, celery and parsnip. CRF was adventive to Ottawa, Canada ca 1885 and now is widespread in North America. Adults small, slender and shiny with head and legs yellow and body dark green or blue-black; Antenna long. CRF adults appear during May and oviposit at base of host plant. Larvae penetrate soil and feed on roots. CRF overwinters as a Pupa within a brown puparium in soil or as larva on roots of host plant. CRF completes 2–3 generations per year. Syn. Carrot Fly. See Psilidae.

CARROT WEEVIL *Listronotus oregonensis* (LeConte); *Listronotus texanus* (Stockton) [Coleoptera: Curculionidae].

CARRUCCIO, ANTONIO (1839–1923) (Lepri 1923, Boll. Ist. zool. R. Univ. Roma 1: 1–4.)

CARRYING CAPACITY The maximum population density a given environment will support for a sustained period. Rel. Ecology.

CARSIDARIDAE Plural Noun. A small Family of Psylloidea (Hemiptera) characterized by a forewing with Vena Spuria. See Psylloidea.

CARSTANJEN, ERNST (1835–1884) (Anon. 1884, Psyche 4: 236.)

CARTAP A nereistoxin-derived compound {S,S'-(2-dimethylamino) trimethylene) bis (thiocarbamate) hydrochloride} used as a contact insecticide and stomach poison for control of aphids, leafhoppers, thrips, leaf-eating insects, leafminers, stem borers, and related insects. Cartap displays some ovicidal activity. Cartap is applied to tree crops, *Citrus*, cotton, grapes, field crops and ornamentals in some countries. Some phytotoxicity noted in cotton and some apples. Compound is not registered for use in USA. Toxic to fishes. Cartap is not mixed with akaline insecticides. Nereistoxin is derived from marine annelid. Trade names include Caldan®, Padan®, Patap®, Sanvex® and Thiobel®. See Carbamate Insecticide.

CARTER, HENRY ROSE (1852–1925) (Musgrave 1932, *A Bibliography of Australian Entomology 1775–1930*. 380 pp. (40), Sydney.)

CARTER, HERBERT JAMES (1858–1940) (A. B. W. 1943, Proc. Linn. Soc. N. S. W. 68: 91–94.)

CARTER, JOHN WILLIAM (1843–1920) (Porritt 1921, Entomol. mon. Mag. 57: 65–68.)

CARTHAEIDAE Common 1966 Plural Noun. A small Family of ditrysian Lepidoptera assigned to Superfamily Bombycoidea and known from Australia. Carthaeids are placed among various Families including Noctuidae, Saturniidae and Geometridae in different classifications. Adult large bodied with wingspan 76–100 mm; head with piliform scales; Ocelli and Chaetosemata absent; male Antenna bipectinate and female Antenna dentate; Proboscis well developed and coiled at repose; Maxillary Palpus small, with three segments; wings broad with prominent eyespots; Epiphysis present; tibial spur formula 0-2-4; Tympanal Organs absent. Eggs are flattened with primary axis horizontal and Micropyle at one end; Chorion smooth, somewhat shiny. Eggs are laid in groups of 2–3 on host plant. Larval head hypognathous; abdominal segments 1–8 with prominent lateral eyespots; abdominal segments 3–6 with Prolegs; segment 8 humped without horn. Pupa strongly sclerotized and head rugose; Cremaster present; loosely constructed cocoon on ground. See Bombycoidea.

CARTHY, JOHN DENNIS (1924–1972) (Butler 1973, Proc. R. ent. Soc. Lond. (C) 37: 56.)

CARTILAGE Noun. (Latin, *cartilago, cratis* = wicker. Pl., Cartilages.) A translucent, elastic tissue that comprises the skeleton of embryos, young vertebrates and which is converted into bone in most higher vertebrates. Cf. Chitin; Resilin.

CARTILAGINOUS Adj. (Latin, *cartilago* = cartilage; *-osus* = possessing the qualities of.) Descriptive of structure resembling cartilage in structure or appearance. Alt. Cartilagineus. See Cartilage.

CARTILAGO ENSIFORMIS See Sternum Collare.

CARTON Noun. (French, *carton* > Latin, *charta* = card. Pl., Cartons.) 1. The paper-like material manufactured by aculeate social Hymenoptera for nest construction. 2. The material forming the walls of Termitaria.

CARUNCLE Noun. (Latin, *caruncula* = small bit of flesh. Pl., Caruncles.) A soft, naked, fleshy excresence or protuberance. Alt. Caruncula.

CARUS, CARL GUSTAV (1789–1869) (Nordenskiold 1935, *History of Biology*. 629 pp. (209), London.)

CARUS, CHARLES GUSTAVUS German academic at Dresden and author of *Memoir on the Circulation of the Larvae of Neuropterous Insects* (1827).

CARUS, JULIUS VICTOR (1823–1903) German zoologist educated in Leipzig, Wurtzburg and Freiburg. Carus was appointed Keeper of Oxford Museum of Comparative Anatomy (1849–1851) and Professor of Comparative Anatomy and Director, Zoological Institute of Leipzig (1853). Important works by Carus include *System der Tierischen Morphologie* (1853), *Handbuch der Zoologie* and *Geschichte der Zoologie*. (Musgrave 1932, *A Bibliography of Australian Entomology 1775–1930*, 380 pp. (43), Sydney.)

CARVEL, MARCEL (1893–1955) (Dupius 1955, Cah. Nat. 11: 97–100, bibliogr.)

CARVIL® See Fenobucarb.

CARYOLITE Noun. (Greek, *karyon* = nut; *lytikos* = loosing; *-ites* = constituent. Pl., Caryolites.) 1. A group of sarcolytes accompanied by a knot of muscular origin (Henneguy). 2. A fragment of muscular fibre (including a nucleus) which is undergoing phagocytosis during insect development (Imms). Rel. Phagocytosis.

CARYOPHYLLEOUS Adj. (Greek, *karyophyllon* = clove tree.) Nut or clove brown.

CARZOL® See Formetanate.

CASAC, OSVALDO HUGO (1931–1971) (Garcia 1971, Mosquito Syst. Newsl. 3: 212–215, bibliogr.)

CASANGES, ABBY H (–1973) (Anon. 1973, Bull. ent. Soc. Am. 19(1): [i].)

CASARETTO, GIOVANNI (1812–) (Papavero 1975, *Essays on the History of Neotropical Dipterology*. 2: 347, São Paulo.)

CASCADE® See Flufenoxuron.

CASE BEARERS See Coleophoridae.

CASE MOTHS See Psychidae.

CASEBEARER MOTHS See Coleophoridae.

CASEIN Noun. (Latin, *caseus* = cheese. Pl., Caseins.) A phosphoprotein available as a fat-free and vitamin-free preparation or substance obtained from cow's milk. Casein is often used as a protein constituent of artificial insect-diets; it may lack some amino acids or have them in insufficient quantities.

CASEMAKING CLOTHES-MOTH A complex of Species resembling *Tinea pellionella* Linnaeus [Lepidoptera: Tineidae]. In older literature probably recognized as *T. pellionella,* a widespread (probably cosmopolitan) larval pest of hair, hides, feathers and some plant material; CCM is more often a pest in domestic homes than commercial facilities. Adult brown with 9–16 mm wingspan. Males are active but female sluggish. Eggshell with longitudinal ridges, sticky and adheres to substrate. Larvae construct a parchment-like case that is carried while feeding; larvae retreats into their case when disturbed. Larvae prefer fabric stained with fruit juices, or containing human sweat or urine for the B-complex vitamins contained therein; larvae can damage but cannot digest cotton or linen. Pupation occurs within a cocoon formed by enlargement of the case. Life cycle: Egg 4–7 days; larva 41–56 days; pupa 9–20 days; adult 42–130 days. CCM is monovoltine in cold-temperate regions and bivoltine elsewhere. Development, biology and anatomy resemble Webbing Clothes Moth but CCM carries web and less common than WCM. Syn Casebearing Clothes-Moth. Cf. Large Pale Clothes-Moth; Tropical Case-Bearing Clothes Moth; Webbing Clothes Moth.

CASEY, THOMAS LINCOLN (1831–1925) American military officer (West Point, 1852) who became Chief of Engineers (1888, Brigadier General) and a distinguished amateur coleopterist (Member, US National Academy of Science 1890). Collected and described more Species of North American Coleoptera than any contemporary. Founding member of Entomological Society of Washington. (Mallis 1971, *American Entomologists*. 549 pp. (260–264), New Brunswick.)

CASPARI, W (–1909) (Anon. 1909, Ent. Rdsch. 26: 81.)

CASSAVA MEALYBUG *Phenacoccus manihoti* Matile-Ferrero [Hemiptera: Pseudococcidae]: A significant pest of cassava, *Manihot esculenta* Crantz, in the western and central Africa areas of cassava growth. CMB was introduced from South America during the early 1970s. Significant biological control of CM was achieved with the encyrtid wasp, *Epidinocarsis lopezi* (DeSantis). See Pseudococcidae.

CASSELBERRY, RAYMOND C (1900–1954) (Remington 1954, Lepid. News 8: 30.)

CASSIDY, THOMAS PART (–1968) (Rainwater & Ewing 1968, J. Econ. Ent. 61: 1475.)

CASSIN, JOHN (1813–1869) (Brewer 1869, Proc. Boston Soc. nat. Hist. 12: 244–248.)

CASSINO, SAMUEL E (1854–1937) (Estate of Cassino 1938, Ent. News 49: 180.)

CASTANEOUS Adj. (Latin, *castanea* = chestnut; Latin, *-osus* = possessing the qualities of.) Chestnut brown or bright red-brown in colour. Alt. Castaneus.

CASTE Noun. (Latin, *castus* = pure. Pl., Castes.) Social Insects: Different groups of individuals of predictable morphological types or behaviour which perform specialized labour within the colony. Caste differences are permanent and not attributed to age. Castes include workers, soldiers and reproductives. The egg and pupal stages are members of the colony but not castes. Syn. Social Class; Social Rank' Social Position. See Brood; Isoptera; Hymenoptera.

CASTE POLYETHISM Social Insects. A division of labour within a colony involving morphological castes that perform specialized social responsibilites. See Polyethism. Cf. Age Polyethism.

CASTE POLYMORPHISM A form of polymorphism expressed within social insects, particularly ants. Several morphologically distinct forms of female reproductives have been noted in Ponerinae, and some socially parasitic Formicinae and Myrmecinae. Several gynomorphic, intermorphic and nearly ergatomorphic specimens are functional queens in *Leptothorax* sp. The Species is functionally monogynous, with only one reproductive female in each colony, including one or several inseminated but non-egg laying potential queens. Intermorphs may have identical offspring, or produce intermorphs and gynomorphs as young potential queens. A genetic mechanism may be responsible for queen polymorphism. See Polymorphism. Cf. Cyclical Polymorphism; Kentromorphism.

CASTEK, JOSEPH (1872–1941) (Tykac 1941, Cas. csl. Spol. ent. 38: 6–7.)

CASTELLANI, OMERO (1903–1974) (Consiglio 1976, Odonatologica 5: 76–78, bibliogr.)

CASTELNEAU, FRANCOIS LOUIS NOMPAR DE CAUMONT DE LAPORTE DE (1810–1880) (Papavero 1971, *Essays on the History of Neotropical Dipterology* 1; 149–159. São Paulo.)

CASTLE, DAVID MACFARLAND (1842–1924) (Wenzel 1924, Ent. News 35: 305–306.)

CASTLE-RUSSELL, SYDNEY GEORGE (1867–1955) (E. A. C. 1955, Entomologist 88: 192.)

CASTNIIDAE Plural Noun. A small Family (ca 30 Species) of ditrysian Lepidoptera forming Superfamily Castnioidea; Species occur in Neotropical, Oriental and Australian Realms. Adult small to large with wingspan 25–120 mm; head with smooth scales; lamellar scales on Vertex directed forward; Frons scales piliform; Ocelli present and Chaetosemata absent; antennal Flagellum with smooth scales and apex clubbed; Proboscis typically large but vestigial in a few Species; Maxillary Palpus small with 2–4 segments; Labial Palpus upturned; fore tibial Epiphysis small and spine-like; tibial spur formula 0-2-4; Ovipositor long and telescopic. Egg fusiform with several longitudinal ridges and Micropyle at one end of long axis. Larva constructs silk-lined tunnels in monocots. Pupation occurs in larval tunnels. Adult is brightly coloured and active diurnally. See Lepidoptera.

CASTRATION Noun. (Latin, *castrare* = to castrate; English, *-tion* = result of an action. Pl., Castrations.) 1. Any process that interferes with or inhibits the production of mature eggs or Spermatozoa in the gonads of an organism. See Emasculation. 2. The removal of the androecium from a flower. Alt. Emasculate; Neuter.

CAT FLEA *Ctenocephalides felis* (Bouché) [Siphonaptera: Pulicidae]: A widespread urban pest of humans and domesticated animals; CF prefers cats but will feed upon ca 50 Species of mammals, including dogs. Adult 1.5–2.5 mm long and laterally compressed; Frons with conspicuous, internal, club-shaped incrassation; Ctenidial Combs horizontal; genal process with small sharp spine; Frontal Tubercle absent; abdominal Terga 2–7 with conspicuous single row of Setae. Female requires blood meal each 12 hours to survive and a blood meal for oviposition; female responds to thermal and visual stimuli. CF eggs are laid on host but fall into carpet or resting place of host; ca 150 eggs are laid during adult life. Larvae remain in host's resting place and feed upon dried blood, exuviae or debris. Pupation occurs within cocoon cemented by debris. Development is temperature and relative-humidity dependent, optimally: eclosion ca 2–4 days; larva 10–11 days; pupa 7–10 days, adult female 11 days and male 7 days when confined, up to 6 weeks when continually on cat; life cycle 20–24 days; development can require 20 months. See Pulicidae. Cf. Dog Flea; Human Flea; Oriental Rat Flea.

CAT FOLLICLE-MITE *Demodex cati* Mégnin [Acari: Demodicidae].

CAT LOUSE *Felicola subrostratus* (Burmeister) [Mallophaga: Trichodectidae]: A pest of cats. Cf. Dog Louse.

CATABENA MOTH *Neogalea esula* (Druce) [Lepidoptera: Noctuidae]: A moth endemic to Mexico; larva feeds on lantana leaves and pupates within cocoons attached to stems. See Noctuidae.

CATALOGUE Noun. (Greek, *katalogos* = a counting up. Pl., Catalogues.) A compendium of taxonomic and nomenclatural information prepared by specialists and arranged by taxonomic categories. Kinds of information taxonomic catalogues include: 1. Original publication for each Genus and Species including author names and dates of publication; 2. synonyms of Genera and Species listed; 3. important taxonomic references; 4. geographical distributions of included Taxa; 5. hosts, host plants and ecological information when known; 6. references to biological data. Alt. Catalog. Cf. Monograph.

CATALPA MIDGE *Contarinia catalpae* (Comstock) [Diptera: Cecidomyiidae].

CATALPA SPHINX *Ceratomia catalpae* (Boisduval) [Lepidoptera: Sphingidae]: A significant pest of Catalpa in USA. CS is univoltine or bivoltine and overwinters as a pupa within soil under Catalpa tree. Adults appear when tree has become in full leaf. Female oviposits white eggs in masses on underside of leaves with as many as 1,000 eggs in a mass; eclosion occurs within 10–14 days. See Sphingidae.

CATALYST Noun. (Greek, *katalysis* = dissolution. Pl., Catalysts.) A chemical compound or element which induces chemical reactions in other substances without itself being affected or changed. Syn. Chemical Reactor; Reactant; Synergist.

CATAPHRACTED Adj. (Greek, *kataphractos* = covered.) Invested with a hard, callous Integument, or with scales closely united. Alt. Cataphractus.

CATAPTERIGIDAE Plural Noun. A small Family of glossatous Lepidoptera assigned to Superfamily Eriocranioidea.

CATATREPSIS Noun. (Greek, *kata* = down; *trepein* = to turn. Pl., Catatrepses.) Embryology: Insect blastokinesis or the phase when an embryo leaves the dorsal aspect to resume its primitive condition on the ventral aspect of the egg (Wheeler). Syn: Katatrepsis. Cf. Anatrepsis.

CATCH Noun. (Old French, *chacier* = to hunt. Pl., Catches.) Collembola: The Retinaculum or Tenaculum.

CATECHU Noun. (Malay, *catechu*.) Any of several astringent, dry, earthy or resin-like substances produced by decoction and evaporation from the wood, leaves or bark of several Species of plants in tropical Asia, most notably *Acacia catechu*. Catechu is used in medicine as an astringent, and in the processes of dyeing and tanning. See Kino.

CATENATION Noun. (Latin, *catena* = chain; *-tion* = an action, a state. Pl., Catenations.) A structure composed of a series of regularly arranged (connected, successive) elements similar in shape and composition (*e.g.* links in a chain). Alt. Catenatus.

CATENIFORM Adj. (Latin, *catena* = chain; *forma* = shape.) Chain-shaped. A term used implicitly to suggest smaller links than 'catenate.' Alt. Catenulate; Catenulatus. See Catenation. Rel. Form 2; Shape 2; Outline Shape.

CATERPILLAR Noun. (Latin, *cattus* = cat; *pilosus* = hairy. Pl., Caterpillars.) 1. An elongate, cylindrical larval stage with well developed head capsule, biting Mandibles, short Antennae, thoracic legs and abdominal Prolegs. 2. Lepidoptera: The polypod or eruciform larva. 3. Mecoptera, symphytous Hymenoptera: The larval stage with abdominal Prolegs. Correct veneral term: An *army* of caterpillars. See Larva. Cf. Grub; Maggot; Slug. Rel. Naiad; Nymph.

CATERVATUM (Latin, *caterva* = in loose order.) By heaps.

CATESBY, MARK (1682–1749) English naturalist who visited America (1712–1719, 1722–1726) and published *Natural History of Carolina, Florida and the Bahama Islands*...Illustrated with 220 plates, including 31 insects (Mark Catesby, London. 1731, 1743.) (Hume 1943, Florida Hist. Quart. 21: 292.)

CATFACING Noun. Injury to a plant which results from the feeding of piercing-sucking insects on developing fruit. Catfacing results in uneven growth and deformation.

CATHREMA Noun. (Greek, *katheter* = to thrust in. Pl., Cathremata; Cathremae.) Lepidoptera: A striated thickened part of the Ejaculatory Duct of male Nepticulidae.

CATTANEO, GIACOMO (1857–1925) (Issel 1926, Riv. Biol. 8: 128–135, bibliogr.)

CATTLE FOLLICLE-MITE *Demodex bovis* Stiles [Acari: Demodicidae].

CATTLE GRUB *Hypoderma lineatum* (de Villers) [Diptera: Oestridae]: A monovoltine pest of cattle in the Holarctic. Adult ca 13 mm long, bee-like, robust and conspicuously setose; Setae on Scutum white anteriad and yellow posteriad; Setae at apex of Abdomen reddish yellow. Adult emergence occurs early in morning and mating occurs soon afterward. Adult cannot feed and lives 3–5 days; female produces ca 300–700 eggs during lifetime and oviposits on legs and lower part of body but does not cause gadding in cattle. Several eggs are deposited in line along shaft of hair and eclosion occurs within 4–7 days. Neonate larvae move down hair shaft, penetrate skin or hair follicle; larvae burrow or migrate through body over a period of several months, temporarily residing in lining of Oesophagus. Later, the larvae continue their migration to back of host, chew small holes in skin and reverse their position. Respiration occurs via Spiracles at posterior end of body at hole in skin; larvae complete their development in place under skin on back of host; mature larvae enlarge their hole in skin, emerge, drop to ground and pupate in soil. See Oestridae. Cf. Northern Cattle Grub.

CATTLE GRUBS The larvae of *Hypoderma* spp. [Diptera: Oestridae] that develop as parasites of cattle. Two notable Species are the Cattle Grub and Northern Cattle Grub. See Hypodermatidae; Oestridae.

CATTLE ITCH-MITE *Sarcoptes bovis* Robin [Acari: Sarcoptidae].

CATTLE TAIL-LOUSE *Haematopinus quadripertusus* Fahrenholz [Anoplura: Haematopinidae].

CATTLE TICK *Boophilus annulatus* (Say), *Boophilus microplus* (Canestrini) [Acari: Ixodidae]: *B. microplus* is a pest in Australia.

CATTLE-BITING LOUSE *Bovicola bovis* (Linnaeus) [Mallophaga: Trichodectidae]: A North American pest of cattle, particularly during winter; CBL attacks base of tail, withers and shoulders. Body ca 1.5 mm long, red with transverse dark stripes across Abdomen. Syn. Red Cattle Louse. See Trichodectidae. Cf. Goat-Biting Louse; Horse-Biting Louse; Sheep-Biting Louse.

CATTLE-POISONING SAWFLY *Lophyrotoma interrupta* (Klug) [Hymenoptera: Pergidae]: A sawfly that is endemic to Australia. Cattle eat mature larvae which accumulate at base of iron-bark trees before pupation See Pergidae.

CATTLEYA FLY See Orchidfly.

CAUCHY, PIERRE AMÉDÉE (1806–1831) (Anon. 1832, Ann. Soc. ent. Fr. 1: 114.)

CAUDA Noun. (Latin, *cauda* = tail. Pl., Caudae.) 1. The tail or any tail-like process. 2. The pointed apex of the Abdomen or extension of the caudal segment or appendage terminating the Abdomen. Alt. Caudulae. See Tail. Cf. Trunk.

CAUDACORIA Noun. (Latin, *cauda* = tail; *corium* = leather. Pl., Caudacoriae.) The indentation continuous with the Notarotaxis between Caudalaria and Scutalaria (MacGillivray).

CAUDAD Adverb (Latin, *cauda* = tail; *ad* = toward.) Toward or in the direction of the posterior or tail-end of the body. See Orientation. Cf. Anteriad; Apicad; Basad; Centrad; Cephalad; Craniad; Dextrad; Dextrocaudad; Dextrocephalad; Distad; Dorsad; Ectad; Entad; Laterad; Mediad; Mesad; Neurad; Orad; Proximad; Rostrad; Sinistrad; Sinistrocaudad; Sinistrocephalad; Ventrad.

CAUDAL Adj. (Latin, *cauda* = tail.) Descriptive of or pertaining to the Cauda or anal end of the insect body.

CAUDAL FAN Mosquito larvae: Feathery bristles arranged fan-like on the ninth abdominal segment.

CAUDAL FILAMENT A long, taperered cuticular process projecting from the posterior end of the body. See Appendix Dorsalis.

CAUDAL GILL A type of Tracheal Gill which occurs in the naiad of some Zygoptera. CGs consist of three external tracheal gills on the naiad, located at the posterior end of the body. One median and two lateral gills are found in early instar naiads; in later instars the gills may become saccate or lamellate. See Gill. Cf. Lateral Abdominal Gill; Rectal Gill. Rel. Cutaneous Respiration.

CAUDAL PLATE 1. Coccids: Sclerites formed by the lateral Pilacerores of segments 5–7; posterior lateral sclerites (MacGillivray). 2. Acarina: See Pygosomal Plate.

CAUDAL POSTANAL SETA Coccids: See Obanal Setae (MacGillivray).

CAUDAL PROCESS Any elongate structure that projects from the apex of the Abdomen of adult or immature stages (*e.g.* Appendix Dorsalis; Caudal Seta).

CAUDAL SETA Long, thread-like processes at the posterior end of the Abdomen in many insects. See Anal Filaments; Postanal Setae.

CAUDAL STYLE See Appendix Dorsalis.

CAUDAL VESSICLE An eversible portion of the hindgut in braconid larvae (Hymenoptera) which functions in respiration.

CAUDALABIA (Latin, *cauda* = tail; *labium* = lip. Pl., Caudalabiae.) Coccids: The Labiae of the Abdomen *sensu* MacGillivray.

CAUDALARIA Noun. (Latin, *cauda* = tail; *alerius* = wing. Pl., Caudalariae.) The Alaria on the posterior part of each lateral margin of the Scutum *sensu* MacGillivray.

CAUDATA Noun. (Latin, *cauda* = tail.) The Order of Salamanders including three extant Suborders, nine Families, 62 Genera and about 360 Species.

CAUDATA CANEGRUB *Lepidiota caudata* Blackburn [Coleoptera: Scarabaeidae]: In Australia, larvae damage pastures by feeding on roots. See Canegrub.

CAUDATE Adj. (Latin, *cauda* = tail.) 1. With tail-like extensions or processes. 2. Hymenoptera (Apocrita): A specialized body form of some endoparasitic ichenumonid larvae, which is typically segmented with long, flexible, caudal appendages. Function of cadual appendages is not established; sometimes they are progressively reduced in later instars and lost in last instar. Alt. Caudatus.

CAUDELL, ANDREW NELSON (1872–1936) (Mallis 1971, *American Entomologists*. 549 pp. (198–200), New Brunswick.)

CAUDO-CEPHALIC (Latin, *cauda* = tail; Greek, *kephale* = head; *-ic* = characterized by.) In a line from head-to-tail. See Orientation. Cf. Caudal; Cephalic.

CAUDULA Noun. (Latin, *cauda* = tail. Pl., Caudulae.) A little tail.

CAUL Noun. (Middle English, *calle* = covering > Middle French, *calotte* = skullcap. Pl., Cauls.) 1. A network covering. 2. An enclosing membrane. 3. The fatty mass of tissue in larvae from which the organs of the future adult were supposed to develop.

CAULFIELD, F B (–1892) (Anon. 1892, Can. Ent. 24: 104.)

CAULICULUS Noun. (Latin, *cauliculus* = stalk of a plant.) 1. The larger of two stalks supporting the Calyx of the mushroom body. 2. Growing on the stem of another plant.

CAULIGASTRIC Adj. (Latin. *caulis* > Greek, *kaul* = stem, stalk; *-ic* = characterized by.) Acarology: Pertaining to a narrow connection between the Prosoma and Opisthosoma. Cf. Pedicel; Petiole.

CAULIS Noun. (Latin, *caulis* = stalk. Pl., Caules.) 1. The Funicle of the Antenna. 2. The corneous basal part of the jaws. 3. Lepidoptera (Tortricidae): The median Apodeme ventrad of the Aedeagus (Juxta + Anellus.) 4. Hymenoptera: The basal portion of male Genitalia consisting of the Gonobase, Gonocoxite and Volsella.

CAVANNA, GUELFO (1850–1920) (Gestro 1922, Memorie Soc. Ent. Ital. 1: 5–7.)

CAVATE Adj. (Latin, *cavus* = hollow.) Hollowed out; cave-like. Alt. Cavatus.

CAVE, C J P (1871–1950) (Wigglesworth 1951, Proc. R. ent. Soc. Lond. (C) 15: 75.)

CAVE WETA See Macropathidae.

CAVEOLAE Plural Noun. (Latin, *cavea* = excavated area. Sl., Caveola.) Microscopic vesicles in the Plasma Membrane which facilitate the passage of molecules. Alt. Caveolae.

CAVERA Noun. (Etymology obscure.) Coccids: Tracheal tube connecting with each spiracle expanded Into a chamber of varying size and shape. See Collar Chamber (MacGillivray).

CAVERNICOLOUS Adj. (Latin, *caverna* = cavern; *colous* = living in or on; *-osus* = possessing the qualities of.) Cave-inhabiting. A term used to characterize subterranean insects that live as troglobites in caves, caverns, lava tubes and other permanent subterranean spaces. Cf.

Agricolous; Aphidicolous; Algicolous; Arenicolous; Caespiticolous; Deserticolous; Ericeticolous; Fimicolous; Fungicolous; Gallicolous; Lapidicolous; Lignicolous; Madicolous; Muscicolous; Nemoricolous; Nidicolous; Paludicolous; Pratinicolous; Ripicolous; Rupicolous; Saxicolous; Silvicolous.

CAVERNOUS Adj. (Latin, *cavernosus* = chambered; *-osus* = possessing the qualities of.) Divided into small spaces or little caverns.

CAVITY Noun. (Latin, *cavitas* = cave; English, *-ity* = suffix forming abstract nouns. Pl., Cavities.) A hollow space or opening.

CAVOLINI (CAOLINIUS), FILIPPO (1756–1810) (Rose 1850, *New General Biographical Dictionary* 6: 146.)

CAYENNE TICK *Amblyomma cajennense* (Fabricius) [Acari: Ixodidae].

CAZIER, MONT A (1911–1995) Curator (American Museum of Natural History 1941–1962) and later Professor of Zoology (Arizona State University); author of 61 publications on behaviour, ecology and systematics of various arthropod groups.

CCC® See Chlormequat.

CCHF See Crimean-Congo Haemorrhagic Fever.

CEANOTHUS SILK-MOTH *Hyalophora euryalus* (Boisduval) [Lepidoptera: Saturniidae].

CEBADILLA See Sabadilla.

CEBALLOS Y FERNANDEZ DE CORDOBA, D GONZALES (1892–1967) (Agenjo 1968, Eos 43: 319–343, bibliogr.)

CECCONI, GAICOMO (1866–1941) (Trotter 1942, Marcellia 30: xllviii.)

CECH, CARL OTTOKAR (1842–1895) (Anon. 1896, Ost. landw. Wbl. 22: 227, 229.)

CECIDIUM Noun. (Greek, *kekidion, kekis* = gall, anything gushing out; Latin, *-ium* = diminutive > Greek, *-idion*. Pl., Cecidia.) A gall.

CECIDOGENOUS Adj. (Greek, *kekis* = oak gall; *genous* = producing; Latin, *-osus* = possessing the qualities of.) 1. Anything producing galls on plants. 2. Diptera: Larvae of Cecidomyiidae that cause and live in galls.

CECIDOMYIDAE See Cecidomyiidae.

CECIDOMYIIDAE Plural Noun. Gall Midges. A cosmopolitan Family of nematocerous Diptera including more than 3,000 nominal Species and about 75 fossil Species. Adult body small to minute and delicate; legs long and without tibial spurs; wings setose with 3–5 longitudinal veins but lacking cross veins. Not all midges induce gall formation in host plants: Larvae of many Species live in galls; some Species are inquilines in galls; some Species are scavengers in decomposing vegetation; some Species are predaceous on small arthropods; some Species are endoparasitic within other insects. Syn. Cecidomyidae; Itonididae. See Chrysanthemum Gall Midge; Clover Seed Midge; Hessian Fly; Sorghum Midge.

CECIDOSIDAE Plural Noun. A small Family of glossatous Lepidoptera assigned to Superfamily Incurvarioidea.

CECIL, RODNEY (1898–1962) (Stone & Campbell 1963, J. Econ. Ent. 56: 424.)

CECROPIA MOTH *Hyalophora cecropia* (Linnaeus) [Lepidoptera: Saturniidae].

CECROPINS Plural Noun. Inducible antibacterial proteins originally isolated and identified from the pupal stage of the moth *Hyalophora cercropia* pupae, and now known to occur in other endopterygote insects. Cercropins cause lysis of bacteria.

CEDAR BEETLES See Rhipiceridae.

CEDAR LEAF-MINER *Blastotere thujella* [Lepidoptera: Argyresthiidae].

CEDARTREE BORER *Semanotus ligneus* (Fabricius) [Coleoptera: Cerambycidae].

CEE Central European Encephalitis. See Tick-Borne Encephalitis.

CEJPA, KARLA (1919–1970) (Pilát 1970, Ceska Mykol. 24: 1–4.)

CEKUDIFOL® See Dicofol.

CEKUFON® See Trichlorfon.

CEKU-GIB® See Gibberellic Acid.

CEKUMETA® See Metaldehyde.

CEKUSAN® See Dichlorvos.

CELERY APHID *Brachycolus heraclei* Takahashi [Hemiptera: Aphididae].

CELERY LEAFTIER *Udea rubigalis* (Guenée) [Lepidoptera: Pyralidae]: A multivoltine pest of celery, beans, spinach and many ornamental plants in North America. Eggs are flattened, scale-like and typically laid in overlapping clutches on the underside of leaves; eclosion occurs within 5–12 days. Neonate larva pale green; older larvae pale yellow with dorsal longitudinal white stripe and dark band in centre of stripe. Larvae consume the lower surface of leaves and upper surface remains intact. Pupation occurs within curled leaf edge held in place by silk threads. In greenhouses, CL may complete eight generations per year. Syn. Greenhouse Leaftier, *Phlyctaenia rubigalis* (Guenée), *Phlyctaenia ferrugialis* Hübner. See Pyralidae.

CELERY LOOPER *Anagrapha falcifera* (Kirby) [Lepidoptera: Noctuidae]. Syn. *Syngrapha falcifera* (Kirby).

CELFUME® See Methyl Bromide.

CELL Noun. (Latin, *cella* = compartment. Pl., Cells.) 1. Any closed area in an insect wing between or bounded by veins. In the Comstock System, cells derive their names from the vein forming the upper margin: *e.g.* all cells just below the Radius are Radial Cells, and they are numbered from the base outward. 2. In a nest or honeycomb: A prepared space or small chamber in which an egg is placed and the larva develops. 3. A unit of biological organization differentiated into cytoplasm and nucleus, from which units plants and animals develop by mitotic division. Rel. Gland; Organ; Parenchyma; Tissue.

CELL CULTURE The growth of cells as independent units (not tissues) suspended within a medium. See Cell Line.

CELL INCLUSION Any of several kinds of naturally occurring, functional Organelles found within the cell, including Centrioles, Ribosomes, Mitochondria, Endoplasmic Reticulum and Golgi Apparatus. See Cell. Cf. Nuclear Inclusions.

CELL JUNCTION A principal mechanism of cell binding. Typically opened osmotically but not through the action of proteolytic enzymes or divalent cation chelators. CJs probably provide intercellular channels for ionic and molecular flow. See Cell. Cf. Continuous Junction; Gap Junction; Septate Junction; Scalariform Junction; Tight Junction. Rel. Desmosome.

CELL LINE Tissue Culture: Cells that originate by subculture of a Primary Culture and which have the capacity to divide *in vitro*. Cell lines may be finite or continuous, depending upon the continuation of subculturing. See Tissue Culture. Cf. Cell Culture; Cell Strain. Rel. Biotechnology.

CELL MOVEMENT Active movement of a cell is achieved with Pseudopodia, Cilia or Flagella. Movement by Pseudopodia is best studied in Amoeba. Insect Spermatozoa move with the aid of Flagella. Cilia and Flagella are extremes of an anatomical continuum and not ultrastructurally distinct organelles. Flagella are typically longer than Cilia and represented by single projections from the cell surface. Cilia are typically short, multiple projections which provide flexural motion composed of asymmetrical movements. Forward (vertical) movement is called effective stroke; rearward (horizontal) movement is called recovery stroke. Cyclical action of a Cilium is more-or-less three dimensional: effective stroke is planar while the recovery stroke swings out of that plane. Characteristically, the wavelength is longer than the organelle. Cf. Sperm Flagellum; Sperm Movement.

CELL REMODELLING One of two processes by which change in Fat Body is maintained during Metamorphosis. The remodelling process sees larval Fat Body dissociated into isolated cells and subsequently reassociated into adult Fat Body. The alternative process (Histolysis) sees larval Fat Body completely destroyed during transformation with simultaneous development of adult Fat Body from undifferentiated stem cells. See Fat Body.

CELL STRAIN Cells derived from a cell line or Primary Culture by selection for a specific property or marker and maintained through subsequent subculture. Selection of a Cell Strain is carried out by cloning. See Cell. Cf. Cell Culture; Cell Line.

CELLAR SPIDERS See Phlocidae.

CELLI, ANGELO (1857–1914) Italian Medical Entomologist who worked on human Malaria in Italy and formed 'La Societa per gli Studii della Malaria.' (Howard 1930, Smithson. Misc. Colln. 84: 483, 487, 491–496.)

CELLOSOLVE Noun. (Etymology uncertain. Pl., Cellosolves.) A clear, liquid compound (ethylene-glycol-monoethyl ether) sometimes useful for clearing histological tissues. Cellosolve does not excessively harden biological tissues, but is highly volatile.

CELLS OF SEMPER Cells of the crystalline eye cone, probably specialized corneagenous cells (Snodgrass).

CELLULAE Plural Noun. (Latin, *cellula* = small cell. Sl., Cellula.) 1. Some lecaniid coccoids: Usually round or more-or-less oval areas. Cellulae frequently are composed of concentric lighter and darker bands, not cells in a histological sense. 2. Dermal cells; dermal pores (MacGillivray).

CELLULAR DEFENCE MECHANISM Methods by which the insect defends itself from pathogens and parasites in the Haemolymph. CDM include Encapsulation, Nodule Formation and Phagocytosis. See Haemolymph. Rel. Pathogen.

CELLULASE Noun. (Latin, *cellula* = small cell. Pl., Cellulases.) A digestive enzyme that breaks up (hydrolyses) cellulose to cellobiose. Syn. Cytase.

CELLULE Noun. (Latin, *cellula* = small cell. Pl., Cellules.) Part of a wing included between veins. A term usually applied to a small area completely enclosed and rarely to interspaces that do not form a closed area.

CELLULOSE Noun. (Latin, *cellula* = small cell. Pl., Celluloses.) The principal component of ligno-cellulose and a structural polysaccharide in green plants which forms in young cell walls. See Ligno-cellulose. Rel. Wood.

CELLUTEC® See Permethrin.

CELMIDE® See Ethylene Dibromide.

CELMONE® See NAA.

CELPHOS® See AlP.

CELYPHIDAE Plural Noun. See Lauxaniidae.

CEMENT GLAND See Colleterial Gland.

CEMENT LAYER A layer of the Cuticle that is not produced by all insects. When present, CL serves as the outermost layer of Integument. CL is produced by Dermal Glands that secrete tanned proteins of a shellac-like substance. CL is thought to provide protection from abrasion; protection is not absolute and if the cement layer is physically abraded, the insect may suffer the rapid loss of water. See Cuticle; Integument. Cf. Cuticulin Layer; Wax Layer.

CENCHRUS Noun. (Greek, *kenchros* = millet. Pl., Cenchri.) Hymenoptera: A pale-coloured, membranous lobe or area on each side of the Metanotum in Symphyta (except Cephidae). The margins are rough and presumably function as a holdfast along the anal region of the forewing at repose. Cf. Hamulus.

CENOGENETIC Adj. (Greek, *kainos* = new; *genesis* = origin; *-ic* = characterized by.) 'Developing sidewise.' 1. Pertaining to insects with secondary adaptations to special ways-of-life as larvae (Comstock). 2. Pertaining to non-phylogenetic processes in the development of the individual. 3. Development of the adaptations found in the early stages of the individual.

CENOGENOUS Adj. (Greek, *kainos* = new; *genous* = producing; Latin, *-osus* = possessing the qualities of.) Producing young at one time oviparously and viviparously at another time. A term sometimes applied to aphids with complex biological characteristics.

CENOSIS See Biocenosis.

CENOZOIC ERA The most recent Era in Geological History which followed Mesozoic Era. EC was initiated about 65 million years ago and presently thought to have been initiated with a massive extinction of Dinosaurs and large vertebrates at end of Mesozoic Era, and adaptive radiation of mammals, snakes, frogs, grasses and flowering plants. CE stratigraphic record is divided into Palaeogene Period and Neogene Period (= Tertiary and Quaternary) which are divided into several Epochs. See Geological Time Scale; Mesozoic Era; Palaeozoic Era. Cf. Palaeogene Period; Neogene Period. Rel. Fossil.

CENTIGRADE Noun. (Latin, *centum* = hundred; *gradus* = step.) The scientific temperature-measurement scale divided into 100 equal parts (degrees) between the melting point of ice and the boiling point of water at sea level (760 mm pressure.) 1°C = 1.8° Fahrenheit.

CENTIMETRE Noun. (Latin, *centum* = hundred; *metrum* = measure. Pl., Centimetres.) A unit of linear measurement. 1 cm = 0.01 metre; 2.54 cm = 1.0 inch. Abbreviated 'cm.'

CENTIPEDES See Chilopoda.

CENTRAD Adv. (Latin, *centrum* = centre ; *ad* = toward.) Toward the centre or interior. See Orientation. Cf. Anteriad; Apicad; Basad; Caudad; Cephalad; Craniad; Dextrad; Dextrocaudad; Dextrocephalad; Distad; Dorsad; Ectad; Entad; Laterad; Mediad; Mesad; Neurad; Orad; Proximad; Rostrad; Sinistrad; Sinistrocaudad; Sinistrocephalad; Ventrad.

CENTRAL BODY Part of the Protocerebrum in the Brain posterior of the Pars Intercerebralis. See Brain. Cf. Protocerebrum.

CENTRAL EUROPEAN ENCEPHALITIS See Tick-Borne Encephalitis.

CENTRAL FOVEOLA See Median Foveola.

CENTRAL NERVOUS SYSTEM Insecta: The Brain and Ventral Nerve Cord. VNC consists of serially arranged, paired Ganglia. CNS groundplan suggests one pair of Ganglia per body segment. Sequential pairs of Ganglia are connected by longitudinal connectives; short, transverse commissures connect members of a pair of ganglia within a segment. Neurosecretory Cells occur in all ganglia of CNS. They are specialized neurons which release hormone into the haemocoel instead of a transmitter substance at a synaptic gap. See Nervous System; Nerve Cell. Cf. Brain; Peripheral Nervous System. Rel. Afferent Neuron; Efferent Neuron; Interneuron; Motor Neuron; Receptor.

CENTRAL REGION The Costal region of the insect wing. See Costa; Wing.

CENTRAL SHEATH See Axoneme.

CENTRAL SYMMETRY SYSTEM Lepidoptera: The median field of a moth wing-pattern bounded basally and distally by the light central line of the transverse anterior and transverse posterior lines respectively (Richards).

CENTRAL TUBULE See Axoneme.

CENTRE OF DISTRIBUTION Ecology: The central area from which any group or particular Species has spread.

CENTRE OF ORIGIN In biogeographical analysis, the endemic focus of related Species or groups of related organisms. CoO determined by (1) an area with the greatest number of Species in the group; (2) an area with the greatest morphological diversity in a group; (3) an area with the most 'primitive' Species in a group; (4) an area with the most 'derived' Species in a group; (5) ecologically optimal areas. See Dispersal Biogeography; Ecological Biogeography; Vicariance Biogeography.

CENTRIFUGAL NEURON See Deutocerebrum.

CENTRIOLE ADJUNCT An organelle of Spermatozoa that occurs in most Orders of insects. CA may be derived from the nucleus. CA is best developed in spermatids and later degenerates or may be lost. Structurally, CA forms a compact sleeve consisting of ribonucleoproteins, as determined by histochemical analysis and autoradiography. Usually, CA assumes form of a granular mass of material (160–320 Å); CA sometimes composed of fibres (*i.e. Gryllus* and *Sciara*). Term first used by Gatenby & Tahmisian (1959). Syn. Flagellar Accessory Structure; Juxtanuclear Body; Postnuclear Body (Gatenby & Wigoder 1929); Pseudoblepharoplast (Sotelo & Trujillo-Cenoz 1958. See Spermatozoon; Sperm Anatomy.

CENTRIONCIDAE See Diopsidae.

CENTRIPETAL Adj. (Latin, *centrum* = centre; *petere* = to seek.) Toward the centre.

CENTRIS Noun. (Latin, *centrum* = centre.) The Sting in aculeate Hymenoptera.

CENTROLECITHAL Adj. (Greek, *kentron* = centre ; *lekthos* = yolk.) Pertaining to eggs with abundant yolk near the centre. This type of yolk distribution is generally restricted to arthropods (except scorpions and a few coelenterates). Centrolecithal eggs undergo meroblastic cleavage. Cf. Alecithal; Teloloecithal.

CENTROPHOBISM Noun. (Greek, *kentron* = centre; *phobos* = fear; English, *-ism* = state. Pl., Centrophobisms.) Avoidance of the centre of a behavioural arena. (Goetz & Biesinger 1985, 156 (3): 319–328.)

CENTROSOME Noun. (Greek, *kentron* = centre; *soma* = body. Pl., Centrosomes.) A spherical body (organelle) in many animals and some plants that appears outside and near the Nucleus of a cell.

CEPALAK, RUDOLF (1886–1972) (Heyrovsky

1974, Zpravy Cesk. spol. ent. 10: 87–88.)

CEPHACORIA Noun. (Etymology obscure.) The indentation continuous with the Notarotaxis, between the Cephalaria and Medalaria (MacGillivray).

CEPHALAD Adverb. (Greek, *kephale* = head; Latin, *ad* = toward.) Toward the anterior part or head-end of the body. See Orientation. Cf. Anteriad; Apicad; Basad; Caudad; Centrad; Craniad; Dextrad; Dextrocaudad; Dextrocephalad; Distad; Dorsad; Ectad; Entad; Laterad; Mediad; Mesad; Neurad; Orad; Proximad; Rostrad; Sinistrad; Sinistrocaudad; Sinistrocephalad; Ventrad.

CEPHALIC Adj. (Greek, *kephale* = head; *-ic* = consisting of.) Belonging to or attached to the head; directed toward the head. See Orientation. Cf. Caudal.

CEPHALIC ARTERY One of the divisions of the Aorta entering the insect head.

CEPHALIC BRISTLE Diptera: Specialized large and relatively coarse Setae that occur on the head.

CEPHALIC FORAMEN The Occipital Foramen of the head through which the Dorsal Vessel, Oesophagus, Salivary Ducts and Ventral Nerve Cord pass from the head to the Prothorax. CE is centrally positioned on posterior aspect of head above mouthparts and below Vertex. Syn. Occipital Foramen.

CEPHALIC HEART Odonata: A pulsating organ that exerts pressure against the egg-shell during eclosion (hatching) and exserts a cap-like Operculum (Imms). See Pulsatile Organ.

CEPHALIC LOBES The head lobes of the embryo, comprising the region of the Prostomium and usually that of the Tritocerebral Somite (Snodgrass).

CEPHALIC POLE Insect egg: The end of an elongated egg that points to the head of the parent while the egg is situated in the ovariole (Imms).

CEPHALIC POSTANAL SETAE Coccids: See Cisanal Setae (MacGillivray).

CEPHALIC SALIVARY GLAND Bees: A pair of glands positioned against the posterior wall of the head (Imms). See Salivary Gland.

CEPHALIC STOMODAEUM Part of the Stomodaeum contained in the head. See Alimentary System; Stomodaeum.

CEPHALIC VESICLE A single sac formed by the union of the larval Pharynx and its diverticula (Imms). See Vesicle.

CEPHALIZATION Noun. (Greek, *kephale* = head; English, *-tion* = result of an action. Pl., Cephalizations.) The concentration of the sense-organs in the anterior part of the body toward the head, or in the head.

CEPHALOCAUDAL SUTURE Vespidae: The median suture dividing the Mesepisternum in some Species (Viereck).

CEPHALOIDAE False Longhorn Beetles. A small Family of polyphagous Coleoptera assigned to Superfamily Tenebrionoidea. Adult moderate sized (8–20 mm long), dark coloured; head diamond-shaped; tarsal formula 5-5-4; superficially resembling Cerambycidae. Adults taken at flowers; larvae inhabit rotten logs.

CEPHALOLATERAL MARGIN Coccids: The margin of the Operculum connecting the inner and outer angles (MacGillivray).

CEPHALOMERE Noun. (Greek, *kephale* = head; *meros* = part. Pl., Cephalomeres.) One of an arthropod's head segments. See Mere; Segmentation.

CEPHALON Noun. (Greek, *kephale* = head. Pl., Cephala.) 1. The head of an insect. 2. The anterior shield of a Triblobite. See Head.

CEPHALO-PHARYNGEAL SKELETON Muscid larva: A heavily chitinized structure withdrawn into the anterior segments; an invaginated portion of the mouthparts.

CEPHALOPHRAGMA Noun. (Greek, *kephale* = head; *soma* = body; *phragma* = fence. Pl., Cephalophragmata.) Orthoptera: A V-shaped partition that divides the head into an anterior and posterior chamber. See Phragma. Cf. Paraphragma.

CEPHALOSOME Noun. (Greek, *kephale* = head; *soma* = body. Pl., Cephalosomes.) The head cover in the pupal stage. See Pupa.

CEPHALOTHORAX Noun. (Greek, *kephale* = head; *thorax* = chest. Pl., Cephalothoraxes.) 1. The anterior body Tagma of Arachnida and Crustacea. The Cephalothorax is analogous, perhaps homologous, with the insect's head and Thorax. 2. The portion of an obtect pupa that covers head and Thorax. 3. Anterior body segments of larvae that have no obviously separated head. 4. Coccids: The fused head and Thorax. See Tagma.

CEPHIDAE Plural Noun. Stem Sawflies. A Family of Symphyta (Hymenoptera). Adult body narrow, 4–18 mm long; Antenna long, thread-like or weakly clubbed, with 16–30 segments; Hypostomal Bridge present; posterior margin of Pronotum nearly transverse; Cenchri absent; fore Tibia with one apical spur; middle Tibia usually with subapical spurs; Abdomen cylindrical or laterally compressed, weakly constricted at distal margin of first segment; Ovipositor exerted and male genitalia orthandrous. Larvae are stem borers. See Symphyta.

CEPHOIDEA Hymenoptera: A monofamilial Superfamily of Apocrita characterized by adults lacking Cenchri and fore tibial spur developed into a Calcar. Larvae are stem borers; some Species inflict economic levels of injury to grasses.

CERAGO Noun. (Latin, *cera* = wax.) Bee-bread. Rel. Apidae.

CERAMBYCIDAE Plural Noun. Longhorned Beetles; Longicorn Beetles. Roundheaded Wood Borers; Roundheaded Borers. A large, cosmopolitan Family of polyphagous Coleoptera assigned to Superfamily Chrysomeloidea. Adult elongate, subcylindrical with compound eye typically emarginate, sometimes completely divided;

Antenna very long, with 11 segments and can be directed rearward; tarsal formula 5-5-5 with fourth segment small and concealed. Adults are taken at flowers while females oviposit in crevices in bark. Eggs hatch and neonate larvae bore into wood to construct tunnels that are circular (round) in cross section. Larvae are elongate, cylindrical and whitish; head retracted into Thorax; legs highly reduced, vestigial or absent. All Species are phytophagous and many Species are signficant economic pests that attack forest and fruit trees; a few Species attack injured trees or fresh-cut logs. See Elm Borer; Locust Borer; Poplar Borer; Roundheaded Apple-Tree Borer. Rel. Buprestidae; Scolytidae.

CERAPHRONIDAE Haliday 1833. Plural Noun. A moderate size Family of parasitic Hymenoptera placed among the Proctotrupoidea by early workers, but generally regarded as an independent lineage of Parasitica or a member of the Stephanoidea. Diagnostic features include very small to small size, dark bodied without metallic coloration, compound eye setose, Antenna with 9–10 segments in female, with 10–11 segments in male, middle Tibia with one spur. Most Species are macropterous, some Species are apterous; Pterostigma is large and Waterston's organ is present.

CERAPHRONOIDEA A Superfamily of Parasitica including the Ceraphronidae and Megaspilidae; sometimes included within Stephanoidea. Ceraphronoids are characterized by a fore Tibia with two apical spurs. See Apocrita. Cf. Chalcidoidea; Cynipoidea; Ephialtitoidea; Evanioidea; Ichneumonoidea; Proctotrupoidea; Stephanoidea; Trigonaloidea.

CERARAN SETAE Coccids: The conical Setae of the Cerari (MacGillivray).

CERARI Noun. (Etymology obscure.) Coccids: See Tricerores (MacGillivray).

CERATHECA Noun. (Greek, *keras* = horn; *theca* = covering. Pl., Cerathecae.) Part of the pupal case that envelops the Antenna. Alt. Ceratotheca. Rel. Theca.

CERATOCANTHIDAE Plural Noun. A small, pantropical Family of polyphagous Coleoptera assigned to the Scarabaeoidea. Adult body 2.5–3.2 mm long, highly convex with head large and deflexed; compound eyes divided; Antenna with 9–10 segments including club with three segments; Scutellum large; tarsal formula 5-5-5; Abdomen with 5–6 Ventrites. Larva with head hypognathous and Stemmata absent; Antenna with four segments; legs with five segments, tarsal claws absent; Urogomphi absent. One Species taken in termite nest. Syn Acanthoceridae. See Coleoptera.

CERATOCOMBIDAE Fieber 1861. Plural Noun. A small, predominantly tropical Family of Heteroptera assigned to the Dipsocoroidea (Hemiptera). Antennal Pedicel twice as long as Scape; Labium with 4 segments, usually extending to hind Coxa. See Heteroptera.

CERATOPHYLLIDAE Plural Noun. Ceratophyllid Fleas. Adult with internatennal suture absent; Genal Comb absent; two rows of bristles on most abdominal Terga. The largest Family of fleas with 143 Species in North America in three Subfamilies, Ceratophyllinae, Leptopsyllinae and Amphipsyllinae. Most Species are ectoparasites of rodents and about 10% are ectoparasites of birds. Notable Species include Common Ground-Squirrel Flea; European Chicken Flea; Northern Rat Flea; Western Chicken Flea. See Siphonaptera.

CERATOPOGONIDAE Plural Noun. (Greek, *keras* = horn; *pogon* = beard.) Biting Midges (Scotland); No-See-Ums (America); Punkies; Sand Flies (Caribbean, Australia). A cosmopolitan, numerically large Family (ca 4,000 Species) of nematocerous Diptera assigned to Superfamily Chironomoidea. Adult flies 1–6 mm long; Ocelli absent; females with biting and sucking mouthparts; male Antenna plumose (sexually dimorphic); wings held scissor-like over Abdomen at repose; Costa ending near wing apex, Median Vein forked, with two branches (M1, M2); wings often pictured; Postnotum usually without longitudinal groove. Ceratopogonids occur in moist habitats, including heavily fertilized areas frequented by livestock. Females of many Species are predatory; Species of Ceratopoginae are haematophagous and entomophagous; blood sucking habits of adult females has a demonstrated role as disease vectors: Ceratoponids transmit Blue-Tongue to sheep, Onchocerciasis to cattle and horses, and Filariasis to humans. Adult ceratopogonids attack vertebrate and invertebrate Species to obtain high-protein meal. Both sexes visit flowers and take nectar; females require a protein-rich meal for egg maturation. Eggs are laid in clutches of varying size to several hundred; individual eggs typically are small, dark, slender, and covered with minute projections (Ansulae). Eggs are deposited in water or moist habitats which periodically become flooded. Larvae with 11 body segments, well developed head capsule but legless. Larvae occur in moist habitats; some larvae live in moist cactus stems. Forcipomyiinae with crawling terrestrial or semiaquatic forms that feed on algae, plant debris or fungi. Dasyheleinae are found in rock pools, tree holes and sap flows; they move by wriggling. Ceratopogoninae larval habits vary from borrowing in moist soil to fully aquatic and free-swimming; they are mostly carnivorous. Leptoconopinae larvae borrow in soil, chiefly in arid areas and on coastal and inland beaches. Syn: Helicidae. See Chironomoidea. Cf. Black Flies; Moth Flies; Sand Flies. Rel. Midges.

CERATUBA Noun. (Greek, *keras* = horn; Latin, *tubus* = tube. Pl., Ceratubae.) Some coccids: An invaginated cuticular tube that forms the terminal outlet of some of the wax glands. Ceratubae vary in size and shape (Comstock).

CERCAL HEART Circulatory Organs reported in stoneflies. See Accessory Circulatory Organ. Cf. Antennal Heart; Leg Heart; Wing Heart.

CERCARIA Noun. (Greek, *kerkos* = tail.) A Basipodite of the Cercus; Cercal Basipodite (MacGillivray).

CERCOBRANCHIATE Adj. (Greek, *kerkos* = tail; *brangchia* = gills.) Nymphal Odonata: With the respiratory apparatus consisting of three terminal lamellate caudal gills.

CERCOPHANIDAE Plural Noun. See Cercophaniidae.

CERCOPHANIIDAE Plural Noun. A small Family of ditrysian Lepidoptera assigned to Superfamily Bombycoidea.

CERCOPIDAE Plural Noun. Froghoppers; Spittlebugs. A numerically large, cosmopolitan Family of Cercopoidea (Hemiptera). Hind margin of Pronotum straight or medially curved; compound eyes circular in dorsal aspect. See Cercopoidea.

CERCOPODA Noun. (Greek, *kerkos* = tail; *pous* = foot. Pl., Cercopodae.) Jointed, foot-like appendages of the last abdominal segment; Cerci.

CERCOPOIDEA Noun. Froghoppers; Spittlebugs. A Superfamily of auchenorrhynchous Hemiptera. Adult head often narrowly produced anteriad; Tentorium complete; Ocelli paired when present; hind Coxa short, conical, laterally dilated. Included Families: Aphrophoridae, Cercopidae, Clastopteridae, Machaerotidae. See Auchenorrhyncha.

CERCUS Noun. (Greek, *kerkos* = tail. Pl., Cerci.) An appendage (generally paired) of the apparent tenth abdominal Tergum (or Terga 10 + 11). Conventionally regarded as a sensory appendage that is typically slender, filamentous and segmented. Unsegmented in Orthoptera, Phasmatodea and Dermaptera. Term is incorrectly applied to appendicular structures on the ninth segment of Coleoptera larvae, or eighth and ninth segments of Hymenoptera. See Appendage. Cf. Pygostylus; Urogomphus. Rel. Appendix Dorsalis.

CEREAL APHID *Rhopalosiphum padi* Linnaeus [Hemiptera: Aphididae].

CEREAL LEAF BEETLE *Oulema melanopus* (Linnaeus) [Coleoptera: Chrysomelidae]: A widespread, monovoltine pest of small grains, particularly barley and oats. CLB is endemic to Turkey, distributed throughout Mediterranean and Europe and was introduced into midwestern USA during 1962. Adult CLB overwinters in field and becomes active in April-May; female oviposits up to 50 eggs during lifetime; eggs are deposited individually on host plant. Eggs are cylindrical, ca 1 mm, yellow when laid but quickly become black. Larvae are predominantly yellow with dark head and legs; mature larva 5–6 mm long. CLB undergoes four larval instars; larval stage requires ca 14 days. Pupation occurs in cell within soil; pupal stage lasts 17–25 days. Adult ca 4 mm long, metallic blue-black with red Prothorax and legs. Larvae and adults chew long strips from leaves and plants appear white-tipped; heavily infested fields display yellow-white patches.

CEREAL WHITEFLY *Aleurocybotus indicus* David & Subrmaniam [Hemiptera: Aleyrodidae].

CEREBELLAR Adj. (Latin, *cerebrum* = brain.) Pertaining to the Cerebellum. See Brain.

CEREBELLUM Noun. (Latin, *cerebrum* = brain. Pl., Cerebella; Cerebellums.) The Suboesophageal Ganglion of insects. See Nervous System.

CEREBRAL GANGLION The Brain of insects, lies between the supporting Apodemes of the Tentorium, just above the Oesophagus, Supra-oesophageal Ganglion. See Brain; Ganglion.

CEREBRUM Noun. (Latin, *cerebrum* = brain. Pl., Cerebrums; Cerebra.) The Supra-oesophageal Ganglion. See Brain; Ganglion.

CEREOUS Adj. (Latin, *cera* = wax; *-osus* = possessing the qualities of.) Wax-like.

CERESA, LEOPOLDO (1901–1957) (Binaghi 1958, Memorie Soc. ent. Ital. 37: 20–22.)

CERNISVITOV, LEV (1902–1945) (Tchernavin 1947, Proc. Linn. Soc. Lond. 158: 129–131.)

CERNUOUS Adj. (Latin, *cernuus* = with face turned downward; *-osus* = possessing the qualities of.) Pertaining to structure with the apex bent downward. Alt. Cernuus.

CEROCOCCIDAE Plural Noun. A small, exclusively tropical Family of sternorrhynchous Hemiptera assigned to Superfamily Coccoidea which includes about 60 Species. See Coccoidea.

CERODECYTE Noun. (Greek, *keros* = wax; *kytos* = hollow. Pl., Cerodecytes.) Cells that aid in forming and conserving wax (Snodgrass). See Oenocyte.

CERONE® See Ethephon.

CEROPHYTIDAE Plural Noun. A small and rarely collected Family of Coleoptera assigned to Superfamily Elateroidea. Adults ca 7–9 mm long, dark coloured and somewhat flattened; hind Trochanter nearly as long as Femur. Inhabit rotten wood and under decaying bark. See Elateroidea.

CERORES Noun. (Sl., Ceroris.) Coccids: The wax-gland openings or pores of the Integument, when not invaginated, through which wax is secreted (MacGillivray).

CERTAMATE® See Proproxur.

CERTAN® A registered biopesticide derived from *Bacillus thuringiensis* var. *aizawai*. See *Bacillus thuringiensis*.

CERUMEN Noun. (Latin, *cera* = wax. Pl., Cerumens.) The nest material of the stingless bees *(Trigona)* and to a lesser extent of *Apis*. Cerumen consists of wax and resin, and is occasionally mixed with other substances. See Apoidea. Rel. Wax Gland.

CERUTTI, NESTOR (1886–1940) (Kuttor 1940, Mitt. schweiz. ent. Ges. 18: 208.)

CERVACORIA Noun. (Latin, *cervix* = neck; *corium*

= leather. Pl., Cervacoriae.) The cephalic membranous end of the Cervix (MacGillivray). See Cervix.

CERVANOTUM Noun. (Latin, *cervix* = neck; Greek, *noton* = back. Pl., Cervanota.) The dorsal part of the Cervix (MacGillivray). See Notum.

CERVAPLEURON Noun. (Latin, *cervix* = neck; Greek, *pleuron* = side. Pl., Cervapleura.) The lateral parts of the Cervix (MacGillivray). See Pleuron.

CERVASTERNANUM Noun. (Latin, *cervix* = neck; Greek, *sternon* = chest. Pl., Cervasternana.) The anterior sclerite of the Cervasternum (MacGillivray). See Sternum.

CERVASTERNELLUM Noun. (Latin, *cervix* = neck; Greek, *sternon* = chest. Pl., Cervasternella.) The posterior sclerite of the Cervasternum (MacGillivray). See Sternum.

CERVASTERNUM Noun. (Latin, *cervix* = neck; Greek, *sternon* = chest. Pl., Cervasterna.) The ventral part of the Cervix (MacGillivray). See Cervix; Sternum.

CERVEPIMERON Noun. (Latin, *cervix* = neck; Greek, *epi* = upon; *meros* = part. Pl., Cervepimera.) The posterior sclerite (Cervasternum) of the Epimeron, when two sclerites are present (MacGillivray). See Epimeron. Cf. Cervepisternum.

CERVEPISTERNUM Noun. (Latin, *cervix* = neck; Greek, *epi* = upon; *sternum* = breast bone. Pl., Cervepisterna.) The anterior sclerite (Cervapleuron), when two sclerites are present (MacGillivray). See Episternum. Cf. Cervepimeron.

CERVICAL Adj. (Latin, *cervix* = neck.) Pertaining to the neck or Cervix.

CERVICAL AMPULLA Some Orthoptera: A cervical membrane of the 'Pronymph' stage. CA is everted by hydrostatic pressure and acts with a ridge or row of teeth on the head to facilitate eclosion. CA also acts in Pronymph emergence from soil or plant tissue.

CERVICAL FORAMEN Coleoptera larvae: The Occipital Foramen.

CERVICAL GLAND Lepidoptera caterpillar: A ventral gland on the Prothorax.

CERVICAL SCLERITES Small chitinous sclerites on the membrane between head and Thorax. See Jugular Sclerites.

CERVICAL SHIELD Lepidoptera larva; The chitinous sclerite on the Prothorax just posterior of the head. Syn: Prothoracic Shield.

CERVICAL TRIANGLE See Epicranial Notch.

CERVICALIA Noun. (Latin, *cervix* = neck.) The cervical Sclerites of the insect (Crampton).

CERVICULATE Adj. (Latin, *cervix* = neck; *-atus* = adjectival suffix.) With a long neck or neck-like portion. Alt. Cerviculatus.

CERVICUM Noun. (Latin, *cervix* = neck.) The membrane between head and Thorax. See Cervix 2.

CERVINUS Adj. (Latin, *cervus* = deer.) Reddish; Deer-gray.

CERVIX Noun. (Latin, *cervix* = neck. Pl., Cervices; Cervixes.) 1. Diptera: Part of Occiput positioned over the junction with the head, (*i.e.* between Vertex and neck). 2. General: A membranous region between head and Thorax. In a more restricted sense, upper part of the neck. Cervix connects Occipital Foramen with Pronotum. Typically, the area is membranous and often concealed. Cervical region has not been studied intensively and disagreement exists over what this region represents. Some morphologists infer an imaginary primary intersegmental line between head and Thorax; some morphologists have called the Cervix a 'Microthorax' and regarded this region as a body segment. Elements of Cervix have been ascribed to the labial segment of the head; other elements of the Cervix have been ascribed to the Thorax. Groundplan of Cervix contains two cervical sclerites in the lateral region on each side of the head; Diptera display up to three pairs of cervical sclerites. Anterior sclerite articulates with Occipital Condyle of head and posterior sclerite articulates with Prothoracic Episternum. Musculature attached to these sclerites increases or decreases the angle between sclerites and thereby creates limited mobility of head. Mantid head has exceptional mobility and can turn because of the musculature associated with the cervical sclerites. Some insects have apparently lost one of the sclerites or posterior sclerite has become fused with Prothoracic Episternum. See Occiput; Postocciput.

CERYLONIDAE Plural Noun. A small Family of polyphagous Coleoptera assigned to Superfamily Cucujoidea. Adult < 3 mm long and somewhat flattened; Antenna with 10 segments including Club of 1–3 segments; Coxae widely separated. Cerylonids are taken beneath bark in decaying wood or leaf litter where they presumably feed upon fungi. See Cucujoidea.

CESAR® See Hexythiazox.

CESPITICOLOUS Adj. (Latin, *caespes* = turf; *colous* = living in or on.) Descriptive of organisms inhabiting grassy places. Rel. Habitat.

CESPITOSE See Caespitose.

CESTIFORM Adj. (Greek, *kestos* = girdle; Latin *forma* = shape.) Girdle-shaped. Rel. Form 2; Shape 2; Outline Shape.

CESTONI, GRACINTO (1637–1718) (Baglioni 1942, Riv. Parassit. 6: 1–12.)

CGA-184699® See Lufenuron.

CGA-215944® See Pymetrozine.

CHABRIER, H DE MONTPELLIER (Swainson 1840, *Taxidermy: With biogr. of zoologists*. 329 pp. (151), London. [Volume 12, *Cabinet Cyclopedia*. Edited by D. Lardner.].)

CHABRIER, J Author of *Essai sur le Vol des Insectes* (1823) which describes in detail the thoracic sclerites, muscles and the sclerites associated with the base of the wing. Chabrier also attempted to explain the phenomenon of flight.

CHADWICK, LEIGH EDWARD (1904–1975) (Storch & Dethier 1975, J. Econ. Ent. 68: 565.)

CHAETA Noun. (Greek, *chaite* = long hair. Pl., Chaetae.) Insects: An ectodermal evagination from the Integument. Form, number and arrangement of Chaetae are sometimes taxonomically useful. Chaetae are functionally diverse and serve as sensory receptors, components of stridulatory devices and diffraction gratings in some insects. Cf. Seta. Rel. Chaetotaxy.

CHAETEESSIDAE Plural Noun. A very small Family of Mantodea with a few Species in Neotropical Realm. Adult body small and macropterous in both sexes; Prothorax subquadrate; raptorial fore legs only with large Setae but spines not present; fore and hindwings similar; Cerci long and 4 segmented. See Mantodea.

CHAETIC Adj. (Greek, *chaite* = long hair; *-ic* = characterized by.) Pertaining to tactile Setae.

CHAETICUM SENSILLUM 1. A tactile sense organ of which the external part is spine or bristle-like (Snodgrass). 2. A modified Trichode that resembles the Trichode Sensillum (appears spine- or hair-like), but has an elevated collar around the base. CS occur on individuals in several groups of insects including the Abdomen of *Notonecta*, *Nepta* and mud-inhabiting tipulid Diptera. CS occur on the Tarsi of the middle and hind legs of *Chironomus*. Two morphological forms of CS are known. Short, Ventral Sensilla Chaeticae (SVSC) have been known since the beginning of 20th century. Under the light microscope these Sensilla appear curved or forked at the apex, but with the scanning electron microscope they are palmate. The SVSC is affected by parasitism (merminthids), and probably serves in contact chemoreception. Long Sensilla Chaetica are less numerous and irregular in distribution. Anatomically they are longer and more blunt than the short ventrals. LSC show a complex morphological form: Thick walled (presumably prevent penetration of chemicals). The Chaeticum Sensillum usually is used for mechanoreception and occasionally for contact chemoreception and auditory reception. See Sensillum. Cf. Trichode Sensillum.

CHAETIFEROUS Adj. (Greek *chaite* = long hair; Latin, *ferre* = to bear; Latin, *-osus* = possessing the qualities of.) Bearing tactile Setae. Syn: Chaetigerous; Setiferous; Setigerous.

CHAETOPARIA Noun. (Greek *chaite* = long hair; Latin, *paries* = wall. Pl., Chaetopariae.) Scarabaeoid larvae: The inner part of the Paria covered with Setae; strongest toward the Pedium and gradually decreasing in size toward the Gymnoparia, or toward the Acanthoparia when the Gymnoparia is absent (Boving). See Paria.

CHAETOPHOROUS Adj. (Greek, *chaite* = long hair; *pherein* = to bear; Latin, *-osus* = possessing the qualities of.) Bristle-bearing. A term applied to Diptera.

CHAETOSEMA Noun. (Greek, *chaite* = long hair; *sema* = sign. Pl., Chaetosemata.) Lepidoptera: Elevated, cuticular patches on the adult head. Chaetosemata bear bristle-like Setae or narrow scales and are typically positioned near the compound eye and behind the Antenna. Present in all butterflies and many Families of moths. Function not understood but connected with the Brain via nerves. Syn. Jordan's Organ; Eltringham's Organ.

CHAETOSOMATIDAE Plural Noun.

CHAETOTAXY Noun. (Greek, *cahaite* = long hair, *taxis* = arrange. Pl., Chaetotaxies.) The arrangement, nomenclature or classification of Setae distributed over the body of insects and other arthropods.

CHAFERS See Scarabaeidae.

CHAFF SCALE *Parlatoria pergandii* (Comstock) [Hemiptera: Diaspididae]: A cosmopolitan pest of *Citrus* and ornamental plants. Adult female's cover greyish, an irregular oval ca 1.5 mm diameter. Female oviposits ca 50 purple eggs; crawlers settle on trunk and mature limbs, two nymphal instars. CS is multivoltine with up to six generations per year, depending upon climate. See Diaspididae.

CHAGAS, CARLOS JUSTINIANO RIBIEROS DO (1879–1934) Pioneer in Medical Entomology best known for his work on Chagas' Disease. Chagas located *Trypanosoma cruzi* in humans and demonstrated the pathogen *Panstrongylus megistus* served as a vector for *T. cruzi*. (Sámano 1935, An. Inst. Biol. Univ. Méx. 6: 2.)

CHAGAS' DISEASE *Trypanosoma (Schisotrypanum) cruzi*. A Neotropical disease of vertebrates that is common and widespread in rural areas and most prevalent in Argentina, Brazil and Venezuela. CD is a common disease of humans with estimates of several million clinical cases per year. Trypanosome also attacks more than 100 Species of mammals including wood rats, dogs, cats and opossums. Disease is transmitted among mammals via excrement of *Triatoma* spp. (Reduviidae). Bug ingests Trypomastigotes circulating in blood of infected mammal; within midgut of bug, Trypomastigotes differentiate into Epimastigotes which later accumulate on the rectal wall. Subsequently, Epimastigotes differentiate into Metacyclic Trypomastigotes that are excreted with faeces when bug feeds on mammal. Chagas' Disease is transmitted when an infective agent (Metacyclic Trypomastigotes) is scratched into the feeding wound and penetrates mucous membranes or skin abrasion. Life cycle of trypanosome in bug requires about three weeks to complete; a bug is infective for life. See Trypanosomiasis.

CHAGAS, EVANDRO SERAFIM LOBO (1905–1940) (Villela 1941, Mems. Inst. Oswaldo Cruz 36: xxxiii–xliii, bibliogr.)

CHAGNON, GUSTAVE (1871–1966) (Fournier *et al.* 1969, Ann. ent. Soc. Quebec 14: 42–46.)

CHAGRINED See Shagreened.

CHAIN TRANSPORT Social Insects: The phenomenon by which food is relayed back to the colony by several workers. See Feeding Strategies.

CHAINSPOTTED GEOMETER *Cingilia catenaria* (Drury) [Lepidoptera: Geometridae].

CHALASTOGASTRA Hymenoptera: More primitive Taxa of the Order and including the sawflies. Syn. Symphyta. See Hymenoptera. Cf. Clistogastra.

CHALASTOGASTROUS Adj. (Greek, *kalastos* = loose; *gaster* = stomach; Latin, *-osus* = possessing the qualities of.) Pertaining to Hymenoptera with a sessile Gaster and without a marked constriction at the Propodeum. Cf. Clistogastrous. Rel. Petiole.

CHALAZA Noun. (Greek, *chalaza* = hail. Pl., Chalazae.) Immatures: A small, mound-like, cuticular elevation (protuberance) that bears a plumose Seta or 1–3 simple Setae. A conical Pinaculum on Lepidoptera larva. See Integument Cf. Pinaculum; Scolus; Verruca.

CHALCEOUS Adj. (Greek, *chalkos* = brass; Latin, *-aceus* = of or pertaining to.) Brassy in colour or appearance. Alt. Chalceus.

CHALCIDIDAE Plural Noun. Chalcidid Wasps; Chalcids. A moderate sized (ca 40 Genera, 1,500 Species), cosmopolitan Family of apocritous Hymenoptera assigned to Superfamily Chalcidoidea. Adult body relatively large (to 10 mm), robust, often sculptured and non metallic; Antenna with 13 segments; Parapsidal Sutures (Notauli) present and complete; macropterous with forewing Marginal Vein short to very long, Stigma short and Postmarginal Vein absent to moderately long; hind Femur enlarged and denticulate on ventral surface; Gaster subsessile to petiolate. All Species parasitic, primary and hyperparasites of Holometabola, particularly Lepidoptera and to lesser extent Diptera, Coleoptera and other Hymenoptera. Hyperparasitic on Tachinidae and parasitic Hymenoptera. Development typically as solitary endoparasites of last instar larva and pupa. Some chalcidids are used in Biological Control (*Brachymeria*).

CHALCIDOIDEA Spinola 1811. A numerically large (ca 20 Families, 17,000 nominal Species), cosmopolitan Superfamily of parasitic Hymenoptera characterized by small body size, reduced wing venation, a Prepectus and 13 or fewer antennal segments. Fossil record extending to Cretaceous Period. Group frequently used in applied biological control of agricultural pests. Included Families: Agaonidae, Aphelinidae, Chalcididae, Elasmidae, Encyrtidae, Eucharitidae, Eulophidae, Eupelmidae, Eurytomidae, Leucospidae, Mymaridae, Mymarommatidae, Ormyridae, Perilampidae, Pteromalidae, Signiphoridae, Tanaostigmatidae, Tetracampidae, Torymidae, Trichogrammatidae. See Hymenoptera. Cf. Ceraphronoidea, Cynipoidea, Ephialtitoidea, Evanioidea, Ichneumonoidea, Proctotrupoidea, Stephanoidea, Trogonalidodea.

CHALCIDOIDS See Chalcidoidea.

CHALCIDS See Chalcididae.

CHALK BROOD A fungal disease of honey bee larvae; disease widespread in temperate regions and is more common among drones than workers. Infected larvae are swollen, fluffy and assume the hexagonal shape of their cells; death occurs ca 2 days after cells have been capped; dead larvae may be white, blue-grey or black. Disease caused by *Ascophaera apis;* spores occur in brown-green spore cysts on the larval Integument; spores ingested by larva with food; spores viable for many years and can infect other Species of bees. Cf. Calcino; Stone Brood. Rel. Pathogen.

CHALYBEATE Adj. (Greek, *chalybes* = iron or steel; *-atus* = adjectival suffix.) Metallic steel-blue in colour. Alt. Chalybeatus; Chalybeous; Chalybeus.

CHAMAEMYIIDAE Plural Noun. A small cosmopolitan Family of cyclorrhaphous Diptera. Placed with Agromyzidae and subsequently removed and called Ochthiphilidae. Family name changed. Adult body chunky with densely grey pollinose; Costa entire, Anal Vein complete with Anal Cell usually apically pointed; Postvertical Bristles convergent or absent; Interfrontal Bristles absent; fore Femora Bristles present but Tibiae lack Preapical Bristles. All Chamaemyiidae are predaceous upon aphids, scales and mealybugs; attacks on scale insects are limited to nondiaspine forms. Eggs are laid singly among prey egg masses or in colonies of prey. Female fly oviposits through or under waxy scale cover of host scale among the eggs of host. Larvae are active feeders and usually consume prey quickly. Pupation among prey remains or open upon surface of substrate. Cocoon formed of coarse threads and incomplete. Puparium coarctate and complete. See Diptera.

CHAMBER Noun. (Old French, *chambre* from Latin, *camera* = chamber; Pl., Chambers.) A segmental dilatation of the insect heart.

CHAMBERLAIN, RALPH (1909–1966) (Pearson 1967, Proc. R. ent. Soc. Lond. (C) 31: 62.)

CHAMBERLIN, JOSEPH CONRAD (1898–1962) (Varshney 1964, Ent. News 75: 55.)

CHAMBERLIN, THOMAS ROSCOE (1889–1958) (Chamberlin 1959, J. Econ. Ent. 52: 181–182.)

CHAMBERS, VICTOR TOUSEY (1831–1883) (Osborn 1946, *Fragments of Entomological History*. Pt. II. 232 pp. (64–65), Columbus, Ohio.)

CHAMPAGNE-CORK ORGANS Ampullaceous Sensilla; so-called from their shape (*sensu* Forel).

CHAMPION, GEORGE CHARLES (1851–1927) (Osborn 1937, *Fragments of Entomological History*. 394 pp. (146), Columbus, Ohio.)

CHAMPION, REGINALD JAMES (1895–1917) (Anon. 1918. Ent. News 29: 80.)

CHAMPLAN, ALFRED B (1882–1957) (Wheeler & Valley 1975, History of Entomology in the Pennsylvania Department of Agriculture, 37 pp. (11–12), Harrisburg.)

CHAMPOLLION, JEAN FRANÇOIS (1790–1831) French Egyptologist. (Rose 1850, *New General Biographical Dictionary* 6: 198–199.)

CHANDLER, HARRY PHYLANDER (1917–1955) Leech 1957, Pan-Pacif. Ent. 33: 31–33, bibliogr.)

CHANELLED Adj. (Middle English, *chanel* > Latin, *canalis* = canal.) 1. Cuticle characterized by displaying conspicuous grooves, Sulci or invaginations. 2. A gutter, groove or furrow. 3. The deeper part of a river, harbour or strait and in which navigation is easier or current flows.

CHANEY, WILLIAM (1828–1906) (Walker 1907, Entomol. mon. Mag. 43: 16–17.)

CHANGA MOLE CRICKET *Scapteriscus didactylus* (Latreille) [Orthoptera: Gryllotalpidae]: A Species of mole cricket endemic to northern South America and the Caribbean. CMC is regarded as a major agricultural pest in Puerto Rico. See Gryllotalpidae. Cf. Imitator Mole Cricket; Tawny Mole Cricket; Short-Winged Mole Cricket.

CHANGAS See Gryllotalpidae.

CHANT, JOHN (–1867) (Newman 1868, Entomologist 4: 106–107.)

CHANTELOT, HENRI (–1963) (Anon. 1963, Bull. Soc. ent. Fr. 68: 49.)

CHAOBORIDAE Plural Noun. Phantom Midge. A small, cosmopolitan Family of nematocerous Diptera assigned to Superfamily Culicoidea and sometimes placed in Culicidae. Adult small, delicate and Ocelli absent; Proboscis short (not forming a biting Proboscis); body displays Setae but lacks scales; Discal Cell absent; wing veins Rs and M each with three branches; scales on wing form marginal fringe. Adults of a few Species feed upon blood as Culicidae. Larvae are aquatic as Culicidae; most Species are aquatic predators; a few chaoborid larvae are filter feeders. Syn. Corethridae. See Culicoidea. Cf. Clear Lake Gnat; Culicidae; Dixidae.

CHAPELIER, ABBÉ (–1861) (A. A. 1863, Mem. Soc. Sci. phys. nat. dept. d'Uect Vilaine 1: 47.)

CHAPERON Noun. (Old French, *chape* from Latin, *cappa* = a covering. Pl., Chaperons.) The Clypeus or Clypeus Anterior.

CHAPIN, EDWARD American taxonomic specialist in Coleoptera at the U. S. National Museum.

CHAPIN, EDWARD ALBERT (1849–1889) (Essig 1931, *History of Entomology* 1029 pp. (567), New York.)

CHAPIN, S F (1839–1889) American physician and amateur Entomologist who was active in pest control in California until his death in an accidental drowning. (Essig 1931, *History of Entomology*. 1029 pp. (567–568), New York.)

CHAPLET Noun. (Latin, *cappa* = garland. Pl., Chaplets.) 1. A small crown or circle of hooks or other small process terminating a segment or appendage. 2. A garland or wreath for the head. 3. A string of beads worn as a necklace.

CHAPMAN, JOHN ARTHUR (1919–1974) (Anon. 1974, Bull. ent. Soc. Can. 6: 138–139.)

CHAPMAN, ROYAL NORTON (1889–1939) (Osborn 1946, *Fragments of Entomological History*. Pt. II. 232 pp. (65–66), Columbus, Ohio.)

CHAPMAN, THOMAS (1816–1879) (Anon. 1880, Scott. Nat. 5: 236.)

CHAPMAN, THOMAS ALGERNON (1842–1921) (Musgrave 1932, *A Bibliography of Australian Entomology 1775–1930*. 380 pp. (44), Sydney.)

CHAPPELL, JOSEPH (1830–1896) (Bailey 1896, Entomol. mon. Mag. 32: 262.)

CHAPUIS, FALICIEN (1824–1879) (Ratzeburg 1847, Forstwissenschaftliches Schriftsteller-Lexicon 1: 110–111, bibliogr.)

CHARACTER Noun. (Greek, *charassein* = to engrave. Pl., Characters.) 1. A quality of physical features (colour, structure or shape), biological characteristics or behavioural attributes of an organism. 2. A general term used in taxonomic Entomology to describe features of insects or insect parts.

CHARACTER POLARIZATION Taxonomy: The process of determining which of two homologous characters is plesiomorphic or apomorphic. Alt. Character Arugmentation. See Character State. Rel. Homology.

CHARACTER STATE Taxonomy: A logical subdivision of characters into several alternative conditions. *e.g.* Eye colour is a character; red, brown, yellow are character states of eye colour. Character states may be used to distinguish among Taxa (Species, Genera, Families) and therefore useful in taxonomy. Character States may be regarded as primitive (ancestral) or derived. See Character. Cf. Attribute.

CHARACTER TRANSFORMATION See Transformation Series.

CHARCOAL BEETLE *Melanophila consputa* LeConte [Coleoptera: Buprestidae].

CHARDINY, LOUIS CURTIUS (1793–1837) (Douzel 1837, Bull. Soc. ent. Fr. 6: xxvi–xxx.)

CHARGE® See Lambda Cyhalothrin.

CHARIPIDAE Dalla Torre & Kieffer 1910. Plural Noun. A small Family of apocritious Hymenoptera assigned to the Cynipoidea, or placed within Cynipidae in some classifications. Adult small bodied (< 2.5 mm long); female Antenna with 13 segments and male Antenna with 14 segments; Pronotum carinate; Vertex, most of Thorax and Gaster polished; Gaster with a ring of pubescence at base of second Tergum; female Gaster laterally compressed, Tergum II or II + III largest. Hyperparasites of braconids and aphelinids attacking aphids. See Cynipoidea.

CHARLETON, WALTER (1619–1707) (Rose 1850, *New General Biographical Dictionary* 6: xxviii–xxlx.)

CHARLEVOIX, PIERRE FRANCOIS XAVIER (1682–1761) (Rose 1850, *New General Biographical Dictionary* 6: 247–248.)

CHARLON, AUGUSTIN (1793–1842) (Douzel 1842, Bull. Soc. ent. Fr. 11: xxviii–xxix.)

CHARPENTIER, M Author of *Horae Entomologicae adjectis tabulis novem colouratis*. (1825).

CHARTRES, S A (–1948) (Wigglesworth 1950, Proc. R. ent. Soc. Lond. (C) 14: 64.)

CHASTER, GEORGE WILLIAM (–1910) (Anon. 1910, Entomol. mon. Mag. 46: 145–146.)

CHATANAY, JEAN (1884–1914) (Berland 1920, Ann. Soc. ent. Fr. 89: 423–425.)

CHATFIELD, ALFRED, F (1816–1900) (Anon. 1900, Ent. News 11: 451.)

CHATHAMIIDAE Plural Noun. A small Family of Trichoptera assigned to Superfamily Limnephiloidea, and known only from Australia and New Zealand. Adult crepuscular, moderate-sized and slender with Ocelli absent; Maxillary Palpus with five segments; middle Tibia longer than Femur; tibial spur formula 2-2-4; female with long Ovipositor. Larvae form portable cases and inhabit marine rock pools where they feed upon algae. See Limnephiloidea.

CHATTERJEE, N C (1899–1950) (M. L. R. R. S. T. 1954, Indian J. Ent. 16: 307.)

CHAUDOIR, BARON MAXIMILIEN STANISSLAVOVITCH DE (1816–1881) Ukranian born Coleopterist specializing in Carabidae. Chaudoir published 108 papers (5,230 pages) in which he described 318 Genera and 2,911 Species. His insect collection is housed in the Muséum National d'Histoire Naturelle, Paris. (Bordas 1921, Insecta 11: 24; Basilewsky 1982, Coleopts Bull. 36 (3): 462–474.)

CHAUL (Greek, *chauliodous*.) With projecting teeth.

CHAULIOPINAE A Subfamily of Lygaeidae.

CHAUSSIER, FRANCOIS (1746–1828) (Rose 1850, *New General Biographical Dictionary* 6: 264.)

CHAVANNES, JACQUES AUGUST (1810–1879) (Blanc 1909, Verh. schweiz. naturf. Ges. 92: 33–39.)

CHAVIGNERIE, J B E DE LA See Bellier de la Chavignerie.

CHAWNER, ETHEL FRANCES (1866–1953) (Buxton 1954, Proc. R. ent. Soc. Lond. (C) 18: 79.)

CHECKERED BEETLES See Cleridae.

CHECKERED WHITE See Southern Cabbageworm.

CHECKMATE CM® See Codelure.

CHECKMATE OFM® See Isomate-M.

CHECKMATE TPW® See Lycolure.

CHECKMATE-PBW® See Gossyplure.

CHEEK Noun. (Anglo Saxon, *ceace* = cheek. Pl., Cheeks.) 1. Hymenoptera: The space between the Mandible socket and ventral margin of the compound eye. 2. Diptera: The Bucca but not extending to the caudal margin of head. See Parafacials. Syn. Gena.

CHEEK GROOVES Diptera: More-or-less distinct impressions below the eyes (Comstock).

CHEESE MITE *Tyrolichus casei* Oudemans [Acari: Acaridae]. Rel. Cheese Skipper.

CHEESE SKIPPER *Piophila casei* (Linnaeus) [Diptera: Piophilidae]: A widespread urban pest of food-production and food-handling facilities; previously a serious pest in cheese, dried meats and dried fish but now less important in refrigerated stored products. CS is endemic to Europe and adventive elsewhere through commerce. Adult CS can be annoying or mechanically transmit disease. CS larvae can be ingested and cause intestinal disturbance, disease or Enteric Pseudomyiasis. Adult 2.5–4.0 mm long with yellow face and black body with metallic bronze colour on Thorax; wings at repose folded flat over Abdomen. Adult feeds on liquid associated with larval food and live ca 3–4 days. Female oviposits on cheese and cured meat; eggs are laid individually or in clutches up to 50; female produces ca 150 eggs during lifetime. Eclosion occurs within 24–36 hours. Larvae (maggots) are legless, yellowish-white in colour and body tapered toward head. Larvae bore into food and often moves with characteristic 'skipping' or jump (hence the common name) by bending body and rapidly straightening. Pupation occurs away from larval food in drier habitat. Life cycle requires 2–6 weeks. See Piophilidae. Rel. Cheese Mite.

CHEESMAN, LUCY EVELYN (1881–1969) (Smith 1969, Entomol. mon. Mag. 105: 217–219. bibliogr.)

CHEETHAM, CHRISTOPHER, A (1875–1954) (Buxton 1955, Proc. R. ent. Soc. Lond. (C) 19: 69.)

CHEIRONYM Noun. (Greek, *cheir* = hand; *onyma* = name. Pl., Cheironyms.) A manuscript name. Rel. Taxonomy; Nomenclature.

CHELA Noun. (Greek, *chele* = claw. Pl., Chelae.) 1. The terminal portion of a limb bearing a lateral, moveable claw like that of a crab. 2. A term specifically applied to the feet in some Parasitica in which the opposable claw forms a clasping structure.

CHELATE Adj. (Greek, *chele* = claw; *-atus* = adjectival suffix.) Pincer-like; bearing a Chela or claw. Term applied when claws can be drawn down or back upon the last tarsal segment. Cf. Achelate; Forcipate.

CHELATION Noun. (Greek, *chele* = claw; English, *-tion* = result of an action. Pl., Chelations.) The combination of organic compounds and metal atoms (*e.g.* Chlorophyll, Cytochrome, Haemoglobin).

CHELICERA Noun. (Greek, *chele* = claw; *keras* = horn. Pl., Chelicerae.) 1. The pincer-like first pair of appendages of adult Chelicerata. Structures regarded as homologous with the second pair of Antennae in Crustacea. 2. Heteroptera: Pinching or grasping claws of Phymatidae. 3. Acarina: Paired preoral appendages, consisting of a basal segment (cheliceral base), blade-like moveable digit (cheliceral blade) and a small fixed digit (Pseudochela). (Goff *et al.* 1982, J. Med. Ent. 19: 222.)

CHELICERAL APODEME Acarina: The sclerotized, internal projection attached to the ventral, posterior angle of the basal segment of the Chelicera (Goff *et al.* 1982, J. Med. Ent. 19: 222.)

CHELICERAL BASE Acarina: The first, immovabale segment of the Chelicera (Goff *et al.* 1982, J. Med. Ent. 19: 222.)

CHELICERAL BLADE Acarina: The optically active, movable digit of the Chelicera. CB is the primary structure for piercing the skin of the host. CB is armed with spines and a cuspid cap. Syn. Chelostyle. (Goff *et al.* 1982, J. Med. Ent. 19: 234.)

CHELIFEROUS Adj. (Greek, *chele* = claw; Latin, *ferre* = to carry; *-osus* = possessing the qualities of.) Bearing or terminating in a very thick forceps or Chela (Kirby & Spence). Alt. Cheliferus.

CHELIFORM Adj. (Greek, *chele* = claw; Latin, *forma* = shape.) Claw-shaped; pertaining to structure shaped as a pincer, claw or chela; descriptive of a digit that articulates with an adjacent digit. See Chela. Cf. Forcipate; Forcipiform Rel. Form 2; Shape 2; Outline Shape.

CHELISOCHIDAE Plural Noun. Black Earwigs. A numerically small, predominantly tropical Family of Dermaptera. See Dermaptera.

CHELOBASE See Cheliceral Base.

CHELONARIIDAE Plural Noun. A small Family of Coleoptera assigned to Superfamily Dryopoidea. Adult 4–5 mm long, dark and oval in outline shape; Scape, Pedicel and basal segments of Funicle held at repose in prosternal groove, remaining segments held along Mesosternum. Larvae are aquatic. See Dryopoidea.

CHELONIFORM Adj. (Greek, *chelys* = tortoise; Latin, *forma* = shape.) Turtle-shaped; characteristically with head concealed and body dorsoventrally compressed (flattened). Rel. Form 2; Shape 2; Outline Shape.

CHEMATHION® Malathion.

CHEMATHOATE® See Dimethoate.

CHEM-FISH® See Rotenone.

CHEMICAL CONTROL A principal approach or pest management strategy that involves use of pesticides or toxic chemicals to include 'Hard Pesticides' and 'Soft Pesticides.' Cf. Biological Control; Chemical Control; Cultural Control; Integrated Pest Management; Natural Control; Regulatory Control.

CHEMICAL DEFENCE SYNDROME. A combination of morphological attributes and/or behavioural characteristics exhibited by chemically protected insects. Features include brachyptery, aposematic coloration, aggregations, sluggishness and related behaviours. (Pasteels *et al.* 1983, Ann. Rev. Ent. 28: 263–289.)

CHEMICAL NAME One of three names given to a pesticide. Naming is based on the chemical structure and follows the rules of the International Union of Pure and Applied Chemistry (IUPAC.) See Approved Common Name. Cf. Trade Name.

CHEMICAL SOIL TREATMENT See Soil-Barrier Treatment.

CHEMIGATION Use of irrigation systems for dispensing pesticides and fertilizers.

CHEMNITZ, JOHANN HIERONYMUS (1730–1800) (Rose 1850, *New General Biographical Dictionary* 6: 268.)

CHEMOPERCEPTION Noun. (English, *chemic* = transmutation; Latin, *recipere* = to receive; English, *-tion* = result of an action. Pl., Chemoperceptions.) Perception through chemical stimuli, *i.e.*, taste, smell.

CHEMORECEPTIVE Adj. (English, *chemic* = transmutation; Greek, *chemeia* = alchemy; Latin, *recipere* = to receive.) Susceptible to chemical stimuli.

CHEMORECEPTOR Noun. (English, *chemic* = transmutation; Greek, *chemeia* = alchemy; Latin, *recipere* = to receive; *-or* = one who does something. Pl., Chemreceptors.) A sense organ (Sensillum) composed of a group of cells sensitive to chemical properties of matter. Chemoreceptors are specialized to respond to 'taste' and 'odour'. Chemoreceptive Sensilla are typically attended by several neurons. These neurons can also serve as mechanoreceptors. Chemoreceptors are important for the location of food, hosts, testing food, initiating escape from noxious compounds and locating oviposition sites. Chemoreceptors are thin walled (used for general olfaction) and thick walled (used in contact chemoreception). Anatomically, the chemoreceptor surface is covered with pores while the mechanoreceptor is not. The Dendrite narrows after leaving the cell body and has a cilium-like appearance and is very short. Above this Cilium the Dendrite thickens again [Dendrite Sheath (Steinbrecht) = Cuticular Sheath (Slifer) = Scolopale (Chapman)]. Pore tubules extend into fluid of the Sensillum's lumen but do not contact the receptor cell membrane. Chemoreceptors operate in olfaction and contact chemoreception. The so-called 'common chemical sense' perceives high concentrations of irritating substances, such as ammonia, but no specific kind of receptor has been identified. Olfactory chemoreceptors perceive low concentrations of a diffuse gas. This is commonly regarded as the sense of smell. Receptors have several sense cells with individual cells in the complex responding to different spectra of odours. Odours can stimulate or inhibit the action of cells. Contact Chemoreceptors perceive chemicals in solution at relatively high concentrations. This is commonly regarded as the sense of taste. Basiconic Sensilla may be contact chemoreceptors because they are thick walled and contain no pore tubules. Instead, filaments extend into an opening at the tip. Nerve fibres react to particular classes of chemical compounds. Contact chemoreceptors are important to feeding and oviposition. They are found on the Labellum of *Tabanus,* Tarsi of *Musca,* and Maxillary and Labial Palpi of *Carabus.* Proboscis eversion in flies is prompted by tarsal contact of sugar solution. See Sensillum. Cf. Mechanoreceptor.

CHEMOSTAT Noun. (Greek, *chemeia* = alchemy;

-*stat* = comb. form meaning stationary. Pl., Chemostats.) Tissue Culture: Continuous culture in which the cells flow out of a fermenter with the effluent. The growth rate of the cells is determined by the rate of supply of a limiting nutrient in the supplied medium.

CHEMOSTERILANT Noun. (Greek, *chemeia* = alchemy; Latin, *sterilis* = barren; -*antem* = an agent of something. Pl., Chemosterilants.) A chemical that induces reproductive sterility.

CHEMOTAXIS Noun. (English, *chemic* = transmutation; *taxis* = arrangement. Pl., Chemotaxes.) A reaction to chemical stimuli. See Orientation. Cf. Aerotaxis; Anemotaxis; Geotaxis; Menotaxis; Osmotaxis; Phototaxis; Phototaxis; Rheotaxis; Rotaxis; Scototaxis; Strophotaxis; Telotaxis; Thermotaxis; Thigmotaxis; Tonotaxis; Tropotaxis.

CHEMOTROPISM Noun. (English, *chemic* = transmutation; *trope* = turn; -*ism* = state. Pl., Chemotropisms.) Reaction to chemical stimuli including smell and taste. See Tropism. Cf. Aeolotropism; Anemotropism; Electrotropism; Galvanotropism; Geotropism; Heliotropism; Hydrotropism; Phototropism; Rheotropism; Stereotropism; Thermotropism; Thigmotropism; Tonotropism. Rel. Taxis.

CHENU, JEAN CHARLES (1808–1879) (Anon. 1880, Am. Nat. 14: 151.)

CHEREPANOV, ALEXEYA IGNAT'EVICH (Kukharchuk 1974, *Fauna and Ecology of Insects from Siberia.* (Ed. N. G. Kolomiets) 205 pp. (1–13, bibliogr.), Novosibirsk. [In Russian.].)

CHERMOCK, FRANKLIN, H (1906–1967) (Masters 1968, Bull. Assoc. Minnesota Ent. 2: 21–23.)

CHERRY Noun. (Latin, *cerasus* = cherry tree. Pl., Cherries.) 1. Holarctic: Any of several Species of *Prunus* (Rosaceae) with globose drupes that enclose a smooth stone. 2. Australia: Any of several trees with fruits resembling cherries.

CHERRY CASEBEARER *Coleophora pruniella* Clemens [Lepidoptera: Coleophoridae]: A pest of cherry in North America. Larvae construct portable cases and feed from their case. See Coleophoridae. Cf. Cigar Casebearer; Pistol Casebearer.

CHERRY FRUIT FLY Adult of *Rhagoletis cingulata* (Loew) [Diptera: Tephritidae]: A monovoltine pest of cherries in the northeastern USA and Canada; CFF also attacks plum and pear. CFF overwinters as a Pupa within soil. Adult CFFs emerge during spring. Adult body ca 3–4 mm long; body black with four white stripes on Abdomen; Tibiae and Tarsi yellow. Adults are attracted to cherry tree, feed by scraping exudates from leaves and fruit. Eggs are yellow, pedicellate and thrust into half-ripe fruit by female's Ovipositor. Neonate larvae feed near pit and later feed in flesh. CFF usually completes development in one fruit. Syn. Cherry Maggot. See Tephritidae. Cf. Black Cherry Fruit Fly; Western Cherry Fruit Fly.

CHERRY FRUIT SAWFLY *Hoplocampa cookei* (Clarke) [Hymenoptera: Tenthredindae]: A monovoltine pest of cherry, plum and prune in western USA. Eggs are shiny, white, kidney-shaped and laid on blossoms. Larvae are crescent-shaped with seven pairs of Prolegs; young larvae enter fruit, eat their way to kernel and exit fruit through hole in side; fruit yellows and drops prematurely. See Tenthredinidae. Cf. European Apple Sawfly; Pear Sawfly.

CHERRY FRUITWORM *Grapholita packardi* Zeller [Lepidoptera: Tortricidae]: A significant pest of blueberry and cherry in North America.

CHERRY LEAF BEETLE *Pyrrhalta cavicollis* (LeConte) [Coleoptera: Chrysomelidae].

CHERRY MAGGOT Larva of *Rhagoletis cingulata* (Loew) [Diptera: Tephritidae]. See Cherry Fruit Fly.

CHESS® See Pymetrozine.

CHESTNUT BLIGHT A fungal disease of Chestnut trees caused by *Endothia parasitica* and vectored by several Species of Coleoptera.

CHESTNUT TIMBERWORM *Melittomma sericeum* (Harris) [Coleoptera: Lymexylidae].

CHESTNUT WEEVIL *Curculio dentipes* (Roelofs) [Coleoptera: Curculionidae].

CHEUX, ALBERT (–1914) (Anon. 1915, Miscnea ent. 22: 50.)

CHEVALIER, LOUIS (1851–1929) (Anon. 1929, Bull. Soc. sci. nat. méd. Seine et Oise 10: 65–71.)

CHEVROLAT, LOUIS ALEXANDRE AUGUSTE (1799–1884) Marseul, S.A. de Abeille (Les ent. et leurs écrits.) 24: 148–164, bibliogr.)

CHEVROLAT, M A Author of *Coleopteres du Mexicque.*

CHEWING MOUTHPARTS Appendages of the head that are opposable in operation and adapted for the mastication of particulate matter or matrix-like material. See Mouthparts. Cf. Piercing-sucking mouthparts.

CHIAPAS AMBER Fossilized resin from Chiapas, Mexico and of Oliogene–Miocene age. CA includes some fossil insects. See Amber; Fossil. Cf. Baltic Amber; Burmese Amber; Canadian Amber; Dominican Amber; Lebanese Amber; Taimyrian Amber.

CHIASMA Noun. (Greek, *chiasma* = cross. Pl., Chiasmata.) An X-like crossing of nerve tracts within a nerve centre .

CHICHERIN, TIKHON SERGEEVICH (1869–1904) (Semenov-Tian-Shansky 1908, Horae Soc. ent. Ross. 38 (4): 1–45, bibliogr.)

CHICKEN BODY-LOUSE *Menacanthus stramineus* (Nitzsch) [Mallophaga: Menoponidae]: A cosmopolitan pest of poultry including chicken, pigeon, turkey and peacock; arguably the most serious pest of chickens. CBL is active and sometimes attains very large numbers on bird; CBL often is found under the wings. Postembryonic development to adult requires 2–3 weeks.

CHICKEN HEAD-LOUSE *Cuclotogaster heterographus* (Nitzsch) [Mallophaga: Philopteridae]: An important pest of poultry that clings to the host's head feathers and is passed from bird-to-bird by contact. CHL is most injurious to young birds. See Philopteridae. Cf. Brown Chicken-Louse; Chicken Body-Louse; Fluff Louse; Shaft Louse; Wing Louse.

CHICKEN MITE *Dermanyssus gallinae* (DeGeer) [Acari: Dermanyssidae].

CHICKEN-DUNG FLY *Fannia pusio* (Wiedemann) [Diptera: Muscidae].

CHIGGER Noun. (French, *chique*. Pl., Chiggers.) Acarina: The parasitic, six-legged larval stage of Trombiculidae. Syn. Bete Mite; Harvest Mite; Itch Mite; Red Mite; Rouget. (Goff *et al.* 1982, J. Med. Ent. 19: 222.)

CHIGOE FLEA See Sand Flea.

CHIKUNGUNYA Noun. (Etymology uncertain.) An arboviral disease in tropical Africa and Asia; assigned to *Alphavirus* of Tongaviridae. Disease is characterized by fever, arthralgia and rash; incubation period 2–6 days and acute phase 3–10 days; Chikungunya is rarely fatal. Probably a disease of other primates and vectored to humans by mosquitoes, including *Aedes aegypti* and *A. africanus*. The disease is often associated with forested areas; sometimes it occurs in urban areas in conjunction with Dengue or Yellow Fever. See Arbovirus; Dengue; Tongaviridae; Yellow Fever. Cf. O'Nyong-Nyong; Ross River Fever; Sindbis.

CHILDION® See Tetradifon.

CHILDREN, JOHN GEORGE (1777–1852) (Gunther 1975, *A Century of Zoology at the British Museum 1815–1915*. 533 pp. (56–62), London.)

CHILDS, LEROY (1888–1963) (Newcomer 1964, J. Econ. Ent. 57: 422.)

CHILLCOTT, JAMES GORDON THOMAS (1929–1967) (Shewell 1967, Can. Ent. 99: 780–783.)

CHILOPOD Noun. (Greek, *cheilos* = lip; *pous* = foot. Pl., Chilopods.) A member of the Class Chilopoda.

CHILOPODA Noun. (Greek, *cheilos* = lip; *pous* = foot.) The Centipedes. A cosmopolitan Class within the Phylum Arthropoda that includes several thousand described Species. Adult body up to 25 cm long, elongate and somewhat dorsoventrally compressed; with one pair of Antennae and numerous body segments each of which has a pair of legs or appendages; 15–173 pairs of legs; first body segment behind head bears modified 'fangs' called Toxicognaths. All Species are nocturnal predators that occur in soil, sand, crevices, beneath rocks, in rotting wood and similar habitats. See Arthropoda. Cf. Diplopoda; Insecta; Pauropoda; Symphyla.

CHIMERA Noun. (Latin, *chimara* = monster. Pl., Chimeras.) One individual composed of tissues from two genotypes. Rel. Gynandromorph; Mosiac.

CHINA, WILLIAM E English Hemipterist and former Keeper of the British Museum.

CHINAGLIA, LEOPOLDO (1890–1916) A promising Acarologist killed during the First World War. (Berlese 1917, Redia 12: 361–366, bibliogr.)

CHINALPHOS® See Quinalphos.

CHINCH BUG *Blissus leucopterus leucopterus* (Say) [Hemiptera: Lygaeidae]: A serious pest of grasses and cultivated grains in USA east of the Rocky Mountains. Preferred host plants include barley, corn, oats, rice, rye and wheat. CB is capable of destroying entire crops. Adult 4.2–5.2 mm long; body black with reddish-yellow legs; black triangular spot on white wing along anterior margin. CBs overwinter as adults in clumped grasses or in field litter and become active during April. Female oviposits ca 200 eggs during one month; eggs are laid on leaf sheaths, in soil, roots or stems near ground. Eggs are elongate, cylindrical and curved, with short projections on the cap; colour white then dark red. Neonate nymphs are red with white strip behind wing pad; mature nymphs are darker in colour and lack a stripe. CB undergoes 4–5 nymphal instars and 2–3 generations per year. Nymphs suck juice from roots or stems; as grains mature, the nymphs migrate in large numbers on foot to more succulent plants. Subspecies include: Hairy Chinch Bug, Southern Chinch Bug, Western Chinch Bug. See Lygaeidae. Cf. False Chinch Bug.

CHINESE GALL A gall produced on sumac by cynipid wasps. CG is high in tannic acid and used for making ink.

CHINESE MANTID *Tenodera aridifolia sinensis* Saussure [Mantodea: Mantidae]: Introduced into North America from the Orient about 189; Species occurs in eastern USA. See Mantidae.

CHINESE OBSCURE-SCALE *Parlatoreopsis cinensis* (Marlatt) [Hemiptera: Diaspididae].

CHINESE ROSE-BEETLE *Adoretus sinicus* Burmeister [Coleoptera: Scarabaeidae]: A serious nocturnal defoliator of many plant Species in Hawaii. Feeding damage is characteristically interveinal on dicots. See Scarabaeidae.

CHINESE WAX-SCALE See Hard Wax-Scale.

CHINESE-JUNK CATERPILLARS *Doratifera* spp. [Lepidoptera: Limacodidae]: Minor pests of ornamental trees, fruit trees and some native *Eucalyptus* and *Tristania* in Australia. The name is inspired by the exotic shape and coloration of the caterpillars. Species also included under other common names: Black Slug Cup-Moth, Four-Spotted Cup Moth, Mottled Cup-Moth and Painted Cup-Moth. See Limacodidae.

CHINO, MITSUSHIGE (1888–1957) (Hasegawa 1967, Kontyû 35 (Suppl.): 52–53, bibliogr. [In Japanese.].)

CHINOMETHIONAT A synthetic, organic-hydrocarbon insecticide {6-methyl-1,3-dithio (4,5-b) quinoxaline-2-one} used as an Acaricide, contact insecticide and Fungicide against Mites, mite eggs, plant-sucking insects and Powdery Mildew. Chinomethonat applied to numerous fruit tree crops, strawberries and ornamentals. Phytotoxic to some apple and pear varieties, and some ornamentals (including alders, some roses, junipers). Toxic to fishes. Syn. Quinomethionate. Trade names include Joust®, Morestan® and Oxythioquinox®. See Chlorinated Hydrocarbon Insecticide.

CHIPCOR® See Ethephon.

CHIPMUNK FLEA *Tamiophila grandis* [Siphonaptera: Ctenophthalmidae]: A parasite of chipmunks in eastern North America; also attacks red squirrel, weasel and rabbit. Adult genal Ctenidium with two spines on each side. Cf. American Mouse Flea.

CHIQUE See Sand Flea.

CHIROMIIDAE See Chyromyidae.

CHIROMYZIDAE See Stratiomyidae.

CHIRONOMIDAE Plural Noun. Midges. A cosmopolitan Family including about 6,000 Species of nematocerous Diptera assigned to Superfamily Chironomoidea. Adult mosquito-like, body elongate, ca 1–10 mm long; Ocelli absent; male Antenna plumose, Flagellum with 11–14 segments in male and 5–14 segments in female; mouthparts non-piercing; head posteriorly flattened; Thorax often projecting above head with median furrow or keel but Mesonotum without 'V' shaped suture; Postscutellum with distinct median longitudinal groove; Wing rather short, Medial Vein not branched. Larvae typically with Prolegs on Prothorax and anal segments, and a pair of Papillae on anal segment. Adults are crepuscular or nocturnal; they do not bite humans or other animals and form mating swarms at dusk. Eggs are laid in gelatinous strings or masses. Immature stages of most Species are aquatic (typically freshwater) with some Species found in peripheral saltwater habitats; a few Species occur in terrestrial habitats that are rich in organic matter. Entomophagous larvae of Tanypodinae, Podonominae and a few Species from other Subfamilies. Entomophagous Species usually feed on other chironomid larvae. Larvae of other Chironomidae are phytophagous. Syn. Tendipedidae. See Chironomoidea. Rel. Biting Midges; Bloodworms.

CHIRONYM See Cheironym.

CHIROTYPE Noun. (Greek, *cheir* = hand; *typos* = pattern. Pl., Chirotypes.) Taxonomy: The type-specimen upon which a manuscript name is based. See Type. Rel. Taxonomy.

CHISELS Psocidae: See Mouth Fork.

CHITIN Noun. (Greek, *chiton* = tunic. Pl., Chitins.) 1. A colourless, nitrogenous polysaccharide, linearly arranged as beta-linked N-acetylglucosamine units. Chitin is widespread in arthropods and plants. Chitin is most abundant component of insect Integument and comprises about 50% of Cuticle's dry weight. Following cellulose, Chitin is the second most abundant natural fibre. Chitin has been identified in Integument of eurypterids (ca 405 MYBP) and may have been available for groundplan of Integument development. Chitin is generally regarded as ubiquitous, but it has not been found in Epicuticle, Tracheae of some insects, or in Lepidoptera scales. Chitin was first isolated from fungi by Braconot during 1811 and called Fungine; it was first isolated from insects (beetle Elytra) in 1823 by Odier. Chitin is a chemical compound intermediate between proteins and carbohydrates; chemical formula (Brach 1912) $(C_3 2H_5 4N_4 O_{21})_x$. Chitin is a noteworthy natural compound adapted for many diverse purposes. Industry uses Chitin to recover toxic metals from factory waste. Medicine uses Chitin to promote wound healing. Chitin also used as a general purpose adhesive and as a treatment of wool by textile industry. Physically, Chitin is colourless as a solid. As other nitrogenous polysaccharides, Chitin has a high molecular weight. Chitin is soluble in concentrated mineral acids and hot alkali solution; Chitin is insoluble in water, dilute alkali, alcohol and organic solvents. Chitin exists in three crystallographic forms, called alpha, beta and gamma. Only alpha form has been found in insects. Chemistry of Chitin explains its remarkable physical properties. N-Acetyl D-glucosamine is functional unit of Chitin; it forms long-chain polymeric bundles about 3 µm in diameter. These bundles are called Micellae; each bundle consists of ca 20 molecules. Micellae align with other Micellae to form fibrils that are oriented in plane of Lamella. Chitin fibres are embedded in a protein matrix. Several proteins may be contained within Integument of one Species of insect, and more than 20 proteins have been isolated from Integument of insects. Bonding between the protein and Chitin is controlled by the epidermal cells. 'Stiffness' is provided by covalent bonds; 'flexibility' is provided by Schiff's bases (CH = N). Alt. Chitine. See Integument. Cf. Entomolin; Sclerotin.

CHITIN SYNTHESIS INHIBITOR An Insect Growth Regulator that inhibits production of Chitin in the Integument of insects and other arthropods. Examples include Cyromazine and Triflumuron. See Insect Growth Regulator. Cf. Juvenile Hormone Analogues.

CHITINIZATION Noun. (Greek, *chiton* = tunic; English, *-tion* = result of an action. Pl., Chitinizations.) The act or process of depositing or filling areas of the Integument with Chitin.

CHITINIZED Adj. (Greek, *chiton* = tunic.) Descriptive of Integument filled with or hardened by Chitin. See Integument.

CHITINOGENOUS Adj. (Greek, *chiton* = tunic; *genous* = producing; Latin, *-osus* = possessing the qualities of.) A term applied to epidermal cells that secrete Chitin.

CHITINOUS CRADLE Coccids: The chitinized arms or bars forming the endoskeleton of the head (MacGillivray).

CHITINOUS PLATE OF HAYES: Scarabaeoid larvae: The two Nesia of Boving.

CHITINOUS Adj. (Greek, *chiton* = tunic; Latin, *-osus* = possessing the qualities of.) Composed of Chitin or like Chitin in texture; as a colour term, amber yellow.

CHITOSAN Noun. (Greek, *chiton* = tunic. Pl., Chitosans.) A product derived from Chitin by treating it with an hydroxide at high temperature.

CHITOSE Noun. (Greek, *chiton* = tunic; *-osus* = full of. Pl., Chitoses.) A decomposition product of Chitin; a glucosamine salt.

CHITTENDEN, FRANK HURLBURT (1850–1929) American government Entomologist (USDA), Coleopterists and Economic Entomologist specializing in truck crops. (Mallis 1971, *American Entomologists*. 549 pp. (100–102), New Brunswick.)

CHITTY, ARTHUR JOHN (1859–1908) (Waterhouse 1908, Proc. ent. Soc. Lond. 1908: xcviii.)

CHLAMYDATE Adj. (Greek, *chlamys* = cloak; *-atus* = adjectival suffix.) Cloaked or bearing a cloak or mantle-like structure. Alt. Chlamydatus.

CHLAMYDOSPORE Noun. (Greek, *chlamys* = cloak; *sporos* = a sowing. Pl., Chlamydospores.) Fungi: A thick-walled resting spore that survives adverse conditions.

CHLORDANE A synthetic, chlorinated hydrocarbon insecticide {1,2, 4, 5, 6, 7, 8, 8-octochloro, 2, 3, 3a,4,7,7a-hexahydoro-4,7-methanoindene} used as a fumigant, contact insecticide and stomach poison against ants, aphids, armyworms, boll weevils, chinch bugs, cockroaches, corn earworms, crickets, earwigs, fleas, flies, grasshoppers, hornworms, Japanese beetles, lygus bugs, mosquitoes, sowbugs, termites, thrips, wasps, wireworms and other insects. Chlordane was developed by Velsicol Chemical Company ca 1945. Chlordane is applied to few crops in some countries, and used in prevention and control of subterranean termites in some countries. Chlordane was removed from registration in USA and many developed countries, in part because it is toxic to bees and fishes and is persistent in soil. Trade names include: Chlor-tox®, Mahatz®, Termex®, Mermiseal®, Termidan®. See Chlorinated Hydrocarbon Insecticide.

CHLORETHEPHON® See Ethephon.

CHLORETHOXYFOS An organic-phosphate (thiophosphoric acid) compound {O,O-diethyl O-(1,2,2,2-tetrachloroethyl) phosphorothioate} used as a broad-spectrum soil insecticide against rootworms, cutworms, wireworms and root maggots. Compound is applied to vegetables, corn, potatoes, sugar beets, sorghum, and turf. Experimental registration in USA. Trade name is Fortress®. See Organophosphorus Insecticide.

CHLORFENVINPHOS An organic-phosphate (phosphoric acid) compound {2-chloro-1-(2,4-dichlorophenyl)vinyl diethylphosphate} used as a contact insecticide for control of cutworms, root-feeding insects, plant-sucking insects, and public-health pests. Compound is applied to vegetables, numerous crops and livestock in many countries, but it is not registered for use in USA. Trade names include: Birlane®, Haptarax®, Haptasol®, Sapecron®, Sedanox®, Stefadone®, Supocade®, Supona®, Unitox®, Vinylphate®. See Organophosphorus Insecticide.

CHLORFLURAZURON A substituted urea compound {N-[4-(3-chloro-5-trifluoromethyl-2-pyridyloxy)-3–5-dichlorophenyl]-N'-(2,6-difluorobenzoxyl)urea} that is used as a Chitin Synthesis Inhibitor for control of Lepidoptera (armyworms, bollworms, budworms, leafworms, loopers). Ovicidal properties and stomach poison effects have been noted in some insects. Compound is applied to cotton, tea and vegetables in some countries. Some phytotoxic effects noted on cabbage and peppers. Chlorflurazuron is not registered for use in USA. Compound shows no effect on mites, aphids or whiteflies, and no systemic activity. Trade names include: Aim®, Atabron®, Helix®, Jupiter®.

CHLORIDE Noun. (Greek, *chloros* = pale green; *-ide* = chemical suffix. Pl., Chlorides.) A salt of hydrochloric acid, the commonest chloride is table salt, NaCl.

CHLORINATED Adj. (Greek, *chloros* = pale green.) Of a greenish-yellow colour.

CHLORINATED HYDROCARBON An organic compound containing chlorine, hydrogen and occasionally oxygen and sulfur. The first widely used synthetic organic insecticides. See DDT (dichlorodiphenyltrichloroethane).

CHLORINATED HYDROCARBON INSECTICIDE An insecticide whose molecule is predominantly carbon and hydrogen, but has had at least one chlorine atom introduced to it that has replaced one of the hydrogen atoms. Syn. Organochlorine. See Insecticide.

CHLORMEPHOS An organic-phosphate (dithiophosphate) compound {S-chloromethyl-O,O-diethyl phosphorodithioate} used as a contact insecticide for control of wireworms, scarab larvae, root flies, crickets, subterranean-nesting wasps and other soil inhabiting pests. Compound is applied to soil, but it is not registered for use in USA. Toxic to fishes. Trade name is Dotan®. See Organophosphorus Insecticide.

CHLORMEQUAT A chlorinated hydrocarbon {2-chloroethyl)-trimethyl-ammonium chloride} used as a Plant Growth Regulator. Trade names include: Adjust®, Arotex-Extra®, Barleyquat-B®, Bettaquat-B®, CCC®, Cycocel®, Cycogan®, Holdup®, Hormocel®, Hyquat®, Hico-Ccl®, Incracel®, Lihochin®, Manipulator®, Minc®, Nainit®, Pentagan®, Siacourt-C®, Stabilene®, Standup®, Titan®. Cf. Insect Growth Regulator.

CHLORO IPC® See Chlorpropham.

CHLOROBENZILATE A chlorinated hydrocarbon {Ethyl 4,4'-dichlorobenzilate} manufactured as a contact acaricide. Compound is phytotoxic to almonds, apples, peaches, plums, prunes and roses. Trade names include Acaraben®, Benzilan® and Kopmite®. See Organochlorine

Insecticide.

CHLOROCRESOL Noun. (Greek, *chloros* = pale green; *kreas* = flesh; Latin, *oleum* = oil.) A crystalline material used to prevent fungal or mould attack of dried insects specimens or soften brittle specimens.

CHLOROCYPHIDAE Plural Noun. A small Family of Odonata assigned to the Calyopterygoidea. Species are restricted to forests of South America. Adults small, robust with Epistome snout-like; Abdomen thick and shorter than wings; wings narrow, petiolate; Pterostigma narrow and elongated; quadrilateral narrow and rectangular. Larvae morphologically are diverse; Antenna with six segments, first segment as long as others combined; two lateral Caudal Gills spike-like, Dorsal Gill vestigial. Syn. Dicteriastidae. See Odonata.

CHLOROFORM Noun. ($CHCl_3$.) A clear, colourless, highly flammable and volatile liquid with an ether-like odour and a burning sweet taste. Chloroform is miscible with alcohol, ether, benzene and oils. Compound is used as a killing agent but stiffens specimens and when stored in dark-coloured glass jars it forms phosgene (carbonyl chloride, $COCl_2$).

CHLOROFOS® See Trichlorfon.

CHLOROLESTIDAE Plural Noun. A Family of Odonata. Adults frequently metallic coloured with body small to moderate sized; Pterostigma rectangular or rhomboidal and second Cubital Vein strongly arched after leaving quadrilateral. Important Genera are: *Chlorolestes* (South Africa), *Chorismagrion* (Australia), *Ecchlorolestes* (South Africa), *Episynlestes* (Australia), *Megalestes* (Asia), *Orolestes* (Asia) and *Synlestes* (Australia). See Odonata.

CHLORONICOTINYL Noun. A group of synthetic insecticides discovered in 1985. Active ingredient is imidacloprid. Registered as a termiticide in USA, Japan, Thailand and Australia. Water-based suspension concentrate; odourless, non-flammable. Symptoms of poisoning include disorientation, lethargy and loss of grooming behaviour; mode-of-action involves nerve cell receptors. Marketed under name 'Premise' by Bayer. See Insecticide. Cf. Botanical Insecticide; Carbamate; Inorganic Insecticide; Organochlorine; Organophosphorus; Synthetic Pyrethroid; Insect Growth Regulators.

CHLOROPERLIDAE Plural Noun. Green Stoneflies. A Family of Plecoptera including 18 Genera and about 125 nominal Species. The Family is best represented in temperate parts of Holarctic; some Genera have restricted distributions. Size and colour varies within Family; difficult to generalize about habitus. Larvae are omnivorous when young and carnivorous when older. Adults do not feed. See Plecoptera.

CHLOROPHANE Noun. (Greek, *chloros* = yellow; *phane* = evident.) An oily, greenish yellow pigment found in insects.

CHLOROPHYLL Noun. (Greek, *chloros* = yellow; *phyllon* = leaf. Pl., Chlorophylls.) The green colouring matter of plants and one of the substances found in the blood of insects. Variants include: chlorophyll a (blue-green chlorophyll) and chlorophyll b (yellow-green chlorophyll).

CHLOR-O-PIC® See Chloropicrin.

CHLOROPICRIN 'Tear Gas.' A highly toxic fumigant {Tricholornitromethane} used against stored product pests, soil insects, nematodes, weeds and rodents. Chloropicrin is non-flammable with a boiling point of 112° C and sometimes it is formulated with Methyl Bromide to act as a warning odour. Compound was first formulated during 1908 in England and used extensively during WW I. Chloropicrin is highly toxic to humans and typically requires a licence for use; it is no longer registered as a grain fumigant in USA. Trade names include Nitrochloroform®, Tear Gas®, Chlor-o-Pic®, Niklor®, Dolochlor®, Tri-Chlor®, HD-Pic®.

CHLOROPIDAE Plural Noun. Eye Flies; Eye Gnats; Frit Flies. A large, cosmopolitan Family (ca 1,000 Species) of brachycerous Diptera assigned to Superfamily Chloropoidea; Species are distributed among 257 Genera and Subgenera within 2–4 Subfamilies. Chloropids are closely related to Milichiidae and Drosophilidae, but separated from these Families by following characters: Wing without an Anal Cell or Anal Vein, vein M1 usually at least slightly bent near middle, Costa broken just before end of vein R1, frontal triangle large, anterior point nearly reaches base of Antenna; Propleuron with lateral part flat and separated from transverse part by a vertical Carina; Postvertical Bristles convergent and never with distinct fronto-orbitals. Family is best known for eye gnats (*Hippelates*) with pestiferous habit and eye disease transmission in man and animals. Chloropids are biologically diverse and typically phytophagous with several members important as crop pests. Some Species are saprophagous; a few Species are predaceous and or parasitic. Larvae of *Pseudogaurax* spp. are predaceous upon egg masses of spiders, notably black widow (*Lactrodectus mactans*), and upon mantid and Lepidoptera eggs. *Chloropisca glabra* is predaceous upon sugarbeet root aphid. Life histories of parasitic forms are fairly uniform: Eggs are deposited near host and first instar larvae actively search for hosts. For Species predaceous upon eggs of spiders: Eggs are laid on outside of sac; neonate larvae burrow through layers of silk and consume contents of sac. Pupation usually occurs near host remains. The flies usually are univoltine but an occasional second generation may develop. Syn. Oscinidae; Siphonellopsidae; Titaniidae.

CHLOROPLAST Noun. (Greek, *chloros* = pallid; *plastos* = moulded, Pl., Chloroplasts.) One of the chlorophyll bodies of the green plant.

CHLOROSIS Noun. (Greek, *chloros* = pallid; *sis* =

a condition or state. Pl., Chloroses.) Plants: Absence of green pigments in plant leaves; caused by disease such as magnesium or iron deficiency.

CHLOROTIC Adj. (Greek, *chloros* = pallid; *-ic* = characterized by.) Pertaining to Chlorosis.

CHLORPROPHAM A carbamate compound {Isopropyl-m-chlorocarbanilate} used as an herbicide outside the USA and a Plant Growth Regulator for potatoes in the USA. Trade names include CIPC®, Chloro IPC®, Decco 276EC®, Keim-Stop®, Pommetrol®, Sprout Nip®, Spud Nic®, Taterpix®. See Carbamate Insecticide. Cf. Insect Growth Regulator.

CHLORPROPYLATE A chlorinated hydrocarbon related to Bromopropylate marked in some countries for control of mites under trade name Rospin® See Organochlorine Insecticide.

CHLORPYRIFOS An organophosphate (thiophosphoric acid) compound {O,O-diethyl-O-(3,5,6-trichloro-2-pyridinyl) phosphorothioate} used as a contact insecticide and stomach poison against mites, thrips, plant-sucking insects, leaf-chewing insects, cutworms, webworms and domiciliary pests. Compound is applied to numerous tree crops, cotton, sugarcane, vegetables, ornamentals, nursery stock, livestock barns and turf; also applied as a public-health insecticide for control of mosquitoes, ticks, fleas, lice, houseflies and ants. Chlorpyrifos does not break down as quickly as other OP compounds. It is used as a chemical barrier treatment against termites. Compound is phytotoxic to some ornamentals including poinsettias, azaleas, camellias and roses. Trade names include: Barral®, Brodan®, Cinch®, Clinch®, Crossfire®, Deter® (Rhone Poulenc), Detmol®, Detmolin®, Dilmor®, Disparo®, Durmet®, Dursban®, Empire®, Eradax®, Estate®, Equity®, Fosban®, Gigant®, Kregan®, Killmaster®, Lentrek®, Lepecid®, Lock-On®, Lorsban® (Dow), Lorvek®, Loxiran®, Nurelle®, Pageant®, Predator®, Pyrinex®, Rimi®, Spannit®, Silrifos®, Talon®, Tenure®, Vexter®, Zidil®. See Organophosphorus Insecticide.

CHLORPYRIFOS-METHYL An organic-phosphate (thiophosphoric acid) compound {O,O-dimethyl-O-(3,5,6-trichloro-2-pyridinyl) phosphorothioate} used as a contact insecticide, stomach poison and fumigant against stored product pests, fruit and vegetable pests and domiciliary pests. Compound is applied to corn, cotton, *Citrus*, rice, oats, vegetables and ornamentals; also applied as a public-health insecticide for control of mosquitoes, fleas and houseflies. Toxic to fishes and Crustacea. Trade names include: Daskor®, Graincote®, Prinex®, Reldan®, Smite®, Zertell®. See Organophosphorus Insecticide.

CHLORTIEPIN® See Endosulfan.

CHLOR-TOX® See Chlordane.

CHOBAUT, ALFRED (1860–1926) (Buysson 1927, Bull. Soc. normande Ent. 3: 51–53.)

CHOKECHERRY LEAFROLLER *Sparganothis directana* (Walker) [Lepidoptera: Tortricidae].

CHOLERA An often fatal disease with symptoms of continuous vomiting and diarrhoea; caused by *Vibrio comma* and transmitted via contaminated water or vectored by housefly through excrement. Also vectored by other Diptera to include *Calliphora vomitoria, C. vicina, Lucilia caesar* and *Sarcophaga carnaria.*

CHOLODKOVSKY, NICOLAS ALEXANDER (1858–1921) Russian M.D, Ph.D., Professor of Zoology at Military Academy of Medicine and well known as translator of English and German poets; author of textbooks on Zoology, Comparative Anatomy and Entomology. Cholodkovsky published extensively on aphids and chermesids. (Howard 1930, Smithson. Misc. Colln. 84: 292, 295, 300, 303, 307, 536.)

CHOPARD, LUCIEN (1885–1971) (Agnilav 1972, Ann. Soc. ent. Fr. 8: 767–861, bibliography.)

CHOPRA, L R (–1942) (Anon. 1942, Indian J. Ent. 4: 97.)

CHORDA Noun. (Greek, *chorde* = string. Pl., Chordata.) Lepidopteran wing: The stem of veins R 4+5 (Turner), the part of vein R4+5 separating the Areole from the basal cell in Lepidoptera (Tillyard).

CHORDOTONAL LIGAMENT A ligamentous structure connecting a scolopophore with the body-wall (Comstock).

CHORDOTONAL NERVE A nerve connecting to a chordontonal or auditory organ (Klots).

CHORDOTONAL ORGAN A sense organ of the scolopophorous type, found between the Tibia and Basitarsus of some Coleoptera, Diptera and Hemiptera. CO is used for the detection of substrate vibrations to enable an insect to 'hear' sounds. The cellular elements forming an elongate structure are attached at both ends to the body wall, but not necessarily containing sense rods or Scolops (Snodgrass). See Johnston's Organ; Subgenual Organ.

CHORDOTONAL Adj. (Greek, *chorde* = chord; *tonos* = tone.) Pertaining to sensory receptors responsive to harmonic vibrations. Term usually is applied to organs adapted for the perception of sound vibrations. See Chordotonal Organ.

CHOREUTIDAE Plural Noun. Silkworm Moths. A cosmopolitan Family of ditrysian Lepidoptera which includes ca 350 Species and is assigned to Superfamily Sesioidea. The Family is most abundant in Orient and Australia. Adult wingspan ca 5–20 mm with head scales smooth and overlapping scales at base of Proboscis; Ocelli are prominent and Chaetosemata are absent; Antenna filiform in male and female; Maxillary Palpus minute with 1–2 segments; Labial Palpus curved with two segments; fore Tibia with Epiphysis; tibial formula 0-2-4; wings often with metallic markings. Species are diurnal and adults walk with jerky gait. Larvae typically feed on leaves surrounded by silken web or folded leaves. Pupation occurs within double-walled silken cocoon. See Sesioidea.

CHORIOGENESIS Noun. (Greek, *chorion* = skin; *genesis* = descent. Pl., Choriogeneses.) The process by which the Chorion of the eggshell is deposited upon the egg, typically occurring in the Vitellarium of the Ovariole.

CHORION Noun. (Greek, *chorion* = skin. Pl., Chorions.) The outer shell (cover) of the insect egg. Initially, the Chorion surrounds the Ooplasm of the insect egg and subsequently it surrounds and protects the insect embryo. Chorion is composed of several layers which are deposited by follicle cells in the Ovariole. The Chorion surface typically bears a Micropyle, Aeroplyes and sometimes a Hydropyle. Surface sometimes smooth, often covered with complex sculptural patterns that can be taxonomically useful. See Egg; Eggshell.

CHORIONIC Adj. (Greek, *chorion* = skin; *-ic* = consisting of.) Descriptive of or pertaining to the Chorion or shell of the insect egg.

CHORIONIN Noun. (Greek, *chorion* = skin; *-in* = chemical suffix = *-ine*. Pl., Chorionins.) The chemical substance of which the Chorion of the insect egg is composed (Tillyard). See Chorion.

CHORIOTHETE Noun. (Greek, *chorion* = skin; *thetikos* = fit for placing. Choriothetes.) A tongue-like organ on the ventral surface of the Uterus in viviparous Diptera used to remove the Chorion of the egg and Exuviae of first instar larva. (Jackson 1948)

CHOROID Noun. (Greek, *chorion* = skin; *eidos* = form. Pl., Choroids.) The black basal membrane of the Ommatidium in the insect eye (Kirby & Spence). Alt. Choroides.

CHOROLOGY Noun. (Greek, *choros* = place; *logos* = discourse. Pl., Chorologies.) Biogeography; Zoogeography.

CHRETIEN, PIERRE (1846–1934) (Dos Passos 1951, J. N. Y. ent. Soc. 59: 160.)

CHRISTENSON, LE ROY DEAN (1906–1968) (Mitchell 1968, Proc. Hawaii ent. Soc. 20: 467–468.)

CHRISTIANI, OLAF GUDMANN (1877–1944) (West 1947, Ent. Meddr. 25: 148.)

CHRISTMAS BEETLES *Anoplognathus* spp. [Coleoptera: Scarabaeidae]: In Australia larvae damage pineapples, strawberries, pasture grasses and sugarcane; adults sometimes defoliate *Eucalyptus*. See Scarabaeidae.

CHRISTMAS-BERRY WEBWORM *Cryptoblabes gnidiella* (Millière) [Lepidoptera: Pyralidae].

CHRISTOPH, HUGO THEODOR (1831–1894) (Anon. 1895, Entomol. mon. Mag. 31: 30.)

CHRISTY, JOHN FELL (–1851) (Westwood 1851, Proc. ent. Soc. Lond. 1: 135.)

CHRISTY, ROBERT MILLER (1861–1928) (P. T. 1928, Proc. Linn. Soc. Lond. 1928: 112–113, bibliogr.)

CHRISTY, WILLIAM MILLER (1863–1939) (Edelsten 1940, Entomologist 73: 24.)

CHRIZOPON® See Indolebutyric Acid.

CHROMATIN Noun. (Greek, *chroma* = colour; Latin, *-in* = an activator.) Minute granules that make up the chromoplasm of a cell nucleus.

CHROMATOCYTE Noun. (Greek, *chroma* = colour; *kytos* = hollow vessel; container. Pl., Chromatocytes.) 1. A pigment-containing cell. Specifically, epidermal cells in which pigments are contained and in which pigments migrate within the cell and display different colours or hues. Alt. Chromocyte. Syn. Color Cell; Pigment Cell.

CHROMATOLYSIS Noun. (Greek, *chroma* = colour; *lysis* = loosening; *-sis* = a condition or state. Pl., Chromatolyses.) A phase of histoloysis in larvae during which the chromatin condenses in the nodules of histolyzing tissue into compact masses (Henneguy).

CHROMATOPHORE Noun. (Greek, *chroma* = colour; *phorein* = to bear. Pl., Chromatophores.) Special cells that contain pigments are widely distributed among animals. Chromatophores are called chromatosomes when they are arranged in groups of cells. Monochromatic chromatophores contain one colour of pigment; dichromatic chromatophores contain two colour pigments; polychromatic chromatophores contain several pigmentary colours. When chromatophores contain more than one coloured pigment granule, the granules may migrate along specified branches called chromorhizae. Typical chromatophores are not found in insects. A rapid colour change by insects is achieved via epidermal cells which induce a pigment granule to migrate. Epidermal cells are not branched chromatophores. Epidermal cells that influence colour changes are best known in stick insects and are controlled by the endocrine gland system.

CHROMATOSOME Noun. See Chromatophore.

CHROMENES Plural Noun. A complex of alkaloid biopesticides extracted from Asteraceae, including Encecalin, Demethoxyencecalin and Chromenes 90 and 91. Chromenes express insecticidal activities by affecting Metamorphosis or destroying the Corpus Allatum of target pests. See Biopesticide; Botanical Insecticide; Natural Insecticide; Insecticide; Pesticides. Cf. Acetogenins; Domoic Acid; Furanocumarins; Mammeins; Polyacetylenes; Terthienyl.

CHROMOPHIL Noun. (Greek, *chroma* = colour; *philein* = to love. Pl., Chromophils.) A rounded or slightly flattened leucocyte blood cell which has a strong affinity for aniline dyes.

CHROMOPHILE LEUCOCYTE See Chromophil.

CHROMORHIZAE. See Chromatophore.

CHROMOSOME Noun. (Greek, *chroma* = colour; *soma* = body. Pl., Chromosomes.) Darkly-staining bodies found within eukaryotic cells and consisting of DNA that encodes genetic information. Typically, a somatic cell contains the diploid number of Chromosomes while a reproductive cell (Oogonium, Spermatozoon) contains a haploid number of Chromosomes. A constant number of Chromosomes occurs in individuals.

See Reproduction. Cf. Heterochromosome. Rel. Chromatin.

CHRONIC BEE-PARALYSIS VIRUS A widespread viral disease of honey bees which displays the symptoms of trembling wings and bodies, bloated Abdomen and an inability to fly. CPV infects Brain and nerve ganglia and can cause mortality in infected workers and colony collapse. See Honey Bee Viral Disease. Cf. Acute Bee-Paralysis Virus. Rel. Pathogen.

CHRONIC POISONING Illness or death from long-term ingestion of, or exposure to, low levels of a toxicant.

CHRYSALIS Noun. (Greek, *chrysallis* = something golden, gold. Pl., Chrysalides.) The stage between larva and adult in butterflies. The Pupa Obtecta of Lepidoptera *sensu* Tillyard. Alt. Chrysalid. See Pupa.

CHRYSANTHEMUM APHID *Macrosiphoniella sanborni* (Gillette) [Hemiptera: Aphididae].

CHRYSANTHEMUM FLOWER-BORER *Lorita abnorana* Busck [Lepidoptera: Cochylidae].

CHRYSANTHEMUM GALL-MIDGE *Rhopalomyia chrysanthemi* (Ahlberg) [Diptera: Cecidomyiidae]: A multivoltine pest of chrysanthemums in greenhouses that is endemic to Europe and was adventive to North America ca 1915. Female CGM lays ca 100 yellow eggs on leaf surface, tips of new growth and stems; eclosion is temperature-dependent and occurs within 3–15 days. Neonate larvae bore into host plants and induce formation of cone-shaped galls on upper surface of leaves and stems. Larvae feed and pupation occurs within galls. CGM completes 5–6 generations per year, depending upon conditions. See Cecidomyiidae.

CHRYSANTHEMUM LACE-BUG *Corythucha marmorata* (Uhler) [Hemiptera: Tingidae]: Widespread in North America and common in glasshouses. Nymphs and adults feed upon plant juices from stems and leaves. Host plants include Chysanthemum, Aster and Goldenrod. See Tingidae.

CHRYSANTHEMUM LEAFMINER *Chromatomyia syngenesiae* Hardy [Diptera: Agromyzidae]: A Species endemic to Europe and adventive elsewhere. Adult with Frons yellow and Mesonotum grey. See Agromyzidae.

CHRYSANTHEMUM STOOL-MINER *Psila nigricornis* Meigen [Diptera: Psilidae]: A minor pest of carrots in Europe and Canada. See Psilidae.

CHRYSANTHEMUM THRIPS *Thrips nigropilosus* Uzel [Thysanoptera: Thripidae].

CHRYSANTHUS, PATER (1905–1972) (Hammen & Van Helsdinge 1972, Ent. Ber., Amst. 32: 145–150, bibliogr.)

CHRYSARGYRUS Adj. (Greek, *chrysos* = gold; *argentos* = silver.) Silvery gilt.

CHRYSIDIDAE Latreille 1802. Plural Noun. Cuckoo Wasps; Gold Wasps; Ruby Wasps. A cosmopolitan, large (ca 4,000 Species) Family of apocritous Hymenoptera assigned to the Chrysidoidea. Diagnostic features include: Adult body heavily sclerotized, sculptured and metallic coloured; sexual dimorphism absent, sexes difficult to identify; Antenna with 13 segments in both sexes but club not apparent; Pronotum not extending to Tegula; wings typically macropterous in both sexes, some amisegine females brachypterous or apterous; wing venation with several closed cells in forewing; hindwing without Jugal Lobe or closed cells; Propodeum large with posterolateral margins denticulate; Metasoma with 3–6 visible Terga, apical segments tubular; Sting considerably reduced (?absent), ventral surface frequently concave. Primary parasites of phasmid eggs and other Hymenoptera; host associations reflect higher classification of Chrysididae: Primitive Subfamilies attack sawflies and phasmid eggs; derived Subfamilies attack aculeate Hymenoptera. Cleptoparasitic forms feed on host and food in cell. First instar larva often planidiform; typically attacks resting prepupal stage of host. Complex behaviour of female in host attack: Chews into host cocoon to oviposit; chews into mud cells and then cocoon, sometimes seal hole with saliva; oviposits on bug nymph and develops on sphecid host. Females often roll into ball for protection (Thanatosis). Chrysidids evolved from an ancestor near Bethylidae (Mesitiinae). Included Subfamilies: Amesiginae, Chrysidinae, Cleptinae, Elapinae and Parnopinae. See Chrysidoidea.

CHRYSIDINAE The nominant Subfamily of Chrysididae (Hymenoptera: Aculeata). Moderately large and cosmopolitan in distribution, the Chrysidinae may be characterized as displaying a forewing with RS stub sclerotized and long, Discoidal and Radial Cells complete; Gaster with three visible Terga, T III with subapical transverse Sulcus or pits, apical margin emarginate, lobate or dentate. Species presumably parasitic on other aculeate Hymenoptera.

CHRYSIDOIDEA Latreille 1802. A primitive or ancestral Superfamily of aculeate Hymenoptera, apparently near the Pompiloidea. Features include: Adult head hypognathous or prognathous; Antenna with 10–40 segments (typically 13); body usually dark or metallic coloured; often dorsoventrally flattened; Pronotum extending to Tegula or separated by a lobe; wing venation typically reduced. All Species primary parasites; midgut and hindgut not connected in the feeding larva. Host spectrum includes Lepidoptera, Coleoptera, Homoptera and other Hymenoptera. Included Families: Bethylidae, Chrysididae, Dryinidae, Embolemidae, Sclerogibbidae and Scolebythidae. Syn. Bethyloidea. See Aculeata.

CHRYSOMELIDAE Plural Noun. Leaf Beetles; Tortoise Beetles; Sweetpotato Beetles; Gold Bugs. A cosmopolitan, Family of polyphagous Coleoptera assigned to Superfamily Chrysomeloidea and including more than 20,000 nominal Species. Chrysomelids are related to Cerambycidae and Bruchidae. Second largest Family of herbivorous insects; largest Subfamily Alticinae (flea beetles). Adult 1–32 mm long with body shape highly variable but often glabrous and brightly coloured; head often with grooves or prominences, sometimes rostrate but Rostrum wider than long; Antenna with 9–11 segments, typically filiform; tarsal formula 4-4-4, Tarsomeres 1–3 with lobes on ventral surface covered with adhesive Setae; fourth segment minute and pseudotetramerous or fused with last segment; Abdomen with 5–6 Ventrites. Larvae highly variable in size and shape, usually weakly sclerotized; Antenna with 1–3 segments; 0–6 Stemmata; Mandible with three or more teeth; leg development variable; Urogomphi present or absent. Larvae and adults are phytophagous and feed on same food plant; chrysomelids rarely feed on wood. See Chrysomeloidea.

CHRYSOMELOIDEA Superfamily of polyphagous Coleoptera, including: Bruchidae, Chrysomelidae, Cerambycidae and Megalopodidae. Adults with pseudotetamerous Tarsi. See Coleoptera.

CHRYSOPA The nominant generic name for Chrysopidae (Neuroptera) and commonly called green lacewings. Agents used in biological control of agricultural pests and currently under the generic name *Chrysoperla*.

CHRYSOPIDAE Plural Noun. Green Lacewings. A Family of Neuroptera.

CHRYSOPOLOMIDAE Plural Noun. A small Family of ditrysian Lepidoptera assigned to Superfamily Zygaenoidea.

CHRYSRON® See Resmethrin.

CHRYZOSAN® See Indolebutyric Acid.

CHRYZOTEK® See Indolebutyric Acid.

CHUN, CARL (1852–1914) (Anon. 1914, Ent. News 25: 335.)

CHURCH, NORMAN STANLEY (1929–1975) (Burgess & Burrage 1975, Bull. ent. Soc. Can. 7: 94.)

CHURCHYARD BEETLES *Blaps* spp. [Coleoptera: Tenebrionidae]: A cosmopolitan Genus with some Species acting as pests on cereals. See Tenebrionidae.

CHURR-WORMS See Gryllotalpidae.

CHYLE Noun. (Latin, *chylus* > Greek, *chylos* = juice > *cheein* = to pour.) The food-mass after it has passed through the 'gizzard' and is mixed with the secretions of the salivary glands and caecal structures, ready to be assimilated.

CHYLIFEROUS Adj. (Greek, *chylos* = juice; Latin, *ferre* = to carry; *-osus* = possessing the qualities of.) Chyle-conducting.

CHYLIFIC STOMACH Insects: The part of the alimentary system in which the Chyle is prepared and digestion begins. See Alimentary System. Alt. Chylific Ventricle.

CHYLIFICATION Noun. (Greek, *chylos* = juice; Latin, *facere* = to make; English, *-tion* = result of an action. Pl., Chylifications.) The process by which Chyle is produced.

CHYME Noun. (Greek, *chymos* = juice. Pl., Chymes.) The semidigested food in the small intestine.

CHYROMYIDAE Plural Noun. A small Family of brachycerous Diptera assigned to Superfamily Heleomyzoidea. Adult small to minute with fronto-orbital bristles reclinate; Vibrissae not differentiated from cheek bristles; vein Cu+1A not reaching margin; vein 1A not broken at junction with cell CuP. Biology poorly known. Syn. Chyromyiidae. See Heleomyzoidea.

CHYROMYIIDAE See Chyromyidae.

CHYZER, KORNELIUS (1836–1909) (Bonnet 1945, *Bibliographia Araneorum* 1: 42.)

CIBARIAL Adj. (Latin, *cibaria* = food.) Descriptive of or pertaining to the Cibarium.

CIBARIAL APPARATUS The mouthparts that manipulate or process food. (Archaic). See Cibarium.

CIBARIAL PUMP A pumping mechanism adapted for imbibing fluids. CP is particularly well developed in the Homoptera.

CIBARIUM Noun. (Latin = *cibaria* = food; *arium* = place of a thing. Pl., Cibaria.) 1. The food pocket of the extraoral or preoral mouth cavity behind the Epipharnyx, above the Hypopharynx and the under surface of the Clypeus (Snodgrass). 2. The food canal; sucking tube *sensu* MacGillivray.

CICADA KILLER *Sphecius speciosus* (Drury) [Hymenoptera: Sphecidae]: A large-bodied, univoltine wasp endemic and widespread in eastern North America during middle to late summer. Female CKs nest in soil and provision their cells with paralysed cicadas. See Sphecidae. Cf. Sand Wasp.

CICADA KILLERS See Sphecidae.

CICADAS See Cicadidae.

CICADELLIDAE Plural Noun. Leafhoppers. The nominant Family of Cicadelloidea (Hemiptera). Cosmopolitan distribution, abundant in tropical regions with many Species regarded as economically important. See Cicadelloidea.

CICADELLOIDEA Leafhoppers; Treehoppers. A numerically large, cosmopolitan Superfamily of auchenorrhynchous Hemiptera. Cicadelloids are characterized by Tentorium reduced, transverse hind Coxa and pretasal claws broad. Included Families: Aetalionidae, Cicadellidae, Eurymelidae, Hylicidae and Membracidae. See Auchenorrhyncha; Hemiptera.

CICADIDAE Plural Noun. Cicadas. A numerically large, cosmopolitan Family of Cicadoidea (Hemiptera).

CICADOIDEA A Superfamily of auchenorrhynchous Hemiptera including about 1,200 Species assigned to Cicadidae and Tetigarctidae. Cosmopolitan but most well represented in tropical and subtropical regions. Antennal Flagellum with five segments that become progressively shorter and tapering distad; three Ocelli; Tentorium complete; wing venation relatively complete; sound production and reception system complex.

CICATRICES Noun. (Etymology obscure.) Coccids: The Labiae *teste* MacGillivray.

CICATRICOSE Adj. (Latin, *cicatrix* = scar; *-osus* = full of.) Pertaining to surface with superficial scars or elevated margins like small-pox on a surface. Alt. Cicatricosus.

CICATRIX Noun. (Latin, *cicatrix* = scar. Pl., Cicatrices.) 1. A scar or elevated ridge; a spot; a scar-like structure. 2. Coccids: The wax threads melted into a compact mass found as a minute ball or dot over the first Exuviae of the scales of many adult females (MacGillivray).

CICINDELIDAE Plural Noun. Tiger Beetles. A numerically large, cosmopolitan Family of adephagous Coleoptera that is sometimes placed in Carabidae. Adult body moderate sized and often brightly coloured. Head deflexed; Labrum large; eyes large and laterally protuberant; legs long and slender. Adults typically are agile predators in open areas such as stream banks and beaches; a few Species are wingless and on tree trunks. Larvae are ambush predators living in vertical tunnels of soil or wood. See Coleoptera.

CIDAL® See Phenthoate.

CIDE-TRAK® See Isomate-M.

CIFERRI, RAFFAELE (1897–1964) (Giacomini 1965, bot. Ital. 72: 705–772, bibliogr.)

CIGAR CASEBEARER *Coleophora serratella* (Linnaeus) [Lepidoptera: Coleophoridae]: A pest of apple and other tree fruit in western North America. Life history resembles Pistol Casebearer. CCB larvae feed from within a portable cigar-shaped case and cause leaf blotches. Syn. Birch Casebearer. See Coleophoridae. Cf. Pistol Casebearer.

CIGARETTE BEETLE *Lasioderma serricorne* (Fabricius) [Coleoptera: Anobiidae]: A cosmopolitan, sporadic pest of stored products (tobacco, milled cereal, grain, dried fruit) and dry-insect collections. Larvae cause most damage through feeding and frass; adults most often seen and cause damage when burrowing through cardboard containers and wrapping. Adult 2–3 mm long, reddish-yellow and oval in outline shape; head drawn under Pronotum when beetle threatened; Antenna with 11 segments, 4–10 serrate; Elytra smooth. CB adults actively fly, particularly at night. Female lays ca 100 eggs during lifetime; eggs are laid on larval food and hatch within 6–8 days at 85ºF. Larvae live 30–60 days and complete 4–6 instars; early instars are active, but late instars are less active. Mature larvae are white, scarabaelform and bore away from feeding sites to construct a pupal cell/cocoon of waste material and food held together by midgut secretion. Pupal stage requires 4–12 days. CB requires humidity >25% to complete development. Adults remain in cocoons ca 3–10 days until sexually mature; females oviposit 6–20 days. Adults take water but do not feed and live ca 25 days. Syn. Tobacco Beetle; Towbug. See Anobiidae. Cf. Drugstore Beetle.

CIIDAE Plural Noun. Minute Tree-Fungus Beetles. A Family of Coleoptera assigned to Superfamily Cucujoidea. Adults <6.0 mm long and dark coloured with body cylindrical; head deflexed and not visible from above; Antenna with 3-segmented club; tarsal formula 4-4-4 with fourth segment longest. Ciids inhabit wood under bark and rotting wood where they feed on fungi. Syn. Cisidae; Cioidae. See Cucujoidea.

CILDON® See Phosphamidon.

CILIA OF POSTERIOR ORBIT Diptera: A row of bristles along the posterior margin of the compound eye (Comstock).

CILIATE Adj. (Latin *cilium* = eyelid; *-atus* = adjectival suffix.) 1. Pertaining to structure fringed or lined with a row of parallel Setae or Cilia (Kirby & Spence). 2. Descriptive of Antennae that resemble Cilia (*e.g.* mantids). Alt. Ciliated; Ciliatus.

CILIUM Noun. (Latin *cilium* = eyelid. Pl., Cilia.) Fringes arranged from series of moderate or thin Setae arranged in tufts or single lines; thin scattered Setae on a surface or margin.

CIMBICIDAE Leach 1817. Plural Noun. Cimbicid Sawflies. A Family of Symphyta (Hymenoptera) assigned to the Tenthredinoidea. Adult body 10–30 mm long; Antenna clubbed with 6–7 segments; Hypostomal Bridge absent but sometimes with membranous Hypostoma; Pronotum short with posterior margin arched; forewing with vein 2r-rs; fore Tibia with a pair of unmodified apical spurs; middle Tibia without subapical spurs; Abdomen dorsally convex, ventrally flattened and laterally carinate; Ovipositor sheaths (Gonostyli) weakly projecting beyond apex of Abdomen; male Genitalia strophandrous. Most Species feed on angiosperms (Betulaceae, Rosaceae, Salicaceae).

CIMICIDAE Plural Noun. Bat Bugs; Bed Bugs; Bird Bugs. A small Family of heteropterous Hemiptera assigned to Superfamily Cimicoidea. Adult body oval, flattened, and often reddish-brown in colour; compound eyes present but Ocelli absent; Clypeus expanded apically; Antenna with 4 segments; beak with three segments; wings reduced to scales; Ctenidia absent; Tarsi with three segments. Male punctures body wall of female to inject sperm into special organ called the Spermalege. Adults and nymphs feed on blood as ectoparasites of mammals and birds but live away from the host in places of concealment. See Cimicoidea. Cf. Bed Bug. Rel. Traumatic Insemination.

CIMICINE Noun. 1. An oily fluid of disagreeable odour secreted by some Heteroptera and presumably used as a defensive secretion. 2. Adj. Descriptive of or pertaining to true bugs.

CIMICOIDEA A Superfamily of Hemiptera including Anthocoridae, Cimicidae, Medocostidae, Nabidae, Plokiophilidae, Polyctenidae and Velocipedidae. Group characterized by haemocoelic insemination with fertilization in the

Vitellarium; embryogenesis occurs before oviposition and the egg lacks a Micropyle.

CIMIER Noun. (Eytmology obscure.) The head crest in pierid chrysalids.

CINCH® See Chlorpyrifos.

CINCTUS Adj. (Latin, *cingulum* = girdle.) With a coloured band; cingulatus.

CINEREOUS COCKROACH *Nauphoeta cinerea* (Oliver) [Blattaria: Blaberidae].

CINEREOUS Adj. (Latin, *cinereus* = ashen; *-osus* = possessing the qualities of.) Ash-coloured, grey tinged with blackish. Alt. Cinereus.

CINERESCENT Adj. (Latin, *cinereus* = ashen.) Ashen in colour or appearance.

CINERIN® See Pyrethrin.

CINEROLONE® See Pyrethrin.

CINGOVSKI, JONCE DIMKO (1926–1983) Yugoslavian custodian at Museum of Natural History of Macedonia, Skopje and student of Symphyta.

CINGULA Plural Noun. (Latin, *cingulum* = girdle.) Coloured bands.

CINGULATE Adj. (Latin, *cingulum* = girdle; *-atus* = adjectival suffix.) Pertaining to structure with one or more Cingula or bands. Alt. Cingulatus. See Cinctus.

CINNABAR MOTH *Tyria jacobaeae* (Linnaeus) [Lepidoptera: Arctiidae]: A moth imported into several countries for biological control of tansy ragwort, *Senecio jacobaea* Linnaeus. CB was imported into Pactific Northwest of USA ca 1960. Adults are active May-June and eggs are deposited on undersides of leaves. Larvae feed on flowers and foliage during summer with plants often defoliated by feeding of larvae. See Arctiidae. Rel. Ragwort Flea Beetle.

CINNABARINE Adj. (Latin, *cinnebaris* > Greek, *kinnabarri*.) Vermilion red; the colour of cinnabar, or red oxide of mercury.

CINNAMOMEOUS Adj. (Latin, *cinnamomum* from Hebrew, *qinnamon* = cinnamon tree; Latin, *-osus* = possessing the qualities of.) Cinnamon brown. Alt. Cinnamomeus.

CINURA See Thysanura.

CIOIDAE See Ciidae.

CIPC® See Chlorpropham.

CIRCA Preposition. (Latin, *crica*.) About; near; around. Often used with a numeral to indicate an approximate time or number. Though anglicized, *circa* should be italicized. Abbrev. *ca.* or *c.* Alt. Circiter.

CIRCADIAN RHYTHM 1. A term proposed by Halberg *et al.* (1959) to describe any endogenous biological phenomenon that is manifest on a 24-hour periodicity. 2. An internal biological clock. 3. A behavioural activity or biological event that occurs around a 24-hour period. See Biological Clock; Diel Periodicity; Photoperiodism. Cf. Endogenous Rhythm; Exogenous Rhythm. Rel. Behaviour.

CIRCINATE Adj. (Latin, *circinatus* = made round.) Ring-shaped; coiled. A term used in Botany to describe structure that is spirally coiled or rolled up on an axis with the apex at the centre of the coil. Examples inlcude a watch-spring, fern frond and butterfly's Proboscis. Alt. Circinal. Adv. Circinately. Rel. Form 2; Shape 2; Outline Shape.

CIRCITER Prep. (Latin, *circus* = circle.) About; round-about. Alt. Circa.

CIRCUITOUS TRANSMISSION Medical Entomology: The complex passage of a pathogen between members of arthropod and vertebrate host-species. See Transmission; Vector. Cf. Horizontal Transmission; Vertical Transmission.

CIRCULAR BLACK SCALE *Chrysomphalus aonidum* (Linnaeus) [Hemiptera: Diaspididae]: A cosmopolitan pest of avocado, banana, *Citrus* and subtropical ornamentals including oleander and palms. CBS resembles California Red Scale in size and habitus; life history is similar to CRS but body of female CBS does not become attached to cover during moulting and CRS produces crawlers. CBS adult female cover is circular, dark reddish brown to black and 2.0 mm diameter; male cover is oval and ca 1.0 mm long. Female oviposits ca 300 eggs under her cover; crawlers settle near mother on leaves and fruit causing leaf drop and fruit disfigurement; CBS completes two nymphal instars and the life cycle requires 6–8 weeks. Species is multivoltine and generations overlap. CBS is controlled with natural enemies including *Aphtyis holoxanthus* DeBach in some regions. Syn. Florida Red Scale. See Diaspididae.

CIRCULAR PURPLE SCALE *Chrysomphalus aonidium* (Linnaeus) [Hemiptera: Diaspididae]:

CIRCULATION Noun. (Latin, *circulatio* = act of circulating; English, *-tion* = result of an action. Pl., Circulations.) The act or process which involves the regular movement (flow) of fluids or gas along specified channels, tubes or paths. See Circulatory System.

CIRCULATORY SYSTEM The CS organizes the movement of fluids through body cavities and appendages of the insect body; CS is derived from embryonic Mesoderm and displays an 'open' plan in which most circulation is achieved outside of tubules. Direction of flow within body is posterior-to-anterior and dorsal-to-ventral; direction of flow within legs is anterior-surface downward and posterior-surface upward; direction of flow within wings in tubular longitudinal veins from base to apex with return along posterior area of wing. Organs and tissues of CS include Dorsal Vessel (Heart and Aorta), Dorsal and Ventral Diaphragms, Alary Muscles, Fat Body, Haemocytes and Oenocytes. Accessory circulatory structures include Pulsatile Organs. CS functions include transport and storage of metabolites, synthesis of intermediate metabolites, sequestering waste products, hydrostatic skeleton, thermoregulation and defence. See Alary Muscles; Dorsal Vessel; Haemocoel; Haemolymph. Cf. Alimentary System; Nervous System; Reproductive System; Respiratory System. Rel. Hydrostatic Skeleton; Thermoregulation.

CIRCULUS Noun. (Latin, *circulus* = circle.) 1. Coccids: Ventralabia of MacGillivray. 2. A glandular structure, the contents of which are discharged internally (Ferris & Murdock).

CIRCUM- Latin prefix meaning 'around, about, on all sides, in a circle.'

CIRCUMANAL Adj. (Latin, *circum* = around; *anus* = anus.) Descriptive of something surrounding the Anus. Cf. Circumoral.

CIRCUMCAPITULAR FURROW Acarina: A transverse furrow, Sulcus or suture separating the Gnathosoma and Idiosoma. Syn. Circum capitular Suture. (Goff *et al.* 1982, J. Med. Ent. 19: 234.)

CIRCUMFERENTIAL LAMELLAE See Lateral Plates (MacGillivray).

CIRCUMFILI Plural Noun. (Latin, *circum* = around; *filum* = thread.) Cecidomyiidae: Looped filaments or tortuous threads on the antennal segments that are arranged in whorls and presumed sensory in function (Imms).

CIRCUMGENITAL GLAND Small circular glands with an excretory orifice at the apex. Diaspididae: CGs are disposed in groups about the genital orifice. Coccidae: The scattered Cerores on the ventral aspect of the Abdomen cephalad and laterad of the Opercula (MacGillivray). Syn. Multilocular Wax Gland. See Exocrine Gland.

CIRCUMGENITAL PORES Diaspididae: Genacerores (MacGillivray).

CIRCUMOCULAR Adj. (Latin, *circum* = around; *oculus* = eye.) Surrounding the eye.

CIRCUMOESOPHAGEAL COMMISSURES Cords or nerve fibres connecting the Suboesophageal Ganglion with the main trunk of the nervous system.

CIRCUMOESOPHAGEAL CONNECTIVES 1. Nervous connectives between the Brain and the Ventral Nerve Cord near the Stomodaeum. In an arthropod groundplan, CCs form the connectives from the Archicerebrum to the first Ventral Ganglion. In insects, the CCs extend between the Tritocerebral Ganglia and the Mandibular Ganglia (Snodgrass). See Tritocerebrum.

CIRCUMOESOPHAGEAL GANGLION See Ventral Nerve Cord.

CIRCUMORAL Adj. (Latin, *circum* = around; *os* = mouth.) Descriptive of something surrounding the mouth. Cf. Circumanal.

CIRCUMSEPTED Adj. (Latin, *circum* = around.) With a vein encompassing the entire wing margin. See Wing Venation.

CIRCUMSPOROZOITE PROTEIN Surface proteins on the sporozoite-stage of *Plasmodium falciparum*, the microorganism responsible for Malaria. The protein is the principal antigen responsible for the development of antibodies to Malaria infection. The development of a vaccine against the Malaria parasite has been slow because circumsporozoite vaccines have not been sufficiently immunogenic.

CIRCUMVERSION Adj. (Latin, *circum* = around; *avertere* = turn away.) A turning around.

CIRDEI, FILIMON (–1971) (Dumont & Kiuta 1972, Odonatologica 1: 165–166, bibliogr.)

CIRILLO, DOMINIQUE (1734–1799) Italian republican, naturalist, Neapolitan physician, Professor of Botany and friend of Buffon. Cirillo was executed rather than swear allegiance to King Ferdinand. Principal writings include *Fundamenta Botanica* (1787) and *Entomologicae Neapolitanae Specimen* (1787). (Weiss 1936, *Pioneer Century of American Entomology.* 320 pp. (103–104), New Brunswick.)

CIRRATE ANTENNA A pectinate Antenna with very long curved lateral branches, sometimes fringed with Setae. See Antenna. Cf. Plumose.

CIRROSE Adj. (Latin, *cirrus* = curl; *-osus* = full of.) Furnished with a fringe of Setae; fringed, having one or more Cirri (Kirby & Spence). Alt. Cirrous; Cirrosus. Cf. Plumose.

CIRRUS Noun. (Latin, *cirrus* = curl. Pl., Cirri.) A curled lock of Setae placed on a thin stalk.

CISANAL SETAE Coccids: The shorter and distal two of four Setae near the caudal ring (MacGillivray). See Setae.

CISIDAE See Ciidae.

CISLIN® See Deltamethrin

CISMONTANE Adj. (Latin, *cis* = on this side; *mons* = mountains.) Pertaining to this side of the mountains; on the hither side of any mountan range. Cf. Transmontane.

CIST, JACOB (1782–1825) (Weiss 1936, *Pioneer Century of American Entomology,* 320 pp. (104), New Brunswick.)

CISTA ACUSTICA An elongate ridge or crest of scolopophores of the integumental type (Imms).

CITATION® See Cyromazine.

CITRAZON® See Benzoximate.

CITRICOLA SCALE *Coccus pseudomagnoliarum* (Kuwana) [Hemiptera: Coccidae]: A moderate pest of *Citrus* which is particularly troublesome in USA, Mexico, Russia, Iran, Australia and Japan. CS also attacks walnut, elm, blackberry and pomegranate. CS produces copious honeydew that promotes sooty mould. Adult female 3–4 mm long, convex and grey-brown; cover with longitudinal ridge extending length of body; immatures are flat and transparent. Females can produce 1,000–1,500 eggs during 4–6 weeks of late spring to early summer; eclosion occurs within 2–3 days. Crawlers settle on leave and twigs; CS passes through two nymphal instars with later stages migrating to twigs. Populations are synchronous and typically univoltine. See Coccidae. Cf. Black Scale; Hemispherical Scale.

CITRINE Adj. (Latin, *citrus* = lemon.) Lemon yellow. Alt. Citrinus.

CITRONELLA OIL An essential oil for histological clearing of tissue; CO is prepared from Oil of Lemon or formed by oxidation of Citronellol (an unsaturated alcohol) or extracted from grass *Cymbopogon nardus.*

CITROPHILOUS MEALYBUG *Pseudococcus calceolariae* (Maskell) [Hemiptera: Pseudococcidae]: A pest of *Citrus*, pome and stone fruits and ornamental plants. CMB is endemic to eastern Australia and after 1910 was adventive to North and South America, South Africa, Europe and New Zealand. Adult female 3–4 mm long, slow-moving, wingless, with mealy white wax covering the body; four apparent dark red longitudinal lines extend along dorsum and four short 'tail' filaments project from caudal end of body; body contains dark red fluid (Haemolymph). Adult male is considerably smaller, more fragile, lacks functional mouthparts and displays two wings and long caudal filaments. Female oviposits ca 500 eggs in specially constructed egg sac; eclosion occurs within 2–3 days; crawlers seek sheltered habitat. Female passes through two post-crawler nymphal instars while the male undergoes a prepupal and pupal stage. Nymphs and adult females produce substantial amounts of honeydew which serves as substrate for sooty mould. CMB is multivoltine with up to four generations per year, depending upon climate. Complete biological control of CMB in California was achieved by imported parasite *Coccophagus gurneyi* Compere. See Pseudococcidae.

CITRUS ANT *Oecophylla smaragdina* (Fabricius) [Hymenoptera: Formicidae]: A large, entomophagous, arbicolous ant that weaves leaves and twigs to form nests in trees. CA has been used for centuries in biological control of *Citrus* pests. CA is endemic to China; elsewhere CA is used for control of other crop pests. In China the ant is called Huang Jin Yi (Yellow Fear Ant) or Huang Gan Yi (Yellow Citrus Ant). See Formicidae.

CITRUS BLACKFLY *Aleurocanthus woglumi* Ashby [Hemiptera: Aleyrodidae]: A widespread, polyphagous pest endemic to south Asia and occurs in Thailand, Burma, Malaya, Sumatra, Java, Phillipines, Singapore and Borneo. Important host plants include *Citrus*, mango, pear, coffee, persimmon. CBF pupal case black, and has black spine-like projections; confused with Orange Spiny Blackfly, *Aleurocanthus spiniferus* (Quaintance), known from Thailand and originally described from Java. CBF eggs are attached by Pedicel to underside of leaves in spiral-formation. Biological control of CBF has been achieved in several regions with imported parasitic Hymenoptera including *Eretmocerus serius*, *Amitus hesperidium* and *Encarsia opulenta*. See Aleyrodidae. Cf. Woolly Whitefly.

CITRUS-BLOSSOM BUG *Austropeplus* sp. [Hemiptera: Miridae]: A minor pest of *Citrus* leaves and flowers; CBB is endemic to eastern Australia and probably feeds on unidentified native host plants. Adult 5 mm long with body predominantly brown, Prothorax with yellow, Scutellum with a 'V' shaped green mark and red spot on forewing, venter of body green. Eggs are laid on plant. CBB undergoes five nymphal instars and completes a life cycle within 6–8 weeks. Species is multivoltine with ca six generations per year. See Miridae.

CITRUS BLOSSOM-MIDGE *Cecidomya* sp. [Diptera: Cecidomyiidae]: A minor pest of *Citrus* blossoms in eastern Australia. CBM is believed endemic to Australia and not known to occur on other host plants. Adult ca 2 mm long and short-lived; female lays ca 50 eggs in *Citrus* blossoms and eclosion occurs 1–2 days after oviposition. Larvae feed ca 7 days. Life cycle requires ca 14 days. CBM is multivoltine with up to 12 generations per year. See Cecidomyiidae.

CITRUS BRANCH-BORER *Uracanthus cryptophagus* Olliff [Coleoptera: Cerambycidae]: A minor and infrequent pest of *Citrus* in Australia. CBB also feeds on other Rutaceae, particularly finger lime *(Microcitrus australasica)*. Adult ca 40 mm long, reddish brown with Antenna as long as body. Eggs are laid in cracks in twigs and small branches; eclosion occurs ca 10 days. Neonate larvae bore into branches to form tunnels and eventually ringbarking to cause loss of entire limbs. CBB is monovoltine. See Cerambycidae.

CITRUS BUD-MITE *Eriophyes sheldoni* (Ewing) [Acari: Eriophyidae].

CITRUS BUTTERFLIES See Large Citrus-Butterfly; Small Citrus-Butterfly.

CITRUS FLAT-MITE *Brevipalpus lewisi* McGregor [Acari: Tenuipalpidae].

CITRUS FLOWER-MOTH *Prays nephelomima* Meyrick [Lepidoptera: Yponomeutidae]: A pest of *Citrus* in southeast Asia and Australia. Adult and larva resemble Lemon Bud Moth and the biology and life cycle of both Species are similar. See Yponomeutidae.

CITRUS FRUIT-BORER *Citripestis sagittiferella* [Lepidoptera: Pyralidae]: A minor pest of *Citrus* in southeast Asia.

CITRUS FRUIT-WEEVIL *Neomerimnetes sobrinus* (Lea) [Coleoptera: Curculionidae]: A minor and sporadic pest of *Citrus*, mainly oranges. CFW is endemic to Australia and only known to feed on *Citrus*. Adult tan coloured and ca 7 mm long. Larvae feed on *Citrus* roots but cause little damage. Adults emerge during spring and damage rind of young fruit. CFF is univoltine. See Curculionidae.

CITRUS GALL-WASP *Bruchophagus fellis* (Girault) [Hymenoptera: Eurytomidae]: A pest endemic to Australia that occurs on all *Citrus* varieties but usually causes serious damage only on grapefruit; endemic host plant is finger lime *(Microcitrus australasica)*. Adult ca 2.5 mm long, black with shiny Abdomen. Female oviposits between bark and wood of young twigs; eclosion occurs within 2–4 weeks; Larvae are thick-set, white and legless; neonate larvae burrow into twigs, feed and form galls after several months. Larvae pupate within galls during late winter; pupal period requires 2–3 weeks and adult lives about 1 week.

CGW is Univoltine. See Eurytomidae.

CITRUS GROUND-MEALYBUG *Geococcus citrinus* Kuwana [Hemiptera: Pseudococcidae].

CITRUS JASSID See Citrus Leafhopper.

CITRUS KATYDID *Caedicia strenua* (Walker) [Orthoptera: Tettigoniidae]: In Australia, a minor and sporadic pest of *Citrus,* attacking Valencias and Washington navels; lemons rarely are affected. Adult green, ca 40 mm long; forewing narrow and opaque; hindwing fan-like and transparent; female with short, recurved Ovipositor. CK damages young fruit and shows preference for thickly-foliaged trees. Natural host plants are not known but blackberries are a common food. See Tettigoniidae. Cf. Angular-Winged Katydid; Broad-Winged Katydid; Crested Katydid; Forktailed Bush-Katydid; Inland Katydid; Mottled Katydid; Philippine Katydid; Spotted Katydid.

CITRUS LEAF-EATING CRICKET *Tamborina australis* (Walker) [Orthoptera: Gryllidae]: A minor pest of *Citrus* in eastern Australia; probably endemic but not recorded on other host plants. Adult pale brown and 35–40 mm long; Antenna twice length of body; hindwing curled and projecting beyond apex of Abdomen; Ovipositor projecting to apex of hindwing. Eggs are laid in green twigs. Nymphs and adults feed nocturnally on foliage by chewing margin of leaves, grazing on surface or chewing holes in leaves; CLEC completes five nymphal instars. Species is multivoltine with 4–5 generations per year. See Gryllidae.

CITRUS LEAF-EATING WEEVIL *Eutinophaea bicristata* Lea [Coleoptera: Curculionidae]: A minor and sporadic pest of *Citrus* which is endemic to Australia. Adult 3 mm long and mottled grey-brown; emerge during spring and feed on young foliage on the lower third of the tree; chew out sections on both leaf surfaces. CLEW larvae feed on roots but create no obvious signs of damage; injury is most serious on lower parts of tree. Pupation occurs in soil. See Curculionidae.

CITRUS LEAFHOPPER *Empoasca smithi* (Fletcher & Donaldson) [Hemiptera: Cicadellidae]: An endemic pest of *Citrus* and castor oil plant in Australia. Adult 3–4 mm long, pale green, broadest across head with body tapering posteriad. Female oviposits eggs individually into midrib of young leaves. Eclosion occurs within one week and CLH undergoes five nymphal instars. CLH life cycle requires 4–5 weeks and completes at least six generations per year. See Cicadellidae.

CITRUS LEAFMINER *Phyllocnistis citrella* Stainton [Lepidoptera: Gracillariidae]: A widespread, serious pest of *Citrus* which is endemic to southeast Asia and spread elsewhere since 1990. Adult ca 2 mm long, delicate with narrow forewing and hingwing with long marginal fringe. Adult female lives ca 7 days and mating occurs at dusk or evening. Female lays ca 50 eggs during lifetime and deposits eggs individually on underside of leaves near midrib. Eggs are oval, ca 0.3 mm long and translucent; eclosion occurs within a day. First instar larvae burrow into leaf to construct mine; larvae are pale-green and consume contents of epidermal cells lacerated by mouthparts. Leaf mines are sinuous, silvery and lined centrally with dark frass. CLM completes three larval instars and completes feeding within 5–6 days; larvae do not leave mine or move from lower to upper surface of leaf. Prepupae are yellowish brown and spin silk to line pupal chamber; prepupal stage is complete in one day; pupal stage requires six days. CLM is multivoltine with up to 15 geneations per year. See Gracillariidae.

CITRUS LEAFROLLER *Psorosticha zizyphi* (Stainton) [Lepidoptera: Oecophoridae]: A pest of *Citrus* in southeast Asia and Australia. Adult ca 6–7 mm long, greyish-brown with black spots on wings. Eggs are pale-green and individually laid on *Citrus* leaves near midrib; up to 300 eggs laid per female during six-week period; eclosion occurs within 3–4 days. Larvae are yellow-orange with dark head. CLR requires five larval instars. Early instar larvae form silk gallery on leaf surface; later instar larvae roll young leaves and feed inside tubes. Mature larvae spin cocoons within leaf roll. CLR life cycle requires 3–4 weeks. Species is multvoltine with ca 10 generations per year. See Oecophoridae.

CITRUS LONGICORN *Skeletodes tetrops* Newman [Coleoptera: Cerambycidae]: A pest of *Citrus* in Australia. Adult ca 12 mm long, brownish-grey with legs unusually long. Female lays eggs in cracks on small branches or where other borers have damaged tree; eclosion occurs within 10 days. Neonate larvae bore into wood and larvae feed under bark. CL is monovoltine. Syn. Spider Longicorn. See Cerambycidae.

CITRUS MEALYBUG PREDATOR *Oligochrysia lutea* (Walker) [Neuroptera: Chrysopidae]. (Australia).

CITRUS MEALYBUG *Planococcus citri* (Risso) [Hemiptera: Pseudococcidae]: A cosmopolitan pest of numerous subtropical and tropical fruits and ornamental plants, including avocado, *Citrus*, cotton, cucumber, potatoes, strawberries, tomatoes and ornamental plants. Adult female ca 3 mm long with mealy white covering and 18 pairs of short wax filaments along margin of body; filaments are shortest at anterior margin of body and become progressively longer posteriad with longest pair at posterior end of body. Male considerably smaller than female, fragile, with non-functional mouthparts, one pair of functional wings and two long filaments at posterior end of body. CMB resembles Longtailed Mealybug but posterior filaments considerably shorter; resembles Citrophilous Mealybug but body contains yellow fluid and lacks red longitudinal stripes. Female CMB oviposits 300–600 pale yellow, oval eggs during a 1–2 week period; eggs are deposited in long, cottony egg sac positioned under and behind female; female dies following oviposition. Eclosion occurs within a week and CMB overwinters in egg stage. Females moult three times; males moult four times, and adult male with pair of wings and pair of Halteres. CMB nymphs and adult females produce copious amounts of honeydew which serves as substrate for sooty mould. CMB is multivoltine with up to 6 generations per year, depending upon host plant and climate. Biological control of CMB has achieved variable results with predatory beetles and parasitic wasps. See Pseudococcidae.

CITRUS PLANTHOPPER *Colgar peracutum* (Walker) [Hemiptera: Flatidae]: A pest of asparagus, *Citrus*, grapes, guava, macadamia and some ornamentals in Australia. Adult ca 8 mm long and pale green to white with small red spot in middle of forewing and pale red stripe along margin of forewing; wing triangular in outline shape from lateral view; wings held tentiform when at repose. Eggs are laid in circular mass comprising about 50 eggs. CPH is multivoltine with life cycle requiring 1–2 months, depending upon climate and ho s plant. See Flatidae.

CITRUS PSYLLA *Trioza eryteae* (Del Guercio) [Hemiptera: Triozidae].

CITRUS RED-MITE *Panonychus citri* (McGregor) [Acari: Tetranychidae]. (Australia).

CITRUS RINDBORER 1. *Adoxophyes templana* (Pag) [Lepidoptera: Tortricidae]: A *Citrus* pest that is endemic to Australia and probably has alternative native host plants. Adult small (ca 5–6 mm long), pale brown with darker markings. Eggs are pale green and laid on leaves or fruit. Larvae are pale yellow, feed on young foliage or penetrate fruits at their points of contact; CRB completes five larval instars. Pupation occurs within silken chamber in folded leaf. Life cycle requires one month and Species is multivoltine with about seven generations per year. 2. *Prays endocarpa* Meyrick [Lepidoptera: Yponomeutidae]: Endemic to southeast Asia where the larva mines the rind of *Citrus.* See Yponomeutidae. Cf. Olive Moth.

CITRUS ROOT-BARK CHANNELER *Pseudomydaus citriperda* (Tryon) [Coleoptera: Curculionidae]. (Australia).

CITRUS ROOT-MEALYBUG *Rhizoecus hondonis* Kuwana [Hemiptera: Pseudococcidae].

CITRUS ROOT-WEEVIL *Pachnaeus litus* (Germar) [Coleoptera: Curculionidae].

CITRUS RUST-MITE. *Phyllocoptruta oleivora* (Ashmead) [Acarina: Eriophyidae]. (Australia).

CITRUS RUST-THRIPS *Chaetanaphothrips orchidii* (Moulton) [Thysanoptera: Thripidae]: A minor and sporadic pest of *Citrus* in Australia, Europe and USA. Adult 1.5 mm long, yellow and wings narrow with long fringe and black marks. Eggs are laid in fruit or leaf tissue. CRT feeding damage results in brown rust marks on rind between touching fruits. CRT resembles Banana Rust Thrips. CRT move rapidly into sheltered positions when disturbed or exposed to light. Pupation occurs in soil. Life cycle requires 3–5 weeks; Species is multivoltine. See Thripidae.

CITRUS SNOW-SCALE See White Louse Scale.

CITRUS SWALLOWTAIL *Papilio xuthus* Linnaeus [Lepidoptera: Papilionidae].

CITRUS THRIPS *Scirtothrips citri* (Moulton) [Thysanoptera: Thripidae]: A pest of *Citrus* in western North America. CT also feeds upon numerous other Species of plants. Eggs are kidney-shaped and inserted into host plant leaves, stems, twigs or fruit; overwinters in egg stage; multivoltine up to 12 generations per year. See Thripidae.

CITRUS TRUNK-BORER *Platyomopsis pulverulens* (Boisduval) [Coleoptera: Cerambycidae]. (Australia).

CITRUS WHITEFLY *Dialeurodes citri* (Ashmead) [Hemiptera: Aleyrodidae]: A widespread, arrhenotokous pest of *Citrus*. CWF is endemic to Old World and adventive to New World by 1880. Eggs are pale-yellow, pedicellate and deposited in underside of leaves. Nymphs move to feeding site, moult and lose legs. CWF produces honeydew which causes sooty mould to drop on fruit. Alternative host plants include ash, chinaberry, gardenias, privet. Long-term biological control attempts have resulted in mixed results. See Aleyrodidae. Cf. Cloudywinged Whitefly.

CIVIL TWILIGHT See Twilight.

CIXIIDAE Plural Noun. A cosmopolitan Family of Hemiptera including about 1,300 Species and assigned to Superfamily Fulgoroidea. Nymphs are root feeders, cavernicolous or myrmecophilous. See Fulgoroidea.

CLAASEN, PETER WALTER (1886–1937) (Osborn 1946, *Fragments of Entomological History.* Pt. II. 232 pp. (66), Columbus, Ohio.)

CLADE Noun. (Greek, *klados* = sprout. Pl., Clades.) Taxonomy: A monophyletic group, *i.e.,* a natural taxon. See Cladism. Rel. Classification.

CLADISM (Greek, *klados* = sprout; English, *-ism* = doctrine.) A philosophical approach to classification which employs logic to develop principles and methods for the purpose of defining taxonomic groups based on genetic relatedness of the members. Relatedness in cladistic principles is defined on the basis of shared, derived characters. In cladistic methodology, derived characters (apomorphies) are distinguished from ancestral characters (plesiomorphies). 'Natural' classifications are regarded as monophyletic; classifications that include members of 'unrelated' lineages are regarded as polyphyletic. Identification of monophyletic groups is made difficult by parallel evolution and character convergence. Cladism was first articulated by Willi Hennig during the 1950s but was not understood, or misunderstood, until critical issues were clarified by workers during the 1970s. See Classification. Cf. Numerical Taxonomy. Rel. Evolution.

CLADISTIC VICARIANCE BIOGEOGRAPHY A philosophical fusion of Cladistic Phylogenetics and Vicariance Biogeography to explain the geographical distribution of organisms. See Vicariance Biogeography. Cf. Panbiogeography.

CLADISTICS (Greek, *klados* = sprout.) Pertaining to Cladism or its methods.

CLADOCEROUS Adj. (Greek, *klados* = sprout; *keras* = horn; Latin, *-osus* = possessing the qualities of.) Pertaining to insects with branched horn-like structures or Antennae.

CLADOGRAM Noun. (Greek, *klados* = sprout; *graphein* = to write. Pl., Cladograms.) A 'branching diagram' that purports to explain or illustrate taxonomic relationship among organisms which are presumed related through common descent. Cladograms are based on origin and sequence of divergence as manifest during long periods of time and when subject to the forces of evolution. Cf. Dendrogram; Phenogram.

CLAES, EDGARD (–1895) (Selys-Longchamps 1895, Ann. Soc. ent. Belg. 39: 241–242.)

CLAIRVILLE, JOSEPH PHILIPPE DE (1742–1830) An Englishman living in Switzerland and author *of Entomolgie Helvetique* (2 vols; 1798, 1806). (Geilinger 1935, Mitt. naturw. Ges. Winterthür. 19: 281–288.)

CLAMBIDAE Plural Noun. A small Family of Coleoptera assigned to Eucinetoidea. Adult oval, convex, 1–1.5 mm long; Antenna with 10 segments, club of two segments; compound eye at least partially divided; hind Coxa with large sclerite that conceals hind Femur; tarsal formula 4-4-4; Abdomen with 5–6 Ventrites. Larval head prognathous; Antenna with three segments and legs with five segments; Urogomphi absent. Clambids inhabit leaf litter and feed upon fungal spores. See Coleoptera.

CLAPAREDE, JEAN LOUIS RENE ANTOINE EDOUARD (1832–1871) (Vogt 1873, J. Zool. Paris 2: 138–159.)

CLAPAREDE'S ORGAN Acarina: A tubular structure projecting between Coxae I and II. CO typically is covered with a protective lid and may function in water balance. (Goff *et al.* 1982, J. Med. Ent. 19: 222.)

CLARK, ALEXANDER HENRY (1837–1911) (Morice 1911, Proc. ent. Soc. Lond. 1911: cxxiii.)

CLARK, AUSTIN HOBART (1880–1954) (Mallis 1971, *American Entomologists*. 549 pp. (336–337), New Brunswick.)

CLARK, BENJAMIN PRESTON (1860–1939) (Usinger 1939, Pan-Pacif. Ent. 15: 90.)

CLARK, BRACY (1771–1860) (Douglas 1860, Proc. ent. Soc. Lond. 5: 148–149.)

CLARK, EDGAR JAMES (–1945) (Carpenter 1945, Entomologist's mon Mag. 81: 143.)

CLARK, GOWAN CRESSWELL CONINGSBY (1888–1964) (C. G. D. 1964, Entomologist's Rec. J. Var. 76: 173–174.)

CLARK, HAMLET (1823–1867) (Musgrave 1932, *A Bibliography of Australian Entomology 1775–1930*. 380 pp. (48), Sydney.)

CLARK, JOHN S (1885–1956) (Brown 1956, Ent. News 67: 197–199.)

CLARK, JOHN ADOLPHUS (1842–1908) (Sequera *et al.* 1909, Entomologist's Rec. J. Var. 21: 22–24.)

CLARK, MARIANNE, E (1880–1963) (Spencer 1963, Proc. ent. Soc. Br. Columb. 60: 53–54.)

CLARKE, ALEXANDER HENRY (1839–1911) (Turner 1911, Entomologist's Rec. J. Var. 23: 256.)

CLARKE, CORA HUIDEKOPER (1851–1911) (Anon. 1916, Ent. News 27: 384.)

CLARKE, EDWARD DANIEL (1769–1822) (Rose 1850, *New General Biographical Dictionary* 6: 341–342.)

CLARKE, WARREN THOMPSON (1863–1929) American academic (University of California) and specialist in economic Entomology who studied life histories of grape leafhopper, peach twig borer and potato tuber moth. (Essig 1931, *History of Entomology*. 1029 pp. (568–570), New York.)

CLASP Noun. (Ango Saxon, *clyppan* = to embrace. Pl., Clasps.) Collembola: See Retinaculum.

CLASPER Noun. (Anglo Saxon, *clyppan* = to embrace. Pl., Claspers.) 1. Genitalia of male Lepidoptera: Term used by most authors as synonymous with Harpe. Sometimes clasper is used to refer to specialized holdfast organs developed on inner face of Harpe (Klots). 2. Hymenoptera: Parameres of external genitalia.

CLASPETTE Noun. (Anglo Saxon, *clyppan* = to embrace. Pl., Claspettes.) Genitalia of male Culicidae: Inner basal lobe of the side-piece.

CLASP-FILAMENT Genitalia of male Culicidae: The articulated appendage or terminal segment of the side-piece or clasp. CF sometimes bears an articulated point or apex and then is the articulated apex (Smith).

CLASS Noun. (Latin, *classis* = division. Pl., Classes.) 1. A division of the animal Kingdom below Phylum and above Order. The Insecta constitute a Class within the Phylum Arthropoda. 2. A set of particular 'things,' considered together through a defining property.

CLASSICAL BIOLOGICAL CONTROL The regulation of a pest population by exotic natural enemies (parasites, predators, pathogens) that have been moved from one area to another (imported) for this purpose. See Biological Control. Cf. Augmentative Biological Control; Conservation Biological Control. Rel. Chemical Control; Integrated Pest Management.

CLASSIFICATION Noun. (Latin, *classis* = division. English, *-tion* = result of an action. Pl., Classifications.) The arrangement (placement) of organisms into taxonomic groups based on characteristics used to define each group. Membership in groups is based on structural features (anatomy), biological characteristics, behavourial attributes or biochemical criteria. Classifications of organisms in biology may be 'artificial' or 'natural'. Artificial classifications are based on features that are convenient to study or easy to observe. Natural classifications are based on features that are shared by members of a group through evolutionary descent. Cf. Systematics; Taxonomy.

CLATHRATE Adj. (Greek, *klethra* = lattice; *-atus* = adjectival suffix.) Reticulate; latticed. Pertaining to structure with elevated ridges crossing each other at right angles. Alt. Clathratus; Clathrose; Clathrosus.

CLAUS, CARL FRIEDRICH WILHELM (1835–1898) (Grabben 1899, Verh. zool.-bot. Ges. Wien 49: 112–116.)

CLAUSEN, CURTIS P American government Entomologist (USDA) during most of his career, working for several years in overseas laboratories (Japan) and later at the University of California (Riverside) (1951–1956). Specialist in Biological Control and the biology of entomophagous insects; author of book *Entomophagous Insects.*

CLAUSSEN, PETER (1804–) (Papavero 1971, *Essays on the History of Neotropical Dipterology.* 1: 89–91. São Paulo.)

CLAUSTRAL COLONY FOUNDING Social Insects: The procedure during which queen ants or royal pairs of termites sequester themselves in a cell and rear the first generation of workers using internal body tissues (fat body, histolysed wing muscles) as the source of nutrition.

CLAUSTRUM Noun. (Latin, *claustrum* = bar. Pl., Clastra.) The device that holds the fore and hindwings together during flight. The device includes hooks of one wing attached to a thickened margin of the other wing, or by a Jugum or Frenum. See Frenum; Hamulus; Jugum.

CLAVA Noun. (Latin, *clava* = club. Pl., Clavata.) 1. Any club-shaped structure. 2. The enlarged apical segments of an Antenna which are differentiated in size and shape from the preceding segments of the Flagellum. Syn. Clavola. See Antenna; Club.

CLAVACERATUBAE Plural Noun. (Latin, *clava* = club; *cera* = wax; *tubus* = tube.) Coccidae: Ceratubae in the Pygidia of Aspidiotini with an enlarged inner end, which gives them a distinct club-shaped or clavate appearance (MacGillivray).

CLAVAL FOLD Hymenoptera: A furrow on the forewing anterior of and parallel with vein 1A, and which extends to the claval notch on the wing margin.

CLAVAL SUTURE Heteroptera: The suture at the base of Hemelytron, separating the Clavus. See Clavus.

CLAVATE HAIRS Collembola: Tenent Setae.

CLAVATE SCALE *Clavaspis herculeana* (Cockerell & Hadden) [Hemiptera: Diaspididae]. (Australia).

CLAVATE SETAE Coccids: Certain marginal Setae with enlarged or dilated ends (MacGillivray). Setae that are short and narrow basally, broadening distally and blunt apically. (See Ferris 1923)

CLAVATE Adj. (Latin, *clava* = club; *-atus* = adjectival suffix.) 1. Clubbed; descriptive of Antennae that gradually or abruptly thicken, expand or dilate toward the apex. Alt. Clavatus; Claviform. See Club. Rel. Antenna. 2. Descriptive of surface sculpture, usually the insect's integument, that displays club-like projections that thicken gradually toward the apex. See Sculpture Pattern. Cf. Alveolate; Baculate; Carinate; Echinate; Favose; Gemmate; Psilate; Punctate; Reticulate; Rugulate; Scabrate; Shagreened; Smooth; Striate; Subclavate; Verrucate.

CLAVICORNIA Noun. A series of beetles the members of which display Antennae more-or-less distinctly enlarged or clubbed at the apex. See Cucujoidea.

CLAVICULA Noun. (Latin, *clavicula* = small key. Pl., Claviculae.) The Coxa of the anterior legs (Kirby & Spence).

CLAVICULAR LOBE Hemiptera: That portion of hindwing posterior of the Anal Veins.

CLAVIFORM Adj. (Latin, *clava* = club; *forma* = shape.) 1. Club-shaped; a descriptor often applied to Antennae or other appendages. 2. Noctuid moths: An elongate spot or mark extending from the transverse anterior line through the submedian interspace, toward and sometimes to the transverse posterior line. Alt. Clavate. See Club. Rel. Form 2; Shape 2; Outline Shape.

CLAVIGERATE ANTENNA An Antenna that terminates in a gradual club. See Antenna.

CLAVOLA Noun. (Latin, *clava* = club. Pl., Clavolae.) The insect Antenna except the first and second segments (Comstock); the Clava *sensu* Smith.

CLAVULUS See Frenulum.

CLAVUS Noun. (Latin, *clavus* = club, nail. Pl., Clavi.) 1. The club of an Antenna; the Clava; Clavola. 2. Hymenoptera: The knob or enlargement at the apex of the Stigmal or Radial Veins of the forewing in some Parasitica. 3. Heteroptera: The sharply pointed anal area of the Hemelytra; adjacent to the Scutellum when folded, in coccids. 4. Genitalia of male Lepidoptera: A rounded or finger-like process, usually setose, arising from the Sacculus of the Harpe (Klots).

CLAW Noun. (Anglo Saxon, *clawu* = claw. Pl., Claws.) 1. A hollow, sharp, multicellular organ, generally paired, at the apex of the insect leg. 2. One (or more) corneous, sharp structures that arm the lobes of the Maxillae (Kirby & Spence). 3. Any sharp, curved process at the apex of an appendage.

CLAYBACKED CUTWORM *Agrotis gladiaria* Morrison [Lepidoptera: Noctuidae]. See Cutworms.

CLAYCOLOURED BILLBUG *Sphenophorus aequalis aequalis* Gyllenhal [Coleoptera: Curculionidae].

CLAYCOLOURED LEAF BEETLE *Anomoea laticlavia* (Forster) [Coleoptera: Chrysomelidae].

CLEARING Adj. (Latin, *clarus* = clear, bright.) A term applied to the histological process by which a transitional fluid replaces absolute ethanol in specimens being prepared for microscopical study. The term 'clearing' is a misnomer; clearing is a consequence of the agent used, not the ultimate objective of the process. Early histologists used essential oils (terpineol, clove oil, cedarwood oil) as transitional fluids and these fluids clear tissue. See Histology. Cf. Maceration.

CLEAR-LAKE GNAT *Chaoborus astictopus* Dyar & Shannon [Diptera: Chaoboridae]: A fly endemic to western USA. Adult brown with pale coloured Setae; active May-October and attracted to lights where they form large swarms; mouthparts short; adults do not feed on blood. Eggs are laid on surface of water and sink to bottom. Larvae are transparent, predaceous and capture prey with prehensile Antenna. See Chaoboridae. Rel. Phantom Larvae.

CLEARMITE® See Azocyclotin.

CLEARWING BORER See Sesiidae.

CLEARWING MOTHS See Aegeriidae.

CLEARWINGED GRASSHOPPER *Camnula pellucida* (Scudder) [Orthoptera: Acrididae]: A pest of range grasses, grains and vegetable crops throughout North America. Adults may swarm but do not migrate long distances. See Acrididae.

CLEARWINGS See Sesiidae.

CLEAVAGE CELLS Cells formed in the egg during segmentation.

CLEAVAGE Noun. (Anglo Saxon = *cleofian* = to cut. Pl., Cleavages.) A series of mitotic divisions that result in transformation of a zygote into an embryo. See Segmentation.

CLEAVAGE LINE A line of weakness in the Integument of the insect body or the Chorion of an insect egg. The line along which emergence from the egg is effected or Ecdysis-Apolysis occurs. CL is variable in form but frequently Y-shaped. See Ecdysial Clevage Line; Epicranial Suture; Operculum.

CLEFT Noun. (Middle English, *clift* = to split > German, *kluft*. Pl., Clefts.) 1. Split; partly divided longitudinally. 2. A space made by a crack, fissure or fracture. 3. Coleoptera: Claws divided such that parts lie one above another.

CLEG Noun. (Old Norse, *klegg*. Pl., Clegs.) 1. A horsefly or gadfly. 2. Specifically, tabanid flies of the Genus *Haematopota* whose eyes display a zig-zag pattern of brilliant colour and speckled wings. Alt. Clegg. Syn. Stouts. See Tabanidae.

CLEGHORN, HUGH FRANCIS CLARKE (1820–1895) (Anon. 1895, J. Botany, Lond. 33: 256.)

CLEIODIC Adj. (Greek, *kleis* = bar; *oon* = egg; *-ic* = of the nature of.) Pertaining to eggs that are enclosed within a membrane.

CLEMATIS BLISTER-BEETLE *Epicauta cinerea* (Forster) [Coleoptera: Meloidae].

CLEMENS, JAMES BRACKENRIDGE (1829–1867) (dos Passos 1951, J. N. Y. ent. Soc. 59: 135.)

CLEMENT, ARMAND LUCIEN (1848–1920) (Iches 1921, Ann Soc. ent. Fr. 90: 66– 68.)

CLEMENTI, VINCENT (1812–1899) (Anon. 1899, Can. Ent. 31: 371.)

CLENCH, HARRY KENDON (1925–1979) (Miller 1980, J. Lepid. Soc. 34: 81–84.)

CLEPTINAE A small Subfamily of aculeate Hymenoptera (Chrysididae), with one Holarctic Genus and and a second Genus in Argentina. Morphologically characterized by a well developed median, longitudinal Sulcus extending ventrad from the median Ocellus; Pronotum with a distinct transverse Sulcus; forewing Discoidal Cell incomplete or absent and Radial Sector short; female with four exposed gastral Terga, Sterna convex and Ovipositor well developed; male with five exposed gastral Terga. Species are parasites of sawfly larvae (Tenthredinidae).

CLEPTOBIOSIS Noun. (Greek, *klepenai* = to steal; *biosis* = a form of life. Pl., Cleptobioses.) A form of commensalism expressed among ant Species in which one Species nests near or on the nests of another Species and feeds upon refuse from middens or steals food from workers or the food stores of the host Species. Syn. Kleptobiosis. See Biosis; Commensalism; Parasitism. Cf. Abiosis; Anhydrobiosis; Antibiosis; Archebiosis; Calobiosis; Hamabiosis; Kleptobiosis; Lestobiosis; Parabiosis; Phylacobiosis; Plesiobiosis; Synclerobiosis; Trophobiosis; Xenobiosis.

CLEPTOPARASITE Noun. (Greek, *klepenai* = to steal; *parasitosis* = parasite; *-ites* = resident. Pl., Cleptoparasites.) 1. A parasite which feeds on the food reserves of another insect. Used primarily in the context of a larva feeding on the food intended for the larva of another Species. The larva for which the food was intended dies, in the absence of food supplies necessary to complete development. See Hyperparasite. Cf. Adelphoparasite; Multiple Parasitism.

CLEPTOPARASITIC Adj. (Greek, *klepenai* = to steal; *parasitosis* = parasite; *-ic* = of the nature of.) Pertaining to a cleptoparasite's behaviour or a cleptoparasite. See Hyperparasitic; Parasitic.

CLEPTOPARASITISM Noun. (Greek, *klepenai* = to steal; *parasitosis* = parasite; *-ism* = condition. Pl., Cleptoparasitisms.) An instance of multiple parasitism in which one parasite preferentially attacks a host already parasitized by another Species of parasite. See Parasitism. Cf. Adelphoparasitism; Hyperparasitism; Multiple Parasitism. Rel. Feeding Strategy.

CLERCK, KARL ALEXANDER (1710–1765) (Bonnet 1945, *Bibliographia Araneorum* 1: 28.)

CLERIDAE Plural Noun. Checkered Beetles. A Family of polyphagous Coleoptera, central group of Cleroidea and including 3,000 nominal Species; clerids are widespread but more abundant in tropical regions. Adult 2–45 mm long; body parallel sided, frequently brightly coloured and conspicuously setose; head somewhat deflexed; Antenna short, with 11 segments, shape variable, usually clubbed; apical palpal segments enlarged (securiform). Fore Coxa large and conical; tarsal formula 5-5-5 or 4-4-4 (basal or penultimate segment often small), segments 1–4 bilobed. Larvae are campodeiform, elongate, cylindrical or somewhat flattened and soft bodied; head well sclerotized; Antenna with three segments; 0–6 Stemmata; legs with five segments, each Pretarsus with one claw; terminal segment well sclerotized; Urogomphi absent or short. Adults and larvae are predaceous and usually attack other insects. Typically, clerids are associated with woody plants (under bark, in tunnels or galls, on plant foliage and dead twigs). Adults typically are diurnal and take (1) carrion *(Necrobia),* (2) pollen *(Trichodes)* or living insects (most clerids); some *Hydnocera* apparently are parasitic and complete development on one host. Larvae are predatory on bees and move from cell to cell devouring bee larvae and provender in cell. Clerids usually lay eggs near their intended prey; eggs hatch and larvae actively search for prey. Larvae that feed on wood boring beetles feed on all stages of prey and inquilines of burrows. See Coleoptera.

CLERK, CHARLES Swedish painter, student of Linnaeus and author of *Icones Insectorum rariorum* (1759–1764).

CLERMAIT® See Azocyclotin.

CLEVELAND, CLARENCE, R (1883–1946) (Anon. 1953, J. Econ. Ent. 46: 538.)

CLICK BEETLES See Elateridae.

CLIDEMIA LEAFROLLER *Blepharomastix ebulealis* (Guenée) [Lepidoptera: Pyralidae].

CLIDEMIA THRIPS *Liothirps urichi* Karny [Thysanoptera: Phlaeothripidae].

CLIMATE Noun. (Latin, *clima* > Greek, *klima* = inclination. Pl., Climates.) 1. 'The sum total of weather at a particular point' (Shelford). 2. The average course of weather conditions at a particular place over a prolonged period of time. Rel. Environment; Temperature; Weather.

CLIMAX AREA Ecology: The most stable and permanent type of biotic environment attainable over a major geographic area (*e.g.* Northern Coniferous Forest). Cf. Subclimax Area.

CLIMAX STAGE The final phase (period, point) of ecological succession that permits a community of organisms to live indefinitely in an area. Rel. Community; Ecology.

CLIMOGRAPH Noun. (Greek, *klima*; *graphein* = to write. Pl., Climographs.) A diagram of climate plotted in the form of a mean monthly wet-bulb temperature and humidity (Shelford).

CLINAL Adj. (Greek, *klinein* = to slant; *-alis* = characterized by.) Descriptive of structure that is inclined or sloped from a horizontal perspective. Cf. Aclinal. Rel. Declivity.

CLINCH® See Chlorpyrifos.

CLINE Noun. (Greek, *klinein* = to slant. Pl., Clines.) A gradual or continuous change in form of a structure over the geographical distribution of a Species possessing the structure. A character gradient. Cf. Ecocline.

CLINTHERIFORM Adj. Plate-shaped. Rel. Form 2; Shape 2; Outline Shape.

CLISTOGASTRA A Suborder of Hymenoptera characterized by members displaying a petiolate Abdomen or a constriction between the Propodeum and third abdominal segment. In such organisms the portion of the Abdomen caudad of the Petiole is termed the Gaster. See Apocrita; Parasitica. Cf. Chalastogastra.

CLISTOGASTROUS Adj. (Greek, *klitos* = cliff; Latin, *-osus* = possessing the qualities of.) Pertaining to Hymenoptera with Gaster separated from Propodeum by a marked constriction or Petiole. See Petiole. Cf. Chalastrogastrous.

CLITHRUM Noun. (Greek, *kleithron* = bar. Pl., Clithra.) Scarabaeoid larvae: A paired, short sclerite in the anterior part of the margin of the Epipharnyx, separating the Corypha and Paria; often absent (Boving).

CLOACA Noun. (Latin *cloaca* = sewer. Pl., Cloacas.) The common chamber into which the Anus and Gonopore open in some insects. Syn. Rectum.

CLOFENTEZINE A tetrazine compound {3,6-bis(2-chlorophenyl)-1,2,4,5-tetrazine} used as a contact acaricide against the egg and larval stages, but not effective against adult mites. Compound is slow-acting, can be used with other pesticides and is non-toxic to bees and predatory mites. Trade names include Acaristop®, Apollo® and Apolo®. See Insecticide.

CLONE Noun. (Greek, *klon* = twig. Pl., Clones.) 1. A group of individuals propagated by mitosis from one ancestor. An apomictic strain. 2. Tissue Culture: A population of genetically identical cells that are derived from an individual cell. Alt. Genetic Copy; Twin.

CLOSED BAND The moveable valvular fold of the inner closing mechanism of a spiracle opposite the valve (Snodgrass).

CLOSED CELL A wing cell completely surrounded by veins (Comstock). Cf. Open Cell.

CLOSED TRACHEAL SYSTEM The anatomical type of Respiratory System in which Spiracles are absent or non-functional. CTS is present in aquatic immature stage of many insect Species in which oxygen enters the Tracheal System over the body surface or through Tracheal Gills. In some Species the Tracheal Plexus lies near the surface of thin Cuticle; in other Species the Tracheal Plexus is restricted to specific areas of the body; in some Species the Tracheal Plexus is restricted to Tracheal Gills. See Tracheal Gill. Cf. Open Tracheal System. Rel. Respiratory System.

CLOSS, ADOLF G (1864–1938) (Anon. 1938, Arb. morph. taxon. Ent. Berl. 5: 352.)

CLOSTEROVIRUS Noun. (Greek, *kloster* = spindle; *virus*. Pl., Closteroviruses.) A member of a group of flexuous-rod plant viruses typically found in the phloem system.

CLOT Noun. (Middle English, *clod* > Anglo Saxon, *clott*. Pl., Clots.) 1. A mass or lump of something such as blood. Cf. Coagulation.

CLOTHES MOTH See Casemaking Clothes Moth; Webbing Clothes Moth.

CLOTHES MOTHS Any of several Species of Lepidoptera; common clothes-moth and casemaking clothes-moth frequently encountered. Adult body pale, small, slender; female wing span 10–12 mm; male slightly smaller. Larvae white to cream coloured with darker head. Adult active at night throughout year; prefer dark and undisturbed areas. Eggs are laid onto food source. Larvae feed on pollen, fur, feathers, skin and hair in natural situations; taken around birds' nests and in carcasses. Larvae spin silken tubes or webbing as protective cover; pupate in cocoons fastened to substrate with silk; cases often attractively patterned. See Tineidae. Cf. Casemaking Clothes-Moth; Webbing Clothes-Moth.

CLOTHING HAIRS The Setae (Chaetae) that invest (cover) the surface of the insect body or its appendages; CH frequently are more-or-less specialized, giving surface a downy appearance.

CLOTHODIDAE Plural Noun. A Family of Embiidina.

CLOUDED PLANT-BUG *Neurocolpus nubilus* (Say) [Hemoptera: Miridae].

CLOUDED SULPHUR *Colias philodice* Godart [Lepidoptera: Pieridae].

CLOUDED WOOD-NYMPHS *Cercyonis pegala* Fabricius *nephele* [Lepidoptera: Satyridae]: Endemic to northeastern North America.

CLOUDLESS SULPHUR *Phoebis sennae eubule* [Lepidoptera: Pieridae]. Alt. Pale Sulphur.

CLOUDY-WING VIRUS A commonly-occurring, widespread viral disease of adult honey bees which is manifest in wings which are not transparent. Viral particles may be airborne and transmitted between bees over short distances. Infected individuals die quickly; heavily infested colonies may be driven to extinction. See Honey Bee Viral Disease. Rel. Pathogen.

CLOUDYWINGED WHITEFLY *Dialeurodes citrifolii* (Morgan) [Hemiptera: Aleyrodidae]: Resembles Citrus Whitefly except susceptible to Aschersonia Fungus and eggs are black. See Aleyrodidae. Cf. Citrus Whitefly.

CLOUET DES PEBRUCHES, LOUIS ARMAND MARIE JOSEPH (1873–1911) (Gaulle 1911, Bull. Soc. ent. Fr. 1911: 341.)

CLOVER APHID *Nearctaphis bakeri* (Cowen) [Hemiptera: Aphididae].

CLOVER APHID PARASITE *Aphelinus lapisligni* Howard [Hymenoptera: Aphelinidae]: A minute internal parasite of aphid nymphs that is potentially useful in applied biological control of some aphid Species. See Aphelinidae.

CLOVER CASEBEARER *Coleophora alcyonipennella* Kollar [Lepidoptera: Coleophoridae]. (Australia).

CLOVER CUTWORM *Discestra trifolii* (Hufnagel) [Lepidoptera: Noctuidae]. See Cutworms.

CLOVER HAYWORM *Hypsopygia costalis* (Fabricius) [Lepidoptera: Pyralidae].

CLOVER-HEAD CATERPILLAR *Grapholita interstinctana* (Clemens) [Lepidoptera: Tortricidae]: A pest of several Species of clover. CHC is endemic to eastern North America with 2–3 generations per year. Eggs are laid on stems, leaves or heads of clover. Larvae are greenish white and feed preferentially on heads, or leaves; larval development requires 30–35 days. Pupation occurs inside silken cocoon on leaves or base of stem. CHC overwinters as larva or pupa in litter. See Tortricidae.

CLOVER-HEAD WEEVIL *Hypera meles* (Fabricius) [Coleoptera: Curculionidae].

CLOVER-LEAF MIDGE *Dasineura trifolii* (Loew) [Diptera: Cecidomyiidae].

CLOVER-LEAF WEEVIL *Hypera punctata* (Fabricius) [Coleoptera: Curculionidae]: A Species endemic to Siberia or Europe and adventive in North America ca 1880. A monovoltine pest of alfalfa and clover; and most evident during spring. Adults appear during summer; neonate adult feeds, then becomes inactive until fall; females oviposit during September-October. Eggs are pale yellow, deposited in plant stems, on stalks or crowns of host plants; oviposition site is sealed with faecal plug. Some eggs hatch during autumn, other eggs hatch during spring. Larvae are green with pale stripe along dorsum, legless with body characteristically curved or curled. Larvae remain concealed at base of plant during day and climb plant stem during night to feed on leaves. CLW overwinters as larva in soil; pupation occurs within coarse brown or greenish cocoon in soil or litter during summer. See Curculionidae. Cf. Alfalfa Weevil; Lesser Clover-Leaf Weevil.

CLOVER LEAFHOPPER *Aceratagallia sanguinolenta* (Provancher) [Hemiptera: Cicadellidae]: Widespread in North America. CLH imbibes fluid from leaves and stems. Overwinters as adult completes 2+ generations per year. Hostplants include alfalfa, clover, lupin, peas, grains and potatoes. See Cicadellidae.

CLOVER LOOPER *Caenurgina crassiuscula* (Haworth) [Lepidoptera: Noctuidae].

CLOVER MITE 1. *Bryobia cristata* (Duges). 2. *Bryobia praetiosa* (C.L. Koch) [Acarina: Tetranychidae]: A pest of fruit and ornamental trees in North America and Australia.

CLOVER ROOT BORER *Hylastinus obscurus* (Marsham) [Coleoptera: Scolytidae]: A monovoltine pest of clover in North America that was introduced from Europe ca 1870. CRB attacks many Species of clover and infested plants typically become discoloured, wilt and die. CRB typically overwinters as an adult in the soil among clover roots. During spring female oviposits a few eggs in the crown of the clover plant; larvae tunnel in the roots and kill the plant. See Scolytidae.

CLOVER-ROOT CURCULIO *Sitona hispidula* (Fabricius) [Coleoptera: Curculionidae]: A bivoltine pest of alfalfa, clover and grasses throughout North America; CRC probably is endemic to Europe. In USA, Species overwinters typically as young larva, or as an egg or adult. Adults are active in May/June and feed actively for 30–45 days; they are less active during midsummer and resume activity during autumn. Eggs are laid on underside of leaves and larvae feed on roots; adults feed on leaves. See Curculioidae.

CLOVER-SEED CHALCID *Bruchophagus platyptera* (Walker) [Hymenoptera: Eurytomidae]: A multivoltine pest of alfalfa and clover in North America. CSC impacts alfalfa seed production but has no effect of hay production. CSC eggs are laid in developing seeds and the larvae consume entire contents of seed within 2 weeks and overwinter as mature larvae. See Euryto-midae. Cf. Alfalfa Seed Chalcid; Clover-Seed Midge.

CLOVER-SEED MIDGE *Dasineura leguminicola* (Lintner) [Diptera: Cecidomyiidae]: A multivoltine pest of red clover in many parts of North America. CSM attacks seed crop but does not affect hay crop, and overwinters as mature larvae within fragile silken cocoons. Adults emerge during spring. Adult mosquito-like, grey to black with bright red Abdomen; Ovipositor as long as body. CSM eggs are pale yellow, attached individually or in clusters to hairs of calyx in clover blossom; female lays ca 100 eggs during lifetime and eclosion occurs within 3–5 days. Neonate larvae move to top of flower and then inside unopened petals where they destroy unopened petals. Mature larvae drop to ground (generally induced by rain) and pupate within soil. See Cecidomyiidae. Cf. Hessian Fly; Sorghum Midge. Rel. Alfalfa Seed Chalcid; Clover-Seed Chalcid.

CLOVER-SEED WEEVIL *Tychius picirostris* (Fabricius) [Coleoptera: Curculionidae].

CLOVER SPRINGTAIL *Sminthurus viridis* (Linnaeus) [Collembola: Sminthuridae]. (Australia).

CLOVER-STEM BORER *Languria mozardi* Latreille [Coleoptera: Languridae]: A pest of clovers in eastern USA. Larvae bore into stems of legumes to cause swelling, cracking, reduced seed production and inferior forage quality. See Languridae.

CLOWN BEETLES See Histeridae.

CLUB Noun. (Middle English, *clubbe* = club > Old Norse, *klubba*. Pl., Clubs.) 1. Insect Antenna: Apical segments of the Flagellum that are enlarged or thickened such that they are anatomically differentiated from the preceding segments. 2. A hand-held tool or weapon, longer than wide, tapering and used to strike objects.

CLUB® See Methiocarb.

CLUB TAILS See Gomphidae.

CLUBBED See Clavate.

CLUBIONIDAE Plural Noun. Vagrant Spiders; Sac Spiders. Spiders with spiracles near Spinnerets; Tarsi with two claws and claw tuft on each leg. Clubionids are predominantly nocturnal spiders that rest during daylight in silken retreats; some Species are active during day. Clubionids do not spin webs; instead, they hunt on ground or vegetation. Female protects her egg sac; some Species show maternal care for spiderlings.

CLUNALIS Adj. (Latin, *clunes* = buttock.) Pertaining to the posterior body parts.

CLUNIS Noun. (Latin, *clunes* = buttock > *cluneum* > *clune*.) Buttock or haunch of an animal.

CLUNIUM Noun. (Latin, *clunius* = rump, buttock; *-ium* = diminutive > Greek, *-idion*. Pl., Clunes.) Fused abdominal Terga (8–10 female, 9–10 male) in some insects, particularly Psocoptera.

CLUSIIDAE Plural Noun. A small Family of acalypterate Diptera of uncertain placement and sometimes assigned to Superfamily Opomyzoidea. Adult 3–4 mm long; wing often with dark spots or infumation. Larvae live in decaying wood or under bark; larvae jump. Syn. Clusiodidae; Heteroneuridae. See Diptera.

CLUSIODIDAE Plural Noun. See Clusiidae.

CLUSIUS, ALFRED (1867–1909) (Horn 1910, Dt. ent. Z. 1910: 113.)

CLUSTER Noun. (Anglo Saxon, *clyster*. Pl., Clusters.) 1. A group of similar objects, parts or organisms close together. 2. A group of bees clinging together to form a solid mass. Cf. Aggregation.

CLUSTER CATERPILLAR *Spodoptera litura* (Fabricius) [Lepidoptera: Noctuidae]: In Australia, a pest of strawberries, tomatoes, melons, cotton and apples. CCs cause damage to leaves, flowers and fruit. Adult moth with wingspan 30–38 mm and body 15–20 mm long. Forewing grey to reddish-brown with a variegated pattern and pale lines along the veins. Male wingbase and apex with bluish areas; hindwing greyish-white with grey margins and dark lines along veins. Adult female lays 1,000–2,000 eggs in batches of 100–300 on underside of leaves; batches are covered in orange-brown to pink scales from apex of female's Abdomen. Eggs 0.6 mm in diameter, spherical and somewhat flattened; egg stage requires 4–12 days depending on temperature. Mature larvae are 40–45 mm long and lack Setae. Young larvae vary in colour from blackish-grey to dark green; mature larvae are reddish-brown or whitish yellow. Sides of body have light and dark longitudinal bands; dorsum of body has crescent-shaped spots on the sides of each segment. Spots on abdominal segments 1 and 8 are larger than other spots and spot on segment 1 interrupts lateral line on side of body. A bright yellow stripe occurs on top of body along entire length. Early instar larvae are gregarious and feed on underside of leaves, leaving the top intact. Late instar larvae disperse and feed at night. The mature larvae spend day in ground under host plant. The larval stage has six instars and lasts 15–23 days. Mature larvae produce earthen cells for pupation; pupae are 15–20 mm long, red-brown and have two small spines on posterior tip; pupal stage lasts 11–13 days. Adult stage is 4–10 days. Life-cycle can be completed in 35 days. Syn. Cotton Leafworm; Tobacco Caterpillar; Tobacco Cutworm. See Noctuidae.

CLUSTER FLY *Pollenia rudis* (Fabricius) [Diptera: Calliphoridae]: An Holarctic, multivoltine urban pest in areas with managed turfgrass containing soils rich in organic matter and earthworms. Adults resemble the housefly but are larger, characterised by Thorax with yellow, curly Setae and Abdomen pollinose and non-metallic. Female CF deposits eggs on moist soil; first instar larvae seek earthworms and penetrate a worm near Clitellum; CF undergoes three larval instars; pupation occurs in soil. Life cycle complete ca 30 days; 3–4 generations per year. Common name is based on habits of adults (last generation during late summer) forming large aggregations in attic or wall void of homes; enter building via small openings but do not enter through doors or windows. Flies remain dormant during winter but may become active during warm periods; fly to lights and windows and die in large numbers on window sills. CF can create significant nuisance but little medical or veterinary importance. See Calliphoridae. Cf. Green Cluster Fly.

CLUTCH SIZE (Middle English, *cleken* = to hatch). The number of eggs deposited during an egg-laying (ovipositional) episode.

CLUTTEN, WILLIAM GEORGE (1886–1935) (Wright 1935, NWest Nat. 10: 50–51.)

CLUTTERBUCK, CHARLES GRANVILLE (1871–1957) (Richards 1959, Proc. R. ent. Soc. Lond. (C) 23: 73.)

CLYPANGULI Plural Noun. (Etymology obscure.) The parts of the Clypeus bearing the Precolia (MacGillivray) at each lateral end of the Postclypeus, antecoxal piece of the Mandible; Paraclypeus; Lateroclypeus (MacGillivray).

CLYPEAL CONSTRICTION. A constriction formed when a surface is drawn in from the sides so as to produce a shield or saddle-like form.

CLYPEAL HEAD A flattish head, with broad flat margins in the Clypeus and Front. Alt. Caput Clypeatum.

CLYPEAL SUTURE The suture marking the boundary between the Clypeus and the Epicranium; the suture separating the Postclypeus and the Preclypeus (MacGillivray).

CLYPEATE Adj. (Latin, *clypeus* = shield; *-atus* = adjectival suffix.) Shield-like in form. Alt. Clypeatus; Clypeiform.

CLYPEIFORM Adj. (Latin, *clypeus* = shield; *forma* = shape.) Shield-shaped; Clypeus-shaped. Alt. Clypeate. See Clypeus; Shield. Rel. Form 2; Shape 2; Outline Shape.

CLYPEOCEPHALIC PROLONGMENT Mayflies: A small, rounded projection arising from the anterior border of the head between the Antennae (Needham).

CLYPEOFRONTAL Adj. (Latin, *clypeus* = shield; *front*.) Descriptive of or pertaining to the Clypeus and the Front. See Clypeus; Head.

CLYPEOFRONTAL SUTURE A flexible junction or suture between the Clypeus and Labrum. Syn. Frontoclypeal Suture.

CLYPEOLUS See Anteclypeus.

CLYPEUS Noun. (Latin, *clypeus* = shield. Pl., Clypeuses.) 1. Anterior sclerite of the insect head below Frons (Face) and above the Labrum. Clypeus is highly variable in size and shape. Dorsally, Clypeus is separated from the Frons by the Frontoclypeal Suture; ventrally, Clypeus is seprated from the Labrum by the Clypeolabral Suture or membrane. The internal surface of the Clypeus provides area for attachment of Cibarial Dilator Muscles. 2. Coccids: The broad band bounding the ventral aspect of the head (Green); not the true Clypeus (MacGillivray). 3. Some Diptera: Clypeus is divided into an Anteclypeus and Pinaculum. See Head. Cf. Anteclypeus; Chaperon; Clypeus Anterior; Epistoma; Nasus; Pinaculum; Prelabrum.

CLYPEUS ANTERIOR See Anteclypeus.

CLYPEUS POSTERIOR See Postclypeus.

CLYPOFRONS Noun. (Latin, *clypeus* = shield; *frons* = forehead.) The transverse line apparently forming the caudal limit of the Clypeus on the external surface of the head, and its invaginated part (MacGillivray).

CLYTHIIDAE See Platypezidae.

CO_2 Chemical formula for carbon dioxide. A colourless, odourless gas that is heavier than air (of which it is a constituent), somewhat soluble in water and not a supporter of combustion. CO_2 is an end-product of animal respiration. See Respiration. Cf. Oxygen.

COACTATAE Adj. (Latin, *coacta* = felt.) Surface texture which is very condensed matt and nearly smooth. The projections forming the surface are typically short and stout. Alt. Coactus.

COADAPTATION Noun. (Latin, *cum* = with; *ad* = to; *aptare* = to fit. English, *-tion* = result of an action. Pl., Coadaptations.) 1. Permanent changes to action, process or structure that are made by unrelated organisms as a result of long-term interactions. Mutual adaptations. Rel. Commensalism. 2. Correlated variation in size, shape or colour of mutually dependent structures or interactive systems. 3. Structures modified from a groundplan which work together harmoniously. Cf. Adaptation. Rel. Evolution.

COADUNATE Adj. (Latin, *coadunare* = unite; *-atus* = adjectival suffix.) 1. Descriptive of structures that have grown together; confluent. 2. Joined together at the base; two or more things joined together. 3. Coleoptera: Pertaining to Elytra permanently united or fused at the suture.

COADUNATION Noun. (Latin, *coadunatio*; English, *-tion* = result of an action. Pl., Coadunations.) The act or process of union of dissimilar substances into one body or mass. See Amalgamation.

COAGULATE Verb. (Latin, *coagulatus*, past part. of *coagulare* from *coagulum* = means of coagulation.) 1. To effect coagulation. 2. To congeal, to clot, to curdle; to change from a fluid to a jelly.

Coagulated; Coagulating.

COAGULATION Noun. (Latin, *coagulatio; cum* = with; *agere* = to drive; English, *-tion* = result of an action. Pl., Coagulations.) 1. The act, process or change of a liquid to a viscous state and ultimately a solid state through chemical process. 2. Coagulation of insect Haemolymph may involve cell agglutination or plasma coagulation. Cell Agglutination involves Coagulocytes and may take several patterns. Cf. Clot.

COAGULIN Noun. (Latin, *coagulum* = rennet. Pl., Coagulins.) Any constituent of blood that aids in the blood-clotting process. Cf. Anticoagulin. Rel. Haemolymph.

COAGULOCYTE Noun. (Latin, *cum* = together; *agere* = to drive; *kytos* = hollow vessel; container. Pl., Coagulocytes.) A basic anatomical form of insect Haemocyte. Coagulocytes are blood cells of variable-size, spherical in shape and hyaline. Coagulocytes display a Plasma Membrane that typically lacks Micropapillae, Filipodia or irregular processes. A Coagulocyte nucleus typically is small, oval, eccentrically positioned; nucleus contains perinuclear Cisternae that give chromatin a cartwheel-like appearance. Coagulocyte cytoplasm contains granular inclusions and numerous Polyribosomes. Coagulocytes are fragile and unstable. Histochemically they show weak PAS-positive reaction. Syn. Cystocyte; Thombocytoid. See Haemocyte. Cf. Adipohemocyte; Granulocyte; Nephrocyte; Oenocyte; Plasmatocyte; Prohaemocyte; Spherulocyte. Rel. Haemolymph.

COAGULUM Noun. (Latin, *coagulum* = rennet. Pl., Coagula.) A clotted mass, usually referring to blood. Rel. Haemolymph.

COALESCE Intrans. Verb. (Latin, *coalescere* > *coalitum* > *co-*; *alescere* = grow.) To grow together; to unite in growth into one body.

COALESCENCE Noun. (Latin, *coalescens.*) Growing together; union; combination.

COALITE Verb (Latin, *coalitus* > *coalescere.*) To unite or associate; fused. Said of any two parts usually separated and distinct (Kirby & Spence). Alt. Coalitus. Syn. Fused.

COALITE STILT PROLEGS Stilt Prolegs united for a part of their length into one organ, which has a bifid apex. See Proleg.

COARCTATE Adj. (Latin, *coarctare* = to press together.) 1. Contracted or compacted. A term applied to a pupal form in which the future adult is concealed by a thickened, usually cylindrical case (cover) that is often hardened larval integument. 2. Compressed such that the Abdomen and Thorax are forced together. 3. Meloidae: The third phase of hypermetamorphic development in which the equivalent of the sixth larval instar is heavily sclerotized, develops rudimentary appendages, but is immobile. Syn Pseudopupa. See Pupa. Cf. Obtect. Alt. Coarctatus.

COARCTATE LARVA A larva that retains the remnants of the previous instar's Integument on the posterior end of the body. See Moulting.

COARCTATE PUPA Brachycerous Diptera: A pupa which remains enclosed in the last larval integument. See Pupa; Puparium. Cf. Obtect Pupa.

COASTAL BROWN-ANT *Pheidole megacephala* (Fabricius) [Hymenoptera: Formicidae]: Body brown; workers 1.5–3 mm long with two morphs, larger morph with large head and functions as soldier; two nodes on abdominal Pedicel; spines on posterior part of Thorax. CBA is pest in dwellings where it prefers food with animal protein. CBA nests in crevices, wall voids, under logs, rocks, pavers and around building foundations. Nests can become very large. In Australian *Citrus*, CBA disrupts natural enemies of other *Citrus* pests. CBA is attracted to honeydew-producing insects (soft scales, mealybugs, flatids, aphids) but under some circumstances prefers proteins and fats over sweets. See Formicidae.

COASTAL FLY *Fannia femoralis* (Stein) [Diptera: Muscidae]. Cf. Latrine Fly; Little House Fly.

COATES, THOMAS JAMES DAGLESS (1920–1972) (Anon. 1972, J. ent. Soc. sth. Afr. 35: 362.)

COBELLI, RUGGERO DE (1838–1921) (Cobelli 1932, Publ. Mus. civ. Stor. Rovereto 59: 1–26, bibliogr.)

COBO, BERNABÉ (1572–1659) (Dusmet y Alonso 1919, Boln. Soc. ent. Esp. 2: 77.)

COCOA THRIPS *Selenothrips rubrocinctus* (Giard) [Thysanoptera: Thripidae]: A widespread pest in tropical and subtropical regions where it inflicts significant damage on cacao, guava, cashew, mango and other tropical crops. Syn. Red-Banded Thrips. See Thripidae.

COCARDES Plural Noun. (French, *cocarde* = distinguishing mark.) Machilidae: Retractile vesicular bodies on each side of the Thorax in some Species. Rel. Vesicle.

COCCIDAE Plural Noun. Soft Scales. A cosmopolitan Family of about 1,000 Species placed within the coccoid Hemiptera. All Species are phytophagous. Some Species are parthenogenetic and known only from females; some Species produce live young called crawlers. Coccids are univoltine or multivoltine, depending upon climate and host plant. Adult thoracic spiracles are similar in size; abdominal spiracles absent. Coccids display a pair of mesially contiguous sclerites at base of anal cleft; some Species produce a soft wax covering, other Species are covered by a convex, hard cover. Crawlers have legs and move to disperse or seek a suitable place to insert their mouthparts into a host plant. Feeding by coccids involves excretion of copious amounts of fluid called honeydew which contains large amounts of sugar; honeydew accumulates on plant and causes sooty mould. Some Species are considered economically important pests through honeydew secretions and sooty mould disease. See Hemiptera. Cf. Diaspididae; Eriococcidae. Rel. Black Scale; Citricola Scale.

COCCIDIVOROUS Adj. (Greek, *kokkos* = berry;

vorare = to devour; Latin, *-osus* = possessing the qualities of.) Pertaining to insects that feed on scale insects or mealybugs (*e.g.* some coccinellid beetles.) See Feeding Strategy. Rel. Aphidivorous.

COCCINELLIDAE Plural Noun. Lady Beetles. A cosmopolitan Family of polyphagous Coleoptera (Cucujoidea) including about 5,000 nominal Species. Adult 1–10 mm long, convex, ovate, frequently bicolourous or spotted; Antenna short, with 7–11 segments and club weakly developed; tarsal formula 3-3-3 or pseudotrimerous; Abdomen with 5–6 Ventrites. Larvae are elongate, oblong or sometimes broadly ovate and body with tubercles, spines, waxy exudate or aposematic coloration; three Stemmata on each side of head; Antenna with 1–3 segments; legs with five segments and each Pretarsus with one apical claw; Urotergites absent. Coccinellids include several Subfamilies: Epilachninae are all phytophagous; Coccinellinae are all predaceous. Epilachninae often are pests of Solanaceae and Curcurbitaceae. Coccinellinae are important in biological control; their prey include aphids, coccids, mites; oviposition near prey. See Cucujoidea.

COCCINEOUS (Latin, *coccinus* = scarlet coloured, from Greek, *keros* = berry, seed; Latin, *-osus* = possessing the qualities of.) Cochineal red, scarlet. Alt. Coccineus.

COCCOID Adj. (Greek, *kokkos* = seed; *eidos* = form.) Coccoid-like; resembling a scale insect in shape (vaguely globular or spherical). Cf. Ovoid; Spheroid. Rel. Eidos; Form; Shape.

COCCOIDEA Noun. Mealybugs; Scale Insects. A cosmopolitan Superfamily of sternorrhynchous Hemiptera that includes about 6,000 nominal Species. Coccoids display profound sexual dimorphism. Adult females are relatively long lived; mouthpart Stylets anchor the body; body frequently protected with wax or a hard cover manufactured by glandular secretions. Antennae, legs and wings are absent, body Tagmata fused. Males short lived, fragile and lack a cover; males lack functional mouthparts and do not feed; winged or wingless, alate forms display two forewings; hindwings called Hamulohalteres. Parthenogenesis is common and complex; reproduction may be Oviparous, Ovoviparous or Viviparous. First instar is called a 'crawler' and possesses legs with which to disperse; subsequent instars may lose the legs and become sedentary. Many Species are regarded as significant pests of agriculture and horticulture. Included Families: Aclerdidae, Asterolecaniidae, Beesoniidae, Cercoccidae, Coccidae, Conchaspididae, Dactylopiidae, Diaspididae, Eriococcidae, Halimococcidae, Kermesidae, Kerriidae, Lecanodiaspididae, Margarodidae, Ortheziidae, Phenacoleachiidae, Phoenicococcidae, Pseudococcidae, Stictococcidae. See Hemiptera.

COCCOPHAGINAE Förster 1878. A Subfamily of Aphelinidae important in biological control. Anatomical features include: Antenna with 8–9 segments including club with three segments; Pronotum and Prepectus complete; forewing broad, uniformly setose, without Linea Calva and with short marginal fringe. Biology: Internal parasites of Coccidae, Aclerdidae, Pseudococcidae, Asterolecanidae. Most important Genus *Coccophagus* Westwood which includes ca 190 Species. See Aphelinidae. Rel. Biological Control.

COCEPHALIC Adj. (Greek, *co-*; *kephale* = head.) Pertaining to a prognathous head in which the Foramen only exists (MacGillivray).

COCHINEAL Noun. (Spanish, *cochinillo* = a little pig.) A commercial dye or histological stain made from the dried bodies of the coccoid *Dactylopius coccus* or related Species.

COCHINEAL INSECT 1. *Dactylopius coccus* Costa [Hemiptera: Dactylopiidae]: A coccid that feeds upon Mexican *Opuntia* spp. (prickly pear cactus) and from which a red dye is obtained. Term 'cochineal' is derived from Spanish 'cochinillo' (a little pig) and given because of gregarious habits of this Species. 2. *Dactylopius ceylonicus*. A coccoid whose significance lies partly in its use as the first successful importation of an insect for biological control of a weed (India 1863 - against prickly-pear cactus). Rel. Biological Control.

COCHINEAL SCALES See Dactylopiidae.

COCHLEARIUM Noun. (Latin, *cochleare* = a spoon with a sharp point for eating snails from the shell > Greek, *kochlias* = snail; *-arium* = place of a thing. Pl., Cochlearia.) Hymenoptera: The Gonostylus.

COCHLEATE Adj. (Greek, *kochlias* = snail; *-atus* = adjectival suffix.) Twisted spirally, like a snail-shell. Alt. Cochleatus; Cochleiform.

COCHLEATE VESICLE A non-musculated, sclerotized, cochliform structure, attached to the base of the Ejaculatory Duct. CV probably is activated hydrostatically and serves as a sperm pump in some male Sciomyzidae.

COCHRAN, JAMES HARVEY (1913–1969) (Hays *et al.* 1969, J. Econ. Ent. 62: 1252.)

COCHYLIDAE Plural Noun. Cochylid Moths. A Family of Lepidoptera.

COCHYLIDIIDAE Plural Noun. See Limacodidae.

COCKAYNE, EDWARD ALFRED (1880–1955) (Worms 1958, Entomologist's Gaz. 9: 73–82. bibliogr.)

COCKCHAFER Noun. (Middle English, *cok* = cock > Old High German, *kevar* = beetle. Pl., Cockchafers.) A common name for the larval stage of scarab beetles, particularly members of *Melolontha* in Europe or *Adoryphorus* and *Aphodius* in Australia. See Scarabaeidae. Cf. Black-Headed Pasture Cockchafer; Little Pasture-Cockchafer; Red-Headed Pasture Cockchafer.

COCKERELL, THEODORE DRU ALISON (1862–1948) English natural historian who spent most of his life as an academic in America (University of Colorado). Cockerell held positions at British Museum (Natural History), in Jamaica and New Mexico State University. He was a prolific writer and author of nearly 4,000 papers on many aspects of natural history and Entomology, including taxonomy of Apoidea and in fossils of Florissant, Colorado. (Mallis 1971, *American Entomologists*. 549 pp. (357–362), New Brunswick.)

COCKLE, GEORGE (–1900) (Verrall 1900, Proc. ent. Soc. Lond. 1900: xiv.)

COCKLEBUR WEEVIL *Rhodobaenus quinquedecimpunctatus* (Say) [Coleoptera: Curculionidae].

COCKLEBUR *Xanthium strumarium* Linnaeus. A widespread rangeland weed.

COCKROACHES Plural Noun. (Spanish, *cucaracha*.) A predominantly tropical group of insects presently comprising about 4,000 Species. Common name for Blattaria. See Blattaria; Blattodea.

COCKSFOOT GRUB *Rhopaea verreauxii* Blanchard [Coleoptera: Scarabaeidae]. (Australia).

COCKSFOOT THRIPS *Chirothrips manicatus* Haliday [Thysanoptera: Thripidae]. (Australia).

COCKTAIL ANT *Crematogaster laeviceps chasei* Forel [Hymenoptera: Formicidae]. (Australia).

COCOA MOTH See Tobacco Moth.

COCOA POD BORER *Conopomorpha cramerella* (Snellen) [Lepidoptera: Gracillariidae]: A major pest of cocoa in southeast Asia. See Gracillariidae.

COCONUT GRASSHOPPER Common name loosely applied to *Sexava nubila* St. and *S. novaeguinae* Branc. [Orthoptera: Tettigoniidae], both of which cause severe defoliation of coconut palms in the Bismarck Archipelago. See Tettigoniidae.

COCONUT LEAFEATING CATERPILLAR *Opisina arenosella* Walker [Lepidoptera: Xyloryctidae]. Syn. *Nephanitis serinopa* Meyrick.

COCONUT LEAF-MINER *Agonoxena argaula* Meyrick [Lepidoptera: Agonoxenidae].

COCONUT LEAFROLLER *Hedylepta blackburni* (Butler) [Lepidoptera: Pyralidae].

COCONUT MEALYBUG *Nipaecoccus nipae* (Maskell) [Hemiptera: Pseudococcidae].

COCONUT RHINOCEROS-BEETLE *Oryctes rhinoceros* [Coleoptera: Scarabaeidae]: A significant pest of coconut palm in palm growing regions of the world. See Scarabaeidae.

COCONUT SCALE *Aspidiotus destructor* Signoret [Hemiptera: Diaspididae].

COCONUT WHITEFLY *Aleurodicus destructor* Mackie [Hemiptera: Aleyrodidae]. (Australia).

COCONUT-SPIKE MOTH *Tirathaba rufivena* Walker [Lepidoptera: Pyralidae].

COCOON Noun. (French = *cocon* = cocoon. Pl., Cocoons.) A covering of silk or other material that surrounds the pupa. Cocoon is spun or constructed by the larval stage and composed (in part or entirely) of silk or other viscid fibres. The cocoons of many Species are the product of glandular secretions. Cocoon is composed of proteins, including fibroin and serecin. Taxonomic distribution of cocoons among holometabolous insects is widespread, but not all Species produce cocoons. All Neuroptera spin a silken cocoon. Cocoons are common among Lepidoptera and are constructed of earth, wood fragments or debris. Cocoons are not common among Coleoptera (exception: chrysomelid *Ophraelia*). Siphonaptera produce a cocoon (during third larval instar) of silk and debris; sometimes flea does not spin a cocoon when ambient humidity is high. Hymenoptera cocoons are taxonomically irregular: Symphyta commonly manufacture a cocoon with Labial Glands; Parasitica rarely produce cocoons (typically with Malpighian Tubules, *Euplectrus, Thysanus*); some Aculeata (Bethylidae) produce cocoons with Labial Glands. Functions of cocoons are diverse: 1. protect pupa from physical damage, 2. deter predation and parasitism, 3. maintain body temperature and 4. maintain water balance. Coccon architecture is highly variable. Cf. Pupa. Rel. Metamorphosis.

COCOON BREAKER Lepidoptera: Structures or processes on the pupa, often on the head, that facilitate adult emergence from the cocoon. Syn. Cocoon-cutter. See Cocoon.

COCOONASE Noun. (French = *cocon* = cocoon; Chemistry *-ase* = enzyme.) An enzyme secreted by some moths which softens or dissolves cocoon silk to facilitate emergence of the adult from the cocoon.

CODELURE A synthetic pheromone {E,E-8,10-dodecadien-1-ol} used to disrupt sex pheromones of the Codling Moth. Compound is applied to fruit trees. Trade names include Codlemone®, Checkmate CM®, Bomate-C® and Rimilure®.

CODINA, ASCENSIO (1877–1932) (Espanol 1932, Butl. Inst. Catalana Hist. nat. 32: 239–242.)

CODLEMONE Noun. The synthetic sex pheromone of the codling moth.

CODLEMONE® See Codelure.

CODLING MOTH *Cydia pomonella* (Linnaeus) [Lepidoptera: Tortricidae]: A global pest of apples, pears, plums and some varieties of European walnuts. CM apparently is endemic in Eurasia and spread elsewhere with infested fruit. First recovered in North America about 1750; CM is currently a major pest of apples and pears in eastern Australia. CM is regarded as most important, persistent and difficult-to-control pest of apples. CM overwinters as mature larva within a thick silk cocoon and pupation occurs during mid-spring. Female deposits ca 50 eggs during lifetime. Eggs are white, flattened and laid individually on upper surface of leaves or twigs; eclosion occurs within 6–20 days, depending upon temperature. Neonate larvae may feed briefly on leaves, then enter young apples at calyx cup, penetrate core and often consume seeds. Pupation occurs within cocoon under loose bark or on ground. Adults do not fly far. CM completes 2–3 generations per year. Early entomological literature published under the generic names *Laspeyresia, Grapholitha* and *Carpocapsa*. See Tortricidae. Cf. Filbertworm; Lesser Appleworm.

CODLING-MOTH GRANULOSIS VIRUS A viral disease of codling moth being advertized and sold in some countries as a stomach poison of codling moth larvae. Trade names include: Carpovirusine®, Cyd-X®, Decyde®, Granupon®, Madex 2®, Madex 3®, Obstamade®, Specific-T®. See Biopesticide; Granulosis Virus. Cf. Summer Fruit-Tortrix Granulosis Virus.

COE, RALPH LEONARD (1906–1968) (Smith 1968, Entomol. mon. Mag. 104: 70–72, bibliogr.)

COECAL Adj. (Greek, *koiloma* = hollow.) 1. Pertaining to tubular structure that ends blindly. 2. Enclosed within a tube or pouch.

COECUM Noun. (Greek, *koiloma* = hollow. Pl., Coeca.) A blind sac or tube. Term applied to a series of appendages opening into the Alimentary Canal at the junction of the Crop and Chylific Ventricle. See Caecum. Rel. Alimentary System.

COELOBLAST Noun. (Greek, *koiloma* = hollow; *blastos* = bud. Pl., Coeloblasts.) The Endoderm *sensu stricto*.

COELOCONIC SENSILLUM A Basiconic Sensillum sunk into a depression or pit in the integument. The pit is called a Sacculus. Several additional pegs may be positioned along the margin of the depression and directed toward the central peg. When the depression is deep and the surface constricted, the receptor is called a Sensillum Ampullaceum. CS are thin or thick walled and best identified when staining with 0.1% silver nitrate in 70% ETOH. Tips appear dark brown when so treated; tips of Trichode Sensilla remain unchanged under such treatment. Coeloconic Sensilla function in olfaction but also may serve as thermoregulators. An anatomical form of CS may serve in hygroreception in the moth *Opisina arenosella* Walker. CS may respond to host plant odour. The third antennal segment of *Sarcophaga* has several sensory pits in which CS occur. One sensory pit occurs in *Drosophila* which has CS. Coeloconic Sensilla are also positioned outside the Sacculus. The Sensilla are double walled with longitudinal channels near the apices and Dendrites do not branch. Proximally the Sensillum is smooth. When positioned within a Sacculus, each Sensillum is innervated by two or three sense cells. Syn. Sensillum Coeloconicum. See Sensilla. Cf. Ampulliform Sensillum.

COELOM Noun. (Greek, *koiloma* = hollow. Pl., Coelomes.) The body cavity.

COELOMIC CAVITY The space between the Viscera and body wall.

COELOM-SAC 1. The cavity containing the Viscera. 2. Embryology: One of a pair of closed sacs arising in the Mesoderm of each segment of the embryo and giving rise to the adult Coelom.

COELOPIDAE Plural Noun. Seaweed Flies. A widespread Family of acalypterate Diptera assigned to Superfamily Sciomyzoidea. Adult moderate sized; thoracic dorsum flattened; body and legs bristly. Occiput typically flat and broadly in contact with Pronotum; Coxae of each pair of legs medially contiguous or nearly so; apical tarsal segment triangular and wider than other Tarsomeres. Adults inhabit seashore in places where seaweed accumulates on beach; larvae live in kelp and sea grass. Syn. Phycodromidae. See Sciomyzoidea; Seaweed Fly.

COENAGRIONIDAE Plural Noun. Narrow-winged Damselflies. A very large, cosmopolitan Family of zygopterous Odonata. Adults are small to medium-sized and slender; Pterostigma short, quadrilateral has lower outer angle acute; superior appendages of male usually not forcipate. Larvae are slender or moderately slender; Antenna with seven segments and third segment typically longest. Three caudal gills long, lamellate and without terminal filaments. Species are found in diverse habitats; adults rest horizontally with wings together over body. See Odonata.

COENAGRIONOIDEA A large Superfamily of Odonata. Adults highly variable in size; wings typically petiolate with Nodus in basal one third of wing; Arculus near Nodus and R4 closer to nodal level than Arculus. Included Families: Coenagrionidae, Lestoideidae, Megapodagrionidae, Platycnemididae, Platystictidae, Protoneuridae and Pseudostigmatidae. See Odonata.

COENOCYTE Noun. (Greek, *koinos* = common; *kytos* – hollow vessel; container. Pl., Coenocytes.) An enlarged Protoplast, the nuclear divisions of which have not been followed by cytoplasmic cleavage (Daubenmire).

COENOGENOUS Adj. (Greek, *koinos* = common; *genous* = producing; Latin, *-osus* = possessing the qualities of.) Oviparous at one season of the year, Ovoviviparous at another season. A term applied to the Aphididae.

COENOSIS See Biocenosis. Alt. Cenosis.

COERULEOUS Adj. (Latin, *caeruleus* = dark blue; Latin, *-osus* = possessing the qualities of.) Sky-blue. Alt. Coeruleus. See Caeruleus.

COFFEE-BEAN WEEVIL *Araecerus fasciculatus* (DeGeer) [Coleoptera: Anthribidae]: A cosmopolitan field pest of Brazil nuts, corn, coffee berries, cacao beans, cassava, groundnuts, various seeds, seed pods and processed foods. CBW also a pest of stored vegetable products in warehouses; direct damage through feeding or contamination of high-value commodities. CBW prefers food with high moisture content and decaying material. CBW is most abundant in tropical and subtropical regions. Adult 3–5 mm long and grey-brown with pale marks on Elytra; Antenna long, clubbed but not geniculate; strong flier. Female lays ca 50 eggs during lifetime; eggs are laid individually on seed, bean or product; incubation requires 5–8 days. Larvae are white, legless and burrow into seed or bean; larval development is dependent upon food. Pupation occurs within seed; pupal stage requires 6–7 days. Adults live ca 130 days; adult male is sexually mature three days after emergence; female is sexually mature six days after emergence; adults mate more than once. Syn. Cacao Weevil; Nutmeg Weevil. See Anthribidae. Cf. Kao Haole Seed Beetle.

COFFEE-BERRY BORER *Hypoihenemus hampei* (Ferrari) [Coleoptera: Scolytidae]: A widespread pest of commercially grown coffee. CBB was first noted in Gabon (1901), Zaire (1903) and Uganda (1908); subsequently it was reported as a pest of coffee elsewhere (Java 1908, Brazil 1926, El Salvador 1969). CBB probably originated on *robusta* coffee in the Congo Basin forests. Adult females burrow into immature coffee berries, feed and lay eggs in the tunnel made within the bean. Larval development requires 2–3 weeks; pupation lasts about one week and occurs within the tunnels. CBB reported attacking beans. See Scolytidae.

COFFEE MEALYBUG *Planococcus kenyae* (LePelly) [Hemiptera: Pseudococcidae].

COFFIN FLY *Conicera tibialis* Schmitz [Diptera: Phoridae]: A fly associated with human corpses after burial. CF is sometimes involved in Forensic Entomology. See Phoridae.

COGHO, AUGUST (1816–1891) (Joseph 1893, Weidmann 24: 75–77.)

COHESION Noun. (Latin, *cohaerere* = to stick together. Pl., Cohesions.) 1. A condition of union of separate parts. Cf. Adhesion. 2. The result of genic integration of a population that forms a tokogenetic plexus and promotes the existence of a unitary lineage. (Frost & Hills 1990, Herpetologica 46: 87–104.) See Lineage; Sexual Plexus.

COHESION SPECIES CONCEPT A Species concept based on intrinsic cohesion mechanisms, such as gene flow and natural selection, that results in Species cohesion. See Species Concepts.

COILA Noun. (Prob. from Old French *coillir* = to gather. Pl., Coilae.) The point of articulation of an appendage upon the body (MacGillivray).

COINCIDENT Ady. (Latin, *co-*; *incidere* = to fall upon.) 1. Descriptive of things that occur in the same time or place; things that are of the same physical properties. 2. Things coinciding, running together or lying in continuation so as to appear as one.

COITION Noun. (Latin, *coire* = to go together; English, *-tion* = result of an action.) The act of copulation. Sexual intercourse. Alt. Coitus.

COLBEAU, JULES ALEXANDRE JOSEF (1823–1881) (Anon. 1881, Leopoldina 17: 101.)

COLCHICINE Noun. (Latin, *colchium* = meadow saffron. Pl., Colchicines.) An alkaloid obtained from *Colchium autumnalis* which, when applied to plant cells, inhibits spindle formation and promotes polyploidy.

COLCORD, MABEL (1872–1954) (Hawes 1955, Proc. ent. Soc. Wash. 57: 88–91.)

COLDEWEY, HENDRIK (1880–1955) (Lempke 1955, Ent. Ber., Amst. 15: 257–259.)

COLE, GEOFFREY ALFRED (1902–1973) (W[orms] Entomologist's Rec. J. Var. 85: 157–159.)

COLEBROOK, HENRY THOMAS (1765–1782) (Rose 1850, *New General Biographical Dictionary* 6: 399.)

COLEMAN, GEORGE ALBERT (1866–1932) (Essig 1932, Mon. Bull. Calif. Dep. Agric. 21: 246–247.)

COLEMAN, LESLIE S L (–1954) (Kevan 1954, Entomologist's Rec. J. Var. 66: 296.)

COLENSO, WILLIAM (1811–1899) (Anon. 1899, Nature 59: 420.)

COLEOPHORIDAE Plural Noun. Casebearer Moths. A Family of ditrysian Lepidoptera assigned to Superfamily Gelechioidea; cosmopolitan in distribution but best represented in Northern Hemisphere. Adult small with wingspan 6–20 mm; head with smooth scales; Ocelli and Chaetosemata absent; Antenna shorter than forewing length; Pecten or dense tuft of scales on Scape; Proboscis densely scaled basad; Maxillary Palpus minute with 1–4 segments; Labial Palpus recurved; fore Tibia with Epiphysis; tibial spur formula 0-2-4; wings lanceolate with long marginal fringe. Egg upright, circular or oval in outline shape; Chorion with radial ribs. First instar larvae mine leaves or rarely flower buds; subsequent instars reside in case formed of plant fragments; cases are diagnostically useful; larvae feed externally on leaves and flowers; some Species form galls or roll leaves. Pupation occurs in the larval case. Adults rest with Antenna directed forward; Coleophorids typically in low populations and not inducing serious damage; some Species are regarded as pests of economically important trees (apple, cherry, elm, pecan, birch), grasses and seed heads. See Gelechioidea.

COLEOPTERA Linnaeus 1758. (Permian-Recent.) (Greek, *koleos* = sheath; *pteryx* = wing.) Beetles. The numerically largest Order of Holometabola with more than 300,000 nominal Species in about 2000 Genera. Body usually strongly sclerotized, to 125 mm long; coloration black to metallic. Head primitively prognathous; Antenna typically with 11 segments and shape highly variable; compound eyes typically present, variable in size, shape, development; euconic Ommatidium usual, exoconic or aconic Ommatidium sometimes developed; Ocelli not common; Mandible well developed with incising teeth, grinding Mola, and/or membranous Prostheca;

Maxillary Palpus with 3–5 segments; Labial Palpus with 1–3 segments. Prothorax well developed, articulated with head and Mesothorax; Mesothorax and Metathorax fused to form Pterothorax; anterior wings modified into protective covers (Elytra); dorsal surface of Elytron called the disc; discs meet in a straight, longitudinal sutural edge dorsally over Abdomen; lateral margin of disc called Epipleuron; disc typically with longitudinal rows of punctures, often in grooves called Striae; spaces between Striae called Interstriae or Intervals; Interstriae numbered from Sutural Edge lateral, and represent primitive veins; Epipleuron = Precostal Vein; Interstriae near Sutural Edge = Cubito-Anal Veins. Posterior wings membranous, folded complexly over Abdomen at repose and beneath Elytra; venation reduced and modified by folding patterns; basic folding patterns include: Adephagan type (Adephaga, Archostemata, Myxophaga) and polyphagan type (Polyphaga). Legs typically are adapted for walking, sometimes adapted for other purposes; Trochantin often exposed and mobile; Trochanter variably developed; Tarsi typically with five segments, often reduced with tarsal formula diagnostically important; Pretarsus typically with two claws, Empodium variable. Abdomen of female with nine segments; Abdomen of male with 10 segments; Sterna more sclerotized than abdominal Terga. Spiracles typically in membrane; groundplan number eight with trend toward reduction. Male genitalia complex. Nervous system typically with three thoracic and eight abdominal ganglia. Pupa adecticous, exarate except some Staphylinidae. Coleoptera are biologically and ecologically diverse: Aquatic to terrestrial; arboreal to hypogaeic; phytophagous to parasitic. Coleoptera are the 'sister group' of Strepsiptera and most closely related to neuropteroid complex, but ordinal affinities not clear. Included Families: Acanthocnemidae, Ademosynidae, Aderidae, Aglycyderidae, Agyrtidae, Alexiidae, Allocorynidae, Amphizoidae, Anobiidae, Anoplodermatidae, Anthicidae, Anthribidae, Archeocrypticidae, Artematopidae, Attelabidae, Belidae, Biphyllidae, Boganiidae, Boridae, Bostrichidae, Bothrideridae, Brachypsectridae, Brentidae, Buprestidae, Byrrhidae, Byturidae, Callirhipidae, Cantharidae, Carabidae, Cavognathidae, Cephaloidea, Cerambycidae, Ceratocanthidae, Cerophytidae, Cerylonidae, Chaetosomatidae, Chalcodryinidae, Chelonariidae, Chrysomelidae, Ciidae, Clambidae, Cleridae, Cneoglossidae, Coccinellidae, Colydiidae, Corylophidae, Cryptophagidae, Cucujidae, Cupedidae, Curculionidae, Cyathoceridae, Dascillidae, Dasyceridae, Dermestidae, Derodontidae, Diphyllostomatidae, Discolomidae, Disteniidae, Drilidae, Dryopidae, Dytiscidae, Elateridae, Elmidae, Endecatomidae, Endomychidae, Erotylidae, Eucinetidae, Eucnemidae, Eulichadidae, Geotrupidae, Glaphyridae, Glaresidae, Gyrinidae, Haliplidae, Helotidae, Heteroceridae, Histeridae, Hobartiidae, Homalisidae, Hybosoridae, Hydraenidae, Hydrophilidae, Hydroscaphidae, Hygrobiidae, Ithyceridae, Jacobsoniidae, Laemophloeidae, Lamingtoniidae, Lampyridae, Languriidae, Lathridiidae, Leiodidae, Limnichidae, Lucanidae, Lutrochidae, Lycidae, Lymexylidae, Melandryidae, Meloidae, Melyridae, Micromalthidae, Micropeplidae, Microsporidae, Monommidae, Mordellidae, Mycetophagidae, Mycteridae, Nemonychidae, Nitidulidae, Nosodendridae, Noteridae, Ochodaeidae, Oedemeridae, Omethidae, Ommatidae, Oxycorynidae, Oxypeltidae, Passalidae, Pedilidae, Perimylopidae, Phalacridae, Phengodidae, Phloeostichidae, Phloiophilidae, Phycosecidae, Plastoceridae, Pleocomidae, Propalticidae, Prostomidae, Protocucujidae, Pselaphidae, Psephenidae, Pterogeniidae, Ptiliidae, Ptilodactylidae, Pyrochroidae, Pythidae, Rhinorhipidae, Rhipiceridae, Rhipiphoridae, Rhizophagidae, Rhysodidae, Salpingidae, Scarabaeidae, Scirtidae, Scraptiidae, Scydmaenidae, Silphidae, Silvanidae, Sphaeritidae, Sphindididae, Staphylinidae, Synchroidae, Synteliidae, Telegeusidae, Tenebrionidae, Tetraphaleridae,Tetratomidae, Throscidae, Torridincolidae, Trachelostenidae, Trachypachidae, Trictenotomidae, Trogidae, Trogossitidae, Urodontidae, Vesperidae and Zopheridae.

COLEOPTERANS See Coleoptera.

COLEOPTEROID Adj. (Greek, *koleos* = sheath; *pteryx* = wing; *eidos* = form.) Beetle-like in shape, form or habits. See Coleoptera. Cf. Neuroid; Orthopteroid; Rel. Eidos; Form; Shape.

COLIN, GABRIEL CONSTANT (1825–1896) (Anon. 1897, Arch. Patol. Anat. Phys. 148: 191.)

COLLAGEN Noun. (Greek, *kolla* = glue; *genous* = produce. Pl., Collagens.) A group of proteins found in white, fibrous connective tissue and as an organic component of bone; Collagen contains high amounts of hydroxyproline. Syn. Collogon. Cf. Chitin.

COLLAGENASE Noun. (Greek, *kolla* = glue; *genous* = produce. Pl., Collagenases.) A proteolytic digestive enzyme. See Enzyme.

COLLAPSE® A registered biopesticide derived from *Bacillus thuringiensis* var. *kurstaki*. See *Bacillus thuringiensis*.

COLLAR Noun. (Middle English, *coler* = collar. Pl., Collars.) 1. General: Any structure between the Head and Thorax of an insect. 2. Hymenoptera: The neck. 3. Diptera: May mean neck, sclerites attached to Prothorax, the Prothorax, or pro-thoracic processes (Antefurca). 4. Coleoptera: The narrow constricted anterior part of Pronotum, and generally margined by a groove (Sulcus). Cf. Cervix.

COLLARE Noun. (Latin, *collare* = collar. Pl., Collares.) A somewhat elevated posterior part of the Collum.

COLLATERAL Adj. (Latin, *cum* = with; *lateralis* = sides.) A lateral branch of an Axon (Snodgrass).

COLLATORIA Noun. The mouth of the Colleterial Duct (MacGillivray).

COLLECTING BASKET An arrangement of Setae, bristles and/or spines on the forelegs of certain insects, in which they collect or hold food while devouring it (Needham).

COLLEDGE, WILLIAM ROBERT (1841–1928) (Anon. 1928, Qd. Nat. 6: 93–94, bibliogr.)

COLLEMBOLA Lubbock 1873. Noun. (Greek, *kolla* = glue; *embolon* = peg.) Springtails. A Class or Order of highly specialized entognathous Hexapoda that is cosmopolitan in distribution and includes about 6,000 nominal Species placed in 500 Genera. Adult body small (2–10 mm long), soft and apterous. Head prognathous in primitive groups but somewhat hypognathous in more derived groups. Compound eye is weakly developed with few Ommatidia; Antenna with 4–6 segments and intrinsic musculature and an olfactory Postantennal Organ (Organ of Tomosvary) generally present. Thoracic Tagma and segmentation are not always well developed; legs with four principal segments (Coxa, Trochanter, Femur, fused Tibiotarsus). Abdomen displaying saltatorial appendages and ventral tube (Collophore) on the first abdominal segment which gives the group its ordinal name; genital aperture on segment five; Cerci absent. Common name (springtail) based on a two-part jumping device found on abdominal Sterna III (Retinaculum) and IV (Furcula). Development by Collembola is ametabolous or epimetabolic. Most Species feed in moist habitats on decaying vegetation, algae, fungi and pollen. Some Species feed on dead or moribund invertebrates in the soil or exuviae. Cannibalism and predation are documented in some groups. Parthenogenesis has been recorded in a few Species. Collembola lack internal fertilization and the male deposits a Spermatophore on substrata where it is captured by the female who takes the Spermatozoa into her genital tract. Adults are epitokous: All Species apparently alternate a 'reproductive' instar with a feeding, 'non-reproductive' instar. Oldest fossil collembolan, *Rhyniella praecursor* (Hirst & Maulik 1926), was taken from Devonian Period beds of Rhynie Chert in England. Included Families: Actaletidae, Brachystomellidae, Coenaletidae, Cyphoderidae, Dicyrtomidae, Entomobryidae, Hypogastruridae, Isotomidae, Mackenziellidae, Microfalculidae, Neanuridae, Neelidae, Odontellidae, Onychiuridae, Oncopoduridae, Paronellidae, Poduridae, Protentomobryidae, Smithuridae, Tomoceridae. See Apterygota.

COLLENCHYMA Noun. (Greek, *kolla* = glue; *engchyma* = infusion. Pl., Collenchymae.) Plant galls induced by Cynipidae in which the layer of tissue positioned directly below the Epidermis has cells with thickened walls. Typically the contents are crystalline, hard and compact (Kinsey).

COLLENETTE, CYRIL LESLIE (1888–1959) (Uvarov 1960, Proc. R. ent. Soc. Lond. (C) 24: 52.)

COLLETERIAL GLAND SECRETIONS Acid mucopolysaccharides, glycoproteins and phospholipids released from Colleterial Glands. Gross inspection shows frothy secretions that harden in Orthoptera, a gelatinous matrix in some Diptera and hyrdophilous silk in some Coleoptera.

COLLETERIAL GLAND Pouch-like accessory glands of the female reproductive system. CG serve several functions including secreting substances that glue eggs together (*e.g.* cockroach and mantid Ootheca) or to the substratum [*e.g.* the noctuid moth *Spodoptera littoralis* (Boisduval)]. The anatomy and ultrastructure of Colleterial Glands are determined by their function and vary among groups of insects. Blattaria display left and right Colleterial Glands that have different gland cell types and secretions. See Accessory Gland; Colleterium. Cf. Cement Gland. Rel. Exocrine Gland.

COLLETERIUM Noun. (Greek = *kolla* = glue; Latin, *-ium* = diminutive > Greek, *-idion*. Pl., Colleteria.) An accessory glandular structure of the Oviduct which manufactures and secretes viscid material used to cement the eggs together. See Ootheca.

COLLETIDAE Plural Noun. Colletid Bees; Plasterer Bees; Yellow-faced Bees. A plesiomorphic Family of aculeate Hymenoptera assigned to Superfamily Apoidea (bees). Adult head with Subantennal Suture below Antennal socket; facial Foveae usually present; Labrum wider than long; apex of Glossa (tongue) short, bilobed or truncate; Flabellum absent; Galea short; Labial Palpus with four subequal, cylindrical segments; forewing second Recurrent Vein s-shaped (Colletinae), Marginal Cell normal, Stigma large, elongate; basitibial sclerite absent; Scopa sometimes present on Tibia but absent from Abdomen; Pygidium absent; Arolia well developed in male and female. Subfamilies include Colletinae, Diphaglossinae, Hylaeinae, Euryglossinae and Xeromelissinae. See Apoidea.

COLLETINAE Plasterer Bees. A Subfamily of Colletidae (Apoidea) consisting of about 15 Genera. Adult body moderate sized, conspicuously setose; forewing with three Submarginal Cells; second Recurrent Vein s-shaped. Females nest in soil; derive common name from habit of lining cell with waterproofing substance.

COLLETT, EDWARD PYEMONT (1863–1937) (Britton 1937, Entomol. mon. Mag. 73: 92.)

COLLETT, ROBERT (1842–1913) (Natig 1944, Norsk. ent. Tidsskr. 7: 28–32, bibliogr.)

COLLICULUM Noun. (Latin, *colliculus* = little hill. Pl., Collicula; Colliculums.) Lepidoptera: A sclerotized plate or thickening near the posterior end of the female's Ductus Bursae.

COLLIER, ALAN EGERTON (–1972) (Lees 1974, Proc. R. ent. Soc. Lond. (C) 38: 58.)

COLLIER, HENRY N (1893?-1975) (Gunn 1976, Proc. R. ent. Soc. Lond. (C) 40: 51.)

COLLIGATE Adj. (Latin, *colligare* = to collect; *-atus* = adjectival suffix.) Attached to any part but not moveable. Alt. Colligatus.

COLLIN, JAMES EDWARD (1876–1968) (Smith 1969, J. Soc. Biblphy nat. Hist. 5: 226–235, bibliogr.)

COLLINS, CHARLES WALTER (1882–1948) (Parker 1949, Proc. ent. Soc. Wash. 51: 84–85.)

COLLINS, DONALD LOUIS (–1973) (Anon. 1974, Rep. N. Y. State Sci. Serv. 1972–1973: 22–24.)

COLLINS, JOSEPH JOYNSON (1865–1942) (A. H. H. 1942, NWest. Nat. 17: 113–114.)

COLLINS, JOSEPH JOYNSON (1865–1942) English curatorial assistant (Hope Museum, Oxford 1905–1935) with interests in Coleoptera.

COLLINS, REGINALD J (1901–1969) (Hinton 1971, Proc. R. ent. Soc. Lond. (C) 35: 52.)

COLLINSON, PETER (1695–1768) (Harper 1943, Trans. Am. phil. Soc. 33: 1–2.)

COLLOGON See Collagen.

COLLOID Noun. (Greek, *kolla* = glue; *eidos* = form. Pl., Colloids.) A glue-like substance of high molecular weight and which does not diffuse through a semipermeable membrane. Physically amorphous and without crystalline structure. See Mucus. Cf. Mucoid.

COLLOPHORE Noun. (Greek, *kolla* = glue; *pherein* = to bear, carry. Pl., Collophores.) Collembola: The ventral tube of the first abdominal segment. A comparatively large, cylindrical, median base that divides into a pair of apical Diverticula in most Species. Diverticula are extensible with hydrostatic pressure. Collophore communicates with Labial Glands via a ventral groove or closed duct which transmits uric acid from labial Nephridia to the Collophore. Collophore functions in osmoregulation in some Species and adhesion after jumping in other Species. Epithelial cells of a Collophore play a role in ion transport, excretion and respiration; may provide legs with a lubricant for grooming.

COLLUM Noun. (Latin, *collum* = neck. Pl., Colla.) 1. The neck or collar of any elongate body that is apically expanded. 2. A slender connection between head and Thorax in Hymenoptera and Diptera. 3. Coleoptera: The posterior, narrow part of the head or to include the Prothorax (Smith).

COLOBATHRISTIDAE Plural Noun. A small Family of Lygaeoidea found in South America, southeast Asia and Australia.

COLON Noun. (Greek, *kolon* = colon. Pl., Colons.) An enlarged part of the Alimentary Canal positioned between the Ileum and Rectum. See Hindgut.

COLONIAL SPIDER *Badumna socialis* (Rainbow) [Araneida: Desidae]. (Australia).

COLONICI Noun. (Etymology obscure.) A development of Gallicolae Migrantes in *Chermes* from the nymphal migrantes (Imms).

COLONIZATION Noun. (Latin, *colonia* = farm; English, *-tion* = result of an action. Pl., Colonizations.) 1. The act of colonizing or state of being colonized. 2. Biological Control: The release of natural enemies (parasites, predators, pathogens) into an area for the control of pests.

COLONY Noun. (Latin, *colonia* = farm. Pl., Colonies.) 1. Individuals of one Species living in close association in space and time. 2. Social Insects: A group of individuals (not a mated pair) that cooperatively construct nests or rear offspring. See Eusocial. Cf. Aggregation.

COLONY FISSION Social Insects: The multiplication of colonies via the departure of reproductives and workers from a parent colony with reproductives and workers remaining in the parent colony. Syn: Hesmosis (ants); Sociotomy (termites); Swarming (honey bees).

COLONY ODOUR Social Insects: Any odour on the body of a social insect that is unique to a colony. Cf. Nest Odour; Species Odour.

COLORADO POTATO-BEETLE *Leptinotarsa decemlineata* (Say) [Coleoptera: Chrysomelidae]: A significant pest of potatoes that is endemic in North America and adventive to Europe. CPB overwinters as an adult buried in the soil. Eggs are yellow or red while laid on-end and grouped in rows on underside of leaves; eclosion occurs within 4–9 days. CPB completes four larval instars. Female lays ca 500 eggs during 4–5 weeks; overwintering adults die after oviposition. Larvae and adults feed on leaves. CPB accidentally was introduced into Eastern Europe sometime after 1945, and has been regarded as a significant pest of potatoes in Russia. CPB prefers potatoes but will consume other plants including peppers, tomatoes, tobacco, eggplant, cabbage and some weeds. Species is bivoltine in North America, except in extreme north and south ends of range. See Chrysomelidae.

COLORADO TICK FEVER An arboviral disease endemic to western North America. CTF is caused by an arbovirus assigned to Reoviridae. Virus is enzootic in wild rodents and transmitted to humans by the tick *Dermacentor andersoni.* Acronym CTF. See Reoviridae. Cf. African Horse-Sickness; Bluetongue Disease; Epizootic Haemorrhagic Disease.

COLORATION Noun. (Middle English, *colour*; *-tion* = result of an action. Pl., Colorations.) 1. The capacity or quality of colour. 2. In insects, an arrangement of colours produced by pigments or structural modification of the integument. Coloration is often used in the identification or description of insects, but can be variable and sometimes an unreliable taxonomic character. Coloration serves insects in several ways, but not all adaptive purposes have been identified. Coloration serves the animal at individual, Species and interspecific levels. Each of these levels may be important to survival of individual or success of population; coloration seen by humans represents a compromise of selection operating at each level. At individual level, a pigment may be biochemically important while same pigment may be important in thermoregulation. (Dark-coloured bodies absorb heat; pale-coloured bodies reflect heat. Insects are poikilothermous; temperature regulation via body colour may be important in metabolic survival and development of individual.) Coloration at intraspecific level may act in visual recognition for members of same Species, particularly recognition of potential mates or potential rivals for mates. A sharp contrast in coloration of drab nocturnal Lepidoptera (moths) compared with conspicuous coloration of diurnal Lepidoptera (butterflies) illustrates this point. When sexual dimorphism in coloration is conspicuous, attraction of mates may be a logical explanation of more conspicuous forms. Recruitment of conspecifics (particularly in social forms) may serve as a function of colour or colour pattern. Coloration at interspecific level provides visual signals that warn predators of distastefulness or danger. Coloration may provide concealment from predators. Deception is a mechanism of selection operating on prey. Coloration creates disguise of size and shape, thereby rendering insect invisible within its environment. Anatomy and colour of prey becomes modified to evade predation. Role of deception probably is more readily appreciated in context of colour, and to a lesser extent shape. Deception is achieved by changing colour patterns or providing a colour pattern that apparently modifies shape. Role played by colour for interspecific recognition is complicated because several factors are involved. Long-term predator-prey relationships have engaged the mechanism of natural selection on both participants. From a morphological viewpoint, anatomy of predator becomes adapted to locate, overpower, kill and consume prey. Alt. Colouration. Rel. Crypsis; Mimicry.

COLUMBIA-BASIN WIREWORM *Limonius subauratus* LeConte [Coleoptera: Elateridae].

COLUMBIAN DEFOLIATOR *Oxydia trachiata* (Guenée) [Lepidoptera: Geometridae]: A multivoltine, polyphagous moth endemic in the neotropics and accidentally introduced into North America. See Geometridae.

COLUMBIAN TIMBER-BEETLE *Corthylus columbianus* Hopkins [Coleoptera: Scolytidae].

COLUMBINE BORER *Papaipema purpurifascia* (Grote & Robinson) [Lepidoptera: Noctuidae].

COLUMBINE LEAF-MINER A complex of Species including *Phytomyza aquilegiana* Frost, *P. aquilegivora* Spenser and *P. columbinae* Seghal [Diptera: Agromyzidae].

COLUMELLA, LUCIUS JUNIUS MODERATUS (5 BC–54 AD) Roman writer on agricultural Entomology (*De re Rustica,* 50) who made recommendation for control of some granary and garden pests. (Rose 1850, *New General Biographical Dictionary* 6: 422.)

COLUMELLA Noun. (Latin, *columella* = little rod. Pl., Columellae.) A little rod or pillar that extends along the central axis of a structure. Rel. Orientation.

COLUMN RAID Army Ants: A raid conducted by branching columns. Terminal elements of columns consist of relatively few workers that lay chemical trails and capture prey. See Formicidae.

COLUMNAR CELLS The most common cell-type in the Midgut. Anatomy of CC is highly variable, hence many names exist by which these cells are identified. CCs of midgut are bound together with Septate Junctions, Gap Junctions and Demosomes. Ultrastructure of CC is influenced by functions of cells and their secretions. CCs store fat, mineral salts and glycogen. CC nucleus typically is positioned near centre of cell. Cytoplasm contains Rough Endoplasmic Reticulum and abundant Golgi material; basal membrane of columnar cells lies adjacent to the Haemolymph and displays a highly convoluted surface of Microvilli facing the lumen of the gut. Apical region of Microvilli lacks organelles but does display Cisternae of Smooth Endoplasmic Reticulum; Mitochondria occur beneath apical region and Pinocytotic Vesicles occur beneath Microvilli. In general, midgut cells are richly supplied with Mitochondria, Endoplasmic Reticulum, Golgi Bodies and Ribosomes (thought to synthesize digestive enzymes). Spherocrystals are composed of concentric layers of mineral salts and originate on the Cisternae of Rough Endoplasmic Reticulum of most insects. In Lepidoptera and Thysanura, Spherocrystals apparently develop from Golgi Vesicles. Syn. Differentiated Cells; Lipophilic Cells; Cuboidal Cells. See Midgut. Cf. Regenerative Cell; Endocrine Cell; Peritrophic Membrane.

COLYDIIDAE Plural Noun. Cylindrical Bark Beetles. A Family of polyphagous Coleoptera including about 1,400 Species; the Family is widespread except on islands of the Pacific Ocean. Colydiids are placed within Cucujoidea but their relationship to other Coleoptera within the group is not certain. Adult body 2–10 mm long, cylindrical, dark coloured, heavily sclerotized and often strongly sculptured. Antenna is retracted beneath frontal ridge and club with 2–3 segments. Fore Coxa globular and closed behind, tarsal segments not bilobed; tarsal formula 4-4-4 with claws simple. Abdomen with five Ventrites. Larva is subcylindrical with short, recurved Urogomphi. Most Species are phytophagous with adults slow-moving and concealed under bark or within burrows in wood constructed by other Coleoptera. Some larvae of colydiids are predaceous on larvae and pupae of wood boring beetles (cerambycids and buprestids). See Cucujoidea. Rel. Anthicidae.

COLYER, CHARLES NORMAN (1908–1970) (Smith 1972, Entomol. mon. Mag. 108: 1–2, bibliogr.)

COMANCHE LACEWING *Chrysopa comanche* Banks [Neuroptera: Chrysopidae].

COMAS, JAIME NONELL (1876–1938) (Sachtleben 1941, Arb. physiol. angew. Ent. Berl. 8: 212.)

COMATE Adj. (Latin, *comatus* = long haired; *-atus* = adjectival suffix.) Setose only on upper part of head or Vertex. Generally setose with long, flexible Setae on upper surface of a structure (Kirby & Spence). Alt. Comatus.

COMB Noun. (Anglo Saxon, *comb* = comb. Pl., Combs.) 1. General: A row or rows of close-set, short Setae, spines, bristles or spindles positioned on the distal end of some leg segments. 2. Hymenoptera: Wax brood-cells constructed by bees that contain individual larvae and store honey. 3. Termite nests: A spongy, dark reddish-brown material made by workers from excreta and used to make fungal beds. 4. Lepidoptera (Lycaenidae): The serrate distal margin of Rostellum of Valva (Tuxen). 5. Diptera (Culicidae): Specialized spicules on each side of abdominal segment VIII (Peterson). 6. Diptera larvae (Culicidae): Setae on the upper surface of a Maxilla, with which brush-like structures are cleaned. See Antenna Cleaner; Ctenidia; Strigilis.

COMB CLAWED BEETLES See Tenebrionidae.

COMB-FOOTED SPIDERS See Theridiidae.

COMBAT® See Hydramethylnon.

COMBINATION COLORATION Colours arising from a combination of pigmentary and structural colours. The phenomenon is believed more common than structural colours (Imms). See Coloration; Pigmentary Colour; Structural Colour. Cf. Advancing Coloration; Alluring Coloration; Anticryptic Coloration; Apetetic Coloration; Aposematic Coloration; Cryptic Coloration; Directive Coloration; Disruptive Coloration; Epigamic Coloration; Episematic Coloration; Procryptic Coloration; Protective Coloration; Pseudepisematic Coloration; Pseudoaposematic Coloration; Seasonal Coloration; Sematic Coloration; Structural Colour. Rel. Crypsis; Mimicry.

COMBINATIONS Plural Noun. Genetics: A 'reshuffling' or reorganization of characters.

COMEAU, NOEL (–1976) (Anon. 1976, Bull. ent. Soc. Can. 8: 10.)

COMITE® See Propargite.

COMMENSAL Noun. (Latin, *cum* = with; *mensa* = table. Pl., Commensals.) A participating member of a commensalistic relationship. Alt. Coenosite. See Commensalism. Cf. Parasite; Symbiont.

COMMENSALISM Noun. (Latin, *cum* = with; *mensa* = table; English, *-ism* = condition. Pl., Commensalisms.) Literally, 'common table.' A term applied to an interspecific relationship in which one Species feeds on the food supply of another Species without destroying or harming the owner of the supply. See Symbiosis. Cf. Inquilinism; Mutualism; Parasitism.

COMMINUTE Adj. (Latin, *comminutus* = broken to bits; crumbled.) To grind exceedingly fine; to abrade or reduce to minute particles.

COMMINUTED Verb. (Latin, *comminutus* = broken to bits; crumbled.) Broken into extremely small fragments.

COMMISSURE Noun. (Latin, *commissura* = seam. Pl., Commissures.) 1. General: The nerve cord connecting any two bodies or structures (*e.g.* tracheal tubes). 2. Heteroptera: The more-or-less lengthened line of union where the Hemelytra meet along the Clavus below the apex of the Scutellum. 3. Coleoptera: The joint in the Costal Nerve of the wing it bends at the transverse fold.

COMMISURAL TRACHEAL TRUNKS Cross tracheal trunks continuous from one side of the body to the other, formed by anastomosis of the dorsal and ventral Tracheae of each side. Syn: Tracheal Commissure (Snodgrass).

COMMODORE® See Lambda Cyhalothrin.

COMMON Adj. (Middle English, *common* > Latin, *communis* = common.) 1. Of frequent occurrence; not rare (*e.g.* common migration). 2. A thing, condition, habit or trait that is shared by many individuals. 3. Anatomy: Formed of or divided into two branches (*e.g.* common oviduct). 4. Lepidoptera: Occurring on adjacent parts; descriptive of a band or Fascia when it crosses the anterior and posterior wings. Ant. Uncommon. Alt. General; Ordinary; Widespread.

COMMON AEROPLANE *Phaedyma shepherdi sheperdi* (Moore) [Lepidoptera: Nymphalidae]. (Australia).

COMMON ALBATROSS *Appias paulina ega* (Boisduval) [Lepidoptera: Pieridae]. (Australia).

COMMON, ALFRED FOGO (1873?–1952) (H.C.H.-1952, Entomologist 85: 216.)

COMMON AMBROSIA-BEETLE *Platyplus parallelus* (Fabricius) [Coleoptera: Curculionidae]. (Australia).

COMMON ANTHELID *Anthela acuta* (Walker) [Lepidoptera: Anthelidae]. (Australia).

COMMON ANTLION *Myrmeleon acer* Walker [Neuroptera: Myrmeleontidae]. (Australia).

COMMON ARMYWORM *Leucania convecta* (Walker) [Lepidoptera: Noctuidae]: A widespread pest of many crops in Australia. Adult wingspan 30–50 mm; forewing grey or reddish brown with two elongate, pointed, white markings; hindwing dark grey. CA adult lives 2–4 weeks; female lays several hundred eggs. Egg are white or pale yellow, round and ca 0.6 mm diam; eggs laid in long rows or clusters preferably in leaf litter; incubation requires 1–3 weeks. Neonate larvae may remain together for ca 1 day; early instar larvae skeletonize or chew holes in leaves; later instar larva consumes leaves from edge; larvae walk in a looping manner. Larvae feed nocturnally and remain concealed during day; when disturbed, larvae drop to ground, curl into ball (head inward) and remain motionless for a few minutes; CA completes six larval instars; mature larva pupates within a cell constructed in soil; overwinters as larva or pupa. CA is multivoltine with 3–4 overlapping generations per year. Syn. *Mythimna convecta*. See Armyworms; Noctuidae. Cf. Southern Armyworm. Rel. Looper.

COMMON AUGER-BEETLE *Xylopsocus gibbicollis* (Macleay) [Coleoptera: Bostrychidae]. (Australia).

COMMON AUSTRALIAN LADY-BEETLE *Coelophora inaequalis* (Fabricius) [Coleoptera: Coccinellidae].

COMMON AUSTRALIAN-ANOPHELINE *Anopheles annulipes* Walker [Diptera: Culicidae]: A pest in Australia and Papua New Guinea. Adult female is speckled grey with spotted wings; Palp as long as Proboscis; apical half of Palp with three white and two black bands; Abdomen with scales on segment VIII only. Female appears to stand on head to bite; bites at night, dusk, dawn or during day in shade. CAA enters buildings and takes blood meals from many hosts including humans. CAA can transmit Malaria and Filariasis. Eggs laid singly on surface of water; larvae usually develop in fresh water pools. See Culicidae. Cf. Australian Malaria Mosquito.

COMMON BANDED-AWL *Hasora chromus chromus* (Cramer) [Lepidoptera: Hesperiidae]. (Australia).

COMMON BANDED-MOSQUITO *Culex annulirostris* Skuse [Diptera: Culicidae]: A widely distributed pest Species taken in Australia, New Guinea, Philippines, Indonesia and the South Pacific. CBM feeds just after sunset on humans, other mammals and birds. CBM is a notable medical concern in that it transmits Filariasis, Murray Valley Encephalitis and Ross River Virus. Proboscis with median white band; Femora mottled and hind tarsal segments with narrow, basal pale bands; Cerci inconspicuous. Eggs are laid in rafts on surface of water; typically breeds in grassy freshwater pools, occasionally brackish water. See Culicidae.

COMMON BLOSSOM THIRPS See Tomato Thrips.

COMMON BLUET *Enallagma ebrium* [Coenagrionidae: Odonata].

COMMON BROWN-BUTTERFLY *Heteronympha merope* (Fabricus) [Lepidoptera: Nymphalidae]. (Australia).

COMMON BROWN-EARWIG *Labidura truncata* Kirby [Dermaptera: Labiduridae]: A predaceous Species found in Australia. Adult body 25–35 mm long, elongate; head bright reddish brown and body dull brown with straw coloured markings; legs, Antenna, plural region of abdominal segments, basal half of Forceps whitish; apex of Abdomen with pair of shiny Forceps. Adult females lay eggs in clusters of 20–30 in soil burrows. Eggs are creamy-white and oval. Female protects eggs and young nymphs within burrow. Nymphs are white when they emerge from egg and gradually darken. Nymphs are similar to adults in appearance, but lack wings. See Labiduridae.

COMMON BROWN-LEAFHOPPER *Orosius argentatus* (Evans) [Hemiptera: Cicadellidae]. (Australia).

COMMON BROWN-RINGLET *Hypocysta metirius* Butler [Lepidoptera: Nymphalidae]. (Australia).

COMMON BROWN-SCORPION *Liocheles waigiensis* (Gervais) [Scorpionida: Buthidae]. (Australia).

COMMON CARPET-BEETLE *Anthrenus scrophulariae* (Linnaeus) [Coleoptera: Dermestidae]: A cosmopolitan pest of animal products including wool, feathers, fur, hair, horns, hides and stored products. Adult oval-shaped, ca 3 mm long with mottled coloration. CCB life cycle duration determined by temperature, relative humidity and food. See Dermestidae. Cf. Australian Carpet Beetle; Black Carpet Beetle; Furniture Carpet Beetle; Varied Carpet Beetle.

COMMON CATTLE-GRUB *Hypoderma lineatus* (de Villers) [Diptera: Oestridae]: A widespread and serious pest of cattle and also known to attack goats, horses, and humans. Adult ca 11–13 mm long, black with longitudinal white stripes on Thorax and tuft of white Setae on lateral part of Prothorax; reddish Setae at apex of Abdomen. Eggs oval, white; laid in rows along hairs on legs and heels of cattle. Upon eclosion, larva penetrates skin and during 6 month period migrates (via Oesophagus and Diaphragm) to back. Larva bores hole in skin, applies posterior spiracles to aperture and completes development within swelling (warble) beneath skin. Mature larva becomes brown or black, burrows through skin, drops to ground and pupates in soil. Syn. Heel Fly; Ox Bot; Ox Warble Fly. See Oestridae. Cf. Northern Cattle Grub.

COMMON CENTIPEDE *Scolopendra morsitans* Linnaeus [Scolopendromorpha: Scolopendridae]. (Australia).

COMMON CLOTHES-MOTH See Webbing Clothes Moth.

COMMON COTTON-LOOPER *Anomis planalis* (Swinhoe) [Lepidoptera: Noctuidae]. (Australia).

COMMON CUTWORM 1. *Agrotis infusa* (Boisduval) [Lepidoptera: Noctuidae]: A bivoltine pest of several vegetable and field crops in Australia. Adult wingspan 30–50 mm; forewing dark brown, to grey-black with a black bar between two pale marks near centre; hindwing greyish to pale brown. Egg is dome-shaped, ca 0.5 mm diam, first white, then red, finally grey-black; eggs laid in clusters 2–3 rows deep; eclosion occurs within 3–7 days. CC typically completes six larval instars. 2. *Agrotis segetum* (Shiffermüller): An Old World pest of root crops including potato. See Cutworms.

COMMON DAMSEL-BUG *Nabis americoferus* Carayon [Hemiptera: Nabidae].

COMMON DART *Ocybadistes flavovittatus flavovittatus* (Latreille) [Lepidoptera: Hesperiidae]. (Australia).

COMMON DUCK-LOUSE *Anotoecus dentatus*

(Scopoli) [Phthiraptera: Philopteridae]. (Australia).

COMMON DUSKY-BLUE BUTTERFLY *Candalides hyancinthinus* (Semper) [Lepidoptera: Lycaenidae]. (Australia).

COMMON EGGFLY *Hypolimnas bolina nerina* (Fabricius) [Lepidoptera: Nymphalidae]. (Australia).

COMMON EUCALYPT-LONGICORN *Phoracantha semipunctata* (Fabricius). A pest of *Eucalyptus* that is endemic to Australia and accidentally established in USA and Europe. CEL larvae bore into trees.

COMMON FURNITURE-BEETLE See Furniture Beetle.

COMMON GARDEN-SNAIL *Helix aspersa* (Müller) [Sigmurethra: Helicidae]: A widespread pest of horticulture.

COMMON GOAT-LOUSE *Bovicola caprae* (Gurlt) [Phthiraptera: Trichodectidae]. (Australia).

COMMON GRASS-BLUE *Zizina labradus labradus* (Godart) [Lepidoptera: Lycaenidae]. (Australia).

COMMON GRASS-YELLOW *Eurema hecabe phoebus* (Butler) [Lepidoptera: Pieridae]. (Australia).

COMMON GREEN-DARNER *Anax junius* (Drury) [Odonata: Aeschnidae].

COMMON GROUND-SQUIRREL FLEA *Diamanus montanus* [Siphonaptera: Ceratophyllidae]: A pest of ground squirrels in western North America.

COMMON HIDE-BEETLE *Dermestes maculatus* DeGeer [Coleoptera: Dermestidae]: A widespread pest of untreated hides and skins, stored products (particularly materials of animal origin) and dried fish (in Asia and Africa). CHB is a major pest of museum collections, including insect collections. CHB is endemic to Europe. Pest as larva and adult; chews holes in products and contaminates food with frass and exuviae. Adult body elongate-oval, 6–10 mm long, dark with black and grey Setae on dorsum. Antenna with 11 segments including club; posterior margin of Elytron serrate with a spine where Elytra meet medially; adults fly readily and are capable of establishing new infestations quickly. Females oviposit within 12 hours of copulation and lay ca 200 eggs during relatively short life (ca 2–3 weeks). Eggs are oval, ca 1.3 mm long, white and deposited in batches of 4–8 within cracks of stored meat or stored products; eggs darken during embryogenesis; incubation requires ca 2 days. Larval stage requires ca 33 days; typically 5–7 moults with as many as 11 instars. Larvae are cannibalistic; mature larvae are white, ca 14 mm long with numerous long Setae and two curved Urogomphi; larvae tunnel into solid substrate for pupation. Pupation occurs within last larval exuvia. Adult emergence within ca 11 days. Syn. Leather Beetle. See Dermestidae. Cf. Hide Beetle; Larder Beetle.

COMMON HOUSE-MOSQUITO See Brown House Mosquito. Syn. Southern House Mosquito.

COMMON HOVER-FLY *Melangyna viridiceps* (Macquart) [Diptera: Syrphidae]: In Australia, adults are important pollinators of plants; larvae are important predators of aphids. See Syrphidae. Cf. Hover Flies.

COMMON, IAN F B Australian taxonomic Entomologist (CSIRO) specializing in Microlepidoptera and author of many technical papers and books dealing with moths.

COMMON IMPERIAL-BLUE *Jalmenus evagoras* (Donovan) [Lepidoptera: Lycaenidae]. (Australia). See Lycaenidae.

COMMON JEZABEL *Delias negrina* (Fabricius) [Lepidoptera: Pieridae]. (Australia). See Pieridae.

COMMON LACEWING See Chrysopidae.

COMMON MALARIA-MOSQUITO *Anopheles quadrimaculatus* Say [Diptera: Culicidae]: A pest widely distributed in eastern North America. Female moderately large bodied, dark brown with four dark spots near centre of wing; Vertex of head with pale scales; Proboscis, Palpus, Thorax and Tarsi with dark scales. Females are crepuscular feeders on blood of humans and domestic animals. See Culicidae.

COMMON MARSUPIAL-TICK. *Ixodes tasmani* Neumann [Acarina: Ixodidae]. (Australia).

COMMON MIGRANT *Catopsilia pyranthe crokera* (W.S Macleay) (Australia).

COMMON MOLE CRICKET *Gryllotalpa pluvialis* Mjoberg [Orthoptera: Gryllotalpidae]: A Species of mole cricket endemic to Australia. See Gryllotalpidae.

COMMON MOONBEAM *Philiris innotata innotata* (Miskin) [Lepidoptera: Lycaenidae]. (Australia).

COMMON NAME The popular or vernacular name given to an insect Order (*e.g.* beetles), Family (*e.g.* dustywings), Genus (*e.g.* carpenter ants) or Species (*e.g.* honey bee). Common names have no authors associated with them or official standing under the rules of Zoological Nomenclature. Synonymy is frequent among common names, and many Species may be known under one common name. See International Code of Zoological Nomenclature. Cf. Scientific Name. Rel. Description.

COMMON NETCASTING-SPIDER *Deinopis subrufa* L. Koch [Araneida: Deinopidae]. (Australia).

COMMON OAKBLUE *Arhopala micale amphis* Waterhouse [Lepidoptera: Lycaenidae]. (Australia).

COMMON PAPER-WASP *Polistes humilis synoecus* Saussure [Hymenoptera: Vespidae]: A Species of social wasp endemic to eastern Australia and adventive to Western Australia ca 1950. Adult worker body ca 15 mm long with markings and bands of brown, black and yellow; Antenna orange-brown. CPW biology and life history are similar to other Species of paper wasps. CPWs are social insects that construct nests consisting of numerous cells fashioned with saliva and macerated plant fibre; nest is attached by a constriction (Petiole) to branch of bush or tree, usually partially concealed by foliage. Nest is established by foundress queen who constructs a few cells then oviposits an egg in each cell; queen provides nectar and prey for developing larvae; larvae pupate within their cells. Adult workers emerge within a few weeks to assume duties of nest construction, foraging and brood care. Late in season, new queens and drones are produced; drones mate with virgin queens and then die; inseminated queens overwinter. CPW colony is multivoltine but nest typically only used during one season. See Vespidae. Cf. European Wasp; Golden Paper-Wasp; Macao Paper-Wasp; Yellow Paper-Wasp.

COMMON PEARL WHITE *Elodian angulipennis* (Lucas) [Lepidoptera: Pieridae]. (Australia).

COMMON PILLBUG *Armadillidium vulgare* (Latreille) [Isopoda: Armadillidiidae]: A widespread minor pest of garden and home that is endemic to Europe and adventive in many parts of world. Adult 10–16 mm long and dark grey to dark purplish (male) or brownish to grey with pale yellow markings (female). Antenna about half as long as body; body with seven pairs of legs; no spines at posterior margin of body; rolls into a ball when disturbed. Female with brood pouch on venter near head; produce 30–100 eggs per reproductive cycle and carries eggs 4–8 weeks; neonate young remain in pouch 1–2 weeks; young gregarious within their cohort; older individuals tend to eat younger individuals; CP undergoes 11 moults to maturity and moults after becoming an adult; egg to adult requires ca six months; females live ca two years and reproduce through most of year (1–4 broods per year). See Native Pillbug. Cf. Slaters.

COMMON POWDERPOST BEETLE *Euvrilletta peltata* [Coleoptera: Anobiidae]: An abundant pest of wood in the eastern USA. CPB typically infests sapwood part of softwoods; old and new sapwood are attacked and infestions continue until sapwood is consumed. Natural populations occur in dead branches and fallen logs; urban populations occur in wood used in construction of buildings. Adults are nocturnal with emergence during spring and early summer; females mate often and lay ca 50 eggs during their reproductive life; eclosion occurs within eight days, first instar larvae bore into wood and tunnel in direction of grain; pupation occurs within feeding chamber near surface of wood. Feeding may require more than a year; pupation requires ca 14 days. See Anobiidae.

COMMON PRICKLY-PEAR *Opuntia inermis* (De Candolle) De Candolle [Cactaceae]: A Species of cactus native to the new world and adventive elsewhere as a significant rangeland pest. In Australia controlled with *Cactoblastis cactorum* (Bergroth) imported from Argentina during 1925. Cf. Prickly Pear. Rel. Biological Control; Cactus Moth.

COMMON RED-EYE *Chaetocneme beata* (Hewitson) [Lepidoptera: Hesperiidae]. (Australia).

COMMON SCORPIONFLIES See Panorpidae.

COMMON SILVER XENICA *Oreixenica lathoniella* (Westwood) [Lepidoptera: Nymphalidae]. (Australia).

COMMON SILVERFISH *Lepisma saccharina* Linnaeus [Thysanura: Lepismatidae]: A cosmopolitan, nocturnal, urban pest of homes, warehouse and stored products. Adult to 15 mm long with body fusiform; Antenna long and filiform; body covered with scales but Setae not barbed or arranged into combs; apterous; Cerci and Appendix Dorsalis shorter than Abdomen. CS food includes paper, starchy material, book bindings, textiles, wallpaper, sizing and linen; CS often colonize roofs and subfloors. Females lay ca 100 eggs during lifetime; egg stage requires 19–43 days; adult lives 3.0–3.5 years. Other principal urban Species called 'silverfish' include: *Acrotelsa collaris* (Fabricius), *Ctenolepisma lineata* (Fabricius) and *Ctenolepisma longicaudata* Escherich. See Silverfish; Thysanura. Cf. Firebrat.

COMMON SKIMMERS See Libellulidae.

COMMON SLATER *Porcellio scaber* Latreille [Isopoda: Porcellionidae]: Apparently endemic to Europe and introduced into Australia where CS is a minor pest of garden and home. Adult 9–15 mm long, dark grey or grey with paler margin; brown or yellow-orange with black spots; Antenna about half as long as body; body with seven pairs of legs and two spine-like appendages at posterior margin. CSs do not roll into ball when threatened or disturbed. See Slaters. Cf. Common Pillbug; Garden Slater; Native Pillbug.

COMMON SPLENDID GHOST-MOTH *Aenetus ligniveren* (Lewin) [Lepidoptera: Hepialidae]. (Australia).

COMMON SPOTTED-LADYBIRD *Harmonia conformis* (Boisduval) [Coleoptera: Coccinellidae]: In Australia, a beneficial predator. Adult 5.0–7.9 mm long, round-ovate and yellow-orange with black markings. Markings on CSL include 18 evenly distributed spots on Elytra and two spots on Thorax. Females lay eggs in batches near larval food. Eggs are yellow, spindle-shaped and laid standing on end. Larvae are elongate, with well defined legs; body black with markings and fleshy spines. CSL completes four larval instars and pupation occurs on plant in last larval exuvia. Adults and larvae are predaceous on aphids and cause no plant damage. Syn. Eighteen-Spotted Ladybird. See Coccinellidae. Cf. Minute Two-Spotted Ladybird; Mite-Eating Ladybird; Striped Ladybird; Three-Banded Ladybird; Three-Spotted Ladybird; Variable Ladybird.

COMMON STONEFLIES See Perlidae.

COMMON SWIFT *Pelopidas agna dingo* Evans [Lepidoptera: Hesperiidae]. (Australia).

COMMON THRIPS See Thripidae.

COMMON TIT *Hypolycaena phorbas phorbas* (Fabricius) [Lepidoptera: Lycaenidae]. (Australia).

COMMON WHISTLING-MOTH *Hecatesia fenestrata* Boisduval [Lepidoptera: Noctuidae]. (Australia).

COMMON WHITE See Southern Cabbageworm.

COMMON WHITE-SPOT SKIPPER *Trapezites petalia* (Hewitson) [Lepidoptera: Noctuidae]. (Australia).

COMMON WOODLOUSE *Armadillidium vulgare* (Latreille) [Isopoda: Armadillidiidae]. (Australia).

COMMON YELLOWJACKET *Vespula vulgaris* (Linnaeus) [Hymenoptera: Vespidae]: A widespread urban pest in the Holarctic. CYJ resembles German Yellowjacket and nests are constructed of macerated wood, typically subterranean with single entrance and nest ca 10 cm below surface; CYJ occasionally is found in wall voids or aerial locations. Colony size may reach 50,000 workers. See Yellowjacket. Cf. Aerial Yellowjacket; German Yellowjacket; Western Yellowjacket.

COMMUNAL Adj. (Latin, *communalis* > *communis* = general, common.) 1. Pertaining to eusocial organisms that live in colonies. 2. Eusocial bees: Females that utilize a composite nest but each female makes and provisions her own cells. Alt. Public; Common; Shared. See Eusocial; Primitively Social; Quasisocial; Semisocial. Subsocial. Cf. Aggregation; Colony; Social.

COMMUNICATION Noun. (Latin, *communicatio* = communicate; English, *-tion* = result of an action. Pl., Communications.) The act or process of imparting or delivering knowledge, opinions, facts or state-of-being from one organism to another. Insects communicate acoustically, chemically, tactually and visually. Alt. Message; Statement.

COMMUNITY Noun. (Latin, *communis* = common; English, *-ity* = suffix forming abstract nouns. Pl., Communities.) 1. Ecology: An assemblage of animals or plants of a specified habitat and which differ from assemblages of other plants or animals. 2. A group of organisms living in the same place and under the same conditions. Alt. Population; Society. 3. A common character or likeness.

COMOSE Adj. (Latin, *comosus* = hairy > *coma* = hair; *-osus* = full of.) 1. Terminating in a tuft of. Setae that form a brush. 2. Bearing a tuft of soft hairs. Alt. Comate.

COMPACTED Adj. (Latin, *compactum,* past participle *compactus* = to make arrangement with.) Descriptive of structures that are concentrated or pressed firmly together; consolidated; closely united.

COMPARATIVE ANATOMY The subdiscipline of anatomy in which structure is compared among Taxa or between sexes of a Taxon. CA was first developed by Aristotle. See Aristotle. Cf. Anatomy; Morphology.

COMPARETTI, ANDREA (1746–1801) (Rose 1850, *New General Biographical Dictionary* 6: 429.)

COMPERE, GEORGE (1858–1928) American agriculturalist and foreign explorer for natural enemies used in Biological Control. (Anon. 1935, Arx. Esc. sup. Agr. Barcelona 1: 293–299 portrait.)

COMPERE, HAROLD (1896–1978) American Entomologist (University of California.) Foreign explorer for natural enemies of *Citrus* pests for California and specialist in taxonomy of parasitic Hymenoptera, particularly Aphelinidae and Encyrtidae. (Gordh, 1994, Pan Pac. Entomol. 70 (3): 188–205.)

COMPETITION Noun. (Latin, *competitio*. Pl., Competitions.) 1. The act of competing. 2. The physical struggle or indirect interactions between individuals for the same object, place or goal. 3. The interaction among individuals for a resource that is used by the individuals or their kin. See Cf. Intraspecific Competition; Interspecific Competition. Rel. Combat.

COMPLANATE Adj. (Latin, *complanatus,* past participle of *complanare* = to make plane; *-atus* = adjectival suffix.) Flattened; make level. Pertaining to structure compressed; flattened above and below; deplanate. Alt. Complanatus.

COMPLEMENTAL REPRODUCTIVE Termitidae: Special sexual forms replace the king (reproductive male) and queen (reproductive female) of a colony when these primary reproductives die. CR are derived from nymphs by special feeding. Alt. Complementary Reproductive. See Termitidae.

COMPLEMENTARY SEX DETERMINATION A genetical mechanism for determination of an individual's sex in which homozygosity or hemizygosity at a specific locus or loci determines a male; heterozygosity at the same locus or loci determines a female.

COMPLETE METAMORPHOSIS Metamorphosis of Holometabola that typically involves four consecutive stages: Embryo (egg), larva, pupa and adult. Each stage is completely different in form and separated in time from other stages of the same individual. Indirect Metamorphosis *sensu* Wardle. See Metamorphosis. Cf. Ametaboly; Heterometaboly; Incomplete Metamorphosis; Paurometaboly.

COMPLEX Adj. (Latin, *complexus,* past participle of *complecti* = to entwine around.) 1. Composite; composed of more than one part. 2. Complicated; intricate. Alt. Compound; Intricate; Multifaceted. Noun. (Latin, *complecti* = to entwine around. Pl., Complexes.) 1. An assemblage of related things. 2. Structure consisting of several differentiated parts. 3. Taxonomy: A group of Species assigned to a Genus that are morphologically or biologically more closely related or similar to one another than to other members of the Genus. 4. Ecology: A grouping or association of organisms.

COMPLICANT Adj. (Latin, *complicans* > *cum* = together; *plicare* = to fold; *-antem* = adjectival suffix.) Overlapping; A condition in which one wing overlaps the other wing.

COMPLICATE Adj. (Latin, *cum* = together; *plicare* = to fold; *-atus* = adjectival suffix.) Longitudinally laid in folds. Intricate or complex; not simple. Alt. Complicatus.

COMPLY® See Fenoxycarb.

COMPONENT Adj. (Latin, *componens,* present participle of *componere* = to put; to set.) 1. Serving to constitute. 2. Constituent.

COMPONENT Noun. (Latin, *componere* = to put; to set. Pl., Components.) 1. A constituent part; an ingredient. 2. Any one part of a combined whole.

COMPOSITE Adj. (Latin, *cum* = together; *ponere* = to place; *-ites* = substance.) Descriptive of structure that is composed of several parts that collectively define that structure or make it functional. Alt. Amalgamated; Compositus; Fused.

COMPOSITE THRIPS *Microcephalothrips abdominalis* (D. L. Crawford) [Thysanoptera: Thripidae].

COMPOUND Adj. (Middle English, *compounen;* Latin, *cum* = together; *ponere* = to place.) 1. Composed of many similar (or dissimilar) parts, elements, things or ingredients. 2. Composed of several individuals joined together. Alt. Complex; Mixture.

COMPOUND ANTENNA Capitate Antenna formed by several segments or Flagellomeres. See Antenna. Cf. Simple Antenna.

COMPOUND EYE Insecta: Paired aggregations of separate visual elements (Ommatidia) that are located on the head. Compound eye is a common anatomical feature among arthropods. CEs are always paired and cyclops-like insects (with one median compound eye) are unknown. Some insects (Ephemeroptera and Gyrinidae) appear to have four CEs, but this condition is the result of extreme modification of the entire eye to service optical needs from above and below the animal. Some highly evolved insects are eyeless, but this condition is derived and represents an adaptation to a specialized environment (caves) or lifestyle (parasitism); ancestors of these eyeless insects had eyes. CE is derived from epidermal cells of integument. CEs typically occur in lateral part of head with Antennae positioned between them. CE may represent appendage-like structures of a primitive segment in groundplan head. This eye-bearing segment occurs behind antennal segment and in front of mandibular segment. Some primitive insects (Protura and Diplura) are blind; Collembola have Ommatidia; Archaeognatha and Thysanura have true CEs. Transition from eyeless to eyed apterygotes suggests that CE developed later in evolutionary history of insects than integument. CE size varies considerably. When CEs are large and meet medially, condition is called holoptic. Holoptic condition occurs in some Diptera (Bombyliidae, Tabanidae, Pipunculidae and Acroceridae). In other Diptera, the holoptic condition is restricted to males that swarm and engage in aerial copulation. When CEs are not as large and do not meet medially, the condition is called dichoptic. CE shape is typically round or oval in outline. CEs may be bean-shaped (reniform) in some insects. An extreme of this condition occurs when CEs are medially emarginate. This emargination is generally an adaptation to accommodate the antennal Scape. Emarginate CEs occur in sphecid wasps and cerambycid beetles. CE is surrounded by a Circumocular Suture in some insects. In beetles, CE is supported by an internal flange called the Occulata. A similar flange is found in some parasitic Hymenoptera, but flange has not been named. The flange supports CE. Plane of vision varies considerably among insects, and seems dependent upon several factors. *Notonecta* views 240° on horizontal and 360° on vertical. Stalk-eyed fly *Cyrtodiopsis whitei* has an extraordinary range of view with considerable binocular overlap along median plane. Overlap ca 140° in frontoventral area of view. See Eye. Cf. Ocellus; Pin-Hole Eye; Simple Eye; Stemma.

COMPOUND EYE ANATOMY See Cone; Lens; Ommatidium; Pigment Cell; Rhabdom.

COMPOUND EYE SENSITIVITY Compound eye is efficient at short range and insects view their nearby environment at distances appropriate for small bodied animals. CE is exceptionally effective at detecting slight movement or motion. Some Species can separate objects within one degree. This feature is important for predators in locating prey and prey in detecting predators. Range of sensitivity extends from ca 300 µ (ultraviolet) to 690 µ (650 µ = Yellow). Ultraviolet light perception has not been studied intensively but insects are probably sensitive to UV. This range is skewed somewhat to human eye from ca 400–790 µ. Range of sensitivity in insect eye is not uniform because when intensity is constant, ERG's show two peaks at 350 µ and 500 µ. Temperature influences spectral sensitivity. Insects are poikilothermous and influence of temperature is apparent on motor and sensory functions, particularly at temperatures lower than 20°C. Sensitivity also varies with amount of time light has impinged upon photoreceptor. So-called light and dark adapted eyes are different morphologically. Cytological changes in eye reflect changes in concentration of mitochondria and movement of screening pigments. Some insects can perceive and discriminate among colours. Colour-blind insects are called deuteranoptic. Colour vision has been inconclusively studied in insects but is known in bees, ants and flies. Anatomical requirements for colour vision include two or more tuned photoreceptors and visual pigments that absorb photons of given wavelength. Polarized light can be perceived by some insects.

Polarized light can be used by ants as a solar compass for orientation and navigation (John Lubbock 1882, aka Lord Avebury). Reflectance pattern of nonpolarized light is circular; reflectance pattern of polarized light is elliptical. Insects can utilize polarization naturally, but man cannot. Vertebrates, including man, cannot see polarized light because rhodopsin molecules float free and the axial orientation is random. Houseflies and fireflies orient to polarized light within a narrow range of intensity. Ability to perceive plane of polarization is lost above 410 nm.

COMPOUND GLAND See Multicellular Gland.

COMPOUND NEST Social Insects: A nest containing colonies of two or more Species. Galleries of the colonies anastomose and adults mingle, but the broods remain separate. See Nest. Cf. Mixed Nest.

COMPOUND OCELLUS An ocellate spot in which the colour is in three or more circles. See Ocellus.

COMPRESSED Adj. (Latin, *cum* = together; *premere* = to press.) Flattened by lateral pressure; flattened laterally. Cf. Depressed. Alt. Condensed; Dense; Squeezed together.

COMPRESSION FOSSIL A type of fossil formed when an organism or its parts are burried under sediment and the weight of the sediment compresses the specimen into a thin film of organic material. When fossil-bearing stone is split, compression fossils appear as dark paint on the exposed surface. CF typical of plant leaves, fish scales and insect wings. See Fossil.

COMPRESSOR OF THE LABRUM A single or paired median muscle, attached on the anterior and posterior walls within the Labrum (Snodgrass). Alt. Labral Compressor. See Labrum.

COMSTOCK, ANNA BOTSFORD (1845–1930) Noted natural history illustrator and wife of J. H. Comstock. (Smith 1976, Ann. Rev. Ent. 21: 1–25, bibliogr.)

COMSTOCK, JOHN ADAMS (1883–1970) (Martin 1971, J. Lepid. Soc. 25: 215.)

COMSTOCK, JOHN HENRY (1849–1931) American academic (Cornell University) and founder of the first Entomology Department. Wrote textbook of Entomology; noted for work on wing classification, spider systematics and many groups of insects. (Herrick 1931, Ann. ent. Soc. Am. 24: 199–204.)

COMSTOCK MEALYBUG *Pseudococcus comstocki* (Kuwana) [Hemiptera: Pseudococcidae]: In eastern North America, a pest of apples, peaches, pears, grapes; a pest of *Cltrus* in some parts of world. Overwinters as egg under bark. See Hemiptera.

COMSTOCK, WILLIAM PHILLIPS (1880–1956) (Mallis 1971, *American Entomologists*. 549 pp. (332–339), New Brunswick.)

CONCA, RAFAEL CHALVER (–) (Albert 1973, In R. C. Conca. La Familia Aphidiidae. (Ins. Him.) en España. 132 pp. (7–9), Valencia.)

CONCATENATE Adj. (Latin, *cum* = together; *catenatus* = chained; *-atus* = adjectival suffix.) Pertaining to structure linked in a chain-like series. Alt. Concatenatus.

CONCAVE Noun. (Latin, *concavus* > *con* = with; *cavus* = hollow.) 1. Hollowed or depressed; the interior of a sphere. 2. Vaulted. Alt. Concavus; Hollow. Ant. Convex.

CONCAVE VEIN 1. A vein that has a tendency to fold downward (Snodgrass). 2. A vein of the insect wing that follows the furrows (Imms). See Vein; Wing Venation. Cf. Convex Vein.

CONCAVO-CONVEX (Latin, *concavus*; *convexus* = vaulted.) Hollowed out or concave on one surface, rounded or convex on the other, like a small segment of a hollow sphere.

CONCENTRATED Adj. (Latin, *cum* = together; *centrum* = centre.) 1. Gathered at one point. 2. Intensified or strengthened by evaporation.

CONCENTRIC Adj. (Medieval Latin, *concentricus* > *cum* = together; *centrum* = centre ; *-ic* = characterized by.) 1. Pertaining to structures with a common centre. 2. Descriptive of circles and spheres of different size and which have a common centre.

CONCHASPIDIDAE Plural Noun. A small Family of scale insects, comprising about 25 Species in three Genera, found in tropical Africa and South America, Orient and Australia. Adult's Antenna and legs well developed, otherwise conchaspids resemble diaspidids and produce a scale cover completely detached from the body; nymphal exuviae not incorporated into adult conchaspid cover. Distal segments of Abdomen fused into a 'Pseudopygidium' in adult female and nymphal instars; most Species have tubular ducts, which alone among the wax-secreting organs may be concerned with cover formation. Conchaspids apparently are related to Diaspididae. See Coccoidea. Cf. Diaspididae.

CONCHATE Adj. (Greek, *kongche* = shell; *-atus* = adjectival suffix.) 1. Shell-like. 2. Orthoptera: A term applied to the shell-like inflation of the Auricle in the anterior Tibia. Alt. Conchatus.

CONCHIFORM Adj. (Latin, *concha* = shell; *forma* = shape.) Shell-shaped; Concha-shaped; shaped as half of a bivalve shell. Alt. Conchate. Cf. Conriform; Heliciform; Mytiliform; Pectiniform; Trochiform. Rel. Form 2; Shape 2; Outline Shape.

CONCHUELA *Chlorochroa ligata* (Say) [Hemiptera: Pentatomidae]: A Species endemic to British Colombia, Colorado south to central Mexico. Eggs first appear pale green then brown, cylindrical in shape and glued in masses to leaves. Nymphs feed gregariously. A minor pest of cotton where the plant and insect are sympatric. See Pentatomidae.

CONCINNE Adj. (Latin, *concinnus* = skilfully joined; neat.) Neat; fine.

CONCOLORATE Adj. (Latin, *concolor* = with the same colour; *-atus* = adjectival suffix.) Of a uniform colour; different parts with the same colour. Alt. Concolorous.

CONCOLORES Adj. (Latin, *concolor* = with the same colour.) Lepidoptera: Descriptive of wings whose upper and lower surfaces are of the same colour.

CONCORD® See Alphacypermethrin.

CONCRESCENCE Noun. (Latin, *concrescere* = to grow together. Pl., Concrescences.) Tissues or structures that grow together.

CONCRETE Adj. (Latin, *concretus* > *concrescere* = grown together.) 1. Grown together to form a mass. 2. Formed by a union of separate parts. Syn. Material; Tangible.

CONCRETION Noun. (French, *concretion* > Latin, *concretion* = grown together; English, *-tion* = result of an action. Pl., Concretions.) 1. The act or process of bring particles or parts together to form a soild structure. 2. A mass or nodule formed by by growing together.

CONCURRENT Adj. (French, *concurrent* > Latin, *concurrens* > *concurrere* = conjoined.) Running together; a term applied to a vein that runs into another vein and does not separate. Ant. Separate. Syn. Coexisting; Concomitant; Parallel; Simultaneous. Cf. Recurrent.

CONCURRENT HOST FEEDING See Host Feeding. Cf. Non-concurrent host feeding; Destructive host feeding; Non-destructive host feeding.

CONDE, OTTO (1905–1944) (Sachtleben 1944, Arb. morph. taxon. Ent. Berl. 11: 141.)

CONDENSATION Noun. (Latin, *condensatio* > *cum* = together; *densare* = to make thick; English, *-tion* = result of an action. Pl., Condensations.) The act or process in which tissue or structure thickens or becomes visible; condensed. Syn. Concentration; Reduction. Cf. Evaporation.

CONDILO SPURIO The mandibular Condyle. A Preartis *sensu* MacGillivray. Syn: Dorsal Condyle; Epicondyle.

CONDITIONED REFLEX The action in the nervous system arising from external conditions.

CONDOR® A biopesticide derived from *Bacillus thuringiensis* EG 2348. See *Bacillus thuringiensis.*

CONDUCTIVITY Noun. (Latin, *conducere* = to lead together; English, *-ity* = suffix forming abstract nouns. Pl., Conductivities.) 1. The property of nervous impulse propagation. 2. Protoplasmic tissue conveying changes in metabolic activity.

CONDUPLICATE Adj. (Latin, *conduplicare* = to double; *-atus* = adjectival suffix.) Doubled or folded together. Alt. Conduplicatus.

CONDYLAR Adj. (Greek, *kondylos* = knuckle.) Descriptive of or pertaining to a Condyle or relating to a joint-surface. Alt. Condylic.

CONDYLE Noun. (Greek, *kondylos* = knuckle. Pl., Condyles.) 1. General: Any process by means of which an appendage is articulated into a pan, cavity or depression. 2. A knob-like process that articulates the base of the Mandible to the head. See Ginglymus. Cf. Acetabulum; Cotyla; Enarthrosis.

CONE Noun. (Latin, *conus* > Greek, *konos* = cone. Pl., Cones.) 1. Any object or structure that resembles a geometric cone. A shape generated by rotation of one of the legs of a triangle around another leg that serves as an axis. The rotating leg forms a circular plane and the area within the hypotenuse creates the cone shape. 2. Thrips: The cone-shaped portion of the head that contains the mouthparts. Cone formed by Labrum and Clypeus above and Labium below; the mouth cone. 3. Compound eye: The usually conical-shaped crystalline body of an Ommatidium. The Cone lies immediately beneath the Lens and constitutes the second element of the light-capturing system. The Cone is formed by Semper Cells. Several anatomical types of Cone have been described in the insect eye. See Pseudoconic Cone; Cf. Aconic. Rel. Compound Eye; Ommatidium.

CONE BEETLES See Curculionidae.

CONE HEADED BUSH-CRICKETS See Conocephalidae.

CONE-NOSE BUGS See Reduviidae.

CONFER Verb. (Latin, *conferre* = to bring together.) 1. To collect; to contribute. 2. To compare; to collate. Italicized and abbreviated *cf., cfr.* Confer, Conferring; Conferred.

CONFERTED Adj. (Latin, *confertus* = crowded.) 1. Pressed together. Descriptive of structures that are closely clustered, crowded or compacted. 2. Botany: Fascicled.

CONFIDOR® See Imidacloprid.

CONFIRM® See Tebufenozide.

CONFLECTS Adj. (Etymology obscure.) Crowded, clustered. Ant. Sparse.

CONFLUENT Adj. (Latin, *confluens* > *confluere* > *con-* together; *fluere* = to flow.) 1. Running together, *e.g.* two spots united in one outline. Alt. Confluens. 2. Tissue Culture: The point of maximum cell density arising through exhaustion of cell-culture medium or the complete cover of existing growth surface.

CONFUSED Adj. (French, *confus* > Latin, *confuses* > *confundere* = to confound.) 1. Pertaining to markings with indefinite outlines or markings running together as lines and spots without definte pattern. Things mingled such that they are not recognized as discrete. 2. Disordered; thrown into disarray. Syn. Perplexed. Ant. Organized.

CONFUSED FLOUR BEETLE *Tribolium confusum* Jacquelin de Val [Coleoptera: Tenebrionidae]: A cosmopolitan pest of plant material and museum collections that is endemic to Africa (perhaps Ethiopia) and moved globally through commerce. CFB larvae prefer milled grains but also feed on cereal products. CFB probably is derived from a bark-inhabiting and fungus-feeding ancestor. CFB adult resembles the Rust-Red Flour-Beetle, but antennal club poorly defined of four segments and CFB is adapted to cooler conditions. Adults with well developed wings but will not fly and are typically seen walking on food. CFB larvae are pests while the adults most often are seen. Larvae and adults occasionally are cannibalistic or facultative predators of their own egg and pupal stages. Eggs are laid directly in food and Species completes 7–8 larval instars; mature larvae emerge from food and pupate exposed (not enclosed in cocoon or pupation chamber). Development is influenced by type of food, temperature and humidity. Egg stage completed within 6–9 days; larval stage requires 16–100 days; pupal stage lasts 6–8 days; adult lives 28–130 days. Syn. Bran Bug. See Tenebrionidae. Cf. Dark Flour Beetle; Rust-Red Flour-Beetle; Yellow Mealworm.

CONGENER Noun. (Latin, *congener* > *con-* = together; *generis* = race. Pl., Congeners.) 1. A Species assigned to the same Genus as another Species. 2. An animal, plant, or microorganism that is morphologically similar to other organisms and on the basis of that similarity assigned together in classification.

CONGENERIC Adj. (Latin, *congener* = of same race; *-ic* = consisting of.) Descriptive of or pertaining to a Species that agrees in all 'generic' characters to other Species with which it is compared. See Genus. Cf. Conspecific. Rel. Classification.

CONGESTED Adj. (Latin, *congestus* > *congere* = to bring together; heaped.) 1. Descriptive of organisms, organelles or things that are heaped together or crowded. 2. Disruptively crowded. Syn. Crowded. Ant. Empty. Cf. Distended.

CONGLOBATE Adj. (Latin, *conglobatus* > *conglobare* = ball-shaped; *-atus* = adjectival suffix.) Ball-shaped; forming a rounded mass, ball or sphere. See Globate. Cf. Lobate. Rel. Shape.

CONGLOBATE GLAND A glandular appendage of male sexual organs in insects, that opens upon one of the external structures. See Gland.

CONGLOMERATE Adj. (Latin, *conglomeratus* > *conglomerare* = to wind into a ball; *cum* = with; *glomerare* = to wind; *-atus* = adjectival suffix.) 1. Descriptive of similar objects congregated or densely massed together. 2. Descriptive of structure or features irregularly grouped into spots. Alt. Conglomeratus. Cf. Adelphous; Aggregate.

CONGO FLOOR MAGGOT *Auchmeromyia luteola* [Diptera: Calliphoridae]: A blood-feeding larval parasite of humans in sub-Saharan Africa and Madagascar. Female lays eggs on ground or bedding material of host. Larvae are nocturnal and feed upon blood of sleeping human. Larvae feed for 15–20 minutes, detaches and moves to a place of concealment; larvae feed repeatedly at night until development has been completed (2–12 weeks). Pupation requires 9–16 days. See Calliphoridae.

CONIC Adj. (Latin, *conus* = cone; *-ic* = of the nature of.) Pertaining to structure with a flat base and tapering to a point. Alt. Conical. See Cone 1. Cf. Cylindrical; Obconic. Rel. Form; Shape.

CONICAL PEG OF GONOSTYLUS Hymenoptera (Pamphilidae): Laminum.

CONICO-ACUMINATE Adj. (Latin, *conus* = cone; *acumen* = point; *-atus* = adjectival suffix.) Descriptive of structure in the form of a long, pointed cone. Cf. Acuminate; Conical.

CONICULUS Noun. (Latin, *conus* = cone. Pl., Conicula.) Acarology: Structure that encloses the preoral cavity.

CONIDIUM Noun. (Greek, *konis* = dust; Latin, *-ium* = diminutive > Greek, *-idion*. Pl., Conidia.) Fungi: Spores that are produced during the asexual phase of the life cycle. See Fungus.

CONIFER SAWFLIES See Diprionidae.

CONIFER SPIDER MITE *Oligonychus coniferarum* (McGregor) [Acari: Tetranychidae].

CONIFEROUS Adj. (Latin, *conus* = cone; *ferre* = bearing; *-osus* = possessing the qualities of.) Bearing cone-like processes. Alt. Coniferus.

CONIOPTERYGIDAE Plural Noun. Aphidwolves; Dustywings. A Family of Neuroptera distantly related to other extant Families. Adult small bodied and covered with wax from wax glands; Antenna short, moniliform, Scape enlarged and Pedicel elongate; Ocelli absent. Wings tentiform at repose and wing venation reduced; Pterostigma absent; wing coupling via Hamuli at base of forewing and hindwing. Adults are arboreal and males do not manufacture a Spermatophore. Larvae are called aphidwolves and act as important predators of aphids and mealybugs. See Neuroptera. Rel. Biological Control.

CONIOPTERYGOIDEA A Superfamily of planipennous Neuroptera including one extant Family, the Coniopterygidae.

CONJOINED Adj. (Latin, *con* = with; *jungere* = to join.) Joined together; united; adnate. Ant. Disjunct. Syn. United.

CONJUGATE Verb. (Latin, *conjugare* = to join together; *-atus* = adjectival suffix.) To bring together in pairs; consisting of one pair.

CONJUGATION Noun. (Latin, *com* = together; *jugare* = to yoke; English, *-tion* = result of an action. Pl., Conjugations.) 1. The physical act of joining, combining or uniting. 2. The state of being joined or united. 3. A biological process similar to fertilization in high organisms that involves the fusion or merging of male and female gametes and which results in a new generation of organisms. Syn. Syngamy. 3. The pairing of Chromosomes. See Reproduction. Cf. Fertilization.

CONJUNCTIVA Noun. (Latin, *conjunctivus, cum* = together; *jungere* = to join. Pl., Conjunctivae.) The membranous, flexible portion of the Integument that connects two segments or sclerites. The Coria *sensu* MacGillivray. Syn. Membrane.

CONJUNCTIVUS Noun. (Latin, *conjunctivus* = connective. Pl., Conjunctivas; Conjunctivae.) A mandibular sclerite between the Molar and the Basalis.

CONJUNCTURA Noun. (Italian, *congiuntura* > *congiunto* = conjoined. Pl., Conjuncturae.) The articulation of a wing to the Thorax.

CONNATE Adj. (Latin, *connatus* > *conasci* > *cum* = together; *gnatus* = born; *-atus* = adjectival suffix.) Pertaining to structures that are fused or immovably united. Term often used to describe sternal sclerites of the beetle Abdomen. Alt. Connatus.

CONNECTIVE Noun. (Latin, *con-*; *nectere* = to bind. Pl., Connectives.) A nerve, Trachea or muscle that connects structures within or between body segments. Example: Longitudinal nerve-cord fibres that connect one ganglion of the nervous system with an adjacent ganglion.

CONNECTIVE TISSUE Any non-specified tissue, meroblastic in origin, that binds together other tissues or organs of the body. CT varies considerably in form, texture and physical properties, including elastic, inelastic, reticulate, fibrous and adipose. See Tissue.

CONNEXIVUM Noun. (Middle French, *connexer* = to connect. Pl., Connexivia.) Heteroptera: The prominent, flattened abdominal margin at the junction of the dorsal and ventral sclerites. See Abdomen. Rel. Sclerites.

CONNIVENT Adj. (Latin, *connivere* = to close the eyes.) Converging; approaching together. A term applied to wings so folded in repose that they unite perfectly at their corresponding margins.

CONNOLD, EDWARD THOMAS (–1909.) (Kuhnt 1910, Dt. ent. Z. 1910: 330.)

CONOCEPHALIDAE Plural Noun. Cone Headed Bush-Crickets. A large Family of ensiferous Orthoptera with about 140 Genera and 1,000 nominal Species, most of which are placed in the Conocephalinae. Adult body typically elongate and green or brownish and head usually subconical to pointed; Antenna longer than the body with scrobal impression weakly margined. Prothoracic Spiracle large and concealed; fore Tibia lacks terminal dorsal spines and auditory organ usually concealed. Tegmina (when present) are long and narrow; Ovipositor long and straight or slightly curved. Species found in low vegetation that is humid. Some Species have a mixed diet while others (Listroscelidinae) are exclusively predatory. Females typically oviposit in plant tissue. See Orthoptera.

CONOESUCIDAE Plural Noun. A small Family of Trichoptera assigned to Superfamily Limnephiloidea. Adult small to moderate-sized with Ocelli absent; female Maxillary Palpus with five segments and male with 1–3 segments; Antenna nearly as long as forewing. Pronotum with a pair of setal warts; Mesoscutum without setal warts; tibial spur formula 2-2-2/3/4. Larvae form cylindrical cases composed of sand, stones, silk or plant material. Conoesucids feed upon alga, moss and plant detritus. See Trichoptera.

CONONOTIDAE See Anthicidae.

CONOPIDAE Plural Noun. Thick-Headed Flies. A widespread Family of brachycerous Diptera assigned to Superfamily Conopoidea and including about 800 described Species. The Family is respresented poorly on Pacific islands. Adult head more-or-less inflated with well developed Ptilinum, Proboscis often geniculate and Antenna elongate. Forewing first Posterior Cell apically narrowed or closed in wing margin; Anal Cell long and apically petiolate. Most Species are marked brightly with yellow and black; Terminalia of male and female are enlarged and conspicuous. Species mimic Vespidae and Syrphidae. Subfamily Myopinae: Species all small, drab in colour with body nearly asetose. Relationships of Conopidae are not precisely established, but may represent a possible sister-group to Schizophora. Biologically, conopids are poorly known; Species have been reared from hosts but little is known about conopid immature stages. Conopinae, Myopinae and Dalmanniinae are solitary internal parasites of aculeate Hymenoptera; Stylogasterinae parasitize calyptrate Diptera, crickets and cockroaches: Females oviposit in adult hosts while in flight. Egg deposition is rapid with insertion directed into Abdomen of host. Following eclosion, concopid larvae remain free in abdominal cavity of host, feeding on body fluids and vital organs. Development of larvae is fairly rapid. Pupation occurs within body cavity of host. Puparium usually is robust, cylindrical, brownish and slightly inflated dorsoventrally. See Conopoidea.

CONOPOIDEA Noun. A Superfamily of schizophorous Diptera that includes the Conopidae. Possible sister group of Schizophora. Wing vein M1 adjacent to or fused with M5 distally, cell CuP long, acute with occasional vestige of Vena Spuria; male Postabdomen symmetrical.

CONQUER® See Esfenvalerate.

CONRADI, ADOLPH (1838–1910) (Osborne 1937, *Fragments of Entomological History*. 394 pp., Columbus, Ohio.)

CONSANI, MARIO (1929–1953) (Lanza 1953, Boll. Soc. ent. ital. 53: 59–60, bibliogr.)

CONSISTENCY INDEX. In cladistical theory, a measure of fit between a character matrix and a cladogram. A numerical index from 0 to 1 which measures 'fit' for a data set and homoplasy within a cladogram. If CI = 1, then homoplasy does not exist; homoplasy increases with a decreasing numerical value in the CI. CI = m/s, where m = number of character states - 1; s = number of observed character state changes (Kluge & Ferris 1969.) See Cladism.

CONSOBRINA CANEGRUB *Lepidiota consobrina* Girault [Coleoptera: Scarabaeidae]: A pest of sugarcane in northern Queensland, Australia. See Canegrub.

CONSOCIES Noun. (Latin, *cum* = together; *socius* = fellow.) 1. Ecology: Groups of plants usually dominated by 1–2 of them and in agreement as to the main features of habitat preference, time of reporduction, reaction to physical factors, *etc.*

(Shelford). 2. A consociation of plants in a developmental stage.

CONSOLIDATED Adj. (Latin, *consolidatus* > *consolidare* > *con-* = with; *solidus* = solid.) 1. Coalescent into one part with any neighbouring, usually detached part. 2. Descriptive of something made hard, compact and firm. Alt. Combined; Fused; Joined.

CONSPECIFIC Noun. (Latin, *con* = together; *species*; *-ic* = of the nature of. Pl., Conspecifics.) Individuals, populations or Subspecies that are assigned to one nominal Species or belong to the same Species. See Species. Cf. Congeneric.

CONSPERSE Adj. (Latin, *conspersus* = sprinkled.) 1. Pertaining to structure that is sprinkled with numerous, minute, irregular spots. 2. Descriptive of dot-like markings or pores.

CONSPERSE STINK BUG *Euschistus conspersus* Uhler [Hemiptera: Coreidae].

CONSPICUOUS Adj. (Latin, *conspicuous* > *conspicere* = to get sight of; *-osus* = possessing the qualities of.) Striking; obvious; readily apparent. Descriptive of something easily seen at first glance. Ant. Obscure. Syn. Prominent; Evident; Striking.

CONSPURCATE Adj. (Latin, *conspersus* = sprinkled.) Confusedly sprinkled with discoloured or dark spots. Alt. Conspurcatus.

CONSTANCY Noun. (Latin, *constantia*.) 1. Something that does not change with time. 2. Permanent. Ant. Vacillation.

CONSTANT Noun. (Latin, *constans*; *-antem* = an agent of something. Pl., Constants.) 1. Genetics: Any unvarying quantity. 2. Mathematics: An invariable or fixed quantity. 3. Biology: A parameter or condition that does not change. Ant. Variable. Syn. Steady; Stable; Invariable.

CONSTANT, ALLEXANDRE (1829–1901) (Walker 1901, Entomologist 34: 212.)

CONSTRICTED Adj. (Latin, *constrictus* = drawn together.) 1. Drawn in; gathered; narrowed medially and dilated toward the extremities. 2. Contracted or compressed at places on the length of a structure. Alt. Constrictus.

CONSTRUCTIONAL MORPHOLOGY An area of research concerned with anatomical features as determined by constituent elements composing the features. See Morphology. Cf. Functional Morphology.

CONSUL® See Hexaflumuron.

CONSULT® See Hexaflumuron.

CONSUMER Noun. (Latin, *consumere* > *con-* = with; *sumere* = to take; *-or* = an agent of action. Pl., Consumers.) 1. An heterotropic organism (animal, fungus, bacterium) that uses organic matter as food. 2. One who uses time or resources. Ant. Producer. See Ecology. Rel. Feeding Strategies.

CONSUTE Adj. (Latin, *consutus* > *consuere* = to sew together.) Zoology: Descriptive of a surface that displays stich-like markings. Alt. Consut.

CONTACT POISON An insecticide that kills an insect by entering the body. CPs are often deployed against insects with piercing-sucking mouthparts. CPs are applied as a spray, dip or dust and their action may be immediate or residual; CPs enter the body through the Integument or respiratory system. See Insecticide; Pesticide. Cf. Fumigant; Stomach Poison.

CONTAGIOUS Adj. (Latin, *contagiosus*.) Descriptive of a disease that is transmitted between organism of the same Species by physical contact or intimate association.

CONTAINMENT Noun. (Latin, *continere* > *con-* with; *tenere* = to hold. Pl., Containments.) Pest Management: A type of regulatory-control programme in which a target pest is confined within a geographical region. The concept of containment is applied in a legal sense against pests established in the area of concern and for which eradication is not practical or applicable. Typically, containment does not attempt to reduce or eliminate pest population in area of infestation. See Regulatory Control. Cf. Eradication; Suppression.

CONTARINI, NICOLO (1780–1849) Italian naturalist and early student of Acarology through reports on several Species. (Goidanich 1969, Memorie Soc. ent. ital. 48: 627–658.)

CONTIGUOUS Adj. (Latin, *contiguus* > *contingere* = to touch on all sides.) 1. Pertaining to adjacent structures; descriptive of structures with points, margins or surfaces which are in contact but which not united or fused. Ant. Separated. Syn. Adjacent; Bordering; Juxtaposed. Cf. Attingent.

CONTINENTAL DRIFT The movement of continents with respect to each other over the earth's surface. CD first was suggested by French geographer Antonio Snider-Pellegrini (1858). Subsequently, the German meteorologist Alfred Wegener (1915) proposed the idea of the Paleozoic supercontinent (Pangaea) which fragmented through the process of Plate Tectonics. The concept of CD widely accepted in Europe but rejected in North America by Zoologists for many years until the conclusion was irrefutable based on geological evidence. See Geological Time Scale. Cf. Gondwana; Laurasia. Rel. Biogeography.

CONTINUOUS VARIATION An expression of anatomical variation in which individuals of a Species or population display variation for a character or feature that is not discrete. See Variation.

CONTORTED Adj. (Latin, *contortus* > *contorquere* > *con-* together; *torquere* = to twist.) Twisted. Obliquely incumbent upon each other. Alt. Knotted; Twisted. Cf. Convoluted.

CONTOUR Noun. (Italian, *contorno* > *contornare* > *con-* = with; *tornare* = to turn. Pl., Contours.) 1. The outline or periphery of a surface or structure. 2. A curve. Alt. Curve; Outline; Shape.

CONTRACT Adj. (Latin, *cum* = together; *trahere* = to draw.) To draw, or drawn, together. To reduce, or reduced, in size by contraction. Alt. Contracted, contractus.

CONTRACTILE Adj. (Latin, *cum* = together; *trahere* = to draw.) 1. Zoology: Capable of being drawn near the body. 2. Drawn together, contracted or having the power of contracting. Ant. Extensile.

CONTRALATERAL Adj. (Latin, *contra* = opposite; *latus* = side.) Bilaterally symmetrical organisms: Descriptive of structure or motions that act in conjunction with similar parts on the opposite side of the body. Cf. Ipsilateral. Rel. Symmetry.

CONTRAST Intrans. Verb. (Italian, *contrasto* > Latin, *contra* = opposite; *stare* = to stand.) 1. Appearing in sharp relief or contrast, as one colour or marking against another. 2. Objects that appear dissimilar when juxtaposed or simultaneously compared.

CONTROL Noun. (French, *contrerole* = verification. Pl., Controls.) The reduction or regulation of populations in an area. See Biological Control; Chemical Control; Cultural Control. Cf. Erradication; Suppression.

CONVERGENCE Noun. (Latin, *convergere* > *con-* = together; *vergere* = to incline. Pl., Convergences.) 1. Coordinated movement of bodies to a fixed point. 2. A coming together of two lineages. 3. A term applied to indicate resemblance between two morphological forms (either animals or their structures) that are derived from different ancestral lineages. Resemblance under conditions of convergence are induced by the adoption of similar habits or through reduction or elimination of original differences (Tillyard). 4. 'The independent development of similar characters in two or more lineages that is not based on inherited genotypic similarity' (Holmes 1980: 49). See Homoplasy; Parallelism; Parallel Evolution. Cf. Divergence.

CONVERGENT Adj. (Latin, *convergere* = to incline together.) Structures or bodies coming together along their long axes. Ant. Divergent.

CONVERGENT EVOLUTION The resemblence of two or more organisms (or structures on two or more organisms) through independent acts of natural or sexual selection.

CONVERGENT LADY-BEETLE *Hippodamia convergens* Guérin-Méneville [Coleoptera: Coccinellidae].

CONVERGING Adj. (Latin, *convergere* = to incline together.) Descriptive of structures that approach each other near the apex. Ant. Diverging.

CONVERSE Adj. (Latin, *conversus* = to turn about.) Turned around; reversed in order. Syn. Obverse; Reverse.

CONVEX Adj. (Latin, *convexus* = vaulted.) The outer, curved surface of a segment of a sphere. Ant. Concave.

CONVEX VEIN A vein that has a tendency to fold upward (Snodgrass). A vein of the insect wing that follows the ridges (Imms). See Vein; Wing Venation. Cf. Concave Vein.

CONVOLUTE Adj. (Latin, *cum* = together; *volvere* = to wind.) Rolled or twisted spirally. A term also applied to wings when they are wrapped around the body. Alt. Convoluted; Convolutus.

CONVOLUTION Noun. (Latin, *cum* = together; *volvere* = to wind; English, *-tion* = result of an action.) 1. A process of winding or folding. 2. A condition in which a structure is rolled or folded on itself.

CONVOLVULUS HAWK-MOTH *Agrius convovuli* (Linnaeus) [Lepidoptera: Sphingidae]. (Australia).

COOK, ALBERT JOHN (1842–1916) American economic Entomologist who studied with Louis Agassiz and H. A. Hagen. Cook became Professor in Michigan until 1894, then Professor of Biology, Pomona College 1894–1911 and subsequently California State Commissioner of Horticulture 1911–1914. Cook was a charter member of American Association of Economic Entomologists. He was a pioneer in fumigation, instrumental in establishing quarantine laws in USA and prolific writer of bulletins and articles of practical importance. (Essig 1931, *History of Entomology*. 1029 pp (578–582); Mallis 1971, *American Entomologists*. 549 pp. (138–141), New Brunswick.)

COOK, BRIAN DIGBY (1939–1960) (Uvarov 1961, Proc. R. ent. Soc. Lond. (C) 25: 49.)

COOK, EDWIN W (1904–1941) (Anon. 1941, Newsl. U. S. Dept. Agric. Bur. ent. Pl., Quar. 8(12): 2–3.)

COOK, FRANK CUMMINGS (–1923) (Bishopp 1923, J. Econ. Ent. 16: 398–399.)

COOK, JOHN HAWLEY (1878–1958) (Fredrick 1958, Lepid. News 12: 207.)

COOK, MELVILLE THURSTON (1869–1952) (Leclerq 1953, Ann. ent. Soc. Am. 46: 172.)

COOK, THOMAS WRENTMORE (1884–1962) (Arnaud & Wale 1973, Pan- Pacif. Ent. 49: 177–181.)

COOK, WILLIAM CARMICHAEL (1895–1967) (Lane & Newcomer 1967, J. Econ. Ent. 60: 1488.)

COOKE, BENJAMIN (1816–1883) (Dunning 1883, Proc. ent. Soc. Lond. 1883: xli-xliii.)

COOKE, BERTRAM HEWETT HUNTER (1874–1946) (Hale-Carpenter 1947, Proc. R. ent. Soc. Lond. (C) 11: 60.)

COOKE, CALEB (1838–1880) (Anon. 1880, Psyche 3: 1–7.)

COOKE, MATTHEW (1829–1887) Irish-American and first economic Entomologist in California. (Essig 1931, *History of Entomology*. 1029 pp. (581–584), New York.)

COOKE, NICHOLAS (1818–1885) (Carrington 1885, Entomologist 18: 175–176.)

COOKE, THOMAS (1814–1885) (Anon. 1888. Entomologist 18: 200.)

COOKSONIA The oldest, simplest and most primitive land vascular plant. Apparently of Upper Silurian age taken from several localities in the Downton Series of Wales. Plants ca 2–3 cm tall, with cutinized Epidermis, triradiate spores and vascular tissue of annular tracheids. Genus consists of two Species, *C. pertoni* and *C. hemispherica.* Named for Isabel Cookson who collected type-material.

COOKTOWN AZURE *Ogyris aenone* Waterhouse [Lepidoptera: Lycaenidae]. (Australia).

COOLEY, ROBERT ALLEN (1873–1968) (Kohb 1969, J. Econ. Ent. 62: 972.)

COOLEY SPRUCE GALL ADELGID *Adelges cooleyi* (Gillette) [Hemiptera: Adelgidae]: Causes galls on blue spruce in North America. Cf. Eastern Spruce Gall Adelgid.

COOLING, LANCELOT E (1893–1924) (Musgrave 1932, *A Bibliography of Australian Entomology 1775–1930*. 380 pp., (55, bibliogr.), Sydney.)

COOLOOLA MONSTER *Cooloola propator* Rentz [Lepidoptera: Cooloolidae]. (Australia).

COOLOOLIDAE Rentz 1980 Plural Noun. Cooloola Monster. A monogeneric Family including two Species found in Queensland, Australia. Superficially resembling Stenopelmatidae, but Cooloolidae are distinguished by very short, 9-segmented Antenna, vestigial eye, sharp and elongate Mandible, and daggerlike Lacinia. Femora short and forelegs adapted for fossorial activities. Mouthpart anatomy suggests that cooloolids are predatory on subterranean prey. (Rentz 1980, Mem. Qlds. Mus. 20: 63–69.)

COOMAN, A DE (–1968) (Kennedy 1969, Proc. R. ent. Soc. Lond. (C) 33: 54.)

COON BUG *Oxycarenus arctatus* (Walker) [Hemiptera: Lygaeidae]. (Australia).

COOPER, ABRAHAM (1829–1911) (Bates 1868, Proc. ent. Soc. Lond. 1868: lvi.)

COOPER, ELLWOOD (1829–1918) (Essig 1931, *History of Entomology*. 1029 pp. (585–587), New York.)

COOPER, JAMES (1792–1879) (Carrington 1879, Entomologist 12: 280.)

COOPER, JAMES GRAHAM (1830–1902) (Essig 1931, *History of Entomology*. 1029 pp. (588), New York.)

COOPER, JOHN ANDERSON (1849–1896) (Anon. 1896, Entomologist 29: 200.)

COOPER, JOSEPH OMER (1893–1972) (A. B. M. W. & J. C. V. H. 1973, J. ent. Soc. sth. Afr. 36: 183–184.)

COOPER, KENNETH American academic (Princeton, Dartmouth, Penn State, Universities of California, Rochester and Florida) with diverse interests in genetics, biology and systematics of Coleoptera, Mecoptera and aculeate Hymenoptera.

COOPEX® See Permethrin.

COORDINATE Noun. (Latin, *co-* = together; *ordinare* = regulate. Pl., Coordinates.) 1. Things equal in rank, value, power or order. 2. Taxonomic Nomenclature: Taxa of the same rank. See Taxon. Rel. Nomenclature.

COOTAMUNDRA WATTLE PSYLLID *Acizzia acaciaebaileyanae* (Froggatt) [Hemiptera: Psyllidae]. (Australia).

COOTIE See Human Louse; Anoplura.

COPAL Noun. (Nauhautl > Spanish, *copalli* = resin. Pl., Copals.) A hard, transparent, citron coloured, shining, odiferous resin produced by several tropical Species of trees from Africa, South America and the Orient. A source of varnishes and some fossil insects. Cf. Amber.

COPE, EDWARD DRINKER (1840–1897) American paleontologist who observed a general trend for the body size of organisms to increase during the evolutionary history of a lineage (Family, Genus). (Anon. 1907, Pop. Sci. Mon. 70: 314–316.)

COPE'S RULE See Law of the Unspecialized. Rel. Dollo's Law.

COPELLO, ANDRES (–1948) (C. A. L. T. 1949, Revta. Soc. ent. argent. 14: 233–234, bibliogr.)

COPEOGNATHA Noun. (Greek, *kope* = oar; *gnathos* = teeth.) Booklice and their allies, the Suborder Psocida (Imms).

COPPER BUTTERFLIES See Lycaenidae.

COPPER JEWEL *Hypochrysops apelles apelles* (Fabricius) [Lepidoptera: Lycaenidae]. (Australia).

COPPERS See Lycaenidae.

COPRA BEETLE See Red-Legged Ham Beetle.

COPRA ITCH A hypersensitivity in humans to stored products caused by the mite *Tyrophagus putrescentiae* (Schrank). Syn. Grocers' Itch.

COPROMORPHIDAE Hampson 1918. Plural Noun. A small, pantropical Family of ditrysian Lepidoptera assigned to Superfamily Copromorphoidea; most Species occur in Asia and Australia. Adult small bodied with wingspan 12–20 mm; head scales smooth, long and slender; Ocelli sometimes present; Chaetosemata absent; antennal form variable; Proboscis without scales; Maxillary Palpus small, with four segments; Labial Palpus larger and upturned; fore tibial Epiphysis present; tibial spur formula 0-2-4; forewing without Pterostigma; hindwing wider than forewing, M vein with three branches. Larval head subprognathous. Pupae stout and sclerotized. Adult with wings folded flat above Abdomen. See Lepidoptera.

COPROMORPHOIDEA Noun. A Superfamily of ditrysian Lepidoptera including Alucitidae, Carposinidae, Epermeniidae and Copromorphidae. See Lepidoptera.

COPROPHAGOUS Adj. (Greek, *kopros* = dung; *phagein* = to eat; Latin, *-osus* = possessing the qualities of.) Pertaining to animals that feed on excrement or decaying vegetable matter of an excrement-like character. Alt. Coprophagus. See Feeding Strategies.

COPROPHAGY Noun. (Greek, *kopros* = dung; *phagein* = to eat. Pl., Coprophagies.) The condition in which organisms feed on excrement or dung. See Feeding Strategies.

COPULA Noun. (Latin, *copula* = bond. Pl., Copulas, Copulae.) 1. Something which connects. 2. The bond or linkage between male and female during copulation in which Semen and other elements are transmitted from male to female. See Copulation. Cf. Phoretic Copulation. Rel. Spermatophore; Spermatozoa.

COPULATE Adj. (Latin, *copulatus*, pp of *copulare* = to couple.) Pertaining to the union during the act of copulation. Transitive, Intrasitive Verb. (Latin, *copulates* > *copulare* = to couple.) To unite in sexual intercourse.

COPULATION Noun. (Latin, *copulatio* = copulate. English, *-tion* = result of an action. Pl., Copulations.) 1. The act of coupling or bonding during sexual intercourse. 2. The state of being joined during sexual intercourse. Syn. Coition.

COPULATION CHAMBER A chamber or cell excavated by certain scolytid beetles in their burrows, in which copulation takes place. Rel. Scolytidae.

COPULATORY OSSICULE Hymenoptera. See Digitus.

COPULATORY POUCH See Bursa Copulatrix.

COPULO (Latin, To couple.) The correct form of the term used to described the act of mating is '*in copulo*', not '*in copula*'.

COQUEBERT, ANTOINE JEAN (1753–1825) French naturalist and author of *Bulletin des Sciences*, and *Illustratio iconographica Insectorum quae in Musaeis Parisinis observavit, J. Chr. Fabricius* (1799–1804). (Musgrave 1932, *A Bibliography of Australian Entomology 1775–1930*. 380 pp., (55), Sydney.)

COQUEREL, CHARLES (1822–1867) (Papavero 1971, *Essays on the History of Neotropical Dipterology* 1:194–195, São Paulo.)

COQUILLETT, DANIEL WILLIAM (1856–1911) (Knight & Pugh 1974, Mosquito Syst. 6: 214–215, 217–219, bibliogr.)

CORACTION Noun. (Latin, *cum* = with; *actio* = action; English, *-tion* = result of an action. Pl., Coractions.) Reciprocal activity among organisms that are closely spaced.

CORAL® See Coumaphos.

CORALLINE Adj. (Greek, *korallion* = coral.) Coral-like in shape or texture. A pale pinkish red. Alt. Corallinus.

CORBEL Noun. (Latin, *corvus* = raven. Pl., Corbels.) 1. A raven. 2. Coleoptera: An ovate area at the distal end of the Tibia, surrounded by a fringe of minute bristles. When the articular cavity is on the side, the Corbel above is closed; Corbel is open when the cavity is at the extreme apex.

CORBET, ALEXANDER STEVEN (1896–1948) (Williams 1949, Proc. R. ent. Soc. Lond. (C)13: 67.)

CORBETT, GEORGE HAMBLIN (1888–1968) (Kennedy 1969, Proc. R. ent. Soc. Lond. (C)33: 54.)

CORBETT, HERBERT HENRY (1856–1921) (Rothschild 1921, Proc. Ent. Soc. Lond. 1921: cxxix.)

CORBICULA Noun. (Latin, *corbicula* = a little basket. Pl., Corbiculae.) Hymenoptera: 1. Apoidea: Typically, a concave, smooth area of the hind Tibia that is surrounded by a fringe of Setae along the margin. Corbicula collects and holds pollen and other materials for transport to the nest. Corbicula differs from the Scopa in the reduction of marginal fringe Setae. 2. Agaonidae: A pollen collecting structure on the legs of female fig wasps. Syn: Pollen Basket. See Scopa. Rel. Leg.

CORBICULATE Adj. (Latin, *corbicula* = a little basket.; *-atus* = adjectival suffix.) Furnished with a brush of strong Setae; having Corbiculae. Alt. Corbiculatus.

CORBIE *Oncopera intricata* Walker [Lepidoptera: Hepialidae]. (Australia).

CORBIN, GEORGE MENTHLEY (1841–1914) (F. V. B. 1914, Entomologist 47: 160.)

CORCULA Noun. (Etymology obscure. Pl., Corculae.) The reservoirs of the dorsal channel through which the insect blood flows, the chambers of the dorsal vessel.

CORDATE Adj. (Latin, *cor* = heart; *-atus* = adjectival suffix.) Heart-shaped. Descriptive of structure that is triangular with basal corners rounded and the apex pointed or bluntly rounded; middle of base not necessarily emarginate. Syn. Cordiform. Cf. Obcordate; Subcordate. Rel. Form 2; Shape 2; Outline Shape.

CORDIER, JULES (1811–1846) (Fairmaire 1947, Bull. Soc. ent. Fr. (2) 5: xv-xvii.)

CORDIFORM Adj. (Latin, *cor* = heart; *forma* = shape.) See Cordate.

CORDLEY, A B (1864–1936) (Hatch 1949, Century of Entomology in the Pacific Northwest. 43 pp. (9–10), Seattle.)

CORDULESGASTRIDAE Plural Noun. Biddies. A small Family of anisopterous Odonata assigned to Superfamily Aeshnoidea and confined to the Palearctic and Oriental regions. Frequently put in its own Superfamily and appears to share features with Aeshnoidea and Libelluloidea. Adult large to very large bodied; compound eyes separated medially; wings typically narrow; Pterostigma moderately elongate and narrow; Antenodal Crossveins not in line in Costal and Subcostal spaces; triangles variable. Naiads are robust, elongate and fusiform; head broad and rather square with a projecting ridge on Frons; eyes prominent and raised; Antenna 7-segmented with third segment longest. See Aeshnoidea.

CORDULIDAE Plural Noun. Green-eyed Skimmers. A Family of aeshnoid Odonata assigned to Superfamily Libelluloidea. Adults vary in size with body generally reddish brown to black with some metallic coloration; compound eyes in contact medially but not broadly and posterior margins with weak projections; Abdomen slender; males with Auricles on segment II and sometimes tuberculate on TX. Nodus closer to apex of wing than base; Pterostigma usually short; anal loop usually well developed; male legs with tibial keels. Naiads are variable in shape with eyes small and raised; Antenna with seven segments and protuberance on head of some Species; Abdomen frequently with hooks laterad and mid-dorsally; mask strongly concave and crenulate. See Libelluloidea.

CORDYLURIDAE See Scathophagidae.

COREIDAE Leach 1815. Plural Noun. Leaf-footed Bugs; Squash Bugs. The nominant Family of Coreoidea (Hemiptera). Coreids are cosmopolitan in distribution, large bodied and elongate with coloration grey or brown. Forewing membrane with numerous veins; Metapleural Scent Gland with Peritreme; legs robust; hind Femur and hind Tibia often enlarged, dilated or flattened with spines or other processes. Species are predominantly phytophagous, with several Species regarded as significant agricultural pests; the Family includes a few predaceous Species. See Hemiptera.

COREINAE A Subfamily of Coreidae.

COREMATA Plural Noun. (Greek, *korema* = broom.) Lepidoptera: A term coined by Janse (1932) for a pair of long, eversible, tubes with glandular cells that occur along the posterior margin (in intersegmental membrane) of various abdominal segments of males. Coremata are typically paired, sometimes are branched and in some Geometridae may occur on the Valvulae. Coremata are invested with Setae along most of the surface, or Setae may be concentrated at the apex of the tube; glandular cells on the Coremata release of sex pheromones. Cf. Hair Pencil. Rel. Exocrine Gland.

COREOIDEA A Superfamily of heteropterous Hemiptera including Alydidae, Coreidae, Hyocephalidae, Rhopalidae and Stenocephalidae. Ocelli are present and the membrane of forewing contains longitudinal Accessory Veins.

CORETHRIDAE See Chaoboridae.

CORETHROGYNE Noun. (Greek, *korema* = broom; *gyne* = female. Pl., Corethrogynes.) Lepidoptera: Scales at the apex of the Abdomen in some females that are shed and used to cover or protect eggs following oviposition.

CORIA Noun. (Latin, *corium* = leather. Pl., Coriae.) The fexible membrane between the body segments or appendage segments. The Conjunctiva or intersegmental membrane (MacGillivray).

CORIACEO-RETICULATE Adj. Descriptive of a surface with impressed reticulations giving a leather-like appearance. See Sculpture Pattern. Cf. Coriaceous; Reticulate.

CORIACEOUS Adj. (Latin, *corium* = leather; *-aceus* = of or pertaining to.) Leather-like; pertaining to thick, tough and somewhat rigid structure. Alt. Coriaceus. See Sculpture Pattern.

CORIAL Adj. (Latin, *corium* = leather.) Descriptive of or pertaining to a Corium. A term specifically applied to the wing of Heteroptera.

CORIARIOUS Adj. (Latin, *corium* = leather; Latin, *-osus* = possessing the qualities of.) Leather-like in sculpture or texture. See Sculpture Pattern.

CORIUM Noun. (Latin, *corium* = leather. Pl., Coria.) 1. Heteroptera: The elongate, middle portion of the Hemelytron that extends from the base to the membrane below the Embolium, if the latter is present. The entire hard part of a wing as distinguished from membrane, exclusive of the Clavus. The outer broader part of the Hemelytra, from the Clavus to the outer edge, or to the Embolium when the latter is present. 2. The membrane between segmental appendages.

CORIXIDAE Leach 1815. Plural Noun. Water Boatmen. A cosmopolitan Family of heteropterous Hemiptera assigned to Superfamily Corixoidea with a fossil record from Jurassic Period. Adult head broad with large, transverse compound eyes that overlap anterior of Prothorax; beak broad, conical and one-segmented; Ocelli absent. Forelegs scoop-like; middle legs long, slender with one tarsal segment; hind legs with fringe of Setae; Tibia and Tarsus with long Setae adapted for swimming. Corixids spend most time submerged, clinging to vegetation with their middle legs. Females attach their eggs to submerged objects near surface of water. Species are predaceous upon aquatic insects, diatoms and algal cells; prey are located with forelegs; some Species can ingest non-liquid food. See Hemiptera.

CORIXOIDEA A Superfamily of Hemiptera including Corixidiae.

CORM Noun. (Greek, *kormos* = trunk. Pl., Corms.) Botany: A short, swollen (enlarged), subterranean stem. Cf. Bulb.

CORMEA Noun. (Etymology obscure. Pl., Cormeae.) Specialized scent-tufts near the apex of the Abdomen of certain male Lepidoptera (Klots).

CORN APHID *Rhopalosiphum maidis* (Fitch) [Hemiptera: Aphididae]. (Australia). See Corn-Leaf Aphid.

CORN BLOTCH LEAF-MINER *Agromyza parvicornis* Loew [Diptera: Agromyzidae].

CORN DELPHACID *Peregrinus maidis* (Ashmead) [Hemiptera: Delphacidae]. Syn. Corn Planthopper.

CORN EARWORM 1. *Helicoverpa zea* (Boddie) [Lepidoptera: Noctuidae]: Endemic to New World and now a cosmopolitan, polyphagous pest of many economically important crops, particularly vegetable and field crops. Adult ca 20 mm long with a wingspan to 40 mm; forewings tan with central spot and dark band and pale fringe near apex; hindwing pale with dark band near fringe. Adults feed, mate and oviposit nocturnally; female lives ca 14 days and lays 1,000–3,000 eggs individually on terminal growth, near flowers and fruit. Eggs are ca 0.5 mm diam, white, dome-shaped with ridges radiating from apex of egg; eclosion occurs within 2–4 days. Neonate and early instar larvae are tan or cream coloured with black spots; mature larva ca 40 mm with head yellow or brown and body yellow, green or black with conspicuous longitudinal stripes along the sides; tubercles with spines. CE larvae prefer flowers and developing fruits; they enter fruit near stem and cause secondary rot at site of penetration; damage to flowers and buds causes abortion or fruit malformation. Larvae pass through five instars within ca 2–5 weeks. Development on tomatoes is solitary and completed on one or more fruit. Pupae are dark brown, brittle, ca 20 mm long; pupation occurs in soil. CE life cycle requires 30–50 days. Adults are strong fliers and capable of long-distance dispersal. Syn. Bollworm; False Budworm; *Heliothis zea* (Boddie); Tomato Fruitworm. Cf. African Bollworm; Egyptian Bollworm; Native Budworm; Pink Bollworm; Spiny Bollworm; Spotted Bollworm. 2. *Helicoverpa armigera* (Hübner) [Lepidoptera: Noctuidae]. (Australia). See Cotton Bollworm.

CORN FLEA-BEETLE *Chaetocnema pulicaria* Melsheimer [Coleoptera: Chrysomelidae]: A pest of maize that transmits Stewart's Disease to plants by depositing excrement on or near feeding wounds. See Chrysomelidae.

CORN-LEAF APHID *Rhopalosiphum maidis* (Fitch) [Hemiptera: Aphididae]: A cosmopolitan pest of maize, wheat, barley, oats, sugarcane, sorghum and other grasses. CLA is found predominantly in tropical and subtropical regions where it is capable of transmitting viral diseases of cereals, including mosaic disease of sugarcane. Corn infested with CLA shows leaf mottling of yellowish or reddish spots. CLA is capable of completing more than 50 generations in one year. See Aphididae.

CORN LEAFHOPPER. *Dalbulus maidis* (DeLong & Wolcott) [Hemiptera: Cicadellidae]: A serious pest of maize in Latin America that transmits stunting pathogens (Maize Rayado Fino Virus and Corn Stunt Spiroplasma).

CORN PLANTHOPPER *Peregrinus maidis* (Ashmead) [Hemiptera: Delphacidae]. Syn. Corn Delphacid.

CORN-ROOT APHID *Anuraphis maidiradicis* (Forbes) [Hemiptera: Aphididae]: A pest of cotton and corn east of the Rocky Mountains in North America; CRA also attacks some grasses and weeds. CRA overwintering eggs are dark green, shiny and collected from corn and stored by Cornfield Ant. Ants move aphid nymphs to ho stplants (smartweed, corn, cotton); nymphs feed and develop into a winged generation that moves to other areas. CRA highly reliant on ants. See Aphididae. Rel. Cornfield Ant; Symbiosis.

CORN ROOT WEBWORM *Crambus caliginosellus* Clemens [Lepidoptera: Pyralidae].

CORN-SAP BEETLE *Carpophilus dimidiatus* (Fabricius) [Coleoptera: Nitidulidae]: A pest of figs in South Africa and dates in other areas. Adult 2.3–3.5 mm long and reddish brown. See Nitidulidae. Cf. Dried-Fruit Beetle.

CORN SILK BEETLE *Metrioidea brunneus* (Crotch) [Coleoptera: Chrysomelidae].

CORN WIRE WORM. *Melanotus communis* (Gyllenhal) [Coleoptera: Elateridae].

CORNALIA, EMILIO (1825–1882) (Stoppani 1884, Atti Soc. ital. Sci. nat. 27: 17–41.)

CORNEA. Noun. (Latin, *corneus* = horny. Pl., Corneas.) The outer, transparent surface of the compound eye or each facet of the compound eye. Syn. Lens. See Compound Eye. Cf. Cone; Pigment Cells; Retina; Rhabdome.

CORNEAGEN LAYER. A layer of colourless, transparent cells over the insect eye beneath the Cornea that secretes and supports the lens (Imms).

CORNEAGENOUS CELLS Epidermal cells that generate the Cornea (Snodgrass). See Corneagen Layer.

CORNEAGENOUS PIGMENT CELL See Primary Pigment Cell.

CORNEAL FACET See Compound Eye.

CORNEAL HYPODERMIS Vitreous Layer; Lentigen Layer; Corneagen Layer (Comstock). See Eye.

CORNEAL LAYER Insect Ocellus: A continuation of Cuticle over the Ocellus. The corneagen layer *sensu* Comstock. See Compound Eye.

CORNEAL LENS The individual lens-like structures of which the Cornea of the compound eye is composed. Syn. Lens. See Compound Eye.

CORNEAL NIPPLE ARRAY Tubercles that project from the Lens of the insect Ommatidium. CANs occur in more than 25 Orders, including Lepidoptera, Neuroptera and Diptera. CNAs serve to reduce light reflection, increase light transmission and serve as an impedance transforming device. See Compound Eye. Cf. Lens. Rel. Vision.

CORNEAL PIGMENT CELL See Primary Pigment Cell.

CORNEATE Adj. (Latin, *corneus* = horny; *-atus* = adjectival suffix.) Horned; pertaining to structure with horns; horn-like structures. Alt. Corneatus.

CORNELIAN *Deudorix epijarbas diovis* Hewitson [Lepidoptera: Lycaenidae]. (Australia).

CORNEOUS Adj. (Latin, *corneus* = horny; *-osus* = possessing the qualities of.) Of a horny or chitinous substance; resembling horn in texture. Rel. Horn.

CORNFIELD ANT *Lasius alienus* (Foerster) [Hymenoptera: Formicidae]: A widespread and endemic pest throughout USA and probably best know for its mutualistic relationship with Corn Root Aphid. CA workers are monomorphic, without antennal club, stingless with anal opening circular and surrounded by Setal fringe; CA utilizes formic acid in defence. CA nests in soil or rotting wood. Foraging workers encounter aphids, bring them to the nest where eggs and aphid nymphs are tended and placed on roots of corn plant. CA is dependent upon honeydew secreted by aphids as nutrition. Nuptial flights of CA occur during August-September. Food of CA includes insects, honeydew, nectar and meat from homes. See Argentine Ant; Black Carpenter-Ant;

Fire Ant; Formicidae; Larger Yellow-Ant; Little Black Ant; Odorous House Ant; Pavement Ant; Pharaoh Ant; Thief Ant. Rel. Corn Root Aphid.

CORNICLE Noun. (Latin, *cornu* = a horn. Pl., Cornicles.) Paired secretory structures on the Abdomen of aphids. Typically, Cornicles occur on the posterior margin of fifth abdominal segment and are less common on sixth segment in some Drepanosiphidae and Greenideidae. Cornicles are variable in shape: Absent or ring-like in Pemphigidae, elongate and cylinder-like in most Families; outer surface ranging from smooth to sculptured; apex with membranous lid and sclerite that serves as a valve manipulated by abdominal musculature. Most Cornicle secretions are triglicerides that act as defensive wax secretions and alarm pheromones but not honey-dew. See Gland. Cf. Siphunculus. Rel. Alarm Pheromone.

CORNICULI Noun. See Corniculum 3.

CORNICULUM Noun. (Latin, *corniculum* diminutive of *cornu* = horn. Pl., Cornicula) 1. Any small horn or horn-like process. 2. Immatures: A small cuticular protuberance or process. Similar to a Chalaza but with an apical Seta. 3. Orthoptera: The small horny apices or parts of the Ovipositor. See Valves.

CORNICULUS Noun. (Latin, *corniculum* = little horn. Pl., Corniculi.) The Cornicles of aphids. See Cornicle.

CORNIFORM Adj. (Latin, *cornu* = horn; *forma* = shape.) 1. Horn-shaped; shaped as the horn of an ox. 2. A long, mucronate or pointed process. See Horn. Cf. Conchiform; Heliciform; Mytiliform; Pectiniform; Trochiform. Rel. Form 2; Shape 2; Outline Shape.

CORNU Noun. (Latin, *cornu* = horn. Pl., Cornua.) 1. A horn or horn-like structure. 2. Odonata (Anisoptera) see Flagella. 3. Orthoptera: See Ancora. 4. Homoptera (Cicadidae): Horn-like process terminating in Vesica. 5. Hymenoptera (Apidae): Projections from the Bursa of the Endophallus. 6. Diptera (adult Chironomidae): See Cornua of Hypophrayngeal Suspensorium (Saether). Larval Muscomorpha: Two horn-like processes in the cephalo-pharyngeal skeleton that point posteriad.

CORNUS Noun. (Latin, *cornu* = horn.) 1. Symphyta: A horn-like or tail-like tergal processes near the apex of the larval body. 2. The dried root-bark of flowering dogwood which contains the bitter tonic Cornin.

CORNUTE Adj. (Latin, *cornutus* = horned.) Having horns or horn-like processes.

CORNUTUS Noun. (Latin, *cornutus* = horned. Pl., Cornuti.) Lepidoptera Genitalia: A slender, heavily sclerotized spine or spines within the male's Ejaculatory Duct. Cornuti penetrate the female's Ductus Bursae and Bursa Copulatrix, and often remain in the Bursa after insemination.

COROLLA Noun. (Latin, *corolla* = little crown. Pl., Corollae; Corollas.) The inner part of the Perianth of a flower. The Corolla is composed of one to several petals. When the petals are united, the condition is called gamopetalous; when the petals are distinct or separate, the condition is called polypetalous. See Flower; Perianth. Cf. Calyx.

CORONA Noun. (Latin, *corona* = crown. Pl., Coronae; Coronas.) Any crown or crown-like process. Genitalia of male Lepidoptera: A specialized row or mass of spines on the Cucullus of the Harpe (Klots).

CORONAL SUTURE A dorsomedian part of the ecdysial cleavage line. A longitudinal suture along the midline of the head, extending from the Epicranial Notch to the apex of the Clypeus. See Epicranial Suture. Syn. Metopic Suture.

CORONATE EGG Any insect egg in which the upper end is surrounded by a circle of spines or spine-like protuberances.

CORONATE Adj. (Latin, *corona* = crown; *-atus* = adjectival suffix.) With a crown-like apex or point of termination. Alt. Coronatus.

CORONET Noun. (Latin, *corona* = crown. Pl., Coronets.) Any small crown-like structure on the head.

CORONULA Noun. (Latin, diminutive of *corona* = crown. Pl., Coronulae.) Literally, a little crown. A circle or semicircle of spines at the apex of the Tibia.

COROTHION® Parathion.

CORPORA Plural Noun. (Latin, *corpus* = body. Sl., Corpus.) Bodies. See Corpus.

CORPORA ALLATA (Sl., *corpus allatum*.) A pair of small, ovoid cellular bodies associated with stomodaeal ganglia behind the Brain on either side of the Oesophagus and connected to Corpora Cardiaca. CA are ductless glands which, according to DuPorte, are entirely epithelial and do not contain nervous tissue. Products of the CA include Juvenile Hormone, Yolk Forming Hormone and a hormone with a general effect on metabolism. Syn. Corpora Incerta (Snodgrass); Ganglia Allata. See Stomatogastric Nervous System. Cf. Central Nervous System; Corpora Cardiaca; Incretory Organs.

CORPORA CARDIACA A pair of small bodies closely associated with the Hypocerebral Ganglion. CC join the Hypocerebral Ganglion by short, transverse commissures and join the posterior part of Protocerebrum via two nerves. CC stores and releases hormones from neurosecretory cells of Brain, activates Prothoracic Gland and regulates heartbeat. CC of pterygotes are typically neurohemal-endocrine organs. Described earlier as paired stomatogastric nerves innervating heart, CC were subsequently identified as endocrine organs. The dual neurohemal-endocrine character was later established by Scharrer (1963). Among apteryogtes, only Zygentoma (= Lepismida) have a CC that is comparable with pterygote CC. CC structurally complex with parenchymal secretory cells, interstitial glial-like cells, Glial Cells, cellular processes of Parenchymal Cells containing secretory granules and extrinsic neurosecretory and ordinary axons. Syn. Oesophageal Ganglion; Pharyngeal Ganglion. See Central Nervous System; Retrocerebral Nervous System. Cf. Brain; Incretory Organs.

CORPORA INCERTA See Corpora Allata.

CORPORA OPTICA The dorsal optic parts of the Brain in some lower insects (Snodrass). See Brain.

CORPORA VENTRALIA (Sl., *Corpus Ventralum.*) The ventral bodies of the Protocerebrum in the ventrolateral parts of the Brain (Snodgrass).

CORPOROTENTORIUM Noun. (Latin, *corpus* = body; *tentorium* = tent. Pl., Corporotentoria.) See Corpotentorium.

CORPOTENDONS Plural Noun. (Latin, *corpus* = body; *tendo* = tendon, from *tendere* = to strech.) The 'tendons' originating on the Corporotentorium (MacGillivray). See Tentorium.

CORPOTENTORIUM Noun. (Latin, *corpus* = body; *tentorium* = tent. Pl., Corpotentoria.) 1. The body of the Tentorium. 2. The prominent bridge that divides the Foramen into two parts; on one side the Alimentary Canal passes and on the other the nervous system. (MacGillivray).

CORPUS Noun. (Latin, *corpus* = body. Pl., Corpora.) 1. The body of a dead animal. 2. The entire body of an insect, including head, Thorax and Abdomen. See Body. 3. Any homogenous structure or mass that forms part of an organ. 4. The collective writings by an individual on a subject, or the collective writings on a subject.

CORPUS ADIPOSUM The mass of fat tissue often found in larval insects. See Fat Body.

CORPUS ALLATUM See Corpora Allata.

CORPUS CARDIACUM See Corpora Cardiaca.

CORPUS CENTRALE The central body of the insect Brain (Snodgrass). See Brain.

CORPUS LUTEUM The mass of degenerating Follicle Cells left in an egg chamber after the discharge of the egg (Snodgrass). See Ovary.

CORPUS PEDUNCULATUM (Pl., Corpora Penduncculata.) 1. Pedunculate or 'Mushroom Bodies' of protocerebral region of insect Brain. Corpora Pendunculata are positioned at sides of Pars Intercerebralis and contain three types of cells and a mass of Neuropile. Kenyon Cells are cell bodies that form a flattened cap over the Neuropil (termed the Calyx). Calyx forms a stalk that divides into two lobes ventrally. An Alpha Lobe is more distal of pair and is invested with sensory neurons. A Beta Lobe is medial and is invested with motor neurons. Alpha and Beta Lobes project downward and rearward in tracts that cross and ultimately connect to Corpora Cardiacum. Size of Corpus Pedunculata appears correlated with complexity of behaviours. CP seems most well developed in social insects and has a role in visual integrations in social Hymenoptera (but not other insects) and sequential organization of behaviour. See Brain; Central Nervous System; Nervous System. Cf. Optic Lobes; Pars Intercerebralis; Protocerebral Bridge. 2. The stalked bodies in the annelid forebrain. 3. Association Centres *sensu* Snodgrass.

CORPUS SCOLOPALE A sense-rod; a Scolopale. A hollow, peg-like structure enclosed by a spindle-shaped bundle of Sensilla. See Scolopale; Scolophore.

CORPUS VENTRALE A paired ventro-lateral mass of the insect Protocerebrum. See Protocerebrum.

CORPUSCLE Noun. (Latin, *corpusculum* = small body. Pl., Corpuscles.) A small cell. Term usually applied to blood cells.

CORREA BROWN *Oreixenica correae* (Olliff) [Lepidoptera: Nymphalidae]. (Australia).

CORRELATED (Latin, *correlatio* = relationship.) Structure, condition or behaviour that is derived from the same ancestral form (state). Correlation is asserted when two or more features or qualities bear a direct or an inverse relation to each other. Correlation does not imply a relation of cause and effect.

CORRELATIVE Adj. (Latin, *correlatio* = relationship.) Of a correlated nature. See Correlated.

CORRIGENDUM (Latin, *corrigens* > *corrigere* = to correct. Pl., *Corrigenda.*) 1. Typographical errors to be corrected. 2. A fault or error that must be corrected.

CORROBOREE CICADA *Macrotristria intersecta* (Walker) [Hemiptera: Cicadidae]. (Australia).

CORRODE Verb. (Latin, *corrodere* > *cor-*; *rodere* = to gnaw.) To eat away gradually, as by rust or decay.

CORRODENT Noun. (Latin, *corrodere* = to gnaw to pieces.) Any member of the Corrodentia.

CORRODENTIA (Latin, *corrodere* = to gnaw to pieces.) See Pscoptera; Psocidae; Booklice.

CORRUGATED Adj. (Latin, *corrugatus,* past participle of *corrugare,* from *cor-* + *rugare* = to wrinkle.) Descriptive of structure that is folded, furrowed or wrinkled; pertaining to a surface with alternate ridges and channels. Syn Wrinkle. Ant. Smooth. See Sculpture Pattern. Cf. Acuductate; Alveolate; Baculate; Clavate; Echinate; Favose; Gemmate; Psilate; Punctate; Reticulate; Rugulate; Scabrate; Shagreened; Smooth; Striate; Verrucate; Wrinkle.

CORSAIR® See Permethrin.

CORSELET Noun. (Latin, *cors* = body. Pl., Corselets.) The Prothorax in Coleoptera. See Prothorax.

CORTEX Noun. (Latin, *cortex* = bark.) 1. The outer or superficial layer of an organ (*e.g.* the insect brain, nerve ganglion). 2. Vascular Plants: The portion of a stem or root that is external of the vascular tissue. 3. The peel of certain fruits (*e.g. Citrus*).

CORTICAL Adj. (Latin, *cortex* = bark.) Relating to the Cortex, outer skin or an outer layer of tissue covering an organ.

CORTICAL CELLS Cells of the Brain's cortex.

CORTICAL CYTOPLASM The peripheral layer of Cytoplasm in the insect egg (Snodgrass).

CORTICINUS Adj. (Latin, *cortex* = bark.) Bark-like sculpture or texture. See Sculpture.

CORTICOLOUS Adj. (Latin, *cortex* = bark; *colous* = living in or on.) Descriptive of organisms that live in or on the bark of plants. See Habitat.

CORVINUS Adj. (Latin, *corvus* = raven.) Crow-black; descriptive of structure coloured a deep, shining black with a greenish lustre.

CORYDALIDAE Plural Noun. Dobsonflies. A numerically small, widespread Family of Megaloptera (Neuroptera). Adult body large with three Ocelli present; fourth tarsal segment unmodified; larva aquatic with eight pairs of abdominal gills. See Megaloptera.

CORYDALOIDIDAE Plural Noun. A Family of Megasecoptera found in the Upper Carboniferous deposits of Commentry, France.

CORYLOPHIDAE Plural Noun. Minute Fungus Beetles. Adults typically minute, oval in dorsal outline-shape; Antenna clubbed; tarsal formula 4-4-4 with third segment small and concealed; hindwing with conspicuous marginal fringe. Corylophids inhabit decaying vegetation. See Coleoptera.

CORYPHA Noun. (Greek, *koryphe* = peak. Pl., Coryphae.) Scarabaeoid larvae: The unpaired anterior region of the Epipharynx between the Clithra and bearing a small number of Setae. Corypha is often merged with the Acropariae into a common apical region when the Clithra are absent (Boving).

CORYPHATE Adj. (Greek, *koryphe* = peak; *-atus* = adjectival suffix.) See Capillatus. Alt. Coryphatus.

CORYSTERIUM Noun. (Greek, *koryphe* = summit; Latin, *-ium* = diminutive > Greek, *-idion*. Pl., Corysteria.) An abdominal glandular structure in some female insects that secretes a glutinous cover for the eggs.

COSBAN® See XMC.

COSMOPOLITAN Adj. (Greek, *kosmos* = world; *polites* = citizen.) 1. Pertaining to organisms occurring throughout the world. 2. Pertaining to a place where organisms from many places may appear. Ant. Insular; Provincial. See Zoogeography. Cf. Local.

COSMOPOLITAN GRAIN PSOCID *Lachesilla pedicularia* (Linnaeus) [Psocoptera: Lachesillidae].

COSMOPOLITAN NEST MITE *Haemolaelaps casalis* (Berlese) [Acarina: Laelapidae]. (Australia).

COSMOPTERIGIDAE Plural Noun. Fringe Moths; Leaf Miner Moths. A cosmopolitan Family (ca 1,200 Species) of ditrysian Lepidoptera assigned to the Gelechioidea. Adult small bodied with wingspan 6–32 mm; head scales smooth; Ocelli sometimes absent; Chaetosemata absent; Antenna shorter than forewing; Pecten sometimes present; Proboscis with scales on basal half; Maxillary Palpus with four segments folded over base of Proboscis; Labial Palpus long, recurved; Epiphysis present, tibial spur formula 0-2-4; hind Tibia with long hair-scales; forewing lanceolate, Pterostigma absent. Pupation occurs within the larval shelter; Pupa with Maxillary Palpus small or absent; Labial Palpus, fore Femur concealed; Cremaster sometimes present with straight or hooked Setae. Larval biology and feeding preferences diverse: larvae bore into stems, seeds and fungi; some Species tie leaves; a few Species form galls; a few Species are predaceous on scale insects. Adults of most Species are nocturnal. Syn. Momphidae. See Gelechioidea.

COSMOTROPICAL Adj. (Greek, *kosmos* = world; *tropikos* = of the solstice.) With a distribution throughout most of the tropics. See Pantropical.

COSSIDAE Leach 1815. Plural Noun. Carpenter-Worm Moths; Carpenter Worms; Goat Moths; Leopard Moths; Wood Moths. A cosmopolitan, primitive Family of ditrysian Lepidoptera containing ca 400 Species and assigned to Superfamily Cossoidea or Tortricoidea. Adult small to large bodied with wingspan 10–240 mm; head with raised scales on Vertex; Ocelli usually absent; Chaetosemata absent; Antenna less than half length of forewing; male Antenna typically bipectinate (some Species filiform), female typically filiform (some Species bipectinate); Proboscis absent or short and without scales; Maxillary Palpus minute or at most with two small segments; Epiphysis typically present, tibial spurs absent or spur formula 0-2-4 or 0-2-2. Eggs are small and upright with vertical and horizontal ribs. Larvae are stout; prothoracic dorsal shield large and heavily sclerotized. Pupae are elongate, cylindrical and sclerotized; head spined; Abdomen moveable. Eggs are laid in crevices of wood or on host plant; larvae bore into wood and often form galleries beneath bark; a few Species tunnel in soil. Pupation occurs in a chamber within tunnel. Cossid life cycle requires 2–3 years. Cossids are considered the most primitive Family of Ditrysia and sister-group of Tortricidae. Syn. Hypoptidae; Zeuxeridae. See Carpenter Worm; Cossoidea; Leopard Moth.

COSSILIDAE Plural Noun. See Fedtschenkidae.

COSSOIDEA A Superfamily of ditrysian Lepidoptera including Cossidae, Dudgeoneidae and Metarbelidae; cossoids are cosmopolitan in distribution but best developed in Orient and Australia.

COSTA Noun. (Latin, *costa* = rib. Pl., Costae.) 1. An elevated ridge that is rounded at its crest. 2. The thickened anterior margin of a wing, typically referring to the forewing. 3. The vein extending along the anterior margin of the wing from base to the point of junction with Subcosta (Comstock). The Costal Vein is not branched and is articulated at its base with the Humeral Plate. See Subcosta. 4. Genitalia of male Lepidoptera: Dorsal part of the Harpe (Klots). Cf. Carina; Keel; Ridge. Rel. Costiform.

COSTA, ACHILLE (1823–1898) (Crovetti 1970, Boll. Soc. sarda Sci. nat. 6: 59– 71.)

COSTA LIMA, ANGELO MOREIRA DA (1887–1964) A Brazilian Entomologist (Escola Nacional de Agronomia e Instituto Oswaldo Cruz) working in taxonomy and biology of parasitic Hymenoptera. (Pearson 1966, Proc. R. ent. Soc. Lond. (C) 30: 62.)

COSTA LIMA'S ORGAN See Metafemoral Spring.

COSTA, ORONZIO GABRIELE (1787–1867) (Salfi 1968, Annuar. Ist. Mus. Zool. Univ. Napoli 18: 1–17.)

COSTAL AREA The area behind Costal Vein. See Costal Field. Rel. Remigium.

COSTAL BRACE Ephemeroptera: A thick veinlet running from Costa to Radius at the base of the wing. The Humeral Veinlet *sensu* Tillyard.

COSTAL CELL The area enclosed between the Costal and Subcostal Veins in the insect wing. In the plural, all cells margined anteriorly by the Costa (Comstock). Hymenoptera (Norton): Includes the 1st, 2nd and 3rd Subcostal of Packard; the 3rd Costal (*i.e.* 2nd Radial). Diptera (Williston): The 2nd Costal *sensu* Smith.

COSTAL CROSS VEINS In many-veined insect wings: Veins that extend between the Costa and Subcosta (Comstock). See Vein; Wing Venation.

COSTAL FIELD Orthoptera: Region of the Tegmina adjacent to the anterior margin or Costa; the Anterior Field.

COSTAL FOLD Lepidoptera males (Hesperiidae): An expansion of the Costal area of the forewing near the margin; folded over or under the wing to form a slit-like pocket containing scales or Setae, that function to release sex pheromone. A membranous flap that may be opened to expose the Androconia (Smith).

COSTAL HINGE The Nodal Furrow.

COSTAL MARGIN The anterior margin of a wing irrespective of Costate.

COSTAL MEMBRANE Hymenoptera: The surface of wing anterior of the Costal Vein.

COSTAL NERVURE The first principal nerve of the insect wing (Jardine); the Costa.

COSTAL REGION Wings of insects: A noncommittal term applied to a more-or-less indefinite region. The area of the wing near the Costa.

COSTAL SCLERITE A more-or-less distinct sclerite at the base of the Costa (Comstock).

COSTAL SPINES Wings of generalized Lepidoptera: A series of slightly curved spine-like Setae on the Costa of the hindwing near the base. Costal spines aid in holding the wings together (Comstock).

COSTAL VEIN Lepidoptera: A vein that extends near and parallel with the Costal margin, extending from base to margin before the apex. The CV is always simple and often absent in the hindwings. Vein 12 of the numerical series on forewings; Vein 8 on hindwings. See Vein; Wing Venation.

COSTALIS Noun. (Latin, *costa* = rib.) The Venella on the cephalic margin of the Rotaxis (MacGillivray).

COSTALLAE Noun. (Latin, *costa* = rib; Sl., Costalla.) The funditae associated with the Costal Vein (MacGillivray).

COSTATE Adj. (Latin, *costa* = rib; *-atus* = adjectival suffix.) Pertaining to structure with Costae or longitudinal, elevated ribs. Alt. Costatus. See Costa.

COSTIFORM Adj. (Latin, *costa* = rib; *forma* = shape.) Rib-shaped; structure in the form of Costae or elevations from a cuticular surface. See Costa. Rel. Form 2; Shape 2; Outline Shape.

COSTORADIAL Adj. (Latin, *costa* = rib; *radius* = ray.) Descriptive of or pertaining to the Radius and the Costa of the insect wing.

COSTULA Noun. (Latin, *costa* = rib. Pl., Costulae.) Hymenoptera: A small ridge dividing the external median metathoracic area into two parts.

COSTULATE Adj. (Latin, *costa* = rib; *-atus* = adjectival suffix.) A diminutive form of Costate; less prominently ribbed than Costate. Alt. Costulatus.

COTHRAN, WARREN American academic (University of California, Davis) who specialized in insect ecology.

COTNION-ETHYL® See Azinphos-Ethyl.

COTNION-METHYL® See Azinphos-Methyl.

COTTAM, ARTHUR (1837–1911) (Barraud, 1912, Entomologist 45: 48.)

COTTAM, CLARENCE (1899–1974) (Tanner 1975, Gt. Basin Nat. 35: 231–239.)

COTTIER, WILLIAM (1905–1964) (Wigglesworth 1965, Proc. R. ent. Soc. Lond. (C) 29: 53.)

COTTLE, JAMES EDWARD (1861–1953) (Leach 1956, Pan-Pacif. Ent. 32: 19–21, 142.)

COTTON APHID *Aphis gossypii* Glover [Hemiptera: Aphididae]: A cosmopolitan pest of many crops incuding cotton, okra, melons, cucumbers, potatoes, tobacco and tomatoes. Adults ca 2 mm long with body globular and colour varies from yellowish-green to blackish-green; Antenna shorter than body; Cornicles tapered, black and darker than Cauda that has 5–7 Setae. In cotton belt of North America, only ovovivparous, parthenogenetic females occur. CA overwinters as winged adults in litter. Spring generations occur on weeds; winged forms appear when plants change; During summer, CA can produce a new generation within five days. On cotton, CA causes seedlings to stunt, leaf drop and premature opening of bolls. Adults and nymphs can produce copious amounts of honeydew that is deposited onto open bolls producing sticky cotton. In some growing regions, sap feeding reduces plant vitality leading to seedling malformation, leaf crumpling and slowed growth. CA can vector plant diseases. CA occurs on oldest leaves near the stem at the base of the plant during fruiting and boll maturation. At the end of the growing season, aphids migrate to the top of the plant. In northern part of distribution, males and oviparious females appear on melon; overwinter in egg stage. Generation can be completed within a week; multivoltine with up to 25 generations per year. Syn. Melon Aphid. See Aphididae. Cf. Green Peach Aphid.

COTTON APHID FUNGUS *Neozygites fresenii* - a disease of CA that can eliminate pest populations within 7–10 days under optimal conditions. Recently killed aphids are covered with velvety white/pale grey pruinescence.

COTTON BLISTER-MITE *Acalitus gossypii* (Banks) [Acari: Eriophyidae].

COTTON BOLLWORM *Helicoverpa armigera* (Hübner) [Lepidoptera: Noctuidae]: Endemic to Australasia and now a cosmopolitan pest of more than 100 Species of plants; a major pest of cotton, tobacco and other crops in southern areas of Europe, Africa and Asia; CB is an important pest of cotton, oilseed, grains and horticultural crops in Australia. Adults probably facultative migrants. Biology resembling *H. zea.* Adult females lay small (0.4–0.6 mm), yellowish-white eggs. The egg laying period 5–24 days depending on region with 500–3,000 eggs produced per female. Eggs adhere to the substrate and display ca 24 longitudinal ridges on the Chorion; eggs turn dark brown before hatching (commonly called the 'brown ring' stage); egg stage lasts 3–10 days depending on temperature. Eggs are most commonly laid on the upper third of the cotton plant, but can be deposited on any structure. First and second instar larvae are yellowish-white to reddish-brown. Prolegs are present on abdominal segments 3–6 and 10. Mature larvae are 30–40 mm long with a brown, mottled head capsule and large white Setae on the first thoracic segment. The larval stage lasts 59–92 days. Mature larvae drop to the ground to pupate in an earthen cell. Pupae 14–18 mm long, mahogany-brown with two parallel spines on the posterior tip. *Helicoverpa armigera* tends to overwinter under late summer crops as pupae in a suspended state of development (diapause). Diapausing *H. armigera* pupae emerge as moths during spring. Damage begins from squaring and continues through to boll maturity. Larvae cut neat round holes into squares and bolls. Young bolls and squares may fall from the plant. Reproductive tissue is favoured. Taxonomically CB is confused with similar Species, including *H. zea* (Hardwick 1965. Canad Ent. Soc. Mem. 40, 247 pp.) Cotton bollworms are 14–20 mm long with a 35–40 mm wingspan. Males greenish-grey and females orange-brown to grey. A band of dark spots along the wing margin and a broad, irregular, brown band occurs on the forewing; hindwings pale with a dark brown band on the apex and a pale patch occurs in the band Antennae long, thread-like (filiform) and covered in fine Setae. Syn. Cotton Earworm; Tobacco Budworm; Tomato Grub; Old World Bollworm. See Corn Earworm; Native Budworm; Nocutidae. Cf. Egyptian Bollworm; Pink Bollworm; Spiny Bollworm; Spotted Bollworm.

COTTON BUD THRIPS See Tomato Thrips.

COTTON FLEAHOPPER *Pseudatomoscelis seriatus* (Reuter) [Hemiptera: Miridae]: A major pest of cotton in the USA. Adults are 3–5 mm long, pale green with minute black spots. Eggs are 0.8 mm long and yellow-white; typically inserted beneath stem surface near apical meristem, and commonly laid in tender weeds including goatweed and horsemint, or under bark of cotton plant. Eclosion occurs within seven days. Adults leave weeds when plants harden and fly to cotton in tender stages of development; adults return to weeds when cotton plants harden. Neonate nymphs are translucent; older nymphs are pale green with red eyes, resemble adults and move rapidly when disturbed. CFH undergoes five instars. Adults and nymphs feed on sap from tender parts of plant (terminals and small squares); squares most susceptible at pinhead-matchead size. CFH overwinters within egg stage and completes 6–8 overlapping generations per year. Syn. *Psallus seriatus* (Reuter). See Miridae.

COTTON HARLEQUIN-BUG *Tectocoris diophthalmus* (Thunberg) [Hemiptera: Scutelleridae]: A pest of cotton in Australia; nymphs feed on cotton bolls and shoots; piercing large bolls causes shedding or discoloration by facilitating entry of rot and mould organisms. Adult semi-circular to oval-shaped, 15–18 mm long and brightly coloured; adults predominantly orange or yellow with metallic green, blue and red patterns. Adult female lays eggs in whorled clusters around stems and twigs. Eggs are white. Adult females protect egg clusters until nymphal emergence. Nymphs resemble adult but are smaller and more brightly coloured; nymphs are predominantly metallic blue or black with red markings. Nymphs are gregarious and occur in large numbers. Adults and nymphs pierce green and maturing bolls with their mouthparts and suck sap. Bolls frequently pierced may fall; bolls become discoloured and boll rotting fungi may enter. Nymphs feed on seeds in small bolls or opened bolls. Adult stage ca 80 days; Life-cycle is completed in 56–84 days. See Scutelleridae.

COTTON LACE-BUG *Corythucha gossypii* (Fabricius) [Hemiptera: Tingidae].

COTTON LEAFHOPPER *Amrasca terraereginae* (Paoli) [Hemiptera: Cicadellidae]: A pest of seedling cotton in Australia. CLH causes severe leaf stippling and retards plant growth. Adult ca 3 mm long, pale green and wedge-shaped; adult with characteristic sideways walk and will hop and fly when disturbed. Adult stage to 60 days. Eggs are 0.7–0.9 mm long, curved and greenish-yellow; eggs are embedded singly in midrib or petiole of a young stem. Egg stage completed within 4–11 days. Nymphs resemble adults but paler and lack wings. Nymphs 0.5–2.0 mm long, pale green and wedge-shaped. Nymphs move sideways in a crab-like movement when disturbed. Nymphal stage requires 7–21 days depending on temperature and food. Leafhoppers insert mouthparts into underside of leaves for feeding. Feeding causes yellowing on underside and reddish-brown coloration on upperside of leaf. Leaves puckered and curl downward. See Cicadellidae.

COTTON LEAFMINER *Stigmella gossypii* (Forbes & Leonard) [Lepidoptera: Nepticulidae].

COTTON LEAF-PERFORATOR 1. *Bucculatrix thurberiella* Busck [Lepidoptera: Lyonetiidae (Bucculatricidae)]: A pest of cotton in the southwestern USA. Eggs ca 0.3 mm diam, bullet-shaped, vertically ribbed, white when laid and becoming reddish when mature. Eggs are deposited singly on leaves, bracts and bolls; eclosion within three days. Neonate larvae are white; older larvae are green-grey with row of black spots along sides of body; mature larva ca 6 mm long. First instar larva forms winding mines in leaves; second instar larva leaves mine and creates holes in leaves (perforations) and forms horseshoe-shaped covers. Pupation occurs within white cocoon attached to leaf petioles and stems of cotton plants. 2. Australia: *Bucculatrix gossypii* Turner. Adult silvery-white with brownish wing tips; hindwing densely fringed with Setae. Eggs are small and brownish-white; Egg stage requires ca 4 days. Larvae are pale green and 6–7 mm long when mature. Mature larvae are pale grey with four black spots behind head. A white, ribbed silken cocoon surrounds a petiole or stem. Pupal stage requires 6–7 days. First three instars mine through leaf Epidermis; fourth instar feeds on underside of leaf and produces clear patches called 'windows'. A silken shelter is produced and caterpillar moults into fifth instar. This instar is most destructive and skeletonizes leaf. See Lyonetiidae.

COTTON LEAF-WORM 1. *Alabama argillacea* (Hübner) [Lepidoptera: Noctuidae]: A significant pest of cotton in the USA; larvae feed on leaves but occasionally attack buds and bolls. Adult ca 15 mm long, pale brown and tinged with green and red; wingspan ca 40 mm; forewing with sinuous dark red-brown lines and distinct spot near centre and white near base. Eggs are blue-green, flattened with ribs; eggs are laid individually on underside of cotton leaves in upper canopy of plant; eclosion within 3–5 days. Larvae are pale-green to black with three narrow, white, longitudinal stripes along dorsum and two parallel rows of black spots surrounded by a white ring. Larvae feed on leaves 15–20 days; mature larva ca 40 mm long. Pupation occurs within a rolled leaf; Pupae are dark brown and ca 20 mm long; adults emerge within 6–8 days. 2. *Spodoptera littoralis* (Boisd.) [Lepidoptera: Noctuidae]. See Cluster Caterpillar.

COTTON LOOPER *Anomis flava* (Fabricius) [Lepidoptera: Noctuidae]: A minor pest of cotton in Australia. Adults with wingspan 28–31 mm; body and forewing reddish-brown and hindwing pale brown. Adult female lays greenish-blue eggs on young leaves and squares. Mature larvae are 28–30 mm long and pale green. Larvae move with a characteristic looping action: Tip of Abdomen drawn toward head creating a loop in the body; front of body lifts from leaf surface, straightens and moves forward. Pupation occurs on underside of bolls, in leaf folds and at base of plant in soil. CL life-cycle is complete within 35 days. Larvae feed mainly on leaves and sometimes will feed on squares and boll surfaces. Larvae prefer older leaves and defoliation progresses upwards on plant. See Noctuidae.

COTTON PLANT-BUG *Aulacosternum nigrorubrum* Dallas [Hemiptera: Coreidae]: A minor pest of cotton in Australia.

COTTON SEED-BUG *Oxycarenus luctuosus* (Montrouzier) [Hemiptera: Lygaeidae]: A pest of cotton in Australia. Adult oval-elongate, 3–5 mm long and dark brown to black; forewing transparent with a dark brown or black spot in the centre; body dark brown and visible through wings when viewed from above. Adults are gregarious and occur in large numbers. Eggs are 1 mm long, oval and creamy-white; eggs are laid in open bolls. Nymphs resemble adults in appearance but lack wings. Nymphs are bright red and occur in clusters among the lint. Adults and nymphs feed on the ripe cotton seeds and can reduce seed weight and germination rates. CSBs pierce seeds from open and partially open bolls and cause lint staining. CSBs occur in cotton during flowering and maintain large populations until harvest. See Lygaeidae.

COTTON SPOTTED BOLLWORM *Earias vittella* (Fabricius) [Lepidoptera: Noctuidae].

COTTON SPRINGTAIL *Entomobrya unostrigata* Stach [Collembola: Entomobryidae]: A pest of cotton in Australia.

COTTON SQUARE BORER *Strymon melinus* Hübner [Lepidoptera: Lycaenidae]: A pest of cotton in Australia.

COTTON STAINER 1. *Dysdercus suturellus* (Herrich-Schäffer) [Hemiptera: Pyrrhocoridae]: A pest of cotton in southeastern North America. Adult 10–20 mm long, reddish orange to tan with white 'collar' and black band on Thorax; Antenna and legs reddish at base and black at apex. Premating period of 2–6 days. Female oviposits up to 60 days. Eggs are smooth, oval, ca 1.5 mm long, white when laid but becoming orange; eggs laid on cotton or on moist ground. Eggs are laid in batches to 100. First instar nymph does not feed; second and third instar nymphs feed on seeds on ground; fourth and fifth instar nymphs and adults wander over plant in search of bolls upon which to feed; during puncture of seeds an exudate stains lint. Adults suck sap from seeds and feeding canals provide site for fungal infection and boll disease. 2. Common name for several Species of Heteroptera in various Families, that cause discoloration of the cotton fibres by piercing immature bolls during feeding. Australia: *Dysdercus cingulatus* (Fabricius). See Pyrrhocoridae.

COTTON STALKGIRDLING BEETLE *Rhyparida australis* (Boheman) [Coleoptera: Chrysomelidae]: A pest of cotton in Australia.

COTTON STEM MOTH *Platyedra subcinerea* (Haworth) [Lepidoptera: Gelechiidae]: A pest of cotton in Australia.

COTTON TIPWORM *Crocidosema plebejana* Zeller [Lepidoptera: Tortricidae]: A pest of cotton in Australia. Adult moth with wingspan of 12 mm; forewing mottled greyish-brown with two small silver V-shaped marks near wing apex. Adult females lay eggs singly among hairs on terminals and leaves of cotton plant. Eggs are small, oval, flattened and transparent blue; as egg develops colour, it becomes whiter and a 'red ring' appears just before hatching; egg stage requires 3–4 days. Larvae are small to medium sized (to 15 mm long) and white with pinkish tinge. Head and first thoracic segment are dark brown. Pupation occurs inside damaged squares that fall to the ground. Larvae feed and tunnel into terminals of young plants destroying the single stem habit and causing multiple branching. Feeding on squares leads to squares dropping to the ground. Cotton tipworm infests many malvaceous plants including cotton. Terminal damage induces secondary branches and delays maturation of plant. See Tortricidae.

COTTON WEBSPINNER *Achyra affinitalis* (Lederer) [Lepidoptera: Tortricidae]: A pest of cotton in Australia.

COTTON WHITEFLY *Bemisia tabaci* (Gennadius) [Hemiptera: Aleyrodidae]. (Australia). Syn. Cassava Whitefly; Silverleaf Whitefly (*Bemisia tabaci* B Biotype); Sweet Potato Whitefly; Tobacco Whitefly. See Sweetpotato Whitefly.

COTTON-PEACH SCALE *Pulvinaria amygdali* Cockerell [Hemiptera: Coccidae]: In North America, a pest of peach.

COTTONSEED BUG *Oxycarenus luctuosus* (Montrouzier) [Hemiptera: Lygaeidae]: A pest of cotton in Australia; CSBs cause reduction in oil content of cotton seeds.

COTTON-SEEDLING THRIPS *Thrips tabaci* Lindeman [Thysanoptera: Thripidae]: A pest of cotton in Australia. Any stage of cotton is attacked but seedling stage vulnerable to loss of vigour, death or unwanted branching. Syn. Sweetpotato Thrips (North America). See Thripidae.

COTTONWOOD BORER *Plectrodera scalator* (Fabricius) [Coleoptera: Cerambycidae].

COTTONWOOD DAGGER MOTH *Acronicta lepusculina* Guenée [Lepidoptera: Noctuidae].

COTTONWOOD LEAF BEETLE *Chrysomela scripta* Fabricius [Coleoptera: Chrysomelidae]: A multivoltine pest of cottonwood, poplar and willow in North America. Eggs reddish and laid in clusters on underside of leaves. Larvae skeletonizes leaves; adults feed on leaves, with only midrib and mainveins remaining. See Chrysomelidae.

COTTONWOOD TWIG BORER *Gypsonoma haimbachiana* (Kearfott) [Lepidoptera: Tortricidae].

COTTONY CAMELLIA SCALE *Pulvinaria floccifera* (Westwood) [Hemiptera: Coccidae]. (Australia).

COTTONY CITRUS SCALE *Pulvinaria polygonata* Cockerell [Hemiptera: Coccidae]: A pest of *Citrus* in southeast Asia and Australia. Adult female yellow-brown, elongate-oval and 3–5 mm long; female produces a cottony sac into which eggs are deposited; ca 200–300 eggs are deposited in one sac. CCS includes two nymphal instars and the life cycle is complete within 2–3 months. Species undergoes 2–3 generations per year. Syn. Pulvinaria Scale (Australia). See Coccidae. Cf. Cottony Cushion Scale.

COTTONY CUSHION SCALE *Icerya purchasi* Maskell [Hemiptera: Margarodidae]: Endemic to Australia, spread to many subtropical and tropical regions after 1860 and now a global pest of *Citrus* where it is grown. CCS attacks acacias and other native trees in Australia. Adult females are 5 mm long, reddish-brown with white mealy covering. Egg sac is attached to body, fluted and ca 10 mm long. CCS resembles Cottony Citrus Scale. CCS is bisexual and females deposit 600–1,000 orange-coloured eggs in specially constructed fluted, egg sac. Crawlers are bright red with yellow tufts on dorsum and margin of body; crawlers prefer midrib on underside of leaves, but later instar and adult shift to twigs and branches. Adult female is mobile until eggsac produced. CCS completes two generations per year. CCS produces abundant amounts of honeydew that serves as substrate for sooty mould. Complete and spectacular biological control of CCS in California and elsewhere has been achieved with dipterous parasite - *Cryptochaetum iceryae* (Williston) and 'vedalia beetle' - a coccinellid predator, *Rodolia cardinalis* (Mulsant). See Margarodidae.

COTTONY FLAVICANS SCALE *Pulvinaria flavicans* Maskell [Hemiptera: Coccidae]. (Australia).

COTTONY GRASS SCALE *Pulvinaria elongata* Newstead [Hemiptera: Coccidae]: A pest of sugar cane in Queensland, Australia.

COTTONY HOPBUSH SCALE *Pulvinaria dodonaeae* Maskell [Hemiptera: Coccidae]. (Australia).

COTTONY LORANTHI SCALE *Tectopulvinaria loranthi* Froggatt [Hemiptera: Coccidae]. (Australia).

COTTONY MAPLE SCALE *Pulvinaria innumerabilis* (Rathvon) [Hemiptera: Coccidae]: Endemic to Europe and widespread in North America. Female CMS produces a large, conspicuous white sac on twigs, oviposits ca 3,000 eggs into sac and dies. Eggs hatch during June-July and first-instar nymphs move to underside of leaves where they feed on sap from midrib and veins. Intermediate-instar female nymphs return to wood, overwinter and complete development during next year; other nymphs complete development on leaves, mate and die. Host plants include shade and fruit trees, maple, pear, grape and gooseberry. See Coccidae.

COTTONY PIGFACE SCALE *Pulvinariella mesembryanthemi* (Vallot) [Hemiptera: Coccidae]. (Australia).

COTTONY SALICORNIA SCALE *Pulvinaria salicorniae* Froggatt [Hemiptera: Coccidae]. (Australia).

COTTONY SALTBUSH SCALE *Pulvinaria maskelli* Olliff [Hemiptera: Coccidae]. (Australia).

COTTONY SYDNEY SCALE *Pulvinaria decorata* Borksenius [Hemiptera: Coccidae]. (Australia).

COTTONY THOMPSON SCALE *Pulvinaria thompsoni* Maskell [Hemiptera: Coccidae]. (Australia).

COTTONY URBICOLA SCALE. *Pulvinaria urbicola* Cockerell [Hemiptera: Coccidae].

COTTY, ERNEST PAUL (1818–1877) (Fauvel 1875, Annu. Ent. 6: 114.)

COTYLA Noun. (Greek, *kotyle* = something hollow.) 1. A cup-like cavity. 2. The articular pan, cup or socket of a ball-and-socket joint. Alt. Cotyle. Cf. Acetabulum.

COTYLOID CAVITY The Acetabulum or coxal cavity in which the Coxa articulates by means of a ball-and-socket joint.

COTYPE Noun. (Latin, *cum* = with; *typus* = image.) 1. Any of several specimens examined by the describer when a Species is described, but no one specimen of which is selected as the Holotype. 2. Any member of the type-series. See Paratype 3. Any other specimen in the type-series, from which series the original description of Species is drawn up. 4. Syntype (Jardine). See Type. Cf. Allotype; Holotype; Paratype; Syntype. Rel. Taxonomy.

COUCALS See Cuculidae.

COUCH FLEA-BEETLE *Chaetocnema australica* (Baly) [Coleoptera: Chrysomelidae]. (Australia).

COUCH MITE *Eriophyes tenuis* (Nalepa) [Acarina: Eriophyidae]. (Australia).

COUCH, JOHNATHAN (1789–1870) (Anon. 1870, Proc. Linn. Soc. Lond. 1869–1879: 99–100.)

COUCHGRASS MITE 1. *Dolichotetranychus australianus* (Womersley) [Acarina: Tenuipalpidae] (Australia). 2. *Eriophyes cynodiniensis* (Sayed) [Acarina: Eriophyidae] (Australia).

COUCHGRASS SCALE *Odonaspis ruthae* Kotinsky [Hemiptera: Diaspididae] (Australia).

COUCHGRASS WEBWORM *Sclerobia tritalis* (Walker) [Lepidoptera: Pyralidae] (Australia).

COUCHTIP MAGGOT *Delia urbana* (Malloch) [Diptera: Anthomyiidae] (Australia).

COUCKE, EDOURD (–1899) (Mik 1900, Wien. ent. Ztg. 19: 88.)

COULEE CRICKET *Peranabrus scabricollis* (Thomas) [Orthoptera: Tettigoniidae]: A minor pest of sagebrush, mustard, field and garden crops in northwestern USA. Eggs are laid individually in grass stools during fall; eclosion occurs during spring. CCs migrate in large numbers and are strongly cannibalistic. Males stridulate. See Tettigoniidae. Cf. Mormon Cricket.

COUMAPHOS An organic-phosphate (thiophosphoric acid) compound {O,O-diethyl-O-(3-chloro-4-methyl-2-oxo-2H-1-benzapyran-7-yl) phosphorothioate} used as a systemic insecticide against livestock pests. Compound is applied to cattle, sheep, goats, horses, swine and dogs for control of lice, screwworms, flies, ticks, mites, mosquitoes and other barnyard pests. As an acaricide, Coumaphos shows potential for control of Varroa Mite that attacks honeybees. Compound is toxic to birds and fishes. Trade names include: Asuntol®, Baymix®, CoRal®, Diolice®, Meldane®, Negashunt®, Perizin®, Resitox®, Umbethion®, Zipcide®. See Organophosphorus Insecticide.

COUMARIN (See Berenbaum 1983, Evolution 37: 163–179.)

COUMARINS 94 See Mammeins.

COUNTERSHADING Noun. (Latin, *contra* = opposite) A physical condition and aspect of mimicry in which an animal is dark dorsally and pale ventrally. When lighting is from above (the generalized condition), the overall coloration of the animal is uniform and thus, presumably, inconspicuous. Countershading tends to reduce shape enhancement. See Coloration; Mimicry. Cf. Disruptive Coloration; Crypsis; Environmental Melanism; Industrial Melanism. Rel. Pigmentary Colour; Structural Colour.

COUPER, WILLIAM (fl 1870) (Paradis 1974, Ann. Soc. ent. Québec 19: 4–15, bibliogr.)

COUROLS See Leptosomatidae.

COURT Noun. (Middle English, *court* > Latin, *cohors* > Greek, *chortos* = an enclosure. Pl., Courts.) Social Bees: A group of workers that surround the queen, antennate her, 'lick' her and sometimes provide food for her.

COURVOISIER, LUDWIG GEORG (1843–1918) (B. 1935, Ent. Z., Frankf. a. M. 49: 113–114.)

COVERING GALL Plant galls characterized by the gall inducer first acting externally to produce hypertrophy of surrounding tissue that quickly results in the inducer becoming enclosed within the gall. An aperture to the outside may be maintained. CG induced by mites, aphids, coccids, cecidomyiids, cynipids. See Gall. Cf. Bud-and-Rosette Gall; Filz Gall; Mark Gall; Pit Gall; Pouch Gall; Roll-and-Fold Gall.

COW KILLER See Mutillidae.

COWAN, FRANK (1844–1905) (Weiss 1936, *Pioneer Century of American Entomology*. 320 pp. (259), New Brunswick.)

COWAN, GEORGE HENRY (1886–1924) (Cooley 1926, Bienn. Rep. Mont. St. Bd. Ent. 6: 14–15.)

COWAN, THOMAS WILLIAM (1849–1926) (H. W.

M. 1927, Proc. Linn. Soc. Lond. 139: 79.)

COWBOY BEETLE *Diaphonia dorsalis* (Donovan) [Coleoptera: Scarabaeidae]. (Australia).

COWLEY, JOHN (1909–1967) (Kennedy 1968, Proc. R. ent. Soc. Lond. (C) 32: 59.)

COWPEA APHID *Aphis craccivora* Koch [Hemiptera: Aphididae].

COWPEA CURCULIO *Chalcodermes aeneus* Boheman [Coleoptera: Curculionidae]: A pest of many kinds of fruit trees in eastern USA. Adults make round feeding punctures in skin of fruit; females cut crescent-shaped flaps in skin and insert their elliptical, white eggs. Larvae feed in fruits for ca 2 weeks. Pupation occurs in soil; adults emerge, feed for about one month and overwinter as adults in trash. CC completes 1–2 generations per year. Feeding and egg punctures permit brown rot. See Curculionidae.

COWPEA WEEVIL See Spotted Cowpea Bruchid.

COWPERTHWAITE, WILLIAM GARDNER (1919–1964) (Denmark 1964, J. Econ. Ent. 57: 422–423.)

COX, B C (–1948) (Williams 1949, Proc. R. ent. Soc. Lond. (C) 13: 67.)

COX, HENRY RAMSAY (1844–1880) (Carrington 1880, Entomologist 13: 248.)

COX, W E (–1948) (Williams 1949, Proc. R. Ent. Soc. Lond. (C) 32: 67.)

COX, WILLIAM SAY ILSTON (1874–1966) (Kennedy 1968, Proc. R. ent. Soc. Lond. (C) 32: 59.)

COXA Noun. (Latin, *coxa* = hip. Pl., Coxae.) The basal (proximal) segment of the insect leg and distal portion of the primitive Coxopodite. Coxae are paired, ventrolateral in position and found on each of thoracic segment. Coxa attaches to the lateroventral part of thoracic wall. Basal part of Coxa is attached to body wall by a Coxal Corium with one (monocondylic), two (dicondylic) or three points (tricondylic) of articulation. Distal part of Coxa attached to Trochanter. Proximal end of Coxa is surrounded by a Basicostal Suture. When BS completely surrounds the Coxa, it forms a narrow basal zone (Basicoxite) and a larger, more distal zone (Disticoxite). Coxa varies in shape, but significance of shape poorly studied. Coxa often forms a cylindrical or truncated cone. Coxa of most neopterous insects is not conspicuously large; Coxa seems unusually long in Strepsiptera and many Mantodea. Coxa of wing-bearing segments of Neuroptera, Mecoptera, Lepidoptera and Trichoptera may be divided into an anterior portion called Coxa Vera (Coxa Genuina) and posterior portion called the Meron. Meron ultimately disassociates from Coxa and incorporates into thoracic Pleuron in some groups of Holometabola (*e.g.* Diptera). See Leg. Cf. Femur; Pretarsus; Tibia; Trochanter. Rel. Basicostal Suture; Coxal Articulations; Trochantin.

COXA GENUINA Some Orders: The apparent anterior subdivision of an enlarged Meron (Snodgrass). Syn. Coxa Vera (Imms). See Coxa. Rel. Leg.

COXA ROTATORIA A Coxa with a monocondylic articulatory point. See Coxa. Rel. Leg.

COXA SCROBICULATA A Coxa with dicondylic articulatory points. See Coxa. Rel. Leg.

COXA VERA The anterior part of the middle and hind Coxae in many insects (Imms). See Coxa. Rel. Leg.

COXACAVA Noun. (Latin, *coxa* = hip. Pl., Coxacavae.)The coxal cavity *sensu* MacGillivray.

COXACOILA Noun. (Latin, *coxa* = hip. Pl., Coxacoilae.) The posterior angle of the Episternum produced into a condylar process (MacGillivray).

COXACORIA Noun. (Latin, *coxa* = hip.) The membrane articulating the legs; the Coria or articulating membrane around each leg-joint. The membrane between the Coxa, Sternum and Pleuron (MacGillivray). Cf. Coxal Cavity.

COXAFOSSA Noun. (Latin, *coxa* = hip; *fossa* = ditch. Pl., Coxafossae.) A depression in the thoracic pleural wall that accommodates the Coxa when at repose. (MacGillivray).

COXAL Adj. (Latin, *coxa* = hip.) Descriptive of or pertaining to the Coxa. Cf. Subcoxal.

COXAL CAVITY The opening or space in which the Coxa articulates. See Acetabulum.

COXAL CORIUM The articular membrane surrounding the base of the Coxa (Snodgrass). See Corium.

COXAL FILE Some aquatic Coleoptera: A series of striations just above the hind Coxa of the male. Regarded as a component of a stridulatory mechanism. See Stridulation.

COXAL GLAND Excretory organ of mites located near the first leg.

COXAL GLANDS Eversible glandular structures at base of legs. Coxal glands are well developed in some Thysanurans and modified variously in higher Orders. See Glands.

COXAL MUSCLE One to several muscles that originate in the Coxa of an appendage and insert into a more distal segment of the appendage. See Muscle.

COXAL PLATES Some aquatic Coleoptera: Plate-like expansions or dilatations of the hind Coxae.

COXAL PROCESS The cuticular projection of the Pleuron located at the ventral end of the Pleural Suture and to which the Coxa articulates.

COXAL STYLETS Thysanura: Short, leg-like, jointed appendages on the Sternum of the abdominal segments. See Stylet 1.

COXARTIS Noun. The primary point of articulation of the Coxa itself to the body (MacGillivray).

COXASUTURE Noun. (Latin, *coxa* = hip. Pl., Coxasutures.) A suture of a leg segment (MacGillivray). See Suture.

COXATA Noun. (Latin, *coxa* = hip. Pl., Coxatae.) The minute sclerite in the Coxacoria on the caudal aspect of the Coxa (MacGillivray).

COXELLA Noun. (Latin, *coxa* = hip. Pl., Coxellae.)

1. The minute sclerite in the Coxacoria near the anterior margin of the Coxa. The accessory or complementary Coxal Sclerite. 2. The Juxtacoxale *sensu* MacGillivray.

COXIFER Noun. (Latin, *coxa* = hip; *ferre* = to carry. Pl., Coxifers.) The pleural pivot of the Coxa (Crampton). See Articulation.

COXITE Noun. (Latin, *coxa* = hip; *-ites* = constituent. Pl., Coxites.) 1. The basal segment of any leg-like appendage, *e.g.,* those of the abdominal sternites (Tillyard). 2. A rudimentary abdominal limb in Thysanura displayed in the form of paired lateral sclerites (Folsom & Wardle). See Coxosternite.

COXOLA Noun. (Pl., Coxolae.) A minute sclerite in the Coxacoria at the apicolateral portion of the Coxa (MacGillivray).

COXOMARGINALE See Basicoxite (Snodgrass).

COXOPLEURE Noun. (Latin, *coxa* = hip; Greek, *pleura* = side. Pl., Coxopleures.) See Antepleuron (MacGillivray).

COXOPLEURITE Noun. (Latin, *coxa* = hip; Greek, *pleura* = side; *-ites* = constituent. Pl., Coxopleurites.) The sclerite of a generalized thoracic Pleuron adjacent to the dorsal margin of the Coxa. The Coxopleurite bears the dorsal coxal articulation, the anterior part of which becomes the definitive Trochantin. See Eutrochantin; Trochantinopleura (Snodgrass). Cf. Coxosternite.

COXOPLURAL Adj. (Latin, *coxa* = hip; Greek, *pleura* = side.) Descriptive of or pertaining to the Coxa and the Pleuron.

COXOPODITE Noun. (Latin, *coxa* = hip; Greek, *pous* = foot; *-ites* = constituent. Pl., Coxopodites.) The primary basal segment of an appendage, representing the primitive limb basis; the Coxa.

COXOPODITE PLATE Thysanura: The fused Coxites and Sternum of each segment (Folsom & Wardle). Alt. Coxopodial Plate.

COXOSTERNITE Noun. (Latin, *coxa* = hip; Greek, *sternon* = chest. Pl., Coxosternites.) The limb base element (or elements) of a Coxosternum or Pleurosternum, commonly called 'Coxite'. Pleurosternite (Snodgrass). Cf. Coxopleurite.

COXOSTERNUM Noun. (Latin, *coxa* = hip; *sternum* = breast bone. Pl., Conxosterna; Coxosternums.) Morphologically, a definitive sternal sclerite that includes the areas of the limb bases; Pleurosternum (Snodgrass).

COXO-TROCHANTERAL JOINT Arthropod leg: One of the two primary flexion points positioned between the Coxa and the Trochanter (Snodgrass). See Articulation; Leg; Joint. Cf. Femoro-Tibial Joint. Rel. Segmentation.

COXO-TROCHANTERAL Adj. (Latin, *coxa* = hip; *trochanter* = runner.) Of or pertaining to the Coxa and Trochanter.

COMPLEMENTARY DNA DNA strand formed from messenger RNA using the enzyme reverse transcriptase. Acronym cDNA.

CRAB LOUSE *Pthirus pubis* (Linnaeus) [Anoplura: Pthiridae]: A cosmopolitan blood-sucking louse that is host-specific on humans. CL eggs are attached to hairs of host. Feeding stages prefer to remain in pubic region or other hirsute areas of the host's body. Nymphs and adults can survive for short periods off the host. See Pthiridae. Cf. Body Louse. Head Louse.

CRAB SPIDERS See Thomisidae.

CRABBE, GEORGE (1754–1832) (Lisney 1960, *Bibliography of British Lepidoptera 1608–1799.* 315 pp. (284–285), London.)

CRABGRASS LEAF BEETLE *Oulema rufotincta* (Clark) [Coleoptera: Chrysomelidae]. (Australia).

CRABHOLE MOSQUITO *Deinocerites cancer* Theobald [Diptera: Culicidae].

CRACK DOWN® See Deltamethrin.

CRADLE Noun. (Anglo Saxon, *cradel*. Pl., Cradles.) Coccids: The ventral aspect of the endoskeleton of the head (Green).

CRAG Noun. (Middle Flemish, *krage* = the neck.) The neck, throat or craw. See Cervix.

CRAGG, FRANCIS WILLIAM (1882–1924) (W. A. H. 1925, J. Econ. Ent. 18: 563.)

CRAIG, CHARLES FRANKLIN (1872–1950) (Anon. 1950, J. Econ. Ent. 43: 965.)

CRAMBIDAE See Pyralidae.

CRAMER, PIETER (–1779) Dutch merchant, amateur Entomologist and author of *Papillons exotiques des trois parties du Monde, l'Asie, l'Afrique, et l'Amerique* (4 vols, 1779–1782.) (dos Passos 1951, J. N. Y. ent. Soc. 59: 140.)

CRAMPTON, GUY CHESTER (1811–1951) (Mallis 1971, *American Entomologists*. 549 pp. (438–439), New Brunswick.)

CRANBERRY *Vaccinium macrocarpon* Aiton [Ericaceae]: An evergreen plant endemic to eastern North America and restricted to acid bogs and swamps; several commercial cultivars produce edible berries.

CRANBERRY FRUITWORM 1. *Rhabdopterus picipes* (Oliver) [Coleoptera: Chrysomelidae]: A pest of cranberry and blueberry in eastern USA. Larvae feed on small roots and adults feed on leaves. See Chrysomelidae. 2. *Acrobasis vaccinii* Riley [Lepidoptera: Pyralidae].

CRANBERRY GIRDLER *Chrysoteuchia topiaria* (Zeller) [Lepidoptera: Pyralidae].

CRANBERRY WEEVIL *Anthonomus musculus* Say [Coleoptera: Curculionidae].

CRANE FLIES See Tipulidae. Cf. European Crane Fly; Range Crane Fly.

CRANIAD Adv. (Latin, *cranium* = skull; *ad* = toward.) Directed toward the head. Alt. Cephalad. See Orientation. Cf. Anteriad; Apicad; Basad; Caudad; Centrad; Cephalad; Dextrad; Dextrocaudad; Dextrocephalad; Distad; Dorsad; Ectad; Entad; Laterad; Mediad; Mesad; Neurad; Orad; Proximad; Rostrad; Sinistrad; Sinistrocaudad; Sinistrocephalad; Ventrad.

CRANIAL Adj. (Latin, *cranium* = skull > Greek, *kranion*.) Descriptive of or pertaining to the Cranium.

CRANIUM Noun. (Latin, *cranium* = skull > Greek, *kranion*. Pl., Crania; Craniums.) The sclerotized portion of the head capsule except the neck (Cervix); sometimes limited to the fixed parts above the Clypeofrontal Suture.

CRAPE-MYRTLE APHID *Tinocallis kahawaluokalani* (Kirkaldy) [Hemiptera: Aphididae].

CRASKE, EDMUND SYDNEY (1873–1943) (Crow 1944, Entomologist 77: 48.)

CRASKE, JOHN CHRISTOPHER BEADNELL (1902–1958) (A.E.C. 1959, Entomologist's Rec. J. Var. 71: 59–60.)

CRASSA Noun. (Latin, *crassus* = thick. Pl., Crassae.) 1. The linear thickenings on each Postgena. 2. Mandibular Apodemes *sensu* MacGillivray.

CRASSUS Adj. (Latin, *crassus* = thick.) Thick; tumid; dense; coarse.

CRAT- Greek (*kratos*) prefix indicating power.

CRATERIFORM Adj. (Latin, *crater* = bowl; *forma* = shape.) Bowl-shaped. Descriptive of structure shaped as a bowl or saucer. Term obsolete, but in older literature usually applied to depressions. Cf. Sulcate; Sulciform. Rel. Form 2; Shape 2; Outline Shape.

CRAW, ALEXANDER (1850–1908) First Los Angeles County Horticultural Commissioner (1881) and member of the California State Horticultural Commission (1888–1904). (Essig 1931, *History of Entomology*. 1029 pp. (593–595, bibliogr.), New York.)

CRAWFORD, FRASER S (1829–1890) (Musgrave 1932, *A Bibliography of Australian Entomology 1775–1930*. 360 pp. (57), Sydney.)

CRAWFORD, JAMES CHAMBERLIN (1880–1950) (Mallis 1971, *American Entomologists*. 549 pp. (371–372), New Brunswick.)

CRAWFORD, WILLIAM MONOD (1876–1942) (Anon. 1941, Irish Nat. J. 7: 336–337; Blair, Proc. R. ent. Soc. Lond. (C) 6: 40.)

CRAWL Noun. (Dutch, *kraal*. Pl., Crawls.) A pen for hogs or slaves. Verb. (Crawled, Crawling. Old Norse, *krafla* = to paw or scrabble with the hands.) To move slowly by drawing the body along the ground, worm-like. To Creep. Note: Legged insects walk; legless larval insects crawl. See Creep.

CRAWLER Noun. (Old Norse, *krafla* = to paw or scrabble with the hands. Pl., Crawlers.) 1. Vernacular name for the ambulatory stage (first nymphal instar) of some scale insects. 2. An adult honey bee incapable of flight due to parasitization by a conopid fly larva. 3. A hellgramite. 4. An organism that crawls.

CRAWLING WATER BEETLES See Haliplidae.

CRAZY ANT 1. *Paratrechina longicornis* (Latreille) [Hymenoptera: Formicidae]: Presumed endemic to India, widespread in tropical-subtropical Australia and a nuisance ant in some urban habitats. Typically, CA is fast-walking with erratic movement and predatory or scavenging feeding habits. 2. *Paratrechina fulva* (Mayr), [Hymenoptera: Formicidae]: Introduced from northern Brazil to northern Colombia in an attempt to suppress poisonous snake populations. The eventual outcome was the disappearance of 36 of 38 Species of ants indigenous to the invaded area as well as at least seven other soil-dwelling insect Species, one Species of snake, and three Species of lizards (de Polania & Wilches 1992). See Formicidae.

CREAM-SPOTTED ICHNEUMON *Echthromorpha intricatoria* (Fabricius) [Hymenoptera: Ichneumonidae]: A parasite of lawn armyworms in Australia.

CREBER Adj. (Latin, *creber* = close.) Closely set.

CREEP Verb. (Middle English, *crepen* > Anglo Saxon, *creopan* > Old Norse, *krjupa* = to creep.) 1. To move with the body prone and near the ground, worm-like or reptile-like. 2. To experience a sensation as of insects creeping over the body. Note: Creep and Crawl are used interchangeably. Crawl seems more common for legless animals. Creep suggests slowness or stealth; crawl suggests servility or abjectness.

CREEPING MYIASIS A type of Myiasis affecting humans that is caused by Species of bot flies that usually infest domesticated animals (horse, mule or donkey). First-instar larvae of the bot fly penetrate human skin and burrow beneath surface; larvae do not complete development and typically die during first instar. Human infestation through close association with true hosts of bot flies. Species most frequently implicated include *Gasterophilus haemorrhoidalis, G. inermis, G. nigricornis* and *G. pecorum*. Syn. Creeping Cutaneous Myiasis. See Bot Flies; Myiasis.

CREEPING WATERBUGS See Naucoridae.

CREIGHTON, WILLIAM STEEL (1902–1973) (Steel 1974, Proc. ent. Soc. Wash. 76: 38.)

CREMASTER Noun. (Greek, *kremastos* = hung. Pl., Cremasters.) 1. The apex of the last segment of the Abdomen. 2. The terminal spine or hooked abdominal process of subterranean pupae that is used to facilitate emergence from the earth. 3. An anal hook by which some Lepidoptera pupae are attached to the silk cocoon or silk pad. Cf. Sustentor.

CREMASTRAL Adj. (Greek, *kremastos* = hung.) Descriptive of or pertaining to the Cremaster.

CRENA Noun. (Latin, *crena* = notch. Pl., Crenae.) A deep groove or Sulcus.

CRENATE Adj. (Latin, *crena* = notch; *-atus* = adjectival suffix.) Scalloped with small, blunt, rounded teeth that typically occur along a margin. Alt. Crenatus.

CRENATION Noun. (Latin, *crenatus* = notched; English, *-tion* = result of an action. Pl., Crenations.) A scalloped margin or rounded tooth.

CRENULATE Adj. (Latin, diminutive of *crena* = little notch.) 1. Descriptive of a structure's margin or a surface with small scallops that are evenly rounded and rather deeply curved. A margin invested with small tooth-like projections. Alt. Crenellated; Crenulated; Crenulatus.

CREOSOTE SPIDER MITE *Psedobryobia drummondi* (Ewing) [Acari: Tetranychidae].

CREP Noun. (Latin, *crepusculum* from *creper* = dusky. Pl., Creps.) A unit of time proposed by Neilsen (1961) that involves twilight and the measurement of photoperiod's influence on organisms. Evening Creps = time of day – time of sunset divided by duration of civil twilight. Morning Creps = time of sunrise – time of day divided by duration of civil twilgiht. See Twilight. Rel. Lux; Photoperiodism.

CREPERA Adj. (Etymology obscure.) A gleam of paler colour on a dark ground.

CREPIDIUM See Ventral Ptyche.

CREPIS Noun. (Greek, *krepis* = foundation. Pl., Crepides.) 1. Scarabaeoid larvae: The thinly sclerotized, anteriorly concave, median cross-bar pertaining to the region Haptolachus. The Crepis is usually asymmetrical and only indicated by a fine line, or completely absent. 2. The transversely, strongly bowed bar of Hayes (Boving).

CREPITACULUM Noun. (Latin, *crepitaculum* = rattle. Pl., Crepitacula.) A stridulating or sound-producing organ, such as found in some Orthoptera and Psocoptera. Rel. Stridulation.

CREPITATE Intransitive Verb. (Latin, *crepitare* = to crackle; *-atus* = adjectival suffix.) A series of short, sharp explosions or crackling sounds, such as those produced by a bombardier beetle or rattlesnake.

CREPITATION Noun. (Latin, *crepitare* = to crackle; English, *-tion* = result of an action. Pl., Crepitations.) 1. A crackling sound or the production of a sharp sound by the discharge of fluid or vapour for defence (*i.e.* sound of the bombardier beetle). 2. Crackling or creaking.

CREPUSCULAR Adj. (Latin, *crepusculum* = dusk.) Pertaining to organisms active at dusk or (more rarely) at daybreak. Cf. Diurnal; Nocturnal. Rel. Circadian Rhythm.

CRESCENT CELL An anucleate, polymorphic cell in the Haemolymph of cockroaches. See Oenocyte.

CRESCENT Noun. (Latin, *crescere* = to grow. Pl., Crescents.) A crescent-shaped structure. Specifically, an object whose outline shape resembles the appearance of the moon between its new and full phases. Rel. Outline Shape.

CRESCENTIC Adj. (Latin, *crescere* = to grow; *-ic* = of the nature of.) Structure that is crescent-shaped or crescent-like in outline shape; meniscoid. Syn. Crescentiform. Cf. Lenticular. Rel. Outline Shape.

CRESCENTIFORM Adj. (Latin, *crescere* = to grow; *forma* = shape.) Crescent-shaped; lunule-shaped. Syn Crescentic. See Crescent. Rel. Form 2; Shape 2; Outline Shape.

CRESCENTMARKED LILY APHID *Neomyzus circumflexus* (Buckton) [Hemiptera: Aphididae].

CRESSON, EZRA TOWNSEND JR (1876–1948) Associate curator of insects at Philadelphia Academy of Sciences and taxonomic specialist in Ephydridae. (Calvert 1949, Ent. News 60: 85–99, bibliogr.)

CRESSON, EZRA TOWNSEND SR (1838–1926) (Mallis 1971, *American Entomologists*. 549 pp. (343–348), New Brunswick.)

CRESSON, GEORGE BRINGHURST (1859–1919) (Weiss 1936, *Pioneer Century of American Entomology*. 320 pp. (238–239), New Brunswick.)

CREST Noun. (Latin, *crista* = crest. Pl., Crests.) 1. A prominent, longitudinal Carina on the upper surface of any part of the head or body. 2. A ridge or elongate prominence on a bone or other hard surface.

CRESTED KATYDID *Alectoria superba* Brunner von Wattenwyl [Orthoptera: Tettigoniidae]. (Australia). Cf. Angular-Winged Katydid; Broad-Winged Katydid; Citrus Katydid; Forktailed Bush-Katydid; Inland Katydid; Mottled Katydid; Philippine Katydid; Spotted Katydid.

CRESTED See Cristate.

CRESUS® See Deltamethrin.

CRETACEOUS PERIOD The third Period (146–65 MYBP) of the Mesozoic Era and named (Latin, *creta* = chalk) for the the soft, fine-grained limestone that accumulated over the late Cretaceous sea floor. Cretaceous System was described in 1822. First insects preserved in amber of Cretaceous Period reveal bees and ants; first parasitic Hymenoptera appear to include Mymaridae, Trichogrammatidae and Tetracampidae. Principal amber deposits include: Taimurian Amber (Siberia, 110 MYBP), Canadian Amber (Cedar Lake, Manitoba and Medicine Hat, 70–90 MYBP), New Jersey Amber (Kinkora, 96–74 MYBP). During the CP, gymnosperms decline while angiosperms originate and radiate; with angiosperm radiation some insect pollinators (Diptera) radiate. CP cataclysmic events include the extinction of dinosaurs, giant swimming reptiles, giant turtles and ammonoids ca 65 MYBP. See Geological Time Scale; Mesozoic Era. Cf. Jurassic Period; Triassic Period. Rel. Fossil.

CRETAVIDAE Sharov 1957. A Family of aculeate Hymenoptera based on a Cretaceous Period fossil (*Cretavus sibiricus* Sharov) from Siberia. Subsequently transferred to Mutillidae. See Hymenoptera. Cf. Plumariidae.

CRETEVANIIDAE Rasnitsyn 1975. A Family of parasitic Hymenoptera assigned to the Evanioidea.

CRETSCHMAR, M (1900–1961) (Werncke 1962, Mitt. dt. ent. Ges. 21: 19–20.)

CREUTZER, CHRISTIAN Author of *Entomlogische Versuche* (1799).

CREVECOEUR, FERDINAND F (1826–1931) (Smith 1931, Trans. Kans. Acad. Sci. 34: 138–144, bibliogr.)

CREWE, H HARPUR (–1883) (Anon. 1883, Entomol. mon. Mag. 20: 118–119.)

CRIBELLATE SPIDERS See Dictynidae.

CRIBELLUM Noun. (Latin, diminutive of *cribum* = seive. Pl., Cribella.) 1. Insects: A sieve-like sclerite that opens via a slit near the upper surface of the Mandibles. 2. Spiders: A flat spinning organ in front of the anterior Spinnerets on the Abdomen of some spider Species. Cribellum consisting of numerous microscopic spigots that issue exceedingly fine threads of silk. Cribellate silk threads combine with larger strands of silk issuing from Spinnerets. Cribellum possibly represents modified anteriomedial pair of Spinnerets or independently derived feature of cribellate spiders. See Calamistrum.

CRIBIFORM PLATES 1. Cuticular pitted or sieve-like Sclerites on the dorsal surface of the Abdomen in some Genera of Asterolecaniinae (MacGillivray). 2. Scarabaeid larvae: Sieve-like openings in the Spiracles.

CRIBRATE Adj. (Latin, *cribrum* = seive; *-atus* = adjectival suffix.) Seive-like in form or function. Descriptive of structure or a surface pierced with closely set, small holes. Alt. Cribratus. See Cribriform; Seive. Cf. Porous; Punctate.

CRIBRATE WEEVIL *Otiorhynchus cribricollis* Gyllenhal [Coleoptera: Curculionidae].

CRIBRIFORM Adj. (Latin, *cribrum* = seive; *forma* = shape.) Sieve-shaped; structure displaying perforations as those of a sieve. See Cribrate. Rel. Form 2; Shape 2; Outline Shape.

CRICKET Noun. (Middle English, *criket*. Pl., Crickets.) See Orthoptera. Proposed veneral term: A crackle of crickets.

CRICKETS See Gryllidae; Saltatoria. Cf. African Mole-Cricket; Australian King-Cricket; Black Field-Cricket; Black-Horned Tree Cricket; Changa Mole Cricket; Citrus Leaf-Eating Cricket; Common Mole Cricket; Dark-Night Mole Cricket; Emma Field Cricket; European Mole Cricket; Field Cricket; Four-Spotted Tree Cricket; Giant Mole Cricket; Greenhouse Stone-Cricket; House Cricket; Imitator Mole Cricket; Indian House-Cricket; Inland Field Cricket; Jerusalem Cricket; Knife Mole-Cricket; Dark-Night Mole Cricket; Mole Cricket; Mormon Cricket; Northern Mole-Cricket; Oceanic Field Cricket; Short-Winged Mole Cricket; Snowy Tree-Cricket; Southern Mole Cricket; Tree Cricket; Two-Spotted Cricket.

CRIDDLE, NORMAN (1875–1933) (Criddle 1975, Manitoba Ent. 8: 5–9.)

CRIMEAN-CONGO HAEMORRHAGIC FEVER An arbovirus of the Bunyaviridae known to occur in the Palearctic, Orient and tropical Africa. CCHF causes moderately high mortality in humans. Incubation period 2–5 days; clinical period 7–10 days; disease causes high fever, headache and bleeding from nose, mouth and vein punctures. CCHF vectored by ca 30 Species of ticks, particularly *Hyalomma* spp. and *Rhipicephalus* spp., with transstadial retention and transovarial transmission in some vectors. See Arbovirus; Bunyaviridiae.

CRINEOUS Adj. (Etymology obscure.) Dark-brown, with a slight admixture of yellow and gray. Alt. Crineus; Crinosus.

CRINITE Adj. (Latin, *crinitus* = with locks of hair; *-ites* = constituent.) Thinly covered with very long flexible Setae.

CRINKLED FLANNEL MOTH *Lagoa crispata* (Packard) [Lepidoptera: Megalopygidae]: A minor pest of apples and berries in eastern North America. Adult yellowish with brown spots or bands on wings; wingspan ca 25 mm. Cocoons typically on twigs, formed of tough silk and provided with a lid. See Megalopygidae.

CRINOPTERYGIDAE Plural Noun. A small Family of heteroneurous Lepidoptera assigned to Superfamily Incurvarioidea. See Incurvarioidea.

CRISODRIN® See Monocrotophos.

CRISP, GEOFFREY (–1960) (Uvarov 1961, Proc. R. ent. Soc. Lond. (C) 25: 49.)

CRISPATE Adj. (Latin, *crispatus* from *crispus* = curly; *-atus* = adjectival suffix.) With a wrinkled or fluted margin. Alt. Crispatus.

CRISPUS Adj. (Latin, *crispus* = curly hair.) Pertaining to a surface with edge disproportionately larger than disc, that causes the margin to have an irregular undulation.

CRISTA Noun. (Latin, *crista* = crest. Pl., Cristae.) 1. 'A cord-like ridge running the full length of the dorsal surface of the egg-capsule in Blatta' (Wheeler). 2. A ridge or crest.

CRISTA ACUSTICA See Siebold's Organ (Comstock).

CRISTATE Adj. (Latin, *cristatus* = crested; *-atus* = adjectival suffix.) Crested; descriptive of surface with a prominent carina or crest on the upper surface of a structure, appendage or segment. Rel. Subcristate.

CRISTER, JOSEPH NEAL (1899–1964) (Curl 1965, J. Econ. Ent. 58: 1040.)

CRISTIFORM Adj. (Latin, *cristatus* = crested; *forma* = shape.) Descriptive of structure shaped as a sharp ridge or crest.

CRISTULA Noun. (Latin, *cristatus* = crested. Pl., Cristulae.) A small crest.

CRISTULATE Adj. (Latin, *cristatus* = crested; *-atus* = adjectival suffix.) With little crescent-like ridges or crests.

CRITICAL HABITAT According to USA federal law, the ecosystems upon which endangered and threatened Species depend. See Ecosystem; Habitat.

CRITICAL POINT DRYING A procedure used to prepare specimens for study under the scanning electron microscope and for preserving soft-bodied specimens traditionally preserved in fluid. Specimens are dehydrated through a graded ethanol series, passed through an intermediate fluid (acetone, ethyl alcohol, Freon), then into a transitional fluid (CO_2, Freon, nitrous oxide) and then subjected to a critical temperature and pressure in a specially constructed chamber. See Desiccate. Cf. Freeze Drying.

CRIVELLI, GIUSEPPE BALSAMO (1800–1974)

(Conci 1967, Atti Soc. ital. Sci. nat. Milan 106: 36–37.)

CROCEA SKIPPER *Neohesperilla crocea* (Miskin) [Lepidoptera: Hesperiidae]. (Australia).

CROCEOUS Adj. (Latin, *croceus* from *crocus* = saffron; *-osus* = possessing the qualities of.) Saffron yellow; yellow with an admixture of red. Alt. Croceus.

CROCHET Noun. (French, *crochet* = small hook. Pl., Crochets.) Lepidoptera larva: Curved spines or hooks on the Prolegs or the Cremaster of pupae. See Proleg. Cf. Biordinal Crochet; Multiordinal Crochet. Rel. Hook.

CROCKER, W (–1949) (Anon. 1951, Trans. Proc. S. Lond. ent. nat. Hist. Soc. 1950–1951: 56.)

CROFT, HARRY HOLMES (1820–1883) (Bethune 1916, Can. Ent. 48: 1–5.)

CROFTON WEED CROWNBORER *Acalolepta argentata* (Aurivillius) [Coleoptera: Cerambycidae]. (Australia).

CROFTON WEED GALL FLY *Procecidochares utilis* Stone [Diptera: Tephritidae]: A biological control agent used against *Eupatorium adenophorum* Sprengel, a noxious weed native to Mexico and spread throughout much of the Pacific Basin. CWGF was discovered in Mexico and established in Hawaii in 1945. Subsequently, it has been established in several other countries and Islands. See Tephritidae.

CROISSANDEAU, JULES (1843–1895) (Ragomot 1895, Bull. Soc. ent. Fr. 64: ccxx.)

CROITES SKIPPER *Croitana croites* (Hewitson) [Lepidoptera: Hesperiidae]. (Australia).

CRONETON® See Ethiofencarb.

CROOK Noun. (Old Norse, *krokr* = hook, bend. Pl., Hooks.) The hook or recurved apex of the Antenna in Hesperidae.

CROP Noun. (Middle English, *croppe* = craw. Pl., Crops.) The dilated portion of the Foregut positioned posterior of the Oesophagus and anterior of the Proventriculus. Some insects (*e.g.* Mecoptera) lack a Crop. Typically the Crop forms a long, slender tube with a bladder-like swelling at the posterior end; in some fluid-feeding Diptera a Diverticulum may branch from the crop that serves as a reservoir. The Crop receives and holds food before its passage to the midgut. The inner wall is often convoluted to accommodate the storage of food. The Crop of the honey ant *Myrmecocystus mexicanus* Wesmael becomes so distended that a replete worker cannot walk. Distended workers become repositories for sugars and lipids for other individuals in the colony. The crop is lined with a thick Cuticular Intima that is generally thought to inhibit digestion. However, in some insects digestion takes place in the Crop when digestive enzymes are passed rearward from the mouth or regurgitated from the midgut. In carabid beetles preoral digestion involves the storage of digestive enzymes in the Crop for periodic release. Enzymes such as trypsin and chymotrypsin are produced in the midgut and emitted to preorally digest protein (*e.g.* caterpillars, slugs). The predigested material is imbibed and later stored in the Crop Syn: Gizzard; Ingluvies; Jabot. See Alimentary System. Cf. Oesophagus, Proventriculus. Rel. Digestion.

CROP MIRID *Sidnia kinbergi* (Stal) [Hemiptera: Miridae]. (Australia).

CROP ROTATION The agricultural practice of seasonally alternating the type of crop planted at a particular location.

CROPP, LUDWIG CHRISTIAN (1718–1796) (Weidner 1967, Abh. Verh. Naturw. Ver. Hamburg Suppl. 9: 39.)

CROSBY, CYRUS RICHARD (1879–1937) (Osborn 1946, *Fragments of Entomological History*. Pt. II. 232 pp. (66–67), Columbus, Ohio.)

CROSS Noun. (Middle English, *crois* > Anglo Saxon, *cros* > Latin, *crux*. Pl., Crosses.) 1. The intersection of two linear, structures of similar form and composition. 2. A hybrid; the offspring of two differing forms.

CROSS NERVE Suboesophageal Commissure.

CROSS, EDWARD WINSLOW (1875–1899) (Anon. 1899, Miscnea ent. 7: 88.)

CROSS, HOWARD BENJAMIN (1889–1921) (Anon. 1922, Am. J. trop. Med. 2: 177.)

CROSS, WILLIAM H (1928–1984) American collector and student of geographical variation.

CROSS, WILLIAM JOHN (1834–1907) (Anon. 1907, Entomologist 40: 96.)

CROSSFIRE® See Chlorpyrifos; Resmethrin.

CROSS-STRIPED CABBAGE WORM *Evergestis rimosalis* (Guenée) [Lepidoptera: Pyralidae].

CROSSVEIN Noun. (Latin, *crux* = cross. Pl., Crossveins.) Insect wing: Typically short veins between the longitudinal veins and their branches. Crossveins typically connect longitudinal veins or project at an angle from longitudinal veins. Crossveins apparently do not contain Tracheae. When an apparent Crossvein (based on position) contains Tracheae, then this vein is regarded as a longitudinal vein that has changed direction. Crossveins show subtle forms of structural diversity. Crossveins may be solid and elliptical or circular or annular. Solid cross veins may form junctions with longitudinal veins in corrugated wings such that the connections form angular brackets or strengthening members. Circular crossveins may connect longitudinal veins at points distant from other points of crossvein contact and in areas of membrane that are not corrugated. Annular crossveins are flexible and occur in areas of wing flexion or folding. Crossveins are numerous in net-veined wings but other insects display relatively few crossveins. Crossveins of the generalized insect wing are variable in position and number. A Humeral Vein connects Costa and Subcosta. A Radial Crossvein connects first branch of Radius (R1) with the anteriormost second branch of Radial Sector (R2). A Sectorial Crossvein connects two medial branches of Radial Sector (R3, R4). A Medial Crossvein connects secondary branches of Median, M2 and M3. A Medio-cubital Crossvein connects posteriormost branch of Median with anteriormost branch of Cubital. See Vein; Wing Venation. Cf. Longitudinal Vein.

CROTALARIA MOTH *Utetheisa lotrix* (Cramer) [Lepidoptera: Arctiidae]. (Australia).

CROTALARIA PODBORER *Argina astrea* (Drury) [Lepidoptera: Arctiidae]. (Australia).

CROTCH, GEORGE ROBERT (1841–1874) (Musgrave 1932, *A Bibliography of Australian Entomology 1775–1930.* 380 pp. (58, bibliogr.), Sydney.)

CROTCH, WILLIAM DUPPA (–1903) (Anon. 1903, Entomol. mon. Mag. 39: 256.)

CROTON BUG See German Cockroach.

CROTON CATERPILLAR *Achaea janata* (Linnaeus) [Lepidoptera: Noctuidae].

CROTON MUSSEL SCALE *Lepidosaphes tokionis* (Kuwana) [Hemiptera: Diaspididae].

CROWLEY, PHILIP (1837–1900) (Anon. 1905, Entomol. mon. Mag. 37: 49.)

CROWN Noun. (Latin, *corona* = crown. Pl., Crowns.) 1. The uppermost portion of the head in Lepidoptera. See Coronet; Corona. Cf. Vertex. 2. Plants: The shortened stem with leaves and axillary buds.

CRUCIATE Adj. (Latin, *crux* = cross; *-atus* = adjectival suffix.) 1. Shaped as a cross. A term applied to wings in which the inner margins lie one over the other, or to incumbent wings that overlie the apex. 2. Diptera: Applied to bristles when they cross in direction. 3. Plants: Cross-shaped leaves or petals. Alt. Cruciatus.

CRUCIATE BRISTLES Diptera: A pair of bristles on the lower part of the Frontal Vitta that are directed inward and forward (Comstock).

CRUCIATO-COMPLICATUS Adj. (Latin, *crux* = cross.) Folded crosswise; incumbent when the inner margins overlap. A term not well distinguished from cruciate. See Cruciate. Rel. Shape.

CRUCIFORM Adj. (Latin, *crux* = cross; *forma* = shape.) Cross-shaped.

CRUICKSHANK, TERTIA SILVIA (1866–1905) (Anon. 1905, Can. Ent. 37: 196.)

CRUISE (Latin, *crux* = cross. Cruises.) 1. Ecology: Jargon used to describe travelling around to count or observe larger animals. 2. The movement of organisms in a prescribed or predictable pattern. 3. A patrol.

CRUISER *Vindula arsinoe ada* (Butler) [Lepidoptera: Nymphalidae]. (Australia).

CRUMENA Noun. (Latin, *crumena* = purse, bag. Pl., Crumenae.) 1. Some Hemiptera: Mandibular and Maxillary Bristles that are much longer than the Beak (Rostrum). 2. A long internal pouch in the head which, in some Species, extends backward into the Thorax and contains mouthpart bristles when they are retracted.

CRUNCH® See Carbaryl.

CRURA CEREBRI Two large cords that connect the Supra- and Suboesophageal Ganglia.

CRURAL Adj. (Latin, *crus* = leg; *-alis* = belonging to, pertaining to.) Descriptive of or pertaining to a leg. See Leg.

CRUS Noun. (Latin, *crura* = leg, shank. Pl., Crura.) A leg, thigh or leg-like structure.

CRUSADE® See Fonofos.

CRUSADER BUG *Mictus profana* (Fabricius) [Hemiptera: Coreidae]: A pest of eucalyptus, cassias, citrus, wattles and ornamental plants in Australia. Adult ca 25 mm long, Antenna predominantly dark-coloured with apical segments yellow and body greyish-brown with cream-coloured 'X' shaped mark on dorsum. CB overwinters as an adult. Eggs are laid in rafts or rows on foliage, fruit or twigs. Nymphal stage completes five instars; first instar with red Abdomen, subsequent instars with orange spots on dorsum of Abdomen and sides of Thorax. CB life cycle requires ca 8 weeks; Species is multivoltine with 3–4 generations per year. See Coreidae.

CRUSTACEA Noun. (Latin, *crusta* = crust; the hard surface of a body or shell.) The Subphylum of the Phylum Arthropoda containing crabs and lobsters; predominantly marine arthropods with more than 44,000 Species. See Arthropoda.

CRUSTACEOUS Adj. (Latin, *crusta* = crust; *-aceus* = of or pertaining to.) 1. Descriptive of structure that is hard, such as the shell of a crab. 2. With the form or physical characteristics of Crustacea.

CRUTTWELL, CHARLES THOMAS (1847–1911) (Morice 1912, Proc. ent. Soc. Lond. 1912: cxxii.)

CRUTTWELL, GEORGE HENRY WILSON (1891–1959) (Newman 1970, Proc. Trans. Brit. ent. nat. Hist. Soc. 3: 50.)

CRUZ, JOSE DA COSTA (1894–1940) (Pacheco 1941, Mems. Inst. Oswaldo Cruz 36: xxv-xxxi, bibliogr.)

CRUZ, OSWALDO GONÇALVES (1872–1917) Brazilian Medical Entomologist who worked on Yellow Fever eradication. The Oswaldo Cruz Institute in Rio de Janerio is named in his honour. (Howard 1930, Smithson. Misc. Colln. 84: 425, 489.)

CRYOGENIC Adj. (Greek, *kryos* = icy cold; *genes* = producing; *-ic* = characterized by.) Pertaining to extremely cold temperatures and their effect on physical properties of matter. Rel. Cryophilic.

CRYOLITE Noun. (Greek, *kryos* = icy cold.) Sodium Aluminofluoride. An inorganic compound ($Na_3Al_1F_6$) that is mined in Greenland. Cryolite is manufactured to function as an acaricide, contact insecticide or stomach poison. Applied to some crops as a dust to control some Lepidoptera, Coleoptera and Thysanoptera. Some phytotoxicity and toxicity to bees has been observed; compound is not used with lime. Trade-names include: Kryocide® and Prokil®. See Inorganic Insecticide.

CRYOPHILIC Adj. (Greek, *kryos* = icy cold; *philos* = loving; *-ic* = of the nature of.) Descriptive of organisms or tissues that are tolerant of extremely cold temperatures. Cf. Cryostatic; Thermophilic.

CRYOPRESERVATION Noun. (Greek, *kryos* = icy cold. Pl., Cryopreservations.) Tissue Culture: The maintenance of cells under extremely cold conditions, such as liquid nitrogen.

CRYOPRESERVATIVE Noun. (Greek, *kryos* = icy cold. Pl., Cryopreservatives.) Compounds such as glycerol and dimethyl sulphoxide (DMSO) that prevent or attenuate cell damage during the freezing and thawing phase of cryopreservation.

CRYOSTAT Noun. (Greek, *kryos* = icy cold; *stasis* = a standing still. Pl., Cryostats.) An apparatus that maintains cold temperatures (< 0°C). Cf. Thermostat.

CRYOSTATIC Adj. (Greek, *kryos* = icy cold; *stasis* = a standing still; *-ic* = of the nature of.) 1. Pertaining to low temperatures. 2. Resultant stoppage of bodily functions due to exposure to cold temperatures. Cf. Cryophilic.

CRYPSIS Noun. (Greek, *kryptos* = hidden. Pl., Crypses.) 1. The phenomenon or capacity for concealment or being hidden within the environment. 2. A type of protective resemblance in which an insect imitates a plant part (leaf, twig, bud), an inanimate object (rock, stone, sand) or animal product (excrement) to avoid detection and predation. Crypsis involving plants is common among insects (*e.g.* Phyllidae 'leaf insects' resemble leaves; Geometridae larvae resemble bark of tree; Phasmidae resemble twigs of trees). Trichoptera larvae in streams have body coloration that promotes concealment and reduces predation; psephenid water beetle (water penny) pupae resemble their distasteful larval stage. Geometrid and other moth larvae resemble bird excrement in shape, size and colour. Resemblance can be as an adult (with wings open imitating 'splashed' dropping or with wings closed around the body if excrement is cylindrical). Body size does not seem to reduce effectiveness of Crypsis. Covering the body with objects is another type of Crypsis (*e.g.* trash collection by tingids, chrysopid larva covers itself with mealybug wax). See Coloration; Mimicry. Cf. Countershading; Disruptive Coloration; Environmental Melanism; Industrial Melanism; Thanatosis. Rel. Pigmentary Colour; Structural Colour.

CRYPT Noun. (Greek, *kryptos* = hidden. Pl., Crypts.) 1. A glandular tube. 2. Minute secretory follicles or cavities; specifically, large gland-like structures between epithelial cells in the chylific ventricle.

CRYPTIC Adj. (Greek, *kryptos* = hidden; *-ic* = characterized by.) 1. Pertaining to a hidden or concealed condition, particularly in relation to protective coloration or camouflage. See Cryptic Coloration. 2. Pertaining to organisms that are very similar in anatomy and/or biology but reproductively isolated. See Cryptic Species. 3. Pertaining to structure, process or action that is concealed or hidden. Alt. Cryptical.

CRYPTIC COLORATION Coloration patterns of animals that make their outline form, shape or body pattern less apparent, and thereby confering a degree of protection from predation. See Coloration. Cf. Advancing Coloration; Alluring Coloration; Anticryptic Coloration; Apetetic Coloration; Aposematic Coloration; Combination Coloration; Diffraction Colour; Directive Coloration; Disruptive Coloration; Epigamic Coloration; Episematic Coloration; Procryptic Coloration; Protective Coloration; Pseudepisematic Coloration; Pseudoaposematic Coloration; Scattering Colour; Seasonal Coloration; Sematic Coloration. Rel. Crypsis; Mimicry.

CRYPTIC COLOURS Pigmentary colours of the insect Integument or animal body that protect the organism from predation. See Crypsis.

CRYPTIC SPECIES Reproductively distinct Species that show little or no outward morphological differences, and thus are difficult to distinguish. Cf. Sibling Species. Rel. Biological Species Concept.

CRYPTO- Greek prefix meaning 'hidden.'

CRYPTOCERATA Plural Noun. A division of the Heteroptera, the members of which display small Antennae that are concealed in a groove under the head and typically shorter than the head. Syn. Adelocerata. See Heteroptera. Cf. Gymnocerata.

CRYPTOCERCIDAE Plural Noun. Brown-hooded Cockroaches. A small Family of cockroaches; one described Species occurs in North America. See Blattaria.

CRYPTOCHAETIDAE Plural Noun. A small Family of acalypterate Diptera assigned to Superfamily Ephydroidea, and including ca 25 Species known mainly from the Ethiopian and Australasian regions. Adult small, stout-bodied, metallic bluish in colour and the first Flagellomere is elongate and without an Arista; Postvertical Bristles convergent and apical pair of Scutellar Bristles not errect. Larvae develop as internal parasites of Margarodidae. Australian Species (*Cryptochaetum iceryae, C. monophlebi*) are important control agents of Cottony Cushion Scale, *Icerya purchasi* Maskell. See Ephydroidea.

CRYPTOGASTRAN Adj. (Greek, *kryptos* = hidden; *gaster* = stomach.) 1. Coleoptera: A type of Abdomen in which the second Sternum is membranous and concealed in the hind Coxal cavity. The third abdominal Sternum is the first continuous Sternum across the Abdomen. The condition is typical of Coleoptera such as Curculionidae. Cf. Haplogastran, Hologastran. 2. With the venter or belly covered or concealed.

CRYPTOGNOMAE Plural Noun. A class of Chelicerata comprised of Anactinotrichida, Ricinulei and the extinct Architarbi.

CRYPTONEPHRIDIUM Noun. (Greek, *kryptos* = hidden; *nephros* = kidney; *-idion* = diminutive. Pl., Cryptonephridia.) The distal surfaces of Malpighian Tubules that envelop the Rectum. This system is common in Coleoptera and Lepidoptera larvae that occur in xeric haitats. Water and salts are taken from the hindgut lumen via rectal epithelium cells. See Excretion.

CRYPTOPENTAMERA Noun. (Greek, *kryptos* = hidden; *pente* = five; *meros* = part. Pl., Cryptopentamerae.) Legs with five small, concealed Tarsomeres.

CRYPTOPHAGIDAE Plural Noun. Silken Fungus Beetles. A small Family of polyphagous Coleoptera, the Species of which feed upon fungus. Adults 1–4 mm long with body pubescent. Individuals occur under bark, on flowers and within bird nests. Species of *Cryptophagus* and *Henoticus* are regarded as pests of stored products. See Coleoptera.

CRYPTOPLEURY Noun. (Greek, *kryptos* = hidden; *pleuron* = side. Pl., Cryptopleuries.) Hymenoptera: A condition in which the Propleuron is concealed by the lateral portion of the Pronotum. Rel. Cryptic.

CRYPTOTETRAMERA Noun. (Greek, *kryptos* = hidden; *tetras* = four; *meros* = part.) Legs with four small, concealed Tarsomeres. See Leg.

CRYPTOTHORAX Noun. (Greek, *kryptos* = hidden; *thorax* = breastplate. Pl., Cryptothoraxes.) An hypothetical thoracic ring between the Mesothorax and Metathorax. See Thorax.

CRYPTOZOIC SCHIZOGONY Medical Entomology: Product of first exoerythrocytic EE-schizonts from the sporozoites that may produce a secondary primary generation. (Huff & Coulston 1944, J. Inf. Dis. 75: 231–249.) See Schizogony. Cf. Exoerythrocytic Schizogony; Metacryptozoic Schizogony.

CRYPTOZOIC Adj. (Greek, *kryptos* = hidden; *zoon* = animal; *-ic* = characterized by.) Pertaining to insects that live in concealed habitats, such as within the soil, beneath bark or under stones. See Habitat. Cf. Hypogaeic; Phanerozoic.

CRYSTAL CELL See Oenocyte.

CRYSTALINE® A registered biopesticide derived from *Bacillus thuringiensis* var. *kurstaki*. See *Bacillus thuringiensis*.

CRYSTALLINE Adj. (Greek, *krystallinos* = crystalline.) Descriptive of structure that is transparent; crystal-like in physical properties (hard and clear). Cf. Hyaline; Opaque.

CRYSTALLINE BODY See Crystalline Cone.

CRYSTALLINE CONE A hard, transparent, refractive structure in euconic eyes. CC is secreted by a group of cells beneath the corneagenous layer of the Cornea. Syn. Crystalline Body; Vitreous Body. See Compound Eye.

CRYSTALLINE HUMOUR The crystalline lens of the eye. See Crystalline Body; Crystalline Cone.

CTENIDAE Plural Noun. Wandering Spiders. A Family of running spiders, mostly tropical in distribution, and confined to the southern states in North America, present throughout the Neotropics and in Australasia. Species in this Family do not form webs, but walk over vegetation or ground searching for prey. Neotropical Species can be very large and many are accidentally introduced into other areas on exported banana bunches. Some Species have highly potent toxin; *Phoneutria fera* is considered one of the most dangerous Species in South America. Members of this Family are similar to Clubionidae, but have three rows of eyes.

CTENIDIUM Noun. (Greek, *ktenos* = comb; *-idion* = diminutive. Pl., Ctenidia.) A comb-like row of short non-innervated spines (bristles) on an insect's body. Ctenidia are most evident on parasitic insects including Diptera (Polyctenidae, Nycteribidae), Coleoptera (Platypsyllidae) and Siphonaptera (most Families). Ctenidia occur over the body of Polyctendidae, but only on the Prothorax and sometimes the Metathorax of Siphonaptera. The rearward orientation of Ctenidia is believed to facilitate movement among hairs on the host's body. The comb-like arrangement of spines in the comb presumably acts as a holdfast by lodging hairs between adjacent spines during the act of grooming by the host. See Projection. Cf. Acantha; Microtrichia; Spine. Rel. Sensillum; Seta.

CTENIZIDAE Plural Noun. Trapdoor spiders. A Family of mygalomorph spiders whose Species excavate deep burrows, line them with silk and use them as retreats. The burrow is topped with a hinged door; prey are ambushed via sensing the vibrations of passing, ground-dwelling insects. Members of this Family have developed comb-like rakes of stiff spines on their Chelicerae used for digging. The burrow is waterproofed with a mixture of saliva and mud to which a lining of silk is applied.

CTENOIDIOBOTHRIUM Noun. (Greek, *ktenos* = comb. Pl., Ctenoidiobothria.) Psocoptera: Long Setae ('bristles') that are arranged in rows along the Tibia. See Tibia. Cf. Bristle; Seta; Trichobothrium.

CTENOPLECTIDAE Plural Noun. A primitive Family of long-tongued bees (Apoidea), presently containing two Genera. See Melittidae (Michener 1980. Zool. J. Linn. Soc. 69: 183–203.) See Apoidea.

CUBAN COCKROACH *Panchlora nivea* (Linnaeus) [Blattaria: Blaberidae].

CUBAN LAUREL THRIPS *Gynaikothrips ficorum* (Marchal) [Thysanoptera: Phlaeothripidae].

CUBE ROOT® See Rotenone.

CUBEROL® See Rotenone.

CUBITAL Adj. (Latin, *cubitalis* = of elbow; *-alis* = pertaining to.) Refering or belonging to the Cubitus. See Wing Venation.

CUBITAL AREA The surface of the wing that lies between the main stem of the Cubitus and the Anal Vein. CA is bounded proximally by the anal crossing (Comstock). Cubtial Areas include all cells that are bounded anteriorly by the Cubitus or its branches (Comstock). Diptera: Radial 3 (Comstock), third Posterior Cell (Loew). Hymenoptera: Radial 3, 4, and 5 (Comstock). See Wing.

CUBITAL CELLULE Insect wing: The cell between the Radial Cell and the vein originating near the extremity of the Cubitus (Jardine). Syn. Cubital Cell.

CUBITAL FORKS Branching or points of separation of the branches of the Cubitus.

CUBITAL NERVE The Cubital Vein. Syn. Cubital Nervure. See Cubitus.

CUBITAL PECTEN In the hindwings of some Tineoidea and Pyraloidea: A comb of stiff, straight Setae on the upper surface that are directed upward and slightly backward on vein Cu2 (Tillyard). See Wing Venation.

CUBITAL SUPPLEMENT Odonata: A midrib-like wing vein that divides the Cubito-Anal loop longitudinally (Tillyard).

CUBITO-ANAL Adj. (Latin, *cubitus* = elbow; *anus*; *-alis* = pertaining to.) Pertaining to the Cubitus and the Anal Vein of the insect wing. See Wing Venation.

CUBITO-ANAL CROSSVEINS Odonata: Wing veins that connect the Cubitus and first Anal Veins (Garman). See Vein; Wing Venation.

CUBITO-ANAL EXCISION In many insects, a notch in the margin of the wing at the point where the anal and preanal areas join (Comstock). See Wing.

CUBITO-ANAL FOLD Insect wings: The continuation of the Cubito-Anal Sulcus, that usually extends to the margin of the wing (Comstock). See Wing.

CUBITO-ANAL LOOP Wings of some Odonata: A loop developed between veins A2 and Cu2; the foot-shaped loop of Needham (Comstock). See Wing Venation.

CUBITO-ANAL SULCUS The deep channel between the anterior and posterior tuberosities of the insect wing (Comstock). See Sulcus; Wing.

CUBITUS Noun. (Latin, *cubitalus* = of elbow. Cubiti.) 1. The Cubital Vein of the insect wing. 2. The fifth longitudinal vein of the insect wing, extending from the base and usually two-branched before reaching the outer margin. 3. Orthoptera: The Internomedian and Ulnar Veins. 4. Neuroptera: The vein positioned between the Media and Anal Veins. 5. Heteroptera: See Brachium.

CUBOIDAL Adj. (Greek, *kyboeides* = cube-like; *-alis* = characterized by.) Cube-shaped; descriptive of structure that resembles a cube.

CUBOR® See Rotenone.

CUCKOO BEES *Coelioxys* spp., *Inquilina* spp., *Nomada* spp., *Thyreus* spp. [Hymenoptera: Anthophoridae]. (Australia). *Sphecodes* spp. [Hymenoptera: Halictidae]. (Australia).

CUCKOO ROLLERS See Leptosomatidae.

CUCKOO SPIT Cercopidae: Liquid in the form of bubbles produced as an anal secretion of immature and used to conceal the body. See Spittle. Cf. Saliva.

CUCKOO WASPS See Chrysididae.

CUCKOOS See Cuculidae.

CUCUJIDAE Plural Noun. Flat Bark Beetles. A cosmopolitan Family of polyphagous Coleoptera assigned to Superfamily Cucujoidea and which includes about 1,000 nominal Species. Adult body to 25 mm long, elongate, parallel sided and dorsoventrally compressed; head large with eyes relatively small; Antenna with 11 segments, usually long and moniliform or nearly so; hind Coxae widely separated, tarsal formula usually 5-5-5, rarely 5-5-4 in male; Tarsomeres not lobed. Abdomen with five Ventrites. Larval head prognathous with up to six Stemmata on each side of head; legs usually with five segments, rarely four; two tarsal claws on each Pretarsus; Urogomphi present or absent. Cucujids are phytophagous with several Species known as pests of stored products, especially *Cryptolestes*. Incl. Passandridae. See Cucujoidea.

CUCUJIFORMIA Coleoptera: A monophyletic series in the Suborder Polyphaga comprised of six Superfamilies, Lymexyloidea, Cleroidea, Cucujoidea, Tenebrionoidea, Chrysomeloidea, and Curculionoidea. See Coleoptera.

CUCUJOIDEA Noun. A Superfamily of polyphagous Coleoptera. Larvae with Pygopod-like tenth abdominal segment. Adult male with ring-type Aedeagus; female lacks 5-5-4 tarsal formula; hindwing may have more than four veins behind MP. Included Families: Biphyllidae, Boganiidae, Bothrideridae, Cavognathidae, Cerylonidae, Coccinellidae, Corylophidae, Cryptophagidae, Cucujidae, Discolomidae, Endomychidae, Erotylidae, Hobartiidae, Laemophloeidae, Lamingtoniidae, Languriidae, Lathridiidae, Nitidulidae, Phalacridae, Phloeostichidae, Propalticidae, Protocucujidae, Rhizophagidae, Silvanidae and Sphindidae. See Coleoptera.

CUCULLATE Adj. (Latin, *cucullus* = hood; *-atus* = adjectival suffix.) 1. Hood-shaped; hood-like. Descriptive of structure with a hood. 2. Descriptive of the Prothorax projecting over the posterior margin of the head in some Orthoptera. Alt. Cucullated; Cucullatus. Cf. Cap; Hood.

CUCULLUS Noun. (Latin, *cucullus* = hood. Pl., Cucullae.) 1. A hood; a hood-shaped covering or structure. 2. Genitalia of male Lepidoptera: The terminal part of the Harpe (Klots). See Capillitium.

CUCUMBER BEETLES See Chrysomelidae.

CUCUMBER FLY *Dacus cucumis* French [Diptera: Tephritidae]: In Australia, a serious pest of cucumber, tomato and papaya grown along the eastern coast of Queensland. See Tephritidae.

CUCUMBER MOSIAC A viral disease of cucumbers and other curcurbits that is transmitted by several Species of aphids and the Cucumber Beetle.

CUCUMBER MOTH *Diaphania indica* (Saunders) [Lepidoptera: Pyralidae]. (Australia).

CUCUMBITATE Adj. (*-atus* = adjectival suffix.)

Shaped as a melon. A term sometimes applied to the shape of insect eggs.

CUCUMIFORM Adj. (Latin, *cucumer; forma* = shape.) Cucumber-shaped.

CUCURBIT LADYBIRD *Epilachna cucurbitae* Richards (Coleoptera: Coccinellidae]. (Australia).

CUCURBIT LONGICORN *Apomecyna saltator* (Fabricius) [Coleoptera: Cerambycidae].

CUCURBIT MIDGE *Prodiplosis citrulli* (Felt) [Diptera: Cecidomyiidae].

CUCURBIT SHIELD BUG *Megymenum insulare* Westwood [Hemiptera: Dinidoridae]: In Australia, a minor pest of cucurbits, mainly pumpkins; CSB feeds on stems, leaf stalks and young fruits. Syn. Pumpkin Bug. *Megymenum affine* Boisduval. (South Australia). See Dinidoridae.

CUCURBIT STEMBORER *Apomecyna histrio* (Fabricius) [Coleoptera: Cerambycidae]. (Australia).

CUCURBIT WILT A bacterial disease of curcurbits caused by *Erwinia tracheiphila* transmitted in the excrement of cucumber beetles.

CUDGEL® See Fonofos.

CUENOT, LUCIEN (1866–1937) (Rostand 1951, Genetics, Princeton 42: 1–6.)

CUILLERON See Alula.

CUISINE, HENRY DE LA (1827–1891) (Grouvelle 1891, Bull. Soc. ent. Fr. 1891: c.)

CULEX Noun. (Latin, *culex* = gnat.) The type-genus of the Family Culicidae. A large, cosmopolitan Genus of mosquitoes. See Culicidae. Cf. *Aedes*; *Anopheles*.

CULICIDAE Plural Noun. Mosquitoes. A numerically large (ca 3,000 Species), cosmopolitan Family of nematocerous Diptera assigned to Superfamily Culicoidea. Higher classiification of mosquitoes is variable and some classifications may recognize several Subfamilies and other Families of Nematocera (*e.g.* Chaoboridae, Dixidae). Adult mouthparts form a Proboscis from an elongate Labium surrounding the Stylets; Proboscis directed forward and as long as head and Thorax; male Antenna plumose (sexually dimorphic); wing with scales along veins, scales forming finge along posterior margin of wing. Adults are crepuscular or nocturnal and engage in aerial copulation. Oviposition occurs in water or moist habitat. Eggs usually are deposited singly on surface of standing water. Eggs hatch within 18 hours and neonate larvae immediately begin foraging for suitable prey. All larvae are aquatic; larvae and pupae are active swimmers; a few Species with larvae that are predaceous upon zooplankton and small crustaceans and/or cannabalistic in pools, ponds, tree holes or crab holes. Larvae of all Toxorhynchitinae are obligate predators of other mosquito larvae. Adult *Toxorhynchites* are large, attractive flies with metallic coloration and mainly tropical distribution; adults feed only on plant juices. Most adult mosquito females are blood-feeders of vertebrates; males do not take blood and many do not feed. Females vector Malaria, Filariasis, Dengue, Myxomatosis and other diseases. Culicids are the preferred and almost exclusive prey of other mosquito larvae. See Chaoboridae; Dixidae. Cf. Asian Tiger Mosquito; Brown House Mosquito; Salt-Marsh Mosquito; Yellow Fever Mosquito. See Nematocera.

CULICOMORPHA A Division of nematocerous Diptera including Ceratopogonidae, (Heleidae), Chaoboridae (Corethridae), Chironomidae (Tendipedidae), Culicidae, Dixidae, Simuliidae (Melusinidae) and Thaumaleidae (Orphnephilidae). See Nematocera.

CULMEN Noun. (Latin, *culmen* = summit. Pl., Culmina; Culmens.) 1. The culmination of a process. 2. The longitudinal Carina of a caterpillar. 3. The dorsal ridge of a bird's bill. Cf. Carina.

CULOT, JULES (1861–1933) (Pictet 1934, Mitt. schweiz. ent. Ges. 16: 129–139, bibliogr.)

CULTELLATE Adj. (Latin, *cultellus* = little knife; *-atus* = adjectival suffix.) Knife-like; descriptive of structure that has a sharp cutting edge and an apical point. Cf. Aciculate.

CULTELLUS Noun. (Latin, *culter* = plowshare, knife; Pl., Cultelli.) 1. A piercing, blade-like mouthpart of blood-sucking flies. 2. The Mandibles of some authors. Cf. Stylet.

CULTRATE Adj. (Latin, *cultratus* = knife-shaped; *-atus* = adjectival suffix.) Shaped as a pruning knife. Alt. Cultriform.

CULTURAL CONTROL A principal-approach or pest-management strategy that involves use of production practices, such as crop spacing, crop rotation, planting and harvest dates, irrigation, pruning and tillage operations, individually or in combination to disrupt a pest's life cycle. See Cultural Management. Cf. Biological Control; Chemical Control; Integrated Pest Management; Natural Control; Regulatory Control.

CULTURAL ENTOMOLOGY The subdiscipline of Entomology that concerns insects as they affect the lifestyle or habits of groups of people. See Entomology.

CULTURAL MANAGEMENT Manipulation of a cropping environment to reduce pest increase and damage. Syn. Cultural Control; Ecological Management.

CULTURE Noun. (Latin, *cultura* = cultivation > *colere* = to till. Pl., Cultures.) 1. The maintenance of organisms in the laboratory.

CULUS Noun. (Latin, *culus* = anus.) The orifice (opening) at the apex of the Anus. See Anus.

CUMMINGS, BRUCE FREDERIC (WILLIAM NERO PONTIUS PILATE BARBELLION) (1899–1919) (Abbott 1973, J. Modern Lit. 3(1): 45–62s.)

CUMULATE Adj. (Latin, *cumulare* = to heap up; *-atus* = adjectival suffix.) Descriptive of structures arranged in groups or heaps.

CUMULUS Noun. (Latin, *cumulus* = heap, mass. Pl., Cumuli.) A group or heap, as of cells within a developing Ovum. Cf. Tumulus.

CUNEATE Adj. (Latin, *cuneatus* = wedge; *-atus* =

adjectival suffix.) Wedge-shaped; descriptive of structure that is elongate-triangular. Term often applied to leaves with an abruptly pointed apex and tapering to the base. Alt. Cuneatus; Cuneiform. See Outline Shape.

CUNEIFORM Adj. (Latin, *cuneus* = wedge; *forma* = shape.) Wedge-shaped.

CUNEUS Noun. (Latin, *cuneus* = wedge. Pl., Cunea.) 1. Heteroptera: A small triangular area along the anterioapical margin of the Corium on the Hemelytron. Cf. Clavus; Corium; Embolium. 2. Odonata: A small triangle of the Vertex between the compound eyes.

CUNI Y MARTORELL, MIGUEL (1827–1902) (Dusmet y Alonso 1919, Boln. Soc. ent. Esp. 2: 188.)

CUNNINGHAM, ALLAN (1791–1839) (Musgrave 1932, *A Bibliography of Australian Entomology 1775–1930*. 380 pp. (59), Sydney.)

CUNNINGHAM, THOMAS H (1838–1916) (Day 1916, Proc. ent. Soc. Brit. Columb. 8: 4.)

CUP Noun. (Middle English, Anglo Saxon *cuppe* > Latin, *cupa* = cask, tub. Pl., Cups.) 1. Wings of Nymphalidae: An odoriferous opening provided with a covering membrane pierced in the centre by a minute pore (Imms). 3. A bowl-shaped vessel designed to hold liquid, often steep sided with a flat bottom. 3. A broad, shallow concave surface; a socket. See Acetabulum; Cotyla.

CUP MOTHS See Limacodidae.

CUPBOARD SPIDER *Steatoda grossa* (C.L. Koch) [Araneida: Theridiidae]. (Australia).

CUPEDIDAE Plural Noun. Reticulated Beetles. A numerically small, widely distributed Family of archostematous Coleoptera. Adult 6–20 mm long, elongate, parallel-sided and moderately to strongly flattened. Antenna inserted dorsally, filiform with 11 segments and longer than head and Prothorax combined; Mandibles with large apical tooth; Maxilla and Labium with elongate, setose apical process. Prosternal process separating fore Coxae; fore Coxa transverse with Trochantin expanded; frontal cavity open and middle Coxae contiguous; Elytra often reticulate; tarsal formula 5-5-5, segment four lobate; Abdomen with five Ventrites. Adults slow-moving. Larvae are eruciform and bore into dead wood attacked by fungi. Syn. Cupesidae; Cupidae. See Archostemata; Coleoptera.

CUPEDOIDEA Syn. Archostemata.

CUPIDINIDAE See Lycaenidae.

CUPINCIDA® See Heptachlor.

CUPOLA ORGAN See Sensillum Campaniformium.

CUPREA ANT-BLUE *Acrodipsas cuprea* (Sands) [Lepidoptera: Lycaenidae]. (Australia).

CUPREOUS (Latin, *cupreus* = copper; *-osus* = possessing the qualities of.) Coppery; Metallic copper red. Alt. Cupreus.

CUPULE Noun. (Latin, *cupula* = little tub. Pl., Cupules.) 1. A cup-shaped organ. 2. A sucker-like process covering the ventral surface of the Tarsi in male Dytiscidae.

CUPULIFEROUS Adj. (Latin, *cupula* = little tub; *ferous* = bearing; *-osus* = possessing the qualities of.) Pertaining to structure with cupules or small cup-like objects.

CUPULIFORM Adj. (Latin, *cupula* = little tub; *forma* = shape.) Cup-shaped; resembling a small cup. Syn. Cyathiform.

CURACRON® See Profenofos.

CURATERR® See Carbofuran.

CURCULIO BEETLE *Otiorhynchus cribricollis* Gyllenhal [Coleoptera: Curculionidae]. (Australia).

CURCULIONIDAE Plural Noun. Snout Beetles; Snout Weevils; Weevils. A Family of polyphagous Coleoptera that forms the largest Family of insects with 45,000 nominal Species and an estimated 85,000 Species (O'Brien & Wimber 1979, Coleop. Bull. 33: 151–166.) Adult 1–60 mm long with body robust, convex, heavily sclerotized and often with Setae or scales. Antenna with 7–11 segments; tarsal formula typically 4-4-4 (pseudotetramerous), rarely 5-5-5. Abdomen with five Ventrites, rarely six Ventrites. Larvae with 0–2 Stemmata; Antenna with one segment; legs absent; Urotergites absent. See Curculionoidea; Polyphaga.

CURCULIONOIDEA Noun. A Superfamily of polyphagous Coleoptera. Adult with head elongate to form a Rostrum, Labrum fused and Mandible reduced; Galea and Lacinea fused; fore coxal and middle coxal cavities closed; Antenna typically clubbed; Rostrum forms a Scrobe for antennal Scape. Curculionid larva typically lacks legs. Most Species are phytophagous; a few Species are mycophagous. Included Families: Anthribidae, Apionidae, Attelabidae, Belidae, Brenthidae, Curculionidae, Nemonychidae, Oxycorynidae and Proterhinidae. See Coleoptera.

CURL GRUB Common name for the larval stage of *Heteronychus arator* (Fabricius) [Coleoptera: Scarabaeidae] in Australia. See African Black Beetle.

CURL, LEO FOWLER (1899–1971) (Spears 1972, J. Econ. Ent. 65: 1219.)

CURLE, RICHARD H P (1884–1968) (Kennedy 1969, Proc. R. ent. Soc. Lond. (C) 33: 54.)

CURLED ROSE SAWFLY *Allantus cinctus* (Linnaeus) [Hymenoptera: Tenthredinidae].

CURLY TOP VIRUS A viral disease of sugarbeets and other vegetables that is transmitted by the Beet Leafhopper. See Beet Leafhopper.

CURRAN, CHARLES H (1895–1972) (Cortes 1973, Revta. chilena Ent. 7: 262.)

CURRANT APHID *Cryptomyzus ribis* (Linnaeus) [Hemiptera: Aphididae]: A multivoltine pest of currant and gooseberry widespread in North America. CAs overwinter as black eggs on canes of host plant. Neonate nymphs appear during spring and feed on underside of leaves. Feeding causes leaves to curl or 'cup' and become bright red near cupping; later instar nymphs feed on underside of leaves within curl or cup. First generations are wingless and parthenogenetic; summer generation produces winged females that migrate to weeds and reproduce sexually during autumn; late autumn generation migrates back to currant and deposits eggs on cane. See Aphididae.

CURRANT BORER *Synanthedon tipuliformis* (Clerck) [Lepidoptera: Sesiidae]: A monovoltine pest of currant, gooseberry, black elder and sumac in North America, Europe and Asia. Eggs are laid on canes and larvae bore into canes to feed on pith and wood. Mature larvae are yellow and ca 12 mm long. CBs overwinter in canes as mature larvae and pupation occurs in tunnels near exit holes. See Sesiidae. Cf. Rhododendron Borer.

CURRANT BUD-MITE *Cecidophyopsis ribis* (Westwood) [Acarina: Eriophyidae]. (Australia).

CURRANT BUD-MOTH *Stathmopoda chalcotypa* Meyrick [Lepidoptera: Oecophoridae]. (Australia).

CURRANT CLEARWING MOTH *Synanthedon tipuliformis* [Lepidoptera: Sesiidae]: A serious pest of black currants on New Zealand (South Island.) Larvae feed in tunnels on pith of canes; damage through destruction of flower buds, cane-tip dieback and cane breakage. See Sesiidae.

CURRANT FRUIT FLY *Epochra canadensis* (Loew) [Diptera: Tephritidae]: A pest of currants and gooseberries in the USA and Canada. See Tephritidae.

CURRANT FRUIT-WEEVIL *Pseudanthonomus validus* Dietz [Coleoptera: Curculionidae].

CURRANT SPANWORM *Itame ribearia* (Fitch) [Lepidoptera: Geometridae].

CURRANT STEM-GIRDLER *Janus integer* (Norton) [Hymenoptera: Cephidae]. See Cephidae. Cf. Imported Currantworm.

CURRIE, ROLLA PATTESON (1875–1960) (Edmunds & Muesebeck 1961, Proc. ent. Soc. Wash. 63: 137–139.)

CURSORIA Plural Noun. Orthopteroids: The series in which the legs are long, slender and adapted for running (cockroaches, *etc.*)

CURSORIAL Adj. (Latin, *cursor* = a runner; *-alis* = characterized by.) Descriptive of legs that are adapted for running. Typically, cursorial legs are long and tapered. See Leg. Cf. Fossorial; Gressorial; Natatorial; Raptorial; Saltatorial; Scansorial. Rel. Locomotion.

CURTAIN WEB SPIDERS See Dipluridae.

CURTIS, (RURICOLA) JOHN (1791–1862) British Entomologist and author of *Illustrations of the Genera of British Insects* and *Guide to an arrangement of British Insects* (1831). (Ordish 1974, *John Curtis and the Pioneering of Pest Control.* 121 pp. Reading, Berks.)

CURTIS, WILLIAM (1746–1799) (Stearn 1969, Introduction to facsimile edition of *History of the Brown Tail Moth.* [1–12]. London.)

CURTIS, WILLIAM PARKINSON (1878–1968) (Kennedy 1969, Proc. R. ent. Soc. Lond. (C) 33: 54.)

CURTONOTIDAE Plural Noun. A small Family of schizophorous Diptera assigned to Superfamily Ephydroidea with about 50 Species in three Genera - *Curtonotum, Axinota* and *Cyrtona.* Curtonids occur mainly in the tropical and subtropical areas of the Ethiopian, Neotropical and Oriental regions. Adults are medium-sized (4–7 mm long), pale brownish with plumose Arista and strongly-humped Thorax; two Mesoplural Bristles present; each Tibia with one preapical, dorsal bristle and longitudinal rows of Setae; wing with vein-like thickening distad from transverse section of CuA. Biology of curtonids is poorly known. One Species is known from Nearctic, *Curtonotum helvum* (Loew). Syn. Cyrtonotidae. See Ephydroidea.

CURVATE Adj. (Latin, *curvus* = curved; *-atus* = adjectival suffix.) Pertaining to structure that is curved along its primary axis or margin, but not appearing broken. Alt. Curvatus. See Shape. Cf. Fractate.

CURVINERVATE Adj. (Latin, *curvus* = curved; *nervus* = nerve; *-atus* = adjectival suffix.) Psocidae: Descriptive of wing veins that are distinctly curved.

CUSHING, EMORY C (1897–1974) (Anon. 1974, J. Econ. Ent. 67: 566.)

CUSHMAN, ROBERT ASA (1880–1957) American taxonomist (U.S.D.A.) specializing in parasitic Hymenoptera, particularly the Ichneumonoidea. (Mallis 1971, *American Entomologists.* 549 pp. (372–373), New Brunswick.)

CUSP Noun. (Latin *cuspis* = a point. Pl., Cusps.) Any pointed process, sometimes at the margin of a wing.

CUSPIDAL Adj. (Latin *cuspis* = a point; *-alis* = characterized by.) Descriptive of structure that ends in a point; pointed. See Shape. Cf. Oblate.

CUSPIDATE Adj. (Latin, *cuspidare* = to make pointed; *-atus* = adjectival suffix.) Prickly pointed. Pertaining to structure that ends in a sharp point. Structure with an acuminate point that ends in a bristle. See Shape. Cf. Acuminate.

CUSTODITE Adj. (Latin, *custos* = guard; *-ites* = inhabitant.) 1. Guarded. 2. Descriptive of a body in an envelope. Alt. Custoditus. Rel. Shroud.

CUTANEOUS Adj. (Latin, *cutis* = skin; *-osus* = possessing the qualities of.) Descriptive of or pertaining to the integument. See Integument.

CUTANEOUS LEISHMANIASIS A kind of human leishmaniasis that causes skin ulcers; known in two forms: Old World and New World. Old World Dry Cutaneous Leishmaniasis is caused by *Leishmania tropica* and vectored by *Phlebotomus* spp.; disease is anthroponotic in urban areas. Old World Wet Cutaneous Leishmaniasis is caused by *L. major* and is zoonotic in rural areas. CL produces lesions that heal over prolonged period of time. NWCL is caused by several Species of *Leishmania* and vectored by *Lutzomyia* spp. Infections can lead to Mucocutaneous Leishmaniasis. See Leishmaniasis. Oriental Sore. Cf. Visceral Leishmaniasis.

CUTANEOUS MYIASIS See Traumatic Myiasis.

CUTANEOUS RESPIRATION The diffusion of oxygen through the surface of the integument. Typical of small-bodied aquatic immature insects with a closed respiratory system (*e.g.* chironomid larvae.) See Tracheae. Cf. Tracheal Gill. Rel. Respiration.

CUTANEOUS WOHLFAHRTIA MYIASIS A type of Myiasis caused by *Wolfahrtia vigil* in Canada and northern USA. Hosts include many domesticated animals and humans. Female fly deposits larvae on or near host; larvae penetrate skin and cause boil-like swelling; larvae complete development, exit skin and pupate in soil. See Traumatic Myiasis.

CUTEREBRIDAE Plural Noun. Robust Bot-Flies. A small Family of parasitic muscoid Diptera regarded as a Subfamily of Oestridae in some classifications. Adults are bee-like, robust-bodied and covered with numerous Setae; head with deep groove on ventral surface and Frontal Suture present; Palps concealed; Scutellum projecting beyond base of Metanotum and Postscutellum not developed; Squamma large; Coxae viewed from ventral aspect contiguous or close-set. Larvae are obligatory dermal parasites of rodents, rabbits and occasionally humans. Most notable Species is Tropical Warble-Fly (Human Bot-Fly or Torsalo). See Tropical Warble-Fly; Rabbit Bot Fly. Cf. Oestridae.

CUTEX Noun. (Latin, *cutis* = skin.) The skin. See Integument.

CUTHBERTSON, ALEXANDER (1901–1942) (Townsend 1942, Revta. Ent., Rio de J. 13: 456–457.)

CUTICLE Noun. (Latin, *cuticula* = thin skin. Pl., Cuticles.) The outer covering of an insect which is formed of several non-cellular layers containing chitin, wax, proteins and secreted by the Epidermal Cells. Cuticle forms most conspicuous component of the Integument and includes ca 50% of dry weight of Integument and 25% of body dry weight. Microscopically, Cuticle is composed of several layers, including an Endocuticle, Exocuticle and Epicuticle. So-called Procuticle comprises area beneath Epicuticle and is represented by chitinous rods embedded in protein matrix. Procuticle functions in support of the Epicuticle, protects Epidermis, stores biomechanical energy and food. Biomechanical properties of Cuticle vary. Physically, Cuticle is solid, rubber-like or Arthrodial Membrane. Solid Cuticle is rigid due to tanning; rubber-like Cuticle is flexible due to resilin; Arthrodial Membrane is flexible and forms contacts between hardened sclerites. Chemically, all three types of Cuticle contain Chitin that is supplemented by various structural proteins (sclerotins), carbohydrates and lipids. See Integument. Cf. Assembly Zone; Basement Membrane; Cement Layer; Cuticulin Layer; Dermal Glands; Epidermal Cells; Endocuticle; Exocuticle; Pore Canals; Wax Layer. Rel. Cutex; Skeleton; Skin.

CUTICULAR Adj. (Latin, *cuticula* = thin skin.) Descriptive of or pertaining to the Cuticle.

CUTICULAR APPENDAGES Outgrowths of Cuticle connected with it by a membranous attachment of two kinds: Setae and spurs (Imms).

CUTICULAR COLOUR Colours contained mostly in the Epidermis, including permanent browns, blacks and yellows (Imms). See Coloration. Cf. Advancing Coloration; Alluring Coloration; Anticryptic Coloration; Apetetic Coloration; Combination Coloration; Cryptic Coloration; Directive Coloration; Disruptive Coloration; Epigamic Coloration; Episematic Coloration; Pigmentary Colour; Procryptic Coloration; Protective Coloration; Pseudepisematic Coloration; Pseudoaposematic Coloration; Scattering Colour; Seasonal Coloration; Sematic Coloration; Structural Colour; Subhypodermal Colour. Rel. Crypsis; Mimicry.

CUTICULAR HYDROCARBONS (See Ann. Rev. Ent. 27: 149–172.)

CUTICULAR NODULES Small, more-or-less conical outgrowths of the Cuticula (Comstock). See Protuberance. Cf. Acantha.

CUTICULAR PROCESSES Outgrowths that are integral parts of the substance of the Cuticle. CPs are rigidly connected with the Cuticle and lack a membranous articulation. CPs are distinct from cuticular appendages (Imms). See Cuticle; Process.

CUTICULARIZATION Noun. (Latin, *cutical* = thin skin; English, *-tion* = result of an action. Pl., Cuticularizations.) 1. Formation of Cuticle. 2. The process of transformation into Cuticle. See Cuticle.

CUTICULIN Noun. (Latin, *cutical* = thin skin. Pl., Cuticulins.) The chemical compound forming the Cuticle of insects and consisting of a mixture of waxes (Wardle).

CUTICULIN LAYER A barrier between the old and new Epicuticle. Cuticulin is abundant in Cuticle that will be tanned to form Sclerotin. The cuticulin layer prevents damage that might occur to the new Cuticle during moulting. Inner Epicuticle consists of tanned lipoproteins; Outer Epicuticle forms a trilaminar lipid membrane plus protein and sets a limit on integumental expansion after moulting. See Integument. Cf. Cement Layer; Wax Layer.

CUTLASS® A registered biopesticide derived from *Bacillus thuringiensis* strain EG 2371. See *Bacillus thuringiensis*.

CUTTER GOLD® See Cyfluthrin.

CUTWORMS Noun. Any of several Species of noctuid moths whose larvae inhabit soil during the day and feed on plants during the night. Cutworms attack numerous vegetable crops, weeds and ornamental plants. Females lay eggs on leaves or on the soil surface near the host plant. Eggs are spherical, ribbed in appearance and ca 0.45 mm high. Egg stage requires 3–6 days. Larvae are 38–50 mm long and vary from light grey to black. Larval stage lasts 25–35 days. Cutworms typically prefer seedling plants and chew through plant stems at ground level. Pupa is overwintering stage. Cutworms pass through several generations per year. See Noctuidae. Cf. Armyworms.

CUVIER, GEORGE LÉOPOLD CHRETIEN FRÉDÉRIC DAGOBERT (1768–1832) French academic, Natural Historian and Comparative Anatomist. Cuvier (1786) popularized the concept of extinction. (Nordenskiöld 1935, *History of Biology*. 629 pp. (331–343), London.)

CYAFORCE® See Hydramethylnon.

CYANE JEWEL *Hypochrysops cyane* (Waterhouse & Lyell) [Lepidoptera; Lycaenidae]. (Australia).

CYANESCENT Adj. (Greek, *kyaneos* = dark blue.) With a deep bluish tinge or shading.

CYANEUS Adj. (Greek, *kyaneos* = dark blue.) Pure dark blue; indigo blue. Alt. Cyaneous.

CYANIDE Noun. (Greek, *kyaneos* = dark blue. Pl., Cyanindes.) A compound of cyanogens that is used in solid form as a killing agent for insects. Common forms include Potassium Cyanide (KCN), Sodium Cyanide (NaCN), and Calcium Cyanide [$Ca(CN)_2$]. As a killing agent, Potassium Cyanide is preferred; Sodium Cyanide is hygroscopic; Calcium Cyanide not widely distributed.

CYANOBACTERIA Plural Noun. (Greek, *kyanos* = blue; *bakterion* = small rod. Sl., Cyanobacterium.) Microscopic organisms that contain chlorophyll A, but lack nuclei and cell inclusions; photosynthesis throughout the cell and not limited to chloroplasts; reproduce by cell fragmentation. Cyanobacteria were the first organisms capable of photosynthesis (ca 3.5 billion years ago) and are extant today. In presence of light, Cyanobacteria obtain hydrogen from H_2O and liberate oxygen, thereby created global atmosphere. Over tens of millions of years, oxygen in atmosphere increased, ozone layer became more dense and filtered ultraviolet radiation thereby making an environment suitable for metazoan life (ca 600 million years ago). Physical evidence of presence of Cyanobacteria is indicated by stromatolite formations in geological deposits. See Stromatolite.

CYANOGENIC Adj. (Greek, *kyanos* = blue; *genesis* = origin; *-ic* = consisting of.) Gas producing. A term applied to repugnatorial glands in myriapods (particularly diplopods) and some insects. See Diplopoda.

CYANOPHOS An organic-phosphate (thiophosphoric acid) compound {O,O-dimethyl O-(4-cyanophenyl) phosphorothioate} used as a foliar insecticide against armyworms, aphids, flea beetles and domiciliary pests. Compound is applied to fruit trees, cabbage, sugar beets, potatoes, soybeans, vegetables and as a public-health insecticide in some countries. Cyanophos is toxic to bees and fishes, and thus is not registered for use in USA. Trade names include Cyanox®, Cyap® and Cynock®. See Organophosphorus Insecticide.

CYANOPHYLLUM SCALE *Abgrallaspis cyanophylli* (Signoret) [Hemiptera: Diaspididae]. (Australia).

CYANOX® See Cyanophos.

CYAP® See Cyanophos.

CYATHIFORM Adj. (Latin, *cyathus* = a cup; *forma* = shape.) Obconical and concave; descriptive of organisms, organs or appendages that are cup-shaped. Syn. Cupuliform. Rel Shape.

CYATHOTHECA Noun. (Greek, *kyathos* = cup; *theke* = case. Pl., Cyathothecae.) The thoracic cover in the pupal stage. See Theca. Rel. Pupa.

CYBOLT® See Flucythrinate.

CYCLAMEN MITE *Phytonemus pallidus* (Banks) [Acarina: Tarsonemidae]: A widespread pest of cyclamen and other ornamental plants, particularly in greenhouses. Infested plants display leaves with purple areas and distorted flowers. Syn. *Tarsonemus pallidus* Banks.

CYCLE Noun. (Greek, *kyklos* = circle. Pl., Cycles.) 1. A recurrent biological act, process or phenomenon (*e.g.* migration, ovulation, respiration). 2. A cycle of development of a population (breeding cycle). 3. The movement of fluid through the body in a programmed and predictable manner (circulation). See Life Cycle. Rel. Phenology.

CYCLICAL PARTHENOGENESIS See Metagenesis.

CYCLICAL POLYMORPHISM A form of polymorphism most fully expressed in Aphididae. CP is exceedingly complex with annual cycles of generations in some Species of aphids that produce as many as 20 morphs. Some of these generations reproduce sexually; other generations reproduce asexually. Polymorphism appears under endocrine control, which in turn is influenced by photoperiod, temperature and population density. Syn. Cyclomorphosis. See Polymorphism. Cf. Caste Polymorphism; Kentromorphism.

CYCLODAN® See Endosulfan.

CYCLODEPSIPEPTIDE Noun. (Pl., Cyclodepsipeptides.) A biopesticide first isolated during 1954 as Destruxin A and B from the fungus *Metarhizium anisopliae.* Subsequently, additional related compounds have been isolated and identified as bearing insecticidal properties (*e.g.* Bassianolide, Enniatin C, Vlinomycin). See Biopesticide. Cf. Alkaloids; Azadirachtin.

CYCLODEVELOPMENTAL TRANSMISSION A form of Biological Transmission in which a pathogen undergoes developmental changes but does not multiply in the body of an arthropod vector. CT typical of helminths that use arthropods as vectors. Cf. Cyclopropagative Transmission; Propagative Transmission.

CYCLODIENE Noun. A chlorinated hydrocarbon insecticide with a carbon-based ring structure. *e.g.* Chlordane.

CYCLODIENE INSECTICIDE See Aldrin.

CYCLOID Adj. (Greek, *kykloiedes* = circular.) 1.

Circular; arranged in or progressing in circles. 2. Pertaining to organisms with concentric lines of growth and a smooth margin. Cf. Deltoid; Trochoid. Rel. Eidos; Form; Shape.

CYCLOLABIA Noun. (Greek, *kyklos* = cycle; Latin, *labium* = lips. Pl., Cyclolabiae.) Dermaptera: The shorter forceps when they are of variable length within a Species. See Forceps. Rel. Dermaptera.

CYCLOMORPHOSIS See Cyclical Polymorphism.

CYCLON® See Hydramethylnon.

CYCLOPEAN Adj. (Greek, *kyklos* = circle; *ops* = eye.) Pertaining to a single median eye. A mutated condition. Alt. Cyclopic.

CYCLOPEAN EAR An unpaired, median auditory organ of mantids. CE is positioned on ventral midline between metathoracic legs, and responds to ultrasound between 25–45 kHz with thresholds of 55–60 decibels. (Yager & Hoy 1986, Science 231: 727.)

CYCLOPOID LARVA Hymenoptera: Hypermetamorphic, endophagous, first-instar larva of some Proctotrupoidea. CL are characterized by a large swollen Cephalothorax, very large sickle-like Mandibles and a pair of bifurcate Caudal Processes. The larva resembles the nauplius larva of crustaceans (Imms). Syn: Cyclopiform Larva. See Larva. Rel. Proctotrupoidea.

CYCLOPROPAGATIVE TRANSMISSION A form of Biological Transmission in which a pathogen multiplies its numbers or undergoes developmental changes in the body of an arthropod vector. CT is typical of Protozoa that use insects as vectors. Only the final stage of the pathogen is infective; intermediate stages of the pathogen cannot be transmitted to the vertebrate host. Cf. Cyclodevelopmental Transmission; Propagative Transmission.

CYCLOPROTHRIN Noun. A synthetic-pyrethroid compound used as a contact and stomach poison; applied to ornamentals, fruits, vegetables and cotton in Australia; also used against sheep blowfly. Not registered in USA. Trade names include: Baclash® and Cyclosal® See Synthetic Pyrethroids.

CYCLOPTERIDAE Martynova 1958. See Cyclopterinidae.

CYCLOPTERINIDAE Carpenter 1986. Plural Noun. Replacement name for Cyclopteridae, a monogeneric fossil Genus of Mecoptera found in the Permian beds of Kuznetz Basin, USSR.

CYCLORRHAPHA Noun. A previous term for the Division of brachycerous Diptera classified into Series Aschiza and Schizophora. Cyclorrhapha are typified by coarctate pupae that are formed from last larval Integument and rotated male Genitalia that are flexed anteriad. Larva typically are acephalic, maggot-like and terrestrial feeding in decaying organic material; some Species are aquatic, predaceous or parasitic. Cyclorrhaphous Diptera differ from other insects in hardening the last (third) larval instar Cuticle to form a puparium. The puparium is barrel-shaped and protects the contained individual. Pupa and adult formation occur within the puparium; the process of puparium formation occurs many hours before pupal formation. The process of puparium formation is distinct from pupa formation and term pupariation is used to described it in cyclorrhaphous Diptera. Adults escape from hardened pupal cases by pushing off an Operculum or covering. Included Families: Acartophthalmidae, Agromyzidae (Phytomyzidae), Anthomyiidae (Anthomyidae, Scopeumatidae), Anthomyzidae, Asteiidae, Aulacigastridae, Boorboridae, Braulidae, Calliphoridae, Camillidae, Canaceidae, Carnidae, Celyphidae, Chloropidae, Chyromyidae, Clusiidae, Clusiodidae, Coelopidae, Conopidae, Cryptochaetidae, Curtonotidae, Cuterebridae, Cypselidae, Cypselosomatidae, Diastatidae, Diopsidae, Drosophilidae, Dryomyzidae, Ephydridae, Fergusonidae, Gasterophilidae, Helcomyzidae, Hippoboscidae, Lauxaniidae (Sapromyzidae), Lonchaeidae, Lonchopteridae, Megamerinidae, Micropezidae, Milichiidae, Muscidae (Glossinidae, Fanniidae), Neottiophilidae, Neriidae, Nycteribiidae, Odiniidae, Oestridae (Hypodermatidae), Opomyzidae, Otitidae, Pallopteridae, Periscelidae, Periscelididae, Phoridae, Piophilidae, Pipunculidae, Platypezidae, Pseudopomyzidae, Psilidae, Pyrgotidae, Rhinotoridae, Ropalomeridae (Rhopalomeridae), Sciadoceridae, Sciomyzidae (Tetanoceridae, Tetanoceratidae), Sepsidae, Sphaeroceridae (Sphaeroceratidae), Streblidae, Syrphidae, Tachinidae, Tanypezidae, Tethinidae, Tetratomyzidae, Thyreophoridae, Trixoscelidae. Syn. Muscomorpha. See Muscomorpha. Cf. Orthorrhapha.

CYCLORRHAPHOUS Adj. (Greek, *kyklos* = circle; *rhaphe* = seam; Latin, *-osus* = possessing the qualities of.) 1. Circular seamed. 2. Pertaining to the Cyclorrhapha.

CYCLOSAL® See Cycloprothrin.

CYCLOTORNIDAE Meyrick 1912 Plural Noun. A small Family of ditrysian Lepidoptera assigned to Superfamily Zygaenoidea and which is endemic to Australia and includes about 40 Species. Adult small with wingspan 10–30 mm and typically grey with spots on wings; head with appressed lamellar scales and two whorls of piliform scales; Ocelli and Chaetosemata absent; Antenna form simple, about half or two-thirds as long as forewing; Proboscis and Maxillary Palpus absent; Labial Palpus short, with three segments; Epiphysis absent; tibial spur formula 0-2-4. Eggs are oval, flattened with longitudinal ridges and transverse ribs; they are laid in large numbers on twigs or bark. Larvae are hypermetamorphic; first instar is parasitic on leafhoppers: Instar feeds, detaches, spins cocoon then moults into second instar; subsequent instars are predatory on ant larvae within ant nest. Pupae are flattened and contained within a white silk cocoon that has a transverse opening at anterior end. See Zygaenoidea.

CYCOCEL® See Chlormequat.

CYCOGAN® See Chlormequat.

CYDARIFORM Adj. (Greek, *kidaris* = a turban; Latin *forma* = shape.) Globose, but truncated at two opposite sides. Alt. Cidariform.

CYDNIDAE Bilberg 1820. Plural Noun. Burrower Bugs; Burrowing Bugs; Cydnid Bugs. A widespread, rather small Family of heteropterous Hemiptera assigned to Superfamily Pentatomoidea; closely related to stink bugs. Body coloration typically black or reddish brown; Scutellum triangular or subtriangular, not covering Hemelytra and Abdomen; Tibiae with long spines; middle and hind Coxae apically with fringe of long Setae. Adults and nymphs of most Species burrow into soil where they feed upon roots, stems and fallen seeds. Sehirinae feed upon plant parts above ground level. Subfamilies include Amnestinae, Cydninae, Clavicorniae and Sehirinae. See Pentatomoidea.

CYD-X® See Codling-Moth Granulosis Virus.

CYFEN® See Fenitrothion.

CYFLEE® See Famphur.

CYFLUTHRIN Noun. A synthetic-pyrethroid compound with contact and stomach-poison activity on insects feeding upon ornamental plants, turf and cotton in USA. Cyfluthrin is used on field crops, ornamentals, vegetables and as public-health control agent in many regions of world. Compound displays minor phytotoxicity on some ornamentals; it is toxic to bees and fishes. Trade names include: Baythroid®, Cutter Gold®, Decathlon®, Insectipen®, Laser®, Optem®, Phthon®, Responsar®, Solfac®, Tempo®. See Synthetic Pyrethroids.

CYGON® See Dimethoate.

CYHALON® See Lambda Cyhalothrin.

CYHEXATIN A triphenyltin compound {Tricyclohexylhydroxystannane} used as a contact Acaricide against mites. Applied to fruit tree crops, grapes, soybeans and other crops in some countries, but compound is not registered for use in USA. Trade names include: Araconol-F®, Metaran®, Oxotin®, Pennstyl®, Plictran®, Silatin®, Triran®. See Acaricide.

CYLINDRACEOUS Adj. (Latin, *cylindrus* = to roll; *-aceus* = of or pertaining to.) Cylindrical; shaped as a cylinder. Alt. Cylindraceus; Cylindrate. Cf. Cylindrical.

CYLINDRACHETIDAE Plural Noun. A small Family of caeliferous Orthoptera assigned to Superfamily Tridactyloidea, and which includes six Species in Australia, one Species in New Guinea and one Species in Patagonia. Adults are moderate-sized with eyes nearly absent and wings absent; legs short, stout, adapted for burrowing; Cercus of one short segment. Cylindrachetids are exclusively subterranean and tunnel in sand. See Tridacytloidea.

CYLINDRICAL Adj. (Latin, *cylindrus* > Greek, *kylindros* > *kylindein* = to roll; *-alis* = characterized by.) Descriptive of structure shaped in the form of a cylinder. An elongate tube whose outline shape is round or oval with walls of equal diameter throughout. Rel. Form; Shape.

CYLINDRICAL AUGER-BEETLE *Xylion cylindricus* Macleay [Coleoptera: Bostrichidae]: A pest of timber in Australia, preferring hardwoods that contain starch. CAB larvae take sapwood only and constructs holes ca 3.0–3.5 mm in diameter. See Bostrichidae.

CYLINDRICAL BARK BEETLES. See Colydiidae.

CYLINDROTOMIDAE Plural Noun. A Family of ca 70 nematocerous Diptera closely related to Tipulidae and sometimes regarded as a Subfamily of them. Adults in marshy habitats; larvae are phytophagous, feeding on mosses and herbaceous plants. See Nematocera.

CYMATOPHORIDAE See Thyatiridae.

CYMBALARIA APHID *Myzus cymbalariae* Stryan [Hemiptera: Aphididae]. (Australia).

CYMBIFORM Adj. (Latin, *cymba* = boat; *forma* = shape.) Boat-shaped; a concave disc with elevated margin; navicular. Rel. Form; Shape.

CYMBIGON® See Cypermethrin.

CYMBUSH® See Cypermethrin.

CYMPERATOR® See Cypermethrin.

CYNIPIDAE Latreille 1802. Plural Noun. Cynipid Gall Wasps. A numerically large, cosmopolitan Family (ca 1,000 nominal Species) of apocritous Hymenoptera assigned to the Cynipoidea. Cynipids are distinguished from other members of Superfamily by a combination of characters including female Antenna with 13–14 segments, male Antenna with 14–15 segments, most Species macropterous, some brachypterous and a few apterous, Mesosoma strongly sculptured, middle and hind Tibia each with two apical spurs, female Gaster laterally compressed, gastral Tergum II or II + III largest. In the restricted sense, the Family is composed exclusively of gall formers or inquilines of galls and host plant specificity is variable. Female wasps display ovipositional site specificity (leaves, stems, flowers, fruits). Galls are induced by salivary secretions from larvae. Larvae are hymenopteriform, and development may be solitary (monothalamous) or gregarious (polythalamous). Life cycles complex in some Species with sexual and asexual generations. See Cynipoidea. Rel. Heterogony.

CYNIPOIDEA Latreille 1802. A cosmopolitan Superfamily of apocritous Hymenoptera consisting of about 3,000 nominal Species. About 30% of the cynipoids are phytophagous, and constitute the largest number of phytophagous forms within the so-called Parasitica. Cynipoids are typically small bodied, dark coloured, forewing with a triangular Radial Cell (Areolet), Pterostigma absent, tarsal formula 5-5-5 and Gaster frequently compressed laterally. Included Families: Archaeocynipidae, Cynipidae, Eucoilidae, Figitidae, Ibaliidae and Lioterpidae.

See Apocrita. Cf. Ceraphronoidea; Chalcidoidea; Ephialtitoidea; Evanioidea; Ichneumonoidea; Proctotrupoidea; Stephanoidea.

CYNOCK® See Cyanophos.

CYNOFF® See Cypermethrin.

CYNONE SKIPPER *Anisynta cynone* (Hewitson) [Lepidoptera: Hesperiidae]. (Australia).

CYNTHIA MOTH *Samia cynthia* (Drury) [Lepidoptera: Saturniidae].

CYPER-ACTIVE® See Cypermethrin.

CYPERKILL® See Cypermethrin.

CYPERMETHRIN Noun. A synthetic-pyrethroid compound with broad-spectrum effectiveness as a contact insecticide and to a lesser extent used as a stomach poison. Cypermethrin is slightly more active than Permethrin. Compound displays minor phytotoxicity but is toxic to bees, fishes and other aquatic animals. Cypermethrin is used on numerous agricultural crops and ornamentals, and is important in control of many significant agricultural pests; it is used as crack or crevice treatment in non-food areas. Trade names include: Agrothrin®, Ammo®, Arrivo®, Avicade®, Barricade®, Barricide®, Cymbigon®, Cymbush®, Cymperator®, Cynoff®, Cyper-Active®; Cyperkill®, Cypersect®, Cypertox®, Daskor®, Demon®, Ectopor®, Ektomin®, Equiband®, Fenom®, Flectron®, Folcard®, Folcord®, Kafil Super®, Kordon®, Kruel®, Lorsban Plus®, Mastor®, Nurelle®, Parzon®, Polytrin®, Prevail®, Ripcord®, Sheerpa®, Siege®, Siperin®, Stockade®, Topple® and Toppel®. See Synthetic Pyrethroids.

CYPERSECT® See Cypermethrin.

CYPERTOX® See Cypermethrin.

CYPHONIDAE See Scirtidae.

CYPHOSOMATIC (Greek, *kyphos* = bent; *soma* = body; *-ic* = characterized by.) Immatures: With the dorsum curved and the venter flattened (Chrysomelidae).

CYPRESS APHID *Cinaria fresia* Blanchard [Hemiptera: Aphididae]. (Australia).

CYPRESS BARK-BEETLE *Phloeosinus cupressi* Hopkins [Coleoptera: Curculionidae]. (Australia).

CYPRESS BARK-WEEVIL *Aesotes leucurus* Pascoe [Coleoptera: Curculionidae]. (Australia).

CYPRESS JEWEL-BEETLE *Diadoxus scalaris* (Laporte & Gory) [Coleoptera: Buprestidae]. (Australia).

CYPRESS LONGICORN *Tritocosmia latecostata* Fairmaire [Coleoptera: Cerambycidae]: A Species endemic to Australia where the larvae tunnel in sapwood of native cypress (*Callitris* spp.); as trees grow tunnels become incorporated in true wood. Attacks by successive generations of CL spoil timber for commercial use. See Cerambycidae.

CYPRESS PINE-APHID *Cinara tujafilina* (del Guercio) [Hemiptera: Aphididae]. (Australia).

CYPRESS PINE-SAWFLY *Zenarge turneri* Rohwer [Hymenoptera: Argidae]. (Australia).

CYPROTUS BLUE-BUTTERFLY *Candalides cyprotus* (Olliff) [Lepidoptera: Lycaenidae]. (Australia).

CYPSELIDAE See Sphaeroceridae.

CYPSELOSOMATIDAE Plural Noun. A small Family of brachycerous Diptera assigned to Superfamily Nerioidea and occuring in Australasia, USA, Central America and the Palearctic. This group contains two North American Species of *Lathetocomyia*, a Genus described in 1956. Only 14 specimens of the two United States Species of *Latheticomyia* are known, and were collected during late twilight at banana-baited traps in Arizona and Utah. Biology of cypselosomatids is poorly known; larvae of at least one Species are associated with bat guano in caves. See Diptera.

CYRIL'S BROWN *Argynnina cyrila* Waterhouse & Lyell [Lepidoptera: Nymphalidae]. (Australia).

CYROMAZINE A trizine compound {N-cyclopropyl-1,3,5-triazine-2,4,6-triamine} used as an Insect Growth Regulator against maggots in manure and leafminers; also a contact insecticide against the maggot (larval stage) of most Diptera. Compound applied to celery, Chinese cabbage, mushrooms, lettuce and peppers in some countries. Cyromazine is also used on poultry for fly control, and appears to act as a selective insecticide on Diptera. Trade names include: Armor®, Citation®, Larvadex®, Neporex®, Trigard®, Vetrazine®

CYRTIDAE See Acroceridae.

CYRTONOTIDAE See Curtonotidae.

CYST Noun. (Greek, *kystis* = bladder. Pl., Cysts.) 1. An abnormal sac or vesicle that contains fluid. 2. A membrane that surrounds a cell. 3. A bladder-like structure. Alt. Capsule; Envelope; Sac; Shroud. Rel. Membrane.

CYSTOBLAST Noun. (Greek, *kystis* = bladder; *blastos* = bud. Pl., Cystoblasts.) Within the apical region of the Ovariole, a daughter cell from a germ-line division that will result in an Oocyte and Follicle Cells. Cf. Cystocyte. Rel. Germ Cell.

CYSTOCYTE Noun. (Greek, *kystis* = bladder; *kytos* = hollow vessel; container. Pl., Cystocytes.) 1. See Coagulocyte. 2. Cells that enclose germ cells in a gonadial tube. Follicle Cells of the Ovary; cyst cells of the Testis (Snodgrass). 3. A cell that arises from the division of a Cystoblast. Cf. Cystoblast. Rel. Germ Cell.

CYTHION® Malathion.

CYTOCHROME Noun. (Greek, *kytos* = hollow vessel; container; *chroma* = colour. Pl., Cytochromes.) An intracellular respiratory pigment of insects (Keilin).

CYTOGENETICS Plural Noun. (Greek, *kytos* = hollow vessel; *genesis*.) The study of genetics at the cellular level. During the 34 year period that Mendel's work on inheritance sat in obscurity, the nature of Chromosomes was being advanced. Upon the rediscovery of Mendel's paper the fusion of inheritance and Chromosome theory became known as Cytogenetics (Srb &

Owen 1953).

CYTOLOGY Noun. (Greek, *kytos* = hollow vessel; container; *logos* = discourse. Pl., Cytologies.) The study of the structure, composition and life history of cells. Rel. Histology.

CYTOLYTIC ENZYMES Hypothetical compounds secreted by one parasite to kill another competing parasite. Term was proposed by Spencer (1926) for interaction among aphid parasites, and it has been suggested by other workers studying other groups of parasites. See Physiological Suppression.

CYTON See Neurocyte.

CYTOPLASM Noun. (Greek, *kytos* = hollow vessel; container; *plasma* = something moulded. Pl., Cytoplasms.) The protoplasm of a cell exclusive of the nucleus; the cell body.

CYTOPLASMIC POLYHEDROSIS VIRUS See Insect Poxvirus. Cf. Granulosis Virus; Nonoccluded-type Virus; Nuclear Polyhedrosis Virus. Rel. Pathogen.

CYTOPLASMIC RESISTANCE Host-plant resistance conferred by mutable substances in Cell Cytoplasm.

CYTROLANE® See Mephosfolan.

CZEKLIUS, DANIEL (1857–1938) (Anon. 1939, Arb. morph. taxon. Ent. Berl. 6: 188.)

CZERNY, LEANDER (1859–1944) (Rankl 1946, Jber, Obergymnas Kremsmünster 89: 3–13.)

CZESCHKA, FRITZ (1857–1910) (Kuhnt 1910, Dt. ent. Z. 1910: 715–716.)

CZIKI, ERNO (1875–1954) (Szekessy 1954, Folia Ent. hung. 7: 1–20, bibliogr.)

CZIZEK, KARL (1871–1925) (Absolon 1933, Cas. morav. Mus. Brne. 1931–32: 28–29, 290–295.)

CZN, C RITSEMA (1873–1916) Curator at Rijksmuseum van Natuurlike in Leiden.

CZWALINA, GUSTAVE (1841–1894) (Seidlitz 1894, Dt. ent. Z. 38: 325–327.)

DA FONSECA, FLAVIO OLIVEIRA RIBEIRO (1900–1963) (Anon. 1964, Revta. bras. Ent. 11 only.)

DACHN- Greek prefix meaning 'to bite' or 'sting.'

DACHNONYPHA Suborder of primitive Lepidoptera. Adults with small body and diurnal activity; fore and hindwings with similar venation; Mandibles present or absent; Maxillary Palpus multi-segmented; middle Tibia spurred. Three included Families are Eriocraniidae, Neopseustidae and Mnesarchaeidae. See Lepidoptera. Cf. Zeugloptera.

DACTYL Noun. (Greek, *daktylos* = finger. Pl., Dactyls.) 1. A finger, toe or digit. 2. Insecta: A tarsal segment following the Basitarsus when that is enlarged, as in bees. 3. Scorpions: The terminal ventral projection of the Pretarsus. Alt. Dactylus. See Tarsus.

DACTYLAR Adj. (Greek, *daktylos* = finger.) Pertaining to digits or finger-like projections, particularly on legs.

DACTYLOPIIDAE Plural Noun. Cochineal Scales; Dactylopiids. A small Family of Coccoidea (Hemiptera). Adult with thoracic spiracles subequal in size; dorsum with Locular Pores; lacking 8-shaped pores; abdominal spiracles absent; anal opening forming crescent-shaped anterior rim. See Coccoidea.

DACTYLOPODITE Noun. (Greek, *daktylos* = finger; *pous* = foot; *-ites* = constituent. Pl., Dactylopodites.) 1. General: A simple, claw-like segment at the distal end of the legs of some Arthropoda (*e.g.* Protura). 2. Crustacea: The claw-like distal leg segment. 3. Spiders: The Metatarsus and Tarsus. 4. The Pretarsus *sensu* Snodgrass. See Tarsus. Cf. Telopodite.

DADANT, CAMILLE PIERRE (1851–1938) (Anon. 1938, Am. Bee. J. 78: 151.)

DADANT, MAURICE G (–1972) (Anon. 1973, Bee Wld 54: 68.)

DADDY LONG-LEGS SPIDER *Pholcus phalangioides* (Fuehustralia).

DADDY LONG-LEGS See Tipulidae.

DAECKE, VICTOR ARTHUR ERICH (1863–1918) (Osborn 1937, *Fragments of Entomological History*. 394 pp. (202–203), Columbus, Ohio.)

DAGGER MARK A mark in the form of a Greek *psi*.

DAGGERTOOTHS See Anotopteridae.

DAHL, ERIC (1887–1944) (Klefbeck 1944, Ent. Tidskr. 65: 214.)

DAHL, FRIEDRICH THEODOR (1856–1929) (Bonnet 1945, *Bibliographia Araneorum* 1: 50.)

DAHL, GEORGE Austrian merchant and author of *Coleoptera und Lepidoptera* (1823).

DAHLBERG, CARL GUSTAV (1721–1781) (Papavero 1971, *Essays on the History of Neotropical Dipterology* 1: 7–8. São Paulo.)

DAHLBOM, ANDERS GUSTAV (1806–1859) (Anon. 1859, Bull. Soc. ent. Fr. (3) 7: cxxix.)

DAHLMAN, JOHN WILLIAM Director of Natural History Museum in Stockholm and author of *Analecta Entomolgica* (1823), *Prodromus Monographiae Castniae* (1825), *Om nagra svenska arter of Coccus* (1826), *Chalcidites* (1820), Synoptical Table of the Butterflies of Sweden (in Mem. l'Acad. Stockholm, 1816), *Ephemerides Entomologicae* (1824) and other works.

DAHLSTRÖM, JULIUS (1834–1907) (Aigner 1907, Rovart. Lap. 14: 13, 183–185.)

DAHM, OSCAR ELIS LEONARD (1812–1883) (Spångberg 1884, Ent. Tidsskr. 5: 73– 79, 94.)

DAKIN, JOHN A (1852–1900) (Calvert 1900, Ent. News 11: 451.)

DAL NERO, VITTORIO (1862–1948) (Ruffo 1948, Atti memorie Accad. agr. Sci. Lett. Verona 125: xlvi-li, bibliogr.)

DALDORFF, DOGOBERT (INGOBERT) CARL DO (–1802) (Zimsen 1964, *The Type Material of I. C. Fabricius*. 656 pp. (12), Copenhagen.)

DALE, CHARLES WILLIAM (1851–1906) (Merrifield 1906, Proc. ent. Soc. Lond. 1906: cxiv.)

DALE, EDWARD ROBERT (–1903) (Poulton 1903, Proc. ent. Soc. Lond. 1903: lxxvii.)

DALE, JAMES CHARLES (1792–1872) (Westwood 1872, Proc. ent. Soc. Lond. 1872: c.)

DALLA FIOR, GIUSEPPE (1884–1967) (Fenaroli 1969, Studi trent. Sci. nat. 46: 9–16, bibliogr.)

DALLAS, WILLIAM SWEETLAND (1824–1890) (Musgrave 1932, *A Bibliography of Australian Entomology 1775–1930*. 380 pp. (61), Sydney.)

DALLA-TORRE, KARL WILHELM VON (1850–1928) (Strand 1929, Int. ent. Z. 22: 337–340.)

DALLS, ERNESTO D (1885–1943) (Lizer y Trelles 1947, Curso de Ent. 1: 48.)

DALLY, E V See Wilson, E. V.

DALMAN, JOHANN WILHELM (1787–1828) (Musgrave 1932, *A Bibliography of Australian Entomology 1775–1930*. 380 pp. (61), Sydney.)

DALMUS, RAYMUND DE (1862–1936) (Bonnet 1945, *Bibliographia Araneorum* 1: 52.)

DALTRY, HAROLD WLLIAM (1887–1962) (Varley 1963, Proc. R. ent. Soc. Lond. (C)27: 50.)

DALTRY, THOMAS WILLIAM (1832–1904) (Porritt 1905, Entomol. mon. Mag. 41: 215.)

DAMAGE Noun. (Latin, *damnum* = damage, fine; *age* > *aticum* = action. Pl., Damages.) 1. Economic Entomology: A measurable loss of commodity value due to insect activity; damage is most often related to loss of yield of the commodity in terms of quantity, quality, or aesthetic appeal. Damage can result from either direct or indirect insect activity. 2. The consequence of injury or impairment of physical action or biological process. See Injury.

DAMBACHER, JACOB JOSEF (1794–1868) (Rudy 1926, Arch. Insektenk. Oberrheingeb 2: 41–52.)

DAMEL'S BLUE BUTTERFLY *Jalmenus daemeli* Semper [Lepidoptera: Lycaenidae]. (Australia).

DAMIANITSCH, RUDOLPH (–1867) (Anon. 1867, Verh. zool.-bot. Ges. Wien 17: 111– 112.)

DAMIN, NARCIS (1845–1905) (Bonnet 1945, *Bibliographia Araneorum* 1: 43.)

DAMINOZIDE A Plant Growth Regulator {Butanedioic acid mono-(2,2 dimethylhydrazide)} used on ornamental plants only in USA. Registration on crop plants has been rescinded in USA. Trade names include: Alar®, B-Nine® and Dazide®. Cf. Insect Growth Regulator.

DAMMERMAN, KAREL WILLEM (1885–1951) (Lieffinck 1952, Treubia 21: 469–480, bibliogr.)

DAMPF, ALFONSO (1884–1948) [Mallis 1971, *American Entomologists*. 549 pp. (440–441)].

DAMPIER, WILLIAM (1652–1715) English navigator and explorer along coast of Australia and New Guinea. (Hagen 1948, *Green World of the Naturalist*. 398 pp., 97–105.)

DAMPING OFF The symptom of a fungal infection in plants characterized by the collapse and subsequent death of seedlings following the formation of stem lesions at soil level. DO also includes the pre-emergent death of seedlings as indicated by poor or erratic emergence. Several pathogenic fungi are causitive agents of DO including *Pythium* spp., *Thanatephorus cucumeris* and *Aphanomyces cochlioides*.

DAMPWOOD BORER *Hadrobregmus australiensis* Pic [Coleoptera: Anobiidae]: A pest of moist and decaying timber in Australia. DWB often attacks softwood subfloors, hardwood bearers and joists. Adults are dark brown, 6–8 mm long; larvae resemble furniture-beetle larva; pupal chambers with smooth walls and darker than surrounding wood. See Anobiidae.

DAMPWOOD TERMITE *Porotermes adamsoni* [Isoptera: Termopsidae]: A minor pest of forest trees and damp timber in Australia. DT can produce pipes and degrade logs. DWT will attack buildings if wood contacts soil; requires decaying wood to initiate attack, but problem disappears in dry wood. See Termopsidae. Cf. Drywood Termite.

DAMPWOOD TERMITES See Hodotermitidae.

DAMSEL BUG *Nabis kinbergii* Reuter [Hemiptera: Nabidae]: An Australian predator frequently taken in Cotton fields. Adults are 8–10 mm long, slender-cylindrical and brown with dark brown markings on Scutellum. Head is elongate with a projection between Antennae; Antennae are long, slender, with four segments; legs long and slender; hind legs longer than fore and mid legs. Adult females deposit eggs into soft plant tissue with circular emergence caps protruding above surface. Nymphs are similar to adults in appearance but lack wings. Adults and nymphs are predatory on moth eggs and larvae and on mites. DB is found on cotton infested with caterpillars and mites. DB adults are collected predominantly during winter. See Nabidae.

DAMSELFLIES See Odonata.

DAMSON HOP APHID *Phorodon humuli* Schrank [Hemiptera: Aphididae].

DANA, JAMES DWIGHT (1813–1895) (Dane 1895, Am. J. Sci. 149: 329–356).

DANADIM® See Dimethoate.

DANAID EGGFLIES *Hypolimnas misippus* (Linnaeus) [Lepidoptera: Nymphalidae]. (Australia).

DANAIDAE Plural Noun. Milkweed Butterflies. A cosmopolitan Family of Lepidoptera that is best represented in tropical regions. Typical adult with atrophied fore leg; Antenna without scales, attached to Vertex and Club moderately developed; male Androconial Scales conspicuous on Cu2 and positioned near outer margin of hindwing. Eggs are cone-shaped with radiating ridges and transverse connections extending between ridges. Larvae are smooth, colourful with transverse stripes and tentacles near anterior and posterior ends of body. Pupae are attached to substrate by Cremaster only. Food plants of danaid larvae contain compounds toxic or distasteful to predators (birds, frogs, lizards). Adults are migratory, sometimes for long distances. See Lepidoptera.

DANBY, WILLIAM HARTLEY (1850–1920) (Hatch 1949, *Century of Entomology in the Pacific Northwest*. 43 pp. (7), Seattle.)

DANCE Noun. (Latin, *de* = in; *ante* = front of. Pl., Dances.) 1. Behaviour: Elaborate or complex and stereotypical movements made by males and/or females during courtship. 2. Social Bees: Movements by a worker on a honey bee comb that provide information to other workers in the nest.

DANCE FLIES See Empidae.

DANCKELMANN, BERNHARD (–1901) (Remele 1901, Z. Forst. -u. Jagdw. 33: 125–135.)

DANDELION GALL-WASP *Phanacis taraxaci* (Ashmead) [Hymenoptera: Cynipidae].

DANDELION THRIPS *Ceratothrips frici* (Uzel) [Thysanoptera: Thripidae]. (Australia).

DANDO, WILHELM (1819–) (Ratzeburg 1874, Forstwissenschaftliches Schriftsteller-Lexicon 1: 28–30.)

DANDRIDGE, JOSEPH (1664–1746) (Bristowe 1967, Entomologist's Gaz. 18: 73–89.)

DANFORTH, STUART T (1900–1938) (Osborn 1946, *Fragments of Entomological History,* Pt. II. 232 pp. (77), Columbus, Ohio.)

DANICUT®. See Amitraz.

DANIEL, KARL (1862–1930) (Heikertinger 1931, Koleopt. Rdsch. 16: 33.)

DANILEVSKII, ALEXANDR SERGEVICH (1911–1969) (Kuznetosov & Falkovitsh 1973, Horae Soc. ent. Ross. 56: 5–7.)

DANIMEN® See Fenpropathrin.

DANITOL® See Fenpropathrin.

DANITRON® See Fenproximate.

DANNATT, WALTER (1863–1940) (Newman 1940, Entomologist 73: 96.)

DANNREUTHER, T (1873–1963) (Wigglesworth 1964, Proc. R. ent. Soc. Lond. (C) 28: 57.)

DARDENNE, M P (–1960) (Anon. 1960, Bull. Soc. ent. Fr. 65: 69.)

DARK CERULEAN *Jamides phaseli* (Mathew) [Lepidoptera: Lycaenidae]. (Australia).

DARK CILIATE BLUE *Anthene seltuttus affinis* (Waterhouse & Turner) [Lepidoptera: Lycaenidae]. (Australia).

DARK DARTER *Telicota ohara ohara* (Plotz) [Lepidoptera: Hesperiidae]. (Australia).

DARK FLOUR BEETLE *Tribolium destructor* Uyttenboogaart [Coleoptera: Tenebrionidae]: A minor pest of stored products in Europe, Afghanistan, Kenya and Ethiopia. Adults of DFB are 5–6 mm long, dark brown or black and resemble Confused Flour Beetle or Rust-Red Flour-Beetle. See Tenebrionidae. Cf. Confused Flour Beetle; Rust-Red Flour-Beetle.

DARK GRASS-BLUE *Zizeeria karsandra* (Moore) [Lepidoptera: Lycaenidae]. (Australia).

DARK MEALWORM *Tenebrio obscurus* (Fabricius) [Coleoptera: Tenebrionidae]: A minor pest of stored products; DM is endemic to Europe and widespread through commerce. DM resembles Yellow Mealworm in appearance and biology. See Tenebrionidae. Cf. Yellow Mealworm.

DARK ORANGE-DART *Ocybadistes ardea heterobathra* (Lower) [Lepidoptera: Hesperiidae]. (Australia).

DARK PURPLE-AZURE *Ogyris abrota* Westwood [Lepidoptera: Lycaenidae]. (Australia).

DARK-HEADED RICEBORER *Chilo polychrysa* (Meyrick) [Lepidoptera: Pyralidae]. (Australia).

DARKLING BEETLES See Tenebrionidae.

DARK-NIGHT MOLE CRICKET *Gryllotalpa monanka* Otte & Alexander [Orthoptera: Gryllotalpidae]: A Species of mole cricket endemic to Australia and for many years confused with the African Mole Cricket. See Gryllotalpidae. Cf. African Mole Cricket.

DARK-SIDED CUTWORM *Euxoa messoria* (Harris) [Lepidoptera: Noctuidae]. See Cutworms.

DARK-SPOTTED TIGER-MOTH *Spilosoma canescens* (Butler) [Lepidoptera: Arctiidae]. (Australia).

DARK-TAIL WASPS *Cameronella* spp. [Hymenoptera: Pteromalidae]. (Australia).

DARK-WINGED FUNGUS GNAT *Bradysia* spp. [Diptera: Sciaridae] (particularly *B. coprophilia* and *B. impatiens*) are recognized as important plant pests in glasshouse and mushroom cellars. DWFG larvae damage seedling plants; larva and adult also promote damage through transmission of fungal diseases via spores of *Verticillium, Cylindrocladium, Pythium* and *Botrytis*. DWFG adults transmit mycopathogens in mushroom production facilities. Adult DWFGs are ca 2–3 mm long, dark-bodied, slender and weakly sclerotized; Antenna moniliform; wing with a distinctive 'Y' pattern not seen in other small Diptera. Adults are weak fliers; females typically are found on underside of leaf or near soil surface; males emerge a day before females and are more conspicuous. Females release a pheromone that causes males to flick their wings and move in a zig-zag path while in pursuit of the female. Female mates once and experience a preoviposition period of 1–2 days. The female dies after oviposition, and the male lives longer than female (ca 3–8 days). Females lay 100–200 eggs on surface of wet or moist soil; eclosion occurs after 3–4 days. Neonate larvae with well developed Mandibles; larval body is worm-like, transparent and the head capsule is black; subsequent larval instars become white or opaque. DWFG completes four larval instars that feed on decaying plant material, fungi, algae and plant roots; larvae can transmit fungal diseases and root rot while feeding. Mature larvae wander on soil surface, orient perpendicular to soil surface in debris and prepare pupal chambers of white silk. DWFG life cycle ca 20–30 days @ 21–24°C; Species is multivoltine with generation overlap. See Sciaridae. Cf. Shore Flies.

DARLING, NOYES (1782–1846) (Weiss 1936, *Pioneer Century of American Entomology*. 320 pp. (167–168), New Brunswick.)

DARLING, SAMUEL TAYLOR (1872–1925) (Hegner 1926, Parasitology 12: 117–119.)

DARNERS See Aeshnidae.

DARNING NEEDLE See Odonata.

DART Noun. (O.F., *dart* = dagger. Pl., Darts.) 1. The Sting (Aculeus) of aculeate Hymenoptera. 2. A pointed structure used to penetrate. See Sting. Cf. Aculeus; Arrow; Needle.

DART® See Teflubenzuron.

DARWIN 755® See Butocarboxim.

DARWIN BROWN-CROW BUTTERFLY *Euploea darchia darchia* (W. S. Macleay) [Lepidoptera: Nymphalidae]. (Australia).

DARWIN, CHARLES ROBERT (1809–1882) English naturalist and proponent of the Theory of Natural Selection. CRD's father was Robert Waring Darwin (M.D.) and grandfather was Erasmus Darwin (philosopher). CRD was educated at Edinburgh University 1825 for study of medicine and Cambridge University 1828 for clerical studies but both programmes were abandoned. CRD married January 1839. CRD was naturalist on voyage of HMS Beagle (December 1831–October 1836). This voyage provided specimens, data and observations seminal to CRD's explanation of Natural Selection. Published: *Origin of Species* (1858), *Descent of Man* (1871), *Expression of the Emotions* (1872). CRD was awarded the Royal Medal (1853), Wollaston Medal (1859) and Copley Medal (1864). (Papavero 1975, *Essays on the History of Neotropical Dipterology*. 2: 233–246. São Paulo.)

DARWIN, ERASMUS (1713–1802) English physician, poet and grandfather of Charles Darwin. ED published *The Botanic Garden* in two parts, *Economy of Vegetation* (1792) and *Loves of the Plants* (1789), and other works. (Nordenskiöld 1935, *History of Biology*. 629 pp. (294–296), London.)

DARWIN RINGLET *Hypocysta adiante* (Hübner) [Lepidoptera: Nymphalidae]. (Australia).

DARWIN'S ANT See Brown House-Ant.

DARWINISM Noun. (Named after Charles R. Darwin.) The Theory of Natural Selection which asserts that speciation by plants and animals is driven in populations by the process in which successful individuals survive and reproduce while less successful individuals die or do not reproduce. Successful individuals pass the genes which confer superiority to their offsping. Cf. Lamarkism. Rel. Evolution.

DARWIN'S TOADS See Rhinodermatidae.

DASCILLIDAE Plural Noun. A Family of polyphagous Coleoptera assigned to the Dascilloidea. Adult body elongate, pubescent and 8–11 mm long; Antenna serrate with 11 segments; Pronotum transverse; tarsal formula 5-5-5; Empodium absent; Abdomen with five Ventrites. Larvae are C-shaped with head hypognathous and Stemmata absent; Antenna apparently with 2–3 segments and legs with five segments, many tarsal claws present; Urogomphi present. Larvae live in soil and feed upon organic matter. See Coleoptera.

DASCILLOIDEA Noun. A Superfamily of polyphagous Coleoptera including Dascillidae, Karumidae and Rhipiceridae.

DASH Noun. (Middle English, *dasshe* > *daschen*. Pl., Dashes.) 1. A spot or short mark of colour on the body or wing which imparts a characteristic display. Rel. Printer's Symbols. 2. A rapid movement of short duration. 3. The striking or breaking of the surface of a liquid with a short, sharp blow.

DASHWOOD-JONES, WILLIAM ARTHUR (1858–1928) (Green 1947, Proc. ent. Soc. Br. Columb. 43: 41–43.)

DASKOR® See Chlorpyrifos-Methyl; Cypermethrin.

DASYCERIDAE Plural Noun. A small Family of Coleoptera assigned to Superfamily Staphylinoidea. Adults are <3 mm long and resemble Lathridiidae except fore Coxae are contiguous and open posteriorly. Elytra cover entire Abdomen and all Sterna are moveable. See Staphylinoidea.

DASYGASTRES Plural Noun. (Greek, *dasys* = hairy, thick; *gastros* = belly, stomach.) Bees with pollen-carrying structures on the Abdomen. Cf. Megachilidae. Rel. Corbicula.

DASYPHILLOUS Adj. (Greek, *dasys* = hairy, *phyllon* = leaf; Latin, *-osus* = abounding in.) 1. Pertaining to leaves that are densely hairy (pubescent). 2. Descriptive of plants with leaves that are dense or thickly set.

DASYPODINAE A Subfamily of Melittidae (Apoidea) consisting of about 10 Genera.

DATA Plural Noun. (Latin, *dare* = to give; *datum* = something given. Sl., Datum.) 1. Information. 2. Collectively, material serving as a basis for discussion and inference. Scientific data typically is stored in the form of numbers or pictures and depicted or presented graphically, pictorially or statistically. Alt. Facts; Records; Staticis. Cf. Database.

DATA ANALYSIS Organizational, mathematical or structural manipulation of data (database) to distil higher levels of order or patterns.

DATA MATRIX Mathematics: Any rectangular arrangement of data in rows and columns.

DATABASE Noun. (Latin, *datum* = something given; base = *basis*. Pl., Databases.) A large collection of data organized such that elements of information can be expanded, updated and retrieved rapidly for various tasks. Alt. Data Base. Cf. Data.

DATEBUG *Asarcopus palmarum* Horvath [Hemiptera: Issidae].

DATE-PALM SCALE *Parlatoria blanchardii* (Targioni-Tozzetti) [Hemiptera: Diaspididae]: A widespread pest of date palms in places that the plant is cultivated. See Diaspididae.

DATE-STONE BORER *Coccotrypes dactyliperda* [Coleoptera: Scolytidae]: A pest of date palm in the Mediterranean region. See Scolytidae.

DATHE, GUSTAV (1813–1880) (Dathe 1892, *Lehrbuch der Bienenzucht*. Pt. 5.)

DATURA LEAF BEETLE *Lema trivittata* Say [Coleoptera: Chrysomelidae]. (Australia).

DAUBE, PIERRE GUSTAV (1807–1872) (Fauvel 1873, Annu. Ent. 1: 106–107.)

DAUGHTER Noun. (Anglo Saxon, *dohto;* Middle English, *doughter*. Pl., Daughters.) 1. A female offspring. 2. Offspring of the first generation without regard to sex (*e.g.* daughter cell, daughter nucleus).

DAUGHTER CELL A cell produced by fission from any antecedent cell. See Cell.

DAUPHIN, ANTOINE (Vallière. 1945, Circ. Sci. Bourb. Cent. Fr. 1940– 44: 19–21.)

DAVAINE, CASIMIR JOSEPH (1812–1882) (Laboulbène 1884, Ann. Soc. ent. Fr. (6) 4: 361–364.)

DAVALL, EDMUND (1793–1860) (Ratzeburg 1847, Forstwissenchaftliches Schriftsteller-Lexicon 1: 136–137.)

DAVENPORT, RACHEL MARGARET (1889–1958) (Scott 1959, Entomol. mon. Mag. 95: 15.)

DAVEY, JAMES THOMAS (1923–1959) (Uvarov 1960, Proc. R. ent. Soc. Lond. (C) 24: 53.)

DAVID, PERE ARMAND (1826–1900) (Anon. 1901, Entomol. mon. Mag. 37: 20.)

DAVIDSON, ANSTRUTHER (1860–) (Essig 1931, *History of Entomology*. 1029 pp. (600–601), New York.)

DAVIDSON, D M N (Hall 1947, Proc. R. ent. Soc. Lond. (C) 11: 60.)

DAVIDSON, JAMES (1885–1945) (Hall 1947, Proc. R. ent. Soc. Lond. (C) 11: 60.)

DAVIDSON, WILLIAM MARK (1887–1961) (Rawson 1962, J. Lepid. Soc. 16: 250.)

DAVIES, EDWIN C H (–1909) (C. M. 1962, Entomologist 42: 168.)

DAVIES, WILLIAM MALDWYN (1903–1937) (Laing 1937, Entomol. mon. Mag. 73: 92–93.)

DAVIP® See Famphur.

DAVIS, ALONZO CLAYTON (1901–1942) (Osborn 1946, *Fragments of Entomological History*. Pt. II, 232 pp. (77), Columbus, Ohio.)

DAVIS, CHARLES ABBOT (1869–1908) (Burgess

1908, J. Econ Ent. 1: 165.)

DAVIS, EDGAR WILLIAM (1895–1954) (McDuffie & Darst 1956, J. Econ. Ent. 49: 285.)

DAVIS, EDWARD MOTT (1888–1943) (Anon. 1944, Fla. Nat. 17: 51–52.)

DAVIS, HARROLD FOSBERG CONSETT (–1944) (Browne 1945, Proc. Linn. Soc. N. S. W. 70: ii.)

DAVIS, JOHN (JACK) JAMES (1933–1970) (Butler 1972, Proc. R. ent. Soc. Lond. (C) 36: 61.)

DAVIS, JOHN JUNE (1885–1965) (Mallis 1971, *American Entomologists*. 549 pp. (441–443), New Brunswick.)

DAVIS, JOSEPH M (1909–1957) (George & Yull 1958, J. Econ. Ent. 51: 415.)

DAVIS, NELSON CARYL (1892–1933) (Shannon 1934, Ent. News 45: 56.)

DAVIS, WILLIAM MORRIS (1850–1934) (Calvert 1934, Ent. News 45: 84.)

DAVIS, WILLIAM THOMPSON (1862–1945) (Mallis 1971, *American Entomologists*. 549 pp. (216–220), New Brunswick.)

DAVY, JOHN (1790–1868) (Anon. 1868, Proc. ent. Soc. Lond. 1868: lxxix–lxxxi.)

DAWSETT, A (–1896) (Anon. 1897, Ent. News 8: 72.)

DAWSON RIVER BLACK FLY *Austrosimulium pestilens* Mackerras & Mackerras [Diptera: Simulidae]. (Australia).

DAWSON, GEORGE (1843–1927) (Anon. Entomologist 60: 144.)

DAWSON, JOHN (–1879) (Dunning 1879, Proc. ent. Soc. Lond. 1897: lix.)

DAWSON, JOHN FREDERIC (1802–1870) (Anon. 1871, Entomol. mon. Mag. 7: 216.)

DAWSON, RALPH WARD (1887–1974) (Telford 1975, Proc. Wash. St. ent. Soc. 36: 388–389.)

DAWSON'S BURROWING BEE *Amegilla dawsoni* (Rayment) [Hymenoptera: Anthophoridae]. (Australia).

DAY DEGREES See Degree Day.

DAY, F H (1875–1963) (Wigglesworth 1964, Proc. R. ent. Soc. Lond. (C) 28: 57.)

DAY, GEORGE O (1854–1942) (Downes 1943, Proc. ent. Soc. Br. Columb. 40: 34–35.)

DAY, WILLIS C (1894–1965) (Edmunds 1966, Pan-Pacif. Ent. 42: 165–167.)

DAY-EYE A type of insect eye, so-called because it is adapted for use in daylight when light is abundant (Comstock). See Vision. Cf. Apposition Eye; Superposition Eye.

DAYFEEDING ARMYWORM *Spodoptera exempta* (Walker) [Lepidoptera: Noctuidae]. (Australia). See Armyworms.

DAYLENGTH Noun. (Old English, *daeg; lengthu*. Pl., Daylengths.) The period or interval during which light is visible. See Photophase.

DAZIDE® See Daminozide.

DAZOMET A soil preplant fumigant {Terahydro-3,5 dimethyl-2H-1,3,5-thiadiazine-2-thione} used for control of weeds, nematodes, soil fungi and soil insects. Trade names include Basamid® and Basolon®

DAZZLE® See Diazinon.

DCIP A chlorinated hydrocarbon used as a nematicide. DCIP is applied to fruit crops, vegetables and ornamentals in some countries but is not registered for use in USA. Tradename is Nemamort®. See Fumigant.

DDT Acronym for Dichloro-diphenyl-trichloroethane, the original synthetic organochloride insecticide. First synthesized in 1873 but insecticidal properties not recognized until the compound was rediscovered in 1939 by Swiss chemist Paul Muller (Noble Prize 1948). DDT first was manufactured by Geigy Chemical Company ca 1940 and widely used from the mid-1940s; numerous formulations are available. DDT is phytotoxic on some vegetables including beans, curcurbits and tomatoes; DDT is toxic to bees and birds. DDT acts as contact insecticide and stomach-poison with long-term persistent, residual action and accumulates in soil. DDT banned from use in USA and most developed countries due to toxic effects on non-target soil-dwelling invertebrates and birds. DDT remains popular in many developing countries for some crop pests and as a public-health insecticide for control of Malaria mosquitoes, and use as a human-body insecticide. Widespread resistance to DDT is expressed in many Species of insects. See Organochlorine Insecticide.

DDVP See Dichlorvos.

DE BACH, PAUL H (1913–1992) American academic (University of California), foreign explorer and specialist on biological control of citrus pests. Author of more than 200 scientific papers and two books. Advocate of the ecological concept of Competitive Displacement and advanced the notion that Species of parasitic Hymenoptera competitively displace related Species that act as ecological homologues in attacking one host.

DE BEAUMONT, J B A L E See Elie De Beaumont.

DE BLOIS GREEN CHARLES (1863–1929) (dos Passos 1951, J. N. Y. ent. Soc. 59: 175.)

DE DOUS, CARL PHILIP EMIL (1866–1936) (Henriksen 1938, Ent. Meddr 20: 102–103.)

DE FILLIPPI, FILIPPO (1813–1867) (Conci 1967, Atti Soc. ital. Sci. nat. 106: 33–34.)

DE GEER, CARL (1720–1778) Swedish aristocrat (Baron) of Dutch origin who studied biology at Utrecht. DeGeer continued the observations of Reaumur under publication *Memoires pour servir a l'histore des insects* (I-VII, 1752–1778). Author of several volumes, the most notable of which were translated by Retzius into Latin under the title *Genera et Species Insectorum* and published in 1783. A version in German was published by Goez. DeGeer contributed a strong taxonomic component to entomological work with pre-Linnean diagnoses; he provided the first descriptions and illustrations of Collembola (1740, 1743), and hypothesized that the Collophore was used to accumulate humidity and adhere to smooth surfaces. (Bonnet 1945, *Bibliographia*

Araneorum 1: 29.)

DE LEON, DONALD (1902–1966) (Frost 1967, Ann. ent. Soc. Am. 60: 869.)

DE MARCH, MARCO (1872–1936) (Corti 1941, Boll. zool. Torino 12: 209–213.)

DE SEABRA, ANTHERO FREDERICO (1874–1952) (Cunha 1954, Mems Estud. Mus. zool. Univ. Coimbra 221: 1–17, bibliogr.)

DEAD-END HOST Medical Entomology: A vertebrate host that is infected with a pathogen, but the incidence of the pathogen is too low to be infective of a blood-feeding arthropod vector. See Parasitism. Cf. Amplifying Host; Resistant Host; Silent Host; Susceptible Host.

DEAD-LEAF CRICKETS See Acridoxenidae.

DEADLINE® See Metaldehyde.

DEALATE Noun. (Latin, *de* = away; *alatus* = winged. Pl., Dealates.) Insects that have shed their wings and become wingless (secondarily apterous). A term applied to reproductive castes of ants and termites. See Caste. Rel. Alate.

DEALATION Noun. (Latin, *de* = away; *alatus* = winged; English, *-tion* = result of an action. Pl., Dealations.) The self-inflicted removal of wings by a winged insect. Typically, dealation occurs in the reproductive castes of ants and termites following nuptial flights or copulation.

DEAN, GEORGE ADAMS (1873–1956) (Mallis 1971, *American Entomologists*. 549 pp. (443–444), New Brunswick.)

DEANDRADE, NUMO FREIRE (1924–1958) (A[lmeida1960, d' Mems Estud. Mus. zool. Univ. Coimbra 262: 1–2, bibliogr.)

DEATH'S-HEAD SPHINX MOTH *Acherontia atropos* (Linnaeus) [Lepidoptera: Sphingidae]: A Species of moth known for its distinctive body markings and its ability to expel air forcibly from the Pharynx to produce a whistling sound. See Sphingidae.

DEATH-WATCH BEETLE *Xestobium rufovillosum* (DeGeer) [Coleoptera: Anobiidae]: An Holarctic pest of structural timber. DWB is endemic to Europe and transported to North America through shipping and commerce. DWB naturally occurs in tree trunks and old and decaying branches of hardwoods As urban pest, DWB infests rafters and other support timbers. DWB was responsible for destruction of many historical medieval buildings of Europe. Adults are 6–7 mm long, reddish-brown to dark brown with yellow-grey Setae on Pronotum and Elytra; Antenna with 11 segments including apical three segments englarged to form a Club. Larvae are pale yellow with black Mandibles. Adult emergence hole is circular, ca 4 mm in diameter. Adults apparently do not fly. Females lay ca 60 eggs in batches of 2–3 on wood surface. Eclosion occurs within ca 20–21 days and first instar larvae penetrate wood. DWB life cycle typically is complete within a year, but can require several years. Common name derived from courtship sound of head tapping on wood; tapping made by both sexes and consists of ca 10 taps during two seconds with intensity increasing during call. Alt. Deathwatch Beetle. See Anobiidae. Cf. Furniture Beetle.

DEAURATE Adj. (Latin, *de* = undoing of an action; *aurum* = colour of gold.) A rubbed or worn gold colour. Alt. Deauratus.

DEBANTIC® See Tetrachlorvinphos.

DEBORD, SARA HOKE (1899–1950) (Heinrich 1950, Proc. ent. Soc. Wash. 52: 157–159.)

DEBRIS Noun. (Old French, *debruisier* = to break. Pl., Debris.) 1. The rubbish, remains or ruins of anything which has decomposed or physically broken down with time. Cf. Detritus; Waste. 2. Organic waste from dead or damaged tissue. Syn. Refuse. Rel. Duff; Humus; Litter; Mor.

DEBRIS BUG *Lyctocoris campestris* (Fabricius) [Hemiptera: Anthocoridae]. (Australia).

DECAMOX® See Thiofanox.

DECAPOPHYSIS Noun. (Greek, *deca* = 10; *apo* = from; *phyein* = to grow; *-sis* = a condition or state. Pl., Decapophyses.) A superior Apophysis of the tenth abdominal segment in female insects. See Apophysis.

DECARY, RAYMOND (1891–1973) (Viette 1974, Bull. Soc. ent. Fr. 78: 297–298.)

DECATHLON® See Cyfluthrin.

DECAUX, FRANÇOIS (–1899) (Bouvier 1899, Bull. Soc. ent. Fr. 1899: 285.)

DECAVALVA Noun. (Greek, *deka* = ten; Latin, *valva* = fold. Pl., Decavalvae.) 1. The dorsal or posterior Ovipositor. 2. Inner medial or posterior Valvulae (MacGillivray).

DECAVALVIFER Noun. (Greek, *deka* = ten; Latin, *valva* = fold. Pl., Decavalvifers.) Orthoptera: The Valvifer (MacGillivray).

DECAY Noun. (Old English, *decayen* > Latin, *de* = down, away; *cadere* = to fall. Pl., Decays.) 1. The physical process of degradation or wearing away of structure. 2. The condition of an organism which has undergone the action of microbial and physical destruction. Cf. Cariose; Putrefication; Rot.

DECAY MOTHS *Barea* spp. [Lepidoptera: Oecophoridae]: Several Species that are distributed along coastal Australia and develop as secondary pests of fungus-infested decaying wood. Adults are 8–12 mm long with wings grey-brown and wingspan 20–25 mm. Larvae are cream-translucent with grey markings at posterior end and covered with stiff, gold-coloured body Setae; mature larva to 30 mm long. Adults are free flying and feed on nectar. Moths oviposit eggs into damp, decaying wood of logs and dead trees. Larvae feed on fungi in wood. Frass of shredded wood particles is webbed with silk and wood fragments to form a loose shelter. Feeding by larvae occurs in shelter and pupation occurs within timber. See Oecophoridae.

DECCO 276EC® See Chlorpropham.

DECEPHALIC Adj. (Latin, *de* = down, away; *cephalicus* = head; *-ic* = characterized by.) Pertaining to an insect with a prognathous head and Occipital Foramen divided into two parts by other structures (MacGillivray).

DECIDUOUS Adj. (Latin, *decdere* = to fall down > *de-* + down, away; *cadere* = to fall; *-osus* = possessing the quality of.) Pertaining to structure that is shed or falls off the body at maturity or at specific periods or development or life (*e.g.* leaves on a tree). Cf. Caducous.

DECIS® See Deltamethrin.

DECIS-PRIME® See Deltamethrin.

DECLINATE Adj. (Latin, *declinatus*, past participle of *declinare* = to turn aside; *-atus* = adjectival suffix.) 1. Descriptive of structure that is bent downward or aside. 2. Structure that is somewhat bent, with the apex downward. A term often applied to Setae. Cf. Inclinate; Proclinate; Reclinate.

DECLIVITY Noun. (Latin, *declivitus* = sloping; English, *-ity* = suffix forming abstract nouns. Pl., Declivities.) A surface that slopes downward. Alt. Declivous; Declivus.

DECOLORATE Adj. (Latin, *decoloratus* = deprived of colour; *-atus* = adjectival suffix.) Faded; of a faded-appearing colour.

DECOMPOSED Adj. (Latin, *de* = away; *cum* = with; *pausare* = to rest.) Pertaining to structure or substance which is not intact or which appears shapeless. Rel. Amorphous.

DECOMPOSER Noun. (French, *decomposer* = decay. Pl., Decomposers.) An heterotrophic organism which utilizes dead organic matter as food and further breaks the material into smaller structural units or chemically less complex substances. Cf. Heterotroph; Producer; Saprogen. Rel. Debris; Detritus.

DECOMPOSITION Noun. (French, *de* = from, down, away; *composer* = to compose. Pl., Decompositions.) The act or process of decay. The state of being decomposed. Cf. Putrefaction.

DECORTICATE Verb. (Latin, *decorticare* = to peel.) To remove the bark or cortex of a plant.

DECOY® See Methiocarb.

DECOY PBW® See Gossyplure.

DECOY TPW® See Lycolure.

DECREPITANS Adj. (Latin, *decrepitare* = to roast so as to cause crackling.) Crackling. Alt. Decrepitant.

DECTICOUS Adj. (Greek, *dektikos* = biting; Latin, *-osus* = possessing the quality of.) Descriptive of holometabolous insects with a pupal stage that possess articulated Mandibles that are capable of being used to free the insect from the pupal cocoon or chamber. See Pupa. Cf. Adecticous. Rel. Metamorphosis.

DECTICOUS PUPA A Pupa with articulated Mandibles that can be used by the pharate adult for defence or extracation from a pupal chamber. See Pupa. Cf. Adecticous Pupa; Obtect Pupa. Rel. Metamorphosis.

DECUMBENT Adj. (Latin, *decumbere* = to lie down.) Bent or deflected downward; bent down at the apex from an upright base. Term applied to Setae which are bent or deflected. Cf. Procumbent; Recumbent.

DECURRENT Adj. (Latin, *decurrere* = to run down.) Pertaining to structure which is closely attached to and extending down another body. Cf. Procurved; Recurrent.

DECURVED Adj. Pertaining to structrure which is bowed, arched or curved downward. Cf. Decumbent; Recurved.

DECUSSATE Adj. (Latin, *decussare* = to cross in the form of an X.) 1. Crossed or intersected at an angle; X-like. 2. Crossed pairs, such as large Setae (bristles) alternately crossed in some Diptera. 3. Descriptive of plants with paired leaves that are crossed.

DECYDE® See Codling-Moth Granulosis Virus.

DEDEVAP® See Dichlorvos.

DEER FLIES Specifically, any member of the Genus *Chrysops* (compound eyes spotted with brilliant colour and banded wings). See Tabanidae. Rel. Horse Flies.

DEER FLY *Chrysops callidus* [Diptera: Tabanidae]: A pest of domestic mammals and man that occurs over most of North America. Adults are 10–14 mm long, predominantly black coloured with eyes green displaying zigzag stripes; Abdomen yellowish with black V-shaped mark on second segment; wings smoky black. Ocelli present; third segment of Antenna with five pseudosegments; hind Tibia with two apical spurs. Adults active June-July near standing water. Eggs black and deposited in masses of ca 100 on vegetation near water. Larva yellowish-greenish; Abdomen with brown rings. See Tabanidae.

DEER FLY FEVER See Tularemia.

DEER KEDS See Hippoboscidae.

DEET An acronym for N,N-diethyl-*m*-toluamide. A synthetic insectide used as a repellant.

DEFARGO, GERARD (1906–1942) (Marietan 1942, Verh. schweiz. naturf. Ges. 122: 290–291, bibliogr.)

DEFAUNATION Noun. (Latin, *de* = away from; *faunus* = god of woods. Pl., Defaunations.) 1. The elimination of symbionts from the gut of an insect. 2. The removal of animal life from an area or habitat. Rel. Extinction.

DEFECATORY Adj. Descriptive of or pertaining to defecation or excretion of faeces.

DEFEND® See Permethrin.

DEFENDER MORPH One of two larval forms of the polyembyronic encyrtid hymenopteran *Copidosomopsis tanytmemus* Caltagirone. DM is an internal parasite of Lepidoptera larvae; DM is highly motile and characterized by a large head and well developed Mandibles. DM attacks and kills or injures larvae of competing Species of parasitic Hymenoptera that also attack Lepidoptera larvae. The DM does not pupate and thus does not become a reproductive adult. See Polyembryony.

DEFENCE GLAND An Exocrine Gland specifically

adapted to provide chemicals which protect insects from predators. Scent may be emitted from the Ostiole of a Defence Gland and contact the body of the prey; the odour of the prey serves to deter the predator. Alternatively, scent may be collected on the prey's Tarsus and applied to the predator. In some systems the scent may be forcefully ejected throught the DG Ostiole and directed at the predator. Single-chemical defence compounds include Butanoic Acid, Isobutyric Acid, Hexanal, Hexyl Acetate and Tridecane Other DG secretions may defend against microorganisms such and fungi. See Exocrine Gland; Gland. Cf. Accessory Gland; Cement Gland; Scent Gland.

DEFENSINS Plural Noun. A small group of cystine-rich cationic proteins that envelop viruses and are active against bacteria and fungi. Insect defensins are similar to vertebrate defensins.

DEFINITIVE ACCESSORY VEINS An arcane term used to characterize Accessory Veins that have attained a position comparable in stability to that of the primitive branches of the principal veins (Comstock). See Vein; Wing Venation.

DEFINITIVE HOST The host (insect or other organism) in which the sexual life of a parasite is passed (Matheson). See Host; Parasitism.

DEFINITIVE RESERVOIR Medical Entomology: A host (insect or other organism) which contains a natural supply of the sexual stage of a parasite (Matheson). See Parasitism.

DEFENSIVE SECRETION Chemical compounds that are regurgitated or emitted from exocrine glands or through intersegmental membranes and which repel aggressive or predatory organisms. See Allomone. Rel. Allelochemical.

DEFLECTED Adj. (Latin, *deflectere* = to bend aside.) Bent downward; pertaining to wings with the inner (posterior) margins overlapping and the outer (anterior) margins declining toward the sides. Cf. Inflected.

DEFLEX Transitive Verb. (Latin, *deflexus* > deflectere. Deflexed; Deflexing.) Abruptly bent downward. Cf. Decumbent.

DEFOLIATOR Noun. (Latin, *de* = down, away; *folium* = leaf; *-or* = one who does something. Pl., Defoliators.) A chewing insect that destroys the leaves of plants. Cf. Leafminer. Rel. Herbivore.

DEFORMED Adj. (Latin, *de* = from; *forma* = shape.) Twisted or set in an unusual form. Specifically, a term applied to some Coleoptera with knotted or twisted Antennae (*e.g.* male Meloidae). Rel. Malformed; Disfigured.

DEFORMED-WING VIRUS A widespread viral disease manifest in adult honey bees. DWV is vectored by Varroa mites to pupal bees. The disease multiplies slowly. Pupae infected at the white-eye stage of development emerge with deformed wings and soon die. See Honey Bee Viral Disease. Cf. Acute Bee-Paralysis Virus. Rel. Pathogen.

DEGENERATE Adj. (Latin, *degeneratus* = decaying; *-atus* = adjectival suffix.) To lose adaptive features; to deteriorate.

DEGENERATION Noun. (Latin, *degenare* = to decay; English, *-tion* = result of an action. Pl., Degenerations.) 1. A progressive deterioration or loss of function of any body part or organ. 2. A transition from a higher form of development to a lower form of development. Alt. Deterioration; Disintegration. Ant. Regeneration. Rel. Evolution.

DEGREE DAY 1. An accumulation of heat units above some threshold temperature for a 24-hour period. DD are related to physiological development of an insect. 2. The number of degrees above a minimum temperature acceptable for growth, multiplied by time in days. DD represent a measure of physiological time through a combination of time and temperature. 3. A linear summation model of development that is the most widely used form of development modeling. Alt. Day Degrees.

DEHISCE Verb. (Latin, *dehiscere* = to gape.) To open for the discharge of contents, such as a seed pod or a Pupa.

DEHISCENCE Noun. (Latin, *dehiscere* = to gape.) 1. A bursting open of a seed capsule or seed pod. 2. The opening of anthers for the release of pollen. 3. The splitting of the pupal Integument during adult emergence in Lepidoptera. Alt. Emergence.

DEHISCENT Adj. (Latin, *dehiscens*.) 1. Characterized by dehiscence; standing open, separating toward the apex; bursting. 2. A term used to characterize ripe fruit or fungus fruiting bodies.

DEINBOLI, PETER VOGELIUS (1783–1874) (Natvig 1944, Norsk. ent. Tidsskr. 7: 10–13.)

DEINOPIDAE Plural Noun. Net-casting Spiders; Ogre-faced Spiders. Many deinopids construct a rectangular web held with the first pair of legs. When prey approach the spider, the spider's web is cast and ensnare the prey.

DEJEAN, PIERRE FRANÇOIS MARIE AUGUST (1780–1845) (Dejean, Compte.) French general and Peer of France. Coleopterist and author of *Catalogue de Coleopteres*. With Latreille, Dejean wrote *Histoire Naturelle et Iconographie des Coleopteres d'Europe*, and with Boisduval, Dejean wrote *Iconographis et Hist. Nat. des Coleop. d'Europe*. (Arnett. 1948, Coleopts Bull. 2: 15–16.)

DEJECTA Plural Noun. (Latin, *dejectus* = to throw or cast down.) 1. All liquid and solid material emanating or shed from the body of an animal. Dejecta from insects include: Setae, silk, pheromones, exuvia, regurgitated food and excrement. Alt. Dejectamenta. 2. Excrement or excretion. See Excreta.

DEJECTAMENTA See Dejecta.

DEL PONTE EDUARDO F (1897–1969) (Begarano 1970, Revta. Soc. ent. argent. 32: 5–8.)

DEL Abbreviation for 'delineavit.' In pre 20th Century illustrated material, the abbreviation was placed in lower left corner of plates to indicate the artist. The names of plate engravers typically was placed in lower right corner of plates.

DELABY, EDMOND (1838–1892) (Anon. 1893, Bull. Soc. linn. Nord Fr. 11: 177–179.)

DELAHAYE, JULES (1826–1889) (Clement 1889, Ann. Soc. ent. Fr. (6)9: 501–504.)

DELALAMDE PIERRE ANTOINE (1787–1828) (Papavero 1971, *Essays on the History of Neotropical Dipterology*. 1: 115–123. São Paulo.)

DELAMINATION Noun. (Latin *de* = from; *lamina* = a thin piece of metal; English, *-tion* = result of an action. Pl., Delaminations.) A splitting or division of structure into layers.

DELAMONTAGNE, ALEXANDRE (1806–1836) (Pierret 1836, Bull. Soc. ent. Fr. 5: lxxx–lxxxii.)

DELAPORTE, M. Author of *Etudes Entomologiques, ou description d'Insectes nouveau, & c.* (1834), *Essai d'une classification systematique de l'ordre des Hemipteres (Heteropteres)*, and other works related to taxonomy of Coleoptera.

DELAROUZEE, CHARLES (1835–1860) (Baron 1861, Ann. Soc. ent. Fr. (4) 1: 259–264.)

DELAYED VOLTINISM An insect life-cycle requiring more than one year for completion. See Voltinism. Cf. Multivoltine; Univoltine. Rel. Aestivation; Diapause; Dormancy.

DELECIA® See AlP.

DELESSE, HUBERT (1914–1972) (Browne 1972, J. Lepid. Soc. 26: 268–274, bibliogr.)

DELETANG, LUIS FRANCISCO ALEJANDRO (1882–1931) (Lizer y Trelles 1947, Curso Ent. 1: 46–47.)

DELFIN® A registered biopesticide derived from *Bacillus thuringiensis* var. *kurstaki*. See *Bacillus thuringiensis*.

DELHI BOIL See Oriental Sore. Syn. Bagdad Boil.

DELICATE SLATER *Porcellionides pruinosus* (Brandt) [Isopoda: Porcellionidae]. (Australia).

DELLA TORRE ETASSO, ALESSANDRO (1881–1937) (Giordani Soika 1937, Boll. Soc Venez. Storia nat. 1: 9–10, 119–200.)

DELMOTTE, ORTOLL (–1963) (Anon. 1963, Mitt. dt. ent. Ges. 22: 44.)

DELONG, DWIGHT MOORE (1892–1984) American academic (Ohio State University) and taxonomic specialist in leafhoppers and insect taxonomy. Author of more than 450 papers and co-author of *An Introduction to the Study of Insects*.

DELPHACID PLANTHOPPERS [Hemiptera: Delphacidae]. (Australia).

DELPHACIDAE Plural Noun. Delphacid Planthoppers. A cosmopolitan Family including ca 1,500 Species assigned to the Fulgoroidea (Hemiptera). Wing polymorphism and brachyptery are common in some Species; hind Tibia with apical, flexible spur. Delphacids predominantly are phloem feeders; a few Species of economic importance. See Fulgoroidea.

DELPINO, GIACOMO GUISEPPE FREDERICO (1833–1905) (Morinii 1905, Rc. Accad. Sci. 1st Bologna 9: 113–145.)

DELTA BT® A registered biopesticide derived from *Bacillus thuringiensis* var. *kurstaki*. See *Bacillus thuringiensis*.

DELTA Noun. (Acronymic: Descriptive Language for Taxonomy.) A computer algorithm designed to write taxonomic keys, generate character matrices and develop taxonomic descriptions. (Dallwitz & Paine, 1986. CSIRO Div. Ent., Rept. 13, 106 pp.)

DELTAMETHRIN A synthetic-pyrethroid compound with broad-spectrum effectivenss as a contact and stomach poison and slightly more active than Permethrin. Deltamethrin has minor phytotoxicity but is toxic to bees, fishes and other aquatic animals. Deltamethrin is used on numerous agricultural crops and for public-health pests in some regions of the world; it is registered in USA against structural pests. Trade names include: Butoflin®, Butoss®, Butox®, Calypso®, Cislin®, Crack Down®, Cresus®, Decis®, Decis-Prime®, K-Obiol®, K-Othrin®; K-K-Otek®, Othrine®, Pasypso®, Thripstik®. See Synthetic Pyrethroids.

DELTANET® See Furathiocarb.

DELTOID Adj. (Greek, *delta*; *eidos* = form.) Triangle-like; descriptive of structure whose outline shape resembles the Greek letter Delta, with apex extended. See Triangle. Cf. Cycloid; Obdeltoid; Trochoid. Rel. Eidos. Rel. Form; Shape.

DELUSORY CLEPTOPARASITOSIS A human mental disorder in which an individual incorrectly believes that insects or mites infest a home or place of habitation. (Grace & Wood 1987, Pan-Pac. Entomol. 63: 1–4.) See Medical Entomology; Parasitology. Cf. Acariasis; Entomophobia.

DELUSORY DERMATITIS A human mental condition in which a sense of itching or a tingling sensation occurs on the skin. The afflicted person believes the sensation is caused by mites, but mites cannot be detected. Imaginary mites are sometimes called 'Paper Mites' or 'Cable Mites' by afflicted person in the belief that electric cables for computers or paper transmit the mites. See Medical Entomology; Parasitology. Cf. Acariasis; Entomophobia.

DELUSORY PARASITOSIS A human mental disorder in which an individual incorrectly believes that insects or mites infest the body (Waldron 1972, Calif. Med. 117: 76–78.) The disorder may quickly develop and continue for years; often seen in late middle age and may be more common among women. Entomologists and parasitologists are often consulted by sufferers when Physicians are indifferent to complaints. See Medical Entomology; Parasitology. Cf. Acariasis; Entomophobia.

DEMAISON, LOUIS (–1937) (Anon. 1937, Bull. Soc. ent. Fr. 42: 161.)

DEMAND® See Lambda Cyhalothrin.

DEMARCATE Verb. (French, *demarquer* = to mark off.) To establish boundaries; set off; to distinguish. See Discriminate.

DEMARCATION Noun. (Spanish, *demarcacion* >

de-; *marcar* = to mark; English, *-tion* = result of an action; Websters: French = *demarcation* > *demarquer*. Pl., Demarcations.) A bounding or limiting of the physical, biological or behavioural margins of a feature (attribute, characteristic) of an organism. Cf. Parameter.

DEME Noun. (Greek, *demos* = in sense; modern usage = a commune. Pl., Demes.) A local population of a Species consisting of potentially interbreeding individuals at a given locality. Cf. Population.

DEMEREC, MILISLAR (1895–1966) (Wagner 1967, Genetics, Princeton 56: 21.)

DEMERSAL Adj. (Latin, *demersus* = submerged; *-alis* = characterized by.) Sinking to the bottom (of a body of water).

DEMERSE Trans. Verb. (Latin, *demersio* = immersion.) To submerge.

DEMERSED Adj. (Latin, *demersus* = submerged.) Botany: Submerged.

DEMERSION Noun. (Latin, *demersio* = immersion. Pl., Demersions.) Submergence.

DEMO® See Primicarb.

DEMOGRAPHIC Adj. (Greek, *demo-*; *graphikos* = to write; *-ic* = characterized by.) Pertaining to demography or parameters of populations. See Demography.

DEMOGRAPHY Noun. (Greek, *demo-*; *graphikos* = to write. Pl., Demographies.) 1. The statistical analysis of populations, including natality and mortality. 2. The rate of growth and age structure of populations and the study of the factors which influence them. Rel. Population.

DEMON® See Cypermethrin.

DEMUTH, GEORGE S (1871–1934) (Anon. 1934, Outdoor Indiana 1(4): 18.)

DENAPON® See Carbaryl.

D-END® See Tralomethrin.

DENDRIFORM Adj. (Greek, *dendros* = tree; Latin, *forma* = shape.) Branch-shaped. Descriptive of outline shape or three-dimensional form that resembles the branching pattern of a tree; arborescent. Alt. Dendritic. See Branch 1. Cf. Fork. Rel. Form 2; Shape 2; Outline Shape.

DENDRITES Noun. (Greek, *dendros* = tree. Pl., Dendrites.) Finely ramifying branches given off from a nerve cell. Alt. Dendrons. See Branch; Nerve Cell.

DENDROBIUM MEALYBUG *Pseudococcus dendrobiorum* Williams [Hemiptera: Pseudococcidae]. A pest of dendrobium orchids in Australia. See Pseudococcidae.

DENDROGRAM Noun. (Greek, *dendros* = tree; *-gram* = drawn. Pl., Dendrograms.) A schematic branching structure which purports to demonstrate degrees of relationship among taxa. See Classification. Cf. Phylogram; Phenogram; Evolutionary Tree. Rel. Cladistics.

DENDROID Adj. (Greek, *dendros* = tree; *eidos* = form.) Branch-like; descriptive of structure with markings that appear to have the branching pattern of a shrub or tree. See Branch. Cf. Dendriform; Neuroid. Rel. Eidos; Form; Shape.

DENDROPHAGOUS Adj. (Greek, *dendron* = tree; *phagein* = to devour; Latin, *-osus* = with the quality of.) Tree-eating; a term used to describe organisms that feed on woody tissues. Alt. Dendrophagus. See Feeding Strategy. Cf. Dendrophilous.

DENDROPHILOUS Adj. (Greek, *dendron* = tree; *philein* = to love; Latin, *-osus* = with the quality of.) Tree-loving; a term used to describe organisms that live on trees or shrubs, or within woody tissue. See Habitat. Cf. Agricolous; Algicolous; Caespiticolous; Fungicolous; Lignicolous; Nemoricolous; Pratinicolous; Silvicolous; Thamnophilous. Rel. Ecology.

DENE See Dune.

DENGUE Noun. (Spanish, *dengue* = prudery, affectation.) *Edhazardia aedis,* an arbovirus with a complex life cycle, involving horizontal (preoral) and vertical (transovarial) transmission through the vector. Dengue virus is replicated in midgut epithelium and transferred to salivary glands by Haemocytes; extrinsic cycle requires 8–14 days; Dengue can be transmitted by sexual contact. Vectors include the mosquitoes *Aedes aegypti* (Linnaeus), *A. albopictus* (Skuse) and *A. scutellaris*.

DENGUE FEVER An arboviral disease whose causative agent belongs to the Family Flaviviridae. DF predominantly is tropical and vectored by several *Aedes* Species; *Aedes aegypti* serves as principal vector in urban areas. DF incubation period 2–7 days in humans, followed by period of high fever (ca 6 days) accompanied by cutaneous eruption, severe pain in the head and limbs; weakness and depression may persist for weeks but the disease is rarely fatal. Mosquitoes can acquire DF from infected persons during febrile period. See Arbovirus; Flaviviridae. Cf. Dengue Haemorrhagic Fever; West Nile; Yellow Fever.

DENGUE HAEMORRHAGIC FEVER A disease of children first recognized in Thailand (1954); DHF currently is widespread in southeast Asia and sporadically a problem among infants and young children in the Caribbean. DHF is vectored by *A. aegypti* in the New World. No vaccine has been developed for DHF; the disease causes an appreciable number deaths among its victims. Cf. West Nile.

DENGUE MOSQUITO See Yellow Fever Mosquito.

DENIER, PEDRO CELESTINO LUIS (1892–1941) (Lizer y Trelles 1947, Curso Ent. 1: 49.)

DENIS, JEAN ROBERT (1893–1969) (Delamare Debouteville *et al.* 1970, Revue Ecol. Biol. Sol 7: 1–10, bibliogr.)

DENIS, MICHAEL (1729–1800) (Rose 1850, *New General Biographical Dictionary* 7: 57.)

DENIS, ROBERT (1893–1969) (Grassé 1969, Bull. Soc. ent. Fr. 75: 221–228, bibliogr.)

DENKAPHON® See Trichlorfon.

DENKAVEPHON® See Dichlorvos.

DENMARK, HAROLD American taxonomic Acarologist and administrator (Florida Department of Agriculture and Consumer Services) specializing in Phytoseiidae.

DENNING, DONALD G (1909–1988) American industry Entomologist and taxonomic specialist on Trichoptera. (Resh 1989, Pan-Pac. Ent. 65: 97–107.)

DENNIS, A W (–1946) (Jackson 1947, Proc. Trans. S. Lond. ent. nat. Hist. Soc. 1946–47: 56.)

DENNIS, GEORGE CHRISTOPHER (–1897) (Porritt 1898, Naturalist 1898: 113–114.)

DENNY, HENRY (1803–1871) (Anon. 1871, Proc. Linn. Soc. Lond. 1870–71: 84–85.)

DENNYS, ARTHUR ALEXANDER (1894–1942) (Buckell 1942, Proc. ent. Soc. Br. Columb. 39: 2–3.)

DENS Noun. (Latin, *dens* = tooth; Pl., Dentes.) 1. A tooth or tooth-like process; the teeth or pointed processes on the medial surface of the Mandible (Smith). 2. Collembola: The long proximal segment of the distal arms of the forks of the Manubrium.

DENSARIA Noun. (Latin, *densus* = thick, close. Pl., Densariae.) 1. Coccids: Distinct thickenings on the ventral aspect associated with the Incisurae (MacGillivray). 2. Thickenings along the margins of the Incisurae (Comstock).

DENSITY DEPENDENCE The change of influence of environmental or physiological factors upon population size as population density increases. Cf. Density Independence.

DENSITY-DEPENDENT FACTOR A population-regulating factor that changes proportionally in intensity with changes in population density. See Density-independent factor.

DENSITY INDEPENDENCE The lack of change in population size as population density changes. Cf. Density Dependence.

DENSITY-INDEPENDENT FACTOR A population-regulating factor that causes mortality and is unrelated to a population's density. See Density-dependent factor.

DENTACERORES Plural Noun. (Latin, *dens* = tooth.) Coccids: The elongated Cerores of the mesial row of Orbacerores which produce irregularities in the membrane surrounding the Anus (MacGillivray); denticulate pores (Smith).

DENTAL SCLERITE Muscid larvae: The small tooth-like sclerite at each side of the base of the mandibular sclerite.

DENTATE Adj. (Latin, *dens* = tooth; *-atus* = adjectival suffix.) Toothed; with tooth-like prominences. Descriptive of structure with acute teeth, the sides of which are equal and the apex is above the middle of the base. Alt. Dentatus. Cf. Mandibulate; Tooth.

DENTATE-SERRATE Adj. (Latin, *dens* = tooth; *serra* = saw.) Descriptive of a structure that is toothed, with the dentations themselves serrated on their edges. Cf. Aculeate-Serrate; Biserrate; Dentate-Sinuate; Multiserrate; Subserrate; Uniserrate.

DENTATE-SINUATE Adj. (Latin, *dens* = tooth; *sinus* = curve.) Descriptive of structure or the margin of a structure that is toothed and indented. Cf. Aculeate-Serrate; Biserrate; Dentate-Serrate; Multiserrate; Subserrate; Uniserrate.

DENTES CANINAE See Canine Teeth.

DENTICLE Noun. (Latin, *denticulus* = small tooth. Pl., Denticles.) A small tooth or cuticular projection. See Projection. Cf. Acantha; Ctenidium; Spine. Rel. Integument; Seta; Sensilla.

DENTICULATE Adj. (Latin, *denticulus* = small tooth; *-atus* = adjectival suffix.) Pertaining to a structure or surface with small teeth, particularly along a margin. Alt. Denticulated; Denticulatus. See Denticle.

DENTICULATE PORES See Dentacerores.

DENTIFORM Adj. (Latin, *dens* = tooth; *forma* = shape.) Tooth-shaped; appearing as a tooth or a surface with many teeth. See Tooth. Cf. Dentate; Digitiform; Papilliform. Rel. Form 2; Shape 2; Outline Shape.

DENTITION Noun. (Latin, *dens* = tooth; English, *-tion* = result of an action. Pl., Dentitions.) The kinds, number and arrangement of teeth or tooth-like structures. The term is usually applied to the Mandible.

DENTON, JOHN M (1829–1896) (Osborn 1937, *Fragments of Entomological History*. 394 pp. (143), Columbus, Ohio.)

DENUDATE Adj. (Latin, *denudatus* > *denudare* = to strip off; *-atus* = adjectival suffix.) Literally, without covering; specifically, structure or a surface without Setae or scales. Alt. Denudated; Denuded. Rel. Asetose.

DENUDE Verb. (Latin, *de* = down; *nudare* = to make bare.) To free from covering; to rub so as to remove the surface covering of scales, Setae, wax, detritus or other vestiture.

DENVIL, HORACE GASKELL (1901–1968) (Wheeler 1969, Proc. Trans. Brit. Ent. nat. Hist. Soc. 2: 52.)

DEODORANT Noun. (Latin, *de* = down; *-antem* = an agent of something. Pl., Deodorants.) A material added to a pesticide formulation to mask unpleasant odours.

DEONG, ELMER RALPH (1882–1966) (Michelbacher 1967, J. Econ. Ent. 60: 897)

DEORSE Adv. (Latin, *deorsum* = downwards. Pl., Deorsus.) Downward, indicating motion. Alt: Deorsum.

DEP® See Trichlorfon.

DEPAUPERATE Adj. (Latin, *depauperare* > *de* = down; *pauparere* = to make poor.) To impoverish.

DEPAUPERATE COLONY Ants: An impoverished colony of ants, bees or other social insects. A colony whose numbers are declining. See Colony. Rel. Eusocial.

DEPENDENT Trans. Verb. (Latin, *depedere* > *de* = down; *pendere* = to hang.) Hanging down. Cf. Suspended.

DEPLANATE See Complanate. Alt. Deplanatus.

DEPLANCHE, EMILE (1824–1875) (Vieillard 1876, Bull. Soc. Linn. Normandie (2) 10: 341–350.)

DEPLUMING MITE *Knemidokoptes gallinae* (Railliet) [Acari: Sarcoptidae]. *Neocnemidocoptes gallinae* (Raillet) [Acari: Knemidokoptidae]. (Australia).

DEPRESSARIIDAE Plural Noun. A small Family of ditrysian Lepidoptera assigned to Superfamily Gelechioidea and which is cosmopolitan in distribution but best represented in temperate regions of Northern Hemisphere. Adults are moderately small with wingspan 10–36 mm and head bearing smooth scales; Ocelli usually absent and Chaetosemata absent; Antenna filiform (bipectinate in males of a few Species), usually shorter than forewing length; Proboscis scaled; Maxillary Palpus folded, with four segments; Labial Palpus recurved; fore Tibia with Epiphysis; tibial formula 0-2-4; forewing broad, without Pterostigma. Larvae with hypognathous head; prothoracic shield and Pinacula present; Prolegs occur on abdominal segments 3–6, 10; crochets are biordinal or triordinal. Larvae feed on leaves joined by silk-and-faecal-pellet galleries. See Gelechioidea.

DEPRESSED Adj. (Latin, *de* = down, away; *premere* = to press.) Descriptive of structure that is flattened by dorsovental pressure; flattened dorsoventrally. Alt. Depressus. Ant. Elevated. Cf. Compressed; Flattened; Impression.

DEPRESSED FLOUR BEETLE *Palorus subdepressus* (Wollaston) [Coleoptera: Tenebrionidae]. (Australia).

DEPRESSOR MUSCLE A muscle employed to lower or depress an appendage. Alt. Depressor. See Muscle. Cf. Elevator Muscle.

DEPTA, PAMIATKE VALERIANA (1885–1963) (Korbel 1964, Sb. slov. narod. Muz. 10: 155–156.)

DEPUISET, LOUIS MARIE ALPHONSE (1822–1886) (Clement 1886, Ann. Soc. ent. Fr. (6) & : 471–474.)

DEPURATORY Adj. (Latin, *depurativus* > *de-* = down, away; *purare* = to purify.) Clean.

DERATOPTERA An obsolete ordinal name for the Orthoptera.

DERBIDAE Plural Noun. A widespread Family of Fulgoroidea (Hemiptera) that includes about 800 nominal Species. Derbids characterized by apical segment of Rostrum not longer than wide and hind Tibia without apical spines. Nymphs feed on fungi associated with bark and adults feed on monocots. See Fulgoroidea.

DERHAM, WILLIAM (1657–1735) English clergy, philosopher and Lepidopterist. (Lisney 1960, *Bibliography of British Lepidoptera 1608–1799*. 315 pp. (65–76, bibliogr.), London.)

DERMA Noun. (Greek, *derma* = skin. Pl., Derms; Dermata.) The inner and usually thicker layer of the Cuticle, and which is laminated in structure, non-pigmented and positioned beneath the Epidermis. Alt. Derm; Dermis; Hypoderma. See Integument. Cf. Cuticle. Rel. Skin.

DERMAL Adj. (Greek, *derma* = skin; *-alis* = belonging to, pertaining to.) Pertaining to the skin (Integument) or outer covering of the body. Alt. Dermis. See Epidermis; Integument; Skin.

DERMAL CELL Coccids: See Cellulae (MacGillivray).

DERMAL GLANDS Modified epithelial cells that are responsible for producing defensive secretions and pheromones. Dermal Gland Cells are often larger or shaped different from the adjacent Epithelial Cells. DGs sometimes form large complexes of cells that project into the Haemocoel to accommodate an increase in size associated with higher metabolic activity. DG ducts project upward and penetrate layers of Cuticle. DG ducts are considerably larger in diameter than pore canals. See Gland; Integument. Cf. Cement Layer; Cuticulin Layer; Epidermal Cells; Epidermis; Endocuticle; Exocuticle; Pore Canals; Wax Layer. Rel. Integument; Skeleton; Skin.

DERMAL LIGHT SENSE A phenomenon involving a diffuse photosensitivity over the body surface. Examples include clam genus *Mya* which retracts its siphon when animal is exposed to light, and anterior segments of earthworm's body that are sensitive to light. Phenomenon is poorly studied in insects but seen in cockroach and may be more widespread. Cf. Dorsal Light Reaction.

DERMAL MYIASIS See Myiasis.

DERMAL PORES Coccids. See Cellulae (MacGillivray).

DERMAPTERA Leach 1815. Noun. (Greek, *derma* = skin; *pteron* = wing.) Earwigs. A cosmopolitan Order of paurometabolous insects whose earliest fossil record dates from the Jurassic Period. Dermaptera consist of about 1,800 nominal Species and the Order is best represented in tropical and warm temperate regions. Dermaptera were placed among Orthoptera in 19th Century classifications. Dermaptera probably are related to grylloblattids. Size to 50 mm long with elongate body and head broad, rather flattened and prognathous; Ecdysial Line on Frons and Vertex; Antenna short, annulate; mouthparts mandibulate; compound eye variable in size and Ocelli absent. Thoracic segments not fused; Prothorax largest in apterous Species, otherwise smallest; two pairs of spiracles in thoracic membrane. Forewing modified into Tegmina; hindwing large, fan-like or circular, composed mostly of Anal Fan; veins anterior of anal region reduced and confined to sclerotized Remigium. Legs relatively short, cursorial with Coxae separated; Tarsus with three segments; tarsal claws long, other pretarsal elements absent. Abdomen elongate with 10 flexible, telescoping segments; eight Sterna present in female; pleural sclerites absent; spiracles found in pleural membrane of segments 1–8; Cerci modified into appendages called Forceps and which usually are larger in males. Male genitalia with Penis apically bifurcate in most Species; one Penis is reduced in some specialized groups. Nervous system with three Thoracic Ganglia and six Abdominal Ganglia. Polytrophic Ovarioles; Arixeniina and Hemimerina with Pseudoplacental Viviparity. Metamorphosis is paurometabolous. Dermaptera typically are nocturnal, thigmotactic and sometimes taken at lights; they occur in forest litter, under bark or stones, within soil, or similar habitats that afford seclusion, darkness and moisture. Eggs lack an Operculum and are laid in batches. Females provide maternal care for their eggs. Earwigs in cooler habitats predominantly are herbivorous; Species in warm temperate and tropical regions predominantly are predaceous. Predatory forms feed on various kinds of insects but seem to prefer soft-bodied larvae. Dermaptera Families include: Anisolabididae, Apachyidae, Arixeniidae, Carcinophoridae, Chelisochidae, Diplatyidae, Forficulidae, Hemimeridae, Karchiellidae, Labiduridae, Labiidae, Pygidicranidae and Spongiphoridae. Syn: Dermatoptera.

DERMATOBLAST Noun. (Greek, *derma* = skin; *blastos* = bud. Pl., Dermatoblasts.) Insect embryo: The outer, thin layer of cells that is segregated from the Ectoderm cells and the Neural Ridges. Dermatoblasts form the body-wall (Imms).

DERMATOPTERA See Dermaptera.

DERMESTID BEETLE *Dermestes frischii* Kugelann [Coleoptera: Dermestidae]: A multivoltine pest of stored products that occurs in the Holarctic, Africa, Madagascar and Australia. DB resembles Common Hide Beetle in anatomy and biology. Adults are 6–9 mm and shining black to dark brown; dorsal surface of body with numerous long recumbent to suberect Setae. Pronotum with dense vestiture of white or yellow Setae along lateral and anterior margin; Elytra with numerous white and black Setae (black Setae more common, white Setae forming patches at base and lateral margin); Antennal club is conspicuous and consists of three segments. Apical margin of Elytra not serrate and inner apices not forming acute spines. Female lays ca 60 eggs in groups of 2–4 during a 10-day period, followed by oviposition pause. Eclosion occurs within 2–3 days at 28–30°C. DB typically completes five larval instars (range 5–9) with number dependent upon food and temperature. Larvae feed upon carcass, bone, carrion, dried skins and stored grain products. Mature larvae may leave food and bore into wood as pupation site; pupal period requires 5–8 days. Life cycle requires 30–50 days. See Dermestidae.

DERMESTIDAE Plural Noun. Carpet Beetles; Dermestid Beetles; Hide Beetles; Skin Beetles. A cosmopolitan Family of polyphagous Coleoptera (Bostrychoidea) with about 750 Species in six Subfamilies. Adult 1–10 mm long; body pubescent or with scales, oblong, compact and typically somber coloured with pattern defined by colour of scales; head deflexed and median Ocellus present. Antenna is short, typically with 11 (rarely nine) segments, clavate and fitting into groove when retracted. Pronotum narrowing anteriad with posterolateral margins acute; fore coxal cavities open posteriorly; tarsal formula 5-5-5 and segments not lobed. Elytra covering Abdomen; Abdomen with five Ventrites. Larvae are eruciform with long Setae and hair pencils; Antenna with three segments; legs with five segments, Pretarsi each with two claws; Tergites present; Urotergites present or absent. Larvae are scavengers on dead plant and animal material, typically with high protein content; larvae may cause dermatitis from urticating Setae. Adults often are taken on flowers where they feed on nectar, pollen or scavenge. Several Species are significant pests of stored products, particularly grains; urticating Setae may be ingested by humans. Predatory Species assigned to *Thaumaglossa*; their larvae eat mantid and Lepidoptera eggs. Syn. Thorictidae; Thylodriidae. See Bostrychoidea.

DERMESTOIDEA In some classifications, a Superfamily of polyphagous Coleoptera including Dermestidae, Derodontidae, Nosodendridae, Sarothriidae and Thoricidae. See Coleoptera. Cf. Bostrychoidea.

DERMOMUSCULAR Adj. (Greek, *derma* = skin; Latin, *musculus* = mouse.) An archane term that pertains to the skin (dermis) and muscles.

DERODONTIDAE Plural Noun. A Family of 10 Species of Coleoptera in the Dermestoidea known only from North America, Europe and Asia. Distinguished from other dermestoids by adults with Ocelli, anterior Coxal cavities closed, hind Coxae extended laterad beyond lateral margin of Metepisternum. Biology poorly known; adult and larva predaceous on *Adelges piceae* (Ratzeburg). Eggs deposited on prey or within the habitat of the prey's eggs. Typically one generation per year. See Dermestoidea.

DERRIN® See Rotenone.

DERRINGER® See Resmethrin.

DERRIS® See Rotenone.

DERRIS ROOT See Rotenone.

DESAREST, EUGENE (1816–1889) (Laboulbène 1890, Bull. Soc. ent. Fr. (6) 10: v–vi.)

DESBROCHERS DES LOGES, JULES (1836–1913) (Deville 1914, Ent. Bl. Biol. Syst. Käfer 10: 64.)

DESCLEROTIZATION Noun. (Latin, *de-*; Greek, *scleros* = hard; English, *-tion* = result of an action. Pl., Desclerotizations.) The act, process of phenomenon involving a loss of sclerotin in normally sclerotized parts or structures. See Sclerotin.

DESCRIPTION Noun. (Latin, *descriptio* > *de-*; *scribere* = to write; English, *-tion* = result of an action. Pl., Descriptions.) Systematic Entomology: The written enumeration of structural features, biological characteristics or behavioural attributes that may be used to separate or distinguish individuals, Species or higher groups from each other. Descriptions serve to identify organisms and to distinguish them from other organisms that appear similar through common descent (close relatives) or convergence (birds and bats; whales and fishes). Descriptions serve as the written foundation of systematic Entomology (Biology). The specimens upon which the descriptions are prepared serve as the concrete basis for the scientific or common name. Rel. Common Name; Scientific Name; Nomenclature.

DESCRIPTIVE ENTOMOLOGY The subdiscipline of Entomology concerned with the enumeration and identification of characters used to distinguish Species from related coordinate taxa. An applied form of comparative anatomy of insects. See Taxonomy. Rel. Classification.

DESECTED See Truncate. Alt. Desectus.

DESERT CORN FLEA-BEETLE *Chaetocnema ectypa* Horn [Coleoptera: Chrysomelidae].

DESERT HARVESTER-ANT *Pogonomyrmex rugosus* [Hymenoptera: Formicidae].

DESERT SPIDER-MITE *Tetranychus desertorum* Banks [Acari: Tetranychidae].

DESERTICOLOUS Adj. (Latin, *desertus* = waste; *colere* = to inhabit.) A term used to characterize organisms that inhabit deserts. See Habitat. Cf. Ammophilous; Arenicolous; Eremophilous; Ericeticolous; Geophilous; Lapidicolous; Psammophilous; Saxicolous; Xerophilous. Rel. Ecology.

DESFONTAINES, RENÉ LOURICHE (1755–1833) (Zimsen 1964, *The Type Material of I. C. Fabricius*. 656 pp. (17), Copenhagen.)

DESGODINS, AUGUSTE (1826–1913) (Oberthür 1916, Etudes de lépidopterologie comparée 11 (Portraits de lépidoptèrologistes (1re Sér.): [3])

DESHPANDE, V G (–1941) (Anon. 1941, Indian J. Ent. 3: 351.)

DESICCATE Verb. (Latin = *desiccatus*, pp of *desiccare* = to dry up.) Transitive: To dry or eliminate moisture from the body. Intransitive: To become dry.

DESICCATION Noun. (Latin = *desiccatus*, past participle of *desiccare* = to dry up; English, *-tion* = result of an action. Pl., Desiccations.) Excessive drying through the natural loss of moisture or by artificial methods. Rel. Transpiration; Perspiration.

DESICCATIVE Noun and Adj. (Latin, *desiccare* = to dry up.) Drying.

DESICCATOR Noun. (Latin, *desiccare* = to dry up; *or* = one who does something. Pl., Desiccators.) A device or apparatus (usually of glass and maintained airtight) for eliminating moisture from specimens or biological tissue. A desiccator usually employs a vacuum and high temperature or chemical agents (*e.g.* calcium chloride). See Critical Point Dry.

DESIDERATUM Noun. (Latin, *desideratus* > *desiderare* = to miss. Pl., Desiderata.) 1. A thing or things that are needed or desired. 2. A list of things regarded as necessary.

DESIGN® A registered biopesticide derived from *Bacillus thuringiensis* var. *aizawai* strain GC-91. See *Bacillus thuringiensis.*

DESJARDINS, JULIEN FRANÇOIS (1799–1840) (Silbermann 1840, Revue Ent. (Silbermann) 5: 351.)

DESLANDES, ANDREW FRANCIS BOUREAU (1690–1757) (Rose 1850, *New General Biographical Dictionary* 7: 68.)

DESLONGCHAMPS, JACQUES AMAND ETUDES (1794–1867) (Grosse *et al.* 1868. J. Conch. Lond. 16: 121.)

DESMAREST, ANSELME GAETAN (1784–1838) (Blanchard 1838, Ann. Soc. ent. Fr. 7: xliii–xlviii.)

DESMAREST, EUGENE (1816–1889) (Laboulbène 1890, Bull. Soc. ent. Fr. (6) 10: v–vi.)

DESMERGATE Noun. (Greek, *desmo* = bond; *ergos* = work; *-ate* = Adj. suffix. Pl., Desmergates.) An ant intermediate between the ordinary worker and the soldier.

DESMOSOME Noun. (Greek, *desmos* = bond; *soma* = body. Pl., Desomosomes.) A principal mechanism by which cells are bound together. Desmosomes are characterized by opposable thickened portions of cell membrane separated by enlarged intercellular spaces and bundles of fine cytoplasmic filaments. See Spot Desmosome; Belt Desmosome. Cf. Cell Apposition; Cell Junction; Chemical Synapse.

DESPAY, RAYMOND (1876–) (Astre. 1950, Bull. Soc. hist. nat. Toulouse 85: 67–76, bibliogr.)

DESQUAMATION Noun. (Latin, *de-*; *squama* = scale; English, *-tion* = result of an action. Pl., Desquamations.) The process of scaling or coming off in Scales. See Squama.

DESTITUTE Adj. (Latin, *destitutus* > *destituere* = to leave alone, to abandon.) Abandoned. Pertaining to a sense of wanting or being without qualities of life or qualities essential for life. Alt. Impoverished.

DESTRUCTIVE HOST FEEDING See Host Feeding. Cf. Concurrent Host Feeding; Non-Concurrent Host Feeding; Non-Destructive Host Feeding.

DESVIGNES, THOMAS (1812–1868) (Newman 1868, Entomologist 4: 108.)

DESVOIDY, ROBINEAU French physician and author of *Recherches sur l'organization vertebrale des Crustaces, des Arachnides, et des Insectes* (1828), *Essai sur la tribu des Culicides* (IN: Mem. Soc. d'Hist. Nat. Paris, vol. 2) and *Sur les Dipteres de la Tribu des Muscides* (IN: Mem. Savants Etrangers de l'Acad. Sci.)

DETERMINATE Adj. (Latin, *determinans* > *determinare* > *de-*; *terminus* = end; *-atus* = adjectival suffix.) With well defined outlines or distinct limits; fixed; marked out. Alt. Determinatus. Cf. Indeterminate.

DETERMINATE EVOLUTION See Orthogenesis.

DETERMINISTIC MODEL See Model. Cf. Stochastic Model.

DETERRENT Noun. (Latin, *deterrens* = to deter. Pl., Deterrents.) A chemical that inhibits feeding or oviposition when being in a place where insects would normally feed or oviposit. Cf. Attractant *sensu* Dethier *et al.* 1960. 53: 136. Alt. Prevention; Restriction.

DETIA® See AIP.

DETIAPHOS® See MGP.

DETMOL® See Chlorpyrifos; Permethrin; Propetamphos.

DETMOLIN® See Chlorpyrifos.

DETONANS Adj. (Latin, *detonare* > *de-*; *tonare* = to thunder.) Exploding; descriptive of something emitting a sudden noise or puff as an explosion. Alt. Detonant.

DETRITIVORE Noun. (Latin, *detritus* = rubbed or worn away; *vorare* = to devour. Pl., Detritivores.) An organism that feeds upon detritus. See Feeding Strategies. Cf. Carnivore; Herbivore; Omnivore; Saprovore. Rel. Ecology.

DETRITIVOROUS Adj. (Latin, *detritus* = rubbed or worn away; *vorare* = to devour; *-osus* = with the quality of.) A term descriptive of organisms that feed upon detritus. Feeding on or eating fur and feather detritus. See Feeding Strategy. Cf. Carnivorous; Omnivorous.

DETRITUS Noun. (Latin, *detritus* = rubbed or worn away.) 1. Material that remains after disintegration, rubbing away or the destruction of structure. 2. Fragmented material. 3. Any disintegrated or broken matter. Cf. Humus; Debris. Duff; Litter; Mor.

DETRITUS MOTH *Opogona omoscopa* (Meyrick) [Lepidoptera: Tineidae]. (Australia).

DETTMER, HEINRICH (1873–1933) (Sala *et al.* 1934, Broteria 30: 5–9, bibliogr.)

DEUBEL, FRIEDRICH (1843–1933) (Holdhaus 1936, Arb. morph. taxon. Ent. Berl. 3: 35–48, bibliogr.)

DEUQUET, CAMILLE FELIX (1882–1972) (Anon. 1973, Circ. Ent. Soc. Aust. (N. S. W.) 236: 24, bibliogr.)

DEUTERANOPTIC Adj. (Greek, *deuteros* = second; *an* = not; *opsis* = sight; *-ic* = of the nature of.)

DEUTEROTOKY Noun. (Greek, *deuteros* = second; *tokos* = birth. Pl., Deuterotokies.) A form of parthenogenetic reproduction in which progeny are male and female. (See Walter 1983). See Parthenogenesis. Cf. Arrhenotoky; Heterogony; Thelytoky.

DEUTEROZOIC Adj. (Greek, *deuteros* = second; *zoe* = life; *-ic* = of the nature of.) A term proposed for the Devonian through Permian Periods of the geological time-scale. See Palaeozoic Era.

DEUTOCEREBRAL Adj. (Greek, *deuteros* = second; Latin, *cerebrum* = brain; *-alis* = belonging to, pertaining to.) Descriptive of or pertaining to the brain's Deutocerebrum. See Deutocerebrum. Cf. Protocerebral; Tritocerebral.

DEUTOCEREBRAL COMMISSURE The connection between the antennal Glomeruli of opposite sides, traversing the lower part of the Brain (Snodgrass). See Brain; Commisure.

DEUTOCEREBRAL REGION The part of the primitive arthropod Brain which innervates the Antennae; the antennular region (Snodgrass). See Brain; Deutocerebrum.

DEUTOCEREBRAL SEGMENT The antennal segment of the insect head. See Head Segmentation.

DEUTOCEREBRUM Noun. (Greek, *deuteros* = second; Latin, *cerebrum* = brain. Pl., Deutocerebrums; Deutocerebra.) A preoral neuromere or middle portion of insect Brain which represents fused Ganglia of hypothetical antennal segment of head. Deutocerebrum innervates Antennae and consists of two Neuropiles: Antennal Lobe and Dorsal Lobe (Antennal Mechansensory and Motor Centre). Both Neuropiles receive primary sensory fibres from receptor cells of Antenna. Antennal Lobe is responsible for processing olfactory information and forms a Neuropile with several classes of central neurons and distinct compartments (Glomeruli). Glomeruli are separated by a layer of Glial Processes (ca 1000 very small glomerulus-like compartments have been counted in *Locusta migratoria*). Sexual dimorphism in insect Antenna probably also involves the Brain's Antennal Lobe. Antennal Mechanosensory and Motor Center (AMMC) is responsible for processing mechanosensory information. Scape and Pedicel contain mechanoreceptors (Bohm's Organ; Janet's Organ; Johnston's Organ). Central projections from these mechanoreceptors concentrate in AMMC. Boundaries of AMMC are difficult to define because Neuropile fuses with adjacent Protocerebrum. Some mechanoreceptors from Flagellomeres bypass Antennal Lobe and terminate in AMMC. Some motor neurons which innervate Antennal muscles have dendritic fields within AMMC. Several types of Neurons are associated with Deutocerebrum. Local Interneurons (Amacrine Cells) with arborizations but lacking a conspicuous Axon are confined to Antennal Lobe. Projection Interneurons (Output Neurons, Principal Neurons) contain Dendrites in Antennal Lobe and Axons that project into Protocerebrum. Centrifugal Neurons contain dendritic arborizations outside Antennal Lobe (in Protocerebrum) and Axons into Antennal Lobe. Primary Afferent Neurons include olfactory receptors (Trichode and Basiconic Sensilla) on flagellar segments which have Axons that project into single Glomeruli of the ipsilateral Antennal Lobe, or are connected to the contralateral Antennal Lobe via Commissures. Receptor neurons on Labial Pit Organs also have Axons that project to Antennal Lobe. Accessory Lobes are typically two in number, ventrolateral in position and connected by transverse fibres. Syn. Antennal Lobe. See Brain; Central Nervous System. Cf. Dorsal Lobe; Protocererbrum; Tritocerebrum.

DEUTONYMPH Noun. (Greek, *deutos* = two; *nymphe* = bride. Pl., Deutonymphs.) Acarina: The active, 8–legged, predaceous, second post-larval instar. Syn. Nymph. (Goff *et al.* 1982, J. Med. Ent. 19: 223.) Cf. Protonymph; Tritonymph; Larva. Rel. Development; Metamorphosis.

DEUTOPLASM Noun. (Greek, *deuteros* = second; *plasma* = mould. Pl., Deutoplasmas.) The yolk or food plasm in the Cytoplasm of an egg or other cell. Alt. Deuteroplasm. Cf. Energid.

DEUTOTERGITE Noun. (Greek, *deuteros* = second; Latin, *tergum* = back; *-ites* = constituent. Pl., Deutotergites.) The secondary dorsal segment of the Abdomen.

DEUTOVUM See Prelarva.

DEVASTATING GRASSHOPPER *Melanoplus devastator* Scudder [Orthoptera: Acrididae].

DEVELOPMENT Noun. (French, *développement* > *développer* = to develop. Pl., Developments.) 1. The change in an organism from one form of structural organization to another form of structural organization. (Seidel 1955, Naturwissenshaften 42: 275–286.) 2. Organisms: The act, process or series of biochemical, physical and biological events that originate with fertilization, proceed through growth, reproduction and senescence, and culminate in death. Development involves predictable change which is genetically programmed and may be altered by nutrition or environmental conditions. See Metamorphosis. Cf. Atrophy; Growth. Rel. Morphology; Ontogeny; Phylogeny.

DEVELOPMENTAL ASYMMETRY The programmed development of the embryo. Radial Cleavage leads to bilateral symmetry; Spiral Cleavage leads to asymmetry in platyhelminths, annelids and molluscs. See Asymmetry; Symmetry. Cf. Absolute Asymmetry; Behavioural Asymmetry; Biological Asymmetry; Morphological Asymmetry; Pattern Asymmetry; Skeletal Asymmetry. Rel. Organization.

DEVELOPMENTAL CYCLE Insects: The period or interval of time between fertilization of the egg and reproduction by the adult. Rel. Life Cycle.

DEVELOPMENTAL THRESHOLD The minimum ambient temperature (cardinal temperature) necessary for physiological functions of insect development to proceed.

DEVILLE, JEAN SAINTE-CLAIRE (1870–1932) (Barthe 1935, Miscnea ent. 36: 66–71, bibliogr.)

DEVIL'S COACH-HORSE *Creophilus erythrocephalus* (Fabricius) [Coleoptera: Staphylinidae]. (Australia).

DEVIL'S ROPE *Opuntia imbricata* (Haw.) DC. [Cactaceae]: A Species of cactus native to the New World and adventive elsewhere as a minor rangeland pest. DR has been controlled with phytophagous insects. See Prickly Pear.

DEVIL'S ROPE PEAR COCHINEAL *Dactylopius tomentosus* (Lamarck) [Hemiptera: Dactylopiidae]. (Australia).

DEVONIAN PERIOD The fourth Period (409–363 MYBP) of the Palaeozoic Era and named after Devonshire, the place where the Devonian Period stratum was first studied. Principal deposits include Rhynie (Scotland), Gilboa (New York) and Alken an der Mosel (Germany). During DP, marine invertebrates flourish, and first isopod Crustacea may have originated. Vertebrates diversify (Devonian called 'Age of Fishes') and included acanthodians, dipnoians and crossopterygians. Lungs developed and paired lateral appendages permitted penetration of terrestrial environment. First amphibians occur during DP. Terrestrial plants became tall with stems bearing well defined sap tubes and transparent Cuticle with stomata; these plants had small, non-vascularized up-turned scale-like leaves that were easy to climb up but difficult to climb down. Leafy plants *(Leclercquia),* first known wood *(Rellimia)* and Chert (an ancient peat) were common. Late Devonian shows lycopods (club mosses), ferns, horsetails, and seed ferns. Invertebrates are well documented during DP: Atelocerata (myriapod), Collembola appear during Middle Devonian and trigonotarbid acarines were fully terrestrial with air breathing book lungs. See Geological Time Scale; Palaeozoic Era. Cf. Cambrian Period; Lower Carboniferous Period; Upper Carboniferous Period; Ordovician Period; Permian Period; Silurian Period. Rel. Fossil.

DEWAR, DAVID ALEXANDER (1872–1957) (Richards 1958, Proc. R. ent. Soc. Lond (C)22: 75.)

DEWDROP SPIDER *Argyrodes antipodianus* Cambridge [Araneida: Theridiidae]. (Australia).

DEWITZ, HERMAN (1848–1890) (Anon. 1886, Festschr. Vers. dt. Naturf. 1886: 227.)

DEWS, SAMUEL CHARLES (1904–1973) (Reed 1974, J. Econ. Ent. 67: 465–466.)

DEXIIDAE See Tachinidae.

DEXIOTORMA Noun. (Greek, *dexios* = right; *tormos* = pivot. Pl., Dexiotormae.) Scarabaeoid larvae: A frequently slender sclerite extending inward from the right hind angle of the Epipharynx, sometimes provided with a heel-shaped Pternotorma (right Torma of Hayes) (Boving).

DEXTRAD Adv. (Latin, *dexter* = right.) Extending or directed toward the right. See Orientation. Cf. Anteriad; Apicad; Basad; Caudad; Centrad; Cephalad; Craniad; Dextrocaudad; Dextrocephalad; Distad; Dorsad; Ectad; Entad; Laterad; Mediad; Mesad; Neurad; Orad; Proximad; Rostrad; Sinistrad; Sinistrocaudad; Sinistrocephalad; Ventrad.

DEXTRAL Adj. (Latin, *dextra* = right-hand side; *-alis* = belonging to, characterized by.) 1. Inclined toward the right side. 2. Pertaining to a coiled structure in which the coil ascends from left to right of the observer. 3. Trichoptera: Applied to the direction of coiling in the case of some Species. Cf. Sinistral. Rel. Orientation.

DEXTRALITY Noun. (Latin, *dextra* = right-hand side. Pl., Dextralities.) 1. The condition or phenomenon of 'handedness' in which one side (right or left) differs from a basic state of appearance (size, shape) or functionality (use, disuse). 2. Right handedness; the condition in which the right hand is used in preference over the left hand.

DEXTRALLY Adv. (Latin, *dexter* = right.) Toward the right, as used in the context or rotation.

DEXTROCAUDAD Adv. (Latin, *dexter* = right, *cauda* = tail, *-ad* = adverbial suffix.) Extending obliquely between dextrad and caudad. See Orientation. Cf. Anteriad; Apicad; Basad; Caudad; Centrad; Cephalad; Craniad; Dextrad; Dextrocephalad; Distad; Dorsad; Ectad; Entad; Laterad; Mediad; Mesad; Neurad; Orad; Proximad; Rostrad; Sinistrad; Sinistrocaudad; Sinistrocephalad; Ventrad.

DEXTROCEPHALAD Adv. (Latin, *dexter* = right, *kephale* = head, *-ad* = adverbial suffix.) Extending obliquely between dextrad and cephalad. See Orientation. Cf. Anteriad; Apicad; Basad; Caudad; Centrad; Cephalad; Craniad; Dextrad; Dextrocaudad; Distad; Dorsad; Ectad; Entad; Laterad; Mediad; Mesad; Neurad; Orad; Proximad; Rostrad; Sinistrad; Sinistrocaudad; Sinistrocephalad; Ventrad.

DEXTRON Noun. (Latin, *dexter* = right. Pl., Dextrons.) The right side of the body. The area to the right of the median longitudinal line. See Orientation. Cf. Levatron.

DEXTROSE Noun. (Latin, *dexter* = right; *-osus* = full of. Pl., Dextroses.) Glucose.

DEYROLLE, ARCHILE (1813–1865) (Pascoe 1866. Proc. ent. Soc. Lond. (3) 2: 139.)

DF See Dengue Fever.

DHF See Dengue Haemorrhagic Fever.

DIACON® See Methoprene.

DIADOCIIDAE See Mycetophilidae.

DIADROMOUS Adj. (Greek, *diadromous* = wandering.) Pertaining to a fan-like nerve network. See Nervous System.

DIAFENTHIURON A thiourea compound {1-*tert*-butyl-3-(2,6-di-isopropyl-4-phenoxyphenyl)-thiourea} used as an acaricide, contact insecticide and stomach poison. Applied to cotton, fruit trees, vegetables and ornamentals in some countries Not registered for use in USA. Toxic to fishes and bees. Properties include slow acting, some ovicidial activity, effective on all stages of mites; may be used with other pesticides. Trade names include Pegasus® and Polo®.

DIAGNOSIS Noun. (Greek, *diagnosis* = discrimination. Pl., Diagnoses.) 1. An abridged or concise description of a Taxon. A statement limited to differential characters that purport to separate a Taxon from other Taxa. 2. A summary of character states that serves to distinguish a Taxon from other Taxa. Cf. Description. Rel. Taxonomy.

DIAGNOSTIC Adj. (Greek, *diagnosis* = discrimination; *-ic* = characterized by.) Pertaining to characteristics or features which purport to separate zoological entities.

DIAKINESIS Noun. (Greek, *dia* = through; *kinesis* = movement. Pl., Diakineses.) The final stage of prophase in meiosis during which the nuclear membrane disappears. See Kinesis. Cf. Allokinesis; Blastokinesis; Klinokinesis; Ookinesis; Orthokinesis; Thigmokinesis. Rel. Taxis; Tropism.

DIAKONOFF, ALEXANDER MIKHAILOVICH (1888–1956) (Lieftinck 1972, Ent. Ber., Amst. 32: 41–52, bibliogr.)

DIALYSIS Noun. (Greek, *dia* = through; *lysis* = loosing; *-sis* = a condition or state. Pl., Dialyses.) 1. Separation of a solution's constituents through a permeable membrane. 2. Separation of the outstanding differential charcters.

DIAMMINAE Subfamily of Thynnidae (Hymenoptera: Aculeata). See Thynnindae.

DIAMOND BEETLE *Chrysolopus spectabilis* (Fabricius) [Coleoptera: Curculionidae]. (Australia).

DIAMONDBACK MOTH *Plutella xylostella* (Linnaeus); *Plutella maculipennis* (Curtis) [Lepidoptera: Tineidae (Plutellidae)]: A cosmopolitan, multivoltine pest of cole crops, particularly cabbage and cauliflower. DBM probably is endemic to Europe and spread elsewhere with commerce. Eggs are yellow-white, laid singly or in groups of 2–3. Larvae wriggle violently when disturbed and often drop from the plant, suspended on a thread of silk. Larvae feed on all plant parts and kill young plants or cause reduced value in crops. Larvae mine leaves and become serious pests of cabbage and radish in Austrialia and southeast Asia. Pupation occurs within a loose, saclike coccon. DBM overwinters as adult in debris. See Tineidae.

DIAMONDBACKED SPITTLEBUG *Lepyronia quadrangularis* (Say) [Hemiptera: Cercopidae]: A monovoltine pest of grasses, shrubs and herbs in North America. Eggs are laid between leaf and mainstem; nymphs produce spittle that surrounds the body. DbSb overwinters in the egg stage. See Cercopidae.

DIAMPHIPNOIDAE Plural Noun. A Family of Plecoptera including two Genera, five Species, known from mountainous areas of Chile. Adults are large bodied and nearly 50 mm long. Larvae are dark green to brown, with branched respiratory gills on first four abdominal segments; second abdominal ganglion connects to metathoracic ganglion; head modified and not 'holognathous' as seen in eustheniids. Larvae are considered primarily herbivorous, but details of their life history are not extensive. See Plecoptera.

DIANA MOONBEAM *Philiris diana diana* Waterhouse & Lyell [Lepidoptera: Lycaenidae]. (Australia).

DIANEX® See Methoprene.

DIAPAUSE Noun. (Greek, *dia* = through; *pausis* = suspending; *diapauein* = to cause to cease. Pl., Diapauses.) 1. A physiological condition or state of restrained development and reduced metabolic activity which cannot be directly attributed to unfavourable environmental conditions. Visual consequences of diapause in postembryonic stages include slowed or suspended growth, differentiation, Metamorphosis or reproduction. In the insect embryo, diapause has been interpreted as a 'resting period' between Anatrepsis and Katatrepsis. The phenomenon of diapause was first experimentally observed by Duclaux (1869); the term diapause was first proposed by Wheeler (1893) for a specific stage in grasshopper embryonic development. Current physiological concept of diapause was proposed by Henneguy (1940). Diapause is regarded as adaptive and thought to increase the probability of survival during environmentally unfavourable periods while maintaining the life cycle in synchrony with seasonal progression. In univoltine insects, diapause is regarded as obligate and is independent of external stimulus. In multivoltine insects, diapause is regarded as facultative and is phase-set by external stimuli. 2. An anticipated, typically long-term, cyclical interruption in growth or development of an organism due to one or more environmental factors that occur well before adverse environmental conditions are manifest (Manisingh 1971, Canad. Entomol. 103: 993). See Aestival Diapause; Hibernal Diapause. Cf. Aestivation; Dormancy; Hibernation; Quiescence. Rel. Development.

DIAPHANOPTERODEA An Order of Palaeoptera and purportedly the most primitive of Palaeozoic Era pterygotes. Characterized by beak-like mouthparts, long Cerci, prominent Ovipositor and wings folded and fluted. Order similar to Megasecoptera in narrow wing base and anterior venation. Wing folding appears similar to Neroptera: Prevents being blown away and facilitates crawling in confined spaces. Diaphanopterodea are regarded as the sister-group of Palaeodictyoptera, Megasecoptera and Permothemistida.

DIAPHANOUS Adj. (Greek, *dia-* = between; *phainein* = to show; Latin, *-osus* = with the quality of.) Descriptive of structure which is clear or semitransparent, but not glass-like (Kirby & Spence). Alt. Diaphanus. Cf. Translucent; Transparent.

DIAPHRAGM Noun. (Greek, *diaphragma* = midriff. Pl., Diaphragms.) 1. Any thin membrane that divides a compartment. 2. The thin membrane separating the cavity containing the Heart from the remainder of the body. See Dorsal Diaphragm. 3. Lepidoptera: The membrane closing the posterior end of the male Abdomen between the bases of the Valvulae. See Ventral Diaphragm.

DIAPRIIDAE Haliday 1833. Plural Noun. A numerically large, cosmopolitan Family of apocritous Hymenoptera assigned to the Proctotrupoidea. Body size variable, typically small, dark, polished; female Antenna with 11–15 segments, male with 13–14 segments; Antenna geniculate, usually clubbed in female, Scape elongate, often on ledge-like prominence; male F I-II sexually dimorphic; forewing with Costal, marginals without Pterostigma; hindwing with 0–1 cells, without Jugal Lobe; Gaster somewhat pedunculate, Ovipositor retracted. Endoparasitic on Diptera pupae within puparia, sometimes gregarious; some Species hyperparasitic on Tachinidae in their hosts, Dryinidae in their hosts; some Species myrmecophilous. See Proctotrupoidea.

DIARACT See Teflubenzuron.

DIARTHROSIS Noun. (Greek, *dia* = through; *arthroun* = to connect by a joint. Pl., Diarthroses.) An articulation that permits motion between adjacent and connected structures. See Articulation; Segmentation. Rel. Acetabulum; Cotyla; Condyle; Enarthrosis; Ginglymus.

DIASPICERA Noun. Coccids: A group of three cells of the Ceratuba (MacGillivray).

DIASPIDIDAE Plural Noun. Armoured Scales. A cosmopolitan Family of about 1,700 Species of Coccoidea (Hemiptera). Adult female not insect-like in habitus and lacks eyes, Antennae and legs; apical segments of Abdomen forming a Pygidium containing glands with tubular ducts and marginal lobes; body cover (the Scale) is composed of glandular secretion, excrement and exuviae. A few Species of Diaspididae form galls and some Species are parthenogenetic. Female oviposits eggs or live young. First instar is called a 'crawler' and bears weakly developed eyes, Antennae and legs. The crawler serves as the agent of dispersal via walking, wind or phoresy. Crawlers insert their mouthparts into a plant, begin to feed and then moult. During the moulting process, the scale cover forms, legs and Antennae are shed and the insect becomes sedentary. Most Species complete three instars but development differs for males and females. Females moult into adults and remain under their scale cover and fixed to the host plant throughout their lives. Males undergo a Pupa-like stage of development, and then moult into a winged adult. Males are attracted to virgin females via sex pheromone, mate and die. Diaspididae may be univoltine, but many Species are multivoltine. Many Species cause serious economic damage to orchard trees, ornamental trees and shrubs. Diaspidids typically do not vector disease to host plant but their sugar-rich excrement (honeydew) accumulates on foliage and serves as substrate for sooty mould. Extensive feeding reduces plant vitality and can result in death. See California Red Scale; Purple Scale; San Jose Scale; Yellow Scale. Cf. Coccidae; Eriococcidae. Rel. Aleyrodidae; Pseudococcidae.

DIASTASE See Amylase.

DIASTATIC Adj. Pertaining to the enzyme diastase.

DIASTATIDAE Plural Noun. A small (seven North American Species) Family of acalypterate Diptera distributed throughout the Holarctic, but extending into the Oriental and Neotropical regions. Variously placed in Superfamily Muscoidea and Ephydroidea; diastatids superfically resemble the Drosophilidae, but are usually dark-coloured. Species are relatively rare and little studied. Adults frequent low deciduous shrubs along the margins of bogs, marshes and other lush forest habitats. Two fossil Species known from amber. See Diptera.

DIASTOLE Noun. (Greek, *diastole* = difference. Pl., Diastoles.) The regular expansion or relaxation of the Heart, through which action, blood is drawn inward or into the Heart. Cf. Systole.

DIASTOLIC Adj. (Greek, *diastole* = dilatation; *-ic* = characterized by.) Pertaining to the diastole. Cf. Systolic.

DIASTOMIAN SCENT GLAND Metathoracic Scent Gland of geocorisian Heteroptera, usually characterized by a widely spaced pair of median ventral orifices near the Metasternal Apophysis (Carayon 1971). See Exocrine Gland. Cf. Omphalian Scent Gland. Rel. Scent Gland System.

DIATIROLOGY Noun. (Pl., Diatirologies.) The science or art of preservation (Beal 1979).

DIATOM RAKE Mayfly nymphs: A configuration of the Galea composed of bristles and pectinated spines (or Setae/spines on Maxillae) used to scrape diatoms from stones or other objects for food (Needham).

DIATOMACEOUS EARTH A material used as a barrier for the control of ants and termites, and in grains for the control of stored product pests; also effective against snails and slugs. DE applied as a slurry, paste or dust; it abrades the Cuticle and causes desiccation. See Inorganic Insecticide.

DIAZIDE® See Diazinon.

DIAZINON An organic-phosphate (thiophosphoric acid) compound {O,O-diethyl-O-(2-isopropyl-6-methyl-5-pyrimidinyl) phosphorothioate} used as an acaricide, contact insecticide and stomach poison against mites, ticks, thrips, plant-sucking insects, leaf-chewing insects and domiciliary pests. Compound is applied to numerous tree crops, vegetables, ornamentals, nursery stock, livestock barns and turf. Diazinon also is applied as a public-health insecticide for control of mosquitoes, fleas, lice, houseflies and ants. Regarded as fast-acting with residual effects and compatible with other insecticides. Compound is phytotoxic to apples, some ornamentals and is not for use on African violets, ferns, gardenia, hibiscus or poinsettia. Trade names include: Basudin®, D.Z.N.®, Dazzle®, Diazide®, Diazital®, Diazol®, Gardentox®, Kayazinon®, Kayazol®, Knox-Out®, Nedcidol®, Nipsan®, Nucidol®, Optimizer®, Patriot®, Sarolex®, Spectracide®, Terminator®. See Organophosphorus Insecticide.

DIAZITAL® See Diazinon.

DIAZOL® See Diazinon.

DIBB, JOHN ROTHWELL (1906–1973) (Lees 1974, Proc. R. ent. Soc. Lond. (C) 38: 58–59.)

DIBETA® See *Bacillus thuringiensis.*

DIBROM® See Naled.

DIBROME® See Ethylene Dibromide.

DICARZOL® See Formetanate.

DICEROUS Adj. (Greek, *dikeras* = two-horned; *-osus* = with the quality of.) With two Antennae (Kirby & Spence). Alt. Dicerus.

DICHAETAE Noun. (Greek, *di* = two; *chaite* = long hair.) A group of brachycerous Diptera with a Proboscis consisting of two parts (muscids, *etc.*)

DICHLOROVOS See Dichlorvos.

DICHLORPHOS® See Dichlorvos.

DICHLORVOS An organophosphorus insecticide (phosphoric acid) compound {2,2-dichlorovinyl dimethyl phosphate} used as a fumigant, acaricide, contact insecticide and stomach poison against pests of pets and livestock; often marketed as strips (Vapona strips, 738 Vapona strips, no-pest strips) or pet collars that create toxic vapor with a fumigant-like action. Compound is used as a surface spray for moderate-term (ca two month) residual control of urban insects (silverfish, cockroaches and some stored grain pests). Compound is applied to domiciles, barns, poultry houses and agricultural premises. In some countries, compound is applied as a foliar spray on crops. Syn. DDVP. Trade names include: Aminatrix®, Atgard®, Canogard®, Cekusan®, Dedevap®, Denkavephon®, Dichlorphos®, Divipan®, Doom®, Equigard®, Herkal®, Hosbit®, Krecalvin®, Lindan®, Mafu®, Marvex®, Nerkol®, Nogos®, Nuvan®, Oko®, Phosvit®, Piran®, Riddex®, Ritron®, Task®, Unifosz®, Vapona®, Vaponite®. See Organophosphorus Insecticide.

DICHOPTIC Adj. (Greek, *dicha* = in two, apart; *opsis* = sight; *-ic* = of the nature of.) With two eyes. Pertaining to insects with compound eyes medially separated by the Vertex, Frons and Face. See Compound Eye. Cf. Holoptic.

DICHOTOMIZE Intransitive Verb. (Greek, *dichotomos* > *dicha* = in two; *temnein* = to cut. Dichotomized; Dichotomizing.) To divide into two repeatedly, as in a stem or root.

DICHOTOMOUS Adj. (Greek, *dicha* = in two, apart; *temnein* = to cut; Latin, *-osus* = with the quality of.) Forked. Descriptive of structure that divides regularly into pairs.

DICHOTOMOUS KEYS A device of logic used in taxonomy to identify taxa. Specifically, dichotomous keys consist of statements about characters or character states which are provided in alternative contrasting statements arranged in couplets. Based on the correctness of the statement with regard to the specimen at hand, the user is directed to other contrasting statements and ultimately the name of the Taxon. Cf. Tabular Key, Pictorial Key. Rel. Identification.

DICHOTOMY Noun. (Greek, *dicha* = in two, apart; *temnein* = to cut. Pl., Dichotomies.) 1. The branching of one ancestral stem into two diverging lineages. Used in phylogeny of ancestral lines of descent. 2. Branches in wing venation of main veins or their branches. 3. A table or key for determining Species or higher groups, in which they are separated by contrasting characters arranged in couplets, two-by-two (Tillyard).

DICHROMATISM Noun. (Greek, *dicha* = in two, apart; *chroma* = colour; English, *-ism* = state. Pl., Dichromatisms.) The expression of two colour forms (morphs) within one Species. Cf. Monochromatic.

DICHTHADIIFORM ERGATOGYNE (QUEEN) Army Ants: A member of an aberrant reproductive caste, characterized by a wingless Alitrunk, large Gaster and expanded Postpetiole. Syn. Dichthadiigyne.

DICHTHADIIGYNE See Eichthadiiform Ergatogyne

DICKERSON, EDGAR (1878–1923) (Weiss 1924, Ent. News 35–38, bibliogr.)

DICKSON, ROBERT C (–1987) American academic and economic Entomologist (University of California, Riverside), specializing in insect vectors of plant pathogens and biology, ecology and taxonomy of aphids. First scientist to demonstrate endogenous periodicity (Circadian Rhythm) in animals.

DICKY RICE WEEVIL See Spinelegged Citrus Weevil.

DICOFEN® See Fenitrothion.

DICOFOL A chlorinated hydrocarbon {2,2,2-trichloro-1,1-bis(4-chlorophynyl) ethanol} manufactured as a contact acaricide and applied to numerous crops and ornamental plants in some countries. Phytotoxicity noted on eggplant, pears and some ornamental plants; toxic to fishes. Trade names include: Acarin®, Carbax®, Cekudifol®, Dicomite®, Difol®, Hilfol®, Kelthane®, Mitigan®. See Organochlorine Insecticide.

DICOMITE® See Dicofol.

DICONDYLIC Adj. (Greek, *di* = two; *kondylos* = knuckle; *-ic* = of the nature of.) 1. Pertaining to structure with two Condyles. 2. Generally, a term used in reference to Mandibles which have two articulatory processes with the Cranium.

DICONDYLIC ARTICULATION A joint with two points of articulation between adjacent segments. Dicondylic articulations are most common in adult insects, and typically anteroposterior, except at the Trochantero-femoral joint (dorsoventral). Alt. Dicondylian; Dicondylous. Syn. Dicondylic Joint *sensu* Snodgrass. See Leg Articulation. Cf. Monocondylic Articulation; Tricondylic Articulation.

DICROTOPHOS An organic-phosphate (phosphoric acid) compound {Dimethyl phosphate of 3-hydroxy-N,N-dimethyl-cis-crotonamide} used for the control of plant-sucking and leaf-eating insects, thrips and mites. Compound is applied to citrus, cotton, cereals, rice, ornamentals and sugarcane in many countries; it may be used for tick control on cattle and is registered on cotton in USA. Dicrotophos is phytotoxic to some grain seeds, and is toxic to bees, birds, fishes and wildlife. Trade names include: Bidrin®, Carbicron®, Ektafos®, Ricron®. See Organophosphorus Insecticide.

DICTYNIDAE Plural Noun. Hackle-band Weavers. The largest Family of the cribellate spiders; widely distributed and common in vegetation, leaf litter, tree bark and fissures in soil. Adults are small bodied (<5 mm long) and construct irregular webs, usually near the tips of plant branches or within crevices.

DICTYOPHARIDAE Plural Noun. A cosmopolitan Family of Fulgoroidea (Hemiptera) including about 600 Species. Head typically prolonged anteriad; terminal segment of beak more than two times longer than wide; median Ocellus absent; claval vein not extending to apex of Clavus; anal area of hindwing without crossveins; second Tarsomere of hindleg with a row of apical spines. Nymph and adult feed on grass. See Fulgoroidea.

DICTYOPTERA Noun. (Greek, *diktyos* = net; *pteron* = wing. Pl., Dictyopterae.) An ordinal name applied to mantids and cockroaches. In early classifications also more generally applied to the Orthoptera (Smith). See Blattaria; Blattodea.

DICTYOSOME Noun. (Greek, *diktyon* = net; *soma* = body. Pl., Dictyosomes.) See Golgi Body.

DICTYOSPERMUM SCALE *Chrysomphalus dictyospermi* (Morgan) [Hemiptera: Diaspididae]: A polyphagous armoured scale insect that is a widely distributed pest of *Citrus*. DS is controlled by parasitic Hymenoptera in some regions. Syn. Spanish Red Scale. See Diaspididae.

DIDACTYLE Adj. (Greek, *di* = two; *dactylos* = finger.) Literally, two-toed. Organisms with two Tarsi or spots touching or confluent. Alt. Didactylus.

DIE Verb. (Middle English, *dien* > Old High German, *touwen* = to die. Died; Died; Dying; Dies). 1. The irreversible termination of the physical processes of life. See Life. 2. To come to an end or pass out of existence. See Death. 3. The collective action of a group to disappear or subside gradually. Ant. Live. See Extinction.

DIE-BACK Plants: The progressive death of roots, shoots or branches; DB typically begins at the tip of shoots or branches.

DIECK, GEORG (1847–1925) (Krausse 1926, Int. ent. Z. 20: 9–10.)

DIEL PERIODICITY Biological phenomena that are expressed on a 24-hour periodicity. A term applied to biological phenomena without regard to whether the periodicity is influenced by external stimuli or internal control. Cf. Biological Clock; Circadian Rhythm; Photoperiod.

DIELDRIN A synthetic, chlorinated hydrocarbon insecticide {1R, 4S, 4aS, 5R, 6R, 7S, 8S, 8aR)- 1, 2, 3, 4, 10, 10-hexachloro-1, 4, 4a, 5, 6, 7, 8, 8a-octahydro-6, 7-epoxy-1, 4, 5, 8-dimethanonaphthalene} used as a contact insecticide and stomach poison against mites, ticks, ants, armyworms, chinch bugs, crickets, cutworms, earwigs, grasshoppers, jap beetles, leaf miners, mosquitoes, snails, slugs, sowbugs, spittlebugs, termites, thrips, wasps and webworms. Dieldrin is used as a public-health insecticide and for control of desert locusts in Africa. The compound is applied to few crops in some countries, but is used in prevention and control of subterranean termites. Dieldrin is toxic to bees and fishes, and is persistent in soil. Compound removed from registration in USA and many developed countries. See Chlorinated Hydrocarbon Insecticide; Organochlorine.

DIELECTRIC WAVEGUIDE HYPOTHESIS Cf. Aeroscepsis.

DIENER, HUGO (1856–1935) (Szekessy 1935, Koleopt. Rdsch. 21: 56–57.)

DIENOCHLOR A synthetic, chlorinated-hydrocarbon Acaricide and Insecticide {Decachloro bis(2,4-cyclopentadiene-1-yl} used against mites and whiteflies in glasshouses on ornamentals. Compound with long residual action and does not affect beneficial insects. Dienochlor is compatible with most pesticides but is toxic to fishes and should not be used on feed for livestock. Trade names include Myten®, and Pentac®. See Chlorinated Hydrocarbon Insecticide.

DIET Noun. (Latin, *diaeta* > Greek, *diaeta* = manner of living. Pl., Diets.) The food necessary for subsistence at the minimum to include activity and reproduction. Cf. Artificial Diet; Chemically Defined Diet; Crude Diet; Semisynthetic Diet; Synthetic Diet.

DIETHION® See Ethion.

DIETL, ALBERT (1849–1901) (Weiss 1943, J. N. Y. ent. Soc. 51: 290.)

DIETRICH, KASPAR (1819–1878) (Anon. 1880, Mitt. schweiz. ent. Ges. 5: 391.)

DIETRICH, KONKARDIA AMALIE (NÉE NELLE) (Weidner 1967, Abh. Ver. naturw. Verh. Hamburg Suppl. 9: 149–153)

DIETZ, HARRY FREDERIC (1890–1954) (Mallis 1971, *American Entomologists*. 549 pp. (444–445), New Brunswick.)

DIETZ, WILLIAM GEORGE (1848–1932) (Hungerford 1933, Ann. ent. Soc. Am. 26: 188.)

DIETZE, KARL (1851–1935) (Seitz 1935, Ent. Rdsch. 52: 177–178.)

DIFENTHOS® See Temephos.

DIFFERENTIAE SPECIFICAE An obsolete phrase for a descriptive method developed by Linnaeus in which a series of descriptive terms was used to distinguish Species.

DIFFERENTIAL GRASSHOPPER *Melanoplus differentialis* (Thomas) [Orthoptera: Acrididae]: A pest of grains, forage crops, cotton, sugar beets, vegetables and fruits. DG is common throughout North America. Female deposits eight pods each containing about 100 eggs. DG typically completes six nymphal instars. Males do not stridulate. See Acrididae.

DIFFERENTIATION Noun. (Latin, *differens* > *differe*.) A genetically-controlled process that yields stage-specific features. Intensification of this differentiation process results in a division of responsibility between stages: The immature (nymph, naiad, larva) feeds and the adult reproduces. During the process of differentiation between immature and adult, a behavioural liaison maintains continuity in an individual's life history. Immatures must utilize suitable food; adult males must know where to find adult females; adult females must know where to lay eggs. A failure to achieve the goals of any stage results in the death of the individual or failure to reproduce. See Metamorphosis. Rel. Development; Growth.

DIFFORMIS Adj. (Latin, *dis-*; *forma* = shape.) Irregular in form or outline shape. Not comparable; anomalous.

DIFFRACTED Adj. (Latin, *dis-*; *fractus* = broken.) Bending in different directions.

DIFFRACTION COLOUR Iridescent colouring due to the presence of a diffraction grating. In insects the grating is structural and composed of a series of parallel, equidistant lines engraved on the Integument. Iridescence by diffraction can only be seen along the path of the light. Thus, depending upon the angle of view, DC may not be visible. See Coloration. Cf. Advancing Coloration; Alluring Coloration; Anticryptic Coloration; Apetetic Coloration; Combination Coloration; Cryptic Coloration; Cuticular Colour; Directive Coloration; Disruptive Coloration; Epigamic Coloration; Episematic Coloration; Interference Colour; Pigmentary Colour; Procryptic Coloration; Protective Coloration; Pseudepisematic Coloration; Pseudoaposematic Coloration; Scattering Colour; Seasonal Coloration; Sematic Coloration; Structural Colour; Subhypodermal Colour. Rel. Crypsis; Mimicry.

DIFFRACTION GRATING Diffraction gratings represent one method of producing iridescent colour. Diffraction gratings consist of a series of parallel, equidistant, closely spaced grooves, lines or ridges on the Integument. Each groove (Sulcus), line or ridge (Carina) will reflect light. Light reflected by different Carinae will travel a different distance. The grooves, lines or parallel ridges are separated by somewhat more than a wavelength of light. Microtrichia are sometimes present on the grating. Spacing of Microtrichia affects colour; direction of Setae or Microtrichia may also affect the colour seen in a manner compared to the pile of a carpet when it is rubbed. Diffraction gratings are common among insects, but rarely observed owing to wavelengths involved (ca. 1–2 millimicrons separating lines optimal for human vision). Diffraction colours are not significant in terms of body heat gain or loss, and therefore are not significant in thermoregulation. Diffraction gratings may provide warning coloration or deceive predators about distance and size; diffraction grating may enhance warning coloration from parts of the body. In another way, distance is confounded by rapid colour or reflectance changes. Distance is perceived by stereoscopic vision (vertebrates) or parallax (vertebrates or invertebrates). If distance cannot be estimated, then size cannot be estimated. See Structural Colour; Iridescent. Cf. Scattering Colour. Rel. Pigmentary Colour.

DIFFUSE Adj. (Latin, *dis-*; *fundere* = to pour.) Spreading out; descriptive of structure of colour without distinct edge or margin. Alt. Diffusus.

DIFFUSION Noun. (Latin, *diffundere*. Pl., Diffusions.) The movement of molecules from an area of high concentration to an area of low concentration. A gradual mixing of two adjacent gases; two liquids, or a gas and a liquid.

DIFFUSION TRACHEAE Cylindrical Tracheae having noncollapsible walls. See Ventilation Tracheae (Snodgrass).

DIFLUBENZURON A chlorinated hydrocarbon {1-(4-chlorophenyl-3-(2,6-difluorobenzoyl urea)} that affects Chitin deposition and hence interrupts the moulting process in insects. Diflubenzuron attacks the larval stage of many insects feeding on crops, forest and livestock. The compound is not phytotoxic and is not taken into plants. Diflubenzuron is not effective on adult insects and does not affect plant-sucking insects but is toxic to Crustacea. Trade names include: Difluron®, Dimilin®, Dudim®, Kitinex®, Larvakil®, Micromite®, Vigilante®. See Organochlorine Insecticide; Insect Growth Regulator.

DIFLURON® See Diflubenzuron.

DIFOL® See Dicofol.

DIGESTION Noun. (Latin, *digestio* = digestion. Pl., Digestions.) The chemical process by which food is broken into units for metabolic assimilation. The digestive process involves three phases: An initial phase characterized by a decrease in the molecular weight of polymeric food molecules by the action of polymer hydrolases (amylase, cellulase, hemicellulase, trypsin), an intermediate phase whose products are dimers or small oligomeres (maltose, cellobiose), dipeptides derived from starch, cellulose and proteins, and a final phase during which the dimers are split into monomers by dimer hydrolases (maltase, cellobiase, dipeptidase). Alt. Absorption. Cf. Extra-oral Digestion. Rel. Alimentary System.

DIGESTIVE CELLS The secretory and absorptive cells of the Ventricular Epithelium as distinguished from the Regenerative Cells (Snodgrass).

DIGESTIVE TRACT See: Alimentary Canal.

DIGGER BEES Members of Subfamily Anthophorinae. Adult body robust and setose. DBs nest in soil and construct cells that are lined with waterproof wax or varnish-like material. Most Species collect pollen; some Species are cleptoparasitic on related Species. See Anthophoridae.

DIGGER WASPS See Sphecidae.

DIGGLE'S BLUE *Hychrysops digglesii* (Hewitson) [Lepidoptera: Lycaenidae]. (Australia).

DIGGLES, SILVESTER (1817–1880) Amateur naturalist who arrived in Queensland, Australia in 1854; he collected many insects in Queensland (particularly Coleoptera and Lepidoptera) that were sent to Francis Walker at the British Museum of Natural History. Diggles helped establish the Queensland Museum and acted as an honorary curator. (Mackerras & Marks 1974, Changing patterns in Entomology. A symposium. 76 pp. (pp. 6–7), Sydney.)

DIGIT Noun. (Latin, *digitus* = finger. Pl., Digits.) 1. Insects: Any finger-like structure. 2. Chelicerates: The distal portion of a Chela or Chelicera. 3. Immature Diptera: Prothoracic Spiracles.

DIGITATE Adj. (Latin, *digitus* = finger; *-atus* = adjectival suffix.) Finger-like. Pertaining to structure divided into finger-like processes. Alt. Digitatus. See Digit. Cf. Digitiform.

DIGITIFORM Adj. (Latin, *digitus* = finger; *forma* = shape.) Finger-shaped. Alt. Digitate. See Digit. Cf. Dentiform; Papilliform. Rel. Form 2; Shape 2; Outline Shape.

DIGITULE Noun. (Latin, *digitus* = finger. Pl., Digitules.) 1. Coccidae: Appendages of the feet that may be broadly dilated or knobbed Setae. 2. Tenent hairs; empodial hairs (MacGillivray).

DIGITUS Noun. (Latin, *digitus* = finger. Pl., Digits; Digitae.) 1. The terminal segment of the Tarsus that bears claws. 2. A small appendage attached to the Lacinia of the Maxilla; rarely present and probably tactile. 3. Genitalia of male Lepidoptera: A finger-like lobe arising from the Costa of the Harpes (Klots).

DIGONEUTIC Adj. (Greek, *di-*; *goneuein* = to generate; *-ic* = characterized by.) Descriptive of insects that are two-brooded. See Bivoltine.

DIGONEUTISM Noun. (Greek, *di-*; *goneuein* = to generate; English, *-ism* = condition. Pl., Digoneutisms.) The capacity to produce two broods in one season. Alt. Bivoltinism.

DIGUET, LÉON (1859–1926) (Calvert 1927, Ent. News 38: 261.)

DIJKSTRA, GERRIT (1910–1971) (Lempke 1971, Ent. Ber., Amst. 31: 21–22.)

DILARIDAE Plural Noun. A Family of Neuroptera comprised of two very rare North American Species that are found along the eastern coast, Cuba and Arizona. Adults rest with their wings apart, females lay eggs under bark and the larvae are predaceous. See Neuroptera.

DILATATE Adj. (Latin, *dilatatus* pp of *dilatare* = to dilate; *-atus* = adjectival suffix.) Widened in some part; dilated. Disproportionately broad.

DILATED Adj. (Latin, *dilatatus* > *dilatare* = to dilate.) 1. Expanded laterally; widened. 2. Flattened.

DILATION Noun. (Latin, *dilatatus* > *dilatare* = to dilate; English, *-tion* = result of an action. Pl., Dilations.) 1. The act or process of dilating. 2. The state or condition of being dilated. 3. An expanded or widened part or structure.

DILATOR Noun. (Latin, *dilatatus* > *dilatare* = to dilate; *-or* = one who does something. Pl., Dilators.) Any muscle that functions to enlarge, expand or open a structure. Cf. Protractor; Retractor; Flexor.

DILATOR BUCCALIS A muscle which opens or dilates the insect mouth. See Mouth Dilator.

DILATOR CIBARII One of a pair of muscles within the Clypeus arising from its anterior wall and inserted in the epipharyngeal surface of the Cibarium (Snodgrass).

DILATOR MUSCLE Any of several muscles extending from the body wall to the Alimentary Canal; suspensory muscles (Snodgrass). Cf. Occlusor Muscle.

DILATOR MUSCLE OF A SPIRACLE A muscle serving to open either the external or the internal closing apparatus of the spiracular Atrium (Snodgrass).

DILATOR PHARYNGEALIS The dilator muscle of the Pharynx.

DILATOR POSTPHARYNGEALIS The dilator muscle of the posterior Pharynx. Where such occurs, the Postpharyngeal Dilator Muscle.

DILATORES PHARYNGIS FRONTALES See Frontal Dilators of the Pharynx.

DILATORES PHARYNGIS POSTFRONTALES See Post-frontal Pharyngeal Dilators.

DILLEN (DILLENIUS) JOANNIS JACOBUS (1684–1747) (Rose 1850, *New General Biographical Dictionary* 7: 82–83.)

DILMOR® See Chlorpyrifos.

DILUENT Noun. (Latin, *diluens* > *diluere* = to wash away. Pl., Diluents.) 1. Diluting. 2. A thing that dilutes or dissolves. 3. Pesticide formulation: Any component added to reduce the concentration of the active ingredient in a pesticide. Oil or Water are common diluents in liquid formulations; talc or clay are common diluents in solid formulations. Cf. Active Ingredient; Inert Ingredient; Solvent; Surfactant; Synergist.

DILUTE Adj. (Latin, *dilutus* > *diluere* > *dis-* = off; *luere* = to wash.) 1. Diluted. 2. Descriptive of a pale or paling colour; thinned. Alt. Dilutus. Abbreviated dil.

DILUTION Noun. (Latin, *diluens* > *diluere* = to wash away. Pl., Dilutions.) Something diluted.

DILUTIOR Adj. (Latin, *diluens* > *diluere* = to wash away; *-or* = a condition.) Much thinned out or diluted.

DIMAZ® See Disulfoton.

DIMBOA (2,4-dihydroxy-7-methoxyl-1,4-benzoxazine-3-one.) A naturally occurring compound found in corn (*Zea mays* L.) that confers temporary resistance against attack by first-generation populations of European Corn Borer.

DIMECRON® See Phosphamidon.

DIMENOX® See Phosphamidon.

DIMERA Noun. (Latin, from Greek *dimeres* = divided in two. Pl., Dimerae.) Forms with two-segmented Tarsi: Specifically applied to some groups of Homoptera.

DIMEROUS Adj. (Greek, *di* = two; *meros* = part.; Latin, *-osus* = with the property of.) 1. Composed of two pieces; pertaining to parts arranged in pairs. 2. Descriptive of insects with only two tarsal segments. See Pretarsus. Cf. Pentamerous.

DIMETHOATE An organic-phosphate (dithio-phosphate) compound {O,O-dimethyl S-(n-methylcarbamoylmethyl) phosphorodithioate} used as an acaricide and systemic/contact insecticide for control of mites, plant-sucking insects (to include aphids, leafhoppers, scale insects, whiteflies), thrips, leaf miners, loopers and grasshoppers. Compound can be used very dilute in a granular formulation by homeowners. Conventional applications include dipteran control in barnyard settings, and other insects and mites on a variety of vegetable, fruit and ornamental crops. Dimethoate is chemically compatible with most other insecticides, fungicides and miticides but not for use with alkaline materials. Compound is applied to alfalfa, artichokes, bananas, cotton, cocoa, coffee, grapes, lettuce, mangoes, melons, soybeans, sugarcane, tea, tree crops, tobacco, vegetables, wheat and ornamentals. Dimethoate is regarded as phytotoxic on beans, chrysanthemums, cotton, figs, peaches, pines, tomatoes, lemons, olives and walnuts; it is also toxic to bees. Trade names include: Chemathoate®, Cygon®, Danadim®, Dimetox®, Dimethogen®, Fosfamid®, Lagon®, Perfekthion®, Rebelate®, Rogor®, Roxion®, Sinoratox®, Stinger®, System®, Tafgor®, Trimethion®, Trounce®, Vitex®. See Organophosphorus Insecticide.

DIMETHOGEN® See Dimethoate.

DIMETOX® See Dimethoate.

DIMIDIATE FASCIA Abbreviated fascia. See Fascia.

DIMIDIATE Adj. (Latin, *dimidius* = half; *-atus* = adjectival suffix.) 1. Descriptive of something divided into two equal halves. Halved. 2. Half-round; extending half-way around or across. 3. Descriptive of something half of the normal condition (*e.g.* dimidiate Elytra that cover only half of the Abdomen). Alt. Dimidiatus.

DIMIDIUS Adj. (Latin, *dimidius* = half.) Of half length.

DIMILIN® See Diflubenzuron.

DIMMOCK, GEORGE (1852–1930) (Emerson 1930, Psyche 37: 299)

DIMORPH Noun. (Greek, *dis* = twice; *morphe* = shape. Pl., Dimorphs) An individual belonging to one of the two forms of the same Species.

DIMORPHIC Adj. (Greek, *dis* = twice; *morphe* = shape; *-ic* = of the nature of.) Pertaining to Dimorphism. Structure that occurs in two distinct forms, typically expressed in coloration, size or shape. Alt. Dimorphous. See: Sexual Dimorphism. Cf. Monomorphic; Polymorphic.

DIMORPHISM Noun. (Greek, *dis* = twice; *morphe* = shape; English, *-ism* = condition. Pl., Dimorphisms.) A genetically controlled (influenced), non-pathological condition in which individuals of a Species are characterized by distinctive or discrete patterns of coloration, size or shape. Dimorphism can be characterized as a seasonal, sexual or geographic manifestation.

See Sexual Dimorphism. Cf. Monomorphism; Polymorphism.

DIMPLING BUG 1. *Campylomma liebknechti* (Girault) [Hemiptera: Miridae]: A pest of apples in Queensland, Australia. Bugs are attracted to apple blossoms and migrate into orchards during warm weather where they feed on developing fruit and cause dimpling of fruit which results in downgrading. Following petal fall, adults migrate from the orchard. Syn. *Campyomma livida* Reuter. 2. *Niastama punctaticollis* Reuter [Hemiptera: Miridae] in Tasmania. See Miridae.

DINERGATANDROMORPH Noun. (Greek, *deinos* = powerful; *ergates* = worker; *aner* = male; *morphe* = form. Pl., Dinergatandromorphs.) A condition in which an ant expresses anatomical features of a soldier and anatomical features of a male. (Donisthorpe 1929, Zool. Anz. 52: 92–96.)

DINERGATE Noun. (Greek, *deinos* = powerful; *ergates* = worker. Pl., Dinergates.) A soldier ant, characterized by a large head and Mandibles.

DINGY BUSHBROWN *Mycalesis perseus perseus* (Fabricius) [Lepidoptera: Nymphalidae]. (Australia).

DINGY CUTWORM *Feltia ducens* Walker [Lepidoptera: Noctuidae]. See Cutworms.

DINGY DART *Suniana lascivia* (Rosenstock) [Lepidoptera: Hesperiidae]. (Australia).

DINGY DARTER *Telicota eurotas eurychlora* Lower [Lepidoptera: Hesperiidae]. (Australia).

DINGY GRASSDART *Taractrocera dolon dolon* (Plotz) [Lepidoptera: Hesperiidae]. (Australia).

DINGY JEWEL *Hypochrysops ignitus olliffi* Miskin [Lepidoptera: Lycaenidae]. (Australia).

DINGY RING *Ypthima arctoa arctoa* (Fabricius) [Lepidoptera: Nymphalidae]. (Australia).

DINGY RINGLET *Hypocysta pseudirius* Butler [Lepidoptera: Nymphalidae]. (Australia).

DINGY SHIELD SKIPPER *Signeta tymbophora* (Meyrick and Lower) [Lepidoptera: Hesperiidae]. (Australia).

DINGY SWALLOWTAIL *Eleppone anctus* (W.S. MacKleay) [Lepidoptera: Papilionidae]. (Australia).

DINIDORIDAE Plural Noun. A small Family of heteropterous Hemiptera assigned to Superfamily Pentatomoidea. Dinidorids are predominantly tropical in distribution with most Species occurring in Asia and tropical Africa; a few Species occur in Australia. Adult 10–15 mm long and dull coloured; Scutellum triangular and not covering Abdomen; Hemelytron reticulately veined; Tibiae lacking strong spines. See Hemiptera.

DINITROPHENOL Noun. A synthetic organic insecticide characterized by nitro groups attached to a phenol ring.

DINNAGE, HARRY (1876–1955) (Hall 1956, Proc. R. ent. Soc. Lond. (C) 20: 75.)

DINNOT, M B (–1969) (Anon. 1969, Bull. Soc. ent. Fr. 74: 7.)

DINOFLAGELLATES A Class of algae (Dinophyceae) important in marine environments as producers of organic matter and which are responsible for the phenomenon called Red Tides.

DINOPIDAE Plural Noun. Net-casting Spiders; Ogre-faced Spiders. A rare Family of small bodied, cribellate spiders. Spiders predominantly tropical in distribution; in USA, they apparently are restricted to southwest. Dinopids are cryptic in coloration and habitus; two extremely large eyes (thus the name ogre-faced). Dinopids are nocturnal in habit and develop a net-like web that is held with their Tarsi. Dinopids hang from vegetation and ensnare or capture passing insects in their web; spiders capture up to nine prey during one night, but must reconstruct a new web after each capture.

DIOECIOUS Adj. (Greek, *di-*; *oikos* = dwelling; Latin, *-osus* = full of.) General: 1. Pertaining to organisms with distinct sexes; male and female sexual structures or glands in different individuals. 2. Botany: Plants with pistillate and staminate flowers on the different plant. Cf. Monoecious.

DIOLICE® See Coumaphos.

DIOPSIDAE Plural Noun. Stalk-Eyed Flies. A small Family of acalypterate Diptera that includes about 150 Species in 13 Genera and assigned to Superfamily Diopsoidea. Adults are small to medium-sized (4–12 mm long) with head subrectangular in frontal aspect and conspicuous eye stalks; Vibrissae absent. Prosternum with Precoxal Bridge and Ovipositor absent. Diopsids are predominantly in tropical Africa, the Oriental region and Malagasy; only a few Species are known in Palaearctic and Nearctic. Biology of diopsids is poorly known. Adult males of some Species form leks; some larvae inhabit wet decomposing vegetation and some known as shoot borers. Syn. Centrioncidae. See Diopsoidea.

DIOPTIDAE Plural Noun. Oakworms. A Family of Lepidoptera assigned to Superfamily Noctuoidea. Dioptids are restricted to New World and are predominantly Neotropical in distribution. Perhaps a Subfamily of Notodontidae. See Noctuoidea.

DIOPTRATE Adj. (Greek, *dia* = through, across; *opsesthai* = to see; *-atus* = adjectival suffix.) 1. A term descriptive of insect eyes that display transverse divisions or partitions. 2. An ocellate spot with the pupil crossed or divided by a transverse line.

DIOPTRIC Adj. (Greek, *dioptrikos* > *dioptra* > *dia* = through; *opsomai* = to see; *-ic* = of the nature of.) Refractive or causing vision by the refraction of light.

DIOPTRIC APPARATUS The outer transparent part of an optic organ, consisting of the Cornea and usually of a subcorneal crystalline body (Snodgrass). See Compound Eye.

DIOPTRIC LAYER A refractive layer of the insect eye. See Compound Eye.

DIOSZEGHY, L (Capuse 1970, Revta. Muzeelor 1970: 124–126)

DIP Noun. (Old English, *dyppan*. Pl., Dips.) 1. An inclination or downward slope from the horizontal. 2. A liquid insecticide or pesticide in water and put into a tank through which livestock are moved. 3. A depression with steep sides that occurs on a surface or path.

DIPEL® A registered biopesticide derived from *Bacillus thuringiensis* var. *kurstaki*. See *Bacillus thuringiensis*.

DIPHAGLOSSINAE A Subfamily of Colletidae (Apoidea) consisting of less than five Genera. See Colletidae.

DIPHAGOUS Adj. (Greek, *dis* = twice; *phagein* = to eat; Latin, *-osus* = with the qualities of.) Pertaining to Species of parastic insects in which both sexes feed on the same host but in different ways. The phenomenon is best known in parasitic Hymenoptera whose males feed externally while females of the same Species feed internally. See Feeding Strategy. Cf. Autoparasitism. Rel. Monophagous.

DIPHASIC ALLOMETRY A polymorphic condition in which the allometric regression line (plotted on a log-log scale) is discontinuous with parts of different slope that connect at an intermediate point not relevant to either part.

DIPLATYIDAE Plural Noun. A small Family of arixeniine Dermaptera assigned to Superfamily Pygidicranoidea. See Dermaptera.

DIPLOGANGLIATA See Arthropoda.

DIPLOGLOSSATA Noun. An ordinal term proposed for Hemimeridae, because of the supposed presence of a second labial segment.

DIPLOID Adj. (Greek, *diploos* = double; *eidos* = form.) Pertaining to organisms with two sets of Chromosomes, or twice the haploid number. See Chromosome. Cf. Haploid.

DIPLOID Noun. (Greek, *diploos* = double; *eidos* = form. Pl., Diploids.) A diploid organism.

DIPLOPOD Noun. (Greek, *diploos* = double; *pous* = foot. Pl., Diplopods.) A member of the Diplopoda. Common name: millipedes. Cf. Cheilopod; Pauropod.

DIPLOPODA Plural Noun. (Greek, *diploos* = lip; *pous* = foot.) Thousand-Legged Worms; Millipedes. A cosmopolitan Class within the Phylum Arthropoda that includes several thousand described Species. Body elongate, cylindrical, with one pair of Antennae, and numerous body segments each of which has two pairs of legs. All diplopod Species are noctural herbivore/scavengers that occur in soil, sand, crevices, beneath rocks, in rotting wood and similar habitats; a few Species are intertidal or maine. Some Species are pests of vegetable crops. See Arthropoda. Cf. Chilopoda; Pauropoda; Symphyla; Insecta.

DIPLOPOLYNEURIDAE Rohdendorf 1961 A Genus of fossil Diptera known from the type-species of Lower Jurassic Period age from Kirghizistan. Placement questioned.

DIPLOPTERA Plural Noun. A Genus of cockroach in the Family Blaberidae (Blattodea); most Species are tropical and occur in leaf litter and debris. Alt. Diplopteryga. See Blaberidae.

DIPLOPTERYGA Noun. Hymenoptera: Wasps in which the wings are longitudinally folded when at rest.

DIPLURA Börner 1904 Noun. (Greek, *diploos* = two; *oura* = tail.) A cosmopolitan Class or Order of entognathous, epimorphic hexapods whose position in relation to the Insecta is questioned. Diplura consist of about 700 nominal Species. Adult body size and shape are highly variable – usually of small size, but some Species up to 50 mm long. Body narrow and eyes absent; Antenna monifilform with each segment containing intrinsic musculature. Legs each with five segments and Pretarsus with a pair of lateral claws and occasionally a median claw. Abdomen with 10 segments, some segments with Styli and eversible vesicles; Cerci usually present but variable in development; Gonostyli and Gonopore between segments 8 and 9; Malpighian Tubules reduced or absent; 7–8 abdominal ganglia; lateral tracheal systems apparently not connected medially. Internal fertilization is not known in Diplura and the males manufacture Spermatophores. Specimens live beneath logs and rocks; a few records associate Diplura with ants and termites. Campodeids are phytophagous; Japygids are predaceous. Included Families of Diplura are: Anajapygidae, Campodeidae, Dinjapygidae, Evalljapygidae, Heterojapygidae, Japygidae, Parajapygidae, Procampodeidae and Projapygidae. See Apterygota. Cf. Protura.

DIPLURIDAE Plural Noun. Funnel-web spiders; Funnel-web Tarantulas. A numerically large, widely distributed Family of hunting spiders related to tarantulas. Diplurid Species in the USA are restricted to the western states. Diplurids are characterized by three claws on each Tarsus, four Spinnerets and sheet-like funnel-shaped webs; spiders usually are situated in vegetation with the apex of the tunnel near the roots or other concealed place. Insects are trapped on the open end of the funnel where the spider takes its prey.

DIPNEUMONE Noun. (Latin, from Greek, *di-*; *pneumon* = lung. Pl., Dipneumones.) Organisms that possess two lungs. A term applied to some spiders.

DIPRIONIDAE Rohwer 1911. Plural Noun. Conifer Sawflies. A small Family of boreal Symphyta (Hymenoptera) assigned to the Tenthredinoidea. Adult body 5–10 mm long; Antenna with 14–32 segments, females serrate, males plumose; Hypostomal Bridge absent; Pronotum short with posterior margin curved; Scutellum without a Carina separating a post-tergum; forewing vein 2R-RS absent; hindwing with RS and M cells closed; fore Tibia with two simple apical spurs; middle Tibia without subapical spur; Ovipositor sheath (Gonostylus) not projecting beyond apex of Abdomen; male genitalia strophandrous. Diprioninae feed on Pinaceae; Monocteninae feed on Cupressaceae. See Symphyta.

DIPSOCORIDAE Dohrn 1859. Plural Noun. The numerically small Family of heteropterous Hemiptera assigned to the Superfamily Dipsocoroidea and closely related to the Schizopteridae. Antennal Pedicel twice as long as Scape; Labium with four segments, extending to fore Coxa; female with two Tarsomeres on all legs; male genitalia asymmetrical. Dipsocorids live within leaf litter and among stones; individuals run rapidly when disturbed but do not jump. See Hemiptera.

DIPSOCOROIDEA A Superfamily of Heteroptera including Ceratocombidae, Dipsocoridae, Hypsipterygidae, Schizopteridae and Stemmocryptidae. See Heteroptera.

DIPTERA Linnaeus 1758. (Permian-Recent.) (Greek, *dis* = twice; *pteron* wing.) Flies. A cosmopolitan Order of holometabolous insects including about 150,000 Species. Diptera are found in all habitats of all Zoogeographical realms. Adults with typically large, manipulable head; compound eyes large and multifaceted; three Ocelli usually present, sometimes two, occasionally anocellate. Antenna highly variable: moniliform and multisegmented to 3-segmented and aristate; mouthparts adapted for sucking, forming Proboscis or Rostrum; piercing-type Proboscis in predatory and blood-sucking Species. Prothorax and Metathorax reduced; membranous mesothoracic wings used in flight; metathoracic wings modified into club-like Halteres used as balancing organs; some apterous Species. Legs are variable in form and structure. Abdomen with 11 segments in ancestral condition, 10 segments in derived condition (10 + 11 fuse to form Proctiger); distal segments sometimes modified into a telescopic Postabdomen; Cerci at apex of segment 10; female Cerci primitively with two segments (most Nematocera and Brachycera), or reduced to one segment (throughout Diptera); male Cerci of one segment on Proctiger; male genitalia complex. Larvae are apodous. Pupae are adecticous and exarate or obtect. Diptera are biologically diverse and medically and ecologically important: Blood sucking Species vectors of many diseases; house flies vector of enteric diseases; larvae of some Species cause Myiasis; tephritid fruit flies serious pests of fruit; leaf miners cause questionable damage to foliage; many Species are beneficial as pollinators and parasites of insect pests. Classified into Suborders Nematocera and Brachycera. See Brachycera; Nematocera.

DIPTEREX® See Trichlorfon.

DIPTEROCECIDIUM Noun. (Greek, *dis* = twice; *pteron* = wing; *kekis* = gall nut; *-idion* = diminutive. Pl., Dipterocecidia.) A gall formed on a plant by a dipterous insect. See Gall.

DIPTEROUS Adj. (Greek, *dis* = twice; *pteron* = wing; Latin, *-osus* = full of.) An organism belonging to the Order Diptera, or having the attributes, characters or features of Diptera.

DIRECT HYPERPARASITOID In parasitical systems involving a host, primary parasite and hyperparasitic Species: A Species of hyperparasitic insect whose adult female stage lays eggs in or near the primary parasite. The host is not attacked or is not the object of oviposition by the hyperparasitic Species. See Hyperparasite. Cf. Indirect Hyperparasitoid. Rel. Parasitism.

DIRECT IMPORTATION In biological control, the movement of an organism from an endemic area into another area and released immediately without propagation in the laboratory. See Biological Control. Cf. Augmentative Biological Control.

DIRECT LOSS Devaluation of a marketable commodity due to the presence of pest insects or insect damage. See Dockage.

DIRECT METAMORPHOSIS See Incomplete Metamorphosis.

DIRECT PINNING The insertion of a standard insect pin directly through the body of an insect. See Insect Pins.

DIRECT WING MUSCLES The Axillary and dorsal muscles of the wings.

DIRECTING TUBE Lepidoptera larvae: Anterior division of the silk-spinning apparatus positioned within the Spinneret.

DIRECTIVE COLORATION Directive marks of colours that tend to divert the attention of any enemy or predator from more vital parts of the organims. See Coloration. Cf. Advancing Coloration; Alluring Coloration; Anticryptic Coloration; Apetetic Coloration; Aposematic Coloration; Combination Coloration; Cryptic Coloration; Disruptive Coloration; Epigamic Coloration; Episematic Coloration; Procryptic Coloration; Protective Coloration; Pseudepisematic Coloration; Pseudoaposematic Coloration; Scattering Colour; Seasonal Coloration; Sematic Coloration. Rel. Crypsis; Mimicry.

DIRPHIA SKIPPER *Motasingha dirphia* (Hewitson) [Lepidoptera: Hesperiidae]. (Australia).

DISC Noun. (Latin, *diskus* = disc. Pl., Discs.) 1. The central upper surface of any anatomical structure or body part; all of the surface area within the margin of a structure. 2. The central area of a wing. 3. Orthoptera: The obliquely ridged outer surface of the hind Femur in Saltatoria. 4. Coleoptera larvae: The abdominal motor processes. Alt. Disk.

DISCA Noun. (Greek, *diskus* = disc. Pl., Discae.) The place of attachment to the body of a large muscle, showing as a disc or ring on the outer surface (MacGillivray).

DISCAL Adj. (Latin, *diskus* = disc.) On or relating to the disc of any surface or structure. Rel. Subdiscal.

DISCAL AREA Wings: The more central portion of the wing or the area covered by the Discal Cell.

DISCAL BRISTLES Diptera: Usually one or more pairs of bristles inserted near the middle of the dorsal wall of the abdominal segments anterior of the hind margin (Comstock).

DISCAL CELL Lepidoptera: The large or Median Cell extending from the base of the wing toward the center; the Radial Cell of Comstock. Diptera: According to Williston, the first Media of Comstock. Odonata: The Discoidal Areolets. Trichoptera: The cell between the forks of the Radial Sector, separated by a crossvein from the second Apical Cell (Smith).

DISCAL CROSSVEIN 1. The vein separating the Discal Cell and second Basal Cell. See Discoidal Crossvein. 2. Vein M of Comstock-Needham system *teste* Curran. See Vein; Wing Venation.

DISCAL ELEVATION Heteroptera (Tingidae): The central area of the forewing raised above the surrounding level.

DISCAL PATCH Some male Hesperidae: The oblique streak of specialized black scales on the disc of the forewing.

DISCAL SCUTELLAR BRISTLES Diptera: See Dorsoscutellar Bristles.

DISCAL SCUTELLARS Diptera: The bristles on the disc of the Scutellum (Curran).

DISCAL SETA Some coccids: A single large Seta on the caudal half of the dorsal surface of the Operculum (MacGillivray).

DISCAL VEIN Lepidoptera: The crossvein closing the Discal or Median Cell, extends from Radius to Media. See Vein; Wing Venation.

DISCALOCA Noun. Coccids: A small round projecting area in the middle of the ventral aspect of the caudal end of the body; also termed Mesodiscaloca (MacGillivray), Vaginal Disc, Vaginal Areola, Ventral Scar or Subcircular Scar (of authors).

DISCIFORM Adj. (Latin, *diskus* = disc; *forma* = shape.) Disc-shaped; descriptive of structure formed or shaped as a disc. See Disc. Cf. Discoid; Operculiform. Rel. Form 2; Shape 2; Outline Shape.

DISCOCELLULAR NERVURE Lepidoptera: The Discal Vein. Alt. Discocellular Vein.

DISCOID Adj. (Greek, *diskos* = disc; *eidos* = form.) Disc-like in form or function. Pertaining to structure that is disc-shaped or plate-like. Alt. Discoidal; Disciform. See Disc. Cf. Patelliform. Rel. Eidos; Form; Shape.

DISCOIDAL AREA 1. The middle area or field of an organ, especially the wings. 2. Orthoptera: The area of the Tegmina between the posterior (Anal) and anterior (Costal) areas. Discoidal field.

DISCOIDAL AREOLETS Odonata: A varying number of rows of cells on the outer side of the triangle between the short sector (M of Comstock) and the upper sector of the triangle (Cu of Comstock); Post-Triangle Cells; Discal Cells.

DISCOIDAL CELL A term applied to some outstanding cells of an insect wing, *e.g.* the Quadrilateral (dragonflies), Median Cell in Diptera (Tillyard). Hymenoptera: (Norton), first Media, Media and Media of Comstock (Smith).

DISCOIDAL CROSSVEIN 1. Diptera: The vein separating the Discal Cell and second Basal Cells. See Discal Crossvein. 2. The M of Comstock-Needham system *teste* Curran. See Vein; Wing Venation.

DISCOIDAL FIELD See Discoidal Area.

DISCOIDAL NERVULE Lepidoptera: The Media *sensu* Comstock.

DISCOIDAL TRIANGLE Odonata: See Triangle.

DISCOIDAL VEIN 1. Diptera: The Media *teste* Schiner; Anterior Intercalary Vein *teste* Loew. 2. Hymenoptera: The Media *teste* Norton. Vein beyond the junction with the Medial Crossvein (Comstock). 3. Orthoptera: The first and largest branch of the Humeral Vein. See Vein; Wing Venation.

DISCOIDEOUS (Greek, *diskos* = disc; *eidos* = form; Latin, *-osus* = full of.) Discoidal. Alt. Discoideus. See Shape.

DISCOLCELLULARS Noun. (Greek, *diskos* = disc; plus cellular.) The collective term applied to the short, more-or-less transverse veins closing the cell of the lepidopterous wing distally (Tillyard).

DISCOLOMIDAE Plural Noun. A Family of Coleoptera assigned to Superfamily Cucujoidea. Adult less than 2 mm long, ovoid in outline shape with body flattened; Antenna with nine segments including apical segment forming a club; Pronotum and Elytra with glandular pores. Larva also are ovoid and flattened with head concealed; club-like Setae around body margin. Discolomids inhabit leaf litter and occur under bark where they feed on fruiting bodies of fungi. See Cucujoidea.

DISCOLOUR Adj. (Latin, *dis* = from; *celare* = to cover; *or* = a condition.) Any colour that appears different from surrounding or adjacent coloured parts of an insect; more-or-less contrasting; not concolorous. Alt. Discolorate; Discolored; Discolorous.

DISCONTINUOUS Adj. (Latin, *dis-*; *continuare* = to continue; *-osus* = with the quality of.) Not continuous; broken off. A term often applied to a line of Setae, wing venation, or band of coloration. Alt. Broken; Fractured; Ruptured. Ant. Continuous.

DISCONTINUOUS DISTRIBUTION Zoogeography: Occurrence of related organisms in widely separated areas without closely related forms in intervening areas. See Zoogeography.

DISCONTINUOUS VARIATION Variation in structure without intermediate forms. A mutation in the broad sense.

DISCOTA Noun. (Pl., Discotae.) Insects in which development of the adults is from imaginal discs in the embryo. See Adiscota.

DISCRETE Adj. (Latin, *discretus,* past participle of *discernere*.) Separate; individually distinct.

DISCRETE Noun. (Latin, *discretus,* past participle of *discernere*.) A unit. Any well defined, well separated structure or unit of organization. Typically applied to parts definitely delimited from other parts. Alt. Distinct; Separate.

DISCUS Noun. (Greek, *diskos* = disc. Pl., Disci; Discuses.) A disc or structure that is circular in outline and somewhat flat in shape.

DISCUS® See Isofenphos.

DISCUS OF MAXILLA 1. The disc (stalk) of the Maxilla. 2. The second part of the Maxilla which joins the insertion.

DISEASE Noun. (Middle English, *disese*. Pl., Diseases.) 1. Any departure from a healthy condition in an organism. 2. An impairment or disruption in the normal course of life processes (activity, vitality, growth, reproduction) in response to environment (malnutrition, climate), infective agents (pathogens) or a combination of factors. Syn. Ailing; Diseased; Ill; Sick. See Epidemiology. Rel. Pathogen.

DISH TRAP See Pitfall Trap.

DISJOINED (Latin, *disiunctus* = separated.) Separated; standing apart. Pertaining to insects with the head, Thorax and Abdomen separated by constrictions. Alt. Disjointed; Disjunct; Disjunctus.

DISJUNCTION Noun. (Latin, *dis-*; *junctum* = to join; English, *-tion* = result of an action. Pl., Disjunctions.) In biogeography, the quality of a taxon's distibution in areas geographically separated (*e.g.* Australia and South America).

DISLOCATED Adj. (Medieval Latin, *dislocatus* > *dislocare* > *dis-* ; *locare* = to place.) 1. Pertaining to something out of order. Not continuous; discontinuous. 2. Pertaining to something out of position; Disjointed; disjunct. 3. Pertaining to striae, bands or lines that are interrupted.

DISMATE® See Gossyplure.

DISPAR SKIPPER *Dispar compacta* (Butler) [Lepidoptera: Hesperiidae]. (Australia).

DISPARLURE A synthetic pheromone {(Z)-7,8-epoxy-2-methyloctadecane} used as an attractant (trap) and to disrupt sex pheromones of the Gypsy Moth. Trade names include Disrupt II® and Gyplure®.

DISPARO® See Chlorpyrifos.

DISPERSAL Adj. (Latin, *dispergere* = to disperse; *-alis* = characterized by, pertaining to.) 1. A non-directional movement of insects within or between habitats. Dispersal sometimes is divided into Passive Dispersal and Active Dispersal. Cf. Migration. 2. In biogeography, an explanation of Taxon distribution due to active movement from one place to another place. Cf. Emigration; Migration.

DISPERSAL BIOGEOGRAPHY An approach to biogeographical analysis that attempts to explain organism distribution based on apparent centres of origin. In dispersal theory related groups of organisms have a centre of origin, then expand their range (diffuse), encounter a barrier, move over the barrier (disperse) and then differentiate in isolation. In dispersal theory, barriers precede disjunction and Taxa are older than the barrier. See Center of Origin. Cf. Ecological Biogeography; Vicariance Biogeography.

DISPERSED (Latin, *dispergere* = to disperse.) With scattered markings, punctures or other small sculptures. Alt. Dispersus.

DISPONS, PAUL (1906–1972) (Carayon 1972, Bull. Soc. ent. Fr. 77: 257–258.)

DISPOSED Adj. (Latin, *dis-*; *poser* = to place.) Arranged or laid out.

DISRUPT II®. See Dispalure.

DISRUPTIVE COLORATION Colour patterns of an animal that obscure the outline of the animal and thereby confer protection from predators. Strongly contrasting adjacent colours tend to break up continuity and thereby obscure shape. Examples of this phenomenon are common: Warships painted with sharply contrasting colour patterns; a fawn covered with light spots; the stabilimentum on the web of banded garden spider. Sarcophagid flies are typically black and grey with stripped or tesselated patterns. This coloration pattern makes the flies difficult to detect on tree trunks, bark or lichens unless they move. These striped contrasting somber colours render the insects invisible in the appropriate habitat. See Coloration. Cf. Advancing Coloration; Alluring Coloration; Anticryptic Coloration; Apetetic Coloration; Aposematic Coloration; Combination Coloration; Cryptic Coloration; Cuticular Colour; Directive Coloration; Epigamic Coloration; Episematic Coloration; Interference Colour; Pigmentary Colour; Procryptic Coloration; Protective Coloration; Pseudepisematic Coloration; Pseudoaposematic Coloration; Seasonal Coloration; Sematic Coloration. Rel. Crypsis; Mimicry.

DISSEPIMENT Noun. (Latin, *dis-*; *saepire* = to hedge in, enclose. Pl., Dissepiments.) 1. A partition or wall. Term applied to the forming Septa separating the Coelom-sacs in the embryo. 2. The thin envelope surrounding members in obtect Pupae.

DISSILIENT Adj. (Latin, *dis-*; *salire* = to leap.) Bursting open elastically.

DISTACALYPTERON Noun. (Pl., Distacalypterons.) The Antosquama (sic); Squamma; Antitegula (MacGillivray).

DISTAD Adv. (Latin, *distare* = to stand apart; *ad* = toward.) Toward the distal end. See Orientation. Cf. Anteriad; Apicad; Basad; Caudad; Centrad; Cephalad; Craniad; Dextrad; Dextrocaudad; Dextrocephalad; Dorsad; Ectad; Entad; Laterad; Mediad; Mesad; Neurad; Orad; Proximad; Rostrad; Sinistrad; Sinistrocaudad; Sinistrocephalad; Ventrad.

DISTADENTES (Pl., Distadentis.) The Dentes of the distal end of the Mandible (MacGillivray).

DISTAL Adj. (Latin, *distare* = to stand apart; *-alis* = pertaining to.) 1. Descriptive of a structure near the free end of an appendage. 2. The portion of a segment farthest from the body or connection with the body. See Orientation. Cf. Apical; Mesial; Proximal.

DISTAL CELL Wings: The cell bounded by the branches of the crossveins. See Wing.

DISTAL PROCESS The peripheral branch or one of several distal branches of a sensory nerve cell (Snodgrass).

DISTAL RETINULA CELLS. Iris Pigment Cells. Densely pigmented cells surrounding the crystalline cone-cells and the corneal Hypodermis (Comstock). See Ommatidium. Rel. Compound Eye.

DISTAL SENSORY AREA Scarabaeoid larvae: The Haptomerum of Boving.

DISTALIA Noun. (Latin, *distare* = to stand apart; *-ia* = noun-forming suffix. Pl., Distaliae.) The collective term for all the segments of the Antenna except the Scape and Pedicel (Tillyard).

DISTANT, WILLIAM LUCAS (1845–1922) English Hemipterist who published extensively and whose work is well illustrated. (Osborn 1937, *Fragments of Entomological History*. 394 pp. (232), Columbus, Ohio.)

DISTANT Adj. (Latin, *distans* > *distare* = to stand apart; *-antem* = adjectival suffix.) Widely separated and indicating the separation of parts from each other by sutures or Sulci.

DISTAPECTINAE Plural Noun. Coccids: Broad Pectinae with the distal end truncate or subtruncate and teeth limited to this end (MacGIllivray).

DISTICARDO Noun. (Latin, *distare* = to stand apart; + *cardo*. Pl., Disticardoes; Disticardos.) The distal region of the Cardo (Crampton).

DISTICHOUS ANTENNA A pectinate Antenna in which the processes originate from the apex of a segment and bend forward at acute angles.

DISTICHOUS Adj. (Latin, *distichus* > Greek, *distichos* > *di-* = two; *stichos* = row; Latin, *-osus* = full of.) Bipartite; descriptive of structure separated into two parts. Alt. Distichus.

DISTIGALEA Noun. (Latin, *distare*, to stand apart; + *galea*. Pl., Distigaleae.) 1. The distal segment of the Galea. 2. The Distagalea *sensu* MacGillivray.

DISTIPHALLUS Noun. (Latin, *distans* = standing apart; Greek *phallos* = penis. Pl., Distiphalli; Distiphalluses.) See Phallus.

DISTIPHARYNX Noun. (Latin, *distans* = standing apart; *pharyngx* = throat. Pl., Distipharnices; Distipharynxes.) The portion of the foregut that connects the Epipharynx and Hypopharynx. See Pharynx. Cf. Basipharynx.

DISTIPROBOSCIS Noun. (Latin, *distans* = standing apart; Greek, *proboskis* = trunk. Pl., Distiproboscises; Distiproboscides.) Muscid flies: The outer third of the Proboscis that bears the Labella. See Proboscis.

DISTIPULVILLUS Noun. (Latin, *distans* = standing apart; *pulvillus* = small cushion. Pl., Distipulvilli; Distipulvilluses.) Heteroptera: A flattened, often lamellate, somewhat cup-like structure projecting from the Basipulvillus. See Pretarsus. Cf. Pulvilus.

DISTITARSAL Adj. (Latin, *distans* = standing apart; Greek, *tarsos* = sole of foot; *-alis* = belonging to, characterized by.) Descriptive of or pertaining to the Distitarsus or the apical tarsal segment. See Leg. Cf. Pretarsus.

DISTITARSUS Noun. (Latin, *distans* = standing apart; Greek, *tarsos* = sole of foot. Pl., Distitarsi.) The distal or apical segment of the Tarsus which bears the Ungues (tarsal claws). See Pretarsus.

DISULFOTON An organic-phosphate (dithio-phosphate) compound {O,O-diethyl-S-2-ethylthioethyl-phosphorodithioate} used as a systemic insecticide and acaricide against mites, aphids, thrips, plant-sucking insects and leaf-chewing insects. Compound is applied to cotton, corn, citrus, grapes, onions, potatoes, rice, tomatoes, vegetables and ornamentals. Trade names include: Dimaz®, Di-Syston®, Disultex®, Ekanon®, Frumin-AL®, Knave®, Oxydisulfoton®, Solvigran®, Solvirex®. See Organophosphorus Insecticide.

DISULTEX® See Disulfoton.

DI-SYSTON® See Disulfoton.

DITHIOMETHON® See Thiometon.

DITOMYIIDAE Plural Noun. See Mycetophilidae.

DITROCHA Noun. (Latin, *di-* = two; + *trocha* = short for trochanter.) Hymenoptera grouped together on the basis of a 2-segmented Trochanter.

DITRYSIA Noun. A Suborder (Series) of Lepidoptera which includes most Species. Female with separate Gonopore and Bursa Copulatrix; S2 with Apodemes; oblique Proboscis muscles; abdominal sex pheromone glands. Included Superfamilies: Calliduloidea, Castnioidea, Copromorphoidea, Cossoides, Bombycoidea, Gelechioidea, Geometroidea, Hesperioidea, Noctuoidea, Notodontoidea, Pterophoroidea, Papilonoidea, Pyraloidea, Sphingoidea, Tineoidea, Tortricoidea, Zygaenoidea and Yponomeutoidea. Cf. Dachnonypha; Monotrysia; Zeugloptera.

DITRYSIAN Adj. Pertaining to female Lepidoptera with two abdominal apertures associated with the reproductive system. An anterior aperture on Sternum VII or VIII receives the Aedeagus during copulation (Bursa Copulatrix, Ostium Bursa); a posterior aperture on Sternum X serves as a Gonopore. Regarded as a derived condition. Cf. Monotrysian; Oviporus.

DITTRICH, RUDOLPH (1850–1922) (Wolf 1924, Jh. Ver. schles. Insektenk. 14: 21– 23.)

DIURESIS Noun. (Greek, *dia* = through; *ouron* = urine. Pl., Diureses.) The phenomenon, act or condition of increased or excessive urine secretion.

DIURETIC Adj. (Greek, *dia* = through; *ouron* = urine.) Pertaining to an increase in secretion of urine.

DIURNAE Day fliers. An obsolete term applied to butterflies.

DIURNAL Adj. (Latin, *diurnae* = pertaining to day; *-alis* = belonging to, characterized by.) Descriptive of activity patterns that occur during daylight only. Cf. Crepuscular; Nocturnal. Rel. Circadian Rhythm.

DIVARICATE Adj. (Latin, *dis-*; *varicare* = to straddle; *-atus* = adjectival suffix.) Forked or divided into two branches, straddling or spreading apart, of wings, lapped at the base and diverging behind. Alt. Divaricatus.

DIVARICATION Noun. (Latin, *divaricare* = to spread apart. Pl., Divarications.) Spread apart or straddling.

DIVEN, EMERSON LISCUM (1899–1919) (Busck 1919, Proc. ent. Soc. Wash. 21: 177–178.)

DIVERGE Verb. (Latin, *dis-*; *vergere* = to bend, incline.) Spreading out from a common base. Cf. Converge.

DIVERGENT Adj. (Latin, *dis-*; *vergere* = to bend, incline.) Descriptive of structural parts that recede farther from each other. Cf. Convergent.

DIVERSE Adj. (Latin, *diversus* = turned up in several directions; OF. *diverser* = to change, vary.) Unequal; of various kinds. Pertaining to structures differing in size or shape. Alt. Sundry; Varied; Various. Ant. Uniform.

DIVERSITY Noun. (Old French, *diversite* = to change or vary > Latin, *diversitas* > *diversus* = diverse; English, *-ity* = suffix forming abstract nouns. Pl., Diversities.) 1. A state or instance of difference. 2. Multiformity. 3. Variety.

DIVERSITY PRINCIPLE In biogeography theory, a principle of the dispersal explanation according to which the direction of movement (dispersal) is indicated by the relative diversity of Taxa in different geographical areas.

DIVERTICULUM Noun. (Latin, *divertere* = to turn away. Pl., Diverticula.) A tube, sac or invagination originating on the wall of a vessel or the Alimentary Canal and closed at the distal end. Cf. Sac; Tube.

DIVIDENS (VENA) Orthoptera: first Anal Vein *sensu* Comstock.

DIVING FLIGHT A form of passive flight in which the insect engages in a fast descent at a large angle between the flight direction and the horizontal. DF is not common among insects but has been noted in Orthoptera and some broad-winged Lepidoptera. See Flight; Cf. Passive Flight.

DIVIPAN® See Dichlorvos.

DIVISION OF LABOUR See Polyethism.

DIXA MIDGES See Dixidae.

DIXEY, FREDERICK AUGUSTUS (1855–1935) English professional Entomologist and curator of Hope Museum (Oxford) ca 1900–1932. (Neave 1936, R. ent. Soc. Lond. President's address. pp. 1–2.)

DIXIDAE Plural Noun. Dixa Midges; Dixid Midges. A sometimes common Family of nematocerous Diptera assigned to Superfamily Culicoidea; dixids are widespread with about 40 Species in North America. Adults are small, slender, mosquito-like with long legs and Antenna; wing lacks scales. Adults lack functional mouthparts and do not bite. Larva are aquatic; three thoracic segments not fused; abdominal segments 1–2 with Pseudopods and segments 5–7 with Ambulacral Combs. Larvae occur at surface of water feeding on algae, microorganisms and decayed organic matter; they move in a horizontal, U-shaped pattern. See Culicoidea. Cf. Chaoboridae; Culicidae.

DIXON, ROBERT DONALD (1938–1974) (Dolinski 1975, Bull. ent. Soc. Can. 7: 12.)

DIXON, ROLAND MAURICE (1858–1910) (Dixon 1911, Zoologist 69: 118–119.)

DIXON® See Phosphamidon.

DIXXON, SAMUEL GIBSON (1851–1919) (Anon. 1918, Ent. News 29: 157.)

DMDT See Methoxychlor.

DNA PROBE Isolated single DNA strands used to detect the complementary strands.

DOANE, RENNIE WILBUR (1871–1942) American academic (Stanford University) who published in Economic and Medical Entomology and Forest Insects. (Osborn 1946, *Fragments of Entomological History*. Pt. II, 232 pp. Columbus, Ohio.)

DOBBINS, TRABER NORMAN (1896–1952) (Flemming 1952, *et al.* J. Econ. Ent. 45: 903.)

DOBREE, NICHOLAS FRANK (1831–1908) (Porritt 1908, Naturalist, Lond. 1908: 48–50, bibliogr.)

DOBREE-FOX, EDWARD CARTERET (–1906) (Bankes 1906, Entomol. mon. Mag. 42: 141.)

DOBSON FLIES See Megaloptera.

DOBSON, GEORGE EDWARD (–1895) (Anon. 1896, Irish Nat. 5: 73.)

DOBSON, H T (–1914) (Anon. 1914, Entomologist's Rec. J. Var. 26: 208.)

DOBSONFLIES See Corydalidae. Cf. Alderflies.

DOBSONFLY *Corydalus cornutus* (Linnaeus) [Neuroptera: Corydalidae]: A Species found in eastern North America. Adults are nocturnal, short lived and non-feeding. Males possess elongate Mandibles that are used in courship; females with short Mandibles. Larvae occur under stones in shallow, fast-flowing water; first seven abdominal segments each with a pair of Tracheal Gills. See Corydalidae. Rel. Hellgrammite.

DOCK APHID *Brachycaudus rumexicolens* (Patch) [Hemiptera: Aphididae]. (Australia).

DOCK SAWFLY *Ametastegia glabrata* (Fallén) [Hymenoptera: Tenthredinidae]: A sawfly endemic to Europe and adventive to North America, Chile and Australia. DS is a multivoltine pest of dock and related plants (Polygonaceae, Chenopodiaceae) in Canada and northern USA. Eggs are laid in leaves. Larvae (Dock False Wireworm) are bright green with white spots; larvae overwinter in hollow stems, softwood or fruit. See Tenthredinidae.

DOCKAGE Noun. (Middle English, *docke* = end of tail; *dock* = to cut off; *age* = *aticum* = cumulative result of. Pl., Dockages.) 1. Residual material from the processing of agricultural commodities including: chaff, broken kernels of grain, grain dust and dirt. 2. Devaluation of a product due to the presence of insects or undesirable foreign matter. See Direct Loss.

DOCTERS VAN LEEUWEN, WILLEM MARIUS (1880–1960) (Anon. 1960, Proc. Linn. Soc. Lond. 172: 131–132.)

DOD, FREDERICK HOVA WALLEY (1871–1919) (dos Passos 1951, J. N. Y. ent. Soc. 59: 175.)

DODD, ALAN PARKHURST (1896–1981) Australian economic Entomologist and son of F. P. Dodd. APD was employed by the Bureau of Sugar Experiment Stations 1912–1921 and responsible for taxonomy and biology of cane grubs. Member, Commonwealth Prickly-Pear Board 1921–1939. APD was responsible for importation of *Cactoblastis cactorum* from Argentina into Australia for biological control of *Opuntia*. Taxonomic specialist on parasitic Hymenoptera, particularly the Proctotrupoidea. He retired as Director, Biological Control Branch, Queensland Department of Lands.

DODD, FREDERICK PARKHURST (1861–1937) Australian self-trained naturalist and insect collector. FPD wrote definitive studies involving several unusual biologies of Australian insects. He developed an extensive collection and business based on material from Australia and New Guinea. (Monteith 1991, *The Butterfly Man of Kuranda*. Mackerras & Marks 1974, Changing patterns in Entomology. A Symposium. 76 pp. (8), Brisbane.)

DODDER GALL WEEVIL *Smicronyx sculpticollis* Casey [Coleoptera: Curculionidae].

DODD'S AZURE *Ogyris iphis* Waterhouse & Lyell [Lepidoptera: Lycaenidae]. (Australia).

DODD'S BUNYIP *Tamasa doddi* (Goding & Froggatt) [Hemiptera: Cicadidae]. (Australia).

DÖDERLEIN, LUDWIG (1855–1936) (Anon. 1936, Zool. Anz. 114: 160.)

DODERO, AGOSTINO (–1937) (Horn 1938, Arb. morph. taxon. Ent. Berl. 5: 72.)

DODGE, EDGAR A (–1933) (Van Duzee 1933, Pan-Pacif. Ent. 9: 52.)

DODGE, HAROLD RODNEY (1913–1973) (Telford & James 1973, Proc. ent. Soc. Wash. 76: 230–232)

DOEBNER, K (–1891) (Anon. 1955, Mitt. naturw. Mus. Aschaffenb. 7: 6.)

DOELLO JURADO, MARTIN (–1948) (Anon. 1949, Revta. Soc. ent. argent. 14: 234–238.)

DOERING, KATHLEEN CLARE (1900–1970) American academic (University of Kansas) who specialized in teaching and study of insect morphology. (Woodruff 1971, J. Kans. Ent. Soc. 44: 3–4, bibliogr.)

DOESBURG, PIETER HENDRIK (1892–1971) (Goot & Doesburg 1971, Ent. Ber., Amst. 31: 205–214, bibliogr.)

DOETS, CORNELIUS (1894–1952) (Lempke 1952, Ent. Ber., Amst. 14: 113–115, bibliogr.)

DOFMEISTER, GEORG (1810–1881) (Holzinger 1883, Mitt. naturw. Ver. Steierm 20: xxvii–xxxiii.)

DOG FLEA *Ctenocephalides canis* (Curtis) [Siphonaptera: Pulicidae]: A pest of dogs, cats, humans, rats, rabbits, squirrels and poultry. DF anatomically is very similar to Cat Flea; adult 1.5–2.5 mm long and laterally compressed; Frons with conspicuous, internal, club-shaped incrassation; Ctenidial Combs horizontal; Genal Process with small sharp spine; Frontal Tubercle absent; abdominal Terga 2–7 with conspicuous single row of Setae. Females lay ca 70 eggs on hair of host or ground of host's habitat. Larvae are whitish, elongate and cylindrical with sparse vestiture of Setae. In Australia, DF is less common than Cat Flea. DF is an intermediate host of the dog tapeworm, *Dipylidium caninum*. See Pulicidae. Cf. Cat Flea; Human Flea; Oriental Rat Flea.

DOG FOLLICLE MITE *Demodex canis* Leydig [Acari: Demodicidae]. (Australia).

DOG LOUSE *Heterodoxus spiniger* (Enderlein) [Phthiraptera: Boopiidae]: A widespread parasite on the domestic dog. DL probably is endemic to Australia with ancestral forms feeding on marsupials; hypothesized that recent ancestor shifted from marsupial to dingo and then to dogs. See Boopiidae; Cf. Dog-Biting Louse; Cat Louse.

DOG MANGE-MITE *Demodex canis* Leydig [Acarina: Demodicidae]. (Australia).

DOG NASAL-MITE *Pneumonyssoides caninum* (Chandler & Ruhe) [Acarina: Halarachnidae]. (Australia).

DOG SUCKING-LOUSE *Linognathus setosus* (Olfers) [Phthiraptera: Linognathidae]. (Australia).

DOG-BITING LOUSE *Trichodectes canis* (DeGeer) [Mallophaga: Trichodectidae]: In North America, a pest of puppies and intermediate host of dog tapeworm, *Dipylidium canium*. Adult ca 1 mm long, broad. Cf. Dog Louse.

DOG-EAR MARKS Apoidea: Small, subtriangular marks of light colour, just below the Antennae (Cockerell).

DOGWOOD BORER *Synanthedon scitula* (Harris) [Lepidoptera: Sesiidae].

DOGWOOD CLUBGALL MIDGE *Resseliella clavula* (Beutenmüller) [Diptera: Cecidomyiidae].

DOGWOOD SCALE *Chionaspis corni* Cooley [Hemiptera: Diaspididae]: A minor pest of deciduous trees in North America. DS resembles Elm Scurfy Scale and Scurfy Scale. See Diaspididae.

DOGWOOD SPITTLEBUG *Clastoptera proteus* Fitch [Hemiptera: Cercopidae].

DOGWOOD TWIG-BORER *Oberea tripunctata* (Swederus) [Coleoptera: Cerambycidae].

DOHANIAN, SENEKERIM MARDIROS (1889–1972) (Leonard 1974, J. Wash. Acad. Sci. 64: 250–251.)

DOHERTY, WILLIAM (1857–1901) (Holland 1902, Ent. News 13: 63–64.)

DOHRN, CARL AUGUST (1806–1892) (Musgrave 1932, *A Bibliography of Australian Entomology 1775–1930*. 380 pp. Sydney.)

DOHRN, FELIX ANTOIN (1840–1909) (Smith 1910, Proc. Linn. Soc. Lond. 1909–1910: 89–90.)

DOHRN, WOLFGANG LUDWIG HEINRICH (1838–1913) (Musgrave 1932, *A Bibliography of Australian Entomology 1775–1930*. 380 pp. (70), Sydney.)

DOI, HIRONOBI (1885–1949) (Hasegawa 1967, Kontyû 35 (Suppl.): 54 [In Japanese].)

DOIDJE, HARRIS (1869–1934) (Anon. 1934, Proc. Somerset archaeol. nat. Hist. Soc. 80: 56–57.)

DOKHTUROV, VLADIMIR SERGHEIVICH (1859–1890) (Musgrave 1932, *A Bibliography of Australian Entomology 1775–1930*. 380 pp. (70), Sydney.)

DOLABRATE Adj. (Latin, *dolabra* = mattock; pick-axe; *-atus* = adjectival suffix.) See Dolabriform. Alt. Dolabratus.

DOLABRIFORM Adj. (Latin, *dolabra* = mattock; pick-ax; *forma* = shape.) Hatchet-shaped; descriptive of compressed structure with a prominent dilated keel and cylindrical base. Syn. Dolabrate. Cf. Securiform. Rel. Form 2; Shape 2; Outline Shape.

DOLESCHALL, CARL LUDWIG (1827–1859) (Bonnet 1945, *Bibliographia Araneorum* 1: 35.)

DOLESCHALL, HENRICH (1855–1936) (Hoffman 1936, Ent. Z., Frankf. a. M. 50: 189.)

DOLICHASTER Noun. (Greek, *dolichos* = long. Pl., Dolichasters.) Ascalaphid larvae: Setal fringe along the lateral segmental processes of the Mandible (Imms).

DOLICHODERINAE A Subfamily of ants. Petiole present; Sting or acidipore or circlet of Setae absent; defensive secretion discharged through slit.

DOLICHOPIDAE See Dolichopodidae.

DOLICHOPODIDAE Plural Noun. Longlegged Flies. A cosmopolitan Family of orthorrhaphous Diptera including ca 150 Genera and 6,000 Species. Dolichopodids occur from high Arctic to southern Chile and several subantarctic islands; the Family is well represented on oceanic islands, and many Species of several Subfamilies occupy coastal or estuarine-intertidal habitats. Dolichopodids are closely related to Empididae, but separated by short Proboscis with fleshy Labellae and different wing venation. Adults are mostly slender, 1–9 mm long, and body often metallic green, sometimes yellow or more rarely brown or black; Frontal Suture absent; wing venation with Discal and second Basal Cells united; Anal Cell shortened or absent; Thorax with strong Setae. Female Abdomen apically pointed and male genitalia bulbous and held beneath Abdomen. Adults and larvae are predaceous. Adults usually are found in moist places; most Species are found on soil or vegetation near streams or moist areas; a few Species are found on tree trunks in deep shade. Some Species rest and feed on surface of water. Species whose larvae feed on adults and larvae of Scolytidae lay eggs in crevices in bark of trees; first instar larvae search out their prey. Larvae of aquatic or semiaquatic Species feed upon various water and soil organisms. Pupation normally occurs in cell. See Diptera.

DOLIOFORM Adj. (Latin, *dolium* = large cask; *forma* = shape.) Barrel-shaped. Cf. Discoid. Rel. Form 2; Shape 2; Outline Shape.

DOLIOLOIDES Noun. Obtect or coarctate Pupae of some insects.

DOLL, JACOB (1847–1929) (Osborn 1937, *Fragments of Entomological History*. 394 pp. (144, 291), Columbus, Ohio.)

DOLLFUSS, ERNEST (1852–1872) (Fauvel 1873, Annu. Ent. 1: 104–106.)

DOLLMAN, HEREWARD CHUNE (1888–1919) (Anon. 1920, Ent. News 31: 30.)

DOLLO PARSIMONY (Farris 1977, Syst. Zool. 26: 77–78.) See Parsimony.

DOLLO'S LAW A controversial concept named after the Belgian palentologist L. Dollo (1893) who titled the concept 'Law of Phylogentical Irreversibility' and noted 'qu'un organisme ne peut retourner meme partiellement, à un étàt antérieur, dejà réalisé dans le série de ses ancetres.' The concept has been taken to mean that evolutionary process is not reversible, or that an organ (structure) lost in the course of evolution will not be developed again through phylogenetical time. Numerous exceptions to Dollo's Law have been recorded and the concept probably should not be regarded as a 'law'.

DOLOCHLOR® See Chloropicrin.

DOMATIUM Noun. (Greek, *domation* = small house; Latin, *-ium* diminutive from Greek, *-idion*. Pl., Domatia.) Non pathological modifications of plant tissue which accomodate ants or mites. Development widespread in tropical dicotyledons; not reported from monocots, gymnosperms or herbs. Typically found at the axis of junctions of second-order veins with midribs. Modifications include pits, depressions, tissue overgrowth and tufts of Setae. Syn. Myrmecodomatia. See Acarodomatium.

DOMBEY, JOSEPH (1742–1796) (Rose 1850, *New General Biographical Dictionary* 7: 105.)

DOME ORGAN See Sensillum Campaniformium.

DOMESTIC CONTAINER-MOSQUITO *Aedes notoscriptus* (Skuse) [Diptera: Culicidae]: A persistent, but usually minor, domestic pest in many places; DCM bites during day, near sunset and at night. Female feeds on humans throughout day but peak feeding at dusk; also feeds on other mammals and birds. Proboscis black with pale band; legs with white bands; wings with dark scales. Female lays eggs individually in moist places near water or at waterline; eggs resistant to desiccation and hatch when immersed in water; natural habitats include tree holes or rock pools with decomposing vegetation; DCM has shifted to urban environments and breeds in roof gutters, plastic containers, car tyres, discarded cans. Experimentally shown to be a poor vector of Yellow Fever and Dengue; good vector of dog heartworm and will support some strains of filariasis worm *Wuchereria bancrofti* but not others.

DOMINANT Noun. (Latin, *dominare* = to dominate; *-antem* = an agent of something. Pl., Dominants.) A character more constant and conspicuous than any other, a type or series occurring in large numbers both as to Genera, Species and individuals and in which differentiation is yet active.

DOMINATOR® See Pirimiphos-Methyl.

DOMINEX® See Alphacypermethrin.

DOMINICAN AMBER Amber of Lower Miocene-Upper Oligocene age from the Dominican Republic which is apparently derived from *Hymenaea* (Leguminosae). See Amber; Fossil. Cf. Baltic Amber; Burmese Amber; Canadian Amber; Chiapas Amber; Lebanese Amber; Taimyrian Amber.

DOMINIQUE, JULES (1838–1902) (Bureau 1903, Bull. Soc. sci. nat. Ouest Fr. (2) 3: 471–491, bibliogr.)

DOMINULA SKIPPER *Anysynta dominula* (Plotz) [Lepidoptera: Hesperiidae]. (Australia).

DOMOIC ACID A biopesticide extracted from seaweed (*Chondria armata, Digenea simplex*). Mode-of-action probably via increased sensitivity of neuromuscular junctions to glutamic acid and DA binds to proctolin receptors. See Biopesticides; Botanical Insecticide; Natural Insecticide; Insecticide; Pesticides. Cf. Acetogenins; Alkaloids; Chromenes; Cyclopepsipeptides; Furanocumarins; Mammeins; Polyacetylenes; Terthienyl.

DONCASTER, LEONARD (1877–1920) (Anon. 1920, Science 52: 11.)

DONCELL, CHARLES DONCKIER (1802–1888) (Dimmock 1889, Psyche 5: 156.)

DONISTHORPE, HORACE ST JOHN KELLY (1870–1951) (Riley 1952, Proc. R. ent. Soc. Lond. (C) 16: 84.)

DÖNITZ, WILHELM (1838–1912) (Knoblauch 1912, Ber. senckenb. naturf. Ges. 4 293–294.)

DONNYSA SKIPPER *Hesperilla donnysa* Hewitson [Lepidoptera: Hesperiidae]. (Australia).

DONOVAN, BESSII (1877–1951) (Baynes 1952, Irish Nat. J. 10: 10.)

DONOVAN, CHARLES (1863–1951) (Riley 1952, Entomologist 85: 120.)

DONOVAN, EDWARD (1768–1837) English naturalist and author or several entomological works including: *An Epitome of the Natural History of the Insects of China*, *An Epitome of the Natural History of the Insects of India*, *An Epitome of the Natural History of the Insects of Asia*, *The Natural History of Insects* and *General Illustrations of Entomology*. (Musgrave 1932, *A Bibliography of Australian Entomology 1775–1930*. 380 pp. (70–71), Sydney.)

DONZEL, HUGO FLEURY (1791–1850) (Mulsant 1853, Opusc. ent. 2: 155–172, bibliogr.)

DONZEL, HYGNES (1810–1879) (Anon. 1880, Naturaliste 2: 159.)

DOODLEBUG Noun. (Etymology obscure. Pl., Doodlebugs.) The immature stage of Neuroptera assigned to the Myrmeleontidae. Doodlebugs live in sandy soil where they excavate a cone-shaped pit that traps ground-dwelling insects. See Myrmeleontidae.

DOOM® See Dichlorvos; Tetramethrin.

DORAMECTIN Noun. A 16-membered macrocyclic-lactone closely related to Abamectin, Moxidectin and Ivermectin that is produced by soil-inhabiting actinomycetes. Doramectin is purported to be effective for control of insects, acarines and gastrointestinal nematodes in yearling cattle.

DOREEN'S PREDATOR MITE *Typhlodromus doreenae* Schicha [Acarina: Phytoseiidae]. (Australia).

DORFMEISTER, GEORG (1810–1881) (Holzinger 1884, Mitt. naturw. Ver. Steierm. 20: xxvii–xxxiii.)

DORIA, GIACOMO (1840–1913) (Capra 1968, Boll. zool. 35: 463–470.)

DORIER, A (1900–1969) (Degrange 1970, Trav. Lab. Hydrobiol. Piscic. Univ. Grenoble 61: 6–16, bibliogr.)

DORILAIDAE See Pipunculidae.

DORMANCY Noun. (Latin, *dormire* = to sleep. Pl., Dormancies.) An imperfectly understood physiological phenomenon that is generally regarded as a state of suspended development or reduced metabolic activity. Dormancy is manifest in plants and animals. Among insects, dormancy may occur in embryonic, immature or adult stages and can be divided into the subcategories Aestivation, Athermopause and Hibernation (Mansingh 1971, Canad. Entomol. 103: 983). See Development. Cf. Aestivation; Athermopause; Hibernation; Diapause.

DORMANT Adj. (Latin, *dormio* = sleeping; *-antem* = adjectival suffix.) Pertaining to organisms during a period of physical inactivity. See Aestivation; Hibernation. Cf. Diapause.

DORMANT OIL A petroleum oil applied to trees only when foliage is not present. DOs are used for suffocating phloem-feeding insects. Highly phytotoxic to all green plant parts. DO are applied to plants to control scale insects and as ovicides. See Summer Oil; See Petroleum Oils.

DORMER, JOHN BAPTISTE JOSEPH (1830–1900) (Anon. 1901, Entomol. mon. Mag. 37: 49.)

DORN, KARL (1884–1973) (Dieckmann 1966, Mitt. dt. ent. Ges. 25: 1–2)

DORPLICA Noun. (Pl., Dorplicae.) The name applied to a tergal plica (MacGillivray).

DÖRRIES, FRITZ (1822–1916) (Weidner 1967, Abh. Verh. naturw. Ver. Hamburg Suppl. 9: 166–168)

DORSAD Adv. (Latin, *dorsum* = back; *ad* = toward.) In the direction of the dorsum or back of an insect. Ant.: Ventrad. See Orientation. Cf. Anteriad; Apicad; Basad; Caudad; Centrad; Cephalad; Craniad; Dextrad; Dextrocaudad; Dextrocephalad; Distad; Ectad; Entad; Laterad; Mediad; Mesad; Neurad; Orad; Proximad; Rostrad; Sinistrad; Sinistrocaudad; Sinistrocephalad; Ventrad.

DORSAL Adj. (Latin, *dorsum* = back; *-alis* = characterized by, pertaining to.) Descriptive of or belonging to the upper surface. See Orientation. Cf. Subdorsal; Lateral; Ventral.

DORSAL ACETABULUM A cavity on the dorsal surface of mandibular base. DA articulates with a Condyle on the Cranium. See Acetabulum; Cotyla.

DORSAL AMNIOSEROSAL SAC The dorsal organ at a later stage of development in the embryos of some insects. DAS is formed from ruptured and contracted Amnion and Serosa (Imms).

DORSAL ARMS OF TENTORIUM Apparent secondary extensions of the anterior arms of the Tentorium; extend from the head wall near the antennal base to the Tentorial Arm. DATs are often slender, tendon-like Apodemes that are sometimes absent. See Tentorium.

DORSAL BLASTODERM The extra-embryonic part of the Blastoderm; the Serosa *sensu* Snodgrass.

DORSAL BLOOD-VESSEL The mesodermal tube in insect embryos. The so-called Heart in adult insects. Syn. Dorsal Vessel. Rel. Aorta.

DORSAL BODY SETA See Dorsal Idiosomal Seta.

DORSAL BRISTLES See Dorso-central Bristles.

DORSAL DIAPHRAGM A Diaphragm that forms the floor of a dorsal compartment and separates the Heart and Perivisceral Sinuses in the Abdomen. DD usually is restricted to Abdomen but can project into Thorax. DD is not a continuous sheet of tissue and contains gaps or fenestrations between Pericardial Sinus and Perivisceral Sinus. See Circulatory System. Cf. Ventral Diaphragm.

DORSAL FLANGE Hymenoptera (Apocrita): An enlarged portion of the dorsolateral angle of each lateral arm of the Labial Sclerite.

DORSAL GLANDS Coccids: See Dorsal Pores (MacGillivray).

DORSAL IDIOSOMAL SETA Acarina: A Seta on the dorsal surface of the Idiosoma, excluding the scutal Setae. (Goff *et al.* 1982, J. Med. Ent. 19: 223.)

DORSAL LAMELLAE Coccids: See Dorsal Plates (MacGillivray).

DORSAL LIGHT REACTION Phenomenon in which an animal, presented with bright light, continues to orient its body in order to maintain the brightest point of illumination on the dorsal part of the body when light is presented from different aspects. DLR occurs in locusts that can fly upside down in a wind tunnel when the floor of the tunnel is lit. Cf. Dermal Light Sense.

DORSAL LINE Lepidoptera larva: A longitudinal line on the middle of the back or dorsum. Cf. Adorsal Line; Lateral Line.

DORSAL LIP Coccids: A transverse, strongly chitinized sclerite which supports the caudal end of the anal tube on the dorsal side (MacGillivray).

DORSAL MUSCLES In insects, the ordinary longitudinal muscles of the back, in which the fibres are typically longitudinal and attached on the intersegmental folds or on the Antecostae of successive Terga (Snodgrass).

DORSAL OCELLI When present, the usual simple eyes of adult insects, normally three in number in some Orders, two in others. See Ocellus. Cf. Anocellate.

DORSAL ORGAN 1. In the insect embryo, the Amnioserosal Sac, sometimes, however, arising from the amnion only (Imms). 2. A mass of cells in the dorsal part of the Embryo apparently produced by the invaginated Serosa (Snodgrass).

DORSAL OSTIOLES Pseudococcidae: The transverse, slit-like openings in the derm of the dorsum, on the Pronotum and the sixth abdominal segment opening into the Haemocoel (Ferris & Murdock). The Labiae *sensu* MacGillivray.

DORSAL PLATE Coccids: The Ceratubae and their external openings (MacGillivray).

DORSAL PORES Coccids: The Ceratubae and their external openings (MacGillivray).

DORSAL SCALE Diaspididae: Part of covering scale that lies above the insect. Cf. Ventral Scale.

DORSAL SETA Coccids: Setae on the dorsal aspect of the Pygidium (MacGillivray).

DORSAL SETAL FORMULA Acari: The arrangement of Setae by rows from anterior to posterior: Humerals, first posthumeral, second posthumeral rows, *etc.* (Goff *et al.* 1982, J. Med. Ent. 19: 223.)

DORSAL SINUS The body cavity containing blood and positioned beneath the abdominal Terga, above the Dorsal Diaphragm and surrounding the dorsal vessel (Heart). Syn Pericardial Sinus *sensu* Snodgrass.

DORSAL SPACE Slug-caterpillars: The area between subdorsal ridges.

DORSAL STYLET Lice: Upper paired Stylet of the mouth.

DORSAL TIBIAL SETAE See Palpal Setae.

DORSAL TRACHEAE The dorsal segmental Tracheae originating at a spiracle.

DORSAL TRACHEAL-COMMISSURE A commissure that crosses above the dorsal blood vessel.

DORSAL TRACHEAL-TRUNK A longitudinal dorsal trunk uniting the series of dorsal Tracheae. See Trachea. Rel. Respiration.

DORSAL TUBERCLES 1. Some coccids: Minute structures resting upon the surface of the Cuticle that resemble a small neckless flask (MacGillivray). 2. Submarginal Tubercles *sensu* Green. See Tubercle. Cf. Acantha; Spine.

DORSAL TUBULAR-SPINNERETS Coccids. See Dorsal Pores (MacGillivray). See Spinneret.

DORSAL VESSEL The 'Insect Heart' or primary channel for blood movement in the insect's body. DV components include a chambered abdominal Heart connected with an unchambered thoracic Aorta and cephalic Aorta. Abdominal Heart also displays Ostial Valves. Anatomy of the Dorsal Vessel is highly variable: Simple straight, parallel-sided tube (Thysanura, Dermaptera, larval Diptera); bulbar organ (Orthoptera, Ephemeroptera, Coleoptera); branched with bilateral extensions in head, Thorax or Abdomen (Mallophaga, Lepidoptera, some Hymenoptera). A 'Primitive' condition (groundplan) may include three thoracic Aorta segments and nine abdominal Heart segments. See Circulation. Cf. Aorta; Heart; Ostial Valve. Rel. Haemolymph.

DORSAL-GLAND ORIFICES Diaspinae: Oval orifices arranged in more or less distinct rows on the surface of the Pygidium, through which is discharged the material of which the dorsal scale is formed. Syn. Dorsal Pores (MacGillivray).

DORSEY, CARL KESTER (1901–1970) (Butler *et al.* 1971, Ann. ent. Soc. Am. 64: 563)

DORSIFEROUS Adj. (Latin, *dorsum* = back; *ferre* = to carry; *-osus* = full of.) Carrying or bearing on the back. A term applied to insects that carry their eggs or immatures (*e.g.* male Belostomatidae).

DORSO-ALAR REGION Diptera: The area between the Transverse Suture (Scuto-scultellar) and the Scutellum on one side and the root of the wing and the dorsocentral region on the other.

DORSOCAUDAL Adv. Toward the back and posterior part of a body or appendage.

DORSOCENTRAL BRISTLES Diptera: A row of bristles on each side, adjacent to and parallel with the Acrostichal Bristles; the Dorsocentrals (Comstock); two or four longitudinal rows on the inner part of the dorsum (Smith).

DORSOCENTRAL REGION Diptera: An area bounded by two imaginary lines drawn from the Scutellar Bridges forward, and coinciding with a space free from bristles that exists on the outer side of the dorsal rows and is often occupied by a dorsal thoracic stripe.

DORSOCENTRALS See Dorsocentral Bristles.

DORSO-HUMERAL REGION Diptera: A space bounded by the anterior end of Thorax and transverse suture on two sides and by the Dorsopleural Suture and dorsocentral region on the two others.

DORSOLUM See Dorsulum; Mesoscutum.

DORSOMESON Noun. (Latin, *dorsum* = back, Greek, *mesos* = middle. Pl., Dorsomesons.) The intersection of the Meson with the dorsal surface of the body (Garman).

DORSOPLEURAL LINE The line of separation between the dorsum and the pleural region of the body, often marked by a fold or groove (Snodgrass).

DORSOPLEURAL SUTURE Diptera: The lateral suture located between the dorsum and pleuron from the Humeri through the wing base. DPS separates the Mesonotum from the Pleuron (Smith).

DORSOSCUTELLAR BRISTLES Diptera: Usually a single pair of bristles on the dorsal part of the Scutellum, one on each side of the median line, slightly behind its middle (Comstock).

DORSOTENTORIA Plural Noun. The dorsal arms of the Tentorium.

DORSOVALVULAE Plural Noun. Dorsal valves of the Ovipositor (Crampton).

DORSOVENTRAL In a line from the upper to the lower surface.

DORSULUM Noun. (Latin, *dorsum* = back. Pl., Dorsula.) The Mesonotum anterior of the Scutellum, to include the wing sockets; specifically, the Mesoscutellum.

DORSUM Noun. (Latin, *dorsum* = back. Pl., Dorsums.) 1. The anatomical upper surface of any structure or body. 2. Odonata: Mespisterna, mesothoracic and metathoracic Terga. 3. Coleoptera: Often confined to Mesothorax and Metathorax. 4. Diptera: Upper surface of Thorax, limited by Dorsopleural Sutures laterally, Scutellum posteriorly and neck anteriorly. 5. Lepidoptera: Lower or inner margin of wing.

DORWARD, KELVIN (1908–1965) (Davis 1966, J. Econ. Ent. 59: 246–247.)

DORYLAIDAE Plural Noun. See Pipunculidae.

DORYLANER Noun. (Greek, *dory* = spear; *aner* = male.) An unusually large form of the male ant occurring in the Dorylinae. Characterized by long and peculiar Mandibles, long cylindrical Abdomen and peculiar genitalia; the typical male in the Subfamily.

DORYLINAE Army Ants. Driver Ants. A Subfamily of Formicidae.

DORYLOPHILE (Greek, *dory* = spear; *philein* = lover.) An obligatory guest of army ants in the Tribe Dorylini. Cf. Ecitophile.

DOS PASSOS, VIOLET (1891–1944) (Comstock 1945, J. N. Y. ent. Soc. 53: 47–48.)

DOTAN®. See Chlormephos.

DOTREPPE, AUGUSTE JOSEPH (–1944) (Berger 1944, Lambillionea 44: 4.)

DOUBLE DRUMMER *Thopha saccata* (Fabricius) [Hemiptera: Cicadidae]. (Australia).

DOUBLE HORN The shape of the Papal Tarsus of higher oribatid mites.

DOUBLE OCELLUS An ocellate spot made up of two such spots.

DOUBLEDAY, EDWARD (1810–1849) (Miller 1973, *That Noble Cabinet* 400 pp. (233) London.)

DOUBLEDAY, HENRY (1809–1875) (Dunning 1877, Entomologist 10: 53–61)

DOUBLEDAY'S SKIPPER *Toxidia doubledayi* (Felder) [Lepidoptera: Hesperiidae]. (Australia).

DOUBLE-DOOR TRAPDOOR SPIDERS [Araneida: Actinopodidae]. (Australia).

DOUBLE-HEADED HAWK MOTH *Coequosa triangularis* (Donovan) [Lepidoptera: Sphingidae]. (Australia).

DOUBLE-SPOTTED CICADA *Cicadetta labeculata* (Distant) [Hemiptera: Cicadidae]. (Australia).

DOUBLE-SPOTTED LINEBLUE *Nacaduba biocellata biocellata* (C. and R. Felder) [Lepidoptera: Lycaenidae]. (Australia).

DOUBLET, CHARLES (–1956) (Anon. 1956, Bull. Soc. ent. Fr. 61: 194.)

DOUBLIER, JEAN THEODORE (1814–1854) (Mulsant 1856, Ann. Soc. Linn. Lyon. 3: 175–186.)

DOUÉ, PIERRE ACHILLE AUGUSTIN (1791–1869) (Buquet 1869, Bull. Soc. ent. Fr. (4) 9: lvi–lviii.)

DOUGLAS, JOHN WILLIAM (1814–1905) (Dale 1906, Entomol. mon. Mag. 42: 16.)

DOUGLAS-FIR BEETLE *Dendroctonus pseudotsugae* Hopkins [Coleoptera: Scolytidae]: A significant, monovoltine pest of Douglas Fir trees in western North America. Overwinters as an adult. Female beetle is monogamous; she bores into tree, develops straight or slightly curved egg galleries in inner bark and score sapwood, emits pheromone and attracts male. See Scolytidae. Cf. Eastern Larch Beetle; Mountain Pine Beetle; Southern Pine Beetle; Red Turpentine Beetle; Western Pine Beetle.

DOUGLAS-FIR CONE MOTH *Barbara colfaxiana* (Kearfott) [Lepidoptera: Tortricidae].

DOUGLAS-FIR ENGRAVER *Scolytus unispinosus* LeConte [Coleoptera: Scolytidae].

DOUGLAS-FIR PITCH MOTH *Synanthedon novaroensis* (Hy. Edwards) [Lepidoptera: Sesiidae]: A pest of Douglas fir, larch and Sitka spruce in North America. Eggs laid in bark crevices or wounds; larvae bore galleries through inner bark and outer wood; overwinter as larvae; pupation in pitch mass. Four years for one generation.

DOUGLAS-FIR TUSSOCK MOTH *Orgyia pseudostugata* (McDunnough) [Lepidoptera: Lymantriidae].

DOUGLAS-FIR TUSSOCK MOTH NPV A naturally-occurring, host-specific multicapsid Nuclear Polyhedrosis Virus used as a biopesticide against larvae of Douglas-Fir Tussock Moth. Trade names include: TM Biocontrol-1® and Virtuss®. See Biopesticide; Nuclear Polyhedrosis Virus. Cf. Beet Armyworm NPV; European-Pine Sawfly NPV; Heliothis NPV; Mamestra brassicae NPV; Redheaded Pine Sawfly NPV; Spodoptera littoralis NPV.

DOUGLAS-FIR TWIG WEEVIL *Cylindrocopturus furnissi* Buchanan [Coleoptera: Curculionidae].

DOUGLASIIDAE Plural Noun. A small Family of ditrysian Lepidoptera assigned to Superfamily Tineoidea or Yponomeutoidea in older classifications; predominantly Palaearctic with a few Species in North America and Australia. Adult body very small with wingspan 7–10 mm and head with smooth, overlapping lamellar Scales; Ocelli large and conspicuous but Chaetosemata absent; Proboscis conspicuous, Antenna shorter than forewing, filiform and Scape lacking Pecten; Proboscis short and without Scales; Maxillary Palpus vestigial; Labial Palpus short, of three segments and pendulous; fore Tibia with Epiphysis; tibial spur formula 0-2-4; hind Tibia with long hair-like Scales; forewing lanceolate. Larvae are fusiform with Prolegs small and crochets absent. Pupation occurs in leaf mines, plant stems or flowers. Biology of douglasiids is poorly known; most Species mine leaves or tunnel stems of Rosaceae and Boraginaceae. See Lepidoptera.

DOUGLASS, BENJAMIN WALLACE (–1939) (Anon. 1944, Outdoor Indiana 11 (10): 4,16.)

DOUGLASS, JAMES ROBERT (1894–1970) (Peay 1972, J. Econ. Ent. 65: 313.)

DOUMERC, ADOLPH JACQUES LOUIS (1802–1868) (Targioni-Tozetti *et al.* 1870, Boll. Soc. ent. ital. 2: 100.)

DOURS, JEAN ANTOINE (1824–1874) (Anon. 1883, Bull. Soc. linn. Nord. Fr. 6: 200–203.)

DOUS, CARL PHILIP EMIL (1866–1936) (Henriksen 1938, Ent. Meddr 20: 102–103.)

DOUSETT, ARTHUR (–1896) (Meldoia 1897, Proc. ent. Soc. Lond. 1897: xciv.)

DOVE, WALTER E (1894–1961) (Anon. 1961, Pyrethrum Post 6: 42.)

DOW, ROBERT PERCY (1865–1936) (Engelhardt 1937, Bull. Brooklyn ent. Soc. 32: 1–4, bibliogr.)

DOWELL, PHILIP (1864–1936) American high-school teacher and amateur Entomologist. (Osborn 1946, *Fragments of Entomological History*. Pt II. 232 pp. Columbus, Ohio.)

DOWNES, WILLIAM (1874–1959) (Andison 1960, Proc. ent. Soc. Br. Columb. 57: 60–62.)

DOWNIE, NORVILLE M (–1994) American Academic (Purdue University) and author of several statistics texts and authority on North American beetles. Coauthor of a 3-volume identification guide, *The Beetles of Northeastern North America*.

DOWNY MILDEW A fungal pathogen of lima beans which is caused *by Phytophthora phaseoli* and vectored by bees.

DRAESEKE, JOHANNES (1892–1970) (Hertel 1972, Ent. Abh. Mus. Tierk. Dresden 37: vii–ix.)

DRAG Noun. (Middle English, *dragge*. Pl., Drags.) Anything that retards motion. Specifically, the retarding force acting on a body (insect) moving through a medium (air, water) parallel and opposite to the direction of motion. Cf. Lift; Thrust. Rel. Airfoil.

DRAGNET® See Permethrin.

DRAGON® See Permethrin.

DRAGONFLIES See Odonata.

DRAKE, CARL JOHN (1885–1965) American academic and student of Hemiptera, particularly the Tingidae. (Mallis 1971, *American Entomologists*. 549 pp. (238–239), New Brunswick.)

DRAPARNAUD, JACQUES PHILIPPE RAYMOND (1772–1805) (Rose 1850, *New General Biographical Dictionary* 7: 138.)

DRAPARNAUD'S GLASS SNAIL *Oxychilus draparnaudi* (Beck) [Sigmurethra: Zonitidae]. (Australia).

DRAPIEZ Belgian chemist and author of *Memoires sur un nouveau genus d'Insectes Coleopteres*, and *Descripton de quelques nouvelles especes d'Insectes* (published in Annales Generales des Sciences Physiques.)

DRASSIDAE Plural Noun. White-Tailed Spiders. Drab-coloured, ground-dwelling spiders with posterior Spinnerets widely spaced, two tarsal claws and claw tuft on each leg; nocturnal hunters which are concealed under rocks and leaf litter during day. Syn. Gnaphosidae.

DRAUDT, MAX (1875–1953) (R. 1953, Z. wein. ent. Ges. 38: 258.)

DRAW-THREAD The silk-producing gland in Lepidoptera.

DRAX® See Boric Acid.

DRAZA® See Methiocarb

DREISBACH, ROBERT R (1888–1964) (Fletcher 1964, J. Econ. Ent. 57: 796)

DREPANIDAE Plural Noun. Hook-tip Moths. A widespread Family of ditrysian Lepidoptera assigned to Superfamily Drepanoidea and including about 700 Species. Adults are small to moderate-sized with wingspan 20–60 mm and body dull coloured. Head with short, smooth scales; Ocelli typically absent and Chaetosemata absent; Antenna filiform, lamellate or bipectinate; Proboscis typically small or absent; Maxillary Palpus typically absent, when present minute and of one segment; Labial Palpus porrect and moderately long; foretibial Epiphysis present or absent; tibial spur formula variable; forewing variable; Abdomen with Tympanal Organ. Eggs are flat, oval or round in outline shape; Chorion with fine micro-sculpture. Larval head hypognathous and abdominal Prolegs on segments 3–6 and 10 (except Drepaninae). Pupae are weakly sclerotized and Cremaster poorly developed. Larvae feed on foliage exposed or in leaf rolls. See Lepidoptera. Cf. Thyatiridae.

DRESCHER, FRIEDERICH CARL (1874–1957) (Lieftinck 1958, Treubia 24: 131–134.)

DRESCHER, RUDOLPH (–1935) (Anon. 1936, Arb. morph. taxon. Ent. Berl. 3: 151.)

DRESNAY, G DU (–1944) (Lucas 1946, Entomologiste 2: 196–197.)

DREWSE, CHRISTIAN (1799–1896) (Henriksen 1925, Ent. Meddr 15: 176–179, bibliogr.)

DREXLER, CONSTANTIN CHARLES (Dos Passos 1951, J. N. Y. ent. Soc. 59: 136.)

DREYFUS, ANDRÉ (1897–1952) (Bier 1952, Revta. bras. Biol. 12: 1–6, bibliogr.)

DRI-DIE® See Silica Compounds.

DRIED-APPLE BEETLES *Araecerus palmeris* (Pascoe) [Coleoptera: Anthribidae]. (Australia).

DRIED-CURRANT MOTH See Almond Moth.

DRIED-FRUIT BEETLE *Carpophilus hemipterus* (Linnaeus) [Coleoptera: Nitidulidae]: A cosmopolitan pest of food, particularly stored dried fruit. DFBs serve as vectors of bacteria and fungi that cause smut, food to spoil and sour. Adults are 2.5–3.0 mm long, oval, flattened, mottled brown, yellow and black; Elytra short (not covering apex of Abdomen) with apical amber-coloured spot and smaller spot at base on side; fly actively. Females lay ca 1000 eggs during a lifetime. Eggs are deposited within decaying or damaged fruit and require incubation period of 2–3 days. Larvae are campodeiform, to 7.0 mm long, yellow with short legs and two pairs of rounded projections at apex of Abdomen. Larvae complete 3–4 instars with larval stage 6–14 days. Pupation occurs in soil or infested stored products and pupal stage requires 5–11 days. DFB overwinters as adult and is multivoltine in warm regions or heated buildings; life cycle requires at least 21 days and typically 1–2 months. See Nitidulidae. Cf. Corn-Sap Beetle.

DRIED-FRUIT BEETLES In Australia, any of several Species of *Carpophilus* [Coleoptera: Nitidulidae].

DRIED-FRUIT DERMATITIS A hypersensitivity in humans associated with stored products and caused by the mite *Carpoglyphus lactis* (Linnaeus). Syn. Grocers' Itch.

DRIED-FRUIT MITE *Carpoglyphus lactis* (Linnaeus) [Acarina: Carpoglyphidae].

DRIED-FRUIT MOTH 1. *Vitula edmandsae serratilineella* Ragnot [Lepidoptera: Pyralidae]. 2. *Ephestia calidella* (Guenée) [Lepidoptera: Pyralidae].

DRI-KILL® See Rotenone.

DRILIDAE Plural Noun. A Family of Coleoptera containing about 80 Species found in the Palaearctic, Ethiopian, and Oriental Zoogeographical Realms. Drilidae are a member of the Cantharoidea. Adults are recognized by flabellate Antenna inserted along lateral margin of Frons anterior of compound eye; females apterous and males possess fully developed Elytra; Prosternum elongate anteriad of fore Coxa. Adults do not feed and larvae are predators of snails. Triunguliform first instar larva actively searches out prey. Completion of life cycle can take several years. See Coleoptera.

DRINOX® See Heptachlor.

DRIVE BEETLES See Elmidae.

DROFIX® See NAA.

DRONE Noun. (Anglo Saxon, *dran* = drone. Pl., Drones.) Hymenoptera: A male bee or ant. Verb. A monotonous, humming sound. Sometimes used to characterize the sound of an insect in flight.

DRONE CELL Apoidea: A cell in which a male bee develops and typically larger than a cell in which a worker bee develops.

DRONE FLIES See Syrphidae.

DRONE FLY *Eristalis texax* (Linnaeus) [Diptera: Syrphidae]: A Species widespread in Europe and North America. Larvae are called rat-tailed maggots, and occur in polluted water rich in rotting organic matter. Adults are Honey Bee-like in habitus and coloration; brown-black with Scutellum yellow; vistis composite flowers. Larvae are cylindrical, grublike with a long respiratory siphon on apex of Abdomen. DF occasionally cause intestinal Myiasis in humans.

DROOPING PRICKLY PEAR *Opuntia vulgaris* Mill. (*O. monacantha* (Willd.) (Haw.) [Cactaceae]: A Species of cactus native to the New World and adventive elsewhere as a minor rangeland pest (Australia). DPP is controlled with *Dactylopius ceylonicus* Green in Australia, India and Sri

Lanka. See Prickly Pear.

DROPLET SPERMATOPHORE A form of Spermatophore found in many apterygote insects and which may be ancestral to the pterygote Spermatophore. Apterygotes engage in indirect Spermatozoa transfer. Typically the droplet Spermatophore consists of a basal adhesive 'foot,' a thread-like stalk and an apical droplet. Some Collembola encyst their Spermatozoa within the droplet. See Spermatophore. Rel. Spermatozoa.

DROSOPHILA A large Genus of the Family Drosophilidae (Diptera) made famous and popularized by geneticists such as Morgan and Sturtevant.

DROSOPHILIDAE Plural Noun. Ferment Flies; Pomice Flies; Small Fruit Flies; Vinegar Flies; Wine Flies. Third largest Family of acalyptrate flies with about 1,000 Species, exceeded only by Chloropidae and Ephydridae. Adults are 2.5–4 mm long and brownish-yellow/black; Subcosta of wing incomplete or vestigial, Costa broken near Humeral Crossvein and near end of vein R1; Postvertical Bristles convergent, front usually with three pairs of Orbital Bristles, anterior pair proclinate others reclinate; wings normally hyaline but may be pictured; Discal Cell and second Posterior Cell confluent; Anal Cell small or absent. Arrangement of bristles on the head and Mesonotum are taxonomically important. Most Species develop in decaying fruits, fungi, sap exudates and similar materials; a few Species are parasitic or predaceous on Homoptera; other Species are primary parasites in spider egg sacs. A few Species develop as commensals or true parasites on land crabs. Adults are strong fliers and often crepuscular. Oviposition by female is upon or near larval food. Eggs hatch within days and larvae are maggot-like, free living and actively search for prey or plant food; life cycle requires 8–14 days. Predators pupate within host remains or in top 5–8 cm of soil. Drosophilids are regarded as urban pests near breweries, wineries, cannaries and vegetable shops; they become a severe nuisance around food-handling facilities and can pose a threat to health through mechanical transmission of disease. If ingested, drosophilid larvae can cause Pseudomyiasis. See Diptera.

DROSOPHILIST Noun. A specialist in the study of *Drosophila*.

DROSOPHILOIDEA A Superfamily of schizophorous Diptera including Braulidae, Camillidae, Canaceidae, Chloropidae (Oscinidae, Titaniidae), Cryptochaetidae, Curtonotidae (Cyrtonotidae), Diastatidae, Drosophilidae, Ephyrididae (Hydrellidae, Notiophilidae), Milichiidae (Phyllomyzidae) and Tethinidae.

DRUCE, HAMILTON HERBERT CHARLES JAMES (1869–1922) (Musgrave 1932, *A Bibliography of Australian Entomology 1775–1930*. 380 pp. (72), Sydney.)

DRUCE, HERBERT (1846–1913) (Musgrave 1932, *A Bibliography of Australian Entomology 1775–1930*. 380 pp. (72), Sydney.)

DRUGSTORE BEETLE *Stegobium paniceum* (Linnaeus) [Coleoptera: Anobiidae]: A cosmopolitan, sporadically serious pest of stored products, granaries, pharmacies and museum collections; high-value commodities easily are ruined by slight damage or contamination. DBs occur indoors in temperate regions but are not as abundant in tropical regions. Larvae feed on broad range of foods including biscuits, chocolates, flour, fruits, nuts, leather and wool; adults do not feed. DB requires humidity >35% to complete development. Adults are 2.0–3.5 mm long, reddish brown and elongate-oval in outline shape; Antenna with loosely defined club of three segments; Elytra with longitudinal striae and lines of fine Setae. Adults live 13–85 days and females lay ca 75 eggs during lifetime. Eggs are laid on dry organic matter and incubation requires 7–12 days. Larvae with 4–6 instars and larval stage requires 35–50 days. Pupation typically occurs within silken cocoon and no cocoon is formed under some conditions; pupal stage 8–20 days. Alt. Drug Store Beetle. Syn. Biscuit Beetle. See Anobiidae. Cf. Cigarette Beetle.

DRUGSTORE BEETLES See Anobiidae. Cf. Deathwatch Beetles.

DRUITT, ALAN (1863–1933) (W. P. C. 1933, Entomologist 66: 96.)

DRURY, DRU (1725–1803) English goldsmith, engraver, amateur Entomologist and author of *Illustrations of Natural History* (3 vols, 1770–1782). (Gambles, R. M. 1976. Odonatologica 5: 2.)

DRY FUNGUS BEETLES See Sphindidae.

DRY, WILLIAM WOODY (1829–1967) (Cancienne 1968, J. Econ. Ent. 61: 885.)

DRYANDER, JONAS (1748–1810) (Rose 1850, *New General Biographical Dictionary* 7: 146–147.)

DRYANDRA MOTH *Carthaea saturnioides* Walker [Lepidoptera: Carthaeidae]. (Australia).

DRYBERRY MITE *Phyllocoptes gracilis* (Nalepa) [Acari: Eriophyidae].

DRYINIDAE Haliday 1833. Plural Noun. Dryinid Wasps. A cosmopolitan, moderate sized (ca 45 Genera, 850 Species) Family of apocritous Hymenoptera assigned to the Chrysidoidea and related to Bethylidae. Adult females with body often ant-like, 1.5–10 mm long and not metallic coloured. Antenna with 10 segments in male and female; male macropterous, females macropterous, brachypterous or apterous; forewing venation typically reduced to Costal Cell and large Stigma; hindwing Marginal Vein without closed cells or Jugal Lobe; female with chelate foretarsi (except Aphelopinae), formed by an enlarged fifth Tarsomere opposable to tarsal claw. Species develop as primary external parasites of auchenorrhynchous Homoptera (Cicadelloidea, Fulgoroidea), particularly Cicadellidae. Females inflict temporary paralysis and sometimes host feed; developing nymph of parasitized host will not moult. First-instar larva Sacciform and later instars Hymenopteriform. Larvae apparently lack Malpighian Tubules; Thalacium from cast larval exuviae (as Rhopalosomatidae); polyembryony is known in one Species and thelytoky is known in another Species. Pupation occurs within a cocoon formed in soil or vegetation. Fossil record of dryinids in Taimyrian, Canadian, and Dominican Amber. See Chrysidoidea.

DRYLAND WIREWORM *Ctenicera glauca* (Germar) [Coleoptera: Elateridae].

DRYOMYZIDAE Plural Noun. A Family of schizophorous Diptera assigned to Superfamily Sciomyzoidea which includes 30 Species in six Genera. Adult flies medium-sized (4–12 mm long), moderately bristly to hairy and most Species apparently are saprophagous. *Oedoparena glauca* (Coquillett) is a predator of barnacles. See Diptera.

DRYOPIDAE Plural Noun. Long-toed Water Beetles. A small Family of polyphagous Coleoptera assigned to Superfamily Byrrhoidea. Adults are <8 mm long, dull coloured and elongate-oval in outline shape. Antenna short with most segments transverse and concealed by prosternal lobe. Adults are aquatic while larvae are vermiform and inhabit soil or decaying wood. See Byrrhoidea.

DRYOPOIDEA A cosmopolitan Superfamily of polyphagous Coleoptera that in some classifications includes Chelonariidae, Dryopidae, Elmidae, Eulichadidae, Eurypogonidae, Heteroceridae, Limnichidae, Psephenidae, and Ptilodactylidae. In other classifications, many of these Families are assigned to Byrrhoidea. See Byrrhoidea.

DRYWOOD TERMITE Any member of *Cryptotermes* (Kalotermitidae). Colonies small, reside in wooden builidings and do not require contact with soil. Colonies often encountered in tropical areas. See Kalotermitidae. Cf. Dampwood Termite.

DUACORIA Noun. (Pl., Duacoriae.) The Coria or membrane of the second abdominal segment (MacGillivray).

DUASPIRACLE. Noun. (Pl., Duaspiracles.) A spiracle of the second abdominal segment of an insect (MacGillivray).

DUBINI, ANGELO (1813–1902) (Giovannola 1939, Riv. Parassit. 3: 1–4.)

DUBOISIA FLEA-BEETLE *Psylliodes parilis* Weise [Coleoptera: Chrysomelidae]. (Australia).

DUBOURGAIS, A (1841?–1906) (Fauvel 1906, Revue Ent. 25: 28.)

DUCHAUSSOY, A French Entomologist who published a few papers on taxonomy of parasitic Hymenoptera during the First World War.

DUCHON, EMANUEL (1865–1944) (Heyrovsky 1945, Cas. csl. spol. ent. 42: 87–88)

DUCKETT, ALLEN BOWIE (1891–1918) (Popenoe *et al.* 1918, Proc. ent. Soc. Wash. 20: 185–186, bibliogr.)

DUCROTAY DE BLAINVILLE, MARIE HENRI. See BLAINVILLE.

DUCT Noun. (Latin, *ductus* = conduit, connecting. Pl., Ducts.) 1. A channel, tube or canal adapted for transporting a secretion from a gland to the point of discharge. Cf. Aperture; Pore. Rel. Gland.

DUCTEOLE Noun. (Latin, *ductus* = conduit, connecting. Pl., Ducteoles.) A ductlet or small duct. Alt. Ductule. See Duct.

DUCTUS BURSAE Lepidoptera: The duct in females which extends from the Ostium Bursa (copulatory aperture) to the Bursa Copulatrix.

DUCTUS EJACULATORIUS A duct or tube in males that is formed by the union of Vas Deferentia from each side. Semial fluid passes through the DE and is discharged into the Vagina. Syn. Ejaculatory Duct.

DUCTUS SEMINALIS Female Lepidoptera: The tube or canal connecting the Bursa Copulatrix with the Common Oviduct (Oviductus Communis, *q.v.* Imms).

DUDA, LADISLAW (1854–1895) (Anon. 1895,. Zool. Anz. 18: 404.)

DUDA, OSWALD (1869–1941) (Hellén 1943, Notul. ent. 23: 60.)

DUDGEONEIDAE Plural Noun. A small Family of ditrysian Lepidoptera that occurs in Africa and Madagascar, and from India to Australia. Placement of dudgeoneids is problematical; some classifications place them within Cossidae; sometimes they are regarded as Family in Cossoidea, Sesioidea or Pyraloidea. Adults are moderate sized with wingspan 25–70 mm and head with somewhat rough scales. Ocelli and Chaetosemata are absent; Antenna about half length of forewing; bipectinate or unipectinate to apex in both sexes; Proboscis absent; Maxillary Palpus minute; Labial Palpus upturned; fore Tibia with Epiphysis; tibial spur formula 0-2-4. Abdomen with paired, sternal Tympanal Organs at base. Larvae apparently bore stems and the Pupae protrude from larval tunnel. See Lepidoptera.

DUDIM® See Diflubenzuron.

DUDLEY, PAUL (1675–1751) (Weiss 1936, *Pioneer Century of American Entomology*. 320 pp. (25–26), New Brunswick.)

DUFF Noun. (Middle English, *dogh* = dough. Pl., Duffs.) The partly decayed vegetable matter on the forest floor. Cf. Debris; Detritus; Humus; Litter; Mor.

DUFFIELD, CHARLES ALBAN WILLIAM (1866–1974) (W[orms] 1975, Entomologist's Rec. J. Var. 87: 127–128.)

DUFOUR, LEON JEAN MARIE (1780–1865) French physician, anatomist and author of *Memoire Anatomique sur une noucelle espece d'Insecte du genre Brachine* (In: Ann. Mus. d'Hist. Nat., vol 18), *Memoires sur l'Anatomie des Coleopteres, des Cigales, des Cicadelles, des Labidoures or Forficulae* (In: Ann. Sci. Nat.), *Recherches Anatomiques et Physiologiques sur les Hemipteres, accompagnees de considerations relatives a l'histoire naturelle et a la classi-*

fication de ces insectes (1833). (Bonnet 1945, Bibliographie Araneorum 1: 32)

DUFOUR'S GLAND Hymenoptera: A single, tube-like accessory gland associated with the female's internal reproductive system. DG is near the venom reservoir and opens near the Sting's base (aculeates) or Ovipositor (Parasitica). DG secretes alkaline elements of poison (venom) which are transmitted to the host/prey/victim by the Sting or Ovipositor. Ultrastructure of DG reveals an inner epithelium layer of large, rounded cells containing lipids and oval, flattened nuclei; outer muscular layer composed of transverse striations. Secretory cells of DG contain numerous free ribosomes and mitochondria concentrated at basal part of cell. Parallel profiles of endoplasmic reticulum also are present. The proximal part of the discharge duct is surrounded by longitudinal and circular muscles. Syn. Alkaline Gland. See Accessory Gland. Cf. Acid Gland. Rel. Exocrine Gland.

DUFRANE, ABEL (1880–1960) (Pierard 1961, Bull. Ann. Soc. r. Ent. Belg. 97: 72–86, bibliogr.)

DUFTSCHMID, GASPARD Author of *Fauna Austriae* (2 vols, 1805–1812).

DUFTSCHMID, KASPAR (Duméril 1823, *Considérations générale sur la classe des insectes*. 272 pp. (266), Paris.)

DUGES, ALFREDO AUGUSTO DELSESCAUT (1826–1910) (Osborn 1937, *Fragments of Entomological History*. 394 pp. Columbus, Ohio.)

DUGES, EUGENE (1826–1895) (Papavero 1971, *Essays on the History of Neotropical Dipterology*. 1: 179. São Paulo.)

DUHAMEL DU MONCEAU, HENRY LOUIS (1700–1782) (Simmons 1929, J. Econ. Ent. 22: 820–821)

DUJARDIN, FELIX (1801–1860) (Nordenskiöld 1935, *History of Biology*. 629 pp. (428–429), London.)

DULFER, ANTON FREDRIK BERNARD (–1954) (Landsmann 1954, Ent. Ber., Amst. 15: 321.)

DULL COPPER *Paralucia pyrodiscus pyrodiscus* (Doubleday) [Lepidoptera: Lycaenidae]. (Australia).

DULL JEWEL *Hypochrysops epicurus* Miskin [Lepidoptera: Lycaenidae]. (Australia).

DULL OAKBLUE *Arthopala centaurus centaurus* (Fabricius) [Lepidoptera: Lycaenidae]. (Australia).

DULOSIS Noun. (Greek, *doulosis* = subjugation. Pl., Duloses.) A relationship in which workers of a parasitic ant Species raid the nest of another ant Species to capture brood that will be reared and become slave nestmates.

DUMDUM FEVER See Kala-azar. Syn. Black Disease; Tropical Splenomagaly.

DUMERIL, ANDRÉ MARIE CONSTANT (1774–1860) French academic at Jardin des Rao and member of the Academy of Science. Dumeril was author of *Dictionnaire des Sciences Nat.*, and *Considerations generales sur la classe des Insectes* (1823). (Anon. 1861, Proc. Linn. Soc. Lond. 1861: xlvi–xlviii, bibliogr.)

DUMIGAN, EDWARD JARRETT (1878–1969) (Anon. 1969, News Bull. ent. Soc. Qd. 55: 8.)

DUMMER, R A (Alexander 1923, Ent. News 34: 192.)

DUMONT D'URVILLE, JULES SÉBASTIEN CÉSAR (1790–1842) (Musgrave 1932, *A Bibliography of Australian Entomology 1775–1930*. 380 pp. (73–74), Sydney.)

DUMOSE Adj. (Latin, *dumosus* = bushy; *-osus* = full of.) Pertaining to structure which is bush-like in appearance.

DUMP FLIES Species of muscid fly-genus *Ophyra*. See Bronze Dump-Fly.

DUNE Noun. (Middle English, *doun*. Pl., Dungs.) A rounded hill of sand formed by the action of wind. Dunes typically occur in or near deserts or along the sea coast. Syn. Dene. Rel. Habitat.

DUNG BEETLES See Geotrupidae; Scarabaeidae.

DUNG FLIES See Scathophagidae; Sphaeroceridae.

DUNLOP, G A (1868–1933) (Anon. 1933, NWest Nat. 1933: 142–145.)

DUNN, GEORG W (1814–1905) (Essig 1931, *History of Entomology*. 1029 pp. (605), New York.)

DUNNING, JOSEPH WILLIAM (1833–1897) (Neave 1933, *Centennial History of the Entomological Society, London*. 224 pp. (137–138), London.)

DUODENUM Noun. (Latin, *duodeni* = twelve each. Pl., Duodena; Duodenums.) The section of the hindgut posteriad of the entrance of the Malpighian Tubules. Syn. Chylific Ventricle. See Alimentary System.

DUPERREY, LOUIS ISIDORE (1786–1865) (Musgrave 1932, *A Bibliography of Australian Entomology 1775–1930*. 380 pp. (75), Sydney.)

DUPION Noun. (French *doupion*. Pl., Dupions.) A cocoon spun collectively by two silk-worms together; also the coarse silk from such a cocoon.

DUPLA SCALE *Duplaspidiotus claviger* (Cockerell) [Hemiptera: Diaspididae]. (Australia).

DUPLAGLOSSA Noun. (Latin, *duplus* = double; Greek, *glossa* = tongue. Pl., Duplaglossae.) Aculeate Hymenoptera: The Glossa when divided or forked into two slender Glossa-like structures (MacGillivray).

DUPLICATE Adj. (Latin, *duplicatus* past participle of *duplicare* = to double; *-atus* = adjectival suffix.) 1. Double. 2. Either one of two identical objects. 3. A copy; something that resembles another object in size, shape and content but which is not of the same source or made in the same manner.

DUPLICATO-PECTINATE Descriptive of an antennal form in which the branches (rami) of a Bipectinate Antenna are alternately long and short. See Antenna. Cf. Pectinate.

DUPONCHEL, PHILOGENE AUGUST JOSEPH (1774–1846) French collaborator on *Histoire Naturelle des Lepidopteres de France* (initiated by M. Godart). Duponchel also published

Monographie du genre Erotyle (IN: Mem. du Museum, vol. 12) (Duméril 1847, Ann. Soc. ent. Fr. (2) 5: 517, bibliogr.)

DUPONT, H See Puech-Dupont, R. H.

DUPORT, MELVILLE E (1891–1981) Born on island of Nevis, West Indies and emigrated to Canada 1910 where he completed a Ph.D. 1921. Duport devoted his entire professional career to faculty of Macdonald College (McGill University). Duport taught, administered and undertook research in morphology. He trained ca 130 students and wrote many papers involving issues of morphology and a popular laboratory manual of insect morphology (DuPorte 1959).

DUPUY, GABRIEL (1846–1913) (Oberthür 1916, Etudes de lépidopterologie comparée 11: (Portraits de lépidoptèristes) (3me Ser.): [5].)

DURACIDE® See Tetramethrin.

DURAKIL® See Fenitrothion.

DURAMITEX® See Malathion.

DURAND, GEORGES (–1963) (Anon. 1963, Bull. Soc. ent. Fr. 69: 213.)

DURAPHOS® See Mevinphos.

DURASCUTELLA Noun. (Pl., Durascutellae.) The infolded longitudinal thickening on each side separating the Mesascutella from Parascutella (MacGillivray).

DURCKHEIM, STRAUS Author of *Considerations generales sur l'Anatomie comparee des animaux articules, auxquelles on joint l'anatomie descriptive du Hanneton* (1828).

DURITAE Noun. (Etymology obscure. Sl., Durita.) 1. The independent Pteraliae interposed between the ends of the wings. 2. The Alariae and Venellae *sensu* MacGillivray.

DURMET® See Chlorpyrifos.

DURRANI, MUHAMMED ZARIF (1930–1967) (Kennedy 1968, Proc. R. ent. Soc. Lond. (C) 32: 59.)

DURRANT, JOHN HARTLEY (1863–1928) (English taxonomic Entomologist (B.M.N.H.) who specialized in Lepidoptera. (V. A. D. 1928, The Times, London 23 Jan 1928.)

DURSBAN® See Chlorpyrifos.

DURUS Adj. (Latin, *dura* = hard.) Hard.

DURY, CHARLES (1847–1931) (Osborn 1937, *Fragments of Entomological History*. 394 pp. (212–213), Columbus, Ohio.)

DUSKE, GEORGII AUGUSTOVICH (–1908) (Kusnezov 1908, Revue Russe Ent. 7: 174.)

DUSKY PASTURE-SCARAB *Sericesthis nigrolineata* Boisduval [Coleoptera: Scarabaeidae]. (Australia).

DUSKY SAP-BEETLE *Carpophilus lugubris* Murray [Coleoptera: Nitidulidae].

DUSKY STINK-BUG *Euschistus tristigmus* (Say) [Hemiptera: Pentatomidae].

DUSKY Adj. (Middle English, *dosk*). Somewhat darkened; pale fuscous.

DUSKYBACKED LEAFROLLER *Archips mortuana* Kearfott [Lepidoptera: Tortricidae].

DUSKY-BIRCH SAWFLY *Croesus latitarsus* Norton [Hymenoptera: Tenthredinidae].

DUST Noun. (Anglo Saxon, *dust* > Greek, *dunst* = vapour. Pl., Dusts.) 1. Earth or other matter finely ground and easily suspended in air. 2. A form of insecticide formulation in which the active ingredient is dry and mixed with a dry diluent (talc or clay). Alternatively, the active ingredient is finely divided and constitutes the dust (Boric Acid). Dust primarily is used in the control of household pests and applied to cracks and crevices. Designated as **'D'** on USA Environmental Protection Agency labels and Material Safety Data Sheet documentation. See Aerosol; Emulsifiable Concentrate; Microencapsulated Concentrate; Suspension Concentrate; Oil Concentrate; Wettable Powder.

DUSTY BROWN BEETLES *Gonocephalum* spp. [Coleoptera: Tenebrionidae]: A moderate-sized Genus of beetles that occurs in tropical and subtropical areas. Some Species are pests in stored products or field crops; some Species are predaceous. See Tenebrionidae.

DUSTYWINGS See Coniopterygidae.

DUTCH ELM DISEASE A highly virulent fungal pathogen [*Ceratostomella ulmi* Schwarz] that affects several Species of elm trees. DED and its vector, European Elm Bark Beetle [*Scolytus multistriatus* (Marsham)], were imported into USA during 1930. Subsequently, DED became responsible for widespread destruction of American Elm trees throughout North America. DED also attacks other kinds of trees including ash and basswood, and is vectored by other Species of bark beetles. See Scolytidae.

DUTFIELD, JAMES (fl 1748) (Lisney 1960, *Bibliography of British Lepidoptera 1608–1719*. 315 pp. (148–151, bibliogr.), London.)

DUTREUX, AUGUSTE (1808–1890) (Koltz. 1891, Fauna Ver. Luxemburg 1: 7–9.)

DUTROCHET, RENÉ JOAQUIM HENRI (1776–1847) (Ratzeburg 1874, Forstwissenschaftliches Schriftsteller-Lexicon 1: 158–160.)

DUTTON, JAMES FAIRCLOUGH (1859–1937) (Anon. 1938, Arb. morph. taxon. Ent. Berl. 5: 186.)

DUURLOO, HANS PETER (1860–1941) (Tuxen 1968, Ent. Meddr 36: 40–41)

DUVAL, P N C J See Jacquelin-Duval, P. N. C.

DUVIVIER, ANTOINE (–1896) (Anon. 1897, Miscnea ent. 5: 15.)

DUZEE, E P VAN See Van Duzee.

DVIGHUBSKII, IVAN ALEXSYEEVICH (1771–1839) (Roulier 1840, Bull. Soc. Nat. Moscou 13: 342–359.)

D-VAC A suction trap that uses a powerful vacuum mounted on a backpack, with a large conical attachment, either fixed or with a flexible tube, for the collection of insects. The D-vac is named for its developer, Everitt Detrick.

DZN® See Diazinon.

DWARF HONEY BEE *Apis florea* Fabricius [Hymenoptera: Apidae]. See Apidae. Cf. European

Honey Bee.

DWIGHT, NATHANIEL (1770–1831) (Weiss 1936, *Pioneer Century of American Entomology*. 320 pp. (70–71), New Brunswick.)

DWINELLE, CHARLES HASCALL (1847–) (Essig 1931, *History of Entomology*. 1029 pp. (504–508), New York.)

DYAD Noun. (Greek, *dyas* = two. Pl., Diads.) A bivalent Chromosome.

DYAR, HARRISON GREY (1866–1929) American Entomologist who specialized in taxonomy of Diptera, particularly Culicidae, and responsible for 'Dyar's Law.' (Knight & Pugh 1974, Mosquito Syst. 6: 11–24, bibliogr. (mosquito papers only).

DYAR'S LAW Lepidoptera larvae: An empirical observation that concludes that increase of headwidth shows a regular geometrical progression in successive instars (Dyar 1890). The observation has been taken to be a predictive tool for estimating instar number for some Species, and has been extrapolated to other groups of insects. Subsequent investigations have shown that DL is not an accurate predictor of instar number and is not generally valid. See Allometry. Cf. Prizbram's Rule; Prizbram's Factor.

DYCARB® See Bendiocarb.

DYE, FRANKLIN (1835–1920) (Weiss 1920, Ent. News 31: 180.)

DYFONATE® See Fonofos.

DYK, VACLAV (1912–) Lobosvarsky 1972, (Zool. Listy 21: 1–2.)

DYLOX® See Trichlorfon.

DYMET A combination of methoxychlor and diazinon used on ornamental plants in some countries. See Organochlorine Insecticide.

DYNAMEX® See Avermectin.

DYNAMIC Adj. (Greek, *dynamis* = power; *-ic* = of the nature of.) Pertaining to motion. Active, as opposed to static. Manifesting power. Alt. Active; Energetic; Vigorous. Ant. Static.

DYNEIN Noun. (Greek, *dynamis* = power.) A protein in Flagella and Cilia that is involved in mechanical work.

DYNEIN ARMS A component of Accessory Doublets in the Sperm Flagellum. Dynein Arms originate on Subfibre A. DAs project from Subfibre A at periodic intervals along the length of the Microtubule. The arms are formed of the protein dynein and contain axoneme's main ATPase activity. Each arm is composed of many polypeptides and ATP-ase high-molecular-weight molecules. Two distinct types of arms have been identified, based on anatomy and chemistry, called an Outer Dynein Arm and an Inner Dynein Arm. The elements are separated by a distance of about 24 nm. Subfibre A temporarily inserts onto Subfibre B of the adjacent doublet via the Dynein Arm. See Microtubule. Cf. Nexin; Radial Arm. Rel. Sperm Flagellum.

DYSDERIDAE Plural Noun. Six-eyed spiders. A numerically small, widely distributed Family of ecribellate spiders. Dysderids are characterized by six eyes that are nearly contiguous and two spiracles posterior of book-lung openings. Dysderids are nocturnal hunters that occur under bark or stones where they construct silken retreats.

DYSENTERY Noun. (Latin, *dysenteria* > Greek, *dys* = bad; *enteron* = intestines. Pl., Dysenteries.) Beekeeping: Adult bees that defecate within or near their colony; dysentery often occurs during late winter after bees have accumulated significant amounts of waste material and ambient temperatures have been too low for flying, or when toxins accumulate in the bee's food. Bees can suffocate or become adversely affected by turmoil and efforts at grooming themselves and the hive. Affected colonies can become infected with pathogens, resulting in colony extinction.

DYSON, DAVID (1823–1856) (Anon. 1867, The Substitute 1: 106–108, 146.)

DYSTROPHIC Adj. (Latin, *dystrophia* > Greek, *dys* = bad; *trephein* = to nourish; *-ic* = of the nature of.) Pertaining to inadequate nutrition. Cf. Eutrophic; Oligotrophic.

DYSTROPHY Noun. (Latin, *dystrophia* > Greek, *dys-* = ill; *trephein* = to nourish. Pl., Dystrophies.) Abnormal or defective nourishment.

DYTISCIDAE Plural Noun. Predaceous Diving Beetles. A cosmopolitan Family of adephagous Coleoptera that includes about 2,000 nominal Species. Adults are predaceous, smooth and boat-shaped with head deeply recessed into Prothorax. Antenna filiform with 11 segments; without sternal keel; hind Coxa very large, lacking Coxal plate; hind leg enlarged, flattened with Tibia and tarsal Setae adapted for swimming; hindlegs stroked simultaneously when swimming; tarsal formula 5-5-5; claws on hind Pretarsus not equal in size. Dytiscids fly after dark and pitch into water. They respire underwater with air stored under Elytra. All Species are aquatic in still and running water. Larvae are campodeiform and predaceous upon other insects, tadpoles and small fish. Dytiscids engage in preoral digestion and imbibe liquified prey via channel along sickle-shaped Mandible. Dytiscids respire through a siphon formed by an elongated eighth abdominal segment. Pupation occurs in moist soil near water. See Coleoptera.

DZIERZON, JOHANNES (1811–1906) Silesian priest and apiculturalist who asserted that male honey bees were produced from unfertilized eggs. He was responsible for the first observation of arrhenotokous parthenogenesis in the Hymenoptera. (Pellet 1929, Am. Bee J. 69: 398–400) Rel. Parthenogenesis.

DZIURZYNSKI, CLEMENS (–1934) (Reisser *et al.* 1935, Z. öst. Ent.Ver. 20: 1–2.)

EALES-WHITE, J C (–1948) (Williams 1949, Proc. R. ent. Soc. Lond. (C) 13: 67.)

EAR Noun. (Anglo Saxon, *ere* > Latin, *auris* > Greek *ous* = an ear. Pl., Ears.) An organ or complex anatomical structure that is adapted for hearing. An 'ear' is well developed in some groups of insects, such as Orthoptera. See Tympanal Organ. Rel. Acoustical Communication.

EAR MANGE MITE *Otodectes cynotis* (Hering) [Acarina: Psoroptidae]. (Australia).

EAR PIERCER Vernacular term for an earwig. See Dermaptera.

EAR TAG A slow release insecticide formulation in which a small plastic tag is impregnated with toxin and is attached to the ears of livestock. The insecticidal vapour from the tag provides extended protection for the head and face from a variety of flying insect pests.

EAR TICK *Otobius megnini* (Dugès) [Acari: Argasidae].

EARDRUM See Tympanum.

EARTHFIRE® See Resmethrin.

EARWIG Noun. (Anglo Saxon, *earwicga*. Pl., Earwigs.) The common name for Dermaptera. In particular, members of the Genus *Forficula*, or sometimes specifically *Forficula auricularis*. Earwig is so named because the insect was believed to penetrate the ear canal of sleeping humans. See Dermaptera.

EARWIG FLIES See Meropeidae.

EAST, ROBERT THEODORE (1923–1967) (Davich 1968, J. Econ. Ent. 61: 886.)

EASTERN BENT-WINGED CICADA *Froggatoides typicus* Distant [Hemiptera: Cicadidae]. (Australia).

EASTERN BLACK-HEADED BUDWORM *Acleris variana* (Fernald) [Lepidoptera: Tortricidae].

EASTERN BROWN-CROW BUTTERFLY *Euploea tulliolus tulliolus* (Fabricius) [Lepidoptera: Nymphalidae]. (Australia).

EASTERN CRANE FLY *Tipula abdominalis* [Diptera: Tipulidae]. See Tipulidae. Cf. European Crane Fly; Range Crane Fly.

EASTERN EQUINE ENCEPHALITIS An arboviral disease of horses and humans localized in eastern USA and South America. Virus is assigned to Genus *Alphavirus* of Tongaviridae. Epizootics have been reported in game birds. EEE is vectored by mosquitoes; EEE is not common but mortality is high in birds, horses and humans. See Arbovirus; Tongaviridae. Cf. Venezuelan Equine Encephalitis; Western Equine Encephalitis.

EASTERN FALSE-WIREWORM *Pterohelaeus darlingensis* Carter [Coleoptera: Tenebrionidae]; A pest of cotton in Australia. Adults are 14.5–17.0 mm long, oval and black; Antenna with loose, flattened, 4-segmented club; Elytra with 18 Striae. Adult females lay eggs in soil. Eggs are 1.2–1.5 mm long, oval, smooth and white. Larvae are elongate-cylindrical, strongly sclerotized and yellow-brown; head and first three segments yellow-brown or with pale brown patches. Pupation occurs in soil. Pupae are elongate, pale brown-cream and 17–23 mm long. EFWs feed on leaf litter and occur in the soil around cotton in Australia. Eggs are laid in summer and autumn; larvae feed on organic matter in the soil until they reach full size during spring. False wireworms prefer dry conditions protected by stubble or weeds. See Tenebrionidae.

EASTERN FIELD-WIREWORM *Limonius agonus* (Say) [Coleoptera: Elateridae].

EASTERN FIVE-SPINED IPS *Ips grandicollis* (Eichhoff) [Coleoptera: Scolytidae].

EASTERN FLAT *Netrocoryne repanda* Felder [Lepidoptera: Hesperiidae]. (Australia).

EASTERN GOLDEN-HAIRED BLOWFLY *Calliphora stygia* (Fabricius) [Diptera: Calliphoridae]: A primary blowfly in Australia but not as important as *C. augur* or *L. cuprina*. See Calliphoridae; Lesser Brown-Blowfly; Sheep Blowfly.

EASTERN GRAPE-LEAFHOPPER *Erythroneura comes* (Say) [Hemiptera: Cicadellidae].

EASTERN HERCULES-BEETLE *Dynastes tityus* (Linnaeus) [Coleoptera: Scarabaeidae].

EASTERN HORN-LERP *Creiis corniculata* (Froggatt) [Hemiptera: Psyllidae]. (Australia).

EASTERN LARCH-BEETLE *Dendroctonus simplex* LeConte [Coleoptera: Scolytidae]: A pest of American larch in eastern North America and Alaska. Adults are 2–4 mm long, reddish brown or black; Pronotum corasely and finely punctate; Elytra striate and punctate. See Scolytidae. Cf. Mountain Pine Beetle; Southern Pine Beetle; Red Turpentine Beetle; Western Pine Beetle.

EASTERN LUBBER-GRASSHOPPER *Romalea guttata* (Houttuyn) [Orthoptera: Acrididae].

EASTERN MOUSE-SPIDER *Missulena bradleyi* Rainbow [Araneida: Actinopodidae]. (Australia).

EASTERN PINE-LOOPER *Lambdina pellucidaria* (Grote & Robinson) [Lepidoptera: Geometridae].

EASTERN PINE-SEEDWORM *Cydia toreuta* (Grote) [Lepidoptera: Tortricidae].

EASTERN PINE-SHOOT BORER *Eucosma gloriola* Heinrich [Lepidoptera: Tortricidae].

EASTERN PINE-WEEVIL *Pissodes nemorensis* Germar [Coleoptera: Curculionidae].

EASTERN RINGED-XENICA *Geitoneura acantha* (Donovan) [Lepidoptera: Nymphalidae]. (Australia).

EASTERN SAND WASP *Bembix spinolae* Lepeletier [Hymenoptera: Sphecidae]: A wasp endemic to eastern North America. Adults are stout, ca 11–15 mm long and predominantly black with white pubescence and pale transverse, interrupted stripes on Gaster; wings hyaline with pale brown venation; Tarsi pale yellow to greenish. Females construct their nests in sand and progressively provision cells with flies. See Sphecidae. Cf. Western Sand Wasp; Sand Wasp.

EASTERN SANDGRINDER *Arenopsaltria nubivena* (Walker) [Hemiptera: Cicadidae]. (Australia).

EASTERN SPRUCE-BUDWORM *Choristoneura*

fumiferana (Clemens) [Lepidoptera: Tortricidae]: A highly destructive forest pest native of North America. Preferred host plants of ESB include spruce, balsam fir, Douglas fir, true firs and lodgepole pine. See Tortricidae.

EASTERN SPRUCE-GALL ADELGID *Adelges abietis* (Linnaeus) [Hemiptera: Adelgidae]: A pest in eastern North America that induces pineapple-shaped growths at the base of young shoots of Norway and white spruce. See Adelgidae. Cf. Cooley Spruce Gall Adelgid.

EASTERN SUBTERRANEAN-TERMITE *Reticulitermes flavipes* (Kollar) [Isoptera: Rhinotermitidae]: A Species endemic to eastern North America. Flights of EST are initiated during spring after warm rain and through summer. EST is a pest of second-growth lumber, wood and wood-products; it consumes soft spring growth and timber along length of grain or consumes all wood. EST colony must maintain contact with ground via earthen tunnels. See Rhinotermitidae. Cf. Western Subterranean Termite.

EASTERN TENT-CATERPILLAR *Malacosoma americanum* (Fabricius) [Lepidoptera: Lasiocampidae]: A univoltine, periodic and sporadic pest of fruit and ornamental trees in eastern North America. ETC overwinters as a mass of several hundred eggs attached to branches of tree. Larvae communally occupy large webs constructed in forks and crotches of trees; larvae do not feed within web but reside there during night and inclement weather. Larvae construct fine silken trail from web to tender leaves that are used as food; communal web is enlarged as feeding progresses; feeding is completed within 30–45 days when mature larvae disperse and pupate within white silken cocoons on tree trunk or nearby object. See Lasiocampidae. Cf. Fall Webworm.

EASTERN YELLOWJACKET *Vespula maculifrons* (Buysson) [Hymenoptera: Vespidae].

EASTON, ALAN MAURICE (1907–1989) English physician and amateur Entomologist specializing in Nitidulidae and Forensic Entomology. (Baccus & Kirk-Spriggs 1990, Antenna 14 (2): 61–63.)

EASTON, NIGEL LEIGH (1902–1963) (Davey 1964, Entomologist's Rec. J. Var. 76: 27–28.)

EASYTEC® See Tecnazene.

EATON, ALFRED EDWIN (1845–1929) (Musgrave 1932, *A Bibliogr. of Australian Entomology 1775–1930*. 380 pp. (76), Sydney.)

EATON, CHARLES BRADFORD (1913–1964) (Yuill & Baker 1964, J. Econ. Ent. 57: 795.)

EAU DE LABARRAQUE A solution of sodium hypochlorite used in bleaching insect structures, somewhat cheaper than Javelle Water and quite as effective.

EBENINE Adj. (Latin, *ebenus* = ebony.) Black in appearance as ebony.

EBICID® See Thiometon.

EBNER, RICHARD (1885–1961) (Beier 1963, Ann. naturh. Mus. Wien 66: 15–16.)

EBOR GRASSGRUB *Oncopera alboguttata* (Tindale) [Lepidoptera: Hepialidae]. (Australia).

EBURNEOUS Adj. (Latin, *eburneus* = of ivory; *-osus* = full of.) Ivory white. Alt. Eburneus.

ECALCARATE Adj. (Latin, *ex* = out of; *calcar* = spur; *-atus* = adjectival suffix.) Without a spur or Calcar. Alt. Ecalcaratus. See Calcar; Spur.

ECARINATE Adj. (Latin, *ex* = out of; *carina* = keel; *-atus* = adjectival suffix.) Without a keel or Carina. Alt. Ecarinatus.

ECAUDATE Adj. (Latin, *ex* = out of; *cauda* = tail; *-atus* = adjectival suffix.) Pertaining to organisms that lack a tail. Alt. Acaudal; Acaudate.

ECAUSATE WING A wing without tail-like processes. See Wing.

ECDYSIAL SUTURE A suture longitudinal on the Vertex that separates epicranial halves of the head. Ecdysial suture may be Y-shaped, U-shaped or V-shaped and is variably developed among insects. Arms of the ES that diverge anteroventrally are called Frontal Sutures (Frontogenal Sutures) but are not present in all insects. The complex of sutures is used by many hemimetabolous and paurometabolous insects to emerge from the old Integument during moulting. Parts of ES are called Coronal Suture, Frontal Suture, Epicranial Suture, Ecdysial Line and Cleavage Line in various groups of insects. See Head; Suture. Cf. Apodeme; Apophysis.

ECDYSIOTROPIN Noun. (Greek, *ekdysai* = to strip > *ek* = out; *dyein* = to enter; Pl., Ecdysiotropins.) Prothoracicotropic hormone is produced in the Brain that stimulates ecdysteroid production by the Prothoracic Gland which ultimately affects moulting and Metamorphosis. Rel. Hormone; Metamorphosis.

ECDYSIS Noun. (Greek, *ekdysai* = to strip > *ek* = out; *dyein* = to enter; *-sis* = a condition or state. Pl., Ecdyses.) The process of shedding the Integument during moulting. See Moulting. Cf. Apolysis.

ECDYSONE Noun. (Greek, *ekdysis* = to strip off; *-one* = chemical suffix for a hormone). A hormone produced by the prothoracic glands upon stimulation by prothoracicotropic hormones (PTTH). The target tissue of ecdysone is the cellular layer of the Integument, and the action is the initiation of Apolysis: cell division, cell elongation and production of moulting fluid. Ecdysone is synthesized in the insect from cholesterol or related steroids obtained from the diet.

ECDYSTEROID Noun. (Greek, *ekdysis* = to strip off; *stereos* = solid; *eidos* = form) Ecdysone, its derivatives such as 20-hydroxyecdysone, or other hormones identified in initiating Apolysis. Collectively these hormones are referred to as ecdysteroids. See Ecdysterone.

ECDYSTERONE Noun. (Greek, *ekdysis* = to strip off; *stereos* = solid; *-one* = chemical suffix for a hormone). Synonym for 20-hydroxyecdysone, a more active form of ecdysone converted by the fat body, Malpighian Tubules and gut in moths. See Ecdysteroid.

ECESIC Adj. (Greek, *oikesis* = act of dwelling; *-ic* = characterized by.) Pertaining to Ecesis.

ECESIS Noun. (New Latin, *ec*; *esis* > Greek, *oikesis* = act of dwelling. Pl., Eceses.) The process of emigrant organisms becoming established in new habitats or localities. Cf. Emigration; Migration.

ECETAM®. See Metham-Sodium.

ECHIDNA TICK *Aponomma concolor* Neumann [Acarina: Ixodidae]. (Australia).

ECHINATE Adj. (Greek. *echinos* = hedgehog; *-atus* = adjectival suffix.) 1. Descriptive of surface sculpture, usually the insect's Integument, that displays dispersed elevations or pustules, each of which is more-or-less spiny at the apex. 2. A body covered with spines and resembling a hedgehog. 3. Covered with prickles. Alt. Echinatus. See Sculpture Pattern. Cf. Alveolate; Baculate; Clavate; Favose; Gemmate; Psilate; Punctate; Reticulate; Rugulate; Scabrate; Shagreened; Smooth; Striate; Verrucate.

ECHINOPHTHIRIIDAE Plural Noun. A small Family of Anoplura parasitic on seals and river otters. Adult body relatively large, densely covered with scales, Setae and peg-like receptors with head relatively small; compound eyes or ocular points absent; Antenna with 3–5 segments; Mandibles absent and mouthparts forming a beak or rounded anteriorly. Fore leg small, slender, with an acuminate claw; middle and hind legs similar in size and shape, and with an elaborate tibial thumb. Abdomen membranous without sclerites; spiracles small, each with long, narrow atrial chamber and long chitinous rod. See Anoplura.

ECHOLOCATION Noun. (Latin, *echo* = echo; *locare* = to place. Pl., Echolocations.) High-frequency sounds emitted by organisms that are used to locate objects. Bats use echolocation to locate moths.

ECHTHRO- Greek prefix meaning 'hate' or 'hateful.' Alt. Echtho-.

ECITOPHILE Noun. (Obscure, *Eciton* = genus of army ant; *philein* = to love. Pl., Ecitophiles.) An obligatory guest of army ants of the Tribe Ecitonini. See Army Ant. Cf. Dorylophile.

ECKERT, JOHN EDWARD (1895–1975) (Anon. 1975, Bee Wld 56: 161.)

ECKLON, CHRISTIAN FRIEDRICH (Weidner 1967, Abh. Ver. naturw. Verh. Hamburg Suppl. 9: 135–136.)

ECKSTEIN, FRITZ (1880–1944) (Weidner 1967, Abh. Ver. naturw. Verh. Hamburg Suppl. 9: 347–348.)

ECKSTEIN, KARL (1859–1939) (Anon. 1940, Z. angew. Ent. 26: 382.)

ECLAHRA® See Fosthiazate.

ECLESIS® See Fosthiazate.

ECLIPSE® See Fenoxycarb.

ECLOSION Noun. (Latin: *e* = out; *clausus* = shut. Pl., Eclosions.) 1. The act (process) of hatching (emerging) from the eggshell. Eclosion is a critical phase of the insect's life. Failure to emerge properly can leave an neonate larva or nymph deformed and will terminate life prematurely. Several anatomical features have been developed by insects to solve problems associated with emergence from the eggshell. So-called hatching organs are common in many insect embryos. These 'organs' come in many forms and are called by many names (*e.g.* Egg Burster, Egg Tooth, Egg Oviruptor, Hatching Spine). See Egg; Eggshell. 2. Emergence of the Imago (adult) insect from the pupal case.

ECNOMINDAE Plural Noun. A cosmopolitan Family of hydropsychoid Trichoptera including about 100 Species. Adult wings dull, mottled with span to 20 mm; Ocelli absent; Antenna shorter than forewing; Maxillary Palpus with five segments. Mesoscutellum and Scutellum each with pair of rounded setal warts; wings narrow, rounded apically. Larvae with thoracic nota sclerotized; fore Trochantin elongate, apically acute; Abdomen without gills. Larvae occupy lentic or lotic water; silken tubes fixed. See Trichoptera.

ECOBIOTIC Adj. (Greek, *oikos* = household; *biosis* = manner of life; *-ic* = characterized by.) Pertaining to the specific mode of life adopted within a specific habitat. See Cf. Abiotic; Rel. Habitat.

ECOCLINE Noun. (Greek, *oikos* = household; *klinein* = slanted. Pl., Ecoclines.) 1. A continuous variation in ecotypes with relation to variation in ecological conditions. 2. A continuous series of forms occuring within a group of organisms that occupy adjacent ecological niches. See Cline; Ecotype.

ECOLOGICAL Adj. (Greek, *oikos* = dwelling; *logos* = discourse.) Descriptive of or pertaining to Ecology. Pertaining to conditions determined by the biotic surroundings.

ECOLOGICAL BIOGEOGRAPHY An approach to biogeographical analysis that attempts to explain the distribution of organisms based on abiotic and biotic needs. See Biogeography. Cf. Dispersal Biogeography; Vicariance Biogeography.

ECOLOGICAL MANAGEMENT See Cultural Management.

ECOLOGICAL SPECIES CONCEPT A Species con-cept based on adaptive zones used by organisms. Cf. Species Concepts.

ECOLOGICAL SUBSPECIES A Subspecies conditioned by habit or habitat (Ferris). See Subspecies.

ECOLOGY Noun. (Greek, *oikos* = dwelling; *logos* = discourse. Pl., Ecologies.) The study of organisms in relation to their environment, including reactions or changes in organisms to the conditions of their existence and modifications of these reactions in relation to changes in environment. Alt. Oecology. See Autecology; Bionomics; Synecology. Cf. Ethology.

ECOMORPHOSIS Noun. (Greek, *oikos* = dwelling; *morphe* = form; *-sis* = a condition or state. Pl.,

Ecomorphoses.) Environmentally induced polymorphism. See Polymorphism.

ECONOMIC DAMAGE The amount of pest-induced injury that justifies the cost of applying pest control measures.

ECONOMIC ENTOMOLOGY The subdiscipline of Entomology concerned with the study of positive and negative economic values of insects as they affect human life and activities. EE is sometimes equated with Agricultural Entomology or the control of agricultural pests. See Entomology. Cf. Agricultural Entomology; Applied Entomology.

ECONOMIC INJURY LEVEL The level of injury to a crop by a pest at which the implementation of control measures becomes cost effective. See Action Level. Acr. EIL.

ECONOMIC POISON Legal classification for a substance used to control, prevent, destroy, repel or mitigate any pest.

ECONOMIC THRESHOLD The pest density at which management action should be taken to prevent an increasing pest population from reaching the economic injury level (EIL). Designated as ET also referred to as Action Threshold. See Economic Injury Level; Threshold. Cf. Action Threshold; Medical Threshold; Social Threshold.

ECOPRO® See Temephos.

ECOSPECIES Singular and Plural Noun. (Greek, *oikos* = dwelling; Latin, *species* = particular kind.) A group of individuals occupying a particular ecological niche, behaving as a Species and whose members are capable of interbreeding without genetical loss. See Species.

ECOSYSTEM Noun. (Greek, *oikos* = dwelling; *systema* = composite whole. Pl., Ecosystems). An ecological system formed of component members (a community) and the abioltic environment surrounding the system. See Ecology. Cf. Agroecosystem; Urban Ecosystem. Rel. Biome.

ECOTECH® A registered biopesticide derived from *Bacillus thuringiensis* var. *kurstaki*. See *Bacillus thuringiensis*.

ECOTECH EXTRA® A biopesticide derived from *Bacillus thuringiensis* strain EG 2424. See *Bacillus thuringiensis*.

ECOTECH PRO® A biopesticide derived from *Bacillus thuringiensis* EG 2348. See *Bacillus thuringiensis*.

ECOTHRIN® See Tetramethrin.

ECOTYPE Noun. (Greek, *oikos* = dwelling; *typos* = pattern. Pl., Ecotypes.) 1. Distinct plant races whose distinguishing characters are related to the selective effects of local environments. Cf. Subspecies. 2. Plants or animals that are adapted to narrowly defined tolerances for various conditions essential to life (*e.g.* temperature, humidity, light, nutrients).

ECTAD Adv. (Greek, *ektos* = outside; Latin, *ad* = toward.) From within and toward the outer surface of the insect body. See Orientation. Cf. Anteriad; Apicad; Basad; Caudad; Centrad; Cephalad; Craniad; Dextrad; Dextrocaudad; Dextrocephalad; Distad; Dorsad; Entad; Laterad; Mediad; Mesad; Neurad; Orad; Proximad; Rostrad; Sinistrad; Sinistrocaudad; Sinistrocephalad; Ventrad.

ECTADENIA Plural Noun. (Greek, *ektos* = outside; *aden* = gland.) The accessory male glands of ectodermal origin and derived through evaginations of the Ejaculatory Duct. Cuticle lines the Ectadenia and a glandular epithelium is not well developed or may be absent. When present, GE may be secondarily derived. Many calypterate Diptera display well developed Ectadenia. Funtionally, Ectadenia are poorly studied and their role is not known. Syn. Ectadene. See Male Accessory Gland. Cf. Mesadenia.

ECTAL Adv. (Greek, *ektos* = outside.) Directed outward or toward the outer surface of the insect body. See Orientation. Cf. Ental.

ECTOBLAST Noun. (Greek, *ektos* = outside; *blastos* = bud. Pl., Ectoblasts.) The outer wall of a cell; the Ectoderm or Epiblast.

ECTODERM Noun. (Greek, *ektos* = outside; *derma* = skin. Pl., Ectoderms.) The outer layer of skin, the outer layer of the Blastoderm. Ectoderm gives rise to the nervous system, Integument, sensory receptors, glands and epithelial structures of the body surface. Cf. Endoderm.

ECTODERMAL Adj. (Greek, *ektos* = outside; *derma* = skin; Latin, *-alis* = pertaining to.) Descriptive of or pertaining to the Ectoderm; arising from the Ectoderm.

ECTODEX® See Amitraz.

ECTOGNATHOUS Adj. (Greek, *ektos* = outside; *gnathos* = jaw; Latin, *-osus* = with the property of.) Pertaining to an organism with exserted mouthparts. See Ectotrophous.

ECTOLABIUM See Labium.

ECTOPARASITE Noun. (Greek, *ektos* = outside; *parasitos* = parasite; *-ites* = inhabitant. Pl., Ectoparasites.) A parasite that develops on the external surface of a host's body. Ectoparasitic insects include fleas (larvae and adults) and lice (nymphs and adults). Vertebrates serve as hosts of ectoparasites. See Ectoparasitoid; Parasite; Parasitism. Cf. Endoparasite. Rel. Predator.

ECTOPARASITIC Adj. (Greek, *ektos* = outside; *parasitos* = parasite; *-ic* = of the nature of.) Pertaining to an Ectoparasite; with the nature of an Ectoparasite. See Parasite. Cf. Endoparasitic.

ECTOPARASITOID Noun. (Greek, *ektos* = outside; *parasitos* = parasite; *eidos* = form. Pl., Ectoparasitoids.) A parasitoid whose larval stage feeds and develops on the external surface of a host's body. Most ectoparasitoids are larval Hymenoptera and Diptera. The hosts of ectoparasitoids include insects and rarely other arthropods. See Endoparasitoid; Parasitoid; Parasitism. Cf. Ectoparasite. Rel. Predator.

ECTOPHAGOUS Adj. (Greek, *ektos* = outside;

phagein = to devour; Latin, *-osus* = with the property of.) A term descriptive of organisms that feed externally or in an exposed situation. Term usually used in the context of parasitism in contrast with organisms (parasites) that feed internally. See Feeding Strategy; Parasitism. Cf. Endophagous.

ECTOPHALLUS Noun. (Greek, *ektos* = outside; *phallos* = penis. Pl., Ectophalluses.) The outer phallic wall as distinct from the Endophallus (Snodgrass).

ECTOPLASM Noun. (Greek, *ektos* = outside; *plasm* = to form or mould. Pl., Ectoplasms.) An external layer of protoplasm in a cell.

ECTOPOR® See Cypermethrin.

ECTOPSOCIDAE Plural Noun. A moderate-sized Family of Psocoptera assigned to Suborder Psocomorpha. Adults with broad wings that are held relatively flat over the Abdomen; Pterostigma nearly parallel-sided and Arolea Postica absent. Species typically occupy leaf litter and may reach high populations. See Psocoptera.

ECTOPTYGMA Noun. (Greek, *ektos* = outside.) The Serosa of the embryo (Graber).

ECTOSKELETAL Adj. (Greek, *ektos* = outside; *skeleton* = mummy; Latin, *-alis* = belonging to, pertaining to.) Referring to something outside the skeleton; exoskeleton; exoskeletal.

ECTOSYMBIONT Noun. (Greek, *ekto-*; *symbionai* = to live with. Pl., Ectosymbionts.) A symbiont that associates with a host colony during at least part of its life cycle in some relationship other than internal parasitism. See Symbiont.

ECTOTRACHEA Noun. (Greek, *ekto-*; Latin, *trachea* = windpipe. Pl., Ectotracheae.) The outer surface-layer of the Trachea.

ECTOTROPHOUS Adj. (Greek, *ektos* = outside; *trophein* = to eat; Latin, *-osus* = with the property of.) Descriptive of organisms with mouthparts exposed or free, and not embedded or concealed within the head. See Feeding Strategy. Cf. Entotrophous.

ECTREPHIDAE See Anobiidae.

ECTRIN® See Fenvalerate.

ECYDONURIDAE See Heptageniidae.

ECYDURIDAE See Heptageniidae.

ED/CT® See Ethylene Dichloride.

EDAPHIC Adj. (Greek, *edaphos* = ground; *-ic* = of the nature of.) Relating to the soil. Pertaining to conditions of soil or substrate as it influences organisms.

EDB® See Ethylene Dibromide.

E-D-BEE® See Ethylene Dibromide.

EDC® See Ethylene Dichloride.

EDDY, BRAYTON (1899–1950) (Anon. 1950, J. Econ. Ent. 43: 573.)

EDDY, CLIFFORD OTIS (1894–1966) (Dickinson 1966, J. Econ. Ent. 59: 1550.)

EDELMANN, HERBERT (1909–1967) (Kaisila 1968, Suomen hyönt. Aikak. 34: 105–106.)

EDELSTEN, HUBERT MCDONALD (1877–1959) (Uvarov 1960, Proc. R. ent. Soc. Lond. (C) 24: 51.)

EDEMA Noun. (Greek, *oidema* = a swelling; Pl., Edemata.) Abnormal accumulation of serous fluid in connective tissues or serous cavities.

EDEMATUS Adj. (Greek, *oidema* = a swelling.) Dull, translucent white. Alt. Edematose; Edematous.

EDENTATE Adj. (Latin, *ex* = without; *dens* = teeth; *-atus* = adjectival suffix.) Pertaining to an organism or structure that lacks teeth; descriptive of a toothless organism. Alt. Edentatus; Edentula; Edentulous. See Dentate.

EDENTULA Adj. (Latin, *ex* = without; *dens* = tooth.) See Edentate.

EDGE Noun. (Anglo Saxon, *ecg* = edge > Middle English, *egge*. Pl., Edges.) 1. The margin; Acies. 2. The cutting side of a blade (knife, lance, sword). 3. The brink or verge of a cliff. 4. A border of a surface or structure. Syn. Margin. 5. An advantage.

EDGREN, PER ADOLF (1802–1891) (Sandahl 1891, Ent. Tidschr. 12: 233.)

EDITUM Noun. (Latin, *editus* > *edere* = to bring forth.) Genitalia of male Lepidoptera: A rounded, usually setose, protuberance that arises more-or-less dorsally from the medial surface of the Harpes (Klots).

EDLESTON, ROBERT SMITH (1819–) (Westwood 1872, Proc. ent. Soc. Lond. 1872: 1.)

EDMONDS, TOMAS HERBERT (1887–1944) (Cockayne 1945, Proc. R. ent. Soc. Lond. (C) 9: 47.)

EDMUNDS, ABRAHAM (1804–1869) (Anon. 1869, Entomol. mon. Mag. 6: 42.)

EDRIZAR® See Amitraz.

EDUFOS® Cadusafos.

EDWARD, THOMAS (1814–1886) (Smiles 1877, Harper, N. Y. 1: 303.)

EDWARDS CLIFFORD (1886–1972) (W[orms] 1973, Entomologist's Rec. J. Var. 85: 69.)

EDWARDS, A M See Milne-Edwards, A.

EDWARDS, EDWIN HUGH (1867–1939) (Osborn 1946, *Fragments of Entomological History*. Pt. II. 232 pp. (78), Columbus, Ohio.)

EDWARDS, FREDERICK WALLACE (1888–1940) (Usinger 1941, Pan-Pacific. Ent. 17: 84.)

EDWARDS, GEORGE (1694–1773) English librarian at the College of Physicians, Naturalist and author of *Gleanings of Natural History*. (Lisney 1960, *Bibliogr. of British Lepidoptera 1608–1799*. 315 pp. (127–144, bibliogr.), London.)

EDWARDS, GEORGE A (1914–1960) (Tompkins 1960, Ann. ent. Soc. Am. 53: 547.)

EDWARDS, H MILNE- See Milne-Edwards, H.

EDWARDS, HENRY (1830–1891) (Osborn 1937, *Fragments of Entomological History,* 394 pp. (162), Columbus, Ohio.)

EDWARDS, JAMES (1856–1928) (Turner 1928, Entomol. mon. Mag. 64: 27.)

EDWARDS, STANLEY (1864–1938) (Turner 1938, Proc. Linn. Soc. Lond. 1937– 38: 311–312.)

EDWARDS, THOMAS GROVES (1880–1958) (S. W. 1958, Proc. Trans. S. Lond. ent. nat. Hist. Soc. 1958: xliv–xlv.)

EDWARDS, WILLIAM HENRY (1822–1909) (Mallis

1971, *American Entomologists*. 549 pp. (288–292), New Brunswick.)

EECKE, R VAN (1886–1975) (Helsdingen 1976, Ent. Ber., Amst. 36: 64.)

EEDLE, THOMAS (1829–1888) (Carrington 1889, Entomologist 22: 52.)

EEE See Eastern Equine Encephalitis.

EFFACED Trans. Verb. (French *es-*, from Latin, *ex-*; *face* = face.) Obliterated; rubbed out; not visible.

EFFECTOR Noun. (Latin, *effectus* > to effect; *-or* = one who does something. Pl., Effectors.) An organ of the body, principally a muscle or a gland, that is activated by nerve stimuli (Snodgrass).

EFFERENT Verb. (Latin, *efferre* = to bear out > *ex* = out of; *ferre* = to carry.) Carrying outward or away from the centre. Cf. Afferent.

EFFERENT NERVE A nerve that conducts from a nerve centre toward the periphery; the Axon of a motor neurone (Snodgrass). Cf. Afferent Nerve.

EFFERENT NEURON A nerve cell that conveys impulses from ganglia outward to muscles or glands, a Motor Neurone. Alt. Efferent Neurone.

EFFLATOUN, HASSAN CHAKER (1893–1957) (Richards 1959, Proc. R. ent. Soc. Lond. (C) 23: 72.)

EFFLUVIUM Noun. (Latin, *effuere* = to flow out; *-ium* = diminutive. Pl., Effluvia.) A foul or unpleasant smell or emanation.

EFITAX® See Alphacypermethrin.

EFLECTED. Adj. (Latin, *ex-*; *flexus* = to bend.) Somewhat angularly bent outward.

EGEST Transitive Verb. (Latin, *egestus* > *egerere* = to carry out.) Evacuate. 1. To cast or expel indigestible matter from the body. 2. To void waste product from the body via respiration, sweating, digestion.

EGESTIA Plural Noun. (Latin, *egestus*.) Something expelled from the body. Syn. Dejecta; Excreta.

EGESTION Noun. (Latin, *egestio*; English, *-tion* = result of an action. Pl., Egestions.) Evacuation.

EGG Noun. (Anglo Saxon, *Aeg* > Islandic, *egg* > Old Norse, *egg*. Pl., Eggs.) 1. A globular, round or ovoid unit of reproduction capable of fertilization and consisting of a cover (eggshell) that surrounds nutrients (yolk) necessary for embryonic development. 2. An Ovum or cell from an Ovary. 3. The first stage of insect development. Correctly: A clutch of eggs. 3. Anything resembling an egg in form or function. The egg is the largest cell of the insect body. Entomologists often refer to the egg as one of the principal stages of life. This characterization is misleading because the egg actually serves as a containment device for the embryonic stage of insect development. This device accommodates, protects and serves physiological needs of the developing embryo. Environmental requirements of an embryo within the eggshell are modest. Most eggs contain all requisites for embryonic development; only oxygen and gas exchange are supplemental. Embryos of some insect Species require water for development; this is particularly evident in Species with aquatic immature stages. Water absorbed during the liquid phase typically increases egg size; water absorbed during the vapour phase does not increase the egg's volume but typically returns the egg to its size prior to desiccation. Some terrestrial insects produce embryos that require nutrients provided by the eggshell (*e.g.* phasmids require calcium in the eggshell). Other insects probably require an external supply of nutrients for development, but this has not been documented. See Ovum. Cf. Eggshell; Egg Shape. Rel. Egg Morphology; Metamorphosis.

EGG BURSTER A point, elevation, ridge or keel on the embryonic head used to mechanically rupture the eggshell during eclosion. The Egg Burster appears common among Hemiptera. EB does not cut the surface of the eggshell. EB is present on a preemergent nymph but does not appear until late in embryonic development. EB is a 'T' shaped sclerite with membranous Cuticle suspended between the arms. The sclerite 'arms' are somewhat arched and form a blunt tetrahedral angle at their point of intersection. Lateral arms are positioned on dorsum along Pronotum's anterior margin; the third arm extends along an imaginary dorso-longitudinal line. The preemergent nymph's head is deflexed such that the 'T' shaped sclerite can apply hydrostatic pressure to the Operculum See Egg. Cf. Egg Tooth; Hatching Spine; Oviruptor. Rel. Eclosion.

EGG CALYX The enlarged portion of the Lateral Oviduct at junction with the Ovariole; EC contains the egg before it passes into the Median Oviduct. See Calyx.

EGG CASE A case (cover, envelope) secreted by a female to contain (bond) a group of eggs laid during one ovipositional episode. Syn. Egg Pod; Ootheca; Ovicapsule.

EGG CHAMBER A compartment or follicle of an ovarial egg tube that forms Follicle Cells that surround an Oocyte (Snodgrass). See Ovariole. Rel. Meroistic Ovary; Panoistic Ovary.

EGG DEFENCE Eggs are vulnerable to predation and parasitism. In response to pressure from egg predators and egg parasites, numerous adaptations have evolved that defend the insect egg from these mortailty agents. Female mosquitoes in Genera *Culex* and *Culiseta* deposit a defensive secretion on their eggs. Sticky filaments on eggs of spined soldier bug, *Podisus maculiventris*, may be an adaptation to deter parasitic Hymenoptera. Eggs of balsam woolly adelgid *Adelges piceae* are covered with small bits of wax and attached to fir tree with a wax thread; wax may deter predation. Other adaptations also deter predation and parasitism. Many insects conceal their eggs. Crypsis can involve several techniques. Spumaline is a froth formed on surface of a liquid by action of effervescence or agitation. Spumaline sometimes is employed to envelope an egg to render it invisible. Eggs of

Bombyliidae are 'dusted' and invisible due to adherent particles of dirt. Spumaline on egg of aphid *Longistigma caryae* also serves to dust egg. Disruptive coloration is an important adaptation used by some insects to protect eggs from predators. Eggs of Lepidoptera (Saturniidae, Lymantriidae Lasiocampidae) display a black spot that resembles an emergence hole made by parasitic wasps and deter predation. Aposematic coloration is important in warning predators of distastefulness, unpalatability or lethality. See Aposematic Coloration; Crypsis; Disruptive Coloration; Repagula; Spumaline.

EGG EPICHORION See Epichorion.

EGG EXTRACHORION A thin outer layer of the egg that contains granules and which envelops a newly deposited egg. (Hartley, Quart. J. Microsc. Sci. 102: 249–255.)

EGG GUIDE Orthoptera: Two small, pointed structures between upper and lower valves on the ventral part of the eighth abdominal segment. The EG is used in oviposition. Cf. Ovipositor.

EGG MORPHOLOGY The study of egg anatomy in relation to function of the component parts of the egg.

EGG POLARITY The insect egg has primary and secondary axes. Entomologists frequently refer to anterior and posterior poles of insect eggs without defining terms. Implicitly, anterior should mean nearest the head; but nearest the head of the female laying an egg or the pole of egg in which embryonic head develops? Topographical features of eggs may not necessarily be used to solve the problem of orientation because many features are not universally present, easily recognized or consistent. Egg orientation in Pentatomidae is such that females attach eggs to the substrata with the primary axis of an egg perpendicular to substrata; posterior pole of an egg emerges from female first, is attached to substrata and is the egg region in which an embryo will develop. Egg orientation among Hymenoptera follows Haller's (1886) Law of Orientation within polytrophic ovariole: Anterior pole is directed toward head of parent female; during oviposition, posterior pole emerges first, which permits regulation of fertilization. Dorsal, ventral and lateral sides of eggs vary within individual female. The embryo remains in original cephalocaudal axis during entire development, but just before eclosion it rotates 180° on longitudinal axis. Some Species of tephritid flies have elongate eggs. Embryo within eggshell rotates 180° before emergence such that head of developing embryo orients toward central axis of the gall or flower head into which the egg is laid. See Egg. Rel. Oviposition.

EGG SHAPE See Acuminate Egg; Encyrtiform Egg; Flat Egg; Hymenopteriform Egg; Macrotype Egg; Membranous Egg; Microtype Egg; Stalked Egg; Pedicellate Egg.

EGG SIZE Insect egg size varies among Species. The smallest insect egg is less than 0.05 mm long; the largest insect egg is several millimetres long. Within that size range, a continuum of sizes is displayed. Physiological, ecological and functional considerations can influence the egg size. Physiological factors are related to female laying an egg. Female size is sometimes correlated with size of egg she lays (*e.g.* stick insect *Extatosoma tiaratum*). In other insects, physiological condition of female may influence size of egg laid (*e.g.* moth *Chilo partellus*). Ecological factors influence egg size but are intrinsically difficult to analyse. Tachinidae are all parasitic as larvae. Tachinid Species that lay macrotype eggs oviposit on hosts; tachinid Species that lay microtype eggs oviposit on vegetation. Significant differences in numbers of eggs laid are also reflected in Ecology. Macrotype eggs are relatively few in number per female (ca 300); microtype eggs laid on vegetation and consumed by the host are relatively abundant (ca 3,000+). Functional considerations can influence egg size. For instance, availability of oxygen subtly influences eggshell anatomy. Complex chorionic structure of eggshell is frequently intended to provide free flow of oxygen and carbon dioxide between embryo and atmosphere. Resisting mechanical stress is another important functional consideration, but is difficult to analyse without knowledge of egg ultrastructure, physical properties and component chemicals that comprise eggshell.

EGG TOOTH See Oviruptor.

EGGAR Noun. (Etymology obscure.) Lepidoptera: Members of *Eriogaster* or *Lasiocampa* (Lasiocampidae). Syn. Egger.

EGGAR MOTHS See Lasiocampidae.

EGGER, HEINRICH (1848–1915) (Wöhlbier. 1931, Mansfelder Lond. (Beil. Eisteber Ztg.) 6: 321–328.)

EGGER, JOHANN (1804–1866) (Schiner 1867, Verh. zool.-bot. Ges. Wien 17: 531–540, bibliogr.)

EGGERS, HANS (1873–1947) (Schedl 1947, Zentbl. Gesamtgeb. Ent. 3: [1–2].)

EGG-FRUIT CATERPILLAR *Sceliodes cordalis* (Doubleday) [Lepidoptera: Pyralidae]. (Australia).

EGGLESTONE, J W (1877–1963) (Wigglesworth 1964, Proc. R. ent. Soc. Lond. (C) 28: 57.)

EGG-POD Orthoptera: A clutch of eggs surrounded by a capsule. Cf. Egg Case; Ootheca.

EGG-POUCH See Ootheca.

EGG-TOOTH Syn. Egg-Burster; Hatching Device. See Oviruptor.

EGG-TUBE See Ovarian Tube.

EGG-VALVE Mayflies: The posterior prolongation of the seventh Sternum (Needham).

EGGPLANT FLEA BEETLE *Epitrix fuscula* Crotch [Coleoptera: Chrysomelidae].

EGGPLANT FRUIT BORER *Leucinoides orbonalis* [Lepidoptera: Pyralidae]: A pest of solanaceous plants in Africa and southeast Asia.

EGGPLANT LACE BUG *Gargaphia solani*

Heidemann [Hemiptera: Tingidae]: In USA, a pest of Linden tree, and members of the Solanaceae. First antennal segment black or dark brown; Paranota angulate or narrowly rounded; Pronotum with vestiture of long, silky Setae. Female guards eggs and early instar nymphs. See Tingidae.

EGGPLANT LEAFMINER *Tildenia inconspicuella* (Murtfeldt) [Lepidoptera: Gelechiidae].

EGGPLANT STEM-BORER *Euzophera osseatella* (Treitschke) [Lepidoptera: Pyralidae]: A pest of solanaceous plants that occur around the Mediterranean Sea.

EGGSHELL Noun. (Anglo Saxon, *aeg;* Middle English, *shelle* > Anglo Saxon > *scell*. Pl., Eggshells.) 1. The typically hard, external covering of the Ovum (egg). See Egg; Ovum. 2. Insects: Collectively, the Chorion and Vitelline Membrane of an insect egg. Pterygote eggshell is produced by follicular cells of the Ovary and forms a biochemically complex structure composed of several layers. Physical and chemical characteristics of each layer are unique. Eggshell may be compared to the Integument: Each layer is responsible for a different function. Vitelline Membrane forms the eggshell's innermost layer. VM surrounds the Oocyte and is a product of the follicular cell epithelium; developmental precursors of VM sometimes are called Vitelline Bodies. A Wax Layer surrounds VM; WL is formed when VM is completed. Lipids are produced by ovarian follicle cells and deposited as plaques. Freeze fracture reveals a smooth-surfaced material arranged longitudinally in multiple layers. Smoothness suggests the material is not protein; longitudinal fractures of layers suggests hydrophobicity. Conventional opinion holds that WL is responsible for waterproofing to resist desiccation and drowning. An Endochorion Layer resides above WL; Endochorion is proteinaceous and of varing thickness and architectural complexity. An Exochorion layer typically surrounds the Endochorion. See Egg. Cf. Chorion; Endochorion; Exochorion; Vitelline Membrane; Wax Layer.

EGGSHELL SCULPTURE The outer surface of insect eggs varies from smooth (polished) to coarse or ornately sculptured. Sculptural patterns consist of elevated ridges. In some insects, the eggshell is elaborately sculptured and taxonomists have developed an extensive nomenclature to describe these features. Rel. Sculpture Pattern.

EGGSHELL TOPOGRAPHY The physical features and surface contours of the eggshell. Topographical features include the surface sculpture, Pedicle, Micropyle, Aeropyle, Hydropyle and Operculum. See Eggshell Sculpture.

EGYPT BEE VIRUS An obscure viral disease of adult honey bees first reported from Egypt. See Honey Bee Viral Disease.

EGYPTIAN ALFALFA WEEVIL *Hypera brunnipennis* (Boheman) [Coleoptera: Curculionidae].

EGYPTIAN BEETLE *Blaps polychresta* Forskal [Coleoptera: Tenebrionidae]. (Australia).

EGYPTIAN BOLLWORM *Earias insulana* (Boisduval) [Lepidoptera: Noctuidae] a widespread pest of cotton in the Old World, particularly Africa, Egypt and India. See Noctuidae. Cf. Cotton Bollworm; Pink Bollworm; Spiny Bollworm. Spotted Bollworm.

EGYPTIAN HOUSE FLY *Musca vicina* Macquart [Diptera: Muscidae]: An urban pest in Egypt. Adults resemble House Fly; EHF prefers to breed in horse and donkey manure, but will also breed in human faeces. See Muscidae. Cf. House Fly; Bush Fly; Face Fly.

EHD See Epizootic Haemorrhagic Disease.

EHF See Epidemic Haemorrhagic Fever.

EHRENBERG, CARL AUGUST (1801–1849) (Papavero 1975, *Essays on the History of Neotropical Dipterology*. 2: 293. São Paulo.)

EHRENBERG, CHRISTIAN GOTTFRIED (1795–1876) (Nordenskiöld 1935, *History of Biology*. 629 pp. (427–428), London.)

EHRENBERG, M Published 'Insects collected in Egypt, Nubia, and Arabia' (In: *Symbolae Physicae*.)

EHRENREICH, PAUL (–1914) (Kuhnt 1914, Dt. ent. Z. 1914: 357.)

EHRET, GEORG DIONYSIUS (1710–1770) (Rose 1850, *New General Biographical Dictionary* 7: 217.)

EHRHORN, EDWARD MACFARLAINE (1862–1941) (Osborn 1946, *Fragments of Entomological History*. Pt. II. 232 pp. (78–79), Columbus, Ohio.)

EHRLICHIEAE A Tribe of rickettsia-like organisms assigned to Family Rickettsiaceae. Ehrlichieae are pathogens of mammals which replicate in cytoplasm of reticuloendothelial cells. Ixodid ticks known as vectors that are trans-stadially transmitted, but not trans-ovarially transmitted. See Rickettsiaceae. Cf. Canine Ehrlichiosis; Heartwater; Potomac Horse Fever.

EHRMAN, GEORGE ALEXANDER (1862–1926) (Hollan 1926, Ent. News 37: 95–96.)

EICHBAUM, ERNER (1883–1966) (Friese 1968, Ent. Berichte, Berl. 1968: 135.)

EICHELBAUM, FELIX (1848–1922) (Weidner 1967, Abh. Verh. naturw. Ver. Hamburg Suppl. 9: 263.)

EICHHORN'S CROW BUTTERFLY *Euploea eichhorni* Staudinger [Lepidoptera: Nymphalidae]. (Australia).

EICHLER, WITOLD (1874–1960) (Sliwinski & Tranda 1963, Polskie Pismo ent. (B) 31–32: 267–269.)

EICHLER'S RULE 1. A hypothesis asserting that numerically large taxonomic groups of hosts will harbour higher taxonomic categories of parasites in association. 2. Groups of hosts with more variation are attacked by more parasites that are taxonomically uniform groups of hosts. Cf. Emery's Rule; Farenholtz' Rule; Manter's Rules;

Szidat's Rule.

EICHOFF, WILHELM JOSEPH (1824–1893) (Wildermann 1894, Jb. naturw. 9: 503.)

EICHWALD, EDUARD VON (1795–1876) (Stricker 1877, Zool. Garten 18: 72.)

EICKWORT, GEORGE C (1940–1994) American academic (Cornell University) and specialist on behaviour, morphology and taxonomy of bees.

EIDEL, KARL (1905–1975) (Tobias & Döhler 1976, Ent. Z. Frankf. A M. 86: 13–15, bibliogr.)

EIDMANN, HERMANN A (–1949) (Brown 1949, Ent. News 60: 262.)

EIDOS Noun. (Greek, *eidos* = shape; form; appearance. Pl., Eidi.) A form, essence or concept. Eidos is used as a combining suffix (-oid) for a Greek noun to construct a word that indicates (without explicit description) the form (three-dimensional shape) of an object (actinoid), substance (albuminoid), texture (malacoid), function (gynecoid), behaviour (parasitoid) or attribute (humanoid) of a structure, body or organism. See Form; Shape; Outline Shape.

EIFFINGER, GEORG (1838–1920) (Pfaff 1934, Festschr. 50 jähr. Beste. Int. Ent. Ver. Frankf. p.7.)

EIGHTEEN SPOTTED LADYBIRD See Common Spotted Ladybird.

EIGHT-SPOTTED FORESTER *Alypia octomaculata* (Fabricius) [Lepidoptera: Noctuidae].

EIGHT-TOOTHED SPRUCE BEETLE *Ips typographus* [Coleoptera: Scolytidae].

EIL See Economic Injury Level.

EIMER, THEODOR (1843–1898) (Carrington 1899, Nat. Sci., Lond. 14: 164–165.)

EISEN, GUSTAVUS AUGUST (1847–1940) (Osborn 1946, *Fragments of Entomological History*. Pt. II. 232 pp. (79), Columbus. Ohio.)

EJACULATE Noun. (Latin, *ejaculatus* = thrown out. Pl., Ejaculates.) Seminal fluid emitted from the body of the male.

EJACULATORY Adj. (Latin, *ejaculare* = to throw out.) Pertaining to ducts that emit fluid from the body.

EJACULATORY BULB Diptera: A strongly-musculated syringe-like structure of the male Ejaculatory Duct (Snodgrass).

EJACULATORY DUCT A slender, flexible tube within the Aedeagus that is introduced into the Ostium and Ductus Bursae of the female during copulation. Spermatozoa pass through the ED from the Vasa Deferentia (Klots). The ED is Ectoderm, lined with Cuticle, sometimes musculated and forms as an invagination between Sterna IX and X. In Thysanura, the median Penis is located between the bases of the Styli of Sternum IX. The ED is homologous with the female's Median Oviduct. In most pterygote insects the duct terminates in membrane between the abdominal segments IX and X. The aperture is called the Gonopore. Presumably, the ED forms as an integumental invagination originating at the Gonopore. Alt. Ductus Ejaculatorius. Cf. Penis

EJACULATORY SAC Any organ that pumps ejaculate from the Vas Deferens through the Ejaculatory Duct to the Aedeagus.

EJACULATORY VALVE In some male moths: A device between the Radix Penis and Glandular Prostatica. The opening of the valve is regulated by the surrounding sphincter, thus impeding the back flow of secretions and seminal fluid within the Radix Penis and resulting in their transport outwards during ejaculation. The musculature of the Ejaculatory Duct and Corpus Penis promotes further transport of these secretions into the female's Bursa Copulatrix. See Vas Deferens. Rel. Copulation.

EKALUX® See Quinalphos.

EKAMET® See Etrimfos.

EKANON® See Disulfoton.

EKATIN® See Thiometon.

EKATOX® Parathion.

EKBLON, AXEL (–1914) (Mjöberg 1914, Ent. Tidskr. 35: 221–222.)

EKELBERG, HANS JONAS (1805–1889) (Sandahl 1889, Ent. Tidskr. 10: 161–164.)

EKSMIN® See Permethrin.

EKSTRÖM, MARTIN (1891–1969) (Svenson 1969, Opusc. ent. 34: 174–175.)

EKTAFOS® See Dicrotophos.

EKTOMIN® See Cypermethrin.

EKWALL, NILS ARVID (1919–1942) (Hanso 1943, Ent. Tidskr. 64: 204.)

EL SEGUNDO BLUE *Euphilotes battoides allyni* (Shields) [Lepidoptera: Lycaenidae].

ELABRATE Adj. (Latin, *e* = without; *labrum* = lip; *-atus* = adjectival suffix.) Descriptive of structure that lacks a Labrum. Alt. Elabratus. See Mouthparts. Cf. Labrum.

ELACATIDAE See Salpingidae.

ELACHISTIDAE Bruand 1850. Plural Noun. Grass Leafminer Moths; Grass Miners. A small Family of ditrysian Lepidoptera assigned to Superfamily Gelechioidea that is cosmopolitan in distribution but most speciose in temperate areas of Northern Hemisphere. Adults are very small, wingspan 5–15 mm and head with smooth, overlapping lamellar scales; Ocelli present or absent; Chaetosemata absent; Antenna shorter than forewing; Scape sometimes with Pecten; Proboscis with scales basad; Maxillary Palpus small, 1–2 segments; Labial Palpus variable; fore Tibia with Epiphysis; tibial spur formula 0-2-4; forewing and hindwing lanceolate; marginal fringe long; Discal cell well developed in hindwing. Eggs are oval or elongate-oval with longitudinal ribs; eggs are laid singly or in small batches on upper surface of leaf near veins or margin. Mature larvae are spindle-shaped with head prognathous and flattened; thoracic legs usually present but sometimes prothoracic legs absent. Most larvae mine leaves throughout stage; a few Species mine stems. Host plants typically are grass and sedge; cause blotch mines in grass. Pupae are spindle-shaped and slightly flattened. Some Species pupate in silken webs; some Species form flimsy cocoons; some Pupae are naked. See Gelechioidea.

ELAEOPHOROSIS Noun. (Greek, *elaion* = olive oil; *phore* = producer; *-sis* = a condition or state. Pl., Elaeophoroses.) The filarial worm *Elaeophora schneideri* that attacks deer, elk and sheep in North America. Parasites move into arteries to cause blindness, antler deformities, muzzle necrosis and possibly death. Tabanid flies serve as vectors for the worms. Syn. Arterial Worm; Blood Worm.

ELAIOSOME Noun. (Greek, *elaion* = oil; *soma* = body. Pl., Elaisomes.) A plastid that forms and stores lipids. Elaiosomes on seeds are attractive to some ants and apparently serve as a device for seed dispersal. Syn. Aril. See Myrmecochory.

ELAMPINAE A moderate-size, cosmopolitan Subfamily of Chrysididae (Hymenoptera: Aculeata) that is numerically most well represented in the Holarctic. Elampines are characterized by forewing with Discoidal Cell weakly developed or absent; Tegula small; Gaster with three exposed Terga and Sterna flat or concave; each tarsal claw with more than one tine (except *Xerochrum*). The Subfamily appears transitional between Subfamilies that attack aculeate Hymenoptera larvae and Subfamilies that have specialized on sawflies and phasmatid eggs. See Chrysididae.

ELASMIDAE Plural Noun. A moderately small Family (ca 1–2 Genera, 200 nominal Species) of parasitic Hymenoptera assigned to the Chalcidoidea, and sometimes placed in the Eulophidae. Elasmid adults typically are small (1.5–3.0 mm) and never metallic coloured; Antenna with nine segments (including Anellus) and male Antenna ramose; forewing wedge-shaped with Marginal Vein long, Stigmal and Postmarginal Veins very short; hind Coxa enlarged and compressed; hind Tibia with zig-zag pattern of dark Setae along posterior margin; Tarsomere formula 4-4-4. Most elasmid Species act as primary, gregarious parasites of Lepidoptera; a few elasmids are hyperparasitic Species attacking Braconidae, Ichneumonidae in cocoons, and Vespidae provisioning cells with Lepidoptera. Females paralyse hosts with venom. The number of eggs deposited on hosts is influenced by host size. First instar larvae are hymenopteriform and segmented; some with thoracic Pseudodpoda (locomotory); four pairs of spiracles on first instar, last instar with nine pairs of spiracles. See Chalcidoidea.

ELASTES Noun. (Greek, *elastikos* = to drive.) Elastic organs of the ventral segments of *Machilis* that assist in the act of leaping.

ELASTIC Adj. (Greek, *elastikos* = to drive; *-ic* = characterized by.) Descriptive of structure that is flexible or pliant.

ELATE Alt. Elatus. See Elevated.

ELATER Noun. (Greek, *elater* = driver. Pl., Elaters.) Collembola: The Furcula.

ELATERIDAE Plural Noun. Click Beetles; Skipjack Beetles; Pithworms; Wireworms. A cosmopolitan Family of about 7,000 described Species that is related to Lampyridae. Adult body length to 60 mm, elongate; head small, with acute ridge between medial margins of compound eyes; head deeply recessed in Prothorax; Antenna not clavate, typically serrate or pectinate. Posterolateral pronotal process corresponding to an anterolateral Mesosternal Sulcus that collectively form a click mechanism. Eggs are often oval to spherical shaped, white and deposited singly. Larvae are elongate, tough bodied, with thoracic legs and head flattened. Larvae are phytophagous or predaceous. Phytophagous Species feed on the roots of many grasses and vegetables including beans, cabbage, cucumber, onions, peas and potatoes; predatory Species live in concealed situations, typically in soil or in wood. Hosts include larvae of xylophagous insects, larvae and Pupae of scarabs, and larvae of Lepidoptera. Cannibalism occurs in some elaterid Species that live in high population densities. Some Species require several years to complete one generation. See Coleoptera.

ELATERIFORM LARVA 1. Any larva that resembles an elaterid (wireworm) larva in form. 2. A slender larva that is moderately heavily sclerotized, lacks elaborate ornamentation or sculpture, and bears three pairs of thoracic legs.

ELATERIFORMIA A Series (*sensu* Lawrence 1988) of Coleoptera that includes Superfamilies Dascilloidea, Buprestoidea, Byrrhoidea and Elateroidea. Adults with streamlined body and pro-mesothoracic coupling device. Most members have a heterogeneous life-cycle with long-lived larvae and short-lived adults.

ELATERIUM Noun. (Greek, *elaterion* > *elaterios* = driving; *-ium* = diminutive. Pl., Elateria.) Hymenoptera: The membranous, flexible portion of the upper surface of the fused second Valvulae that forms the Ovipositor shaft (Smith 1969; Quicke *et al.* 1994, J. Nat Hist 28: 635). The Elaterium varies in size and shape but may contain Resilin and is thought to enable expansion of the paired second Valvulae during passage of the egg through the egg canal during oviposition. See Ovipositor.

ELATEROIDEA Noun. A Superfamily of polyphagous Coleoptera of diverse composition. Includes Cantharoidea in some sense. Brachypsectridae, Cantharidae, Cebronidae, Cerophytidae, Elateridae, Eucnemidae, Lampyridae, Lycidae, Perothopidae, Rhinorhiphidae, Throscidae (Trixagidae). Adults lack mandibular Mola, transverse Metasternal Suture; most with four Malpighian Tubules; larvae with one Stemma on each side of head, mouthparts adapted for liquid feeding. See Coleoptera.

ELATTOSTASE Noun. (Etymology obscure.) Acarology: A stage of development in which the mouthparts atrophy, but the individual remains mobile.

ELBOWED ANTENNA See Geniculate Antenna.

ELCAR® See *Heliothis* NPV.

ELDER SHOOT BORER *Achatodes zeae* (Harris) [Lepidoptera: Noctuidae].

ELECTROANTENNOGRAM Noun. (Greek, *elektro*; *antenna*; *graphein* = to write. Electroantennograms.) An apparatus used to measure the electric potential of an insect Antenna stimulated by volatile compounds.

ELECTROPHORESIS Noun. (Greek, *elektro*; *pherein* = bear, carry; *-sis* = a condition or state. Pl., Electrophoreses.) A process by which gene products of an individual organism are separated by an electrical field in a gel medium and then stained so that they may be identified and classified. Electrophoresis is used to infer genotypes in populations and relative gene frequencies. Various techniques include Horizontal Starch Gel Electrophoresis, Cellulose Acetate, Polyacrylamide Electrophoresis and Isoelectric Focusing.

ELECTROTOMIDAE Rasnitsyn 1977. Plural Noun. A Family of Symphyta (Hymenoptera) assigned to the Tenthredinoidea. See Symphyta.

ELECTROTROPISM See Galvanotropism.

ELEMENT Noun. (Latin, *elementum* = first principle element. Pl., Elements.) 1. A component or constituent part of something. 2. An ingredient of several substances mixed together. Cf. Member.

ELEOCYTE See Spherulocyte.

ELEPHANT BEETLE *Xylotrupes gideon* (Linnaeus) [Coleoptera: Scarabaeidae]. (Australia).

ELEPHANT BEETLES See Scarabaeidae.

ELEPHANT FLIES Members of the Genus *Tabanus* with eyes unmodified or with horizontal bands and pale wings. See Tabanidae.

ELEPHANT WEEVIL *Orthorhinus cylindrirostris* (Fabricius) [Coleoptera: Curculiondiae]: A pest of cultivated and native trees in Australia, particularly dying or felled rainforest trees. EW hosts include apples, citrus, *Eucalyptus*, grape and stonefruit. Adults are ca 20 mm long, grey with darker markings and Proboscis 5–7 mm long; fore legs longer than middle or hind legs; Elytra with four apical protuberances. Females lay eggs in niches in bark and neonate larvae penetrate wood to form tunnels (circular in cross section, usually along grain) that become tightly packed with fine fibrous material. Tunnels made by each larva can extend ca 0.6 m. Pupation occurs within cells at ends of tunnels. Adult beetles emerge and bore directly to exterior. EW is monovoltine and the life cycle requires ca one year. See Curculionidae.

ELEPHANTIASIS Noun. (Greek, *elephas* = elephant; *-iasis* = suffix for names of diseases. Pl., Elephantiases.) In humans, Elephantiasis is a chronic affection of the skin commonly affecting the legs and genitals with hypertrophy of cellular tissue due to the obstruction in lymphatic circulation and blood vessels by an infestation of the filarial worm, *Wucheria bancrofti* or *Brugia malayi*. Filarial worms are passed from one mammalian host to another by the feeding of their intermediate hosts, mosquitoes in the Genera *Aedes*, *Anopheles* and *Culex*. The disease is Pan-tropical. Syn. Yava Skin; Barbados Leg. Cf. Filariasis. Rel. Parasitism.

ELEUTHERATA Noun. (Greek, *eleutheros* = free.) Coleoptera: All forms with free, separated Maxillae.

ELEUTHEROTOGONY Noun. (Greek, *eleutheros* = free; *-gony* = to be born. Eleutherotogonies.) A condition expressed during embryonic development when the back of the insect is formed without participation of the membranes (Henneguy).

ELEVATED Adj. (Latin, *elevatus* > *elevare* = to lift up; *-atus* = adjectival suffix.) Descriptive of a structure or part that is higher than its surroundings. Alt. Elevate; Elevatus. Ant. Depressed.

ELEVATOR Noun. (Latin, *elevatus* > *elevare* = to lift up; *-or* = one who does something. Pl., Elevators.) A muscle that, when contracted, elevates or moves an appendage or structure upward. Cf. Depressor; Rotator.

ELEVEN-SPOTTED LADYBIRD *Coccinella undecimpunctata* Linnaeus [Coleoptera: Coccinellidae]. (Australia).

ELIA BROWN *Nesoxenica leprea* (Hewitson) [Lepidoptera: Nymphalidae]. (Australia).

ELIE DE BEAUMONT, JEAN BAPTISTE ARMAND LOUIS LÉONÉE (1798–1874) (Bertrand 1877, Mém. Acad. Sci. Inst. Fr. 39: lx–xxxvi.)

ELIENA SKIPPER *Trapezites eliena* (Hewitson) [Lepidoptera: Hesperiidae]. (Australia).

ELIMINATION Noun. (Latin, *eliminare* > *ex-*; *limen* = threshold; English, *-tion* = result of an action. Pl., Eliminations.) The physiological discharge of any waste or useless substances from the body tissues. Rel. Ingestion.

ELINGUATE Adj. (Latin, *elinguatus* = to deprive of the tongue; *-atus* = adjectival suffix.) Without a tongue; descriptive of organisms in which the Maxillae are connate with the Labium. Syn. Synista. Cf. Tongue.

ELIOT, SAMUEL LEWELL (1844–1889) (Packard 1889, Entomologica Amer. 5: 83–84.)

ELIPSOCIDAE Plural Noun. A numerically small, cosmopolitan Family of Psocoptera assigned to the Suborder Psocomorpha. Antenna with 13 segments and lacking secondary annulations; forms with 2- and 3-segmented Tarsi. Member Taxa are associated mostly with trees. See Psocoptera.

ELISA Enzyme-Linked Immunosorbent Assay. An immunological method employed to identify and quantify the presence of an antigen or antibody. The principle of ELISA involves the detection of an antigen with enzyme-labelled antibodies. The enzyme causes a colorimetric reaction that is proportional to the number of antigen molecules bound to the enzyme-antibody conjugate. Three common ELISA protocols include: Competitive ELISA, direct double-antibody sandwich ELISA, and indirect antibody sandwich ELISA.

ELITE Social Insects: A colony member that displays great initiative and activity. Cf. Altruism.

ELLERTON, JOHN (1910–1971) (A. M. E. 1972, Proc. Brit. ent. nat. Hist. Soc. 5: 74–75.)

ELLERTSON, FLOYD ELROY (1919–1963) (Jones & Richter 1963, J. Econ. Ent. 57: 423.)

ELLIIDAE Krzeminska *et al.* 1993 Plural Noun. A small Family of fossil Diptera known from Mesozoic deposits in central Asia. See Diptera.

ELLIOTT, D C (1909–1943) (Vige 1945, Ann. ent. Soc. Am. 38: 140–141.)

ELLIOTT, ERNEST A (–1936) (Imms 1937, Proc. R. ent. Soc. Lond. (C) 1: 55.)

ELLIPSOIDAL See Elliptical.

ELLIPTICAL Adj. (Greek, *elleiptikos* = elliptical, defective; Latin, *-alis* = characterized by.) Oblong-oval; descriptive of structure whose ends are equally rounded and together forming an even ellipsoid. Alt. Ellipticum. See Outline Shape.

ELLIS, HENRY, WILLOUGHBY (1869–1943) (Cockayne 1944, Proc. R. ent. Soc. Lond. (C) 8: 69.)

ELLIS, JOHN (1710–1776) (Rose 1850, *New General Biographical Dictionary* 7 227.)

ELLISON, GEORGE (1862–1941) (Ellison 1944, NWest Nat. 19: 23–26.)

ELLISON, R D (–1959) (Uvarov 1960, Proc. R. ent. Soc. Lond. (C) 24: 53.)

ELLISON, ROBERT ELDON (1895–1959) (Worms 1960, Entomologist 63: 92.)

ELLISOR, LEWIE OWEN (191O-1939) (Osborn 1946, *Fragments of Entomological History*. Pt. II. 232 pp. (79–80), Columbus, Ohio.)

ELLISTON-WRIGHT, FREDERICK ROBERT (1879–1966) (Pearson 1967, Proc. R. ent Soc. Lond. (C) 31: 62.)

ELM APHID *Tinocallis platani* (Kaltenbach) [Hemiptera: Aphididae]: A monophagous aphid endemic to the Palearctic that was introduced into North America where high populations and consequent honeydew are a nuisance. See Aphididae.

ELM BARK-BEETLE *Scolytus multistriatus* (Marsham) [Coleoptera: Curculionidae]. (Australia). See Smaller European-Elm Bark-Beetle.

ELM BORER *Saperda tridentata* Oliver [Coleoptera: Cerambycidae]: A pest of slippery elm and white elm in eastern North America. EB attacks trees of lowered vitality and can kill them. Eggs are deposited in bark fissures of twigs and branches; eclosion occurs during autumn and neonate larvae penetrate inner bark. EBs overwinter as early-instar larvae in bark and sapwood; larvae resume feeding during spring and form irregular channels within inner bark and sapwood. Pupation occurs within cells formed in sapwood. Adults are active during early summer; EB typically are monovoltine but may require two years to complete development under adverse conditions. See Cerambycidae. Cf. Linden Borer; Locust Borer; Poplar Borer.

ELM CALLIGRAPHA *Calligrapha scalaris* (LeConte) [Coleoptera: Chrysomelidae].

ELM CASEBEARER. *Coleophora ulmifoliella* McDunnough [Lepidoptera: Coleophoridae].

ELM COCKSCOMB GALL-APHID *Colopha ulmicola* (Fitch) [Hemiptera: Aphididae]: A pest of red elm throughout USA that induces plants to produce unsightly galls on leaves which appear as rooster's comb. See Aphididae.

ELM FLEA-BEETLE *Altica carinata* Germar [Coleoptera: Chrysomelidae].

ELM LACE-BUG *Corythucha ulmi* Osborn & Drake [Hemiptera: Tingidae].

ELM LEAF-BEETLE *Pyrrhalta luteola* (Müller) [Coleoptera: Chrysomelidae]: A highly destructive pest of Elms that was introduced into the USA from Europe during the 1830s. Adults are ca 7–8 mm long with body finely pubescent, yellow-green in coloration with three dark spots on Pronotum, and dark stripe along side of Elytra. ELB overwinters as an adult in places of concealment. Eggs are orange-yellow, spindle shaped and laid on end in clusters on underside of leaves. Neonate larvae are black, becoming yellow with two dark stipes on dorsum of head but legs and tubercles remain black. Larvae feed gregariously on underside of leaves and may aggregate in large numbers on trunk. Pupation occurs in bark fissures. ELBs overwinter as adults. ELB is the most important defoliator of urban elm trees; infested trees display yellow, skeletonized foliage. Syn. *Xanthogaleruca luteola*. See Chrysomelidae. Cf. Larger Elm-Leaf Beetle.

ELM LEAFHOPPER *Ribautiana ulmi* (Linnaeus) [Hemiptera: Cicadellidae]. (Australia).

ELM SAWFLY *Cimbex americana* Leach [Hymenoptera: Cimbicidae]: A widespread pest of broad-leaved trees in North America. ES eggs are oval, flattened, transparent and inserted into leaves. Larvae feed on leaves and remain coiled at rest. ES overwinters as a larva within a cocoon on ground; pupation occurs during following spring. Adults girdle bark on twigs. See Cimbicidae.

ELM SCALE *Gossyparia spuria* (Modeer) [Hemiptera: Eriococcidae].

ELM SCURFY-SCALE *Chionaspis americana* Johnson [Hemiptera: Diaspididae]: A pest of hackberry in North America. ESS resembles Scurfy Scale and Dogwood Scale. See Diaspididae.

ELM SPANWORM *Ennomos subsignarius* (Hübner) [Lepidoptera: Geometridae]. A defoliator of hardwood forests in the eastern United States.

ELM SPHINX *Ceratomia amyntor* (Geyer) [Lepidoptera: Sphingidae].

ELMIDAE Plural Noun. Riffle Beetles. A Family of

polyphagous Coleoptera assigned to Byrrhoidea or Dryopoidea. Adults are 1–6 mm long and body with tomentose tracts on venter that serve as a Plastron. Antenna with 11 segments and legs long; tarsal formula 5-5-5, Pretarsi each with two long claws; Abdomen with five Ventrites. Larva onisciform and head prognathous; Antenna with three segments; legs with five segments and each Pretarsus with one claw; Urogomphi absent. Larvae respire through anal gills. Most adults and all larvae live under water in association with rocky bottoms where they feed on decaying vegetation and algae. Elimid larvae leave water to pupate. Syn. Elminthidae; Helminthidae. See Coleoptera.

ELM-LEAF APHID *Tinocallis ulmifolii* (Monell) [Hemiptera: Aphididae].

ELM-LEAF MINER *Fenusa ulmi* Sundevall [Hymenoptera: Tenthredinidae]: An Holarctic pest of European and American elms. The ELM larva mines leaves. See Tenthredinidae.

ELOCRIL® See Phoxim.

ELONGATE Adj. (Latin, *elongare* = to prolong; *-atus* = adjectival suffix.) Drawn out; structure that is much longer than wide. Ant. Shortened; Transverse.

ELONGATE ANTENNA An Antenna that is as long as the body.

ELONGATE FLEA BEETLE *Systena elongata* (Fabricius) [Coleoptera: Chrysomelidae].

ELONGATE HEMLOCK SCALE *Fiorinia externa* Ferris [Hemiptera: Diaspidae].

ELONGATED PORES Coccids: See Dorsal Pores (MacGillivray).

ELSAN® See Phenthoate.

ELSTER, GEORGE RUDOLPH (1897–1940) (Osborn 1946, *Fragments of Entomological History*. Pt. II. 232 pp. (80), Columbus, Ohio.)

ELTHAM COPPER *Pralucia pyrodiscus lucida* Crosby [Lepidoptera: Lycaenidae]. (Australia).

ELTRINGHAM, HARRY (1873–1941) English businessman and amateur lepidopterist interested in mimicry. Eltringham served as Curator of Insects at the Hope Museum, Oxford (1924–1937). (Riley 1942, Entomologist 75: 25–27.)

ELTRINGHAM ORGAN A structure that occurs along the basal portion of the hindwing posterior margin in some male Myrmeleontidae. EO facilitates dispersal of volatile thoracic gland secretions (Lofqvist & Bergstom 1980. Ins. Biochem. 10: 1–10.)

ELUTE Adj. (Latin, *eluere* from *ex-*; *luere* = wash.) With indistinct markings. Rel. Sculpture.

ELWES, HENRY JOHN (1846–1922) (Riley 1924, Proc. S. Lond. ent. nat. Hist. Soc. 1923–24: 76.)

ELY, CHARLES RUSSELL (1870–1939) (Osborn 1946, *Fragments of Entomological History*. Pt. II. 232 pp. (80), Columbus, Ohio.)

ELYTRAL LIGULA A tongue-like process on the inner face of the side margin of the Elytra. EL serves to perfect the union with the ventral segments in Dytiscidae. See Elytron; Ligula.

ELYTRIFORM Adj. (Greek, *elytron* = sheath; Latin, *forma* = shape.) Shaped or appearing as an Elytron.

ELYTRIN See Chitin.

ELYTRON Noun. (Greek, *sheath*, from *elyein* = to roll around. Pl., Elytra.) The forewings of Coleoptera. Elytra are not membranous, instead, they have become modified into leathery or chitinous covers that protect the hindwings. Typically, Elytra are not used in active flapping flight. At repose, Elytra typically meet and form a straight line along the middle of dorsum. The Elytron is divided into several regions. The Disc forms the surface and is marked with longitudinal striae that contribute to the Elytron's structural rigidity. The basic number of Striae is nine, and by convention they are numbered from the Sutural Edge laterad. The number of Striae in beetles range from 0 to about 25. The Interstriae (interstices) are areas between Striae. Odd-numbered Interstriae correspond to primitive wing veins. Frequently Interstriae are externally different in size, convexity and pilosity. The Sutural Edge is represented by the adducting medial surfaces of the Elytra. The Epipleuron forms the lateral, incurved border of the Elytron. Two tubercles are formed at the base that articulates with the Thorax to rotate the Elytron forward in preparation for flight. The Epipleuron is not present in Curculionidae or Meloidae. The Elytra have several locking mechanisms. Sutural Edges lock by a flange in a groove (tongue-and-groove, mortise-and-tenon). Elytra are fused in many vestigially winged Taxa. The lock affords protection to the soft abdominal Terga and tightly cover the abdominal spiracles. Elytra are not always fused and sometimes they are just locked. Sometimes, the Epipleural Flange locks with the abdominal Sternum. Alt. Elytrum. See Wing Modification. Cf. Hemelytra; Tegmina.

ELYTROPTERA See Coleoptera.

EMA, TEIZIRO (1867–) (Hasegawa 1967, Kontyû 35 (Suppl.): 19, bibliogr. [In Japanese.])

EMANDIBULATA Noun. (Latin, *ex* = without; *mandibulum* = jaw.) Insects that lack functional mandibules in all stages. *e.g.* butterflies and moths. Term applied in any life stage. Alt. Emandibulatus. See Mandible.

EMARGINATE Adj. (Latin: *ex* = out; *marginare* = to limit; *-atus* = adjectival suffix.) Pertaining to the margin of a sclerite or structure that displays an incision, dent or notch. Alt. Emarginatus. See Cf. Entire. Adv., Emarginately.

EMARGINATION Noun. (Latin, *ex* = out; *marginare* = to delimit; English, *-tion* = result of an action. Pl., Emarginations.) A notch or an incised area at an edge, margin, or apex.

EMBA-FUME® See Methyl Bromide.

EMBALMING FLUID A liquid intended to preserve insects, particularly large, soft-bodied insects. Components include: Toluene (or Xylene) 60 ml; Tertiary Butyl Alcohol (TBA) 25 ml; Ethyl Alcohol

15 ml; Phenol 5 gr; Naphthelene 20 gr; Canada Balsam in Xylene 10 drops.

EMBATHION® See Ethion.

EMBIIDAE Plural Noun. A Family of Embiidina.

EMBIIDINA Enderline 1903. Web Spinners. Foot Spinners. The correct Ordinal name for insects that have been called Embioptera. A widespread group of about 200 nominal Species and an indeterminant number of undescribed Species. Fossil record begins with specimens in Oligocene Baltic Amber. Web spinners are well represented in tropical and warm temperate regions. Adults with body small and elongate; Head prognathous; Antenna filiform with 12–32 segments; mandibulate mouthparts and Gula present; compound eye reniform, large in male, small in female; Ocelli absent. Females are wingless and males are winged or wingless; both conditions may occur in males of one Species; wing venation is reduced when wings are present. Legs each with three Tarsomeres. Adult male and female and all nymphal instars spin silk; fore Basitarsus contains silk glands; silk issues from ducts on ventral surface of Basitarsus and second Tarsomeres. Abdomen with ten apparent segments and segment 11 reduced; Cercus tactile with two segments; panoistic Ovarioles in female. Nervous system with three thoracic ganglia and seven abdominal ganglia. All Species are gregarious with nymphs and adults occupying silken galleries. Rapid rearward movement is enabled by large depressor muscle of hind Tibia. Species feed on bark, moss, lichen, dead leaves and other plant material. Cannibalism may occur and males sometimes are eaten by females after copulation. Eggs and early instar nymphs are guarded by female. Eggs are elongate, curved and operculate; eggs are laid on walls of tunnels where they are guarded by the female. Included Families: Anisembiidae, Australembiidae, Clothodidae, Embiidae, Embonychidae, Notoligotomidae, Oligotomidae, Tetratembiidae. See Embioptera.

EMBIOPTERA Shipley 1904 (Greek, *embio* = lively; *pteron* = wing.) An emendation of Embiidina for uniformity of Ordinal name endings. See Embiidina.

EMBOLAR Adj. (Greek, *embolos* = wedge.) Heteroptera: Descriptive of or pertaining to the Embolium of the wing.

EMBOLEMIDAE Förster 1856. Plural Noun. A widespread, numerically small, rare Family of apocritious Hymenoptera assigned to the Chrysidoidea that includes about 15 nominal Species with strong sexual dimorphism. Both sexes with Antenna 10-segmented; Torulus on prominence and distant from Clypeus; when head viewed in lateral aspect, Mandible posterior of level of compound eye; male typically macropterous; female frequently brachypterous or apterous with habitus ant-like and cone-shaped head. Reported hosts include crickets and Fulgoridae but biology of Species is poorly known. Apparently, eggs are deposited internally and hosts show no sign of paralysis. Embolemid larvae develop externally near wing base in a sac. Larvae are gregarious, do not form masses of cocoons and disperse before pupation. Placement of the Family is problematical, but probably near Bethylidae and Dryinidae. See Chrysididae.

EMBOLIUM Noun. (Greek, *embolos* = wedge, something interposed; Latin, *-ium* = diminutive > Greek, *-idion*. Pl., Embolia.) The differentiated Costal part of the Corium in the forewing (Hemelytron) of some Heteroptera (Tillyard). The special enlargement at the base of the forewing that fits into the cavity in which the wing is moved (Smith).

EMBONYCHIDAE. Plural Noun. A small Family of Embiidina (Embioptera).

EMBOSSED Adj. (Middle English, *embosen*.) Ornamented with raised figures.

EMBRY, B (1898–1963) (Wigglesworth 1964, Proc. R. ent. Soc. Lond. (C) 28: 57.)

EMBRYO Noun. (Greek, *embryon* > *en* = in; *bryein* = to swell. Pl., Embryos.) 1. The animal or plant in early stages of development. 2. Animals: An organism during the period between cleveage and birth (hatching, eclosion). The insect embryo displays features adaptive to that stage only. Unique embryonic features include amniotic folds for protection, Trophamnion for nutrition and a tooth on the embryo's head used for eclosion. These features are useful to the embryo but not subsequent stages. 3. Seed-bearing plants: The sporophyte. See Differentiation. Cf. Foetus; Rel. Adult; Immature.

EMBRYOGENESIS Noun. (Greek, *embryon* = embryo; *genesis* = descent. Pl., Embryogeneses.) The second phase of the developmental process experienced by insects. The embryonic phase of develop begins inside the eggshell with germ-band formation and ends with dorsal closure. See Development. Cf. Metamorphosis; Oogenesis. Rel. Histology.

EMBRYOLOGY Noun. (Greek, *embryon* = embryo; *logos* = discourse. Pl., Embryologies.) The study of plants and animals that is concerned with growth, differentiation and development between the time of fertilization and birth (emergence, eclosion, germination). See Development; Ontogeny. Cf. Gerantology.

EMBRYONAL Adj. (Greek, *embryon* = embryo; Latin, *-alis* = belonging to, pertaining to.) Descriptive of or pertaining to the embryo.

EMBRYONIC Adj. (Greek, *embryon* = embryo; *-ic* = characterized by.) 1. Found in, or relating to the embryo. 2. An organism in an undeveloped state or condition.

EMBRYONIC PERIOD The interval of the life-cycle between fertilization and eclosion from the egg.

EMDEN, FRITZ ISIDORE VAN (1898–1958) (Richards 1959, Proc. R. ent. Soc. Lond. (C) 23:

72.)

EMENDATION Noun. (Latin, *emendatio*. Pl., Emendations.) 1. A change in practise, procedure or principle. 2. A critical change to correct an error of fact or interpretation. 3. Taxonomy: An intentional modification of the spelling of a previously published scientific name.

EMENDATUS Latin, amended. Abbreviated 'emend.' See Emendation 2.

EMERALD COCKROACH WASP *Ampulex compressa* (Fabricius) [Hymenoptera: Sphecidae].

EMERGENCE Noun. (Latin, *emergere* from *ex*-; *mergere* = to plunge. Pl., Emergences.) The escape of an adult winged insect from its cocoon, pupal case or nymphal Integument. Cf. Eclosion; Hatching. Rel. Metamorphosis.

EMERTON, JAMES HENRY (1847–1930) (Bonnet 1945, *Bibliographia Araneorum* 1: 46.)

EMERY, CARLO (1848–1925) (Musgrave 1932, *A Bibliogr. of Australian Entomology 1775–1930*. 380 pp. (79), Sydney.)

EMERY, JOHN W (1918–1972) (Butler 1973, Proc. R. ent. Soc. Lond. (C) 37: 55.)

EMERY'S RULE Social Insects: The notion that social parasites are very similar to their hosts and presumably closely related phylogenetically. Cf. Eichler's Rule; Fahrenholtz Rule; Manter's Rules; Szidat's Rule.

EMEX WEEVIL *Perapion antiquum* (Gyllenhal) [Coleoptera: Brentidae]. (Australia).

EMGE, JOSEPH (1840–1895) (Anon. 1895, Societas ent. 9: 181.)

EMIGRATION Noun. (Latin, *emigratio* > *emigrare* = to remove; *ex* = away, *migrare* = to move English, *-tion* = result of an action. Pl., Emigrations.) The movement of individuals, groups or populations from one place to another place. Cf. Dispersal; Migration.

EMMA FIELD CRICKET *Teleogryllus emma* (Ohmachi & Matsuura) [Orthoptera: Gryllidae].

EMMATOS® Malathion.

EMMET Noun. (Middle English, *emete* = ant. Pl., Emmets.) An ant.

EMMONS, EBENEZER (1799–1863) (Osborn 1937, *Fragments of Entomological History*. 394 pp. (141), Columbus, Ohio.)

EMPEROR GUM MOTH *Opodiphthera eucalypti* (Scott) [Lepidoptera: Saturniidae]. (Australia).

EMPEROR MOTHS See Saturniidae.

EMPERORS See Lethrinidae.

EMPIDAE Plural Noun. Dance Flies. A cosmopolitan, orthorrhaphous Family of Diptera with about 6,000 Species, with most Species found in north temperate and montane regions. Empids are closely related to Bombyliidae and Dolichopodidae, and show a fossil record dating to the Middle Cretaceous Period. Empids display piercing/sucking mouthparts, and some Species have a short Proboscis with fleshy Labellum; wing venation with Anal Cell shortened, reduced or absent; two or three posterior cells. Adults are predaceous with characteristic mating flights or dances; males of many Genera form mating swarms that are sometimes elaborate. Males frequently utilize captured insect prey or bubbles of a frothy secretion as a lure for females. Habits of immature stages are poorly known. Larvae are aquatic or terrestrial in leaf litter, soil, rotting wood. Larvae of a few Genera have been reared from cow dung. Most females oviposit; females of Holarctic *Ocydromia glabricula* (Fallen) larvaposit on dung. Larvae presumed predaceous on immature stages of other arthropods. See Diptera; Empidoidea.

EMPIDIDAE See Empidae.

EMPIDOIDEA Noun. A Superfamily of orthorrhaphous Diptera including Dolichopodidae (Dolichopidae) and Empidae (Empididae). Wing with shortened CuA that usually is recurved to apex in basal half of A1; genitalia complex. Adults and larvae typically are predaceous; larvae are aquatic or terrestrial. (See Waters 1989, A Cretaceous Period dance fly (Diptera: Empidae) from Botswana. Syst. Ent. 14 (2): 233–241, regarding phylogeny.)

EMPIRE® See Chlorpyrifos.

EMPIRICAL Adj. (Greek, *empeirikos* = experienced.) Descriptive of something that is learned though experience or observation as opposed to theoretical knowledge. Cf. Heuristic; Theoretical.

EMPODIAL HAIR A digitule; a tenent Seta on the Tarsus or Tibia of coccids (MacGillivray).

EMPODIUM Noun. (Greek, *en* = in; *pous* = feet; *empodion* = an obstacle; Latin, *-ium* = diminutive > Greek *idion*. Pl., Empodia.) 1. A structure of variable size and shape located between the pretarsal claws. 2. A process of the Unguitractor (Crampton), often erroneously applied to a Pulvillus or Onychium. The Empodium in Heteroptera forms paired, spine-like divergent Setae called Parempodia. 3. The single pad-like or filiform median structure often present in the insect claws, either between the paired Pulvilli or alone without such Pulvilli – a term proper to this structure only (Imms). 4. Acarina: The thin, usually claw-like structure arising medially, between the claws, from the Pretarsus (Goff *et al.* 1982, J. Med. Ent. 19: 223.) See Pretarsus. Cf. Arolium; Median Unguitractor Plate. Rel. Leg.

EMPSON, DAVID WADSWORTH (1921–1968) (Kennedy 1969, Proc. R. ent. Soc. Lond. (C) 33: 54.)

EMPUSIDAE Plural Noun. A small Family of Mantodea, including eight Genera, found in Africa, the Mediterranean and Asia. Adult body typically large, slender and unusually formed; Frons and Clypeus carinate with a median cranial process; compound eyes round; male Antenna bipectinate; fore Femur with 4–5 discoidal spines and five outer spines; Abdomen laterally lobate and supra-anal sclerite transverse.

EMULSIFIABLE CONCENTRATE A form of insecticide formulation in which an emulsifying agent is added to the active ingredient and oil-based solvent (toluene) to dilute the material with water. EC = active ingredient + solvent + emulsifier. EC diluted in water becomes an emulsion in which globules of insecticide and solvent are uniformly distributed. EC is designated as 'EC' or 'E' on USA Environmental Protection Agency label or Material Safety Data Sheet documentation. See Aerosol; Dust; Microencapsulated Concentrate; Oil Concentrate; Suspension Concentrate; Wettable Powder.

EMULSIFIER Noun. (Latin, *emulsum* = to milk out. Pl., Emulsifiers.) A type of surfactant added to pesticides that affects the mixing of oil-based liquids in water. Emulsifiers cause insecticides and oil to form small globules and disperse uniformly in water. See Pesticide; Wetting Agent.

EMULSION Noun. (Latin, *emulsio* > *emulgere* > *emulsum* = to drain out. Pl., Emulsions.) A temporary suspension of microscopic droplets of one liquid in a second liquid with which the first does not mix. *e.g.* oil and water shaken together form an emulsion.

EMULSIONIZE See Emulsify.

ENAMELLED SPIDER *Araneus bradleyi* (Keyserling) [Araneida: Araneidae]. (Australia).

ENANTIOMER Noun. (Greek, *enantios* = opposite; *meros* = part. Pl., Enantiomers.) An optical isomer.

ENANTIOMORPHIC Adj. (Greek, *enantios* = opposite; *morphe* = form; *-ic* = characterized by.) Pertaining to objects that are mirror images.

ENARTHROSIS Noun. (Greek, *en* = in; *arthron* = joint; *-sis* = a condition or state. Pl., Enarthroses.) An articulation such as a ball-and-socket joint. See Articulation. Cf. Acetabulum; Diarthrosis; Condyle; Cotyla; Ginglymus.

ENCAPSULATION Noun. (Greek, *en* = in; Latin, *capsa* = box; English, *-tion* = result of an action. Pl., Encapsulations.) A Cellular Defence Mechanism (immune response) of insects in which moderate-sized foreign material, microorganisms (pathogens) or parasitoids in Haemolymph of an insect are surrounded or enveloped by Phagocytes (Plasmatocytes) or other types of Haemocytes. Encapsulation occurs around objects too large for Haemocyte Phagocytosis. Encapsulation process involves cooperation between haemocytes and is mediated by cytokines and adhesion molecules. See Haemolymph. Cf. Nodule Formation; Phagocytosis. Rel. Coagulation.

ENCEPHALIC Adj. (Greek, *engkephalos* = brain; *-ic* = characterized by.) Pertaining to the brain.

ENCEPHALITIS Noun. (Greek, *enkephalos* = brain; English, *-tion* = result of an action. Pl., Encephalitides.) 1. Any of several arboviral diseases that affect the Brain of humans and which are often fatal. 2. Diseases carried in other animals and transmitted to humans by mosquitoes. Most notable encephalitides of human concern include: St Louis Encephalitis, Murray Valley Encephalitis, West Nile Encephalitis and Japanese Encephalitis. See Arbovirus. Cf. Japanese Encephalitis; Murray Valley Encephalitis; St Louis Encephalitis; West Nile Encephalitis.

ENCEPHALIZATION Noun. (Greek, *enkephalos* = brain; English, *-tion* = result of an action. Pl., Encephalizations.) The process of Brain formation.

ENCEPHALON Noun. (Greek, *enkephalos* = brain. Pl., Encephalons.) 1. The brain. 2. The portion of the head containing the brain. Alt. Encephalum. See Brain; Cephalon.

ENCIRCLED Adj. (Greek, *en-*; *kyklon* = circle.) Ringed; margined round about.

ENCYRTIDAE Plural Noun. Encyrtid Wasps. A cosmopolitan, large (ca 500 Genera, 3,000 nominal Species) Family of parasitic Hymenoptera assigned to the Chalcidoidea. Diagnositic features include adult body robust (0.75–5.0 mm long), often metallic or dark coloured; female Antenna usually with 11 segments (5–13 segments, without Anelli), male Antenna usually with nine segments (5–10 segments) sometimes ramose; Mesoscutum transverse, Notauli or Parapsidal Sutures usually absent, rarely present and invariably incomplete; Mesopleuron enlarged, conspicuous, convex; forewing usually macropterous, rarely apterous, Marginal Vein very short (punctiform), Postmarginal and Stigmal Veins short, subequal; middle tibial spur enlarged, spindle-like; middle Basitarsus with pegs or blunt spines; Tarsomeres formula 5-5-5, rarely 4-4-4; Pygostyli located anteriad of Gaster apex. All Species are endoparasites and some Species are hyperparasitic. Host spectrum of Encyrtidae is broad, but predominantly parasitic of Homoptera. Ovarian eggs are encyrtiform with the deposited egg stalked. Polyembryony is known in many Species. Encyrtidae presumably are related to Tanaostigmatidae and Eupelmidae. Encyrtids are important in biological control of mealybugs and scale insects. See Chalcidoidea.

ENCYRTIFORM EGG A form of egg characteristic of parasitic Hymenoptera Families Encyrtidae and Tanaostigmatidae. Within Ovary, eggs are shaped as two spheres connected by a narrow cylinder. Eggs are deposited into host and one sphere collapses with narrow cylinder projecting as a stalk outside the body of the host. An aeroscopic sclerite, used for embryonic and larval respiration, usually is found on stalk and sometimes projects onto the body of the egg. The radical change in shape after oviposition is noteworthy.

END CHAMBER The Germarium of a gonadial tube (Snodgrass). See Ovariole; Testis.

END HOOK Dragonflies: A tooth along the medial border of the lateral lobe.

END ORGAN Nerve: The peripheral receptors for external stimuli, at the end of a nerve fibre.

ENDANGERED SPECIES Any Species that is 'in

danger of extinction throughout all or a significant portion of its range.' (United States Environmental Species Act of 1973 as amended through the 100th Congress, Sec. 3, United States Fish and Wildlife Service, Department of the Interior, Washington DC.)

ENDEMIC Adj. (Greek, *endemos* = native; *-ic* = of the nature of.) Pertaining to organisms in a specific geographical region or ecological habitat; organisms native to a region and not introduced. Term applied to Species or higher taxonomic groups. See Zoogeography. Cf. Adventive; Introduced; Native. Rel. Pandemic.

ENDEMIC TYPHUS See Murine Typhus.

ENDEMISM Noun. (Greek, *endemos* > *en-*; *demos* = the people; English, *-ism* = state. Pl., Endemisms.) Biogeography: The quality of having unique resident Species of plants and animals.

ENDERLEIN, GUENTER (1872–1968) (Rohlfien 1975, Beitr. Ent. 25: 291 only.)

ENDITE Noun. (Greek, *endon* = within; *-ites* = constituent. Pl., Endites.) A mesial lobe of any limb segment (Snodgrass). Cf. Exite.

ENDOBLAST Noun. (Greek, *endon* = within; *blastos* = bud. Pl., Endoblasts.) An inner layer formed by the invagination of the Blastoderm.

ENDOCARDIUM Noun. (Greek, *endon* = within; *kardia* = heart; Latin, *-ium* = diminutive > Greek, *-idion*. Pl., Endocardia.) The membrane that forms the inner lining of the heart. Cf. Mycardium.

ENDOCEL® See Endosulfan.

ENDOCHORION Noun. (Greek, *endon* = within; *chorion* = skin. Pl., Endochorions.) An architecturally complex layer of the Chorion of the insect eggshell that is positioned between the Wax Layer and Exochorion. The Endochorion is predominantly protein. During Endochorion development, spaces between Endochorion 'pillars' may be filled with polysaccharides or glycoproteins. When eggshell is completed, this material dries and cavities form between pillars. These cavities contain air and serve in respiration, or the innermost part of the Endochorion forms an Aerostatic Layer a few micra thick. AL may be distributed uniformly throughout the egg's inner lining. See Eggshell. Cf. Exochorion; Vitelline Membrane; Wax Layer.

ENDOCIDE® See Endosulfan.

ENDOCRANIUM Noun. (Greek, *endon* = within; *kranion* skull; Latin, *-ium* = diminutive > Greek, *-idion*. Pl., Endocrania.) The inner surface of the Cranium.

ENDOCRINE CELLS One of the three types of cells found in the Midgut. EC have not been identified in all insects, but do occur in Orthoptera, Blattaria, Heteroptera, Coleoptera, Diptera and Lepidoptera. Endocrine cells have also been reported in the Apterygota. ECs are secretory in function and microscopically characterized by pale Cytoplasm and granular inclusions. The apices of Endocrine Cells can maintain contact with the midgut lumen. Based on immunological and histochemical tests, endocrine cells synthesize and release polypeptide hormones. Syn. Clear Cells; Granular Cells; Secretory Cells. See Alimentary System; Midgut. Cf. Columnar Cell; Peritrophic Membrane; Regenerative Cell.

ENDOCRINE GLAND Glands that provide secretions used inside the body. See Hormone. Cf. Exocrine Gland. Rel. Gland.

ENDOCUTICLE Noun. (Greek, *endon* = within; Latin, *cuticula* = diminutive of skin. Pl., Endocuticles.) The innermost layer of the Cuticle and typically forms the thickest layer of the Integument. Endocuticle is not tanned and not sclerotized. Functions of Endocuticle include protecting the epidermal cells, providing nutrients during starvation and providing flexibility during movement or stress. Microscopically, Endocuticle consists of several horizontal layers. Desert Locust ultrastructure shows that Endocuticle is deposited in circadian growth cycles. Layers of Endocuticle deposited at night consist of lamellae; layers deposited during the day do not contain lamellae. These cycles of Endocuticle growth persist during the adult stage. Many insects deposit Endocuticle layers on a circadian or 24-hour periodicity; some Coleoptera apparently deposit layers of Endocuticle on a periodicity other than 24 hours. Each lamella is composed of elongate microfibrils; microfibrils of each lamella are oriented in one 'preferred' direction or in a helicoidal manner. In preferred direction, successive lamina display microfibrils oriented in same direction. If microfibrils of consecutive lamella are angled with respect to previous lamella such that preferred direction constantly rotates in one direction, then arrangement is called helicoidal. See Integument. Cf. Assembly Zone; Basement Membrane; Cement Layer; Cuticulin Layer; Demal Glands; Epidermal Cells; Exocuticle; Pore Canals; Wax Layer. Rel. Skeleton; Skin.

ENDODERM Noun. (Greek, *endon* = within; *derma* = skin. Pl., Endoderms.) The inner layer of the Blastoderm in the embryo, giving origin to the mid-intenstine and other visceral organs. See Entoderm.

ENDODERMAL Adj. (Greek, *endon* = within; *derma* = skin; Latin, *-alis* = belonging to, pertaining to.) Descriptive of Endoderm; pertaining to Endoderm. Cf. Ectodermal.

ENDOGENOUS Adj. (French, *endogene*; Latin, *-osus* = with the quality of.) Pertaining to process or action that originates from within a cell, organ or body. Descriptive of something mainfest from internal causes and not from external causes. Cf. Exogengous. Rel. Circadian Rhythm.

ENDOGENOUS RHYTHM Behavioural, biological or physiological phenomena that periodically express themselves and which occur independent of external stimuli such as photoperiod. See Photoperiodism. Cf. Exogenous Rhythm. Rel.

Biological Clock; Circadian Rhythm; Diel Periodicity.

ENDOGNATHOUS Adj. (Greek, *endon* = within; *gnathos* = jaw; Latin, *-osus* = with the quality of.) Pertaining to mouthparts (Mandibles and Maxillae) that are concealed by the lateral margins of the Labium or cranial folds. The condition is best exemplified in Protura. See Feeding. Cf. Ectognathous. Rel. Mandible.

ENDOLABIUM Noun. (Greek, *endon* = within; *labium* = lip; *-ium* = diminutive > Greek *idion*. Pl., Endolabia.) 1. The inner surface of the Labium. 2. The Hypopharynx when that structure is well developed. See Hypopharnyx; Labium.

ENDOMESODERM Noun. (Greek, *endon* = within; *mesos* = middle; *derma* = skin. Pl., Endomesoderms.) The inner layer formed by an invagination of the middle portion of the primitive band of the embryo, and from which the Endoderm and Mesoderm are subsequently differentiated.

ENDOMYCHIDAE Plural Noun. Handsome Fungus Beetles. A small Family of polyphagous Coleoptera assigned to Superfamily Cucujoidea. Adults are 1–8 mm long; habitus resembling coccinellids with body ovate to elongate and globose; black with red or yellow coloration; head visible from dorsal aspect; Frontoclypeal Suture present; Antenna long with loose club of four segments. Adults and larvae typically occur under bark, in rotting wood, in fungi or decaying fruit; HFB feed upon fungus. Syn. Merophysiidae. See Cucujoidea.

ENDOPARAMERE Noun. (Greek, *endon* = within; *para* = beside; *meros* = part. Pl., Endoparameres.) The lateral end of the Peronea produced internally as a stout Paramere (MacGillivray). See Paramere.

ENDOPARASITE Noun. (Greek, *endon* = within; *parasitos* = parasite; *-ites* = constituent. Pl., Endoparasites.) Any organism that develops as a parasite within the body of another organism at the expense and to the detriment of the 'host'. See Ectoparasite; Parasitism. Cf. Endoparasitoid. Rel. Predator.

ENDOPARASITOID Noun. (Greek, *endon* = within; *parasitos* = parasite; *eidos* = form. Pl., Endoparasitoids.) A parasitoid insect whose larval stage feeds and develops inside the body of the host. Endoparasitoids are common among some groups of Hymenoptera and Diptera. See Ectoparasitoid; Parasitoid; Parasitism. Cf. Endoparasite. Rel. Predator.

ENDOPHAGOUS Adj. (Greek, *endon* = within; *phagein* = to devour; Latin, *-osus* = with the property of.) 1. Pertaining to parasitic animals that feeding inside their hosts. 2. Medical Entomology: Pertaining to a disease vector (female mosquito) that enters dwellings to feed on a host. See Endophilic; Feeding Strategy. Cf. Exophagous.

ENDOPHALLUS Noun. (Greek, *endon* = within; *phallos* = Penis. Pl., Endophalluses.) The inner chamber of the Phallus invaginated at the apex of the Aedeagus, into which the Ejaculatory Duct opens. Typically an eversible sac or tube; sometimes a permanently internal phallic structure (Snodgrass).

ENDOPHILIC Adj. (Greek, *endon* = within; *philein* = to love; *-ic* = characterized by.) Medical Entomology: Pertaining to a disease vector (female mosquito) that enters a dwelling to rest or effect transmission. Alt. Endophily. See Endophagous. Cf. Exophilic; Exophillous.

ENDOPHYTIC Adj. (Greek, *endon* = within; *phyton* = plant; *-ic* = characterized by.) Living within plant or tree tissue, as borers or miners. See Feeding Strategies. Cf. Saprophytic.

ENDOPHYTIC OVIPOSITION A form of oviposition in dragonflies with elongated eggs, that are inserted by an Ovipositor into leaves or stems of water plants (Imms). See Oviposition.

ENDOPLASM Noun. (Greek, *endon* = within; *plasma* = mould. Pl., Endoplasms.) The inner or central part of the Cytoplasm.

ENDOPLASMIC RETICULUM A cellular meshwork of double membrane found within the cytoplasm that is continuous with the cell and nuclear membranes. Frequently called rough ER when lined with Ribosomes or smooth ER when not lined with Ribosomes. See Cell Inclusions.

ENDOPLEURAL RIDGE See Plural Ridge.

ENDOPLEURITE Noun. (Greek, *endon* = within; *pleura* = side. Pl., Endopleurites.) A cuticular invagination between pleurites. See Apodeme.

ENDOPODITE Noun. (Greek, *endon* = within; *pous* = foot. Pl., Endopodites.) 1. General: The mesial branch of a biramous appendage. 2. The main shaft of the limb beyond the Basipodite (Snodgrass). 3. The second part of the Maxilla.

ENDOPTERYGOTA Noun. (Greek, *endon* = within; *pterygion* = little wing.) Insects with complex Metamorphosis and which develop wings and genitalia internally. Endopterygote Metamorphosis is complete, and consists of the egg, larva, Pupa and adult stages. Some entomologists argue that Endopterygota evolved from Exopterygota; others entomologists believe that Ametabola were ancestral to the Endopterygota. During the 18th century conventional opinion held that insects arose *de novo*. Syn. Holometabola; Oligoneoptera; Oligoneuroptera. Cf. Exopterygota. Rel. Metamorphosis.

ENDOPTERYGOTE Adj. (Greek, *endon* = within; *pterygion* = little wing.) Pertaining to insects whose wings develop internally. Cf. Exopterygote.

ENDOSKELETON Noun. (Greek, *endon* = within; *skeletos* = dried. Pl., Endoskeletons.) A cuticular invagination into the body cavity from the body wall (Integument) that provides structural support and serves for points of muscle attachment. See Integument. Cf. Hydrostatic Skeleton. Rel. Apodeme; Furca; Sulcus; Tentorium.

ENDOSMOSIS Noun. (Greek, *endon* = within;

osmos = impulse; *-sis* = a condition or state. Pl., Endosmoses.) Osmotic diffusion toward the inside of a cell or vessel.

ENDOSTERNAL RIDGE Y-shaped union of convergent ridges from the bases of an Apophyses (Snodgrass).

ENDOSTERNAL Adj. (Greek, *endon* = within; Latin, *sternum* = sternum; *-alis* = belonging to, pertaining to.) Descriptive of or pertaining to the Endosternum. See Sternum. Rel. Apodeme; Furca.

ENDOSTERNITE Noun. (Greek, *endon* = within; Latin, *sternum* = Sternum; *-ites* = constituent. Pl., Endosternites.) Part of an Apodeme arising from the Intersternal Membrane.

ENDOSTERNUM Noun. (Greek, *endon* = within; *sternon* = chest. Pl., Endosterna.) The inner part of the Sternum. See Sternum.

ENDOSULFAN A synthetic, chlorinated hydrocarbon insecticide {hexachlorohexahydromethano-2,4,3-benzodioxathiepin-oxide} used as an Acaricide, contact insecticide and stomach poison against Cyclamen mites, armyworms, aphids, boll weevils, bollworms, corn earworms, cutworms, green vegetable bug, leafhoppers, loopers, psyllids, stemborers, thrips and other insects. Endosulfan applied to numerous fruit tree crops, vegetable crops and field crops. Toxic to fishes; corrosive to iron. Trade names include: Benzoepin®, Beosit®, Chlortiepin®, Cyclodan®, Endocel®, Endocide®, Hexasulfan®, Hildan®, Malix®, Melophen®, Phaser®, Thifor®, Thimul®, Thiodan®, Thionex®, Thiosulfan®, Tionel®. See Chlorinated Hydrocarbon Insecticide.

ENDOTERGITE Noun. (Greek, *endon* = within; Latin, *tergum* = back; *-ites* = constituent. Pl., Endotergites.) 1. A transverse infolding between adjacent Tergites of the Endothorax. 2. A Phragma.

ENDOTHECA Noun. (Greek, *endon* = within; *theke* = case. Pl., Endothecae.) The inner wall of the Phallotheca (Snodgrass).

ENDOTHORAX Noun. (Greek, *endon* = within; *thorax* = chest. Pl., Endothoraxes.) The internal framework or processes of the Thorax.

ENDOTOKY Noun. (Greek, *endon* = within; *tokos* = birth. Pl., Endotokies.) A form of reproduction in which eggs are developed within the body of the mother. See Exotoky.

ENDOTRACHEA Noun. (Greek, *endon* = within; Latin, *trachia* = windpipe. Pl., Endotracheae.) The inner surface or lining of the Trachea. See Intima.

ENDRIN A synthetic, chlorinated hydrocarbon insecticide {1R, 4S, 4aS, 5S, 6S, 7R, 8R, 8aR)-1, 2, 3, 4, 10, 10-hexachloro-1, 4, 4a, 5,6,7,8,8a-octahydro-6,7-epoxy-1,4:5,8-dimethano-naphthalene} used as a contact insecticide and stomach poison against ants, aphids, armyworms, bollworms, chinch bugs, corn borers, crickets, cutworms, grasshoppers, hornworms, leaf hoppers, leaf miners, lygus bugs and thrips. Endrin is applied to few crops (cotton, rice, sugarcane) in some countries. Endrin has been removed from registration in USA and many developed countries. The compound is toxic to bees and fishes, and is persistent in soil. Trade name is Hexadrin®. See Chlorinated Hydrocarbon Insecticide.

ENDROMIDAE Plural Noun. A small Family of ditrysian Lepidoptera assigned to Superfamily Bombycoidea. See Lepidoptera.

ENEOPTERIDAE Plural Noun. (Greek, *eneos* = dumb, stupid, silent.) Bush crickets (not 'Bush-crickets' as used for Tettigonioidea.) A cosmopolitan Family of ensiferous Orthoptera in Superfamily Grylloidea. Currently about 80 Genera and 500 Species in four Subfamilies (Podoscirtinae, Eneopterinae, Itarinae, and Prognathogryllinae). Prognathogryllinae is restricted to Hawaiian Islands; Podoscirtinae inhabit New World; Itarinae inhabit Old World. Body sometimes with green markings and head (including eyes) small to moderate sized; second tarsal segment depressed and rather heart-shaped, hind Tibia spinose and serrated between spines; Tegmina usually projecting beyond apex of Abdomen. Male stridulatory apparatus reduced. Female Ovipositor long and straight. Most Species live on trees and shrubs, and some Species are mimetic. Biology of eneopterids is poorly known.

ENERVIS Adj. (Latin, *ex-*; *nervis* = sinew.) Without veins of any kind. A term applied to insect wings. See Wing Venation.

ENGEL, ERICH OTTO (1866–1944) (Rohlfien 1975, Beitr. Ent. 25: 290 only.)

ENGELHARDT, GEORGE PAUL (1871–1942) (Osborn 1946, *Fragments of American Entomology*. Pt. II. 232 pp. (80–82), Columbus, Ohio.)

ENGELHART, CHRISTIAN (1857–1919) (Tuxen 1968, Ent. Meddr 36: 67 only.)

ENGELMANN SPRUCE WEEVIL *Pissodes strobi* (Peck) [Coleoptera: Curculiondiae]. Syn. Sitka Spruce Weevil; White Pine Weevil.

ENGELMANN, WILHELM (1808–1878) (Ehlers 1879, Z. wiss. Zool. 32: 1–12.)

ENGERLING SPILZ® See *Beauveria brongniartii.*

ENGLISH GRAIN-APHID *Sitobion avenae* (Fabricius) [Hemiptera: Aphididae]: In North America, a widespread pest of small grains, wild grasses and cultivated grasses. Aphids accumulate on bracts of grain head; their feeding causes kernels to shrivel. See Aphididae.

ENGLISH WASP *Vespula vulgaris* (Linnaeus) [Hymenoptera: Vespidae]. (Australia).

ENGLISH, JAMES LAKE (1820–1888) (Carrington 1888, Entomologist 21: 72.)

ENGRAMELLE, MARIE DOMINIQUE JOSEPH (1727–1780) French cleric and author of *Papillons d'Europe* (8 vols, 1779–1793).

ENGRAVED Adj. (Greek, *en* = in; Anglo Saxon, *grafan* = to dig.) Ornate patterns of sculpture on the Integument of insects. See Sculpture. Cf.

Exsculptus.

ENGRAVER BEETLES See Curculionidae.

ENGRAVERS Plural Noun. (Greek, *en* = in; Anglo Saxon, *grafan* = to dig.) 1. Members of the Genus *Ips* (Scolytidae) that typically feed upon the cambium of weakened, dying or recently fallen pine trees. Engravers are typically multivoltine with 2–5 generations per year, overwinter as adults and attack the tree trunk and larger branches with thin bark. Egg galleries are kept clean and radiate from the nuptial chamber to form distinctive patterns; eggs are laid in niches along sides of the gallery; eggs hatch and larvae construct mines directed away from gallery. See Scolytidae; California Five-Spined Ips; Monterey-Pine Engraver; Pine Engraver; Six-Spined Ips. Cf. Ambrosia Beetle; Bark Beetle.

ENHANCED MICROBIAL DEGRADATION Unusually rapid breakdown of soil pesticides by microorganisms.

ENICOCEPHALIDAE Stål, 1860, Plural Noun. Gnat Bugs; Unique-headed Bugs. A small, cosmopolitan Family of heteropterous Hemiptera assigned to Superfamily Enicocephaloidea. Adult body 2–5 mm long; head constricted at base and between compound eyes and slightly swollen between constrictions; Pronotum formed into three regions by Sulci; fore leg raptorial with Tarsi opposable to Tibia. Male genitalia reduced. Enicocephalids typically occur under bark of dead trees, under rocks in wooded areas and under leaves. Some Species form swarms in sunlight. See Enicocephaloidea

ENICOCEPHALOIDEA A Superfamily of Heteroptera including the Aenictopecheidae and Enicocephalidae. Representatives are predominantly tropical or subtropical. Adults are small to moderate size; resemble reduviids; tarsal formula of nymphs 1-1-1 and adult usually 1-2-2 or rarely 1-1-1 or 2-2-2. Species are predators in leaf litter, rotting logs and under stones. See Hemiptera.

ENLARGE Trans. Verb. (Old French, *en* = in; *large* = wide.) To make larger. Cf. Reduce.

ENNATON Noun. (Greek, *ennea* = nine.) The ninth segment in insects.

ENOCK, FREDERICK (1845–1916) English preparator of microscopical slides and specialist in taxonomy of Mymaridae. (Bethune-Baker 1916, Entomologist's Rec. J. Var. 29: 218–219.)

ENSATE Adj. (Latin, *ensis* = sword; *-atus* = adjectival suffix.) Sword-shaped; two-edged. Pertaining to structure large at the base and tapering toward the apex. See Anceps. Alt. Ensatus; Ensiform.

ENSHEATHE Transitive Verb. (Greek, *en* = in; Middle English, *shethe.*) To cover; to enclose with a sheath.

ENSIFERA (Latin, *ensis* = sword, *ensifer* = sword bearing.) Crickets; Katydids; Long-horned Grasshoppers. A cosmopolitan Suborder of Orthoptera containing about 1,500 Genera and 8,500 Species with three Superfamilies recognized: Gryllacridoidea, Tettigonioidea, and Grylloidea. Higher classification of Ensifera sometimes is disputed: Grylloptera and Ensifera as Orders; Stenopelmatoidea and Schizodactylidea sometimes recognized as distinct from Gryllacridoidea. Ensifera are predominantly tropical and found in arboreal habitats. Most Species are nocturnal and stridulate (a characteristic acquired independently several times); fore Tibia used for sound reception in some groups and hind Femur modified to form a Strigil. Mandibles symmetrical with acute teeth, adapted for biting and chewing; pleural portion of Thorax mostly concealed by pronotal lobes; prothoracic spiracles frequently modified for sound perception; large hind legs adapted for jumping; Tarsi with 3–4 segments. Some groups are omnivorous, some are carnivorous; phytophagous habits probably evolved secondarily within Ensifera. Copulatory stance is noteworthy with female superior to male (venter-to-venter with male or female mounting male); copulatory stance is reversed in Caelifera and male Phallus is not intromittent; accessory glands of female open independently at base of Ovipositor and do not form a tube common to Gonopore and Spermathecal Duct. Spermatophore typically is pyriform, flask-shaped or subspherical, and extruded from female who eats part of it. Eggs are laid in batches but deposited separately in plant tissue, and not bound together in Ootheca. Ensifera are rarely gregarious although some migratory flights involve millions of individuals. See Gryllacrididea; Gryllidea; Schizodactyloidea; Stenopelmatidea; Tettigoniidea.

ENSIFORM Adj. (Latin, *ensis* = sword; *forma* = shape.) Sword-shaped; pertaining to structures sword-like in shape; specifically flattened structure with sharp edges and tapering to a slender point. See Sword. Cf. Acuminate; Dolabriform; Falcate; Remiform. Rel. Form 2; Shape 2; Outline Shape.

ENSIGN SCALES See Ortheziidae.

ENSIGN WASPS See Evaniidae.

ENSTAR® See Kinoprene.

ENSTARII® See Kinoprene.

ENTAD Adv. (Greek, *entos* = within; Latin, *ad* = toward.) Extending inward from the outer margin. See Orientation. Cf. Anteriad; Apicad; Basad; Caudad; Centrad; Cephalad; Craniad; Dextrad; Dextrocaudad; Dextrocephalad; Distad; Dorsad; Ectad; Laterad; Mediad; Mesad; Neurad; Orad; Proximad; Rostrad; Sinistrad; Sinistrocaudad; Sinistrocephalad; Ventrad.

ENTAL Adj. (Greek, *entos* = within; Latin, *-alis* = characterized by.) Descriptive of something projecting inward from the external surface of the insect body. A term applied to the structures within the body. See Orientation. Cf. Ectal. Rel. Apodeme.

ENTELECHY Noun. (Greek, *en* = in; *telos* = end;

echein = to hold. Pl., Entelechies.) 1. A metaphysical concept in which a cause leads to an end as a purpose; a condition in which actuality follows on potentiality. 2. An expression encountered in philosophical biology that is used to indicate a cause that leads to an effect predetermined by the inner constitution of the cause. Entelechy is, in effect, a principle in nature that conditions, influences and directs organisms to purposeful ends.

ENTENSION SOLE The pad-like Pulvillus that may be extended by the extension plate through the pressure plate.

ENTERIC Adj. (Greek, *enterikos* = intestine; *-ic* = of the nature of.) Relating to the digestive canal (Enteron). See Alimentary System.

ENTERIC ACARIASIS The accidental ingestion of mites. Implicitly, a form of Enteric Arthropodiasis. See Acariasis.

ENTERIC ARTHROPODIASIS The accidental ingestion of living arthropods. See Arthropodiasis. Cf. Enteric Acariasis.

ENTERIC CAECA A sac-like Diverticulum positioned at the oesophageal end of the stomach. See Gastric Caeca.

ENTERIC CANTHARIASIS The accidental ingestion of beetles, particularly Meloidae.

ENTERIC EPITHELIUM A layer of cells on the wall of the digestive tract resting on a basement membrane (Snodgrass). See Midgut. Rel. Alimentary System.

ENTERIC MYIASIS See Myiasis. Cf. Canthariasis.

ENTEROKINASE Noun. (Greek, *enterikos* = intestine; *kinase* = an enzyme.) An enzyme produced in the lining cells of the insect gut that activates trypsin. See Alimentary System.

ENTERON Noun. (Greek, *enteron* = an intestine. Pl., Enterons.) The alimentary canal. A general term used to refer to the entire digestive canal.

ENTEX® See Fenthion.

ENTIRE Adj. (Latin, *integer* = untouched.) Structure with an even, smooth, unbroken margin. A term applied to wings when their margin is not divided, notched or incised. Cf. Emarginate.

ENTITY Noun. (Latin, *entitas* > *entis* = thing > *esse* = to be; English, *-ity* = suffix forming abstract nouns.) 1. An individual in the philosophical sense. 2. Being; essence.

ENTOCRANIUM Noun. (Greek, *entos* = within; *kranion* = skull. Latin, *-ium* = diminutive > Greek, *idion*. Pl., Entocranium.) The internal skeletal structure (Apodemes) of the insect's head including the tentorial complex. See Tentorium.

ENTODERM Noun. (Greek, *entos* = within; *derma* = skin. Pl., Endoderms.) The innermost germ layer of the embryo. Entoderm produces the Epithelium of the Alimentary Canal and accessory structures Syn. Endoderm; Hypoblast. Cf. Ectoderm.

ENTODORSUM See Entotergum.

ENTOGNATHA See Entognathata.

ENTOGNATHATA Henning 1953. A Subclass of Insecta characterized by Mandibles and Maxillae that are concealed by facial Integument, eyes that are reduced or absent and Malpighian Tubules that are reduced. Included Taxa: Diplura, Collembola and Protura.

ENTOGNATHOUS Adj. (Greek, *entos* = within, inner; *gnathos* = jaw; Latin, = *-osus* = with the property of.) Pertaining to the Entognatha. Insects with mouthparts recessed within the head. A condition sometimes regarded as a feature typical of a primitive ancestral condition. See Head; Mouthpart. Cf. Entotrophous.

ENTOLOMA Noun. (Greek, *entos* = within, inner; *loma* = hem, fringe.) The inner margin of the wings.

ENTOMIASIS (Greek, *entomon* = insect; *-iasis* = suffix for names of diseases.) Lesions produced by insects in general on the tissues of living animals. Cf. Myiasis.

ENTOMOBRYIDAE Plural Noun. A numerically large, geographically widespread Family of arthropleonian Collembola. Entomobryids are characterized by well developed Furcula, body elongate with scales, Pronotum reduced and thoracic and most abdominal segments separated by conspicuous membrane. Entomobryids are found in leaf litter, moss and under bark. See Collembola.

ENTOMOGENOUS Adj. (Greek, *entomon* = insect; *genes* = born; Latin, *-osus* = with the property of.) Pertaining to organisms that develop in or on an insect. Examples include entomogenous fungi and nematodes. See Feeding Strategy.

ENTOMOGENOUS FUNGI Fungi that act as pathogens of insects, particularly *Beauveria* spp., *Metarhizium* spp., *Nosema* spp., or *Verticillium* spp. and are used as biopesticides for the control of some insect pests. EF often have broad host ranges. Common agents include *Beauveria bassiana* (Balsamo) Vuillemin, *Metarhizium anisopliae* (Metschnikoff) Sorokin, *Nomuraea rileyi* (Farlow) Samson and *Paecilomyces fumosoroseus* (Wize) Brown & Smith. Syn. Entomopathogenic Fungi. See Fungus. Rel. Biopesticide; Microbial Pesticide.

ENTOMOGENOUS NEMATODES Nematodes that act as pathogens of insects, particularly *Steinernema* spp. and *Heterorhabditis* spp. EN are used as biopesticides for the control of some pests, notably soil-inhabiting insects to include cutworms, white grubs, wireworms, weevils, armyworms and mole crickets; some nematodes are Species-specific in control of pests and kill larva within 2 days. Trade names include: Bioflea®, Biosafe®, Biosafe-N®, Biovector®, Exhibit®, Fightagrub®, NC All®, Millenium®, Nemsys®, Otinem®, Pianbiot®, Sanoplant®, Scanmask®, Terrix®, Vector®. See Biopesticide.

ENTOMOGRAPHY Noun. (Greek, *entomon* = insect; *graphein* = to write. Pl., Entomographies.) The description of an insect or of its life history (Archaic).

ENTOMOLIN Noun. (Greek, *entomon* = insect; *-in*

= chemical suffix.) Chitin (Archiac). Alt. Entomoline.

ENTOMOLOGIST Noun. (Greek, *entomon* = notch, insect; *logos* = discourse. Pl., Entomologists.) 1. A person that collects and/or studies insects. 2. A person that studies problems caused by insects and attempts to solve these problems. See Entomology. Cf. Acarologist. Rel. Agronomist; Botanist.

ENTOMOLOGY Noun. (Greek, *entomon* = notch, insect; *logos* = discourse. Pl., Entomologies.) A subdiscipline of Zoology that involves the study of insects. The term comes from Aristotle's *Entoma* which was the name for 'bloodless' animals (insects, arachnids, myriapods, worms) in his system of classification. The word 'entomon' refers to the notched appearance of the body which is created by Tagmata and areas of articulation between the sclerites. Entomology is divided into several professional areas of research including: Agricultural Entomology, Applied Entomology, Economic Entomology, Forensic Entomology, Insect Behaviour, Insect Physiology, Insect Morphology, Insect Pathology, Medical Entomology, Taxonomic Entomology (Systematics), Urban Entomology and Veterinary Entomology. See each term for an explanation. Cf. Acarology; Agronomy; Botany.

ENTOMOPALYNOLOGY Noun. (Greek, *entomon* = insect; *palynein* = to scatter; *logos* = discourse. Pl., Entomopalynologies.) The study of the ways that insects use pollen and how pollen becomes attached to an insect's body. See Pollen. Cf. Melissopalynology; Palynology. Rel. Spore.

ENTOMOPATHOGENIC FUNGI See Entomogenous Fungi.

ENTOMOPHAGOUS Adj. (Greek, *entomon* = insect; *phagein* = to devour; Latin, *-osus* = with the property of.) 1. Feeding upon insects. 2. Sometimes applied to wasps that feed their larvae with immature stages of other insects. See Feeding Strategy.

ENTOMOPHAGY Noun. (Greek, *entomon* = insect; *phagein* = to devour. Pl., Entomophagies.) The consumption of insects by other organisms.

ENTOMOPHILOUS Adj. (Greek, *entomon* = insect; *philein* = to love; Latin, *-osus* = full of.) Insect-loving; a term applied to plants especially adapted for pollination by insects. See Habitat. Cf. Agricolous; Aphidicolous; Entomophytous; Zoophilous. Rel. Ecology. Ant. Entomophobic.

ENTOMOPHOBIA Adj. (Gr. *entomon* = insect; *phobos* = terror.) A persistent, excessive and irrational fear of insects. Cf. Arachnophobia; Delusory Dermatitis; Delusory Parasitosis; Illusions of Parasitosis (Waldron 1972. Calif. Med. 117: 76–78.)

ENTOMOPHYTOUS Adj. (Greek, *entomon* = insect; *phyton* = plant; Latin, *-osus* = with the property of.) Pertaining to plants that feed upon insects. See Entomogenous; Habitat. Cf. Entomophilous; Fungicolous; Phytophilous; Silvicolous; Zoophilous. Rel. Ecology.

ENTOMOSIS Noun. (Greek, *entomon* = insect; *-osis* = denotes disease. Pl., Entomoses.) A disease caused by a parasitic insect.

ENTOMOTAXY Noun. (Greek, *entomon* = insect; *taxis* = arrange. Pl., Entomotaxies.) The preservation and preparation of insects for study. See Entomology. Cf. Curation.

ENTOMOTOMIST Noun. (Greek, *entomon* = insect. Pl., Entomotomists.) 1. A person who studies insect structure. 2. A person skilled in insect dissection. See Entomology. Cf. Anatomist; Morphologist. Rel. Acarologist.

ENTOMOTOMY Noun. (Greek, *entomon* = insect; *tomos* = cutting. Pl., Entomotomies.) The discipline involving internal structure of insects; insect anatomy.

ENTOMUROCHROME Noun. (Greek, *entomon* = insect; *ouron* = urine; *chroma* = colour. Pl., Entomurochromes.) The colouring matter of the Malpighian tubes.

ENTOPARASITE Noun. (Greek, *entos* = within; *parasitos* = parasite. Pl., Entoparasites.) An internal parasite; a parasite that feeds within the body of the host. See Parasite. Cf. Ectoparasite.

ENTOPLEURON Noun. (Greek, *entos* = within; *pleuron* = side. Pl., Entopleura.) Cuticular arm extending into the body-cavity from each Pleuron. Syn. Apodema; Lateral Apodeme; Pleural Arm; Pleural Ridge (MacGillivray). See Pleuron.

ENTOPTERARIA See Entopterygota.

ENTOPTYGMA Noun. (Greek, *entos* = within. Pl., Entoptygmata.) The Amnion in the insect embryo (Graber).

ENTOSTERNUM Noun. (Greek, *entos* = within; *sternon* = chest. Pl., Entosterna.) The internal processes from the Sternum. Syn. Furca. See Sternum.

ENTOTERGUM Noun. (Greek, *entos* = within; *tergum* = back. Pl., Entoterga.) A large, V-shaped ridge on the undersurface of the Notum with its apex directed forward (Snodgrass). See Tergum.

ENTOTHORAX Noun. (Greek, *entos* = within; *thorax* = breastplate. Pl., Entothoraces; Entothoraxes.) The apodemes or processes extending inward from the sternal Sclerites. See Apophysis; Furca.

ENTOTROPHOUS Adj. (Greek, *entos* = within; *trephein* = to nourish; Latin, *-osus* = with the property of.) Descriptive of mouthparts that are 'buried' in the head. See Feeding Strategy. Cf. Ectotrophous; Entognathous.

ENTOZOA Plural Noun. (Greek, *entos* = within; *zoe* = life. Pl., Entozoae.) Animals that live within the body of other animals. Rel. Symbiosis.

ENVELOPE Noun. (French, *enveloppe*. Pl., Envelopes.) An enclosing membrane, shell or covering. Cf. Cover; Cyst; Mantle; Sac; Shroud.

ENVELOPING CELL The intermediate cell of a sense organ, or of one of the component sensory units of the organ, probably corresponding to the Trichogen of a Seta (Snodgrass).

ENVENOMATION Noun. (Middle English, *envenimen* > Latin, *en* = in; *venim* = poison. Pl., Envenomations.) The poisonous effects caused by the bites, stings or secretions of insects and other arthropods.

ENVIRONMENT Noun. (Old French, *en* = in; *viron* = circle. Pl., Environments.) The total of the natural conditions under which animals live, including climatic, geographic, physiographic and faunal conditions.

ENVIRONMENTAL FATE The activity and movement of a substance from introduction into the environment, its destiny and that of its degradation products.

ENVIRONMENTAL MELANISM See Crypsis.

ENVIRONMENTAL RESISTANCE The physical and biological restraints that prevent a Species from realizing its biotic potential.

ENZONE A thiocarbonate compound {Sodium Tetrathiocarbonate} used as a fumigant for control of nematodes and as a contact insecticide. Compound is used as a soil treatment for Grape Phylloxera, soil insects, soil diseases and nematodes. Enzone is registered on citrus and grapes. Trade name is GY-81.

ENZOOTIC Adj. (Greek, *en* = in; *zoion* = an animal; *-ic* = characterized by.) Pertaining to a disease affecting animals. Cf. Endemic; Epizootic

ENZYME Noun. (Greek, *enzymos* = leavened. Pl., Enzymes.) A complex organic substance produced by living cells and causing chemical changes in organic substances by catalytic action. In writing, the names of enzymes are indicated by the suffix '-ase.'

EOCENE EPOCH The interval of the Geological Time Scale (57 MYBP–34 MYBP) that represents the second Epoch in the Palaeogene Period. EE first formally recognized by Charles Lyell during 1833. Principal deposits include Green River Shales (Wyoming) and Baltic Amber (Upper Eocene) that include many Hymenoptera. Frogs, penguins, pennipeds probably orginated in EE. See Cenozoic Era; Geological Time Scale. Cf. Palaeocene Epoch. Rel. Fossil.

EONE BLUE *Pseudodipsas eone iole* Waterhouse & Lyell [Lepidoptera: Lycaenidae]. (Australia).

EONYMPH Noun. (Greek, *eo* = dawn; *nymphe*. Pl., Eonymphs.) The last-instar larva in Sawflies. The Prepupa of Sawflies. See Metamorphosis. Cf. Larva; Nymph.

EOPLECIIDAE Rohdendorf 1946 Plural Noun. A small Family of fossil Diptera known from the Jurassic Period of Germany, Russia and China. See Diptera.

EOPOLYNEURIDAE Rohdendorf 1962. Plural Noun. A Family of fossil Diptera known only from the type-species of two Genera preserved in Lower Jurassic Period deposits at Kirghizistan. See Diptera.

EOSENTOMIDAE Plural Noun. A Family of Protura containing eight Species in the Genus *Eosentomon* in North America. Protura are characterized by tracheae present, Thorax with two pairs of spiracles and abdominal appendages 2-segmented with a terminal vesicle. See Protura.

EPACME Noun. (Greek, *epi* = upon; *akme* = prime. Pl., Epacmies.) The phase of development in a taxon's lineage immediately preceding its highest point (acme). Cf. Acme; Paracme.

EPALLAGIDAE Plural Noun. A small Family of Odonata assigned to the Calopterygoidea. Adults are small to moderate sized, robust and head robust with large, rounded eyes; wings long, narrow, not petiolate; venation dense with numerous Antenodal Crossveins and Arculus well developed; Nodus midway between base and apex of wing; outer part of wing with numerous supplementary crossveins; Anal Vein long and curved at origin; legs short and, stout; wings frequently iridescent; Abdomen elongate. Larvae are robust and resemble Amphipterygidae; Abdomen with Pseudopodia; Caudal Gills swollen and setose. See Odonata.

EPALPATE Adj. (Greek, *e* = without; *palpus*; *-atus* = adjectival suffix.) Descriptive of an insect that lacks Palpi.

EPANDRIUM Noun. (Latin, *-ium* = diminutive > Greek, *idion*. Pl., Epandria.) The ninth abdominal Tergum of the male insect.

EPAULET Noun. (French, *epaulette* > *epaule* = shoulder > Latin, *spatula* = broad piece > Greek, *spathe* = broad rib. Pl., Epaulets.) 1. Odonata: compressed dorsal expansion on the side of the Prothorax. 2. Diptera: The first setose scale at the base of the Costa, followed by a bare 'scale,' the Basicosta (Curran).

EPERMENIIDAE Plural Noun. A small Family of ditrysian Lepidoptera assigned to Superfamily Epermenioidea in some classifications and Copromorphoidea in other classifications. Globally the Family includes about 75 Species and is represented in North America by 11 Species. Adults are small bodied with wingspan 8–20 mm and head with smooth scales; Ocelli and Chaetosemata absent; Antenna 0.5–0.8 times as long as forewing; Proboscis small, without scales; Maxillary Palpus small to minute with three segments; Labial Palpus upturned; fore Tibia with Epiphysis; tibial spur formula 0-2-4; middle and hind Tibia with whorls of apical bristles; forewing lanceolate, without Pterostigma. Eggs are flattened and ovoid in outline shape; Chorion with ridges. Eggs are laid individually. Larval head is hypognathous with body densely and finely spinulose. Larvae mine leaves or feed upon seeds, fruits and flower heads. *Epermenia pimpinella* Murtfeldt forms leaf mines in parsley. Pupae are sclerotized and pupation occurs within fine silken cocoon in leaf litter, on ground or on host plant. See Copromorphoidea.

EPHEBIC Adj. Adj. (Greek, *ephebos* > *epi* = at; *hebe* = early manhood; *-ic* = characterized by.) Relating to the winged, adult stage.

EPHEMERELLIDAE Plural Noun. A widespread Family of Ephemeroptera whose members are strong fliers. Nymphs are taken under rocks in cool streams and lakes. See Ephemeroptera.

EPHEMERIDA Leach 1817. (Upper Carboniferous-Recent.) Syn. Ephemeroptera.

EPHEMERIDAE Plural Noun. A Family of Ephemeroptera whose members are 10–25 mm long, with two or three caudal filaments, and the wings of some Species infuscate. Nymphs burrow into sand or silt on lake or stream bottoms. See Ephemeroptera.

EPHEMERINA See Ephemeridae.

EPHEMEROIDEA A Superfamily of Ephemeroptera including Behningiidae, Caenidae, Ephemeridae and Neoephemeridae. See Ephemeroptera.

EPHEMEROPTERA (Greek, *ephemeros* = lasting a day; *pteron* = wing.) Mayflies; Shadflies. Palaeopterous insects related to Palaeodictyoptera and regarded as most primitive Order of extant insects; fossil record begins in Permian. Adults are elongate, soft-bodied, short-lived near water and take no food. Compound eyes often divided; three Ocelli; Antenna filiform, multi-segmented and shorter than head; mandibulate mouthparts vestigial and not strongly sclerotized. Prothorax reduced; Mesothorax and Metathorax fused; legs modified, forelegs sometimes vestigial, foreleg of male sometimes adapted for grasping female; wings membranous, always fluted and held upright in repose. Abdomen 10-segmented with postabdominal segment fused to segment 10; Cerci elongate and multi-segmented; Appendix Dorsalis usually present; segments 1–8 with spiracles. Male often with segmented Gonostyli and double Penis with independent Vas Differentia. Females with paired Gonopores along posterior margin of Sternum 7; panoistic Ovarioles leading to parallel Oviducts; Common Oviduct absent; accessory glands absent. Mouthparts of naiadal stages adapted for chewing. Naiad aquatic with gills along sides of Abdomen. Metamorphosis simple with unique developmental stage (Subimago) that is the initial winged form. Naiad found in fast-flowing streams or still waters of ponds where some Species burrow into bottom muck. Adults usually are seen near water on vegetation and other objects; occasionally they are attracted in large numbers to lights. Included Families: Ameletopsidae, Ametropodidae, Baetidae, Baetiscidae, Behningiidae, Caenidae, Coloburiscidae, Ephemerellidae, Euthyplociidae, Ephemeridae, Heptageniidae, Isonychiidae, Leptohyphidae, Leptophlebiidae, Neoephmeridae, Oligoneuriidae, Oniscigastridae, Palingeniidae, Potamanthidae, Polymitarcyidae, Prosopistomatidae, Siphlonuridae, Teloganodidae, Tricorythidae. See Insecta.

EPHIALTITIDAE Plural Noun. Handlirsch 1906. A Family of parasitic Hymenoptera assigned to the Ephialtitomorpha.

EPHIALTITOMORPHA Handlirsch 1906. A Superfamily of Hymenoptera including the Ephialtitidae, Karataidae and Karatavitidae.

EPHYDRIDAE Plural Noun. Brine Flies; Shore Flies. A widespread Family of acalyptrate Diptera including ca 1,700 Species and 115 Genera. Placed in Superfamily Ephydroidea and regarded as near the Diastatidae. Adults typically are found in association with aquatic and semiaquatic habitats including freshwater marshes, coastal marine habitats, highly alkaline or saline wetlands and thermal springs. Ephydrid feeding ecology is diverse: Some adults feed on Haemolymph of ants without killing victim. Ephydrid larvae usually are found in water or within stems or shoots of aquatic plants where they ingest algae, bacteria, Cyanobacteria and detritus; larval stage of some Species are scavengers, a few Species are leaf miners and a few are predators. See Ephydroidea.

EPHYDROIDEA A Superfamily of cyclorrhaphous Diptera. Adults are characterized by Postvertical Bristles absent, convergent or replaced by divergent Postocellar Bristles; Vibrissae usually present. Antennal segment two more-or-less cap-like with dorsal slit on distal portion; segment three with basodorsal prominence concealed in recess on segment two. Prosternum with Precoxal Bridges. Costa interrupted at end of Sc; CuA+1A short and arising preapically from posterior side of cell CuP. Included Families: Camillidae, Campichoetidae, Cryptochetidae, Curtonidae, Distatidae, Drosophilidae and Ephydridae. See Diptera.

EPIBLAST Noun. (Greek, *epi* = upon; *blastos* = bud. Pl., Epiblasts.) The outer germ layer of the Gastrula in embryonic development. Syn. Ectoblast.

EPICARANIAL NOTCH A V-shaped dorsomedial area margined laterally by the larger cranial Sclerites. Syn. Cervical Triangle; Vertical Triangle.

EPICHORION Noun. (Greek, *epi* = upon; *chorion*. Pl., Epichorions.) Acrididae: A more-or-less temporary covering of the egg that originates from follicular cells of the Ovariole nearest the Oviduct (Extrachorion *sensu* Hartley). See Eggshell; Cf. Endochorion; Exochorion; Vitelline Membrane; Wax Layer.

EPICNEMIS Noun. (Greek, *epi* = upon; *knemis* = tibia.) An accessory joint at the base of the Tibia in some arachnids; apparently without motion.

EPICNEMIUM Noun. (Greek, *epi* = upon; *knemis* = tibia; *-ium* = diminutive, > Greek *idion*. Pl., Epicnemia.) Hymenoptera: The Prepectus.

EPICOPEIIDAE Plural Noun. A small Family of ditrysian Lepidoptera of uncertain placement. See Lepidoptera.

EPICRANIAL ARM A divarication or branch of the Epicranial Suture (MacGillivray).

EPICRANIAL LOBE Lepidoptera larva: A lateral, superior, convex lobe of the head. See Cranium; Lobe.

EPICRANIAL NOTCH See Epicranial Suture.

EPICRANIAL PLATE Some larvae: A plate-like structure forming the Epicranium.

EPICRANIAL STEM See Coronal Suture.

EPICRANIAL SUTURE 1. The dorso-longitudinal suture that separates epicranial halves of the head. Typically, a Y-shaped, U-Shaped, or V-shaped suture on the dorsal surface of the head, the arms of which diverge anteroventrally. 2. The line of junction of the two procephalic lobes *sensu* Smith. 3. Lepidoptera larva: A 'V' or 'Y'-shaped indentation in the dorsomedial area of the head capsule. Syn. Cleavage Line; Ecdysial Line; Epicranial Notch.

EPICRANIAL Adj. (Greek, *epi* = upon; *kranion* = skull; Latin, *-alis* = belonging to, characterized by.) Relating or pertaining to the Epicranium.

EPICRANIUM Noun. (Greek, *epi* = upon; *kranion* = skull; Latin *-ium* = diminutive, > Greek, *-idion*. Pl., Epicrania.) The upper part of the head from the front to the neck; often used to include front, Vertex and Genae.

EPICUTICLE Noun. (Greek, *epi* = upon; Latin, *cuticula* = thin skin. Pl., Epicuticles.) A layer of the Cuticle positioned above the Exocuticle and typically less than a few microns thick. Structurally Epicuticle is also laminate and consists of several layers, including a cuticulin layer, wax layer and cement layer. Epicuticle covers the entire insect body except chemoreceptors and end apparatus of gland cells. Epicuticle lines the inner layer of foregut and hindgut, but it is not found in midgut. Epicuticle has a high tensile strength. Chemically, Epicuticle lacks Chitin and consists of tanned proteins. Epicuticle serves several functions including waterproofing, protection from abrasion, protection from microorganisms, and it produces iridescence in many insects; Epicuticle is partly responsible for determining the size of a growing insect. See Cuticle. Rel. Integument.

EPICUTICULA Noun. (Greek, *epi* = upon; Latin, *cuticula* = skin.) 1. The nonchitnous, structureless, external film-like layer of Cuticle covering the Exocuticula. See Epicuticle.

EPIDEICTIC PHEROMONE 1. A multifunctional substance or 'factor' that plays a role in regulating the spatial density of conspecific individuals (Orber 1972). Alt. Spoor Factor (Flanders 1951). 2. An intraspecific chemical messenger that elicits dispersal behaviour from potentially crowded food sources.

EPIDEME See Articulatory Epideme.

EPIDEMIC Adj. (Greek, *epidemos* = among the people; *-ic* = characterized by.) Medical Entomology: Pertaining to any disease that is manifest and prevalent in an area during a short period of time. A widespread, severe outbreak of an infectious disease. Cf. Pandemic.

EPIDEMIC HAEMORRHAGIC FEVER An arbovirus in the Genus *Hatnavirus* (Bunyaviridae), that occurs in Russia, China, Korea and Japan. Viron is enzootic in rodents and vectored to humans by the trombiculid mite *Leptotrombidium scutellare*. EHF transmitted by contact with saliva or excrement of infected rodents, and larval feeding of mite; Transovarial Tranmission has been demonstrated. Syn. Korean Haemorrhagic Fever; Nephropathia Epidemica. See Arbovirus; Bunyviridae.

EPIDEMIC RELAPSING FEVER 1. A highly virulent, louse-borne disease of humans caused by the spirochaete *Borrelia recurrentis*. ERF is vectored by *Pediculus humanus* in epidemic proportion in Africa, eastern Europe and Russia during 1910–1945 and periodically in Sudan and Ethiopia since 1960. First episode of disease ca six days, followed by nine days of remission and then a two-day relapse; mortality to 40% in untreated cases. Spirochaete is ingested by louse in blood of human host, penetrates midgut epithelium and multiplies in Haemocoel; Spirochaete is found in muscles and CNS of vector, but not Ovaries or Salivary Glands. Disease is transmitted to human when louse is crushed on host and Spirochaetes penetrate skin or are scratched into wound; no transovarian transmission or transmission by bite of louse. 2. A form of ERF transmitted by *Borrelia duttoni* in eastern, central and southern Africa. Form of disease only known in humans and transmitted by nymphal instars of the synanthropic tick *Ornithodoros moubata.* Tick ingests Spirochaete with blood of infested person, Spirochaete penetrates midgut, passes into Haemocoel and replicates in CNS, Salivary Glands, Coxal glands and ovarian tissue of tick; Spirochaete is transmitted by tick feeding, skin abrasion or penetration of skin. Spirochaete is transmitted trans-ovarially and trans-stadially for several generations within ticks. See Relapsing Fever. Cf. Lyme Disease.

EPIDEMIC TYPHUS A sometimes fatal rickettsial disease [*Rickettsia prowazeki* Da Rocha-Lima] transmitted to humans by the human louse, *Pediculus humanus humanus.* ET is experimentally transmitted by head louse and crab louse. ET is common in temperate regions where considerable clothing is worn but uncommon in tropical countries where little clothing is worn. Epidemics are associated with crowded conditions, lack of sanitation and warfare. Louse contracts ET when feeding on an infected human. Infection is derived from scratching louse or its excrement into a wound; disease is not transmitted by feeding of louse. *Rickettsia* multiplies in midgut epithelium of louse; cells rupture and pass disease with faeces; *R. prowazeckii* kills louse within 10 days of contracting organism; *R. prowazeckii* can survive more than 60 days in dried faeces of louse. Syn. Brill-Zinnser Disease; Jail Fever; War Fever. Cf. Murine Typhus; Tick-Borne Spotted Fever.

EPIDEMIOLOGY (Greek, *epidemios* = among the people; *logia* = discourse.) The study of the distribution of diseases and the factors responsible for establishing, maintaining and moving disease through populations of organisms. See Disease.

EPIDERMA Noun. (Greek, *epi* = upon; *derma* = skin. Pl., Epidermae.) The outer layer of cortex in Fungi.

EPIDERMAL Adj. (Greek, *epi* = upon; *derma* = skin; Latin, *-alis* = pertaining to.) Descriptive of structure pertaining to the Epidermis; resembling the Epidermis. Alt. Epidermic; Epidermidal. Rel. Subepidermal.

EPIDERMAL LAYER 1. General: The single layer of cells forming the Epidermis or skin of plants and animals. 2. The outer covering of a cynipid gall, including the Epidermis and abnormal developments from it; typically bare or most with stellate Setae (Kinsey).

EPIDERMATA Noun. (Greek, *epi* = upon; *derma* = skin. Pl., Epidermatae.) Abnormal excrescences or growths projecting outward from the Integument.

EPIDERMIS Noun. (Greek, *epi* = upon; *derma* = skin. Pl., Epidermises.) The cellular layer of the Integument that lies beneath and secretes the Cuticle. The Epidermis is immediately above the Basement Membrane and forms a living core of the Integument. Epidermis serves many functions including manufacture and transfer of chemical constituents of the Cuticle. Epidermis also forms external sensory receptors, including chemoreceptors and tactile receptors. Epidermis is one-cell layer thick, but the density of this layer varies with stage of development, area of body and Species. Epidermal cells are bound together by Belt Desmosomes, Gap Junctions and Septate Junctions; thickness of Epidermis varies with mitotic activity, starvation and phase in moulting cycle. Epidermal cells invested with Pore Canals that project outward through the Cuticle. See Integument. Cf. Assembly Zone; Basement Membrane; Cement Layer; Cuticulin Layer; Dermal Glands; Epidermal Cells; Endocuticle; Exocuticle; Pore Canals; Wax Layer. Rel. Skeleton; Skin.

EPIDIDYMIS Noun. (Greek, *epi* = upon; *didymos* = testicle. Pl., Epididymides.) A highly convoluted portion of the Vas Deferens in the male internal reproductive system. The Epididymis is not present in all male insects; it is frequently massed at the posterior part of the Testes. See Testes. Cf. Vas Deferens; Seminal Vesicle. Rel. Gonad.

EPIGAMIC COLORATION Combination of colours or colour patterns displayed by animals during the attraction of mates or courtship. See Coloration. Cf. Advancing Coloration; Alluring Coloration; Anticryptic Coloration; Apetetic Coloration; Aposematic Coloration; Combination Coloration; Cryptic Coloration; Directive Coloration; Disruptive Coloration; Episematic Coloration; Procryptic Coloration; Protective Coloration; Pseudepisematic Coloration; Pseudoaposematic Coloration; Seasonal Coloration; Sematic Coloration. Rel. Crypsis; Mimicry; Sexual Dimorphism; Sexual Selection.

EPIGAMIC Adj. (Greek, *epi* = upon; *gamos* = marriage; *-ic* = characterized by.) Pertaining to a colour, size, shape, chemical or behaviour that attracts members of the opposite sex.

EPIGASTRIUM Noun. (Greek, *epi* = upon; *gaster* = stomach; Latin, *-ium* = diminutive, > Greek, *idion*. Pl., Epigastra.) The first entire ventral sclerite of the Abdomen. See Sternum.

EPIGEAL Adj. (Greek, *epi* = upon; *ge* = earth; Latin, *-alis* = belonging to.) Living on or foraging upon the surface of the earth. Alt. Epigaeic. Cf. Hypogeal.

EPIGENESIS Noun. (Greek, *epi* = upon; *genesis* = descent. Pl., Epigeneses.) A philosophical doctrine or biological hypothesis that asserts that development occurs through the processes of growth ahd differentiation of tissue from undifferentiated gametes. Epigenesis opposes the concept of Preformation. See Preformation.

EPIGENETIC PERIOD Period after union of male and female elements, during which organs are forming.

EPIGLOSSA See Epipharynx.

EPIGLOTTIS See Epipharynx.

EPIGUSTA Noun. (Greek, *epi* = upon; Latin, *gustare* = to taste. Pl., Epigustae.) Part of the Propharynx posterior of the Epipharynx and Tormae (MacGillivray).

EPIGYNUM Noun. (Greek, *epi* = upon; *gyne* = woman. Pl., Epigynums.) Arachnids: The external genitalic opening in females. 2. A sclerite covering the genitalic opening in arachnids. Syn. Vulva. Alt. Epigynium; Epigyne. Cf. Gonopore.

EPILABRUM Noun. (Greek, *epi* = upon; Latin, *labrum* = lip. Pl., Epilabra.) A sclerite at each side of the Labrum, specifically applied to myriapods. See Labrum.

EPILOBE Noun. (Greek, *epi* = upon; *lobos* = lobe. Pl., Epilobes.) Carabidae: Part of the Mentum, corresponding to a partially divided Ligula; a lateral appendage of a bilobed Mentum.

EPIMASTIGOTE Noun. (Greek, *epi* = upon; *mastigos* = a whip. Pl., Epimastigotes.) An anatomical form manifest at a specific phase in the complex life cycle of some parasitic Protozoa (*Leishmania* and *Trypanosoma*). Epimastigote displays a flagellar base anterior of nucleus and Flagellum emerges laterally to form an undulating membrane toward anterior end of cell. See Leishmania; Trypanosoma. Cf. Amastigote; Opisthomastigote; Promastigote; Trypomastigote.

EPIMERAL PARAPTERUM The posterior Basalar Sclerite positioned just behind the Pleural Wing Process and above the Epimeron.

EPIMERAL SUTURE The caudal part of the Sternopleural Suture of authors (MacGillivray).

EPIMERE Noun. (Greek, *epi* = upon; *meros* = part. Pl., Epimeres.) A dorsal process of the Phallobase (Snodgrass).

EPIMERON Noun. (Greek, *epi* = upon; *meros* = part. Pl., Epimera.) The posterior division of a thoracic Pleuron adjacent to the Coxa and posterior of the Pleural Suture. Typically the smaller sclerite of the thoracic Pleuron that is narrow or triangular. See Episternum.

EPIMORPHIC Adj. (Greek, *epi* = upon; *morphe* = form; *-ic* = characterized by.) Pertaining to oranisms that maintain the same form during growth. Cf. Anamorphic.

EPIMORPHOSIS Noun. (Greek, *epi* = upon; *morphe* = shape; *-sis* = a condition or state. Pl., Epimorphoses.) Development or a condition in which the postembryonic immature stage has all body segments completely developed at eclosion. This condition is found in virtually all pterygote insects. Retention of body shape with an increase in body size between moults is seen in anamorphic and epimorphic development and may be regarded as an ancestral characteristic. See Metamorphosis. Cf. Anamorphosis; Morphogenesis. Rel. Development; Growth.

EPINEURAL SINUS Embryo: The space produced mainly by the separation of the yolk from the embryo, over the region of the ventral nerve cord. The beginning of the permanent body-cavity *teste* Imms.

EPINEURIUM Noun. (Greek, *epi* = upon; *neuron* = nerve; Latin, *-ium* = diminutive > Greek, *idion*. Pl., Epineuria.) The membrane that invests a nerve ganglion (Imms).

EPINGLE® See Pyriproxyfen.

EPINOTAL SPINES Myrmicinae: The spines on the Epinotum that protect the Pedicel (Wheeler).

EPINOTUM Noun. (Greek, *epi* = upon; *notum* = back. Pl., Epinota.) Hymenoptera: 1. The Propodeum (Comstock). 2. The dorsal aspect of the Pronotum (MacGillivray).

EPIOPHLEBIIDAE Plural Noun. A small Family of anisozygopterous Odonata found in Japan and the Himalayas. Adults are small and robust with a slender Abdomen. Wings are narrow and petiolate; eyes widely separated. Pterostigma short and thick. Adults fly quickly. Larvae are small and robust (as Anisoptera) but the Antenna has five segments. See Odonata.

EPIOPTICON Noun. (Greek, *epi* = upon; *opsis* = sight. Pl., Epiopticons.) The external medullary mass of the optic segment of the insect Brain (Hickson) or second ganglionic swelling of the Optic Tract.

EPIPHALLUS Noun. (Greek, *epi* = upon; *phallus* = Penis. Pl., Epiphalluses.) Orthoptera: A sclerite in the floor of the genital chamber proximal of the base of the Phallus; A Pseudosternite *teste* Snodgrass. See Phallus.

EPIPHARYNGEAL Adj. (Greek, *epi* = upon; *pharyngx* = throat. Latin, *-alis* = pertaining to.) Descriptive of structure or function relating to the Epipharynx.

EPIPHARYNGEAL SCLERITES Bees: A pair of strap-like sclerites extending backward from the sides of the base of the Epipharynx. See Hypopharyngeal Sclerites.

EPIPHARYNGEAL WALL The membranous inner surfaces of the Clypeus and Labrum (Snodgrass).

EPIPHARYNX Noun. (Greek, *epi* = upon; *pharyngx* = throat. Pl., Epipharnices.) 1. An organ, probably of taste, attached to the inner surface of the Labrum and supposed to correspond to the palate of higher animals. 2. The Epiglossa or Epiglottis: In scarabaeoid larvae, the complex buccal area forming an inner or under lining of the Labrum and extending below the Clypeus (Boving). 3. The Epipharnyx forms a roof of the food pump in Heteroptera, and is positioned along the medial surface of Labium and Clypeus. Generally, an Epipharnyx is not visible externally but often it is complex morphologically and adapted for mastication and filtering of particulate matter. In some Species, an Epipharnyx is visible as an apical projection from the Labrum. In most groups, the projections secure Maxillary and Mandibular Stylets in the Medial Labial Groove See Mouthparts. Cf. Hypopharnyx. Rel. Labium.

EPIPHYSIS Noun. (Greek, *epi* = upon; *phyein* = to grow. Pl., Epiphyses.) Lepidioptera: A subapical, flexible, spur-like or leaf-like process on the medial surface of the anterior Tibia. The Epiphysis is used for grooming the Antennae and Proboscis. Cf. Calcar. Rel. Tibia.

EPIPHYTE Noun. (Greek, *epi* = upon; *phyton* = plant. Pl., Epiphytes.) A plant growing non-parasitically upon other plants (*e.g.* ferns, orchids, *etc.*) and often upon the branches of trees (Tillyard).

EPIPLEURITE Noun. (Greek, *epi* = upon; *pleuron* = side, rib; *-ites* = constituent. Pl., Epipleurites.) The upper pleural sclerite when a pleuron is horizontally divided into a Basalar Sclerite and Subalar Sclerite in a wing-bearing thoracic segment. Differentiated from the upper ends of the Episternum and Epimeron, respectively. The Parapteron *sensu* Snodgrass. See Pleuron.

EPIPLEURON Noun. (Greek, *epi* = upon; *pleura* = side, rib. Pl., Epipleura.) 1. Coleoptera: The deflexed or inflexed portion of the Elytron. Immediately beneath the edge, the inflexed part of the Pronotum sometimes is called Epipleura. More correctly, the entire deflected margin of an Elytron (Kirby & Spence 1826). 2. Hymenoptera: The lateral margin of a gastral Tergum (Thomson 1878: 251). Alt. Epipleurum.

EPIPLOON Noun. (Greek, *epiploon*. Pl., Epiploa.) The Fat-Body of insects. Rare. See Caul.

EPIPLURAL FOLD The elevated lower edge of the Epipleura. See Hypomera.

EPIPODITE Noun. (Greek, *epi* = upon; *pous* = foot; *-ites* = constituent. Pl., Epipodites.) An outer lobe of the Coxopodite, often a gill-bearing organ (Snodgrass).

EPIPROCT Noun. (Greek, *epi* = upon; *proktos* = anus. Pl., Epiprocts.) The supra-anal sclerite or Pygidium *sensu* Needham. The dorsal part of the eleventh segment in insects (Snodgrass).

EPIPROSOMA Noun. (Greek, *epi* = upon; *pro* =

front; *soma* = body. Pl., Epiprosmata.) Acarology: A division of the body comprised of the Gnathosoma and the Aspidosoma. The Prosoma without the Podosoma.

EPIPSOCIDAE Plural Noun. A small Family of Psocoptera with three Species represented in the United States. See Psocoptera.

EPIPYGIUM Noun. (Greek, *epi* = upon; *pyge* = rump; Latin, *-ium* = diminutive, > Greek, *idion*. Pl., Epipygnia.) The dorsal arch of the last abdominal segment.

EPIPYROPIDAE Plural Noun. A small, widespread Family of ditrysian Lepidoptera assigned to Superfamily Zygaenoidea and which resembles Australian Cyclotornidae. Epipyropidae consist of about 10 Genera and 30 Species, most of which occur in the Indo-Australian Realm. Adults are small with wingspan 5–15 mm; eyes small; Ocelli and Chaetosemata absent; Antenna less than half as long as forewing; bipectinate to apex in male and female; Proboscis and Maxillary Palpus absent; Labial Palpus minute; fore Tibia lacking Epiphysis; tibial spurs absent; forewing broadly triangular and lacking Pterostigma; Eggs are discoid or ovate with a thin Chorion; a female can lay 3000 eggs in one day; eggs of clutch hatch over prolonged period. First instar larvae are hypermetamorphic (Triungulinid) with head small, body stout, Prolegs reduced and piercing Mandibles. Triungulinids actively search for hosts; all instars are external parasites of Homoptera of several Families; larvae attach to host Cuticle with hooked claws, attach to a few strands of silk to host's Abdomen to support attachment. Epipyropids complete 4–5 larval instars. When host is killed, then replete larva detaches from host and spins cocoon on twig or grass stem. Cocoon is heavily invested with wax from larval body; cocoon with broad slit at aperture. Adults are active after dusk; fly rapidly and are attracted to lights. Epipyropids are univoltine. See Zygaenoidea.

EPISEMATIC COLORATION Recongition colours in the theory of mimicry. See Coloration. Cf. Advancing Coloration; Alluring Coloration; Anticryptic Coloration; Apetetic Coloration; Aposematic Coloration; Combination Coloration; Cryptic Coloration; Directive Coloration; Disruptive Coloration; Epigamic Coloration; Episematic Coloration; Procryptic Coloration; Protective Coloration; Pseudepisematic Coloration; Pseudoaposematic Coloration; Seasonal Coloration; Sematic Coloration. Rel. Crypsis; Mimicry.

EPISITE Noun. (Greek, *epi* = upon; *-ites* = inhabitant. Pl., Episites.) A predatory insect that lives by directly attacking and consuming tissues of a succession of prey to complete its life-cycle. Syn. Predator. Cf. Parasite.

EPISTASIS Noun. (Greek, *epi* = upon; *stasis* = standing. Pl., Epistases.) Genetics: The dominance of one gene over another non-allelomorphic gene.

EPISTEMOLOGY Noun. (Greek, *episteme* = knowledge; *logos* = discourse. Pl., Epistemologies.) The study of knowledge and the conditions of knowning.

EPISTERNAL Adj. (Greek, *epi* = upon; *sternon* = chest; Latin, *-alis* = pertaining to.) Pertaining to the Episternum.

EPISTERNAL LATERAL See Preepisternum (MacGillivray).

EPISTERNAL PARAPTERA The two anterior Basalar Sclerites positioned immediately above the Episternum.

EPISTERNAL SUTURE Anterior part of the Sternopleural Suture of authors (MacGillivray).

EPISTERNITES Plural Noun. 1. Orthoptera: The upper pair of horn-like appendages forming the Ovipositor in grasshoppers. 2. A sclerotic subdivision of the Episternum.

EPISTERNO-PRECOXAL Adj. Descriptive of or pertaining to the Episternum and Precoxa combined.

EPISTERNUM Noun. (Greek, *epi* = upon; *sternon* = chest. Pl., Episterna.) 1. A thoracic sclerite anterior of the Pleural Suture and sometimes adjacent to the Coxa. Typically, the Episternum is largest lateral thoracic sclerite between the Sternum and the Notum. 2. The anterior sclerite of the Pleuron *sensu* Imms. 3. The Episternum and Epimeron of many insects have become subdivided into several secondary sclerites with attendant secondary sutures. In its simplest form, the Episternum is subdivided into a dorsal Anepisternum and a ventral Katepisternum. Similarly, the Epimeron is subdivided into a dorsal Anepimeron and a ventral Katepimeron. Some insects show modification of the Episternum into a Preepisternum and Prepectus. Prepectus of Orthoptera such as *Romalea* is 'U' shaped and continuous medially. See Thorax. Cf. Epimeron.

EPISTOMAL RIDGE An internal projection of the Epistomal Suture. Cf. Apodeme.

EPISTOMAL SUTURE 1. A suture uniting the anterior ends of the Subgenal Sutures across the face, and separating the Frons and Clypeus. The ES forms a strong, internal ridge (Sulcus). The ES usually is transverse and straight, but arched in some groups of insects and absent in other groups. Syn. Frontoclypeal Suture.

EPISTOME Noun. (Greek, *epi* = upon; *stoma* = mouth. Pl., Epistoma; Epistomes.) 1. General: The lower face between the Buccal Cavity and compound eyes. That sclerite immediately behind or above the Labrum, whether it is Clypeus or an indeterminant sclerite. 2. Odonata: Clypeus, Hypostoma. 3. Rhynchophorous Coleoptera: The reduced frontoclypeal region. 4. Diptera: Part of the face between the front and the Labrum; the oral margin and an area of indefinite size immediately contiguous; Peristoma. Alt. Epistomis; Epistomum.

EPITHELIAL Adj. (Greek, *epi* = upon; *thele* = nipple; Latin, *-alis* = pertaining to.) Descriptive of or pertaining to the Epithelium.

EPITHELIAL LAYER 1. Any layer of cells, one surface of which serves as a boundary of a space (*e.g.* epithelial layer of the Integument). The cellular lining of the gut surrounding the Enteron (Klots).

EPITHELIAL SHEATH Some insects: An outside covering of flat cells of the Tunica Intima; the wall of a testicular tubule, sometimes two-layered (Snodgrass).

EPITHELIUM Noun. (Greek, *epi* = upon; *thele* = nipple; Latin, *-ium* = diminutive > Greek, *idion*. Pl., Epithelia.) Any layer of cells that covers a surface or lines a cavity.

EPITOKOUS Adj. (Greek, *epi* = upon; *tokos* = birth; Latin, *-osus* = with the property of.) Pertaining to Epitoky.

EPITOKY Noun. (Greek, *epi* = upon; *tokos* = birth. Pl., Epitokies.) 1. Collembola: The act, process or phenomenon involved in the alternation of a feeding, non-reproductive instar and a reproductive instar during the adult stage. Instars separated by Ecdysis and some polymorphism between the instars. Cf. Moult. 2. The heteronereid stage of some polychaetes.

EPITORMA Noun. (Greek, *epi* = upon; *tormos* = a hole. Pl., Epitormae.) Scarabaeoid larvae: An elongate sclerite or bar extending from the Zygum toward the Clithrum on the right side of the Epipharynx. In many Species present even when the Clithrum is absent; in others embodied in the Tylus or absent (Boving).

EPIZOOTIC HAEMORRHAGIC DISEASE A minor viral disease of deer, cattle and buffalo in Australia, USA and Japan. EHD is caused by arbovirus assigned to Reoviridae. *Culicoides* spp. probably vector the disease. Acronym EHD. See Reoviridae. Cf. African Horse Sickness; Bluetongue Disease; Colorado Tick Fever.

EPIZOOTIC Adj. (Greek, *epi* = upon; *zoa* = animal; *ticos* = denotes fitness; *-ic* = characterized by.) The widespread, rapid occurrence of a disease affecting many individuals or a large proportion of an animal population at the same time. Plagues are sometimes epizootic. Cf. Enzootic; Pandemic.

EPIZOOTIC Noun. (Greek, *epi* = upon; *zoa* = animal; *ticos* = denotes fitness. Pl., Epizootics.) 1. An epizootic disease. 2. An outbeak of an epizootic disease. See Plague.

EPN An organic-phosphate (thiophosphoric acid) compound {O-ethyl O-4-nitrophenyl phenylphosphorothioate} used as an acaricide and insecticide against mites, thrips, plant-sucking insects, leaf-chewing insects and bollworms. Applied to cotton, rice, vegetables and fruit trees in some countries. No longer registered in USA. Some phytotoxicity on apples. Toxic to fishes. Trade name is Santox®.

EPOMIA Noun. (Pl., Epomiae.) Hymenoptera: The elevated margin of an oblique furrow in the Propleura for the reception of the anterior Femora. Ichneumonidae: A Carina on the lateral margin of the Pronotum, obliquely crossing a Sulcus in the lateral face of the Pronotum.

EPON 812 An epoxy resin developed as an embedding medium for use in microtomy and electron microscopy. (Luft 1961, J. Biophys. Biochem. Cytol. 9: 409.) Cf. Araldite.

EPONYM Noun. (Greek, *epi* = by; Aeolic, *onyma* = name. Greek, *eponymia* = a surname given after a person or thing. Pl., Eponyms.) The name of a person used to designate a zoological entity, biological phenomenon, morphological feature or physical principle. Typically the person's name has become a designation of that thing or phenomenon Examples include: Achilles' Tendon; Darwinella; Dolo's Law; Doufour's Gland; Mithridatism; Oedipism. Cf. Acronym; Patronym. Rel. Taxonomy; Nomenclature.

EPONYMIC Adj. (Greek, *eponymia* = a surname given after a person or thing; *-ic* = characterized by.) Pertaining to an eponym. Alt. Eponymous.

EPOXY RESIN A category of plastic embedding media used in histology. ER display superior thermal stability and low polymerization damage to specimen. See Histology. Cf. Methacrylates; Polyesters. Rel. Embedding.

EPPELSHEIM, EDUARD (1837–1896) (Anon. 1896, Ent. News 7: 256.)

EPPELSHEIM, FRIEDRICH (–1900) (Wilderman 1901, Jb. Naturw. 16: 501.)

EPUPILLATE Adj. (Latin, *ex-*; *pupilla* = pupil of the eye; *-atus* = adjectival suffix.) Destitute of a pupil or central spot; a term applied to ocellate spots. Cf. Ocellate.

EQUAL Adj. (Latin, *aequalis* = equal; *-alis* = characterized by.) Pertaining to structures with similar length, size, shape, texture or composition; structures without superficial inequalities. Ant. Inequal. Rel. Subequal.

EQUATE Verb. (Latin, *aequatus* from *aequus* = level, equal.) To represent or express as equal.

EQUIBAND® See Cypermethrin.

EQUIDISTANT Adj. (Latin, *equi-*; *distans* = distant; *-antem* = adjectival suffix.) Equally distant from any two or more points.

EQUIGARD® See Dichlorvos.

EQUILIBRIUM Noun. (Latin, *aequilibrium* > *aequus* = equal; *libra* = balance; *-ium* = diminutive, > Greek, *idion*. Pl., Equilibria.) Chemistry: The state reached when a reversible reaction is completed and no further reaction takes place.

EQUINE BABESIOSIS A widespread form of the disease Babesiosis expressed in horses primarily by *Babesia caballi* and *B. equi*. Disease sometimes is called Biliary Fever or Equine Babesiosis. EB is not known to occur in USA, Great Britain, Germany, Switzerland, Austria and Japan; disease agent is transmitted by ixodid ticks, including *Dermacentor* spp., *Hyalomma* spp. and *Rhipicephalus* spp. Disease is transstadially transmitted by some vectors and transovarially transmitted by other vectors. Incubation period is 10–30 days, followed by fever, anaemia, malaise and weight loss; disease sometimes is fatal. See Babesiosis. Cf. Bovine Babesiosis; Canine Babesiosis.

EQUINE ONCHOCERCAS A filarial disease of horses caused by *Onchocerca cervocalis* or *O. reticulata.* Microfilariae are ingested by blackflies with blood meal. Cf. Bovine Onchocercas.

EQUINE PIROPLASMOSIS See Equine Babesiosis.

EQUITANT Adj. (Latin, *equitare* = overlapping; *-antem* = adjectival suffix.) Pertaining to structures laminated, overlapping or folded one upon the other.

EQUITY® See Chlorpyrifos.

ERADAX® See Chlorpyrifos.

ERADICANT FUNGICIDE Any fungicide that kills fungi within a host and has a curative effect. See Fungicide. Cf. Protectant Fungicide.

ERADICATION Noun. (Latin, *eradicare* from *ex-*; *radix* = root; English, *-tion* = result of an action.) A type of regulatory-control program in which a target pest is eliminated from a geographical region. Term usually is applied against pests recently introduced but not established in the area of concern. See Control; Regulatory Control; Quarantine Action. Cf. Containment; Suppression.

ERB, HERMAN G (1870–1940) (Anon. 1940, Bull. Brooklyn ent. Soc. 35: 143.)

ERBE, BENJAMIN TRAUGOTT (1802–1866) (Schmidt 1969, Jber. Ges. Freunden Naturw. Gera 12: 45.)

ERBER, JOSEF (–1882) (Anon. 1882, Wien Ent. Ztg 1: 104.)

ERCOLANI, GIOVANNI BATTISTA (1817–1883) Italian physician and founder of veterinary sciences in Italy. Ercolani described mites as parasites of domestic animals. (Anon. 1883, Bull. Sci. méd. Bologna (6) 12: (4–5): i–vii.)

ERDÖS, JOSEF (1900–1971) Hungarian amateur Entomologist and specialist in taxonomy of chalcidoid wasps, particularly the Encyrtidae. (Szelenyi 1972, Folia ent. hung. 25: 179–186, bibliogr.)

ERECT Adj. (Latin, *erigere* = to raise up.) Standing upright, not necessarily perpendicular. Alt. Erectus. Rel. Suberect.

ERECTILE Adj. (Latin, *erigere* = to raise up.) Capable of being erected; a term applied to an appendage, Seta, process, or any tissue that may be distended and made rigid.

ERECTO-PATENT Hesperidae: When at rest, having the forewings erect and the hindwings horizontal.

EREISMA Noun. (Etymology obscure.) The Furcula in *Sminthurus* (Kirby & Spence).

EREMIAPHILIDAE Plural Noun. A small Family of Mantodea including two Genera that are confined to arid deserts of North Africa and Asia. Body thickset, sand-coloured with a coarse Integument; ocellar tubercles absent; both sexes brachypterous; middle and hind legs long and slender; fore Tibia with double-ranged spines. See Mantodea.

EREMOCHAETUS Adj. (Greek, *eremos* = solitary, desert; *chaetus* = hair.) Diptera: With a general absence of bristles.

EREMOPHILOUS Adj. (Greek, *eremos* = desert; *philein* = to love; Latin, *-osus* = full of.) Desert-loving; a term applied to Species of plants and animals that live in deserts or arid regions. See Habitat. Cf. Ammophilous; Deserticolous; Ericeticolous; Geophilous; Hypogenous; Lapidicolous; Psammophilous; Xerophilous. Rel. Ecology.

ERGATANDROMORPH Noun. (Greek, *ergates* = worker; *aner* = male; *morphe* = form.) A genetical condition in which an ant expresses morphological characteristics of a male and worker. (Donisthorpe 1929, Zool. Anz. 52: 92–96.)

ERGATANDROUS Adj. (Greek, *ergates* = worker; *andros* = male; Latin, *-osus* = with the property of.) Formicidae: Ant Species with worker-like males.

ERGATANER Noun. (Greek, *ergates* = worker; *aner* = male. Pl., Ergataners.) Social Insects: A male ant without wings, resembling the ergate or worker.

ERGATE Noun. (Greek, *ergates* = worker. Pl., Ergates.) A worker ant.

ERGATOGYNANDROMORPH Noun. (Greek, *ergates* = worker; *gyne* = female; *aner* = male; *morphe* = form. Pl., Ergatogynandromorphs.) A genetical condition in which an ant expresses morphological features of a worker and morphological features of a queen. (Berndt & Kremer 1983, Ins. Soc. 30: 461–465.)

ERGATOGYNE Noun. (Greek, *ergates* = worker; *gyne* = female. Pl., Ergatogynes.) Social Insects: Any form anatomically intermediate between a worker and a queen; a female or queen ant without wings, resembling the worker.

ERGATOGYNOUS Adj. (Greek, *ergates* = worker; *gyne* = female; Latin, *-osus* = with the property of.) Social Insects: Ants with worker-like queens.

ERGATOID REPRODUCTIVE (Greek, *ergates* = worker; *eidos* = form.) Termites: A reproductive termite with rounded head, without wing buds and nymph-like in form.

ERGATOTELIC TYPE Social Insects: A group in which only the secondary instincts are manifested in the queen and the worker retains all primary instincts (Wheeler).

ERGONOMIC STAGE Social Insects: An intermediate stage of colony development in which only workers are produced. The ES follows the Founding Stage and precedes the Reproductive Stage.

ERGONOMIC Adj. (Greek, *ergon* = work, as combining form erg-, ergo-; *nomos* = law.) 1. Pertaining to theory involving social insects which proposes that a society should maximize reproductive production at the least cost in terms of sterile worker production. 2. Pertaining to the functional efficiency of structures in motion or operation. Cf. Bionomic.

ERGONOMIC Noun. (Greek, *ergon* = work, as combining form *erg-, ergo-; nomos* = law; *-ic* = characterized by. Pl., Ergonomics.) The quantitative study of work, performance and efficiency.

ERGOT Noun. (Old French, *argot* = a spur.) A fungal disease of grasses, including barley, rye and wheat that is caused by *Claviceps purpurea* and transmitted by bees and flies. Grains are replaced with black, club-shaped bodies called ergots.

ERI SILKWORM *Samia cynthia ricini* (Donovan) [Lepidoptera: Saturniidae].

ERIAPHYTINAE Hayat 1978. A small Subfamily of Aphelinidae. Morphological features include: Antennal formula 11032; Maxillary Palpus with two segments; Labial Palpus with one segment; Pronotum not medially divided; Mesoscutum densely setose; Scutellum large, convex, nearly hexagonal with about eight pairs of Setae; Axillae produced anteriad as *Tetrastichus;* forewing broad, Speculum present, Marginal Vein shorter than Submarginal Vein, Stigmal Vein long, Postmarginal Vein present; Gaster with Syntergum; Tarsomere formula 5-5-5. Biology: Gregarious parasites of Asterolecaniidae. See Aphelinidae.

ERIAPORINAE A small Subfamily of Aphelinidae. Morphological features include: Antenna with nine segments excluding two ring segments; Scutellum with three pairs of Setae. Ethiopian and Oriental in distribution. See Aphelinidae.

ERICETICOLOUS Adj. (Latin, *ericetum* = a heath; *colere* = living in or on.) Descriptive of organisms that inhabit or frequently visit sandy or gravelly places. See Habitat. Cf. Ammophilous; Arenicolous; Deserticolous; Geophilous; Nemoricolous; Psammophilous; Xerophilous. Rel. Ecology.

ERICHSON, WILHELM (GUILLAUME) FERDINAND (1808–1849) (Papavero 1971, *Essays on the History of Neotropical Dipterology* 1: 110. São Paulo.)

ERICSON, ISAAC BIRGER (1847–1920) (Uyhenboogaart 1921, Ent. Ber., Amst. 5: 329–330.)

ERIGERON ROOT APHID *Aphis middletonii* (Thomas) [Hemiptera: Aphididae].

ERINNIDAE See Xylophagidae.

ERIOCEPHALIDAE See Micropterygidae.

ERIOCOCCIDAE Plural Noun. Eriococcid Scales; Felt Scales. A cosmopolitan Family of coccoid Hemiptera that includes about 500 Species. Abdominal spiracles absent; terminal segments of Abdomen not forming a Pygidium; anal lobes well developed; Anal Ring setose, with a single row of pores. Some Species form galls; some Species are regarded as significant agricultural pests; Species are found on woody shrubs, trees and grasses. See Coccoidea.

ERIOCOCCUS CATERPILLAR *Stathmopoda melanochra* Meyrick [Lepidoptera: Oecophoridae]. (Australia).

ERIOCOTTIDAE Plural Noun. A small Family of ditrysian Lepidoptera assigned to Superfamily Tineoidea. Adult body small to medium sized; wingspan 15–50 mm; head with coarse piliform scales; Antenna filiform, Pecten well developed, each flagellar segment with lamellar Setae on lateral surface; Proboscis without scales; Maxillary Palpus typically with four segments or reduced; Labial Palpus with three segments, porrect with long appressed scales and lateral bristles; tibial spur formula 0-2-4. Pupal abdominal Terga each with transverse row of spines. Biology, eggs and larvae unknown. See Tineoidea.

ERIOCRANIIDAE Plural Noun. Small Family of moths (ca 30 Species) found in Holarctic. Eriocranids are regarded as the most primitive of Glossata and representative of first moths with functional Proboscis. Adult with Chaetosemata; Mandibles distinct but small; mandibular articulations weak. Epiphysis present or absent; tibial spur formula 0-1-4; wing venation homoneurous; wing-coupling jugate; forewing-metathoracic locking device present. Adults are diurnal; eggs are inserted into leaf tissue; larvae mine leaves. Pupation occurs within cocoon in soil. Pupae are decticous and exarate with hypertrophied Mandibles. Alt. Eriocranidae. See Lepidoptera.

ERIOCRANIOIDEA Noun. A Superfamily of dacnonyphous Lepidoptera including Eriocranidae, Mnesarchaeidae and Neopseustidae. See Lepidoptera.

ERLANGER, KARL VON (–1904) (Pagenstecher 1905, Jb. nassau Ver. Naturk. 58: xiii.)

ERMINE MOTH *Yponomeuta padella* (Linnaeus) [Lepidoptera: Yponomeutidae]: A Palearctic pest feeding on Rosaceae, including *Crataegus* spp. and *Prunus* spp. In North America, EM larvae form communal web on apple and cherry. See Yponomeutidae.

ERMINE MOTHS See Arctiidae; Yponomeutidae.

ERMISCH, KARL (1898–1970) (Rohlfien 1975, Beitr. Ent. 25: 267.)

ERNST, ADOLPH (1832–1899) (John 1932, Boln. Soc. venez. Cienc. nat. 9: 317–380 bibliogr.

ERODED Adj. (Latin, *erosus* = gnawed off.) 1. Gnawed, typically at a margin or edge. Rel. Suberoded. 2. Pertaining to a margin with irregular teeth and emarginations.

EROTYLIDAE Plural Noun. A moderate sized Family of polyphagous Coleoptera assigned to the Cucujoidea. Adults are 5–25 mm long, ovate to elongate and somewhat flattened; Antenna with 11 segments including 3–5 segments in club; tarsal formula 5-5-5; Abdomen with five Ventrites. Larvae are elongate, subcylindrical with head hypognathous; 5–6 Stemmata on each side of head; legs with five segments; two tarsal claws on each Pretarsus; Urogomphi present or absent. Adults and larvae are mycophagous. See Cucujoidea.

ERSCH, JOHANN SAMUEL (1766–1828) (Rose 1850, *New General Biographical Dictionary* 7: 251.)

ERSCHOFF, NICOLLAS GRIGOREVICH (1837–1896) (Alpheraky 1897, Horae Soc. ent. Ross. 31: xi–xix, bibliogr.)

ERUCA Noun. (Latin, *eruca* = caterpillar. Pl., Erucae.) A larva; a caterpillar. See Caterpillar. Cf. Erucina.

ERUCIC STOMATITIS The disorder resulting from the ingestion of caterpillar Setae (urticating hairs) by humans.

ERUCIFORM Adj. (Latin, *eruca* = caterpillar; *forma* = shape.) 1. Shaped as a caterpillar. 2. An anatomical form characterized by a well developed head capsule, thoracic legs and abdominal Prolegs. Eruciform larvae are seen in many Species of Lepidoptera and some Coleoptera, Mecoptera and Hymenoptera (Symphyta). See Caterpillar. Cf. Hymenopteriform; Vermiform.

ERUCINA Noun. (Latin, *eruca* = caterpillar. Pl., Erucinae.) The caterpillar-like larvae of sawflies. See Larva. Cf. Eruca.

ERUCIVOROUS Adj. (Latin, *eruca* = caterpillar; *vorare* = to devour; Latin, *-osus* = with the property of.) Pertaining to predators or parasites of caterpillars. See Feeding Strategy.

ERUPTIVE CELL See Spherulocyte.

ERXLEBEN, JOHANN CHRISTIAN POLYCARP (1744–1777) (Rose 1850, *New General Biographical Dictionary* 7: 253–254.)

ERYTHRAEOUS Adj. (Greek, *erythranos* = red; Latin, *-osus* = with the property of.) Red; nearly arterial blood-red. Alt. Erythraeus.

ERYTHRINUS Adj. (Greek, *erythranos* = red.) Deep brick-red; tending to blood-red. Alt. Erythrine.

ERYTHROIC Adj. (Greek, *erythros* = red; *-ic* = characterized by.) Reddish; reddened. Alt. Erythrous.

ERYTHROPSIN Noun. (Greek, *erythros* = red; *ops* = eye. Pl., Erythropsins.) A colouring substance found in the eyes of night-flying insects that impregnates the Retinula (Imms).

ESAKI, TEISO (1899–1957) (Richards 1958, Proc. R. ent. Soc. Lond. (C) 22: 74.)

ESBEN-PETERSEN, PETER (1869–1942) (Wesenberg-Lund 1943, Vidensk. Meddr dansk. naturh. Foren 106: 144–145.)

ESBIOTHRION®. See Allethrin.

ESCHERICH, KARL LEOPOLD (1871–1951) (Willenstein 1971, Anz. Schädlinsk. Pflanenzschutz 44: 127–128.)

ESCHER-KÜNDIG, JAKOB (1842–1930) (Rebel 1946, Vjschr. naturf. Ges. Zurich 91: 83–84.)

ESCHOLTZ, JOHANN FRIEDRICH (1793–1831) (Papavero 1971, *Essays on the History of Neotropical Dipterology* I: 51–52. São Paulo.)

ESCUTCHEON Noun. (French, *escuchon* > Latin, *scutus* = shield. Pl., Escutcheons.) 1. A shield or object shaped as a shield. 2. Coleoptera: The Scutellum. See Shield.

ESCUTELLATE Adj. (Latin, *ex* = without; *scutellum* = little shield; *-atus* = adjectival suffix.) Descriptive of insects that lack a Scutellum. Alt. Escutellatus; Excutellate; Exscutellatus.

ESFENVALERATE Noun. A synthetic-pyrethroid compound formed as the alpha-isomer of fenvalerate. Some phytotoxicty noted on oranges and vegetables; toxic to fish. Esfenvalerate acts as a contact pesticide and stomach poison that is substantially more active than fenvalerate. Trade names include: Asana®, Conquer®, Halmark®, Sumi-alpha®, Sumidan® and Sumiton®. See Synthetic Pyrethroid.

ESMARK, LAURITZ (1806–1884) (Natvig 1944, Norsk. ent. Tidsskr. 7: 17–19.)

ESODERMA Noun. (Greek, *eso-* = within; *derma* = skin. Pl., Esodermata.) A fibrous Cuticle lining the Exoderma. See Cuticle.

ESPER, HUGEN JOSEPH CHRISTOPH (1742–1810) German academic and author of *Europaische Schmetterlinge* (5 vols, 1777–1794). (Anon. 1858. *Accentuated list of British Lepidoptera*. 118 pp. Oxford & Cambridge Entomological Societies, London.)

ESPUNDIA Noun. (Etymology obscure.) See American Mucocutaneous Leishmaniasis. Syn. Uta.

ESSELBAUGH, CHARLES ORRIS (1898–1952) (J[ames] 1953, Ann. ent. Soc. Am. 46: 172–173, bibliogr.)

ESSENTIAL CHARACTER See Specific Character.

ESSENTIALISM Noun. (Latin, *essentia* = essence; English, *-ism* = doctrine or principle. Pl., Essentialisms.) The doctrine that 'kinds' are determined by intrinsic, essential properties. See Typology.

ESSIG, EDWARD OLIVER (1884–1964) American academic (University of California, Berkeley) and author of several notable publications including *History of Entomology* and *Insects of Western North America*. (Mallis 1971, *American Entomologists*. 549 pp. (161–165), New Brunswick.)

ESSIG'S APHID-FLUID A fluid used to preserve aphids and other small, soft-bodied insects. Components include: Lactic acid 20 parts; Glacial Acetic Acid 4 parts; Phenol (saturated aqueous solution) 2 parts; Distilled Water 1 part.

ESTATE® See Chlorpyrifos.

ESTER Noun. (Latin, *aether* = ether. Pl., Esters.) An organic salt formed when an alcohol and an organic acid react.

ESTERASE Noun. (Greek, *aether* = ether; *saeure* = acid; *-ase* = chemical suffix for an enzyme). Any of a group of enzymes that when present accelerates the hydrolysis of esters.

ESTIVATION Noun. (Latin, *aestivare* = summer; English, *-tion* = result of an action. Pl., Estivations.) See Aestivation.

ESTRIATE Adj. (Latin, *e* = out of; *striatus* = grooved; *-atus* = adjectival suffix.) Descriptive of a surface or structure that does not display parallel grooves or ridges. See Sculpture Pattern. Syn. Smooth.

ESTRUP, PEDER JUNGERSEN (1797–1830) (Henriksen 1926, Ent. Meddr. 15: 222–223.)

ESTUARINE Adj. (Latin, *aestuarium* = estuary.) Pertaining to objects or organisms that inhabit an estuary or an estuary type of habitat. See Estuary. Cf. Benthic; Littoral; Pelagic.

ESTUARY Noun. (Latin, *aestuarium* > *aestus* = tide. Pl., Estuaries.) An inlet on the sea; a place where a river meets the sea and creates a unique habitat.

ET ID GENUS OMNE Latin phrase meaning 'and everything of the sort.'

ETALENE® See Fenitrothion.

ETHANOX® See Ethion.

ETHEPHON An organic phosphorus compound {2-chloroethyl} phosphonic acid used as a Plant Growth Regulator on numerous fruit and nut trees, barley, coffee, cotton, cucumbers, melons, ornamentals, peppers, pineapples, squash, tobacco and wheat. Applications include fruit loosening before harvest. Trade names include: Arvest®, Bromoflor®, Campostan®, Cerone®, Chipcor®, Chlorethephon®, Etherfon®, Etheverse®, Ethotaf®, Ethrel®, Flordimenx®, Florel®, Nu-Tomatonone®, Pistil®, Pluck®, Prep®, Refon®, Stance®, Tomathrel®, Xtargro®. See Organophosphorus Insecticide.

ETHER (Diethyl Ether, C_2H_5-O-C_2H_5) A liquid with characteristic burning sweetish taste, highly inflammable and explosive. Ether is slightly soluble in water, miscible with alcohol, benzene and oils. Ether has been employed in Entomology as a killing agent and used by museum curators to clean and degrease specimens.

ETHERFON® See Ethephon.

ETHERIDGE, ROBERT (1847–1920) (Dun 1926, Records Austr. Mus. 15: 1–27, bibliogr.)

ETHEVERSE® See Ethephon.

ETHIOFENCARB A carbamate compound {2-((ethylthio) methyl)phenyl) methylcarbamate} used as a contact/systemic insecticide and stomach poison for control of aphids. Ethiofencarb is applied to tree crops, field crops and ornamental plants in some countries. The compound is not registered for use in USA; it is toxic to fishes but not bees. Trade names include Arylmate® and Croneton®. See Carbamate Insecticide.

ETHIOL® See Ethion.

ETHION An organic-phosphate (dithiophosphate) compound {O,O,O',O'-tetraethyl S,S'-methylene biphosphorodithioate} used as an acaricide and broad-spectrum insecticide against mites, plant-sucking insects (to include aphids, psyllids, scale insects, leafhoppers, lygus bugs), seed maggots, thirps, leaf miners and Mexican bean beetles. Compound is applied to citrus and many crops outside USA. Ethion is toxic to bees, fishes and wildlife. Trade names include: Diethion®, Embathion®, Ethanox®, Ethiol®, Ethiosul®, Ethodan®, Hylemox® and Rhodocide®. See Organophosphorus Insecticide.

ETHIOPIAN REALM The faunal region including Africa south of the Sahara, southern Arabia and Madagascar. See Zoogeography.

ETHIOPIAN Adj. (Greek, *aethiops* = burned, burnished.) Pertaining to the Ethiopian Realm.

ETHIOSUL® See Ethion.

ETHMIIDAE Plural Noun. A Family containing ca 250 Species of ditrysian Lepidoptera that has been recognized as a Subfamily of Oecophoridae in some classifications. Adults often are brightly coloured with black spots on forewing with wingspan 9–40 mm and head with overlapping lamellar scales; Ocelli and Chaetosemata absent; Antenna shorter than forewing, and Scape typically lacking Pecten. Proboscis is scaled; Maxillary Palpus typically with four segments that fold over base of Proboscis; Labial Palpus strongly recurved, smooth scaled; fore Tibia with Epiphysis; tibial spur formula 0-2-4; wings rather broad, Pterostigma absent. Eggs are oval in outline shape, flattened in some Species and deposited individually or in small groups on underside of leaves. Larvae are cylindrical and head prognathous. Host plants include Boraginaceae and Hydrophillaceae. Adults of some Species are diurnal. See Oecophoridae.

ETHNOLOGY Noun. (Greek, *ethnos* = nation; *graphein* = to write. Pl., Ethnologies.) The study of human races.

ETHOCLINE Noun. (Greek, *ethos* = custom; *klinein* = slope. Pl., Ethoclines.) A graded series of similar behaviours among Species presumably related through descent. See Transformation Series.

ETHODAN® See Ethion.

ETHOGRAM Noun. (Greek, *ethos* = custom; *gramma* = anything written. Pl., Ethograms.) A complete description of the behaviourial repertoire of an individual, caste or Species.

ETHOLOGY Noun. (Greek, *ethos* = custom; *logos* = discorse. Pl., Ethologies.) The study of animal behaviour.

ETHOPROP An organic-phosphate (thiophosphoric acid) compound {O-ethyl S,S-dipropyl phosphorothioate} used as a contact insecticide and nematicide against nematodes, wireworms, rootworms, webworms, Jap beetles and other pests. Compound is applied to bananas, beans, cabbage, cucumbers, potatoes, mushrooms, ornamentals and tobacco. Compound is toxic to wildlife and fishes. Trade name is Mocap®. See Organophosphorus Insecticide.

ETHOTAF® See Ethephon.

ETHREL® See Ethephon.

ETHYL ACETATE A killing agent (CH_3CO_2-C_2H_5) used in Entomology. Popular because it is relatively non-toxic to humans, stuns quickly, kills slowly, and leaves specimens in a relaxed condition.

ETHYL ALCOHOL See Alcohol-Ethyl.

ETHYLENE DIBROMIDE A fumigant {1,2-dibromoethane} used as a soil fumigant and on stored products. Fumigated commodities (outside USA) include bananas, beans, cantaloupe, cherries, corn, cucumbers, grains, guavas, mangoes, papayas, peppers, pineapples, plums and zucchini. EB is used as soil fumigant on vegetables, peanuts, peppers, potatoes, strawberries, soybeans in some countries. EB is not registered for use in USA. Trade names include: EDB®, E-D-Bee®, Bromo-fume®, Celmide®, Dibrome®, Nephis®. See Fumigant. Cf. Sulphuryl Fluoride.

ETHYLENE DICHLORIDE A fumigant {1,2-dichloroethane} that is used against commodity pests. Fumigated commodities include grains in some countries. No longer registered for use in USA. Compound is flammable and corrosive to aluminium and magnesium. Trade names include: EDC®, Granosan® and ED/CT®. See Fumigant.

ETHYLENE OXIDE A highly toxic fumigant used in vacuum chambers for the treatment of packaged food, museum objects and medical supples. EO boils at 11°C and typically is formulated with carbon dioxide or freon. EO is applied for insect, fungus and bacterial control. EO is a hazardous material that requires special licence or training.

ETIELLA MOTH *Etiella behrii* (Zeller) [Lepidoptera: Pyralidae]. (Australia).

ETILON® Parathion.

ETIOLATE Adj. (Old French, *esteule* = straw, from Latin, *stupula* = stubble; *-atus* = adjectival suffix.) Whitened or bleached from lack of sunlight or from disease.

ETIOLATION Noun. (French = *etioler* = to blanch; English, *-tion* = result of an action. Pl., Etiolations.) The development of a plant that is tall and spindly.

ETIOLOGICAL AGENT The parasite causing disease in plants or animals. See Parasite.

ETOFENPROX A chlorinated hydrocarbon {2-(4-ethoxyphenyl)-2-methylpropyl 3-phenoxybenzyl ether} marketed as a contact insecticide and stomach poison in some countries, but not registered in USA. Trade names include: Lenatop®, Permit®, Trebon® and Vectron®. See Organochlorine Insecticide.

ETRIMFOS An organic-phosphate (thiophosphoric acid) compound {O-6-ethoxy-2-ethyl-pyrimidin-4-yl-O,O-dimethyl-phosphorothioate} used as a contact insecticide and stomach poison against aphids, leaf-eating Lepidoptera and Coleoptera and stem borers. Compound is applied to alfalfa, citrus, corn, crucifers, fruit trees, grapes, olives, potatoes, rice, tobacco, and vegetables in some countries; Etrimfos also is used for stored product pests and processed food pests, but is not registered for use in USA. Etrimfos is toxic to bees. Trade names include Ekamet® and Satisfar®. See Organophosphorus Insecticide.

ETROFOLAN® Isoprocarb.

ETTINGHAUSEN, CONSTANTIN VON (1826–1871) (Hoernes 1897, Mitt. naturw. Ver. Steierm. 34: 76–106, bibliogr.)

EUCALYPT-DEFOLIATING SAWFLY *Pergagrapha bella* (Newman) [Hymenoptera: Pergidae]. (Australia).

EUCALYPT FLIES [Diptera: Fergusoninidae]. (Australia).

EUCALYPT KEYHOLE-BORER *Xyleborus truncatus* (Erichson) [Coleoptera: Curculionidae]. (Australia).

EUCALYPT LEAF-BEETLES *Chysophtharta* spp., *Paropsis* spp. [Coleoptera: Chrysomelidae]. (Australia).

EUCALYPT LEAF-GALL SCALE *Opisthoscelis subrotunda* Schrader [Hemiptera: Eriococcidae]. (Australia).

EUCALYPT PINWORM *Atractocerus kreuslerae* Pascoe [Coleoptera: Lymexylidae]. (Australia).

EUCALYPT PSYLLID See Red Gum Sugar Lerp.

EUCALYPT RING-BARKING LONGICORN *Tryphocoria mastersi* Pascoe [Coleoptera: Cerambycidae]. (Australia).

EUCALYPT-SHOOT PSYLLID *Blastopsylla occidentalis* Taylor [Hemiptera: Psyllidae]. (Australia).

EUCALYPTUS LONGHORNED BORER *Phoracantha semipunctata* (Fabricius) [Coleoptera: Cerambycidae]: A native of Australia and a significant pest of *Eucalyptus* Species in California and other areas of the world. See Cerambycidae.

EUCALYPTUS THRIPS *Thrips australis* (Bagnall) [Thysanoptera: Thripidae]. (Australia).

EUCALYPTUS TORTOISE-BEETLES *Chrysophtharta* spp., *Parapsis* spp. [Coleoptera: Chrysomelidae]. (Australia).

EUCALYPTUS WEEVIL *Gonipterus scutellatus* (Gyllenhal) [Coleoptera: Curculionidae]. (Australia).

EUCEPHALOUS Adj. (Greek, *eu* = good; *kephale* = head; Latin, *-osus* = with the property of.) With a well developed head that bears normal appendages. A term applied to some dipterous larvae. See Head.

EUCHARITIDAE Plural Noun. A cosmopolitan, moderately small (ca 45 Genera, 350 Species) Family of apocritous Hymenoptera assigned to Chalcidoidea. Adult body to 10 mm long, often metallic coloured with head and Pronotum small; Mandibles often falcate. Antenna typically with 11–13 segments, without ring segments or defined club but male Antenna often ramose. Prepectus fused to Pronotum and Mesosoma globose; Notaulices well developed and complete; Scutellum frequently with apical spine or prolongation; forewing Marginal Vein long, postmarginal and Stigmal Veins short. Petiole elongate; Gaster small and evaniid-like. Euchartidids are related to Perilampidae, or placed with Pteromalidae in some classifications. Eucharitids are typified by a complex biology that includes larval Heteromorphosis. Larvae develop as ectoparasites or endoparasites of ant larvae. Male eucharitids sometimes swarm over the crown of an ant nest waiting for conspecific females to emerge; copulation occurs soon after females emerge from the ant nest. Adult females are proovigenic and may contain several thousand eggs. Eggs are deposited on vegetation where they hatch into a Planidium-type first-instar larvae. The Planidium body consists of 12 segments and a caudal sucker. Planidia are transported to ant nest by workers and ant Prepupae serve as host. Pupation occurs within the ant's cocoon or exposed in the nest's brood chamber. See Chalcidoidea.

EUCINETIDAE Plural Noun. A Family of Coleoptera assigned to Superfamily Eucinetoidea. Adult body fusiform, less than 3 mm long with head deflexed; Antenna with 11 segments, incrassate or filiform; hind Coxa very large and partially concealing Femur; tarsal formula 5-5-5; Abdomen with 5–7 Ventrites. Larval Antenna with three segments; legs with five segments; Urogomphi absent. Eucinetids feed on slime-mould spores and fruiting bodies of fungi. See Eucinetoidea.

EUCINETOIDEA A Superfamily of polyphagous Coleoptera including Clambidae, Eucinetidae and Scirtidae (Cyphonidae, Helodidae). See Coleoptera.

EUCNEMIDAE Plural Noun. A cosmopolitan Family of Coleoptera consisting of about 1,100 Species. Eucnemids resemble elaterids but the head is strongly deflexed and Labrum absent; the click mechanism of eucnemids is poorly developed. Biology of eucnemids is poorly known. They are found in dead wood or beneath bark, and presumably are carnivorous. See Coleoptera.

EUCOILIDAE Thomson 1862. Plural Noun. A cosmopolitan Family of apocritous Hymenoptera assigned to Superfamily Cynipoidea. Eucoilids are the largest parasitic Family in Cynipoidea with more than 1,000 nominal Species. Adult female Antenna with 13 segments and enlarged distad; male Antenna 15 segments with segments III, IV modified; Scutellum with a raised, flat 'plate', cup or disc; forewing Radial Cell usually closed, some open; gastral Terga II, III fused. Eucoilids develop as ectoparasites of schizophorous Diptera larvae; females oviposit in young host larvae where the parasite completes 3–5 larval instars. Pupation occurs within a host Puparium; hypermetamorphosis probably is universal. Instar I is eucoiliform with paired lateral thoracic lobes for emergence; median caudal projection for emergence from egg; instar I lacks spiracles; instar II with spiracles; respiratory system gradually develops (eight spiracles in Instar V). The caudal projection is gradually lost with successive moults. See Cynipoidea.

EUCOILIFORM LARVA Hymenoptera: An hypermetamorphic larval form diagnostic of Eucoilidae. Primary larval form displays three pairs of long thoracic appendages but lacks cephalic process and girdle of Setae of teleaform larva. Subsequent instars display polypodeiform larval form. EL also is seen in Charipidae and Figitidae. See Hypermetamorphosis; Larva.

EUCONE Noun. (Greek, *eu* = good; *konos* = cone. Pl., Eucones.) A light-concentrating component of the compound eye produced by Semper Cells and positioned beneath the lens and above the Rhabdom. See Eucone Eye.

EUCONE EYE A compound eye with elements as described by Grenacher (1879). Semper Cells produce a crystalline, cylindrical cone consisting of four cells. Distal surface of cone typically flat or concave, body of cone tapered with proximal surface elliptical; hour-glass or pear-shaped in some Species. The Eucone is not birefringent in polarized light and displays a strongly graded refractive index. Composition of Eucone is predominantly protein, lacks Chitin and sometimes displays a sleeve of glycogen. Eucone is regarded as the groundplan condition among pterygotes. According to Superposition Theory, Lens and Crystalline Cone form an afocal lenspair. Distal part of Crystalline Cone and Lens focus an intermediate image on plane of cone waist, proximal tip of cone projects image onto Retina. See Compound Eye. Cf. Aconic Eye; Exocone Eye. Rel. Superposition Eye.

EUCONIC Adj. (Greek, *eu* = good; *konos* = cone; *-ic* = characterized by.) Pertaining to compound eyes with a Euconic Cone.

EUCOXA Noun. (Greek, *eu* = good; Latin, *coxa* = hip. Pl., Eucoxae; Eucoxas.) The anterior division of the insect Coxa (Crampton).

EUDERMAPTERA Noun. A Superfamily of the Order Dermaptera that contains the single Family Apachyidae, with only one Genus (Burr). See Dermaptera.

EUDIAPAUSE Noun. (Greek, *eu* = good; *diapauein* = to cause to cease. Pl., Eudiapauses.) Dormancy in an organism that is initiated and terminated by different environmental factors (Muller 1965, Zool. Anz. 29 suppl: 182). In some insects photoperiod is thought to induce dormancy and temperature is thought to terminate dormancy. See Aestivation; Dormancy; Hibernation. Cf. Diapause; Quiescence. Rel. Oligopause; Parapause.

EUGENIA CATERPILLAR *Targalla delatrix* Guenée [Lepidoptera: Noctuidae].

EUGENITAL Adj. (Greek, *eu* = good; *genitalis*; *-alis* = pertaining to.) Descriptive of the primary genital opening of Actinotrichida mites.

EUGLENIDAE See Aderidae.

EUHOLOGNATHA Noun. A Suborder of Plecoptera. Naiads are active nocturnally and generally phytophagous; they often eat detritus, especially during first instar. Adults of some Species emerge during winter; adults feed on algae and other plants, flowers and pollen. See Plecoptera. Cf. Systellognatha.

EUKARYON Noun. (Greek, *eu* = good; *karyon* = nut. Pl., Eukaryons.) The Nucleus of a eukaryotic organism. See Nucleus. Cf. Prokaryon.

EUKARYOTE Noun. (Greek, *eu* = good; *karyon* = nucleus. Pl., Eukaryotes.) 1. Organisms whose Chromosomes or DNA are contained within a nucleus. 2. Organisms other than blue-green algae and bacteria. Cf. Prokaryote.

EUKARYOTIC Adj. (Greek, *eu* = good; *karyon* = nucleus; nut; *-ic* = characterized by.) Pertaining to eukaryotes. See Eukaryote.

EULABIUM Noun. (Greek, *eu* = good; Latin, *labium* = lip. Pl., Eulabia.) The distal-most division of a Labium. See Mouthparts. Cf. Labium.

EULOPHIDAE Plural Noun. Eulophid Wasps. A large (ca 325 Genera 3,100 nominal Species), cosmopolitan Family of parasitic wasps assigned to the Superfamily Chalcidoidea. Eulophids are typically small to very small (1–6 mm), often weakly sclerotized with coloration variable, usually dark or metallic; Antenna with 5–9 segments including up to four funicular segments; male Antenna sometimes ramose (branched funiculars); four-segmented Tarsi; fore tibial spur not bent; Axilla often produced anteriad of Transscutal Suture; Parapsidal Sutures sometimes developed; Scutellum sometimes with parallel, longitudinal grooves. Eulophids predominantly act as primary parasites with some hyperparasitic and a few phytophagous Species. Solitary or gregarious development as endoparasite or ectoparasite. Eulophids display a broad host spectrum including all immature stages attacked; most Species are parasitic or hyperparasitic, attacking concealed Lepidoptera and Diptera. Some larvae behave as predators. Eulophids are important in some biological control programmes involving sawflies on pine and leaf miners. Included Subfamilies Entedontinae, Euderinae, Eulophinae and Tetrastichinae. See Chalcidoidea.

EUMASTACIDAE Plural Noun. Monkey Grasshoppers. A predominantly tropical Family of caeliferous Orthoptera assigned to Superfamily Eumasticoidea. Adults are wingless with short Antennae. Eumastacids occur in bushes and small trees. See Eumasticoidea.

EUMASTICOIDEA Noun. A Superfamily of caeliferous Orthoptera including Biroellidae, Chorotypidae, Episactidae, Eruciidae, Eumastacidae, Euschmidtiidae, Gomphomastacidae, Mastacideidae, Miraculidae, Morabidae and Thericleidae. See Orthoptera.

EUMECOPTERA Noun. A Suborder of Mecoptera including Apteropanorpidae, Bittacidae, Boreidae, Choristidae, Eomeropidae, Nannochoristidae and Panorpidae. See Mecoptera. Cf. Protomecoptera.

EUMENIDAE Plural Noun. Potter Wasps; Mason Wasps. A large (ca 3,000 Species), cosmopolitan Family of aculeate Hymenoptera assigned to Superfamily Vespoidea, or sometimes placed as Subfamily of Vespidae. Female Antenna with 12 segments, male Antenna with 13 segments; medial margin of compound eye notched; Mandibles elongate and overlapping apicad; Pronotum projecting to Tegula; middle Coxae medially juxtaposed; middle Tibia with one spur; wings macropterous and folded longitudinally at repose, forewing with elongate Disoidal Cell; hindwing with closed cell and Jugal Lobe; first gastral Tergum and Sternum partially fused with a constriction between first and second gastral segments. Species usually are solitary, rarely gregarious, with female constructing nests of mud, clay, in hollow plant stems or insect burrows; nest architecture frequently is complex. Eggs are deposited in cell, then cell is provisioned with paralysed Lepidoptera or Coleoptera larvae. Cells often are mass-provisioned. Eumeninae are solitary wasps with three Submarginal Cells in the forewing, filiform Antenna, middle Tibia with one apical spur, and toothed or otherwise modified tarsal claws. Massarinae are solitary wasps that provision mud cells with pollen and nectar; they are separated from other vespoids by presence of two Submarginal Cells in forewing and clubbed Antenna. See Vespoidea. Cf. Vespidae.

EUNOTUM Noun. (Greek, *eu* = good; *notum* = the back. Pl., Eunota.) The wing-bearing notal sclerite anterior of the Postnotum (Crampton).

EUONYMUS ALATUS SCALE *Lepidosaphes yanagicola* Kuwana [Hemiptera: Diaspididae].

EUONYMUS SCALE *Unaspis euonymi* (Comstock) [Hemiptera: Diaspididae]: A pest of *Euonymus* and *Pachysandra* in the eastern USA. An occasional pest of *Citrus* in nurseries, and overwinters as mature female. See Diaspididae.

EUORTHOPTERA Noun. (Greek, *eu* = good; *orthoptera*.) Orthoptera excluding the Dermaptera. See Dermaptera; Orthoptera.

EUPATHIDIUM Noun. (Greek, *eu* = good; Latin, *-ium* = diminutive > Greek, *-idion* = diminutive. Pl., Eupathidia.) Acarina: A tapering, nude, specialized Seta on Tarsi I and II; frequently the palpal Tarsus, composed of an optically active actinopiline sheath surrounding a core of protoplasm (Goff *et al.* 1982, J. Med. Ent. 19: 223.) Cf. Palpal Seta; Pretarsala; Subterminala.

EUPATORIUM GALL FLY *Procecidochares utilis* Stone [Diptera: Tephritidae]. See Crofton Weed Gall Fly.

EUPELMIDAE Plural Noun. Eupelmid Wasps. A moderate sized (ca 60 Genera, 750 Species), cosmopolitan Family of parasitic Hymenoptera assigned to the Chalcidoidea. Important features include female body elongate, frequently metallic coloured and often 'U'-shaped in dried specimens; female Antenna with 11–13 segments including Anellus; male Antenna with nine segments, frequently ramose; Mesoscutum elongate, frequently with depression; Mesopleuron enlarged, convex, without furrow; middle Coxa near hind Coxa, tibial spur robust; Basitarsus with pegs or blunt spines; forewing usually macropterous, sometimes brachypterous; Marginal Vein typically long; Postmarginal, Stigmal Veins shorter; Paratergites absent; Pygostyli not advanced. Family predominantly parasitic with some facultatively hyperparasitic Species. Development typically is solitary with larvae feeding ectoparasitically or endoparasitically. Eupelmids show a broad host spectrum with all progenitive strategies demonstrated, and some larvae feed as predators of eggs and larvae. Eupelmids are related to Encyrtidae and Tanaostigmatidae. See Chalcidoidea.

EUPLANTULA Noun. (Greek, *eu* = good; Latin, *planta* = sole of foot. Pl., Euplantulae.) Diptera, Orthoptera: A pad-like structure on the ventral surface of Tarsomeres or the Pretarsus. Syn. Tarsal Pulvilli. See Pretarsus. Cf. Arolium; Empodium; Rel. Leg.

EUPLEURON Noun. (Greek, *eu* = good; *pleura* = rib. Pl., Eupleura; Pl., Eupleurons.) The dorsal arch of the primitive pleural region (Crampton). See Anapleurite; Pleuron.

EUPLEXOPTERA Noun. (Greek, *eu* = good; *plexus* > *plectere* = plaited; *optera* = wing.) With beautifully folded wings. An ordinal term applied to the earwigs. Syn. Dermaptera.

EUPTEROTIDAE Plural Noun. A Family of ditrysian Lepidoptera assigned to Superfamily Bombycoidea that includes ca 400 Species with most occurring in Ethiopian and Oriental realms. Adults are variable in size with wingspan 20–140 mm and head displaying piliform scales; Antenna bipectinate to apex in male and female; Ocelli, Chaetosemata and Maxillary Palpus absent; Proboscis weakly developed or absent; Labial Palpus short-minute or concealed by scales of Frons; Epiphysis present or absent; tibial spur formula 0-2-2; hindwing as broad as forewing. Larvae of some Species are gregarious. Pupation occurs within cocoon of silk and larval Setae. Adults are nocturnal and visit lights. See Bombycoidea.

EUPYRENE SPERM A type of Spermatozoa produced by Lepidoptera during Spermatogenesis. ES are characterized by correctly formed Chromatin with Spermatozoa remaining in bundles during transfer to female at copulation. Eupyrene Spermatozoa fertilize eggs. See Spermatozoa. Cf. Apyrene Sperm; Hyperpyrene Sperm; Oligopyrene Sperm.

EUPYRENE Adj. (Greek, *eu* = good, well; *pyren* = fruit stone.) Pertaining to Eupyrene Spermatozoa with Chromosomes. Cf. Apyrene; Hyperyrene; Oligopyrene.

EURASIAN PINE-ADELGID *Pineus pini* (Macquart) [Hemiptera: Adelgidae].

EUROPEAN ALDER LEAFMINER *Fenusa dohrnii* (Tischbein) [Hymenoptera: Tenthredinidae]: An Holarctic pest of alder. See Tenthredinidae.

EUROPEAN APPLE SAWFLY *Hoplocampa testudinea* (Klug) [Hymenoptera: Tenthredinidae]: A minor pest of apples in the eastern USA. Adults inhabit blossoms; larvae mine fruit. See Tenthredinidae. Cf. Cherry Fruit Sawfly; Pear Sawfly.

EUROPEAN BIRCH APHID *Euceraphis betule* (Koch) [Hemiptera: Aphididae]. (Australia).

EUROPEAN BLUEBOTTLE *Calliphora vicina* Robineau-Desvoidy [Diptera: Calliphoridae]. (Australia).

EUROPEAN BROOD DISEASE See European Foulbrood Disease. Cf. American Brood Disease.

EUROPEAN CABBAGEWORM *Pieris rapae* (Linnaeus) [Lepidoptera: Pieridae]. See Imported Cabbageworm.

EUROPEAN CHAFER *Rhizotrogus (Amphimallon) majalis* (Razoumowsky) [Coleoptera: Scarabaeidae]: A significant pest of turfgrass in some regions of North America.

EUROPEAN CHERRY FRUIT-FLY *Rhagoletis cerasi* Linnaeus [Diptera: Tephritidae].

EUROPEAN CHICKEN FLEA *Ceratophyllus gallinae* (Schrank) [Siphonaptera: Ceratophyllidae]: A pest of birds that is endemic to Europe and adventive to eastern North America. ECF hind Femur with 4–6 bristles along medial surface. See Ceratophyllidae. Cf. Western Chicken-Flea.

EUROPEAN CORN BORER *Ostrinia nubilalis* (Hübner) [Lepidoptera: Pyralidae]: An European Species of moth first discovered in the United States near Boston during 1917. ECB is regarded as the most signifcant pest of *Zea mays* Linnaeus (corn) in North America; it attacks more than 200 kinds of plants, including peppers, potatoes, tomatoes and snap beans. Adult females are pale yellow-brown with sinuous, dark lines on forewing; male darker with olive-brown marks on wings. Adults are strong fliers with wingspan ca 25 mm. Females lay eggs in clusters of 15–30 in overlapping scale-like form on underside of leaves; eggs are pellucid. Early larval instars feed on leaves, tassels and stems; later larval instars bore into stalk, leaf stem or ear. ECB overwinters as mature larvae inside burrows. Mature larvae are tan with small, circular dark spots. Pupae are smooth and brown; pupation occurs in loose cocoon during spring. See Pyralidae.

EUROPEAN CRANE FLY *Tipula paludosa* (Meigen) [Diptera: Tipulidae]: A Species endemic to Europe and adventive to Canada during 1955. Adult females are 18–24 mm long and males are 15–18 mm long. ECF adults emerge during August-September and fly during early morning and evening. Eggs are shiny black and put in soil. Young larvae are pinkish in coloration and feed on roots; older larvae with black head and greyish body, migrate at night to feed on plants. Larvae are called leatherjackets and constitute a serious pest of lawns, grains (barley, oats, wheat, rye), sod, turnips, strawberries, corn and peas. ECFs overwinter as larvae and pupation occurs during May. Syn. Marsh Crane Fly. See Tipulidae. Cf. Eastern Crane Fly; Range Crane Fly.

EUROPEAN DUNG MITE *Macrocheles glaber* (Muller), *Macrocheles perglaber* Filippone & Peggazano [Acarina: Macrochelidae]. (Australia).

EUROPEAN EARWIG *Forficula auricularia* Linnaeus [Dermaptera: Forficulidae]: A pest endemic to Europe and now widespread in temperate regions to include North America and Australia. Adults are brown, elongate, 12–15 mm long with body soft or pliant. Antenna is thread-like with 12 segments and extending to base of Abdomen. Pronotum is shield-like; forewing (tegmen) small and smooth; hindwing large, membranous, fan-shaped and folded under Tegmen. Abdomen is longer than head and Thorax combined; Cerci are large, conspicuous and forcipate; Cerci of males curved; Cerci of females nearly straight. Eggs are laid in small cells constructed in soil, ca five cm from surface; broods consist of ca 30 individuals. EEs overwinter as eggs in nests or as adults under stones; females lay two clutches of eggs per year (one in spring, second in early summer). Eclosion occurs within 2–3 weeks and female guards her eggs in cell and sometimes remains with early instar nymphs. EEs complete 5–6 nymphal instars. Species are Monovoltine and complete life cycle ca 70 days. EEs are regarded as a pest of fruits, vegetables and flowers in gardens; they inhabit garbage and also feed upon other insects. EEs may enter houses and feed upon starch, sugar and meat. See Forficulidae.

EUROPEAN ELM SCALE *Gossyparia spuria* (Modeer) [Hemiptera: Eriococcidae]: A widespread, univoltine pest of elms in North America. Females are surrounded by flimsy cocoon and deposit eggs beneath their bodies. Eclosion occurs within one hour of oviposition. Young nymphs feed on underside of leaves and older nymphs (second instar) migrate to limbs or trunk during autumn. EES overwinters as second instar nymphs in bark fissures. See Eriococcidae.

EUROPEAN FOULBROOD DISEASE A widespread bacterial disease that affects honey bee larvae. EFB is caused by *Melissococcus pluton,* a gram-positive, lanceolate bacterium that occurs singly, in chains or in clusters within bee larvae. Bacteria are ingested with food, invade midgut of young larvae, rapidly multiply and typically kill 4–5 day-old larvae. Dead bee larvae become flaccid and change colour to yellow-brown. EFB only affects bee larvae in open cells; disease often is transmitted among brood by nurse bees. EFB typically occurs during summer and infested colonies may recover rapidly. The bacterium has been discovered in other Species of *Apis*. Syn. *Bacillus pluton*; *Streptococcus pluton*. Cf. American Foulbrood Disease.

EUROPEAN FRUIT LECANIUM *Parthenolecanium corni* (Bouché) [Hemiptera: Coccidae]: A monovoltine pest of numerous shade, fruit and and ornamental trees in North America. Females lay several hundred pearl-white eggs during June; eggs hatch and nymphs migrate to underside of leaves along midrib and veins where they feed through summer. During late summer nymphs migrate to limbs. Alt. Brown Apricot Scale. See Coccidae.

EUROPEAN FRUIT SCALE *Quadraspidiotus ostreaeformis* (Curtis) [Hemiptera: Diaspididae]: A pest of deciduous fruit trees in North America. EFS resembles San Jose Scale. See Diaspididae.

EUROPEAN GALL APHID *Pemphigus spyrotheca* [Hemiptera: Aphididae].

EUROPEAN GARDEN SPIDER *Dysdera crocota* C. L. Koch [Araneidae: Dysderidae]. (Australia).

EUROPEAN GRAIN MOTH *Nemapogon granella* (Linnaeus) [Lepidoptera: Tineidae].

EUROPEAN HONEY BEE *Apis mellifera* Linnaeus [Hymenoptera: Apidae]: A domesticated bee endemic to the Old World tropics (Africa) and distributed globally as a source of honey and providing an important role in the pollination of plants. Adult workers are 12–15 mm long and body covered with long yellowish Setae, underlying colour black on head and Mesosoma; Metasoma predominantly reddish brown or yellow with transverse black stripes; compound eyes with long, conspicuous Setae. Forewing with three Submarginal Cells; hindwing Jugal Lobe more than half as long as Vannal Lobe. Middle Tibia with one apical spur; hind Tibia and Basitarsus strongly compressed; hind Tibia lacking apical spur. Sting well developed in workers. EHB larval stage completes five instars. EHB are highly eusocial with queen, drone and workers; queen takes nuptial flight and mates aerially with several males (drones); males lose genitalia during copulation and subsequently dies. EHB nest is exposed or in cavities and composed of wax secreted by sternal glands of workers; nest cells are hexagonal-cylindrical in cross-section and typically horizontal forming vertical combs of two layers of cells that open in opposite directions. Male cells are larger than worker cells and both cell types are used interchangeably for pollen and honey storage and larval development. Queen-producing cells usually hang from brood combs. Eggs are laid individually in unprovisioned cells; eggs are oriented vertically in the bottom of each cell; hundreds-of-thousands of cells are active simultaneously in large colonies. Larvae are fed progressively and cells are not closed until larva completes development. EHB is affected by diseases (American Foul Brood; European Foul Brood), insects (Small Hive Beetle, Wax Moth) and mites (Tracheal Mite; Varroa Mite). See Apidae. Cf. Dwarf Honey Bee; Eastern Honey Bee. Rel. American Foul Brood; European Foul Brood; Small Hive Beetle; Tracheal Mite; Varroa Mite; Wax Moth. Rel. Eusocial.

EUROPEAN HONEYSUCKLE LEAFROLLER *Ypsolophus dentella* (Fabricius) [Lepidoptera: Plutellidae].

EUROPEAN HORNET *Vespa crabro* Linnaeus [Hymenoptera; Vespidae]: A common urban pest in Europe and North America; adults are ca 20–30 mm long and cause damage to ornamental plants and fruit trees. Males and new queens are produced during late summer, leave nest and mate. Nests are founded during spring by mated, overwintered queens; nests sometimes are relocated as season progresses and size increase demands movement. Colonies typically contain fewer than 800 workers. Workers take nectar, fruit and protein; they can forage during night at lights. See Vespidae. Cf. Hornet; Yellowjacket.

EUROPEAN HOUSE SPIDERS *Tegenaria* spp. [Araneidae: Agelenidae]. (Australia).

EUROPEAN HOUSE-DUST MITE *Dermatophagoides pteronyssinus* (Trouessart) [Acarina: Pyroglyphidae]. (Australia).

EUROPEAN MANTID *Mantis religiosa* Linnaeus [Mantodea: Mantidae]: Endemic to the Palearctic and introduced into North America during 1899. EM occurs throughout eastern United States and eastern Canada. Syn. Praying Mantis. See Mantidae.

EUROPEAN MOLE CRICKET *Gryllotalpa gryllotalpa* (Linnaeus) [Orthoptera: Gryllotalpidae]: A Species endemic to Europe that was accidentally introduced into the USA ca 1913 and now established in several eastern states. EMC fore Tibia with four dactyls (digging claws) at apex; base of fore Femur with a short, bladeless process; hind Femur longer than Pronotum; dorsomedial margin of hind Tibia with five spines and apex with seven spines. See Gryllotalpidae; Mole Crickets. Cf. African Mole Cricket.

EUROPEAN MOUSE FLEA *Leptopsylla segnis* (Schönherr) [Siphonaptera: Ceratophyllidae (Leptopsyllidae)] is distributed worldwide and sometimes common in port cities. Adult sometimes attacks humans and rats but typically acts as a parasite of house mouse (*Mus musculus*). See Ceratophyllidae.

EUROPEAN PEACH SCALE *Parthenolecanium persicae* (Fabricius) [Hemiptera: Coccidae].

EUROPEAN PINE SAWFLY *Neodiprion sertifer* (Geoffroy) [Hymenoptera: Diprionidae]: Endemic to Europe and accidentally introduced into USA during 1925 where it is a pest of pines and spruce. Eggs are inserted into slits in needles of the current year's growth. EPS overwinters in egg stage. Larvae feed gregariously on needles, starting at tip and consuming entire needle. Pupation occurs within coccon on ground. See Diprionidae.

EUROPEAN PINE SAWFLY NPV A naturally-occurring, host-specific Nuclear Polyhedrosis Virus used as a biopesticide against larvae of European-Pine Sawfly (*Neodiprion sertifer*). The pathogen attacks the stomach and kills larvae within 10–20 days. Trade names include: Ns NPV®, Sertan® and Virox®. See Biopesticide; Nuclear Polyhedrosis Virus. Cf. Beet Armyworm NPV; Douglas-Fir Tussock Moth NPV; Heliothis NPV; Mamestra brassicae NPV; Redheaded Pine Sawfly NPV; Spodoptera littoralis NPV.

EUROPEAN PINE-SHOOT MOTH *Rhyacionia buoliana* (Denis & Schiffermüller) [Lepidoptera: Tortricidae]: A Palearctic pest of pines accidentally imported and first detected in North America during 1914. Eggs are flat, yellowish and laid individually or in small clusters on needles, twigs or buds. EPSM larvae spin a protective web then bore into base of needle; they overwinter as larvae within the bud. Pupation occurs inside shoot. See Tortricidae.

EUROPEAN RABBIT FLEA *Spilopsyllus cuniculi* (Dale) [Siphonaptera: Pulicidae]: A Species imported into Australia for control of rabbits. ERF serves as a vector for Myxomatosis. See Pulicidae. Rel. Myxomatosis.

EUROPEAN RAT FLEA *Nosopsyllus fasciatus* (Bosc) [Siphonaptera: Ceratophyllidae]: A widespread urban pest. Threshold temperature for egg development ca 5°C. See Ceratophyllidae. Cf. Oriental Rat Flea.

EUROPEAN RED MITE *Panonychus ulmi* (Koch) [Acari: Tetranychidae]: A multivoltine pest of fruit and shade trees in Europe and North America, and apples in Australia. Nymphs and adults suck contents of leaf cells; upper surface of leaves discoloured by feeding while extensive feeding damage causes bronzing and defoliation. ERM overwinters as a fertilized female or egg. Fruit from damaged trees fails to colour properly; fruit from defoliated trees becomes sunburned. See Tetranychidae.

EUROPEAN SPRUCE BEETLE *Dendroctonus micans* (Kugelann) [Coleoptera: Scolytidae.]

EUROPEAN SPRUCE SAWFLY *Gilpinia hercyniae* (Hartig) [Hymenoptera: Diprionidae]: Endemic to Europe-Eurasia and accidentally introduced into Canada during 1922. In Canada and USA, ESS attacks spruce. Eggs are pale green and inserted in slits made in needles. Larvae are green with longitudinal white stripes; larvae feed gregariously on needles and prefer needles at least a year old in upper crown. ESSs overwinter as larvae within a cocoon in litter, and complete 1–2 generations per year; some larvae spend three winters in cocoon. See Diprionidae.

EUROPEAN STRAWBERRY WEEVIL *Otiorhynchus sulcatus* (Fabricius) [Coleoptera: Curculionidae]. (Australia).

EUROPEAN VINE MOTH *Lobesia botrana* Denis & Schiffermüller [Lepidoptera: Tortricidae].

EUROPEAN WASP *Vespula germanica* (Fabricius) [Hymenoptera: Vespidae]: A predatory wasp of European origin that was accidentally introduced into North American and Australia. EW is a significant agricultural, urban and ecological pest where adventive and established. In Europe, colonies are annual with workers dying each autumn; in Australia workers can overwinter with colonies and reach very large numbers (100,000 individuals). Workers ca 12–15 mm long; queen and male ca 20 mm long; Antenna black and body black with yellow markings resembling Common Paper-Wasp; legs not held suspended from body while in flight. Flight patterns are rather direct and movement is rapid. Workers sting repeatedly and actively defend their nest. Males die during autumn while the fertilized queen overwinters and establishes a new colony during spring. Nests are always concealed in ground, tree stumps, wall cavities and roof voids; nests consist of several tiers of individual cells with tiers supported by short columns. An entire nest is surrounded by envelope of macerated wood particles and plant fibre; nest size 15 cm to 5 m. Workers are attracted to food and drink, creating nuisance and medical threat through stings to mouth, throat and neck. Workers are a problem for domestic animals, particularly dogs. See Vespidae. Cf. Common Paper-Wasp; Yellow Paper-Waps.

EUROPEAN WHEAT-STEM SAWFLY *Cephus pygmaeus* (Linnaeus) [Hymenoptera: Cephidae]: A pest of wheat that is endemic in Europe and adventive to eastern North America. Syn. Wheat-Stem Borer. See Cephidae. Cf. Black Grain-Stem Sawfly; Hessian Fly; Wheat-Stem Sawfly.

EUROPEAN WHIRLIGIG MITE *Anystis baccarum* (Linnaeus) [Acarina; Anystidae]. (Australia).

EUROPEAN WOOD WASP *Sirex noctilio* Fabricius [Hymenoptera: Siricidae]: An Holarctic secondary forest pest of *Pinus* spp. EWW accidentally was introduced into New Zealand about 1900, in Victoria ca 1940s and Tasmania ca 1950s. EWW typically is monovoltine. Adult females are metallic blue, ca 25 mm long and oviposit through bark into sapwood. Larvae display a sharp, hard, brown spine at apex of Abdomen. Pupation occurs in wood. Fungus is introduced into sapwood by wasps with their egg. Fungus blocks water-transporting tracheids, eventually killing a tree. See Siricidae.

EURYBRACHIDAE Plural Noun. A Family of fulgoroid Homoptera consisting of less than 200 Species. Most abundantly represented in Africa, the Orient and Australia. Characterized by a rather broad, dorsoventrally depressed habitus.

EURYCANTHINAE Plural Noun. A Subfamily of Phasmatidae [Phasmatodea], mainly found in Papua New Guinea with a few Species in northern Queensland.

EURYCINIDAE See Lycaenidae.

EURYGAMOUS Adj. (Greek, *eurys* = wide; *gamos* = marriage; Latin, *-osus* = with the property of.) Species that require large spaces for mating. A term sometimes applied to mosquitoes. Cf. Stenogamous.

EURYGLOSSINAE A Subfamily of Colletidae (Apoidea) consisting of about 15 Genera.

EURYHALINE Adj. (Greek, *eurys* = wide; *hallnos* = saline.) Pertaining to organisms that adapt to a wide range of salinity. Cf. Stenohaline. Rel. Habitat.

EURYHYGRIC Adj. (Greek, *eurys* = wide; *hygros* = wet; *-ic* = of the nature of.) Pertaining to organisms that adapt to wide ranges of relative humidity. Cf. Stenohygric. Rel. Habitat.

EURYMELIDAE Plural Noun. A numerically small Family of Cicadelloidea (Hemiptera), restricted to New Guinea, Australia and New Caledonia. Adult face is broadly flattened with Ocelli anterior of diamond-shaped Frontoclypeus; Aedeagus not associated with Parameres. Eurymelids typically are associated with *Eucalyptus*; Species are myrmecophilous with gregarious nymphs. See Cicadelloidea.

EURYMERODESMIDAE Causey 1951. Plural Noun. A Family of polydesmoid millipeds consisting of two Genera and about 25 Species found in the south-central USA. Eurymeodesmids are characterized by unmodified, caudally projecting gonopodal telolpodites, long prefemoral Setae, laterally oriented gonopods with aperture setose and Coxal Cannula lateral of the telopodite.

EURYPHAGOUS Adj. (Greek, *eurys* = wide; *phagein* = to eat; Latin, *-osus* = with the property of.) 1. Pertaining to organisms that exhibit euryphagy. 2. Pertaining typically to a herbivorous insect that feeds on many Species of food plants, a predator that feeds on many Species of prey or a parasite that feeds on many Species of host. 3. Pertaining to an omnivorous diet. Syn. Oligophagous. See Feeding Strategy. Cf. Monophagous; Oligophagous; Stenophagous.

EURYPHAGY Noun. (Greek, *eurys* = wide; *phagein* = to eat. Pl., Euryphagies.) 1. The lowest level of specialization in food or least limited range of dietary requirements. 2. An unspecialized diet. 3. An unlimited dietary range. See Omnivorous. Syn. Polyphagy. Cf. Monophagy; Oligophagy; Stenophagy.

EURYPOGONIDAE See Artematopidae.

EURYTOMIDAE Plural Noun. Jointworms; Seed Chalcids. A moderately large (ca 75 Genera, 1,100 Species), cosmopolitan Family of apocritous Hymenoptera assigned to the Chalcidoidea. Adult body 3–15 mm long, robust, dark, non-metallic with head and Thorax frequently coarsely punctate. Antenna with 13 segments, sometimes with Anelli; male Antenna with verticellate whorls of Setae. Pronotum rectangular (subquadrate) from dorsal aspect; Parapsidal Sutures usually present and complete; hind Coxa not enlarged; two apical hind tibial spurs; Tarsomere formula 5-5-5; female Metasoma usually laterally compressed; gastral Terga I-II polished or minutely punctate. Development typically is solitary and rarely gregarious; some Species act as ectoparasites with larval combat; some Species are endoparasitic of cecidogenic insects; some Species are phytophagous in galls, other Species form galls; some Species are egg predators; some Species are parasites of phytophages in seeds, stems and galls; parasites of other phytophages; hyperparasites. Some eggs are spinose and some with micropylar projections. First instar larva is Hymenoteriform, often with five pairs of spiracles. Host spectrum and feeding strategies are diverse: Attack Coleoptera, Hymenoptera, Diptera, Lepidoptera. See Chalcidoidea.

EUSOCIAL Adj. (Greek, *eu* = good; Latin, *sociare* = to associate; *-alis* = characterized by.) 1. Pertaining to insects that display social characteristics including cooperation among individuals in the care of young, contact between generations and reproductive division of labour. Eusocial insects include termites, ants, most bees, many wasps and some other insects. Eusocial insects display several levels of complexity in interaction. 2. Eusocial Bees: Colonies that are Family-groups composed of two generations (mothers and daughters) and typically consist of one mother (queen) and her daughters (workers). Division of labour and development of castes are well developed in eusocial bees. Levels of association between members of one Species nesting in an area include: Aggregation; Communal; Highly Social; Primitively Social; Quasisocial; Semisocial; Solitary; Subsocial. Cf. Communal; Social.

EUSTERNUM Noun. (Greek, *eu* = good; *sternon* = chest. Pl., Eusterna; Eusternums.) 1. The intrasegmental ventral sclerite of a thoracic segment exclusive of the Spinasternum, but usually including the Sternopleurites (Snodgrass). 2. A large sclerite of variable shape frequently extending laterally and upward into the pleural region (Imms). 3. The anterior sternite of the Sternum (Wardle).

EUSTHENIIDAE Plural Noun. A primitive Family of Plecoptera including six Genera and 17 nominal Species assigned to Superfamily Eusthenioidea and found in Australia, New Zealand and South America. Adult body large, brightly coloured with head 'holognathous' and Submentum exceedingly long. Wings with numerous cross veins in all areas; tibial spurs modified, one apical spur apparently migrated to midlength of Tibia and other remains apical but reduced in size. Larvae bear 5–6 pairs of lateral, segmental respiratory gills. Larvae are carnivorous. See Eusthenioidea.

EUSTHENIOIDEA Noun. A Superfamily of Plecoptera roughly equivalent to Archiperlaria. The Eusthenioidea includes most plesiomorphic Taxa in Plecoptera. Characters include larval gills on lateral portion of first 4–6 abdominal segments, large body size (wing span to 110 mm), hindwing anal fan with 8–9 Anales. See Plecoptera.

EUSTIPES See Stipes (MacGillivray).

EUSTIS, HENRY W (1877–1951) (Harris 1952, Lepid. News 6: 77.)

EUSYNANTHROPE Noun. (Greek, *eu* = good; *syn* = together; *anthropos* = human. Pl., Eusynanthropes.) An insect (or other organism) that is completely associated with humans and totally dependent upon humans and/or the human environment for its reproduction, development and continued existence. *e.g.* Human Louse. See Synanthrope. Cf. Hemisynanthrope.

EUSYNANTHROPIC Adj. (Greek, *eu* = good; *syn* = together; *anthropos* = human.) Pertaining to an organism that acts as a Eusynanthrope.

EUSYNANTHROPY Noun. (Greek, *eu* = good; *syn* = together; *anthropos* = human. Pl., Eusynanthropies.) An extension of the concept of synanthropy that recognizes an association with insects (or other organisms) in which the insects are totally dependent upon humans or the human environment. See Eusynanthrope. Cf. Hemisynanthropy.

EUTRACHEATA Noun. (Greek, *eu* = good; *trachelos* = neck. Pl., Eutracheatae.) A term applied to articulates (arthropods including insects) with a well developed tracheal system. Rel. Trachea.

EUTROCHANTIN Noun. (Greek, *eu* = good; *trochilia* = articular. Pl., Eutrochantins.) The ventral arch of the primitive pleural region (Crampton). See Coxopleurite.

EUTROPHIC Adj. (Greek, *eu* = good; *trephein* = to eat; *-ic* = characterized by.) Descriptive of an organism in a well nourished condition. Cf. Dystrophic; Oligotrophic.

EUZOONOSIS Noun. (Greek, *eu* = well; *zoon* = animal; *nosos* = disease. Pl., Euzoonoses.) A zoonotic disease of wild or domestic animals that is transmitted to humans and for which a human is an accidental or terminal host for the pathogen. See Zoonosis; Parazoonosis.

EVACUATE Transitive Verb. (Latin, *evacuatus*, past part. of *evacuare* = to empty out.) 1. To make empty. 2. To eliminate waste products of digestion. 3. To eliminate a gas by evacuation or ventilation, or a liquid through pumping.

EVACUATION Noun. (Latin, *evacuare* = to empty; English, *-tion* = result of an action. Pl., Evacuations.) 1. The act of eliminating waste products from the digestive system. 2. The rapid, organized movement of a population from one place to another place.

EVAGINATE Verb. (Latin, *evaginare* = to unsheath.) To turn inside out or to cause a concealed organ or part within the body to protrude or project externally and conspicuously.

EVAGINATION Noun. (Latin, *e* = out; *vagina* = sheath; English, *-tion* = result of an action. Pl., Evaginations.) Protrusion or eversion of an internal surface or structure by turning it outward. Evagination effected through hydrostatic pressure and/or muscular action.

EVANESCENT Adj. (Latin, *evanescere* = to vanish.) Disappearing; fading. Pertaining to structures or features that become gradually less apparent.

EVANIIDAE Plural Noun. Ensign Wasps. A widespread (predominantly tropical) Family of parasitic Hymenoptera consisting of about 400 Species, and assigned to various Superfamilies in early classifications, but generally is regarded as the nominant Family of the Evanioidea. Adult body size moderate, Antenna with 13 segments; Mesosoma large in relation to the Metasoma; Petiole cylindrical, elongate and positioned high on Propodeum; hindwing with a Jugal Lobe;

Gaster rather small, laterally compressed and ovate in lateral aspect. Ovipositor not exserted. All Species apparently develop as endoparasites of cockroach Oothecae. See Evanioidea.

EVANIOIDEA Latreille 1802. A Superfamily of apocritious Hymenoptera characterized by the Petiole attached high on the Propodeum and functional gastral spiracles found only on Tergum VIII. The Taxon may be polyphyletic, and representative Families have been placed in the Ichenumonoidea and Proctotrupoidea by some classifications. Extant representatives of the Evanioidea are biologically diverse. Included Families are: Anomopterellidae, Aulacidae, Baissidae, Cretevaniidae, Evaniidae, Gasteruptiidae, Kotujellidae and Praeaulacidae. See Apocrita; Hymenoptera.

EVANS, ALWEN MYLANWY (–1937) (Anon. 1938, Arb. morph. taxon. Ent. Berl. 5: 72.)

EVANS, G H (–1948) (Williams 1949, Proc. R. ent. Soc. Lond. (C) 13: 67.)

EVANS, HOWARD ENSIGN American academic (Kansas State University, Cornell University, Harvard University, Colorado State University) and specialist in taxonomy and ethology of aculeate Hymenoptera. Author of many popular books and monographic studies.

EVANS, JOHN WILLIAM (1906–1990) Australian Entomologist and director of Australian Museum. Evans was born in Jabalpore, India and educated Wellington College, Berkshire, England. He was a student of R. J. Tillyard, employed by CSIR (CSIRO) and the first Australian taxonomic specialist on Homoptera.

EVANS, SAMUEL G (1839–1929) (Davis 1931, Proc. Indiana Acad. Sci. 41: 52–54.)

EVANS, WILLIAM (1851–1922) (Bonnet 1945, *Bibliographia Araneorum* 1: 51.)

EVANS, WILLIAM HARRY (1876–1956) (Riley 1957, Nature 179: 127.)

EVAPORATION Noun. (Latin, *evaporare* > *ex-*; *vapor* = steam. Pl., Evaporations.) The act or process by which a substance is converted from a liquid or solid into a vapour and divested of its original solution or matrix. Cf. Condensation.

EVAPORATIVE COOLING A thermoregulatory process by which body fluids (Haemolymph) and/or imbibed liquid (water, nectar) are regurgitated onto mouthparts or expelled from Anus and held in droplet form. Droplet temperature is lowered by action of evaporation and droplet is then imbibed or retracted back into the body. Fluid thus drawn into the body aids in lowering the insect's internal temperature. See Thermoregulation. Cf. Basking. Rel. Circulatory System.

EVE-CHURRS See Gryllotalpidae.

EVELATE Adj. (Latin, *e* = out; *velatus* = veiled; *-atus* = adjectival suffix.) Pertaining to structure that lacks a veil or velum.

EVENING BROWN BUTTERFLY *Melanitis leda bankia* (Fabricius) [Lepidoptera: Nymphalidae]. (Australia).

EVENIUS, JOACHIM (1896–1973) (Anon. 1974, Bee Wld 55: 31.)

EVERCIDE® See Permethrin.

EVERGREEN Adj. (Old Engish, *aefre*; *green*.) Descriptive of plants that do not loose or shed their leaves and appear green during the entire year.

EVERGREEN CROP SPRAY® See Pyrethrin.

EVERMANN, BARTON WARREN (–1932) (Van Duzee 1933, Pan-Pacif. Ent. 9: 10.)

EVERSIBLE BLADDER Thysanoptera: A large, protrusible sac at the Pretarsus apex that serves as an adhesion organ. See Leg. Cf. Pretarsus.

EVERSIBLE Adj. (Latin, *eversio* > *eversion*; *ible*.) 1. Capable of being everted. 2. Descriptive of something turned outward or inside out. Cf. Protrusible.

EVERSMANN, EDUARD (–1861) (Anon. 1858. *Accentuated List of British Lepidoptera*. 118 pp. Oxford & Cambridge Entomological Societies, London.)

EVERT Verb. (Latin, *evertere* = to turn.) To turn outward or inside out. (Everted; Everting.)

EVERTS, EDOUARD JACQUES GUILLAUME (1849–1932) (Oudemans 1933, Tijdschr. Ent. 76: 1–46, bibliogr.)

EVERY, ROBERT WILSON (1910–1970) (Crowell 1971, J. Econ. Ent. 64: 778.)

EVIDENT Adj. (Latin, *evidens* > *ex-* = out; *videre* = to see.) Something that is easily seen, sensed, recognized or apparent. Syn. Apparent; Obvious; Unmistakable.

EVOLUTION Noun. (Latin, *evolvere* = to unroll; English, *-tion* = result of an action. Pl., Evolutions.) A unifying Theory of biological organization proposed by Charles Darwin in 1858 to explain the modification in structure and habit of biological systems (organisms) through descent. See Regressive Evolution. Cf. Genesis. Rel. Darwinism; Lamarkism.

EVOLUTIONARY SPECIES A genetical lineage evolving independent of other genetical lineages and which displays its own unitary evolutionary role and tendencies (Simpson 1961). Cf. Agamospecies; Morphospecies.

EVOLUTIONARY TAXONOMY Taxonomy or classification of living organisms according to their descent or relationship to other organisms. See Classification; Taxonomy.

EWALD (EWALDT), BENJAMIN (1674–1719) (Rose 1852, *New General Biographical Dictionary* 7: 279.)

EWAN, HERBERT GEORGE (1929–1963) (Dickerman 1964, J. Econ. Ent. 57: 184.)

EWERS, WILLIAM (1871–1935) (Mickell 1937, Ann. ent. Soc. Am. 30: 182.)

EWING, HENRY ELLSWORTH (1883–1951) (Anon. 1951, Proc. ent. Soc. Wash. 53: 147–149.)

EWING, KY PEPPER (1898–1974) (Rainwater & Parencia 1974, J. Econ. Ent. 67: 568–569.)

EX LARVA Latin phrase meaning 'from the larva' or 'out of the larva.' Term usually is applied to adult specimens that have been reared from larvae held for observation.

EX NOMINE Latin phrase used in taxonomy to mean 'under that name.'

EX OVUM (Pl., Ex Ova.) Latin phrase meaning 'from the egg' or 'out of the egg.' Term applied to specimens that have been reared from the egg stage.

EX PARTE 1. Latin phrase meaning 'on one side only.' 2. A phrase used in taxonomy to mean 'in part.'

EXAMINIUM See Ventral Ptyche.

EXARATE PUPA A 'free' Pupa. A form of pupal development in which body appendages (Antenna, legs and wings) are not immobile or 'glued' to the body. Characteristic of lower Endopterygota, but not restricted to that group. See Pupa. Cf. Obtect Pupa; Pupa Exarata. Rel. Metamorphosis.

EXARATE Adj. (Latin, *exaratus* = plowed; *-atus* = adjectival suffix.) Pertaining to Integument the surface of which displays Sulci or sculpture patterns. See Sculpture Pattern.

EXARTICULATE Adj. (Latin, *ex* = without; *articulus* = a small joint; *-atus* = adjectival suffix.) Without distinct joints. Alt. Exarticulatus. See Articulation.

EXARTICULATE ANTENNA An Antenna with one segment. See Antenna.

EXASPERATE Adj. (Latin, *exasperare* = to irritate; *-atus* = adjectival suffix.) Rough with irregular elevations. Alt. Exasperatus.

EXCALCARATE Adj. (Latin, *ex-* = without; *calcar* = spur; *-atus* = adjectival suffix.) Without spurs. Alt. Excalcaratus.

EXCALIBER® See Lambda Cyhalothrin.

EXCAUDATE Adj. (Latin, *ex* = out of; *cauda* = tail; *-atus* = adjectival suffix.) See Ecaudate.

EXCAVATION Noun. (Latin, *excavatus* = hollow; *-tion* = result of an action. Pl., Excavations.) 1. A cavity or depression made by cutting, digging or scooping. 2. A pit, groove or depression in the Integument. 3. The removal of superposed material from the site of a previous inhabitation, culture or civilization. Alt. Excavatus.

EXCEL® See AIP.

EXCENTRIC Adj. (Latin, *ex-*; *centrus* = center; *-ic* = characterized by.) Not in the centre; revolving or arranged about a point that is not central.

EXCIND Adj. (Unknown origin.) Having an angular notch on an end. Alt. Excindate.

EXCISED Adj. (Assumed Latin, *accensare* = to tax.) Pertaining to structures with a deep cut or notch, usually along the margin.

EXCISION Noun. (Latin, *excisio* = cutting. Pl., Excisions.) 1. The act or process of excising. 2. A deep cut, notch or incision in the surface of a body. 3. A surgical removal of structure or body part. Cf. Incision.

EXCITATION Noun. (Latin, *excitare* = to arouse; English, *-tion* = result of an action. Pl., Excitations.) The act of producing or increasing stimulation.

EXCITE-R® See Pyrethrin.

EXCORIATE Trans. Verb. (Latin, *excoriare* = to skin.) 1. To strip, wear away or abrade the skin, hide or covering of an organism. 2. To remove the Cuticle of an insect or the bark of a tree.

EXCREMENTACEOUS Adj. (Latin, *excrementum* = to grow out; Latin, *aceus* = of or pertaining to.) Made up of or resembling excrement or faeces. Alt. Excrementitious; Excrementous. See Habitat. Rel. Ecology.

EXCRESCENCE Noun. (Latin, *excrescentia* from *excrescere* = an outgrowth. Pl., Excrescenses.) An outgrowth or elevation, sometimes considered abnormal.

EXCRETA Plural Noun. (Latin, *excretum* = separated.) The waste products of digestion that are eliminated from the body. Syn. Excrement. See Faeces. Cf. Dejecta.

EXCREMENT See Excreta.

EXCRETION Noun. (Latin, *ex* = out; *cernere* = to sift; English, *-tion* = result of an action. Pl., Excretions.) 1. The act of eliminating waste product from the body. 2. Any material or substance produced by any secretory glands or structures and which is voided or otherwise sent out from them. 3. The elimination of metabolic waste products from cells. Rel. Frass; Honeydew; Meconium.

EXCRETORY ORGANS Structures that collect, process and transport waste products from the body. See Alimentary System. Cf. Cryptonephridium; Malpighian Tubules; Ureter. Rel. Colon; Hindgut; Ileum; Rectum.

EXCURRENT OSTIUM (Pl., Ostia.) Paired openings in lateral wall of Dorsal Vessel that permit blood to escape from Dorsal Vessel into Pericardial Sinus. EO variable in size, shape and number. EO do not occur in all insects: Thysanura with midventral EO on abdominal segments 1–5; Embioptera with midventral Ostia on Mesothorax, Metathorax and first abdominal segment; Orthoptera with EO laterally developed as 'holes' in heart. EO typically lack valves; Papillae may surround EO and expand when Heart contracts so Haemolymph passes outward; papillae contract when Heart relaxes so Haemolymph is prevented from flowing into Heart. See Dorsal Vessel; Ostium. Cf. Incurrent Ostium. Rel. Circulation.

EXCURRENT Adj. (Latin, *ex* = out; *currere* = to run.) Attenuate; narrowly prolonged.

EXCURVATE Adj. (Latin, *ex* = out; *curvare* = to curve; *-atus* = adjectival suffix.) Curved outward.

EXEMPLI GRATIA A Latin phrase used to mean 'by way of example.' Abbreviated and italicized: *e.g.* Cf. *id est.*

EXFLAGELLATION Noun. (Latin, *ex* = out; *flagellum* = a small whip; English, *-tion* = result of an action. Pl., Exflagellations.) The protrusion of thread-like daughter cells from the Microgametocyte of *Plasmodium*.

EXHIBIT® See Entomogenous Nematodes.

EXITES Plural Noun. (Greek, *exo* = outside.) An outer lobe of any limb segment (Snodgrass). Cf.

Endites.

EXNER, SIGMUND (1846–1926) (Anon. 1926, Science 63: 227.)

EXOCHORION Noun. (Greek, *exo* = outside; *chorion* = skin. Pl., Exochorions.) The layer of insect eggshell enveloping the Endochorion. Exochorion is occasionally absent (*e.g.* silkmoth). Exochorion is composed of polysaccharides and mucopolysaccharides that are apparently secreted through Follicle Cell microvilli. Exochorion may be pierced by Aeropyles, many of which correspond to Endochorion pillars. The pillars may protrude beyond the Exochorion surface. See Eggshell. Cf. Endochorion; Epichorion; Vitelline Membrane; Wax Layer.

EXOCONE EYE A type of compound eye in which the crystalline Cone is replaced by one of extracellular cuticular origin; Exocone appears as a deep ingrowth from the inner aspect of the corneal Facet (Imms). The Exoconic Cone occurs in malacoderms and polyphagous Coleoptera. The Exocone is an extension of the Lens and is composed of Chitin and protein arranged in stacks of lamellae. Electron micrographs of cross-sections show that the Lens spirals. Longitudinal sections show a laminate composition. Nuclei of cells project to basement membrane as fine filaments called Cone Cell Threads. See Compound Eye. Cf. Aconic Eye; Euconic Eye.

EXOCORIUM Noun. (Greek, *exo* = without; Latin, *corium* = leather; *-ium* = diminutive > Greek, *-idion* = diminutive. Pl., Exocoria.) Heteroptera: A narrow marginal part of the Hemelytra; the outer margin of the Corium. See Hemelytra.

EXOCRINE GLAND Glands that provide secretions used outside the body. Ectodermal Exocrine Glands are grouped into two categories: Simple glands that are composed of one cell or aggregations of uniform cells, and Compound glands that contain several types of cells, including secretory cells, reservoir cells, duct cells and an aperture. See Allelochemical; Pheromone. Cf. Endocrine Gland.

EXOCUTICLE Noun. (Greek, *exo* = without; Latin, diminutive of *cutis* = skin. Pl., Exocuticles.) 1. The outer chitinous layer of the Cuticula that contains sclerotic deposits of the Cuticula when the latter are present (Snodgrass). 2. The middle, structureless amber-coloured layer of the Cuticle, composed of Cuticulin; a protein and Chitin compound (Wardle). 3. Exocuticle lies immediately above the Endocuticle and more-or-less homogeneous compared with layered Endocuticle. Exocuticle is thinner and stronger than Endocuticle, and laid before Ecdysis. Physically, Exocuticle is hard, contains Chitin and the tanned protein sclerotin. Ultrastructurally, Exocuticle contains lamellae and Chitin microfibrils. The microfibrils are parallel and perpendicular to the plane of section, and embedded in a protein matrix. Exocuticle stores potential energy and provides rigidity to the Integument. Alt. Exocuticula. See Cuticle; Integument. Cf. Endocuticle.

EXODERMA Noun. (Greek, *exo* = without; *derma* = skin. Pl., Exodermata.) The outer skin. Alt. Exoderma.

EXOERYTHROCYTIC SCHIZOGONY Medical Entomology: The period of Sporozoite development in the tissue of a vertebrate host (James & Tate 1937, Nature 139: 546.) Subsequently divided into two types: Primary and Secondary (Davey 1946, Trans. R. Soc. Trop. Med. Hyg. 40: 171–182.) See Malaria; Schizogony. Cf. Cryptozoic Schizogony; Metacryptic Schizogony.

EXOLOMA Noun. (Greek, *exo-* = outside; *loma* = hem, fringe. Pl., Exolomata.) The apical margin of the wings.

EXOPHAGOUS Adj. (Greek, *exo* = outside; *phagein* = to devour; Latin, *-osus* = with the property of.) 1. Pertaining to parasitic animals that feed externally on their hosts. 2. Medical Entomology: Pertaining to a disease vector (female mosquito) that feeds on a host outside dwellings. See Exophilic. Cf. Endophagous.

EXOPHILIC Adj. (Greek, *exo* = outside; *philein* = to love; *-ic* = characterized by.) Medical Entomology: Pertaining to a disease vector (female mosquito) that does not enter dwellings to rest. See Exophagous. Cf. Endophilic.

EXOPHYTIC Adj. (Greek, *exo* = outside; *phyton* = plant.) Relating to the outside of plant tissue. Cf. Endophytic; Saprophytic.

EXOPHYTIC OVIPOSITION A form of oviposition in Anisoptera in which the rounded eggs are either dropped freely into the water or attached superficially to water plants (Imms).

EXOPODITE Noun. (Greek, *exo* = outside; *pous* = foot; *-ites* = constituent. Pl., Exopodites.) 1. The segment on the outer aspect of the Basipodite in Crustacea. 2. The third segment of a Maxillary Palpus.

EXOPORIAN Adj. (Greek, *exo* = outside; *poros* = hole.) Pertaining to female Lepidoptera that display separate apertures for copulation and oviposition, and an external groove along which sperm pass from the Copulatory Bursa to the Vagina.

EXOPTERARIA See Exopterygota.

EXOPTERYGOTA Noun. (Greek, *exo* = outside; *pterygion* = little wing.) The Heterometabola (Hemimetabola). A division of Pterygota including insects that pass through simple Metamorphosis in which wing pads develop externally on the nymph's body. Some exopterygotes are wingless and their ancestors lost wings during transitional phases of evolution. Ephemeroptera have two winged instars called the Subimago and Imago. A pupal stage has been reported in only a few forms (*e.g.* some Thysanoptera) See Hemimetabola; Metamorphosis; Paurometabola. Cf. Endopterygota.

EXOPTERYGOTE Adj. (Greek, *exo* = outside; *pterygion* = little wing.) Pertaining to insects whose wings develop externally in all stages; insect with incomplete Metamorphosis. Cf. Endopterygote.

EXOPTERYGOTOUS Adj. (Greek, *exo* = outside; *pterygion* = little wing; Latin, *-osus* = with the property of.) Descriptive of or pertaining to the Exopterygota. The Heterometabola with wings that develop externally. Cf. Endopterydotous.

EXOSKELETAL PLATES Sclerties that form the external skeleton of insects. See Sclerites.

EXOSKELETON Noun. (Greek, *exo* = outside; *skeletos* = hard. Pl., Exoskeletons.) The body wall of an arthropod. Muscles attach to the inner surface. See Integument. Cf. Endoskeleton.

EXOTIC Adj. (Greek, *exotikos* = foreign; *-ic* = characterized by.) 1. Pertaining to organisms not native of a particular place or geographical area. Noun. (Greek, *exotikos* = foreign. Pl., Exotics.) 1. An introduced Species. 2. A Species taken in an area beyond the natural, known limits of that Species. Cf. Endemic.

EXOTIC NATURAL ENEMY In classical biological control, the regulation of a pest population by natural enemies imported for this purpose. See Biological Control. Cf. Augmentative Biological Control.

EXOTOKY Noun. (Greek, *exo-*; *tokos* = birth. Pl., Exotokies.) A form of reproduction in which eggs are developed outside the body of the mother and without care by the mother. Cf. Endotoky.

EXPALPATE Adj. (Latin, *ex* = without; *palpare* = to feel; *-atus* = adjectival suffix.) Without Palpi. Syn. Epalpate.

EXPANDED Applied to adult winged insects prepared for storage in a museum cabinet with the wings spread.

EXPANSE Noun. (Latin, *ex-* = without; *pandere* = to spread out. Pl., Expanses.) The distance between the apices or other widest point of the wings when fully spread.

EXPANSIO ALARUM Latin phrase meaning 'wing stretch' or 'wing spread.' See Expanse.

EXPAR® See Permethrin.

EXPARARTIS Noun. (Etymology obscure.) The arm of the Parartis, articulating on the external surface of the Paracoila (MacGillivray).

EXPIRATORY Adj. (Latin, *ex-* = out; *spirare* = to breathe.) Relating to the act of expiration. A phase of respiration when the Abdomen is contracted and air contained within abdominal Tracheae is expelled. Cf. Inspiratory.

EXPLANATE Adj. (Latin, *ex* = out; *planare* = to make plain; *-atus* = adjectival suffix.) Spread and flattened. A term sometimes applied to the margin of structure, such as an explanate Pronotum of adult lampyrid beetles.

EXPLANT Noun. (Latin, *ex* = cut; + plant; Pl., Explants.) Tissue Culture: A Tissue removed from a plant or animal and transferred to a culture medium.

EXPLICATE Adj. (Latin, *explicare* = to unfold; *-atus* = adjectival suffix.) Unfolded, open; without folds or Plica.

EXPLORATORY TRAIL Social Insects: See Recruitment Trail.

EXPONENTIAL PHASE 1. Tissue Culture: The period or interval of time after the Lag Phase and before the Stationary Phase. 2. Cells: During the Exponential Phase, cell populations grow exponentially. See Growth Curve. Cf. Lag Phase; Stationary Phase.

EXPOSED Adj. (Latin, *ex-*; *poser* = to place.) Visible. Cf. Concealed.

EXQUEMELEN, ALEXANDER OLIVER (fl 1590) (MacGillavry 1947, Ent. Ber., Amst. 12: 175–178.)

EXSCULPTATE Adj. (Latin, *exculptum* = to carve; *-atus* = adjectival suffix.) Surface with superficial, irregular, more-or-less longitudinal depressions, as if carved or etched. Alt. Exsculptus. See Sculpture.

EXSCUTELLATE Adj. (Latin, *ex* = out; *scutum* = shield; *-atus* = adjectival suffix.) Descriptive of insects that lack a Scutellum. Alt. Exscutellatus. See Scutellum.

EXSERTED Adj. (Latin, *exserere* = to stretch out.) Protruded; structure projecting beyond the body or beyond a given point. Alt. Exsertum.

EXSERTILE Adj. Latin, *exserere* = to stretch out.) Possible to exsert or extrude.

EXSERTION Noun. (Latin, *exserere* = to stretch out; English, *-tion* = result of an action. Pl., Exsertions.) A protrusion or an extension of a line or other ornamentation beyond its ordinary course.

EXSICCATAE Plural Noun. (Latin, *exsiccare* = dried.) A collection of dried herbarium specimens.

EXTENDED Adj. (Latin, *ex-* = out; *tendere* = to stretch.) Spread out; not lying one upon the other.

EXTENSACUTA Noun. (Etymology obscure.) A small sclerite in the Mandacoria connecting the head and the Mandible (MacGillivray).

EXTENSE Adj. (Latin, *ex-* = out; *tendere* = to stretch.) Extended, expanded. Alt. Extensus.

EXTENSILE Adj. (Latin, *extendere* = stretch.) Descriptive of tissue or structure that is capable of being stretched or drawn out. Alt. Extensible. Ant. Contractile.

EXTENSION PLATE A structure at the base of the Pulvillus that functions to extend it.

EXTENSOR Noun. (Latin, *extendere* = stretch; *or* = one who does something. Pl., Extensors.) General: A muscle that serves to straighten an appendage or part. Insects: A muscle that moves an appendage away from the body. A term used in connection with many muscles. Cf. Flexor; Retractor.

EXTENSOR MUSCLE Any muscle that extends or straightens an appendage or structure. Cf. Flexor Muscle.

EXTENSOR ROW Diptera: A row of bristles on the upper surface of the Femur (Comstock). Cf. Flexor Row.

EXTENSOR TENDON The Tendon to which the extensor muscles are attached (MacGillivray).

EXTENUATE Verb. (Latin, *extenuare* = to make thin.) To make or to become weak, thin or slender.

EXTERIOR EDGE The edge of the insect wing extending from the base to the apex.

EXTERIOR MARGIN The outer margin, sometimes used for Costal Margin. See Margin.

EXTERIOR PALPI The Maxillary Palpi.

EXTERIOR REGION The costal region of the wing.

EXTERNAL AREA Hymenoptera: The upper of the three cells or areas of the Metanotum, between the median and lateral longitudinal Carinae. The first lateral basal area.

EXTERNAL BEE-MITE *Acarapis dorsalis* Morgenthaler, *Acarapis externus* Morgenthaler [Acari: Tarsonemidae]. (Australia).

EXTERNAL CHIASMA Optic lobe of insects: A structure formed by the nerve fibres from the eye, that cross one another very complexly after leaving the ganglionic sclerite (Needham).

EXTERNAL GENITALIA Organs involved specifically with copulation and ovipositon. See Adaegus; Ovipositor. Cf. Internal Genitalia.

EXTERNAL MEDIAN AREA Hymenoptera: The median of the three cells (areas) between the median and lateral longitudinal Carinae; the second lateral area.

EXTERNAL MEDULLARY MASS The Medulla Externa; Epiopticon.

EXTERNAL PARAMERA All the genital appendages (except the internal) of the male.

EXTERNAL RESPIRATION The process of transferring the respiratory gases through the body wall, taking place in insects through thin areas of the Ectoderm, either at the body surface or in the walls of evaginations (gills) or invaginations (Tracheae) (Snodgrass).

EXTERNOMEDIAL VEIN 1. Hymenoptera: The Radius of Comstock *teste* Norton. 2. Orthoptera: The Media of Comstock. See Vein; Wing Venation.

EXTERNOMEDIAN NERVE The Humeral and Discoida Veins together.

EXTEROCEPTOR Noun. (Latin, *exterus* = on the outside; *receptor* = receiver; *or* = one who does something. Pl., Exteroceptors.) A sense organ located externally for the perception of external stimuli.

EXTINCTION Noun. (Latin, *extinctio*.) The act or process of destroying life or eliminating it. Syn. Annihilation; Destruction; Extermination. Rel. Defaunation

EXTRA ANTENNAE Antennae inserted on the outside of the eyes, or set very distant from the eyes.

EXTRA FIELD See Serosa; Dorsal Blastoderm.

EXTRA-CAVITY Preoral cavity. See Mouth Cavity.

EXTRA-EMBRYONIC Lying outside of the embryo.

EXTRA-EPICARDIAL Without an Epicardium; outside the Epicardium.

EXTRA-FLORAL NECTARIES Nectar-secreting plant structures other than flowers.

EXTRA-INTESTINAL Outside of the intestine.

EXTRA-OCULAR Remote from or beyond the eyes.

EXTRA-ORAL Outside of or beyond the mouth.

EXTRA-ORAL DIGESTION The chemical treatment of food before ingestion that enhances nutrient quality or accessibility. EOD is widespread and common among arachnids, chilopods and insects. Broadly divided into EOD type I (chemical liquefaction of prey within prey's body) and type 2 (mechanical disassembly of prey and chemical digestion outside a predator's mouth, but within the sphere of influence of mouthparts.) Type I is divided into refluxers (repeatedly pump enzymes in-and-out of prey) and nonrefluxers (enzymes pumped into prey, drawn into predator's gut.) Syn. Pre-oral Digestion.

EXTRAX® See Rotenone.

EXTREMITY Noun. (Latin, *extremitas* = limit; English, *-ity* = suffix forming abstract nouns. Pl., Extremities.) The point most remote from the base. An appendage or distal part of an appendage.

EXTRIN® See Fenvalerate.

EXTRINSIC Adj. (Latin, *extrinsecus* = on outside; *-ic* = characterized by.) Extraneous. External; not in or of a body. Pertaining to musculature that originates in another segment of the body. See Orientation. Cf. Intrinsic.

EXTRINSIC ARTICULATION A type of articulation in which the articulating surfaces are areas of contact on the outside of the skeletal parts (Snodgrass).

EXTRORSE Adj. (Latin, *extrorsus* = outwardly.) Pertaining to structure turned toward the outside.

EXTRUDE Verb. (Latin, *ex-* = out; *trudere* = thrust.) To turn or force out.

EXTRUSION Noun. (Latin, *extrusus*, past part. *extrudere* = to thrust out. Pl., Extrusions.) The act of pushing or forcing out.

EXUDATE Noun. (Latin, *exudare* = to sweat. Pl., Exudates.) Any substance given through the process of exudation.

EXUDATION Noun. (Latin, *exudare* = to sweat; English, *-tion* = result of an action. Pl., Exudations.) The process of discharge from a cell, an organ or organism through a membrane, incision, pore, gland or orifice. *e.g.* Mammal sweat; plant resins; insect defensive chemicals.

EXUDATORIA Plural Noun. Pseudomyrminae (ants) and some termites: Special Papillae or finger-like projects on larvae that exude a secretion highly attractive to their adult worker nurses.

EXUDE Transitive Verb. (Latin, *exudate* = to sweat out.) To secrete, emit, seep or discharge slowly through small openings.

EXUVIAE Plural Noun. (Latin, *exuere* = to strip off.) 1. General: The portion of the Integument of a larva, nymph or naiad that is shed from the body during the process of moulting. 2. Diaspididae: Nymphal Integument that is shed from the body and incorporated into the Scale Cover. Incorrectly given as 'exuvium' when a singular form of the noun is intended. Alt. Exuvia.

EXUVIAL GLAND Epidermal glands that secrete moulting fluid which facilitates Ecdysis (moulting). Alt. Moulting Glands.

EXUVIATE Verb. (Latin, *exuere* = to strip off.) To cast the skin; to moult.

EXUVIATION Noun. (Latin, *exuere* = to strip off; English, *-tion* = result of an action. Pl., Exuviations.) The act or physical process of shedding the cast-off 'skin' or Exuviae. See Moult. Rel. Integument.

EYE Noun. (Anglo Saxon, *eage* = eye. Pl., Eyes.) An organ of sight or vision that is capable of interpreting or processing information in the visual band of the electromagnetic spectrum. An insect eye is composed of numerous Ommatidia arranged in well defined groups on each side of the head. The term is properly applied to compound eyes only, but is sometimes used to designate the so-called simple eye or Ocellus. See Cf. Simple Eye; Compound Eye. Rel. Facet; Ommatidium; Stemmata.

EYE FLIES See Chloropidae.

EYE GNATS *Hippelates* spp. [Diptera: Chloropidae]: A New World Genus of flies (adults ca 1.5–2.5 mm long with apical spur on hind Tibia). EGs frequently and persistently visit eyes, sores and wounds of mammalian hosts. EGs are considered a nuisance and implicated in the mechanical transmission of human diseases such as pinkeye and yaws. See Chloropidae. Cf. Siphunculina.

EYE OF COLOUR An ocellate spot.

EYE SPOT 1. A rudimentary or evanescent Ocellus. 2. A pigmentary spot on the Integument that superficially resembles an eye. A coloured eye-like spot on the wing or body of insects, typically ringed with one or more contracting colours. ES is considered part of a startle response of some insects that is used as an adaptation to evade predators.

EYE-CAP Some Lepidoptera: The greatly-enlarged basal segment of the Antenna (Comstock).

EYED CLICK BEETLE *Alaus oculatus* (Linnaeus) [Coleoptera: Elateridae]: A Species common in rotting logs and stumps of eastern USA. Larvae are predaceous upon wood-boring beetles. Syn. Big-eyed Click Beetle; Eyed Elater. See Elateridae.

EYED ELATER See Eyed Click Beetle.

EYE-LIKE GLANDS Coccidoidea: Labiae *sensu* MacGillivray.

EYE-SPOTTED BUD MOTH *Spilonota ocellana* (Denis & Schiffermüller) [Lepidoptera: Tortricidae]: A pest of apples in northwestern USA and Europe. Eggs are transparent, discoid and laid on underside of leaves. Larvae feed beneath silken web; overwintered larvae attack blossoms and the summer brood feeds on underside of webbed leaves. ESBMs overwinter as larvae within cocoons attached to twig. Host plants include apple, cherry, pear and plum. See Tortricidae.

EYRE PENINSULA FUNNELWEB SPIDER *Hadronyche eyrei* (Gray) [Araneidae: Hexathelidae]. (Australia).

EYSELL, ADOLF (1846–1934) (Anon. 1935, Arb. physiol. angew. Ent. Berl. 2: 99.)

FABACEAE Legumes. A large, cosmopolitan and economically important Family of plants. Representatives include beans, peas, lentils, soybeans, peanuts and alfalfa.

FABER, HORST (1918–1974) (Anon. 1974, Prakt. SchädlBekämpf. 26: 108.)

FABRE, JEAN-HENRI (1823–1915) French naturalist, student of insect biology and observational behaviourist. Fabre wrote *Souveniers Entomologiques*. (Gerin 1974, Ann. Soc. ent. Fr. 10: 667–673, bibliogr.; Delange 1981, Fabre, L'Homme qui aimait les insectes. Paris: Jean-Claude Lattes, 354 pp.)

FABRICIUS, HIERONYMUS Early modern embryologist at Padua. Teacher of William Harvey and author of *De Formatione Ovi et Pulli* and *De Formato Foetu*.

FABRICIUS, JOHNANN CHRISTIAN (7 January 1745, Tonder, Duchy of Slesvig – 3 March 1808, Kiel, Germany) Danish academic, natural historian, and systematist. Fabricius was the son of a physician and a student of Linnaeus (1762–1764). Fabricius became Professor first at Copenhagen (1770) and then Kiel (1775) until his death. Fabricius was author of nearly 10,000 scientific names for insects and developed a classification based on mouthpart structures. His *Philosophia Entomologica* (1778) is generally regarded as the first textbook of Entomology; other works by Fabricius include: *Systema Entomologiae*, *Systema Insectorum classes, ordines, genera, et species* (1775), *Genera Insectorum* (1776), *Species Insectorum* (2 vols, 1781), *Mantissa Insectorum, sistens eorum species nuper detectas* (2 vols), *Entomologia systematica emendata et aucta* (4 vols, 1792–1796), *Systema Eleutheratorum* (2 vols, 1801), *Systema Rhyngotorum* (1801), *Systema Piezatorum* (1804), *Systema Antliatorum* (1805), *Systema Glossatorum* (1807). (Burgess 1973, *Portraits of doctors & scientists in the Wellcome Institute of the history of medicine*. 458 pp., London.)

FABRICIUS, OTHO (1744–1822) Missionary in Greenland (1768–1774) and author of *Fauna Groenlina* (1780). (Henriksen 1922, Ent. Meddr. 15: 81–83.)

FABRICIUS, PHILIPPE-CONRAD (1714–1774) (Anon. 1889, Abeille (Les ent. et leurs écrits) 26: 273, bibliogr.)

FAC® See Prothoate.

FACE Noun. (Latin, *facies* = form, shape, face. Pl., Faces.) 1. The general appearance of a structure or organism as presented to a viewer. 2. The upper or outer surface of any body part or appendage. 3. A generalized term used to describe the anteromedial portion of the head that is bounded dorsally by insertion of Antennae, laterally by medial margins of compound eyes and ventrally by Frontoclypeal Suture. In some insects, the Face is coincident with some, most or all of the Frons. Among most Diptera, lateral margins of the Face are determined by Frontogenal Sutures that originate at Anterior Tentorial Pits and extend dorsad toward the base of Antennae. The sclerite between Frontogenal Sutures is called the Facial Plate in Diptera. Diptera with a Ptilinum possess Ptilinial Sutures that extend ventrad lateral of Frontogenal Sutures. The area between compound eye and Ptilinial Suture is called the Parafacial. 4. Ephemeroptera: A fusion of the Front and Vertex. 5. Hymenoptera: Generally, the area between Antennae and Clypeus; in bees the Face extends between the eyes to base of Antennae.

FACE FLIES See Muscidae.

FACE FLY *Musca autumnalis* DeGeer [Diptera: Muscidae]: A widespread urban pest in houses, stables and barns. FF is primarily a pest of animals transmitting eyeworm *(Thelazia rhodesii)* and pinkeye in cattle *(Mordax bovis)*. Adults resemble the house fly, but are larger bodied; female Abdomen black and male Abdomen orange; Puparium off-white. FF is endemic in Africa, Asia and Europe; it was introduced into North America during 1952, and now is widespread in Canada and most of USA. Females oviposit in fresh cow manure; egg stage requires about one day. Larvae feed 2–4 days; generation time ca 14 days. FFs overwinter as unmated adults. See Muscidae. Cf. Bush Fly; House Fly.

FACET Noun. (French, *facette* = small face. Pl., Facets.) 1. Any smooth, flat, small surface of a structure. 2. Lens-like divisions of the compound eye. See Ommatidium. Cf. Lens. 3. Any sharply defined aspect of a problem, process or subject.

FACETED EYE See Compound Eye.

FACIAL Adj. (Latin, *facies* = face; *-alis* = pertaining to.) 1. Descriptive of structure or process associated with the 'face.' 2. Pertaining to the anterior, vertical surface of the hypognathous head. See Face. Cf. Frontal; Genal; Occipital.

FACIAL ANGLE The angle formed by the junctions of the Face and Vertex. See Head.

FACIAL BRISTLES Diptera: A series of Setae on each side of the middle part of the face, above the Vibrissae and along the Facialia.

FACIAL CARINA (Pl., Carinae.) 1. Carinae of the Frontal Costa and the accessory (lateral) Carinae of the Face. 2. Orthoptera: Usually restricted to the accessory Carinae. 3. Diptera: The single median ridge in the Face separating the antennal grooves. See Carina.

FACIAL DEPRESSION 1. An antennal Fovea. 2. Hymenoptera: The scrobal impressions near the place of attachment for the Antennae. See Head. Cf. Scrobe; Fovea. Rel. Pit; Sulcus.

FACIAL ORBIT Hymenoptera: Part of the head adjacent to the inner margin of the compound eye.

FACIAL RIDGES Diptera: Elevated lateral borders of the antennal grooves (Smith). Syn. Vibrissal Ridges (Comstock).

FACIAL TUBERCLE Diptera: A median convexity below middle of Face. Facial Carinae.

FACIALIUM Noun. (Latin, *facies* = face; Latin, *-ium* = diminutive > Greek, *-idion*. Pl., Facialia.) Diptera: Part of the Face between the lower part of the Frontal Fissure and the Antennal Foveae. See Vibrissal Ridges.

FACIES Plural Noun. (Latin, *facies* = face.) The typical aspect or appearance of a structure, body form, Species, Genus or group of insects. See Habitus. Cf. Form. Rel. Subfacies.

FACIES FOSSILS A geological term used in stratigraphy to refer to a group of beds differing in lithologic character and/or fossil content from beds of the same age. See Fossil. Cf. Compression Fossil.

FACILITATION Noun. (Latin, *facilitas* = facility, from *facilis* = easy; English, *-tion* = result of an action. Pl., Facilitations.) The increased ease of performance of a behavioural action with the continued successive application of a necessary stimulus. Ant. Inhibition. Cf. Fatigue; Habituation.

FACIO-ORBITAL BRISTLES Diptera: Bristles on the facial region adjacent to the compound eye. Syn. Fronto-Orbital Setae.

FACKLER, H L (1894–1935) (Bentley 1936, J. Econ. Ent. 29: 224–225.)

FACTITIOUS HOST 1. Biological Control: A plant or animal host Species used in the mass culture of a natural enemy. Factitious hosts are relatively simple or inexpensive to culture in large numbers and typically are not true hosts of the natural enemy. 2. A Species of host that is attacked by parasites or parasitoids under laboratory conditions but not attacked in the field or under natural conditions.

FACULTATIVE Adj. (Latin, *facultas* = faculty, ability.) 1. Descriptive of conditions that permit alternative responses under different conditions. 2. Pertaining to organisms capable of living under conditions unusual for life or development. The term is often applied to parasites.

FACULTATIVE DIAPAUSE A type of diapause that can be induced or terminated by change in photoperiod, temperature or both. The condition may be characteristic of multivoltine Species. See Diapause. Cf. Obligatory Diapause.

FACULTATIVE HYPERPASITISM A type of hyperparasitism in which the immature hyperparasite can complete feeding and development as a primary parasite or use a primary parasite as a host. See Hyperparasitism. Cf. Obligatory Hyperparasitism.

FACULTATIVE MULTIPLE PARASITISM Hymenoptera: The random, occasional occurrence of more than one Species of parasite simultaneously developing on a host. See Parasitism. Cf. Obligatory Multiple Parasitism. Rel. Mutualism.

FACULTATIVE MYIASIS A type of Myiasis that includes parasitic fly larvae that are not dependent upon vertebrate hosts for completion of a phase of their development. Fly larvae involved in FM feed upon hosts as a consequence of opportunity. See Myiasis. Cf. Obligatory Myiasis; Pseudomyiasis.

FACULTATIVE PARASITE 1. An organism that exists as a free-living or parasitic individual. 2. Hymenoptera: A primary parasite that is capable of developing on plant tissues (such as galls) when insect hosts are not available. 3. Fungi: Species that live on a host plant only during part of the life cycle. See Parasite; Parasitism. Cf. Obligatory Parasite.

FACULTATIVE PREDATION An aspect of predation in which a predator feeds on prey that do not represent an essential part of a predator's diet. FP may be adaptive in that it can result in population growth. See Predation.

FACULTATIVELY Adv. (Latin, *facultatas* = capability.) In a facultative manner or way.

FAECAL Adj. (Latin, *faeces* = dregs; *-alis* = characterized by, pertaining to.) Descriptive of or pertaining to faeces (excrement).

FAECES Plural Noun. (Latin, *faeces* = dregs.) Excrement; the eliminated wastes of the digestive process. See Digestion. Cf. Frass; Meconium. Alt. Feces.

FAECULA Noun. (Latin, *faecula* = burnt tartar. Pl., Faeculae.) The excrement of insects.

FAECULENT Adj. (Latin, *faeculentia*.) Foul or covered with excrement.

FAESTER, K (1887–1966) (Tuxen 1966, Ent. Meddr. 34: 183–186, bibliogr.)

FAFAI® See Permethrin.

FAGGOT BAGWORM *Clania ignobilis* (Walker) [Lepidoptera: Psychidae]. (Australia).

FAGGOT CASE-MOTH *Clania ignobilis* (Walker) [Lepidoptera: Psychidae]. (Australia).

FAHRAEUS, OLAF LIMMANUEL (1796–1884) (Sandahl 1884, Ent. Tidskr. 5: 111–114, 209–210, bibliogr.)

FAHRENHOLZ, HEINRICH (1882–1945) (Viets 1951, Abh. naturw. Ver. Bremen 32: 457–461, bibliogr.)

FAHRENHOLTZ'S RULE An hypothesis that asserts that phylogenetic trends within a taxonomic group of parasites reflect phylogenetic trends within the taxonomic group that serve as their hosts. (Stammer 1957, Zool. Anz. 159: 255–267). Cf. Eichler's Rule; Emery's Rule; Manter's Rule; Szidat's Rule. Rel. Systematics.

FAHRINGER, JOSEF (1876–1950) (Strouhel 1955, Annln naturh. Mus. Wien. 60: 7.)

FAIRCHILD, ALEXANDER GRAHAM BELL (1907–1994) Research Entomologist (University of Florida) and grandson of Alexander Graham Bell.

FAIRMAIRE, LÉON (1820–1906) (Musgrave 1932, *A Bibliography of Australian Entomology 1775–1930*, 380 pp. (88–89), Sydney.)

FAIRY FLIES See Mymaridae.

FAIRY WASPS See Mymaridae.

FAIRY YELLOW *Eurema daira* [Lepidoptera: Pieridae].

FALCATE Adj. (Latin, *falcatus* = bent or curved, from *falx* = sickle; *-atus* = adjectival suffix.) Sickle-shaped; convexly curved. Descriptive of structure that is long, narrow, tapered and curved. *e.g.* Leaves of many plants; Mandibles of some parasitic Hymenoptera larvae (Braconidae) or adults (Eucharitidae). Alt. Falcatus. Syn. Falciform. Cf. Ensiform; Remiform; Subfalcate. Rel. Shape.

FALCES Plural Noun. (Latin, *falces* = sickles.) Lycaenid butterflies: Long, strongly curved, sharp, sclerites of the genitalia that probably are modifications of the Uncus (Klots).

FALCIFORM, Adj. (Latin, *falx* = sickle; *forma* = shape.) Sickle-shaped; descriptive of structure that is long and curved as a sickle. Syn. Falcate. See Sickle. Cf. Dolabriform; Ensiform. Rel. Dolabrate. Rel. Form 2; Shape 2; Outline Shape.

FALCONER, HUGH (1908–1965) (Anon. 1867, Proc. R. Soc. 15: xiv–xx.)

FALCONER, WILLIAM (1862–1943) (Cockayne 1945, Proc. R. ent. Soc. Lond. (C) 9: 47.)

FALCOZ, LOUIS (1870–1938) (Roman 1938, Bull. Soc. ent. Fr. 43: 197.)

FALDERRMANN, FRANZ (1799–1838) (Marseul 1884, Abeille 22: (Les ent. et leurs écrits.) 140.)

FALKENSTRÖM, GUSTAF ADOLF (1867–1942) (Leech 1944, Can. Ent. 76: 211.)

FALL ARMYWORM *Spodoptera frugiperda* (J. E. Smith) [Lepidoptera: Noctuidae]: A pest of middle-to-late season cotton in USA and maize in some areas such as Nicaragua; a sporadic pest of wheat. Eggs are ca 0.4 mm in diameter, pale pink, radially striped and becoming grey before eclosion. Eggs are deposited in masses on underside of leaves low on plant; an egg mass displays a waxy, translucent covering. Eclosion occurs within 3–4 days. Neonate larvae are creamy-white with black head and black Setae, and y-shaped suture on head. Mature larvae are ca 45 mm long, pale tan or green to nearly black with brown or black spots and two darker bands on either side of dorsum, and three white lines on first thoracic sclerite. Larval development requires ca 15–25 days. Pupation occurs in soil and requires 7–10 days; Pupae are dark brown, 14–17 mm long with two small spines forming a 'V' at apex of Abdomen. Adults are ca 25 mm long with wingspan of 35 mm; forewings are mottled and hindwings whitish with grey margin. Adult females live ca 12 days and deposit ca 1,000 eggs in clutches of 100. Syn. *Laphygma frugiperda* Smith. See Armyworms; Noctuidae. Cf. African Armyworm; Beet Armyworm; Southern Armyworm.

FALL CANKERWORM *Alsophila pometaria* (Harris) [Lepidoptera: Geometridae]: A sporadic defoliator of fruit and shade trees in North America. FC Overwinters in egg stage on trunk or branches of tree. Larvae with three pairs of prolegs; they feed ca 30 days on foliage and then move to soil and pupate. Adults appear during late autumn; females are wingless and males are winged. Females climb trees, mate with males and deposit masses of eggs in single layer. FC complete one generation per year. See Geometridae. Cf. Spring Cankerworm.

FALL WEBWORM *Hyphantria cunea* (Drury) [Lepidoptera: Arctiidae]: A widespred pest of ornamental, fruit and woodland trees in North America and China. FC overwinters as Pupae within silken cocoons on ground, under trash or under bark. Adults appear during spring, mate and oviposit eggs in masses on leaves; eggs are covered with layer of scales from female's body. Larvae enclose foliage on branches within loosely constructed, off-white web; webs become contaiminated with black faecal pellets of larvae and feeding is completed within 30–45 days; pupation follows. Adults emerge during late summer and females oviposit eggs of a second generation. See Arctiidae. Cf. Eastern Tent Caterpillar.

FALL, HENRY CLINTON (1862–1939) (Mallis 1971, *American Entomologists*. 549 pp. (266–269), New Brunswick.)

FALLEN, CHARLES FREDERICK Swedish academic (University of Lund) and author of *Diptera Suecica* (1814–1817), *Monographica Scandinaviae monographia tractati et iconibus illustrati* (1832), and other works.

FALLOPIAN TUBES See Oviduct.

FALLOU, JULES FERDINAND (1812–1895) (Meldola 1895, Proc. ent. Soc. Lond. 1895: lxxiii.)

FALLOW Adj. (Middle English, *falow* = arable land.) Land unseeded and usually clean cultivated during a growing season.

FALSADENTES Diptera: Prestomal teeth (MacGillivray).

FALSE BLISTER-BEETLES See Oedemeridae.

FALSE BLOSSOM A viral disease of cranberries transmitted by leafhoppers. Rel. Cicadellidae.

FALSE BUDWORM See Corn Earworm.

FALSE CELERY-LEAFTIER *Udea profundalis* (Packard) [Lepidoptera: Pyralidae].

FALSE CHINCH-BUG *Nysius raphanus* Howard [Hemiptera: Lygaeidae]: A widespread pest of agriculture in North America. FCB hostplants include alfalfa, grasses, grains, truck crops, deciduous fruits, citrus, grapes and berries. See Lygaeidae. Cf. Chinch Bug.

FALSE CLICK BEETLES See Eucnemidae.

FALSE CLOWN BEETLES See Sphaeritidae.

FALSE CODLING-MOTH *Cryptophlebia leucotreta* (Meyrick) [Lepidoptera: Tortricidae]: A pest of citrus fruits, cotton bolls and macadamia nuts in sub-Saharan Africa. See Tortricidae. Cf. Codling Moth.

FALSE DARKLING-BEETLES See Melandryidae.

FALSE FIREFLY BEETLES See Drilidae.

FALSE GERMAN-COCKROACH *Blattella lituricollis* (Walker) [Blattodea: Blattellidae].

FALSE LADYBIRD BEETLES See Tenebrionidae.

FALSE LEGS Spurious legs; Prolegs. Cf. Pseudopodia.

FALSE LONGHORN BEETLES See Cephaloidae

FALSE LONGHORNED BEETLES See

Cephaloidae.

FALSE MANTID See Mantispidae.

FALSE MELON-BEETLE *Atrachya menetriesii* (Feldermann) [Coleoptera: Chrysomelidae].

FALSE METALLIC WOODBORING BEETLES See Throscidae.

FALSE ORIENTAL FRUIT-FLY *Dacus opilae* Drew & Hardy [Diptera: Tephritidae]. (Australia).

FALSE OVIPOSITOR Trichoptera: The female's ninth abdominal segment when it is capable of retraction and concealment.

FALSE POTATO-BEETLE *Leptinotarsa juncta* (Germar) [Coleoptera: Chrysomelidae].

FALSE POWDER-POST BEETLES See Bostrychidae.

FALSE SCORPIONS See Pseudoscorpionida.

FALSE SKIN BEETLES See Biphyllidae.

FALSE SPIRACLES Nepidae: Pairs of sieve-like structures on the Connexivum of Sterna 3–5 near the spiracles. FS are not connected with the tracheal system and their function is not determined.

FALSE STABLE-FLY *Muscina stabulans* (Fallén) [Diptera: Muscidae]: A widespread, multivoltine pest on poultry farms. Adults are larger and more robust than House Fly, with body grey and Thorax with four dark longitudinal stripes and pale spot at apex; wings transparent but faintly embrowned. FSF eggs are deposited in fresh manure, human excrement or decaying vegetable matter; females may also oviposit on food in homes. Eclosion occurs within 12–18 hours. Larvae feed in decaying vegetable matter, manure and human excrement, and become carnivorous or cannibalistic late in development. Larval stage requires 15–25 days. Pupation occurs in drier part of manure. Larvae can cause Intestinal Myiasis in humans; adults serve as mechanical vectors of enteric diseases. Syn. Non-Biting Stable Fly. See Muscidae. Cf. Stable Fly; House Fly; Lesser House Fly.

FALSE STAINER *Aulacosternum nigrorubrum* Dallas [Hemiptera: Coreidae]. (Australia).

FALSE WIREWORM In Australia, the common name of several beetle Species, to include: *Celibe* sp., *Gonocephalum misellum* (Blackburn), *Gonocephalum walkeri* (Champion) and *Isopteron punctatissimus* (Pascoe) [Coleoptera: Tenebrionidae].

FALSE WIREWORMS Significant pests of native grasses, oats, beans, corn, cotton, alfalfa and other plants in North America. Adults are dark-bodied, elongate, <24 mm long and flightless. When disturbed, FWW elevate the posterior portion of body and insect appears to 'stand on its head.' Eggs are laid in soil. Larvae resemble wireworms; body cylindrical and elongate, with Antenna and legs apparent. Larvae feed in soil on roots and underground parts of plant stems. See Tenebrionidae. Cf. Wireworms.

FALSE-HEMLOCK LOOPER *Nepytia canosaria* (Walker) [Lepidoptera: Geometridae].

FALSIFORMICIDAE Rasnitsyn 1975. A small Family of aculeate Hymenoptera assigned to the Scolioidea.

FALX Noun. (Latin, *falx* = sickle. Pl., Falces.) 1. Siphonaptera: An internal thickening of the Integument that extends over the Vertex between antennal scrobes (Comstock). 2. Lepidoptera: A pair of slender, curved processes in the male genitalia that extend caudad and ventrad from the Tegumen on each side of the base of the Uncus. The Falx probably is homologous with the Gnathos *sensu* Klots.

FALZONI, ADOLFO (1875–1945) (Porte 1946, Memorie Soc. ent. ital. 25: 74.)

FAMFOS® See Phosphamidon.

FAMILY Noun. (Latin, *familia* = household. Pl., Families.) In zoological classification, a level in the taxonomic hierarchy below the Order and above the Genus. All zoological Family names end in the suffix '-idae'; all zoological Subfamily names end in the suffix '-inae'. The Family-level taxon must include a type-genus that holds a type Species. See Classification. Cf. Subfamily. Rel. Taxonomy.

FAMOPHOS® See Famphur.

FAMPHUR An organic-phosphate (thiophosphoric acid) compound {O,O-dimethyl O-p-(dimethylsulphamoyl) phenyl phosphorothioate} used as a systemic livestock insecticide against bot flies, lice and other pests. Compound is applied to livestock or placed in feed. Trade names include Cyflee®, Davip®, Famophos®, Fanfos® and Warbex®. See Organophosphorus Insecticide.

FAMULUS Noun. (Latin, *famulus* = attendant. Pl., Famuli.) 1. A medieval scribe. 2. Acarina: A small, hollow Seta on Tarsus I and II. Syn. Microtarsala. (Goff *et al.* 1982, J. Med. Ent. 19: 223).

FAN Noun. (Anglo Saxon, *fann* = fan, from Latin, *vannus* = winnow. Pl., Fans) 1. Anatomy: A broad, flattened object or surface in the outline form of a sector of a circle. 2. A device used to propel an air current (honeybee wings used to fan the hive to keep it cool.) Cf. Flabellum.

FANFOS® See Famphur.

FANG Noun. (Ango Saxon, *fang* = grip. Pl., Fangs.) 1. A long, conical, apically pointed tooth. 2. A distal portion of the Chelicera in arachnids.

FANNIIDAE Latrine Flies. See Muscidae.

FANTHAM, HAROLD BENJAMIN (1876–1937) (Roubaud 1938, Bull. Soc. Path. exot. 31: 178.)

FARA, LODISLAV (1902–1962) (Heyrovsky 1962, Cas. csl. Spol. ent. 59: 293.)

FARCTATE Adj. (Latin, *farctus* = stuffed, pp of *farcire* = to stuff.) Pertaining to structure that is fully filled. Ant. Hollow.

FARIMOS® See *Bacillus sphaericus.*

FARINA Noun. (Latin, *farina* = meal, flour.) 1. Any substance with the texture or consistency of flour. 2. The fine powder-like material on the body of some insects.

FARINACEOUS Adj. (Latin, *farina* = meal, flour; *aceus* = of or pertaining to; *-osus* = with the property of.) 1. Containing, consisting of, or made from flour or meal. 2. Mealy. A term applied to powdery looking wings and surfaces. Alt. Farinaceus.

FARINOSE Adj. (Latin, *farinosus* = mealy, from Latin, *farina* = meal, flour; *-osus* = full of.) 1. Dotted with many single, flour-like spots. 2. Mealy. Pulverulent; Pollinose. Adj. Farinosus.

FARMATOX® See Fenpropathrin.

FARN, ALBERT BRIDGES (1841–1921) (Jenkinson 1922, Entomol. mon. Mag. 58: 20–22.)

FARNESENE A synthetic pheromone {E-beta-farnesene} used to mimic alarm pheromone signals in aphids. Compound causes feeding disruption, increased movement and more effective contact with insecticides. Farnesene is not registered for use in USA. Trade name is Panic 24 EC®.

FARNESOL Noun. An alcohol ($C_{15}H_{25}OH$) with juvenile hormone activity in insects.

FARR, WILLIAM BERRY (–1890) (Anon. 1890, Entomologist 23: 208.)

FARREN, WILLIAM (1836–1887) (Carrington 1888, Entomologist 21: 71–72.)

FARWELL, CHRISTOPHER GUY (1886–1971) (R. W. W. 1972, Entomol. mon. Mag. 84: 55–56.)

FASCIA Noun. (Latin, *fascia* = band, sash. Pl., Fasciae.) 1. Anatomy: A thin layer of connective tissue that covers, supports, binds or connects muscles or body organs. 2. Taxonomy: A transverse band or broad line. In older literature, a term commonly applied when the fascia crosses both wings or wing covers.

FASCIATE Adj. (Latin, *fasciatus* > *fasciare* = to wrap with bands; *-atus* = adjectival suffix.) Pertaining to structure with a broad transverse stripe or band. Alt. Fasciated.

FASCICULA Noun. (Latin, *fascuculus* = a bundle. Pl., Fasciculae.) A bundle of muscle fibres.

FASCICULATE Adj. (Latin, *fascuculus* = a bundle; *-atus* = adjectival suffix.) Bundled; clustered as in a bundle. Tufted; superficially covered with bundles of long Setae. Alt. Fasciculatus.

FASCICULATE ANTENNA An Antenna in which each segment has a distinct pencil or fasicule of long Setae. See Antenna.

FASSL, ANTON HEINRICH (HERMANN) (1876–1922) (Rohlfien 1975, Beitr. Ent. 25: 288 only.)

FASSNIDGE, WILLIAM (1888–1949) (Wigglesworth 1949, Proc. R. ent. Soc. Lond. (C)14: 64.)

FASTAC® See Alphacypermethrin.

FASTIGIAL FOVEOLAE Orthoptera: A pair of pits on the Fastigium of the Vertex near the anterior margin or ventral of it.

FASTIGIAL FURROW Orthoptera: A narrow, deep Sulcus along the apex of the vertexal Fastigium.

FASTIGIAL Adj. (Latin, *fastigium* = a slope, roof; *-alis* = pertaining to.) Descriptive of or pertaining to a Fastigium.

FASTIGIATE Adj. (Latin, *fastigatus* = sloping to a point; *-atus* = adjectival suffix.) Tapering to an apex and thereby forming a conical structure. See Shape.

FASTIGIUM Noun. (Latin, *fastigium* = a slope, roof; Latin, *-ium* = diminutive. Pl., Fastigia.) 1. Orthoptera: The extreme point or front of the Vertex. 2. Some insects: A prominent angle between the Vertex and Face (Imms).

FAT BODY Structurally pleiomorphic cells that occur in immature and adult insects. FB gross anatomy and distribution are variable among insect groups. FB is categorized into two types based upon location and function (Perivisceral Fat Body and Peripheral Fat Body). Most FB cells are called Trophocytes (Adipocytes). Anatomically, FB typically forms thin lobes of yellow or white tissue suspended in Haemolymph; FB is tracheated, surrounds internal organs and is distributed through the body cavity. FB near the Epidermis is called Peripheral Fat Body or Subcuticular Fat Body; FB surrounding the gut is called Perivisceral (Visceral) Fat Body. Functionally, FB is the principal organ involved in intermediary metabolism and storage organ for Haemolymph proteins, lipids and carbohydrates. In Diptera, larval FB produces larval serum proteins; adult FB produces yolk polypeptides. FB is typically associated with specific types of cells, depending upon Species under consideration: Adipocytes (Trophocytes), Mycetocytes (in Collembola, Thysanura, Blattaria), Oenocytes and Urocytes. Syn. Corpus Adiposum. See Peripheral Fat Body; Perivisceral Fat Body. Cf. Haemocyte; Mycetocyte; Oenocyte; Trophocyte; Urocytes. Rel. Haemocyte; Nephrocyte.

FAT CELL A component cell of the Fat Body. See Mycetocyte; Trophocyte; Urocyte.

FATIGUE Noun. (French from Latin, *fatigare* = to weary. Pl., Fatigues.) Behaviour: Exhaustion induced from stimuli that follow each other too rapidly to permit complete recovery. Cf. Facilitation; Habituation.

FATIO, VICTOR (1838–1906) (Anon. 1906, Revue suisse Zool. 14.)

FATISCENT Adj. (Latin, *fatiscens* > *fastiscere* = to open in chinks.) An arcane term used to describe structure containing cracks, crevices or openings. Cf. Fissate.

FATTIG, PERRY WILBUR (1881–1953) (Remington 1954, Lepid. News 8: 30.)

FAUNA Noun. (Latin, *fauna* = Roman goddess, sister of Faunus. Pl., Faunas; Faunae.) The collective animal life of any region, place or area.

FAUNAL TURNOVER Episodes in the fossil record that depict the synchronous appearance and disappearance of Species from a community.

FAURE, JACABUS CHRISTIAN (1891–1973) (Lees 1974, Proc. R. ent. Soc. Lond. (C) 38: 57.)

FAURIELLIDAE Plural Noun. A numerically small Family of Thysanoptera assigned to Terebrantia.

FAUSSEK, VICTOR (1861–1910) (Kusnezov 1910, Revue Russe Ent. 10: 251– 261, bibliogr.)

FAUST, JOHANNES (1812–1903) (Musgrave 1932, *A Bibliography of Australian Entomology 1775–*

1930. 380 pp., (89), Sydney.)

FAUVEL, CHARLES ADOLPHE ALBERT (1840–1921) (Buysson 1925, Bull. Soc. Normandie ent. 1: 1–22, bibliogr.)

FAVARCQ, L J (–1900) (Anon. 1900, Exchange 16: 9–10.)

FAVOSE Adj. (Latin, *favus* = honeycomb; *-osus* = full of.) Descriptive of surface sculpture (usually the insect's Integument) that displays relatively large, deep pits such that the surface resembles the cells of a honeycomb. Alt. Favosus; Favous. See Sculpture Pattern. Cf. Alveolate; Baculate; Clavate; Echinate; Gemmate; Psilate; Punctate; Reticulate; Rugulate; Scabrate; Shagreened; Smooth; Striate; Verrucate.

FAVUS Noun. (Latin, *favus* = honeycomb. Pl., Favi.) An hexagonal cell forming part of a matrix that resembles a honeycomb.

FAY RICHARD WILLIAM (1912–1972) (Schoof 1973, J. Econ. Ent. 66: 828– 829.)

FAZ, ALFREDO (1863–1931) (Rohlfien 1975, Beitr. Ent. 25: 267.)

FEA, LEONARDO (1852–1903) (Gestro 1921, Annali Mus. civ. Stor. nat. Genova 50: 30–31, 35.)

FEATHER CHEWING-LICE See Philopteridae.

FEATHER MITE *Megninia cubitalis* (Mégnin) [Acari: Analgidae].

FEATHER-WINGED BEETLES See Ptiliidae.

FEATHER-WINGED CRICKETS See Pteroplistidae.

FEATHER-WINGED MOTHS See Alucitidae.

FEBROGENESIS Noun. (Latin, *febris* = fever; *gegnere* = orgin; *-sis* = a condition or state. Pl., Febrogeneses.) A behavioural fever manifest in an increase in thermal preference by an animal.

FECUNDATE Verb. (Latin, *fecundare* = to make fruitful.) To pollinate a plant, fertilize an egg; to impregnate.

FECUNDATION Noun. (Latin, *fecundus* = fruitful; English, *-tion* = result of an action. Pl., Fecundations.) Fertilization, as an egg by a Spermatozoon.

FECUNDITY Noun. (Latin, *fecunditas* = fruitfulness. Pl., Fecundities.) 1. The number of eggs produced by a female during her lifetime. 2. The innate capacity for increase in individuals of a population or Species. Cf. Fertility.

FEDCHENKO, ALEKSEII PAVOLOVICH (1844–1873) (Regel 1874, Garten Flora 1874: 3–7.)

FEDDAN A unit of land measure in Egypt. One feddan equals 1.038 acres.

FEDDERSEN, FRIEDRICH CHRISTIAN (1835–1898) (Anon. 1899, Forst u. Jagdwes. 31: 233–234.)

FEDDERSEN, TUGE (1898–1959) (Anon. 1959, Ent. Meddr. 29: 167–169.)

FEDERLEY, HARRY (1879–1951) (Soumalainen. Scientia genet. 4: 201–204.)

FEDOROV, STEPAN MITROFANOVICH (1888–1961) (Bei-Bienko 1962, Ent. Obozr. 41: 241–244, bibliogr. (Translation: Ent. Rev. 41: 147–149.).)

FEDRIZZI, GIACENTO (1858–1879) (Moltoni 1940, Riv. ital. orit. (2) 10: 63–72, bibliogr.)

FEDSCHENKIIDAE See Sapygidae.

FEDTSCHENKIINAE A monogeneric Subfamily of small, fossorial wasps found in western USA, Iran, Lebanon, Turkmenistan and Uzbeckistan. North American Species is a parasite of the ground-nesting eumenid wasp, *Pterochilus trichogaster* Bohart. See Aculeata; Sapygidae.

FEEBLE Adj. (Latin, *febilis* = lamentable, to weep over.) 1. Lacking in strength, endurance or vigour. 2. Deficient in qualities that indicate vigour or power. 3. Pertaining to structure that is slight or scarcely noticeable. Alt. Feebly. Ant. Vigorous.

FEEDING STRATEGY The complex of anatomical, behavioural and physiological adaptations that are responsible for the acquisition of food and processing of nutrients by an organism. At the organism level, feeding strategies are broadly categorized as saprophagy, phytophagy, predation and parasitism. Within each of these categories are subcategories that are modifications of the basic feeding patterns. Special terms are used to describe feeding preferences with respect to the range or kinds of Species used as food: Monophagy, Oligophagy, Polyphagy and Stenophagy. See Haematophage; Phytophagy; Predation; Parasitism; Saprophagy. Cf. Monophagy; Oligophagy; Polyphagy; Stenophagy. Rel. Filter Feeding; Leaf Feeding; Vortex Feeding.

FEEDING TUBE A cylindrical device created by adult female parasitic Hymenoptera with the aid of the Ovipositor. FT is used to feed upon a host's Haemolymph in concealed habitats or places in which the body of the host cannot be contacted by mouthparts of the adult parasite. The parasite's Ovipositor penetrates the host's Integument and induces Haemolymph to exude from the puncture. Haemolymph escapes from the host's body, surrounds the parasite's Ovipositor and dries. Dried Haemolymph thereby forms a tube from which fluid can be imbibed (by capillary action) to the parasite's mouthparts. (Fulton 1933, Ann. ent. Soc. Am. 26: 536–553). Rel. Host Feeding.

FEELER Noun. (Middle English, *felen* = to feel. Pl., Feelers.) 1. A tactile process of an organism. 2. A sensory appendage, typically an Antenna. Arcane: A term applied to an Antenna or Palpus (mouth-feeler). See Antenna.

FEET Plural Noun. (Greek, *podus* = foot. Sing. Foot.) The apical-most part of paired legs or appendages of locomotion. Feet make contact with the substrate and are necessary for walking, running and jumping. Feet are not used for crawling. Insects: One pair of feet is attached to each thoracic segment. The Abdomen of larvae often have paired appendages used for locomotion. See Appendage. Cf. Proleg.

FEIGE, CURT (1880–1962) (Anon. 1962, Mitt. dt.

ent. Ges. 21: 66.)

FEIJO, JOAO DA SILVA (1760–1824) (Papavero 1971, *Essays on the History of Neotropical Dipterology*. 1: 48–49. São Paulo.)

FEIJOO, BENITO GERONIMO (1676–1764) (Dusmet y Alonso 1919, Boln. Soc. ent. Esp. 2: 78.)

FEISTHAMEL, JOACHIM FRANÇOIS PHILIBERTO DE (1791–1851) (Ratzeburg 1874, Forstwissenschaftliches Schriftsteller-Lexicon 1: 175–178.)

FELDER, CAJETOAN VON (1814–1894) (Musgrave 1932, *A Bibliography of Australian Entomology 1775–1930*. 380 pp. (90), Sydney.)

FELDER, RUDOLPH (1842–1871) (Schiner 1872, Verh. zool.-bot. Ges. Wien 22: 41–50, bibliogr.)

FELDER'S LINEBLUE *Prosotas felderi* (Murray) [Lepidoptera: Lycaenidae]. (Australia).

FELDMAN, HENRY (1814–1881) (Osborn 1937, *Fragments of Entomological History*. 394 pp. (231), Columbus, Ohio.)

FELDTMANN, EDUARD (1875–1960) (Weidner 1967, Abh. Verh. naturw. Ver. Hamburg Suppl. 9: 191.)

FELLER, GUSTAV (–1960) (Anon. 1960, Mitt. dt. ent. Ges. 19: 45.)

FELT CHAMBER See Stigmatic Chamber.

FELT, EPHRAIM PORTER (1868–1943) (Mallis 1971, *American Entomologists*. 549 pp. (399–402), New Brunswick.)

FELT LINE Hymenoptera: A dense, elongate cluster of Setae within a cuticular depression along the lateral margin of Metasomal Tergum 3 in mutillid wasps. The FL is associated with a glandular secretion.

FELT SCALES See Eriococcidae.

FELTED GRASS-COCCID *Antonina graminis* (Maskell) [Hemiptera: Pseudococcidae]. (Australia).

FELTED PINE-COCCID *Eriococcus araucariae* Maskell [Hemiptera: Eriococcidae]. (Australia).

FEMALE PRONUCLEUS The Nucleus within an egg after maturation of the Oocyte.

FEMALE Noun. (Latin, *femina* = woman. Pl., Females.) The organism in which eggs are developed, or the individual that possesses reproductive organs adapted for the production of female gametes. Cf. Male.

FEMINA (Latin, *femina* = woman.) The female; an individual of the female sex.

FEMORAL Adj. (Latin, *femur* = thigh; *-alis* = characterized by, pertaining to.) Pertaining to structure or action associated with the Femur.

FEMORALA Noun. (Latin, *femur* = thigh. Pl., Femoralae.) Acarina: A specialized Seta (Solenidium) on the Femur (Goff *et al.* 1982, J. Med. Ent. 19: 225).

FEMORATE Adj. (Latin, *femur* = thigh; *-atus* = adjectival suffix.) Pertaining to abnormal or unusually developed Femora. Alt. Femoratus.

FEMORO-ALARY ORGANS Orthoptera: The sound-producing apparatus of some Species that rub the hind Femora against the Tegmina (Folsom & Wardle). Rel. Crickets.

FEMORO-TIBIAL Adj. (Latin, *femur* = thigh; *tibia* = shin bone; *-alis* = pertaining to.) Descriptive of structure or action involving the Femur and Tibia or to the articulation between them.

FEMORO-TIBIAL JOINT One of the two primary flexion ponts of the arthropod leg, and positioned between the Femur and Tibia (Snodgrass). See Articulation; Leg; Joint. Cf. Coxo-Trochanteral Joint. Rel. Segmentation.

FEMORO-TROCHANTERIC Adj. (Latin, *femur* = thigh; *trochanter* = ball upon which hip-bone rotates; *-ic* = consisting of.) Descriptive of or pertaining to the Femur and the Trochanter.

FEMUR Noun. (Latin, *femur* = thigh. Pl., Femora; Femurs.) The third segment of the insect leg. Articulated with body through Trochanter and Coxa, and bearing a Tibia at its distal margin. Femur is typically the largest leg segment and the most variable in shape. A terete Femur is cylindrical and tapering; other insects display a sulcate Femur that is furrowed or grooved; an incrassate Femur is thickened and generally associated with burrowing or digging. Femur is often modified for leaping (Orthoptera), capture of prey (Mantoidea) and grooming. See Leg. Cf. Coxa; Pretarsus; Tibia; Trochanter. Rel. Meropodite.

FEN Noun. (Old English, *fenn* = marsh. Pl., Fens.) A plant community that occurs on wet peat. Cf. Bog; Marsh; Swale; Swamp. Rel. Community.

FENAMIPHOS An organic-phosphate (phosphoric acid) compound {Ethyl-3-methyl-4-(methylthio)phenyl (l-methylethyl) phosphoramidate} used as a nematicide, and contact/systemic insecticide to control mites, nematodes, plant-sucking insects, thrips and mole crickets. Compound is applied to apples, peaches, cherries, citrus, cotton, vegetables, bananas, ornamentals, strawberries and turf. Trade name is Nemacur®. See Organophosphorus Insecticide.

FENAZAQUIN A quinazoline compound {4-[[4-(1,1-dimethylene) phenyl]ethoxy] quinazoline} used as an acaricide in some countries with experimental registration in USA. Fenazaquin is applied to vegetables, stone fruit, cotton, grapes and ornamentals. Trade name is Magister®.

FENAZOX A diphenyl compound that acts as an acaricide and contact insecticide. Compound is toxic to fish and not registered in USA. Trade name is Fentoxan®.

FENBUTATIN-OXIDE A synthetic, organic-tin compound {Hexakis-(2-methyl-2-phenylpropyl)-distannoxane} used as a contact Acaricide against mites on fruit trees, strawberries and grapes. Compound is toxic to fishes. Syn. Hexakis. Trade names include Bendex®, Neostanox®, Novran®, Osadan®, Torque® and Vendex®.

FENDONA® See Alphacypermethrin.

FENESTRA Noun. (Latin, *fenestra* = window. Pl.,

Fenestrae.) 1. A window. 2. Any transparent glassy spot or mark, such as a pellucid mark in a vein. 3. Blattaria: A small, pale, membranous area at the base of the Antenna. 4. Isoptera: Window-like perforations in a membrane. Cf. Fontanel.

FENESTRATE Adj. (Latin, *fenestra* = window; *-atus* = adjectival suffix.) Pertaining to structure with transparent or window-like spots, such as in the wings of some Lepidoptera. Alt. Fenestrated; Fenestratus.

FENESTRATE MEMBRANE Compound eye: The membrane at the base of the Ommatidia at their junction with the optic nerve. See Retina.

FENESTRATE OCELLUS An ocellate spot with a transparent area in the eye.

FENESTRELLA Noun. (Latin, *fenestra* = window. Pl., Fenestrellae.) A transparent eye-like spot in the anal area of the Tegmina of some Orthoptera (Kirby & Spence).

FENITROTHION An organic-phosphate (thiophosphoric acid) compound {O,O-dimethyl O-(4-nitro-m-totyl phosphorothioate} used as a contact insecticide and stomach poison against mites, thrips, plant-sucking insects, leaf-chewing insects and bollworms. Compound is applied to numerous crops in many countries, ornamentals, nursery stock and forest trees. Compound also is applied as a public-health insecticide for control of mosquitoes and bed bugs, and used as a surface spray for moderate-term (ca two month) residual control of stored-grain pests and cockroaches. Fenitrothion is regarded as comparable with Parathion but is safer to handle; non-systemic, fast-acting with residual effects and some ovicidal activity. Fenitrothion is phytotoxic to cotton, brassicas, and some fruit crops. Trade names include: Cyfen®, Dicofen®, Durakil®, Etalene®, Fenstan®, Kaleit®, Metathion®, Micromite®, Novathion®, Nuvanol®, Pestroy®, Senthion®, Sumanone®, Sumithion®, Verthion®. See Organophosphorus Insecticide.

FENKILL® See Fenvalerate.

FENN, CHARLES (–1925) (Adkin 1926, Entomologist 59: 120.)

FENN, ELLENOR (1744–1813) (Lisney 1960, *Bibliography of British Lepidoptera 1608–1799*. 315 pp. (290–291, bibliogr.), London.)

FENNAH, RONALD GORDON (1910–1987) English government Entomologist, administrator (Commonwealth Institute of Entomology) and specialist in Fulgoroidea.

FENNEL APHID *Dysaphis foeniculus* (Theobald) [Hemiptera: Aphididae]. (Australia).

FENNOTOX® See Heptachlor.

FENOBUCARB A carbamate compound {2-sec-butylphenylmethylcarbamate} used as a contact insecticide for control of plant-sucking insects on cotton, cereals, rice, sugarcane, tea and vegetables in some countries. Fenobucarb is not registered for use in USA, in part because it is toxic to fishes. Trade names include Bassa®, BPMC®. Carvil® and Hopcin®. See Carbamate Insecticide.

FENOM® See Cypermethrin.

FENOTHRIN® See Phenothrin.

FENOXYCARB A carbamate compound {Ethyl[2-(4-phenoxyphenoxy)ethyl] carbamate} used as a contact insecticide, stomach poison and Insect Growth Regulator for control of domestic pests and plant-sucking insects. Fenoxycarb is applied to ornamentals and stored products. Phytotoxicity to some varieties of grapes, citrus, pears and many ornamentals. Fenoxycarb is toxic to fishes. Trade names include Award®, Comply®, Eclipse®, Hurricane®, Insegar®, Logic®, Precision®, Preclude®, Regulator®, Sustain®, Torus® and Varikill®. See Carbamate Insecticide.

FENPROPAR® See Propargite.

FENPROPATHRIN A synthetic-pyrethroid compound with repellency and contact activity as an acaricide and selective insecticide in many countries. The compound is used on ornamental plants and cotton in USA. Fenpropathrin displays minor phytotoxicity but toxic to fishes. Trade names include: Danimen®, Danitol®, Farmatox®, Herald®, Kilumal®, Lody®, Meothrin®, Methrin®, Randal®, Rody®, Smash®, Sumirody®; Tame®. See Synthetic Pyrethroids.

FENPROXIMATE An oxime compound {(E)-1,1-dimethylethyl-4-[[[[(1,3-dimethyl-5-phenoxy-1H-pyrazol-4-yl) methylene]amino] oxy]methyl]benzoate} used as an acaricide, contact insecticide and stomach poison to control mites. The compound is effective in all stages, is non-systemic and gives quick knockdown. Fenproximate is applied to apples, citrus, stone fruits and ornamentals in some countries. Toxic to fishes. Trade names include: Danitron®, Kendo®, Nja®, Ortus®.

FENPYRAD A pyrrole compound {N-(4-t-butylbenzyl)-4-chloro-3-ethyl-1-methyl-pyrazole-5-carboxamide} used as a contact acaricide. Compound is applied to apples, pears, citrus, grapes, cotton, vegetables and ornamentals in some countries. Compound apparently is effective against plant-sucking insects (aphids, psyllids, whiteflies), thrips and related insects. Fenpyrad is under experimental registration in USA. Fenpyrad is phytotoxic to cucumber, Japanese pear and some varieties of roses. Compound demostrates quick knockdown, residual activity and is compatible with other pesticides but is not compatible with alkaline compounds. Syn. Tebufenpyrad. Trade names include Masai® and Pyranica®.

FENSTAN® See Fenitrothion.

FENSUL® See Fenvalerate.

FENTHIOCARB A carbamate compound {S-(4-phenoxybutyl)-N,N-dimethyl thiocarbamate} used as an acaricide for control of mites on apples, citrus, cotton, vegetables and other crops in some countries. Compound is phytotoxic to some apple varieties, cotton, peaches and melons. Fenthiocarb is not registered for use in USA. Trade names include Panocon® and Panosin®. See Carbamate Insecticide.

FENTHION An organic-phosphate (thiophosphoric acid) compound {O,O-dimethyl-O-[4-(methylthio)-m-tolyl] phosphorothioate} used as an acaricide, contact insecticide and stomach poison against mites, thrips, plant-sucking insects, armyworms, leaf-miners, bollworms, crickets and public-health pests. Fenthion is used as a surface spray for moderate-term (ca 2 month) residual control of spiders, fleas, flies and cockroaches. Compound is applied to vegetables, cereals, corn, tea, cotton, tobacco, rice, citrus, fruit trees, olives and grapes in many countries. Fenthion is registered for mosquito control in USA, but is toxic to bees, fishes and Crustacea. Trade names include: Baycid®, Baytex®, Entex®, Lebaycid®, Queletox®, Spotton®, Tiguvon®. See Organophosphorus Insecticide.

FENTOXAN® See Fexazox.

FENVALERATE A synthetic-pyrethroid compound that acts as a selective insecticide through contact or as a stomach poison; fast knockdown with residual effect but no systemic activity. Fenvalerate is applied to numerous fruit and vegetable crops; it is used to control domiciliary cockroaches and is toxic to bees and fish. Trade names include: Aqmatrine®, Belmark®, Bovaid®, Ectrin®, Extrin®, Fenkill®, Fensul®, Fenvalethrin®, Moscade®, Pydrin®, Pyrid®, Sanvalerate®, Sumibac®, Sumicidin®, Sumifleece®, Sumifly®, Sumitick®, Sumitok®, Tirade® and Tribute®. See Synthetic Pyrethroid.

FENVALETHRIN® See Fenvalerate.

FENYES, ADELBERT (1863–1937) (Blackwelder 1942, Pan-Pacif. Ent. 18: 17–22.)

FERAL Adj. (Latin, *fera* = wild animal; *-alis* = characterized by.) 1. Descriptive of animals or plants that are not under domestication. 2. Descriptive of organisms that have escaped domestication and reverted to a natural form.

FERDINAND, (JOZEF HUYSKINS) BROEDER (1896–1965) (Schepdael. 1967, Linneana Belg. 3: 84, 90.)

FEREDAY, RICHARD WILLIAM (–1899) (Anon. 1899, Zool. Anz. 22: 464.)

FERGUSON, EUSTACE WILLIAM (1884–1927) (Musgrave 1932, *A Bibliography of Australian Entomology 1775–1930*. 380 pp. (90–92), Sydney.)

FERGUSONINIDAE Hennig 1958. Plural Noun. A monogeneric Family of Diptera in the Australasian region. Larvae live in close association with gallicolous nematodes of Myrtaceae (esp. *Eucalyptus*).

FERGUSSON, A (–1949) (Wigglesworth 1950, Proc. R. ent. Soc. Lond. (C) 14: 64.)

FERMENT FLIES See Drosophilidae. (Australia).

FERMENT Noun. (Latin, *fermentum* = ferment. Pl., Ferments.) See Enzyme.

FERMENTATION Noun. (Latin, *fermentum* = ferment; English, *-tion* = result of an action. Pl., Fermentations.) The process by which microorganisms (yeast, bacteria) chemically convert raw material such as glucose into products such as alcohol.

FERN APHID *Idiopterus nephrelepidis* Davis [Hemiptera: Aphididae].

FERN FLIES See Teratomyzidae. (Australia).

FERN MOTH *Callopistria floridensis* (Guenée) [Lepidoptera: Noctuidae].

FERN SCALE 1. *Pinnaspis carisis* Ferris [Hemiptera: Diaspididae]. (Australia). 2. *Pinnaspis aspidistrae* (Signoret) [Hemiptera: Diaspididae].

FERNALD, CHARLES HENRY (1838–1921) (Mallis 1971, *American Entomologists*. 549 pp. (141–150), New Brunswick.)

FERNALD, HENRY TORSEY (1866–1952) (Mallis 1971, *American Entomologists*. 569 pp. (141–150), New Brunswick.)

FERNALD, MARIA E (1839–1919) (Osborn 1946, *Fragments of Entomological History*. 394 pp. (83–84), Columbus, Ohio.)

FERNANDEZ, R P AMBROSIO (1882–1953) (Agenjo 1954, Graellsia 12: 1–19, bibliogr.)

FERNEX® See Pirimiphos-Ethyl.

FERNOS® See Primicarb.

FERNOW, BERNARD EDUARD (1851–1923) (Anon. 1923, U. S. Dept. agric. Off. Rec. 2(7): 5.)

FERRANT, VICTOR (1856–1942) (Sachtleben 1942, Arb. physiol. angew. Ent. Berl 9: 285.)

FERRANTE, GIOVANNI (1858–1945) (Lusena 1946, Bull. Inst. Egypte 27: 151–165.)

FERRARI, JOHANN ANGELO (1806–1876) (Katter 1877, Ent. Kal. 2: 71.)

FERRARI, PIETRO MANSUETO (1823–1893) (Wilderman 1894, Jb. naturw. 9: 503.)

FERREIRA, TITO MANUEL PAZ (Dias 1970, Revta. cienc. Vet. (Ser. A) 3: 337–340.)

FERREOUS Adj. (Latin, *ferreus* = made of iron; *-osus* = with the property of.) The metallic grey of polished iron. Alt. Ferreus.

FERRER, ASCENSIO CODIAN (1876–1932) (Navas 1932, Boln. Soc. ent. Esp. 15: 120–124.)

FERRIERE, CHARLES Swiss entomologist specializing in taxonomy of Chalcidoidea.

FERRIS, GORDON FLOYD (1893–1958) American academic (Stanford University) and specialist in taxonomy of Coccoidea and insect morphology. Ferris published more than 200 scientific papers and was Founder and Editor of *Microentomology*. (Mallis 1971, *American Entomologists*. 549 pp. (169–173), New Brunswick.)

FERRUGINO-TESTACEOUS Adj. (Latin, *ferruginus* = rusty; *testa* = a piece of burned clay; Latin, *aceus* = of or pertaining to.) A rusty yellow-brown.

FERRUGINOUS Adj. (Latin, *ferruginus* = rusty; *-osus* = with the property of.) Rusty red-brown. Alt. Ferrugineus; Ferruginosus; Ferruginous.

FERTILE Adj. (Latin, *fertilis* = fertile.) 1. Producing viable gametes, spores, or offspring. 2. Eggs or seeds that are capable of development. Ant. Infertile. Cf. Fecund.

FERTILITY Adj. (Latin, *fertilis* = fertile; English, *-ity* = suffix forming abstract nouns.) The ability to reproduce or produce viable offspring. Cf. Fecundity.

FERTILIZATION Noun. (Latin, *fertilis* = fertile; English, *-tion* = result of an action. Pl., Fertilizations.) The union of male and female gametes. In insects, fertilization is characterized by mature sperm moving through the Micropyle of an egg to unite with the female Pronucleus. A term loosely and incorrectly applied to the act of copulation or its completion. See Reproduction. Cf. Conjugation. Rel. Coitus; Copulation; Insemination.

FERTILIZE Trans. Verb. (French, *fertilize*. Ferilized, Fertilizing.) To inseminate via the introduction of Spermatozoa to the egg.

FERTON, CHARLES (1856–1921) (Anon. 1924, Revue zool. Agr. appliq. 23: 99–100.)

FESTA, ENRICO (1868–1939) (Beaux 1940, Annali Mus. civ. Stor. nat. Giacomo Doria 60: [1].)

FESTIVUS Adj. (Latin, *festivus* = gay.) Variegated with bright colours.

FESTOON Noun. (French, *feston* = garland > Italian, *festone*. Pl., Festoons.) Marginal structures arranged in loops or garlands as if pendulously suspended.

FETID Adj. (Latin, *fetidus* = stinking; an offensive odour.) Malodorous; substance with a disagreeable odour.

FETID GLAND Orthoptera: A gland that secretes a malodorous fluid. See Gland. Cf. Defensive Gland; Repugnatorial Gland.

FETTIG, FRANÇOISE JOSEPHE (1824–1906) (Bourgeois 1908, Mitt. naturhist. Ges. Colmar. 9: 199–239, bibliogr.)

FEVER FLIES See Bibionidae.

FEYTAUD, JEAN (1881–1974) (Anon. 1974, Bull. Soc. ent. Fr. 78: 297.)

FIBONACCI SERIES An unending numerical sequence in which each term is defined as the sum of the two preceding numbers in the sequence (1, 1, 2, 3, 5, 8, 13, 21, 34, 55, 89...)

FIBRE Noun. (Latin, *fibra* = band. Pl., Fibres.) Any thread-like structure, typically muscle, nerve or connective tissue. Cf. Fibril.

FIBRIL Noun. (Latin, *fibrilla,* diminuitive of *fibra* = small fibre. Pl., Fibrillae.) 1. A small fibre or root hair. 2. A component or element of a muscle fibre, *i.e.* a muscle sarcostyle or the finer fibrous structure of a muscle or nerve. Cf. Fibre.

FIBRILLAR THEORY An hypothesis of the late 19th Century that asserted Cytoplasm is composed of fibrils in individual strands (filar theory) or a network (reticular theory) within a homogeneous ground substance.

FIBRILLATED Adj. (Latin, *fibrilla* = small fibre.) Formed or consisting of Fibrillae.

FIBRIN Noun. (Latin, *fibra* = band. Pl., Fibrins.) An insoluble protein compound, derived from fibroinogen, that makes up a large part of the muscular tissue. Fibrin is found in blood and other body liquids.

FIBROBLAST Noun. (Latin, *fibra* = band; Greek, *blastos* = bud. Pl., Fibroblasts.) Tissue Culture: A cell type found in connective tissue associated with Collagen. Flattened connective tissue cell.

FIBROIN Noun. (Latin, *fibra* = band. Pl., Fibroins.) 1. Larval insects: A chemical compound found in silk produced from Fibroinogen. 2. Lepidoptera silk: Fibroin is the protein filament surrounded by a cementing protein (Sericin). See Silk. Cf. Sericin. Rel. Bave.

FIBROINOGEN Noun. (Latin, *fibra* = band; Greek, *genos* = birth. Pl., Fibroinogens.) A blood protein involved in the production of Fibrin. Rel. Silk.

FIBULA Noun. (Latin, *fibula* = buckle. Pl., Fibulae.) 1. The Jugal Lobe of the insect wing. 2. The structure holding together the fore and hindwings of some insects. See Jugum.

FICALBI, EUGENIO (1858–1922) (C[hiarugi] 1922, Monitore zool. Ital. 33: 145–146.)

FICAM® See Bendiocarb.

FICHT, GEORGE AUGUSTUS (1900–1941) (Osborn 1946, *Fragments of Entomological History*. Pt. II. 232 pp. (84), Columbus, Ohio.)

FICK, WILHELM (1872–1930) (Weidner 1967, Arb. Verh. naturw. Ver. Hamburg Suppl. 9: 91.)

FICKERT, KARL RUDOLPH DIETRICH (1849–1904) (Linden 1904, Leopoldina 40: 52–54.)

FIDDLER BEETLE *Eupoecila australasiae* (Donovan) [Coleoptera: Scarabaeidae]. (Australia).

FIDE Latin term meaning 'on authority of.' In taxonomy, a term employed to indicate an author has not seen a work or a specimen cited. Term used in brackets.

FIDELIIDAE A small Family (ca five Genera) of aculeate Hymenoptera assigned to Superfamily Apoidea or regarded a Subfamily of Megachilidae or Apidae in some classifications. Fideliids occur in South Africa and Chile. Head with one Subantennal Suture and Facial Foveae absent; Labrum longer than wide; apex of Galea acute or rounded; Flabellum present; Glossa short basal of Palpus and long apical of Palpus; Labial Palpus segments 1–2 elongate and flattened; Preepisternal Suture absent below Scrobal Suture; forewing Marginal Cell normal; Stigma well developed; Basitibial Plate present in female, variable in male; Pygidium present in male and female; Arolia present. See Apoidea.

FIDELITY Noun. (Latin, *fidelitas* = faithfullness; English, *-ity* = suffix forming abstract nouns. Pl., Fidelities.) The tendency of an insect to remain in a habitat, visit a particular host plant or behave in a predictable manner.

FIEBER, FRANZ XAVIER (1807–1872) (Osborn 1937, *Fragments of Entomological History*. 394 pp. (portrait only), Columbus, Ohio.)

FIELD CRICKET *Teleogryllus* spp. [Orthoptera: Gryllidae] in Australia. Adults are 20–25 mm long with body cylindrical and dark brown to black; head small and rounded; Antenna longer than body and thread-like (filiform). Foreleg and middle leg are similar in shape with hindleg adapted for jumping. Forewing small and leathery with hindwing long, fan-like and folded beneath forewing with apical portion rolled and projecting rearward. Multi-segmented Cerci at apex of Abdomen; Cerci taper apically and about as long as hind Tibia and Tarsus. Female Ovipositor ca 1.5 times as long as Cerci. Males stridulate to attract females. Black field-crickets are multivoltine with 2–3 generations per year. Adult females lay ca 1,000 eggs. Eggs are banana-shaped, 2.5–3.0 mm long, white to pale yellow and laid singly in soil 1–4 cm deep. Egg stage requires 1–3 weeks. Nymphal stage involves 8–12 moults but typically 9–10. Nymphs are similar in appearance to adults but lack wings; wing buds appear in last nymphal instar. Adults and late instar nymphs feed on leaves and stems of young seedlings. Damage only caused when crickets are in plague numbers. FC found early in season at base of plants. See Gryllidae.

FIELD, GEORGE HAMILTON (1850–1937) (Abbot 1937, Ent. News 48: 270–272.)

FIELD, HERBERT HAVILAND (1868–1921) (Ward 1921, Science 54: 424–428.)

FIELDE, ADELE MARION (1839 –1916) (Stevens 1918, Field Mem. Comm. 1918: 377.)

FIERY HUNTER *Calosoma calidum* (Fabricius) [Coleoptera: Carabidae]: A predatory beetle widespread in North America. Larva and adult feed on cutworms and armyworms but do not climb trees to search for prey. See Carabidae.

FIERY JEWEL *Hypochrysops ignitus ignitus* (Leach) [Lepidoptera: Lycaenidae]. (Australia).

FIERY SKIPPER *Hylephila phyleus* (Drury) [Lepidoptera: Hesperiidae].

FIFTH LONGITUDINAL VEIN Diptera: The Media of Comstock *teste* Williston. See Vein; Wing Venation.

FIG BARK-BEETLE *Aricerus eichhoffi* (Blandford) [Coleoptera: Curculionidae]. (Australia).

FIG BEETLE See Fig Leaf Beetle.

FIG FRUITBORER *Phycomorpha prasinochroa* (Meyrick) [Lepidoptera: Copromorphidae]. (Australia).

FIG LEAF-BEETLE 1. *Poneridia australis* (Boheman) [Coleoptera: Chrysomelidae]: In Australia adults and larvae cause minor defoliation of fig trees. 2. *Poneridia semipullata* (Clark) [Coleoptera: Chrysomelidae]. (Australia).

FIG-LEAF MOTH *Talanga tolumnialis* (Walker) [Lepidoptera: Pyralidae]. (Australia).

FIG LEAFHOPPER *Dialecticopteryx australica* Kirkaldy [Hemiptera: Cicadellidae]. (Australia).

FIG LONGICORN *Acalolepta vastator* (Newman) [Coleoptera: Cerambycidae]: A pest of *Citrus*, grapes and figs in Australia. Adults are ca 30 mm long, brownish-grey with prominent spine on side of Thorax. Females lay eggs in cracks on small branches or where other borers have damaged tree. Eclosion occurs within 10 days and neonate larvae bore into wood. Larvae inhabit trunk and heavier limbs and bore deep into wood. Monovoltine. See Cerambycidae.

FIG MITE *Eriophyes ficus* Cotte [Acari: Eriophyidae].

FIG PSYLLID See Moreton Bay Fig Psyllid.

FIG SCALE *Lepidosaphes conchiformis* (Gmelin) [Hemiptera: Diaspididae].

FIG WASP *Blastophaga psenes* (Linnaeus) [Hymenoptera: Agaonidae]: Endemic to Mediterranean and introduced elsewhere for pollination of the edible fig, *Ficus carica* Linnaeus. Larvae develop within syconia of wild (inedible) caprifigs. Adults pollinate Calimyrna figs. See Agaonidae. Rel. Caprifig; Calimyrna Fig.

FIGHTAGRUB® See Entomogenous Nematodes.

FIGIAN GINGER-WEEVIL *Elytroteinus subtruncatus* (Fairmaire) [Coleoptera: Curculionidae].

FIGITIDAE Hartig 1840. Plural Noun. A moderate sized, cosmopolitan Family of parasitic Hymenoptera assigned to Cynipoidea. Adults typically are small bodied and macropterous; female and male Antenna with 13 segments; Mesosoma with sculpture and Scutellum sometimes with an apical spine. Hosts include Neuroptera and Diptera larvae. Eggs are deposited in neuropteran Haemocoel. Larvae are solitary and develop as internal parasites until host spins a cocoon, then figitid migrates outside host and feeds externally. Figitids that attack Diptera prefer Syrphidae and saprophagous cyclorrhaphans. See Cynipoidea. Cf. Eucoilidae.

FILA See Filum.

FILACEOUS Adj. (Latin, *filum* = thread; *aceus* = of or pertaining to.) Pertaining to structure composed of filaments or threads.

FILAMENT Noun. (Latin, *filum* = thread. Pl., Filaments) 1. A thread-like slender process of uniform diameter. 2. An elongated appendage.

FILAMENT BEARER *Nematocampa limbata* (Haworth) [Lepidoptera: Geometridae].

FILAMENT PLATE Early stage of insect embryo: A differentiated sheet of cells that connects the genital rudiment with the Heart rudiment of the same side of the body (Imms).

FILAMENTOUS Adj. (Latin, *filum* = thread; *-osus* = with the property of.) 1. Filament-like; thread-like. 2. Pertaining to or descriptive of long, thin structures or appendages. Term typically is applied to Antenna.

FILAMENTOUS VIRUS A virus that causes disease of honey bee. FV multiplies in Fat Body and ovarian tissue of adults; the Haemolymph of severely infected bees becomes milky-white. The disease was first noted by Clark (1978) in the USA. See Honey Bee Viral Disease. Rel. Pathogen.

FILARIA Noun. (Latin, *filum* = thread. Pl., Filariae.) Any of several thread-like nematodes whose adults live as parasites within the lymphatic system of man and other animals. Filaria produce changes that may cause Elephantiasis and other diseases. The larval stage develops within biting insects.

FILARIAL WORMS Members of the Class Nematoda (Superfamily Filarioidea) that are responsible for human and veterinary diseases vectored by blood-sucking Diptera. Principal Families of worms include Filariidae and Onchocercidae.

FILARIASIS Noun. (Latin, *filum* = thread; Greek, *-iasis* = suffix for names of diseases. Pl., Filariases.) A disease, transmitted by mosquitoes or other nematocerous Diptera, caused by the presence of microscopic nematodes (Filaria). Syn. Filariosis.

FILARICIDE Noun. (New Latin, *filaria* = worm; *-cide* > *caedere* = to kill. Pl., Filaricides.) A chemical agent that kills filarial nematodes. Cf. Acaricide; Herbicide; Insecticide; Nematicide; Pesticide.

FILATE Adj. 1. (Latin, *filum* = thread; *-atus* = adjectival suffix.) Thread-like; thin as a thread. 2. Separated from the disc by a channel, which produces a very slender thread-like margin. Alt. Filiate. See Filiform.

FILATE ANTENNA Diptera: A simple Antenna, without lateral Seta or dilation.

FILATOR Noun. (Latin, *filum* = thread; *-or* = one who does something. Pl., Filators.) 1. The Spinnerets of most caterpillars. 2. The silk spinning structure of silkworms that regulates the size of silk fibres. See Spinneret 2.

FILBERT APHID *Myzocallus coryli* (Goetze) [Hemiptera: Aphididae].

FILBERT BUD-MITE *Phytocoptella avellanae* (Nalepa) [Acari: Nalepellidae].

FILBERT WEEVIL *Curculio occidentis* (Casey) [Coleoptera: Curculionidae]. A pest of cultivated filbert nuts along the Pacific Coast of USA. See Curculionidae. Cf. Hazlenut Weevil.

FILBERTWORM *Cydia latiferreana* (Walsingham) [Lepidoptera: Tortricidae]: A pest of walnuts and filberts in western USA. Eggs are laid individually near host fruit; first white, then transparent. Larvae feed within gall or nut. Pupation occurs within cocoon at feeding site. FW undergoes 2.5 generations per year; some larvae can span 2–3 winters. Host plants include acorns, almonds, hazelnuts and walnuts. Syn. *Melissopus latiferreanus* (Walsingham). See Tortricidae. Cf. Codling Moth.

FILE Noun. (Old High German, *fila.* Pl., Files.) 1. Crickets: A diagonal, ridged vein near the Tegmina base that is used in stridulation. 2. General: Any corrugated (ribbed) part of a stridulatory device that creates sound when rubbed against a membrane or opposable device. File is located on any appendage or area of the insect body. See Stridulation. Cf. Strigil; Plectrum. Rel. Sound.

FILE BEETLES See Bostrichidae.

FILE OF PHANERES Acarology: Row of Phaneres.

FILHO, JOSÉ OTAVIO (1906–1964) (Anon. 1964, Studia Ent. 7: 487.)

FILICIFORM Adj. (Latin, *filix* = fern; *forma* = shape.) Fern-shaped; shaped as the frond of a fern. Cf. Adiantiform; Bipectinate; Penniform; Pinnate. Rel., Form 2; Shape 2; Outline Shape.

FILICORNIA Noun. Insects with thread-like Antennae, *e.g.* Coleoptera - the Carabidae.

FILIFORM Adj. (Latin, *filum* = thread; *forma* = shape.) Thread-shaped; descriptive of a structure that is long, slender, cylindrical and of uniform diameter or parallel-sided. The term is descriptive of antennal shape in some insects. See Antenna. Cf. Flagelliform; Moniliform; Papiliform; Vermiform. Rel. Form 2; Shape 2; Outline Shape.

FILIFORM ANTENNA An Antenna that is thread-like and of uniform thickness for its entire length. See Antenna.

FILIP'EV, NIKOLIYA NAKOLAEVICH (1882–1942) (Koshanchikov 1953, Ent. Obozr 33: 363–368, bibliogr.)

FILIPPI, NATALE (1896–1959) (Anon. 1960, Boll. Assoc. romana Ent. 15: 1.)

FILIPPI'S GLANDS Lepidoptera larvae: Paired accessory glands associated with the silk glands.

FILITOX® See Methamidophos.

FILLET Noun. (Latin, *filum* = thread. Pl., Fillets.) 1. Lepidoptera: A transverse, raised structure between the Antennae. 2. A band of white matter in the Midbrain and Medulla Oblongata. Alt. Lemniscus.

FILOSE Adj. (Latin, *filum* = thread; *-osus* = full of.) Pertaining to structure that ends in a thread-like process. Cf. Filamentous.

FILTER APPARATUS Finely-branched processes of the atrial wall of some spiracles that often form two thick but air-pervious mats just within the atrial orifice (Snodgrass).

FILTER CHAMBER A part of the alimentary system in Homoptera in which the two ends of the Ventriculus and the beginning of the intestine are bound together in a membranous and muscular sheath (Snodgrass).

FILTER FEEDING A method of feeding seen in insects such as Helodidae that filter food on the bottom of ponds. FF involves sifting mud with Maxillary Combs provided by Apical Mandibular Combs that move food toward the mouth. Mandible serves a dual function in that the Molar region of the appendage grinds the food into manageable units for ingestion. Mosquito and midge larvae also filter feed. See Feeding Strategy.

FILTH FLIES See Milichiidae.

FILUM Noun. (Latin, *filum* = thread. Pl., Fila.) 1. A thread. 2. A slender, paired, unjointed, filiform, Anal Organ in *Machilis.*

FILUM SPINOSUM A row of rather short, dark, blunt Setae that extend obliquely from the Stigma toward the posterior margin of the forewing of some chalcidoid wasps. FS is immediately laterad of the Linea Calva. See Linea Calva.

FILUM TERMINALE The media Cercus; Pseudocercus.

FILZ GALL Plant galls characterized by the presence of hairs or hairy outgrowths of the Epidermis. FG typically are induced on the undersides of leaves with gall-inducer living external on the plant; many mites induce and inhabit FG. See Gall. Cf. Bud-and-Rosette Gall; Covering Gall; Mark Gall; Pit Gall; Pouch Gall; Roll-and-Fold Gall.

FIMBRIA Noun. (Latin, *fimbria* = fringe. Pl., Fimbriae.) 1. Any fringe-like structure. 2. A fringe-like vestiture of Setae at the apex of any structure. Alt. Lacinea; Pileus. See Fringe. Cf. Anal Fimbria.

FIMBRIATE Adj. (Latin, *fimbriatus* = fringed; *-atus* = adjectival suffix.) Structure displaying marginal Setae of irregular length or non-parallel Setae or bristles. Alt. Fimbriatus.

FIMBRIATE ANTENNA A setaceous Antenna, each segment of which bears one lateral Seta. See Antenna.

FIMBRIATE PLATES Coccids: See Pectinae (MacGillivray).

FIMICOLOUS Adj. (Latin, *fimus* = dung; *colere* = to inhabit, dwell.) Pertaining to organisms that inhabit dung or develop upon dung. See Habitat. Cf. Stercoraceous. Rel. Ecology.

FIN Noun. (Anglo Saxon, *finn* = fin. Pl., Fins.) Elongate, flattened body structures used in locomotion by aquatic organisms.

FINDAL, J KRISTIAN (1871–1938) (Hansen 1940, Ent. Meddr. 20: 585–586.)

FINGER Noun. (Old English, *finger*. Pl., Fingers.) 1. Any digit-like process from the body. 2. The Digitus of the Maxilla.

FINGERED LERP *Cardiaspina maniformis* Taylor [Hemiptera: Psyllidae]. (Australia).

FINGERLING, MAX (1844–1904) (Möbius 1943, Dt. ent. Z. Iris 57: 4.)

FINIPHOS® See Mevinphos.

FINLAY, CARLOS JUAN (1833–1915) Cuban physician who (1880) advanced the notion that mosquitoes transmit Yellow Fever. His experiments were not conclusive. (Hoffman 1944, Revta. Med. trop. Parasit. Habana 10: 125–128.)

FINLAY, JOHN (1835–1897) (Anon. 1897, Entomologist 30: 228.)

FINLAYSON, L R (1903–1965) (Beirne 1965, Proc. ent. Soc. Ont. 96: 129)

FINOT, PIERRE ADRIEN PROSPER (1838–1908) (Kheil 1911, Int. ent. Z. 5: 185, 190–191, 197–199, 203–204, 213–215.)

FINTELMANN, LOUIS (1802–) (Ratzeburg 1874, Forstwissenschaftliches Schriftsteller -Lexicon 1: 478, footnote.)

FINTESCO, GEORGE (1875–1948) (Popescu-Gorg. 1948, Notat. biol Buc. 6: 184–186, bibliogr.)

FINZI, BRUNO (1897–1941) (Menozzi 1941, Memorie Soc. ent. ital. 20: 190–191, bibliogr.)

FIORI, ANDREA (1854–1933) (Montanaro 1934, Atti Soc. nat. Modena 13: 22–25, bibliogr.)

FIORI, ATTILIO (–1958) (Anon. 1958, Boll. Soc. ent. ital 88: 130.)

FIORINIA SCALE *Fiorinia fioriniae* (Targioni Tozzetti) [Hemiptera: Diaspididae] (Australia).

FIPRONIL A phenyl pyrazole compound {5-amino-1-[2,6-dichloro-4-(trifluoromethyl) phenyl-4-(trifluoromethyl) sulphinyl]-1H-pyrazola-3-carbonitrate} used as a contact insecticide and stomach poison for control of grasshoppers, root and stem-boring insects and thrips. Fipronil is applied to corn, cotton, sugar beets, bananas, potatoes and alfalfa in many countries, but not registered for use in USA. Trade names include MB 46030® and Regent MC®.

FIR CONE-LOOPER *Eupithecia spermaphaga* (Dyar) [Lepidoptera: Geometridae].

FIR ENGRAVER *Scolytus ventralis* LeConte [Coleoptera: Scolytidae]: A monovoltine, significant pest of fir trees in western North America that can complete a partial second generation in warm areas and one generation each two years in cold areas. Adults are 2–4 mm long, black, punctate with Elytra striated. Females construct nuptial chambers across grain of wood, then construct egg gallery horizontal into bark and sapwood. Male aids female in removal of debris from gallery. Eggs are laid in niches along sides of galleries. Larvae feed separately in individual mines excavated at right angle to gallery (with grain). A brown fungal stain is associated with larval feeding areas. FE overwinters as a mature larva in cell at end of tunnel. Pupation occurs during following summer. See Scolytidae. Cf. Douglas-Fir Engraver; Shothole Borer; Smaller European-Elm Bark Beetle. Rel. Engravers.

FIR-SEED MOTH *Cydia bracteatana* (Fernald) [Lepidoptera: Tortricidae].

FIR TREE BORER *Semanotus litigiosus* (Casey) [Coleoptera: Cerambycidae].

FIRE ANTS Members of the Genus *Solenopsis* (typically), or *Ochetomyrmex*. See Formicidae. Cf. Imported Fire Ant; Little Fire Ant; Native Fire Ant; Southern Fire Ant.

FIRE BEE *Trigona (Oxytrigona) mellicolor* [Hymenoptera: Apidae].

FIRE BEETLE *Merimna atrata* (Laporte & Gory) [Coleoptera: Buprestidae]. (Australia).

FIRE BEETLES See Phengodidae.

FIRE BLIGHT A bacterial disease of apple, pear and quince. FB is caused by *Erwinia amylovora* and vectored by many Species of bees, wasps, flies and bugs. FB was the first plant disease shown to be transmitted by insects (1892).

FIRE-BLIGHT BEETLE *Pyrgoides orphana* (Erichson) [Coleoptera: Chrysomelidae]. (Australia).

FIRE COLOURED BEETLES See Pyrochroidae.

FIRE FLIES See Lampyridae.

FIREBAR SWORDTAIL *Graphium aristeus parmatum* (Gray) [Lepidoptera: Papilionidae]. (Australia).

FIREBRAT *Lepismodes inquilinus* Newman [Thysanura: Lepismatidae]: A cosmopolitan pest of paper, starchy foods, book bindings, textiles, wallpaper, sizing, glue, cotton, linen and silk. Setae are barbed and arranged in combs; Maxillary Palpus with six segments; abdominal Terga each with a dorsal and lateral pair of setal combs; scales on dorsum pale and dark brown; Cerci and Appendix Dorsalis longer than Abdomen. Typically Firebrats occur where temperature is 32–44°C and relative humidity is high. The Species is an urban problem in bakeries, kitchens, stoves, boiler rooms and hot-water storage facilities. Egg stage requires 14–18 days but no eclosion at <20°C. Nymphal stage requires 330 days at 27°C. Adults live to 2.5 years and moult throughout life. Syn *Thermobia domestica* Packard. See Lepismatidae.

FIRETAIL *Cicadetta denisoni* (Distant) [Hemiptera: Cicadidae]. (Australia).

FIREWEED Noun. *Senecio madagascariensis* an asteraceous plant native of Madagascar and pest of coastal pastures in eastern Australia.

FIRST ANAL VEIN The wing-vein next following the Cubitus (Comstock). See Vein; Wing Venation.

FIRST AXILLARY SCLERITE A pterothoracic sclerite which morphologists believe has detached from the median lateral Notum. FAS articulates with the Anterior Notal Process and is associated with the base of the Subcostal Vein. Syn. Notal Ossicle *sensu* Crampton; Mesoptero *sensu* Berlese; Condylophore and Anterior Arm Anterobasale *sensu* Jordan. See Wing Articulations. Cf. Second Axillary Sclerite.

FIRST BASAL CELL Diptera: A cell positioned between the 1st, 2nd, 3rd and 4th longitudinal veins on the basal half of the wing (Curran).

FIRST CLYPEUS See Postclypeus.

FIRST GONAPOPHYSIS See Volsella.

FIRST GRUB The second phase of hypermetamorphic larval development in Meloidae. The first instar triungulin moults into a scarabaeiform larval shape which is retained in instars 2–5.

FIRST INCISURA Coccoidea: The distinct indentation or notch of the margin of the Pygidium on the meson between the median pair of lobes (MacGillivray).

FIRST INNER APICAL NERVURE Hymenoptera (Norton): The cubitus; from Media to First Anal (Comstock).

FIRST JUGAL VEIN See Vena Arcuata; Vein; Wing Venation.

FIRST LATERAL SUTURE Odonata: A suture originating beneath the forewing base behind the Humeral Suture and meets it behind the middle Coxa (Smith).

FIRST LONGITUDINAL VEIN Diptera: The Radius (Comstock). See Vein; Wing Venation.

FIRST MAXILLAE Second pair of appendages of gnathal region of head; in insects called 'the Maxillae' (Snodgrass).

FIRST MEDIAN PLATE A small sclerite of variable shape positioned in the angle between the second Axillary Sclerite and the distal arm of the third Axillary Sclerite at the base of the mediocubital field of the wing; accessory to the third Axillary in function, and usually attached to it (Snodgrass).

FIRST RADIO-MEDIAL CROSSVEIN Neuroptera: Hindwing vein arising from the Radial Sector and extending toward the base of the wing; joining the Media near its base (Comstock). See Vein; Wing Venation.

FIRST SEGMENT Any segmented appendage: The segment nearest the point of attachment with the body.

FIRST SPECIES RULE A practice among some early taxonomic workers who regarded the first Species named in a new Genus to be the type of that Genus (International Code). The rule was applied in general to the works of older authors where no generic type was designated. The concept is not regarded as applicable under the current rules of Zoological Nomenclature.

FIRST SUBMARGINAL CROSS-NERVURE Hymenoptera: Part of Media and the Radio-Medial Crossvein (Comstock).

FIRST THORACIC SPIRACLE The mesothoracic spiracle; often displaced into the posterior part of the Prothorax (Snodgrass).

FIRST TROCHANTER The first segment of the Telopodite; the Basipodite *sensu* Snodgrass.

FIRST VEIN Diptera: The vein positioned immediately behind the Auxiliary Vein, or when absent, immediately behind Costa, R and R1 (Curran). See Vein; Wing Venation.

FIRTH, JOHN (1832–1885) (Dimmock 1888, Psyche 5: 35.)

FISCHBACH, CARL (Fricke 1902, Forst. Jagdwes. 34: 41–43.)

FISCHER VON RÖSLERSTAMM, JOSEPH EMANUEL (1783–1866) (Manns 1866, Verh. zool. -bot. Ges. Wien 16: 51–54.)

FISCHER VON WALDHEIM, GOTTHELF (1770–1853) Director of the Imperial Museum of Moscow. Author of *Descriptions de quelques Insectes, dans les Memoires des Naturalistes de Moscou* (1820), *Entomographia Imperii Russici* (3 vols), *Notice sur une Mouche carnivore, nommee Medetere* (1819), *Notice sur l'Argas de Perse* (1823), and *Elateroids* (1824). (Essig 1931, *History of Entomology*. 1029 pp. (631–632), New York.)

FISCHER, AUGUST (1864–1931) (Joest 1931, Ent. Z., Frankf. a. M. 45: 171.)

FISCHER, CECIL ERNEST CLAUDE (1874–1950) (Burkill 1952, Indian Forester 78: 50.)

FISCHER, EMIL (1868–1954) (Sachtleben 1956, Beitr. Ent. 6: 200–201.)

FISCHER, F (Essig 1931, *History of Entomology*. 1029 pp. (630–631), New York.)

FISCHER, FRANS CHRISTIAN JOHAN (1902–1973) (Botosaneanu 1974, Nouv. Revue Ent. 4: 87–88)

FISCHER, HEINRICH LEOPOLD (1817–1886) (Keller 1928, Arch. Insektenk. Oberrheingeb. 2: 217–224.)

FISCHER, JOHANN BAPTIST (1804–1932) (Gistl. 1832, Faunus 1: 50.)

FISCHER-SIGWART, HERMANN (1842–1925)

(Lautharat 1925, Tätber. naturf. Ges. Baselland 7: 163–171)

FISH FLIES See Megaloptera.

FISHER, CYRIL EDMUND (1931–1958) (Richards 1959, Proc. R. ent. Soc. Lond. (C) 23: 73.)

FISHER, WARREN SAMUEL (1878–1971) (Wheeler & Valley 1975, History of Entomology in Pennslvania Department of Agriculture. 37 pp. (12), Harrisburg.)

FISH-TOX® See Rotenone.

FISKE, WILLIAM FULLER (1831–) (Anon. 1936, Proc. ent. Soc. Wash. 38: 114.)

FISON, ALBERT JAMES (1840–1912) (Wheeler 1912, Entomologist's Rec. J. Var. 24: 280.)

FISSATE Adj. (Latin, *fissilis,* from *fissus* = split; *-atus* = adjectival suffix.) Divided or split; structure with fissures or cracks. Cf. Fatiscent.

FISSILE Adj. (Latin, *fissilis,* from *fissus* = split.) Pertaining to structure capable of being split in the direction of the grain of wood. Descriptive of Cleft or divided wings (plume-moths).

FISSIPAROUS Adj. (Latin, *fissus* = split; *parere* = to bring forth; *-osus* = with the property of.) Reproducing by fission; a form of asexual reproduction in which the parent divides with each part becoming a new individual. See Reproduction.

FISSURE Noun. (Latin, *fissura* . *fissus* = split. Pl., Fissures.) 1. A crevice or narrow longitudinal opening of undefined length and relatively significant depth. 2. A slit. 3. A crack in the surface of the Integument. Syn. Break, Crack; Fracture; Split. Cf. Sulcus.

FISTULA Adj. (Latin, *fistula* = pipe, tube. Pl., Fistulae.) 1. A slender tube. 2. Lepidoptera: The channel formed by the union of the two parts of the Proboscis. 3. An abnormal passage from an abscess to the surface; a passage between two hollow organs.

FISTULAR Adj. (Latin, *fistula* = pipe, tube.) Structure resembling a slender, cylindrical tube.

FISUS Adj. (Latin, *fissus* = split.) Cleft; pertaining to structure longitudinally divided from apex to base.

FITCH, ASA (1809–1897) Born in New York; farmer, physician, student of natural history and New York State Agricultural Society Entomologist (1854–1873). Fitch was author of many practical reports on insects and has been called 'Father of Economic Entomology in North America'. (Mallis 1971, *American Entomologists.* 549 pp. (326–327), New Brunswick; Barnes 1988, *Asa Fitch and the Emergence of American Entomology.* 120 pp.)

FITCH, EDWARD ARTHUR (1854–1912) (Morley 1912, Entomologist 45: 235–236.)

FITZ, ERNST (1869–1939) (Schawada 1939, Z. wien. Ent. Ver. 24: 161–162)

FITZGERALD, FRANCIS (fl 1789) (Lisney 1960, *Bibliography of British Lepidoptera 1608–1799.* 315 pp. (226–228), London.)

FITZGERALD, L D F V See Vesey-Fitzgerald, L. D. F.

FITZINGER, LEOPOLD JOSEPH (1802–1884) (Anon. 1884, Psyche 4: 236.)

FIVE-SPINED BARK BEETLE *Ips grandicollis* (Eichhoff) [Coleoptera: Scolytidae]: A small, dark brown to black, cylindrical beetle ca 3 mm long with apex of body truncated; excavated above and bears five spines (one very small) on margin of excavation. Larvae are white, slightly curved, with reddish brown head. Adult males tunnel through bark, into cambium and excavate a nuptial chamber. Males produce sex pheromone, attract females and other males to log or tree. Males may attract 1–7 females and copulation occurs in nuptial chamber. Each female bores a tunnel away from nuptial chamber and deposits eggs in cavities along sides. Frass is pushed along tunnel by female and cleared through entry tunnel by male. Eggs hatch within seven days and larvae tunnel through inner bark. Larval period a few weeks in summer, a few months in winter. Mature larvae form cells at end of galleries and pupate. Pupal stage requires one week. Neonate adults feed on inner bark before boring emergence holes similar to holes through which parents entered. Main damage to timber not in tunnelling, but rather timber-staining caused by fungus. FSBB is native to North America and accidentally introduced into South Australia ca 1943 via importation of pine timber from USA; beetle was discovered in Queensland 1982 and established in *Pinus* growing areas. FSBB attacks recently felled trees and logging debris; it will infest and kill nearby standing trees when beetle populations are high. Injured trees are particularly susceptible to attack. Timber-staining fungus may aid in killing trees and degrade logs. Attacks only *Pinus*, with all Species in Genus susceptible. See Scolytidae.

FIXATION Noun. (Latin, *fixus* = fixed; English, *-tion* = result of an action. Pl., Fixations.) An histological procedure intended to terminate life processes quickly with a minimum distortion to cytological detail. Fixation also must prevent autolysis, prevent microbial action and increase the refractive index of tissue. See Histology.

FIXED HAIRS See Microtricha (Kellogg).

FLAB Noun. (Middle English, *flappe*. Pl., Flabs.) Diptera: The Labella. Lobes at the apex of the mouthparts.

FLABBY Adj. (Related to Middle English *flappe*.) Soft, pliant, limp or easily yielding to touch; not rigid. Cf. Turgid.

FLABELLAE Adj. Coccids: Flabelliform marginal Setae.

FLABELLATE Adj. (Latin, *flabellare* = to fan; *-atus* = adjectival suffix.) Fan-shaped; structure with long, thin processes lying flat on each other, and resembling the folds of a fan. Flabellate is characteristic of some antennal configurations seen in Coleoptera, Lepidoptera and Hymenoptera. Alt. Flabelliform. See Antenna.

FLABELLIFORM Adj. (Latin, *flabellum* = fan; *forma* = shape.) Fan-shaped. Alt. Flabellate. See Fan.

Cf. Pectiniform. Rel. Form 2; Shape 2; Outline Shape.

FLABELLIFORM MARGINAL HAIRS Coccids: Flattened, scale-like marginal Setae that are broadly oval in outline shape in some Genera.

FLABELLUM Noun. (Latin, *flabellum* = fan. Pl., Flabella.) 1. A fan. 2. A leaf-like or fan-like structure. 3. Apoidea: The translucent lobe at the apex of the Glossa. 4. Part of a Flagellum.

FLACCID Adj. (Latin, *flaccidus* > *flaccus* = flabby.) 1. Descriptive of structure that yields to pressure because it is not firm (rigid). 2. Something lacking vigour or force. 3. Feeble; limber; lax. Syn. Limp; Pliant; Soft. Ant. Turgid.

FLACH, KARL (1856–1920) (Anon. 1955, Mitt. naturw. Mus. Aschaffenburg 7: 6.)

FLAGELLAR ACCESSORY STRUCTURE See Centriole Adjunct.

FLAGELLATE Adj. (Latin, *flagellum* = whip; *-atus* = adjectival suffix.) Whip-like; pertaining to structure with a Flagellum or whip-like feature. Alt. Flagellatus.

FLAGELLIFORM Adj. (Latin, *flagellum* = whip; *forma* = shape.) Whip-shaped; descriptive of structure that is elongate, slender and sometimes tapered apicad. See Flagellum. Cf. Filiform; Moniliform; Palpiform; Vermiform. Rel. Form 2; Shape 2; Outline Shape.

FLAGELLOMERE Noun. (Latin, *flagellum* = whip; Greek, *meros* = part. Pl., Flagellomeres.) A superficially discrete or identifiable unit of the Antenna distad of the Pedicel. Flagellomeres are often erroneously referred to as Antennal segments. Flagellomeres are not true segments in the morphological sense because they lack intrinsic musculature and they are serially homodynamous with adjacent Flagellomeres. See Antenna. Cf. Flagellum. Rel. Segmentation.

FLAGELLUM Noun. (Latin, *flagellum* = whip. Pl., Flagella.) 1. The portion of the Antenna distad of the Pedicel, including Funicle and Club. Flagellum segmentation varies between and among groups of insects; groundplan number of Flagellomeres in nematocerous Diptera is 14; primitive Brachycera have eight Flagellomeres; Asilomorpha have three Flagellomeres and Muscomorpha have four Flagellomeres. Shape of Flagellum among insects is exceedingly plastic. Within a Species, Genus or Family the number of segments and their shapes can serve as reliable diagnostic characters. Some generally recognized shapes include moniliform, filiform, aristate, geniculate, clavate, capitate, ramose, setaceous, pectinate, flabellate, lamellate. Primitive dipteran Flagellum is often thread-like or filiform. Sometimes, Flagellomeres bear whorls of long Setae, giving Antenna a plumose appearance. In higher Diptera first Flagellomere is large and remaining Flagellomeres are reduced or modified to form an Arista or Stylus. Arista is typically hair-like, long and flexible; Stylus is variable in length but usually inflexible. Antenna bearing an Arista is called aristate; Antenna bearing a Stylus is called stylate. See Antenna. 2. Any long, thin whip-like process. 3. The tail-like process of a spermatozoon. See Sperm Flagellum. Cf. Aflagellar.

FLAGRO® See Gibberellic Acid.

FLAME SKIPPER *Hesperilla idothea* (Miskin) [Lepidoptera: Hesperiidae]. (Australia).

FLAMINIO, RUIZ P (1883–1942) (Beltram 1960, Revta. univ. Santiago 44–45: 131–145, bibliogr.)

FLAMMATE Adj. (Latin, *flagrare* = to burn; *-atus* = adjectival suffix.) Flaming or fiery red. Alt. Flammeus.

FLANDERS, STANLEY E (1894–1984) American academic (University of California, Riverside), specializing in Biological Control and biology of parasitic Hymenoptera. Author of more than 300 scientific papers.

FLANGE Noun. (Old French, *flangir* = to bend, turn. Pl., Flanges.) 1. A projecting rim or edge that provides structure support and mechanical strength. Cf. Rim. 2. A part that spreads out like a rim. Syn. Projection; Protrusion.

FLANK Noun. (Old French, *flanc* = loin, hip. Pl., Flanks.) 1. The side or lateral aspect of the insect Thorax. 2. The Pleuron. See Thorax.

FLANKING SETA Psocoptera: A Seta near the hyaline cone of the distal paraproctal margin.

FLANNEL MOTHS See Megalopygidae.

FLAPPING FLIGHT One of two basic functional forms of insect flight. FF is distinguished by wingbeats that create active movement through air. FF is the principal form of flight by insects; FF displays different forms including hovering flight and forward flight. See Flight. Cf. Passive Flight.

FLARING Adj. (Unknown origin.) Descriptive of structure that expands like the bell of a trumpet.

FLASK-LIKE SENSE-ORGAN See Sensillum Ampullaceum.

FLAT BARK-BEETLES See Cucujidae.

FLAT BEETLES See Cucujidae.

FLAT BROWN-MILLIPEDE *Brachydesmus superus* Latzel [Polydesmida: Polydesmidae] (Australia).

FLAT BUGS See Aradidae.

FLAT FLIES See Hippoboscidae.

FLAT GRAIN-BEETLE *Cryptolestes pusillus* (Schonherr) [Coleoptera: Cucujidae]: A cosmopolitan pest of grain and cereal products. FGB is common in humid tropics. Adults and larvae cannot penetrate intact grain but do access small defects in seed coat. Adults and larvae are cannibalistic. Adults are 1.5–2.0 mm long, pale brown with long Antennae. Larvae are small, whitish, elongate, with appendages at apex of Abdomen. Life cycle requires 7–9 weeks and adults may live one year. Females lay ca 240 eggs which are deposited singly in furrows or crevices of kernel. FGB completes four larval instars; last instar produces cocoon with silk from prothoracic gland. See Cucujidae.

FLAT GRAIN-BEETLES 1. *Cryptolestes* spp.

[Coleoptera: Cucujidae (Laemophloeidae)]. 2. In Australia; Taxa assigned to the Silvanidae.

FLAT HUNTSMAN-SPIDER *Delena cancerides* Walckenaer [Araneida: Heteropodidae]. (Australia).

FLAT SCALE *Paralecanium expansum* (Green) [Hemiptera: Coccidae]. (Australia).

FLAT-FOOTED FLIES See Platypezidae.

FLATHEADED APPLE-TREE BORER *Chrysobothris femorata* (Olivier) [Coleoptera: Buprestidae]: A significant pest of apple and other deciduous fruit and ornamental trees throughout North America. Eggs are laid in fissures or cracks in bark and neonate larvae enter cracks, to feed on sapwood of young trees and on bark of older trees. During autumn, larvae burrow into wood and construct chambers for overwintering. See Buprestidae. Cf. Bronze Birch Borer; Pacific Flathead Borer; Roundheaded Apple-Tree Borer.

FLATHEADED BORERS Members of the buprestid beetle Genus *Chrysobothris*. See Buprestidae. Cf. Roundheaded Borer.

FLATHEADED CONE BORER *Chrysophana placida conicola* Van Dyke [Coleoptera: Buprestidae]: A minor pest of pine cones in western North America. See Buprestidae.

FLATHEADED FIR BORER *Melanophila drummondi* (Kirby) [Coleoptera: Buprestidae].

FLATHEADED PASTURE WEBWORM *Oncopera mitocera* (Turner) [Lepidoptera: Hepialidae] (Australia).

FLATHEADED WOOD BORERS See Buprestidae.

FLATID PLANTHOPPERS See Flatidae.

FLATIDAE Plural Noun. Flatid Planthoppers. A numerically large, predominantly tropical Family of auchenorrhynchous Hemiptera assigned to Superfamily Fulgoroidea. Adults with body wedge-shaped and colour often green but sometimes sombre coloured; forewing broad and strongly tectiform with numerous crossveins in Costal region; hind Tibia without large, articulated spur at apex but spine at apex of second Tarsomere. Flatids typically feed on vines, shrubs and trees in wooded habitats. See Fulgoroidea; Hemiptera.

FLAVESCENT Adj. (Latin, *flavescere* = to turn yellow.) Descriptive of structure with a somewhat yellow colour; verging on yellow. Flavescent is used incorrectly in Odonata to indicate a slightly smoky colour.

FLAVID Adj. (Latin, *flavidus* = yellow.) Descriptive of colour that is golden yellow or sulphur yellow. See Yellow. Cf. Helvolus.

FLAVIVIRIDAE Plural Noun. A Family of viruses that contains about 60 arboviruses; *Flavivirus* generally is regarded as the most important Genus of arboviruses in terms of impact on humans. Virons are small (ca 40–50 nm dia), spherical with lipoprotein envelope and positive-stranded RNA. Most *Flavivirus* virons are vectored by mosquitoes; a few are vectored by ticks. Viruses produce several diseases in humans and livestock, including Yellow Fever, Dengue Fever, Classical Swine Fever (Hog Cholera), Border Disease, St Louis Encephalitis and Tick-borne Encephalitis. See Arbovirus. Cf. Bunyaviridae; Tongaviridae; Reoviridae; Rhabdoviridae.

FLAVONES Plural Noun. Plant pigments that are found in some insects and produce the red in *Lygaeus* and *Leptocoris*. See Pigmentary Colour. Cf. Carotenoids; Melanins; Ommochromes; Pterines (Pteridines); Quinones. Rel. Structural Colour.

FLAVOTESTACEOUS Adj. (Latin, *flavus* = yellow; *testa* = piece of burnt clay; *aceus* = of or pertaining to.) Light yellow- brown, almost luteous or clay colour.

FLAVOUS (Latin, *flavus* = yellow; *-osus* = full of.) Adj. Pure, clear yellow.

FLAVOVIRENS Adj. (Latin, *flavus* = yellow; *virens* = green.) Green verging upon yellow.

FLAX BOLLWORM *Heliothis ononis* (Denis & Schiffermüller) [Lepidoptera: Noctuidae].

FLAX TOW Short, broken fibres removed during scutching or hackling. FT is used in the textile industry to produce twine, yarn and stuffing.

FLEA BEETLES Members of the chrysomelid beetle Subfamily Alticinae (Galerucinae). Many Species are economically important pests (Potato Flea Beetle); a few Species are used in control of weeds (Ragwort Flea Beetle). Name ascribed to the capacity of adults to catapult their bodies into the air in a propulsive jump via a Metafemoral Spring. See Chrysomelidae. Cf. Alligator-Weed Flea Beetle; Couch Flea-Beetle; Pale-Striped Flea Beetle; Potato Flea-Beetle; Ragwort Flea-Beetle; Red-Legged Flea Beetle; Striped Flea Beetle; Submetallic Flea Beetle; Three-Spotted Flea Beetle; Toothed Flea Beetle.

FLEA COLLAR A slow-release insecticide formulation for control of ectoparasitic fleas and ticks on domesticated dogs and cats. The collar is a plastic strip impregnated with a toxin worn around the neck. The insecticidal vapour from the collar provides extended protection for the body.

FLEA TYPHUS See Murine Typhus.

FLEAS See Siphonaptera.

FLECTRON® See Cypermethrin.

FLEE® See Bifenthrin.

FLEECEWORM See Green Blow Fly.

FLEISCHER, ANTON (1850–1934) (Obenberger 1950, Cas. csl. Spol. ent. 47: 215–217)

FLEISCHER, JOSEPH (1866–1940) (Obenberger 1940, Cas. csl. Spol. ent. 37: 16–17)

FLEISCHMANN, ALBERT (1862–1942) (Stammer. 1952, Sber. phys.-med. Soz. Erlangen 75: xx–xxxv.)

FLEISCHMANN, FRIEDRICH (1874–1909) (Rebel 1909, Verh. zool.-bot. Ges. Wien 59: [240–241].)

FLEMING, GEORGE (1856–1927) (Anon. 1927, Entomologist 60: 143–144.)

FLEMING, JOHN (1785–1857) (Bryson 1861, Trans. R. Soc. Edinb. 22: 655–680.)

FLESH FLIES See Sarcophagidae. Adults are 6–14 mm long, typically with sombre coloration combining grey with darker longitudinal stripes on thoracic dorsum and spots on Abdomen. FFs are active during warm, sunny weather and are attracted to food source (human food, excrement, decaying waste). Females larviposit on meat, excrement and garbage. Eclosion occurs within 24 hours for egg-laying Species. Larvae feed for several days and then pupate away from food in drier habitat. FF life cycle requires 2–4 weeks, depending upon Species and conditions. See Sarcophagidae.

FLETCHER, JAMES (1852–1908) (Mallis 1971, *American Entomologists*. 549 pp. (115–116), New Brunswick.)

FLETCHER, JOHN EDWARD (1836–1902) (Fowler 1902, Proc. ent. Soc. Lond. 1902: lviii.)

FLETCHER, JOSEPH JAMES (1850–1926) (Musgrave 1932, *A Bibliography of Australian Entomology 1775–1930*. 380 pp. (93–94), Sydney.)

FLETCHER, ORLIN KENYON (1908–1975) (Lund 1975, Georgia ent. Soc. 10: 189.)

FLETCHER, ROBERT KEMBLE (1884–1956) (Gaines 1957, J. econ Ent. 50: 229.)

FLETCHER, THOMAS BAINBRIGGE (1878–1950) (Sen 1952, Indian J. Ent. 14: 87–90.)

FLETCHER, WILLIAM HOLLAND BALLETT (1853–1941) (Blair 1942, Proc. R. ent. Soc. Lond. (C) 6: 41.)

FLETCHER'S SCALE *Parthenolecanium fletcheri* (Cockerell) [Hemiptera: Coccidae]. In North America, a pest of greenhouse plants, arborvitae and yew. Alt. Arborvitae Scale.

FLEURY (–1967) (Anon. 1967, Bull. Soc. ent. Fr. 72: 8.)

FLEURY, ARTHUR C (1888–1938) (Anon. 1939, Bull. Calif. Dep. agric. 28: 130–131.)

FLEUTIAUX, EDMOND JEAN BAPTISTE (1858–1951) (Anon. 1952, Beitr. Ent. 2: 328.)

FLEX Trans. verb. (Latin, *flexus* > *flectere* = to bend.) To bend; to curve back.

FLEXIBILITY Noun. (French, *flexibilité* > Latin, *flexibilitas* = bendable; English, *-ity* = suffix forming abstract nouns. Pl., Flexibilities.) The state or quality of being flexible.

FLEXIBLE Adj. (Latin, *flexibilis* > *flexus*.) 1. Structure capable of being bent, turned, bowed or twisted without breaking. 2. Anything with the capacity for change or modification and adaptation to new conditions or applications. Syn. Elastic; Resilient.

FLEXOR Noun. (Latin, *flexus* = bending; *or* = one who does something. Pl., Flexors.) 1. General: A muscle that serves to bend or deflect an appendage or body part. 2. Insects: A muscle that draws an appendage toward the body. A term used to describe the action of many muscles. Cf. Extensor.

FLEXOR MEMBRANE Membranous areas between the tarsal claws. FMs apparently transfer tension of the apodeme on the Unguitractor to the claws, thereby forcing them upward.

FLEXOR MUSCLE A muscle that bends any jointed structure. Cf. Extensor Muscle.

FLEXOR ROW Diptera: One or more rows of bristles along the lower surface of the Femur (Comstock). Cf. Extensor Row.

FLEXUOSE Adj. (Latin, *flectere* = to bend; *-osus* = full of.) Almost zig-zag, without acute angles but more acute at angles than undulating. Flexuose differs from sinuate in being alternately bent and nearly straight. Alt. Flexuosus; Flexuous.

FLEXURE Noun. (Latin, *flexura* > *flexus*. Pl., Flexures.) 1. A flexing or bending; a state of being flexed or bent. 2. A turn, bend or fold.

FLIES See Diptera.

FLIGHT Intrans. Verb. To take flight in numbers, migrate. Noun. (Anglo Saxon, *fliht* = a flying. Pl., Flights.) The movement of adult insects with the aid of wings. Flight is divisible into categories on the basis of functional aspects (Flapping Flight, Passive Flight) and behavioural aspects (Mass Flight, Migratory Flight, Swarming Flight, Trivial Flight). See each term for explanation.

FLIGHT INTERCEPT TRAP An insect trap that consists of a vertical sheet of fine-mesh cloth or plastic maintained between two poles. A rectangular sheet of plastic serves as a roof suspended above the vertical sheet. A rectangular gutter filled with preservative rests at the base of the vertical sheet. Flying insects encounter the vertical sheet and fall into the gutter. The trap is effective at collecting Coleoptera and small Hymenoptera. See Trap. Cf. Malaise Trap.

FLINDERS FUNNEL-WEB SPIDER *Hadronyche flindersi* (Gravy) [Araneida: Hexathelidae]; In Australia, a highly venomous large-bodied, black mygalomorph spider with distribution restricted to the eastern coast of Australia south to Tasmania.

FLINT, WESLEY PILLSBURY (1883–1943) (Mallis 1971, *American Entomologists*. 549 pp. (445–467), New Brunswick.)

FLIT GUN A hand-held, pump-action sprayer that is used for the application of liquid insecticide.

FLOCCULENT Adj. (Latin, *flocculus* = lock of wool.) 1. Descriptive of structure consisting of soft flakes. 2. Descriptive of structure covered with waxy material that gives the impression of wool.

FLOCCULUS Noun. (Latin, *flocculus* = lock of wool. Pl., Flocculi.) Hymenoptera: A setose or bristly appendage on the hind Coxa of some Species.

FLOCCUS Noun. (Latin, *flocculus* = lock of wool. Pl., Floccuses.) 1. A tuft of wool or wool-like Setae. 2. Any tuft-like structure. 2. A mass of fungal filaments or mycelia.

FLÖGEL, JOHANN HEINRICH LUDWIG (1834–1918) (Weidner 1967, Abh. Verh. naturw Ver. Hamburg Suppl. 9: 274–279.)

FLOHR, JULIUS (1837–1896) (Anon. 1896, Ent. News 7: 192.)

FLOODWATER MOSQUITO *Aedes sticticus*

(Meigen) [Diptera: Culicidae].

FLOOR Noun. (Middle English. *Flor*. Pl., Floors.) 1. Insect anatomy: The lower interior surface of any cavity. 2. The lower inside surface of any hollow structure. Cf. Ceiling; Roof; Wall.

FLOR, GUSTAV VON (–1883) (Anon. 1883, Entomol. mon. Mag. 20: 72.)

FLORA Noun. (Latin, *flos* = flower; *flora* = Roman goddess of flowers. Pl., Floras.) 1. A comprehensive taxonomic treatment of plants or plant communities in any region. 2. The plants or plant-life characteristics of a particular geographical region or geological time. Cf. Fauna.

FLORALTONE® See Gibberellic Acid.

FLORBAC®. A registered biopesticide derived from *Bacillus thuringiensis* var. *aizawai*. See *Bacillus thuringiensis*.

FLORDIMENX® See Ethephon.

FLOREL® See Ethephon.

FLORIDA CARPENTER-ANT *Camponotus floridanus* (Buckley) [*Camponotus abdominalis* (Fabricius)] [Hymenoptera: Formicidae]: An endemic and the most common *Camponotus* Species in southeastern USA. FCA infests beehives and the wood in buildings. See Formicidae. Cf. Black Carpenter-Ant; Brown Carpenter-Ant; Red Carpenter-Ant.

FLORIDA FERN-CATERPILLAR *Callopistria floridensis* (Guenée) [Lepidoptera: Noctuidae]: A pest of ornamental ferns in North America. FFC eggs are deposited on underside of leaves and eclosion occurs within 5–7 days. Larvae feed at night and consume leaflets of old growth or consume entire new growth. See Noctuidae.

FLORIDA-GLADES MOSQUITO *Aedes Psorophora confinnis* [Diptera: Culicidae]: A pest in the gulf coastal states of the USA. Females are sooty black with white rings on Proboscis and Tarsi. FGMs bite aggressively during night and are particularly abundant in grassy areas. See Culicidae.

FLORIDA HARVESTER-ANT *Pogonomyrmex badius* (Latreille) [Hymenoptera: Formicidae]. See Formicidae. Cf. California Harvester-Ant; Red Harvester Ant; Harvester-Ant; Western Harvester-Ant.

FLORIDA RED-SCALE See Circular Black Scale.

FLORIDA WAX-SCALE *Ceroplastes floridensis* (Comstock) [Hemiptera: Coccidae]: A pest of citrus in the New World, Hawaii, southeast Asia and Israel. FWS resembles Pink Wax Scale and Hard Wax Scale. Adult females are 2–4 mm long, pale pink to white, globular, smooth, with a depression at centre of cover's top and depressions along margin. Females deposit ca 1,400 brick-red eggs beneath their scale covers; FWS completes two nymphal instars and 1–2 generations per year depending upon climate and host plant. FWS produces large amounts of honeydew that serves as substrate for sooty mould and other fungi. See Coccidae.

FLORISSANT SHALE Fossil-bearing geological deposits near Florissant, Colorado. Regarded by early workers (including T. D. A. Cockerell) as Miocene and now believed to represent material of Lower Oligocene.

FLOSCULIFEROUS Adj. (Latin, *flosculus* = floweret; *-osus* = full of.) Bearing a Flosculus.

FLOSCULUS Adj. (Latin, *flosculus* = flowerlet.) 1. Flowery. 2. Noun. Fulgoridae: A small, tubular, lunate, anal organ with a central style. Rare.

FLOUR BEETLES See Dermestidae.

FLOUR MITE *Acarus siro* Linnaeus [Acari: Acaridae] (Australia).

FLOURINATED SULFONAMIDE A novel synthetic insecticide used in baits for ants and cockroaches. FS acts to uncouple oxidative phosphorylation. See Insecticide.

FLOURY BAKER *Abricta curvicosta* (Germar) [Hemiptera: Cicadidae]. (Australia).

FLOWABLE Noun. A pesticide formulation in which the active ingredient is wet-milled with a clay diluent and water to produce a gel-like compound that can be measured as a liquid and mixed with water for spraying. Designated as 'F' or 'L' on U.S. Environmental Protection Agency label or Material Safety Data Sheet documentation.

FLOWER Noun. (Middle English, *flour* > French, *fleur* > Latin, *floris*. Pl., Flowers.) 1. The reproductive organs of a seed-bearing plant. Flowers are typically conspicuous through their size, colour and fragrance. 2. Anatomically, a flower consists of a Perianth (outer covering) divided into Calyx and Corolla, an Androecium (with one or more stamens) and a Gynoecium or Pistil (with one or more Carpels bearing ovules).

FLOWER BEETLES *Carpophilus* spp. [Coleoptera: Nitidulidae]: In Australia, Adults are ca 3.2 mm long and body black with amber-brown to red-brown legs. Elytra do not cover apex of Abdomen and expose ca one third of Abdomen; Elytra with prominent amber-brown spots at base and apex. Adult females lay ca 1,000 eggs. Eggs are white and the egg stage requires ca 2 days. Larvae are white with brown head and apex of Abdomen. Larvae ca 6 mm in length and larval stage lasts 7–14 days. Pupae overwinter in soil; pupal stage requires at least 5–14 days. FBs are found on cotton feeding on pollen, but are not known to cause damage. FBs sometimes abundant in cotton at the flowering stage. See Nitidulidae.

FLOWER BUGS See Anthocoridae.

FLOWER FLIES See Syrphidae.

FLOWER SPIDERS See Thomisidae.

FLOWER THRIPS *Frankliniella tritici* (Fitch) [Thysanoptera: Thripidae]: A pest of grasses, alfalfa and garden flowers in North America. FTs eggs are minute, white and deposited in plant tissue. Nymphs feed about 6 days, moult twice, drop to ground, moult twice and enter pseudopupal stage. FTs cause deformation of flowers and petals to appear flecked. See Thripidae. Cf. Tobacco Thrips; Western Flower Thrips.

FLOWER WASPS See Tiphiidae.

FLOWER-LOVING FLIES See Apioceridae.

FLOYD, BAYARD FRANKLIN (1882–1945) (Watson 1945, Fla. Ent. 28: 39, 41.)

FLUCTUATING VARIATIONS The neo-Darwinian concept of unstable changes in the characteristics of a Species, that appear or disappear from generation to generation, or which depend on genetic or somatic influences conditioned by known or unknown factors. FVs eventually, by natural selection, tend to crystallize and become permanent, a characteristic of nascent Species in plastic aggregates (Kinsey).

FLUCYCLOXURON A benzoylurea compound {1-[alpha-(4-chloro-alpha-cyclopropylbenzylideneamino-oxy)-p-tolyl]- 3-(2,6-difluorobenzoyl) urea} used as a Chitin Synthesis Inhibitor against rust mites, spider mites, Lepidoptera larvae (armyworms, bollworms, leafrollers), some Coleoptera and mosquitoes. Flucycloxuron is applied to vegetables, field crops, nuts, tea, pome and citrus fruits and ornamentals in some countries. Compound is not registered for use in USA. Flucycloxuron is toxic to Crustacea but non-toxic to beneficial insects. Trade name is Andalin®. See Insect Growth Regulator.

FLUCYTHRINATE Noun. A synthetic-pyrethroid compound with broad-spectrum activity which is used as a contact and stomach poison on vegetables, fruits and cotton in some countries. Flucythrinate is not registered in USA; it displays minor phytotoxicity and is toxic to fishes and other aquatic animals and bees. Trade names include: Cybolt®, Gardian®, Guardian®. See Synthetic Pyrethroids.

FLUFENOXURON A urea compound {1-[4-(2-chloro-alpha, alpha, alpha-trifluoro-p-tolyloxy)-2-fluoro-phenyl]-3-(2,6-difluorobenzoyl) urea} used as an Insect Growth Regulator against mites. Flufenoxuron is applied as a foliar treatment to numerous crops in some countries. Not registered for use in USA. Trade name is Cascade®. See Insect Growth Regulator.

FLUFF LOUSE *Goniocotes gallinae* (DeGeer) [Mallophaga: Philopteridae]: A pest of poultry which resides on fluffy portion of feathers where it feeds on barbs and barbules. Syn. Lesser Chicken Louse. See Philopteridae. Cf. Brown Chicken-Louse; Chicken Body-Louse; Shaft Louse; Wing Louse.

FLUITER, HENDRICK JACOB DE (1907–1970) (Tenhouten 1970, Entomophaga 15: 125–128)

FLUKE, CHARLES LEWIS (1891–1959) (Mallis 1971, *American Entomologists*. 549 pp. (404), New Brunswick.)

FLUME Noun. (Latin, *flumen*, from *fluere* = to flow. Pl., Flumes.) 1. A channel for the transportation of water. The Flume is sometimes used as a collecting source by Entomologists.

FLURY, FERDINAND (1877–1947) (Anon. 1948, Anz. Schädlinsk. 21: 44–45.)

FLUTED Adj. (Middle English, *floute* = flute.) Decorated with flutes. Descriptive of a surface with channels or grooves.

FLUTED SCALES See Margarodidae.

FLUVALINATE An acaricide used to control Varroa mite, *Varroa jacobsoni* Oudemans, a significant pest of honeybees. The material is marketed under the tradename Apistan®. In some places (Italy, USA), the mite has become resistant to this compound. See Varroa Mite.

FLUVIATILE Adj. (Latin, *fluviatilis* > *fluvius* = river; *-atilis*.) Pertaining to organisms that inhabit the margins of running streams.

FLY Noun. (Middle English, *flien* = fly. Pl., Flies.) See Diptera.

FLY-BELT The area infested by tsetse flies in Africa.

FLY-BLOWS Eggs or young maggots of flesh flies. Meat is called fly-blown when eggs or larvae have been deposited on it.

FLY PAPER A sticky trap made from a thin film or paper strips that are yellow in colour and coated with a thick, sticky substance. Adult flies are attracted to the yellow colour and become trapped in the coating when they alight. Fly paper traps are hung in areas of intense fly activity such as barns or other livestock shelters.

FLYING-HAIRS Very long, slender surface Setae whose bases are set in cuticular punctures (pits).

FLYLURE® See Muscalure.

FODDER MITE *Lepidoglyphus destructor* (Schrank) [Acarina: Glycyphagidae]. (Australia).

FOERSTER, ARNOLD (1810–1884) (Wackerzapp 1886, Verh. naturh. Ver. preuss. Rheinl. 43: 33–41.)

FOESTER, KUND (1887–1966) (Tuxen 1968, Ent. Meddr. 36: 105)

FOGELQVIST, GUSTAF (1881–1946) (Lindroth 1946, Ent. Tidskr. 67: 233.)

FOIL® A biopesticide derived from *Bacillus thuringiensis* strain EG 2424. See *Bacillus thuringiensis.*

FOLBEX® See Bromopropylate.

FOLCARD® See Cypermethrin.

FOLDED MEMBRANE Cicadidae: The membrane in the anterior wall of the ventral cavity of the Chordontonal Organ (Comstock).

FOLEY, FRANCIS B (1906–1959) (Campbell 1959, J. Econ. Ent. 52: 1034)

FOLIACEOUS Adj. (Latin, *folia* = leaf; *-aceus* = of or pertaining to.) 1. Leaf-like; resembling a leaf. 2. Descriptive of thin, laminate structures. Alt. Foliaceus; Foliate.

FOLIAFUME® See Rotenone.

FOLIATE Adj. (Latin, *folium* = leaf; *-atus* = adjectival suffix.) Leaf-like; resembling a leaf. Alt. Foliaceous.

FOLIDOL® Parathion.

FOLIMAT® See Omethoate.

FOLIOLES Adj. Leaf-like processes from a margin or protuberance. Alt. Foliolae.

FOLLIAS, ALEXIS RUPERT (1813–1873) (Buquet 1873, Bull. Soc. ent. Fr. (5) 3: cv–cvi.)

FOLLICLE Noun. (Latin, *folliculus* = small sac. Pl., Follicles.) 1. Any cellular sac or tube, such as as a gland. 2. The male's sperm tube which represents a structural homologue of the female's Ovariole and functional unit of the Testis. Histologically, the male Follicle is divided into several regions including Germarium, Zones of Growth, Maturation and Transformation. The Germarium forms the distal, blind end of the tubule and contains a nutritive apical cell (Versonian Cell, Trophocyte) that is surrounded by primary Spermatogonia. The zone of growth is immediately distad of the Germarium and contains Spermatogonia that divide by Mitosis and become encysted by somatic mesodermal cells. The Zone of Maturation is an area in which Spermatocytes undergo two meiotic divisions to form spermatids. The zone of transformation is an area in which spermatids transform into the definitive shape of Spermatozoa. Follicles form a compact body and may be serially arranged. Follicles vary in number among groups of insects: Adephagous beetles may have one Follicle; lice have two Follicles; Acrididae (grasshoppers) may have more than 100 follicles. In some Diptera, the follicles may become fused into a large, common sac. See Testes. Cf. Ovariole; Germarium; Vas Efferens. Rel. Gonad. 3. A cocoon. See Incunabulum.

FOLLICLE CELL Cells of the female reproductive system that are derived from Prefollicular Cells in the Germarium and that surround the Primary Oocyte. Follicle Cells develop synchronously or continuously during each gonadotrophic cycle and sequester yolk protein precursors. See Ovariole; Vitellogenesis. Cf. Trophocyte. Rel. Ovary; Reproductive System.

FOLLICLE MITE *Demodex follicularum* (Simon) [Acari: Demodicidae].

FOLOGNE, EGIDE (–1919) (Lameere 1919, Ann. Soc. ent. Belg. 59: 123–124.)

FOLOSAN® See Tecnazene.

FOLSOM, JUSTUS WATSON (1871–1936) (Mallis 1971, *American Entomologists*. 549 pp. (446–447), New Brunswick.)

FOLTIN, HANS (Hörleinsberger 1969, Z. wien. ent. Ges. 54: 70–71.)

FONOFOS An organic-phosphate (dithiophosphate) compound {O-ethyl-S-phenyl ethylphosphorodithioate} used as a soil insecticide for control of soil-inhabiting pests, aphids, corn rootworm, and wireworms. Toxic to birds, wildlife and fishes. Trade names include: Admiral®, Capfos®, Crusade®, Cudgel®, Dyfonate®, Mailstay®, Metro®, Tycap®. See Organophosphorus Insecticide.

FONSCOLOMBE, ETIENNE LAURENT JOSEPH HIPPOLYTE BAYER DE (1772–1853) (Mulsant 1853, Opusc. ent. 2: 129–144, bibliogr.)

FONSECA, FLAVIO R (1901–1963) (Wigglesworth 1964, Proc. R. ent. Soc. Lond. (C) 28: 57.)

FONTANA, FELICE (1730–1805) (Ambrois 1889, Bull. Soc. ven.-trent. Sci. nat. (Padova) 4: 145–149.)

FONTANA, PRADA PIETRO (1875–1948) (Panzera 1967, Boll. Soc. ticin. Sci. nat. 58: 51–66.)

FONTANELLE Noun. (French, *fontanelle* = little fountain) Isoptera: A pale, shallow, circular depression of the surface of the head between the Compound Eyes. The Fontanelle serves as a reservoir for defensive secretions emitted from a glandular pore opening in termite Families Rhinotermitidae and Termitidae. See Frontal Gland. Alt. Fontanelle.

FOOD CANAL Hemiptera: A longitudinal groove on the medial surface of the Maxillary Stylets that forms a tube for the passage of food from the feeding site to the head. FC is anterior of the Salivary Canal. Cf. Salivary Canal.

FOOD CHAIN A trophic path or succession of populations through which energy flows as a result of feeding.

FOOD CHANNELS Delicate parallel grooves of the lateral and ventral surfaces of the sensory lobes in the Pupae of certain Diptera.

FOOD MEATUS Food channel. (Archaic)

FOOD PLANT A plant on which an insect habitually feeds. Not to be confused with a host plant on which an insect lives, since certain predacious forms haunt particular plants, that are the food-plants of their prey.

FOOD RESERVOIR Lepidoptera: A blind sac or diverticulum from the posterior part of the Oesophagus positioned in the Abdomen dorsal of the stomach.

FOOD WEB A complex of food chains that connect populations in an ecosystem.

FOOT SPINNER See Embiidina.

FOOT Noun. (Greek, *podos* = foot. Pl., Feet.) The Tarsus. A term improperly used interchangably with 'leg'. See Feet; Tarsus.

FOOTE, BENJAMIN A (1928–) American academic (Kent State University) and specialist in ecology and systematics of acalyterate Diptera, particularly the immature stages.

FOOTMEN See Arctiidae.

FOOT-SHAPED LOOP Wings of Odonata. See Cubito-anal Loop.

FOOT-SHIELD Lepidoptera larva: The chitinous sclerite on the outer side of the abdominal feet.

FOOT-STALK Of the Maxilla; the stipes.

FORAGE LOOPER *Caenurgina erechtea* (Cramer) [Lepidoptera: Noctuidae].

FORAGING ANT See Army Ant.

FORAMACORIA Noun. (Latin, *foramen* = hole; *corium* = leather. Pl., Coramacoriae.) The Coria or membrane of a Foramen (MacGillivray).

FORAMEN Noun. (Latin, *foramen* = hole. Pl., Foramina; Foramens.) 1. An opening in the body wall for the passage of a vessel or nerve. 2. Any opening at the apex of a structure. 3. The opening of a cocoon that permits the emergence of the adult. 4. Small openings in the body wall. 5. Orthoptera: The auditory organ on the fore Tibia.

FORAMEN MAGNUM See Occipital Foramen.

FORAMEN OCCIPITALE See Occipital Foramen.

FORAMINAL Adj. (Latin, *foramen* = hole; *-alis* = pertaining to.) Descriptive of or pertaining to a Foramen.

FORAY® A registered biopesticide derived from *Bacillus thuringiensis* var. *kurstaki*. See *Bacillus thuringiensis.*

FORBES, JAMES (1749–1819) (Rose 1850, *New General Biographical Dictionary* 7: 412.)

FORBES SCALE *Quadraspidiotus forbesi* (Johnson) [Hemiptera: Diaspididae]: A pest of deciduous fruit trees (apple, apricot, cherry, currant, pear, plum, and quince) in North America. FS resembles San Jose Scale in habitus and biology. FS completes 1–3 generations per year, depending upon host plant and climate. See Diaspididae.

FORBES, STEPHEN ALFRED (1844–1930) (Mallis 1971, *American Entomologists*. 549 pp. (55–60), New Brunswick.)

FORBES, WILLIAM ALEXANDER (1855–1883) (Dunning 1883, Proc. ent. Soc. Lond. 1883: xlii.)

FORBES, WILLIAM TROWBRIDGE MERRIFIELD (1886–1968) (Franclemont 1969, Ent. News 80: 51.)

FORBIVOROUS Adj. (Greek, *phorbe* = food, fodder; *vorare* = to devour.) Pertaining to insects that feed on any herb other than grass.

FORBUSH, EDWARD HOWE (May 1928, Proc. Boston Soc. nat. Hist. 39 (2): 33–72, bibliogr.)

FORCE® See Tefluthrin.

FORCEPS Noun. (Latin, *forceps* = tongs, pincers. Pl., Forcipes.) 1. Hook-like or pincer-like appendages at the apex of the Abdomen of many insects. Forceps are used in copulation, defence, predation or sensory reception. 2. Protura: Periphallus; Ephemeroptera: Genostyles; Blattaria: Phallus; Dermaptera: Cerci; Auchenorrhyncha: Styles; Heteroptera: Parameres. Psyllidae: Appendages on posterior portion of Hypandrium; Lepidoptera: Process of Sacculus (Valvae); Coleoptera: Aedeagus; Diptera: Gonopods (Cerci); Hymenoptera (various combinations).

FORCEPS BASE Mayflies: See Styliger Plate.

FORCIPATE Adj. (Latin, *forceps* = tongs, pincers; *-atus* = adjectival suffix.) Bearing Forceps or similar structures. Alt. Forcipated; Forcipatus. See Chelate.

FORCIPIFORM Adj. (Latin, *forceps* = tongs, pincers; *forma* = shape.) Forcep-shaped. Descriptive of structure with the form of forceps or pincers. See Forceps. Cf. Cheliform. Rel. Form 2; Shape 2; Outline Shape.

FORD, A (1871–1943) (Turner 1943, Entomologist's Rec. J. Var. 55: 70.)

FORD, LEONARD TALMAN (1880–1961) (Brown 1961, Trans. Proc. So. Lond. ent. nat. Hist. Soc. 1959: xxxv.)

FORD, W K (1909–1972) (Butler 1973, Proc. R. ent. Soc. Lond. (C) 37: 56.)

FORDHAM, WILLIAM JOHN (1882–1942) (Walsh 1943, Entomol. mon. Mag. 79: 48.)

FORE Adj. (Anglo Saxon, *for*; Old High German, *fora*; Sanskrit, *pura* = in front, before in time or space.) Anterior.

FOREBRAIN Noun. (Anglo Saxon, *for* = in front; brain. Pl., Forebrains.) See Protocerebrum.

FOREGUT Noun. (Anglo Saxon, *for* = in front; Old English, *guttas*. Pl., Foreguts.) One of the three principal divisions of an insect's Alimentary System. Foregut originates at the mouth (Os, Buccal Cavity) and terminates at the so-called Gizzard. Foregut is largely confined to the head. Transverse section through foregut shows an innermost layer called the Cuticular Intima. CI is relatively thick in many insects, surrounds the foregut lumen and is shed at each moult. Depending upon the region of the foregut, cuticular projections may extend into the lumen. A layer of epithelium cells surround the CI and a Basement Membrane surrounds the Epithelium. In many areas, the CI and Epithelium are thrown into longitudinal folds that provide areas for immediate expansion to accommodate ingested food. Foregut is enveloped in a layer of longitudinal muscles that in turn are covered with a layer of circular muscles. Circular muscles do not attach to the Epithelium. Longitudinal muscles can attach to the Epithelium or Circular Muscles. See Alimentary System. Cf. Midgut, Hindgut.

FOREHEAD Noun. (Anglo Saxon, *for* = in front; Middle English, *hed*. Pl., Foreheads.) Mallophaga: The head anterior of the Mandibles and Antennae.

FOREIGN GRAIN-BEETLE *Ahasverus advena* (Waltl) [Coleoptera: Cucujidae]: A cosmopolitan pest of stored products including cereal, cocoa, copra, grain, peanuts, dried fruits, herbs, spices and roots. FGB is a problem in mouldy produce at high humidity and often feeds on fungal mycelia. Adults are active, strong fliers, 2–3 mm long with Prothorax quadrate and single apical tooth. Females display discrete oviposition cycles. Eggs typically are laid singly or sometimes in small clusters; incubation requires 4–5 days. Larval development requires 11–19 days and 4–5 larval instars. Pupation occurs within a cell of food fragments cemented with anal secretion. Life cycle is complete within 30 days. FGB does not reproduce at <65% RH.

FOREINTESTINE Noun. (Anglo Saxon, *for* = in front; Latin, *intestinium* > *intestinum*. Pl., Foreintestines.) See Foregut.

FOREL AUGUSTE HENRI (1848–1931) (Handschin 1958, Mitt. schweiz. ent. Ges. 31: 109–120.)

FOREMOST FORMATION A stratum of Upper Cretaceous Period (Campanian) age found in Manitoba near Cedar Lake that contains fossil insects preserved in amber. Absolute age ca 74 MYBP.

FORENSIC ENTOMOLOGY The study of insects associated with human corpses. FE information typically is used to determine time of death, place of death and other issues of medical or legal importance. See Entomology.

FOREST DAY-MOSQUITO *Aedes albopictus* (Skuse) [Diptera: Culicidae].

FOREST FLIES See Hippoboscidae.

FOREST TENT-CATERPILLAR *Malacosoma disstria* Hübner [Lepidoptera: Lasiocampidae]: A serious defoliator of ornamental and forest trees in North America. FTC overwinters in egg stage and eclosion occurs during spring. Biology and life history resemble Eastern Tent Caterpillar. Larvae do not construct a tent, but aggregate in clusters on twigs, branches and trunk when not feeding. Coccon is formed of leaves webbed together. See Lasiocampidae. Cf. Eastern Tent Caterpillar.

FOREST TREE-TERMITE *Neotermes connexus* Snyder [Isoptera: Kalotermitidae].

FORESTERS Procridinae [Lepidoptera: Zygaenidae]. (Australia).

FOREST FLIES See Nemouridae.

FORESTUS, PETRUS (1522–1597) (Rose 1850, *New General Biographical Dictionary* 7: 415.)

FOREWING Noun. (Sanskrit, *puras* = in front; Middle English, *winge*, from Old Norse, *vaengr*. Pl., Forewings.) The anterior wing of an insect which is attached to the Mesothorax. See Wing. Cf. Hindwing. Rel. Elytron; Hemelytron; Tegmen.

FORFEX Noun. (Latin, *forfex* = shears.) 1. A pair of shears. 2. A paired, anal structure that opens and shuts transversely, as a pair of scissors.

FORFICATE Adj. (Latin, *forfex* = shears; *-atus* = adjectival suffix.) Deeply notched. Alt. Forcipate.

FORFICULIDAE Plural Noun. (Latin, *forfex* = shears.) European Earwigs. A numerically large, cosmopolitan Family of Dermaptera that includes the common European earwig, *Forficula auricularia* Linnaeus. See Forficuloidea.

FORFICULOIDEA A Superfamily of Dermaptera containing Families Spongiphoridae, Chelisochidae and Forficulidae. See Dermaptera.

FORGET-ME-NOT *Catochrysops panormus platissa* (Herrich-Schaffer) [Lepidoptera: Lycaenidae]. (Australia).

FORK Noun. (Latin, *furca* = fork. Pl., Forks.) The division of a structure or body into branches. The branches are of the same material as the main body and are subordinate in size and function. See Furca. Cf. Branch. Rel. Dendriform; Dendritic; Dendroid.

FORKED FUNGUS-BEETLE *Bolitotherus cornutus* [Coleoptera: Tenebrionidae].

FORKTAILED BUSH-KATYDID *Scudderia furcata* Brunner von Wattenwyl [Orthoptera: Tettigoniidae]. Cf. Angular-Winged Katydid; Broad-Winged Katydid; Citrus Katydid; Crested Katydid; Inland Katydid; Mottled Katydid; Philippine Katydid; Spotted Katydid.

FORM Noun. (Latin, *forma* = shape. Pl., Forms.) 1. An image or likeness. 2. The characteristic three-dimensional composition and shape of an inorganic substance or biological structure. See Outline Shape; Shape. 3. A taxonomically neutral term for incompletely analysed Phena (Taxa) of questionable taxonomic status or placement. 4. Any representative of a Species that differs from the normal 'type' in some uniform character; seasonal form when occurring at a period different from the type; dimorphic form under alternation of generations or when two colour patterns occur; sexual form when members of one sex differ uniformly from those of the other sex. The term 'form' is often used as a suffix (-form) for a Latin noun to form a word that indicates (without explicit description) the shape of an object (*e.g.* aciniform; setiform). See Shape; Outline Shape; Eidos. Cf. Anatomy; Morphology; Taxonomy. Rel. Monomorphic; Polymorphic.

FORM AND FUNCTION The influence of form and function on Anatomy (Morphology) has not been resolved after centuries of debate. Aristotle believed that structure is caused by function and structures differentiate (change in form) only after the influence of function. In contrast, the German biologist Bernhard Rensch argues that function is caused by structure. Other workers in early 20th century believed that 'form' is morphology and 'function' is physiology. The change in form of many anatomical features appears to enhance the capacity of an organism to utilize mechanical energy. The functional significance of change in form of other features is not always apparent; form is three dimensional while function is four dimensional. Form involves shape, size and any set of physical properties of a structure, appendage or body. Function is the motion of change of form. Form and function may be studied inductively and deductively. Inductive study involves analysis of part to whole, particular to general or specific to universal. In distinction, deductive study involves analysis of general to particular or general to general. In a sense, deduction may be viewed as the opposite of induction. That is, conclusion necessarily follows from premise. Logical principles must be applied to scientific research See Hypothetico-deductive method in science (Popper 1959).

FORMA Latin term meaning 'form.' Abbreviated 'f.'

FORMALDEHYDE A 40% solution in water forming a clear, colourless liquid with a pungent odour. Histologists use formaldehyde as a preservative and for hardening tissues.

FORMANER, ROMUALD (1853–1927) (Heikertinger 1927, Koleopt. Rdsch. 13: 243–245, bibliogr.)

FORMATENDON Noun. (Latin, *forma* = shape; *tendo* > *tendere* = to strech. Pl., Formatendons.) The tendon-like apodeme of the Tentorium (MacGillivray).

FORMATION Noun. (Latin, *forma* = shape; English, *-tion* = result of an action. Pl., Formations.) Ecology: Groups of physiologically similar associations (Shelford).

FORMATIVE CELL 1. See Prohaemocyte. 2. A Trichogen or modified Epidermis cell that produces

Trichia or Setae.

FORMETANATE A carbamate compound {N,N-dimethyl-N'-(3(((methylamino) carbanoyl)-oxy)phenyl)methanimide-monohydrochloride} used as an acaricide and contact insecticide for control of mites, thrips, leaf hoppers, leaf miners, stinkbugs, and related Species. Formetanate is applied to tree crops, alfalfa and other crops in some countries. Formetanate is toxic to bees and fishes. Trade names include Carzol® and Dicarzol®. See Carbamate Insecticide.

FORMIC Adj. (Latin, *formica* = ant; *-ic* = of the nature of.) Of, pertaining to, or derived from ants.

FORMIC ACID 1. In general, the irritating agent in the Sting. 2. A component of the venom (poison) of Hymenoptera. 3. The acid secreted by some ants.

FORMICARY Noun. (Latin, *formica* = ant. Pl., Formicaries.) An ant nest; an ant-hill; an artificial nest for ants used in the laboratory. Alt. Formicarium.

FORMICIDAE Stephens 1829. Plural Noun. Ants. A cosmopolitan, large Family containing about 8,800 Species and 300 Genera of aculeate Hymenoptera assigned to the Vespoidea or forming a unique Superfamily (Formicoidea). Perhaps 15,000–20,000 Species exist. Subfamilies include Dolichoderinae, Formicinae, Formiciinae (fossil), Myrmeciinae, Myrmicinae. Formicid fossil record begins in Cretaceous Period. Diagnostic features of formicids include male with 10–13 antennal segments, female with 10–12 antennal segments; Scape elongate and Antenna geniculate in both sexes. Wings usually present only for a short period following emergence as an adult, then individual dealate; hindwing with 1–2 closed cells and without anal lobe. Metapleural Gland usually is present with orifice above hind Coxa. Gaster petiolate with abdominal segments II or II and III forming scale-like nodular processes; Sting typically is present but reduced in some Species, or replaced by repugnatorial glands in Subfamilies Formicinae and Dolichoderinae. Formicids are eusocial insects with three castes (male, worker, queen) and a complex division of labour between worker and queen; a few parasitic Species have lost the worker caste. Nests are formed in soil, wood or plant debris; males are seasonally produced, short lived and inseminate virgin females in air (phoretic copulation) or ground. Inseminated female sheds wings, sequester selves or in small groups (pleometrotsis), begin egg laying and development of new colony. Colony is sustained on many forms of plant and animal food. Complex chemical communication system has been developed by most Species. Ants serve as hosts for many inquilines. See Aculeata. Cf. Bees; Wasps. Rel. Inquilinism; Myrmecophily.

FORMICIINAE A Subfamily of ants characterized from Taxa taken in Eocene deposits in USA, England, Germany. See Formicidae.

FORMICINAE Lutz 1986. A Subfamily of ants. Sting absent, apex of Gaster surrounded by circlet of Setae; waist of one segment; petiolar node typically in form of scale. See Formicidae.

FORMICOIDEA Latreille 1802. A monofamilial Superfamily of aculeate Hymenoptera comprising the ants. Characterized by development of worker caste, division of labour, contact between generations, Metapleural Gland, and Gaster separated from Propodeum by one or two nodiform segments. See Aculeata. Cf. Apoidea; Chrysidoidea; Pompiloidea; Sphecoidea.

FORMOSAN SUBTERRANEAN TERMITE *Coptotermes formosanus* Shiraki [Isoptera: Rhinotermitidae]: A Species endemic to China and adventive to subtropical and coastal areas throughout the world. Worker Mandible sword-like without teeth; Fontanelle on anterior margin of head; Pronotum flat without anterior lobe; four tarsal segments; Cerci of two segments. FST is regarded as a very serious structural pest in USA (Hawaii and southeast) and South Africa. In continental USA, FST was first noted during 1965 in Texas but may have been introduced after WWII; detected in coastal states after 1965; evidence indicates FST can displace endemic *Reticulitermes* spp. Classified among subterranean termites but often forms 'aerial infestations' (not connected with ground) in homes. A mature colony may contain eight million individuals, forage up to 100 m radius and be capable of destroying homes within a few years if left untreated. Queens can deposit 2,000 eggs per day. See Rhinotermitidae. Cf. West Indian Drywood Termite.

FORMOTHION An organic-phosphate (dithiophosphate) compound {S-[2-foromylmethylamino]-2-oxyethyl]-O,O-dimethyl phosphorodithioate} used as an acaricide and contact/systemic insecticide for control of mites, plant-sucking insects (to include aphids, mealybugs, scale insects, whiteflies), thrips, leaf miners, flies and mosquitoes. Compound applied to numerous crops and ornamentals in many countries. Not registered for use in USA. Formothion is toxic to bees and fishes. Trade name is Anthio®. See Organophosphorus Insecticide.

FORMULATION Noun. (Latin, *formula,* diminutive of *forma* = shape. Pl., Formulations.) A mixture of active and inert ingredients designed to promote the safe and effective delivery of a toxicant into the environment. The product available to the public for use against pests. Alt. Composition. Cf. Adjuvant.

FORNICATE Adj. (Latin, *fornicatus* = vaulted; *-atus* = adjectival suffix.) Pertaining to structures that are arched or vaulted (concave inside, convex outside).

FORSIUS, RUNAR (1884–1935) (T. H. S. 1936, Norsk. ent. Tidsskr. 4: 135–136.)

FORSKAL, PEHR (1732 Helsingfors–1763) Student of Linnaeus. Forskal died on the 'Arabian

Journey,' a Danish government-sponsored scientific expedition to Egypt and Arabia. (Zimsen 1964, *The Type Material of I. C. Fabricius.* 656 pp. (17), Copenhagen.)

FORSSELL, ARVID LEONARD (1879–1950) (Luther 1952, Memo Soc. Fauna Flora fenn. 27: 182.)

FORSSELL, NILS EDVARD (1821–1883) (Sandahl 1883, Ent. Tidskr. 4: 97–100, 120–122.)

FORSSLUND, KARL HERMAN (1900–1973) (Lindroth 1974, Ent. Tidschr. 94: 137–139.)

FORSSTROM, JOHANN ERIC (1775–1824) (Papavero 1971, *Essays in the History of Neotropical Dipterology* 1: 103. São Paulo.)

FORSTER MOTHS See Agaristidae.

FORSTER, ARTHUR H (–1946) (Hall 1947, Proc. R. ent. Soc. Lond. (C) 11: 61.)

FÖRSTER, CARL (1854–1923) (Arndt 1924, Jh. Ver. schles. Insektenk. 14: 25–27.)

FORSTER, HAROLD W (1909–1974) (Mackechnie-Jarvis 1975, Proc. Brit. ent. nat. Hist. Soc. 8: 60.)

FORSTER, JOHANN REINHOLD (1729–1798) Prussian by birth and companion of Captain Cook on his voyages of exploration. Forster wrote *Insectorum Centuria* (1771) and *Catalogue of British Insects* (1770). (Hoare 1976, The Tactless Philosopher. J. R. Forster. Melbourne.)

FORSTER, JOHN (1766–1770) (Lisney 1960, *Bibliography of British Lepidoptera 1608–1799.* 315 pp. (199–200), London.)

FOR-SYN® See Resmethrin.

FORSYTH JOHNSTONE J (–1944) (Cockayne 1945, Proc. R. ent. Soc. Lond. (C) 9: 48.)

FORSYTH, MARGUERITE SHEPARD (–1951) (Klots 1952, Lepid. News. 6: 76–77.)

FORSYTH, WILLIAM (1737–1804) (Rose 1850, *New General Biographical Dictionary* 7: 418.)

FORTI, ARCHILLE (1878–1937) (Trotter 1937, Marcellia 29: 102–104.)

FORTRESS® See Chlorethoxyfos.

FORTUNE, ROBERT (1813–1880) (Anon. 1880. Naturaliste 2: 223.)

FORWERG, HUGO BERNHARD (1835–1905) (Anon. 1905, Insektenbörse 22: 186.)

FORZA® See Tefluthrin.

FOSBAN® See Chlorpyrifos.

FOSDAN® See Phosmet.

FOSFAMID® See Dimethoate.

FOSSA Noun. (Latin, *fossa* = ditch. Pl., Fossae.) A relatively large pit, deep Sulcus or significant depression on the body surface.

FOSSIL Noun. (Latin *fossilis* = dug out, dug up. Pl., Fossils.) Any traces, impressions, animals or plants that are preserved in the Earth's crust. Study of the process of fossilization is called Taphonomy. Fossils are not limited to material preserved in stony form. Most fossils are preserved in rock or amber. Fossils constitute one line of evidence used in establishing classifications. See Compression Fossil; Facies Fossil. Rel. Amber; Classification; Resins.

FOSSILIZATION Noun. (French, *fossile* from Latin, *fossilis* = dug up; -tion. Pl., Fossilizations.) The process of preservation, the rate of which depends upon available oxygen, temperature, moisture, depth of burial and elements in the surrounding sediments.

FOSSORIA Noun. (Latin, *fossor* = digger.) Organisms adapted for burrowing and living within burrows (*e.g.* Orthoptera: Gryllotalpidae; Hymenoptera: Sphecidae)

FOSSORIAL Adj. (Latin, *fossor* = digger; *-alis* = characterized by.) 1. Descriptive of structures modified for digging or burrowing. Rel. Subfossorial. 2. Pertaining to burrowing or digging habits. See Leg. Cf. Cursorial; Gressorial; Natatorial; Raptorial; Saltatorial; Scansorial. Rel. Locomotion.

FOSSULA Noun. (Latin, diminutive of *fossa* = ditch. Pl., Fossulae.) 1. A somewhat long and narrow depression or gutter. 2. An elongated shallow groove (Kirby & Spence). Term specifically applied to grooves on the head or sides of the Prothorax in which Antennae are concealed (Smith). Alt. Fossulet.

FOSSULATE Adj. (Latin, diminutive of *fossa* = ditch; *-atus* = adjectival suffix.) With superficial oblong depressions or fossulae. Alt. Fossulatus.

FOSTER, EDWARD (1863–1930) (Holloway 1930, J. Econ. Ent. 23: 1017–1018.)

FOSTER, SHIRLEY WATSON (1884–1923) (Anon. 1924, J. Econ. Ent. 17: 162.)

FOSTER, WILLIAM S (1868–1899) (dos Passos 1951, J. N. Y. ent. Soc. 59: 164.)

FOSTHIAZATE An organic-phosphate (thiophosphoric acid) compound {(RS)-S-sec-butyl O-ethyl 2-oxo-1,3-thiazolido-3-ylphosphonothioate} used as a nematicide and systemic insecticide against nematodes, mites, thrips, aphids and other insects. Compound is applied to vegetables, tobacco, potatoes, bananas, citrus, peanuts and other crops in some countries. Compound is under experimental registration in USA. Trade names include: Eclahra®, Eclesis®, IKI-1145®, Nemathorin®. See Organophosphorus Insecticide.

FOSTION® See Prothoate.

FOUDRAS, ANTOINE CASIMIR MARGUERITE EUGENE (1783–1859) (Galcoz 1859, Exchange 28: 19–22, 27–29, 36–38, 43–45, 51–52, 61–63, 66–68.)

FOUNDER EFFECT The principle that the founders of a new population carry only a random fraction of the genetic diversity found in the larger, parent population.

FOUNDER MODEL A rapid speciation scenario in which a small, isolated population (such as a few colonists on an island) undergoes rapid divergence from its parent population. Syn. Quantum Speciation.

FOUNDING STAGE Social Insects: The earliest period in colony development. FS is typified by the queen rearing the first brood. The FS is succeeded by the Ergonomic Stage.

FOUNTAINE, MARGARET ELIZABETH (1862–1940) (Sheldon 1940, Entomologist 73: 193–195, bibliogr.)

FOUR-BANDED LEAFROLLER *Argyrotaenia quadrifasciana* (Fernald) [Lepidoptera: Tortricidae].

FOUR-BAR SWORDTAIL *Protographium leosthenes* (Doubleday) [Lepidoptera: Papilionidae]. (Australia).

FOUR-JAWED SPIDERS *Tetragnatha* spp. [Araneida: Tetragnathidae]. (Australia).

FOUR-SPOTTED CUP MOTH *Doratifera quadriguttata* (Walker) [Lepidoptera: Limacodidae]: A bivoltine, minor pest of fruit and ornamental trees in Australia. Females oviposit on leaves in rows of 8–12 eggs per row; eggs are bright yellow, flattened and covered with short, dark Setae from female. Caterpillars are reddish green with double row of large, black-and-white coloured markings on dorsum at anterior and posterior ends of body, and two pairs of stellate tubercles composed of defensive spines. Syn. Chinese-Junk Caterpillar. See Limacodidae.

FOUR-SPOTTED SPIDER MITE *Tetranychus canadensis* (McGregor) [Acari: Tetranychidae].

FOUR-SPOTTED TRAPDOOR SPIDER *Aganippe sutristis* Cambridge [Araneida: Idiopidae]. (Australia).

FOUR-SPOTTED TREE CRICKET *Oecanthus quadripunctatus* Beutenmüller [Orthoptera: Gryllidae].

FOURCROY, ANTOINE FRANÇOIS DE (1755–1809) (Nordenskiöld 1935, *History of Biology*. 629 pp. (370–371), London.)

FOURLINED PLANT-BUG *Poecilocapsus lineatus* (Fabricius) [Heteroptera: Miridae]: A monovoltine pest of gooseberry, currant, rose, chrysanthemum, aster and sunflower in eastern North America. Eggs are laid within slits in stems during fall. FLPB overwinters in egg stage and nymphs emerge during spring. See Miridae.

FOURLINED SILVERFISH *Ctenolepisma lineata pilifera* [Thysanura: Lepismatidae]. A minor urban pest in southwestern USA. See Silverfish.

FOURNIER DE HORRACK (–1952) (Riley 1953, Proc. R. ent. Soc. Lond. (C) 17: 72.)

FOURNIER, OVILA (1899–1974) (Anon. 1975, Proc. ent. Soc. Ont. 105: 3.)

FOURNIER, PIERRE (–1957) (Anon. 1957, Bull. Soc. ent. Fr. 62: 221.)

FOURTH AXILLARY A wing-base sclerite that articulates proximally with the Posterior Notal Process and distally with the third Axillary Sclerite (Imms).

FOURTH LONGITUDINAL VEIN Diptera: The longitudinal vein originating near the base of the wing which separates the two basal cells and borders the Discal Cell anteriorly (medial, M1,2,3). The Discoidal *sensu* Curran. See Vein; Wing Venation.

FOVEA Noun. (Latin, *fovea* = depression. Pl., Foveae.) A pit or depression in the Integument with well defined walls. Cf. Apodeme; Apophysis.

FOVEOLA Noun. (Latin, *foveola* = small depression. Pl., Foveolae.) A small pit or depression in the Integument. Alt. Foveolet.

FOVEOLATE Adj. (Latin, *foveola* = small depression; *-atus* = adjectival suffix.) Pertaining to a surface with pits or depressions (Foveae). Alt. Foveolatus; Foveate; Foveatus.

FOWL TICK *Argas persicus* (Oken) [Acari: Argasidae].

FOWLER, WILLIAM WEEKS (1849–1923) (Prior 1972, Bull. amat. ent. Soc. 31: 116–117.)

FOX, CARROLL (1875–1936) (Anon. 1937, Arb. morph. taxon. Ent. Berl. 4: 161.)

FOX, CHARLES LOUIS (1869–1928) (Van Duzee 1928, Pan-Pacif. Ent. 4: 192.)

FOX, E C DOBRÉE See Dobrée-Fox.

FOX, HENRY (1875–1951) (Rehn 1952, Ent. News 63: 113–119)

FOX, RICHARD MIDDLETON (1911–1967) (Kennedy 1969, Proc. R. ent. Soc. Lond. (C) 33: 54.)

FOX, WILLIAM HENRY (1857–1921) (Wade 1936, Proc. ent. Soc. Wash. 38: 115.)

FOX-WILSON, GEORGE (1896–1951) (Riley 1952, Proc. R. ent. Soc. Lond. (C) 16: 88.)

FOXGLOVE APHID *Aulacorthum solani* (Kaltenbach) [Hemiptera: Aphididae]. A pest of potatoes in western North America. See Aphididae.

FRACTATE Adj. (Latin, *fractus* = broken; *-atus* = adjectival suffix.) Pertaining to structure abruptly bent at an angle and appearing broken. Cf. Curvate.

FRACTATE ANTENNA An Antenna with one very long segment and with distal segments attached to it at an angle. Cf. Elbowed; Geniculate.

FRACTURE Noun. (Latin, *fractus* = broken. Pl., Fractures.) Heteroptera: The suture or indentation in the Hemelytron that separates Cuneus from Corium.

FRACTURED Adj. (Latin, *fractus* = broken.) Pertaining to structure interrupted in continuity; often applied to wing venation or sclerite. See Broken; Discontinuous; Ruptured.

FRACTUS Adj. (Latin, *fractus* = broken.) 1. Broken. 2. A term also applied to a geniculate Antenna.

FRAENKEL, GOTTFRIED, S (1912–) (Friedman *et al.* 1972, Israel J. ent. 7: i-v, 157–168, bibliogr.)

FRAGILE Adj. (Latin, *fragilis* = fragile.) Easily breakable; thin and brittle. Cf. Gracile.

FRAMBESIA See Yaws.

FRANCESCHINI, FELICE (1845–1928) (Grandori 1930, Boll. Lab. zool. Ist. agr. Milan 1: 5–8.)

FRANCHINI, GUISEPPE (1879–1938) (Roubaud 1938, Bull. Soc. Path. exot. 31: 533–534.)

FRANCILLON, JOHN (1744–1816) (M. J. 1976, Proc. Brit. ent. nat. Hist. Soc. 8: 99.)

FRANCIS, WILLIAM (1817–1904) (Anon. 1904, Proc. Linn. Soc. Lond. 1903–04: 31.)

FRANCK, GEORGE (1839–1923) (Bather 1924, Bull. Brooklyn ent. Soc. 19: 103.)

FRANCK, PAUL A R (1874–1936) (Weidner 1967, Abh. Ver. naturw. Verh. Hamburg Suppl. 9: 263.)

FRANCLEMONT, JOHN G (1912–) American academic (Cornell University) and taxonomic specialist in Lepidoptera, particularly noctuid moths of the New World.

FRANCOIS, PHILIPPE (1859–1908) (Caullery *et al.* 1909, Bull. scient. Fr. Belg. 42: lxxxv–xciv, bibliogr.)

FRANK VON FRANKENAU, GEORG (1644–1704) (Rose 1850, *New General Biographical Dictionary* 7: 438.)

FRANK, ALBERT BERNHARD (1839–1900) (Schröder 1900, Illte Z. Ent. 5: 390–391.)

FRANK, HANS (1884–1950) (Reisser 1950, Z. wein. ent. Ges. 35: 123–214.)

FRANKENBERGER, ZDENEK (1892–1966) (Balthasar 1966, Acta ent. bohemoslovaca 63: 241–243, bibliogr.)

FRANKLIN, BENJAMIN (1706–1790) (Rose 1850, *New General Biographical Dictionary* 7: 440–442.)

FRANKLIN, HENRY JAMES (1883–1958) (Mallis 1971, *American Entomologists*. 549 pp. (447–448), New Brunswick.)

FRANTZIUS, ALEXANDER VON (1821–1877) (Ecker 1878, Arch. Anthrop. 10: 343.)

FRANZEN, LUDWIG (1878–1945) Amateur entomologist in Queensland, Australia who published life histories on many Species of Hemiptera, Neuroptera and Lepidoptera.

FRASER, FREDERIC CHARLES (1880–1963) (Wigglesworth 1964, Proc. R. ent. Soc. Lond. (C)28: 57.)

FRASER, GERALD DE COURCY (–1953) (Worms 1953, Entomologist 86: 58.)

FRASS Noun. (German; *fressen* = to devour; MHG *vraz* = to devour.) 1. Solid larval insect excrement (Torre-Bueno). 2. Macerated plant material fashioned by wood-boring insects that is often combined with excrement (Borror *et al.*). See Excreta; Excretion. Cf. Honeydew; Meconium. Rel. Digestion.

FRAUENFELD, GEORG RITTER VON (1807–1873) (Papavero 1975, *Essays on the History of Neotropical Dipterology*. 2: 286–288. São Paulo.)

FRAYED SETAE Coccids: Clavate Setae in which the margin of the distal part is indented or toothed (MacGillivray).

FRAZIUS, WOLFGANG (1564–1628) (Rose 1850, *New General Biographical Dictionary* 7: 442.)

FREAR, DONALD, E E (1906–1973) (Coon 1974, J. Econ. Ent. 67: 314.)

FREDERICI, ENRICO (1901–1925) (Cannezzi 1925, Riv. biol. Rome 7: 607–609, bibliogr.)

FREE Adj. (Middle English, *fre* = free.) 1. Unrestricted in movement. 2. Detached, not firmly joined with or united to any other part. A term used to describe Pupae when all the parts and appendages are separately encased (*e.g.* Coleoptera).

FREE PUPA Pupae in which the appendages and limbs are not fused to the outer covering. See Pupa. Cf. Obtect Pupa.

FREE WATER Water in any chemical compound (such as protoplasm) that may be evaporated without changing the composition or formula of the compound. Cf. Bound Water.

FREEBORN, STANLEY BARRON (1891–1960) (Usinger 1960, J. Econ. Ent. 54: 1069–1070.)

FREEMAN, FRANCIS FORD (1847–1908) (Waterhouse 1908, Proc. ent. Soc. Lond. 1908: cvi.)

FREEMAN, THOMAS N (1911–1975) (McGuffin 1975, Bull. ent. Soc. Can. 7: 95)

FREESE, ANTON (1866–1923) (Weidner 1967, Abh. Verh. naturw. Ver. Hamburg Suppl. 9: 293.)

FREEZE DRYING A procedure for preserving specimens in a dry condition that would otherwise be preserved in fluid. The insect is frozen in a natural position and then dehydrated under vacuum in a desiccator kept at –4 to –7°C. The frozen condition prevents distortion while drying. Time required to complete drying is variable, with a few days for small specimens and about a week for large specimens. When dry, specimens can be brought to room temperature and pressure and permanently stored in a collection. Cf. Critical Point Dry.

FREIGIUS (FREY) JOHANN THOMAS (1543–1583) (Rose 1850, *New General Biographical Dictionary* 7: 449.)

FREMLIN, HEAVER STUART (1865–1952) (Riley 1953, Proc. R. ent. Soc. Lond. (C) 17: 72.)

FRENATAE Noun. (Latin, *frenare* = to bridle.) The series of Lepidoptera in which a more or less well marked Frenulum occurs.

FRENATE Adj. (Latin, *frenare* = to bridle; *-atus* = adjectival suffix.) Pertaining to a Frenulum.

FRENCH ANYSTIS-MITE *Anystis wallacei* Otto [Acarina: Anystidae]. (Australia).

FRENCH, CHARLES (1840–1933) (Meiners 1948, Lepid. News 2: 40.)

FRENCH, GEORGE HAZEN (1841–1935) (dos Passos 1951, J. N. Y. ent. Soc. 59: 162.)

FRENCH'S CANEGRUB *Lepidiota frenchi* Blackburn [Coleoptera: Scarabaeidae]: In Australia, larvae attack sugar cane throughout Queensland. See Canegrub; Scarabaeidae.

FRENULAR HOOK Male frenate Lepidoptera: A hook or fold into which the Frenulum is fitted.

FRENULUM Noun. (Latin, *frenulum*, diminutive of *frenum* = bridle. Pl., Frenula.) 1. Lepidoptera: One or a group of fused or closely appressed Setae (bristles) (simple in males, compound in females) arising from the Costal base of the hindwing in many taxa. Frenular bristles project toward the forewing and engage the Retinaculum to unite wings during flight. 2. Cicada: The triangular lateral sclerite on the Mesonotum that connects with the Trochlea, the anal area of the hindwings, and thus the Tendo.

FRENUM Noun. (Latin, *frenum* = bridle. Pl., Frena.) 1. A device that holds things together. 2. A lunate or triangular portion at the inner and posterior base of the wing in Odonata and Trichoptera. (See Tendo). 3. Heteroptera: The lateral groove in the upper margin of the Scutellum that catches the channelled locking device on the lower edge of the hemelytral Clavus.

FRERS, ARTURO GERMAN (1900–1925) (Lizer y Trelles 1947, Curso Ent. 1: 50–51.)

FREUND, LUDWIG (1878–1953) (Hase 1953, Anz. Schädlingsk. 26: 189.)

FREY, HEINRICH (1822–1890) (Anon. 1958, Mitt. schweiz. ent. Ges. 31: 112.)

FREY, J T See Freigius.

FREY, RICHARD HJALMAR (1886–1965) (Hackman 1966, Arsb. Vuosil. Soc. sci. fenn. 44 C: 1–7.)

FREY, WALTER (1911–1973) (Weidner 1974, Anz. Schädlingsk. Pflanzen- Umbelts 47: 30–31.)

FREYER, JOHN (1886–1948) (Gimingham 1948, Nature 162: 986–987.)

FREY-GESSNER, EMILE (1826–1919) (Handschin 1958, Mitt. schweiz. ent. Ges. 31: 109–120.)

FREYREISS, GEORG WILHELM (1789–1825) (Papavero 1971, *Essays in the History of Neotropical Dipterology* 1: 57–60. São Paulo.)

FRIEB, HERMANN (–1948) (Horion 1949, Koelopt. Z. 1: 78.)

FRIEDENFELS, EUGEN DROTLEFF VON (1819–1885) (E. A. B. 1886, Verh. Mitt. siebenbürg Ver. Naturw. 36: 1–5.)

FRIEDERICHS, KARL (1879–1969) (Schmidt 1969, Mitt. dt. ent. Ges. 28: 39.)

FRIEDLANDER, JULIUS (1827–1882) (Anon. 1882, Leopoldina 18: 210.)

FRIEDRICH, OTTO (1800–1880) (Geiser 1937, *Naturalists of the Frontier*. 341 pp. (323), Dallas.)

FRIEND, ROGER BOYNTON (1896–1962) (Mallis 1971, *American Entomologists*. 549 pp. (448–449), New Brunswick.)

FRIGATE-BIRD FLY *Olfersia spinifera* (Leach) [Diptera: Hippoboscidae].

FRIGYES, WAGNER (1873–1938) (Szent-Ivany 1938, Folia ent. hung. 4: 65–66.)

FRINGE Noun. (Middle English, *frenge*. Pl., Fringes.) 1. Marginal Setae, scales or other processes that project beyond the edge of the wing membrane. 2. Coccoidea: The marginal band of glassy threads of wax excreted from the Octacerores in the Asterolecaniinae. 3. The pygidial fringe *sensu* MacGillivray. Syn. Marginal Fringe.

FRINGE MOTHS See Cosmopterygidae.

FRINGED BLUE-BUTTERFLY *Neolucia agricola* (Westwood) [Lepidoptera: Lycaenidae]. (Australia).

FRINGED NETTLEGRUB *Natada nararia* Mo. [Lepidoptera: Limacodidae]: A significant pest of tea in parts of southeast Asia. Multivoltine with life cycle in 11–12 weeks. Eggs are laid on upper surfaces of old leaves while young larvae feed on undersurface of leaves. Cocoons are beneath bush or under stones; adult female lays eggs on day following emergence with ca 400 eggs per female.

FRINGED-ORCHID APHID *Cerataphis orchidearum* (Westwood) [Hemiptera: Aphididae].

FRINGED PLATES Coccids: Pectinae *sensu* MacGillivray.

FRINGED SETAE Coccids: The transverse row of Setae on the ventral aspect of the caudal end of the anal tube (MacGillivray).

FRINGS, KARL (1874–1931) (Pfaff *et al.* 1934, Feschr. Bestehen. Int. Ent. Ver. Frankf. 50: 7.)

FRIOMET, CHARLES (–1945) (Berger 1945, Lambillionea 45: 77.)

FRIPP, HENRY EDWARD (1816–1880) (Anon. 1892, Proc. Bristol Soc. nat. Hist. 7: 1–3, bibliogr.)

FRISCH, JOHANN LEONHARD (1666–1743) German teacher and author of *Beschreibung von allerley Insekten in Teutschland* (13 parts, 1720–1738), a well illustrated account of the life history and damage caused by ca 300 Species endemic to Germany. (Howard 1930, Smithson. misc. Collns. 84: 202.)

FRISEN DAHL, AXEL (1890–1919) (Lundblad 1919, Ent. Tidskr. 40: 53–61, bibliogr.)

FRISON, THEODORE HENRY (1895–1945) (Mallis 1971, *American Entomologists*. 549 pp. (449–451), New Brunswick.)

FRIT FLIES See Chloropidae.

FRIT FLY *Oscinella frit* (Linnaeus) [Diptera: Chloropidae].

FRITILLARIES See Nymphalidae.

FRITILLARY Noun. (Pl., Fritillaries.) Common name for some nymphalid butterflies with checkered or spotted pattern of coloration on wings.

FRITSCH, KARL (1812–1879) (Anon. 1880, Psyche 3: 71.)

FRITZ, AUGUST (1901–1967) (Wyniger 1967, Mitt. ent. Ges. Basel 17: 144.)

FRIVALD, EMÉRIC FRIVALDSZKY VON (1799–1870) (Wallace 1871, Proc. ent. Soc. Lond. 1871: liii.)

FRIVALDSZKY, JANOS (1822–1895) (Csiki 1903, Rovart. Lap. 10: 23–25.)

FROG CICADA *Venustria superba* Goding & Froggatt [Hemiptera: Cicadidae]. (Australia).

FROG HOPPERS See Cercopidae.

FROGGATT, JOHN LEWIS (1891–1975) (Monteith 1975, News Bull. ent. Soc. Qd 3: 46–47.)

FROGGATT, WALTER WILSON (1858–1937) (Walton 1942, Proc. Linn. Soc. N. S. W. 67: 77–81.)

FROGGATT'S CANEGRUB *Lepidiota froggatti* Macleay [Coleoptera: Scarabaeidae]. (Australia). See Canegrub.

FROGHOPPERS See Cercopidae.

FROHAWK, FREDERICK WILLIAM (1861–1946) (Riley 1947, Entomologist 80: 25–27.)

FROHBERG, ERICH (1910–1974) (Grueber 1974, Prakt. SchädlBekämpF. 26: 103.)

FRÖLICH, FRANC A G (Anon. 1858, *Accentuated*

list of British Lepidoptera. 118 pp. (xxii), Oxford & Cambridge Entomological Societies, London.)

FRÖMMING, EWALD (–1960) (Zimmerman, Mitt. dt. ent. Ges. 19: 75.)

FRONS Noun. (Latin, *frons* = forehead. Pl., Frontes.) 1. General: An unpaired head sclerite positioned between the arms of the Epicranial Suture. Frons typically bears the median Ocellus (Imms). Frons of most insects is limited ventrally by the Frontoclypeal Suture (Epistomal Suture), a transverse suture beneath the antennal sockets. Frons of most Schizophora Diptera has become modified to form a Ptilinum. 2. General: Upper/anterior portion of the head capsule, usually a distinct sclerite between Epicranium and Clypeus (Tillyard). 3. Auchenorrhynchous Hemiptera: A conventional area of the head, not definitely delineated from Epicranium and Clypeus (Tillyard). 4. Siphonaptera: Part of the anterior dorsal wall of the head anterior of the Antennal groove and the Falx (Comstock). Syn: Front. See Head Capsule. Cf. Clyeus; Face.

FRONTAL Adj. (Latin, *frons* = forehead; *-alis* = pertaining to.) 1. Descriptive of structure or process associated with the 'front' of the head. 2. Pertaining to the anterior aspect of any structure. Rel. Subfrontal. Cf. Facial; Genal; Occipital.

FRONTAL AREA 1. Generalized insect head: Typically the facial region between the Antennae or the Frontal Sutures when present, and base of the Labium (Snodgrass). 2. Hymenoptera: A head region located between the scrobal impressions, frontal crest and ocellar furrow. 3. Specifically Ants: A small, demarked, usually triangular space in the middle line just above or behind the Clypeus (Wheeler).

FRONTAL BRISTLES (FRONTALS) Diptera: Setae along the medial margin of the Parafrontals (Curran).

FRONTAL COSTA Orthoptera: Prominent, vertical ridge of the head which is median or lateral. See Median Carina, Lateral Carina.

FRONTAL CREST Hymenoptera: An elevation extending across the head just above the antennal sockets. FC usually is limited on each side by the Antennal furrows (Sulcus), but sometimes extending across the Antennal furrows nearly to the margin of the Compound Eyes. Frequently interrupted by the median Fovea, when it is said to be broken.

FRONTAL DILATORS OF PHARYNX One or more pairs of slender muscles arising on the Frons and inserted on the anterior part of the Pharynx (Snodgrass).

FRONTAL DISC Muscid larvae: A prominent histoblast upon which the rudiment of an Antenna is developed (Comstock).

FRONTAL FASTIGIUM Orthoptera: A facial process that extends dorsad between the Antennae and meeting or nearly meeting the Fastigium of the Vertex in Tettigonidae.

FRONTAL FISSURE Diptera: The impressed line extending from the Frontal Lunule to the border of the mouth.

FRONTAL GANGLION CONNECTIVES Connectives between the Tritocerebrum and the Frontal Ganglion. See Tritocerebrum.

FRONTAL GANGLION A small, triangular, median ganglion positioned above Oesophagus and anteriad of Brain. FG is anteriormost incretory organ, positioned anterior or ventral of Tritocerebrum and connected to it via a pair of Frontal Ganglionic Connectives. See Incretory Organs; Tritocerebrum.

FRONTAL GLAND A large, median gland beneath the Integument of head in certain soldier-termites. FG produces milky secretions for defence and opens via a frontal pore. See Gland.

FRONTAL LOBE Psyllidae: Two lobes (swellings) more-or-less completely divided by a suture in which an Ocellus is situated.

FRONTAL LUNULE Diptera: An oval or crescentic space above the base of the Antennae; in Cyclorrhapha, bounded by the Frontal Suture.

FRONTAL NERVE Nerve arising from the Frontal Ganglion.

FRONTAL OF THE TENTORIUM Basally fused anterior arms of the Tentorium which form a broad sclerite in Blattidae (Imms).

FRONTAL ORBIT See Facial Orbit (MacGillivray).

FRONTAL ORBITS Diptera: The space contiguous to the eyes on the Front (Curran). The genovertical plates *sensu* Comstock.

FRONTAL PITS The invaginations of the anterior arms of the Tentorium (Crampton).

FRONTAL PLANE Any one of several mutually parallel planes that pass through a body at a right angle to the transverse plane and sagittal plane. Syn. Horizontal Plane. Cf. Parasagittal Planes; Sagital Planes; Transverse Plane.

FRONTAL PROCESSES Diptera: The middle of the Front when it is membranous or discoloured. Syn. Vitta Fronalis.

FRONTAL SUTURE 1. The suture between the Front and the Clypeus. 2. Diptera: FS separates the Frontal Lunule from the part of the head above it. 3. Coleoptera: Clypeal Suture or the suture formed by the arms of the Epicranial Suture (Crampton). See Epistomal Suture.

FRONTAL TRIANGLE Diptera: The triangular space in males, between the eyes and limited ventrally by a line drawn through base of the Antennae. The triangle in holoptic flies, or the area bounded above by the eyes and below by the Antennae (Curran). See Frons.

FRONTAL TUBERCLE 1. Isoptera: A prominent, often horn-like, structure in soldiers, through which the Frontal Gland duct opens. Aphididae: The raised structures upon which Antennae are placed (Smith).

FRONTAL VESICLE Odonata: A swelling between the Compound Eyes, bearing the Ocelli (Garman).

FRONTAL VITTA Diptera: The softer area between

the rows of frontal bristles or Setae extending from the Antennae to the Ocelli (Curran).

FRONTALIA Noun. Latin, *frons* = forehead.) Diptera: The central stripe of the front, the Frontal Vitta (Curran).

FRONTISPIECE Noun. (Latin, *frontispicium* = beginning. Pl., Frontispieces.) The part that meets the eye first.

FRONTOCLYPEAL AREA Part of the insect head, typically the facial region between the Antennae or the Frontal Sutures when present, and the base of the Labium (Snodgrass).

FRONTOCLYPEAL SUTURE Suture between the front and the Clypeus. See Epistomal Suture.

FRONTOCLYPEAL Adj. Latin, *frons* = forehead; *clypeus* = shield;; *-alis* = pertaining to.) Descriptive of or pertaining to the combined Frons and Clypeus.

FRONTOCLYPEUS Noun. (Latin, *frons* = forehead; *clypeus* = shield.) 1. Diptera: The combined Front and Clypeus when the suture between them is obosolete. 2. The Epistoma *sensu* Comstock.

FRONTOGENAL SUTURE See Subantennal Suture.

FRONTO-ORBITAL BRISTLES Diptera: Setae on each side of the Front, ventrad of the vertical bristles (Smith).

FROST, CHARLES ALBERT (1872–1962) (Darlington 1963, Psyche 70: 3–6, bibliogr.)

FROST, HAROLD LOCK (1875–1940) (Higgins 1942, Fernald Cl. Mass Sta. Coll. Ybk. 11: 1–3.)

FROST PROTECTION Thermoregulatory processes by which insects avoid freezing by the presence of cryoprotectants in the Haemolymph. Frost Protection may involve Supercooling or Frost Resistance. Supercooling (avoidance of freezing) is due to the absence of nucleation sites for the formation of ice crystals in the Haemolymph. Glycerol, Sorbitol, Mannitol and Threitol are polyols in the Haemolymph which promote Supercooling; thermal-hysteresis proteins may also promote Supercooling. The mechanism of Frost Resistance (tolerance of freezing) is poorly understood. See Thermoregulation. Rel. Circulatory System.

FROST, S W (Anon. 1958, Ent. News 69: 67–78, bibliogr., only.)

FROSTED SCALE *Eulecanium pruinosum* (Coquillett) [Hemiptera: Coccidae]. (Australia).

FROTH GLANDS Cercopidae: See Glands of Batelli (Comstock).

FRUACARB® See Carbofuran.

FRUGIVOROUS Adj. (Latin, *frux* = fruit; *vorae* = to devour.) Cf. Carpophagous.

FRUHSTORFER, HANS (1863–1922) (Rohlfien 1975, Beitr. Ent. 25: 268)

FRUIT BORERS See Agonoxenidae.

FRUIT FIX® See NAA.

FRUIT FLIES See Drosophilidae; Tephritidae. Cf. Vinegar Flies.

FRUITING BODY Fungi: A general term for spore-bearing structures.

FRUITONE-N® See NAA.

FRUITPIERCING MOTH *Othreis fullonia* (Clerck), *Othreis materna* (Linnaeus), *Eudocima salaminia* (Cramer) [Lepidoptera: Noctuidae]: Pests of many tropical and subtropical fruits including banana, carambola, fig, guava, mango, papaya and persimmon; FPM is widespread in western Pacific, southeast Asia and Africa. Adults are large and stout-bodied, mottled brown; wingspan ca 100 mm; hindwing predominantly yellow with large black spots and margin with row of white spots. Larvae are black with two large white-and-black spots on segments 6 and 7 and undergo five instars. In Australia larvae feed on native vines (Menispermaceae); adults are nocturnal, use their Proboscis to pierce rind of developing fruit and create a feeding hole ca 2 mm diameter that causes premature decay, discolouring, fruit drop and is a source of entry for other fruit-feeding insects. Pupation occurs within silken chamber in webbed leaves. Life cycle requires a few months. FPM is multivoltine with several generations per year, the precise number determined by climate.

FRUIT-SPOTTING BUG *Amblypelta nitida* Stål [Hemiptera: Miridae]: A multivoltine pest of numerous horticultural crops in Australia that damages most fruits and macadamia nuts by feeding activities. FSB overwinters as an adult; females lay a few eggs per day but about 150 eggs during lifetime. Eggs are pale green-opalescent and triangular with rounded corners; eggs are laid singly on fruit or foliage. Nymphs and adults feed on nuts/kernels and cause large dark brown spots on green husks; necrotic spots cause fruit fall. See Miridae.

FRUIT-TREE BARK BORER See Shothole Borer.

FRUIT-TREE BORER *Maroga melanostigma* (Wallengren) [Lepidoptera: Oecophoridae]: A minor pest of *Citrus* in Australia. See Oecophoridae.

FRUIT-TREE LEAFROLLER *Archips argyrospila* (Walker) [Lepidoptera: Tortricidae]: In North America, a widespread, monovoltine pest of deciduous fruit trees and periodic pest of apples. FLT overwinters in the egg stage; eggs are laid in masses of 30–100 on tree trunk or branches and covered with 'cement'; the masses are cryptically coloured; eclosion occurs during spring when trees are beginning to bud. Larvae feed on leaves, buds and small fruit which are webbed together; feeding requires about a month. Pupation occurs in rolled leaves and webbing within a flimsly cocoon. Adults emerge during July with oviposition soon thereafter. See Tortricidae. Cf. Red-Banded Leafroller.

FRUIT-TREE PINHOLE BORER *Xyleborus saxesen* (Ratzeburg) [Coleoptera: Scolytidae]. (Australia).

FRUIT-TREE ROOT WEEVIL *Leptopius squalidus* (Boheman) [Coleoptera: Curculionidae]. (Australia).

FRUIT-WORM BEETLES See Byturidae.

FRUMIN-AL® See Disulfoton.

FRY, ALEXANDER (1821–1905) (Merrifield 1906, Proc. ent. Soc. Lond. 1905: lxxxvi.)

FRYER, ALFRED (1826–1912) (Bennett 1912, Proc. Linn. Soc. Lond. 124: 46–47.)

FRYER, HERBERT FORTESCUE (1854–1930) (Fryer 1930, Entomol. mon. Mag. 66: 114.)

FRYER, JOHN CLAUD FORTESCUE (1886–1948) (Williams 1949, Proc. R. Ent. Lond. (C) 13: 67.)

FUCHS, AUGUST (1830–1904) (Anon. 1943, Dt. ent. Z. Iris 57: 6.)

FUCHS, CHARLES (CARL) (1839–1914) (Osborn 1937, *Fragments of Entomological History*. 394 pp. (144), Columbus, Ohio.)

FUCHS, ROBERT (–1934) (Anon. 1936, Koleopt. Rdsch. 22: 121.)

FUCHS, RUDOLPH WILHELM (–1955) (Weidner 1967, Abh. Verh. naturw. Ver. Hamburg Suppl. 9: 293.).

FUEGLISTALLER-REINHOLD, HANS (– 1965?) (Hunziker 1966, Mitt. ent. Ges. Basel 16: 14)

FUGACIOUS Adj. (Latin, *fugacis* = fleeting; *-osus* = full of.) Botany: Pertaining to structures falling off a plant unusually early when compared with similar structures on other plants. Cf. Caducous; Marcescent.

FUGITIVE Adj. (Latin, *fugio* = to flee.) Short lived; soon disappearing; not permanent. Sometimes applied to Species.

FÜGNER, KARL (1842–1916) (Rosse 1934, Int. ent. Z. 28: 434.)

FUHR, JOSEF (1964–1941) (Sachtleben 1942, Arb. morph. taxon. Ent. Berl. 9: 66.)

FUHRMANN, OTTO (1871–1945) (Delachaux 1945, Verh. schweiz. naturf. Ges. 125: 366–367.)

FULCRAL Adj. (Latin, *fulcrum* = support; *-alis* = characterized by.) Descriptive of or pertaining to the fulcrum of a lever or similar mechanism.

FULCRAL PLATES Aculeate Hymenoptera: See Triangular Plates.

FULCRANT TROCHANTER A Trochanter which continues along the Femur but not intervening between it and the Coxa, as in Carabids. See Second Trochanter.

FULCRUM Noun. (Latin, *fulcrum* = support. Pl., Fulcra.) 1. General: Any structure that serves as a support to another structure. 2. Heteroptera: The Trochanter. 3. Diptera and Hymenoptera: The chitinous envelope at the base of the mouth covering the beginning of the Oesophagus or *os hyoideum*.

FULDA, OSCAR (–1945) (Anon. Bull. Brooklyn ent. Soc. 40: 71.)

FÜLDNER, JOHANN MORITZ (1818–1873) (Arnett 1874, Arch. Ver. Freunde Naturg. Mecklen. 28: 143–147.)

FULFILL® See Pymetrozine.

FULGID Adj. (Latin, *fulgeo* = to shine.) A bright, fiery red colour. Alt. Fulgidus.

FULGORIDAE Plural Noun. Fulgorid Planthoppers. A predominantly tropical Family of auchenorrhynchous Hemiptera assigned to Superfamily Fulgoroidea including about 700 described Species. Adults are moderate to large bodied (to 30 mm); Clypeus with lateral Carina; forewing Clavus long; hindwing with numerous cross-veins between Anal Veins; hind Tibia without large, articulated spur at apex. Nymphs and adults feed on shrubs and trees, inserting mouthparts through bark. See Fulgoroidea.

FULGOROIDEA Planthoppers. A cosmopolitan Superfamily of auchenorrhynchous Hemiptera with about 10,000 nominal Species that are best represented in tropical regions. Adult head carinate, not with sutures; Clypeus not extending between Compound Eyes; antennal socket beneath eye; Pedicel large with numerous Sensilla; typically two Ocelli, rarely three; Tegula on Mesothorax; middle Coxa long; females and nymphs often produce wax. Fulgoids are predominantly phloem feeders on angiosperms. Included Families: Acanaloniidae, Achilidae, Achilixiidae, Cixiidae, Delphacidae, Derbidae, Dictyopharidae, Eurybrachyidae, Flatidae, Fulgoridae, Gengidae, Hypochthonellidae, Issidae, Kinnaridae, Lophopidae, Meenoplidae, Nogodinidae, Ricaniidae, Tettigometridae and Tropiduchidae. See Hemiptera.

FULIGINOUS Adj. (Latin, *fuligo* = soot; *-osus* = full of.) Sooty or smoky brown. Alt. Fuliginosus.

FULLAWAY, DAVID TIMMINS (1880–1965) (Pemberton 1965, J. Econ. Ent. 58: 593–594.)

FULL-BAC® A registered biopesticide derived from *Bacillus thuringiensis* var. *kurstaki*. See *Bacillus thuringiensis*.

FÜLLEBORN, FRIEDRICH (1866–1933) (Stunkard 1934, J. Parasit. 20: 203, 206.)

FULLER ROSE-BEETLE *Asynonychus godmani* Crotch [Coleoptera: Curculionidae]. *Pantomorus cervinus* (Boheman). See Fuller's Rose Weevil.

FULLER, ANDREW SAMUEL (1828–1896) (Anon. 1896, Ent. News 7: 192.)

FULLER, CLAUDE W (1872–1928) (Anon. 1929, S. Afr. J. nat. Hist 6: 236–240, bibliogr.)

FULLER, FREEMAN M (1924–1959) (Potts 1959, J. Econ. Ent. 52: 788)

FULLER, HENRY SHEPARD (1917–1964) Krombein *et al.* 1964, Proc. ent. Soc. Wash. 66: 197–204, bibliogr.)

FULLER'S ROSE-WEEVIL *Asynonychus cervinus* (Boheman) [Coleoptera: Curculionidae]: Probably native of South America and a sporadic but important pest of tree crops, particularly citrus in many countries. Adults are ca 8 mm long, grey-brown with crescent-shaped white mark on the Elytra. FRW is parthenogenetic and males are unknown; females cannot fly and walk up trunk of trees to feed and oviposit. Eggs are yellow and laid in masses of 20–30 under stones, under bark, under Calyx lobes on fruit, or in curled dead leaves; females lay ca 160 eggs during life. Larvae are yellowish, legless and C-shaped; larvae feed on roots. Pupation occurs in earthen cells. FRW development requires about a year; and Species is monovoltine or perhaps completes two generations per year. See Curculionidae. Cf. Fuller Rose Beetle.

FULMEK, L (1883–1969) (Anon. 1970, Annln naturh. Mus. Wien 74: 667–670.)

FULTON, BENTLEY HALL (1889–1960) (Smith 1961, J. Econ. Ent. 54: 613–614.)

FULTURAE Plural Noun. (Latin, *fultura* = prop.) Myriapods: Sclerites that support the Hypopharynx. See Suspensorium of Hypopharynx.

FULVESCENT Adj. (Latin, *fulvus* = tawny.) Shining brown; of tawny lustre.

FULVID Adj. (Latin, *fulvus* = tawny.) Tawny, brownish yellow. Alt. Fulvous; Fulvus.

FULVO-AENEOUS Adj. (Latin, *fulvus* = tawny; *aeneous* = bronze; *-osus* = full of.) Brazen; with a touch of brownish yellow.

FULVOUS Adj. (Latin, *fulvus* = tawny; *-osus* = with the property of.) Tawny; a dull brownish-yellow.

FUMATE Adj. (Latin, *fumus* = smoky; *-atus* = adjectival suffix.) Smoky grey. Alt. Fumatus.

FUMEUS Adj. (Latin, *fumus* = smoky.) Smoke-coloured.

FUMI-CEL® See MGP.

FUMIGANT Noun. (Latin, *fumigare* = to fumigate; *-antem* = an agent of something. Pl., Fumigants.) 1. Volatile chemicals that act as poisonous gases in confined areas. Fumigants exist as small molecules that can move between fibres of wood and aggregate particles of concrete blocks. Fumigants are highly toxic compounds and include: Aip, Carbon Disulphide, Carbon Tetrachloride, Chloropicrin, Dazomet, DCIP, Hydrogen Cyanide, Enzone, Ethylene Dibromide, Ethylene Dichloride, Ethylene Oxide, MGP, Metham-Sodium, Methyl Bromide, Phosphine, Sulphuryl Fluoride and Vikane. 2. A volatile insecticide that enters an insect via the respiratory system.

FUMIGATION Noun. (Latin, *fumigare* = to fumigate; English, *-tion* = result of an action. Pl., Fumigations.) The act of fumigating or exposing something to fumes in order to disinfect or eliminate pests. See Fumigant. Rel. Methyl Bromide; Sulphuryl Fluoride.

FUMI-STRIP® See MGP.

FUMITE® See Tecnazene.

FUMITOXIN® See AlP.

FUMOSE Adj. (Latin, *fumus* = smoky; ; *-osus* = full of.) Smoky.

FUNCTION Noun. (Latin, *functio* > *functus* > *fungi* = to perform; English, *-tion* = result of an action. Pl., Functions.) 1. The work which an appendage, structure or organ normally performs. 2. The action for which an individual, structure or device is specifically responsible or adapted. Alt. Duty; Role; Responsibility. 3. An action that fits within a group of actions which is necessary for operation or completion of a task. Rel. Operation. Cf. Form and function.

FUNCTIONAL MONOGYNY Social Insects: A condition in which several inseminated queens coexist, but only one queen produces the reproductive brood.

FUNCTIONAL MORPHOLOGY Study of anatomical structure in relation to the way in which components operate. See Anatomy. Cf. Constructional Morphology.

FUNDAMENT Noun. (Middle English, *foundation* > Latin, *fundamentum* > *foundement* > *fundare* = foundation. Pl., Fundaments.) 1. The base of a structure (Obsolete). 2. The Anus. (Obscure). 3. The beginning or foundation of a structure in the embryo (Folsom & Wardle). 4. Primordium; Anlage.

FUNDARIMA Noun. (Etymology obscure.) The Fissure, line or furrow marking the line of fusion of the Stipulae (MacGillivray).

FUNDATRIGENIA Noun. (Latin, *fundus* = foundation. Pl., Fundatrigeniae.) Aphididae: Apterous, viviparous, parthenogenetic females which live on primary hostplants. Fundatrigeniae are the progeny of the Fundatrix. Cf. Alienicola; Fundatrix; Migrantes; Sexupara.

FUNDATRIX SPURIA Aphididae: A winged agamic form or migrant. See Fundatrix.

FUNDATRIX Noun. (Latin, *fundator* > *fundare* = to establish. Pl., Fundatrices.) Aphididae: Apterous, viviparous, parthenogenetic females that emerge during spring from overwintered eggs. Morphologically Fundatrices are characterized by smaller eyes, legs and other body parts. The Fundatrix typically forms a new colony by laying eggs on a leaf. Syn: Stem Mother. See Aphididae. Cf. Alienicola, Fundatrigenia, Migrantes, Sexupare, Sexuales.

FUNDITAE Plural Noun. (Latin, *fundus* > bottom. Sing., Fundita.) Chitinized areas between the Venellae and the Duritae (MacGillivray).

FUNDUS Noun. (French, *fond* > Latin, *fundus* = bottom, foundation, ground.) 1. The bottom of a tubular structure or pit. 2. The bottom of the inner surface of a hollow organ.

FUNETÉ, JOSÉ MARIA DE LA (1855–1932) (Navas 1932, Boln. Soc. ent. Esp. 15: 125–127)

FUNG-FU® See Lambda Cyhalothrin.

FUNGICIDE Noun. (Latin, *fungus* = mushroom; *-cide* > *caedere* = to kill. Pl., Fungicides.) A chemical compound which kills or inhibits development of fungi. Broadly categorized as Protectant Fungicides and Eradicant Fungicides. See Pesticide. Cf. Acaricide; Filaricide; Herbicide; Insecticide; Nematicide; Pesticide.

FUNGICOLOUS Adj. (Latin, *fungus* = mushroom; *colere* = living in or on.) Pertaining to organisms that live in or on fungi. Syn. Fungivorous. See Habitat. Cf. Caespiticolous; Dendrophilous; Entomophilous; Entomophytous; Fimicolous; Madicolous; Mymecophilous. Rel. Ecology.

FUNGINE Noun. (Latin, *fungus* = mushroom. Pl., Fungines.) The chitinous substance forming the cell wall of some fungi. Fungine was first isolated in 1811 by Odier.

FUNGIVORE Noun. (Latin, *fungus* = mushroom;

vorare = to devour. Pl., Fungivores.) Alt. Mycetophage.

FUNGIVORIDAE See Mycetophilidae; Sciaridae.

FUNGIVORITIDAE See Pleciofungivoridae.

FUNGIVOROUS Adj. (Latin, *fungus* = mushroom; *vorare* = to devour; *-osus* = with the property of.) Fungus-eating or devouring. A dietary habit apparently widespread in primitive groups of Coleoptera. Syn: Fungicolous; Mycophagous. See Feeding Strategy. Cf. Cannibalistic; Carnivorous; Carpophagous; Frugivorous; Parasitic; Predaceous; Phytophagous; Saprophagous; Xylophagous. Rel. Ecology.

FUNGOID Adj. (Latin, *fungus* = mushroom; Greek, *eidos* = form.) Fungus-like in physical appearance, structural texture or biological properties. See Fungus. Cf. Agaricoid. Rel. Eidos; Form; Shape.

FUNGUS Noun. (Latin, *fungus* = mushroom; Pl., Fungi.) A Kingdom of organisms the members of which are characterized by hyphae aggregated into mycelia and which lack chlorophyll. About 100,000 Species of fungi are known; the group is predominantly saprophytic, with some parasitic Species and other Species living as mycorrhizal symbionts. Fungal life cycle is complex and includes asexual (anamorphic) and sexual (telomorphic) phases. Fungi are the most common and frequently most important cause of plant disease. Sometimes categorized as Facultative and Obligate Parasites. See Mycorrhiza.

FUNGUS GARDEN 1. A chamber in ant nests in which ants cultivate fungi for food. 2. Beds in which many of the higher termites (Termitidae) grow fungi.

FUNGUS GNAT See Dark-Winged Fungus Gnat.

FUNGUS GNATS See Mycetophilidae.

FUNGUS WEEVILS See Anthribidae.

FUNGUS-EATING LADYBIRD *Illeis galbula* (Mulsant) [Coleoptera: Coccinellidae]. (Australia).

FUNICLE Noun. (Latin, *funiculus* = small cord. Pl., Funicles.) 1. The Antennal segments located between ring segments (Anelli) and the club (Clavus). 2. A small cord; a slender stalk. 3. Hymenoptera: A slender ligament or tendon connecting the Propodeum to the Petiole dorsally. Alt. Funicule; Funiculus.

FUNICULAR JOINT Hymenoptera: Any segment of the Funicle.

FUNICULATE Adj. (Latin, *funiculus* = small cord; *-atus* = adjectival suffix.) Whip-like; structures which are long, slender, composed of many flexible joints. Alt. Funiculatus.

FUNIS Noun. (Latin, *funis* = a rope.) 1. Diptera: The bar on which the Pseudotracheae terminate (MacGillivray). 2. The umbilical cord.

FUNKHAUSER, WILLIAM D (1881–1948) (Kopp & Yonke 1974, Ent. News 85: 131–145, bibliogr. and list of names proposed.)

FUNNEL Noun. (Medieval Latin, *fundibulum* > Latin, *infundibulum* > *infundere* = to pour in. Pl., Funnels.) 1. A structure in the shape of a hollow cone with a tube attached to the base. 2. Insects: The Peritrophic Membrane. (Arcane). Rel. Conduit; Cylinder; Tube.

FUNNEL ANT *Aphaenogaster pythia* Forel [Hymenoptera: Formicidae]: A pest of sugarcane grown on granitic gravel loams and sandy clay loams in Australia. Ants construct mounds of loose pellets of earth and subsoil in cane rows. Burrowing and soil inversion cause weakening of soil binding roots; root drying caused by increased aeration.

FUNNEL-WEB SPIDERS See Dipluridae; Hexathelidae.

FUNNEL-WEB TARANTULAS See Dipluridae.

FUR BEETLE *Attagenus pellio* (Linnaeus) [Coleoptera: Dermestidae]. (Australia).

FURACON® See Aminofuracarb.

FURADAN® See Carbofuran.

FURANOCOUMARINS. Plural Noun. A complex of light-activated alkaloid biopesticides extracted typically from Rutaceae and Umbelliferae but which also occur in other plant Families. Best known from the umbelliferous wild parsnip [*Dacus carota* Linnaeus] and rutaceous *Thamnosama montana.* Active compounds include Angelican, Bergapten, Isopsoralen, Psoralen and Xanthotoxin. See Biopesticide; Botanical Insecticide; Natural Insecticide; Insecticide; Pesticides. Cf. Polyacetylenes; Terthienyl.

FURATHIOCARB A carbamate compound {butyl 2,3-dihydro-2,2-dimethylbenzofuran-7-yl-N',N'-dimethyl-N',N'-thiodicarbamate} used as a nematicide, contact/systemic insecticide and stomach poison for control of nematodes, mites, rootworms, wireworms and aphids. Furathiocarb is applied to corn, cotton, cereals, rice, soybeans, sunflowers, sugar beet and vegetable crops in some countries. Compound is not registered for use in USA in part because it is toxic to fishes. Trade names include Deltanet® and Promet®. See Carbamate Insecticide.

FURCA Noun. (Latin, *furca* = fork. Pl., Furcae; Furcas.) 1. Collembola: A fork-like appendage on the fourth abdominal Sternum. Furca consists of a median, basal projection (Manubrium) that divides into a pair of Dentes (Dens), each of which may bear an apical segment called the Mucro. The Furca forms a spring-like device that engages hooks of the third Sternum and is used to catapult the animal into the air. Syn. Furcula. 2. Thysanura: A fork or anal appendage used for leaping. See Furcula. 3. Pterygota: The forked ental (inward projecting) processes of the Sternum; an endosternite. 4. A term generally applied to the Sternellum (Snodgrass). Cf. Antefurca; Apophysis; Entosternum; Entothorax; Medifurca; Postfurca; Episterne (MacGillivray). 5. Genitalia of male Lepidoptera: A structure, often consisting of paired halves, that arises as an elaboration of the ventral part of the Juxta (Klots). See Branch; Fork. Cf. Antefurca; Apodeme; Spina.

FURCAE MAXILLARES The Paragnaths in certain Corrodentia (Comstock).

FURCAL ARMS Sternal Apophyses (internal rod-like projections) of the Integument. Furcal Arms form an internal skeleton or surface for muscle attachment. See Furca. Cf. Apodeme. Rel. Skeleton.

FURCAL ORIFICE See Furcina. Sternal Orifice.

FURCAPECTINAE Plural Noun. (Latin, *furca* = fork; *pecten* = comb.) 1. Coccoidea: Pectinae with a slender shaft that displays 2–3 small inconspicuous teeth at the distal end. 2. Furcate sclerites of systematists (MacGillivray). Rel. Furca; Pectin.

FURCASTERNUM Noun. (Latin, *furca* = fork; *sternum* = breast bone. Pl., Furcasterna.) A distinct part of the Sternum in some insects bearing the Furca; a term generally applied to the Sternellum. See Sternum.

FURCATE Adj. (Latin, *furca* = fork; *-atus* = adjectival suffix.) Forked. Alt. Furcated; Furcatus. See Furca. Cf. Branched; Dendritic.

FURCATE PLATES Coccids: See Furcapectinae (MacGillivray).

FURCATE SETAE Coccids: frayed Setae.

FURCELLA Noun. (Latin, *furcula* = little fork. Pl., Furcellae.) The Spina or Monapophysis of the Sternellum (MacGillivray). See Apophysis; Furca.

FURCELLINA Noun. (Latin, *furcula* = little fork. Pl., Furcellinae.) The pit or thickening on the outer surface of the Sternellum marking the invagination of the Furcella (MacGillivray).

FURCINA Noun. (Latin, *furca* = fork. Pl., Furcinae.) 1. The point on the outer surface of an insect where a Epicranium has been invaginated. 2. The pit or thickening on the outer surface of the Sternum, marking the point of invagination of a Epicranium (MacGillivray).

FURCULA SUPRE-ANALIS See Supranal Fork.

FURCULA Noun. (Latin, diminutive *furca* = fork; little fork. Pl., Furculae.) 1. Any small forked process. 2. Collembola: See Furca. 3. Orthoptera: A pair of caudally directed appendages that overlie the base of the supra-anal sclerite in a more-or-less forked position.

FURCULAR Adj. (Latin, diminutive *furca* = fork; little fork.) Descriptive of or pertaining to the Furcula. See Furca.

FURNITURE BEETLE *Anobium punctatum* (DeGeer) [Coleoptera: Anobiidae]: A serious pest of timber, but not a major pest in warmer climates. FB is endemic to Europe and has become established elsewhere as a pest of pine, willow, maple and beech used in cabinet woods, shelving, flooring, housing timbers and imported antique furniture. Baltic pine *Pinus sylvestris*, New Zealand white pine *Podocarpus dacrydioides*, rimu *Dacrydium cupressinum*, hoop pine, kauri pines *Agathis* spp. are severely affected in Australia. FB also attacks hardwoods such as English oak *Quercus robur* and spotted gum. FB infestations usually are encountered in timbers in-service at least 20 years. Adults are 2.5–4.0 mm long, cylindrical, dark brown with reddish brown legs and head concealed beneath cowl-like Prothorax; Antennal club of 3 segments. Body with fine yellowish Setae and longitudinal rows of pits on Elytra. Females lay ca 30–80 ovoid eggs singly or rows of 2–4 in cracks, crevices and abraded areas of timber. Eclosion occurs within ca 2–5 weeks and neonate larvae bite part of eggshell on which female has deposited yeast cells. Symbiotic yeast enables the larvae to convert cellulose to protein in gut and the process is assisted by an enzyme. Larval head is gold-brown, Mandibles chestnut brown and body greyish white, covered with fine Setae. Larvae are hook-shaped and curl into a tight ball when removed from tunnel. Larvae tunnel in wood 1–5 years; tunnels are fashioned with grain of wood and across grain; tunnels are loosely packed with frass. Frass consists of cigar-shaped pellets containing chewed wood. When rubbed into palm of hand, frass feels coarse and gritty. Mature larvae move near surface to construct pupal chamber and the pupal stage requires 3–4 weeks. Adults emerge through round flight hole ca 2 mm in diameter and beetles live four weeks. Females shelter in flight holes periodically after emergence with many opportunities for reinfesting attacked wood; the natural spread of infestation is slow. Infested timber which is removed from service should be burned. Fumigation of valuable antiques is effective but does not prevent reinfestation. See Anobiidae. Syn. Common Furniture Beetle.

FURNITURE BEETLES Members of Anobiidae; ca 1,100 Species world-wide. FB feed upon relatively dry matter ranging from dried breakfast cereals to antique furniture; FB Species typically utilize only a narrow range of materials as food. Most larvae bore in wood or bark of dead trees. See Anobiidae.

FURNITURE CARPET-BEETLE *Anthrenus flavipes* LeConte. [Coleoptera: Dermestidae]: A cosmopolitan pest of wool and animal skins, upholstered furniture, carpets and some stored foods. FCB is endemic to Oriental Realm and distributed globally through commerce. Larvae are called 'woolly bears' because their bodies are covered with long Setae. Adults are 2.0–3.5 mm long, oval in outline shape mottled yellow, white and black and apex of Abdomen with a cleft where Elytra meet. Larvae are ca 5 mm long, dark with body widest toward anterior end, narrows toward apex of Abdomen. Larvae often feed in limited area and exuviae accumulate in one place. FCB is not regarded as serious a pest as Black and Varied Carpet Beetles. See Dermestidae. Cf. Australian Carpet Beetle; Black Carpet Beetle; Common Carpet Beetle; Varied Carpet Beetle.

FURRED Adj. (Middle English, *furre* = fur.) Covered with short, dense decumbent Setae; resembling fur.

FÜRSTENAU, JOHANN HERMANN (1688–1756) (Rose 1850, *New General Biographical Dictionary* 7: 463.)

FURY® See Zeta Cypermethrin.

FUSAREX® See Tecnazene.

FUSARIUM WILT A fungal disease of cotton which is caused by *Fusarium vasinfectum* and transmitted by grasshoppers.

FUSCESCENT Adj. (Latin, *fuscus* = dark, dusky; *-escent* = adjectival suffix, starting to be.) Becoming brown; structure with a brown shading.

FUSCO-FERRUGINOUS Adj. (Latin, fusco = dark; *ferruos* = rusty; *-osus* = full of.) Brownish; rust red.

FUSCO-PICEOUS Adj. (Latin, *fusco* = dark; *piceus* = pitch; *aceus* = of or pertaining to.) Pitch black with a brown tinge or admixture.

FUSCO-RUFOUS Adj. (Latin, *fusco* = dark; *rufus* = red; *-osus* = full of.) Red-brown; approaching liver brown.

FUSCO-TESTACEOUS Adj. (Latin, *fusco* = dark; *testa* = burnt piece of clay; *-aceus* = of or pertaining to.) Dull reddish brown.

FUSCOUS Adj. (Latin, *fusco* = dark; *-aceus* = of or pertaining to.) Dark brown, approaching black; a plain mixture of black and red. Alt. Fuscus.

FUSED Transitive Verb (Latin, *fusus*, past part. *fundere* = to pour out.) Combined; as when two normally separated spots become confluent and share a common outline.

FUSI Plural Noun. (Latin, *fusus* = spindle.) Spiders: Spinning organs that consist of two retractile processes that issue from the Mammulae and form silk threads. Cf. Spinneret.

FUSI PILIFORMIS Coccids: The sclerites (MacGillivray).

FUSI SPINIFORMIS Coccids: The sclerites (MacGillivray).

FUSIFORM Adj. (Latin, *fusus* = spindle; *forma* = shape.) Spindle-shaped; any structure which is elongate, broad near the middle of the long (primary) axis and tapering toward the ends. Adj. Fusiformate. See Spindle. Cf. Clavate; Filiform; Lentiform; Subfusiform; Vermiform. Rel. Form 2; Shape 2; Outline Shape.

FUSION Noun. (Latin, *fusio* > *fundere* > *fusum* = to melt, to pour. Pl., Fusions.) 1. A mass, body or group formed by bring together different parts or elements. 2. An amalgamation of adjacent, ordinarily discrete, parts or sclerites of the insect body. Cf. Amalgamation; Composite.

FUSS, CARL ADOLF (1817–1874) (Teutsch 1876, Verh. Mitt. siebenb. Ver. Nat. 26: 11–16.)

FUSS, HERMAN (1824–1915) (Hubenthal 1915, Dt. Ent. Z. 1915, 577–578.)

FÜSSLEY, JOHN CASPAR (1743–1786) (Rose 1850, *New General Biographical Dictionary* 7: 458.)

FUSULAE Plural Noun. Spiders: Minute tubules comprising the Spinnerets.

FUTURA® A registered biopesticide derived from *Bacillus thuringiensis* var. *kurstaki*. See *Bacillus thuringiensis*.

FUTURA® See Carbofuran.

FYFANON® Malathion.

FYLES, THOMAS W (1832–1921) (Bethune 1921, Can. Ent. 53: 262–264.)

GAAB, WILLIAM MORE (1839–1878) (Essig 1931, *History of Entomology*. 1029 pp. (638), New York.)

GÄBLER, HELLMUT (1904–1969) (Bassus 1970, Mitt. dt. ent. Ges. BRD 29: 1–2.)

GABRIEL, ALFRED GEORGE (1884–1968) (Kennedy 1969, Proc. R. ent. Soc. Lond. (C) 33: 55.)

GABRIEL, JOSEPH JOHANN CONRAD (1841–1937) (Anon. 1937, Arb. morphol. taxon. Ent. Berl. 4: 351.)

GAD FLIES Common name for adult flies of Genus *Hypoderma,* particularly *H. bovis,* which cause cattle to become restless and stampede to escape adult oviposition. See Tabanidae; Oestridae. Cf. Cattle Grubs.

GADAMER, HERMANN FRIEDRICH RUDOLF HEINRICH (1818–1885) (Wallengren 1885, Ent. Tidskr. 6: 177–178, 219–220.)

GADDIS, BEVY MARSHALL (1891–1949) (Annand & Richmond 1950, J. Econ. Ent. 43: 571–572.)

GADEAU DE KERVILLE, HENRI (1858–1940) (Regnier 1941, Bull Soc. Amis. Sci. nat Rouen (9) 76–77: 1–36, bibliogr.)

GAEDE, H M Author of *Beytrage zur Anatomie der Insekte*' (1815).

GAEDEKEN, CARL GEORG (1832–1900) (Englehart 1918, Ent. Meddr. 12: 22–23.)

GAGLIARDI, ALDO (1883–1969) (Martelli 1970, Memoria Soc. ent. ital. 49: 156–158, bibliogr.)

GAGNE, WAYNE C (1942–1988) American systematic Entomologist (Bernice P. Bishop Museum) specializing in Miridae. (Beardsley 1990, Proc. Haw. Ent. Soc. 30: 21–22.)

GAHAN, ARTHUR BURTON (1880–1960) American systematic Entomologist (U.S. Department of Agriculture) specializing on taxonomy of Chalcidoidea (Hymenoptera). (Cory & Muesebeck 1960, Proc. ent. Soc. Wash. 62: 198–204, bibliogr.)

GAHAN, CHARLES JOSEPH (1862–1939) (Anon. 1939, Museums J. 38: 586–587.)

GAILLARDOT-BEY, CHARLES (1814–1883) (Chevrolat 1883, Bull. Soc. ent. Fr. (6) 3: cx.)

GAIMARD, JOSEPH PAUL (1793–1858) (Musgrave 1932, *A Bibliogr. of Australian Entomology 1775–1930*. 380 pp. (117), Sydney.)

GAIN THRESHOLD The beginning point of economic damage by a pest to an agricultural product that is expressed in terms of the amount of harvestable produce. Concurrently, the point when the cost of suppressing insect injury is equal to the money to be gained from avoiding the damage, assuming the control is effective.

GAJL, KAZIMIERZ (1896–1934) (Jaczewski 1934, Polskie Pismo ent. 13: 215–217.)

GALACTICIDAE Plural Noun. A small Family of ditrysian Lepidoptera assigned to Superfamily Tineoidea. Galactidae are endemic to Old World with a few Species adventive to New World. Adults are small, wingspan 10–15 mm and head smooth with lamellar scales. Ocelli and Chaetosemata absent; Antenna scaled, filiform or bipectinate, usually with Pecten; Proboscis long, unscaled; Maxillary Palpus small, of two segments; Labial Palpus short, porrect; fore Tibia with Epiphysis; tibial spur formula 0-2-4; wings held tectiform over Abdomen at repose. Larvae are communal in webs during early instars; later development in enlarged web or individually between webbed leaves. Pupae with Maxillary Palpus and most abdominal segments with a transverse row of spines. Pupation occurs within silken cocoon inside the larval web.

GALACTOSE Noun. (Greek, *gala* = milk; *-osus* = full of.) A hexose sugar differing slightly from glucose in its molecular arrangement. Formed with an equal quantity of glucose when lactose is hydrolysed.

GALAN Y RUIZ, ALFONZO (–1919) (Arras 1919, Boln. R. Soc. esp. Hist nat. 19: 122–123.)

GALARASTRA Noun. (Latin, *galea* = helmet; *raster* = rake.) The Rastra on the Galea. The Galeafimbrium *sensu* MacGillivray.

GALASSI, RENATO (1896–1972) (Berio 1973, Mem. Soc. ent. ital 52: 59–60.)

GALEA Noun. (Latin, *galea* = helmet. Pl., Galeae.) 1. Generalized Maxilla: The lateral sclerite attached to the distal margin of the Stipes. Galea sometimes appears 2-segmented and often hood-like. Galea is subject to great modifications in Hymenoptera and Diptera; it forms the elongate, tubular Proboscis of Lepidoptera. See Maxilla. Cf. Lacinia; Palpifer. Rel. Subgalea. 2. Acarina: Paired, anterolateral, lamellate projections extending from the Gnathobase and curl dorsally around the Chelicerae. A Seta is present on each anterior surface. Syn. Outer lobe of palpal base (Goff *et al.* 1982, J. Med. Ent. 19: 225).

GALEA PALPIFORMIS The Galea when it is distinct from the Lacinia and composed of several cylindrical segments.

GALEAFIMBRIUM Noun. (Latin, *galea* = helmet. *fibra* = band; Latin, *-ium* = diminutive > Greek, *-idion*. Pl., Galeafimbria.) The Galarastra *sensu* MacGillivray.

GALEALA Noun. (Latin, *galea* = a helmet. Pl., Galealae.) Acari: Setae on the anterolateral surface of the Galea. Syn. Galeal Seta; Protorostral Seta (Goff *et al.* 1982, J. Med. Ent. 19: 225).

GALEARIA Noun. Hymenoptera: The lobe of the Proxagalea *sensu* MacGillivray.

GALEARIS Noun. Lepidoptera: The closely appressed Galeae *sensu* MacGillivray.

GALEATE Adj. (Latin, *galeatus* = helmeted.) Helmet-shaped; hooded. Alt. Galeiform. See Helmet; Hood.

GALEAZZI, GIACOMO (–1869) (Targioni-Tozzetti 1869, Boll. Soc. ent. Ital. 1: 256.)

GALEIFORM Adj. (Latin, *galea* = helmet; *forma* = shape.) Helmet-shaped. See Helmet; Hood. Cf. Galeate. Rel. Form 2; Shape 2; Outline Shape.

GALEN (ca 130–200 AD) Greek physician, anatomist and developmental biologist.

GALEOTHECA Noun. (Latin, *galea* = helmet; *theca* = sheath. Pl., Galeothecae.) Part of the pupal case that covers the Galea.

GALGULIDAE Billberg 1820. See Gelastocoridae. Name *Galgulus* Fabricius in Insecta preoccupied by *Galgulus* Pliny in Aves.

GALL Noun. (Anglo Saxon, *gealla* = gall. Pl., Galls.) 1. An abnormal growth or swelling of plant tissue induced by insects (gall wasp, gall midge), mites, bacteria, parasitic fungi or diseases of the plant. 2. Any deviation in the normal pattern of plant growth produced by a specific reaction to the presence and activity of a foreign organism (Bloch 1965). 3. An overgrowth of plant tissue. Küster (1911) recognized two types of galls: Organoid gall and histioid gall. More recent classifications recognize several types: Bud-and-Rosette Gall, Covering Gall, Filz Gall, Mark Gall, Pit Gall, Pouch Gall, Roll-and-Fold Gall. See each term for explanation. Alt. Cedidium.

GALL APHIDS See Phylloxeridae.

GALL FLIES See Tephritidae.

GALL INSECT Insects that induce a developmentally regulated proliferation of tissues on the plants they infest. Gall anatomy is often Species-specific. Gall-forming insects are referrable to Upper Carboniferous (Labanderia, Nature 1995.) See Gall.

GALL MIDGES See Cecidomyiidae.

GALL WASP 1. Any member of the hymenopterous Genus *Cynips* which causes gall formation in plants. 2. In general, any hymenopterous insect that induces tumorous growth on a plant.

GALLA Noun. (Anglo Saxon, *gealla* = gall. Pl., Gallae.) A plant gall.

GALLARDO, ANGEL (1867–1934) (Bruck 1934, Rev. Soc. ent. argent. 6: 235–242, bibliogr.)

GALLED Adj. (Anglo Saxon, *gealla* = gall.) Pertaining to a plant which has galls.

GALLEGOS, JOSÉ MARIA (–1925) (Van Duzee 1925, Pan-Pacif. Ent. 2: 96.)

GALLFLIES See Cynipidae.

GALLI, VALERIO BRUNO (1867–1943) (Roux 1938, Schweizer/Arch. Tierheilk. 80: 57–59.)

GALLICOLA Noun. (Anglo Saxon, *gealla* = gall; Latin, *colere* = to dwell. Pl., Gallicolae.) Leaf-gall formers in the Genus *Phylloxera*.

GALLICOLAE MIGRANTES Winged gall-making forms that fly to an intermediate host, *e.g.* *Chermes* of spruce, these hibernate as first stage nymphs (Imms).

GALLICOLOUS Adj. (Anglo Saxon, *gealla* = gall; Latin, *colere* = to dwell.) Pertaining to organisms that live within galls as producers of galls or as inquilines of galls produced by other organisms. See Gall; Habitat. Cf. Aphidicolous. Rel. Ecology.

GALLINIPPER Noun. (Anglo Saxon, *gealla* = gall; English, *nip* = to pinch. Pl., Gallinippers.) *Psorophora ciliata* [Diptera: Culicidae]: A pest of eastern USA. Females are large-bodied with wingspan ca 14 mm and median, longitudinal band of yellow scales on Mesonotum flanked by bare spot. Hind Tibia and Tarsus with dense tuft of long Setae. Larvae inhabit flood waters and rain pools; predaceous upon other mosquito larvae. See Culicidae.

GALLIPHAGOUS Adj. (Anglo Saxon, *gealla* = gall; Greek, *phagein* = to devour; *-osus* = with the property of.) Pertaining to organisms that feed upon galls or gall tissue. Alt. Gallivorous. See Feeding Strategy.

GALLMAKING MAPLE-BORER *Xylotrechus aceris* Fisher [Coleoptera: Cerambycidae].

GALLOIS, JEAN (1632–1707) (Rose 1850, *New General Biographical Dictionary* 7: 475.)

GALTON, FRANCIS (1822–1911) (Anon. 1911, Entomol. mon. Mag. 47: 72.)

GALVAGNI, EGON (1874–1955) (Stronhal 1955, Annl. naturh. Mus. Wien 60: 17–19.)

GALVANOTROPISM Noun. (After Luigi Galvani; *-tropism*, comb. form = a tendency to turn; English, *-ism* = state. Pl., Galvanotropisms.) Electrotropism. An observable reaction by organisms to electric currents. See Tropism. Cf. Aeolotropism; Anemotropism; Chemotropism; Electrotropism; Geotropism; Heliotropism; Hydrotropism; Phototropism; Rheotropism; Stereotropism; Thermotropism; Thigmotropism; Tonotropism. Rel. Taxis.

GAMASIDA Plural Noun. A group of Anactiotrichid mites. Syn. Mesostigmata.

GAMBIAN SLEEPING SICKNESS See Sleeping Sickness.

GAMERGATE Noun. (Pl., Gamergates.) Social Insects: A mated, egg-laying worker.

GAMETE Noun. (Greek, *gametes* = spouse. Pl., Gametes.) The mature egg within the female reproductive system; or sperm-cell. Gametes typically contain the haploid number of Chromosomes. Gametes typically unite during the process of fertilization and produce a Zygote with the diploid complement of Chromosomes. See Fertilization. Cf. Ova; Sperm. Rel. Chromosome.

GAMETOCYTE Noun. (Greek, *gametes* = spouse; *kytos* = hollow vessel; container. Pl., Gametocytes.) A sex cell in *Plasmodium*, the malarial parasite, developed from the Merozoite.

GAMMA HCH See Lindane.

GAMMA TAXONOMY Taxonomy: Studies of the biology, interspecific interactions and rates of evolution of Taxa. See Taxonomy. Cf. Alpha Taxonomy; Beta Taxonomy.

GAMMALIN See Lindane.

GAMMA-MEAN See Lindane.

GAMMAPHEX See Lindane.

GAMMA-SOL See Lindane.

GAMMEX See Lindane.

GAMMEXANE See Lindane.

GAMOGENESIS Noun. (Greek, *gametes* = spouse; *genos* = descent. Pl., Gamogeneses.) Reproduction through fertilization. See Agamogenesis.

GAMOGENETIC EGGS Aphididae: The eggs of the

true sexual forms, as distinguished from the so-called Pseudova of the agamic females (Comstock).

GANDARA, ALVARO FRANCO (–1958) (Azevedo 1959, Anais Inst. Med. trop. Lisb. 16: 763–768, bibliogr.)

GANDOLPHE, PAUL LOUIS (–1889) (Anon. 1890, Bull. Soc. ent. Fr. (6) 9: ccxxx.)

GANGLBAUER, LUDWIG (1856–1912) (Anon. 1912, Entomol. mon. Mag. 48: 217–218.)

GANGLIA ALLATA See Corpora Allata.

GANGLION VENTRICULARE The unpaired ganglion anterior of the Brain. Syn. Frontal Ganglion.

GANGLION Noun. (Greek, *ganglion* = little tumor. Pl., Ganglia.) Groups of nerve cells that form a mass called a Ganglion. A Ganglion consists of cell bodies along the periphery and terminal arborizations, collaterals of Motor Neurons and a dense mass of nerve fibrils. Collectively the central part of the Ganglion forms a Neuropil (Neuropile). Groups of terminal fibrils form compact bodies within the Neuropil called Glomeruli. An Ectodermal supporting tissue within the ganglion is composed of Glial Cells. See Nervous System. Cf. Nerve Cell; Neuropile. Rel. Abdominal Ganglion; Brain; Thoracic Ganglion.

GANGLIONATE Adj. (Greek, *ganglion* = little tumor.) Pertaining to structure with ganglia or nerve-masses.

GANGLIONIC CELLS Nerve cells of the ganglia.

GANGLIONIC CENTRE A coalescence of two or more ganglia of adjoining segments.

GANGLIONIC COMMISSURE The short nerve cord connecting any two adjacent ganglia in series (Wardle).

GANGLIONIC LAYER Ganglionic sclerite.

GANGLIONIC PLATE A kind of screen placed across the path of the postretinal fibres of the insect Brain. Composed of a layer of cylindrical columns of nervous tissue with many nuclei in the outer margin and a single row of nuclei near its inner margin (Needham). The Periopticon.

GANIN, MITROFAN STEPHANOVICH (–1839) (Bogdanov 1841, Izv. imp. obshch. Lyub. Estest. Antrop. Etnogr. imp. Mosc. Univ. 55: [73–74], bibliogr.)

GANOGENE CELLS. Coccoids: Two cells of the Ceratuba, one on each side of the wax cell and attached to the Bulla (MacGillivray).

GARBE, ERNST WILHELM (1853–1925) (Dó 1931, Revta. Mus. paul. 17: 567–570.)

GARBIGLIETTI, ANTONIO (1807–1887) (Lessona 1887, G. Accad. Med. Torino 35: 31–37, bibliogr.)

GARCIA LOPEZ, ANGEL (1888–1937) (Cañizo 1941, Boln. Patol. veg. Ent. agric. 10: 361–363, bibliogr.)

GARCIA MERCET, RICARDO (1860–1933) Spanish pharmacist and Entomologist specializing in taxonomy of parasitic Hymenoptera, particularly the Chalcidoidea. Author of numerous publications and two books. (Dusmet 1933, Resen. cient. R. espan. Hist. nat. 8: 113–123, bibliogr.)

GARCIA, FABIAN (1871–1948) (Eyer 1948, J. Econ. Ent. 41: 1000–1001.)

GARDCIDE® See Tetrachlorvinphos.

GARDE, PHILIP LE HARDY DE LA (–1913) (Anon. 1913, Ent. Rec. J. Var. 25: 205.)

GARDEN FLEAHOPPER *Haltictus bractatus* (Say) [Heteroptera: Miridae]: A multivoltine, sporadic pest of alfalfa and clover in eastern North America. Females are dimorphic (macropterous and brachypterous); males are monomorphic. GFH overwinters as an adult. Eggs are inserted into mouthpart punctures on stems and leaves. Nymphs suck sap from plants. See Miridae.

GARDEN MAGGOT *Bibio imitator* Walker [Diptera: Bibionidae]. (Australia).

GARDEN MILLIPEDE *Oxidus gracilis* Koch [Polydesmida: Paradoxomatidae].

GARDEN ORBWEAVING-SPIDER *Eriophora biapicata* (L. Koch); *Eriophora transmarina* (Keyserling) [Araneida: Araneidae]. (Australia).

GARDEN PEBBLE-MOTH *Evergestis forficalis* (Linnaeus) [Lepidoptera: Pyralidae].

GARDEN SLATER *Porcellionides pruinosus* [Isopoda: Porcellionidae]. Body ca 10 mm long and bluish-grey. GS is common in gardens, under bricks, stones, flower pots, timber and bark. GS occasionally is found in houses, probably is endemic to Europe and adventive to Australia. See Common Slater; Slaters. Cf. Pillbugs.

GARDEN SOLDIER-FLY *Exaireta spinigera* (Wiedemann) [Diptera: Stratiomyidae]. (Australia).

GARDEN SPRINGTAIL *Bourletiella hortensis* (Fitch) [Collembola: Bourletiellidae].

GARDEN SPRINGTAILS *Bourletiella* spp. [Collembola: Sminthuridae]. (Australia).

GARDEN SYMPHYLAN *Scutigerella immaculata* (Newport) [Symphyla: Scutigerellidae].

GARDEN WEBWORM *Achyra rantalis* (Guenée) [Lepidoptera: Pyralidae]: A multivoltine pest of numerous garden crops and weeds in North America. GWW is endemic to New World with wide distribution in North America. Adults appear during early spring and eggs are deposited in masses (2–3, 20–50) on host-plant leaves. Eclosion occurs within 3–7 days and Larvae feed on underside of host plant leaves within protective silken web; when threatened, larvae drop to ground or move into their tubular silken retreats. During periods of starvation, GWW larvae may behave as armyworms and move in large numbers. GWW overwinter as Pupae in soil around plants where the larvae fed. See Pyralidae. Cf. Alfalfa Webworm.

GARDEN WEEVIL *Phlyctinus callosus* Boheman [Coleoptera: Curculionidae]. (Australia).

GARDEN WOLF-SPIDER *Lycosa godeffroyi* L.Koch [Araneida: Lycosidae]. (Australia).

GARDENIA BUD-MITE *Colomerus gardeniella* (Keifer) [Acari: Eriophyidae].

GARDENTOX® See Diazinon.

GARDIAN® See Flucythrinate.

GARDNER, JAMES CLARK MOLESWORTH (1894–1970) (Hinton 1972, Proc. R. ent. Soc. Lond. (C) 35: 53.)

GARDNER, JOHN (1842–1921) (Rothschild 1921, Proc. ent. Soc. Lond. 1921: cxxix.)

GARDNER, WILLOUGHBY (1860–1953) (Buxton 1954, Proc. R. ent. Soc. Lond. (C) 18: 79.)

GARDONA® See Tetrachlorvinphos.

GARD-STAR® See Permethrin.

GARGISH, BHAGWAN DAS (–1950) (Batra 1950, Indian J. Ent. 12: 256–257.)

GARIAL® See Amitraz.

GARLEPP, GUSTAV (1862–1907) (Papavero 1975, *Essays on the History of Neotropical Dipterology*. 2: 293–295. São Paulo.)

GARLIC SNAIL *Oxychilus alliarius* (Miller) [Sigmurethra: Zonitidae]. (Australia).

GARMAN, PHILIP (1891–1972) (Johnson 1973, Odonatologica 2: 333–334.)

GARMAN, WILLIAM HARRISON (1858–1944) (Osborn 1946, Pt. II. 232 pp. (85) Columbus, Ohio.) American economic Entomologist, educated at Johns Hopkins University, worked at Kentucky Agricultural Experiment Station and became president of the Association of Economic Entomologists (1905).

GARNEYS, WILLIAM (1831–1881) (Anon. 1881, Entomol. mon. Mag. 18: 163–164.)

GARNIER, JAQUES (1808–1888?) (Anon. 1889, Bull. Soc. linn. Nord Fr. 9: 67–70.)

GARRETA, LÉON (1887–1914) (Berland 1920, Ann. Soc. ent. Fr. 89: 425–427, bibliogr.)

GARRETT, ANDREW (1823–) (Weidner 1967, Abh. Verh. naturw. Ver. Hamburg Suppl. 9: 153–154, 141.)

GARRETT, GARET (1868–1896) (Garrett 1942, Trans. Suffolk nat. Soc. 5: 1012.)

GARVOX® See Bendiocarb.

GAS EXCHANGE The movement of gas from one part of the body to another part of the body or from a tissue to the ventilatory system. See Ventilation. Rel. Respiration.

GASPERINI, FERDINAND DE (1791–1883) (Anon. 1834, Bull. Soc. ent. Fr. 3: lxxii.)

GASSNER, IGNAZ (1806–1890) (Reitter 1890, Wien. Ent. Ztg. 9: 184.)

GAST, ROBERT THEODORE (1923–1967) (Davich 1968, J. Econ. Ent. 62: 886.)

GASTALDI, BARTOLEMO (1818–1879) (Seller 1879, Atti Accad. Lincei (3) 3: 82–92, bibliogr.)

GASTER Noun. (Greek, *gaster* = stomach, Latin, *gaster* = belly. Pl., Gasters.) Apocritous Hymenoptera: The posterior 7–8 segments of the Abdomen, behind the constricted second segment (Petole). In Symphyta the Abdomen is broadly attached to the Abdomen and a Gaster is not apparent as separated from the Thorax by a constriction. In more highly evolved Hymenoptera (Parasitica and Aculeata), the first abdominal segment has become 'separated' from the remainder of the Abdomen by the second segment which takes the form of a sclerotized, ring-like constriction (Petiole). The first (anteriormost) abdominal segment is called the Propodeum; the second segment (which forms the constriction) is called the Petiole; the remaining segments of the Abdomen are collectively referred to as the Gaster. See Tagma. Cf. Mesosoma; Metasoma. Rel. Tagmatization.

GASTEROCOELE Noun. (Greek, *gaster* = stomach; *kele* = tumour. Pl., Gasterocoeles.) The gastrulation cavity or Archenteron *sensu* Snodgrass.

GASTEROPHILIDAE Plural Noun. (Greek, *gaster* = stomach; *philein* = to love.) Horse Bot-Flies. A small Family of parasitic muscoid Diptera endemic to Africa and Asia; regarded as Subfamily of Oestridae in some classifications. Adults with vestigial mouthparts; Postscutellum not developed; wing vein M1 not curved forward to meet R4+5; Squamma small. Most notable Genus is *Gasterophilus*, the Species of which attack horses. Eggs are laid on horse hair and neonate larvae enter the horse's mouth to develop in alimentary system; larvae pass from host in faeces during defecation. Pupation occurs in soil. Other Genera of gasterophilids parasitize African and Indian Elephant *(Cobboldia, Platycobboldia, Rodhainomyia),* African and Asiatic Rhinoceros *(Gyrostigma)* and Zebra. See Horse Bot-Fly; Oestridae.

GASTEROPHILUS ENTERIC MYIASIS Myiasis of the donkey, elephant, horse, mule, rhinoceros and zebra. Aetiological agents include ca 10 members of Genus *Gasterophilus* [Oestridae]. Larvae infest hosts throughout year; heavy infestations cause gastritis with resulting loss of condition in host; death of animal is rare. See Bot Flies; Myiasis.

GASTEROPHILUS INERMIS Brauer [Diptera: Oestridae]: A significant pest of horses in the Old World. Females lay 300–500 eggs during lifetime; eggs are laid on cheeks of host and neonate first-instar larvae penetrate skin and burrow to mouth. Burrowing causes 'Summer Dermatitis' in horses. Third instar larvae are attached to the Rectum near the Anus. See Gasterophilus Enteric Myiasis. Cf. Horse Bot-Fly; Nose Bot-Fly; Throat Bot-Fly.

GASTEROPHILUS PECORUM [Diptera: Oestridae]: Largest Species of Bot Fly affecting horses, and the most serious pest with death of host sometimes occurring; GP is an Old World Species that is not found in USA. Females lay ca 1,300–2,400 eggs during lifetime; eggs are glossy black and laid in batches of 10–15 on grass; eclosion occurs when eggs are ingested by a horse. First and second-instar larvae may attach to tongue or soft palate of the host; third instar passes to stomach of host. Mature larvae pass from body of host with faeces and pupation occurs in soil. See Gasterophilus Enteric Myiasis. Cf. Horse Bot-Fly; Nose Bot-Fly; Throat Bot-Fly.

GASTEROTHECA Noun. (Greek, *gaster* = stomach; *theke* = case. Pl., Gasterothecae.) The abdominal case or that part of the Theca or pupal-case which encloses the Abdomen. See Pupa; Theca.

GASTERUPTIIDAE Ashmead 1900. Plural Noun. A numerically moderate sized (ca 500 Species) and geographically widespread Family of apocritious Hymenoptera assigned to the Evanioidea. Adults are medium sized; male Antenna with 13 segments, female Antenna with 14 segments; Propleuron elongate and forming a neck; forewing plicated with extensive venation and hindwing with reduced venation; gastral Petiole attached high on Propodeum. Gaster elongate and enlarging caudad, Ovipositor exserted. Adults visit flowers and rotting logs. Developmental biology is transitional between cleptoparasitism and ectoparasitism of bees and wasps. Some Species lay eggs in cells of bees or wasps; eggs hatch and parasite larvae attack the host larvae; in other Species the parasite egg hatches and the larva consumes provisions within the host cell. Third-instar larvae void excrement, spin weakly developed cocoons and the mature larvae overwinter; pupation occurs during the following year. See Evanioidea.

GASTION® See AIP.

GASTOXIN® See AIP.

GASTRAL Adj. (Greek, *gaster* = stomach, Latin, *gaster* = belly; *-alis* = pertaining to.) Pertaining to the Gaster. See Gaster.

GASTRIC CAECA Caecum. See Enteric Caeca.

GASTRIC MYIASIS See Gasterophilus Enteric Myiasis.

GASTRIC Adj. (Greek, *gaster* = stomach; *-ic* = characterized by.) Pertaining to the stomach. Ants: Pertaining to the Gaster.

GASTROCELUS Noun. (Greek, *gaster* = stomach; *kele* = tumor. Pl., Gastrocoeli.) Hymenoptera: A lateral impression on the second gastral Tergum of Ichneumonidae which frequently contains a sensory area (Thyridium). See Thyridium.

GASTRO-ILEAL FOLD In insects, the boundary between the intestine and the Chylific Stomach (Ventricle), which forms a valve.

GASTRULA Noun. (Greek, *gaster* = stomach; *ula* = diminutive suffix. Pl., Gastrulae; Gastrulas.) 1. Insect embryo: The stage resembling a sac, with an outer layer of Epiblastic Cells and an inner layer of Hypoblastic Cells. 2. The Embryo after Gastrulation. Cf. Bastula; Morula.

GASTRULATION Noun. (Greek, *gaster* = stomach; English, *-tion* = result of an action. Pl., Gastrulations.) 1. Embryonic development: The act or process of forming a Gastrula. 2. The act or process of forming the Endoderm.

GATHERING HAIRS The flexible, flattened and often hooked Setae that occur on the 'tongue' of bees and other Hymenoptera. Syn. Hooked Hairs. See Seta.

GAUCHO® See Imidacloprid.

GAUDICHAUD-BEAUPRE, CHARLES (1789–) (Papavero 1971, *Essays on the History of Neotropical Dipterology* 1: 124–217. São Paulo.)

GAULE, JULES DE (–1922) (Joannis 1922, Bull. Soc. ent. Fr. 1922: 281–283.)

GAUSE'S RULE Ecology: The concept which asserts that no two Species with identical ecological requirements can coexist in the same place at the same time. Cf. Competitive Displacement. Rel. Ecological Homologue.

GAUTARD, VICTOR DE (–1870) (Anon. 1871, Petites Nouv. Ent. 3: 117.)

GAUTHIER DES COTTES, BARON (– 1875) (Anon. 1875, Petites Nouv. Ent. 3: 117.)

GAUTHIER, GEORGES (1901–1972) (Maltais & Paradis 1973, Ann. Soc. Ent. Québec 18: 103–107, bibliogr.)

GAY, CLAUDIO (1800–1873) (Porter 1902, Revta. chil. hist. Nat. 6: 109–132, bibliogr.)

GAYNER, FRANCIS (1870–1933) (Anon. 1933, Br. med. J. 2: 1097.)

GAZELLA DUNG-BEETLE *Onthophagus gazella* (Fabricius) [Coleoptera: Scarabaeidae]: A Species that was introduced into northern Australia from Africa for habitat management. GDB feeds on and buries cattle dung; availability of breeding sites for bush fly and buffalo fly may be reduced by activities of this beetle.

GAZELLE® See Carbosulfan.

GEBIEN, HANS (1874–1947) (Martini 1950, Verh. Ver. naturw. Heimatforsch 30: vii–xv, bibliogr.)

GEBLER, FREDERICK AUGUSTUS (1782–1850) (Marseul 1883, Abeille 21 (Les ent. et leurs écrits): 71–72.)

GEDDES, GAMBLE (1850–1896) (Anon. 1896, Can. Ent. 28: 117–118, bibliogr.)

GEDOELST, LOUIS M (1861–1927) (Mesnil 1927, Bull. Soc. Path. exot. 20: 205–206.)

GEDYE, A F J (–1963) (Wigglesworth 1964, Proc. R. ent. Soc. Lond. (C) 28: 58.)

GEE, GEORGE FREDERICK (1873–1945) (Hall 1947, Proc. R. ent. Soc. Lond. (C) 11: 61.)

GEE, NATHANIEL GIST (1876–1937) (Graben 1938, Peking Nat. Hist. Bull 12: 167–168.)

GEGENBAUER, CARL (1826–1903) (Nordenskiöld 1925, *History of Biology,* 629 pp. (499–504), London.)

GEHIN, JOSEPH JEAN BAPTISTE (1816–1889) French Entomologist and specialist in Coleoptera who published a series of papers on the life history of insects of the Department of the Moselle. (Anon. 1890, Bull. Soc. ent. Fr. 1890: ccxxxi.)

GEHRING, JOHN GEORGE (1857–1932) (Hungerford 1933, Ann. ent. Soc. Am. 26: 188.)

GEIB, ARTHUR F (1911–1974) (Bickley 1975, Mosquito News 35: 107.)

GEIGY, R (1902–1962) (Wyniger 1962, Mitt. ent. Ges. Besel 12: 81–82.)

GEINTIZ, HANS BRUNO (1814–1900) (Anon. 1900, Geol. Mag. 7: 477–478.)

GEKUGIB® See Gibberellic Acid.

GELASTOCORIDAE Kirkaldy 1897. Plural Noun. Toad bugs; Gelastocorids. A widespread Family of semiaquatic Hemiptera consisting of less than 100 Species, placed in Subfamilies Gelastocorinae Champion and Nerthrinae; Gelastocorids do not occur in Europe. Fossil record of gelastocorids is unknown and their phylogeny is vague. Taxonomically, gelastocordis are placed near Saldidae, Ochteridae and Naucoridae; currently placed in Ochteroidea. Adult body moderate-sized, robust, ovoid in outline shape and dorsoventrally compressed; Rostrum with four segments; Antenna short, typically concealed, with four segments; Ocelli present in most Species; compound eyes bulging. Pronotum large; Scutellum large, triangular in outline, somewhat elevated; anterior leg raptorial; front and middle Tarsi of one segment; hind Tarsi of three segments; typically flightless although sometimes winged. All Species are presumed predaceous but details of development, life history and behaviour are poorly documented. Gelastocorinae characteristically are found along the sandy banks of streams and ponds where they merge imperceptibly with the substrate; Nerthrinae frequently are removed from aquatic habitats. Syn. Galgulidae Billberg 1820. See Hemiptera.

GELATINOUS Adj. (Latin, *gelare* = to congeal; Greek, *genes* = producing; Latin, *-osus* = full of.) Pertaining to substance with a jelly-like texture or consistency. Cf. Viscid.

GELBER, FRANCIS Russian physician, naturalist and author of *Observationes Entomologicae* and various works in Mem. Doc. Nat. Moscou.

GELECHIIDAE Plural Noun. Gelechiid Moths. A cosmopolitan Family (ca 4,000 Species) of ditrysian Lepidoptera assigned to the Gelechioidea. Adults are small with wingspan 8–35 mm and head scales typically smooth with long lamellar scales on Frons; Ocelli often present but Chaetosemata absent; Antenna shorter than forewing and Pecten usually absent; Maxillary Palpus small with 4–3 segments folded over base of Proboscis; Labial Palpus recurved; Epiphysis present; tibial spur formula 0-2-4; hind Tibia with long hair scales; forewing shape variable and lacks Pterostigma. Eggs are oval or elongate-oval, apically rounded with chorionic sculpture weak or absent. Eggs typically are laid individually on host plant leaves. Larval head prognathous. Pupae with Maxillary Palpus, Labial Palpus and fore Femur concealed; Cremaster sometimes present. Pupation occurs in silken cocoon or on ground in detritus. Early instars of larvae often are leaf miners; later instars fold or tie leaves or bring leaves together with silk; a few Species form galls or tunnel in stems; some larvae feed on seeds. Adults of most Species are nocturnal, concealed during day and rest with wings folded over Abdomen and Antennae on wings; some Species are diurnal. Several Species are regarded as major agricultural pests.

GELECHIOIDEA A Superfamily of ditrysian Lepidoptera including Agonoxenidae, Anomologidae, Blastobasidae, Coleophoridae, Cosmopterigidae, Elachistidae, Ethmiidae, Gelechiidae, Lecithoceridae, Metachanididae, Oecophoridae, Physoptilidae, Pterolonchidae, Scythridae, Stathmopodidae, Stenomidae, Strepsimanidae, Timridae and Xyloryctidae. See Lepidoptera.

GELIN, HENRI (1848–1923) (Anon. 1924, Bull. Soc. ent. Fr. 1924: 7–8.)

GELONUS BUGS *Gelonus* spp. [Hemiptera: Coreidae]. (Australia).

GEMIGNANI, EMILIO V (1897–1949) (Anon. 1949, Revta. Soc. ent. argent. 14: 240–243, bibliogr.)

GEMINATE Adj. (Latin, *gemini* = twins; *-atus* = adjectival suffix.) Arranged in pairs composed of two similar parts; doubled; twinned. Alt. Geminatus; Geminous; Geminus.

GEMMA Noun. (Latin, *gemma* = bud. Pl., Gemmae.) A bud or bud-like organic growth.

GEMMATE Adj. (Latin, *gemmare* = to bud.) 1. Descriptive of surface sculpture, usually the insect's Integument, that displays elements as wide or wider than tall and constricted at the base. See Sculpture Pattern. Cf. Alveolate; Baculate; Clavate; Echinate; Favose; Psilate; Punctate; Reticulate; Rugulate; Scabrate; Shagreened; Smooth; Striate; Verrucate. 2. Bud-like. 3. Reproducing by buds (Gemmae). 4. Marked with metallic or bright coloured spots. Alt. Gemmatus.

GEMMATION Noun. (Latin, *gemma* = bud; English, *-tion* = result of an action. Pl., Gemmations.) Budding or the process of bud formation. Reproduction by budding.

GEMMINGER, MAX (1820–1887) (Will 1887, Ent. Nachr. 13: 237–238.)

GEMMIPAROUS Adj. (Latin, *gemma* = bud; *parere* = to produce; *-osus* = with the property of.) A form of asexual reproduction in which new individuals arise as buds from the germ body of the parent. See Reproduction.

GEMSTAR® A selective biopesticide used for control of Corn Earworm, Cotton Bollworm, Tobacco Budworm and Tomato Fruitworm. Active ingredient: Polyhedral Occlusion Bodies formed by NPV of *Heliothis virescens.* See Nuclear Polyhedrosis Virus. Cf. Spod-X®. Rel. Biopesticide.

GENA Noun. (Latin, *gena* = cheek. Pl., Genae.) 1. Trilobites: The anterolateral part of the Prosoma. 2. Generalized Insects: The so-called cheek or sclerotized area on the side of the head below the compound eye and extending to the Gular Suture. 3. Odonata: Area between compound eyes, Clypeus and mouthparts. 4. Diptera: Space between the lower border of the compound eye and oral margin, merging into the face at the front and limited by the occiptial margin behind. See Head Capsule. Cf. Postgena.

GENACERORES Noun. Some Diaspididae: The Cerores located on the ventral aspect of the Pygidium near the Meson (MacGillivray).

GENAL Adj. (Latin, *gena* = cheek; *-alis* = pertaining to.) Descriptive of structure or process associated with the Gena. See Head. Cf. Facial; Frontal; Occipital.

GENAL BRISTLES Diptera: Setae on the cheeks near lower corner or eye.

GENAL COMB Siphonaptera: A row of powerful spines on the latero-ventral border of the head. See Ctenidia.

GENAL ORBIT Part of an orbit adjacent to the ventral margin of a compound eye (MacGillivray).

GENAPONTA Noun. (Latin, *gena* = cheek; *pons* = bridge. Pl., Genapontae.) 1. The bridge formed by the fusion of the elongated Postgenae which divides the Cervacoria in specialized insects. 2. The Gula (MacGillivray). Cf. Hypostomal Bridge; Postgenal Bridge.

GENATASINUS Noun. The genital pouch *sensu* MacGillivray.

GENCOR® See Hydroprene.

GENE Noun. (Greek, *genos* = descent. Pl., Genes.) The unit of heredity carried on the chromosomes. A segment of DNA which codes for a specific protein. Rel. Chromosome.

GENE, CARLO GUISEPPE (1800–1847) (Sismonda 1851, Mem. acad. Sci. Torino 11: 1–19, bibliogr.)

GENE SPLICING Manipulating genes in order to tie them together. See Genetic Engineering. Cf. Biotechnology.

GENE-FOR-GENE RELATIONSHIP The concept of a resistant or susceptible allele at a gene locus in a particular plant cultivar corresponding to a susceptible or virulent allele at the same locus in a pest. Essentially, a one-to-one relationship between host and pest genotypes.

GENERAL EQUILIBRIUM POSITION A population's long-term average density.

GENERAL HOMOLOGY An abstract concept in which the structural resemblance of organs, structures or parts of an organism are compared with a conceptual or hypothetical ancestral form. See Homology. Cf. Serial Homology; Special Homology.

GENERALIZED Adj. (Latin, *generalis* = of one kind.) Biology: A comparative term used to indicate an ancient, primitive or long-standing character when compared with a relatively recently evolved character or 'specialized' (caenogenetic) character. An archaic, primitive or generalized 'organism' is one in which ancient characters predominate (Tillyard). Cf. Derived; Specialized.

GENERAL-USE PESTICIDE One of two categories for pesticide handling established by the U.S. Environmental Protection Agency. General-use pesticides contain a percentage of active ingredient that is considered safe for public use without previous training or qualification. Pesticides that can be sold/purchased and used without a licence or permit. See Pesticide.

GENERATION Noun. (Latin, *generatio* = reproduction; English, *-tion* = result of an action. Pl., Generations.) A cohort of individuals comprising a population with common ancestors and which demonstrate collective dynamic properties including birth, maturation, reproduction and death.

GENERATION TIME 1. The time interval between consecutive cell divisions. 2. The time interval during which individuals of a population complete one life-cycle.

GENERIC NAME The name of a Genus. The GN is one word (simple or compound), written with a capital initial letter and employed as a substantive (noun) in the nominative singular (International Code of Zoological Nomenclature).

GENE'S ORGAN Acarology: A Subscutal or Capitular Gland that secretes a viscous substance that is used when transferring eggs from the Gonopore to the dorsal surface of the body of ovipositing ticks.

GENESIS Noun. (Latin > Greek, *genesis* > *gignesthai* = to be born. Pl., Geneses.) 1. The origin or coming into being of any organism (through development) or lineage of organisms (through descent). 2. The act or process of creation of life. 3. Life's process or physical conditions. Rel. Evolution.

GENESIS® See Thiodicarb.

GENETIC CONTROL A method of pest control that employs genetically altered individuals which are released into a 'wild' population, successfully copulate with members of that population and produce sterile or inviable progeny. See Biological Control.

GENETIC DISTANCE A method of treating electrophoretic data which is an alternative to more traditional character–character/ state data in systematic studies. See Nei's Genetic Distance Index.

GENETIC ENGINEERING Technology used to isolate genes from an organism, manipulate the genes in the laboratory and insert them into another organism. Rel. Biotechnology.

GENETIC HERMAPHRODITE An insect in which the gonads of both sexes are present in the same individual.

GENICULAR ARC Orthoptera: A curved dark mark or stripe on the posterior knee-joint.

GENICULATE Adj. (Latin, *geniculatus, geniculum* = little knee; *-atus* = adjectival suffix.) Descriptive of a structure (appendage) that is 'knee jointed' or abruptly bent in an obtuse angle (*e.g.* geniculate Sntenna). Alt. Elbowed; Geniculatus. Rel. Subgeniculate. Adv. Geniculately.

GENICULUM Noun. (Latin, *geniculum* = little knee. Pl., Genicula.) A small knee or bend in structure.

GENISTA CATERPILLAR *Uresiphita reversalis* (Guenée) [Lepidoptera: Pyralidae].

GENITAL Adj. (Latin, *gignere* = to beget; *-alis* = pertaining to.) Pertaining to the structure or function of the genitalia (external reproductive organs). See Genitalia. Rel. Subgenital.

GENITAL ARMATURE Collectively the structures concerned with copulation in the male (*e.g.* Aedaegus, Endophallus, Parameres *etc.*) Cf. Genital Chamber 1.

GENITAL CHAMBER 1. Female: Primarily an invagination or cavity behind the eighth abdominal Sternum which contains the Gonopore and spermathecal aperture. The GC is often converted into a pouch-like or tubular Vagina (Uterus), and in some insects opening secondarily on or behind the ninth Sternum. 2. Male: An invagination (cavity) behind (above) the ninth Sternum that contains the intromittent organ (Penis) (Snodgrass).

GENITAL FOSSA Odonata (Dragonflies): A median depression on the ventral wall of the second Sternum in which the copulating organs are lodged. Rel. Odonata.

GENITAL HAMULE 1. A small hook or sclerite covering the anal cavity of the male; the Supra-anal or Genital Hook. 2. Lepidoptera: The Uncus. Odonata: 1–2 pairs of lateral processes of the male genitalia on the ventral surface of the second abdominal segment.

GENITAL HOOK See Genital Hamule.

GENITAL PAPILLA Some Sminthuridae: A tubercular elevation upon which the genital aperture opens.

GENITAL PORES Ephemeroptera: Openings of the sperm ducts at the end of the Penis (Needham).

GENITAL POUCH Some Diptera: A sac into which the Aedeagus is thrust in repose.

GENITAL RIDGES Insect embryo: Thickenings of the splanchnic wall of the Mesoderm in the abdominal region of the body. The cell groups of which are the rudiments of the Testes or Ovaries (Snodgrass).

GENITAL SEGMENTS 1. Male: The ninth abdominal segment. 2. Female: The eighth and ninth abdominal segments. Syn. Gonosomites. Modifications of genital sclerites can be considerable. Adult Pterygota have well developed GS to include organs of copulation and oviposition. Dual function has resulted in considerable differentiation of associated segments and contributed to difference of opinion regarding Homology of genitalic parts. Among Pterygota, male genitalia are generally positioned on segment nine. Female genitalia are generally positioned on segments eight and/or segment nine. The ninth sternum is called a Hypandrium in some insects; tenth sternum of Ephemeroptera is called an Hypandrium. Fusion of segments nine and ten in Psocoptera results in a structure called the Clunium. See Abdomen. Cf. Gonopore.

GENITAL SPIKE Diaspididae: The sheath of Penis which takes the form of a long mucronate spike. Cf. Mucron; Spike 4.

GENITAL STYLES The unmodified outer Gonapophysis, as in the males of *Periplaneta* and the Isoptera. Cf. Stylus.

GENITAL TUFT Lepidoptera: An expansible tuft of fine Setae associated with the apical segments of the adult Abdomen and which typically produce pheromones. Cf. Brush Organ; Hair Pencil.

GENITAL VALVE Odonata: A pair of sclerites, one on either side of the Ovipositor which are derived from the ninth abdominal Sternum.

GENITALIA Plural Noun. (Latin, *gegnere* = to beget.) 1. Genitalia are external components of the reproductive system which are derived from Ectoderm. During embryological development, the Polypod shows limb buds on all abdominal segments. These buds are regarded as homologous. The embryo then passes from the Polypod to the Oligopod stage during which most of the limb buds are lost but Genitalia remain in a different form. 2. Collectively, Genitalia include all genitalic structures. 3. The Gonapophyses *senu* Comstock. See External Genitalia.

GENLIS, STÉPHANIE FELICITATÉ DUCREST DE SAINT AUBIN DE (Rose 1850, *New General Biographical Dictionary* 7: 502.)

GENOHOLOTYPE Noun. (Greek, *genos* = race; *holos* = entire; *typos* = pattern. Pl., Genoholotypes.) 1. The Species upon which a Genus is founded, whether unique (monotypic). 2. One Species within a series of Species. 3. A Species named as generic type by the author (Obsolete). See Type. Cf. Allotype; Cotype; Genolectotype; Genosyntype; Genotype; Holotype; Lectotype; Neotype; Syntype; Topotype. Rel. Nomenclature; Taxonomy.

GENOLECTOTYPE Noun. (Greek, *genos* = race; *lektos* = chosen; *typos* = pattern. Pl., Genolectotypes.) The Species, of a series of Species, within a Genus that is selected as the 'type' of that Genus. The designation occurs subsequent to the description of the Genus and placement of several Species within the Genus. See Type. Cf. Allotype; Cotype; Genoholotype; Genosyntype; Genotype; Holotype; Lectotype; Neotype; Syntype; Topotype. Rel. Nomenclature; Taxonomy.

GENOCLINE Noun. (Greek, *genos* = race; *kleinen* = incline.) Cf. Ecocline.

GENOME Noun. (Greek, *genos* = offspring. Pl., Genomes.) 1. All of the genes carried by an individual; the complete set of genes in an organism. 2. The total genetic characteristics of a cell.

GENOSYNTYPE Noun. (Greek, *genos* = race; *syn* = with; *typos* = pattern. Pl., Genosyntypes.) One of a series of Species upon which a Genus is established but no Species in the series has been mentioned as Genotype by the author. See Type. Cf. Genosyntype; Genotype; Holotype; Lectotype; Neotype; Syntype. Rel. Nomenclature; Taxonomy.

GENOTYPE Noun. (Greek, *genos* = race; *typos* = pattern. Pl., Genotypes.) A taxonomic concept in which a Species is designated as the type-Species of a Genus, and the Species upon which the Genus is established. See Type. Cf. Holotype; Lectotype; Neotype. Rel. Nomenclature; Taxonomy.

GENOVERTICAL PLATES Diptera: The so-called 'Front' of writers on Chaetotaxy; the Parafrontalia *sensu* Comstock.

GENTROL® See Hydroprene.

GENU Noun. (Latin, *genu* = knee. Pl., Genua.) Acarina: A knee; the articulation between Femur and Tibia. Syn. Patella. Cf. Moula.

GENUALA Noun. (Latin, *genu* = knee. Pl., Genualae.) Acari: Specialized nude Setae on Genua of legs I–III (Goff *et al.* 1982, J. Med. Ent. 19: 226.)

GENUS Noun. (Latin, *genus* = race. Pl., Genera.) 1. An assemblage of Species agreeing in one character or a series of characters. The concept is considered as arbitrary and opinionative, although some taxonomists regard the Genus as a natural assemblage. 2. 'A concept or idea that has no material counterpart anywhere in objective nature, a category or class on which we set purely arbitrary limits' (J. C. Chamberlain). Rel. Subgenus.

GENUS NOVUM Latin phrase meaning 'New Genus.' An epithet used to identify newly proposed names for previously undescribed Taxa. Abbreviated gen. nov., g. n., or n. g.

GEODEPHAGOUS Adj. (Greek, *ge* = earth; *adephagos* = voracious from *aden* = to one's fill; *phagein* = to eat; Latin, *-osus* = with the property of.) Adephagous.

GEODROMICA Noun. (Greek, *ge* = earth; *dromikos* = good at running.) The terrestrial Heteroptera in which the Antennae are not concealed.

GEOFFROY SAINT-HILAIRE, ETIÉNE (1772–1844) French physician and entomologist who published *Histoire abregee des insectes qui se trouvent aux environs de Paris* (1762). (Beltran 1944, Revta. Soc. mex Hist. nat. 5: 155–166.)

GEOFFROY SAINT-HILAIRE, ISIDORE (1805–1861) (Nickles 1862, Am. J. Sci. 34: 122–123.)

GEOFFROY, ETIENNE LOUIS (1727–1810) French physician and naturalist. Author of *Historie abregee des Insectes des Environs de Paris* (2 vols, 1764.) (Rose 1850, *New General Biographical Dictionary* 7: 566.)

GEOGRAPHIC CONGRUENCE In biogeography, the agreement of the interrelationships among geographic areas, as displayed by the cladograms of different Taxa.

GEOGRAPHIC EPIDEMIOLOGY See Amer. Sci. 75: 252–259; 280–284.

GEOGRAPHIC INFORMATION SYSTEM A computerized system of organizing and analysing any spatial array of data and information. Acronym GIS.

GEOGRAPHIC ISOLATION Isolation by geographic barriers, as oceans (in the case of islands), mountains, or similar obstacles to diffusion of Species.

GEOGRAPHIC SUBSPECIES A Subspecies conditioned by geographic factors (Ferris).

GEOLOGICAL HISTORY The historical record of the earth's existence as preserved in stratigraphy and the fossil record. The Geological Record is framed in Eras (Archaeozic, Palaeozoic, Mesozoic, Cenozoic) and Periods within Eras. Each Era or Period is defined by an interval of time (millions of years) and characterized by geological events, weather conditions and biotic diversity. Transitions between consecutive Eras and Periods are typified by massive changes in biota or conditions. See Cenozoic; Mesozoic; Palaeozoic.

GEOLOGICAL TIME SCALE A chronological reference system that measures the age of stratigraphic deposits in the earth's crust and correlates forms of life with intervals of time. GTS is hierarchical. The highest level consists of three Eons: Archean (4.6–2.5 BYBP), Proterozoic (2.5 BYBP–570 MYBP) and Phanerozoic (570 MYBP –Present). Phanerozoic Eon divided into three subordinate Eras: Palaeozoic (570 MYBP–245 MYBP), Mesozoic (245 MYBP–65 MYBP) and Cenozoic (65 MYBP–Present). Palaeozoic Era divided into six Periods: Cambrian, Ordovician, Silurian, Devonian, Carboniferous, Permian; Mesozoic Era divided into three Periods: Triassic Period, Jurassic Period, Cretaceous Period; Cenozoic Era divided into two Period: Palaeogene, Neogene. The sequence of Eons, Eras and Periods is accepted as constant; the relative ages of each interval is periodically adjusted. See each Eon, Era or Period for explanation.

GEOMAGNETIC RECEPTOR A sense organ involved in the perception of the Earth's magnetic fields.

GEOMET® See Phorate; Terbufos.

GEOMETERS See Geometridae.

GEOMETRID Adj. Measuring Worm. The larvae of Species assigned to the Family Geometridae. Name Measuring Worm is derived from the characteristic manner of walking in which a geometrid larva elevates the middle segments of the body, attaches the caudal appendages to the substrate, then moves head and Thorax forward and captures the substrate with thoracic appendages. The caudal appendages release their attachment and the posterior end of the body is drawn forward and in the process the middle segments of the body are arched. The sequence repeats and the larvae 'walk' forward. See Geometridae. Cf. Rectigrad.

GEOMETRIDAE Leach 1815. Plural Noun. Canker Worms; Geometers; Geometrid Moths; Measuring Worms. A cosmopolitan Family of ditrysian Lepidoptera assigned to Superfamily Geometroidea and including more than 20,000 described Species. Adult size is highly variable with wingspan 10–120 mm and head typically with smooth scales. Some Species with a conical prominence on the Frons; Ocelli nearly always absent and Chaetosemata present above the eye; Antenna simple in female but male form variable; Proboscis not scaled and typically coiled when not in use; Maxillary Palpus small with 1–2 segments; Labial Palpus variable in form; Thorax typically with lamellar scales; fore Tibia with Epiphysis; tibial spur formula typically 0-2-4; males of some Species with hair pencil on hind Tibia. Eggs are highly variable in shape and sculpture but typically flat type; eggs are laid individually, in small groups or in heaps. Larvae typically are long and slender with body often twig-like or leaf-like and head hypognathous; abdominal Prolegs on segments 3–6, often only on segment 5; anal Prolegs large. Larvae typically feed in exposed habitat, rarely concealed and many Species are economically important. Pupae typically are well sclerotized; Maxillary Palpus absent; Cremaster present; dorsal spines absent. Pupation occurs within a flimsy cocoon, in soil or leaf litter. Adults are nocturnal (some Larentiinae diurnal) and typically cryptically coloured with pattern corresponding to preferred resting substrate; diurnal Species are conspicuously coloured. Most Species rest with wings spread away from body, closely pressed to substrate, but a few Species hold wings as butterflies over Thorax. See Geometroidea.

GEOMETROIDEA A Superfamily of ditrysian Lepidoptera including Axiidae, Drepanidae, Epilemidae, Geometridae, Sematuridae, Thyatriridae and Uraniidae. See Lepidoptera.

GEOPHAGOUS Adj. (Greek, *ge* = earth; *phagein* = to devour; Latin, *-osus* = with the property of.) Feeding on earth that contains much organic matter. See Feeding Strategy.

GEOPHILOUS Adj. (Greek, *ge* = earth; *philein* = to love; Latin, *-osus* = with the property of.) Ground-loving; living on the ground. A term applied to Species that live on the surface of the ground or frequently come into contact with it. See Habitat. Cf. Hypogenous; Nidicolous; Psammophilous; Saxicolous. Rel. Ecology.

GEOPHOS® See Phorate; Terbufos.

GEORG, WILHELM (1817–1869) (Ratzeburg 1874, Forstwissenschafliches Schriftsteller-Lexicon 1: 782–783.)

GEORGE, DOROTHY CHANCEY (1887–1938) (Anon. 1939, Phytopathology 29: 389.)

GEORGI, RUDOLF OTTO (1879–1956) (Schimistchek 1956, Anz. Schädlinsk. 29: 148.)

GEORYSSIDAE Plural Noun. Minute Mud-loving Beetles. A very small Family of polyphagous Coleoptera assigned to Superfamily Hydrophiloidea. North American fauna comprised of two small and rare Species. Species occur in mud along the banks of lakes and streams, apparently feeding on algae. Georyssids are considered a Subfamily of Hydrophilidae in Australia and represented by one Genus, *Georissus*. Adult <2 mm long; black; oval in outline shape; inhabit mud along the bakcs of streams and lakes; apparently feed upon algae. See Hydrophilloidea.

GEOTAXIS Noun. (Greek, *ge* = earth; *taxis* = arrangement. Pl., Geotaxes.) Orientation or reaction to the earth or ground. See Orientation. Cf. Aerotaxis; Anemotaxis; Chemotaxis; Menotaxis; Osmotaxis; Phototaxis; Rheotaxis; Rotaxis; Scototaxis; Strophotaxis; Telotaxis; Thermotaxis; Thigmotaxis; Tonotaxis; Tropotaxis.

GEOTROPISM Noun. (Greek, *ge* = earth; *trope* = turn; English, *-ism* = state. Pl., Geotropisms.) Reaction toward the earth or ground. See Tropism. Cf. Aeolotropism; Anemotropism; Chemotropism; Electrotropism; Galvanotropism; Heliotropism; Hydrotropism; Phototropism; Rheotropism; Stereotropism; Thermotropism; Thigmotropism; Tonotropism. Rel. Taxis.

GEOTRUPIDAE Plural Noun. A Family of polyphagous Coleoptera assigned to Scarabaeoidea. Adults with body stout, strongly convex, to 24 mm long and head prognathous with horn; Antenna with 11 segments including club of three segments; Pronotum often with horn; tarsal formula 5-5-5; Abdomen with 6–7 Ventrites. Larvae with prognathous head not strongly sclerotized; Stemmata absent; Antenna with three segments; legs with three segments; tarsal claws absent; Urogomphi absent. Adults are attracted to lights, and females lay eggs in deep burrows and provide larvae with fungi, decaying vegetable matter or dung as food. See Scarabaeoidea.

GERANIOL Noun. (Pl., Geraniols.) An alcohol formed from geranial. A constituent of oil of orange-rind, cheap oil of lemon-grass, and oil of citron.

GERASIMOV, ALEKSEY MAKSIMOVICH (1904–1942) (Kozhanchikov 1948, Ent. obozr. 30: 165–167, bibliogr.)

GERBER, HENRI See IMHOFF-GERBER, H.

GERGER, EDUARD (1838–1907) (Csiki 1907, Rovart. Lap. 14: 139–140.)

GERHARD, WILLIAM J (1873–1958) (Neuzel 1959, Ann. ent. Soc. Am. 52: 338, 340.)

GERHARDT, JULIUS (1827–1912) (Henke 1913, Ent. Bl. Biol. Syst. Käfer 9: 1–8, bibliogr.)

GERM Noun. (French, *germe*, from Latin, *germen* = a bud. Pl., Germs.) 1. A microscopic organism capable of causing disease. 2. Something which serves as the origin of development.

GERM BALL Reproductive cells in larvae from which young may develop as buds.

GERM BAND The area of thickened cells on the ventral side of the Blastoderm which becomes the embryo. Embryonic rudiment; germ disc; primitive streak (Snodgrass). Alt. Germinal Band.

GERM CELLS Early-stage undifferentiated reproductive cells in the Ovariole which are destined to become Ova or Spermatozoa. Germ Cells become differentiated from Somatic Cells in the Ovarioles during cleavage. Syn. Germ Line. Cf. Somatic Cells. Rel. Cystoblast; Cystocyte. Alt. Germinal Cells.

GERM GLAND See Gonad.

GERM PLASM The part of the germ cell that carries the hereditary characteristics. Cf. Somatoplasm.

GERM THEORY OF DISEASE Proposed by Pasteur (1877).

GERM TRACT The cytoplasmic area of the Blastula which contains germ cells; posterior Polar Plasm (Snodgrass).

GERMAIN, RILIBERTO (1827–1913) (Porter 1917, Acta Soc. scient. Chil. 24: 1–9, bibliogr.)

GERMAN COCKROACH *Blatella germanica* (Linnaeus) [Blattodea: Blattellidae]: A cosmopolitan, urban pest. Adults are 10–15 mm long, pale amber brown with two longitudinal dark stripes on Pronotum; winged but does not fly. Adults live 4–6 months and females deposit 5–8 Oothecae, each containing 25–40 eggs. Oothecae are carried during embryonic development of contained eggs and released just before eclosion; GC nymphs undergo 6–7 moults and Species completes 3–4 generations per year. GC probably is most common domiciliary cockroach globally; common in kitchens, pantries, storerooms, and prefers warmth and moisture of food preparation areas. Nymphs and adults feed on most organic matter but thrive on human foods. Syn Croton Bug; Shiner; Steamfly; Steamer. See Blattodea. Cf. Asian Cockroach; Brown-Banded Cockroach.

GERMAN YELLOWJACKET *Paravespula germanica* (Linnaeus) [Hymenoptera: Vespidae]: A widespread, serious urban pest in campgrounds, public areas and gardens. GYJ is endemic to Europe and adventive elsewhere. In Europe, GYJ typically builds subterranean nests while in North America it typically constructs nests in wall voids and attics. Nest may contain 15 combs and several thousand workers; colonies usually are annual. Adults are 15 mm long, coloration yellow and black, and body with numerous long, black Setae forming a diffuse covering; first gastral Tergum transverse (truncate) along anterior margin; hindwing with anal lobe. Syn. *Vespula germanica*. See Vespidae. Cf. Aerial Yellowjacket; Common Yellowjacket; Western Yellowjacket.

GERMAR, E. FRANCIS German academic (mineralogist) and author of *Magazin der Entomologist* (4 vols, 1813, 1821), and *Insectorum Species novae aut minus cognitae & c* (1824). Germar continued to publish Ahrens' *Fauna Insectorum Europae* (which was a continuation of *Fauna Insectorum Germanicae Initia*). Germar's primary entomological interest involved taxonomy of Coleoptera.

GERMAR, ERNST FRIEDRICH (1786–1853) German naturalist and author of *Dissertatio sistens Bombycum Species*. Also continued publication of Illiger's *Magazin für Insectenkunde*. (Newman 1853, Proc. ent. Soc. Lond. 1852–1853: 149–150; Marseul 1883, Abeille 21: (Les ent. et leurs écrits): 112–117, bibliogr. only.)

GERMARIUM Noun. (Latin, *germen* = bud; *-arium* = place of a thing. Pl., Germaria.) Reproductive system: The distal portion of an Ovariole or Testicular Follicle. The Germarium contains the primary Oogonia or Spermatogonia (Cystoblasts) that differentiate into Oocytes or Spermatocytes and nurse cells. The Germarium is divided into three zones: Zone I is characterized by intense mitotic activity and distinct cell boundaries. Zone II is a transitional area of Trophocyte differentiation, indistinct cell boundaries, and Trophocyte nuclei clustered with a common cytoplasm. Zone III contains nucleoli and nuclei considerably larger than in Zone II and peripherally arranged around a central trophic core. Giant nuclei, found in the trophic core, are formed by the fusion of migratory nuclei. Subsequently, giant nuclei and migratory nuclei break down and release DNA into the trophic core. See Ovariole. Cf. Vitellarium.

GERMINAL DISC See Germ Band.

GERMINAL EPITHELIUM Epithelial tissue which develops into gametes (sex-cells). See Epithelium.

GERMINAL LAYERS Three layers of cells in the early development of the embryo: The outer layer (Ectoderm), middle layer (Mesoderm) and inner layer (Endoderm). Syn. Germ Layers.

GERMINAL VESICLE The nucleus of the insect egg (Imms).

GERONTOGEIC Adj. (Greek, *geron* = old man; *ge* = earth; *-ic* = consisting of.) Pertaining to organisms or lineages of organisms which originated in the Old World. Alt. Gerontogaeous. See Neogeic.

GERRIDAE Leach 1815. Plural Noun. Water Striders; Pond Skaters; Water Skaters; Wherrymen. A cosmpolitan Family of heteropterous Hemiptera assigned to Superfamily Gerroidea. Adult body size is variable but covered with dense vestiture of hydrophobic Setae; legs expceptionally long, thin; fore Coxa distantly separated from middle Coxa; middle Femur wider than hind Femur; hind Tibia projecting beyond apex of Abdomen; tarsal claws preapical on apical Tarsomere of each leg; Metasternum with single, median scent-gland oriface (Omphalium); abdominal scent gland orifice absent. Most Species of gerrids are associated with fresh water; a few Species are found on salt water (*Halobates* spp.) distant from land. Gerrid nymphs and adults are predatory and 'skate' on surface of water; eggs are deposited on floating objects. See Gerroidea.

GERROIDEA Noun. A Superfamily of Hemiptera including Hermatobatidae, Gerridae and Veliidae. See Hemiptera.

GERSTAECKER, CARL EDUARD ADOLPH (1828–1895) (Meldola 1895, Proc. ent. Soc. Lond. 1895: lxxi–lxxii.)

GERVAIS, FRANÇOIS LOUIS PAUL (1816–1879) (Anon. 1879, Bull. Soc. ent. Fr. (5) 9: xxx–xxxi.)

GESIN, HERMANN (1917–1967) (Anon. 1967, Mitt. ent. Ges. Basel 17: 145–146.)

GESNER, CONRAD (1516–1565) Swiss Physician and Natural Historian. Gesner has been regarded by many historians of science as the most outstanding naturalist of the period between Aristotle and 1800. (Wellisch 1975, Soc. Biblphy nat. Hist. 7: 151–247, bibliogr.)

GESSNER, JOHANN (1709–1796) (Rudio 1896, Festschr. naturf. Ges. Zurich 1746–1896: 58–64.)

GESTATION Noun. (Latin, *gestatio* = bear, carry; English, *-tion* = result of an action. Pl., Gestations.) 1. Pregnancy; The period from egg fertilization to oviposition during which the embryo matures in the body of the female parent. Cf. Fertilization.

GESTRO, RAFFAELLO (1845–1935) (Invrea 1939, Memorie Soc. ent. ital 17: 241–252, bibliogr.)

GETZENDANER, CHARLES WILBUR (1899–1970) (Landis *et al.* 1971, J. Econ. Ent. 64: 777.)

GEYER, G GUYULA (JULIUS) (1828–1906) (Aigner-Abafi 1907, Rovart. Lap. 14: 47–49.)

GHIESBREGHT, AUGUST B (1810–1893) (Papavero 1971, *Essays on the History of Neotropical Dipterology* 1: 176–178. São Paulo.)

GHIGI, ALESSANDRO (1875–1970) (Pasquini 1971, Boll. Zool. 38: 89–95.)

GHILIANI, VICTOR (1812–1878) (Papavero 1975, *Essays on the History of Neotropical Dipterology* 2: 342–343. São Paulo.)

GHOSH, SHREE C C (1881–1949) (Bose 1949, Indian J. Ent. 11: 227–228.)

GHOST ANT *Tapinoma melanocephalum* (Fabricius) [Hymenoptera: Formicidae]. (Australia).

GHOST MOTHS See Hepialidae.

GIACOMELLI, EUGENIO (–1941) (Anon. 1949, Revta. Soc. ent. argent. 14: 223–227, bibliogr.)

GIACOMINI, ERCOLE (1864–1944) (Pasquini 1945, Rc. Sess. Accad. Sci. Ist. Bologna 1944–1945: 15–45, bibliogr.)

GIANELLI, GIACENTO (1840–1932) (Turanti 1932, Memorie Soc. ent. ital. 11: 106–108, bibliogr.)

GIANT AFRICAN-SNAIL *Achatina fulica* Bowdich [Stylommatophora: Achatinidae]. Endemic to Africa and spread through the Pacific by Japanese soldiers during WWII. A significant pest of agriculture and horticulture; an object of quarantine regulations.

GIANT BARK-APHID *Longistigma caryae* (Harris) [Hemiptera: Aphididae].

GIANT BURROWING-COCKROACH *Macropanesthia rhinoceros* Saussure [Blattodea: Blaberidae]. (Australia).

GIANT CELLS See Teratocyte.

GIANT CENTIPEDE *Ethmostigmus rubripes* (Brandt) [Scolopendromorpha: Scolopendridae]: An urban pest in Australia that occurs in moist leaf litter and under stones. Adults are ca 25 cm long with body reddish brown and legs and Antenna amber-coloured. GCs are nocturnal predators that can cause pain and swelling through its bite. See Chilopoda. Cf. House Centipede; Ribbon Centipede.

GIANT CICADA KILLER See Cicada Killer.

GIANT COCKROACHES See Blaberidae.

GIANT EARWIG *Titanolabis colossea* (Dohrn) [Dermaptera: Labiduridae]. (Australia).

GIANT FISHKILLERS *Lethocerus* spp. [Hemiptera: Belostomatidae]. (Australia).

GIANT FUSIFORM CELL See Plasmatocyte.

GIANT GRASSHOPPER *Valanga irregularis* (Walker) [Orthoptera: Acrididae]: A univoltine Australian pest of garden trees and shrubs that can cause damage to commercial fruit and nut trees. Adults are 90 mm long; pale brown typically with narrow longitudinal white to green stripe; nymphs are pale green. Eggs are laid in pods deposited in soil. Nymphs and adults feed on foliage. GG completes seven nymphal instars. See Acrididae.

GIANT HAWAIIAN-DRAGONFLY *Anax strenuus* Hagen [Odonata: Aeschnidae].

GIANT LACEWINGS See Polystoechotidae.

GIANT MOLE CRICKET *Gryllotalpa major* Saussure [Orthoptera: Gryllotalpidae]: A Species endemic to North America and not regarded as a significant pest. Apex of fore Tibia with four Dactyls; hind Femur longer than Pronotum; apex of hind Tibia with seven spines; dorsomedial margin of hind Tibia not armed. See Gryllotalpidae; Mole Crickets.

GIANT NORTHERN-TERMITE See Giant Termite.

GIANT PINE-WEEVIL *Eurohamphus fasciculatus* Shuckard [Coleoptera: Curculionidae]. (Australia).

GIANT PREDATORY CLICK-BEETLE *Paracalais gibboni* (Newman) [Coleoptera: Elateridae]. (Australia).

GIANT SILKWORM-MOTHS See Saturniidae.

GIANT STAG-BEETLE *Lucanus elaphus* Fabricius [Coleoptera: Lucanidae].

GIANT SWALLOWTAIL See Orange-Dog.

GIANT TADPOLE-KILLERS *Lethocerus* spp. [Hemiptera: Belostomatidae]. (Australia).

GIANT TERMITE *Mastotermes darwiniensis* [Isoptera: Mastotermitidae]: A very primitive termite that is regarded as a serious pest of wood in tropical Australia, including Queensland, Northern Territory and Western Australia. GT attacks wood in contact with ground. Queens are ca 20 mm long; workers are ca 15 mm long. Queens are not physiogastric and lay clutches of 20 eggs contained within pods. Colonies are small (to a few thousand individuals) and nests are subterranean, in tree stumps or root crowns of trees. Colonies can be formed by 'breaking off' with developing reproductives used as neotenic queens. Syn. Giant Northern Termite. See Mastotermitidae.

GIANT THRIPS *Idolothrips spectrum* Haliday [Thysanoptera: Phlaeothripidae] (Australia).

GIANT TROPICAL-ANT *Paraponera clavata* [Hymenoptera: Formicidae].

GIANT WATER BUGS See Belostomatidae.

GIANT WATER-BUG *Lethocerus americanus* (Leidy) [Heteroptera: Belostomatidae]: A widespread and common Species in North America. Adults attracted to lights. Eggs operculate, striped at apical end. See Belostomatidae.

GIANT WATER-BUGS See Belostomatidae.

GIANT WETAS See Mimnermidae.

GIANT WOOD-MOTH *Xyleutes cinerus* (Tepper) [Lepidoptera: Cossidae]: In Australia a moth common along Queensland coast. Adults are grey with dark spot on Thorax. Female wingspan ca 25 cm and body weight 25 g; male half the size of female. Mature larvae are white, ca 15 cm long and three cm in diameter. Females may contain ca 20,000 eggs, deposit eggs in crevices in bark of living trees and cover eggs with glutinous secretion. Neonate larvae lower themselves on silken thread and are dispersed by wind. GWM presumably spends early life feeding on roots underground. Later, 25 mm larvae bore into trunk of smooth-barked eucalypt and excavate flat, oval chamber in sapwood. Entrance to bore is sealed with hard plug of fine, tightly compacted sawdust mixed with silk and flush with outer bark surface. Growing larvae continue to enlarge burrows that extend vertically upward. Sawdust produced by activity is not eaten but is ejected through hole in entrance plug. Larval food: Soft callus tissue that the tree produces in cambium in an attempt to seal larval burrow. GWM Larvae grow for two years. Before Pupation, larvae gnaw top of turning chamber through outer bark to create exit hole. Larvae then retreat up tunnel to pupate behind plug of silk and sawdust. When Pupae are ready to hatch, they cut a way out using tooth on head, wriggle along burrow and partially through exit hole. In that position, moth emerges and flies. In mature trees, tunnelling is not important damage but exit hole may allow entry of water and fungi. Holes seal within about a year but tunnelling may weaken small trees. See Cossidae.

GIARD, ALFRED (1846–1908) (Dentec 1909, Bull. scient. Fr. Belg. 42: lii–lxxiii, bibliogr.)

GIARDINA, ANDREA (1875–1948) (Mariani 1948, Naturalista sicil. (3) 3: 113–139, bibliogr.)

GIBB, DOUGLAS GORDON (1923–1970) (Burton 1971, Proc. Bristol nat. Soc. 32: 10.)

GIBB, LACHLAN (1852–1922) (Adkin 1922, Entomologist 55: 95–96.)

GIBB (Latin, *gibber* = bent, hunched.) Humpbacked in appearance.

GIBBA Noun. (Latin, *gibbus* = humped.) A rounded protuberance or prominence. See Gibbose.

GIBBERELLIC ACID An important plant hormone {3S, 3aR, 4S, 4aS, 7S, 9aR, 9bR, 12S)-7,12-dihydroxy-3-methyl-6-methylene-2-oxoperhydro-4a, 7-methano-9b,3-propenoazuleno[1,2-b] furan-4-carboxylic acid} that is used as a Plant Growth Regulator. GA is registered on numerous crop plants including artichokes, beans, blueberries, carrots, celery, cherries, cocoa, cole crops, cotton, cucumbers, grapefruit, grapes, hops, lemons, melons, mustard, onions, oranges, ornamentals, peanuts, peppers, prunes, rhubarb, rye, rice, potatoes, prunes, strawberries, spinach, squash, sugarcane, tomatoes, turnips, wheat and other crops. Trade names include: Activol®, Brellin®, Floraltone®, Gekugib®, Ceku-Gib®, Flagro®, Gibberellin®, Gib-Gro®, Gib-Sol®, Gib-Tabs®, Gibrel®, Grocel®, Release®, Ryz-Up®, Uvex®. See Insect Growth Regulator; Plant Growth Regulator.

GIBBERELLIN® See Gibberellic Acid.

GIBBINS, ERNEST GERALD (1900–1942) (Cockayne 1944, Proc. R. ent. Soc. Lond. (C) 8: 70.)

GIBBONS, CHARLES (1841–1927) (Musgrave 1932, *A Bibliogr. of Australian Entomology 1775–1930*. 380 pp. (120), Sydney.)

GIBBONS, WILLIAM (1781–1845) (Weiss 1936, *Pioneer Century of American Entomology*. 320 pp. (132–133), New Brunswick.)

GIBBOSE Adj. (Latin, *gibbus* = humped; *-osus* = full of.) Humpbacked; very convex. A surface presenting one or more large elevations. Structure that is elevated in a curve which is not the segment of a circle, like the moon when more than half full. A term frequently applied to a Macula or spot. Alt. Gibbosus; Gibbous; Gibbus.

GIBBS, ARTHUR ERNEST (1859–1917) (Turner 1917, Entomologist's Rec. J. Var. 29: 91–92.)

GIB-GRO® See Gibberellic Acid.

GIBREL® See Gibberellic Acid.

GIB-SOL® See Gibberellic Acid.

GIBSON, ARTHUR (1875–1959) (Mallis 1971, *American Entomologists*. 549 pp. (454–455), New Brunswick.)

GIBSON, FREDERICK ALLEN CLARENCE (1915–1968) (Chadwick 1969, J. ent. Soc. Aust. (N.S.W.) 5: 62–63.)

GIBSON, LESTER ERNEST (1889–1942) (Osborn 1946, *Fragments of Entomological History,* Pt II. 232 pp. (85) Columbus, Ohio.)

GIB-TABS® See Gibberellic Acid.

GIEBEL, CHRISTOPH GOTTFRIED ANDREAS (1820–1881) (Anon. 1881, Z. ges. Naturw. 54: 631–637.)

GIEMSA, GUSTAV (–1948) (Anon. Anz. Schädlingsk. 21: 61.)

GIFFARD WHITEFLY *Bemesia giffardi* (Kotinsky) [Hemiptera: Aleyrodidae]. Alt. Giffard's Whitefly (Australia).

GIFFARD, WALTER M (1856–1929) (Van Duzee 1929, Pan-Pacif. Ent. 6: 46–47.)

GIGANT® See Chlorpyrifos.

GIGASIRICIDAE Plural Noun. A fossil Family of Symphyta (Hymenoptera.)

GIGLIOLI, ENRICO HILLYER (1845–1909) (Rose 1909, Boll. Soc. ent. Ital. 41: 19–27.)

GIGLIOLI, GEORGE (–1975) (Anon. 1975, The Times, London 27.1.1975.)

GIGLIO-TOS, ERMANNO (1865–1926) (Zavattari 1926, Memorie Soc. ent. ital. 5: 35–41, bibliogr.)

GILBERT, LOUISE FITZ RUDOLPH (–1900) (Anon. 1900, Ent. News 11: 484.)

GILBERT'S BLUE-BUTTERFLY *Candalides gilberti* Waterhouse [Lepidoptera: Lycaenidae]. (Australia).

GILBOA DEPOSIT Devonian fossil-bearing geological deposts near Gilboa, New York. GD consists of fine-grained mudstone ca 376–379 MYBP that is predominantly terrestrial with the lycopod *Leclercquia complexa* representing a dominant botanical element. Fossil arthropods include trigonotarbid acarines and collembolans. Cf. Rhynie Chert; Alken an der Mosel.

GILL Noun. (Middle English, *gile*, of Scandanavian origin. Pl., Gills.) A respiratory organ of aquatic immature insects. Gills are usually hollow, thin-walled, lamellar, filament-like projections from various regions of the body. Oxygen and other gasses in solution are passed from the water into the body through gills. Cf. Branchia; Tracheal Gill. Abdominal Gill; Rectal Gill. Rel. Respiration.

GILL TUFT A group of filamentous gills, generally lateral.

GILL, THEODORE NICHOLAS (1837–1914) (Wade 1936, Proc. ent. Soc. Wash. 38: 116–117.)

GILL, WALTER BATTERSALL (–1900) (Anon. 1900, Entomol. mon. Mag. 36: 66–67.)

GILLES, WILLIAM SETTEN (1876–1938) (Huggins 1939, Entomologist 72: 80.)

GILLET, JOSEPH (–1937) (Frennet 1937, Bull. Ann. Soc. ent. Belg. 77: 165.)

GILLETTE, CLARENCE PRESTON (1859–1941) (Mallis 1971, *American Entomologists*. 549 pp. (225–226), New Brunswick.)

GILLIAT, FRANCIS TRANTHAM (1865–1955) (Hall 1956, Proc. R. ent. Soc. Lond. (C) 20: 75.)

GILLIATT, FREDERICK COURTNEY (1889–1938) (Kelsall 1938, Can. Ent. 70: 197–198, bibliogr.)

GILLMER, MAX (1857–1923) (Wiedner 1967, Abh. Verh. naturw. Ver. Hamburg Suppl. 9: 191.)

GILLO, ROBERT (–1891) (Anon. 1891, Entomol. mon. Mag. 27: 200.)

GILLOTT, ARTHUR GEORGE MALIN (1868–1927) (Riley 1928, Entomologist 61: 168.)

GILSON, GUSTAVO (1859–) (Anon. 1937, Annali. Pont. Accad. Sci. 1: 341–348, bibliogr.)

GILSON'S GLANDS Trichoptera: Metameric thoracic glands regarded as organs of excretion (Imms).

GILVOUS Adj. (Latin, *gilvus* = pale yellow; *-osus* = with the property of.) Pale yellow. Alt. Gilvus.

GIMINGHAM, CONRAD THEODORE (1884–1957) (Thomas 1958, Ann. appl. Biol. 46: 124–125.)

GIMMERTHAL, BENJAMIN AUGUST (1779–1848) (Nesse 1849, KorrespBl. NaturfVer. Riga 3: 117–123.)

GIN TRAP A device consisting of opposable sclerites on the dorsum of some insect larvae. The GT serves as a pincer to deter penetration of the larva by the Ovipositor of parasitic Hymenoptera.

GINANNI, FRANCESCO (1716–1766) (Briosi 1914, Atti Ist. bot. Univ. Lab. crittogam. Pavie 13: I–v.)

GINER, MARI JOSÉ (1901–1946) (Ignacio Sela 1947, Revta. iber. 116: 1–7.)

GINGER MAGGOT *Eumerus figurans* Walker [Diptera: Syrphidae].

GINGLYMUS (Greek, *ginglumos*.) 1. A ball-and-socket joint. 2. A hinged joint permitting flexion only in one plane. See Articulation. Rel. Acetabulum; Cotyla; Condyle.

GINKGO Noun. (Jap. *Ginko*. Pl., Ginkgoes.) A monotypic genus of relictual gymnosperm (Ginkoaceae) that originated during the late Palaeozoic/early Triassic Period in an area now called eastern China. Ginkgoes are characterized by fan-like needles and yellow drupe-like fruit. Syn. Maidenhair Fern.

GIOVANNOLA, ARNOLDO (1901–1939) (Carradetti 1939, Riv. Parassit, 3: 345–350, bibliogr.)

GIRARD, MAURICE JEAN AUGUSTE (1822–1886) (Poujade 1886, Ann. Soc. ent. Fr. (6) 6: 475–480, bibliogr.)

GIRAUD, JOSEPH ETIENNE (1808–1877) (Fairmaire 1877, Ann. Soc. ent. Fr. (5) 7: 389–396, bibliogr.)

GIRAULT, ALEXANDRE ARSENE (1884–1941) American born and educated taxonomist specializing in parasitic Hymenoptera (Chalcidoidea). Early in his career, Girault was employed by the USDA but moved to Australia during WWI. In Australia, Girault intermittently worked in economic Entomology and published privately printed papers on the taxonomy of chalcidoids. (Gordh *et al.*, 1979, Mem. Amer. Ent. Inst. 27: 1–400.)

GIRDLE Noun. (Middle English, *gurdel* = encircle. Pl., Girdles.) 1. A structure that girds, encircles, confines or restrains. 2. A cut around a tree trunk, branch or stem through the outer bark and cortex resulting in loss of vascular flow and tissue death.

GIRSCHNER, ERNST (1860–1913) (Hetschenko 1914, Wien. Ent. Ztg. 33: 230–234, bibliogr.)

GIS Acronym: Geographical Information System.

GISIN, HERMANN (1917–1967) (Szeptycki 1968, Polskie Pismo ent. 38: 673–674.)

GISTEL (GISTL) JOHANNES VON NEPOMUK FRANZ XAVER (1803–1873) (Strand 1919, Arch. naturgesch. 83A: 124–149, bibliogr.)

GIUSTINO, AGOSTINO DODERO DU (–1937) (Horn 1938, Arb. morph. taxon. Ent. Berl. 5: 72.)

GIUSTIZIERE® See Pirimiphos-Methyl.

GIZZARD Noun. (Middle English, *giser* from Latin, *gigeria* = cooked entrails of poultry. Pl., Gizzards.) The Proventriculus or pouch-like structure between the Crop and the Chylific Stomach, furnished with chitinous teeth (Acanthae) or sclerites in which the food is masticated by grinding or sifting. See Cardia.

GLABROUS Adj. (Latin, *glaber* = smooth, bald; Latin, *-osus* = with the property of.) A term used to characterize Integument which is smooth, Asetose and without punctures or structures. Alt. Glaber; Glabrate. Ant. Hairy.

GLADIOLUS THRIPS *Taeniothrips simplex* (Morrison) [Thysanoptera: Thripidae]: A Species found throughout North America where gladiolus, iris and lilies are grown. GT damage is characterized by 'blasting' of foliage caused by feeding punctures, deformation of flowers and corms become sticky from exudate. GT is parthenogenetic; eggs are kidney-shaped and inserted into growing plant tissue or corms; eclosion occurs within a week. Host plants include gladiolus, iris and lily. Infested leaf sheaths become brown, leaves become silvered, reduced size and colour of flowers. See Thripidae. Cf. Iris Thrips.

GLAIRY Adj. (Middle English, *glayre* from Latin, *clarus* = clear.) Having the aspect of or resembling glair or the white of an egg.

GLAND Noun. (Latin, *glans, glandis* = an acorn. Pl., Glands.) One or more cells which synthesize a chemical product and release it into the body (Endocrine Gland) or outside the body (Exocrine Gland) as a characteristic product (hormone, pheromone, saliva, silk, venom, wax). Based on gross anatomy and simple form, a gland consists of a secretory, vessel-like part, an elongate duct and an aperture. See Organ; Parenchyma; Tissue. Cf. Endocrine Gland; Exocrine Gland.

GLAND ORIFICE An external opening or aperture through which glandular products are released. Cf. Duct; Pore.

GLAND SPINES 1. Coccidae, spiny appendages, each of which is supplied with a single gland whose opening is at the apex. 2. The plates *sensu* MacGillivray.

GLAND-BEARING PROMINENCE Diaspididae: A prominence on the margin, bearing a gland opening in the dorsal surface.

GLANDIFEROUS SPINES Ortheziidae: The Pilacerores (MacGillivray).

GLANDS OF BATELLI Cercopidae: Epidermal glands that produce spittle. Syn. Froth Gland (Comstock).

GLANDS OF FILIPPI Lepidoptera larvae: A pair of small, accessory, Acinous Glands that open into the ducts of the silk-press (Snodgrass). See Filippi's Glands.

GLANDUBA Hymenoptera: Presumably ectocrine glands surrounded by sclerotized rings on the Integument of larval Symphyta.

GLANDULAE MUCOSA Accessory glands at the base of the Vasa Deferentia, whose secretion mixes with the Semen and forms the seminal packets (Packard).

GLANDULAR Adj. (Latin, *glans* > *glandis* = an acorn.) Having the character or function of a gland; pertaining to glands. A term sometimes used to describe specialized Setae, spines or other processes.

GLANDULAR BRISTLE Stout and rigid glandular Setae, such as the 'urticating hairs' of some Lepidoptera.

GLANDULAR CHAETA The tubular outlets of Epidermal Glands. Syn. Glandular Bristles (Wardle).

GLANDULAR HAIR Lepidoptera larvae: Hollow, smooth Setae filled with a poisonous secretion.

GLANDULAR OPAQUE SPOT Some Heteroptera: 2–3 spots on the sides of the Abdomen, near the spiracles of the third and fourth abdominal segments.

GLANDULAR PORE Coccids: The Spiracerores (MacGillivray).

GLANDULAR SETA Tubular Setae that function as the outlet for the secretions of Epidermal Glands.

GLANVILLE, ELEANOR (ELIANOR ELIZABETH) (1654–1709?) (Wilkinson 1966, Entomologist's Gaz. 17: 149–158.)

GLANVILLE, M E (–1888) (Dimmock 1888, Psyche 5: 156.)

GLASENAPP, SERGEJ PAVLOVICH (1848–1937) (Anon. 1937, Arb. physiol. angew. Ent. Berl. 4: 246.)

GLASER, LUDWIG (1818–1898) (Glaser 1898, Ent. Nachr. 24: 217–223, bibliogr.)

GLASER, RUDOLF WILLIAM (1888–1947) (Weiss 1948, Ent. News 59: 29–30.)

GLASGOW, HUGH (1884–1948) (Hartzell 1948, J. Econ. Ent. 41: 837–838.)

GLASS-HOUSE STRIPED SCIARID *Bradysia tritici* (Coquillett) [Diptera: Sciaridae]. (Australia).

GLASS-HOUSE BLACK SCIARID *Bradysia impatiens* (Johannsen) [Diptera: Sciaridae]. (Australia).

GLASSWING *Acraea andromacha andromacha* (Fabricius) [Lepidoptera: Nymphalidae]. (Australia).

GLASSY CUTWORM *Apamea devastator* (Brace) [Lepidoptera: Nocutidae]: A monovoltine, subterranean pest of sod-crops in most of USA. Larvae with glassy-texture, not granulated, body greenish-white with reddish head. See Nocutidae. Cf. Cutworms.

GLASSY-WINGED SHARPSHOOTER *Homolodisca coagulata* (Say) [Hemiptera: Cicadellidae]: A Species adventive in North America and which attacks a wide range of economically important plant Species including citrus and oleander; GWSS is a vector of Oleander blight, Pierce's Disease. See Cicadellidae.

GLAUCOUS Adj. (Greek, *glaukos* > Latin, *glaucus* = bluish green; Latin, *-osus* = with the property of.) A term used to describe colour as sea-green or pale bluish green. Alt. Glaucus.

GLAZONOV, DIMITRI KONSTANTINOVICH (1869–1913) (Semenov-Tianshansky 1914, Rev. Russe Ent. 14: lxxxvii–xcv, bibliogr.)

GLEDITSCH, JOHANN GOTTLIEB (1714–1786) (Rose 1850, *New General Biographical Dictionary* 8: 40.)

GLEICHEN (RUSSWORM), WILHELM FRIEDRICH VON (1717–1783) (Rose 1850, *New General Biographical Dictionary* 8: 40.)

GLENN, PRESSLEY ADAMS (1867–1938) (Osborn 1946, *Fragments of Entomological History,* Pt. II. 232 pp. (87), Columbus, Ohio.)

GLIAL CELL (Greek, *glia* = glue.) Nonexcitable cells adjacent to or surrounding the neuron and isolating Axons. Some insect neurons are not invested with Glial Cells. Typically, GCs outnumber nerve cells in Central Nervous System by 10–50 times in vertebrates and 1.5–8 times in invertebrates. GCs undergo mitotic division in Central Nervous System of insects and are bound together by Desmosomes and Tight Junctions. GCs are called a Mesaxon when forming a spiral arrangement around an Axon. GCs anchor nerve cells, limit movement, are responsible for Blood-Brain Barrier and may pass nutrients to neurons. Perineurium stores glycogen, which passes to underlying Glia and then to Neuron. GCs protect nerve cells from physical damage and insulate individual nerve cells from adjacent electrical discharges. Syn. Glial Tissue; Gliacyte. See Nervous System; Nerve Cell. Cf. Axon; Cell Body; Dendrite. Rel. Central Nervous System.

GLIDING FLIGHT A form of passive flight which follows full speed in fast flapping flight during which the insect engages prolonged forward movement without beating its wings. GF is possible in insects with large wings and low wing loading (low ratio of body mass to wing area). Seen in insects with long, narrow wings such as anisopterous Odonata and Lepidoptera (Danaidae, Nyphalidae, Papilionidae). See Flight; Passive Flight.

GLISTENING BLUE *Sahuna scintillata* (Lucas) [Lepidoptera: Lycaenidae]. (Australia).

GLITZ, CHRISTIAN THEODOR (1818–1889) (Staudinger 1890, Stettin ent. Ztg. 51: 8–10.)

GLOBATE Adj. (Latin, *globus* = globe; *-atus* = adjectival suffix.) Spherical or perfectly round in all directions. Alt. Globosus; Globular. Syn. Globose. See Lobe. Cf. Lobate. Rel. Form 2; Shape 2; Outline Shape.

GLOBE FLIES See Celyphidae.

GLOBOSE SCALE *Sphaerolecanium prunastri* (Boyer de Fonscolombe) [Hemiptera: Coccidae].

GLOBOSE Adj. (Latin, *globus* = globe.) Descriptive of structure which is spherical or globular in shape. Rel. Subglobose.

GLOBULAR Adj. (Latin, *globus* = globe.) See Outline Shape.

GLOBULAR SPIDER-BEETLE *Trigonogenius globulum* Solier [Coleoptera: Ptinidae].

GLOBULAR SPRINGTAILS [Collembola: Sminthuridae]. (Australia).

GLOBULI CELLS Specialized association cells of the Brain, usually distinguished by their small size, poverty of cytoplasm, and richly chromatic nuclei (Snodgrass).

GLOBULIN Noun. (Latin, *globus* = globe. Pl., Globulins.) Any of several simple proteins including albumenoid protein compound in insect Haemolymph, Fibrinogen and Myosin.

GLOCHIS Noun. (Greek, *glochis* = an arrow point. Pl., Glochines.) A barbed point or spine, such as the small spines found on some cactus.

GLOMERULATE Adj. (Latin, diminutive of *glomus* = ball; *-atus* = adjectival suffix.) Congregated or massed together in a cluster. Alt. Glomeratus.

GLOMERULE Noun. (New Latin, *glomerulus*. Pl., Glomerules.) An inflorescence composed of a head-like cyme.

GLOMERULOUS Adj. (Latin, diminutive of *glomus* = a ball; *-osus* = with the property of. Pl., Glomeruli.) Descriptive of or formed by Glomerules.

GLOMERULUS Noun. (Latin, diminutive of *glomus* = ball. Pl., Glomeruli.) 1. A small, compact, convoluted mass of intermingled terminal arborizations of nerve fibres within a nerve centre (Snodgrass). 2. A convoluted mass of small capillaries. 3. The convoluted portion of the ducts of an excretory gland.

GLOOMY SCALE *Melanaspis tenebricosa* (Comstock) [Hemiptera: Diaspididae].

GLOSSA Noun. (Greek, *glossa* = tongue. Pl., Glossae.) 1. The median lobe formed by the fusion of two Paraglossae and two Funicles (Imms). 2. The paired, medial lobes or gnathobases of the Labium of fused second Maxilla (Tillyard). 3. A term loosely used as a synonym for the tongue; especially applied to the coiled structure of the Lepidoptera. See Ligula. Rel. Subglossa.

GLOSSARIA Noun. (Greek, *glossa* = tongue.) A chitinized area of the Latarima of the Glossa (MacGillivray).

GLOSSARIUM Noun. (Greek, *glossa* = tongue; Latin, *-ium* = diminutive > Greek, *-idion*. Pl., Glossaria.) Diptera: The slender, pointed Glossa (Labrum-Epipharnyx) of some Species.

GLOSSATA Noun. (Greek, *glossa* = tongue.) 1. A Fabrician term for Lepidoptera. 2. A Suborder of Lepidoptera including most of the extant Taxa. Adults posess a Proboscis, lack functional Mandibles in post-pharate condition; larvae with articulated Spinnerets. See Lepidoptera.

GLOSSATE Adj. (Greek, *glossa* = tongue; *-atus* = adjectival suffix.) Pertaining to insects furnished with a spiral tongue.

GLOSSELYTRODEA Martynov 1938. An extinct Order of Neopteran insects (Permian Period - Lower Jurassic Period) represented by one or two fossil specimens from Australia.

GLOSSINID Adj. (Greek, *glossa* = tongue.) Any fly closely related or superficially similar to the Genus *Glossina*.

GLOSSINIDAE Plural Noun. Tsetse Flies. A small, monogeneric Family of calypterate Diptera assigned to Superfamily Muscoidea. Glossinids are sometimes regarded as a Subfamily of Muscidae. *Glossina* includes ca 25 Species restricted to tropical Africa and Saudi Arabia. Adult body moderate to rather large sized and brownish coloured; Proboscis long (5–15 mm) and directed anteriad; Palps as long as Proboscis and forming a sheath for Proboscis; antennal Arista plumose on dorsum of segment; wings folded at repose and projecting slightly beyond apex of Abdomen. Female Ovary consists of two polytrophic Ovarioles. Eggs are fertilized and retained in Uterus; egg incubation requires four days at 25°C; eclosion achieved with labral tooth. Larvae feed on secretion from 'milk' gland within Uterus; larvae are held in place within Uterus by Choriothete during instars 1–2; three larval instars with development requiring nine days at 25°C. Females are viviparous and deposit mature larvae late in afternoon; pupariaton occurs in soil. Flies are teneral during first 10–14 days of adult life. Both sexes require a vertebrate-blood meal. Sex ratio near 1:1; males copulate many times; females probably inseminated once under natural conditions. Repoduction is extremely low and glossinds produce fewer than 10 offspring with one mature larva produced at 10-day intervals following a post-emergence 10 day teneral-period. Females can live six months, probably average 3–4 months; males live 2–6 weeks. Species are important vectors of trypanosomes that cause Sleeping Sickness in man and Nagana in domestic livestock. See Muscoidea. Rel. Nagana; Sleeping Sickness; Tsetse Fly.

GLOSSOTHECA Noun. (Greek, *glossa* = tongue; *theke* = box. Pl., Glossothecae.) Part of the pupal cover which protects the Proboscis. Syn. Tongue-case.

GLOSSY SHIELD BUG *Cermatulus nasalis* (Westwood) [Hemiptera: Pentatomidae]: In Australia, a predator of pests in cotton. Adults are 8.5–15.0 mm long, elliptical or ovoid with body brass coloured or light red with brown punctations; head with a pale coloured line down centre and Antenna black; mouthparts extend to the hind Coxa; lateral projections from Pronotum rounded with two dark spots. Hemelytron projects beyond apex of Abdomen; Scutellum has a distinctive Y-shaped marking and two spots occur on Hemelytron. GSB often is misidentified as Brown Shield Bug but that Species has a matt brown surface, shorter head and smaller eyes. Black eggs are laid in 'rafts' of 50 or more and have short, white spines around the rim. Nymphs are dark red and brown with early instars bright red. Nymphs lack wings. Nymph and adult are predatory on moth and sawfly larvae, predatory on loopers and *Helicoverpa* larvae in cotton. See Pentatomidae. Cf. Brown Shield Bug. Rel. Predatory Shield Bug.

GLOVER, ALFRED K (–1928) (Calvert 1928, Ent. News 39: 317.)

GLOVER, TOWNSEND (1813–1883) Born Rio de Janerio, Brazil, of British parents; educated in England and later Germany as a still-life artist under Mattenheimer. Glover travelled to America ca 1836 to paint and became involved in pomology. First American Entomologist of USDA as Special Agent/Entomologist (1853) and early worker in economic Entomology. (Mallis 1971, *American Entomologists*. 549 pp. (61–69), New Brunswick.)

GLOVER'S SCALE *Lepidosaphes gloverii* (Packard) [Hemiptera: Diaspididae]: A cosmopolitan pest of citrus, coconut, mango and palms. GC resembles Purple Scale in shape, general appearance and life history. Adult female's cover is brown, ca 6 mm long, slender and parallel-sided. Females oviposit ca 50–100 white eggs in two rows beneath cover and eclosion within two weeks; crawlers settle in sheltered places, often on mature leaves, GS life cycle is completed within 6–8 weeks. GS is multivoltine with up to six generation per year, depending upon climate. See Diaspididae. Cf. Purple Scale.

GLOWWORMS *Arachnocampa* spp. [Diptera: Mycetophilidae], best known for cave-dwelling *Arachnocampa luminosa* (Skuse) in New Zealand. Light organ formed from enlarged apex of Malpighian Tubules. Glowworm life cycle requires 10–11 months. Eggs are deposited on cave ceiling with incubation requiring 20–24 days. Five larval instars; larval stage requires several months, depending upon food supply. Larvae use salivary gland secretion (mucous) and silk to form a tunnel to 10 cm long; larva suspends ca 70 threads from ceiling of cave; threads ca 15 cm long with periodic droplets of sticky substance to catch prey. Food includes insects or other arthropods caught in thread; cannibalism is common among crowded larvae, and includes Pupae or adults trapped in web. Pupal stage requires 12–13 days; Pupae are luminescent intermittently; female Pupae are luminescent until emergence; male luminescence is lost 2–3 days before emergence. Adults live 1–4 days; both sexes intermittently luminescent; male probably is attracted to female by luminosity. Females lay ca 130 eggs along margin of existing glow of flies. See Mycetophilidae.

GLOWWORMS See Lampyridae.

GLUCOSAMINE Noun. (Greek, *glykes* = sweet; *ammoniakon* = gum. Pl., Glucosamines.) An amino sugar based on glucose, found in plants and animals, which forms part of the Chitin molecule.

GLUCOSE Noun. (Greek, *glykes* = sweet; *-osus* = full of.) A hexose sugar found within all cells and the primary end-product of carbohydrate digestion. Digestible sugar produced from starches by action of saliva. Syn. Blood Sugar; Dextrose; Grape Sugar.

GLUCOSIDASE Noun. (Greek, *glykes* = sweet; *-osus* = full of.) A glucose-splitting enzyme.

GLUCOSIDE Noun. (Greek, *glykes* = sweet; chemistry *-ide* = a compound of two elements. Pl., Glucosides.) Any compound which, on hydrolysis, yields a monosaccharide (usually glucose) together with one other substance.

GLUME Noun. (Latin, *gluma* = a hull or husk, from *glubere* = to cast off the shell. Pl., Glumes.) Anatomy: Sensory ridges found on antennal Flagellomeres of many parasitic Hymenoptera. See Antenna. Cf. Multiporous Plate Sensilla; Tyloids.

GLUTAMINIC ACID A decomposition product of the sclero-proteins of the animal body, a dibasic monoamino-acid, present in casin in a relatively large amount.

GLUTINOSE Adj. (Latin, *gluten* = glue; *-osus* = full of.) Slimy; viscid. Alt. Glutinous.

GLUTOPIDAE Krivosheina 1971. A Family of brachycerous (Orthorrapha) Diptera.

GLYCERINE Noun. (Greek, *glykes* = sweet.) A colourless, dense liquid obtained from the saponification of animal fats. Glycerine is miscible with water, oils and alcohol; it is used in dissecting and for microscopic preparations; when added to alcohol for preservation of biological specimens, glycerine prevents them from drying out.

GLYCOGEN Noun. (Greek, *glykes* = sweet; *genes* = producing. Pl., Glycogens.) 1. Animal starch; a branch-chained polysaccharide in the form in which carbohydrates are stored in animals. 2. A food storage substance in the liver and muscles of vertebrates; also found within invertebrates and some plants.

GLYCOSLYATION Noun. (Pl., Glycoslyations.) The process of adding carbohydrate groups to a protein immediately following its synthesis in a eukaryotic cell.

GLYMMAE Plural Noun. (Etymology obscure.) Hymenoptera: Convolutions of the Integument anteriad of the Petiolar Spiracle. Term used predominantly in taxonomy of Ichneumonoidea, and infrequently in other Hymenoptera.

GLYPHIPTERIGIDAE Plural Noun. Glyphipterigid Moths. A small, cosmopolitan Family of ditrysian Lepidoptera assigned to Superfamily Yponomeutoidea, containing about 350 Species and best represented in temperate regions. Adults are small with wingspan 5–20 mm and head with smooth lamellar scales; Ocelli present but Chaetosemata absent; Antenna filiform and shorter than forewing, Scape without Pecten; Proboscis present and lacking scales; fore Tibia with Epiphysis, tarsal formula 0-2-4. Eggs are oval in outline shape with fine sculpture on Chorion; eggs are laid individually on host plant. Larvae are spinose with head hypognathous; larvae bore into buds, stems and seeds of monocotyledons. Pupae with small projections on head; pupation occurs away from host plant in a loose, poorly developed cocoon. Adults are diurnally active in damp habitats; they raise and lower wings while standing. Syn. Choreutidae; Hemerophilidae. See Yponomeutoidea.

GLYPHOSATE An herbicide marketed under the trade name Roundup® (Dow Chemical).

GLYPTOL Noun. (Pl., Glyptols.) A synthetic medium used for ringing temporary microscopic slide mounts that are prepared in Hoyers Medium and similar water-based mountants. Cf. Zuts. Rel. Histology.

GMELIN, JOHANN FRIEDERICH (1748–1804) (Rose 1850, *New General Biographical Dictionary* 8: 44.)

GNAPHOSIDAE Plural Noun. See Drassidae.

GNAT BUGS See Enicocephalidae.

GNATHAL REGION The first of three anatomical regions into which the embryonic trunk segments become segregated. The appendages of the GR become structures responsible for processing food. See Gnathocephalon.

GNATHAL Adj. (Greek, *gnathos* = jaw; Latin, *-alis* = pertaining to.) Pertaining to the 'jaws' or feeding appendages including Mandibles and Maxillae.

GNATHAL SEGMENTS Embryo: The segments that develop the Mandibles and first and second Maxillae (Snodgrass).

GNATHITE Noun. (Greek, *gnathos* = jaw; *-ites* = constituent. Pl., Gnathites.) 1. A jaw, Mandible or jaw-like appendage which is used for processing food. 2. The functional mouthparts.

GNATHOBASE Noun. (Greek, *gnathos* = jaw; *basis* = base. Pl., Gnathobases.) 1. A lobe or projecting portion of one of the basal segments of an appendage positioned near the mouth. The Gnathobase is used in the process of feeding. The insect Maxillae normally displays an inner and an outer Gnathobase (Tillyard). 2. Acari: A single sclerite formed by the ventral fusion of the palpal Coxae. Gnathobase bears a pair of branched Setae (Interpalpal Setae). Syn. Basis Capituli; Capitular Sternum (Goff *et al.* 1982, J. Med. Ent. 19: 226.)

GNATHOCEPHALON Noun. (Greek, *gnathos* = jaw; *kaphalos* = head. Pl., Gnathocephalons.) Portion of the head formed by gnathal segments and bearing Mandibles, Maxillae and Labium. In modern insects, head segments are completely united with no trace of an intersegmental line, except the Postoccipital Suture. The Postoccipital Suture is thought to lie between second and third gnathal segments (*i.e.* first and second Maxillae). See Head.

GNATHOCHILARIUM Noun. (Greek, *gnathos* = jaw; *-arium* = place of a thing. Pl., Gnathochilaria.) A sclerite formed by the labial structures.

GNATHO-IDIOSOMATIC ARTICULATION Acarology. Movable joint of the Gnathosoma and Idiosoma.

GNATHOPOD Noun. (Greek, *gnathos* = jaw; *pous* = foot. Pl., Gnathopods.) Any arthropod appendage in the head region which may be used to manipulate or process food.

GNATHOS Noun. (Greek, *gnathos* = jaw.) Genitalia of male Lepidoptera: A pair of appendages which arise from the posterior margin of the ninth abdominal Tergum somewhat below the base of the Uncus and extending ventrad and caudad along the sides of the Anus; often fused mesially below or around the Anus and supporting it; forming the structure termed 'Scaphium' or Subscaphium by some workers (Klots).

GNATHOSOMA Noun. (Greek, *gnathos* = jaw; *soma* = body. Pl., Gnathosomata, Gnathosomas.) Acari: The portion of the body anterior of the Circumcapitular Furrow. Appendages of the Gnathosoma include Chelicerae and Palpi. Syn. Capitulum; Hypostoma. (Goff *et al.* 1982, J. Med. Ent. 19: 226.)

GNATHOTHORACIC Adj. (Greek, *gnathos* = jaw; *thorax* = chest; *-ic* = characterized by.) Insect embryo: Pertaining to the combined gnathal and thoracic regions of the developing body (Snodgrass).

-GNATHOUS A Greek combining form to indicate mouthparts or mouthpart orientation. Cf. Hypognathous.

GNATROL® A biopesticide derived from *Bacillus thuringiensis* serotype H-14. See *Bacillus thuringiensis.*

GOANNA TICK *Aponomma undatum* (Fabricius) [Acari: Ixodidae]. (Australia).

GOAT FOLLICLE-MITE *Demodex caprae* Railliet [Acari: Demodicidae].

GOAT MOTHS See Cossidae.

GOAT WEED See Saint-Johns Wort.

GOAT-BITING LOUSE *Bovicola caprae* (Gurlt) [Mallophaga: Trichodectidae]: Probably the most common louse pest of goats. Cf. Cattle-Biting Louse; Horse-Biting Louse; Sheep-Biting Louse.

GOAT-SUCKING LOUSE *Linognathus stenopsis* (Burmeister) [Phthiraptera: Linognathidae]. (Australia).

GOBLET CELL See Calyciform Cell.

GOBRYIDAE McAlpine 1997. Hinge Flies. A small Family of acalypterate Diptera assigned to Superfamily Diopsoidea. Gobryids occur in Taiwan, Philippines, Malaysia, Indonesia and New Guinea; Family contains one Genus and <10 Species. Adults are elongate and chaetotaxy reduced with fronto-orbital bristles absent; Parafacial Suture obsolete; head wider than Thorax, somewhat depressed; Arista not segmented; Postalar Bristle absent; abdominal Tergum I with deep membranous sinuation along anterior margin. Common name 'hinge fly' alludes to the articulation between Thorax and Abdomen.

GODART, JEAN BAPTISTE (1775–1825) French lepidopterist who contributed to *Encyclopedie Methodique* and was author of volumes 1–5 of *Historire Naturelle des Lepidopteres ou Papillons de France* (1822). (Duponchel 1982, Hist. nat. Lepid. Fr. 6: 5–10.)

GODEFFROY, JOHANN CESAR (1813–) (Weidner 1967, Abh. Verh. naturw. Ver. Hamburg Suppl. 9: 139–142.)

GODFREY, EDWARD JOHN (–1933) (Anon. 1933, J. Siam Soc. 9: 264.)

GODING, FREDERICK WEBSTER (1858–1933) (Osborn 1934, J. N. Y. ent. Soc. 42: 443–449, bibliogr.)

GODMAN, FREDERICK DU CANE (1834–1919) English gentleman-naturalist who conducted survey of natural history of Mexico and Central America (1857–1874, with Osbert Salvin). Salvin and Godman published *Biologia Centrali-Americana* (63 vols, 1879–1915). (Godman 1915, Biol. Centrali Americana 1: 1–12. (Autobiogr.))

GODRON, DOMINIQUE ALEXANDRE (1807–1880) (Bonnet 1880, Naturaliste 2: 310–311, bibliogr.)

GOECKE, HANS (–1963) (Anon. 1963, Mitt. dt. ent. Ges. 22: 44.)

GOEDART, JEAN (1620–1668) Dutch painter and natural historian; Goedart observed the Metamorphosis of ca 140 Species of insects and was author of *Metamorphosis et Historia Naturalis Insectorum* (3 vols., 1662–1667). (Swainson 1840, *Taxidermy: With Biogr. of Zoologists*. 392 pp. (202–203), London. (Vol. 12, *Cabinet*, D. Lardner, ed.)

GOELDI, EMILIO AUGUSTO (1859–1917) (Studer 1917, Verh. dt. schweiz. Naturf. Ges. 1917: 1–24.)

GOETGHEBUER, MORICE (1876–1962) (Kiriakoff 1963, Bull. Ann. Soc. R. ent. Belg. 98: 240–247, bibliogr.)

GOETHE, JOHANN WOLFGANG VON (1749–1832) German poet, philosopher and natural historian. Goethe is best known for his literary works. Moreover, Goethe was the first person to use the term 'Morphology' to embrace the study of shape as related to comparative anatomy. Goethe was instrumental in forming the Natur Philosophishe Schule. (Nordenskiöld 1935, *History of Biology*. 629 pp. (279–285), London.)

GOETSCH, WILHELM (1887–1957) (Hellmich 1960, Verh. dt. zool. Ges. Suppl. 24: 540–543.)

GOETSCHMANN, THEODOR (1852–1912) (Dittrich 1912, Jh. Ver. schles. Insektenk 5: xxii–xxv.)

GOETZE, JOHANN AUGUST EPHRAIM (1731–1793) (Anon. *Accentuated list of British Lepidoptera*. 18 pp. (xxiv), Oxford & Cambridge Entomological Societies, London.)

GOFF, CARLOS CLYDE (1905–1939) (Watson 1939, Fla. Ent. 22: 12–13, bibliogr.)

GOFF, DOROTHY CLARA SCHULTZ (1907–1937) (Anon. 1937, Fla. Ent. 20: 5–6.)

GOFFE, EDWARD RIVENHALL (1887–1952) (Riley 1953, Proc. R. ent. Soc. Lond. (C) 17: 22.)

GOFFERED Adj. (French, *gaufrer* = honeycomb, waffle.) Pertaining to structure with regular, close-set impressions separated by narrow, parallel ridges; reticulated. Alt. Gauffered; Gauffred.

GOLD BUGS See Chrysomelidae.

GOLD WASPS See Chrysididae.

GOLDEN ANT *Polyrhachis ornata* Mayr [Hymenoptera: Formicidae]. (Australia).

GOLDEN APPLE BEETLES See Chrysomelidae.

GOLDEN BUPRESTID *Buprestis aurulenta* Linnaeus [Coleoptera: Buprestidae].

GOLDEN CRICKET-WASP *Liris aurulentus* (Fabricius) [Hymenoptera: Sphecidae].

GOLDEN DRUMMER *Thopha colorata* Distant [Hemiptera: Cicadidae]. (Australia).

GOLDEN EMPEROR *Anapsaltodea pulchra* (Ashton) [Hemiptera: Cicadidae]. (Australia).

GOLDEN MEALYBUG *Nipaecoccus aurilanatus* (Maskell) [Hemiptera: Pseudococcidae]: A minor pest of *Citrus* in Australia.

GOLDEN MOSQUITO *Coqillettidia xanthogaster* (Edwards) [Diptera: Culicidae] (Australia).

GOLDEN MYSTERY-SNAIL *Pila canaliculata* (Lamarck) [Monotacardia: Ampullariidae] (Australia).

GOLDEN OAK-SCALE *Asterolecanium variolosum* (Ratzeburg) [Hemiptera: Asterolecaniidae]: A pit-making scale insect that became a pest in New

Zealand and Australia before the WW II but controlled with an encyrtid wasp imported from North America. See Coccidae. Syn. *Asterodiaspis variolosa* (Ratzeburg).

GOLDEN ORB-WEAVERS *Nephila* spp. [Araneida: Araneidae] (Australia).

GOLDEN PAPER-WASP *Polistes fuscatus aurifer* Saussure [Hymenoptera: Vespidae]: (North America.) See Vespidae. Cf. Common Paper Wasp; Macao Paper Wasp; Yellow Paper Wasp.

GOLDEN SPIDER-BEETLE *Niptus hololeucus* (Faldermann) [Coleoptera: Ptinidae]: A pest of stored products; endemic to western Asia and adventive to Europe. Adults are 3.0–4.5 mm long and covered with numerous golden Setae. See Ptinidae.

GOLDEN STAG-BEETLE *Lamprima aurata* Latreille [Coleoptera: Lucanidae] (Australia).

GOLDEN SUN-MOTH *Synemon plana* Walker [Lepidoptera: Castniidae] (Australia).

GOLDEN TORTOISE-BEETLE *Metriona bicolor* (Fabricius) [Coleoptera: Chrysomelidae].

GOLDEN TRAPDOOR-SPIDER *Arbanitis* sp. [Araneida: Idiopidae] (Australia).

GOLDEN-EYED LACEWING *Chrysopa oculata* Say [Neuroptera: Chrysopidae].

GOLDEN-GLOW APHID *Dactynotus rudbeckiae* (Fitch) [Hemiptera: Aphididae].

GOLDEN-HAIRED BARK BEETLE *Hylurgus ligniperda* (Fabricius) [Coleoptera: Curculionidae] (Australia).

GOLDEN-TIPPED TUBULAR-THRIPS *Haplothrips gowdeyi* (Franklin) [Thysanoptera: Phlaeothripidae] (Australia).

GOLDFUSS, GEORG AUSUST (1782–1848.) (Ratzeburg 1874, Forstwissenschaftliches Schriftsteller-Lexicon 1: 48.)

GÖLDI-BRAUN, ROBERT (1861–1940) (Jögren 1940, Bitidningen 39: 229–230.)

GOLDING, FREDERICK DENNIS (1877–1961) (Varley 1962, Proc. R. ent. Soc. Lond. (C) 26: 53.)

GOLDSCHMIDT, RICHARD B (1878–1958) (Beltran 1958, Revta. Soc. méx. Hist. nat. 20: 185–193.)

GOLDSMITH, OLIVER (1728–1774) (Rose 1950, *New General Biographical Dictionary* 8: 57–58.)

GOLGI APPARATUS (After Camillo Golgi, Italian physician.) An organelle within the Cytoplasm of eukaryotic cells that consists of smooth-surfaced, double-membraned Vesicles whose function is not certainly established. GA may secrete giant molecules and play a role in metabolite transport.

GOLGI BODY An individual element of the Golgi Apparatus. Syn. Dictyosome.

GOLIATH BEETLES See Scarabaeidae.

GÖLLER, ERNST (–1938) (Pillel 1938, Ent. Z. Frankf. a. M. 52: 117–118.)

GOLLMER, JULIUS (–1861) (Papavero 1975, *Essays on the History of Neotropical Dipterology* Vol. 2. 217–446 (275), São Paulo.)

GOLOVYANKO, ZINOVIYA STEPANOVICH (1876–1953) (Gusev 1956, Ent. Oboz. 35: 230–236, bibliogr.)

GOLTZ, HANS FREIHERR (1864–1941) (Cailless 1934, Int. ent. Z. 28: 145–154; 169–175; 180–182.)

GOLUBATZ FLY *Simulium colombaschense* Fabricius [Diptera: Simuliidae]: A serious pest of domestic livestock and feral animals in middle and southern Europe. High mortality is associated with salivary toxins injected into host at feeding. Massive attacks by GF result in increased permeability of animal's capillaries which cause fluid from circulatory system to ooze into body cavities and tissue spaces.

GOMELIN® A registered biopesticide derived from *Bacillus thuringiensis* var. *kurstaki*. See *Bacillus thuringiensis*.

GOMES, FRANCISO AGOSTINHO (1769–1842) (Papavero 1971, *Essays on the History of Neotropical Dipterology.* 1: 48. São Paulo.)

GOMES, JOÃO FLORENCIO DE SALLES (1886–1919) (Taunay 1919, Revta. Mus. paul. 11: 563–575, bibliogr.)

GOMPHIDAE Plural Noun. Club Tails. A large, cosmopolitan Family of Odonata. Adults are variable in size, body green or yellow and marked with brown or black and head broad with eyes widely separated. Abdomen typically slender but often with inflated subterminal segments. Triangles of wing subequal in shape. Females lack an Ovipositor. Larval shape is variable with Antenna short, four segmented with third segment long and thick and apical segment minute. Legs robust and Tarsi of foreleg and middle leg each with two segments. See Odonata.

GONAD Noun. (Greek, *gonos* = begetting. Pl., Gonads.) The Ovary, Testis, or embryonic rudiment of reproductive tissue. The gonad (Testis) is the basic internal component of the male reproductive system and is derived from Mesoderm. The development of Gonads is brief during embryonic life and resumes during the nymphal stage of Paurometabola, the naiadal stage of Hemimetabola and the larval stage of Holometabola. Gonad is formed by splanchnic Mesoderm cells enveloping the germ cells (Snodgrass). Syn. Germ Gland; Testis. Cf. Ovary. Rel. Genitalia.

GONADIAL Adj. (Greek, *gone* = seed, *idion* = small; *-alis* = characterized by.) Descriptive of or pertaining to the gonads.

GONADIAL TUBE Any of the fine tubes lined with germinal epithelium that constitutes the gamete-forming part of a gonad or sex-organ (Klots).

GONAPOPHYSES Plural Noun. (Greek, *gone* = generation; *apo* = away; *phyein* = to grow. Sl., Gonapophysis.) 1. The four sclerotized structures which form the Ovipositor. Gonapophysis VIII forms the two lower valves of the Ovipositor; Gonapophysis IX forms the two upper valves of the Ovipositor. 2. The appendages surrounding the Gonopore (Tillyard). 3. Odonata: The leaf-like processes of the Ovipositor (Garman). See

Ovipositor.

GONDWANALAND Noun. Name proposed in 1885 by Austrian Geologist Edward Suess for southern landmass of supercontinent Pangaea. Syn Gondwana. See Pangaea. Cf. Laurasia.

GONGYLIDIA Plural Noun. (Greek, *gongylos* = round; *-idion* = small.) Social Insects: An enlarged hyphal apex of symbiotic fungi cultured by attine ants and used as food. See Staphyla. Alt. Gongylidium.

GONIOTHALAMICIN See Acetogenins.

GONOBASE Noun. (Greek, *gonos* = offspring; *basis* = base. Pl., Gonobases.) 1. Hymenoptera: A ring-like sclerite at the base of the gonocoxites. 2. Neuroptera: See Gonarcus. Syn. Basal Ring; Gonocardo.

GONOCARDO Noun. (Greek, *gonos* = offspring; Latin, *cardo* = hinge. Pl., Gonocardines.) Hymenoptera: See Gonobase.

GONOCOXA Noun. (Greek, *gone* = seed; Latin, *coxa* = hip. Pl., Gonocoxae.) 1. The basal sclerite of a insect Gonopod. 2. Hymenoptera: The sclerite which serves as the base of the Ovipositor sheaths. Syn. Gonocoxite.

GONOCOXAL APODEME 1. Hymenoptera: Anterior apodeme(s) which project into concave portions of the Gonobase. 2. Diptera: Internal process of Gonocoxite. See Apodeme.

GONOCOXAL ARMS Hymenoptera: Ventrobasal extensions of the Gonocoxites that form a ventral gonocoxal bridge when the GA are fused.

GONOCOXITE See Gonocoxa.

GONODUCT Noun. (Greek, *gonos* = birth; Latin, *ductus* = led. Pl., Gonoducts.) 1. The duct leading from a gonad to the genital opening. 2. The female's Oviduct; the male's Vasa Deferentia.

GONOFORCEPS Plural Noun. (Greek, *gonos* = birth; Latin, *forecps* = tongs.) 1. Hymenoptera: See Gonocoxa. 2. Heteroptera: Parameres.

GONOLACINIA Hymenoptera: See Digitus.

GONOMACULA Hymenoptera: See Cupping Disc.

GONOPLAC See Ovipositor Sheaths.

GONOPOD Noun. (Greek, *gonos* = begetting; *pous* = foot. Pl., Gonopods.) An appendage of the genital segment or associated segment that is modified for copulation, intromission or oviposition (Snodgrass).

GONOPORE Noun. (Greek, *gone* = seed; *poros* = channel. Pl., Gonopores.) 1. The aperture of the Common Oviduct through which the egg passes during oviposition. 2. Male Insects: The external aperture of the median Ejaculatory Duct, usually concealed in the Endophalus, or an apertures of the paired genital ducts. See Reproductive System. Cf. Epigynum.

GONOSICULUS Noun. (Greek, *gone* = seed; *sikua* = gourd-shaped; *-ulus* = dim.) Hymenoptera: See Digitus.

GONOSQUAMA Noun. (Greek, *gone* = seed; Latin, *squama* = a scale.) Hymenoptera: See Gonostylus.

GONOSTIPES Noun. (Greek, *gone* = seed; *stipes* = tree trunk.) 1. Mecoptera: See Gonocoxite. 2. Diptera: See Basistylus. 3. Hymenoptera: See Gonocoxite.

GONOSTIPITAL ARMS Hymenoptera: Gonocoxal Arms.

GONOSTYLUS Noun. (Greek, *gonos* = begetting; *stylus* = pillar. Pl., Gonostyli.) 1. General: The stylus on the Gonocoxa of some insects. See Saw Guides. 2. Styli of the ninth abdominal segment. When present, the Gonostylus is generally modified to form clasping organs called Harpagones.

GONOTREME Noun. (Greek, *gonos* = offspring; *trema* = hole. Pl., Gonotremes.) 1. The external opening of the Bursa Genitalis in either sex. 2. The Vulva of the female (Snodgrass).

GONYCORIA Noun. (Greek, *gony* = knee; *corium* = leather. Pl., Gonycoriae.) The part of the Tibiacoria associated with the articulation of the Tibia (MacGillivray).

GONYODON Noun. (Greek, *gony* = knee. Pl., Gonyodons.) Noctuidae: A tooth-like articulated process at the apex of the Femur in some Species.

GONYTHECA Noun. (Greek, *gony* = knee; *theca* = case. Pl., Gonythecae.) The apical part of the Femur which articulates with the Tibia. See Leg.

GOOD, HENRY GEORGE (1897–1964) (Arant 1964, J. Econ. Ent. 57: 1016–1017.)

GOODMAN, OLIVER RICHARDSON (1877–1929) (Turner 1929, Entomologist's Rec. J. Var. 41: 52.)

GOODRICH, EDWIN STEPHEN (1868–1946) (Beer 1946, Nature 157: 184–185.)

GOODRICH, SAMUEL GRISWOLD (1793–1860) (Weiss 1936, *Pioneer Century of American Entomology*. 320 pp. (134), New Brunswick.)

GOODSIR, JOHN (1814–1867) (Anon. 1868, Proc. R. Soc. London 16: xiv-xvi.)

GOODSON, LAWRENCE GRAHAM HEMBLETON (1920–1967) (Kennedy 1968, Proc. R. ent. Soc. Lond. (C) 32: 59.)

GOODWIN, CHARLES CYRIL RAY (–1972) (Butler 1973, Proc. R. ent. Soc. Lond. (C) 37: 56.)

GOODWIN, EDWARD (1867–1934) (Frohawk 1935, Entomologist 68: 48.)

GOOSE BODY-LOUSE *Trinoton anserinum* (Fabricius) [Phthiraptera: Menoponidae]. (Australia).

GOOSE WING-LOUSE *Anaticola* spp. [Phthiraptera: Philopteridae]. (Australia).

GOOSEBERRY FRUITWORM *Zophodia convolutella* (Hübner) [Lepidoptera: Pyralidae].

GOOSEBERRY SAWFLY *Pristophora pallipes* (Lep.) [Hymenoptera: Tenthredinidae].

GOOSEBERRY WEEVIL *Ecrizothis inaequalis* Blackburn [Coleoptera: Curculionidae]. (Australia).

GOOSEBERRY WITCHBROOM-APHID *Kakimia houghtonensis* (Troop) [Hemiptera: Aphididae].

GOOSEY, WILLIAM (–1878) (Pratt 1879, Entomologist 12: 64.)

GOOSSENS, THÉODORE (1827–1889) (Mabille

1889, Ann. Soc. ent. Fr. (6) 9: 499–500.)

GOPHER-TORTOISE TICK *Amblyomma tuberculatum* Marx [Acari: Ixodidae].

GORDON, GEORGE (1801–1893) (Trail 1894, Ann. Scot. nat. Hist. 1894: 65–71.)

GORDON, JOHN DOUGLAS (1901–1959) (Uvarov 1960, Proc. R. ent. Soc. Lond. (C) 24: 54.)

GORDON, RUPERT MONTGOMERY (1874–1961) (Anon. 1961, Annals Trop. med. Parasit. 55: 261.)

GORDON, WILLIAM M (–1944) (Anon. 1944, Ent. News 55: 86.)

GOREAU, C C (1790–1879) French soldier (Colonel), student of insects injurious to plants and author of *Injurious Insects* (1862, 250 pp.). Goreau provides early documetation of the role of parasites that attack leaf miners.

GORGAS, WILLIAM CRAWFORD (1854–1920) (Martin 1924, Gorgas Mem. Inst. Chicago 1924: 1–76.)

GORGERET Noun. (French, *gorge* = neck.) The barbed Sting of the honey bee. Cf. Aculeus 1.

GORHAM, HENRY STEPHEN (1939–1920) (Tomlin 1920, Entomol. mon. Mag. 56: 112–113.)

GORI® See Permethrin.

GÖRNITZ, KARL (Kirchberg 1965, Mitt. dt. ent. Ges. 24: 1–2.)

GORRIZ Y MUNOZ, RICARDO JOSÉ (–1916) (Dusmet y Alonso 1919, Boln. Soc. ent. Esp. 2: 162.)

GORSE SEED-WEEVIL 1. *Apion ulicis* (Forster) [Coleoptera: Curculionidae]. 2. *Exapion ulicis* (Forster) [Coleoptera: Brentidae]. (Australia).

GORSE THRIPS *Odontothripiella australis* (Bagnall) [Thysanoptera: Thripidae]. (Australia).

GORSE WEEVIL see Gorse Seed-Weevil.

GORTANI, MICHELE (1883–1966) (Desio 1969, Attit. Accad. Sci. Lett. Udine (7) 7: 85–124, bibliogr.)

GORY, HIPOOLYTE LOUIS (–1852) (Westwood 1853, Proc. ent. Soc. Lond. (2) 2: 53.)

GORY, M Author of *Monographie des Cetiones* (with M. Percheron) and *Monographie du genre Sisyphe*.

GOSS, HERBERT (–1908) (Anon. 1908, Entomologist's Rec. J. Var. 20: 70–71.)

GOSSARD, HARRY ARTHUR (1868–1925) (J. Econ. Ent. 19: 193–199.)

GOSSE, PHILIP HENRY (1810–1888) (Gosse 1896, *The Naturalist of the Seashore. The life of Philip Henry Gosse*. 387 pp. London.)

GOSSYPLURE A synthetic pheromone {1:1 Z,Z:Z,E 7,11-hexadecadienyl acetate} used to monitor populations and disrupt sex pheromone signals between male and female Pink Bollworm attacking cotton. Trade names include Agricon-G®, Attract 'N' Kill PBW®, Checkmate-PBW®, Decoy PBW®, Dismate®, Nomate PBW-MEC®, PB-Ropes®, Pectamone®, Pectone®, Rimilure-PBW®, Silibate-PBW®, Sirene®, Stirrup-PBW®. See Synthetic Pheromone. Rel. Parapheromone.

GOSSYPOL Noun. (Pl., Gossypols.) A complex terpenoid aldehyde produced by cotton plant glands that has insecticidal or feeding-deterrent properties. See Natural Insecticide.

GOTHNION® See Azinphos-Methyl.

GOTWALD, ADAM (1881–1970) (Koehler 1970, Polskie Pismo ent. 49: 877–878.)

GOUGELET, JEAN SCIPION (1798–1892) (Desmarest 1872, Ann. Soc. ent. Fr. (5) 2: 511–512.)

GOULD, AUGUSTUS ADDISON (1805–1866) (Anon. 1868, Proc. Amer. Acad. Arts. Sci. 7: 300–304.)

GOULD, JOHN (1804–1881) English naturalist, taxidermist and painter of mammals and birds. Gould was among the most prolific painters of birds in Asia, Australia and the Pacific. (Musgrave 1932, *A Bibliogr. of Australian Entomology 1775–1930*. 380 pp. (128). Sydney.)

GOUNELLE, PIERRE EMILE (1850–1914) (Millot 1920, Ann. Soc. ent. Fr. 89: 109–112, bibliogr.)

GOUREAU, CLAUDE CHARLES (1790–1879) (Miot 1879, Ann. Soc. ent. Fr. (5) 9: 389–400, bibliogr.)

GOUT FLIES See Chloropidae.

GOWDEY, CARLTON CRAIG (1884–1928) (Brown & Heineman 1972. *Jamaica and its Butterflies*. 478 pp. (17–18), London.)

GOZIS, MAURICE PERROTT DES (1852?-1909) (Horn 1909, Dt. ent. Z. 1909: 687.)

GRABER, VEIT (1844–1892) (Jaworowski 1892, Wien. ent. Ztg. 11: 253–262, bibliogr.)

GRABER'S ORGAN Diptera: An elaborate, flame-shaped structure on the Abdomen of tabanid larvae. GO is everted through a tube positioned dorsally between the apical two Terga. The function of GO remains unknown (Imms).

GRABOW, CARL WILHELM LOUIS (1790–1859) (Serhagen 1901, Allg. Z. Ent. 6: 241–242.)

GRABOWSKI, MARJAN (1864–1930) (N. 1930, Polski Pismo Ent. 9: 292–293.)

GRACIE, W MACAULEY (–1958) (Richards 1959, Proc. R. ent. Soc. Lond. (C) 23: 73.)

GRACIL- Latin prefix meaning 'thin' or 'slender'.

GRACILE Adj. (Latin, *gracilus* = slender.) Slender; graceful. A term descriptive of structures, bodies or appendages that are long, thin, flexible or in some way indefinably attractive. Cf. Fragile.

GRACILITIPULIDAE Hong & Wand 1990. A monotypic Family of Diptera known from the Laiyang Jurassic Period deposits of China. Apparently near the Tipulidae. See Diptera.

GRACILLARIIDAE Stainton 1854. Leaf-blotch Miners. A cosmopolitan Family of ditrysian Lepidoptera assigned to Tineoidea and including about 1,600 Species. Adults are small bodied, slender with wingspan 5–20 mm and head typically with smooth, lamellar Scales; Ocelli and Chaetosemata absent; Antenna filiform and as long or longer than forewing; Scape slender, sometimes forming an eye cap, usually without Pecten; Proboscis without scales; Maxillary Palpus typically with four segments, sometimes reduced in number; Labial Palpus with three segments; usually porrect; fore Tibia with Epiphysis; tibial spur formula 0-2-4; wings narrow with long marginal fringe. Eggs are circular or oval in outline shape with dorsal surface convex and finely sculptured; ventral surface flat. Larvae are heteromorphic with early instars dorsoventrally flattened, Mandible falcate, thoracic legs and Prolegs absent; later instars are cylindrical with thoracic and Prolegs present. Pupae with long Antenna and Proboscis; Abdomen with dorsal spines. Pupation typically occurs within oval, flattened cocoons inside mines. At least early instar larvae feed on sap; later instar larvae mine leaves, stems, petioles and twigs or form galls; some Species with a transitional non-feeding instar. Syn Phyllocnistidae. See Tineoidea.

GRADATE Noun. (Latin, *gradus* = step; *-atus* = adjectival suffix. Pl., Gradates.) One grade or step at a time; arranged in a series and blending to merge one into the other *e.g.* colours. Archaic Gradatim.

GRADATE VEINS A transverse series of crossveins in the wing of megalopterous insects; each crossvein forms part of a graded series. See Vein; Wing Venation.

GRADE Noun. (Latin, *gradus* = step. Pl., Grades.) 1. A stage in a process. 2. A position on a scale; a rank. 3. A degree of height, level or size. 4. A taxonomic category between Kingdom and Phylum. 5. An assemblage or organisms with similar features but which are not directly related.

GRADIENT Noun. (Latin, *gradus* = step. Pl., Gradients.) A Grade or regular ascent or descent (actual or arithmetical) or other graphic rise and fall.

GRADUAL METAMORPHOSIS See Paurometabolous.

GRADUALISM Noun. (Medieval Latin, *gradualis* > *gradus* = step; English, *-ism* = principle or doctrine. Pl., Gradualisms.) An hypothesis which attempts to explain the way evolution operates. Cf. Punctuated Equilibrium.

GRADWELL, GEORGE RONALD (1926–1974) (Lees 1975, Proc. R. ent. Soc. London (C) 39: 55.)

GRAEF, EDWARD LOUIS (1842–1922) (Englehardt 1922, Bull Brooklyn ent. Soc. 17: 43–45.)

GRAELLS DE LA AGÜERA, MARIANO DE LA PAZ (1809–1898) (Anon. 1943, Graellsia 1: 7–21.)

GRAENICHER, SIGMUND (1855–1937) (Cockerell 1937, Science 86: 364–365.)

GRAESER, LUDWIG (LOUIS) CARL FRIEDRICH (1840–1913) (Weidner 1967, Abh. Verh. naturw. Ver. Hamburg Suppl. 9: 130–134.)

GRÄFFE, EDUARD (1833–1916) (Soldanski 1916, Dt. ent. Z 1916: 605, bibliogr.)

GRAHAM, SAMUEL ALEXANDER (1891–1967) (Knight 1968, Michigan Ent. 1: 287–288.)

GRAIN APHID *Sitobion avenae* (Fabricius) [Hemiptera: Aphididae].

GRAIN MITE *Acarus siro* Linnaeus [Acari: Acaridae].

GRAIN RUST-MITE *Abacarus hystrix* (Nalepa) [Acari: Eriophyidae].

GRAIN THRIPS *Limothrips cerealium* (Haliday) [Thysanoptera: Thripidae].

GRAIN WEEVIL See Granary Weevil.

GRAINCOTE® See Chlorpyrifos-Methyl.

GRAM POD-BORER *Helicoverpa armigera* (Hübner) [Lepidoptera: Noctuidae].

GRAMANN, AUGUST (1876–1936) (R. W. 1937, Mitt. schweiz. ent. Ges. 17: 186–187.)

GRAMINACEOUS Adj. (Latin, *gramen* = grass; *-aceus* = of or pertaining to.) Grass-coloured, of the green of grass. Alt. Graminaceus; Gramineous; Graminelus.

GRAMINIVOROUS Adj. (Latin, *gramen* = grass; *vorein* = to devlour; *-osus* = with the property of.) Grain-eating. A term used to characterize many stored product pests. See Feeding Strategy.

GRANARY WEEVIL *Sitophilus granarius* (Linnaeus) [Coleoptera: Curculionidae]: A widespread, very serious pest of stored products, particularly grains (wheat, barley, oats, maize, rye). GW is distributed throughout world via commerce; less common in tropical and subtropical areas. GW is anatomically very similar to Rice Weevil and Maize Weevil; GW adults are cylindrical, dark brown to black, 2.5–4.0 mm long with elongate-oval pits on Pronotum; Rostrum ca 0.25 times as long as body; Elytra lack colour patches; beetles flightless. GW tolerates cooler temperatures than RW or MW and is found in temperate climates. Females typically chew holes in kernels and oviposit within endosperm. Larva to 4.0 mm, white with dark head. Life history resembles the rice weevil but adults cannot fly. Life cycle requires 4–6 weeks and adult may live 8 months. Syn. Grain Weevil. See Curculionidae. Cf. Rice Weevil; Maize Weevil.

GRANDI, GUIDO (1886–1970) Italian Entomologist and specialist on fig wasps. (Goidanachi 1973, Memorie Soc. ent. ital. 52: 5–34, bibliogr. (Reprinted from Atti Accad. naz. ital. Ent. Rc.))

GRANDORI, REMO (1885–1956) (Domenichini 1955, Boll. Zool. agric. Bachic 21: 71–80, bibliogr.)

GRANDSLAM® See Methiocarb.

GRANGER, CHARLES (–1972) (Kelner-Pillault 1972, Bull. Soc. ent. Fr. 77: 177–178.)

GRANNY MOTH *Dasypodia selenophora* Guenée [Lepidoptera: Noctuidae]. (Australia).

GRANNY'S CLOAK MOTH *Speiredonia spectans* (Guenée) [Lepidoptera: Noctuidae]. (Australia).

GRANOSAN® See Ethylene Dichloride.

GRANOSE Adj. (Latin, *granum* = grain; *-osus* = full of.) Descriptive of structure which appears as a string of beads. Cf. Moniliform.

GRANT, CHARLES E (1850–1936) (Anon. 1936. Can. Ent. 68: 212.)

GRANT, FREDERICK THOMAS (1870–1959) (Southwood 1959, Proc. Trans. S. London. ent. nat. Hist. Soc. 1959: xliii–xliv.)

GRANT, HAROLD JOHNSON (1921–1966) (Phillips 1966, Ent. News 77: 233–238, bibliogr.)

GRANT, PERCY (–1974) (Smith 1975, Austral. ent.

Soc. News Bull. 11: 47.)

GRANULAR Adj. (Latin, *granum* = grain.) Covered with small grains. Alt. Granulate; Granulatus. A type of pesticide formulation prepared by applying liquid insecticide to coarse particles of porous inorganic material such as clay or organic material such as macerated walnut shells or oatbran. Designated as 'G' on U.S. Environmental Protection Agency label or Material Safety Data Sheet documentation.

GRANULAR LEUCOCYTE A Leucocyte with the granular structure of the Cytoplasm that is temporarily phagocytic (Snodgrass). See Granulocyte. Cf. Leucocyte. Rel. Circulatory System.

GRANULAR SPHERES Phagocytes distended with the inclusion of tissues they have attacked within the Pupa during Histolysis (Imms).

GRANULATE CUTWORM *Agrotis subterranea* (Fabricius) [Lepidoptera: Noctuidae]. See Cutworms.

GRANULE Noun. (Latin, *granum* = grain. Pl., Granules.) A little grain or a minute grain-like elevation.

GRANULOCYTE Noun. (Latin, *granulum* = small grain; Greek, *kytos* = hollow vessel; container. Pl., Granulocytes.) A basic anatomical form of insect Haemocyte. Variable-sized Haemocytes whose Plasma Membrane is papillate or with irregular processes; nucleus small, round-oval and centrally positioned; cytoplasm granular. Granulocytes of Holometabolous insects generally are smaller than Granulocytes of Paurometabolous or Hemimetabolous insects. Syn. Amoebocyte; Cystocyte; Hyaline Cell; Phagocyte; Pycnoleucocyte. See Haemocyte. Cf. Adipohemocyte; Coagulocyte; Nephrocyte; Oenocyte; Plasmatocyte; Prohaemocyte; Spherulocyte.

GRANULOSE Adj. (Latin, *granum* = grain; *-osus* = full of.) 1. Pertaining to a granular surface or surface roughened. 2. The surface of a structure composed of distinct grains. See Texture.

GRANULOSIS VIRUS Insect-pathogenic viruses of the Baculoviridae. See Nuclear Polyhedrosis Virus; Nonoccluded-type Virus. Cf. Cytoplasmic Polyhedrosis Virus; Insect Poxvirus.

GRANUM TINCTORIUM See Kermes.

GRANUPON® See Codling Moth Granulosis Virus.

GRANUTOX® See Phorate; Terbufos.

GRAPE BERRY-MOTH *Polychrosis viteana* (Clemens) [Lepidoptera: Tortricidae]: A bivoltine pest of grapes in eastern USA and Canada. GBMs overwinter as Pupae within cocoons attached to trash or grape bark. Adults appear during late spring and females oviposit flat, circular, cream-coloured eggs upon all exposed parts of plant. Larvae web berries together and cause them to drop from vine prematurely; surface of nearly ripe grapes are scarred by feeding. Syn. American Vine Moth; *Endopiza viteana*. See Tortricidae.

GRAPE-BLOSSOM MIDGE *Contarinia johnsoni* Felt [Diptera: Cecidomyiidae]: A pest of grapes in North America. GBM larvae induce buds to swell and discolour. See Cecidomyiidae. Cf. Sorghum Midge; Pear Midge.

GRAPE BORER *Xylotrechus pyrrhoderus* Bates [Coleoptera: Cerambycidae].

GRAPE-CANE GALLMAKER *Ampeloglypter sesostris* (LeConte) [Coleoptera: Curculionidae].

GRAPE COLASPIS *Colaspis brunnea* (Fabricius) [Coleoptera: Chrysomelidae]: In North America, a widespread, univoltine pest of grapes, strawberries, beans, clover, melons, potatoes and cowpeas. GC overwinter as early instar larvae. Adults feed on foliage and deposit eggs on or near roots of the host plants. Pupation occurs within earthen cells. See Chrysomelidae.

GRAPE CURCULIO *Craponius inaequalis* (Say) [Coleoptera: Curculionidae]: A pest of grapes in eastern USA. Eggs are laid in punctures made within berries during July-August; larvae feed on seeds and berries. Pupation occurs within earthen cell; adults emerge during autumn and overwinter in field trash. See Curculionidae.

GRAPE ERINEUM-MITE *Colomerus vitis* (Pagenstecher) [Acari: Eriophyidae].

GRAPE FLEA-BEETLE *Altica chalybea* Illiger [Coleoptera: Chrysomelidae]: A pest of apple, grape, plum and quince in eastern North America; GFB also attacks beech, elm and Virginia creeper. Adults are ca 12 mm long, metallic dark bue-green in colour, overwinter in or near grape plantings and become active during spring. Adults feed on buds of host plants as they open; feeding damage gives opening foliage a ragged, tattered appearance. GFB females oviposit on bark of canes or upper side of leaves; eggs are pale yellow. Larvae feed on leaves and blossom buds. Mature larvae drop to ground and pupate within soil. See Chrysomelidae.

GRAPE LEAF BLISTER-MITE *Colomerus vitis* (Pagenstecher) [Acari: Eriophyidae]. (Australia).

GRAPE LEAF RUST-MITE *Calepitrimerus vitis* (Napela) [Acari: Eriophyidae]. (Australia).

GRAPE LEAF-FOLDER *Desmia funeralis* (Hübner) [Lepidoptera: Pyralidae]: A widespread pest of grapes in North America. Eggs are laid on underside of leaves and early instar larvae feed gregariously on leaves loosely webbed together; later instars feed individually in pencil-sized leaf rolls. Pupation occurs in leaf rolls and GLF completes three generations per year. Insect causes reduced photosynthesis in affected plants. See Pyralidae.

GRAPE-LEAF SKELETONIZER *Harrisina americana* (Guérin) [Lepidoptera: Zygaenidae].

GRAPE LEAFHOPPER *Erythroneura elegantula* Osborn [Hemiptera: Cicadellidae]: A pest of grape and apple in North America; alternative host plants include Virginia creeper, Boston Ivy and blackberry. Adults are ca 3–4 mm long, pale yellow with red markings on forewings. GLHs overwinter as adults in ground trash and become active in spring when grape leaves are half developed; they then fly to leaves and feed on underside. Adults feed 2–3 weeks before oviposition. Eggs are laid in leaf tissue; Nymphs requires 20–35 days for development. Nymph and adult feeding cause leaves to mottle, blotch and drop or reduce chlorophyll. GLH completes 2–3 generations per year, depending upon climate. Some control of this pest is provided by a mymarid wasp, *Anagrus epos* Girault. See Cicadellidae. Cf. Beet Leafhopper.

GRAPE MEALYBUG *Pseudococcus maritimus* (Ehrhorn) [Hemiptera: Pseudococcidae]: An occasional pest of table grape varieties in western North America. Several internal parasites (Encyrtidae) have been recovered from the pest. See Pseudococcidae.

GRAPE PHYLLOXERA *Daktulosphaira vitifoliae* (Fitch) [Hemiptera: Phylloxeridae]: A pest of grapes endemic to North America. Adults are minute, with four forms. GP overwinter as eggs on grape canes or phylloxerans within gall nodules on grape roots. CP completes 5–8 generations per year. Eggs hatch during spring and phylloxerans migrate to leaves where they feed and form half-pea-sized galls. Mature parthenogenetic females produce nymphs within galls that form new galls. Leaf-inhabiting phylloxerans drop to ground, form galls on roots. Subsequently, winged forms appear during autumn, emerge from galls on roots, mate and lay eggs on leaves. GP was a serious pest of grapes introduced from North America into Europe during late 1880s. See Phylloxeridae.

GRAPE PLUME-MOTH *Pterophorus periscelidactylus* Fitch [Lepidoptera: Pterophoridae]: Common name from the adult whose wings are feather-like. GPM is a pest of grape vines in eastern North America. Eggs are laid on branches and larvae feed on terminal leaves drawn together with silk. GPM pupa is attached to plant with silk and is suspended as a butterfly chrysalis. See Pterophoridae.

GRAPE ROOT-BORER *Vitacea polistiformis* (Harris) [Lepidoptera: Sesiidae]: A moth whose larvae burrow into grape roots and thereby cause damage to grape plants. See Sesiidae.

GRAPE ROOTWORM *Fidia viticida* Walsh [Coleoptera: Chrysomelidae]: A monovoltine pest of grape in eastern North America. GRWs overwinter as larvae deep within soil; as soil warms during spring, larvae move toward surface. Pupation occurs within cells ca 2–5 cm from surface during May-June. Adults feed on upper surface of foliage and cause chain-like rows of holes. Eggs are yellowish, deposited in masses of 20–30 on canes under loose bark. Eclosion occurs within a few days and neonate larvae drop to ground, enter soil and feed on grape roots. See Chrysomelidae. Cf. Western Grape Rootworm.

GRAPE SAWFLY *Erythraspides vitis* (Harris) [Hymenoptera: Tenthredinidae].

GRAPE SCALE *Diaspidiotus uvae* (Comstock) [Hemiptera: Diaspididae].

GRAPE-SEED CHALCID *Evoxysoma vitis* (Saunders) [Hymenoptera: Eurytomidae].

GRAPE THRIPS *Drepanothrips reuteri* Uzel [Thysanoptera: Thripidae].

GRAPE TRUNK-BORER *Clytoleptus albofasciatus* (Laporte & Gory) [Coleoptera: Cerambycidae].

GRAPE-VINE APHID *Aphis illinoisensis* Shimer [Hemiptera: Aphididae]: A pest on succulent terminal shoots of grape in North America. See Aphididae.

GRAPE-VINE HAWK-MOTH *Hippotion celerio* (Linnaeus) [Lepidoptera: Sphigindae]. (Australia).

GRAPE-VINE LOOPER *Eulithis diversilineata* (Hübner) [Lepidoptera: Geometridae].

GRAPE-VINE MOTH *Phalaenoides glycinae* Lewin [Lepidoptera: Noctuidae]. (Australia).

GRAPE-VINE SCALE *Parthenolecanium persicae* (Fabricius) [Hemiptera: Coccidae]: In Australia, a pest of grape that sucks sap from foliage and canes; sooty mould fungus develops which makes fruit unusable. See Coccidae.

GRAPE WHITEFLY *Trialeurodes vittata* (Quaintance) [Hemiptera: Aleyrodidae].

GRAPENTIEN, HUGO (1860–1935) (A. S. 1935, Mitt. schweiz. ent. Ges. 16: 611–612.)

GRASLIN, ADOLPHE DE (1802–1882) (Mabille 1883, Ann. Soc. ent. Fr. (6) 3: 561–564, bibliogr.)

GRASS Noun. (Middle English, *gras* > Anglo Saxon, *graes* > Latin, *gramen*. Pl., Grasses.) Green herbage that taxonomically is assigned to the Cyperaceae, Juncaceae and Poaceae. Grasses collectively constitute a dominant form of terrestrial vegetation. Grass architecture is simple without complex branching structure and leaves modified to form narrow, spear-shaped blades; secondary thickenings are not present and growth proceeds at intercalary meristem, protected by hard leaf sheaths. Grasses typically lack secondary defensive compounds which deter herbivores. As a consequence, many Species of insects feed upon grasses. See Plant. Cf. Bush; Tree; Shrub.

GRASS ANTHELID *Pterolocera amplicornis* Walker [Lepidoptera: Anthelidae]. (Australia).

GRASS BLUE-BUTTERFLY *Zizina labradus labradus* (Godart) [Lepidoptera: Lycaenidae]. (Australia).

GRASS CATERPILLAR *Herpetogramma licarsisalis* (Walker) [Lepidoptera: Pyralidae]. (Australia).

GRASS COCCID *Symonicoccus australis* (Maskell) [Hemiptera: Coccidae]. (Australia).

GRASS FLEAHOPPER *Halticus chrysolepis* Kirkaldy [Hemiptera: Miridae].

GRASS FLIES See Chloropidae.

GRASS ITCH-MITE *Odontacarus australiensis* (Hirst) [Acari: Trombiculidae]. (Australia).

GRASS JEWEL *Freyeria putli* (Kollar) [Lepidoptera: Lycaenidae]. (Australia).

GRASS LEAFMINER-MOTHS See Elachistidae.

GRASS MITE *Siteroptes graminum* (Reuter) [Acari: Siteroptidae].

GRASS MOTHS See Crambidae; Elachistidae; Pyralidae.

GRASS SAWFLY *Pachynematus extensicornis* (Norton) [Hymenoptera: Tenthredinidae].

GRASS SCOLYTID *Hypothenemus pubescens* Hopkins [Coleoptera: Scolytidae].

GRASS SHARPSHOOTER *Draeculacephala minerva* Ball [Hemiptera: Cicadellidae].

GRASS SHEATH-MINER *Cerodontha dorsalis* (Loew) [Diptera: Agromyzidae].

GRASS THRIPS *Anaphothrips obscurus* (Müller) [Thysanoptera: Thripidae].

GRASS TUNNEL-MOTH *Philobota chionoptera* Meyrick [Lepidoptera: Oecophoridae]. (Australia).

GRASS WEBBING-MITE *Olygonychus araneum* Davis; *Olygonychus digitatus* Davis [Acari: Tetranychidae]. (Australia).

GRASS WEBWORM 1. *Calamotropha leptogrammella* (Meyrick) [Lepidoptera: Pyralidae]. (Australia). 2. *Herpetogramma licarsisalis* (Walker) [Lepidoptera: Pyralidae]: An important pest of turfgrasses in Hawaii. See Pyralidae.

GRASS-CROWN MEALYBUG *Antonina graminis* (Maskell) [Hemiptera: Pseudococcidae]. (Australia).

GRASSHOPPER Noun. (Middle English, *gras*; *hoppen* = to hoop. Pl., Grasshoppers.) A common name applied to Orthoptera, particularly Acrididae, that do not migrate. See Acrididae. Cf. Locust; Long-horned Grasshopper; Short-horned Grasshopper. Correct veneral term: A cloud of grasshoppers.

GRASSHOPPER BEE-FLY *Systoechus vulgaris* Loew [Diptera: Bombyliidae].

GRASSHOPPER EGG-PARASITES *Scelio* spp. [Hymenoptera: Scelionidae]. (Australia).

GRASSHOPPER MAGGOTS *Blaesoxipha* spp. [Diptera: Sarcophagidae]: Internal parasites of grasshoppers. Female fly larviposits on grasshopper and GMs penetrate host where they develop in thoracic region of body. See Sarcophagidae.

GRASSHOPPER SPORE® See *Nosema locustae*.

GRASSHOPPERS See Acrididae; Saltatoria.

GRASSI, GIOVANNI BATTISTA (1854–1925) (Fedele 1927, Boll. Soc. nat. Napoli 38: (append): 59–93, bibliogr.)

GRASSMAN, PETER CHASE (1925–1944) (Essig 1946, Ann. ent. Soc. Am. 39: 344–345.)

GRASSROOT MEALYBUG *Rhyzoecus rumicis* (Maskell) [Hemiptera: Pseudococcidae]. (Australia).

GRATZ, HEINRICH (1886–1966) (Muller 1968, Ent. Ber., Berlin 1968: 135–136.)

GRAVENHORST JEAN LOUIS LUIS CHARLES (1777–1857) German Entomologist and author of numerous works including: *Coleoptera microptera Brunsvicensia* (1802), *Monographia Coleopterorum micropterorum* (1806), Monographie du genre Ichneumon (1814), *Monographia Ichneumonum Pedemontae regionis* (volume 24 of Mem. l'Acad. Sci., Turin), *Monographie des Ichneumons Apteres*, *Conspectus generum es familiarum Ichneumonidum* (with Nees ab Eisenbeck), and *Beitrage zur Entomologie*. Perhaps his most know work is titled *Ichneumonologia Europaea* (3 vols, 1829). (Letzner 1857, Jber. schles. Ges. vaterl. Kult. 1857: 111–115.)

GRAVES, PHILIP PERCIVAL (1876–1953) (Buxton 1954, Proc. R. ent. Soc. Lond. (C) 18: 79.)

GRAVID Adj. (Latin, *gravidus* = loaded.) Pertaining to females with eggs in the reproductive system which are suitable for oviposition, or larvae in the Uterus which are suitable for larviposition.

GRAY, ASA (1810–1888) American physician (Fairfield Medical College, 1831) who was appointed Fisher Professor of Natural History (Harvard University, 1842) and regarded as a dominant individual in 19th century Botany. Gray was the first professional botanist in North America and author of *The Botanical Text-book* (1842) which emphasized plant structure and classification. He is best remembered for *Manual of the Botany of the Northern United States* (1848) (Hewitson 1889, Proc. Calif. Acad. Sci 1: 309–312.)

GRAY, DAVID EDGAR (1913–1972) (Eidt 1973, Bull. Ent. Soc. Can. 5: 11.)

GRAY, GEORGE ROBERT (1808–1872) (Anon. 1872, Proc. Linn. Soc. Lond. 1871–1872: lxii-lxiv.)

GRAY, JOHN (–1878) (Somerville 1910, Glasgow Nat. 2: 68–69.)

GRAY, JOHN (–1881) (Anon. 1882, Entomol. mon. Mag. 18: 190–191.)

GRAY, JOHN EDWARD (1800–1875) (Anon. 1875, Ann. Mag. nat. Hist (4) 15: 281–285; Gunther 1975, *A Century of Zoology at the British Museum 1815–1914.* 533 pp. (1–210), London.)

GRAY, PERCY HEATH HOBART (1891–1971) (Sheppard 1972, J. Lepid. Soc. 26: 66–67, bibliogr.)

GREAT ASH-SPHINX *Sphinx chersis* (Hübner) [Lepidoptera: Sphingidae].

GREAT DORSAL-RECTI MUSCLES The wider set of longitudinal dorsal muscles which lie near the Heart in insects (Comstock & Kellogg). See Muscle.

GREAT FIG-TREE BORER *Batocera boisduvali* (Hope) [Coleoptera: Cerambycidae]. (Australia).

GREAT STRIPED GARDEN-SLUG *Limax maximus* Linnaeus [Sigmurethra: Limacidae]. (Australia).

GREAT YELLOW-SLUG *Lehmannia flava* (Linnaeus) [Sigmurethra: Limacidae]. (Australia).

GREAT-BASIN WIREWORM *Ctenicera pruinina* (Horn) [Coleoptera: Elateridae]. A pest of potatoes in the Great Basin and Pacific Northwest USA. See Elateridae.

GREATER COCONUT-SPIKE MOTH A complex of Species whose larvae feed on the inflorescence of coconuts grown throughout tropical Asia and

Oceana. Species include *Tirathaba complexa* (Butler) and *T. rufivena* (Walker) [Lepidoptera: Pyralidae]. See Pyralidae.

GREATER DATE MOTH *Arenipses sabella* Hampson [Lepidoptera: Pyralidae].

GREATER EUROPEAN-SPRUCE BEETLE *Dendroctonus micans* (Kugelmann) [Coleoptera: Scolytidae]: A scolytid beetle endemic to Palearctic coniferous forests. Primarily attacks *Picea* spp. See Scolytidae.

GREATER OCELLARS Diptera: See Ocellar Bristles (Comstock).

GREATER SANDY DUNG-BEETLE *Euoniticellus africanus* (Harold) [Coleoptera: Scarabaeidae]. (Australia).

GREATER WAX-MOTH *Galleria mellonella* (Linnaeus) [Lepidoptera: Pyralidae]: A pest of the honey bee that generally inhabits comb not occupied by bees. Male moth releases a pheromone that calls a female to vicinity of hive where mating may occur. Females enter hive to lay eggs at night; moths are attacked if detected by worker bees. Larvae form silken burrows, feed on wax of comb and can destroy hive but do not attack bees. GWM completes 6–7 larval instars and pupation occurs within a tough cocoon on side of hive. See Pyralidae. Cf. Lesser Wax Moth.

GREDLER, VINCENZ MARIA PATER (1823–1912) (Ohaus 1912, Dt. ent. Z. 1912: 375–376.)

GREEDY SCALE *Hemiberlesia rapax* (Comstock) [Lepidoptera: Tortricidae].

GREEN Noun. (Middle English, *grene* > Anglo Saxon, *grene*. Pl., Greens.) 1. One of the six psychological primary colours perceived when energy of wavelengths 495–515 is presented as a stimulus to the eye. Specifically, the psychological hue seen at wavelength 505.5 millimicrons when presented as a stimulus to the human eye. 2. A colour, the hue of which is between yellow and blue. See Colour. Cf. Black; Blue; Red; Yellow; White.

GREEN AWL *Hasora discolor mastusia* Fruhstorfer [Lepidoptera: Hesperiidae]. (Australia).

GREEN BARON *Psaltoda magnifica* Moulds [Hemiptera: Cicadidae]. (Australia).

GREEN BLOW-FLY *Phaenicia sericata* (Meigen) [Diptera: Calliphoridae]: A cosmopolitan, multivoltine pest of sheep and significant pest in Britain. GBF is adventive to Australia and New Zealand where it is a minor pest of sheep. GBF adults are 11–14 mm long, metallic green without markings and anatomically very similar to *L. cuprina,* another pest of sheep. Eggs are deposited in clusters ca 150–200 in decomposing garbage or dead animals. GBF larvae (fleeceworms) attack soiled sheep wool. GBF overwinters as mature larvae in soil and pupation occurs in soil. Larvae are used in medical treatment of nonhealing wounds and osteomyelitis. Syn. *Phaenicia sericata;* Green Bottle Fly. See Calliphoridae. Cf. Sheep Blow-Fly. Rel. Maggot Debridement Therapy.

GREEN BOTTLE FLIES *Lucilia* spp. [Diptera: Calliphoridae].

GREEN BUDWORM *Hedya nubiferana* (Haworth) [Lepidoptera: Tortricidae].

GREEN BUNYIP *Tamasa rainbowi* Ashton [Hemiptera: Cicadidae]. (Australia).

GREEN CARAB-BEETLE *Calosoma schayeri* Erichson [Coleoptera: Carabidae]: An Australian Species. Adults are 22–25 mm long, elongate and metallic green and black with a constriction between Pronotum and Elytra; mouthparts occur at apex of head and consist of strong, sharp Mandibles; Antenna long and eyes large. Eggs are laid in soil or leaf litter. Larvae are elongate, lightly sclerotized with strong mouthparts adapted for seizing prey. Pupation occurs in soil or leaf litter. Nocturnal larvae wander over plants in search of caterpillar prey. Adults prey on cutworms and other soil-dwelling caterpillars and may be attracted to lights at night. GCB may be found on cotton throughout Australia. If handled roughly, adult beetles give off an unpleasant odour. See Carabidae.

GREEN CARPENTER-BEES *Lestis* spp. [Hymenoptera: Anthophoridae] (Australia).

GREEN CLOVERWORM *Plathypena scabra* (Fabricius) [Lepidoptera: Noctuidae]: A multivoltine, sporadic pest of leguminous crops in the United States with larval feeding on foliage giving plant a ragged appearance. Adult wingspan 30 mm, body brown with black spots or mottling and triangular outline shape when at repose. Eggs are ca 0.5 mm with polar ridges, and translucent green becoming brownish with red spots before eclosion. Eggs are deposited individually on underside of leaves of host plant; eclosion occus within three days. Mature larvae are ca 25 mm long, pale green with two narrow white stripes along sides of body; larvae display three pairs of abdominal prolegs and pair of anal prolegs. Pupae are brown and within silken cocoons. GCWs overwinter as Pupae or adults, or are active throughout year in tropical climate. GCW typically completes 3–4 generations per year. See Noctuidae.

GREEN CLUSTER FLY *Dasyphora cyanella* [Diptera: Muscidae]: An urban pest of houses in Europe. Adults overwinter in attics and roof spaces of buildings. Biology and life history similar to Cluster Fly. See Muscidae. Cf. Cluster Fly.

GREEN COFFEE SCALE *Coccus viridis* (Green) [Hemiptera: Coccidae]: A widespread pest of citrus, coffee and ornamental plants. Adult females are pale yellow-green, 3–4 mm long and oval to elongate in outline shape with flattened cover; gut visible through scale cover as black spots or U-shaped line. GCS is parthenogenetic and females produce live young or eclosion occurs immediately after oviposition; life cycle requires ca 6–9 weeks and Species completes 3–4 generations per year. GCS produces honeydew which serves as substrate for sooty mould and other fungi. See Coccidae.

GREEN CRAB-SPIDER *Hedana* spp. [Araneida: Thomisidae]. (Australia).

GREEN CUTWORM *Neumichtis saliaris* (Guenée) [Lepidoptera: Noctuidae]. (Australia). See Cutworms.

GREEN-EYED SKIMMERS See Corduliidae.

GREEN FAIRY *Baeturia flava* (Goding & Froggatt) [Hemiptera: Cicadidae]. (Australia).

GREEN FRUIT-WORM *Lithophane antennata* (Walker) [Lepidoptera: Noctuidae].

GREEN FRUIT-WORMS The larval stage of several Species of noctuid moths that eat holes in the fruit of deciduous forest trees and fruit trees. See Nocutidae.

GREEN GARDEN-LOOPER *Chrysodeixis eriosoma* (Doubleday) [Lepidoptera: Noctuidae].

GREEN GLANDS The highly modified Nephridia in Crustacea.

GREEN GUAVA-MEALYBUG *Chloropulvinaria psidii* (Maskell) [Hemiptera: Pseudococcidae].

GREEN-HEAD ANT *Rhytidoponera metallica* (Smith) [Hymenoptera: Formicidae]: Workers are 4–6 mm long, head metallic-green, metallic black-green Thorax and Gaster. Species with a particularly painful sting. In Australia, nest in small colonies under paths or in ground and rarely invade homes. Nests are relatively large and conspicuous, with multiple entrance holes that are often decorated with twigs, small pebbles and crumbs of soil. This is one of the few ponerines that will forage in trees, and workers may commonly be seen on tree-trunks. GHA is an urban problem of lawns and sports fields. Feed on vegetable matter. See Formicidae.

GREEN HEADS See Tabanidae.

GREEN JUNE-BEETLE *Cotinis nitida* (Linnaeus) [Coleoptera: Scarabaeidae]: A pest of fruit trees, vegetables and lawns in southern USA. Adults feed on fruit and foliage; larvae feed on roots. See Scarabaeidae. Cf. Fig Beetle; Japanese Beetle.

GREEN LACEWING *Mallada* spp. [Neuroptera: Chrysopidae]: In Australia. Adults are 5–12 mm long with wingspan 7–16 mm and body elongate, pale green, sometimes brown or with brown markings on head and Thorax. Head triangular; Antenna as long as forewing; eye large and may be pigmented; wings transparent with lace-like venation. Eggs are white, oval, stalked and laid on underside of leaves. Eggs are placed in groups, sometimes in a ring or U-shape. Larvae are elongate and tapered distally; larvae appear humped and covered in long spines with prey items lodged on them for camouflage. Mouthparts sickle-shaped, large and positioned at apex of head. Pupation occurs on foliage within a cocoon that is camouflaged with prey. Only larvae are predaceous and attack aphids, mites, soft-bodied insects and eggs. Larvae inhabit foliage where they seek prey. See Chrysopidae. Cf. Brown Lacewing.

GREEN LACEWINGS See Chrysopidae.

GREEN LEAFHOPPER 1. *Nephotettix malayanus* Ishihara & Kawase; *Nephotettix nigromaculatus* (Motschulsky) [Hemiptera: Cicadellidae]. (Australia). 2. *Nephotettix* spp. [Hemiptera: Cicadellidae]. 2. *Tettigella viridis* Linnaeus [Hemiptera: Tettigoniidae].

GREEN MANTID *Orthodera ministralis* (Fabricius) [Mantodea: Mantidae]. (Australia). See Mantidae.

GREEN MIRID *Creontiades dilutus* (Stål) [Hemiptera: Miridae]: A sap-sucking pest of stone fruits and an important secondary pest of Cotton in Australia. Adults are 2–6 mm long, yellow-green to green and elongate-oval with head and Pronotum triangular; second antennal segment longer than other segments. Adult stage lasts up to 28 days. Adult females lay eggs singly within plant tissue with only an oval egg cap exposed above leaf or petiole surface. Egg stage requires 7–10 days. Nymphs are similar to adult but lack wing buds. Nymphs are 1–6 mm long, green to yellow-green with long Antennae. See Miridae. Cf. Brown Mirid.

GREEN MUSCARDINE-FUNGUS *Metarhizium anisopliae* (Metchnikoff) Sorokin: An infection similar to *Beauveria bassiana*. Green muscardine fungus was discovered in Russia by Metchnikoff in 1879 infecting larvae of the wheat cockchafer. The fungus also causes sporadic destruction of canegrub populations. Infected larvae become covered with white fungal growth that later turns grey-green when fungal spores are produced. In North America at least 75 insects have been found to be infected with *Metarhizium* but its use in biological control has been limited.

GREEN PEACH-APHID *Myzus persicae* (Sulzer) [Hemiptera: Aphididae]: A cosmopolitan pest that feeds on many tree and vegetable crops, and a significant pest of flue-cured tobacco in USA; endemic to Europe and adventive through global commerce. GPA transmits Beet Mosaic, Lettuce Mosaic, Beet Yellows and Potato Leaf Roll. Feeding by large numbers of individuals may induce wilting. Adults are 1.5–2.5 mm long, globular and yellowish-green with a dark blotch on Abdomen; apex of Abdomen (Cauda) is constricted near middle and has six Setae protruding from the side. Adult females are fertilized and lay eggs on the bark of the winter host plant (peach, plum, apricot, ornamental shrubs and flowering plants). Eggs are shiny black. First-instar nymphs are similar to adult but lack wings; nymphs are pale yellowish-green with three dark lines on dorsum of Abdomen. Females from this generation are parthenogenetic and produce live young. Individuals from second and third generations have wings and migrate from winter host plant to summer host plant during spring. Many generations are produced parthenogenetically until autumn. Adult females then migrate back to the winter host plant and produce offspring capable of sexual reproduction. Males and females mate and the female lays eggs to repeat the cycle. In North America, overwintering eggs typically are laid on peach, plum and cherry; nymphs emerge during spring and feed; later generations shift to vegetables. GPA completes as many as 30 generations per year. Syn. Spinach Aphid; Tobacco Aphid. See Aphididae. Cf. Cotton Aphid.

GREEN PLANTHOPPER *Siphanta acuta* (Walker); *Siphanta hebes* (Walker) [Hemiptera: Flatidae]: Pests of numerous ornamental native plants in Australia; *S. acuta* also occurs in New Zealand and Hawaii. Adults are ca 8 mm long, typically pale green with small yellow spots and triangular in outline shape from lateral view; wings held tentiform when at repose. Eggs are laid in a circular mass comprising about 50 eggs. GPH is multivoltine with a life cycle requiring 1–2 months, depending upon climate and host plant. See Flatidae.

GREEN POTATO-BUG *Cuspicona simplex* Walker [Hemiptera: Pentatomidae]: A minor pest of potato in Australia where it attacks tender shoots and growing points. See Pentatomidae.

GREEN RICE-LEAFHOPPER *Nephotettix virescens* (Distant) [Hemiptera: Cicadellidae]. Alt. Green Leafhopper.

GREEN ROSE-CHAFER *Dichelonyx backi* (Kirby) [Coleoptera: Scarabaeidae].

GREEN SCALE *Coccus viridis* (Green) [Hemiptera: Coccidae].

GREEN SCARAB-BEETLE *Diphucephala colaspidoides* (Gyllenhal) [Coleoptera: Scarabaeidae]. (Australia).

GREEN SEMILOOPER *Plusia eriosoma* (Doubleday) [Lepidoptera: Noctuidae].

GREEN SHIELD-SCALE *Pulvinaria psidii* Maskell [Hemiptera: Coccidae]. (Australia).

GREEN SNAIL *Helix aperta* (Born) [Sigmurethra: Helicidae]. (Australia).

GREEN SOLDIER-BEETLE *Chauliognathus pulchellus* W. S. Macleay [Coleoptera: Cantharidae]: An Australian Species. Adults are 8–12 mm long, elongate and pale yellow-green with black head and legs; black markings occur on Pronotum and Elytra; Antenna long; Elytra apically rounded. Larvae are elongate, creamy-white with flattened, sclerotized head. Larvae occur in soil or leaf litter. Pupation occurs in soil or leaf litter. Adults are found on foliage and flowers; they feed on foliage, but are not classified as pests. Larvae are predaceous on other insects. GSBs sometimes occur in flying swarms. See Cantharidae.

GREEN SPHINX *Tinostoma smaragditis* (Meyrick) [Lepidoptera: Sphingidae].

GREEN SPRING-BEETLE *Diphucephala edwardsii* Waterhouse [Coleoptera: Scarabaeidae]. (Australia).

GREEN SPRUCE-APHID *Cinara fornacula* Hottes [Hemiptera: Aphididae].

GREEN STINK-BUG 1. *Acrosternum hilare* (Say) [Hemiptera: Pentatomidae]: A widespread pest of vegetable crops and soybeans. Adults and nymphs damage crop with piercing-sucking mouthparts. Adults are ca 15 mm long, shield-shaped with body green; Antenna with five segments, banded with black coloration. GSB overwinters as adults in concealed habitats; adults may feed intermittently during winter in subtropical or warm-temperate climates. Eggs are barrel-shaped and laid in clusters on host plant; eclosion occurs within 12 days. Nymphs are predominantly green with black Pronotum and vertical black-and-yellow stripes on Abdomen; neonate nymphs remain in aggregations and do not feed for several days. GSBs undergo five instars. Biology similar to other Species of economically important stink bugs: GSB prefers tender plant tissue and immature seeds as food but may attack stems, foliage and blooms. In Australia, GSB sometimes invades bean crops in large numbers and damages pods by sucking juices; Species may also damage tomatoes, silverbeet and grapevines. 2. *Plautia affinis* (Dallas) [Hemiptera: Pentatomidae]. (Australia). See Pentatomidae. Cf. Brown Stink Bug; Green Vegetable Bug.

GREEN STONEFLIES See Chloroperlidae.

GREEN TREE-ANT *Oecophylla smaragdina* (Fabricius) [Hymenoptera: Formicidae]. (Australia).

GREEN TREEHOPPER *Sextius virescens* (Fairmaire) [Hemiptera: Membracidae]. (Australia).

GREEN TRIANGLE-BUTTERFLY *Graphium macfarlanei macfarlanei* (Butler) [Lepidoptera: Papilionidae]. (Australia).

GREEN VEGETABLE-BUG *Nezara viridula* (Linnaeus) [Hemiptera: Pentatomidae]: A cosmopolitan, polyphagous pest of horticultural, oilseed and legume crops. Host records include about 15 Species of monocots and 180 Species of dicots. GVB is a notable pest of field and vegetable crops. In Australia, it acts as a major pest of grain legumes, beans and oilseed crops and minor pest of many fruit and vegetables. GVB prefers to feed on fruit and seeds but may also damage foliage; now partially controlled by an egg parasite. GVB presumably originated around Mediterranean or Africa. Adults are green, shield-shaped, 12–17 mm long with 3–5 yellow-white spots on anterior margin of Scutellum. GVBs emit malodorous chemical when disturbed. Feeding preferences of GVB include immature fruit and seed. Damage by adult and late-instar nymph involves reduction in seed, fruit quality or fruit yield. Eggs are barrel-shaped, 1.3 mm tall and laid in groups of 30–120 on underside of leaves. Females may lay 200–600 eggs in several clutches during lifetime. GVB completes five nymphal instars; first instar nymph round, reddish-brown with varied colour patterns of black, green, yellow and red; first instar is gregarious and does not feed; early nymphal instars remain in aggregations; later nymphal instars disperse. Mature nymph is 12–13 mm long, dark brown to black with wing buds and Thorax has yellow margins and Abdomen with 2 rows of white spots on either side. Egg-to-adult in 4–12 weeks and 3–4 overlapping generations per year. Adults overwinter, active in flight and can disperse long distances. Syn. Southern Green Stink Bug. See Pentatomidae.

GREEN VEGETABLE-BUG EGG-PARASITE *Trissolcus basalis* (Wollaston) [Hymenoptera: Scelionidae]. (Australia).

GREEN WHIZZER *Macrotristria intersecta* (Walker) [Hemiptera: Cicadidae]. (Australia).

GREEN, EDWARD ERNEST (1861–1949) (Laing 1949, Entomol. mon. Mag. 85: 215–216.)

GREEN, JOHN WAGNER (1889–1968) (Leech 1968, Ent. News 79: 284.)

GREENBOTTLES See Calliphoridae.

GREENBUG *Schizaphis graminum* (Rondani) [Hemiptera: Aphididae]: A cosmopolitan pest of alfalfa, sorghum, wheat, and wild grasses. Grains and sorghum are injured by piercing-sucking mouthparts that cause mechanical damage and affect plant uptake of nutrients; toxins also may be injected during feeding that can cause chlorosis (redding and yellowing) and necrosis (browning) of plant tissues. Greenbug also transmits viral diseases to small grains (Barley Yellow Dwarf Virus) and sorghum (Maize Dwarf Mosaic Virus). Greenbug first reported in North America during 1882 in Virginia. In northern USA, Greenbug eggs are shiny black and overwintering eggs produce females; several generations of winged and wingless females produce live young. Autumn generation produces male and female offspring. In southern USA, only parthenogenetic females produce live nymphs. Nymphs are ca 1.5–2.0 mm long, pale green with longitudinal dark green stripe on dorsum; legs and Cornicles green with apices black. Adult female can produce ca 80 offspring in 25 days; Species can complete 5–14 generations per year, depending upon local conditions. See Aphididae. Cf. Russian Wheat Aphid.

GREENE, CHARLES TULL (1879–1958) (Fisher *et al.* 1959. Proc. ent. Soc. Wash. 61: 187–190, bibliogr.)

GREENE, JOSEPH (1824–1906) (Anon. 1906, Entomologist's mon. Mag 42: 66–67.)

GREENGROCER *Cyclochila australasiae* (Donovan) [Hemiptera: Cicadidae]. (Australia).

GREENHEADED SPRUCE-SAWFLY *Pikonema dimmockii* (Cresson) [Hymenoptera: Tenthredinidae].

GREENHOUSE LEAF-TIER 1. *Udea rubigalis* (Guenée) [Lepidoptera: Pyralidae]. 2. *Anagrapha falcifera* (Kirby) [Lepidoptera: Noctuidae]: A pest of many ornamental plants and soft-leaved greenhouse plants in North America. Eggs are flattened, scale-like and deposited on underside of leaves; eclosion occurs within 5–12 days. Syn. Celery Looper. See Noctuidae.

GREENHOUSE ORTHEZIA *Orthezia insignis* Browne [Hemiptera: Ortheziidae]: A pest of greenhouse ornamental plants in some countries, including Australia. GO resembles mealybugs in general appearance and habits. Nymphs are ca 1–3 mm diameter, green with waxy plates over dorsum. Females drag a fluted, white wax egg sac behind body. See Ortheziidae.

GREENHOUSE STONE-CRICKET *Tachycines asynamorus* Adelung [Orthoptera: Gryllacrididae].

GREENHOUSE THRIPS *Heliothrips haemorrhoidalis* (Bouché) [Thysanoptera: Thripidae]: A cosmopolitan pest outdoors in tropical and subtropical areas and glasshouse production in temperate areas. Host plants include avocados, carnations chrysanthemums, cucumbers, mangoes and roses. Adult are 1.5 mm long, black with yellow legs; larvae white or pale yellow. GT apparently are parthenogenetic but males of this Species have been reported in Peru. Eggs are 0.3 mm long, inserted individually into leaves or fruit; feeding stages cause stippling of leaves which curl and wither apically; quiescent stage (Propupa or Pseudopupa) remains on host plant. Life cycle requires five weeks; GT is multivoltine with several generations per year. See Thripidae. Cf. Gladiolus Thrips; Onion Thrips.

GREENHOUSE WHITEFLY *Trialeurodes vaporariorum* (Westwood) [Hemiptera: Aleyrodidae]: A widespread, pest of vegetables and ornamentals in glasshouses and gardens in warm environments. Adults are white, oval and 1.5–2.0 mm long; wings covered in a white waxy powder. Adults are very active and can live for 30–40 days. Adult females lay 100–150 eggs on the underside of leaves. Eggs are yellowish-green, elliptical/conical, Pedicelate and laid in a small ring. Eggs turn brownish-black a few days after oviposition. The egg stage lasts 6–15 days depending on temperature. Nymphs are ca 0.8 mm long, oval, flat and pale green to transparent. All instars have fine, long and short, white waxy filaments radiating from body. Nymphs settle on leaf near where eggs were laid and insert mouthparts to feed. Nymphs remain in the same position until they become adults. Nymphal period lasts ca 28–30 days. GHW overwinters as a nymph. Damage is induced by phloem consumption and accumulation of honeydew on foliage and fruit which promotes growth of sooty moulds (*Cladosphaerospermum* spp.) Nymphs vector many plant diseases including Lettuce Yellow and Squash Silver Leaf. Control of whitefly is achieved under some conditions with the parasite *Encarsia formosa* Gahan [Hymenoptera: Aphelinidae]. See Aleyrodidae. Cf. Cotton Whitefly.

GREENHOUSE-WHITEFLY PARASITE *Encarsia formosa* Gahan [Hymenoptera: Aphelinidae].

(Australia).

GREENING, LINNAEUS (1855–1927) (Dallman 1928, Proc. Linn. Soc. Lond. 1827–28: 118–119.)

GREENISH DARTER *Telicota ancilla ancilla* (Herrich-Schaffer) [Lepidoptera: Hesperiidae]. (Australia).

GREEN'S MEALYBUG *Pseudococcus citriculus* Green (Hemiptera: Pseudococcidae). A widespread pest of citrus which has been controlled in some regions with natural enemies. See Pseudococcidae.

GREENSLADE, RALPH MICHAEL (–1975) (Gunn 1976, Proc. R. ent. Soc. Lond. (C) 40: 51.)

GREEN-SPOTTED TRIANGLE BUTTERFLY *Graphium agamemnon ligatum* (Rothschild) [Lepidoptera: Papilionidae]. (Australia).

GREENSTRIPED GRASSHOPPER *Chortophaga viridifasciata* (DeGeer) [Orthoptera: Acrididae].

GREENSTRIPED MAPLEWORM *Dryocampa rubicunda* (Fabricius) [Lepidoptera: Saturniidae].

GREEN-TREE-ANT MIMICKING SPIDER *Amyciaea albomaculata* Cambridge [Araneida: Thomisidae]. (Australia).

GREER, ALBERT H (1891–1943) (Lyle 1944, Ann. ent. Soc. Am. 37: 132.)

GREER, THOMAS (1874–1949) (Beirne 1949, Irish Nat. J. 9: 327–328.)

GREGARINES Plural Noun. (Latin, *grex* = flock.) Large sized Protozoa (Telosporidia) that are parasitic of invertebrates; adult stage of gregarines develops extracellularly.

GREGARIOUS Adj. (Latin, *grex* = flock; *-osus* = full of.) 1. Pertaining to organisms that live in societies or communities, but not social. 2. Parasites of one Species that develop simultaneously on one host. Cf. Solitary.

GREGARIOUS GALL WEEVIL *Strongylorhinus clarki* Marshall; *Strongylorhinus ochraceus* Schonherr [Coleoptera: Curculionidae]. (Australia).

GREGARIOUS OAK LEAFMINER *Cameraria cincinnatiella* (Chambers) [Lepidoptera: Gracillariidae].

GREGARIOUS PARASITE A parasite which simultaneously develops in or on a host with other members of the same Species. Syn. Gregarious Parasitoid. See Parasitism. Cf. Solitary Parasite.

GREGARIOUS PARASITISM A phenomenon in which several parasite larvae complete development on or in one host. GP is common among parasitic Hymenoptera and some Diptera. See Parasitism. Cf. Solitary Parasitism.

GREGE Noun. (French *grege* = raw silk.) 1. Raw silk, including the gummy outer layer as spun by a caterpillar. Cf. Bave. 2. The colour beige. 3. The colour nutria.

GREGOR, FRANTISEK (1896–1942) (Kratochvil 1946, Sborn. Prir. klubu. Trebici 4: 16–21, bibliogr.)

GREGSON, CHARLES STUART (1817–1899) (Anon. 1899, Entomol. mon. Mag. 35: 96–97.)

GREINER, JOHANN (1863–1938) (Anon. 1939, Arb. morph. taxon. Ent. Berl 6: 69.)

GREMMINER, ALOIS (–1959) (Anon. 1959, Z. wien. ent. Ges. 44: 216.)

GRENIER, AUGUSTE JEAN FRANÇOIS (1814–1890) (Bonvouloir 1890, Ann. Soc. ent. Fr. (6) 10: 563–566.)

GRENSTED, LAWRENCE WILLIAM (1885–1964) English academic (Professor of Religion) and amateur Entomologist. (Wigglesworth 1965, Proc. R. ent. Soc. Lond. (C) 29: 53.)

GRENTLENBERG, ROBERT (1823–1886) (Anon. 1893, Schr. naturf. Ges. Danzig 8: 104.)

GRES Noun. (French = a type of sandstone) The gummy layer surrounding the silk thread spun by a caterpillar. See Silk. Cf. Bave; Grege.

GRESSITT, JUDSON LINSLEY (1914–1982) American biogeographer (B. P. Bishop Museum) and specialist on insects of the Pacific. (Int. J. Entomol. 25 (1): 1–10; 25 (2–3): 87–102.)

GRESSORIA Noun. (Latin, *gressus* = to walk.) A taxonomic category that includes Mantidae and Phasmatidae in some classifications. Cf. Saltatoria.

GRESSORIAL Adj. (Latin, *gressus* = to walk; *-alis* = characterized by.) 1. A term used to describe legs adapted for walking. 2. Lepidoptera with anterior legs aborted, but middle and hind legs used for walking. Alt. Gressorius. See Leg. Cf. Cursorial; Fossorial; Natatorial; Raptorial; Saltatorial; Scansorial. Rel. Locomotion.

GRETSCHMER, MAX (–1961) (Gleichauf 1962, Verh. dt. zool. Ges. (Zool. Anz.) Suppl. 26: 717–718.)

GREVILLEA LOOPER *Oenochroma vinaria* Guenée [Lepidoptera: Geometridae]. (Australia).

GREVILLEA MEALYBUG *Australicoccus grevilleae* (Fuller) [Hemiptera: Pseudococcidae]. (Australia).

GREW, NEHEMIAH (1641–1712) (Rose 1850, *New General Biographical Dictionary* 8: 118–119.)

GREY ALBATROSS *Appias melania* (Fabricius) [Lepidoptera: Pieridae]. (Australia).

GREY-BACK CANE BEETLE *Dermolepida albohirtum* (Waterhouse) [Coleoptera: Scarbaeidae]: In Australia, a serious pest of sugarcane in northern Queensland; early instar larvae cause most damage to cane roots. See Scarabaeidae.

GREY-BACK CANEGRUB *Dermolepida albohirtum* (Waterhouse) [Coleoptera: Scarabaeidae]. (Australia). See Canegrub.

GREY-BANDED LEAF WEEVIL *Ethemaia sellata* Pascoe [Coleoptera: Curculionidae]. (Australia).

GREY BLISTER-BEETLE *Epicauta aethiops* (Latreille) [Coleoptera: Meloidae]: A minor crop pest in northeast Africa.

GREY CHRISTMAS-BEETLE *Trioplognathus griseopilosus* (Ohaus) [Coleoptera: Scarabaeidae]. (Australia).

GREY CLUSTER-BUG *Nysius clevelandensis* Evans [Hemiptera: Lygaeidae]: In Australia, a minor pest of Cotton; primary hosts include sunflowers, potatoes, stone fruits, citrus and weed. Adults are ca 5 mm long, slender-ovate and greyish-brown with dark markings; males are smaller and darker than female; wings nearly transparent; Pronotum and legs with fine Setae. Adult stage lasts ca 28 days during summer, but adults overwinter at end of autumn. Overwintering adult females lay eggs during spring. Eggs are small, creamy-white and laid either in the soil or on the terminals of plants. Eggs are laid in batches of 2–7. The egg stage lasts 5–8 days. Nymphs resemble adults but lack wings; nymphs pear-shaped and reddish. See Lygaeidae. Cf. Rutherglen Bug.

GREY DUNG-BALL ROLLER *Sisyphus spinipes* (Thunberg) [Coleoptera: Scarabaeidae]. (Australia).

GREY-FURROWED ROSE CHAFER *Trichaulax philipsii* (Schreibers) [Coleoptera: Scarabaeidae]. (Australia).

GREY LAWN-LEAFHOPPER *Exitianus exitiosus* (Uhler) [Hemiptera: Cicadellidae].

GREY LOOPER *Cleora inflexaria* Snellen [Lepidoptera: Geometridae]. (Australia).

GREY PINEAPPLE-MEALYBUG *Dysmicoccus neobrevipes* Beardsley [Hemiptera: Pseudococcidae].

GREY PLANTHOPPER *Anzora unicolor* (Walker) [Hemiptera: Flatidae]. (Australia).

GREY PYRALID *Pyralis manihotalis* Guenée [Lepidoptera: Pyralidae]: A pest of stored products to include dried meat, seeds, hides and bones. GP is endemic to South America, now pantropical and periodically recovered in Europe. Adults are pale brown with pale sinuous bands on wings and wingspan 25–38 mm. See Pyralidae. Cf. Painted Meal Moth.

GREY SCALE *Lindingaspis rossi* (Maskell) [Hemiptera: Diaspididae]. (Australia).

GREY SNOUT-BEETLE *Eremnus setulosus* Boheman [Coleoptera: Curculionidae] (South Africa). Cf. Speckled Snout-Beetle.

GREY-STRIPED MOSQUITO *Aedes vittiger* (Skuse) [Diptera: Culicidae]: A vicious day-biting mosquito and minor pest in Australia. Adults are grey with four longitudinal black stripes on Scutum; Palpus ca 0.25 as long as Proboscis; wing scales dark; apex of Abdomen tapering and with scales. Eggs are laid singly in mud and withstand desiccation. GSM breeds in temporary pools with sunlight and emergent vegetation; larval and pupal stages require ca 6 days during mid summer. GSM is not known to vector disease. Cf. Scotch Grey. See Culicidae.

GREY SUGARCANE-MEALYBUG *Dysmicoccus boninsis* (Kuwana) [Hemiptera: Pseudococcidae]. (Australia).

GREY SUNFLOWER-SEED WEEVIL *Smicronyx sordidus* LeConte [Coleoptera: Curculionidae]: A pest of sunflower in parts of North America. Adults aggregate on developing seed heads, feed on pollen and oviposit in developing achenes. See Curculionidae. Cf. Red Sunflower Seed Weevil.

GREY WILLOW LEAF-BEETLE *Pyrrhalta decora decora* (Say) [Coleoptera: Chrysomelidae].

GREY WOLF-SPIDER *Lycosa simsoni* (Simon) [Araneida: Lycosidae]. (Australia).

GREYBACKS See Petaluridae.

GREYBANDED LEAF-ROLLER *Argyrotaenia mariana* (Fernald) [Lepidoptera: Tortricidae].

GREY, THOMAS DE See Walsingham.

GREY, WILLIAM (1827–1896) (Anon. 1897, Ent. News 9: 32.)

GRIBODO, GIOVANNI (1846–1924) (Invrea 1925, Memorie Soc. ent. ital 4: 223–228, bibliogr.)

GRIDELLI, EDUARDO (1895–1958) (Invrea 1960, Annali Mus. civ. stor. nat. Giacomo Doria 71: 435–449, bibliogr.)

GRIEBEL, JULIUS (–1919) (Hedicke 1919, Dt. ent. Z. 1919: 233.)

GRIEP, ERDMANN (1899–1957) (Anon. 1957, Ent. Bl. Biol. Syst. Käfer 53: 122–134, bibliogr.)

GRIFFINI, ARCHILLE (1895–1932) (Anon. 1932, Boll Soc. ent. Ital. 64: 109.)

GRIFFITH, ARTHUR FORSTER (–1933) (A[tkin] 1934, Entomologist 67: 48.)

GRIFFITH, HORACE GREELEY (–1899) (Anon 1899, Ent. News 10: 252.)

GRIFFITHS, GEORGE CHARLES (1852–1924) (Bartlett 1925, Entomologist 58: 71–72.)

GRILL, CLAES (1851–1919) (Trägardh 1920, Ent. Tisdschr. 41: 41–45, bibliogr.)

GRIMES, DILLARD WISTER (1891–1943) (Osborn 1946, *Fragments of Entomological History,* Pt. II. 232 pp. (87–88), Columbus, Ohio.)

GRIMM, ADAM (1824–1876) (Pellet 1929, Am. Bee. J. 69: 608–609.)

GRIMSHAW, PERCY HALL (1869–1939) (Smart 1945, Soc. Biblphy. nat. Hist. 2: 29–42, bibliogr.)

GRINDING TEETH Dentes Molares; Molars.

GRINNEL, FORDYCE (1882–1943) (Anon. 1943, Bull. Soc. Calif. Acad. Sci. 42: 136–137.)

GRIPOPTERYGIDAE Plural Noun. A Family of Plecoptera containing 37 Genera, about 125 nominal Species, and which occurs in high mountains of South America, Australia and New Zealand. NZ with 21 endemic Genera. Distribution Amphinoitic. Adults are typically small bodied and dark coloured; eggs are relatively unmodified, lack collars, adhesive organs and other holdfasts noted in other stoneflies. All Species apparently are herbivores. See Plecoptera.

GRISCENT Adj. (French, *gris* = gray.) Ashen grey.

GRISEA CANEGRUB *Lepidiota grisea* Britton [Coleoptera: Scarabaeidae]. (Australia). See Canegrub.

GRISEOUS Adj. (Medieval Latin, *griseus* from Greek, *greis* = gray; Latin, *-osus* = with the property of.) Pale gray. Alt. Griseus.

GRIST, CHARLES (1863–1952) Riley 1953, Proc. R. ent. Soc. Lond. (C) 17: 72.

GRISWOLD, GRACE HALL (1872–1946)

(Johannsen 1946, J. Econ. Ent. 39: , 423–424.)

GROCEL® See Gibberellic Acid.

GROCERS' ITCH A hypersensitivity in humans to stored products caused by the mite *Glycyphagus domesticus* (DeGeer). Sometimes GI is a collective term for hypersensitivity caused by reactions to other mites. See: Bakers' Itch; Copra Itch; Dried-fruit Dermatitis; Wheat-Pollard Itch.

GRODHAUS, GAIL (1928–1987) American public health officer and specialist in Chironomidae. (Sublett *et al.* 1989, JKES 62: 294–296.)

GRÖGL, FRITZ (–1966) (Anon. 1966, Z. wien. ent. Ges. 51: 96.)

GROLL, H W (1819–1900) (Anon. 1900, Tijdschr. Ent. 43: 33.)

GRÖNBLOM, THORWALD (1885–1971) (Sotavalta 1972, Suomen hyönt. Aikak 38: 109–111.)

GRÖNLIEN, NILS (1847–1939) (Knaben 1940, Norsk. ent. Tidsskr. 5: 192–194, bibliogr.)

GROOM Transitive Verb. (Middle English, *grom* = a servant. Groomed; Grooming.) Cleaning the body. Syn. Preening.

GROSCHKE, FRANZ (–1956) (Lindner 1956, Jh. Ver. caterl. Naturk. Würt. 111: 10, 103.)

GROSIER, JEAN BAPTISTE GABRIEL ALEXANDRE (1743–1823) (Rose 1850, *New General Biographical Dictionary* 8: 131.)

GROSS, JULIUS (–1933) (Anon. 1933, Zool. Anz. 102: 339.)

GROSSBECK, JOHN ARTHUR (1883–1914) (Davis 1914, J. N. Y. ent. Soc. 22: 271–275, bibliogr.)

GROSSINGER, JAMES BAPTIST (1728–1803) (Aigner 1887, Rovart. Lap. 4: 69–71.)

GROSSUS Adj. (Old French, *gros* = large.) Descriptive of structure or mass which is large or big in size or bulk; gross.

GROSVENOR, GEORGE HERBERT (1880–1912) (Morice 1912, Proc. ent. Soc. Lond. 1912, clxv–clxvi.)

GROTE, AUGUST RADCLIFFE (1841–1903) (Strecker 1878, *Butterflies and Moths of North America*, (pp 232–236, bibliogr. (part); Anon. 1903, Ent. News 14: 277–278; Mallis 1971, *American Entomologists*. 549 pp. (304–308), New Brunswick.)

GROTH, CLAUS HEINRICH (1859–1930) (Zirk 1930, Verh. Ver. naturw. Heimatforsch 22: xix–xx.)

GROTH, KURT (1884–1957) (Petersen 1958, Ent. Meddr. 28: 187–189, bibliogr.)

GROUND BEETLES See Carabidae.

GROUND BUGS See Lygaeidae.

GROUND CRICKETS See Gryllidae.

GROUND MEALYBUG *Rhizoecus falcifer* Künckel d'Herculais [Hemiptera: Pseudococcidae].

GROUND PEARL The outer, pearly covering of preadult nymphal stage of *Margarodes*. GPs are strung for necklaces, particularly in South Africa and the Bahamas. A pest of Bermuda grass and grapevine roots in southwestern USA. See Margarodidae.

GROUND SHIELD BUG *Choerocoris paganus* (Fabricius) [Hemiptera: Scutelleridae].

GROUND SPIDERS See Lamponidae.

GROUNDNUT BEETLE *Caryedon serratus* (Oliver) [Coleoptera: Bruchidae]: A widespread pest of groundnuts (peanuts) and legume pods in warmer regions. GNB is endemic to Asia. Adults are 4–7 mm long, reddish brown with dark spots on Elytra; eyes not emarginate; hind Femur with a comb of spines; hind Tibia curved. Eggs stick to pod or groundnut. Larvae bore into pods or shells and feed on seed; mature larvae leave the pod to pupate in thin cocoon on outer surface. GNB life cycle requires 40 days. See Bruchidae.

GROUNDNUT LEAF-WEBBER *Aproaerema modicella* Deventer [Lepidoptera: Gelechiidae].

GROUNDPLAN Noun. (Middle English, *ground* > Anglo Saxon, *grund* = foundation; French, *plan* > Latin, *planta* = sole of food. Pl., Groundplans.) 1. An hypothetical anatomical form from which an existing adaptation has been derived. 2. A basic form from which subsequent morphological adaptation is based. A groundplan may refer to a specific anatomical part, an entire organism, or any contrived taxonomic grouping. Primitive organisms (considered part of a basal clade) may show more groundplan features than their more derived (= apomorphic) relatives. Often, the groundplan for a particular anatomical feature is presumed because a fossil record has not been reported. Advanced (derived) organisms retain many groundplan features, with only few features appearing modified or derived. Ideas regarding a groundplan may come from fossils, embryology, comparative anatomy or molecular data. Evidence from all disciplines in a specific entomological problem is rarely available because knowledge about any particular group of insects is very limited. Intermediate stages of development between a groundplan and an existing form constitute a Transformation Series. Members of TS may constitute an hypothetical series of events that depict the major features or structural changes in a structure or organism through evolution. Syn. Archetype; Bauplan. Cf. Anagenesis; Katagenesis; Morphological Transformation; Transformation Series. Rel. Classification; Evolution; Phylogeny.

GROUNDSELBUSH GALL-FLY *Rhopalomyia californica* Felt [Diptera: Cecidomyiidae]. (Australia).

GROUNDSELBUSH LEAF-BEETLE *Trirhabda baccharidis* (Weber) [Coleoptera: Chrysomelidae]: A monovoltine pest endemic to USA whose larval and adult stages feed on foliage and pupation occurs in soil. See Chrysomelidae.

GROUNDSELBUSH LEAF-WEBBING CATERPILLAR *Aristotelia* sp. [Lepidoptera: Gelechiidae]. (Australia).

GROUNDSELBUSH STEMBORER *Megacyllene mellyi* (Chevrolat) [Coleoptera: Cerambycidae].

GROUP Noun. (French, *gruppe*. Pl., Groups.) 1. An assemblage of objects based on resemblance or common features. 2. An assemblage of organisms based on natural relationships. 3. An indefinite division of classification used for a series of allied Species, Genera or larger assemblages. Cf. Taxon. Rel. Classification; Taxonomy.

GROUP EFFECT A change in behaviour or physiology induced by stimuli not directed by space or time. Social Facilitation.

GROUP PREDATION Social Insects: The search, capture and consumption of living animals through cooperation by members of one Species. GP is seen in Army Ants.

GROUP TRANSPORT Social Insects: The cooperative movement of food by members of one Species. GT is performed by workers.

GROUPED GLANDS Diaspididae: The Genacerores *sensu* MacGillivray. See Circumgenital Glands.

GROUPED ORIFICES OF SPIRACLES The Spiracerores *sensu* MacGillivray.

GROUT, ROY (–1974) (Anon. 1975. Bee Wld 56: 54.)

GROUVELLE, ANTOINE HENRI (1843–1917) (Desbroides 1919, Ann. Soc. ent. Fr. 88: 345–360, bibliogr.)

GROUVELLE, PHILIPPE (–1923) (Rabaud 1923, Bull Soc. ent. Fr. 1923: 185–186.)

GROWTH CURVE A graphical display which plots population size against time. The points on the graph generate a Growth Curve which explains population-size change in terms of numbers of individuals over time. The Growth Curve is often divided into several phases including a Lag Phase, Log (Exponential) Phase and Stationary Phase. Rel. Population.

GROZDANIC, SIMEUN (1896–1972) (Vasic 1972, Acta ent. jugosl. 8: 135–137.)

GRUB Noun. (Middle English, *grobbe* = grub. Pl., Grubs.) An elongate, whitish insect larva. The term is loosely applied to all insects, but more specifically applied to the larvae of Coleoptera (Scarabaeiform) and some Hymenoptera (Apoidea). Cf. Larva; Maggot. Verb. (Grubbed, Grubbing.) (Middle English, *grubben* = grub.) Intransitive: To dig in or under the soil for an object difficult to remove. Transitive: To break or clear the surface of the soil; to remove roots by digging.

GRUBE, ADOLPHE EDOUARD (1812–1880) (Zaddach 1900, Schr. Phys.-öken Ges. Köningsb. 21: 113–130, bibliogr.)

GRUBER, FRITZ (–1958) (Anon. 1958, Z. wiener ent. Ztg. 43: 184.)

GRUBWORM See White Grubs.

GRUDE-NIELSEN, MARTIN ADOLF (1894–1961) (Opheim 1962, Norsk. ent. Tidsskr. 12: 69.)

GRUNACK, ALBERT (1842–1907) (Anon. 1907, Ent. Z. Frankf. a. M. 21: 90, 102, 160.)

GRUNDELL, JULIUS GEORGE (1857–1933) (Van Dyke 1934, Pan-Pacif. Ent. 10: 48.)

GRUNERT, JULIUS THEODOR (1809–) (Ratzeburg 1874, Forstwissenschaftlisches Scriftsteller-Lexicon 1: 208–213.)

GRUT, FERDINAND (1820–1891) (McLachlan 1891, Entomol. mon. Mag. 27: 251–252.)

GRYLLACRIDIDAE Stål 1874. Plural Noun. Wingless Long-horned Grasshoppers. The largest Family in the Stenopelmatoidea, or Gryllacridoidea, including about 75 Genera and 550 nominal Species. Gryllacridids are characterized by the head not disproportionately large and anterior Coxa spinose, but fore Tibia not spinose and tibial auditory organ absent. Wings are fully developed but lacking stridulatory device. Cerci not annulated, and male posterior abdominal segments inflate to form a 'scrotum' for Testes. Ovipositor long, narrow and concealed by Styli. Gryllacriids are found most commonly in the tropics of Indomalaysia and America. Only one Species is found in North America. Most Species are arboreal and conceal themselves in rolled leaves during the day. They are predominantly predatory or scavengers on insects (particularly aphids) and other invertebrates. See Gryllacridoidea.

GRYLLACRIDOIDEA A primitive Superfamily of ensiferous Orthoptera that is cosmopolitan in distribution and consists of about 1,000 Species. Member Taxa are characterized by forewings absent or not tegminized and lacking stridulatory devices; fore and middle Tibiae with conspicuous articulated spines; Tarsi four-segmented; Antenna considerably longer than body. Representative Families include Gryllacrididae, Schizodactylidae, Stenopelmatidae, Rhaphidophoridae and ?Cooloolidae. See Orthoptera.

GRYLLIDAE Saussure 1894. Plural Noun. Ground Crickets; Field Crickets; Short-tailed Crickets. A cosmopolitan Family of ensiferous Orthoptera containing about 100 Genera and 800 nominal Species that are assigned to three Subfamilies: Nemobiinae (ground crickets), Gryllinae (field crickets), and Brachyturpinae (short-tailed crickets). Features include head globular; Antenna inserted high on head; hind Tibia strongy spinose without denticles between spines; second tarsal segment compressed. See Orthoptera.

GRYLLOBLATTARIA Bruner 1915. Syn. Grylloblattodea.

GRYLLOBLATTIDAE Walker 1914. Plural Noun. Ice Crawlers. A primitive, numerically small Family of Grylloblattodea apparently restricted to alpine areas of Northern Hemisphere. Individuals are apterous with prognathous heads; they are active during cold weather and inhabit caves (Korea), crevices or interface between soil and stone. Grylloblattids are active at temperatures of 0°C but killed by formation of ice within body at –6°C; their Haemolymph lacks high concentrations of glycerol, sorbitol and erythritol. Eggs are large, black and laid individually (not in Ootheca) over a period of several months or years. Nymphs are subterranean and lack Pulvilli (Euplantulae). Species undergo eight nymphal instars with a life cycle up to 7 years. Placed in the Order Notoptera or Grylloblattodea; affinities with Orthoptera or Dermaptera have been suggested. See Grylloblattodea.

GRYLLOBLATTODEA Brues & Melander 1932. (Greek, *Grylloblatta*; *eidos* = form.) Rock Crawlers. No fossil record. An Order consisting of one Family and about 20 cryptozoic Species found in cold, wet habitats of North America, Japan and Russia. Features include: Soft-bodied, elongate, slender, mandibulate insects from 15 to 30 mm long. Prognathous head; compound eyes reduced or absent; Ocelli absent; long; filiform Antenna with less than 45 segments. Prothorax large; apterous; cursorial legs with large Coxae. Abdomen with 11 segments; Cerci long and segmented, Ovipositor strongly exserted. Grylloblattodoids typically are found in rotting logs, beneath stones on talus slopes near snow. Their diet includes moss and other insects. Syn. Notoptera Crampton 1915. Grylloblattoidea Brues & Melander 1915. See Grylloblattaria.

GRYLLOIDEA The largest Superfamily of crickets with about 325 Genera and 2,300 Species. Higher classification of the Superfamily varies with 2–13 Families recognized as valid. Grylloidea are cosmopolitan, predominantly tropical; inhabit all terrestrial habitats from deserts to swamps. Some Species are cavernicolous, others are aquatic. Includes Cachoplistidae, Eneopteridae, Gryllidae, Mogoplistidae, Myrmecophilidae, Oecanthidae, Pentacentridae, Phalangopsidae, Pteroplistidae, Scleropteridae and Trigonidiidae. See Orthoptera.

GRYLLOTALPIDAE Brunner 1882. Plural Noun. Changas; Churr-worms; Eve-churrs; Mole Crickets. Gryllotalpids consist of two Subfamilies: Gryllotalpinae (five Genera about 50 Species) and Scapteriscinae (one Genus about 10 Species). Gryllotalpinae are cosmopolitan in distribution; Scapteriscinae are Neotropical and in India. Gryllotalpids are nocturnal; males and females can stridulate at dusk, become silent during night. Females lay eggs in burrows or specially constructed brood chambers; they probably require more than a year to complete life cycle. Adult and nymphs occur in same habitat; adults of some Species show care of eggs or nymphs; eject a sticky, black, malodorous material from the Anus when disturbed. All Species probably are omnivores and can eat soil arthopods and subterranean plant parts. See Gryllotalpoidea. Cf. African Mole Cricket.

GRYLLOTALPOIDEA (Latin, *gryllus* = cricket, *talpa* = mole.) Changas; Churr Worms; Eve-churrs; Mole Crickets. A Superfamily of ensiferous Orthoptera containing about five Genera and 50 Species. Typically, adult body large, densely setose with head small and conical; two Ocelli present; Antenna very short. Species are fossorial or cavernicolous with legs modified for burrowing and wings frequently absent, when present wings form stridulatory organs on male, but lack characteristic mirror of related Grylloidea; Ovipositor is weakly developed or absent. See Orthoptera.

GRYLLUS Noun. (Greek, *gryllos* = cricket.) Variation of Gryllos. In Greco-Roman art, Gryllos was a comic combination of animals or of animal and human form.

GRYOPIDAE Plural Noun. A small Family of amblycerous Mallophaga occuring in New World. Antenna clubbed and concealed in groove; Maxillary Palpus present; Tarsi with one claw or claw absent. Gryopids are parasitic on guinea pigs. See Mallophaga.

GRZEGORZEK, ADDLBERT (1819–1890) (Mik 1890, Wien ent. Zig 9: 160, bibliogr.)

GSCHWANDNER, ROBERT (1875–1927) (Robert 1928, Annin naturh. Mus. Wien 42:1.)

GUANINE Noun. (Peruvian, *huanu* = dung.) A white, amorphous compound that occurs in transparent areas of some wings and displays a milky tinge; also found in the photogenic organs of Lampyridae. Guanine is an excretory substance in Varroa mites; composition: C H N O (von Furth).

GUAR MIDGE *Contarinia texana* (Felt) [Diptera: Cecidomyiidae].

GUARDED PUPA. See Pupa Custodiata.

GUARDIAN® See Flucythrinate.

GUAVA MEALYBUG *Ferrisia virgata* (Cockerell) [Hemiptera: Pseudococcidae]: A widespread, polyphagous pest of many crops that is found in tropical and subtropical areas. Syn. Striped Mealybug. See Pseudococcidae.

GUAVA MOTH *Coscinoptycha improbana* Meyrick [Lepidoptera: Carposinidae]

GUBIN, ALEXANDER FEDOROVICH (1898–1956) (Khalifman 1958, Ent. Obozr. 37: 476–485, bibliogr.; Translation: Ent. Rev. Wash. 37: 407–417, bibliogr.)

GUDMAN, FREDERIK CARL JULIUS EMIL (1869–1932) (Jense 1932, Flora og Fauna 1932: 37–39.)

GUENÉE, ACHILLE (1809–1880) (Essig 1931, *History of Entomology*. 1029 pp. (641–641), New York.) Correct: Guenée.

GUERIN, FRANCOIS ETINNE Editor of '*Magazin de Zoologie*' and author of several articles published in the *Encyclopedie Methodique*. Author of *Iconographie de Regne Animal*.

GUERIN-MENEVILLE, FELIX EDOUARD (1799–1874) Probably the most prolific French entomological author of the 19th century who published more than 400 papers. He was a taxonomic specialist of Coleoptera and produced extensive work in economic Entomology with reports on insects attacking wheat, barley, clover, olives; his later work involved sericulture. (Anon. 1874, Entomol. mon. Mag. 10: 233–234; Marseul 1885, Abeille (Les ent. et leurs écrits)

21: 75–100.) Correct: Guérin-Méneville.

GUEST Noun. (Middle English, *gest*. Pl., Guests.) 1. An insect which lives in the nest, gall or other dwelling place of another Species without adversely affecting the primary occupant or host. 2. An inquiline or social symbiont. See Inquiline. Cf. Cleptobiosis.

GUETTARD, JEAN ETIENNE (1715–1756) (Rose 1850, *New General Biographical Dictionary* 8: 146.)

GUICHARD, ALBERTO FRAGA (1883–1969) (Pino 1973, Revta. Chil. Ent. 7: 262–263.)

GUIDE FOSSILS See Index Fossil.

GUIGLIA, D Italian museum curator (Genoa) and specialist on aculeate Hymenoptera.

GUIGNON, CHANOINE (1856–1933) (Anon. 1934, Lambillionea 34: 25.)

GUILD Noun. (Middle English, *gylde*. Pl., Guilds.) Groups of insects that share a spatially or temporally definable niche or its resources. Typically guilds are identified in conjunction with herbivorous insects. Examples of guilds include: Seed feeders, leaf miners, stem borers, gall formers, and root feeders.

GUILDING, LANSDOWN (1797–1831) (Papavero 1975, *Essays on the History of Neotropical Dipterology* 2: 219–221. São Paulo.)

GUILLEBEAU, FRANCISQUE (1821–1897) (Anon. 1898, Miscnea ent. 5: 134–136.)

GUIMARAES, JOSÉ GEMINIAMO GOMES (–1935) (Anon. 1837, Campo 7: 23.)

GUINEA ANT *Tetramorium bicarinatum* (Nylander) [Hymenoptera: Formicidae].

GUINEA FEATHER LOUSE *Goniodes numidae* Mjberg [Mallophaga: Philopteridae].

GUINEA PIG MANGE MITE *Trixacarus caviae* Fain [Acari: Sarcoptidae].

GUINEA WORM A disease of humans caused by the nematode *Dracunculus medinensis* (Linnaeus). The nematode uses several Species of cyclopoid copepods as intermediate hosts and ultimately contacts humans that drink contaminated water.

GUITERAS, JUAN (1852–1925) (Tovar 1944, Rev. Med. trop. Parasit. Habana 10: 49–63.)

GULA Noun. (Latin, *gula* = throat. Pl., Gulae.) 1. A form of sclerotic closure along the posterior aspect of head which separates Occipital Foramen and Labium. Gula typically is developed in prognathous axial orientation of the head in which Posterior Tentorial Pits migrate anteriad. The median sclerite (Gula) apparently forms *de novo* in the membranous neck region between the lateral extensions of the Postocciput. Cf. Gulamentum; Hypostomal Bridge; Postgenal Bridge. 2. A median sclerite (typically elongate) that forms the central part of the head ventrum. Gula occurs in some Coleoptera, Neuroptera, Isoptera, and other insects displaying a prognathous head. Gula originates adjacent to the Posterior Tentorial Pits, extends from the Submentum to the posterior margin, and laterally bounded by Genae. Syn. Gular Plate. See Head Capsule. Cf. Gulamentum.

GULAMENTAL PLATE A large basal sclerite of the Labium; the fused Gula and Submentum in termites and some other insects (Crampton).

GULAMENTUM Noun. (Latin, *gula* = throat; *mentum* = chin. Pl., Gulamenta.) 1. The Gula and Submentum fused into one sclerite. 2. A sclerite containing the gular and submental regions (Crampton). See Head Capsule. Cf. Gula; Hypostomal Bridge; Postgenal Bridge.

GULAR Adj. (Latin, *gula* = throat.) Descriptive of or pertaining to the Gula (throat).

GULAR PEDUNCLE Coleoptera: The Submental Peduncle.

GULAR PIT An invagination of a posterior arm of the Tentorium (Crampton). Syn. Posterior Tentorial Pit.

GULAR PLATE The Gula *sensu* MacGillivray.

GULAR SUTURE The line of division between the Gula (throat) and Genae (cheeks). See Suture.

GULDE, JOHANNES (1872–1929) (Sach 1930, Natur. Mus. Frankf. 60: 48.)

GULF STRIP See Semitropical.

GULF WIREWORM *Conoderus amplicollis* (Gyllenhal) [Coleoptera: Elateridae].

GULF-COAST TICK *Amblyomma maculatum* Koch [Acari: Ixodidae].

GULLET Noun. (Middle English, *golet* = throat. Pl., Gullets.) The Oesophagus.

GULLIVER, GEORGE JAMES (–1931) (Sheldon 1931, Entomologist 64: 192.)

GULOMENTAL Adj. (Latin, *gula* = throat; *mentum* = chin.) Referring or pertaining to the region covered by the Gula and Mentum.

GUM Noun. (Middle English, *gomme* from Latin, *gummi* = gum. Pl., Gums.) Polysaccharides produced within plants as a result of bacterial infection. Cf. Resin; Sap.

GUM DAMAR A natural exudate from the plant *Shorea wiesneri* that is not miscible with water or alcohol but is miscible with benzene or toluene. GD is used in histological preparations. See Histology. Cf. Canada Balsam.

GUMILLA, JOSE (ca 1690–ca 1758.) (Dusmet y Alonso 1919, Boln. Soc. ent. Esp. 2: 78–79.)

GUMLEAF SKELETONIZER *Uraba lugens* Walker [Lepidoptera: Noctuidae]: A pest of *Eucalyptus* spp. in Australia. See Noctuidae.

GUMMOSIS Noun. (Latin, *gummosus* = gummy; *-osis* = disease. Pl., Gummoses.) The production of gum by a plant as a consequence of disease or disorder. Syn. Gumming.

GUMPPENBERG, CARL VON (–1893) (Anon. 1893, Leopoldina 29: 159.)

GUMTREE HOPPERS *Eurymela* spp. [Hemiptera: Eurymelidae].

GUMTREE SCALE *Eriococcus coriaceus* Maskell [Hemiptera: Eriococcidae].

GUMTREE SCALE LADYBIRD *Rhyzobius ventralis* (Erichson) [Coleoptera: Coccinellidae].

GUNDER, JEANE DANIEL (1888–1948) (Martin 1948, Lepid. News 2: 105.)

GUNDERSON, HAROLD (1913–1971) (Stockdale 1972, J. Econ. Ent. 65: 937.)

GUNDLACH, JUAN (JOHANNES CHRISTOPHER) (1810–1896) (Ramsden 1915, Ent. News 26: 241–260, bibliogr.)

GUNKEL, WOLFGANG (–1964) (Anon. 1964, Mitt. dt. ent. Ges. 23: 62.)

GÜNTHER, KLAUS (1907–1975) (Urich 1975, Zool. Beitr. 2: 347–361, bibliogr.)

GUNTHERT, HANS (1895–1931) (Aue 1931, Ent. Anz. 11: 203–204.)

GUNTON, HERBERT CHARLES (–1953) (Buxton 1954, Proc. R. ent. Soc. Lond. (C) 18: 80.)

GURLST, ERNST FRIEDRICH (1794–1882) (Anon. 1882, Zool. Anz. 5: 556.)

GURNEY, ASHLEY (–1988) American government Entomologist (U.S.D.A. at U.S. National Museum) specializing in taxonomy of Orthoptera.

GURNEY, WILLIAM BUTLER (1882–1939) (Holmes 1940, Proc. Linn. Soc. N.S.W. 65: i–ii.)

GUSATHION® See Azinphos-Methyl.

GUSATHNION® See Azinphos-Ethyl.

GUSE, CARL AUGUST HERMANN (–1915?) (Möller 1915, Z. Forst. Jagdwes. 47: 324–330.)

GUSMANN, PAUL (1866–1942) (Benick 1947, Verh. Ver. naturw. Heimatforsch 29: xiv–xv, bibliogr.)

GUSSAKOVSKY, VSEVOLOD VLADIMIROVICH (1904–1948) (Narzikulov & Pulavskii 1970, Ent. Obozr. 49: 502–507, bibliogr.; Translation: Ent. Rev., Wash. 49: 302–305.)

GUSTATORY Adj. (Latin, *gustare* = to taste.) Relating to the sense of taste.

GUSTATORY NERVES The nerves of the organs of taste.

GUSTATORY ORGAN Nymphs of Mayflies: A row of Sensillae on the underside of the triangular Labrum (Needham).

GUSTATORY SENSE Taste. Perception of flavours through the mouth in general, bound up with the sense of smell.

GUT-BUSTER® A registered biopesticide derived from *Bacillus thuringiensis* var. *kurstaki*. See *Bacillus thuringiensis.*

GUTHEIL, A (1832–1886) (Anon. 1887, Insekten-Wlt. 3: 112.)

GUTHION® See Azinphos-Methyl.

GUTHRIE, JOSEPH EDWARD (1871–1935) (Anon. 1935, Proc. Iowa Acad. Sci. 42: 21–24, bibliogr.)

GUTIERREZ, RAMON (1917–1952) (Looser 1953, Revta. univ. Santiago 38: 13–16, bibliogr.)

GUTTA PERCHA Noun. (Malay.) A substance nearly white to brown resembling rubber, but containing more resin, from the latex of several Malaysian trees.

GUTTA Noun. (Latin, *gutta* = drop. Pl., Gutttae.) A circular, coloured spot, particularly on the wing of an insect.

GUTTATE Adj. (Latin, *guttatus*, from *gutta* = a drop; *-atus* = adjectival suffix.) 1. With guttae; covered with small spots. 2. Small drops. Alt. Guttatus.

GUTTIFORM Adj. (Latin, *gutta* = a drop; *forma* = shape.) Drop-shaped. Cf. Ampulliform; Lageniform; Pyriform. Rel. Form 2; Shape 2; Outline Shape.

GUYON, GEORGE (1825–1878) (Anon. 1878, Entomol. mon. Mag. 14: 263–264.)

GWATKIN-WILLIAMS, RUPERT STANLEY (1875–1949) (Cockayne 1949, Entomologist's Rec. J. Var. 61: 92.)

GY-81® See Enzone.

GYLEK, LUDWIG (1864–1923) (Hoffman 1923, Ent. Anz. 3: 101–102.)

GYLLENHAL, LEONHARD (1752–1840) Swedish Entomologist and author of *Insecta Suecica* (1808–1827). (Swainson 1840, *Taxidermy; with Biogr. of Zoologists*. 392 pp. (208), [Vol. 12 of *Cabinet*, D. Lardner, ed.] London.)

GYMNOCERATA Noun. (Greek, *gymnos* = naked; *keras* = horn.) Insects with freely movable, conspicuous Antennae. Specifically, a division of the Heteroptera which includes Species that have Antennae longer than the head. Gymnocerata include most terrestrial bugs and water striders. See Heteroptera. Cf. Cryptocerata.

GYMNODOMOUS Adj. (Greek, *gymnos* = naked; *domos* = house; Latin, *-osus* = with the property of.) Hymenoptera: Pertaining to wasp nests with exposed combs (*Polistes*).

GYMNOGASTRA Noun. (Greek, *gymnos* = naked; *gaster* = stomach. Pl., Gymnogastrae.) Hymenoptera: Species in which the venter is visible. See Cryptogastra.

GYMNOPARIA Noun. (Greek, *gymnos* = naked; Latin, *paries* = wall. Pl., Gymnopariae.) Scarabaeoid larvae: The exposed part of the Paria between the Acanthoparia and the Chaetoparia and behind the Acroparia; not always present or distinct (Boving).

GYMNOPTERA Noun. (Greek, *gymnos* = naked; *pteros* = wing. Pl., Gymnopterae.) Species with membranous wings that are not covered with scales.

GYMNOSPERM Noun. (Greek, *gymnos* = naked; *sperma* = seed. Pl., Gymnosperms.) A member of the Gymnospermae. See Gymnospermae. Cf. Angiosperm.

GYMNOSPERMAE Plural Noun. (Greek, *gymnos* = naked; *sperma* = seed.) The smaller of the two major divisions of the seed-bearing plants. Gymnospermae include members that lack flowers. Gymnosperm seeds are exposed or develop on a bract (sporophyll), most often in a cone. Major lineages include cycads, conifers and ginkgos. Cf. Angiospermae.

GYNAECANER Noun. (Greek, *gyne* = female; *aner* = male. Pl., Gynaecaners.) A male ant which resembles the female and has the same number of antennal segments.

GYNAECOTELIC TYPE Social insects: A group in which the female is the complete prototype of the sex, with all the primary instincts of the sex, including those of the worker caste (Wheeler).

GYNANDROMORPH Noun. (Greek, *gyne* = female; *aner* = male; *morphe* = form. Pl., Gynandromorphs.) 1. An insect that has some features which are clearly male in form or function and some structural features which are clearly female in form or function. 2. A sexual mosaic. Gynandromorphs appear often in Orthoptera (particularly Acrididae). Gynandromorphs are not common in Plecoptera. Gynandromorphs have been reported in the principal groups of Holometabola, including Diptera, Lepidoptera, Coleoptera and Hymenoptera. Virtually all major taxonomic groups of Hymenoptera display gynandromorphs, including Symphyta (sawflies), Parasitica and Aculeata. Cf. Bilateral Gyandromorph; Intersex.

GYNANDROMORPHIC Adj. (Greek, *gyne* = female; *aner* = male; *morphe* = form; *-ic* = of the nature of.) 1. Pertaining to gynandromorph. An individual that displays gynandromorphism. 2. Pertaining to the condition of gynandromorphism.

GYNANDROMORPHISM Noun. (Greek, *gyne* = female; *aner* = male; *morphe* = form; *-ism* = condition. Pl., Gynandromorphisms.) A condition in which male and female features are displayed in one individual. Gynandromorphs are frequently called intersexes and occur in many groups of insects. Some groups of insects may lack profound sexual dimorphism or sexual dimorphism may be limited to internal features and therefore go undetected. Several factors contribute to the expression of gynandromorphism. Genetic causes include: (1) partial fertilization in which sperm is delayed from penetrating Ooplasm until cleavage commences. Fertilized daughter nuclei become females; unfertilized daughter nuclei become male. (2) Dispermy (polyspermy) occurs when some eggs have two nuclei; male Spermatozoa unite with female nucleus to produce females. (3) Chromosomal elimination involves a normal female zygote that divides and a daughter nuclei loses the X-chromosome. Subsequent daughter nuclei of the X-less strain become male; the normal chromosomal complement strain becomes female. Parasitism can also result in a gynandromorph (Homoptera); male chironomid midges become female due to merminthid parasites. Hymenoptera infested with Stylopidae can undergo alteration of the reproductive organs. Environmental Conditions such as high temperature can produce gynandromorphs in parasitic Hymenoptera including *Trichogramma* and *Ooencyrtus*.

GYNE Noun. (Greek, *gyne* = woman. Pl., Gynes.) Social Hymenoptera (ants, bees, wasps): A female member of a reproductive caste; a potential or actual queen. The term is typically applied to potential queens.

GYNE CELL Social Hymenoptera: A cell in a nest in which a Gyne is reared. See Gyne.

GYNECOID Adj. (Greek, *gyne* = woman; *eidos* = form.) 1. Female-like in appearance, behaviour or function. 2. Social Insects; Queen-like; a worker with typically queen features (*e.g.* an enlarged Gaster). 3. An egg-laying worker ant. See Gyne. Cf. Arrhenoid; Androgyne; Android. Rel. Eidos; Form; Shape.

GYNERGATANDROMORPH Noun. (Greek, *gyne* = female; *ergates* = worker; *aner* = male; *morphe* = form. Pl., Gynergatandromorphs.) Social Insects: A genetical condition in which an ant displays morphological characteristics of a queen and morphological characteristics of a worker. (Berndt & Kremer, 1983. Ins. Soc. 30: 461–465.)

GYNERGATE Noun. (Greek, *gyne* = woman; *ergates* = worker. Pl., Gynergates.) Social Insects: A female with structures or features of the queen and worker. See Gynergatandromorph.

GYNETYPE Noun. (Greek, *gyne* = woman; *typos* = pattern. Pl., Gynetypes.) Taxonomy: A female type (Banks & Caudell). See Type. Cf. Androtype. Rel. Nomenclature; Taxonomy.

GYNOCHROMATYPIC Adj. (Greek, *gyne* = woman; *chroma* = colour; *typos* = pattern; *-ic* = of the nature of.) Pertaining to the typical female coloration pattern in a sexually dimorphic population. Syn. Heterochromatic; Heterochrome; Heteromorphic; Heteromorphous. Cf. Androchromatypic. (Hilton 1987, Ent. News 98: 221–223.) Rel. Sexual Dimorphism.

GYNOGENESIS Noun. (Greek, *gyne* = female; *genesis* = descent. Pl., Gynogeneses.) Unisexual reproduction in which unreduced gametes are activated by sperm penetration of the egg. The paternal genome is not incorportated into the Zygote. Syn. Pseudogamy. See Reproduction. Cf. Parthenogenesis.

GYÖRFI, JANOS (1905–1966) (Kadlubowski 1968, Polski Pismo Ent. 38: 943–945.)

GYPLURE® See Dispalure.

GYPSY MOTH *Lymantria dispar* (Linnaeus) [Lepidoptera: Lymantriidae]: A polyphagous moth endemic to the Palearctic; imported into North America (Massachussets) during 1868 by French Astronomer (Leopold Trouvelot) employed at Harvard Observatory. Trouvelot conducted hybridization studies of silk-producing caterpillars for development of resistance to Pebrine Disease which was threatening French sericulture industry. GM escaped culture in Medford, MA where Trouvelot lived; moth remained unnoticed until 1889 when it reached plague proportions. Subsequently, GM became a serious and chronic pest of evergreen and deciduous trees. GM is a particular problem on ornamental trees and defoliates hardwoods in northeastern USA. GM is monovoltine and overwinters in the egg stage. Females are unable to fly and lay their eggs in clusters of ca 400 eggs, covered by body scales. Egg clusters are deposited on ground, tree bark or objects such as automobile tyres. All attempts at control of GM have been unsuccessful. See Lymantriidae. Cf. Brown-tail Moth; White-Tailed Tussock Moth.

GYRI-CEREBRALES Lobes of the Oesophageal Ganglion of the embryo that are connected with the primary lobe. The stalked bodies.

GYRINIDAE Plural Noun. Whirligig Beetles. A cosmopolitan, Family of aquatic, adephagous Coleoptera including about 500 Species. Adult body small to moderate sized and oval in outline shape; Antenna very short and compact; Compound Eyes divided by integumental stripe into dorsal and ventral regions. Fore leg long, slender and raptorial; middle and hind legs short, compressed and paddle-like with marginal fringe of swimming Setae. Eggs are deposited on emergent aquatic vegetation. Larvae are bottom-feeding predators with biting-type mouthparts; larvae respire via feather-like abdominal gills and swim via undulatory motion of entire body. Larvae leave water to pupate in cocoon; pupation site above high-water mark. Adults are scavengers or predators on water surface; gregarious, probably nocturnal and diurnal; swim on surface of water in tight circles aided by Pygidial Gland surfactant. See Coleoptera.

GYROPIDAE Plural Noun. Rodent Chewing Lice. A Family of amblycerous Mallophaga predominantly found in the neotropics. Gyropids are parasitic on guinea pigs. Diagnostic features include Mandibulate mouthparts; Antenna typically capitate, with four segments, concealed in grooves; Tarsi with one claw. See Mallophaga.

HAAG-RUTENBERG, JOHANN GEORG (1830–1897) (Kraatz 1880, Dt. ent. Z. 24: 231–235, bibliogr.)

HAAP, OTTO (1864–1922) (Hedicke 1922, Dt. ent. Z. 1922: 241–242, bibliogr.)

HAAR, DIRK (1860–1905) (Oudemans 1906, Tijdschr. Ent. 49: 1–7, bibliogr.)

HAASE, ERICH (1857–1894) (Steinert 1894, Dt. ent. Z. Iris 7: 364–366.)

HAASE, JOSEF (1891–1971) (Utech 1971, Ent. Ber., Berlin 1971: 78–80,bibliogr.)

HABATSU, KANKICHI (1878–1914) (Hasegawa 1967, Kontyû 35 (Suppl.): 68.)

HABENA Noun. (Latin, *habena* = strap. Pl., Haneae.) A fascia on the Thorax.

HABER, VERNON RAYMOND (1887–1947) (Frost 1947, J. Econ. Ent. 40: 613–614, bibliogr.)

HABERLANDT, FREDERICO (1826–1878) (Bolle 1875, Atti Socl. agric. Gorizia 17: 97–99.)

HABIT Noun. (Latin, *habere* = to have. Pl., Habits.) 1. A predictable pattern of behaviour. 2. A pattern of behaviour that is acquired with repetition. See Learning. Cf. Habitat; Habitus.

HABITAT Noun. (Latin, *habitare* = to inhabit. Pl., Habitats.) 1. Ecology: The area within which an organism naturally occurs, but not a particular location (Shelford). 2. Habitat is defined by the totality of biotic (plants, animals, microorganism) and abiotic conditions (temperature, humidity, rainfall) in that area. Abbreviated 'hab.' Habitats that pertain to chemical, soil or biochemical conditions: Acidophilous; Nitrophilous. Habitats that relate to plants or plant communities: Agricolous; Algicolous; Caespiticolous; Corticolous; Dendrophilous; Fungicolous; Lignicolous; Madicolous; Muscicolous; Nemoricolous; Paludicolous; Phytophilous; Pratinicolous; Silvicolous; Thamnophilous; Xylophilous. Habitats that relate to abiotic factors in the environment: Ammophilous; Arenicolous; Cavernicolous; Deserticolous; Eremophilous; Ericeticolous; Geophilous; Halophilous; Hydrophilous; Hypogenous; Lapidicolous; Psammophilous; Rhophilous; Ripicolous; Rupicolous; Saxicolous; Xerophilous. Habitats that relate to animals or animal products: Aphidicolous; Entomophilous; Entomophytous; Fimicolous; Gallicolous; Malacophilous; Myrmecophilous; Nidicolous; Ornithophilous; Stercoraceous; Zoophilous. See Ecology.

HABITUATION Noun. (Latin, *habituare* = to promote a habit; English, *-tion* = result of an action. Pl., Habituations.) A form of learning in which a cell or organism responds less frequently, less rapidly or less intensively to repeated stimulation by a stimulus that elicits a typical response. Cf. Associative Learning; Fatigue; Facilitation; Imprinting.

HABITUS Singular and Plural Noun. (Latin, *habitus* = appearance.) The form, shape or superficial appearance of an organism. See Form; Shape. Cf. Habit; Habitat.

HABU, MICHIYA (–1904) (Hasegawa 1967, Kontyû 35 (Suppl.): 68.)

HACKBERRY ENGRAVER *Scolytus muticus* Say [Coleoptera: Scolytidae].

HACKBERRY LACE-BUG *Corythucha celtidis* Osborn & Drake [Heteroptera: Tingidae].

HACKBERRY NIPPLE-GALL MAKER *Pachypsylla celtidismamma* (Fletcher) [Hemiptera: Psyllidae]: In North America, a pest of hackberry which causes mammiform galls on the underside of leaves. Adults swarm during autumn. See Psyllidae.

HACKER, HENRY (1876–1973) Australian entomological curator at Queensland Museum (1911–1929) and economic Entomologist with Department of Agriculture and Stock who published on native bees and Hemiptera of Australia. (Marks & Dahms 1974, Mem. Qd Mus. 1974, 191–194.)

HACKWITZ, GUSTAF OLAF DAVID VON (1835–1914) (Kemner 1915, Ent. Tidskr. 36: 74–75.)

HADWEN, S A (1877–1947) (Williams 1948, Proc. R. ent. Soc. Lond. (C) 12: 64.)

HADZI, JOVAN (1884–1972) (Carnelutti 1972, Acta. Ent. jugosl. 8: 139–140.)

HAECKEL, ERNST HEINRICH PHILIPP AUGUST (1834–1919) German Zoologist and developmental biologist. (Anon. 1919, Ent. Bl. Biol. Syst. Käfer 15: 255.)

HAECKEL'S LAW An hypothesis that states 'ontogeny recapitulates phylogeny.' That is, the evolution of a group of animals can be seen in the embryology of group members. This observation may hold in a general sense for some organisms, but the 'Law' is now generally discredited.

HAEGER, ERICH (Urbshn 1965, Mitt. schweiz. ent. Ges. 24: 61.)

HAEGERLE, ROWLAND WELLS (1892–1940) (Shill 1940, J. Econ. Ent. 33: 708.)

HAEMAL Adj. (Greek, *haema* = blood; *-alis* = belonging to.) Descriptive of, pertaining to, or connected with blood or the circulatory system. Rel. Haemocoel; Haemolymph.

HAEMATOCYTE See Haemocyte.

HAEMATOMYZIDAE Plural Noun. A monogeneric Family of Phthiraptera assigned to Suborder Rhyncophthirina and consisting of two ectoparasitic Species, one that lives on Indian and African Elephants and the other lives on Wart Hogs. See Phthiraptera.

HAEMATOPHAGE Noun. (Greek, *haima* = blood; *phagein* = to eat; Pl., Haematophages.) An organism that feeds on blood. See Feeding Strategies. Cf. Solenophage.

HAEMATOPHAGOUS Adj. (Greek, *haima* = blood; *phagein* = to eat; Latin, *-osus* = with the property of.) 1. Descriptive of blood feeding or blood-feeding characteristics. 2. Pertaining to insects that feed upon blood. See Feeding Strategies. Cf. Sanguinivorous; Saprophagous; Solenophagous.

HAEMATOPINIDAE Plural Noun. Wrinkled Sucking Lice. A moderately large Family of Anoplura which contains many Species that parasitize domestic livestock. Head relatively small; compound eyes absent, Ocular Points present; Mandibles absent, mouthparts forming a Beak or rounded anteriorly. See Anoplura.

HAEMATOZOON Noun. (Greek, *haima* = blood; *zoon* = animal. Pl., Haematozoons.) Any organism that develops parasitically within the blood of another organism.

HAEMOCOEL Noun. (Greek, *haima* = blood; *koilos* = hollow. Pl., Haemocoels.) 1. The fluid-filled cavity or cavities of the embryo between the Mesoderm and other germ layers; probably a remnant of the Blastocoel (Snodgrass). 2. The large body cavity in which Haemolymph flows. In postembryonic stages of insects the Haemocoele consists of the internal portion of Head, Pericardial Sinus, Perivisceral Sinus and Perineural Sinus. Alt. Haemocoel. See Circulatory System. Cf. Pericardial Sinus; Perivisceral Sinus; Perineural Sinus. Rel. Haemolymph.

HAEMOCOELIC Adj. (Greek, *haima* = blood; *koilos* = hollow; *-ic* = characterized by.) Descriptive of or pertaining to the Haemocoel or insect's blood cavity.

HAEMOCOELIC INSEMINATION Cimicidae males have hypodermic-like Penes which can be inserted at many sites along the female body. Sperm are injected into the Haemocoel and migrate to the Ovary where they accumulate in large masses. The phenomenon is called 'Haemocoelic Insemination.' In some groups, females have developed a new genital system with a Spermatheca which opens on the dorsum. In *Cimex lectularius*, HI occurs with Spermatozoa activated by accessory gland secretions. HI may provide additional nutrients for the female. In the ultimate case of paternal deception, male cimicids sometimes inject Spermatozoa into other males *(Xyleocoris)*. The Spermatozoa make their way to the inseminated male's sperm ducts and then into the female. Syn. Traumatic Insemination.

HAEMOCOELIC VIVIPARITY A kind of viviparous reproduction in which Oviducts are absent and eggs occur within the Haemocoel. Eggs lack a Chorion and nourishment is obtained through a trophic membrane. Eggs hatch inside the mother's body, larvae consume her internal tissue and emerge through the body wall. HV occurs in Strepsiptera and some Cecidomyiidae (Diptera). See Viviparity. Cf. Adenotrophic Viviparity; Pseudoplacental Viviparity.

HAEMOCYANIN Noun. (Greek, *haima* = blood; *kyanos* = blue. Pl., Haemocyanins.) A colourless, oxygen-carring respiratory pigment found in the blood of arthrodpods. Analogous in function and chemical structure to Haemoglobin in other animals. Haemocyanin contains copper (not iron) and turns blue upon oxidation in the blood of insects. Alt. Hemocyanin.

HAEMOCYTE Noun. (Greek, *haima* = blood; *kytos* = hollow vessel; container. Pl., Haemocytes.) Nucleated cells, typically free-floating and circulating within or associated with the insect's Haemolymph. Some Haemocytes may be the physiological equivalent of white blood cells in vertebrates. Schwammerdam (1758) discovered Haemocytes within louse Haemolymph and called them 'transparent globules.' Haemocytes are highly pleiomorphic and have been given many names based on shape; some Haemocytes change shape during their existence and thereby confuse their classification. Seven anatomical types of Haemocytes are currently recognized: Adipohemocyte, Coagulocyte, Granulocyte, Oenocyte, Plasmatocyte, Prohaemocyte and Spherulocyte. Most common Haemocytes in Haemolymph include: Granulocyte, Nephrocyte, Oenocyte, Prohaemocyte and Plasmatocyte. Haemocyte functions include metabolism, nutrient synthesis and storage, hormone transport, cellular immunity, poison detoxification, connective tissue and Basement Membrane synthesis, and injury repair. Alt. Haemocyte. See Circulatory System; Haemolymph. Cf. Adipohemocyte; Coagulocyte; Granulocyte; Oenocyte; Plasmatocyte; Prohaemocyte; Spherulocyte.

HAEMOGLOBIN Noun. (Greek, *haima* = blood; Latin, *globos* = sphere. Pl., Haemoglobins.) A red haemochrome pigment that consists of two pairs of polypeptide chains. Haemoglobin occurs in vertebrate blood and functions as an agent of oxygen transport. Haemoglobin is present in few insects (chironomid larvae), displays lower molecular weight and diminished capacity to transport oxygen. Abbreviated Hb; Hgb.

HAEMOID Adj. (Greek, *haima* = blood; *eidos* = form.) Blood-like. Pertaining to matter which is blood-like in appearance, substance or function. See Haemolymph; Haematocyte. Rel. Eidos; Form; Shape.

HAEMOLYMPH Noun. (Greek, *haima* = blood; *lympha* = water. Pl., Haemolymphs.) 1. The watery blood-like or lymph-like nutritive fluid of lower invertebrates. 2. Insects: The extracellular circulating fluid which fills the Haemocoel. Haemolymph includes plasma (fluid) and inclusions (hormones, metabolic products, cells, microorganisms and parasites). Inorganic ions in solution include: Sodium, potassium, calcium, chloride, phosphate, bicarbonate and magnesium. Trehalose is the primary sugar in Haemolymph; proteins and amino acids also occur in Haemolymph. Cells within the Haemolymph are called Haemocytes. As a physical element of the Circulatory System, Haemolymph is the agent of thermoregulation and frost protection, the agent of protection from physical (mechanical) injury and the agent of defence from disease, parasitism and predation. Physical protection from injury involves Coagulation; Cellular Defence Mechanisms and Lysozymes protect against microorganisms and parasites; Reflex Bleeding protects against predation. Syn. Insect Blood. See Cellular Defence Mechanisms; Circulatory System; Lysozyme; Reflex Bleeding. Cf. Coagulation; Haemocyte; Plasma; Thermoregulation. Rel. Haemocoel.

HAEMOLYTIC Adj. (Greek, *haima* = blood; *lysis* = loosing; *-ic* = characterized by.) Pertaining to something that causes Haemolysis; the breakdown or destruction of red blood cells. Alt. Hemolytic

HAEMOPARASITE Noun. (Greek, *haima* = blood; *parasitos* = parasite; *-ites* = inhabitant. Pl., Haemoparasites.) Parasitic organisms which live some or all of their lives in the blood of a host. See Parasite.

HAEMOPHAGOUS Adj. (Greek, *haima* = blood; *phagein* = to devour; Latin, *-osus* = with the property of.) Blood-eating. A descriptive term used to characterized the feeding habits of some nematocerous flies, fleas, lice, ticks and mites. See Feeding Strategy.

HAEMOPOIETIC ORGAN Tissue in the Haemocoel which produces Haemolymph cells. See Haemolymph.

HAEMORRHAGE Noun. (Greek, *haima* = blood; *-rrhagia* comb. form from *rhagnynai* = to burst; *age* > *aticum* = cumulative result of. Pl., Haemorrhages.) 1. Any discharge of blood from a ruptured vessel. 2. Insects: The loss of Haemolymph through a rupture of the Integument. Alt. Hemorrhage. Rel. Haemolymph.

HAEMOTOXIC Adj. (Greek, *haima* = blood; *toxikon* = poison; *-ic* = characterized by.) Pertaining to toxins that affect the blood. See Toxin. Cf. Neurotoxic.

HAEMOTOXIN Noun. (Greek, *haima* = blood; *toxikon* = poison. Pl., Haemotoxins.) A toxin that produces Haemolysis or destruction of the functional role(s) of blood or components of the blood. A characteristic of the venom of some spiders, some snakes and other venomous organisms. Alt. Haemopathin. See Toxin. Cf. Neurotoxin.

HAEMOXANTHINE Noun. (Greek, *haima* = blood; *xanthos* = red.) A dissolved albuminoid which has a respiratory and nutritive function in insect blood.

HAEMOZOIN Noun. (Greek, *haima* = blood; *zoon* = animal. Pl., Haemozoins.) A dark-coloured pigment (Melanin) produced through the digestion of red blood cells by *Plasmodium*. Syn. Haemozoin Granules. See Malaria.

HAFERKORN, ARTHUR (1858–1919) (Möbius 1944, Dt. ent. Z. Iris 57: 7.)

HAG MOTH *Phobetron pithecium* (J. E. Smith) [Lepidoptera: Limacodidae].

HAGEDORN, JULIUS MAX (1852–1914) (Anon. 1915, Ent. News 26: 384.)

HAGEMANN, AXEL OTTO CHRISTIAN (1856–1909) (Natvig 1944, Norsk. Ent. Tidsskr 7: 47, bibliogr.)

HAGEN, HAROLD R (1886–1960) (Creighton 1960, J. N. Y. ent. Soc. 68: 63–65.

HAGEN, HERMAN AUGUST (1817–1893) (Henshaw 1894, Proc. Am. Acad. Arts Sci. 29: 419–423.)

HAGENBACH JOHN JAMES (–1826) Conservator of the Royal Museum of Leyden and probably the author of *Symbola Faunae Insectorum Helvetiae*.

HAGEN'S GLAND Abdominal exocrine gland(s) with ducts on the gastral Terga or intersegmental membrane of some male Braconidae. Function of HG has not been determined. (Buckingham & Sharkey 1988: 199–242.) See Gland. Cf. Waterston's Organ.

HAGGAR, GEORGE (1816–1892) (Douglas 1892, Entomol. mon. Mag. 28: 112.)

HAGGART, JAMES CREVAR (1874–1934) (Williams 1935, Entomologist's Rec. J. Var. 47: 11.)

HAGLIDAE Plural Noun. A small Family of ensiferous Orthoptera assigned to Superfamily Tettigonioidea and not know to occur in Australia or North America. See Prophalangopsidae.

HAGLUND, CARL JOHAN EMIL (1837–1901) (Sjöstedt 1902, Ent. Tidschr. 23: 41–46, bibliogr.)

HAHN, CARL WILHELM (1786–1836) (Bonnet 1945, *Bibliographia Araneorum* 1: 32.)

HAHN, JAROSLAV (1897–1933) (Anon. 1933, Cas. csl. Spol. ent. 30: 11, bibliogr.)

HAHNEL, PAUL (1843–1887) (Staudinger 1890, Dt. ent. Z. Iris 3: 128–132.)

HAIARI® See Rotenone.

HAIDENTHALER, LEOPOLD (–1961) (Anon. 1961, Z. wien. ent. Ges. 46: 120.)

HAIJ, JULIUS BERNHARD (1859–1920) (Bengtssen 1921, Ent. Tidskr. 42: 120–124, bibliogr.)

HAIMBACH, FRANK (1859–1930) (Wilhems *et al.* 1930, Ent. News 41: 281–284.)

HAINES, FREDERICK HASLEWOOD (1864–1946) (F. C. F. 1946, Entomol. mon. Mag. 82: 96.)

HAIR Noun. (Anglo Saxon, *haer* = hair. Pl., Hairs.) 1. An epidermal outgrowth unique to mammals, consisting of distal, cornified epidermal cells that grow by cell division within a basal follicle. 2. Any slender, flexible filament largest at the base and often tapering toward the apex. 3. A term almost universally and incorrectly applied to the hair-like Setae of insects. See Seta.

HAIR FOLLICLE MITE *Demodex folliculorum* (Simon) [Acari: Demodicidae].

HAIR PENCIL Tufts or compact groups of long Setae on the body of male Lepidoptera and some Acanthaclisinae Neuroptera. HP are positioned on many parts of the body and legs, most frequently on anterior part of the Abdomen, and associated with exocrine glands. HP are expanded and displayed during courtship and are responsible for dispersal of sex pheromone. Syn. Brush Organ. See Seta. Cf. Coremata.

HAIR PLATES Groups of hair-like cuticular projections positioned at various places on the insect body, but more often at points of articulation, or concentrations of Setae such as the face of some Diptera. HP function in proprioreception. See Proprioreceptor; Sensillum. Cf. Hair Plate.

HAIR-FIELDS The 'Spinules' *sensu* Jardine. Syn. Hair-Scales.

HAIR-SCALES See Hair-Fields.

HAIRLESS FLOWER THRIPS *Pseudanaphothrips achaetus* (Bagnall) [Thysanoptera: Thripidae].

HAIRSTREAK BUTTERFLIES See Lycaenidae.

HAIRSTREAKS See Lycaenidae.

HAIRY Adj. (Middle English, *heer* > Anglo Saxon, *haer* = hair.) Descriptive of insects covered, invested or clothed with Setae. Ant. Glabrous. See Seta. Rel. Vestiture.

HAIRY ANT *Paratrechina longicornis* (Latreille) [Hymenoptera: Formicidae].

HAIRY CHINCH-BUG *Blissus leucopterus hirtus* Montandon [Hemiptera: Lygaeidae]: Subspecies of Chinch Bug found in northeastern North America as a pest of lawns. See Chinch Bug; Lygaeidae.

HAIRY CICADA *Tettigarcta crinita* Distant [Hemiptera: Tettigarctidae].

HAIRY FLOWER-WASPS See Scoliidae.

HAIRY FUNGUS-BEETLE *Typhaea stercorea* (Linnaeus) [Coleoptera: Mycetophagidae]: A cosmopolitan pest of mouldy cereals and other foods; HFB is most abundant in tropical regions. See Mycetophagidae.

HAIRY IMPERIAL-SPIDER *Ordgarius furcatus* (Cambridge) [Araneida: Araneidae].

HAIRY LEAF-EATING CATERPILLAR *Xanthodes congenita* (Hampson) [Lepidoptera: Noctuidea].

HAIRY LINE-BLUE BUTTERFLY *Erysichton lineata lineata* (Murray) [Lepidoptera: Lycaenidae].

HAIRY POWDERPOST-BEETLE *Minthea rugicollis* (Walker) [Coleoptera: Bostrichidae].

HAIRY RICE-CATERPILLAR *Celama taeniata* (Snellen) [Lepidoptera: Noctuidae].

HAIRY RICE-STEMBORER *Niphadoses palleucus* Common [Lepidoptera: Pyralidae].

HAIRY ROVE-BEETLE *Creophilus maxillosus* (Linnaeus) [Coleoptera: Staphylinidae].

HAIRY SCARAB *Saulostomus villosus* Waterhouse [Coleoptera: Scarabaeidae].

HAIRY SPIDER-BEETLE *Ptinus villiger* (Reitter) (Coleoptera: Ptinidae].

HAIRY-MARY CATERPILLAR *Anthela nicothoe* (Boisduval) [Lepidoptera: Anthelidae].

HAIRY-MAGGOT BLOWFLY *Chrysomya rufifacies* (Macquart) [Diptera: Calliphoridae]: In Australia, the most important secondary sheep blowfly. HMB maggots infest lesions caused by primary blowflies to compete with and prey upon primary blowfly maggots. See Calliphoridae.

HAJNEHO, FRANTISKA (1901–1972) (Okal 1974, Ent. problémy 12: 274.)

HALASZ, ARPOD (1857–1904) (Anon. 1904, Rovart. Lap. 11: 194.)

HALBERD-SHAPED Descriptive of structure that is triangular in shape and hollowed along the sides and base.

HALBERT, JAMES NATHANIEL (1872–1948) (Beirne 1948, Irish Nat. J. 9: 168–171.)

HALBHERR, BERNARDINO (1844–1934) (Heikertinger 1935, Koleopt. Rdsch. 21: 196–197.)

HALDE, JEAN BAPTISTE DU (1674–1743) (Rose 1850, *New General Biographical Dictionary* 8: 175.)

HALDEMAN, SAMUEL STEHMAN (1812–1880) (Mallis 1971, *American Entomologists*. 549 pp. (33–36), New Brunswick.)

HALE, MATTHEW (1609–1676) (Rose 1850, *New General Biographical Dictionary* 8: 175–176.)

HALFORDIA FRUIT FLY *Dacus halfordiae* Tryon [Diptera: Tephritidae].

HALICTIDAE Plural Noun. Sweat Bees. A widespread, moderately large Family of aculeate Hymenoptera assigned to Superfamily Apoidea (bees). Adults are small bodied, often metallic coloured and bear one Subantennal Suture; Facial Foveae absent; Labrum wider than long, subtriangular or oval; apex of Glossa acute or rounded; Flabellum absent; Galea typically longer basal than apical of Palpus; Labial Palpus segments typically subcylindrical and subequal; Pre-Episternal Suture nearly complete; forewing Basal Vein curved and not straight; posterior part of second Recurrent Vein not arcuate; Marginal Cell normal, Stigma large and elongate; Pygidium present, sometimes retracted; Arolia well developed in male and female. Halictids are primitively social bees with most Species nesting in soil. Nesting females form aggregations with each female constructing an individual nest. Nests consist of a vertical shaft and horizontal tunnels from shaft. Tunnels terminate in an enlarged cell which accommodates a larva and its provisions. Halictids are important in pollination of plants; a few Species are cleptoparastic upon other bees. Subfamilies include Dufoureinae, Halictinae, Nomiinae and Rophitinae. See Apoidea.

HALICTINAE The nominant Subfamily of Halictidae (Hymenoptera: Apoidea), and consisting of about 60 Genera. See Halictidae.

HALIDAY, ALEXANDER HENRY (1807–1870) (Anon, 1870. Nature 2: 240.)

HALIK, JAROSLAV (1894–1973) (Anon. 1973, Revta. bras. Ent. 17: viii.)

HALIPLIDAE Plural Noun. Crawling Water Beetles. A cosmopolitan Family of about 100 Species of adephagous Coleoptera, and presumably related to Gyrinidae. Adults are small, convex and boat shaped; Antenna filiform with 11 segments. Elytra seriately punctate; Scutellum absent; fore Coxa globose; frontal cavity externally open, internally closed. Tibiae and Tarsi with long Setae adapted for swimming; tarsal formula 5-5-5; hind Coxal plates very large. Abdomen with six Ventrites, basal three Ventrites connate. All Species are aquatic and usually reside in ponds, lakes and slow-flowing streams. Haliplids are associated with vegetation; adults and larvae feed upon stoneworts and other green algae. See Coleoptera.

HALIZAN® See Metaldehyde.

HALL SCALE *Nilotaspis halli* (Green) [Hemiptera: Diaspididae].

HALL, CHRISTOPHER GEORGE (1842–1890) (Anon. 1890, Entomol. mon. Mag. 26: 304–305.)

HALL, GAYLORD CROSSETTE (1871–1954) (Webb 1954, J. N. Y. ent. Soc. 62: 153–159.)

HALL, MAURICE CROWTHER (1881–1938) (Chapin *et al.*1938, Proc. ent. Soc. Wash. 40: 147–149.)

HALL, WILFRED JOHN (1893–1965) (Pearson 1966, Proc. R. ent. Soc. Lond. (C) 30: 62–63.)

HALLER, ALBRECHT VON (1708–1777) (Nordenskiöld 1935, *History of Biology*. 629 pp. (234–238), London.)

HALLER, GUSTAV (1853–1886) German Zoologist. (Dimmock 1888, Psyche 5: 35.)

HALLER, HERBERT (1894–1972) (Hall & Reed 1973, Proc. ent. Soc. Wash. 75: 260–263.)

HALLER'S ORGAN Ticks: A small vesicle contaning sensory receptors on the Tarsus of the first pair of legs; olfactory in function. HO was named after Gustav Haller.

HALLETT, HOWARD MOUNTJOY (1878–1958) (Richards 1959, Proc. R. ent. Soc. Lond. (C) 23: 72.)

HALLEZ' LAW An observation (not a Law) which asserts that the primordial head of an insect embryo is orientated anteriad within the Ovariole.

HALLINEN, JOSEPH EDWARD (1859–1932) (Hungerford 1934, Ann. ent. Soc. Am. 27: 122.)

HALLMARK® See Lambda Cyhalothrin.

HALMARK® See Esfenvalerate.

HALOPHILE Noun. (Greek, *hals* = salt; *philein* = to love. Pl., Halophiles.) An organism that lives and reproduces in water with relatively high salt concentrations. Example: Brine Flies. Alt. Halophilic. See Habitat. Cf. Mesophile; Osmophile; Psychrophile; Thermophile. Rel. Ecology.

HALOPHILOUS Adj. (Greek, *hals* = salt; *philein* = to love; Latin, *-osus* = with the property of.) Salt-loving; descriptive of organisms that live in or near salt marshes, marine habitats or habitats near the seashore. Alt. Halophilic. See Habitat. Cf. Algicolous; Hydrophilous; Madicolous; Paludicolous; Pratinicolous; Rhophilous; Ripicolous. Rel. Ecology.

HALOPHITE Noun. (Greek, *hals* = salt; *phyton* = plant; *-ites* = substance. Pl., Halophytes.) A plant that lives and reproduces in water with relatively high salt concentrations. See Halophile. Cf. Aerophyte; Psammophyte.

HALOPHOBE Noun. (Greek, *hals* = salt; *phobos* = fear. Pl., Halophobes.) Any organism, particularly plants, that cannot tolerate high concentrations of salt. Cf. Halophile.

HALTERATA Noun. The Diptera.

HALTERE Noun. (Greek, *halter* = weight. Pl., Halteres.) The structurally and functionally modified hindwing of Diptera. The Haltere forms a complex orientation device. Anatomically, the Haltere consist of a base (Scabellum), an elongate pedicel and an apically enlarged knob (Capitulum). The base of the Haltere is invested with campaniform Sensilla, Hicks' Papillae and chordotonal organs. The Haltere is a balancing organ used to maintain stability during flight. Halteres vibrate at the same frequency as the forewings but at antiphase. Halteres vibrate in vertical motion, not 8-shaped as the forewings. The Halteres' centre-of-gravity is located near the knob. Campaniform Sensilla on the Haltere respond to changes in stress developed from changes in the inertia of harmonic motion of the oscillating Haltere. Dipterists speculate that the Haltere originated as an adaptation for aerial swarming. Alt. Balancing Organs. See Wing; Wing Modification.

HALTERIPTERA Noun. The Diptera.

HALYZIA SKIPPER *Mesodina halyzia* (Hewitson) [Lepidoptera: Hesperiidae].

HAMA, YSUGIO (1907–1939) (Anon. 1939, Kontyû 35 (Suppl.) 69, bibliogr.)

HAMABIOSIS Noun. (Greek, *hama* = together; *bios* = life; *-sis* = a condition or state. Pl., Hamabioses.) A form of symbiosis in which two Species of insects live in association without obvious motive or advantage to either. See Biosis; Commensalism; Parasitism. Cf. Abiosis; Anhydrobiosis; Antibiosis; Archebiosis; Calobiosis; Cleptobiosis; Kleptobiosis; Lestobiosis; Parabiosis; Phylacobiosis; Plesiobiosis; Synclerobiosis; Trophobiosis; Xenobiosis.

HAMADENS Plural Noun. (Latin, *hamatus* = hooked; *dens* = tooth.) A tooth-like projection of the Lacinia proximad of the Maxadentes (MacGillivray).

HAMATE Adj. (Latin, *hamatus* = hooked; *-atus* = adjectival suffix.) Barbed; pertaining to structures furnished with hooks or barbs; hook-like; bent as a hook. Alt. Hamiform. See Barb; Hook. Cf. Aduncate; Ankistroid; Ankyroid; Barbed; Hastiform; Uncate; Uncinate. Rel. Form 2; Shape 2; Outline Shape.

HAMBLETON, JAMES CHASE (1863–1938) (Osborn 1946, *Fragments of Entomological History*, Pt. II. 232 pp. (8), Columbus, Ohio.)

HAMBLETON, JAMES ISAAC (1895–1969) (Campbell 1969, Proc. ent. Soc. Wash. 71: 593–597.)

HAMER, JAMES (1841–1887) (Dimmock 1888, Psyche 5: 35.)

HAMET, H (1815–1889) (Am. 1889, Bull. Insect agric. Paris 14: 143–147.)

HAMFELT, HENRY KRISTIAN (1892–1938.) (Kemner 1939, Opusc. ent. 4: 112.)

HAMIDOP® See Methamidophos.

HAMIFORM Adj. (Latin, *hamus* = hook; *forma* = shape.) Hook-shaped; descriptive of structure that is hooked. Alt. Hamate. See Barb; Hook. Cf. Hastiform; Unciform; Uncinate. Rel. Form 2; Shape 2; Outline Shape.

HAMILTON, CLYDE CARNEY (1890–1938) (Filmer 1960, J. Econ. Ent. 53: 1142.)

HAMILTON, JOHN (1827–1897) (Osborn 1937, *Fragments of Entomological History*. 394 pp. (144), Columbus, Ohio.)

HAMLIN, AMOS (1766–1843) (Weiss 1936, *Pioneer Century of American Entomology*. 320 pp. (69–70), New Brunswick.)

HAMLIN, JOHN CALHOUN (1896–1943) (Osborn 1946, *Fragments of Entomological History*. Pt. II. 232 pp. (88–89), Columbus, Ohio.)

HAMLYN, EDWARD TREMAYNE (–1953) (Buxton 1954, Proc. R. ent. Soc. Lond. (C) 18: 80.)

HAMM, ALBERT HARRY (1861–1951) English printer and amateur Entomologist interested in Lepidoptera and courtship behaviour of Empididae. Hamm was employed as an assistant at Hope Museum (Oxford) (1897–1931) and awarded an honorary M.A. (Oxford, 1942). (Carpenter 1952, Proc. Linn. Soc. Lond. 163: 83–84.)

HAMMAR, ALFRED GOTTLIEB (1880–1913) (Riley 1914, J. Econ. Ent. 7: 155–157.)

HAMMARSTRÖM, EDWARD RUDOLF (1863–1928) (Forsius 1928, Notul. Ent. 8: 62–63.)

HAMMEL, AROID DAID Author of *Quelques Observations sur la Blatte Germanique* (1821) and several numbers of *Essais Entomologique* (1827).

HAMMER See Lindane.

HAMMOCK Noun. (Spanish, *hamaca*. Pl., Hammocks.) The hammock-like covering of caterpillars of certain moths.

HAMMOND, HAROLD EDWARD (1902–1963) (Smith 1964, Entomologist 97: 49–51.)

HAMNER, ARTHUR LEE (1901–1970) (Green & Wilson 1971, J. Econ. Ent. 64: 997.)

HAMPE, CLEMENS (1802–1884) (Reitter 1885, Wien ent. Ztg. 4: 1–2.)

HAMPSON, GEORGE FRANCIS (1860–1936) (Imms 1937, Proc. R. ent. Soc. Lond. (C) 1: 55–56.)

HAMULA Noun. (Latin, *hamulus* = little hook. Pl., Hamuli) The Retinaculum. Collembola: The fused ventral process used for jumping. See Retinaculum.

HAMULAR PROCESS See Hamuli.

HAMULOHALTERE Noun. (Latin, *hamulus* = little hook; Greek, *halter* = weight. Pl., Hamulohalteres.) The reduced form of the hindwing in male Coccoidea.

HAMULUS Noun. (Latin, *hamulus* = little hook. Pl., Hamuli.) 1. Any small hook or hook-like process. 2. Odonata: A fork of the Hamule (Garman). Forked appendage of the second segment of the male. Syn. Genital Hamule; Haules; Hamuli Anteriores; Hamuli Posteriores; Posterior Lamina (Garman). 3. Orthoptera: A hook-like processes of tree-cricket genitalia (Smith). 4. Siphonaptera: Moveable sclerites from the lateral walls of aedeagal Palliolum. 5. Hymenoptera and some Trichoptera: Small hook-like structures along the anterior margin of the hindwing which attach with the posterior margin of the forewing during flight. 6. Acari: Triple-hooked, retrorse, apicoventral processes on the palpal Femur (Goff *et al.* 1982, J. Med. Ent. 19: 226.) Rel. Wing.

HANBURY, DANIEL (1825–1875) (Behn 1877, Leopoldina 13: 1.)

HANBURY, FREDERICK JANSON (1851–1875) (Wallis 1939, Proc. Linn. Soc. Lond. 1938–1939: 318–319.)

HANCOCK, G L R (–1940) (Gedge 1940, J. E. Afr. Uganda nat. Hist. Soc. 15: 83–84.)

HANCOCK, JOSEPH LANE (1864–1922) (Osborn 1937, *Fragments of Entomological History*. 394 pp. (231), Columbus, Ohio.)

HAND See Manus.

HANDLIRSCH, ADAM PETER JOSEF (1864–1890) (Brauer 1890, Verh. zool-bot. Ges. Wien 40: 515–516, bibliogr.)

HANDLIRSCH, ANTON (1865–1935) Trained as Pharmacist but became assistant to Friedrich Brauer at Vienna Naturhistorisches Museum, and ultimately became Director. Handlirsch is best known from *Die Fossilen Insekten* (1906–1908) and volume 3 of Schröder's *Handbuch der Entomologie* (1920–1925). (Esaki 1936, Mushi 10: 152–155.)

HANDSCHIN, EDUARD (1894–1962) (Portmann 1962, Verh. natuf. Ges. Basel 73: 332–339, bibliogr.)

HANDSOME FUNGUS BEETLES See Endomychidae.

HANELD, WILHELM (–1910) (Kuhut 1910, Dt. ent. Z. 1910, 329–330.)

HANG FLIES See Mecoptera; Bittacidae.

HANGING FLIES See Mecoptera; Bittacidae.

HANHAM, ABDIEL WILLIAM (1857–1944) (Downes 1945, Proc. ent. Soc. Br. Columb. 42: 27, bibliogr.)

HANITSCH, KARL RICHARD (1860–1940) Born and educated in Germany, Hanitsch became a Naturalized English citizen who worked in Singapore at Raffles Library and Museum (1895–1919) and was awarded an honorary M.A. (Oxford, 1935). (Imms *et al.* 1940, Nature 146: 360.)

HANITSCH, RITCHARD (1860–1940) See Hanitsch, K. R.

HANNA, G DALLAS (1887–1970) (Miller 1962, Proc. Calif. Acad. Sci. 32: 1–40; Arnaud 1972, Pan-Pacif. Ent. 48: 59–61.)

HANSEN, HANS JACOB (1855–1936) (Calman 1937, Proc. Linn. Soc. Lond. 1936–1937: 193–195.)

HANSEN, JAMES (1874–1934) (Hungerford 1936, Ann. ent. Soc. Am. 29: 184.)

HANSEN, PEDRO JÖRGENSEN (1865–1937) (Lolfino 1938, Physis 12: 407–408.)

HANSEN, VICTOR (1889–1974) (Tuxen 1975, Ent. Meddr. 42: 97–108. bibliogr.)

HANSON, BROR HENRIK (1906–1966) (Landin 1966, Opusc. ent. 32: 3–6.)

HANSSEN, HANS KRISTIAN (1860–1948) (Strand

1950, Norsk. ent. Tidsskr. 8: 136–139, bibliogr.)

HANSTRÖM, BERTILL (1891–1966) (Lindroth 1969, Opusc. ent. 34: 171.)

HAPLODIPLOIDY Noun. (Greek, *haploos* = simple; *diploos* = double; *eidos* = form. Pl., Haplodiploidies.) A genetical condition seen in Hymenoptera, Acari and other arthropods wherein males are hemizygous and possess half the chromosomal complement of the diploid female. See Diploid; Haploid. Cf. Arrhenotoky; Deuterotoky; Thelytoky. Rel. Parthenogenesis.

HAPLOGASTRAN Adj. Coleoptera: A type of Abdomen in which the second Sternum is modified to form two small, lateral sclerites. The third abdominal Sternum is the first continuous Sternum across the Abdomen. Typical of Coleoptera such as Scarabaeidae. Cf. Cryptogastran; Hologastran.

HAPLOID Adj. (Greek, *haploos* = simple; *eidos* = form.) 1. Pertaining to the Chromosome number contained within an organism's gamete. 2. The typical number of Chromosomes observed after meiosis in a gametic cell. See Chromosome. Cf. Diploid; Polyploid.

HAPLOMETROSIS Noun. (Greek, *haploos* = simple; *metros* = mother; *-sis* = a condition or state. Pl., Haplometroses.) Social Hymenoptera: Foundation of a colony (nest) by one reproductive female (queen). Cf. Pleometrosis.

HAPLOMETROTIC Adj. (Greek, *haploos* = simple; *metros* = mother; *-ic* = of the nature of.) Pertaining to Haplometrosis.

HAPTARAX® See Chlorfenvinphos.

HAPTASOL® See Chlorfenvinphos.

HAPTOLACHUS Noun. (Greek, *hapto* = to fasten, bind; *lachos* = portion. Pl., Haptolachi.) Scarabaeoid larvae: The posteromedial region of the Epipharynx behind the Pedium and below the Clypeus. Haptolachus is composed of the Nesia, a number of Sensilla, and the Crepis. Syn: Proximal Sensory Area *sensu* Hayes (Boving).

HAPTOMERUM Noun. (Greek, *hapto* = to fasten, bind; *meros* = part. Pl., Haptomeri.) Scarabaeoid larvae: The anteromedial region of the Epipharynx anterior of the Pedium and posterior of the Corypha, or behind the apical region consisting of the united Acropariae and Corypha. Haptomerum is composed of the Zygum, various Sensilla and a series of Crepis. Syn. Proximal Sensory Area *sensu* Hayes (Boving).

HAPTOTYPE Noun. (Greek, *haptein* = to touch; *typos* = pattern. Pl., Haptotypes.) Botany: An Icotype collected with the Holotype, but possibly taken from another plant. See Type. Cf. Holotype; Icotype. Rel. Nomenclature; Taxonomy.

HARA, KANESUKE (1885–1962) (Hasegawa 1967, Kontyû 35 (Suppl.): 70.)

HARBISON, JOHN STEWART (1826–1912) (Pellett 1929, Am. Bee J. 69: 556–558.)

HARD MAPLE BUDMINER *Obrussa ochrefasciella* (Chambers) [Lepidoptera: Nepticulidae].

HARD PESTICIDE A chemical used in control of pests which tends to have a long residue life and is harmful to the environment and beneficial organisms. Example: Organophosphates. See Pesticide. Cf. Soft Pesticide.

HARD TICKS See Ixodidae.

HARD WAX SCALE *Ceroplastes sinensis* Del Guercio [Hemiptera: Coccidae]: Endemic to South America and now a widespread pest of *Citrus* and ornamental plants. Adult females are 7 mm, globular with a dirty white, hard-wax cover. HWS is parthenogenetic with males rarely encountered; females may produce ca 5,000 orange-coloured eggs. Crawlers often prefer to settle on midrib of leaves and after several weeks shift to twigs and small branches. HWS completes two nymphal instars. HWS produces large amounts of honeydew which serves as substrate for sooty mould and other fungi. Syn. Chinese Wax Scale. See Coccidae.

HARD-HITTER® See Permethrin.

HARDONK, MARTINUS (1896–1959) (Lempke 1959, Ent. Ber., Amst. 19: 169–170.)

HARDPAN Noun. (Pl., Hardpans.) A compact soil with a high concentration of clay and an accumulation of salts, conditions which result in poor drainage and restricted root growth.

HARDS, CHARLES HENRY (1896–1964) (Wigglesworth 1965, Proc. R. ent. Soc. Lond. (C) 29: 53.)

HARDY, GEORGE AUSTIN (1888–1966) (Carl 1966, J. ent. Soc. Br. Columb. 63: 43–44.)

HARDY, GEORGE HUDLESTON HURLSTONE (1882–1966) Australian Entomologist specialising in taxonomy of brachycerous Diptera and author of 173 papers (1914–1967). (Chadwick 1966, J. ent. Soc. Austral. (N.S.W.) 3: 48–57, bibliogr.)

HARDY, JAMES (1815–1898) (Ferguson 1898, Hist. Berwicksh. Nat. Club 16: 341–372, bibliogr.)

HARDY, JOHN RAY (1844–1921) (Britten 1922, Rep. Trans. Manchester ent. Soc. 19: 71–72.)

HARDY–WEINBERG EQUILIBRIUM Population genetics: The concept that gene (allelic) frequencies remain constant from generation to generation in a population. The HWE stipulates several assumptions about the population: The population is infinitely large, it is randomly interbreeding, that no selection occurs, migration does not occur, and mutation does not occur. HWE used for the calculation of expected and observed allelic frequencies.

HARE, E. J. (1884–1969) (Worms 1969, Entomologist's Gaz. 81: 151–152.)

HARFORD, W G W (1825–1911) (Essig 1931, *History of Entomology*. 1029 pp. (650–651), New York.)

HARGREAVES, ERNEST (1899–1974) (Gunn 1976, Proc. R. ent. Soc. Lond. (C) 40: 52.)

HARKER, JAMES ALLEN (1848–1894) (Anon. 1895, Proc. Linn. Soc. Lond. 1894–1895: 32.)

HARLEQUIN BUG 1. *Murgantia histrionica* (Hahn) [Hemiptera: Pentatomidae]: In North America, a pest of many Species of vegetables and cole crops. HB was introduced from Mexico into the southeast after the Civil War. In warmer areas, HB is active and reproductive throughout year; in more temperate climates, HB overwinters in refuse. Eggs are barrel-shaped and deposited in clusters on underside of leaves of host plant; eclosion is temperature-dependent. Nymphs undergo five instars. Nymphs and adults suck sap from host plants. HB typically completes three generations. See Pentatomidae. 2. *Dindymus versicolor* (Herrich-Schäffer) [Hemiptera: Pyrrhocoridae]: In Australia, a pest of deciduous fruits including citrus; HB occurs in southern states. See Pyrrhocoridae.

HARLEQUIN COCKROACH *Neostylopyga rhombifolia* (Stoll) [Blattaria: Blattidae].

HARLEQUIN FLIES See Bibionidae.

HARMANN, SAMUEL WILLARD (1894–1948) (Hartzell 1948, J. Econ. Ent. 41: 838–839.)

HARNED, ROBEY WENTWORTH (1884–1968) (Ewing & Rainwater 1968, J. Econ. Ent. 61: 1133–1134.)

HAROLD, EDGAR VON (1830–1886) (Anon. 1886, Berl. ent. Z. 30: 149–150.)

HARPE Noun. (Latin, *harpago* = hook. Pl., Harpes.) 1. Lepidoptera: Genitalic claspers of the male. One of the pair of clasping organs developed as appendages of the ninth abdominal Sternum, articulating with the caudal margin of the Vinculum (Klots). In British usage, specialized growths of stiff Setae or bristles on the ninth Sternum in male Lepidoptera (Tillyard). 2. Diptera: See Diststylus; Harpagones; Parameres.

HARPER, ALFRED (1813–1884) (Anon. 1884, Entomologist 17: 264.)

HARPER, GEOFFREY WITHINGTON (1902–1973) (Lees 1974, Proc. R. ent. Soc. Lond. (C) 38: 59.)

HARPER, PHILIP HENRY (–1883) (Dunning 1883, Proc. ent. Soc. Lond. 1883: xlii.)

HARRINGTON, WILLIAM HAGUE (1852–1918) (Gibson 1918, Can. Ent. 50: 181–187, bibliogr.)

HARRIOT, THOMAS (1560–) (Weiss 1936, *Pioneer Century of American Entomology*. 320 pp. (3–4), New Brunswick.)

HARRIS THADDEUS MASON (1768–1842) (Weiss 1936, *Pioneer Century of American Entomology*. 320 pp. (41–42, 53), New Brunswick.)

HARRIS, EDWARD DOUBLEDAY (1839–1919) (Leng 1919, J. N. Y. ent. Soc. 27: 237–240, bibliogr.)

HARRIS, HERBERT GEORGE (–1942) (Russell 1942, Entomologist's Rec. J. Var. 54: 147–148.)

HARRIS, JOHN THOMAS (–1892) (Godman 1892, Proc. ent. Soc. Lond. 1892: lv-lvi.)

HARRIS, MOSES (1731–1788) English painter of miniatures, engraver and naturalist. Principal works include *An Exposition of English Insects* (1781) and *The Aurelian or Natural History of English Insects, namely, Moths and Butterflies* (1766). Harris also wrote *The English Lepidoptera, or the Aurelian's Pocket Companion* (1775). (Weiss 1926, Science Mon. 23: 560–564.)

HARRIS, PERCIVAL FREDERIC (1892–1951) (H. C. H. 1951, Entomologist 84: 120.)

HARRIS, THADDEUS WILLIAM (1795–1856) (Scudder 1860, Proc. Boston Soc. nat. Hist. 7: 72, 213–222, bibliogr.)

HARRISIA CACTUS LONGICORN *Alcidion cereicola* Fisher [Coleoptera: Cerambycidae].

HARRISIA CACTUS MEALYBUG *Hypogeococcus pungens* (Granara de Willink) [Hemiptera: Pseudococcidae].

HARRISIA CACTUS WEEVIL *Eriocereophaga humeridens* O'Brien [Coleoptera: Curculionidae]

HARRISON, ALBERT (1860–1911) (Turner 1911, Entomologist's Rec. J. Var. 23: 282–283.)

HARRISON, JOHN (1693–1776) (Rose 1950, *New General Biographical Dictionary* 8: 218.)

HARRISON, JOHN (1834–1907) (Waterhouse 1907, Proc. ent. Soc. Lond. 1907, xcvii.)

HARRISON, JOHN WILLIAM HESLOP (1881–1967) (Peacock 1968, Biogr. Mem. Fellows R. Soc. 14: 243–270, bibliogr.)

HARRISON, LAUNCELOT (1880–1928) (Musgrave 1932, *A Bibliography of Australian Entomology 1775–1930*. 380 pp. (145–146, bibliogr.), Sidney.)

HART, CHARLES ARTHUR (1859–1918) (Malloch 1918, Ent. News 29: 157–159.)

HART, ESTHER HASTINGS (1862–1940) (Walton 1940, J. Econ. Ent. 33: 820.)

HARTERT, ERNST JOHANN OTTO (1959–1933) (Anon. 1934, J. Bombay nat. Hist. Soc. 37: 474–475.)

HARTIG, ROBERT (1839–1901) (Denglar 1902, Forst.-Wiss. Zbl. 24: 1–5.)

HARTIG, THEODOR (1805–1880) (Anon. 1880, Leopoldina 16: 70–71, bibliogr.)

HARTLEY, EDWIN A (1893–1926) (Osborn 1937, *Fragments of Entomological History*. 394 pp. (218), Columbus, Ohio.)

HARTMANN, AUGUST (?1807–1880) (Kriechbaumer 1880, Ent. NachBl. 6: 148–149.)

HARTMANN, JOHANN DANIEL (1793–1862) (Jäggli 1950, Mitt. schweiz, ent. Ges. 31: 201.)

HARTMANN, KARL FRIEDRICH (1859–1932) (Günther 1943, Dt. ent. Z. Iris 1943: 6–9.)

HARTWIEG, FRITZ (1877–1962) (Anon. 1963, Z. wien. ent. Ges. 48: 88.)

HARTWIG, WILHELM (1889–1915) (Soldanski 1916, Dt. ent. Z. 1916: 1–2.)

HARTZELL FREDERICK ZELLER (1879–1958) (Chapman 1960, J. Econ. Ent. 53: 180–181.)

HARTZELL, ALBERT (1891–1972) (Torgeson 1972, J. Econ. Ent. 65: 935–936.)

HARUKAWA, CHUKIDU (1887–1968) (Yasue 1969, Kontyû 37: 463–465.)

HARVEST MITE *Neotrombicula autumnalis* (Shaw) [Acari: Trombiculidae]: An european Species whose larval stage commonly attacks rabbits.

HARVESTER ANT *Messor barbarus* Linnaeus; *Pogonomyrmex salinas* Olsen [Hymenoptera: Formicidae]. See Formicidae. Cf. California Harvester-Ant; Florida Harvester-Ant; Red Harvester

Ant; Western Harvester-Ant.

HARVESTER ANTS Members of the Genus *Pogonomyrmex*. See Formicidae. Cf. California Harvester-Ant; Florida Harvester-Ant; Red Harvester Ant; Harvester-Ant; Western Harvester-Ant.

HARVESTERS TERMITES *Drepanotermes* spp. [Isoptera: Termitidae].

HARVESTERS See Lycaenidae; Opiliones.

HARVESTING ANTS Ant Species which predominantly feed upon seeds and store them in the nest. Harvesting behaviour has developed independently in several ant lineages. See Formicidae.

HARVESTMEN Plural Noun. The common name for Opiliones. Sometimes specifically members of the opilionid Genera *Equitius* [Triaenonychidae], *Spinicrus* [Phalangiidae] or *Triaenobunus* [Triaenonychidae].

HARVEY, FRANCIS LEROY (1850–1900) (Calvert 1900, Ent. News 11: 451–452.)

HARVEY, JOHNSTON (1881–1951) (Anon. 1952, Proc. Linn. Soc. N.S.W. 77: v.)

HARVEY, WILLIAM (1578–1657) English embryologist probably best known for his description of the circulation of blood. Author of *Exercitatio Anatomica de Motu Cordis et Sanguinis in Animalibus* (1628) and the exceptionally important *Exercitationes de Generatione Animalium* (1651) which first introduced the concept of Metamorphosis and Epigenesis. Harvey hypothesized that all life came from eggs and noted that the chick embryo originated as the Blastoderm.

HARWOOD, BERNARD SMITH (1876–1933) (Anon. 1933, Entomol. mon. Mag. 69: 256–257.)

HARWOOD, PHILIP (1882–1957) (Brown 1958, Entomologist's mon.; Mag. 94: 60.)

HARWOOD, WILLIAM HENRY (1840–1917) (Anon. 1918, Entomol. mon. Mag. 54: 40–41.)

HASE, ALBRECHT (–1962) (Herfs 1963, Anz. Schädlingsk. 36: 44.)

HASEBROEK, KARL (1860–1941) (Weidner 1967, Abh. Verh. naturw. Ver. Hamburg Suppl. 9: 280–284.)

HASEMAN, LEONARD (1884–1969) (Enns 1969, J. Econ. Ent. 62: 973.)

HASHIMOTO, SAGORO (1866–1952) (Hasegawa 1967, Kontyû 35 (Suppl.): 67, bibliogr.)

HASSALL, ALBERT (1862–1942) (Price *et al.* 1943, J. Parasit. 29: 233–235.)

HASSELQUIST, FREDERICK (1722–1752) (Rose 1850, *New General Biographical Dictionary*. 8: 226–227.)

HASSELT, ALEXANDER WILLEM MICHIEL VAN (1814–1902) (Leesberg 1903, Tijdschr. Ent. 46: 71–77.)

HASTATE Adj. (Latin, *hasta* = spear; *-atus* = adjectival suffix.) Arrow-shaped; structure pointed apically and with spreading lobes or flanges. Alt. Hastiform. See Arrow. Cf. Hamiform.

HASTATE PUPIL In an ocellate spot, when the pupil is hastate or halberd-shaped.

HASTIFORM Adj. (Latin, *hasta* = spear; *forma* = shape.) Hastate-shaped; arrow-shaped with flaring barbs. Syn. Hastate. See Arrow. Cf. Hamiform; Unciform. Rel. Form 2; Shape 2; Outline Shape.

HASTISETA Noun. ((Latin, *hasta* = spear; *seta* = bristle. Pl., Hastisetae.) One of three setal types found on the body of larval Dermestidae. Typically, Hastisetae are filiform, barbed and spear-tipped in form. Hastisetae are prone on the body when the larva is calm, but become erect when the larva is irritated. Hastisetae are loosely anchored and easily disengaged by vibration. Hastisetae detach from the body in response to attack by predators; they accumulate, adhere and interlock to form a tangled mass which is difficult to groom. See Sensillum. Cf. Nudiseta; Spiciseta.

HASWELL WILLIAM AITCHESON (1854–1925) (Carter 1928, Proc. Linn Soc. N.S.W. 53: 485–498.)

HATAKEYAMA, HISASHIGE (1880–1959) (Hasegawa 1967, Kontyû 35 (suppl.): 67–68.)

HATCH ACT United States of America legislation signed into law by President Grover Cleveland on 2 March 1887 establishing Agricultural Experiment Stations at Land Grant Colleges. Named for William H. Hatch, a congressman from Missouri. See Morrill Act.

HATCHET Noun. (Middle Engish, *hachet* = hatchet. Pl., Hatchets.) A weapon devised by humans and used for cutting. Specifically, hatchets are constructed of metal, with a handle of wood or metal, and a head which is broad, flat with a sharp edge. Hatchet is used as a comparative descriptor in some entomological terms that describe shape, form or function. Cf. Ensiform; Falcate. Rel. Sickle; Spear; Sword.

HATCHET WASPS See Evaniidae.

HATCHING Noun. (Middle English, *hacche*. Pl., Hatchings.) Surface marks, lines or scratches which are numerous, short and close-set. Alt. Hatched. Cf. Crosshatching. Verb. (Middle English, *hacchen*.) The breaking of the eggshell by an insect embryo during the process of emergence. See Eclosion.

HATCHING MEMBRANE A membranous sheath (not the amnion) surrounding the immature insect within the egg at the time of hatching (eclosion). HM probably is an embryonic exuvial Cuticula, shed while hatching or soon after hatching (Snodgrass).

HATCHING SPINE Any cuticular projection on the head of first instar larvae that may be used to facilitate emergence from the eggshell. *e.g. Myodopsylla insignis* (Rothschild) larvae have a dorsomedial projection that is lost after the first moult (Smith & Clay 1988). See Egg Burster; Egg Tooth; Oviruptor. Rel. Eclosion.

HATCHING STRUCTURE See Eclosion.

HATTORI, TORU (–1908) (Hasegawa 1967, Kontyû

35 (suppl.): 68.)

HAU LEAFMINER *Philodoria hauicola* (Swezey) [Lepidoptera: Gracillariidae].

HAUDER, FRANZ (1860–1923) (Christel 1958, Wien. ent. Ges. 43: 187–206.)

HAUG, ALBERT (1866–1937) (Anon. 1937, Ent. Z. Frankf. a. M. 51: 14.)

HAUPT, HERMANN (1873–1959) (Anon. 1959, Ent. News 70: 244.)

HAUPT, R (–1916) (Soldanski 1916, Dt. ent. Z. 1916: 227.)

HAUSCHILD, LOUIS MARCUS (1840–1913) (Kryger 1918, Ent. Meddr. 12: 25–29.)

HAUSER, FRIEDRICH (1853–1932) (Ross 1935, Int. ent. Z. 28: 484.)

HAUSER, GUSTAV (–1935) (Ross 1935, Int. ent. Z. 28: 484–485.)

HAUSTELLUM Noun. (Latin, *haurire* = to drain; *haurio* = to suck. Pl., Haustella) Lepidoptera: The elongate Galeae which form a tube adapted for sucking liquid food. Syn. Mediproboscis; Proboscis.

HAUY, RENÉ JUST (1743–1822) (Rose 1850, *New General Biographical Dictionary* 8: 230.)

HAVELKA, JAROMIR (1911?–1971) (Heyrovsky 1972, Acta ent. Bohemoslovaca 69: 133, bibliogr.)

HAVILAND, M D See Brindley.

HAWAIIAN ANTLION *Eidoleon wilsoni* (McLachlan) [Neuroptera: Myrmeleontidae].

HAWAIIAN BEET-WEBWORM *Spoladea recurvalis* (Fabricius) [Lepidoptera: Pyralidae].

HAWAIIAN BUD-MOTH *Helicoverpa hawaiiensis* (Quaintance & Brues) [Lepidoptera: Noctuidae].

HAWAIIAN CARPENTER-ANT *Camponotus variegatus* (F. Smith) [Hymenoptera: Formicidae].

HAWAIIAN FLOWER-THRIPS *Thrips hawaiiensis* (Morgan) [Thysanoptera: Thripidae].

HAWAIIAN GRASS-THRIPS *Anaphothrips swezeyi* Moulton [Thysanoptera: Thripidae].

HAWAIIAN PELAGIC WATER-STRIDER *Halobates hawaiiensis* Usinger [Heteroptera: Gerridae].

HAWAIIAN SPHINX *Hyles calida* (Butler) [Lepidoptera: Sphingidae].

HAWK MOTHS See Sphingidae.

HAWKESWORTH, JOHN (1715–1773) (Rose 1850, *New General Biographical Dictionary*. 8: 232.)

HAWKINS, CHARLES N (–1970) (Hinton 1972, Proc. R. ent. Soc. Lond. (C) 35: 53.)

HAWKINS, KENNETH (1890–1946) (Anon. 1946, J. Econ. Ent. 39: 681.)

HAWKSHAW, JOHN CLARKE (1841–1921) (Scott 1921, Entomol. mon. Mag. 57: 94.)

HAWKSHAW, O (1869–1949) (Wigglesworth 1950, Proc. R. ent. Soc. Lond. (C) 14: 64.)

HAWLEY, IRA MYRON (1894–1966) (Knowlton *et al.* 1968, J. Econ. Ent. 61: 887.)

HAWLEY, WILLIAM C B (–1948) (Williams 1949, Proc. R. ent. Soc. Lond. (C) 13: 67.)

HAWORTH, ADRIAN HARDY (1767–1833) English author of *Lepidoptera Britannica* (1803). (Anon. 1901, Trans. Hull scient. Fld. Nat. Club 1: 229–232.)

HAWTHORN LACE-BUG *Corythucha cydoniae* (Fitch) [Hemiptera: Tingidae]: A pest endemic to North America. Nymphs and adults feed upon leaf fluids of Hawthorn, Cotoneaster, Pyracantha and Japanese quince. HLB causes leaves to discolour or drop prematurely. HLB overwinters as an egg or adult. See Tingidae.

HAYASHI, JUNNOSUKE (1898–1966) (Hasegawa 1967, Kontyû 35 (Suppl.): 69–70, bibliogr.)

HAYASHI, KEI (1914–1962) (Hasegawa 1967, Kontyû (Suppl.) 35: 70. bibliogr.)

HAYNES, HARRY (1864–1951) (Worms 1951, Entomologist 84: 48.)

HAYWARD, ARTHUR RUSHER (1878–1939) (Rippon 1939, Entomologist 72: 296.)

HAYWARD, HAROLD CARLYLE (1876–1935) (Shelden 1935, Entomologist 68: 144.)

HAYWARD, KENNETH J (1891–1972) (Pena 1973, Revta. chil. Ent. 7: 263–264.)

HAYWARD, ROLAND (1865–1906) (Henshaw 1906, Psyche 13: 101–103, bibliogr.)

HAZARD Noun. (Middle English, *hasard*. Pl., Hazards.) 1. A thing or condition that creates danger or peril. 2. The danger that injury will occur with use of a particular pesticide, depending on active ingredient toxicity and exposure duration.

HAZEL APHID *Myzocallis coryli* (Goeze) [Hemiptera: Aphididae].

HAZELNUT WEEVIL 1. *Curculio obtusus* (Blanchard) [Coleoptera: Curculionidae]: A pest of hazelnut that is widespread in North America. Cf. Filbert Weevil. 2. *Curculio nucum* Linnaeus [Coleoptera: Curculionidae]: A pest of hazelnut in Europe whose Larvae feed on kernels, then chew holes in shells and pupate in soil. See Curculionidae.

HD-PIC® See Chloropicrin.

HEAD Noun. (Anglo Saxon, *heafod* = head. Pl., Heads.) The first (anteriormost) Tagma of the insect's body. Head articulates with Thorax, and bears eyes, Ocelli, mouthparts and Antennae. Head is the result of consolidation (fusion) of several segments that were extant in primitive ancestors. Archetypal segments include: 1. Ocular (Protocerebral) Segment; 2. Antennal (Deutocerebral) Segment; 3. second Antennal (Tritocerebral) Segment; 4. Mandibular Segment; 5. Superlingual Segment; 6. Maxillary Segment; 7. Labial (second maxillary) Segment. Each head region is innervated by nerves from the Brain. Superficially, the relationships seem uncomplicated: Protocerebrum innervates eyes, Deutocerebrum innervates Antennae and Tritocerebrum innervates postantennal appendages. Head regions each correspond to a 'primitive' segment of an ancestral insect's body. A similar argument prevails for head and thoracic segmentation among all arthropods. Unfortunately, general structure of the insect head is not consistent with cephalic nervous system. Mouth lies anterior of the Tritocerebral Somite; Protocerebral and

Deutocerebral Ganglia appear preoral. Syn: Prosoma; Cranium. See Tagma. Cf. Abdomen; Thorax. Rel. Gnathocephalon; Head Orientation.

HEAD CAPSULE 1. The fused sclerites of the head which form a hard compact case (Imms). 2. Sclerites or hardened regions of the head include the Vertex, Frons, Face, Gena, Postcranium, Occiput and Cervix. See Head. Cf. Gnathocephalon.

HEAD LOUSE *Pediculus humanus capitis* (DeGeer) [Phthiraptera: Pediculidae]: A cosmopolitan blood-sucking parasite of humans. Adults are 2.5–3.5 mm long and anatomically indistinguishable from Body Louse, but generally taken on human head. HL eggs are attached to hair on head of host. HL is not a vector of Epidemic Typhus. See Pediculidae. Cf. Body Louse. Rel. Epidemic Typhus.

HEAD ORIENTATION Axial position or posture of the head in its resting condition. HO is important in providing definitions regarding the head. Axial position in insects falls into three basic categories based on HO, including hypognathous, prognathous and opisthognathous. Some Species do not conform to any convenient definition. Axial position may be helpful in determining topographical features, but only within a limited extent because axial positions have evolved several times in diverse lineages and are subject to different selective pressures. See Head.

HEAD VESICLE Diptera: See Ptilinum.

HEADLEE, THOMAS JEFFERSON (1877–1946) (Peairs *et al.* 1946, J. Econ. Ent. 39: 681–683.)

HEALEY, IAN NEVILL (1941–1972) (Butler 1973, Proc. R. ent. Soc. Lond. (C) 37: 56.)

HEALY, JOHN L (1864–1926) (Wyatt 1926, Ent. News 37: 128.)

HEARDER, GEORGE JONATHON (–1894) (Anon. 1894, Entomol. mon. Mag. 30: 65–66.)

HEARING Noun. (Middle English, *heringe* = to hear. Pl., Hearings.) The capacity or ability to perceive sound, or vibrations in the air or substrate. See Sound.

HEARLE, ERIC (1893–1934) (Hungerford, 1935, Ann. ent. Soc. Am. 28: 179.)

HEARSEY, JOHN BUNNET (–1865) (Pascoe 1866, Proc. ent. Soc. Lond. (3) 2: 138–139.)

HEART Noun. (Anglo Saxon, *heorte* = heart. Pl., Hearts.) The posterior part of the Dorsal Vessel which is divided into chambers and confined to the Abdomen. The Heart propels Haemolymph and controls the rate, force and general direction of circulation within the body. See Dorsal Vessel. Cf. Aorta. Rel. Circulatory System; Haemolymph.

HEART CHAMBER A segmental swelling of the Heart (Snodgrass). See Heart.

HEARTBEAT Noun. (Anglo Saxon, *heorte* = heart. Old English, *beatan*. Pl., Heartbeats.) A complete pulstation of the Heart.

HEARTWATER Noun. (Anglo Saxon, *heorte* = heart. Old English, *waetar*. Pl., Heartwaters.) A rickettsia-like disease of sheep, goats and cattle in southern Africa. Heartwater also infects native ruminants. Aetiological agent *Cowdria ruminantium* assigned to Tribe Ehrlichieae of Rickettsiaceae. Vectors of disease include *Amblyomma* spp. See Rickettsiaceae. Cf. Canine Ehrlichiosis.

HEATH SPITTLEBUG *Clastoptera saintcyri* Provancher [Hemiptera: Cercopidae].

HEATH, EDWIN FIRMSTONE (1840–1914) (A. F. 1914, Can. Ent. 46: 299–300.)

HEAUTOTYPE See Autotype.

HEBARD, MORGAN (1887–1946) (Rehn 1948, Ent. News 59: 57–69.)

HEBENSTREIT, JOHANN ERNST (1703–1751) (Rose 1850, *New General Biographical Dictionary* 8: 242.)

HEBERDEN, WILLIAM (1710–1801) (Rose 1850, *New General Biographical Dictionary* 8: 244.)

HEBRIDAE Amyot & Serville 1843. Plural Noun. Velvet Water Bugs. A small, widespread Family of heteropterous Hemiptera assigned to Gerromorpha or Superfamily Hebroidea depending upon classification. Adult body <3.5 mm long, oblong in outline shape and covered with dense pubescence; Ocelli present; Antenna with four or five segments; Beak of four segments; Scutellum present; legs each with two Tarsomeres; tarsal claws apical; macropterous, brachypterous or apterous; macropterous individuals with membrane lacking veins. Adults and nymphs live on surface of ponds or in vegetation along the shore of streams; a few Species inhabit tropical rainforest. Hebrids are regarded as predaceous. See Hebroidea.

HEBROIDEA A Superfamily of Hemiptera including the Hebridae. See Hemiptera.

HECKEL, PAUL (1856–1935) (Horn 1935, Arb. morph. taxon. Ent. Berlin 2: 121.)

HECKENDORN, FRITZ (1879–1963) (Allenspach 1963, Mitt. ent. Ges. Basel 13: 92–93.)

HECQUET, PHILIPP (1661–1737) (Rose 1850, *New General Biographical Dictionary* 8: 245–246.)

HEDEMANN, WILHELM VON (1836–1903) (Rebel 1903, Verh. zool.-bot. Ges. Wien 53: 421–423, bibliogr.)

HEDGE GRASSHOPPER *Valanga irregularis* (Walker) [Orthoptera: Acrididae].

HEDGEHOG FLEA *Archaeopsylla erinacei* [Siphonaptera: Pulicidae].

HEDGEHOG SLUG *Arion intermedius* Normad [Sigmurethra: Arionidae].

HEDGEHOG TICK *Ixodes hexagonus* Leach [Arari: Ixodidae].

HEDGES, ALFRED VANDER (1893–1957) (Richards 1959, Proc. R. ent. Soc. Lond. (C) 23: 73.)

HEDGREN, GUSTAF WILHELM (1894–1942) (Lindroth 1944, Ent. Tijdskr. 65: 36.)

HEDICKE, HANS (1911–1970) (Konigsmann 1971, Dt. ent. Z. 18: 387–400, bibliogr.)

HEDICKE, HANS (–1949) (Kerrich 1949, Entomol. mon. Mag. 85: 264.)

HEDWIG, KARL (1875–1965) (Weidner 1965, Nachr. naturw. Mus. Aschaffenb. 72: lxiv, bibliogr.)

HEEGER, ERNST (–1866) (Frauenfeld 1866, Verh. zool.-bot. Ges. Wien 16: 102–103.)

HEEL FLIES See Hypodermatidae.

HEEL FLY See Ox Warble Fly.

HEELS Hymenoptera: The Spinulae (Torre Bueno).

HEER, OSWALD (1809–1883) Swiss palaeontologist, entomologist, theologian and Director of the Botanical Garden in Zürich. (Scudder 1883, Science 2: 583–586.)

HEFNAWY, TAWFIK (1940–1975) (Hoogstraal 1976, J. Med. Ent. Honolulu 13: 48, bibliogr.)

HEGNER, ROBERT (1880–1942) (Beltran 1942, Rev. Soc. méx. Hist. nat. 3: 183–192.)

HEIDEMANN, OTTO (1842–1916) (Howard 1916, Proc. ent. Soc. Wash. 18: 201–205, bibliogr.)

HEIDENREICH, ERNST (1880–1964) (Anon. 1964, Mitt. dt. ent. Ges. 23: 43.)

HEIDER, KARL (1856–1935) (Ulrich 1960, Stadium Berolinense 1960, 868–897.)

HEIKERTINGER, FRANZ (1876–1953) (Strouhal 1955, Annln naturh. Mus. Wien 60: 20–35, bibliogr.)

HEINEMANN, HERMANN VON (1812–1871) (Blasius 1887, Jber. Ver. naturw. Braunschweig 5: 120–125.)

HEINIG, HERMANN (1858–1970) (Montfort 1971, Beitr. naturk. Forsch. SüdwDtl. 30: 169–170.)

HEINK, CHARLES L (1869–1940) (Meiners 1941, Ent. News 52: 119–120.)

HEINRICH, CARL (1880–1955) (Wade 1955, Proc. ent. Soc. Wash. 57: 249–255, bibliogr.; Gates-Clarke 1974, J. Lepid. Soc. 28: 186–189.)

HEINRICH, RUDOLF (1859–1939) (Anon. 1939, Arb. morph. taxon. Ent. Berl. 6: 69.)

HEISER, RUDOLF (1839–1905) (Anon. 1907, Ent. News 18: 112.)

HEISTER, LORENZ (1683–1758) (Rose 1850, *New General Biographical Dictionary* 8: 251.)

HELARION® See Metaldehyde.

HELBING, HERMANN (–1939) (DeBeaux 1940, Annali Mus. Civ. Stor. nat. Giacomo Doria 60: (1)-(2).)

HELCODERMTUS SPINES Holometabola: Spines on the Pupa that are adapted for boring or tearing. See Spine.

HELCODERMTUS Adj. (Greek, *helcos* = festering wound; *derma* = skin.) Pertaining to a surface with ulcer-like depressions.

HELCOMYZIDAE Plural Noun. Seabeach Flies. A small Family of cyclorrhaphous Diptera assigned to Superfamily Sciomyzoidea. Adults resemble Sciomyzidae but differ in lacking Femoral Bristles and R1 ending beyond middle of wing; Palpus lacking apical bristles. Three Species occur along northwestern coast of North America; larvae inhabit decomposing seaweed. See Sciomyzoidea.

HELEIDAE Plural Noun. See Ceratopogonidae.

HELENA BROWN BUTTERFLY *Tisiphone helena* (Olliff) [Lepidoptera: Nymphalidae]

HELENA MOTH *Opodiphthera helena* (White) [Lepidoptera: Saturniidae].

HELENITA BLUE BUTTERFLY *Candalides helenita helenita* (Semper) [Lepidoptera: Lycaenidae].

HELEOMYZIDAE Plural Noun. A moderate sized Family of cyclorrhaphous Diptera assigned to Superfamily Heleomyzoidea. Adult body moderate size, typically brown with well developed oral Vibrissae; Postvertical Bristles convergent; Costa spinose; Adults frequent moist, shady habitats. Larvae occur in fungi or inhabit decaying plant and animal matter. Syn. Helomyzidae; Trixoscelidae. See Heleomyzoidea.

HELEOMYZOIDEA A Superfamily of schizophorous Diptera including Chyromyidae (Chyromiidae), Coelopidae (Phycodromidae), Heleomyzidae (Trichoscelidae, Trixoscelidae), Somatiidae and Sphaeroceridae (Boorboridae, Cypselidae, Sphaeroceratidae). See Schizophora.

HELGRAMMITE Noun. (Etymology obscure. Pl., Helgrammites.) The aquatic larval stage of *Corydalus cornutus* or other Corydalidae (Megaloptera). Typically found in swift water of stony creeks. Alt. Hellgrammite. See Corydalidae.

HELICIFORM Adj. (Greek, *helix* = spiral; Latin, *forma* = shape.) Snail-shaped. Descriptive of structure in the form or shape of a spiral snail shell. A term applied to describe the cases of some Trichoptera. Syn. Helicoid. Cf. Chonchiform; Conriform; Mytiliform; Pectiniform; Trochiform. Rel. Form 2; Shape 2; Outline Shape.

HELICONIANS See Heliconiidae.

HELICONIIDAE Plural Noun. Heliconians. A predominantly tropical Family of butterflies. Forewings usually are long, narrow and projecting beyond hindwings; forelegs atrophied. Eggs are barrel-shaped. Larvae with Setae which become branched spines after moulting. Host plants include passion vine (*Passiflora*). Chrysalids are irregular in shape with numerous sharp projections, and are suspended by Cremaster. Heliconians are extensively used as a 'model' in mimentical systems owing to their distasteful-lethal features. Heliconiidae sometimes are regarded as a Subfamily of Nymphalidae. See Nymphalidae.

HELIOCHARITIDAE Plural Noun. A small Family of Odonata assigned to Calopterygoidea. Adults are small to medium sized with Abdomen slender. Wings are long, narrow and petiolate; numerous Antenodal Crossveins present; Pterostigma elongate and quadrilateral rectangular; Anal Vein well developed. See Calopterygoidea.

HELIOCOPSYCHIDAE Plural Noun. Snailcase Caddisflies. A small Family of Trichoptera assigned to Superfamily Sericostomatoidea. Adults are 5–8 mm long with body pale coloured with wings mottled. Head dorsum with large warts; Ocelli absent; Antenna shorter than forewing; Maxillary Palpus with less than five segments; mesoscutal Setae usually confined to warts; tarsal segments with spines arranged irregularly. Common name refers to snail-shell shape of larval case; larvae occur on sandy bottoms of streams; case constructed of sand grains. See Sericostomatoidea.

HELIODINIDAE Plural Noun. Heliodinid Moths. A small Family (ca 400 Species) of ditrysian Lepidoptera assigned to Superfamily Yponomeutoidea. Adult body very small with wingspan 7–15 mm and head with smooth scales, metallic reflections; Ocelli present and Chaetosemata absent; Antenna nearly as long as forewing, Scape lacks Pecten; Proboscis lacks scales; Maxillary Palpus very small and Labial Palpus variable; fore Tibia with Epiphysis; tibial spur formula 0-2-4; Tympanal Organs absent. Adults are diurnal; when standing, hind legs elevated or pressed against Abdomen. Eggs are oval in outline shape and Chorion conspicuously punctate or sculpture pattern reticulate. Feeding habits of larvae are diverse; some Species feed beneath webbing on leaves or fruit; some Species mine grass; at least one Species feeds as an internal parasite of *Kermes* spp. *Stathmopoda auriferella* (Walker) resembles *Sitotroga cerealella*; larvae feed on sorghum in Nigeria. See Yponomeutoidea.

HELIOPHOBIC Adj. (Greek, *helios* = sun; *phobos* = fear, terror; *-ic* = of the nature of.) Pertaining to an aversion to sunlight; negatively heliotropic. Cf. Phototropic.

HELIOTACTIC Adj. (Greek, *helios* = sun; *taxis* = arrangement; *-ic* = of the nature of.) Seeking sunlight. Pertaining to organisms that seek sunlight.

HELIOTHIS NPV A naturally-occurring Nuclear Polyhedrosis Virus used as a biopesticide against larvae of *Heliothis* spp. and *Helicoverpa* spp. (bollworms) in some countries. NPV must be ingested to be effective; kills larvae within 2–10 days. Trade names include: H-NPV® and Elcar®. See Biopesticide; Nuclear Polyhedrosis Virus. Cf. Beet Armyworm NPV; Douglas-Fir Tussock Moth NPV; European-Pine Sawfly NPV; Redheaded Pine Sawfly NPV; *Spodoptera littoralis* NPV.

HELIOTROPE FLEA-BETTLE *Longitarsus albinens* (Fondras) [Coleoptera: Chrysomelidae].

HELIOTROPE MOTH *Utetheisa pulchelloides* Hampson [Lepidoptera: Arctiidae].

HELIOTROPIC Adj. (Greek, *helios* = sun; *trope* = a turning; *-ic* = of the nature of.) Pertaining to heliotropism.

HELIOTROPISM Noun. (Greek, *helios* = sun; *tropos* = a turn; English, *-ism* = condition. Pl., Heliotropisms.) A tropism in response to the stimulus of sunlight. See Tropism. Cf. Aeolotropism; Anemotropism; Chemotropism; Electrotropism; Galvanotropism; Geotropism; Hydrotropism; Phototropism; Rheotropism; Stereotropism; Thermotropism; Thigmotropism; Tonotropism. Rel. Taxis.

HELIOZELIDAE Plural Noun. Shield Bearers. A small, cosmopolitan Family of heteroneurous Lepidoptera assigned to Superfamily Incurvarioidea. Adult body small, often metallic with wingspan 3–9 mm and head with smooth, overlapping scales, often shining; Ocelli and Chaetosemata absent; Labrum typically without spinose Pilifer; Antenna filiform, considerably shorter than forewing; Proboscis short but longer than Labial Palpus, with overlapping scales basad; Maxillary Palpus short, not folded, with five or fewer segments; Labial Palpus short and pendulous; fore Tibia with Epiphysis; tibial spur formula 0-2-4; wings lanceolate; hindwing without Discal cell. Eggs are inserted individually beneath bark, in petiole or leaf of host plant. Larvae are without thoracic legs; Prolegs sometimes with Crochets on segments 4–6; two Stemmata on each side of head. Pupae are exarate; Maxillary Palpus absent; abdominal Terga with fine spines. Adults are diurnal, fly in sunshine and rest on flowers; some Species visit lights during night. Biology is poorly known; larvae of most Species are leaf and petiole miners; last instar larvae cut oval case which drops to ground; flat part of case is attached with silk to leaf litter at pupation site. See Incurvarioidea.

HELIX® See Chlorflurazuron.

HELLEN, WOLTER Finnish Entomologist specializing in taxonomy of Coleoptera and parasitic Hymenoptera.

HELLER, KARL MARIA JOSEF (1864–1945) (Draeseke 1957, Abh. Berl. Mus. Tierk. Dresden 23: 257–268.)

HELLERS, GUSTAF ADOLF (1889–1940) (Anon. 1939, Bitidningen 38: 19–20; Anon. 1940, Bitidningen 39: 21.)

HELLGRAMMITE Noun. (Etymology obscure. Pl., Hellgrammites.) The aquatic, predaceous larvae of the Nearctic megalopteran *Corydalus cornutus*. Alt. Helgrammite. See Megaloptera.

HELLIESEN, TOR AMBROSIUS (1855–1914) (Natvig 1915, Ent. Bl. Biol. Syst. Käfer 11: 128, bibliogr.)

HELLINS, JOHN (1829–1887) (Anon. 1887, Psyche 24: 20.)

HELLM, OTTO (1826–1902) (Lakowitz 1928, Ber. westpreuss. bot.-zool. Ver. 50: 22–23.)

HELLWEGER, MICHELE (1865–1930) (Schuler 1933, Öst. entVer. 18: 21–23.)

HELMET FLEAS See Stephanocircidae.

HELMET Noun. (Old High German, *helm* = helmet. Pl., Helmets.) 1. A hard, protective covering of the head. 2. The Galea. Cf. Hood.

HELMHOLTZ, HERMAN LUDWIG FERDINAND (1821–1894) (A. W. R. 1896, Proc. R. Soc. Lond. 59: xvii–xxx.)

HELMINTHIDAE See Elmidae.

HELMS, RICHARD (1842–1914) (Weidner 1967, Abh. Verh. naturw. Ver. Hamburg Suppl. 9: 164–166.)

HELOCEROUS Adj. (Greek, *helos* = nail; Latin, *-osus* = with the property of.) Pertaining to insects with clavate Antennae. See Antenna.

HELODIDAE Plural Noun. A cosmopolitan Family of Coleoptera with about 400 nominal Species. Adults characteristically less than 10 mm long, head invisible from above and with a longitudinal Carina beneath eye; Antenna somewhat elongate and compressed with Scape stout; tarsal formula 5-5-5 with fourth segment bilobed and tarsal claws simple. Larvae are aquatic with anal Gills and functional Spiracle on eighth abdominal segment. Helodids are remarkable among endopterygotes in that larval Antennae are multisegmented. Adults can swim, but apparently rarely enter water. Pupation occurs in mud along waterline. Helodidae are sometimes considered synonym of Scirtidae. See Coleoptera.

HELORIDAE Förster 1856. Plural Noun. A widespread Family of apocritous Hymenoptera assigned to the Proctotrupoidea and consisting of one Genus with eight nominal Species. Adults typically are small, dark bodied and polished; Antenna are not geniculate with 15 segments with an Anellus; Antenna are not inserted on prominent elevation; forewing with Pterostigma and 2–4 closed cells; Trochantellus free on some legs. Larvae develop as solitary, internal multivoltine parasites of Chrysopidae. Eggs are deposited in host larvae, free within Haemocoel of host. Polypodeiform first-instar larvae remain in first instar until host spins cocoon; rapid development of 2–3 instars follows with pupation outside host. Adult parasite emerges from cocoon. See Proctotrupoidea.

HELOTHION® See Sulprofos.

HELOTREPHIDAE Plural Noun. Aquatic Heteroptera in lakes and ponds that feed on vegetation.

HELUS Noun. (Etymology obscure. Pl., Heli.) Scarabaeoid larvae: A coarse, fixed spine without a cup, belonging to the region of the Haptomerum (Boving).

HELVOLUS Adj. (Latin, *helvus* = light yellow.) Descriptive of colour that is tawny or dull reddish-yellow. See Tawny; Yellow. Cf. Flavus.

HELVUS Adj. (Latin, *helvus* = light yellow.) Descriptive of colour that is honey-yellow. See Yellow. Cf. Flavus.

HEMELYTRON Noun. (Greek, *hemi* = half; *elytron* = sheath. Pl., Hemelytra, Hemielytra.) 1. Hemiptera–Heteroptera: The anterior (fore) wing whose basal half is thickened and apical part is membranous. 2. Termed by some authors 'Elytron' or 'Tegmen'; sometimes applied to Orthopteran Tegmen. Alt. Hemielytron. See Wing Modification. Cf. Elytron; Tegmen.

HEMEROBIIDAE Plural Noun. Brown Lacewings. A cosmopolitan, moderate-sized Family of hemerobioid Neuroptera. Hemerobiids are small bodied and delicate in appearance; Antenna are moniliform and as long as wing; wings typically are subequal in size with a frenulum-type wing coupling mechanism in macropterous forms; forewing with at least two Rs veins arising from R. Adults are predaceous in trees and shrubs and commonly display thanatotic behaviour; females are long-lived with high fecundity. Larvae are fusiform, predaceous and do not carry debris. See Hemerobioidea. Cf. Chrysopidae.

HEMEROBIOIDEA A Superfamily of plannipenous Neuroptera including Chrysopidae and Hemerobiidae. See Neuroptera.

HEMEROCALLIS APHID *Myzus hemerocallis* Takahashi [Hemiptera: Aphididae].

HEMI- Greek prefix meaning 'half.'

HEMICELLULOSE Noun. (Greek, *hemi* = half; Latin, *cellula* = little cell. Pl., Hemicelluloses.) A principal component of lignocellulose; alkali-soluble polysaccharides composed of sugars (arabinose, xylose, mannose, galactose) forming part of the cell wall of plants and the food store of some seeds. Hemicellulose is interspersed and covalently linked with lignin to form a matrix surrounding cellulose microfibrils. See Lignocellulose.

HEMICEPHALOUS Adj. (Greek, *hemi* = half; *kephale* = head; Latin, *-osus* = with the property of.) 1. Pertaining to insect larvae with a head reduced in size and structure. 2. Diptera: A type of larvae in an intermediate condition between the eucephalous and the acephalous types.

HEMIMERINA A Suborder of Dermaptera (or a group excluded from that Order) that consists of ca 10 nominal Species in tropical Africa. Species are flattened, blind and apterous with Tarsi adapted for gripping hair and Cerci rod-like. Biology is poorly known. Species are associated with rats (*Cricetomys*) and mice (*Beamys*); may be parasitic.

HEMIMEROPTERA An obsolete term for Hemiptera.

HEMIMETABOLA Noun. (Greek, *hemi* = half; *metabole* = change.) A Division of the Heterometabola including aquatic immatures that differ from adults in the presence of provisional organs. Representatives include Plecoptera, Ephemeroptera, Odonata and some Homoptera. See Metaboly; Metamorphosis. Cf. Ametabola; Paurometabola; Holometabola.

HEMIMETABOLOUS Adj. (Greek, *hemi* = half; *metabole* = change; Latin, *-osus* = full of.) 1. Pertaining to the Hemimetabola. 2. Insects with an incomplete Metamorphosis and developing with gradual changes in size and form from a first instar nymph to an adult. Hemimetabolous development is seen in exopterygote insects and refers to organisms with Metamorphosis in which anatomical changes between immature and adult are made in one radical, conspicuous transition. Hemimetabolous development can be regarded as a specialized transitional phase associated with aquatic immatures since most hemimetabolous insects are aquatic as immatures. Examples include Plecoptera, Ephemeroptera, Odonata and some Homoptera. See Development; Metamorphosis. Cf. Ameabolous; Holometabolous; Paurometabolous.

HEMIMETABOLY Noun. (Greek, *hemi* = half; *metabole* = change.) The condition of being Hemimetabolous. Hemimetaboly is a derived condition among pterygote insects. See Metamorphosis. Cf. Ametaboly; Holometaboly; Paurometaboly. Rel. Groundplan.

HEMIPEPLIDAE See Mycteridae.

HEMIPHLEBIIDAE Plural Noun. A monotypic Family of Odonata (*Hemiphlebia* Selys, 1869; *Hemiphlebia mirabilis* Selys 1869) occurring in Australia. This Species is a rare, small, green damselfly that may be the most ancient Species of Odonata and may have been in Australia for ca 280 million years; *H. miribilis* possesses primitive and reduced anatomical features and is considered the sister Taxon of Odonata (Trueman *et al.* 1992; Watson & O'Farrell 1991). See Odonata.

HEMIPHLEBIOIDEA A Superfamily erected for the rare, Australian damselfly, *Hemiphlebia mirabilis* (Hemiphlebiidae).

HEMIPNEUSTIC RESPIRATION A type of insect respiration in which pairs of spiracles are closed. HR is prevalent among larval insects. See Hypopneustic Respiration.

HEMIPNEUSTIC Adj. (Greek, *hemi* = half; *phein* = to breathe; *-ic* = consisting of.) 1. Pertaining to insects that have fewer than 10 pairs of functional Spiracles in the Tracheal System. HTS is subdivided into Apneustic and Hypopneustic. 2. Pertaining to a type of insect Tracheal System in which pairs of Spiracles are closed. HTS is regarded as prevalent among larvae. See Tracheal System; Spiracle. Cf. Apneustic; Hypopneustic; Oligopneustic; Polypneustic; Propneustic. Rel. Respiratory System.

HEMIPSOCIDAE Plural Noun. A numerically small Family of Psocoptera assigned to the Suborder Psocomorpha. Antenna with 13 segments, without secondary annulations; Tarsi with 2–3 segments; forewing with M vein forked, Areola Postica connected to M by crossvein; some Species with maculated wings. Species found in dried leaves and leaf litter. See Psocoptera.

HEMIPTERA Noun. (Greek, *hemi* = half; *pteron* = winged.) A cosmopolitan Order of Neoptera and the largest Order of Exopterygota, divided into Suborders Auchenorrhyncha, Sternorrhyncha and Heteroptera. Fossil record of Hemiptera extends from the Lower Permian to Recent. Adi;t body to 115 mm long and exceptionally diverse in form. Head prognathous in most Heteroptera, and opisthognathous in Sternorrhyncha and Auchenorrhyncha; Antenna with 3–10 segments; Clypeus large, often divided into Postclypeus (Frons) and Anteclypeus; Labrum small, triangular, covering base of mouth; mouthparts adapted for suction feeding with long, thin Mandibles and Maxillae forming stylets enveloped by a segmental beak (Rostrum); Mandibular Stylets lateral, Maxillary Stylets medial; longitudinal grooves on medial aspect of Maxillary Stylets form anterior Food Canal and posterior Salivary Canal; Maxillary and Labial Palpi absent; mouthparts absent from all male and some female Coccoidea and sexuales of some Aphidoidea; Compound Eyes usually present, sometimes absent; two Ocelli usually present, three Ocelli in some groups. Thorax variously modified with spines, lobes, processes and lateral extensions; Mesonotum usually large with primary groundplan sclerites; Sterna typically reduced; spiracles present on Mesothorax and Metathorax. Legs typically cursorial, forelegs raptorial in some predatory groups; hindlegs with flattened Tibiae in some aquatic groups; apex of Pretarsus with various structural elements of ambiguous Homology; Tarsi with 1–3 segments in Auchenorrhyncha and Heteroptera, 1–2 segments in Sternorrhyncha. Tarsi typically with two claws, one claw in Coccoidea, some Enicocephaloidea and Nepomorpha, claw absent on some legs of Gerromorpha. Wings in repose tectiform, flat and partially overlapping on Abdomen, or Coleoptera-like in some groups. Forewing texture coriaceous to membranous in most homopterous groups, thickened basally and membranous apically in most Heteroptera (Hemelytron). Hindwing membranous, reduced or absent; wing coupling mechanisms absent or variously developed; Autotomy known in some Gerromorpha. Abdomen usually broadly attached to Thorax; 10 segments typically present with apical segment forming anal tube (Proctiger); eight pairs of spiracles present in most Hemiptera, aquatic groups with specialized mechanisms of respiration; abdominal nota of Heteroptera modified into sclerotized dorsal Terga and lateral Connexivia. Diverse wax and scent gland systems developed; sound producing and sound reception systems developed in some groups; Peritrophic Membrane absent; 1–4 nerve Ganglia comprise Ventral Nerve Cord; accessory glands present in male and female; acrotrophic Ovarioles; 0–3 Spermathecae present, simple to complex. Paurometabolous development typical, some groups with Pupae in male; nymphs lack Ocelli, tarsal and antennal segmentation sometimes reduced in nymph. Predominantly plant feeding insects; phytophagous Heteroptera feed upon parenchyma, seeds, pollen; Sternorrhyncha and Fulgoroidea predominantly phloem feeders; Cicadelloidea, Cicadoidea, Cercopoidea predominantly xylem feeders; predation and parasitism developed in Heteroptera. Included Families: Sternorrhyncha: Aclerdidae, Adelgidae, Aleyrodidae, Aphididae, Asterolecaniidae, Beesoniidae, Calophyidae, Carsidaridae, Cercoccidae, Coccidae, Conchaspididae, Dactylopiidae, Diaspididae, Eriococcidae, Halimococcidae, Homotomidae, Kermesidae, Kerriidae, Lecanodiaspididae, Margarodidae, Ortheziidae, Phacopteronidae, Phenacoleachiidae, Phoenicococcidae, Phylloxeridae, Pseudococcidae, Psyllidae, Stictococcidae and Triozidae. Auchenorrhyncha: Acanaloniidae, Achilidae, Achilixiidae, Aetalionidae, Aphrophoridae, Cercopidae, Cicadellidae, Cicadididae, Cixiidae, Clastopteridae, Delphacidae, Derbidae, Dictyopharidae, Eurybrachyidae, Eurymelidae, Flatidae, Fulgoridae, Gengidae, Hylicidae, Hypochthonellidae, Issidae, Kinnaridae, Lophopidae, Machaerotidae, Meenoplidae, Membracidae, Nogodinidae, Ricaniidae, Tettigarctidae Tettigometridae, Tropiduchidae. Heteroptera: Acanthosomatidae, Aenictopecheidae, Aepophilidae, Alydidae, Anthocoridae, Aradidae, Belostomatidae, Berytidae, Canopidae, Ceratocombidae, Cimicidae, Colobathristidae, Coreidae, Corixidae, Cydnidae, Dinidoridae, Dipsocoridae, Enicocephalidae, Gelastocoridae, Gerridae, Hebridae, Helotrephidae, Hermatobatidae, Hyocephalidae, Hydrometridae, Hypsipterygidae, Idiostolidae, Joppeicidae, Largidae, Leptopodidae, Lestoniidae, Lygaeidae, Macroveliidae, Malcidae, Medocostidae, Megarididae, Mesoveliidae, Microphysidae, Miridae, Nabidae, Naucoridae, Nepidae, Notonectidae, Omaniidae, Pachynomidae, Paraphrynoveliidae, Peloridiidae, Pentatomidae, Philoeidae, Piesmatidae, Plataspididae, Pleidae, Plokiophilidae, Polyctenidae, Progonocimicidae, Pyrrhocoridae, Reduviidae, Rhopalidae, Saldidae, Schizopteridae, Scutelleridae, Stemmocryptidae, Stenocephalidae, Termitaphididae, Tessaratomidae, Thaumastellidae, Thaumastocoridae, Tingidae, Veliidae, Velocipedidae, Vianaididae. Fossils: Archescytinidae, Thyreocoridae, Urostylidae. See Insecta.

HEMISPHERIC Adj. (Greek, *hemi* = half; *sphaira* = globe; *-ic* = characterized by.) Shaped as the half of a globe or sphere; semiglobose. Alt. Hemispherical; Hemisphaericum.

HEMISPHERICAL SCALE *Saissetia coffeae* (Walker) [Hemiptera: Coccidae]: In North America, a pest of ferns, greenhouse and house plants. In Australia, a minor and sporadic pest of citrus that produces honeydew which causes sooty mould. See Coccidae. Cf. Black Scale.

HEMISYNANTHROPE Noun. (Greek, *hemi* = half; *syn* = together; *anthropos* = humans. Pl., Hemisynanthropes.) An insect (or other organism) which is not completely associated with humans or not totally dependent upon humans and/or the human environment for its reproduction, development or continued existence. *e.g.* Housefly. Adj. Hemisynanthropic. See Synanthrope. Cf. Eusynanthrope.

HEMISYNANTHROPY Noun. (Greek, *hemi* = half; *syn* = together; *anthropos* = humans. Pl., Hemsynanthropies.) An extension of the concept of Synanthropy which recognizes an association with insects (or other organisms) in which the insects are not totally dependent upon humans or the human environment. *e.g.* Housefly. See Synanthrope; Synanthropy. Cf. Eusynan-thropy.

HEMITERGITE Noun. (Greek, *hemi* = half; Latin, *tergum* = back; *-ites* = constituent. Pl., Hemitergites.) Embiidae: Two parts of Tergum 10 in the male (Tillyard). Hymenoptera: Left and right sclerites of any abdominal Tergum which has been divided longitudinally. See Tergum; Tergite.

HEMLOCK BORER *Melanophila fulvoguttata* (Harris) [Coleoptera: Buprestidae].

HEMLOCK LOOPER *Lambdina fiscellaria fiscellaria* (Guenée) [Lepidoptera: Geometridae].

HEMLOCK SAWFLY *Neodiprion tsugae* Middleton [Hymenoptera: Diprionidae].

HEMLOCK SCALE *Abgrallaspis ithacae* (Ferris) [Hemiptera: Diaspididae].

HEMLOCK WOOLLY-ADELGID *Adelges tsugae* [Hemiptera: Adelgidae].

HEMMING, ARTHUR FRANCIS (1893–1964) An autocratic English commissioner of zoological nomenclature who was instrumental in the enactment of many poor decisions regarding the names of insects. (Riley 1964, J. Lepid. Soc. 18: 237–239.)

HEMOCYANIN See Haemocyanin.

HEMOSIS Noun. (Etymology obscure. Pl., Hemoses.) Hymenoptera: Colony fission as experienced in some Species of ants (Formicidae). See Hesmosis.

HEMP Noun. (Middle English, *hemp*. Pl., Hemps.) *Cannabis sativa* Linnaeus. An herbaceous annual originating from Asia. Fibres used in production of rope and sailcloth; flowers and leaves used as a drug. Also know as Marijuana.

HEMP SAWFLY *Trichiocampus cannabis* Xiao & Huang [Hymenoptera: Symphyta].

HEMPEL, ADOLPH American born and educated (Rawlins College, Fla; Univ Illinois) who emigrated to Brazil during 1895 to work with H. von Ihring at Museu Paulista. Hempel served as State Entomologist for Sao Paulo and was a specialist on taxonomy of scale insects, particularly Coccidae, and control of coffee pests.

HENDEL, FRIEDRICH GEORG (1874–1936) (Lindner 1936, Konowia 15: 240–247, bibliogr.)

HENDER, LESLIE (–1970) (Butler 1972, Proc. R. ent. Soc. Lond. (C) 36–61.)

HENDERSON, JOHN LOFTUS (1884–1965) (Buck 1966, Proc. S. Lond. ent. nat. Hist. Soc. 1966, 35–36.)

HENDERSON, MATTHEW (–1890) (Wright 1894, Nat. Hist Trans. Northumb. 11: 28–29.)

HENDERSON, WILLIAM WILLIAMS (1879–1944) (Tanner 1944, Gt. Basin Nat. 5: 23–24, bibliogr.)

HENLE, FRIEDRICH GUSTAV JACOB (1809–1885) (Nordenskiöld 1935, *History of Biology*. 629 pp. (397–398), London.)

HENNEBELLE, AUGUSTINE (–1857) (Hamet 1858, Apicultur 2: 97.)

HENNEGUY, LOUIS FELIX (1850–1928) (Dupont 1928, Bull Soc. ent. Fr., 1928: 33–34.)

HENNERT, CARL WILHELM (1739–1806) (Ratzeburg 1874, Forstwissenschaftliches Schriftsteller Lexicon 1: 235–237.)

HENNIG, WILLI (1913–1976) German Dipterist and systematic theoretician who forged his ideas into an operational method of analysing and interpreting genealogical relationships for groups of organisms. Hennig's approach to developing classifications is called Cladistics or Cladistical Methodology. See Cladistics.

HENNIN, DE BOUSSO-WALCOURT, DON GUY DE (–1945) (Collart 1945, Bull. Ann. Soc. ent. Belg. 31: 77.)

HENNING, JOHANN FRIEDRICH (–1831) (Kühlwin 1832, Bull. Soc. nat. Mosc. 4: 40.)

HENNINGS, OTTO (–1948) (Horion 1949, Koleopt. Z. 1: 78.)

HENOPIDAE See Acroceridae.

HENRICH, CARL (1850–1920) (Jickeli 1921, Verh. Mitt. siebenb. Ver. Naturw. 71: 47–57.)

HENRICK, GEORGE HAMILTON (1850–1939) (Anon. 1939, Arb. morph. taxon. Ent. Berl. 6: 349.)

HENRIKSEN, KARL LUDWIG (1888–1940) (Saales 1940, Suomen hyönt Aikak 6: 77–79.)

HENROT, HENRI (1913–1973) (Balazue 1973, Entomologiste 30: 178–180.)

HENSCHEL, GUSTAV A O (1835–1895) (Anon. 1895, Oesterr. Forst.-Ztg. 13: 100–102.)

HENSEL, H A (1833–1874) (Kraatz 1875, Dt. ent. Z. 19: 8.)

HENSEL, REINHOLD FRIEDRICH (1826–1881) (Martens 1882, Leopoldina 18: 19–21.)

HENSEL, WALTER (1889–1921) (Hensel 1924, Jh. Ver. schles. Insektenk. 14: 27–28.)

HENSHAW, SAMUEL (1852–1941) (Jackson 1941, Science 93: 342–343.)

HENTZ, NICHOLAS MARCELLUS (1797–1856) (Burgess 1875, Occ. Pap. Boston Soc. nat. Hist. 2: 5–13, bibliogr.)

HEPATIC Adj. (Greek, *hepar* = liver; *-ic* = of the nature of.) 1. Liver-coloured or liver-brown. 2. Descriptive of or pertaining to the liver (the Nephridia of insects.) 3. Botanical jargon for liverworts.

HEPATIC CAECA Diverticula or side chambers of the midgut. Syn. Gastric Caeca. See Caecum.

HEPATIC CELLS See Nephrocytes.

HEPATIC POUCHES Caecal Pouches.

HEPATICOLOUR Adj. (Greek, *hepar* = liver; Latin, *colour* > *colare* = to conceal.) Liver-brown.

HEPIALIDAE Latreille 1809. Plural Noun. Ghost Moths; Swift Moths. *Sensu stricto*, a cosmopolitan Family containing ca 500 Species of primitive, heteroneurous Lepidoptera which is the nominant Family of Hepialoidea. Adult body is medium to very large with wingspan 20–250 mm and head with Piliform Scales; Ocelli absent; Antenna short with flagellar segments angular and highly variable in form; Proboscis vestigial or absent; Maxillary Palpus minute and 1–2 segments; Labial Palpus short and usually with three segments; wings held tectiform at repose. Eggs are small and spherical with Chorion nearly smooth. Eclosion is achieved through splitting of Chorion, not neonate larva chewing. Females lay large numbers of eggs during lifetime, perhaps 40,000 in extreme instance. Females are unusual among Lepidoptera in apparently dropping eggs while in flight. Larvae are elongate, cylindrical and head with six Stemmata. Thorax with three pairs of legs; Abdomen with Prolegs on segments 3–6, bearing multiserial Crochets in later instars. Pupae are long, cylindrical with small Mandibles; Cremaster absent. Larvae feed in concealed places; diet is broad with some Species mycophagous and most phytophagous; larvae of many Australian Species are subterranean and feed upon roots of *Acacia* and *Eucalyptus*, hepialid larvae are used as food by aboriginal peoples. Adults do not feed. See Hepialoidea.

HEPIALOIDEA A small Superfamily of monotrysian Lepidoptera including Hepialidae, Neotheoridae (*Neotheora* of Brazil), Palaeosetidae and Prototheoridae (Anomosetidae, *Anomoses* of Australia). Adult body size is highly variable; Antenna with dense mat of cuticular projections; Ocelli and Chaetosemata absent; Mandible small, non-functional; Proboscis small or absent; wings usually with Microtrichia.

HEPNER, LEON (1915–1995) American academic (Fort Hays State College 1946–1958, Mississippi State University (1958–1980); A.B., A.M., Ph.D. (1946) University of Kansas. Medical Entomologist in Algeria and Italy 1943–45. Taxonomic specialist on Cicadellidae, especially *Erythroneura*.

HEPP, ALBERT (1883–1941) (Sachtleben 1941, Arb. morph. taxon. Ent. Berl. 8: 287.)

HEPTACHLOR A synthetic, chlorinated hydrocarbon insecticide {1, 4, 5, 6, 7, 8, 8-heptachloro,3a,4,7,7a-tetrahydro-4,7-methanoindene} used as a contact insecticide and stomach poison against ants, boll weevils, cornborers, cutworms, grasshoppers, Jap beetles, rootworms, termites, thrips, wireworms and other insects. Heptachlor developed by Velsicol Chemical Company ca 1948. Applied as a seed treatment in some countries; used as wood treatment for control of termites in some contries. Heptachlor removed from registration in USA and many developed countries. Toxic to bees and fishes; persistent in soil. Trade names include: Biarbinex®, Cupincida®, Drinox®, Fennotox®, Heptox®. See Chlorinated Hydrocarbon Insecticide; Organochlorine.

HEPTAGENIIDAE Plural Noun. A large, widespread Family of Ephemeroptera assigned to Superfamily Heptagenioidea. Forewing with Cubital Intercalaries parallel; hindwing present; posterior margin of female's ninth Sternum with a median notch. Naiad occurs on the underside of stones in streams and ponds. Syn. Ecyduridae; Ecydonuridae. See Heptagenioidea.

HEPTAGENIOIDEA A Superfamily of Ephemeroptera including Heptageniidae. See Ephemeroptera.

HEPTAPOSOGASTERIDAE Plural Noun. Family of chewing lice (Phthiraptera: Ischnocera).

HEPTENOPHOS A synthetic, organic phosphate insecticide {7-chloro-bicyclo-(3.2.0)-hepta-2,6-dien-6yl)-dimethyl phosphate} used as a contact/systemic insecticide against ticks, plant sucking insects (aphids, mealybugs, psyllids, whiteflies), fleas, lice and flies. Heptenophos applied to numerous fruit tree crops, vegetable crops, field crops, pets and livestock in some countries. Not registered for use in USA. Toxic to fishes. Trade names include Hostaquick® and Ragadan®. Cf. Organophosphate Insecticide.

HEPTOX® See Heptachlor.

HERALD® See Fenpropathrin.

HERB Noun. (Latin, *herba* = green crop. Pl., Herbs.) In a botanical sense, any seed-bearing plant which has a non-woody stem. Cf. Bush; Shrub;

HERBARIUM Noun. (Latin, *herba* = herbage; *-arium* = place of a thing. Pl., Herbaria.) A place in which dried, identified and catalogued plants and their parts are preserved for scientific study. Cf. Museum.

HERBERT, G K P (–1952) (Buxton 1954, Proc. R. ent. Soc. Lond. (C) 18: 80.)

HERBICIDE Noun. (Latin, *herba* = green crop; *caedere* = to kill. Pl., Herbicides.) A chemical compound that kills or inhibits plant development. Common herbicides include Glyphosate (Roundup® Dow), Oxyflurofen (Goal®, ICI), Simazine (Princep®, Ciba Geigy). See Mycoherbicide; Pesticide. Cf. Fungicide; Insecticide; Nematicide.

HERBIVORE Noun. (Latin, *herba* = green crop, *vorare* = to devour. Pl., Herbivores.) 1. Any animal that feeds on plant tissue. 2. Specifically, a plant feeding insect or a phytophagous insect. Syn. Phytophage. See Feeding Strategies. Cf. Carnivore; Omnivore; Saprovore. Rel. Ecology.

HERBIVOROUS Adj. (Latin, *herba* = green crop; *vorare* = to devour.) 1. Descriptive of organisms (plants or animals) that feed upon plant tissue. 2. Leaf feeding insects. See Cannibalistic; Parasitic; Predaceous. Cf. Carnivorous; Phytophagous; Saprophagous.

HERBIVORY Noun. (Latin, *herba* = green crop; *vorare* = to devour. Pl., Herbivories) 1. The act or process of being herbivorous. 2. A type of feeding strategy that requires plant tissue as a source of nutrition. See Feeding Strategies; Digestion; Extra-oral Digestion. Cf. Cannibalism; Parasitism; Predation; Necrophily. Rel. Carnivory; Phytophagy; Saprophagy.

HERBST, CARL FRIEDRICH (Geiser 1939, Field Lab. 7: 38.)

HERBST, JOHN FREDERICH WILLIAM (1743–1807) German cleric and author of *Natursystem aller bekanten in und Auslandischen Inseckten & c. von Carl Gustaf Jablonsky forgesetz von J. F. W. Herbst* (10 vols, 1785–) (Marseul 1887, Abeille 24 (Les ent. et leurs écrits): 181–187, bibliogr.)

HERBST, PAUL (1861–1927) (Porter 1929, Revta. chil. Hist. nat. 33: 77–80, bibliogr.)

HERCULES MOTH *Coscinocera hercules* (Miskin) [Lepidoptera: Saturniidae].

HEREDITARY Adj. (Latin, *hereditas* = heirship.) Inherited; genetically determined features or characteristics transmitted from parent to progeny.

HEREDITY Noun. (Latin, *hereditas* = heirship; English, *-ity* = suffix forming abstract nouns. Pl., Heredities.) The transmission of organized protoplasm and its innate capacities from parent to offspring. Inheritance from progenitors; organic resemblance based on commonness of descent (Conklin).

HEREMETABOLA Noun. With slight or incomplete Metamorphosis, but with a resting stage at the end of nymphal life. An arcane term specifically applied to the Cicadidae.

HERETIEN, FRÉDÉRIC (–1857) (Hemet 1858, Apiculteur 2: 96–97.)

HERFS, ADOLF (–1975) (Anon. 1975, Prakt. SchädlBekampf 27: 95.)

HERIARD-DUBREUIL, (–1955) (Anon. 1955, Bull. Soc. ent. Fr. 60: 34.)

HERING, EDUARD (1842–1911) (Dohrn 1911, Stettin ent. Ztg. 72: 383–384.)

HERING, EDUARD VON (1799–1881) (Rueff 1882, Jh. Ver. vaterl. Naturk. Württ. 38: 42–50.)

HERING, ERICH MARTIN (1893–1967) (Hanneman 1967, J. Lepid. Soc. 22: 123–124.)

HERING, HERMANN CONRAD WILHELM (1800–1886) (Schleich 1886, Stettin. ent. Ztg 47: 178–182.)

HERISSANT, LOUIS ANTOINE PROSPER (1745–1769) (Rose 1850, *New General Biographical Dictionary* 8: 292.)

HERKAL® See Dichlorvos.

HERKLOTS, JANUS ADRIAN (1820–1872) (Anon. 1872, NachrBl. dt. malak. Ges. 5: 89–90.)

HERMAN, OTTO (1835–1914) (Csiki 1915, Rovart. Lap. 22: 61–67, 138, bibliogr.)

HERMANN (HERMANNUS), PAUL (1646–1695) (Rose 1850, *New General Biographical Dictionary* 8: 293.)

HERMANN, JOHANN (1738–1800) (Rose 1850, *New General Biographical Dictionary* 8: 292–293.)

HERMAPHRODITE Noun. (Greek, *hermaphroditos* = combining both sexes; *-ites* = resident. Pl., Hermaphrodites.) An individual in which the characters or features of both sexes appear. Cf. Gynandromorph.

HERMAPHRODITIC Adj. (Greek, *hermaphroditos* = combining both sexes; *-ic* = of the nature of.) Of the nature of an hermaphrodite.

HERMAPHRODITISM Noun. (Greek, *hermaphroditos* = combining both sexes; English, *-ism* = condition. Pl., Hermaphroditisms.) The real or apparent condition of being a hermaphrodite. An organism being of both sexes; bisexuality in the individual.

HERMATOBATIDAE Plural Noun. A small Family (ca 10 Species) of marine heteropterous Hemiptera. Adult body ca 3–4 mm long, elongate-ova and head widened, declivant with compound eye granular and with reduced number of Ommatidia; adult with dorsal abdominal scent glands; fore tarsal claws preapical, middle and hind tarsal claws apical. See Hemiptera.

HERMINIER, FELIX LOUIS (1779–1833) (Urban 1903, Symbolae Antillanea 3: 74–75.)

HERMINIIDAE Plural Noun. A small Family of ditrysian Lepidoptera assigned to Superfamily Noctuoidea and most abundant in Neotropical Realm. Adults are small and dull-coloured; Labial Palpus large; Tibia of foreleg very short and modified in male; Metathorax beneath Tympanal Organ enlarged. Larval Integument Granular and Stellar Chalazae with Setae apically expanded. Larvae with four pairs of Prolegs and feed on dead and decaying leaves. See Noctuoidea.

HERMS, WILLIAM BRODBECK (1876–1949) American academic (University of California, Berkeley) specializing in Medical Entomology. (Furman 1949, Pan-Pacif. Ent. 25: 192.)

HERNANDEZ, FRANCISCO GARCIA (1514–1578) (Rose 1950, *New General Biographical Dictionary* 8: 295.)

HEROLD, HANS (1877–1937) (Anon. 1938, Carinthia II 128: 135–136.)

HERON, FRANCIS ARTHUR (1864–1940) (A. B. G. 1942, Entomologist 75: 23–24.)

HERRERO, PAULINE J (1865–1941) (Del Canizo 1941, Boln. Patol. veg. Ent. agric. 10: 367–368.)

HERRICH-SCHAEFFER, GOTTLIEB AUGUST WILHELM (1799–1874) German taxonomist and author of *Nomenclator Entomologicus, De generation insectorum partibusque ei inservientibus* (1821), and *Verzeichniss der europaischen Insecten; zur Erleichterung des Tauschverkehrs mit Preisen versehen* (Heft I 1835, Heft II 1844.) Friedrich Pustet. Regensburg. See: (Hofman 1874, Stettin ent. Ztg. 35: 277–284.)

HERRICH-SCHAEFFER, GUSTAV ADOLF (1836–1903) (Fürnohr 1903, Ber. naturw. Ver. Regensburg 9: 129.)

HERRICK, EDWARD CLAUDIUS (1811–1862) (Anon. 1862, Am. J. Sci. 34: 159–160.)

HERRICK, GLENN WASHINGTON (1870–1965) (Rawlins 1965, J. Econ. Ent. 58: 809–810.)

HERRING, JON LAMAR (1922–1985) American government Entomologist (Systematic Entomology Laboratory, USDA, Washington) specializing in taxonomy of aquatic Hemiptera. (Henry & Froescher 1987, Proc. ent. Soc. Wash. 89: 384–388.)

HERSILIIDAE Plural Noun. Two-tailed Spiders. A numerically small and rare Family of ecribellate spiders whose members display distinctively long Spinnerets. Hersiliids are distributed mainly in the New World tropics with one Species in North America (restricted to Texas) and a Genus in Australia. These spiders do not form webs; instead, they hunt prey on vegetation and rocks.

HERTWIG, OSCAR (1849–1922) (Weissenberg 1959, Lebensdarst dt. NatForsch 7: 1–62, bibliogr.)

HERVE, ERNEST (–1914) (Alluaud 1914, Bull Soc. ent. Fr. 1914, 496.)

HERVEY, GEORGE EDWARD ROMAINE (1894–1962) (Chapman 1963, J. Econ. Ent. 56: 424–425.)

HERZ, ALFRED OTTO (1853–1905) (Kusnezov 1905, Revue Russe Ent. 5: 311–312, bibliogr.)

HESLOP, IAN ROBERT PENICUICK (1904–1970) (Burton 1971, Proc. Bristol Nat. Soc. 32: 10–11.)

HESLOP, ROBERT (1851–1880) (Anon. 1880, Entomol. mon. Mag. 17: 71.)

HESLOP-HARRISON, GEORGE (1911–1964) (Pearson 1965, Proc. R. Ent. Soc. Lond. (C) 29: 53.)

HESMOSIS Noun. (Etymology obscure. Pl., Hesmoses.) Social Insects: The foundation of new colonies by reproductives and workers from an established colony. Syn. Colony Swarming. See Hemosis.

HESPERIIDAE Plural Noun. Skippers. A cosmopolitan Family of ditrysian Lepidoptera assigned to Hesperioidea. Adult head with short, slender scales and compound eye surrounded by row of small Ommatidia; Ocelli absent; Chaetosemata present; Antennal Scape with scale-tuft, Flagellum dilated apically, clubbed with apex typically hooked; Proboscis without scales; stalked veins not present in wings; Epiphysis present, tibial spur formula 0-0-4. Larvae with numerous short, fine, secondary Setae; constricted behind head. Larvae typically feed on monocotyledons, a few Species feed on dicotylendons. Larvae feed in silk-lined shelter of leaves or rolled leaves. Pupae are attached to larval habitat by Cremaster and Central Girdle. See Hesperioidea.

HESPERIOIDEA A Superfamily of ditrysian Lepidoptera including Hesperiidae and Megathymidae. See Lepidoptera.

HESS, RICHARD ALEXANDER (1835–1916) (Hillerick 1916, Forstwiss. ZentBl. 38: 543–546.)

HESSE, JOHAN WILHELM (–1815) (Zimsen 1964, *The Type Material of I. C. Fabricius*, 656 pp. (13), Copenhagen.)

HESSEL, SIDNEY A. (–1974) (Anon. 1974, News Lepid. Soc. 1974 (5): 9.)

HESSIAN FLY *Mayetiola destructor* (Say) [Diptera: Cecidomyiidae]: A Species endemic to the Caucasus of Russia and a serious pest of wheat, barley and rye in North America; oats are never infested and native grasses are rarely attacked. Adults are ca 3 mm long with body sooty black and Abdomen of female reddish; live ca 3–4 days; passively dispersed moderate distances (ca few miles) by wind. Eggs reddish, elongate, deposited end-to-end in lines of 2–15 on upper side of leaves; female deposits ca 250–300 eggs during lifetime; eclosion temperature-dependant (3–10 days). Neonate larva reddish; older larva white; larval stage causes all damage to host plant by feeding on the lower part of the stem; larva feeds externally between base of sheath and stem; larva does not enter stem. Typically HF produces two generations per year and overwinters as mature larvae inside puparia; pupation occurs during spring. Heavily infested plants die during winter and the straws of infested plants typically break when the heads begin to fill. Common name refers to first detection in US during 1779 when Hessian soldiers introduced the insect with straw used as bedding. See Cecidoimyiidae. Cf. Wheat Jointworm; Wheat-Stem Sawfly.

HETERO- Greek prefix meaning 'unequal' or 'different from.'

HETEROBATHMIIDAE Plural Noun. A Family of one Genus and a few Species of Lepidoptera found in temperate South America, placed within its own Suborder and Superfamily. Adults are small with wingspan about 10 mm; compound eye relatively small; external Ocelli and Chaetosemata present; Antennal Flagellum filiform; mouthparts as Micropterygidae. Epiphysis on foreleg; tibial formula 0-0-4. Wings elongate, narrow with a single layer of lamellar scales; wing coupling jugate; forewing-metathoracic locking device present. Adults are active during late winter and early spring, when they presumably feed on pollen of *Nothofagus.* Eggs are laid individually in jelly-like substance on underside of leaves of *Nothofagus;* larvae mine leaves of *Nothofagus.* Pupation occurs in soil within strong silken cocoon. See Heterobathmioidea.

HETEROBATHMIOIDEA Superfamily of Lepidoptera containing only Heterobathmiidae.

HETEROCERA Lepidoptera whose Antennae are not apically clubbed. See Lepidoptera. Cf. Rhopalocera.

HETEROCERIDAE Plural Noun. A cosmopolitan Family of polyphagous Coleoptera including about 150 nominal Species. Adult body small, pubescent and dark coloured; Antenna short and compact; Tibiae spinose laterally and tarsal formula 4-4-4. Adults and larvae inhabit mud and feed on organic matter; some adults taken at lights. See Coleoptera.

HETEROCHROMATIC (Greek, *heteros* = other; *chroma* = colour; *-ic* = of the nature of.) Of different colour. A term applied to Species which show two colour forms of one sex, one of which (homoeochrome) resembles the opposite sex. A condition seen in some Odonata and Lepidoptera.

HETEROCHROMOSOME Noun. (Greek, *heteros* = other; *chroma* = colour; *soma* = body. Pl., Heterochromosomes.) Syn. Allosome. See Chromosome.

HETEROCHRONIC Adj. (Greek, *heteros* = other; *chronos* = time; *-ic* = of the nature of.) Pertaining to Heterochrony.

HETEROCHRONY Noun. (Greek, *heteros* = other; *chronos* = time. Pl., Heterochronies.) An irregular development in point of time; a later stage becoming evident before one that is earlier in ordinary course. Heterochronic conditions create problems in phylogenetic analysis. See Paedomorphosis.

HETERODACTYL Adj. (Greek, *heteros* = other; *dactyl* = toe.) Descriptive of insects with Ungues that are different in size and shape. See Ungues. Cf. Homodactyl.

HETEROECY Noun. (Greek, *heteros* = other; *oikos* = house. Pl., Heteroecies.) The seasonal alternation of host plant Taxa as demonstrated by many Species of aphids. In heteroeceous Species the egg, Sexual and Fundatrix generations live on primary host plant Species, while the female generations live on secondary, taxonomically unrelated, host plant Species. See Aphididae.

HETEROGAMY Noun. (Greek, *heteros* = other; *gamos* = offspring. Pl., Heterogamies.) An alternation of generations which involves a sexual and a parthenogenetic generation. Heterogamy commonly is expressed in gall-forming Cynipidae (Hymenoptera). See: Heterogony. Cf. Arrhenotoky; Deuterotoky; Thelytoky.

HETEROGENEOUS Adj. (Greek, *heteros* = other; *genos* = a kind; Latin, *-osus* = with the property of.) 1. Pertaining to structure that is composed of dissimilar parts. 2. Descriptive of things differing in kind or quality.

HETEROGENESIS Noun. (Greek, *heteros* = other; *genesis* = descent. Pl., Heterogeneses.) Abiogenesis. The Alternation of Generations. Alt. Heterogeny. See Abiogenesis.

HETEROGONY Noun. (Greek, *heteros* = other; *gonos* = birth. Pl., Heterogonies.) 1. A botanical phenomenon in which flowers have stamens of several lengths. 2. The term is sometimes used interchangeably with Heterogamy to mean cyclical Parthenogenesis by insects (a mechanism of reproduction in which Sexual and Parthenogenetic generations are produced in a Species). Cf. Metagenesis.

HETEROGYNA Noun. (Greek, *heteros* = other; *gyne* = woman.) The ants; referring to the different kinds of females - queens and workers - as distinguished from males.

HETEROGYNIDAE Nagy 1969. A small Family of primitive aculeate Hymenoptera with Species found in the eastern Mediterranean, Africa and Madagascar. Sometimes placed in the Sphecidae (Sphecoidea) or the Plumariidae. Morphologically similar to Embolemidae (Chrysidoidea). Syn. Heterogynaidae.

HETEROGYNOUS Adj. (Greek, *heteros* = other; *gyne* = woman; Latin, *-osus* = with the property of.) A condition pertaining to social insects with different morphs or behaviourally different kinds of females (queens and workers).

HETEROMERA Noun. (Greek, *heteros* = other; *meros* = part. Pl., Heteromerae.) Coleoptera in which the anterior and middle Tarsi are 5-segmented and the posterior Tarsi are 4-segmented. Tarsal formula 5-5-4.

HETEROMEROUS Adj. (Greek, *heteros* = other; *meros* = part; Latin, *-osus* = with the property of.) Pertaining to insects with an unequal number of tarsal segments on the legs of a specimen. See Tarsal Formula. Cf. Homomerous.

HETEROMETABOLA Noun. (Greek, *heteros* = other; *metabole* = change.) Insects with an incomplete or direct Metamorphosis, applied to members of the lower Orders which pass through a simple Metamorphosis. There is no pupal stage and the immature insects are known as nymphs (Imms). See Metamorphosis. Cf. Ametabola; Hemimetabola; Holometabola; Paurometabola.

HETEROMORPHA Noun. (Greek, *heteros* = other; *morphe* = form. Pl., Heteromorphae.) A division of the Subclass Pterygogenea (*sensu* Brauer) which includes the Holometabola or Endopterygota.

HETEROMORPHIC Adj. (Greek, *heteros* = other; *morphe* = form; *-ic* = characterized by.) Descriptive of organisms that have different forms, shapes or coloration at distinct life-stages. Frequently the forms are profoundly different and association between stages of the same Species is difficult.

HETEROMORPHOSIS Noun. (Greek, *heteros* = other; *morphe* = shape; *-sis* = a condition or state. Pl., Heteromorphoses.) The replacement or regeneration of an organ, structure or part by a different organ, structure or part following injury or removal. Alt. Xenomorphosis. See Metamorphosis. Cf. Larval Heteromorphosis. Rel. Homoeosis.

HETEROMORPHOUS Adj. (Greek, *heteros* = other; *morphosis* = shaping; Latin, *-osus* = with the property of.) 1. Pertaining to insects with Metamorphosis complete and in abrupt stages, rendering the larva and adult dissimilar. 2. Pertaining to presumably related structures with different form, appearance or construction. See Metamorphosis. Cf. Isomorphous.

HETERONEMIIDAE Plural Noun. Common Walking Sticks. An numerically small Family of Phasmida with seven Genera distributed in USA. Adult apterous and distinctly stick-like in appearance; the longest insect in USA is *Megaphasma dentricrus* (Stål) reaching a length of 150–180 mm. See Phasmida.

HETERONEURA An Infraorder (Division Monotrysia) of Lepidoptera, the members of which display hindwing venation reduced with R2–R5 absent. See Lepidoptera.

HETERONEURIDAE See Clusiidae.

HETERONOMOUS Adj. (Greek, *heteros* = other; *nomos* = law; Latin, *-osus* = with the property of.) Pertaining to structure with two parts which, when compared with each other, are of different quality. Pertaining to comparable structures that differ in development or function.

HETERONOMY Noun. (Greek, *heteros* = other; *nomos* = law. Pl., Heteronomies.) Different development. Species in which males and females develop in fundamentally different ways. Heteronomy is seen in some parasitic Hymenoptera in which the males and females develop on different host Species. (Walter 1988).

HETEROPALPI Plural Noun. (Greek, *heteros* = other; *palpus* = soft palm of the hand.) Trichoptera: Palpi with a different number of segments in the male and the female.

HETEROPARTHENOGENESIS Noun. (Greek, *heteros* = other; *parthenos* = virgin; *genesis* = descent. Pl., Heteroparthenogeneses.) Cyclical Parthenogenesis. A condition in which generations of a population are exclusively female with a generation composed of males and females occasionally produced. A condition seen in Aphididae and Cyniipidae. Syn. Heterogony. See Parthenogenesis.

HETEROPHAGA Noun. (Greek, *heteros* = other; *phagein* = to eat.) An alternative name for the Parasitica (Hymenoptera). The term 'Heterophaga' may be used to draw attention to diverse, often non-parasitic, feeding habits of this taxonomic assemblage. See Hymenoptera.

HETEROPHYTE Noun. (Greek, *heteros* = other; *phyto* = plant. Pl., Heterophytes.) A plant that receives its food from living or dead plants or animals, or their products. See Feeding Strategies. Cf. Autotrophyte; Parasite; Saprophyte.

HETEROPLASTIC Adj. (Greek, *heteros* = other; *plastos* = molded; *-ic* = of the nature of.) 1. Pertaining to grafts that combine individuals of different Species or Genera. 2. Pertaining to grafts from one body region to another body region. See Autoplastic. Cf. Homoplastic.

HETEROPODIDAE Plural Noun. Huntsmen. Giant Crab Spider. A Family of hunting spiders resembling thomisid crab spiders. Heteropodids occur in Australia, the Neotropics and the southern states of the USA. Species live under bark, in crevices and on plant foliage.

HETEROPTERA Latreille 1810. (Permian-Recent.) (Greek, *heteros* = other; *pteron* = wing.) A numerically large, cosmopolitan Suborder of Hemiptera. Adult's head is proganthous with Clypeus reduced; Antenna with up to five segments, Flagellum not aristate; Pedicel not heavily invested with Sensilla; Labium distantly removed from Prosternum and separated by Gula; two Ocelli at most. Thoracic Scent Glands well developed and dorsal in nymph, ventral in adult; anterior wings (Hemelytra) held flat over Abdomen, apically overlapping with base sclerotized forming a Corium and apex membranous; Tarsi with up to three segments. Phytophagous Species feed on parenchyma; predaceous Species feed on homogenized tissue and Haemolymph; aquatic Species are common with legs adapted for predation and swimming. See Hemiptera.

HETEROPTEROUS Adj. (Greek, *heteros* = other; *pteron* = wing; Latin, *-osus* = with the property of.) Descriptive of wings with different texture in different parts; pertaining to the Order Heteroptera.

HETERORHABDITIS SPP. See Entomogenous Nematodes.

HETEROSIS Noun. (Greek, *heteros* = other; *-sis* = a condition or state. Pl., Heteroses.) See Hybrid Vigor.

HETEROSYNTHESIS. Noun. (Greek, *heteros* = other; *syntithenai* = to put together. Pl., Heterosyntheses.) The condition in which an Oocyte receives yolk from an extra-ovarian source. Heterosynthesis is regarded as a derived mechanism of supplying yolk to ovarian Follicles. Cf. Autosynthesis. Rel. Nutrition.

HETEROTHRIPIDAE Plural Noun. Heterothripids. A numerically small Family (70 Species in four Genera) of Thysanoptera assigned to Suborder Terebrantia. Two Genera occur in the USA (*Heterothrips, Oligothrips)* and most Species feed in flowers. Antennae with nine segments and sensoria occurring as a small band around the third and fourth segments, or as a blunt cone. See Thysanoptera.

HETEROTROPH Noun. (Greek, *heteros* = other; *trophe* = nourishment. Pl., Heterotrophs.) A organism that cannot use CO_2 or carbonates and inorganic nitrogen compounds for metabolic synthesis. *e.g.* Animals and most plants which cannot photosynthesize. Cf. Autotroph.

HETEROTROPHIC Adj. (Greek, *heteros* = other; *trophe* = nourishment; *-ic* = of the nature of.) Pertaining to heterotrophs.

HETEROTYPICAL Adj. (Greek, *heteros* = other; *typos* = pattern; *-alis* = appropriate to, characterized by.) Descriptive of a Species in a Genus, including more than one Species, all Species of which differ in essential features of structure. Cf. Polytypic.

HETSCHKO, ALFRED (1854–1933) (Heikertinger 1933, Wien. ent. Ztg. 50: 1–5, bibliogr.)

HEURISTIC Adj. (Greek, *heuriskein* = to discover; *-ic* = characterized by.) Descriptive of something providing aid in the solution of a problem but itself not justified or incapable of justification. A term applied to problem-solving techniques that utilze self-education as a feed-back mechanism. Cf. Empirical.

HEURN, WILLEM CONELIS VAN (1887–1972) (Holthuis & Hussou 1973, Zool. Bijdr. 16: 1–67, bibliogr.)

HEWITSON, WILLIAM CHAPMAN (1806–1878) (Embleton 1880. Trans. nat. Hist. Soc. Northumb. 7: 223–235, bibliogr.)

HEWITT, CHARLES GORDON (1885–1920) Canadian medical Entomologist who significantly contributed to scientific understanding of the housefly. (Gibson *et al.* 1920, Can. Ent. 52: 97–105, bibliogr.)

HEWITT, JOHN (1881–1961) (Varley 1962, Proc. R. ent. Soc. Lond. (C) 26: 53.)

HEXACHAETOUS Adj. (Greek, *hex* = six; *chaite* = hair; Latin, *-osus* = with the property of.) Diptera in which the mouth structures possess six piercing Setae.

HEXADRIN® See Endrin.

HEXAFLUMURON A urea compound {N-(((3,5-dichloro-4-(1,1,2,2-tetrafluoroethoxy) phenyl) amino) carbonyl)-2,6-difluoro benzamide} used as an Insect Growth Regulator and stomach poison against Coleoptera, Diptera, Lepidoptera and Isoptera. Hexaflumuron has some ovicidal activity. Compound is applied to fruit crops, forestry, potatoes and cotton in some countries. Experimental registration in USA for control of structural pests. Trade names include: Carpocapsa®, Consul®, Consult®, MAF-46®, Redale®, Sentricon®, Trueno®. See Insect Growth Regulator.

HEXAMETHYLDISILIZANE (HMDS) A chemical used to prepare soft insect tissue for scanning electron microscopy. (Nation 1983. Int. J. Ins. Morphol. Embryol. 12: 201–210.) Cf. Osmium Tetroxide.

HEXANEPHRIC Adj. (Greek, *hex* = six; *nephros* = kidney; *-ic* = consisting of.) Organisms with six kidneys, or excretory structures used as kidneys. Rel. Malpighian Tubule.

HEXAPOD Adj. (Greek, *hex* = six; *pous* = foot.) Any six-legged arthropod. A term often intended to mean an insect.

HEXAPODA Noun. (Greek, *hex* = six; *pous* = foot.) (Devonian-Recent.) A Superclass of Arthropoda and sister-group of the Myriapoda. Both groups possess Malpighian Tubules and Tömösvary Organs but have lost the mandibular Palpus. Hexapods are Tracheate arthropods with head, Thorax and Abdomen forming distinct Tagmata and six legs (one pair on each thoracic segment) in the adult stage. Included Taxa: Collembola, Protura, Diplura, Archaeognatha, Thysanura and Pterygota.

HEXAPODOUS Adj. (Greek, *hex* = six; *pous* = foot; Latin, *-osus* = with the property of.) Pertaining to six-legged arthropods. Alt. Hexapodal; Hexapodus.

HEXASULFAN® See Endosulfan.

HEXAVIN® See Carbaryl.

HEXHAM GREY See Scotch Grey.

HEXYTHIAZOX A carboxamide compound {4-RS, 5-RS-5-(4-chlorophenyl)-N-cyclohexyl-4-methyl-2 oxothiazolidinone-3-carboxamide} used as a contact and stomach poison for mites. Hexythiazox is applied to tree crops, citrus, coffee, grapes, rice, strawberries, tea, vegetables, ornamentals in some countries. Registered on pears in USA. Compound is not effective on adult mites but does show ovicidal activity. Trade names include Acariflor®, Acorit®, Calibre®, Cesar®, Metacar®, Nissolan®, Nissorun®, Salibre®, Savey®, Stopper®, Zeldox®. See Carbamate Insecticide.

HEYDEN, CARL HEINRICH GEORG VON (1793–1866) (Kirschbaum 1866, Jb. nassau. Ver. Naturk 19–20: 511–516.)

HEYDEN, LUCAS FRIEDRICH JULIUS DOMINICUS VON (1838–1915) (Reitter 1908, Ent. Bl. Biol. Syst. Käfer 4: 85–88.)

HEYDENREICH, GUSTAV HEINRICH (–1897) (Anon. 1897, Leopoldina 33: 113.)

HEYMONS, ALBRECHT (–1960) (Anon. 1960, Mitt. dt. ent. Ges. 19: 94.)

HEYMONS, RICHARD (1867–1943) (Anon. 1941, Biologie 10: 417–418.)

HEYNE, ALEXANDER (1869–1927) (Korchefsky 1929, Dt. ent. Z. 1929: 73–74, bibliogr.)

HIANS Adj. (Latin, *hiare* = to gape.) Gaping.

HIATUS Noun. (Latin, *hiare* = to gape.) Any large, conspicuous gap or opening.

HIBERNACULUM Noun. (Latin, *hibernaculum* = winter quarters. Pl., Hibernacula.) 1. A tent-like structure or sheath constructed from a leaf or other material, in which a larva hides or hibernates. 2. Any place of hibernation or extended refugium.

HIBERNATE Noun. (Latin, *hibernus* = wintry. Pl., Hibernates.) To pass the winter in an arrested physiological condition. See Hibernation. Cf. Aestivate.

HIBERNATION Noun. (Latin, *hibernus* = wintry; English, *-tion* = result of an action. Pl., Hibernations.) 1. Winter Dormancy. 2. A period of suspended development in organisms that occurs during seasonally low temperatures. 3. Inactivity in organisms during winter in temperate climates. 4. A physiological condition of growth retardation or arrest, primarily designed to overcome lower than optimum temperatures (Mansingh 1971, Canad. Entomol. 103: 991). See Development. Cf. Aestivation; Athermopause; Hypnody. Rel. Diapause; Oligopause; Quiescence.

HIBISCUS FLOWER BEETLE *Aethina* (*Olliffura*) *concolor* (Macleay) [Coleoptera: Nitidulidae].

HIBISCUS LEAFMINER *Philodoria hibiscella* (Swezey) [Lepidoptera: Gracillariidae].

HIBISCUS MEALYBUG 1. *Nipaecoccus viridis* (Newstead) [Hemiptera: Pseudococcidae]. 2. *Maconellicoccus hirsutus* (Green) [Hemiptera: Pseudococcidae]: A mealybug endemic to the Orient and a pest in Egypt and Australia. Female oval, yellow-orange and covered in white wax. Mealybugs do not resemble other insects and lack apparent body segmentation. Legs and Antenna are only visible with a hand lens. Adults feed by placing mouthparts into a plant and sucking sap. Adults are sedentary and feed in one place. Male are rarely found. Females lay small round, yellow eggs into wax secreted from body. First instar nymphs are called 'crawlers' and are an active stage in life-cycle. Crawlers move from wax mass to locate a place to feed. Once settled, crawlers moult into sedentary nymphs. Nymphs resemble adults in appearance, but are smaller. HMB honeydew produces sticky cotton. See Pseudococcidae. Cf. Pink Hibiscus Mealybug; Stripped Mealybug.

HIBISCUS WHITEFLY *Singhius hibisci* (Kotinsky) [Hemiptera: Aleyrodidae].

HIBROM® See Naled.

HICKORY BARK-BEETLE *Scolytus quadraspinosus* Say [Coleoptera: Scolytidae]: A monovoltine, significant pest of Hickory in the eastern USA, and also attacks pecan and butternut. Adults are 2–4 mm long, dark brown-black; male head and Pronotum fringed with long Setae; Pronotum finely punctate; Elytra striae coarsely punctate, intervals finely punctate. Eggs are laid in niches along sides of vertical galleries parallel with the wood's grain. Each larva constucts its personal tunnel perpendicular to gallery. HBB overwinters as mature larvae in cells at end of tunnels; pupation occurs during spring. See Scolytidae. Cf. Fir Engraver; Shothole Borer; Smaller European-Elm Bark Beetle. Rel. Engravers.

HICKORY HORNED-DEVIL *Citheronia regalis* (Fabricius) [Lepidoptera: Saturniidae]. Syn. Regal Moth.

HICKORY LEAFROLLER *Argyrotaenia juglandana* (Fernald) [Lepidoptera: Tortricidae].

HICKORY PLANT-BUG *Lygocoris caryae* (Knight) [Hemiptera: Miridae].

HICKORY SHUCKWORM *Cydia caryana* (Fitch) [Lepidoptera: Tortricidae]: A serious pest of pecan and hickory in southern USA. HSW larvae create long tunnels in shuck which prevent separation of nut. HSW overwinters as mature larvae in shucks. See Tortricidae.

HICKORY TUSSOCK-MOTH *Lophocampa caryae* (Harris) [Lepidoptera: Arctiidae].

HICKS, CHARLES HENRY (–1941) (Anon. 1942, Ann. ent. Soc. Am. 35: 124.)

HICKS' BOTTLE Hymenoptera: A campaniform Sensilla which is a flask-shaped pit or depression on antennal Flagellomeres of bees and ants. HB was regarded as an organ of hearing by early workers. See Organ of Hicks. Cf. Organ of vom Rath.

HICO-CCL® See Chlormequat.

HIDE Noun. (Middle English, *hide* > Anglo Saxon, *Hyd* > Latin, *cutis* > Greek, *skytos* = skin. Pl., Hides.) 1. The skin or tough outer covering of large animals. Cf. Integument; Pelt; Skin. 2. A measure of land in the Domesday Book, ca 120 acres.

HIDE BEETLE *Dermestes ater* DeGeer [Coleoptera: Dermestidae]: A cosmopolitan larval pest of bones, carcasses, wool, wood and cork. HB is endemic to Americas and prefers food with high protein content; HB is an occasional structural pest of homes; softwoods are selected as site for pupation, not as source of food. Adults are 7–9 mm long and uniformly dark; Elytron without apical serrations or spines at sutural angle.

Eggs are laid singly or in small clusters in concealed habitats; incubation requries 2–6 days and is not affected by humidity. Mature larvae ca 15 mm long and very setose. Larval instar number is variable, but increases with lowering of temperature and food quality. Larval stage requires 35–238 days, and is influenced by humidity. Pupation occurs within chambers constructed by mature larvae; pupation site often creates economic damage. Pupal stage requires 5–30 days, but is not affected by humidity. Adults mate upon emergence and complete a preoviposition period lasting 7–13 days. Fecundity is correlated with mating, temperature, food, water. Females lay ca 300–400 eggs. Adult stage 42–48 days and life cycle complete within 2–9 months. HB reputedly found on Egyptian mummies in 1834 by F. W. Hope. See Dermestidae. Cf. Common Hide Beetle; Larder Beetle.

HIEROGLYPHIC Adj. (Greek, *hieros* = sacred; *glyphein* = to carve; *-ic* = consisting of.) Marked with a pattern of characters resembling hieroglyphics. Alt. Hieroglyphiicus.

HIERONYMUS, GEORGE HANS ENIO WOLFGANG (1846–1921) (Trotter 1920, Marcellia 19: xxxv.)

HIEROXESTIDAE See Tineidae.

HIGGINS, EDMUND THOMAS (Anon. 1880, Bull. Soc. ent. Fr. (5) 10: xcvii.)

HIGH PLAINS GRASSHOPPER *Dissosteira longipennis* (Thomas) [Orthoptera: Acrididae]: A Species found at elevations of 4,000–6,000 feet in Colorado, New Mexico, Oklahoma and Texas. HPG pod contains about 65 eggs; pods are massed in 'egg beds'. Nymphs with 5–6 instars; adults become gregarious. HPG occasionally becomes a serious pest of range grasses. See Acrididae.

HILCRON® See Monocrotophos.

HILDAN® See Endosulfan.

HILDRETH, SAMUEL PRESCOTT (1783–1863) (Weiss 1936, *Pioneer Century of American Entomology*. 320 pp. (115), New Brunswick.)

HILFOL® See Dicofol.

HILL Noun. (Middle English, *hil*. Pl., Hills.) The dome-shaped mound or crater surmounting a Formicary; perforated with cavities which serve as incubators for immature stages of ants (Wheeler).

HILL, JOHN (1716–1775) (Rose 1850, *New General Biographical Dictionary* 8: 321–322.)

HILL, JOHN (1843–1916) (Anon. 1916, Entomologist 49: 96.)

HILL, RICHARD (1795–1782) (Weiss 1936, *Pioneer Century of American Entomology*. 320 pp. (237–238), New Brunswick.)

HILL, WILLIAM W (1833–1888) (Smith 1888, Entomologica Am. 3: 236.)

HILLER, MAX (1900–1970) (Tauber 1970, Ent. NachrBl. 17: 158.)

HILLER, REINHOLD (–1903) (Kraatz 1904, Dt. ent. Z. 48: 7.)

HILLER, ROBERT (–1941) (Sachtleben 1942. Arb. morph. taxon. Ent. Berl. 9: 65.)

HILL'S BROWN BLOWFLY *Calliphora hilli* Patton [Diptera: Calliphoridae].

HILPMANN, GEORG (1830–1888) (Eisinger 1888, Ent. Z. Frankf. a. M. 2: 94.)

HIMALAYAN HAEMORRHAGIC DISEASE A presumed tick-borne disease that first became apparent among Indian soldiers during 1962.

HIMSL, FERDINAND (1868?-1907) (Christle 1908, Z. wien. ent. Ges. 43: 187–206.)

HINCKS, WALTER DOUGLAS (1906–1961) (Steele 1964, Coleopt. Cat. (Suppl.) 35 (3): lii–iv.)

HIND ANGLE 1. Lepidoptera: The point where forewing inner and outer margins meet. 2. Anal angle of the hindwing.

HIND BODY See Abdomen.

HIND HEAD Mallophaga: Part of head behind the Mandible and Antenna.

HIND INTESTINE. Syn: Hind Gut.

HINDGUT Noun. (Old English, *hinder* = behind; Old English, *gyte*. Pl., Hindguts.) The hindgut is the posterior region of the Alimentary System, attached to midgut at Pyloric Valve and typically confined to the Abdomen. Hindgut and foregut both are lined with Integument but hindgut Integument is thinner than corresponding layer of foregut, or modified in other ways. Typically, hindgut musculature is poorly developed; when present, longitudinal muscles are external to circular muscles and resemble the condition in midgut. Hindgut is responsible for excretion of waste products and conservation of water. The primary excretory product is Uric Acid (NH). HG includes Malpighian Tubules and Anal Glands. GH is divided into several anatomical regions: Ileum, Colon, Rectum and Anus. Syn. Proctotaeum. See Alimentary System. Cf. Foregut; Midgut. Rel. Digestion.

HINDLE, EDWARD (1886–1973) (Garnham 1974, Biogr. Mem. Fellows. R. Soc. 20: 217–234, bibliogr.)

HINDS, WARREN ELMER (1876–1936) (Thomas 1936, J. Econ. Ent. 29: 225–226; Bailey 1948, Fla. Ent. 31: 43–44, bibliogr. only.)

HINDWING Noun. (Middle English *hinder*, *winge* > Old Norse, *vaengr*. Pl., Hindwings.) The posterior wing of an insect, attached to the Metathorax. Some insects have profoundly modified hindwings (*e.g.* Haltere of Diptera). See Wing; Posterior Wing. Cf. Forewing.

HINE, JAMES STEWART (1866–1930) (Kennedy 1931, Ent. News 42: 177–180.)

HINGE FLIES See Gobryidae.

HINGE OF MAXILLA 1. The Cardo. 2. The point of articulation of any moveable joint. See Maxilla.

HINGSTON, RICHARD WILLIAM GEORGE (1887–1966) (Pearson 1967, Proc. R. ent. Soc. Lond. (C) 31: 62.)

HINKLEY, HOLMES (–1891) (Anon. 1891, Psyche 6: 60.)

HINTON, HOWARD EVEREST (1912–1977) English academic (University of Bristol) who was born in Mexico as son of mining engineer and completed a tertiary education in United States (UC Berkeley). Hinton was an exceptionally prolific writer in physiology, evolution and functional morphology. Major works include a three volume treatise on the insect egg.

HIPPO FLIES Specifically, members of the Genus *Tabanus* with eyes unmodified or with horizontal bands and pale wings. See Tabanidae.

HIPPOBOSCIDAE Plural Noun. Bird Ticks; Flat Flies; Keds; Louse Flies; Tick Flies; Wallaby Flies. A Family of ca 200 Species of blood-sucking muscoid Diptera. Adults ca 5–6 mm long, with flattened, leathery body; mouthparts porrect; wing with well developed veins in anterior region; legs robust; tarsal claws long and recurved. Adults live as ectoparasites on birds and mammals. Females larviposit mature larvae that are ready to pupate, usually away from host. See Diptera; Sheep Ked. Cf. Nycteribiidae.

HIPS See Coxa.

HIRAYAMA, SHUJIRO (1887–1954) (Hasegawa 1967, Kontyû 35 (Suppl.) 70–71, bibliogr. pl. 4 fig. 8.)

HIRST, ARTHUR STANLEY (1883–1930) (Anon. 1930, Science 72: 7.)

HIRST'S MARSUPIAL TICK *Ixodes hirsti* Hassall (Acari: Ixodidae].

HIRSUTE Adj. (Latin, *hirsutus* = shaggy.) Clothed with long, strong, shaggy Setae.

HIRSUTISCULUS Adj. (Latin, *hirsutus* = shaggy; *sicula* dim. of *sica* = dagger.) Somewhat hairy, entirely or in part. See Hirsute. Rel. Vestitute.

HISPID Adj. (Latin, *hispidus* = rough.) Pertaining to structure bristly or sparsely set with short, stiff Setae.

HISTER BEETLE *Hister nomas* Erichson [Coleoptera: Histeridae].

HISTER BEETLES See Histeridae.

HISTERIDAE Plural Noun. Hister Beetles. A cosmopolitan Family of polyphagous Coleoptera containing about 3,500 nominal Species in about 260 Genera. Histeridae occupies a central position in Histeroidea, or is assigned to Superfamily Hydrophiloidea. Adult body 1–15 mm long, round to oval, compact and polished; Antenna with 8–11 segments, short, geniculate, clubbed. Head deeply inserted; Pronotum keeled; Elytra abbreviated with 1–2 exposed apical abdominal Terga; fore Coxa transverse; fore Tibia spinose or dentate; tarsal formula 5-5-5, rarely 5-5-4. Abdomen with five Ventrites. Larvae are campodeiform and superficially resemble carabid larvae; legs with four segments and one tarsal claw on each leg; Urogomphi with two segments. Histerids are exclusively predaceous as adults and larvae, but occupy diverse habitats. Some Species are termitophilous, some Species are myrmecophilous, some Species are associated with carrion, and some Species are found in and around animal waste. Dorsoventrally compressed histerids are taken beneath bark of dying or dead trees; other histerids are found in nests of various warm and cold blooded vertebrates, including birds, gophers and foxes. See Histeroidea.

HISTEROIDEA A Superfamily of polyphagous Coleoptera including Histeridae, Sphaeritidae and Synteliidae. See Coleoptera.

HISTOBLAST Noun. (Greek, *histos* = tissue; *blastos* = bud. Pl., Histoblasts.) 1. A small group of cells not affected by the Histolysis of old tissue (Comstock). 2. The morphological unit or cell characteristic of a particular tissue (Smith). 3. The imaginal bud in the Histolysis of the larval structures during pupation. 4. Imaginal disc.

HISTOBLASTIC Adj. (Greek, *histos* = tissue; *blastos* = bud; *-ic* = characterized by.) 1. Descriptive of or pertaining to Histoblasts. 2. Of the nature of Histoblasts or Imaginal Buds.

HISTOGENESIS Noun. (Greek, *histos* = tissue; *genesis* = descent. Pl., Histogeneses.) 1. The development and formation of tissue. 2. Holometabolous insects: The formation and development of adult tissue from histolysed larval organs during the pupal period. Among holometabolous insects, Histogenesis involves the formation and development of adult tissue and internal organs from histolysed larval organs during the pupal period. See Development.

HISTOGENETIC Adj. (Greek, *histos* = tissue; *genes* = producing; *-ic* = characterized by.) Descriptive of or pertaining to Histogenesis.

HISTOLOGICAL Adj. (Greek, *histos* = tissue; *logos* = discourse; *-alis* = characterized by.) Descriptive of or relating to Histology. See Histology.

HISTOLOGY Noun. (Greek, *histos* = tissue; *logos* = discourse. Pl., Histologies.) The study of an organism's tissues at the cellular and subcellular levels. At the tissue level of organization, Histology involves two phases: Histolysis and Histogenesis. Tissue preparation involves fixation, embedding, sectioning and staining. See Morphology. Cf. Organography; Histogenesis; Histolysis. Rel. Cytology; Ultratructure.

HISTOLYSIS Noun. (Greek; *histos* = tissue; *lyein* = to dissolve; *-sis* = a condition or state. Pl., Histolyses.) The act, process or phenomenon of degeneration and dissolution of tissue (internal structure) by action of various cellular components of Haemolymph in a deliberate, programmed manner. See Metamorphosis. Cf. Histogenesis. Rel. Bacteriolysis.

HISTOLYTIC Adj. (Greek; *histos* = tissue; *lyein* = to dissolve; *-ic* = characterized by.) Descriptive of or pertaining to Histolysis; concerning the nature of Histolysis.

HITCHCOCK, EDWARD (1793–1864) (Weiss 1936, *Pioneer Century of American Entomology*. 320 pp. (129), New Brunswick.)

HIVE Noun. (Middle English, *hive*. Pl., Hives.) A man-made structure which serves as the home of a honey bee colony. See Honey Bee.

HIVE ODOUR Hymenoptera: Unique nest odours.

HLISNIKOVSKY, JOSEF (1905–1972) (Jelinck 1973, Acta ent. bohemoslovaca 70: 430–432, bibliogr.)

HNATECK, JOHANN SAMUEL (1801–1880) (Frey 1880, Mitt. schweiz. ent. Ges. 5: 557–560.)

H-NPV® See Heliothis NPV.

HOARY (Obs.) Adj. Greyish-white.

HOBART BROWN *Argynnina hobartia* (Westwood) [Lepidoptera: Nymphalidae].

HOBBY, BERTRAM MAURICE (–1983) English academic, specialist in insect biology and editor of Entomologist's Monthly Magazine (1964–1981).

HOBSON, R P (Thomas 1941, Nature 147: 793.)

HOC TEMPORE Latin phrase meaning 'at this time.'

HOCH, KARL (Horion 1965, Ent. Bl. Biol. Syst. Käfer 61: 129–133; Cymorek 1966, Ent. Bl. Biol. Syst. Käfer 62: 65–67.)

HOCHHUTCH, JOHANN HEINRICH (–1872) (Marseul 1887, Abeille 24 (Les ent. et leurs écrits): 192, bibliogr.)

HOCHSTEIN, ANTON (1829–1911) (Weiss *et al.* 1925, Ent. News 38: 1–4.)

HOCKEN, MELVILLE (1888–1949) (Morley 1949, Nature 163: 792.)

HOCKING, BRIAN (1914–1974) (Anon. 1974, Bull Ent. Soc. Can. 6: 61; Lees 1975, Proc. R. ent. Soc. Lond. (C) 39: 55–56.)

HODGKINSON, JAMES B. (1823–1897) (Anon. 1893, Brit. Nat. 3: 101–103.)

HODGSON, GERALD GEORGE (1861–1911) (Anon. 1911, Entomol. mon. Mag. 47: 72–73.)

HODOTERMITIDAE Plural Noun. Dampwood Termites. A Family of Isoptera which lacks a worker caste and individuals lack a Fontanelle. Hodotermitids typically are found in dead wood. They do not require contact with the soil, but do require moisture. See Isoptera.

HOE 084498® See Silafluofen.

HOEFNAGEL, GEORGE (1545–1617?) (Felts 1854. Bull. Acad. r. Belg. 21: 978–1012.)

HOEFNAGEL, JORIS (Ca 1600–1650) Renaissance painter; *Diversae insectorum volatilium icones ad vivum depictae* (1630–1646) depicts his better illustrations. Also, *Mira Calligraphia Monumenta* (Model Book of Calligraphy), inscribed by Georg Bocska, 1561–62, Illuminated by Hoefnagel for Emperor Rudolph II ca 1590–96, Ms. 20, fol 7. 86 MV, 572. (Wilson 1835. A Treatise on Insects, p. 5.)

HOEG, NEILS (1876–1951) (Hansen 1949, Ent. Meddr. 25: 459–462, bibliogr.)

HOFENEDER, KARL (1878–1951) (Strouhal 1954, Annln naturh. Mus. Wien 60: 8.)

HOFFER, EDUARD (1841–1915) (Günter 1961, Mitt. naturw. Ver. Steierm. 52: 1–12, bibliogr.)

HOFFMAN, M A (–1967) (Anon. 1967, Bull. Soc. ent. Fr. 72: 285.)

HOFFMAN, WILLIAM ALBERT (1894–1943) (Rozeboom 1943, J. Parasit. 29: 301–302.)

HOFFMANN, ADOLPHE (1899–1967) (Balachowsky 1967, Bull. Soc. ent. Fr. 73: 156–165, bibliogr.)

HOFFMANN, CARLOS CHRISTIAN (1876–1942) (Hoffmann 1962, An. Inst. Biol. Univ. Méx. 33: 387–398, bibliogr.)

HOFFMANN, PAUL (1850–1921) (Caliess 1922, Int. ent. Z. 15: 185–188.)

HOFFMANN, R WOLFGANG (–1940) (Sachtleben 1942, Arb. morph. taxon. Ent. Berl. 9: 134.)

HOFFMANSEGG, JOHANN CENTURIUS (1766–1849) (Lichtenstein 1850, Protokoll Ges. Natur. Heilk. Dresden 1849: 5–9.)

HOFMAN, KARL (–1926) (Oches 1926, Ent. Bl. Biol. Syst. Käfer 22: 192.)

HOFMANN, CHRISTOPH (1907–1942) (Escherich 1942, Z. angew. Ent. 30: 153–156, bibliogr.)

HOFMANN, ERNST (1837–1892) Curator of Entomology at Staatliches Museum fur Naturkunde, Stuttgart. (Hofmann 1892, Dt. ent. Z. Iris 5: 459–463.)

HOFMANN, FRIEDRICH (1798–1870) (Hofmann 1871, Stettin. ent. Ztg 31: 264.)

HOFMANN, OTTMAR (1835–1900) (Escherich 1900, Illte Z. Ent. 5: 140–143, bibliogr.)

HÖFNER, GABRIEL (1842–1921) (J. T. 1921, Ent. Z. Frankf. a. M. 35: 9.)

HOG FOLLICLE MITE *Demodex phylloides* Csokor [Acari: Demodicidae].

HOG LOUSE *Haematopinus suis* (Linnaeus) [Anoplura: Haematopinidae]: The largest Species of blood-sucking lice. HL is resident in North America and typically found during winter on ears and folds of skin. See Haematopinidae.

HOGARTH, A M (1877–1947) (Proc. R. ent. Soc. Lond. (C) 12: 64.)

HÖGE, C F (Weidner 1967, Jbh. Verh. naturw. Ver. Hamburg; Suppl. 9: 178–179.)

HOGENTOBLER, HANS ST GALLEN (Wädenswil 1967, Ent. Bl. Biol. Syst. Käfer 63: 123)

HOGG, HENRY ROUGHTON (1850–1923) (Bonnet 1945, *Bibliographia Araneorum* 1: 51.)

HOLARCTIC REALM A faunal region comprising the whole of Europe, Northern Africa as far as the Sahara, Asia, down to the Himalaya Mountains and North America down to Mexico. See Zoogeography. Cf. Palaearctic; Nearctic.

HOLARCTIC Adj. (Greek, *holos* = whole; *arktos* = great bear; *-ic* = consisting of.) Descriptive of or pertaining to the Holarctic Realm.

HOLDAWAY, FREDERICK GEORGE (1902–1965) Australian academic (University of Adelaide), subsequently government Entomologist (CSIRO), and student of Lepidoptera. (Hodson 1965, J. Econ. Ent. 58: 594.)

HOLDEM® A combination of Ethoprop and Phorate used in granular form for the control of soil insect pests and nematodes attacking potatoes. See Ethoprop.

HOLDHAUS, KARL (1883–1975) (Anon. 1976, Alm. öst. Akad. Wiss. 1975: 77.)

HOLDUP® See Chlormequat.

HOLIDIC DIET An artificial diet which consists entirely of known chemicals. HD theoretically possible and conceptually ideal but practically impossible owing to contaminants. Many diets claim holidic status based on known ingredients, but do not account for unknown contaminants. See Artificial Diet. Cf. Meridic Diet; Oligidic Diet.

HOLL, NICOLAS JOSEF EUGENE (1855–) (Oberthür 1916, Etudes de Lépidopterologie comparée 11: (Portraits de lépidoptèristes. 3me Sér): (6), 1858–1861: 94–96.)

HOLLAND, GEORGE PEARSON (1911–1985) Canadian Entomologist specializing in Siphonaptera and Director of Canadian Entomology Research Institute. (Arnaud & Leech, 1986. Pan-Pac. Ent. 62: 167.)

HOLLAND, WILLIAM (1845–1930) English shoemaker and amateur Entomologist who was employed by Hope Museum (Oxford) as an assistant (1893–1913). (J. E. T. Entomologist 63: 263–264.)

HOLLAND, WILLIAM JACOB (1848–1932) (Gundner 1929, Ent. News 40: 205–217.)

HOLLAR, (HOLLARD), WENZEL (1607–1677) (Rose 1850, *New General Biographical Dictionary* 8: 347–348.)

HÖLLMER, FRANZ (1820–1885) (Dimmock 1888, Psyche 5: 36.)

HOLLOWAY, JAMES KEEVER (1900–1964) (Haeussler 1965, J. Econ. Ent. 58: 808–809.)

HOLLRUNG, MAX (1858–1937) (Anon. 1927, Arb. physiol. angew. Ent. Berl. 4: 246.)

HOLLY LEAFMINER *Phytomyza ilicis* Curtis [Diptera: Agromyzidae].

HOLLY SCALE *Dynaspidiotus britannicus* (Newstead) [Hemiptera: Diaspididae].

HOLLYHOCK PLANT-BUG *Brooksetta althaea* (Hussey) [Hemiptera: Miridae].

HOLLYHOCK WEEVIL *Apion longirostre* Oliver [Coleoptera: Curculionidae].

HOLMBERG, EDUARDO LADISLAO (1852–1937) (Bonnet 1945, *Bibliographia Araneorum* 1: 46–47.)

HOLMBOE, FREDERICK VOGT (–1954) (L. R. N. 1954, Ent. Tidskr. 9: 263–264.)

HOLMERZ, CONRAD GEORG GOTTFRID (1839–1907) (Trybonn 1907, Ent. Tidskr. 28: 119–121.)

HOLMGREN, AUGUSTE EMIL (1829–1888) (Spångberg 1889, Ent. Tidskr. 10: 165–174, bibliogr.)

HOLOBLASTIC DIVISION The type of cleavage in which the entire egg is divided (Snodgrass).

HOLOCRINE Adj. (Greek, *holos* = entire; *krinein* = to separate.) Descriptive of or pertaining to holocriny.

HOLOCRINY Noun. (Greek, *holos* = entire; *krinein* = to separate. Pl., Holocrinies.) The passage of an enzyme into the lumen of a gut or salivary duct through disintegration of the parent cell.

HOLOCYCLIC SPECIES Aphididae: Species in which the winter host plants are relatively few, reproduction is always sexual and does not migrate to summer host plants (Wardle).

HOLOGASTRAN Adj. Coleoptera: A type of Abdomen in which the second Sternum is the first continuous Sternum across the Abdomen. Typical of Coleoptera such as Cantharidae. Cf. Cryptogastran; Haplogastran.

HOLOGNATHA Noun. (Greek, *holos* = entire; *gnathos* = jaw. Pl., Holognathae.) A Suborder of Plecoptera containing the Families Grypoteridae, Capniidae and Nemouridae. See Plecoptera.

HOLOMETABOLA Noun. (Greek, *holos* = entire; *metabole* = change.) The higher pterygote insects which pass through a complete and complex Metamorphosis and which have an egg, larva, Pupae, and adult (Imago) stage. A taxonomic equivalent of the Endopterygota. Representative Taxa include male Coccoidea, Megaloptera, Neuroptera, Mecoptera, Diptera, Siphonaptera, Trichoptera, Lepidoptera, Coleoptera, Strepsiptera, Hymenoptera. Cf. Ametabola; Paurometabola; Hemimetabola.

HOLOMETABOLAN Adj. (Greek, *holos* = entire; *metabole* = change.) A holometabolous insect.

HOLOMETABOLOUS Adj. (Greek, *holos* = entire; *metabole* = change; Latin, *-osus* = with the property of.) Insects with a complete transformation during Metamorphosis with egg, larval, pupal and adult stages distinctly separated by profoundly different morphs at each stage. See Metamorphosis. Cf. Hemimetabolous; Paurometabolous.

HOLOMETABOLY Noun. (Greek, *holos* = entire; *metabole* = change.) The condition of being Holometabolous. Holometaboly is a derived condition among pterygote insects. See Metamorphosis. Cf. Ametaboly; Hemimetaboly; Paurometaboly. Rel. Groundplan.

HOLOPELTID Adj. Acarology: A Prodorsal Shield not subdivided by transverse sutures.

HOLOPHYLY Cf. Monophyly; Polyphyly.

HOLOPNEUSTIC Adj. (Greek, *holos* = entire; *pnein* = to breathe; *-ic* = consisting of.) Pertaining to a respiratory system displaying two thoracic and eight abdominal spiracles, all of which are open and functional. The Holopneustic condition is regarded as the primitive type of respiratory system among insects. See Respiratory System. Cf. Apneustic; Hemipneusitic; Metapneustic.

HOLOPTIC Adj. (Greek, *holos* = entire; *optikos* = pertaining to sight; *-ic* = of the nature of.) Pertaining to insects with enlarged compound eyes which are medially contiguous and provide an extensive field of vision. See Vision. Cf. Dichoptic. Rel. Compound Eye.

HOLOSERICEOUS Adj. (Greek, *holos* = entire; Latin, *sericeus* = silken; *-aceus* = of or pertaining to.) Pertaining to structure with a short, dense, decumbent, silky pile of hair, which gives the surface a satiny luster. Alt. Holosericeus.

HOLOTYPE Noun. (Greek, *holos* = entire; *typos* = pattern. Pl., Holotypes.) 1. The single specimen selected by the author of a Species as its type or the only specimen known at the time of description. 2. The namebearer of a Species. 3. The 'Type' in the sense of taxonomy and nomenclatural rules. See Type. Cf. Allotype; Cotype; Lectotype; Neotype; Paratype; Syntype; Topotype. Rel. Nomenclature; Taxonomy.

HOLSTEBROE, HERTZ OTTO (1857–1940) (West 1942, Ent. Meddr. 22: 276–277.)

HÖLZEL, EMIL (1894–1973) (Demelt 1974, ArbGem. öst. Ent. 25: 66–68, bibliogr.)

HOMALISIDAE Plural Noun. A small Family of polyphagous Coleoptera assigned to Superfamily Elateroidea. Strictly Old World in distribution. See Elateroidea.

HOMBERG, WILLIAM (1652–1715) (Rose 1850, *New General Biographical Dictionary* 8: 353.)

HOME, EVERARD (1736–1832) (Rose 1850, *New General Biographical Dictionary* 8: 355.)

HOMELYTRA Plural Noun. (Greek, *homoios* = alike; *elytron* = sheath. Pl., Homelytrae.) Elytra of similar or equal size, shape, colour or texture.

HOMEOSTASIS Noun. (Greek, *homoios* = alike; *stasis* = standing. Pl., Homeostases.) 1. The maintenance of a constant internal environment through physiological mechanisms. 2. The maintenance of an equilibrium between organisms and their environment through self-regulation. Alt. Homoeostasis.

HOMEOTYPE See Homoeotype.

HOMEYER, ALEXANDER VON (1834–1903) (Pagenstecher 1904, Jb. nassau. Ver. Naturk. 57: xiii.)

HOMILOPSOCIDEA Plural Noun. A Superfamily designation in the Psocomorpha (Psocoptera), which contains 10 Families. Taxonomic relationships within this Superfamily are not clear and most taxonomists regard the grouping as 'convenient' pending further study. See Psocoptera.

HOMOCHRONIC HEREDITY The appearance of a variation in offspring at whatever age it first appeared in the parent. Alt. Homochronous Heredity.

HOMOCHRONOUS Adj. (Greek, *homos* = same; *chronos* = time; Latin, *-osus* = with the property of.) Occurring at the same time. Changes in an organism which appear in the offspring at the same age at which they appeared in the parent. Rel. Synchronous.

HOMODACTYL Adj. (Greek, *homos* = same; *dactyl* = toe.) Pertaining to Ungues that are similar in size and shape. See Ungues. Cf. Heterodactyl.

HOMODYNAMIC Adj. (Greek, *homos* = same; *dynamis* = power; *-ic* = of the nature of.) Developing without resting stages; insects that do not require diapause. Pertaining to Homodynamism.

HOMODYNAMOUS Adj. (Greek, *homos* = same; *dynamis* = power; Latin, *-osus* = with the property of.) 1. Structures or body segments that are serially homologous. 2. Pertaining to the Serial Homology of Metameres.

HOMOEOTHERMAL Adj. (Greek, *homoios* = alike; *thermos* = warmth; *-alis* = characterized by.) With regard to animals, the characteristic of warm-bloodedness. Alt. Homeothermal. Cf. Poikilothermal.

HOMOEOCHROMATISM Noun. (Greek, *homoios* = alike; *chroma* = colour; English, *-ism* = state. Pl., Homoeochromatisms.) A phenomenon involving regional colour variation, as when over a given region many butterflies tend to vary similarly as regards colour.

HOMOEOCHROME (Greek, *homoios* = alike; *chroma* = colour.) Of the same colour. Cf. Heterochrome.

HOMOEOSIS Noun. (Greek, *homoiosis* = likeness; *-osis* = a condition or state. Pl., Homeoses.) 1. The assumption of one structure or organ likeness by another. 2. The substitution of homotypes or serial homologues for one another. *e.g.* Modification of an Antenna into a leg. See Homology. Cf. Heteromorphosis.

HOMOEOTYPE Noun. (Greek, *homoios* = alike; *typos* = pattern. Pl., Homoeotypes.) A specimen compared with the Holotype (Lectotype, Neotype) and considered conspecific by someone other than the author of a Species (Banks & Caudell). Syn. Homeotype; Homotype. See Type. Cf. Plesiotype. Rel. Nomenclature; Taxonomy.

HOMOGENEOUS Adj. (Greek, *homos* = same; *genos* = kind; Latin, *-osus* = with the property of.) 1. Of the same kind or nature. 2. Pertaining to structure similar in texture or parts. 3. Pertaining to organisms which are similar in features through descent. Cf. Heterogeneous.

HOMOGENY (Greek, *homos* = same; *genos* = race.) Correspondence of parts due to common descent of organisms. See Lankester (1870, Ann. Mag. Nat. Hist. 6: 35–43.) See Homoplasy.

HOMOIOPTERIDAE Handlirsch. A Family of Palaeodictyoptera that bears hollow tergal projections from the Thorax and Abdomen. The projections are regarded as homologous with transverse abdominal Carinae in extant Odonata. See Palaeodictyoptera.

HOMOLOGIZATION Noun. (Greek, *homologous* = agreeing; English, *-tion* = result of an action. Pl., Homologizations.) The act of homologizing or of being identical in structure and origin. See Homology.

HOMOLOGOUS Adj. (Greek, *homologous* = agreeing; Latin, *-osus* = with the property of.) Pertaining to organs or structures which have a common origin, but the insects which bear them may be distantly related as reflected in classification. Homologous organs or structures may appear modified or adapted for different purposes among lineages. See Analogous; Paralogous.

HOMOLOGUE Noun. (Greek, *homologia* = agreement. Pl., Homologues.) A term popularized by English vertebrate anatomist Richard Owen (1843) to express fundamental similarity in organs among organisms: 'The same organ in different animals under every variety of form and function' (Owen 1843). Any 'thing' or part homologous to another part. See Homology. Cf. Analogue; Paralogue.

HOMOLOGY Noun. (Greek, *homologia* = agreement. Pl., Homologies.) The central concept of anatomy and the essence of perfecting biological classifications which reflect natural (evolutionary) relationships. 1. Structures which occur in different lineages and which are acquired through common ancestry. Structures related, but not necessarily identical in function or construction. 2. Structural correspondence between two or more characteristics of organisms caused by continuity of information (Van Valen 1982). Sometimes subdivided into Historical Homology and Repetitive Homology. Criteria used to establish Homology include (a) similarity in relative position and connection of corresponding parts, (b) similarity in structural organization of parts and (c) similarity in development of parts. See General Homology; Iterative Homology; Latent Homology; Operational Homology; Phylogenetic Homology; Serial Homology (Homotypy *sensu* Owen); Special Homology. Cf. Analogy; Convergent Evolution; Paralogy.

HOMOMEROUS Adj. (Greek, *homos* = same; *meros* = part; Latin, *-osus* = with the property of.) Descriptive of legs with an equal number of tarsal segments. Syn: Isomerous. Cf. Heteromerous.

HOMOMORPHA Noun. (Greek, *homos* = same; *morphe* = shape.) Insect Species in which the larvae resemble the adults.

HOMOMORPHOUS Adj. (Greek, *homos* = same; *morphe* = shape; Latin, *-osus* = with the property of.) Pertaining to organisms or parts of organisms with similar external appearance or form.

HOMONOMOUS Adj. (Greek, *homos* = same; *nomos* = development; Latin, *-osus* = with the property of.) 1. Pertaining to structure with the same substance or texture. 2. Pertaining to Homology of parts arranged on a transverse axis, similarly developed and of equal function.

HOMONYM Noun. (Greek, *homos* = same; *onyma* = name. Pl., Homonyms.) One and the same name for two or more different things (International Code). Rel. Taxonomy; Nomenclature.

HOMONYMOUS Adj. (Greek, *homos* = same; *onyma* = name; Latin, *-osus* = with the property of.) One name applying to different concepts.

HOMOPHONOUS Adj. (Greek, *homos* = same; *phone* = voice; Latin, *-osus* = with the property of.) Of words, orthographically different but indistinguishable in sound. Term applied to different conceptions.

HOMOPLASTIC Adj. (Greek, *homos* = same; *plastos* = mould; *-ic* = consisting of.) 1. Pertaining to organs similar in situation and purpose, but not in structural identity. 2. Pertaining to organs similar in shape or structure but not of common origin. Alt. Homoplasious. See Heteroplastic. Cf. Autoplastic.

HOMOPLASY Noun. (Greek, *homos* = same; *plastos* = mould. Pl., Homoplasies.) Character state reversals and convergence. The concept of structurally similar characters originating independent of one another. Homoplasy is demonstrated to exist when two Taxa possessing a character have a common ancestor which does not possess that character. Homoplasy is attributed to analogy, parallelism or convergence. (Lankester 1870, Ann. Mag. Nat. Hist. 6: 35–43.) See Homology. Cf. Parallelism; Convergence.

HOMOPTERA Leach 1815. (Upper Carboniferous-Recent.) An ordinal term applied to Hemiptera (*sensu lato*) in which the forewing is of the uniform consistency. See Heteroptera. Cf. Auchenorhyncha; Sternorrhyncha.

HOMOTENE Adj. (Greek, *homos* = same; Latin, *teneo* = to hold.) Retaining a primitive form.

HOMOTENOUS Adj. (Greek, *homos* = same; Latin, *teneo* = to hold; Latin, *-osus* = with the property of.) Pertaining to the primitive form. A term sometimes applied to hexapods without Metamorphosis or with an incomplete Metamorphosis.

HOMOTYPE See Homeotype.

HONEGGER-ROSENMUND, HERMANN (1845–1927) (Anon. 1928, Mitt. schweiz. ent. Ges. 14: 85–86.)

HONEY Noun. (Middle English, *hony* = honey. Pl., Honeys; Honies.) The thickened and partly digested nectar of flowers produced by adults of various bees and used as food by the larval stage of the bees.

HONEY ANT *Myrmecocystus mexicanus hortideorum* (McCook) [Hymenoptera: Formicidae]: A Species that occurs in Mexico and from southern California to Colorado. Workers are 5–10 mm long, nocturnal and feed on honeydew secreted by Homoptera. HA workers consume honeydew, store nectar in their Crops and their bodies becomes distended ('replete'); engorged workers hang suspended by Tarsi from ceiling of nest chambers and discharge honeydew when solicited by other workers of colony. HAs consumed by North American indians as food. See Formicidae. Cf. Larger Yellow Ant.

HONEY BEE See European Honey Bee.

HONEY BEE MITE Any of several Species of mites that infest and develop parasitically or phoretically upon honey bees. Most Species assigned to Family Tarsonemidae; other Species assigned to Families Erythraeidae, Laelaptidae and Varroidae. See *Acarapis*; Tracheal Mite; Varroa Mite.

HONEY BEE VIRAL DISEASE Any of several viral diseases that affect honey bee larvae, Pupae and/or adults. Known viral diseases include: Acute Bee-Paralysis Virus; Apis Iridescent-Virus; Arkansas Bee-Virus; Bee Virus-X; Bee Virus-Y; Black Queen-Cell Virus; Chronic Bee-Paralysis Virus; Cloudy-Wing Virus; Deformed-Wing Virus; Egypt Bee-Virus; Filamentous Virus; Kashmir Bee-Virus; Sacbrood; Slow-Paralysis Virus. Cf. American Foulbrood; *Nosema apis*; Honey Bee Mite; Chalk Brood. Rel. Pathogen.

HONEY BEETLE *Heteronyx flavus* Blackburn [Coleoptera: Scarabaeidae].

HONEY STOMACH Apoid Hymenoptera: The thin-walled, distensible Crop of workers. HS serves as a reservoir for liquids, particularly nectar which is transported back to the nest and stored in cells as honey.

HONEY TUBES See Cornicles; Siphonets; Siphuncles.

HONEY YELLOW A clear, light golden yellow.

HONEYBROWN BEETLE *Echnolagria grandis* (Gyllenhal) [Coleoptera: Tenebrionidae].

HONEYCOMB Noun. (Anglo Saxon, *hunigcamb*. Pl., Honeycombs.) 1. The aggregation of hexagonal cells made of wax, formed by the honey bee, as breeding cells for the larvae and as storage places for honey. 2. Any structure composed of hexagonal cells, or any material of large pores, with a distant resemblance to the bee comb structure.

HONEYDEW GLANDS Coccids: The rectal wax pores (MacGillivray).

HONEYDEW Noun. (Pl., Honeydews.) 1. The sugar-rich, honey-like secretion of many Homoptera, particularly aphids. Obtained from the phloem fluid of plants by insects through feeding, and passed through the Anus as excrement. Honeydew used as a source of nutrition by many insects, particularly ants. 2. An exudate from the surface of some plant galls. See Excreta; Excretion. Cf. Frass; Meconium. Rel. Digestion.

HONEYLOCUST PLANT-BUG *Diaphnocoris chlorionis* (Say) [Hemiptera: Miridae].

HONEYLOCUST SPIDER-MITE *Platytetranychus multidigitali* (Ewing) [Acari: Tetranychidae].

HONEYPOT Noun. (Pl., Honeypots.) Social Insects: The members of a physiological caste of ants in the Genus *Myrmecocystus* which store large volumes of honeydew in their hyperdistended crops. See Formicidae.

HONEYPOT ANT *Melophorus bagoti* Lubbock [Hymenoptera: Formicidae].

HONEYSUCKLE Common name for *Lonicera* spp. (Caprifoliaceae), about 150 Species of shrub-like plants common in the Northern Hemisphere. Also common name for *Aquilegia canadensis, Justicia californica* and *Rhododendron prinophyllum.*

HONEYSUCKLE APHID *Hyadaphis foeniculi* (Passerini) [Hemiptera: Aphididae].

HONEYSUCKLE LEAFMINER *Swezeyula lonicerae* Zimmerman & Bradley [Lepidoptera: Elachistidae].

HONEYSUCKLE SAWFLY *Zaraea inflata* Norton [Hymenoptera: Cimbicidae].

HÖNNE, HERMAN (1883–1963) (Mannheims 1963, Bonn. zool. Beitr. 14: 246–247.)

HONRATH, EDUARD G. (1837–1893) (Anon. 1893, Insektenbörse 10: 81.)

HOOD Noun. (Old High German, *huot* = helmet. Pl., Hoods.) 1. Any covering of the head, neck and sometimes the shoulders. Often soft, pliant and flexible (membrane-like) or hard and inflexible (helmet-like). 2. A covering of straw over a beehive. Cf. Cap; Cucullus.

HOOD OF THE MAXILLA 1. The Maxilla and Galea. 2. Tingitidae: The elevated anterior part of the Prothorax, often covering the head. 3. Lepidoptera: An evaginated area of the lateral part of the first abdominal segment extending anteriorly to cover partially the tympanal cavity (Richards).

HOOD, J DOUGLAS (O'Neill 1974, Proc. ent. Soc. Wash. 76: 297–309, bibliogr. only, 1948–1960, with list of new Taxa.)

HOOK TIP MOTHS See Drepanidae.

HOOKE, ROBERT (1635–1730) English Professor of Geometry (London), early microscopist and author of *Micrographia* (1667). (Rose 1850, *New General Biographical Dictionary* 8: 363–364.)

HOOKED HAIRS Setae which are apically curved and believed responsible for gathering or holding appendages together. Cf. Hamulus.

HOOKER, CHARLES WORCESTER (1883–1913) (Fernald 1913, J. Econ. Ent. 6: 334–335.)

HOOKER, WILLIAM JACKSON (1785–1865) (Anon. 1867, Proc. R. Soc. Lond. 15: xxv–xxx.)

HOOKLET See Hamulus.

HOOK-TIP MOTHS See Drepanidae.

HOOKTIPPED BROWN LACEWING *Drepanacra binocula* (Newman) [Neuroptera: Hemerobiidae].

HOOP-PINE BARK BEETLE *Hylurdrectonus piniarius* Schedl [Coleoptera: Curculionidae]: In Australia, adults construct brood galleries in the phloem layers beneath bark of trees; larvae cause damage by tunneling. See Curculionidae.

HOOP-PINE BORER *Pachycotes australis*; *Pachycotes clavatus* Schedl [Coleoptera: Curculionidae].

HOOP-PINE BRANCHCUTTER *Strongylurus decoratus* (Mckeown) [Coleoptera: Cerambycidae].

HOOP PINE JEWEL BEETLE *Prospheres aurantiopictus* (Laporte & Gory) [Coleoptera: Buprestidae]: In Australia, a significant pest of hoop pine that also attacks bunya pine *Araucaria bidwilli* and *Pinus* spp. in Queensland. HPJB occurs along coastal Queensland from New South Wales to Cape York, in all areas where hoop pine grows. Adults are shiny black, torpedo-shaped and 15–20 mm long, with short saw-like Antenna; often with 10 red-orange spots on Elytra. Larvae are elongate, white with Thorax distinctly flattened and expanded. Eggs are laid on exposed wood, in bark crevices, on freshly felled or dying trees and on tree wounds. Neonate larvae penetrate wood; tunnels often follow growth rings. Pupate in wood; neonate adult emerges through oval-shaped hole 3–5 mm diam. HPJB life cycle usually 1–2 years but in dry timber may be longer (record 17 years). Emergent adults do not reinfest timber; females lay eggs on unseasoned timber only. Syn. Cobra Borer; Flathead Borer. See Buprestidae.

HOOP-PINE LONGICORN *Diotimana undulata* (Pascoe) [Coleoptera: Cerambycidae]. An Australian Species that attacks standing dead trees and freshly felled logs of hoop pine. HPL recorded from slash pine *Pinus elliottii*. Larvae tunnel mostly in cambial region, bore into log to pupate; plug opening with shreds of wood. See Cerambycidae.

HOOP-PINE SEED MOTH *Hieromantis ephodophora* Meyrick [Lepidoptera: Oecophoridae].

HOOP-PINE STITCH BEETLE *Hyleops glabratus* Schedl [Coleoptera: Curculionidae]

HOOS, EMMA (–1874) (Hehl 1974, Prakt. SchädlBekämpf. 26: 154.)

HOP APHID *Phorodon humuli* (Schrank) [Hemiptera: Aphididae].

HOP FLEA-BEETLE *Psylliodes punctulata* Melsheimer [Coleoptera: Chrysomelidae].

HOP LOOPER *Hypena humuli* (Harris) [Lepidoptera: Noctuidae].

HOP PLANT-BUG *Taedia hawleyi* (Knight) [Hemiptera: Miridae].

HOP VINE-BORER *Hydraecia immanis* (Guenée) [Lepidoptera: Noctuidae]: A pest of strawberries in Nova Scotia. See Noctuidae.

HOP Noun. (Middle English, *hoppen,* > *hoppian*. Pl., Hops.) A short leap. Toadbugs hop with their hind legs; Grasshoppers jump with their hind legs. The jump is achieved by the rapid extension of the metathoracic Tibiae. A small hop or defensive kick can be executed with the Tibia starting from a partially extended position, but for the jump to be maximally developed, the Tibia must be fully flexed initially. Hopping and jumping by insects typically involves only one pair of legs, but not always the same pair of legs. Jumping usually involves the the metathoracic legs, but exceptions occur, such as some parasitic Hymenoptera (Encyrtidae and Eupelmidae). Jumping is a more rapid form of locomotion than hopping and more distance is traversed. [See Brown 1963, Jumping Arthropods, Times Science Reviews, Summer 1963, pp 6–7]. Cf. Jump. Transitive Verb. (Middle English, *hoppen,* > *hoppian*.) To move by a short leap or series of leaps with all feet off the ground. See Locomotion. Cf. Crawl; Jump; Run; Walk.

HOPCIN® See Fenobucarb.

HOPE, FREDERICK WILLIAM (1797–1862) English cleric with financial independence that enabled him to pursue interests in natural history. Hope was a Fellow of the Linnean Society (1822) and Royal Society (1834), a founding member of the Entomological Society of London (1833) and its President (1835, 1839, 1845). His financial donations established the Hope Department of Zoology and Museum at Oxford University.

HOPFFER, CARL HEINRICH (1810–1876) (Westwood 1876, Proc. ent. Soc. Lond. 1876: xliv.)

HOPFFGARTEN, GEORG MAXIMILIAN VON (1825–1904) (Haupt 1905, Dt. ent. Z. 49: 174–176, bibliogr.)

HOPFINGER, JOHN CARL (1888–1961) (Eft 1962, J. Lepid. Soc. 16: 147–149.)

HOPKINS BIOCLIMATIC LAW The observational principle that 'other conditions being equal, the variations in time of occurrence of a given periodic event in life activity in temperate North America is at the general average rate of four days to each degree of latitude, 5° off longitude and 400 feet of altitude, later northward, eastward and upward in spring and early summer and the reverse in late summer and autumn' (Hopkins).

HOPKINS HOST SELECTION PRINCIPLE An hypothesis which contends that a female insect capable of breeding on more than one host Species will prefer to oviposit on the host Species from which that female developed.

HOPKINS, ANDREW DELMAR (1857–1948) (Stemple 1966, Bull. ent. Soc. Am. 12: 25–28, bibliogr.)

HOPKINS, GEORGE HENRY EVANS (1899–1973) (Lees 1974, Proc. R. ent. Soc. Lond. (C) 38: 60.)

HOPLEY, EDWARD WILLIAM JOHN (–1869) (Anon. 1869, Entomol. mon. Mag. 6: 18.)

HOPLOPLEURIDAE Plural Noun. Small-Mammal Sucking Lice. A moderately small Family of Anoplura whose Species parasitize rodents. Head relatively small; compound eyes and Ocular Points absent; Mandibles absent, mouthparts forming a beak or rounded anteriorly; head and Thorax with a few Setae; hind legs largest; Abdomen with Paratergites. See Anoplura.

HOPPE, DAVID HENRY English apothecary and author of *Enumeratio Insectorum Elytratorum Indigenorum* (1795).

HOPPER STOPPER® See *Nosema locustae*.

HOPPERBURN Noun. (Etymology obscure. Pl., Hopperburns.) A virus-like disease which causes stippling, curling and browning of leaves of potato and eggplant. Disease is transmitted by Potato Leafhopper. Rel. Potato Leafhopper.

HOPPIN, WASHINGTON (1827–1867) (Weiss 1925, Ent. News 36: 257–267.)

HOPPING, GEORGE REDSTONE (1899–1974) (Reid 1974, Bull. ent. Soc. Can. 6: 96.)

HOPPING, RALPH (1868–1941) (Blaisdell *et al.* 1942, Pan-Pacif. Ent. 18: 1–3, bibliogr.)

HORA, SUNDER LAL (1896–1955) (O'Donoghue 1956, Nature 177: 358–359.)

HOREHOUND BUG *Agonoscelis rutila* (Fabricius) [Hemiptera: Pentatomidae]: A minor pest in Australia. See Pentatomidae.

HORI, MATSUGI (1896–1938) (Tagashi 1970, Kontyû 38: 391–392.)

HORII, EIKICHI (–1920) (Hasegawa 1967, Kontyû 35 (Suppl.): 73.)

HORION, ADOLF (Freude 1968, NachrBl. bayer. Ent. 17: 33.)

HORISMOLOGY Noun. (Greek, *horismos* = a marking by bounds; *logos* = discourse. Pl., Horismologies.) The art of naming the venation of insect wings. By convention, wing veins are named from anterior to posterior and from proximal to distal. Because confusion exists over wing vein terms, identify veins from the wing base and not associations with Axillary Sclerites. Wing veins are convex or concave, based on their tendency to fold upward or downward when the wing is pulled over the body. Lameer (1922) and Martynov (1924) named branches of veins as anterior and posterior based on their profile: Anterior veins are convex and posterior veins are concave. Entomologists attempt to name wing veins with respect to Homology. Perfecting a system of vein nomenclature has been difficult because many Species of insects exist with many patterns of wing venation; veins are sometimes reduced or lost. The task has further been complicated by a failure to correctly identify homologous veins between Orders and Families of insects. Establishing Homology between veins based upon the fossil record is not possible because the fossil record is incomplete. Further, intermediary conditions or transitional phases of wing-vein development are very weakly indicated in the existing record. The geological record suddenly reveals wings in well developed form with complex venation within several lineages. As a consequence of the controversy over Homology in wing venation, several systems have been proposed for naming wing veins. The Comstock Needham is widely used, but this system of venation should be credited to Redtenbacher (1886) who developed the system. Comstock (1918) acknowledged Redtenbacher, but this credit is generally ignored. See Orismology; Vein.

HORIZON Noun. (Middle English, *orisonte*. Pl., Horizons.) See Stratum.

HORIZONTAL Adj. (Middle English, *orisonte*; *-alis* = characterized by.) Parallel with the horizon in terms of position or motion. A term applied to the wings of an insect when held in repose parallel to the horizon.

HORIZONTAL BORER *Austroplatypus incompertus* (Schedl) [Coleoptera: Curculionidae].

HORIZONTAL PLANE See Frontal Plane.

HORIZONTAL TRANSMISSION Medical Entomology: Passage of a pathogen (virus, bacteria, protozoa, *etc.*) through an arthropod host Species (vector) to a vertebrate host Species. Syn. Circuitous Transmission. See Transmission; Vector. Cf. Vertical Transmission.

HORKEL, JOHANN (1796–1846) (Ratzeburg 1874, Forstwissenschaftliches Schriftsteller-Lexikon 1: 251–254.)

HORMEX ROOTING-POWDER® See Indolebutyric Acid.

HORMOCEL® See Chlormequat.

HORMODIN® See Indolebutyric Acid.

HORMOFIX® See NAA.

HORMOLIGOSIS Noun. (Greek, *hormaein* = to excitel *oligos* = little; *-sis* = a condition or state. Pl., Hormoligoses.) Reproductive stimulation by sublethal doses of pesticides.

HORMONE Noun. (Greek, *hormaein* = to excite. Pl., Hormones.) A chemical, produced by body organs, specialized cells, tissues or Endocrine glands, released into the body and which mediates behavioural or physiological actions. Hormones are typically named on the basis of the process they control. Cf. Semiochemical. Rel. Allelochemical; Endocrine Gland; Gland.

HORMOZAKI, CONSTANTIN VON (1863–1937) (Anon. 1937, Arb. morph. taxon. Ent. Berl. 4: 241.)

HORN FLY *Haematobia irritans* (Linnaeus) [Diptera: Muscidae]: A pest of cattle throughout Northern Hemisphere that feeds to some extent on sheep and horses. HF causes annoyance and irritation resulting in milk loss to dairy cattle and weight loss to beef cattle; HF acts as a mechanical vector of pathogens experimentally. Adults ca 4 mm long; resemble Stable Fly in habitus and features of mouthparts and wing. Adult HFs are continuously active on host and both sexes take blood from cattle; female leaves host only to oviposit; feeds with body oriented head-downward. Female HF oviposits beneath fresh manure on grass or ground and deposits groups of 4–6 eggs; eclosion requires one day. Female lays ca 400 eggs during lifetime. HF larvae burrow through manure, aided by spine-like processed on ventral surface of segments 6–12; larval development is completed within 4–8 days. Pupation occurs in feeding medium or soil; pupal stage requires 6–8 days. HFs overwinter as Pupae in cold climates. HF life cycle requires ca 14 days. See Muscidae. Cf. Buffalo Fly; House Fly; Stable Fly.

HORN Noun. (Greek *keras* > Latin *cornu* > Anglo Saxon *horn*. Pl., Horns.) 1. A conspicuous, apically tapered or pointed process on the head. 2. Any process resembling an animal's horn. 3. Archane: An insect Antenna.

HORN, GEORGE HENRY (1840–1897) (Anon. 1898, Ent. News 9: 1–3.)

HORN, WALTHER (1871–1939) (Porter 1939, Revta. chil. Hist. nat. 43: 195–198, bibliogr.)

HORNBECK, HANS BALTAZAR (1800–1870) (Papavero 1971, *Essays on the History of Neotropical Dipterology* 1: 107. São Paulo.)

HORNE, ARTHUR (1861–1922) (S[heldon] 1922, Entomologist 55: 288.)

HORNE, CHARLES (1824–1872) (Anon. 1972, Enomologist's mon. Mag. 8: 295.)

HORNE, FRANK ROBERT (–1975) (Anon. 1975, Bee Wld. 56: 161.)

HORNED PASSALUS *Odontotaenius disjunctus* (Illiger) [Coleoptera: Passalidae].

HORNED SQUASH BUG *Anasa armigera* (Say) [Hemiptera: Coreidae].

HORNED TREEHOPPERS See Membracidae.

HORNER, ARTHUR CLAYPON (–1893) (Elwes 1893, Proc. ent. Soc. Lond. 1893: lvi.)

HORNET Noun. (Anglo Saxon, *hyrnet*, influenced by Greek, *horniss* = horn. Pl., Hornets.) Aculeate social wasps of the Genus *Vespa* that includes ca 25 Species; Genus is predominantly Asian with a few Species in Europe and North America. Workers typically construct arboreal nests; colonies contain a few hundred to several thousand workers. Food consists of insects, spiders, nectar and fruit. See Vespidae. Cf. European Hornet; Yellowjacket.

HORNET MOTH *Sesia apiformis* (Clerck) [Lepidoptera: Sesiidae].

HORNETS See Vespidae.

HORNIG, HERMANN (1858–1942) (Poole 1942, Ent. News 53: 238–239.)

HORNIG, JOHANN VON (1819–1886) (Rebel 1887, Verh. zool.-bot. Ges. Wien 37: 42–46, bibliogr.)

HORNTAILS See Siricidae.

HORNUNG, SOFUS (1877–1971) (Tuxen 1972, Ent. Meddr. 40: 29–30.)

HORSE-BITING LOUSE 1. *Bovicola equi* (Denny) [Mallophaga: Trichodectidae]: A pest of horse, mule and donkey in North America. Attacks host around neck and base of tail. See Trichodectidae. Cf. Cattle-Biting Louse; Goat-Biting Louse; Sheep-Biting Louse. 2. *Werneckiella equi* (Denny) [Phthiraptera: Trichodectidae].

HORSE BOT-FLIES See Gasterophilidae.

HORSE BOT-FLY *Gasterophilus intestinalis* (DeGeer) [Diptera: Oestridae]: A monovoltine pest of horses endemic to Palaearctic but now widespread in association with horses; most common bot of horses in USA and Australia. HBF is an aetiological agent of Enteric Myiasis. Adults are 15–17 mm long, brown, very setose and resembling bumble bee; wing faintly dusky; does not take food, lives 3–20 days. Female lays ca 700–1,000 eggs during lifetime. Eggs are pale yellow and glued to hair on horse, typically front legs. Embryonic development is complete within 2–4 days and eclosion occurs 3–6 days later. Horses lick their legs and first-instar larvae attach to mucous membrane of tongue then quickly pass into Oesophagus. Second and third-instar larvae feed in-place attached to Oesophagus or cardiac region of stomach. Mature larvae pass from body of host with faeces. Pupation occurs within soil and requires 3–5 weeks. See Oestridae. Cf. Nose Bot-Fly; Throat Bot-Fly. Rel. Gasterophilus Enteric Myiasis.

HORSE FLIES Specifically, members of the Genus *Tabanus* with eyes unmodified or with horizontal bands and pale wings. See Tabanidae.

HORSE FOLLICLE-MITE *Demodex equi* Railliet [Acari: Demodecidae].

HORSE STINGER See Odonata.

HORSE SUCKING-LOUSE *Haematopinus asini* (Linnaeus) [Phthiraptera: Haematopinidae].

HORSERADISH FLEA-BEETLE *Phyllotreta armoraciae* (Koch) [Coleoptera: Chrysomelidae].

HORSESHOE CRAB BEETLES See Ptiliidae.

HORSFIELD, THOMAS (1773–1859) English physican, naturalist and author of *Catalogue of Lepidopterous Insects* (1828–1829). (Gray 1861, Trans. ent. Soc. Lond. 1858–1861: 94–96.)

HORTICULTURE Noun. (Latin, *hortus* = garden; *culture*. Pl., Horticultures.) 1. The practice of cultivating plants. 2. The science involved with growing vegetables, fruit and ornamental plants. Cf. Agriculture; Agronomy. Rel. Apiculture.

HORTON, EDWARD (1816–1871) (Anon. 1871, Entomol. mon. Mag. 7: 215–216.)

HORVATH, GEZA de (1847–1937) Hungarian Hemipterist who published extensively on Heteroptera. (Lundblad 1984, Ent. Tidskr. 69: 5–7.)

HOSBIT® See Dichlorvos.

HOST Noun. (Latin, *hospes* = stranger, guest, host. Pl., Hosts.) An organism which supplies nutrition or protection essential for the development of another organism, termed a parasite. See Parasitism. Cf. Primary Host; Secondary Host. Rel. Prey.

HOST DISCRIMINATION 1. The ability of a female parasite to determine whether a potential host has been parasitized, and to reject or accept the host as a site for oviposition based on that determination. 2. The ability of a female parasite to distribute her eggs among hosts in a non-random manner. The phenomenon is widespread among parasitic Hymenoptera. See Parasitism. Cf. Multiple Parasitism; Superparasitism.

HOST FEEDING A phenomenon in which an adult parasitic insect (parasitoid) feeds upon the body or body fluids of a potential host for the larval stage. See: Concurrent Host Feeding; Non-Concurrent Host Feeding; Destructive Host Feeding; Non-Destructive Host Feeding. Rel. Feeding Tube.

HOST HABITAT LOCATION The initial phase of host selection, characterized by a female parasite detecting the site, or area in which a potential host may reside.

HOST PLANT 1. A plant invaded by a pathogen and from which the pathogen obtains nutrients. 2. A plant used by an insect as a place to live or feed. Predaceous Heteroptera: HP determined by the preferred prey which lives on it.

HOST PLANT RESISTANCE The relative amount of heritable qualities possessed by a plant that reduces the degree of damage done to the plant by the activity of a pest.

HOST RANGE The kinds of organisms attacked by a specific group of pathogens or parasites. HR determined by taxonomy, habitat, distribution or chemistry of the host organism.

HOST SELECTION A complex process, subdivided into a subordinated series of events or decisions, responsible for successful propagation of a parasitic insect. Traditionally divided into (1) host habitat-location, (2) host location, (3) host acceptance, (4) host suitability, and (5) host regulation. See these terms for explanation.

HOSTAQUICK® See Heptenophos.

HOSTATHION® See Triazophos.

HOTTA, MASAZO (1891–1964) (Hasegawa 1967, Kontyû 35 (Suppl.) 74.)

HOUARD, CLODOMIR ANTONY VINCENT (1873–1943) (Trotter *et al.* 1948, Marcellia 30: 252–268.)

HOUGH, GARRY DE NORD (1861–1903?) (Aldrich 1903, Ent. News 14: 245–247, bibliogr.)

HOUGHTON, ALBERT (–1897) (Moberly 1897, Entomologist's Rec. J. Var. 9: 124–125.)

HOULBERT, CONSTANT VINCENT (1857–1947) (Jolivet 1948, Miscnea ent. 45: 103–108, bibliogr.)

HOUSE Noun. (Anglo Saxon, *hus.* Pl., Houses.) 1. A place of human habitation. 2. Any gelatinous substance or secretion from an insect or other invertebrate which is used by as a place in which to reside. Cf. Habitat.

HOUSE CENTIPEDE 1. *Allothereua maculata* (Newport) [Scutigeromorpha: Scutigeridae]: A minor urban pest in Australia that occurs in damp places such as subfloor areas, glasshouse and fern houses. Adults are 20–25 mm long, pale brown with dark markings and head with long, slender Antenna; eyes reduced; first pair of legs modified into poison claws; body with 15 segments (8 apparent from dorsal aspect) and 15 pairs of exceptionally long legs; legs shed easily under stress or trauma; a pair of caudal appendages resemble Antennae in length. HC is predaceous upon insects and other arthropods and generally regarded as not harmful. See Chilopoda. Cf Giant Centipede; Ribbon Centipede. 2. *Scutigera coleoptrata* (Linnaeus) [Scutigeromorpha: Scutigeridae].

HOUSE CRICKET *Acheta domesticus* (Linnaeus) [Orthoptera: Gryllidae]: A Species widespread throughout North America and similar to Field Cricket. HCs live outdoors during spring and summer, enter homes during autumn. Eggs laid in ground during summer and hatch during following spring. HCs are a minor pests of wool and carpets. See Gryllidae.

HOUSE DUST MITES Any one of several Species of mites that inhabit houses and affect man. HDM are typically found in carpet, furniture and mattress. Common Species include *Dermatophagoides pteronyssinus, D. farinae, Euroglyphus maynei* and *Glycyphagus domesticus.* See European House Dust Mite.

HOUSE FLIES See Muscidae.

HOUSE FLY MITE *Macrocheles muscaedomesticae* (Scopoli) [Acari: Macrochelidae].

HOUSE FLY *Musca domestica* Linnaeus [Diptera: Muscidae]: A cosmopolitan, synanthropic, multivoltine pest; most significant urban pest inside dwellings; less common in hot tropical areas. HF primarily is diurnal or less active under artificial light; adults live about 2–4 weeks and life cycle requires 2–4 weeks. HC completes as many as 12 generations per summer. HF can be problem in cattle feedlot, dairy farms and poultry houses. Adults are 4–8 mm long with body nonmetallic and predominantly greyish with four longitudinal stripes on Thorax; males with two yellow patches on base of Abdomen. Adult females release sex pheromone (muscalure) which attracts males; females typically mate once and oviposit batches of 100–150 eggs in moist organic matter (40–70% water). Females lays ca 700–900 eggs per lifetime; eggs are white, elongate and ca 1 mm long. Eclosion occurs within eight hours to three days. Larvae are typically cyclorrhaphous with three instars; larvae feed in moist, warm, organic matter and decaying vegetation; mature larvae are 10–12 mm long. Pupation occurs in habitat drier than feeding site ca 5–60 cm below surface; period requires 3–30 days. Adults actively fly in temperature range 11–32°C and inactive at <7°C and killed at <0°C. Adults are highly annoying but more significant as mechanical vector of many pathogenic human diseases including food-poisoning bacteria *Salmonella* and diarrhoea bacteria *Shigella.* Adults are attracted to disease-containing organic material and human food; adults are transported through commerce. See Muscidae. Cf. Bush Fly; Egyptian House Fly; Face Fly; Lesser House Fly.

HOUSE ITCH MITE *Glycyphagus domesticus* (De Greer) [Acari: Glycyphagidae].

HOUSE LONGHORN BEETLE *Hylotrupes bajulus* (Linnaeus) [Coleoptera: Cerambycidae]: A minor pest of conifers in Europe and South Africa. See Cerambycidae.

HOUSE MITE *Glycyphagus domesticus* (DeGeer) [Acari: Glycyphagidae].

HOUSE MOSQUITO *Culex quinquefasciatus* Say [Diptera: Culicidae]: Common name in North America. See Culicidae. Cf. Brown House Mosquito.

HOUSE MOUSE MITE *Liponyssoides sanguineus* (Hirst) [Acari: Macronyssidae].

HOUSE SPIDERS See Desidae.

HOUSEHOULD CASEBEARER *Phereoeca uterella* Walsingham [Lepidoptera: Tineidae].

HOUSER, JOHN SAMUEL (1881–1947) (Parks 1947, J. Econ. Ent. 40: 611–613.)

HOUSKA, JAROSLAV (1885–1946) (Heyrovsky 1949, Cas. csl. Spol. ent. 46: 74–76.)

HOVER FLIES Diptera belonging to the Family Syrphidae. Adults are 5–15 mm long, elongate and black with yellow and brown markings. Coloration pattern, body size and adult habit of feeding on flower nectar and pollen, give hover flies the appearance of bees and wasps. Hover flies are distinguished from wasps by their characteristic hovering over flowers (similar to dragonflies) and single pair of wings. Hover flies have very large eyes and a bristle on an enlarged third Antennal segment. Adult females lay white, oval eggs near colonies of aphids. Larvae are maggot-like, vary in colour from creamy-white to brown and are active predators on aphids. Pupae are brown, tear-shaped and located under leaves or in leaf litter. Adults feed on flower nectar and pollen. See Syrphidae.

HOWARD, CHARLES WALTER (1882–1928)

(Hoffman 1927, Lingnan J. Sci. 5: 293–299, bibliogr.)

HOWARD, JOHN OLIVER TUNLEY (1905–1961) (Worms 1962, Entomologist's Rec. J. Var. 74: 27–28.)

HOWARD, LELAND OSSIAN (1857–1950) American economic Entomologist employed by U.S. Department of Agriculture. Howard was a specialist in taxonomy of parasitic Hymenoptera early in his career and later became an Administrator of the Entomology Branch of the USDA over the major portion of his career. (Gahan *et al.* 1950, Proc. ent. Soc. Wash. 52: 224–233.)

HOWE, FREDERICK JOHN (Butler 1972, Proc. R. ent. Soc. Lond. (C) 36: 61.)

HOWE, REGINALD HEBER (1875–1932) (Anon. 1932, Ent. News 43: 40.)

HOWES, FRANK NORMAN (–1973) (Anon. 1974, Bee Wld 55: 31–32.)

HOWES, WILLIAM GEORGE (1879–1946) (Miller 1946, Trans. Proc. R. Soc. N. Z. 76: 262–263, bibliogr.)

HOWLETT, FRANK MILBURN (1877–1920) (Scott 1920, Entomol. mon. Mag. 56: 234, 262.)

HOY, PHILO ROMAYNE (1816–1892) (Schorger 1944, Passenger Pigeon 6: 55–59, bibliogr.)

HOYER'S MEDIUM A water-based temporary mountant used for the preservation of microscopic slide preparations. Chloral hydrate 2 parts; Water 50 parts; Gum arabic (granules) 50 parts.

HOZAWA, SANJI (1885–1947) (Hasegawa 1967, Kontyû 35 (Suppl.): 73.)

HRUBY, KAREL (1910–1962) (Moucha 1963, Cas. csk. spol. ent. 60: 264–265, bibliogr.)

HUARD, VICTOR A (1853–1929) (Maheux 1930, Naturaliste Can. 57: 5–10.)

HUBBARD, GEORGE E (Davis 1931, Proc. Indiana Acad. Sci. 41: 55.)

HUBBARD, HENRY GUERNSEY (1850–1899) (Smith 1899, Ent. News 10: 80–83.)

HUBENTHAL, WILHELM (1871–1949) (Anon. 1941, Arb. morph. taxon. Ent. Berl. 8: 287.)

HUBER, ALBERT (1890–1972) (Wittmer 1972, Mitt. ent. Ges. Basel 22: 102.)

HUBER, FRANZ (1750–1831) Swiss naturalist and author of *Nouvelles observations sur les Abeilles* (2 vols, 1792). Although blind, Huber made contributions to honey bee biology through his son and valet by asking questions of their observations. (Decandolle 1832, Notice sur la vie et les écrits de F. Huber. Paris.)

HUBER, JAKOB (1868–1914) (Anon. 1914, Ent. News 25: 288.)

HUBER, JEAN PIERRE (1777–1840) Swiss naturalist, son of Franz Huber and author of *Recherches sur les Moeurs des Fourmis indigenes* (1810) *and Observations sur les Bourdons* (vol. 6, Linnean Transactions). (Duméril 1823, *Considérations générale sur la classe des insectes*. 272 pp. (265–266), Paris.)

HÜBNER, JACOB (1761–1826) (Geyer 1827, *In* Thon's Archiv. Ent. 1: 28–31, bibliogr.)

HÜBNER, JOHANN GOTTFRIED (1746–1812) Austrian painter and author of Lepidoptera of Europe. Hübner collected and sent specimens to Fabricius. Hübner's personal collection was sold to H. v. Minkwitz and the Coleoptera and Lepidoptera subsequently were sold to E. Fr. Germar. (Germar 1913, Mag. Ent. 1: 140.)

HUDAK, EDE AGOST (1822–1902) (Aigner-Abafi 1902, Rovart. Lap. 9: 13, 131–133.)

HUDD, ALFRED EDMUND (1845–1920) (Griffiths 1921, Entomol. mon. Mag. 56: 262–263.)

HUDSON, ERIC JOHN (1899–1968) (Kennedy 1969, Proc. R. ent. Soc. Lond. (C) 33: 55.)

HUDSON, GEORGE VERNON (1867–1946) (Anon. 1946, Proc. N. Z. Inst 76: 264–266.)

HUDSON, JOSHUA (–1974) (Gunn 1976, Proc. R. ent. Soc. Lond. (C) 40: 52.)

HUDSONIAN ZONE Part of the Boreal Region comprising the northern part of the transcontinental coniferous forests: In the eastern United States restricted to the cold summits of the highest mountains, from northern New England to western North Carolina: In the West it covers the higher slopes of the Rocky and Sierra-Cascade systems.

HUE Noun. (Middle English, *hoew* = shape, colour, form. Pl., Hues.) 1. One of three attributes of colour which may be assigned to several categories (red, green, yellow, blue). Hue separates colours from greys of the same brilliance. See Coloration. Cf. Brilliance; Saturation. 2. The form, figure or appearance of a structure or body.

HUEMER, HANS (1858–1935) (Muller 1935, Z. österr. Ent. Ver. 20: 45–47.)

HUFNAGEL, GEORGE (1546–1600) See HOEFNAGEL, G.

HUGENTOBLER, HANS (1901–1967) (Florin 1967, Mitt. ent. Ges. Basel 17: 95–96.)

HUGGINS, ETHEL A (–1963) (H. C. H. 1963, Entomologist's Rec. J. Var. 75: 228.)

HUGHELS, A W See McKenny-Hughes.

HUGHES, GRIFFITH (Rose 1850, *New General Biographical Dictionary* 8: 403.)

HUGO, RUDOLPH (1874–1938) (Anon. 1938, Arb. morph. taxon. Ent. Berlin 5: 296.)

HUGUENIN, JULIUS CAESAR (1840–1926) (Van Duzee 1927, Proc. Pacif. Cst. ent. Soc. 2: 95–96.)

HUJIMAKI, YUKIO (1883–1949) (Hasegawa 1967, Kontyû 35 (suppl.): 71.)

HUKAYA, TORU (1888–1916) (Hasegawa 1967, Kontyû 35 (suppl.): 71.)

HUKKINEN, YRJÖ ARMAS (1886–1947) (Kamervo 1946, Memo. Soc. Fauna Flora Fenn. 23: 240–241.)

HUKUDA, JIRO (1911–1962) (Hasegawa 1967, Kontyû 35 (Suppl.): 71–72.)

HULBIRT, EDWIN RAY (1886–1965) (Comstock 1965, J. Lepid. Soc. 19: 243–244.)

HULKONEN, OLAVI (1905–1937) (Horn 1937, Arb. physiol. angew. Ent. Berlin 4: 159.)

HULL Noun. (Anglo Saxon, *helan* = to cover. Pl.,

Hulls.) The outer covering of cereal seeds. Alt. Husk.

HULL, JOHN B (1900–1949) (Dove 1949, J. Econ. Ent. 42: 869.)

HULST, GEORGE DURYEA (1846–1900) (Weeks *et al.* 1900, J. N. Y. ent. Soc. 8: 248–254, bibliogr.)

HULTGREN, JOHAN ALBERT (1845–1938) (Kemner 1939, Opusc. ent. 4: 111.)

HUMAN BOT-FLY See Tropical Warble-Fly.

HUMAN FLEA *Pulex irritans* (Linnaeus) [Siphonaptera: Pulicidae]: A cosmopolitan pest of humans and domesticated animals. Adults with interantennal ridge present and Frontal Tubercle absent; small spine on Gena below eye; Pronotal and Genal Ctenidia absent; Metanotum at least half as long as T1. Adults lack the strong ctenidial combs characteristic of Dog Flea and Cat Flea. Females lay ca 500 eggs on host or host's habitat during lifetime. Life cycle requires 15–65 days depending upon temperature. HF is capable of jumping 20 cm horizontal and 13 cm vertical. HF is most commonly associated with domesticated pigs in agricultural environment; *Pulex* is a New World Genus and most Species are associated with porcines. HF is a notable pest of humans in some developing countries, but replaced by Cat Flea as the dominant Species in developed urban habitations. HF is not regarded as a vector of plague bacillus and is not implicated in transmission of disease to humans. HF can serve as an intermediate host of dog tapeworm, *Dipylidium caninum*. See Pulicidae. Cf. Cat Flea; Dog Flea; Oriental Rat Flea.

HUMAN ITCH MITE *Sarcoptes scabei* (DeGeer) [Acari: Sarcoptidae]. A serious pest of humans which causes allergies by the mites burrowing and feeding in the Epidermis. Mites prefer soft skin in which to burrow and feed; several weeks may pass between initial infestation and the appearance of symptoms (itching and dermatitis). Female burrows in skin, oviposits in tunnels; eclosion occurs within 5 days; nymphs move over surface of skin; within 6 days nymphs become males or immature females. Adults live 2–6 weeks and the life cycle requires 14 days. HIM is transmitted by direct contact. Alt. Scabies; Norwegian Itch. Cf. Mange Mite.

HUMAN LICE See Pediculidae. Cf. Body Louse; Head Louse; Pubic Louse.

HUMAN ONCHOCERCIASIS See Onchocerciasis.

HUMAN TRYPANOSOMIASIS A protozoan disease that causes African Sleeping Sickness in humans. Flagellate protozoan Genus *Trypanosoma* infests vertebrates; an intermediate form of trypanosome is acquired in blood taken by fly from infected human. The intermediate form passes into midgut and is contained within Peritrophic Membrane; intermediate form differentiates into Procyclic Trypomastigote which divides and develops within endoperitrophic space; Procyclic Trypomastigote later migrates to salivary glands; division within the salivary glands results in a metacyclic form of Trypanosome which is injected into another vertebrate host while Tsetse Fly feeds. See Sleeping Sickness; Tsetse Fly. Cf. Leishmaniasis; Nagana.

HUMBOLT, FRIEDERICH HEINRICH ALEXANDER VON (1769–1859) German aristocrat (Baron), traveller and naturalist. With permission of King of Spain, Humbolt and his friend Aime Bonpland, travelled throught South America for five years (1799–1804) and wrote *Personal Narrative of Travels*. Later in life, Humbolt travelled extensively in Russia, India and China. During his life, Humbolt published many scientific works but is best known for his *Cosmos* (1845–1858) which was recognized as the most important scientific work of its time.

HUMERAL Adj. (Latin, *humeralis* > *humerus* = shoulder; *-alis* = characterized by.) Descriptive of structure relating to the shoulder or Humerus. See Humerus.

HUMERAL ANGLE 1. The angle at the base of the Costal margin of the wing. 2. Orthoptera: The obtusely rounded angle formed by the deflection of the sides of the Pronotum from the dorsum. 3. Lepidoptera: The angle of the wings at the base of the Costa, near the point of attachment to the body. 4. Coleoptera: The outer anterior angle of the Elytra.

HUMERAL BRISTLES Diptera: One or more bristles on the Humeral Callus (Comstock).

HUMERAL CALLUS Diptera: Each of the anterior angles of the Prescutum of the Mesothorax, usually a more-or-less rounded tubercle (Comstock). The Humeri.

HUMERAL CARINA Coleoptera: An elevated ridge or keel on the outer anterior angle of the Elytra.

HUMERAL CROSSVEIN A crossvein that extends between the Costa and Subcosta near the wing base (Comstock). See Vein; Wing Venation.

HUMERAL LOBE The area of contact between the hindwing and the forewing.

HUMERAL NERVE Insect wings: A transverse nerve extending between the Costa and Subcosta. Alt. Humeral Nervure.

HUMERAL PIT Diptera: See Pseudosutural Fovea (Curran).

HUMERAL PLATE The anterior Preaxillary Sclerite of the wing base which supports the Costal Vein. HP very large and conspicuous in Odonata; HP of most insects is rather small, inconspicuous or absent. See Wing Articulation.

HUMERAL STRIPE Odonata: A stripe which covers the Humeral Suture.

HUMERAL SUTURE Odonata: A suture extending from anterior of the forewing base to edge of the middle Coxa. See Basal Suture.

HUMERAL VEINLET Tillyard: See Costal Brace.

HUMERAL VEINS Lepidoptera: Secondary veins on the hindwings that strengthen the humeral angle. See Vein; Wing Venation.

HUMERALIS Coleoptera: An angulated projection along the margin at the base of the Elytra.

HUMERUS Noun. (Latin, *humerus* = shoulder. Pl., Humeri.) 1. Literally, the shoulder. 2. Orthoptera: The Femur of the foreleg. 3. Heteroptera: The lateral angle of the Prothorax. 4. Coleoptera: The basolateral angle of the Elytra. 5. Diptera: The anterior angles of the Mesonotum. 6. Hymenoptera: The Subcostal Vein in some groups; the anterior part of the Pronotum.

HUMID Adj. (Latin, *humere* = to be moist.) Pertaining to moisture. Climatically, descriptive of regions in which the normal rainfall is sufficient to produce ordinary farm crops without irrigation. Cf. Arid.

HUMMEL, ARVID DAVID (–1836) (Anon. 1858, *Accentuated List of British Lepidoptera*. 118 pp. (xxvi), Oxford & Cambridge Entomological Societies, London.)

HUMP Noun. (Middle Low German, *hump*; Greek, *hump* = lump, knoll. Pl., Humps.) A rounded protuberance. 2. A fleshy protuberance on the back of an animal. See Process 1, 2. Cf. Bump.

HUMPBACKED FLIES See Phoridae.

HUMPBACKED FLY *Megaselia aletiae* [Diptera: Phoridae]: A Species widespread in the USA. Adults are 3–5 mm long, body predominantly yellow, eyes dark; Thorax reddish and Abdomen darker with yellow bands. Eggs are deposited on caterpillar host. Larvae enter their host via its Anus. See Phoridae.

HUMPED SPIDERS See Uloboridae.

HUMP-WINGED CRICKETS See Prophalangopsidae.

HUMUS Noun. (Latin, *humus* = earth. Pl., Humes.) Dark organic material which is complex in structure and the product of decomposing plant and animal remains. Humus is the organic part of soil. See Soil.

HUNGERFORD, HERBERT BARKER (1885–1963) American academic (University of Kansas) and student of aquatic Hemiptera. (Woodruff 1963, J. Kansas ent. Soc. 36: 197–199, bibliogr.)

HUNTER, JOHN (–1893) (Wheeler 1929, N. E. J. Med. 200: 810–823.)

HUNTER, WALTER DAVID (1875–1925) (Howard 1925, Proc. ent. Soc. Wash. 27: 169–181, bibliogr.)

HUNTER, WILLIAM (1718–1783) English physician and anatomist. Hunter collected some of the Species described by Fabricius. (Zimsen 1964, *The Type Material of I. C. Fabricius*. 656 pp. (15–16), Copenhagen.)

HUNTING BILLBUG *Sphenophorus venatus vestitus* Chittenden [Coleoptera: Curculionidae].

HUNTINGTON, EDGAR IRVING (–1962) (Anon. 1963, J. N. Y. ent. Soc. 71: 58.)

HUNTSMAN SPIDERS Heteropodidae (Sparassidae).

HURICHI, JOICHIRO (1828–1898) (Hasegawa 1967, Kontyû 35 (Suppl.): 72–73.)

HURRICANE® See Fenoxycarb.

HURTIG, HENRY (1918–1973) (Anon. 1974, Bull. ent. Soc. Can. 6: 24.)

HUSK Noun. (Middle English, *huske* = husk. Pl., Husks.) The outer coating or covering of seeds. Rel. Walnut Husk Fly.

HUSSEY, ROLAND F American adcademic (University of Florida), nomenclaturist and bibliographer of Hemiptera. Hussey published a catalogue of Pyrrhocoridae.

HUTCHINGS, CLARENCE BASDEN (1877–1945) (Gibson 1946, Can. Ent. 77: 235–236, bibliogr.)

HUTCHINS, ROSS E (1912–1983) American Academic (Mississippi State University) and specialist in insect photography.

HUTCHINSON, CLAUDE MACKENZIE (–1941) (Butler 1941, Nature 148: 367.)

HUTCHINSON, EMMA SARAH (1820–1905) (Bankes 1906, Entomol. mon. Mag. 42: 43, 274, 275.)

HUTCHINSON, WILBUR LAURIN (–1933) (Hungerford 1934, Ann. ent. Soc. Am. 27: 122.)

HUTTON, FREDERICK WOUASTON HUTTEN (1836–1905) (Howard 1930, Smithson. misc. Collns. 84: 400–401.)

HUXLEY, JULIAN (1887–1975) Biologist and humanist, Professor of Zoology at Keys College, London, and formulated 'Evolutionary Humanism', an ethical theory based on natural selection. Works include: *Essays of a Biologist* (1923); *Religion Without Revelation* (1927); *The Tissue-Culture King* (1927); *Animal Biology* (with J. B. S. Haldane - 1927); *The Science of Life* (with H. G. Wells - 1931); *Evolution: the Modern Synthesis* (1942); *Evolutionary Ethics* (1943); *Biological Aspects of Cancer* (1957); *Towards a New Humanism* (1957).

HUXLEY, THOMAS HENRY (1825–1895) (Anon. 1895, Entomologist's Rec. J. Var. 7: 24.)

HUYGHE, FERNALD (1887–1916) (Berland 1920, Ann. Soc. ent. Fr. 89: 427–428, bibliogr.)

HUYSKENS, J See FERDINAND, B.

HYACINTHINE Adj. (Latin, *hyacinthinis* > Greek, *hyakinthinos*.) The purple blue of the hyacinth.

HYALINE Adj. (Latin, *hyalinus* > Greek, *hyalinos* = glassy.) A term used to describe structure that is optically clear, transparent or partly so (*e.g.* membrane). Glass-like or descriptive of water-like in colour. Alt. Hyalinus. Cf. Crystalline; Opaque.

HYALINE CELL See Granulocyte; Spherulocyte.

HYALINE GRASS BUG *Liorhyssus hyalinus* (Fabricius) [Hemiptera: Rhopalidae].

HYALINE SWIFT *Parnara amalia* (Semper) [Lepidoptera: Hesperiidae].

HYALOPLASM Noun. (Greek, *hyalos* = glass; *plasm* = anything formed or molded. Pl., Hyaloplasms.) 1. The clear, non-contractile matter in which the Fibrillar Spongioplasm of insect muscle is imbedded (Wardle). 2. The ground substance of Protoplasm within a cell. Alt. Hyaloplasma.

HYATT, ALPHEUS (1838–1902) (Benjamin 1890, Harper's Weekley 34: 925–926.)

HYATT, JAMES (1817–1904) (Schoonhoven 1904, Science 19: 635–636.)

HYBLAEIDAE Plural Noun. A small, monotypic

Family of ditrysian Lepidoptera which forms the only element of Superfamily Hyblaeoidea. Taxonomy is problematic with hyblaeids regarded as a Subfamily of Noctuidae in some classifications or a Family of Pyraloidea or Sesioidea. Adults are medium-sized, stout-bodied with wingspan 25–40 mm and head small with smooth scales; Ocelli small and Chaetosemata absent; Antenna filiform, less than half as long as forewing, with scales on dorsal surface only; Proboscis without scales; Maxillary Palpus with 3–4 segments, porrect and densely scaled; Labial Palpus porrect and densely scaled; Thorax without Tympanal Organs, with smooth scales; fore Tibia with Epiphysis; tibial spur formula 0-2-4 in female, 0-2-0 in male; middle and hind Tibiae with dense vestiture of long, piliform scales; Tarsi spined along ventral surface; male hind Tibia with conspicuous hair pencil; forewing without Pterostigma; hindwing broader than forewing. Eggs are translucent, flattened with sides parallel and ends rounded with Chorion striated. Eggs are laid individually on underside of young leaves of host plant. Mature larvae are stout with head subprognathous and body spinulose. Larvae feed within shelters formed of leaves. Pupae are stout and strongly sclerotized; Pupation occurs within larval shelter or foliage shelter near ground on in leaf litter. See Pyraloidea.

HYBOSORIDAE Plural Noun. A small Family of polyphagous Coleoptera assigned to the Scarbaeoidea. Adult with body shining, dark, 5–15 mm long; Antenna with 10 segments including club of three segments; tarsal formula 5-5-5; Abdomen with 6–7 segments. Larval head hypognathous and Antenna with three segments; legs with five segments, two tarsal claws; Urogomphi absent. Species feed on carrion. See Scarabaeoidea.

HYBRID Noun. (Latin, *hybrida* = the offspring of a tame sow and a wild boar > Greek, *hybris* = violation, outrage. Pl., Hybrids.) A genetic cross between two Species which displays a greater or lesser degree of ancestral characters of both lines.

HYBRIDOMA Noun. (Greek, *hybris* = violation, outrage; *-oma* = tumour. Pl., Hybridomas.) A unique fused cell that produces quantities of a specific antibody and which reproduces endlessly. Rel. Biotechnology.

HYBRIZONTIDAE Marshall 1872. A small Family of apocritous Hymenoptera assigned to the Ichneumonoidea. See Paxylommatidae.

HYDRADEPHAGA Noun. Aquatic, predatory, pentamerous beetles with filiform Antennae. See Adephaga.

HYDRADEPHAGOUS Adj. (Greek, *hydro* = water; + *adephaga*; Latin, *-osus* = with the property of.) Relating to or like Hydradephaga.

HYDRAENIDAE Plural Noun. Minute Moss Beetles. A numerically small Family of aquatic Coleoptera. Adults are elongate-oval, minute and dark-bodied; Antenna inconspicuous, concealed in grooves beneath head and between Prosternum and pronotal Hypomera; Maxillary Palpus well developed and sometimes longer than Antenna; Abdomen with 6–7 Ventrites; Plastron-type respiration via hydrofugal Setae. Larvae are elongate, campodeiform with legs well developed; Urogomphi articulated and pair of hooks on segment 10. Hydraenids inhabit vegetation along margin of streams and seashore where adults and larvae feed on algae. (Perkins 1980, Quaest. Ent. 16: 3–554.) See Coleoptera.

HYDRAMETHYLNON An hydrazone compound {Tetrahydro-5,5-dimethyl-2(1H)-pyrimidinone (3-(4-(trifluoromethyl)phenyl)-1(2-(4-(trifluoro-methyl) phenyl) ethenyl)-propenylidene) hydrazone} used as a stomach poison for cockroaches and fire ants. Hydramethylnon is applied to pasture, range, lawns, turf and non-crop areas. Compound is slow acting (2–3 months for cockroaches, 2–8 weeks for ants) and toxic to fishes. Trade names include: Amdro®, Blatex®, Combat®, Cyaforce®, Cyclon®, Impact®, Matox®, Maxforce®, Pyramdron®, Seige®, Wipeout®.

HYDRANGEA SCALE *Pulvinaria hyndrangeae* Steinweden [Hemiptera: Coccidae]

HYDRELLIDAE See Ephydridae.

HYDRO- Greek combining form relating to water and used as a prefix.

HYDROBIOSIDAE Plural Noun. A widespread Family of rhyacophiloid Trichoptera including about 150 Species. Adult features include Ocelli present, Antenna about as long as forewing; Maxillary Palpus with five segments. Mesoscutum and Scutellum without setal warts; wings dark, sometimes mottled; forewing to 30 mm long. Larvae are free-living with head prognathous; predatory in fast moving cold water with a few Species in standing water; foreleg chelate or modified to grasping prey; abdominal gills absent; anal Prolegs long and stout; Pupal chamber made of small stones and Pupae contained within silken cocoon. See Trichoptera.

HYDROCHIDAE Plural Noun. See Hydrophilidae.

HYDROCHLORIC ACID HCl is a colourless, fuming liquid when pure, and commercially available as a 25% solution in distilled water, when it shows no fuming, HCl mixes with water in any proportion and is used in histology as a tissue-hardener. Rel. Histology.

HYDROFLOURIC ACID DIGESTION A technique for extracting arthropod fossils from mineral deposits. (Grierson 1976, Amer. J. Botany 63: 1184.)

HYDROGEN CYANIDE A highly toxic fumigant with penetration capacity less than Methyl Bromide, and thus used in empty buildings. Use of HC on decline in favour of other fumigants, and has been banned as a fumigant in some countries.

HYDROGEN ION Chemistry: Any free, electrically charged particle of hydrogen arising from the breaking up of a compound of hydrogen in solution (ionization), indicated by turning blue litmus pink or by an acid taste, characteristic of all acids in solution, symbol, pH.

HYDROLYSE Verb. (Greek, *hydro* = water; *lyein* = to dissolve.) To decompose into two or more compounds, with the fixation of the elements of water or of some hydroxide.

HYDROLYSIS Noun. (Greek, *hydro* = water; *lyein* = to dissolve; *-sis* = a condition or state. Pl., Hydrolyses.) The chemical decomposition of a compound by water, causing formation of a new compound.

HYDROMETRIDAE Billberg 1820. Plural Noun. Marsh Treaders; Water Measurers; Water Treaders. A numerically small, cosmopolitan Family of heteropterous Hemiptera assigned to Superfamily Hydrometroidea. Adult body ca 5–8 mm long, somber-coloured, elongate, slender and stick-like; legs thread-like; compound eyes laterally protuberant and positioned halfway between anterior and posterior margins of head; wings reduced or absent; Hydrometrids inhabit vegetation associated with ponds, streams and rivers where they behave as somewhat sluggish predators of mosquito larvae, mosquito Pupae and other organisms that live on the surface of water. Eggs are glued individually on objects near water. See Hydrometroidea.

HYDROMETROIDEA A Superfamily of Hemiptera including the Families Hydrometridae, Macroveliidae and Paraphyrynoveliidae. See Hemiptera.

HYDROPHILIDAE Plural Noun. Scavenger Water Beetles. A cosmopolitan Family of polyphagous Coleoptera with about 1,600 nominal Species. Adults with body 2–40 mm long, convex and oval; Antenna short, loosely clavate, with 7–9 segments; Maxillary Palpus long and filiform; Clypeus large. Middle and hind legs modified for swimming; hind legs stroked alternately; tarsal formula 5-5-5. Larvae are campodeiform and thoracic legs short with four segments; metapneustic respiratory system with eighth spiracles functional and positioned in respiratory atrium. Adults are phytophagous or saprophagous; larvae are exclusively predaceous. Hydrophilids are predominantly aquatic but include a few terrestrial Species. Syn. Hydrochidae; Spercheidae. See Hydrophiloidea.

HYDROPHILOIDEA A Superfamily of polyphagous Coleoptera including Georyssidae, Histeridae, Hydrochidae, Hydrophilidae and Spercheidae. Adults with relatively short Antenna, long Scape, 3–segmented club; aquatic forms with segment preceding club transverse, concave and used to replentish oxygen supply; fore Coxa very large, legs spinose or dentate; wings almost always with R-M loop. Larvae are predaceous; Mandible falcate without Mola; Labrum fused; Abdomen membranous; Urogomphi relatively small and articulated. See Coleoptera.

HYDROPHILOUS Adj. (Greek, *hydro* = water; *philein* = to love; Latin, *-osus* = with the property of.) Moisture-loving; pertaining to organisms that frequently visit or live in damp places. See Habitat; Hydrophobic. Cf. Algicolous; Halophilous; Madicolous; Paludicolous; Pratinicolous; Rhophilous; Ripicolous. Rel. Ecology.

HYDROPHOBIC Adj. (Greek, *hydro* = water; *phobos* = fear; *-ic* = characterized by.) The physical or chemical characteristic of repelling water. See Hydrophilous; Hydropyle. Rel. Water Balance.

HYDROPHYTE Noun. (Greek, *hydro* = water; *phyton* = plant. Pl., Hyrodphytes.) A plant adapted to live in water or under conditions of extreme moisture. Cf. Halophyte.

HYDROPIC Adj. (Greek, *hydro* = water; *skopein* = to regard; *-ic* = of the nature of.) Descriptive of insect eggs which contain little or no yolk. Hydropic eggs occur in some parasitic Species and swell and derive nutrient for Embryogenesis when placed in the body of a host. Cf. Anhydropic.

HYDROPRENE A synthetic organic compound {Ethyl (E,E)-(R,S)-3,7,11-trimethyldodera-2,4-dienoate} used as an Insect Growth Regulator for cockroach control within non-food areas of buildings. Compound acts as a juvenile hormone analogue and does not affect adults; causes anatomical abnormalies (deformed wings, darkened Integument) manifest during last nymphal instar and induces sterility. Applied several times per year. Trade names include Gencor®, Gentrol® and Mator®. See Internal Growth Regulator.

HYDROPSYCHIDAE Plural Noun. Net-spinning Caddisflies. An numerically large Family of Trichoptera. Larvae commonly occur in small streams where the current is strongest and where they construct a case-like shelter of sand or other debris. Near the case, hydropsychids also construct a small cup-shaped net oriented to catch prey items floating down stream. See Hydropsychoidea.

HYDROPSYCHOIDEA A Superfamily of Trichoptera (caddisflies) including Arctopsychidae, Dipseudopsidae, Ecnomidae, Hydropsychidae, Philopotamidae, Polycentropodidae, Psychomyiidae, Stenopsychidae and Xiphocentronidae. See Trichoptera.

HYDROPTILIDAE Plural Noun. Microcaddisflies. A cosmopolitan Family with about 700 Species of rhyacophilous Trichoptera. Hydroptilids are the smallest caddisflies and adults are very setose with mottled colourpattern and wing to 12 mm long; Ocelli present or absent; Maxillary Palpus with five segments; Antenna shorter than forewing; Mesoscutum without setal warts. First four larval instars are free-living and subsequent instar forms silken case; anal claws small. Pupal cases are anchored to substrate. Adults aggregate on vegetation near water and may be attracted to lights. See Trichoptera.

HYDROPYLE Noun. (Greek, *hydro* = water; *pyle* = gate. Pl., Hydropyles.) A structure (organelle) adapted for the absorption of the liquid phase of water. Hydropyles occur on the eggs of some aquatic insects (Belostomatidae). A relationship exists between the onset of water absorption by eggs and embryonic growth. Three types of Hydropyles have been reported: Serosal Hydropyles, Serosal Cuticle Hydropyles and Chorionic Hydropyles. Serosal Hydropyles occur on eggs of insects such as *Pteronarchys proteus* and *Melanoplus differentialis.* Chorionic and Serosal Hydropyles occur on eggs of *Nepa cinerea.* A Columnar Serosa is found in many Heteroptera (*e.g.* Gerridae, Saldidae, Ochteridae and Notonectidae). A Basal Hydropyle on egg of *Belostoma lautarium* occurs in basal area of egg and is composed of an outer area with many fine excrescences that surround a central core of tightly packed filaments that form a group. See Eggshell; Egg. Cf. Aeropyle; Capitulum; Micropyle; Operculum. Rel. Water Balance.

HYDROSCAPHIDAE Plural Noun. A numerically small Family in the Suborder Myxophaga (Coleoptera) commonly known as skiff beetles. Hydroscaphids occur in North America and Australia. Adults and larvae are aquatic and can be found in diverse habitats including hot springs. Most commonly they occur in filamentous algae growing on rocks in freshwater streams. The adults bear a ventral Plastron and larvae possess spiracular or tracheal gills. Hydroscaphids are represented by a single Species in North America, *Hydroscapha natans* LeConte, that occurs in southwestern states. See Coleoptera.

HYDROSTATIC SKELETON A fluid mechanism that employs contractile elements (muscles) of the body to antagonize and localize contractions of body regions to increase Haemolymph pressure. Component parts of the HS include an incompressible fluid contained within a Coelome (cavity) and surrounded by muscles that extend in different directions. HS is distinct from the skeletal-muscle system because no muscle has a specific antagonist; in HS system one muscle's contraction makes it antagonistic to all other muscles. In HS system, if no bulk movement of fluid from one place to another is effected, then the muscles typically are arranged in a longitudinal and circular manner. If bulk movement of fluid from one place to another is effected, then the muscle arrangement is otherwise. Hydrostatic Skeleton is used in moulting, egg and pupal eclosion, ptilinum activation and wing expansion; localized HS activities include reflex bleeding, Proboscis eversion, genital exsertion, Haemolymph circulation in appendages. See Circulatory System. Cf. Resilin. Rel. Integument; Jumping Mechanism; Accessory Circulatory Organs.

HYDROSTATIC Adj. (Greek, *hydro* = water; *statikos* = causing to stand; *-ic* = characterized by.) 1. Descriptive of or pertaining to the pressure of water. 2. Air-filled sacs used by aquatic insects as organs of flotation. Cf. Aerostatic.

HYDROTROPISM Noun. (Greek, *hydro* = water; *trepein* = to turn; English, *-ism* = condition. Pl., Hydrotropisms.) The reaction or response of an organism to water. See Tropism. Cf. Aeolotropism; Anemotropism; Chemotropism; Electrotropism; Galvanotropism; Geotropism; Heliotropism; Phototropism; Rheotropism; Stereotropism; Thermotropism; Thigmotropism; Tonotropism. Rel. Taxis.

HYDROXYCUMARIN A secondary plant compound. (Berenbaum 1983, Evolution 37: 163–179.)

HYERES, CYBOD (Crombie 1952, Endeavour 11 (44.))

HYGRIC Adj. (Greek, *hygros* = wet; *-ic* = characterized by.) Pertaining to moist conditions or high relative humidity. Cf. Mesic; Xeric.

HYGROBIIDAE Plural Noun. A widespread Family of adephagous Coleoptera consisting of five Species. Adults are stout, oval and 8–12 mm long. Antenna are filiform with 11 segments; eyes strongly protuberant. Mesosternum with transverse suture; fore Coxa globose; frontal cavity open; middle Coxa narrow; Tibiae and Tarsi with Setae adapted for swimming; tarsal formula 5-5-5. Abdomen displays six Ventrites and basal three Ventrites that are connate. Adults and larvae are aquatic; adults are slow-moving bottom-feeding predators in stagnant water. Adults swim, stridulate and store air beneath Elytra. See Coleoptera.

HYGROMETABOLISM Noun. (Greek, *hydro* = water; *metabole* = change; English, *-ism* = state. Pl., Hygrometabolisms.) The dependence of metabolism on humidity.

HYKU, V (1924–1945) (Anon. 1946, Cas. csk. spol. Ent. 43: 89–90.)

HYLAEINAE A Subfamily of Colletidae (Apoidea) consisting of about 18 Genera.

HYLEMOX® See Ethion.

HYMEN Noun. (Greek, *hymen* = membrane. Pl., Hymens.) A thin, flat membrane serving as a partition.

HYMENOID Adj. (Greek, *hymen* = membrane; *eidos* = form.) Membrane-like in form or appearance.

HYMENOPODIDAE Plural Noun. A numerically small, little-known Family of Mantodea occurring in northern Australian rainforests. See Mantodea.

HYMENOPTERA Linnaeus 1758. (Triassic Period-Recent.) Noun. (Greek, *hymen* = membrane, *pteron* = wing.) Ants; Bees; Sawflies; Wasps. A numerically large, (ca 125,000 described Species), cosmopolitan Order of endopterygote holometabolous insects. Hymenoptera are subdivided into Suborders Symphyta (sawflies) and Apocrita (bees, wasps, ants, parasitica). Adults with head mobile, usually hypognathous (sometimes prognathous) and not fused with Pronotum; compound eyes typically large and multifaceted and reduced in some Taxa; three Ocelli usually present, occasionally absent; Antenna typically geniculate with segmentation variable and frequently sexually dimorphic; Funicle sometimes ramose in male and apical segments sometimes differentiated into club. Mandibles are present and mouthparts usually are adapted for biting-chewing; some bees are adapted for chewing-sucking. Adults bear four membranous wings; forewing usually significantly larger than hindwing with pair connected by Hamuli during flight; forewing venation complex and extensive in large bodied Species, drastically reduced in small to minute parasitic Species; hindwing venation with a few cells at most, otherwise reduced; apterous Species throughout Order; reproductive ants shed wings after nuptial flight. Apocrita with first abdominal segment incorporated into thoracic region and called a Propodeum. Propodeum and Thorax called 'Mesosoma' or 'Alitrunk' (ants). Propodeum usually separated from remainder of Abdomen by constricted segment 2 (Petiole) or segments 2 and 3 (Postpetiole of ants). Abdomen behind Petiole called 'Metasoma' or 'Gaster' (ants). Abdomen with 10 segments in primitive condition; first segment (Propodeum) and second segment (Petiole) typically without sternal element; female with appendicular Ovipositor (sawflies, parasitica) or Sting (aculeata). Cerci consisting of 1-segmented or modified into two setose patches (Pygostyli); male genitalia complex. Larvae are caterpillar-like (Symphyta) or apodous and vermiform (Apocrita). Instar number is variable and larval habitus diverse. Some Species of Hymenoptera undergo hypermetamorphic development. Apocritous larval gut not functionally connected with the Anus until feeding completed. Pupae are adecticous, usually exarate, or sometimes obtect. Biology of Species is diverse: Phytophagy, parasitism, predation, gall forming; polyembryonic development in some Species; eusociality has evolved several times. Parthenogenesis is universal within Order; females are diploid and males are haploid. Hymenoptera are highly beneficial in terms of pollination of crops and parasitism of agricultural pests. Oldest fossil Hymenoptera sawflies (Xyelidae) referrable to Late Triassic Period from Ipswich deposits in Queensland, Australia and Kirghisia, Russia. Subsequent records from Jurassic Period show 10 Families of Symphyta and Apocrita. Hymenoptera found in Baltic amber with ca 370 Species known from this deposit. Families include Agaonidae, Agriotypidae, Alloxystidae, Ampulicidae, Anacharitidae, Anaxyelidae, Andrenidae, Anthophoridae, Aphelinidae, Aphidiidae, Apidae, Apozygidae, Argidae, Aulacidae, Austroniidae, Austroserphidae, Bethylidae, Blasticotomidae, Braconidae, Bradynobaenidae, Cephidae, Ceraphronidae, Chalcididae, Charipidae, Chrysididae, Cimbicidae, Colletidae, Ctenoplectridae, Cynipidae, Diapriidae, Diprionidae, Dryinidae, Elasmidae, Embolemidae, Encyrtidae, Eucharitidae, Eucoilidae, Eulophidae, Eumenidae, Eupelmidae, Eurytomidae, Evaniidae, Figitidae, Formicidae, Gasteruptiidae, Halictidae, Heloridae, Hybrizontidae, Ibaliidae, Ichneumonidae, Leucospidae, Liopteridae, Loboscelidiidae, Masaridae, Megachilidae, Megalodontidae, Megalryidae, Megaspilidae, Melittidae, Monomachidae, Mutillidae, Mymaridae, Mymarommatidae, Ormyridae, Orussidae, Oxaeidae, Pamphiliidae, Paxylommatidae, Pelecinidae, Peradeniidae, Pergidae, Perilampidae, Platygasteridae, Plumariidae, Pompilidae, Proctotrupidae, Pteromalidae, Rhopalosomatidae, Roproniidae, Rotoitidae, Sapygidae, Scelionidae, Sclerogibbidae, Scoliidae, Scolobythidae, Sierolomorphidae, Signiphoridae, Siricidae, Sphecidae, Stenotritidae, Stephanidae, Syntexidae, Tanaostigmatidae, Tenthredinidae, Tetracampidae, Tiphiidae, Torymidae, Trichogrammatidae, Trigonalyidae, Vanhorniidae, Vespidae, Xiphydriidae and Xyelidae. Cf. Clistogastra; Chalastogastra. See Aculeata, Parasitica. Syn: Phleboptera; Piezata.

HYMENOPTERIFORM EGG (Greek, *hymen* = membrane, *pteron* = wing; Latin, *forma* = shape.) The generalized (hypothetical) ancestral form (shape) of Hymenoptera egg. HE is typically sausage-like with rounded poles and the outline shape is several times longer than wide. This egg form is expressed by most Hymenoptera and it is also found in some Diptera (Nemestrinidae, Bombyliidae, Cecidomyiidae). See Egg. Cf. Acuminate; Encyrtiform; Pedicellate. Rel. Shape.

HYMENOPTERIFORM LARVA The generalized (hypothetical) larval body form of apocritous Hymenoptera. HL is typically featureless with a pale to translucent body, head capsule weakly developed or absent, body spindle-shaped, and without thoracic legs. Cf. Eruciform.

HYOCEPHALIDAE Bergroth 1906. A small Family of Hemiptera known from Australia. Placement questionable but presently assigned to the Coreoidea and probably related to Stenocephalidae. (Stys 1964, Acta Zool. 10: 229.) See Hemiptera.

HYOID Adj. (Greek, *hyoeides* = Y-shaped.) Shaped with the form of the Greek upsilon.

HYOID SCLERITE Diptera: A small U-shaped sclerite in the pharyngeal wall between the lower end of the Fulcrum and the base of the Labrum-Epipharynx. The Hyoid Sclerite maintains the lumen of the Pharynx in a distended condition. See Sclerite.

HYPANDRIUM Noun. (Greek, *hypo* = beneath, *aner* = male; Latin, *-ium* = diminutive > Greek, *-idion*. Pl., Hypandria.) 1. The apical abdominal sternal sclerite of many insects (Sternum 10 of Ephemeroptera, Sternum 9 of Psocoptera). 2. The Hypoproct of Needham; Subgenital Plate of Snodgrass. 3. A projecting sclerite or extension over a sclerite (Tillyard). 4. The ninth abdominal Sternum of male insects (Crampton).

HYPENA MOTH *Hypena laceratalis* Walker [Lepidoptera: Noctuidae].

HYPERCEPHALIC Adj. (Greek, *hyper* = above; *kephale* = head; *-ic* = characterized by.) Descriptive of pertaining to heads of some Diptera (Richariidae, Otitidae, Platystomidae, Tephritidae, Micropezidae, Periscelididae, Sepsidae, Diopsidae, Drosophilidae). Hypercephalic males have extremely broad heads that are disproportionately large and an example of sexual dimorphism. See Head. Cf. Macrocephalic.

HYPERCEPHALY Adj. (Greek, *hyper* = above; *kephale* = head.) Diptera: A form of sexual dimorphism in which males have broad head.

HYPERGAMESIS Noun. (Greek, *hyper* = above; *gamos* = marriage. Pl., Hypergameses.) 1. The process by which excess (supernumerary) Spermatozoa are absorbed or utilized by the female. The phenomenon has been recognized in several unrelated groups of insects, which suggests that Hypergamesis has evolved several times and probably is adaptive. *e.g.* Bed bug [*Cimex lectularius* Linnaeus] and sheep blowfly [*Lucilia cuprina* (Wiedemann)]. 2. The utilization of excess Spermatozoa by the female *Cimex* as a source of nutrition.

HYPERMETAMORPHOSIS Noun. (Greek, *hyper* = above; *meta* = after; *morphosis* = shaping. Pl., Hypermetamorphoses.) 1. Endopterygote insects whose larvae change form, shape or substance during successive instars as a normal consequence of development. Examples are found in (but not restricted to) Coleoptera (Meloidae), Strepsiptera, Diptera (Acroceridae, Bombyliidae), Lepidoptera (Epipyropidae), and Hymenoptera (Eucharitidae, Perilampidae). See Larval Heteromophosis. 2. Insects which pass through a larger than expected number of stages with supernumerary stages interpolated between the mature larva and adult. Rel. Parasitism; Planidium; Triungulin; Triungulinid.

HYPERPARASITE Noun. (Greek, *hyper* = above; *para* = beside; *sitos* = food; *-ites* = resident. Pl., Hyperparasites.) An organism which develops as a parasite of another parasite. Syn: Secondary Parasite. See Parasite. Cf. Tertiary Parasite.

HYPERPARASITISM Noun. (Greek, *hyper* = above; *para* = beside; *sitos* = food; English, *-ism* = condition. Pl., Hyperparasitisms.) A progenetive strategy in which individuals of one Species behave as parasites in relation to individuals of another Species that is developing as a parasite of a free-living organism. Within the Insecta, Hyperparasitism seems restricted (or most common) among Hymenoptera, Coleoptera and Diptera. See Parasite. Cf. Facultative Hyperparasitism; Obligatory Hyperparasitism; Adelphoparasitsm; Tertiary Hyperparasitism.

HYPERPARASITOIDISM Noun. (Greek, *hyper* = above; *para* = beside; *sitos* = food; English, *-ism* = state or condition. Pl., Hyperparasitoidisms.) A condition in which one parasitoid is in turn parasitized by another parasitoid. Syn: Hyperparasite. See Parasitism. Cf. Direct Hyperparasitoid; Indirect Hyperparasitoid. Rel. Host Discrimination.

HYPERPNEUSTIC Adj. (Greek, *hyper* = above; *pneuma* = breathe; *-ic* = characterized by.) Pertaining to insects with supernumerary Spiracles, as in Thysanura and the Thorax of some Diplura (Snodgrass).

HYPERPYRENE Adj. (Greek, *hyper* = above; *pyren* = fruit stone.) Pertaining to Hyperpyrene Sperm.

HYPERPYRENE SPERM Spermatozoa that display a normal length, but their heads are larger and the Flagella are disproportionately thicker than normal Spermatozoa. See Spermatozoa. Cf. Apyrene Sperm; Eupyrene Sperm; Oligopyrene Sperm.

HYPERSENSITIVITY Noun. (Greek, *hyper* = above; *sentire* = to feel; English, *-ity* = suffix forming abstract nouns. Pl., Hypersensitivities.) 1. A condition of being excessively sensitive to a stimulus. 2. A pathological condition in which an obligatory parasite (pathogenic fungus) penetrates a host plant, causes surrounding plant cells to die, induces conditions promoting death of the invading parasite, and thereby terminates the fungal infection. (Wood 1967, Physiological Plant Pathology.)

HYPERTELY Noun. (Greek, *hyper* = above; *telos* = end. Pl., Hypertelies.) 1. Beyond the bounds of the useful. 2. A condition in which forms resemble other objects more closely than needful, or without apparent purpose. 3. Excessive imitation in coloration or coloration pattern.

HYPERTRIGONAL SPACE Supreatriangular space.

HYPERTROPHIDAE Plural Noun. A small Family of ditrysian Lepidoptera assigned to Superfamily Gelechioidea and which includes ca 50 described Species, all from Australia and New Guinea. Adults are small with brightly coloured scales and a wingspan of 8–35 mm; head with smooth scales; Ocelli and Chaetosemata absent; Antenna filiform, about 0.75 times as long as forewing; Scape without Pecten; Proboscis scaled; Maxillary Palpus with four segments, folded over Proboscis; Labial Palpus short, recurved; fore Tibia with Epiphysis; tibial formula 0-2-4. Larvae feed on leaves of Myrtaceae and occupy short, tubular, open-ended shelters on underside of leaves; faecal material woven into shelter. Pupae are twig-like with Maxillary Palpus exposed and Labial Palpus concealed; Proleg scars on abdominal segments 4–6; pupation in diverse places including at leaf tips, on vegetation and within leaf litter. Adults probably are diurnal and walk with waddling gait; wings held tectiform over Abdomen at repose.

HYPERTROPHY Noun. (Greek, *hyper* = above; *trophe* = nourishment. Pl., Hypertrophies.) Abnormal enlargement or excessive development of a structure, organ or body part. Cf. Atrophy.

HYPHA Noun. (Greek, *hyphe* = web. Pl., Hyphae.) The thread-like projections or filaments from the fungal body. See Fungus. Cf. Mycelium.

HYPISTOMA Noun. The Hypopharynx.

HYPNODY Adj. (Greek, *hypnos* = sleep; *eidos* = form.) Lethargy. A condition similar to or identical with hibernation. A term proposed by Kunckel as a substitute for Hypermetamorphosis. Cf. Aestivation; Athermopause; Hibernation.

HYPNOTHECA Noun. (Greek, *hypnos* = sleep; *theke* = case. Pl., Hypnothecae.) The Pseudonymph stage of Meloidae (Kunckel). See Hypermetamorphosis.

HYPO- (Greek, *hypo* = under.) Prefix meaning 'under,' 'beneath,' 'less' or 'down.'

HYPOBLAST Noun. (Greek, *hypo* = under; *blastos* = bud. Pl., Hypoblasts.) The inner germ layer of the Gastrula which becomes Entoderm, and to some extent Mesoderm and Endoderm. Alt. Endoblast.

HYPOCEREBRAL GANGLION A median nerve mass positioned ventrad of Aorta, above Pharynx and posteriad of Tritocerebrum. HG is connected to Frontal Ganglion by Anterior Recurrent Nerve HG is sometimes fused with Corpora Cardica and continues rearward as Posterior Recurrent Nerve to Ingluvial Ganglion. Ingluvial Ganglion (Ventricular Ganglion, Stomachic Ganglion) is positioned on gut at junction of Foregut and Midgut. Imms (1930) notes that in saltatorial Orthoptera, two Ingluvial Ganglia exist and are serviced independently by a pair of Posterior Recurrent Nerves. See Incretory Organs; Occipital Ganglion; Tritocerebrum.

HYPOCHONDRIUM Noun. (Greek, *hypo* = under; *chondros* = cartilage; Latin, *-ium* = diminutive > Greek, *-idion*. Pl., Hypochondria.) Paired partial segments which come between the first entire ventral segment and the posterior part of the Postpectus in some Coleoptera.

HYPOCRATERIFORM Adj. (Latin, *hypo* = less; *crater* = bowl; *forma* = shape.) Shaped as a shallow bowl or dish. See Crateriform. Rel. Form 2; Shape 2; Outline Shape.

HYPODACTYLE Noun. (Greek, *hypo* = under; *daktylos* = finger. Pl., Hypodactyles.) The so-called Labium of Hemiptera.

HYPODERMAL Adj. (Latin, *hypo* = under; *dermis* = skin; *-alis* = characterized by.) Descriptive of or relating to the Hypodermis or outer skin. Alt. Hypodermatic; Hypodermic.

HYPODERMAL COLOURS Colours lodged in cells of the insect Hypodermis (Epidermis) in the form of granules or fat droplets. HC include red, orange, yellow or green and are very evanescent after death (Imms).

HYPODERMATIDAE Plural Noun. Cattle Grubs; Heel Flies; Warble-Flies. A small Family of parasitic muscoid Diptera regarded as a Subfamily of Oestridae in some classifications. Adult Postscutellum large; Squamma large. Larvae with reduced mouth-hooks and develop as dermal parasites, principally of rodents, rabbits, cattle and deer. See Diptera. Cf. Oestridae.

HYPODERMIC ENVELOPE See Peripodal Sac.

HYPODERMIN Noun. A peptide first isolated and identified from the larva of a warble fly *Hypoderma lineatum*. Hypodermin displays little action on small molecule substrates of trypsin, chymotrypsin, elastase or microbial collagenases, and is a homologue of chymotrypsin.

HYPODERMIS Noun. (Latin, *hypo* = under; *dermis* = skin. Pl., Hypodermata.) 1. The cellular layer beneath and secreting the chitinous cuticule of the Integument. Syn. Epidermis. 2. The lining membrane of Elytra and Hemelytra. Alt. Hypoderm.

HYPOGEAL Adj. (Greek, *hypo* = under; *geos* = earth; *-alis* = characterized by.) Descriptive of organisms with a subterranean existence or habitat. Used to characterize some wasps (Mutillidae) and beetles (Scarabaeidae) which live in the soil. Alt. Hypogean; Hypogaeic; Hypogaeous. Cf. Cryptobiotic.

HYPOGENOUS Adj. (Greek, *hypo* = under; *geos* = earth; Latin, *-osus* = with the property of.) Descriptive of organisms that grow or develop beneath the surface of the soil. See Habitat. Cf. Geophilous. Rel. Ecology.

HYPOGLOSSIS Noun. (Greek, *hypo* = under; *glossa* = tongue; *-sis* = a condition or state. Pl., Hypoglosses.) The ventral surface of the tongue. The Hypoglottis. See Tongue.

HYPOGLOTTIS Noun. (Greek, *hypo* = under; *glotta* = tongue. Pl., Hypoglottises.) A sclerite inserted between the Mentum and Labium in many Coleoptera.

HYPOGNATHOUS Adj. (Greek, *hypo* = under; *gnathos* = jaw; Latin, *-osus* = with the property of.) 1. General zoological useage: Descriptive of animals with the lower jaw slightly longer than the upper jaw. 2. Entomological useage: Descriptive of insects (immature or adult) with the head vertically oriented and the mouth directed ventrad. Most insects with an hypognathous condition display an Occipital Foramen near centre of head's posterior surface. The hypognathous condition is considered by most insect morphologists to represent the primitive or generalized condition. The hypognathous position is evident in most major groups of insects and can be seen in the grasshopper, housefly and honeybee. Other conditions are probably derived from ancestors with an hypognathous head. See Head. Cf. Opisthognathous; Orthognathous; Prognathous.

HYPOGRAPHOUS Adj. (Greek, *hypo* = under; *graphein* = to write; Latin, *-osus* = with the property of.) Shaded. A term applied to a Fascia that becomes gradually darker.

HYPOGYNIUM Noun. (Greek, *hypo* = under; *gyne* = female; Latin, *-ium* = diminutive > Greek, *-idion*. Pl., Hypogynia.) The eighth or apical abdominal Sternum of the female insect.

HYPOMERE Noun. (Greek, *hypo* = under; *meros* = part. Pl., Hypomeres.) A ventral process of the Phallobase *sensu* Snodgrass.

HYPOMERON Noun. (Greek, *hypo* = under; *meros* = part. Pl., Hypomera.) Coleoptera: The inflexed edge of the Pronotum (Pronotal Hypomera), and the elevated lower margin of the Epipleura (Elytral Hypomera). See Epiplural Fold.

HYPONYM Noun. (Greek, *hypo* = under, *onyma* = name. Pl., Hyponyms.) 1. A generic name not founded upon a type Species. 2. A provisional name for a specimen. Rel. Nomenclature; Taxonomy.

HYPOPARATYPE Noun. (Greek, *hypo* = under; *para* = beside; *typos* = pattern. Pl., Hypoparatypes). See Type. Rel. Nomenclature; Taxonomy.

HYPOPHARYNGEAL Adj. (Greek, *hypo* = under; *pharyngx* = gullet; Latin, *-alis* = pertaining to.) Pertaining to the Hypopharynx.

HYPOPHARYNGEAL GLAND Social Hymenoptera: Paired glands in the head of adult workers who produce food for the brood. HG opens via ducts into the Hypopharynx. Syn. Food Glands; Pharyngeal Glands.

HYPOPHARYNGEAL SCLERITES Bees: A pair of strap-like sclerites that extend from the Hypopharynx to the Mentum. See Epipharyngeal Sclerite.

HYPOPHARYNGEAL SKELETON See Tentorium.

HYPOPHARYNGEAL SUSPENSORIUM See Suspensorium of Hypopharynx.

HYPOPHARYNX Noun. (Greek, *hypos* = under; *pharyngx* = Pharynx. Pl., Hypopharynices.) 1. A tongue-like, sensory structure projecting from the oral cavity. 2. The upper surface of Labium that serves as an organ of taste, or a true 'tongue.' 3. The Maxillula and Lingua *sensu* MacGillivray. The Hypopharynx forms a floor of the functional mouth and appears a product of several gnathal segments. Hypopharynx is well developed in many Paurometabola, including Hemiptera, Psocoptera and Mallophaga. In cockroach *Periplaneta americana,* Hypopharynx forms a tongue-like lobe that is reinforced by two lingual sclerites (Matsuda 1965: 100). Hypopharynx is not well developed or extensively represented in many groups of Holometabola. Salivary Gland ducts open onto the Hypopharynx. See Mouthparts; Pharnyx. Cf. Epipharnyx.

HYPOPLEURAL BRISTLES Diptera: Bristles on the Hypopleura, usually in a vertical row (Curran).

HYPOPLEURAL ROW Diptera: See Hypopleural Bristles.

HYPOPLEURITE Noun. (Greek, *hypos* = under; *pleuron* = side; *-ites* = constituent. Pl., Hypopleurites.) 1. The lower pleural sclerite formed when the Pleuron is divided horizontally into two parts. 2. Symphyta larvae: Enlarged portion of a Proleg ventrad of Postepipleurite. See Pleuron. Rel. Sclerite.

HYPOPLEURON Noun. (Greek, *hypos* = under; *pleuron* = side. Pl., Hypopleura.) 1. Diptera: The lower part of the mesothoracic Epimeron (Katepimeron) (Comstock). 2. A space over the middle and hind Coxae, between the Metapleura and Pteropleura. 3. The side of the Metasternum (Smith).

HYPOPNEUSTIC Adj. (Greek, *hypo* = under; *pnein* = to breathe.) Pertaining to a type of Tracheal System in which insects display fewer than 10 pairs of functional Spiracles. See Tracheal System; Spiracle. Cf. Apneustic; Hemipneustic; Holopneustic; Oligopneustic; Polypneustic; Propneustic. Rel. Respiratory System.

HYPOPTERA Noun. (Greek, *hypo* = under; *pteron* = wing. Pl., Hypopterae.) The Tegula. Alt. Hypoptere

HYPOPUS Noun. (Greek, *hypo* = under; *pous* = foot. Pl., Hypopi.) The phoretic, non-feeding deutonymphal stage of some astigmatid mites. 'Anal suckers' consist of an adhesive that attaches the Hypopus to an insect which transports the mite from one place to another place. Rel. Phoresey.

HYPOPYGIAL SETAE Some coccids: The two longitudinal rows of Setae on the ventral aspect of the outer wall of the anal tube (MacGillivray).

HYPOPYGIAL SPINE *Cynips*: A spinous structure of the Hypopygium.

HYPOPYGIDIUM Noun. (Greek, *hypo* = under; *pyge* = rump; Latin, *-ium* = diminutive > Greek, *-idion*. Pl., Hypopygidia.) Hymenoptera: Sternum IX. Syn: Subgential Plate.

HYPOPYGIUM Noun. (Greek, *hypo* = under; *pyge* = rump; Latin, *-ium* = diminutive > Greek, *-idion*. Pl., Hypopygia.) 1. Adults: The posterior portion of the Abdomen; the Anus; the ventral sclerite of the anal opening. 2. Diptera: The male sexual organs and terminal segments of the Abdomen. The Propygium. 3. Coleoptera: The last segment behind the Elytra (MacGillivray).

HYPOSTERNUM Noun. (Greek, *hypo* = under; *sternon* = chest. Pl., Hyposterna.) Collectively, the Peristernum, Preepisternum, Katopleura, Prepectus and Praesternum *sensu* MacGillivray.

HYPOSTIGMAL CELL 1. Neuroptera Planipennia: A greatly elongated cell behind the fusion of Sc and R . 2. The Hypostigmatic Space *sensu* Tillyard.

HYPOSTIGMATIC CELL or SPACE See Hypostigmal Cell.

HYPOSTOMA Noun. (Greek, *hypo* = under; *stoma* = mouth. Pl., Hypostomata; Hypostomatae) 1. Diptera: Part of the head between the Antennae, compound eyes and mouth. 2. Hemiptera: The lower part of face. 3. Ticks: A dart-like structure arising from the median ventral surface of the Basis Capituli (Matheson). See Gnathosoma. 4. Crustacea: The upper lip or Labrum. Alt. Hypostome.

HYPOSTOMAL AREAS Sclerites set off by the Hypostomal Sutures.

HYPOSTOMAL BRIDGE One of three fundamental closures on the posterior aspect of the insect Cranium with the resultant separation of Occipital Foramen from the oral cavity. Characteristically, the HB is developed in adult heads displaying a hypognathous axial orientation, and includes members of the Diptera, Hymenoptera and Heteroptera. HB is formed by the medial extension and fusion of the hypostmal lobes (Hypostoma). In Diptera, the Hypostomal Bridge has been called the Pseudogula. According to Crampton (1942) the Hypostomal Bridge is not present in some Tanyderidae and Anisopodidae. See Head Capsule. Cf. Gula; Postgenal Bridge.

HYPOSTOMAL SCLERITE Muscid larvae: Each of two irregularly shaped lateral sclerites connected by a ventral, bar-like sclerite, to which the mandibular sclerite articulates.

HYPOSTOMAL SUTURE Part of the Subgenal Suture posterior of the Mandible. Often, the HS is obsolete or suppressed (Snodgrass).

HYPOTENUSIS Noun. Odonata: The simple or broken crossvein between Media and Cubitus which forms the outer boundary of the triangle.

HYPOTHESIS Noun. (Greek, *hypotithenai* = to propose, to suppose; *-sis* = a condition or state. Pl., Hypotheses.) 1. An assumption made as the beginning or basis of a course of logical inquiry. *e.g.* All men are created equal. 2. A proposition tentatively accepted and which undergoes modification through testing. Facts of process (evolution), procedure (courtship behaviour) or action (pentose shunt) that emerge through critical analysis of hypothesis. 3. A stage in the hierarchical assembly of facts that ultimately leads to the explanation of universal truths: Observation leads to Hypothesis development; Hypothesis testing leads to formulation of Theory; Theory testing leads to the formulation of Law. Law is the universal truth. Cf. Law; Theory. Rel. Falsifable Hypothesis; Verifyable Hypothesis.

HYPOTHETICAL Adj. (Greek, *hypothetickos*; Latin, *-alis* = characterized by.) Descriptive of a logical or conditional hypothesis. See Hypothesis. Cf. Theoretical.

HYPOTHETICAL TAXONOMIC UNIT (HTU) See Taxonomy.

HYPOTHETICO-DEDUCTIVE REASONING (METHOD) A step-wise process that involves speculation based on observation to generate an hypothesis, followed by experimentation to test that hypothesis. See Mayr (1982), Weisman (1892: 303) Hull (1973) Ruse (1975, 1979.) Rel. Hypothesis.

HYPOTOME Noun. (Greek, *hypo* = under; *temnein* = to cut. Pl., Hypotomes.) Hymenoptera: Sternum IX; structure in bumble bees between the Volsella and the Penis (Dufour).

HYPOTYPE Noun. (Greek, *hypo* = under; *typos* = pattern. Pl., Hypotypes.) A specimen (not the Holotype) upon which a subsequent or supplementary description or figure is based. See Type. Cf. Apotype; Plesiotype. Rel. Nomenclature; Taxonomy.

HYPOVALVAE Noun. (Greek, *hypo* = under; Latin, *valva* = fold.) Valvular processes of the eighth abdominal Sternum (Crampton).

HYQUAT® See Chlormequat.

HYSLOP, JAMES AUGUSTUS (1884–1953) (O'Neill 1953, Proc. ent. Soc. Wash. 55: 153–156.)

HYSTEROSOMA Noun. (Greek, *hysteros* = after; *soma* = body. Pl., Hysterosomata.) Acarology: A division of the body posteriad of the Sejugal Furrow, composed of the Opisthosoma and the two posterior segements of the Podosoma. See Idiosoma.

HYSTEROTHELY Noun. (Greek, *hysteros* = after; *telos* = completion. Pl., Hysterothelies.) 1. An error expressed during Metamorphosis in which larval characters are present in the Pupae or pupal characters are present in the adult. Hysterotely is meditated by environmental characteristics (temperature), parasites (merminthid nematodes) or chemicals (internal growth regulators). The condition has been reported in Coleoptera and Lepidoptera. 2. The expression of sexual maturity during the larval stage (Torre Bueno). 3. A condition in which sexual immaturity is prolonged indefinitely. See Teratology. Cf. Metathetely; Prothetely.

HYSTRICHOPSYLLIDAE Plural Noun. Hystrichopsyllid Fleas. The second-largest Family of Siphonaptera. Head with 2–3 rows or bristles; Interantennal Suture typically present; three thoracic Terga collectively longer than first abdominal Tergum; Pronotal Comb present; Metanotum without marginal spines; fore Femur with lateral apical bristle larger than medial apical bristle; hind Tibia with a short apical tooth; abdominal Terga 2–7 typically with two rows of bristles. Adults live as parasites of small rodents, insectivores (shrews) and carnivores. See Siphonaptera.

HYTEC® See Tecnazene.

HYTHERGRAPH Noun. A temperature-rainfall diagram or graph.

HYTOX® Isoprocarb.

IACCHOIDES SKIPPER *Trapezites iacchoides* Waterhouse [Lepidoptera: Hesperiidae].

IACCHUS SKIPPER *Trapezites iacchus* (Fabricius) [Lepidoptera: Hesperiidae].

-IASIS Greek suffix indicating a morbid medical condition; suffix for names of diseases.

IBA® See Indolebutyric Acid.

IBALIIDAE Thomson 1862. Plural Noun. Small Family (2 Genera, about 15 Species) of apocritous Hymenoptera assigned to the Cynipoidea. *Ibalia* is endemic to Northern Hemisphere; *Heteribalia* is endemic to southeast Asia. Adult body to 20 mm long; female Antenna with 13 segments and male Antenna with 15 segments; large spur on hind Tarsomere II; Gaster compressed; Tergum VI largest; Ovipositor concealed. Ibaliids display a complex biology (solitary egg-larval parasite); females locate hosts boring in wood or host's eggs; females oviposit into eggs or early larvae of Siricidae; first instar larvae are polypodeiform; later instars are hymenopteriform. Early larval instars are endoparasitic; late larval instars are ectoparasitic; fourth instar with large Mandibles but does not feed. Pupation occurs in tunnel of host. Ibaliids are used in Tasmania and New Zealand for biological control of siricids in pines. See Cynipoidea. Cf. Liopteridae.

ICE CRAWLERS See Grylloblattidae.

ICEPLANT SCALE *Pulvinariella mesembryanthemi* (Vallot); *Pulvinaria delottoi* Gill. [Hemiptera: Coccidae].

ICHES, M L (–1960) (Anon. 1960, Bull. Soc. ent. Fr. 65: 5.)

ICHNEUMON FLIES See Ichneumonidae.

ICHNEUMON WASPS See Ichneumonidae.

ICHNEUMONIDAE Latreille 1802. Plural Noun. Ichneumonflies; Ichneumon Wasps. The Largest Family of parasitic Hymenoptera, containing about 25 Subfamilies, 1,250 Genera and 20,000 nominal Species. Adult body size is variable; Antenna not geniculate, typically with more than 13 segments and not apically clubbed; Mandible typically bidentate. Pronotum extending to Tegula; typically macropterous, Pterostigma present, two Recurrent Veins, sometimes brachypterous or apterous. Most abundant in tropical-subtropical regions. All Species are parasitic, host spectrum focused on larval or pupal Holometabola, particularly Lepidoptera, Symphyta and phytophagous Coleoptera. Most Ichneumonidae are solitary internal parasites, with five larval instars; sometimes gregarious; sometimes hyperparasitic. Most Species spin cocoons and pupate where the host died or pupate within the cocoon or Pupae of host. Ichneumonidae are usually arrhenotokous; some Species are thelytokous. See Ichneumonoidea.

ICHNEUMONOIDEA Latreille 1802. A large, cosmopolitan Superfamily of generalized apocritous Hymenoptera with long Antenna of more than 13 segments, and a Trochantellus. Considered by some workers as among the most primitive groups of the Parasitica. Included Families: Agriotypidae, Aphidiidae, Apozygidae, Braconidae, Ichneumonidae, Paxylommatidae (= Hybrizontidae), Praeichneumonidae, and in some classifications the Megalyridae and Stephanidae. See Apocrita; Hymenoptera. Cf. Ceraphronoidea; Chalcidoidea; Cynipoidea; Ephialtitoidea; Evanioidea; Proctotrupoidea; Stephanoidea; Trigonaloidea.

ICHTHYOPHAGOUS (Greek, *ichthus* = fish; *phagein* = to eat; Latin, *-osus* = with the property of.) Fish-eating. See Feeding Strategy.

ICILIUS BLUE BUTTERFLIES *Jalmenus icilius* Hewitson [Lepidoptera: Lycaenidae].

ICON® See Lambda Cyhalothrin.

ICON SPECIES See Indicator Species.

ICONOREX One of a Family of clay-coated papers used in publishing high-quality glossy prints and for producing illustrations with pen-and-ink. The clay-coating allows for scraping of inked parts to add the effect of highlights where needed. Iconorex is a brand name of Tomasetti Papers located in Brisbane, Australia.

ICOTYPE Noun. (Greek, *eidos* = resemble; *typos* = pattern. Pl., Icotypes.) A specimen that is part of the type-series, has been examined by taxonomic specialists and serves for the purpose of identification of the Species. See Type. Rel. Nomenclature; Taxonomy.

ICTINUS BLUE BUTTERFLY *Jalmenus ictinus* Hewitson [Lepidoptera: Lycaenidae].

IDENTIFICATION Noun. (Latin, *idem* = the same; English, *-tion* = result of an action. Pl., Identifications.) The process of determining the name of the biological form in hand. Identifications are made using taxonomic keys, comparison with descriptions, or comparison with other specimens.

IDEOTYPE Noun. (Greek, *idein* = to see; *typos* = pattern. Pl., Ideotypes.) A specimen named by the author after comparison with the type, but not also a Topotype. See Type. Cf. Holotype; Topotype; Syntype. Rel. Nomenclature; Taxonomy.

IDIOBIOLOGY Noun. (Greek, *idios* = personal; *bios* = life; *logos* = discourse. Pl., Idiobiologies.) The biology of an individual organism. Syn. Autobiology. Cf. Autecology.

IDIOBIONT Noun. (Greek, *idios* = distinct, personal; *bion* = living. Pl., Idiobionts.) Protelean parasites (parasitoids) which kill, permanently impair or paralyse their hosts after oviposition and thereby prevent further development of the hosts. Typically, Idiobionts are ectoparasites that attack hosts in concealed situations and express a broad host spectrum (generalists). See Parasitism. Cf. Koinobiont. (See Askew & Shaw, pp. 225–265 in: *Insect Parasitoids*. Waage & Greathead, eds. Academic Press.)

IDIOGASTRA The hymenopterous Suborder Orussoidea in some classifications. See Hymenoptera. Cf. Chalastogastra.

IDIOSOMA Noun. (Greek, *idios* = distinct, *soma* = body. Pl., Idiosomae.) Acarology: A division of the body posteriad of the Circumcapitular Furrow, *i.e.,* the body without the Gnathosoma.

IDIOSTOLIDAE Plural Noun. A small Family of heteropterous Hemiptera assigned to Superfamily Idiostoloidea, known from Australia and southern South America. Idiostolids are ground-dwelling Species but their biology is poorly known; Species are asssociated with *Notho-fagus*. See Idiostoloidea.

IDIOSTOLOIDEA A Superfamily of Hemiptera including Idiostolidae. See Hemiptera.

IGNITUS Adj. (Latin, *ignus* = fire.) Fire-red.

IGR Acronym. See: Insect Growth Regulator.

IHERING, HERMANN VON (1850–1930) (Chiarelli 1931, Physis 10: 339–342.)

IHERING, RODOLPHO THEODORE GASPAR WILHELM (1883–1939) (Urban 1840, *In* Martius, Flora Brasiliensis 1: 34.)

IIZUKA, AKIRA (1862–1938) (Hasegawa 1967, Kontyû 35 (Suppl.): 5–6.)

IKEDA, SAKUJIRO (1860–1938) (Hasegawa 1967, Kontyû 35 (Suppl.): 9–10.)

IKI-1145® See Fosthiazate.

IKUMA, YOICHIRO (–1916) (Hasegawa 1967, Kontyû 35 (Suppl.): 6.)

ILEAL VALVE A strong constriction in the Intestine near the eighth abdominal segment (Needham).

ILEOCOLON Noun. (Latin, *ileum* = groin, *colon* = the Large Intestine.) The anterior portion of the hindgut which extends from the midgut to the Rectum. The term is applied when the hindgut is not distinctly differentiated into Ileum and Colon.

ILEUM Noun. (Greek, *Ile* = roll, twist; Latin, *ileum* = groin.) The anterior part of the Anterior Intestine, between the Pylorus and the Colon; the Small Intestine *sensu* Snodgrass. An undifferentiated tube connected to the Pylorus in most insects; in some insects the Ileum is convoluted. Functionally, convolution increases surface area without substantially increasing the abdominal volume it occupies. The Ileum has different functions in various groups of insects: Pouches for flagellates in termites; removes water from excretory product in Heteroptera; produces hormones in some Diptera *(Ostrinia),* and excretes ammonia in blowflies. The Ileum forms a fermentation chamber in Scarabeidae. See Excretion; Hindgut. Cf. Colon; Malpighian Tubule; Pyloric Valve; Rectum.

ILIAC GLANDS Paired tubular glands attached to the hindgut of larval parasitic Hymenoptera. IG perhaps are modified Malpighian Tubules.

ILIMA LEAFMINER *Philodoria marginestrigata* (Walsingham) [Lepidoptera: Gracillariidae].

ILIMA MOTH *Amyna natalis* (Walker) [Coleoptera: Curculionidae].

ILLIDGE, ROWLAND (1846–1929) Naturalist in Australia who studied life histories of many insects. Butterflies of his collection went to the South Australian Museum. (Tryon 1929, Qld. Nat. 7: 13–19, bibliogr.)

ILLIDGE'S ANT-BLUE *Acrodipsas illidgei* (Waterhouse & Lyell) [Lepidoptera: Lycaenidae].

ILLIGER, JOHANN CARL WILHELM (1775–1825) German academic who completed Kugelann's catalog called *Verzeichniss der Kaefer Preussens* (1798). Illiger also published *Magazin fur Insectenkunde* (7 vols, 1801–1807), and *Systematiches Verzeichniss von den Schmetterlingen der Wiener gegend* (2 vols, 1801), and continued Rossi's *Fauna Etrusca.* (Marseul 1887, Abeille 24 (Les ent. et leurs écrits): 188–190, bibliogr. only.)

ILLUSIONS OF PARASITOSIS A human mental condition in which physical discomfort attributed to insects or mites is actual and due to environmental conditions (static electricity, *etc.*) Cf. Delusions of Parasitosis; Delusory Parasitosis; Entomophobia.

ILTSCHEV (ILCHEV), DIELCHO (1885–1925) (Anon. 1935, Ent. News 36: 224.)

IMAGE, SELWYN (1849–1930) (Sheldon 1930, Entomologist 63: 238–239.)

IMAGINAL Adj. (Latin, *imago* = image; *-alis* = characterized by.) Descriptive of conditions or pertaining to the adult (Imago).

IMAGINAL BUDS Holometabolous insects: Embryonic cells around and from which the organs and appendages of the future adult develop. See Imaginal Disc.

IMAGINAL CELLS Cells that will produce adult structure.

IMAGINAL DISC A mass of tissue in the larva which will ultimately become a structure or ogan in the adult.

IMAGINATION Noun. (Latin, *imago* = image, likeness; English, *-tion* = result of an action.) The process of becoming an Imago or adult insect (Packard).

IMAGINE Noun. (Latin, *imago* = image, likeness.) The Imago or full grown insect. An adult.

IMAGO Noun. (Latin, *imago* = image, likeness. Pl., Imagines; Imagos.) The adult stage or sexually developed insect. Cf. Subimago.

IMAMURA, SHIGEMOTO (1904–1936) (Hasegawa 1967, Kontyû 35 (Suppl.): 14.)

IMBRICATE Adj. (Latin, *imbrex* = tile; *-atus* = adjectival suffix.) Pertaining to similar anatomical structures that partly overlap. Structures arranged or appear as the scales on a fish or the shingles on a roof. Archane, Imbricatus. Cf. Valvate.

IMBRICATED SNOUT WEEVIL *Epicaerus imbricatus* (Say) [Coleoptera: Curculiondiae]: A monovoltine pest of apple, strawberries and other plants. ISW eggs are laid on leaves of plants. Larvae feed on stems and roots of host plants while the adults feed on leaves and buds.

IMHOF, OTHMAR EMIL (1855–1936) (Schinz 1937, Vjschr. naturf. Ges. Zürich 82: 472–474.)

IMHOFF, LUDWIG (1801–1868) (Bischof-Ehinger 1872, Mitt. schweiz, ent. Ges. 3: 73–81, bibliogr.)

IMHOFF-GERBER, HENRI (1879–1959) (Beuret

1959, Mitt. ent. Ges. Basel 9: 61–65, partial bibliogr.)

IMIDACLOPRID A nitroqualdine compound {1-(chloro-3-pyridylmethyl)-N-nitroimidazolidin-2-ylidiniamine} developed as a synthetic analogue of Nicotine. Imidacloprid is used as a systemic insecticide and stomach poison used to control plant-sucking insects, thrips and leaf-feeding beetles. Applied as a soil drench or foliar spray to ornamentals, cereals, corn, rice, potatoes, vegetables and cotton in some countries. Imidacloprid may be used in combination with entomopathogenic fungus on some vegetables. Compound with long-lasting effects and antifeeding activity; can be used with other pesticides. Trade names include: Admire®, Bay-NTN-33893®, Confidor®, Gaucho®, Marathon®, Merit®, Premier®, Premise®, Provado®. Marketed under name Premise® by Bayer for control of termites. Cf. Nicotine.

IMIDAN® See Phosmet.

IMITATOR MOLE CRICKET *Scapteriscus imitatus* Nickle & Castner [Orthoptera: Gryllotalpidae]: A Species of mole cricket morphologically smiliar to the Changa and separable only by a combination of measured features. IMC occurs in Puerto Rico, Venezuela, Guyana, Ecuador and Brazil. See Mole Crickets. Cf. Changa Mole Cricket; Tawny Mole Cricket; Short-Winged Mole Cricket.

IMMACULATE Adj. (Latin, *immaculatus* from *im* = not; *maculare* = to stain; *-atus* = adjectival suffix.) Destitute of spots or marks. Archane: Immaculatus.

IMMARGINATE Adj. (Latin, *im* = not; *margo* = border; *-atus* = adjectival suffix.) Without an elevated rim or margin. Archane: Immarginatus.

IMMERSED GERM BAND A germ band that sinks into the yolk of the egg (Folsom & Wardle).

IMMERSED Adj. (Latin, *immersus* from *im* = in; *megere* = to plunge.) Inserted, imbedded or concealed within. Archane: Immersus.

IMMIDAE Heppner 1977 Plural Noun. A small Family of ditrysian Lepidoptera assigned to Superfamily Immoidea. Family consists of ca 250 Species most of which occur in Orient and Australia; a few Species in neotropics, Africa and Japan. Adults are small with wingspan 15–35 mm and head with smooth scales; Ocelli nearly always absent and Chaetosemata nearly always present; Antenna less than 0.75 times as long as forewing, typically filiform; Proboscis without scales; Maxillary Palpus minute with 1–2 segments; Labial Palpus curved upward; fore Tibia with Epiphysis; tibial spur formula 0-2-4. Egg stage is unknown. Biology is poorly known. See Lepidoptera.

IMMIGRANT ACACIA WEEVIL *Orthorhinus klugi* Boheman [Coleoptera: Curculionidae].

IMMIGRANT ORGANISMS See Adventive. Cf. Introduced Organisms.

IMMIGRATION Noun. (Latin, *immigratus* > *immigrare* = to go; to remove.) Movement of people from one country into another country. Cf. Dispersal; Migration.

IMMOBILE Adj. (Latin, *im* = not; *mobilis* = to move.) Descriptive of ogranisms that do not have motion.

IMMOBILIZED ENZYME An enzyme that is localized in a defined region and which can be reused in a continuous process.

IMMORTALITY Noun. (Latin, *im* = not; *mortalis* = mortal; English, *-ity* = suffix forming abstract nouns.) Genetics: The concept of continuing and continuous life of the germplasm.

IMMORTALIZATION Noun. (Latin, *im* = not; *mortalis* = mortal; English, *ion* = suffix.) Tissue Culture: The transformation of a cell with a finite life-span into a cell posessing an indefinite life-span. Rel. Biotechnology.

IMMS, AUGUSTUS DANIEL (1880–1949) English Entomologist perhaps best known for his *Textbook of Entomology*. (Wigglesworth 1949, Obit. Not. Fellows. R. Soc. Lond. 6 (18): 463–470, bibliogr.)

IMMUNE RESPONSE See Cellular Defence Mechanism. Rel. Antibody; Antigen.

IMMUNITY Noun. (Latin, *immunitas* > *immunis*; English, *-ity* = suffix forming abstract nouns. Pl., Immunities.) 1. A condition or capacity of an organism to resist disease or infection from microorganisms. Immunity is manifest in several ways: Active Immunity; Acquired Immunity; Natural Immunity; Passive Immunity. 2. A high resistance to feeding by insects as shown in some plants.

IMPACT® See Hydramethylnon.

IMPEL® See Inorganic Borate.

IMPERATOR® See Permethrin.

IMPERFECT Adj. (Latin, *imperfectus*.) 1. Descriptive of structure or process not being perfect in form or function. 2. A term sometimes applied to the immature insects in contradistinction to perfect or the adult stage.

IMPERFORATE Adj. (Latin, *im-* = not; *perforatus* = boring through.) Not perforated.

IMPERIAL BLUE BUTTERFLY *Jalmenus evagoras* (Donovan) [Lepidoptera: Lycaenidae].

IMPERIAL MOTH *Eacles imperialis* (Drury) [Lepidoptera: Saturniidae].

IMPERIAL WHITE BUTTERFLY *Delias harpalyce* (Donovan) [Lepidoptera: Pieridae].

IMPLICATE Adj. (Latin, *im* = in; *plica* = fold; *-atus* = adjectival suffix.) Folded inward.

IMPORTED CABBAGEWORM *Pieris rapae* (Linnaeus) [Lepidoptera: Pieridae]: Endemic to Europe and introduced into North America ca 1860. A widespread, significant pest of crucifers. ICW overwinters as Pupae and adults emerge during spring, mate and oviposit on underside of host plant leaves. Eggs are yellow, bullet-shaped with ridges. Larvae are green, feed on leaves and complete development within 15 days. ICW with three generations per year in parts of its range. Alt. Cabbage Butterfly; European

Cabbageworm. See Pieridae.

IMPORTED CRUCIFER WEEVIL *Baris lepidii* Germar [Coleoptera: Curculionidae].

IMPORTED CURRANTWORM *Nematus ribesii* (Scopoli) [Hymenoptera: Tenthredinidae]: A pest of currant and gooseberry. ICW is endemic to Europe and was adventive to North America ca 1857. ICW overwinters as larvae or Pupae within cocoons on ground. Adults are ca 8–10 mm long, predominantly black with pale marks on Abdomen. Eggs are white, elongate, flattened and laid end-to-end along veins or midrib on underside of leaves. Larvae are green with black spots; they feed along edges of leaves, sometimes stripping plant. When threatened, larvae elevate anterior and posterior parts of their bodies from leaf and remain attached via abdominal prolegs on middle segments of body. ICW completes 1–3 generations per year, depending upon climate. Syn. *Pteronidea ribesii.* See Tenthredinidae. Cf. Currant Stem-Girdler.

IMPORTED FIRE ANT *Solenopsis saevissima richteri* Forel [Hymenoptera: Formicidae]. A noxious pest in the southeastern USA that was adventive to Alabama ca 1920. IFA workers are polymorphic, 2.5–6 mm long and variable in colour (reddish to reddish brown); Gaster blackish with basal yellow band and Mandible with four well-defined teeth. Nests are characterized by mounds of variable height. IFAs feed upon other insects and occasionally seeds. Workers sting readily and cause pustules on skin. See Formicidae. Cf. Little Fire Ant; Native Fire Ant; Southern Fire Ant. Rel. Argentine Ant; Black Carpenter-Ant; Fire Ant; Larger Yellow-Ant; Little Black Ant; Odorous House Ant; Pavement Ant; Pharaoh Ant; Thief Ant.

IMPORTED LONGHORNED WEEVIL *Calomycterus setarius* Roelofs [Coleoptera: Curculionidae].

IMPORTED WILLOW LEAF BEETLE *Plagiodera versicolora* (Laicharting) [Coleoptera: Chrysomelidae].

IMPREGNATE (Latin, *impraegnatus*, pp of *impregnare* = to make pregnant.) Verb. To make fertile or pregnant, to fertilize.

IMPRESSED Adj. (Latin, *impressus* > *imprimere* > *im* = in; *premere* = to press.) Descriptive of structure or surface containing shallow depressed areas or markings. Alt. Impressus. Cf. Depressed.

IMPRESSION Noun. (Latin, *impressio*. Pl., Impressions.) 1. An indentation, mark, stamp or depression made through physical contact with a surface. *e.g.* Marks in wood made by wood feeding insects. 2. An influence or effect made on an organism as measured by a change in behaviour. Cf. Depression.

IMPRINTING Noun. (Latin, *imprimere* = to imprint.) A form of learning in which a newborn recognizes that an object has a particular function. See Learning. Cf. Associative Learning; Facilitation; Habituation.

IMPUBIS Adj. (Latin, *in* = not; *pubescere* = maturity.) Descriptive of structure without Setae (Archaic).

IMPUNCTATE (Latin, *in* = not; *punctum* = point; *-atus* = adjectival suffix.) Not punctate or marked with punctures. Alt. Impunctatus. Cf. Smooth; Polished; Punctate.

IN VITRO (Latin, *in vitro* = in glass.) Biological process, growth or development of organisms under artificial conditions.

IN VIVO (Latin, *in vivo* = in something alive.) Biological process, growth or development of organisms under natural conditions.

INA GRASSDART *Taractrocera ina* Waterhouse [Lepidoptera: Hesperiidae].

INAEQUALIS Adj. Latin term meaning 'not equal'.

INAMURA, SOZO (1890–1945) (Hasegawa 1967, Kontyû 35 (Suppl.): 13.)

INANITION Noun. (Latin, *inanition* = emptiness; English, *-tion* = result of an action.) 1. The state of emptiness. 2. Starvation or exhaustion from a lack of food or non-assimilation of nutrients.

INARTICULATE Adj. (Late Latin, *inarticulatus* = not articulate.) Not jointed or segmented.

INAURATE Adj. (Latin, *inaurere* = to gild; *-atus* = adjectival suffix.) Golden yellow. Alt. Inauratus.

INCANUS Adj. (Unknown origin.) White with a slight admixture of black; hoary; the colour of a grey head (Kirby & Spence).

INCASED PUPA See Pupa Folliculata.

INCENSE CEDAR SCALE *Xylococculus macrocarpae* (Coleman) [Hemiptera: Margarodidae], an insect commonly taken on Incense Cedar along the western coast of North America.

INCENSE CEDAR WASP *Syntexis libocedrii* Rohwer [Hymenoptera: Anaxyelidae].

INCERTAE SEDIS Latin phrase used in Taxonomy to mean 'of uncertain position' – pertains to organisms whose taxonomic placement is not clear.

INCH Noun. (Middle English, *inche* from Latin *uncia* = the 12th part.) The English and American standard of length in insect measurement. One inch = 12 lines, ca. 25 mm; usually expressed in units and hundredths, as 1.01.

INCHBALD, PETER (1815–1896) (Anon. 1896, Entomol. mon. Mag. 32: 164.)

INCHMEN *Myrmecia* spp. [Hymenoptera: Formicidae].

INCIPIENT SPECIES 1. An aberrant form or group of forms showing plasticity in the direction of fluctuating changes in structure, of specific value if stabilized (Ferris). 2. Variations from the norm of a Species, which appear to be stable and differentiable to a greater or less degree from the usual form of a Species (Kinsey).

INCIPIENT Adj. (Latin, *incepere* = to begin.) 1. Beginning to be; to appear. 2. Species of animals in the process of arising.

INCISED Adj. (Latin, *incedere* = to incise.) Notched; descriptive of structure that is deeply cut. Alt.

Incisus.

INCISED NOTCHES Coccids: The Incisurae *sensu* MacGillivray.

INCISION Noun. (French *incision*; from Latin, *incisio* = cutting into.) 1. Any cut into the margin of a structure or through a surface. 2. The marginal slits or notches in Coccidae. Cf. Excision.

INCISOR Noun. (Latin, *incisus* = cut into; *or* = one who does something. Pl., Incisors.) 1. A sharp tooth adapted for cutting. 2. Any structure that produces an incision. See Mandible. Cf. Molar; Tooth.

INCISOR LOBE Mandibles of some insects: The toothed, distal lobe which is used for biting. See Mandible. Cf. Molar; Tooth.

INCISURA AXILLARIS The axillary incision of the insect wing.

INCISURA Noun. (Latin, *incidere* = to cut into. Pl., Incisurae.) 1. A notch or indentation in a smooth or flat surface. 2. Coccids: Incisions in the margin of the Pygidium (MacGillivray).

INCISURE Noun. (Latin, *incidere* = to cut into. Pl., Incisures.) An impressed line marking the junction of two segments. An incision.

INCLINATE Adj. (Latin, *inclinus* = bending; leaning; *-atus* = adjectival suffix.) Inclined. Pertaining to structure which is leaning or tilting. Cf. Delinate; Prolinate; Reclinate.

INCLUSIVE FITNESS Social Insects: The fitness of the individual and the individual's influence on the fitness of relatives other than direct descendants. See: Individual Selection; Kin Selection.

INCLUSUS Adj. (Latin, *inclusus* = confined; included; shut-up.) Pertaining to structure entirely or partly concealed.

INCONSPICUOUS Adj. (Latin, *inconspicuus; -osus* = with the property of.) Not attracting attention or quickly noticeable.

INCRACEL® See Chlormequat.

INCRASSATE Adj. (Latin, *incrassare* = thickened; *-atus* = adjectival suffix.) Thickened. Pertaining to a structure that is expanded or swollen at some one point, especially near the apex. Alt. Incrassated; Incrassatus.

INCRETORY ORGANS Specialized groups of cells within a Head that synthesize and secrete hormones necessary for growth and Metamorphosis. IOs are not part of the Brain, but are intimately associated with it. Incretory Organs include Frontal Ganglion, Hypocerebral Ganglion, Ingluvial Ganglion, Corpora Cardiaca and Corpora Allata. See Central Nervous System. Cf. Corpora Allata; Corpora Cardiaca; Frontal Ganglion, Hypocerebral Ganglion, Ingluvial Ganglion.

INCUBATE (Latin, *incubare* = to lie on; *-atus* = adjectival suffix.) To brood; to cause to develop, as an egg.

INCUBATION Noun. (Latin, *incubare* = to lie on; English, *-tion* = result of an action.) 1. The embryonic period of development. 2. The period between fertilization and eclosion. Rel. Development.

INCUMBENT Adj. (Latin, *incumbere* = to lie upon.) Lying one over another. Pertaining to wings when they cover the dorsum horizontally. Alt. Incumbens.

INCUMBENT WINGS Wings that lie horizontal over the dorsum when the insect is at rest.

INCUNABULUM Adj. (Latin, *incunabula* = swaddling clothes, craddle. Pl., Incunabula.) Folliculus; cocoon.

INCURRENT Adj. (Latin, *in* = into; *currere* = to run.) Leading into. Cf. Excurrent.

INCURRENT OSTIUM (Pl., Ostia.) Paired openings in lateral wall of the Heart that admit Haemolymph from the Pericardial Sinus into the Heart. When the Heart is bulbar or chambered, the Incurrent Ostium usually is located behind the imaginary midline which bisects the Chamber; IO sometimes are positioned near the posterior end of the Chamber. See Dorsal Vessel. Cf. Excurrent Ostium. Rel. Circulation.

INCURVARIIDAE Plural Noun. Incurvariid Moths. A widespread Family of heteroneurous Lepidoptera and nominant Family of Superfamily Incurvarioidea; most abundant in temperate regions. Adults are small bodied with wingspan 5–18 mm and head with elevated hair-like scales; Ocelli absent; Antenna filiform, about half as long as forewing; Proboscis short, without scales and probably functional; Maxillary Palpus with five segments, folded; Labial Palpus with three segments, pendulous; tibial spur formula 0-2-4; Ovipositor strongly sclerotized, flattened, triangular. Larvae display six Stemmata and thoracic legs sometimes are absent. Pupae are exarate and Maxillary Palpus present. Larvae feed as leaf miner on trees and shrubs; some Species are leaf miners only during first instar; late instars of some Species are casebearers. See Incurvarioidea.

INCURVARIOIDEA A Superfamily of monotrysian Lepidoptera including Adelidae, Crinopterygidae, Cecidostidae, Heliozelidae, Incurvariidae, Prodoxidae and Tscheriidae. Body small to very small; Ocelli and Chaetosemata absent; Antenna often with Pecten; Proboscis absent, short or long; wings typically with Microtrichia; Ovipositor adapted for piercing plant tissue. Larvae are apodous or with short Prolegs; Crochets in short rows; Anal Prolegs and Crochets usually absent. Larvae are leaf miners, leaf skeletonizers, gall formers or borers in twigs and stems. See Lepidoptera.

INCURVATE Adj. (Latin, *incurvus* = bent; *-atus* = adjectival suffix.) Bowed or curved inwards. Alt. Incurvatus; Incurved.

INDENTATION Noun. (Latin, *indentatus* = indented; English, *-tion* = result of an action.) A dent or dimple; a depression in the surface of Cuticle or body part. See Invagination. Cf. Evagination.

INDEPENDENT VEIN 1. Lepidoptera: A Vein arising from the Crossvein closing the cell, and which does not branch directly from any Vein reaching the base. 2. The Media *sensu* Comstock. 3. Vein number 5 of the series of Smith. See Vein; Wing Venation.

INDETERMINATE Adj. (Latin, *indeterminatus* = not determined; *-atus* = adjectival suffix.) Not defined or well marked; obscure. Pertaining to structure without constant form or shape. Arcane: Indeterminatus.

INDEX FOSSIL Taxa (typically Genera or Species) that are used to correlate spatially separate stratigraphic sections. IF are easily distinguished from related Taxa, geographically widespread, occur in many kinds of sedimentary rocks and are limited to narrow stratigraphic intervals. Syn. Facies Fossil; Guide Fossils. Rel. Trace Fossil.

INDIAN HONEY BEE *Apis cerana* [Hymenoptera: Apidae]. Syn. Asian Honey Bee. Cf. European Honey Bee.

INDIAN HOUSE-CRICKET *Gryllodes sigillatus* (Walker) [Orthoptera: Gryllidae].

INDIAN MEAL-MOTH *Plodia interpunctella* (Hübner) [Lepidoptera: Pyralidae]: A cosmopolitan pest of grains, nuts, dried fruits, cereals in kitchens, stores and granaries; IMI is endemic to Europe and distributed through commerce. Household infestations sometimes are attributed to dried pet food; late instar larvae can penetrate polyethylene packages and earlier instar larvae follow. Females lay 50–400 eggs within a few days of emergence. Eggs are coppery, laid individually or in masses of 10–30 on larval food; eclosion occurs within 3–14 days, depending upon temperature. Larvae cover their food with silken web while feeding, and leave food clinging to web. IMM completes 5–7 larval instars; larval development requires 6–8 weeks. Pupation occurs in silken cocoon distant from food source; pupal period requires 1–3 weeks depending upon temperature. Life cycle is completed in eight weeks and Species can complete 1–2 generations per year in temperate regions or eight generations in tropical regions. See Pyralidae. Cf. Mediterranean Flour Moth.

INDIAN NEEM TREE *Azadirachta indica* A. Jussieu [Meliaceae]. A fast-growing tree native to arid India and Burma that has been planted in Africa, Madagascar, Mauritius, Indonesia, Thailand, Australia and Caribbean. Neem is widely grown for seeds that contain the limonoid compound Azadirachtin which is used as a Botanical Insecticide. See Azadirachtin; Neem. Cf. Philippine Neem Tree; Thai Neem Tree; Marrango.

INDIAN WEED CATERPILLAR *Heliothis rubrescens* (Walker) [Lepidoptera: Noctuidea].

INDIAN WHITE WAX SCALE *Ceroplastes ceriferus* (Fabricius) [Hemiptera: Coccidae].

INDICATOR SPECIES A Species used as a gauge for the condition of a particular habitat, community or ecosystem. A characteristic or surrogate Species for a community or ecosystem. Syn. Icon Species; Sentinel Species.

INDIGENOUS Adj. (Latin, *indigena* = native; Latin, *-osus* = with the property of.) 1. Native to a geographical region or ecological habitat. See Autochthonous. Rel. Ecology. 2. Occurring or living naturally in an area; not introduced. Related terms: Exotic; Adventive; Native.

INDIGO FLASH *Rapala varuna simsoni* (Miskin) [Lepidoptera: Lycaenidae].

INDIGOTIC Adj. (Spanish, *indigo* = blue dye; *-ic* = consisting of.) A very deep indigo blue.

INDIRECT Adj. (Latin, *indirectus* = not direct.) As applied to Metamorphosis, complete.

INDIRECT HYPERPARASITOID In parasitical systems involving a host, primary parasite and hyperparasitic Species: A Species of hyperparasitic insect whose adult female stage lays eggs in or near the host Species. IHs attack of the host is made without regard to the presence or absence of the primary parasite. See Hyperparasitism. Cf. Direct Hyperparasitoid.

INDIRECT WING MUSCLES The dorsal and tergosternal muscles. Cf. Direct Wing Muscles. See Flight Muscles.

INDIVIDUAL Noun. (Latin, *in* = not; *dividuus* = divisible; *-alis* = pertaining to.) An entity, person, zooid or particular thing, localized in space and time, and possessing unique properties. Cf. Population.

INDOLE BUTYRIC® See Indolebutyric Acid.

INDOLEBUTYRIC ACID An organic compound {Indole-3-butyric acid} used as a Plant Growth Regulator on plant cuttings to promote and accelerate rooting. Trade names include: Chrizopon®, Chryzosan®, Chryzotek®, Hormex Rooting-Powder®, Hormodin®, IBA®, Indole Butyric®, Jiffy Grow®, Oxyberon®, Rhizopon AA®. Cf. Plant Growth Regulator.

INDUCED RESISTANCE A type of host-plant resistance in which a particular plant condition or environmental state makes a plant more resistant to pest activity than under other circumstances.

INDUMENTUM Adj. (Latin, *indumentum* = covering.) Noun. A covering of Setae, scales or tufts.

INDURATED Adj. (Latin, *indurescere* = to harden.) Hardened; pertaining to structure becoming harder. Alt. Induratus; Indurescent.

INDUSIUM Noun. (Latin, *induere* = to put on; *-ium* = diminutive > Greek, *-idion*. Pl., Indusia.) 1. A third envelope of the embryonic insect which appears as a disc-like thickening of the Blastoderm anterior of the future head in Locustidae (Imms). 2. An epidermal plant covering which protects the sorus of ferns. 3. The case made by the insect larva (Smith).

INDUSTRIAL MELANISM A form of crytpic coloration in which natural selection, operating through selective predation, modifies the coloration of prey to changing ground colour patterns produced by industrial pollutants. A phenomenon observed for some Lepidoptera. See Crypsis.

INDUVIAE Plural Noun. (Latin, *induviae* = garments.) A chitinous, concentrically layered structure symmetrically placed between the compound eyes, posterior of the Brain and above the Oesophagus. Induviae also found in the ventral side of the eighth and ninth abdominal segments (Needham).

INEQUAL Adj. (Latin, *in* = not; *aequus* = equal; *-alis* = characterized by.) Superficially with irregular elevations and depressions.

INERMIS Adj. (Latin, *inermis* = unarmed.) Not armed; without Setae, spines or any other sharp process. See Mutic; Muticus. Alt. Inermous.

INERT INGREDIENT A chemical in a pesticide formulation that has no direct effect on killing organisms. See Pesticide. Cf. Active Ingredient; Diluent; Solvent; Surfactant; Synergist.

INFANTILE Adj. (Latin, *infantilis, in* = not; *fari* = to speak.) 1. Pertaining to the actions, physical condition or appearance of an infant. See Development. Cf. Adulthood; Juvenile.

INFECTION Noun. (Latin, *inficere* = to taint; English, *-tion* = result of an action. Pl., Infections.) The establishment of a pathogen or parasite within a host. See Disease. Cf. Injury.

INFECTIVITY Noun. (Latin *infecere* = to taint; English, *-ity* = suffix forming abstract nouns.) Cf. Pathogenicity.

INFERICORNIA Hemiptera: Taxa whose Antennae appear inserted well down on the sides of head, *e.g.* Lygaeidae.

INFERIOR Adj. (Latin, inferior = lower; *or* = a condition.) A term of position indicating beneath, below or behind. Cf. Superior.

INFERIOR APPENDAGE Male Odonata: The lower one or two of the terminal abdominal parts used to clasp the female in copulation.

INFERIOR LOBE See Lobus Inferior.

INFERIOR ORBIT See Genal Orbit (MacGillivray).

INFERIOR REGION The Costal region of the wing.

INFERIOR WING The hindwing. See Secondaries.

INFERIORS Odonata: The lower anal appendages (Garman).

INFERO-POSTERIOR Positioned below and behind.

INFERTILITY Noun. (Latin, *in* = in; *fertilis* from *ferre* = to bear, carry; English, *-ity* = suffix forming abstract nouns.) An inability to reproduce. Cf. Fertility.

INFESTATION Noun. (Latin, *infestare* = to attack, disturb.) 1. The act or process of being infested. 2. The condition or state of being infested by large numbers of internal or external parasites so as to be harmful or bothersome.

INFILTRATION Noun. (French, *infiltration*; English, *-tion* = result of an action.) The act or process of filtering or permeating. Implicitly, a slow or gradual process.

INFLATED (Latin, *inflatus* past part., *inflare* = to inflate.) Adj. Blown up; distended; bladder-like. Cf. Bullate; Tumid.

INFLECTED Adj. (Latin, *inflectere* = to bend.) Bent inward at an angle. Cf. Deflected.

INFLECTION Noun. (Latin, *inflexio* > *inflectere* = to bend; English, *-tion* = result of an action. Pl., Inflections.) An inward bending or flexion.

INFLUENT Noun. (Latin, *influere* = to flow into. Pl., Influents.) Ecology: An animal or plant having an influence on other living forms. Cf. Subinfluent.

INFRA-ANAL FLAPS Podical Sclerites.

INFRA-ANAL LOBE Lepidoptera larva: A thick, conical, fleshy lobe, often ending in chitinous point, positioned beneath the vent.

INFRA-ANAL PLATES See Podical Plates.

INFRABRUSTIA Noun. Brustia limited to the ventral or caudal aspect of the Mandible (MacGillivray).

INFRABUCCAL CAVITY Adult Formicidae: A cavity on the floor of the mouth, opening by a short, narrow canal. IBC presumably is used for storage and compaction of indigestible material that will be eliminated. Syn. Infrabuccal Chamber; Infrabuccal Pocket.

INFRACAPITULUM Noun. Acarology: Inferior part of the Gnathosoma bearing the lips and Palpi and containing the mouth and Pharnyx.

INFRACERCAL PLATES Orthoptera: Generally inconspicuous paired sclerites that underlie in part the Cerci and in part the lateral portion of the Supra-anal Sclerite.

INFRACLYPEUS Noun. (Latin, *infra* = below; *clypeus* = shield.) Anteclypeus; Rhinarium.

INFRACOXAL Adj. (Latin, *infra* = below; *coxa* = hip; *-alis* = pertaining to.) Lying below the Coxa.

INFRACTED Adj. (Latin, *infractus* = breaking.) Structure directed abruptly inward, as if broken. Alt. Infractus.

INFRAEPIMERON Noun. (Latin, *infra* = below; *epimeron*.) The Sclerite of the Epimeron; the Katepimeron. See Thorax; Sternum. Cf. Infraepisternum.

INFRAEPISTERNUM Noun. (Latin, *infra* = below; *episternum*. Pl., Infraepisterna.) The Katepisternum or lower Sclerite (division) of the Episternum. See Thorax; Sternum. Cf. Infraepimeron.

INFRAGENITAL Adj. (Latin, *infra* = below; genital; *-alis* = pertaining to.) Below the genital opening or process.

INFRAMARGINAL Adj. (Latin, *infra* = below; marginal.) Positioned below or behind the Marginal Cell of the wing, or behind or below any margin.

INFRAMEDIAN VEIN Orthoptera: Ulnar Vein. See Vein; Wing Venation.

INFRAOCULAR Adj. (Latin, *infra* = below; ocular.) Positioned below and between the eyes.

INFRAOESOPHAGEAL Adj. (Latin, *infra* = below; esophagus; *-alis* = characterized by.) Positioned below the Oesophagus. See Suboesophageal.

INFRARED Adj. (Latin, *infra* = below; red.) Beyond the visible wavelengths of light at the red end of the spectrum.

INFRASPECIFIC Adj. (Latin, *infra* = below; *species*; *-ic* = characterized by.) A taxonomic category below the level of Species. Includes Subspecies, varieties. Cf. Conspecific.

INFRASQUAMAL SETULAE Diptera: Fine Setae

positioned below the point of attachment of the Squamae (Curran).

INFRASTIGMATAL Adj. (Latin, *infra* = below; *stigmata*; *-alis* = characterized by.) Positioned below the Stigmata or spiracles.

INFRINGING Intrans. Verb. (Latin, *in* = in; *frangere* = to break.) Encroaching upon.

INFUMATED Adj. (Latin, *infumatus* = smoking.) Smoke-coloured; clouded, as with smoke. Alt. Infumatus.

INFUNDA Noun. 1. Hemiptera: The chitinized globular swelling on which the Salivos is located. The Salivary Pump, Salivary Injector (MacGillivray). 2. Coccoids: The Hypopharynx of Berlese & Green (MacGillivray).

INFUNDIBULATE Adj. (Latin, *infundibulum* = a funnel; *-atus* = adjectival suffix.) Shaped as a funnel. Alt. Infundibuliform.

INFUNDIBULUM Noun. (Latin, *infundibulum* = a funnel.) Any funnel-shaped organ or structure.

INFUSCATE Adj. (Latin, *in* = in; *fuscare* = to make dark; *-atus* = adjectival suffix.) Smokey grey-brown, with a blackish tinge. Alt. Infuscated; Infuscatus.

INGENITZKII, IVAN (1862–1900) (Anon. 1900, Zool. Anz. 23: 384.)

INGENS Adj. (Origin unknown.) Unusually large or disporportionate in size.

INGEST Verb. (Latin, *ingerere* = to put in.) To eat.

INGESTION Noun. (Latin, *ingestio* = to pour in; English, *-tion* = result of an action. Pl., Ingestions.) Taking in of food; eating.

INGESTIVE Adj. (Latin, *ingerere* = to put in.) Pertaining to ingestion or the swallowing of food.

INGESTIVE INFECTION Medical Entomology: That form in which parasites are received at the time of feeding. The infective stage of a parasite taken through the mouth (Matheson).

INGHAM, CHARLES HENRY (1904–1957) (Henne 1957, Lepid. News 11: 169–170.)

INGLIS, JAMES DALGLEISH (1909–1933) (Gregson 1948, Proc. Ent. Soc. Br. Columb. 44: 35–36.)

INGLUVIAL Adj. (Latin. *ingluvies* = a maw or crop; *-alis* = pertaining to.) Descriptive of or pertaining to the Ingluvies (Crop) of insects.

INGLUVIAL GANGLION A paired ganglion of the Stomodaeal Nervous System in some insects. IG is positioned on the side of the Crop. The Gastric Ganglion *sensu* Snodgrass. See Nervous System.

INGLUVIES See Crop.

INGPEN, ABEL (1796–1854) (Anon. 1855, Proc. Linn. Soc. Lond. 2: 425–426.)

INGURGITATION Noun. (Latin, *ingurgtare* = to pour in; English, *-tion* = result of an action.) The act of swallowing.

INITIATORIN Noun. A secretion of the posterior Glandula Prostatica of some moths which acts as a sperm-activating factor within the inner and outer matrices of the Spermatophore.

INJURY Noun. (Middle English, *injurie*; from Latin, *injuria* = wrong, an injury. Pl., Injuries.) The effects of pest activities that are deleterious on host plant condition or physiology. See Damage.

INLAND ARMYWORM *Persectania dyscrita* Common [Lepidoptera: Noctuidae]. See Armyworms.

INLAND FIELD CRICKET *Lepidogryllus parvulus* (Walker) [Orthoptera: Gryllidae].

INLAND KATYDID *Caedicia simplex* (Walker) [Orthoptera: Tettigoniidae]: A pest of citrus in Australia. Adults are ca 50 mm long and green with yellow stripe on anterior margin of wing. Eggs are disc-like, ca 5 mm long and laid in parallel rows of 10–12 on tree trunks or limbs. Nymphs feed on young foliage, flowers and fruit; Species completes five instars and is Monovoltine. See Tettigoniidae. Cf. Angular-Winged Katydid; Broad-Winged Katydid; Citrus Katydid; Crested Katydid; Forktailed Bush-Katydid; Mottled Katydid; Philippine Katydid; Spotted Katydid.

INLAND ROBUST SCORPION *Urodacus yaschenkoi* (Birula) [Scorpionida: Scorpionidae].

INNER CLASPERS 1. Hymenoptera: See Volsella. 2. Auchenorrhyncha: See Styles.

INNER DORSOCENTRAL BRISTLES Diptera: See Acrostichal Bristles.

INNER LOBE 1. Generally referring to the Maxilla. 2. The Lacinia.

INNER MARGIN 1. The line extending along the lower or interior edge of the wing from the base to the hind or Anal Angle. 2. Coccids: The margin connecting the inner angle and the apex of the Operculum (MacGillivray).

INNER PARAMERE Hymenoptera: Volsella.

INNER PLATE Hymenoptera: See Second Valvifers.

INNER SQUAMA Diptera: The larger of the two Squamae arising from the posterior scutellar margin of the wing-bearing segment, and forming a protective hood-like canopy over the Haltere.

INNER SURFACE OF LEG 1 The medial surface of the leg when it is extended ventral from the body. 2. The ventral surface of the insect leg when it is extended at right angles to the body (Snodgrass). See Orientation.

INNERVATE Trans. Verb. (Latin, *in* = in; *nervus* = sinew; *-atus* = adjectival suffix.) To supply with nerves.

INNES, WALTER FRANCIS (1858–1937) (Anon. 1937, Bull. Soc. ent. Egypte 21: 173–174.)

INNOVATE Adj. (Latin, *inovare* = to renew; *-atus* = adjectival suffix.) Without markings.

INOCELLIIDAE Plural Noun. Snakeflies. A very small Family of Rhaphidoidea with about thirty nominal Species. Inocelliids are widespread in Holarctic. Adults lack Ocelli. See Rhaphidoidea.

INOCULAR Adj. (Latin, *in* = into; *oculus* = eye.) Descriptive of an Antenna or structure originating near the eye.

INOCULAR ANTENNA An Antenna whose base (Scape) is partly or entirely surrounded by the compound eye. See Antenna.

INOCULATION Noun. (Latin, *inoculare* = to engraft. Pl., Inoculations.) 1. Medicine: The introduction (usually through injection) of microorganisms into a body or tissue in order to produce antibodies. 2. The introduction by insects (usually through feeding or oviposition) of pathogenic organisms into plants or animals.

INOCULATIVE Adj. Medical Entomology: The manner of infection of the organism when it takes place through the act of biting by the carrier, the organism being inoculated through the feeding process (Matheson).

INOCULATIVE RELEASE In biological control, the repeated release of relatively small numbers of a natural enemy into an area with the intent of building a field population or establishing the natural enemy. Cf. Augmentative Release; Inundative Release.

INOCULUM Noun. (Latin, *inoculare* = to ingraft. Pl., Inocula.) A pathogen or pathogen parts capable of infecting a host plant. Inoculua include fungal spores, virus particles and bacterial cells. Cf. Pathogen.

INOMATA, SHUJIRO (1893–1958) (Hasegawa 1967, Kontyû 35 (Suppl.): 13–14.)

INOPEPLIDAE See Salpingidae.

INORATE SCALE *Aonidiella inornata* McKenzie [Hemiptera: Diaspididae].

INORGANIC BORATE Disodium Octaborate Tetrahydrate, an inorganic compound ($Na_2B_8O_{13}$-$4H_2O$) which occurs in borate minerals. Applied to wood to kill wood-feeding insects and fungi. Compound phytotoxic; treatment rapidly penetrates wood to deposit microscopic crystals; IB permanent, odourless, non-staining and non-corrosive. Trade names include: Bora-Care®, Impel® and Tim-Bor®. See Inorganic Insecticide.

INORGANIC INSECTICIDE A category of insecticides whose members are of mineral origin and which do not include carbon. II typically exist in the form of crystalline solids; as insecticides, II used in baits and dusts. Member compounds vary in mammalian toxicity from moderate to very high and are stable over long periods of time. Examples include: Arsenious Trioxide, Boric Acid, Cryolite, Diatomaceous Earth, Inorganic Borate, Liquid Nitrogen, Silica Compounds, Sodium Fluoride, Sodium Hexafluorosilicate. See Insecticide. Cf. Botanical Insecticide; Natural Insecticide. Rel. Acaricide; Herbicide; Nematicide; Pesticide.

INOSCULATE Verb. (Latin, *in* = in; *osculari* = to kiss.) To unite or to join and make one, such as vessels or capillaries.

INOUS BLUE BUTTERFLY *Jalmenus inous* Hewitson [Lepidoptera: Lycaenidae].

INOVAT® See Phosmet.

INQUILINE Noun. (Latin, *inquilinius* = tennant. Pl., Inquilines.) 1. A Species that lives as a 'guest' of another Species. Typically, an insect that habitually lives within the nests of other Species (bees, ants, wasps and termites) but does no harm to the nesting Species. 2. An insect that lives within a gall developed by another Species of insect. Cf. Commensal; Parasite.

INQUILINISM Noun. (Latin, *inquilinius* = tennant; English, *-ism* = condition. Pl., Inquilinisms.) A commensal relationship in which a socially parasitic Species spends its entire life within the nest of a host Species. See Symbiosis. Cf. Commensalism; Parasitism.

INQUILINOUS Adj. (Latin, *inquilinius* = tennant; Latin, *-osus* = with the property of.) 1. Living as guests in the nests of other Species. 2. Living within the galls of other insects.

INSECT Noun. (Latin, *insectum* > *insectus* > *insecare* = to cut in. Pl., Insects.) Any member of the Class Insecta (Hexapoda). The name apparently is derived from the appearance of the body which appears cut or incised.

INSECT BEHAVIOUR The subdiscipline of Entomology concerned with the application of principles of animal behaviour as those principles relate to insects. See Entomology.

INSECT GROWTH REGULATOR A category of insecticidal compounds which function as Juvenile Hormone Analogues that interrupt growth and development or act as Chitin Synthesis Inhibitors. IGRs display low mammalian toxicity and are effective on immature insects only. Examples include Buprofezin, Chlorflurazuron, Cyromazine, Flucycloxuron, Flufenoxuron, Hexaflumuron, Hydroprene, Kinoprene, Lufenuron, Methoprene, Novaluron, Pyriproxyfen, Tebufenozide and Teflubenzuron. Acronym: IGR. See Chitin Synthesis Inhibitor; Insecticide; Juvenile Hormone Analogue. Cf. Botanical Insecticide; Carbamate; Inorganic Insecticides; Organochlorine Insecticide; Organophosphorus Insecticide; Plant Growth Regulator; Synthetic Pyrethroids.

INSECT MORPHOLOGY The subdiscipline of Entomology concerned with the study of insect anatomy and the function of anatomical parts. See Anatomy. Cf. Insect Physiology.

INSECT PHYSIOLOGY The subdiscipline of Entomology concerned with the study of the principles of animal physiology as they relate to insects. See Physiology. Cf. Insect Morphology.

INSECT PIN A needle-like piece of metal thrust through the bodies of dead insects and used to impale them. The point of the pin on which the insect is impaled is then stuck in the bottom of a box or carton with a soft bottom. Numerous specimens are accumulated in boxes, specialized drawers or cabinets to form a collection. Standard insect pins are 38 mm long and range in size from 000–7. Heads are commonly made of nylon. 'Upset' is an integral head made by mechanically squeezing the end of the pin, or a small piece of metal pressed onto the pin. Size 2 pins (0.46 mm in diameter) are most useful. Pins of larger diameter (size 3–7) are used for pinning large insects. Standard insect pins are currently made of ordinary spring steel (called 'black') or stainless steel with a blued or a lacquered (japanned) finish.

INSECT POXVIRUS See Cytoplasmic Polyhedrosis Virus. See Granular Virus; Nonoccluded-type Virus; Nuclear Polyhedrosis Virus.

INSECT SOCIETY See Eusocial Insect.

INSECTA Noun. (Latin, *insectum*.) The largest Class of the Phylum Arthropoda and which contains about 1.8 million nominal Species and estimates of 10–30 million extant Species. The Class is cosmopolitan in distribution and representatives have been extant since the Devonian Period. The Insecta represent a dominant life form from the viewpoint of numbers of individuals, numbers of Species, length of existence and diversity of lifestyle. Anatomical features include; Mandibulate tracheates with three body Tagma (head, Thorax, Abdomen), multisegmented Antenna, Thorax composed of three segments, wings usually present or secondarily lost and associated with middle and posterior segments; each pair of legs articulated with and consecutively arranged on each of the thoracic segments. Characterized by a complex life-cycle involving Metamorphosis. Included Orders: Archaeognatha, Blattodea, Coleoptera, Collembola, Dermaptera, Diplura, Diptera, Embioptera (Embiidina), Ephemeroptera, Grylloblattodea, Hemiptera, Hymenoptera, Isoptera, Lepidoptera, Mantodea, Mecoptera, Megaloptera, Neuroptera, Odonata, Orthoptera, Phasmatoidea, Phthiraptera, Plecoptera, Protura, Psocoptera, Raphidioptera, Siphonaptera, Strepsiptera, Thysanoptera, Thysanura, Trichoptera, Zoraptera.

INSECTABAN® See Permethrin.

INSECTAN Adj. (Latin, *insectum*.) Pertaining to or characteristic of insects in general. Alt. Insectean

INSECTARY Noun. (Latin, *insectum*. Pl., Insectaries.) 1. A place, usually a specially constructed building, where insects are propagated. 2. Often, a building where beneficial insects are produced for mass release as natural enemies against pests. Alt. Insectarium.

INSECTICIDAL NATURAL PRODUCTS See Botanical Insecticides; Biopesticides.

INSECTICIDE Noun. (Latin, *insectum* > *insecare* = to cut in; *–cide* > *caedere* = to kill. Pl., Insecticides.) Chemical compounds used to kill insects. Broadly categorized into Botanical Insecticide, Carbamate, Cyclodiene Insecticide; Inorganic Insecticide, Natural Insecticide, Organochlorine Insecticide, Organophosphorus Insecticide, Synthetic Pyrethroid and Insect Growth Regulator. See each term; Pesticide. Cf. Acaricide; Filaricide; Fungicide; Herbicide; Nematicide.

INSECTICIDE FORMULATION The form in which 'technical material' is provided to an applicator/consumer/pest control operator by an insecticide manufacturer. IFs vary in complexity and substance; IFs intended to improve material effectiveness, simplify on-site preparation or application, improve safety in handling, improve aspects of storage. Principal types of IF for urban pest control: Aerosols, Dusts, Emulsifiable Concentrates, Microencapsulated Concentrates, Oil Concentrates, Suspension Concentrates, Wettable Powders.

INSECTIPEN® See Cyfluthrin.

INSECTIVORE Adj. (Latin, *insectum*; *vorare* = to eat.) An organism which eats insects. Cf. Carnivore; Herbivore; Saprovore.

INSECTIVOROUS Adj. (Latin, *insectum*, *vorare* = to eat; Latin, *-osus* = with the property of.) Insect-eating; pertaining to organisms subsisting on insects. See Feeding Strategy; Entomophagous. Cf. Herbivorous; Predaceous; Phytophagous.

INSEGAR® See Fenoxycarb.

INSEMINATE Trans. Verb. (Latin, *inseminatus* > *inseminare* = to sew. Inseminated, Inseminating.) To fertilize with Semen. Rel. Copulation; Fertilization.

INSEMINATION REACTION See Asada & Kitagawa 1988, Jap. J. Genet. 63: 137–148.

INSERTED Adj. (Latin, *insertus* = thrusting in.) 1. A condition in which the base of one part is set into an adjacent part. 2. Attached by natural growth.

INSERTED HEAD The head of an insect that is entirely or partly concealed within the Thorax.

INSERTIO MAXILLAE The lowest part of the Maxillae adjoining the head (Jardine).

INSERTIO LABII The lowest part of the Labium adjoining the head (Jardine).

INSERTION Noun. (Latin, *in* = in; *serere* = to join; English, *-tion* = result of an action. Pl., Insertions.) 1. The point of attachment of a moveable part. 2. The part of attachment of a muscle to a moveable part. Alt. Insertio. Cf. Origin. Rel. Muscle.

INSIDIOUS FLOWER BUG *Orius insidiosus* (Say) [Heteroptera: Anthocoridae]: A beneficial Species widespread in North America. Prey of IFB include chinch bug, thrips, corn earworm, insect eggs and mites. See Anthocoridae.

INSOLATION Noun. (Latin, *insolare* = to expose to the sun; English, *-tion* = result of an action. Pl., Insolations.) Exposure to the sun's rays.

INSPIRATORY Adj. (Latin, *inspiratus* = to blow into.) Descriptive of or pertaining to inspiration or the intake of air during the process of breathing. Rel. Respiration.

INSTAR Noun. (Latin, *instar* = form. Pl., Instars.) 1. The postembryonic, immature insect between moults (Fischer 1853). 2. 'The arthropod between two moults embracing a portion of the somatic growing phase, eclosion from the successive egg being purposely equated with a moult' (Carlson 1983). In the larval stage, instars are numbered to designate the various periods. The first instar larva is the stage between the egg and first moult. Moulting is the equivalent of Ecdysis. Some physiologists including Wigglesworth argue that instar is defined as the animal from Ecdysis to Ecdysis. Other physiologists argue that instar is defined from Apolysis to Apolysis. The process of Cuticle separation from the Epidermis (Apolysis) is obscure and does not necessarily occur simultaneously over the entire animal; defining an instar is difficult and sometimes impossible. See Apolysis; Ecdysis. Cf. Stadium; Stage.

INSTAR® See Beet Armyworm NPV.

INSTINCT Noun. (Latin, *instinctus* = impulse. Pl., Instincts.) A chain of coordinated reflex actions, each taking place when the previous one is completed. Any series of reactions not depending on antecedent experience.

INSTINCTIVE BEHAVIOUR Behaviour which is a chain of reflex actions (Wardle).

INSTITIA Noun. (Unknown origin. Pl., Institiae.) Striae or furrows of equal width on the surface of a structure.

INSTRUMENTA CIBARIA The mouthparts or Trophi.

INSTRUMENTA CIBARIA LIBERA Unencumbered mouthparts; biting mouthparts; Instrumenta Cibaria Mordentia.

INSTRUMENTA CIBARIA MORDENTIA The biting organs (Mandibles) of an insect.

INSTRUMENTA MASTICANDI The chewing organs of the mouth.

INSTRUMENTA SUCTORIA The mouthparts of a Haustellate insect.

INSULA See Islet.

INTEGER Noun. (Latin, *integer* = entire.) Entire; without incisions, as a margin.

INTEGRATED CONTROL An approach to pest control which embraces techniques including Biological Control, Chemical Control, Cultural Control. See Biological Control; Chemical Control; Cultural Control; Natural Control; Regulatory Control. Syn. Integrated Pest Management.

INTEGRATED PEST MANAGEMENT See Integrated Control. Abbreviated as IPM. A programatic approach to pest control that has as its foundation the use of biological control methods, plant breeding, and the judicious application of pesticides, especially selective pesticides.

INTEGUMENT Noun. (Latin, *integumentum* = covering. Pl., Integuments.) The sclerotized Cuticle covering the insect body. Specifically, a multiple-layered, composite organ which defines body shape, size and colour. Integument is composed of living Epidermal Cells and secretory product of those cells. Each layer is of a different thickness, chemical composition and displays physical properties different from surrounding layers. Integument is largest body organ in weight and volume. Each layer is responsible for a different aspect of organ's collective responsibility. Insect Integument represents a compromise between 'form' and 'function'. If skeleton is structurally weak, then it will fail in buckling or compression. If skeleton is too strong, then it will be cumbersome due to an increased thickness and weight. A very strong Integument may be biochemically expensive and thus a liability to animal which possesses it. A biomechanical compromise exists in maintaining an external skeleton. The insect body is a hollow cylinder that renders it stronger than a solid rod of same material and cross-sectional area. Typically Integument would be about three times stronger in resisting buckling, bending and compression. The amount of benefit which can be biomechanically derived is limited. An isometrically thicker Integument covering an hypothetical insect with body size of an elephant would be impossible because thickened Integument would shatter. See Skeleton. Cf. Basement Membrane; Dermal Glands; Epidermal Cells; Epidermis; Endocuticle; Exocuticle; Cuticulin Layer; Wax Layer; Cement Layer; Pore Canals. Rel. Bone; Cartilage; Chitin; Cutex; Skin.

INTEGUMENTAL SCOLOPHORE A scolophore (scolopophore) in the Integument of which the nerve-ending is attached to the body wall (Comstock). Alt. Scolopophore.

INTEGUMENTARY Adj. Descriptive of or pertaining to the Integument.

INTERALAR SPACE Odonata: The Terga of the Mesothorax and Metathorax.

INTERANTENNAL SETAE Coccids: Several Setae on the head ventral and caudad of the antennal articulation (MacGillivray).

INTERANTENNAL SULCUS A shallow protuberance between the Antennal sockets. See Torulus.

INTERANTENNAL Adj. (Latin, *inter* = between; + antennal.) Pertaining to structure or space between the basal segments of the Antenna.

INTERARTICULAR Adj. (Latin, *inter* = between; + articular.) Descriptive of something positioned between joints or segments.

INTERBIFID GROOVES 1. Diptera: The pseudotracheal openings. 2. The Mentum *sensu* Snodgrass.

INTERCALARY Adj. (Latin, *intercalaris* = inserted.) 1. Descriptive of something additional and interpolated. 2. Something inserted between two structures (*e.g.* a wing vein).

INTERCALARY APPENDAGES The rudimentary postantennal or premandibular appendages (Snodgrass).

INTERCALARY PLATES 1. Some coccids: The longitudinal row of sclerites on each side between the mesial sclerites and the lateral sclerites. 2. The Supplementary Scales of Newstead (MacGillivray).

INTERCALARY SEGMENT The premandibular segment of the insect head, in the embryo. See Tritocerebral Segment.

INTERCALARY VEIN 1. Any added or supplementary wing vein. 2. Ephemeridae: Any one of several longitudinal veins between the Anal Vein and the first Axillary, but not a branch of either. 3. Diptera: Anterior intercalary of Loew. 4. Discoidal Vein; Posterior Intercalary. 5. The Cubitus of Comstock *teste* Curran. 6. The posterior branch of the fourth vein in cases where its base partly closes the Discal Cell. 7. M of Comstock. See Vein; Wing Venation.

INTERCALATE Trans. Verb. (Latin, *intercalatus* = inserting; *-atus* = adjectival suffix.) To interpolate or insert into any serial arrangement between any two consecutive members.

INTERCELLULAR Adj. (Latin, *inter* = between; cellular.) Pertaining to structure positioned between and among cells.

INTERCHELICERAL GLAND Acarology: Unpaired Prosomatic Glands of unknown function.

INTERCOSTAL Adj. (Latin, *inter* = between; *costa* = a rib; *-alis* = pertaining to.) Between veins or Costae, usually in the narrow grooves between veins in the Costal region of a wing.

INTERCOSTAL VEIN Hymenoptera: Subcosta. See Vein; Wing Venation.

INTERCOSTULA Noun. (Latin, *inter* = between, *costa* = a rib; *ula* = diminutive form.) Small, vein-like structures between the normal veins which are visible on a wing margin but lost toward the disc.

INTERCOXAL PROCESS Coleoptera: A median protrusion of the basal abdominal segment between the hind Coxae.

INTERFACETAL HAIR Setae that develop on the compound eye between Ommatidia (facets).

INTERFERENCE COLOUR One of three forms of structural (physical) colour. IC produced by optical interference of reflections from laminae or a series of ribs. IC is produced by thin films, such as those on soap bubbles and oil films. On the insect body, interference is a structural colour produced through optical interference of reflections from laminae or alternating thin-layers of Cuticle which have different refractive indices. Typically more laminae produce greater purity of colour. IC include the corneal interference filters in the compound eye of some Diptera. See Coloration. Cf. Iridescence; Scattering Colour; Structural Colour; Subhypodermal Colour. Rel. Bioluminescence; Diffraction; Pigmenary Colour; Scattering.

INTERFRONTAL BRISTLES Diptera: Enlarged Setae (bristles) on the Frontal Vitta (Curran). Syn. Interfrontal Hairs.

INTERFRONTALIA Diptera: The Frontal Vitta (Curran).

INTERGANGLIONIC Adj. (Latin, *inter* = between; *ganglion* = tumour; *-ic* = consisting of.) Pertaining to structures located between Ganglia.

INTERGANGLIONIC CONNECTIVES See Abdominal Ganglion.

INTERGANGLIONIC NERVE CORD The fused Commissures of the Ganglia forming the ganglionic centres (Wardle).

INTERIOR EDGE The boundary of the inner margin of the wing. Alt. Inner Edge.

INTERIOR PALPI The Labial Palpi.

INTERLOBULAR INCISIONS Coccids: Incisurae *sensu* MacGillivray.

INTERMAXILLAIRE Noun. (Latin, *inter* = between; *maxilla* = jaw.) The Maxillary Lobe.

INTERMEDIATE Adj. (Latin, *intermediatus*, pp of *intermediare* = to come between.) Lying between others in position, or possessing characters between two other forms.

INTERMEDIATE FIELD In Tegmina, the discoidal field.

INTERMEDIATE HOST Medical Entomology: The host (insect or otherwise) in which the asexual stages of a parasite are passed (Matheson).

INTERMEDIATE HOST RESERVOIR Medical Entomology: Hosts (insect or otherwise), in which a natural supply of the asexual stages of a parasite occur (Matheson).

INTERMEDIATE MESENTERON RUDIMENT A median strand of cells of the ventral Endoderm remnant taking part in the regeneration of the Mesenteron in some insects (Snodgrass).

INTERMICELLAR Adj. (Latin, *inter* = between; + *micellae*; *-ar* = adj. suffix.) Pertaining to something occurring along the Micellae of Chitin.

INTERNAL AREA Hymenoptera: The posterior of the three areas between median and lateral longitudinal Carina on the Metanotum; the third lateral area.

INTERNAL CELL Hymenoptera: The second Anal Cell of Comstock (*teste* Smith).

INTERNAL CHIASMA The optic segment of the insect Brain (Packard).

INTERNAL GENITALIA Female Insects: The Ovaries and their component parts (Ovarioles, Pedicel, Calyx, Lateral Oviduct) including the Spermatheca, Accessory Glands, Genital Chamber and the Vagina (Snodgrass). Cf. External Genitalia.

INTERNAL MAXILLARY PALPI See Palpi Maxillares Interni.

INTERNAL MEDULLARY MASS The Opticon *sensu* Hickson.

INTERNAL PARAMERA 1. The inner pair of the genital appendages of the male. 2. The functional Penis.

INTERNAL RESPIRATION The process of oxidation accompanying metabolism in the cells of the body tissue (Snodgrass).

INTERNAL TRIANGLE Odonata: See triangle.

INTERNAL VEINS 1. Lepidoptera: 1–3 veins that are free and extend from the base to the outer margin of the wing near the hind angle and are never branched. 2. Veins 1a to 1c in the numerical series. 3. The Anal Veins of Comstock *teste* Smith. See Vein; Wing Venation.

INTERNAL WING CELL Odonata: The triangular wing cell posterior and proximad of the triangle (Garman).

INTERNATIONAL CODE See International Rules of Zoological Nomeclature.

INTERNATIONAL RULES OF ZOOLOGICAL NOMENCLATURE The guidelines developed for naming animal Taxa adopted by the Fifth International Congress of Zoology, Berlin 1901. Alt. Règles Internationales de la Nomenclature

Zoologique. The French text is the official version. See Plenary Powers.

INTERNEURAL Adj. (Latin, *inter* = between; Greek, *neuron* = nerve, Latin, *-alis* = pertaining to.) Pertaining to structure, space or colour between the nerves (or veins) of wings.

INTERNEURON Noun. (Latin, *inter* = between; Greek, *neuron* = nerve. Pl., Interneurons.) Unipolar Neurons that lie exclusively in the Central Nervous System and function as a switchboard to direct nerve impluses along the correct path. Syn. Adjustor Neurons. See Nervous System; Nerve Cell. Cf. Afferent Neuron; Motor Neuron; Efferent Neuron. Rel. Central Nervous System.

INTERNIDIAL Adj. (Latin, *inter* = between; *nidus* = nest; *-alis* = characterized by.) Descriptive of interactions between nests. Pertaining to phenomena involving more than one nest. Cf. Intranidial.

INTERNODE Adj. (Latin, *inter* = between; *nodus* = knot. Pl., Internodes.) Between nodes. Part of a plant stem between two successive nodes.

INTERNOMANDIBULAR GLAND Hymenoptera: Paired Salivary Glands in bees near the medial surface of Mandible base (Smith). See Mandibular Gland.

INTERNOMEDIA VEIN Orthoptera: The Cubitus *sensu* Comstock. See Vein; Wing Venation.

INTERNUNCIAL NEURONE See Association Neurone.

INTEROCELLAR FURROW Hymenoptera: A short depression or Sulcus extending from the middle of the Ocellar Furrow to the median Ocellus. The IF flares out adjacent to the median Ocellus and is frequently a depressed area surrounding the Ocellus.

INTEROCEPTOR Noun. (Modern Latin, internal and receptor. Pl., Interoceptors.) An internal sense organ for the perception of internal stimuli.

INTEROCULAR Adj. (Latin, *inter* = between; ocular.) Structures located between the eyes.

INTEROCULAR ANTENNA Antennae that are set between the compound eyes.

INTERPLEURITE Noun. (Latin, *inter* = between; Greek, *pleura* = rib; *-ites* = constituent. Pl., Interpleurites.) One of the Intersegmentalia between the Pleurites (Imms).

INTERPLICAL Adj. (Latin, *inter* = between; *plica* = fold; *-alis* = characterized by.) 1. Lying between folds. 2. Orthoptera; A term specifically applied to the alternate ridges and grooves in the anal area of the hindwings.

INTERPLURAL SUTURE 1. Odonata: The suture between the Mesopleuron and Metapleuron (Garman). 2. The reduced part of the Metacoria between the Pleural Sclerites (MacGillivray).

INTERPOSED SECTORS Odonata: The shorter longitudinal vein in the wings of some Species between the chief veins; supplementary sectors.

INTERRADIAL NEXUS Wings of Myrmeleonidae: A conjuction of adjacent veins which joins the apex of the first and second branches of the Radial Sector (Comstock).

INTERRUPTED Adj. (Latin, *interrumpere* = to interrupt.) Broken in continuity, but with the apices of the broken parts in line with each other. Alt. Interruptus.

INTERSEGMENTAL Adj. (Latin, *inter* = between; *segmentum*; *-alis* = characterized by.) Descriptive of structure, process or condition that occurs between segments; interarticular. Cf. Intrasegmental.

INTERSEGMENTAL MEMBRANE The flexible infolded part of the Cuticle or Conjunctiva between two secondary adjacent segments. IM is typically the nonsclerotized posterior part of a primary segment and provides freedom of movement of the body.

INTERSEGMENTAL PLATES The Cervical Sclerites *sensu* Crampton.

INTERSEGMENTALIA Noun. (Latin, *inter* = between; *segmentum*.) Small detached sclerites between adjacent segments of the insect body.

INTERSEX Noun. (Latin, *inter* = between; *sexus* = sex. Pl., Intersexes.) 1. An insect with its general sexual characteristics intermediate between male and female. See Gynandromorph. 2. Intersexes have been called 'phenotypic mosaics in time.' The individual begins development as one sex but later changes to the other sex or displays parts of the other sex. Intersexes are symmetrical. The phenomenon is common among invertebrates, and has been reported in vertebrates (frogs and the sturgeon). Among vertebrates, control of secondary sexual characteristics is under hormonal control mediated by the gonads. Within the Insecta the picture is not clear, and some erroneous cases are attributable to faulty moulting.

INTERSPACE Noun. (Latin, *inter* = between; *spatium*.) 1. Coleoptera: The plane surface between Elytral Striae. 2. Lepidoptera: spaces between wing veins not included in closed cells. 3. Orthoptera: A deep incision or Sulcus on the posterior margin of the Metasternum (Walden).

INTERSPACIAL Adj. (Latin, *inter* = between; *spatium*; *-alis* = characterized by.) Occurring in the interspaces between two wing veins or two Elytral Striae.

INTERSPECIFIC COMPETITION The interaction among individuals of more than one Species for a resource used by the individuals. See Competition. Cf. Intraspecific Competition.

INTERSTERNITE Noun. (Latin, *inter* = between; *sternum* = breast bone; *-ites* = constituent. Pl., Intersternites.) Primary intersegmental sclerites of the venter that become the Spinasterna of the Thorax (Snodgrass).

INTERSTERNUM Noun. (Latin, *inter* = between; *sternum* = breast bone.) 1. Odonata: A large sclerite of the thoracic Sternum immediately anterior of the Abdomen (Garman). 2. The Postpleurite or Opisthopleurite *sensu* MacGillivray.

INTERSTICE Noun. (Latin, *inter* = between; *sistere* = to stand.) A space between two lines, whether striate or punctate. Alt. Interstitium.

INTERSTITIAL LINE The elevated ridge between two striae or series of punctures. See Elytron.

INTERSTRIAE See Elytron.

INTERTERGITE Noun. (Latin, *inter* = between; *tergum* = back; *-ites* = constituent. Pl., Intertergites.) One of the Segmentalia between the Tergites (Imms).

INTERTIDAL Adj. (Latin, *inter* = between; Anglo Saxon, *tid* = tide; *-alis* = characterized by.) Pertaining to something or phenomenon occurring on the beach between high-water and low-water levels.

INTERTIDAL DWARFBUGS See Omannidae.

INTERTIDAL TRAPDOOR SPIDER *Idioctis yerlata* Churchill & Raven [Araneida: Barychelidae].

INTERVAL Adj. (Latin, *inter* = between; *vallum* = wall; *-alis* = characterized by.) 1. The space between two structures or sculptural elements. 2. The time periods between successive stages of development. See Stadium.

INTERVALVULA Noun. (Latin, *inter* = between; *valvula* diminutive of *valva*. Pl., Valvulae.) 1. Sternal sclerites in the venter of the ninth abdominal segment between the second Valvifers (Snodgrass). 2. Intermediate Valves of the Ovipositor (Crampton).

INTERVENTRICULAR VALVE The inner opening between the chambers of the Heart (Smith).

INTERVENTRICULAR VALVULE A small Heart valve anterior of the Semilunar Valve.

INTERVENULAR Adj. (Latin, *inter* = between; *vena* = blood vessel.) Pertaining to structure between any two veins or the wing or body.

INTESTINA PARVA The small intestine.

INTESTINAL CAECUM The part of the Large Intestine anterior of the junction with the Small Intestine. See Alimentary System; Caecum.

INTESTINAL MYIASIS See Myiasis.

INTESTINE Noun. (Latin, *intestinus* = internal. Pl., Intestines.) The part of the Alimentary Canal through which the food passes from the Stomach, and in which absorption is completed and the excretions are formed for expulsion. See Alimentary System. Cf. Foregut; Hindgut; Midgut.

INTIMA Noun. (Latin, *intimos* = innermost.) 1. The internal lining of the foregut and hindgut in insects. The Intima is continuous with the Cuticula of the body-wall (Matheson). 2. The membranous lining of the Tracheae and Endotrachea. See Foregut; Hindgut. Rel. Integument.

INTIMA OF THE TRACHEA The innermost chitinous lining of the respiratory system. See Intima. Cf. Trachea. Rel. Respiratory System.

INTORT Trans. Verb. (Intorted, Past Part. Latin, *intorquere* = to twist.) Turned or twisted inwardly. Alt. Intortus.

INTRA VITAM Latin phrase meaning 'within life' and refers to microscopic preparations of insect staining with very dilute stains while the insect is alive, by direct injection or by ingestion of food. See Histology.

INTRA-ALAR BRISTLES Diptera: A row of 2–3 bristles between the Supraalar and Dorsocentral groups.

INTRACELLULAR Adj. (Latin, *intra* = within; + cellular.) Occurring within the cell or in a cell.

INTRACRANIAL Adj. (Latin, *intra* = within; Greek, *kranion* = skull; *-alis* = characterized by.) Within the Cranium (head).

INTRAEPICARDIAL Adj. (Latin, *intra* = within; *epi* = upon; Greek, *kardia* = heart; *-alis* = characterized by.) Within or inside the Epicardium.

INTRAEPIDERMAL Adj. (Latin, *intra* = within; *epi* = upon; Greek, *derma* = skin; *-alis* = characterized by.) Within or inside the Epidermis.

INTRAGANGLIONIC Adj. (Latin, *intra* = within; Greek, *ganglion* = a mass of nerve cells; *-ic* = of the nature of.) Within or inside Ganglia.

INTRAHUMERAL BRISTLES Diptera (Calyptrata). Setae located immediately in front of the thoracic suture, between the Humeral Calus and the Presutural Depression. See Presutural Bristles.

INTRANIDIAL Adj. (Latin, *intra* = within; *nidus* = nest; *-alis* = characterized by, pertaining to.) Within a nest. A term applied to phenomena occuring within one nest. Cf. Interndidal.

INTRAOCULAR Adj. (Latin, *intra* = within; *oculus* = eye.) Positioned within the eye, actually or apparently.

INTRAOCULAR ANTENNA An Antenna inserted below the eyes.

INTRAORGANIC Adj. (Latin, *intra* = within; *organicus* = organ; *-ic* = of the nature of.) Within an organ. Term applied to endoparasitic forms which live within the organs of their hosts.

INTRAPULMONARY RESPIRATION A type of respiration which does not involve movements of the outer body wall and is confined to the respiratory organs.

INTRASEGMENTAL Adj. (Latin, *intra* = within; *secare* = to cut; *-alis* = pertaining to.) Descriptive of a condition, process or structure that occurs within a segment. Cf. Intersegmental.

INTRASPECIFIC Adj. (New Latin, *intra* = within, *species; -ic* = characterized by.) Pertaining to interactions among members of the same biological Species or population.

INTRASPECIFIC COMPETITION The interaction among individuals of one biological Species for a resource used by the individuals. Cf. Interspecific Competition.

INTRAUTERINE Adj. (Latin, *intra* = within; *uterus* = womb.) Applied to development in which eggs hatch within the Vagina of the mother (some Diptera).

INTRICATE Adj. (Latin, *intricare* = to entangle; *-atus* = adjectival suffix.) With depressions or elevations so run into each other as to be difficult to see (Kirby & Spence). Confused; Irregular. Alt. Intricatus.

INTRINSIC Adj. (Latin, *intrinsecus* = inwards; *-ic* = of the nature of.) Inward; inherent. Pertaining to structure inside a larger body.

INTRINSIC ARTICULATION A type of articulation in which the points making contact are sclerotic prolongations within the Articular Membrane (Snodgrass).

INTRINSIC MUSCULATURE Insect muscles which have their origin and insertion within a body segment.

INTRINSIC RATE OF INCREASE Ecology. The fraction, symbolized by *r*, by which a population is growing at any instant in time.

INTRODUCED See Adventive.

INTRODUCED DADDY-LONGLEGS *Nelima doriae* (Canestrini) [Opiliones: Phalangiidae].

INTRODUCED GREY HOUSE SPIDER *Achaearanea tepidariorum* (C.L. Koch) [Araneida: Theridiidae].

INTRODUCED ORGANISMS See Adventive. Cf. Immigrant organisms.

INTRODUCED PINE SAWFLY *Diprion similis* (Hartig) [Hymenoptera: Diprionidae].

INTROMITTENT Adj. (Latin, *intro* = within; *mittere* = to send.) Pertaining to structure adapted for entering or insertion into a sac, bursa or pouch.

INTROMITTENT ORGAN The male Phallus; the Penis.

INTRORSE Adj. (Latin, *introsus* = inwards.) Directed inwards or toward the body.

INTRUSUS Adj. (Latin, *intrusus* = thrusted.) Intruded; opposite of protruded.

INTUMESCENT Adj. (Latin, *intumescere* = to swell.) Enlarged; swollen; expanded.

INTUSSUSCEPTION Noun. (Latin, *intus* = within; *suscipere* = to receive; English, *-tion* = result of an action.) Deposition of new particles of formative material in a tissue or structure.

INUNDATIVE RELEASE Biological Control: The organized release of relatively large numbers of a natural enemy into an area with the intent of reducing a pest population quickly. See Biological Control. Cf. Augmentative Release; Innoculative Release.

INVAGINATE Verb. (Latin, *in* = into; *vagina* = sheath; *-atus* = adjectival suffix.) To turn inward or to retract within the body wall. Cf. Intorse.

INVAGINATION Noun. (Latin, *in* = into; *vagina* = sheath; English, *-tion* = result of an action. Pl., Invaginations.) A pouch or sac formed by an infolding or depression of an outer surface. See Indentation. Cf. Evagination. Rel. Sculpture.

INVAGINATION OF THE EMBRYO The direct infolding of the embryo into the egg (Snodgrass).

INVALID NAME (Latin, *invalidus* = not strong.) The scientific name of a Subspecies, Species or Genus that is without standing under the rules specified by the International Code of Zoological Nomenclature. See Scientific Name; Common Name. Cf. Valid Name. Rel. Nomenclature; Synonym.

INVERMAY BUG *Nysius turneri* Evans [Hemiptera: Lygaeidae].

INVERTASE A digestive enzyme which brings about the hydrolysis of sucrose (cane sugar) and converts it into a mixture of glucose and fructose (invert sugars). See Sucrase.

INVERTEBRATA Plural Noun. (Latin, *in* = not; *vertebra* = turning point.) The large category of animals without a vertebral column or backbone. Invertebrata include the Arthropoda. See Arthropoda. Cf. Vertebrata.

INVERTEBRATE Adj. (Latin, *in* = not; *vertebra* = turning point; *-atus* = adjectival suffix.) Without a backbone or vertebral column. Alt. Invertebral.

INVEST Verb. (Latin, *investire* = to clothe.) To cover.

INVESTITUS Adj. (Latin, *investius* > *investire* = to invest.) An arcane term referring to Integument without scales or Setae. Unclothed.

INVESTMENT Noun. (Latin, *in* = in; *vestire* = to clothe.) A covering, typically the body wall.

INVOLUCRATE Adj. (Latin, *involucrum* = covering; *-atus* = adjectival suffix.) Pertaining to structure involving or resembling an involucre. Alt. Involucratus.

INVOLUCRE Noun. (Latin, *involucrum* = covering.) Botany: A whorl of distinct or united leaves or bracts subtending a flower or inflorescence.

INVOLUCRUM Noun. (Latin, *involucrum* = covering.) A sheet or sheets of Cerumen surrounding the brood chamber in nests of most stingless bees

INVOLUCRUM ALARUM Dermaptera: A flap of the Metanotum.

INVOLUTE Adj. (Latin, *involvere* = involve.) Rolled inward spirally. Alt. Involuted, Involutus.

INVOLUTI Noun. Butterflies whose larvae live in a folded leaf, *e.g.* Hesperiidae.

INVOLUTION OF THE EMBRYO Invagination of the embryo accompanied by a revolution and final reversal of position in the egg (Snodgrass).

INVOLVULUS Adj. (Latin, *involvare* = to wrap.) An arcane term used to describe larvae that wrap themselves in leaves.

INVREA, FABIO (1884–1968) (Conci 1975, Memorie Soc. ent. ital. 48: 1(a) (1969): lxxxv–xcv. bibliogr.)

IO MOTH *Automeris io* (Fabricius) [Lepidoptera: Saturniidae].

IOTERIUM (Etymology obscure.) Hymenoptera: See: Poison Gland. (Kirby & Spence).

IPM See Integrated Pest Management.

IPS Noun. A Genus of scolytid beetles that is injurious to some trees. See Scolytidae.

IPSILATERAL Adj. (Latin, *ipse* = same; *latus* = side; *-alis* = characterized by.) Descriptive of structures or appendages on the same side of the body. Cf. Contralateral.

IPSO FACTO Latin phrase meaning 'in the fact itself'.

IRIDESCENCE Noun. (Latin, *iridis* = rainbow. Pl., Iridescences.) 1. The rainbow-like play of interference colours reflected from an object (*e.g.* soap bubble, peacock feathers, mother of pearl, body or Scales of some insects). Iridescence is produced by interference and diffraction; scattering cannot produce iridescence. With insects, the body glitters of different colours when viewed from different angles, due to diffraction of light from closely ribbed or corrugated surfaces. Non-iridescent objects appear the same from any angle of view. Depending upon the insect, iridescence probably comes from different layers of the Integument. Layers form interference reflectors that consist of alternating layers of high and low refractive index material whose thickness is similar to the wavelength of light. A certain proportion of incident light is reflected at each interface. Depending on layer thickness and wavelength, the reflections constructively or destructively interfere. Constructive interference occurs for wavelengths that are twice the sum of the optical thickness of the electron-lucent and electron-dense layers. See Structural Colour. Cf. Interference Colour; Scattering Colour. Rel. Integument; Scales.

IRIDOMYRMEX Mayr 1862. A large, economically important Genus of ants. See Formicidae.

IRIS Noun. (Greek, Latin, *iris* = rainbow.) 1. The dark pigment surrounding the dioptric apparatus of an eye (Snodgrass). 2. The coloured circle enclosing the pupil of an ocellate spot.

IRIS BORER *Macronoctua onusta* Grote [Lepidoptera: Noctuidae].

IRIS CELLS Slender spindle-shaped pigment cells surrounding the inner tapering end of the crystalline cone cells and the distal end of the retinal cells in the Ommatidia (Needham).

IRIS PIGMENT Iris Tapetum (Smith).

IRIS PIGMENT CELLS Cells containing Iris pigment of the eye (Snodgrass); Iris Cells. See Secondary Pigment Cell. Syn. Outer Pigment Cells; Secondary Iris Cell.

IRIS SKIPPER *Arhenes dschilus iris* (Waterhouse) [Lepidoptera: Hesperiidae].

IRIS TAPETUM The pigment layer of the compound eye just below the Crystalline Cone.

IRIS THRIPS *Iridothrips iridis* (Watson) [Thysanoptera: Thripidae].

IRIS WEEVIL *Mononychus vulpeculus* (Fabricius) [Coleoptera: Curculionidae].

IRIS WHITEFLY *Aleyrodes spiraeoides* Quaintance [Hemiptera: Aleyrodidae].

IRISED Adj. (Latin, *iris* = rainbow.) With rainbow colours.

IRMSCHER, EMIL BERNARD (1862–1912) (Möbius 1943, Dt. ent. Z. Iris 57: 8.)

IRON Noun. (Middle English, *iren*.) A grey metallic element present in the Haemoglobin of the blood and necessary in food stuffs to all red-blooded animals, the oxygen-carrying component of animal tissues. Chem Abbr: Fe.

IRONBARK BEETLE *Zopherosis georgei* White [Coleoptera: Zopheridae].

IRONBARK BORER See Ironbark Beetle.

IRONBARK LACE LERP *Cardiaspina vittaformis* (Froggatt) [Hemiptera: Psyllidae].

IRONBARK SAWFLY *Lophyrotoma anlis* (Costa) [Hymenoptera: Pergidae].

IRONOMYIIDAE Plural Noun. A small Family of acalypterate Diptera assigned to Superfamily Platypezoidea. See Diptera.

IRREGULAR Noun. (Latin, *in* = not; *regula* = rule.) Unequal, curved, bent or otherwise twisted or modified without order or apparent symmetry. Syn. Atypical; Unusual. Cf. Regular.

IRRI International Rice Research Institute. An international centre devoted to rice research and training. Founded by Ford and Rockefeller Foundations in 1960 and located in Los Banos, Philippines.

IRRITABILITY Noun. (Latin, *irritare* = to provoke; English, *-ity* = suffix forming abstract nouns.) The capacity of an organism to receive external stimulus and respond to that stimulus.

IRRITANT Noun. (Latin, *iritare* = to provoke; *-antem* = an agent of something. Pl., Irritants.) Any external stimulus that provokes a behavioural response.

IRRORATE Adj. (Latin, *irrorare* = to bedew; *-atus* = adjectival suffix.) Freckled; covered with minute spots or coloured markings. A term sometimes applied to the colour patterns of Lepidoptera wings. Alt. Irrorated; Irroratus.

ISAAK, JULJUSZ (1870–1924) (Eichler 1923, Polskie Pismo Ent. 2: 157–158.)

ISABELLINE Adj. (French, *Isabella* = feminine name; *-ine*.) Pale yellow with some red and brown. Alt. Isabellinus.

ISAZOFOS An organic-phosphate (thiophosphoric acid) compound {O-(5-chloro-1-methylethyl)IH-1,2,4-triazol-3yl)O,O-diethyl phosphorothioate} used as a nematicide, contact insecticide and stomach poison against nematodes, rootworms, leaf hoppers, mole crickets, stem borers, maggots, and other soil pests. Toxic to fishes. Trade names include: Brace®, Miral®, Triumph®, Victor®. See Organophosphorus Insecticide.

ISCHIA Noun. (Greek, *ischion* = hip.) The Pleura.

ISCHIOPODITE Noun. (Greek, *ischion* = hip; *pous* = foot; *-ites* = constituent. Pl., Ischiopodites.) 1. The third segment of a generalized limb, or second segment of the Telopodite. 2. The second Trochanter or Prefemur *sensu* Snodgrass. Alt. Ischium.

ISCHNOCERA Noun. A small Suborder or Division of Phthiraptera (Mallophaga in part). Included Families: Heptapsogasteridae, Philopteridae and Trichodectidae. Species feed exclusively on the feathers and scurf of birds. Hosts include penguins, albatrosses, petrels, shearwaters and all other Families of marine birds. See Mallophaga. Cf. Amblycera. Rel. Parasite.

ISCHNOPSYLLIDAE Plural Noun. Bat Fleas. A small Family of Siphonaptera assigned to Superfamily Ceratophylloidea. Adult Genal Comb of two broad spines; eyes reduced or absent; middle Coxa with apodeme; hind Tibia with an apical tooth; dorsal surface of Sensillum flat; female with one Spermatheca. Adults parasitize bats. See Siphonaptera. Cf. Ceratophyllidae.

ISE, HIDEO (1911–1938) (Hasegawa 1967, Kontyû

35 (Suppl.): 11.)

ISELY, DWIGHT (1887–1974) (Miner 1976, J. Econ. Ent. 69: 298–299.)

ISELY, FREDERICK B (1873–1947) (Alexander 1949, Ent. News 60: 29–30.)

ISENSCHMID, H MORITZ (1850–1878) (Perty 1880, Mitt. Schwiez. ent. Ges. 5: 488–492.)

ISERT, PAUL ERDMANN (1756–1787) (Zimsen 1964, *The Type Material of I. C. Fabricius*. 656 pp. (13), Copenhagen.)

ISHIDA, MASATA (1877–1940) (Hasegawa 1967, Kontyû 35 (Suppl.): 9.)

ISHII, OMOKAGE (1894–1959) (Hasegawa 1967, Kontyû 35 (Suppl.): 7–8.)

ISHII, SHIGANI (1883–1933) (Hasegawa 1967, Kontyû 35 (Suppl.): 7.)

ISHII, T Japanese taxonomist specializing in parasitic Hymenoptera.

ISHIKAWA, CHIYOMATSU (1860–1935) (Hasegawa 1967, Kontyû 35 (Suppl.): 8.)

ISHIMORI, NAOTO (1890–1961) (Hasegawa 1967, Kontyû 35 (Suppl.): 10–11.)

ISHIMURA, KIYOSHI (1910–1956) (Hasegawa 1967, Kontyû 35 (suppl.): 10.)

ISHITANI, FUKUMOBU (1915–1946) (Hasegawa 1967, Kontyû 35 (Suppl.): 9–10.)

ISHIWADA, SHIGETANI (1868–1941) (Hasegawa 1967, Kontyû 35 (Suppl.): 11.)

ISLAND Noun. (Old English, *igland* > Old Norse, *eyland* > Latin, *insula* = island. Pl., Islands.) 1. A body of land surrounded by water. 2. Morphology: An isolated part within the boundaries of any structure or body.

ISLAND FRUIT FLY *Dirioxa pornia* (Walker) [Diptera: Tephritidae].

ISLAND PINHOLE BORER *Xyleborus perforans* (Wollaston) [Coleoptera: Scolytidae]: A widespread pest of living trees via injuries or diseased bark; IPB often is found in dying or dead trees, green logs or newly sawn timber. IPB is not size-selective; constructs tunnel with numerous irregular branches but no brood chambers. Eggs are laid in parent tunnels and larvae feed on ambrosia fungus in tunnels. Life cycle requires 3 months during summer. See Curculionidae.

ISLE OF WIGHT DISEASE A disease of honey bees first noted during 1906 on the Isle of Wight, and subsequently recorded on a few occasions at that place before 1920. Symptoms of the disease include adult bees walking and dying on the ground near their hives during the active season, with spectacular colony extinction. Numerous colony extinctions in other countries have occurred under similar circumstances but diagnosis has not been confirmed. Disease remains of unknown origin. Rel. Honey Bee.

ISLET Noun. (Old French, *islette*, diminutive of *isle*. Pl., Islets.) 1. A small mass of tissue isolated within another type of tissue. 2. A spot of a different colour included within a larger spot. Cf. Ocellate.

ISOELECTRIC FOCUSING The process of separating proteins on a membrane set across an electrical gradient. Proteins move to an isoelectric point based on their relative size, and thus, charge. Abbr: IEF.

ISOFENPHOS An organic-phosphate (phosphoric acid) compound {1-methylethyl 2-((ethoxy((1-methylethyl)amino)phosphinothioyl)-oxy)benzoate} used as a contact insecticide against root-feeding insects, thrips, termites and leaf-feeding insects. Compound is applied to turf, ornamentals and nursery stock, and vegetables in some countries. Toxic to fishes and earthworms. Trade names include: Amidocid®, Discus®, Le-Mat®, Oftanol®. See Organophosphorus Insecticide.

ISOLATE Noun. (Italian *isola* from Latin, *insula* = island.) 1. Separate from other things of a similar origin or nature. 2. Something occurring alone.

ISOLATION Noun. (Latin, *insula* = island; English, *-tion* = result of an action.) Complete separation from similar forms, in biology, may be geographic, climatic, *etc*. Cf. Aggregation.

ISOMATE TPW® See Lycolure.

ISOMATE-M A synthetic pheromone {(Z-8) and (E-8)-dedecenyl acetate} used for attraction of Oriental Fruit Moth, Macadamia Nut Borer and Koa Seedworm. Formulations include ropes and laminated dispensers which are applied to peaches and nectarines. Trade names include Checkmate OFM®, Cide-Trak®, Quant 5® and Rak-5®.

ISOMERA Noun. (Greek, *isos* = equal; *meros* = part.) A series of Coleoptera whose members have an equal number of Tarsomeres on all feet. See Tarsomere.

ISOMEROUS Adj. (Greek, *isos* = equal; *meros* = part; Latin, *-osus* = with the property of.) With an equal number of tarsal segments (Tarsomeres) on all feet. Syn. Homoeomerous. Cf. Heteromerous.

ISOMETOPIDAE Jumping Tree Bugs. A small Family of heteropterous Hemiptera assigned to Miroidea and sometimes recognized as a Subfamily of Miridae. Adult body less than 3 mm long; Ocelli present; forewing with Cuneus and 1–2 cells in Membrane. Isometopids are frequently taken on bark or limbs of dead trees. Some Species are predaceous. See Miridae.

ISOMETRY Noun. (Greek, *isos* = equal; *metron* = measure.) The condition in which the relative size of two body structures or parts remain constant when the entire body changes in size. Cf. Allometry.

ISOMORPHOUS Adj. (Greek, *isos* = equal; *morphe* = shape; Latin, *-osus* = with the property of.) Pertaining to structures with the same form, appearance or construction. Cf. Heteromorphous.

ISOMURA, JUNICHI (1890–1932) (Hasegawa 1967, Kontyû 35 (Suppl.): 11–12.)

ISOPALPI Plural Noun. (Latin, *iso* = equal; *palpare* = to stroke.) Palpi with the same number of segments in both sexes, *e.g.* Trichoptera.

ISOPARTHENOGENESIS Noun. (Greek, *isos* =

equal; *parthenos* = virgin; *genesis* = descent.) An archane term to characterized the 'typical' form of Parthenogenesis, such as seen in bees (Henneguy).

ISOPHENE Noun. (Greek, *isos* = equal; *phainein* = to show. Pl., Isopheres.) A contour line of equal-event dates of rhythmic periodic phenomena at the same altitude.

ISOPODA Lateille 1817. (Middle Pennsylvanian–Recent.) Woodlice. (Greek, *isos* = equal; *pous* = foot.) A cosmopolitan Order of Crustacea found in all ecological habitats, ranging from terrestrial to marine benthic and bathypelagic. Present classifications recognize nine Suborders and about 100 Family-level Taxa. Suborders include Anthuridea Leach 1814, Asellota Latreille 1803; Calabozoidea Van Lieshout 1983; Epicaridea Latreille 1831; Flabellifera Sars 1882; Gnathiidea Leach 1814; Oniscidea Latreille 1803; Phreatoicidea Stebbing 1893 and Valvifera Sars 1882. Anatomically, isopods are characterized by exposed pereonites (lacking a carapace over pereopods); eyes, when present, not stalked; thoracic limbs (pereopods) without exites; pleopods usually broad and lamellar. Internal fertilization is almost universal in isopods, and young are retained in brood pouches formed on basal segments of the pereopods.

ISOPODS See Isopoda.

ISOPROCARB A carbamate compound {2-(1-methylethyl phenyl methylcarbamate} used as a stomach poison. Isoprocarb is applied to tree crops, cotton, coffee, rice, vegetables, peanuts, potatoes, ornamentals in some countries. Effective against plant-sucking insects, thrips and related Species. Trade names include: Etrofolan®, Hytox®, Mipc®, Mipcin®. See Carbamate Insecticide.

ISOPROPYL ALCOHOL (IUPAC: 2-propanol; $CH_3CHOHCH_3$) A member of the lower alcohols, with a distinct odoUr, a relatively high boiling point (82.4°C), and complete solubility in water. Isopropyl alcohol is used as a drying agent in insect histology, and is used infrequently to store insect specimens.

ISOPTERA Brulle 1832. (Eocene-Recent.) (Greek, *isos* = equal; *pteron* = wing.) Termites; White Ants. A cosmopolitan Order including about 2,300 Species of social insects. Isoptera are best represented in tropical and subtropical regions. Adults are soft-bodied, polymorphic and cryptozoic; Antenna moniliform or filiform; mandibulate mouthparts; primary reproductives with four membranous, similar sized, net-veined wings shed following dispersive flight. Abdomen with 10 apparent segments; Cerci with 1–5 segments; sclerotized external genitalia absent from most Species. Metamorphosis is paurometabolous. Primary food includes cellulose broken down by symbiotic bacteria or flagellate protozoans that occur in the termite's gut; incidental food includes exuviae, dead nestmates and excrement. Symbionts are passed to newly moulted individuals via proctodaeal feeding (trophallaxis). Nests are arboreal, subterranean and within Termitaria. Reproductives swarm; royal pair (king and queen) excavate nuptial chamber, copulate and establish new colony. Royal pair are often long-lived, perhaps to 50 years in some Species. Physogastry of female reproductives is typical in some groups. Castes include: Primary reproductives (king, queen), secondary reproductives (neoteinics), soldiers (male, female), workers (male, female) or worker-like individuals (pseudergates). Economically significant destruction of wood and wood products. Isoptera probably are related to Blattaria. Included Families: Hodotermitidae, Kalotermitidae, Mastotermitidae, Rhinotermitidae, Serritermitidae, Termitidae and Termopsidae. See Insecta.

ISOTHERM Noun. (Latin, *iso* = the same; *therme* = heat.) In chemistry, an isothermal line or process.

ISOTOMIDAE Plural Noun. A numerically moderate, widely distributed Family of arthropleonian Collembola. Isotomids are characterized by an elongate body with separate thoracic and abdominal segments, lacking Scales and Furcula often reduced. Isotomids are found in diverse moist habitats including soil, debris and pools. See Collembola.

ISOTONIC Adj. (Greek, *isos* = the same; *tonos* = stretching, tension; *-ic* = characterized by.) Tissue Culture: Pertaining to a culture medium in which the osmotic pressure is the same as intercellular fluid.

ISOTOX See Lindane.

ISOTROPIC Adj. (French, *isotropique* > Latin; *iso* = equal; Greek, *tropikos* = turn.) The pale part of the Sarostyle.

ISOTYPICAL GENUS A Genus described from more than one Species, all of which are congeneric (Smith).

ISOXATHION An organic-phosphate (thio-phosphoric acid) compound {O,O-diethyl O-(5-phenyl 3-isoxazolyl) phosphorothioate} used as a contact insecticide and stomach poison against mites, plant-sucking insects, cutworms, armyworms and chewing insects. Compound applied to orchards trees, forage crops, citrus, rice, ornamentals and vegetables in some countries. Not registered for use in USA. Toxic to fish. Trade name in Karphos®. See Organophosphorus Insecticide.

ISOZYME Noun. (New Latin.) Multiple forms of an enzyme. These are not isomers, as a rule, but distinct protein molecules with unique metabolic functions. Alt. Isoenzyme.

ISSEL, RAFFAELE (1878–1936) (Remotti 1936, Boll. Musei Lab. Zool. Anat. comp. Univ. Genova 16: 1–22, bibliogr.)

ISSIDAE Plural Noun. Issid Planthoppers. A large, widespread, predominantly tropical Family of auchenorrhynchous Hemiptera assigned to

Superfamily Fulgoroidea. Typically dark-coloured; many Species with weevil-like snout; forewing convex, small, often Elytra-like; Clavus not granulated; apex of second Tarsomere of hind leg with a spine. Biology poorly known; presumably phloem feeders on angiosperms. See Fulgoroidea.

ITALIAN PEAR SCALE *Epidiaspis leperii* (Signoret) [Hemiptera: Diaspidiae].

ITCH MITE *Sarcoptes scabiei* (DeGeer) [Acari: Sarcoptidae].

ITCH MITES See Sarcoptidae.

ITERATIVE HOMOLOGY Correspondence between characteristics (parts, structures) within an individual. See Homology. Cf. Phylogenetic Homology.

ITEROPAROUS Adj. (Latin, *iterare* = to repeat; *parere* = to bear; Latin, *-osus* = with the property of.) Pertaining to adult insects which are relatively long lived (compared to the developmental period) and whose eggs are laid singly or in small clutches throughout adulthood. Cf. Semelparous.

ITHONIDAE Plural Noun. Moth Lacewings. A Family of primitive Neuroptera consisting of three Genera and about 15 Species in Australia, one Genus and a few Species in North America. Habitus moth-like; body stout; Antenna not clubbed; foreleg not raptorial; wings broad at base, large with complex venation and Nygmata; female Ovipositor plowshare-like; male with large genitalic claspers (Parameres). Adults nocturnal or crepuscular; eggs large; larvae subterranean, perhaps feeding on decaying plant material. See Neuroptera.

ITHONOIDEA A Superfamily of plannipenous Neuroptera including Ithonidae. See Neuroptera.

ITO, TOKUTARO (1865–1941) (Hasegawa 1967, Kontyû 35 (Suppl.): 12.)

ITOGA, AKIRA (1894–1941) (Hasegawa 1967, Kontyû 35 (Suppl.): 12–13.)

ITONIDIDAE Plural Noun. See Cecidomyiidae.

IVASCHINZOFF, MARTIN (1901–1934) (Hellen 1935, Notul. ent. 15: 103–104.)

IVERMECTIN Noun. A commercially available compound introduced in 1981 for control of parasitic organisms in livestock, companion animals and humans. Ivermectin is highly effective against nematode and arthropod parasites and has been used to control human Onchocerciasis (river blindness). See Avermectin; Abamectin.

IVES, BENJAMIN HALE (1806–1837) (Weiss 1936, *Pioneer Century of American Entomology*. 320 pp. (133), New Brunswick.)

IVY APHID *Aphis hederae* Kaltenbach [Hemiptera: Aphididae].

IVY LEAFROLLER *Cryptoptila immersana* (Walker) [Lepidoptera: Tortricidae].

IVY SCALE *Aspidiotus nerii* Bouché [Hemiptera: Diaspididae]

IWAKAWA, TOMOTARO (1855–1933) (Hasegawa 1967, Kontyû 35 (Suppl.): 14.)

IWASAKI, TAKUYA (1869–1937) (Hasegawa 1967, Kontyû 35 (Suppl.): 14–15.)

IWASE, TARO (1906–1970) (Hinton 1971, Proc. R. ent. Soc. Lond. (C) 35: 53.)

IXODIDA Plural Noun. One of four groups of anactiotrichid mites.

IXODIDAE Plural Noun. Hard Ticks. See Ixodida. Cf. Argasidae; Nuttalliellidae.

IZQUIERDO, VICENTE (1850–1926) (Porter 1926, Revta. chil. Hist. nat. 30: 184–186, bibliogr.)

JABLONOWSKI, JOSEPH (1863–1943) (Escherich 1944, Z. angew. Ent. 30: 492–494.)

JABOT Noun. (French, origin uncertain. Pl., Jabots.) Originally intended to mean the ruffle on a man's shirt beneath the neck, but used in an entomological context to mean the Crop of the foregut. See Crop.

JACK PINE BUDWORM *Choristoneura pinus pinus* Freeman [Lepidoptera: Tortricidae]: A pest of Christmas tree plantations in northeastern USA and Canada. JPB will attack several Species of pine and spruce. See Tortricidae. Cf. Spruce Budworm.

JACKSON, ARTHUR RANDELL (1877–1944) (Dallman *et al.* 1944, NWest. Nat. 19: 182–192, bibliogr.)

JACKSON, CHARLES HERBERT NEWTON (–1955) (Hall 1956, Proc. R. ent. Soc. Lond. (C) 20: 75.)

JACKSON, CHARLES THOMAS (1805–1880) (Weiss 1936, *Pioneer Century of American Entomology*. 320 pp. (222–223), New Brunswick.)

JACKSON, DOROTHY JEAN (1893–1973) (Lees 1974, Proc. R. ent. Soc. Lond. (C) 38: 57.)

JACKSON, GEORGE (–1899) (Dutton 1899, Naturalist Hull 1899: 288.)

JACKSON, REGINALD (REX) ANDREW (1890–1969) (Worms 1970, Entomologist's Rec. J. Var. 82: 59–60.)

JACKSON, THOMAS HERBERT ELLIOT (1903–1968) (Carcasson & Clench 1969, J. Lepid. Soc. 23: 131–134.)

JACOB, JOHN KENNETH (–1941) (Spencer 1941, Proc. ent. Soc. Br. Columb. 38: 4–5.)

JACOBS, JEAN CHARLES (1821–1907) (Fologne 1908, Mém. Soc. r. ent. Belg. 15: 1–5.)

JACOBSEN, OLUF (1846–1921) (Kryger 1923, Ent. Meddr. 14: 120–124, bibliogr.)

JACOBSON (YAKOBSON), GHEORGII GHEORGHIEVICH (1871–1926) (Anon. 1928, Revue Russe Ent. 22: 1–28, bibliogr.)

JACOBSON, EDWARD (1870–1944) (De Meijère 1946, Ent. Ber., Amst. 12: 2–4, bibliogr.)

JACOBSON, HELMUT (1914–1942) (Sachtleben 1943. Arb. morph. taxon. Ent. Berl. 10: 174–175.)

JACOBSONIIDAE Plural Noun. A small Family of polyphagous Coleoptera assigned to Superfamily Bostrichoidea. Jacobsoniids are not represented in North America; a few Species occur in Australia. Adults are narrow and elongate; Antenna moniliform with 10–11 segments; tarsal formula 3-3-3, appearing with two segments on legs of some Species; Metasternum as long as combined length of abdominal Ventrites. Larvae are elongate, somewhat flattened with dorsum sclerotized; Antenna and Urgomphus short. Syn. Sarothriidae. See Bostrichoidea.

JACOBY, MARTIN (1842–1907) (Anon. 1908, Entomologist 41: 25–26.)

JACOT, ARTHUR PAUL (1890–1939) (Anon. 1941, Fla. Ent. 24: 43–47, bibliogr.)

JACQUELIN DUVAL, PIERRE NICOLAS CAMILLE (1828–1862) (Migneaux 1862, Ann. Soc. ent. Fr. (4) 2: 617–619.)

JACQUET, ERNEST (1842–1888) (Chobault 1888, Echange 4: 3.)

JACQUIN, NICHOLAS JOSEPH (1727–1817) (Rose 1850, *New General Biographical Dictionary* 8: 473.)

JACZEWSKI, TADEUSZ ANTONI FRANCISCEK (1899–1974) (Lees 1975, Proc. R. ent. Soc. Lond. (C) 39: 56.)

JAEGER, BENEDICT (1787–1869) (Weiss 1922, Proc. N. J. Hist. Soc. 7: 196–207.)

JAEGER, JULIUS (1834–1922) (Porritt 1922, Entomol. mon. Mag. 58: 114–115.)

JAESCHKE, GUSTAV (1834–1923) (Weidner 1967, Abh. Verh. naturw. Ver. Hamburg Suppl. 9: 192.)

JAFFUEL, FELIX (1874–1939) (Porter 1939, Revta chil. Hist. nat. 43: 124–126, bibliogr.)

JÄGER, GUSTAV (1832–1916) (Schenkling 1933, Ent. Bl. Biol. Syst. Käfer 29: 28–30.)

JÄGERSKIÖLD, L A (–1945) (Lindblad 1945, Ent. Tidskr. 66: 155–156.)

JAHN, ALFRED (1882–1964) (Anon. 1964, Mitt. dt. ent. Ges. 23: 84.)

JAIL FEVER See Epidemic Typhus.

JAKOBSON, G G See Jacobson.

JAKOVLEV, ALEXANDRE IVANOVICH (1863–1909) (Horn 1909, Dt. ent. Z. 1910: 170.)

JAKOVLEV, VASILII EGOROVICH (1839–1908) (Bogdanov 1891, Izv. imp. obshch. Lyub. Estest. Antrop. Etnogr. imp. Mosc. Univ. 77: [203–205]. bibliogr.)

JAMES, HEDLEY GORDON (1902–1975) (Maw 1976, Bul. ent. Soc. Canada 8: 19.)

JAMES, RUSSEL (1875–1942) (Cockayne 1944, Proc. R. ent. Soc. Lond. (C) 8: 70.)

JAMES, SYDNEY PRICE (1870–1946) (Christophers 1947, Obit. not. Fellows R. Soc. 5: 507–523, bibliogr.)

JAN, GEORG (1791–1866) (Silbenrock 1901, Bot. Zool. Oesterr. 1850–1900 Festschr, pp 445–446.)

JANCKE, OLDWIG (1901–1960) (Ehrenbarat & Leib 1961, Anz. Schädlingsk. 34: 59–60.)

JANDER, ALBERT (1840–1920) (Hedwig 1921, Jh. Ver. schles. Insektenk. 13: 21–23.)

JANET, CHARLES (1849–1932) (Berland 1932, Ann. Soc. ent. Fr. 101: 157–164, bibliogr.)

JANET'S ORGAN A Chordotonal Organ of the insect Antenna. (Janet 1911, C. R. Acad. Sci. 152: 110–112.) See Antenna. Cf. Bohm's Organ; Johnston's Organ; Multiporous Plate Sensilla.

JANGIUS, JOACHIM (1587–1657) (Weidner 1967, Abh. Verh. naturw. Ver. Hamburg Suppl. 9: 17–22.)

JÄNICHEN, ROBERT (–1912) (Belling 1912, Int. ent. Z. 6: 204.)

JANNONE, GIUSEPPE (1907–1971) (Boll. Lab. Ent. agr. Filippo Silvestri 29: 326–357, bibliogr.)

JANSCHA, ANTON (1734–1773) (Fraser 1951, Janscha on the Swarming of Bees. 28 pp. Royston, Herts.)

JANSE, ANTONIE JOHANNES THEODORUS

(1877–1970) (Lea 1957, J. ent. Soc. sth. Afr. 20: 3–9, bibliogr.)

JANSON, EDWARD WESTEY (1822–1891) (Anon. 1891, Ent. News 2: 188.)

JANSON, OLIVER ERICHSON (1850–1926) (Calvert 1927, Ent. News 38: 260–261.)

JANSSENS, ANDRÉ (1906–1954) (Collart 1954, Bull. Inst. r. nat. Belg. 31 (1): 1–7, bibliogr.)

JANSSON, ANTON (1880–1963) (Lindroth 1963, Opusc. ent. 28: 162.)

JANTHINE Adj. (Greek, *ianthinos* from *ion* = violet; *anthos* = flower.) Violet coloured.

JANVIER, HYPPOLYTE (1892–1986) French Entomologist specializing in Hymenoptera.

JAPANESE BEETLE *Popillia japonica* Newman [Coleoptera: Scarabaeidae]: An exceptionally destructive scarab beetle endemic to Japan and accidentally introduced into the eastern USA about 1916. Adults skeletonize foliage, consume flowers and gouge fruit of more than 300 Species of plants, including most deciduous fruit and shade trees. Larvae are serious pests through feeding on roots of many Species of plants, particularly grasses. In North America, JB completes one generation per year with larva overwintering in soil and pupating during spring. Adult emerges during May - October. Most effective natural enemy appears to be *Bacillus popilliae* Dutky which produces Milky Spore Disease in the larva. Syn. Jap Beetle. See Scarabaeidae. Rel. Milky Spore Disease.

JAPANESE ENCEPHALITIS An arboviral disease whose agent belongs to the Family Flaviviridae. JE endemic to Japan, spread to other parts of Asia and now the most common epidemic encephalitis globally. JE vectored by *Culex* Species; *Culex tritaeniorhynchus* serves as principal vector. Mosquitoes acquire JE from infected aquatic birds (herons, egrets), transmit viron to pigs where disease multiplies; mosquitoes then transmit JE to humans and horses. Mortality in humans age-dependent but considerable. See Arbovirus. Flaviviridae. Cf. St Louis Encephalitis. Murray Valley Encephalitis.

JAPYGIDAE Plural Noun. A numerically small, widespread Family of Diplura. Adults to 50 mm long; Mandible without Prostheca; Labial Palpus present; Thorax with four pairs of spiracles; Abdomen with Vesicles; Cerci forcipate, consisting of one segment. Biology of Japygidae is poorly known. See Diplura.

JAQUES, HARRY EDWIN (1880–1963) (Millspaugh 1964, Ann. ent. Soc. Am. 57: 265–266.)

JAROSCHEWSKY, A See Yaroschevski.

JARRAH LEAFMINER *Perthida glyphopa* Common [Lepidoptera: Incurvariidae]: A pest feeding on the leaves of *Eucalyptus marginata* in Australia. Alt. Jarrah Leafminer.

JARRIGE, JEAN (1904–1975) (Ruter 1975, Entomologiste 31: 230–232.)

JARVIS' FRUIT FLY *Dacus jarvisi* (Tryon) [Diptera: Tephritidae].

JAVELIN® A broad-spectrum biopesticide used for control of Lepidoptera larvae attacking vegetables, fruit tree and vine crops. Active ingredient *Bacillus thuringiensis* var. *kurstaki*. See *Bacillus thuringiensis*. Cf. Agree®; Teknar®. Rel. Biopesticide.

JAVELLE WATER A solution of postassium hypochlorite used in bleaching insect structures. Also called Eau de Javelle.

JAW-CAPSULE The case which contains the mouth structures in those dipterous larvae in which the head is differentiated.

JAYEWICKREME, SAMUEL HUBERT (1908–1953) (Buxton 1954, Proc. R. ent. Soc. Lond. (C) 18: 80.)

JAYNE, HORACE (1859–1913) (Calvert 1913, Ent. News 24: 383–384.)

JAYNES, HAROLD ANDRUS (1900–1971) (Allen & Leonard 1974, J. Econ. Ent. 67: 143.)

JE See Japanese Encephalitis.

JEANNEL, RENÉ (1879–1965) (Deboutteville & Paulian 1966, Ann. Soc. ent. Fr. 2: 1–37, bibliogr.)

JEDLICKY, ARNOST (Maran 1968, Acta. ent. bohemoslovaca 65: 250–251.)

JEFFERY, HUGH GEOFFREY (–1943) (Wakely 1948, Entomologist's Rec. J. Var. 60: 68.)

JENISON-WALWORTH, WILHELM VON (1796–1853) (Heyden 1855, Stettin. ent. Ztg 16: 15–16.)

JENKINSON, FRANCIS JOHN HENRY (1853–1923) (Scott 1923, Entomol. mon. Mag. 39: 261–262.)

JENNER, JAMES HERBERT AUGUSTUS (1849–1924) ([Adkin 1924], Entomologist 57: 215.)

JENNINGS, ALLAN HINSON (1866–1918) (Pierce *et al.* 1919, Proc. ent. Soc. Wash. 21: 61–63, bibliogr.)

JENNISON, HARRY MILLIKEN (1885–1940) (Anon. 1841, Mass. St. Coll. Fernald Clb. Yrbk. 10: 42–43.)

JENNY, FRED KEISER See Keiser-Jenny, F.

JENÖ, GYÖRFFY (1882–1970) (Endrodi 1972, Folia ent. hung. 25: 5–26, bibliogr.)

JENSEN, ADOLF S (1866–1953) (Tuxen 1954, Ent. Meddr. 27: 46–48, bibliogr.)

JENSEN, FRITZ (–1961) (Natvig 1962, Norsk. ent. Tidskr. 12: 57–59, bibliogr.)

JENSEN, LARS PETER (1869–1934) (Wolff 1935, Ent. Meddr. 19: 183–186, bibliogr.)

JENSEN, OTTO GORDIUS (1833–1905) (Engelhardt 1918, Ent. Meddr. 12: 20.)

JENSEN-HAARUP, ANDERS CHRISTIAN (1863–1934) (Esben-Petersen 1936, Flora Fauna Sillberg 42: 36–39.)

JENYN, LEONARD See Blomefield, L.

JEPSON, F P (–1950) (A. W. R. 1950, Trop. Agric. 106: 130.)

JERDON, THOMAS CLAVERHILL (1811–1872) (Anon. 1876, Hist. Berwicksh. nat. Clb. 1873–1875: 143–151.)

JERUSALEM CRICKET *Stenopelmatus fuscus* [Orthoptera: Stenopelmatidae]: Endemic to western North America and sometimes common along Pacific Coast. Eggs are white and oval; masses are laid in holes in ground. Nymphs feed on other insects, decaying organic matter, plant roots and tubers. JC typically is nocturnal and slow moving; adults are cannibalistic. Syn. Earth Child; Potato Bug. See Stenopelmatidae.

JERUSALEM CRICKETS See Stenopelmatidae.

JESTER, FRIEDRICH ERNST (1743–1822) (Ratzeburg 1874, Forstwissenschaftliches Schriftsteller-Lexicon 1: 270–272.)

JET STREAMS A permanent, high-altitude, long, narrow, meandering current of high-speed winds near the tropopause. JRs move generally west to east at altitudes of 15 to 25 kilometres (10 to 15 miles) and often exceed a speed of 402 kilometres (250 miles per hour). JSs affect the development and movement of weather systems.

JEUNE CELL Jeune Globule; Jeune Leucocyte. See Prohaemocyte.

JEWEL BEETLES Members of Buprestidae, named for brilliant colour on adult. Adults feed primarily on flower nectar while larvae usually feed in wood or roots of trees or shrubs. Most Species are not economically important, despite the feeding habits. See Buprestidae.

JEWEL BUGS See Scutelleridae.

JEWEL SPIDERS *Gasteracantha* spp. [Araneida: Araneidae].

JEWETT, HOWARD HERMAN (1884–1959) (Townsend 1960, J. Econ. Ent. 53: 18.)

JIFFY GROW® See Indolebutyric Acid.

JIGGER See Sand Flea.

JIROVEC, OTTO (1907–1957) (Anon. 1957, Zool. Listy 20: 84–94, bibliogr.)

JOANNIS, JOSEPH DE (1864–1932) (Viette 1949, Lepid. News 3: 77.)

JODFENPHOS See Phoxim.

JOHANNSEN, OSKAR AUGUSTUS (1870–1961) (Crook 1973, Ent. News 84: 101–102.)

JOHANSEN, JOHAN PETER (1844–1915) (Engelbert 1918, Ent. Meddr. 12: 18–19.)

JOHANSON, CARL HANS (1828–1908) (Lampa 1908, Ent. Tisskr. 29: 279–281.)

JOHN, OSKAR (1875–1935) (Horn 1935, Arb. morph. taxon. Ent. Berl. 2: 63.)

JOHNSON, CHARLES WILLISON (1863–1932) (Brooks 1932, Bull. Bost. Soc. nat. Hist. 65: 3–5, bibliogr.)

JOHNSON, J PETER (1899–1973) (Turner 1974, J. Econ. Ent. 67: 315.)

JOHNSON, JAMES SMITH (1936–1920) (H. S. 1921, Ent. News 32: 63–64.)

JOHNSON, ORSON BENNETT (1848–1917) (Anon. 1917, Ent. News 28: 338.)

JOHNSON, WILLIAM FREDERICK (1852–1934) (Walker 1934, Entomol. mon. Mag. 70: 164–165.)

JOHNSON, WILLIS GRANT (1866–1908) (Felt *et al.* 1908, J. Econ. Ent. 1: 163–164.)

JOHNSTON, FREDERICK ANDREW (1887–1941) (Anon. 1942, Ybk Mass. Sta. Coll. Fernald Clb. 11: 35–36.)

JOHNSTON, H BENNETT (–1974) (Lees 1975, Proc. R. ent. Soc. Lond. (C) 39: 56.)

JOHNSTON, T HARVEY (1881–1951) Australian academic (University of Queensland) and founder of academic Entomology in Queensland. (Mackerras & Marks 1974, Changing Patterns of Entomology. A symposium, 76 pp., (8), Brisbane.)

JOHNSTONE, DOUGLAS CHARLES (1890–1932) (Frohawk, 1932. Entomologist 65: 120.)

JOHNSTON'S ORGAN A scolopophorous auditory-organ located within the second antennal segment (Pedicel) of insects representing most larger Orders of insects. Alt. Johnstonian Organ (Johnston 1855, Quart. J. Microscop. Sci. 3: 97–102.) See Antenna. Cf. Bohm's Organ; Janet's Organ; Multiporous Plate Sensilla.

JOHOW, FREDERICO (–1833) (Porter 1933, Revta chil. Hist. nat. 37: 57–58.)

JOICEY, JAMES JOHN (1871–1932) (Riley 1932, Entomologist 65: 142–144.)

JOINT Noun. (O.F., *joindre* from Latin, *jungere* = to joint. Pl., Joints.) 1. A point or area of articulation between two sclerites. 2. The area of fusion between limb segments. 3. Non-sclerotized Cuticle between adjacent sclerotized regions of the Integument (Snodgrass).

JOINTWORMS See Eurytomidae.

JÖKER, ASTRID (1883–1946) (Larsson 1947, Ent. Meddr. 25: 149, bibliogr.)

JOLICOEUR, HENRI (–1895) (Anon. 1895, Leopoldina 31: 58.)

JONAS, FREDERICK MAURICE (1851–1924) (Esaki 1956, Kontyû 24: 115–117.)

JONES, ALBERT HUGH (–1924) (S[heldon] 1924, Entomologist 57: 95–96.)

JONES, CHARLES MALCOLM (1908–1966) (Pearson 1967, Proc. R. ent. Soc. Lond. (C) 31: 62.)

JONES, EDMOND PRICE (1877–1957) Amateur Australian lepidopterist who collected in Queensland.

JONES, EDWARD WALLEY (1904–1955) (Gibson & Douglas 1957, J. Econ. Ent. 50: 230.)

JONES, ELMER THOMAS (1892–1970) (Knutson 1971, J. Econ. Ent. 64: 344.)

JONES, ELWYN PARRY (1907–1965) (Pearson 1966, Proc. R. ent. Soc. Lond. (C) 30: 63.)

JONES, FRANK MORTON (1869–1962) (McDermott 1963, Ent. News 74: 29–36, bibliogr.)

JONES, HUGH (fl1671–1701) (Wilkinson 1974, Gt. Lakes Ent. 7: 129–131.)

JONES, HUGH PARRY (1893–1937) (Carr 1937, Entomol. mon. Mag. 73: 93–94.)

JONES, J R J LLEWELLYN (–1954) (Buxton 1955, Proc. R. ent. Soc. Lond. (C) 19: 69.)

JONES, MERLIN PERRY (1895–1963) (Sherman 1963, J. Econ. Ent. 56: 545.)

JONES, SIDNEY CARROL (1898–1974) (Crowell 1975, J. Econ. Ent. 68: 568.)

JONES, THOMAS HENRY (1885–1951) (Hyslop 1941, Proc. ent. Soc. Wash. 43: 61–62, bibliogr.)

JONES, THOMAS RYMER (1810–1880) (Anon. 1881, Am. Nat. 15: 175.)

JONES, WILLIAM (–1818) (Poulton *et al.* 1934, Trans. Soc. Br. Ent. 1: 139–155.)

JONSTON, JOH (1603–1675) (Eiselt 1836, *Geschichte, Systematik und Literature der Insektenkunde*. 255 pp. (20), Leipzig.)

JOPPEICIDAE Plural Noun. A small Family of heteropterous Hemiptera assigned to Superfamily Tingoidea. See Tingoidea.

JORDAN, HEINRICH ERNST KARL (1861–1959) (Anon. 1960, Trans. R. Ent. Soc. Lond. 109: 1–9.)

JORDAN, KARL HERMANN CHRISTIAN (1875–1972) (Schwenke 1968, Z. angew. Ent. 62: 240–241.)

JORDAN, ROBERT COANE ROBERTS (1825–1890) (Warren 1890, Entomologist 23: 238–240, bibliogr.)

JORDAN'S ORGAN See Chaetosemata.

JORDIS, CARL (1845–1903) (Anon. 1903, Societas Ent. 18: 109.)

JÖRGENSEN, HANS NICOLAY LAVRIDS (1865–1937) (Sonderup 1938, Ent. Meddr. 20: 107–108.)

JÖRGENSEN, PETER (1870–1937) (Esben-Petersen 1938, Ent. Meddr. 20: 105–106.)

JOSEPH, EDWIN GEORGE (1887–1975) (Gunn 1976. Proc. R. ent. Soc. Lond. (C) 40: 52.)

JOUGL, HANS A (1862–1910) (Kheil 1911, Int. ent. Z. 4: 243–244).

JOURHEUILLE, CAMILLE (1830–1909) (Mabille 1909, Ann. Soc. ent. Fr. 78: 575–577, bibliogr.)

JOUSSEAUME, FELIX (–1921) (Anon. 1921, Bull. Mus. natn. Hist. nat. Paris 27: 482–483.)

JOUST® See Chinomethionat.

JOUTEL, LOUIS HIPPOLYTE (1858–1916) (Davis 1916, J. N. Y. ent. Soc. 24: 239–242.)

JOWLS Diptera: The cheeks behind the depressed anterior part. The Peristoma of Curran.

JOY, ERNEST EDWARD COOPER (1869–1940) (W. R. S. 1940. Entomologist 73: 264.)

JOY, NORMAN HUMBERT (1874–1953) (Anon. 1953, Entomologist's Rec. J. Var. 65: 96.)

JUBATE Adj. (Latin, *jubatus* = with a maine.) Descriptive of a surface fringed with long, pendent Setae. Cf. Setose.

JUCCI, CARLO (1897–1963) (Lerma 1964, Atti Accad. Naz. ital. Ent. Rc. 12: 17–23.)

JUDAY, CHANCEY (1871–1944) (Noland *et al.* 1945, Limnol. Soc. Amer. Spec. publ. 16: 1–9, bibliogr.)

JUDEICH, JOHANN FREIDRICH (1828–1894) (Anon. 1894, Suisse Econ. Forest 1894: 131–132.)

JUGAL Adj. (Latin, *jugum* = yoke; *-alis* = pertaining to.) Acarology: Pertaining to the furrow separating the Prosoma and Opisthosoma.

JUGAL BRISTLES Large Setae along the margin of the Jugal Lobe.

JUGAL FOLD See Plica Jugals.

JUGAL LOBE 1. An area of the forewing projecting posteriad and contacting the hindwing of the insect. Syn. Jugal Region; Fibula. 2. The small projecting piece at the base of the forewing's posterior margin in some Lepidoptera and Trichoptera (Tillyard).

JUGAL REGION 1. A posterior basal lobe or area of the wing set off from the Vannal Region by Plica Jugalis, containing Vena Arcuata and Vena Cardinalis when these veins are present (Snodgrass). 2. The Neala *sensu* Martynov. See Wing. Cf. Anal Region; Axillary Region; Remigium.

JUGATAE Noun. (Latin, *jugum* = yoke.) A series of Lepidoptera in which the Jugum (not Frenulum) unites the wings in flight.

JUGATE Adj. (Latin, *jugum* = yoke; *-atus* = adjectival suffix.) Lepidoptera having a Jugum wing-coupling apparatus; any Lepidoptera having a Jugum. Cf. Frenulum; Retinulaculum.

JUGOFRENATE Adj. (Latin, *jugum* = yoke; *frenum* = bridle; *-atus* = adjectival suffix.) Insect wings having both a Jugum and a Frenulum as a coupling apparatus. Seen in Lepidoptera, such as the Micropterigidae.

JUGULAR Adj. (Latin *jugulum* = collar bone.) Descriptive of or pertaining to the throat.

JUGULAR SCLERITES Small sclerites in the membrane connecting the head with the Thorax. See Cervical Sclerites.

JUGULUM Noun. (Latin *jugulum* = collar bone.) 1. A sclerite just behind the Submentum. 2. The Gula; the cavity of the posterior part of the head to which the neck is joined. 3. The lateral and ventral parts of the Prothorax (Smith).

JUGUM Noun. (Latin, *jugum* = yoke. Pl., Juga.) 1. Heteroptera: Paired lateral lobes of the head, one on each side of the Tylus. 2. Some Lepidoptera and Trichoptera: A lobe, process or finger-like projection at the base of the forewing which overlaps the hindwing and holds them together during flight. Syn. Fibula.

JULICH, WILHELM (1839–1893) (Anon. 1894, Ent. News 5: 32.)

JULIEN'S ORGAN Satyridae: The Corema (Klots).

JULLIEN, JOHN (1873–1928) (Pictet 1928, Bull. Soc. lepidopt. Genève 6: 17–18, 45–62, bibliogr.)

JUMP Intrans. Verb. (Unknown origin.) To spring away from the ground (substrate) with the aid of muscular action of appendages. Flea beetles (Chrysomelidae) jump with their hind legs. Energy for the jump is stored in a Metafemoral Spring. The propulsive force is considerable and some Species can jump about 100 times their body length. The spring is a scroll-shaped organ inside the hind Femur that is held in place by Tibial Extensor Muscle insertions and attached to the hind Tibia base by a ligament-like structure. Special energy storage mechanism of the Metafemoral Spring differ from resilin because the amino acid composition in the protein is not like that of resilin. A Metafemoral Spring has evolved independently within several lineages of

Coleoptera, including Chrysomelidae, Curculionidae, Buprestidae and Bruchidae. See Locomotion. Cf. Crawl; Flight; Hop; Run; Walk. Rel. Pleural Arch; Resilin.

JUMPER ANTS *Myrmecia* spp. [Hymenoptera: Formicidae]. See Bulldog Ants.

JUMPING BEAN Seeds of Euphorbiaceae *(Sebastiania, Sapium)* which move or 'jump' owing to the movement of insect larvae within the seeds.

JUMPING BRISTLETAILS See Machilidae.

JUMPING MECHANISM Biomechanical devices which facilitate jumping by insects. Common JM include middle tibial extension of the leg and hydrostatic pressure in the Abdomen. See Furca; Hydrostatic Skeleton; Resilin; Tibial Extension.

JUMPING PLANTLICE See Psyllidae.

JUMPING SPIDERS See Salticidae.

JUMPING TREE BUGS See Isometopidae.

JUNCKEL, GUSTAV (1849–1919) (Möbius 1944, Dt. ent. Z. Iris 57: 8.)

JUNCO Y REYES D (1890–1970) (Mingo 1970, Graellsia 25: 241–243, bibliogr.)

JUNE BEETLES See Scarabaeidae; White Grubs.

JUNE BUGS See Scarabaeidae.

JUNGHANS, HANS HORST (1835–1908) (Möbius 1944, Dt. ent. Z. Iris 57: 8.)

JUNIOR HOMONYM Taxonomic nomenclature: The most recently published of two or more identical names for the same or different taxonomic categories. See Taxonomy. Cf. Senior Homonym.

JUNIOR SYNONYM Taxonomic nomenclature: The most recently published of two or more available synonyms for the same taxonomic category. See Taxonomy. Cf. Senior Synonym.

JUNIPER APHID *Cinara juniperi* (De Greer) [Hemiptera: Aphididae].

JUNIPER SCALE *Carulaspis juniperi* (Bouché) [Hemiptera: Diaspididae]: A pest of arborvitae, cypress, juniper and red cedar in North America. See Diaspididae.

JUNIPERUS Linnaeus. A Genus of coniferous plants widespread in the Northern Hemisphere. *Juniperus* is second in size to *Pinus*, and includes about 80 Species-level Taxa, assigned to three Sections (Caryocedrus, Oxycedrus and Sabina).

JUNK, WILHELM (1866–1942) (Riley 1944, Proc. R. ent. Soc. Lond. (C) 9: 19.)

JUPITER® See Chlorflurazuron.

JURADO, MARTIN DOELLO (–1948) (Anon. 1949, Revta Soc. ent. argent. 14: 235–238.)

JURAPRIIDAE Rasnitsyn 1983. A Family of apocritous Hymenoptera assigned to Proctotrupoidea. See Proctotrupoidea.

JURASSIC PERIOD 'Age of Cycads.' The second Period (208–146 MYBP) of the Mesozoic Era and named after the Jura Mountains on the border of Switzerland and France where Jurassic deposits were first studied. Important deposits include the Upper Jurassic Solenhofen Beds (Germany): An estuary-type habitat of Lithographic Stone in which *Archaeopteryx*, Dragonflies and *Pseudosirex* (Hymenoptera) were preserved. Other Jurassic deposits include British Lias (Somerset to Gloucestershire), Lias of Dobbertin (Mecklenberg), Montsech (Spain) and East Karatau, Turkestan (Slates). Terrestrial plants included Ginkgos, conifers, cycads, ferns and tree-ferns common. Great amphibians died out and were replaced with frogs and toads. Reptiles were dominant on land, including Pterosaurs (which soared and did not fly). Four Orders of mammals are represented in the JP. Insects include dragonflies (most common insects in estuary-type habitats), grasshoppers, termites, beetles and flies. Middle Jurassic is poorly known but the Late Jurassic shows a rich hymenopteran fauna. Representatives included the Praeaulacidae (extinct), Ephialtididae (extinct), Mesoserphidae (extinct), Heloridae, Megalyridae, Anaxyelidae and Xyelidae; *Pseudosirex* reveals body length 85 mm, wing expanse 60 mm and a long, conspicuous Ovipositor. See Geological Time Scale; Mesozoic Era. Cf. Cretaceous Period; Triassic Period.

JURECEK, STÉPHAN (1877–1940) (Obenberger 1940, Cas. csl. Spol. ent. 37: 1–2.)

JURESONG® See Lambda Cyhalothrin.

JURINE, LOUIS (1751–1819) Swiss academic, surgeon and author of several entomological works including *Nouvelle Methode de classer les Hymenopteres et les Dipteres* (1807), *Observations sur les Xenos Vesparum* (1816) and *Observations sur les ailes des Hymenopters* whch appears appeared in Volume 24 of Memoires de l'Academie de Turin. (Eiselt 1836, *Geschichte, Systematik und Literatur der Insektenkunde.* 255 pp. (105–106), Leipzig.)

JUSTA- Latin prefix for 'juxta' meaning 'near'.

JUTE LOOPER *Anomis involuta* (Walker) [Lepidoptera: Noctuidae].

JUTE STEM WEEVIL *Apion corchori* [Coleoptera: Curculionidae].

JUVABIONE Noun. (Etymology obscure.) A juvenile hormone analogue.

JUVENILE Adj. (Latin, *juvenilis* from *juvenis* = young.) 1. Descriptive of an organism which is immature, young or undeveloped physically, biologically or behaviourally. 2. Descriptive of organisms, organs or structures that remain in an immature or young stage of development. See Development; Metamorphosis. Cf. Adult; Infantile. Rel. Growth.

JUVENILE HORMONE Noun. A sesquiterpene hormone produced by the Corpora Allata. JH is best known for its inhibition of development to the adult stage; also involved in the regulation of many physiological functions including diapause, heartbeat, migration, moulting and oogenesis.

JUVENILE HORMONE ANALOGUE See Insect Growth Regulator.

JUVENOID Noun. (Latin, *juvenilis* from *juvenis* = young; Greek, *eidos* = condition. Pl., Juvenoids.) Juvenile Hormone analogue. Chemical compounds which act as Internal Growth Regulators to mimic Juvenile Hormone and disrupt the insect's endocrine system. Ingestion or topical application induces supernumerary instars, retardation of development to normal Pupae and/or adult, and loss of fertility. Examples include: Methoprene, Hydroprene and Fenoxycarb. Juvenoids are not related to nerve poisons; Juvenoids are highly selective insecticides, display very low mammalian toxicity and show no residue problems. See Insect Growth Regulator; Insecticide.

JUXTA Male Lepidoptera: A sclerite beneath the Aedeagus and to which it may be hinged or fused; part of the Fultura Inferior. See Anellus.

JUXTANUCELAR BODY See Centriole Adjunct.

KAAD Acronym for Kerosene, Acetic Acid and Dioxane. A fixative for immature insects, consisting of: Kerosene (1 part), Glacial Acetic Acid (2 parts), Dioxane (1 part) and Ethanol (10 parts).

KADETHRIN A synthetic-pyrethroid compound used as a contact insecticide against flies, mosquitoes, midges and cockroaches in some regions of world. Highly phytotoxic and toxic to bees and fishes. Kadethrin is not available in USA. Trade names include: Spray-Tox®. See Synthetic Pyrethroids.

KAEMPFER, ENGELBERT (1651–1716) (Rose 1850, *New General Biographical Dictionary* 9: 73–74.)

KAESER, WALTER (1917–1975) (Wahl 1975, Apidologie 6: 190.)

KAESTNER, ABRAHAM GOTTHELF (1719–1800) (Rose 1850, *New General Biographical Dictionary* 9: 74.)

KAFIL® See Permethrin.

KAFIL SUPER® See Cypermethrin.

KAHL, PAUL HUGO ISADOR (1859–1941) (Osborn 1946, *Fragments of Entomological History,* Pt. II. 232 pp. (93–94), Columbus, Ohio.)

KAIROMONE (Greek, *kairos* = opportunistic; *hormaein* = to excite. Pl., Kairomones.) An interspecific chemical 'messenger' emitted by one Species and received by a second Species. The chemical messenger is beneficial only to the receiver Species. See Allomone; Semiochemical. Cf. Pheromone; Synomone. Rel. Hormone.

KAISER, OSKAR (1896–1935) (Anon. 1935, Arb. morph. taxon. Ent. Berl. 6: 69.)

KALA-AZAR A widespread form of Visceral Leishmaniasis produced by *Leishmania donovani* (Laveran & Mesnil). Disease localizes in reticuloendothelial cells of humans and causes progressive enlargment of spleen and liver; Kala-Azar sometimes is fatal. Syn. Black Disease; Dumdum Fever; Tropical Splenomagaly. See Leishmaniasis; Visceral Leishmaniasis. Cf. Cutaneous Leishmaniasis.

KALANDADTZE, LEONID POLIYEUKTOVICH (1898–1968) (Batiashvil 1969, Ent. Obozr. 48: 703–710. (Translation: Ent. Rev. Wash.48: 444–449).

KALCHBERG, ADOLF VON FREIHERR (1841–1899) (Anon. 1900, Jber. wien. ent. Ver. 10: 25–26.)

KALEIT® See Fenitrothion.

KALIS, JORINUS PIETER ADRIANUS (1899–1949) (Roepke 1949, Ent. Ber., Amst. 12: 425–427.)

KALLOGRAMMATIDAE Handlirsch 1906.

KALM, PEHR (1715–1779) (Brendel 1879, Am. Nat. 13: 755–756.)

KALOTERMITIDAE Plural Noun. Drywood Termites; Powderpost Termites. A Family of termites found in drywood and dampwood habitats. Species lack a worker caste, while individuals lack a Fontanelle and do not construct earthen tubes. See Isoptera.

KALSHOVEN, LOUIS GEORGE EDUMUND (1892–1970) (Vecht 1970, Ent. Ber., Amst. 30: 89–90, bibliogr.)

KALTENBACK, JOHANN HEINRICH (1807–1876) (Anon. 1876, Ent. NachrBl. 2: 112–114.)

KALUGINA, NADESHDA SERGEEVNA (–1922) Russian specialist in fossil nematoceran Diptera.

KAMBERSY, OTTO (1859–1907) (Reitter 1907, Wien. ent. Ztg. 26: 325–326.)

KAMEHAMEHA BUTTERFLY *Vanessa tameamea* Eschscholtz [Lepidoptera: Nymphalidae].

KÄMPFER, ENGELBERT (1651–1716) (Müldener 1891, Natur 20: 111–112, 118–120, 137–140.)

KANDA, SHIGEO (1898–1961) (Hasegawa 1967, Kontyû 35 (Suppl.): 30, bibliogr.)

KANE, WILLIAM FRANCIS DE VISMES (1840–1918) (Gardner 1918, Entomol. mon. Mag. 54: 254–255.)

KANERVA, NIILO (1898–1942) (Kivirikko 1943, Suomen hyönt. Aikak 9: 1–3.)

KANGAROO BEETLE *Sagra papuana* Jacoby [Coleoptera: Chrysomelidae].

KANGAROO BOT FLY *Tracheomyia macropi* (Froggatt) [Diptera: Oestridae].

KANGAROO HARD TICK *Amblyomma triguttatum* C.L. Koch [Acari: Ixodidae].

KANGAROO LOUSE *Heterodoxus longitarsus* (Piaget) [Phthiraptera: Boopidae].

KANGAROO SOFT TICK *Ornithodoros gurneyi* Warburton (Acari: Argasidae].

KANGAS, E (Weidner 1968, Z. angew. Ent. 61: 350–353.)

KANI, TOKACHI (1908–1944) (Hasegawa 1967, Kontyû 35 (Suppl.): 28, bibliogr.)

KANO, TADAO (1906–1945) (Hasegawa 1967, Kontyû 35 (Suppl.): 28–29, bibliogr. pl. 4, fig. 8.)

KANT, IMMANUEL (1724–1804) A Prussian philosopher of humble origins, Kant was educated at Collegium Fredericianum and the University of Konigsberg. He served as a private tutor (1746–1755), teacher at University of Konigsberg (1755–1770) and Professor and chair of logic and metaphysics (1770–1797). Among Kant's most influential works are: *Kritik der reinen Vernunf* (1781); *Kritik der practischen Vernunft* (1788) and *Kritik der Urtheilskraft* (1790.) Kant was educated in the Rationalism of Wolff, and inclined to the Empiricism of Hume. Kant ultimately developed the Transcendental Method of Philosophy which contends that knowledge does not exist without experience. Experience is a compound of matter given in sensation and form with principles of arrangement and synthesis which come from the mind.

KANZAWA, TSUNEO (1889–1954) (Hasegawa 1967, Kontyû 35 (Suppl.): 29–30, bibliogr.)

KAO HAOLE SEED BEETLE *Araecerus levipennis* Jordan [Coleoptera: Anthribidae]: A minor pest of stored products. See Anthribidae. Cf. Coffee Bean Weevil.

KAP® See Phenthoate.

KAPPA Noun. (Greek, k.) Diptera: The Palpiger

sensu MacGillivray.

KARAMAN, ZORA (1907–1974) (Serafimovski 1975, Zastita Bilja 25: 279–280.)

KARATAIDAE Rasnitsyn 1977. A Family of Hymenoptera assigned to the Ephialtitomorpha.

KARATAVITIDAE Rasnitsyn 1963. Plural Noun. A fossil Family of Symphyta (Hymenoptera) assigned to the Ephialtitomorpha.

KARATE® See Lambda Cyhalothrin.

KARAVAEV, VLADIMIR AFANASSIEVICH (1864–1939) (Paramonov 1941, Trav. Mus. zool. Kiev 24: 3–8, bibliogr.)

KARBASPRAY® See Carbaryl.

KARBATION® See Metham-Sodium.

KARBOFOS® Malathion.

KARNEY, HEINRICH HUGO (1886–1939) (Sachtleben 1939, Arb. physiol. angew. Ent. Berl. 6: 315–316.)

KARPHOS® See Isoxathion.

KARSCH, FERDINAND ANTON FRANZ (1853–1936) (Bonnet 1946, *Bibliographia Araneorum* 1: 48–49, bibliogr.)

KARSTEN, HERMAN GUSTAV WILHELM KARL (1817–1877) (Anon. 1877, Leopoldina 13: 130, 162–163, bibliogr.)

KARUMIIDAE Plural Noun. An exceedingly small Family of Coleoptera with a disjunct distribution known only from South America, southwest Asia and Afghanistan. Karumiids are members of the Cantharoidea, and probably most closely related to the Drilidae and Phengodidae. Females are larviform, flightless, termitophilous and probably rarely leave termite nests. Krumariids probably are not predators of the termites. See Cantharoidea.

KARYOLOGY Noun. (Greek, *karyon* = nucleus; *logos* = discourse.) The study of chromosomes and nuclear cytology.

KARYON Noun. (Greek, *karyon* = nucleus. Pl., Karya.) The cell nucleus.

KARYOTHECA Noun. (Greek, *karyon* = nucleus; *theke* = covering. Pl., Karyothecae.) The nuclear membrane.

KARYOTYPE Noun. (Greek, *karyon* = nucleus; *typos* = pattern.) 1. The complement of chromosomes in a cell. 2. Individuals with the same Chromosome number and similar linear arrangement of genes in homologous chromosomes.

KASCHKE, KARL (1852–1935) (Rupp, 1936. Ent. Z. Frankf. a. M. 50: 25–27.)

KASHMIR BEE-VIRUS A viral disease first discovered in *Apis cerana* from Kashmir and India; KBV subsequently was detected in *Apis mellifera* from Australia and New Zealand. KBV multiplies rapidly and kills adult honey bees. See Honey Bee Viral Disease.

KATABOLIC Adj. (Greek, *kata* = down; *ballein* = to throw; *-ic* = of the nature of.) Pertaining to katabolism or of the nature of katabolism. Cf. Anabolic.

KATABOLISM Noun. (Greek, *kata* = down; *ballein* = to throw; English, *-ism* = condition.) Metabolic processes which break down or destroy proteins, fats and carbohydrates.

KATAPLEURE See Preepisternum.

KATATREPSIS Noun. (Greek, *kata* = down; *-sis* = a condition or state.) During blastokinesis, the passage of the embryonic insect in from the dorsal aspect of the Ovum to its original position on the ventral aspect (Wheeler, see Henneguy). Cf. Anatrepsis. Rel. Blastokinesis.

KATEPIMERON See Infraepimeron.

KATEPISTERNUM Noun. (Greek, *kata* = down, *epi* = upon; *sternon* = chest. Pl., Katepisterna.) The lower part of the Episternum when it is divided by a suture or cleft into two parts. The lower part of the Episternum (Comstock). See Infraepisternum; Pleuron. Cf. Anepisternum.

KATIE'S SPRINGTAIL *Katianna australis* Womersley [Collembola: Sminthuridae].

KATIPO *Latrodectus hasselti* (Thorell) [Araneae: Theridiidae] (New Zealand).

KATO, MASAYO (1898–1967) (Ishikura 1968, Kontyû 36: 203–205.)

KATO, SHIZUO (1906–1962) (Hasegawa 1967, Kontyû 35 (Suppl.): 27, bibliogr.)

KATSOMATA, KANAME (1896–1945) (Hasegawa 1967, Kontyû 35 (suppl.): 26–27, bibliogr.)

KATSUKI, KIYOSHI (–1937) (Anon. 1938, Arb. physiol. angew. Ent. Berl. 5: 79.)

KATYDIDS See Tettigoniidae; Saltatoria.

KAUFMANN, OTTO (1896–1944) (Sachtleben 1944, Arb. physiol. angew. Ent. Berl. 11: 157–158.)

KAUFMANN-JAN, WILLY (–1947) (Vogel 1948, Mitt. schweiz. ent. Ges. 21: 296–297.)

KAUP, JOHANN JAKOB (1803–1873) (Anon. 1873, Leopoldina 9: 18–20.)

KAURI COCCID *Conifericoccus agathidis* Brimblecombe [Hemiptera: Margarodidae].

KAURI MOTHS See Agathiphagidae.

KAUTZ, HANS (1870–1954) (Anon. 1956, Z. wien. ent. Ges. 41: 241–245.)

KAVAN, OTAKAR (1888–1956) (Heyrovsky 1956, Cas. csk. Spol. ent. 53: 227.)

KAWAKAMI, TIKIYA (1871–1915) (Hasegawa 1967, Kontyû 35 (Suppl.): 29, bibliogr.)

KAWALL, JOHANN HEINRICH CARL (1799–1881) (Kraatz 1881, Dt. ent. Z. 25: 340.)

KAWAMURA, TAMIJI (1883–1964) (Hasegawa 1967, Kontyû 35 (Suppl.): 29.)

KAYAFUME® See Methyl Bromide.

KAYAPHOS® See Propaphos.

KAYAZINON® See Diazinon.

KAYAZOL® See Diazinon.

KAYE, WILLIAM JAMES (1875–1967) (Kennedy 1968, Proc. R. ent. Soc. Lond. (C) 32: 59.)

KEA, J W (1911–1936) (Anon. 1937, Fla. Ent. 19: 54.)

KEARFOTT, WILLIAM DUNHAM (1864–1917) (Anon. 1918, Ent. News 29: 1–3, bibliogr.)

KEDS See Hippoboscidae.

KEEL Noun. (Old Norse, *kjölr* = keel.) An elevated ridge or Carina. See Carina. Rel. Sculpture.

KEELEY, R G (–1874) (Anon. 1874, Entomol. mon. Mag. 11: 70.)

KEFERSTEIN, GEORG ADOLF (1793–1884) (Kolbe 1885, Berl. ent. Z. 29: 173–180.)

KEILEM, ALFRED (–1963) (Heyrovsky 1963, Cas. csl. Spol. ent. 60: 267.)

KEIM-STOP® See Chlorpropham.

KEISER-JENNY, FRED (1895–1969) (Eglin-Dederding 1969, Mitt. ent. Ges. Basel 19: 121–123, bibliogr.)

KELER, STEPHEN VON (1897–1967) (Eichler 1973, Lounais-Hameen Luonto 46: 1–46, bibliogr.)

KELLER, GEORGE J (1873–1926) (Buckholtz 1926, J. N. Y. ent. Soc. 34: 293.)

KELLER, JOHN CARLOS (1918–1971) (Fluno 1971, J. Econ. Ent. 64: 996–997.)

KELLERMAN, KARL FREDERICK (1879–1934) (Taylor 1934, Science 80: 373–374.)

KELLICOTT, DAVID SIMONS (1842–1898) American academic (University of Buffalo; Ohio State University) and specialist in Odonata. (Webster 1898, Can. Ent. 30: 166–167.)

KELLNER, AUGUST (1794–1883) (Anon. 1883. Wien. ent. Ztg. 2: 184.)

KELLNER-PILLAULT, SIMONE (1925–1985) (Caussanel 1986. Ann. Ent. Soc. Fr. 22: 311–312.) Curator of Entomology (Hymenoptera) at the Museum National d'Histoire Naturelle, Paris.

KELLOGG, VERNON LYMAN (1867–1937) (McClung 1938, Science 87: 158–159.)

KELLY, EDWARD GUERRANT (1880–1949) (Anon. 1949, J. Econ. Ent. 42: 162–163.)

KELP FLIES See Coelopidae.

KELSALL, ARTHUR (1892–1974) (Eidt 1975, Bull. ent. Soc. Canad. 7: 97.)

KELSALL, THOMAS (–1903) (Bailey 1904, Entomol. mon. Mag. 40: 18–19.)

KELTHANE® See Dicofol.

KEMEROVO TICK FEVER A tick-borne viral disease of humans in Siberia which is vectored by *Ixodes persulcatus;* etiological agent not known.

KEMNER, NILS VIKTOR ALARIK (1887–1948) (Tuxen 1949, Ent. Meddr. 25: 330–331.)

KEMOLATE® See Phosmet.

KEMP, STANLEY (–1945) (Ashe 1945, Entomol. mon. Mag. 81: 240.)

KEMPER, HEINRICH (Döhring 1967, Mitt. dt. ent. Ges. 26: 1–3.)

KEMPNY, PETER (1862–1906) (Navas 1906, Boln. Soc. aragon. Cienc. nat. 5: 182–185, bibliogr.)

KENDERESSY, DÉNES (1846–1881) (Bordon 1897, Rovart Lap. 4: 25–28.)

KENDO® See Fenproximate.

KENNEDY, CLARENCE HAMILTON (1870–1952) (Riley 1953, Proc. R. ent. Soc. Lond. (C) 17: 72.)

KENNEL, JULIUS (1854–1939) (Strand 1940, Folia zool. hydrobiol. (Riga) 10: 364–368, bibliogr.)

KENNICOTT, ROBERT W (1835–1866) (Dos Pasos 1951, J. N. Y. ent. Soc. 59: 137.)

KENOFURAN® See Carbofuran.

KENRICK, GEORGE HAMILTON (1850–1939) (Bethune-Baker 1939, Entomologist's Rec. J. Var. 61: 116.)

KENTROMORPHIC Adj. (Greek, *kentron* = point; *morphe* = form; *-ic* = characterized by.) Descriptive of or pertaining to Kentromorphism.

KENTROMORPHISM Noun. (Greek, *kentron* = point; *morphe* = form; *-ism* = condition. Pl., Kentromorphisms.) 1. A type of environmentally induced polymorphism found in some Orthoptera, some Lepidoptera and a few other insects. One form (phase) is manifest at low population densities; another form (phase) is manifest at high population densities; so-called intermediate forms occur at changing population densities. Kentromorphism is best illustrated in locusts that have a solitary and migratory phase (Haskell 1962). 2. A term applied to distinguish 'phases' (*sensu* Uvarov) from seasonal and temperature-induced pigmentary phases of insects (Key & Day 1954, Aust. J. Zool. 2: 309–339.) See Polymorphism. Cf. Caste Polymorphism; Cyclical Polymorphism.

KENWAY, HAROLD CECIL (1872–1952) (Riley 1953, Proc. R. ent. Soc. Lond. (C) 17: 72–73.)

KENYA TICK TYPHUS A minor rickettsial disease in Kenya and caused by the aetiological agent *Rickettsia conorii.* Elsewhere, disease reported under other names including South African Tick-Bite Fever, Marseilles Fever and Pimply Fever. KTT poorly studied with few deaths reported; ticks apparently vector disease to humans from rodents, rabbits and mammals endemic to regions. See Rickettsiaceae; Tick-Borne Spotted Fever. Cf. Rickettsial Pox; Rocky Mountain Spotted Fever.

KENYAN DUNG BEETLES *Copris fallaciosus* Gillet [Coleoptera: Scarabaeidae].

KEPPEN, FEDOR PETROVICH (1833–1908) (Adelung 1908, Revue Russe Ent. 8: xv–xviii.)

KERATIN Noun. (Greek, *keras* = horn.) The nitrogenous, sulphur containing fibrous protein that forms ungulate horns, hair, nails and feathers. Analogous to Chitin in insects.

KERRIIDAE Plural Noun. Lac Scales. A Family of Hemiptera that is predominantly tropical/subtropical in distribution. Females are globular in shape, legless and live within resinous cells. The Indian lac insect [*Laccifer lacca* (Kerr)] is of economic importance and produces large quantities of high-quality lac used to make varnish; ca. 4 million pounds are produced annually. See Hemiptera.

KERKLOIS, JEAN ADRIEN (1820–1872) (Vollenhoven 1872, Petites Nouv. Ent. 4: 201.)

KERLEE, ROY (1905–1928) (Anon. 1928, Bienn. Rep. Mo. St. Bd. Ent. 7: 9.)

KERMES Noun. (Arabic, *qirmiz* = crimson.) A red dye prepared from the dried females of the coccid *Kermes ilicis.*

KERMESINUS Adj. (Arabic, *qirmiz* = crimson.) Dark red, with much blue.

KERNBACH, KURT (1912–1968) (Schaeffer 1968, Mitt. dt. ent. Ges. 27: 13–14.)

KERNEL GRUB *Assara seminivale* (Turner) [Lepidoptera: Pyralidae].

KERNER VON MARILAUN, ANTON RITTER (1831–1898) (Mik 1898, Wien. ent. Ztg. 17: 184.)

KEROPLATIDAE Plural Noun. A cosmopolitan Family consisting of about 700 Species and 100 Genera of Diptera. Adults occur in dark, damp habitats; larvae occur in caves, under logs and bracket fungi where they spin silken webs; larvae of some Species are predaceous. See Diptera.

KERREMANS, CHARLES (1847–1915) (Semenov-Tian-Shansky, Revue Russe Ent. 15: 683; Fenyes 1916, Ent. News 27: 48.)

KERSEY, ROBERT HOBART (1890–1951) (Riley 1953, Proc. R. ent. Soc. Lond. (C) 17: 73.)

KERSHAW, JAMES ANDREW (1866–1946) (R. A. K. 1947, Mem. natn. Mus. Victoria 15: 180–181.)

KERSHAW, SIDNEY HARDINGE (1881–1964) (P.B.M.A. 1964, Entomologist's Rec. J. Var. 76: 265.)

KERSHAW'S BROWN *Oreixenica kershawi* (Miskin) [Lepidoptera: Nymphalidae].

KERTÉSZ, KALMAN (1867–1922) (Cresson 1923, Ent. News 34: 128.)

KESSEL, EDWARD LUTHER (1904–) American academic (University of San Francisco) and Associate Curator (California Academy of Science). Kessel was a specialist in biology and systematics of Platypezidae. (Arnaud 1989, Myia 4: vii-xxvii, 16 figs.)

KESSLER, HERMANN FREIDRICH (1816–1897) (Ackermann 1889, Ent. Nachr. 13: 76–78, bibliogr.)

KESSLER, PAUL (–1964) (Bros 1964, Mitt. ent. Ges. Basel 14: 164.)

KETTNER, FRIEDRICH WILHELM (Weidner 1971, Ent. Mitt. zool. St. Inst. zool. Mus. Hamburg 4: 205–206.)

KEVAN, DOUGLAS KEELY (1896–1968) (Kennedy 1969, Proc. R. ent. Soc. Lond. (C) 33: 55.)

KEY Noun. (Middle English, *key* = key. Pl., Keys) A taxonomic device by which objects are identified based on suites of characters or character states. Cf. Tabular Key; Pictorial Key; Dichotomous Key. Rel. Identification; Taxonomy.

KEY PEST A perennial, severe pest of crop production; a pest that dominates the development and implementation of pest control activities. See Secondary Pest; Subeconomic Pest.

KEYHOLE WASP *Pachydynerus nasidens* (Latreille) [Hymenoptera: Vespidae].

KEYS, JAMES H (1855–1941) (Cameron 1941, Entomol. mon. Mag. 77: 60–61.)

KEYSERLING, EYGENE VON (1833–1889) (Mar 1889, Entomologica Amer. 5: 159–160.)

KEYSTONE SPECIES Species that have a disproportionately large effect on other Species in a community.

KFD See Kyasanur Forest Disease.

KHAPRA BEETLE *Trogoderma granarium* Everts [Coleoptera: Dermestidae]: A serious international pest of stored grains and dried vegetables. KB is endemic to India and is now pantropical. KB was adventive to USA in 1953 but eradicated. KB lives indoors in temperate climates and outdoors in hot-dry tropical-subtropical climates. KB only exclusively phytophagous Species of dermestid; it prefers whole grain and cereals. KB larvae excavate grain and contaminate product with exuviae; larvae cannot penetrate sound grain until fourth instar; adult rarely feeds. Adults are 2–3 mm long, oval, mottled black and brown; median Ocellus present; Antenna with club of 3–5 segments; wings are present but adults do not fly. KB displays low reproductive potential (ca 35 eggs per female). Eggs are laid singly among grains with an incubation period of 4–10 days, depending upon temperature but not humidity. Larval stage requires 4–8 instars and supernumerary instars at lower temperatures; females complete an additional instar. Larvae can survive years without food and larval diapause may last six years; mature larvae are ca 5 mm long and lacks Urogomphi. Pupation occurs within last larval Integument; pupal stage requires 3–6 days. Teneral adults remain in last larval Integument ca 1–2 days and mating occurs after emergence. Adult females undergo a preoviposition period of 2–3 days at 25°C. Adults are short-lived (4–13 days) and fecundity not dependent upon food or moisture. See Dermestidae. Cf. Common Hide Beetle; Hide Beetle; Larder Beetle.

KHEIL, NAPOLEON MANUEL (1849–1923) (Vavra 1924, Acta ent. Mus. Prag. 2: 3–4, bibliogr.)

KIAER, HANS (1865–1929) (Anon. 1892, Ent. Tidskr. 13: 69–70.)

KIAWE BEAN WEEVIL *Algarobius bottimeri* Kingsolver [Coleoptera: Bruchidae].

KIAWE FLOWER MOTH *Ithome concolorella* (Chambers) [Lepidoptera: Cosmopterygidae].

KIAWE ROUNDHEADED BORER *Placosternus crinicornis* (Chevrolat) [Coleoptera: Cerambycidae].

KIAWE SCOLYTID *Hypothenemus biramnus* (Eichhoff) [Coleoptera: Scolytidae].

KIDD, JOHN (1775–1851) (Westwood 1852, Trans. ent. Soc. Lond. 1: 137.)

KIDNEY Noun. (Middle English, *kidenei* = kidney. Pl., Kidneys.) In vertebrates, paired organs of excretion that collect and eliminate urea, uric acid and other waste products. The human kidney is bean-shaped. Cf. Malpighian Tubule.

KIDNEY-SHAPED See Reniform; Nephroid.

KIEFFER, JEAN JACQUES (1856–1925) French priest, educator and taxonomic specialist on parasitic Hymenoptera (Bethyloidea, Proctotrupoidea) and Cecidomyiidae (Diptera). (Trotter 1925, Marcellia 22: 130–133; Travares 1926, Broteria (Zool.) 23: 126–148, bibliogr.)

KIELLERUP, CARL EMIL (1822–1908) (Papavero 1975, *Essays on the History of Neotropical Dipterology* 2: 364. São Paulo.)

KIESENWETTER, ERNST AUGUST HELLMUTH

VON (1820–1880) (Anon. 1880, Entomol. mon. Mag. 16: 67–70, bibliogr.)

KIKUYU GRASS BUG *Halticus chrysolepis* Kirkaldy [Hemiptera: Miridae].

KILLER BEE See Africanized Honey Bee.

KILLIAS, EDUARD (1829–1891) (Anon. 1892, Mitt. schweiz. ent. Ges. 8: 373–375.)

KILLINGTON, FREDERICK JAMES (1894–1956) (G. J. K. 1957, J. Soc. Br. Ent. 5: 230.)

KILLMASTER® See Chlorpyrifos.

KILLZONE® See Proproxur.

KILMAN, LEROY N (1884–1954) (Remington 1954, Lepid. News 8: 30.)

KILUMAL® See Fenpropathrin.

KILVAL® See Vamidothion.

KILVAR® See Vamidothion.

KIN RECOGNITION Social Insects: The recognition and discrimination among various categories of related individuals.

KIN SELECTION Social Insects: 1. The selection of genes as a result of individuals in a population (Species) affecting the survival and reproduction of relatives who possess the same genes by common ancestry. 2. An ecological phenomenon in which individual behavioural actions benefit the population, but not necessarily the individuals comprising the population. Cf. Altruism.

KINALUX® See Quinalphos.

KINCAID, TREVOR (1872–1949?) (Hatch 1950, Studies honoring T. Kinkaid, 10 pp., bibliogr.; Kincaid 1962. *Autobiogr. - the adventures of an Entomologist.* 2987 pp.)

KINDERMANN, ALBERT (1810–1860) (Lederer 1860, Wien. ent. Monatschr. 4: 251–255.)

KINESIS Noun. (Greek, *kinesis* = motion. Pl., Kineses.) A non-orientational movement exhibited by an organism in response to an external stimulus. See Orientation. Cf. Allokinesis; Blastokinesis; Diakinesis; Klinokinesis; Ookinesis; Orthokinesis; Thigmokinesis. Rel. Taxis; Tropism.

KING Noun. (Anglo Saxon, *cyng* = king.) Isoptera: The sexually developed male.

KING CHRISTMAS BEETLE *Anoplognathus viridiaeneus* (Donovan) [Coleoptera: Scarabaeidae].

KING CRICKETS See Mimnermidae; See Stenopelmatidae.

KING STAG BEETLE *Phalacrognathus muelleri* (Macleay) [Coleoptera: Lucanidae].

KING, ALBERT FREEMAN AFRICANUS (1841–1914) (Howard 1915, Pop. Sci. Mon. 87: 175.)

KING, EDMUND (Rose 1850, *New General Biographical Dictionary* 9: 100.)

KING, GEORGE B (1848–1916) (Fernald 1916, J. Econ. Ent. 9: 572.)

KING, HAROLD (1888–1956) (H. S. 1956, Entomologist's Rec. J. Var. 68: 120.)

KING, JAMES JOSEPH FRANCIS XAVIER (1855–1933) (Anon. 1933, Anon. Entomol. mon. Mag. 69: 166.)

KING, JOSEPH LYONEL (1888–1952) (Anon. 1953, J. Econ. Ent. 46: 189.)

KING, PHILIP PARKER (1793–1855) (Papavero 1975, *Essays on the History of Neotropical Dipterology* 2: 229–230. São Paulo.)

KING, ROBERT LETHBRIDGE (1823–1897) (Musgrave 1932, *A Bibliogr. of Australian Entomology 1775–1930,* viii + 380 pp. (175), Sydney.)

KING, VERNON (1886–1918) (Anon. 1918, J. Econ. Ent. 11: 390–391.)

KINGDOM Noun. (Anglo Saxon, *cyningdom* = kingdom. Pl., Kingdoms.) A taxonomic category stemming from the primary division of natural objects, living and non-living; Three Kingdoms recognized by older workers: Animal Kingdom (containing all animals), Vegetable Kingdom, (containing all plants), Mineral Kingdom (containing all inorganic forms of matter.) Five Kingdoms currently are recognized by most biologists.

KINNMARK, FOLKE (1898–1951) (Lindroth 1952, Opusc. ent. 17: 105–106.)

KINO Noun. (Mandingo, *kino.*) A dark red or brown or blackish resin-like tanniferous substance produced by several Species of tropical and subtropical trees. African kino from *Pterocarpus erinaceus*, Malabar kino from *P. marsupium,* Australian kino from *Eucalyptus* spp., Bengal kino from *Butea monospermae,* Jamacian kino from *Coccoloba uvifera.* Used by the trees in defence, in medicine as an astringent, and in the processes of dyeing and tanning. See Catechu.

KINOPRENE An organic compound {2-propynyl (E,)-3,7,11-trimethyl-2,4-dodecadienoate} used as an Insect Growth Regulator for control of plant-sucking insects (aphids, mealybugs, scale insects, whiteflies), fungus gnats and related insects. Kinoprene applied to glasshouse plants, indoor ornamentals and vegetable seed crops. Phytotoxicity on some varieties of roses and poinsettas has been observed. Compound not for use on crop plants. Trade names include Altodel®, Enstar® and EnstarII®.

KINOPRENE Noun. A juvenile hormone analogue.

KINOPSIS Noun. (Greek, *kinesis* = motion; *-sis* = a condition or state.) Social Insects: Alarm communication (recruitment) only through movement of nestmates.

KINOSHITA, SHUTA (1884–1955) (Hasegawa 1967, Kontyû 35 (Suppl.): 30–31, bibliogr.)

KINSEY, ALFRED C (1894–1955) (Christy 1956, Proc. Ind. Acad. Sci. 66: 30–31.)

KIPP, FRIEDRICH (1814–1869) (Anon. 1969, Verbl. West. fal.-Rhein. Ver. Bienen Seidenzucht 20: 17–18.)

KIPSIN® See Methomyl.

KIRBY, HAROLD (1900–1953) (Buxton 1954, Proc. R. ent. Soc. London (C) 18: 80.)

KIRBY, WILLIAM (1759–1850) English cleric and entomologist. Kirby was author of *Monographia Apum Angliae* (2 vols, 1802), and several other works; he used comparison of wing venation in classification. Kirby was coauthor with William Spence of *Introduction to Entomology* (4 vols, 1815–1826). (Spence 1850, Proc. ent. Soc. Lond. 1850, 33, bibliogr.)

KIRBY, WILLIAM FORSELL (1844–1912) (Kirby 1912, Entomologist's Rec. J. Var. 24: 314–317.)

KIRCH, ARTUR (1891–1969) (Evers 1970, Ent. Bl. biol. syst. Käfer 66: 65.)

KIRCH, CHRISTFRIED (1694–1740) (Rose 1850, *New General Biographical Dictionary* 9: 118.)

KIRCHBERG, ERICH (1914–1968) (Döhring 1969, Mitt. dt. ent. Ges. 28: 1–3.)

KIRCHER, ATHANASE (1602–1680) (Percheron 1837, Bibliographia Entomologica 1: 207.)

KIRCHHOFFER, OTTO (1863–1914) (Anon. 1914, Dt. ent. Z. 1914: 649.)

KIRCHMAIER, GEORG CASPAR (1635–1700) (Rose 1950, *New General Biographical Dictionary* 9: 119.)

KIRCHNER, LEOPOLD ANTON (–1879) (Fitch 1880, Entomologist 13: 118–119.)

KIRITSHENKO, ALEXEYA NIKOLAEVICH (1884–1971) (Kerzhner & Stackelberg 1971, Ent. Oboz. 50: 719–729, bibliogr.)

KIRK, FLORENCE JAN (1897–1935) (Donisthorpe 1935, Entomologist's Rec. J. Var. 47: 56.)

KIRKALDY WHITEFLY *Dialeurodes kirkaldyi* (Kotinsky) [Hemiptera: Aleyrodidae].

KIRKALDY, GEORGE WILLIS (1873–1910) Scotch Hemipterist who for many years lived in Hawaii and published a world catalogue on Pentatomoidea. (Bueno 1910, Ent. News 21: 240–242.)

KIRKLAND, ARCHIE HOWARD (1873–1931) (Burgess 1941, Yb. Mass. Sta. Coll. Fernald Cl. 10: 1–3.)

KIRKPATRICK, THOMAS WINFRID (–1971) Entomologist with the Kenya Department of Agriculture. (Butler 1972, Proc. R. ent. Soc. Lond. (C) 36: 61–62.)

KIRSCH, THEODOR FRANZ WILHELM (1818–1889) (Meyer 1889, Abh. zool. Mus. Dresden 2: 1–7, bibliogr.)

KIRSCHBAUM, CARL LUDWIG (1812–1880) (Koch 1880, Jb. nassu Ver. Naturk. 31–32: 324–334.)

KIRSTEN (KIRSTENIUS), GEORG (1613–1660) (Rose 1850, *New General Biographical Dictionary* 9: 120.)

KIRTLAND, JARED POTTER (1793–1877) (Anon. 1878, Proc. Amer. Acad. Sci. 13: 452–453.)

KISS VON ZILAH, ANDREAS (1873–1931) (Müller 1933, Verh. Mitt. siebenb. Ver. Naturw. 81–82: 1–4.)

KISSING BUGS See Reduviidae.

KISSINGER, JOHN B (–1909) (Horn 1909, Dt. ent. Z. 1909: 583–584.)

KITE, VITAE (–1940) (Meiners 1941, Ent. News 52: 120.)

KITINEX® See Diflubenzuron.

KITRON® See Acephate.

KITTNER, THEODOR (–1906) (Schmeichler 1906, Verh. naturf. Ver. Brünn 45: 41.)

KITZBERGER, IVAN F (1880–1927) (Stepanek 1927, Cas. csl. Spol. ent. 24: 44.)

KIVIRIKKO, KAARLO EEMELI (1870–1947) (Saalas 1947, Suomen Hyönt. Aikak 13: 61–63.)

KJAER, EJVIND (1885–1951) (Wolff 1952, Ent. Meddr. 26: 247–249.)

K-K-OTEK® See Deltamethrin.

KLÄGER (–1916) (Rose 1934, Int. ent. Z. 27: 525.)

KLAGES HENRY G (1860–1936) (Avinoff 1937, Ent. News 48: 29–30.)

KLAGES, ELIZA (1835–1919) (Klages 1919, Ent. News 30: 180.)

KLAGES, FREDERICK W (1859–1886) (Anon. 1886, Entomologica Am. 2: 56.)

KLAMATH WEED See Saint-Johns Wort. Syn. Goat Weed.

KLAMATH-WEED BEETLE *Chrysolina quadrigemina* (Suffrain) [Coleoptera: Chrysomelidae]: Endemic to Palearctic and an effective control agent of Saint-Johns Wort in Australia and North America. Adults and larvae defoliate plants. See Chrysomelidae.

KLÄMPFEL, GABRIEL (–1831) (Gistl 1832, Faunus 1: 50–51.)

KLAPALEK, FRANTISEK (1863–1919) (Anon. 1924, Jubil. Sborn. Cas. ent. Spol. 1903–1924: 1–2, bibliogr.)

KLARTAN® See Tau-Fluvalinate.

KLATT, BERTHOLD (1885–1958) (Weidner 1967, Abh. Verh. naturw. Ver. Hamburg Suppl. 9: 354–358.)

KLEBS, RICHARD HERMANN ERDMANN (1850–1911) (Tornquist 1911, Schr. phys.-ökon. Ges. Konigsb. 25: 31–37, bibliogr.)

KLEE, HEINRICH (1905–1943) (Sachtleben 1943, Arb. physiol. angew. Ent. Berl. 10: 259–260.)

KLEE, WALDEMAR G (1853–1891) (Essig 1931, *History of Entomology*. 1029 pp. (672–673), New York.)

KLEEMAN, CHRISTIAN FREDERICH KARL (1735–1789) (Eisinger 1925, Ent. Z. Frankf. a. M. 39: 66–67, 74–75.)

KLEFBECK, EINAR (1888–1963) (Lindroth 1963, Opusc. ent. 28: 163–164.)

KLEIN, J T (1846–1904) (H. H. A. 1904, Ent. News 15: 353.)

KLEIN, JACOB THEODOR (1685–1759) (Rose 1850, *New General Biographical Dictionary* 9: 122.)

KLEIN, PETER ADOLF (1853–1889) (Englehart 1918, Ent. Meddr. 12: 17.)

KLEINE, GEORG (1806–1894) (Anon. 1888, Dt. Bienenfr. 1888: 1.)

KLEINE, RICHARD (1874–1948) (Sachtleben 1949, Koleopt. Z. 1: 89–92.)

KLEINSCHMIDT, JOHANN THEODOR (Weidner 1967, Abh. Verh. naturw. Ver. Hamburg Suppl. 9: 156.)

KLENE, HEINRICH (1845–1933) (Carmer 1933, Natuurh. Maandbl. 22: 143–145.)

KLEPTOBIOSIS Noun. (Greek, *klepenai* = to steal; *bios* = life form; *-sis* = a condition or state. Pl.,

Kleptobioses.) A life style in which one organism lives by theivery. Kleptobiosis is seen in some Species of ants. See Biosis; Commensalism; Parasitism. Cf. Abiosis; Anhydrobiosis; Antibiosis; Archebiosis; Calobiosis; Cleptobiosis; Hamabiosis; Lestobiosis; Parabiosis; Phylacobiosis; Plesiobiosis; Synclerobiosis; Trophobiosis; Xenobiosis.

KLEPTON Noun. (Greek, *kleptenai* = to steal.) (Dubois & Gunther, 1982. Zool. Jhrb. Syst. 109: 290–305.) Cf. Synklepton.

KLEPTOPARASITE Noun. (Greek, *kleptenai* = to steal; *parasitos* = parasite; *-ites* = resident. Pl., Kleptoparasites.) A parasitic insect (parasitoid) which consumes the food of another parasite and in so doing often kills that parasite.

KLETKE, PAUL (1835–1917) (Anon. 1919, Jh. Verh. schles. Insektenk 10–12: 22–23.)

KLICKA, LADISLOV (–1937) (Obenberger 1937, Cas. csl. Spol. ent. 34: 97–100.)

KLIJNSTRA, BONNO HYLKO (1880–1953) (Brouerius 1959, Ent. Ber., Amst. 193.)

KLIMA, ANTON (1871–1941) (Sachtleben 1942, Arb. morph. taxon. Ent. Berl. 9: 65.)

KLIMESCH, JOSEF (1884–1935) (Reisser 1962, Z. wien. ent. Ges. 47: 57–58.)

KLIMSCH, EDGAR (1878–1939) (Puschnig 1939, Carinthia II 49: 125.)

KLINGER, HEINZ (1911–1940) (Sachtleben 1940, Arb. physiol. angew. Ent. Berl. 7: 168.)

KLINGSTED, TORSTEN HOLGER (1919–1947) (Palmgren 1948, Meddn Soc. Fauna Flora fenn. 24: 251–253.)

KLINKOWSKI, MAXIMILIAN (1904–1971) (Bercks 1971, Anz. Schädlingsk. Pflanzen 44: 175.)

KLINOKINESIS Noun. (Greek, *klinein* = to lean; *kinesis* = motion.) See Kinesis. Cf. Allokinesis; Blastokinesis; Diakinesis; Ookinesis; Orthokinesis; Thigmokinesis. Rel. Taxis; Tropism.

KLITSCHKA, THEODOR (–1938) (Guhn 1939, Ent. Z. Frank. a. M. 52: 317–318.)

KLOCKER, ALBERT (1862–1923) (Henriksen 1925, Ent. Meddr. 14: 449–453, bibliogr.)

KLUG, FRANCIS German physician and director of the Berlin Museum. Klug was author of *Monographia Siricum Germaniae &c.* (1803), *Entomologische Monographien* (1824), and *Jahrbucker der Insecten Kunde* (1834).

KLUG, JOHANN CRISTOPH FRIEDRICH (1775–1856) German natural historian and monographer principally interested in Coleoptera and Hymenoptera. (Gerstaecker 1856, Stettin ent. Ztg. 17: 225–237; Marseul 1887, Abeille 24 (Les ent et leurs écrits) 181–186, bibliogr.)

KLUG'S XENICA *Geitoneura klugii* (Guérin-Méneville) [Lepidoptera: Nymphalidae].

KLUK, KRZSZTOF (1739–1796) Polish intellectual, natural historian and driving force behind the 'new agriculture' movement. Author of a book on domestic and wild animals and an encyclopedia of natural history in Poland.

KNAB, FREDERICK (1865–1918) (Caudell 1919, Proc. ent. Soc. Wash. 21: 41–52, bibliogr.)

KNABEN, NILS (–1969) (Opheim 1969, Norsk. ent. Tidsskr. 16: 61–62.)

KNABL, HERMANN (1880–1940) (Heikertinger 1940, Koleopt. Rdsch. 26: 92.)

KNAGGS, HENRY GUARD (1832–1908) (Walker 1908, Entomol. mon. Mag. 44: 49–50.)

KNAPP, ROYCE BURTON (1924–1955) (App 1956, J. Econ. Ent. 49: 574.)

KNATZ, JOHANN LUDWIG (1831–1892) (Anon. 1892, Ber. Ver. Naturk. Cassel 38: 10.)

KNAUS, WARREN (1858–1937) (Dean 1938, J. Kans. ent. Soc. 11: 1–3.)

KNAUSS, KAUFMANN OTTO (1889–1965) (Steinig 1965, Mitt. dt. ent. Ges. 24: 78–79.)

KNAUTH, JOHANNES (1843–1905) (Daniel 1906, Münch. koleopt. Z. 2: 394.)

KNAVE® See Disulfoton; See Quinalphos.

KNEE Noun. The point of junction between the Femur and Tibia of an insect leg.

KNER, RUDOLPH (1810–1869) (Korbell 1870, Sber. bayer. Akad. Wiss. 1: 417–418.)

KNETZGER, AUGUST (1867–1940) (Meiners 1941, Ent. News. 52: 119.)

KNIEPHOFF, JOHANNES (1865–1940) (Anon. 1941, Ent. Bl. Biol. Syst. Käfer 37: 93.)

KNIFE Noun. (Middle English, *Knif.* Pl., Knives.) An instrument (typically of glass, steel or other durable material) with a sharpened edge used under pressure for cutting. The edge is composed of a bevel angle, facet angle, and clearance angle. Little is known about the cutting process (see Wachtel *et al.* 1966 in A. W. Pollister, Physical Techniques in Biological Research, Vol. IIIA). An important device in histology and microtomy. See Ralph Knife; Latta and Hartmann Knife.

KNIFE MOLE-CRICKET *Gryllotalpa cultriger* Uhler [Orthoptera: Gryllotalpidae]: A Species of 'four-toed' mole cricket endemic to southwestern USA. See African Mole Cricket; European Mole Cricket.

KNIGHT, HARRY HAZELTON American academic (Iowa State University) and student of Hemiptera, particularly Miridae. Demonstrated the usefulness of male genitalia in the taxonomy of Miridae.

KNIGHT, HUGH (1877–1943) (Quayle 1944, J. Econ. Ent. 37: 330–331.)

KNIGHT, THOMAS ANDREW (1759–1838) (Rose 1850, *New General Biographical Dictionary* 9: 126.)

KNIPLING, EDWARD F (1909–2000) American government entomologist (USDA) (B.A. Texas A& M University; M.S., Ph.D. Iowa State University). Director USDA Entomology Division (1953–1971). Pioneered sterilized-male technique to protect livestock from screwworms; awarded Japan Prize from Science and Technology Foundation of Japan; National Medal of Science (1966); President's Award for Distinguished Federal Civilian Service (1971) and USDA Distinguished Service Award.

KNISCH, ALFRED (1884–1926) (Heikertinger 1927, Koleopt. Rdsch. 13: 86–88, bibliogr.)

KNOCH, AUGUST WILHELM (1742–1818) (Zinken 1818, Magazin Ent. (Germar) 3: 458–460.)

KNOCH, JULIUS (1828–1893) (Anon. 1893, Leopoldina 29: 162–163.)

KNOCHE, ERNST (1867–1939) (Sachtleben 1939, Arb. physiol. angew. Ent. Berl. 6: 315.)

KNOCK, AUGUST WILLIAM German academic at Brunswick and author of *Neuebeytrage zur Insectenkunde* (1801).

KNOCKDOWN Adj. Jargon: Descriptive of the mode of action of certain insecticides that rapidly stun or narcotize insects. The term commonly refers to pyrethrum and pyrethroid compounds. The knockdown effect does not necessarily kill the insect because many Species can metabolize the toxin and recover from its effects.

KNOOP, ROELOF (1905–1964) (Lempke 1964, Ent. Ber., Amst. 24: 81.)

KNÖRLEIN, JOSEF (1806–1883) (Munganast 1883, Wien. ent. Ztg. 2: 80.)

KNORR, GEORG WOLFGANG (1705–1761) (Wisinger 1930, Ent. Z. Frankf. a. M. 43: 241–248, bibliogr.)

KNOTH, MAX FRIEDRICH AUGUST (1863–1937) (Anon. 1937, Arb. morph. taxon. Ent. Berl. 4: 160.)

KNOT-HORN MOTHS Members of the Phycitinae [Pyralidae]. A large group, predominantly tropical in distribution and including several Species that are important pests of stored products.

KNOWER, HENRY MCELDERRY (1868–1940) (Anon. 1940, Ent. News 51: 51.)

KNOWLES, ROBERT (1883–1936) (Anon. 1936, J. Parasit. 22: 550–551.)

KNOX, ROBERT (1638–1720) (Rose 1850, *New General Biographical Dictionary* 9: 131.)

KNOX-OUT® See Diazinon.

KNULL, JOSEF NISSLEY (1891–1975) (White 1976, Proc. ent. Soc. Wash. 78: 116.)

KNUTH, PAUL (1854–1899) (Ludwig 1899, Illte Z. ent. 4: 356–367, bibliogr.)

KO, HO JE (–1986) Korean Entomologist at Forestry Research Institute, Seoul with interest in Symphyta.

KOA BUG *Coleotichus blackburniae* White [Hemiptera: Pentatomidae].

KOA HAOLE SEED WEEVIL *Araecerus levipennis* Jordan [Coleoptera: Anthribidae].

KOA MOTH *Scotorythra paludicola* (Butler) [Lepidoptera: Geometridae].

KOA SEEDWORM *Cryptophlebia illepida* (Butler) [Lepidoptera: Tortricidae].

KOBA, YASUKI (1888–1946) (Hasegawa 1967, Kontyû 25 (Suppl.): 36.)

KOBAKHIDZE, DAVID NESTROCICH (1911–1970) (Anon. 1971, Ent. Oboz. 50: 472–479, bibliogr.)

KOBERT, EDUARD RUDOLF (1854–1918) (Anon. 1919, Ent. News 30: 210.)

K-OBIOL® See Deltamethrin.

KOCH, CARL LUDWIG (1778–1851) (Bonnet 1945, *Bibliographia Araneorum* 1: 32.)

KOCH, CARLOS (1804–1870) (Frey 1970, Ent. Arb. Mus. Georg Frey 21: 1–2.)

KOCH, GABRIEL (1807–1881) (Anon. 1881, Entomol. mon. Mag. 17: 240.)

KOCH, LUDWIG CAROL CHRISTIAN (1825–1908) (Dittrich 1911, Jh. Ver. schles. Insektenk. 4: xxii–xv, bibliogr.)

KOCH, MANFRED (1901–1972) (Ebert & Klausnitzer 1973, Ent. Nachr., Dresden 17: 25–32, bibliogr.)

KOCH, ROBERT (1843–1910) (Libbertz 1910, Ber. senckenb. naturf. Ges. 41: 306–318.)

KOCH, VALDEMAR (1852–1902) (Englehart 1918, Ent. Meddr. 12: 19.)

KOCHER, LOUIS (1894–1972) (Bailly-Choumara 1973, Bull. Soc. sci. nat. phys. Maroc 52: 1–10, bibliogr.)

KOEBELE, ALBERT (1852–1924) German emigrant to America; USDA explorer in Australia during 1888–1889 for natural enemies of cottony cushion scale, imported Vedalia Beetle and other natural enemies to California with spectacular success; VB sent to Hawaii during 1890 with similar results; employed by Hawaiian Sugar Planter's Association and Territorial Government (1893–1910) as Entomologist and explorer for insect pests and Lantana. Perkins & Swezey 1925, Hawaii. Plt. Rec. 29: 359–376.

KOEHN, HENRY (1892–1963) (Weidner 1967, Abh. Verh. naturw. Ver. Hamburg, Suppl. 9: 294.)

KOELREUTER, JOSEPH GOTTLIEB (1733–1806) (Nordenskiöld 1935, *History of Biology*. 629 pp. (254–257) London.)

KOENIG, JOHAN GERHARD (1728–) (Zimsen 1964, *The Type Material of I. C. Fabricius.* 656 pp. (12–13), Copenhagen.)

KOENIG, SAMUEL (1712–1751) (Rose 1850, *New General Biographical Dictionary* 9: 134.)

KOEPCKE, MARIA (P. G. A. F. 1972, Revta. peru. Ent. 15: iii.)

KOEPPEN, FEDORA PETROVICH (1833–1908) (Adelung 1908, Revue Russe Ent. 8: xv–xviii.)

KOFOID, A C (1874–1949) (West 1949, Ent. Meddr. 25: 331–332.)

KOH Potassium Hydroxide. A chemical compound used for macerating insect specimens and appendages for microscopical study. Solutions of 10% KOH in distilled water typically are used.

KOHL, FRANZ FRIEDRICH (1851–1924) (H. F. 1925, Konowia 4: 89–96, bibliogr.)

KÖHLER, JOHANN CHRISTIAN GOTTLIEB (1759–1833) (Schummel 1858, Z. Ent. 12: 15–24.)

KÖHLER, JOHANNES AUGUST ERNST (1829–1903) (Möbius 1943, Dt. ent. Z., Iris 57: 9.)

KOHL'S SEABIRD TICK *Ixodes kohlsi* Arthur [Acari: Ixodidae].

KOHLS, GLEN MILTON (1905–1986) American government acarologist specializing in ticks and tickborne diseases. (Clifford & Keirans, 1986. Proc. ent. Soc. Wash. 89: 375–383.)

KOINOBIONT Noun. (Pl., Koinobonts.) Protelean parasites (parasitoids) which do not kill, permanently impair or paralyse their hosts after oviposition and thereby do not prevent further development of the hosts. Typically endoparasitic parasites which attack hosts in exposed situations and thereby demonstrate a limited host range (specialists.) See Parasitism. Cf. Idiobont. (See Askew & Shaw, pp. 225–265 in *Insect Parasitoids.* Waage & Greathead, eds. Academic Press.)

KOIZUMI, MAKATA (1882–1952) (Hasegawa 1967, Kontyû 35 (Suppl.): 33–34, bibliogr.)

KOKUEV, NIKITTA RAFAILOVICH (1848–1914) (Semenov-Tian-Shansky 1916, Revue Russe Ent. 16: lv-lxx, bibliogr.)

KOLA NUT WEEVIL *Balanogastris kolae* (Desbrochers) [Coleoptera: Curculionidae]: A tropical pest attacking nuts of *Kola acuminata.* Larva feeds inside nuts and taken into stores where they complete development.

KOLBE, HERMANN JULIUS (1855–1939) (Kuntzen 1925, Dt. ent. Z. 1925: 439–40.)

KOLBE, WILHELM (1852–1929) (Hinke 1919, Ent. Bl. Biol. Syst. Käfer 15: 172–180.)

KOLENATI, FRÉDÉRIC A (1813–1864) (Anon. 1888, Abeille (Les ent et leurs écrits) 25: 220–223, bibliogr.)

KOLLAR, VINCENZ (1797–1860) (Schiner 1860, Wien. ent. Monastschr. 4: 222–224.)

KÖLLER, HERMANN (1885–1968) (Müller 1968, Ent. Berichte 1968: 81–92.)

KÖLLIKER, RUDOLPH ALBERT VON (1817–1905) German Academic (Professor of Anatomy and Zoology, Wurtzburg.) Published comparative study of development in insects and vertebrates (1843). (Taschenberg 1906, Leopoldina 42: 75–82, 87–91, 103–116, bibliogr.)

KOLTZE, WILHELM (1839–1914) (Horn 1915, Ent. Mitt. 4: 1–3.)

KOMAREK, JULIUS M (1892–1955) (Kratochvil 1952, Zool. Ent. Listy 15: 211–216, bibliogr.)

KOMET® See Tefluthrin.

KOMP, WILLIAM H WOOD (1893–1955) (Wake *et al.* 1956, Proc. ent. Soc. Wash. 58: 47–55, bibliogr.)

KONAKOV, NIKOLAI NIKOLAIVITCH (1900–1947) (Grunin 1948, Ent. Obozr. 30: 168–170, bibliogr.)

KÖNIG, GOTTLOB (1779–1849) (Ratzeburg 1874, Forstwissenschaftliches Schriftsteller-Lexicon 1: 288–290.)

KONIGSBERGER, JACOB CHRISTAAN (1867–1951) (Roepke, 1951. Ent. Ber., Amst. 13: 257–258.)

KONISHI, GINSHICHI (1900–1944) (Hasegawa 1967, Kontyû 35 (Suppl.): 35–36, bibliogr. (pl.4, fig. 8.)

KONO, HIROMICHI (1905–1963) (Hasegawa 1967, Kontyû 35 (Suppl.): 34–35, bibliogr. (pl. 4, fig. 7.)

KONO, TEIZO (1818–1871) (Hasegawa 1967, Kontyû 35 (Suppl.): 34.)

KONOW, FRIEDRICH WILHELM (1842–1908) (Horn 1908, Dt. ent. Z. 1908: 428.)

KONOWIELLIDAE Bischoff 1914. Plural Noun. Family-group name of aculeate Hymenoptera. Cf. Plumariidae.

KOONS, BENJAMIN FRANKLIN (1844–1903) (Anon. 1904, Ent. News 15: 112.)

KOORNNEEF, JAN (1868–1955) (Fischer 1955, Ent. Ber., Amst. 15: 521–523.)

KOPEC, STEFAN (Anon. 1941, Nature 148: 655.)

KOPMITE® See Chlorobenzilate.

KOPONEN, JUHANA SAMALI VALDIMAR (1882–1948) (Valle 1948, Suomen hyönt. Aikak 14: 49.)

KORB, MAXIMILIAN (1851–1933) (Arnold 1933, Mitt. Münch. ent. Ges. 23: 103–107.)

KORB, ROSINA (–1911) (Seidlitz 1911, Mitt. Münch. Ent. Ges. 2: 22–38.)

KORDON® See Cypermethrin.

KOREAN HAEMORRHAGIC FEVER See Epidemic Haemorrhagic Fever.

KOREISHI, TAKAHASHI (1906–1965) (Hasegawa 1967, Kontyû 35 (Suppl.): 37.)

KORLEVIC, ANTON (1851–1915) (Csiki 1915, Rovart. Lap. 22: 43–44.)

KOROTNEV, ALEKSYEI ALEKSEEVICH (–1915) (Semenov-Tian-Shansky 1915, Revue Russe Ent. 15: 682–683.)

KORSCHELT, EUGENE (1858–1946) (Koller 1948, Naturw. Rdsch. Stuttg. 1: 182–183.)

KORTEBOS, HERMAN HENDRIK (1879–1950) (Lempke 1950, Ent. Ber., Amst. 13: 113.)

KOSCHABEK, FRANZ (1884–1961) (Reisser 1961, Z. wein. ent. Ges. 46: 172–173.)

KOSCHEVNIKOV'S GLAND A small group of glandular cells located near the Sting chamber at the apex of the Metasoma in queen bees. Apparently a pheromone producing gland whose secretion attracts worker bees.

KOSOBUTZKOGO, MEFODIYA IL'YA (1896–1964) (Sazonov 1964, Ent. Obozr. 43: 737–742, bibliogr. (Translation: Ent. Rev. Wash. 43: 377–380, bibliogr.)

KOSSMANN, MAX THEODOR (1840–1902) (Anon. 1903, Z. Ent. 28: xxxiv–xxxv.)

KOSTER'S CURSE *Clidemia hirta* (Linnaeus) D. Don, a perennial shrub native to the West Indies, Central and South America which became accidentally introduced into Fiji and controlled with the imported thrips *Liothrips urichi* Karney.

KOTAKE, HIROSHI (1869–1923) (Hasegawa 1967, Kontyû 35 (Suppl.): 35, bibliogr.)

KOTHE, ALBERT (1828–1885) (Dimmock 1888, Psyche 5: 36.)

K-OTHRIN® See Deltamethrin.

KOTINSKY, JACOB (1873–1928) (Osborn 1937, *Fragments of Entomological History.* 394 pp. (235) Columbus, Ohio.)

KOTUJELLIDAE Rasnitsyn 1975. A Family of parasitic Hymenoptera assigned to the Evanioidea, sometimes placed within the Gasteruptiidae.

KOTULA, BOLESLAV (1849–1898) (Kulczynzki 1899, Spraw. Kom. Fizyogr. Kraju 34: xx–xxvii.)

KOU LEAFWORM *Ethmia nigroapicella* (Saalmüller) [Lepidoptera: Oecophoridae].

KOVALEV, VLADIMIR GRIGOREVICH (1942–1987) Russian Dipterist specializing in fossil Brachycera and extant Empidoidea.

KOWARZ, FERDINAND (1838–1914) (Becker 1915, Dt. ent. Z. 1915, 1–3, bibliogr.)

KOYAMA, UMITARO (1870–1953) (Hasegawa 1967, Kontyû 35 (Suppl.): 36–37, bibliogr.)

KOZHANCHIKOV, IGORYA VASIL'EVICH (1904–1958) (Schtakel'berg 1959, Ent. Obozr. 38: 243–251, bibliogr. Translation: Ent. Rev. Wash. 38: 221–230, bibliogr.)

KRAATZ, ERNST GUSTAV (1831–1909) (Horn 1910, Dt. ent. Z. 1910: 109–112.)

KRAEPELIN, KARL (1848–1915) (Weidner 1967, Abh. Verh. Naturw. Ver. Hamburg Suppl. 9: 197–207).

KRAFFT, LUDWIG (1811–1908) (Anon. 1917, Pfälz. Heimatk. 13: 8.)

KRAMAR, JAROSLAV (–1960) (Chalupsky 1970, Acta. ent. bohemoslovaca 67: 277–278, bibliogr.)

KRAMER, HEINRICH (1872–1935) (Anon. 1935, Arb. morph. taxon. Ent. Berl. 2: 121.)

KRANCHER, PAUL OSKAR (1857–1936) (Anon. 1936, Ent. Z. Frankf. a. M. 50: 13–14.)

KRANZL, ERWIN (–1955) (Anon. 1955, Z. wien. ent. Ges. 40: 176, 244.)

KRASY, THEODOR (1875–) (Heyrovsky 1945, Cas. csl. Spol. ent. 42: 19–21, bibliogr.)

KRATOCHVIL, JOSEPH (Rosicky 1968, Zool. Listy 31: 295–298)

KRAUSE, ERNST (1839–1903) (Insektenbörse 20: 306. Pseudonym: Carus Sterne.)

KRAUSE, FRANZ (1833–1898) (Anon. 1898, Leopoldina 34: 172.)

KRAUSE'S MEMBRANE A transverse septum in the middle of each clear band or zone (Sarcomere) of the insect muscle, to which the Sarcostyles are joined. KM consists of a network of radially distributed threads which cut across the muscle fibre (Imms).

KRAUSS, FERDINAND V (1812–1890) Collected Hymenoptera in Natal.

KRAUSS, HERMAN, AUGUST (1848–1937?) (Anon. 1938, Arb. morph. taxon. Ent. Berl. 5: 352.)

KRAUSS, WILLIAM CHRISTOPHER (Anon. 1896, Am. mon. micr. J. 17: 1–4.)

KRAUSSE, ANTON HERMANN (1878–1929) (Anon. 1919, Arch. Naturgesch. 85 A (12): 60–77, bibliogr.)

KRAUTSCHNEIDER, HUGO (–1965) (Anon. 1965, Mitt. schweiz. ent. Ges. 24: 23.)

KREBS, LUDWIG (1792–1844) (Urban 1903, Symbollae Antillanae 3: 69.)

KRECALVIN® See Dichlorvos.

KREFFT, JOHANN LOUIS GERHARD (1830–1880) (Whitley 1932, Rec. Aust. Mus. 18: 328.)

KREFFT'S DARTER *Telicota augias krefftii* (W. J. Macleay) [Lepidoptera: Hesperiidae].

KREGAN® See Chlorpyrifos.

KREITHNER, EDUARD (1858–1888) ('Rghf' 1888, Wien ent. Ztg. 7: 116.)

KREKICH-STRASSOLDO, HANS (1864–1929) (Heberdey 1931, Koleopt. Rdsch. 16: 29–32, bibliogr.)

KREYE, HERMANN (1856–1940) (Sachtleben 1940, Arb. morph. taxon. Ent. Berl. 7: 76.)

KRICHELDORFF, ADOLPH (1880–1962) (Harms 1961, Mitt. dt. ent. Ges. 20: 1–2.)

KRIECHBAUMER, JOSEPH (1819–1902) (Anon. 1902, Entomol. mon. Mag. 38: 288–289.)

KRIEG, HANS (1891–1975) (Weidner 1975, Anz. Schädlingsk. Pflanzenschutz Umwet. 48: 144.)

KRIESCH, JANOS (1834–1888) (Anon. 1889, Jb. Naturw. Ver. Magdeb. 4: 551.)

KRISHNA, AYYAR, P N (1894–1946) (Lal 1947, Indian J. Ent. 8: 137–138.)

KRISTOFA, PAMIATKE JOZEFA (1925–1974) (Jasenák 1976, Entom. problémy 13: 174.)

KRIZEK, ALEXANDER (1851–1906) (Dittrich 1907, Z. Ent. 32: xlix–l.)

KRÖBER, OTTO (1882–) German dipterist with world collection of Tabanidae, Therevidae, Conopidae and Scenopinidae. (Weidner 1967, Abh. Verh. naturw. Ver. Hamburg Suppl. 9: 229–232.)

KROGERUS, ROLF (1882–1965) (Lindberg 1966, Arsb/Vuosik. Soc. Scient. Fenn. 44 C: 3–11.)

KROGH, AUGUST (1874–1949) (Weis-Fogh 1949, Ent. Meddr. 25: 435–445, bibliogr.)

KRÖYER, HENRIK, NICOLAJ (1799–1870) (Henriksen 1926, Ent. Meddr. 15: 204–206.)

KRUEL® See Cypermethrin.

KRUG, CARL WILHELM LEOPOLD (1833–1898) (Papavero 1975, *Essays on the History of Neotropical Dipterology* 2: 296–297. São Paulo.)

KRÜGER, EDGAR (Weidner 1967, Abh. Verh. naturw. Ver. Hamburg Suppl. 9: 267–268.)

KRÜGER, ERMANNO GIORGIO (1871–1940) (Prosdocimo 1940, Annali Mus. libico Stor. nat. 2: 335–336, bibliogr.)

KRÜGER, FRIEDRICH WILHELM CARL (1899–1945) (Thienemann 1945, Arch. Hydrobiol. 41: 430–434, bibliogr.)

KRÜGER, LEOPOLD (1861–1938) (Anon. 1939, Stettin. ent. Ztg 100: 61.)

KRULIKOVSKII, LEONID KONSTANTINOVICH (1864–1930) (Sheljuzhoko 1931, Revue Russe Ent. 24: 236–245, bibliogr.)

KRUTZSCH, KARL LEBRECHT (1772–1852) (Ratzeburg 1874, Forstwissenschaftliches Schriftsteller-Lexicon 1: 274 footnote.)

KRYGER, JENS PETER (1874–1951) (Tuxen 1952, Ent. Meddr. 26: 231–243, bibliogr.)

KRYNICKY, JOHANN (1797–1838) (Kaleniczenko 1839, Bull. Imp. Soc. Moscou 12: 25–33, bibliogr.)

KRYOCIDE® See Cryolite.

K-SALT FRUIT FIX 200® See NAA.

KUBARY, JOHANN STANISLAUS (Weidner 1967, Abh. Ver. naturw. Ver. Hamburg Suppl. 9: 154–155, 141.)

KUBE, INOKICHI (1874–1939) (Hasegawa 1967,

Kontyû 35 (Suppl.): 32.)

KUBES, P AUGUSTIN (1862–1924) (Sustera 1924, Cas. csl. Spol. ent. 21: 41.)

KUDAS, KARL (1900–1974) (Aspock 1975, Arbeitsem Ost. Ent. 26: 113–117.)

KUDLICKA, EDUARD (1866–1940) (Heyrovsky 1940, Cas. csl. Spol. ent. 37: 81–82.)

KUDO, RICHARD ROKSABRO (1886–1967) (Fisher 1967, Trans. Illinois Acad. Sci. 60: 329–336, bibliogr.)

KUDOS® See Permethrin.

KUHLMAN, LUDWIG (1857–1928) (Seitz 1928, Ent. Rdsch. 45: 17.)

KÜHN, HEINRICH (1860–1906) (Hartert 1907, Novit. Zool. 14: 340–341.)

KÜHN, JULIUS (1825–1910) (Könnedcke & Klinkowski 1960. Kühn-Arch. 74: 1–18.)

KUHNT, PAUL (–1934) (Ross 1934, Int. ent. Z. 27: 526–527.)

KUIK® See Methomyl.

KUJAU, MAXIMILIAN (–1939) (Weidner 1967, Abh. Verh. naturw. Ver. Hamburg Suppl. 9: 294.)

KULAGIN, NIKOLAI MICKALOVICH (1860–1940) (Boghdanov-Katkov 1940, Bull Plant Prot. (Leningr.) 1–2: 3–7.)

KULZER, HANS (1889–1974) (Frey 1975, Ent. Arb. Mus. Frey 26: 363.)

KUMM, PAUL (1866–1927) (Anon. 1928, Ber. westpreuss. bot. -zool. Ver. 50: 48–49, bibliogr.)

KÜNCKEL, D'HERCULAIS, JULES PHILIPPE ALEXANDRE (1843–1918) (Birabin 1908, An. Soc. cient. argent 1900: 1–97, bibliogr.)

KÜNNEMAN, GEORG (1866–1922) (Ross 1934, Int. ent. Z. 27: 527–528.)

KÜNOW, GOTTHOLD (1840–1909) (Hoen 1909, Dt. ent. Z. 1909: 468–469.)

KUNTZE, ALBERT FRIEDRICH ARTHUR (1842–1933) (Heller 1935, Arb. morph. taxon. Ent. Berl. 2: 276–282.)

KUNTZE, HEINRICH (1838–1900) (Dittrich 1900, Z. Ent. 25: 27–28.)

KUNTZE, ROMAN (1902–1944) (Kapascencki 1948, Polskie Pismo ent. 18: 129–141, bibliogr.)

KUNZE, GUSTAV (1793–1851) (Kiesenwetter 1851, Stettin ent. Ztg. 12: 257–260.)

KUPER, CHARLES AUGUSTUS FREDERICK (1806–1887) (McLachlan 1887, President's address, ent. Soc. Lond. 1887: 8.)

KURENTZOV, ALEKSEYA IVANOVICH (1896–1975) (Krivolutzkaya 1975, Ent. Obozr. 54: 926–935, bibliogr.)

KURIUMOFF, N V (1882–1917) (Borodin 1921, J. Econ. Ent. 14: 377–380, bibliogr.)

KUROSAWA, MIKIO (1903–1967) (Hasegawa 1967, Kontyû 35 (Suppl.): 32, bibliogr., pl. 4, fig. 6.)

KURRAJONG LEAF-TIER *Lygropia clytusalis* (Walker) [Lepidoptera: Pyralidae].

KURRAJONG POD BEETLE *Idaethina froggatti* Kirejshuk & Lawrence [Coleoptera: Nitidulidae]

KURRAJONG PSYLLID See Kurrajong twig psyllid.

KURRAJONG SEED WEEVIL *Tepperia sterculiae* Lea [Coleoptera: Curculionidae].

KURRAJONG STAR PSYLLID *Protyora sterculiae* (Froggatt) [Hemiptera: Carsidaridae].

KURRAJONG TWIG PSYLLID *Aconopsylla sterculiae* (Froggatt) [Hemiptera: Psyllidae].

KURRAJONG WEEVIL *Axionicus insignis* Pascoe [Coleoptera: Curculionidae].

KUSCHAKEWITSCH, JACOB ALEXANDER (1826–1866) (Kuschakewitsch 1869, Trudy Soc. Ent. Ross 4: 3–7.)

KUSDAS, KARL (1900–1974) (Christl 1958, Z. wien. ent. Ges. 43: 187–206.)

KUSNEZOV, VICTOR N (–1932) (Horn 1934, Arb. morph. taxon. Ent. Berl. 1: 309.)

KÜSTER, HEINRICH CARL (1807–1876) (Meyer 1876, NachrBl. malakozool. Ges. 8: 81–86.)

KUWANA, SHINKAI INOKICHI (1872–1933) (Essig 1933, J. Econ. Ent. 26: 1185–1188, bibliogr.)

KUWAYAMA, SHIGERU (–1912) (Patch 1912, Ent. News 23: 288.)

KUWERT, AUGUST FERDINAND (1828–1894) (Anon. 1894, Insektenbörse 11: 191–192.)

KUZNETZOV, NIKOLAI YAKOVELEVICH (1873–1948) (Navlovskii 1949, Ent. Obozr. 30: 171–180, bibliogr.)

KWIAT, A K See Wyatt.

KYASANUR FOREST DISEASE An arboviral disease whose aetiological agent belongs to the Flaviviridae. KFD first reported in Karnataka State, India during 1957. KFD epizootic in monkeys and transmitted to humans by ticks (*Haemaphysialis* spp.), particularly *H. spinigera*. Viron recovered from rodents, shrews and macaque; trans-stadial transmission demonstrated but not transovarial transmission. Incubation period in humans 3–8 days; patient may experience more than one episode with meningoencephalitis; some mortality manifest from KFD. See Arbovirus; Encephalitis; Flaviviridae. Cf. Omsk Haemorrhagic Fever; Tick-borne Encephalitis.

KYPFOS® Malathion.

LA PLAT WEEVIL *Sphenophorus brunnipennis* (Germar) [Coleoptera: Curculionidae].

LABACORIA Plural Noun. (Latin, *labium* = lip; *corium* = leather.) The membrane between the Labium and the margin of the head; the Basimaxillary Membrane (MacGillivray).

LABELLAR Adj. (Latin, *labellum* = small lip.) Diptera: Descriptive of or pertaining to the Labella.

LABELLUM Noun. (Latin, *labellum* = small lip. Pl., Labella.) 1. Diptera: Sensitive, ridged, apical mouth structures of some Species. 2. Coleoptera and Hemiptera: A prolongation of the Labrum covering the base of Rostrum (Smith). 3. Honey bee: A small spoon-shaped lobe at the apex of the Glossa. Syn. Bouton (Imms).

LABIA Plural Noun. (Latin, *labium* = lip.) 1. Parts of a Peritreme bounding a spiracle; lips of the Peritreme (MacGillivray). 2. Coccoidea: Four peculiar paired structures, two on the head and two on the Coria between the sixth and seventh abdominal segments (MacGillivray).

LABIAL Adj. (Latin, *labium* = lip; *-alis* = pertaining to.) Descriptive of structure or action associated with the Labium.

LABIAL GLAND The most highly developed glands of head appendages and open via a median duct between the Hypopharynx base and Labium, or on the Hypopharynx (Snodgrass). Ducts from LG unite and deposit secretory product into the Salivarium. In its simplest design, Salivarium forms a pocket at base of Hypopharnyx. This area becomes a silk press in Lepidoptera larvae, and a salivary syringe in Hymenoptera, Diptera and Homoptera. Labial Glands occur in all primary orders of insects except Coleoptera. Lepidoptera and Hymenoptera larvae may produce silk; blood-sucking Diptera may produce an anticoagulant with Labial Gland secretions. Syn. Salivary Gland. See Salivary Gland; Silk Gland.

LABIAL PALPUS (Pl., Labial Palpi.) The paired, segmented, sensory appendage of the Labium. LP originates on the Palpiger and is shorter than the Maxillary Palpus and commonly displays three segments. Syn. Anterior Palpus.

LABIAL SEGMENT The seventh segment of the head which bears the Labium. The second maxillary segment. Cf. Maxillary Segment.

LABIAL SUTURE A Suture between the Labium and Mentum. See Suture.

LABIATE Adj. (Latin, *labium* = lip; *-atus* = adjectival suffix.) Lip-like or having lip-like sutures. Alt. Labiatus.

LABIDURIDAE Plural Noun. Striped Earwigs. A cosmopolitan Family of Dermaptera assigned to Superfamily Anisolabidoidea. Adult 10–45 mm long; Antenna with 25–30 segments; apterous or macropterous; second tarsal segment not projecting beneath base of third tarsal segment. Some Species are nocturnal and predaceous.

LABIELLA Noun. (Latin, *labium* = lip.) 1. A mouthpart of Myriapoda. 2. The Hypopharynx *sensu* Packard.

LABIIDAE Plural Noun. See Spongiphoridae.

LABILE Adj. (Latin, *labilis* = disposed to slip.) Frequently or readily changing in form or composition and therefore unstable in nature. Ant. Stable.

LABILLARDIERE, JACQUES JULIEN HOUTON DE (1755–1834) (Müller 1886, Pap. Proc. R. Soc. Tasm. 1885: 334–335.)

LABIOIDEA See Spongiphoroidea.

LABIOMAXILLARY Adj. Pertaining to the Labium and Maxilla collectively.

LABIOSTIPITES Noun. (Latin, *labium* = lip; *stipes* = stalk.) The Prementum, or part of the Labium formed by the Stipes of the labial (second Maxillary) appendages (Snodgrass).

LABIPALP Noun. (Latin, *labium* = lip; *palpare* = to feel. Pl., Labipalpi.) An arcane term for a Labial Palpus.

LABIS Noun. (Greek, *labis* = forceps. Pl., Labides.) Genitalia of male Lepidoptera: One of a pair of setose appendages, usually rounded terminally, which arise from the dorsal posterior margin of the Tegumen in Lycaenidae; possibly they represent the Uncus or Socii (Klots). Alt. Labipalpus.

LABIUM Noun. (Latin, *labium* = lip. Pl., Labia.) A compound, bilaterally-symmetrical sclerite that forms the 'lower lip' or floor of the mouth in mandibulate insects, and sometimes called the 'tongue'. Labium is positioned behind the first Maxilla and opposed to the Labrum. Labium is regarded as the 'second Maxilla' and seems serially homologous with second Maxillary Sclerites of crustacean mouthparts. The insect Labium consists of laterally paired structures in the embryonic insect which become sclerotized and fuse in postembryonic life. Labium exhibits considerable variation in development among mandibulate insects. In lower pterygotes Labium consists of a Postmentum (attached to Cranium) and a more distal Prementum (attached to apical margin of Postmentum). In Hemiptera, Labrum is represented by a short flap opening at the base of Rostrum. The Labrum is usually shorter than the Labium and not segmented. Three types of Labrum in Hemiptera: Transverse, flap-like without epipharyngeal projections, elongate without epipharyngeal projections and transverse with epipharyngeal projections. Controversey prevails over naming labial parts. In some insects, Postmentum is divided into a proximal sclerite called Submentum and a distal sclerite called Mentum. The Mentum is frequently lost. Prementum (Prelabium, Ligula) forms the distal (moveable) part of Labium and is homologous with Stipes of Maxilla. Palpiger contains musculature attached to the base of Prementum. Labial Palpus is usually shorter than the Maxillary Palpus and typically displays fewer segments. See Mouthparts. Cf. Mandible; Maxilla; Rostrum.

LABIUM SUPERIUS An arcane Latin name for the Labrum.

LABLER, KARL (1882–1935) (Obenberger 1935, Cas. Csl. Spol. ent. 32: 97–98.)

LABOISSIERE, VICTOR (1875–1942) (Jolivet 1948, Miscnea ent. 45: 113–117.)

LABOUCHERE, FRANCIS ANTHONY (1870–1951) (Riley 1952, Proc. R. ent. Soc. Lond. (C) 16: 85.)

LABOULBENE, JEAN JOSEPH ALEXANDRE (1825–1898) (Fairmaire 1906, Ann. Soc. ent. Fr. 75: 63–66.)

LABRAL Adj. (Latin, *labrum* = lip; *-alis* = pertaining to.) Descriptive of structure or action associated with the Labrum.

LABRAL NERVE The element of the labro-frontal nerve which passes to the Labrum.

LABRAL SUTURE A suture between the Labrum and Clypeus. Syn. Clypeolabral Suture.

LABRAM, JONAS DAVID (1785–1852) (Jäggli 1950, Mitt. schweiz. ent. Ges. 31: 199–200.)

LABRARIA Noun. Hymenoptera: The prominent, closely folded, transverse lobe or lip of the Epigusta (MacGillivray). Syn. Epipharynx

LABRARIS Noun. Hymenoptera: An arcane term for the tube formed by the Glossae in bees (MacGillivray).

LABRECULA Noun. Hymenoptera: An arcane term for the small, transverse lip at the entrance to the Basipharynx (MacGillivray).

LABREY, BEEBEE BOWMAN (1817–1882) (Anon. 1882, Entomol. mon. Mag. 19: 22.)

LABROFRONTAL LOBES The Tritocerebrum.

LABROFRONTAL NERVE A short nerve trunk arising anteriorly from the Tritocerebrum (Snodgrass). See Brain.

LABRUM Noun. (Latin, *labrum* = lip. Pl., Labra.) The 'upper lip' of the insect head which covers the base of the Mandible and forms the roof of the mouth. Labrum traditionally has been viewed as a preoral feature that is not homologous with segmental appendages. Labrum articulates with the cranial capsule via membrane (Clypeolabral 'Suture') or is fused to the cranial capsule and immobile. Apical margin of Labrum of some insects is unusually shaped or invested with Setae. For instance, digitate Labrum of parasitic Hymenoptera Eucharitidae and Perilampidae. Digits of Labrum mesh with Setae on labiomaxillary complex and form a sieve-like apparatus. Functionally, this apparatus may exclude pollen when nectar feeding. See Head. Cf. Clypeus; Frons; Gena. Rel. Mouthparts.

LABRUM-EPIPHARYNX Diptera: The median, unpaired Lancet of Species with piercing mouthparts.

LABYRINTH Noun. (Latin, *labyrinthus* = labyrinth. Pl., Labrynths.) Any convoluted structure.

LAC Noun. (Persian, *lak* = lacquer.) The yellowish or reddish-brown resinous substance produced from the Epidermal Glands of *Tacchardia lacca*. Lac is a resin with many industrial and scientific uses. Lac Production originated in India but was adopted in China by the 7th century for drugs and dyes. See Seed Lac; Shellac; Stick Lac. Cf. Pe-La.

LAC DYE A red colouring substance produced from the females of the lac insect, *Tacchardia lacca*.

LAC GLAND Coccoid Hemiptera: The gland which secretes Lac.

LAC INSECTS See Kerriidae.

LACAZE-DUTHIERS, FELIX JOSEF HENRI DE (1821–1901) (Pruvot *et al.* 1902, Archs. zool. exp. gén. (3) 10: 1–78, bibliogr.)

LACCAIC ACID A organic acid in lac.

LACCATE Adj. (Italian, *lacca* = varnish; *-atus* = adjectival suffix.) Descriptive of structure with a varnish-like appearance.

LACE BUGS See Tingidae.

LACENE, ANTOINE MARIE ETIENNE (1769–1859) (Mulsant 1860, Opusc. ent. 12: 201–227, bibliogr.)

LACER Noun. (Latin, *lacerare* = to tear.) A lappet. A term applied to a structural margin with irregular, broad and deep emarginations that leave lappet-like intervals.

LACERATED Adj. (Latin, *lacerare* = to tear.) Describing structure with a ragged or torn appearance. See Lacer.

LACERDA, ANTONIO DE (1834–1885) (Dimmock 1888, Psyche 5: 36.)

LACERTIFORM Adj. (Latin, *lacerta* = lizard; *forma* = shape.) Lizard-shaped; descriptive of a structure or a body shaped as a lizard. Rel. Form 2; Shape 2; Outline Shape.

LACEWING Noun. Common name for Neuroptera. See Chrysopidae; Hemerobiidae.

LACEY, LIONEL (1873–1962) (Varley 1963, Proc. R. ent. Soc. Lond. (C) 27: 51.)

LACHESILLIDAE Plural Noun. Psocids. A numerically small Family of Psocoptera assigned to the Suborder Psocomorpha. Antenna with 13 segments but lacking secondary annulations; Tarsi with two segments; wings without Setae. Species inhabit dried leaves and leaf litter.

LACHRYPHAGOUS Adj. (Latin, *lacrima* = tear; Greek, *phagein* = to devour.) Pertaining to an organism that feeds upon tears or the secretions of Lachrymal Glands. Alt. Lacriphagous.

LACHRYPHAGY Noun. (Latin, *lacrima* = tear; Greek, *phagein* = to devour.) The behaviour, habit or phenomenon of consuming tears.

LACINARASTRUM Noun. (Pl., Lacinarastra.) The Rastrum of the Lacinea; the Laciniafimbrium *sensu* MacGillivray.

LACINELLA Noun. (Latin, *lacinea* = flap. Pl., Lacinellae.) The lateral lobe of the Lacinea when it is two-lobed (MacGillivray).

LACINIA Noun. (Latin, *lacinea* = flap. Pl., Laciniae.) 1. A blade-like, mesial sclerite attached to the distal margin of the Maxillary Stipes. Often, Lacinia is articulated with Stipes and bears brushes of Setae or spines. See Maxilla. Cf. Cardo; Galea; Palpifer; Palpiger; Stipes; 2. Diptera: A nonsegmented, flat, lancet-like sclerite adapted for piercing. 3. Psocoptera: The styliform appendage of Ribaga. See Maxilla. Cf.

Galea.

LACINIA CONVOLUTA The Haustellum rolled or coiled like the spring of a clock and suspended beneath the head.

LACINIA CORIARIA A long, leather-like, flexible Lacinea.

LACINIA EXTERIORIS and INTERIORIS. Aphididae: The Palpiger and Paraglossa. An arcane term for the Maxillary Galea and Lacinea.

LACINIA FALCATA An acute, sickle-shaped Lacinea.

LACINIA MOBILIS A small, plate-like appendage near the extremity of the Mandible in *Campodea* and *Anajapyx* (Imms). Ephemeroptera: A small appendage on the Mandible between the Molar surfaces and the Canines. LM often curved, sometimes blunt at the apex or minutely toothed apically, frayed as a long brush of fine Setae (Needham).

LACINIA OBTUSA A Lacinia rounded and not produced or acute.

LACINIAFIMBRIUM (Latin, *lacinia* = flap; *-ium* = diminutive > Greek, *-idion*. Pl., Laciniafimbria.) The Laciniarastra *sensu* MacGillivray.

LACINIATE Adj. (Latin, *lacinia* = flap; *-atus* = adjectival suffix.) Pertaining to structure cut into irregular, unequal and deep segments; jagged. Alt. Laciniated; Laciniatus.

LACINOIDEA Noun. (Latin, *lacinea* = lip; Greek, *eidos* = form.) The mesial lobe of the Lacinea when it is bilobed (MacGillivray).

LACKEY CATERPILLAR *Malacosoma neustrium* [Lepidoptera: Lasiocampidae].

LACKSCHEWITZ, PAUL (1865–1936) (Knorre 1937, KorrespBl. naturfVer. Riga 62: 20–29, bibliogr.)

LACORDAIRE, JEAN THÉODORE (1801–1870) French entomologist and founding member of Entomological Society of France. Author of *Essai sur les Coleopteres de la Guayane Francaise* and *Introduction a l'Entomologie* (1834). (Candèze 1872, Ann. Acad. Belg. 38: 139–160, bibliogr.)

LACQUER Noun. A varnish produced from lac.

LACROIX, J G L (–1939) (Anon. 1939, Lambillionea 39: 141.)

LACTASE Noun. (Latin, *lac* = milk.) A digestive enzyme, which hydrolyses lactose or milk sugar.

LACTATE Adj. (Latin, *lac* = milk; *-atus* = adjectival suffix.) Descriptive of structure which is milk-white.

LACTEAL Adj. (Latin, *lac* = milk; *-alis* = pertaining to.) Relating to milk; milky in appearance.

LACTEOUS Adj. (Latin, *lac* = milk; Latin, *-osus* = with the property of.) Milky colour. Alt. Lacteus.

LACTESCENT Adj. (Latin, *lactescere* = to turn to milk.) Secreting or yielding a milky fluid.

LACTIC ACID An acid produced by the fermentation of milk, or by heating sugars with alkalis; a thick, sour liquid, miscible with water, alcohol and ether.

LACUNA Noun. (Latin, *lacuna* = cavity. Pl., Lacunae.) A pit, a gap, an empty space; irregular impressions or cavities, specifically, the non-walled cavities of the body.

LACUNOSE Adj. (Latin, *lacuna* = cavity; Latin, *-osus* = with the property of.) Pertaining to structure with scattered, irregular, broad, shallow cavities; structure which is furrowed or pitted. Alt. Lacunosus.

LACUSTRINE Adj. (Latin, *lacus* = lake.) Pertaining to lakes, or near lakes.

LADEBURG, CARL F (1897–1956) (Hunter 1957, Fla Ent. 40–43.)

LADY BEETLES See Coccinellidae.

LADYBIRD BEETLES See Coccinellidae.

LADYBIRDS See Coccinellidae.

LADYBUGS See Coccinellidae.

LAEMOBOTHRIIDAE Plural Noun. A small, cosmopolitan Family of Amblycera. Head elongate with strong swelling of Gena (anterior of eyes); mandibulate mouthparts; Antenna with four segments, typically forming club and concealed in grooves. Species parasitic on water birds (rails, storks) and predatory birds (hawks).

LAEMODIPODIFORM Shaped as the nymph of a walking stick. Rel. Form 2; Shape 2; Outline Shape.

LAEOTORMA Noun. (Greek, *laios* = left; *tormos* = socket.) Scarabaeoid larvae: A transverse sclerite from the left hind angle of the Epipharynx; usually provided with Pternotorma, often with Epitorma or a part of the Epitorma and, more rarely with Apotorma (Boving).

LAEOTROPIC Adj. (Greek, *laios* = left; *trope* = turning; *-ic* = characterized by.) Turning to the left, or inclined to turn to the left.

LAET, JOANNES DE (1593–1649) (Rose 1950, *New General Biographical Dictionary* 9: 153.)

LAETUS Adj. (Latin, *laetus* = cheerful, glad. From Sanskrit root.) Very bright in colour. Alt. Laete.

LAEVIS Adj. (Latin, *laevis* = left.) Smooth, shining and without superficial elevations. Alt. Laevigatus; Levigate.

LAFERTE-SENECTERE, F THIBAULT DE LA CARTE DE (1808–1886) (Marseul 1887, Abeille (Les ent. et leurs écrits) 24: 173–174, bibliogr.)

LAFFERTY, BOYD M (1917?–1973) (Jobbins 1973, Mosq. News 33: 520.)

LAFFOON, JEAN LUTHER (1922–1973) (Robinson 1973, J. Econ. Ent. 66: 831.)

LAFITOLE, MARQUIS DE (–1881) (Anon. 1881, Naturaliste 1: 494.)

LAFORGE, FREDERICK D (1882–1958) (Busbey 1959, J. Econ. Ent. 52: 180.)

LAG PHASE Tissue Culture: The period following sub-culture of cells and before the Exponential Growth Phase. Cells become acclimated to their new environment during the Lag Phase. See Growth Curve. Cf. Exponential Phase; Stationary Phase.

LAGENIFORM Adj. (Latin, *lagena* = flask; *forma* = shape.) Bottle-shaped; Flask-shaped; descriptive of structure with a large, cylindrical body that becomes apically constricted into a narrow neck. Alt. Lagenoid. See Ampulliform; Adeniform; Pyreniform. Rel. Form 2; Shape 2; Outline Shape.

LAGON® See Dimethoate.

LAGRIIDAE See Tenebrionidae.

LAHARPE, JEAN JAQUES CHARLES (1802–1877) (Anon. 1878, Verh. schweiz. naturf. Ges. 60: 293–304.)

LAHILLE, FERNANDO (1861–1940) (Porter 1940, Revta chil. Hist. nat. 44: 266–268, bibliogr.)

LAICHARTING, JOHANN NEPOMUK (1745–1797) Austrian academic and author of *Verzeichniss der Tyrolen Insecten* (2 vols 1781–1784). (Anon. 1934, Beitr. Gesch. Statist. Naturk. Innsbruck 8: 186–224, bibliogr.)

LAIDLAW, F F (1876–1963) (Dance 1964, J. Conch. Lond. 25: 288–291.)

LAJONQUIERE, ETIENNE DE (–1975) Bustillo 1975, Shilap 12: 239.

LAKE Noun. (French, *laque*, from Persian *lak*. Pl., Lakes.) 1. Specifically, a purplish red pigment from cochineal or lac produced by precipitation of the colouring matter with metallic compounds. 2. Generally, variously coloured insoluble metallic compounds of dyes (*e.g.* brazilwood lake, yellow lake). Lake pigments are prepared by precipitation from solutions of dyes with salts of aluminium or tin. During the dyeing process, the metallic salts are called mordants. See Dye; Mordant.

LAKIN, CHARLES ERNEST (1878–1972) (Butler 1973, Proc. R. ent. Soc. Lond. (C) 37: 56.)

LALLEMAND, VICTOR (1880–1965) (Verstraeten 1966, Bull. rech. agron. Gembloux 1: 324–331, bibliogr.)

LAMACERATUBAE Plural Noun. Diaspididae: The long slender Ceratubae in the Pygidia (MacGillivray).

LAMARCK, JEAN BAPISTE PIERRE ANTOINE DE MONET DE (1744–1829) French aristocrat and botanist and evolutionary biologist; developed an improved classification of plants that was articulated in his *Flore Francaiaem ou Descpription succincte de toutes les Plantes qui croissent naturellement den France* (1780). Following the French Revolution, Lamarck became Professor of Zoology in Jardines des Plantes. Lamarck preceded Darwin in developing explanations for organic evolution. Most important work *Philosophie Zoologique*. Essential points in Lamark's hypothesis regarding evolution were called Lamarkism (Anon. 1902, Am. Nat. 36: 495–497.) Cf. Charles Darwin.

LAMARKISM Noun. Principles (factors) expressed by Jean Lamark and taken to form the basis of his hypothesis of evolution. Lamark's principles are called 'Laws' in older publications and include. 1. Frequent and sustained use of organs strengthens and enlarges them; lack of frequent and sustained use of organs causes them to weaken and atrophy. 2. The conditions of organs in an individual are trasmitted to offspring, providing the conditions are common to both sexes. See Darwinism.

LAMB, CHARLES GEORGE (1861–1941) (Blair 1942, Proc. R. ent. Soc. Lond. (C) 6: 41.)

LAMBDA CYHALOTHRIN Noun. A synthetic-pyrethroid compound used as a contact and stomach poison against insects attacking cotton, turf, ornamentals and public-health. Toxic to fishes. Trade names include: Battle®, Charge®, Commodore®, Cyhalon®, Demand®, Excaliber®, Fung-Fu®, Hallmark®, Icon®, Juresong®, Karate®, Matador®, Perimpak®, Saber®, Samourai®, Scimitar®, Sentinel®. See Synthetic Pyrethroids.

LAMBERS, D. HILLE RIS (1909–1984) Specialist on Aphidoidea.

LAMBERT, CLAUDE FRANCOIS (1705–1765) (Rose 1950, *New General Biographical Dictionary* 9: 170.)

LAMBIN, CHARLES (1822–1885) (Dimmock 1888, Psyche 5: 36.)

LAMBORN, WILLIAM ALFRED STEDWELL (1877–1960) (Uvarov 1961, Proc. R. ent. Soc. Lond. (C) 25: 49.)

LAMBOTTE, HENRI ANTOINE (1816–1873) (Denis 1873, Ann. Soc. malac. Belg. 8: i–xxiv.)

LAMEERE, ALFRED LUCIEN GASTON (1864–1942) (Hale-Carpenter 1946, Proc. R. Ent. Soc. Lond. (C) 10: 52.)

LAMEERE, M AUGUSTE (1864–1942) (Anon. 1943, Bull. Cerc. Zool. congol. 17: (51–52).)

LAMELLA Noun. (Latin, diminutive of *lamina; lamella* = small plate; small metal plate. Pl., Lamellae.) 1. A thin sclerite. 2. A plate-like or leaf-like process. 3. The Parademe (MacGillivray); Coccids: The plates (MacGillivray).

LAMELLATE ANTENNA An Antenna with the club formed of closely opposed leaf-like surfaces, the concealed surface set with sensory pits.

LAMELLATE Adj. (Latin, *lamella* = small plate; *-atus* = adjectival suffix.) Sheet-like; leaf-like; pertaining to structure composed of, or covered with, laminae or thin sheets. Alt. Lamellatus; Laminate; Laminatus. See Lamelliform.

LAMELLES Noun. Coccids: Lobes (MacGillivray).

LAMELLICORN BEETLES. See Scarabaeidae.

LAMELLICORNIA Noun. Coleoptera in which the Antennae terminate in a lamellate or leaf-like club.

LAMELLIFORM Adj. (Latin, *lamella* = small plate; *forma* = shape.) Leaf-shaped; descriptive of structure that is broad and flat; structure resembling leaves, blades or lamellae. Alt. Lamellate. See Lamella; Leaf. Cf. Laminiform; Phylliform; Squamiform. Rel. Form 2; Shape 2; Outline Shape.

LAMELLOCYTE See Plasmatocyte.

LAMINA Noun. (Latin, *lamina* = plate. Pl., Laminae; Laminas.) 1. A thin, flat, chitinous scale-like sclerite. 2. Any of several broad, flat anatomical structures. 3. Siphonaptera: The corneous sclerite on each side of the mouth (Jardine).

LAMINA EXTERNA The Paraglossa.

LAMINA GANGLIONARIS 1. One of two ganglionic bodies of the optic lobe in brachipod Crustacea. 2. The Periopticon in insects (Snodgrass).

LAMINA INFRA-ANALIS Anisopteran Odonata nymphs: One of two lateroventral sclerites in the walls of the circular fold which contains the Anus (Snodgrass).

LAMINA INTERNA The Ligula.

LAMINA SUBGENITALIS 1. The Subgenital Sclerite. 2. Cockroaches: The seventh ventral sclerite of the female and ninth ventral sclerite of the male.

LAMINA SUPRA-ANALIS Anisoptera larvae (Odonata): A small dorsal sclerite in the walls of the circular fold which contains the Anus (Snodgrass); Suranal Sclerite.

LAMINAR Adj. (Latin, *lamina* = plate.) Descriptive of a composite structure consisting of several broad, flat, leaf-like parts. Alt. Laminate. See Lamina. Cf. Laminiform.

LAMINAR FLOW CABINET Tissue Culture: A cabinet or enclosed space in which filtered air circulates. LFC maintain a sterile environment for the propagation of cell cultures.

LAMINATE Adj. (Latin, *lamina* = plate; *-atus* = adjectival suffix.) 1. Consisting of lamina or laminae. 2. A composite structure formed of thin flat layers of variable thickness, texture and composition.

LAMINATO-CARINATE With an elevated ridge or keel, formed of thin sclerites.

LAMINIFORM Adj. (Latin = *lamina* = plate; *forma* = shape.) Scale-shaped; structure with the appearance of an Lamina. See Lamina. Cf. Lamelliform; Phylliform; Squamiform. Rel. Form 2; Shape 2; Outline Shape.

LAMINITENTORIUM Noun. (Latin = *lamina* = plate; *tentorium* = tent. Pl., Laminitentoria.) The frontal sclerite of the Tentorium (MacGillivray).

LAMITENDONS Noun. Tendons attached to the cephalic part of the Laminitentorium (MacGillivray).

LAMNADENS Noun. (Latin = *lamina* = plate; *dens* = tooth.) Diptera: The tooth-sclerite bearing the Prestomal Teeth (MacGillivray).

LAMPA, SVEN (1839–1914) (Palmen 1915, Meddn. Soc. Fauna Flora Fenn. 41: 104–105.)

LAMPRECHT, HERBERT (1890–1969) (Lindroth 1969, Opusc. ent. 34: 175.)

LAMPYRIDAE Plural Noun. Fireflies; Glowworms; Lightningbugs. A cosmopolitan Family of polyphagous Coleoptera with ca 1,800 nominal Species; assigned to Elateroidea. Adult 4–12 mm long; elongate, soft-bodied, somewhat flattened; head concealed by Pronotum when viewed from above; Antenna short, filiform with 11 segments; male with large eyes; Labrum frequently indistinct; Pronotum explanate; tarsal formula 5-5-5; females wingless in some Species; winged females not known to fly; males always winged, crepuscular; Abdomen with six Ventrites. Adults with luminescent organs on female abdominal Sternum 5 and male abdominal Sterna 5 or 6. Adults typically do not feed, some take water and some take nectar from flowers; adults of *Photinus*, *Lucidota*, *Pyractomena* may be herbivorous. Larva campodeiform; body covered with tough, shield-like sclerites dorsally; head small, prognathous; each side of head with one Stemma; Mandibles channelled; legs with five segments, each with two tarsal claws; Urogomphi absent; entomophagous, possess Acinose Glands at anterior end of foregut; glands produce paralytic and proteolytic enzymes; digestion preoral in most lampyrids.

LANA Noun. (Latin, *lana* = wool.) Long abdominal Setae of some Lepidoptera.

LANAI BUCKET A container used for the placement of irradiated fruitfly Pupae in the field which permits emergence of the flies. LB used as an alternative or complement to ground and aerial release of irradiated flies.

LANATE Adj. (Latin, *lana* = wool; *-atus* = adjectival suffix.) Descriptive of structure densely covered with long, fine, somewhat curled Setae; resembling wool. Alt. Lanatus.

LANCE FLIES See Lonchaeidae.

LANCEOLATE Adj. (Latin, *lanceola* = little lance; *-atus* = adjectival suffix.) Lance- or spear-shaped; oblong and tapering apicad. Alt. Lanceolatus. See Outline Shape.

LANCEY, WILLIAM (–1853) (Newman 1854, Zoologist 12: 428.)

LANDHOPPERS A group of peracarid crustaceans including ca 70 Species '*Talitrus* spp.' [Amphipoda: Talitridae] which have colonized terrestrial habitats including mangrove, leaf litter and soil of wet forests in Australia, islands of the Pacific and Caribbean. Primarily nocturnal and feed upon angiosperm leaf litter and debris; occasionally invade homes. Body smooth, laterally compressed; head with small, sessile eyes; two pairs of Antenna with pair 1 small and pair 2 sometimes as long as forelegs; Thorax (Peraeon) of seven segments; appendages near head modified for manipulating or processing food; five pairs of walking legs (Peraeoprods); Abdomen (pleon) with reduced appendages adapted for swimming; female with brood pouch on thoracic Sternum.

LANDIS, BIRELY (1904–1974) (Butt 1975, J. Econ. Ent. 68: 568.)

LANDOIS, HERMANN (1835–1905) (Recker 1905, Jber. westf. ProvVer. Wiss. Kunst. 33: 9–17, bibliogr.)

LANDWEHR, FRIEDRICH (1866–1911) (Roc 1911. Sber. naturh. Ver. preuss. Rheinl. Westfal 1911, 180–181.)

LANE, JOHN (1905–1963) (Del Ponte 1963, Revta Soc. ent. argent. 25: 1–2.)

LANG, HENRY CHARLES (–1909) (Anon. 1910, Entomol. mon. Mag. 46: 39.)

LANGE METALMARK *Apodemia mormo langei* Comstock [Lepidoptera: Riodinidae].

LANGE, B (1910–1969) (Anon. 1969, Anz.

Schädlinsk. Pflanzenschutz 42: 156.)

LANGHOFFER, AUGUST (1861–1940) (Sachtleben 1940, Arb. morph. taxon. Ent. Berl. 7: 169–170.)

LANGSDORFF, GEORG HEINRICH VON (1774–1852) (Carvalho 1918, Revta Mus. paul 10: 877–883.)

LANGURIIDAE Plural Noun. Lizard Beetles. A widespread Family of polyphagous Coleoptera (Cucujoidea). Adult 1.5–10 mm long, elongate, subcylindrical to flattened; head large, Ommatidia with coarse appearance; elytral Epipleura well developed, complete; tarsal formula 5-5-5; Tarsomeres with setose lobes on ventral surface; Abdomen with five Ventrites. Larva elongate, cylindrical to flattened; head prognathous; Stemmata present or absent; Antenna with three segments; legs with five segments; Pretarsi each with two claws; Urogomphi present. Larval feeding diverse: pollen, decaying vegetation, leaf litter, stem borers.

LANIARII Noun. (Latin, *laniarius* = pertaining to a butcher.) Very sharp, usually long conical teeth. Alt. Laniary.

LANKESTER, EDWIN (1814–1874) (J. F. P. 1875, J. micros. sci. 15: 59–62.)

LANKIALA, ACRO (1877–1959) (Nordström 1959, Opusc. ent. 24: 78.)

LANKTREE, PATRICK ADRIEN DESMOND (1919–1975) (Gunn 1976, Proc. R. ent. Soc. Lond. (C) 40: 52.)

LANNATE® See Methomyl.

LANOX® See Methomyl.

LANSBERGE, JOHAN WILHELM (1830–1909) (Bos 1888, Tijdschr. Ent. 31: 201–234.)

LANTANA *Lantana camara* Linnaeus [Verbenaceae]: An hairy, sometimes prickly shrub growing to 1.5 m. Endemic to tropical and subtropical America; moved elsewhere as an ornamental plant (India 1809, Australia 1841, Hawaii 1885); considered a serious weed pest throughout world. Leaves and seeds toxic to livestock; overwintering site and alternative host plant for agricultural pests; reduces carrying capacity of pasture. Lantana supports an exceptionally large complex of herbivorous insects; more than 30 natural enemies tried in 27 countries, only provide partial control. Syn. Yellow Sage.

LANTANA BUG See Greenhouse Orthezia.

LANTANA CERAMBYCID *Plagiohammus spinipennis* Thomson [Coleoptera: Cerambycidae].

LANTANA-DEFOLIATOR CATERPILLAR *Hypena strigata* (Fabricius) [Lepidoptera: Noctuidae].

LANTANA FLOWER-CATERPILLAR *Epinotia lantana* (Busck) [Lepidoptera: Tortricidae]: Endemic to Central America; introduced into India, Australia and South Pacific for biocontrol of Lantana. Adult oviposits into flowers; larvae mine shoots, fruits, flowers; pupation in seed receptacle.

LANTANA GALL-FLY *Eutreta xanthochaeta* Aldrich [Diptera: Tephritidae].

LANTANA HISPID *Uroplata girardi* Pic [Coleoptera: Chrysomelidae]: Endemic to South America; introduced into many countries for control of Lantana; performance erratic. Multivoltine with overlapping generations; eggs laid individually in leaf; larvae mine leaves; pupation within mine. Adults cause feeding scars on upper surface of leaves. Species prefers cool, semi-shaded conditions; overwinters as adult. Alt. Lantana Leaf Beetle (Australia)

LANTANA LACE-BUG 1. *Teleonemia elata* Drake [Hemiptera: Tingidae]: Endemic to Brazil; nymphs and adults suck leaf juices of Lantana. 2. *Teleonemia scrupulosa* Stål [Hemiptera: Tingidae]: Endemic to Central America; introduced into many countries for biocontrol of Lantana. Multivoltine with overlapping generations; eggs laid in groups along main veins of leaves; young nymphs feed in groups on underside of leaves; old nymphs, adults feed on leaves, flowers and buds; cause shoot dieback and defoliation; important when plants under stress from drought. 3. *Leptobyrsa decora* Drake [Hemiptera: Tingidae]: Endemic to Peru; nymphs and adults suck leaf juices of lantana; eggs laid in batches of 40–50 on underside of leaf mid-rib.

LANTANA LEAF-BEETLE *Octotoma scabripennis* Guérin-Méneville [Coleoptera: Chrysomelidae]: Endemic to Mexico; introduced into many countries for biocontrol of Lantana. Biology resembles Lantana Hispid; larva develops mine near main vein of leaf; Species prefers hot exposed conditions; complements action of LH.

LANTANA LEAF-FOLDING CATERPILLAR *Anania haemorrhoidalis* (Guenée) [Lepidoptera: Pyralidae]: Endemic to Central America; Larvae feed on lantana, web and roll leaves.

LANTANA LEAFMINER 1. *Cremastobombycia lantanella* (Schrank) [Lepidoptera: Gracillariidae]. 2. *Uroplata girardi* Pic [Coleoptera: Chrysomelidae]: Endemic to Brazil; Larvae mine and pupate in leaves; adults feed on upper Epidermis and mesophyll of leaves. See Lantana Leaf Beetle.

LANTANA LEAF-MINING BEETLE See Lantana Leaf Beetle.

LANTANA LEAF-MINING FLY *Calcomyza lantanae* (Frick) [Diptera: Agromyzidae]: Endemic to tropical America; introduced into Australia and South Africa for biocontrol of Lantana. Adults feed from punctures made by females; females oviposit into leaves; larvae form blotch mines leaves; pupation in soil.

LANTANA LEAF-TIER *Salbia haemorrhoidalis* Guenée [Lepidoptera: Pyralidae].

LANTANA PLUME-MOTH *Lantanophaga pusillidactyla* (Walker) [Lepidoptera: Pterophoridae]: Endemic to Central America; larvae feed on flowers.

LANTANA-SEED FLY *Ophiomyia lantanae* (Froggatt) [Diptera: Agromyzidae]: Endemic to Central America; introduced into numerous countries for biocontrol of Lantana. Transferred from

Hawaii to New Caledonia during 1908–1909 and Fiji during 1911. Female lays individual egg in green berry; larva feeds on outer fruit; pupation in ripe berry or peduncle.

LANTANA STICK CATERPILLAR *Neogalea sunia* (Guenée) [Lepidoptera: Noctuidae].

LANTERN FLIES See Fulgoridae.

LANTZ, DAVID ERNEST (1855–1918) (Anon. 1918, Ent. News 29: 400.)

LANUGINOSE Adj. (Latin, *lanugo* = down; *-osus* = full of.) Covered with long, fine, flexible Setae that give a down-like appearance. See Crinitus. Alt. Langinosus; Lanuginous.

LANUGO Noun. (Latin, *lanugo* = down.) Slender, fine, individual Setae.

LAPAROSTICT Noun. (Greek, *lapero-* comb. form = flank; *stictos* = to prick.) Any lamellicorn beetle with abdominal spiracles connecting membrane between dorsal and ventral rings.

LAPIDICOLOUS Adj. (Latin, *lapis* = stone; *colere* = living in or on.) Descriptive of organisms that live under stones or deeply imbedded among stones. See Habitat. Cf. Rupicolous. Rel. Ecology.

LAPORTE, FRANCOIS LOUIS See CASTELNAU.

LAPOUGE, GEORGES VACHER DE (1854–1936) (Barthe 1937, Miscnea ent. 38: 29–31, bibliogr.)

LAPPET MOTH *Phyllodesmia americana* (Harris) [Lepidoptera: Lasiocampidae].

LAPPET MOTHS See Lasiocampidae.

LAPSUS CALAMI (Latin, Pl., *Lapsus calamorum*.) A Latin phrase meaning 'slip-of-the-pen.' In taxonomic publications, an inadvertent error in the publication of a scientific name.

LAPSUS LINGUAE (Latin.) A slip of the tongue.

LARCH APHID *Cinara laricis* (Walker) [Hemiptera: Aphididae].

LARCH CASEBEARER *Coleophora laricella* Hübner [Lepidoptera: Coleophoridae]: A moth endemic to central Europe found in alpine habitats on *Larix decidua* Mill. First detected in North America during 1896, the moth spread rapidly through Canada and the northern USA.

LARCH SAWFLY *Pristiphora erichsonii* (Hartig) [Hymenoptera: Tenthredinidae]: A sawfly endemic to Europe and a sporadic pest in Canadian forests. Status in Nearctic unclear: Apparently four biotypes in Canada; perhaps two biotypes endemic; two biotypes Eurasian, migrated across Bering Land bridge; possibly adventive biotype to Canada during 1880. Eggs translucent, laid in rows under bark on terminal twigs; egg slits may cause twig curl. Larvae gregarious, grey-green, with seven pairs of Prolegs; feed on foliage in upper crowns, strip branches. Overwinter as larvae within tough cocoon in litter. Monovoltine; some larvae spend two winters in cocoons. Some control achieved with ichneumonid *Mesoleius tenthredinis* Morley.

LARDER BEETLE *Dermestes lardarius* Linnaeus [Coleoptera: Dermestidae]: A cosmopolitan pest of stored animal products including hair, horn, feathers, fur and carcasses; not a problem in hot tropical areas. Adult and larva pests by biting holes in skin and furs and contaminating stored products with frass. Adult 7–10 mm long, nearly black with transverse yellow stripe and dark spots across basal part of Elytra; larva to 15 mm, brown, conspicuously setose, apex of Abdomen with two curved processes. Life cycle typically 2–3 months but as long as nine months; adult may live 1.5 years. Preoviposition period ca 10 days following emergence; female oviposits 200–800 eggs during lifetime. Eggs ca 2 mm long, laid individually in crevices or cracks near animal products; egg stage 2–12 days. Larval instars variable in number, typically 5–6 with more in female than male; larvae bristly with two posteriorly directed terminal Urogomphi; mature larva ca 12–15 mm long; stage 15–80 days. Pupal stage 8–15 days; may pupate in timber or other concealed habitat away from site of infestation; exposed Pupae remains in last larval Integument but concealed. Syn. Bacon Beetle. See Dermestidae. Cf. Common Hide Beetle; Hide Beetle; Khaphra Beetle.

LAREYNIE, PHILIPPE (1826–1857) (Fairmaire 1859, Ann. Soc. ent. Fr. (3) 7: 261–266.)

LARGAIOLLI, VITTORIO (1868–1951) Italian acarologist.

LARGE AMBROSIA BEETLE *Platypus froggatti* Sampson [Coleoptera: Curculionidae].

LARGE ANT-BLUE *Acrodipsas brisbanensis* (Miskin) [Lepidoptera: Lycaenidae].

LARGE ASPEN-TORTRIX *Choristoneura conflictana* (Walker) [Lepidoptera: Tortricidae].

LARGE AUGER-BEETLE *Bostrychopsis jesuita* (Fabricius) [Coleoptera: Bostrychidae]. In Australia, a univoltine pest of *Eucalyptus*, wattles, orchard and ornamental plants.

LARGE BAGWORM *Oiketicus elongatus* Saunders [Lepidoptera: Psychidae].

LARGE BANDED AWL *Hasora khoda haslia* Swinhoe [Lepidoptera: Hesperiidae].

LARGE BIGEYED-BUG *Geocoris bullatus* (Say) [Hemiptera: Lygaeidae].

LARGE BROWN AZURE *Ogyris idmo* Hewitson [Lepidoptera: Lycaenidae].

LARGE BROWN HOUSE-MOTH See Granny Moth.

LARGE BROWN MANTIDS *Archimantis* spp. [Mantodea: Mantidae]. See Mantidae.

LARGE BROWN-SPIDER *Heteropoda venatoria* (Linnaeus) [Araneae: Sparassidae].

LARGE CARPENTER-BEES *Xylocopa* spp. [Hymenoptera: Anthophoridae].

LARGE CHESTNUT-WEEVIL *Curculio caryatrypes* (Boheman) [Coleoptera: Curculionidae]: A pest of chestnuts in eastern North America. Female drills hole in nut to deposit eggs; larva feeds on chestnut until it falls; pupation in soil.

LARGE CHICKEN-LOUSE *Goniodes gigas* (Taschenberg) [Mallophaga: Philopteridae].

LARGE CITRUS-BUTTERFLY *Princeps aegeus* (Donovan) [Lepidoptera: Papilionidae]: Endemic to Australia and a minor but frequent pest of citrus. Caterpillars feed on young foliage and chew large areas from leaf edge. Severe attack can reduce shoots to bare twigs; usually not harmful to mature trees. Adult wingspan ca 130 mm; male forewing predominantly black with white markings; hindwing of female with white, orange and blue markings. Eggs yellow, spherical, 2–3 mm diameter. Caterpillar to 70 mm long, with Osmeterium, brown to olive green with reddish band along anterior part of body. Five larval instars. Pupation in upright position on twig and held in place by girdle. Life cycle 2–3 months; multivoltine with at least three generations per year. Cf Small Citrus-Butterfly.

LARGE COTTONY-SCALE *Pulvinaria mammeae* Maskell [Hemiptera: Coccidae].

LARGE CROSSVEIN Diptera: The crossvein closing the Discal cell; posterior crossvein (m and M) (Curran). See Vein; Wing Venation.

LARGE DARTER *Telicota anisodesma* Lower [Lepidoptera: Hesperiidae].

LARGE DINGY SKIPPY *Toxidia peron* (Latreille) [Lepidoptera: Hesperiidae].

LARGE DUCK-LOUSE *Trinoton querquedulae* (Linnaeus) [Phthiraptera: Menoponidae].

LARGE FALSE WIREWORM *Pterohelaeus alternatus* Pascoe [Coleoptera: Tenebrionidae]. Adult 16.5–19.0 mm long, oval and black; Antenna with loose, flattened, 4-segmented club. Elytra with 18 striae. Adult females lay eggs in soil. Eggs 1.2–1.5 mm long, oval, smooth and white. Larvae elongate-cylindrical, strongly sclerotised and yellow-brown; head and first three segments yellow-brown or with light brown patches. Pupation occurs in soil. Pupae elongate, light brown-cream and 17–23 mm long. Eggs laid in summer and autumn; larvae feed on organic matter in the soil until they reach full size in spring. False wireworms prefer dry conditions protected by stubble or weeds. See Eastern False Wireworm.

LARGE FERN WEEVIL *Syagrius fulvitarsus* Pascoe [Coleoptera: Curculionidae].

LARGE GREEN JASSID *Batracomorphus angustatus* (Osborn) [Hemiptera: Cicadellidae].

LARGE GREEN SAWFLY *Perga affinis insularis* Riek [Hymenoptera: Pergidae].

LARGE GREEN-BANDED BLUE *Danis danis serapis* Miskin [Lepidoptera: Lycaenidae].

LARGE HEN LOUSE *Goniodes gigas* (Taschenberg) [Phthiraptera: Philopteridae].

LARGE HUNTSMAN SPIDERS *Isopeda* spp. [Araneida: Heteropodidae].

LARGE INTESTINE The Colon. See Alimentary System.

LARGE KISSING-BUG *Triatoma rubrofasciata* (DeGeer) [Hemiptera: Reduviidae].

LARGE LEAF-EATING LADYBIRD *Epilachna guttatopustulata* (Fabricius) [Coleoptera: Coccinellidae]: In Australia a pest of duboisia; adults and larvae eat leaf surface and cause severe damage if infestation is heavy.

LARGE MANGO TIPBORER See Mango Shoot Caterpillar.

LARGE MILKWEED-BUG *Oncopeltus fasciatus* (Dallas) [Hemiptera: Lygaeidae]: A pest of milkweed throughout North America, but most common east of the Rocky Mountains. Eggs elongate, red and deposited on milkweed. LMB overwinter as adults on trees; may congregate aroung buildings on warm winter days.

LARGE ORANGE-SULPHUR *Phoebis agarithe* [Lepidoptera: Pieridae].

LARGE PALE CLOTHES MOTH *Tinea pallescentella* Stainton [Lepidoptera: Tineidae]: A cosmopolitan pest of animal products including wool, hair and feathers. Cf. Casemaking Clothes-Moth; Tropical Case-Bearing Clothes Moth; Webbing Clothes Moth.

LARGE POND SNAIL *Lymnaea stagnalis* (Linnaeus) [Basommatophora: Lymnaeidae].

LARGE SHOOT WEEVIL *Rhinaria concavirostris* Lea [Coleoptera: Curculionidae].

LARGE SULPHUR *Phoebis philea* [Lepidoptera: Pieridae].

LARGE TURKEY-LOUSE *Chelopistes meleagridis* (Linnaeus) [Mallophaga: Philopteridae]: A significant pest of turkeys. Cf. Slender Turkey-Louse.

LARGE WAX-MOTH *Galleria mellonella* (Linnaeus) [Lepidoptera: Pyralidae].

LARGE YELLOW UNDERWING *Noctua pronuba* (Linnaeus) [Lepidoptera: Noctuidae]: A pest of root crops in the Palearctic.

LARGE-LEGGED THRIPS See Merothripidae.

LARGER BLACK FLOUR-BEETLE *Cynaeus angustus* (LeConte) [Coleoptera: Tenebrionidae]: A minor pest of farm-stored grains in North America. Prefers damaged kernels, but can feed upon intact kernels. Eggs laid singly in concealed places; incubation 3–4 days. Larva with 9–11 instars; development density dependent, ca 22–92 days at 30°C. Pupation in a cell constructed of kernels; duration 4–6 days. Adults live ca one year; preoviposition period 5–7 days; females lay 350–450 eggs.

LARGER CANNA-LEAFROLLER *Calpodes ethlius* (Stoll) [Lepidoptera: Hesperiidae].

LARGER ELM-LEAF BEETLE *Monocesta coryli* (Say) [Coleoptera: Chrysomelidae]: A monovoltine pest of elm, hawthorn, hazelnut, pecan in eastern USA. Adult ca 10–12 mm long; yellow with greenish-blue on base and apex of Elytra. Egg lemon like in colour and shape, deposited in masses to underside of leaves during summer. Larvae metallic reddish-brown coloured, feed gregariously; three instars; overwinter as mature larvae within a cell in soil. Pupation during spring. Cf. Elm Leaf Beetle.

LARGER GRAIN-BORER *Prostephanus truncatus* (Horn) [Coleoptera: Bostrichidae]: A minor pest of corn in Central America and major pest of cassava in east Africa; feeds on many softer grains; in Americas reproduces only on corn. LGB endemic to Central America; adventive to South

America, North America and Africa (Tanzania, Kenya, Togo). Adult 3–4 mm long, dark, cylindrical; Prothorax large, hood-like and conceals head, conspicuously tuberculate; Antenna of 10 segments with apical three segments large and form loose club; Elytra rectangular in outline shape, tuberculate, posterior declivity steep, flat with posterior margin carinate; readily fly. Adult lives 40–60 days, bores into grain or cassava. Female lays eggs 50–225 eggs during 14 days; eggs laid in small chambers created from tunnels in grain or cassava; incubation 3–7 days. Larva white, scarabaeiform with thoracic segments large; 4–6 instars, typically 4; mature larva 4 mm long. Pupation inside case of frass or food; pupal stage ca five days. Cf. Lesser Grain-Borer.

LARGER PALE-TROGIID *Trogium pulsatorium* (Linnaeus) [Psocoptera: Trogiidae].

LARGER SHOT-HOLE BORER *Scolytus mali* (Bechstein) [Coleoptera: Scolytidae].

LARGER YELLOW-ANT *Acanthomyops interjectus* (Mayr) [Hymenoptera: Formicidae]: Workers 2.5–4.0 mm long; yellow; emanates a lemon odour when bodies are crushed. LYA typically nests within large mounds in open areas, but can also be taken in wall voids and subfloors; nocturnal nuptial flights. Can be found in large numbers around houses, but does not take human food; can tend aphids and mealybugs on plant roots. See Formicidae. Cf. Smaller Yellow Ant. Rel. Argentine Ant; Black Carpenter-Ant; Cornfield Ant; Fire Ant; Little Black Ant; Odorous House Ant; Pavement Ant; Pharaoh Ant; Thief Ant.

LARGIDAE Amyot & Serville 1843. Plural Noun. Largid Bugs. A small Family of Hemiptera including about 15 Genera and placed in Superfamily Pyrrhocoroidea; Includes Subfamilies Larginae Hussey 1929 and Physopeltinae Hussey 1929. Largidae are related to Lygaeidae. Largid Species typically stout with four-segmented beak; Pronotum lateral margin not expanded; forewing membrane with 7–8 branching veins arising from two basal cells. Species are plant feeders on ground on low weeds; some Species are ant mimics.

LARKSPUR LEAFMINER *Phytomyza* spp. [Diptera: Agromyzidae].

LARRIMER, WALTER HARRISON (1889–1970) (Poos 1971, J. Econ. Ent. 64: 345.)

LARROUSE, FERNAND (1888–1937) (Brumpt 1937, Ann. Parasit. hum. comp. 15: 552–556, bibliogr.)

LARSEN, CARL C R (1846–1920) (Wolff 1953, Ent. Meddr. 26: 474–481, bibliogr.)

LARSON, ANDREW OLOF (1887–1948) (Campbell *et al.* 1949, J. Econ. Ent. 42: 165.)

LARTIGUE, HENRI (1830–1884) (Léveillé 1884, Ann. Soc. ent. Fr. (6) 4: 364–366.)

LARVA Noun. (Latin, *larva* = mask, ghost. Pl., Larvae.) 1. An immature stage of an holometabolous insect. 2. The stage of development following the egg stage, preceding the pupal stage and differing fundamentally from the adult. 3. In a strict zoological sense: The immature form of animals which undergo Metamorphosis (Imms). The term larva (*larva* = mask, *i.e.* adult identity concealed behind a larval form) is difficult to define because it has broad application in zoology. Larvae are found in Crustacea and Insecta, but the two are not ontogenetically equivalent. The insect larva hatches with definitive body segmentation; appendages and mouthparts are usually different in size, shape and structure from the adult. A compromise definition for larva: The immature, postembryonic stage which has acquired adaptive characteristics that an adult ancestor did not possess and the forms or expressions of which are not carried into the adult stage. 2. Acari: The active, immature, six-legged, parasitic stage between prelarva and protonymph (Goff *et al.* 1982, J. Med. Ent. 19: 226.) See Caterpillar; Grub; Maggot; Slug. Cf. Nymph; Naiad. Rel. Metamorphosis; Pupa.

LARVA ACULEATA A larva with a dense covering of fur-like Setae.

LARVA CORNUTA A larva with fleshy horn-like processes.

LARVA FURCIFERA A larva with furcate processes.

LARVA URSINA A setose caterpillar; a woolly bear.

LARVADEX® See Cyromazine.

LARVAKIL® See Diflubenzuron.

LARVAL COMBAT Overt aggression between larvae of one or more Species. Characteristic of some endoparasitic Hymenoptera which must compete for nutritional resources (the host) and typically results in the death of one or more combatants. Syn: Physical Combat. Cf. Physiological Suppression.

LARVAL HETEROMORPHOSIS Endopterygote larvae morphologically adapted to different forms for different lifestyles during successive instars (Snodgrass 1954). The phenomenon is characteristic of (but not restricted to) parasitic insects. See Hypermetamorphosis. See Development; Metamorphosis. Cf. Heteromorphosis. Rel. Planidium; Triungulin; Triungulinid.

LARVAL PELLICLE Coccids: The first cast skin or exuviae of the larva.

LARVAL PELT Acari: The larval exuvium (Goff *et al.* 1982, J. Med. Ent. 19: 226.)

LARVAL PROLEG See Proleg.

LARVAL THERAPY See Maggot Therapy.

LARVAPOD Noun. (Latin, *larva* = ghost; *pous* = foot. Pl., Larvapods.) A larval proleg (MacGillivray).

LARVARIUM Noun. (Latin, *larva* = ghost; *arium* = place of a thing.) A tube or case made by a larva as a shelter or retreat.

LARVATA Masked; an arcane term applied to coarctate and obtect Pupae. Alt. Larvated.

LARVATROL® A registered biopesticide derived from *Bacillus thuringiensis* var. *kurstaki*. See *Bacillus thuringiensis.*

LARVIFORM Adj. (Latin, *larva* = ghost; *forma* = shape.) Larva-shaped; pertaining to a body shaped as a larva; resembling a larva. A term of little value because the range of shapes and forms in larval insects is considerable. See Larva. Cf. Campodeiform; Cylindrical; Limaciform. Rel. Form 2; Shape 2; Outline Shape.

LARVIN® See Thiodicarb.

LARVINA Noun. (Latin, *larva* = mask. Pl., Larvinae.) A maggot; specifically, a dipterous larva without legs or a well defined head.

LARVIPAROUS Adj. (Latin, *larva* = ghost; *parere* = to produce; *-osus* = full of.) Reproduction in which living larvae emerge from the mother. A mode of reproduction manifest in some Diptera. See Reproduction. Cf. Oviparous; Ovoviparous; Puparous.

LARVIPOSITOR Noun. (Latin, *larva* = ghost; *parere* = to produce; *or* = one who does a specified thing. Pl., Larvipositors.) Larviparous Diptera: The modified Ovipositor which is adapted to deposit larvae but not eggs.

LARVO-BT® A registered biopesticide derived from *Bacillus thuringiensis* var. *kurstaki*. See *Bacillus thuringiensis*.

LARVULE Noun. (Latin, *larvula* = small larva. Pl., Larvules.) The early stage of ephemerid development when the naiad has no developed respiratory, circulatory or nervous systems.

LASER SCANNING CONFOCAL MICROSCOPY A technique for obtaining high-resolution images and three-dimension reconstruction of biological specimens. Light from a laser is focused by an objective lens on a fluorescent specimen; the reflected light is focused on a photodetector (photomultiplier) via a beam splitter. A confocal aperture (pinhole) is placed in front of the photodetector; light reflected from points on the specimen not within the focal plane of the laser is defocused and out-of-focus information is reduced. The spot focused on the centre of the pinhole is the confocal spot.

LASER® A registered biopesticide derived from *Bacillus thuringiensis* var. *kurstaki*. See *Bacillus thuringiensis*; See Cyfluthrin.

LASHED Adj. (Middle English, *laschen* = a whip.) Structure with a more-or-less complete fringe of stiff Setae or bristles along the margin of the compound eye.

LASIOCAMPIDAE Plural Noun. Lappet Moths; Tent Caterpillars. A widespread Family of ditrysian Lepidoptera (ca 1,500 Species, 150 Genera) assigned to Bombycoidea; lasiocampids do not occur naturally on New Zealand and Solomon Islands. Adult stout-bodied; sexual dimorphism pronounced with females larger than males; wingspan 2–170 mm; head with small piliform scales concealing Frons and base of mouthparts; Ocelli and Chaetosemata absent; eyes often setose; Antenna bipectinate to apex in male and female; Proboscis vestigial or absent; Maxillary Palpus absent; Labial Palpus porrect with Chaetosemata-like sense organ on base; Epiphysis in male, absent or reduced in female; tibial spurs very short, formula 0-2-2. Adult short lived (ca 1–2 weeks); male fast flying (some diurnal) female sluggish (nocturnal); female emerges with eggs available for oviposition; nocturnal oviposition. Egg flattened, oval in outline shape, Chorion patterned but not sculptured; micropyle on upper surface; eggs usually laid individually, sometimes in small clusters or rarely in large masses. Larva cylindrical or flattened, sometimes with urticating Setae or dorsal protuberance, verrucae secondary Setae, not with branched Setae; head hypognathous; Thorax often with lateral lobes (lappets); abdominal Prolegs on segments 3–6, 10. Larva solitary or gregarious, sometimes living in silken 'tent'. Pupae stout, rounded anteriorly; Epicranial Suture present; Maxillary Palpus absent; Labial Palpus exposed; Antenna short, tapering; Cremaster absent; cocoon stiff, white, parchment-like found in foliage of host plant. Family collectively polyphagous; many Species restricted to one host plant Family or Genus; most Species rare but a few Species of significant pest status (*Malacosoma* spp.) Subfamilies include Archaeopachinae, Chondrosteginae, Gastropachinae, Gonometinae, Lasiocampinae and Micromphaliinae.

LASKA, FRANTISEK (1904–1965) (Hrabe 1966, Zool. Listy 15: 91–92, bibliogr.)

LASPEYRES, JACOB HEINRICH German bureaucrat and author of *Sesiae Europeae iconibus et descriptionibus illustratae* (1801) and '*Critical Observations on the Systematic Catalogue of the Lepidoptera of the environs of Vienna*' (In: Illiger's Magazin). Anon. 1858, *Accentuated list of British Lepidoptera*. xliv + 118 pp. Oxford & Cambridge Entomological Societies, London.)

LASS, HERMANN (1859–1938) (Anon. 1938, Arb. morph. taxon. Ent. Berl. 5: 296.)

LASUREUS Adj. (Latin, *lazureus* = blue.) The dark blue colour of lapis lazuli.

LASZLOROC, SZALAY (1887–1970) (Allodiatoris 1973, Folia ent. hung. 26: 5–16, bibliogr.)

LATACORIA Noun. The Coria between a Sternum and a Tergum (MacGillivray).

LATADENTES Plural Noun. Coccids: The lobe-like projections which bound the indentations in the Lateris; the 'lateral teeth' of Marlatt *teste* MacGillivray.

LATADISCALOCA Noun. Coccids: The structure located on each of the mesial Discaloca (MacGillivray).

LATAGENACERORES Noun. Some coccids: The Pregenacerores and Postgenacerores combined to form a elongated group on each side of the Vulva (MacGillivray).

LATALLAE Noun. (Sl., Latalla.) Small chitinized areas or sclerites in the Latacoriae (MacGillivray).

LATANIA SCALE *Hemiberlesia lataniae* (Signoret) [Hemiptera: Diaspididae].

LATAPECTINAE Noun. Coccids: Pectinae with a pointed distal end providing a triangular outline with teeth arranged along both sides of the shaft

(MacGillivray).

LATARIMA Noun. The fissure separating the Glossa from the Paraglossa (MacGillivray).

LATASTE, FERNALD (1847–1934) (Porter 1934, Revta chil. Hist. nat. 38: 53–56.)

LATASUTURE Noun. A Latacoria when reduced to a suture (MacGillivray).

LATATERGUM Noun. The side of a Tergum folded into its lateral or ventral aspect in a distinct narrow longitudinal area (MacGillivray).

LATENT Adj. (Latin, *latens* = hidden.) Pertaining to structures or actions which are hidden or concealed.

LATENT INFECTION Infection of a host by a pathogen without the development of symptoms.

LATERAD Adv. (Latin, *latus* = side; *ad* = toward.) Toward the side and away from the median line. See Orientation. Cf. Anteriad; Apicad; Basad; Caudad; Centrad; Cephalad; Craniad; Dextrad; Dextrocaudad; Dextrocephalad; Distad; Dorsad; Ectad; Entad; Mediad; Mesad; Neurad; Orad; Proximad; Rostrad; Sinistrad; Sinistrocaudad; Sinistrocephalad; Ventrad.

LATERAL Adj. (Latin, *latus* = side; *-alis* = characterized by.) Descriptive of structure or movement relating to the side. Cf. Medial.

LATERAL ABDOMINAL GILL A type of Tracheal Gill which occurs in the naiad of some primitive calopterygid dragonflies and Ephemeroptera and the larvae of Megaloptera. LAG are filamentous lamellae which originate on the ventral surface of abdominal segments 2–8 and project laterally. Lamellae may be single or double; some lamellae may have basal brachial filaments. See Gill. Cf. Caudal Gills; Rectal Gill. Rel. Cutaneous Respiration.

LATERAL APODEME An invagination of the thoracic segments for the attachment of muscles; derived from the Pleurite (Needham). A process of the thoracic endoskeleton; an Endopleurite; Entopleuron (MacGillivray).

LATERAL AREAS Hymenoptera: The three spaces on the Metanotum, between the median and lateral long Carinae.

LATERAL BRISTLES Diptera: Bristles on or near the lateral margins of the abdominal segments (Comstock).

LATERAL CALLIS Some coccids: The Callis laterad of the Mesial Callis (MacGillivray).

LATERAL CARINAE Orthoptera: Cuticular ridges on the head extending downward from the front margin of the eyes; ridges on Prothorax extending along each lateral margin of the dorsum.

LATERAL CERARI Coccids: All the Cerari except the eighteenth pair or Anal Cerari (MacGillivray).

LATERAL FACIAL BRISTLES Diptera: 1–2 bristles sometimes present on the sides of the head below the eyes (Comstock).

LATERAL FILAMENTS Long, tapering appendages on the margins of the Abdomen in certain aquatic larvae (Comstock & Kellogg).

LATERAL FOVEA Hymenoptera: A pit on each side of the head near the Antennal Sockets and sometimes connected with them by a short furrow; these represent the Antennal Furrows between the vertical furrows and the Antennal Sockets. LF vary considerably in size, shape and location when not completely obsolete.

LATERAL FOVEOLAE Orthoptera: Foveate depressions on the margins of the Vertex near the front border of the eye.

LATERAL KEEL Odonata: A ridge on the side of the Abdomen (Garman). Coccids: The Lateral pilacerores, (MacGillivray).

LATERAL LINE Case-bearing trichopterous larvae: A delicate longitudinal cuticular fold beset by fine Setae on each side of the Abdomen (Imms). Caterpillars: A line at the margin of the dorsum between the Subdorsal and Suprestigmatal Lines (Smith).

LATERAL LOBE 1. Odonata: The part of the Labium which corresponds to the Paraglossa with the Palpiger and Palpus (Gerstaecker), or more probably, to the Palpus alone (Butler). An expansion borne by the Squama at the side (Garman); the deflexed part of the Pronotum which covers the sides of the Prothorax in many Orthoptera. 2. Hymenoptera: The parts on each side of the Parapsidal Furrows of the Mesoscutum, the Scapula (Smith).

LATERAL LONGITUDINAL AREA Hymenoptera: An area extending between the median and pleural Carinae of Metanotum.

LATERAL MUSCLES In insects, muscles which are typically dorsoventral and both intrasegmental and intersegmental (Snodgrass).

LATERAL NERVE CORDS The lateral strands of nerve tissue produced from the Ventral neuroblasts (Snodgrass).

LATERAL NOTCH Coccids: The notch on the lateral side of a lobe (MacGillivray).

LATERAL OCELLI The grouped, simple eyes at the sides of the head of holometabolous larvae. Variable in number and located and generally thought capable of detecting but not effective at resolving images. The Stemmata *sensu* Snodgrass. See Ocellus. Cf. Compound Eye.

LATERAL ORBAERORES Coccids: The outer and shorter row of Cerores of the anal ring (MacGillivray).

LATERAL OVIDUCT In most insects, Lateral Oviducts are a pair of tubes attached basally to the Calyx and distally to the Median Oviduct. LO derived from embryonic Mesoderm, but in some higher groups (*e.g.* Holometabola) LO are replaced by Ectodermal branches from the Median Oviduct. LO typically form simple tubes but in some acridid grasshoppers a gland may form at the anterior end. The cell wall of the Lateral Oviduct is surrounded by circular muscles, longitudinal muscles, or both. Syn. Oviductus Lateralis. Cf. Median Oviduct.

LATERAL PHARYNGEAL GLAND Bees: A pair of large glands at the sides of the pharynx (Imms).

LATERAL PILACERORES Coccids: The series of sclerites which extends around the margins of the body and forms the lateral sclerites of the test, lateral sclerites, circumferential lamellae, marginal sclerites (MacGilliray).

LATERAL PLATES Embryo: The lateral areas of the germ band after differentiation of the middle plate (Snodgrass), in coccids the lateral pilacerores (MacGillivray).

LATERAL RIDGE Slug caterpillars: A raised line along the lateral series of abdominal tubercles.

LATERAL SCALE One of the lateral processes of the Ovipositor in Cynipidae positioned within and below the anal scale.

LATERAL SETAE Odonata: Setae of the proximal segment of the Labial Palpi (Garman).

LATERAL SPACE Slug caterpillars: The area on each side of the body between the subdorsal and lateral ridges.

LATERAL SPINAE The Spinae one on each side of the Median Spina of the Stigmatic Clefts (MacGillivray).

LATERAL SPINES Odonata: Spines at the caudal end of the Lateral Keel (Garman).

LATERAL SUTURES Odonata: Sutures on the sides of Thorax.

LATERAL TRACHEAL TRUCK The usual longitudinal tracheal trunk on each side of the body closely connected with the lateral spiracles (Snodgrass).

LATERAL TUBERCLE Lepidoptera larva: A tubercle on the thoracic and abdominal segments. See Tubercle.

LATERICEOUS Adj. (Latin, *later* = brick; *-aceus* = of or pertaining to.) Brick-red. Alt. Latericius; Lateritious; Lateritius.

LATERIS Noun. (Pl., Lateres.) Coccids: The part of the lateral pygidial margin extending on each side, forming the cephalic margin of the Pygidium (MacGillivray).

LATEROCERVICALIA Noun. The lateral neck sclerites (Crampton).

LATERODORSAL Adj. (Latin, *latus* = side; + dorsal; *-alis* = pertaining to.) Descriptive of structure or action toward the side and back.

LATERO-OPITHOSIMA GLAND Paired integumental glands in the laterodorsal aspect of the Opithosoma.

LATEROPHARYNGEAL Adj. (Latin, *latus* = side; + pharyngeal.) Descriptive of something positioned on the side of the Pharynx.

LATEROPOSTNOTUM Noun. The Postalar Bridge of the thoracic Notum.

LATEROSTERNAL Adj. (Latin, *latus* = side; + sternal.) To the side of the Sternum. Descriptive of something positioned on a laterosternite.

LATEROSTERNITE Noun. (Latin, *latus* = side; + sternum; *-ites* = constituent. Pl., Laterosternites.) 1. The lateral part of a definitive thoracic Sternum, apparently derived from the Sternopleurite of the Subcoxa. 2. A small sclerite occurring in the pleural region of the Abdomen (actually a pleurite, in the sense that it lies in the pleural region) (Snodgrass). 3. Separate lateral sclerites of the sides of the Eusternum in Isoptera and Dermaptera (Imms). See Sternum. Cf. Laterotergite.

LATEROSTIGMATAL Adj. (Latin, *latus* = side; + stigma.) Descriptive of or pertaining to the side, immediately above the spiracle.

LATEROTERGAL Adj. (Latin, *latus* = side; + terga.) Descriptive of structure or process on the side of the Tergum.

LATEROTERGITE Noun. (Latin, *latus* = side; + terga; *-ites* = constituent. Pl., Laterotergites.) 1. A lateral sclerotization of the dorsum distinct from a principal median Tergum. The paratergite *sensu* Snodgrass. See Tergum. Cf. Laterosternite. 2. A sclerotized area of the Postnotum, divided into a dorsal anatergite and ventral katatergite. See Postnotum 2.

LATEROVENTRAL Adj. (Latin, *latus* = side; + ventral.) Descriptive of structure or pertaining to something which is toward the side and the venter. Cf. Dorsoventral. Rel. Orientation.

LATEROVENTRAL AMBULATORY APPENDAGES 1. Generalized or hypothetical appendages which reflect a groundplan of primitive legs. 2. Any evaginations in the lower animals which function as structures for walking.

LATEROVENTRAL METATHORACIC CARINA Odonata: The dividing line between the Metepimera and the Metasternum.

LATEROVERTED Adj. Displaced toward the side of the body; laterally displaced.

LATESCENT Adj. (Latin, *latescens*, present particple of *latescere* = to be concealed.) Becoming obscure or hidden.

LATEX Noun. (Latin, *latex* = milky.) A milky exudation or sap of numerous, taxonomically unrelated plants; consists of gum resins, waxes, oils and compounds toxic to animals. Produced by Asclepiadaceae, Apocynaceae, Euphorbiaceae, Moraceae, Sapotaceae. Primary commercial products of latex include rubber, gutta-percha and chicle. Cf. Catechu; Kino; Resin.

LATHRIDIIDAE Plural Noun. Plaster Beetles. Minute Brown Scavenger Beetles. A cosmopolitan Family of Coleoptera consisting of about 500 Species. Morphologically similar to Ciidae, distinguished by minute to small size (1–3 mm), tarsal formula 3-3-3, Coxal cavities closed posteriad, Antenna 8–11 segmented with a 2–3 segmented club, and Abdomen with 5–6 visible Sterna. Predominantly phytophagous; feeds on moulds and related material. Included Subfamilies: Lathridiinae and Corticariinae.

LATHROTELIDAE Clark 1971. A Family of ditrysian Lepidoptera including one Species taken from Rapa Island. Regarded as an aberrant Pyralidae by some workers.

LATHY, PERCY I (1870–1943) (Talbot 1943, Entomologist 76: 263–264.)

LATIERS, HENRY J H (–1929) (Cremers 1929, Naturh. Maandbl. 18: 53–54.)

LATIFOLIATE Adj. (Latin, *latus* = wide; *folium* = leaf; *-atus* = adjectival suffix.) Descriptive of plants with broad leaves. Cf. Angustifoliate.

LATIGASTRIC Adj. (Latin, *latus* = wide; *gaster* = stomach; *-ic* = characterized by.) Acarology: Having a broad juncture between the Prosoma and Opisthosoma as in the Chelicerata.

LATREILLE, PIERRE ANDRÉ (1762–1833) French zoologist and dominant figure in science during the first part of the 19th century. First entomological work entitled *Precis des characteres generiques des Insectes, disposes dans un order naturel* (1796). Later dominant work *Histoire generale et particuliere des Crustaces et des Insectes* (14 vols, 1802–1805). (*Genera Crustaceorum et Insectorum secundum ordinem naturalem in familias desposita, iconibus exemplisque plurimis explicata*. Parisiis, Argentoruat, A. Konig, Paris). Other noteworthy publications include *Considerations generales sur l'ordre naturel des animaux composant les classes des Crustaces des Arachnides, et des Insectes, avec un tableau methodique de leurs genres distribues en familles* (1810), and *Regne Animal distribue d'apres son organization & c.* (4 vols, 1817). (Percheron 1837, Bibliogr. Ent. 1: 225–234, bibliogr.)

LATREILLE'S SEGMENT Hymenoptera: The Propodeum. The first abdominal segment of Apocrita in which the segment has dissociated from the functional Abdomen (Gaster) and become fused with the Thorax. See Propodeum.

LATRINE FLIES See Fanniidae.

LATRINE FLY *Fannia scalaris* (Fabricius) [Diptera: Muscidae]: A cosmopolitan synanthropic, urban pest near dwellings; not common in tropics and absent from Arctic. Adult 3–6 mm long; resembles Lesser House Fly but larger with bluish-black Thorax with four brown stripes; legs black; wings hyaline; Halteres yellow. Adult diurnal; lives about 2–3 weeks. Larva with simple spines on lateral aspect of thoracic segments and forked spines on abdominal segments. Female prefers to oviposit in human excrement over excrement of other animals or decaying plant material. Vector of human diseases. See Muscidae. Cf. House Fly; Lesser House Fly; Coastal Fly.

LATRODECTISM Noun. A medical condition in humans caused by bites of spiders, *Latrodectus* spp. Latrodectism is common in warmer areas of the world, particularly in agricultural areas; problem persistent and may be locally epidemic in some seasons or under ideal conditions. Spider's bite generally not felt, but soon develops swelling and pain at site of bite; more severe cases result in general ache, rigidity, tonic spasms; intense pain in inguinial region, shock, fever, nausea, headache, sweating and difficulty in breathing; fatality rate ca 5%. See Black-Widow Spider; Brown-Widow Spider; Redback Spider. Cf. Loxoscelism.

LATTA and HARTMANN KNIFE A triangular-shaped glass knife used in ultramicrotomy and histology. See Ralph Knife. (Latta & Hartmann 1950, Proc. Soc. Exp. Biol. Med. 74: 436).

LATTER, OSWALD H (1864–1948) (Williams 1949, Proc. R. ent. Soc. Lond. (C) 13: 69.)

LATTICED Adj. (Middle English, *latis* = lath.) Cancellate. Marked to resemble lattice or crosswork.

LATTIN, GUSTAV DE (–1968) (Cleve 1968, Mitt. dt. ent. Ges. 27: 51.)

LATUS Noun. (Latin, *latus* = the side.) The side of the insect body. Broad.

LATUSCULA Noun. 1. The facets of the compound eye (Smith). 2. Diptera: The Notopleural or Dorsopleural Suture *sensu* MacGillivray.

LATZEL, ROBERT (–1956) (Anon. 1956, Z. Wein ent. Ges. 41: 184).

LAUBE, GUSTAV CARL (1839–1923) (Mobius 1943, Dt. ent. Z. Iris 57: 10)

LAUBERT, RICHARD (1870–1952) (Anon. 1953, Nachrbl. Dt. Pflanzen Braunsch. 5: 15–16.)

LAURACEAE A preferred host plant Family for papilionid butterflies, particularly the Papilionini.

LAURASIA Noun. Name proposed by South African geologist Alexander du Toit for northern landmass of supercontinent Pangaea. See Pangaea. Cf. Gondwanaland.

LAURENT, CHARLES (1821–1870) (Chevrolet 1871, Ann. Soc. ent. Fr. (5) 1: 119–121, bibliogr.)

LAURENT, PHILIP (–1942) (Anon. 1942, Ent. News 53: 227.)

LAUROP, CHRISTIAN PETER (1772–1858) (Ratzeburg 1874, Forstwissenschaftliches Schriftsteller-Lexicon 1: 291–295.)

LAUTERBORN, ROBERT (1869–1952) (Marker 1953, Verh. dt. zool. Ges. 1952: 572–575, bibliogr.)

LAUTNER, JULIUS (1897–1972) (Allenspach 1972, Mitt. schweiz. ent. Ges. 45: 230.)

LAUXANIIDAE Plural Noun. A numerically fairly large Family of cyclorrhaphous Diptera assigned to Superfamily Lauxanioidea. Small, robust flies that vary in colour; some Species display patterned wings. Lauxaniids are distinguished from other acalypterates by an complete Subcosta, absence of Oral Vibrissae, converging Postvertical Bristles and Preapical Tibial Bristles. Larvae inhabit decaying vegetation; adults inhabit moist, shady places. Syn. Sapromyzidae.

LAVALLETE, M A (–1954) (Anon. 1954, Bull. Soc. ent. Fr. 59: 33.)

LAVERAN, ALPHONSE (1845–1922) Pioneer in Medical Entomology who discovered (1880) the human malarial parasite *Plasmodium malariae* in the red blood cells. (Dorald 1936, Clinical med. Surg. 43: 511–572.)

LAVERNINDAE See Momphidae.

LAVERS, CHARLES HENRY (–1941). (Lavers 1945, Trans. Am. microsc. Soc. 64: 228.)

LAW Noun. (Middle English, *lawe*. Pl., Laws.) 1. A succinct statement in words or equation that express an order or relationship of phenomena that has no exception. *e.g.* Second Law of Thermodynamics. Although many phenomena in Biology carry the word 'Law,' these terms are inappropriate as the principles they serve to characterized have not been rigorously tested or have exceptions. *e.g.* Dolo's Law. Syn. Rule. Cf. Principle. Rel. Hypothesis; Theory.

LAW OF DISHARMONY A mathematical principle in which the logarithm of the dimension of the part is proportional to the logarithm of the whole, or y = Kx.

LAW OF PRIORITY A principle in taxonomic nomenclature which provides that the valid name of a Genus or Species can be only that name under which it was first designated, under certain conditions prescribed in the International Code of Zoological Nomenclature.

LAW OF THE UNSPECIALIZED An observation which asserts that major successful lineages are derived from generalized ancestors with small body size (Cope 1896).

LAWN ARMYWORM *Spodoptera mauritia* (Boisduval) [Lepidoptera: Noctuidae]: A serious pest of turf, lawn and pasture in eastern Australia and Hawaii. Adult wingspan 30–45 mm; forewing grey or grey-brown with wavy lines and dark mark near centre; hindwing grey with dark margin. Female lays ca 700 eggs; egg greenish-pale brown, slightly flattened, ca 0.5 mm diam; laid in compact mass of 50–200 with mass covered by long Setae from body of female; oviposition sites seemingly random; eclosion within 4–10 days. Neonate larvae remain together, disperse within hours; larva prefers grass; early instars feed during day, later instars feed at night; neonate larva skeletonizes leaf, later instars eat holes in upper leaves, older larvae feed from edge of leaves; when disturbed larva drops to ground, curls in a ball (head inward) and remains motionless for several minutes; larval development ca 27–30 days with 7–8 instars. Pupation within cell constructed in soil; pupal stage 7–14 days. Generation time 42 days at 25°C. See Armyworms; Noctuidae.

LAWN FLY *Hydrellia tritici* Coquillett [Diptera: Ephydridae].

LAWN LEAFHOPPER *Deltocephalus hospes* Kirkaldy [Hemiptera: Cicadellidae].

LAWSON, J A (–1948) (Williams 1949, Proc. R. ent. Soc. Lond. (C) 13: 68.)

LAWSON, JOHN (–1712) (Weiss 1936, *Pioneer Century of American Entomology*. 320 pp. (17–18), New Brunswick.)

LAWSON, PAUL BOWEN (1888–1954) (Hungerford 1954, J. Kans. ent. Soc. 27: 81, bibliogr.)

LAYARD, E LEOPOLD (1824–1900?) (Anon. 1900, Zoologist 4: 47–48.)

LAZEAR, JESSE WILLIAM (1866–1900) (Thayer 1900, Johns Hopkins Hosp. Bull. 11: 290–291.)

LC_{50} The concentration of a toxicant in some medium that kills 50% of the test organisms exposed; generally expressed as milligrams per kilogram of body weight. See LD_{50}.

LD_{50} The dose of a toxicant that will kill 50% of the test organisms to which it is administered; generally expressed as milligrams of toxicant per kilogram of body weight. See LC_{50}.

LE CONTE, JOHN EATTON (1784–1860) (Anon. 1861, Am. J. Sci. (2) 31: 303, 462, 463.)

LE GRICE, F (–1902) (Anon. 1902, Entomologist's Rec. J. Var. 14: 109.)

LE ROI, OTTO (–1916) (Hedicke 1917, Dt. ent. Z. 1917: 326.)

LEA, ARTHUR MILLS (1868–1932) Australian entomologist who served in several government positions, including government entomologist in Tasmania (1899–1911). Primary entomological interest was with Coleoptera. (Hale 1932, Rec. So. Austral. Mus. 4: 411–432, bibliogr.)

LEA, H A F (1907–1989) South African entomologist and world authority on locust control. (Brown 1991, Antenna 15: 61–66.)

LEA, ISAAC (1792–1886) (Anon. 1887, Leopoldina 23: 53–54.)

LEACH, EDWIN RALPH (1878–1971) (Arnaud 1971, Pan-Pac. Ent. 47: 312.)

LEACH, FRANK A (–1929) (Davis 1931, Ann. ent. Soc. Am. 24: 187.)

LEACH, JOHN ARTHUR (1870–1929) (Croll *et al.* 1930, Emu 29: 230–233.)

LEACH, WILLIAM ELFORD (1790–1836) English physician and naturalist whose entomological works were published in Shaw's Zoological Miscellany (3 volumes) and the Linnean Transactions. (Hope 1837, Bull. Soc. ent. Fr. 6: xxxiv–xxxv.)

LEAD CITRATE A heavy-metal solution (salt) used in conjunction with uranyl acetate as a post-fixative stain for transmission electron miscroscopy. These stains provide increased contrast of biological tissues, thus added clarity and visibility, due to their affinity for a variety of tissue types.

LEAD-CABLE BORER *Scobicia declivis* (LeConte) [Coleoptera: Bostrichidae].

LEADER, BENJAMIN JOHN (1914–1942) (N. A. W. 1943, Entomologist 76: 88.)

LEAF Noun. (Middle English, *leef,* Old Norse, *lauf* = leaf. Pl., Leaves.) One of the lateral outgrowths of a stem and plant part which constitutes a part of the foliage and which is involved in photosynthesis. Leaves provide the principal or exclusive type of food for many Species of insects. See Feeding Strategy. Leaf outline-shape terms have been extensively used in taxonomic Entomology to describe anatomical structural shapes of insects. See Outline Shape. Cf. Bark; Root; Wood.

LEAF BAGWORM *Hyalarcta huebneri* (Westwood) [Lepidoptera: Psychidae] (Australia).

LEAF BEETLES See Chrysomelidae.

LEAF-BLISTER SAWFLY *Phylacteophaga froggatti* Riek [Hymenoptera: Pergidae]: A sawfly endemic to Australia that causes serious damage to *Eucalyptus* spp. Larvae mine leaves, consume leaf tissue under the surface layer and induce large silver-white blisters. See Pergidae.

LEAF BLOTCH MINERS See Gracillariidae.

LEAF BUGS See Miridae.

LEAF CASE MOTH *Hyalarcta heubneri* (Westwood) [Lepidoptera: Psychidae]: A minor pest of citrus in Australia.

LEAF CRUMPLER *Acrobasis indigenella* (Zeller) [Lepidoptera: Pyralidae]: A monovoltine pest of apple, cherry, pear, plum and quince in Ontario to Upper Mississippi Valley in North America. LC overwinters as a larva within a case; during spring, case is loosened and larva moves to new buds and binds several leaves together with silk; pupation occurs during early summer. Eggs are laid on new leaves and eclosion occurs within 15–20 days. Larvae construct portable, tough, horn-like cases of silk and debris, they feed on shoots and leaves; cases are attached to twigs during fall. Syn. *Mineola indigenella* (Zeller). See Pyralidae.

LEAF-CUTTING ANTS Ants of the Genera *Acromyrmex* and *Atta*. See Formicidae. Cf. Texas Leaf-Cutting Ant.

LEAF-CUTTING BEES See Megachilidae.

LEAF FEEDING A specialized type of herbivory that is widespread among many groups of insects. Leaf-feeding insects display well developed Mandibles of generalized form (*e.g.* Orthoptera, Dermaptera). Many Coleoptera are leaf-feeding, including the Chrysomelidae. Leaf toughness often affects Mandible wear and tough leaves erode the cutting surface of Mandibles more than tender leaves. Insects with eroded Mandibles feed slower (less efficiently) than insects with less Mandible wear. Leaf toughness may influence feeding patterns in herbivorous insects. See Feeding Strategy. Rel. Ecology.

LEAF-FOOTED BUG *Leptoglossus phyllopus* (Linnaeus) [Hemiptera: Coreidae].

LEAF-FOOTED PINE SEED BUG *Leptoglossus corculus* (Say) [Hemiptera: Coreidae].

LEAF HOPPERS See Cicadellidae.

LEAF INSECTS See Phasmatodea; Phasmida; Phyllidae.

LEAF ROLL A viral disease of potatoes which is transmitted by aphids.

LEAF ROLLERS See Tortricidae.

LEAF-ROLLING CRICKETS See Gryllacrididae.

LEAF-ROLLING SAWFLIES See Pamphiliidae.

LEAF-ROLLING WEEVILS See Attelabidae.

LEAF-SKELETONIZER MOTHS See Zygaenidae.

LEAF TIERS See Tortricidae.

LEAFCURL PLUM APHID *Brachycaudus helichrysi* (Kaltenbach) [Hemiptera: Aphididae].

LEAFCURLING SPIDER *Phonognatha graeffei* (Keyserling) [Araneida: Araneidae].

LEAFCUTTER MOTHS See Incurvariidae.

LEAFFOLDER *Cnaphalocrocis medinalis* (Guenée) [Lepidoptera: Pyralidae].

LEAFHOPPER ASSASSIN-BUG *Zelus renardii* Kolenati [Hemiptera: Reduviidae].

LEAFHOPPERS See Cicadellidae.

LEAFMINER Noun. An insect (typically the larva stage) that uses its mouthparts to create a tunnel between the upper and lower surfaces of a leaf. See Feeding Strategies. Cf. Defoliator. Rel. Agromyzidae.

LEAFMINERS See Agromyzidae; Nepticulidae.

LEAFMINER FLIES See Agromyzidae.

LEAFMINER MOTHS See Cosmopterygidae.

LEAFROLLER MOTHS See Tortricidae.

LEAN, O B (–1973) (Anon. 1973, Locust News. F. A. O. 25: 14–15.)

LEARNING Noun. (Middle English, *lernen* = to learn.) A persistent change in insect behaviour caused by experience of association (conditioning), imprinting or habituation. See Associative Learning; Imprinting. Cf. Fatigue; Facilitation; Habituation.

LEATHER BEETLES See Dermestidae.

LEATHERJACKET See Tipulidae.

LEBANESE AMBER Amber deposits from the Albian to Aptian, Lower Cretaceous Period (120 MYBP). LA contains some of the oldest insect inclusions preserved in Amber. See Amber; Fossil. Cf. Baltic Amber; Burmese Amber; Canadian Amber; Chiapas Amber; Dominican Amber; Taimyrian Amber.

LEBARON, WILLIAM (1814–1876) (Goding 1885, Entomologica am. 1: 122–125, bibliogr.)

LEBAYCID® See Fenthion.

LEBBECK MEALYBUG *Nipaecoccus viridis* (Maskell) [Hemiptera: Pseudococcidae].

LEBEDEV, ALEXANDER G (1874–1936) (Anon. 1936, Arb. morph. taxon. Ent. Berl. 3: 151.)

LEBERT, HERMANN (1813–1878) (Anon. 1878, Zool. Anz. 1: 228.)

LEBOEUF, CHARLES (1859?-1884) (Anon. 1884, Feuille jean. Nat. 14: 103.)

LECERF, FERDINAND (1881–1945) (Fletcher 1945, Entomologist's Rec. J. Var. 57: 76.)

LECITHAL Adj. (Greek, *lekithos* = egg yolk; *-alis* = characterized by.) Descriptive of eggs that contain large amounts of yolk. Cf. Alecithal.

LECITHOCERIDAE Plural Noun. A Family of ditryisan Lepidoptera assigned to Superfamily Gelechioidea. Old World distribution and most abundant in tropical and subtropical areas. Lecithocerids are regarded as Subfamily of Gelechiidae in some classifications. Lecithocerids are differentiated from Gelechiidae on basis of wing-locking Setae on fore and hindwings of both sexes, musculature of male genitalia, upright egg and larval chaetotaxy. Adults typically are nocturnal; a few brightly coloured Species apparently are diurnal. Moths rest with head directed downward, Antennae extended forward and wings folded flat above Abdomen. Biology of lecithocerids is poorly known. The larvae apparently feed upon dead plant material; pupation occurs in silken cocoons covered with detritus. Syn. Timyridae. See Gelechioidea.

LECLERC, EUGENE ALEXANDRE (–1868) (Liegard 1868, Bull. Soc. Linn. Normandie (2) 3: 285–289.)

LECLERC, GEORGES LOUIS (1707–1788) See Buffon, Georges Louis Le Clerc.

LECONTE, JOHN LAWRENCE (1825–1883) American Coleopterist. (Horn 1883, Science 2: 783–786.)

LECONTVIRUS® See Redheaded Pine Sawfly NPV.

LECTOTYPE Noun. (Greek, *lektos* = chosen; *typos* = pattern. Pl., Lectotypes.) In the case of a nominal Species that did not have a Holotype designated at the time of its description: A specimen selected as the name bearer of a nominal Species from a series of Syntypes. See Type. Cf. Cotype; Holotype; Neotype; Syntype. Rel. Nomenclature; Taxonomy.

LEDER, HANS (1843–1921) (Hwtschko 1922, Wien ent. Ztg. 39: 95–96, bibliogr.)

LEDERER, JULIUS (1821–1870) (Anon. 1871, Stettin. ent. Ztg 32: 179–183.)

LEE, ARTHUR BOLLES (1849–1927) (Anon. 1927, Nature 119: 432.)

LEECH, DANIEL HERBERT (1878–1941) (Leech 1947, Proc. ent. Soc. Br. Columb. 44: 36–38.)

LEECH, JOHN HENRY (1862–1900) (Anon. 1901, Entomologist 34: 33–38, bibliogr.)

LEED, HENRY ATTFIELD (1873–1958) (Q. 1959, Entomologist 92: 22.)

LEEFMANS, S (1884–1954) (Kalshoven 1954, Indones. J. nat. Sci. 110: 129–142, bibliogr.)

LEEK MOTH *Acrolepiopsis assectella* (Zeller) [Lepidoptera: Acrolepiidae].

LEES, FRANK HENRY (1883–1973) (Worms 1974, Entomologist's Rec. J. Var. 86: 171–173.)

LEESBURG, ANTONIUS FRANCISCUS ADOLPHUS (1848–1906) (Everts 1907, Tijdschr. Ent. 50: 117–120.)

LEEUWENHOEK, ANTON VAN (1632–1723) Dutch businessman, microscopist and naturalist. Leeuwenhoek is generally credited with invention of optical microscope. His improvements to lenses enabled magnifications ca 270 diameters and permitted descriptions of minute anatomical detail. Leeuwenhoek provided the first descriptions of Spermatozoa; he dissected and described the compound eye of beetles and postulated the mosiac appearance of images; he reported parthenogenesis and ovoviviparity in aphids. Leeuwenhoek's entomological observations were important but not organized into discrete publications. Instead, his scientific observations were frequently published in the form of letters to scientists in Europe and summarized in *Arvana Naturae Detectae Ope Microscopiorum* (4 vols, 1695–1722). (Miall 1912, *Early Naturalists, Their Lives and Work.* 396 pp. (200–223), London.)

LEFEBURE DE CERISY, LOUIS CHARLES (1789–1864) (Caieu 1967, Mem. Soc. Emul. Abbeville 10: 699–704.)

LEFEBVRE, ALEXANDER LOUIS (1798–1867) Musgrave 1932, *A Bibliogr. of Australian Entomology 1775–1930.* 380 pp. (197), Sydney.)

LEFEVRE, EDOUARD (1839–1894) (Fairmaire 1895, Ann. Soc. ent. Fr. 64: 121–126, bibliogr.)

LEFROY, HAROLD MAXWELL (1877–1925) (Anon. 1925, Entomologist 58: 279–280.)

LEFT-HANDED POND SNAIL *Physa acuta* (Daprarnaud) [Basommatophora: Physidae].

LEG Noun. (Middle English, *legge* = leg. Pl., Legs.) Insects: Cuticular appendages that project from the ventrolateral portion of Thorax. Six legs are present in nymphal Paurometabola and adult insects; leg number varies in larvae, depending upon taxonomic group. Legs are nearly always present in adult Pterygota; some nymphal sternorrhynchous Homoptera and some larval Holometabola are legless; a legless condition is called 'apodous.' McNamara & Trewan (1993) report that a Silurian fossil suggests that insects arose from an extinct lineage of euthycarcinoid arthropods by paedomorphic loss of legs. This interpretation is highly speculative but points to alternative explanations regarding the origin of insects and their appendages. Traditional viewpoint suggests that primitive (groundplan) thoracic leg contained six segments; Kukalová-Peck (1987) suggests that groundplan leg contained at least 11 segments. Presumed Homologies of segments between legs of insects and other arthropods must account for additional subdivisions. Basal portion of groundplan leg called a Coxopodite; distal portion of groundplan leg called a Telopodite. Groundplan leg also possessed lobes; some fossils suggest that lobes were sometimes annulated. Lobes are named by their relative position: Medial processes called Endites or Endite Lobes; lateral processes called Exites or Exite Lobes. Lobes probably were used as tactile devices, for respiration and/or swimming aids in aquatic Species. A leg is divided into component parts, including Coxa, Trochanter, Femur, Tibia, Tarsus and Pretarsus. A leg is primarily an appendage for locomotion, but other adaptions have evolved, including prey capture, courtship and excavation. Legs resemble Maxillary and Labial Palps in general features such as position (ventrolateral), segmentation, intrinsic musculature and shape; resemblance of legs and Mandibles is less obvious because Mandibles have become highly modified. Geological record shows oldest fossil terrestrial hexapods date to Lower Devonian and had well developed, segmented legs. Most primitive extant hexapods include Archaeognatha and Thysanura; examining these Taxa may provide insight into changes in the leg. The micro-coryphian (bristletail) *Machilis* sp. has three pairs of undifferentiated legs. These animals jump with aid of abdominal musculature and not modifications of the leg. See Coxa; Femur; Gonytheca;

Pretarsus; Tarsus; Tibia and Trochanter. Rel.

LEG ARTICULATION Each segment of the insect leg is relatively strongly sclerotized and separated by an Articular Corium or membrane. The sclerotized segment provides points for muscle attachment and strength for the structure. The membrane between segments provides flexibility to the entire leg and provides points of articulation that define planes of motion for the leg segments. Leg articulations are Monocondylic, Dicondylic or Tricondylic. See Leg; Joint. Cf. Dicondylic Articulation; Monocondylic Articulation; Tricondylic Articulation. Rel. Segmentation.

LEG FUNCTION The primary function of the leg involves locomotion (walking, running, jumping). Numerous secondary functions have evolved. See Cursiorial; Fossorial; Gressorial; Natatorial; Raptorial; Saltatorial; Scansorial.

LEG HEART Circulatory organs found in Odonata, Hemiptera and Homoptera. LHs are formed from specialized skeletal muscles located in the upper part of the Tibia and below the femurotibial articulation. See Accessory Circulatory Organ. Cf. Antennal Circulatory Organ; Cercal Heart: Wing Heart.

LEG MUSCULATURE The insect leg contains many muscles (*e.g.* leg of cockroach *Periplaneta americana* contains 368 postcoxal muscles). Musculature of the generalized leg is classified as intrinsic when it originates within a Telopodite or the segment served. Intrinsic musculature of the Telopodite consists of two muscles per segment. One muscle of the pair is called the elevator and is responsible for raising the segment. The other muscle of the pair is called the depressor and is responsible for lowering the segment. Apodemes provide structural support for appendages during movement. Leg musculature is called extrinsic when a muscle originates in one area (segment) and inserts in a different region. Extrinsic leg musculature originates on body wall and inserts on a Coxopodite or Telopodite. Six extrinsic muscle groups are recognized. Tergal promotors are inserted on a Trochantin or Coxa and move Coxa forward. Tergal remotors move Coxa rearward. Sternal promotors move Coxa in anterior rotation. Sternal remotors move Coxa in posterior rotation. Sternal adductors move Coxa toward the midventral line. Pleural abductors move a Coxa away from the body.

LEGIEST, LEON (–1946) (Lambillionea 1946, 46: 117.)

LEGION Noun. (Latin, *legio* = legion. Pl., Legions.) A taxonomic grouping of Genera. A Legion is used in the classification of some animals, but is not widely employed.

LEGIONARY ANT See Army Ant.

LEGNUM Noun. (Unknown origin.) The margin of a Squama.

LEGULA Noun. (Pl., Legulae.) Hymenoptera: The distal portion of Gonapophysis IX when the distal Rachis is separated from the gonopod.

LEGUME WEBSPINNER *Lamprosema abstitalis* (Walker) [Lepidoptera: Pyralidae].

LEHMANN, AUGUST (1802–1868) (Ratzeburg 1874, Forstwissenschaftliches Schriftsteller-Lexicon 1: 299–301.)

LEHMANN, JOHANN GEORG (1792–1860) (Weidner 1967, Abh. Verh. naturw. Ver. Hamburg Suppl. 9: 97–101.)

LEHMANN, MARTIN CHRISTIAN GOTTLIEB (1775–1856) German-Dutch sensory physiologist and author of *De antennis insectorum* (1799, 1800) which could not conclude the function of the insect Antenna. (Henriksen 1926, Ent. Meddr. 15: 218–220.)

LEICHARDT, FRIEDRICH WILHELM LUDWIG (1813–1848) (Musgrave 1930, Aust. Zool. 6: 197.)

LEICHHARDT'S GRASSHOPPER *Petasida ephippigera* White [Orthoptera: Pyrgomorphidae].

LEIDY, JOSEPH (1823–1891) (Chapman 1892, Proc. Acad. nat. Sci. Philad. 1891: 342–388, bibliogr.)

LEIODIDAE Plural Noun. Moderately small Coleoptera assigned to Staphylinoidea. Body 1–5 mm long; adult Antenna with 10–11 segments including club of 3–5 segments; segments 7, 9, 10 with internal sensory vesicles; legs often spinose; tarsal formula 5-5-5 to 3-3-3; Abdomen concealed by Elytra. Larva campodeiform; head prognathous; Urogomphi articulated at base. Species common in decaying vegetation, carrion, associated with fungi; some scavengers, some myrmecophiles; some cavernicolous. Syn. Anisotomidae, Catopidae, Cholevidae.

LEISHMANIA Noun. A Genus of protozoan parasites causing Leishmaniasis in humans. Parasite assumes distinctive forms at different parts of life cycle. Amastigote form lacks Flagellum, with flagellar base and Kinetoplast anterior of the nucleus; Promastigote Form with Flagellum at anterior end of body, flagellar base and Kinetoplast anterior of nucleus; Opisthomastigote Form with Flagellum base posterior of nucleus and with long flagellar pocket projecting to anterior end of cell; Epimastigote Form with flagellar base anterior of nucleus and Flagellum emerging laterally to form an undulating membrane toward anterior end of cell; Trypomastigote Form with flagellar base posterior of nucleus from which Flagellum emerges laterally and forms long, undulating membrane. See Leishmaniasis. Cf. Trypanosomiasis.

LEISHMANIASIS Noun. (After Gen. Sir William Boog Leishman [1865–1926]; *-iasis* = suffix for names of diseases.) A disease of humans caused by parasitic Protozoa of Genus *Leishmania*. Disease manifest in several clinical forms and is vectored by sandflies *Phlebotomus* spp. in Old World and *Lutzomyia* spp. in New World. Old World manifestation of disease involves reservoir hosts (zoonotic); New World manifestation of disease does not involve reservoir hosts (anthroponotic). Inside vector, *Leishmania* extracellular and motile; inside humans, *Leishmania* amastigote form resident within reticuloendothelial cells. See Cutaneous Leishmaniasis; Mucocutaneous Leishmaniasis; Visceral Leishmaniasis. Cf. Trypanosomiasis.

LEISHMANN, WILLIAM BOOG (1865–1926) (C. J. M. 1928, Proc. R. Soc. London (B) 102: ix-xviii.)

LEITGEB, BUHERT (1835–1888) (Wettstein 1901, Bot. Zool. Oesterr. 1850–1900. Festschr. p 200.)

LEIVERS, ABRAHAM (1874–1953) (Richards 1959, Proc. R. ent. Soc. Lond. (C) 23: 73.)

LEK Noun. (Swedish, *lek* = game. Pl., Leks.) 1. An aggregation of males at a site where conspecific females occur. 2. A site in which males compete for females. 3. The area in which lekking behaviour occurs.

LEKKING Noun. (Swedish, *lek* = game.) An overt courtship display.

LELONG, B M (1858–1901) (Anon. 1901, Pacif. rural press 61: 290.)

LEMANN, FREDERICK C (–1908) (Anon. 1908, Entomologist's Rec. J. Var. 20: 96.)

LEMARCHAND, S (1876–1953) (Anon. 1953, Lepid. News. 7: 166.)

LE-MAT® See Isofenphos.

LEMBERK, FRANK (1894–1942) (Silhavy 1946, Sborn. Pfir. Klubu Trebici 4: 22–24.)

LEMBERT, JOHN B (1840–1896) (Bethune 1896, Can. Ent. 28: 217–218.)

LEMERY, NICOLAS (1645–1715) (Rose 1850, *New General Biographical Dictionary* 9: 234.)

LEMMER, FREDERICK (1876–1941) (Engelhardt 1942, Bull Brooklyn ent. Soc. 37: 4–5.)

LEMMON, HELEN LEE (1916–1967) (Pyle 1968, J. Lepid. Soc. 22: 196.)

LEMNISCATE Adj. (Greek, *lemniskos* = ribbon; *-atus* = adjectival suffix.) Ribbon-like; pertaining to structure in the form of an 8.

LEMOINE, VICTOR (–1897) (Grouvelle 1897, Bull. Soc. ent. Fr. 1897: 129.)

LEMON BUD MOTH *Prays parilis* Turner [Lepidoptera: Yponomeutidae]: A minor and sporadic pest of citrus in the Philippines, southeast Asia and Australia. Adult small (ca 6–7 mm long), greyish-brown. Eggs hemispherical and laid on citrus buds; up to 20 eggs laid per bud; upon eclosion larva bores into bud; only one larva completes development in a bud. Larva yellow in early instars becoming green and reddish-brown in late instars; mature larva forms lace-like cocoon on curled leaf or flower. Life cycle 3–4 weeks Multivoltine with 6–7 generations per year. See Yponomeutidae.

LEMON BUTTERFLY *Papilio demoleus* Linnaeus [Lepidoptera: Papilionidae].

LEMON MIGRANT *Catopsilia pomona pomona* (Fabricius) [Lepidoptera: Pieridae].

LEMONIIDAE Plural Noun. A small Family of ditrysian Lepidoptera assigned to Superfamily Bombycoidea.

LEMORD, EUGENE (1841–1892) (Anon. 1892, Leopoldina 28: 156.)

LEMURIA REALM The faunal region restricted to the Island of Madagascar.

LENATOP® See Etofenprox.

LENG, CHARLES WILLIAM (1859–1941) (Davis 1941, J. N. Y. ent. Soc. 49: 189–192.)

LENGERKEN, HANNS VON (1889–1966) (Hüsing 1959, Mitt. dt. ent. Ges. 18: 54–55.)

LENGERSDORF, FRANZ (1880–1965) (Mannheims 1965, Mitt. schweiz. ent. Ges. 24: 41–43.)

LENITIC Adj. See Lentic.

LENKO, KAROL (1914–1975) (Reichardt 1975, Studia ent. 18: 619–622, bibliogr.)

LENNOM, WILLIAM (1818–1899) (Service 1900, Ann. Scot. nat. Hist. 1900: 134–136.)

LENS Noun. (Latin, *lentil* = lens. Pl., Lenses.) A transparent, cuticular structure that forms the outer (distalmost) component of an Ommatidium. Lens is usually colourless and contains a special type of Cuticle; Lens of some male whiteflies (*Aleyrodes*) contains yellow pigment. A Lens focuses rays of light on the Rhabdom. Lens typically appears dome shaped when viewed externally; in cross section, Lens is usually biconvex; sometimes convex-concave and occasionally plano-convex. Convexity of external surface influences amount of light captured; a highly convex Lens captures more light than a flat Lens. Lens surface typically is smooth or sometimes displays projections called tubercles. A smooth Lens occurs in toadbug *Gelastocoris occulatus*. Tubercles also called a Corneal Nipple Array. Lens size is correlated with behaviour in beetles: Diurnal beetles posses relatively small Facets and nocturnal beetles have relatively large Facets. Light is captured over Lens surface. Light rays travelling through Lens move at slightly different speeds because they impinge on surface at different points and travel different distances through the Lens. The substance of a Lens contains different radial gradients of refractive index. These are low at the periphery of the Lens and high at its core. The area beneath Lens is transparent, birefringent and without pore canals; it reflects middle and high UV (high energy damages photoreceptors). Ultrastructure shows Lens is lamellate or helix-like. Syn. Cornea. See Compound Eye. Cf. Cone; Corneal Nipple Array; Pigment Cells; Retina; Rhabdome.

LENTIC Adj. (Latin, *lentus* = slow; *-ic* = characterized by.) Pertaining to standing water, such as lakes, ponds and swamps. An ecological term applied to those organisms whose normal habitat is streams (Needham). Alt. Lenitic. Cf. Lotic.

LENTICEL Noun. (French, *lenticelle*, dim. of *lenticule* = lens-shaped. Pl., Lenticels.) A minute hole in a fruit or stem which permits gas exchange. Cf. Stoma.

LENTICULAR Adj. (Latin, *lenticularis* = lentil.) 1. Lens-shaped, or more specifically resembling a double-convex Lens. 2. Descriptive of structure in the size or shape of a lentil. Alt. Lenticulatus.

Syn. Lentiform.

LENTICULATE Adj. (Latin, *lenticule* = lentil; *-atus* = adjectival suffix.) Pertaining to an object whose sides taper and meet in a sharp point.

LENTIFORM Adj. (Latin, *lens* = lentil; *forma* = shape.) Lentil-shaped; descriptive of structure which is bi-convex. Syn. Lenticular. See Lentil. Cf. Amygdaliform; Fusiform. Rel. Form 2; Shape 2; Outline Shape.

LENTIGEN LAYER The corneagen layer.

LENTIL Noun. (Latin, *lenticula* = diminutive of lens. Pl., Lentils.) 1. The flattened seeds of the lentil plant, *Lens culinaris*. 2. A thin geological stratum enclosed and surrounded by strata of different composition. Arcane: Lentile.

LENTREK® See Chlorpyrifos.

LENZ, HARALD OTHMAR (1799–1870) (Anon. 1870, Zool. Garten 11: 100.)

LEO AFRICANUS, JOHN (–1526) (Rose 1850, *New General Biographical Dictionary* 9: 246.)

LEONARD, MORTIMER DEMAREST (1890–1975) American economic entomologist. (Russell 1975, Proc. ent. Soc. Wash. 505–507.)

LEONARD, ROBERT EDWARD (?1897–1972) (Conney 1976, Ann. Rep. Proc. Lancs. Cheshire ent. Soc. 95–96: 90–91.)

LEONARDI, GUSTAVO (1869–1918) (Silvestri 1918, Boll. Lab. Zool. gen agr. R. Scuola Agric. Portici 11: 291–298.)

LEONHARD, OTTO (1853–1929) (Horn 1929, Ent. Bl. Biol. Syst. Käfer 25: 113–118.)

LEONI, GUISEPPE (1866–1928) (Grandi 1929, Memorie Soc. ent. ital. 8: 5–7, bibliogr.)

LEOPARD *Phalanta phalantha araca* (Waterhouse & Lyell) [Lepidoptera: Nymphalidae]. (Australia).

LEOPARD MOTH *Zeuzera pyrina* (Linnaeus) [Lepidoptera: Cossidae]: A moth endemic to Europe and adventive to North America ca 1879. LM is a pest of deciduous and shade trees in northeastern USA. Eggs oval, pinkish and deposited in bark fissures; eclosion within ca 10 days. LM larvae bore into heartwood of crown of trees; overwinters as larva and requires 2–3 years to complete development. Pupation occurs in larval tunnels. See Cossidae. Cf. Carpenter Worm.

LEOPARD MOTHS See Cossidae.

LEOTICHIIDAE Plural Noun. A small, Old World Family of aquatic Hemiptera (Heteroptera) assigned to Superfamily Leptopodoidea. Related to Saldidae and sometimes placed within Leptopodidae.

LEPECID® See Chlorpyrifos.

LEPELETIER DE ST FARGEAU, AMÉDEE LOUIS MICHEL (1770–1845) (Serville 1846, Ann. Soc. ent. Fr. (2) 4: 193–200.)

LEPELLETIER DE ST FARGEAU Author of *Monographie des Chrysis des environs de Paris* (in Ann. Mus d'Hist. Nat., 58) and *Monographia tenthredinetarum Synonymia extricata* (1823). Coauthored with M. de Serville the *Entomologie de l'Encyclopedie Methodique*.

LEPESME, PIERRE (1913–1957) (Richards 1958, Proc. R. ent. Soc. London (C) 22: 75.)

LEPICERIDAE Plural Noun. A rarely collected, South African Family of aquatic Coleoptera assigned to the Myxophaga. Taxa include *Lepicerus inaequalis* and *Lepicerus bufo*.

LEPID® A registered biopesticide derived from *Bacillus thuringiensis* var. *kurstaki*. See *Bacillus thuringiensis*.

LEPIDIC ACID A yellow pigment obtained from certain butterfly scales, a derivative of uric acid. See Lepidopteric Acid.

LEPIDIC Adj. (Greek, *Lepis* = scale; *-ic* = consisting of.) Comprised of scales, as the wing of a moth. Consisting of scales; covered with scales.

LEPIDOCIDE® A registered biopesticide derived from *Bacillus thuringiensis* var. *kurstaki*. See *Bacillus thuringiensis*.

LEPIDOID Adj. (Greek, *Lepis* = scale; *eidos* = form.) Scale-like in shape. Resembling a scale.

LEPIDOPSOCIDAE Plural Noun. A numerically small, infrequently collected Family of Psocoptera placed in the Suborder Trogiomorpha. Most Species resemble Microlepidoptera, inhabit leaf litter, bark; Antenna with more than 20 segments; wings, body covered with scales; Tarsi with three segments, claws with preapical tooth.

LEPIDOPTERA Linnaeus 1758. (Upper Triassic Period? Jurassic Period-Recent). (Greek, *lepis* = scale; *pteron* = wing.) Butterflies, Skippers, Moths. Cosmopolitan, panorpoid (mecopteroid) Order of Holometabola including about 140,000 nominal Species; sister group of Trichoptera. Adult body and appendages with scales; head free; mouthparts 'haustellate' with a spirally coiled Proboscis formed from elongate, medially-locked Maxillary Galeae; Mandibles typically absent (vestigial in Micropterygidae); Maxillary Palps small or absent; Labial Palps well developed; Compound Eyes large and multifaceted; Ocelli typically absent or lateral Ocelli present; Antennae elongate, Flagellum with numerous Flagellomeres, diverse in appearance. Thorax agglutinate; Prothorax small, membranous often with pair of articulated sclerites (Patagia); Mesothorax large, Metathorax somewhat smaller. Fore and hindwings typically large, membranous, scale-covered; hollow scales contain pigment to provide colour; solid scales provide structural colour (iridescence); amplexiform, jugal or frenal wing-coupling mechanisms. Legs typically adapted for walking; Tarsi with five segments; fore Tibia with apical Epiphysis for grooming; forelegs reduced in some Papilionoidea; hindlegs reduced in some Hepialidae, Geometridae; all legs reduced in some female Psychidae. Abdomen with 11 segments, usually 10 visible segments; spiracles on segments 1–7, rarely on segment 8; Cerci absent; Anus of male behind segment 9; Anus of female at posterior end of Abdomen; eversible pheromone glands common near apex of female Abdomen in many Families; hair pencils

(Coremata) common in males; male genitalia complex; female with one genital opening (monotrysian) or two genital openings (ditrysian). Larva (caterpillar) with head well sclerotized; each thoracic segment bearing pair of legs containing five segments; each leg terminating in single claw; Abdomen 10 segmented with Prolegs usually on segments 3–6 and 10; Prolegs apically truncate (the Planta) with Crochets (sclerotized hooks). Pupae decticous, exarate or adecticous, obtect Divided into Suborders Zeugloptera, Dachnonypha, Monotrysia (a few Species each) and Ditrysia (nearly all Species). See each for discussion. Included Families: Acanthopteroctetidae, Adelidae, Aganaidae, Agathiphagidae, Agonoxenidae, Alucitidae, Anthelidae, Anomosetidae, Apatelodidae, Apoprogonidae, Arctiidae, Argyresthiidae, Arrhenophanidae, Axiidae, Batrachedridae, Blastobasidae, Blastodacnidae, Bombycidae, Brachodidae, Brahmaeidae, Bucculatricidae, Callidulidae, Carposinidae, Carthaeidae, Castniidae, Catapterigidae, Cecidosidae, Cercophanidae, Choreutidae, Chrysopolomidae, Coleophoridae, Copromorphidae, Cosmopterigidae, Cossidae, Crinopterygidae, Cyclotornidae, Dalceridae, Depressariidae, Dioptidae, Douglasiidae, Drepanidae, Dudgeoneidae, Elachistidae, Endromidae, Epermeniidae, Epicopeiidae, Epipyropidae, Eriocottidae, Eriocraniidae, Ethmiidae, Eupterotidae, Galacticidae, Gelechiidae, Geometridae, Glyphipterigidae, Gracillariidae, Hedylidae, Heliodinidae, Heliozelidae, Hepialidae, Herminiidae, Hesperiidae, Heterobathmiidae, Heterogynidae, Hibrildidae, Holcopogonidae, Hyblaeidae, Hypertrophidae, Immidae, Incurvariidae, Lasiocampidae, Lathrotelidae, Lecithoceridae, Lemoniidae, Limacodidae, Lophocoronidae, Lycaenidae, Lymantriidae, Lyonetiidae, Megalopygidae, Metarbelidae, Micropterygidae, Mimallonidae, Mnesarchaeidae, Momphidae, Neopseustidae, Neotheoridae, Nepticulidae, Noctuidae, Notodontidae, Nymphalidae, Oecophoridae, Opostegidae, Oxytenidae, Palaeosetidae, Palaephatidae, Papilionidae, Pieridae, Plutellidae, Prodoxidae, Prototheoridae, Pseudarbelidae, Psychidae, Pterophoridae, Pterothysanidae, Pyralidae, Ratardidae, Roeslerstammiidaee, Saturniidae, Schreckensteiniidae, Sematuridae, Sesiidae, Somabrachyidae, Sphingidae, Symmocidae, Scythrididae, Thaumetopoeidae, Thyretidae, Thyrididae, Tineidae, Tineodidae, Tischeriidae, Tortricidae, Uraniidae, Urodidae, Yponomeutidae, Ypsolophidae, Zygaenidae.

LEPIDOPTERAN Adj. A butterfly or moth. A member of the Lepidoptera.

LEPIDOPTERIC ACID A green pigment obtained from the wing scales of Lepidoptera; a derivative of uric acid. See Lepidotic Acid.

LEPIDOPTERIN Noun. A protein decomposition product found in Lepidoptera.

LEPIDOPTEROUS Adj. (Greek, *lepsis* = a scale; *pteron* = wing; Latin, *-osus* = with the property of.) Descriptive of or pertaining to the Order Lepidoptera; the moths and butterflies.

LEPIDOTRICHIDAE Plural Noun. Primitive Bristletails. A rare Family of large bodied Thysanura. One Species found in California related to Species in Baltic Amber.

LEPIS Noun. (Greek.) A scale.

LEPISMA Noun. The type Genus of Lepismatidae; a silverfish.

LEPISMATIDAE Plural Noun. A Family of Thysanura some Species of which are domicillary and commonly encountered, including the silverfish *Lepisma saccharina* Linnaeus and firebrat *Thermobia domestica* (Packard). Species feed on starch found in books, paper and fabrics.

LEPNEVA, SOF'YA GHEGHOR'EV (1883–1966) (Shul'tova 1967, Ent. Obozr. 46: 259–264, bibliogr. Transl. Ent. Rev. Wash. 46: 152–156, bibliogr.)

LEPREA BROWN *Nesoxenica leprea* (Hewitson) [Lepidoptera: Nymphalidae].

LEPRIEUR, CHARLES EUGENE (1815–1893) (Saulcy 1894, Ann. Soc. ent. Fr. 1894: 453–458, bibliogr.)

LEPRIEUR, FRANÇOIS RENÉ MATHIAS (1799–1870) (Papavero 1971, *Essays on the History of Neotropical Dipterology*. 1: 147. São Paulo.)

LEPROUS Adj. (Greek, *lepis* = scale; Latin, *-osus* = with the property of.) Descriptive of structure with loose, irregular scales.

LEPT- Greek (*leptos*) derived prefix suggesting small, weak or thin.

LEPTIDAE See Rhagionidae.

LEPTIFORM Adj. See Campodeiform.

LEPTINIDAE Plural Noun. Mammal Nest Beetles. Adults small, flattened beetles with recumbent Setae; compound eyes, wings reduced; Antenna with 11 segments, often clubbed; Elytra fused medially; ectoparasitic on rodents and insectivores. Larvae resemble staphylinids, body depressed, Antenna with three segments; eyes absent. Three larval instars.

LEPTOCERIDAE Plural Noun. Longhorned Caddisflies. A cosmopolitan Family of limnephilous Trichoptera with about 1,000 Species. Ocelli absent; Antenna filiform, to three times longer than wing; Maxillary Palpus with five segments. Mesoscutum elongate, setal warts absent, two longitudinal bands of setigerous puctures present; Scutellum small with a few setigerous punctures; wings slender, to 40 mm long. Larval legs slender, hind Femur divided; abdominal gills single, branched or absent. Inhabit swift, cold streams to temporary pools; construct tubes of rock or plant material; some omnivorous, some predatory.

LEPTOCORIS BUG *Leptocoris mitellata* Bergroth [Hemiptera: Rhopalidae].

LEPTOPERLOIDEA A Superfamily of Plecoptera including Austropelidae (Penturoperlidae) and

Gripopterygidae.

LEPTOPHLEBIIDAE Plural Noun. Widespread Family of Ephemeroptera containing Subfamilies Atalophlebiinae, Leptophlebiinae and Mesonetinae (fossil only). The Family contains more than 100 nominal Genera and 500 Species. Most abundant mayfly Family in Australia. Highly variable in size; upper portion of male eye with large facets, lower portion of male eye with small facets. Hindwing reduced or rarely absent; male tarsal formula 5-4-4; female tarsal formula 4-4-4. Male forceps with four segments. Adult males of some Species form dense swarms during day.

LEPTOPHLETIOIDEA A Superfamily of Ephemeroptera including Ephemerellidae and Leptophlebiidae.

LEPTOPODIDAE Amyot & Serville 1843. Plural Noun. Spiny Shore Bugs. A small Family of heteropterous Hemiptera assigned to Superfamily Leptopodoidea; share affinities with Saldidae. Family is Old World with one Species introduced into New World. Adult >2.0 mm long; body slender; compound eyes large, protuberant and often with conspicuous spines; Ocelli close set on tubercle; Hemelytra with overlapping membranes; Labium short, with spines, apex not projecting over Prosternum. Leptopodids live near streams, often on stones or rock walls and sometimes in drier habitats. Species are predaceous on insects and other small arthropods.

LEPTOPOIDEA A Superfamily of Hemiptera including Leptopodidae and Omaniidae.

LEPTOPSYLLINAE Plural Noun. Mouse Fleas. A numerically small Subfamily of Ceratophyllidae (Siphonaptera) whose members parasitize rodents. Adults characterized by an Interantennal Suture and Genal Comb. See European Mouse Flea.

LEPTOS Adj. (Greek.) Small, fine.

LEPTOX® A registered biopesticide derived from *Bacillus thuringiensis* var. *kurstaki*. See *Bacillus thuringiensis*.

LEREBOULLET, AUGUSTE (1804–1865) (Anon. 1866, Am. J. Sci. (2) 41: 110.)

LERP Noun. (Aboriginal, *lerp* = sweet. Pl., Lerps.) A sweet-tasting secretion produced by homopterous insects feeding on *Eucalyptus* spp. in continental Australia, Tasmania and New Zealand.

LERP INSECTS See Psylloidea.

LERUTH, ROBERT (1912–1940) (Maréchae 1942, Bull. Mus. r. Hist. nat. Belg. 18: 1–14, bibliogr.)

LESION Noun. (Latin, *laesio* = injury. Pl., Lesions.) Localized area of diseased tissue. Cf. Scab.

LESNE, PIERRE (1871–1949) (Anon. 1951, Ann. Soc. ent. Fr. 120: 1–16 bibliogr.)

LESPEDEZA WEBWORM *Tetralopha scortealis* (Lederer) [Lepidoptera: Pyralidae].

LESPES, PIERRE GABRIEL CHARLES (1827–1872) (Fauvel 1873, Annu. ent. 1: 106.)

LESSE, HUBERT DE (1914–1972) (Brown 1972, J. Lepid. Soc. 26: 268–274, bibliogr.)

LESSER APPLEWORM *Grapholita prunivora* (Walsh) [Lepidoptera: Tortricidae]: In North America, a pest of apple and prunes. Biology and habits similar to Codling Moth.

LESSER ARMYWORM *Spodoptera exigua* (Hübner) [Lepidoptera: Noctuidae]: Adult moths with wingspan of ca 32 mm; forewing greyish-brown with pale spot in middle of front margin; hindwing white with a dark margin on top of wing; body ca 10 mm long. Adult female lays eggs in early spring and deposits egg masses of ca 80 eggs; egg masses covered in Setae or scales from female's body; a female may lay up to 600 eggs in 4–10 days; egg stage 2–5 days. Young larvae skeletonize leaves; larvae feed ca 21 days and then spin a light silken web over foliage; larvae remain under webbing and pass through five instars. Mature larvae are green with prominent dark lateral stripes. Pupation in soil cells produced by gluing soil particles, leaf litter and larval secretion together. Life-cycle 24–36 days depending on temperature. Pupal stage is overwintering stage in cold climates and adult is overwintering stage in warmer climates. See Noctuidae; Armyworms.

LESSER AUGER-BEETLE *Heterobostrychus aequalis* (Waterhouse) [Coleoptera: Bostrichidae].

LESSER BLADDER-CICADA *Cystosoma schmeltzi* Distant [Hemiptera: Cicadidae].

LESSER BROWN-BLOWFLY 1. *Calliphora augur* (Fabricius) [Diptera: Calliphoridae]: A primary blowfly distributed in eastern Australia; regarded as second in importance to *L. cuprina* in causing Cutaneous Myiasis in sheep. Adult with bluish Thorax and golden Abdomen with longitudinal bluish stripe along dorsum. 2. *Calliphora dubia* (Macquart) [Diptera: Calliphoridae]. See Eastern Goldenhaired Blowfly; Sheep Blowfly; Calliphoridae.

LESSER BROWN-SCORPION *Isometrus maculatus* DeGeer [Scorpiones: Buthidae].

LESSER BUDMOTH *Recurvaria nanella* (Denis & Schiffermüller) [Lepidoptera: Gelechiidae].

LESSER BULB-FLY *Eumerus tuberculatus* Rondani [Diptera: Syrphidae]: A widespread, bivoltine pest of bulbous plants including onion, shallot, amaryllis, carrot, ginger and potato. Life history and biology resembling Narcissus Bulb Fly. See Syrphidae. Cf. Narcissus Bulb Fly.

LESSER CANNA-LEAFROLLER *Geshna cannalis* (Quaintance) [Lepidoptera: Pyralidae].

LESSER CLOVER-LEAF WEEVIL *Hypera nigrirostris* (Fabricius) [Coleoptera: Curculionidae]: A pest of red clover in western USA. Overwinters in woodlands as adult; during spring, adults become active and feed on clover leaves, female oviposits in stems or terminal buds of clover plant. Neonate adult brown, then green; female lays 200–300 eggs during April-May; eclosion ca 14–21 days. Neonate larva white, then whitish brown, legless with curved posture;

larval stage ca 20–25 days; mature larva with black head; body white with dark transverse stripe behind head. Pupal cocoon round-oval, transparent or whitish in colour; pupal stage 5–12 days. See Curculionidae. Cf. Alfalfa Weevil; Clover-Leaf Weevil.

LESSER CORN-STALK BORER *Elasmopalpus lignosellus* (Zeller) [Lepidoptera: Pyralidae]: A widespread pest of beans, peanuts, peas, turnips and wheat in the USA, and major pest of peanuts in Texas. LCB overwinters as larva, Pupae or adult, depending upon locality and climate; LCB bivoltine in southern USA. Adult wingspan ca 25 mm; male forewing brownish-yellow with dark spots and grey margin; female forewing nearly black. Adult active in early spring, female deposits eggs on leaves and stems. Egg pale greenish and eclosion within 7 days. Larva lives in silken tubes in soil; tunnels to feed on roots and stems; feeding period 15–20 days. Mature larva ca 20 mm long, pale bluish-green with transverse brown band around each body segment. Pupae brownish; pupation within silken cocoon in crop trash; pupal period ca 15–20 days. See Pyralidae.

LESSER DUNG FLIES See Sphaeroceridae.

LESSER ENSIGN-WASP *Szepligetella sericea* (Cameron) [Hymenoptera: Evaniidae].

LESSER FOLLICLE-MITE *Demodex brevis* Bulanova [Acari: Demodicidae].

LESSER GHOST-MOTH *Fraus simulans* Walker [Lepidoptera: Hepialidae].

LESSER GRAIN-BORER *Rhyzopertha dominica* (Fabricius) [Coleoptera: Bostrichidae]: A serious pest of grains, particularly stored grains and cereals; larvae cannot develop on spices or oilseeds. Tropical origin; cosmopolitan distribution; occurs within heated facilities in temperate regions. Adult 2.5–3.0 mm long, reddish brown, body cylindrical; head concealed beneath hood-like, tuberculate Prothorax; Antenna 10-segments with loosely defined club of three large segments; Elytra with rows of setigerous punctures; active fliers but seldom seen flying near infestations. Female lays ca 200–500 eggs over prolonged period (to four months); eggs laid individually or in small groups on loose grain or frass. Larvae white, parallel-sided, with small head and prominent legs; first instar campodeiform, with prominent posterior, median spine. Larva moves actively to feed on grain, bores into damaged grain; second instar less active; third instar scarabaeiform; 3–7 larval instars vary with temperature and grain. Pupation within grain; period ca three days. Life cycle 3–6 weeks, ca a month in tropical regions; development influenced by temperature and moisture content of food. Syn Australian Wheat Weevil. See Bostrichidae. Cf. Larger Grain-Borer.

LESSER HORNED CITRUS-BUG *Vitellus antemna* Breddin [Hemiptera: Pentatomidae].

LESSER HOUSE-FLY *Fannia canicularis* (Linnaeus) [Diptera: Muscidae]: A synanthropic, cosmopolitan pest near dwellings; attacks decaying produce, dung, fungi; LHF a problem in poultry houses and farms. Adult 4–6 mm long; smaller, more slender than House Fly; body grey, Thorax with three dark longitudinal stripes; rests with wings overlapping. LHF diurnal; lives about 2–3 weeks; life cycle 3–4 weeks. Adults hover in entryways and shaded areas; spend considerably more time in flight than House Fly. Female prefers to oviposit in semi-liquid such as moist excrement or decaying plant material; eggs laid in batches of ca 50; eggs with 'floating' appendages; eclosion within two days. Larva somewhat flattened with numerous lateral spine-like processes bearing Setae on each body segment used to propel larva through medium; feed 8–10 days in moist habitat (chicken manure, dog excrement); mature larva ca 9 mm long, moves into drier habitat and pupate. Pupation not deep in soil; pupal stage 1–4 weeks. Vector of human diseases; larvae cause Intestinal Myiasis when ingested. See Muscidae. Cf. False Stable Fly; House Fly; Latrine Fly.

LESSER LAWN-LEAFHOPPER *Graminella sonora* (Ball) [Hemiptera: Cicadellidae].

LESSER MEALWORM *Alphitobius diaperinus* (Panzer) [Coleoptera: Tenebrionidae]: A cosmopolitan pest of damp, mouldy grains and cereals, and structural pest in poultry houses. Endemic to Africa where it inhabits bird nests and bat caves. Adult and larva feed on dead birds; larva may feed on moribund birds. Harbours and transmits diseases in poultry, including Avian Leucosis. Adult 6–7 mm long, shiny black; larva to 20 mm, dark yellow with brown bands. Life cycle 9–12 months. Eggs glued to substrate, often in groups; egg incubation 3–10 days; 6–11 larval instars. Adult lives ca 400 days, maximum 700 days; oviposits many eggs per day through adult life, ca 1,300 eggs during lifetime. See Tenebrionidae. Cf. Dark Mealworm; Yellow Mealworm.

LESSER MULBERRY-PYRALID *Glyphodes pyloalis* (Walker) [Lepidoptera: Pyralidae].

LESSER OCELLAR BRISTLES Diptera: 3–12 pairs of bristles, in two, sometimes four, parallel lines beginning very close to the greater ocellars. The post-ocellars *sensu* Comstock.

LESSER ORCHID-WEEVIL *Orchidophilus peregrinator* Buchanan [Coleoptera: Curculionidae].

LESSER PEACH-TREE BORER *Synanthedon pictipes* (Grote & Robinson) [Lepidoptera: Sesiidae (Aegeriidae)]: A pest of cherry, peach and plum in North America. Overwinters as larva within silken cocoon attached to bark of trunk or branches. Larvae active during spring and feed on inner bark of tree; larval feeding causes accumulation of masses of gum exudation from wounds in bark. Pupal cocoon spun within feeding burrows. Adults emerge May-June; oviposit in cracks and crevices in bark; eclosion within

10 days; neonate larvae penetrate bark to begin feeding. LPTB completes 1–2 generations per year, depending upon host plant and climate. See Sesiidae. Cf. Peach-Tree Borer; Peach-Twig Borer.

LESSER QUEENSLAND-FRUIT-FLY *Dacus neohumeralis* Hardy [Diptera: Tephritidae].

LESSER ROSE-APHID *Myzaphis rosarum* (Kaltenbach) [Hemiptera: Aphididae].

LESSER SHOOT-WEEVIL *Rhachiodes dentifer* Boheman [Coleoptera: Curculionidae].

LESSER WANDERER BUTTERFLY *Danaus chrysippus petilia* (Stoll) [Lepidoptera: Nymphalidae].

LESSER WATERBOATMEN See Corixidae.

LESSER WAX-MOTH *Achroia grisella* (Fabricius) [Lepidoptera: Pyralidae]: A minor pest of honey bee hives. Larvae of LWM chew the cap of sealed brood which induces worker honey bees to remove the remainder of the cap. The exposed cell induces a condition called 'bad brood.' Some adults emerging from uncapped cells display deformed legs and wings. See Pyralidae. Cf. Greater Wax Moth.

LESSER, FRIEDRICH CHRISTIAN (1692–1754) (Rose 1850, *New General Biographical Dictionary* 9: 256.)

LESSERT, ROGER DE (1875–1945) (Revilliod 1946, Verh. schweiz. naturf. Ges. 126: 383–387, bibliogr.)

LESSON, RENÉ PRIMEVERE (1794–1849) (Swainson 1840, *Taxidermy; with Biogr. of Zoologists.* 392 pp. (241–243), London. [Volume 12 of *Cabinet Cyclopedia*. Edited by D. Lardner.]

LESSONA, MICHELE (1823–1894) (Camerano 1894, Boll Musei Zool. Anat. comp. R. Univ. Torino 9 (188): 1–72, bibliogr.)

LESTAGE, JOHANNES ANTOINE (1879–1945) (Carpentier 1947, Bull. Mus. r. Hist. nat. Belg. 23 (3): 1–23, bibliogr.)

LESTIDAE Plural Noun. Spread-winged Damselflies. A Family of Odonata. Adults small to medium-sized and slender. Pterostigma rectangular; Veins R3 and R4 originate near Arculus. Species found in swamps; individuals rest vertically on vegetation with wings partly spread.

LESTINOIDEA A Superfamily of Odonata. Adults typically small or medium-sized with wings petiolate or narrowed basad; wing bears a Nodus at basal third and Pterostigma usually long and rectangular; Arculus positioned about midway between base of wing and Nodus; two Antenodal Crossveins. See Odonata.

LESTOBIOSIS Noun. (Greek, *lestes* = plunderer; *bios* = lifestyle; *-sis* = a condition or state. Pl., Lestobioses.) Social Insects: A relationship in which a colony of small Species lives within the nest of larger Species and preys upon the brood and steals the food of the larger Species. See Biosis; Commensalism; Parasitism; Social Parasitism. Cf. Abiosis; Anhydrobiosis; Antibiosis; Archebiosis; Calobiosis; Cleptobiosis; Hamabiosis; Kleptobiosis; Parabiosis; Phylacobiosis; Plesiobiosis; Synclerobiosis; Trophobiosis; Xenobiosis. Rel. Thief Ants.

LESTOIDEIDAE Plural Noun. A monogeneric Family of Odonata known only from forested creeks of Queensland, Australia. Adult features common with Lestidae, Amphipterygidae and some Coenagrionionoidea. Adult body small; Pterostigma elongate; quarilateral nearly rectangular; Cubital Vein reduced. Larva with trapezoidal head; second segment longer than first; Thorax robust; Abdomen short; Femora thick; three caudal filaments apically plumose.

LESTONIIDAE Plural Noun. A small Family of heteropterous Hemiptera endemic to Australia and assigned to Superfamily Pentatomoidea. Head and Pronotum with lateral margin laterally expanded to give flange-like appearance; Scutellum large, convex, and covering Abdomen and wings at repose; Tarsi with two segments. Adult and nymph taken on native *Callitris.*

LETH, KARL OTTO (1910–1962) (Kaiser 1962, Ent. Meddr. 31: 335–337, bibliogr.)

LETHARGIC Adj. (Greek, *lethargo* = forgetful; *-ic* = characterized by.) Torpid, inactive.

LETHIERRY, LUCIEN FRANÇOIS (1830–1894) (Puton 1894, Rev. Ent. 13: 118–119.)

LETISIMULATE Noun. (Latin, *letum* = death; *simulare* = to feign. Pl., Letisimulates.) Death feigning. (Weir 1889, *Dawn of Reason*). See Thanatosis.

LETTSOM, JOHN COAKLEY (1744–1815) (Lesney 1960, *Bibliogr. of British Lepidoptera 1608–1799,* viii + 315 pp. (207–209), London.)

LETTUCE ROOT-APHID *Pemphigus bursarius* (Linnaeus) [Hemiptera: Aphididae].

LETZNER, KARL WILHELM (1812–1889) (Dittrich 1890, Z. ent. Breslau 15: 1–18, bibliogr.)

LEUCAENA PSYLLID *Heteropsylla cubana* Crawford [Hemiptera: Psyllidae]: A small psyllid native to Central and South America which feeds on the legume *Leucaena leucocephala* (Lam.) deWit and other members of the Genus *Leucaena.* A significant pest of *Leucaena* spp. in southeast Asia, India, Australia and islands of the Pacific Ocean. Defoliation is caused by nymphs and adults which feed on sap from young shoots, leaves and inflorescences. See Psyllidae.

LEUCINE Noun. (Greek, *leukos* = white. Pl., Leucines.) A white crystalline compound, the product of animal decomposition, found in the Malpighian Tubules. As a colour, Leucine is cheesy white.

LEUCKART, KARL GEORG FRIEDRICH RUDOLPH (1822–1898) (Blanchard 1898, Archiv. Parasit. 1: 185–190.)

LEUCKART, RUDOLF (1822–1898) (Petzsch 1973, Anz. Schädlinsk. Pflanzen-umwelts 46: 63–64.)

LEUCOCYTE Noun. (Greek, *leukos* = white; *kytos* = hollow vessel; container. Pl., Leucocytes.) Blood cells: Colourless, sometimes ameboid cells that occur within the Haemolymph. Alt.

White Blood Cell. See Amoebocyte; Plasmatocyte. Cf. Circulatory System.

LEUCOSPIDAE Walker 1834 Plural Noun. A numerically small (about 150 Species), widespread (predominantly tropical, subtropical) Family of parasitic Hymenoptera assigned to the Chalcidoidea. No fossil record. Apparently related to Chalcididae. Robust, comparatively large bodied, Antenna with 13 segments, club undifferentiated; Pronotum large, often transversely carinate; hind Femur enlarged, dentate; Gaster subsessile; Ovipositor projecting over Gaster when at repose. Primary parasites of solitary wasps and bees.

LEUCTRIDAE Plural Noun. Needleflies. A Family of Plecoptera including about 12 Genera and 200 Species. Some Genera are regional in distribution. Small and slender as adults; wings curiously roll around Abdomen at repose; Cerci small, with one segment; male Cerci heavily sclerotized.

LEUNIS, JOHANNES (1802–1873) (Keese 1873, Leopoldina 8: 82–85.)

LEUSSLER, RICHARD A (1866–1943) (Anon. 1943, Ent. News 54: 242.)

LEUTHHARDT, FRANZ (1861–1934) (E. R. 1935, TätBer. naturf. Ges. Baselland 10: 199–209.)

LEVANDER, KAARLO MAINIO (1867–1943) (Luther 1945, Arbo/Vuosik. Soc. Scient. fenn. 23: 1–17, bibliogr.)

LEVATOR Noun. (Latin, *levare* = to raise; *or* = one who does something. Pl., Levators.) Muscles which elevate a structure. Cf. Depressor.

LEVATOR MUSCLE Any muscle used to raise an appendage.

LEVEILLE, ALBERT (–1911) (Janet 1911, Bull Soc. ent. Fr. 1911: 25–26.)

LEVETT, CHARLES (1846–1981) (Anon. 1918, Entomologist 51: 96.)

LEVIGATE Adj. (Latin, *levigatus* = smoothed; *-atus* = adjectival suffix.) Smooth, polished sometimes shiny; pertaining to surface without elevations or depressions. Alt. Levigatus; Laevis; Levis.

LEVOITURIER, J ALEXANDRE (1814–1896) (Anon. 1898, Bull Soc. sci. nat. Elbeuf 16: 95–97.)

LEVRAT, JEAN NICHOLAS BARTHELMY GUSTAVE (1823–1859) (Mulsant 1859, Opusc. ent. 11: 69–80, bibliogr.)

LEVULOSE Noun. (Latin, *laevus* = left; *-ule* = diminutive; *-osus* = full of.) Fruit sugar, an 'invert' sugar (a levarotatory enantiomer, as opposed to dextrarotatory) found in honey and many fruit juices, formed by hydrolysis of cane sugar; fructose.

LEWES, GEORGE HENRY (1817–1878) (Anon. 1878, Nature 19: 106–107.)

LEWIN, JOHN WILLIAM (1770–1819) Son of William Lewin. English painter and author of *Natural History of Lepidopterous Insects of New South Wales* (1805). (Musgrave 1930, Aust. Zool. 6: 192–193.)

LEWIN, WILLIAM (fl 1791–1795) Author of *The Papilios of Great Britain systematically arranged and painted from Nature.* (1795) (Lisney 1960, *Bibliogr. of British Lepidoptera 1608–1799.* 315 pp. (286–289), London.)

LEWIN'S BAG-SHELTER MOTH *Panacela lewinae* (Lewin) [Lepidoptera: Eupterotidae].

LEWIS, CLARENCE WATERMAN (1882–1936) (Hyslop 1936, J. Econ. Ent. 29: 1030.)

LEWIS, GEORGE (1839–1926) (Edwards 1927, Proc. Linn. Soc. Lond. 1926–27: 89–90.)

LEWIS, WILLIAM ARNOLD (–1877) (Anon. 1877, Entomologist 10: 264.)

LEXICON Noun. (Greek, *lexis* = phrase, word. Pl., Lexicons.) 1. A systematic collection of words or phrases with definitions. 2. A dictionary.

LEY, HELLMUT (1909–1973) (Anon. 1874, Prakt. SchädlBefämpf. 26: 46.)

LEYBOLD, FRIEDRICH (1827–1879) (Anon. 1887, Leopoldina 23: 208–210, bibliogr.)

LEYDIG, FRANZ VON (1821–1908) (Taschenburg 1909, Leopoldina 45: 37–44, 47–52, 70–76, 82–88, bibliogr.)

LEZINIDAE Plural Noun. A Family of ensiferous Orthoptera containing *Lezina*, a stout-bodied, apterous Genus found in xeric habitats of North Africa to Central Asia. Species lack an Ovipositor; Cerci long, slender, segmented. See Ensifera.

L'HOMME, LÉON (1867–1949) (Anon. 1949, Revue franç. Lép. 12: 65–67.)

LI See Louping Ill.

LIANA Noun. (Latin, *ligare* = to bind.) Any climbing plant with roots in the ground. Such plants are characteristic of, but not restricted to, tropical rainforests.

LIBAVIUS, ANDREAS (1560–1616) (Rose 1850, *New General Biographical Dictionary* 9: 266.)

LIBELLULIDAE Plural Noun. Common Skimmers. The largest Family of Odonata which is cosmopolitan in distribution and occupies all habitats frequented by Odonata. Adults are highly variable in size with Abdomen sometimes short. Auricles absent; eyes broadly contiguous or shortly contiguous in crepuscular forms; posterior edge of eye usually lacks a small projection. Tibiae of male lacks a keel and hindwing not angled. Antenodal crossveins tend to be in line in Costal and Subcostal spaces. Antennal loop generally moderate-sized to long, and anal appendages of male simple. Female lacks an Ovipositor. Larvae are variable in shape and size, but eyes usually prominent and elevated. Abdomen with lateral spines and sometimes with mid-dorsal hooks. Legs normally are shorter than in Corduliidae. See Odonata.

LIBELLULOIDEA A Superfamily of Odonata whose adults are variable in size but the eyes are in contact medially. Wings vary in shape but Pterostigma narrow; antenodal cossveins usually more-or-less coincident in Costal and Subcostal spaces; Discoidal Cells variable in shape, usually trianglular; anal loop well developed. The larval mask is deeply concave and setose. See Odonata.

LIBYTHEIDAE Plural Noun. Snout Butterflies. A numerically small Family of Lepidoptera assigned to the Papilionoidea. The Family contains 10 Species, but is cosmopolitan in distribution, and closely related to Nymphalidae. Adults are small bodied, brownish with long labial palps; males with front legs reduced. Larvae feed on plants in Ulmaceae. Some texts treat this group as a Subfamily of Nymphalidae. See Papilionoidea.

LICATUS (LICETI), FORTUNEUS (1577–1651) (Rose 1850, *New General Biographical Dictionary* 9: 267.)

LICE See Phthiraptera

LICHEN CASE MOTH *Cebysa leucotelus* Walker [Lepidoptera: Psychidae].

LICHEN-EATING CATERPILLAR *Manulea replana* (Lewin) [Lepidoptera: Arctiidae].

LICHENOPHAGOUS Adj. (Greek, *leichein* = to lick; *phagein* = to eat; Latin, *-osus* = with the property of.) Pertaining to oganisms that feed on lichen. See Feeding Strategy.

LICHENSTEIN, A A H German philologist, academic and author of 'A Dissertation on Two Natural Genera hitherto confounded under the name of Mantis.' (Published in Linnean Transactions, volume 6.)

LICHT, SOL FELTY (1886–1947) (Bullock 1947, Science 106: 483–484.)

LICHTENSTEIN, ANTON AUGUST HEINRICH (1753–1816) (Weidner 1967, Abh. Verh. naturw. Ver. Hamburg Suppl. 9: 43–50.)

LICHTENSTEIN, MARTIN HEINRICH CARL (1780–1851) (Weidner 1967, Abh. Verh. naturw. Ver. Hamburg Suppl. 9: 50–54.)

LICHTENSTEIN, WILLIAM AUGUSTE JULES (1818–1886) (Mayet 1887, Ann. Soc. ent. Fr. (6) 7: 49–58, bibliogr.)

LICHTWARDT, BERNHARD (1857–1943) (Sachtleben 1943, Arb. morph. taxon. Ent. Berl. 10: 174.)

LICHTWERK, JULIUS (1890–1963) (Weidner 1967, Abh. Verh. naturw. Ver. Hamb. Suppl. 9: 295.)

LICKERISH, LESLIE ARTHUR (–1971) (Butler 1972, Proc. R. ent. Soc. Lond. (C) 36: 62.)

LIDGETT, JAMES (–1941) (Miller 1941, Victorian Nat. 58: 47.)

LIEBECK, CHARLES (1863–1947) (Richmond 1947, Ent. News 58: 165–168.)

LIEBIG, JUSTUS VON (1803–1873) (Ratzeburg 1874, Forstwissenschaftliches Schriftsteller-Lexicon 1: 312–317.)

LIEBMANN, FREDERIK MICHAEL (1813–1856) (Papavero 1975, *Essays on the History of Neotropical Dipterology*. 2: 365. São Paulo.)

LIEBMANN, MORITZ (1858–1920) (Spröngerts 1921, Dt. ent. Z. Iris 35: 52–53, bibliogr.)

LIEBMANN, WALTER (Pfitzer-Lohse 1965, Ent. Bl. Biol. Syst. Käfer 61: 65–66.)

LIEGEL, EMANUEL (1859–1894) (Anon. 1894, Ent. News 5: 160.)

LIENARD, VALERE (1856–1886) (Plateau 1886, C. R. Soc. ent. Belg. 30: cxlix-cli, bibliogr.)

LIENIG, FRIEDERIKE (–1858) (Reichenbach 1846, Allg. dt. Naturh. Ztg. 1: 303–304.)

LIENIG, MINE (–1885) (Anon. 1858, *Accentuated list of British Lepidoptera*. 118 pp. (xxix). Oxford & Cambridge Entomological Societies, London.)

LIEU, K O VICTORIA (–1956) (Hall 1957, Proc. R. ent. Soc. Lond. (C) 21: 66.)

LIEURY, JEAN-BAPTISTE (1818–1888) (Anon. 1889, Bull. Soc. Amis. sci. nat. Rouen (3) 24: 347–349.)

LIFE CYCLE 1. The time interval between fertilization of an egg and the death of the individual which develops from that egg. 2. The time interval between egg-deposition and attainment of sexual maturity as shown by egg-laying. 3. The time interval between hatching from the egg and emergence of the adult from the Pupae. 4. The successive stages of reproduction, growth and development of an organism between the appearance and reappearance of the same stage. See Phenology.

LIFE FORM The characteristic body shape of an organism at different phases of development.

LIFE HISTORY See Phenology.

LIFE STAGE In insects, several periods in life which are radically different from each other in appearance and functional responsibility. See: Egg; Larva; Pupa; Holometabola; Nymph; Hemimetabola; Paurometabola.

LIFE ZONE See Biome.

LIFT Noun. The action or process of moving a body from a relatively low level to a higher level. *e.g.* The movement of wings cause Lift which results in the body of the insect being elevated into the air. Cf. Drag; Thrust. Rel. Airfoil.

LIGAMENT Noun. (Latin, *ligamentum* = bandage. Pl., Ligaments.) Tough, fibrous tissue between parts or segments of a body.

LIGASE Noun. An enzyme used by genetic engineers to join cut ends of DNA stands. Rel. Biotechnology; Genetic Engineering.

LIGHT BROWN FLEA BEETLE *Longitarsus victoriensis* Blackburn [Coleoptera: Chrysomelidae].

LIGHT MICROSCOPE See Scanning Electron Microscope; Transmission Electron Microscope.

LIGHTBROWN APPLE MOTH *Epiphyas postvittana* (Walker) [Lepidoptera: Tortricidae]: A pest of orchards, grapes, ornamentals, pasture crops and vegetables; endemic to Australia and introduced into New Zealand, Hawaii and England. Adult male moths are 10–15 mm long, pale brown with dark brown markings on the forewing from wing apex to centre of wing. Female moths are 20–25 mm long, pale brown with indistinct markings on forewing. Adults are active at dusk and rest during the day on vegetation. Wings are folded forming a bell-shape outline at rest. Adult stage lasts 25–30 days. Eggs are green, flattened and laid in overlapping rows on leaves, stems and fruit; clusters of eggs resemble fish scales. Adult females lay 100–200 eggs in batches of 20–25 on the host plant; egg stage lasts 7–14 days. All instars produce webbing for cover when feeding. Larvae are 15–25 mm long, green with a dark brown head capsule; larval stage lasts 24–60 days depending on food and temperature. Larvae feed within silken tubes and move vigorously backward when disturbed or sometimes hang from silk thread outside tunnel. Early instar larvae are pale yellow-green, late-instar larvae with a dark green stripe and brown head; six larval instars. Larvae produce webs that causes leaf to curl by folding leaf onto itself, by attaching two leaves together, or by attaching a leaf to a fruit. Damage through larval feeding malforms seedlings and injures fruit crops. Pupation occurs within silken chambers in rolled leaves or plant debris. Life cycle 3–4 months. Multivoltine with 3–4 generations per year. See Tortricidae.

LIGHTNING BUGS See Lampyridae.

LIGNAC, JOSEPH ADRIAN DE LARGE DE (–1762) (Rose 1850, *New General Biographical Dictionary* 9: 271.)

LIGNEOUS Adj. (Latin, *lignum* = wood; *-osus* = with the property of.) Wood-like; woody; made of wood; resembling wood in structure; wood-brown. Alt. Ligneus; Ligniform.

LIGNICOLOUS Adj. (Latin, *lignum* = wood; *colere* = living in or on.) Pertaining to organisms that live in wood. See Habitat. Cf. Agricolous; Dendrophilous; Nemoricolous. Rel. Ecology.

LIGNIN Noun. (Latin, *lignum* = wood. Pl., Lignins.) A principal component of lignocellulose which occurs as a structural polymer in the cell walls and middle lamella of most higher plants. Lignin is interspersed and covalently linked with hemicellulose to form the matrix that surrounds cellulose microfibrils. See Lignocellulose. Cf. Tannin. Rel. Chitin; Collagen.

LIGNIVOROUS Adj. (Latin, *lignum* = wood; *vorare* = to devour; Latin, *-osus* = with the property of.) Pertaining to organisms which feed upon wood or woody tissues. See Feeding Strategy.

LIGNOCELLULOSE Noun. (Latin, *lignum* = wood; *cellula* = little cell. Pl., Lingocelluloses.) The woody tissue produced through photosynthetic fixation of carbon dioxide. Constituents include cellulose, hemicellulose and lignin. Decomposition of lignocellulose is achieved primarily through the action of microorganisms and augmented by various arthropods including termites. See Cellulose; Hemicellulose; Lignin.

LIGULA Noun. (Latin, *ligula* = little tongue. Pl., Ligulae.) 1. A small spoon, strap or tongue. 2. The median sclerite of the Labium which inserts onto distal margin of Prementum, and forms the terminal lobe of the Labium. Sometimes the Ligula is fused to form a single sclerite but more often it is subdivided into four lobes, or modifications of four lobes. These correspond to the Galea and Lacinea of the first Maxillae. Often used synonymously with 'Glossa' and 'tongue.' See Elytral Ligula. Cf. Glossa; Lacinea; Paraglossa.

LIGULATE (Latin, *ligula* = little tongue; *-atus* = adjectival suffix.) Strap-shaped; tongue-like in shape or function. Pertaining to structure which is linear and much longer than broad. See Ligula. Cf. Linguiform. Rel. Form 2; Shape 2; Outline Shape.

LIHOCHIN® See Chlormequat.

LIK'YANOVICH, FEDORA KONSTANTINOVICH (1904–1942) (Richter 1951, Ent. Obozr. 31: 315–320, bibliogr.)

LILAC BORER *Podosesia syringae* (Harris) [Lepidoptera: Sesiidae]: A monovoltine pest of lilac, ash, mountain ash and privet in eastern North America. Eggs are laid in bark near base of tree. Neonate larvae bore into trees and overwinter as half-grown larvae; mature larvae are 15–18 mm long, cream coloured with brown head. Pupation occurs within tunnels. Syn. Ash Borer. See Sesiidae. Cf. Ash Borer; Currant Borer; Peach-Tree Borer. See Sesiidae.

LILAC LEAFMINER *Caloptilia syringella* (Fabricius) [Lepidoptera: Gracillariidae]: A pest of lilac, azalea and ash; larvae mine leaves, then roll edges or web several leaves together and feed on surface. See Gracillariidae. Cf. Azalea Leaf Miner.

LILACEOUS Adj. (Latin, *lillium* = lily; Latin, *-aceus* = of or pertaining to.) Lilac colour. Alt. Lilaceus; Liacinous; Lilacinus.

LILY APHID *Aulacorthum circumflexum* (Buckton) [Hemiptera: Aphididae].

LILY BULB THRIPS *Liothrips vaneeckei* Priesner [Thysanoptera: Phlaeothripidae].

LILY CATERPILLAR *Spodoptera picta* (Guérin-Méneville) [Lepidoptera: Noctuidae].

LILY LEAF-BEETLE *Lilioceris lilii* (Scopoli) [Coleoptera: Chrysomelidae].

LILY PILLY PSYLLID *Trioza eugeniae* Froggatt [Hemiptera: Triozidae].

LILY WEEVIL *Agasphaerops nigra* Horn [Coleoptera: Curculionidae].

LIMA-BEAN POD BORER *Etiella zinckenella* (Treitschke) [Lepidoptera: Pyralidae]: Widespread throughout North America; a pest of lima bean, wild peas and lupin. Eggs are white, elliptical and laid individually on pods of lima beans or lupines. Larvae bore into pod to feed; holes in pods seal making detection of larvae impossible. Pupation occurs in soil and LBPBs overwinter as mature larvae within cocoon in soil. Species completes one generation per year on annual plants; 2–4 generations per year on lima bean. See Pyralidae.

LIMA-BEAN VINE BORER *Monoptilota peregratialis* (Hulst) [Lepidoptera: Pyralidae].

LIMACIFORM Adj. (Latin, *limax* = slug; *forma* = shape.) Slug-shaped; descriptive of or in the form of *Limax*. A term often applied to legless larvae, particularly Symphyta (Hymenoptera) and

Limacodidae (Lepidoptera). See Slug. Rel. Form 2; Shape 2; Outline Shape.

LIMACODIDAE Plural Noun. Nettlegrubs; Slug-Caterpillar Moths; Cup Moths. A widespread, but predominantly tropical, Family of ditrysian Lepidoptera which includes ca 800 Species assigned to Superfamily Zygaenoidea in most classifications but sometimes placed in Cossoidea. Adults are stout-bodied with wingspan 14–74 mm and with smooth scales on head; Ocelli and Chaetosemata absent; male Antenna at least partially bipectinate; female Antenna dentate or filiform; Proboscis and Maxillary Palpus absent or minute; head retractile; Antenna about half as long as long forewing; foreleg lacking Epiphysis; tibial spur formula 0-2-4 or 0-2-2; Thorax, Femora and Tibiae with dense vestiture of lamellar scales. Eggs are typically scale-like, oval in outline shape, with Chorion translucent; eggs usually are deposited in overlapping arrangement with dense cover of piliform Setae from female's Abdomen; some Species lay eggs singly on foliage of host plant. Caterpillars are stout, slug-like; thoracic legs small to minute and Prolegs absent. Development is heteromorphic with first instar broad, dorsoventrally flattened and displaying spines on each body segment; subsequent instars are slug-like with urticating Setae. Cocoon is hard, smooth with projecting, urticating spines. Pupal cocoon is oval or pyriform and located in diverse places: On host plant, fissure of bark, leaf litter and in soil. Adults of most Species are nocturnal and fly erratically. Caterpillars feed exposed on foliage; some Species are serious defoliators of ornamental and crop trees; faecal material of caterpillar is cup-shaped, hence common name. See Zygaenoidea.

LIMATOX® See Metaldehyde.

LIMB Noun. (Anglo-Saxon, *lim*. Pl., Limbs.) 1. Any appendage to include a leg or wing 2. The circumference, the edge or rim (obsolete). 3. Cicada: The area along the outer and posterior margin of the wing beyond the closed cells (Smith). Alt. Limbus.

LIMB BASIS The primary basal segment of an appendage supporting the Telopodite, sometimes subdivided into a proximal Subcoxa (Pleuropodite or Pleuron), and a distal Coxa. The Coxopodite *sensu* Snodgrass.

LIMBATE Adj. (Latin, *limbus* = border; *-atus* = adjectival suffix.) With a margin or limb of a different colour. Alt. Limbatus.

LIME APHID *Eucallipterus tiliae* [Hemiptera: Aphididae].

LIMINOID Noun. (Spanish, *lima* = lemon; *-oid* = like. Pl., Liminoids.) Plant-produced terpenoids that discourage herbivory by insects. Syn. Tetranortriterpenoids. See Botanical Insecticide; Natural Insecticide; Insecticide; Pesticides.

LIMNEPHILIDAE Plural Noun. Northern Caddisflies. A numerically large, widespread Family of Trichoptera assigned to Superfamily Limnephiloidea. Adults are more than 6 mm long; Ocelli present; female Maxillary Palpus with five segments; male Maxillary Palpus with three segments; Antenna about as long as forewing; forewing Discoidal Cell closed; Median Cell absent; Anal Veins fused. Larvae with prosternal horn; abdominal gills branched with lateral fringe; feed on fine particulate matter in streams. See Limnephiloidea.

LIMNEPHILOIDEA A Superfamily of Trichoptera (caddisflies) including Anomalopsychidae, Antipodoeciidae, Atriplectididae, Barbarochthonidae, Beraeidae, Brachycentridae, Calamoceratidae, Calocidae, Chathamiidae, Conoesucidae, Goeridae, Helicophidae, Helicopsychidae, Hydrosalpingidae, Kokiriidae, Lepidostomatidae, Leptoceridae, Limnephilidae, Limnocentropodidae, Molannidae, Odontoceridae, Oeconesidae, Petrothrincidae, Philorheithridae, Phryganeidae, Phryganopsychidae, Plectrotarsidae, Sericostomatidae, Tasimiidae and Uenoidae. See Trichoptera.

LIMNICHIDAE Plural Noun. A small Family of polyphagous Coleoptera assigned to Superfamily Byrrhoidea. Adults are 1–4 mm long; convex and oval in outline shape. Larvae inhabit sand and leaf-litter where they presumably feed on decaying organic matter. See Byrrhoidea.

LIMNOLOGY Noun. (Greek, *limne* = marshy lake; *logos* = discourse.) A research discipline concerned with the study of the fauna of waters. See Ecology; Habitat. Rel. Zoogeography.

LIMONIIDAE Plural Noun. A large, cosmopolitan Family of nematocerous Diptera sometimes included with Tipulidae. Limoniidae consists of about 11,000 Species. Immature limoniids are associated with aquatic and damp habitats where larvae feed upon fungi, algae and decomposing plant material. See Tipulidae.

LIMOPHAGOUS Adj. (Latin, *limus* = mud; Greek, *phagein* = to eat; Latin, *-osus* = with the property of.) Descriptive of organisms that consume mud. See Feeding Strategies.

LIMPAGA (Etymology obscure.) Common name for Philippine Neem Tree in Sabah. See Philippine Neem Tree.

LIMPET Noun. (Anglo Saxon, *lempedon* = limpet. Pl., Limpets.) Close-fitting to stones or other objects in water, after the fashion of the small mollusc of that name.

LIMPID Adj. (Latin, *limpa* = water.) Clear and transparent. A term applied to wings and ornamentation.

LIMULODIDAE See Ptiliidae.

LINACERATUBAE Coccoidea: Long, slender Ceratubae in the Pygidia in which the sides are parallel or nearly parallel (MacGillivray).

LINAVERTEG Noun. The Vertex when reduced to a linear area between the Front and a compound eye; Parafrons *sensu* MacGillivray.

LINCECUM, GIDEON (1792–1894) (Anon. 1875, Amer. Nat. 9: 191.)

LINCK, ERNST (1874–1963) (Allenspach 1963, Mitt. ent. Ges. Basel 13: 91–92.)

LINDACOL See Lindane.

LINDAFOR See Lindane.

LINDAN® See Dichlorvos.

LINDANE A synthetic, chlorinated hydrocarbon insecticide {Gamma isomer of 1,2,3,4,5,6-hexachloro-cylohexane} used as a contact insecticide, stomach poison and fumigant against Mange mites, ants, aphids, boll weevils, cockroaches, flea beetles, flies, grasshoppers, leaf miners, lygus bug, mosquitoes, psyllids, spittlebugs, termites, thrips, wireworms and related insects. Lindane is applied to numerous vegetable crops and field crops, tobacco, cotton, sudan grass, sugar beets and agricultural buildings. Lindane is toxic to bees, wildlife and fishes and persists in soil for more than a year; it is not for use on poultry or poultry houses. Syn. Gamma HCH. Trade names include: Ambrocide®, BHC®, Benesan®, Gammaphex®, Gamma-Mean®, Gamma-Sol®, Gammalin®, Gammex®, Gammexane®, Hammer®, Isotox®, Lindacol®, Lindafor®, Lintox®, Novigam®, Silvanol®. See Chlorinated Hydrocarbon Insecticide.

LINDBERG, HAKAN (1898–1966) (Lindroth 1976, Opusc. ent. 32: 3–4.)

LINDBERG, HARALD (1871–1963) (Lindroth 1963, Notul. ent. 43: 68.)

LINDBERG, KNUT (1892–1962) (Lindroth 1963, Opusc. ent. 27: 127–128.)

LINDEMAN, KARL (1844–1929) (Bogdanov 1930, Plant Prot. Leningr. 6: 823–831, bibliogr.)

LINDEMANS, JOHANNUS (1874–1944) (Zollner 1947, Ent. Ber., Amst. 12: 97–100.)

LINDEN APHID *Eucallipterus tiliae* Linnaeus [Hemiptera: Callaphididae].

LINDEN BORER *Saperda vestita* Say [Coleoptera: Cerambycidae]: A pest of linden in eastern North America. Eggs are laid near ground in small clutches within notches cut into bark by adult. Larvae mine under bark near ground, work into wood and move downward toward roots. During midsummer, adults feed on green bark, veins and stems of leaves. See Cerambycidae. Cf. Elm Borer; Locust Borer; Poplar Borer.

LINDEN LOOPER *Erannis tiliaria* (Harris) [Lepidoptera: Geometridae].

LINDEN, MARIA GRÄFIN VON (1869–1936) (Horn 1936, Arb. morphol. taxon. Ent. Berl. 3: 300.)

LINDEN, PL L VANDER Belgian academic and author of several minor entomological works including *Observations sue les Hymenopteres d'Europe, de la famille des Fouisseurs* (1819).

LINDENBAUER, MAURITZ (–1863) (Anon. 1963, Z. wien. ent. Ges. 48: 144.)

LINDER, JULES (1830–1869) (Deyrolle 1869, Petites Nouv. Ent. 1: 28.)

LINDER, STIG (1901–1965) (Stenram 1966, Opusc. ent. 31: 150.)

LINDLEY, JOHN (1799–1865) (Anon. 1867, Proc. ent. Soc. Lond. 1866–1867: xxx-xxvii.)

LINDROTH, HARALD (–1963) (Lindroth 1963, Opusc. ent. 28: 166.)

LINDSEY, ARTHUR WARD (1894–1963) (Voss 1963, J. Lepid. Soc. 17: 181–190, bibliogr.)

LINE Noun. (Latin, *linea* = line. Pl., Lines.) 1. A narrow streak or stripe. 2. A unit of measurement, one-twelfth of an inch, used by some 19th century English and American taxonomists.

LINE GRASS YELLOW *Eurema laeta sana* (Butler) [Lepidoptera: Pieridae].

LINE OF WEAKNESS Cuticular features which are used at moulting. Lines of weakness are frequently and incorrectly named as if they were sutures. For instance the so-called Ecdysial Cleavage Line is a line of weakness that is sometimes considered as synonymous with the Epicranial Suture. The two features are similar in position and appearance, but structurally they may have been derived from different conditions. Cf. Apodeme; Sulcus; Suture.

LINEA Noun. (Latin, *linea* = line.) A line, line-like structure, mark or narrow stripe.

LINEA CALVA An oblique 'hairless' streak on the forewing of some parasitic Hymenoptera. LC extends obliquely from the Stigma toward the posterior margin of the wing. Syn. Speculum. See Filum Spinosum.

LINEAGE Noun. (Old French, *ligne* = line of descent; *age* = *aticum* = cumulative result of. Pl., Lineages.) Descent in a genealogical line from a common ancestor or Taxon.

LINEAR Adj. (Latin, *linea* = line.) Straight; in the form of a straight line. See Outline Shape.

LINEAR BUG *Phaenacantha australiae* Kirkaldy [Hemiptera: Colobathristidae].

LINEATE Adj. (Latin, *linea* = line; *-atus* = adjectival suffix.) Pertaining to structure with linear marks including longitudinally striped or parallel lines; raised or depressed lines. Alt. Lineated; Lineatus.

LINED CLICK-BEETLE *Agriotes lineatus* (Linnaeus) [Coleoptera: Elateridae].

LINED SPIDERS *Miturga* spp. [Araneida: Miturgidae].

LINED SPITTLEBUG *Neophilaenus lineatus* (Linnaeus) [Hemiptera: Cercopidae].

LINED STALK-BORER *Oligia fractilinea* (Grote) [Lepidoptera: Noctuidae].

LINELL, MARTIN LARSSON (1849–1897) (Schwarz 1901, Proc. ent. Soc. Wash. 4: 177–180, bibliogr.)

LINEOLATE Adj. (Latin, *Linea* = line; *-atus* = adjectival suffix.) A delicate, fine line. Alt. Lineolet.

LINGNAU, WERNER AUGUST (1904–) (Rohlfien 1975, Beitr. Ent. 25: 270.)

LINGUA Noun. (Latin, *lingua* = tongue.) 1. A median lobe of the Hypopharynx in Apterygota. 2. Hymenoptera: The Ligula. 3. Lepidoptera and Diptera: The Maxillary structures; the Hypopharynx.

LINGUA SPIRALIS The spiral or coiled tongue of Lepidoptera. See Glossa.

LINGUACUTA Noun. (Latin, *lingua* = tongue; Latin, *acutus* = sharp.) The slender sclerite articulating against the extension of each Paralingua (MacGillivray).

LINGUAL Adj. (Latin, *lingua* = tongue; *-alis* = characterized by.) Connected with the tongue. Descriptive of or pertaining to the tongue. Rel. Sublingual.

LINGUAL GLANDS 1. Psocoptera: A pair of chitinized sclerites on the ventral aspect of the Hypopharynx, apparently without glandular structure. 2. Mallophaga: A pair of ovoid sclerites with rod-like stalks, associated with the Hypopharynx, presumably of a glandular nature (Imms).

LINGUATENDON Noun. (Latin, *lingua* = tongue; *tendere* = to strech. Pl., Linguatendons.) The tendon attached to the Linguacuta (MacGillivray).

LINGUIFORM Adj. (Latin, *lingua* = tongue; *forma* = shape.) Tongue-shaped; an elongate structure with the apical margin obtusely rounded. See Tongue. Cf. Ligulate; Lobiform. Rel. Form 2; Shape 2; Outline Shape.

LINGULA Noun. (Latin, *lingula* = little tongue.) 1. Aleyrodidae: A more-or-less slender tongue or strap-shaped organ attached cephalad within the vasiform orifice. 2. A term proposed by Leuckart for the Ligula of bees (Smith). 3. The sclerite on each side supporting the caudal half of each lateral area of the Hypopharynx (MacGillivray).

LINKAGE Noun. (Anglo Saxon, *hlince* = link; *age* = *aticum* = action. Pl., Linkages.) 1. Genetics: The tendency of groups of characters in chromosomes to keep together instead of assorting freely and haphazardly.

LINKAGE DISEQUILIBRIUM (Weir 1979, Biometrics 35: 235–254).

LINKE, MAX (1876–1963) (Dorn 1960, Mitt. dt. ent. Ges. 19: 73–75.)

LINKOLA, JUSSI (1911–1938) (Savas 1938, Suomen hyönt. Aikak. 4: 186–187.)

LINNAEUS (LINNE), CAROLUS VON (12 May 1707–10 Jan 1778) 'He found biology a chaos; he left it a cosmos.' Linnaeus was a Swedish Natural Historian, physician to the King of Sweden and Professor at the University of Upsala. He began as a Botany student at University of Lund and became Supervisor of Botanical Garden at Upsala (1728); M.D. degree University of Harderwyk (1737). At Upsala, Linnaeus was Professor of Medicine (1740) and Professor of Botany (1741). Under Linnaeus lectureship, enrolment at Upsala increased from 500 to 1,500 students. He enunciated principles of systematic biology, defined Genera and Species and established binomial nomenclature. Linnaeus described ca 2,900 Species of insects, in addition to many more Species of plants and animals. His best know zoological publication is *Systemae Naturae* (1735), with its 10th edition (1758) taken as the starting date for modern Zoological Nomenclature (except Spiders, see LeClerq *Svenska Spindlar* 1757). Other notable works by Linnaeus include *Fundamenta Botanica* (1736), *Genera Plantarum* (1737), *Flora Lapponica* (1737), *Classes Plantarum* (1738), *Philosophia Botanica* (1751) and *Species Plantarum* (1751). (Fabricius 1780, Dt. Mus. 1: 431–441; Fée 1831, Mém. Soc. sci. Agric. Lille 1: 1–37, bibliogr.; Jardin 1848, Nats. Libr. 6: i–xxxiii, 27–92, bibliogr.)

LINNELL, JOHN (–1906) (Saunders 1907, Entomol. mon. Mag. 43: 69–70.)

LINOGNATHIDAE Plural Noun. Smooth Sucking Lice. A moderately small Family of Anoplura whose Species parasitize domestic mammals. Head relatively small; compound eyes and ocular Points absent; Mandibles absent, mouthparts forming a Beak or rounded anteriorly; head and Thorax with a few Setae; fore Coxae widely separated; Abdomen without Paratergites. See Anoplura.

LINTNER, JOSEPH ALBERT (1822–1898) (Goding 1891, Trans. N. Y. Agric. Soc. 1890: 361–363; Weiss 1936, *Pioneer Century of American Entomology*. 320 pp. (248), New Brunswick.)

LINTOX See Lindane.

LION BEETLE *Ulochaetes leoninus* LeConte [Coleoptera: Cerambycidae].

LIOPTERIDAE Ashmead 1895. Plural Noun. A numerically small, pantropical Family of apocritious Hymenoptera assigned to the Cynipoidea. Morphologically similar to the Ibaliidae but separated by relatively high placement of Petiole and gastral Terga IV–VI largest. Biology unknown. See Cynipoidea. Cf. Ibaliidae.

LIOY, PAULO (1834–1911) (Schmitz 1926, IIIth int. ent. Kongr. Zurich: 39–40.)

LIP Noun. (Middle English, *lippe* > Anglo Saxon, *lippa*.) 1. Soft, pliant lobes that surround the oriface of the mouth. 2. Acarology: Protuberances anterior of the mouth that enclose the preoral cavity and are a continuation of the pharyngeal wall.

LIPASE Noun. (Greek, *lipos* = fat.) A digestive enzyme which splits or breaks fats in food. See Enzyme.

LIPOCHROMOUS Adj. (Greek, *lipos* = fat; *chroma* = colour; Latin, *-osus* = with the property of.) Descriptive of structure that lacks colour. See Coloration.

LIPOID Adj. (Greek, *lipos* = fat; *eidos* = form.) A fat or fat-like substance utilized by living organisms; insoluble in water but soluble in ether and other fat-solvents.

LIPOLYTIC Adj. (Greek, *lipos* = fat; *lyein* = to dissolve; *-ic* = characterized by.) Fat reducing; pertaining to the enzyme lipase; having the action of lipase.

LIPOPHORINS Noun. (Greek, *lipos* = fat; *phoros* = bearing.) A major category of proteins that carry lipids to various body tissues and Oocytes. Lipophorins are classified into low-denisty lipoproteins, high-density lipoproteins and very high-density lipoproteins. Cf. Vitelogenin.

LIPOPTERA The Order (or Suborder) Mallophaga; specifically, biting lice or bird lice. See Mallophaga.

LIPOSCELIDAE Plural Noun. Booklice. A cosmopolitan Family of troctomorphous Psocoptera including the book louse, *Liposcelis divinatorius* (Müller). Adult body small and dorsoventrally compressed; Antenna with fewer than 20 segments and with secondary annulations; hind Femur enlarged; Tarsi with three segments; winged and wingless forms. Several Species are domicillary and associated with stored products. See Psocoptera.

LIQUEFACTION Noun. (French, *liquefier* > Latin, *liquere* = to be liquid; English, *-tion* = result of an action.) The process of turning solids into liquids.

LIQUID NITROGEN A compound used in wall voids as a treatment for drywood termites. See Inorganic Insecticide.

LIQUI-STIK® See NAA.

LIRIOPEIDAE See Ptychopteridae.

LIRO, JOHAN (JSI) IVAR (1872–1943) (Salaas 1943, Suomen hyönt. Aikak. 9: 217–218.)

LISLE, (DELISLE), GUILLAUME DE (1675–1726) (Rose 1850, *New General Biographical Dictionary* 7: 50.)

LISNEY, DOROTHY ELIA (1910–1973) (Lees 1975, Proc. R. ent. Soc. Lond. (C) 39: 57.)

LIST, GEORGE M (1885–1956) (Daniels & Palmer 1957, J. Econ. Ent. 50: 524.)

LISTER, JOSEPH JACKSON (1857–1927) (Collins 1927, Proc. ent. Soc. Lond. 2: 104–105.)

LISTER, MARTIN (1638–1712) (Duméril 1823, *Considérations générale sur la classe des insectes.* 272 pp. (245–246), Paris.)

LISTO, JAAKO (1900–1935) (Tullgren 1935, Ent. Tijdschr. 56: 190–191.)

LITCHI ERINOSE MITE *Eriophyes litchii* Kieffer [Acari: Eriophyidae].

LITCHI FRUIT-MOTH *Cryptophlebia ombrodelta* (Lower) [Lepidoptera: Tortricidae].

LITCHI MITE *Eriophyes litchii* Keifer [Acari: Eriophyidae].

LITCHI STINK BUG *Lyramorpha rosea* Westwood [Hemiptera: Tessaratomidae].

LITERATE Adj. (Latin, *litera* = a letter; *-ate* = to put into action; *-atus* = adjectival suffix.) Ornamented with characters like letters. Alt. Literatus.

LITERATURE Noun. (Latin, *literatura* = a writing.) The collective major and minor publications in all languages pertaining to any science. Not construed as exemplifying a classic style in any one of them.

LITHOCHROA BLUE BUTTERFLY *Jalmenus lithochroa* Waterhouse [Lepidoptera: Lycaenidae].

LITHURGINAE A Subfamily of Megachilidae (Apoidea) consisting of about five Genera.

LITTAC® See Alphacypermethrin.

LITTER Noun. (Middle English, *liter* = bed. Pl., Litters.) The uppermost layer of partially decayed organic matter on the forest floor. Cf. Debris; Detritus; Humus; Mor.

LITTLE BLACK-ANT *Monomorium minimum* (Buckley) [Hymenoptera: Formicidae]: Endemic to North America, omnivorous and common in urban habitats. Workers are monomorphic, ca 1.5–3.0 mm long, slender bodied, dark brown or black and shining with sparse pubescence; Antenna with 12 segments including well developed club of three segments. Colonies nest out-of-doors and form conical craters; occasionally nest in homes. LBA workers forage for fruits, sweets and sometimes meats. See Formicidae. Cf. Argentine Ant; Black Carpenter-Ant; Cornfield Ant; Fire Ant; Larger Yellow-Ant; Odorous House Ant; Pavement Ant; Pharaoh Ant; Thief Ant.

LITTLE CARPENTER BEE *Ceratina dupla* Say [Hymenoptera: Xylocopidae (Anthophoridae)]: A bivoltine, polylectic bee widespread in North America. Adults are metallic blue; female excavates nest from pith of dead stems and twigs; nests consist of cells in twigs that are partitioned by cemented fragments of stem. One female constructs several cells and waits in her personal cell at the top of the nest for progeny emergence; development of progeny is synchronous and daughters fly away with mother; one daughter returns to nest and establishes residency. See Anthophoridae. Cf. Carpenter Bee; Mountain Carpenter Bee. Rel. Bumble Bee.

LITTLE CARPENTER-WORM *Prionoxystus macmurtrei* (Guérin) [Lepidoptera: Cossidae].

LITTLE EARWIGS See Spongiphoridae.

LITTLE FIRE-ANT *Ochetomyrmex auropunctatus* (Roger) [Hymenoptera: Formicidae].

LITTLE GREEN-LEAFHOPPER *Balclutha incisa hospes* (Kirkaldy) [Hemiptera: Cicadellidae].

LITTLE HOUSE-FLY *Fannia canicularis* (Linnaeus) [Diptera: Muscidae]: A widespread, synanthropic pest that replaces the House Fly in boreal habitats. LHF resembles House Fly but smaller and more slender; Arista not plumose; thoracic dorsum with three dark longitudinal stripes. Adults often hover and dart in midair. Larvae are flattened with conspicuous lateral spines and feed in decaying organic matter, including excrement. Larvae cause Myiasis in humans. See Muscidae. Cf. House Fly; Latrine Fly; Coastal Fly.

LITTLE MARBLED SCORPION *Lychas marmoreus* (C.L. Koch) [Scorpionida: Buthidae].

LITTLE PASTURE-COCKCHAFER *Aphodius frenchi* Blackburn [Coleoptera: Scarabaeidae].

LITTLE SULPHURS Any member of the pierid butterfly Genus *Eurema*.

LITTLE YELLOW-ANT *Plagiolepis alluaudi* Emery [Hymenoptera: Formicidae].

LITTLER, FRANK MERVYN (1880–1922) (Musgrave 1932, *A Bibliogr. of Australian Entomology 1775–1930*. 380 pp. (202), Sydney.)

LITTLEWOOD, FRANK (1882–1949) (Anon. 1949, Rep. Raven ent. nat. Hist. Soc. 1949: 37.)

LITTORAL Adj. (Latin, *litus* = seashore; *-alis* = characterized by.) Descriptive of a habitat that pertains to life along the sea-coast or in the shore debris; strictly, between tide marks. Alt. Litoral. Cf. Benthic; Estuarine; Pelagic.

LITTORAL ZONE The biogeographic marine realm near the shoreline and involving life that occurs between high and low tide. See Habitat. Cf. Abyssal Zone; Neritic Zone; Pelagic Zone; Photic Zone.

LITTORALIA Noun. (Latin, *litus* = seashore.) Heteroptera that live on the margins of streams or bodies of water.

LITURA Noun. (Latin, *litura* = a blur. Pl., Liturae.) An indistinct spot with pale margins; a spot which appears blotted.

LITURATE Adj. (Latin, *litura* = a blur; *-atus* = adjectival suffix.) Marked with Litura. Alt. Lituratus.

LIVELY ANT-GUEST BEETLES See Tenebrionidae.

LIVETT, H W (–1907) (Fowler 1901, Proc. ent. Soc. Lond. 1901: xxxv.)

LIVID (Latin, *lividus* = bluish.) Adj. Lead-colour; liver-colour. A combination of black with blue, a pale purplish brown (Kirby & Spence).

LIVINGSTONE, CLERMONT (–1907) (Harvey 1907, Quart. Bull. Br. Columb. ent. Soc. 8:1.)

LIZARD BEETLES See Languridae.

LIZER Y TRELLES, CARLOS A (1887–1959) (B[iraben] 1959, Neotropica 5: 55.)

LJUNGDAHL, DAVID (1870–1940) (Sachtleben 1942, Arb. morph. taxon. Ent. Berl. 9: 65.)

LJUNGH, SVEN INGEMAR (1757–1828) (Anon. 1829, svenska VetenskAkad. Handl. 1828: 279–282.)

LLANSO, JAINE (–1942) (Anon. 1942, Revta Soc. ent. argent. 11: 382.)

LLOYD, BERTRAM (1881–1944) (Hayward 1944, NWest nat. 19: 192–193.)

LLOYD, CHARLES THOMAS (1883–1951) (Riley 1952, Proc. R. ent. Soc. Lond. (C) 16: 86.)

LLOYD, ROBERT WYLIE (1868–1958) (Anon. 1858, Entomol. mon. Mag. 94: 97–99.)

LOA LOA Noun. (Congo dialect.) A filarial worm, *Filara loa* (Guyot), in African rainforests that causes Loiasis in humans. Microfilarial stage of worm develops in fat-body of tabanid fly by moulting twice to become infective larva; larvae then migrate into head of fly, accumulate and penetrate Hypopharynx to be passed into vertebrate while fly is feeding. Disease symptoms include Calabar Swellings. Alt. Loa. See Loiasis.

LOBATE Adj. (Greek, *lobos*; Latin, *lobus* = lobe; *-atus* = adjectival suffix.) Lobe-like. Descriptive of structure with lobes; structure divided by deep, undulating and successive incisions. Alt. Lobatus. Syn. Lobiform. Cf. Conglobate; Globate.

LOBE Noun. (Greek, *lobos*; Latin, *lobus* = lobe. Pl., Lobes.) 1. Any prominent, rounded process or excrescence along a margin or the surface of a structure. 2. Lateral expansions of the abdominal segments. 3. A wing area (posterior) defined by two convex margins separated by a notch. See Notch; Anal Lobe; Claval Lobe. 4. Diaspididae: The rounded, tooth-like processes on the margin of the Pygidium. Coccidae: The broad semi-oval projections of the pygidial fringe (MacGillivray).

LOBELET Noun. (Greek, *lobos*; Latin, *lobus* = lobe. Pl., Lobelets.) Coccids: Any of the subdivisions of a Lobe (MacGillivray).

LÖBERBAUER, RUDOLF (–1967) (Kusdas 1967, Wien. Ent. Ges. 52: 107–108, bibliogr.)

LOBES OF PRONOTUM Orthoptera: The spaces or areas formed by three transverse impressions on the Pronotum.

LOBIFORM Adj. (Latin, *lobus* = lobe' *forma* = shape.) Lobe-shaped; a rounded process. Syn. Lobate. See Lobe. Cf. Linguiform; Pyreniform. Rel. Form 2; Shape 2; Outline Shape.

LOBLOLLY-PINE SAWFLY *Neodiprion taedae linearis* Ross [Hymenoptera: Diprionidae].

LOBOSCELIDIIDAE Yoshimoto & Maa 1961. Plural Noun. A small Family of apocritous Hymenoptera of problematical placement: Originally in Proctotrupoidea, later transferred to Chrysiodidea as an independent Family, and more recently placed as a Subfamily within Chrysididae. Loboscelidiids consist of about 20 nominal Species found in the Oriental and Indoaustralian Realms. Adults are strongly sexually dimorphic with 13 antennal segments, forewing with Humeral Vein and Jugal Lobe; tibial spur formula 1-2-2; Occipital Flanges present. Loboscelidiid biology is uncertain: Purportedly myrmecophilous and one Species reported from phasmatid eggs. See Chrysididae.

LOBULATE Adj. (Latin, *lobus* > Greek, *lobos* = lobe; *-atus* = adjectival suffix.) 1. Pertaining to any margin that is incised. 2. Divided into, or with may small lobes or lobules.

LOBULE Noun. (Latin, *lobus* > Greek, *lobos* = lobe. Pl., Lobules.) 1. A small lobe. 2. Coccids: Lobes *sensu* MacGillivray.

LOBULUS Noun. (Latin, *lobus* = lobe. Pl., Lobuli.) The partly separated portion of the wings of some flies and of the hindwings in some Hymenoptera. See Alula.

LOBUS INFERIOR The lower lobe off the Maxillae covered by the upper or Superior Lobus.

LOBUS MAXILLAE An arcane term for the Maxillary Lobe or the Galea.

LOBUS SUPERIOR An arcane term for the Galea.

LOCAL INTERNEURONS See Deutocerebrum.

LOCHHEAD, WILLIAM (1864–1927) (Leopold 1927, Rep. ent. Soc. Ont. 58: 86–91, bibliogr.).

LOCK-AND-KEY HYPOTHESIS A long standing hypothesis which argues that the complex anatomy of genitalia observed in many Species maintains specificity via reproductive isolation through physical barriers to copulation. Rel. Biological Species Concept.

LOCK-ON® See Chlorpyrifos.

LOCO CITATO Latin phrase meaning 'a place cited' and used in taxonomic literature to indicate the publication and page for a description or reference pertaining to the object under discussion. Abbrev., *l.c.*

LOCOMOTION Noun. (Latin, *locus* = place; *motivus* = capable of moving. Pl., Locomotives.) 1 Progressive movement. 2. The act, ability or capacity to move from one place to another. Insect locomotion involves the action of legs or wings, or the rapid contraction of muscles to strike a part of the body against the substrate and propel the insect. See Leg; Wing. Cf. Crawl; Fly; Hop; Jump; Run; Walk. Rel. Flight; Cursorial; Fossorial; Gressorial; Raptorial; Saltatorial; Scansorial.

LOCOMOTOR Adj. (Latin, *locus* = a place; *motio* = motion; *or* = one who does something.) Descriptive of or pertaining to locomotion or movement from place to place; serving for locomotion. Alt. Locomotory.

LOCUCIDE® See *Nosema locustae*.

LOCUST Noun. (Latin, *locusta* = locust. Pl., Locusts.) Highly destructive Species of Orthoptera, typically Acrididae, that periodically occur in dense, strongly migrating plagues. The term 'locust' is incorrectly applied to cicadas. Correctly: A plague of locusts. See Orthoptera. Cf. Grasshopper.

LOCUST-BEAN MOTH See Carob Moth.

LOCUST BORER *Megacyllene robiniae* (Forster) [Coleoptera: Cerambycidae]: A monovoltine, significant pest of black locust in eastern North America. Adults are ca 12–18 mm long, velvet-like with pubescence forming black and golden crossbars on dorsum of body; venter uniformly black; Antenna and legs red. Eggs are elongate, white and laid within bark fissures; eclosion occurs within 14 days. LB overwinters as early-instar larvae within innerbark or sapwood of host plant; feeding larvae bore through sapwood and heartwood; feeding damage often causes death of tree. Pupation occurs inside cells formed within wood. Adults become active during September. See Cerambycidae. Cf. Elm Borer; Painted Hickory-Borer; Poplar Borer.

LOCUST LEAFMINER *Odontota dorsalis* (Thunberg) [Coleoptera: Chrysomelidae].

LOCUST LEAF-ROLLER *Nephopterix subcaesiella* (Clemens) [Lepidoptera: Pyralidae].

LOCUST TWIG-BORER *Ecdytolopha insiticiana* Zeller [Lepidoptera: Tortricidae].

LOCUSTAKININ A signalling peptide isolated from locusts by use of myotropic assays. Myotropic peptide Families are associated with myotropic-, allatostatic-, pheromonotropic and physiological activities, diapause induction, stimulation of cuticular melanization, and diuresis. Some Locustakinins may be neurotransmitters present in nerves innervating the oviduct, salivary glands, male accessory glands, and the Heart; other peptides are stored in neurohemal organs. Other locust signalling peptides include locusta-myotropins, locustapyrokinins, locustatachykinins, Locusta accessory gland myotropins, locusta-sulfakinin and Locusta cardioactive peptide.

LOCY, WILLIAM ALBERT (1857–1924) (Calvert 1924, Ent. News 35: 386.)

LODEIZEN, J A F (–1950) (Stärcke 1950, Ent. Ber., Amst. 13: 225.)

LODGE, GEORGE EDWARD (1860–1954) (Buxton 1955, Proc. R. ent. Soc. Lond. (C) 19: 69.)

LODGEPOLE-CONE BEETLE *Conophthorus ponderosae* Hopkins [Coleoptera: Scolytidae].

LODGEPOLE NEEDLE-MINER *Coleotechnites milleri* (Busck) [Lepidoptera: Gelechiidae]: A widespread pest of lodgepole pine in North America. Syn. *Evagora milleri* (Busck).

LODGEPOLE-PINE BEETLE *Dendroctonus murrayanae* Hopkins [Coleoptera: Scolytidae].

LODGEPOLE SAWFLY *Neodiprion burkei* Middleton [Hymenoptera: Diprionidae].

LODGEPOLE TERMINAL-WEEVIL *Pissodes terminalis* Hopping [Coleoptera: Curculionidae].

LÖDING, HENRY PETER (1869–1942) (Engelhardt 1942, Bull Brooklyn ent. Soc. 37: 50–51.)

LODY® See Fenpropathrin.

LOEB, JACQUES (1859–1924) (Levene 1924, Science 59: 427–430.)

LOEBEL, FRIEDRICH (–1960) (Anon. 1960, Z. wien ent. Ges. 45: 160.)

LOEFLING, PEHR (1729–) (Papavero 1971, *Essays on the History of Neotropical Dipterology*. I: 4–7. São Paulo.)

LOELIGER, ROBERT (–1952) (Benz 1952, Mitt. ent. Ges. Basel 2: 54–55.)

LOEW, F HERMANN (1807–1878) German taxonomist with major contribution involving Diptera found in Baltic amber. (Kowarz 1879, Verh. zool.-bot. Ges. Wien 29: 45–47.)

LÖFGREN, GUSTAF (1848–1910) (S. B. 1911, Ent. Tidskr. 37: 108–109.)

LÖFGREN, VAINÖ ARMAS (1894–1940) (Kangas 1940, Suomen hyönt. Aikak 6: 79–80.)

LOFTHOUSE, T A (Hale-Carpenter 1946, Proc. R. ent. Soc. Lond. (C) 10: 54.)

LOFTIN, ULPHIAN CARR (1890–1946) (Heinrich *et al.* 1946, Proc. ent. Soc. Wash. 48: 240–243, bibliogr.)

LOG PHASE Tissue Culture: The period after the Lag Phase and before the Stationary Phase. Cells. During the Log Phase, cell populations grow exponentially. See Growth Curve. Cf. Exponential Phase; Lag Phase.

LOGAN, ROBERT FRANCIS (1827–1887) (Anon. 1887, Entomol. mon. Mag. 24: 92–93.)

LOG-CABIN CADDISFLIES See Limnephilidae.

LOGIC® See Fenoxycarb.

LOHMANDER, HANS (1896–1961) (Lindroth 1961, Opusc. ent. 26: 1–3.)

LOHMANN, HANS (1863–1933?) (Schnakenbeck 1933, Biologie 2: 316–317.)

LOHWAG, KURT (1913–1970) (Weindlmayer 1970, Anz. Schädlingsk. Pflandenschultz 43: 125.)

LOIASIS Noun. (From Loa; *-iasis* = suffix for names of diseases.) A filarial disease caused by infection with *Loa loa* and affects humans in tropical

Africa. Symptoms include 'Calabar swellings' that periodically appear on wrists and ankles in response to antigenic material released by worms in the body. Adult female worms are ca 50–70 mm; adult male worms are ca 30–35 mm. Adults move under skin through connective tissue; microfilariae occur in blood of host. Disease seems restricted to forested habitats. Two forms of disease are recognized; human form with microfilariae active in blood during day and vectored by day-biting tabanid flies of Genus *Chrysops,* particularly *C. dimidiata* and *C. silacea.* Second form of disease occurs in monkeys.

LOKAY, EMANUEL (1822–1880) (Obenberger 1928, Cas. csl. Spol. ent. 25: 127–130, bibliogr.)

LOLOC® See *Nosema locustae.*

LOMBARDI, MASSIMILIANO (–1947) (Anon. 1947, Boll Soc. ent. Ital. 7: 1.)

LOMBARDINI, GIACONDO (1886–1965) (Anon. 1966, Bol. Soc. ent. Ital. 69: 16.)

LOMNICKI, JEROSLAW (–1931) (Kinee 1932, Polskie Pismo ent. 11: 1–16, bibliogr.)

LOMNICKI, MARTIN VON (1844–1915) (Soldanski 1916, Dt. ent. Z. 1916: 88.)

LONA, CARLO (1885–1971) (Anon. 1972, Boll Soc. ent. ital. 104: 21.)

LONCHAEIDAE Plural Noun. A cosmopolitan Family of about 500 Species. Species often are rare in nature and scarce in collections. Lonchaeids are closely related to Lauxaniidae and superficially some Species resemble anthomyiids or piophilids. Adults are usually shiny or metallic blue-black or black in colour. Females possess long piercing Ovipositors similar to Otitidae; setose Frons; one reclinate Fronto-orbital Bristle, Mesopleuron setose with one or more bristles; Haltere black. Larvae develop in fruits and vegetables; larvae usually are secondary invaders in association with other phytophagous insects. Some Species spend larval lives under bark of dead and dying trees. With one exception, Species are saprophytic and not beneficial except in the role they play in breakdown of rotting fruit and vegetables. Cf. Lauxaniidae.

LONCHOPTERIDAE Plural Noun. Spear-Winged flies. A small, monogeneric Family of cyclorrhaphous Diptera assigned to Superfamily Lonchopteroidea. Adults are less than 5 mm long, slender and yellowish with wings pointed apically. Males are extremely rare and some Species may be parthenogenetic. SWF can be common in moist shady areas where larvae inhabit decaying plant material. Pupae coarctate. Syn. Muscidoridae. See Lonchopteroidea.

LONCHOPTEROIDEA A Superfamily of cyclorrhaphous Diptera included within the Series Aschiza and containing the Lonchopteridae (Musidoridae). See Diptera.

LONE-STAR TICK *Amblyomma americanum* (Linnaeus) [Acari: Ixodidae].

LONG BROWN-SCALE *Coccus longulus* (Douglas) [Hemiptera: Coccidae]. See Long Soft-Scale.

LONG SOFT-SCALE *Coccus longulus* (Douglas) [Hemiptera: Coccidae]: A cosmopolitan pest of citrus, carambola, custard apple, lychee, leucanena, fig and ornamental plants. Adult females are grey-brown, oval in outline shape and 4–6 mm long; eye spots are visible. Females produce crawlers over a 1-month period. LSS completes two nymphal instars and the life cycle requires ca 2 months. LSS is multivoltine with 4–6 generations per year. LSS produces large amounts of honeydew which serves as substrate for sooty mould and other fungi. See Coccidae.

LONG-HEADED FLOUR BEETLE *Latheticus oryzae* Waterhouse [Coleoptera: Tenebrionidae]: A widespread pest of stored barley, corn, flour, rice and wheat; LHFB is a minor pest of flour and processed grains in Australia but serious pest of stored cereals in southeast Asia and other topical regions. LHFB damage is through feeding and persistent odour from defensive gland secretions. Adults are 2.5–3.0 mm long, slender, flattened and pale yellow-brown; apical antennal segment smaller than other segments; head disproportionately long. Low reproduction by female with 5–10 eggs laid over three days; incubation requires 3–4 days. LHFB usually completes 6–7 larval instars, with regressive moults at low temperature and poor diet. Larval development requires ca 15 days, pupation requires 3–4 days and life cycle is complete within 4–6 weeks. See Tenebrionidae. Cf. Broad-horned Flour Beetle; Slender-horned Flour Beetle.

LONG-HORN BEETLES See Cerambycidae.

LONG-HORNED CADDIS *Triplectides australis* Navas [Trichoptera: Leptoceridae].

LONGHORNED CADDISFLIES. See Leptoceridae.

LONGHORN MOTHS See Adelidae.

LONGHORNED BEETLES See Cerambycidae.

LONGHORNED GRASSHOPPERS See Tettigoniidae.

LONGICORN Noun. (Pl., Longicorns.) 1. Insects with an Antenna as long or longer than the body. 2. A common name typically applied to cerambycid beetles. See Cerambycidae.

LONGICORN BEETLE Coleoptera of the Family Cerambycidae. Name from adult's long, many-segmented Antenna. LB are forest pests that attack stressed, dying or damaged standing trees and freshly felled logs. Adults usually are elongate-rectangular 5–85 mm long. Larvae generally are white, thick and fleshy, somewhat cylindrical in shape tapering towards apex of Abdomen. LB with expanded Thorax and brown-black head and mouthparts. Larvae bore and tunnelling does not structurally weaken timber but reduces value. When infested timber is used in a building or furniture, emerging adults may damage lining or covering materials. See Cerambycidae. Syn. Longhorned Beetles.

LONGICORNIA Noun. The Cerambycidae or longhorned beetles.

LONGIPENNATE Adj. (Latin, *longus* = long; *penna*

= wing; *-atus* = adjectival suffix.) Long-winged.

LONGITUDINAL Adj. (Latin, *longitudo* from *longus* = long; *-alis* = characterized by.) A structure or appendage oriented in the direction of the long axis of the body.

LONGITUDINAL VEINS Wing veins that typically extend lengthwise through the wing directly from the base or as branches of veins that originate at the wing base and extend toward the apex. Longitudinal veins that originate at the base of the wing often contain Tracheae. Longitudinal vein-like structures that are solid may be called Pseudoveins and their Homology with true veins is drawn into question. Pseudoveins occur in some Heteroptera whose longitudinal veins are solid. Biomechanically, longitudinal veins may act as trusses while crossveins act as struts. Tubular veins are cylindrical in cross-section; the cross-sectional area, shape and thickness of longitudinal wing veins differ, which suggests dynamical properties of veins beyond reinforcement and containment of wing membrane. A perfectly circular vein of uniform diameter along its length would resist forces of compression and tension equally along the vein. An elliptical vein increases rigidity and resists bending in the long axis of the vein cross-section. The short axis of the cross-section can increase its strength in resisting bending by periodic contact with cross veins. Other outline shapes of longitudinal veins are possible and point to more sophisticated mechanisms of strength. Longitudinal veins often branch or subdivide along their course toward the apex of the wing. Three basic patterns of branching are recognized. A dichotomous branching pattern is most common, and consists of periodic forks or divisions of the longitudinal veins. A pectinate branching pattern consists of a serial arrangement of branches that lie along a main longitudinal vein. The branches form in the same direction and display a similar angle of divergence from the main longitudinal vein. A triadic branching pattern is seen in palaeopterous insects that display a secondary longitudinal vein between pairs of branches of main longitudinal veins. The intercalated (secondary) vein is of opposite sign to the adjacent longitudinal veins. A triad is called positive when the intercalated vein is negative and adjacent veins are positive; a triad is called negative when the intercalated vein is positive and adjacent veins are negative. Longitudinal veins of the generalized insect wing: Precosta is a small vein at the wing base of some fossil insects but it is not present in modern insects as a distinct vein. Costal Vein is the dominant vein near the anterior margin of the wing in modern insects. Sometimes the costal vein is confluent with the anterior margin of the wing. An Ambient Vein may be continuous with the apex of the Costal Vein and continue around the wing margin. When present, the ambient vein probably stiffens the margin of the wing. The Subcostal Vein is the second vein of the wing, and articulates with the anterior process of the First Axillary Sclerite. The Subcostal Vein is forked once distad to form two short branches (Sc1, Sc2). Radial Vein is the third vein of the wing and generally the most well developed vein. RV articulates with the Second Axillary Sclerite at the wing base, forks near the middle of the wing and forms an anterior R1 and posterior Radial Sector. Distally, the RS branches twice to form four subveins R2–5. The Medial Vein is the fourth vein of the wing. Hypothetically the MV branches into a Media Anterior and Media Sector The Media Anterior has two further branches; the MS has four terminal branches. Most extant insect Species (Neoptera) have lost the Anterior Media; the Posterior Media remains four-branched. (In contrast Odonata have lost the Media Sector. Media of other insects is attached to or continuous with the Distal Median Plate). The Cubital Vein is the fifth vein of wing with two primary basal branches (Cu1, Cu2). Cu1 has variable number of secondary branches, but typically the number is two; Cu2 has been mistaken for an Anal Vein by some workers. Proximally, Cubitus is associated with the Distal Median Plate; Cubitus is fused to M in ephemeropteroid orders and fused to Postcubitus in Planoneoptera. Postcubitus is the first Anal Vein in the Comstock-Needham classification, but is considered a separate vein because it is not attached to Anal Veins basally and is associated with the Cubital Vein proximally rather than the Third Axillary Sclerite. Typically Postcubitus is not branched. Anal veins (Vannal Veins) are associated with the Third Axillary Sclerite. Anal Veins vary in number (0–16) and are responsible for flexing the wing. Distally, Anal Veins are branched or not branched. Jugal Veins are variable in number (0–2) and typically short. Jugal veins are not responsible for support of the proximal wing region. Primary veins responsible for support of wing surface are: Costa, Subcosta and Radius, at least in proximal region. See Wing; Vein. Cf. Crossvein. Rel. Pseudovein.

LONG-JOINTED BEETLES See Tenebrionidae.

LONG-LEGGED ANT *Anoplolepis longipes* (Jerdon) [Hymenoptera: Formicidae].

LONG-LEGGED FLIES See Dolichopodidae.

LONG-LEGGED HARVESTMEN See Phalangiidae.

LONG-LIPPED BEETLES See Telegeusidae.

LONG-NOSED CATTLE LOUSE *Linognathus vituli* (Linnaeus) [Phthiraptera: Linognathidae].

LONGJAWED SPIDERS *Tetragnatha* spp. [Araneida: Tetragnathidae].

LONGLEAF-PINE SEEDWORM *Cydia ingens* (Heinrich) [Lepidoptera: Tortricidae].

LONGSTAFF, GEORGE BLUNDELL (1849–1921) (Image *et al.* 1921, Entomol. mon. Mag. 57: 157–161.)

LONG-TAILED FRUIT-FLY PARASITE *Opius longicaudatus* Ashmead [Hymenoptera: Braconidae].

LONGTAILED MEALYBUG *Pseudococcus longispinus* (Targioni Tozzeti) [Hemiptera: Pseudococcidae]: A cosmopolitan pest of avocado, citrus, grape, mango, pear and numerous ornamental plants. Adult females are oval in outline shape, 3–4 mm long, slow-moving, covered with white mealy wax and filaments along margin of body and two caudal filaments as long and body. Females do not produce egg sac, but produce living crawlers that accumulate among wax threads under her body; 200–300 crawlers are produced during 2–3 week interval. All stages prefer sheltered habitats including curled leaves, under fruit calyx, and fissures in bark. Females moult three times; males moult four times. Nymphs and adult females produce copious amounts of honeydew that serves as substrate for sooty mould and other fungi. Complete biological control of LTM has been achieved on Avocado in California with parasitic Hymenoptera. See Pseudococcidae.

LONG-TAILED ORB-WEAVING SPIDER *Argiope protensa* L. Koch [Araneida: Araneidae].

LONG-TOED WATER BEETLES See Dryopidae.

LONG-TONGUED BEES Hymenoptera: Any member of the Anthophoridae, Apidae, and Megachilidae in which the adult Glossa is elongate and apically pointed. See Bee. Cf. Short-Tongued Bee.

LÖNNEBERG, EINAR (1865–1942) (Anon. 1944, Annali Mus. Civ. Stor. nat. Giacomo Doria 62: (5)-(6).)

LOOMIS, HENRY (1839–1920) (Anon. 1920, Ent. News 31: 240.)

LOOP Noun. (Middle English, *loupe* = bend. Pl., Loops.) The structure at the base of the inner side of the forewings into which the Frenullum of male moths is fitted. See Retinaculum.

LOOPED Adj. Elongated structures with a closed or nearly closed curve at the end, like a loop.

LOOPER Noun. (Pl., Loopers.) A common name for geometrid or other lepidopterous caterpillars that lack some or all the middle abdominal legs. Without these legs, movement is accomplished by bringing tail to Thorax and forming a loop of the middle segments of the body. The shape assumed by the larva in locomotion provides impetus for the common name. See Geometridae. Cf. Armyworm.

LOOPER CATERPILLAR *Chrysodeixis argentifera* (Guenée); *Chrysodeixis eriosoma* (Doubleday); *Chrysodeixis subsidens* (Walker) [Lepidoptera: Noctuidae].

LOOS, KURT (1859–1933) (Springer 1933, Natur. Heimat. 4: 65–67.)

LOPEZ, ALONZO WILLIAM (1900–1932) (Essig 1933, J. Econ. Ent. 26: 306–307, bibliogr.)

LOPEZ-SEOANE, VICTOR (1834–1900) (Dusmet y Alonso 1916, Boln. Soc. ent. Espan. 2: 188.)

LOPHOCATERIDAE Plural Noun. A small Family of Coleoptera, including Siamese Grain Beetle. See Coleoptera.

LOPHOCORONIDAE Common 1973. Plural Noun. A Family of ditrysian Lepidoptera endemic to Australia, assigned to Superfamily Lophocoronoidea and containing fewer than 10 Species. Adults are very small, wingspan 10–15 mm and head with erect, long, hair-like scales; Ocelli and Chaetosemata absent; Antenna filiform; Labrum without Pilifers; Mandibles lobate, not musculated; Proboscis short and weakly developed; Maxillary Palpus of five segments; Labial Palpus of three segments. Epiphysis absent; tibial spur formula 0-2-4; wings held tectiform over Abdomen when at rest; female with piercing Ovipositor. Immature stages are unknown. See Lepidoptera.

LOPHOPIDAE A small Family (ca 150 spp) of auchenorrhynchous Hemiptera assigned to Superfamily Fulgoroidea. Antennal Scrobes below compound eyes; Pedicel large, with conspicuous Sensilla; Frons carinate, longer than wide; Clypeus with lateral Carinae; hind Tibia without large, moveable spur; apex of hind second Tarsomere without spines. Biology of Lophopidae is poorly known and presumably Species are phloem feeders on angiosperms. See Fulgoroidea.

LORD AVEBURY See Lubbock, John.

LORD HOWE ISLAND CICADA *Psaltoda insularis* Ashton [Hemiptera: Cicadidae].

LORD HOWE ISLAND WAX SCALE *Ceroplastes insulanus* De Lotto [Hemiptera: Coccidae].

LORD, JOHN KEAST (1817–1872) (Newman 1873, Entomologist 6: 296.)

LORENZEN, IWER THOR (Weidner 1967, Abh. Verh. naturw. Ver. Hamburg Suppl. 9: 269–270.)

LOREY, TUISKO (1845–1901) (Wimmenauer 1901, Allg. Forst. Jagdztg 78: 113–118.)

LORITZ, JEAN (1891–1965) (Fiammengo & Klinzig 1965, Bull. Soc. ent. Mulhouse 1965 (Suppl. Oct.): 1–8, bibliogr.)

LORQUIN, PIERRE JOSEPH MICHEL (1797–1873) (Boisduval 1869, Ann. Soc. ent. Belg. 12: 5–10.)

LORSBAN PLUS® See Cypermethrin.

LORSBAN® See Chlorpyrifos.

LORUM Noun. (Latin, *lorum* = thong; Pl., Lora.) 1. A sclerotized band connecting the Submentum and Cardo of the Maxilla (Comstock). 2. The Submentum or 'small cords' upon which the base of the Proboscis is seated (Say). 3. The anterior part of the Genae at the edge of the mouth. 4. A sclerite in the lateral wall of the Cranium. Lorum is a lateral extension of the Hypopharnyx *teste* Snodgrass. Syn. Mandibular Plate. 5. Hemiptera: A small sclerite lateral of the Clypeus and front extending toward the Gena; the Mandibular Sclerite (Plate). Heteroptera: A sclerite laterad of the Jugum and anterior of the eyes. 6. Some Diptera: A corneous process attached to the mouth's which flexor muscles; the Maxillary Palpifer. 7. Honeybee: A flexible, transverse band supporting the base of the Submentum, with its extremities attached to the distal ends of the Cardines; the Lora of some authors (Imms). See Mouthparts. Cf. Rostrum.

LORVEK® See Chlorpyrifos.

LOSY, JOSEF (1874–1916) (Csiki 1917, Rovart. Lapok 24: 124–125, 132, bibliogr.)

LOTHARALBER-ROEDER (1902–1972) (de Bros 1972, Mitt. ent. Ges. Basel 22: 99–100.)

LOTIC Adj. (Latin, *lotus* = washing; *-ic* = consisting of.) Pertaining to organisms living in fast-moving waters. A term applied to organisms that normally inhabit rivers (Needham). Cf. Lentic.

LOTIS BLUE *Lycaeides argyrognomon lotis* (Lintner) [Lepidoptera: Lycaenidae].

LOUDON, JOHN CLAUDIUS (1783–1843) (Rose 1850, *New General Biographical Dictionary* 9: 317–318.)

LOUNSBURY, CHARLES PUGSLEY (1872–1955) (Mossop 1955, J. ent. Soc. sth. Afr. 18: 144–148.)

LOUPING ILL A tick-borne arboviral disease closely related to Central European Encephalitis. Typically a disease of sheep in Scotland and Ireland that rarely occurs in humans. LI is enzootic in deer and birds; vectored by sheep tick, *Ixodes ricinus;* trans-stadial transmission occurs but not trans-oviarial transmission. See Arbovirus. Cf. Tick-Borne Encephalitis.

LOUSE Noun. (Middle English, *lous.*) Minute, ectoparastic, wingless insects that feed on warm blooded animals. Two Suborders are currently recognized under the Ordinal name of Phthiraptera: The true or sucking lice (Anoplura) and the biting lice (Mallophaga).

LOUSE FLIES See Hippoboscidae.

LOVE BUGS See Bibionidae.

LOVELL, JOHN HARNEY (1860–1939) (Anon. 1939, Am. Bee J. 79: 568–570.)

LØVENDAL, EMIL ADOLF (1839–1901) (Hansen 1901, Ent. Tidskr. 22: 177–183; Haiso 1901, Ent. Tidschr. 22: 177–183, bibliogr.)

LØVENSKIOLD, MIKAEL HERMAN (1751–1807) (Zimsen 1964, *The Type Material of I. C. Fabricius*. 656 pp. (14), Copenhagen.)

LOVETT, ARTHUR LESTER (1885–1924) (Anon. 1924, J. Econ. Ent. 17: 501–504.)

LÖW, FRANZ (1829–1889) (Rogenhofer 1890, Verh. zool.-bot. Ges. Wien 40: 165–167.)

LOW, HUGH (1824–1905) (Cowan 1968, J. Soc. Biblphy nat. Hist. 4: 327–343, bibliogr.)

LOWE, FRANK EDWARD (1854–1917) (Wheeler 1918, Entomologist's Rec. J. Var. 30: 59–60.)

LOWE, VICTOR HUNT (1869–1903) (Osborn 1937, *Fragments of Entomological History,* 394 pp. (70), Columbus, Ohio.)

LOWE, WILLIAM HENRY (–1900) (Verrall 1900, Proc. ent. Soc. Lond. 1900: xliii.)

LOWER AUSTRAL ZONE The faunal area comprising the southern part of USA from Chesapeake Bay to the great interior valley of California, interrupted by the continental divide in eastern Arizona and west New Mexico and divided according to conditions of humidity into an eastern or Austroriparian and western or lower part.

LOWER CARBONIFEROUS PERIOD The fifth Period (363–323 MYBP) of the Palaeozoic Era, named for extensive deposition of coal generated by the tropical swamp-forest community. Terrestrial plants included scale ferns (lycopods) and seed ferns in swamps. Most abundant lycopods include Genera *Lepidodendron* and *Sigillaria.* Marine invertebrates continued to flourish and adaptively radiate; cephalopods (molluscs) were particularly abundant but graptolites died out during this Period. Insect Infraclass Neoptera (with the ability to flex their wings) invaded new niches and became wind blown. Orders in the LCP fossil record include Diaphanopteroidea, Megasecoptera and Palaeodictyoptera. Principal LCP insect deposits include Commentary (France), Mazon Creek, (Illinois) and Saar (Germany). Syn. Early Carboniferous; Mississippian. See Geological Time Scale; Palaeozoic. Cf. Cambrian Period; Devonian Period; Upper Carboniferous Period; Ordovician Period; Permian Period; Silurian Period.

LOWER MARGIN of TEGMINA (Thomas). The Costal or anterior margin of other authors.

LOWER RADIAL VEIN Lepidoptera: The Media of Comstock. See Vein; Wing Venation.

LOWER SECTOR OF TRIANGLE Odonata: Cubitus of Comstock.

LOWER SONORAN FAUNAL AREA That which comprises the most arid deserts of North America, beginning west of lat. 98° in Texas, with narrow arms going into southern New Mexico, interrupted by the Continental Divide, it covers a large part of western and southern Arizona, south western Nevada, southwestern California, a part of central California, and most of Lower California, these areas are irregular and incapable of brief definition.

LOWER, OSWALD BERTRAM (1863–1925) (Carter 1926, Proc. Linn. Soc. N. S. W. 5: iii–iv.)

LOWER'S DARTER *Telicota mesoptis mesoptis* Lower [Lepidoptera: Hesperiidae].

LOWLAND TREE TERMITE *Incisitermes immigrans* (Snyder) [Isoptera: Kalotermitidae].

LOWREY, P F J (–1891) (Frohawk 1891, Entomologist 24: 200.)

LOWRY, PHILIP ROSEMONT (1896–1931) (Davis 1932, Ann. ent. Soc. Am. 25: 250.)

LOWTHER, RICHARD CHARLES (1884–1950) (Riley 1952, Proc. R. ent. Soc. Lond. (C) 16: 86.)

LOXIRAN® See Chlorpyrifos.

LOXOSCELISM Noun. A medical condition in humans caused by bites of the spider *Loxosceles reclusa* Gertsch & Mulaik in North America and *L. laeta* (Nicolet) in South America. In USA, *L. reclusa* is endemic to eastern USA and common in warmer states. Spiders live several years and populations inhabit buildings and may remain undetected until moving house, furniture or other large objects. Spider's bite generally not felt, but causes gangrenous necrosis near site of bite; severe cases result in persistence of open wound for weeks or months with skin grafts required; fatalities recorded at low rate. See Brown-Recluse Spider. Syn. Necrotic Arachnidism. Cf. Latrodectism.

LOZENGED Adj. (Old French, *losenge* from Latin, *lausa* = stone or slab.) Pertaining to structure that is lozenge-shaped or rhomboidal; a regular rhombus form.

LOZOVOJ, D I (1904–1970) (Zaitzev & Ter-Minassian 1971, Ent. Obozr. 50: 240–245, bibliogr.)

LUBBER GRASSHOPPER *Brachystola magna* (Girard) [Orthoptera: Acrididae].

LUBBOCK, JOHN WILLIAM (LORD AVEBURY) (1834–1913) (Duff (ed) 1924, *Life and Work of Lord Avebury*. 261 pp. London.)

LUBRICATE Verb. (Latin, *lubricare* = to lubricate.) Slippery, as if oily; covered with a sippery mucus. Alt. Lubricatus; Lubricous.

LUCANAL® See Naled.

LUCANIDAE Plural Noun. Stag Beetles. A cosmopolitan Family of polyphagous Coleoptera assigned to the Scarabaeoidea. Adult body to 60 mm long, heavily sclerotized with head prognathous; Antenna lamellate with 10 segments including 3–7 segmented club; Mandibles dimorphic, male Mandible often antler-like, female Mandible pincer-like; fore Tibia spinose or dentate; tarsal formula 5-5-5; Abdomen with five Ventrites. Larvae are scarabaeiform and head hypognathous; at most with one Stemma on side of head; Antenna with 3–4 segments; legs with five segments; stridulatory Pars Stridens on middle Coxa, Plectrum on hind Trochanter; Urogomphi absent. Species prefer moist habitats. Adults typically are nocturnal and most Species probably do not feed; some feed on nectar. Larvae inhabit and feed upon rotten wood. See Scarabaeoidea.

LUCANTE, A (1850–1889) (Anon. 1889, Feuille jeun. Nat. 19: 155.)

LUCAS, ARTHUR HENRY SHAKESPEARE (1853–1936) (Doley 1936, Victorian Nat. 53: 54–55.)

LUCAS, GRACE ELLEN (1877–1950) (Baynes 1952, Irish Nat. J. 10: 10.)

LUCAS, PIERRE HIPPOLYTE (1814–1899) (Anon. 1899, Entomol. mon. Mag. 35: 276.)

LUCAS, THOMAS PENNINGTON (1843–1917) (Gurney 1918, Proc. R. Soc. Qld. 30: 2, bibliogr.)

LUCAS, WILLIAM JOHN (1858–1932) (Killington 1932, Entomologist 65: 25–27, bibliogr.)

LUCCA, CARMEL DE (–1971) (Butler 1973, Proc. R. ent. Soc. Lond. (C) 37: 56.)

LUCERNE APHID PARASITE *Aphidius ervi* Haliday [Hymenoptera: Braconidae].

LUCERNE BUD MITE *Eriophyes medicaginis* (Keifer) [Acari: Eriophyidae].

LUCERNE CROWNBORER 1. *Corrhenes stigmatica* (Pascoe) [Coleoptera: Cerambycidae]: A pest of alfalfa in Australia. 2. *Zygrita diva* Thomson [Coleoptera: Cerambycidae]: A pest of alfalfa in Australia.

LUCERNE FLEA *Sminthurus viridis* (Linnaeus) [Collembola: Sminthuridae].

LUCERNE LEAF-EATING BEETLE *Colaspoides foveiventris* Lea [Coleoptera: Chrysomelidae].

LUCERNE LEAFHOPPER *Austroasca alfalfae* (Evans) [Hemiptera: Cicadelidae]: A pest of alfalfa in Australia.

LUCERNE LEAFROLLER *Merophyas divulsana* (Walker) [Lepidoptera: Tortricidae]: A pest of alfalfa in Australia.

LUCERNE LOOPER *Zermizinga indocilisaria* Walker [Lepidoptera: Geometridae].

LUCERNE SEED WASP *Bruchophagus roddi* (Gussakovski) [Hymenoptera: Eurytomidae].

LUCERNE SEED WEB MOTH *Etiella behrii* Zeller [Lepidoptera: Pyralidae].

LUCHT, HENDRIK (1884–1951) (Roepke 1951, Ent. Ber., Amst. 13: 286.)

LUCID Adj. (Latin, *lucere* = to shine.) Bright, clear, shining, mirror-like. Alt. Lucidate; Lucidatus; Luciidus.

LUCIFERASE Noun. (Latin, *lux* = light; *ferre* = to carry.) An enzyme in the luminous organs of light-producing beetles.

LUCIFERIN Noun. (Latin, *lux* = light; *ferre* = to carry.) A substance in the Haemolymph of luminous beetles which, when brought into contact with luciferase, produces light.

LUCIFEROUS Adj. (Latin, *lux* = light; *ferre* = to carry; *-osus* = with the property of.) Light emitting.

LUCIFUGOUS Adj. (Latin, *lucifuga* = avoiding light; *-osus* = with the property of.) Pertaining to organisms which avoid light. Term applied to nocturnal forms or those that live in concealment.

LÜDDEMANN, ANDREAS (1858–1938) (Horn 1938, Arb. morph. taxon. Ent. Berl. 5: 296.)

LÜDEKE, OSCAR (1854–1942) (Sachtleben 1942, Arb. morph. taxon. Ent. Berl. 9: 206.)

LUDLOW, CLARA SOUTHMAYD (1852–1924) (Knight & Pugh 1974, Mosquito Syst. 6: 214–217, bibliogr.)

LUDOLF (LEUTHOLF), HIOB (JOB) (1624–1704) (Rose 1850, *New General Biographical Dictionary* 9: 348.)

LUDWIG, AUGUST (1867–1951) (Hüsinger 1952, Neue Brehm Bücherei 31: 3.)

LUDWIG, BARON KARL VON (1784–1847) German aristocrat and natural historian who collected in South Africa.

LUDY, FRIEDRICH (–1896) (Anon. 1896, Wien Ent. Ztg. 15: 124.)

LUEDERWALDT, HERMANN (1865–1934) (Taunay 1937, Revta Mus. paul. 21: 31–47, bibliogr.)

LUETGENS, AUGUST (1838–1908) (Anon. 1908, J. N. Y. ent. Soc. 16: 52.)

LUFENURON A benzamide compound {N-[[[2,5-dichloro-4-(1,1,2,3,3,3-hexafluoropropoxy) phenyl] amino]=carbonyl]-2,6-difluorobenzamide} used as an Insect Growth Regulator (Chitin Inhibitor) against Colorado Potato Beetle, Lepidoptera larvae (armyworms, bollworms, leaf rollers, loopers), thrips, whiteflies and related insects. Lufenuron does not affect adult insects but some ovicidal activity and stomach poison activity have been observed. Compound is applied to fruit crops, forestry, cotton, citrus, grapes tea and vegetables in some countries. Experimental registration only in USA. Trade names include: Axor®, CGA-184699®, Match®, Sorba®. See Insect Growth Regulator.

LUFF, WILLIAM AMBRIDGE (1851–1910) (Marquand 1910, Rep. Trans. Guernsey Soc. nat. Hist. 6: 147–154, bibliogr.)

LUGGER, OTTO (1844–1901) (Howard 1901, Ent. News 12: 222–224.)

LUGINBILL, PHILIP (1917–1973) (App 1973, Proc. ent. Soc. Wash. 75: 490.)

LUIGIONI, PAOLO (1873–1937) (Anon. 1937, Annuar. pontif. Accad. Sci. 1: 520–527, bibliogr.)

LUKASSEN, VALK (–1939) (De Beaux 1940, Annali mus. civ. Stor. Nat. Giacomo Doria 60: (3).)

LUMEN Noun. (Latin, *lumen* = light. Pl., Lumina.) The enclosed space or cavity of any hollow or vesicular organ or structure.

LUMHOLTZ, CARL SOFUS (1851–1922) (L. R. 1944, Norsk. ent. Tidsskr. 7: 44–45.)

LUMIALA, ONNI VEIKKO (1910–1944) (Kotilainen 1945, Memo Soc. Fauna Flora fenn. 21: 233–238.)

LUMINAL Adj. (Latin, *lumen* = light; *-alis* = pertaining to.) Pertaining to a lumen.

LUMINESCENCE Noun. (Latin, *lumen* = light.) The light of fire-flies, as a substitute for phosphorescence.

LUMPER Noun. (Jargon. Pl., Lumpers.) A taxonomist who, in describing Species or Genera, recognizes only prominent or obvious morphological characters to the exclusion of minor colour or variable characters of maculation or structure. A philosophy of classification in which reductionism has influence. Cf. Splitter.

LUMSDEN SUCTION TRAP A collecting device for biting insects. (Bull. Ent. Res. 52: 233–238.) See Trap.

LUNA MOTH *Actias luna* (Linnaeus) [Lepidoptera: Saturniidae].

LUNALA Noun. (Latin, *luna* = moon. Pl., Lunulae.) 1. A small lunate mark or crescent that sometimes is coloured. 2. Hymenoptera: One of the crescent-shaped marks near the compound eye (Smith). See: Lunule. Alt. Lunulet.

LUNARDONI, AGOSTINO (1857–1933) (Anon. 1933, Studi trent. Sci. nat. 14: 52–53.)

LUNATE Adj. (Latin, *luna* = moon; *-atus* = adjectival suffix.) Crescent-shaped. Alt. Lunare; Lunaris.

LUND, GUSTAV BUDDE See Budde-Lund.

LUND, NIELS TÖNDER (1749–1809) (Heinriksen 1923, Ent. Meddr. 15: 106–109).

LUND, PETER WILHELM (1801–1880) (Anon. 1880, Ibis (4) 4: 483–484.)

LUNDBECK, WILLIAM (1863–1941) (Kryler 1942, Ent. Meddr. 22: 268–273, bibliogr.)

LUNDBERG, JONAS (Lindroth 1967, Opusc. ent. 32: 7.)

LUNDSTROM, CARL AUGUST (1844–1914) (Frey 1915, Luonnon Ystävä 1: 12–15, bibliogr.)

LUNG BOOKS 1. The respiratory pouches of Arachnida. Book lungs are so-called because their walls are produced into parallel, lamellate folds (Snodgrass). 2. The invaginated brancheae *sensu* Imms. Syn: Book Lungs.

LUNG MITE *Sternostoma tracheacolum* Lawrence [Acari: Rhinonyssidae].

LUNULATE Adj. (Latin, *luna* = moon; *-atus* = adjectival suffix.) Pertaining to structure that is crescentic in shape or composed of a series of small lunules, as a line. Obsolete: Lunulatus

LUNULE Noun. (Latin, *lunula* = small moon.) Diptera: A small, crescent-shaped sclerite between the Ptilinal Suture and the base of the Antennae. See Lunala.

LUQUE, ANTONIO VAREA DE (–1975) (Bustillo 1975, Shilap 12: 239–240.)

LURECTRON® See Muscalure.

LURID Adj. (Latin, *luridus* = pale-yellow.) Dirty brown bluish tinge. A term also used to indicate an obscuring of bright colours. Alt. Luridus.

LUSUS NATURAE Latin phrase meaning a sport or freak of nature. A mutation.

LUTEOTESTACEOUS Adj. (Latin, *luteus* = brownish; *testa* = piece of burnt clay; *-aceus* = of or pertaining to.) Dark clay-yellow in coloration. Cf. Luteous; Testaceous.

LUTEOUS Adj. (Latin, *luteus* = brownish; *-osus* = full of.) Clay-colour, brownish-yellow; deep yellow with a tint of red (Kirby & Spence). Alt. Luteus.

LUTESCENT Adj. (Latin, *luteus* = brownish.) Becoming or approaching clay yellow in coloration.

LUTOSE Adj. (Latin, *luteus* = brownish; *-osus* = full of.) Apparently or actually covered with clay or dirt. Alt. Lutosus.

LUTSCHNIK (LUTSNIK) VIKTORA NIKOLAEVICH (1892–1936) (Plavilstshikov 1938, Revue Russe Ent. 27: 267–280, bibliogr.)

LUTZ, ADOLFO (1855–1940) (Anon. 1940, Brasil Med. 54: 855–858, 874–878, bibliogr.)

LUTZ, FRANK EUGENE (1879–1943) (Weiss *et al.* 1944, J. N. Y. ent. Soc. 52: 63–73, bibliogr.)

LUTZAN, KARL VON (1850–1923) (Kupffer 1924, KorrespBl. NaturVer. Riga 58: 7–8, bibliogr.)

LUX Noun. (Latin, *luxatus* > *luxare*. Luxes.) A unit of measurement of light, based on the spectral sensitivity of the human eye. A lux is the illuminance of a surface receiving one lumen per square metre. Alternatively, 1 Lux = 0.093 foot candles.

LYCAENIDAE Plural Noun. Blues (Plebejinae); Coppers (Lycaeninae); Hairstreaks (Theclinae); Harvesters (Gerydinae). A numerically large, cosmopolitan Family of butterflies. Adults with compound eye frequently emarginate along medial margin; Maxillary Palpus absent; Labial Palpus ascending; foreleg of male often reduced; foreleg of female normal; Epiphysis absent; Tibial Spur formula 0-2-2, 0-1-1 or 0-0-0; hindwing with posterolateral margin developed into 1–2 tails. Larvae are onisciform with head typically retractile; Abdomen with dorsal medial gland on segment 7 (Newcomer's Organ), paired eversible glands on segment 8 (Tentacular Organ) and Perforated Cupoplas near spiracles. Pupae with central silken girdle and lacking Cremaster. Larvae frequently are associated with ants. Lycaenids are parasitic or predaceous upon ants and other inquilines. Syn. Cupidinidae; Erycinidae; Nemeobiidae; Riodinidae; Ruralidae. See Lepidoptera.

LYCHEE STINK-BUG *Tessaratoma papillosa* [Hemiptera: Tessaratomidae].

LYCIDAE Plural Noun. Net-winged Beetles. A Family of polyphagous Coleoptera assigned to Elateroidea and including about 3,000 nominal Species considered and many Genera or groups of Genera. Adult body 5–20 mm long, depressed with Integument soft and leathery; head usually triangular; Antenna usually with 11 flattened, serrate or pectinate segments. Elytra thin with deep, lattice-like sculpture; tarsal formula 5-5-5. A Species with larviform adult females. Larvae are elongate, somewhat flattened and heavily sclerotized; head prognathous; Stemma absent; Antenna with two segments; Mandibles blade-like, non-opposable and longitudinally divided into two parts; legs with five segments, each with two tarsal claws; Urogomphi typically are absent. Lycids typically are found in warm, forested areas. Adults are day-flying, aposematically coloured; reflexive bleeding confers chemical protection. Feeding habits are controversial: Predatory or not predatory; larvae are lignivorous. Typically, larvae live beneath bark or within soil. Species with larviform adult females do not pupate; other Species pupate in habitat of larva. See Elateroidea.

LYCOLURE A synthetic pheromone {(E)-4-tridecenyl acetate; (Z)-4-tridecenyl acetate} used to disrupt sex pheromone of Tomato Pin Worm. Trade names include Checkmate TPW®, Decoy TPW®, Nomate TPW SPIRAL®, Isomate TPW® and Pin-Down®. See Synthetic Pheromone. Rel. Parapheromone.

LYCORIIDAE Plural Noun. See Sciaridae.

LYCOSIDAE Plural Noun. Wolf Spiders. Moderate to large-sized, drab-coloured spiders. Wolf spiders 10–40 mm long; body oval-elongate and vary in colour from grey-black to dark brown and red-brown. A 'Union Jack' pattern is often evident on carapace; Abdomen patterned with a bell and V-shaped lines. Eye placement characteristic: two large eyes positioned centrally on front of carapace; a row of four small eyes beneath two large eyes and pair of small eyes are situated above and lateral of large eyes. Antenna and wings are absent; all legs long, similar in shape and Setae. Adult females deposits eggs within a sac which is carried on Spinnerets. The neonate spiderlings carried on back of female for several weeks. Spiderlings similar in appearance to adults. Adults live for 1–2 years. Wolf spiders are ground-dwelling spiders which actively hunt their prey on ground, but may climb plants. Most Species build burrows for retreats. Cf. Lynx Spiders.

LYCTIDAE Plural Noun. Powderpost Beetles. A Family of Coleoptera.

LYELL, CHARLES (1797–1841) English geologist and author of *Principles of Geology*, an influential 19th century text.

LYELL'S SWIFT *Pelopidas lyelli lyelli* (Rothschild) [Lepidoptera: Hesperiidae].

LYGAEIDAE Plural Noun. Lygaeids; Seed Bugs. A cosmopolitan Family of Hemiptera consisting of more than 4,000 Species. Body form and size diverse; frequently with red or yellow; Antenna with four segments and arising from area below imaginary line bissecting eyes; Ocelli absent and brachypterous or Ocelli present; forewing membrane with five veins. Biology unusual in that most Species feed on seeds dispersed in ground litter; some Species sap-sucking, predaceous or feeding upon vertebrate blood.

LYGAEOIDEA A Superfamily of Hemiptera including Berytidae, Colobathristidae, Lygaeidae, Largidae, Malcidae, Pyrrhocoridae.

LYGISTORRHINIDAE Plural Noun. A small Family of Diptera related to Mycetophilidae.

LYGUS BUG *Lygus* spp. [Heteroptera: Lygaeidae]: Significant pests of cotton and other plants in North America.

LYLE, CLAY (1894–1971) (Colmer 1972, J. Econ. Ent. 65: 1532–1533.)

LYLE, GEORGE TREVOR (1873–1930) (W. J. F. 1930, Naturalist 1930, 434–435.)

LYMAN, HENRY HERBERT (1854–1914) (Bethune C. J. S. 1914, Can. Ent. 46: 221–226, bibliogr.)

LYMAN, THEODORE (1833–1897) (Anon. 1899, Proc. Am. Acad. Arts Sci. 34: 656–663.)

LYMANTRIIDAE Plural Noun. Gypsy Moths; Tussock Moths. A Family of ditrysian Lepidoptera assigned to Superfamily Noctuoidea; includes 2,500 Species, cosmopolitan distribution but best represented in Old World tropics. Adult variable in size; wingspan 15–100 mm; head with piliform scales; Ocelli and Chaetosemata absent; Antenna bipectinate in male and usually in female, often with few long Setae on terminus of each branch; Proboscis typically absent or vestigial; Maxillary Palpus absent or of one segment; Labial Palpus small to medium sized, upturned or porrect. Thorax with long, piliform scales; metathoracic Tympanal Organs directed obliquely rearward; fore tibial Epiphysis present in male, absent or reduced in female; tibial spur formula 0-2-4, rarely 0-2-2; female sometimes apterous or brachypterous; female Abdomen with dense tuft of deciduous scales used to cover egg mass. Egg upright, circular, globular or with flattened apex and base; eggs laid in masses covered with scales. Larval head hypognathous with numerous secondary Setae; body with dense tuft of secondary Setae; urticating Setae in some Species; gland on dorsum of abdominal segments 6 and usually 7; abdominal Prolegs on segments 3–6, 10. Pupae stout; Maxillary Palpus absent; Labial Palpus typically exposed; Cremaster present with short hooked Setae at apex; cocoon composed of larval Setae, typically concealed under bark or within crevices. First instar larva often dispersed on strands of silk; larva feeds on foliage of trees and woody shrubs, less often on herbage. Adult typically nocturnal, visit lights; a few diurnal Species; Wings held tectiform over Abdomen or wings pressed to substrate with legs exposed. Includes economically important Brown-Tail Moth and Gypsy Moth. Syn. Liparidae.

LYME DISEASE An infectious, tick borne disease occurring in USA and Europe. Named from Old Lyme, Connecticut where the disease was first recognized in 1975. Disease caused by a spirochete (spiral-shaped bacterium), *Borrelia burgdorferi* which is transmitted by *Ixodes pacificus* in the western USA. Early symptoms include rash, fever and aches; arthritis is a common long-term symptom of the disease. Natural history of disease is poorly understood.

LYMEXYLIDAE Plural Noun. Ship Timber Beetles. A small, widespread Family of polyphagous Coleoptera assigned to Superfamily Lymexyloidea. Adults elongate, narrow, 9–13 mm long; head projecting downward and narrowed behind the eyes; Antenna filiform or serrate; Tarsi with five segments. Larvae inhabit dead logs and stumps. European Species are known pests of ship timbers. In the USA, STB cause much of the pin hole damage observed in chestnut (*Aesculus* spp.)

LYMEXYLOIDEA A numerically small Superfamily of polyphagous Coleoptera including Lymexylidae.

LYMPHATIC Adj. (Latin, *lympha* = water; *-ic* = consisting of.) Producing, carrying or relating to the lymph or Haemolymph.

LYMPHOCYTE See Plasmatocyte.

LYNCH-ARRIBALZAGA, ENRIQUE (1856–1935) (Birabin 1935, Revta Mus. La Plata 83: 85, bibliogr.)

LYNCH-ARRIBALZAGA, FÉLIX (1854–1928) (Dallas 1928, Revta Soc. ent. argent. 2: 5–12.)

LYNX SPIDERS See Oxyopidae.

LYOCYTOSIS (Greek, *lyein* = to loosen; *kytos* = a hollow; *-sis* = a condition or state.) According to Anglas, a kind of extracellular digestion which brings about the destruction of tissues in phagocytosis of larvae (Imms).

LYON, FRANCIS HAMILTON (1886–1964) (Pearson 1966, Proc. R. ent. Soc. Lond. (C) 30: 63.)

LYON, MARCUS WARD (1875–1942) (Just 1942, Am. Midl. Nat. 27: (3): ii-iv, bibliogr.)

LYONET, PIERRE (1707–1789) Dutch lawyer and insect anatomist. Author of *Traite Anatomique de la Chenille de Saule* (1762); noted for precise attention to detail and accurate descriptions of dissections. Posthumous work published by de Haan of Leyden and entitled *Recherches sur l'Anatomie et les Metamorphoses de differentes especes d'Insectes* (1832). (Snellen von Vollenhoven 1880, Album Natur 1: 1–14.)

LYONETIID MOTHS See Lyonetiidae.

LYONETIIDAE Stainton 1854. Plural Noun. Lyonetiid Moths. A Family of ditrysian Lepidoptera placed in Superfamily Yponomeutoidea or Tineoidea in older classifications. Adult very small; wingspan 4–12 mm; head usually with smooth scales, tuft of scales between Antennae; Ocelli and Chaetostema absent; Antenna about 0.6 times as long as forewing; scale forming eye cap or Pecten; Proboscis short or vestigial and without scales; Maxillary Palpus absent, short or minute with one segment; Labial Palpus variable; foretibial Epiphysis present; tibial spur formula 0-2-4; forewing lanceolate, venation reduced, Pterostigma absent. Egg oval in outline shape; usually laid individually on leaf, sometimes inserted into leaf. Larva flattened; head prognathous; early instars typically lack Prolegs; late instars with Prolegs on abdominal segments 3–6, 10 and Crochets in uniordinal arrangement; larva mine leaves or develop in web between leaves. Pupation in cocoon on host plant or suspended by a few strands of silk. Alt. Lyonetidae.

LYOPHILIZE (Greek, *lyein* = loosen; *philos* = loving.)

LYPOR® See Temephos.

LYRATE Adj. (Greek, *lyra* = lyre; *-atus* = adjectival suffix.) Lyre-shaped or cut into several transverse segments and gradually enlarging towards the apex. Alt. Lyriform. See Lyre; Outline Shape.

LYRE Noun. (Greek, *lyra* = lyre. Pl., Lyres.) 1. The upper wall or border of the spinning tube of caterpillars. 2. A stringed muscial instrument with characteristic shape. See Lyrate.

LYRIFISSURE Noun. Acarology: a sense organ whose opening appears as a cuticular fissure. Hypothesized to detect deformations of the exoskeleton.

LYSHOLM, BJARNE (1861–1939) (Strand 1940, Norsk. ent. Tidsskr. 5: 189–190, bibliogr.)

LYSIS Noun. (Greek, *lysis* = loosing. Pl., Lyses.) 1. The physical process or act of disintegration of biological tissue. 2. Cell Destruction.

LYSOGENESIS Noun. (Greek, *lysis* = loosening; *genesis* = descent. Pl., Lysogeneses.) 1. The production of lysins. 2. The phenomenon of Lysis.

LYSOZYME Noun. (Greek, *lysis* = loosing; *zyme* = leaven. Pl., Lysozymes.) Enzymes in the Haemolymph which lyse the cell wall of Gram-Positive Bacteria to act as a defence mechanism. See Haemolymph. Cf. Cellular Defence Mechanism.

LYSTER, MARTIN (–1711) English physician to Queen Anne and naturalist. Author of *Cum Scarabaeorum Anglicanorum quibusdam tabulis mutis* (1685).

LYTHRACEAE A cosmopolitan Family of Myrtales containing ca 30 Genera and 550 Species. Related to the Onagraceae; speciose in subtropical areas. Largest Genus *Cuphea* (260 New World Species) is studied as source of lauric acid which is used in chemical, food and detergent industries.

MAAS, FRANZ (1850–1929) (Rapp 1930, Ent. Bl. Biol. Syst. Käfer 26: 1–4.)

MAAS, OTTO (–1916) (Soldanski 1916, Dt. ent. Z. 1916: 228.)

MAASEN, J PETER (1810–1890) (Anon. 1890, Entomologist 23: 328.)

MABILLE, PAUL (1835–1923) (Anon. 1923, Ent. News 34: 256.)

MACADAMIA Noun. (Etymology obscure.) A Genus of Proteaceae including about 10 Species; Genus is endemic from Madagascar to Australia. *Macadamia integrifolia* Maiden & Betche is endemic to Queensland, Australia but first was grown commercially in Hawaii and widely cultivated for edible nuts.

MACADAMIA CUP-MOTH *Mecytha fasciata* (Walker) [Lepidoptera: Limacodidae].

MACADAMIA FELTED-COCCID *Eriococcus ironsidei* Williams [Hemiptera: Eriococcidae]: An apparently host plant-specific pest of *Macadamia* spp. in Australia. MFC infests all above-ground parts of *Macadamia* and causes stunting, yellowing and spotting of older leaves; heavy infestations kill young plants and reduce yield of mature plants. Life cycle displays an egg, two crawler stages and an adult female enclosed within a felt sac; adult male winged. See Eriococcidae.

MACADAMIA FLOWER-CATERPILLAR *Cryptoblabes hemigypsa* Turner [Lepidoptera: Pyralidae]: A pest of *Macadamia* spp. in eastern Australia. MFC not host plant specific but reproduces on many Proteaceae and migrates to *Macadamia* from other host plants. MFC eggs are white-yellow, oval, 0.5 mm long and laid beneath small bracts; young larvae live within buds, later cover racemes with webbing; old larvae eat buds and flowers. See Pyralidae.

MACADAMIA LACE BUG *Ulonemia concava* Drake [Hemiptera: Tingidae].

MACADAMIA LEAFMINER *Acrocerops chionosema* Turner [Lepidoptera: Gracillariidae]: A moth endemic to Australia. Larvae are major, sporadic pests of lush foliage of macadamia; mines are serpentine-like during early development and blotch-like during late development; both leaf surfaces may be mined. Heavy infestations of MLM may cause dieback; larvae abandon mines to pupate. See. Gracillariidae.

MACADAMIA MUSSEL SCALE *Lepidosaphes macadamiae* Williams [Hemiptera: Diaspididae].

MACADAMIA NUT-BORER *Cryptophlebia ombrodelta* (Lower) [Lepidoptera: Tortricidae]: A pantropical pest of macadamia whose larvae feed on many exotic ornamental Species. Eggs are small, scale-like and laid on or near macadamia husk. Larvae enter husks, bore into kernel. MNB reduces kernel quality and yield. See Tortricidae.

MACADAMIA TWIG-GIRDLER *Xylorycta luteotactella* (Walker) [Lepidoptera: Oecophoridae]: A pest endemic to Australia that feeds upon numerous Proteaceae. As a pest of macadamia, MTG larvae girdle twigs, attack foliage and tunnel into nuts; larvae feed beneath webs. Heavy infestations of MTG check growth or kill trees. See Oecophoridae.

MACADAMIA WHITE-SCALE *Pseudaulacaspis brimblecombei* Williams [Hemiptera: Diaspididae].

MACAO PAPER WASP *Polistes macaensis* (Fabricius) [Hymenoptera: Vespidae]. See Vespidae. Cf. Common Paper Wasp; Golden Paper Wasp; Yellow Paper Wasp.

MACBAL® See XMC.

MACBRIDE, ERNEST WILLIAM (1866–1940) (Calman 1941, Obit. Not. Fell. R. Soc. Lond. 10: 447–759, bibliogr.)

MACCOLLOCH, JOHN (1773–1835) (Rose 1850, *New General Biographical Dictionary* 9: 379–380.)

MACDOUGALL, R STEWART (1862–1947) (Williams 1948, Proc. R. ent. Soc. Lond. (C) 12: 65.)

MACERATE Adj. (Latin, *macerare* = to soften; *-atus* = adjectival suffix.) Macerated. Verb (Latin, *macerare* = to soften.) 1. Transitive: To make lean. Intransitive: To soften and wear away by seeping, digestion or other means.

MACERATION Noun. (Latin, *maceratus* = soft, weak; *-tion* = result of an action. Pl., Macerations.) An histological technique that renders the insect Integument transparent. Maceration enables clear observation of internal structure without dissection. Popular macerating agents include Potassium Hydroxide and Sodium Hydroxide. See Histology. Cf. Clearing. Rel. Fixation.

MACFIE, J W SCOTT (1879–1948) (Anon. 1948, Entomologist 82: 21–22.)

MACGILLIVRAY, ALEXANDER DYAR (1868–1924) (Riley 1924, Ent. News 35: 224–288.)

MACGILLIVRAY, D (1869–1951) (Wiel 1951, Ent. Ber., Amst. 13: 241–243.)

MACHATSCHKE, JOHANN W (1912–1974) (Delkeskamp 1975, Ent. Germanica 1: 38.)

MACHIDA, JIRO (1885–1964) (Hasegawa 1967, Kontyû 35 (Suppl.): 77.)

MACHILIDAE Plural Noun. Jumping Bristletails. A primitive Family of Archaeognatha found predominantly in Northern Hemisphere and placed with Thysanura in older classifications. Characteristics include a Monocondylic Mandible and Styli on Coxae. Machilids are active and jump when disturbed. See Archaeognatha. Cf. Meinertellidae.

MACHIN, WILLIAM (1882–1894) (Tutt 1894, Entomologist's Rec. J. Var. 5: 209–210.)

MACHULKA, VACLAV (1889–1949) (Anon. 1950, Cas. csl. Spol. ent. 47: 3–6, bibliogr.)

MACK, GEORGE (1900–1963) (Pearson 1966, Proc. R. ent. Soc. Lond. (C) 30: 63.)

MACKENZIE, COMPTON (–1972) (Campbell 1973, Entomologist's Rec. J. Var. 85: 107–108.)

MACKERRAS, MABEL JOSEPHINE (1896–1971) Daughter of T. L. Bancroft and taxonomist, collector and parasitologist. MJM worked with muscoid Diptera, blackflies and cockroaches. (Marks 1972, Proc. R. Soc. Qd 38: 103.)

MACKIE, DAVID BARCLAY (1882–1944) (Keifer 1945, Bull. Calif. Dep. Agric. 34: 3.)

MACKWOOD, F M (1843–1931) (Eltringham 1932, Proc. ent. Soc. Lond. 6: 107.)

MACKWORTH-PRAED, CYRIL WYNTHROP (1891–1974) (Worms 1975, Entomologist's Rec. J. Var. 87: 29–30.)

MACLEAY, ALEXANDER (1767–1848) (Mackecknie-Jarvis 1976, Proc. Brit. ent. nat. Hist. Soc. 8: 92–93.)

MACLEAY, WILLIAM JOHN (1820–1891) (Haswell 1891, Proc. Linn. Soc. N. S. W. (2) 6: 707–716.)

MACLEAY, WILLIAM SHARP (1792–1865) Author of *Horae Entomologicae*, or *Essays on the Annulose Animals* (1819–1821), and several articles in the Linnean Transactions and Zoological Journal. Macleay was the first Entomologist to describe the Mediterranean Fruit Fly as a pest of citrus. (Fletcher 1929, Proc. Linn. Soc. N. S. W. 45: 584, 591–629.)

MACLEAY'S GRASS YELLOW *Eurema herla* (W. S. Macleay) [Lepidoptera: Pieridae].

MACLEAY'S SWALLOWTAIL *Graphium macleayanum macleayanum* (Leach) [Lepidoptera: Papilionidae].

MACLOSKIE, GEORGE (1834–1920) (Anon. 1920, Ent. News 31: 89–90, bibliogr.)

MACLURE, WILLIAM (1763–1840) Born in Scotland and emigrated to America; author of *The Geology of the United States*. Maclure was a participant in the New Harmony, Indiana 'experiment' and sometimes has been regarded as 'Father of American Geology.' (Morton 1841, Mem. Acad. nat. Sci. Phil. 1841: 1–5, bibliogr.)

MACNEILL, NIALL (1899–1969) (Asahina 1971, Tombo 14: 11.)

MACQUART, JEAN PIERRE MARIE (1778–1855) French worker interested in Diptera and insects associated with trees, shrubs and herbaceous plants of Europe. Macquart was author of *Insectes dipteres du Nord de la France* (in: Mem. Soc. R. Sci, 1826–1829), and *Histoire Naturelle des Insectes (Dipteres)* (2 vols, 1834–1835). (Crosskey 1971, Bull. Br. Mus. nat. Hist., Entomol. 25: 256–263.)

MACQUER, PIERRE JOSEPH (1718–1784) (Rose 1850, *New General Biographical Dictionary* 9: 395.)

MACRANER Noun. (Greek, *makros* = large; *aner* = male. Pl., Macraners.) The larger size-class in Species of ants with more than one size-class of males. Cf. Micraner. Rel. Allometry.

MACRERGATE Noun. (Greek, *makros* = large; *ergates* = worker. Pl., Macrergates.) A worker ant of unusually large size.

MACROCEPHALIC Adj. (Greek, *makros* = large; *kephale* = head.) 1. Pertaining to a disproportionately large head. 2. Descriptive of an organism with a disproportionately large head. Macrocephaly is a condition in which overall size of the head is disproportionately large in relation to body size. The condition may be restricted to one sex. Macrocephalic heads appear among several groups of insects including some eusocial Hymenoptera. Syn. Macrocephalous. Cf. Hypercephalic; Phragmotic.

MACROCHAETA Noun. (Greek, *makros* = large; *chaite* = hair. Pl., Macrochaetae.) The long bristles that occur singly on the body of Diptera. See Microchaeta.

MACROEVOLUTION Noun. (Greek, *macros* = large; *evolutio* = an unrolling; English, *-tion* = result of an action. Pl., Macroevolutions.) Evolution above the Species level over protracted periods of geological time. Elements include speciation, morphological change and extinction. See Gradualism; Punctuated Equilibrium.

MACROGAMETE Noun. (Greek, *makros* = large; *gametes* = spouse. Pl., Macrogametes.) The mature female Gametocyte of the malarial protozoan, *Plasmodium*.

MACROGAMETOCYTE Noun. (Greek, *makros* = large; *gametes* = spouse; *kytos* = hollow vessel; container. Pl., Macrogametocytes.) The female sex cell developed from the Merozoite of the malarial protozoan.

MACROGEOGRAPHICALLY Cf. Microgeographically.

MACROGYNE Noun. (Greek, *makros* = large; *gyne* = woman. Pl., Macrogynes.) The larger class size of Species of ants with two size-classes of queens. Cf. Microgyne.

MACROLABIUM Noun. (Greek, *makros* = large; Latin, *labium* = lip. Pl., Macrolabia.) Dermaptera: The longer form of Forceps.

MACROMIIDAE Plural Noun. Belted Skimmers; River Skimmers. A small Family of Odonata assigned to Superfamily Libelluloidea. Placed in Corduliidae in some classifications. Hindwing wider than forewing; triangles in front and hindwings not similar in shape. Species frequently encountered near lakes, streams and marshy habitats. See Libelluloidea.

MACRONEUCLEOCYTE See Prohaemocyte.

MACROPATHIDAE Plural Noun. Cave Weta. A Family of ensiferous Orthoptera sometimes placed in Rhaphidophoridae but differing in having paired dorsal spines on apex of Metatarsus. Family circumantarctic (New Zealand, Southern Chile, Patagonia, Falkland Islands) and includes 20 Genera and 50 Species. Most Species live in caves or crevices and are predatory or scavenge other invertebrates. See Orthoptera. Cf. Rhaphidophoridae.

MACROPHAGOUS Adj. (Greek, *makros* = large; *phagein* = to eat; Latin, *-osus* = with the property of.) Descriptive of organisms feeding on large particles of food.

MACROPORE Noun. (Greek, *makros* = large; *poros*

= channel. Pl., Macropores.) Coccids: See Oraceratubae (MacGillivray).

MACROPSYLLIDAE Plural Noun.

MACROPTEROUS Adj. (Greek, *makros* = large; *pteron* = wing; Latin, *-osus* = with the property of.) Long or large winged. A term descriptive of wings that are not reduced in size and presumably fully functional in active flight. See Apterous; Polymorphism. Cf. Brachypterous; Micropterous; Subapterous.

MACROSOMITE Noun. (Greek, *makros* = large; *soma* = body; *-ites* = constituent. Pl., Macrosomites.) 1. Primordial regions of the primitive band of the insect embryo. 2. Any inordinately large segment of an arthropod's body.

MACROTRICHIA Noun. (Greek, *makros* = large; Latin, *-trichia,* > Greek, *thrix* > *trichos* = hair; *ia* = a quality. Pl., Macrotrichiae.) Diptera: Comparatively larger microscopic Setae on the surface of wings (Curran). Cf. Microtrichium.

MACROTYPE EGG Characteristically, a large egg with a thick, opaque dorsal surface and thin, flat, transparent ventral surface. Macrotype eggs are oblong in dorsal aspect and semicircular in lateral aspect. Surface features which may be present include a flange margin for ventral surface, and spumaline for adhesion to host. Macrotype eggs typically have an extensive chorionic respiratory system. Marcrotype eggs are restricted to the Tachinidae and were subdivided into dehiscent and indehiscent forms. See Egg. Cf. Membranous Egg; Microtype.

MACROZAMIA BORER *Tranes internatus* Pascoe [Coleoptera: Curculionidae].

MACSWAIN, JOHN WINSLOW (–1970) (Anon. 1970, Pan-Pacif. Ent. 46: 308.)

MACULA Noun. (Latin, *macula* = spot. Pl., Maculae.) 1. A coloured mark or spot whose shape is indeterminate. 2. A shallow depression in a surface. 3. A cuticular tubercle.

MACULAR FASCIA A Fascia composed of distinct spots.

MACULATE Adj. (Latin, *macula* = spot.) Structure that is spotted or marked with figures of any shape, or colour differing from the ground or base colour. Alt. Maculated; Maculatus.

MACULATE CURCULIO *Desiantha diversipes* (Pascoe) [Coleoptera: Curculionidae].

MACULATE LADYBIRD *Harmonia octomaculata* (Fabricius) [Coleoptera: Coccinellidae].

MACULATION Noun. (Latin, *maculare* = to spot; English, *-tion* = result of an action. Pl., Maculations.) The ornamentation or pattern of markings on a plant or animal.

MACULOSE Adj. (Latin, *maculosus* = spotted.) Spotted; with many marks or spots.

MADAGASCAR Globally, the Fourth largest island and separated from continental Africa during Jurassic Period (160–165 MYBP). Surface dimensions 1,600 km long, to 370 km wide and 587,000 square kilometres; situated between 12° and 25° 30' south in latitude; separated from Africa by Mozambique Channel. Climatically diverse and inhabited by humans for ca 2000 years (Verin 1967, Bull. Madagascar 259: 947–976.)

MADARA, JINDRICHA (1876–1971) (Maran 1971, Acta ent. bohemoslovaca 68: 427–428, bibliogr.)

MADEIRA ANT *Pheidole megacephala* (Fabricius) [Hymenoptera: Formicidae].

MADEIRA COCKROACH *Leucophaea maderae* (Fabricius) [Blattaria: Blaberidae]: A tropical and subtropical urban pest. Adult 40–50 mm long.

MADELUNG, HJALMAR (–1946) (Sonderup 1947, Flora Fauna Silkeburg 53: 38.)

MADER, LEOPOLD (1886–1961) (Janezyk 1963, Ann. naturh. Mus. Wien 66: 17.)

MADESCENT Adj. (Latin, *madescere* = to become moist.) Pertaining to structure becoming moist or damp.

MADEX 2® See Codling-Moth Granulosis Virus.

MADEX 3® See Codling-Moth Granulosis Virus.

MADICOLOUS Adj. (Latin, *madidus* = moist, wet; *colere* = living in or on.) Moisture loving. Pertaining to insects that prefer moist or wet habitats. See Habitat. Cf. Algicolous; Halophilous; Hydrophilous; Madicolous; Paludicolous; Pratinicolous; Rhophilous; Ripicolous. Rel. Ecology. Rel. Ecology.

MADREPORE Noun. (French, *madrepore* from the Latin, *mater* = mother; Greek, *poros* = friable stone. Pl., Madrepores.) Reef-building coral that is branched and covered with small apertures.

MADREPORIFORM BODIES Coccids: See Cribriform Plates (MacGillivray).

MADSEN, HAROLD F (1921–1987) American academic (University of California, Berkeley) and subsequently Head, Entomology Section, Agriculture Canada Research Station, Summerland, British Colombia. Primary research interest in pest management.

MADWAR, S (–1953) (Buxton 1954, Proc. R. ent. Soc. Lond. (C) 18: 80.)

MAERIANUM Noun. (Etymology uncertain.) 'That segment of the post-pectus situated one on each side behind the acetabulum and paraplerum, it supports the posterior feet' (Say). See Meriaeum.

MAF-46® See Hexaflumuron.

MAFU® See Dichlorvos.

MAGENTA Adj. (Italian). Fuchsine dye from Magent, Italy, in allusion to a battle fought there at the time the dye was discovered. A brilliant dark red with a purplish cast.

MAGESTRETTI, LUIGI (–1958) (Anon. 1958, Boll. Soc. ent. ital. 88: 69.)

MAGGI, LEOPOLDINA (1840–1905) (Cattaneo 1905, Monitore zool. ital. 16: 78–84, bibliogr.)

MAGGOT Noun. (Middle English, *magot* = grub. Pl., Maggots.) The legless larva of Diptera, typically without a distinct head and body usually pointed anteriorly and blunt posteriorly. See Larva. Cf. Caterpillar; Grub; Slug. Rel. Naiad; Nymph.

MAGGOT DEBRIDEMENT THERAPY The use of live fly larvae in medical treatment for cleaning wounds that have not healed. William Baer

(Johns Hopkins University) was the first physician in the U.S. to actively promote MDT (1932). MDT was successfully and routinely performed in over 300 hospitals, until the mid-1940's, when its use was supplanted by the new antibiotics and surgical techniques. Currently, only *Phaenicia sericata* (green blow fly) larvae are used in clinical work, because it has been used successfully in maggot therapy. See Maggot; Green Blow Fly. Cf. Myiasis. Rel. Biotherapy.

MAGGOT THERAPY The medicinal use of sarcophagid fly larvae to clean open sores or slow-to-heal wounds for the prevention of gangrene. Larval feeding is limited to dead or decaying tissues and facilitates rapid healing of the wound. MT treatment can be superior for patients with poor circulation or sensitivity or allergies to medication. Syn. Larval Therapy.

MAGGS, P (1876–1962) (Wigglesworth 1964, Proc. R. ent. Soc. Lond. (C) 28: 58.)

MAGIS Adv. (Latin, anomalous form of *magnus*, *magis* = in a higher degree, more.) More.

MAGISTER® See Fenazaquin.

MAGNAM® A registered biopesticide derived from *Bacillus thuringiensis* var. *kurstaki*. See *Bacillus thuringiensis.*

MAGNESIUM Noun. (Latin, *magnesia*.) A silvery white metallic element found in chlorophyll and necessary for the growth of all common plants.

MAGNETIC RESONANCE IMAGING (MRI.) An imaging technique involving tissue samples placed in a strong magnetic field causing protons to align with and precess within the applied field at the Larmor frequency. A radiofrequency pulse at the same frequency is absorbed by the precessing protons and reradiated yielding an nuclear magnetic resonance (NMR) signal. By application of a series of radiofrequency pulses and magnetic field gradients, the signal can be mapped from the sample into a digital array yielding an image of a magnetically selected portion of the tissue sample.

MAGNETIC RESONANCE MICROSCOPY Nuclear Magnetic Resonance signals spatially localized by application of magnetic gradients. (Lauterbur 1973. Nature 242: 190.) See Microscopy.

MAGNETIC TERMITE *Amitermes meridionalis* (Froggatt) [Isoptera: Termitidae].

MAGNIFICENT SPIDER *Ordgarius magnificus* (Rainbow) [Araneida: Araneidae].

MAGNOLIA SCALE *Neolecanium cornuparvum* (Thro) [Hemiptera: Coccidae]: Among the largest scale insects at 1.3 cm long. MS overwinters as a first-instar nymph. A pest of Magnolia in North America. See Coccidae.

MAGNUM® See Thiodicarb.

MAGNUS-GOTHUS, OLAUS (1490–1568) (Rose 1850, *New General Biographical Dictionary* 9: 412.)

MAGPIE MOTH *Nyctemera amica* (White) [Lepidoptera: Arctiidae].

MAGRETTI, PAOLO (1854–1913) (Senna 1913, Boll. Soc. ent. ital. 45: 245–247, bibliogr.)

MAGTOXIN® See MGP.

MAHATZ® See Chlordane.

MAHETA SKIPPER *Trapezites maheta* (Hewitson) [Lepidoptera: Hesperiidae].

MAHOGONY BARK-WEEVIL *Macrocopturus floridanus* (Fall) [Coleoptera: Curculionidae].

MAHOGONY LEAFMINER *Phyllocnistis meliacella* Becker [Lepidoptera: Gracillariidae].

MAHOGONY WEBWORM *Macalla thyrsisalis* Walker [Lepidoptera: Pyralidae].

MAHONY, EUGENE (1899–1952) (Stelfox 1952, Irish Nat. J. 10: 229–230.)

MAIDENHAIR FERN See Ginkgo.

MAIDENHAIR FERN APHID *Idiopterus nephrelepidis* Davis [Hemiptera: Aphididae].

MAIDENHAIR FERN WEEVIL *Neosyagrius cordipennis* Lea [Coleoptera: Curculionidae].

MAIDL, FRANZ (1887–1951) (Beier 1953, Annln naturh. Mus. Wien 59: 1–4.)

MAILLE, ARSEN (1784–1839) (Serville 1839, Ann. Soc. ent. Fr. 8: 603–606.)

MAILLOT, GASTON (–1974) (Anon. 1974, Bull. Soc. ent. Fr. 79: 270.)

MAILSTAY® See Fonofos.

MAIMETSHIDAE Rasnitsyn 1975. A Family of parasitic Hymenoptera assigned to the Stephanoidea.

MAIN, HUGH (–1948) (Blair 1948, Entomol. mon. Mag. 84: 96.)

MAINDRON, MAURICE (1857–1911) (Desbordes 1912, Ann. Soc. ent. Fr. 80: 503–510, bibliogr.)

MAINDRONIIDAE Plural Noun. A small, obscure Family of Thysanura, whose members lack exertile Vesicles on abdominal Sterna. See Thysanura.

MAIZE BILLBUG *Sphenophorus maidis* Chittenden [Coleoptera: Curculionidae]: In North America, a pest of corn which tunnels and feeds inside stalks. See Curculionidae.

MAIZE LEAFHOPPER *Cicadulina bimaculata* Evans [Hemiptera: Cicadellidae].

MAIZE PLANTHOPPER *Peregrinus maidis* (Ashmead) [Hemiptera: Delphacidae].

MAIZE STALK-BORER *Chilo partellus* (Swinhoe) [Lepidoptera: Pyralidae].

MAIZE WEEVIL *Sitophilus zeamais* Motschulsky [Coleoptera: Curculionidae]: A cosmopolitan pest of cereal grains, particularly in subtropical and tropical climates; the name suggests that MW may prefer maize. Adult MW is anatomically nearly identical with Rice Weevil, except genitalia characters of male and female; body 2.5–4.0 mm long, reddish brown with four pale spots on Elytra; Pronotum pitted; larva to 4.0 mm, white with dark head. Life cycle requires 4–6 weeks; adults fly and live about 6 months. See Curculionidae. Cf. Granary Weevil; Rice Weevil.

MAJOR, JOHN DANIEL (1634–1693) (Rose 1850, *New General Biographical Dictionary* 9: 435.)

MAJOR WORKER Social Insects: A member of the subcaste of the largest workers. In ants, usually specialized for defence. Syn. Soldier. Termites:

A large worker in a nest. Cf. Media Worker; Minor Worker.

MAKI, MOICHIRO (1886–1959) (Hasegawa 1967, Kontyû 35 (Suppl.): 75–76.)

MÄKLIN, FRIEDRICH WILHELM (1821–1883) (Sandahl 1883, Ent. Tidsskr. 4: 6–8, 51–52, bibliogr.)

MALA Noun. (Latin, *mala* = cheek, jaw. Pl., Malae.) 1. Insects; A lobe, ridge or grinding surface of the Maxilla or Mandible. See Mola. 2. Myriapods: The 3rd segment of the Mandible.

MALA MANDIBULARIS An arcane term for the grinding surface or area of a Mandible.

MALA MAXILLAE An arcane Latin term for either of the lobes of the Maxilla; specifically, the Galea and Lacinia.

MALACHIIDAE See Melyridae.

MALACOID Adj. (Greek, *malakos* = soft; *eidos* = form.) Mollusc-like; pertaining to structure which is soft, pliant or of undetermined shape. See Mollusc. Cf. Amoeboid; Helicoid. Rel. Eidos; Form; Shape.

MALACOLOGY Noun. (Greek, *malakos* = soft; *logos* = discourse. Pl., Malacologies.) The study of molluscs, including snails.

MALACOPHAGOUS Adj. (Greek, *malakos* = soft; *phagein* = to eat.) Descriptive of animals that feed on soft-bodied organisms, particularly snails (*e.g.* Sciomyzidae, Drilidae and some Staphylinidae).

MALACOPHILOUS Adj. (Greek, *malakos* = soft; *philein* = to love.) 1. Mollusc-loving; pertaining to an organism associated with molluscs, particularly snails. 2. Pertaining to plants that are pollinated by molluscs. See Habitat. Rel. Ecology.

MALAISE TRAP A complex, tent-like sampling device that intercepts and captures flying insects. MT was developed by René Malaise. MT consists of a vertical net that serves as a baffle, end nets, and a sloping canopy leading upward to a collecting device. The collecting device may be a jar with a solid or evaporating killing agent or a liquid in which the insects drown. The original design is unidirectional or bidirectional with the baffle in the middle; more recent types include a nondirectional type with cross baffles and with the collecting device in the centre. (Steyskal, G. C. 1981. Bibliogr. of the Malaise trap. Proc. ent. Soc. Wash. 83: 225–229.) See Trap.

MALAMAR® Malathion.

MALAPHOS® Malathion.

MALAPOPHYSIS Noun. (Latin, *mala* = cheek; Greek, *apo* = away; *phyein* = to grow.) Acarology: Paired anterior region of the Infracapitulum. Dorsally the continuation of the Cervix; ventrally the continuation of the Mentum.

MALAR Adj. (Latin, *mala* = cheek bone.) Pertaining to the cheek.

MALAR SPACE Hymenoptera: The lateral area on the head included between the proximal end of the Mandible and the ventral margin of the compound eye. The Gena *sensu* MacGillivray. See Genal Orbit.

MALARIA Noun. (Italian, *mala* = bad; *aria* = air.) An acute or chronic (sometimes fatal), pan-tropical disease of humans caused by sporozoan parasites (*Plasmodium* spp.) and transmitted by various insects. More than 400 Species of anopheline mosquitoes are recognized as vectors. Malaria is the most important, widespread and persistent disease that affects humans. Estimates report that more than 300 million cases of Malaria exist with 1–3 million deaths per year; disease most common in Africa. Common forms include Falciparum Malaria and Vivax Malaria. Other animals are also affected by Malaria. Cf. Benign Tertian Malaria; Malignant Tertian Malaria; Quartan Malaria.

MALASPRAY® Malathion.

MALATHION An organic-phosphate (dithiophosphate) compound {O,O-dimethyl phosphorodithioate ester of diethyl mercaptosuccinate} used as a insecticide and acaricide against mites, thrips, plant-sucking insects, leaf-chewing insects, armyworms, bollworms, cutworms, webworms and domiciliary pests. Compound applied to cotton, cucumbers, cherries, grapes, onions, peppers, potatoes, tomatoes, strawberries, vegetables, ornamentals; also applied as a public-health insecticide for control of mosquitoes, ticks, fleas, lice, houseflies and ants. Compound phytotoxic to some apples, some grapes, pears, cucurbits, beans, and some ornamentals. Compound corrosive to iron and other heavy metals. Toxic to bees and fishes. Trade names include: Carbophos®, Chemathion®, Cythion®, Duramitex®, Emmatos®, Fyfanon®, Karbofos®, Kypfos®, Malamar®, Malaphos®, Malaspray®, Malphos®, Maltox®, Mercaptothion®, MLT®, Rion®, Zithiol®. See Organophosphorus Insecticide.

MALAXATE Transitive Verb. (Greek, *malassein* = to soften; Latin, *malaxatus* > *malaxare* = to soften.) To soften or knead to softness.

MALAXATION Noun. (Greek, *malassein* = to soften; English, *-tion* = result of an action. Pl., Malaxations.) A behaviour of some adult wasps (*e.g.* Bethylidae, Pompilidae) in which the Integument of paralysed prey (hosts) (spiders or caterpillars) is chewed, kneaded or made pliant by chewing action of the Mandibles. Malaxation presumably makes the prey (host) more suitable as food for developing wasp progeny or restricts the motion of the prey.

MALAYAN POWDER-POST BEETLE *Minthea rugicollis* (Walker) [Coleoptera: Bostrichidae]: A pest of rainforest hardwood timbers in southeast Asia. See Bostrichidae. Cf. Powder-Post Beetle; Small Powder-Post Beetle.

MALAYSIAN FRUIT-FLY *Dacus latifrons* (Hendel) [Diptera: Tephritidae]: A fly originating in southeast Asia which is regarded as a potential pest of many solanaceous plants, including tomato, pepper and eggplant. See Tephritidae.

MALCIDAE Plural Noun. A small Family of

heteropterous Hemiptera assigned to Superfamily Lygaeoidea. (Stys 1967, Acta Ent. Mus. Nat. Prague 37: 351–516.)

MALDISON An organophosphorus insecticide used as a surface spray space spray (mist) for residual control of mosquitoes and flies. See Organophosphorus Insecticide.

MALE Noun. (French, from Latin, *masculus*, dim. of *mas* = male. Pl., Males.) In biparental Species, the sex with organs for production of Spermatozoa or male gametes.

MALE ACCESSORY GLAND Ectodermal pouches or blind ducts that branch from the Vas Differens, Seminal Vesicle or Ejaculatory Duct. Anatomically, AG consists of an epithelial layer with a central margin that bears microvilli which project into the lumen of the gland. Many insects possess two pairs of AGs. One pair (Mesadenia) are mesodermal and open into the Vas Deferens; other pair (Ectadenia) are ectodermal and open into the Ejaculatory Duct. Conspicuous accessory glands are not evident in many calypterate muscoid Diptera, but anterior part of Ejaculatory Duct is thickened and glandular. AGs of Blattaria are not paired; Orthopteran AGs typically are multilobed. Male AG secretions form a Spermatophore, facilitates Spermatozoa transfer from male to female, activate Spermatozoa, provides nutrients for inseminated female, stimulates female genital tract to move Spermatozoa, affect female reproductive and oviposition behaviour. Male AG secretions stimulate vitellogenin synthesis in female mosquitoes. AG also manufacture mating plugs. See Accessory Gland. Cf. Female Accessory Gland.

MALE GENITAL CHAMBER The conjunctival membrane behind the ninth Sternum invaginated within the ninth segment in the male (Snodgrass).

MALEK, IVAN (–1959) (Kratochvil 1959, Zool. Listy 8: 288–289.)

MALEZIEU, NICHOLAS DE (1650–1727) (Rose 1850, *New General Biographical Dictionary* 9: 445.)

MALICE® See Bensultap.

MALIGNANT TERTIAN MALARIA The most serious form of human malaria with a high rate of mortality. MTM found in warmer regions of world. Caused by the protozoan *Plasmodium falciparum* whose Merozoite stage attacks red blood cells of all ages; Schizogony occurs in capillaries of internal organs and interrupts their blood supply. Blackwater Fever and Haemolytic Anemia associated with MTM. See Malaria. Cf. Benign Tertian Malaria.

MALINDEVA SKIPPER *Hesperilla malindeva* Lower [Lepidoptera: Hesperiidae].

MALINKOWSKI, AUGUST LUDWIG VON (1809–1862) (Anon. 1863, KorrespBl. Ver. Naturk. Pressburg 2: 109–111.)

MALIPEDES Noun. (Latin, *mala* = cheek, *pedes* = foot.) The third and fourth pair of 'footjaws' in centipedes (Packard).

MALIX® See Endosulfan.

MALLEE MOTHS See Oecophoridae.

MALLEE WITCHETYGRUG *Cnemoplites blackburni* Lameere; *Cnemoplites edulis* Newman [Coleoptera: Cerambycidae].

MALLEOLI Noun. The Halteres.

MALLOCH, JOHN RUSSELL (1875–1963) (Sabrosky 1963, Ann. ent. Soc. Am. 56: 565.)

MALLOPHAGA Nitzsch 1818. (Recent.) (Greek, *mallos* = wool; *phagein* = to devour.) Biting Lice; Bird Lice; Chewing Lice; Wool-Eaters. An Order or Suborder name applied to biting or chewing lice. Small, wingless, flattened external parasites of birds and mammals (mostly of birds); mandibulate insects with thoracic segments similar in size and shape; Metamorphosis simple. Immatures resemble the adults, and all stages occur on the host; lice rarely found away from host. Diet consists of feathers or mammalian skin scurf and blood of various hosts; some biting lice prey on the eggs and moulting nymphs of other lice, including members of their own Species. Included Families: Boopiidae, Gyropidae, Heptapsogasteridae, Laemobothriidae, Menoponidae, Philopteridae, Ricinidae, Trichodectidae and Trimenoponidae. Six Families commonly found in North America; Philopteridae on poultry, Trichodectidae on cattle, horses and dogs. Species of Mallophaga are not known to attack man. See Amblycera; Ischnocera. Cf. Anoplura. Rel. Parasite. Syn. Lipoptera.

MALLY, FREDERICK WILLIAM (1868–1939) (Sasscer 1939, J. Econ. Ent. 32: 601.)

MALM, AUGUST WILHELM (1821–1882) (Spongberg 1882, J. Ent. Tidsskr. 3: 157–159, 161–162.)

MALMAZET, JEAN ANDRÉ (1808–1858) (Mulsant 1878, Ann. Soc. Linn. Lyon 25: 75–82.)

MALMIGNATTE Noun. (Italian, *malminatta*.) A small black spider with red spots of the Genus *Latrodectus* of southern Europe, reputed to be deadly.

MALOCH, JOHANN (JAN) (1825–1911) (Uzel 1913, Beitr. Insekt. Fauna Böhems. 8: 1–18.)

MALPHOS® Malathion.

MALPIGHAMOEBA MELLIFICAE A protozoan (Order Sarcodina) that infects the Malpighian Tubules of adult honey bees and causes a disease that results in the atrophy of tubules. Syn. *Vahlkampfia mellificae*.

MALPIGHI, MARCELLO (1628–1694) Italian academic (University of Bologna) and physician. Regarded as founder of Embryology and Microscopy. Author of *De Anatome Plantarum;* important works on chick embryology published through the Royal Society of London. Important entomological work entitled *Dissertatio de Bombyce* (1669), written at encouragement of British Royal Society. Malpighi's work separated animal anatomy from medicine. (Rose 1850, *New General Biographical Dictionary* 9: 450.)

MALPIGHIAN TUBULES The organs of excretion and osmoregulation in insects; MT typically attached to hindgut at junction with midgut. MT usually float free in Haemolymph to decontaminate themselves; MT also contact fat body. Distal surface of MT sometimes envelops the Rectum and forms a Cryptonephridium. MT not found in Collembola, Diplura and Aphididae. In Collembola, Labial Glands probably serve in an excretory role. MT weakly developed in proturan *Acerentomon* and papillate in Protura, Diplura and Strepsiptera. MT number varies among insect groups: scale insects (Coccidae) usually two, Orthoptera more than 100; most Diptera two pairs. Generally, MT short when numerous and long when fewer in number. When paired, MT join and form a common excretory duct called the Ureter. MT consist of three ultrastructurally distinct regions: proximal, middle and distal (name refers position relative to midgut). All cell types possess ultrastructural features characteristic of ion-transporting cells. Basal and apical membranes elaborated and associated with mitochondria. Proximal and distal regions short, heavily tracheated and composed of one distinct cell type; middle region longest portion of MT and is composed of two distinct cell types, primary and secondary (both types binucleate). Primary cells more numerous, contain large nuclei, laminate concretions in membrane-bound vacuoles, and posses large microvilli that contain mitochondria; Secondary cells do not contain laminate concretions. MT one cell layer thick and composed of epithelium cells that show structural polarity typical of cells which have a transport function. Most textbooks maintain two discrete regions of MT: Proximal area which is absorptive and a distal area which is secretive; Na^+, K^+. Cl^- serve as osmolytes. Cellular components of MT epitheial cells are complex and cross section of epithelium cell shows three regions: basal region (contacts Haemolymph) includes a Basement Membrane, basal infoldings and mitochondria. Intermediate region with diverse organelles (vesicles, nucleoli and ampulla-like bodies); nucleus occurs in intermediate region and is frequently polyploid. Apical region projects into lumen of MT (site of ion exchange.) Anatomical features of apical region include a brush border and endoplasmic reticulum. Secondary Ureter forms common duct for pairs of tubules which open into hind gut. Ureter surrounded by muscle cells which contract by peristalsis and move urine into hindgut. Tracheoles between muscles and epithelium cells facilitate gas exchange that is aided by muscle contraction. Structure of uterine cells changes proximally to distally. Distal cells similar to main tube; proximal cells similar to epithelium cells of midgut. Musculature with MT varies considerably; some insects lack muscles; some insects show muscles. Circular muscles may surround entire tubule. When muscle cells are present, peristaltic contraction brings urine into lumen of tubule. See Alimentary System; Hindgut. Cf. Fat Body. Rel. Excretion; Kidney.

MALTASE Noun. (Anglo Saxon, *mealt* = malt.) A digestive enzyme by which maltose or malt sugar is hydrolysed and converted into glucose.

MALTOSE Noun. (Anglo Saxon, *mealt* = malt.) A reducing sugar (disaccharide) produced by hydrolysis of starch. Produced by germinating barley and hydrolysed to glucose by maltase and dilute acids.

MALTOX® Malathion.

MALYSHEV, SERGEI IVANOVICH (1884–1967) (Stel'nikov 1968, Ent. Obrozr. 47: 688–693, bibliogr. (Trans.: Ent. Rev. Wash. 47: 421–424, bibliogr.) Russian Entomologist specializing in Hymenoptera.

***MAMESTRA BRASSICAE* NPV** A naturally-occurring Nuclear Polyhedrosis Virus used as a biopesticide against larva of *Mamestra brassicae* and other Lepidoptera in some countries. Must be ingested to be effective; kills larva within 2–10 days. Trade names include: Mb-NPV®, Mamestrin® and Virin KS®. See Biopesticide; Nuclear Polyhedrosis Virus. Cf. Beet Armyworm NPV; Douglas-Fir Tussock Moth NPV; European-Pine Sawfly NPV; *Heliothis* NPV; Redheaded Pine Sawfly NPV; *Spodoptera littoralis* NPV.

MAMESTRIN® See *Mamestra brassicae* NPV.

MAMMAL CHEWING LICE See Trichodectidae.

MAMMAL NEST BEETLES See Leptinidae.

MAMMEINS Plural Noun. A complex of alkaloid biopesticides known from seed of the evergreen tree *Mammea americana* and roots of *M. longifolia*. Mammeins include Coumarins 94 and Surangin B. See Biopesticide; Botanical Insecticide; Natural Insecticide; Insecticide; Pesticides. Cf. Acetogenins; Azadirachtin; Chromenes; Cyclodepsipeptides; Furanocumarins; Polyacetylenes; Terthienyl.

MAMMILLATE Adj. (Latin, *mamma* = breast; *-atus* = adjectival suffix.) Pertaining to structure with nipple-like protuberances or processes. Alt. Mammillated; Mammillatus.

MAMMULAE Plural Noun. (Latin, *mamulla,* diminutive of *mamma* = breast.) 1. Any small, conical Papilla. 2. Spiders: The Spinnerets that issue silk. Cf. Spinneret.

MANASE, AI (1851–1888) (Hasegawa 1967, Kontyû 35 (Suppl.): 80–81.)

MANCINI, CESARE (1881–1967) (Tamanini 1968, Memorie Soc. ent. ital.47: 5–10, bibliogr.)

MANDACORIA Plural Noun. (Latin, *mandere* = to chew; *corium* = skin.) The membrane on the cephalic or dorsal aspect between the Mandibularia of the head and a Mandible, and between the Preartis and Postartis (MacGillivray).

MANDERS, NEVILLE (1857–1915) (Rowland-Brown 1915, Entomol. mon. Mag. 51: 317–319.)

MANDEVILLE, BERNARD DE (1670–1733) (Rose 1850, *New General Biographical Dictionary* 9: 454–456.)

MANDIBLE Noun. (Latin, *mandibulum* = jaw. Pl., Mandibles.) 1. The anterior-most pair of oral appendages on the insect head. Mandibles are lateral appendages, immediately behind Labrum and may represent segmental appendages of fourth segment of head groundplan. Early in their evolutionary history, insects perfected Mandible as an appendage for processing food and it represents one of the hardest parts of insect's Integument (ca 3.0 on Mohs Scale). Mandibles vary considerably in size, shape and apical geometry. Mandibles are stout and highly modified in form, but not showing signs of segmentation. Mandible shape is strongly influenced by function: Tooth-like in chewing insects and needle-like or sword-shaped in piercing-sucking insects. Mandibles are not always used for feeding. Some Holometabola use their Mandibles to exit Puparium or area of pupal confinement (*e.g.* exodont Mandible of alysiine braconid wasps). Some bees and wasps use Mandibles to construct nests in soil, wood and other hard material (*e.g.* bee, *Perdita opuntae* excavates a nest from stone). Mandible is useful for processing matrices of varying structural complexity, chemical composition and physical hardness. Complex plant fibres of differing degrees of hardness require a Mandible that is harder than fibres under process and a Mandible with complex surface features. The origin of the insect and myriapod Mandible has been considered by many morphologists. 2. The mouth hooks of muscoid Diptera. See Mouthpart. Cf. Molar; Prostheca; Tooth; Rel. Cultellus; Labium; Maxilla.

MANDIBULAR ARTICULATION Points of contact between the Mandible and cranial wall. The hypothetical ancestor of insects is thought to possess a Mandible with one point of articulation. Later, insects acquired a second point of articulation. The basis of this assumption comes from a survey of the Hexapoda. Modern Apterygota have a monocondylic Mandible; Pterygota have a dicondylic Mandible. See Articulation. Cf. Dicondylic; Monocondylic. Rel. Acetabulum; Cotyla; Condyle.

MANDIBULAR GANGLIA Nerve masses which control the Mandibles.

MANDIBULAR GLAND Paired glands that open mesially at base or near apex of insect Mandible. Mandibular Glands occur in the Apterygota, Isoptera, Orthoptera, Coleoptera, Hymenoptera, and larval Lepidoptera. Within taxonomic groups that express MGs, development of glands is not uniform (*e.g.* within Sphingidae, *Celerio* larvae have Labial and Mandibular Glands while *Protoparce* has neither). MGs usually are well developed in Lepidoptera larvae where they form tubes that project into the body. Tubular salivary glands in Thorax on each side of foregut, communicate with mouth via a pore at base of Mandible. MGs occur in many Hymenoptera including wasps, bees and ants. Functions of MGs are diverse: Ant *Calomyrmex* MG have strong antimicrobial activity. Weaver Ant MGs are large and discharge secretions when disturbed; glandular compounds include Mellein. Queens differ from workers quantitatively but not qualitatively. Ants shows strong sexual dimorphism in MG secretions. Male MG of *Camponotus thoracicus* contain sex-specific compound, 2,4 dimethylhexan-5-olide. Methyl 6-methylsalicylate also occurs in this Species and in many others; compound stimulates seasonal swarming behaviour. Among ponerine ants, compound does not occur in male, but occurs in MG secretions of worker. In *Gnamptogenys pleurodon* compound acts as an alarm pheromone. Mutillidae produce short-chained ketones and unidentified compounds; ketones are alarm pheromones in ants that are predators of mutillid wasps. MG secretions are probably defensive compounds. Bees: MG a sac-like gland opening at inner angle of each Mandible, larger in queen than worker and poorly developed in drone (Imms). Many male bees use MG secretions to mark territory. Ground-nesting bees use MG secretion to waterproof their cells. Honey bee queen secretes sex pheromone (trans-9-keto-2-decenoic acid) from MG. MG provide an alarm function and repugnance to insectivorous mammals. See Gland. Cf. Labial Gland; Maxillary Gland.

MANDIBULAR MUSCULATURE The groups or individual muscles which bring the Mandibles together or separate the Mandibles and process food or masticate other materials. Mandibular Musculature is arranged according to the type of Mandible which the insect displays. The monocondylic Mandible represents the primitive (plesiomorphic) condition and corresponds to the dorsal articulation of a walking appendage (leg). This development is seen in simple Crustacea and all apterygotes except Lepismatidae. The monocondylic Mandible contains one point of articulation, and four muscles for each Mandible. Muscles with a dorsal origin insert on the anterior and posterior face of the mandibular base. These muscles are the anterior and posterior remotor muscles. Muscles with a ventral origin include posterior muscle that originate on the Sternal Apophysis or Hypopharyngeal Sclerite (future Anterior Tentorial Arm of pterygotes); anterior muscles fuse medially to form a median ligament (Common Zygomatic Adductor). Collectively, these muscles bring the Mandibles together. The dicondylic Mandible shows a derived (apomorphic) condition and is found in the Lepismatidae and Pterygota. The dicondylic Mandible has secondarily acquired an articulation point anterior of the first point in the monocondylic Mandible. This develops a long plane of attachment. The monocondylic Mandible has no plane of attachment and the Mandibles may move forward or rearward when the remotor muscles contract. Two points of articulation create a plane of movement which restricts the direction of Mandible movement. The posterior muscle becomes the Dorsal Adductor, and is responsible for closing the Mandibles because its point of insertion is proximad of the line between the two points of articulation. The anterior muscle becomes the Dorsal Abductor, and is responsible for opening the Mandibles because its point of insertion is distad of the line between the two points of articulation. See Muscle. Cf. Maxillary Musculature.

MANDIBULAR NERVES Nerves of the Mandibles.

MANDIBULAR PALP Mayflies: See Lacinia Mobilis.

MANDIBULAR POUCH Thysanoptera: An invagination of the head containing the single functional Mandible (Snodgrass).

MANDIBULAR SCAR Some Coleoptera: A round or oval area on the Mandible which serves as a support for the deciduous provisional Mandibles of the Pupa (Imms).

MANDIBULAR SCLERITE 1. Muscid larvae: A heavily chitinized sclerite with a broad base and bearing a pair of mouth hooks. A part of the cephalo-pharyngeal skeleton. 2. Heteroptera: The Juga, Lora and Fulcra (Comstock).

MANDIBULAR SCROBE Coleoptera: A broad, deep groove on the lateral aspect of Mandible in some Taxa.

MANDIBULAR SEGMENT The fourth segment or Mandible-bearing segment of an insect's head.

MANDIBULAR STYLET Modified Mandibles in several groups of insects with piercing-sucking mouthparts, best exemplified in Hemiptera-Heteroptera. Two pairs of stylets exist: The lateral pair are modified Mandibles and the median pair are modified Maxillae. A cross section through Mandibular and Maxillary Stylets reveals that each Stylet contains one neural canal that extends the length of the Stylet. Mandibular Stylets sometimes envelop Maxillary Stylets. Maxillary Stylets interconnect dorsally and ventrally along a longitudinal tongue-in-groove connection. The medial surfaces of Maxillary Stylets are convex and form two longitudinal channels or tubes. The ventral channel is called the Salivary Duct and forms a tube through which digestive enzymes and other fluids pass. The dorsal channel is called the Food Canal, a tube responsible for drawing liquified (or particulate food) into the body. The Food Canal may have Acanthae or bristles that filter larger particles of solid food. Fluid conduction is facilitated by the Cibarial Pump. The apical margin of one Mandibular Stylet is toothed for cutting into host plant Cuticle, vertebrate skin, or arthropod Integument. Teeth are linearly arranged and number of teeth varies from a few to more than 20; predaceous and hematophagous Species display more teeth than phytophagous Species. Size and shape of teeth also varies among Species; teeth may consist of small, unformly sized serrations or teeth may be considerably larger. In some Species, teeth are harpoon-like and may be used as an anchor. The other Mandibular Stylet displays an apical rasp that makes the apical modifications of the Mandibular Stylets asymmetrical. Functionally, Stylet Teeth cut or penetrate tissue; the rasp pulverizes or shreds tissue so that small amounts of tissue can be imbibed through the Food Canal. Maxillary Stylets splay apart at the apex; this action probably increases the effectiveness of the Maxillary Stylets in probing into appendages such as legs. The apical margin of Maxillary Stylets are spoon-shaped and sharpened in *Geocoris punctipes*. This also probably facilitates the ingestion of food.

MANDIBULAR TEETH The distalmost part of the Mandible is frequently invested with conical projections called teeth (incisors). Teeth differ in size, shape and arrangement. Some insects show the distal part of the Mandible modified into a broad, chisel-like cutting surface called a truncation. Other insects display a bidentate Mandible characterized by two apical teeth. Some insects display a tridentate Mandible with three apical teeth. Each pattern of dentition represents a solution to problems presented by differences in the matrix being processed by the Mandibles for feeding, burrow construction, emerging from pupal containment or other biological constraints.

MANDIBULARIA Noun. (Latin, *mandibulum* = jaw.) 1. The Trochantin of the Mandible. 2. The Basimandibula *sensu* MacGillivray.

MANDIBULATA Noun. (Latin, *mandibulum* = jaw.) Insects in which the adults have functional Mandibles used for biting. Cf. Chelicerata.

MANDIBULATE MOTHS See Micropterygidae.

MANDIBULATE Adj. (Latin, *mandibulum* = jaw; *-atus* = adjectival suffix.) 1. Pertaining to organisms with a lower jaw. 2. Pertaining to arthropods with biting jaws. 3. Hymenoptera: Segmented apocritous larvae with a sclerotized and unusually large head, large falcate Mandibles and body tapered posteriad. Endoparasitic and ectoparasitic larvae are mandibulate. Alt. Mandibulated; Mandibulatus. See Larva.

MANDIBULIFORM Adj. (Latin, *mandibulum* = jaw; *forma* = shape.) Mandible-shaped; descriptive of structure which resembles a Mandible or is jaw-shaped. Ideally, the structure is heavily sclerotized with two ball-shaped processes at the base, an elongate body and apical conical or pointed processes. Alt. Mandibuliformis. See Mandible. Cf. Maniform. Rel. Form 2; Shape 2; Outline Shape.

MANDIBULRIS (Latin, *mandibulum* = jaw.) Aphaiptera: The sucking tube, formed by the Mandibles (MacGillivray).

MANDOGENAL SUTURE A suture separating the Trochantin of the Mandible from the Front and Vertex (MacGillivray).

MANDORIS Noun. (Etymology uncertain.) The mouth-like slit between the Prepharynx and Postpharynx (MacGillivray).

MANDUCATE Adj. (Latin, *manducare* = to chew; *-atus* = adjectival suffix.) A term descriptive of Mandibles that are capable of biting and chewing.

MANEVAL, JEAN CHARLES HENRI (1892–1942) (Pitau 1943, Revue Sci. nat. Auvergne 9: 1–4, bibliogr.)

MANGANARO, ANA GIORDANO (–1921) (Trotter 1920, Marcellia 19: xxxv.)

MANGE MITE *Sarcoptes scabei canis* (DeGeer) [Acari: Sarcoptidae]. A Species of mite which infests the Epidermis of dogs causing blisters and inflammation. MM can affect humans closely associated with dogs. Cf. Human Itch Mite.

MANGO *Mangifera indica* Linnaeus [Anacardiaceae]: Evergreen tree, native to India, grown for edible fruit in drier tropics.

MANGO BARK-BEETLE *Hypocryphalus mangiferae* (Stebbing) [Coleoptera: Scolytidae].

MANGO BUD-MITE *Eriophyes mangiferae* (Sayed) [Acari: Eriophyidae]. Alt. Mango Bud Mite.

MANGO FLOWER-BEETLE *Protaetia fusca* (Herbst) [Coleoptera: Scarabaeidae].

MANGO FLY *Dacus frauenfeldi* Schiner [Diptera: Tephritidae].

MANGO GREY-SCALE *Genaparlatoria pseudaspidiotus* (Ldgr.) [Hemiptera: Diaspididae]: An Old World, tropical pest of mango and occasionally found on citrus. See Diaspididae.

MANGO LEAF-WEBBER *Orthaga exvinacea* Hamps. [Lepidoptera: Pyralidae].

MANGO NUT-WEEVIL *Sternochetus gravis* (Fabricius) [Coleoptera: Curculionidae].

MANGO PLANTHOPPER *Colgaroides acuminata* (Walker) [Hemiptera: Flatidae]: A pest of mango and citrus in Australia. Adults are ca 13 mm long; pale green to white, with small red spot in middle of forewing and pale red stripe along margin of forewing; triangular in outline shape from lateral view; wings held tentiform when at repose. Eggs are laid in circular mass comprising about 50 eggs. MPH is multivoltine with life cycle requiring 1–2 months, depending upon climate and host plant. See Flatidae.

MANGO SCALE *Aulacaspis tubercularis* Newstead; *Pseudaulacaspis cockerelli* (Cooley) [Hemiptera: Diaspididae].

MANGO SEED WEEVIL *Sternochetus mangiferae* (Fabricius) [Coleoptera: Curculionidae]: A pest of mango in the Old World tropics. See Curculionidae.

MANGO SHOOT-CATERPILLAR *Penicillaria jocosatrix* Guenée [Lepidoptera: Noctuidae].

MANGO SPIDER-MITE *Oligonychus mangiferus* (Rahman & Sapra) [Acari: Tetranychidae].

MANGO TIPBORER *Chlumetia euthysticha* (Turner) [Lepidoptera: Noctuidae].

MANGO WEEVIL *Cryptorhynchus mangiferae* (Fabricius) [Coleoptera: Curculionidae].

MANGO WHITE-SCALE *Parlatoria crypta* MacKenzie [Hemiptera: Diaspididae].

MANGOLD APHID *Rhopalosiphoninus staphyleae* (Koch) [Hemiptera: Aphididae].

MANGROVE CICADA *Aruntia interclusa* (Walker) [Hemiptera: Cicadidae].

MANICA Noun. (Latin, *manicatus* = sleeved.) Lepidoptera: The membranous sheath of the Penis (Tillyard).

MANICATE Adj. (Latin, *manicatus* = sleeved; *-atus* = adjectival suffix.) Fur-like; covered with irregularly arranged or matted Setae.

MANIFORM Adj. (Latin, *manus* = hand; *forma* = shape.) Hand-shaped; descriptive of structure which is broad, flat with conical projections from the apical margin. Alt. Maniformis. See Hand. Cf. Actiniform; Rotuliform. Rel. Form 2; Shape 2; Outline Shape.

MANIPULATOR® See Chlormequat.

MANITOBA TRAP A canopy-type trap for horse flies (Diptera: Tabanidae). The trap forms an inverted cone set on a tripod with a black sphere suspended beneath the canopy. Adult flies are attracted to the black sphere. tend to fly upward and are subsequently contained at the apex of the cone in a collecting jar.

MANITRUNK Noun. (Latin comb. form, from *manus* = hand; *truncus* = the main stem.) Part of a trunk that bears the anterior legs; the Prothorax. Alt. Manitruncus. See Prothorax.

MANK, EDITH WEBSTER (1892–1945) (Weiss 1946, Ann. ent. Soc. Am. 39: 4, bibliogr.)

MANLY, GEORGE B (1890–1967) (Anon. 1955, Entomologist's Rec. J. Var. 67: 216.)

MANN, BENJAMIN PICKMAN (1848–1926) (Anon. 1926, Ent. News 37: 192.)

MANN, JOHN (1905–1994) Australian specialist in biological control of weeds. Member, Commonwealth Prickly Pear Board and instrumental in control of Prickly Pear in Australia. Awarded MBE for contribution to rural industry. Founding member of Queensland Entomological Society (1923).

MANN, JOSEPH JOHANN (1804–1889) (Rogenhofer 1889, Wein ent. Ztg. 8: 241–244.)

MANN, WILLIAM M (1886–1960) (Snyder 1961, Proc. ent. Soc. Wash. 63: 69–73, bibliogr.)

MANNEHEIMS, BERNHARD (–1971) (Neithammov 1971, Bonn. zool. Beitr. 22: 1–3.)

MANNERHEIM, CARL GUSTAV VON (1804–1854) Councillor of the Tsar of all the Russias. Author of *Eucnemis insectorum genus* (1823), *Observations sur le genre Megalope* (vol 10, Mem. Imp. Acad. Sci. St. Petersburg (1824), *Description de quarante nouvelles especes de Scarabaeides du Brazil.* (Newman 1854, Proc. ent. Soc. Lond. 1854: 54; Motschulsky 1855, Etude Ent. 4: 5–7; Sahlberg 1919, Finska Tidsskr. 87: 76–100.)

MANNING CANYON SHALE FORMATION A Late Mississippian - Early Pennsylvanian (Namurian A and B) formation found in central Utah. Deposits consist of shales, siltstones, orthoquartzites and interbedded limestones. Plants are among the best known for Carboniferous deposits in western North America and suggest a swamp-like, moist lowland with summer-like conditions. Type-locality for *Brodioptera stricklani* Nelson & Tidwell [Megasecoptera]; insect presumably with beak adapted for sucking and feeding on cones of lycopods, cordaiteans and pteridosperms. See Megasecoptera.

MANNING, FRANCIS JOSEPH (1912–1966) (Pearson 1967, Proc. R. ent. Soc. Lond. (C) 31: 62.)

MANSON, PATRICK (1844–1922) English medical entomologist and first person (1878) to realize development of the nematode *Wuchereria bancrofti* in the mosquito *Culex quinquefasciatus*. (Wood 1922, Am. J. trop. Med. 2: 361–368.) See Bancroftian Filariasis.

MANTA® See Methoprene.

MANTERO, GIACOMO (1878–1949) (Invrea 1951, Annali Mus. civ. Stor. nat. Giacomo Doria 64: 335–338, bibliogr.)

MANTER'S RULES A series of generalizations regarding parasites and their hosts: 1. Parasites evolve more slowly than their hosts; 2. The longer the period of association between parasites and hosts, the more pronounced the host specificity; 3. The largest number of parasites attacking a host Species will be found in the area of endemicity of the host. Cf. Eichler's Rule; Emery's Rule; Farenholtz' Rule; Szidat's Rule.

MANTIDAE Plural Noun. Soothsayers. The largest Family of Mantodea including 263 Genera in 21 Subfamilies; cosmopolitan in distribution. Body size from small to very large. Large Species can capture and eat small lizards and frogs. Subfamilies include Amelinae, Angelinae, Caliridinae, Choeradodinae, Compsothespinae, Deroplatyinae, Haaniinae, Iridopteryginae, Liturgusinae, Mantinae, Oligonychinae, Orthoderinae, Oxyothespinae, Photininae, Phyllotheliinae, Schizocephalinae, Sibyllinae, Tarachodinae, Toxoderinae, Thespinae and Vatinae. See Mantodea. Cf. Australian Mantid; Black-Barrel Mantid; Burmeister Mantid; Carolina Mantid; Chinese Mantid; European Mantid; Green Mantid; Large Brown Mantids; Narrow-Winged Mantid; Small Brown Mantids.

MANTIDFLIES See Mantispidae.

MANTIDS See Mantodea.

MANTISPIDAE Plural Noun. Mantidflies; Mantispids. A moderate sized, widespread Family of Neuroptera. Adult body is small to moderate-sized. Distinctive mantid-like adults with raptorial foreleg, elongate Pronotum; elongate fore Coxa, wings narrow and forewing with Pterostigma. Larvae are hypermetamorphic and parasitic on spiders and social wasps. High fecundity is reported in some Species. See Neuroptera.

MANTISPIDS See Mantispidae.

MANTISPOIDEA A Superfamily of plannipenous Neuroptera including Berothidae and Mantispidae.

MANTLE CELLS In the insect eye, corneagenous cells that ensheath the Retina (Snodgrass).

MANTODEA Burmeister 1838. (Triassic Period-Recent.) Praying Mantids. An insect Order placed with the Orthoptera in some classifications. Terrestrial Neoptera including about 1,800 Species and 320 Genera; cosmopolitan but predominantly tropical or subtropical. Adult body size usually moderate to large; head usually hypognathous, triangular in frontal aspect and tapering towards mouthparts, shortened when viewed in dorsal and lateral aspect; head articulates freely with Thorax; mouthparts typically orthopteroid; Maxillary Palpus with five segments; Labial Palpus with three segments; compound eyes large; three Ocelli (larger in male) but sometimes lost; Antenna multisegmented, ciliate, longer in male than female. Prothorax narrow, elongate (in primitive Families frequently leaf-shaped and articulates with Pterothorax); Mesothorax and Metathorax spiraculate, similar in size, shape and sclerotization; fore leg raptorial, fore Coxa elongate, cylindrical, highly mobile; fore Femur elongate, robust, spined along ventral surface; fore Tibia elongate, cylindrical spined along ventral surface with spines opposable to ventral spines of Femur; middle and hind legs cursorial with Coxae large and mobile; tarsal formula 5-5-5; forewing elongate, relatively narrow, hardened into Tegmina, overlapping abdominal Terga in repose; hindwing membranous, fanlike, fluted, folded under Tegmina at repose. Abdomen dorsoventrally compressed, with 11 segments, Sternum 1 usually reduced or absent; spiracles in pleural membrane of segment 1, in lateral margin of Terga 2–8; Cercus segmented, inserted immediately behind Tergum 10. Female with panoistic Ovarioles; Common Oviduct with large accessory glands whose secretions form Ootheca; size, shape and site of Ootheca Species-specific. Male genitalia asymmetrical. Digestive tract short, straight with large Crop; 6-ribbed proventriculus; midgut with eight Caecae; Malpighian Tubules numerous; Thorax with three Ganglia; Abdomen with seven Ganglia. Immature and adult predaceous. Defensive mechanisms include attack, Crypsis, Aposematic Coloration and Mimicry. Included Families Amorphoscelidae, Chaeteessidae, Empusidae, Eremlaphilidae, Hymenopodidae, Mantidae, Mantoididae and Metallyticidae. See Insecta.

MANTOIDIDAE Plural Noun. A monogeneric Neotropical Family of Mantodea which holds only a few Species. Adult body small, delicate, with females occasionally brachypterous; Pronotum quarate or square; raptorial forelegs armed only with short delicate spines; Tibiae with stronger apical claws; Cerci long. See Mantodea.

MANUBRIUM Noun. (Latin, *manubrium* = handle. Pl., Manubria.) 1. Coleoptera: That part of the Mesosternum in Elateridae which forms the process for fitting into the cavity of the Prothorax. 2. Collembola: The large median base of the Furcula.

MANUS Noun. (Latin, *manus* = hand.) A term formerly applied to the anterior Tarsus.

MANUSCRIPT NAME A scientific name applied to a Taxon in speech, correspondence or report which is supposed new to science but which has not been validated by the criteria set forth in the International Code of Zoological Nomenclature. MS can apply to many levels of the taxonomic hierarchy (Family, Genus, Species), but most commonly encountered at the Species level. Cf. Nomen Nudum; Nomen Dubium.

MANY-PLUMED MOTHS See Alucitidae.

MAOMETABOLA Noun. Insects with a slight or gradual Metamorphosis and without a resting stage, *e.g.* the Orthoptera. Cf. Paurometabola.

MAORI MITE *Phyllocoptruta oleivora* (Ashmead) [Acari: Eriophyidae] (Australia).

MAPLE APHID *Periphyllus californiensis* Shinji [Hemiptera: Aphididae].

MAPLE BLADDER-GALL MITE *Vasates quadripedes* Shimer [Acari: Eriophyidae].

MAPLE CALLUS-BORER *Synanthedon acerni* (Clemens) [Lepidoptera: Sesiidae].

MAPLE LEAFCUTTER *Paraclemensia acerifoliella* (Fitch) [Lepidoptera: Incurvariidae].

MAPLE PETIOLE-BORER *Caulocampus acericaulis* (MacGillivray) [Hymenoptera: Tenthredinidae].

MAPLE TRUMPET-SKELETONIZER *Epinotia aceriella* (Clemens) [Lepidoptera: Tortricidae].

MAPLE, JOHN DINWIDDIE (1910–1945) American student of parasitic Hymenoptera interested in eggshells of Encyrtidae. Killed in airplane crash on Okinawa during World War II. (Flanders 1946, Ann. ent. Soc. Am. 39: 4–5.)

MAPOSOL® See Metham-Sodium.

MARALDI, GIACOMO FILIPPI (1665–1729) (Rose 1850, *New General Biographical Dictionary* 9: 470.)

MARAN, JOSEPH (1905–1965) (Moucha 1960, Cas. csl. Spol. ent. 57: 407–408, bibliogr.)

MARATHON® See Imidacloprid.

MARBLE GALL Hymenoptera: The hard, spherical gall of *Andricus kollari* (Cynipidae). See Oak Apple.

MARBLED Adj. (Greek, *marmaros* = stone.) Irregularly mottled, gray and white, like marble. Alt. Marmorate; Marmoratus.

MARBLED BLUE BUTTERFLY *Erysichton palmyra tasmanicus* (Miskin) [Lepidoptera: Lycaenidae].

MARBLED SCORPION *Lychas variatus* Thorell [Scorpionida: Buthidae].

MARCESCENT Adj. (Latin, *marcescere* = to wither.) Botany: Withering but persisting; not falling off readily, as a calyx after fertilization. Cf. Caducous; Fugacious.

MARCGRAVE, GEORGE (JORGE) (1610–1644) (Gudger 1912, Pop. Sci. Mon. 81: 250–274.)

MARCH FLIES See Bibionidae; Tabanidae.

MARCHAL, PAUL (1862–1942) (Vayssiere 1942, Ann. Inst. Nat. Agron. 33: 5–33; Vayssiere 1942, Ann. Soc. ent. Fr. 111: 149–165, bibliogr.)

MARCHAND, HENRI (1899–1956) (Straub 1956, Mitt. ent. Ges. Basel 6: 41–43, bibliogr.)

MARCHAND, M S LE (–1953) (Anon. 1953, Bull. Soc. ent. Fr. 58: 137.)

MARCHESONI, VITTORIO (1912–1963) (Giacomini 1964, Lav. Soc. ital. Biol. 1964: 219–223.)

MARCHIAPAVOR, ETTORE (1847–1935) (Missiroli 1936, Riv. Malar 15: 185–189.)

MARCHOUX, EMILE (1862–1943) (Roubaud 1943, Bull. Soc. Path. exot. 36: 319–324.)

MARCRESCENT Adj. (Latin, *marcresco* = to fade.) Withering; pertaining to structure in the act of withering. Cf. Caducous, Fugacious.

MARELLI, CAROLOS A (–1966) (Orfila 1966, Revta Soc. ent. argent. 29: 42.)

MAREY, C French physiologist and first to report figure-8 motion of the insect wing-stroke during flapping flight (1869, Mémoire sur le vol des insectes et des oiseaux. Ann Sci. Nat. Zool. 5 (12): 49–150.)

MARGANOTUM Noun. (Pl., Marganota.) A thin ridge, sometimes a fine Carina, separating the Epinotum and Pleuranotum (MacGillivray).

MARGARITACEOUS Adj. (Greek, *margarites* = a pearl; Latin, *-aceus* = of or pertaining to.) Decriptive of structure that shines or resembles mother of pearl; nacre. Alt. Margaritaceus.

MARGARODIDAE Plural Noun. Margarodid Scales. A primitive Family of Coccoidea. Some males with compound eyes; some adult females without functional mouthparts; thoracic and abdominal spiracles present; anal opening simple. Included Subfamilies: Margarodinae, Matsucoccinae, Xylococcinae. See Coccoidea.

MARGATERGUM Noun. (Pl., Margaterga.) A sharp ridge separating a Latatergum from the mesial part of a Tergum (MacGillivray).

MARGHERITIS, AURELIO EMILIO (1921–1974) (Serantes de Gonzalez 1974, Revta Soc. ent. argent. 34: 196.)

MARGIN Adj. (Latin, *margo* = edge.) The more-or-less narrow part of a surface within the edge, bounded on the inner side by the submargin. Alt. Margo.

MARGINAL ACCESSORY VEINS Short, twig-like branches resulting from bifurcations of veins that have not extended far back from the margin of the wings (Comstock). See Vein; Wing Venation.

MARGINAL AREA Orthoptera: See Mediasinal Area.

MARGINAL BRISTLES Diptera: Setae inserted on the posterior margin of the abdominal tergal segments (Comstock).

MARGINAL CELL 1. Diptera (Williston): The Subcostal Cell of Schiner, the Radial Cell of Comstock. 2. Hymenoptera: Radial 1 and 2 *sensu* Comstock. 3. In general, the wing cell beyond the Stigma.

MARGINAL CELLULE See Radial Cellule.

MARGINAL FIELD The Costal Field in Tegmina.

MARGINAL FRINGE The edge of the insect wing.

Some insects (Thysanoptera) display narrow wings with a long MF. Hymenoptera possess a marginal fringe which is inconspicuous. The MF reaches elaborate expression among Mymaridae (parasitic wasps) and Ptinidae (beetles) whose Species have reduced the wing blade to a narrow strip and the MF is several times longer than the wing's width. See Wing.

MARGINAL GLAND OPENINGS Coccids: See Dorsal Pores.

MARGINAL LUNAR PORES Coccids: See Dorsal Pores.

MARGINAL PLATES Coccids: See Lateral Pilacerores.

MARGINAL PORES Coccids: See Dorsal Pores.

MARGINAL SCUTELLAR BRISTLES Diptera: Usually a distinct row of large bristles on margin of Scutellum (Comstock). Syn. Marginal Scutellars.

MARGINAL SETAE Coccids: Fringing Setae on the prominent margin (MacGillivray).

MARGINAL VEIN 1. General: The vein forming the Marginal Cell. 2. Orthoptera: The Costa *sensu* Comstock. 3. Hymenoptera: The Radius of Comstock *teste* Norton. Alt. Marginal Nervure. See Vein; Wing Venation.

MARGINATE Adj. (Latin, *margo* = edge; *-atus* = adjectival suffix.) Bounded by an elevated or attenuated margin; with the margin edged by a flat border. Alt. Marginatus; Margined.

MARGINED BLISTER-BEETLE *Epicauta pestifera* Werner [Coleoptera: Meloidae].

MARGOSAN-O® A biopesticide product developed by Vikwood Ltd and consisting of an ethanolic extract of Neem seed (20–25% neem oil and 0.3% azadirachtin). Technology licensed to W. R. Grace & Co. and used on ornamental and landscape plants. See Azadirachtin. Rel. Neem.

MARIANI, MARIO (1898–1965) (Romano 1967, Mem. Soc. ent. ital. 46: 18–26, bibliogr.)

MARICOPA HARVESTER-ANT *Pogonomyrmex maricopa* Wheeler [Hymenoptera: Formicidae].

MARIETTI, BERNARDO (1786–1844) (Conci 1967, Atti Soc. ital. Sci. nat.106: 41, 54.)

MARIGOLD APHID *Neotoxoptera oliveri* (Essig) [Hemiptera: Aphididae].

MARINE Adj. (Latin, *marinus,* from *mare* = sea; *inus* = ine.) Descriptive of organisms associated with salt-water (sea) habitats. Marine habitats (zones) include abyssal zone, littoral zone, neritic zone, oceanic zone, pelagic zone and photic zone.

MARINE SPIDER *Desis* spp. [Araneida: Desidae].

MARK GALL Plant galls characterized by eggs deposited on tissue or larvae boring into tissue. Larvae completely enclosed at initiation of gall development. MG typically produced on stems or sometimes on leaves by tenthredinid sawflies, tephritid and cecidomyiid flies. See Gall. Cf. Bud-and-Rosette Gall; Covering Gall; Filz Gall; Pit Gall; Pouch Gall; Roll-and-Fold Gall.

MARKEL, AUGUST (1837–1897) (Dietz 1897, Ent. News 8: 184.)

MÄRKEL, JOHANN CHRISTIAN FRIEDRICH (1790–1860) (Anon. 1889, Abeille (Les ent. et leurs écrits) 26: 279–280, bibliogr.)

MARKET VALUE The amount of money that a seller can expect to obtain for a commodity.

MARLATE® See Methoxychlor.

MARLATT, CHARLES LESTER (1863–1954) American economic Entomologist with USDA. (Cory *et al.* 1955, Proc. ent. Soc. Wash. 37: 37–43.)

MARLIN® See Methomyl.

MARMONT, LINDSAY EDGAR (1860–1949) (Glendenning 1949, Proc. ent. Soc. Br. Columb. 45: 32.)

MARMORACEOUS Adj. (Latin, *marmor* = marble; *-aceus* = of or pertaining to.) Marble-like in colour and markings; variegated. Alt. Marmorate; Marmoratus.

MARNIX, PHILIPPE DE (1538–1598) (Rose 1850, *New General Biographical Dictionary* 9: 498.)

MARNO, ERNST (–1883) (Rogenhofer 1884, Verh. zool.-bot. Ges. Wien 33: 21–22.)

MARQUAND, ERNEST DAVID (1848–1918) (Rowswell 1918, Trans. Guernsey Soc. nat. Sci. 8: 83–90, bibliogr.)

MARQUARDT, MAX (–1960) (Anon. 1960, Mitt. dt. ent. 19: 94.)

MARQUES AZEVEDO, LUIS AUGUSTO (–1939) (Anon. 1939, Revta Ent. Rio de J. 10: 483.)

MARQUESAS ISLANDS Twelve Islands situated between latitudes 7° 50' and 10° 35' south, and longitudes 138° and 50' and 140° 50' west. Hawaiian islands located about 3220 kilometers north-northwest. Marquesas discovered by American Captain Joseph Ingraham on ship Hope of Boston.

MARQUET, CHARLES (1820–1900) (Trutat 1900, Bull. Soc. Hist. nat. Toulouse 33: 182–187, bibliogr.)

MARRANGO Noun. See Philippine Neem Tree.

MARSCHALL, AUGUST FRIEDRICH (1804–1887) (Rogenhofer 1887, Verh. zool.- bot. Ges. Wien 37: 62–63.)

MARSDEN-JONES, ERIC (1887–1960) (Varley 1961, Proc. R. ent. Soc. Lond. (C) 25: 50.)

MARSEILLES FEVER See Kenya Tick Typhus.

MARSEUL, SYLVIAN AUGUSTIN DE (1812–1890) (Perraudière 1890, Ann. Soc. ent. Fr. (6) 10: 421–428, bibliogr.)

MARSH Noun. (Middle English, *mersch* = meadowland, Anglo Saxon, *mersc*. Pl., Marshes.) An ecological habitat characterized by plants living together on wet, non-peaty ground. Syn. Swamp. Cf. Bog; Fen. Rel. Community.

MARSH BEETLES See Scirtidae.

MARSH FLIES See Sciomyzidae; Ephydridae.

MARSH TREADERS See Hydrometridae.

MARSH, DUDLEY GRAHAM (1891–1969) (Anon. 1969, Entomologist's Rec. J. Var. 81: 312.)

MARSH, HAROLD OSCAR (1885–1918) (Craighead 1918, J. Econ. Ent. 11: 438–439, bibliogr.)

MARSHALL® See Carbosulfan.

MARSHALL, ARTHUR MILNES (1852–1894) (F. E. W. 1893, Trans. Rep. Manchester microscop. Soc. 1893: 71–72.)

MARSHALL, GUY ANSTRUTHER KNOX (1871–1959) (Thomson 1969, Biogr. Mem. Fellows R. Soc. 6: 169–181, bibliogr.)

MARSHALL, JOHN FREDERICK (1874–1959) (Shute 1950, Nature 165: 16–17.)

MARSHALL, PATRICK (1869–1950) (Riley 1952, Proc. R. ent. Soc. Lond. (C) 16: 86.)

MARSHALL, THOMAS ANSELL (1827–1903) (Bignell 1903, Entomologist's Rec. J. Var. 15: 190–191.)

MARSHALL, WILLIAM (1845–1907) (Schaufuss 1907, Ent. Wbl. 24: 174–175.)

MARSHAM, THOMAS (–1819) English Entomologist and author of *Entomologia Britannica sistens Insecta Britanniae indigena secundum methodum Linnaenam disposita* (1802). (Fletcher 1920, Proc. Linn. Soc. N. S. W. 45: 570–571, 574.)

MARSHY Adj. (Middle English, *mersch* = meadow land > Anglo Saxon, *mersc*.) Pertaining to habitat that resembles a marsh or swamp. See Habitat.

MARSIGLI, (MARSILIUS), LUIGI FERNANDO (1658–1730) (Rose 1850, *New General Biographical Dictionary*. 9: 504–505.)

MARSTON, LEON CHESTER (1905–1937) (Mickel 1938, Ann. ent. Soc. Am. 31: 121.)

MARSUPIAL COCCID See Greenhouse Orthezia.

MARSUPIUM Noun. (Latin, *marsupium* = pouch; Latin, *-ium* = diminutive > Greek, *-idion*.) Some Coccidae: A pouch, formed from Anal Sclerites, in which females carry eggs and young.

MARTELLI, GIOVANNI MARIA (1877–1954) (Roberti 1954, Boll. Lab. Ent. agr. 'Filippo Silvestri' 13: 303–311.)

MARTEN, WERNER (–1974) (Anon. 1974, Shilap 2 (6): 96.)

MARTENS, GEORGE MATHIAS (1788–1872) (Anon. 1872, Leopoldina 7: 91.)

MARTIN, C D (–1950) (Anon. 1950, Rep. Raven ent. nat. Hist. Soc. 1950: 43.)

MARTIN, CHARLES JACOB (1835–1929) (Englehart 1930, Bull. Brooklyn ent. Soc. 25: 39.)

MARTIN, HENRY NEWELL (1848–1896) (Anon. 1897, Ir. Nat. J. 6: 103.)

MARTIN, JAMES OTIS (1870–1951) (Linsley 1952, Pan-Pacif. Ent. 23: 71–74, bibliogr.)

MARTIN, JOANNY (–1905) (Bouvier 1905, Bull. Soc. ent. Fr. 1905: 221–222.)

MARTIN, JOHANNES KARL LUDWIG (1851–1942) (Altena 1946, Nature 1946, 157: 866.)

MARTIN, JOHN (1925–1957) (Richards 1958, Proc. R. ent. Soc. Lond. (C) 22: 75.)

MARTIN, LUDWIG (1858–1924) (Roslin 1925, Dt. ent. Z. Iris 39: 5–10, bibliogr.)

MARTIN, MATTHEW (1748–1838) (Lisney 1960, *Bibliogr. of British Lepidoptera 1608–1799*. 315 pp. (222–223), London.)

MASON BEE *Hoplitis producta* (Cresson) [Hymenoptera: Megachilidae]: A Species widespread in North America and which includes numerous Subspecies. See Megachilidae.

MASS EXTINCTION The disappearance of Taxa of plants or animals from the fossil record during relatively short intervals of geological time. Examples: Late Cambrian mass extinction of Trilobites, late Permian mass extinction of marine invertebrates; Late Cretaceous Period mass extinction of dinosaurs. Cf. Adaptive Radiation. Rel. Geological Time Scale.

MASS FLIGHT Large numbers of conspecific individuals taking flight simultaneously, often at specific times of the day, typically for mating, migration or dispersal. See Flight. Cf. Trivial Flight.

MASTOPARAN Noun. (Greek, *mastos* = breast.) Peptides that occur in wasp venoms that are analogous to Melittin in honeybee venom. The mode of action or functional significance of Mastoparan in toxic action remains unknown.

MASTOTERMITIDAE Plural Noun. A monotypic Family of termites generally regarded as being among the most primitive Taxa of termites and probably closely related to cockraches. The described Species, *Mastotermes darwiniensis,* is known from tropical Australia and New Guinea; does not build mound but typically nests in the boles of trees. Adult alate body size large; compound eyes relatively large; Ocelli present; Tarsi with five segments; Arolium present; Antenna with 29–32 segments; hindwing with Anal Lobe but lacking Basal Suture. Eggs are deposited in pods. Physogastric neoteinics are unknown; apterous neoteinics are common in mature colonies. See Isoptera.

MAULIK'S ORGAN See Metafemoral Spring.

MAXADILAN A polypetide in the saliva of phlebotomine sand flies that can enhance the infection of *Leishmania* in vertebrates presumably by inhibiting macrophage activity against the pathogen. M also has vasodilation activity in vertebrates.

MAXILLA Noun. (Latin, *maxilla* = jawbone. Pl., Maxillae.) Paired, lateral Accessory 'jaws' located immediately posterior of the Mandibles. Maxilla is structurally more complex than Mandible. (Mandible is not an apparently segmented structure in insects and not obviously derived from a segmented appendage). Maxilla demonstrates more clearly a condition of generalized Homology with an appendage. Maxillary appendage components include: Coxopodite (Cardo and Stipes) and Telopodite (Maxillary Palpus). Maxilla typically is elongate with one point of articulation on Cranium. Cardo (basal segment that articulates with Cranium) is attached to Stipes. Cardo is variable in size and shape. Stipes forms second segment of Maxilla. Stipes is broadly attached to Cardo basally, bears a movable Palpus laterally, and is attached to Galea and Lacinea distally. Stipes is modified into a piercing device in some Diptera and into a lever for flexing Proboscis in Diptera. See Mouthparts. Cf. Cardo Galea; Lacinea; Palpifer; Palpiger; Stipes. Rel. Labium; Mandible.

MAXILLAE SETOSAE Bristled or 'hairy' Maxillae.

MAXILLARIA Noun. (Latin, *maxilla* = jawbone. Pl., Maxillariae.) The paired, narrow, ribbon-like sclerites along the mesial margin of the Postgena. Maxillary Pleurites or Maxillifers *sensu* MacGillivray.

MAXILLARY Adj. Attached or pertaining to the Maxilla.

MAXILLARY GLAND A relatively uncommon type of gland in insects. MG has been reported among Collembola, Protura, Heteroptera and sporadically in Holometabola (larval Trichoptera, Neuroptera, Coleoptera and Hymenoptera). Two groups of glandular cells of Maxillary segment are placed mesially above Buccal Tube and Infrabuccal Cavity. MG usually are small, inconspicuous and open at base of Mandible. See Gland. Cf. Labial Gland; Mandibular Gland.

MAXILLARY LOBES See Lobus Maxillae.

MAXILLARY MUSCULATURE Muscles associated with the Maxilla and responsible for processing food or other substances that will be ingested. Maxillary musculature is complex compared with Mandible musculature. A Maxilla is monocondylic with the head via the basal Cardo; Stipes is distad of Cardo and displays a medial Lacinea and lateral Galea. Dorsal musculature includes anterior and posterior muscles that originate on head wall and insert onto the Cardo's upper margin. An additional anterodorsal muscle inserts onto the apex of the Stipes or base of the Lacinea (called Cranial Flexor of Lacinea). Ventral Muscles are anterior and posterior adductors that originate on Tentorium and insert (one each) onto Cardo and Stipes. Muscles manipulating the Lacinea and Galea have their origin on Stipes. These are called Stiptial Flexor of Lacinea and Stipital Flexor of Galea. The Maxillary Palpus is manipulated by Levator and Depressor Muscles that have their origin on Stipes (not Palpifer). Each palpal segment usually has one intrinsic muscle.

MAXILLARY NERVES Nerves of the Maxillae; Nervi Maxillarum.

MAXILLARY PALP (Pl., Palpi.) The Palpus carried by the Stipes on its outer apical surface; consisting of 1–7 segments. MPs typically are sensory in function (Imms). Alt. Maxillary Palpus; Maxipalp; Maxipalpus.

MAXILLARY PLATE Hemiptera: The sclerite posterior of the Lorum and continuous dorsally with the cranial wall.

MAXILLARY PLEURITES The lateral sclerites (Epimera and Episterna) of the Maxillary segment (Smith).

MAXILLARY SEGMENT The sixth segment of the head and which bears the Maxillae. Cf. Labial Segment.

MAXILLARY STYLET 1. Hemiptera: The inner pair of stylets of the mouth. 2. The thin, distal part of the swollen maxillary sclerite. See Mandibular Stylet.

MAXILLARY TENDON 1. Two slender rod-like sclerites in the basal third of the muscid Proboscis. 2. The remnant of the Palpifer, to which muscles for flexing the Proboscis are attached. See Lora.

MAXILLARY TENTACLE Female *Pronuba yuccasella*: A long, curled, spinose appendage of the Maxilla adapted for the collection of *Yucca* pollen (Comstock).

MAXILLIPED Noun. (Latin, *maxilla* = jaw; *pes* = foot. Pl., Maxillipeds.) The three pairs of appendages in Crustacea following the second Maxillae. The first pair sometimes (Amphipoda) united to form a Labium-like structure attached to the head (Snodgrass).

MAXILLOLABIAL Adj. (Latin, *maxilla* = jaw; *labium* = lip; Latin, *-alis* = pertaining to) Descriptive of structure or action associated with the Maxilla and Labium. See Mouthparts.

MAXILLULAE Noun. (Latin, *maxilla* = jaw) 1. The Paragnatha *sensu* Snodgrass. 2. The Superlinguae (Imms deemed it an undesirable term). 3. Thysanura: A pair of appendages between the Mandibles and first Maxillae (Smith).

MAXWELL LEFROY, H M See Lefroy, H. M. Maxwell.

MAY, A. W. A. (1916–1966) (Chadwick 1968, J. Aust. ent. Soc. (N.S.W.) 6: 41–42.)

MAY BEETLES See Scarabaeidae.

MAY, HARRY H. (1878–1943) (Wynn *et al.* 1944, Entomol. mon. Mag. 80: 43.)

MAY, JAN (–1959) (Anon. 1960, Cas. csl. Spol. ent. 57: 94–95, bibliogr.)

MAY, JOHN WILLIAM (1814–1902) (Anon. 1960, Entomol. mon. Mag. 38: 186.)

MAYER, ARNOLD (1913–1959) (Mafel 1959, Anz. Schädlingsk. 32: 139.)

MAYER, HELMUT (1920–1954) (Beier 1954, Ann. naturh. Mus. Wien 60: 5–6.)

MAYER, KAREL (1912–1939) (Sachtleben 1940, Arb. morph. taxon. Ent. Berl. 7: 336.)

MAYER, KARL (–1970) (Anon. 1970, Mitt. dt. ent. Ges. 29: 25–26.)

MAYER, PAUL (1848–1923) (Péterfi 1924, Z. wiss. Mikrosk. 41: 145–154, bibliogr.)

MAYET, VALÉRY (1839–1909) (Horn 1909, Dt. ent. Z. 1909: 583.)

MAYFLIES See Ephemeroptera.

MAYNARD, CHARLES JOHNSON (1845–1929) (Batchelder 1951, J. Soc. Biblphy. nat. Hist. 2: 227–260, bibliogr.)

MAYNE'S HOUSE DUST MITE *Euroglyphus maynei* Fain [Acari: Pyroglyphidae].

MAYR, GUSTAV (1830–1908) (Dalla Torre 1908, Wien ent. Ztg. 27: 255–271, bibliogr.)

MAYRIAN FURROW When viewed in dorsal aspect, the Y-shape groove in the Mesonotum of certain male ants.

MAZARAKII, VIKTOR VIKTOROVICH (1857–1912)

(Jacobson 1912, Revue Russe Ent. 12: xxix–xxxiii, bibliogr.)

MAZARREDO, CARLOS (–1910) (Dusmet y Alonso 1919, Bol. Soc. ent. Esp. 2: 172.)

MAZON CREEK A geological site in Illinois, USA noted for its fossil deposits of Upper Carbonifeorus insects.

MAZZA, SALVADOR (–1946) (Biraben 1947, Revta. Soc. ent. argent. 13: 341–343.)

MB 46030® See Fipronil.

Mb-NPV® See *Mamestra brassicae* NPV.

McARTHUR, HARRY (1846–1910) (South 1910, Entomologist 43: 103–104.)

McATEE, WALDO LEE American biologist living in Washington, D.C. and student of Hemiptera

McCAMPBELL, SAM CORNELIUS (1895–1966) (Davis 1967, J. Econ. Ent. 60: 194.)

McCATHIE, JAMES GORDON (1918–1968) (Kennedy 1969, Proc. R. ent. Soc. Lond. (C) 33: 55.)

McCLANAHAN, HOWARD SAMUEL (–1952) (Anon. 1952, Psyche 35: 90.)

McCLAY, A T (1899–1970) (Anon. 1970, Coleopt. Newsl. 3: 13.)

McCLUNG, CLARENCE ERWIN (1870–1946) (Wenrich 1946, Science 103: 551–552.)

McCOLLOCH, JAMES WALKER (1889–1929) (American Entomologist. (Parker *et al.*, 1930. J. Kans. ent. Soc. 3: 51–52, bibliogr.)

McCONELL, HAROLD SLOAN (1893–1958) (Anderson *et al.* 1959, Proc. ent. Soc. Wash. 61: 36–38.)

McCONNELL, WILBUR ROSS (1881–1920) (Walton 1920, J. Econ. Ent. 13: 371–373.)

McCOOK, HENRY CHRISTOPHER (1837–1911) (Calvert 1911, Ent. News 22: 433–438; Ludwig 1911, J. Presbyt. Hist. Soc. 6: 97–146 bibliogr.)

McCOY, FREDERICK (1823–1899) (H. W. 1905, Proc. R. Soc. London 75: 43–45.)

McCULLOUGH, ALLAN RIVERSTON (1885–1925) (Anderson *et al.* 1926, Rec. Aust. Mus. 15: 141–148, bibliogr.)

McDANIEL SPIDER MITE *Tetranychus mcdanieli* McGregor [Acari: Tetranychidae].

McDERMOTT, FRANK ALEXANDER (1855–1966) (Buck 1968, Ent. News. 79: 49–54, bibliogr.)

McDUFFIE, WILLIAM CARL (1910–1967) (Fluno & Weidhaas 1968, J. Econ. Ent. 61: 590.)

McDUNNOUGH, JAMES HALLIDAY (1877–1962) (Ferguson 1962, J. Lepid. Soc. 16: 209–222, bibliogr.)

McELHOSE, ARTHUR (1890–1957) (Remington 1954, Lepid. News 8: 30.)

McGLASHAN, CHARLES FAYETTE (1847–1931) (Essig 1931, Pan-Pacif. Ent. 7: 97–99.)

McGREGOR, MALCOLM EVAN (1889–1933) (Horn 1934, Arb. physiol. angew. Ent. Berl. 1: 305.)

McGREGOR, RICHARD CRITTENDON (1871–1936) (Alexander 1937, Philipp. J. Sci. 63: 359–361.)

McILROY, WILLIAM DENMARK (1906–1932) (Hungerford 1934, Ann. ent. Soc. Am. 27: 122.)

McINDOO, NORMAN EUGENE (1881–1956) (Siegler 1957, Proc. ent. Soc. Wash. 59: 43–44.)

McKELLAR, HUGH (1849–1929) (Criddle 1929, Can. Ent. 61; 288.)

McKENNY-HUGHES, ALFRED WESTON (1895–1970) (Hinton 1971, Proc. R. ent. Soc. Lond. (C) 35: 52.)

McKENZIE, HOWARD LESTER (1910–1968) American academic (University of California, Davis) and taxonomic specialist on Homoptera, particularly mealybugs and scale insects. (Miller *et al.* 1969, Pan-Pacif. Ent. 45: 245–259, bibliogr.)

McKINNEY, KENNETH BARBEE (1890–1946) (Caffery & White 1947, J. Econ. Ent. 40: 283.)

McLACHLAN, ROBERT (1837–1904) (Calvert 1904, Ent. News 15: 226–228.)

McLAINE, LEONARD SEPTIMUS (1887–1943) (Fernald 1943, J. Econ. Ent. 36: 946–947; Keenan *et al.* 1944, Can. Ent. 76: 1–4, bibliogr.)

McLEMORE, JOHN ANDERSON (Cockerham 1926, J. Econ. Ent. 19: 418–419.)

McLEOD, MURDOCH (–1950) (Wigglesworth 1951, Proc. R. ent. Soc. Lond. (C) 15: 76.)

McMILLAN, H ELLIS (–1934) (Anon. 1936, Rep. ent. Soc. Ont. 66: 5.)

McNAB, WILLIAM RAMSAY (1844–1889) (Anon. 1890, Entomol. mon. Mag. 26: 26.)

McNAY, EVETT, J (1910–1939) (Anon. 1939, Am. Bee J. 79: 76.)

McPHAIL TRAP [See History and use of the McPhail Trap. Fla. Ent. 60: 11–16.] See Trap. Cf. Light Trap; New Jersey Trap; Wilkinson Trap.

MEABROME® See Methyl Bromide.

MEAD, EDITH EDWARDS (1852–1927) (dos Passos 1951, J. N. Y. ent. Soc. 59: 157.)

MEAD, TEODOR LUQUEER (1852–1936) (Brown 1955, Lepid. News. 9: 185–190.)

MEADE, RICHARD HENRY (1814–1899) (Verrall 1899, Proc. ent. Soc. Lond. 1899: xxxvii–xxxix.)

MEADE-WALDO, GEOFFREY (1884–1916) (Soldanski 1916, Dt. ent. Z. 1916: 364–365.)

MEADOW ARGUS BUTTERFLY *Junonia villida calybe* (Godart) [Lepidoptera: Nymphalidae].

MEADOW PLANT BUG *Leptopterna dolabrata* (Linnaeus) [Hemiptera: Miridae].

MEADOW SPITTLEBUG *Philaenus spumarius* (Linnaeus) [Hemiptera: Cercopidae]: A pest of alfalfa and clover in northeastern and north central states of the USA. See Cercopidae.

MEAL MOTH *Pyralis farinalis* (Linnaeus) [Lepidoptera: Pyralidae]: A cosmopolitan pest of bran, flour and potatoes in storage facilites. Adult wingspan 25–30 mm with wings held flat, laterally extended and Abdomen curled upward over Thorax. Larvae are grey, ca 25 mm long and live in silken gallery attached to substrate; development can require two years. See Pyralidae.

MEAL WORM BEETLES See Tenebrionidae.

MEALWORMS See Tenebrionidae.

MEALY Adj. (Latin, *molere* = to grind.) Farinose; with a flour-like dusting.

MEALY BUGS See Coccoidea.

MEALY PLUM APHID *Hyalopterus pruni* (Geoffroy) [Hemiptera: Aphididae].

MEALYBUG DESTROYER *Cryptolaemus montrouzieri* Mulsant [Coleoptera: Coccinellidae]: A beetle extensively used in biological control of pseudococcoid pests. Syn. Mealybug Ladybird. See Coccinellidae.

MEALYBUG PARASITE *Leptomastix dactylopii* Howard [Hymenoptera: Encyrtidae].

MEALYBUGS See Pseudococcidae. Rel. Armoured Scales; Softscales; Whiteflies.

MEASURING WORMS See Geometridae.

MEAT ANT *Iridomyrmex purpureus* (F. Smith) [Hymenoptera: Formicidae]: Endemic to Australia and a minor pest in temperate areas; bites savagely but does not sting. Workers 13–14 mm long, typically red and black. MAs nest in large colonies in ground and produce mounds topped with gravel; pests of tennis courts. MAs feed on meat and animal tissue; take honeydew from Homoptera; enter dwellings and feed on sweets. See Formicidae.

MEAT ANTS *Iridomyrmex* (*purpureus* gp.) spp. [Hymenoptera: Formicidae].

MEATUS Noun. (Latin, *meatus* = passage.) A channel or duct.

MECAGLOSSA Mecoptera: Reduced Glossae and Paraglossae, that are fused with each other and the Stipulae (MacGillivray).

MECAPTERA See Mecoptera.

MECARBAM An organic-phosphate (dithio-phosphate) compound {S-(N-methylcarbamoylmethyl) diethyl phosphorodithioate} used as an acaricide, contact insecticide and stomach poison for control of plant-sucking insects (to include aphids, leafhoppers, scale insects, whiteflies), thrips, leaf miners, root fly, carrot fly and codling moth. Compound is applied to citrus, cotton, rice, stone fruit, pome fruit and rice in some countries, but is not registered for use in USA. Mecarbam is regarded as phytotoxic on egg plant and is toxic to bees. Trade names include Afos® and Murotox®. See Organophosphorus Insecticide.

MECHANICAL TRANSMISSION Any object that facilitates the passage of a plant or animal pathogen from one host to another host, but is not essential for transmission of the pathogen. Mechanical vectors are not intermediate hosts for the pathogens; when infected, a mechanical vector can immediately transmit a pathogen to another host. Agricultural equipment can create abrasions on infected plants during cultivation, pathogen-laden fluids are moved on the equipment and transmitted to uninfected plants.

MECHANICAL VECTOR An arthropod that facilitates the passage of a pathogen from one vertebrate host to another host, but is not essential for transmission of the pathogen. Mechanical vectors are not intermediate hosts for the pathogens; when infected, a mechanical vector can immediately transmit a pathogen to another vertebrate host. Eliminating a mechanical vector will not eliminate transmission of the pathogen. Example: Houseflies serve as mechanical vectors of enteric diseases to humans. See Transmission. Cf. Biological Vector.

MECHANISTIC THEORY Behaviour: A theory that all animal action depends only on stimuli received through or from a nerve from purely external or objective sources.

MECHANORECEPTOR Noun. (Latin, *mechanicus*; *or* = one who does something. Pl., Mechanoreceptors.) Sensory receptors that respond to mechanical deflection of a sensillum's surface. Deflection may involve wind current, tactile manipulation, harmonic vibrations of air, water, substrate or mechanical distortion of the body. Typically, mechanoreceptors are formed with one or two neurons. Most Trichogens are capable of being bent or deflected when pressure is exerted from any direction. Some mechanoreceptors can provide directional sensitivity or respond to mechanical stimulus which comes from one direction. Mechanoreceptors are involved with coordinated movement of the body Tagma and appendages (wing, leg, Antenna, Maxillary and Labial Palpi). Mechanoreception is achieved with the aid of Campaniform Sensilla, Proprioreceptors (Stretch Receptors), Statocysts, Hair Plates and Chordotonal Organs. See Sensillum. Cf. Chemoreceptor. Rel. Campaniform Sensilla; Chordotonal Organs; Hair Plates; Proprioreceptors; Statocysts; Stretch Receptors.

MECKEL, JOHANN FRIEDERICK (1781–1833) (Nordenskiöld 1935, *History of Biology*. 625 pp. (355–359), London.)

MECKLENBORG-STREBLITZ, ADOLF FRIEDRICH (1882–1918) (Reha 1918, Ent. News 29: 159–160.)

MECONEMATIDAE Plural Noun. Oak crickets. A Family of ensiferous Orthoptera including about 30 Genera and 200 Species, most of which are found in the Palearctic and Ethiopian Realms. Adults are usually small bodied, green or green-yellow with head rounded. Antenna inserted between compound eyes, Prosternum spineless, and anterior Tibia sometimes with raptorial spines along ventral surface. Wing development is variable. Insects are arboreal, nocturnal and predatory. Eggs of one Species are deposited in cynipid galls. See Orthoptera.

MECONIUM Noun. (Latin, *meconium* > Greek, *mekonion* = poppy juice, > *mekon* = a poppy; Latin, *-ium* = diminutive > Greek, *-idion*. Pl., Meconia.) 1. The liquid substance excreted from the Anus of some holometabolous insects after emergence of the Adult from the Chrysalis or Pupa (Torre-Bueno, Imms). 2. The first excrement of new-born infant (Brown). 3. Parasitic Hymenoptera: The pellet-like excrement of larvae or prepupae (Askew). See Excreta; Excretion. Cf. Frass; Honeydew. Rel. Digestion.

MECOPODIDAE Plural Noun. A Family of

ensiferous Orthoptera including 57 Genera and about 130 Species, most of which are confined to the Old World tropics. Mecopodids typically are large bodied, head rounded with Antenna inserted between compound eyes; Scrobes margined, Prothoracic Spiracles large, elongate and partly concealed by Pronotum. First two tarsal segments grooved laterad. Tegmina are leaf-like when present and males with well developed basal Stridulatory Organ; Ovipositor recurved. Mecopodids occur in forests or arboreally on open bushes. All Species apparently are phytophagous. See Orthoptera.

MECOPTERA Packard 1886. (Greek, *meco* = long; *pteron* = wing.) Hang flies; Hangingflies; Scorpion flies. A cosmopolitan Order of about 500 Species with a fossil record from the Lower Permian. Adults are moderate-sized, slender bodied and long-winged; head hypognathous, prolonged into a Rostrum beneath compound eyes; mouth mandibulate; Ocelli are present or absent; Antenna filiform with 14–60 segments; Cervix membranous; Pronotum saddle-like; four wings typically present, not folded, similar in size, shape, venation; Mesothorax, Metathorax subequal in size; Metathorax fused with first abdominal Tergum; middle and hind Coxae typically with well developed Meron. Abdomen with 11 segments; Cerci with two segments in female, one segment in male or rarely absent. Metamorphosis is complete. Larvae are eruciform, scarabaeiform or elongate; some larvae with compound eyes. Pupae are decticous and exarate. Males are predaceous; females do not take living prey. Mecoptera are divided into Suborders Protomecoptera and Eumecoptera. Included Families: Apteropanorpidae, Bittacidae, Boreidae, Choristidae, Eomeropidae, Meropeidae, Nannochoristidae, Notiothaumidae, Panorpidae and Panorpodidae. See Insecta.

MEDACORIA Noun. The lateral notal emargination (MacGillivray).

MEDALARIA Noun. The Alaria on the anterior part of each lateral margin of the Scutum (MacGillivray).

MEDALIFERA Noun. The Alifera set between the Prealifera and the Pleuralifera, Posterior Basalare, second Parapteron (MacGillivray).

MEDENBACK, ALEXANDER BENJAMIN DE ROOY VAN (1841–1878) (Kraatz 1878, Dt. ent. Z. 22: 226.)

MEDER, OSKAR (1877–1944) (Heydemann 1944, Ent. Z. 58: 33–34.)

MEDFLY See Mediterranean Fruit Fly.

MEDIA Noun. (Latin, *medius* = middle.) 1. The fourth of the longitudinal veins. Media extends from base through middle of wing; not more than four branches, the branches numbered on margin from one nearest the apex to four nearest the Anal Angle. 2. Ants: Forms intermediate between the large-headed (majors) and the small-headed (minors) worker-ants (Wheeler).

MEDIA WORKER Social Insects: A member of the medium-sized subcaste in ants with three or more subcastes of workers. Cf. Major Worker; Minor Worker.

MEDIAD Adv. (Latin, *medius* = middle; *ad* = to.) Toward the median plane or middle of a body, structure or appendage. See Orientation. Cf. Anteriad; Apicad; Basad; Caudad; Centrad; Cephalad; Craniad; Dextrad; Dextrocaudad; Dextrocephalad; Distad; Dorsad; Ectad; Entad; Laterad; Mesad; Neurad; Orad; Proximad; Rostrad; Sinistrad; Sinistrocaudad; Sinistrocephalad; Ventrad.

MEDIAL Adj. (Latin, *medius* = middle; *-alis* = pertaining to.) Descriptive of something toward the middle of a structure or toward the midline of a body; extending toward the middle. See Orientation. Cf. Distal; Lateral; Mesial; Proximal.

MEDIAL CELL 1. Wing cells anteriorly bounded by the Media or its branches. 2. Hymenoptera: The Medial Cell includes the Median and Cubital Cells of Comstock (*teste* Norton).

MEDIAL CROSSVEIN The Crossvein which extends from Media to Media (Comstock). See Vein; Wing Venation.

MEDIALE Noun. The second Axillary Sclerite of the insect wing (Crampton).

MEDIAN Adj. (Latin, *medius* = middle.) In or at the middle. Descriptive of or pertaining to the middle.

MEDIAN AREA 1. Wings in Orthoptera: Area between Radial and Ulnar Veins, Radius and Media (Comstock). 2. Metathorax of Hymenoptera: Middle of dorsum, divided into three spaces or cells, 1st (Basal Area), 2nd (Upper Media, Areola), 3rd (Apical or Petiolar Area).

MEDIAN CARINA 1. Any keel-like Carina set medially on a part of an insect. 2. Orthoptera head: A median dorsal Carina. 3. Any Carina that extends down middle of Front from Fastigium.

MEDIAN CELL 1. Lepidoptera: The closed area formed by a line extending from the end of Subcostal to end of Median Veins. 2. The Radial Cell of Comstock.

MEDIAN CERCUS A segmented sensory outgrowth of the 11th abdominal Tergum (Wardle). See Pseudocercus.

MEDIAN CLAW Any single unpaired claw, not to be confused with those cases in which one of a pair disappears at the last moult of the nymph in Hemiptera.

MEDIAN CORD Insect embryo: A chain of cells separated from Ectoderm lining the Neural Groove (Imms).

MEDIAN CROSSVEINS Odonata: Crossveins that cross the median space (Smith). See Vein; Wing Venation.

MEDIAN FLEXOR PLATE See Median Unguitractor Plate.

MEDIAN FORKS 1. Orthoptera: The forks of the Median. 2. Some Heteroptera: The indentation which separates the Embolium from remainder of Corium.

MEDIAN FOVEA Hymenoptera: A rounded or angular pit near the ventral margin of the Frontal Crest. Syn. Antennal Fovea.

MEDIAN FOVEOLA Orthoptera: The foveate depression of the Vertex between the eyes. Syn. Central Foveola.

MEDIAN INCISURA Some Coccidae: The Incisura in the Meson between the median pair of lobes (MacGillivray).

MEDIAN LAMELLAE Coccidae: Dorsal plates *sensu* MacGillivray.

MEDIAN LINES Forewings of many moths: The first (transverse anterior), which crosses about one-third from base. The second (transverse posterior), which crosses beyond the outer third and is usually sinuate.

MEDIAN LOBE Coccidae: One of the lobes on each side of an Incisura (MacGillivray).

MEDIAN LOBE OF LABIUM Odonata: The Mentum (Garman).

MEDIAN LONGITUDINAL CARINAE Hymenoptera: Cuticular ridges positioned parallel to the imaginary midline of a structure.

MEDIAN NERVE CORD The median strand of nerve tissue produced from the ventral Neuroblasts (Snodgrass).

MEDIAN NERVES Unpaired nerves arising from the Ganglia of the Ventral Nerve Cord between the roots of the connectives (Snodgrass).

MEDIAN NERVULES Lepidoptera (Holland): The 1st Cubitus of Comstock; the 2nd Cubitus of Comstock; the 3rd Media of Comstock.

MEDIAN NEXUS Wings of Myrmeleonidae: A conjunction of adjacent veins that joins the apex of vein M + with the vein on each side of it (Needham).

MEDIAN NOTCH Coccidae: A Sulcus near the margin of the Pygidium, at the posterior extremity of the body. The Mesial Notch *sensu* MacGillivray.

MEDIAN OVIDUCT Reproductive System: In most insects the Median Oviduct is secondarily developed from ectodermal invaginations of intersegmental membrane. The MO is well musculated and, like the Lateral Oviduct, is often enveloped by circular and longitudinal muscles. The Median Oviduct is often relatively thick walled and forms a fertilization chamber in insects such as Diptera. The MO is also called the Vagina or Bursa Copulatrix when it receives the male Aedaegus; MO of viviparous Diptera is called the Uterus. Syn. Oviductus Communis; Common Oviduct.

MEDIAN PLANE A vertical plane which divides animals into right and left parts.

MEDIAN PLATE 1. Insect embryo: The middle sclerite. 2. Hymenoptera Sessiliventres: The dorsal sclerite connecting the Thorax and Abdomen (Smith). 3. Sclerites called Median Plates are at the base of the wing lateral of the second Axillary Sclerite. Their shape is not distinctive, but they are important in wing flexion. The Proximal Median Plate lies distad of the second Axillary Sclerite and is probably a subdivision of it. The Distal Median Plate is not always developed. When present, it is separated from the Proximal Median Plate by a fold of membrane. Alternatively the Distal Median Plate may be represented by a vague sclerotization of the area at the base of the Median Vein and Cubital Vein.

MEDIAN PLATES OF WINGBASE See First Media Plate; Second Media Plate.

MEDIAN SECTOR Odonata: The Media of Comstock.

MEDIAN SEGMENT Clistogastrous Hymenoptera: The basal segment of the Abdomen which becomes broadly attached to the Metathorax. See Propodeum. Syn. Latreille's Organ. Formicidae: Epinotum.

MEDIAN SHADE Lepidoptera: A linear mark which crosses at or about middle of wings. Alt. Median Line.

MEDIAN SPACE 1. Lepidoptera: The area between the Media Lines. 2. Odonata: the Cubital Cell (Comstock); the Basilar Space of the wing, q.v. (Garman); the space at the base of the wing between Radius and first Anal; Medial Cell of Comstock (Selys 1896, and authors).

MEDIAN SPINA The usually longer Spina of the Stigmatic clefts in the middle of the series, when there are three or more (MacGillivray). See Spina.

MEDIAN SUPERIOR ANAL APPENDAGES Male Anisoptera: Appendages of the eleventh Somite, positioned above the Anus (Imms).

MEDIAN SUSPENSORY LIGAMENT The ligaments of the two Ovaries when they are combined into one compact structure (Snodgrass).

MEDIAN SUTURE A longitudinal suture on the middle line of the Terga and Sterna (Comstock).

MEDIAN UNGUITRACTOR PLATE Some insects: A ventral, median, sclerotized structure between the Ungues at the base of the Pretarsus. The Pretarsal Depressor Muscle, or its Apodeme, attaches to the MUP and depresses the Ungues. Syn. Median Flexor Plate. See Pretarsus. Cf. Empodium. Rel. Leg.

MEDIAN VEIN 1. Odonata and Lepidoptera: The Radius of Comstock. 2. Hymenoptera: The 3rd vein from the Costal margin (Smith). See Vein; Wing Venation.

MEDIASTINAL Adj. (Latin, *medius* = middle; *-alis* = pertaining to.) Descriptive of structure relating to an imaginary longitudinal median line or area. See Orientation.

MEDIASTINAL AREA Orthoptera: The area between Median (Mediastinal) Vein and the costal or front margin; marginal area.

MEDIASTINAL VEIN 1. Orthoptera and Diptera: The Subcosta of Comstock. 2. Diptera: The Auxiliary Vein of Meigen. See Vein; Wing Venation.

MEDICAL ENTOMOLOGY The subdiscipline of Entomology concerned with the study or control of insects affecting human health. See Entomology. Cf. Applied Entomology; Veterinary Entomology.

MEDICAL THRESHOLD See Threshold. Cf. Economic Threshold; Social Threshold.

MEDIELLA Noun. (Pl., Mediellae.) The Funditae associated with the Submedia, of which they may be a part (MacGilliivray).

MEDIFURCA Noun. (Latin, *medius* = middle; *furca* = fork. Pl., Medifurcae.) 1. A forked process of the anterior surface of the Medipectus. 2. Flat Apodemes that diverge and bridge the commissure internally in the Thorax (Packard).

MEDIOCUBITAL CROSSVEIN The vein connecting the series M and C (Comstock). See Vein; Wing Venation.

MEDIOCUBITAL Adj. (Latin, *medius* = middle; *cubitus*; *-alis* = pertaining to.) Descriptive of something associated with the Media and Cubitus of the wing.

MEDIOTERGITE Noun. (Latin, *medius* = middle; *tergum* = the back; *-ites* = constituent. Pl., Mediotergites.) The mesial region of the Postscutellum (Crampton).

MEDIOVENTRAL LINE Lepidoptera larva: An imaginary line of reference projecting along the middle of the ventral surface of the body.

MEDIPECTUS Noun. (Latin, *medius* = middle; *pectus* = chest.) The ventral surface of the Mesothorax. Syn. Mesosternum.

MEDIPROBOSCIS Noun. (Latin, *medius* = middle; Greek, *proboskis* > *pro* = before; *boskein* = to graze. Pl., Mediproboscises.) 1. Diptera (Muscidae): The strongly chitinized middle third of the Proboscis. 2. The Haustellum *sensu* MacGillivray. Syn. Proboscis.

MEDITERRANEAN FLOUR-MOTH *Anagasta kuehniella* (Zeller) [Lepidoptera: Pyralidae]: A cosmopolitan pest of flour mills, granaries and homes. MFM perhaps native to Europe or Central America and most common in temperate regions; within North America. MFM was first reported in Canada during 1889. Success of MFM in temperate urban situations has been attributed to survival at low relative humidity (<25%). MFM reproduction involves nocturnal courtship, copulation and oviposition. Eggs are laid in clutches ca 100–500 near flour, meal or grain waste; eclosion occurs within 3–16 days, depending upon temperature. Larvae can feed on many foods, but is primarily a pest of flour. MFM is common in mills, bulk and storage facilities; larva webs food and feeds within web. Pupation occurs within silken cocoon and complete within 8–12 days. Life cycle requires 8–10 weeks, depending upon temperature and humidity. Males are sterile when reared under high temperatures (ca 30°C) or continuous light. Syn. *Ephestia kuehniella* Zeller. See Pyralidae. Cf. Almond Moth; Tobacco Moth; Raisin Moth; Indian Meal Moth.

MEDITERRANEAN FRUIT FLY *Ceratitis capitata* (Wiedemann) [Diptera: Tephritidae]: A very serious, widespread agricultural pest which attacks fruits, vegetables and nuts of more than 200 plant Species. MFF is a major concern to quarantine globally. MFF is endemic to east-central Africa and was introduced into Mediterranean region during late 18th century, probably via Arab trade routes importing coffee from Ethiopia. MFF was first described as a serious pest of *Citrus* from the Azores by W. S. MacLeay (1829); MFF was distributed around globe by shipping and established on every continent and major island by mid-1900s; MFF was established in Western Australia (1895), discovered in Hawaii (1910), Florida (1929) and California (1979); repeatedly intercepted in California and Florida. Adults are 4–5 mm long, Thorax black with silver or grey spots, wings mottled, eyes pale-green and Abdomen brown with two pale-coloured rings. Female feeds on microbes, honeydew and protein on surface of fruit and leaves for egg development; MFF develops ca 1000 eggs during lifetime. Eggs are white, ca 1 mm long and banana shaped; female punctures fruit to lay 2–30 eggs under surface of ripening fruit; eclosion requires ca 2–3 days. Neonate larva bores into fruit to feed; MFF completes three larval instars and requires 10–14 days for larval development; mature larvae are ca 8 mm long, and drop to ground to pupate in soil. Female can live six months; life cycle requires 4–15 weeks, MFF is multivoltine with 3–10 generations per year, depending upon climate and host plant. Syn. Medfly. See Tephritidae. Cf. Apple Maggot; Melon Fly; Mexican Fruit Fly; Oriental Fruit Fly; Papaya Fruit Fly; Queensland Fruit Fly.

MEDITHORAX Noun. See Mesothorax.

MEDITRUNCUS Noun. See Mesothorax.

MEDIUS Adj. (Latin *medius.*) Middle; Incorrectly used for the Media of the insect wing.

MEDLURE® See Trimedlure.

MEDULLA Noun. (Latin, *medulla* = marrow, pith.) 1. Brachiopod Crustacea: The proximal of the two ganglionic bodies (Snodgrass). 2. The central area of a Ganglion (Wardle).

MEDULLA EXTERNA See Epiopticon.

MEDULLA INTERNA See Opticon.

MEDULLARY SUBSTANCE The dense fibrous mass of nerve terminals forming the interior of a Ganglion or Neuropile (Snodgrass).

MEDULLARY TISSUE See Medullary Substance.

MEDVIDRIN® See Mevinphos.

MEEHAN, THOMAS (1826–1901) (Howard 1930, Smithson. misc. Collns. 84: 15.)

MEEN® See Azadirachtin.

MEES, ADOLF (–1915) (Soldanski 1916, Dt. ent. Z. 1916: 227.)

MEGACHILIDAE Plural Noun. Leaf-cutting Bees. A large Family of aculeate Hymenoptera assigned to Superfamily Apoidea (bees). Body moderate sized, stout; one Subantennal Suture present; Facial Foveae absent; Labrum longer than wide, subquadrangular; Flabellum present;

Galea short basal of Palpus, long apical of Palpus; Labial Palpus segment 1–2 elongate, flattened; Pre-episternal Suture absent; forewing with two Submarginal Cells subequal in size, Marginal Cell normal, Stigma subquadrate; Metasomal Sternum modified into a Scopa for pollen transport; Basitibial Suture absent; Pygidium absent. Megachilids nest in soil, natural cavities or in wood; nest cells lined with leaf material cut from living plants. Subfamilies include the Fideliinae, Megachilinae and Lithurginae. See Apoidea.

MEGACHILINAE The nominant Subfamily of Megachilidae and consisting of about 85 Genera.

MEGALODONTIDAE Konow 1897. Plural Noun. A Family of Symphyta (Hymenoptera) assigned to the Pamphilioidea.

MEGALODONTOIDEA Hymenoptera: Superfamily of Symphyta containing Megalondontidae. See Symphyta. Cf. Pamphilioidea; Xyeloidea.

MEGALOPTERA Latreille 1802. (Greek, *megalo* = large; *pteron* = wing.) Alderflies; Dobsonflies. Megaloptera are regarded as very primitive Endopterygota with a fossil record dating from the Permian Period. Megaloptera are considered a Suborder of Neuroptera or distinct Order (depending upon classification) and a sister-group of Rhaphidoptera. About 300 nominal Species have been described, and most of them occur in temperate regions. Adults are moderate to large bodied, with head prognathous and generally dorsoventrally compressed; compound eyes bulge laterally; Mandibles are well developed; Maxillary Palpus with 5–4 segments; Labial Palpus with 3–4 segments. Thoracic segments are articulate; Pronotum large, subtrapezoidal or subrectangular; wings are large, membranous, subequal in size; longitudinal veins with reduced end-twigging; tarsal formula is 5-5-5. Abdomen soft, pliant with spiracles on segments 1–8; nervous system with three thoracic ganglia and seven abdominal ganglia. Larvae are predaceous, elongate, dorsoventrally compressed and aquatic; larvae display biting-type mouthparts with powerful Mandibles; abdominal gills are present on most segments. Pupae are exarate and decticous. Eggs are laid in masses above water on vegetation or rocks; the neonate larvae drops into water. Included Families are Corydalidae and Sialidae. See Insecta. Cf. Planipennia; Rhaphidoidea.

MEGALOPYGIDAE Plural Noun. Flannel Moths. A small Family of ditrysian Lepidoptera assigned to Superfamily Zygaenoidea with about 10 Species known in North America. Adults with a wool-like appearance owing to the mixture of typical scales interspersed with curled Setae. Larvae bear urticating spines, five typical pairs of Prolegs and two pairs of Prolegs which are sucker-like and lack Crochets. One Species (Crinkled Flannel Moth) is a minor pest. See Zygaenoidea.

MEGALRYIDAE Schletterer 1889. Plural Noun. A small Family of parasitic Hymenoptera, sometimes assigned to Ichneumonoidea, sometimes placed in Stephanoidea, and sometimes as an unassigned Taxon. Family is distributed through South America, South Africa, Australia and Orient. Adult Antennal Flagellum with 12 segments, head with Subantennal Sulcus, Mesoscutum with median longituditunal Sulcus, spiracle on Mesothorax surrounded by pronotal Cuticle; hindwing venation reduced. See Ichneumonoidea.

MEGALUROTHRIPS *Megalurothrips kellyanus* (Bagnall) [Thysanoptera: Thripidae]: A pest of *Citrus* in Australia. Adults are 2–3 mm long, black body and legs, wings dark; larvae are yellow. Development includes two larval instars, propupa and pupa. Species is multivoltine with six generations per year. See Thripidae.

MEGALYROIDEA Schletterer 1889. Hymenoptera. Superfamily of Apocrita containing Megalyridae and Stephanidae. Characterized by adults with Subantennal Sulcus and hind tibial spurs unmodified. See Stephanoidea.

MEGAMERINIDAE Plural Noun. A small Family of acalypterate Diptera known from the Palearctic and Oriental Realms. Adult body elongate with bristles absent from anterior part of Thorax and wing narrow basally; Metapleuron and Metasternum forming base for attachment of hind Coxa; Metathorax with deep Postcoxal Bridge; hind Femur enlarged and ventrally spinose; Abdomen with narrow attachment to Thorax. Megamerinidae sometimes is assigned to Superfamily Diopsoidea, more recently placed within Nerioidea (Micropezoidea). Syn. Megameridae Hendel 1916. See Diptera.

MEGANISOPTERA Martynov 1932. An Order of insects extant during the Upper Carboniferous to the Lower Triassic Period. See Insecta.

MEGANOMIINAE A Subfamily of Melittidae (Apoidea) consisting of about five Genera.

MEGAPODAGRIONIDAE Plural Noun. A moderately large Family of Odonata found in tropical regions. Adults resemble letids; wings strongly petiolate; Pterostigma with several supplementary long veins along outer part of wings (usually); 2–3 Antenodal Crossveins present; Quadrilateral Vein rectangular; vein R4 near Nodus. Male anal appendage forcipate. Larvae are small, robust with prominent eyes; Antenna with seven segments, 3rd segment elongate; three Caudal Gills foliate and held horizontal, not vertical as most Zygoptera. See Odonata.

MEGASECOPTERA Brongniart 1893 An extinct Order of haustellate paleopterous insects, known from the Upper Carboniferous to Upper Permian and consisting of several Families. Megasecopterans characterized by beak-like mouthparts, Cerci long, Ovipositor prominent and wings folded and fluted; wing folding is similar to Neuroptera and presumably was adapted to prevent being blown about and permit crawling in confined spaces. Related to Diaphaneropterodea and Palaeodictyoptera based on beaks presumably adapted for sucking the contents of fructifications and cones of lycopods, pteridosperms or other plants. Megasecoptera differs from other contemporaneous Orders in a few features: Body typically smaller, more slender; wings petiolate; Thorax enlarged. See Paleoptera. Cf. Diaphanopteroidea; Paleodictyoptera.

MEGASPILIDAE Ashmead 1893. Plural Noun. A moderate sized Family of parasitic Hymenoptera assigned to the Ceraphronoidea or Stephanoidea. Diagnostic features include Antenna with 11 segments in male and female, fore and middle Tibia each with two spurs and Waterston's Organ absent. Species are ectoparasites of diverse Taxa including Coccidae, Mecoptera, Neuroptera and Diptera. Some Species develop as hyperparasites of aphidiids that attack Aphididae; other Species are myrmecophiles and probably attack Diptera. See Ceraphronoidea.

MEGATHYMIDAE Plural Noun. A small Family of Lepidoptera assigned to Superfamily Hesperioidea. Megathymids are regarded in contemporary classifications as a Subfamily of Hesperiidae.

MEGERLE VON MÜHLFELD, JOHANN CARL (1765–1832) Collection apparently in Naturhistorische Museum, Wien. (Gistle 1832, Faunus 1: 55.)

MEGGIOLARO, GUISEPPE (1931–1967) (Bucciarelli 1970, Mem. Soc. ent. Ital. 49: 27–32, bibliogr.)

MEHES, GYULAROL (1881–1961) (Bela 1966, Folia ent. hung. 19: 1–8.)

MEHRING, JOHANNES (–1878) (Anon. 1879, VerBl. Westf.-Rhein. Ber. Bienen Seidenzicht 30: 35.)

MEIDELL, OUE (1903–1942) (Holgersen 1942, Norsk. ent. Tidsskr. 6: 226–228, bibliogr.)

MEIER, BERNARD (1894–1973) (Anon. 1973, Bull. Soc. ent. Mulhouse (Suppl.) Apr.1973: 1.)

MEIG, JUAN (Agenjo 1969, Graellsia 24: 289–304.)

MEIGEN, JOHN WILHELM (1764–1845) German naturalist and author of *Beschreibung der Europaischen Zweyflugeligen Insecten*, 6 vols, 1818–1830. (Steyskal 1974, Mosquito Syst. 6: 79–87. Translation of Förster, Stettin. ent. Ztg. 7: 66–74, 130–141.)

MEIJERE, JOHANNES CORNELIS HENDRICK DE (1866–1947) (Barendrecht & Krüsemann 1949, Tijdschr. Ent. 90: 1–15, bibliogr.)

MEIKLE, AGNES ADAM (–1951) (Riley 1952, Proc. R. ent. Soc. Lond. (C) 16: 86.)

MEINERS, EDWIN PAUL (1893–1960) (Remington 1962, J. Lepid. Soc. 16: 71–75, bibliogr.)

MEINERT, FREDERIK VILHELM AUGUST (1833–1912) (Henriksen 1927, Ent. Meddr. 15: 253–262, bibliogr.)

MEINERTELLIDAE Plural Noun. A derived Family of apterygotes assigned to the Order Archaeognatha; most meinertellids occur in the Southern Hemisphere. See Machilidae.

MEISNER, KARL FRIEDRICH (1765–1825) (Brunner 1825, Annln schweiz. Ges. Naturw. 2: 241–253.)

MEISSNER, GEORG (1829–1905) (Weiss 1904, Naturw. Rdsch. Stutt. 20: 349–351.)

MEIXNER, ADOLF (1883–1965) (Anon. 1966, Z. wiener ent. Ges. 51: 96.)

MELANDER, AXEL LEONARD (1878–1962) American academic. (Spieth 1966, Ann. ent. Soc. Am. 59: 235–237.)

MELANDRYIDAE Plural Noun. False Darkling Beetles (Coleoptera). Adults are dark coloured and hard bodied; Antenna filiform with 11 segments; Palpi large; tarsal formula 5-5-4. Adults and larvae occur under tree bark or within dry wood and fungi. Larvae are phytophagous or carnivorous. Syn. Tetratomidae. See Coleoptera.

MELANIC Adj. (Greek, *melas* = black; *-ic* = characterized by.) Pertaining to structure with a blackish surface.

MELANINS Plural Noun. (Greek, *melas* = black.) Any one of a group of organic pigments that produce black, amber and dark brown colours by deposition in the Cuticle (Wardle). Melanins are among the oldest pigments known and are phylogenetically important. Melanins are found in many microganisms, fungi and all animal groups. Typically, organisms that deposit melanin can synthesize it. Melanins usually produce brown and black in the Integument, but may produce other colours. For instance, structural features in bird feathers can interact with melanins to produce unusual metallic colours. Melanins serve many functions and can occur in granular form. Melanins shield the body from harmful rays of light. Melanins create cryptic coloration which visually blends the insect's body into the environment. Melanins are involved in some internal defence reactions. Melanization is associated with encapsulation of parasitic larvae within the body of some insects, such as caterpillars. Melanin is derived from tyrosine by way of the phenolase reactions. Some insect physiologists believe that the substrate and enzyme are physically separated within certain blood cells normally, but injury causes the reaction to proceed. See Pigmentary Colour. Cf. Carotenoids; Flavinoids; Ommochromes; Pterines (Pteridines); Quinones. Rel. Structural Colour.

MELANISM Noun. (Greek, *melas* = black; English, *-ism* = condition. Pl., Melanisms.) An abnormal or unusual darkening: A suffusion with blackish.

MELANISTIC Adj. (Greek, *melas* = black; *-ic* = characterized by.) Dark or blackish.

MELANOCHROIC Adj. (Greek, *melas* = black; *chros* = colour; *-ic* = consisting of.) Dark coloured or tending to blackness.

MELASTOMA BORER *Selca brunella* (Hampson) [Lepidoptera: Noctuidae].

MELBOURNE TRAPDOOR SPIDER *Stanwellia grisea* (Hogg) [Araneida: Nemesiidae].

MELDANE® See Coumaphos.

MELDOLA, RAPHAEL (1849–1915) (Distant 1916, Entomologist 49: 23–24.)

MELICHAR, LEOPOLD (1856–1924) (Rambousek 1925, Cas. csl. Spol. ent. 22: 1–3, bibliogr.)

MELIN, DOUGLAS (1895–1946) (Lundblad 1946, Ent. Tidskr. 67: 215–217.)

MELIPHAGOUS Adj. (Greek, *meli* = honey; *phagein* = to eat, devour; Latin, *-osus* = with the property of.) Pertaining to animals that feed on honey; honey-eating. Alt. Mellivorous. See Feeding Strategy.

MELIPONINAE A Subfamily of Apidae (Apoidea) consisting of about 10 Genera.

MELIS, ANTONIO (1871–1963) (Zocchi 1963, Redia 48: i-xii, bibliogr.)

MELISSAEUS Adj. (From Latin, *Melissa* = a small Genus of Old World plants, *M. officianalis* - cultivated balm.) Balm-scented.

MELISSOPALYNOLOGY Noun. (Greek, *meli* = honey; *pale* = pollen; *logos* = discourse. Pl., Melissopalynologies.) The study of the ways that honeybees use pollen and nectar; the microscopic study of honey. Cf. Entomopalynology. Palynology. Rel. Pollen.

MELITTIDAE Plural Noun. A small Family of aculeate Hymenoptera assigned to Superfamily Apoidea. Adults are small-bodied, dark-coloured with one Subantennal Suture present and Facial Foveae absent; Labrum wider than long, subtriangular or oval; apex of Glossa acute or rounded; short-tongued; Galea usually short basal of Palpus; Labial Palpus segments cylindrical and subequal; forewing second Recurrent Vein not arcuate; hindwing Jugal Lobe shorter than Submedian Cell; Basitibial Plate present in female; Arolia well developed in male and female; Pygidium present. Melittids are regarded as primitive bees that superficially resemble andrenids but lack Facial Foveae and the Scopa is restricted to posterior Tibia and Basitarsus. Subfamilies include Dasypodinae, Meganomiinae and Melittinae. See Apoidea.

MELITTIN Noun. (Latin, *melis* = bee.) A small basic, polypeptide toxin derived from the venom of the European honey bee, *Apis mellifera* L., that inhibits protein-kinases and cell-membrane lytic factor. Melittin is the most prevalent toxin derived from honeybee venom and is a potent anti-inflammatory agent, 100 times more potent than hydrocortisol. The peptide is a 26-amino acid chain with no disulfide bridge; the NH_2 terminal part of the molecule is predominantly hydrophobic and the C terminal part is hydrophilic and strongly basic.

MELITTINAE Noun. The nominant Subfamily of Mellitidae, consisting of about five Genera. See Mellitidae.

MELLIFERA Adj. (Latin, *mel* = honey; *ferre* = to carry.) Honey-makers. A term that is applied to bees as a whole.

MELLIFEROUS Adj. (Latin, *mel* = honey; *ferre* = to carry; *-osus* = with the property of.) Honey-producing; producers of honey.

MELLIN, ALFRED (1859–1920) (Merschner 1921, Jh. Ver. schles. Insektenk. 13: 23–24.)

MELLISUGOUS Adj. (Latin, *mel* = honey; *sugere* = to suck; *-osus* = with the property of.) Honey-sucking: Feeding on honey.

MELLO-LETAO, CANDIDO FIRMINO DE (1886–1948) (Anon. 1938, Boln Esc. nac. Agron 1: 71–96, bibliogr.)

MELLOWS, W T (–1950) (Anon. 1951, Trans. S. Lond. ent. nat. Hist. Soc. 1950–1951: 56.)

MELLY, ANDRE (1802–1851) (Schaum 1852, Stettin ent. Ztg. 13: 67–71.)

MELNIKOV, NIKOLAI MIKHALOVICH (Bogdanov 1841, Izv. imp. obshch. Lyub. Estest. Antrop. Etnogr. imp. Mosc. Univ. 55: (61), bibliogr., pl. 9.)

MELOCEPHALIC Adj. Having a pseudohypognathous type of head (MacGillivray). Cf. Hypognathous; Opisthognathous; Prognathous.

MELOIDAE Plural Noun. Blister Beetles. A cosmopolitan Family of polyphagous Coleoptera (Tenebrionoidea) consisting of about 2,800 Species distributed among 75 Genera. Classifications including 2–5 Subfamilies are recognized by different workers and meloids are placed among cucujoids or tenebrionoids. Adults are 7–25 mm long with body elongate, convex and glabrous, and weakly sclerotized with soft Integument. Head is deflexed and constricted behind eyes to form pronounced neck; Antenna with 11 segments that are filiform or submoniliform. Prothorax without lateral Carinae and narrows anteriorly; legs slender; fore Coxa elongate; tarsal formula 5-5-4; claws pectinate; Abdomen with 5–6 exposed abdominal Ventrites. Larval Antenna with one or three segments; Stemmata are present or absent; legs are present or absent; Urogomphi absent. Family is most abundant in arid areas. Adults are phytophagous with a few Species of some economic importance; a defensive secretion (Cantharidin) causes blistering upon contact with skin. Meloids display hypermetamorphic larvae; first instar larva triungulin, second caraboid, third–fourth scarabaeoid, fifth coarctate, sixth scolytoid. First instar larvae attach to adult stage of host and are transported to nest; fifth instar apparently is not a feeding stage and mouthparts are not functional. Sixth instar feeds. Larvae are predaceous on eggs of grasshoppers or parasitic on larvae of bees (megachilids and andrenids). Eggs of host are consumed and larval development is completed on nest provisions. Oviposition typically occurs at a location removed from host or prey. Pupation by bee parasites and predators usually occurs in cells of the host. Adult females lay clutches of eggs which number into thousands. Meloids developing on grasshopper egg pods lay smallest number of eggs per clutch. Oviposition usually occurs in soil with female using Mandibles and forelegs to create depression in which eggs are deposited; eggs are covered after oviposition. Some Species oviposit under rocks, in or near nests of bees, on flower heads. Species attacking egg pods of grasshoppers pupate near remains of prey. See Coleoptera.

MELOLONTHOID Adj. (Greek, *Melolonthe* = cockchafer; *eidos* = form.) Melolontha-like; resembling a June Beetle or Chafer in appearance or habits. Alt. Scarabeioid. Rel. Eidos; Form; Shape.

MELON APHID *Aphis gossypii* Glover [Hemiptera: Aphididae]. Syn. Cotton Aphid.

MELON FLY *Bactrocera cucurbitae* (Coquillett) [Diptera: Tephritidae]: A widespread, multivoltine pest of curcurbits; MF is native to Southeast Asia, but widely distributed by commerce and considered the most serious pest of melons and squash. Females oviposit on any part of plant; larvae are capable of developing on most plant parts. Larval feeding causes stunting of buds, stems and developing fruit; mature fruit can become heavily infested with larvae and unsuitable for consumption. See Tephritidae. Cf. Apple Maggot; Cherry Fruit Fly; Mediterranean Fruit Fly; Melon Fly; Mexican Fruit Fly; Oriental Fruit Fly; Papaya Fruit Fly; Queensland Fruit Fly.

MELON THRIPS *Thrips palmi* Karny [Thysanoptera: Thripidae].

MELON WEEVIL *Baris traegardhi* Auriv. [Coleoptera: Curculionidae]: A minor pest of melons in northeastern Africa. See Coleoptera.

MELON WORM *Diaphania hyalinata* (Linnaeus) [Lepidoptera: Pyralidae]: A widespread pest of pumpkins, melons and other cucurbits. MW biology resembles pickleworm but feeds more extensively on foliage. See Pyralidae. Cf. Pickleworm.

MELOPHEN® See Endosulfan.

MELSHEIMER, FREDERICK ERNST (1782–1873) (Arnett 1948, Coleopts Bull. 2: 27–28.)

MELSHEIMER, FREDERICK VALENTINE (1749–1814) (Schwarz 1894, Proc. ent. Soc. Wash. 3: 134–138.)

MELSHEIMER, JOHN F (1780–1829) (Fox 1901, Ent. News 110–113, 138–141, 173–177, 203–205, 233–236, 281–283, 314–316.)

MELUSINIDAE Plural Noun. See Simuliidae.

MELVILL, JAMES COSMO (1845–1929) (Weiss 1931, Proc. Linn. Soc. Lond. 142: 211–213.)

MELYRIDAE Plural Noun. Soft-winged Flower Beetles. A cosmopolitan Family of polyphagous Coleoptera assigned to Cleroidea and including about 4,000 nominal Species. Disagreement over higher placement: Cleroidea or Cantharoidea. Adults are 11–10 mm long, rather elongate, somewhat flattened and soft-bodied; Antenna with 10–11 segments and shape variable; front Coxae prominent; tarsal formula 5-5-5, rarely 4-5-5 in male; Elytra not striate and with long Setae; Abdomen with six Ventrites. Larvae with head prognathous, body elongate, slighlty flattened and weakly sclerotized; 1–5 Stemmata on each side of head; Antenna with three segments; legs with 5 segments, each Pretarsus with one tarsal claw; Urogomphi present. Most Species are entomophagous; some adults frequent flowers and feed on pollen; some larvae are scavengers; most larvae and adults feed on invertebrates, including worms, molluscs and larval insects. *Collops* Species are predaceous on aphids, leafhoppers and other soft-bodied insects. Most predators feed on exuded fluids of victims rather than consuming body. Syn. Malachiidae.

MELZER, JULIUS (1878–1934) (Borgmeier 1935, Recta Ent. Rio de J. 5: 89–90.)

MEMBER Noun. (Old French, *membre*, from Latin, *membrum* = a limb. Pl., Members.) 1. A limb or appendage of a body. 2. An organism within a group. Particularly, an individual belonging to a Species; a Species belonging to a Genus; a Genus belonging to a Family; a Family belonging to Order. 3. In general, any element of a subset belonging to set within a hierarchical organization. Cf. Element. Rel. Classification; Taxonomy.

MEMBRACIDAE Plural Noun. Treehoppers. A numerically large Family of auchenorrhynchous Hemiptera assigned to Superfamily Cicadelloidea. Membracids are most abundant in the neotropics. Adults are 8–14 mm long; Pronotum considerably enlarged, projecting over head and abdomen, and typically hump-backed, often with distinctive ornamental shape or feature to include horns, keels, or spines. All Species are phytophagous. Most Species feed on trees and shrubs; the nymphs of some Species feed on grasses and may be attended by ants; many Species are relatively specialized in host plant associations. A few Species are pests of ornamental plants. Typically, membracids are univoltine or bivoltine and overwinter in the egg stage. See Cicadelloidea.

MEMBRANA FENESTRATA Insects: The Basement Membrane of the compound eye (Snodgrass). See Compound Eye; Ommatidium.

MEMBRANA RETINENS The streched part of the membrane around the Rectum of butterfly larvae and used in the transformation to the Chrysalis.

MEMBRANACEOUS See Membranous.

MEMBRANE Noun. (Latin, *membrana* = membrane. Pl., Membranes.) 1. Any thin, transparent, flexible body tissue. Specifically the wing tissue between the veins. 2. Heteroptera: The thin, transparent or translucent apex of the Hemelytra, as distinguished from the thickened basal part, the Corium. Alt. Membrana.

MEMBRANIZATION Noun. (Latin, *membrana* = membrane; English, *-tion* = result of an action.) Process by which tissue is changed into a membrane.

MEMBRANOUS Adj. (Latin, *membrana* = membrane; *-osus* = with the property of.) Pertaining to tissue which is thin and semi-transparent; like a membrane. A tissue of a thin, pliable texture. Alt. Membranaceus.

MEMBRANOUS EGG Typically variable in size, Chorion thin, transparent and appears membranous. Surface reticulation pattern and pliancy provide an impression of membrane. An egg typically is ejected from the female which contains a mature embryo ready to emerge; eclosion occurs soon after oviposition. Eggs often are glued to the host and site specificity has been suggested. The distinction between macrotype and membranous eggs is sometimes lost. Membranous egg shape is representative of Diptera (Tachinidae, Sarcophagidae). See Egg. Cf. Macrotype Egg.

MEMBRANULE Noun. Anisopteran Odonata: The small, veinless darker expansion at the base of the wings. See Anal Membrane.

MEMININI, GIANFRANCO (1935–1959) (Anon. 1959, Mem. Soc. ent. ital. 38: 152–153.)

MEMPEL, ADOLPH (1870–1949) (Anon. 1950, Boln. Soc. bras. Ent. 1: 30–31.)

MENDENHALL, EUGENE WARREN (1874–1944) (Barringer 1945, Ann. ent. Soc. Am. 38: 142.)

MENDENHALL, WILFRED THOMAS (1901–1966) (Davis 1967, J. Econ. Ent. 60: 898.)

MENDES AZEVEDO, CANDIDO (1874–1944) (Luisier 1944, Broteria 40: 43–48, bibliogr.)

MENDES, DARIO (1892–1963) (Anon. 1963, Studia Ent. 6: 586.)

MENETRIES, EDOUARD (1802–1861) (Marseul 1884, Abeille 22: 137–139, bibliogr.)

MENGE, FRANZ ANTON (1808–1880) (Bonnet 1945, *Bibliographia Araneorum* 1: 35–36.)

MENGEL, LEVI WALTER SCOTT (1868–1941) (Anon. 1941, Ent. News 52: 178–180.)

MENISCOIDAL Adj. (Greek, *meniskos* > *menes* = little moon; *-alis* = characterized by.) Crescent-shaped; descriptive of structure that is 'concave-convex' with one side convex and the other side concave.

MENITE® See Mevinphos.

MENOGNATHA Noun. (Greek, *meno-* = to remain; *gnathos* = jaw.) Insects in which both young and adults feed by Mandibles, *e.g.* the Orthoptera. See Feeding Strategies. Cf. Menorhyncha; Metagnatha.

MENOPONIDAE Plural Noun. Poultry Body Lice. The largest Family of Amblycera. Head broadly triangular and expanded behind eyes; Antenna typically capitate, with four segments, concealed in grooves; mandibulate mouthparts; Maxillary Palpus with 2–3 segments; Tarsi with paired claws. Cosmopolitan parasites of birds. See Amblycera. Cf. Chicken Body Louse; Shaft Louse.

MENORHYNCHA Noun. (Greek, *meno-* = to remain; *rhynchos* = snout.) Forms in which both young and adult take food by suction, *e.g.*, Hemiptera. See Feeding Strategies. Cf. Menognatha; Metagnatha.

MENOTAXIS Noun. (Greek, *meno-* = to remain; *taxis* = arrangement.) Partial or indefinite orientation. See Orientation. Cf. Aerotaxis; Anemotaxis; Chemotaxis; Geotaxis; Notarotaxis; Osmotaxis; Phototaxis; Rheotaxis; Rotaxis; Scototaxis; Strophotaxis; Telotaxis; Thermotaxis; Thigmotaxis; Tonotaxis; Tropotaxis.

MENOZZI, CARLOS (1892–1943) (Grandi 1943, Memorie Soc. ent. ital. 22: 118–124, bibliogr.)

MENSIK, EMANUEL (1851–1913) (Dittrich 1914, Jh. Ver. Schles. Insektenk. 7: xxiv–xxv.)

MENTACORIA Noun. The oral part of the Cervacoria, adjacent to the Labium (MacGillivray).

MENTAL SETAE Odonata: Setae on the inner surface of the Mentum (Garman).

MENTAL SUTURE A cuticular line between the Submentum and Gula.

MENTAL Adj. (Latin, *mentum* = chin; *-alis* = characterized by.) Descriptive of or pertaining to the Mentum.

MENTASUTURE Noun. (Latin, *mentum* = chin; *sutura*.) 1. The suture-like reduced Mentacoria. 2. A suture between the Submentum and Gula (MacGillivray).

MENTIGEROUS Adj. (Latin, Mentum = chin; *gerere* = to carry; Latin, *-osus* = with the property of.) Bearing or having a Mentum.

MENTUM Noun. (Latin, Mentum = chin.) 1. The distal sclerite of the insect's Labrum bearing the moveable parts, and attached to (sometimes fused with) the Submentum. The Mentum corresponds to the (united) Stipes of the Maxillae. 2. Coleoptera: The so-called Mentum is the Submentum. 3. Diptera: The posterior oral margin. See Thyroid. 4. Hymenoptera: Part of the 'tongue,' the second segment bearing the Labial Palpi, Paraglossae and Ligula. Cf. Submentum. 5. Acarology: Unpaired part of the ventral surface of the Infracapitulum. See Head.

MENZEL, AUGUST (1810–1878) Anon. 1879. Mitt. schweiz. ent. Ges. 5: 492–494, bibliogr.)

MEOBAL® See MPMC.

MEOTHRIN® See Fenpropathrin.

MEPHOSFOLAN An organic-phosphate (dithiophosphate) compound {2-diethoxyphosphinylimino)-4-methyl-1,3-dithiolane} used as a contact/systemic insecticide, stomach poison and acaricide against spider mites, thrips, plant-sucking insects, leaf-chewing insects and bollworms. Compound applied to fruit, cotton, corn, citrus, grapes, onions, potatoes, rice, tomatoes, vegetables and ornamentals in some countries. Not registered for use in USA. Compound is toxic to bees and fishes. Trade names include: Cytrolane®. See Organophosphorus Insecticide.

MEQUIGNON, A (–1958) (Anon. 1958, Bull. Soc. ent. Fr. 63: 157.)

MERA, ARTHUR WILLIAM (1849–1930) (Turner

1930, Entomologist's Rec. J. Var. 42: 143–144.)

MERAL PLATE Diptera: The Meron.

MERCAPTOTHION® Malathion.

MERCET, GARCIA RICARDO (1860–1932) Spanish pharmacist who in his later years became a taxonomic specialist of parasitic Hymenoptera (Aphelinidae & Encyrtidae). (Dusmet 1933, Boln. Soc. ent. Esp. 16: 112–113.)

MERCHANT GRAIN-BEETLE *Oryzaephilus mercator* (Fauvel) [Coleoptera: Cucujidae]: A widespread pest of household cereals, dried fruits, nuts and spices, but not stored grain; MGB is most common in tropical and subtropical regions. Species is very similar to Saw-Toothed Grain Weevil anatomically and biologically but MGB prefers oilseed and more restricted in distribution. See Cucujidae. Cf. Saw-toothed Grain Weevil.

MERCK, PAUL (1793–1849) (Pierret 1849, Bull. Soc. ent. Fr. (2) 7: lxviii–lxix.)

MERDAFOS® See Sulprofos.

MERDIVOROUS Noun. (Latin, *merda* = excrement; *vorare* = to eat.) Feeding upon dung or excrement. See Feeding Strategies. Cf. Scatophagous.

MERE Noun. (Greek, *meros* = a part. Pl., Meres.) A part or subdivision of a sclerite (*e.g.* Pleuromere) or segment (*e.g.* Flagellomere).

MERE, ROBIN MARCUS (1909–1966) (Messenger 1966, Proc. S. Lond. Ent. nat. Hist Soc. 1966, 93–94.)

MEREDITH, LOUISA ANNE (1812–1895) (Swann 1929, Proc. R. Aust. Hist. Soc. 15: 1–29.)

MERIAEUM Noun. 1. The sclerites of the sockets of the posterior legs, behind the Acetabulum and Parapleuron (Knoch). 2. Coleoptera: The posterior inflected part of the Metasternum (Smith).

MERIAN, MADAME MARIA SIBYLLA (FRAU J A GRAFF) (1647–1717) Dutch noblewoman of German extraction, and author of several entomological works including *Der Raupen wunderbare Verwandlung* (2 vols, 1679, 1683), *Metamorphosis Insectorum Surinamensis* (1705) and *Erucarum Ortus, alimentum et paradoxa metamorphosis* (1718). Merian was a superb artist who prepared her own watercolour illustrations that were engraved on copper plates. (Erlanger 1976, Insect Wld. digest 3: 13–21.)

MERIDIC DIET An artificial diet which consists mostly of defined chemicals but includes undefined components (*e.g.* Casein). MD often complex and contain specific components (amino acids, sterols, vitamins, minerals) combined with agar and water. Most artificial diets for insects are meridic. See Artificial Diet. Cf. Oligidic Diet; Holidic Diet.

MERINO, GONZALO R (1890–1969) (Cendana 1969, Philipp. Ent. 1: 258–259.)

MERIT® See Imidacloprid.

MERKEL, AUGUST (1837–1897) (Dietz 1891, Ent. News 8: 184.)

MERKEL, RUDOLF (–1948) (Anon. 1948, Anz. Schädlingsk. 21: 96.)

MERMISEAL® See Chlordane.

MERMITHANER Noun. (Greek, *mermis* = cord; *aner* = male. Pl., Mermithaners.) A male ant parasitized by the nematode *Mermis*.

MERMITHERGATE Noun. (Greek, *mermis* = cord; *ergates* = worker. Pl., Mermithergates.) A worker ant parasitized by the nematode *Mermis*.

MERMITHOGYNE Noun. (Greek, *mermis* = cord; *gyne* = female. Pl., Mermithogynes.) A female or queen ant parasitized by the nematode *Mermis*.

MEROANDRY Noun. (Greek, *meros* = part; *aner* = male.) The condition of reduced testicular number.

MEROBLASTIC DIVISION The type of egg cleavage in which only the Nucleus and the nuclear cytoplasm are divided (Snodgrass).

MEROBLASTIC Adj. (Greek, *meros* = part; *blastos* = bud; *-ic* = characterized by.) Pertaining to eggs which undergo partial cleavage.

MEROCRINE Adj. (Greek, *meros* = part; *krinein* = to separate.) Descriptive of or pertaining to Merocriny.

MEROCRINY Noun. (Greek, *meros* = part; *krinein* = to separate. Pl., Merocrinies.) The passage of a digestive enzyme into the lumen of gut or salivary duct through the free borders of the parent cells (Wardle).

MEROISTIC Adj. (Greek, *meros* = part; *oon* = egg; *-ic* = characterized by.) Pertaining to Ovarioles that contain Trophocytes (Nurse Cells).

MEROISTIC EGG TUBE A derived type of Ovary which contains Trophocytes (nutritive cells, nurse cells) which provide macromolecules (including nucleic acid) that are utilized by the developing Oocyte. The function of Trophocytes was recognized by the English Peer Sir John Lubbock (Lord Avebury) in 1859. Two types of Meroistic Ovary have been described: Telotrophic and Polytrophic. See Ovariole. Cf. Telotrophic Ovariole; Polytrophic Ovariole. Rel. Panoistic Egg Tube.

MEROISTIC OVARIOLE A type of Ovariole in which nutritive cells are present (Imms). Syn. Meroistic Egg Tube (Snodgrass). See Ovariole. Cf. Panoistic Ovariole; Telotrophic Ovariole. Rel. Oogenesis.

MERON Noun. (Greek, *meros* = upper thigh. Pl., Mera.) A lateral, postarticular, basal area of the Coxa. Meron can be a sclerite dissociated from the Coxa and incorporated into the Pleuron. Meron is typically large and conspicuous in panorpoid and neuropteroid insects. Meron of Diptera forms a separate sclerite in the thoracic Pleuron.

MEROPACHYDINAE A Subfamily of Coreidae.

MEROPEIDAE Plural Noun. Earwigflies. A small Family of Mecoptera with Species known from North America and Western Australia. One Species occurs in eastern North America; one Species from Western Australia. Meropeids are characterized by the head opisthognathous and partially concealed by Pronotum, Tarsus not raptorial, each Tarsus with two claws; Ocelli absent; wings macropterous, broad with reticulate venation; wings folded over abdomen at repose; thoracic sclerites not fused; male genital claspers with long and slender basal and apical segments. Species probably are phytophagous. See Mecoptera.

MEROPHYSIIDAE Plural Noun. See Endomychidae.

MEROPLEURITE Noun. (Greek, *meros* = upper thigh; *pleura* = side; *-ites* = constituent. Pl., Meropleurites.) A composite sclerite including elements of the Meron, Coxa and lower region of the Epimeron (Crampton). See Meron; Epipleuron.

MEROPODITE Noun. (Greek, *meros* = upper thigh; *pous* = foot; *-ites* = constituent. Pl., Meropodites.) 1. Chelicerata: The Femur. 2. The fourth segment of a generalized limb; the Femur. Cf Telopodite.

MEROTHRIPIDAE Plural Noun. Large-legged Thrips. The most primitive extant Family of Thysanoptera, and including about 15 nominal Species assigned to Terrebrantia; merothripids predominantly are Neotropical with some Species cosmopolitan through trade. Adult Pronotum with two longitudinal sutures; front and hind Femora enlarged; female SVIII with a pair of overlapping lobes; TX with pair of apical Trichobothria. Merothripids are fungivorous and live in leaf litter. See Thysanoptera.

MEROZOITE Noun. (Greek, *meros* = part; *zoon* = animal; *-ites* = resident. Pl., Merozoites.) The third stage in the asexual development of the protozoan which causes Malaria. The Merozoite is derived from the Schizont, one of the liberated spores of the *Plasmodium* Schizont (Comstock).

MERRET, CHRISTOPHER (1614–1695) (Rose 1850, *New General Biographical Dictionary*. 10: 106.)

MERRIAM, CLINTON HART (1856–1942) (Daubenmire 1938, Q. Rev. Biol. 13: 327–332, bibliogr.)

MERRICK, FRANKLIN A (1844–1912) (Anon. 1913, Can. Ent. 45: 170.)

MERRICK, HARRY DUNCAN (1869–1907) (Anon. 1907, Ent. News 18: 320.)

MERRIFIELD, FREDERICK (1831–1924) (Poulton 1924, Entomologist 57: 239–240.)

MERRILL, GEORGE BATES (1886–1971) (Denmark 1971, Fla. Ent. 54: 314.)

MERRILL, JOSEPH HENRY (1881–1946) (Metcalf 1947, J. Econ. Ent. 40: 141.)

MERRIN, JOSEPH (1820–1904) (Anon. 1904, Entomol. mon. Mag. 40: 112.)

MERYCIDAE Plural Noun. A small Family of Coleoptera assigned to Tenebrionoidea and synonymous with Colydiidae or Zopheridae.

MESA Noun. (Spanish.) A large, flat-topped hill. Cf. Butte.

MESAD Adv. (Latin, *medius* = middle; *ad* = toward.) Toward or in the direction of the median plane of the insect body, or Meson. See Orientation. Cf. Anteriad; Apicad; Basad; Caudad; Centrad; Cephalad; Craniad; Dextrad; Dextrocaudad; Dextrocephalad; Distad; Dorsad; Ectad; Entad; Laterad; Mediad; Neurad; Orad; Proximad; Rostrad; Sinistrad; Sinistrocaudad; Sinistrocephalad; Ventrad.

MESADENIA Plural Noun. (Greek, *mesos* = middle; *aden* = gland.) Mesodermal accessory glands of the male insect which are derived from evaginations of the Vasa Deferentia. Composed of a secretory epithelium on cell-layer thick and conspicuous Basement Membrane. Secretory cells are Apocrine or Merocrine; secretory products protein of mucoprotein; secretory activities apparently under control of hormones. Secretions produce Spermatophore, seminal fluid, activate Spermatozoa and affect female behaviour. See Male Accessory Gland. Cf. Ectadenia.

MESAL Adj. (Latin, *medius* = middle; *-alis* = pertaining to.) Descriptive of structure that is positioned on or in the median plane of the body. See Orientation. Cf. Distal; Mesial; Proximal.

MESAL CALIS Coccids: One of the Calles adjacent to the Meson (MacGillvary).

MESAL CERARI Coccids: The single dorsal row of Cerari (MacGillivray).

MESAL LOBES Coccids: The median lobes *sensu* MacGillivray.

MESAL MARGIN Coccids: See Inner Margin (MacGillivray).

MESAL NOTCH Coccids: The notch on the mesial side of a lobe (MacGillivray).

MESAL ORBACERORES Coccids: The inner and mesial row of Orbacerores surrounding the Anal Ring (MacGillivray).

MESAL PLATES 1. Some coccids: The plates of wax located on the Meson between the dorsal sclerites, limited to the Mesothorax, Metathorax and first abdominal segments. 2. The wedge-shaped sclerites *sensu* MacGillivray.

MESANEPISTERNUM Noun. (Greek, *mesos* = middle; *an* = both; *epi* = upon; *sternon* = chest. Pl., Mesanepisterna.) Odonata: The Anepisternum of the Metathorax (Garman).

MESARIMA Noun.(Etymology obscure.) The fissure separating the Glossae (MacGillivray).

MESASCUTELLA Noun. (Greek, *mesos* = middle; *scutellum*.) The middle area of the Mesoscutellum (MacGillivray).

MESAXON Glial Cells that which form a spiral around an Axon. See Nervous System; Nerve Cell. Cf. Glial Cell. Rel. Central Nervous System.

MESENCHYMA Noun. (Greek, *mesos* = middle; *engchein* = to pour in.) Mesoblastic tissue formed of loosely connected or scattered cells (Snodgrass). Alt. Mesenchyme.

MESENTERON Noun. (Greek, *mesos* = middle; *enteron* = gut.) The midgut, stomach or Chylific Ventricle. The middle portion of the primitive intestinal canal, lined with Entoderm. See Midgut.

MESENTERON RUDIMENTS The groups of Endoderm cells that regenerate the Mesenteron, including an anterior, a posterior, and sometimes an intermediate rudiment (Snodgrass).

MESEPIMERON Noun. (Greek, *mesos* = middle; *epi* = upon; *meros* = upper thigh. Pl., Mesepimera.) Odonata: The sclerite between Humeral and first Lateral Suture. The Epimeron of the Mesothorax.

MESEPISTERNUM Noun. (Greek, *mesos* = middle; *epi* = upon; *sternon* = chest. Pl., Mesepisterna.) Odonata: The oblique lateral sclerites of the Mesothorax that meet dorsally in a ridge (Smith). Ants: The Mesothoracic Sternum.

MESIAL Adj. (Latin, *medius* = middle; *-alis* = pertaining to.) Descriptive of something near an imaginary line dividing a body into left and right halves; the median plane of the body. See Orientation. Cf. Distal; Medial; Proximal.

MESIALLY Adj. (Latin, *medius* = middle.) At or to the middle or midline of a structure.

MESIC Adj. (Greek, *mesos* = middle; *-ic* = of the nature of.) Descriptive of habitat or environmental conditions that are moderate, *i.e.* not too dry, not too wet, not too hot and not too cold. Cf. Hygric; Xeric.

MESINFRAEPISTERNUM Noun. (Greek, *mesos* = middle; *infra* = below; *epi* = upon; *sternon* = chest. Pl., Mesinfraepisterna.) A sclerite formed between Propleuron, Mesepisternum, Mesepimeron and second Coxa.

MESITHONIDAE Panfilov 1980.

MESKE, OTTO VON (1837–1890) (Anon. 1890, Entologica am. 6: 180.)

MESNIL, FELIX (1868–1938) (Roubaud 1938, Bull Soc. Path. exot. 31: 173–177.)

MESOBLAST Noun. (Greek, *mesos* = middle; *blastos* = sprout. Pl., Mesoblasts.) The middle germ layer of the embryo; the Mesoderm.

MESOBLASTIC Adj. (Greek, *mesos* = middle; *blastos* = sprout.) Descriptive of or pertaining to the Mesoblast.

MESOBLASTIC SOMITES Segmental divisions of the embryonic insect Mesoderm.

MESOCEPHALIC PILLARS Bees: Two large, oblique, strongly chitinous bars that form a brace between the anterior and posterior walls of the head (Snodgrass). Cf. Tentorium.

MESOCHRYSOPIDAE Handlirsch 1906.

MESOCORIA Noun. The Coria anterior of the Mesothorax (MacGillivray).

MESOCOXA Noun. (Greek, *mesos* = middle; *coxa* = hip. Pl., Mesocoxae.) The middle Coxa or basal segment of the middle leg of and insect. See Leg.

MESOCUTICLE Noun. (Greek, *mesos* = middle; + cuticle.) A transitional cuticular region, consisting of several layers associated with the outer portion of Endocuticle. Mesocuticle is not a distinct layer but gradually merges with Exocuticle. As the Endocuticle, Mesocuticle is not tanned and is not sclerotized. See Cuticle; Integument.

MESODERM. Noun. (Greek, *mesos* = middle; *derm* = skin. Pl., Mesoderms.) The embryonic layer of tissue lying between the Ectoderm and Endoderm. Alt. Mesoblast. See Embryology. Cf. Ectoderm; Endoderm.

MESODISCALOCA Coccids: See Discaloca (MacGillivray).

MESODONT Male Lucanidae: Mandibles intermediate in size between the Teleodont (large) and the Priodont (small) Mandibles. Cf. Amphiodont.

MESOEPISTERNUM Noun. (Greek, *mesos* = middle; *epi* = upon; *sternon* = chest. Pl., Mesoepisterna.) The Episternum of the Mesothorax. Alt. Mesepisternum.

MESOFACIAL PLATE Diptera: See Face.

MESOFURCA Noun. (Greek, *mesos* = middle; Latin, *furca* = fork.) The Furca of the Mesothorax. Syn. Middle Furca.

MESOGENACERORES Coccids: The median group of Genacerores (Comstock).

MESOMEROS The second to fifth abdominal segments in Lepidoptera.

MESON Noun. (Greek, *meson* = middle.) An imaginary middle plane dividing the insect body into right and left parts.

MESONOTUM Noun. (Greek, *mesos* = middle; *noton* = back. Pl., Mesonota.) The primitive upper surface of the second or middle thoracic segment (Mesothorax) of the insect body.

MESOPARAPTERON Noun. Formicidae: The Prescutellum (Wheeler).

MESOPEDES Plural Noun. (Greek, *mesos* = middle; *podos* = foot.) The middle legs.

MESOPHILE Noun. (Greek, *mesos* = middle; *philein* = to love. Pl., Mesophiles.) An organism that lives and reproduces optimally at relatively moderate temperatures. Cf. Halophile; Osmophile; Psychrophile; Thermophile.

MESOPHRAGMA Noun. (Greek, *mesos* = middle; *phragmos* = fence.) An internal prolongation of the Metapraesucutum, which provides points for attachment to some of wing muscles.

MESOPHYLL Noun. (Greek, *mesos* = middle; *phyllon* = leaf. Pl., Mesophylls.) The internal Parenchyma of a leaf, positioned between the upper and lower Epidermis and usually photosynthetic. See Parenchyma.

MESOPHYTE Noun. (Greek, *mesos* = middle; *phyton* = plant. Pl., Mesophytes.) A plant adapted to live under moderate conditions of moisture. Cf. Xerophyte; Halophyte.

MESOPLEURON Noun. (Greek, *mesos* = middle; *pleura* = side. Pl., Mesopleura.) The lateral part of the Mesothorax. In winged insects, postioned ventral of the fore-wing base. dorsal of the middle Coxa and divided by the Plural Suture. Diptera: The upper portion of the Episternum of the Mesothorax (Comstock). Syn. Mesopleurum. See Anepisternum; Anepimeron; Katepisternum; Katepimeron. Cf. Mesosternum.

MESOPLEURUM Noun. See Mesopleuron.

MESOPLURAL Adj. (Greek, *mesos* = middle; *pleura*

= side.) Descriptive of or pertaining to the Mesopleura.

MESOPLURAL BRISTLES Diptera: Large Setae inserted in the angle formed by the Dorsopleural and Mesopleural Sutures.

MESOPLURAL ROW Diptera: A posterior row of bristles or large Setae located on the Mesopleuron (Comstock).

MESOPLURAL SUTURE A suture on each side separating the Episternum and the Epimeron of the Mesothorax (Comstock).

MESOPSOCIDAE Plural Noun. A numerically small, widespread Family of eupscoious Psocoptera. Family best represented in Old World. Adult rather large; brachypterous or macropterous. Most Species inhabit tree branches.

MESOSCIOPHILIDAE Blagoderov 1994. Plural Noun. A Family of fossil Diptera known from several Genera and Species taken in Mesozoic deposits in Russia.

MESOSCUTELLUM Noun. (Greek, *meso* = middle; *scutellum* = little shield. Pl., Mesoscutella.) The Scutellum of the Mesothorax.

MESOSCUTUM Noun. (Greek, *meso* = middle; *scutum* = shield. Pl., Mesoscuta.) 1. Heteroptera: The anterior part of the Mesothorax, positioned under the edge of the Prothorax which is sometimes exposed but usually not visible. 2. Hymenoptera: The Scutum of the Mesothorax. 3. Diptera: See Mesonotum (Curran).

MESOSERIES Noun. (Greek, *mesos* = middle; *series*. Pl., Mesoserieses.) The arrangement of Crochets of the larval Proleg in a single, inner or mesial, longitudinal band.

MESOSERPHIDAE Kozlov 1970. A Family of apocritous Hymenoptera assigned to the Proctotrupoidea.

MESOSOMA Noun. (Greek, *mesos* = middle; *soma* = body. Pl., Mesosomata.) 1. The middle region of the insect body which holds the legs and wings. 2. Higher Hymenoptera: The middle portion of the body which includes the Thorax and Propodeum. Syn. Alitrunk. See Propodeum. Cf. Prosoma; Metasoma; Opisthosoma. 3. Lepidoptera: A sclerite connecting genitalic Valvulae that support the Aedaegus.

MESOSPIRACLES Noun. Coccids: The anterior pair of spiracles on the Mesothorax (MacGillivray).

MESOSTERNAL CAVITY Elateridae: The opening into which the Prosternal Spine or Mucro is fitted.

MESOSTERNAL EPIMERA Coleoptera: The narrow sclerites separating the Mesosternal and Metasternal Episterna.

MESOSTERNAL EPISTERNA Coleoptera: Sclerites on each side of the Mesosternum along the anterior border and Epimera, and typically separated by a distinct suture.

MESOSTERNAL LOBES Orthoptera: The Mesosternellum.

MESOSTERNELLUM Noun. 1. General: The Sternellum of the Mesothorax. 2. Two median lobes of Mesosternum, one on each side of the deep median notch. See Sternellum.

MESOSTERNUM Noun. (Greek, *mesos* = middle; *sternon* = chest. Pl., Mesosterna; Mesosternums.) The ventral surface of the Mesothorax postioned between the Mesopleura. Syn. Medipectus. See Sternum. Cf. Mesopleuron.

MESOSTETHIDIUM Noun. (Greek, *mesos* = middle; *stethos* = chest; *-idion* = diminutive. Pl., Mesostethidia.) 1. The Mesothorax. 2. Acarology: An unsclerotized median part of the Prodorsum.

MESOSTETHIIUM Noun. (Greek, *mesos* = middle; *stethos* = chest; *idion* = diminutive.) The middle portion of the ventrum of the Metathorax, positioned between the middle and hind legs.

MESOSTIGMA Noun. (Greek, *mesos* = middle; Stigma = pricking.) Odonata: small sclerites surrounding the Mesothoracic spiracles. See Caudal Plate (Garman).

MESOSUBSCUTELLUM Noun. (Greek, *mesos* = middle, Latin, *sub* = below; *scutum* = shield.) The middle region of the Subscutella (MacGillivray).

MESOSULCUS Noun. (Greek, *mesos* = middle; Latin, *sulcus* = furrow. Pl., Mesosulci.) Hymenoptera: A median, longitudinal Sulcus or furrow of the Mesosternum.

MESOSUTURE Noun. (Greek, *mesos* = middle; Latin, *sutura* = seam. Pl., Mesosutures.) The suture of the Mesothorax (MacGillivray).

MESOTARSUS Noun. (Greek, *mesos* = middle; Latin, *tarsus* = ankle.) The Tarsus of the middle leg. See Tarsus.

MESOTERGUM Noun. (Greek, *mesos* = middle; Latin; *tergum* = back. Pl., Mesoterga.) The Tergum of the Mesothorax. The Mesonotum. See Tergum.

MESOTHORACOTHECA Noun. (Greek, *mesos* = middle; *thorax* = breast plate; *theke* = case.) The pupal covering of the Mesothorax.

MESOTHORAX Noun. (Greek, *mesos* = middle; *thorax* = breast plate.) The second or middle thoracic segment which bears the middle legs and the anterior wings. Cf. Metathorax; Prothorax.

MESOVELIIDAE Douglas & Scott 1867. Plural Noun. Water Treaders. A numerically small Family of heteropterous Hemiptera assigned to Superfamily Mesovelioidea. Adult body ca 2–5 mm long, slender, greenish to yellow in colour; Ocelli present in macropterous individuals but absent from apterous individuals of Species; tarsal claws apical. Mesoveliids occur on emergent vegetation or along the margin of water. Nymphs and adults are active predators of mosquito larvae, mosquito Pupae and other organisms that live on or near the surface of water. See Mesovelioidea. Cf. Veliidae.

MESOVELIOIDEA A Superfamily of Hemiptera–Heteroptera including the Mesoveliidae. See Hemiptera.

MESOZOIC ERA 'Age of Dinosaurs.' The interval of the Geological Time Scale (245–65 MYBP) characterized by the origin, evolution, adaptive radiation and extinction of dinosaurs and flying reptiles. Birds originated and persisted during ME. The gymnosperms radiated into ginkgos, cycadeoids, cycads and conifers; angiosperms appear late in ME. ME stratigraphic record divided into Triassic Period, Jurassic Period and Cretaceous Period. The Supercontinent (Pangaea) separated into many fragments via plate tectonics and continental drift. See Geological Time Scale; Cenozoic Era; Palaeozoic Era. Cf. Cretaceous Period; Jurassic Period; Triassic Period. Rel. Fossil.

MESSA, GIUSEPPE (1872–1932) (Anon. 1932, Boll. Soc. ent. ital. 64: 173.)

MESUROL® See Methiocarb.

META- Greek prefix used to designate any posterior (generally third) part of structure (*e.g.*, Metapedes, Metapleura).

META® See Metaldehyde.

META-ANEPISTERNUM Odonata: The Anepisternum of the Metathorax (Garman).

METABLASTIC Adj. (Greek, *meta* = after; *blastos* = sprout; *-ic* = characterized by.) Relating to the Ectoblast or Metablast or Ectoderm.

METABOLA Noun. (Greek, *meta* = change; *ballein* = to throw.) Insects that pass through a complete Metamorphosis, to include a series of stages from the just-hatched (neonate) young to the adult, with each stage differing from each other in varying, though obvious degrees (Wardle).

METABOLIC ACTIVITIES All forms of animal activity concerned with metabolism, such as digestion, respiration and excretion.

METABOLISM Noun. (Greek, *metabolikos* = change; *-ism* = condition.) The entire process or series of changes of food into tissue and cell-substance and of these latter into waste products. The first of these changes are anabolic (constructive) and the second are katabolic (destructive).

METABOLITE Noun. (Greek, *metabolikos* = change; *-ites* = substance. Pl., Metabolites.) 1. The product of biochemical activity. 2. Metabolized substances; in general any of the products of metabolism.

METABOLOUS Adj. (Greek, *metabolikos* = change; Latin, *-osus* = with the property of.) Undergoing Metamorphosis or transformation. See Development; Metamorphosis. Cf. Ametabolous; Hemimetabolous; Holometabolous; Paurometabolous.

METABUME® See Methyl Bromide.

METACAR® See Hexythiazox.

METACENTRIC Adj. (Greek, *meta* = beside; *kentron* = center; *-ic* = characterized by.) Pertaining to chromosomes with a Centromere not positioned on the end. Cf. Acentric; Telocentric.

METACEPHALON Noun. (Greek, *meta* = after; *kephale* = head.) Diptera: The area behind the mouth extending up toward the neck (Curran).

METACORIA Noun. (Greek, *meta* = beside; Latin, *corium* = leather.) The Coria anterior of the Metathorax (MacGillivray).

METACOXACORIA Noun. (Greek, *meta* = beside; *coxa*; Latin, *corium* = leather.) The Coxacoria of a metathoracic leg (MacGillivray).

METACOXAL PLATE Coccinellidae: Part of the first ventral segment included above the ventral lines visible on that segment.

METACRATE® See Metolcarb.

METACRYPTOZOIC SCHIZOGONY Medical Entomology: Metacryptozoites which directly parasitize blood cells and initiate the erythorocytic phase of infection. (Huff & Coulston 1944. J. Inf. Dis. 75: 231–249.) See Schizogony. Cf. Cryptozoic Schizogony; Exoerythrocytic Schizogony.

METAEPISTERNUM Noun. (Greek, *meta* = after; *epi* = upon; *sternon* = chest.) The Episternum of the Metathorax.

METAFEMORAL SPRING Coleoptera: A sclerotized spring-like extension of the metatibial tendon which contains a mixture of alpha-chitin and protein (not Resilin). MS serves as an energy-storing device used in jumping by alticine Chrysomelidae, diverse Curculionidae, Buprestidae and Bruchidae. MS gives Flea Beetles their common name. See Tibial Extension. Syn. Costa Lima's Organ; Maulik's Organ.

METAGENESIS Noun. (Greek, *meta* = change; *genesis* = birth.) Alternation of generations. Syn. Cyclical Parthenogenesis. Cf. Heterogony.

METAGNATHA Noun. (Greek, *meta* = change; *gnathos* = jaw.) Insects that feed with Mandibles when young and by suction, with tubular mouths, when mature (*e.g.* Lepidoptera). See Feeding Strategies. Cf. Menognatha; Menorhyncha.

METAGONIA Noun. The posterior or anal angle of a wing.

METALA Noun. (Etymology obscure.) The hindwing.

METALDEHYDE A pesticide compound {2,4,6,8-tetramethyl-1,3,5,7-tetraoxacyclooctane} that acts as a stomach poison on snails and slugs. Compound applied in a bait to garden soils, vegetables, tree crops and on ornamentals. Care should be exercised in areas where children play. Phytotoxic to orchids and when in contact with foliage of plants. Toxic to dogs and poultry but not toxic to fishes. Trade names include: Antimilace®, Ariotox®, Cekumeta®, Deadline®, Halizan®, Helarion®, Limatox®, Meta®, Metarex®, Metason®, Mifaslug®, Namekil®, Optimol®, Slugit®. See Pesticide. Cf. Insecticide.

METALLIC FLEA BEETLES Members of the Genus *Altica* spp. [Coleoptera: Chrysomelidae]. See Chrysomelidae.

METALLIC-GREEN TOMATO FLY *Lamprolonchaea brouniana* (Bezzi) [Diptera: Lonchaeidae].

METALLIC SHIELD BUG *Scutiphora pedicellata* (Kirby) [Hemiptera: Scutelleridae].

METALLIC WOOD BORERS See Buprestidae.

METALLIC WOODBORING BEETLES See Buprestidae.

METALLYCTIDAE Plural Noun. A monogeneric

Family with a few Species found in the Indomalayan Realm. Adult body compact with metallic green or blue coloration; Pronotum outline shape nearly square; anal lobe of hindwing moderate sized; forewing indurate; raptorial forelegs strongly spined and proximal outer spine of fore Femur elongate.

METALMARKS See Lycaenidae; Riodinidae.

METALNIKOV Nere 1946, Ann. Inst. Pasteur, Paris 72: 860–861.

METALOMA Noun. (Greek, *meta* = after; *loma* = fringe, margin. Pl., Metalommae.) The sutural or inner margin of the forewings.

METAM® See Metham-Sodium.

METAMERE Noun. (Greek, *meta* = after; *meros* = part. Pl., Metameres.) A primary body segment, Somite or Athromere.

METAMERIC Adj. (Greek, *meta* = after; *meros* = part; *-ic* = of the nature of.) Dividing the body into primary body segments or Metameres. Descriptive of or pertaining to metamerism.

METAMERIC SACS See Osmeteria.

METAMERISM Adj. (Greek, *meta* = after; *meros* = part; English, *-ism* = state.) 1. The condition of a body in which the segments are more-or-less alike. 2. Division of the insect body into primary body segments or Metameres. Alt. Segmentation. Cf. Metameric Segmentation; Zonal Symmetry. Rel. Oligomerization.

METAMORPHOSIS Noun. (Greek, *meta* = change of; *morphe* = form; *sis* = a condition or state.) Third phase of the developmental process experienced by insects. Metamorphosis includes transformation in shape, form or substance during successive stages of development. Metamorphosis is an intricate phenomenon associated exclusively with postembryonic insects. Increase in size and change in shape are important aspects of Metamorphosis; the phenomenon is difficult to define in a way satisfactory to all zoologists. Entomologists typically use the term in a restrictive sense to mean morphological and physiological changes from immature to adult. A more inclusive definition would be 'individual form change.' Depending upon perspective, the study of Metamorphosis can be approached in several ways. Traditionally, Metamorphosis has been divided into three domains: (1) external shape at different stages of development, (2) internal organs with their changes over time and (3) attendant physiological processes associated with physical changes over time. Alternatively, since metamorphic changes are initiated at the cellular level, histological changes may represent a more objective approach to the study of Metamorphosis. See Anamorphosis; Epimorphosis; Heteromorphosis; Histology; Hypermetamorphosis; Holometabolous Metamorphosis; Hemimetabolous; Paurmetabolous Metamorphosis. Larval Heteromorphosis. Cf. Instar; Stadium; Stage. Rel. Development; Growth; Oogenesis; Embryogenesis.

METANOTAL GLAND Males of Oecanthus: A large gland indicated externaly by a deep depression in the Metatergum (Imms).

METANOTAL SLOPES Diptera: Swellings on the sides of the Metanotum or its sloping sides (Pleurotergites) (Curran).

METANOTUM Noun. (Greek, *meta* = after; *noton* = back. Pl., Metanota.) 1. The primitive upper surface of the third or posterior thoracic ring (Metathorax). 2. Diptera: The oval arched part behind, beneath the Scutellum, best developed in flies with a long, slender Abdomen (*e.g*, Tipulidae).

METAPARAPTERON Noun. (Pl., Metaparaptera.) Ants: The Postscutellum (Wheeler).

METAPEDES Noun. (Greek, *meta* = after; *podos* = foot.) The hind legs.

METAPHRAGMA Noun. (Greek, *meta* = after; *phragma* = fence. Pl., Metaphragmata.) The posteriormost internal thoracic septum.

METAPLANTA Noun. (Greek, *meta* = after; *planta* = foot. Pl., Metaplantae.) The second tarsal segment.

METAPLEURAL BRISTLES Diptera: A fan-like row of bristles on the Metapleuron in certain Families (Comstock).

METAPLEURAL GLAND Formicidae: A paired gland with aperture on the posterolateral wall of the Propodeum. MG is a diagnostic feature of ants; found in all forms except male Dorylinae and some parasitic Species. Secretions include phenylacetic acid and Mellein which have fungicidal and bactericidal properties. Syn. Metasternal Gland; Metathoracic Gland. See Gland Cf. Scent Gland.

METAPLEURAL SUTURE A suture separating the Episternum and the Epimeron of the Metathorax in Odonata (Garman).

METAPLEURON Noun. (Greek, *meta* = behind; *pleura* = side. Pl., Metapleura.) 1. Diptera: The Pleuron of the Metathorax. 2. Hymenoptera: The piece behind and below the insertion of the hindwings. 3. In general, the lateral area of the Metathorax. Alt. Metapleurum.

METAPNEUSTIC Adj. (Greek, *meta* = after; *pneuma* = breathe; *-ic* = consisting of.) 1. Diptera: Pertaining the respiratory system of larvae in which only the last abdominal pair of spiracles is open. See Respiration. Cf. Oligopneustic; Polypneupstic.

METAPNYSTEGA Noun. The circular area of the Metanotum behind the Postscutellum.

METAPODEON Noun. (Greek, *meta* = after; *podeon* = neck. Pl., Metapodea.) Hymenoptera: The portion of the Abdomen behind the Podeon (Petiole). See Abdomen. Cf. Gaster; Metasoma.

METAPODOSOMA Noun. (Greek, *meta* = after; *pous* = foot. Pl., Metapodosomata.) Acarology: A part of the Podosoma posterior of the Sejugal Furrow.

METAPOSTSCUTELLUM Noun. (Greek, *meta* = after; Latin, *post* = after; *scutellum* = small shield.)

Pl., Metapostscutella.) Hymenoptera: The Postscutellum of the Metathorax.

METAPYGIDIUM Noun. (Greek, *meta* = after; *pygidion* = rump; *-idion* = diminutive. Pl., Metapygidia.) Dermaptera: The second segment of the Supraanal Sclerite (Tillyard).

METARAN® See Cyhexatin.

METARBELIDAE Plural Noun. A small Family of ditrysian Lepidoptera assigned to Superfamily Cossoidea.

METAREX® See Metaldehyde.

METARHIZIUM ANISOPLIAE A fungal pathogen whose spores are used as a biopesticide for the control of Black Vine Weevil (*Otiorhnychus sulcatus*) and other soil-inhabiting Coleoptera in some countries. Applied in glasshouses and nurseries for ornamental plants and turf. Regarded as non-phytotoxic and not toxic to warm-blooded animals. Trade name Bio 1020®. See Biopesticide; Fungicide. Cf. *Beauveria brongniartii*; *Nosema locustae*; *Verticillium lecanii*.

***METARHIZIUM ANISOPLIAE* ESF1®** A fungal pathogen whose spores are used as a biopesticide for the control of domicillary cockroaches, termites, flies, aphids and whiteflies in some countries. Material is applied in glasshouses for control of pests on ornamental plants and vegetables; special infection chambers are placed in buildings for contamination with spores. Material is regarded as non-phytotoxic and not toxic to warm-blooded animals. Trade names include: Back Off-1®, Bioblast®, Biopath Fly Control®; Biopath Roach Control®. See Biopesticide; Fungicide. Cf. *Beauveria brongniartii*; *Nosema locustae*; *Verticillium lecanii*.

METASCUTELLUM Noun. (Greek, *meta* = after; Latin, *scutellum* = small shield. Pl., Metascutella.) The Scutellum of the Metathorax.

METASON® See Metaldehyde.

METASPIRACLE Noun. (Greek, *meta* = after; Latin, *spiraculum* = air hole. Pl., Metaspiracles.) 1. A spiracle of the Metathorax (MacGillivray). 2. Coccids: The posterior pair of spiracles on the Metathorax (MacGillivray).

METASTERNAL Adj. (Greek, *meta* = after; *sternon* = chest; Latin, *-alis* = pertaining to.) Descriptive of structure or process associated with the Metasternum.

METASTERNAL EPIMERA (Sl., Metasternal Epimeron.) Small sclerites separating the Metasternal Episterna from the ventral segments.

METASTERNAL EPISTERNA Sclerites positioned on each side of the Metasternum, immediately behind the Mesosternal Epimera.

METASTERNAL GLAND See Metapleural Gland.

METASTERNAL WING Some aquatic Coleoptera: A leaf-like expansion above the coxal sclerites.

METASTERNELLUM Noun. (Greek, *meta* = after; Latin, *sternellum* = diminutive of *sternum* = breastbone. Pl., Metasternella.) The Sternellum of the Metathorax.

METASTERNUM Noun. (Greek, *meta* = after; *sternon* = chest. Pl., Metasterna.) The ventral surface of the Metathorax.

METASTETHIDIUM Noun. (Greek, *meta* = after; *stethos* = chest; *-idion* = diminutive. Pl., Metastethidia.) Acarology: The unsclerotized posterior part of the Prodorsum. See Metathorax.

METASTETHIUM Noun. (Greek, *meta* = after; *stethos* = chest; *-idion* = diminutive. Pl., Metastethia.) The Metasternum or ventral surface of the Metathorax.

METASTIGMATA Noun. (Greek, *meta* = after; *stigma*.) The posterior spiracles of the Metathorax (Needham).

METASTOMA Noun. (Greek, *meta* = behind; *soma* = body. Pl., Metastomata.) 1. The posterior most subdivision of the insect body; the Abdomen of most insects. Higher Hymenoptera: Segments of the Abdomen posterior of the Propodeum. See Abdomen; Gaster; Propodeum. Cf. Mesosoma; Prosoma. 2. Orthoptera: The Hypopharynx.

METASUTURE Noun. (Greek, *meta* = after; *sutura*.) The suture of the Metathorax (MacGillivray).

METASYSTEMOX® See Oxydemeton-Methyl.

METASYSTOX-R® See Oxydemeton-Methyl.

METATARSUS Noun. (Greek, *meta* = after; Tarsus = sole of foot. Pl., Metatarsi.) See Basitarsus, Sarothrum.

METATENTORINA Noun. A Gular pit *sensu* MacGillivray.

METATENTORIUM Noun. (Greek, *meta* = after; Latin, *tendere* = to strech; *-ium* = diminutive > Greek, *-idion*. Pl., Metatentoria.) 1. The 'arms' of the Tentorium invaginated on the ventral aspect of the head adjacent to the Foramen and articulation of the Maxillae. 2. The 'Posterior Tentorial Arms' *sensu* MacGillivray.

METATERGUM Noun. (Greek, *meta* = after; *tergum*. Pl., Metaterga.) See Metanotum.

METATHETELY Noun. (Greek, *metathesis* from *metatithenai* = to place differently, to transpose.) A retardation and reduction of the formation of wing pads in otherwise normal insect Metamorphosis (Folsom & Wardle). 2. Larval characters in Pupa or pupal characters in adult (Strickland). See Teratology. Cf. Hysterothetely; Paedogenesis.

METATHION® See Fenitrothion.

METATHORACIC ACCESSORY GLAND Adult Heteroptera: An Exocrine Gland associated with the Metathoracic Scent Gland. See Scent Gland System. Cf. Abdominal Scent Gland; Metapleural Gland.

METATHORACIC SCENT GLAND Adult Heteroptera: An Exocrine Gland that opens between the Mesothoracic and Metathoracic legs and which produces defensive secretions. MSG anatomically complex with branches, accessory glands, secretory tubules, reservoirs and tracheloles. Classified into Omphalian and Diastomian Glands based on anatomy of the glandular orifice. See Diastomian Scent Gland; Omphalian Scent Gland; Scent Gland System. Cf. Abdominal Scent Gland; Metapleural Gland.

METATHORACIC SCUTUM The Scutum of the Metathorax; the Metascutum.

METATHORACOTHECA Noun. The pupal covering of the Metathorax.

METATHORAX Noun. (Greek, *meta* = behind; *thorax* = chest. Pl., Metathoraxes.) The 3rd or posterior-most thoracic segment which bears the hind legs and posterior pair of wings. Segment variable: Distinct, sometimes closely united with the Mesothorax and sometimes appearing as a part of the Abdomen. See Thorax. Cf. Mesothorax; Prothorax.

METATYPE Noun. (Greek, *meta* = after; *typos* = pattern. Pl., Metatypes.) A specimen compared by the author of a Species with the type and determined by him as conspecific with it (Banks & Caudell). Some authors also require the specimen to be topotypic. See Type. Cf. Holotype; Lectotype. Rel. Nomenclature; Taxonomy.

METAZOA Noun. (Greek, *meta* = after; *zoon* = animal. Pl., Metazoans.) In classification, the higher, many-celled animals, including insects.

METAZOIC Adj. (Greek, *metazoa*; *-ic* = characterized by.) Pertaining to the Metazoa or many-celled animals.

METAZONA Noun. 1. Orthoptera: The portion of the Pronotum posterior of the principle transverse Sulcus. 2. The posterior part of the Pronotum. Cf. Prozona.

METCALF, CLELL LEE (1888–1948) (Balduf 1948, J. Econ. Ent. 41: 997–998.)

METCALF, JOHN WILLIAM (1872–1952) (Riley 1953, Proc. R. ent. Soc. Lond. (C) 17: 73.)

METCALF, ZENO PAYNE (1885–1956) (Smith 1956, Ann. ent. Soc. Am. 49: 302.)

METCHNIKOFF (MECHNIKOV), ILYA ILICH (1845–1916) (Plá 1916, Mems Soc. cubana Hist. nat. 'Filipe Poey' 2: 228–234, bibliogr.)

METELKA, FERENCZ (1814–1885) (Vángel 1885, Rovart Lap. 2: 129–133.)

METEPIMERON Noun. (Greek, *meta* = after; *epi* = upon; *meros* = part. Pl., Metepimera.) Odonata: The part of the Epimeron behind the second Lateral Suture, extending ventrally to the Sternum (Smith). The part of the Epimeron on the dorsal side of the Epimeral Suture, Pteropleura, Aepimeron, Pteropleurite (MacGillivray).

METEPISTERNUM Noun. (Greek, *meta* = after; *epi* = upon; *sternon* = chest. Pl., Metepisterna.) Odonata: The sclerite between the first and second lateral thoracic sutures.

METEX® See Amitraz.

METHACRYLATES Plural Noun. A category of plastic compounds used as embedding media for microtomy and electron microscopy. Typically employed as mixtures of methyl and butyl methacrylates, with medium hardness determined by relative proportions of constituent elements; pure methyl methacrylate is the hardest. Advantages of methacrylates include rapid infiltration and ease of sectioning; disadvantages include distortion due to shrinkage. Methacrylates are useful for examining lipid-soluble substances. See Histology. Cf. Epoxy Resin; Polyesters. Rel. Embedding.

METHAMIDOPHOS An organic-phosphate (thiophosphoric acid) compound {O,S-dimethyl phosphoramidothioate} used as an acaricide and systemic insecticide against mites, thrips, armyworms, loopers, cutworms and plant-sucking pests. Compound is applied to cotton, potatoes, soybeans, sugar beets and vegetables. Methamidophos is toxic to bees, wildlife and fishes. Trade names include: Filitox®, Hamidop®, Monitor®, Nitofol®, Nuratron® Patrole®, Rometa®, Swipe®, Tam®, Tamanox®, Tamaron®, Vetaron®. See Organophosphorus Insecticide.

METHAM-SODIUM A soil fumigant {Sodium N-methyldithiocarbamate dihydrate} used for control of bacteria, fungi, nematodes, weeds and soil insects. Target pests include Verticillium, *Pythium*, *Phytophthora*, Grape Phylloxera and wireworms. Compound is toxic to fishes but is not flammable. Trade names include: Vapam®, VPM®, S.M.D.C.®, Metam®, Soil-Prep®, Karbation®, Maposol®, Polefume®, Vaporooter®, Busan 1020®, Trimaton®, Sistan®, Unifume®, Sectagon II®, Nemasol®, Ecetam® See Fumigant.

METHAVIN® See Methomyl.

METHIDATHION An organic-phosphate (dithiophosphate) compound {O,O-dimethyl phosphorodithioate S-ester with 4-(mercaptomethyl)-2-methoxy-1,3,4-thiadiazolin-5-one} used as an acaricide and contact insecticide for control of mites, plant-sucking insects (to include aphids, scale insects, spittle bugs), thrips, leaf-eating insects and borers. Compound is applied to fruit trees (apples, almonds, apricots, cherries, citrus, mangoes, peaches, pears, pecans, plums, prunes, walnuts) and ornamentals in USA and other crops in many countries. Methidathion is phytotoxic on some plums, apples, almonds, sorghum and tobacco, and is toxic to bees, fishes and wildlife. Trade names include: Supracide®, Supracidin®, Suprathion®, Ultracide®. See Organophosphorus Insecticide.

METHIOCARB A carbamate compound {3,5-dimethyl-4-(methylthio) phenyl methylcarbamate} used as an Acaricide, Molluscicide, contact insecticide and stomach poison. Methiocarb is applied to non-bearing fruit tree crops, ornamentals, turf and citrus in USA; also applied to grapes, cotton, corn, vegetables and ornamentals in some countries. Compound is effective against mites, plant-sucking insects (aphids, scale insects, leafhoppers, psyllids, whitelfies), mosquitoes, grasshoppers, thrips, and related Species. Trade names include: Club®, Draza®, Decoy®, Grandslam®, Mesurol®,

Metmercapturon® See Carbamate Insecticide.

METH-O-GAS® See Methyl Bromide.

METHOMYL A carbamate compound {S-methyl N-(methylcarbamoyl) oxy) thioacetimidate} used as an Nematicide, contact insecticide and stomach poison against aphids and foliage feeding insects; some ovicidal activity. Metomyl is applied to non-bearing fruit tree crops, cotton, corn, field crops, grapes, ornamentals, vegetables and turf. Methomyl is phytotoxic to some varieties of apples and some vegetables under specific conditions. Compound is applied as a bait for the control of adult flies and is toxic to bees, fishes and birds. Trade names include: Kipsin®, Kuik®, Lannate®, Lanox®, Marlin®, Methavin®, Metox-900®, Nudrin®, Stimukil®.

METHOPRENE An organic compound {Isopropyl (2E,4E)-11-methoxy-3,7,11-trimethyl-2,4-dodecadienoate} marketed as an Insect Growth Regulator for control of glasshouse pests (cucumber beetle, leaf miners, leaf hoppers) and stored product pests (ants, cigarette beetle, tobacco moth) and related insects. Methoprene also is used as a public health insecticide for control of urban ants, fleas, lice, mosquito larvae and maggots in manure. Methoprene used in warehouses, food processing plants and grain storage facilities. Compound also applied to rice, pasture, mosquito breeding sites, and fed to livestock for control of manure-infesting flies. Effective at low concentrations and particularly effective on flies; effective on ants when used in a bait. Methoprene is toxic to fishes and Crustacea; non-persistent and should not be combined with other insecticides. Trade names include: Altosid®, Apex®, Diacon®, Dianex®, Manta®, Minex®, Ovitrol®, Pharorid®, Precor®. See Insect Growth Regulator.

METHOXCIDE® See Methoxychlor.

METHOXO® See Methoxychlor.

METHOXYCHLOR A chlorinated hydrocarbon {1,1,1-trichloro-2,2-bis (4-ethoxyphenyl) ethane} insecticide closely related to DDT. Phytotoxic on Chinese elm, some maples and redbuds; toxic to bees and fishes. Compound is used on many agricultural crops and as a public-health pesticide against flies, lice and ticks. Syn. DMDT. Trade names include: Marlate®, Methoxcide®, Methoxo®. See Organochlorine Insecticide.

METHRIN® See Fenpropathrin.

METHYL ALCOHOL See Alcohol, Methyl.

METHYL BROMIDE A fumigant (CH_3Br) used against structural pests and in the soil for agricultural pest control. MB is considered highly effective due to small molecules with effective penetration and very quick action. Fumigated commodities include: Fruits, grains, nuts, sorghum, strawberries and vegetables. Space-fumigation for agricultural builidings, boxes, grainaries and greenhouses. MB is non-flammable, non-explosive, odourless liquid that boils at 4°C. Sometimes mixed with chloropicrin (as a warning odour) if not exposed to food or drugs. Linked to health problems and ozone depletion from the atmosphere; a chronic poison that causes blistering in humans. Material composed of sponge rubber, foam rubber, reclaimed rubber, fur, horeshair, feathers, leather goods, woollens, angora and viscose rayon should not be treated with methyl bromide. The compound displays extreme phytotoxicity. Trade names include: Bromomethane®, Celfume®, Kayafume®, Profume®, Meth-o-Gas®, Tri-Brome®, Emba-Fume®, Meabrome®, Metabume®. See Fumigant. Cf. Sulphuryl Fluoride.

METHYL EUGENOL A parapheromone used to monitor and attract Oriental Fruit Fly. See Pheromone.

METHYLLYCACONITINE An alkaloid biopesticide extracted from the seeds of *Delhinium*. See Biopesticide. Cf. Nereistoxin; Nicotine; Rotenone; Pyrethrin; Pyrroles; Ryania; Sabadilla; Stemofoline.

METHYL-PARATHION See Insecticide.

METICULOSE Adj. (Latin, *meticulosus* from *metus* = a fear; *-osus* = full of.) Flame-like. Term applied to a maculation in the form of a series of coloured flames. Alt. Meticulosus.

METINFRAEPISTERNUM Noun. (Greek, *meta* = after; *infra* = below; *epi* = upon; *sternon* = chest.) Odonata: The sclerite just above the base of the hind Coxa, below the Metepisternum and anterior of the Metepimeron.

METMERCAPTURON® See Methiocarb.

METOCHY Noun. (Greek, *metoche* = sharing.) 1. The relationship between a 'neutral' guest and its host. 2. The symbiotic relationship with ants in which the tolerated guests (inquiline) in an ant nest demands and receives nothing from the ants. See Synecthry; Symphily.

METOL See Alcohol; Methyl.

METOLCARB A carbamate compound {3-m-tolyl-n-methylcarbamate} used as a contact/systemic insecticide with some vapour activity. Metolcarb is applied to citrus, cotton, rice and vegetables in some countries. Compound is not registered for use in USA. Metolcarb is effective against plant-sucking insects (aphids, mealybugs, scale insects, plant hoppers, leaf hoppers, psyllids, whitelfies) thrips and codling moth. Trade names include Metacrate®, MTMC® and Tsumacide®. See Carbamate Insecticide.

METOPIC SUTURE See Coronal Suture.

METOPIDIUM Noun. (Greek, *metopion* = forehead; *-idion* = diminutive. Pl., Metapodia.) Membracidae: The anterior declivous surface of the Prothorax.

METOX-900® See Methomyl.

METRE Noun. (Greek, *metron* = measure. Pl., Metres.) The standard of length in the metric system; equivalent to 39.37 inches. See Centimetre. Millimetre. Alt. Metre.

METRO® See Fonofos.

METROSIS Noun. (Greek, *meter* = mother; *-osis* =

condition. Pl., Metroses.) Social Insects: The number of queens which establish a new colony. See Haplometrosis; Pleometrosis.

METZGER, ANTON (1832–1914) (Rebel 1914, Verh. zool.-bot. Ges. Wien 64: 164–167, bibliogr.)

METZKY, WITBURG (–1951) (Reisser 1951, Z. wien. ent. Ges. 36: 2–3.)

MEUCHE, ALFRED (1913–1942) (Sachtleben 1943, Arb. physiol. angew. Ent. Berl. 10: 260.)

MEULEN, GREELT SIJBRANDI ANTON VAN DER (1883–1970) (Lempke 1970, Ent. Ber., Amst. 30: 209–210.)

MEUNIER, FERNAND ANATOLE (1868–1926) Belgian entomologist and curator at Antwerpener Tiergartens. Meunier described many insects found in Baltic amber. (Anon. 1926, Bull. Soc. ent. Fr. 1926: 65.)

MEUNIER, STANISLAUS (1843–1925) (Dollfus 1925, Bull. Soc. Sci. nat. Reims 44: 175–179.)

MEVES, FRIEDRICH WILHELM (1814–1891) (Sandahl 1891, Ent. Tidskr. 12: 81–86.)

MEVES, JULIUS SEELHORST (1844–1926) (Aurivillius 1926, Ent. Tidskr. 47: 248–251, bibliogr.)

MEVINPHOS An organic-phosphate (thiophosphoric acid) compound {(2-carbomethyoxy-1-methylvinyl dimethyl phosphate} used as an acaricide and contact insecticide with systemic activity against mites, thrips, plant-sucking insects, armyworms, loopers, cutworms, leaf-miners, grasshoppers and other insects. Compound applied to vegetables, watermelons, grapes and strawberries. Rapid-acting and does not accumulate in soil. Toxic to bees, wildlife and fishes. Trade names include: Duraphos®, Finiphos®, Medvidrin®, Menite®, Phosdrin®, Phosfene®. See Organophosphorus Insecticide.

MEXICAN BEAN BEETLE *Epilachna varivestis* Mulsant [Coleoptera: Coccinellidae]: A multivoltine pest of many legume crops in the southeastern USA. Adult and larval stages defoliate plant with heavy infestations causing damage to stems and pods. Adults are 6–9 mm, oval in outline-shape with body yellow to copper-coloured and three longitudinal rows of dark spots on Elytra. MBB overwinters as an adult in crop trash, fissures in soil and beneath rocks; Species becomes active during spring. Females lay ca 500 eggs during lifetime; eggs are deposited in clusters of ca 50 on underside of leaves. Eggs are yellow-orange, ca 1 mm and elliptical. Larvae undergo four instars while feeding on leaves; mature larvae ca 8 mm long, yellow and oval; dorsum with six longitudinal rows of black-tipped spines. Pupation occurs on underside of undamaged leaves; Pupa attach to leaf surface, partially invested in last larval Integument with exposed area of pupal Integument yellow, smooth and rounded; pupal stage requires ca 10 days. Syn. *Epilachna corrupta* Mulsant. Bean Lady Beetle. See Coccinellidae.

MEXICAN BEAN WEEVIL *Zabrotes subfasciatus* (Boheman) [Coleoptera: Bruchidae]: A widespread pest of stored legumes; endemic to South and Central America and adventive in tropical Africa, Indian and Mediterranean regions. Adults are 2.0–2.5 mm long; oval in outline shape; Antenna long; hind Femur without spines. Eggs adhere to pods or testas of beans. Larva bores into pods and consumes seeds. MBW life cycle requires ca 25 days. See Bruchidae.

MEXICAN BLACK SCALE *Saissetia miranda* (Cockerell & Parrott) [Hemiptera: Coccidae].

MEXICAN CORN LEAFHOPPER *Dalbulus elimatus* (Ball) [Hemiptera: Cicadellidae]: A significant pest of maize in parts of Latin America where it transmits stunting pathogens. See Cicadellidae.

MEXICAN CORN ROOTWORM *Diabrotica virgifera zeae* Krysan & Smith [Coleoptera: Chrysomelidae]: An important pest of corn (*Zea mays* Linnaeus) in Mexico and southern Texas; MRC is closely related to the Western Corn Rootworm. Adults are ca 9 mm long, yellow to pale green with black spots on Elytra. Adults oviposit in soil during summer and autumn; eclosion occurs during April. Larvae are indistinguishable from Northern Corn Rootworm; mature larvae ca 12 mm long with dark spot on dorsum of apical abdominal segment. Larvae feed upon corn roots; adults prefer corn silks, but will forage on other plant parts. See Chrysomelidae.

MEXICAN FRUIT FLY *Anastrepha ludens* (Loew) [Diptera: Tephritidae]: The most important pest of fruit in Mexico and a major pest of citrus and other tree crops in Mexico, Belize and Guatemala. MFF is the object of eradication in Mexico and quarantine in North America. See Tephritidae. Cf. Apple Maggot; Cherry Fruit Fly; Mediterranean Fruit Fly; Melon Fly; Oriental Fruit Fly; Papaya Fruit Fly; Queensland Fruit Fly.

MEXICAN LEAFROLLER *Amorbia emigratella* Busck [Lepidoptera: Tortricidae].

MEXICAN MEALYBUG *Phenacoccus gossypii* Townsend & Cockerell [Hemiptera: Pseudococcidae].

MEXICAN PINE BEETLE *Dendroctonus approximatus* Dietz [Coleoptera: Scolytidae].

MEXICAN RICE BORER *Eoreuma loftini* (Dyar) [Lepidoptera: Pyralidae].

MEXICAN SULPHUR *Eurema mexicana* [Lepidoptera: Pieridae].

MEXIDE® See Rotenone.

MEYER, CHRISTIAN ERICH HERMANN VON (1801–1869) (Zittel 1872, J. Zool., Paris 1: 95–96.)

MEYER, FELIX (1853–1926) (Pfaff *et al.* 1934, 50 Jahr Bestchan. Int. ent. Verh. Frankf. (Festschrift) p 7.)

MEYER, GEORG FERDINAND (1860–1947) (Albers 1954, Verh. naturw. Heimatforsch 31: xiii–xiv.)

MEYER, GUSTAV (–1958) (Anon. 1958, Z. Wein. ent. Ges. 43: 136.)

MEYER, JULIUS E (dos Passos 1951, J. N. Y. ent.

Soc. 59: 138.)

MEYER, NICOLAS (1734–1775) (Füessly 1779, Magazin Liebh. Ent. 2: 51–65.)

MEYER, PAUL (1876–1951) (Scheerpeltz 1952, Ent. Bl. Biol. Syst. Käfer 47–48: 113–119, bibliogr.)

MEYER, WILHELM (1854–1935) (Urbahn 1935, Stettin. ent. Ztg 96: i–iv.)

MEYER-DARCIS, GEORGES (–1914) (Doebeli 1914, Mitt. schweiz. ent. Ges. 12: 313–316.)

MEYER-DÜRR, L. RUDOLF (1812–1885) (Stierlin 1885, Mitt. schweiz. ent. Ges. 7: 170–181, bibliogr.)

MEYERINCK, HEINRICH EUGENE VON (1786–1848) (Ratzeburg 1874, Forstwissenschaftliches Schriftsteller -Lexicon 1: 357–360.)

MEYRICK, EDWARD (1854–1938) English educator and insect taxonomist specializing in Microlepidoptera. Meyrick spent his early professional years in Australia and New Zealand. He described ca 15,000 Species of Lepidoptera and his collection is housed in BMNH. (Hull 1939, Obit. not. Fell. R. Soc. Lond. 3: 531–548, bibliogr.; Robinson 1986, 20: 359–367.)

MEZIRIDAE A small Family of Hemiptera placed within the Aradidae in some classifications.

MGP A stored-product fumigant (Mg_3P_2 + $6H_2O$ $3Mg(OH)_2$ + $2PH_3$) used as a space-fumigant in warehouses, mills and elevators for the control of stored-product beetles and moths. Trade names include Fumi-Cel®, Fumi-Strip®, Magtoxin®, Detiaphos®. See Fumigant.

MIALL, LOUIS COMPTON (1843–1921) (Calvert 1921, Ent. News 32: 191–192, bibliogr.)

MICANS Adj. (Latin, *micans* from *mico* = to sparkle.) Shining.

MICELLAE The elongate, submicroscopic, crystalline parts of the Chitin fibres lying parallel with their axes (Snodgrass).

MICHAEL, OTTO (1859–1934) (Wrede 1934, Ent. Z. Frank. a. M. 48: 137–138, bibliogr.)

MICHAELSEN, WILHELM (1860–1937) (Anon. 1937, Der Biologie 6: 144.)

MICHAILOVITCH, NICOLAS (–1919) (Moreau 1919, Bull. Soc. ent. Fr. 1919: 69.)

MICHEL, JOSEPH (1890–1963) (Z. Wien. Ent. Ges. 49: 67–68.)

MICHELBACHER, ABRAHAM EZRA (1899–1991) American academic (University of California, Berkeley) and author of 280 articles involving taxonomy of *Symphyla* and insect pest control. (Chemsak *et al.* 1992, 68 (2): 225–242, bibliogr.)

MICHELET, JULES (1795–1874) (Anon. 1874, Nature 2: 214–215.)

MICHELI, LUCIO (1887–1951) (Anon. 1951, Boll. Soc. ent. ital. 81: 1.)

MICHENER, CHARLES DUNCAN (1919–) American academic (University of Kansas) specializing in biology, social behaviour, and systematics of Apoidea. Watkins Distinguished Professor of Systematics and Ecology. Elected Member, United States National Academy of Sciences.

MICRALIFERA Noun. The minute sclerite located posterior of the Postalifera (MacGillivray).

MICRANER Noun. (Greek, *mikros* = small; *aner* = male. Pl., Micraners.) Social Insects: The smallest size-class in Species with more than one size-class of male. A dwarf male ant. Cf. Macraner.

MICRERGATE Noun. (Greek, *mikros* = small; *ergates* = worker. Pl., Micrergates.) Social Insects: The smallest size class in Species with more than one size class of workers. A dwarf worker ant. Alt. Microergate.

MICROBIAL INSECTICIDE 1. A biological preparation of viruses, bacteria, fungi or other microorganism (or their products) applied and used in ways similar to conventional chemical insecticides and intended to kill, eliminate or destroy pest plants, invertbrates or vertebrates. Cf. Biopesticide; Microbial Pesticide. 2. Toxins, typically proteins, produced by microorganisms formulated into insecticides. Cf. *Bacillus thuringiensis*. Alt. 'Bt'.

MICROCADDIS FLIES See Trichoptera.

MICROCADDISFLIES See Hydroptilidae.

MICROCARB® See Carbaryl.

MICROCARRIER Noun. Tissue Culture: A microscopic particle capable of supporting attachment and growth in cells which require anchorage.

MICROCHAETA Noun. (Greek, *mikros* = small; *chaete* = hair. Pl., Microchaetae.) Any small bristle or Setae. Cf. Macrochaetae.

MICROCORIA Noun. (Greek, *mikros* = small; Latin, *corium* = skin.) Coccids: The Coria of the Microthorax (MacGillivray).

MICROCORYPHIA Noun. (Greek, *mikros* = small; *corpyphia* = head.) See Archaeognatha.

MICROENCAPSULATED CONCENTRATE A form of insecticide formulation in which the active ingredient (AI) is incorporated into a permeable covering (small spheres). See Aerosol; Dust; Emulsifiable Concentrate; Microencapsulated Concentrate; Oil Concentrate; Suspension Concentrate; Wettable Powders.

MICROENCAPSULATION Noun. (Greek, *mikros* = small; + encapsulation.) A formulation technique for fertilizer or pesticides in which chemical-containing, permeable microscopic spheres or capsules are created to permit release at a slow and consistent rate. Microencapsulation process was first developed by Pennwalt Corporation to increase residual activity of produce while reducing volatility and mammalian toxicity. Process involves mechanical dispersion of hydrophobic pesticide in an aqueous medium. Pesticide contains a soluble monomer (*e.g.* diacid chloride); when droplets reduced to appropriate size (ca 35–50 µ), a water soluble monomer (diamine) is added and polymerization occurs at droplet interface. Polymer wall can be altered through cross-linking with polyfunctional isocyanate. Release rate can be varied by the nature of the monomer, degree of cross-linking, thickness of the capsule wall and capsule size. Rel. Insecticide; Pesticide.

MICROGAMETE Noun. (Greek, *mikros* = small; *gametes* = spouse.) The small linear bodies (male elements) derived from the civrogametocytes of the malarial protozoan.

MICROGAMETOCYTE Noun. (Greek, *mikros* = small; *gametes* = spouse; *kytos* = hollow vessel; container.) The small-sized gametocyte of *Plasmodium*; the male sex-cell developed from the merozoite of the malarial protozoan.

MICROGENUALA See Microseta.

MICROGERMIN-F® See *Verticillium lecanii.*

MICROGERMIN-G® See *Verticillium lecanii.*

MICROGYNE Noun. (Greek, *mikros* = small; *gyne* = woman. Pl., Microgynes.) The smallest size class of queen in ants that have more than one size class of queen. A dwarf queen or female ant. See Gyne. Cf. Macrogyne; Micraner.

MICROLEUCOCYTE Noun. (Greek, *mikros* = small; *leuco* = brownish; *kytos* = hollow vessel; container. Pl., Microleucocytes.) A small form of leucocyte found in the insect Haemolymph. See Leucocyte. Rel. Haemolymph.

MICROMALTHIDAE Plural Noun. A monotypic, rarely collected Family of Coleoptera of uncertain placement that has been assigned to the Archostemata or Polyphaga. One Species (*Micromalthus debilis* LeConte) occurs in North America and inhabits decaying logs; the Species is paedogenic with larvae reproducing parthenogenetically. The Species has been transported to other areas. See Coleoptera.

MICROMITE® See Diflubenzuron; Fenitrothion.

MICRON Noun. (Greek, *mikros* = small. Pl., Micra.) A unit of microscopic measurement, 0.001 mm; represented by symbol μ; symbol μμ represents 0.001 of a micron.

MICRONUCLEOCYTE See Plasmatocyte.

MICROPEZIDAE Plural Noun. Stilt-Legged Flies. A widespread Family of schizophorous Diptera assigned to Superfamily Micropezoidea (Nerioidea). Adults are moderate sized and elongate; Ocellar Bristles minute or absent; legs exceptionally long; Costa unbroken; CuA straight; first Posterior Cell of wing (R5) apically narrow. Adults inhabit moist habitats; larvae live in excrement, decaying wood or other plant material. Syn. Calobatidae; Trepidariidae; Tylidae. See Micropezoidea.

MICROPEZOIDEA A Superfamily of schizophorous Diptera including Cypselosomatidae, Megamerinidae, Micropezidae (Calobatidae, Tylididae), Neriidae and Pseudopomizidae.

MICROPHAGOUS Adj. (Greek, *mikros* = small; *phagein* = to devour; Latin, *-osus* = with the property of.) Descriptive of organisms feeding on small particles. See Feeding Strategy.

MICROPHAGY Noun. (Greek, *mikros* = small; *phagein* = to devour.) The act of feeding on microorganisms.

MICROPHYSIDAE Dohrn 1859. Plural Noun. A small Family of heteropterous Hemiptera assigned to Superfamily Miroidea; Microphysidae include about 30 described Species. Adult body very small, oval-oblong and dorsoventrally flattened; head projects forward; Ocelli present; wings broad with lateral margin expanded; Cuneus present; wing membrane with small cell but lacking veins; Tarsi with two segments. See Miroidea.

MICROPORE Noun. (Greek, *mikros* = small; *poros* = channel.) Coccids: Oraceratuba (MacGillivray).

MICROPTERISM Noun. (Greek, *mikros* = small; *pteron* = wing; English, *-ism* = condition. Pl., Micropterisms.) 1. The state of having small wings. 2. The tendency to produce small wings. Alt. Microptery.

MICROPTEROUS Adj. (Greek, *mikros* = small; *pteron* = wing; Latin, *-osus* = with the property of.) Small winged. A condition in which the wings of an insect are disproportionately small and presumably less effective as instruments of flight. Alt. Microptery. See Wings. Cf. Brachypterous; Macropterous.

MICROPTERYGIDAE Herrich-Schaeffer 1855. Plural Noun. Mandibulate Moths. A cosmopolitan Family of primitive Lepidoptera, consisting of about 120 Species. Adult body small; head with raised piliform scales; Ocelli and Chaetosemata present; Antenna moniliform to filiform, with ascoid Sensilla on flagellar segments; Mandibles present, dicondylic, functional; Paraglossae, premental lobes on Labium; tibial spur formula 0-0-4; wing venation homoneurous, Sc and R1 somtimes forked; jugate wing coupling, hindwing-body lock present. Larvae are slug-like with body rugose and hexagonal in cross section; head prognathous and retractable; Spinneret absent; 5–6 Stemmata; Antenna with three segments; thoracic legs with 3–4 segments, each with apical claw; Prolegs non-musculated, without Crochets, often on abdominal segments 1–8. Pupation occurs within tough, silken cocoon; Pupae decticous, exarate; Mandibles used to escape cocoon. Adults inhabit forests; diurnal, feed on pollen and spores; larval diet diverse, feed on moss, liverwort, fungal hyphae, leaf litter. Earliest fossil from Lebanese amber (Lower Cretaceous Period). Syn. Eriocephalidae; Micropterigidae.

MICROPTERYGOIDEA Micropterigoidea. A Superfamily of zeuglopterous Lepidoptera containing only Microperygidae.

MICROPYLAR Adj. (Greek, *mikros* = small; *pyle* = gate.) Descriptive of or pertaining to the micropyle of the insect egg.

MICROPYLE Noun. (Greek, *mikros* = small; *pyle* = gate. Pl., Micropyles.) A minute pore or aperture in the Chorion of the insect eggshell, through which Spermatozoa pass during fertilization. Micropyle's position on eggshell surface is variable, but often it occurs on anterior pole of egg. In some insects, Micropyle can assume different shapes among different Taxa of insects. In housefly, Micropyle is funnel-shaped and surrounded by an elevated collar. Micropyle of beetle *Dineutes horni* is stalked in a depression at anterior end of egg. Micropyle can form a complex of apertures in eggshell. Complex can consist of three parts: Micropyle (aperture) through which Spermatozoa migrate; Sperm Guide (depression) in Chorion's surface that typically lacks sculpture and facilitates Spermatozoon penetration of Micropyle; and Micropylar Canal (Tube) leading from Micropyle into egg. See Sperm Guide. Cf. Aeropyle; Hydropyle; Operculum. Rel. Fertilization.

MICROSATELLITE Noun. (Greek, *mikros* = small; Old French from Latin, *satelles* = an attendant; *-ites* = resident. Pl., Microsatellites.) Molecular Biology: Genetic markers which consist of tandemly repeated short nucleotide sequences which often show unusual G + C composition.

MICROSCULPTURE OF THE TEGUMENT Acarology: fine, superficial markings of the Tegument.

MICROSETA Noun. (Greek, *mikros* = small; Latin, *seta* = bristle. Pl., Microsetae.) 1. Acari: Minute, distodorsal Seta on Genu and Tibia I of Trombiculidae and Genu II of Leuwenhoekiinae. Optically active under cross-polarized light; function unknown. Syn. Microgenuala. (Goff *et al.* 1982, J. Med. Ent. 19: 227.) 2. Diptera: Extremely short, dense Setae on the legs. See Seta. Cf. Bristle.

MICROSOMITE Noun. (Greek, *mikros* = small; *soma* = body. Pl., Microsomites.) Any small, secondary ring or Somite of a Macrosomite in the embryo, which subsequently becomes a body segment. See Somite.

MICROSPORIDAE Plural Noun. A small, widespread Family of myxophagous Coleoptera. Adult less than 1 mm long, globose, shiny; Antenna with 11 segments including Club with three segments; fore Coxa globose; frontal cavity open; middle Coxae contiguous; tarsal formula 3-3-3; Abdomen with three Ventrites. Syn. Sphaeriidae.

MICROSPORIDIA Plural Noun. (Greek, *mikros* = small; *sporos* = seed. Sl., Microsporidium.) A group of microscopic, unicellular, spore-forming organisms sometimes placed within Protozoa or Fungi. Many Microsporidia develop are symbiotic in insects while others are parasitic and induce chronic diseases in insects. Cf. *Nosema*; Pebrine.

MICROTARSALA See Famulus.

MICROTHORAX Noun. (Greek, *mikros* = small; *thorax* = breast plate. Pl., Microthoraxes.) Odonata: A minute division of the cephalic end of the Thorax (Garman). A term for the neck (Cervix) of insects, so employed in the view that the Cervix is a reduced body segment (Snodgrass).

MICROTIBIALA See Microseta.

MICROTOME Noun. (Greek, *mikros* = small; *tome* = cutting. Pl., Microtomes.) A precision instrument designed for cutting matter into thin sections for microscopical examination. Term proposed by Chevalier (1839) for the device invented by Cummings and originally called a 'cutting engine'. Modifications of the principal device include: Rocking Microtome, Rotary Microtome, Sliding Microtome and Ultra Microtome.

MICROTOMY Noun. (Greek, *mikros* = small; *tome* = cutting. Pl., Microtomies.) The cutting of tissue, cells or structures into thin sections (< 10 μ) for microscopic analysis.

MICROTRICHIUM Noun. (Greek, *mikros* = small; *-trichia,* combining form, Greek, *thrix, trichos* = hair. Pl., Microtrichia.) Small, sclerotized non-innervated cuticular projects on the body and wings of insects; Microtrichae are also found in the Tracheae. 1. Thysanoptera: Hair-like structures on the Abdomen. 2. Mecoptera and Diptera: Minute, hair-like structures found on the wings which resemble small covering Setae, but lack the basal articulating socket (Tormogen). 3. Lepidoptera: minute hair-like spines on the wing or Cuticle of primitive moths. Syn. Fixed Setae; Aculei. 4. Plecoptera: Cuticular projections of unknown function (Kapoor 1985, 63: 1360–1367.) See Seta. Cf. Macrotrichia. Rel. Acantha; Ctenidium.

MICROTUBULE Noun. A complex of microscopic tubes inside a sperm tail, including Central Tubules, Accessory Tubules (Doublets) and Peripheral Tubules (Singlets). Axoneme Microtubules typically are parallel to one another and the Flagellum's primary (long) axis. Exceptions to this arrangement are known in unrelated groups of insects. Tubules form an extension of Centriole's Basal Body. Each Microtubule consists of two principal components: subfibre A and subfibre B. Subfibre A is synthesized first and is composed of 13 protofilaments that comprise wall of each Microtubule. Subfibre B is synthesized later and is composed of 10–11 protofilaments that comprise its wall. Distance between doublets ca 22 nm. Other components of Axoneme Microtubules include dynein arms and Radial Spokes. Central Tubules, Accessory Doublets and Peripheral Singlets of axoneme have been numbered and a formula describes Microtubule configuration. Formula '9+2' means that axoneme contains nine Accessory Doublets and two Central Tubules. Formula '9+0' means that Spermatozoon contains nine Accessory Doublets but lacks Central Tubules. Peripheral Singlets also are noted in formula (*e.g.* 9+9+2). Most apterygotes display '9+2' axonemal pattern which is basic to this ancestral group. Notable exceptions to this arrangement occur in Protura which contain 12 doublets and lack central tubules (12 + 0, *Acerentulus*) or 14 accessory doublets without central tubules (14 + 0, *Acerentomon*). Formula '9+9+2' generally regarded as groundplan of Axoneme structure in pterygote insects. This pattern was first noted by Rothschild (1955) for the honeybee, Yasuzumi (1956) for *Drosophila* and Kaye (1970) for the cricket. Variations in groundplan appear sporadically throughout Pterygota and ATs are also absent from tailed

Spermatozoa of Mecoptera, some Trichoptera and Thysanoptera. See Axoneme; Sperm Flagellum. Cf. Centriole Adjunct; Dynein Arms; Radial Spokes.

MICROTYPE EGG Typically minute, variable in shape, with dorsal and lateral surfaces thick and dark, ventral surface thin and membranous. Embryonic development occurs in the Uterus. Egg must be consumed by the host if development is to proceed. Stimulus for hatching is unknown. Microtype eggs are widely distributed among Tachinidae. See Egg. Cf. Macrotype Egg.

MIDDENDORFF, ALEXANDER THEODOR (1815–1894) (Ratzeburg 1874, Forstwissenschaftliches Schriftsteller -Lexicon 1: 360–370, bibliogr.)

MIDDLE APICAL AREA See Internal Area.

MIDDLE APICAL FIELD See Discoidal Field.

MIDDLE APICAL LOBES Orthoptera: See Lobes.

MIDDLE APICAL PLATE The median strip of cells in the germ band between the lateral sclerites (Snodgrass).

MIDDLE APICAL PLURAL AREA Hymenoptera: The median of the three areas between the lateral and pleural Carinae, the second pleural area.

MIDDLE CLASPER Hymenoptera: See Digitus.

MIDDLETON, BERTRAM LINDSAY (–1950) (Anon. 1951, Proc. Linn. Soc. N. S. W. 76: v.)

MIDDORSAL THORACIC CARINA A ridge or elevated line at the meeting of the Mesepisterna in Odonata.

MIDGES See Chironomidae. Cf. Biting Midges.

MIDGULAR SUTURE The single Mesial Suture when the Gula is infolded and concealed. MS formed through the fusion of the two Gular Sutures (MacGillivray).

MIDGUT Noun. (Middle English, *middel; gut*. Pl., Midguts.) The region of the Alimentary System between the Foregut and Hindgut. Adult insects typically display a midgut that is long, straight and confined to the thoracic region. The midgut of many insects often contains differentiated, blind pouches (Caecae) where food is stored and symbiotic microorganisms reside, and other regions where food is subjected to digestive enzymes and absorption. The midgut of haematophagous insects usually initiates Vitellogenesis and Oogenesis. The midgut may be involved in intermediary metabolism and also serves a role in excretion by breaking down Haemoglobin and discharging it into the midgut lumen. The anterior part of the midgut in Acridoidea (grasshoppers) contains six Caecae at the junction with the foregut. Each Caecum includes an anterior arm which projects forward along the foregut and a posterior arm which extends rearward. Typically, the Epithelium of the anterior arms is thrown into several longitudinal folds which are important in digestion. The Epithelium of the posterior arms is not folded and therefore contributes less surface area to the Caecum. Some Species have a specialized 'pocket region' at the base of the posterior arm. The midgut lacks the Cuticular Intima found in the foregut. The Epithelium of the midgut consists of Columnar Cells (differentiated, lipophilic, cuboidal cells), Regenerative Cells and sometimes Endocrine Cells (clear, granular, secretory cells.) The pH of the midgut typically ranges between 6.0–7.5; Scarabaeidae, Nematocera, higher Isoptera and Lepidoptera have extraordinarily alkaline midgut with a pH of 9–12, necessary for the extraction of hemicellulose from plant cell walls. Cyclorrhaphous Diptera midgut pH ranges from 2.8–3.5; Heteroptera midgut pH ranges from 4.5–5.0. Syn. Mesenteron. See Alimentary System. Cf. Foregut; Hindgut. Rel. Digestion.

MIDINTESTINE See Midgut.

MIDRIB Noun. (Middle English, *middel; ribbe*. Pl., Midribs.) The central longitudinal rib of a leaf.

MIEG, JUAN (1779–1859) (Dufour 1861, Ann. Soc. ent. Fr. (4) 1: 17–20, bibliogr.)

MIERS, L (1789–1879) (Anon. 1879, Ann. Mag. nat. Hist. (5) 4: 469–471.)

MIFASLUG® See Metaldehyde.

MIGEOT, DE BARAIN (1830–1864) (Anon. 1888, Abeille 25: (Les ent. et leurs écrits) 193., bibliogr.)

MIGNEAUX, JULES (–1898) (Bouvier 1898, Bull. Soc. ent. Fr, 1898: cxxi–cxxii.)

MIGRANT Noun. (Latin, *migrans* = wandering; *-antem* = an agent of something. Pl., Migrants.) A wandering organism. Syn. Emigrant. See. Migration.

MIGRANTE Noun. (Latin, *migrans* = wandering. Pl., Migrantes.) A form of aphids with a complex life history, characterized by winged, parthenogenetic and viviparous females that develop from later generations of Fundatrigeniae. Migrantes develop on the primary host plant Species and subsequently migrate to the secondary host plant. Cf. Fundatrix; Fundatrigenia; Alienicola; Sexupara; Sexuale. Rel. Migration.

MIGRATION Noun. (Latin, *migration* > *migratio* > *migratus* = a removal, transfer; English, *-tion* = result of an action. Pl., Migrations.) 1. The physical act of moving from one place to another. 2. A group or population of insects that are physiologically, morphologically or behaviourally predisposed to long-range dispersal, either away from a declining resource or as part of a seasonal cycle. Cf. Dispersal; Emigration; Immigration; Trivial Flight. Rel. Autochthonous.

MIGRATORY Adj. (Latin, *migrare* = to transfer.) Migrating; pertaining to organisms subject to migration or passing from one place to another.

MIGRATORY FLIGHT A behavioural category for insect flight which includes individuals of populations involved in steady flight, generally in one direction and not interrupted by mating. MF sometimes divided into Active Migration and Passive Migration. See Flight. Cf. Mass Flight. Swarming Flight. Trivial Flight.

MIGRATORY GRASSHOPPER *Melanoplus sanguinipes* (Fabricius) [Orthoptera: Acrididae]: A Species endemic in North America from south-central Canada to northern Mexico. Female deposits about 20 egg pods each containing about 15 eggs; nymphs and adults feed on all crops and rangeland plants. Spectacular flights in central states during 1938–1940.

MIGRATORY LOCUST *Locusta migratoria* (Linnaeus) [Orthoptera: Acrididae]: A widespread pest of numerous crops. Found in Africa and Australia.

MIGRATORY OOKINETE A stage in *Plasmodium* produced by the union of male and female gametes in the Alimentary Canal of *Anopheles* (Comstock).

MIHI Latin term meaning 'of me' or 'belonging to me'. A term used by taxonomists in the sense of scientific names for organisms. Abbreviated 'm'.

MIK, JOSEF (1839–1900) (Brauer 1901, Wien ent. Ztg. 20: 1–7, bibliogr.)

MIKAN, JOHANN CHRISTIAN (1769–1844) (Eiselt 1836, *Geschichte, Systematik und Literatur der Insektenkunde* 255 pp. (65), Leipzig.)

MIKLOS, NATTAN (1910–1970) (Laszlo 1970, Folia ent. hung. 24: 279–280.)

MILES, WILLIAM HENRY (1863–1930) (Adkin 1930, Entomologist 63: 264.)

MILES-MOSS, A (1873–1948) (Williams 1948, Proc. R. ent. Soc. Lond. (C) 12: 65.)

MILICHIIDAE Plural Noun. A widespread Family of about 100 Species, closely related to the Drosophilidae. Not found in New Zealand. Small, inconspicuous, usually black and often dull but on occasion highly polished; Proboscis usually long and geniculate; lower frontal orbitals incurved or reclinate, Postvertical Bristles convergent or parallel; Mesonotum has one row of Acrostichal Bristles; Costa broken twice; Anal Vein shorter than Anal Cell; Ovipositor large, broadly oval and compressed. Reared from manure, compost heaps, decaying plant material and bird and ant nests. Adults insect feeders and feeding habits apparently transitional between scavenger and predator. Adults of *Desmometopa* associated with robber flies. Attach themselves to adult fly and remain until prey captured, then loose their hold and feed upon prey of asilid. Other Species feed upon insects trapped in spider webs. Syn. Phyllomyzidae.

MILK GLAND A gland in the female reproductive system of some hipoboscid flies (*Melophagus*). Consists of two pairs of branched glands which secrete a fluid for the nourishment of the developing larva. See Viviparity.

MILKWEED BEETLES Members of the chrysomelid Genus *Tetraopes*.

MILKWEED BUGS *Spilostethus* spp. [Hemiptera: Lygaeidae].

MILKWEED BUTTERFLIES See Danaidae; Nymphalidae.

MILKWEED BUTTERFLY *Danaus plexippus* (Linnaeus) [Lepidoptera: Danaidae]. See Monarch Butterfly.

MILKY DISEASE A bacterial disease of Japanese beetle caused by *Bacillus popilliae* or *Bacillus lentimorbis*.

MILKY SPORE DISEASE A bacterial pathogen (*Bacillus popilliae*) restricted to scarab beetles. MSD was the first pathogen of insects registered in USA (1948) and one for which successful control of the Japanese beetle (*Popilla japonica*) has been achieved. Syn. Milky Disease.

MILLENIUM® A biopesticide used for the control of some ornamental pests (Blackvine Weevil, Strawberry Weevil), cutworms, sod webworms and fleas. Active ingredient the nematode *Steinernema carpocapsae*. See Entomogenous Nematodes. Cf. Biovector®. Rel. Biopesticide.

MILLER, DAVID (1890–1973) (Salmon 1975, Proc. R. Soc. N. Z. 103: 123–125, partial bibliogr.)

MILLER, ELLEN ROBERTSON (–1931) (Osborn 1946, *Fragments of Entomological History,* Pt. II. 232 pp. (103), Columbus, Ohio.)

MILLER, JOHN (1715–1790) (Lisney 1960, *Bibliogr. of British Lepidoptera 1608–1799,* 315 pp., (176–179, bibliogr.), London.)

MILLET DE LA TURTAUDIERE, PIERRE AIME (1783–1873) (Anon. 1867, Ann. Inst. Prov. (2) 9: 524–531, bibliogr.)

MILLET, LOUISE CÉSAR AUGUST (1810–1884) (Ramé 1884, Bull. Insect. Agric. 9: 137–140.)

MILLIERE, PIERRE (1811–1887) (des Gozis 1887, Revue Ent. 6: 248–253, bibliogr.)

MILLIMETRE Noun. (Latin, *mille* = one thousand; *meter* = unit of measure. Pl., Millimetres.) A unit of linear measurement; 1 millimetre equals 0.001 metre, 0.01 centimetres, or 0.04 of an inch. Abbreviated 'mm', not 'mm.'

MILLIN DE GRANDMAISON, AUGIN LOUIS (1759–1818) (Rose 1850, *New General Biographical Dictionary* 10: 143.)

MILLIPEDES See Diplopoda.

MILLS, HARLOW BURGES (1906–1971) (Stannard 1971, Ann. ent. Soc. Am. 64: 1476–1477.)

MILMAN, EDWIN JOHN (1853–1940) (P. P. M. 1940, Entomologist 73: 70–71.)

MILMAN, PHILIP P (1878–1952) (Entomologist's Rec. J. Var. 65; 95–96.)

MILNE-EDWARDS, ALPHONSE (1835–1900) (Anon. 1900, Bull. Soc. Hist. nat. Autun 13: 371–404.)

MILNE-EDWARDS, HENRI (1800–1885) French zoologist and physician (Paris 1823); elected to French Academy of Sciences (1838) and succeeded Cuvier; Professor of Entomology of Museum in Paris (1941) and held various chairs in University of Paris; Director of Museum (1864); Milne-Edwards was not an advocate of evolutionary theory, but was the author of numerous books and scientific papers, including *Resume d'Entomologie, au d'Historie des Animaux Articules* (2 vols, 1829) with Jean Audouin. (Quatrefages 1886, Arch. zool. exp. gén. (2) 4: 1–16.)

MILNER, WILLIAM (1821–1867) (Newman 1867, Ent. Mag. 3: 248.)

MIMESIS Noun. (Greek, *mimesis* = imitation; *sis* = a condition or state. Pl., Mimeses.) See Mimicry.

MIMETIC Adj. (Greek, *mimikos* = imitating; *-ic* = characterized by.) Imitative.

MIMETIDAE Plural Noun. Robber Spiders. Pirate Spiders. Non-web-builiding spiders that enter the webs of other spiders to prey upon the resident.

MIMEUR, JEAN MARIE (1898–1945) (Warner 1947, Bull. Soc. Sci. nat. Maroc 25–27: 26–27.)

MIMIC Verb. (Greek, *mimikos* = imitating; *-ic* = of the nature of.) To assume or resemble another organism closely in colour, size, shape or habits.

MIMIC® See Tebufenozide.

MIMIC BEETLES See Histeridae.

MIMICRY Noun. (Greek, *mimikos* = imitating. Pl., Mimicries.) 1. The resemblance in form, colour or habits of two or more unrelated organisms that live in the same area or habitat. 2. An animal's resemblance to plants and inanimate objects. 3. The physical or behavioural resemblance of two or more organisms for the benefit of one or more of the organisms. The associated organisms form a mimetical complex. Components of the mimetical complex include a model, a mimic and an intermediary organism that tries to distinguish between the model and mimic. The model has properties that are objectionable, distasteful or potentially lethal to the intermediary organism (typically a predator). Objectionable features of a model include anatomical adaptations (spines, claws and other cuticular projections). Chemical adaptations include venoms and secretions of repugnatorial gland. A model's distastefulness is generally attributed to chemical repugnance. For instance, milkweed plants of Genus *Asclepias* are characterized by toxic compounds such as cardiac glucosides that plants use to deter feeding by herbivores. (Cardiac glucosides cause an emetic response in birds; birds learn to avoid insects they recognize as distasteful). Some insects have breached the repellent aspect of host-plant chemistry and feed with impunity on distasteful plants such as milkweed (*e.g.* milkweed butterfly *Danaus plexippus*). The mimic is an individual of another Species that has, through evolutionary time, changed its shape, coloration or behaviour to resemble features displayed by the model. In making these changes, the mimic has diverged from features of closely related, non-mimetic Species. For instance, nymphalid and papilionid butterflies mimic danaid butterflies. (Mimetic nymphalids and papilionids superficially resemble danaids more than other, non-mimetic nymphalids and papilionids.) The intermediary organism (Species) is usually a predator and observes other members of the complex. The mimetical complex relies heavily on the intermediary learning to avoid noxious (distasteful or objectionable) Species. Insects have relied on some colours more than others to convey this kind of information. Visual warnings involve aposematic colours (typically conspicuous reds and yellows). Colour is not mimicry, but when colour is used by a mimic to achieve an advantage (confuse an intermediary), the colour serves mimicry. Modifications to organisms in the complex are adaptations that reduce frequency of encounter with predators or incidence of predation. Mimicry typically involves organisms that are not closely related. Mimicry usually involves two or more Species assigned to different Families, Orders or Classes. Organisms within a mimetical system possess very dissimilar genomes yet they are capable of constructing morphs with striking resemblance in size, shape, colour and behaviour. More examples of mimicry have been provided from Insecta than all other animals combined. Mimicry has been observed, documented and experimentally manipulated in many Orders of insects. Taxonomic distribution, intensity of association and scope of mimicry among insects can raise interesting questions which relate to morphology. Syn. Mimesis. See Batesian Mimicry; Müllerian Mimicry. Cf. Aggressive Mimicry; Crypsis. Rel. Aposematic Coloration.

MIMNERMIDAE Plural Noun. King Crickets; Giant Wetas. Ensiferous Orthoptera (about 40 Genera and 150 nominal Species) characterized by an enlarged head with Pronotum expanded anteriad, male Mandible sometimes enlarged significantly, anterior Coxa spinose, anterior Tibia spinose with Auditory Organ, wings brachypterous usually apterous, Ovipositor long and the Cerci are not ringed. Mostly predaceous, found on ground, tree trunks and under wood. One Species, *Dienacrida heteracantha*, has been called an invertebrate mouse because it occupies the mouse's ecological niche on New Zealand.

MIMOSA FLOWERBUD WEEVIL *Coelocephalapion aculeatum* (Fall) [Coleoptera: Brentidae]

MIMOSA LEAF BEETLE *Chlamisus mimosae* Karren [Coleoptera: Chrysomelidae]

MIMOSA WEBWORM *Homadaula anisocentra* Meyrick [Lepidoptera: Plutellidae].

MINACIDE® See Promecarb.

MINC® See Chlormequat.

MINCHIN, EDWARD ALFRED (1866–1915) (Woodcock 1925, Parasitology 25: 157–162.)

MINE Noun. (Middle English. Pl., Mines.) Galleries, burrows or tunnels between the upper and lower surfaces of a leaf made by insect larvae. Mines are constructed by the feeding activities of some larval Diptera, Coleoptera, Lepidoptera and Hymenoptera. Linear mines are narrow and only a little sinuous; serpentine mines are curved or coiled, becoming gradually larger to a head-like end; trumpet mines start small and enlarge rapidly toward the apex; blotch mines are irregular blotches; tentiform mines are blotch-like mines which throw the leaf into a fold on one side.

MINEX® See Methoprene.

MINGAZZINI, PIO (1864–1905) (Giacomini 1905, Monitore zool. ital. 16: 171–176, bibliogr.)

MINGEE, WILLIAM MALCOLM (Cockerham 1926, J. Econ. Ent. 19: 418–419.)

MINIATE Adj. (Latin, *miniatus* = coloured with vermillion; *-atus* = adjectival suffix.) Of the colour of red lead. Alt. Miniatus.

MINIM Noun. (Latin, *minimus* = least. Pl., Minims.) Social Insects: A minor worker ant. The smallest worker typically seen in founding colonies; the smallest workers in colonies of strongly polymorphic Species. See Minor Worker.

MINING BEE 1. *Andrena carlini* Cockerell [Hymenoptera: Andrenidae]. 2. *Anthophora occidentalis* Cresson [Hymenoptera: Anthophoridae].

MINING BEES Members of the anthophorid Genus *Anthophora* which typically nest in large aggregations in sand or clay banks of streams and rivers. MB construct 'chimneys' of mud pellets at tunnel entrances and which curve downward along the slope of the bank.

MINING SCALE *Howardia biclavis* (Comstock) [Hemiptera: Diaspididae].

MINKIEWICZ, STAINISLAW (1877–1944) (Pruffer *et al.* 1949, Polskie Pismo ent. 19: 3–22, bibliogr.)

MINOR WORKER Social Insects: A member of the smallest worker subcaste. Termites and ants: A small worker in one nest. See Major Worker; Media Worker.

MINOT, CHARLES SEDGEWICK (1852–1914) (Calvert 1915, Ent. News 26: 47–48.)

MINT APHID *Ovatus crataegarius* (Walker) [Hemiptera: Aphididae].

MINUTE BEETLES See Clambidae.

MINUTE BLACK SCAVENGER FLIES See Scatopsidae.

MINUTE BROWN SCAVENGER BEETLES See Lathridiidae.

MINUTE EGG-PARASITE *Trichogramma minutum* Riley [Hymenoptera: Trichogrammatidae]: An egg parasite of Lepidoptera sometimes used in biological control programmes; endemic to North America often confused with other Species of *Trichogramma*.

MINUTE FALSE WATER BEETLES See Limnichidae.

MINUTE FUNGUS BEETLES See Corylophidae.

MINUTE HOUSE-ANT *Plagiolepis alluaudi* Emery [Hymenoptera: Formicidae]: In Australia, a fast-moving ant which nests in walls and ceilings; a prevalent pest in houses.

MINUTE MOSS-BEETLES See Hydraenidae.

MINUTE MOULD BEETLE *Corticaria* spp. [Coleoptera: Lathridiidae]

MINUTE MUD-LOVING BEETLES See Georyssidae.

MINUTE PINE-BARK BEETLE *Cryphalus fulvus* Niijima [Coleoptera: Scolytidae].

MINUTE PIRATE-BUG *Orius tristicolour* (White) [Hemiptera: Anthocoridae]: A beneficial insect widespread in North America. Predaceous as nymph and adult. Eggs deposited on host plant of prey. Prey include chinch bugs, thrips, corn earworm, insect eggs and mites.

MINUTE SCAVENGER BEETLE *Cortinicara hirtalis* (Broun) [Coleoptera: Lathridiidae].

MINUTE TREE-FUNGUS BEETLES See Ciidae.

MINUTE TWO-SPOTTED LADYBIRD *Diomus notescens* (Blackburn) [Coleoptera: Coccinellidae]: An Australian Species. Adult 1.5–2.5 mm long, round-ovate and brown-black with two orange-red spots in centre of Elytra; Elytra setose. Adult females lay eggs in batches near larval food. Eggs yellow, spindle-shaped and laid standing on end. Larva elongate, with well defined legs; body black with markings and fleshy spines; four larval instars. Pupal stage 8–9 days. Minute two-spotted ladybirds are predators of aphids, mealybugs and *Helicoverpa* eggs and larvae on cotton. Adult and larva occur on cotton feeding on aphids and *Helicoverpa* eggs. Cf. Mite-Eating Ladybird. See Coccinellidae.

MIOCENE EPOCH The interval of the Geological Time Scale (23 MYBP–5 MYBP) represented by the first Epoch within the Neogene Period. ME was characterized by Charles Lyell (1833) (Greek, *meio* = smaller), to imply fewer modern types of animals than today; predaceous cats evolved. European Alps and Himalayan Mountains formed, accompanied by intense volcanic activity. Principal insect deposits of ME occur in amber: Bitterfield Amber (Germany, Lower Miocene); Chiapas Amber (Mexico); Dominican Amber (Dominican Republic, low mountains between Santiago and Puerto Plata). Florissant Shales of Colorado reveal modern insect fauna. See Geological Time Scale; Neogene Period. Cf. Pliocene Epoch; Pleistocene Epoch; Recent Epoch. Rel. Fossil.

MIPC® Isoprocarb.

MIPCIN® Isoprocarb.

MIRAL® See Isazofos.

MIRIDAE Hahn 1833. Plural Noun. Leaf Bugs; Plant Bugs. A numerically large, cosmopolitan Family of heteropterous Hemiptera assigned to Superfamily Miroidea. Body typically fragile, coloration highly variable from green to brown to cryptically coloured; compound eyes large; Ocelli absent; Antenna and Rostrum with four segments; Mesofemur and Metafemur with Trichobrothria; Tarsi with three segments; forewing membrane with 1–2 cells; Cuneus well developed; a few Species brachypterous. Predominantly phytophagous and suck plant fluids (*Lygus*), with many Species of economic importance; some Species predaceous on soft-bodied insects, insect eggs; a few Species mycetophagous.

MIROIDEA A Superfamily of heteropterous Hemiptera which includes Microphysidae and Miridae. Labium with four segments; forewing cuneate.

MIRROR Noun. (Latin, *mirari* = to wonder; *or* = one

who does something. Pl., Mirrors.) 1. Any polished or smooth surface which forms images by reflection of rays of light. 2. Cicadidae: A tense, mica-like membrane in the posterior wall of the ventral cavity; a part of the chordotonal structure in cicadas. See Specular Membrane.

MISCHOPTERIDAE Plural Noun. A Family of Paleozoic insects found in the Upper Carboniferous deposits of Commentry.

MISHIMA, YATARO (1867–1919) (Hasegawa 1967, Kontyû 35 (Suppl.): 81.)

MISKIN SWIFT *Sabera dobboe autoleon* (Miskin) [Lepidoptera: Hesperiidae].

MISKIN'S JEWEL *Hypochrysops miskini* (Waterhouse) [Lepidoptera: Lycaenidae].

MISRA, RAI BUHADUR CHANDRA SHEKAR (1880–1939) (Bose 1940, Indian J. Ent. 2: 114–115.)

MISSION BLUE *Icaricia icarioides missionensis* (Hovanitz) [Lepidoptera: Lycaenidae].

MISSROLI, ALBERTO (1883–1952) (Mosna 1952. Riv. Parassit. 13: 3–15, bibliogr.)

MISTLETOE BROWNTAIL MOTH *Euproctis edwardsii* Newman [Lepidoptera: Lymantriidae].

MITAC® See Amitraz.

MITCHELL, JOSEPH DANIEL (1850–1922) (Bailey 1923, J. mammal. 4: 48–49.)

MITCHELL, SAMUEL LATHAM (1764–1831) (Weiss 1936, *Pioneer Century of American Entomology,* 320 pp. (56–57), New Brunswick.)

MITE Noun. (Anglo Saxon, *mite*. Pl., Mites.) Any individual of the Order Acari. Typically small to minute size, body sac-like, adults with four pairs of legs, larvae with three pairs of legs, Antennae and wings absent. Biologically and ecologically diverse; parthenogenesis common. See Acari.

MITE-EATING LADYBIRDS *Stethorus* spp. [Coleoptera: Coccinellidae]: In Australia, adults and larvae important predators of various mite Species on avocados, strawberries, bananas. Adult 1–2 mm long, oval and black-brown with yellow legs; Antenna and body covered with fine Setae. Adult overwinters and emerges in spring to lay ca 1,000 orange eggs in batches of 10–50. Egg stage 2–6 days. Larva elongate, dark brown to black with unpigmented patches; larval stages ca 20 days. Pupal stage ca 3–10 days. Adults live 3 months to 1 year. Cf. Minute Two-Spotted Ladybird; Mite-Eating Ladybird; Striped Ladybird; Three-Banded Ladybird; Three-Spotted Ladybird; Variable Ladybird.

MITIGAN® See Dicofol.

MITIS, HEINRICH VON (1845–1905) (Rebel 1905. Verh. zool. -bot. Ges. Wien 55: 267–269, bibliogr.)

MITOCHONDRIAL DERIVATIVES Highly modified Mitochondria found in Spermatozoa. Typical mitochondria occur in Spermatozoa of ancestral groups such as Collembola, Diplura, Machilidae and primitive Pterygota. In most insects, Mitochondria typically form elongate structures that lie adjacent to the Axoneme. Dermaptera and Embioptera Spermatozoa display large Mitochondria that are partially or entirely filled with a crystalloid substance. Modified Mitochondria with a crystalline core are found in Lepidoptera, Hemiptera and many other pterygote insects. Typically two Mitochondrial Derivatives are found in each spermatozoon and flank the Axoneme, sometimes for its entire length. Only one Mitochondrial Derivative has been found in Spermatozoa of Mecoptera and Trichoptera; in these cases, Spermatozoon Mitochondria fuse directly into a single mass. See Axoneme; Sperm Flagellum. Rel. Centriole; Centriole Adjunct.

MITOCHONDRIAL DNA (mtDNA) A closed, circular, duplex molecule that ranges in size from 15,000–20,000 base pairs.

MITOCHONDRIAL DNA ANALYSIS A sensitive method for examining the evolutionary divergence among animals, because of several unusual properties, including ease of isolation and purification, it is a rapidly evolving molecule, and it is maternally inherited.

MITOCHONDRION (New Latin, from Greek, *mito* = thread; *chondros* = grit, cartillage. Pl., Mitochondria.) Small (0.2–3.0 millimicrons in diameter), polymorphic, granular organells or thread-like (moniliform) filaments in the cytoplasm of nearly all eukaryotic cells that use oxygen. Usually rod or granular shaped, but shape change can be induced by Hydrogen ion concentration or Osmotic Pressure. Basic structure consisting of Outer Mitochondrial Membrane, Inner Mitochondrial Membrane (including Folds or Cristae which penetrate the interior), and a Mitochondrial Matrix. Number of Cristae seems positively correlated with oxidative capacity of mitochondrion. Mitochondria are variable in number and distribution within a cell. Functions include enzymatic actions, aerobic respiration at cellular level, and protein synthesis by Transamination.

MITOGEN Noun. A chemical capable of inducing mitosis.

MITOSIS Noun. (Greek, *mitos* = thread; *sis* = a condition or state. Pl., Mitoses.) The process by which one cell is made into two in division of the fertilized cell. See Amitosis.

MITOSOMA Noun. (Greek, *mitos* = thread; *soma* = body. Pl., Mitosomae.) The Spindle Filament of secondary Spermatocytes which forms a connecting piece and tail envelope of a developing spermatozoon.

MITOXUR® See Proproxur.

MITSUHASHI, SHINJI (1878–1953) (Hasegawa 1967, Kontyû 35 (Suppl): 81–82.)

MITSUKURI, KAKICHI (1857–1909) (Horn 1910, Dt. ent. Z. 1910: 212.)

MITTEN, GEORGE (1826–1916) (Davis 1931, Proc. Indiana Acad. Sci. 41: 52.)

MIXAN® See Tetradifon.

MIXED FUNCTION OXIDASE (MONOOXYGENASES) Oxygenases function in forming essential metabolites such as sterols, prostaglandins, and active derivatives of vitamin D. Oxygenases are classified as either dioxygenases or monooxygenases (= hydroxylases.) A typical monooxygenase reaction is the hydroxylation of an alkane to an alcohol.

MIXED NEST Social Insects: A nest containing colonies of two or more Species in which the broods and adults are mixed. See Compound Nest.

MIYAJIMA, MIKINOSUKI (1872–1944) (Hasegawa 1967, Kontyû 35 (Suppl): 83.)

MIYAKE, TSUNEKATA (1880–1921) (Hasegawa 1967, Kontyû 35 (Suppl): 82–83. (pl. 3, fig. 5.)

MLOKOSEVICH, LUDWIG FRANTSOVICH (–1909) (Semenov-Tian-Shansky 1909, Rev. Russe Ent. 9: 344.)

MLT® Malathion.

MNESARCHAEIDAE Plural Noun. A small Family of Exoporia including one Genus and 14 Species known from New Zealand. Adult Frons and Vertex covered with narrow, upright scales; Ocelli and Chaetosema absent; Antenna filiform; Mandibles small, lobe-like, not articulated; Maxillary Palpus with three segments; Labial Palpus with three segments; vom Rath's Organ absent. Wing coupling jugate; forewing-metathoracic coupling device present. Eggs oval, smooth; placed on substrate but not glued or scattered. Larva with four instars; found on mosses, liverworts, rotten logs, tree trunks; feed on periphyton layer. Pupae adecticous; Mandibles small not articulated; appendages fused to one another, but not Abdomen; Pupation in broad, spindle-shaped cocoon. Adults diurnal, active on bright humid days; found on floor of moist forest.

MNESZECH, GEORGES VANDALIN VON (–1881) (Anon. 1882, Am. Nat. 16: 65.)

MOBBING Adj. Hymenoptera: Agonistic social behaviour demonstrated by aculeates in which potential predators are impeded by several potential prey. A group-defence tactic. Cf. Mauling.

MOBILE Adj. (Latin, *mobilis* = moveable, loose.) Pertaining to structure which is moveable or with the power of motion.

MÖBIUS, KARL (1825–1908) (Weidner 1967, Abh. Verh. naturw. Ver. Hamburg Suppl. 9: 123–126.)

MOCAP® See Ethoprop.

MOCHI, ALESSANDRO (1920–1995) Italian physician and amateur hymenopterist specializing in aculeate wasps. Personal collection in Museo Regionale de Scienze Naturali in Torino, Italy.

MOCQUERYS, EMILE (1825–1916) (Anon. 1916, Bull. Soc. ent. Fr. 1916: 277.)

MOCQUERYS, SIMON (1792–1879) (Fauvel 1880, Annu. Ent. 1880: 121–122.)

MOCSARY, ALEXANDER (SANDOR) (1841–1915) (Mocsáry 1912, Rovart Lap. 19: 81–113, 127–128.)

MOCZAR, MIKLOS (1884–1971) (Szelenyi 1972, Folia ent. hung. 25: 169–178, bibliogr.)

MOCZARSKI, EMIL (1879–1945) (Schedl 1946, Zentbl. Gesamtgeb. Ent. 1: 95.)

MODE OF ACTION The means by which a toxin affects the anatomy, physiology, or biochemistry of an organism.

MODEL Noun. (Latin, *modulus* = a small measure. Pl., Models.) 1. Plans for a structure-to-be-built or figures-to-scale of an extant structure. A groundplan; an archetype. 2. Anything which serves as an example for imitation, *i.e.* a model in a mimicry complex. 3. That which resembles something; a copy. 4. A mathematical explanation applied to biological phenomena; a stochastic or deterministic programme designed to predict the outcome of a process given certain conditions. See Mimicry. Deterministic Model. Stochastic Model.

MODERATE Adj. (Latin, *moderatus* pp of *moderare* = to keep within bounds.) In proportion; neither large nor small. Ant. immoderate; excessive.

MODIFICATION Noun. (Latin, *modus* = limit; *ficare* = to make.) Genetics: widespread differences resulting from nurture (environment) rather than nature (inheritance); modifications may or may not be inheritable.

MODIGLIANI, ELIO (1860–1932) (Vinciguerra 1932, Annali Mus. civ. Stor. nat. Giacomo Doria 56: 122–129, bibliogr.)

MODIOLIFORM Adj. (Latin, *modiolaris* = mediolus of the ear; *forma* = shape.) Shaped as the nerve or hub of a wheel; descriptive of structure that is more-or-less globular with truncated ends. Rel. Form 2; Shape 2; Outline Shape.

MOE, NIELS GREEN (1812–1892) (Schöyen 1892, Ent. Tidskr. 13: 275–279.)

MOESER, FRANK E (1869–1914) (Bird 1914, Can. Ent. 46: 268.)

MOFFAT, CHARLES BETHUNE (1859–1945) (Kennedy *et al.* 1946, Irish Nat. J. 8: 349–370.)

MOFFAT, JOHN ALSTON (1825–1904) (Bethune 1904, Rep. ent. Soc. Ont. 35: 109–110, bibliogr.)

MOFFETT THOMAS (1553–1604) English physician and naturalist. Produced an early treatise on insects entitled *Insectorum sive minimorum Animalium Theatrum*, published posthumously (1634). Recommended bathing and regular change of clothes to avoid plague. Variants on spelling: Moufet, Mouffet, Moffet, Mofet. (Swann 1973, Bull. Brit. Arach. Soc. 2: 169–173.)

MOGGRIDGE, JOHN TRAHERNE (–1874) (Simon 1875, Ann. Soc. ent. Fr. (5) 5: 5–8.)

MOGOPLISTIDAE Plural Noun. (Greek, *mogos* = trouble, travail.) Scaly crickets. A small Family (15 Genera, 200 Species) of ensiferous Orthoptera. Divided into two Subfamilies (Bothriophylacinae and Mongolplistinae). Adult body small with silvery scales and head and Pronotum yellowish or reddish-brown; head with short, blunt Rostrum between Antennae, second tarsal segment minute and compressed; hind Tibia serrated and spinose. Species are apterous or males display very reduced Tegmina; Ovipositor straight and needle-like. Individuals live on shrubs, within ground litter or crevices. Biology of mogoplistids is poorly understood. See

Orthoptera.

MOHL, HUGO VON (1805–1872) (Anon. 1874, Leopoldina 10: 34–39.)

MOHR, ERNA (1894–1968) (Kratochvil 1969, Zool. Listy 18: 142.)

MOILLIET, THEODORE ALBERT (1883–1935) (Moilliet 1947, Proc. ent. Soc. Br. Columb. 43: 43–45.)

MOKRZECKI, ZYGMUNT ATANAZY (1865–1935) (Minkiewicz 1935, Na Jubilensz 70 Leia Urodzin Prof. Z. Mokrzeckiego. 31 pp. bibliogr.)

MOLA Noun. (Latin, *mola* = millstone.) The ridged or roughened grinding surface of the Mandibles. When the Mandible is compound, the Molar corresponds to the Subgalea of Maxilla. Syn. molar.

MOLAR LOBE A proximal lobe on the Mandible of some insects which is used for chewing or grinding.

MOLAR Adj. (Latin, *molere* = to grind.) Pertaining to a structure adapted for grinding. Noun. 1. A tooth adapted for grinding. 2. A surface of a tooth adapted for grinding. The Molar of the Mandible is subapical and medial. When Mandibles are drawn together medially, the opposable Molar regions serve as a grinding surface. The Molar is well developed in groups such as Orthoptera; Mandibles of Species may be characterized based on adaptation for certain foraging patterns: Graminivorous, herbivorous, ambivorous, omnivorous and grater. See Mandible. Cf. Incisor; Prostheca.

MOLE CRICKET Minor pests of crops and pastures; anatomically adapted for living underground; feed on roots and other insects. See Gryllotalpidae.

MOLE CRICKETS Members of the Gryllotapidae. Common Species include: African Mole Cricket; Changa Mole Cricket; European Mole Cricket; Giant Mole Cricket; Imitator Mole Cricket; Knife Mole Cricket; Northern Mole Cricket; Short-Winged Mole Cricket; Southern Mole Cricket; Tawny Mole Cricket. See Saltatoria.

MOLESCHOTT, JACOB ALBERT WILLIBRORD (1822–1893) (Nordenskiöld 1935, *History of Biology*. 625 pp. (449–450), London.)

MOLEYRE, LOUIS (1886) (Bourgeois 1886, Bull. Soc. ent. Fr. (6) 6: xvii.)

MOLINA, JUAN IGNACIO (1740–1829) (Porter *et al.* 1929, Revta chil. Hist. nat. 33: 7–14, 169–170, 214–216, 223–225, 428–488.)

MOLINARI, ENRIQUE (–1927) (E. D. D. 1927, Boln. Soc. ent. argent. 3: 21.)

MOLLENKAMP, WILHELM (1846–1913) (Kuhnt 1913, Dt. ent. Z. 1913, 226.)

MÖLLER, ALFRED (1860–1922) (Falk 1927, Hausschwamm-Forschungen 9: 1–11.)

MOLLER, C A (1845–1912) (Tuxen 1968, Ent. Meddr. 36: 18 only.)

MÖLLER, CARL AUGUST (1845–1912) (Engelhart 1918, Ent. Meddr. 12: 16–17.)

MOLT See Moult.

MOMPHIDAE Plural Noun. Momphid Moths. A Family of Lepidoptera assigned to Superfamily Gelechioidea. Nearly 40 Species occur in North America. Elements of Family sometimes placed within Cosmopterygidae. Adult body small; wings long, narrow, apically pointed. Syn. Lavernidae.

MONARCH BUTTERFLIES See Nymphalidae.

MONARCH BUTTERFLY *Danaus plexippus* (Linnaeus) [Lepidoptera: Nymphalidae]: A large-bodied, multivoltine, New World butterfly which develops on milkweed (*Asclepias* spp.) Adults in North America migrate southward into Mexico during August, and northward during the following March. The migration can involve up to 2,000 miles. Larvae sequester glucosides ingested from *Asclepias* which are maintained in adult; adults inedible to birds and serve as model for Viceroy butterfly in a mimetical scheme. Alt. Milkweed Butterfly. See Nymphalidae.

MONARSENOUS Adj. (Greek, *monarchus* from *monos* = alone; *archein* = to be first; Latin, *-osus* = with the property of.) A condition among insects where one male exists among many females. Cf. Polygamous; Monandrous.

MONASTERO, SALVATORE (1900–1972) (Delucchi 1972, Entomophaga 17: 355–356.)

MONBEIG, JAN THÉODORE (1875–1914) (Oberthür 1916, Etudes de Lépidopterologie comparée 11 (portraits de lépidoptèristes 3me Sér.): [12].)

MONCHADSKY, ALEKSANDR SAMOILOVICH (1897–1974) (Gutsevich 1975, Ent. Obozr. 54: 681–683, bibliogr.)

MONCREIFFE, THOMAS (1822–1879) (White 1879, Scott. Nat. 5: 145–148.)

MONELL, JOSEPH TARRIGAN (1859–1915) (Davis 1915, Ent. News 26: 380–383, bibliogr.)

MONGENIC RESISTANCE Resistance of a host plant to pest insect damage resulting from expression of a single gene.

MONILIFORM Adj. (Latin, *monil* = necklace; *forma* = shape.) Bead-shaped; descriptive of structure which resembles a necklace composed of beads; an elongate structure constricted at regular intervals. See Bead. Cf. Filiform; Flagelliform; Vermiform. Rel. Form 2; Shape 2; Outline Shape.

MONILIFORM ANTENNA An Antenna composed of bead-like segments or Flagellomeres. A characteristic or diagnositic antennal shape for some Species of insects.

MONITOR® See Methamidophos.

MONKEY GRASSHOPPERS See Eumastacidae.

MONKEY LUNG MITE *Pneumonyssus simicola* Banks [Acari: Halarachnidae].

MONKEYPOD MOTH *Polydesma umbricola* Boisduval [Lepidoptera: Noctuidae].

MONKEYPOD ROUNDHEAD-BORER *Xystrocera globosa* (Oliver) [Coleoptera: Cerambycidae].

MONOBASIC Adj. (Greek, *monos* = one; *basis* = base; *-ic* = consisting of.) A term applied to Genera that originally were established with one included Species only. See Genus; Type 2.

MONOCANTHA COCHINEAL *Dactylopius ceylonicus* (Green) [Hemiptera: Dactylopiidae].

MONOCHROMATIC Adj. (Greek, *monos* = one; *chromos* = colour; *-ic* = characterized by.) Of one colour throughout.

MONOCIL® See Monocrotophos.

MONOCLONAL Adj. (Greek, *monos* = one; *klon* = a twig; Latin, *-alis* = pertaining to.) Descriptive of a cell population derived from one cell.

MONOCLONAL ANTIBODY Immunology: The use of antibodies to detect and isolate specific molecules. Monoclonal antibodies are developed by cloning a single, antibody-secreting B lymphocyte to obtain uniform antibodies, each with an identical antigen-binding site, in large quantities. See Antibody. Rel. Biotechnology.

MONOCONDYLIC ARTICULATION A single point of articulation between the adjacent segments. Monocondylic articulations are common in immature insects (except Neuroptera and Trichoptera), and typically dorsal in position on the leg. The leg can be elevated, but its strength is limited. Syn. Monocondylic Joint *sensu* Snodgrass. See Leg Articulation. Cf. Dicondylic Articulation; Tricondylic Articulation.

MONOCONDYLIC Adj. (Greek, *monos* = one; *kondylos* = knuckle; *-ic* = consisting of.) Descriptive of an appendage that has one Condyle or point of articulation. Cf. Dicondylic; Tricondylic.

MONOCRON® See Monocrotophos.

MONOCROTOPHOS An organic-phosphate (phosphoric acid) compound {Dimethyl (E)-1-methyl-2-(methylcarbamoyl) vinyl phosphate} used as a systemic and contact insecticide against bollworms, loopers and mites. Compound applied to citrus, grapes, fruits, vegetables, olives, tomatoes, sugar beets, cotton, tobacco, rice and ornamentals in many countries. Registration rescinded in USA. Phytotoxic to apples, almonds, peaches, cherries and some sorghum varieties. Highly toxic to birds, bees, and wildlife. Trade names include: Aimocron®, Alphate®, Azodrin®, Bilobran®, Carbicron®, Crisodrin®, Hilcron®, Monocil®, Monocron®, Monophos®, Nuvacron®, Pandar®, Phillardrin®, Sufos®, Susvin®. See Organophosphorus Insecticide.

MONODACTYLE Noun. (Greek, *monos* = one; *daktylos* = finger.) With a single, moveable claw which closes on the apex of other leg structures, as in the louse Genus *Pediculus*. Alt. Monodactylus.

MONODOMOUS Adj. (Greek, *monos* = one; *domos* = house; Latin, *-osus* = with the property of.) Ants: having one nest only for each colony. Cf. Polydomous.

MONOECIOUS Adj. (Greek, *monos* = one; *oikos* = dwelling; Latin, *-osus* = with the property of.) General: Pertaining to organisms possessing both sexual elements or glands in one individual. Botany: Plants with pistillate and staminate flowers on the same plant. Cf. Dioecious.

MONOEMBRYONIC Adj. (Greek, *monos* = one; *embryon* from *en* = in; *bryein* = to swell; *-ic* = characterized by.) Descriptive of a condition characterized by monoembryony or pertaining to monoembryony, *i.e.,* one embryo in one egg.

MONOEMBRYONY Noun. The production of one embryo only from the fertilized Ovum or egg.

MONOGAMOUS Adj. (Greek, *monos* = one; *gamos* = marriage; Latin, *-osus* = with the property of.) Pertaining to a union where a female is fertilized by one male only.

MONOGAMY Noun. (Greek, *monos* = one; *gamos* = marriage.) The condition of being monogamous or of having one mate only. Cf. Polygamy.

MONOGRAPH Noun. (Greek, *monos* = one; *graphein* = to write. Pl., Monographs.) A comprehensive taxonomic treatise examining a limited subject (Genus, Tribe, Family) in detail. Cf. Revision; Review; Survey.

MONOGYNE FORM Social ant colonies which have one inseminated, functional, egg-laying queen. Cf. Polygyne Form.

MONOGYNOUS Adj. (Greek, *monos* = one; *gyne* = female; Latin, *-osus* = with the property of.) 1. A condition manifest in colonial eusocial Hymenoptera wherein one fecundated female (queen) maintains progeny production. 2. A condition in which a male inseminates only one female. See Polygynous. Cf. Monandrous; Pleometrotic.

MONOGYNY Noun. (Greek, *monos* = one; *gyne* = female.) 1 Social Insects: The presence of only one functional queen in a nest. Monogyny may be dividided into Primary Monogamy and Secondary Monogamy. 2. A condition in which a male insect will mate (court, copulate and inseminate) only one female in a breeding population, only one female during a breeding cycle. Cf. Polygyny.

MONOLEPTA BEETLE See Red-Shouldered Leaf-Beetle.

MONOMACHIDAE Szepligeti 1889. Plural Noun. A small Family (ca 15 Species) of apocritous Hymenoptera assigned to the Proctotrupoidea, with a disjunct austral distribution (Australia, South America). Female adult Antenna with 15 segments; Maxillary Palpus with three segments; Labial Palpus with five segments. Pronotum without reentrant declivity, posterior margin overlapping Mesoscutum; Scutellum with a transverse row of punctures posteriad. Forewing Rs not forked; Radial Cell closed posteriad, Rs + M absent; Median Cell absent. Hindwing Basal vein and M + Cu tracheate. Trochantellus present on all legs; tibial spur formula 1-2-2. Petiole formed by T1 and S1, separated by Suture. Gaster slender, tapering, not laterally compressed. Gastral Terga and Sterna broadly overlapping. Apical Sternum not medially slit. Cercus plate-like. Ovipositor internal. Male Antenna with 14 segments; Flagellomeres not modified. Cercus digitiform. Adults found in moist, cool, forest habitats. Only known host the mature larva or Puparia of Stratiomyidae.

MONOMERI (Greek, *monos* = one; *meros* = part.) Insects with 1-segmented Tarsi.

MONOMEROUS Adj. (Greek, *monos* = one; *meros*

= part; Latin, *-osus* = with the property of.) Descriptive of an appendage or multi-segmented body part displaying only one segment for that part.

MONOMIAL Adj. (Greek, *monos* = one; Latin, *nomos* = name; *-alis* = characterized by.) Descriptive of an organism that has one name only. Syn. Uninomial. See Taxonomy. Cf. Binomial; Polynomial.

MONOMMATIDAE See Monommidae.

MONOMMIDAE Plural Noun. A numerically small, widespread Family of Coleoptera assigned to Superfamily Tenebrionoidea. Adult 5–12 mm long, black convex dorsally, flattened ventrally; compound eyes large and nearly contiguous medially; Antenna with 11 segments including 2–3 segments forming Club, Flagellum concealed at repose within Sulcus on ventral aspect (Hypomeron) of Prothorax; Epimeron moderately wide and complete. Larva elongate, paralell sided and somewhat compressed dorsoventrally; Urogomphi separated by pit. Tarsal formula 5-5-4. Adult inhabits leaf litter; larva lives in rotten wood. Syn. Monommatidae.

MONOMORPHIC Adj. (Greek, *monos* = one; *morphe* = form; *-ic* = characterized by.) Descriptive of or pertaining to a deme, population, race, or Species that displays only one form. Cf. Allomorphic; Dimorphic; Polymorphic.

MONOMORPHISM Noun. (Greek, *monos* = one; *morphe* = form; English, *ism* = condition. Pl., Monomorphisms.) 1. Social Insects: The existence of only one worker subcaste within a Species or colony. Cf. Dimorphism; Polymorphism.

MONOPHAGOUS Adj. (Greek, *monos* = one; *phagein* = to devour; Latin, *-osus* = with the property of.) Descriptive of an organism that exhibits monophagy. Pertaining typically to an herbivorous organism that feeds on one Species of food plant, a predator that feeds on one Species of prey or a parasite that feeds on one Species of host. Alt. Monophagus. Cf. Euryphagous, Oligophagous, Polyphagous, Stenophagy. Rel. Diphagous.

MONOPHAGY Noun. (Greek, *monos* = one; *phagein* = to devour. Pl., Monophagies.) 1. The highest level of specialization in food or most limited range of dietary requirements. Typically shown in an herbivorous organism which feeds on one Species of food plant, a predator which feeds on one Species of prey or a parasite which feeds on one Species of host. Cf. Euryphagy; Oligophagy; Polyphagy; Stenophagy.

MONOPHASIC ALLOMETRY Polymorphism in which the allometric regression line has one slope. See Allometry.

MONOPHOS® See Monocrotophos.

MONOPHYLETIC Adj. (Greek, *monos* = one; *phyle* = tribe; *-ic* = consisting of.) In taxonomic classification, a group of organisms all members of which are derived from a common ancestor and not excluding from an historical group any descendants of the ancestor of that group. See Cladistics. Cf. Holophyletic; Paraphyletic; Polyphyletic. Rel. Phylogeny.

MONOPHYLY A basic characteristic of a Taxon required for the examination of that Taxon by means of phylogenetic analyses. There are several competing definitions which are discussed by E. O. Wiley (1981). Included: 'A taxon is monophyletic if its members are descendants of a common ancestor'; 'Monophyly is the derivation of a taxon through one or more lineages, from one immediately ancestral taxon of the same or lower rank.' Cf. Holophyly, Polyphyly.

MONOPIS MOTHS *Monopis* spp. [Lepidoptera: Tineidae].

MONOTHALAMOUS Adj. (Greek, *monos* = one; *thalamos* = chamber; Latin, *-osus* = with the property of.) Pertaining to single chambered galls found on plants, particularly those induced by gall-forming insects. See Cynipidae. Cf. Polythalamous.

MONOTHELIOUS Adj. (Greek, *monos* = alone; *thelys* = female; Latin, *-osus* = with the property of.) Pertaining to monothely.

MONOTHELY Noun. (Greek, *monos* = alone; *thelys* = female.) A phenomenon in which one female copulates with or is fertilized by many males.

MONOTROCHA Noun. (Greek, *monos* = one; *trochos* = wheel.) Hymenoptera with one Trochanter per leg.

MONOTROCHOUS Adj. (Greek, *monos* = one; *trochos* = wheel.) A condition in which the Trochanter is 1-segmented.

MONOTRYSIA Borner 1939. A small Suborder of Lepidoptera. Composition varies among classifications. In the restricted sense Monotrysia includes the Hepialoidea, Incurvarioidea and Nepticuloidea. Apparently a primitive group under any classification. Variable in body size, lacking Mandibles, wings generally aculeate with similarly reduced venation on forewing and hindwing, female with 1–2 genital openings on Sternum 9–10, Pupae adecticous and obtect. Cf. Dachnonypha, Ditrysia, Zeugloptera.

MONOTRYSIAN Adj. (Greek, *monos* = one; Middle English, *tryst* = a place for waiting.) Pertaining to female Lepidoptera with a single, terminal abdominal aperture which receives the Aedeagus during copulation and serves passage of the egg from the body during oviposition. Regarded as a primitive condition. Cf. Ditrysian.

MONOTYPE Noun. (Greek, *monos* = one; *typos* = type.) The Holotype of a Species that is based on a single specimen (Banks & Caudell). See Type. Cf. Lectotype; Paratype; Syntype.

MONOTYPIC Adj. (Greek, *monos* = one; *typos* = type; *-ic* = consisting of.) Pertaining to a named Species that is based on one 'type,' form or characteristic.

MONOTYPICAL GENUS A Genus based on one Species only, with no other Species known at the time the Genus was characterized (described). See Type by Original Description; Isotypical; Heterotypical.

MONOVOLTINE Adj. (Greek, *monos* = one; Italian, *volta* = time.) See Univoltine.

MONOXENOUS Adj. (Greek, *monos* = one; *xenos* = foreigner; Latin, *-osus* = with the property of.) A parasite which has one Species of host. See Parasitism. Cf. Polyxenous; Pleioxenous. Rel. Host Specificity.

MONRO, HECTOR ALEXANDER URQUHART (1906–1970) (Bond 1970, J. Econ. Ent. 63; 1721.)

MONROS, FRANCISCO (1922–1958) (Halffter 1958, Ciencia Méx. 18: 152–153.)

MONTANDON, ARNOLD LUCIEN (–1922) (Anon. 1923, Entomol. mon. Mag. 59: 39.)

MONTANO, BENITO ARIAS (1527–1598) (Dusmet y Alonso 1919, Boln. Soc. Esp. 2: 76.)

MONTE, OSCAR (1895–1948) (Borgmeier 1948, Revta Ent. Rio de J. 19: 589–590.)

MONTEALEGRE, ABRAHAM (1883–1928) (Porter 1928, Revta chil. Hist. nat. 32: 345–347, bibliogr.)

MONTELL, JUSTUS ELIAS (1869–1954) (Nordmann 1955, Notul ent. 35: 33–34.)

MONTEREY PINE-CONE BEETLE *Conophthorus radiatae* Hopkins [Coleoptera: Scolytidae].

MONTEREY PINE-RESIN MIDGE *Cecidomyia resinicoloides* Williams [Diptera: Cecidomyiidae].

MONTEREY-PINE WEEVIL *Pissodes radiatae* Hopkins [Coleoptera: Curculionidae].

MONTGAUDRY, BARON DE (–1857) (Hamet 1859, Apiculteur 2: 97.)

MONTGOMERY, EDMUND DUNCAN (1835–1911) (Geiser 1931, SWest Rev. 16: 200–235.)

MONTGOMERY, J H (1875–1940) (E. W. B. 1940, Fla. Ent. 23: 25–26.)

MONTGOMERY, ROBERT EUSTACE (1880–) (Roubaud 1938, Bull. Soc. Path. exot. 31: 178–179.)

MONTGOMERY, THOMAS HARRISON (1873–1912) (Bonnet 1945, *Bibliographia Araneorum* 1: 55–56.)

MONTI, RINA STELLA (1861–1937) (Baldi 1938, Riv. Biol. 25: 347–361, bibliogr.)

MONTICELLI, FRANCESCO SAVERIO (1863–1927) (Zirpolo 1929, Boll. Soc. Nat. Napoli 41: 301–336, bibliogr.)

MONTILLOT, LOUIS (–1902) (Anon. 1903, Leopoldina 39: 86.)

MONTORO, SAVERIO PATRIZI (–1957) (Anon. 1957, Boll. Ist. ent. Univ. Bologna 22: 203.)

MONTROOZIER, XAVIER (1821–1897) (Grouvelle 1897, Bull Soc. ent. Fr. 1896: 233.)

MONURA Sharov 1957. An extinct Order of Insecta (Subclass Ectognathata Hennig 1953; Infraclass Archentomata Boudreaux 1979) known from Upper Carboniferous and Lower Permian deposits. Compound eyes well developed; Ocelli imperceptible in imprints; Postoccipital Suture well developed; prothoracic Tergum reduced. Abdominal Styli present on abdominal segments 1–4; abdominal Tergum X enlarged; Appendix Dorsalis as long as body; Cerci apparently absent.

MONZEN, KOTA (1883–1960) (Hasegawa 1967, Kontyû 35 (Suppl.): 86–87, bibliogr.)

MOORE, FREDERIC (1830–1907) (Fruhstorfer 1907, Ent. Wbl. 24: 151–152.)

MOORE, G A (1878–1966) (Vickery 1965, Proc. ent. Soc. Ont. 96: 128.)

MOORE, HARRY (1857–1949) (Wigglesworth 1950, Proc. R. ent. Soc. Lond. (C) 14: 65.)

MOORE, J W (–1948) (Williams 1949, Proc. R. ent. Soc. Lond. (C) 13: 68.)

MOORE, JOHN LEE (1877–1939) (Anon. 1945, Trans. Suffolk Nat. Soc. 5: xcvi-xcvii.)

MOORE, KENNETH MILTON (1912–1990) Australian forest Entomologist and specialist of psyllids associated with dying trees.

MOORE, WILLIAM (1887–1972) (Campbell 1973, J. Econ. Ent. 66: 1009–1010.)

MOQUIN-TANDON, HORACE BENEDICT ALFRED (1804–1863) (Clos 1864, Mém. Acad. Sci. Inscript. Toulouse (6) 2: 5–46, bibliogr.)

MOR Noun. (Danish. Pl., Mors.) A layer of humus consisting of organic decaying material above mineral-containing soil. Cf. Debris; Detritus; Humus; Litter.

MORAVOHYMENIDAE Kukalova-Peck 1972, Plural Noun. Family of Paleopterous insects known from a wing found in the Lower Permian of Czechoslovakia. Similar to Bardohymenidae.

MORAWITZ, FERDINAND (1827–1896) (Semenov 1897, Horae Soc. ent. Ross 31: i–x., bibliogr.)

MORDANT Adj. (Latin, *mordere* = to bite; *-antem* = adjectival suffix.) 1. Biting, caustic, corrosive. 2. Acting as a mordant in dyeing or histology. Noun. (Old French, pp of *mordre* = to bite; *-antem* = an agent of something. Pl., Mordants.) 1. Any substance combined with a dyestuff to form an insoluble compound, or lake, which serves to produce a fixed colour in a tissue or textile. 2. A corroding substance used in etching.

MORDELLIDAE Plural Noun. (Latin, *mordere* = to bite.) Pin Tail Beetles; Tumbling Flower Beetles. A numerically large, cosmopolitan Family of polyphagous Coleoptera (Cucujoidea.) Adult 1.5–15 mm long; body wedge-shaped, 'hunchbacked,' smooth or pubescent. Head deflexed; Antenna short with 11 segments, terminal segments serrate; legs long, thin with spurs, tarsal formula 5-5-4; apical abdominal Terga exposed beyond Elytra, caudally tapered to a long, slender point; five abdominal Ventrites. Larva elongate, eruciform, head hypognathous; Stemmata present or absent; legs degenerate; Urogomphi present or absent. Adults common on flowers; most larvae parasitic or predaceous, some mine stems, leaves or bore into decayed wood.

MORDVILKO, ALEXANDER KONSTATIN (1867–1938) (Borodon 1940, Ann. Soc. ent. Am. 33: 487–494, bibliogr.)

MORE, ALEXANDER GOODMAN (1830–1895) (Anon. 1895, J. Bot. Lond. 33: 225–227.)

MORES Ecology: Groups of organisms in full agreement as to physiological life-histories, usually belonging to a single Species, but possibly including more than one Species (Shelford).

MORESTAN® See Chinomethionate.

MORETON BAY FIG PSYLLID *Mycopsylla fici* (Tryoni) [Hemiptera: Homotomidae].

MORETON BAY FIG WASP *Pleistodontes froggatti* Mayr [Hymenoptera: Agaonidae].

MOREY, FRANK (1858–1925) (J. G. 1926, Proc. Linn. Soc. Lond. 1925–1926: 91–92.)

MORGAN, ALFRED COOKMAN (1876–1931) (Anon. 1931, J. Econ. Ent. 24: 1114.)

MORGAN, HARCOURT ALEXANDER (1867–1950) (Anon. 1950, J. Econ. Ent. 43: 964–965.)

MORGAN, THOMAS HUNT (1866–1945) (Sturtevant 1946, Am. Nat. 80: 22–23.)

MORGAN, WILLIAM LONGWORTH (1904–1968) (Chadwick 1969, J. ent. Soc. Aust. 5: 60–61, bibliogr.)

MORGANTHALER, OTTO (1886–1973) (Schneider 1973, Mitt. schweiz. ent. Ges. 46: 155–156.)

MORI, TAMEZO (1884–1962) (Hasegawa 1967, Kontyû 35 (Suppl.): 85–86.)

MORIBAETIS A small Genus of Neotropical Baetidae whose larvae inhabit the splash-zone of streams (Waltz & McCafferty 1985, 87: 239–251.)

MORIBUND Adj. (Latin, *moribundus* from *moriri* = to die.) An organism which is diseased or dying.

MORICE, FRANCIS DAVID (1849–1926) English cleric and amateur Entomologist interested in Hymenoptera. (Laing 1926, Entomol. mon. Mag. 62: 268–269.)

MORINO, ISAKU (1897–1962) (Hasegawa 1967, Kontyû 35 (Suppl.): 86.)

MORITZ, JOHAN WILHELM KARL (1797–1866) (Papavero 1975, *Essays on the History of Neotropical Dipterology.* 2: 297–298. São Paulo.)

MORLEY, ARTHUR MACDONELL (1879–1972) (Butler 1973, Proc. R. Soc. ent. Lond. (C) 37: 56.)

MORLEY, BENJAMIN (1892–1932) (Bayford 1932, Naturalist, No. 906: 223–225.)

MORLEY, CLAUDE (1874–1951) (Riley 1951, Entomologist 85: 121–122.)

MORMON CRICKET *Anabrus simplex* Haldeman [Orthoptera: Tettigoniidae]: A Species endemic to western North America and a perioidic pest of rangeland grasses, fruit and vegetable crops, alfalfa and wheat. Eggs elongate, laid in midsummer with eclosion during following spring; eggs deposited individually or in groups to 100 within sun-exposed soil. MC undergoes seven nymphal instars. Nymphs and adults have tendency to migrate. Males stridulate and cannibalism is common. See Tettigoniidae. Cf. Coulee Cricket.

MORMOTOMYIIDAE Plural Noun. A small Family of schizophorous Diptera assigned to Superfamily Muscoidea.

MORNINGGLORY LEAFMINER *Bedellia somnulentella* (Zeller) [Lepidoptera: Lyonetidae].

MORO, GIAN BATTISTA (1899–1971) (Guiglia 1972, Memorie Soc. ent. Ital. 51: 197–198, bibliogr.)

MORODER, SALA EMILIO (Anon. 1940, Ciencia 1: 121.)

MOROFSKY, WALTER F (1899–1965) (Guyor 1965, J. Econ. Ent. 58: 805.)

MORPHO BUTTERFLIES See Nymphalidae.

MORPHOGENESIS Noun. (Greek, *morphe* = form; *genesis* = descent. Pl., Morphogeneses.) 1. The development of shape or form. Morphogenesis is anatomical change of structure during development. 2. The development of organs or parts of organisms. Morphogenesis may be gradual, such as the transitions of structures within a stage (nymph, naiad or larva) or radical. Anatomical change is achieved through moulting. See Growth; Metamorphosis. Cf. Anamorphosis; Epimorphosis. Rel. Moulting.

MORPHOLOGICAL Adj. (Greek, *morphe* = shape; *logos* = discourse; Latin, *-alis* = characterized by.) Descriptive of concepts relating to form and structure of an organism. See Morphology. Cf. Anatomical; Physiological.

MORPHOLOGICAL ASYMMETRY A type of asymmetry seen at several levels in insects. Bilateral and mosaic gynandromorphs frequently behave in ways that show subtle differences in functional morphology. Genitalia of many insects are asymmetrical (*e.g.* bedbug *Cimex* and male Grylloblattoidea). Mandibles of some insects are asymmetrical (*e.g.* Thysanoptera lack a right Mandible; carabid *Chlaenius* left Mandible long and straight while right Mandible shorter and curved). See Asymmetry; Symmetry. Cf. Absolute Asymmetry; Behavioural Asymmetry; Biological Asymmetry; Developmental Asymmetry; Pattern Asymmetry; Skeletal Asymmetry. Rel. Organization.

MORPHOLOGICAL CORRESPONDENCE (Woodger 1945, On biological transformations, pp 95–120. *Essays on Growth and Form presented to D'Arcy Wentworth Thompson.* Oxford University Press.) See Bauplan.

MORPHOLOGICAL TRANSFORMATION A concept concerned with documenting anatomical change within a phyletic lineage over evolutionary time. Given an anatomical structure in an hypothetical (groundplan) organism or fossil, a morphological transformation traces the change in structure through evolutionary time and documents change to structure. See Phylogeny. Cf. Transformation Series. Rel. Groundplan.

MORPHOLOGY Noun. (Greek, *morphe* = form, shape; *logos* = discourse. Pl., Morphologies.) The term proposed by Goethe for a discipline within Biology that is concerned with form and function of an organism's structure. Morphology is divided into several subdisciplines: Anatomy (structure), Histology (ultrastructure), Constructional Morphology, and Functional Morphology. See each term for explanation. Morphology centres on interpretation of anatomical facts based upon analytical study. Anatomy of an insect is an expression of organic evolution. Biologists seek to explain the processes and consequences of organic evolution. Morphology represents one line of investigation concerned with providing explanations for the organic diversification which we observe and whose process of change is 'Evolution.' Before 1940, insect morphology focused on naming and describing anatomical structure. The need for this activity has not diminished, but theory leaps over data and new ideas are proposed more rapidly than they are evaluated. As a consequence, Morphology as a discipline remains unfulfilled. See Organism.

MORPHOMETRICS Plural Noun. (Greek, *morphe* = form; *metros* = measure.) The quantitative description and analysis of shape (form) and variation in shape. Cf. Allometry.

MORPHOSPECIES A typological Species recognized on the basis of morphological difference. Cf. Phenon.

MORPHOTYPE Noun. (Greek, *morphe* = form; *typos* = pattern. Pl., Morphotypes.) 1. The type-specimen of one form of a polymorphic Species. See Polymorphism; Type. 2. See Baupan.

MORREN, CHARLES FRANÇOIS ANTOINE (1807–1858) (Morren 1860, Ann. Acad. Sci. Belg. 26: 167–251, bibliogr.)

MORREN, CHARLES JACQUES EDOUARD (1833–1886) (Crépin 1886, Ann. Acad. Sci. Belg. 53: 419–452, bibliogr.)

MORRIL LACEBUG *Corythucha morrilli* Osborn & Drake [Hemiptera: Tingidae].

MORRILL, AUSTIN WINFIELD (1880–1954) (Mallis 1971, *American Entomologists*. 549 pp. (482–483), New Brunswick.)

MORRILL LAND-GRANT COLLEGE ACT United States legislation signed into law by President Abraham Lincoln during 1862. The Act provided public domain to establish colleges in which to educate students in agriculture and engineering. Circumstances required a Second Morrill Act during 1872 which was passed into law in 1890 which gave a direct annual appropriation in support of each land-grant college. See Hatch Act.

MORRIS, FRANCIS ORPEN (1810–1893) (Morris 1897, *Francis Orpen Morris, A Memoir*, 323 pp. London.)

MORRIS, JOHN GOODLOVE (1803–1895) (Anon. 1895, Ent. News 6: 273–274.)

MORRIS, MARGARETA HARE (1797–1867) (Weiss 1936, *Pioneer Century of American Entomologists,* 320 pp. (136–138), New Brunswick.)

MORRISON, HAROLD (1890–1963) American economic entomologist and specialist in Coccoidea. (Russell 1963, Proc. ent. Soc. Wash. 65: 311–313.)

MORRISON, HERBERT KNOWLES (1854–1885) (Anon. 1885, Science 5: 532.)

MORRISON, HUGH ENGLE (1905–1967) (Crowell 1969, J. Econ. Ent. 62: 281.)

MORRISON, LEWIS (1895–1968) (Watt 1970, N. Z. Ent. 4: 87–91.)

MORS, HANS (–1941) (Wellenstein 1943, Z. angew. Ent. 30: 157–159, bibliogr.)

MORS, LOUIS AUGUSTE REMACLE (1826–1884) (Fairmaire 1884, Ann. Soc. ent. Fr. (6) 4: 367–368.)

MORSE, ALBERT PITTS (1863–1936) (Dow 1937, Psyche 44: 1–11, bibliogr.)

MORTALITY Noun. (Latin, *mortalitas* from *mortis* = to be mortal, death. Pl., Mortalities.) The properties or factors which cause death of individuals and a subsequent decrease in population size. Cf. Natality.

MORTIMER, CHARLES HENRY (–1932) (Blair *et al.* 1932, Entomol. mon. Mag. 68: 279.)

MORTON, ALEXANDER (?1855–1907) (Anon. 1907, Proc. R. Soc. Tasm. 1906–1907: xlvii–xlix.)

MORTON, EMILY L (1841–1920) (Newcomb 1917, Ent. News 28: 97–101.)

MORTON, KENNETH J (1858–1940) (Fraser 1940, Ent. News 51: 237–240.)

MORULA Noun. (Latin, *morula* diminutive of *morus,* from *morum* = mulberry. Pl., Morulae.) 1. The globular mass of cells formed by holoblastic cleavage of the egg (Snodgrass). 2. A cluster of developing male germ cells.

MOSAIC Noun. (Latin, *mosaicus* = spotted. Pl., Mosaics.) A patchy variation in normal green colour of leaves. A symptom of viral infection of plants. Cf. Mottle.

MOSAIC THEORY OF VISION An explanation of the functioning of the insect compound eye, in which each Ommatidium conveys a single point of light to each Retinula, whose points, as a whole, combine to produce the single erect optical image. See Compound Eye.

MOSCADE® See Fenvalerate.

MOSCHATE Adj. (Greek, *moschos* = musky; *-atus* = adjectival suffix.) With a musky odour. Alt. Moschatus.

MÖSCHLER, HEINRICH BENNO (1831–1888) (Christoph 1889, Berl. ent. Z. 33: 193–196, bibliogr.)

MOSELEY, HENRY NOTTIDGE (1844–1891) (Lankester 1891, Nature 45: 79–80.)

MOSELY, MARTIN EPHRAIM (1867–1948) (Kimmins 1948, Entomol. mon. Mag. 84: 240.)

MOSER, JULIUS (–1929) (Ross 1934, Int. ent. Z. 27: 538.)

MOSHER, EDNA (1879–1972) (Zimmerman 1973, Bull. ent. Soc. Can. 5: 143–145.)

MOSHER, FRANKLIN HERBERT (1861–1925) (Burgess 1925, J. Econ. Ent. 18: 562.)

MOSKITOCID® A registered biopesticide derived from *Bacillus thuringiensis* var. *israelensis*. See *Bacillus thuringiensis.*

MOSOSERIES Lepidoptera larva: A uniserial circle in which more than half of the Crochets are absent from the Proleg. See Proleg.

MOSQUITO HAWK See Odonata.

MOSQUITO NET A fine-weave net-like fabric that is suspended above a bed and which hangs around the bed or the hem of which may be fitted beneath the mattress. The MN is used to provide an enclosure that protects sleeping people from attack by mosquitoes and other small-bodied, blood-feeding arthropods.

MOSQUITOES See Culicidae. Cf. Asian Tiger-Mosquito; Australian Malaria-Mosquito; Black Salt-Marsh Mosquito; Black-Striped Mosquito; Brown House-Mosquito; Brown Saltmarsh-Mosquito; California Salt-Marsh Mosquito; Common Banded-Mosquito; Common House-Mosquito; Common Malaria-Mosquito; Crabhole Mosquito; Dengue Mosquito; Domestic Container-Mosquito; Floodwater Florida Glades Mosquito; Forest Day-Mosquito; Golden Mosquito; Grey-Striped Mosquito; House Mosquito; Northern House-Mosquito; Northwest Coast-Mosquito; Pitcher-Plant Mosquito; Salt-Marsh Mosquito; Salt-Water Mosquito; Southern House-Mosquito; Southern Saltmarsh Mosquito; Vexans Mosquito; Western Treehole-Mosquito; Yellow Fever Mosquito.

MOSS BUGS See Peloridiidae.

MOSS, FRANK HUMPHREY (1906–1965) (R.P.H. 1905, Entomologist 98: 94–95.)

MOSSYROSE GALL WASP *Diplolepis rosae* (Linnaeus) [Hymenoptera: Cynipidae].

MOTE, DON CARLOS (1887–1972) (Ferguson *et al.* 1974, J. Econ. Ent. 67: 570.)

MOTH Noun. (Anglo Saxon, *mothe*. Pl., Moths.) A common name applied to nocturnal Lepidoptera whose Antennae are not clubbed. Cf. Butterfly; Skipper.

MOTH BALL Noun. See Naphthalene.

MOTH BUTTERFLY *Liphyra brassolis* Rothchild [Lepidoptera: Lycaenidae].

MOTH FLIES Small, moth-like flies sometimes regarded as urban pests near bathrooms, greenhouses and other moisture-laden habitats with decaying organic matter. In domestic situations, MF are often associated with slime in pipes and drains. Adults 2–4 mm long, sombre coloured, densely setose; sucking mouthparts but most do not bite; poor fliers with jerky movements; wings and body densely covered with Setae giving moth-like appearance; wings at repose held vaulted over Abdomen. Larvae with well developed, sclerotized head capsule and Mandibles to feed on decaying vegetation and animal matter; spiracles at posterior end of cone-like body remain in contact with air. Adults typically nocturnal; most Species associated with forested environments near water. Female oviposits in polluted water; eggs hatch within 24 hours; larvae aquatic or semiaquatic, feed 4–15 days; pupation near or within food. MF may cause allergies through inhalation of flies or their body parts. See Psychodidae. Cf. Owl Midges; Sand Flies. Rel. Fungus Gnats.

MOTH FLY See Trickling Filter Fly.

MOTH LACEWINGS See Ithonidae.

MOTHER GENUS The original Genus from which other Genera have been derived by nomenclatural splitting.

MOTHS See Lepidoptera.

MOTILE Adj. (Medieval Latin, *motivus* = moving.) Moving or being able to move.

MOTOR NERVE A nerve that controls motion.

MOTOR NERVOUS SYSTEM The part of the nervous system lying entirely within the body and which transmits stimuli to the motor elements (muscles).

MOTOR NEUROCYTE The Neurocyte of a Motor Neuron.

MOTOR NEURON A Neuron (typically monopolar) whose Ganglia and Axons pass to effectors (muscles and glands). Syn. Motor Neurone; Efferent Neurone. Syn. Efferent Nerve; Efferent Neuron. See Nervous System. Cf. Afferent Neuron; Interneuron; Sensory Neuron.

MOTSCHULSKY (MOCHULSKY), VICTOR IVANOVICH (1810–1871) (Solsky 1868, Horae Soc. ent. Ross. 6 (Suppl.): 1–118, bibliogr.)

MOTTLE Adj. (Middle English, *motteley*.) An irregular pattern of pale and dark green areas on leaves. Cf. Mosiac.

MOTTLED CUP-MOTH *Doratifera vulnerans* (Lewin) [Lepidoptera: Limacodidae]: A bivoltine, minor pest of fruit and native ornamental trees in eastern Australia. Female oviposits on leaves in rows of 8–12 eggs per row; eggs bright yellow, flattened and covered with short, dark Setae from female. Caterpillar green-brown with yellow-white patches, bordered with orange and purple grey in middle of dorsum, and two pairs of stellate tubercles composed of defensive spines anterior and posterior ends of body. Constructs a cup-like cocoon. Syn. Chinese-Junk Caterpillar.

MOTTLED FLOWER SCARAB *Protaetia fusca* (Herbst) [Coleoptera: Scarabaeidae]: In Australia, a pest of cultivated garden plants and field crops; can seriously defoliate maize and destroy mango blossom.

MOTTLED KATYDID *Ephippitytha trigintiduoguttata* (Serville) [Orthoptera: Tettigoniidae]. Cf. Angular-Winged Katydid; Broad-Winged Katydid; Citrus Katydid; Crested Katydid; Forktailed Bush-Katydid; Inland Katydid; Philippine Katydid; Spotted Katydid.

MOTTLED TORTOISE BEETLE *Deloyala guttata* (Oliver) [Coleoptera: Chrysomelidae].

MOUCHA, JOSEPH (1930–1972) (Carneluffi 1972, Acta ent. jugosl. 8: 138, bibliogr.)

MOUFFET, THOMAS (1550–1599) English physician and author of *Insectorum sive Minorum Animalium Theatrum* published posthumously (1634.) Work also includes numerous unpublished notes by Conrad Gesner and Edward Wooton. Also spelled Moffett.

MOUFFLET, ALFRED (1821–1866) (Deyrolle 1866, Ann. Soc. ent. Fr. (4) 6: 607–610.)

MOULA Noun. (Latin, *moula* = knee ball. Pl., Moulas; Moulae.) The proximal portion of a leg's Tibia, typically bent and forming a ball with articulatory processs for the reception of the Femur. See Tibia. Rel. Leg.

MOULD MITE *Tyrophagus putrescentiae* (Schrank) [Acari: Acaridae].

MOULT Intransitive Verb. (Latin, *moutare* = to change.) The act by which insects cast off elements of the Integument during postembryonic growth. Noun. (Latin, *moutare* = to change. Pl., Moults.) 1. The period of transformation when the larva, nymph, naiad or subimago changes from one instar to another. Moulting serves to accommodate growth in size and change in shape. The number of moults varies considerably among insects. Apterygotes do not show a constant number of moults. The collembolan *Willowsia jacobsoni* has an average number of moults to adulthood of five in males and six in females. Apterygotes often moult after attaining adulthood, and the average number of moults in this Species during lifetime is ca 30 in males and 29 in females. Pterygota show no fixed number of moults, but most Species have a typical number of moults. Cyclorrhaphous Diptera have three larval instars. Many higher taxonomic groups show a tendency toward reduction in moult number in more highly evolved or specialized forms. The last larval instar of many insects (including bees) does not feed and does not moult. This instar is sometimes called the prepupa or pharate Pupae. Morphologists determine number of instars by direct observation, frequency-distribution graphs and numerical techniques such as linear regression. 2. The cast portion of the Integument (exuvia) resulting from the processing of moulting. Alt. Moult. See: Apolysis; Ecdysis. Rel. Epitoky; Instar.

MOULTING A periodic process of loosening and discarding the Cuticule, accompanied by formation of a new Cuticle, and often by structural changes in body wall and other organ (Snodgrass). Moulting consists of two phases, Apolysis and Ecdysis. Apolysis involves the physical separation of old Cuticle from epidermal cells of Integument. Ecdysis involves the shedding of the old Cuticle from the body. Moulting is a complex phenomenon involving integument's structure. Moulting is a physiochemical adaptation that solves several problems associated with a terrestrial existence, including desiccation, locomotion, and structural support. Growth is a principal problem for insects during postembryonic development. Growth is fundamentally different between vertebrates and invertebrates. Vertebrates display continuous growth through most of life but in later years undergo regressive growth. In contrast, invertebrates must moult periodically because the Integument limits cuticular expansion. Growth is limited by extensibility of the outer Epicuticle. Beneath the old Cuticle, new Epicuticle is highly convoluted. This arrangement increases integument's surface area without significantly increasing the volume that it occupies. Rel. Integument.

MOULTING FLUID See Moulting Liquid.

MOULTING GLANDS See Exuvial Glands (Snodgrass).

MOULTING HORMONE See Ecdysone; Ecdysteroid.

MOULTING LIQUID A fluid secreted abundantly by growing insects in the act of moulting, by dermal glands of the body surface.

MOULTON, DUDLEY (1878–1951) (Bailey 1951, Pan-Pacif. Ent. 27: 147–147.)

MOULTON, JOHN CONEY (1886–1926) (Anon. 1926. Entomol. mon. Mag. 62: 242.)

MOUND NEST Social Insects: A nest which is constructed in part of soil or carton material that projects above the surface of the ground. The architecture of the mound is often specific, frequently elaborate and sometimes involved in maintaining temperature within the nest.

MOUNTAIN BLUE BUTTERFLY *Neolucia hobartensis* (Miskin) [Lepidoptera: Lycaenidae].

MOUNTAIN CARPENTER BEE *Xylocopa orpifex* Smith [Hymenoptera: Anthophoridae]: A Species consisting of several named Subspecies that occur in western North America. Constructs tunnels in sound wood, including redwood and Douglas fir. See Anthophoridae. Cf. Carpenter Bee; Little Carpenter Bee. Rel. Bumble Bee.

MOUNTAIN KATYID *Acripeza reticulata* Guérin-Méneville [Orthoptera: Tettigoniidae].

MOUNTAIN LEAFHOPPER *Colladonus montanus* (Van Duzee) [Hemiptera: Cicadellidae].

MOUNTAIN MIDGES See Blephariceridae.

MOUNTAIN PINE BEETLE *Dendroctonus ponderosae* Hopkins [Coleoptera: Scolytidae]: A significant pest of several pine tree Species in western North America and frequent cause of tree mortality. Monovoltine; overwinters as egg larva or adult. Adult ca 2–4 mm long; body black with Elytra finely punctate and striated. Adults emerge June–September; eggs deposited in niches on alternate sides of gallery; larvae tunnel at right angles to gallery; pupation occurs at end of larval tunnels. See Scolytidae. Syn. Black Hills Beetle, *Dendroctonus monticolae*. Cf. Southern Pine Beetle; Eastern Larch Beetle; Red Turpentine Beetle; Western Pine Beetle.

MOUNTAIN PINE CONEWORM *Dioryctria yatesi* Mutuura & Munroe [Lepidoptera: Pyralidae].

MOUNTAIN PINHOLE BORER *Platypus subgranosus* Schedl [Coleoptera: Curculionidae].

MOUNTAIN SKIPPER *Anisynta monticoloe* (Olliff) [Lepidoptera: Hesperiidae].

MOUNTAIN SPOTTED SKIPPER *Oreisplanus perornatus* (Kirby) [Lepidoptera: Hesperiidae].

MOUNTAIN-ASH SAWFLY *Pristiphora geniculata* (Hartig) [Hymenoptera: Tenthredinidae].

MOUNTANT Noun. (French, *montant* ppv of *monter* = to mount. Pl., Mountants.) Any substances in which a specimen is placed for observations, usually beneath a coverslip for microscopical observation. Mountants are classified as temporary or permanent. Mountants may be solid (plastic), semisolid (gelatin), liquid (glycerine) or gas (air). Mountants miscible in water include Hoyers Medium, Lactic Acid-PVA and glycerin jellies. Mountants miscible in alcohol include resins, gum mastic, Venice terpentine and Gum Sandarac. Mountants miscible in aromatic hydrocarbons include natural resins (Canada Balsam) synthetic resins (Permount, Euparol, Histoclad, Piccolyte, Hyrax) and mineral oils (Nujol).

MOURNINGCLOAK BUTTERFLY *Nymphalis antiopa* (Linnaeus) [Lepidoptera: Nymphalidae]: A widespread Holarctic Species. Favoured host plants of the larva include elm, poplar, willow. Bivoltine; overwinters as adult. Alt. Camberwell Beauty (UK); larva called 'Spiny Elm Caterpillar'.

MOUSE FLEA *Leptopsylla segnis* (Schonherr) [Siphonaptera: Leptopsyllidae].

MOUSE FLEAS See Leptopsyllidae.

MOUSE SPIDERS See Actinopodidae.

MOUTH Noun. (Middle English. Pl., Mouths.) The anterior opening of the Alimentary Canal, located on the head and the place where feeding appendages are positioned and in which the food is prepared for ingestion. Cf. Oriface. Rel. Aperture; Buccal Cavity.

MOUTH BEARD Asilidae: A prominent tuft of Setae in the front of the head. Syn. Mystax.

MOUTH CAVITY Space enclosed by the Labrum and the mouth appendages, Syn. Preoral Cavity, Extraoral Cavity (Snodgrass).

MOUTH CONE 1. General: Rostrum; Proboscis; Prostomium. 2. Pediculus: A small protractile snout-like tube terminating the elongate head. 3. Thysanoptera: The united Labrum, Labium and Galeae, which contains the Stylets. Cf. Rostrum.

MOUTH DILATORS A pair of muscles originating on the Clypeus and inserting on the Stomodaeum near the mouth. Syn. Dilatores Buccales (Snodgrass).

MOUTH FORK 1. Psocidae: A pair of long, slender, styliform appendages near, but not connected with, the Maxillae. MFs are capable of considerable extrusion from the mouth and used for gouging tree bark or cutting fungal mycelia. 2. Chisels *sensu* Tillyard.

MOUTH HOOKS Muscoid larvae: Solid (not hollow), cuticular claw-like structures secondarily developed on on either side of the mouth opening. MH are the substitute 'jaws' of larvae and often erroneously called Mandibles (Snodgrass).

MOUTHPARTS Plural Noun. Cranial appendages specifically adapted for the acquisition and processing of food. Principal mouthparts include Mandible, Maxilla and Labium; Each appendage is subdivided into component parts of varing complexity and functional interaction among different groups of insects. The Labrum and Antenna sometimes act as accessory feeding structures. Insects have modified their mouthparts in many ways as adaptation for biting, chewing, piercing and sucking. In the context of evolution, these modifications are viewed as strategies for processing food items of different physical complexity and chemical properties. Food is utilized by insects as a liquid, a solid, or a solid suspended in a liquid. Syn. Trophi. See Feeding Strategies. Cf. Labium; Labrum; Mandible; Maxilla. Rel. Labial Glands; Mandibular Glands; Maxillary Glands.

MOVABLE HOOK Dragonflies: A small tooth on the inner border of the lateral lobe, not fixed, as its name implies (Imms).

MOVEMENT OF FLIGHT In the insect wing, the movements that make flight possible, consisting of an upstroke, a downstroke, a forward movement, a rearward movement and a partial rotation of each wing on its axis.

MOXIDECTIN A 16-membered macrocyclic-lactone, closely related to Abamectin, Doramectin and Ivermectin, that is produced by soil-inhabiting actinomycetes. Moxidectin is purported to be effective for control of nematodes, insects, other arthropods and as a dewormer for horses Cf. Abamectin; Doramectin; Ivermectin.

M-PERIL® The delta-endotoxin of *Bacillus thuringiensis* var. *kurstaki* which has been killed and encapsulated in *Pseudomonas fluorescens*. Developed and marketed by Mycogen Corporation (1990.) See *Bacillus thuringiensis*.

MPMC A carbamate compound {3,4-dimethylphenylmethylcarbamate} used as a contact insecticide for control of plant-sucking insects (aphids, leaf hoppers, scale insects, mealybugs) in some countries. MPMC applied to fruit trees, rice and tea in some countries. Not registered for use in USA. Toxic to fishes. Trade names include Xylylcarb® and Meobal®. See Carbamate Insecticide.

MRAZ, JARO (?1880–1927) (Obenberger 1927, Cas. csl. Spol. ent. 24: 77–80.)

MRCIAK, MILAN (1923–1975) (Dusbábek 1975, Acta ent. bohemoslovaca 72: 428–429.)

MSR® See Oxydemeton-Methyl.

MTMC® See Metolcarb.

M-TRAK® A biopesticide derived from *Bacillus thuringiensis* var. *tenebrionis*. See *Bacillus thuringiensis*.

MU A unit of land-area measurement in China. The traditional mu is about 675 square metres (800 square yards). In modern China, the mu is 1/15 hectare (666 square metres or 797 square yards). Alt. Mou.

MUCILAGINOUS Adj. (Late Latin, *mucilage* = a musty juice; *-osus* = with the property of.) Gummy, gum-like, sticky, like gum or mucilage.

MUCIN Noun. (Latin, *mucus*. Pl., Mucins.) An acidic glycoprotein, insoluble in water, but soluble in alkaline solutions. When dissolved a mucin has a glutinous appearance and is not coagulated by boiling. Secreted by the mantle of some bivalve molluscans onto the outer surface of the shell and to which sand grains adhere.

MUCOCUTANEOUS LEISHMANIASIS A New World form of leishmaniasis resulting from a previous case of Cutaneous Leishmaniasis and which involves nasopharyngeal lesions; ML can result in death. See Leishmaniasis.

MUCOID Adj. (Latin, *mucosus* = mucus.) Mucus-like in form or substance; descriptive of moist, vicid material. Alt. Mucoidal. See Mucus. Cf. Colloid. Rel. Eidos, Form, Shape.

MUCOREOUS Adj. (Latin, from *Mucor*, a genus of fungi; *-osus* = with the property of.) Appearing as if mouldy, superficially covered with small, fringe-like processes. Alt. Mucoreus

MUCOUS See Mucus.

MUCRO Noun. (Latin, *mucro* = sharp point. Pl., Mucrones.) 1. Any short, stout, sharp, apically pointed process (Kirby & Spence). 2. Collembola: The short terminal segment of a fork of the Manubrium, variously shaped and borne upon the Dens. 3. Coleoptera: Prosternal Process in Elateridae. 4. The terminal spine or process of an obtect Pupae. 5. Median posterior part of the Epigastrium when differentiated by elevation. 6. Hymenoptera: Apex of Gonapophysis IX used as a wedge.

MUCRONATE Adj. (Latin, *mucro* = sharp point; *-atus* = adjectival suffix.) Descriptive of a structure which terminates abruptly in a sharp point or spine. Alt. Mucronatus.

MUCUS Noun. (Greek, *myxa* = nasal mucus. Pl., Mucuses.) A viscid, slippery, mucin-rich glandular secretion manufactured by animals and plants. Cf. Colloid; Slime. Rel. Mucoid.

MUD CELL Hymenoptera: Any cell fashioned of mud and used for the containment of prey. Typically holds the immature stage of the wasp which constructs the cell.

MUD DAUBERS *Sceliphron* spp. [Hymenoptera: Sphecidae]: Solitary, large-bodied yellow-and-black mud-nesting wasps. Females provision nest with spiders; eggs laid on prey. See Sphecidae. Cf. Organ Pipe Mud Dauber.

MUD-EYES Common name of some Odonata naiads.

MUESEBECK, CARL FREDERICK WILLIAM (1894–1987) American government Entomologist (USDA) and administrator. Head, USDA Division of Insect Identification (1935–1954). Taxonomic specialist of parasitic Hymenoptera, particularly the Braconidae. Author of 137 scientific publications and principal editor of first North American Hymenoptera Catalog. President, Entomological Society of America (1946), Honorary Member ESA (1959), President, Entomological Society of Washington (1940), Honorary President (1971–1987), USDA Distinguished Service Award (1951), ESA L. O. Howard Distinguished Achievement Award (1978). (Krombein & Marsh 1988, Proc. ent. Soc. Wash. 90 (4): 513–523.)

MUGA SILKWORM *Antheraea assama* Westwood [Lepidoptera: Saturniidae].

MÜGGE, AMANDUS (1870–1949) (Weidner 1967, Abh. Verh. naturw. Ver. Hamburg (Suppl.) 9: 296–297.)

MÜGGENBURG, FRIEDRICH HANS (1865–1901) (Lichtwardt 1902, Z. syst. Hymenopt. Dipterol. 2: 1.)

MUHAIGAWA, YUSAKU (1883–1927) (Hasegawa 1967, Kontyû 35 (Suppl.): 83.)

MÜHL, ADOLF (1834–1911) (Anon. 1968, Mitt. dt. ent. Ges. 27: 1–2.)

MÜHLEN, MAX VON (1850–1918) (Schneider 1927, KorrespBl. NaturfVer. Riga 59: (5)-(6.)

MÜHLIG, JOHANN GOTTFRIED GOTTLIEB (1813–1884) (Anon. 1884, Psyche 4: 236.)

MUIR, FREDERICK ARTHUR GODFREY (1872–1931) (Williams 1932, Proc. Hawaii. ent. Soc. 8: 13–15, 141–152, bibliogr.)

MUIR, JOHN (1837–1914) (Grinell 1915, Ent. News 26: 95–96.)

MUIZON, JOSEPH DE (1890–1960?) (Marcou 1960, Mém. Inst. fr. Afr. noire 59: 7.)

MUIZON, M J J DE (–1958) (Anon. 1958, Bull. Soc. ent. Fr. 63: 158.)

MUKAIGAWA, Y Japanese amateur Entomologist interested in Thrips.

MULBERRRY SILKWORM *Bombyx mori* (Linnaeus.) [Lepidoptera: Saturniidae].

MULBERRY PYRALID MOTH *Margaronia pyloalis* Walker [Lepidoptera: Pyralidae].

MULBERRY TIGER MOTH *Spilarctia imparilis* Butler [Lepidoptera: Arctiidae].

MULBERRY WHITEFLY *Tetraleyrodes mori* (Quaintance) [Hemiptera: Aleyrodidae].

MULDER, CLAAS (1796–1867) (Ermerines 1867, Jaarb. K. Akad. Wet. Amst. 1867: 1–21, bibliogr.)

MULDER, JOHN FREDERICK (1840–1921) (Anon. 1922, Victorian Nat. 38: 138.)

MULDER, RUDOLPH HERMAN (1914–1992) Dutch born Australian amateur collector of Insects with special interests in Coleoptera. (Chadwick 1993, Myrmecia 29 (4): 8–9.)

MULGA ANT *Polyrhachis macropus* Wheeler [Hymenoptera: Formicidae].

MULLAN, J P (–1957) (Richards 1958, Proc. R. ent. Soc. Lond. (C) 22: 75.)

MULLEIN THRIPS *Haplothrips verbasci* (Osborn) [Thysanoptera: Phlaeothripidae].

MÜLLER, A JULIUS (–1926) (Hedicke 1926, Dt. ent. Z. 1926: 359.)

MÜLLER, ARNOLD (1884–1934) (Ebner 1935, Konowia 14: 8.)

MULLER, ERNST (1864–1937) (Michalk 1937, Ent. Z. Frankf. a. M. 51: 1–2.)

MÜLLER, FRIEDRICH AUGUST CLEMENS (1829–1902) (Kraatz 1903, Dt. ent. Z. 47: 173–176.)

MÜLLER, FRITZ (JOHANNES FRIEDRICH THEODOR) (1822–1897) (Roquette-Pinto 1929, Bol. Mus. nac. Rio de J. 5: 1–23.)

MÜLLER, GEORG (1864–1946) (Rapp 1935, Beitr. Fauna Thürungen 1: 52.)

MÜLLER, GIUSEPPE (1880–1964) (Invrea 1966, Memorie Soc. ent. ital. 45: 135–148, bibliogr.)

MÜLLER, GUSTAF FREDERIK (1826–1889) (Neren 1889, Ent. Tidskr. 10: 181–190.)

MULLER, H J (Dunn 1947, Nature 215: 108–109.)

MÜLLER, HERMANN (1829–1883) (Anon. 1883, Science 2: 487–488.)

MÜLLER, JOHANN KARL AUGUST (1818–1899) (Taschenberg 1899, Jber. naturf. Ges. Graübundens 42: xvi-xxvi.)

MÜLLER, JOHANNES PETER (1801–1858) (Lohe 1902, Archs Parasit. 5: 95–116, bibliogr.)

MÜLLER, OTTO FRIEDRICH (1730–1784) Author of *Fauna Insectorum Fredrichsdaliana* (1764). (Eiselt 1836, *Geschichte, Systematik und Literateur der Insektenkunde*. 255 pp. (52), Leipzig.)

MÜLLER, PAUL (1897–1957) (Anon. 1957, Ent. Bl. Biol. Syst. Käfer 53: 129–130.)

MÜLLER, PHILIP LUDWIG STATIUS (1725–1776) (Anon. 1776, Beschäft. berlin Ges. naturf. Fr. 2: 582–592.)

MULLER, PHILIP WILBRAND JACOB (1772–1851) (Westwood 1853, Proc. ent. Soc. Lond. 1852–53: 53–54.)

MULLER, ROBERTO (1859–1932) (Hoffman 1932, An. Inst. Biol. Univ. Méx. 3: 133–148.)

MULLERIAN ASSOCIATION An association of Species assigned to distantly related Taxa which have similar colours, possess more or less distasteful qualities and live in the same locality. See Mullerian Mimicry.

MULLERIAN BODIES Bodies produced at the base of petioles of *Cercropia* trees and which are consumed by *Azteca* ants. Cf. Beccarian Bodies; Beltian Bodies.

MULLERIAN MIMICRY A type of Mimicry in which the model and mimic are distasteful, inedible or harmful. Mullerian Mimicry was named after German zoologist Fritz Müller (1822–1897) who collected amazonian butterflies and published the results of his investigations in 1878. In Mullerian Mimicry, two or more distasteful or harmful Species are involved in the system as models. They resemble one another, and one or more of the participants gains advantage from the resemblance. See Mimicry. Cf. Batesian Mimicry. Rel. Crypsis.

MÜLLER-KNUCHEL, AUGUST (1874–1954) (Anon. 1955, Mitt. ent. Ges. Basel 5: 14–15.)

MÜLLER-RUTZ, JEAN (1854–1944) (Th. 1944, Mitt. schweiz. Ent. Ges. 19: 204–207, bibliogr.)

MULLER'S ORGAN 1. In the insect ear, a group of scolophores forming a swelling applied to the inner surface of the Tympanum (Imms). 2. The swollen termination of the insect's auditory nerve (Folsom & Wardle).

MULLER'S THREAD The common terminal thread of all the ovarian tubules. Syn. Terminal Filament, Terminal Ligament.

MULSANT, ETIENNE (1797–1880) (Essig 1931, *History of Entomology*, 1029 pp. (715–717, bibliogr.), New York.)

MULTAMATA® See Bendiocarb.

MULTANGULATE Adj. (Latin, *multi* = many; *angulatus* past part. *angulare* = to make angular; *-atus* = adjectival suffix) Pertaining to structure with many angles.

MULTI- Latin prefix meaning 'many'.

MULTIARTICULATE Adj. (Latin, *multi* = many; *articulatus* past part. *articulare* = to divide into joints; *-atus* = adjectival suffix.) Pertaining to structure or appendages with many joints or segments.

MULTICELLULAR Adj. (Latin, *multi* = many; *cellula* from *celare* = to hide.) Consisting, comprised or composed of two or more cells.

MULTICELLULAR GLAND A gland consisting of more than one cell. Typically Exocrine Multicellular Gland cells form invaginations of the Epidermal cells of the body wall. Anatomical components of the MG include a Secretory Cell, Cuticular Intima, Reservoir, Duct, and Pore (Aperture.) MGs' forms include tubular, convoluted, branched and acinose (racemose). Syn. Compound Gland. See Exocrine Gland; Gland. Cf. Unicellular Gland.

MULTICELLULAR PROCESSES Integument: Cuticular structures which are hollow projections from the body wall; MP usually large and spine-like in form and may arise from any part of the Integument (Snodgrass).

MULTICIDE CONCENTRATE® See Phenothrin.

MULTICOLONIAL Adj. (Latin, *multi* = many; *colonia* from *colonus* = farmer, *colere* = to cultivate, dwell; *-alis* = characterized by.) Social Insects: Pertaining to a population that is divided into colonies and whose members recognize nest boundaries. See Social Insect; Colony. Cf. Unicolonial. Rel. Aggregation.

MULTICUSPID CAP See Cheliceral Blade.

MULTIFID Adj. (Latin, *multifidus* from *multi* = many; *findere* = to split.) Cut into many segments. Alt. Multifidous.

MULTILOCULAR WAX GLAND Pseudococcidae: An uncommon type of Wax Gland apparently restricted to the ventral surface of the last five body segments. In part because of their position, these glands have sometimes been called Circumgenital Glands in scale insects. MWG consists of 12 cells as the tubular wax gland but displays an aperture which is substantially larger than other types of Wax Glands. The MWG aperture is positioned in a shallow depression on a low Papilla. The aperture consists of a circle of ten triangular openings surrounding a larger central opening. The wax forms 'irregular masses.' Cf. Triangular Wax Glands; Tubular Wax Gland; Quinquelocular Wax Glands.

MULTILOCULAR Adj. (Latin, *multi* = many; *oculus* = eye.) With many large cells, spaces or cavities.

MULTINUCLEATE Adj. (Latin, *multi* = many; + nucleus; *-atus* = adjectival suffix.) Having many nuclei. Term applied to cells.

MULTIORDINAL CROCHET Lepidoptera larvae: Descriptive of Proleg Crochets that display several sizes. See Crochet. Cf. Biordinal Crochet; Uniordinal Crochet.

MULTIPARASITISM Noun. See Multiple Parasitism.

MULTIPARTITE Adj. (Latin, *multipartitus* from *multi* = many; *partitus* = divided; *-ites* = constituent.) Divided into many parts.

MULTIPLE PARASITISM Hymenoptera: The oviposition of eggs on or in a host by more than one Species of parasite. Syn. Multiparasitism. See Parasitism. Cf. Facultative Multiple Parasitism; Obligatory Multiple Parasitism. Rel. Superparasitism; Host Discrimination.

MULTIPLE RELEASE In augmentative biological control, a Species (biotype, race, strain, variety) imported and released in an area on more than one occasion. See Augmentative Biological Control. Cf. Single Release.

MULTIPLICATE Adj. (Latin, *multi* = many; *plica* = fold; *-atus* = adjectival suffix.) Descriptive of structure with many longitudinal folds or lines of plication. Alt. Polyplicate. See Plicate. Cf. Uniplicate.

MULTIPOLAR CELL A nerve cell with more than two Neurons proceeding from the cell body. See Neuron; Nerve Cell; Nervous System. Cf. Bipolar Cell; Unipolar Cell. Rel. Muscle.

MULTIPOROUS PLATE SENSILLUM Hymenoptera: A form of Sensilla Placodea on the Antenna of Apocrita. MPS are elongate, elevated ridges above the surface of the Cuticle and subparallel to the long axis of the funicular segment. The Sensilla vary in number and distribution, often they occur on all Flagellomeres. MPS consist of two cuticular channels which form three invaginations that extend the length of the sensillum. Pores may not be evident with SEM or LM but are evident with TEM. The Sensilla are innervated with as many as 50 bipolar neurons, most of which are positioned at the proximal portion of the median channel. Syn. Glume; Rhinaria; Tyloids. See Pore Plate Sensillum; Sensillum.

MULTISEGMENTAL Adj. (Latin, *multus* = many; *segmentum* = a slice, a zone; *-alis* = characterized by.) Descriptive of a body, body region, appendage or structure composed of more than one segment (*e.g.* antennal Flagellum with many segments). See Segmentation. Cf. Unisegmental. Alt. Polysegmental.

MULTISEPTATE Adj. (Latin, *multus* = many; *septum* = hedge; *-atus* = adjectival suffix.) Descriptive of structure which has more than one septum or internal partition with a cell, cavity or coelome. See Septum. Cf. Uniseptate. Alt. Polyseptate.

MULTISERIAL BANDS Lepidoptera larva: Two transverse bands formed when the Proleg Crochets are absent from the mesial and lateral parts of the circle (Imms). Cf. Uniserial Circle.

MULTISERIAL CIRCLE Lepidoptera larva: The arrangement of the Proleg Crochets in several concentric circles. See Crochet.

MULTISERRATE Adj. (Latin, *unus* = one; *serra* = saw; *-atus* = adjectival suffix.) Descriptive of structure with several rows of serrations along a margin. Cf. Aculeate-Serrate; Biserrate; Dentate-Serrate; Subserrate; Uniserrate. Alt. Polyserrate.

MULTISPINOSE Adj. (Latin, *multi* = many, *spina* = spine; *-osus* = full of.) Descriptive of structure with may spines.

MULTIVARIATE DISCRIMINANT ANALYSIS (See Neff & Smith 1979, Syst. Zool. 28: 176–196.)

MULTIVOLTINE Adj. (Latin, *multi* = many; Italian, *volta* = turn.) Pertaining to organisms with many generations in a year or season. Term often applied to Lepidoptera, Diptera and other insects of economic importance. Cf. Univoltine.

MUMIA Noun. (Latin, *mumia* = mummy.) The Pupa.

MUMIA PSEUDONYMPHA Lamarck's name for a pupae which has some degree of locomotion.

MUMMIFY Intransitive Verb. To dry and shrivel as a mummy.

MUMMY Noun. (Latin, *mumia*. Pl., Mummies.) 1. An aphid in the advanced stages of parasitism as attached by an internal parasitic hymenopteran. 2. A dried and shrivelled fruit.

MUNAKATA, TETSUZO (1886–1913) (Hasegawa 1967, Kontyû 35 (Suppl.): 84.)

MUNDELLA, R C (–1942) (Osborn 1946, *Fragments of Entomological History,* Pt. II. 232 pp. (103–104), Columbus, Ohio.)

MUNG MOTH *Maruca testulalis* (Geyer) [Lepidoptera: Pyralidae]: A pantropical pest of legumes. See Pyralidae.

MUNGANAST, EMIL (1848–1914) (Reitter 1914, Wien. ent. Ztg. 33: 210.)

MUNGER CELL An artificial cell of container in which insects are cultured or are maintained.

MUNITE Adj. (Latin, *munitus.*) Armed; provided with an armature.

MUNRO, J W (1888–1968) (Herford 1968, J. stored Prod. Res. 4: 99.)

MUNSTER, THOMAS GEORG (1855–1938) (Natvig 1938. Norsk ent. Tidsskr. 5: 49–54, bibliogr.)

MURAI, TEIKO (1885–1963) (Hasegawa 1967, Kontyû 35 (Suppl.): 84.)

MURAT, MARÉ (1909–1940) (Maire 1941, Bull. Soc. Hist. nat. Afr. N. 32: 155–157.)

MURATA, TOSHICHI (1879–1945) (Hasegawa 1967, Kontyû 35 (Suppl.): 84–85.)

MURFITE® See Tetradifon.

MURICATE Adj. (Latin, *muricatus*, having sharp points; *-atus* = adjectival suffix.) Superficially covered with sharp, thick, but not close, elevated rigid points (Kirby & Spence).

MURINE Adj. (Latin, *murinus* = mouse-grey.) Mouse-grey with a yellowish cast. Alt. Murinus

MURINE TYPHUS A cosmopolitan rickettsial disease principally of rats. Aetiological agent, *Rickettsia typhi* [= *T. mooseri*] most prevalent rickettsial disease of humans and widespread in ports and rural areas. MT incubation period in humans ca 6–14 days; clinical period ca 12 days. MT transmitted from rats [*Rattus rattus, R. norvegicus*] to humans by fleas [*Xenopsylla cheopis, Leptopsylla segnis, Nosopsyllus fasciatus*] and Spined Rat Louse [*Polyplax spinulosa*]. *Rickettsia* multiplies in midgut epithelium of flea or louse and liberated without rupturing cells; fleas pass disease with faeces within 10 days of infection. MT does not harm flea, louse or rat; low mortality associated with disease in humans, except in persons over 50. Syn. Endemic Typhus; Flea Typhus; Rat Typhus; Shop Typhus. Cf. Epidemic Typhus; Tick-Borne Spotted Fever.

MURKY MEAL CATERPILLAR *Aglossa caprealis* (Hübner) [Lepidoptera: Pyralidae].

MURKY MEAL MOTH *Aglossa caprealis* (Hübner) [Lepidoptera: Pyralidae]: A widespread, omnivorous, hyrgophilous pest of stored products, particularly grains, cereals and beans. See Pyralidae.

MUROTOX® See Mecarbam.

MURRAY VALLEY ENCEPHALITIS An arboviral disease whose agent belongs to the Family Flaviviridae. MVE is endemic in southeastern Australia and now widespread in Australia. Viron occurs in birds and is vectored to humans by *Culex annulirostris.* See Arbovirus; Encephalitis; Flaviviridae. Cf. Japanese Encephalitis; Murray Valley Encephalitis; St Louis Encephalitis; West Nile.

MURRAY, ANDREW (1812–1878) (Anon. 1878, Entomol. mon. Mag. 14: 215–216.)

MURRAY, G H (Williams 1948, Proc. R. ent. Soc. Lond. (C)12: 65.)

MURRAY, JAMES (1872–1942) (Day 1942, Entomol. mon. Mag. 78: 120.)

MURRAY, WILLIAM (– 1885) (Anon. 1886, Rep. ent. Soc. Ont. 16: 23.)

MURRAY-AARON, EUGENE (1852–1940) (Anon. 1940, Ward's nat. Sci. Bull. 14 (3–4): 12.)

MURTFELDT, MARY ESTHER (1848–1913) (Osborn 1946, *Fragments of Entomological History*. 394 pp. (165–166), Columbus, Ohio.)

MURVIN® See Carbaryl.

MUSCALURE A synthetic pheromone {Z-9-tricosene} that acts as an aggregation pheromone for Diptera. Muscalure is used in fly baits combined with insecticide for the attraction and elimination of fly pests. Trade names include Flylure®, Lurectron® and Muscamone®. See Synthetic Pheromone. Rel. Parapheromone.

MUSCAMONE® See Muscalure.

MUSCARDINE Noun. (French.) A fungal disease of caterpillars. See Calcino.

MUSCICOLOUS Adj. (Latin, *muscus* = moss; *colere* = living in or on.) Pertaining to organisms that live or grow among or on mosses. Alt. Muscicoline. See Habitat. Cf. Agricolous; Algicolous; Caespiticolous; Fungicolous; Madicolous. Rel. Ecology.

MUSCIDAE Plural Noun. House Flies; Stable Flies. A numerically large, biologcially diverse, cosmopolitan Family of cyclorrhaphous Diptera that is the nominant Family of Muscoidea. Adults of some Species are blood feeders with prestomial teeth of Labellum used for piercing skin. Adults and larvae of some Species are predaceous; some Species are scavengers; other Species feed on decaying vegetable or animal matter. Many Species are economically or medically important. Muscids that are serious pests include Buffalo Fly, Face Fly, Horn Fly, House Fly, Stable Fly and Tsetse Fly. See Muscoidea.

MUSCIDIAN Adj. Descriptive of or pertaining to the muscid Diptera.

MUSCIDORIDAE See Lonchopteridae.

MUSCLE Noun. (Latin, *musculus* = muscle. Pl., Muscles.) Body cells that are highly modified and organized into fibres of the insect body that contract and move appendages, internal structures and body organs. Each muscle has two points of attachment with the body, one point at the origin of the muscle and the other point at the insertion of the muscle into the appendage or stucture that is moved by contraction of the muscle. Rel. Abductor; Adductor; Rotator.

MUSCOID Adj. (Latin, *muscus* = moss; Greek, *eidos* = form.) 1. Moss-like; mossy. 2. 'Fly-like' in habitus or habits; term often applied to a larva which displays a cylindrical, leg-less body that tapers to a conical anterior end and truncate posterior end. A term loosely applied to a large group of Diptera resembling or allied to the Genus *Musca*. See Muscoidea. Rel. Eidos; Form; Shape.

MUSCOIDEA A Superfamily of schizophorous Diptera including Anthomyiidae (Anthomyidae, Scopeumatidae), Calliphoridae, Gasterophilidae, Hippoboscidae, Mormotomyiidae, Muscidae (Fannidae, Glossinidae), Nycteribiidae, Oestridae (Hypodermatidae), Sarcophagidae, Scatophagidae, Streblidae and Tachinidae. See Diptera.

MUSCULARIS Noun. (Latin, *musculus* = muscle.) A muscular sheath investing all parts of an insect's Alimentary Canal (Snodgrass).

MUSCULATED (Latin, *musculus* = muscle.) Furnished or supplied with muscles.

MUSCULATURE Noun. (Latin, *musculus* = muscle. Pl., Musculatures.) 1. The system, arrangement or classification of muscles within an organism (Species). 2. The collective (complete) muscular structure of an organism.

MUSCULI DILATORES SPIRACULORUM An arcane Latin term describing dilator muscles of the respiratory spiracles.

MUSCULI DORSALES An arcane Latin term describing Dorsal Muscles. Cf. Musculi Laterales.

MUSCULI DORSALES EXTERNI LATERALES An arcane Latin term describing Lateral External Dorsal Muscles.

MUSCULI DORSALES EXTERNI MEDIALES An arcane Latin term describing Median External Dorsal Muscles.

MUSCULI DORSALES INTERNI LATERALES An arcane Latin term describing Lateral Internal Dorsal Muscles.

MUSCULI DORSALES INTERNI MEDIALES An arcane Latin term describing Median Internal Dorsal Muscles.

MUSCULI DORSALIS EXTERNI An arcane Latin term describing External Dorsal Muscles.

MUSCULI DORSALIS INTERNI An arcane Latin term describing Internal Dorsal Muscles.

MUSCULI LATERALES An arcane Latin term describing Lateral Muscles.

MUSCULI LATERALES EXTERNI An arcane Latin term describing External Lateral Muscles.

MUSCULI LATERALES INTERNI An arcane Latin term describing Internal Lateral Muscles.

MUSCULI OCCLUSORES SPIRACULORUM An arcane Latin term describing Occlusor Muscles or muscles that close the spiracles. See Occulsor.

MUSCULI PARATERGALES An arcane Latin term describing Paratergal Muscles.

MUSCULI SPIRACULORUM An arcane Latin term describing Spiracular Muscles or muscles associated with the spiracles.

MUSCULI TRANSVERSALES An arcane Latin term describing Transverse Muscles.

MUSCULI TRANSVERSI DORSALES An arcane Latin term describing Dorsal Transverse Muscles.

MUSCULI TRANSVERSI VENTRALES An arcane Latin term describing Ventral Transverse Muscles.

MUSCULI VENTRALES An arcane Latin term describing Ventral Muscles.

MUSCULI VENTRALES INTERNI An arcane Latin term describing Internal Ventral Muscles.

MUSCULI VENTRALES INTERNI LATERALES An arcane Latin term describing Lateral Internal Ventral Muscles.

MUSCULI VENTRALES INTERNI MEDIALES An arcane Latin term describing Median Internal Ventral Muscles.

MUSCULIS ANTLIA An arcane Latin term describing muscles associated with the Proboscis of Lepidoptera.

MUSEUM BEETLE *Anthrenus museorum* (Linnaeus) [Coleoptera: Dermestidae].

MUSGRAVE, ANTHONY (1895–1959) (Whitley 1959, Proc. Linn. Soc. N.S.W. 1958–59: 9–20, bibliogr.)

MUSHROOM BODIES Two stalked nerve structures in the Protocerebral Lobes supposed to be the principal motor and psychic centers of the insect Brain. The stalked bodies, Corpora Pedunculata. See Brain; Protocerebrum.

MUSHROOM FLIES See Mycetophilidae; Phoridae.

MUSHROOM PHORID *Megaselia halterata* (Wood) [Diptera: Phoridae].

MUSHROOM RED PEPPER MITE *Siteroptes mesembrinae* (Canestrini) [Acari: Pygmephoridae].

MUSHROOM SCIARID *Lycoriella auripila* (Fitch); *Lycoriella mali* (Fitch) [Diptera: Sciaridae].

MUSHROOM-SHAPED GLAND The large compact mass formed by the male accessory genital glands (attributed to Huxley by Imms).

MUSHROOM SPRINGTAIL *Hypogastrura denticulata* (Bagnall) [Collembola: Hypogastruridae].

MUSHROOM WHITE-CECID *Heteropeza pygmaea* Winnertz [Diptera: Cecidomyiidae].

MUSHROOM YELLOW-CECID *Mycophila barnesi* Edwards [Diptera: Cecidomyiidae].

MUSIDORIDAE See Lonchopteridae.

MUSIDOROMIMIDAE A monogeneric Family of Diptera known only from the fossil type-Species taken in Lower Jurassic Period deposits of Kirghizistan.

MUSPRATT, VERA MOLESWORTH (1887–1962) (Varley 1963, Proc. R. ent. Soc. Lond. (C) 27: 51.)

MUSSEL SCALE See Purple Scale.

MUSSEL SCALE PARASITE *Aphytis lepidosaphes* Compere [Hymenoptera: Aphelinidae].

MUSSON, CHARLES TUCKER (1856–1928) (Musgrave 1932, *A Bibliogr. of Australian Entomology 1775–1930*. 380 pp. (235), Sydney.)

MUSTARD APHID *Lipaphis erysimi* (Kalt.) [Hemiptera: Aphididae].

MUSTARD SAWFLY *Athalia lugens proxima* [Hymenoptera: Tenthredinidae].

MUTAGEN Noun. (Latin, *mutare* = to change; *genys*. Pl., Mutagens.) A compound capable of inducing mitosis.

MUTANT Noun. (Latin, *mutare* = to change; *-antem* = an agent of something. Pl., Mutants.) An individual with hertiable characteristics different from those of the parents.

MUTATE Verb. (Latin, *mutare* = to change.) To undergo or exhibit a mutation.

MUTATING Adj. (Latin, *mutare* = to change.) A mutant during the process of mutation.

MUTATION Noun. (Latin, *mutare* = to change; English, *-tion* = result of an action. Pl., Mutations.) A change in the amount or structure of an individual's genetic material such that characteristics or features of the organism are changed.

MUTATION THEORY An hypothesis concerning the origin of new forms in organisms that arise from the unchanged parent abruptly, regardless of environment and without transitional forms.

MUTATO NOMINE Latin phrase meaning 'the name being changed.'

MUTCH, JOHN PRATT (1855–1934) (Sheldon 1934, Entomologist 67: 263–264.)

MUTE Adj. (Latin, *mutus* > *muttire* = to mutter.) Silent; without power to produce audible sound.

MUTIC Adj. (Latin, *muticus* = docked; *-ic* = of the nature of.) Unarmed; lacking processes where such usually occur. Alt. Muticus.

MUTICI Noun. (Latin, *muticus* = maimed.) Acridiids without a prosternal spine.

MUTILATE Adj. (Latin, *mutilare* = to maim; *-atus* = adjectival suffix.) Cut off; mutilated. Term applied to abbreviated or incomplete structures.

MUTILATION Noun. (Latin, *mutilare* = to maim; English, *-tion* = result of an action.) The loss of an appendage or structure through physical separation.

MUTILLIDAE Plural Noun. Velvet Ants; Cow Killers. A cosmopolitan, moderate sized Family (ca 5,000 Species) of aculeate Hymenoptera assigned to Superfamily Scolioidea. A few fossil mutillids have been reported, including Species from Baltic amber and Dominican amber. Adults are moderate-sized with strong sexual dimorphism; female typically heavily sclerotized, apterous, densely pubescent with ant-like habitus and 'felt line' on Metasoma Tergum III; male winged; posterolateral portion of Pronotum extending to Tegula. Females are often long-lived (ca 1 year); males are short lived. Larva with well developed Mandible; forms cocoon for pupation. Larvae develop as primary external parasites of concealed immatures of various insects, typically ground-nesting aculeate Hymenoptera, Diptera, Coleoptera and Lepidoptera. Most larvae are solitary; gregarious parasitism is unusual; presumably canibalism occurs when more than one larva attempts to develop on a host. Wingless females often dig or enter small spaces to oviposit; wings prone to damage and thus maladaptive. Many males larger than conspecific females; some may carry females before or during copulation; a small but significant number of Species with wingless males. Male relatively short-lived; female relatively long-lived, some live more than a year. Females spend large proportion of time underground; males spend little time underground. Subfamilies include Apterogyninae, Mutillinae, Myrmillinae, Myrmosinae, Sphaerophthalminae, Typhoctinae.

MUTTKOWSKI, RICHARD ANTHONY (1887–1943) (Calvert 1943. Ent. News 54: 173–174.)

MUTUALISM Noun. (Latin, *mutuus* = exchanged, borrowed, lent; English, *ism* = condition. Pl., Mutualisms.) A form of symbiotic relationship in which both Species or partners in the relationship derive benefit from the association. See Symbiosis; Commensalism. Cf. Parasitism.

MUTUALISTIC Adj. (Latin, *mutuus* = exchanged, borrowed, lent; *-ic* = of the nature of. consisting of. characterized by.) Pertaining to mutualism. Cf. Symbiotic.

MUTZELL, MAX (1818–1887) (Dimmock 1888, Psyche 5: 36.)

MVE See Murray Valley Encephalitis.

MVP® A registered biopesticide derived from *Bacillus thuringiensis* var. *kurstaki*. See *Bacillus thuringiensis*.

MYBP An acronym meaning 'Millions of Years Before Present.' See Geological Time Scale. Cf. BYBP.

MYCANGIA Plural Noun. Coleoptera: Cuticular depressions in the Integument that serve as a repository for carrying ambrosia fungus from one host tree to another by an adult ambrosia-feeding Ipidae (Francke-Grosmann 1963, Ann. Rev. Ent. 8:421). In *Xylosandrus germanus* (Blandford), Mycangia are a pair of fused, partly sclerotized pouches between prothoracic and mesothoracic nota. Mycangia occur exclusively in female beetles and their distribution and form may have taxonomic implications. Mycangia categorized by Batra as (1) Prothoracic Plural Mycangia, (2) Pro-mesonotal Mycangia, (3) Elytral Mycangia, (4) Prosternal-subcoxal Mycangia, and (5) Oral Mycangia. Cf. Acarinaria.

MYCELIUM Noun. (Greek, *mykes* = fungus; Latin, *-ium* = diminutive > Greek, *-idion*. Pl., Mycelia.) Thread-like filaments or hyphae that form a vegetative network of fungi. Alt. Hyphostroma.

MYCETOBIIDAE See Anisopodidae.

MYCETOCYTE Noun. (Greek, *mykes* = fungus; *kytos* = hollow vessel; container. Pl., Mycetocytes.) Amoeboid cells within the insect's Haemocoel which contain microorganisms. Mycetocytes occur between the Ovariole Sheath and Tunica Propria. Mycetocytes also are found in other parts of the body, such as the Fat Body of cockroaches.

MYCETOME Noun. (Greek, *mykes* = fungus; *tome* = to cut. Pl., Mycetomes.) Aphids: The Pseudovitellus.

MYCETOPHAGE Noun. (Greek, *mykes* = fungi; *phagein* = to eat; *age* = *aticum* = action. Pl., Mycetophages.) An organism which feeds upon fungi.

MYCETOPHAGIDAE Plural Noun. Hairy Fungus Beetles. A Family of 200 Species of Coleoptera. Adult 1–5 mm long, black or brown with spots on Elytra, body densely pubescent; resemble Dermestidae. Feed upon fungus; some Species found in mouldy produce stored under damp conditions. See Hairy Fungus Beetle.

MYCETOPHAGOUS Adj. (Greek, *mykes* = fungus; *phagein* = to eat; Latin, *-osus* = with the property of.) Insects which feed exclusively on fungi (after Francke-Grosmann 1963, Ann. Rev. Ent. 8: 421.) See Feeding Strategy.

MYCETOPHILIDAE Plural Noun. Fungus Gnats; Mushroom Flies. A cosmopolitan Family of nematocerous Diptera containing more than 3,000 Species. Also abundant in fossil record with more than 260 described Species. Adults recognized by strong thoracic and tibial bristles, long Coxae, filamentous Antenna and characteristic wing venation in which r-m Crossvein absent and wing membrane covered with Microtrichia. Adults found in humid areas, especially moist woodlands and caves. Larvae fungus feeders, but Species belonging Keroplatinae feed on small arthropods. Larva of *Planaivora insignis* Hickman recorded as ectoparasite in land planarians. Predatory Species, as larvae, spin webs on surface of rocks, under loose bark on rotting logs, on roofs of caves. Webs coated with gelatinous liquid or covered with droplets of same liquid. Small insects entrapped in webs.

MYCOLOGIST Noun. (Greek, *mykes* = fungus; *logos* = discourse. Mycologists.) A person who studies fungi.

MYCOLOGY Noun. (Greek, *mykes* = fungus; *logos* = discourse. Pl., Mycologies.) The study of fungi.

MYCOPHAGOUS Adj. (Greek, *mykes* = fungus; *phagein* = to eat; Latin, *-osus* = with the property of.) Pertaining to organisms that feed on fungi. See Feeding Strategy.

MYCOPLASMA Noun. (Greek, *mykes* = fungus; *plasma* = form. Pl., Mycoplasmas.) An organism resembling a Gram-negative bacterium but lacking a rigid cell wall and varying in shape. Mycoplasmas reside within cell cytoplasm and are transmitted by sap-sucking leafhoppers to infect phloem tissue of plants. Typical diseases include 'aster yellows,' 'little leaf' and 'big bud.' Cf. Spiroplasmas. Rel. Pathogens.

MYCORRHIZA Noun. (Greek, *mykes* = fungus; *rhiza* = root. Pl., Mycorrhizae.) Symbiotic fungal mycelia associated with the roots of plants and which aid in nutrient and water uptake. Two types of mycorrhiza are known: Endomycorrhiza which penetrate the host-plant's root and Ectomycorrhiza which form a mantle around the host-plant's root. Alt. Mycorhiza.

MYCORRHIZAL Adj. (Greek, *mykes* = fungus; *rhiza* = root; Latin, *-alis* = pertaining to.) Pertaining to Mycorrhiza.

MYCOTAL® See *Verticillium lecanii.*

MYCOTOXIN Noun. (Greek, *mykes* = fungus; *toxikon* = poison. Pl., Mycotoxins.) 1. Any of several compounds produced by fungi and which are toxic to animals. 2. A fungal metabolite produced on seed, food or feed which causes illness or death when consumed by humans (*e.g.* aflatoxins).

MYCTERIDAE Plural Noun. A numerically small, geographically widespread Family of Coleoptera assigned to Superfamily Tenebrionoidea. Adults of some Species elongate, slender, flattened but other Species robust; body length 2–13 mm; Antenna with 11 segments, filiform or serrate; tarsal formula 5-5-4. Adults and larvae predaceous; occur under bark, rocks and within vegetation. Syn. Hemipeplidae.

MYDAS FLIES See Mydidae.

MYDASIDAE See Mydidae.

MYDIDAE Plural Noun. Mydas Flies. A Family of Diptera that includes about 340 Species which occur throughout the warmer parts of world, mostly from Afrotropical and Neotropical/southern Nearctic regions. Adult body elongate, wasp-like, medium sized to very large (length of 60 mm); Vertex of head generally broadened, somewhat depressed; one Ocellus on anterior margin; Antenna with four segments, elongate, club-shaped; antennal Flagellomere divided into two segments, basal segment typically cylindrical, longer than head length and apical segment distinctly swollen a little shorter; wing venation similar to Apioceridae with 1–2 branches of M curved forward parallel with the posterior branches of R and hind margin of wing; Tarsi without Empodium. Flower-feeding adults and entomophagous larvae; mouthpart structure of adults indicates they probably feed on nectar. Adults often perch on vegetation or on ground. Larvae found in soil, rotten logs, stumps, nests of fungus ants (*Atta* spp.), usually in close association with beetle larvae (especially Scarabaeidae) upon which they probably feed.

MYELIN Noun. (Greek, *myelos* = marrow. Pl., Myelins.) The protective sheath of nerve fibres. See Nerve.

MYELINATED Adj. Of nerve fibres, surrounded by fatty tissue.

MYERS, JOHN GOLDING (1897–1942) (China 1942, Nature 149: 406.)

MYERS, PAUL REVERE (1888–1925) (Walton *et. al.* 1925, Proc. ent. Soc. Wash. 27: 65–67, bibliogr.)

MYERSIIDAE Plural Noun. A Family-level name in parasitic Hymenoptera proposed by Viereck (1912, Proc USNM 43: 575.) Junior synonym of Ichneumonidae.

MYIASIS Noun. (Greek, *myia* = fly; *-iasis* = suffix for names of diseases. Pl., Myiases.) The infestation of living vertebrates (including humans) with Diptera larvae which feed on the vertebrate's tissue (healthy or necrotic), fluids (blood or other liquid) or ingested food. Myiasis broadly classified as: Dermal (Subdermal) Myiasis, Creeping Myiasis, Nasopharyngeal Myiasis, Enteric (Intestinal, Gastric) Myiasis, Rectal Myiasis and Urinogenital Myiasis. Dermal Myiasis involves larvae penetrating unbroken skin, wounds or lesions; the term Traumatic Myiasis is sometimes used for larvae penetrating wounds. Creeping Myiasis involves larvae burrowing beneath the skin. Nasopharyngeal Myiasis involves larvae penetrating the nasal passage, eye sockets or pharyngeal cavity. Enteric Myiasis involves larvae that enter the mouth or Rectum to live in the gut of the host. Urinogenital Myiasis involves larvae which enter the urinogenital system of the host. See Bot Flies. Cf. Apimyiasis; Entomiasis; Maggot Debridement Therapy.

MYMARIDAE Haliday 1833. Plural Noun. Fairy Flies. Moderately large (ca 95 Genera, 1,200 Species), cosmopolitan Family of parasitic Hymenoptera assigned to Chalcidoidea. Adult body small to minute (0.2 mm long), nonmetallic coloured; Antenna long with 8–13 segments, without Anelli; male Antenna typically filamentous or thread-like; female Antenna with 1–3 apical club segments; Head with Trabeculae, Toruli distantly positioned from Clypeus and widely separated; Scutellum usually with transverse Sulcus; wings sometimes brachypterous or absent; macropterous forewing long, narrow, venation reduced to short Marginal and Stigmal veins, long marginal fringe evident; macropterous hindwing exceedingly narrow and petiolate; legs long, slender with 4–5 Tarsomeres;

Gaster conspicuously petiolate or sessile. More than half mymarid Genera monotypic; most Species assigned to *Polynema*, *Anaphes* and *Gonatocerus*. Mymarids display an extensive fossil record in Cretaceous Period (90–110 MYBP) which includes five Genera and 20 Species. All Species develop as endoparasites of insect eggs; typically solitary, rarely gregarious; prefer host eggs in concealed habitats such as in plant tissue, under bark and in soil. Mymarids not host specific but seem to prefer Auchenorrhyncha (Homoptera). Aquatic females of some Species parasitize submerged Dytiscidae eggs. Females can 'swim', mate underwater and remain submerged for 15 days. Larval instar number varies from 2–4; first instar sacciform or mymariform, with six segments, lobate projections or Setae; second instar lacks segments, projections, Setae. Tracheal system not located in any larval instar of any mymarid Species. Pupation inside host egg. Some Species successful agents in biological control programs.

MYMARIFORM LARVA Hymenoptera: An apocritous larva whose head and caudal end each bears a conical process anterad. The Abdomen of some Species is segmented. The larval form is found in Mymaridae and Trichogrammatidae. See Larva. Cf. Teleaform Larva.

MYMAROMMATIDAE Debauche 1948. Plural Noun. A small (1 Genus, ca 20 Species), widespread Family of parasitic Hymenoptera assigned to Superfamily Chalcidoidea or Superfamily Mymarommatoidea. Adult minute, non-metallic coloured; exodont Mandible; geniculate Antenna without Multiporous Plate Sensilla; Prepectus apparently absent; forewing without venation but displaying reticulations on Remigium; hindwing very narrow, petiolate; Metasomal Petiole composed of two segments; Ovipositor concealed. Biology of the Family unknown. A few Species have been recovered from Cretaceous Period amber, thereby establishing the lineage as ancient. Regarded as sister group of Chalcidoidea when placed in separate Superfamily.

MYOBLAST Noun. (Greek, *myos* = muscle; *blastos* = bud. Pl., Myoblasts.) A cell which produces muscular tissue.

MYOCARDIUM Noun. (Greek, *myos* = muscle; *kardia* = heart; Latin, *-ium* = diminutive > Greek, *-idion*. Pl., Myocardia.) The muscles that form the Heart.

MYODOCHIDAE See Lygaeidae. (Servadei 1951, Redia 36: 171–220.)

MYODYNAMIC Adj. (Greek, *myos* = muscle; *dynamis* = power; *-ic* = of the nature of.) Pertaining to musclular contraction, or muscular force.

MYOFIBRILLA Noun. (Greek, *myos* = muscle; Latin, *fibrilla* = fibre. Pl., Myofibrillae.) A contractile fibre of muscle tissue. Alt. Sarcostyles.

MYOGENESIS Noun. (Greek, *myos* = muscle; Latin, *genesis* = origin. Pl., Myogeneses.) The development of muscles.

MYOGENIC Adj. (Greek, *myos* = muscle; *gennaein* = to produce; *-ic* = characterized by.) Pertaining to muscle contractions which are spontaneous and independent of nervous stimulation. See Muscle. Cf. Neurogenic.

MYOGLOBIN Noun. (Greek, *myos* = muscle; Latin, *globos* = globe. Pl., Myoglobins.) Respiratory protein in muscle which is involved in oxygen storage and transport.

MYOGLYPHIDES Coleoptera: The muscle notches in the posterior margin of the Collum.

MYOHAEMATIN Noun. (Greek, *myos* = muscle; *haima* = blood. Pl., Myohaematins.) An iron-containing pigment of insect muscle (Folsom & Wardle).

MYOLOGY Noun. (Greek, *myos* = muscle; *logos* = discourse. Pl., Myologies.) The branch of anatomy involved with the study of muscles.

MYOPSOCIDAE Plural Noun. A numerically small Family of Psocoptera assigned to Suborder Psocomorpha. Antenna with 13 segments, without secondary annulations; forewing asetose, mottled; Tarsi with three segments. Species found on bark or fences where they feed on algae and fungi.

MYOTOME Noun. (Greek, *myos* = muscle; *tome* = to cut. Pl., Myotomes.) Embryology: A division of the body muscles corresponding to a Metamere (Snodgrass).

MYOTROPIC PEPTIDES Myotropic peptides have been isolated from several types of insects, mostly *Locusta migratoria* and *Schistocerca gregaria*. MPs are grouped into peptide families: insulin, tachykinin, CRF, gastrin/CCK, and vasopressin. These peptides are associated with myotropic-, allatostatic-, and pheromonotropic physiological activities, diapause induction, stimulation of cuticular melanization, and diuresis. Some may be neurotransmitters present in nerves innervating the oviduct, salivary glands, male accessory glands, and the Heart; other peptides are stored in neurohemal organs. See Locustakinin; Cercropin.

MYRIAPODA Noun. (Greek, *myrios* = numberless; *pous* = feet.) The Class of Arthropoda including Chilopoda, Diplopoda, Symphyla, Pauropoda. Head with one pair of Antennae followed by serially repetitive, homodynamous, leg-bearing segments.

MYRIAPODS See Myriapoda.

MYRMECIINAE A numerically small Subfamily of ants found in New Caledonia and Australia. Malar area reduced, Mandible base adjacent to compound eye; Ocelli present; palpal formula 6-4; Petiole and Postpetiole present; sting powerful. Cf. Myrmicinae.

MYRMECIOID COMPLEX A major taxonomic grouping of ants recognized in some classifications. Cf. Ponerioid Complex.

MYRMECOCHORE Noun. (Greek, *myrmex* = ant; *chore* = farm. Pl., Myrmecochores.) An oily seed which attracts ants that transport it.

MYRMECOCHORY Noun. (Greek, *myrmex* = ant;

chore = farm. Pl., Myrmecochories.) Dispersal of seeds (myrmecochore) by ants.

MYRMECOCLEPTY Noun. (Greek, *myrmex* = ant; *klepto* = thief.) The form of symbiosis exhibited by the ant-guest *Atelura*, in which the symbiont steals regurgitated honey during its passage from the mouth of one ant to another.

MYRMECOLOGY Noun. (Greek, *myrmex* = ant; *logos* = discourse. Pl., Mrymecologies.) The study of ants. Cf. Apidology; Melittology.

MYRMECOMORPHOUS Adj. (Greek, *myrmex* = ant; *morphe* = form.) Ant-like.

MYRMECOMORPHY Noun. (Greek, *myrmex* = ant; *morphe* = form.) The phenomenon of displaying an ant-like appearance or habitus.

MYRMECOPHAGOUS Adj. (Greek, *myrmex* = ant; *phagein* = to devour.) Descriptive of or pertaining to an organism that preys on ants. Cf. Myrmecophile.

MYRMECOPHAGY Noun. (Greek, *myrmex* = ant; *phagein* = to devour.) The act or habit of feeding upon ants.

MYRMECOPHIL Noun. (Greek, *myrmex* = ant; *philos* = loving. Pl., Myrmecophils.) A guest in the nest of ants.

MYRMECOPHILE Noun. (Greek, *myrmex* = ant; *philos* = loving. Pl., Myrmecophiles.) A commensal or parasite of ants which inhabits ant nests. Some myrmecophiles are tended by the ants; others prey upon the ants or their brood (Wheeler). *e.g.* Some Staphylinidae (Dorylomimini.) Alt. Myrmecophil. Cf. Myrmecophag.

MYRMECOPHILIDAE Plural Noun. (Greek, *myrmex* = ant; *philein* = to love.) Ant Crickets. Ant-loving Crickets. A Family of ensiferous Orthoptera including 2–5 Genera and ca 50 Species. Worldwide in distribution but absent from Neotropical and Ethiopian rainforests. Considered by some taxonomists as a Subfamily or closely related to bothriophylacine Mongo-plistidae. Body minute, oval and rarely more than 2.0 mm long; compound eyes reduced to a few Ommatidia; Antenna not much longer than body; hind Femur short but considerably enlarged; both sexes apterous; Ovipositor short. Species always associated with ants; some Species parthenogenetic.

MYRMECOPHILOUS Adj. (Greek, *myrmex* = ant; *philos* = loving; Latin, *-osus* = with the property of.) Ant-loving; a term descriptive of insects or other organisms that live in ant nests. See Habitat; Inquiline. Cf. Entomophilous; Termitophilous. Rel. Commensalism; Mutualism; Parasitism.

MYRMECOTROPHY Noun. (Greek, *myrmex* = ant; *trophe* = nourishment.) Stimulation of growth in plants by nutrients carried to plants by ant associates.

MYRMECOXENES See Symphile.

MYRMELEONTIDAE Plural Noun. Antlions; Doodlebugs. The numerically largest Family of Neuroptera, assigned to Superfamily Myrmeleontoidea; geographically widespread and best represented in arid and semiarid areas. Adult elongate, soft-bodied; head relatively small and eyes relatively large; Antenna long with weakly developed club ('knobbed'); wings large, similar in size, shape and reticulate venation. Larvae with sickle-like mouthparts; Stemmata loosely grouped on tubercle; hind Tibia and Tarsus fused. Larvae of all Species predaceous but differing in habits: Some Species wait at the bottom of conical pits in sandy or fine-grain soils; some Species wait under surface; some Species wait on surface of ground. Pupation occurs within silken cocoon in soil. Syn. Myrmeliontidae.

MYRMELEONTIFORMIA A Superfamily of plannipenous Neuroptera including Ascalaphidae, Myrmeleontidae, Nemopteridae, Nymphidae, Psychopsidae and Stilbopterygidae.

MYRMICIIDAE Plural Noun. A fossil Family of Symphyta (Hymenoptera).

MYRMICINAE A Subfamily of ants with Petiole and Postpetiole; workers rarely with Ocelli; pretarsal claws simple. Cf. Myrmeciinae.

MYSTACINE Adj. (Greek, *mystax* = moustache.) Bearded; with a setose fringe above mouth or on the Clypeus. Alt. Mystacinous.

MYSTAX Noun. (Greek, *mystax* = moustache.) Diptera: A patch of Setae above the mouth, on the lower part of the Hypostoma above the Vibrissae; common in Asilidae.

MYTEN® See Dienochlor.

MYTILIFORM Adj. (Greek, *mytilos* = sea mussel; Latin, *forma* = shape.) Shell-shaped; mussel-shell shaped; descriptive of swimming legs in some aquatic Hemiptera. Cf. Chonchiform; Conriform; Heliciform; Pectiniform; Trochiform. Rel. Form 2; Shape 2; Outline Shape.

MYXOMA Noun. (Greek, *myxa* = slime; *oma* = tumor. Pl., Myxomata.) Soft, gelatin-like tumours composed of mucous-type connective tissue.

MYXOMATOSIS Noun. (Greek, *myxa* = nasal mucus; *-osis* = a disease. Pl., Myxomatoses.) A viral disease condition characterized by Myxomata in the body. Specifically, a viral disease of rabbits, endemic to New World and moved elsewhere. Principal vector of Myxomatosis in England is European rabbit flea, *Spilopsyllus cuniculi.* Elsewhere vectors include other Species of fleas, mosquitoes, blackflies, lice and mites. Rabbit myxomatosis not virulent in New World *Sylvilagus*; but highly virulent in Old World *Oryctolagus.* Disease produces lesions of skin, lymph glands, testicle, spleen and lung. Disease develops rapidly; death within 1–2 weeks. Disease deliberately introduced into Australia for biocontrol of European rabbit *Orcytolagus cuniculus.*

MYXOPHAGA (Greek, *myxa* = slime; *phagein* = to devour.) A small Suborder of Coleoptera related to Archostemata. Prothorax without Notoplural Sutures; Hypomeron joined to Sternum; Propleuron reduced and concealed. Species found in aquatic and riparian habitats. Included Families: Cyathoceridae, Hydroscaphidae, Microsporidae, Torridincolidae.

N Symbolically, in Chemistry, N stands for Nitrogen or 'normal' with reference to solutions.

NAA An organic acid (1-Naphthaleneacetic acid) used as a Plant Growth Regulator on fruit trees and ornamentals to thin fruit and prevent preharvest drop. Also used to promote rooting of plant cuttings. Trade names include Aperdex®, Betaral®, Celmone®, Drofix®, Fruit Fix®, Fruitone-N®, Hormofix®, K-Salt Fruit Fix 200®, Liqui-Stik®, NAA-800®, Nafusaku®, Olive Stop®, Phyomone®, Planofix®, Primacol®, Prinacol®, Rhizopon-B®, Rhodofix®, Stafast®, Strike®, Tekkam®, Thin-N-Stop-Drop®, Tipoff®, Tip-Off®, Trecut®, Tre-Hold®. Cf. Insect Growth Regulator.

NAA-800® See NAA.

NABIDAE Costa 1853. Plural Noun. Damsel Bugs. A cosmopolitan Family of heteropterous Hemiptera assigned to Superfamily Cimicoidea. Adult typically dull-brown in colour, 3–11 mm long, slender bodied with elongate head; compound eyes present; Labium actually four-segmented with three apparent segments; Ctenidia absent; Pronotum two-lobed; fore Femur slightly enlarged; Hemelytron without Cuneus. Adults and nymphs active, aggressive plant or ground-roaming predators; female inserts eggs into plant tissue such as grass stems.

NACRÉ Noun. (Medieval Latin, *nacchara;* Italian, *nacchara*. Pl., Nacres.) The iridescent inner surface of mollusc shells consisting of thin, overlapping layers of calcium carbonate and conchiolin. See Iridescence. Rel. Mother-of-Pearl.

NACREOUS Adj. (Latin, *nacara* = a drum; French, *nacre* = with a pearly lustre; Latin, *-osus* = with the property of.) Resembling mother-of-pearl, or the iridescent inner layer of oyster shell. Alt. Nacrous; Nacry. See Iridescent; Opalescent. Cf. Matte; Mother-of-Pearl.

NACRINE Noun. (Arabic, *nakir* = hollowed. Pl., Nacrines.) The iridescent hue of nacre; Mother-of-Pearl coloured. See Iridescence.

NADIG, ADOLF (1877–1960) (Gasche 1960, Verh. schweiz. naturf. Ges. 1960: 229–236, bibliogr.)

NAFUSAKU® See NAA.

NAGANA Noun. (Zulu.) A disease of domestic livestock in Africa that is transmitted principally by Tsetse Fly with other flies serving as mechanical vectors. Disease agents of Nagana include *Trypanosoma brucei, T. congolense, T. evansi* and *T. vivax*. See Trypanosomiasis; Tsetse Fly. Cf. Sleeping Sickness; Surra.

NAGANO, KIKUJIRO (1868–1919) (Hasegawa 1967, Kontyû 35 (Suppl.): 57–58, bibliogr.)

NAGARAJAN, SUBRAMANIA (1917–1964) (Pearson 1965, Proc. R. ent. Soc. Lond. (C) 29: 54.)

NAGASAWA, SHOBEI (1865–) (Hasegawa 1967, Kontyû 35 (Suppl.): 57.)

NAGEL, PAUL (1859–1924) (Wolf 1924, Jh. Ver. schles. Insektenk. 14: 23–24.)

NÄGELI, ALFRED (1863–1935) (V.A. 1935, Mitt. schweiz. ent. Ges. 16: 613–614.)

NÄGELI, WERNER (1900–1971) (Maksymov 1972, Anz. Schädlingsk. Pflanzenshutz 45: 63.)

NAIAD Noun. (Greek, *naias* = water nymph. Pl., Naiads.) The aquatic immature stage of the hemimetabolous insects (Odonata, Ephemeroptera, Plecoptera) that displays gills and does not resemble the adult in shape or general appearance. The term naiad was proposed by Comstock to distinguish between forms living in water and terrestrial nymphs of Paurometabola. See Nymph. Cf. Caterpillar; Grub; Maggot; Slug. Rel. Metamorphosis.

NAIL Noun. (Anglo Saxon, *naegel* = nail. Pl., Nails.) A tarsal claw. Specifically the stout, apically pointed claws in predatory Heteroptera. See Unguis.

NAINIT® See Chlormequat.

NAIROBI SHEEP DISEASE An arbovirus of the Bunyaviridae known to affect sheep and goats in Africa. Exceedingly high mortiality in livestock; rarely affects humans. NSF vectored by the tick *Rhipicephalus appendiculatus* with transovarial transmission. See Arbovirus; Bunyaviridiae.

NAJA® See Fenproximate.

NAKABAYASHI, HYOJI (1876–1934) (Hasegawa 1967, Kontyû 35 (Suppl.): 58, bibliogr.)

NAKAMURA, MASAO (1867–1943) (Hasegawa 1967, Kontyû 35 (Suppl.): 58–59.)

NAKAMURA, YAMATO (1896–1948) (Hasegawa 1967, Kontyû 35 (Suppl.): 59.)

NAKATA, SETSUKO (1930–1971) (Gressitt & Higa 1972, Proc. Hawaii. ent. Soc. 21: 295–297)

NAKAYAMA, SHONOSUKE (Okamoto 1969, Kontyû 37: 460–462.)

NAKED Adj. (Anglo Saxon, *nacod* = naked.) 1. Not clothed; without a covering. An insect's body or sclerite which lacks a vestiture or covering of Setae. 2. A term used to describe the Pupa when it is not enclosed in a cocoon or other covering.

NAKEGAWA, HISAKAZU (1859–1921) (Hasegawa 1967, Kontyû 35 (Suppl.): 56–57.)

NALED Noun. An organic-phosphate (phosphoric acid) compound {1,2-dibromo-2,2-dichloroethyl dimethyl phosphate} used as an acaricide, contact insecticide and stomach poison. Compound applied to numerous crops, vegetables and ornamentals for control of mites, plant-sucking insects and leaf-eating insects. Also used as a public-health insecticide for control of cockroaches, mosquitoes and flies; surface spray for residual control. Phytotoxic to fruit trees and some ornamentals. Trade names include Dibrom®, Hibrom® and Lucanal®. See Organophosphorus Insecticide.

NAMBOUR CANEGRUB *Antitrogus mussoni* (Blackburn) [Coleoptera: Scarabaeidae]. See Canegrub.

NAME Noun. 1. A word or phrase by which a thing is known. 2. Taxonomic Jargon: The correct scientific name or binomen applied to a valid Taxon. See Systematic Name. Cf. Manuscript Name.

Rel. Pseudonym.

NAMEKIL® See Metaldehyde.

NAMIE, MOTOKISHI (1854–1918) (Hasegawa 1967, Kontyû 35 (Suppl.): 59–60. Bibliogr. pl. 1, fig. 6.)

NANIOT, LOUIS (–1907) (Anon. 1907. Entomologist's Rec. J. Var. 19:146.)

NANITIC WORKER Social Insects: Dwarf workers (Hymenoptera and Isoptera) produced by starved broods.

NANOID Adj. (Greek, *nanos* = dwarf; *eidos* = form.) Dwarf-like; descriptive of exceptionally small structures (bodies) in comparison with related structures. Cf. Nanos. Rel. Eidos; Form; Shape.

NANOPSOCETAE A Superfamily of trogiomorph Psocoptera that includes the Families Liposcelidae, Pachytroctidae and Sphaeropsocidae.

NANOS Adj. (Latin, *nanus* = dwarf.) Dwarf-like; descriptive of a body proportionately smaller than the norm. Cf. Nanoid.

NANTUCKET PINE TIP MOTH *Rhyacionia frustrana* (Comstock) [Lepidoptera: Tortricidae]: A pest of *Pinus* spp. endemic in eastern North America. Larva hollows tips of new pine shoots and buds. Pupation occurs within the tips of shoots. Cf. Pitch-Pine Tip Moth.

NaOH Sodium hydroxide. A chemical compound used to macerate delicate or weakly sclerotized specimens or appendages for microscopical study. See KOH.

NAPIFORM Adj. (Latin, *napus* = turnip; *forma* = shape.) Turnip-shaped; pertaining to structure that is large and globular above and tapering below. Generally applied to roots. See Turnip. Cf. Adeniform; Ampulliform; Calyciform; Campaniform. Rel. Form 2; Shape 2; Outline Shape.

NAPHTHALENE Noun. (Greek, *naphtha* = bitumen; plus *al*cohol and *ene* = chemical suffix). A white, crystalline, aromatic hydrocarbon produced in the fractional distillation of coal tar and used for repelling closet-storage pests such as moths and dermestids. See Moth Ball.

NARCISSUS BULB FLY *Merodon equestris* (Fabricius) [Diptera: Syrphidae]: A monovoltine pest of amaryllis, galtonia, hyacinth and narcissus; NBF endemic to Europe and adventive to North America. Adult resembles bumble bee; body black with yellow or orange-coloured Setae. Eggs laid on neck of bulb, base of leaves or on soil. Neonate larvae bore into bulb with mouth hooks. Puparia develop in bulb or in soil. See Syrphidae. Cf. Lesser Bulb Fly; Onion Bulb Fly.

NARCISSUS JEWEL *Hypochrysops narcissus narcissus* (Fabricius) [Lepidoptera: Lycaenidae].

NARCOSIS Noun. (Greek, *narke* = numbness; *sis* = a condition or state. Pl., Narcoses.) A state of stupor induced by a narcotic.

NARCOTIC Noun. (Greek, *narke* = numbness; *-ic* = characterized by. Pl., Narcotics.) 1. Any drug which produces numbness or a state of unconciousness. 2. A chemical that causes reversible depression of the nervous system. Often used as a fumigant in structural pest control.

NARCOTIZE Verb. (Greek, *narke* = numbness.) To place under the influence of a narcotic or sleep-producing agent.

NARDO, GIOVANNI DOMENCIO (1802–1877) (Rirona 1878, Atti Ist. veneto Sci. (5)4: 785–850, bibliogr.)

NARES Plural Noun. (Latin, *nares* = nostrils.) Nostrils.

NARROW-HORNED FLOUR BEETLE See Slender-Horned Flour Beetle.

NARROW-NECKED GRAIN BEETLE *Anthicus floralis* (Linnaeus) [Coleoptera: Anthicidae].

NARROW-WAISTED BARK BEETLES See Salpingidae.

NARROW-WINGED DAMSELFLIES See Coenagrionidae.

NARROW-WINGED DAMSELFLY *Lestes unguiculatus* [Lestidae: Odonata].

NARROW-WINGED MANTID *Tenodera augustipennis* Saussure [Mantodea: Mantidae]. See Mantidae.

NARROW-WINGED PEARL WHITE *Elodina padusa* (Hewitson) [Lepidoptera: Pieridae].

NARUTO, YOSHITAMI (1833–1913) (Hasegawa 1967, Kontyû 35 (Suppl.): 60, bibliogr.)

NASAL CARINA Ephemeroptera: A longitudinal ridge on the head anterior of the middle Ocellus which often appears nose-like when viewed in profile.

NASAL SUTURE See Clypeal Suture.

NASCENT Adj. (Latin, *nascens* pres. part. of *nasci* = to be born.) 1. Beginning to exist; beginning to grow. 2. The act of being born.

NASO Noun. Acarology: Unpaired protuberance at the rostral extremity of the Idiosoma.

NASONOV, NIKOLAI VIKTOROVICH (1855–1939) (Uvarov 1939, Nature 143: 549.)

NASONOV'S GLAND A group of glandular cells on the dorsum of the seventh abdominal Tergum of worker bees. The gland is not found in queens or males. NG may serve several functions but commonly secretes a pheromone that attracts other bees when wings are fanned. The odour secreted by Nasonov's Gland is not race or colony specific.

NASOPHARYNGEAL MYIASIS See Myiasis.

NASUS Noun. (Latin, *nasus* = nose. Pl., Nasi.) 1. General: The so-called 'nose' or part of the insect head which articulates with the Labrum. 2. Odonata: The Clypeus or a modification of it. The upper part of the Clypeus. See Supraclypeus; Postclypeus. 3. Isoptera: A snout-like integumental modification of the head used to eject defensive secretions at nest intruders. 4. Diptera: A projection from the Rostrum. 5. Hymenoptera: A nose-like, anterior termination of the face in certain Species. 6. Acari: An anteromedian projection of the Sternellum (Goff *et al.* 1982, J. Med. Ent. 19: 228.)

NASUTE Noun. (Latin, *nasutus* = large nosed. Pl., Nasutes.) A type of soldier caste manifest in some 'higher' termite Families (Termitidae, Rhinotermitidae). Anatomically, Nasutes display a head with well developed horn-like median frontal projection (Nasus) and reduced Mandibles. The head contains a large Frontal Gland which produces defensive fluid that is ejected through the Nasus. See Soldier.

NATAL Adj. (Latin, *natalis* = birth; *-alis* = pertaining to.) Pertaining to birth.

NATAL FRUIT FLY *Ceratitis rosa* Karsch [Diptera: Tephritidae]: Named for the Natal region of South Africa.

NATALITY Noun. (Latin, *natalis* = birth.) 1. The birth rate of a population. 2. The properties or factors which promote survival of individuals at birth and a consequent increase in population size. Cf. Mortality.

NATATORIAL Adj. (Latin, *natatorius* = to swim; *-alis* = characterized by.) Descriptive of swimming or adaptations for locomotion in water. Alt. Natant; Natatory; Natatorius. See Legs.

NATATORY LAMELLAE Tridactylinae: Long slender sclerites on the hind Tibia (Comstock).

NATHAN, LEONARD (1883–1960) (Uvarov 1961, Proc. R. ent. Soc. Lond. (C)25: 50.)

NATIVE Adj. (Latin, *natus* = born.) Pertaining to the natural distribution of an organism or Species (plant or animal) in that the organism currently lives or its ancestors originated or evolved in a specified area. See Zoogeography. Cf. Adventive; Endemic; Introduced.

NATIVE BUDWORM *Helicoverpa punctigera* (Wallengren) [Lepidoptera: Noctuidae]: An important pest of cotton, oilseed, grains and horticultural crops in Australia. Adults 14–20 mm long with a 35–40 mm wingspan. Males greenish-grey; females orange-brown to grey. Forewings with a band of dark spots along the margin and a broad, irregular, brown band. Hindwings pale with a dark brown band on the apex. A pale patch does not occur in the band as in *H. armigera*. Antennae long, thread-like (filiform) and covered in fine Setae. Adult females lay small (0.4–0.6 mm), yellowish-white eggs; oviposition period 5–24 days depending on region. Egg number 500–3,000 per female; eggs adhere to substrate and display 24 longitudinal ridges on the Chorion; eggs become dark brown before hatching (commonly called the 'brown ring' stage.) Egg stage 3–10 days depending on temperature; eggs commonly laid on upper third of the cotton plant, but can be deposited on any structure. First and second instar larvae yellowish-white to reddish-brown. Prolegs present on abdominal segments 3–6 and 10. Intermediate instars with dark saddle-shaped pigments on the fourth segment. Mature larvae are 30–40 mm long with a brown mottled head capsule and large black Setae on the first thoracic segment. Larval stage 59–92 days. Mature larvae drop to the ground to pupate in an earthen cell. Pupae 14–18 mm long, mahogany brown with two parallel spines on the posterior tip. The Pupa is the overwintering stage. Damage begins from the onset of squaring and continues through to boll maturity. Larvae cut neat circular holes into squares and bolls. Young bolls and squares may fall from the plant. NB is endemic to Australasia; adults are highly migratory and may be obligatory migrants. See Cotton Bollworm; Noctuidae. Cf. Rough Bollworm.

NATIVE DRYWOOD TERMITE *Cryptotermes primus* (Hill) [Isoptera: Kalotermitidae]

NATIVE ELM BARK BEETLE *Hylurgopinus rufipes* (Eichhoff) [Coleoptera: Scolytidae] A pest of elm trees endemic to eastern North America. Adult ca 2.5 mm long; dark coloured, without ventral declivity and spine; head and Pronotum densely punctate; Elytra with rows of dense punctures. Female constructs an egg gallery under bark and across the wood's grain; larval mines are with the grain. NEBB overwinters as a larva or adult in tunnels in bark. Adults emerge during May and serve as vectors of Dutch Elm Disease. See Scolytidae.

NATIVE FIG MOTH *Lactura caminaea* Meyrick [Lepidoptera: Zygaenidae].

NATIVE FIRE ANT *Solenopsis geminata* (Fabricius) [Hymenoptera: Formicidae]: A significant pest endemic to North America, ranging from Carolinas to Costa Rica. Workers polymorphic; 2–4 mm long; coloration yellowish, brownish to blackish; head large; Mandibles incurved and usually without teeth. NFA is an important predator of other insects. NFA typically nests in ground around clumped vegetation, or in rotting wood. See Formicidae. Cf. Imported Fire Ant; Little Fire Ant; Southern Fire Ant.

NATIVE HOLLY LEAFMINER *Phytomyza ilicicola* Loew [Diptera: Agromyzidae].

NATIVE PILLBUG *Australiodillo bifrons*. Resembles Common Pillbug. Adult ca 10 mm long; grey; body with longitudinal keel-like ridge on dorsum and transverse rows of protrusions on each Tergum; fold when disturbed. See Common Pillbug. Cf. Slaters.

NATTERER, JOHANN (1787–1843) (Goeldi 1895, Bol. Mus. Paraense 1: 189–217.)

NATTERER, LEOPOLD (–1942) (Jarvis 1940, Entomol. mon. Mag. 82: 160.)

NATURAL CLASSIFICATION 1. A system of classifying, sorting, or arranging groups of organisms based on putative genetical relationship. A theoretical concept developed in conjunction with generally held views of organic evolution, or some variant of it. 2. A classification of organisms that is based upon taxonomic characters (anatomical, behavioural, biological) that are considered in relation to their phylogenetic significance. See Classification. Cf. Artificial Classification. Rel. Cladistics; Phylogeny; Taxonomy.

NATURAL CONTROL The collective action of climate, edaphic factors, competition and natural enemies upon a population to determine population size. Cf. Biological Control; Chemical Control; Cultural Control; Microbial Control; Regulatory Control.

NATURAL ENEMY An organism which causes the death or impairment of another Species of organism. The concept of natural enemy is used in a positive sense by biological control workers interested in management of pests. See Biological Control. Rel. Parasite; Pathogen; Predator.

NATURAL INSECTICIDE A poorly-defined category of chemicals that are produced by blue-green algae, fungi and plants. The chemicals typically are not related to the basal metabolism of the producer and confer some degree of toxicity to insects and vertebrate herbivores or deter them from feeding upon the alga, fungus or plant. NI include Botanical Insecticides in part. Insecticidal properties of the chemicals probably result from coevolution of the producer and insect. See Insecticide. Cf. Botanical Insecticide; Synthetic Insecticide.

NATURAL SELECTION The principle or hypothesis advanced by Charles Darwin which asserts that in nature any organism not able to endure surrounding conditions is eliminated. Under Darwin's concept of NS, only 'fit' individuals survive and reproduce. See Selection; Evolution. Cf. Artificial Selection; Orthoselection; Sexual Selection. Rel. Genetics.

NAUCORIDAE Leach 1815. Plural Noun. Creeping Water Bugs. A widespread Family of heteropterous Hemiptera assigned to Superfamily Naucoroidea. Adults 9–13 mm long; body brownish coloured; oval in outline shape and slightly flattened; Ocelli absent; fore legs raptorial; all Tarsi of two segments with long tarsal claws; middle and hind legs lack long marginal fringe and not adapted for swimming; forewing lacks veins in membrane. Naucorids are aquatic and typically occur in ponds, springs and streams in association with submerged vegetation. Species are predaceous upon insects and other small aquatic animals.

NAUCOROIDEA A Superfamily of Hemiptera including Naucoridae.

NAUFOCK, ALBERT (1878–1937) (Reisser 1937, öst. EntVer. 22: 53–55, bibliogr.)

NAUPLIFORM LARVA See Cyclopoid Larva.

NAUPLIIFORM Adj. (Latin, *nauplius* = shellfish; *forma* = shape.) 1. Resembling the form or shape of the nauplius stage of Crustacea. 2. Hymenoptera: First instar larva of Platygasteridae. 3. A larva with large, falcate Mandibles and a pair of bifurcate caudal appendages. Alt. Naupliform. See Nauplius. Cf. Cyclopoid larva. Rel. Form 2; Shape 2; Outline Shape.

NAUPLIUS Naun. (Greek, *nauplios* = shellfish. Pl., Nauplii.) The first stage of a crustacean larva, characterized by three pairs of appendages (Antennules, Antennae, Mandibles), a median eye and reduced body segmentation.

NAUTICAL TWILIGHT See Twilight.

NAVAJAS, EDUARDO (1905–1962) (Anon. 1964, Revta. bras. ent. 11 (portrait only.)

NAVARRO DE ANDRADE, EDMUNDO (1881–1941) (Borgmeier 1942, Revta. Ent., Rio de J. 13: 182–188.)

NAVAS, LONGINOS (1858–1938) (Porter 1934, Revta. chil. Hist. nat. 38: 208–213, bibliogr.)

NAVEL ORANGEWORM *Amyelois transitella* (Walker) [Lepidoptera: Pyralidae]: A pest of *Citrus,* almonds and walnuts in the USA and Mexico. A pest of almonds in California since the 1940s. NOW eggs laid in navel of oranges or cracked husks of nuts. Larva acts as a scavenger on injured oranges or enters nuts and spins a web; pupation occurs within a silken cocoon inside mummified fruit or nuts. See Pyralidae.

NAVICULA Noun. (Latin, diminutive: *navis* = ship. Pl., Naviculas.) The fourth Axillary Sclerite (MacGillivray).

NAVICULAR Adj. (Latin, diminutive: *navis* = ship.) Boat-shaped. Cf. Cymbiform; Scaphoid. Rel. Form 2; Shape 2; Outline Shape.

NAVLOVSKII, EVGENII NIKANOVICH (Perfil'ev 1955, Ent. Obozr. 34: 3–8.)

NAWA, UMEHICHI (1847–1945) (Hasegawa 1967, Kontyû 35 (Suppl.): 60–61, bibliogr.)

NAWA, YASUSHI (1857–1926) (Hasegawa 1967, Kontyû 35 (Suppl.): 61, bibliogr. pl. 1, fig. 9.)

NC ALL® See Entomogenous Nematodes.

NE NPV® See Redheaded Pine Sawfly NPV.

NEALA Noun. (Greek, *ne* = not; Latin, *ala* = wing. Pl., Nealeae.) A membranous lobe often developed at the base of the wing proximal of the Vannus. The jugal region or Jugum of the wing *sensu* Snodgrass.

NEALLOTYPE. Noun. (Greek, *neo* = new; *allos* = other; *typos* = pattern. Pl., Neallotypes.) An Allotype described or designated after publication of the original description. See Type. Cf. Holotype. Rel. Taxonomy.

NEALOGY Noun. (Greek, *neales* = youthful; *logos* = discourse. Pl., Nealogies.) The study of immatures or young animals. Cf. Gerontology; Palaeontology.

NEANIC Adj. (Greek, *neanikos* = youthful; *-ic* = of the nature of.) Pertaining to immature stages in the life history of an individual animal. Term sometimes used in reference to the pupal stage.

NEARCTIC REGION (Greek, *neos* = new; *arktos* = giant bear.) One of the principal zoogeographical regions of the world. The part of the Holarctic Realm occupying almost the entire continent of North America, and including Greenland and northern Mexico. Syn. Nearctic Realm. Cf. Holarctic Region; Palaearctic Region; Neotropical Region; Ethiopian Region; Oriental Region; Oceanic Region; Australian Region.

NEAVE, B W (–1923) (Riley 1924, Proc. S. Lond. ent. nat. Hist. Soc. 1923–24: 75.)

NEAVE, SHEFFIELD AIREY (1879–1961) (Hall 1964, In: Nomenclator Zoologicus 6: i–ii.)

NEBEL, LUDWIG (1861–1911) (Heidenreich 1911, Dt. ent. Natn Biblthk 2: 135–136.)

NEBENKERN Noun. (German, *neben* = near; *kern* = nucleus. Pl., Nebenkerns.) A structure in the cell cytoplasm composed of concentric layers of Endoplasmic Reticulum.

NEBULA Noun. (Latin, *nebula* = smoke, vapour. Pl., Nebulae.) 1. A cloud. 2. A vague, indefined, dusky shading.

NEBULIN® See Tecnazene.

NEBULOSE Adj. (Latin, *nebulosus* = dark, clouded.) Cloudy; pertaining to structure without definite form or outline. Alt. Nebulosus; Nebulous.

NEBULOUS VEIN Hymenoptera: A more-or-less pigmented wing vein without a tubular structure and displaying ill-defined edges. NV is visible by transmitted light and reflected light. (Mason 1986, 88: 2). See Vein; Wing Venation. Cf. Spectral Veins; Spurious Vein; Trace Vein; Tubular Vein. Rel. Horismology.

NECK Noun. (Middle English, *necke* = neck. Pl., Necks.) The Cervix or slender connecting structure between head and Thorax of insects which display the head free of the Prothorax. Any contraction of the head at its juncture with the Thorax.

NECKAM, ALEXANDER (1157–1217) (Weiss 1927, J. N. Y. ent. Soc. 35: 417–418.)

NECRON Noun. (Greek, *nekros* = dead. Pl., Necrons.) Plant material that is dead but which has not decomposed.

NECROPHAGOUS Adj. (Greek, *nekros* = dead; *phagein* = to devour; Latin, *-osus* = with the property of.) Death-feeding; Pertaining to organisms that feed on dead or decaying matter. Term typically applied to insects that feed on dead animals or dead insects. Habit well developed among some Diptera (Sarcophagidae) and Coleoptera (Silphidae). Alt. Necrophagus. See Feeding Strategies. Cf. Saprophagous.

NECROPHILOUS Adj. (Greek, *nekros* = dead; *philein* = to love; Latin, *-osus* = with the property of.) Death-loving; a term descriptive of organisms that are associated with dead and decaying plants or animals. Example: Carrion Beetles. See Habitat. Cf. Rel. Ecology.

NECROPHORESIS Noun. (Greek, *nekros* = dead; *pherein* = bear, carry; *-sis* = a condition or state. Pl., Necrophoreses.) Social Insects: The act of removing dead colony members from the nest.

NECROPHORIC Adj. (Greek, *neckros* = dead; *pherein* = to carry; *-ic* = of the nature of.) Pertaining to Necrophoresis; the transport of dead bodies. Alt. Necrophorous.

NECROPSY Noun. (Greek, *nekros* = corpse, dead tissue; *-opsis* = sight, view. Pl., Necropsies.) The examination of a dead body. Cf. Autopsy; Biopsy. Rel. Pathology.

NECROSIS Noun. (Greek, *nekros* = deadness; *-osis* = a condition or state. Pl., Necroses.) 1. The condition of decay. 2. The sudden death of cells or tissues (typically through a breakdown of the Plasma Membrane) followed by cell swelling and rupture. 3. An unspecified diseased condition of plant tissues that causes them to become black and then decay.

NECROTAULIIDAE Plural Noun. A Family of Trichoptera known from the Triassic Period of Kirgizia.

NECROTIC Adj. (Greek, *nekrosis* = deadness; *-ic* = of the nature of.) Tissue in a dead and decayed condition. Descriptive of or pertaining to Necrosis.

NECROTIC ARACHNIDISM See Loxoscelism.

NECSEY, STEFAN (1870–1902) (Argner 1903, Rovart. Lap. 10; 1–9.)

NECTAR Noun. (Greek, *nektar* = death overcoming; drink of gods. Pl., Nectars.) Sweet, frequently scented, substances secreted by flowers, nectaries or other plant structures. Nectar provides nutrition for many insects, particularly Hymenoptera, Lepidoptera and some Diptera and serves as an attractant for pollination.

NECTAR GUIDES Lines, marks or colours on flowers which direct bees to the nectaries.

NECTAR SCARABS *Phyllotocus* spp. [Coleoptera: Scarabaeidae]

NECTARY Noun. (Greek, *nektar* = nectar. Pl., Nectaries.) 1. Aphids: Honey-tubes, Cornicles or Siphuncles. 2. Plants: Spurs or sacs of a flower which secrete a sweetened liquid termed nectar; any plant structure with the same function, termed an extra-floral nectary.

NECTOPOD Noun. (Greek, comb. form *nectos* = swimming; *podos* = foot. Pl., Nectopods.) A limb or appendage adapted for swimming. Cf. Swimmeret.

NEDCIDOL® See Diazinon.

NEEDHAM, JAMES GEORGE (1868–1957) American Entomologist, educated Knox College (Galesburg, Ill.) and Johns Hopkins (1893). Professor of Biology at Lake Forest University. (Mallis 1971, *American Entomologists*. 549 pp. (174–178), New Brunswick.)

NEEDHAM, JOHN TURBERVILLE (1713–1781) English Naturalist and Roman Catholic priest who served as Director of the Academy of Sciences, Brussels. Published *New Microscopical Discoveries* (1745) and *Idée sommaire, ou Vue Générale du Système physique et métaphysique sur la Générátion* (1780). (Rose 1850, *New General Biographical Dictionary* 10: 298.)

NEEDLE Adj. (Middle English, *nedle* from Anglo Saxon, *needl.*) Any slender pointed object. Syn. Acicular.

NEEDLE BUG *Ranatra dispar* Montandon [Hemiptera: Nepidae].

NEEDLEFLIES See Leuctridae.

NEEM BENEFIT® See Azadirachtin.

NEEM See Indian Neem Tree; Philippine Neem Tree; Thai Neem Tree.

NEEMAZAL-T/S® A neem insecticide with 1.0% azadirachtin as the active ingredient. Marketed by Trifolio-M GmbH in Germany.

NEEMISIS® See Azadirachtin.

NEEMIX® A broad-spectrum biopesticide that can

be applied to cole and bulb vegetables, citrus, pome and stone fruits, legumes and fruiting vegetables, root and tuber vegetables, and other crops. Active ingredient 0.25% azadirachtin. Also marketed as Neemix 4.5 with active ingredient at 4.5% azadirachtin. See Azadirachtin.

NEES VON ESENBECK, CHRISTIAN GOTTFRIED DANIEL (1776–1858) German natural philosopher, physician and academic [Professor of Botany at Erlangen (1818), Bonn (1819), Breslau (1831), Berlin (1848–1852).] Wrote several important works including *Naturphilosophie* (1841), *Hymenopterorum Ichneumonibus affinium Monographiae, Genera Europaa et species illustrantes. Scripsit Christ. Godofr. Nees ab Esenbeck*. Stuttgartia, Tubinga, sumptibus, (Kieser 1860, Nova Acta Acad. Caesar. Leop. Carol. 27: xxxv–xcvii, bibliogr.)

NEGASHUNT® See Coumaphos.

NEGATIVE PHOTOTAXIS A behavioural reaction (typically in the form of movement or body orientation) in which an organism responds negatively to the stimulus of light. See Phototaxis. Cf. Positive Phototaxis; Scototaxis. Syn. Photopathy.

NEGATIVE PHOTOTROPISM Reaction or orientation of a body away from light. See Tropism. Cf. Photopathy.

NEGATIVE TROPISM A behavioural reaction in which a stimulus acts to repel (Wardle). See Tropism.

NEGATORIA CANEGRUB *Lepidiota negatoria* Blackburn [Coleoptera: Scarabaeidae]. See Canegrub.

NEGI, PRATAP SINGH (1899–1961) (Mehra 1960, Indian J. Ent. 22: 73–74.)

NEGRO BUG *Corimelaena (Thyrecoris) pulicaria* (Germar) [Hemiptera: Cydnidae]: Widely distributed throughout North America and southward to Guatemala. NB regarded as a significant pest of celery, berries and grasses. Overwinter as an adult; body 2.5–3.5 mm long, black, corial apex rounded. Eggs laid individually on leaves; eclosion occurs within 14 days. Nymphs and adults feed on host plants and secretions/excretion leave a foul taste to plant. See Thyreocoridae.

NEGRO BUGS See Thyreocoridae.

NEGUVON® See Trichlorfon.

NEI'S GENETIC DISTANCE INDEX A mathematical measure of the genetic distance between populations. (Nei 1972, Am. Nat. 106: 283–292; Nei *et al.* 1983, J. Mol. Evol. 19: 153–170.) An approach to systematic problems of relationship which serves as an alternative to classical data such as morphological characters and character states.

NEIDIDAE See Berytidae.

NEISWANDER, CLAUDE REVERE (1893–1974) (Rings & Treece 1974, J. Econ. Ent. 67: 702.)

NEIVA, ARTURO (1880–1943) (Lent 1943, Revta. bras. Biol. 3: 273–291, bibliogr.)

NEJEDLY, Z (1878–1962) (Anon. 1962, Cas. csk. Spol. Ent. 59: i–iii.)

NEKOS® See Rotenone.

NEKTON Noun. (Greek, *nektos* = swimming. Pl., Nektons.) The consocies or group of animals that swim freely in the surface layers of an aquatic habitat. The aggregate of active forms living in the surface layers of water. Cf. Neuston; Plantkton; Seston.

NEKTONIC Adj. (Greek, *nekton* = swimming; *-ic* = of the nature of.) Free-swimming; pertaining to organisms that swim in the surface layers of an aquatic habitat. Cf. Planktonic.

NELL, PHILIP (1857–1923) (Skinner 1924, Ent. News 35: 35.)

NELSON JAMES ALLEN (1875–1941) (Phillips 1942, Ent. News 53: 59–60.)

NELSON, W G F (–1954) (Buxton 1955, Proc. R. ent. Soc. Lond. (C)19: 69.)

NEMACUR® See Fenamiphos.

NEMAMORT® See DCIP.

NEMASOL® See Metham-Sodium.

NEMATHORIN® See Fosthiazate.

NEMATICIDE Noun. (Greek, *nema* = thread; Latin, *-cide* > *caedere* = to kill. Pl., Nematicides.) A chemical compound which kills nematodes. See Pesticide. Cf. Acaricide; Biopesticide; Filaricide; Fungicide; Herbicide; Insecticide; Pesticide.

NEMATID Adj. (Greek, *nematos* = thread.) Thread-like. See Nematode.

NEMATOCERA A Suborder of Diptera whose members display long Antennae each with at least six segments. Regarded as the most primitive group of Diptera with fossil record dating from Jurassic Period. Higher classifications in debate but several Divisions recognized. Included Families: Anisopodidae, Axymyiidae, Bibionidae, Blephariceridae, Canthyloscelidae, Cecidomyiidae, Ceratopogonidae (Heleidae), Chaoboridae (Corethridae), Chironomidae (Tendipedidae), Culicidae, Deuterophlebiidae, Dixidae, Hyperoscelidae (Corynoscelidae), Mycetophilidae (Bolitophilidae, Diadociidae, Ditomyiidae, Ceroplatidae, Macroceratidae, Fungivoridae, Sciophilidae), Nymphomyiidae, Pachyneuridae, Perissommatidae, Psychodidae, Ptychopteridae (Liriopeidae), Scatopsidae, Simuliidae (Melusinidae), Tanyderidae, Thaumaleidae (Orphnephilidae), Tipulidae and Trichoceridae. Syn. Nemocera See Diptera. Cf. Bibionomorpha, Blephariceromorpha, Culicimorpha, Nymphomyiomorpha, Psychodomorpha, Ptychopteromorpha, Tanyderomorpha, Tipulimorpha.

NEMATOCEROUS. Adj. (Greek, *nema* = thread; *keras* = horn.) Pertaining to Diptera with long, thread-like Antennae. Cf. Brachycerous.

NEMATOCYTE See Plasmatocyte.

NEMATODE Noun. (Greek, *nema* = thread; *eidos* = form. Pl., Nematodes.) Small, worm-like animals which are parasitic on plants or other animals, or free-living in soil or living in water.

NEMATOLOGY Noun. (Greek, *nema* = thread; *logos* = discourse. Pl., Nematologies.) The study of nematodes to include taxonomy, morphology and the application of the discipline to agricultural and horticultural pest problems.

NEMEC, BOHUMIL (1873–1966) (Maresquelle 1966, Marcellia 33: 193–196, bibliogr.)

NEMEOBIIDAE See Lycaenidae.

NEMESIIDAE Plural Noun. Trapdoor Spiders.

NEMESTRINIDAE Plural Noun. Tangle-Veined Flies. A small Family of Diptera which occurs in all zoogeographical regions but poorly represented in the tropics. Family often placed between Asilidae and Stratiomyidae. Body lacks bristles but has many fine hairs; Thorax and Abdomen often banded or striped; wing venation with Radius and Medius Branches converging and ending before wing apex; composite Diagonal Vein extending from first basal cell to posterior margin of wing. Species are obligate parasites of grasshoppers and scarabaeid beetles. Eggs deposited in holes or crevices in posts, tree trunks and other elevated objects near grasshopper hosts. Female lays 4,000–5,000 eggs during lifetime. First instar larva planidiform and actively searches for its host; larva enters host's Haemocoel through intersegmental membrane on the Thorax or Abdomen, or through Trachea via the spiracles; larva moults to second instar after entering the host's Haemocoel; larva develops respiratory tube attached at point of entry of first instar larva; larva feeds on fat body and ovarian tissue. Host frequently survives for a short period after emergence of the parasite larva but cannot reproduce. Larva leaves host through intersegmental membrane and enters soil for pupation. Adults visit flowers. Cf. Bee Flies.

NEMOGLOSSATA. Noun. (Greek, *nema* = thread; *glossa* = tongue.) Bees with a thread-like tongue.

NEMOLT See Teflubenzuron.

NEMONYCHIDAE Plural Noun. Pine-Flower Snout Beetles. A small, cosmopolitan Family of polyphagous Coleoptera assigned to Superfamily Curculionoidea. Adult <6.0 mm long, elongate and somewhat compressed with moderate vestiture of decumbent Setae; Rostrum elongate; Antenna inserted near midline of Rostrum, with 11 segments and loose club of three segments; Labrum articulate, not fused to Clypeus; Palps flexible. Adult and larva feed on pollen.

NEMOPTERIDAE Plural Noun. A small Family of myrmeleontoid Neuroptera. Head rostrate; Antenna at most tapering slightly apicad; Trichosors absent; hindwing extremely long and narrow. Larva with long neck and legs; predaceous in sandy habitats under ledges and in twilight zone of caves.

NEMORICOLOUS Adj. (Latin, *nemoris* = grove; *colere* = living in or on.) Descriptive of insects that inhabit open and sunny wooded habitats. See Habitat. Cf. Agricolous; Ammophilous; Arenicolous; Caespiticolous; Madicolous; Paludicolous; Silvicolous. Rel. Ecology.

NEMOURIDAE Plural Noun. Forestflies. A Family of Plecoptera including 18 Genera, ca 400 nominal Species and well represented in all areas of the Holarctic. Adult with 'X' shaped area on Notum near base of forewings. All Species are herbivorous.

NEMOUROIDEA The largest Superfamily of Plecoptera. Adult mouthparts are simple, somewhat small and dull coloured. Taxa are restricted to the Northern Hemisphere, except Notonemouridae. All Species are herbivorous.

NEMSYS® See Entomogenous Nematodes.

NEOANNONIN See Acetogenins.

NEODIPRION Noun. A symphytan Genus in the Family Diprionidae whose larvae feed on coniferous trees. Species of *Neodiprion* are important economic pests of forests especially in Canada and the northern USA.

NEOEPHEMERIDAE Plural Noun. A small Family of Ephemeroptera assigned to Superfamily Caenoidea and which includes a few Species in North America. Species resemble Ephemeridae but forewing veins M and Cu1 strongly divergent at base and Costal Crossveins reduced; hindwing outer fork absent, hindwing M1 and M2 divergent beyond middle of wing, acute costal projection near base and three tails in both sexes; hind Tarsi with four segments. Naiads found in streams and rivers.

NEOGAEA Noun. (Greek, *neos* = new; *gaia* = earth.) The Zoogeographical Region comprising the Western Hemisphere or New World.

NEOGAEIC Adj. (Greek, *neos* = new; *gaia* = earth; *-ic* = of the nature of.) Descriptive of organisms which occur in the Western Hemisphere or New World. See Gerontogaeic.

NEOGALLICOLAE-GALLICOLAE Dimorphs of the phylloxeran Fundatrigeniae, which will become Gallicolae (Imms).

NEOGALLICOLAE-RADICOLAE Dimorphs of the phylloxeran Fundatrigeniae which pass to the root and become Radicolae (Imms).

NEOGENE PERIOD The interval of the Geological Time Scale (23 MYBP–Present) within the Cenozoic Era and which includes the Miocene, Pliocene, Pleistocene and Recent Epochs. During NP, grasses spread and humans evolved from hominid ancestors; Rocky Mountains and Himalayas arose; extensive episodic glacial activity occurs in Northern Hemisphere. See Geological Time Scale; Cenozoic Era. Cf. Palaeogene Epoch. Rel. Fossil.

NEOLEPIDOPTERA Noun. (Greek, *neos* = new; *lepid* = scale; *pteron* = wing.) All haustellate Lepidoptera, except the generalized Micropterygidae. Included Taxa lack functional Mandibles and the Pupa is incomplete or obtect. See Palaeolepidoptera; Protolepidoptera.

NEOPITROID® See Permethrin.

NEOPSEUSTIDAE Plural Noun. A small Family of primitive moths with disjunct distribution (Chile, Argentina, northeastern India, Burma, Taiwan and southwest China). Adult broad-winged, weakly scaled; Ocelli absent; Chaetosemata present; Mandibles large, triangular in shape, not functional; Galeae united to form short Proboscis with double food-canal; Maxillary Palpus with five segments; Labial Palpus with three segments. Epiphysis present; tibial spur formula 0-2-4; wing venation homoneurous; Microtrichia densely cover wing surface; wing scales perforated; Jugum present. Adults are crepuscular or nocturnal; immatures are unknown.

NEOPTERA Noun. (Greek, *neos* = new; *pteron* = wing. Pl., Neopterae.) A taxonomic category of insects whose members display incomplete Metamorphosis; category includes Paraneoptera and Polyneoptera.

NEO-PYNAMIN® See Tetramethrin.

NEORON® See Bromopropylate.

NEOSOMATIC Adj. (Greek, *neos* = new; *soma* = body; *-ic* = characterized by.) Pertaining to Neosomy.

NEOSOME Noun. (Greek, *neos* = new; *soma* = body. Pl., Neosomes.) An organism which has undergone Neosomy.

NEOSOMULE Noun. (Greek, *neos* = new; *soma* = body; *-ule* = diminutive. Pl., Neosomules.) Acarina: A new external structure resulting from Neosomy. (Goff *et al.* 1982, J. Med. Ent. 19: 228.)

NEOSOMY Noun. (Greek, *neos* = new; *soma* = body. Pl., Neosomies.) 1. Radical intrastadial Metamorphosis associated with arthropod symbioses. (Audy *et al.* 1972, J. Med. Ent. 9: 487–494.) 2. Acari: The addition of new external anatomical features in part due to the secretion of new Cuticle during the single active stadium of a chigger which typically alters external form only through a moult (Goff *et al.* 1982, J. Med. Ent. 19: 228.)

NEOSTANOX® See Fenbutatin-Oxide.

NEOTEINIA See Neoteny. Alt. Neoteiny.

NEOTENIC Adj. (Greek, *neos* = new; *teinein* = to stretch; *-ic* = characterized by.) Isoptera: 'secondary' reproductives which developed in a termite colony when the 'Primary' Reproductives (King and Queen) are lost to the colony. Anatomically, Neotenics are characterized by various combinations of the following characters: Compound Eyes small or absent, less sclerotization and less pigmentation on body than primary reproductives, wings absent (aptery) or display rudimentary wing buds (brachyptery). Alt. Neoteinic; Neotenous. Syn. Secondary Reproductive. Cf. Primary Reproductive. Rel. Social Insect.

NEOTENY Noun. (Greek, *neos* = new; *teinein* = to stretch. Pl., Neotenies.) 1. Retarded somatic maturation (Bruce 1979, Evolution 33: 998–1000.) Cf. Metathetely. 2. The expression of sexual maturity during the larval stage. 3. The persistence of larval characteristics (anatomical, biological or behavioural) in the adult. Alt. Neotenia; Neoteinia. Cf. Paedogenesis.

NEOTROPICAL CORNSTALK BORER *Diatraea lineolata* (Walker) [Lepidoptera: Pyralidae].

NEOTROPICAL REGION The Zoogeographical area embracing South America, Central America, the West Indies and southern and coastal Mexico. Mexico forms a transitional area between the Neotropical and the Nearctic Regions.

NEOTTIOPHILIDAE See Piophilidae.

NEOTYPE Noun. (Greek, *neos* = new; *typos* = pattern. Pl., Neotypes.) A plesiotype or specimen selected to represent the name-bearer when the original type or holotype is lost or destroyed. See Type. Cf. Holotype; Plesiotype. Rel. Nomenclature; Taxonomy.

NEOXANTHIN Noun. (Greek, *neos* = new; *xanthos* = yellow. Pl., Neoxanthins.) A xanthophyllic carotenoid pigment that occus in algae. See Pigment. Cf. Carotin.

NEPHIS® See Ethylene Dibromide.

NEPHRIC Adj. (Greek, *nephros* = kidney; *-ic* = of the nature of.) Pertaining to the kidney or excretory organ of insects.

NEPHRIDIAL Adj. (Greek, *nephros* = kidney; *idion* = diminutive; Latin, *-alis* = pertaining to.) Descriptive of structure or function of Nephridia.

NEPHRIDIUM Noun. (Greek, *nephros* = kidney; *-idion* = diminutive. Pl., Nephridia.) Tubular structures functioning as kidneys in annelids and molluscs. A term incorrectly used for the Malphighian Tubules of insects (Smith).

NEPHROCYTE Noun. (Greek, *nephros* = kidney; *kytos* = hollow vessel; container. Pl., Nephrocytes.) Cells ensheathed beneath the Basement Membrane and arranged singly or in groups within various parts of body. Nephrocytes are abundant in the Pericardial Sinus (as Pericardial Cells) along the Heart and Aorta, and beneath the Oesophagus; Nephrocytes form a chain (Garland Cells) between salivary glands in blowfly larvae. Nephrocytes maintain Haemolymph homeostasis by taking in (endocytosis) nonparticulate colloids and release molecules into the Haemolymph through exocytosis; Nephrocytes store waste products in Cytoplasm, regulate Haemolymph composition and share responsibility with Haemocytes. Nephrocytes are distinguished from sessile Haemocytes by enclosure within Basement Membrane and pinocytotic behaviour (not phagocytic as Haemocytes). Syn. Hepatic Cells. See Haemocyte. Cf. Garland Cell; Pericardial Cell; Phagocyte. Rel. Fat Body.

NEPHROID Adj. (Greek, *nephroeides* = as a kidney; *eidos* = form.) Kidney-like in shape or function. See Kidney. Cf. Reniform; Oval. Rel. Eidos; Form; Shape.

NEPHROPATHIA EPIDEMICA See Epidemic Haemorrhagic Fever.

NEPIDAE Latrielle 1802. Plural Noun. Water Scorpions. A small Family of aquatic heteropterous Hemiptera assigned to Superfamily Nepoidea. Fore legs raptorial with Femur and Tibia adapted for prey capture; all legs with one tarsal segment; middle and hind legs slender, not flattened or fringed; Cerci modified into and elongate respiratory tube (siphon) which is often as long as the body. Females insert eggs into tissue of aquatic plants. Nepids inhabit ponds, lakes and streams where they are predatory upon insects and small fish.

NEPIONIC Adj. (Greek, *nepios* = infant; *-ic* = consisting of.) Pertaining to the stage of development immediately following the embryonic stage; postembryonic. Proposed as a substitute for the term 'larval.'

NEPIONOTYPE Noun. (Greek, *nepios* = infant; *typos* = pattern. Pl., Nepionotypes.) The type-specimen as the larval stage of a Species. See Type.

NEPOIDEA A Superfamily of Hemiptera including Belostomatidae and Nepidae. See Hemiptera.

NEPOREX® See Cyromazine.

NEPTICULID MOTHS See Nepticulidae.

NEPTICULIDAE Stainton 1859. Plural Noun. Nepticulid Moths; Leaf Miners. A widespread Family of heteroneurous Lepidoptera with ca 400 described Species. Adult very small, dark coloured; wingspan 3–6 mm; head with erect piliform scales, paired Chaetosemata and collar; Ocelli absent; Antenna about half as long as forewing; Scape forming an eye cap; flagellar segments with unique, branched Sensilla; Proboscis short but functional in some Species; foreleg without Epiphysis; tibial formula 0-2-4; forewing with M and R coalescent; male wings with specialized scales. Eggs convex, typically oval in outline shape. Larva with pair of Stemmata; without segmented thoracic legs; Pseudolegs (Calli) often present on Mesothorax, Metathorax, abdominal segments 1–7; Crochets absent. Pupal appendages exarate; abdominal segments 2–7 moveable. Adults often fly during day; attracted to lights during night; egg deposited on leaves; neonate larva of most Species burrows into leaf and forms mine with mine shape often diagnostic; some Species mine Petiole and a few feed on bark or fruit. Mature larva exits mine, bark or fruit and pupation occurs on ground, in leaf litter or soil; cocoon dense, oval in outline shape and slightly flattened. Syn. Stigmellidae.

NEPTICULOIDEA A Superfamily of monotrysian Lepidoptera including Nepticulidae and Opostegidae. Ocelli and Chaetosemata absent; Antennal Scape expanded, concave, forming an eye cap; foretibial Epiphysis absent, tibial spur formula 0-2-4; wings with reduced venation; Ovipositor non-piercing.

NEREISTOXIN Noun. An alkaloid biopesticidide extracted from the marine annelid *Lumbriconereis heteropoda* Marenz. See Biopesticide. Cf. Methyllycaconitine; Nereistoxin; Nicotine; Rotenone; Pyrethrin; Pyrroles; Ryania; Sabadilla; Stemofoline.

NEREN, CARL HARALD (–1901) (Nordenström 1902, Ent. Tidskr. 23: 195–197, bibliogr.)

NERESHEIMER, JULIUS (1880–1943) (Sachtleben 1944, Arb. morph. taxon. Ent. Berl. 11: 57.)

NERIIDAE Plural Noun. Proposed as Neriades (Westwood 1840).

NERIKI, KIZO (1850–1910) (Hasegawa 1967, Kontyû 35 (Suppl.): 64–65, bibliogr. pl. 1, fig. 4.)

NERITIC ZONE The biogeographic marine realm along coastline between low tide and a depth of 100 metres. See Habitat. Cf. Abyssal Zone; Littoral Zone; Pelagic Zone; Photic Zone.

NERKOL® See Dichlorvos.

NERVATE Adj. (Latin, *nervus* = sinew; *-atus* = adjectival suffix.) Pertaining to a body, structure or appendage with nerves or veins.

NERVE Noun. (Latin, *nervus* = sinew. Pl., Nerves.) A thread-like structure, composed of delicate filaments of tissue and which transmits sensations or stimuli to or from a Ganglion.

NERVE CELL The element of the Nervous System responsible for receiving, propagating and transmitting an electrical stimulus from one part of the body to another. The nerve cell and its branches are called a Neuron (Neurone). Typically the Neuron is elongate with one or more branches. The primary branch from the cell body is called the Axon and is typically long, tapered and contains neural filaments. The Axon directs the electrical impulse away from the cell body. Lateral branches from the Axon are called Collateral Branches The distal portion of the Axon and Collateral Branches form a series of smaller fibres called terminal arborizations. The cell body displays a series of small branches called Dendrites. The Dendrite is typically shorter than the Axon and is directly connected to the cell body (Neurocyte). The Dendrite also displays Terminal Arborizations but not Collateral Branches. The Dendrite receives the stimulus from adjacent cells. The cell body is an enlarged portion of the nerve cell. Cytoplasm contained within the nerve cell-body is called the Perikaryon. The term 'Perikaryon' also is applied to the cell body of the nerve cell. The cell body contains large nuclei and Endoplasmic Reticulum. Basophilic material called Ergatoplasm forms part of the Endoplasmic Reticulum; Ergatoplasm is basophilic because it stains with Safranin, Gentian Violet and Toluidine Blue. The Endoplasmic Reticulum is a lace-like double membrane with RNA granules associated. See Nervous System; Neuron. Syn. Neurocyte.

NERVE FIBRE The Axon or other branches of a Neurocyte (Snodgrass).

NERVE LABII See Labial Nerve.

NERVE MANDIBULARIUM See Mandibular Nerve.

NERVE MAXILLARUM See Maxillary Nerve.

NERVE OCELLARII See Ocellar Nerve.

NERVE SHEATH A membrane-like envelope which surrounds entire Central Nervous System, including Brain, connectives and peripheral nerves. NS is non-cellular, provides support and is probably derived from mesodermal connective tissue. NS consists of mucopolysaccharides, mucoproteins and collagen-like fibres that form narrow, oriented filaments and a thicker layer of fibrils. Syn. Neurolemma; Neural Lamella. See Nerve Cell; Nervous System. Cf. Axon; Cell Body; Dendrite; Glial Cell. Rel. Central Nervous System.

NERVE TRACT A strand of nerve fibres; usually applied to tracts within a nerve centre (Snodgrass).

NERVE TRUNK A bundle of nerve fibres, motor, sensory or both, in the peripheral system.

NERVELLUS Noun. Hymenoptera: A wing vein in Ichneumonidae.

NERVOUS SYSTEM All higher forms of animal life possess a system by which information is processed in a systematic, timely manner. The insect nervous system is derived from Ectoderm and is bilaterally symmetrical. Basic elements of the Nervous System include Afferent Sensory Nerves which proceed inward from the sensory receptor, a Central Nervous System (Interneurons) which recombines information and Efferent Sensory Nerves (Motor Neurons) which transmit information to the area of the body requiring action. Interneurons are interpolated between sensory and motor neurons. See Afferent Nerve; Efferent Nerve; Interneuron; Motor Neuron. Cf. Brain; Nerve Cell; Ventral Nerve Cord. Rel. Alimentary System; Circulatory System; Reproductive System; Respiratory System.

NERVULATION Noun. (Latin, *nervus* = sinew; English, *-tion* = result of an action. Pl., Nervulations.) 1. Arrangement of nerves in the body. 2. Arrangement of veins in the wing. Alt. Nervuration.

NERVULE Noun. (Latin, *nervus* = sinew. Pl., Nervules.) 1. Any of the rod-like, rib-like or vein-like structures which support the wing membranes. 2. A vein or veinlet. Alt. Nervure.

NERVURA COSTALIS The costal nerve (costal nervure) of the wing; the wing-nerve nearest the upper margin of the wing.

NERVUS ANTENNALIS The Antennal nerve. See Deutocerebrum. Rel. Nervous System.

NERVUS GANGLII OCCIPITALIS The nerve of the Occipital Ganglion (Snodgrass).

NERVUS LABROFRONTALIS The labro-frontal nerve.

NERVUS LATERALIS The lateral nerve. A slender nerve in Lepidoptera larvae (Snodgrass).

NERVUS OPTICUS Optic Nerve of the compound eye in insects (Snodgrass). See Optic Lobe.

NERVUS POSTANTENNALIS The Postantennal Nerves, the principal nerves of the Tritocerebral Ganglia in Crustacea (Snodgrass).

NERVUS SUBPHARYNGEALIS The Subpharyngeal Nerve.

NERVUS TEGUMENTALIS The Tegumentary Nerve.

NESIUM Noun. (Etymology obscure; Latin, *-ium* = diminutive > Greek, *-idion*. Pl., Nesia.) Scarabaeoid larvae: A sclerotized, more or less projecting mark in the space between the inner ends of the Tormae, anterior to the Crepis, usually 1–2 Nesia present, right Nesium, the chitinous plate of Hayes, left Nesium, the sense cone of Hayes (Boving).

NESSLING, ELIEL (1871–1941) (Hellén 1943, Notul. ent. 23: 58–59.)

NEST Noun. (Middle English. Pl., Nests.) 1. Social insects: A place or structure fashioned or made by ants, bees, wasps and termites which is used to receive and incubate eggs, rear immatures, store food and serve as the coordinating centre of social organization. 2. Non-social immature insects: Habitat or food fashioned into an area of concealment, refuge or protection for and by an individual and used while feeding self and sometimes pupation. Nests sometimes communal constructions, as in some caterpillars (*e.g.* Lymantriidae).

NEST AURA See Nest Odour.

NEST FLIES See Neottiophilidae.

NEST ODOUR Social Insects: The distinctive, characteristic or unique odour of an insect nest perceived by its inhabitants to distinguish that nest from those of other social insects or the environment. Syn. Nest Aura. See Colony Odour.

NEST ROBBING See Cleptobiosis.

NEST USURPATION Social insects: Behaviour expressed by some bees and wasps in which a female nest occupant (provisioner) is displaced by another female.

NET-CASTING SPIDERS See Deinopidae.

NET-SPINNING CADDISFLIES See Hydropsychidae.

NET-WINGED BEETLES See Lycidae.

NET-WINGED MIDGES See Blephariceridae.

NETOLITZSKY, FRITZ (1875–1945) (Jarvis 1946 Entomol. mon. Mag. 82: 160.)

NETTLE TREE LEAF BEETLE *Hoplostines laporteae* (Weise) [Coleoptera: Chrysomelidae].

NETTLEGRUBS See Limacodidae.

NEUMANN, CASPAR (1683–1737) (Rose 1850, *New General Biographical Dictionary* 10: 316.)

NEUMOEGEN, BERTHOLD (1845–1895) (Anon. 1895, Ent. News 6: 65–66.)

NEURAD Adv. (Greek, *neuron* = nerve; Latin, *ad* = toward.) Toward a nerve; toward the central nervous system. See Orientation. Cf. Anteriad; Apicad; Basad; Caudad; Centrad; Cephalad; Craniad; Dextrad; Dextrocaudad; Dextrocephalad; Distad; Dorsad; Ectad; Entad; Laterad; Mediad; Mesad; Orad; Proximad; Rostrad; Sinistrad; Sinistrocaudad; Sinistrocephalad; Ventrad.

NEURAFORAMEN Noun. (Greek, *neuron* = nerve; Latin, *foramen* = hole.) An arcane term used to describe the opening in the Occipital Foramen through which the nervous system passes from head to Thorax (MacGillivray).

NEURAL Adj. (Greek, *neuron* = nerve; Latin, *-alis* = pertaining to.) Descripitve of structure associated with the nervous system. Descriptive of or pertaining to the nerves or to the nervous system of an animal.

NEURAL CANAL An incomplete tunnel on the floor of the Mesothorax and Metathorax. The NC is formed by fusion of Apodemes, serves for the reception and protection of the Ventral Nerve Cordand provides attachment sites for muscles.

NEURAL GROOVE The furrow in the primitive layer of the embryo in which the nerve cord is formed.

NEURAL RIDGES The two longitudinal ventral ridges of the Ectoderm of the germ band of the embryo in which are formed the lateral cords of Neuroblasts (Snodgrass).

NEURATION Noun. (Greek, *neura* = nerve; English, *-tion* = result of an action.) The venation of the insect wing.

NEURILEMMA Noun. (Greek, *neuron* = nerve; *lemma* = skin. Pl., Neurilemmae.) The external sheath of a nerve fibre. The nucleated sheath of nerve tissue covering the ganglia, nerve trunks and terminal branches (Snodgrass). Adj. Neurilemmal.

NEURITE Noun. (Greek, *neuron* = nerve; *-ites* = constituent. Pl., Neurites.) The cylindrical process along the primary axis of the nerve cell. The Axon. See Axon.

NEUROBLAST Noun. (Greek, *neuron* = nerve; *blastos* = bud. Pl., Neuroblasts.) Insect embryo: The inner layer of cells which forms the nervous tissue; tissue derived from ectodermal cells forming the neural ridges (Imms).

NEUROCHAETIDAE McAlpine 1978. Plural Noun. Upside-down flies.

NEUROCYTE Noun. (Greek, *neuron* = nerve; *kytos* = hollow vessel; container. Pl., Neurocytes.) The cell body of a neuron, usually termed the 'nerve cell.'

NEUROGLOEA Plural Noun. (Greek, *neuron* = nerve; *glia* = glue.) The gelatinous substance which holds together the fine twigs of the Axons of the ganglion cells.

NEUROHORMONE Noun. (Greek, *neuron* = nerve; *hormaein* = to excite. Pl., Neurohormones.) Any member of a class of secretory products which are synthesized in Brain and endocrine glands (Corpus Cardiacum). Neurohormones are released from Neurohaemal Organs to control physiological, behavioural and metabolic events (diuresis, heartbeat, lipid metabolism, moulting, eclosion). Neurohormones are not neuro-transmitters.

NEUROID Adj. (Greek, *neuron* = nerve; *eidos* = form.) Nerve-like in appearance or function. See Nerve. Cf. Dendroid. Rel. Eidos; Form; Shape.

NEUROLOGY Noun. (Greek, *neuron* = nerve; *logos* = discourse. Pl., Neurologists.) The study of the structure and function of the nervous system.

NEUROMERE Noun. (Greek, *neuron* = nerve; *meros* = part. Pl., Neuromeres.) Insect embryo: Paired swellings of the neural ridges at the bases of the embryonic appendages (Imms). The embryonic rudiment of a ganglion. The segmental nervous system of annelids and arthropods.

NEUROMUSCULAR Adj. (Greek, *neuron* = nerve; Latin, *musculus* = muscle.) Pertaining to the complex and intimate association between nerves and muscles. Alt. Neuromyal.

NEURON Noun. (Greek, *neuron* = nerve. Pl., Neurons.) A functional nerve cell, including the Neurocyte, Axon, Dendrites and associated Arborizations. A neuron which has one primary branch is called monopolar; a neuron which has two primary branches is called bipolar; a neuron which has several primary branches is called multipolar. Sensory neurons are associated with receptors of sense organs and are bipolar or multipolar. Most insect sensory axons do not have Synapses. Afferent Neurons connect sensory neurons with Motor Neurons (Efferent Nerves). Motor Neurons are typically monopolar; their ganglia and axons pass to effectors (muscles and glands). Interneurons (Adjustor Neurons) are unipolar and lie exclusively in the Central Nervous System. They act as a switchboard to direct nerve impluses along the correct path. Syn. Neurone. See Nerve Cell.

NEUROPILE Noun. (Greek, *neuron* = nerve; *pilos* = felt. Pl., Neuropiles.) The medullary substance or mass of fibrous tissue within a ganglion (Snodgrass). Alt. Neuropil; Neuropilema.

NEUROPORE See Trichospore.

NEUROPTERA Linnaeus 1758. (Lower Permian-Recent.) (Greek, *neuro* = nerve; *pteron* = wing.) Alderflies, Antlions, Dobsonflies, Fishflies, Lacewings, Owlflies, Snakeflies. A cosmopolitan Order including ca 5,000 nominal Species and which is predominantly tropical. Endopterygote Neuroptera; small to very large, soft bodied; head freely mobile; two compound eyes well developed; Ocelli usually absent; Antenna long, multisegmented; mandibulate, biting-type mouthparts; Maxillary Palpus five segments; Labial Palpus usually three segments. Prothorax moveable; Thorax loosely organized; four wings membranous, usually large, subequal in size, tentiform at repose; venation abundant, net-like with terminal bifurcations (end-twigging); coupling mechanism simple; legs typically cursorial; tarsal formula 5-5-5. Abdomen 10-segmented (except Chrysopidae); Cerci absent; three thoracic and seven abdominal ganglia. Metamorphosis complete. Typically three larval instars, occasionally more; larva with piercing-sucking mouthparts from Mandible and Maxilla forming elongate tube on either side of head; midgut and hindgut not connected in larva. Pupation in cocoon; cocoon silk from Malpighian Tubules, secreted from Anus; Pupae exarate, decticous, Mandibles well developed. Larvae typically predaceous, some parasitic, a few phytophagous; adults predominantly predaceous, some highly beneficial to agriculture. In its older use, the term applied to all net-veined insects irrespective of Metamorphosis or thoracic structure. Classification *sensu* *lato*: Suborder Megaloptera: Corydalidae, Sialidae; Suborder Planipennia: Coniopterygoidea: Coniopterygidae; Hemerobioidea: Chrysopidae, Hemerobiidae; Polystoechotidae, Psychopsidae; Ithonoidea: Ithonidae; Mantispoidea: Berothidae, Mantispidae; Myrmeleontoidea: Ascalaphidae, Myrmeleontidae, Nemopteridae, Nymphidae, Stilbopterygidae; Osmyloidea: Dilaridae, Neurothidae, Osmylidae, Sisyridae. Incertae Sedis: Brucheiseridae, Rapismatidae; Suborder Rhaphidoidea: Inocelliidae, Rhaphidiidae. See Megaloptera; Rhaphidiodea.

NEUROPTEROID Plural Noun. Like the Neuroptera in the broad sense. A term applied to extinct neuropteroid forms that have a general resemblance to Neuroptera.

NEUROPTEROUS Adj. (Greek, *neuron* = nerve; *pteron* = wing.) Descriptive of or pertaining to the Order Neuroptera.

NEUROSECRETORY Adj. (Greek, *neuron* = nerve; Latin, *secernere* = to separate.) Pertaining to nerve cells which secrete hormones. Alt. Neurocrine.

NEUROSECRETORY CELLS See Central Nervous System.

NEUROSENSORY Adj. (Greek, *neuron* = nerve; Latin, *sensus* = sense.) Pertaining to sensory receptors in the Integument.

NEUROSOME Noun. (Greek, *neuron* = nerve; *soma* = body. Pl., Neurosomes.) Mitochondria of the nerve cell.

NEUROSPONGIUM Noun. (Greek, *neuron* = nerve; *spongia* = sponge; Latin, *-ium* = diminutive > Greek, *-idion*. Pl., Neurospongia.) 1. A medullary substance formed of the fine twigs of the axons of the ganglion cells held together by neurogloea (Imms). 2. The transparent medullary part of a nerve ganglion (Folsom & Wardle). 3. A granular matrix in the Periopticon of the insect eye (Smith).

NEUROSYNAPSE Noun. (Greek, *neuron* = nerve; *synapsis* = union. Pl., Neurosynapses.) The synapse or area of contact between adjacent nerve cells.

NEUROTHIDAE Plural Noun. A numerically small Family of osmyloid Neuroptera. Anocellate; wings asetose; forewing Costal Vein forked, usually with Nygmata; fore leg not raptorial. Adults occur near streams and in forests; larvae are aquatic or inhabit moist leaf litter.

NEUROTOXIC Adj. (Greek, *neuron* = nerve; *toxikon* = poison; *-ic* = characterized by.) Pertaining to substances (poisons) which adversely affect the nervous system. See Toxin. Cf. Haemotoxic.

NEUROTOXIN Noun. (Greek, *neuron* = nerve; *toxikon* = poison.) A chemical compound which affects nerve cells or the nervous system. A characteristic of the venom of some wasps, reptiles, fishes and other venomous organisms. See Toxin. Cf. Haemotoxin.

NEUROTRANSMITTER Noun. (Greek, *neuron* = nerve; Latin, *transmittere* = to send across.) A chemical compound (acetylcholine, noradrenaline) secreted at a nerve ending and which enables a nervous impulse to pass over a synapse to an adjacent nerve cell.

NEUROTROPHIC Adj. (Greek, *neuron* = nerve; *trephein* = to nourish; *-ic* = of the nature of.) Pertaining to nourishing the nerves or nervous system.

NEUSTETTER, HEINRICH (–1958) (Anon. 1958, Z. wien. ent. Ges. 43: 64.)

NEUSTON Noun. (Greek, *neustos* = floating.) Organisms which float or swim in the surface layers of water. Cf. Nekton; Plankton; Seston.

NEUTER Noun. (Latin, *neuter* = neither.) Social Hymenoptera: A term used to characterize workers (genetically female) or reproductively undeveloped females. In publications, neuter individuals are indicated by an imperfect form of the astronomical sign for Venus.

NEUTRAL Adj. (Latin, *neuter* = neither; *-alis* = characterized by.) pH = 7.

NEUWYLER (NEUVILLIER), MELCHIOR (1819–1845) (Wolf 1845, Bull. Soc. ent. Fr. 1845: cv–cvi.)

NEVADA SAGE GRASSHOPPER *Melanoplus rugglesi* Gurney [Orthoptera: Acrididae].

NEVERMANN, WILLIAM HEINRICH FERDINAND (GUILLERMO ENRIQUE FERNANDO) (1881–1938) (Anon. 1938, Ent. News 49: 239–240.)

NEVEU-LEMAIRE, MAURICE (1872–1951) (Lavier 1951, Ann. Parasit. hum. comp. 26: 271–273)

NEVIANI, ANTONIO (1857–1946) (Vitolo 1948, Memorie Soc. tosc. Sci. Nat. 55: 1–9, bibliogr.)

NEVINSON EDWARD BONNEY (1858–1927) (Gardner 1928, Entomologist's mon Mag. 64: 117–118.)

NEVINSON, BASIL GEORGE (1852–1909) (Fowler 1910, Entomol. mon. Mag. 46: 93–94.)

NEVSKY, VALERIANA PAVLOVOVICH (1893–1951) (Narzikulov Ent. Obozr. 40: 460–462, bibliogr. Translation: Ent. Rev., Wash. 40: 241–242, bibliogr.)

NEW GUINEA SUGARCANE WEEVIL *Rhabdoscelus obscurus* (Boisduval) [Coleoptera: Curculionidae].

NEW HOUSE BORER *Arhopalus productus* (LeConte) [Coleoptera: Cerambycidae].

NEW JERSEY MECHANICAL TRAP A device for collecting mosquitoes. Syn. New Jersey Trap. See Trap.

NEW SPECIES 1. A Species 'recently' described, purportedly unknown to science and which does not have a scientific name (binomen). 2. A term appended to the description of a formally described and nomenclaturally validated Species-level entity. See Taxonomy.

NEW WORLD SCREW-WORM *Cochliomyia hominivorax* (Coquerel) [Diptera: Calliphoridae]: An obligatory parasite of domestic livestock and humans in the USA and Neotropical Region to northern Chile and Argentina; does not develop on cold-blooded vertebrates. Adult NWSW attracted to cuts, sores and wounds; etiological agent of nasopharyngeal Myiasis in humans. Adult body green-blue metallic coloration; face red, yellow or orange; Thorax with three dark stripes on dorsum. Female lays batches of 10–400 eggs on host; can lay 2,800 eggs during lifetime; eclosion occurs in less than 24 hours. Larval development 3–5 days in host; larvae can develop in carcass or decaying meat but female only oviposits on warm-blooded hosts. Pupation in soil; pupal stage ca seven days. Cannot overwinter in cold climates. Species the object of extensive eradication attempts by USA and first insect eradicated from an area through Sterile Insect Technique. Syn. Primary Screw-Worm. A functional homologue of Old World Screw-Worm. See Calliphoridae. Cf. Old World Screw-Worm; Secondary Screw-Worm.

NEW YORK WEEVIL *Ithycerus noveboracensis* (Forster) [Coleoptera: Curculionidae].

NEW ZEALAND CATTLE TICK *Haemaphysalis longicornis* Neumann [Acari: Ixodidae].

NEW ZEALAND SEABIRD TICK *Ixodes eudyptidis* Maskell [Acari: Ixodidae].

NEWBERRY, EMANUEL AUGUSTUS (1845–1927) (Anon. 1928, Entomol. mon. Mag. 64: 15–16.)

NEWCOMB, SIMON (1835–1909) (Howard 1910, Bull. phil. Soc. Wash. 15: 133–167.)

NEWCOMER'S ORGAN Lepidoptera: An exocrine gland, regarded as a myrmecophilous organ, on the seventh abdominal segment of many lycaenid larvae. Ants take secretion from the gland. First noted by Newcomer (1912). (Kitching & Luke 1985. J. Nat. Hist. 19: 259–276.) Syn. Newcomer's Gland, Dorsal Nectary Organ. Cf. Perforated Cupola Organ. Tentacular Organ.

NEWELL, IRVIN American academic (University of California, Riverside) specializing in taxonomy of acarines and biological control.

NEWELL, WILMON (1878–1943) (Creighton 1943, J. Econ. Ent. 36: 947–949.)

NEWLAND, GORDAN (–1956) (Hall 1957, Proc. R. ent. Soc. Lond. (C) 21: 66.)

NEWMAN FLY *Dacus newmani* (Perkins) [Diptera: Tephritidae].

NEWMAN, EDWARD (1801–1876) (Anon. 1876, Entomol. mon. Mag. 13: 45–46.)

NEWMAN, LEONARD WOODS (1873–1949) (Cockayne 1949, Entomologist's Rec. J. Var. 61: 80–81.)

NEWMAN, LESLIE JOHN (1878–1938) (Musgrave 1932, *A Bibliogr. of Australian Entomology* 1775–1930, 380 pp. (238–240), Sydney.)

NEWPORT, GEORGE (1803–1854) (Anon. 1854, Proc. Linn. Soc. Lond. 2: 309–312, bibliogr.)

NEWSTEAD, ROBERT (1859–1947) (Anon. 1947, Nature 159: 428–429.)

NEWTON, ARTHUR HENRY (1885–1966) (Kennedy 1968, Proc. R. ent. Soc. Lond. (C) 32: 59.)

NEWTON, J HARRY (1893–1956) (Daniels 1957, J. Econ. Ent. 50: 524–525.)

NEX® See Carbofuran.

NEXIN Noun. Thin, elastic filaments of protein which extend at regular intervals between adjacent Accessory Doublets that comprise the Axoneme of a Spermatozoon. Points of fibre attachment are near the Radial Spoke. Functionally, Nexin fibres apparently regulate the amount of shear displacement between adjacent doublets during microtubule movement. Structurally, Nexin fibres preserve the spatial relationship of the doublets forming the Axoneme during the 'resting' phase. See Axoneme; Microtubule. Cf. Dynein Arm; Radial Spoke. Rel. Sperm Tail.

NEXTER® See Pyridaben.

NEXUS Noun. (Latin, *nexus* = tying. Pl., Nexuses.) A connection between individuals of a group or series.

NIACIN Noun. Nicotinic Acid.

NIBLETT, MONTAGUE (1877–1967) (Parmenter 1967, Entomologist's Rec. J. Var. 79: 321–322.)

NICELLE, GRAF G. VON (Anon. 1858, *Accentuated list of British Lepidoptera*. xliv + 118 pp. (xxxv). Oxford & Cambridge Entomological Societies, London.)

NICEVILLE, CHARLES LIONEL AUGUSTUS DE (1852–1901) (Fowler 1901, Proc. ent. Soc. Lond. 1901: xxxiv–xxxv.)

NICHE Noun. (French *niche,* from Italian *nicchia* = wall recess.) Ecology: The total range of climatic conditions, edaphic factors and food necessary for a population to exist and reproduce.

NICHINO® See Buprofezin.

NICIPPE YELLOW *Eurema nicippe*. [Lepidoptera: Pieridae]. Alt. Sleepy Orange.

NICKERL, FRANZ ANTON (1813–1871) (Howard 1930, Smithson. misc. Collns. 84: 309.)

NICKERL, OTTOKAR (1838–1920) (Varia 1923, Cas. csl. Spol. ent. 1: 3–12, bibliogr.)

NICODAMIDAE Plural Noun. Red-and-black spiders.

NICOLAS, HECTOR (1834–1899) (Chobaut 1899, Mém. Acad. Vaucluse 18: 347–544, bibliogr.)

NICOLAY, ALAN S (–1950) (Blackwelder 1950, Coleopts. Bull. 4:2.)

NICOLET, HERCULE (–1872) (Bonnet 1945, *Bibliographia Araneorum* 1:35.)

NICOLETIIDAE Plural Noun. A moderate-sized, cosmopolitan Family of Thysanura. Body to 20 mm long; scales present (Atelurinae) or absent (Nicoletiinae), tapered with Abdomen narrower than Thorax; compound eyes absent; Proventriculus absent; abdominal Sterna often subdivided into medial sternites and lateral coxites; Exertile Vesicles present on Sterna 2–7 of some Species; Appendix Dorsalis present; male Spermatozoa contained in mucoid spermatolophids. Species of Nicoletiinae are cavernicolous and often occur under stones, within leaf litter, or in mammal burrows; Species of Atelurinae occur in ant and termite nests; a few Species are regarded as plant pests.

NICOTINE Noun. (French, from Jacques Nicot,

French ambassador at Lisbon, introduced tobacco to France in 1560. Pl., Nicotines.) An alkaloid derived typically from *Nicotiana tabacum* L. or *N. rustica* L. (Solanaceae) and present in several other Families of plants. Probably most well known and widely used alkaloid-type insecticide; used as a control agent for aphids and soft-bodied insects before the advent of synthetic insecticides. Nicotine is most commonly used to control garden pests and characterized by high mammalian toxicity, limited spectrum of effectiveness and instability in light. Accidental poisoning has resulted in many deaths, and nicotine has been used as an agent of suicide. Chronic exposure causes various reproductive and cardiac conditions. See Biopesticide; Botanical Insecticide; Natural Insecticide; Insecticide. Cf. Methyllycaconitine; Pyrethrin; Rotenone; Ryania; Sabadilla. Rel. Imidacloprid.

NICOULINE® See Rotenone.

NICTITANT Adj. (Latin, *nictare* = to wink; *-antem* = adjectival suffix.) Pertaining to an Ocellus with a lunar spot of another colour.

NIDAMENTAL Adj. (Latin, *nidamentum* = nest material; *-alis* = pertaining to.) Pertaining to a glandular secretion that covers the eggs of some insects.

NIDAMENTUM Noun. (Latin, *nidamentum* = nest material. Pl., Nidamenta.) The gelatinous mass within which *Chironomous* sp. lays its eggs (Henneguy).

NIDICOLOUS Adj. (Latin, *nidus* = nest; *colere* = to dwell.) Descriptive of life within a nest after eclosion from the egg stage. See Habitat. Rel. Ecology

NIDIFICATE Adj. (Latin, *nidus* = nest; *facere* = to make; *-atus* = adjectival suffix.) A term applied when eggs are placed in a nesting area.

NIDIFICATION Noun. (Latin, *nidus* = nest; *facere* = to make.) The process of constructing a nest. The behaviour associated with constructing a nest.

NIDIFUGOUS Adj. (Latin, *nidus* = nest; *fugere* = to flee; *-osus* = with the property of.) Pertaining to an organism that leaves its nest soon after eclosion from the egg.

NIDUS Noun. (Latin, *nidus* = nest. Pl., Nidi.) Centres of development of the large, secreting cells of the midgut epithelium (Comstock). In general, a definite cell-group.

NIELSE, CÉSARE (1898–) (Bucciarelli 1973, Odonatologica 2: 65–67, bibliogr.)

NIELSEN, EMIL (1876–1938) (Braendegaard 1938, Ent. Meddr. 20: 185–187.)

NIELSEN, JENS CHRISTIAN (1881–1918) (Kryger 1919, Ent. Meddr. 13: 1–11, bibliogr.)

NIELSEN, PEDER (1893–1975) (Tuxen 1975, Ent. Meddr. 43: 137–143, bibliogr.)

NIELSEN, PEDER KRISTIAN (1882–1965) (Wolff 1965, Ent. Meddr. 34: 172–174.)

NIELSON, ANKER (1907–) (Tuxen. 1968, Ent. Meddr. 36: 109.)

NIEMEYER, C H ROBERT (1862–1907) (Weidner 1967, Abh. Verh. naturw. Ver. Hamburg Suppl. 9: 192.)

NIEMEYER, LUDWIG (–1974) (Anon. 1975, Prakt. SchädlBekämpf. 27: 99.)

NIEPELT, WILHELM (1862–1936) (Röber 1932, Int. ent. Z. 26: 327–333, bibliogr.)

NIEREMBERG, JOHANN EUSEBIUS VON (1595–1658) (Rose 1850, *New General Biographical Dictionary* 10: 341.)

NIESIOKOWSKI, WITOLD (1867–1954) (Bielewicz 1955, Polskie Pismo ent. 25: 5–8, bibliogr.)

NIETO, JOSE APOLINARIO (1810–1874) (Sallé 1874, Ann. Soc. ent. Fr. (5) 4: 359–361.)

NIEZABITOWSKI, LUBICZ EDWARD (–1947) (Anon. 1948, Polskie Pismo ent. 18: 3.)

NIGER Adj. (Latin, *niger* = black.) Black or the colour of lamp-black.

NIGGER *Orsotraena medus moira* Waterhouse & Lyell [Lepidoptera: Nymphalidae].

NIGGERHEAD TERMITE *Nasutitermes graveolus* (Hill); *Nasutitermes walkeri* (Hill) [Isoptera: Termitidae].

NIGHT EYES Insect eyes adapted for seeing at night when light is dim (Comstock). See Superposition Eye; Apposition Eye.

NIGHTFEEDING SUGARCANE ARMYWORM *Leucania loreyi* (Duponchel) [Lepidoptera: Noctuidae]. See Armyworms.

NIGHTSTALKING SPIDERS *Cheiracanthium* spp. [Araneida: Clubionidae]: Adult 10–15 mm long, pale yellow with a faint grey mark on Abdomen; Carapace and Abdomen elongate; fangs and Chelicerae elongate. Eyes in two rows of four on the front margin. Female constructs a silken retreat (or sac) under leaves to protect egg sac and to hide in during day. Spiderlings similar in appearance to adults. Nightstalking spiders active predators at night. Adult nightstalking spiders largest spiders regularly found on foliage. Cf. Lynx Spider; Wolf Spider.

NIGIDIUS JEZABEL *Delias ennia nigidius* Miskin [Lepidoptera: Pieridae].

NIGRA SCALE *Parasaissetia nigra* (Nietner) [Hemiptera: Coccidae]: A cosmopolitan pest of avocado, citrus, custard apples, guava and hibiscus. Adult female dark brown to black, elongate-oval in outline shape, 3–4 mm long; cover smooth to slightly wrinkled. Immatures sometimes confused with soft brown scale. Parthenogenetic; female oviposits ca 800 eggs beneath scale cover; eclosion occurs within two weeks; two nymphal instars; life cycle ca 2 months during summer. See Soft Brown Scale.

NIGRESCENT Adj. (Latin, *nigrescere* = to turn black.) Descriptive of structure whose colour verges on black; turning black; blackish. Alt. Nigricans; Nigricante. See Black. Cf. White. Rel. Colour.

NIIJIMA, YOSHINAO (1871–1943) (Hasegawa 1967, Kontyû 35 (Suppl.): 61–62, bibliogr.)

NIIMURA, TARO (1917–1951) (Hasegawa 1967,

Kontyû 35 (Suppl.): 62 bibliogr.)

NIISIMA (NIIJIMA) Y (–1944) (Remington 1946, Ann. ent. Soc. Am. 39: 449.)

NIKLAS, OTTO FRIEDRICH (1912–1971) (Franz 1972, Anz. Schädlingsk. Pflanzenschutz 45: 44–45.)

NIKLOR® See Chloropicrin.

NIKOLSKAYA, MARIA NIKOLAEVNA (1896–1969) Russian taxonomic specialist on parasitic Hymenoptera, particularly the Chalcidoidea (Zoological Institute, Soviet Academy of Sciences.) (Trjapitsyn 1970, Ent Obozr. 49: 496–501, bibliogr. (Translation: Ent. Rev., Wash. 49: 298–301.)

NIMITEX® See Temephos.

NINNI, ALESSANDRO PERICLE (1837–1892) (Pavesi 1892, Boll. Soc. ven.-trent. Sci. nat. 5: 70–78.)

NINOMIYA, MOTOTAKA (1891–1928) (Hasegawa 1967, Kontyû 35 (Suppl.): 64.)

NIPSAN® See Diazinon.

NIRVAL® See Thiodicarb.

NISHIMURA, MASATSUGU (1879–1943) (Hasegawa 1967, Kontyû 35 (Suppl.): 62–63.

NISHIYA, JUNICHIRO (1890–1961) (Hasegawa 1967, Kontyû 35 (Suppl.): 63.)

NISSEN, BENDIX (1844–1917) (Wagner 1920, Verh Ver. naturw. Unterh. Hamb. 16: lxvii–lxviii.)

NISSOLAN® See Hexythiazox.

NISSORUN® See Hexythiazox.

NISUS Noun. (Latin, *nisus* = effort.) Strong effort such as the forceful explusion of eggs, young or excrement from the body.

NIT Noun. (Anglo Saxon, *hnitu*. Pl., Nits.) 1. The egg of a louse or other parasitic insect. Often used in reference to the egg when attached to a hair. Sometimes used in reference to the immature parasitic insect. 2. The egg of a louse (Phthiraptera), often attached to an avian feather or mammalian hair by means of accessory gland secretions from the ovipositing female.

NITIDOUS Adj. (Latin, *nitidus* = shining; *-osus* = with the property of.) Descriptive of structure that is shiny and glossy; reflecting light. Alt. Nitid. See White. Cf. Iridescent; Niveous; Opalescent. Rel. Colour.

NITIDULIDAE Plural Noun. Sap Beetles. A cosmopolitan Family of polyphagous Coleoptera including about 2,200 nominal Species assigned to Cucujoidea. Adults less than 6 mm long, oblong to oval, strongly convex to flattened, uniformly dark coloured or with red or yellow markings; head typically constricted at base of Clypeus; Labrum frequently bilobed and separated from remainder of head; Frontoclypeal Suture usually absent; Antenna with 11 segments including club of three segments; procoxal cavities transverse; Metacoxae grooved; tarsal formula 5-5-5, rarely 4-4-4; Abdomen with 5–6 Ventrites. Larval head prognathous; 2–4 stemmata on each side of head; Antenna with three segments; legs with five segments, two tarsal claws on each Pretarsus; Urogomphi present or absent. Divided into five Subfamilies. Occupy diverse habitats; most Species phytophagous, but Species of *Cybocephala* and *Pithophagus* predaceous. *Cybocephala* a conspicuous element in predatory complex attacking coccoid scale insects; *Pithophagus* reported feeding on scolytids. Syn. Smicripinidae.

NITOBE, INAO (1883–1915) (Hasegawa 1967, Kontyû 35 (Suppl.): 63–64, bibliogr.)

NITOFOL® See Methamidophos.

NITRATE Noun. (Latin, *nitrum* = native soda; chemistry, *-ate* = salt formed from acid. Pl., Nitrates.) A salt of nitric acid, HNO.

NITRITE Noun. (Latin, *nitrum* = native soda; chemistry, *-ites* = salt formed from acid. Pl., Nitrites.) A salt of nitrous acid, hence any compound containing the NO-radical.

NITROCHLOROFORM® See Chloropicrin.

NITROGEN Noun. (French, *nitrogene*.) A colourless, odourless and tasteless gas that is active at ordinary temperatures. Nitrogen is present in alkaloids and nearly all organic matter in the form of proteins. Nitrogen forms about 80% of air. Cf. Oxygen.

NITROGENOUS Adj. (Greek, *nitron* = soda; *genos* = descent; ; Latin, *-osus* = with the property of.) Descriptive of substance containing nitrogen.

NITROPHILOUS Adj. (Greek, *nitron* = soda; *philein* = to love.) Pertaining to organisms that require or prefer nitrogen; descriptive of organisms that inhabit nitrogenous soils. See Habitat. Cf. Acidophilous; Halophilous. Rel. Ecology.

NITSCHE, HEINRICH (1845–1902) (Heller 1903, Sber. naturw. Ges. Isis Dresden 1902: v–xi, bibliogr.)

NITSCHE, JOSEF (1873–1901) (Sachtleben 1942, Arb. morph. taxon. Ent. Berl. 9: 134.)

NIVEOUS Adj. (Latin, *niveus* = snow; *-osus* = with the property of.) Snow-white; descriptive of structure whose colour is pure white with an azure under-tint. Alt. Niveus. See White. Cf. Nitidous. Rel. Colour.

NIXON, GILBERT E J (1905–1987) English taxonomic Entomologist specializing in Hymenoptera.

NOBBLE® A mixture of several salts applied as a pesticide for control of slugs and snails in England. Cf. Metaldehyde.

NOBIS Adj. Latin term meaning 'belonging to me'. Applied in some taxonomic works to Species and abbreviated *nob*.

NO-BRAND GRASS YELLOW *Eurema brigitta zoriade* (Felder & Felder) [Lepidoptera: Pieridae].

NOCTUIDAE Latreille 1809. Plural Noun. Owlet Moths; Underwings; Armyworms; Cutworms; Semiloopers. A cosmopolitan Family of ditrysian Lepidoptera assigned to Superfamily Noctuoidea; noctuids are the largest Family of moths and comprise about 25,000 Species. Adult small to large-bodied; wingspan 10–170 mm; head with scales of variable length, tufted or forming conical projection anteriad; Ocelli present in most

Subfamilies, absent from Hypenodinae and Nolinae; Chaetosemata absent; antennal form highly variable; Proboscis typically well developed, coiled at repose; Maxillary Palpus with one minute segment; Labial Palpus variable in size and shape, ascending. Thorax typically with dense vestiture of long, piliform scales; fore tibial Epiphysis present; tibial spur formula 0-2-4; Tibia and Tarsi sometimes spined; Metathorax with paired Tympanal Organs; abdominal base with Counter-tympanal Cavity. Egg upright; usually circular in outline shape; Micropyle centrally placed on top with radiating ribs extending toward bottom of egg and less distinct cross-ribs connecting cross-ribs; eggs typically laid individually, sometimes in small groups and sometimes covered with deciduous scales from female's Abdomen. Larval head hypognathous; body typically with uniformly-arranged primary Setae but lacking secondary Setae; often large-bodied with longitudinal stripes; Prolegs variable in number and development; common names often reflect feeding habits: cutworms, fruitworms, leafworms. Pupae typically well sclerotized; Proboscis and middle Tarsus typically extend to wing tip; wing projecting to abdominal segment four or slightly beyond; Cremaster sometimes present; apex of Abdomen with two straight spines and hooked Setae; pupation within silken cocoon or cell in soil. Predominantly phytophagous, feeding on fruits, flowers and foliage of living plants, including many significant agricultural crops; a few Species feed on dead leaves or detritus; a few predaceous Species attack coccoids. Adults of most Species nocturnal, cryptically patterned and visit lights.

NOCTUIDS See Noctuoidea.

NOCTUOIDEA A Superfamily of ditrysian Lepidoptera including Agaristidae, Amatidae, Arctiidae, Herminiidae, Hypsidae, Lymantriidae, Noctuidae, Nolidae and Thyretidae.

NOCTURNAL Adj. (Latin, *nox* = night; *-alis* = characterized by.) Descriptive of organisms that are active during the night. A term applied to insects that fly or are active at night. Cf. Crepuscular; Diurnal. Rel. Circadian Rhythm.

NODAL FURROW Odonata: A transverse suture which originates at a point in the costal margin corresponding to the Nodus and extends toward the inner margin.

NODAL SECTOR Odonata: The Media of Comstock which arises from the upper sector of the Arculus near the Nodus and extends to the outer margin.

NODDER, FREDERICK (fl 1773–1801) (Lisney 1960, *Bibliogr. of British Lepidoptera 1608–1799*, 315 pp. (229–245), London.)

NODE Noun. (Latin, *nodus* = knob. Pl., Nodi; Nodes.) 1. A pathological swelling or enlargement of structure or tissue; a knob, knot or rounded protuberance. 2. A distinctive form of tissue surrounded by another form of tissue. 3. Odonata: (Nodus) the stout crossvein near the middle of the costal border of the wing which joins the Costa, Subcosta and Radius (Needham). 4. Hymenoptera: The small segment(s) between the Propodeum and Gaster, most evident in ants. Syn. Petiole; Postpetiole. Archane: Nodus.

NODICORN Adj. (Latin, *nodus* = knob; *cornu* = horn.) Descriptive of Antennae with the apex of each segment swollen.

NODIER, CHARLES (1783–1884) (Desmarest 1845, Ann. Soc. ent. Fr. 3 (2): 18–20.)

NODIFORM Adj. (Latin, *nodus* = knob; *forma* = shape.) Knot-shaped; descriptive of structure In the form of a knot, knob or node. See Node. Cf. Form 2; Shape 2; Outline Shape.

NODOSE Adj. (Latin, *nodus* = knob; *-osus* = full of.) 1. Knotted; knotty. A term descriptive of slender parts which have knots or swellings in the joints or segments. 2. Surface sculpture with almost isolated knots or protuberances. Cf. Granular. Alt. Nodosus. See Node.

NODULAR Adj. (Latin, *nodulus* = little knob.) Knotty. Pertaining to structure with small knots, protuberances or swellings. Sometimes applied to integumental surface sculpture and glandular structure. Alt. Nodulate; Nodulose; Nodulosus; Nodulous.

NODULAR SCLERITE Lepidoptera: A small sclerite bounding the anterior end of the tympanal membrane; anatomically a detached prong of the Metepimeron (Richards).

NODULE Noun. (Latin, *nodulus* = diminutive of *nodus* = knob. Pl., Nodules.) A small knot or swelling.

NODULE FORMATION A Cellular Defence Mechanism in which many small foreign objects or a few relatively large-sized microorganisms or foreign material are surrounded or enveloped. NF is effective when Phagocytosis is not effective. NF is presumably a two-stage process involving rapid coagulation by Granulocytes/Coagulocytes and sheath formation by Phagocytes (Plasmatocytes). See Haemolymph. Cf. Encapsulation; Phagocytosis. Rel. Coagulation.

NODULIFEROUS Adj. (Latin, *nodulus* = little knob; *-osus* = with the property of.) Bearing small knob-like structures.

NODULUS Noun. (Latin, *nodulus* = little knob. Pl., Nodules.) A small Nodus or Node.

NODUS Noun. (Latin, *nodus* = knob. Pl., Nodi) 1. Odonata: An indentation near the middle of the costal margin of the wing. 2. A swelling in the insect wing. Rel. Subnodus.

NOE, GIOVANNI (JUAN) CREVANI (1877–) (Anon. 1932, Revta. chil. Hist. nat. 36: 183–187.)

NÖGGERATH, JOHANN JACOB (1788–1877) (Dechan 1877, Verh. naturh. Ver. preuss. Rheinl. 34: 79–97.)

NOGOS® See Dichlorvos.

NOGUCHI, TUKUZO (1897–1965) (Hasegawa 1967, Kontyû 35 (Suppl.): 65.)

NOHIRA, AKIO (1892–1966) (Hasegawa 1967,

Kontyû 35 (Suppl.): 66, bibliogr.)

NOLIDAE Plural Noun. See Noctuidae.

NOLOBAIT® See *Nosema locustae*.

NOMAD Noun. (Greek, *nomas* = roaming about for pasture. Pl., Nomads.) An individual which wanders.

NOMADIC PHASE The period in an Army Ant activity cycle during which the colony actively forages and changes bivouac sites frequently. Queens do not reproduce during the NP and most of the brood is in the larval stage. Cf. Statary Phase.

NOMADISM Noun. (Greek, *nomados* = roaming; English, *ism* = condition.) The tendency of a colony or entire population to move from one area to another. See Migration.

NOMATE PBW-MEC® See Gossyplure.

NOMATE TPW SPIRAL® See Lycolure.

NOMEN CONSERVANDUM (Latin, Pl., *Nomina Conservanda.*) An otherwise unacceptable scientific name which has been conserved by the International Commission of Zoological Nomenclature.

NOMEN DUBIUM (Latin, Pl., *Nomina Dubia.*) The scientific name of a nominal Species for which available evidence is insufficient to permit recognition of the taxonomic Species to which the name was applied.

NOMEN INQUIRENDUM (Latin, Pl., *Nomina Inquirenda.*) A name which which does not have nomenclatural priority, but which should be preserved and not cast into junior synonomy, as ruled by the International Commission on Zoological Nomenclature. A name to be inquired into or whose status is subject to investigation.

NOMEN NOVUM (Latin, Pl., *Nomina Nova.*) A new scientific name. Typically abbreviated nom. nov., or n. n.

NOMEN NUDUM (Latin, Pl., *Nomina Nuda.*) A scientific name which fails to satisfy the rules of availability under the Code of Zoological Nomenclature.

NOMEN OBLITUM (Latin, Pl., *Nomina Oblita.*) A phrase used in conjunction with a scientific name which has not been used in the primary zoological literature for more than fifty years.

NOMEN TRIVIALE (Latin, *Nomina Trivialae.*) A phrase used in pre-1960 editions of the International Code of Zoological Nomenclature. The second word of a binomen (Species) or the third word of a trinomen (Subspecies.)

NOMENCLATURE Noun. (Latin, *nomen* = name; *calare* = to call. Pl., Nomenclatures.) Biology: The application of a grammatically correct name to any biological Taxon. The study of scientific names. See Taxonomy.

NOMINAL SPECIES Named entities that are recognized in taxonomy but may not be valid names of 'Species' in the sense of the 'Biological Species Concept'.

NOMOLOGY Noun. (Greek, *nomos* = law; *logos* = discourse. Pl., Nomologies.) The study and formulation of principles or laws in science.

NOMOLT See Teflubenzuron.

NOMURA, HIKOTARO (1861–) (Hasegawa 1967, Kontyû 35 (Suppl.): 66–67.)

NON-BITING MIDGES See Chironomidae.

NON-BITING STABLE FLY See False Stable-Fly.

NON-CELLULAR OUTGROWTH Cuticular outgrowths of the body-wall produced by epidermal cell secretions. Syn. Noncellular Process.

NONCHITINOUS Adj. (Latin, *non, noenum* = not one; Greek, *chitin* = tunic; Latin, *-osus* = abounding in.) Pertaining to structure not formed of Chitin or not containing Chitin.

NON-CONCURRENT HOST FEEDING See Host Feeding. Cf. Concurrent host feeding; Destructive host feeding; Non-destructive host feeding.

NONELLCOMAS, JAIME (1876–1938) (Anon. 1940, Boln. Patol. veg. Ent. agric. 9: 298–302, bibliogr.)

NON-GRANULAR SPINDLE CELL See Oenocyte.

NON-NUCLEATE Adj. Without a nucleus.

NON-OCCLUDED TYPE VIRUS Insect-pathogenic viruses of the Baculoviridae. See Granulosis Virus; Nuclear Polyhedrosis Virus. Cf. Cytoplasmic Polyhedrosis Virus; Insect Poxvirus.

NON-PREFERENCE Adj. Resistance conferred to a plant due to characteristics that insects will discriminate against during assessment for suitability as a resource. Syn. Antixenosis. Rel. Host Plant Resistance, Induced Resistance, Behavioural Resistance.

NON-REVERTIBLE THELYTOKY A form of thelytokous parthenogenesis found in some parasitic Hymenoptera. Female wasps maintain all-female progeny production without fertilization of female gametes. NRT cannot be 'cured' by treatment of females with diet-induced antibiotics or high temperature. See Parthenogenesis; Thelytoky. Cf. Revertible Thelytoky.

NOOGOORA BURR LONGICORN *Nupserha vexator* (Pascoe) [Coleoptera: Cerambycidae]: A beetle endemic to India and Pakistan and imported into Australia to feed upon Noogoora Burr. NBL eggs laid in stems and petioles; larvae tunnel up to apex of stem and then down to root; pupation in soil.

NOOGOORA BURR SEED FLY *Euaresta aequalis* (Loew) [Diptera: Tephritidae].

NORDENSKIÖLD, ADOLPH ERIK (1832–1885) (Ross 1934, Int. ent. Z. 27: 539.)

NORDENSKIÖLD, ERIC (1872–1933) (Palmgren 1934, Mem. Soc. Fauna Flora fenn. 9: 209–211, 227, 229.)

NORDENSTRÖM, HENNING (–1919) (Roman 1920, Ent. Tidskr. 41: 139–141, bibliogr.)

NORDIN, ALBAN EMANUEL (1853–1939) (Borgvall 1952, Opusc. ent. 17: 21–22.)

NÖRDLINGER, HERMAN (1818–1897) (Anon. 1892, Zentbl. ges. Forstw. 23: 137–145, bibliogr.)

NORDMAN, ALEXANDER VON (1803–1866) (Hjelt 1871, Acta soc. Sci. fenn. 9 (2): 1–40, bibliogr.)

NORDSTROM, A (1882–1965) (Hellen 1965, Notul.

ent. 45: 96.)

NORDSTROM, FRITHIOF (1882–1971) (Krogerus 1972, Notul. ent. 52: 32.)

NORFOLK ISLAND CICADA *Kikihia convicta* (Distant) [Hemiptera: Cicadidae].

NORGATE, FRANK (Rowland-Brown 1919, Entomologist 52: 119.)

NORM Adj. (Latin, *norma* = rule.) A rule, guideline or type.

NORMAL Adj. (Latin, *norma* = rule; *-alis* = belonging to.) 1. Descriptive of process, structure or substance that exhibits the usual 'form' (type, kind) when compared with the condition shown by similar entities. 2. Not exceptional. Cf. Abnormal.

NORMAL SALT SOLUTION For work in insect histology or anatomy, a solution containing 1% NaCl in distilled water. See Physiological Saline; Ringer's Solution.

NORMAN, GEORGE (1823–1882) (T.S. 1900, Trans. Hull scient. Fld nat. Club 1: 105–112, bibliogr.)

NORRIS, M J See Richards, M. J.

NORTH QUEENSLAND DAY MOTH *Alcides zodiaca* (Butler) [Lepidoptera: Uraniidae].

NORTHEASTERN SAWYER *Monochamus notatus* (Drury) [Coleoptera: Cerambycidae].

NORTHERN ARMYWORM *Leucania separata* (Walker) [Lepidoptera: Noctuidae]. See Armyworms.

NORTHERN AUGER BEETLE *Xylothrips religiosus* (Boisduval) [Coleoptera: Bostrichidae].

NORTHERN BROWN HOUSE MOTH *Dasypodia cymatodes* Guenée; *Speiredonia spectans* (Guenée) [Lepidoptera: Noctuidae].

NORTHERN CADDIS FLIES See Trichoptera; Limnephilidae.

NORTHERN CATTLE GRUB *Hypoderma bovis* (Linnaeus) [Diptera: Oestridae]: A monovoltine pest of cattle in the Holarctic; periodically introduced into South America but not established. Adult ca 15 mm long, bee-like, robust, conspicuously setose with Setae at apex of Abdomen reddish yellow. Biology and behaviour similar to Cattle Grub except female lays eggs while in flight or after landing on host; eggs attached singly to hairs of rump or hindleg; adult flies cause gadding in cattle; larvae temporarily reside in epidural fat of spinal column before migrating to back to complete development. Syn. Ox Bot; Ox Warble Fly; Heel Fly. See Oestridae. Cf. Common Cattle Grub.

NORTHERN CHERRYNOSE *Macrotristria sylvara* (Distant) [Hemiptera: Cicadidae].

NORTHERN CITRUS BUTTERFLY *Princeps fuscus canopus* (Westwood) [Lepidoptera: Papilionidae].

NORTHERN CORN ROOTWORM *Diabrotica barberi* Smith & Lawrence [Coleoptera: Chrysomelidae]: A monovoltine pest of corn and grasses in the central USA; Neonate and early instar larvae feed on root hairs and outer root tissue; older-instar larvae feed on inner tissues and vascular bundle; adults feed on silks and pollen; high populations of NCR cause significant reduction in yield due to poor pollination. Egg white, ca 0.1 mm long, football-shaped; female lays 200–1,000 eggs during year; eggs deposited in soil near roots of host plant during late summer and autumn; NCR overwinters in egg stage with eclosion during May and June. Larvae burrow through soil, feed on roots; undergo three larva instars and Prepua. Mature larva ca 10–12 mm long, head black, body white with dark dorso-apical spot at apex of abdomen; larva indistinguishable from Mexican Corn Rootworm. Pupae white, fragile; pupation in cell within soil, requires 5–10 days. Adult ca 6 mm long, neonate pale brown and becoming pale green with age; Neonate adult emerges from soil, active, associated with flowers and tumble from flowers when disturbed. Syn. *Diabrotica longicornis.* See Chrysomelidae.

NORTHERN DINGY DART *Suniana lascivia* (Rosenstock) [Lepidoptera: Hesperiidae].

NORTHERN DOUBLE DRUMMER *Thopha sessiliba* Distant [Hemiptera: Cicadidae].

NORTHERN FALSE WIREWORM *Gonocephalum carpentariae* (Blackburn) [Coleoptera: Tenebrionidae].

NORTHERN FOWL-MITE *Ornithonyssus sylviarum* (Canestrini & Fanzago) [Acari: Macronyssidae].

NORTHERN GRASSDART *Taractrocera ilia ilia* Waterhouse [Lepidoptera: Hesperiidae].

NORTHERN GREENGROCER *Cyclochila virens* Distant [Hemiptera: Cicadidae].

NORTHERN HOUSE-MOSQUITO *Culex pipiens* Linnaeus [Diptera: Culicidae]: A significant pest endemic to northern USA states that occurs above 36ºN and widely in Old World. Significant vector of human filarial worm and St Louis Equine Encephalitis. Closely related to Brown House-Mosquito. NHM most common night-biting mosquito; invades homes readily. Feeds upon blood of birds, domestic animals and humans. Female oviposits rafts of eggs upon surface of water; egg stage 1–2 days; larval stage 10–14 days; pupal stage two days. Syn. Rain-Barrel Mosquito. See Culicidae. Cf. Brown House-Mosquito; Southern House-Mosquito.

NORTHERN IMPERIAL BLUE BUTTERFLY *Jalmenus eichhorni* Staudinger [Lepidoptera: Lycaenidae].

NORTHERN JEZABEL *Delias argenthona argenthona* (Fabricius) [Lepidoptera: Pieridae].

NORTHERN MASKED-CHAFER *Cyclocephala borealis* Arrow [Coleoptera: Scarabaeidae].

NORTHERN MOLE-CRICKET *Neocurtilla hexadactyla* (Perty) [Orthoptera: Gryllotalpidae]: A common pest of gardens, nurseries and orchards; endemic and widely distributed throughout North America. Foretibia with four Dactyls (digging claws); apex of hind Tibia with eight spines. Eggs deposited in chamber at end of burrow; female guards eggs and first-instar nymphs. See Gryllotalpidae. Cf. Mole Crickets; Southern Mole Cricket.

NORTHERN PEARL-WHITE *Elodina perdita perdita* Miskin [Lepidoptera: Pieridae].

NORTHERN PITCH TWIG-MOTH *Petrova albicapitana* (Busck) [Lepidoptera: Tortricidae].

NORTHERN RAT-FLEA *Nosopsyllus fasciatus* (Bosc) [Siphonaptera: Ceratophyllidae]: A pest of rodents and humans; endemic to Holarctic, now cosmopolitan through association with Norway rat. NRF serves as a minor vector of plague and intermediate vector of tapeworms infesting rats and mice. See Ceratophyllidae. Cf. European Mouse Flea.

NORTHERN RINGLET *Hypocysta irius* (Fabricius) [Lepidoptera: Nymphalidae].

NORTHERN RIVERS FUNNELWEB SPIDER *Hadronyche formidabilis* (Rainbow) [Araneida: Hexathelidae].

NORTHERN ROUGH BOLLWORM *Earias vitella* (Fabricius) [Lepidoptera: Noctuidae]: A minor pest on cotton in northern Australia. NRB is widely distributed throughout Asia and is an important pest of cotton in India. Adult moth 12–15 mm long with a wingspan of 18–22 mm; forewing greenish-beige with a dark wedge-shape band in the centre. (Green transverse lines on the forewing of rough bollworm distinguish it from northern rough bollworm); hindwing white with greenish-beige markings on the margin; Thorax dark beige; Abdomen white. Antenna long and slender (filiform). Labial palp prominent. Adult females lay small (0.5 mm in diameter), light blue-green eggs on young leaves and stems; eggs with 30 longitudinal ridges with alternate ridges projecting upwards to form a crown. Mature larva 15–20 mm long; larva spiny with mottled coloration of brown, orange, yellow and grey. Pupation occurs on the plant; Pupa ca 13 mm long and brown in a greyish silken cocoon of irregular shape; life-cycle ca 28–42 days. Larvae enter terminals and tunnel down stems; larvae can enter squares and feeding may cause square to fall. Larvae may also tunnel down main stem causing destruction of primary growing point at any growth stage including seedlings. Larvae attack cotton throughout all stages of plant development. See Noctuidae; Rough Bollworm.

NORTHERN SANDY DUNG BEETLE *Euoniticellus intermedius* (Reiche) [Coleoptera: Scarabaeidae].

NORTHERN TERRITORY FRUIT FLY *Dacus aquilonis* (May) [Diptera: Tephritidae].

NORTHWEST COAST-MOSQUITO *Aedes aboriginis* Dyar [Diptera: Culicidae].

NORTON, EDWARD (1823–1894) (Anon. 1894, Ent. News 5: 161–163, bibliogr.)

NORWAY-MAPLE APHID *Periphyllus lyropictus* (Kessler) [Hemiptera: Aphididae].

NORWEGIAN ITCH See Human Itch Mite.

NOSE BOT-FLY *Gasterophilus haemorrhoidalis* (Linnaeus) [Diptera: Oestridae]: A pest of horses. HBF an aetiological agent of Enteric Myiasis; widespread and generally not as common as Horse Bot or Throat Bot-Flies but considered more pernicious. Adult with red Setae at apex of abdomen. Life history similar to Horse Bot-Fly except female lays ca 160 eggs during lifetime. Egg with long, corrugated pedicle, attached to hair on lips of host. Larval development in fundus of stomach; mature larva attaches to intestine near Anus before dropping to ground to pupate in soil. See Oestridae. Cf. Horse Bot-Fly; Throat Bot-Fly. Rel. Gasterophilus Enteric Myiasis.

NO-SEE-UMS See Ceratopogonidae.

NOSEMA APIS A cosmopolitan, microsporidian fungal pathogen of adult honey bees; larvae are not affected. The pathogen develops exclusively within the midgut epithelium of adult bees. Host specificity and tissue specificity have been debated; records have been published which assert other Species of bees and other tissues are affected. Spores are ingested and pass to the midgut where they develop and multiply in the epithelial-cell cytoplasm; spores replicate and are transmitted to other midgut cells or passed into the hindgut inside sloughed midgut cells and passed from the body with excrement; spores are ingested by young workers when they clean contaminated comb. Disease significantly shortens the life of infected individuals. Cf. Pebrine; Microsporidia. Rel. Pathogen.

NOSEMA LOCUSTAE A microsporidian fungal pathogen whose spores are used as a biopesticide in the control of grasshoppers. Applied to rangeland, pasture and agricultural crops. Combined with wheat bran bait and must be consumed to be effective; regarded as non-phytotoxic and not toxic to warm-blooded animals. Slow-acting and affecting the fat-body of infected individuals; transmitted transovarially and via cannibalism among grasshoppers. First commercially developed by Reuter Labs and Evans Biocontrol (ca 1979); currently marketed by several companies. Trade names include: Grasshopper Spore®, Hopper Stopper®, Locucide®, Nolobait®, Loloc®, Semapore® and Trojan-10® See Biopesticide. Cf. *Beauveria brongniartii*; *Metarhizium anisopliae* ESF 1®; *Verticillium lecanii*.

NOSKIEWICZ, JAN (1890–1963) (Smreczynski 1964, Polskie Pismo ent. 34: 1–18, bibliogr.)

NOSODENDRIDAE Plural Noun. A widespread Family of Coleoptera with about 30 Species in one Genus (*Nosodendron*). Nosodendrids have not been found in South America or Africa. Placement of the Family is disputed: In Byrrhoidea based on adult characters; in Dermestoidea based on adult and larval characters. Nosodendrid adult body is compact, convex and ovoid; legs at repose positioned in cavities on ventral surface of body; antennal club large, well defined with three segments. Biology of Family is poorly known but Species are suspected predators of Diptera larvae. Adult and larva taken from beneath bark, in tree holes and in tree wounds. Rel. Anthicidae.

NOSTRIL Noun. (Anglo Saxon, *nosthyrl* = nostril. Pl., Nostrils.) See Rhinarium.

NOTACORIA Noun. A distinct area in the Mesothorax and Metathorax (MacGillivray).

NOTAL COMB Siphonaptera: A row of conspicuous spines along posterior margin of Prothorax.

NOTALIA Noun. A projection covering part of Mesonotum and Mesopleura. Formed by infolded posterior part of Pronotum. The posterior notal ridge (*sensu* MacGillivray).

NOTAROTAXIS Noun. The layer of Cuticle continuous with the dorsal surface of the wing Notum (MacGillivray).

NOTASUTURE Noun. A Suture of the Notum (MacGillivray).

NOTATE Adj. (Latin, *notatus* = marked; *-atus* = adjectival suffix.) Descriptive of structure marked by lines, stripes or spots, or a surface with a series of depressed marks as a sculpture.

NOTAULUS Noun. (Greek, *notos, noton* = back; *aulix* = furrow. Pl., Notauli.) Hymenoptera: Paired, longitudinal, cuticular furrows on the Mesoscutum that converge posteriad and divide the Mesoscutum into a medial and lateral areas. Alt. Notaulices; Sl., Notaulix. See Mesopraescutum. Cf. Parapsidal Sutures; Parapsides.

NOTCH Noun. (Origin obscure. Pl., Notches.) 1. An indentation, cut or incision in a surface. 2. A point where the fold of a wing meets the wing margin. See Lobe.

NOTCHED Adj. Indented, cut or nicked. A term usually applied to the margin of a structure.

NOTCHED PLATES Coccids: Pectinae (MacGillivray).

NOTCHES Plural Noun. Coccids: Incisurae (MacGillivray).

NOTEPISTERNUM Noun. (Greek, *noton* = back; Episternum. Pl., Notepisterna.) Part of the Episternum dorsad of the Episternal Suture. Rel. Anepisternum; Mesopleura.

NOTERIDAE Plural Noun. Burrowing Water Beetles. A widespread, small Family (ca. 25 Species) of caraboid Coleoptera. Adult 1–6 mm long; Antenna filiform with 11 segments. Scutellum absent; fore Coxa globose; hind Coxal plates longitudinally oriented; tarsal formula 5-5-5; hind tarsal claws equal in size. Abdomen with six Ventrites and basal three Ventrites connate. Species inhabit margin of shallow ponds, aquatic as immatures and adults. Presumably feed on detritus. Larva obtain oxygen from plants via spiracle-bearing siphon at apex of Abdomen. Pupa resides in air-filled cocoon attached to roots of aquatic plants. Related to the Dytiscidae.

NOTHYBIDAE Plural Noun. A small Family of acalypterate Diptera assigned to Superfamily Diopsoidea.

NOTODONT Noun. (Greek, *noton* = back; *odous* = tooth.) With toothed backs, one of a series of moths whose larvae are more or less conspicuously humped on the dorsal surface.

NOTODONTID MOTHS See Notodontidae.

NOTODONTIDAE Plural Noun. Notodontid Moths; Prominents. A Family of ditrysian Lepidoptera assigned to Superfamily Notodontoidea or Noctuoidea; cosmopolitan distribution with more than 2,000 Species known. Adult medium to large-bodied; wingspan 35–135 mm; head typically with rough scales or scale crests; Ocelli and Chaetosemata absent; male Antenna typically bipectinate, female Antenna filiform or bipectinate; Maxillary Palpus small, of two segments; Labial Palpus short, porrect, ascending Frons; Proboscis typically well developed, lacking scales and coiled at repose. Thorax typically with dense Setae; metathoracic Tympanum directed ventrad; Epiphysis present in male, absent in female; tibial spur formula 0-2-4 or 0-2-2. Egg upright, hemispherical with base broad and flat or nearly spherical; Micropyle at apex; Chorion with little or no sculpturing; eggs typically deposited individually on host plant. Larval head Hypognathous; body with spines or other protuberances; Prothorax sometimes with osmeterium-like structure beneath head which secretes formic acid and ketones; Prolegs on segments 3–6; anal legs often reduced or modified. Pupae strongly sclerotized; Maxillary Palpus absent; thoracic dorsum and Abdomen often punctate; Cremaster present; pupation in soil of strong cocoon. Larvae feed or trees and woody shrubs, in arboreal habitat exposed on leaves. Adult nocturnal, typically rests with wings tectiform over Abdomen; a few Species hold wings flat. Syn. Ceruridae.

NOTODONTOIDEA A Superfamily of ditrysian Lepidoptera including Dioptidae, Notodontidae, Thaumetopoeidae and Thyretidae. Regarded as synonymous with Noctuoidea in some classifications.

NOTOGAEA Noun. (Greek, *noto* = south; *gaia* = earth.) One of three proposed zoological realms to include the Australian, Polynesian and Hawaiian regions. See Zoogeography.

NOTOLIGOTOMIDAE Plural Noun. A small Family of Embiidina that occurs in Old World and New World tropics. Notoligotomids are typically winged; abdominal Sternum IX with posterior margin developed into sclerotized lobe (Hypandrium).

NOTONECTAL Adj. (Greek, *notos* = back; *nektos* = swimming; Latin, *-alis* = characterized by.) Descriptive of organisms that swim as Notonectidae (*e.g.* with the back directed downward and venter upward).

NOTONECTIDAE Latreille 1802. Plural Noun. Backswimmers. A numerically large, cosmopolitan Family of heteropterous Hemiptera assigned to Superfamily Notonectoidea. Dorsum highly convex; eyes large, reniform; foreleg raptorial; hind leg oar-like; hind Tibia flattened with a fringe of long Setae; each leg with one claw; Abdomen with ventral, median keel. Species are aquatic and swim upside-down, and most common in ponds and lakes; some Species occur in slow-moving streams. Notonectids are predaceous and attack other insects, tadpoles and small fish.

NOTONECTOIDEA A Superfamily of predaceous, heteropterous Hemiptera. Body strongly arched; forewing membrane short or absent. Included Families Helotrephidae, Notonectidae, Pleidae.

NOTONEMOURIDAE Plural Noun. A Family of Plecoptera including 16 Genera and ca 60 nominal Species that displays a Gondwanaland distribution. Each Genus is endemic and restricted to the zoogeographical realm in which it occurs. Subgenital sclerite of females is modified into an Ovipositor. All Species apparently herbivorious; little known about life history. Early classifications place notonemourids as a Subfamily of Capniidae.

NOTOPLEURAL Adj. (Greek, *notos* = back; *pleuron* = side.) Descriptive of or relating to the Notopleuron.

NOTOPLEURAL SUTURE Dorsopleural Suture.

NOTOPLEURON Plural Noun. (Greek, *noton* = back; *pleuron* = side. Pl., Notopleura.) Diptera: A triangular depression positioned anterior of the transverse suture and posterior of the Humeri (Curran).

NOTOPLURAL BRISTLES Diptera: Usually two bristles inserted immediately above the Dorsopleural Suture between Humeral Callus and root of wing, on the Notopleuron (Comstock).

NOTOPTERA Crampton 1915. 1. An Ordinal name proposed for the inclusive Family Grylloblattidae. Syn: Grylloblattodea. 2. Noun. (Greek, *notos* = back; *pteron* = wing.) Coleoptera: The parallel ridges or thickenings of the posterior part of the Metascutum (MacGillivray).

NOTOPTERALE Noun. The first Axillary Sclerite of the wing (Crampton).

NOTOPTERARIA Noun. Coleoptera: The median notal groove (MacGillivray).

NOTOTHECA Noun. (Greek, *notos* = back; *theke* = case. Pl., Notothecae.) The portion of the the pupal case over upper surface of the Abdomen.

NOTUM Noun. (Greek, *notos* = back. Pl., Nota.) The dorsal or upper surface of a body segment, particularly of the Thorax. Syn. Tergum. Ant. Sternum. See Dorsum. Rel. Orientation; Segmentation.

NOUALHIER, MARTIAL JEAN MAURICE (1860–1898) (Bouvier 1898, Bull. Mus. Hist. nat. Paris 4: 229–232.)

NOUGARET, RAYMOND LOUIS (1866–1933) (Essig 1933, J. Econ. Ent. 26: 239–741, bibliogr.)

NOUN MOTH *Lymantria monacha* [Lepidoptera: Lymantriidae].

NOVA SPECIES See New Species. Abbreviated, n.s., n. sp., nov. sp.

NOVABAC-3® A registered biopesticide derived from *Bacillus thuringiensis* var. *kurstaki*. See *Bacillus thuringiensis*.

NOVAK, GIAM-BATTISTA (–1893) (Anon. 1894, Wien. ent. Ztg. 13: 195.)

NOVAK, PETAR (1879–1968) (Miksic 1969, Boll. Assoc. Romana Ent. 24: 80.)

NOVALURON A urea compound {1-[3-chloro-4-(1,1,2-trifluoro-methoxyethoxy) phenyl]-3-(2,6-difluorobenzoly) urea} used as an Insect Growth Regulator against mosquitoes, whiteflies, psyllids, bollworms and related insects. Novaluron has broad-spectrum activity. Compound is applied to fruit crops, cotton, corn, vegetables and forestry. Experimental registration in USA only. Trade names is GR-572®. See Insect Growth Regulator.

NOVATHION® See Fenitrothion.

NOVAVALVAE Plural Noun. The dorsal lateral Valvulae of the female insect (MacGillivray).

NOVIGAM See Lindane.

NOVODOR®. A biopesticide derived from *Bacillus thuringiensis* var. *tenebrionis*. See *Bacillus thuringiensis*.

NOVRAN® See Fenbutatin-Oxide.

NOVUM Noun. (Latin, *novus* = new. Pl., Nova.) New. A term used in taxonomy to refer to a Taxon new to science. Abbreviated *nov.*

NOVUM GENUS See new Genus. Abbreviated, *n.g., gen. nov.*

NOWICKI, MAXIMILIAN SILA (1826–1890) (Wierzejski 1891, Wien. ent. Ztg. 10: 17–30, bibliogr.)

NOWOTNY, HANS (1897–1971) (Voigt 1971, Beitr. naturk. Forsch. SüdwDtl 30: 165–167, bibliogr.)

NOXFIRE® See Rotenone.

NOXFISH® See Rotenone.

NOXIA CANEGRUB *Lepidiota noxia* Britton [Coleoptera: Scarabaeidae]. See Canegrub.

NPV See Nuclear Polyhedrosis Virus.

NUBILACID® A registered biopesticide derived from *Bacillus thuringiensis* var. *kurstaki*. See *Bacillus thuringiensis*.

NUCHA Noun. (Latin, *nucha* = nape of neck. Pl., Nuchae.) 1. The upper surface of the neck connecting head and Thorax. 2. The posterior dorsomedial portion of the Propodeum of some parasitic Hymenoptera; well developed and taxonomically important in some Families such as the Pteromalidae.

NUCIDOL® See Diazinon.

NUCIFEROUS Adj. (Latin, *nux* = nut; *-osus* = with the property of.) Nut-bearing.

NUCIVOROUS Adj. (Latin, *nux* = nut; *vorare* = to devour; *-osus* = with the property of.) Nut-eating.

NUCLEAR Adj. (Latin, *nucleus* = kernel.) Pertaining to the Nucleus.

NUCLEAR CYTOPLASM The small mass of egg cytoplasm containing the egg nucleus.

NUCLEAR POLYHEDROSIS VIRUS Insect-pathogenic viruses assigned to the Baculoviridae and typified by proteinaceous nuclear occulsions (inclusions) in which progeny virons are embedded at a late stage of infection. NPV commonly attack Lepidoptera larvae, but also occur in larvae of other Orders. Some NPVs are governmentally registered, industrially produced and commercially formulated for sale to control insect pests. Acronym NPV. See Granulosis Virus; Nonoccluded-type Virus. Cf. Cytoplasmic Polyhedrosis Virus; Insect Poxvirus. Rel. Biopesticide.

NUCLEI OF SEMPER Nuclei of the crystalline cone cells of the compound eye (Imms).

NUCLEIFORM Adj. (Latin, *nucleus* = kernel; *forma* = shape.) Nucleus-shaped. Pyrenoid. See Nucleus. Cf. Pyrenoid. Rel. Form 2; Shape 2; Outline Shape.

NUCLEOLAR Adj. (Latin, *nucleus* = kernel.) Pertaining to the nucleolus.

NUCLEOLUS Noun. (Latin, *nucleolus* = little kernel. Pl., Nucleoli.) A small, spherical body in the nucleus of most cells which contains RNA and is associated with the nucleolar organizer.

NUCLEOPROTEIN Noun. (Latin, *nucleus* = kernel; A compound of one or more protein molecules in combination with nucleic acid. An acid of unknown composition containing phosphorus and nitrogen, but usually no sulphur.

NUCLEUS Noun. (Latin, *nucleus* = kernel. Pl., Nuclei.) 1. A well defined, differentiated, round/ ovoid body imbedded within cell contents. The Nucleus consists of a nuclear membrane which surrounds an hyaline ground substance and genetic material (chromosomes) in insect cells. The Nucleus is essential for cellular metabolism, growth and reproduction. 2. Honeybees: A small mass of bees and combs of brood used in forming a new colony of rearing queens. 3. A centre for growth, increase in size or development. 4. The focus of a process or principle.

NUDE Noun. (Latin, *nudus* = naked.) A term used to describe Integument devoid of Setae, scales or surface ornamentation. Alt. Nudus.

NUDISETA Noun. (Latin, *nudus* = naked; *seta* = bristle. Pl., Nudisetae.) One of three anatomical types of Setae found on the body of larval Dermestidae. Nudisetae are smooth, sparse and generally distributed over the body. Their function is unknown, but Nudisetae may be formed in response to parasites. See Sensillum. Cf. Hastiseta; Spiciseta.

NUDITAS (Latin, *nudus* = naked.) Nakedness; the condition of being without covering or vestiture. Alt. Nudity.

NUDRIN® See Methomyl.

NUDUM Noun. (Latin, *nudus* = naked.) Bare areas on a surface which is surrounded by Setae.

NUGENT, FRITZ (–1966) (Anon. 1966, Z. wien. ent. Ges. 51: 96.)

NU-GRO® See Pirimiphos-Methyl.

NULLARBOR CAVES COCKCROACH *Troglo-blattella nullarborensis* Mackerras [Blattodea: Blattellidae].

NULLIPAROUS Adj. (Latin, *nullus* = null; *parere* = to give birth). Descriptive of a female that has not produced offspring. Cf. Virgin.

NUMBAT TICK *Ixodes vestitus* Neumann [Acari: Ixodidae].

NUMERICAL TAXONOMY A philosophical approach to classification which employs mathematical techniques to provide 'decisions' for the limits of Taxa and groupings of Taxa. NT prefers to employ the concept of 'Operational Taxonomic Unit' (OUT) over the concept of 'Species.' Proponents argue that NT is an 'operational' approach to producing classification whereas other systems have inherent bias through Species concepts or preconceived notions of relationship. NT was first tested by the Botanist Adanson during the early part of the 19th century and made popular by Robert Sokal and colleagues during the 1960s. Cf. Cladism.

NUNENMACHER, FREDERICK WILLIAM (1870–1946) (Leach Pan-Pacif. Ent. 24: 1–5, bibliogr.)

NUPTIAL FLIGHT Social Insects: The mating flight of winged male (drone or king) and queens.

NUPTIAL GIFT A substance produced or obtained by a male insect which is presented to a prospective conspecific female as part of a courtship ritual. NGs sometimes provides nutrition for the female which is necessary for reproduction. Regurgitated NGs are a form of trophallaxis seen in some Diptera, including Sciomyzidae, Empidae, Platystomidae, Asteidae and Tephritidae. NGs are highly developed among some Diptera (Sciomyzidae, Tephritidae) and Mecoptera. Cf. Trophallaxis.

NURATRON® See Methamidophos.

NURELLE® See Chlorpyrifos; See Cypermethrin.

NURSE Noun. (Middle English, *nourse*. Pl., Nurses.) Social Insects: A worker ant or worker bee that cares for brood (eggs, larvae, Pupae) but does not forage. See Social Insect. Cf. Drone; Queen; Worker.

NURSE CELLS Cells in the ovarioles of some insects which furnish nutrients to the developing eggs (Comstock). The Trophocytes of the Ovary or Testis (Snodgrass). See Meroistic Ovary.

NURSE, CHARLES GEORGE (1862–1933) (Riley 1934, Entomologist 67: 23–24.)

NURSE, EUSTON JOHN (1865–1945) (Anon. 1945, Trans. Suffolk Nat. Soc. 5: c.)

NURSERY-WEB SPIDERS See Pisauridae.

NÜSSLIN, OTTO (–1915) (Fuchs 1911, Ent. Bl. Biol. Syst. Käfer 7: 1–5, bibliogr.)

NUT BORERS See Agonoxenidae.

NUTANS Adj. (Latin, *nutare* = to nod.) Nodding; drooping; with apex bent downward. Alt. Nutant.

NUTGRASS ARMYWORM *Spodoptera exempta* (Walker) [Lepidoptera: Noctuidae]. See Armyworms.

NUTGRASS BILLBUG *Sphenophorus cariosus* (Oliver) [Coleoptera: Curculionidae].

NUTGRASS BORER MOTH *Bactra venosana* (Zeller) [Lepidoptera: Tortricidae].

NUTGRASS WEEVIL *Athesapeuta cyperi* Marshall [Coleoptera: Curculionidae].

NUTMEG WEEVIL See Coffee-Bean Weevil.

NU-TOMATONONE® See Ethephon.

NUTRIENT Adj. (Latin, *nutrire* = to nourish.) Nourishing; Nutritive.

NUTRITION Noun. (Latin, *nutrire* = to nourish; *-tion* = result of an action.) See Diet.

NUTRITIVE CHAMBER An enlarged portion of the ovariole which is filled with granular nutritive material used in egg development.

NUTRITIVE LAYER Cynipid galls: The innermost tissue of a gall which lines the larval cell (Kinsey).

NUTTALL BLISTER-BEETLE *Lytta nuttalli* Say [Coleoptera: Meloidae].

NUTTALL, GEORGE HENRY FALKINER (1862–1937) (Graham Smith *et al.* 1938, Parasitology 30: 403–418, bibliogr.)

NUTTALLIELLIDAE Plural Noun. A Family of ticks known from one Species that occurs in tropical Africa. See Ixodidea. Cf. Argasidae; Ixodidae.

NUVACRON® See Monocrotophos.

NUVAN® See Dichlorvos.

NUVANOL® See Fenitrothion.

NUVANOL-N® See Phoxim.

NYCTERIBIIDAE Plural Noun. Bat Flies. A numerically small, geographically widespread Family of schizophorous Diptera assigned to Superfamily Muscoidea or Hippoboscoidea. Body arachniform, dorsoventrally compressed and leathery; compound eyes reduced or absent; Ocelli absent; head rather small, narrow and folding at repose into a Sulcus on Mesonotum; Basistarsi exceptionally long; Middle and Hind Coxae widely separated; apterous. Adults live as ectoparasites in the fur of bats where they feed on host's blood; females larviposit on walls of bat roosts. Adult emergence typically is synchronized by contact or warmth of roosting bats. Cf. Streblidae.

NYGMATA Noun. (Greek, *nygmatos* = puncture, prick, sting. Sl., Nygma.) Small sensory areas or spots on wings of some Trichoptera, Neuroptera and Hymenoptera. Neuroptera: Between posterior two branches of apparent Rs and/or basally between Rs and M.

NYLANDER, WILLIAM (1822–1899) (Anon. 1899, Entomol. mon. Mag. 35: 148.)

NYLAR® See Pyriproxyfen.

NYMPH Noun. (Greek, *nymphe* = chrysalis. Pl., Nymphs.) 1. General: The immature stage between egg and adult of non-holometablous insects without distinction for habitat or habitus. 2. Restricted: The terrestrial immature stage of paurometablous insects (Orthoptera, Blattaria, *etc.*) that usually resembles the adult in shape and general appearance. See Metamorphosis. Cf. Larva; Naiad. 3. Termites: Immatures with enlarged gonads, external wingpads and the capacity to develop into functional reproductives. Nymphs are immature insects that emerge from the egg in a relatively advanced stage of development. Nymphs differ from adults in displaying incomplete wings and genitalia. Historically, there has been inconsistent application of the term 'nymph.' North American and English entomologists typically refer to insects without a pupal stage as nymphs. European, particularly French, entomologists use the term nymph to refer to the Pupa.

NYMPHA INCLUSA (Latin.) A Coarctate Pupa.

NYMPHALIDAE Plural Noun. Brushfooted Butterflies. A cosmopolitan Family of ditrysian Lepidoptera assigned to Papilionidea. Largest Family of butterflies; moderate to large size, strong fliers; active at midday. Adult Antenna typically as long as body, densely setose, tricarinate, some antennal segments with two ventral grooves; elongate Chaetosemata parallel to compound eye margin; Maxillary Palpus with one segment; Labial Palpus ascending Frons; forelegs reduced, female foreleg not used for walking, male foreleg with long scales, without Pretarsus and fewer than five Tarsomeres; Epiphysis absent; tibial spur formula 0-2-2 or 0-0-0; hindwing with Discal Cell open along distal margin. Eggs barrel-shaped with vertical ridges or reticulate sculpture. Larva with horned head or branching spines, long, paired filaments, secondary Setae and bifid anal segment. Chrysalis with numerous projections, suspended by Cremaster. Heliconiidae sometimes regarded as Subfamily of Nymphalidae. Syn. Argyreidae.

NYMPHIDAE Plural Noun. A numerically small Family of myrmeleontoid Neuroptera found in Australia and New Guinea. Large bodied; wings broad, hyaline; Trichosors along cubital and anal margin. Eggs large, stalked, arranged in U-shape. First instar larva typically of group; subsequent instar larvae concealed in rotting wood and sometimes covered with trash; pupate within cocoon in sand.

NYMPHIPARA Insects that bear living young in an advanced stage of development. See Pupipara.

NYMPHOCHRYSALIS See Protonymph.

NYMPHOMYIIDAE Plural Noun. A very small Family of nematocerous Diptera of problematic placement and assigned to monotypic Superfamily Nymphomyioidea in some classifications. Adult small, ca 1–2 mm; slender, pale, weakly sclerotized; compound eyes separated dorsally but contiguous ventrally; antennal Flagellum elongate, club present; wings narrow and apically pointed; veination reduced with no closed cells; marginal fringe long. Adults associated with streams in North America; larvae aquatic and associated with moss on stones in streams.

NYSA JEZABEL *Delias nysa nysa* (Fabricius) [Lepidoptera: Pieridae].

O'NYONG-NYONG A non-fatal arboviral disease endemic in Malawi, Uganda and Tanzania which is caused by *Alphavirus* of Tongaviridae. Mosquito vectors include *Anopheles* spp. Disease incubation period ca 8 days followed by arthralgia, rash and headache for several days. See Arbovirus; Tongaviridae. Cf. Chikungunya; Ross River Fever; Sindbis.

OAK APHID *Myzocallis castanicola* Baker; *Tuberculatus annulatus* (Hartig) [Hemiptera: Aphididae].

OAK-APPLE A gall formed on oak (*Quercus* spp.) by the cynipid wasp *Biorhyzia pallida*. See Marble Gall. Cf. Oak Root-Gall.

OAK BUTTON-GALL A small, flat, circular gall induced by the cynipoid wasp *Neuroterus quercusbaccarum*.

OAK CLEARWING MOTH *Paranthrene asilipennis* (Boisduval) [Lepidoptera: Sesiidae].

OAK CRICKETS See Meconematidae.

OAK LACEBUG *Corythucha arcuata* (Say) [Hemiptera: Tingidae].

OAK LEAFMINER *Phyllonorycter messaniella* (Zeller) [Lepidoptera: Gracillariidae].

OAK LEAFROLLER *Archips semiferana* (Walker) [Lepidoptera: Tortricidae].

OAK LEAFTIER *Croesia semipurpurana* (Kearfott) [Lepidoptera: Tortricidae].

OAK LECANIUM *Parthenolecanium quercifex* (Fitch) [Hemiptera: Coccidae].

OAK ROOT-GALL A gall formed on roots of oak (*Quercus* spp.) by the cynipid wasp *Biorhyzia pallida*.

OAK SAPLING BORER *Goes tesselatus* (Haldeman) [Coleoptera: Cerambycidae].

OAK SKELETONIZER *Bucculatrix ainsliella* Murtfeldt [Lepidoptera: Lyonetiidae].

OAK TIMBERWORM *Arrhenodes minutus* (Drury) [Coleoptera: Brentidae]: A Species found in eastern North America that occurs beneath bark of oak, beech, maple and poplar. Females use their long, slender beak to drill hole in wood; male assists in extraction of female's beak and the female lays eggs in holes. Larvae feed upon wood and fungi; adults feed upon fungi, tree exudations and other insects. See Brentidae.

OAK WEBWORM *Archips fervidana* (Clemens) [Lepidoptera: Tortricidae].

OAKWORMS See Dioptidae.

OAT APHID *Rhopalosiphum padi* (Linnaeus) [Hemiptera: Aphididae].

OBANAL SETAE Coccids: The two longest of four Setae, nearest the Postanal Ring (MacGillivray).

OBCONIC Adj. (Latin, *ob* = against; *conus* = cone.) Cone-shaped but attached at the apex; structure in the form of a reversed cone, *i.e.,* with the apex as a base and the base apical. Alt. Obconical. See Cone. Cf. Obcordate; Obdeltoid; Oblanceolate; Oblate; Obovate. Rel. Form 2; Shape 2; Outline Shape.

OBCORDATE Adj. (Latin, *ob* = against; *cor* = heart; *-atus* = adjectival suffix.) Inversely heart-shaped; structure with an apical notch and basal point. Term descriptive of outline shape of some leaves. Alt. Obcordatus; Obcordiform; Obdeltoid; Oblanceolate; Obovate. See Cordate; Cf. Obdeltoid; Oblanceolate. Rel. Form 2; Shape 2; Outline Shape.

OBDELTOID Adj. (Latin, *ob* = against; Greek, *delta*; *eidos* = form.) Triangular with apex at point of attachment. Inversely deltoid. A term describing the outline shape of some leaves. See Deltoid. Cf. Obcordate; Oblanceolate. Rel. Form 2; Shape 2; Outline Shape.

OBENBERGER, JAN (1892–1964) (Pfeffer 1965, Acta ent. bohemoslovaca 62: 70–73.)

OBERT, IVAN STANISLOVICH (1809–1900) (Jacobson 1901, Horae Soc. ent. Ross. 35: xxxvii–xxxix.)

OBERTHÜR, CHARLES (1845–1924) (Houlbert 1924, Ann. Soc. ent. Fr. 93: 163–178, bibliogr.)

OBERTHÜR, RENE (1852–1944) (Cockayne 1945, Proc. R. ent. Soc. Lond. (C) 9:47.)

OBESE Noun. (Latin, *obesus* = eaten away.) Fat; unnaturally large and distended. A term usually applied to the Abdomen. Alt. Obesus.

OBICULARIS Adj. (Latin, *orbis* = orb.) Pertaining to muscles which surround an oriface or aperture.

OBLANCEOLATE Adj. (Latin, *ob* = toward; *lancea* = lance; *-atus* = adjectival suffix.) Inversely lanceolate. Descriptive of structure that is widest near the apex, with the apex pointed. Term descriptive of outline shape of some leaves. See Lanceolate. Cf. Obconic; Obcordate; Obdeltoid; Oblate; Oblong; Obovate; Obpyriform. Rel. Form 2; Shape 2; Outline Shape.

OBLATE Adj. (Latin, *oblatus* = stretched.) Flattened. A term applied to a spheroid of which the diameter is shortened at two opposite ends. Cf. Obcordate; Obdeltoid; Oblanceolate; Oblong; Obpyriform; Prolate. Rel. Form 2; Shape 2; Outline Shape.

OBLIGATORY DIAPAUSE A condition of suspended development that is an integral part of the life cycle of an insect. OD coincides with unfavourable conditions (temperature, moisture) and may be characteristic of univoltine Species. See Diapause. Cf. Facultative Diapause.

OBLIGATORY HYPERPARASITISM A form of hyperparasitism in which the immature hyperparasite must complete feeding and development using a primary parasite as host. See Hyperparasitism. Cf. Facultative Hyperparasitism.

OBLIGATORY MULTIPLE PARASITE Hymenoptera: Occurence of more than one Species of parasite on or in a host and through which one Species derives benefit. Cf. Facultative Multiple Parasite.

OBLIGATORY MYIASIS A category of Myiasis which includes parasitic fly larvae that are dependent upon vertebrate hosts for completion of a phase of their development. See Myiasis. Cf. Facultative Myiasis.

OBLIGATORY PARASITE An organism which lives

exclusively at the expense of another organism. See Parasitism. Cf. Facultative Parasite; Host Specific.

OBLIQUE Adj. (Latin, *obliquus* = slanting.) Slanting; any direction between perpendicular and horizontal.

OBLIQUE BANDED LEAF-ROLLER *Choristoneura rosaceana* (Harris), *Choristoneura lafauryana* (Rag.) [Lepidoptera: Tortricidae].

OBLIQUE STERNALS Very short muscles connecting adjacent edges of the abdominal Sterna.

OBLIQUE TERGALS Short muscles connecting edges of the abdominal Terga.

OBLIQUE VEIN 1. Odonata: An apparent crossvein positioned between M and Rs, distad to the level of the Nodus and inclined obliquely, the basal part of Rs, the Radial Sector. 2. Myrmeleonidae: A branch of the Media (Comstock). See Vein; Wing Venation.

OBLITERATE Adj. (Latin, *obliteratus* = erased; *-atus* = adjectival suffix.) Nearly rubbed out; indistinct; term applied to colour or structure on the insect body. Alt. Obliterated; Obliteratus.

OBLOMOVISM Noun. (Russian.) A collective term for animal behaviour displayed in Hibernation and Aestivation. Coined after Ivan Goncharov's character Ilya Oblomov. (Boss 1974, Trans. Amer. Microsc. Soc. 93 (4): 460–481.)

OBLONG Adj. (Latin, *oblongus* = rather long.) Rectangular in outline shape with adjacent sides unequal. Descriptive of structure longer than wide, with the longitudinal diameter more than twice the transverse. Cf. Obconic; Obcordate; Obdeltoid; Oblanceolate; Oblate; Obovate; Obpyriform. Rel. Form 2; Shape 2; Outline Shape.

OBLONG PLATES Aculeate Hymenoptera: innermost or posterior pair of sclerites associated with the Sting, representing the divided ninth Sternum (Imms). See Second Valvifers.

OBLONGUM Noun. Coleoptera: A closed cell in the hindwing (Tillyard).

OBOVATE Adj. (Latin, *ob* = against; *ovum* = egg; *-atus* = adjectival suffix.) Inversely egg-shaped (ovate) with the narrow end basal. Term descriptive of outline shape of some leaves. See Ovate. Cf. Obconic; Obcordate; Obdeltoid; Oblanceolate; Oblate; Oblong; Obpyriform. Rel. Form 2; Shape 2; Outline Shape.

OBPYRIFORM Adj. (Latin, *ob* = against; *pyrum* = pear; *forma* = shape.) Inversely pyriform; descriptive of structure inversely pear-shaped. See Pyriform. Cf. Obconic; Obcordate; Obdeltoid; Oblanceolate; Oblate; Oblong; Obovate; Obpyriform. Rel. Form 2; Shape 2; Outline Shape.

OBRATZTOV, NIKOLAUS SERGEYEVICH (1905–1966) (Agenjo 1967, Graellsia 23: 35–53, bibliogr.)

OBSCURE Adj. (Latin, *obscures* = covered.) Dark; not readily seen; pertaining to structure not well defined. Alt. Obscurus.

OBSCURE-AENEOUS Adj. (Latin, *obscures* = covered; *aeneus* = bronze; *-osus* = with the property of.) An arcane descriptor of an indistinct bronze colour. Alt. Obscure-aeneus. See Bronze; Copper.

OBSCURE MEALYBUG *Pseudococcus affinis* (Maskell) [Hemiptera: Pseudococcidae].

OBSCURE ROOT WEEVIL *Sciopithes obscurus* Horn [Coleoptera: Curculionidae].

OBSCURE SCALE *Melanaspis obscura* (Comstock) [Hemiptera: Diaspididae]: In North America, a pest of elm, hackberry, oak, pecan. Overwinters as nymph.

OBSITE Adj. (Latin, past part. of *obsero* = to cover over; *-ites* = constituent.) Superficially covered with equal scales or other bodies. Alt. Obsitus.

OBSOLESCENT Adj. (Latin, *obsolescere* = to wear out.) Pertaining to structure in the process of disappearing or of becoming non-functional.

OBSOLETE Adj. (Latin, *obsolescere* = to wear out.) Pertaining to structure which is indistinct, not well developed or almost absent. A comparative term best applied when discussing a character or character state in transition or when comparing the same structure in a coordinate group of Taxa. Alt. Obsoletus.

OBSTAMADE® See Codling-Moth Granulosis Virus.

OBSUBULATE Adj. (Latin, *ob* = against; *subula* = awl; *-atus* = adjectival suffix.) Reverse awl-shaped; narrow and tapering from apex to base. Cf. Acuminate; Aciculate. Rel. Form 2; Shape 2; Outline Shape.

OBTECT Adj. (Latin, *obtectus* = covered over.) Covered; pertaining to structure within a hard covering. Alt. Obtected.

OBTECT PUPA A Pupa with appendages fused or 'glued' to the body by a hardening of the exoskeleton. Termed 'theca' in older works. OP occurs in Nematocerous Diptera, some Lepidoptera, some Hymenoptera. See Decticous Pupa; Adecticous Pupa. Cf. Exarate Pupa; Coarctate Pupa.

OBTUSE Adj. (Latin, *obtusus* = blunt, dull.) Not pointed; descriptive of a surface with a blunt or rounded end. Cf. Acute. Rel. Form 2; Shape 2; Outline Shape.

OBTUSE-ANGULATE Adj. Forming an obtuse angle, as markings or angles.

OBTUSILINGUES Adj. (Latin, *obtusi* = a combining form meaning obtuse; *lingua* = tongue.) Bees with short tongues apically obtuse or bifid. See Acutilingues.

OBUMBRANT Adj. (Latin, *obumbrans* = shaded; *-antem* = adjectival suffix.) 1. Overhanging. 2. Darkened as if in a shadow. Alt. Obumbrans.

OBVERSE Adj. (Latin, *obvertere* = to turn around.) 1. Pertaining to a surface facing the observer. 2. Bilateral symmetry: The surface of structures which complement or face one another, such as the dorsal surface of wings. 3. Pertaining to appendages: The basal portion of the structure narrower than the distal portion. Syn. Converse; Reverse. Noun. 1. The principally viewed surface of any structure; the front. 2. The opposite side of reverse. Cf. Converse.

OCCAM'S RAZOR See Ockham.

OCCASIONAL PEST A pest with a general equilibrium position substantially below the Economic-Injury Level (EIL). The highest pest population fluctuations occasionally and sporadically exceed the Economic-Injury Level, thus requiring intervention.

OCCIPITAL Adj. (Latin, *occiput* = back of head; *-alis* = characterized by.) Descriptive of structure or process associated with the Occiput or posterior part of the head. Cf. Facial; Frontal; Genal.

OCCIPITAL ARCH The area of the head between the Occipital and Postoccipital Sutures. OA dorsal (medial) part is the Occiput proper; the ventral (lateral) part is the Postgena (Snodgrass).

OCCIPITAL CILIA See Occipital Fringe.

OCCIPITAL CONDYLE A process on the margin of the Postocciput to which the lateral neck-sclerites articulate (Snodgrass).

OCCIPITAL FORAMEN An opening (aperture, hole) on posterior surface of the head which is opposed to a similar opening in the Prothorax. The OF provides a passage for the dorsal vessel, ventral nerve cord, tracheal system and other structures that extend from the head into the Thorax. In pterygotes such as Orthoptera, the OF and mouth are not separated. More highly evolved insects have developed sclerotized separations between mouthparts and OF. At least three types of closure between mouthparts and OF have been identified: Hypostomal Bridge, Postgenal Bridge and Gula. Most insects with an hypognathous condition display an Occipital Foramen near the centre of the posterior surface of the head. Insects with a prognathous condition display an Occipital Foramen near the vertex. Syn. Foramen Magnum; Alaforamen. See Head Capsule. Cf. Occiput. Rel. Gula; Hypostomal Bridge; Postgenal Bridge.

OCCIPITAL FRINGE Diptera: The fringe of fine Setae behind the compound eyes (Curran).

OCCIPITAL GANGLION 1. A single or paired postcerebral ganglion of the Stomodaeal Nervous System. 2. The Pharyngeal, Oesophageal, or Hypocerebral Ganglia (Snodgrass).

OCCIPITAL HORN Odonata: A chitinous horn just below the occipital ridge on each side of the head (Garman).

OCCIPITAL LOBE Hymenoptera: The posterolateral surface of the head.

OCCIPITAL MARGIN Mallophaga: The posterior margin of the head.

OCCIPITAL ORBIT The part of the head adjacent to the posterior or lateral margin of a compound eye. Syn. Posterior Orbit (MacGillivray).

OCCIPITAL RIDGE Odonata: A ridge extending between the compound eyes on the head's posterodorsal margin (Garman).

OCCIPITAL SPINE Odonata: A spine on the posterodorsal surface of the head, between the compound eyes (Garman).

OCCIPITAL SUTURE 1. A transverse groove sometimes present on the posterior surface of the head and that ends ventrally anterior of the Mandible's posterior articulation (Snodgrass). 2. The Suture separating the Occiput and Postgena from the Alavertex. 3. The Hypostomal Suture *sensu* MacGillivray. See Head Capsule. Cf. Postoccipital Suture.

OCCIPITO-ORBITAL BRISTLE Diptera: Any of the coarse, rather large Setae positioned along the posterior margin of the compound eye.

OCCIPUT Noun. (Latin, *occiput* = back of head. Pl., Occiputs.) 1. General: The posterior portion of the head between the Vertex and the Foramen Magnum. The Occiput is rarely present as a distinct sclerite or clearly demarcated by 'benchmark' sutures. 2. Diptera: The entire posterior surface of the head (*sensu* Smith). See Head Capsule. Cf. Postocciput.

OCCLUSOR MUSCLE 1. Any muscle responsible for closing an aperture or opening of the body. 2. A muscle attached to or near a respiratory spiracle which, by contraction, closes the spiracle or prevents communication between the spiracular Atrium and the ventillatory Trachea. Cf. Dilator Muscle.

OCCULT Adj. (Latin, *oculere* = to cover.) Hidden; concealed from view. Alt. Occultus.

OCEANIC Adj. (Greek, *okeanos* = ocean; *-ic* = characterized by.) Descriptive of or pertaining to the ocean. A term applied to animals that inhabit the open sea; pelagic. *e.g. Hyalobates*.

OCEANIC AREA The biogeographic marine realm containing the Neritic Zone, Photic Zone and above Abyssal Zone. See Habitat. Cf. Abyssal Zone; Littoral Zone; Neritic Zone; Pelagic Zone; Photic Zone.

OCEANIC BURROWER BUG *Geotomus pygmaeus* (Dallas) [Hemiptera: Cydnidae].

OCEANIC EMBIID *Aposthonia oceania* (Ross) [Embiidina: Oligotomidae].

OCEANIC FIELD CRICKET *Telogryllus oceanicus* (LeGuillou) [Orthoptera: Gryllidae].

OCELIFORM Adj. (Latin, *ocellus* = little eye; *forma* = shape.) Pertaining to structure which resembles an Ocellus. Syn. Ocelloid.

OCELLAE See Ocellus.

OCELLALAE Grouped simple eyes in nymphs and adults of Collembola (MacGillivray).

OCELLANAE Noun. The simple eyes in the nymphs and adults of Exopteraria (MacGillivray).

OCELLAR BASIN Hymenoptera: A concave area occupying the median portion of the frontal area of the head. OB shape and size differ among taxonomic groups.

OCELLAR BRISTLES Diptera: Coarse or rather large Setae near the Ocelli, usually directed forward.

OCELLAR CENTRES Brain centres in the outer

ends of the ocellar Pedicels (Snodgrass).

OCELLAR FURROW Hymenoptera: A transverse furrow, groove or depression which extends between the ends of the vertical furrows near the dorsal margin of the lateral Ocelli. OF frequently confluent with the space around the lateral Ocelli.

OCELLAR PAIR Diptera: Ocellar Bristles (Comstock).

OCELLAR PEDICELS The long, slender, nerve stalks which connect the facial Ocelli with Protocerebrum (Snodgrass).

OCELLAR PLATE See Ocellar Triangle.

OCELLAR RIBBON A crescent-shaped, smooth, thin stripe across the eye region in butterfly chrysalids.

OCELLAR RIDGE Odonata: A ridge behind the Ocelli (Garman).

OCELLAR STRIPE Odonata: A pale stripe behind the Ocelli (Garman).

OCELLAR TRIANGLE Diptera: Ocellar sclerite; vertical triangle (Comstock).

OCELLAR TUBERCLE Diptera: An elevated portion of Vertex which contains the Ocelli. The OT is common in Asilidae.

OCELLARAE Plural Noun. (Latin, *ocellus* = little eye.) The simple eyes in larval Entopteraria (MacGillivray).

OCELLATE Adj. (Latin, *ocellus* = little eye; *-atus* = adjectival suffix.) 1. A head with Ocelli or simple eyes. 2. Structure displaying spots ringed with another colour. Alt. Ocellated; Ocellatus.

OCELLI See Ocellus; Stemmata.

OCELLIFORM Adj. (Latin, *ocellus* = little eye; *forma* = shape.) Ocellus-shaped; dome-shaped; descriptive of structure which resembles an Ocellus. Syn. Ocelloid. See Ocellus. Cf. Campaniform. Rel. Form 2; Shape 2; Outline Shape.

OCELLIGEROUS Adj. (Latin, *ocellus* = little eye; English *gerous* from Latin *ger* = bear; *-osus* = with the property of.) Furnished with or bearing Ocelli. Alt. Ocelligerus.

OCELLOID Adj. (Latin, *ocellus* = little eye; Greek *eidos* = form.) Ocellus-like; See Ocelliform. Rel. Eidos; Form; Shape.

OCELLUS Noun. (Latin, *ocellus* = little eye. Pl., Ocelli.) 1. The 'simple eye' of many adult insects, positioned on the Vertex and between compound eyes. Most holometabolous insects display three Ocelli; 1–2 Ocelli occur in some insects or may be absent in other Species. 2. Lateral simple eyes in larval holometabolous insects (MacGillivray). 3. Optical structures on the head of most adult pterygote insects. Anatomically, an Ocellus consists of a biconvex lens on the Vertex. Light collected by lens is cast onto sense cells (Rhabdom). Usually, an Ocellus is circular in outline but in some insects (*e.g.* Odonata and some bumblebees) it is bilobed. An Ocellus does not form a visual image because light collected by lens is focused beneath sensory cells, but it is sensitive to low intensities of light. Functions of Ocelli are diverse: involved in entrainment to light cycles and mediates a general stimulatory effect on insect. For instance, honeybee forages earlier in morning and longer in day if it's Ocelli are intact. Ocelli may orientate insect toward linearly polarized light, modulate phototactic behaviour or orient insect toward edges and certain objects. Ocelli probably constitute part of visual groundplan system because they occur in most insects. The anterior (Median) Ocellus probably represents fusion of two Ocelli because it is innervated from both sides of Deutocerebrum. Some termites lack Ocelli; other termites have two lateral Ocelli but lack a median Ocellus. When Ocelli are absent, condition is termed anocellate; most Lepidoptera are anocellate. Ocelli may be correlated with other anatomical features. Wasp *Sclerodermus* shows apterous and macropterous individuals of same Species: Macropterous individuals display Ocelli and apterous individuals lack Ocelli. Thysanoptera: Ocelli are present in winged adults only. See Ommatidium; Vision. Cf. Compound Eye; Stemma. Rel. Anocellate; Dermal Light Sense.

OCELLUS COECUS See Blind Ocellus.

OCELLUS SIMPLEX See Simple Ocellus.

OCHRACEOUS Adj. (Greek, *ochros* = yellow; Latin, *-aceus* = of or pertaining to.) Descriptive of structure that is yellow with a slight tinge of brown. Alt. Ochraceus; Ochraeus; Ochreous; Ochreus. See Yellow. Cf. Helvolus; Helvus.

OCHROLEUCUS Adj. (Greek, *ochros* = yellow; *leukos* = white.) Yellow-white. Descriptive of structure that appears dilute ochraceous; a whitish ochre-yellow. See Yellow.

OCHS, GEORG (Avers 1966, Ent. Bl. Biol. Syst. Käfers 62: 67–68.)

OCHSENHEIMER, FERNAND (–1822) German author of *Schmetterlinge von Europa* (4 vols, 1806–1816). After his death the work was continued by M. Treitschke, who added three volumes through 1829.

OCHTERIDAE Kirkaldy 1902. Plural Noun. Velvety Shore Bugs. A numerically Family of heteropterous Hemiptera assigned to Superfamily Ochteroidea. Body 3–5 mm long, oval in outline shape, dark coloured, velvety in texture; Ocelli present; Antenna short with four segments and mostly concealed beneath head; Labium (Rostrum) long, four segments and projecting rearward between hind legs; front and middle Tarsi of two segments; hind Tarsus of three segments. Ochterids occur on mud, sand flats and in grass along margin of ponds and streams. Species are predatory upon other insects.

OCKENDEN, GEORGE RICHARD (1868–1906) (Halert 1907, Novit. Zool. 14: 341–342.)

OCKHAM'S RAZOR *Pluralitas non est ponenda sine necessitate.* A maxim attributed to William of Ockham, 14th Century English Philosopher. Literally taken to mean 'Plurality must not be posited without necessity'. Generally taken to mean that the simplest explanation of facts is probably the correct interpretation. Or, if several hypotheses are available, then the simplest is probably correct.

OCTACERORES Noun. Asterolecaniinae: The paired Cerores which resemble a figure 8 in outline (MacGillivray).

OCTAPOPHYSIS Noun. An inferior apophysis of the structures of the eighth abdominal segment in female insects (MacGillivray).

OCTAVALVAE Noun. The ventral or anterior Valvulae, or anterior Ovipositor (MacGillivray).

OCTAVALVIFER Noun. Orthoptera: The sclerite on the proximal part of the lateral surface of an Octavalva (MacGillivray).

OCTOON Noun. (Greek, *octo* = eight. Pl., Octoons.) The eighth segment of the Abdomen.

OCTOPAMINE Noun. A biogenic amine which functions as a neuromodulator, neurotransmitter and neurohomone in insects.

OCULAR Adj. (Latin, *oculus* = eye.) Descriptive of or pertaining to the eyes. Rel. Subocular.

OCULAR EMARGINATION Mallophaga: A lateral emargination of the head in which the eye is received posteriorly.

OCULAR FLECK Mallophaga: A small, intensely black spot of pigment in the eyes.

OCULAR FRINGE Mallophaga: Closely set, small Setae on the posterior half of an ocular emargination; OF sometimes extends to the temporal margin.

OCULAR LOBES The Protocerebrum of the insect Brain.

OCULAR NEUROMERE Insect embryo: The primitive cephalic ganglion from which the optic lobe of the Brain arises.

OCULAR REGION See Protocerebral Region.

OCULAR SCLERITE 1. The first (Protocerebral) segment of the head (Smith). 2. A ring of hardened Integument surrounding each compound eye.

OCULAR SUTURE The line (inflection) on the head surrounding the compound eye and forming an internal circumocular ridge (Snodgrass). Ocular Suture is not present in all insects, and is difficult to see in some insects unless head is cleared and observed with a microscope. When present, the OS probably provides strength and prevents deformation of the compound eye. See Head Capsule. Cf. Subocular Suture.

OCULAR TUBERCLES Aphids: A group of prominent facets on the posterior portion of the compound eye.

OCULARIUM Noun. (Latin, *oculus* = eye; *arium* = place of a thing. Pl., Ocularia.) The more-or-less elevated area bearing the grouped simple eyes in larvae (MacGillivray). Alt. Ocularum.

OCULATA Noun. A narrow ring-like area surrounding each compound eye and the apodeme or cuticular flange projecting into the head cavity (MacGillivray).

OCULI Plural Noun. (Sl., Oculus.) Eyes; in particular, the Compound Eyes.

OCULOCEPHALIC Adj. (Latin, *oculus* = eye; Greek, *kephale* = head; *-ic* = characterized by.) 1. Descriptive of or pertaining to the eyes and head. 2. Hymenoptera: Imaginal buds that produce the cephalic region.

OCULOMALAR SPACE The area separating the inferior angle of the eye from the insertion of the Mandible (Needham).

OCULOMOTOR Noun. (Latin, *oculus* = eye; *motor* = a mover.) A nerve centre of the muscle which moves the eye in decapod Crustacea (Snodgrass).

ODA, FUJIO (1895–1943) (Hasegawa 1967, Kontyû 35 (Suppl.): 25.)

ODA, KAZUMA (1882–1956) (Hasegawa 1967, Kontyû 35 (Suppl.): 24–25, bibliogr.)

ODAKA, TOMOO (1899–1956) (Hasegawa 1967, Kontyû 35 (Suppl.): 25.)

ODD BEETLE *Thylodrias contractus* Motschulsky [Coleoptera: Curculionidae]: A cosmopolitan larval pest of silk, wool and museum specimens. Adult female larviform, wingless. Egg stage 23–30 days; larval stage 242–338 days; pupal stage 7–14 days; adult stage 9–50 days.

ODD, DONALD ALEXANDER (1905–1970) (Hinton 1971, Proc. R. ent. Soc. Lond. (C) 35: 53.)

ODINIIDAE Plural Noun. A very small Family of schizophorous Diptera assigned to Superfamily Opomyzoidea and apparently related to Agromyzidae as assigned to that Family in some classifications. Adult with Postvertical Bristles divergent; Arista not plumose; Mesoplural Bristles absent; wing patterned, Costa not broken; Preapical tibial bristles present; Ovipositor poorly developed. Syn. Odinidae. Adults associated with fungi on decaying tree trunks.

ODONA Toothed. A term applied to Odonata by Fabricius because of the long teeth on the Maxilla and Labium (Smith).

ODONATA Fabricius 1792. (Greek, *odous* = tooth.) (Lower Permian-Recent.) Damselflies; Darners; Darning Needles; Horse Stinger; Mosquito Hawk. A Palaeopterous Order divided into several Suborders but only three survive today: Anisoptera (dragonflies), Zygoptera (damselflies) and Anisozygoptera. Head large, mobile; Clypeus large with transverse Sulcus dividing structure into Postclypeus and Anteclypeus; three Ocelli on Vertex; mandibulate mouthparts adapted for predation; Antenna short, bristle-like, less than eight segments. Thorax agglutinate; wings similar in size, complex reticulate-venation. Abdomen with 10 segments; spiracles on segments 1–8; accessory copulatory organs of male developed on abdominal Sterna 2–3, distantly removed from testes. Metamorphosis simple. Naiads usually aquatic and respire with gills. Naiads display elongate, scoop-like, prehensile Labium which can be rapidly extended forward to grasp prey. Naiads taken from most freshwater habitats and occasionally from brackish water.

Some Naiads tolerate exposure to humid air, members of at least one Genus in Hawaii terrestrial. Metamorphosis usually occurs at night or early morning. Adults fast fliers and most easily caught when resting. Adults typically diurnal, sometimes crepuscular and occasionally nocturnal. Adults use sight to locate prey; prey captured in flight. Included Families: Aeshnidae, Amphipterygidae, Calopterygidae, Chlorocyphidae, Coenagrionidae, Cordulegastridae, Corduliidae, Dicteriastidae, Epiophlebiidae, Euphaeidae, Gomphidae, Hemiphlebiidae, Isostictidae, Lestidae, Lestoideidae, Libellulidae, Megapodagrionidae, Neopetaliidae, Perilestidae, Petaluridae, Platycnemididae, Platystictidae, Polythroidae, Protoneuridae, Pseudolestidae, Pseudostigmatidae, Synlestidae.

ODONATE Adj. (Greek, *odontos* = tooth; *-atus* = adjectival suffix.) Pertaining to or descriptive of dragonflies or a toothed condition resembling dragonflies.

ODONTOIDEA A triangular projection on each side of the Foramen not covered by the Cervacoria and serving as a point of articulation for the sclerites of the Cervapleura. In a generalized condition, borne by the Maxillariae (MacGillivray).

ODONTUS See Palpal Claw.

ODORATUS Adj. See Odoriferous.

ODORIFEROUS GLANDS Heteroptera: Glands which produce strong, odorous, volatile secretions; some odours pleasant, others disgusting in the extreme (to humans). OG in the nymph open dorsally on the Abdomen and in the adult open by Ostioles with corrugated evaporative Peritremes, near the middle legs on the Sternum.

ODORIFEROUS HOUSE ANT See Odorous House Ant.

ODORIFEROUS Adj. (Latin, *odorifer* from *odor* = odour; *ferre* = to carry; *-osus* = with the property of.) Scented; diffusing an odour.

ODOROUS HOUSE ANT 1. *Tapinoma sessile* (Say) [Hymenoptera: Formicidae]: A Species endemic and widespread in USA as an urban pest. Worker 3.0–3.5 mm long; monomorphic; brownish grey with velvet-like appearance; relatively soft-bodied; Petiole consisting of a Node, concealed by a relatively broad Gaster. OHA typically nests out-of-doors in soil or under wood or stones and within building foundations. A colony may include several queens and several thousand workers. Workers form trails and are attracted to honeydew from homopterous insects on plants. When crushed, workers smell of rotten coconut; odour derived from anal-gland ketones. Syn. Odoriferous House Ant. 2. *Tapinoma minutum* Mayr. In Australia a widespread Species in coastal and forested habitats and which includes several named Subspecies. Australian OHA is facultatively polygynous and polydomous as an opportunistic cavity dweller. See Formicidae. Cf. Argentine Ant; Black Carpenter-Ant; Cornfield Ant; Fire Ant; Larger Yellow-Ant; Little Black Ant; Pavement Ant; Pharaoh Ant; Thief Ant.

ODOUR TRAIL Social Insects: A chemical deposited on the substrate by one insect and followed by another insect. Cf. Trail Pheromone; Trail Substance.

OECOLOGY Noun. Archaic form of the word Ecology.

OECOPHORID MOTHS See Oecophoridae.

OECOPHORIDAE Baruand 1849. Plural Noun. Oecophorid Moths. A numerically large, cosmopolitan Family of ditrysian Lepidoptera assigned to Superfamily Gelechioidea. Adults are variable in size with wingspan 8–75 mm and head typically with smooth, overlapping, lamellar scales; Ocelli typically absent and Chaetosemata absent. Antenna filiform, ciliate or pectinate with length variable and Scape often with Pecten; Epiphysis present; hind Tibia with long piliform scales; Proboscis variable in size; when present always with overlapping, lamellate scales basad; Maxillary Palpus typically with four segments and folded; Labial Palpus typically with three segments and recurved; fore Tibia with Epiphysis; tibial spur formula 0-2-4. Egg flattened, oval in outline shape and Chorion sculptured; eggs typically deposited individually and often concealed. Larva cylindrical or dorsoventrally flattened; thoracic legs present; Prolegs on abdominal segments 3–6, 10 and Crochets variable in arrangement. Pupae cylindrical or dorsoventrally flattened; Maxillary Palpus usually well developed; Labial Palpus concealed; pupation typically in larval habitat and lined with silk or a formed cocoon. Biology diverse, but larva concealed; tunnel in wood, stems, branches, flowers or galls; a few mine leaves; a few tunnel in soil. Syn. Depressariidae; Ethmiidae; Stathmopodidae; Stenomatidae; Tinaegeridae.

OEDAGUS Noun. A variant spelling of Aedeagus.

OEDEMERIDAE Plural Noun. False Blister Beetles. A widespread Family of polyphagous Coleoptera. Adult body slender, small to moderate sized; palpal segments long; Antenna usually long, filiform; tarsal formula 5-5-4. Larva slender, cylindrical; head large; legs well developed; Urogomphi absent. Adults diurnal, contain cantharidin; larvae bore into wood.

OEDOEAGUS See Aedeagus.

OENOCYTE Noun. (Greek, *oinos* = wine; *kytos* = hollow vessel; container. Pl., Oenocytes.) A basic anatomical form of insect Haemocyte. Oenocyte cells are variable-sized, oval-elongate or crescent-shaped; Plasma Membrane typically lacks Micropapillae, Filipodia or irregular processes; the Nucleus typically is small, round-elongate and eccentrically positioned; some cells binucleate; Cytoplasm thick, homogeneous with rod-, needle- or plate-like inclusions and an elaborate system of filaments. Oenocytes are very fragile, lyse quickly and eject non-phagocytic material into Haemolymph. Oenocytes typically show cycles of development with the moulting cycle in immature insects. Oenocytes are arranged in clusters in Lepidoptera and Orthoptera; Oenocytes are associated with Fat Body in Homoptera, Hymenoptera, some Diptera; Oenocytes reside between epidermal cells and Basement Membrane in Heteroptera, Odonata, Ephemeroptera. Oenocyte function is poorly understood but they probably synthesize wax, lipoproteins and ecdysteroids. Syn. Crescent Cell; Crystal Cell; Crystaloid Cell; Non-granular Spindle Cell; Non-phagocytic Giant Haemocyte. See Fat Body; Haemocyte. Cf. Adipohemocyte; Coagulocyte: Granulocyte; Nephrocyte; Plasmatocyte; Prohaemocyte; Spherulocyte.

OENOCYTOID Noun. (Greek, *oinos* = wine; *kytos* = hollow vessel; container; *eidos* = form. Pl., Oenocytoids.) Oenocyte-like; specifically, large, non-phagocytic Haemocytes which are spherical in shape and contain a strongly acidophilic cytoplasm. Named from their resemblance to oenocytes; a rounded or spherical Leucocyte. See Haemocyte. Cf. Oenocyte. Rel. Eidos; Form; Shape.

OERSTED, A S (1816–1872) (Papavero 1975, *Essays on the History of Neotropical Dipterology*. 2: 365–366. São Paulo.)

OERTZEN, EBERHARD (1856–1909) (Kalbe 1910, Berl. ent. Z. 54: 81–88, 229–231.)

OESOPHAGEAL Adj. (Greek, *oisophagos* = gullet; Latin, *-alis* = pertaining to.) Descriptive of structure or process associated with the Oesophagus; near the Oesophagus. Alt. Esophageal. See Foregut.

OESOPHAGEAL BONE Psocidae: A sclerite below the anterior part of the Oesophagus.

OESOPHAGEAL BULB Some adult Tephritidae: An evagination of the Oesophagus anteriad of the Brain. OB does not possess musculature and its function remains unknown (Dean 1933). Syn. Subclypeal Pump.

OESOPHAGEAL COMMISSURE One of the paired nerve cords which connect the Suboesophageal Ganglion with the insect Brain (Folsom & Wardle).

OESOPHAGEAL DIVERTICULA Food reservoirs more generally applied to any saclike structure connected with the Gullet.

OESOPHAGEAL GANGLION See Occipital Ganglion.

OESOPHAGEAL LOBES The posterior part of the Brain; the Tritocerebrum.

OESOPHAGEAL SCLERITE Mallophaga: A thickening of the chitinous lining of the anterior part of the Oesophagus.

OESOPHAGUS Noun. (Greek, *oisophagos* = gullet.) The portion of Foregut between Pharnyx and Crop. Typically an undifferentiated tube responsible for food conduction. In some insects a Diverticulum from the Oesophagus may develop. Adults of the apple maggot, *Rhagoletis pomonella* Walsh display an Oesophageal Bulb which originates on the dorsal surface of the Oesophagus anterior of the Brain. The bulb is packed with bacteria used in food breakdown. See Alimentary Canal. Cf. Pharnyx; Crop. Rel. Digestion. Alt. Esophagus.

OESTRIDAE Plural Noun. Bot Flies; Gad Flies; Stomach Flies; Warble Flies. A small Family of muscoid Diptera, related to Tachinidae. Currently includes several Subfamilies which are recognized as Families in some classifications: Hypodermatidae (Warble Flies); Gasterophilidae (Bot Flies, Stomach Flies); Cuterebridae (Gad Flies.) Adult large, stout-bodied, resemble bees; mouthparts vestigial and not adapted for feeding. Larvae obligate parasites of mammals, infesting skin, nasal, digestive and respiratory passage of hosts. Family *(sensu lato)* includes many pests of economic or veterinary importance; notable pests in Oestrinae include Sheep Bot-Fly and Ox Warble-Fly. See Bot Flies; Diptera; Myiasis. Cf. Cuterebridae; Gasterophilidae; Hypodermatidae.

OETTEL, AUGUST (1839–1905) (Daniel 1906, Münch. koleopt. Z. 2: 390–391.)

OETTINGE, HEINRICH VON (1878–1956) (Anon. 1956, Mitt. dt. ent. Ges. 15: 55.)

OETTL, JOHANN NEPOMUK (1801–1866) (Zacke 1868, Jber. Ver. Bienenzucht Böhmens 1868: 57–63.)

OFFENBURG, FRANZ (–1969) (Anon. 1969, Mitt. dt. ent. Ges. 28: 50.)

OFFICIAL LIST OF GENERIC NAMES IN ZOOLOGY A record of generic names which have been validated, conserved, or stabilized by the International Commission on Zoological Nomenclature through the use of its Plenary Powers or rendering of an official Opinion [Bull. Zool. Nomencl. 4: 269–271.] See Plenary Powers.

OFFICIAL LIST OF SPECIFIC TRIVIAL NAMES IN ZOOLOGY A record of trivial names of Species or Subspecies which have been validated, conserved or stabilized by the International Commission on Zoological Nomenclature through the use of its Plenary Powers or the rendering an official Opinion [Bull. Zool. Nomencl. 4: 269–271.]

OFNACK® See Pyridafenthion.

OFNAK® See Pyridafenthion.

OFTANOL® See Isofenphos.

OFUNACK® See Pyridafenthion.

OGIER DE BAULNY, FERDINAND MARIE (1839–1870) (Simon 1871, Ann. Soc. ent. Fr. 1 (5): 122–124.)

OGLOBLIN, ALEJANDRO A. (1891–1967) Russian emigrant to Argentina and taxonomic specialist in parasitic Hymenoptera. (De Santis 1967, Revta. Soc. ent. argent. 30: 37–38.)

OGLOBLIN, DIMITRA ALEXEEVITSCH (1893–1942) (Stackelberg 1945, Revue Russe Ent. 28: 131–134, bibliogr.)

OGRE-FACED SPIDERS See Dinopidae.

OGUMA, TAROKICHI (1874–1938) (Hasegawa 1967, Kontyû 35 (Suppl.): 22–23, bibliogr.)

OHAUS, FRIEDRICH (1864–1946) (Schunk 1948,

Mainz. Kal. 1948: 1–19, bibliogr.)

OHAYASHI, KANZO (1915–1967) (Hasegawa 1967, Kontyû 35 (Suppl.): 21, bibliogr.)

OHF See Omsk Haemorrhagic Fever.

OHLERT, GUSTAV HEINRICH EMIL (1807–1871) (Bonnet 1945, *Bibliographia Araneorum* 1: 35.)

OHOSHIMA, MASAMITSU (1884–1965) (Hasegawa 1967, Kontyû 35: 19–20, bibliogr.)

OIL BEETLES See Meloidae.

OIL CONCENTRATE A form of insecticide formulation in which the active ingredient is dissolved in an oil. OC may be further diluted by end users with an oil based solvent or diluent such as kerosene or diesel oil. See Aerosol; Dust; Emulsifiable Concentrate; Suspension Concentrate; Wettable Powder.

OINOPHILIDAE See Tineidae.

OKADA, JYUZO (1874–1936) (Hasegawa 1967, Kontyû 35 (Suppl.): 22, bibliogr.)

OKADA, TADAO (1871–1920) (Hasegawa 1967, Kontyû 35 (Suppl.): 22, bibliogr.)

OKADA, TORAJIRO (1872–1920) (Hasegawa 1967, Kontyû 35 (Suppl.): 22, bibliogr.)

OKAJIMA, GONJI (1875–1955) (Hasegawa 1967, Kontyû 35 (Suppl.): 21, bibliogr.)

OKAMOTO, HANJERO (1882–1960) (Hasegawa 1967, Kontyû 35 (Suppl.): 23, bibliogr.)

O'KANE, WALTER COLLINS (1877–1973) (Northon 1974, J. Econ. Ent. 67: 144–145.)

OKEN, LUDWIG LORENZ (1779–1851) (Nordenskiöld 1935, *History of Biology*. 625 pp. (287–289) London.)

OKLAND, FRIDTHJOF (1893–1957) (Lindroth 1957, Opusc. ent. 22: 122–123.)

OKO® See Dichlorvos.

OKUMURA, OOTADA (1887–1923) (Hasegawa 1967, Kontyû 35 (Suppl.): 24.)

OKUNI, AKIRA (–1957) (Hasegawa 1850, Kontyû 35 (Suppl.): 19, bibliogr.)

OLAFSSON, EGGERT (1726–1768) (Henriksen 1922, Ent. Meddr. 15: 79.)

OLANE AZURE *Ogyris olane* Hewitson [Lepidoptera: Lycaenidae].

OLD HOUSE BORER *Hylotrupes bajulus* (Linnaeus) [Coleoptera: Cerambycidae] a widespread, significant pest of structural and ornamental wood; endemic to Atlas Mountains of North Africa and moved through commerce during 17th century. Adult 15–25 mm long; colour variable; larva cream coloured with brown Mandibles and three conspicuous Ocelli on each side of head. Adults emerge in spring, mate; female mates once and lays ca 170 eggs in small batches during 5 days; eggs deposited in crevices; eclosion occurs ca 9 days following oviposition. First instar larva penetrates wood and tunnels parallel to grain near surface; larva feeds in sapwood of fir, pine and spruce with moisture content of 10–20%; larva can digest cellulose and can feed on seasoned softwood for several years; frass somewhat granular consisting of irregular-shaped indigestible particles and barrel shaped faecal pellets. Larval stage 2–10 years with most rapid development at warmer temperatures and higher moisture content of wood; do not live in heartwood or decayed wood; feeding activity of second and third instar larvae audible. Mature larva cuts oval-shaped exit hole in surface of wood, retreats into wood, packs space with wood fibres and constructs pupal chamber. Pupal stage ca 20 days; neonate adult remains in chamber for several days before emerging. Pheromone in frass apparently influences oviposition by females; low pheromone concentrations attractive while high pheromone concentrations unattractive as sites for oviposition. OHB probably among most highly adapted insects to human environment. Feral populations almost unknown and biology of Species well adapted to feeding on structural timber. See Cerambycidae.

OLD LADYMOTH *Dasypodia selenophora* Guenée [Lepidoptera: Noctuidae].

OLD WORLD BOLLWORM See Cotton Bollworm.

OLD WORLD CABBAGE WEBWORM *Hellula undalis* (Fabricius) [Lepidoptera: Pyralidae]: A pest of crucifers which is endemic to Old World tropics and adventive to Hawaii. Cf. Cabbage Webworm.

OLD WORLD SCREW-WORM *Chrysomya bezziana* (Villeneuve) [Diptera: Calliphoridae]: A significant pest of livestock in Africa, India, southeast Asia, Indonesia and New Guinea; a functional homologue of New World Screwworm. Female resembles Oriental Latrine Fly; female with frontal stripe on head broad and parallel-sided; male compound eye facets small with those in upper part of eye slightly larger than others; Squamae white. Female lays 150–500 eggs on edge of wounds; eclosion occurs in ca 24 hours; larva penetrates deeply into tissue with only posterior respiratory spiracles exposed; larval stage ca 6–7 days. Pupation in soil; pupal stage ca 7 days in tropical areas. Adult female sexually receptive after four days and inseminated once (monogamous). See Calliphoridae. Cf. New World Screw-Worm; Oriental Latrine Fly.

OLDENBERG, LORENZ (1863–1931) (Rohlfien 1975, Beitr. Ent. 25: 270–271.)

OLDHAM, CHARLES (1868–1942) (Cash 1942, NWest. Nat. 17: 12–13.)

OLEANDER APHID *Aphis nerii* Boyer de Fonscolombe [Hemiptera: Aphididae].

OLEANDER BUTTERFLY *Euploea core corinna* (W.S. MacLeay) [Lepidoptera: Nymphalidae].

OLEANDER HAWK MOTH *Daphnis nerii* (Linnaeus) [Lepidoptera: Sphingidae].

OLEANDER PIT SCALE *Asterolecanium pustulans* (Cockerell) [Hemiptera: Asterolecaniidae].

OLEANDER SCALE *Aspidiotus nerii* Bouché [Hemiptera: Diaspididae]: A pest of oleander in North America.

OLEOCELLOSIS Noun. (New Latin, *oleo* = oil; *cella* = room; *sis* = a condition or state.) A spotting of citrus fruits by oil liberated from the oil glands of the rind.

OLEUTHREUTIDAE See Tortricidae.

OLFACTION Noun. (Latin, *olfacere* = to smell; English, *-tion* = result of an action.) Behaviour and physiology: The perception of odours; the sense of smell.

OLFACTOMETER Noun. A device for testing the behavioural response of insects to odours. Early models of olfactometers consisted of a 'T' or 'Y'-shaped glass tube into which an odour or odour-bearing object was introduced at one terminus. Air was drawn through the apparatus and the insect introduced at the base. Modern devices are more complicated.

OLFACTORY Adj. (Latin, *olfacere* = to smell.) Pertaining to the sense of smell.

OLFACTORY CONE See Sensillum Basiconicum.

OLFACTORY LOBES Paired, prominent swellings on the anteroventral aspect of the insect Brain; the Deutocerebrum. See Antennary Lobes.

OLFACTORY PORES See Campaniform Sensilla.

OLIFF, ARTHUR SIDNEY (1865–1895) (Guthrie 1876, Agric. Gaz. N.S.W. 7: 1–4, bibliogr.)

OLIGIDIC DIET An artificial diet which consists of crude, natural plant material and liver or other animal product. OD used in mass-rearing programmes for insects when components are easy to obtain, inexpensive and simple to prepare. See Artificial Diet. Cf. Meridic Diet; Holidic Diet.

OLIGOCENE EPOCH The interval of the Geological Time Scale (34 MYBP–23 MYBP) which is the third Epoch of the Palaeogene Period. OE first formally recognized by Heinrich Ernst von Beyrich (1854). Notable deposits include Dominican Amber (lower Miocene–upper Oligocene; Florissant, Colorado. (Lower Oligocene); Chiapas Amber. See Geological Time Scale; Cenozoic. Cf. Palaeocene Epoch; Eocene Epoch. Rel. Fossil.

OLIGOGYNY (Greek, *oligos* = few; *gyne* = woman.) Social Insects: The presence of more than one functional queen in a nest. Workers tolerate more than one queen, but queens are antagonistic to one another and must be separated within the colony.

OLIGOLECTIC Adj. (Greek, *oligos* = few; *lektos* = chosen; *-ic* = of the nature of.) Descriptive of insects, particularly bees, that forage and collect pollen from one or a few plant Species. Term applied to Species, not individuals of a Species. Cf. Monolectic; Polylectic.

OLIGOMERIZATION Noun. (Greek, *oligos* = few; *meros* = parts; *-tion* = an act or process. Pl., Oligomerizations.) An evolutionary phenomenon involving the organizational process of the insect body through which regions of the body assume different functional responsibilities. *e.g.* Each Tagma of the insect body (head, Thorax, Abdomen) has developed specific functional responsibilities. See Morphology. Cf. Tagmatization. Rel. Groundplan.

OLIGONEPHROUS Adj. (Greek, *oligos* = few; *nephros* = kidney; Latin, *-osus* = with the property of.) Descriptive of insects with few Malpighian Tubules. Alt. Oligonephric.

OLIGONEURA Noun. (Greek, *oligos*, = few; *neuron* = nerve.) Wings with few veins; a term specifically applied to Cecidomyiidae (Diptera).

OLIGONEURIIDAE Plural Noun. A small Family of Ephemeroptera. Nymphs found on sandy bottoms of rapidly flowing streams.

OLIGOPAUSE Noun. (Greek, *oligos* = few; *pausi* = ending. Pl., Oligopauses.) 1. Dormancy that is initiated and terminated by the same intensity of one environmental factor (*e.g.* photoperiod) (Müller 1965, Zool. Anz. 29 suppl: 182.) 2. An anticipated, typically long-term, cyclical interruption in growth or development of an organism due to one or more environmental factors and which occurs shortly before adverse environmental conditions are manifest (Manisingh 1971, Canad. Entomol. 103: 993.) See Aestivation; Dormancy; Hibernation. Cf. Diapause; Quiescence. Rel. Eudiapause; Parapause.

OLIGOPHAGOUS Adj. (Greek, *oligos* = few; *phagein* = to eat; Latin, *-osus* = with the property of.) Pertaining to an organism which exhibits oligophagy. Pertaining typically to an herbivorous organism which feeds on a few Species of food plants, a predator which feeds on a few Species of prey or a parasite which feeds on a few Species of host. By implication, the dietary items are closely related. Alt. Stenophagous. Cf. Euryphagous; Monophagous; Polyphagous.

OLIGOPHAGY Noun. (Greek, *oligos* = few; *phagein* = to eat.) 1. A limited dietary range. 2. A high level of specialization in food or a limited range of dietary requirements. Typically shown in an herbivorous organism which feeds on a few Species of food plant, a predator which feeds on a few Species of prey or a parasite which feeds on a few Species of hosts. Syn. Stenophagy. Cf. Monophagy, Polyphagy.

OLIGOPNEUSTIC Adj. (Greek, *oligos* = few; *pneustokos* = breath; *-ic* = consisting of.) Pertaining to a system of respiration characterized by 1–2 pairs of spiracles. OS subdivided into Amphipneustic (one mesothoracic and one postabdominal spiracles, as in psychodids) and Metapneustic (one postabdominal spiracle only, as in mosquito larvae.) Cf. Apneustic; Polypneustic; Propneustic.

OLIGOPOD Noun. (Greek, *oligos* = few; *pous* = feet. Pl., Oligopods.) 1. Animals with few appendages, legs or feet. 2. Insect larvae with thoracic limbs well developed and functional, but abdominal legs not developed. Cf. Protopod; Polypod.

OLIGOPOD LARVA 1. An active larval stage with well developed functional limbs. 2. Thysanuriform or campodeiform larvae of older texts (Folsom & Wardle).

OLIGOPOD PHASE Insect embryo: A phase in which the individual has reached an advanced condition of development (Imms).

OLIGOPYRENE Adj. (Greek, *oligos* = few; *pyren* = fruit stone.) Pertaining to Oligopyrene Sperm.

OLIGOPYRENE SPERM Spermatozoa with a reduced number of chromosomes. See Spermatozoa. Cf. Apyrene Sperm; Eupyrene Sperm; Hyperpyrene Sperm.

OLIGOTOKOUS Adj. (Greek, *oligos* = few; *tokous* = offspring; Latin, *-osus* = with the property of.) Descriptive of females that bear few offspring. Cf. Arrhenotoky; Thelytoky.

OLIGOTOMIDAE Plural Noun. Webspinners. A numerically large, cosmopolitan Family of Embiidina (Embioptera) characterized by Mandible with teeth, male left Cercus terminal segment not modified, winged males with R4+5 not forked.

OLIGOTROPHIC Adj. (Greek, *oligos* = little; *trophe* = nourishment; *-ic* = of the nature of.) Descriptive of a condition which has provided insufficient nutrition. Term often applied to aquatic habitats and the availabiity of nutrients. Cf. Eutrophic; Dystrophic.

OLIGOXENOUS Adj. (Greek, *oligos* = few; *xenos* = host; Latin, *-osus* = with the property of.) Descriptive of a parasite Species that has few Species of hosts. Alt. Stenophagous. Cf. Monoxenous; Pleioxenous; Polyxenous. See Parasitism.

OLIM Adj. (Latin.) Formerly.

OLIVACEOUS Adj. (Latin, *oliva* = olive; *-aceus* = of or pertaining to.) Descriptive of structure that is olive-green; the colour of green olives. Alt. Olivaceus. See Green. Cf. Aeruginous. Rel. Colour.

OLIVE FRUIT FLY *Dacus oleae* (Gmelin) [Diptera: Tephritidae]: A serious pest of olives around the Mediterranean basin. See Tephritidae. Cf. Apple Maggot; Cherry Fruit Fly; Mediterranean Fruit Fly; Melon Fly; Mexican Fruit Fly; Oriental Fruit Fly; Papaya Fruit Fly; Queensland Fruit Fly.

OLIVE KNOT A bacterial disease of olives caused by *Pseudomonas savastanoi* and transmitted by the Olive Fruit Fly. Disease transmitted through infected eggs of OFF and adult female's oviposition.

OLIVE LACE BUG *Froggattia olivinia* Froggatt [Hemiptera: Tingidae].

OLIVE MOTH *Prays oleae* (Fabricius) [Lepidoptera: Yponomeutidae]. A pest endemic to the Mediterranean region where the larva bores into olives. Cf. Citrus Rind Borer.

OLIVE PARLATORIA SCALE See Olive Scale.

OLIVE PSYLLA *Euphyllura phillyreae* Förster [Hemiptera: Aphalaridae].

OLIVE SCALE *Parlatoria oleae* (Colvée) [Hemiptera: Diaspididae]. A pest of olives in California but controlled by combined action of parasitic Hymenoptera *Aphytis paramaculcornis* DeBach & Rosen and *Coccophagoides utilis* Doutt. Alt. Olive Parlatoria Scale

OLIVE STOP® See NAA.

OLIVEIRA, EMMANUEL PAULINUS D' (1837–1899) (Anon. 1913, Broteria 11: 5–14, bibliogr.)

OLIVER, G B (1876–1966) (Anon. 1966, Entomologist's Rec. J. Var. 78: 271.)

OLIVER, WALTER REGINALD BROCK (1883–1957) (J.T.S. 1958, Proc. R. Soc. N.Z. 85: 60–64, bibliogr.)

OLIVI, GIUSEPPE (1769–1795) (Camerano 1905, Boll. Musei Zool. Anat. comp. R. Univ. Torino 20: 1–6.)

OLIVIER, ANTOINE GUILLAUME (1756–1814) French academic, naturalist and author of *Voyage dans l'Empire Ottoman, l'Egypte et al. Perse* (3 vols, 1807), Insectes [in: *Encyclopedie Methodique* (1789, 4 vols) and *Histoire Naturelle des Coleopteres* (6 vols, 1789–1808.)] (Rose 1850, *New General Biographical Dictionary* 10: 392–393.)

OLIVIER, JOSEPH ERNEST (1844–1914) (Pic 1914, Ann. Soc. ent. Fr. 83: 443–457, bibliogr.)

OLIVIERO, MANOEL LOPEZ DE FILHO (1870–1938) (Borgmeier 1938, Revta. Ent., Rio de J. 8: 441–442.)

OLLIVER, C W (–1963) (Anon. 1963, Bull. Soc. ent. Fr. 68: 9.)

OLSEN, CHRIS EMIL (1879–1965) (Ruckes 1965, J. N.Y. ent. Soc. 73: 243.)

OLSEN, EMIL (1857–1931) (Kryger 1932, Ent. Meddr. 18: 205–208.)

OLSEN, SVEND HEROLD (1909–1974) (Tuxen 1975, Ent. Meddr. 43: 115–117.)

OLSSON, AXEL (1888–1963) (Lindroth 1963, Opusc. ent. 28: 165–166.)

OMALOPTERA The pupiparous flies.

OMANIIDAE Cobben 1970. Intertidal Dwarf Bugs. A small Family of intertidal heteropterous Hemiptera resembling Saldidae; known from Indopacific and Red Sea. Adult small (1–2 mm long), dorsum convex, Vertex collar-like; Hemelytra shield-like without membrane; wings absent; Scutellum reduced; Abdomen short; Ovipositor plate-like. Tijdschr. Ent. 113: 61–90.

OMBONI, GIOCANNI (1829–1910) (Dal Piaz 1910, Boll. Soc. geol. ital. 39: xcvi–cvi, bibliogr.)

OMER-COOPER See J. Cooper.

OMETHOATE An organic-phosphate (thiophosphoric acid) compound {O,O-dimethyl S-(n-methylcarbamoylmethyl) phosphorothioate} used as a systemic insecticide and acaricide against mites, codling moth, gypsy moth, plant-sucking insects, thrips and other pests. Compound applied to ornamentals, forests, deciduous fruits, citrus, grapes, cereals, sugarcane, cotton, potatoes and vegetables in some countries. Not registered for use in USA. Compound toxic to bees. Trade name is Folimat®. See Organophosphorus Insecticide.

OMIA Noun. 1. The shoulders: Lateral anterior angles of an agglutinated Thorax when they are distinct. 2. Coleoptera: A corneous sclerite to which anterior coxal muscles are attached. 3. The lateral margin of the Prothorax or lateral margin of the Scutellum in carabids and dytiscids.

See Umbone.

OMITE® See Propargite.

OMMATA Noun. (Greek, *omma* = eye.) Individual facets or elements of the compound eye in some insect larvae (Tillyard).

OMMATEUM Noun. (Greek, *ommation* = little eye.) The compound eye.

OMMATIDAE Plural Noun. A small Family of archostematous Coleoptera. Taxa occur in Australia, South America and Italy. Adult body 5–25 mm long; Antenna filiform with 11 segments; fore Coxa globose with Trochantin expanded; frontal cavity open, middle Coxae contiguous; tarsal formula 5-5-5; Abdomen with five Ventrites.

OMMATIDIAL Adj. (Greek, *ommation* = little eye; Latin, *-alis* = pertaining to.) Descriptive of structure or process associated with the Ommatidia. See Ommatidium. Rel. Compound Eye.

OMMATIDIUM Noun. (Greek, *ommation* = little eye, *ommato* = to furnish with eyes; *-idion* = diminutive. Pl., Ommatidia.) The basic visual element which forms the compound eye. Elements of the Ommatidium include a Lens, Cone, Rhabdom, and Pigment Cells. The number of Ommatidia that form the compound eye varies considerably among Species. Some insects are eyeless; workers of some ants and fleas have one Ommatidium; *Drosophila* adults have about 700 Ommatidia; the cockroach *Periplaneta americana* and many dragonflies have about 2,000 Ommatidia in each compound eye; the compound eye of the bollworm *Helicoverpa armigera* (Hübner) contains ca 8,900 Ommatidia. Sizes and shapes of Ommatidia also vary considerably among Species. Dimensions of an Ommatidium typically vary from 17–22 μ in width and 70–125 μ in length. In some insects, Ommatidia in dorsal part of compound eye are larger than Ommatidia in ventral part of eye. In Diptera such as some Blephariceridae and Axymiidae, compound eyes are divided into dorsal and ventral parts. See Compound Eye. Cf. Cone; Facet; Lens; Pigment Cell; Rhabdom.

OMMOCHROMES Plural Noun. (Greek, *omma* = eye; *chroma* = colour.) Pigments derived from tryptophan and widely distributed in insects. Within the Epidermis, Ommochromes can form yellow, reds and browns. They are also associated with Pterines, particularly within the pigment cells of the compound eye. See Pigmentary Colour. Cf. Carotenoids; Flavinoids; Melanins; Pterines (Pteridines); Quinones. Rel. Structural Colour.

OMNAGENACERORIS Some coccids: A large U-shaped group of Genacerores evidently arising from a fusion of all five groups of Genacerores (MacGillivray).

OMNIVOROUS Adj. (Latin, *omnis* = all; *vorare* = to devour; Latin, *-osus* = with the property of.) A term descriptive of organisms that feed on any plant or animal. See Feeding Strategy. Cf. Carnivorous; Detritovorous.

OMNIVOROUS LEAF-ROLLER *Platynota stultana* Walshingham [Lepidoptera: Tortricidae]: Widespread in USA and Mexico. Eggs flat, overlapping laid in clusters. Larva damages several field crops, walnut, citrus and grapes. Biology similar to Orange Tortrix.

OMNIVOROUS LEAF-TIER *Cnephasia longana* (Haworth) [Lepidoptera: Tortricidae].

OMNIVOROUS LOOPER *Sabulodes aegrotata* (Guenée) [Lepidoptera: Geometridae].

OMNIVOROUS PINHOLE BORER *Crossotarsus omnivorous* Lea [Coleoptera: Curculionidae].

OMNIVOROUS TUSSOCK MOTH *Acyphas leucomelas* (Walker) [Lepidoptera: Lymantriidae].

OMPHALIAN SCENT GLAND Metathoracic Scent Gland of hydrocorisian Heteroptera (surface aquatic bugs, Dipsocoridae, Enicocephalidae, Omanniidae, Saldidae) usually characterized by a median orifice or occasionally by a closely spaced pair of median ventral orifices. OSG also characterized by three mechanisms of occlusion and different physiological function among Families possessing it (Carayon 1971). See Exocrine Gland. Cf. Diastomian Scent Gland. Rel. Scent Gland System.

OMPHALIUM Noun. (Greek, *omphalos* = navel; Latin, *-ium* = diminutive > Greek, *-idion*. Pl., Omphalia.) Gerridae and Veliidae: The elevated aperture of the median gland in the Metasternum; development variable among Species. See Omphalian Scent Gland.

OMPHRALIDAE See Scenopinidae.

OMSK HAEMORRHAGIC FEVER A tick-borne arboviral disease whose agent belongs to the Family Flaviviridae. OHF affects humans in western Siberia, typified by two cycles of infection. Disease naturally occurs in muskrats and vectored to humans by *Dermacentor pictus;* disease also transmitted through contaminated water and contact with infected animals. Acronym OHF. See Arbovirus; Flaviviridae. Cf. Kyasanur Forest Disease; Tick-bourne Encephalitis.

ONCHOCERCIASIS Noun. (Greek, *onkos* = tumour; *kerkos* = tail; *-iasis* = suffix for diseases.) An human disease primarily in tropical Africa and to a lesser extent in Central America, South America and Mexico; typically associated with life along swift-flowing streams. Disease caused by the filarial worm *Onchocera volvulus* (Leuckart) and transmitted only by simuliid flies: *Simulium damnosum* (West Africa), *S. neavei* (Kenya), *S. ochraceum, S. metallicum, S. callidum* (Central America.) Biology of disease agent resembles Bancroftian Filariasis. Microfilariae in skin of vertebrate ingested by blackfly with blood meal; Microfilariae develop to infective larvae in flight muscles of blackfly. Infective stage migrates to head, penetrates Labium and passed to new vertebrate host when blackfly feeds. Adult worms in subcutaneous tissue or encapsulated in nodules; large female ca

50 cm long and male slightly smaller. Disease manifest as subcutaneous nodules, dermatitis or eye lesions that cause blindness. Syn. River Blindness; Human Onchocerciasis; Robles' Disease (Guatemala); Sowda (Yemen). See Bovine Onchocercas. Equine Onchocercas.

ONCODIDAE Plural Noun. See Acroceridae.

ONCOGENE Noun. (Greek, *onkos* = tumour; *genos* = descent. Pl., Oncogenes.) A gene which is active a transformed cell. Many tumour-causing viruses carry one or several genetic loci known as oncogenes. These loci are responsible for neoplastic transformation of the host cell, one pathway by which viruses disturb the control of cell growth and division. Another pathway involves integration of viral DNA into the host genome (= insertional mutagenesis).

ONCOL® See Aminofuracarb.

ONCUS Noun. (Greek, *onkos* = tumour. Pl., Onci.) 1. A welt. 2. A term applied to welt-like ridges on caterpillars.

O'NEILL, WILLIAM JOHN (1900–1974) (Telford 1974, J. Econ. Ent. 67: 569)

ONE-SPOTTED STINK BUG *Euschistus variolarius* (Palisot de Beauvois) [Hemiptera: Pentatomidae].

ONIC® See Alanycarb.

ONION APHID *Neotoxoptera formosana* (Takahashi) [Hemiptera: Aphididae].

ONION BODY An organelle found in chemoreceptive Sensilla, consisting of several sets of double-membrane Cisternae bounded by a single membrane and with Ribosomes sometimes on the outer membrane.

ONION BULB FLY *Eumerus strigatus* (Fallén) [Diptera: Syrphidae]: A pest of onions endemic to Europe and adventive to North America. See Syrphidae. Cf. Narcissus Bulb Fly; Lesser Bulb Fly.

ONION FLY *Delia antiqua* (Meigen) [Diptera: Anthomyiidae]: Adult stage of onion maggot.

ONION MAGGOT *Delia antiqua* (Meigen) [Diptera: Anthomyiidae] larva: A sporadic pest of onions in northern USA and Canada. Overwinters as mature larva or Pupa in Puparia within soil; emerge during spring, mate and oviposit. Eggs elongate, white, deposited on base of onion plant or within soil; eclosion occurs ca 2–7 days. Neonate larvae crawl behind leaf sheath and penetrate bulb; larval development completed ca 15–20 days; Pupation in soil ca 15–20 days. Third generation damage causes onion rot in storage. See Anthomyiidae. Cf. Cabbage Maggot; Seed-Corn Maggot.

ONION PLANT BUG *Lindbergocapsus allii* (Knight) [Hemiptera: Miridae].

ONION THRIPS *Thrips tabaci* Lindeman [Thysanoptera: Thripidae]: A cosmopolitan pest of onions; OT also attacks cauliflower, cabbage, spinach, beets, turnips and cotton; occasional pest of greenhouse tomatoes and cucumbers. Feeding causes 'white blast' or 'silver top' while leaves become twisted or crinkled. OT can transmit spotted wilt from infected plants to greenhouse tomatoes. Adult 1.0–1.2 mm long, elongate-slender and yellow; dark brown patches occur on Thorax and Abdomen. Males rare and lack wings (apterous.) Females with four wings, narrow with a long fringe of Setae along margins; Tarsi lack claws but end with a small bladder. Adult female uses Ovipositor to saw into leaf tissue and deposits eggs in slit. Eggs laid in leaves or stems; eggs white and kidney-shaped; egg stage 5–10 days. Nymphs resemble the adults in appearance and habits. Nymphs lack wings and are paler than adults. Nymphs pass through four instars, two occur in soil and are non-feeding. All life stages occur on plant and overlap. Multivoltine with 5–10 generations per year. Adults and nymphs feed on the growing tips of young seedlings; damage causes underside of leaves to become silvery with dark droplets of excrement and leaf shape becomes malformed. Adults puncture leaves and stems to feed on exuding sap. Syn. Melon Thrips; Potato Thrips; Tobacco Thrips. See Thripidae. Cf. Greenhouse Thrips.

ONISCIFORM Adj. (*Oniscus* = generic name of isopod; *forma* = shape.) Shaped as a wood-louse of the genus *Oniscus*. Structure of body which is oval or elliptical in outline shape, convex above and flat or concave below. A term used to describe the shape of some caterpillars (Lycaenidae) and some beetle larvae (Byrrhidae). See Slaters. Rel. Form 2; Shape 2; Outline Shape.

ONISCIGASTRIDAE Plural Noun. A small Family of baetoid Ephemeroptera that includes a few Species in Australia.

ONO, MAGUSABURO (–1914) (Hasegawa 1967, Kontyû 35 (Suppl.): 26, bibliogr. pl. 1, fig. 7.)

ONSLOW, VICTOR A H HUIA (1890–1922) (Cockayne 1922, Entomologist's Rec. J. Var. 34: 148.)

ONTOGENETIC Adj. (Greek, *on* = being; *genesis* = descent; *-ic* = of the nature of.) Relating to or descriptive of the development of the individual.

ONTOGENY Noun. (Greek, *on* = being; *genesis* = descent. Pl., Ontogenies.) 1. The sequence of developmental events which collectively characterize an individual as distinguished from development of the Species. 2. The development of a colony of social insects from its inception through maturity to senescence. Syn. Ontogenesis. See Development. Cf. Phylogeny. Rel. Embryology.

ONTOLOGY Noun. (Greek, *on* = being; *logos* = discourse. Pl., Ontologies.) A branch of metaphysics concerned with the nature and relationships of beings.

ONUKI, SHINTARO (1869–1910) (Hasegawa 1967, Kontyû 35 (Suppl.): 25, bibliogr. pl. 3, fig. 2.)

ONYCHES Noun. (Greek, *onyx* = nail.) Claws. Typically referring to tarsal claws or the Pretarsus.

ONYCHII Noun. (Greek, *onyx* = nail.) The Pulvilli

(Comstock).

ONYCHIUM Noun. (Greek, *onyx* = nail; *idion* = diminutive. Pl., Onychia.) Diptera: A nail, claw or small processes between the Tarsal Claws. See Empodium. Coleoptera: A more-or-less retractile process on the Pretarsus. Hymenoptera: The apical Tarsal Segment bearing the claws. See Arolium; Pulvillus. 'Used in such a variety of ways that it appears impossible to define satisfactorily' (Holway). The Calcanea prolonged into a long slender spine-like projection (MacGillivray).

ONYCHIURIDAE Plural Noun. A small Family of arthropleonian Collembola. Body elongate, pale, delicate; head bearing Pseudocelli; thoracic and abdominal segments separate. One Species reproducing parthenogentically.

ONYCHOPHORA (Greek, *onyx* = nail; *phora* = producing.) A Phylum of about 80 Species most of which inhabit the Southern Hemisphere. Onychophorans are elongate, non-segmented, worm-like animals usually less than 6 cm long. The head bears three pairs of appendages: Antennae, jaws and Slime Papillae. The body bears 14–44 pairs of legs or lobopods which terminate in a claw with several spines. Onychoporans are predaceous and capture prey with a sticky substance emitted from the Papillae. Species are found in moist habitats such as rotting wood and leaf litter. The group includes *Peripatus*, the presumed connecting link between the Annelida and Arthropoda of older workers (Snodgrass). Viewed as an independent lineage by many contemporary taxonomists.

ONYCHOTRICHES Plural Noun. (Greek, *onyx* = nail; *trichos* = hair.) Acari: Minute, hair-like projections from the lateral surfaces of pretarsal claws and/or Empodium (Goff *et al.* 1982, J. Med. Ent. 19: 228.)

OOBLAST Noun. (Greek, *oon* = egg; *blastos* = bud. Pl., Ooblasts.) The primitive germinal nucleus of an egg.

OOCYST Noun. (Greek, *oon* = egg; *kystis* = bladder. Pl., Oocysts.) Sporozoa: A Cyst surrounding two conjugating gametes.

OOCYTE Noun. (Greek, *oon* = egg; *kytos* = hollow vessel; container. Pl., Oocytes.) A female gamete before formation of the first Polar Body. The egg cell differentiated from the Oogonium before maturation (Snodgrass).

OOGENESIS Noun. (Greek, *oon* = egg; *genesis* =descent. Pl., Oogeneses.) Egg-maturation; the formation of polar bodies in the egg. Oogenesis is the first phase of insect development; it begins with incorporation of proteins and nutrients within the Oocyte that resides in the Germarium of the Ovariole. Subsequently, an eggshell is fabricated. See Development. Cf. Embryogenesis; Metamorphosis.

OOGONIUM Noun. (Greek, *oon* = egg; *gonos* = begetting; Latin, *-ium* = diminutive > Greek, *-idion*. Pl., Oogonia.) The first stage in the differentiation of a egg cell from a primary female germ cell (Snodgrass). See Egg.

OOID Adj. (Greek, *oon* = egg; *eidos* = form.) Egg-like; oval. See Egg. Cf. Adeniform; Oval; Oviform; Ovoid. Rel. Eidos; Form; Shape.

OOKINESIS Noun. (Greek, *oon* = egg; *kinesis* = movement. Pl., Ookineses.) The mitotic stages of nuclear maturation and fertilization of an egg. See Kinesis. Cf. Allokinesis; Blastokinesis; Diakinesis; Klinokinesis; Orthokinesis; Thigmokinesis. Rel. Taxis; Tropism.

OOKINETE Noun. (Greek, *oon* = egg; *kinein* = to move. Pl., Ookinetes.) An elongate, worm-like mobile form produced by the fusion of the Macrogametocyte, with the thread-like daughter cells of the Microgametocyte in the life-cycle of *Plasmodium*.

OOLEMMA Noun. (Greek, *oon* = egg; *lemma* = husk. Pl., Oolemmae.) The cell wall of an egg. The Zona Pellucida. See Vitelline Membrane.

OOPHAGY Noun. (Greek, *oon* = egg; *phagein* = to devour. Pl., Oophagies.) 1. Egg eating. 2. Social Insects: The consumption of eggs by nestmates.

OOSPORE Noun. (Greek, *oon* = egg; *sporos* = seed. Pl., Oospores.) Fungi: Thick-walled resting spores formed during sexual reproduction in some Species.

OOTHECA Noun. (Greek, *oon* = egg; *theke* = case. Pl., Oothecae.) The covering or case over an egg mass, as in some Orthoptera and Blattaria. Syn. Ovicapsule. See Egg Case.

OOTHECAL MEMBRANE See Ootheca.

OOTHECAL PLATES Two chitinous sclerites at the external opening of the Cloacal Chamber, used by the female insect to cut the oothecal membrane at oviposition (Jardine).

OOZE Noun. (Anglo Saxon, *wos* = juice, sap. Pl., Oozes.) The soft mud or slime at the bottom of bodies of water.

OPACOUS Adj. (Latin, *opacus* = opaque; *-osus* = with the property of.) Arcane term meaning 'without surface lustre'; not transparent. Alt. Opaque. Cf. Translucent; Transparent.

OPALESCENT Adj. (Latin, *opalios* > Sanskrit, *upala* = precious stone.) 1. Descriptive of colour with a bluish white luster; not transparent. 2. Reflecting an iridescent colour. Syn. Iridescent.

OPALINE Adj. (Latin, *opalios* from Sanskrit *upala* = precious stone.) Opal-like; translucent and milky. Descriptiive of changeable colour as the opal. Alt. Opalinus; Opalizans.

OPAQUE Adj. (Middle English, *opake* > Latin, *opacus* = shady.) Descriptive of structure that is not transparent; not permitting light to pass through structure; not shining or reflecting light.

OPEN CELL A wing 'cell' or area bounded by veins which extends to the margin of the wing (Comstock). Cf. Closed Cell.

OPEN TRACHEAL SYSTEM The anatomical type of Respiratory System in which Spiracles are open, exposed and functional. OTS seen in terrestrial insects and some aquatic insect Species. In OTS oxygen enters the Tracheal System through Spiracles. Various modifications of the OTS are adaptations to conserve water in xeric-adapted terrestrial insects or prevent flooding in aquatic insects. See Spiracle; Tracheae; Tracheole. Cf. Closed Tracheal System. Rel. Respiratory Muscles; Respiratory System.

OPERANT BEHAVIOUR Behaviour tending to produce effects. Behaviour acting on the environment to produce reinforcing effects.

OPERARIA Noun. (Latin, *operari* = operate.) Workers in eusocial Hymenoptera.

OPERATIONAL HOMOLOGY Features which 'are very much alike in general and in particular'. (Sokal & Sneath 1973: 70.) See Homology.

OPERCULARIA In Species of coccids deeply imbedded in wax, the deeply chitinized caudal part of the body prolonged into a horn-or handle-like structure with the anal cleft located at the caudal end of the horn (MacGillivray).

OPERCULATE Adj. (Latin, *operculum* = lid; *-atus* = adjectival suffix.) Pertaining to a lid-like covering. See Operculiform.

OPERCULIFORM Adj. (Latin, *operculum* = lid; *forma* = shape.) Operculum-shaped; structure with the shape or form of a lid or cover. Syn. Operculate. See Operculum 1. Cf. Disciform. Rel. Form 2; Shape 2; Outline Shape.

OPERCULUM Noun. (Latin, *operculum* = lid. Pl., Opercula; Operculums.) 1. A lid (cover) implicitly circular or oval in outline shape and flat or convex. 2. The lid-like portion of an insect eggshell (typically over the anterior pole) which partly separates and elevates from eggshell 'body' to permit escape of a Nymph, Naiad or Larva during eclosion. An Operculum has been noted in many groups of insects, including Phasmida, Embiidina, Mallophaga and Anoplura; it is probably best studied among the Hemiptera. Southwood (1956) defined Operculum for Hemiptera as separated from the remainder of the Chorion by a 'sealing bar.' Hemiptera eggs 'glued' on vegetation typically display an Operculum over the anterior pole. Operculum of some Pentatomidae may not be functional and eggshell Chorion splits independent of the Operculum. The Operculum typically remains attached to remainder of Eggshell at one point, but is not hinged. There appears to be no point for tearing action. Physical features are not always employed or sufficient to facilitate emergence, and sometimes other methods are used by the Larva or Nymph to leave its Eggshell. See Capitulum; Egg; Eggshell; Cf. Eclosion. 3. Aleyrodidae: The lid-like structure covering the Vasiform Orifice. 4. Cicadidae: One of the paired sclerites covering the Timbals. 5. Lecaniine coccids: The anal sclerites taken together or singly (MacGillivray). 6. Diptera: The chitinous lower part of muscid mouth; the Labrum-Ephipharynx of Dimmock; the Scutes covering the Metathoracic Stigmata.

OPERCULUM GENITALIS Hymenoptera: Sterna VIII and IX.

OPERE CITATO Latin phrase meaning 'work cited.' Abbreviated *op. cit., op. c.* and italicized.

OPHTHALMIC Adj. (Greek, *ophthalmos* = eyes; *-ic* = of the nature of.) Relating to the eye.

OPHTHALMOTHECA Noun. (Greek, *ophthalmos* = eyes, *theke* = case.) The eye-case or part of the Pupa that covers the eyes.

OPILIOACARIDA Plural Noun. One of four groups of anactinotrichid mites.

OPISTHOGNATHOUS Adj. (Greek, *opisthe* = behind; *gnathos* = jaw; Latin, *-osus* = with the property of.) 1. General zoological useage, pertaining to animals with retreating jaws. 2. Entomological useage, pertaining to one of the three principal orientations of the insect cranium in relation to the body. Characterized by a posteroventral position of the mouthparts resulting from a deflection of the facial region. Seen in most Homoptera (Snodgrass). See Head. Cf. Hypognathous; Prognathous.

OPISTHOGONEATE Adj. (Greek, *opisthe* = behind; *gonia* = angle; *-atus* = adjectival suffix.) Myriapoda: One of a group in which the genital opening is at the posterior end of the body. Cf. Progoneate.

OPISTHOGONIA Noun. (Greek, *opisthe* = behind; *gonia* = angle.) The anal angle of the hindwings.

OPISTHOMASTIGOTE Noun. (Greek, *opisthe* = behind; mastigote = protozoa with Flagella.) An anatomical form manifest at a specific phase in the complex life cycle of some parasitic Protozoa (*Leishmania* and *Trypanosoma*). Opisthomastigote displays Flagellum base posterior of nucleus with long flagellar pocket projecting to anterior end of cell. See Leishmania. Trypanosoma. Cf. Amastigote; Promastigote; Epimastigote; Trypomastigote.

OPISTHOMERE Noun. (Greek, *opisthe* = behind; *meros* = part. Pl., Opisthomeres.) The collective name applied to the three segments of the Supra-anal Sclerite in Dermaptera (Tillyard).

OPISTHOSOMA Noun. (Greek, *opisthe* = behind; *soma* = body. Pl., Opisthosomata.) Acarology: A division of the body posterior of the Disjugal Furrow. See Idiosoma.

OPIZ, PHILIPP MAXIMILIAN (1787–1858) (Dvovsky 1858, Lotus 8: 152–158.)

OPOMYZIDAE Plural Noun. A numerically small, Holarctic Family of schizophorous Diptera assigned to Superfamily Opomyzoidea. Adult small to minute-bodied; ocellar triangle small; Oral Vibrissae present; wing basally constricted; Costa broken, not spinose; Anal Cell present. Flies occur among grasses where their larvae feed on the stems. See Opomydoidea.

OPOMYZOIDEA A Superfamily of schizophorous Diptera including Acartophthalmidae, Clusiidae (Clusiodidae, Heteroneuridae), Fergusonidae, Lonchaeidae, Neottiophilidae, Odiniidae (Odinidaea), Opomyzidae, Pallopteridae and Piophilidae. See Diptera.

OPOSTEGIDAE Plural Noun. A small, widespread

Family of nepticuloid Lepidoptera, including less than 100 Species. Adult small bodied; wingspan 4–16 mm; head with long hair-like smooth Setae between Antennae; Antenna shorter than forewing length, forming eye-caps; Ocelli absent; Proboscis short and probably not functional; Maxillary Palpus with five segments and folded; Labial Palpus short and pendulous; fore Tibia without Epiphysis and spur; forewing lanceolate, venation reduced, with 4–5 unbranched veins; hindwing marginal fringe very long. Larva apodous; body long and slender; head prognathous, flattened, wedge-like; one pair of stemmata. Pupal Antenna nearly covering compound eye. Biology poorly known; larvae of some Species apparently leaf miners and a few mine bark or cambium.

OPPERMANN, HENRY (–1974) (Hehl 1974, Prakt. SchädlBekämpf. 26: 22.)

OPTEM® See Cyfluthrin.

OPTIC Adj. (Greek, *optikos* = sight; *-ic* = characterized by.) Relating to vision.

OPTIC CENTRES The Brain centres of the compound eye, positioned in the optic lobes (Snodgrass).

OPTIC DISC Muscid larvae: A disc-like thickening near the Brain in each of the so-called Brain appendages. A Histoblast which develops into the compound eye of the adult (Comstock).

OPTIC GANGLIA See Optic Lobes.

OPTIC LOBES Lateral lobes of Protocerebrum that contain nerves supplying organs of vision (eyes). OLs develop as lateral extensions of Protocerebrum and form ganglionic centres of compound eyes. Size of OL is closely associated with development of compound eyes. Insects which display large and well developed compound eyes, have large optic lobes. Each ganglion is triangular in cross section, wide beneath eye and narrow at its junction with Brain. Composition of OL appears relatively constant throughout Insecta. Medulla Interna represents a large Neuropile farthest from compound eye and separated from Medulla Externa by an Inner Chiastma. Medulla Externa is a neuropile between Medulla Interna and outermost layer (Lamina Ganglionaris). Lamina Ganglionaris is separated from Medulla Externa by Outer Chiastma; Lamina Ganglionaris is part of optic lobe nearest compound eye and receives Axons from Retinula Cells (termed post-retinal fibres). Axons within Lamina Ganglionaris synapse with monopolar neurons. Syn. Optic Ganglia; Optic Tract. See Protocerebrum. Cf. Lamina Ganglionaris; Medulla Externa; Medulla Interna; Pars Intercerebralis; Protocerebral Bridge. Rel. Central Nervous System; Nervous System; Nerve Cell.

OPTIC SEGMENT Segment of head which contains Protocerebrum including Optic Lobes.

OPTIC TRACT See Optic Lobes.

OPTICON Noun. The proximal part of the Optic Lobe in the Protocerebrum of the insect Brain. Syn. Medulla Interna.

OPTIMIZER® See Diazinon.

OPTIMOL® See Metaldehyde.

OPTIMUM Noun. (Latin, *optimus* = best.) The most favourable condition or conditions for the survival, growth and development of an organism. Optimal conditions involve all biotic factors in the environment, including light intensity, photoperiod, heat, temperature, cumulative temperature, moisture, relative humidity, food, shelter and habitat.

ORA Noun. 1. A border, specifically in some Coleoptera the wide lateral margin of the Prothorax. 2. Plural of Os.

ORA COLEOPTERORUM The margin of the Elytra in beetles.

ORACERATUBAE Coccids: The external opening of a Ceratuba, located in the external Cuticle (MacGillivray).

ORACERORIS Coccids: The opening of the Ceroris through which the wax is poured (MacGillivray).

ORACONARIS The mouth cone in the Physapoda (MacGillivray).

ORAD Adv. (Latin, *os* = mouth; *ad* = toward.) Toward the mouth. See Orientation. Cf. Anteriad; Apicad; Basad; Caudad; Centrad; Cephalad; Craniad; Dextrad; Dextrocaudad; Dextrocephalad; Distad; Dorsad; Ectad; Entad; Laterad; Mediad; Mesad; Neurad; Proximad; Rostrad; Sinistrad; Sinistrocaudad; Sinistrocephalad; Ventrad.

ORAL Adj. (Latin, *os* = mouth; *-alis* = pertaining to.) Descriptive of structure or process associated with the mouth or any structure connected with the mouth. See Foregut. Cf. Anal.

ORAL CAVITY The mouth; the buccal cavity.

ORAL FOSSA Mallophaga: A furrow positioned anterior of the Mandibles.

ORAL HOOKS Larval Diptera: See Mouth Hooks.

ORAL SEGMENT The ring or segment which bears the mouth.

ORAL SUCKER Diptera: A broad disc formed by the large soft pads of the Labella when spread outward from the end of the stipital stalk (Snodgrass).

ORANGE AEROPLANE *Pantoporia consimilis consimilis* (Boisduval) [Lepidoptera].

ORANGE BLACKBOY *Gudanga browni* (Distant) [Hemiptera: Cicadidae]

ORANGE BUSHBROWN *Mycalesis terminus terminus* (Fabricius) [Lepidoptera: Nymphalidae].

ORANGE CATERPILLAR PARASITE *Netelia producta* (Brullé) [Hymenoptera: Ichneumonidae]: An Australian Species. Adult wingspan 6–20 mm; body elongate and reddish-orange; head reddish-orange with Antenna longer than head and Thorax. Female Ovipositor short and can sting to inflict pain. Eggs laid within the host. Eggs with a long, curled tube that enables gaseous exchange with the external environment of the host. Larvae live within host feeding on organs.

Pupation within host. Cocoons black, ovoid with black fuzzy strands surrounding them. Adults do not emerge until after host has pupated. Orange caterpillar parasites use larval Lepidoptera as their host.

ORANGE DART *Suniana sunias rectivitta* (Mabille) [Lepidoptera: Hesperiidae].

ORANGE-DOG *Papilio cresphontes* Cramer [Lepidoptera: Papilionidae]: The largest butterfly in North America and an occasional pest of citrus in Florida. Overwinters as Pupa; 2–3 generations per year. Host plants include prickly ash, citrus, hop tree. Alt. Giant Swallowtail. Caterpillar: Orange Puppy.

ORANGE FAIRY *Cicadetta sulcata* (Distant) [Hemiptera: Cicadidae].

ORANGE FRUITBORER *Isotenes miserana* (Walker) [Lepidoptera: Tortricidae]: A multivoltine, minor but sporadically serious pest, especially on Washington navel in coastal Australia. Feeds on all varieties of citrus, *Cupressus* and new growth of camphor laurel and fruits. Feeding on citrus foliage is uncommon. Eggs pale green laid in clusters on leaves or fruit; eggs overlap like fish scales. Larvae bore into fruit or calyx. Moths nocturnal, fly weakly from tree-to-tree. Larval stage can require several months. Pupate in dead rolled leaf, mass of flower debris or webbed foliage. Pupal stage 2–3 weeks. See Tortricidae.

ORANGE GRASSDART *Taractrocera anisomorpha* (Lower) [Lepidoptera: Hesperiidae].

ORANGE-HUMPED MAPLEWORM *Symmerista leucitys* Franclemont [Lepidoptera: Notodontidae].

ORANGE JEZABEL *Delias aruna inferna* Butler [Lepidoptera: Pieridae].

ORANGE LACEWING *Cethosia penthesilea paksha* Fruhstorfer [Lepidoptera: Nymphalidae].

ORANGE MIGRANT *Catopsilia scylla etesia* (Hewitson) [Lepidoptera: Pieridae].

ORANGE PALMDART *Cephrenes augiades sperthias* (Felder) [Lepidoptera: Hesperiidae].

ORANGE PUPPY See Orange-Dog.

ORANGE RINGLET *Hypocysta adiante* (Hübner) [Lepidoptera: Nymphalidae].

ORANGE SPINY WHITEFLY *Aleurocanthus spiniferus* (Quaintance) [Hemiptera: Aleyrodidae]: A widespread pest of citrus which originated in the Orient. Attempts at biological control with parasitic Hymenoptera have witnessed variable results.

ORANGE-STRIPED OAKWORM *Anisota senatoria* (J. E. Smith) [Lepidoptera: Saturniidae].

ORANGE-TAILED POTTER WASP *Delta latreillei petiolaris* (Schulz) [Hymenoptera: Vespidae].

ORANGE TIGER BUTTERFLY *Danaus affinis alexis* (Waterhouse & Lyell) [Lepidoptera: Nymphalidae].

ORANGE TIPS See Pieridae.

ORANGE TORTRIX *Argyrotaenia citrana* (Fernald) [Lepidoptera: Tortricidae]: Widespread in North America; 2–4 generations per year; overwinters as larva; pupates in nest. Eggs laid in shingle-like masses; larva nests among buds and blossoms; larva feeds on leaves, buds and blossoms. Pest of apples, apricots, citrus, grapes, pears, prunes. Larva feeding on grape clusters create conditions favourable for rot. Cf. Omnivorous leaf-roller.

ORANGE WHITE-SPOT SKIPPER *Trapezites heteromacula* Meyrick & Lower [Lepidoptera: Hesperiidae].

ORANGEFOOTED CENTIPEDE *Cormocephalus aurantiipes* (Newport) [Scolopendromorpha: Scolopendridae].

ORB (Latin, *orbis* = circle.) A globe; a circle.

ORBACERORES Noun. Coccids: The Cereores surrounding the anal ring (MacGillivray).

ORBICULA Noun. Hymenoptera: A small, dorsal sclerite at the base of the Arolium and distad of the Unguifer. The chitinized area on the dorsal aspect of the Articularis (*sensu* MacGillivray).

ORBICULAR Adj. (Latin, *orbis* = orb.) Circular in outline, flat with a central attachment. Alt. Orbicularis; Orbiculate.

ORBICULAR SPOT A circular, round or oval spot in the Discal Cell in the forewing of some noctuid moths (Comstock).

ORBICULATE Adj. (Latin, *orbicultus* = rounded; *-atus* = adjectival suffix.) Orb-shaped; pertaining to structures orb-shaped or circular in outline. See Outline Shape.

ORBIGNY, ALCIDES DESSALINES D' (1802–1857) (Papavero 1971, *Essays on the History of Neotropical Dipterology.* 1: 136–144. São Paulo.)

ORBIGNY, CHARLES DESSALINES D' (1806–1876) (Musgrave 1932, *A Bibliogr. of Australian Entomology 1775–1930,* viii + 380 pp. (247), Sydney.)

ORBIGNY, HENRI D' (1845–1915) (Wheeler 1933, Q. Rev. Biol. 8: 325–330.)

ORBIT Noun. (Latin, *orbis* = eye socket. Pl., Orbits.) An imaginary border around the insect eye, the narrow part of the Vertex adjacent to the margin of the compound eye (MacGillivray).

ORBITAL Adj. (Latin, *orbis* = eye socket; *-alis* = characterized by.) Descriptive of structure or substance associated with an orbit, specifically, the orbit of the compound eye.

ORBITAL BRISTLES Diptera: Coarse Setae or bristles, usually proclinate or divergent, positioned on the Parafrontals between the Frontals and Orbits.

ORBITAL SCLERITE A narrow sclerite encircling the compound eyes.

ORBWEAVER See Araneidae.

ORB-WEAVING SPIDERS See Araneidae.

ORCEPHALIC With a hypognathous head having the Foramen divided into two parts by other structures (MacGillivray).

ORCHARD BUTTERFLY *Princeps aegus* (Donovan) [Lepidoptera: Papilionidae].

ORCHEX® See Petroleum Oils.

ORCHID APHID 1. *Cerataphis orchidearum* (Westwood) [Hemiptera: Aphididae]. See Yellow Orchid Aphid. 2. *Macrosiphum luteum* (Buckton) [Hemiptera: Aphididae].

ORCHID BEETLE *Stethopachys formosa* Baly [Coleoptera: Chrysomelidae].

ORCHID DUPE *Lissopimpla excelsa* (Costa) [Hymenoptera: Ichneumonidae]: Adult wingspan 7–18 mm; body elongate, reddish-brown; head pale with Antenna longer than Abdomen. Wings transparent with brownish coloration; Abdomen polished black with yellow spots on first four segments. Female Ovipositor very long; tarsal claws on legs enlarged. Females fly over grass probing for host stage of Lepidoptera. Males pollinate orchids. The orchid *Cryptostylis leptochila* is a wasp mimic and attracts male wasp. Males attempt to mate with flower and thus transfer pollen. Eggs laid within host. Larva lives within host feeding on organs. Pupation occurs within host. A beneficial parasite of lawn armyworm in Australia. See Ichneumonidae.

ORCHID PARLATORIA SCALE *Parlatoria proteus* (Curtis) [Hemiptera: Diaspididae].

ORCHID SCALE *Diaspis boisduvalii* Signoret [Hemiptera: Diaspididae].

ORCHID THRIPS *Chaetanaphothrips orchidii* (Moulton) [Thysanoptera: Thripidae].

ORCHID WEEVIL *Orchidophilus aterrimus* (Waterhouse) [Coleoptera: Curculionidae].

ORCHIDFLY *Eurytoma orchidearum* (Westwood) [Hymenoptera: Eurytomidae]: A cosmopolitan pest of orchids in greenhouses. Larva legless, feeds on bulbs, stems, leaves and buds. Syn. Cattleyafly.

ORCHYMONT, ARMAND D' (1881–1847) (Collart 1950, Bull. Inst. r. Sci. nat. Belg. 26 (37): 1–20, bibliogr.)

ORD, GEORGE (1781–1899) (Rhoads 1908, Cassinia 12: 1–8)

ORDER Noun. (Latin, *ordo* = order. Pl., Orders.) 1. One of the primary taxonomic divisions of the Class Insecta. The Order was originally based largely on wing structure with ordinal names usually ending in *-ptera*. Most of the currently accepted ordinal names were proposed by Linnaeus in *Systema Naturae*. Shipley (1903) proposed modifications to some Order names to add *-ptera* (Siphonaptera, Dermaptera). 2. Order is energy (Myerhoff 1916). 3. The organization of physical matter based on an accepted group of principles. Order is omnipresent and fundamental to morphological groundplans. Riedl (1978) recognized several patterns of Order, including a Standard Part Order, Hierarchical Order, Interdependent Order and Traditive Order. The functional unit of Morphology is the standard part. See Anatomy; Morphology; Organism; Organization.

ORDINAL Adj. (Latin, *ordo* = order; *-alis* = characterized by.) Descriptive of taxonomic status relating to an Order.

ORDINARY CROSSVEIN Diptera: Anterior or small crossvein; r-m of Schiner *teste* Curran. See Vein; Wing Venation.

ORDINATE Adj. (Latin, *ordinatus* = arranged; *-atus* = adjectival suffix.) With markings arranged in rows. Arranged in rows, as spots, punctures, *etc.*

ORDOVICIAN PERIOD The second Period (510–439 MYBP) of the Palaeozoic Era; named for *Ordovices* (a Celtic tribe) in the part of Wales where OP was first investigated. OP noted for adaptive radiation of seaweeds (red, green, and other algae), many Classes and Orders of marine animals (graptolites, conodonts and articulate brachiopods); first vertebrates appear (Ostracoderms). Late Ordovician stratigraphy indicates that most skeletonized animals lived on surface of sediment. Late OP sees mass extinction of brachiopods, bryozoans and elimination of 100 Families of marine animals. See Geological Time Scale; Palaeozoic Era. Cf. Cambrian Period; Devonian Period; Lower Carboniferous Period; Upper Carboniferous Period; Permian Period; Silurian Period. Rel. Fossil.

ORDURE Noun. (Old French, *ord* = filthy from Latin, *horridus* = horrid. Pl., Ordures.) Excrement: The term is usually applied in a negative context, such as foul or offensive.

OREGON FIR SAWYER *Monochamus scutellatus oregonensis* (LeConte) [Coleoptera: Cerambycidae].

OREGON WIREWORM *Melanotus longulus oregonensis* (LeConte) [Coleoptera: Elateridae].

ORESTE® Pyridafenthion.

ORFILA, RICARDO NESTOR (1909–1967) (Pastrina 1968, Revta. Soc. ent. Argent. 30: 131–132.)

ORGAN Noun. (Old English, *organa,* Latin, *organum,* Greek, *organon* = tool, instrument. Pl., Organs.) A differentiated structure of plants or animals, composed of various cells and tissue, that is adapted for particular function or functions. Organs are grouped to form organ systems. Cf. Gland; Parenchyma; Tissue.

ORGAN CULTURE Tissue Culture: The *in vitro* maintenance or growth of an organ or tissue. Cell differentiation and tissue architecture are preserved through Organ Culture.

ORGAN OF BERLESE *Cimex: A* small, usually unpaired, rounded body, which lies on the right side of the Abdomen in close association with a small longitudinal incision on the fourth Sternum, it functions as a copulatory pouch to receive Spermatozoa (Imms). Cf. Organ of Ribaga.

ORGAN OF HICKS A Campaniform Sensillum. See Sensillum Campaniformium.

ORGAN OF JOHNSTON See Johnston's Organ.

ORGAN OF RIBAGA. Lice: The copulatory cavity between Sterna 5–6 as a notch to right of midline. Cf. Organ of Berlese.

ORGAN OF TOMOSVARY Collembola: See Postantennal Organ.

ORGAN OF VOM RATH Lepidoptera: A flask-shaped or pit-shaped invagination on the 3rd segment of the Labial Palpus. Structure first observed by Rath (1887, Zool. Anz. 10: 627–631, 645–649) and subsequently named in his honour. The invagination contains Sensilla and is sensitive to carbon dioxide. The structure is regarded as unique (autapomorphic character) to Lepidoptera and diagnostic for the Order. Cf. Hicks Bottle.

ORGANELLE Noun. (Greek, *organon* = instrument. Pl., Organelles.) The cellular inclusions which have specific functions (*e.g.* Golgi Apparatus, Mitochondria). See Cell.

ORGANIC Adj. (Latin, *organicus,* Greek, *organikos* = instrument, organ; *-ic* = of the nature of.) 1. Pertaining to an organ or system of organs in a plant or animal. 2. Pertaining to matter derived from living organisms. 3. Demonstrating characteristics or features of living organisms. 4. Descriptive of plants cultivated without the use of synthetic pesticides.

ORGANISM Noun. (Greek, *organon* = instrument; English, *-ism* = state. Pl., Organisms.) 1. A self-contained, organic entity capable of sustaining life processes (growth, differentiation, reproduction). 2. An animal or plant. Cf. Virus; Viroid. 3. A mosaic of anatomical features in various stages of complexity. Anatomical features are called 'characters' and their stages of expression are called 'character states.' Some anatomical features reflect the so-called primitive condition or are early in a transformational series. Contemporary evolutionary taxonomists and morphologists call primitive conditions 'plesiomorphic.' Anatomical features which are more modified or elaborate reflect a derived condition which may be advanced in a transformation series. These features are called 'apomorphic.' A complete fossil record for any group of organisms would clearly show the process of evolutionary change (transformation) applied to Anatomy (Morphology). Categorizing organisms as primitive (plesiomorpic) or derived (apomorphic) is erroneous because the anatomical features (characters) that comprise an organism become modified at different rates. See Transformation Series.

ORGANIZATION Noun. See Asymmetry; Symmetry.

ORGANOCHLORINE INSECTICIDE A group of synthetic insecticides whose members vary in mammalian toxicity from moderate to high and which are stable in the soil over long periods of time. Effective against structural urban pests such as termites. Long-term detrimental effects on environment discourage general application. Symptoms of human poisoning include diarrhoea, vomiting, numbness of hands and feet, restlessness, excitability and muscular twitching. Treatment includes: If conscious and poison ingested, then induce vomiting; milk and oil should not be ingested. If skin contamination, then remove clothing, wash skin thoroughly with soap and water. Consult medical advice rapidly. Examples include: Bromopropylate, Chlorobenzilate, Chlorpropylate, DDT, Dicofol, Diflubenzuron, Dymet, Etofenprox, Methoxychlor, Tetradifon. See Insecticide. Cf. Botanical Insecticide; Carbamate; Inorganic Insecticide; Insect Growth Regulator; Organophosphorus Insecticide; Synthetic Pyrethroid.

ORGANOGENESIS Noun. (Greek, *organon* = organ; *genesis* = descent. Pl., Organogeneses.) Embryology: The formation of three germ layers and subsequent formation of different tissues and organs within the embryo. Syn. Organogeny.

ORGANOGENY See Organogenesis.

ORGANOPHOSPHATE Noun. An organic compound containing phosphorus; three main derivatives are based on the oxygen, carbon, sulphur and nitrogen make-up of the ester functional group: 1. Aliphatic, 2. Heterocyclic, and 3. Phenyl.

ORGANOPHOSPHORUS INSECTICIDE A group of synthetic insecticides derived from phosphoric acid and first discovered by Schrader (1937); OPs also known from cultures of *Streptomyces antibioticus* DSM 1951. OPs are chemically unstable compounds used as surface sprays that are effective for days to a few months. OPs function as nerve poisons which inhibit cholinesterase activity. Generic compounds include: Acephate, Azamethiphos, Azinphos-Ethyl, Azinphos-Methyl, Cadusafos, Chlorethoxyfos, Chlorfenvinphos, Chlormephos, Chlorpyrifos, Chlorpyrifos-Methyl, Coumaphos, Cyanophos, Diazinon, Dichlorvos, Dicrotophos, Dimethoate, Disulfoton, EPN, Ethion, Ethoprop, Etrimfos, Famphur, Fenamiphos, Fenitrothion, Fenthion, Fenitrothion, Fonofos, Formothion, Fosthiazate, Isazofos, Isofenphos, Isoxathion, Malathion, Maldison, Mecarbam, Mephosfolan, Methamidophos, Methidathion, Mevinphos, Monocrotophos, Naled, Methoate, Oxydemeton-Methyl, Parathion, Phenthoate, Phosalone, Phorate, Phosdiphen, Phosmet, Phosphamidon, Phoxim, Pirimiphos-Ethyl, Pirimiphos-Methyl, Profenofos, Propaphos, Propetamphos, Prothoate, Prothiofos, Pyraclofos, Pyridafenthion, Quinalphos, Sulprofos, Tebupirimfos, Temephos, Terbufos, Tetrachlorvinphos, Thiometon, Triazophos, Trichlorfon, Vamidothion. Members vary in mammalian toxicity from moderate to very high. Symptoms of human poisoning include headache, fatigue, giddiness, salivation, sweating, blurred vision and pin-point pupils, tightness of chest, difficulty in breathing, rapid heartbeat, nausea, convulsions and paralysis. Treatment includes: if swallowed then induce vomiting; if skin contaminated then remove clothes and wash thoroughly with soap and water; if respiration has failed then attempt resuscitation. Seek medical advice rapidly. See Insecticide; Pesticide. Cf. Botanical Insecticide; Carbamate; Inorganic Insecticide; Insect Growth Regulator; Organochlorine Insecticide; Synthetic Pyrethroid; Systemic Insecticide.

ORGAN-PIPE MUD-DAUBER *Trypoxylon politum* (Say) [Hymenoptera: Sphecidae]: A North American, solitary wasp that builds elongate, usually vertical, mud tubes on exposed surfaces or plant stems. Female provisions cells of tube with live, paralysed spiders; egg deposited on spider. Wasps black bodied or black with red markings.

ORIBATIDA Plural Noun. A group of Actinotrichid mites.

ORICHALCEOUS Adj. (Greek, *oros* = mountain; *chalkos* = brass; Latin, *-aceus* = of or pertaining to.) A colour between gold and brass; aurichalceous, aurichalceus. Alt. Orichalceus.

ORICHORA BROWN *Oreixenica orichora* (Meyrick) [Lepidoptera: Nymphalidae].

ORIENT Noun. (Latin, *oriri* = to rise. Pl., Orients.) To find, or set in, a given direction.

ORIENTAL ARMYWORM *Pseudaletia separata* (Walker) [Lepidoptera: Noctuidae]. See Armyworms.

ORIENTAL BEETLE *Anomala orientalis* Waterhouse [Coleoptera: Scarabaeidae].

ORIENTAL COCKROACH *Blatta orientalis* Linnaeus [Blattodea: Blattidae]: A cosmopolitan urban pest; possibly endemic to Africa or south Russia. Adult dark brown to black, 20–27 mm long; female micropterous; male wings extending nearly to apex of Abdomen but does not fly. Adult lives 3–6 months; female lays 8–14 Oothecae; Oothecae contains 12–16 eggs; Oothecae usually glued to substrate; nymphs undergo 7–10 moults; nymphal development 6–18 months. Most common urban cockroach in Britain; prefers cooler areas; found at or below ground level in buildings. Nymphs and adults feed on decaying organic matter around garbage disposal areas; also feed on starch and fermented food. Syn. Black Beetle. Waterbug. See Blattodea. Cf. American Cockroach.

ORIENTAL COCKROACHES See Blattidae.

ORIENTAL COWPEA BRUCHID *Callosbruchus chinensis* (Linnaeus) [Coleoptera: Bruchidae]: A widespread pest of pulse crops; endemic to Asia by now found in warmer regions of world. Anatomy and biology very similar to Spotted Cowpea Bruchid. See Pulse Beetle. Cf. Spotted Cowpea Bruchid.

ORIENTAL FRUIT FLY *Dacus dorsalis* Hendel [Diptera: Tephritidae]: A Species complex containing up to 65 Species widely distributed through the African and Australasian realms. OFF significant pests of fruits throughout tropical and subtropical areas. Control by means of introduced natural enemies has had mixed success, probably owing to the complex taxonomy. See Tephritidae. Cf. Apple Maggot; Cherry Fruit Fly; Mediterranean Fruit Fly; Melon Fly; Mexican Fruit Fly; Oriental Fruit Fly; Papaya Fruit Fly; Queensland Fruit Fly.

ORIENTAL FRUIT MOTH *Grapholita molesta* (Busck) [Lepidoptera: Tortricidae]: A polyphagous pest of apple, apricot, peach, plum, pear and quince; a major pest of late-season peaches and nectarines. Economic damage caused by larvae tunnelling in fruit and young shoots. OFM is native to northwest China and now distributed through stonefruit growing areas including Europe, North and South America, North Africa, Middle East, New Zealand and Australia. OFM established in USA ca 1913 via imported nursery stock. Larva pinkish or cream-coloured with brown head; overwinters as mature larva within cocoon on bark of tree or rubbish on ground. Pupation occurs during spring for first generation; other generations follow as a cocoon on bark or on ground. Adults emerge during early spring and oviposit on young foliage. Eggs white, laid on underside of leaves near tip of twigs. Early generations bore into tender twigs and cause dieback; later generations penetrate fruit, give no sign of entry. Multivoltine with 6–7 generations per year. See Tortricidae. Cf. Codling Moth; Peach Tree Moth.

ORIENTAL GRASSROOT APHID *Tetraneura nigriabdominalis* (Sasaki) [Hemiptera: Aphididae].

ORIENTAL HOUSE FLY *Musca domestica vicina* Macquart [Diptera: Muscidae].

ORIENTAL LATRINE FLY *Chrysomya megacephala* (Fabricius) [Diptera: Calliphoridae]: An Old World synanthropic fly of medical and veterinary importance; most noteworthy in Oriental and Australasian realms. Resembles Old World Screwworm but male compound eye facets in upper part of eye significantly larger than other facets; Squamae brown. OLF a common pest in markets where it contaminates meat, fish, fruits and vegetables. Larvae feed in carrion, garbage and excrement; an OW homologue of Secondary Screwworm. See Calliphoridae; Secondary Screwworm.

ORIENTAL MOTH *Cnidocampa flavescens* (Walker) [Lepidoptera: Limacodidae].

ORIENTAL RAT FLEA *Xenopsylla cheopis* (Rothschild) [Siphonaptera: Pulicidae]: A cosmopolitan pest of rodents and humans, and of significant medical importance found in cities and rural areas between 35°N and 35°S. Adult genal margin without spine; genal and pronotal Ctenidia absent; interantennal ridge and frontal tubercle absent; Frons not heavily sclerotized; 6–8 large Setae along posterior margin of head; mesothoracic rods present; Metanotum longer than half length of T1; abdominal Terga 2–7 each with one row of Setae. Egg 0.3–0.5 mm oval, white, sticky; threshold temperature for development 12°C. Larvae require high humidity and B vitamin to complete development; three larval instars; requires more than 30 days to complete development. Pupal stage 25–35 days, depending upon temperature. Adult female emerges 3–4 days before male; copulation requires ca 10 minutes; can live one year. Adults capillary feeders; Maxillae penetrate skin, Epipharynx penetrates capillary; saliva contains anticoagulant. Fed adults negatively phototactic; unfed adult positively phototactic. Adult can jump ca 30 cm horizontally. ORF primary vector of Plague (*Yersinia pestis*) and a secondary vector of Epidemic Typhus (*Rickettsia prowazeckii*). ORF can serve as an intermediate host of rodent tapeworm (*Hymenolepis diminuta*.) See Pulicidae. Cf. Cat Flea. Dog Flea. Human Flea. Rel. Chigoe. Alt. Tropical Rat Flea.

ORIENTAL REALM Zoogeographical term used by Wallace to indicate that part of the earth's surface including Asia east of Indus River, south of Himalayas and Yangtse-kian watershed, Sri Lanka, Sumatra, Java and Philippines. See Indoaustralian Realm.

ORIENTAL SCALE *Aonidiella orientalis* (Newstead) [Hemiptera: Diaspididae]

ORIENTAL SCALE PARASITE *Comperiella lemniscata* Compere & Annecke [Hymenoptera: Encyrtidae]

ORIENTAL SCALE PREDATOR *Chilocorus baileyi* Blackburn; *Telsimia* sp. [Coleoptera: Coccinellidae].

ORIENTAL SORE A form of Cutaneous Leishmaniasis caused by *Leishmania tropica* (Wright). Disease widespread and vectored by blood-sucking *Phlebotomus* spp. See Leishmaniasis. Syn. Bagdhad Boil; Delhi Boil.

ORIENTAL STINK BUG *Plautia stali* Scott [Hemiptera: Pentatomidae].

ORIENTAL TEA TORTRIX *Homona magnanima* Diakonoff [Lepidoptera: Tortricidae]: A pest of tea in the Orient.

ORIENTAL TOBACCO BUDWORM *Heliothis assulta* Guenée [Lepidoptera: Noctuidae].

ORIENTATION Noun. (Latin, *orientis*, past part. of *oriri* = to rise; English, *-tion* = result of an action. Pl., Orientations.) 1. The sense of direction, arrangement or organization demonstrated by structures, bodies or organisms. 2. Taxonomy and Morphology: A frame-of-reference that typically involves descriptors of position or aspect (*e.g.* anterior, posterior, dorsal, ventral, proximal; distal). Cf. Aspect. 3. The change of position by organisms in response to external stimuli (light, heat, sound). Cf. Kinesis; Taxis; Tropism. Descriptive terminology associated with anatomical features, direction and movement must be accurate, concise and unambiguous. Taxonomic description reports the aspect or viewpoint from which observations are made. Most insects are bilaterally symmetric, horizontally oriented and move forward through the environment. Most organisms, including insects, display an elongate body. Elongation creates one major (primary) and two minor (secondary) axes. These axes are mutually perpendicular. Most animals are horizontal in their natural posture; humans who describe such things are naturally vertical. A structure characterized as dorsal for a human would be very differently placed than a dorsal structure for an insect. Similar ambiguities are created for structures that are moveable and have several positions relative to the body. The wing of an insect is such a structure. Some workers would describe points or structures on the wing as if the wing were horizontal and at a right angle to the body; other workers would describe the same point or structure as if the wing were horizontal and parallel to the body. Adverbial forms of words involving orientation include: Anteriad; Apicad; Basad; Caudad; Centrad; Cephalad; Craniad; Dextrad; Dextrocaudad; Dextrocephalad; Distad; Dorsad; Ectad; Entad; Laterad; Mediad; Mesad; Neurad; Orad; Proximad; Rostrad; Sinistrad; Sinistrocaudad; Sinistrocephalad; Ventrad. With few exceptions, adjectivial forms of the same words typically include the suffix '-al.' (*e.g. apical, basal, caudal*). See Aspect; Axis.

ORIFICE Noun. (Latin, *os* = mouth; *facere* = to make. Pl., Orifices.) Any opening (aperture) for a tube, sac or duct. *e.g.* a mouth or an Anus. Alt. Orificium. Cf. Aperture; Entrance; Mouth; Opening.

ORIFICIAL CANAL Ostiolar canal.

ORIFICIUM Noun. (Latin, *orificium*. Pl., Orficula.) An archaic term for the anal or genital opening. See Orifice. Cf. Anus.

ORIGIN Noun. (Latin, *origio* from *oriri* = rise and become visible. Pl., Origins.) 1. The ancestry of a genealogical lineage. See Phylogeny. Cf. Primitive; Derived. 2. The larger or more stationary attachment of a muscle. Cf. Insertion. 3. The primary source or cause of a disease, problem, action or behaviour.

ORIGINAL TYPE 1. The specimen from which a published description is prepared. 2. The type by original designation. See Type. Cf. Holotype; Primary Type. Rel. Nomenclature; Taxonomy.

ORION® See Alanycarb.

ORISMOLOGY Noun. (Greek, *horismos* = definition; *logos* = discourse. Pl., Orismologies.) The defining of scientific or technical terms. See Horismology. Alt. Orismologia.

ORMANCEY, PEDRO (1811–1852) (Mulsant 1853, Ann. Soc. linn. Lyon 1: 77–80, bibliogr.)

ORMAY, SANDOR (ALEXANDOR) (1855–1938) (Zoltan 1939, Folia ent. hung. 4: 90–92, bibliogr.)

ORMEROD ELEANOR ANNE (1828–1901) (Howard 1930, Smithson. misc. Collns. 84: 221, 232, 236, 371, 372, 375, 393, 541.)

ORMEROD, GEORGIANA ELIZABETH (1823–1896) (Ormorod 1896, Entomologist 29: 310–311.)

ORMYRIDAE Plural Noun. A cosmopolitan, small Family (ca three Genera, 60 Species) of parasitic Hymenoptera assigned to the Chalcidoidea. Sometimes placed in Torymidae or Pteromalidae. Diagnositic features include body 1–7 mm long, robust, metallic, strongly sclerotized and sculptured; Antenna short, with 13 segments, including 1–3 Anelli; Notauli shallow; forewing macropterous, marginal vein long, Stigmal and postmarginal veins short; Tarsomeres formula 5–5–5; Metasoma coarsely punctate, not petiolate; Ovipositor not exserted. Species parasitize gall-forming Diptera, Cynipidae and Eurytomidae in seeds.

ORNAMENTATION Noun. (Latin, *ornare* = to adorn. Pl., Ornamentations.) 1. Acarina: Markings on the sclerotized regions of the Integument, usually consisting of a pattern of pits or puncta. (Goff *et al.* 1982, J. Med. Ent. 19: 229.) 2. General: Markings, colour, sculpture or adornment associated with insect structure.

ORNATE APHID *Myzus ornatus* Laing [Hemiptera: Aphididae].

ORNATE KANGAROO TICK *Amblyomma triguttatum* C.L. Koch [Acari: Ixodidae].

ORNITHOPHILOUS (Greek, *ornithos* = bird; *philein* = to love; Latin, *-osus* = with the property of.) Bird-loving; pertaining to some Species of flowers fertilized by birds. Alt. Ornithopilous. See Habitat. Rel. Ecology.

ORNITHOPHILY Adj. (Greek, *ornithos* = bird; *philein* = to love.) Medical Entomology: Pertaining to Species or organisms (especially insects and ticks) that bite or feed upon birds. Cf. Anthropophily; Zoophily.

ORO See Oropouche Virus.

OROGENESIS Noun. (Greek, *oros* = mountain; *genesis* = creation.) The process of mountain building. Rel. Biogeography.

OROGENEY Noun. (Greek, *oros* = mountain; *genesis* = creation.) A particular episode of orogenesis.

OROPOUCHE VIRUS An arbovirus of the Bunyaviridae known to occur in Brazil and Trinidad. Acronym ORO; incubation period 4–8 days; causes fever with associated aches and pains; disease non-fatal. Vectors of ORO include mosquitoes (*Culex quinquefasciatus*) and biting midges (*Culicoides paraensis*). See Arbovirus; Bunyaviridae. Cf. Rift Valley Fever; Sandfly Fever.

ORPHNEPHILIDAE Plural Noun. See Thaumaleidae.

ORR, HUGH LAMONT (–1913) (Anon. 1913, Irish Nat. J. 22: 115.)

ORR, JAMES (1879–1946) (J.A.S.S. 1946, Irish Nat. J. 8: 433–434.)

ORSINI, ANTONIO (1786?-1870) (Villa 1870, Atti Soc. ital. Sci. nat. 13: 144–145.)

ORSTADIUS, ERNST TEODOR (–1939) (Kemner 1939, Opusc. ent. 4: 111.)

ORTALIDAE See Otitidae.

ORTALIDIDAE See Otitidae.

ORTAN® See Acephate.

ORTHANDRIA A Series of symphytan Hymenoptera in which the male genitalia do not rotate 180° prior to eclosion. Included Superfamilies are Cephoidea, Megalodontoidea, Siricoidea and Xyeloidea. Cf. Strophandria.

ORTHANDROUS COPULATION The conventional copulatory stance expressed by Hymenoptera in which the male mounts the dorsum of the female and curves the apex of his Metasoma under the apex of the female Metasoma to achieve genitalic contact.

ORTHENE® See Acephate.

ORTHEZIA LADY-BEETLE *Hyperaspis jocosa* (Mulsant) [Coleoptera: Coccinellidae].

ORTHEZIIDAE Plural Noun. Ensign Scales. A primitive, widespread Family of Homoptera with about 80 Species in six Genera. Adult male with compound eyes; adult female secretes wax plates on dorsum, with abdominal spiracles, setiferous and poriferous anal ring; eggs deposited in wax Ovisac.

ORTHOGENESIS Noun. (Latin, from Greek, *orthos* = straight; *genesis* = development, creation. Pl., Orthogeneses.) 1. An hypothesis which asserts that variation of organisms in successive generations occurs along predetermined lines which results in 'progressive' evolutionary trends that are independent of Natural Selection. Syn. Determinate Evolution. Cf. Natural Selection. 2. A racial tendency toward a particular line of development that is uncontrollled by external conditions.

ORTHOGENETIC SELECTION Eimer's theory to account for the origin of Species or variations as the result of a control along definite directions, independently.

ORTHOGNATHOUS Adj. (Greek, *orthos* = straight; *gnathos* = jaw; Latin, *-osus* = with the property of.) Literally, straight jaws. Pertaining to a head whose primary axis is at a right angle to the primary axis of the body. Cf. Hypognathous; Opisthognathous; Prognathous.

ORTHOGRAPHIC VARIANT A spelling of a scientific name or epithet which is not identical with the original spelling in the description of the Taxon. See Original Spelling.

ORTHOKINESIS Noun. (Greek, *orthos* = straight; *kinesis* = movement. Pl., Orthokineses.) The velocity of movement by an organism in response to the intensity of an environmental stimulus. See Kinesis. Cf. Allokinesis; Blastokinesis; Diakinesis; Klinokinesis; Ookinesis; Thigmokinesis. Rel. Taxis; Tropism.

ORTHOPTERA Oliver 1789 Upper Carboniferous. Crickets; Grasshoppers; Katydids; Locusts. (Greek, *orthos* = straight; *pteron* = wing.) A numerically large (ca 20,000 Species), cosmopolitan Order of exopterygote neopterous insects. Head usually hypognathous with posterior margin enveloped by Pronotum; Vertex sometimes fastigiate with face reflexed; Antenna length and segmentation variable; Mandible well developed, often asymmetrical; Maxillary Palpus with 5–6 segments and Labial Palpus with three segments; compound eyes usually well developed but reduced in cavericolous Species. Pronotum large, collar-like and relatively immobile; fore and middle legs gressorial; hind legs saltatorial. Wings present or absent, when present, forewings thickened to form Tegmina. Spiracles in Mesothorax and Metathorax. Orthoptera are divided into the Suborders Caelifera and Ensifera. Ensiferous Orthoptera include the Superfamilies: Gryllacridoidea (Gryllacrididae, Mimnermidae), Gryllotalpoidea (Gryllotalpidae), Grylloidea (Cachoplistidae, Eneopteridae, Gryllidae, Mogoplistidae, Myrmecophilidae, Oecanthidae, Pentacentridae, Phalangopsidae, Pteroplistidae, Sclerопteridae, Trigonidiidae), Rhaphidophoroidea (Macropathidae, Rhaphidophoridae), Schizodactyloidea (Schizodactylidae), Stenopelmatoidea (Cooloolidae, Lezinidae, Stenopelmatidae), and Tettigonioidea (Acridozenidae, Bradyporidae, Conocephalidae, Meconematidae, Mecopodidae, Phaneropteridae, Phasmodidae, Phyllophoridae, Prophalangopsidae, Pseudophyllidae, Tettigoniidae, Tympanophoridae). Caeliferous Orthoptera include the Superfamilies: Acridoidea (Acrididae, Lathiceridae, Lentulidae, Ommexechidae, Pamphagidae, Pamphagodidae, Pauliniidae, Pyrgomorphidae, Romaleidae, Tristiridae), Eumasticoidea (Biroellidae, Chorotypidae, Episactidae, Eruciidae, Eumastacidae, Euschmidtiidae, Gomphomastacidae, Mastacideidae, Miraculidae, Morabidae, Thericleidae), Tetrigoidea (Batrachideidae, Tetrigidae, Tridactyloidea, Cylindrachetidae, Ripipterygidae, Tridactylidae), Trigonopterygoidea (Borneacrididae, Trigonopterygidae) and Xyronotoidea (Pneumoridae, Tanaoceridae, Xyronotidae). Syn. Deratoptera

ORTHOPTEROID Adj. (Greek, *orthos* = straight; *pteron* = wing; *eidos* = form.) 1. Orthoptera-like; descriptive of features or habitus that resemble an orthopteran. 2. A member of the Orthoptera or related lineage. See Orthoptera. Rel. Eidos; Form; Shape.

ORTHORRHAPHA Noun. A Division of Diptera. Wing venation relatively primitive, Rs with three branches and M with as many as four branches; Pupa obtect. Adults and larvae of many Species predaceous or parasitic. Included Families: Acroceridae (Acroceratidae, Cyrtidae, Henopidae, Oncodidae), Apioceridae (Apioceratidae), Athericidae, Asilidae, Bombyliidae, Dolichopodidae (Dolichopidae), Empididae (Empidae), Hilarimorphidae, Mydidae (Mydasidae, Mydaidae), Nemestrinidae, Pantophthalmidae, Pelecorhynchidae, Rhagionidae (Leptidae), Scenopinidae (Omphralidae), Stratiomyidae (Stratiomyiidae, Chiromyzidae), Tabanidae, Therevidae, Vermileonidae, Xylomyidae (Xylomyiidae), Xylophagidae (Erinnidae, Coenomyiidae, Rachiceridae.) See Diptera. Cf. Cyclorrhapha.

ORTHORRHAPHOUS Adj. (Greek, *orthos* = straight; Latin, *-osus* = with the property of.) Pertaining to Orthorrhapha Diptera or with the characteristics of Orthorrhapha.

ORTHOTYPE Noun. (Greek, *orthos* = straight; *typos* = pattern. Pl., Orthotypes.) In nomenclature, the type Species of a Genus that was originally designated as the 'type.' See Type. Cf. Genotype. Rel. Nomenclature; Taxonomy.

ORTIZ, CAROLOS STUARDO (1895–1962) (Cortes 1962, Publnes Cent. Estud. Ent. Univ. Chile 4: 69–72.)

ORTONEDA, VICENTE (Porter, Revta. chil. Hist. nat. 34: 394.)

ORTRAN® See Acephate.

ORTRIL® See Acephate.

ORTSTREUE Social Insects: The tendency of workers to return to the same site for food or guard duty.

ORTUS® See Fenproximate.

ORUSSIDAE Newman 1834. A small, cosmopolitan Family of Symphyta (Hymenoptera) assigned to the Orussoidea or Siricoidea in different classifications. Family with ca 70 Species, 15 Genera; most abundant in tropical regions. Body 3–15 mm long; female Antenna with 10 segments, penultimate segment elongate and swollen, apical segment peg-like; male Antenna with 11 segments; Antenna inserted below apparent Clypeus; female fore Tibia swollen, fore Tarsus with three segments; Ovipositor very long and coiled within Abdomen at repose; male genitalia orthandrous. Larva with Antenna of one segment; thoracic legs reduced; Abdomen with 10 segments, without Supranal Process. Larval stage parasitic on wood-boring insects, particularly Coleoptera. Eggs oviposited in frass-filled tunnels; first instar larvae may feed on frass. *Guiglia schauinslandi* first instar larva feeds externally on *Sirex* sp.; remaining instars may feed internally.

ORUSSOIDEA Newman 1834. A Superfamily of Symphyta including the Orussidae and Paroryssidae. May be related to Siricoidea.

OS Noun. (Latin, *os, osseous* = bone. Pl., Ossa.) A bone.

OSADAN® See Fenbutatin-Oxide.

OSBORN, HENRY FAIRFIELD (1851–1935) (Grabau 1936, Peking nat. Hist. Bull. 10: 165–166.)

OSBORN, HERBERT (1856–1954) (Mallis 1971, *American Entomologists*. 549 pp. (156–161), New Brunswick.)

OSBURN, RAYMOND CARROLL (1872–1955) (Melander 1955, Ann. ent. Soc. Am. 48; 422.)

OSBURNE, HUMPHREY ERNEST (1899–1973) (Mulder 1973, Circ. ent. Soc. Aust. (N.S.W.) 238: 31.)

OSCINIDAE Plural Noun. See Chloropidae.

OSCULA See Osculum.

OSCULANT Adj. (Latin, *osculans* = kissing; *-antem* = adjectival suffix.) 1. Closely adherent. 2. An intermediate character state or transitional feature. Alt. Osculate.

OSCULAR Adj. (Latin, *osculum* = small mouth.) An Osculum.

OSCULATI, CAJETANO (1808–) (Papavero 1975, *Essays on the History of Neotropical Dipterology*.

2: 343–347. São Paulo.)

OSCULUM Noun. (Latin, *osculum* = small mouth. Pl., Oscula.) 1. A small mouth. 2. An excurrent aperture of a sponge.

OSHANIN, VASILII FEDOROVICH (1844–1917) (Kirichenko 1940, Moskva obshch. Isp'talelii prirod. (Hist. Ser.) 5: 1–30, bibliogr.)

OSIMA, GINKICHI (1868–1925) (Hasegawa 1967, Kontyû 35 (Suppl.): 24.)

OSMATERIUM See Osmeterium.

OSMATIC Adj. (Greek, *osme* = smell; *-ic* = of the nature of.) Empowered with the sense of smell.

OSMETERIAL Adj. (Greek, *osme* = smell; Latin, *-alis* = pertaining to.) Descriptive of structure or substance associated with the Osmeterium.

OSMETERIUM Noun. (Latin, from Greek, *osme* = smell; *terein* = to keep; Latin, *-ium* = diminutive > Greek, *-idion*. Pl., Osmateria.) An eversible, tubular, V-shaped gland on the Prothorax of papilionid caterpillars, or other body parts of other insects. Typically, Osmeterium is inverted but when the larva is disturbed, two glandular tubules are everted by hydrostatic pressure. Numerous Papillae are distributed over the tubules and apparently secrete strongly odorous (presumably defensive) compounds including isobutyric acid and 2-methylbuteric acid. Syn: Osmaterium. See Defensive Gland. Cf. Ptilinum. Rel. Hydrostatic Pressure.

OSMICS Adj. (Greek, *osme* = smell.) The study of the sense of smell (olfaction) and odiferous structures and substances.

OSMIUM Noun. (Latin > Greek, *osme* = smell – in allusion to chlorine-like odour of osmium tetroxide; Latin, *-ium* = diminutive > Greek, *-idion*.) A metallic element of the platinum group and the heaviest substance known (sp. gravity as crystalline: 22.48, atomic wt. 190.2.) Hard in substance, grey-whitish to blue-whitish in colour, insoluble in all acids and melting at 2700°C.

OSMIUM TETROXIDE A noncoagulant fixative used in conjunction with glutataraldehyde for electron microscopy. Preserves cellular and cytoplasmic detail but toxic and should be used under a fume hood. Molecular weight 254.2, melting point 41°C, boiling point 131°C.

OSMONT, AUSGUSTE E (–1894) (Fauvel 1894, Revue Ent. 13: 287.)

OSMOPHILE Noun. (Greek, *osmos* = impulse; *phelein* = to love. Pl., Osmophiles.) An organism which lives and reproduces in relatively high concentrations of sugar or salt. Cf. Halophile; Mesophile; Psychrophile; Thermophile.

OSMOPHORE Noun. (Greek, *osme* = smell; *pherein* = to bear, carry.) A chemical group or radical (hydroxyl *etc.*) that is part of a molecule and which gives the compound its characteristic odour.

OSMORECEPTOR Noun. (Greek, *osme* = smell; Latin, *receptor*. Pl., Osmoreceptors.) A sensillar form on the gills of Plecoptera which may have an osmoregulatory function. These receptors are found on the gills of the plecopteran nymph *Thaumatoperla alpina* Burns & Neboiss, other Eustheniidae, Gripopterygidae and Austroperlidae. Anatomically, these receptors appear intermediate between a Placode-type Sensilla of aphids and a thin-walled Basiconic Sensillum. These putative osmoreceptors form a mushroom-like structure consisting of a large dome on a stalk. Upon the dome is a palisade of 20 curved spines. Morphogenetically a sensillum develops into a more complex form with successive moults. The structure is innervated by one neuron whose dendrites fan out to fill the cavity below the dome. See Sensillum. Cf. Chemoreceptor; Hygroreceptor; Mechanoreceptor.

OSMOREGULATION Noun. (Greek, *osmos* = impulse; Latin, *regulatus* = regulated; English, *-tion* = result of an action. Pl., Osmoregulations.) The regulation of osmotic pressure in a body by controlling the relative amounts of salts and water in the body.

OSMOSIS Noun. (Greek, *osmos* = impulse; *sis* = a condition or state. Pl., Osmoses.) Diffusion of a solvent, typically water, through a semipermeable membrane from a dilute solution to a concentrated solution, or from a pure solvent to a solution.

OSMOTAXIS Noun. (Greek, *osmos* = impulse; *taxis* = arrangement. Pl., Osmotaxes.) A behavioural movement in response to changes in osmotic pressure. See Orientation. Cf. Aerotaxis; Anemotaxis; Chemotaxis; Geotaxis; Menotaxis; Phototaxis; Rheotaxis; Rotaxis; Scototaxis; Strophotaxis; Telotaxis; Thermotaxis; Thigmotaxis; Tonotaxis; Tropotaxis.

OSMOTIC Adj. (Greek, *osmos* = impulse; *-ic* = characterized by.) Having the property of osmosis.

OSMYLIDAE Leach 1815. Plural Noun. A numerically moderate-sized, widespread Family of Neuroptera placed in Osmyloidea. Not found in North America. Body moderate to large sized; Antenna filamentous, setose, relatively short; three Ocelli present but not always well defined; wings without coupling mechanism; Nygmata present; males of some Species with eversible scent glands between abdominal segments 8–9; female with two Spermathecae. Eggs laid in short rows. Larvae elongate, predaceous. Subfamilies include Eidoporisminae, Kempyninae, Porisminae, Spilosmylinae, Stenosmylinae.

OSMYLOIDEA A Superfamily of plannipenous Neuroptera including Neurothidae, Osmylidae, Sisyridae and Rapismatidae.

OSSEOUS Adj. (Latin, *osseus* = bony; *-osus* = with the property of.) A small nodule of Chitin resembling a bone.

OSSICLE Adj. (Latin, diminutive of *os* = little bone. Pl., Ossicles.) A small bone or bone-like sclerite.

OSSICULA Noun. (Latin, diminutive of *os* = little bone. Pl., Ossiculae.) Small corneous pieces that serve in the articulation of the wings to the Thorax.

OSSOWSKI, L L J (1905–1960) (Jahn 1961, Anz.

Schädlingsk. 34: 44, bibliogr.)

OSTEN-SACKEN, CHARLES ROBERT (1828–1906) (Bogdanov 1891, Izv. imp. obshch. Lyub. Estest. Antrop. Etnogr. Univ. Mosc. 57: [207–208], bibliogr. (pl. 24.)

OSTERBERGER, BULLION ALPHONSE (1901–1939) (Eddy 1939, J. Econ. Ent. 32: 893–894.)

OSTHELDER, JUDWIG (1877–1954) (Reisser. 1956, Z. Wien. ent. Ges. 41: 212–213.)

OSTIA See Ostium.

OSTIAL Adj. (Latin, *ostium* = door; *-alis* = pertaining to.) Descriptive of structure or function associated with an Ostium.

OSTIAL VALVE Valve-like pouches of the Heart wall containing the Ostia at their inner ends. Typically OV are slit-like and serve as a passage for Haemolymph into the Dorsal Vessel (Incurrent Ostia) or flow of Haemolymph out of the Dorsal Vessel (Excurrent Ostia). See Heart. Cf. Excurrent Ostium; Incurrent Ostium. Rel. Dorsal Vessel.

OSTIOLA Noun. (Latin, *ostiolum* = little odour. Pl., Ostiolae.) A small aperture. In Hemiptera: The opening of the scent glands; placed near the Coxa in the adult, paired and dorsal in the nymph.

OSTIOLAR CANAL A furrow leading from an Ostiole, specifically in Hemiptera.

OSTIOLAR PERITREME In the Hemiptera: The thickened and sometimes furrowed or corrugated border around the Ostiole.

OSTIUM Noun. (Latin, *ostium* = door. Pl., Ostia.) 1. General: An aperture, entrance or mouth-like opening of a body, tube, organ or enclosed anatomical structure. 2. A slit-like opening in the insect Heart. Number of pairs of Ostia varies among Taxa: Cockroach 12, Crickets 10, Honey bee 4, Housefly 3. See Heart; Ostial Valve. Cf. Dorsal Vessel. Rel. Circulation. 3. Lepidoptera: The external genitalic opening of the female which is used for the reception of Sperm, and which leads to the Ductus Bursae and Bursa Copulatrix (Klots).

OSTIUM BURSAE The opening or aperture of the Bursa Copulatrix in Lepidoptera; OB equivalent to the Vulva of female insects displaying the gential opening on the eighth segment (Snodgrass).

OSTROGOVICH, ANDRIANO (1870–1957) (Anon. 1957, Trav. Mus. Hist. nat. Gr. Antipa 1: 375–377.)

OTHRINE® See Deltamethrin.

OTIN, H (–1970) (Anon. 1970, Bull. Soc. ent. Fr. 75: 165.)

OTINEM® See Entomogenous Nematodes.

OTITIDAE Plural Noun. Picture-winged Flies. A cosmopolitan Family of about 500 Species of Diptera assigned to Superfamily Tephritoidea. Superficially resemble Tephritidae. Adult often with pictured wings; Costa entire, not broken near apex of Subcosta; Subcosta complete, free from first vein and ending in Costa; first Posterior Cell open in wing margin; Anal Cell apically produced or angulate; oral Vibrissae absent. Predominantly phytophagous and scavengers in decaying plant materials. Little known of immature stages. Nothing known of egg deposition or pupation sites. Syn. Ortalidae; Ortalididae; Ulidiidae.

OTOCYST Noun. (Greek, *oto* = ear; *kystis* = bladder.) In invertebrates: The auditory organ containing a fluid and Otoliths.

OTOLITH Noun. (Greek, *oto* = ear; *lithos* = stone.) A calcareous concretion in the Otocyst.

OTSUKA, YOSHINARI (1862–1925) (Hasegawa 1967, Kontyû 35 (Suppl.): 20–21, bibliogr.)

OTTANDER, ALEX (1892–1958) (Lindroth 1959, Opusc. ent. 24: 10.)

OTTO, ANTON (1877–1949) (R[eisser]. 1950, Z. Wein. ent. Ges. 35: 124.)

OTTO, PAUL VON (1869–1939) (Hartweig 1939, Ent., Z. Frankf. a. M. 53: 65–66.)

OTTOLENGUI, RODRIGUEZ (–1937) (Hartweig 1939, N. Y. Times 13. 7. 1937.)

OUDEMANS, ANTHONIE CORNELIUS (1858–1943) (Eyndhoven 1943, Tijdschr. Ent. 86: 1–56, bibliogr.)

OUDEMANS, JOHANNES THEODORUS (1862–1934) (Meijere 1934, Tijdschr. Ent. 77: 167–174, bibliogr.)

OUELLET, JOSEPH (1869–1952) (Larochelle 1976, Fabreries 2 (2): 13–16, bibliogr.)

OUTER ANGLE Coccids: The angle at the lateral margin of an Operculum (MacGillivray).

OUTER CLASPER Hymenoptera: The Gonocoxite and Gonostylus.

OUTER LOBE In the Maxilla, the galea.

OUTER MARGIN 1. The outer edge of the wing, between the apex and the hind angle. 2. In coccids: The margin connecting the apex and the lateral margin of the Operculum (MacGillivray).

OUTER PIGMENT CELL Syn. Iris Pigment Cell. Syn. Secondary Iris Cell.

OUTER PLATE 1. Heteroptera (Nepidae): The Gonocoxite of abdominal segment VIII. 2. Hymenoptera (Aculeata): Ninth Hemiterigite.

OUTER RAMUS OF STIPES Hymenoptera: See gonostylus.

OUTER SQUAMA In Diptera: The Squama arising from the wing base behind the 3rd Axillary Vein, evidently representing the Jugal Lobe of other insects (Snodgrass).

OUTER SURFACE OF THE LEG The dorsal surface when the insect leg is extended at a right angle to the body (Snodgrass).

OUTFLANK® See Permethrin.

OUTGROUP COMPARISON (See Maddison *et al.* 1984, Syst. Zool. 33: 83–103.)

OUTLINE SHAPE The shape defined by a line that conforms to the boundary (margin) of a figure (silhouette). Silhouettes are two-dimensional (length and width) with a contour line (margin) that varies. See Shape. Cf. Form. Common outline forms include: Acerate; Acuminate (terminating in a sharp point); Aciculate (needle-like);

Cordate (heart-shaped); Cuneate; Ellipical; Ensiform; Fusiform (spindle-like); Globular; Hastate; Hemispherical; Hypocrateriform (salver-shaped); Lanceolate; Limaciform (slug-like); Linear; Lyrate; Obovate; Orbiculate; Ovaliform (oval-shaped); Ovate; Palmate; Pandurate; Peltate; Pyriform (pear-shaped); Reniform; Runcinate; Sagittate (Triangular); Securiform (triangular-compressed; ax-shaped); Semicircular. (half circle); Semi-cordate (partly heart-shaped); Semicoronate (partly surrounded by a margin of spines or hooks and resembling a crown); Semicoronet (Semicoronate); Semicylindrical (half a cylinder); Semiglobate (half a globe); Semilunar (half a crescent); Spatulate; Spherical; Tectiform (roof-like).

OUTPUT NEURON See Deutocerebrum.

OVA See Ovum.

OVA FAVOSA A Latin phrase meaning 'eggs deposited in closed cells.'

OVA GALLATA A Latin phrase meaning 'eggs deposited in galls.'

OVA GLEBATA A Latin phrase meaning 'eggs deposited or concealed in dung.'

OVA GUMMOSA A Latin phrase referring to eggs covered with a sticky substance which adheres them to surfaces or other objects.

OVA IMPOSITA A Latin phrase referring to eggs deposited in the substrate that is to serve as food for the larve. See Parasitica.

OVA NUDA Literally, naked eggs. A Latin phrase referring to eggs broadcast into the environment in a seemingly random manner, *e.g.* Odonata, Phasmida.

OVA PILOSA A Latin phrase referring to eggs covered with Setae, usually from the Abdomen of the female, *e.g. Lymantria* (Gypsy Moth).

OVA SOLITARIA A Latin phrase referring to eggs deposited singly rather than in groups or clusters.

OVA SPIRALITER DEPOSITA A Latin phrase referring to eggs deposited on a substrate in a spiral pattern. See Aleyrodidae (*Aleyrodes dispersus*).

OVAL GUINEAPIG LOUSE *Gyropus ovalis* Burmeister [Mallophaga: Gyropodidae].

OVAL Adj. (Latin, *ovum* = egg; *-alis* = characterized by.) Egg-shaped in outline; a body in the outline shape of the longitudinal section of an egg; ellipsoidal. See Outline Shape. Cf. Circular.

OVAL PORES See Dorsal Pores.

OVALIFORM Adj. (Latin, *ovum* = egg; *forma* = shape.) Oval-shaped. Term applied to two-dimensional objects or outline forms. See Oval. Cf. Adeniform; Ooid; Oviform; Ovoid; Rel. Form 2; Shape 2; Outline Shape.

OVARIAL LIGAMENT A ligamentous strand attaching the Terminal Filaments of an Ovary to the Dorsal Diaphragm or to the body wall, sometimes united with the opposite side into a Median Ligament attached to the ventral wall of the Dorsal Blood Vessel (Snodgrass).

OVARIAL Adj. (Latin, *ovarium* = ovary; *-alis* = pertaining to.) Descriptive of structure or function associated with an Ovary. See Ovary.

OVARIAN TUBE A tubular structure that contains cells that form Ova. See Germarium. Rel. Reproductive System.

OVARIOLE Noun. (Latin, *ovarium* = ovary. Pl., Ovarioles.) Tapering, elongate, apically closed tubules that collectively comprise the Ovary. Oocytes are produced and develop into eggs within the Ovariole. Anatomically, each Ovariole is composed of a Terminal Filament, Germarium, Vitellarium and Pedicel. Functionally, three types of Ovarioles have been described in insects: Panoistic, Polytrophic and Telotrophic. See Ovary; Testis. Cf. Germarium; Pedicel; Terminal Filament; Vitellarium. Rel. Panoistic Ovariole; Polytrophic Ovariole; Telotrophic Ovariole.

OVARY Noun. (Latin, *ovarium* = ovary. Pl., Ovaries.) The enlarged basal portion of the female reproductive system. Ovaries are compact, composite bodies within the female Abdomen and are positioned on either side of the Alimentary System. Ovaries are paired in most insects, but female butterflies and moths typically have one Ovary. Each Ovary is composed of several Ovarioles. See Reproductive System. Cf. Ovariole. Rel. Testis.

OVASYN® See Amitraz.

OVATE Adj. (Latin, *ovum* = egg; *-atus* = adjectival suffix.) Descriptive of structure that is egg-shaped. Traditionally ovate has been interpreted as shaped as the egg of a bird. However, the comparison between insect and bird egg shape is inappropriate because insect-egg shapes are highly variable. Alt. Ovatus. See Egg.

OVATE-ACUMINATE Ovate with an acuminate point.

OVATE-LANCEOLATE Ovate with a lanceolate point.

OVATE-OBLONG Ovate, elongate along the primary axis, bluntly rounded on the ends.

OVER-TIME® See Permethrin.

OVICAPSULE Noun. (Latin, *ovum* = egg; *capsula* = small box. Pl., Ovicapsules.) An accessory gland secretion which covers or envelopes a clutch of eggs developed during one ovipositional episode. Ovicapsule displayed in Mantodea, Blattaria and some Orthoptera. Syn. Ootheca. See Egg; Cf. Ootheca. Rel. Glands; Reproductive System.

OVICIDE Noun. (Latin, *ovum* = egg; *caedere* = to kill. Pl., Ovicides.) The destruction of the eggs on or in a host by female parasitic Hymenoptera.

OVIDIP® See Propetamphos.

OVIDREX® See Amitraz.

OVIDUCAL Adj. (Latin, *ovum* = egg; *ducere* = to lead; *-alis* = characterized by.) Descriptive of structure or function involving the Oviduct.

OVIDUCT Noun. (Latin, *ovum* = egg; *ducere* = to lead. Pl., Oviducts.) The distal tubular portion of the female reproductive system which transmits the egg outside the body.

OVIDUCTUS COMMUNIS The 'Common Oviduct' or Median Oviduct.

OVIDUCTUS LATERALIS The Lateral Oviduct.

OVIEDO Y VALDES, GONZALO FERNANDEZ DE (1478–1557) (Hagen 1948, *Green World of the Naturalist.* xvii + 398 pp. (16–30), London.)

OVIFORM Adj. (Latin, *ovum* = egg; *forma* = shape.) Egg shaped. Term applied to three-dimensional objects that are typically oval in outline shape. See Oval. Cf. Ovoid; Ooid. Rel. Form 2; Shape 2; Outline Shape.

OVIGEROUS Adj. (Latin, *ovum* = egg; *genere* = to carry; *-osus* = with the property of.) Carrying eggs; a term applied to the fertilized female. Alt. Oviferous.

OVIPARITY Noun. (Latin, *ovum* = egg; *parere* = to bring forth; English, *-ity* = suffix forming abstract nouns.) 1. Conventional form of development in most insects. The egg is fertilized within the body of the female with embryonic development resulting in the development of one individual from one egg which has been fertilized by one sperm. Embryogenesis usually proceeds after the egg has been oviposited, but in some instances considerable embryogenesis may occur within the body of the mother. 2. Cockroaches: Female produces an Ootheca containing eggs and then deposits the Ootheca on the substate; found in all Families of cockroaches except Blaberidae. Cf. Ovovivparity; Viviparity.

OVIPAROUS Adj. (Latin, *ovum* = egg; *parere* = to bring forth; *-osus* = with the property of.) The process of egg laying. Cf. Ovoviviparity; Viviparity.

OVIPORUS Noun. (Latin, *ovum* = egg; *porus* = hole.) Lepidoptera: The posterior opening of the Vagina in most groups, and which serves only for the discharge of eggs when two gential apertures are present (Snodgrass). See Lepidoptera. Cf. Monotrysian; Ditrysian.

OVIPOSIT Verb. (Latin, *ovum* = egg; *ponere* = to place.) The action of laying eggs. Cf. Larviposit.

OVIPOSITION Noun. (Latin, *ovum* = egg; *ponere* = to place; English, *-tion* = result of an action. Pl., Ovipositions.) 1. The act or process of ovipositing. The movement of the egg from the Median Oviduct through an aperture (Gonopore) and outside the female's body. Some insect Species bear a specialized tubular device (Ovipositor) attached to the Gonopore and through which the egg is further extended into plant tissue, places of concealment or within hosts. Small apterygote collembolan *Willowsia jacobsoni* can lay about 360 eggs during its lifetime. Fecundity is probably greatest in social insects. The honeybee, *Apis mellifera,* can produce about 220,000 eggs during a 12 month period (= 602 eggs per day, 25 eggs per hour or an egg laid every 25 seconds). The termite *Kalotermes flavicollis* has been estimated to lay 10 million eggs per year (27,400 eggs per day). Non-social insects typically lay fewer eggs but number can be substantial. Among Neuroptera, Chrysopidae can lay 600 eggs during a lifetime; Mantispidae can lay 8,000 eggs per lifetime. Meloid beetles can lay 4,500 eggs during a lifetime. Pyralid moth *Chilo partellus* female lives about five days as an adult and lays about 500 eggs. See Egg. Cf. Larviposition. Rel. Egg Polarity; Ovipositor; Reproductive System.

OVIPOSITION SITE Places where eggs are left by ovipositing females. OSs are numerous and ecologically diverse; diversity of OSs in response to needs of embryo or postembryonic immature insect. Groundplan for OS difficult to establish, obscured by millions of years of evolution and little fossil information relating to insect eggs or oviposition habits. Phasmida eggs are broadcast (widely dispersed) where food is abundant. This may reflect a primitive condition in paurometabolous, phytophagous, terrestrial insects, but alternative explanations for broadcasting involve derived biological motivations. Broadcasting eggs in some Taxa may be an adaptation to confound predators or parasites from efficiently locating eggs. In other instances, broadcasting eggs may be an adaptation to reduce cannibalism among larvae or nymphs; broadcasting eggs may be an adaptation to utilize ephemeral resources. Most insects lay their eggs in terrestrial habitats. Females of some 'primitive' terrestrial insects deposit eggs in discrete groups. An Ootheca (Ovicapsule) is developed in Mantodea, Blattaria and some Orthoptera. Some plant-feeding insects (*e.g.* Cicadellidae), insert eggs into plant tissue including young or tender shoots and leaves of host plants. Some Aleyrodidae insert their egg pedicel into host plant stomata. Social Insects (bees and wasps) put eggs in specially constructed cells; eggshells of these social insects are not thick, elaborately sculptured or unusually shaped. Some insects deposit eggs in an aquatic medium that evaporates. Some midge and caddisfly eggs have complex respiratory systems like those seen in terrestrial forms. Other aquatic insects embed their eggs in a jelly-like, hygroscopic matrix produced by the female. This 'jelly' may conserve water and discourage predation. Eggs in aquatic habitats are sometimes glued to substrata (plant or rock) to hold eggs in place (*e.g.* libellulid dragonflies). Some corduliids and gomphids lay eggs in masses or in long strings. Saldidae insert their eggs into plant stems near the surface or into algal mats. Eggs are also broadcast on water surface individually or in small clusters (some Libellulidae); eggs of mosquitoes are sometimes deposited in rafts. Females of a few aquatic insects oviposit on the male; this behaviour is seen in some Belostomatidae (*Abedus, Belostoma* and *Sphaerodema*) while others (*Lethocerus*) oviposit on vegetation. Parasitic insects show different constraints on oviposition site. Females whose larvae develop as parasites may place eggs on a host, within a host, or on a substrata which will contact a host. Each of these circumstances may provide conditions which influence the egg. Bethylid wasps paralyse hosts to glue their eggs on paralysed body. Another parasitic wasp, *Euplectrus*, attaches its eggs via a corkscrew-like apparatus on one pole of the egg. The fly *Dermatobia hominis* attaches its eggs to the mosquito *Psorophora* which carries the *D. hominis* egg until a suitable host is encountered. When the mosquito settles on a vertebrate host to feed, *Dermatobia* egg hatches and parasite larva bores into the host. The louse *Pediculus capitus* attaches its egg to hairs of vertebrate host. Internal parasites show other strategies for egg placement. Rel. Egg.

OVIPOSITOR Noun. (Latin, *ovum* = egg; *ponere* = to place. Pl., Ovipositors.) An egg-laying tube of a female insect's Abdomen. An extension of the common oviduct; sometimes rigid and fixed in length; sometimes flexible and telescopic. Not present in all insects, but well developed in some groups such as Thysanura, some Orthoptera, sawflies and parasitic Hymenoptera. Syn. Oviscapte.

OVIPOSITOR SHEATH Heteroptera (Miridae, Nabidae): Part of the body wall and genitalia which contains the Ovipositor. Homoptera (Psyllina): Apical portion of fused inner Valvulae. Higher Diptera: Modification of abdominal segment VII. Hymenoptera: Gonostylus. Syn. Gonoplac.

OVIRUPTOR Noun. (Latin, *ovum* = egg; *ruptor* = a breaker; *or* = one who does something. Pl., Oviruptors.) A cuticular structure (usually on the head) in the form of a spine on the pre-emergent first instar larva. Used for rupturing the eggshell's Chorion to aid eclosion. The tooth or spine is often lost after eclosion and not present on subsequent instars. See Egg Tooth; Hatching Device; Egg-Burster. Rel. Eclosion.

OVISAC Noun. (Latin, *ovum*; *saccus* = bag. Pl., Ovisacs.) Coccoidea: A receptacle for eggs.

OVISCAPTE Noun. (Latin = *ovum* = egg; *captare* = to conduct.) An Ovipositor. Alt. Oviscape (Diptera).

OVITROL® See Methoprene.

OVIVALVULE Noun. (Latin, *ovum* = egg; diminutive of *valva*.) In Ephemeridae: An appendage of the female reproductive organs.

OVOID Adj. (Latin, *ovum* = egg; Greek, *eidos* = form.) Egg-like in shape or appearance. See Egg. Cf. Oviform. Rel. Eidos; Form; Shape.

OVOIDAL Adj. (Latin, *ovum* = egg; Greek, *eidos* = form; *-alis* = characterized by.) Descriptive of an Ovoid shape. See Oultine Shape.

OVOVIVIPARITY Noun. (Latin, *ovum* = egg; *vivus* = living; *parere* = to bring forth; English, *-ity* = suffix forming abstract nouns. Pl., Ovoviviparities.) 1. A method of reproduction in which eggs are maintained in the Common Oviduct (vagina) until eclosion, or eclosion occurs soon after oviposition. Represented in most Orders of insects; a common form of development in some Diptera (Tachinidae), some Aphidoidea, Coccoidea and large-bodied Thysanoptera. Typically, wall of Vagina extensively tracheated to provide oxygen for the developing embryo. Trend toward reduction in number of eggs laid per female in some groups which utilize Ovoviviparity. 2. Cockroaches: Development in which eggs are retained in the Uterus for embryogenesis and before oviposition. Divided into False Ovoviviparity (most Blaberidae, some Blatellidae; Ootheca formed, extruded then drawn into the Uterus) and True Ovoviviparity (some Genera of panestheine Blaberidae; Ootheca not formed; eggs extruded into Uterus and not surrounded by membrane). Cf. Oviparity; Polyembryony; Parthenogenesis; Viviparity.

OVOVIVIPAROUS Adj. (Latin, *ovum* = egg; *vivus* = living; *parere* = to bring forth; *-osus* = with the property of.) Pertaining to Ovoviviparity. See Ovoviviparity.

OVSYANNIKOV, FILIPP VASILEVICH (1827–1906) (Bogdanov 1889, Izv. imp. obshch. Lyub. Estest. Antrop. Entrnogr. imp. Mosc. Univ. 57: [203–206], bibliogr. (pl. 2.)

OVUM Noun. (Latin, *ovum* = egg. Pl., Ova.) 1. A female gamete that corresponds to the male gamete (Spermatozoa). 2. The cell produced in the female reproductive system and which is capable of developing first into an embryo and ultimately into another individual of the same kind. 3. An Egg. See Egg. Rel. Development; Metamorphosis.

OWEN, FRANCIS (–1880) (Carrington 1880, Entomologist 13: 312.)

OWEN, RICHARD (1804–1892) English vertebrate zoologist and anatomist, who among other contributions coined the terms 'Homology' and 'Analogy.' (Canright 1958, Proc. Indiana Acad. Sci. 67: 268–273) See Analogy; Homology.

OWL FLIES See Ascalaphidae.

OWL MIDGES See Psychodidae.

OWLET MOTHS See Noctuidae.

OX WARBLE FLY *Hypoderma bovis* (Linnaeus) and *H. lineatus* (de Villers) [Diptera: Oestridae]. See Common Cattle-Grub; Northern Cattle Grub. Cf. Sheep Bot Fly.

OXAEIDAE Plural Noun. A small Family of aculeate Hymenoptera assigned to Superfamily Apoidea, or considered Subfamily of Andrenidae in some classifications; consists of two Genera which occur in Neotropical Realm and southwestern USA. Adult with two Subantennal Sutures; Facial Foveae present in female; Labrum slightly longer than wide; apex of Glossa acute; Flabellum present; Galea short; Labial Palpus segments short, cylindrical, first segment longest; Pre-episternal Sutures absent; Marginal Cell slender, elongate, Stigma nearly absent; Basitibial Plate present in female; Scopa present on hind Tibia and second abdominal Sternum;

Pygidium well developed in female, poorly developed in male; Arolia absent in female, present in male. Species nest deep in soil. See Apoidea.

OXALIS RED-SPIDER-MITE *Petrobia harti* (Ewing) [Acari: Tetranychidae].

OXALIS WHITEFLY *Aleyrodes shizuokensis* Kuwana [Hemiptera: Aleyrodidae].

OXAMYL A carbamate compound {S-methyl-N',N'-dimethyl-N-(methyl carbamoyloxy)oxy)-1-thaoxamimidate} used as an acaricide, nematicide and contact/systemic insecticide for control of nematodes, mites, plant-sucking insects (aphids, leaf hoppers, scale insects, mealybugs), leaf miners, thrips and leaf-eating insects. Oxamyl applied to non-fruit bearing trees, citrus, cotton, peanuts, peppers, potatoes, soybeans, tobacco, tomatoes, some vegetables and ornamentals. Oxamyl expresses phytotoxicity on some varieties of strawberries. Toxic to bees, fishes and wildlife. Trade names include Blade®, Thioxamyl® and Vydate®. See Carbamate Insecticide.

OXIDASE Noun. (Chemistry, *oxy* = oxygen; *-ase* = enzyme.) An enzyme which catalyses oxido-reductions using molecular oxygen as an acceptor. Cf. Reductase.

OXIDATION Noun. (French, *oxygene*; English, *-tion* = result of an action.) To combine with oxygen, or deprive a compound of hydrogen as by the action of oxygen. Cf. Reduction.

OXOTIN® See Cyhexatin.

OXYBERON® See Indolebutyric Acid.

OXYCANUS GRASSGRUB *Oxycanus antipoda* (Herrich-Schaffer) [Lepidoptera: Hepialidae].

OXYDEMETON-METHYL An organic-phosphate (thiophosphoric acid) compound {S-(2-ethylsulphinyl)ethyl) O,O-dimethyl phosphorothioate} used as an acaricide, systemic and contact insecticide against mites, plant-sucking insects, thrips and other insects. Compound applied to alfalfa, beans, broccoli, citrus, corn, crucifers, fruit trees, grapes, olives, and vegetables in many countries. Not registered for use in USA. Toxic to bees, fishes and wildlife. Trade names include Metasystemox® Metasystox-R® and MSR®. See Organophosphorus Insecticide.

OXYDISULFOTON® See Disulfoton.

OXYGEN Noun. (French, *oxyene* > Greek, *oxys* = sharp; *gen* = to be born.) An element that occurs as a colourless, odourless, tasteless gas in the atomosphere; 23% by weight and 21% by volume of the total. Oxygen is an essential component of aerobic respiration in animals. Symbolized by O; atomic weight 16.0; boiling point −183°C; melting point −218.4°C. See Respiration. Cf. CO_2.

OXYOPIDAE Plural Noun. Lynx Spiders. Adult body size to 15 mm; Abdomen narrow and pointed; body usually pale brown to green and striped; legs with numerous characteristic erect spines. Eyes arranged in hexagonal pattern on anterior margin of head. Adults bind leaves with silk in a tent-like fashion to protect egg sac. Spiderlings similar in appearance to adults. Lynx spiders eat small larvae, leafhoppers, aphids and vinegar flies. Species with well developed sight and specialized to live on vegetation. Females do not build webs but bind leaves to form a shelter for their egg sacs. Young disperse by 'ballooning' from tall plants where silk is played out into breeze until pull lifts and carries them away. Cf. Lycosidae; Nightstalking Spiders.

OXYTHIOQUINOX® See Chinomethionate.

OYSTERSHELL SCALE 1. *Lepidosaphes ulmi* (Linnaeus) [Hemiptera: Diaspididae]: A widespread, polyphagous pest of many ornamental trees in North America. One or two generations per year. OS overwinters in egg stage under female's scale-cover; ca 40–150 eggs deposited per female. Crawlers settle on bark within two days, moult, shed appendages, produce scale cover. OS undergoes two nymphal instars. Adult dark brown, ca 3 mm long, ovoid. OS economic problem on apple, apricot, pear, plum and grape. Trees loose vigour, foliage becomes speckled yellow, and death of tree can result from heavy infestation. Syn. Apple Oystershell Scale. 2. *Quadraspidiotus ostreaeformis* (Curtis) [Hemiptera: Diaspididae]. See Diaspididae.

OZONE Noun. (Greek, *ozein* = to smell. Pl., Ozones.) A gaseous allotropic form of oxygen present in the atmosphere, especially in the upper regions, and a powerful oxidizer. Symbolized by O_3.

P Symbolically, represents the number 400 in Roman Numerals; represents the element Phosphorus in chemistry; represents the Greek Letter PI in mathematics (indicating a continued product, pi, the ratio of a circumference of a circle to its diametre.)

PAASCH, ALEXANDER (1813–1882) (Anon. 1882, Berl. ent. Z. 26: ii.)

PABST, MORTIZ (1833–1908) (Möbius 1944, Dt. ent. Z. Iris 57: 14–15.)

PABULAR Adj. (Latin, *pablum* = nourishment.) Pertaining to food.

PABULATION Noun. (Latin, *pabulo* = to feed; English, *-tion* = result of an action.) The act of feeding or obtaining food.

PABULUM Noun. (Latin, *pablum* = nourishment.) Food.

PACHER, DAVID (1816–1902) (Joabornegg 1902, Carinthia 92: 92–98.)

PACHYNEURIDAE Plural Noun. A small Family of fossil Diptera known from Mesozoic deposits of Mongolia, Siberia and Kazakhstan. Taxonomic position and composition has been debated.

PACHYNOMIDAE Plural Noun. See Linnavouri 1974, Ann. Ent. Fenn. 42:116–138. A small Family of heteropterous Hemiptera assigned to Superfamily Reduvioidea (Cimicomorpha).

PACHYTROCTIDAE Plural Noun. A numerically small, infrequently collected Family of Psocoptera assigned to Suborder Troctomorpha. Body frequently ornately sculptured; Antenna with fewer than 20 segments and secondary annulations; Tarsi with three segments. Inhabit bark and debris.

PACIFIC BEETLE-COCKROACH *Diploptera punctata* (Eschscholtz) [Blattaria: Blaberidae].

PACIFIC-COAST HUMID AREA The faunal area of the transition zone comprising the western parts of Washington and Oregon between the Coast Mountains and Cascade range, parts of northern California and most of the coast region from near Cape Mendocino south to the Santa Barbara Mountains, to the south and east it passes into the arid transition and in places into the upper Sonoran.

PACIFIC-COAST TICK *Dermacentor occidentalis* Marx [Acari: Ixodididae].

PACIFIC-COAST WIREWORM *Limonius canus* LeConte [Coleoptera: Elateridae].

PACIFIC COCKROACH *Euthyrrhapha pacifica* (Coquebert) [Blattaria: Polyphagidae].

PACIFIC DAMPWOOD-TERMITE *Zootermopsis angusticollis* (Hagen) [Isoptera: Termopsidae].

PACIFIC FLATHEAD-BORER *Chrysobothris mali* Horn [Coleoptera: Buprestidae]: A monovoltine pest of deciduous fruit and ornamental trees in North America. PFB overwinters as larva (grub) in mines (burrows) or main trunk or large branches of tree; mature larvae 2.5–5.5 mm deep in wood from bark while younger larvae are closer to the bark. Larvae typically mine wood on sunny side of trunk; mines are packed with fine-grained frass. Female oviposits May-August; eggs deposited in fissures of bark near darmaged or sunspot areas in bark; eggs yellow, wrinkled and disk-like. See Cerambycidae. Cf. Flatheaded Apple-Tree Borer; Roundheaded Apple-Tree Borer.

PACIFIC KISSING-BUG *Oncocephalus pacificus* (Kirkaldy) [Hemiptera: Reduviidae].

PACIFIC MEALYBUG *Planoccus minor* (Maskell) [Hemiptera: Pseudococcidae].

PACIFIC PELAGIC WATER-STRIDER *Halobates sericeus* Eschscholtz [Hemiptera: Gerridae].

PACIFIC SPIDER-MITE *Tetranychus pacificus* McGregor [Acari: Tetranychidae]: A pest of deciduous tree crops and ornamental trees in the pacific states of USA. Adult female resembles Two-Spotted Spider Mite. PSM overwinters as adult female under bark or within field trash; feed on new growth during spring. Female lays ca 50 eggs; Syn. Pacific Mite.

PACIFIC TENT-CATERPILLAR *Malacosoma constrictum* (Hy. Edwards) [Lepidoptera: Lasiocampidae].

PACIFIC WILLOW LEAF-BEETLE *Pyrrhalta decora carbo* (LeConte) [Coleoptera: Chrysomelidae].

PACK, HERBERT JOHN (1892–1930) (Osborn 1937, *Fragments of Entomological History,* 394 pp. (218–219), Columbus, Ohio.)

PACKARD, ALPHEUS SPRING (1839–1905) (Cockerell 1920, Biogr. Mem. Natn. Acad. Sci. 9: 181–236, bibliogr.)

PACKARD, CLYDE MONROE (1889–1971) (Vance *et al.* 1972, J. Econ. Ent. 65: 1531.)

PACKARD GRASSHOPPER *Melanoplus packardii* Scudder [Orthoptera: Acrididae].

PACK-BERESFORD, DENIS ROBERT (1864–1942) (Praeger *et al.* 1942, Ir. Nat. J. 8: 38–40.)

PACYNEURIDAE Plural Noun. A small Family of nematocerous Diptera assigned to Superfamily Bibionoidea. Two Species in Nearctic; one Species occurs in vegetation near streams in mountains.

PAD Noun. (variant of *pod.* Pl., Pads.) 1. A cushion. 2. A mass of anything soft. 3. Coleoptera: The Pulvillus or part of it capable of extension and retraction.

PADAN® See Cartap.

PADDLE Noun. (Middle English, *padell* = small spade. Pl., Paddles.) Aquatic Heteroptera: The flattened or compressed Tarsomeres of the hind legs.

PADDOCK, FLOYD B (1889–1973) (Anon. 1972, Amer. Bee. J. 112: 434.)

PADDY BUG *Leptocorisa acuta* (Thunberg) [Hemiptera: Alydidae].

PADDY STEM-BORER *Scirpophaga incertulas* Walker [Lepidoptera: Pyralidae].

PAEDOGENESIS Noun. (Greek, *pais* = child; *genesis* = descent. Pl., Paedogenesis.) Reproduction in which ovaries become functional during the larval stage and eggs develop parthenogenetically. Some of the developing eggs result in larvae which themselves become parthenogenetic and other developing eggs result in normal adults. Paedogenesis is manifest in some Diptera (Cecidomyiidae), Coleoptera (*Micromalthus*) and some mites (*Pyemotes ventricosus*). Alt. Pedogenesis. Cf. Neoteny. Rel. Gynandromorphism.

PAEDOGENETIC Adj. (Greek, *pais* = child; *genesis* = descent; *-ic* = of the nature of.) Reproduction by juvenile or otherwise immature animals. Cf. Neotenic.

PAEDOMORPH Noun. (Greek, *pais* = child; *morphosis* = form. Pl., Paedomorphs.) Individuals that are paedomorphic.

PAEDOMORPHIC Adj. (Greek, *pais* = child; *morphosis* = form; *-ic* = of the nature of.) Pertaining to Paedomorphosis.

PAEDOMORPHOSIS Noun. (Greek, *pais* = child; *morphe* = form; *sis* = a condition or state.) An evolutionary change in which primitive structures or juvenile charateristics are retained in the adult. (See: Sullivan 1980, J. Herpet. 14: 79–80.)

PAEDOPARTHENOGENESIS Noun. (Greek, *pais* = child; *parthenos* = virgin; *genesis* = descent.) Parthenogenesis during the larval stage. A phenomenon encountered in some Diptera. See Paedogenesis.

PAGAST, FELIX (–1954) (Peus 1954, Arch. Hydrobiol. 49: 423–426, bibliogr.)

PAGE PRECEDENCE The sequence in which two Species or Genera appear in a publication. PP was a principle used to establish the priority of disputed names in early rules of the Code of Zoological Nomenclature. Not a valid principle under existing Rules of Nomenclature.

PAGE, HERBERT E (1868–1945) (Anon. 1945, Entomologist's Rec. J. Var. 57: 48.)

PAGEANT® See Chlorpyrifos.

PAGENSTECHER, ARNOLD (1873–1913) (Dreyer 1913, Jb. nassau. Ver. Naturk. 66: v-xvi, bibliogr.)

PAGENSTECHER, HEINRICH ALEXANDER (–1889) (Anon. 1958, Mitt. Hamb. zool. Mus. 56)

PAGET, CHARLES JOHN (1811–1844) (Paget 1895, Trans. Norfolk Norwich Nat. Soc. 6: 74–76.)

PAGET, JAMES (1814–1897) (Bloomfield 1900, Entomol. mon. Mag. 36: 89.)

PAGINA INFERIOR The lower surface of a wing. (Archaic.) Cf. Converse surface of a wing. Cf. Obverse.

PAGINA Noun. (Latin, *pagina* = leaf, page.) 1. The surface of a wing. 2. Orthoptera: external flattened surface of the hind Femora.

PAGIOPODA Noun. (Greek, *pagos* = something firmly set; *podos* = foot.) Heteroptera: Posterior Coxae not globose and articulation is a hinge joint. Cf. Trochalopoda.

PAGIOPODOUS Adj. (Greek, *pagos* = something firmly set; *podos* = foot; Latin, *-osus* = with the property of.) Heteroptera with a coxal hinge on the hind leg.

PAHVANT VALLEY PLAGUE See Tularemia.

PAILLOT, ANDRÉ (1885–1944) (Anon. 1949, Revue Zool. agric. Gironde 46: 1–15.)

PAINTED APPLE-MOTH *Teia anartoides* Walker [Lepidoptera: Lymantriidae].

PAINTED BEAUTY *Vanessa virginiensis* (Drury) [Lepidoptera: Nymphalidae].

PAINTED CUP-MOTH *Doratifera oxleyi* (Newman) [Lepidoptera: Limacodidae]: A bivoltine, minor pest of fruit and native ornamental trees in eastern Australia. Female oviposits on leaves in rows of 8–12 eggs per row; eggs bright yellow, flattened and covered with short, dark Setae from female. Caterpillar reddish-green with double row of black-and-white patches along middle of dorsum and row of black-and-white oval spots along side of body; two pairs of stellate tubercles composed of defensive spines anterior and posterior ends of body. Although named 'cup moth,' Species constructs oval cocoon. Syn. Chinese-Junk Caterpillar.

PAINTED HICKORY-BORER *Megacyllene caryae* (Gahan) [Coleoptera: Cerambycidae]: A pest of black walnut, hickory, mulberry and osage orange in North America. Anatomically very similar to Locust Borer. See Cerambycidae. Cf. Elm Borer; Locust Borer; Poplar Borer.

PAINTED LADY *Vanessa cardui* (Linnaeus) [Lepidoptera: Nymphalidae].

PAINTED LEAFHOPPER *Endria inimica* (Say) [Hemiptera: Cicadellidae].

PAINTED MAPLE-APHID *Drepanaphis acerifoliae* (Thomas) [Hemiptera: Aphididae].

PAINTED MEAL MOTH *Pyralis pictalis* (Curtis) [Lepidoptera: Pyralidae]: A pest of stored products, primarily grain; endemic to Asia and now widespread. Adult forewing black basally, with two sinuous pale lines, reddish brown apically; wingspan 15–35 mm. Cf. Grey Pyralid.

PAINTED PINE-MOTH *Orgyia australis* Walker [Lepidoptera: Lymantriidae].

PAINTED SKIPPER *Hesperilla picta* (Leach) [Lepidoptera: Hesperiidae].

PAINTED VINE-MOTH *Agarista agricola* (Donovan) [Lepidoptera: Noctuidae].

PAINTER, REGINALD HENRY (1901–1968) (Knutson 1969, J. Kans. ent. Soc. 44: 3–4.)

PALA Noun. (Latin, *palus* = a stake. Pl., Palae.) 1. Dilated anterior tarsal segment in Corixoideae. Palar pegs on anterior legs are used to maintain secure contact between male and female corixids during copulation (Bakonyi 1984, Acta Zool. Hung. 30: 249–255.) 2. Generally, a scoop-like or shovel-shaped structure.

PALAEOBOTANY Noun. The study of fossil plants.

PALAEOCENE EPOCH The interval of the Geological Time Scale (65 MYBP–57 MYBP) that represents the first Epoch in the Palaeogene Period. Tropical and temperate conditions were more widespread than today. Grasses originate near wooded areas or swamps. Rodents appear. See Geological Time Scale; Palaeogene Period. Cf. Eocene Epoch; Oligocene Epoch. Rel. Fossil.

PALAEODICTYOPTERA Goldenberg 1854. An extinct Order of insects whose Species lived during the Carboniferous and Permian. Families include the Homoiopteridae. Palaeodictyoptera are among the most abundant fossil pterygotes taken in some carboniferous sites. Ancestral to modern insects; features include beak-like piercing mouthparts; wings perpendicular to body; wing span 32–430 mm; hingwings orthopteroid; prothoracic lobes. Cf. Diaphanopteroidea; Megasecoptera.

PALAEOENTOMOLOGY Noun. (Greek, *palaios* = ancient; *entomon* = insect; *logos* = discourse. Pl., Palaeoentomologies.) The branch or subdiscipline of Palaeontology concerned with extinct forms, particularly fossil insects.

PALAEOGENE PERIOD The interval of the Geological Time Scale (65 MYBP–23 MYBP) within the Cenozoic Era and which includes the Palaeocene, Eocene and Oligocene Epochs. PP characterized by appearance of major lineages of mammals; grasses originate; See Geological Time Scale; Cenozoic Era. Cf. Neogene Period. Rel. Fossils.

PALAEOLEPIDOPTERA Haustellate Lepidoptera: Mandibles distinct and Pupa is free. Includes Micropterygidae. See Protolepidoptera, Neolepidoptera.

PALAEONTOLOGY Noun. (Greek, *palaios* = ancient; *onta* = existing things; *logos* = discourse. Pl., Palaeontologies.) The study of life during past geological periods; the study of extinct or fossil animals.

PALAEOPLECIIDAE Rohdendorf 1962. Plural Noun. A Family of Diptera based on fossil material of a monotypic Genus preserved in Lower Jurassic Period deposits from Kirghizistan.

PALAEOSETIDAE Turner 1922 Plural Noun. A numerically small, widespread, diverse Family of exoporous Lepidoptera assigned to Hepialoidea. Species known from Taiwan, Australia, India, South America. Adult body small; wingspan 14–18 mm; head with broad lamellar scales and scattered hair-like or smooth lamellate scales; Antenna less than half as long as forewing, covered with lamellar scales on proximal third, remainder of Antenna without scales; Mandibles and Maxillae absent; Labial Palps short, of 2–3 segments; Epiphysis absent or vestigial; tibial spurs absent; wings held tectiform at repose.

PALAEOZOIC ERA The interval of the Geological Time Scale (570–245 MYBP) characterized by the inferred origin, evolution and adaptive radiation of multicellular plants, arthropods and marine vertebrates. PE divided into Cambrian Period, Ordovician Period, Silurian Period, Devonian Period, Early Carboniferous Period, Late Carboniferous Period and Permian Period. See Geological Time Scale; Phanerozoic Eon. Cf. Cenozoic Era; Mesozoic Era. Rel. Fossil.

PALAEPHATIDAE Davis 1986. A Family of monotrysian Lepidoptera consisting of about five Genera and 30 Species found in southern Argentina, Chile and Australia; assigned to Superfamily Palaephatoidea. Adult small bodied; wingspan 8–36 mm; head with raised piliform scales; Ocelli absent from most Taxa; Chaetosemata absent; Antenna filiform, shorter than forewing length; Mandibles vestigal; Pilifers present; Proboscis short; Maxillary Palpus with five (rarely 4) segments; Labial Palpus with three segments, curved upward slightly; Epiphysis typically present; tibial spur formula 0-2-4; wings heteroneurous; Microtrichia on all wing surfaces; female with 2–4 Frenular Bristles, male with a Frenular Bristle; wings held tectiform over Abdomen at repose. Biology poorly known.

PALAEPHATOIDEA A Superfamily of heteroneurous Lepidoptera containing Palaephatidae.

PALATE See Hypopharynx.

PALCHEVSKII, NIKOLAI ALEXANDROVICH (–1909) (Seminov-Tian-Shansky 1909, Revue Russe Ent. 9: 344.)

PALE APHID-LEAFROLLER *Pseudoexentera mali* Freeman [Lepidoptera: Tortricidae].

PALE CERULEAN *Jamides cytus claudia* (Waterhouse & Lyell) [Lepidoptera: Lycaenidae].

PALE CHRYSANTHEMUM APHID *Coloradoa rufomaculata* (Wilson) [Hemiptera: Aphididae].

PALE CILIATE BLUE *Anthene lycaenoides godeffroyi* (Semper) [Lepidoptera: Lycaenidae].

PALE COTTON-STAINER *Dysdercus sidae* Montrouzier [Hemiptera: Pyrrhocoridae]: Endemic to northern Australia and a pest of cotton. Adult oval-elongate, 12–14 mm long, dark brown with pale reddish-brown wings and head; leathery part of forewing pale reddish-brown with a dark brown or black spot in centre; membranous part of forewing black. Antenna long with four segments. Head triangular with prominent lateral eyes; mouthparts adapted for piercing and sucking plant material. Mating adults will remain *in copula* for several days before female oviposits. Adult females lay eggs in batches of 100 in soil and leaf litter. Eggs elongate-oval, 1.5 mm long and creamy-white when freshly laid but turn orange as embryo develops. Egg stage 7 days. Early instar nymphs orange-red to red and black; late instars with white in colour pattern. Nymphs gregarious and cluster in bolls feeding on seeds. Nymphal stage 21–30 days. Adult and nymphs pierce plant tissue and suck sap. PCS prefer open and partially open bolls for feeding. Feeding leads to staining of lint plus germination and seed oil yield reductions. Boll-rotting fungi may gain entry through feeding damage; migrates to cotton late in the growing season. Damage generally occurs only in Autumn. Early season hosts include hibiscus, bottlebrush, kurrajong and weeds. See Pyrrhocoridae.

PALE DAMSEL-BUG *Nabis capsiformis* Germar [Hemiptera: Nabidae].

PALE DARTER *Telicota colon argeus* (Plotz) [Lepidoptera: Hesperiidae].

PALE JUNIPER-WEBWORM *Aethes rutilana* (Hübner) [Lepidoptera: Cochylidae].

PALE LEAFCUTTING-BEE *Megachile concinna* Smith [Hymenoptera: Megachilidae].

PALE LEAF-SPIDER *Chiracanthium mordax* Koch [Araneae: Clubionidae].

PALE LEGUME-BUG *Lygus elisus* Van Duzee [Hemiptera: Miridae].

PALE ORANGE DART *Ocybadistes hypomeloma* Lower [Lepidoptera: Hesperiidae].

PALE SULPHUR See Cloudless Sulphur.

PALE TUSSOCK-MOTH *Halysidota tessellaris* (J. E. Smith) [Lepidoptera: Arctiidae].

PALE WESTERN-CUTWORM *Agrotis orthogonia* Morrison [Lepidoptera: Noctuidae]: A monovoltine, subterranean feeder on alfalfa, beets and small grains in North America. Eggs spherical, white, laid individually in soil. Larva grey, Integument with pavement-like granulations. See Cutworms.

PALEACEUS Adj. (Latin, *palea* = chaff, straw; *aceus* = of or pertaining to.) Like chaff; pertaining to structure with chaff-like scales (Say). Alt. Paleace; Paleaceous.

PALEARCTIC REALM Part of the Holarctic Realm including Europe, Africa north of the Sahara and Asia as far south as the southern edge of the Yangtse-kiaang watershed and the Himalayas and west to the Indus River. Alt. Palearctic Region.

PALEBROWN SAWFLY *Pseudoperga lewisii* (Westwood) [Hymenoptera: Pergidae].

PALEGREEN TRIANGLE BUTTERFLY *Graphium eurypylus lycaon* (C. & R. Felder) [Lepidoptera: Papilionidae].

PALEODICTYOPTERA See Palaeodictyoptera.

PALEOENTOMOLOGY See Palaeoentomology.

PALEONTOLOGY See Palaeontology.

PALE-SIDED CUTWORM *Agrotis malefida* Guenée [Lepidoptera: Noctuidae]. See Cutworms.

PALE-STRIPED FLEA BEETLE *Systena blanda* Melsheimer [Coleoptera: Chrysomelidae]: In North America, a widespread pest of corn. Larva attacks corn seed and feeds on germ such that seed does not sprout; Adult attacks many Species of plants including weeds. Adult ca 4–5 mm long; Elytron with white median stripe and margin pale brown; legs reddish. Larva pale white; head brown; body tapering toward head; six legs. See Flea Beetles.

PALES WEEVIL *Hylobius pales* (Herbst) [Coleoptera: Curculionidae].

PALEY, WILLIAM (1743–1805) (Rose 1850, *New General Biographical Dictionary* 10: 450–451.)

PALIARDI, ANTON ALOIS (1799–1873) (Anon. 1874, Leopoldina 9: 98–99.)

PALIDIUM Noun. (Latin, *palus* = stake; *-ium* = diminutive > Greek, *-idion*. Pl., Palidia.) Scarabaeoid larvae: A group of Pali (Sl., Palus) arranged in one or more rows. Medially placed across the venter in front of the lower anal lip; paired and extending forward and inward from one of the ends of the anal slit; paired and extending straight, arcuately or obliquely forward from inside of one of the ends of the anal slit. Pali usually recumbent with apices directed toward the septula. The Palidium may be monostichous, distichous, tristichous or polystichous, according to whether there are 1, 2, 3 or many rows of Pali (Boving).

PALISOT, BARON DE BEAUVOIS (1755–1820) Author of Afrique et en Amerique & c. (1805.)

PALISOT DE BEAUVOIS, AMBROISE MARIE FRANÇOIS JOSEPH (1752–1820) (Merrill 1937, Proc. Am. phil. Soc. 76: 899–920.)

PALLA Noun. (Pl., Pallae.) Coccoidea: One of the lobes. (MacGillivray).

PALLAS, PETER SIMON (1741–1811) German-born zoologist who spent his life under Russian czarist patronage. Author of *De insectis viventibus intra viventia* (1760) in Berlin; 1867 went to St Petersburg and sent by Catherine II on collecting trips throughout Russia. Pallas published *Icones Insectorum, praesertim Rossiae Siberiaeque peculiarum* (1781–1806) as a product of collections made on his travels. (Jardine 1848, Nats Libr. 18: 17–76, bibliogr.)

PALLESCENT Adj. (Latin, *pallescens* pres. part. of *pallescare* = to grow pale.) Becoming pale; light in colour or tint.

PALLETHRINE® See Pyrethrin.

PALLETTE Noun. (French, *palette*; from Latin, *pala* = spade.) 1. Dytiscidae: Disc-like structure on anterior legs of males composed of three tarsal segments. 2. Coccids: One of the lobes. (MacGillivray).

PALLID Adj. (Latin, *pallidus* = pale.) Pale; very pale. Alt. Pallidus.

PALLIDE-FLAVENS Adj. (Latin, *pallidus* = pale; *flavescentis* = becoming yellow.) Pale or whitish-yellow.

PALLIUM Noun. (Latin, *pallium* = mantle, cover; *-ium* = diminutive. Pl., Pallia.) An erectile membrane partially closing the open cavity formed by the walls of the subgenital sclerite in Melanopli.

PALLOPTERIDAE Plural Noun. A Family of about 30 Species of medium-sized brachycerous acalyptrate Diptera. Found in north temperate regions of the Nearctic, temperate South America and New Zealand. Usually placed near Lonchaeidae and Otitidae. Taxonomically close to Lauxanidae. Recognized by absence of oral Vibrissae, Costa of wing with a Subcostal Break, Postocellar Bristles divergent, Mesonotum with one or more Presutural Dorsocentral Bristles and Anal Cell without an extension. Larvae mostly phytophagous; a few suspected of being entomophagous on scolytid or cerambycid larvae found under bark of felled trees. Eggs laid in cracks and crevices of bark; first instar larvae actively search for hosts. Pupation under bark.

PALM APHID *Cerataphis lataniae* (Boisduval); *Cerataphis variabilis* Hille Ris Lambers [Hemiptera: Aphididae].

PALM, CHARLES (1836–1917) (Anon. 1917, J. N. Y. ent. Soc. 25: 237–238.)

PALM LEAF BEETLE *Brontispa longissima* (Gestro) [Coleoptera: Chrysomelidae].

PALM-LEAF SKELETONIZER *Homaledra sabalella* (Chambers) [Lepidoptera: Coleophoridae].

PALM MEALYBUG *Palmicultor palmarum* (Erhorn) [Hemiptera: Pseudococcidae].

PALM SEEDBORER *Coccotrypes dactyliperda* (Fabricius) [Coleoptera: Curculionidae].

PALM WEEVIL BORER *Diocalandra frumenti* (Fabricius) [Coleoptera: Curculionidae].

PALMA Noun. (Latin, *palma* = palm. Pl., Palmae.) The basal segment of the anterior Tarsus when it is broadened or specifically modified; sometimes furnished with a Strigilis.

PALMATE Adj. (Latin, *palma* = palm; *-atus* = adjectival suffix.) Palm-like; resembling the palm of the hand with finger-like processes. Alt. Palmated; Palmatus. See Outline Shape.

PALMEN, JOHAN AXEL (1845–1919) (Lavander 1919, Mem. Soc. Fauna Flora fenn. 45: 227–233.)

PALMEN, JOHN AXEL (1845–1919) See Palmen, Johan Axel.

PALMEN'S ORGAN A statocystic organ on the head of larval and adult Ephemeroptera. Cuticular node posterodorsal of the eyes at the junction of four tracheoles. See Sensilla. Cf. Statocyst.

PALMER BODIES See Induvia.

PALMER WORM *Dichomeris ligulella* Hübner [Lepidoptera: Gelechiidae].

PALMER, EDWARD (1821–1911) (Stafford 1911, Pop. Sci. Mon. 78: 341–354.)

PALMER, EDWARD GILLETT WORCESTER (–1914) (Dan 1915, Proc. Linn. Soc. N.S.W. 40: viii.)

PALMER, KENNETH L (1889–1956) (Hall 1957, Proc. R. ent. Soc. Lond. (C)20: 75.)

PALMER, MERVYN GROVE (1879–1955) (Hall 1956, Proc. R. ent. Soc. Lond. (C) 20: 75.

PALMFLY *Elymnias agondas australiana* Fruhstorfer [Lepidoptera: Nymphalidae].

PALMIRANI, ANGELO (–1941) (Castellani 1945, Boll. Assoc. romana Ent. 1: 32.)

PALMKING BUTTERFLIES See Nymphalidae.

PALMONI, YAACOV (1897–1971) (Lulav 1972, Israel J. Zool. 21: 55–60.)

PALMULA Noun. (Latin, *palma* = palm.) 1. The Pulvillus, *sensu lato*. 2. The sensory or adhesive structure between the tarsal claws.

PALP Noun. (Latin, *palpare* = to feel, to touch. Pl., Palps.) See Palpus.

PALPAL Adj. (Latin, *palpare* = to feel, to touch; *-alis* = pertaining to.) Descriptive of structure belonging, relating or attached to a Palpus.

PALPAL CLAW Acarina: A claw arising from the distal end of the palpal Tibia. PC lacks musculature and is optically active; possibly derived from a seta. PC single, notched or forming a central axial prong from which 1+ accessory prongs arise. (Goff *et al.* 1982, J. Med. Ent. 19: 229.) Syn. Accessory Prong.

PALPAL FORMULA The number of segments in Maxillary and Labial Palpi. PF sometimes diagnostically important and provided in taxonomic keys or descriptions. Conventionally given as two numbers (separated by a dash, comma or hyphen) with the Maxillary number first. Cf. Tarsal Formula.

PALPAL SETA Acari: Seta on the palpal segments that may be branched, nude or forked. (Goff *et al.* 1982, J. Med. Ent. 19: 229.)

PALPAL THUMB-CLAW PROCESS Acari: The Palpal Tarsus which arises subapically from the Tibia and appears to oppose the Palpal Claw (Goff *et al.* 1982, J. Med. Ent. 19: 229.)

PALPARIUM Noun. (Latin, *palpare* = to feel; *arium* = place of a thing. Pl., Palparia.) Coleoptera: The membranous base of the Labial Palpi which permits extension not possible when Palpi are fused to the Labium.

PALPATE Adj. (Latin, *palpus* = caress, soft palm of the hand; *-atus* = adjectival suffix.) With a palpus. Trans. Verb (Latin, *palpatio,* from *palpare* = to stroke.) To examine by touch. During courtship, the contact of Palpi of one animal on the body of a potential mate.

PALPATION Noun. (Latin, *palpare* = to stroke; English, *-tion* = result of an action.) The act by which insects touch a substrate or other organisms with the Maxillary or Labial Palpi, presumably to explore or communicate. Cf. Antennation; Tarsation.

PALPI TURGIDI Palpi in which the last segment is turgid or swollen.

PALPICORN Adj. With long, slender, Antenna-like Palpi.

PALPIFER Noun. (Latin, *palpare* = to stroke; *ferre* = to carry. Pl., Palpifers.) 1. The apparent first segment or basal attachment of the first Maxillary Palpus. Palpifer is articulated with Stipes and contains intrinsic musculature; may be analogous with the antennal Scape. See Maxilla. Cf. Palpiger; Palpus.

PALPIFEROUS Adj. (Latin, *palpare* = to stroke; *ferre* = to carry; *-osus* = with the property of.) Bearing a Palpus. Alt. Palpigerous.

PALPIFORM Adj. (Latin, *palpare* = to stroke; *forma* = shape), Palpus-shaped; descriptive of structure which is short, cylindrical and segmented. See Palpus. Cf. Antenniform; Flagelliform; Papilliform. Rel. Form 2; Shape 2; Outline Shape.

PALPIGER Noun. (Latin, *palpare* = to stroke; *gerere* = to carry. Pl., Palpigers.) A Palpus-bearing sclerite. Specifically, the Palpus-bearing structure of the Mentum. Cf. Palpifer.

PALPIGEROUS STIPES Coleoptera larvae: Palpifer.

PALPIGRADI Noun. Whipscorpions. An Order of arthropods assigned to the Class Arachnida. Palpigrades typically 1–2 mm long, Prosoma with three unequal-sized dorsal sclerites and abdomen with long, slender, multisegmented

Flagellum at apex. Whipscorpions reside in caves, moist soil and intertidal habitats; they are infrequently collected.

PALPIMACULA Noun. (Latin, *palpare* = to stroke; *macula* = spot. Pl., Palpimaculae.) A sensory area on the Labial Palpus of some insects.

PALP-LIKE APPENDAGES Aculeate Hymenoptera: The dorsal structures of the oblong sclerites. (Imms).

PALPOGNATH Noun. (Latin, *palpare* = to stroke; Greek, *gnathos* = jaw.) The second Maxilla in centipedes (Comstock).

PALPULUS Noun. (Latin, *palpare* = to stroke. Pl., Palpuli.) The Maxillary Palpus in Lepidoptera when visibly developed.

PALPUS Noun. (Latin, *palpare* = to stroke. Pl., Palpi.) A paired, digitiform appendage of the Maxilla and Labium. The Maxillary Palpus is a multisegmented telopodite; each segment of the Palpus bears intrinsic musculature; this feature is also regarded as an indication of primitive segmentation. Receptors on the palpus are tactile or chemosensory in function. See Maxilla; Labium.

PALUDICOLOUS Adj. (Latin, *palus* = marsh; *colere* = to inhabit.) Descriptive of organisms that inhabit or visit marshes or marsh-like habitats. See Habitat. Cf. Agricolous; Algicolous; Caespiticolous; Fungicolous; Pratinicolous; Silvicolous. Rel. Ecology.

PALUMBO, AUGUSTO (1842–1896) (Sciascia 1896, Naturalista sicil. 1: 199–202.)

PALUS Noun. (Latin, *palus* = stake. Pl., Pali.) 1. A straight, pointed spine. 2. A component of the Palidium *sensu* Boving.

PALYNOLOGY Noun. (Greek, *palynein* = to scatter, *pale* = pollen; *logos* = discourse. Pl., Palynologies.) The study of pollen and spores. See Pollen; Spore. Cf. Entomopalynology; Melissopalynology.

PAMPANA, EMILIO (1895–1973) (Cambournac 1973, An. Inst. hyg. Med. trop. 1: 379–381.)

PAMPEL'S FIXATIVE A fluid fixative used by some Orthopterists to preserve colour patterns before long-term preservation of specimens in ETOH. Contents include: 3 parts Glacial Acetic Acid, 15 parts 95% ETOH, 30 parts distilled water and 6 parts formalin.

PAMPHILIIDAE Cameron 1890. Plural Noun. Web-Spinning Sawflies; Leaf-Rolling Sawflies. A small Family of Holarctic Symphyta (Hymenoptera) assigned to the Pamphilioidea or Megalodontoidea. Insects 7–15 mm long; dorsoventrally compressed; Antenna thread-like with 18–24 segments; Pronotum long with posterior margin nearly transverse; Forewing Sc free; 2r-rs present. Middle and hind Tibiae with subapical spurs; Abdomen laterally carinate; first and second Terga medially incised; Ovipositor short and slightly projecting from apex of Abdomen; male genitalia orthandrous. Cephaliciinae associated with pines; Pamphiliinae associated with angiosperms (Betulaceae, Rosaceae, Salicaceae.)

PAMPHILIOIDEA Cameron 1890. In some classifications a Superfamily of Symphyta [Hymenoptera] including the Families Megalodontidae, Pamphiliidae and Praesiricidae. See Meglodontoidea.

PANBIOGEOGRAPHY A methodology first developed by Croziat to explain the geographical distribution of organisms. See Vicariant Biogeography. Cf. Cladistic Vicariance Biogeography.

PANCERI, PAULO (1833–1877) (Cornalia 1877, Rc. 1st Lombardo Sci. lett. (2) 10: 445–480, bibliogr.)

PANDANUS MEALYBUG *Laminicoccus pandani* (Cockerell) [Hemiptera: Pseudococcidae].

PANDAR® See Monocrotophos.

PANDELLE, LOUIS (1824–1905) (Gobert 1905, Ann. Soc. ent. Fr. 74: 287–288.)

PANDEMIC Adj. (Greek, *pan* = all; *demos* = people; *-ic* = characterized by.) Medical Entomology: Pertaining to any disease which is manifest and prevalent simultaneously over a country, a continent, or widespread throughout the world during a limited period of time. Plague pandemics of Europe during Middle Ages serve as an example. Cf. Epidemic.

PANDORA MOTH *Coloradia pandora* Blake [Lepidoptera: Saturniidae].

PANDURATE Adj. (Latin, *pandura* = a bandore; *-atus* = adjectival suffix.) See Panduriform.

PANDURIFORM Adj. (Greek, *pandoura* = lute; Latin, *forma* = shape.) Violin-shaped; lute-shaped; descriptive of structure oblong with rounded ends and medially constricted. Cf. Calceiform; Soleaform; Unguliform. Rel. Form 2; Shape 2; Outline Shape.

PANGAEA Noun. (Greek, *pan* = all, *ge* = earth.) The supercontinent of geological history (Palaeozoic-Mesozoic) which subsequently split and formed Gondwana. See Gondwana.

PANGOLA GRASS APHID *Schizaphis hypersiphonata* (Basu) [Hemiptera: Aphididae].

PANIC24 EC® See Farnesene.

PANMICTIC Adj. (Greek, *pan* = all; *miktos* = mixed; *-ic* = characterized by.) 1. Pertaining to panmixia; 2. Pertaining to structures no longer of use or supposedly lost through Panmixsis.

PANMIXIA Noun. (Greek, *pan* = all; *mixis* = mixing.) Interbreeding not under the influence of natural selection.

PANNEWITZ, JULIUS (1788–1867) (Grunert 1861, Forstliche Blätter 2: 192–195; Ratzeburg 1874, Forstwissenschaftliches Schriftsteller-Lexicon 1: 390–393.)

PANOCON® See Fenthiocarb.

PANOISTIC OVARIOLE (Greek, *pan* = all; *oon* = egg.) 1. An Ovariole which lacks Trophocytes (Nurse Cells). The Oocytes are nourished by the follicular Epithelium. Panoistic oogenesis occurs in Thysanura, Siphonaptera, Odonata, and all orthopteroid Orders except Dermaptera. 2. A type of egg tube in which the Vitellarium contains eggs only (*sensu* Snodgrass). 3. Presumably the most primitive type of Ovariole in which Nutritive Cells are absent (Imms), or one in which the germ cells occur without interruption from one end to the other (Tillyard). Syn. Panoistic Egg Tube; Panoistic Ovary. See Reproductive System. Cf. Meroistic Ovary. Rel. Polytrophic Ovariole; Telotrophic Ovariole.

PANOISTIC OVARY An Ovary composed of Panoistic Ovarioles. Cf. Panoistic Ovariole.

PANORPATAE See Panorpidae.

PANORPIDAE Plural Noun. Common Scorpionflies. A moderate sized Family of Mecoptera. Adults with spotted or striped wings; male with enlarged apical genital bulb projecting above Abdomen. The male's habitus vaguely resembles a scoprion. Panorpids occur on low, broad-leafed vegetation along the margin of woods where the adults and larvae feed on dead insects. Cf. Panorpodidae.

PANORPODIDAE Short-faced Scorpionflies. A small Family of Mecoptera with two Species in North America. Males with enlarged genital bulb, but not carried over dorsum of Abdomen; females of some Species are brachypterous. Adults scrape vegetation for food. Cf. Panorpidae.

PANOSIN® See Fenthiocarb.

PANTEL, JOSÉ (1853–1920) (Dusmet y Alonso 1919, Boln. Soc. ent. Esp. 2: 166–167; Foulquier 1920, Science 52: 266–267; Navas 1920, Boln. Soc. ent. Esp. 3: 105–108, bibliogr.)

PANTHERINE Adj. (Latin, *pantherinus* = with spots like a panther.) 1. A colour almost like cervinus. 2. Spotted like a panther.

PANTHION® Parathion.

PANTON, E STUART (1866–1962) (Brown & Heinemann 1972, *Jamaica and its Butterflies*. xv + 478 pp. (15), London.)

PANTON, J HOYES (–1898) (Anon. 1898, Can. Ent. 30: 77–78.)

PANTOPHAGOUS Adj. (Greek, *panto* = all; *phagein* = to devour; Latin, *-osus* = with the property of.) Pertaining to organisms with a broad-spectrum diet; omnivorous. See Feeding Strategy.

PANTOPHTHALMIDAE Plural Noun. A Family of orthorrhaphous Diptera assigned to Superfamily Tabanoidea. Adult large-bodied, robust; Species occur in Neotropics.

PANURGINAE A Subfamily of solitary, ground-nesting, oligolectic bees (Andrenidae) found in all zoogeographical realms except Australia. Consists of about 35 Genera, most of which are found in the New World.

PANZER, GEORG WOLFGANG FRANZ (1755–1829) German Entomologist and author of many significant works dealing with Hymenoptera. Publications include: *Deutschlands Insecten (Faunae Insectorum Germanicae initia)* (published in cahiers at irregular intervals), *Entomologischen versuch uber die Turineschen Gattungen der Linneischen Hymenoptern* (1806) and *Index Entomologicus, pars prima, Eleutherata* (1813). (Duméril 1823, Considérations générales sur la classe des insectes. xii + 272 pp. (261–262), Paris.) Swainson, 1840, *Taxidermy; with Biography of Zoologists*. 320 pp. (287, bibliogr.), London. [Volume 12 of *Cabinet Cyclopedia*. Edited by D. Lardner.]; Eisinger 1919, Int. ent. Z. 13: 89–92, bibliogr.)

PAOLI LUIGIONI, RICORDO DI (1873–1937) (F.T. 1967, *Boll. Soc. romana Ent.* 22: 49–50.)

PAOLI, GUIDO (1881–1947) Italian student of Berlese and active researcher in Acarology and applied Entomology. Director of Agricultural Entomology Station in Florence. (Binaghi 1947, Memorie Soc. Ent. Ital. 26: 3–19, bibliogr.)

PAPAYA FRUIT-FLY 1. *Bactrocera papayae* Drew & Hancock [Diptera: Tephritidae] a member of the Oriental Fruit Fly complex and major international pest. PFF endemic in SE Asia from Thailand to Kalimantan and Sumatra; introduced into Irian Jaya (1989), Papua New Guinea (1992) and Queensland, Australia (1995.) Host plant range includes 160 Species in 50 Families; infests nearly all edible fruits, some vegetables and some cucurbit crops; a particular pest of mango, banana, citrus and papaya; larvae infest greener stages than Queensland Fruit Fly. Adult resembles QFF; with dark T-shaped mark on Abdomen dorsum. Strong flies and capable of long-distance movement in short period of time. Female feeds on microbes, honeydew and protein on surface of fruit and leaves for egg development; eggs white, banana-shaped, ca 1 mm long; 10–12 eggs deposited under surface of fruit; eclosion occurs ca 2–3 days; neonate larvae burrow into fruit; mature larvae pupate in soil. Multivoltine with number of generations depending upon climate and host plant. See Oriental Fruit Fly. Queensland Fruit Fly. 2. *Toxotrypana curvicauda* Gerstaecker [Diptera: Tephritidae]: Endemic to new world tropics and minor pest of papaya seeds. Adult mimics ichneumon-flies; female with long, curved Ovipositor. Syn. Pawpaw Fruit Fly.

PAPE, PAUL (1859–1933) (Rohlfien 1975, Beitr. Ent. 25: 274.)

PAPER MITE See Delusory Dermatitis.

PAPER WASP See Macao Paper-Wasp.

PAPER WASPS See Papernest Wasps.

PAPERBACK CICADA *Cicadetta hackeri* (Distant) [Hemiptera: Cicadidae].

PAPERNEST WASP 1. *Ropalidia revolutionalis* (Saussure) [Hymenoptera: Vespidae]: A Species of social wasp common in eastern Australia. Adult 10–12 mm long; wing length 7.0–7.5 mm; body dark brown with narrow yellow band or spots at base of Gaster; Clypeus coarsely punctured, 1.5 times wider than tall; dorsal tooth of Mandible blunt or truncate; first gastral Tergum apically not strongly convex; pronotal keel strong, but not sharp and distinctly sinuate; Metanotum with smooth posterior area extending to anterior margin; Propodeum coarsely punctate, with dense white pubescence and angles not strongly striate. Species widespread in Queensland and active throughout year except July. Nests common in garden shrubs and under eaves of buildings; nest consists of two rows of cells suspended from the Pedicel attached at the top; nest not long and sometimes consists of several adjacent combs. 2. *Ropalidia romandi* (Saussure) [Hymenoptera: Vespidae]: A Species of social wasp common in eastern Australia. Adult 6–8 mm long; wing 4.5–6.0 mm long; body predominantly yellow and ferruginous; Mesoscutum with two yellow longitudinal stripes; Mesopleuron predominantly yellow; Gaster yellow or ferruginous; punctures on head and Thorax weakly developed; second gastral Tergum longer than wide; Clypeus weakly and sparsely punctured, Mesepisternum smooth anteriorly, coarsely punctured posteriorly. Species common in Queensland and active throughout year except July. Nests common in trees well above ground level; nest may be more than a metre long and consist of many combs suspended free from one another and surrounded by a thin envelope with and entrance at the bottom; nest may contain tens of thousands of workers.

PAPERNEST WASPS Globally, representatives of the Subfamily Polistinae (Vespidae); in Australia, wasps of the Genera *Polistes* and *Ropalidia*. Polistinae all eusocial; nests constructed of plant material; cooperative care of brood; workers progressively feed larvae with masticated prey; cells used more than once; workers sterile. PNW regarded as pests in some urban habitat because people may be stung; regarded as beneficial insects in that adults collect caterpillars. See Common Paper Wasp.

PAPILIOFORM Adj. (Latin, *papilio* = butterfly; *forma* = shape.) Shaped as a butterfly wing. Rel. Form 2; Shape 2; Outline Shape.

PAPILIONACEOUS Adj. (Latin, *papilio* = butterfly; *-aceus* = of or pertaining to.) Butterfly-like in shape or aspect.

PAPILIONIDAE Plural Noun. Swallowtail Butterflies; Swallowtails; Parnassians. A cosmopolitan Family of ditrysian Lepidoptera. Adult large bodied; Antenna short; Maxillary palp minute; foreleg well developed, Epiphysis present; tibial spur formula 0-2-2; Pulvilli and arolium reduced; forewing R4 and R5 typically stalked. Larva stout, Thorax often humped; Prothorax with eversible Osmeterium. Pupa exposed, attached by Cremaster and central silken girdle. (See Hancock 1983, Smithersia 2: 1–48.)

PAPILIONOIDEA A highly derived Superfamily of ditrysian Lepidoptera. Included Families Libytheidae, Lycaenidae, Nymphalidae, Papilionidae, Pieridae and Riodinidae. Ocelli absent; Chaetosemata prominent; Antenna clubbed distally without hook; Proboscis without scales; tympanal organs absent. Pupa usually exposed; frequently with central silk girdle.

PAPILLA Noun. (Latin, *papilla* = nipple. Pl., Papillae.) 1. Any small, soft or pliant projection. 2. Any conical cuticular projection. 3. The modified Ligula in silk-spinning caterpillars.

PAPILLAE ANALES Lepidoptera: A pair of lobes at the apex of the female Abdomen which are used in oviposition; AP in some Species sclerotized and used to pierce plant tissue; in some Species invested with Setae and presumably sensory.

PAPILLARY Adj. (Latin, *papilla* = nipple.) Descriptive of structure with nipple-like processes and the apices rounded.

PAPILLATE Adj. (Latin, *papilla* = nipple; *-atus* = adjectival suffix.) With small surface elevation, porous at the apex. Alt. Papillatus.

PAPILLIFORM Adj. (Latin, *papilla* = nipple; *forma* = shape.) Papilla-shaped; descriptive of structure attached to a surface and projecting to resemble a Nipple, Tubercle or Papilla. See Papilla; Tubercle. Cf. Dentiform; Digitiform; Tuberculate. Rel. Form 2; Shape 2; Outline Shape.

PAPILLOSE Adj. (Latin, *papilla* = nipple; *-osus* = full of.) Pimply; superficially covered with raised circular spots, pimples or papillae. Alt. Papillosus. Papillous.

PAPILLULATE Adj. (Latin, *papilla* = nipple; *-atus* = adjectival suffix.) Like Papillae; with depressions or elevations or a small elevation in the centre; beset with many Papillules (Kirby & Spence).

PAPILLULE Noun. (Latin, *papilla* = nipple. Pl., Papillules.) Any Tubercle or Variole with a nipple-like elevated projection in the middle.

PAPPOSE Adj. (Latin, *pappus* = down; *-osus* = full of.) Downy; covered with pappus.

PAPPUS Adj. (Latin, *pappus* = down.) Pertaining to a fine down.

PAPYRACEOUS Adj. (Greek, *papyros* = reed-paper; Latin, *-aceus* = of or pertaining to.) Paper-like; papery. Papyrus-like in texture, appearance or physical properties.

PARABIOSIS Noun. (Greek, *para* = beside; *biosis* = manner of life. Pl., Parabioses.) A symbiotic relationship expressed by ants in which different Species use common nest galleries but colonies maintain distinct broods and do not unite or fuse. See Biosis; Commensalism; Symbiosis. Cf. Abiosis; Anhydrobiosis; Antibiosis; Archebiosis; Calobiosis; Cleptobiosis; Hamabiosis; Kleptobiosis; Lestobiosis; Phylacobiosis; Plesiobiosis; Synclerobiosis; Trophobiosis; Xenobiosis.

PARACARDO See Subcardo (MacGillivray).

PARACEPHALIC SUTURE Laterocephalic Suture; Frontogenal Suture; Epicranial Arm (MacGillivray).

PARACERATUBAE Plural Noun. Coccoidea: A type of Cerores with six openings arranged in a circle around the periphery of a central area (MacGillivray).

PARACHUTING FLIGHT A form of passive flight in which the insect engages in a slow, vertical descent; typified in male mayflies and male longhorn moths during swarming. See Flight; Passive Flight.

PARACLYPEAL LOBES Heteroptera: The Juga.

PARACLYPEAL PIECE A part on each side of the Maxillary Palpi in Pupae of some generalized Families of Lepidoptera.

PARACLYPEUS Noun. (Greek, *para* = beside; Latin, clypeus = shield. Pl., Paraclypeuses.) Lepidoptera larva: A narrow sclerite bordering the Clypeus at the sides.

PARACME Noun. (Greek, *parakme* = decadence. Pl., Paracmies.) The decline phase in the lineage of a Taxon after reaching its highest point of development. Cf. Acme; Epacme.

PARACOILA Noun. (Pl., Paracoilae.) The articulation of the Maxilla on the ventral aspect (MacGillivray).

PARACOPRID Adj. A group of coprophagous scarabeid beetles.

PARADEME Noun. (Greek, *para* = beside; *demas* = body.) A secondary inflection of the Integument which provides a surface for muscle attachment; a Phragma (MacGillivray).

PARADENSAE Noun. Coccoidea: More-or-less distinct thickenings on the ventral aspect of the Pygidium in some Species; resembling the Calles but extending longitudinally (MacGillivray).

PARADERM Noun. (Greek, *para* = beside; *derma* = skin. Pl., Paraderms.) The limiting membrane enclosing the Pronymph of Muscidae.

PARADERMAPTERA A Superfamily of the Order Dermaptera with an excessively flattened, conspicuously coloured, body. Contains only the Family Apachyidae.

PARADICHLOROBENZENE A white solid crystal with an oily surface and mothball-like odour. PDB is used in museum cabinets as an insect repellent or to repel wasps that nest in wall voids. Inhalation of PDB may result in headache, nausea, and throat/eye irritation. Prolonged contact may result in skin irritation and allergies. Ingestion of PDB may result in nausea, vomiting, diarrhea, liver and kidney damage. PDB has lower acute toxicity than naphthalene, which is also commonly used as a commercial repellent.

PARADIGM Noun. (Greek, *paradeigma* = example > *para* = beside; *deiknyai* = to show.) An example, model or pattern that demonstrates a principle or mode-of-action for a process. Alt. Model.

PARADORSAL MUSCLE A longitudinal muscle or group of longitudinal fibres positioned on the lateral part of the dorsum above the line of the spiracles; pleural muscles.

PARAFACIAL Noun. (Greek, *para* = beside; Latin, *facies* = face; *-alis* = pertaining to.) 1. Descriptive of structure or substance associated with the Parafacial region of the head. 2. Diptera: Part of the face between Facial Ridges and eyes (Curran); so-called Genae of writers on chaetotaxy (Comstock). Alt. Parafacialia.

PARAFRONS Noun. (Greek, *para* = beside; Latin, *frons* = forehead.) The Vertex when reduced to a linear area between the front and a compound eye (MacGillivray).

PARAFRONTAL Noun. (Greek, *para* = beside; Latin, *frons* = forehead; *-alis* = pertaining to. Pl., Parafrontals.) Diptera: The part of the front laterad of the Frontal Bristles (Curran).

PARAGLOSSA Noun. (Greek, *para* = beside; *glossa* = tongue. Pl., Paraglossae.) A paired, labial structure positioned at either side of the Ligula; sometimes connected with Ligula, sometimes free and two-segmented; corresponds with Maxillary Galea.

PARAGNATHA Noun. (Greek, *para* = beside; *gnathos* = jaw. Sl., Paragnath or paragnathus.) 1. Appendage-like organs between the Mandibles and Maxillae of Thysanura and Collembola. 2. Paralinguae of Folsom; 3. Maxillae of Hansen (Comstock); 4. two small processes developed at the sides of the Hypopharynx, in insects sometimes called the Maxillulae (Tillyard).

PARAGULA Noun. (Greek, *para* = beside; Latin, *gula* = gullet. Pl., Paragulae; Paragulas.) Part of a Postgena along the lateral margin of a Gular Suture (MacGillivray). See Gula.

PARALECTOTYPE Noun. (Greek, *para* = beside; *lektos* = chosen; *typos* = pattern. Pl., Paralectotypes.) All members of the type-series remaining after a Lectotype has been selected. See Type. Cf. Cotype; Holotype; Paratype; Syntype. Rel. Nomenclature; Taxonomy.

PARALINGUA Noun. (Greek, *para* = beside; Latin, *lingue* = tongue. Pl., Paralinguae.) A short sclerite near the anterior end of each Pharyngea (MacGillivray).

PARALLELISM Noun. (Greek, *parallelos* from *para* = beside; *allelon* = of one another; English, *-ism* = condition. Pl., Parallelisms.) Systematic theory: 'The separate development of similar characters in two or more relatively closely related lineages on the basis of genotypic similarity inherited from a common ancestor' (Holmes 1980, 49). See Homoplasy. Cf. Convergence; Reversal.

PARALOGOUS Adj. (Greek, *para* = beside; *logos* = reason, discourse.) Descriptive of or pertaining to paralogues.

PARALOGUES (Greek, *para* = beside; *logos* = reason, discourse.) Plural Noun. Anatomically similar parts or organs which are not necessarily similar in function and whose bearers are not related through descent.

PARALOGY Noun. (Greek, *para* = beside; *logos* = reason, discourse. Pl., Paralogies.) 'A part or organ in one animal similar in anatomical or microanatomical structure to a part or organ in a different animal.' (Hunter 1964, Nature 204: 604.) See Operational Homology; Morphological Correspondence. Cf. Analogy; Homology.

PARAMAR®. Parathion.

PARAMERA See Paramere.

PARAMERE Noun. (Greek, *para* = paired; *meros* = body. Pl., Parameres.) 1. Half of a bilaterally symmetrical structure. 2. Coleoptera: One of paired lateral processes or lobes of the Phallobase; also applied to the Gonapophyses (Snodgrass). 3.

Paired processes on Sternum IX, arising near the base of the Penis in insects other than the Endopterygota (Tillyard). 4. The smaller medial pair of male Gonapophyses that are closely associated with the Aedeagus (Imms). See Endoparamere.

PARAMETER Noun. (Late Latin, *parametrum*.) A variable.

PARAMETRIC STATISTICS See Non-Parametric Statistics.

PARAMONOV, SEREJ (ALEXEJ) (–1967) (Under pseudonym S. Lesnoi.) (Liepa 1969, J. ent. Soc. Aust. (N.S.W.) 5: 3–22, bibliogr.)

PARAMORPH Noun. (Greek, *para* = beside; *morphe* = form. Pl., Paramorphs.) Any anatomical form which can be attributed to environmentally induced factors and not genetically predetermined factors.

PARAMUTUALISM Noun. (*para* = beside; *mutuus* = exchanged; English, *-ism* = condition. Pl., Paramutualisms.) See Facultative Symbiosis.

PARANAL Adj. (Greek, *para* = beside; *anus* = anus; *-alis* = pertaining to.) Descriptive of something near or adjacent to the Anus; pertaining to anal structures.

PARANAL FORKS Lepidoptera larva: Two lateral, bristle-like structures in some caterpillars; PF used to throw frass pellets a distance from the body.

PARANAL LOBES Podical sclerites, in the broad sense.

PARANAL PLATES Coccids: The tenth pair of dorsal sclerites on the sixth abdominal segment (MacGillivray).

PARANEUROPTERA See Odonata.

PARANOTAL EXPANSION The lateral expansions of the Notum, under the form which insect wings have arisen (Tillyard).

PARANOTAL HYPOTHESIS See Paranotal Theory.

PARANOTAL LOBES 1. Lateral cuticular projections of the Pronotum in certain fossil insects. 2. Hypothetical lobes of the Mesonotum and Metanotum. Regarded by some Entomologists as precursors of wings. 3. Insect embryo: Wing rudiments of the Thorax in winged insects (Snodgrass).

PARANOTAL THEORY More correctly termed 'Paranotal Hypothesis.' An explanation for the origin of insect wings which asserts that wings arose first as lateral expansions, or Paranota, positioned along the sides of the thoracic Terga (Crampton). Cf. Tracheal Gill Theory.

PARANOTUM Noun. (Greek, *para* = beside; *noton* = back. Pl., Paranota.) Lateral expansions of the thoracic tergal region (Wardle). Heteroptera (Tingidae): Flattened or laminate, more-or-less downwardly bent sides of the Pronotum. Lateral lobe; Parapsidis; Scapula or Justascutellum *sensu* MacGillivray.

PARAOESOPHAGEAL Adj. Positioned near or adjacent to the Oesophagus.

PARAPAMPHILIIDAE Plural Noun. A fossil Family of Symphyta (Hymenoptera).

PARAPATRIC Adj. (Greek, *para* = beside; *pater* = father; *-ic* = characterized by.) Pertaining to populations whose geographical distributions have a very narrow zone of overlap. Cf. Sympatric.

PARAPATRIC SPECIATION (See Bush 1975, Ann. Rev. Syst. Ecol. 6: 339–364.) See Allopatric Speciation; Stasipatric Speciation.

PARAPAUSE Noun. (Greek, *para* = beside; *pausi* = ending. Pl., Parapauses.) Dormancy in an organism that is initiated and terminated by different intensities of the same environmental factor (*e.g.* photoperiod) (Müller 1965, Zool. Anz. 29 suppl: 182.) See Aestivation; Dormancy; Hibernation. Cf. Diapause; Quiescence. Rel. Eudiapause; Oligopause.

PARAPHARYNX Noun. (Greek, *para* = beside; *pharnyx* = gullet. Pl., Parapharynices.) Ventral portion of the Prepharynx (MacGillivray).

PARAPHEROMONE Noun. (Greek, *para* = beside; *phereum* = to carry; *hormone* = to excite. Pl., Parapheromones.) Specific chemical compounds produced by plants and which mimic the effect of insect pheromones. Parapheromones may be synthesized and used to detect, monitor, mass trap or disrupt the mating of target insect Species. Cf. Pheromone.

PARAPHRAGMA Noun. (Greek, *para* = beside; *phragma* = fence. Pl., Paraphragmata.) The Phragma between the Sternellum and Scutellum (MacGillivray). See Phragma. Cf. Cephalophragma.

PARAPHRAGMINA Noun. (Greek, *para* = beside; *phragma* = fence. Pl., Paraphragminae.) The transverse external opening (aperture) or thickening marking the entrance to the Paraphragma (MacGillivray).

PARAPHYLETIC Adj. (Greek, *para* = beside; *phylon* = race; *-ic* = characterized by.) Pertaining to a group of Taxa whose most recent common ancestor has given rise to one or more excluded Taxa or monophyletic groups of excluded Taxa of which the sister group is completely included in the group. (Oosterbroek 1987, Syst. Zool. 36 (2): 103–108.) Cf. Polyphyetic; Monophyletic.

PARAPHYLY Noun. See Monophyly; Polyphyly.

PARAPHYSIS Noun. (Greek, *para* = beside; *physis* = growth; *sis* = a condition or state. Pl., Paraphyses.) The chitinized thickenings or marginal projections on the Pygidium of scale insects. (Densaria *sensu* MacGillivray).

PARAPLECIIDAE Hong 1983. A monogeneric Family of fossil Diptera known only from the type Species taken in Middle Jurassic Period deposits of Liaoning, China.

PARAPLECOPTERA Martynov 1938. (Carboniferous-Triassic Period.) An Order of insects; Sister group of Plecoptera, Plecoptera differ from Paraplecoptera in that former have three tarsal segments, fusion of MP and CU1 in both wings and fusion of MA with RS in the base of the hindwing. See Plecoptera.

PARAPLEURON Noun. (Greek, *para* = beside; *pleuron* = side. Pl., Parapleura.) The undivided Pleura of the Thorax in some Coleoptera, positioned ventrally on each side of the Sterna (Tillyard). Alt. Parapleurum

PARAPODIUM Noun. (Greek, *para* = beside; *pous* = foot; Latin, *-ium* = diminutive > Greek, *-idion*. Pl., Parapodia.) Annelida: Primitive feet, abdominal false-legs, or Pseudopods (Packard). Specifically, segmented abdominal processes of Symphyla (Smith); a slender cylindrical process of the proximal leg-joint in Symphyla (Comstock).

PARAPROCT Noun. (Greek, *para* = beside; *proktos* = anus. Pl., Paraprocts.) One of the two lobes formed by the ventrolateral parts of the Epiproct (Snodgrass); a pair of lobes bordering the Anus laterally (Tillyard); the lateral part of the vestigial eleventh segment in Mayflies (Needham); Parapodial Sclerites (Crampton). A lateral lobe of the eleventh abdominal Tergum positioned laterad of the Anus.

PARAPSIDAL FURROWS Hymenoptera: Longitudinal grooves on each side of the Mesoscutum of Proctotrupidae that separate Parapsides from the middle lobe. Syn. Parapsidal Grooves (Chalcidoidea); Parapsidal Suture (Formicoidea). See Notaulus.

PARAPSIDAL SUTURE Formicidae: A suture which separates the median area of the Mesonotum from the Parapsis of each lateral area of the Mesonotum. Bethylidae: See Notaulus.

PARAPSIDAL Adj. (Greek, *para* = beside; *apsis* = arch; *-alis* = characterized by.) Descriptive of structure associated with the Parapsides.

PARAPSIS Noun. (Greek, *para* = beside; *apsis* = arch; *sis* = a condition or state. Pl., Parapsides.) Hymenoptera: Lateral part of the Scutellum separated from the mesial part by the Parapsidal Furrow, Suture, or Groove. Chalcidoidea: The sides of the Scutellum; the lateral portion of the Scutellum of the Mesothorax when it is divided into three parts by longitudinal sutures.

PARAPTERON Noun. (Greek, *para* = beside; *pteron* = wing. Pl., Paraptera.) A small sclerite, articulated on the dorsal extremity of the Episternum just below the wings. The Parapteron is absent from the Prothorax. Epipleurite in the broad sense.

PARAPULVILLUS Noun. (Greek, *para* = beside; Latin, *pulvillus* = small cushion. Pl., Parapulvilli.) The pad ventral of the insect pretarsal claw.

PARAQUAT Noun. A toxic, broad-spectrum herbicide used on aquatic weeds.

PARARCHEXYELIDAE Plural Noun. A fossil Family of Symphyta (Hymenoptera).

PARARHOPHITIDAE Plural Noun. A monogeneric Family of Apoidea (Hymenoptera).

PARASAGITTAL PLANES Any of several planes that pass parallel to the midline in the same plane as the Sagittal Section. If the insect body is cut into an infinite number of thin sections, then only one Sagittal Section would exist with an infinite number of Sarasagittal Sections. See Axis; Aspect; Orientation. Cf. Frontal Plane; Sagittal Plane; Transverse Plane.

PARASCUTELLUM Noun. (Greek, *para* = beside; Latin, *scutellum* = small shield. Pl., Parascutella.) 1. The lateral part of the Mesosubscutella between the Mesascutella and an Epimeron (MacGillivray). 2. Areas on each side of the Scutellum (Crampton).

PARASCUTULES Noun. (Sl., Parascutulis.) Hemiptera: The lateral parts of the Scutellum covered by the wings (MacGillivray).

PARASITA See Parasitica.

PARASITE Noun. (Greek, *parasitos* = one who eats at the table of another > *para* = beside, *sitos* = food; Latin, *parasitus* > *para* = beside; *sitos* = food; *-ites* = inhabitant. Pl., Parasites.) An organism that lives on or in another organism, or at the expense of another organism. The concept of parasitism embraces many groups of plants, invertebrate and vertebrate organisms. Parasitic insects are distinguished from parasitoid insects in several features. Parasitic insects include Siphonaptera (fleas), Mallophaga (biting and chewing lice), some Diptera and Species scattered in other Taxa. The hosts of 'parasitic' insects are typically vertebrates. Parasitic insects are 'parasitic' as adults and larvae or nymphs. Parasitic insects are significantly smaller than their hosts; parasitic insects feed upon all stages of their hosts; ideally, parasitic insects do not kill their hosts. See Primary Parasite; Secondary Parasite; Hyperparasite; Facultative Parasite; Obligatory Parasite; External Parasite; Internal Parasite. Cf. Autophyte; Episite; Heterophyte; Host; Idiobont; Parasitoid; Protelean Parasite; Saprophyte. Rel. Predator; Pathogen; Symbiont; Commensal.

PARASITE FLIES See Tachinidae.

PARASITGMATIC PORES Coccids: The spiracerores (MacGillivray).

PARASITIC Adj. (Greek, *parasitikos*; Latin, *parasiticus; -ic* = of the nature of.) Pertaining to organisms that live as parasites. Living on or in another animal in such a way as to derive nourishment from the tissues of the host. Alt. Parasitical. See Parasitism.

PARASITIC GRAIN WASP *Cephalonomia waterstoni* Gahan [Hymenoptera: Bethylidae]: A small, black wasp which develops as an external parasite of beetle larvae infesting grain.

PARASITIC WOOD WASPS See Orussidae.

PARASITICA Noun. 1. The Anoplura or sucking lice *sensu* Torre Bueno and older workers. 2. Hymenoptera: The subdivision of Apocrita which includes the Ichneumonoidea, Evanioidea, Ceraphronoidea, Proctotrupoidea, Chalcidoidea, and Cynipoidea, and aberrant members of these groups which are sometimes classified as independent Superfamilies. Characterized by adult females which possess an appendicular, tube-like Ovipositor and immatures which typically develop as parasites of other insects and spiders or as gall-formers on many groups of plants. Cf. Apocrita.

PARASITISM Noun. (Greek, *parasitos* = parasite; English, *-ism* = condition.) 1. The act or state of being a parasite. 2. A form of symbiosis involving at least two unrelated Species. One symbiont (the parasite) lives at the expense of the other symbiont (the host), provides no benefit to the host and eventually destroys the host. See Ectoparasitism; Endoparasitism; Solitary Parasitism; Gregarious Parasitism. See Symbiosis. Cf. Commensalism; Hyperparasitism; Inquilinism; Predation; Social Parasitism; Temporary Social Parasitism.

PARASITIZE Transitive Verb. To attack, as a parasite.

PARASITOID Noun. (Greek, *parasitos* = parasite; *eidos* = form. Pl., Parasitoids.) An organism which resembles a parasite. Within the Insecta, a condition intermediate between the idealized concepts of predation and parasitism. Parasitoid insects are distinguished from parasitic insects in several features: 1. Parasitoids appear taxonomically restricted to the Hymenoptera and Diptera. 2. The hosts of parasitoids are typically other insects or rarely other arthropods. 3. Parasitoids are 'parasitic' as larvae only (adults of some parasitoids may host feed). 4. Parasitoids are smaller than the hosts upon which they develop but are within the same order-of-magnitude in size. 5. Typically, parasitoids only attack one stage of host (egg/larva/nymph/Pupa/adult) with a few Species exhibiting variants on this scheme (egg-larva parasitoids, egg-Pupa parasitoids, larva-Pupa parasitoids.) In a few instances the parasitoid larva is parasitic during the early stages and episitic during later development (*e.g.,* Tachinidae.) 6. Parasitoid larvae kill their hosts and death of the host transpires near the end of the parasitoid's larval development. See Symbiont; Commensal. Cf. Bacteroid; Hyperparasitoid; Viroid. Rel. Parasite; Predator; Pathogen.

PARASITOLOGY Noun. (Greek, *parasitos* = parasite; *logos* = discourse. Pl., Parasitologies.) The study of parasitic organisms. The social emphasis of Parasitology is typically focused on parasites living in or on animals and humans.

PARASOCIAL Adj. (Greek, *para* = beside; *sociare* = to associate; *-alis* = characterized by.) Descriptive of a behavioural condition involving a colony of bees in which adult females of one generation live together. A condition less than Eusocial in which individuals display care for young or reproductive division of labour or overlapping generations which contribute to colony labour. Categories under Parasocial include: Communal, quasisocial and semisocial. See Eusocial. Cf. Presocial; Subsocial.

PARASTIGMA See Pterostigma.

PARASTIGMATIC GLANDS Small, circular glands, which secrete a waxy powder, sometimes present around the spiracles of Coccidae, the Spiracerores of MacGillivray).

PARASTIPES Noun. (Greek, *para* = beside; Latin, *stipes* = stalk.) The Subgalea or sclerite mesally bordering the Stipes (Crampton). See Stipes.

PARATELI CORCULUM The twelfth cardinal chamber.

PARATELUM Noun. (Greek, *para* = beside; *telos* = end. Pl., Paratella.) An obscure term for the twelfth (penultimate) segment in the hypothetical (generalized) insect body plan. See Groundplan.

PARATELY Noun. (Greek, *para* = beside; *telos* = end. Pl., Paratelies.) Evolution from material unrelated to that of 'type', but resulting in superficial resemblance. See Evolution.

PARATERGITE Noun. (Greek, *para* = beside; *tergum* = back; *-ites* = constituent. Pl., Paratergites.) 1. The lateral marginal region of the Notum (Crampton). 2. Chalcidoidea (Encyrtidae): Paired, narrow sclerites located symmetrically lateral of Syntergum IX in Tetracneminae; Paratergites indicate that primarily tergite IX was undivided. See Laterotergite.

PARATHENE® Parathion.

PARATHION An organic-phosphate (thiophosphoric acid) compound {O,O-diethyl-O-4-nitrophenyl phosphorothioate} used as an acaricide, contact insecticide and stomach poison against mites, thrips, leaf miners, plant-sucking insects, leaf-chewing insects, crickets and borers. Compound applied to alfalfa, barley, corn, cotton, rice, soybeans, sunflowers and wheat; also applied as a public-health insecticide for mosquito control. Compound first manufactured by Bayer AG in Germany ca 1947; subsequently produced by numerous formulators. Trade names include: Alkron®, Alleron®, Bladan®, Corothion®, Ekatox®, Etilon®, Folidol®, Panthion®, Paramar®, Parathene®, Parawet®, Phoskil®, Soprathion®, Strathion®, Thiophos®. See Organophosphorus Insecticide.

PARATIS Noun. 1. A cuticular enlargement at the proximal end of the Subcardo or Cardo. 2. The Artis articulating against a Parcoila (MacGillivray).

PARATORMA Noun. (Greek, *para* = beside, *tormos* = socket. Pl., Paratormae.) Diptera: A strongly chitinized sclerite connecting the lateral margins of the Pharynx and Tormae (MacGillivray).

PARATYPE Noun. (Greek, *para* = beside; *typos* = pattern. Pl., Paratypes.) Nomenclature: Any specimen in a series from which a description has been prepared, other than the one specified as the Type-specimen or Holotype of the Species. Incorrectly, a specimen which has been compared with the Type (Jardine). See Type. Cf. Cotype; Holotype; Syntype. Rel. Nomenclature; Taxonomy.

PARAUPSILON Noun. Diptera: The arm of the Y of the upsilon extending obliquely across the lateral surface of the Mediproboscis; Furca-2 (MacGillivray).

PARAWET® Parathion.

PARAXYMYIIDAE Rohdendorf. Plural Noun. A Family of fossil Diptera known from two monotypic Genera preserved in Jurassic Period deposits of central Asia and China.

PARAZOONOSIS Noun. (Greek, *para* = beside; *zoon* = animal; *nosos* = *disease*. Pl., Parazoonoses.) A zoonotic disease of wild or domestic animals transmitted to humans and for which man is an essential host for the pathogen. See Euzoonosis; Zoonosis.

PARCE Adj. (Latin, *parcus* = sparing, frugal.) Sparse or sparsely.

PARCIDENTATE Adj. (Latin, *parcus* = sparing; *dens* = tooth.) Pertaining to animals with few teeth.

PARECIUM Noun. (Greek, *para* = beside; *oikos* = house; Latin, *-ium* = diminutive > Greek, *-idion*.) Termitidae: The air space surrounding the fungal garden.

PARELCANIDAE Carpenter 1966. Plural Noun. See Anelcanidae.

PAREMPODIUM Noun. (Greek, *para* = beside; *en* = in; *pous* = foot; Latin, *-ium* = diminutive > Greek, *-idion*. Pl., Parempodia.) Heteroptera: Bristle-like appendages of the Empodium (Holway 1935.) See Pretarsus. Cf. Paronychia. Rel. Leg.

PARENCHYMA Noun. (Greek, *parenkein* = to pour in beside. Pl., Parenchymas.) 1. Soft, thin-walled undifferentiated plant cells that vary in structure and function, and which are capable of mitosis when mature. 2. The essential and distinctive (characteristic) tissue of an organ. 3. The abnormal tissue of an gland or organ. 4. The soft, jelly-like interstitial connective tissue in flatworms and some other invertebrates. Alt. Parenchyme. See Mesophyll; Pith. Cf. Aerenchyma. Rel. Organ; Tissue.

PARENCHYMATOUS Adj. (Greek, *para* = beside; *engchyma* = infusion; Latin, *-osus* = with the property of.) Composed of soft cellular and connective tissue.

PARENT, BENOIT (1922–1976) (Paradis 1976, Bull. ent. Soc. Can. 8: 11.)

PARENT, OCTAVE (1882–1942) (Sachtleben 1942, Arb. morph. taxon. Ent. Berl. 9: 133–134.)

PARENTI, ALBERTO (–1965) (Tassi 1965. Boll. Assoc. romana Ent. 20: 21–23.)

PARFIN, S I (1918–1966) (Gurney & Walkley 1967, Proc. ent. Soc. Wash. 69: 190–192, bibliogr.)

PARFITT, EDWARD (1820–1892) (Anon. 1893, Entomol. mon. Mag. 29: 73.)

PARIA Noun. (Latin, *paries* = wall. Pl., Pariae.) Scarabaeoid larvae: A lateral, paired region of the Epipharynx extending from the Clythrum, Epizygum and Haptomerum (or in their place, the Tylus) back to the parietal elements (Dexiotorma and Laeotorma) and delineated intero-laterally from the Pedium by bristles or Asperities of the subregion Chaetoparia and the Phobae. The lateral lobe of Hayes (Boving).

PARIETAL Adj. (Latin, *paries* = wall; *-alis* = pertaining to.) Descriptive of structure or subtance associated with the wall (lining) of a cavity of the body or of an organ.

PARIETALIA Noun. (Latin, *paries* = wall.) The dorsal sclerites of the head between the frontal and occipital regions (Crampton).

PARIETALS Noun. (Latin, *paries* = wall. Sl., Parietal.) 1. Lateral areas of the head between the Frons and Occiput. Parietal sclerites are separated dorsally by the Coronal Suture. Each area bears an Antenna, lateral Ocellus and compound eye (Snodgrass). 2. Adfrontals; the Vertex of the insect head (MacGillivray).

PARIETES Noun. (Latin, *parietes*. = wall.) 1. General: Walls. 2. The perpendicular sides of a honey-comb. 3. Inner walls of any body-cavity (Archaic).

PARIGENITALS Plural Noun. Coccoidea: The Genacerores (MacGillivray).

PARIS, AUGUSTE SIMON (1794–1869) (Reiche 1869, Ann. Soc. ent. Fr. (4) 9: 599–600, bibliogr.)

PARIS GREEN Copper acetoarsenite. An arsenical dust which functions as a stomach poison to control some insects.

PARISH, HERBERT SIMPSON (1875–1957) (Alexander 1959, Ent. News 70: 29–32.)

PARISI, BRUNO (1884–1957) (Moltoni 1957, Atti Soc. ital. sci. nat. 96: 211–222, bibliogr.)

PARITY Noun. (Latin, *paria* = equal things; *paritas* > *par* = equal; *itas* = a state or degree.) 1. The equality in quality, state or condition of two or more things. 2. An equivalence in a farmer's purchasing power and the purchasing power for a period of time covered by government financial support for agricultural commodity price. 3. A ratio between agricultural and nonagricultural prices at a specified past time. 4. The number of offspring born to a female.

PARK, ORLANDO (–1969) (Pearce 1969. Entomol. mon. Mag. 105: 150.)

PARKER, FRANK HENRY (1910–1984) American amateur Entomologist specializing in Coleoptera of Arizona. (Werner 1986, Pan-Pacif. Ent. 62: 1–5, bibliogr.)

PARKER, H L U. S. Bureau of Entomology and Plant Quarantine Entomologist specializing in biological control of insects.

PARKER, JOHN ROBERT (1884–1972) (Blickenstaff & Cowan 1973, J. Econ. Ent. 66: 588.)

PARKER, RALPH LANGLEY (1892–1968) (Smith & Knutson 1968, Ann. ent. Soc. Am. 61: 1631.)

PARKER, SAMUEL (1759–1825) (Rose 1850, *New General Biographical Dictionary* 10: 480 481.)

PARKER, WILLIAM B (1885–1974) (Gardner & Michelbacher 1975, J. Econ. Ent. 68: 281.)

PARKES, WALTER RANDELL (1905–1932) (Worms 1932, Entomologist 65: 96.)

PARKS, THADDEUS HEDGES (1887–1971) (Goleman, 1971, J. Econ. Ent. 64: 1578.)

PARLATORIA DATE-SCALE *Parlatoria blanchardi* (Targioni-Tozzetti) [Hemiptera: Diaspididae].

PARLODION The trade name for a collodoin of nitrocellulose. A spread film used to coat copper grids bearing specimens or thin sections for examination by transmission electron microscopy.

PARMAN, DANIEL CLEVELAND (1885–1967) (Lindquist 1968, J. Econ. Ent. 61: 589.)

PARMENTER, LEONARD (1903–1969) (Payne 1970, Lond. Nat. 49: 130–131.)

PARNASSIANS Plural Noun. (Greek, *parnassos* = mountain in Greece.) See Papilionidae. Papilionid butterflies which lack tail on hindwing; female abdominal venter with pouch formed of wax by male following copulation. Eggs turbinate; larvae flattened, leech-like; pupate on ground among leaves. In North America typically alpine in western states.

PARNOPINAE A small Subfamily of Chrysididae (Hymenoptera: Aculeata). One Genus is found in the New World; two Genera are found in the Old World. Characterized by females with the gastral Sterna flat or concave and three exposed Terga; male with four exposed Terga; a long, exserted tongue, extending to middle Coxae. Species are parasites of sandwasps (Bembicinae).

PARONYCHIUM Noun. (Greek, *para* = beside; *onyx* = nail; Latin, *-ium* = diminutive > Greek, *-idion*. Pl., Paronychia.) A bristle-like structure on the Pulvillus of the insect foot.

PARORYSSIDAE Martynov 1925. Plural Noun. A fossil Family of Symphyta (Hymenoptera) assigned to the Orussoidea.

PAROUS RATES The rate of offspring production (birth rate, natality rate), usually associated with synanthropic Diptera.

PARRENIN, DOMINIQUE (1665–1742) (Rose, 1850, *New General Biographical Dictionary* 10: 486–487.)

PARROTT, P J American Entomologist with New York State Experimental Station.

PARRY, FREDERICK JOHN SIDNEY (1810–1885) (McLachlan 1885, Proc. ent. Soc. Lond. 1885: xli; Musgrave 1932, *A Bibliography of Australian Entomology 1775–1930,* viii + 380 pp. (249–250), Sydney.)

PARRY, THOMAS (–1872) (Anon. 1872, Entomol. mon. Mag. 9: 292.)

PARS Noun. (Latin, *pars* = part.) A part of a structure.

PARS BASALIS See Cardo.

PARS INTERCEREBRALIS The dorsomedian part of the Protocerebrum. Cells in anterior part connected with ocellar nerves; also contains Neurosecretory Cells; other cells connect with Protocerebral Bridge. See Brain; Protocerebrum; Central Body.

PARS STIPITALIS LABII See Prementum.

PARS STRIDENS The rasp (file) portion of a stridulatory device composed of tubercles. PS occurs on different parts of the insect body. Syn. Stridulatory File. See Stridulation. Cf. Plectrum.

PARSHLEY, HAROLD MADISON (1884–1953) American academic (Smith College) and student of Aradidae. Parshley published an influential '*Bibliogr. of North American Heteroptera.*'(Mallis 1971, *American Entomologists*. 549 pp. (234–235), New Brunswick.)

PARSIMONIOUS Adj. (Latin, *parsimonia* fr. *parcere* = to save; to spare; *-osus* = with the property of.) Pertaining to parsimony. Economical. See Parsimony.

PARSIMONY Noun. (Latin, *parsimonia* fr. *parcere* = to save; to spare. Pl., Parsimonies.) In systematics theory: Events or modifications of structure which are the result of the fewest intervening steps or processes. Parsimony exists at several levels in theoretical development of systematics. 1. Evolutionary Parsimony argues that evolution proceeds along the most economical course with the minimum number of 'steps' (Kluge 1984, Cladistics: Perspectives on the Reconstruction of Evolutionary History, Duncan & Stuessy eds.) 2. Methodological Parsimony is a procedural approach to determining the relationship among Taxa being classified. Several methods have been proposed: Camin-Sokal Method (Camin & Sokal, 1965); Dollo Parsimony (Farris 1977); Wagner Parsimony (Kluge & Farris, 1969); Polymorphism Parsimony. See each for details.

PARSLEY APHID *Dysaphis apiifolia* (Theobald) [Hemiptera: Aphididae].

PARSLEYWORM *Papilio polyxenes asterius* Stoll [Lepidoptera: Papilionidae] larva. Syn. Black Swallowtail (adult).

PARSNIP SEED WASP *Systole* sp. [Hymenoptera: Eurytomidae].

PARSNIP WEBWORM *Depressaria pastinacella* (Duponchel) [Lepidoptera: Oecophoridae]: A pest of parsnip, celery and wild carrot in northeastern USA and southern Canada. Overwinters as an adult in leaf litter and under bark.

PARSONS, CARL T (1914–1973) (Lawrence 1975, Coleopts Bull. 29: 355–356, bibliogr.)

PARSONS, RUDOLF E R (–1967) (Kennedy 1969, Proc. R. ent. Soc. Lond. (C) 33: 55.)

PARTES ORIS An archaic term referring to the mouthparts of insects, including the Mandibles, Maxillae and Labium (second Maxillae).

PARTHENIUM STEM-GALLING MOTH *Epiblema strenuana* (Walker) [Lepidoptera: Tortricidae].

PARTHENOGAMY Noun. (Greek, *parthenos* = virgin; *gamos* = marriage. Pl., Parthenogamies.) See Parthenomixis.

PARTHENOGENESIS Noun. (Greek, *parthenos* = virgin; *genesis* = descent. Pl., Parthenogeneses.) Reproduction without fertilization in which development of a Zygote from Ova occurs without fertilization by a male gamete. Parthenogenesis is a reproductive phenomenon common in some groups of arthropods, including the Acari and Insecta. See Arrhenotoky; Deuterotoky; Heterogony; Thelytoky. Cf. Gynogenesis.

PARTHENOGENETIC Adj. (Greek, *parthenos* = virgin; *genesis* = descent.) Pertaining to organisms developed through Parthenogenesis. See Asexual; Parthenogenesis. Cf. Deuterotoky; Heterogony; Thelytoky.

PARTHENOGONE Noun. (Greek, *parthenos* = virgin; *gonos* = offspring. Pl., Parthenogones.) Any organism produced through Parthenogenesis.

PARTHENOTE Noun. (Greek, *parthenos* = virgin. Pl., Parthenotes.) A parthenogenetically produced haploid organism.

PARTIAL CLAUSTRAL COLONY FOUNDING Social Insects: A phenomenon in which a founding queen sequesters herself in a chamber but occasionally leaves the chamber to forage for food. Phenomenon manifest in some Species of ants.

PARTIALLY RESISTANT HOST Medical Entomology: A vertebrate host which is infected with pathogen for a long period of time before recovering or being overcome by the pathogen. Cf. Amplifying Host; Dead-End Host; Resistant Host; Silent Host; Susceptible Host.

PARTICLE FILMS Microscopic mineral particles (kaolin) which are modified in size and shape and applied to plants to deter feeding or oviposition by insects and mites. Cf. Biopesticide.

PARTICOLOURED AUGER BEETLE *Mesoxylion collaris* (Erichson) [Coleoptera: Bostrichidae].

PARTIM (Latin.) Part.

PARTITE Adj. (Latin, *partitus* = divided; *-ites* = constituent.) Pertaining to a division, such as the divided eyes of Gyrinidae.

PARTURITION Noun. (Latin, *partuire* = to bring forth; English, *-tion* = result of an action. Pl., Parturitions.) 1. The process of birth. Cf. Eclosion. 2. The period during which the female is producing eggs or larvae.

PARVIS, GUILIO CESARE (–1952) (Anon. 1952, Boll. Soc. ent. ital. 82: 17.)

PARVOVIRIDAE Plural Noun. A small Family of viruses the members of which are characterized by relatively small size and non-enveloped, single-stranded DNA. Parvoviruses cause several diseases of humans and other animals, including human parvovirus B19, porcine parvovirus (PPV) and canine parvovirus (CPV.) B19 is associated with several human disorders; PPV is a major cause of reproductive failure in swine; CPV causes enteritis in dogs. Cf. Baculoviridae.

PARVULA SKIPPER *Toxidia parvula* (Plotz) [Lepidoptera: Hesperiidae].

PARZON® See Cypermethrin.

PASCOE, FRANCES POLKINGHORNE (1813–1893) (McLachlan 1893, Entomol. mon. Mag. 29: 194–196.)

PASKEM, VLADIDAVEM (–1954) (Patocka 1955, Zool. ent. Listy 18: 104–106, bibliogr.)

PASPALUM WHITEGRUB *Lepidiota laevis* Arrow [Coleoptera: Scarabaeidae].

PASQUALE, GIUSEPPE ANTONIO (1820–1893) (Paladino 1893, Atti Accad. pontan. 23: 1–16, bibliogr.)

PASSALIDAE Plural Noun. Bess Beetles; Passalids. A cosmopolitan Family of polyphagous Coleoptera assigned to the Scarabaeoidea. Best represented in tropical regions. Body black, large to very large, elongate; strongly sclerotized; constricted between Prothorax and Elytra. Head prognathous, sometimes with horn; Antenna with 10 segments including club of 3–6 segments; curved but not geniculate; Scutellum not visible; Elytra striate; tarsal formula 5-5-5; Abdomen with five Ventrites; stridulate with Plectrum on hindwing against Pars Stridens in abdominal dorsum. Larva elongate; head prognathous; stemmata absent; Antenna with two segments; fore and middle legs well developed, each with five segments; hind legs reduced; stridulate with hind leg moved over middle Coxa; Urogomphi absent. Adults and larvae gregarious or subsocial, feed on rotten wood. Stridulation apparently maintains colonial structure. See Coleoptera.

PASSANDRIDAE Plural Noun. A Family of Coleoptera consisting of about 100 described Species; widespread in distribution but not recorded from Europe or New Zealand. Higher classification controversial, but probably most closely related to Cucujidae. Separated from cucujids on basis of confluent Gular Sutures, contiguous anterior tendons of adult and Maxillae concealed by a corneous, porrect sclerite projecting from Genae. Biologically poorly studied. Adults live in tunnels of wood-boring Coleoptera; larvae ectoparasitic on wood boring larvae.

PASSAVANT, PHILIPP THEODOR (1804–1893) (Reichenbach 1893, Ber. senckenb. naturf. Ges. 1893: cxxvii–cxxviii.)

PASSERINI, CARLO (1793–1857) (V.A. 1857, Annali Mus. fis. Stor. Nat. Firenze 1857: 209–222; Toni 1893, Boll. R. Ist. bot. Univ. Parma 1892–93: 5–16, bibliogr.)

PASSERINI, GIOVANI (1816–1893) (Toni 1893, Boll. R. Ist. bot. Univ. Parma 1892–93: 5–16, bibliogr.)

PASSERINI, NAPOLEONE (1862–1952) (Conci 1952, Boll. Soc. ent. ital. 82: 65.)

PASSIONVINE BUG *Fabrictilis gonagra* (Fabricius) [Hemiptera: Coreidae]: A pest of passionfruit, cucurbits, papaya and citrus in Australia.

PASSIONVINE HOPPER *Scolypopa australis* (Walker) [Hemiptera: Ricaniidae]: A minor pest of citrus and cultivated plants in Australia; PVH also occurs in New Zealand. Adult moth-like, ca 8 mm long; body brown; wings mottled brown with hyaline areas; wings project beyond apex of Abdomen and held flat over body when feeding. Nymphs with white, waxy, feather-like tail filaments. Females oviposit into slits cut in bark and twigs; five nymphal instars; possibly multivoltine. Adults and nymphs aggregate to feed on twigs, midrib vein of leaves, fruit stems. PVH produces honeydew which serves as substrate for sooty mould.

PASSIONVINE MEALYBUG *Planococcus minor* (Maskell) [Hemiptera: Pseudococcidae].

PASSIONVINE MITE *Brevipalpus phoenicis*

(Geijskes) [Acari: Tenuipalpidae].

PASSIVE FLIGHT One of two basic functional forms of insect flight. PF is distinguished by an absence of wingbeat, the wings do not create thrust and movement is not generated through inertia as in flapping flight. PF is restricted to heavy, large-bodied insects. PF displays different forms including diving, gliding, soaring and parachuting. Syn Gliding Flight, *sensu latu*. See Flight. Cf. Flapping Flight. Rel. Aerodonetics.

PASSIVE SUCTION VENTILATION A form of Tracheal Ventilation in which a partial vacuum is developed in the Trachea and Oxygen is drawn into the system through a partially open Spiracle. Spiracles are typically closed to prevent loss of water vapour and gas exchange (carbon dioxide and oxygen); spiracle periodically opens to permit intake of oxygen and escape of carbon dioxide. The interval between openings may be hours and is dependent upon temperature, age and stage of development. During spiracular closure, carbon dioxide accumulates in form of bicarbonate. PSV occurs in fleas, Pupae and some small-bodied insects. PSV may involve collapse of Trachea during period when Spiracles are closed; inflation of Trachea by action of first pair of abdominal spiracles. See Tracheal Ventilation. Cf. Active Ventilation. Rel. Respiratory System.

PASSIVE TRANSMISSION The invasion of pathogenic microorganisms (bacteria, fungi, *etc.*) into plants or animals through mechanical damage caused by the feeding or oviposition caused by insects. Cf. Biological Vector.

PASSOS, CYRIL FRANKLIN dos (1887–1986) American lawyer, amateur Lepidopterist and active Research Associate with the American Museum of Natural History.

PASTEUR, LOUIS (1822–1895) (Carrington 1895, Sci. Gossip 2: 197–198.)

PASTURE DAY MOTH *Apina callisto* (Angas) [Lepidoptera: Noctuidae].

PASTURE MITE *Bryobia repensi* Manson [Acari: Tetranychidae].

PASTURE SNOUT MITE *Bdellodes lapidaria* (Kramer) [Acari: Bdellidae].

PASTURE TUNNEL MOTH *Philobota productella* (Walker) [Lepidoptera: Oecophoridae].

PASTURE WEBWORM *Hednota crypsichoroa* Lower, *Hednota longipalpella* (Meyrick), *Hednota panteucha* (Meyrick), *Hednota pedionoma* [Lepidoptera: Pyralidae].

PASTURE WHITEGRUBS *Rhopaea* spp. [Coleoptera: Scarabaeidae].

PASYPSO® See Deltamethrin.

PASZTOR, ISTVAN (1874–1909) (Anon. 1909, Ent. Rdsch. 26: 45.)

PATAGIUM Noun. (Latin, *patagium* = border; Latin, *-ium* = diminutive > Greek, *-idion*. Pl., Patagia.) One of the paired, small, dorsolateral processes or thin lobe-like erectile expansions of the Prothorax (not to be confused with Tegulae) (Imms). Lepidoptera: Lobe-like prothoracic structures covering the base of the forewings; often used synonymously with Tegula or Squamula; homologized with the Paraptera of the Mesothorax (Smith). Culicidae: A sausage-shaped body on each side of the Prothorax in front of the first pair of spiracles.

PATAP® See Cartap.

PATCH, EDITH MARION (1876–1954) (Adams & Simpson 1955, Ann. ent. Soc. Am. 48: 313–314.)

PATE, VERNON SENNOCK LYONESSE LIANCOUR (1903–1958) (Kempf 1961, W.W. Studia ent. 4: 542–545; Krombein 1961. Ent. News 72: 1–5.)

PATELLA Noun. (Latin, *patella* = small pan, diminutive of *patina* = pan. Pl., Patellae; Patellas.) 1. A thick, flattened, triangular, moveable bone that forms knee cap of humans. 2. Dytiscidae: The modified joints of the anterior Tarsi; plate-like horny structures on the ventral surface of the tarsal segments. 3. Ticks: The Tibia *sensu* Matheson. 4. Chelicerata: A segment of the leg between the Femur (Meropodite) and the Tibia (Carpopodite).

PATELLAR Adj. (Latin, *patella* = small pan.) Pertaining to the knee-joint or cap.

PATELLIFORM Adj. (Latin, *patella* = pan; *forma* = shape.) Disc-shaped with a narrow rim. See Patella. Cf. Acetabuliform; Ampulliform; Calathiform; Discoid. Rel. Form 2; Shape 2; Outline Shape.

PATELLULA Noun. (Late Latin, dim. of *patella* = pan.) A Patella with ring-like openings. A tarsal suction device to adhere to surfaces.

PATENCY Noun. (Latin, *patens* = lying open.) A process during the early phase of Vitellogenesis during which interfollicular channels enlarge and is correlated with increased uptake of Vitellogenin by the Ovary.

PATENS Adj. (Latin, *patens* = lying open.) Open, diverging, spreading apart.

PATENT-LEATHER BEETLES See Passalidae.

PATEOLLO-TIBIAL Adj. Descriptive of or pertaining to the Patella and Tibia.

PATERNITY ASSURANCE Behavioural or mechanical methods employed by males to assure that their sperm (rather than another male's sperm) fertilize egg. Paternity assurance behaviours include mate-guarding, copulation of long duration and transmission of behaviourally modifying accessory gland secrections. PA mechanical methods include Spermatophores, sperm plugs or forceful removal of another male's sperm.

PATERSON, GUILLERMO (–1946) (Hayward 1947, Revta Soc. ent. argent. 13: 343.)

PATHOGEN Noun. (Greek, *pathos* = suffering; *genes* = producing. Pl., Pathogens.) Any disease-producing microorganism. Principal pathogens include bacteria, viruses, fungi and nematodes. Pathogen represents one of three principal categories of Natural Enemies used in applied biological control. See. Bacterium; Fungus; Nematode; Virus. Cf. Predator; Parasite. Rel. Biological Control.

PATHOGEN-DERIVED RESISTANCE The use of gene sequences from a pathogen to protect a host from the effects of a pathogen (Sanford & Johnson 1985.)

PATHOGENIC Adj. (Greek, *pathos* = suffering; *genes* = producing; *-ic* = of the nature of.) Disease-causing or disease-producing; a term applied to organisms which cause, carry or transmit disease.

PATHOGENICITY Noun. (Greek, *pathos* = suffering; *genes* = producing; English, *-ity* = suffix forming abstract nouns. Pl., Pathogenicities.) The capacity of a pathogen to cause disease. Cf. Infectivity.

PATHOLOGICAL Adj. (Greek, *pathos* = suffering; *logos* = discourse; Latin, *-alis* = characterized by.) A diseased or abnormal condition; unhealthy or arising from unhealthy conditions.

PATHOLOGIST Noun. (Pl., Pathologists.) A student of pathology. Correct veneral term: A gross of pathologists. Cf. Anatomist; Morphologist.

PATHOLOGY Noun. (Greek, *pathos* = suffering; *logos* = discourse. Pl., Pathologies.) The study of diseases. Cf. Autopsy; Biopsy; Necropsy.

PATHOVAR Noun. (Comb. Pl., Pathovars.) Bacteria: strains that infect only plants within a particular Genus or Species.

PATRIA Noun. (Latin, native country.) Country or home.

PATRIN® See Carbaryl.

PATRIOT® See Diazinon.

PATRIZI, MONTORO SAVERIO (1902–1957) (Tortonese 1957, Annali Mus. civ. Stor. nat. Giacomo Doria 69: 370–374, bibliogr.)

PATROLE® See Methamidophos.

PATROLLING Social Insects: The act of periodically inspecting the interior of a nest and the surrounding area, presumably to detect predators and parasites.

PATRONYM Noun. (Greek, *pater* = father; *onoma* = name.) A patronymic. Rel. Taxonomy; Nomenclature.

PATRONYMIC Adj. (Greek, *pater* = father; *onoma* = name; *-ic* = consisting of.) Pertaining to a patronym. Patronymous.

PATRONYMIC Noun. (Greek, *pater* = father; *onoma* = name.) The addition of suffix or prefix to a name which thereby gives indication of a patrilineal relationship. (*e.g.* Ivanovich; Johnson; MacDonald.) Cf. Acronym; Eponym.

PATTEN, WILLIAM (1861–1932) (Gerould 1932, Science 76: 481–482.)

PATTERN ASYMMETRY The asymmetrical colour patterns on the body of insects which facilitates camouflage. Sometimes pattern is subtle (spot patterns on a butterfly wing). Asymmetry in colour pattern must be achieved without affecting aerodynamical considerations of glide. Wing overlap in insects is another example of asymmetry. Wing overlap is usually a specific sequence. Although overlap arrangement is sometimes apparently trivial, it is vital. Orthopteran stridulatory device on one wing (the Plectrum) must interface with a component on other wing (the Mirror). Insect cannot stridulate if the wing overlap sequence is reversed. See Asymmetry; Symmetry. Cf. Absolute Asymmetry; Behavioural Asymmetry; Biological Asymmetry; Developmental Asymmetry; Morphological Asymmetry; Skeletal Asymmetry. Rel. Organization.

PATTERSON, ALICE MCDOUGALL (–1935) (Spencer 1938, Proc. ent. Soc. Br. Columb. 34: 63.)

PATTERSON, JOHN ELLIOTT (1887–1962) (Easton & Struble 1964, Pan-Pacif. Ent. 40: 14.)

PATTERSON, JOHN P (1835–1898) (Anon. 1898, Ent. News 9: 104.)

PATTERSON, ROBERT (1802–1872) (Anon. 1872, Nature 5: 332.)

PATTERSON'S CURSE *Echium lycopsis* Linnaeus [Boraginaceae] (syn. *E. plantgineum* Linnaeus) An errect biennial plant endemic to the Mediterranean and adventive to Australia. Regarded as a pest to graziers and beneficial plant to beekeepers. Object of legal action in Australia which affected quarantine regulations and biological control importations. Syn. Salvation Jane.

PATTON, WALTER SCOTT (1867–1960) (Anon 1960, Ann. trop Med. Parasit. 54: 2.)

PATULOSE Adj. (Latin, *patulus* = standing open; -*osus* = full of.) Spreading open; expanding; pertaining to structure with a opening which expands distad. Alt. Patulent; Patulosus; Patulous.

PAUL Y AROZARENA, MANUEL DE (1852–1930) (Anon. 1930, Boln. Patol. veg. Ent. Agric. 5: 223–224.)

PAUL, JOSEPH JOHN (1912–1968) (Lund 1968, J. Econ. Ent. 61: 1132.)

PAUL, MORITZ (–1898) (Anon. 1898, Mitt. schweiz. ent. Ges. 10 136.)

PAULINO D'OLIVEIRO, MANOEL (1837–1899) (Nobre 1901, Anais Sci. nat. 7: 173–175, bibliogr.)

PAULMIER, FREDERICK CLARK (1873–1906) (E.B.W. 1906, Science 23: 556.)

PAULSEN, FERNANDO (1842–1908) (Porter 1930, Revta. chil. Hist nat. 34: 114–115, bibliogr.)

PAULY, AUGUST (1850–1914) (Röhrl 1914, Ent. Bl. Biol. Syst. Käfer 10: 129–135, bibliogr.)

PAUNCH Noun. (Latin, *pantex* = paunch.) 1. A crop-like accessory pouch in some Mallophaga. 2. Any pouch-like appendage of the Alimentary Canal.

PAUP Phylogenetic Analysis Using Parsimony. A computer program developed to analyse comparative data to produce the most parsimonious pattern or tree (See Swofford & Begle 1993). See Phylogeny.

PAUROMETABOLA Noun. (Greek, *pauros* = little; *metabole* = change.) A division of the Heterometabola characterized by a gradual development in which the young resemble the adults in general form and mode of life. Examples include Dermaptera, Orthoptera, Embioptera, Isoptera, Zoraptera, Corrodentia, Mallophaga, Anopleura, Heteroptera, most Homoptera. See Metamorphosis. Cf. Hemimetabola; Holometabola.

PAUROMETABOLOUS Adj. (Greek, *pauros* = little; *metabole* = change; Latin, *-osus* = with the property of.) 1. Pertaining to organisms with Metamorphosis in which the changes of form are gradual and inconspicuous. Paurometaboly is an early transitional phase or condition that shows biological and structural improvement over Ametaboly. Examples of paurometabolous insects include Dermaptera, Orthoptera, Embioptera, Isoptera, Zoraptera, Corrodentia, Mallophaga, Anoplura, Heteroptera and most Homoptera. Aphidoidea (Phylloxeridae, Aphididae, Adelgidae) are more complex because development is mixed. The stem mother (Fundatrix) is apterous and alate females are paurometabolous. Sexuales (sexual males, females) are holometabolous. Texts incorrectly report that sexually mature males and females hatch from the egg (Pergande 1904; Whitehead & Eastep 1937; Caldwell & Schuder 1979). See Development; Metamorphosis. Cf. Ametabolous; Hemimetabolous; Holometabolous.

PAUROMETABOLY Noun. (Greek, *pauros* = little; *metabole* = change.) The condition of being paurometabolous. Paurometaboly is a derived condition among pterygote insects that have developed patterns of Metamorphosis. See Metamorphosis. Cf. Ametaboly; Hemimetaboly; Holometaboly. Rel. Groundplan.

PAUROPODA Plural Noun. (Greek, *pauros* = little; *podos* = foot.) A Class of worm-like segmented animals belonging to the Phylum Arthropoda. Body ca 1 mm long; head present; 12 body segments; 8–9 pairs of legs, typically covered with six large dorsal sclerites. Most Species are terrestrial and occur in decaying humus; some Species are intertidal. Cf. Symphyla.

PAVAN'S GLAND Formicidae: A gland which opens near or on the sixth abdominal Sternum. Produces a trail pheromone in some ants (Dolichoderinae).

PAVEL, JANOS (1842–1901) (Lajos 1901, Rovart. Lap. 8: 132–136)

PAVEMENT ANT *Tetramorium caespitum* (Linnaeus) [Hymenoptera: Formicidae]: An Holarctic urban pest. Workers monomorphic, ca 2–3 mm long, robust, hard-bodied, pale brown to black with pale legs and Antennae; antennal club absent; gastral Pedicel of two nodes; head and Thorax with parallel grooves. Winged sexuals occur throughout year, but most abundant during summer. Colonies occur under stones, along pavement, in grass; colonies outdoors most of year and move indoors during winter. Colony refuse consists of grit, small wood fibres, fragments of insect bodies. PA omnivorous and forage throughout year; seek sweets and grease indoors; seek plant material outdoors; feeds on roots and seeds. See Formicidae. Cf. Argentine Ant; Black Carpenter-Ant; Cornfield Ant; Fire Ant; Larger Yellow-Ant; Little Black Ant; Odorous House Ant; Pharaoh Ant; Thief Ant.

PAVESI, PIETRO (1844–1907) Italian academic (Professor of Zoology, University of Genoa) and specialist in Arachnology. (Pavesi 1903, Memorie Accad. Agiati 1903: 832–834; Parona 1907, Monitore zool. ital. 18: 250–253, bibliogr.)

PAVILION Noun. (Middle English, *pavilion* = tent; Latin, *papilio* = butterfly; tent. Pl., Pavilions.) 1. A large tent, usually with a peaked top. 2. A building or part of a building used for exhibitions or displays. 3. An artificial enclosure constructed by ants as a shelter for groups of aphids tended by the ants.

PAVLOVICH, PETERSON LAIMON (1912–1974) (Prieditich 1976, Trudy Latvaan agric. Akad. 100: 3–8, bibliogr.)

PAVLOVSKII, EUGENE NIKANOVICH (1883–1965) (Bei-Bienko 1965, Trudy zool. Inst. Leningr. 35: 3–15, bibliogr.)

PAWLITSCHER, ALFRED (1857–1931) (Hormuzaki 1936, Verh. zool.-bot. Ges. Wien 85: 132–134.)

PAWLOWSKY'S GLANDS *Pediculus*: The pair of glands opening into the Stylet Sac.

PAWPAW FRUITFLY See Papaya Fruitfly.

PAXILLA Noun. (Latin, *paxillus* = peg. Pl., Paxillae.) 1. A small stake or peg. 2. A bundle of spicular processes.

PAXILLATE Adj. (Latin, *paxillus* = peg.) Pertaining to groups of small pegs, spines or spicular processes.

PAXYLOMMATIDAE Förster 1862. Plural Noun. A Holarctic Family of apocritous Hymenoptera assigned to the Ichenumonoidea and consisting of seven Species within one Genus (*Hybrzon*). Individuals lack a second Recurrent Vein in forewing resulting in placement in the Braconidae or in the Ichneumononidae under some classifications. Biology poorly known, but believed internal parasites of ant larvae. Synonym: Hybrizontidae.

PAYKULL, GUSTAVUS VON (1757–1826) Swedish bureaucrat and author of *Fauna Suecica (Insecta)* (3 vols, 1800) and *Monographia Histeroidum* (1811). His collection of insects is housed in the Naturhistoriska Riksmuseet, Stockholm. (Duméril 1823, *Considérations générales sur la classe des insectes.* xii + 272 pp. (265). Paris; Anon. 1827, K. svenska VetenskAkad. Handl. 1826: 350–358, bibliogr.; Zimsen 1964, *The Type Material of J. C. Fabricius.* 656 pp. (17), Copenhagen.)

PAYLOAD® See Acephate.

PAYNE, HENRY T (–1930) (Anon. 1930. *Lond. Nat.* 1930: 39.)

PB-ROPES® See Gossyplure.

PCA See Principal Component Analysis.

PCR See Polymerase Chain Reaction.

PEA APHID *Acyrthosiphon pisum* (Harris) [Hemiptera: Aphididae]: A sporadic pest of legumes throughout North America. When present in large numbers, PA causes wilt in peas and alfalfa. Overwinters as egg or ovoviviparous stem mothers. Winged forms appear at high population densities. Multivoltine to 20 generations per year. See Aphididae.

PEA BEETLE Rel. Pea Weevil.

PEA BLUE BUTTERFLY *Lampides boeticus* (Linnaeus) [Lepidoptera: Lycaenidae].

PEA FLY *Kleinschmidtimyia pisi* (Kleinschmidt) [Diptera: Agromyzidae].

PEA GALL A pea-like gall induced by the wasp *Diplolepis nervosa* on rose leaves. See Gall. Cf. Bedeguar.

PEA LEAF WEEVIL *Sitona lineatus* (Linnaeus) [Coleoptera: Curculionidae].

PEA LEAFMINER *Liriomyza huidobrensis* (Blanchard) [Diptera: Agromyzidae].

PEA MOTH *Laspeyresia nigricana* (Fabricius) [Lepidoptera: Tortricidae]: Endemic to Europe; introduced into North America ca 1900 where it is a pest of peas. PM adult a weak flier, active in afternoon. Eggs white, flat, laid anywhere on plant. Larva bores into pea pods, spins web, partially consumes seeds. Pupation occurs within cocoon in soil; overwinters as mature larva within cocoon; 1–2 generations per year. See Tortricidae.

PEA STEM-FLY *Ophiomyia phaseoli* (Tryon) [Diptera: Agromyzidae].

PEA WEEVIL *Bruchus pisorum* (Linnaeus) [Coleoptera: Bruchidae]: A cosmopolitan, monovoltine pest of peas. Native to southeastern Europe; adventive to North America, Australia, Africa. Eggs oval-elongate, orange and laid individually on pods with 1–15 per pod; eclosion occurs within 5–18 days depending upon temperature. Larvae crescent-shaped, cream coloured; first-instar larva spinose, long legged; penetrates pod, loses spines and legs shorten in later instars; larva feeds in pod until seeds develop, penetrates pea to cause damage to pea seeds and reduces yield; one larva per pea; larva feeds 4–6 weeks. Pupation occurs within pea when mature larva applies mucoid oral secretion on walls of chamber; pupal stage requires two weeks. PW overwinters in adult stage. See Bruchidae. Cf. Bean Weevil; Spotted Cowpea Bruchid. Rel. Pea Aphid.

PEA WEEVILS See Bruchidae.

PEACH, ALEC HAMILTON (1885–1960) (H.W.T. 1961, Proc. Bristol Nat. Soc. 30: 1–3.)

PEACH BARK-BEETLE *Phloeotribus liminaris* (Harris) [Coleoptera: Scolytidae]: A bivoltine pest of peach trees, cherry and other stone fruit in eastern North America; PBB does not attack pome fruits. PBB overwinters as an adult in pupal cell. Adults emerge during spring and are attracted to trees in poor condition or trees with dying branches; male and female adult excavate 'Y'-shaped gallery across a branch and within inner bark; female oviposits along sides of gallery. Neonate larvae construct individual burrows perpendicular to adult gallery and parallel to grain of wood. See Scolytidae. Cf. Shothole Borer; Peach-Tree Borer.

PEACH SILVER MITE *Aculus cornutus* (Banks) [Acari: Eriophyidae].

PEACH WHITE SCALE *Pseudaulacaspis pentagona* (Targioni Tozzetti) [Hemiptera: Diaspididae].

PEACH YELLOWS A viral disease of peaches that is transmitted by the Plum Leafhopper during feeding.

PEACH-TREE BORER *Synanthedon exitiosa* (Say) [Lepidoptera: Sesiidae (Aegeriidae)]: A monovoltine pest of apricot, peach, plum and prune in North America. PTB overwinters as larva, but size of individuals varies considerably (ca 3–12 mm long). During spring larvae resume feeding; pupation occurs within a cocoon of silk and bark near burrow entrance; larvae feed on cambium and inner bark near ground line. Adults active diurnally July-October; female lays 200–800 eggs during life; eggs laid individually or in groups on leaves, on tree trunks or in fissures in soil; eclosion occurs ca 10 days. PTB serves as vector for Plum Wilt. See Sesiidae. Cf. Lesser Peach Tree Borer; Peach Bark Beetle; Western Peach-Tree Borer.

PEACH-TWIG BORER *Anarsia lineatella* Zeller [Lepidoptera: Gelechiidae]: A widespread pest of fruit trees in Europe, Asia and North America; known in USA since 1860. Host plants include almond, apricot, peach, plum and prune. Overwinters as larva within hibernaculum of faecal pellets, bark and silk; silken cocoon attached to bark of trunk or branches. Larvae become active during spring and enter new growth, later fruits; larva may enter more than one twig before completing development; pupation occurs in dark-brown cocoon on bark. Adults oviposit on twigs, bark, leaves or fruit. PTB completes 1–4 generations per year, depending upon host plant and climate. See Gelechiidae. Cf. Oriental Fruit Moth; Peach-Tree Borer; Lesser Peach-Tree Borer.

PEACOCK AWL *Allora doleschallii doleschallii* (Felder) [Lepidoptera: Hesperiidae].

PEACOCK JEWEL *Hypochrysops pythias euclides* Miskin [Lepidoptera: Lycaenidae].

PEAIRS, LEONARD MARION (1886–1956) (Cory 1956, J. Econ. Ent. 49: 430–431.)

PEALE, TITIAN RAMSEY (1797–1885) Early American naturalist and painter. Member of Long's expedition and contributed to Thomas Say's *American Entomology.* (Anon. 1883, Appleton's Cyclopedia of American Biogr. 4: 691; Dimmock 1885. Psyche 4: 266; Anon. 1913, Ent. News 24: 1–3.)

PEANUT MITE *Paraplonobia* sp. [Acari: Tetranychidae].

PEANUT SCARAB 1. *Heteronyx piceus* Blanchard. 2. *Heteronyx rugosipennis* Macleay [Coleoptera: Scarabaeidae].

PEANUT TRASH BUG *Elasmolomus sordidus*

(Fabricius) [Hemiptera: Lygaeidae].

PEAR-AND-CHERRY SLUG *Caliroa cerasi* (Linnaeus) [Hymenoptera: Tenthredinidae]: In Australia: Larvae feed on upper leaf surfaces and produce a skeletonizing effect; attack hawthorns, apples, quinces, fruit-bearing and flowering pears, cherries and plums. Syn. Pear Slug; *Eriocampoides limacina* Retzius.

PEAR FRUIT CHAFER *Euphoria inda* [Coleoptera: Scarabaeidae].

PEAR MIDGE *Contarinia pyrivora* (Riley) [Diptera: Cecidomyiidae]: A pest of pears in North America that causes deformed fruit and premature drop. PM eggs deposited among blossoms and larvae infest fruit. See Cecidomyiidae. Cf. Grape Blossom Midge; Sorghum Midge.

PEAR PLANT-BUG *Lygocoris communis* (Knight) [Hemiptera: Miridae].

PEAR PSYLLA *Cacopsylla pyricola* Foerster [Hemiptera: Psyllidae]: In North America, a widespread, multivoltine pest of pear. PP is endemic to Europe and adventive to USA ca 1832. Species overwinters as an adult under bark or among litter and completes 3–5 generations per year. PP eggs pear-shaped, orange coloured, laid at base of fruit on leaf bud or twigs during spring; egg attached to plant via pedicel and displays a thread-like filament from apical portion of egg; eclosion occurs within 15–30 days. PP undergoes five nymphal instars; instars 2–4 immersed in honeydew; final instar mobile as 'hardshell.' Feeding by PP causes leaves to yellow while honeydew promotes fungus.

PEAR ROOT APHID *Eriosoma pyricola* Baker & Davidson [Hemiptera: Aphididae].

PEAR RUST-MITE *Epitrimerus pyri* (Nalepa) [Acari: Eriophyidae].

PEAR SAWFLY *Caliroa cerasi* (Linnaeus) [Hymenoptera: Tenthredinidae]: In Europe and widespread in North America; a minor pest of pears in eastern USA. Host plants include cherry, hawthorn, pear, plum, quince and mountain ash. PS overwinters as a Pupa within a coccoon formed in soil. Adults appear during spring when cherries develop leaves. Female PS inserts her Ovipositor into leaves and oviposits. Eggs white, oval, inserted under leaf Epidermis; eclosion occurs within a few days. PS larva slug-like with slimy secretion on body; skeletonizes upper surface of leaves; larvae feed 15–20 days. Species bivoltine with first brood thelytokous and second brood arrhenotokous. Syn. Pear Slug; Pear and Cherry Slug. Cf. Cherry Fruit Sawfly; European Apple Sawfly.

PEAR SCALE *Quadraspidiotus pyri* (Lichtenstein) [Hemiptera: Diaspididae].

PEAR SLUG See Pear Sawfly.

PEAR THRIPS *Taeniothrips inconsequens* (Uzel) [Thysanoptera: Thripidae]: A monovoltine pest of pears, plums, prunes and cherries in USA, South America and Europe; other host plants include apples, apricots, grapes and some ornamental trees. PT endemic to Europe and adventive to USA ca 1902. PT overwinter as pharate adults (quiescent Pseudopupa) in cells within soil. Adults black bodied, emerge during spring and feed on bud scales. Eggs deposited in stems of fruit and foliage during early spring; eclosion occurs within 15 days. Nymphs white, complete feeding within 30 days; enter soil during summer. PT causes leaves and young flowers to brown. See Thripidae.

PEARL BODY Social Insects: Lipid-rich pearl-like bodies used by plants to attract and maintain ants. Syn. Bead Gland.

PEARL, RAYMOND (1879–1940) (Anon. 1940, Q. Rev. Biol. 15 (4); Reed 1940, Science 92: 595–597.)

PEARLACEOUS Adj. (Middle English, *perle* = pearl; Latin, *-aceus* = of or pertaining to.) Having the appearance of pearl in colour.

PEARLEAF BLISTER-MITE *Phytoptus pyri* (Pagenstecher) [Acari: Eriophyidae].

PEARMAN, JOHN VICTOR (1887–1970) (Anon. 1971, Proc. Bristol Nat. Soc. 32: 11; Clay & Smithers 1971. Entomol. mon. Mag. 107: 65–66, bibliogr.)

PEARMAN'S ORGAN An apparent stridulatory device on the medial surface of the hind Coxa of some Psocoptera. PO consists of a rugose dome-like Sclerite and adjacent membrane.

PEARSON, ERIC OMAR (1906–1968) (Anon. 1968, Rev. Appl. Ent. (B) 56 (8): 149; Kennedy 1969, Proc. R. ent. Soc. Lond. (C) 33: 55.)

PEAY, WALTER EDWIN (1907–1975) (Nielsen *et al.* 1975, J. Econ. Ent. 68: 567)

PEBRINE Noun. (French.) A microspordian fungal disease of silkworm, *Bombyx mori.* Disease caused by *Nosema bombycis.* See Fungus. Cf. *Nosema*; Microsporidia.

PECAN BUD MOTH *Gretchena bolliana* (Slingerland) [Lepidoptera: Tortricidae].

PECAN CARPENTERWORM *Cossula magnifica* (Strecker) [Lepidoptera: Cossidae].

PECAN CIGAR-CASEBEARER *Coleophora laticornella* Clemens [Lepidoptera: Coleophoridae].

PECAN LEAF SCORCH-MITE *Eotetranychus hicoriae* (McGregor) [Acari: Tetranychidae].

PECAN LEAF-CASEBEARER *Acrobasis juglandis* (LeBaron) [Lepidoptera: Pyralidae]: A univoltine pest of pecans, hickory and walnuts in southeastern United States. Overwinter as larvae, feed on buds and twigs during early spring. Eggs laid in clusters on underside of leaves; young larvae spin winding cases for protection; larvae spin 'hibernacula' during autumn.

PECAN LEAFMINER *Stigmella juglandifoliella* (Clemens) [Lepidoptera: Nepticulidae].

PECAN LEAF-PHYLLOXERA *Phylloxera notabilis* Pergande [Hemiptera: Phylloxeridae].

PECAN LEAF-ROLL MITE *Eriophyes caryae* Keifer [Acari: Eriophyidae].

PECAN NUT CASE-BEARER *Acrobasis nuxvorella*

Neunzig [Lepidoptera: Pyralidae].

PECAN PHYLLOXERA *Phylloxera devastrix* Pergande [Hemiptera: Phylloxeridae]. Syn. Pecan Stem Phylloxera.

PECAN SERPENTINE LEAFMINER *Stigmella juglandifoliella* (Clemens) [Lepidoptera: Nepticulidae].

PECAN SPITTLEBUG *Clastoptera achatina* Germar [Hemiptera: Cercopidae].

PECAN WEEVIL *Curculio caryae* (Horn) [Coleoptera: Curculionidae]: A pest of pecans and hickory in southestern USA. Female drills hole in shell, deposits 2–4 eggs/hole; larva full grown within month, emerges from shell, forms cell in soil and remains for 1–2 years before pupation during autumn. Adults remain in soil until following summer. Life cycle requires 2–3 years; monovoltine.

PECCHIOLI, VICTOR (–1870) (Anon. 1870, Boll. Com. geol. ital. 1: 317–218.)

PECIRKA, JAROMIREM (–1933) (Obenberger 1933. Cas. csl. Spol. ent. 30: 9–10.)

PECK, GEORGE W (1837–1909) (Anon. 1909, Can. Ent. 41: 220.)

PECK, WILLIAM DANDRIDGE (1763–1822) American academic (Harvard University) presented some of the first lectures in Entomology. Wrote several papers on insects of economic importance. (Anon. 1822, Boston Daily Advertiser 35 (81) No.2902. 8 October 1822; Quincy 1840, History Harvard Coll. 2: 329–330; Howard 1930, Smithson. misc. Collns. 84: 11, 30, 72.)

PECKHAM, GEORGE WILLIAM (1845–1914) (Anon. 1904, National cyclopedia. American biogr. 12: 347; Russell *et al.* 1913, Bull. Wisc. nat. Hist. Soc. 11: 109–112, bibliogr.)

PECTAMONE® See Gossyplure.

PECTEN Noun. (Latin *pecten* = comb. Pl., Pectens) 1. Any comb-like structure. 2. Hymenoptera: Curved, rigid Setae on the base of the Maxilla and Labium; compact rows of tibial spines on legs of fossorial wasps; rows of spines on the Tarsomeres of pollen-gathering bees; any series of bristles arranged like a comb. 3. Mosquito larvae: Comb-like teeth on the breathing tube. 4. Lepidoptera: Stiff, scales or Setae arranged comb-like on anterior surface of Antennal Scape or vein CuA on the upper surface of the hindwing; Antenna Scape-scales sometime concave on ventral surface and serve as an eye cap. See Cubital Pecten.

PECTINA Noun. (Latin, *pectinatus* > *pecten* = comb. Pl., Pectinae.) Coccoidea: A broad, fringed sclerite of the Pygidium (Comstock); one of several thin projections with apical teeth or tines located in the Incisurae between the lobes and upon the Lateres (MacGillivray).

PECTINATE Adj. (Latin, *pectinatus* > *pecten* = comb; *-atus* = adjectival suffix.) 1. Comb-like. A term applied to structures (especially Antennae and Tarsi) with processes that resemble the teeth of a comb. Alt. Pectinated; Pectinatus. See Antenna; Tarsus. 2. Pectiniform.

PECTINATELY Adv. (Latin, *pectinatus* > *pecten* = comb.) In a pectinate or feathered manner.

PECTINATO-FIMBRIATE Adj. Descriptive of structure with pectinations that are fringed with Setae.

PECTINE Noun. (Latin, *pecten* = comb. Pl., Pectines.) Two movable processes below the posterior legs and fixed to the Metasternum.

PECTINIFER Noun. (Latin, *pecten* = comb; *fere* = bear, carry. Pl., Pectinifers.) Lepidoptera: A comb of Setae or Sensilla on the valva of some male Nepticuloidea and Incurvarioidea.

PECTINIFORM Adj. (Latin, *pecten* = a comb.) 1. Scallop-shell shaped. See Scallop. Cf. Chonchiform; Conriform; Flabelliform; Heliciform; Mytiliform; Pecteniform. 2. Pectinate. Rel. Form 2; Shape 2; Outline Shape.

PECTINIFORM ANTENNA An Antenna with processes on each side of a Flagellomere such that the Antenna resembles a comb. Syn. Distichous Antenna.

PECTONE® See Gossyplure.

PECTORAL PLATE Coleoptera: The Sternum.

PECTORALIS Adj. Relating to the breast.

PECTUNCULATE Adj. (New Latin, dim. of Latin, *pecten* = comb; *-atus* = adjectival suffix.) With a row of minute appendages like the teeth of a comb, *e.g.,* some maxillary structures.

PECTUS Noun. (Latin, *pectus* = breast. Pl., Pectuses.) 1. The ventral part of the Thorax, variably applied. Coleoptera: The entire Meso- and Metathorax, also to the Pro- and Mesosternum. 2. Diptera: The inferior surface of the Thorax between the legs.

PEDAL Adj. (Latin, *pes* = foot; *-alis* = pertaining to.) Pertaining to the Tarsomeres or legs of an insect.

PEDAL LINE Lepidoptera larva: A line extending along the base of the feet.

PEDAL NERVE The nerve controlling the insect leg.

PEDAL TUBERCLE Cuticular protuberances on the thoracic and abdominal rings of caterpillars or on the anterior side of leg base and correspondingly, on apodal segments, VII of the Abdomen, where it consists of three Setae, VI of the Thorax where the Setae are not numbered, constant (Dyar).

PEDALIAN Adj. (Latin, *pedalis* = of the foot.) An arcane term pertaining to the foot or leg.

PEDAMINA Plural Noun. (Latin, *pes* = foot.) The aborted forelegs of nymphalid butterflies. See Nymphalidae.

PEDATE Adj. (Latin, *pes* = foot; *-atus* = adjectival suffix.) Foot-bearing; descriptive of organisms with feet. Cf. Apodal.

PEDDLER Noun. (Middle English, *peddere* = peddler. Pl., Peddlers.) 1. The larva of some cassid beetles (*Cassida*) that carry their excrement and cast skins on an Anal Fork.

PEDES See Pedis.

PEDES NATATORII Swimming legs. (Archaic). See

Natatorial.

PEDES RAPTORII Legs adapted for seizing prey. (Archaic.) See Raptorial.

PEDES SPURII Spurious legs; the Prolegs.

PEDICEL Noun. (Latin, *pediculus* = small foot. Pl., Pedicels.) 1. Generally, a stalk or stem supporting an organ or other structure. 2. The second segment of the insect Antenna, with intrinsic musculature and forming the pivot between Scape and Funicle. 3. Apocritious Hymenoptera: The second basal segment of the Abdomen, or first (Petiole) and second segment (Postpetiole) in ants. Pedicel is characteristically cylindrical in shape and reduced in size; Pedicel of ants is nodiform or bearing an erect or inclined Scale. 4. A narrow duct or constriction at the base of the Ovariole. A plug forms between the Vitellarium and the Pedicel in immature insects. A plug also forms between ovipositional episodes of mature female insects. 5. Insect Egg: An elongate, prominent structure or stalk that serves as an anchor for the body of the egg when attached to a plant (Psylloidea; Chrysopidae), to a host insect (Eulophidae) or the substrata (Hydrometridae). The Pedicel may form an adaptation for respiration. Cf. Attachment Organ; Polar Cap.

PEDICELLATE Adj. (Latin, *pediculus* = small foot; *-atus* = adjectival suffix.) Supported by a Pedicel; stalked, or on stalk.

PEDICELLATE EGG An apparent variation of the stalked egg in which one end is modified to anchor egg to Integument or seta of host. Typically deposited externally on host; internal eggs attached to host via ventral surface of egg. Pedicel may originate from stalk, from body of egg or from modified micropylar structure. Egg widely distributed among parasitic Hymenoptera, including Chalcidoidea, Ichneumonoidea and Diptera (Cecidomyiidae, Conopidae, Tachinidae). See Egg.

PEDICULIDAE Plural Noun. Human Lice. A moderately small Family of Anoplura whose Species parasitize primates. Head relatively small; compound eyes present; Mandibles absent, mouthparts forming a Beak or rounded anteriorly; forelegs smaller than middle or hind legs; Abdomen longer than basal width. See Crab Louse. Body Louse.

PEDICULOSIS Noun. (Latin, *pediculus* comb. form = louse; *sis* = a condition or state.) Infested with lice; the presence of lice on any part of the body. Cf. Phthiriasis.

PEDICULOUS Adj. (Latin, *pediculus* comb form = louse; *-osus* = with the property of.) Lousy; infested with lice.

PEDIGEROUS Adj. (Latin, *pes* = foot; *gerere* = to carry; Latin, *-osus* = with the property of.) Feet-bearing.

PEDILIDAE See Anthicidae.

PEDIPALP Noun. (Latin, *pes* = foot; *palpare* = to feel, Pl., Pedipalpi; Pedipalps.) Chelicerata: The second pair of appendages on the cephalothorax; used in crushing prey, corresponding to the Mandibles in Mandibulata. Obs. Pedipalpus

PEDIS Noun. (Latin, Pl., Pedes.) Foot or leg of an insect. Alt. Pes.

PEDIUM Noun. (Latin, *pedis* = foot; *-ium* = diminutive > Greek, *-idion*. Pl., Pedia.) Scarabaeoid larvae: The bare central region of the Epipharynx, extending between the Haptomerum and Haptolachus and limited laterally by the interolateral features of the right and left Paria, sometimes marked on the left by the Epitorma (Boving).

PEDOGENESIS See Paedogenesis.

PEDUNCLE Noun. (New Latin, *pedunculus* dim of *pedis* = foot. Pl., Peduncles.) 1. A stalk, stem or Petiole; any stalk-like structure supporting an organ or other structure. 2. Hemiptera: The basal segment of the Antenna. Syn. Pedicel. 3. The large stalk of the Mushroom Bodies of the insect Brain. 4. Botany: The stalk of an inflorescence or solitary flower. Alt. Pedunculus. See Petiole.

PEDUNCULAR Adj. (New Latin, *pedunculus* diminutive of *pedis* = foot.) Descriptive of or pertaining to a Peduncle. Cf. Petiolar.

PEDUNCULATE Adj. (New Latin, *pedunculus* diminutive of *pedis* = foot; *-atus* = adjectival suffix.) Structure positioned on a stalk or peduncle; attached by a slender stalk or neck; petiolate. Alt. Pedunculated; Pedunculatus.

PEDUNCULATED BODY The mushroom body of the insect Brain.

PEEBLES, H M (–1944) (Cockayne 1945, Proc. R. ent. Soc. Lond. (C) 9: 48.)

PEG BEETLES See Passalidae.

PEGASUS® See Diafenthiuron.

PEHLKE, ERNST (1875–1933) (Papavero 1975, *Essays on the History of Neotropical Dipterology*. 2: 299. São Paulo.)

PEICHARDT, AKSELYA NIKOLAEVICH (1891–1942) (Shtakel'berg 1953, Ent. Obozr. 33: 369–375, bibliogr.)

PEILE, HARRY DIAMONT (1872–1959) (Hutchinson 1959, Entomologist 92: 244.)

PE-LA Noun. A commercial Chinese wax, produced since the 13th century in China, and used as a drug and for candles. Wax secreted by the coccid *Ericerus pela*. Cf. Lac.

PELAGIC Adj. (Greek, *pelagos* = sea; *-ic* = characterized by.) Descriptive of organisms inhabiting the open sea; oceanic. See Biogeography. Cf. Benthic; Estuarine; Littoral.

PELAGIC ZONE The biogeographic marine realm characterized by open sea (away from the Littoral Zone) and above the Abyssal Zone. See Habitat. Cf. Abyssal Zone; Littoral Zone; Neritic Zone; Photic Zone.

PELARGONIUM APHID *Acyrthosiphon malvae* (Mosley) [Hemiptera: Aphididae].

PELECINIDAE Haliday 1840. Plural Noun. Pelecinid Wasps. A monogeneric Family of apocritious Hymenoptera assigned to the Proctotrupoidea and restricted to the New World. Placement among the Parasitica, including an independent Superfamily, or within the Proctotrupoidea. A highly distinctive habitus with large, elongate, black body, shiny coloration, extremely long, compressed Gaster and compressed hind Tibia. Primary internal parasites of Coleoptera and often taken near or on decaying logs.

PELECORHYNCHIDAE Plural Noun. A numerically small Family of brachycerous Diptera assigned to Superfamily Tabanoidea. Most Species occur in cool temperature habitats including southern Chile and Australia; a few occur in North America. Adults robust bodied; antennal Flagellum with eight segments; Torulus above middle of head; Labellum dolabriform; Calypters small or vestigial. Adults taken on flowers. Larva resembles rhagionid larva; predaceous in wet or damp habitats such as swamp or stream bank.

PELLET SPIDER *Stanwellia nebulosa* (Rainbow & Puleine) [Araneida: Nemesiidae].

PELLET, PETRI (–1877) (Anon. 1878, Abeille 16: 44.)

PELLICLE Noun. (Latin, *pellis* = skin. Pl., Pellicles.) 1. A thin skin or membrane; Plasma Membrane. 2. The Exuviae or cast larval skins of many insects. 3. Coccidae: Especially applied to the hardened nymphal skin attached to Diaspinae.

PELLIT Adj. (Latin, *pellis* = skin.) Covered with long, drooping Setae, irregularly placed. Alt. Pellitus.

PELLUCID Adj. Transparent, whether clear or coloured. Alt. Pellucidate; Pellucidus.

PELORIDIIDAE Plural Noun. A small Family (ca 30 Species) of heteropterous Hemiptera assigned to Superfamily Peloridioidea; Species occur in southern South America, eastern Australia, Lord Howe Island, New Caledonia and New Zealand. Sister group of Heteroptera with features referrable to Homoptera and Heteroptera. Adult small bodied (2–5 mm long); dorsoventrally compressed; Ocelli typically absent; Antenna with 3 segments, ventrally positioned; Tentorium complete; Rostrum with propleural keels; Gula present; most Species brachypterous; 8 pairs of abdominal spiracles. Species occur in moist, cool habitats associated with moss, sphagnum or leaf litter.

PELORIDIOIDEA A Superfamily of Heteroptera including the Peloridiidae.

PELOTON Noun. (Spanish, *pelota* = a ball. Pl., Pelotons.) The balls of fine Tracheae in larvae, developed to supply the adult organism.

PELOTTAE Noun. (Obscure origin.) Arolia.

PELT Noun. (Middle English from French, *peleterie*). The outer covering of a smaller animal, typically with long hair. Cf. Hide; Integument; Skin.

PELTATE Adj. (Latin, *pelta* = shield; *-atus* = adjectival suffix.) Target-shaped (Say); shield-shaped; descriptive of structure with an outline margin that is round or circular. Alt. Scutate. See Outline Shape.

PELTIDIUM Noun. (Latin, *pelta* = shield; *-ium* = dim. > Greek, *-idion*.) Acarology: Prodorsal Shield.

PELTOPERLIDAE Plural Noun. Roachlike Stoneflies; Roachflies. A Family of Plecoptera including 13 Genera and 35 described Species; Holarctic and Oriental in distribution. Immatures always dark brown, cockroach-like in appearance due in part to disproportionately large Thorax and overlapping segments; adults less cockroach-like, vary in colour from dark brown to yellow.

PEMBERTON, C W Economic Entomologist (Hawaiian Sugar Planters' Association.)

PENAL CLASPERS Proctotrupidae: Lateral fringed processes of the male genitalia.

PENAL SHEATH The hard outer covering of the Penis.

PENCIL Noun. (Middle English, *pencel*. Pl., Pencils.) 1. A small, elongate, brush of Setae. 2. Diptera: A group of sensory receptors on the antennal Flagellum.

PENCILLATE MAXILLA Maxilla in which the entire superior surface of the upper jaw is clothed with Setae.

PENCILLED BLUE BUTTERFLY *Candalides absimilis* (Felder) [Lepidoptera: Lycaenidae].

PENCILLIFORM Adj. (Latin, *penicillus* = a pencil or small brush; *forma* = shape.) 1. Pencil-shaped (long, thin, parallel-slided). 2. Resembling a small painter's brush or pencil. Alt. Penicilliformis. See Pencillate. Cf. Fusiform. Rel. Form 2; Shape 2; Outline Shape.

PENDENT Adj. (Latin, *pendens* = hanging down.) Hanging.

PENDLEBURY, HENRY MAURICE (1893–1945) (Hale-Carpenter 1946, Proc. R. ent. Soc. Lond. (C) 10: 54.)

PENDLEBURY, W J DE MONTÉ (1889–1951) (Riley 1952, Proc. R. ent. Soc. Lond. (C) 16: 87.)

PENDULOUS Adj. (Latin, *pendere* = to hang; Latin, *-osus* = with the property of.) Drooping; suspended; attached to one end only. Alt. Pendulus.

PENEAU, JOSEPH (–1971) (Anon. 1971, Bull. Soc. ent. Fr. 76: 49.)

PENELLIPSE Noun. (Latin, *penes* = within; *ellipses* = fall short.) Lepidoptera larvae: The figure formed when a part less than the half of a uniserial circle of Crochets is absent from the Proleg.

PENES Noun. (Latin, *penes* = within.) Open, slit-like structures of the seminal vesicles to the outer surface in Euplectoptera (Smith).

PENICILLATE Adj. (Latin, *penicillus* = a pencil or small brush; *-atus* = adjectival suffix.) With a long flexible brush or pencil of Setae, often at the end of a thin stalk. Alt. Pencilatus.

PENICILLUM Noun. (Latin, *penicillus* = a pencil or small brush; Pl., Penicilli.) 1. A narrow, elongate pencil-like brush of Setae. 2. Male Noctuidae: Attached at the end of a stalk as long as the brush and folded into a lateral groove. 3. Neuroptera: See Pleuritoquamae. 4. Hymenoptera: Paired small style- or Cerci-like structures at the apex of the eighth Tergum of the Abdomen, or of the genitalia (Smith). Alt. Penicillus.

PENICULUS Noun. (New Latin.) Genitalia of male Lepidoptera: A setose process arising from the caudal edge of the Tegumen, dorsad of the Harpe (Klots).

PENILE PAPILLAE The external opening of the vas deferens in male Isopoda.

PENIS Noun. (Latin *penis* = penis. Pl., Penes.) The flexible, partially membranous, intromittent, copulatory organ of the male insect. See Aedeagus; Phallus. Syn. Vesica.

PENIS BULB Hymenoptera: A peculiar oval body carried within the upper part of the Penis by drone bees during the nuptual flight.

PENIS FUNNEL See Ring Wall.

PENIS POUCH See Penis Sheath.

PENIS ROD Siphonaptera: See Virga Penis. Hymenoptera: See Penis Valves.

PENIS SHEATH The layer of membrane which surrounds the Aedeagus.

PENIS VALVE 1. Diptera: See Parameral Lobe; Chironomidae: See Phallapodeme. 2. Mecoptera: see Parameres. 3. Hymenoptera: A paired Gonapophysis of ninth abdominal segment which is associated with the Penis and collectively forming the Aedeagus.

PENIS VESICLE Dragonflies: A small sac communicating with the genital fossa (Imms).

PENISFILUM Noun. (Latin, *penis* = penis; *filum* = thread. Pl., Penisfila.) A thread-like extension of the insect Penis (Tillyard).

PENNACEOUS Adj. (Latin, *penna* = feather; *-aceus* = of or pertaining to.) Pertaining to structure with a feathered appearance. Alt. Pennaceus; Pennate; Pennatus.

PENNANT, THOMAS (1726–1798) English naturalist and traveller. Principal works include *Arctic Zoology, British Zoology, Indian Zoology, History of Quadrupeds* and *A Tour of Scotland.* (Fée 1831, Mem. Soc. Sci. Agric. Lille 1: 185–186; Rose 1850, *New General Biographical Dictionary* 11: 23–25.)

PENNATE Adj. (Latin, *pennatus* = winged.) Feather-like; descriptive of structure resembling a feather. Alt. Penniform; Pinnate. See Feather.

PENNIFORM Adj. (Latin, *penna* = feather; *forma* = shape.) Feather-shaped; resembling a feather in appearance. Alt. Pennate; Pinnate. See Feather. Cf. Aristiform; Filiciform; Plumose; Scopiform. Rel. Form 2; Shape 2; Outline Shape.

PENNINGTON, KENNETH M (1897–1974) (C. B. C. 1974, Entomologist's Rec. J. Var. 86: 250–252, bibliogr.)

PENNSTYL® See Cyhexatin.

PENNY, DONALD D (1894–1975) (Gardner & Michelbacher 1975, Pan-Pacif. Ent. 5: 177.)

PENNY, THOMAS (c1532–c1588) (Gardner 1931, Rep. Lancs. Chesh. ent. Soc. 1928–30: 1–24; Wilkinson 1973, Gt. Lakes Ent. 6: 16–18.)

PENTAC® See Dienochlor.

PENTACENTRIDAE Plural Noun. (Greek, *pente* = Five; *kentron* = a point.) Five spined crickets. A small Family of ensiferous Orthoptera consisting of 15 Genera and 50 Species distributed through four Subfamilies (Pentacentrinae, Lissotrachelinae, Aphemogryllinae and Nemobiopsidinae.) Most Species inhabit the Indo-Australian Realm. Similar to Gryllidae but differ in that the head is less flattened and the Antenna is inserted below middle of face. Individuals live on the ground or in caves.

PENTAGAN® See Chlormequat.

PENTAGON Noun. (Greek, *pente* = five; *gonia* = angle; Latin, *pentagonum.* Pl., Pentagons.) A 5-sided figure with five equal or unequal angles.

PENTAGONAL Adj. (Greek, *pente* = five; *gonia* = angle; Latin, *-alis* = characterized by.) Descriptive of structure with margin 5-sided. See Outline Shape.

PENTAMERA Noun. (Greek, *pente* = five; *meros* = part.) Coleoptera with 5-segmented Tarsi.

PENTAMERAL SYMMETRY See Radial Symmetry.

PENTAMEROUS Adj. (Greek, *pente* = five; *meros* = part; Latin, *-osus* = with the property of.) Pertaining to legs displaying five Tarsomeres. See Pretarsus. Cf. Dimerous.

PENTASTOMIDAE Plural Noun. Tongue worms. A Class of worm-like, segmented animals of the Phylum Arthropoda.

PENTATOMIDAE Plural Noun. Stink Bugs. A numerically large, cosmopolitan Family of Hemiptera. Body oval or elliptical in outline shape; Antenna with five segments; Ocelli present; head tapering and narrower than Pronotum; Pronotum with large, rounded, Postlateral Lobes; Mesonotum and Metanotum visible in lateral aspect; Scutellum large, convex, U-shaped, covering Abdomen and wings at rest. Most Species are phytophagous and live above ground on host plants; some Species are predaceous and regarded as beneficial predators of pest insects. See Pentatomoidea.

PENTATOMOIDEA Leach 1815. A Superfamily of Hemiptera, Suborder Heteroptera, and characterized by Antenna with four segments in nymphal stage and five segments in adult stage; Scutellum large, triangular or elliptical; trichobothria on Abdomen (a common feature with Coreoidea, Lygaeoidea, Pentatomoidea). Included Families: Acanthosomatidae, Canopidae, Cydnidae, Dinidoridae, Lestoniidae, Megarididae, Pentatomidae, Philoeidae, Plataspididae, Scutelleridae, Tessaratomidae, Thaumastellidae, Thyreocoridae and Urostylidae. See Hemiptera.

PENUCI Noun. (Sl., Penuncus.) A pair of lateral processes arising from the Penis in some Entopterygota (Tillyard).

PENULTIMATE Adj. (Latin, *paene* = almost; *ultimus* = last; *-atus* = adjectival suffix.) 1. Descriptive of structure adjacent to the terminal segment of a body or appendage. 2. The next-to-the-last segment. The segment second from the end. Cf. Antepenultimate; Ultimate.

PEPPER-AND-SALT MOTH *Biston betularia cognataria* (Guenée) [Lepidoptera: Geometridae].

PEPPER MAGGOT *Zonosemata electa* (Say) [Diptera: Tephritidae]: A pest of hot cherry peppers in eastern USA; not a pest on sweet peppers or bell peppers. See Tephritidae.

PEPPER WEEVIL *Anthonomus eugenii* Cano [Coleoptera: Curculionidae]: A multivoltine pest of pepper, potato and eggplant in the southern USA. Eggs laid in punctures made in buds or immature pods. Larva feeds on buds or among seeds within pods; pupation occurs within bud or pod. See Curculionidae.

PEPPER, BAILEY BREAZEALE (1906–1970) (Hansens 1970, J. Econ. Ent. 63: 1579.)

PEPPERGRASS BEETLE *Galeruca browni* Blake [Coleoptera: Chrysomelidae].

PEPPERMINT LOOPER *Paralaea beggaria* (Guenée) [Lepidoptera: Geometridae].

PEPSIN Noun. (Greek, *pepsis* = digestion. Pl., Pepsins.) A protein-splitting enzyme found in pepsinogen and active in the stomach when it comes in contact with dilute hydrochloric acid.

PEPTIDASE Noun. (Greek, *peptein* = to digest. Pl., Peptidases.) A proteolytic enzyme which aids in digestion of complex proteins.

PEPTONE Noun. (Greek, *peptein* = to digest. Pl., Peptones.) A soluble protein compound produced by digestion of albumenoid food substances.

PERADENIIDAE Naumann & Masner 1985. Plural Noun. A small Family (ca two Species) of apocritous Hymenoptera assigned to the Proctotrupoidea, known only from south-eastern Australia. Morphologically bearing strongly pedunculate Gaster, medial margins of Compound Eyes convergent ventrad, unidentate Mandible, female with 13 Antennal segments, male with 12 Antennal segments, Pronotum strongly declivous. Biology unknown. Sometimes included in Heloridae.

PERCARDIAL CHAMBER An open space or sinus around the Heart or Dorsal Vessel.

PERCHERON, ACHILE RÉMI (1797–1869) (Swainson 1840, *Taxidermy with Biography of Zoologists.* 392 pp. (291, bibliogr.) London. [Volume 12 of *Cabinet Cyclopedia.* Edited by D. Lardner.]; Marseul 1889. Abeille 26 (Les ent. et leurs écrits): 277–278, bibliogr.)

PERCHERON, M French author of *Monographie des Passales* and *Genera des Insects* (with M. Guerin).

PERCHLOROMETHANE® See Carbon Tetrachloride.

PERCIPIENT Adj. (Latin, *percipiens* pres. part. of *percipere* = to perceive.) With the power of perceiving.

PERCOLATE Trans. Verb. (Latin, *percolare* = to strain; *-atus* = adjectival suffix.) 1. To cause a liquid to pass through a permeable substance. 2. To pass through or by a porous substance; to filter.

PERCURRENT Adj. (Latin, *percurrens* = running throughout.) 1. Pertaining to substance or colour extending the length of a structure; continuous. 2. Extending from base to apex.

PERDITA F. Smith 1853. A numerically large Genus of small-bodied, ground-nesting bees (Andrenidae, Pangurinae) restricted to southern Canada, the United States and Mexico. Presumably monophyletic and most closely related to *Nomadopsis* and related forms. *Perdita* is predominantly xeric adapted, more than 750 Species-level names have been proposed. Generally regarded as the most speciose Genus of bees in the Nearctic.

PEREION See Prothorax.

PEREIPODA Noun. (Greek, *pereion* = to go about; *podos* = foot.) The 2nd and 3rd pair of thoracic legs of larvae and the 2nd pair in adults.

PERENNIAL Adj. (Latin, *per* = through; *annus* = year; *-alis* = characterized by.) Descriptive of a process that continues or plant that lives for several years. Cf. Annual.

PERENNIAL CANKER A fungal disease of apples cause by *Gloeosporium perennans* and vectored by the Woolly Apple Aphid. WAA feeds on apple, causes cracking of callus which permits entry of fungus.

PERENNIAL COLONY Eusocial Hymenoptera: A colony that lasts more than one season.

PEREYASLAVTSEVA, SOFYA MIKHAILOVNA (1851–1903) (Kuznezov 1903, Revue Russe Ent. 3: 422.)

PEREZ ARCAS, LUREANO (1824–1894) (Martinez y Saeg 1894. An. Soc. esp. Hist. nat. 23: 278–296, bibliogr.)

PEREZ, CHARLES (Horn 1936, Arb. physiol. angew. Ent. Berl. 3: 70.)

PEREZ, JEAN (1833–1914) (Alluaud 1914, Bull. Soc. ent. Fr. 1914: 434; Anon. 1916, Ann. Soc. ent. Fr. 85: 355–366, bibliogr.)

PERFECT INSECT An archane term for the Imago or adult stage of the insect.

PERFEKTHION® See Dimethoate.

PERFILEV, BORIS VASIL'EVICH (1891–1969) (Kuznetsov 1970, Arch. Hydrobiol. 67: 276 281, bibliogr.)

PERFOLIATE Adj. (Latin, *per* = through; *folum* = leaf; *-atus* = adjectival suffix.) Divided into leaf-like plates. Applied to Antennae with disc-like expansions connected by a stalk passing nearly through their centres, also to any part possessing a well developed leaf-like or plate-like expansion. Alt. Perfoliatus.

PERFORATE Adj. (Latin, *perforare* = to bore through; *-atus* = adjectival suffix.) Applied to Antennae in which a part of each segment is dilated or flattened and the remaining part cylindrical so the Antenna appears like a thread on which the segments are strung. Alt. Perforata.

PERFORATED CUPOLA ORGAN Minute, epidermal exocrine organs characterized by a depressed, porous central area (sieve plate). PCOs often found in association with the spiracles of larval Lycaenidae and Riodinidae and secrete amino acids in some Species. Regarded as homologous to Setae and may be responsible for maintaining associations with ants. Cf. Newcomer's Organ; Tentacular Organ.

PERFORATOR Noun. (Latin, *perforare* = to bore through. Pl., Perforators.) A structure adapted for boring through tissue.

PERFUSION (Latin, *per* = through; *fundere* = to pour.) Tissue Culture: Continuous culture of cells in a fermenter.

PERGAMENOUS Adj. (Latin, *pergamena* = parchment; *-osus* = with the property of.) Pertaining to tissue or Integument which is thin and partly transparent; resembling parchment; parchment-like. Alt. Pergamentaceous.

PERGANDE, THEODORE (1840–1916) (Anon. 1916, Ent. News 27: 240, 291, bibliogr.; Anon. 1916, Science 43: 492; Howard 1930, Smithson. misc. Collns. 84: 57, 85–90, 293.)

PERGID SAWFLIES See Pergidae.

PERGIDAE Ashmead 1898. Plural Noun. A Family of Symphyta (Hymenoptera) assigned to the Superfamily Tenthredinoidea. Pergids are abundant in Australia and South America; poorly represented in Nearctic and Orient. Adult 3–30 mm long; Antenna with 5–24 segments; Hypostomal Bridge present or absent; forewing without crossvein 2r; fore Tibia with 1–2 apical spurs. Larva with Stemmata, legs with 3–5 segments and claws; abdominal Prolegs on segments 2–6, 2–8 or 3–8 and 10.

PERIANTH Noun. (New Latin, *perianthum* > Greek, *peri* = around; *anthos* = flower. Pl., Perianths.) The external envelope of a flower, which is often differentiated into a Calyx and Corolla. See Flower.

PERICARDIAL Adj. (Greek, *peri* = around; *kardia* = hear; Latin, *-alis* = characterized by.) Around the Heart; pertaining to the Pericardium. See Heart; Pericardium.

PERICARDIAL CELLS Specialized cells (Nephrocytes) that lie along both sides of the Heart and function to purify the blood; Nephrocytes of the dorsal sinus.

PERICARDIAL CORD A longitudinal thread-like filament which is in the position of the dorsal pulsatory vessel in the Protura (Berlese).

PERICARDIAL DIAPHRAGM A delicate, membranous tissue attached to the ventral surface of the Heart and laterally to the body wall; the Dorsal Diaphragm.

PERICARDIAL SINUS See Dorsal Sinus.

PERICARIDAL CAVITY See Dorsal Sinus.

PERICARP Noun. (Greek, *peri* = around; *karpos* = fruit.) The wall of a fruit which has developed from the Ovary wall.

PERIGEN® See Permethrin.

PERI-INTESTINAL Surrounding or near the Alimentary Canal.

PERIKARION Noun. (Greek, *peri* = around; *karyon* = nucleus; Pl., Perikaryia.) The cytoplasm surrounding the nucelus of a nerve cell. The body of a nerve cell. Alt: Pericaryon.

PERILAMPIDAE Plural Noun. A small (ca 25 Genera, 200 Species), cosmopolitan Family of parasitic Hymenoptera assigned to the Chalcidoidea. Fossil record Lower Oligocene; classification problematical, near Eucharitidae or Chrysolampinae a Subfamily of Pteromalidae. Diagnostic features: Body robust, moderately large, metallic coloured; Antenna short with 13 segments including one Anellus; Labrum 'flap-like', with setal comb along margin; head and Thorax frequently coarsely punctate; Notauli well developed; forewing Marginal Vein long, Postmarginal short; Petiole sometimes well developed; Gaster triangular outline; Terga I, II often fused. Biology poorly understood, apparently primary and secondary parasitic habits, primary parasites attack xylophagous beetles, hyperparasites attack braconids, ichneumonids, tachinids on Symphyta, Lepidoptera larvae. Perilampinae: A few Species may be obligatory hyperparasites; some females proovigenic, oviposit near host; egg pedunculate, planidial first instar larva; Planidium with 12 segments + Caudal Sucker; Planidium penetrates Integument of primary host, searches and penetrates primary host larva; remains inactive until primary host pupates, exits primary host and feeds externally. Chrysolampinae: Ovarian eggs are not pedunculate; first-instar larvae hymenopteriform, with 13 segments, tapered posteriad, segments 2–9 with cup-shaped structure, terminal segment bilobed; ectoparasites of Coleoptera larvae, adults; feed at intersegmental grooves.

PERILESTIDAE Plural Noun. A Family of Odonata. Body small to medium sized with a long, slender Abdomen. Principal diagnostic characters include short Pterostigma, Anal Vein leaves posterior margin from base of wing and below outer end of quadrilateral.

PERILLENE A monoterpene with lemon-like odour first isolated from essential oils in the leaves of *Perilla citriodora* Makino. Subsequently found in other plants and mandibular gland secretions of ants and the anal secretions of thrips. Believed to serve as an alarm pheromone in thrips.

PERIMPAK® See Lambda Cyhalothrin.

PERIMYLOPIDAE Plural Noun. A small Family of polyphagous Coleoptera assigned to Superfamily Tenebrionoidea.

PERINEURAL SINUS The space in the body-cavity of an insect limited by the Ventral Diaphragm.

PERINEURAL Adj. (Greek, *peri* = around; *neuron* = nerve; Latin, *-alis* = pertaining to.) Surrounding a nerve, as the body cavity immediately surrounding the nervous system.

PERINEURIUM Noun. (Greek, *peri* = around; *neuron* = nerve; Latin, *-ium* = diminutive > Greek, *-idion*. Pl., Perineuria.) A tubular sheath of nerve fibres bound together by Septate Desmosomes and Tight Junctions. Perineurium offers no resistance to diffusion and may serve a trophic function or may act as a brain/blood barrier. Barrier involves active transport of ions between Haemolymph and fluid surrounding Axons. Removal of Perineurium results in impairment or prevention of stimulus transmission. See Nervous System; Nerve Cell. Cf. Axon; Cell Body; Dendrite; Neurolemma; Perineurium. Rel. Central Nervous System.

PERINGUEY, LOUIS ALBERT (1924) (Anon. 1924, Ent. News 35: 190, 262; Anon. 1924, Nature 113: 541; Anon. 1925, S. Afr. J. nat. Hist. 5: 1–8, bibliogr.)

PERINUCLEAR Adj. (Greek, *peri* = around; *nucleus* = kernel.) Near or surrounding the cell nucleus.

PERIODICAL Adj. (Greek, *periodos* = a period of time; *-alis* = pertaining to.) Pertaining to cyclical phenomena that recur at regular intervals. Cf. Circadian Rhythm.

PERIODICAL CICADA Species of *Magicicada* which represent 13- and 17-year life cycle forms of three morphologically and behaviourally distinct cicadas. Historically, *M. septendecim* (Linnaeus) [Hemiptera: Cicadidae], a Species endemic to eastern North America. Eggs laid May-June within slits in bark or wood, with 12–20 eggs per slit. Eclosion occurs within 6–8 weeks; first instar nymphs drop to ground, burrow into soil, construct cell and feed on root xylem fluids. Nymphs feed 13 or 17 years, fifth instar nymph emerges ·rom soil, climbs trunk of tree, moults into adult. Adult active 4–6 weeks. Cf. Dogday Cicada; Orchard Cicada.

PERIOPOD Noun. (Greek, *peri* = around; *pous* = feet. Pl., Periopods.) Legs, limbs or appendages used for walking by arthropods (Snodgrass).

PERIOPTICON Noun. (Greek, *peri* = around; *opsis* = sight.) The 3rd ganglionic swelling of the optic nerve in insects. Composed of many cylindrical masses of Neurospongium arranged side-by-side (Packard). The outermost tract of the optical lobe of the insect brain (Imms).

PERIPATUS See Onychophora.

PERIPHALLIC ORGANS Peripheral genital structures on the ninth segment of the male. POs include clasping structures which develop independent of the phallic lobes. This condition occurs in Thysanura, Ephemeroptera, Dermaptera and Orthoptera. Alternatively, primary phallic lobes produce the entire male genitalia, including Aedaegus and clasping structures. In this system, Periphallic Organs are usually absent. This condition is observed in hemipteroids and the Endopterygota. See Genitalia. Cf. Primary Phallic Lobes.

PERIPHERAL Adj. (Greek, *peripherein* = to move around; Latin, *-alis* = characterized by.) 1. Descriptive of structure that is associated with the outer margin of a circular object; distant from the centre and near the circumference. 2. Descriptive of process that is not central to a concept, plan or function. Cf. Central.

PERIPHERAL FAT BODY One of two principal types of Fat Body; PFB is typically diffuse and suspended in the Haemolymph and located beneath the Integument. See Fat Body. Cf. Visceral Fat Body.

PERIPHERAL NERVOUS SYSTEM A delicate network of nerve fibres and multipolar nerve cells in the Integument below the Epidermis. PNS is connected with sensory receptors (Setae, Sensilla). See Central Ganglia. Cf. Central Nervous System; Sympathetic Nervous System.

PERIPHERAL PAD The persistent part of the Peripodal Sac (Bugnion); Annular Zone (Kunckel).

PERIPHERIA Noun. (Greek, *peripherein* = to move around.) The entire outline of the body.

PERIPHERY Noun. (Greek, *peripherein* = to move around.) The outer margin of a body, organ, appendage or structure.

PERIPLASM Noun. (Greek, *peri* = around; *plasma* = form.) A bounding layer of protoplasm in the insect egg which lies just beneath the Vitelline Membrane and completely surrounds the egg (Imms).

PERIPNEUPSTIC Adj. (Greek, *peri* = around; *pneustikos* = breathing; *-ic* = consisting of.) Pertaining to spiracles arranged in a row along the body.

PERIPNEUSTIC RESPIRATORY SYSTEM Insect larvae: The prevalent type RS with spiracles in a row on each side of the body; spiracles of the wing-bearing segments closed.

PERIPODAL CAVITIES Pouches in the embryo in which the rudiments of the future legs and wings are developed.

PERIPODIAL MEMBRANE The cell layer or wall surrounding the peripodal cavities, continuous with the Integument.

PERIPODIAL SAC The membrane enclosing the imaginal bud (Packard); hypodermic envelope.

PERIPROCT Noun. (Greek, *peri* = around; *proktos* = anus.) The primitive terminal segment of the body containing the Anus, anterior of which the true Somites are formed in animals with ring-like segments; Telson (Snodgrass).

PERIPSOCIDAE. Plural Noun. A numerically small, cosmopolitan Family of Psocoptera assigned to the Suborder Psocomorpha. Antenna with 13 segments, without secondary annulations; forewing Areola Postica absent; Tarsi with two segments. Species found on bark and dried foliage.

PERISCELIDAE Plural Noun. A small, widespread Family of acalypterate Diptera assigned to Superfamily Asteioidea. Arista plumose; Postverticals short, diverging; Costa not broken; Radial Sector without bristles. Adults taken near sap flows on wounded trees; larvae of a few Species reared from sap. Syn. Stenomicridae.

PERISSOMMATIDAE Plural Noun. A small Family of nematocerous Diptera including extant Species from Chile and Australia; fossil Species known from Siberia and Mongolia.

PERISTAETIUM PERIISTETHIUM The Mesosternum (Smith).

PERISTALITIC Adj. (Greek, *peristaltikos; peri + stellein* = to send around; *-ic* = characterized by.) Pertaining to Peristalsis.

PERISTALSIS Noun. (Greek, *peristaltikos; peri + stellein* = to send around; *sis* = a condition or state. Pl., Peristalses.) The series of successive waves of contraction along the gut and other body tubes which move digested food toward the anal extremity and fluid through the circulatory system. Cf. Antiperistalsis.

PERISTIGMATIC GLANDS Glands near the spiracles which secrete hydrophobic substances that prevent wetting of the spiracles (Imms).

PERISTOME Noun. (Latin, *peristoma*. Pl., Peristomes.) Diptera: The border of the mouth or oral margin, sometimes used as equivalent to Epistoma. 2. A membranous tissue surrounding the mouthparts at the base, forming the true ventral wall of the head. 3. Anellida: The anteriormost, mouth-bearing segment. Alt. Peristoma; Peristomium.

PERISTOMIAL Adj. (Greek, *peri* = around; *stoma* = mouth; Latin, *-alis* = pertaining to.) Descriptive of structure or process associated with the Peristomium.

PERITHECIUM Noun. (Latin, *peri* = around; *thecium* = case; Latin, *-ium* = diminutive > Greek, *-idion*. Pl., Perithecia.) Fungi: A microscopic, globular or flask-shaped structure containing sexual spores. Cf. Pycnidium; Theca.

PERITONEAL Adj. (Greek, *peritonos* = stretched around; Latin, *-alis* = pertaining to.) Descriptive of structure or process associated with the Peritoneum.

PERITONEAL ENVELOPE See Epithelial Layer (Comstock).

PERITONEAL MEMBRANE See Peritoneal Sheath.

PERITONEAL SHEATH 1. The covering of the entire insect Ovary in young stages and sometimes in the adult. PS consisting of an envelope of adventitious connective tissue. 2. The outer covering of the Testicle (all Follicles) may surround each Testis or surround both Testes. The later condition is seen in some Hymenoptera.

PERITONEUM Noun. (Greek, *peritonos* = stretched across. Pl., Peritonea; Peritoneums; Peritonaea.) 1. The membranous lining of the body cavity of some invertebrates. 2. Insects: The featureless connective tissue forming the outermost layer of the digestive tract (Folsom & Wardle). Alt. Peritonaeum. See Alimentary Canal.

PERITRACHEAL Adj. (Greek, *peri-*; *trachelos* = neck; Latin, *-alis* = pertaining to.) Surrounding the Trachea.

PERITREME Noun. (Greek, *peri* = around; *trema* = hole. Pl., Peritremes.) 1. The margin of a shell opening. 2. The cuticular margin which surrounds a spiracle. 3. Archaic: small sclerite under the forewing through which the spiracle opens (Audouin). 4. Any sclerotic plate surrounding any body opening. Alt. Peritrema. See Spiracle.

PERITROPHIC Adj. (Greek, *peri-*; *trephein* = to nourish; *-ic* = of the nature of.) Descriptive of or pertaining to the membranous envelope within the Ventriculus.

PERITROPHIC MEMBRANE Typically, a delicate membrane which forms a cylindrical envelope surrounding food in the midgut. Structurally, the PM is formed by a network of Chitin and a protein/carbohydrate matrix. PM is impermeable to particulate material, but freely permeable to molecules in solution. Functionally, the PM serves to physically and chemically protect midgut cells, provide a physical barrier against microorganisms entering the Haemocoel and conserve digestive enzymes by preventing their excretion. Most insects bear a Peritrophic Membrane irrespective of the physical nature of the food they ingest. PM has not been found in Mecoptera or many Heteroptera. PM poorly developed in most Diptera, but some exceptions have been suggested. In carabid beetles, the Peritrophic Membrane remains fluid anteriad and is very delicate posteriad and may be overlooked. PM is formed in several ways. In Orthoptera, Odonata, Coleoptera and Hymenoptera PM formed as a delamination of the entire midgut. A fluid precursor of PM typically condenses to form a definitive membrane. In most insects PM is secreted along the entire length of midgut. In some insects, PM remains in a water-soluble fluid state. Diptera are an exception to the generalized pattern of PM formation. The PM is secreted by a band of Cardia Cells as a viscous fluid at the anterior end of the midgut and forced through a mould at an invagination of the midgut valve. The cells producing the PM are rich in Rough Endoplasmic Reticulum and secretory granules. See Alimentary System; Midgut. Cf. Filter Chamber. Rel. Digestion.

PERIVISCERAL Adj. (Greek, *peri-*; Latin, *visceris* = internal organ; *-alis* = pertaining to.) Descriptive of structure surrounding or within the cavity containing the Alimentary Canal and its appendages. See Alimentary Canal.

PERIVISCERAL FAT BODY One of two principal types of Fat Body, typically surrounding the digestive tract. See Fat Body. Cf. Peripheral Fat Body.

PERIVISCERAL SINUS The principal part of the body cavity, between the Dorsal Diaphragm and the Ventral Diaphragm if the latter is present (Snodgrass).

PERIZIN® See Coumaphos.

PERKINS, F A Australian academic Entomologist and founder of Entomology Department, University of Queensland.

PERKINS, LILLY (Brown & Heinemann 1972, *Jamaica and its Butterflies*. xv + 478 pp. (16), London.)

PERKINS, ROBERT CYRIL LAYTON (1866–1955)

Economic Entomologist (Hawaiian Sugar Planter's Association.) (Benson 1955, Entomol. mon. Mag. 19: 289–291; Scott 1956, Biogr. mem. Fellows R. Soc. 2: 215–236, bibliogr.)

PERKINS, VINCENT ROBERT (1831–1922) (Anon. 1922, Ent. Rec. J. Var. 34 : 115; Perkins 1922, Entomol. mon. Mag. 58:110–111.)

PERLATE Adj. (Prob. French, *perle* = pearl; *-atus* = adjectival suffix.) Beaded; bearing relieved, rounded points in series.

PERLIDAE Plural Noun. Common Stoneflies. The largest Family of Plecoptera with 48 Genera and about 400 nominal Species. Family name refers the head which is pearl-like in roundness and brilliance. Representatives in all Zoogeographical Realms except Australia. Adult apical segment of Palpi and two basal segments of Tarsi reduced. Larvae with thoracic and abdominal gills; larval stage lasts several years. All Species carnivorous.

PERLOIDAE Plural Noun. Springflies; Stripetails. A Family of Plecoptera with 40 Genera and about 250 nominal Species. Predominantly Holarctic; one Genus found in Oriental Realm. Most larvae lost segmental respiratory gills, but *Oroperla* Species have gills on Thorax and Abdomen. Larvae carnivorous; adults do not feed and die shortly after emergence.

PERLOPSIDAE Plural Noun. A Family of fossil Plecoptera known from Permian deposits of Russia.

PERMANDINE® See Permethrin.

PERMECTRIN® See Permethrin.

PERMETHRIN Noun. A synthetic-pyrethroid compound with broad-spectrum effectiveness and residual action as a pesticide; acts as contact and stomach poison. Among pesticides, notable with lowest oral and dermal toxicity to mammals; minor phytotoxicity on some ornamentals. Used on numerous agricultural crops and ornamentals; important in control of many significant agricultural pests; used as a dust or wettable powder to control silverfish, cockroaches, bedbugs, ants, fleas, flies and clothes moths. Trade names include: Adion®, Ambush®, Astro®, Atroban®, Biothrin®, Brunol®, Cellutec®, Coopex®, Corsair®, Defend®, Detmol®, Dragnet®, Dragon®, Eksmin®, Evercide®, Expar®, Fafai®, Gard-Star®, Gori®, Hard-Hitter®, Imperator®, Insectaban®, Kafil®, Kudos®, Neopitroid®, Outflank®, Over-Time®, Perigen®, Permandine®, Permectrin®, Permit®, Perthrine®, Picket®, Pounce®, Pramex®, Pynoset®, Qamlin®, Residroid®, Rondo®, Sanbar®, Stockade®, Stomoxin®, Talcord®, Tornade® and Torpedo®. See Synthetic Pyrethroids.

PERMIAN PERIOD The seventh Period (290–245 MYBP) of the Palaeocene Era and named for the Province of Perm in the Ural Mountains in which the PP was first studied. Principal deposits include Lower (Early) Permian beds in Elmo, Kansas; Moravia; Obara, Czechoslovakia; Kuznetsk Basin, Tschekarda, Siberia) and Upper Permian (New South Wales, Australia; Natal, South Africa). PP characterized by mass-extinction twice the magnitude of end-Ordovician episode. Lycopods that formed coal-swamps, sphenopsid and cordaite trees all eliminated; 20 Families of therapsids, 45% of marine Families eliminated and 96% of marine Species eliminated. The Permian Extinction marks the end of trilobites, eurypterids and the marine community dominated by sessile, filter-feeding brachiopods, bryozoan and crinoids. Insect radiation reflected in Ordinal Classification of modern fauna: Odonata, Plecoptera, Thysanoptera, Heteroptera, Coleoptera, Rhaphidioidea, Megaloptera, Neuroptera, Mecoptera, Diptera and Trichoptera. See Geological Time Scale; Palaeozoic. Cf. Cambrian Period; Devonian Period; Lower Carboniferous Period; Upper Carboniferous Period; Ordovician Period; Silurian Period. Rel. Fossil.

PERMINERALIZATION Noun. (Latin, *per* = through; *minerale* = mineral; English *-tion* = result of an action.) A type of fossilization process in which the cells and spaces in a dead organism are replaced with mineral matter and the organic matter remains intact or unchanged. Cf. Petrification. See Fossilization.

PERMIT® See Etofenprox; See Permethrin.

PERMOTANYDERIDAE Plural Noun. A Family containing two Genera (*Choristotanyderus* Riek, *Permotanyderus* Riek) of Upper Permian age in Australia. Assigned to Order Protodiptera with Permotipulidae and a possible link between Diptera and Mecoptera. Taxa with two pairs of wings; forewings with venation resembling primitive Diptera.

PERMOTIPULIDAE Plural Noun. A Family containing two Genera (*Permotipula* Tillyard, *Permila* Williams) of Upper Permian age in Australia and Russia. Assigned to Order Protodiptera with Permotanyderidae and a possible link between Diptera and Mecoptera. Taxa with two pairs of wings but hindwings considerably reduced.

PERON, FRANCOIS (1775–1810) (Iredale 1929, Aust. Mus. Mag. 3: 357–359.)

PERONEA Noun. A clasper of the genitalia (MacGillivray).

PEROPAL® See Azocyclotin.

PERRAULT, CLAUDE (1613–1688) (Rose 1850, *New General Biographical Dictionary* 11: 42–43.)

PERRET-MAISONNEUVE, A (1866–1937) (Lesne 1938, Bull. Soc. ent. Fr. 43: 8.)

PERRIER, ALFRED (1890–) (Morière 1867, Bull. Soc. linn. Normandie (2) 2 : 161–171.)

PERRIER, EDMOND (1844–1921) (de la Torre. 1921, Mems Soc. cub. Hist. nat. 'Filipe Poey' 4: 29–32.)

PERRIN, ELEAZAR ABEILLE DE See Abeille De Perrin.

PERRIN, JOSEPH (1864–1936) (McLaine 1937, Can. Ent. 69: 19–26.)

PERRIS, JEAN PIERRE OMER ANNE EDOUARD

(1808–1878) (Mulsant 1878, Ann. Soc. linn. Lyon 25: 85–110, bibliogr.; Laboulbène 1879, Ann. Soc. ent. Fr. (5) 9: 373–388, bibliogr.)

PERRONCITO, EDOARDO (1847–1936) (Ghislein 1936, Annali Accad. Agric. Torino 79: 213–22.)

PERROUD, BENOIT PHILIBERT (1796–1878) (Musgrave 1932, *A Bibliography of Australian Entomology 1775–1930,* viii + 380 pp. (255), Sydney.)

PERSICINUS Adj. (Latin, *perse* = peach; *aneus* = like.) Describing colour as the red of peach blossoms.

PERSIMMON BORER *Sannina uroceriformis* Walker [Lepidoptera: Sesiidae].

PERSIMMON PSYLLA *Trioza diospyri* (Ashmead) [Hemiptera: Psyllidae].

PERSISTENT Adj. (Latin, *persistens* pres. part. of *persistere* from *per-*; *sistere* = to stand fixed.) Remaining constantly; always present.

PERSONAL PROTECTION EQUIPMENT The minimum equipment required for the safe handling and application of pesticides. PPE standards are developed by the U.S. Environemental Protection Agency and appear on the label and Material Safety Data Sheet. Acr. PPE.

PERSONATE Adj. (Latin, *personatus* = masked; *-atus* = adjectival suffix.) 1. Masked; disguised. Alt. Personatus. 2. A form different from the typical adult form.

PERTHRINE® See Permethrin.

PERTY, JOSEPH ANTON MAXIMILIAN (1804–1884) (Anon. 1884, Psyche 4: 236.)

PERTY, M Author of *Delectus animalium atriculatorum quae in itinere, & c.* The work describes Species collected in Brazil by Spix and Martius.

PESCATORE, GUSTAV (–1916) (Soldanski 1916, Dt. ent. Z. 1916: 227.)

PESCHET, RAYMOND (1880–1940) (Méquignon 1941, Ann. Soc. ent. Fr. 110: 369–372.)

PESSELLA Noun. Male Cicadidae: Two small, acute processes one in each socket of the hind leg; presumably intended to supress the Opercula (Kirby & Spence).

PEST Noun. (Latin, *pestis* origin obscure. Pl., Pests.) 1. Any organism that reduces quality or yield of crops or other products. Cf. Beneficial Organism. 2. A Species or organism that interferes with human health, activities or property, or is objectionable.

PEST MANAGEMENT The theory and practice of controlling organisms that pose a threat to humans, livestock or plants. Pest management takes several forms, including Biological Control, Chemical Control and Integrated Pest Management.

PEST STATUS The ranking of a pest relative to the economics of controlling it.

PESTICIDE Noun. (Pl., Pesticides.) A substance (active ingredient) that is used to kill 'undesirable' organisms including animals, plants and fungi. Pesticides kill organisms in different ways, but often are non-selective in killing and may affect non-target organisms. See Chemical Control. Cf. Acaricide; Biopesticide; Herbicide; Filaricide; Fungicide; Herbicide; Insecticide; Nematicide. Rel. Pest Management.

PESTICIDE ACT Legislation enacted which affects the registration, packaging, labelling, sale, storage and disposition of toxic substances used as pesticides.

PESTIS MINOR A form of plague caused by the bacterium *Yersinia pestis* and vectored by fleas. Disease regarded as minor. See Plague.

PESTROY® See Fenitrothion.

PETAGNA, LUIGI (1779–1823) (Vulpes 1834, Atti 1st. Incoragg. Sci. nat. 5: 287–310.)

PETAGNA, VINCENT Neapolitan author of *Specimen Insectorum Ulteriorus Calabriae* (1787.)

PETAGNA, VINCENZIO (1734–1810) (Anon. 1818, Atti 1st Incoragg. Sci. nat. 2: 340–342, bibliogr.)

PETALURIDAE Plural Noun. Graybacks. A primitive Family of Odonata. Adult large to very large, robust; head broad with widely separated eyes and elongate Abdomen; wings long and narrow with male hindwing anal margin curved; Pterostigma exceptionally long and curved; triangles slightly dissimilar in fore and hindwings; anal loop poorly developed; superior anal appendages of male foliate. Female with an Ovipositor. Naiad body soft, pale and elongate; eyes small and elevated; Antenna with seven segments, second segment longest. Naiads inhabit swampy soil and forage nocturnally.

PETERS, HERMANN (–1928) (Ross 1934, Int. ent. Z. 27: 539.)

PETERS, WILHELM CARL HARTWIG (1825–1883) (Anon. 1883, Psyche 4: 59.)

PETERSDFORFF, EMIL (1836–1915) (Belling 1917, Dt. ent. Z. 1917: 322–324.)

PETERSEN, AXEL (1877–1964) (Anon. 1964, Ent. Meddr 32: 453–455.)

PETERSEN, JOHANNES (1900–1966) (Hansen. 1968, Ent. Meddr 36: 407–408.)

PETERSEN, P E See Esben-Petersen.

PETERSEN, WILHELM C (1854–1933) (Anon. 1933, Entomologist's Rec. J. Var. 45: 142.)

PETERSON, ALVAH (1888–1972) (Cambell 1973, J. Econ. Ent. 66: 1243–1244)

PETERSON, HAROLD O (1906–1963) (Roberts 1963, J. Econ. Ent. 56: 907–908)

PETERSON, JOHN B (–1944) (Lellie 1973, Mosquito News 4: 128.)

PETHEN, ROBERT W (1877–1944) (S. A. 1944, Lond. Nat. 24: 67, bibliogr.)

PETIOLAREA See Petiolar Area.

PETIOLAR Adj. (Latin, *petiolus* = small foot.) Descriptive of or pertaining to a Petiole. Cf. Peduncular.

PETIOLAR AREA Hymenoptera: The apical or posterior of the three median, metanotal cells; 3rd median area. Syn. Apical Area; Petiolarea.

PETIOLATA Noun. Hymenoptera in which there is a slender stalk between the Thorax and Abdomen. See Apocrita.

PETIOLATE Adj. (Latin, *petiolus* = small foot; *-atus* = adjectival suffix.) Stalked; placed upon a stalk. Alt. Petioltus. See Petiole. Cf. Petioliform.

PETIOLE Noun. (Latin, *petiolus* = small foot. Pl., Petioli.) 1. A stem or stalk-like projection. 2. Some Diptera and apocritous Hymenoptera: The slender or narrow tubular segment between the Thorax and Abdomen. Hymenopteran Pedicel formed of one segment, or two-segmented in ants. 3. Plants: The slender stalk of a leaf.

PETIOLIFORM Adj. (Latin, *petiolus* = small foot; *forma* = shape.) Petiole-shaped; descriptive of structure that is short and cylindrical. Alt. Petiolate. See Petiole. Rel. Form 2; Shape 2; Outline Shape.

PETIOLULE Noun. A small Petiole.

PETIVER, JAMES (c 1660–1718) (Lisney 1960, *Bibliography of British Lepidoptera 1608–1799*, 315 pp. (42–44, bibliogr.), London.)

PETRI, KARL (1852–1932) (Müller 1931, Verh. mitt. sieben. Ver. naturw. 80–81: 6–10, bibliogr.)

PETRIFICATION Noun. A form of the fossilization process in which all organic matter in a dead organism is replaced with mineral matter. Cf. Permineralization. See Fossilization.

PETROCCHI, JUANA MIGUELA (1893–1925) (Anon. 1925, Physis B. Aires 7: 417–420, bibliogr.)

PETROLEUM OILS Hydrocarbons applied to plants which function as insecticides, acaricides and ovicides. Kerosene first used as a PO ca 1900. Many commercial products marketed globally. Common names include Dormant Oils and Summer Oils. Trade names include: Orchex, Saf-t-side, Scalecide, Stylet, Sun-Spray, Supreme Oils, Volck Oils. See Insecticide; Pesticide.

PETROVITZ, RUDOLF (1906–1974) (Miksic 1976, Arbeitsgem. Ost. Ent. 27: 120–124, bibliogr.)

PETRUNKEVITCH, ALEXANDER (1875–) (Hutchinson 1945, Trans. Conn. Acad. Arts Sci. 36: 9–15, bibliogr.)

PETTIT, JOHNSON (–1898) (Anon. 1898, Ent. News 9: 184.)

PETTIT, RUFUS HIRAM (1869–1898) (Anon. 1946, Rec. Mich. Sta. Coll. 57 (3): 15.)

PETZ, JOSEF (1866–1926) (Heikertinger 1926, Wien. ent. Ztg 43: 47.)

PETZCH, HANS (–1974) (Anon. 1974, Prakt. SchädlBekämpf. 26: 168.)

PEUCER, GASPARD (CASPAR) (1525–1603) (Rose 1850, *New General Biographical Dictionary* 11: 72.)

PEYERIMHOFF DE FONTENELLE, PAUL MARIE (1873–1957) (Bernard & Pierre, Ann. Soc. ent. Fr. 127: 1–8.)

PEYERIMHOFF, MARIE ANTOINE HERCULE HENRI (1838–1877) (Anon. 1877, Bull. Soc. ent. Fr. (5) 7: lxxvi–lxxvii.)

PEYR, JOSEF (1862–1936) (Anon. 1937, Arb. morph. taxon. Ent. Berl. 4: 242.)

PFAFF, PAUL (1897–1974) (Gruber 1974, Prakt. SchädlBekämpf 26: 142.)

PFAFF, SVEND WILLEMOES (1920–1939) (Wolf 1940, Ent. Meddr 20: 586–587.)

PFANKUCH, KARL (1871–1924) (Alfken 1924, Dt. ent. Z. 1924: 549–553, bibliogr.)

PFAU, R JOHANNES (1881–1966) (Cleve 1967, Mitt. dt. ent. Ges. 26: 3.)

PFEFFER, ANTON (1904–1964) (Heyrovsky 1964, Cas. csl. Spol. ent. 61: 73–78, bibliogr.)

PFEIFFER, ERNST (1893–1955) (Anon. 1955, Mitt. Münch. ent. Ges. 44–45: 531–536.)

PFEIFFER, LUDWIG (1878–1926) (A. H. 1926, Ent. Z. Frankf. a. M. 40: 269–271, bibliogr.)

PFEIL, FRIEDRICH WILHELM LÉOPOLD (1783–1859) (Grunet 1861, Forstliche. Blätter. 1: 41.)

PFITZNER, RUDOLF (1864–1921) (Seitz 1921, Ent. Rdsch. 38: 15.)

PFLEGER, KAREL (1900–1951) (Havelka 1951, Cas csl. Spol. ent. 48: 141–142.)

PFLUG, PAUL GOTTFRIED (1741–1789) (Papavero 1971, *Essays on the History of Neotropical Dipterology*. 1: 22. São Paulo.)

PFLUGFELDER, OTTO (–1968) (Frank 1968, Mitt. dt. ent. Ges. 27: 61–63.)

pH A measure of Hydrogen ion concentration. pH scale ranging from 0–14; pH = 0–7 acidic, pH 7 = neutral; pH 7–14 alkaline. Numbers 1–14 represent the negative logarithm of the hydrogen ion concentration.

PHAEISM Noun. (Greek, *phaics* = dusky; English, *-ism* = condition.) A duskiness in butterflies occurring in a limited region.

PHAGE See Bacteriophage.

PHAGOCYTE Noun. (Greek, *phagein* = to eat; *kytos* = hollow vessel; container. Pl., Phagocytes.) 1. A blood cell (Haemocyte) that consumes, engulfs or absorbs noxious organisms or foreign material including the immature stages of parasitic organisms. See Plasmatocyte. 2. See Granulocyte.

PHAGOCYTIC ORGANS The splenic organs; clumps of Leucocytes found along the sides of the dorsal vessel in insects (Wardle).

PHAGOCYTIC Adj. (Greek, *phagein* = to eat; *kytos* = hollow vessel; container; *-ic* = of the nature of.) Descriptive of or pertaining to Phagocytes or their action.

PHAGOCYTOSIS Noun. (Greek, *phagein* = to eat; *kytos* = hollow vessel; container; *sis* = a condition or state.) 1. A Cellular Defence Mechanism which destroys small-sized microorganisms or foreign material by the action of Plasmatocytes. 2. The process by which parasitic insect larvae are destroyed by their host. 3. The incorporation of solid particles into a Vesicle within the cell. See Haemolymph. Cf. Autophagocytosis; Encapsulation; Nodule Formation. Rel. Coagulation.

PHAGOSTIMULANT Noun. (Greek, *phagein* = to eat; Latin, *stilus* = stake; *-antem* = an agent of something. Pl., Phagostimulants.) A chemical compound or physical feature that is touched, tasted or ingested and which affects the feeding behaviour of an insect.

PHALACRIDAE Plural Noun. Shining Flower Beetles. A cosmopolitan Family of about 500 Species that are assigned to Cucujoidea; Family may be related to nitidulids and Cucujidae (Passandrinae). Adult typically minute to small with strongly compact, convex bodies oval in outline; tarsal formula 5-5-5 with Tarsomeres 1–3 ventrally lobed and segment four reduced in size; Abdomen with five visible Sterna and Terga concealed by entire Elytra. Biologically poorly known. Adults frequently taken at flowers, particularly Asteraceae.

PHALAENAE A Linnaean term embracing most of the heterocerous Lepidoptera, more specifically applied to the Geometridae.

PHALANGOPSIDAE Plural Noun. (Greek, *phalangion* = spider.) Spider Crickets, True Cave Crickets. The Family contains about 60 Genera and 230 Species included in two Subfamilies (Phalangopsinae and Luxarinae). Head short and vertical, Antenna inserted above middle of face, body with long, slender legs, hind Femur enlarged basad and hind Tibia serrated between the spines. Phalangopsidae are tropical and found in humid situations near water or in association with caves.

PHALANX Noun. (Greek, *phalangx* = like of battle. Pl., Phalanxes.) 1. A segment of the Tarsus. 2. A division of classification of uncertain value, similar to Tribe.

PHALERATED PHALERATUS Adj. (Latin, *phaleratus* = ornamented.) Beaded.

PHALLIC Adj. (Greek, *phallus* = penis; *-ic* = characterized by.) Descriptive of or pertaining to the Phallus.

PHALLIC ORGAN The median male intromittent apparatus of the 9th abdominal segment in male insects (Snodgrass). See Phallus.

PHALLOBASE Noun. (Greek, *phallus* = penis; base = step. Pl., Phallobases.) 1. The proximal part of the male Phallus. Phallobase is characterized by highly variable development: Sometimes sclerotized and supporting the Aedeagus; sometimes forming a sheath for the Aedeagus (Phallotheca); often contains an Apodeme that may provide support or a point for muscle attachment. See Phallus. Cf. Aedeagus; Endophallus; Phallotheca.

PHALLOCRYPT Noun. (Greek, *phallus* = penis; *kryptos* = hidden.) An Atrium of the Phallobase or of the genital chamber wall that contains the base of the Aedeagus (Snodgrass).

PHALLOMERE Noun. (Greek, *phallos* = penis; *meros* = part. Pl., Phallomeres.) Genital lobes formed at the sides of the Gonopore in the ontogeny of some insects. In most cases they unite to form the Phallus, but in Blattidae and Mantidae they develop separately into complex genital organs of the adult (Snodgrass).

PHALLOSOME Noun. (Greek, *phallos* = penis; *soma* = body. Pl., Phallosomes.) The Aedeagus or intromittent portion of the external male genitalia. See Aedeagus; Phallus. Cf. Phallobase.

PHALLOTHECA Noun. (Greek, *phallos* = penis; *theke* = box.) A fold or tubular extension at the apex of the Phallobase which connects to the base of the Aedeagus. See Phallus. Cf. Aedeagus; Phallobase. Rel. Endotheca.

PHALLOTREME Noun. (Greek, *phallos* = penis; *trem* = a hole. Pl., Phallotremes.) The distal opening of the Endophallus, usually at the apex of the Aedeagus. See Phallus. Cf. Aedeagus; Endophallus. Rel. Gonopore.

PHALLUS Noun. (Greek, *phallos* = penis. Pl., Phalli.) The intromittent reproductive organ of the male. The Phallus consists of a conical, tubular structure of variable complexity. Primitive insects may not display differentiate parts, and the entire structure may be long, sclerotized and tapering apicad. In a groundplan condition for Pterygote insects, the Phallus shows a basal sclerotized region (Phallobase) and distal sclerotized region (Aedeagus); the Phallobase and Aedeagus are connected by a membranous Phallotheca. The external (sclerotized) walls of the Phallobase and Aedeagus are called the Ectophallus. The Gonopore is positioned at the apex of the Ejaculatory Duct and is concealed within the Phallobase; the Gonopore is connected to the apex of the Aedeagus via a membranous tube called the Endophallus. In some insects the Endophallus is permanently concealed within the Aedeagus or Phallobase; in other insects the Endophallus may be everted through the Aedeagus. The circular aperture at the apex of the Aedeagus is called the Phallotreme. In some insects the Endophallus and Gonopore may be everted through the Phallotreme and into the female's Bursa Copulatrix. See Aedaegus; Phallobase.

PHANERE Noun. Acarology: Any obvious integumental formation.

PHANEROPTERIDAE Plural Noun. The largest Family in the Tettigoniodea (Ensifera: Orthoptera), including about 300 Genera and 2,000 Species within two Subfamilies. Cosmopolitan in distribution but predominantly tropical. Round head with face not flattened or slanted, compound eye small, Antenna longer than body, scrobal impression between eyes, Prosternum not spinose. Anterior Tibia with covered Acoustical Tympana, Tegmina leaf-like and sometimes resemble fungal damage, Ovipositor short and sickle shaped. Most Species in the nominate Subfamily are phytophagous and nocturnal. Many Species mimic other insects and spiders when immature.

PHANEROZOIC EON The third interval in the earth's Geological Time Scale (570 MYBP–Present). PE includes Palaeozoic, Mesozoic and Cenozoic Eras. See Geological Time Scale. Cf. Archaean Eon; Proterozoic Eon.

PHANTOM® See Primicarb.

PHANTOM CRANE FLIES See Ptychopteridae.

PHANTOM HEMLOCK LOOPER *Nepytia*

phantasmaria (Strecker) [Lepidoptera: Geometridae].

PHANTOM LARVAE See Clear Lake Gnat.

PHANTOM MIDGES See Chaoboridae.

PHARAOH ANT *Monomorium pharaonis* (Linnaeus) [Hymenoptera: Formicidae]: A cosmopolitan urban pest originally described from Egypt (hence the name); probably endemic to India or North Africa. Worker small bodied (1.5–2 mm long), head and Thorax reddish-yellow with Gaster darker; Antennal club distinct, of three segments; Pedicel of two nodes. Nest unstructured and crevices used as nesting sites. Queens cannot fly; mate in nest; colonies may contain hundreds of queens with each laying <4,500 eggs. Colonies may contain several hundred thousand workers; new colonies formed by Sociotomy; colonies often maintain contact without aggression between workers of different colonies. Workers omnivorous and forage during day or night; Sting modified for pheromone trail marking and trails used for worker recruitment. Workers do not sting but acid gland produces chemical repellant. Significant urban pest in homes, office buildings and hospitals. In hospitals, PA can serve as mechanical vector for diseases including *Pseudomonas, Salmonella* and *Staphylococcus*. See Formicidae. Cf. Argentine Ant; Black Carpenter-Ant; Cornfield Ant; Fire Ant; Larger Yellow-Ant; Little Black Ant; Odorous House Ant; Pavement Ant; Thief Ant.

PHARATE Adj. (Greek, *pharos* = loose mantle; *-atus* = adjectival suffix.) 1. A term applied to the instar after Apolysis and before Ecdysis (Jenkin & Hinton 1966). 2. The concealed instar enclosed within the cast, old Cuticle of the preceding stage of Cuticle formation. See Metamorphosis; Pupa. Cf. Exarate.

PHARMACEOPHAGOUS Adj. (Greek, *pharmakon* = medicine; *phagein* = to eat; Latin, *-osus* = with the property of.) Pertaining to drug-eating behaviour.

PHARNYX Noun. (Greek, *pharyngx* = gullet.) The portion of the Foregut located posterior of the Buccal Cavity and anterior of the Brain connectives (Tritocerebrum) and Suboesophageal Ganglion. Dilator muscles which originate on the Tentorium and Frons insert onto the Pharynx. Region well developed in nectar feeding and sucking insects and sometimes anatomically divided into an anterior and posterior Pharynx. In some insects, Pharynx represents a slight enlargement at beginning of the Oesophagus. Diptera: sometimes restricted to space between Hypopharynx and Subclypeal Pump and then equivalent to the Subclypeal Tube. In some Diptera, the proximal part of Pharynx is termed the Basipharynx and is continuous with the Oesophagus. See Alimentary System. Cf. Buccal Cavity; Oesophagus. Rel. Digestion.

PHARORID® See Methoprene.

PHARYNGARIS Noun. Diptera: The sucking tube formed by the Labrum-Epipharynx and the Hypopharynx together (MacGillivray).

PHARYNGEA A pharyngeal sclerite (MacGillivray).

PHARYNGEAL DUCT Hemiptera: A narrow continuation of the Pharynx into the labral region.

PHARYNGEAL GANGLION One of the ganglia which lie on the Oesophagus just behind the Brain, each joined with the Hypocerebral Ganglion (Imms). See Occipital Ganglion.

PHARYNGEAL MUSCULATURE Several groups of muscles in the head and associated with the Pharnyx, including Compressor Muscles, Dilator of Cibarium, Anterior and Posterior Labral Muscles and Hypopharyngeal Muscles.

PHARYNGEAL PUMP The sucking pump of some fluid-feeding insects.

PHARYNGEAL SCLERITES Muscoid larvae: Two lateral irregular sclerites united at their dorsoanterior ends by a transverse sclerite, to the heavily chitinized anterior portion of which (the Dorsopharyngeal Sclerite) the Hypostomal Sclerite articulates.

PHARYNGEAL SKELETON Muscoid larvae: The conspicuous sclerotic structure positioned in the anterior end of the maggot and formed of the strongly sclerotized lateral walls of the pump and the walls of the clypeal wings leading back to the antenno-ocular pouches; bucco-pharyngeal armature (Snodgrass).

PHARYNGEAL TUBE Lice: A pair of structures each in the form of a half tube, arising from the floor of the first chamber of the Pharynx and fitting closely together to form a tube (Imms).

PHASER® See Endosulfan.

PHASIC RECEPTOR A type of mechanoreceptive sensillum in which the action potential is generated due to bending or straightening the shaft of the receptor. PR occur on the Antenna, Tarsi, other appendages of the insect body. See Sensillum. Cf. Tonic Receptor.

PHASMATIDAE Karney 1923. Plural Noun. (Greek, *phasma* = apparition, ghost.) Winged Walkingsticks. A numerically large, predominantly tropical Family of Phasmida.

PHASMATIDS Phasmatidae.

PHASMATODEA Stick Insects. See Phasmida.

PHASMIDA Leach 1815. (Lower Triassic Period–Recent.) (Greek, *phasma* = apparition, ghost.) A cosmopolitan Order of large bodied, elongate, terrestrial, phytophagous, exopterygote Neoptera. About 2,500 described Species; most abundant in tropical regions. Body often spinose; head typically prognathous; Antenna with 8–100 segments; Ocelli usually absent, sometimes present in winged morphs; Mandible well developed; Maxillary Palpus with five segments; Labial Palpus with three segments. Prothorax moveable on Mesothorax, sometimes with repugnatorial gland; Mesothorax longest; Thoracic Spiracles anterior of Mesothoracic and Metathoracic Episterna. Wings absent from most Species; winged Species most common in tropics;

when present often restricted to male; forewing modified into Tegmina; hindwings broad with uniform branching pattern. Legs long, slender, gressorial, often spined; Coxae small and separated; tarsal formula typically 5-5-5 except when legs regenerated. Abdomen with 11 segments; Sternum 1 absent or evanescent; spiracles on segments 1–8; Cerci not segmented, sometimes long and clasper-like in male. Nervous system with three thoracic ganglia and seven abdominal ganglia. Ovarioles panoistic; Bursa Copulatrix forms independent aperture; accessory glands present; 1–2 Spermathecae present. Parthenogenesis common. Eggs structurally diverse, operculate, diagnostically useful. Nymphs add antennal segments with successive moults; wings and genitalia develop with successive moults. Elaborate defence mechanisms include thanatosis, autotomy, regurgitation, repugnatorial secretions, aposematic coloration. Included Families: Phasmatidae, Phylliidae.

PHASMODIDAE Plural Noun. Stick crickets. A Family of ensiferous Orthoptera (also called Zaprochilidae) with fewer than five described Species found in Australia. Body elongate with elongate prognathous head, Prosternum spineless, legs are long and thin, fore Tibial Auditory Organ reduced or absent, hind Tibia lacks apical spines, wing development variable. Species apparently anthophilous and feed on pollen and nectar.

PHAULOPTERA An obsolete ordinal term for the scale insects (Laporte 1835).

PHENETIC Adj. (Greek, *phainein* = to show; *-ic* = consisting of.) Cf. Cladistic.

PHENETIC ANALYSIS Provides estimates of overall similarity and does not necessarily imply phylogeny.

PHENGODIDAE Plural Noun. A Family of Coleoptera presently known from fewer than 60 Species, restricted to western hemisphere. Placed in Lampyridae based on presence of luminous organs in adult and larval characteristics. Separated from lampyrids by relatively short Elytron and biplumose Antenna of male. Profound sexual dimorphism exists with female larviform. Biologically poorly known; larva feeds on myriapods.

PHENOGRAM Noun. A branching diagram which represents similarity between terminal units on the diagram. Cf. Cladogram.

PHENOL See Carbolic Acid.

PHENOLOGY Noun. (Greek, *phainein* = to show; *logos* = discourse. Pl., Phenologies.) The study of periodic or cyclical biological phenomena (reproductive cycles, host availability, life cycles) in relation with edaphic factors, climate and weather changes. See Life History.

PHENOMENOLOGY See Phenology.

PHENON A phenotypically uniform sample or group of specimens.

PHENOTHRIN Noun. A synthetic-pyrethroid compound used as broad-spectrum contact insecticide with fast knockdown for control of flies, mosquitoes, midges, fleas, lice, cockroaches, wasps and other urban pests. Toxic to bees and fishes. Trade names include: Fenothrin®, Multicide Concentrate® and Sumithrin®. See Synthetic Pyrethroid.

PHENOTYPE Noun. (Greek, *phainein* = to show; *typos* = pattern. Pl., Phenotypes.) The physical features of an organism as expressed by environment and genotype. See Type. Cf. Genotype.

PHENTHOATE An organic-phosphate (dithiophosphate) compound {O,O-dimethyl S(alpha-ethoxycarbonylbenzyl)-phosphorodithioate} used as an acaricide, contact insecticide and stomach poison for control of mites, plant-sucking insects (to include aphids, mealybugs, psyllids, scale insects, whiteflies), thrips, leaf miners, flea beetles, cabbage worms and borers. Compound applied to cereal, corn, citrus, coffee, cotton, fruit trees, sunflower, sugarcane and vegetables in many countries. Phytotoxic on some grapes, peaches, figs and some apples. Not registered for use in USA. Toxic to bees and fishes. Trade names include: Aimsan®, Cidal®, Elsan®, Kap®. See Organophosphorus Insecticide.

PHEROMONE Noun. (Greek, *pherein* = to carry; *hormein* = excite, stimulate. Pl., Pheromones.) Term proposed by Karlson & Butenandt (1959, 183: 155–156) for a chemical compound secreted by an animal which mediates behaviour of another animal belonging to the same Species. An agent of intraspecific chemical communication. Pheromones are subdivided into several types based on the nature of the interaction. Sex Pheromones are intraspecific chemicals released by members of one sex to attract sexual partners. Sex pheromones are among the most carefully studied insect glandular secretions, in part because of their potential usefulness in the control of some pests. Alarm Pheromones are intraspecific chemical messages which alert individuals to danger. The glandular cells which produce and secrete alarm pheromones are found on various parts of the insect body. Aggregation Pheromones are intraspecific chemical messages which attract individuals to a small area. Aggregation pheromones have been identified in several groups of insects, including many groups of social insects. Trail Pheromone is produced by Pavan's Gland. Syn. Ectohormone (*sensu* Beth 1932, Naturwissenschäften 20: 177–183.) See Allelochemical; Kairomone; Semiochemical. Cf. Aggregation Pheromones; Alarm Pheromones; Sex Pheromones; Trail Pheromone. Rel. Parapheromone.

PHIAL Noun. (Greek, *phiale* = vessel or bottle. Pl., Phials.) A small sac which receives fluid for the purpose of increasing the weight of the wing (Kirby & Spence). Alt. Phialum.

PHIGALIA SKIPPER *Trapezites phigalia* (Hewitson)

[Lepidoptera: Hesperiidae].

PHIGALIOIDES SKIPPER *Trapezites phigalioides* Waterhouse [Lepidoptera: Hesperiidae].

PHILIP, CORNELIUS BECKER (1900–1987) American medical Entomologist specializing in disease transmission by biting flies and ticks. Employed at Rocky Mountain Laboratory in Hamilton, Montana for more than 40 years.

PHILIPPI, FREDERICO (–1910) (Moore 1910, Boln. Mus. nac. Hist. nat. Chile 2: 264–298.)

PHILIPPI, R A Published a work entitled *Orthoptera Berolinensis*.

PHILIPPI, RUDOLPH AMANDO (1808–1904) (Porter 1904, Revta chil. Hist. nat. 8: 174–177, bibliogr.)

PHILIPPIEV, VICTOR IVAN (1857–1906) (Kusnezov, Revue Russe Ent. 6: 383–384.)

PHILIPPINE KATYDID *Phaneroptera fucifera* Stål [Orthoptera: Tettigoniidae]. Cf. Angular-Winged Katydid; Broad-Winged Katydid; Citrus Katydid; Crested Katydid; Forktailed Bush-Katydid; Inland Katydid; Mottled Katydid; Spotted Katydid.

PHILIPPINE NEEM TREE *Azadirachta excelsa* (Jack) Jacobs [Melliaceae]: A tree whose seed kernels contain chemical compounds (Azadirachtin, Marrangin) which display insecticidal properties. Marrango tree ca 40–50 m with open and uneven crown; trunk cylindrical in young trees, buttressed in mature tall trees; bark smooth and pink in young trees, fissured and scales in long flakes in mature trees; branches ascend, spread obliquely but do not droop; leaves to 1 m long with 7–15 alternate leaflets, ca 30 large leaves clustered at the end of long, slanting branchlets. Flowers white, fragrant, hang from axillary panicles; typically five petals, five sepals and 10 anthers. Ovary comprises three cells each with two ovules. Fruits ovoid, green when young and yellow when ripe; pericarp and endocarp connected; endocarp contains one seed kernel covered by think, brown testa. Seed emits garlic-like odour when cut. 1,000 seed weight 1,140 g; 1,000 kernel weight 584 g. Old trees may produce ca 50 kg of seeds per year. Tree endemic to Borneo, Thailand, Philippines and Myanmar; planted in Dominican Republic, Ecuador and West Africa. Other common names include: Ranggu (Sarawak); Limpaga, Ranggau (Sabah); Sentang, Sentang, Setan (West Malaysia); Sadao Chang (Thailand); Marango, Philippine Neem Tree (Philippines); Tamargalay (Myanmar.) See Neem. Cf. Indian Neem Tree, Thai Neem Tree.

PHILIPTSCHENKO, JURIUS (1882–1930) (Anon. 1931, Ent. News 42: 95–96, bibliogr.)

PHILLARDRIN® See Monocrotophos.

PHILLIPS, EVERETT FRANKLIN (1878–1951) (Mallis 1971, *American Entomologists*. 549 pp. (486–487), New Brunswick.)

PHILLIPS, LEONARD STEVENS (1908–1968) (Irwin 1968, J. Lepid. Soc. 22: 21–26.)

PHILLIPS, RICHARD HENRY (1866–1938) (Robinson 1975, *Macrolepidoptera of Fiji and Rotuma* 361 pp. (3–5), London.)

PHILLIPS, ROBERT ALLEN (1866–1945) (A.W.S. 1946, Irish Nat. J. 8: 391–394.)

PHILOBOTA *Philobota productella* (Walker) [Lepidoptera: Oecophoridae].

PHILONIST Noun. (Greek, *philos* = lover; *noos* = mind.) Seeker of knowledge.

PHILOPATRIC (Greek, *philos* = lover; *patria* = native country; *-ic* = characterized by.) See Klahn 1979, Behav. Ecol. Sociobiol. 5: 417–424.

PHILOPOTAMIDAE Plural Noun. A Family of Trichoptera known from the Triassic Period of Kirgizia.

PHILOPTERIDAE Plural Noun. Feather Chewing Lice. The largest Family of ischnocerous Mallophaga. Cosmopolitan parasites of birds. Head relatively large; Antenna not concealed, with five segments; Mandibles vertical; Maxillary Palpus absent; Mesothorax, Metathorax fused; tarsal claws paired.

PHILOTARSIDAE Plural Noun. A numerically small Family of Psocoptera assigned to the Suborder Psocomorpha. Antenna with 13 segments, without secondary annulations; macropterous in both sexes; wings setose; nymphs with glandular Setae. Species found on foliage. (Thornton 1981, Syst. Zool. 6: 413–452.)

PHILPOTT, ALFRED (Eltringham, Proc. ent. Soc. Lond. 6: 108.)

PHINCO-TZZ® See Tetramethrin.

PHIPPS, CLARENCE RITCHIE (1895–1933) (Patch 1933, J. Econ. Ent. 26: 920–922, bibliogr.)

PHLAEOTHRIPIDAE Plural Noun. Phlaeothripids. A large Family of Thysanoptera; only Family of Suborder Tubulifera. Wings without longitudinal veins, Microtrichia; marginal fringe Setae not socketed; abdominal segment 10 tubular; Ovipositor eversible. Some Species feed on fungi, some are predaceous, some are economic pests.

PHLEBOPTERA Noun. See Hymenoptera.

PHLEBOTOMINAE See Psychodidae.

PHLOEIDAE Plural Noun. A small Family of Heteroptera assigned to Superfamily Pentatomoidea.

PHLOEM Noun. (Greek, *phloios* = inner bark.) Vascular Plants: A complex tissue that forms tubes through which sugars and some amino acids are conducted from leaves down the stem. Often used as a source of nutrition for plant-feeding fluid-consuming insects, particularly Hemiptera. Cf. Xylem.

PHLOIOPHILIDAE Plural Noun. A small Family of polyphagous Coleoptera assigned to Superfamily Cleroidea. Species are mycohpagous and occur in the Palearctic.

PHLOX PLANT BUG *Lopidea davisi* Knight [Hemiptera: Miridae].

PHOENICOCOCCIDAE Plural Noun. Phoenicococcid Scale; Red Date Scale. A small Family of sternorrhynchous Hemiptera assigned to Superfamily Coccoidea. One Species occurs on date palm trees in southwestern USA. Adult female wingless, body not covered by scale cover; apical segments of Abdomen not fused into Pygidium; Adult male with Ocelli and one pair of wings; body fragile.

PHOLCIDAE Plural Noun. Daddy-Long-Legs Spiders; Long-Legged Spiders; Cellar Spiders. A Family of ecribellate spiders. Adult 2–6 mm long; eyes in two groups, each containing three simple eyes; Chelicerae fused at base; Spiracles absent; legs long and slender; six Spinnerets equally spaced. Phlocids resemble opilionids but construct irregular or sheet-like web in cellars and similar habitats. Adult hangs suspended from web as thereidiids.

PHONORECEPTOR Noun. (Greek, *phone* = sound; Latin, *receptor* = receiver; *or* = one who does something.) A sense organ or Sensilla which receives sound waves and transduces them into electical signals sent to the central nervous system. Cf. Chemoreceptor; Photoreceptor.

PHORATE An organic-phosphate (dithiophosphate) compound {O,O-diethyl-S-((ethylthio)methyl) phosphorodithioate} used as an acaricide, contact/systemic insecticide and fumigant against mites, plant-sucking insects (to include aphids, leafhoppers, psyllids, whiteflies), thrips, cutworms, rootworms, wireworms, leafminers, shootfly, and nematodes. Compound applied to beans, coffee, corn, cotton, peanuts, potatoes, sorghum, sugarbeets, sugarcane and ornamentals. Phytotoxic to apples and tobacco. Toxic to bees. Compound flammable and should not be used with alkaline compounds. Trade names include: Geomet®, Geophos®, Granutox®, Rampart®, Terrathion®, Thimet®, Timet® and Volphor®. See Organophosphorus Insecticide.

PHORESY Noun. (Greek, *pherein* = to bear, to carry.) A form of symbiosis in which one organism is carried on the body of another larger-bodied organism; the former does not feed on the latter. Phoresy commonly is manifest as mites on insects and insects on insects. Alt. Phoresia. See Dispersal; Migration. Rel. Advectitious.

PHORETIC COPULATION A copulatory act undertaken by insects in flight. Often, one sex (typically female) is wingless and the other sex is winged. PC presumably serves as a mechanism of dispersal. Seen in some aculeate Hymenoptera, including Mutillidae, Bethylidae and Tiphiidae.

PHORETOMORPHY Noun. (Greek, *pherein* = to bear, to carry; *morphe* = form.) Environmentally induced intraspecific polymorphism. Found in acarine Superfamily Pygmephoroidea (Pyemotidae, Moser & Cross 1975, Ann. ent. Soc. Am. 68: 820–822.)

PHORIDAE Plural Noun. Scuttle Flies; Humpbacked Flies. Cretaceous Period-Recent. A cosmopolitan, Family containing about 750 Species of aschizous Diptera, assigned to Superfamily Platypezoidea and closely related to Sciadoceridae and Ironomiidae; separated by Subcostal Vein ending in Radial Vein, apical part of Subcosta absent; Radial Sector simple with two short apical branches which end in Costa, or wings vestigial. Phorids distinguished by humpbacked appearance; head with strong reclinate bristles; Antenna appears 1–segmented, with first two segments minute; antennal Arista usually long and pubescent; wings folded flat over Abdomen at repose; Costa, Subcosta and Radial Veins strong and end near middle of wing; posterior veins simple and very weak; hind femora enlarged, stout, laterally compressed. Some females with aborted or vestigial wings, also lack Halteres. Biologically varied life styles: Live in fungi or decayed plant tissues, some carrion feeders, inquilines in ant and termite nests and others parasitic or predaceous in habit. Parasitic Species attack ants, coccinellid Pupae, lepidopterous larvae and Pupae, dipterous larvae, adult bees, crickets and myriapods. Predaceous Species feed on spider and locust eggs. Usually gregarious; some Species solitary.

PHOROIDEA Noun. A Superfamily of cyclorrhaphous Diptera included within the Series Aschiza and contianing the Phoridae, Platypezidae (Clythiidae) and Sciadoceridae.

PHORONT Noun. (Greek, *phora* = producing; *ontos* = being.) Holotricha: The encysted stage produced by a tomite and leading to formation of a trophont in the life cycle.

PHOSALONE An organic-phosphate (dithiophosphate) compound {S-[(6–chloro-2-oxo-3 (2HO-benzoxazoyl) methyl] O,O-diethyl phosphorodithioate} used as an acaricide, contact insecticide and stomach poison for control of mites, plant-sucking insects (to include aphids, leafhoppers, psyllids), codling moth and borers. Compound applied to fruit and nut crops, grapes, rice, potatoes, tea, ornamentals and vegetables in many countries. Phytotoxic on many plants. Toxic to bees and fishes. Trade names include: Azofene®, Benzofos®, Rubitox®, Zolone®. See Organophosphorus Insecticide.

PHOSDIPHEN An organic-phosphate (phosphoric acid) compound {bio(2,4-dichlorophenyl)ethyl phosphate} used as a contact insecticide. Not registered in USA. Trade name is Baron®. See Organophosphorus Insecticide.

PHOSDRIN® See Mevinphos.

PHOSFENE® See Mevinphos.

PHOSFINON® See AIP.

PHOSKIL® Parathion.

PHOSMET An organic-phosphate (dithiophosphate) compound {N-mercaptomethyl) phthalimide S-(O,O-dimethyl phosphorodithioate} used as an acaricide and contact insecticide for control of mites, plant-sucking insects (to include aphids, mealybugs, psyllids, scale insects, whiteflies), thrips, leaf-eating insects and borers. Compound applied to fruit trees (apples, almonds, apricots, cherries, peaches, plums, prunes, pears) in USA and other crops and livestock in many countries. Phytotoxic on some pines. Toxic to bees, fishes and wildlife. Trade names include: Appa®, Fosdan®, Imidan®, Inovat®, Kemolate®, Phthalophos®, PMP®, Prolate®, Safidon®. See Organophosphorus Insecticide.

PHOSPHAMIDON An organic-phosphate (phosphoric acid) compound {2-chloro-2-diethylcarbamoyl-1-methyl-vinyl-dimethylphosphate} used as an acaricide and systemic stomach poison to control plant-sucking insects, thrips, leaf-chewing insects and mites. Used on numerous crops in many countries but no longer registered in USA. Phytotoxic on fruit trees; highly toxic to bees; low toxicity to fishes. Trade names include: Cildon®, Dimecron®, Dimenox®, Dixon®, Famfos®, Phos-Sul®, Phosron®, Umecron®. See Organophosphorus Insecticide.

PHOSPHINE Noun. An insecticidal fumigant (hydrogen phosphide) used to control rodents and pests in vehicles, buildings and stored food. First used as grain fumigant in Germany during 1930s. Excellent penetration, acts as potent inhibitor of oxidative phosphorylation through inhibition of cytochrome-c oxidase; may also inhibit acetylcholinesterase; kills within 2–3 days. Hydrogen phosphide (phosphine, PH_3) generated by action of atmospheric moisture with aluminium phosphide or magnesium phosphide; usually not mixed with other gasses; releases a garlic or ammonia-like odour. Active ingredient formulated with inert ingredient in form of tablet, pellet or sachet; very slightly soluble in cold water. See Fumigant. Cf. Methyl Bromide.

PHOSPHORESCENT Adj. (Greek, *phosphoros* = bringing light.) Shining or glowing in the dark, like phosphorus.

PHOSRON® See Phosphamidon.

PHOS-SUL® See Phosphamidon.

PHOSTEK® See AlP.

PHOSTOXIN® See AlP.

PHOSVIT® See Dichlorvos.

PHOTIC ZONE The biogeographic marine realm characterized by the area in which light penetrates and above the Abyssal Zone. See Habitat. Cf. Abyssal Zone; Littoral Zone; Neritic Zone; Pelagic Zone.

PHOTOCHEMICAL Noun. (Greek, *phos* = light; Medieval Latin, *alchymista* = alchemist. Pl., Photochemicals.) A chemical compound or substance which by its composition is acted upon by light (*e.g.,* nitrate of silver, $AgNO_3$); of or relating to any reaction depending upon the action of light on the chemical constitution of matter.

PHOTOGEN Noun. (Greek, *phos* = light; *genes* = producing. Pl., Photogens.) A light-producing organ or light-producing substance.

PHOTOGENIC Adj. (Greek, *phos* = light; *genes* = producing; *-ic* = characterized by.) Light producing; pertaining to organisms which produce a phosphorescent glow.

PHOTOGENY Noun. (Greek, *phos* = light; *genes* = producing. Pl., Photogenies.) Production of light.

PHOTOPATHY Noun. (Greek, *phos* = light; *pathos* = feeling. Pl., Photopathies.) A movement away from light; an aversion to light expressed by some organisms. See: Negative Phototaxis; Negative Phototropism.

PHOTOPERIOD Noun. (Greek, *phos* = light; *periodos* = circuit. Pl., Photoperiods.) 1. The entire cycle of illumination and darkness to which an organism is exposed. Conventionally expressed as **L:D** and totaling 24 on a day length (*e.g.* L8:D16; L14:D10). 2. The period of exposure to light in a light-dark cycle. See Diel Periodicity; Daylength. Rel. Photophase; Scotophase; Twilight.

PHOTOPERIODISM Noun. (Greek, *phos* = light; *periodos* = circuit; English, *-ism* = condition. Pl., Photoperiodisms.) 1. The study of biological phenomena that are influenced by periods of exposure or entrainment to light or darkness. Cf. Biological Clock; Circadian Rhythm; Diel Periodicity. 2. The response of organisms to relative day length. This response is manifest in growth, sexual activity, seasonal migration and other periodic phenomena.

PHOTOPHASE Noun. (Greek, *phos* = light; *phainein* = to show. Pl., Photophases.) The daylight portion of the photoperiod. Under natural conditions, the photophase would be some interval of time less than 24 hours (*e.g.* L8:D16; L12:D12; L16:D8). Under experimental conditions the photophase may extend beyond 24 hours (*e.g.* L36:D6; L48:D6). See Photoperiod. Cf. Scotophase. Rel. Circadian Rhythm; Diel Periodicity.

PHOTOPHILIC Adj. (Greek, *phos* = light; *philein* = to love; *-ic* = characterized by.) Pertaining to an attraction to light; light-loving. Sometimes interpreted as positive phototaxis. Ant. Photophobic.

PHOTOPHOBIC Adj. (Greek, *phos* = light; *phobos* = fear; *-ic* = of the nature of.) Pertaining to an aversion to light; light-avoiding. Ant. Photophilic.

PHOTORECEPTOR Noun. (Greek, *phos* = light; Latin, *receptor* = receiver; *or* = one who does something. Pl., Photoreceptors.) A sense organ responsive to light. Typically the Ommatidium of a compound eye or the Ocellus.

PHOTOTACTIC Adj. (Greek, *phos* = light; *taxis* = arrangement.) Pertaining to phototaxis.

PHOTOTAXIS Noun. (Greek, *phos* = light; *taxis* = arrangement. Pl., Phototaxes.) A behavioural reaction, typically in the form of movement or body orientation, by an organism to the stimulus of light. See Orientation. Cf. Aerotaxis; Anemotaxis; Chemotaxis; Geotaxis; Menotaxis; Osmotaxis; Rheotaxis; Rotaxis; Scototaxis; Strophotaxis; Telotaxis; Thermotaxis; Thigmotaxis; Tonotaxis; Tropotaxis.

PHOTOTROPIC Adj. (Greek, *phos* = light; *trepein* = to turn; *-ic* = of the nature of.) Descriptive of or pertaining to phototropism. Cf. Heliotropic.

PHOTOTROPISM Noun. (Greek, *phos* = light;

trepein = to turn; English, *ism* = state. Pl., Phototropisms.) A tropism in response to the stimulus of any form of light. See Tropism. Cf. Aeolotropism; Anemotropism; Chemotropism; Electrotropism; Galvanotropism; Geotropism; Heliotropism; Hydrotropism; Rheotropism; Stereotropism; Thermotropism; Thigmotropism; Tonotropism. Rel. Taxis.

PHOXIM An organic-phosphate (thiophosphoric acid) compound {O-(2,5-dichloro-4-iodophenyl) O,O-dimethyl phosphorothioate} used as a contact insecticide and stomach poison against public health pests. Not registered for use in USA. Toxic to bees. Syn. Jodfenphos. Trade names include Elocril® and Nuvanol-N®. See Organophosphorus Insecticide.

PHRAGMA Noun. (Greek, Phragma = fence. Pl., Phragmata.) A partition, dividing membrane or structure; a transverse partition of the endoskeleton; a projecting structure or internal ridge from the Endocuticula to which a muscle is attached, occurring at the junction of the tergites, pleurites and sternites, especially in the Thorax (Wardle).

PHRAGMANOTUM See Postnotum.

PHRAGMATAL Adj. (Greek, *phragma* = fence.) Descriptive of or pertaining to a Phragma.

PHRAGMINA Noun. (Greek, *phragma* = fence.) The line or path of cuticular invagination of a Phragma (MacGillivray).

PHRAGMOCYTTARES Noun. (Greek, *phragma* = fence; *kyttarous* = honeycomb cell.) Social wasps in which the combs of the nest are supported by the covering envelope. Cf. Poecilocyttares; Stelocyttares.

PHRAGMOSIS Noun. (Greek, *phragmos* = fence; *-sis* = a condition or state. Pl., Phragmoses.) Social Insects: A behavioural response by soldier termites or ants which block the entrance to a nest with their bodies. Morphological adaptations for Phragmosis include a flattened head or truncate Abdomen (Wheeler 1927). Cephalotini (a Tribe of arboreal ants restricted to New World tropics) genus *Zacryptocerus* display soldiers and queens with a head typified by a cephalic disc that is rimmed and saucer shaped. This head shape represents an adaptation for blocking nest entrance with a minimum of vulnerability to defender's body. Some phragmotic heads accumulate dirt; soil-binding properties enable insects to resemble debris thus serving as camouflage. Phenomenon called Cryptic Phragmosis. *Zacryptocerus* with odd-shaped Setae, in association with shallow pits in the disc, accumulate dirt. Pits are invested with many minute secretory pores that apparently produce a binding substance. Some termite soldiers also display a heavily sclerotized phragmotic head that is used to block tunnels in the termite nest.

PHRAGMOTIC Adj. (Greek, *phragmos* = fence; *-ic* = of the nature of.) Pertaining to Phragmosis.

PHRYGANEIDAE Plural Noun. Large Caddisflies. A small Family of Trichoptera assigned to Superfamily Limnephiloidea (Phryganeoidea.) Adult ca 15–25 mm long; Ocelli present; male Maxillary Palpus with four segments; female Maxillary Palpus with five segments; wings typically mottled with grey and brown; middle Tibia with two preapical spurs. Larvae occur in lakes and marshes; cases cylindrical and open at both ends.

PHRYNEIDAE See Anisopodidae.

PHTHALOPHOS® See Phosmet.

PHTHALTHRIN® See Tetramethrin.

PHTHIRAPTERA (Greek, *phthir* = lice; *aptera* = wingless.) Lice. In some classifications, the Ordinal name used to embrace the Anoplura, Mallophaga and Rhynchophthirina. A cosmopolitan group of about 3,000 Species. Body dorsoventrally compressed; Antenna short, 3–5 segmented; compound eyes reduced or absent; Ocelli absent; mouthparts mandibulate (Amblycera, Ischnocera) or piercing-sucking; Maxillary Palpus absent (Ischnocera) or with 2–4 segments (Amblycera). Prothorax usually free with Mesothorax and Metathorax fused (Mallophaga), or all thoracic segments fused (Anoplura); apterous; legs scansorial with 1–2 tarsal segments adapted for grasping. Abdomen 8–10 segments; Cerci absent; Ovipositor absent. Paurometabolous development. Eggs operculate, typically laid on host; most Species with three nymphal instars. Adults relatively short lived, remain on host, feed on blood (Anoplura), or feather, cast skin or blood (Mallophaga). Lice are disease vectors to humans, notably transmitting Epidemic Typhus. Derived from psocopteroid ancestors. Included Families: Boopiidae, Echinophthiriidae, Gyropidae, Hematopinidae, Heptapsogasteridae, Hoplopleuridae, Laemobothriidae, Linognathidae, Menoponidae, Pediculidae, Philopteridae, Polyplacidae, Pthiridae, Ricinidae, Trichodectidae, Trimenoponidae. See Anoplura, Mallophaga, Rhynchophthirina.

PHTHIRIASIS (Greek, *phthir* = lice; *-iasis* = suffix for names of diseases.) 1. A disease condition of the skin caused by sucking lice. 2. A crab louse infestation characteristic of adult humans. Cf. Pediculosis.

PHTHISANER Noun. (Greek, *phthisis* = wasting; *aner* = male.) Ants: A pupal male in which wings are suppressed and head, Thorax, legs and Antennae remain abortive owing to the extraction of body fluids during the larval or prepupal stage by an *Orasema* larva [Chalcidoidea: Eucharitidae]. (Wheeler).

PHTHISERGATE Noun. (Greek, *phthisis* = wasting; *ergates* = worker.) Ants: A pupal worker parasitized by *Orasema* [Chalcidoidea: Eucharitidae], which is unable to pass to the adult stage; an infra-ergatoid form (Wheeler).

PHTHISOGYNE Noun. (Greek, *phthisis* = wasting; *gyne* = woman.) Ants: A form arising from a female larva under the same conditions as a phthisaner (Wheeler).

PHTHON® See Cyfluthrin.

PHYCODROMIDAE See Coelopidae.

PHYCOLOGIST Noun. (Greek, *phykos* = seaweed; *logos* = discourse.) A person who studies algae.

PHYCOLOGY Noun. (Greek, *phykos* = seaweed; *logos* = discourse.) The study of algae.

PHYCOSECIDAE Plural Noun. A small Family of polyphagous Coleoptera assigned to the Cleroidea. Adult 1.5–3.5 mm long, ovate, convex; Antenna with 10 segments; head prognathous; tarsal formula 4-4-4; Abdomen with five Ventrites. Larva with head prognathous, six stemma on each side of head; legs with five segments, one tarsal claw on each Pretarsus; Urogomphi present. Larva and adult scavenge dead birds and fish.

PHYLACOBIOSIS Noun. (Greek, *phylax* = guard; *biosis* = lifestyle.) Ants: A form of symbiosis exhibited by *Camponotus mitarius* which nests in the hills of termites and seems to be on friendly terms with them (Wasmann). See Biosis; Commensalism; Symbiosis. Cf. Abiosis; Anhydrobiosis; Antibiosis; Archebiosis; Calobiosis; Cleptobiosis; Hamabiosis; Kleptobiosis; Lestobiosis; Parabiosis; Plesiobiosis; Synclerobiosis; Trophobiosis; Xenobiosis.

PHYLIP See Phylogenetic Inference Package. Cf. Paup.

PHYLLIFORM Adj. (Greek, *phyllon* = leaf; Latin, *forma* = shape.) Leaf-shaped. See Leaf. Cf. Lamelliform; Laminiform; Squamiform. Rel. Form 2; Shape 2; Outline Shape.

PHYLLIIDAE Karny 1923. Plural Noun. Leaf Insects. A numerically small Family of Phasmida found in southeast Asia, New Guinea and tropical Australia. Body dorsoventrally flattened with lateral expansion of abdominal Terga. Habitus strongly leaf-like and conveys cryptic colour and form.

PHYLLOCNISTIDAE See Gracillariidae.

PHYLLOMYZIDAE See Milichiidae.

PHYLLOPHAGOUS Adj. (Greek, *phyllon* = leaf; *phagein* = devour; Latin, *-osus* = with the property of.) Feeding upon leaf tissue. See Feeding Strategy.

PHYLLOPHORIDAE Plural Noun. Giant Leaf Crickets. A Family of ensiferous Orthoptera with about 10 Genera and 60 Species found in the Indo-Australian Region. Morphologically characterized by large body, round head, Antenna inserted between eyes, scrobal impression weakly margined, Prosternum with a pair of spines, Metasternum with pair of dorsally denticulate lobes which extend below corresponding Coxa. Tympanal Organs of anterior Tibia present and sometimes covered; wings macropterous. Biology of phyllophorids not well understood, but apparently Species are arboreal and phytophagous. Included in the Family are the largest insects.

PHYLLOXERIDAE Plural Noun. Phylloxerans. A Holarctic Family of sternorrhynchous Hemiptera assigned to Superfamily Aphidoidea. Body sometimes covered with powder but not wax filaments; Antenna of Apterae with three segments, Antenna of Alatae with three or four segments; Wings held horizontally at repose; Hindwings without oblique veins; Cornicles and Ovipositor absent. About 70 Species known, all of which feed on deciduous trees and vines; many Species form galls on the leaves or roots of plants. See Grape Phylloxera; Pecan Leaf-Phylloxera; Pecan Phylloxera.

PHYLOGENETIC Adj. (Greek, *phylon* = race; *genesis* = descent; *-ic* = of the nature of.) Relating to tribal or stem development.

PHYLOGENETIC CLASSIFICATION A system of classification which purports to be natural in that all subordinant Taxa are members of a higher category (Genus, Family, Order) by genetical relatedness or descent through evolutionary process. Cf. Artificial Classification.

PHYLOGENETIC HOMOLOGY Correspondence of characteristics (features, parts, structures) between Taxa. See Homology. Cf. Iterative Homology.

PHYLOGENETIC SPECIES CONCEPT A Species concept based on branching, or cladistic relationships among Species or higher Taxa. PSC presents an hypothesis of the 'true' genealogical relationship among Species, based on the concept of shared, derived characters (synapomorphies). Cf. Species Concepts.

PHYLOGENY Noun. (Greek, *phylon* = race; *genesis* = descent. Pl., Phylogenies.) The development of taxonomic groupings through evolutionary studies, comparative anatomy and related phenomena. Cf. Ontogeny.

PHYLOPTERA Noun. 1. A superordinal name proposed to include all the net-veined Orders. 2. The Orthoptera and Dermoptera.

PHYLUM Noun. (Greek, *phylon* = race. Pl., Phyla.) A stem-group category used in biological classification. Occupies a position below the Kingdom-level and above the Class-level.

PHYMATIDAE Laporte 1832. Plural Noun. Ambush Bugs. A small Family of Hemiptera often regarded as a Subfamily of Reduviidae. Adult stout-bodied, coloration yellow, brown, black, to pale green. Fore leg raptorial with enlarged Femur and opposable Tibia; Tarsi reduced or absent. Phymatidae reside in flower heads and are often cryptically coloured. All Species are predaceous and feed upon bees, flies and wasps which visit flowers.

PHYOMONE® See NAA.

PHYSCINAE Yasnosh 1976. A small Subfamily of Aphelinidae (Hymenoptera). Morphological features include: Female Antenna with seven segments incuding club with two segments; Antenna often bicolorous; male Antenna with eight segments; Pronotum and prepectus not divided; forewing without Linea Calva, uniformly setose, with a short marginal fringe; middle tibial spur stout; Sternum VII not reaching apex of Gaster. Biology: Internal parasites of Diaspididae.

PHYSIOGRAPHIC BARRIERS Dividing lines between distinct groups of organisms arising from the formation of the land, *e.g.*, the Rocky Mountains.

PHYSIOGRAPHY Noun. Physical geography, the surface structure of the earth, as mountains, oceans.

PHYSIOLOGICAL SPECIES A group of a given Species differentiated from other groups apparently structurally identical by its life-history and physiological activities (habitat, food, plants).

PHYSIOLOGICAL SALINE See Ringer's Solution.

PHYSIOLOGICAL SUPPRESSION The elimination of supernumerary larval parasites by competitors without physical combat. Mechanisms of supression poorly understood. Encapsulation of moribund parasite by host phagocytes may result in death via anoxia. Alternatively, older parasitic larva may consume nutrients essential for development of younger larvae and death results from inanition. Cf. Larval Combat.

PHYSOGASTRIC Adj. (Greek, *physan* = to blow up; *gaster* = belly; *-ic* = of the nature of.) Pertaining to females with a swollen or abnormally distended Abdomen filled with eggs or immatures.

PHYSOGASTRY Noun. (Greek, *physan* = to blow up; *gaster* = belly.) 1. General: A pathological enlargement of the Abdomen associated with the development of immatures within the body. Females give birth to larvae or adults via 'birthing openings' or rupture of the Integument. A condition characteristic of pygmephoroid mites. 2. Insects: An abdominal enlargement or swelling due to increase in size of fat body and/or Ovarioles. Seen in highly evolved termites, ants and parasitic wasps reared on artificial diet.

PHYSOPELTINAE Hussey 1929. A Subfamily of Largidae.

PHYSOPODA Bladder-footed; the Thysanoptera.

PHYTOALEXIN Noun. A type of phenolic compound produced by plants that become diseased or injured by insects; the compound confers resistance to plants from further attack by insects.

PHYTOGEOGRAPHICAL Adj. (Greek, *phyton* = plant; *ge* = earth; *graphein* = to write; Latin, *-alis* = belonging to.) Descriptive of things pertaining to the distribution of plants. Cf. Zoogeographical.

PHYTOGEOGRAPHY Noun. (Greek, *phyton* = plant; *ge* = earth; *graphein* = to write. Pl., Phytogeographies.) 1. Plant geography; distribution of plants over the earth's surface. 2. The study of distribution and distributional patterns and association or groupings of plants in space, realm or region. See Biogeography. Cf. Zoogeography. Rel. Realm.

PHYTOMETER Noun. (Greek, *phyton* = plant; *metron* = measure.) A plant biometer which indicates conditions through variations in body-size. Commonly referred to as yield, seed production, germination survival and other factors (Shelford).

PHYTOMYZIDAE See Agromyzidae.

PHYTOPHAGA Noun. (Greek, *phyton* = plant; *phagein* = to eat.) Plant feeding beetles characterized by 4th and 5th tarsal segments ankylosed and 3rd tarsal segment lobed. See Coleoptera.

PHYTOPHAGE Noun. (Greek, *phyton* = plant; *phagein* = to eat.) 1. Any organism that feeds on plant tissue. Syn Herbivore. See Feeding Strategies. Cf. Carnivore; Herbivore; Omnivore; Saprovore. Rel. Ecology.

PHYTOPHAGOUS Adj. (Greek, *phyton* = plant; *phagein* = to eat; Latin, *-osus* = with the property of.) Descriptive of organisms (plants or animals) that feed upon plant tissue. Most phytophagous insects show restricted feeding habits in terms of number of Taxa utilized as raw materials for food and plant parts that are acceptable nutritionally. About 90% of phytophagous insects feed on 1–3 plant Families; about 70% of phytophagous insects feed on 1–3 Genera of host plants. Specialization (oligophagy, stenophagy, monophagy) appears selectively advantageous for most phytophagous insects. Alt. Phytophagus. Syn. Herbivorous. See: Feeding Strategies. Cf. Cannibalistic; Carnivorous; Fungivorous; Parasitic; Predaceous; Saprophagous; Xylophagous. Rel. Ecology.

PHYTOPHAGY Noun. (Greek, *phyton* = plant; *phagein* = to eat.) The condition, act or process of being phytophagous. A generalized feeding strategy adopted by insects and other animals that feed upon plants. Included within phytophagy are leaf feeding, phloem feeding, xylem feeding, wood feeding, bark feeding, fruit feeding and root feeding. Feeding may be restricted to a certain Species of plant (monophagy, host-specific), a narrow range of plants (stenophagy) or a wide range of plants (polyphagy). See Feeding Strategies; Digestion; Extra-oral Digestion; Phytosuccivorous; Xylophagous. Cf. Cannibalism; Parasitism; Predation; Necrophily. Rel. Carnivory; Saprophagy.

PHYTOPHILOUS Adj. (Greek, *phyton* = plant; *philos* = to love; Latin, *-osus* = with the property of.) Plant loving; a term applied to organisms that live on, in or near plants. Alt. Phytophilus. See Habitat. Cf. Anthophilous. Rel. Ecology.

PHYTOPHILY Adj. (Greek, *phyton* = plant; *philos* = to love.) 1. The love of plants. 2. Descriptive of organisms that feed on plants. 3. A term used to describe ants that visit or live in or on certain plants (Wheeler). Alt. Phytophilous. See Feeding Strategy. Rel. Commensalism; Mutualism; Symbiosis.

PHYTOSCOPIC Adj. (Greek, *phyton* = plant; *skopein* = to look at; *-ic* = characterized by.) 1. Displaying a form of protective larval coloration derived from the superficial colour of a leaf rather than from its contained pigment (Wardle). 2. Pertaining to colours of larvae produced by light or illumination (Smith).

PHYTOSUCCIVOROUS Adj. (Greek, *phyton* =

plant; Latin, *succus* = juice, sap; *vorare* = to devour; *-osus* = with the property of.) Sap-sucking; sap-eating. Descriptive of plant-feeding organisms that take their nutrients from the sap of plants. Alt. Succivorous. See Feeding Strategy. Cf. Phytophagy; Xylophagous.

PHYTOTELMATA Noun. Structural modifications of plants for water collection used to provide a habitat for animal life. (Varga 1928, Biol. Zbl. 48: 143–162)

PHYTPHTHIRA Noun. (Greek, *phyton* = plant; *phthir* = lice.) The plant lice; some authors include scale insects. (Archaic.)

PIAGET, EDOUARD (1817–1910) (Musgrave 1932, *A Bibliography of Australian Entomology 1775–1930,* viii + 380 pp. (255), Sydney.)

PIAN See Yaws.

PIANBIOT® See Entomogenous Nematodes.

PIC, MAURICE (1890–1957) (Schmidt 1958, Ent. Bl. Biol. Syst Käfer 54: 189.)

PICARD, FRANCOIS (1879–1939) French academic. (Berland 1939, Ann. Soc. ent. Fr. 108: 173–181, bibliogr.)

PICARD, PIERRE FRANCOIS (1810–1936) (Feisthamel 1839, Ann. Soc. ent. Fr. 8: 587–594.)

PICARD-CAMBRIDGE, ARTHUR WALLACE English arachnologist and coleopterist.

PICCIOLI, FERDINANDO (1821–1900) (Bargagli 1900, Boll. Soc. ent. ital. 32: 217–228.)

PICEOUS Adj. (Latin *piceous* = pitch; *-aceus* = of or pertaining to.) Pitchy black; black with reddish tinge (Kirby & Spence).

PICINUS Adj. (Latin, *piceous* = pitch.) Black; structure with bluish or bluish-oily luster. Alt. Picine.

PICK Noun. (Anglo Saxon, *pycan* = to pick. Pl., Picks.) Corrodentia: A moveable sclerite attached to Maxilla.

PICKARD-CAMBRIDGE, ARTHUR (1873–1952) (Riley 1953, Proc. R. ent. Soc. Lond. (C) 17: 73.)

PICKARD-CAMBRIDGE, FREDERICK OCTAVIUS (1861–1905) (Bankes 1905, Entomol. mon. Mag. 41: 97.)

PICKARD-CAMBRIDGE, OCTAVIUS (1828–1917) (Turner 1917, Entomologist's Rec. J. Var. 29: 89–91.)

PICKEL, BENTO JOSÉ (1890–1963) (Anon. 1963, Studia Ent. 6: 585.)

PICKERELWEED BORER *Bellura densa* (Walker) [Lepidoptera: Noctuidae].

PICKERING, CHARLES (1805–1878) (Scudder 1891, Psyche 6: 57–60, 121–124, 137–141, 169–172, 185–187, 297–298, 345–346, 357–358.)

PICKET® See Permethrin.

PICKLEWORM *Diaphania nitidalis* (Stoll) [Lepidoptera: Pyralidae]: A widespread, multivoltine pest of cucumbers, melons and squash. Eggs deposited in small clutches on buds, new leaves, stems or fruits. Larvae penetrate fruits, stems, and buds. Cf. Melonworm.

PICTET, (DE LA RIVE), FRANCOIS JULES (1809–1872) Author of *Recherches pour servir a l'Historie et a l'Anatomie des Phryganides* (1834). (Anon. 1872, Entomol. mon. Mag. 8: 294–295.)

PICTET, A EDOUARD (1835–1879) (Anon. 1879, Entomol. mon. Mag. 16: 24.)

PICTET, ARNOLD (1869–1948) (L.B. 1948, Lambillionea 48: 49.)

PICTICOLLIS CANEGRUB *Lepidiota picticollis* Lea [Coleoptera: Scarabaeidae]. See Canegrub.

PICTURE-WINGED FLIES See Otitidae; Platystomatidae; Tephritidae.

PIE-DISH BEETLES *Helea* spp. [Coleoptera: Tenebrionidae].

PIEL, O (1876–1945) (Hall 1947, Proc. R. ent. soc. Lond. (C) 11: 61.)

PIEPERS, MURINUS CORNELIS (–1919) (Martin 1913, Dt. ent. Z. Iris 33: 134–135.)

PIERANTONI, UMBERTO (1876–1959) (Salfi 1959, Archo zool. ital. 44: iii–xv, bibliogr.)

PIERCE, FRANK NELSON (1861–1943) (Cockayne 1944, Proc. R. ent. Soc. Lond. (C) 8: 69.)

PIERCE, W DWIGHT American Economic Entomologist (U.S. Bureau of Entomology). Employed 1927–1930 by North Negros Sugar Company and studied pests of sugarcane in Philippines. Editor of *Sanitary Entomology - The Entomology of Disease, Hygiene and Sanitation.* Boston.

PIERCE, WILLIAM CLINTON (1906–1965) (Baker 1965, J. Econ. Ent. 58: 1039.)

PIERIDAE Plural Noun. Whites; Sulphur Butterflies; Orange Tips. A cosmopolitan Family of ditrysian Lepidoptera assigned to Papilionoidea. Adult body moderate sized; Maxillary Palp absent; Labial Palp ascending Frons; foreleg fully developed, Tibia spined, Epiphysis absent; tibial spur formula 0-2-2. Larva cylindrical; body with numerous short, fine secondary Setae; sometimes gregarious. Pupa angular, sometimes with spines or ridges; cremaster and central silken girdle present. Syn. Asciidae.

PIERRET, ALEXANDRE (1814–1850) (Doné 1850, Ann. Soc. ent. Fr. (2) 8: 351–360.)

PIERSIG, GUSTAV RICHARD (1857–1906) (Kranker 1908, Ent. Jb. 17: 204.)

PIERSON, HENRY BYRON (1894–1973) (Nash 1973, J. Econ. Ent. 66: 1360–1361.)

PIESBERGEN, FRANZ (1860–1928) (Reihlen 1928, Jh. Ver. vaterl. Naturk. Württ. 84: xxxiv.)

PIESMATIDAE Spinola, Amyot & Serville 1843. Plural Noun. Ash-Grey Plant Bugs; Ash-Grey Leaf Bugs; Piesmatids. A numerically small, geographically widespread Family of heteropterous Hemiptera assigned to Superfamily Piesmatoidea. Body less than 5 mm long; Ocelli present; Pronotum with five longitudinal ridges; Corium and Clavus with irregular network of cells; Hemelytra covering Abdomen; Juga anterior of Tylus; Tarsi with two segments.

PIESMATIDS See Piesmatidae.

PIESMATOIDEA A Superfamily of Hemiptera including Piesmatidae.

PIESZCZEK, ADOLF (–1928) (Hepp 1929, Ent. Z. Frankf. a. M. 43: 29.)

PIEZA Noun. (Greek, *piezein* = to press.) The combined biting and sucking mouth of the Hymenoptera.

PIEZATA Noun. The Fabrician term for Hymenoptera.

PIFFARD, ALBERT (–1909) (Morley 1910, Entomologist 43: 127–128.)

PIFFARD, BERNARD (1832–1916) (Lyle 1916, Entomologist 49: 143–144.)

PIG FOLLICLE-MITE *Demodex phylloides* Csokor [Acari: Demodicidae].

PIG LOUSE *Haematopinus suis* (Linnaeus) [Phthiraptera: Haematopinidae].

PIG MANGE-MITE *Demodex phylloides* Csokor [Acari: Demodicidae].

PIGEON FLY *Pseudolynchia canariensis* (Macquart) [Diptera: Hippoboscidae]: A parasite of pigeons in tropical and warm temperate regions throughout the world. Female takes blood of host; larviposits on host; mature larvae roll off host into nest. Pupal stage ca 30 days. PF vector of blood protozoan *Haemoproteus columbae*.

PIGEON LOUSE *Colpocephalum turbinatum* Denny [Phthiraptera: Menoponidae].

PIGEON TREMEX *Tremex columba* (Linnaeus) [Hymenoptera: Siricidae]. A pest of deciduous hardwoods in North America. Eggs drilled into bark and wood; larva constructs tunnel in sapwood and heartwood. Two-year life cycle.

PIGFACE SCALE See Cottony Pigface Scale.

PIGG, THOMAS (–1901) (Anon. 1902, Entomologist's Rec. J. Var. 14: 27.)

PIGMENT Noun. (Latin, *pigmentum* > *pingere* = to paint. Pl., Pigments.) 1. Any colouring matter or material that gives a colour appearance. 2. Any of various colouring matters that occur in plants and animals.

PIGMENT CELL Insect Ommatidium: An accessory cell loaded with pigment positioned between the Retinulae, or a visual cell containing pigment (Imms); Iris Pigment Cells; Retinal Pigment Cells. See Basal Pigment Cell; Primary Pigment Cell; Secondary Pigment Cell.

PIGMENT LAYER Insect eye: Accessory cells with pigment surrounding the visual and tapetum layers, forming the iris (Needham).

PIGMENTARY COLOUR Any of several colours that are derived from substances of a definite chemical composition. PC is produced by pigments in plants and animals. Pigments may serve an insect in intermediary metabolism or in other physiological ways. PC absorb some light wavelengths and reflect other light wavelenths as colours. PC substances are the products of metabolism and are excretory in nature (Imms). Pigmentary colour changes in the body may be slow (morphological pigment) and involve physical synthesis or destruction of pigments. Alternatively, the action of pigments may be rapid (physiological pigment) and involve migration of pigment granules or a change in shape of cells containing a pigment. Common biological colours include colloids of suspensions and solutions. Most common biological pigments in solution are carotenoids (in fats), urochrome (in urine) and anthocyanins (in plant-cell sap). Pigmentary colour is an artefact of molecular configuration of some chemicals, especially those with double bonds of carbon and carbon connecting oxygen, or nitrogen bound to itself or carbon. Particularly important are NH_2 and Cl_2 radicals which shift absorptive wavelengths and tend to absorb longer wavelengths. A colour-producing molecule is called a chromophore bound to a protein to form a chromoprotein. A visible coloration of pigments is not simply a function of absorbed wavelength but also depends upon: (1) absorption spectrum of the pigment, (2) ultrastructure of environment surrounding the pigment, (3) position of the pigment molecules, (4) mixture of reflected wavelengths from different levels in pigment and (5) chemical state of pigments (granules or amorphous). See Aphins; Carotenes; Flavones; Melanin; Ommochromes; Pterines; Quinones; Tetrapyrroles. Cf. Structural Colour. Rel. Chromatophores; Chromatosomes; Coloration.

PIGMENTATION Noun. (Latin, *pingere* = to paint; English, *-tion* = result of an action.) The deposition of pigment in a body, thereby giving it colour.

PIGMENTED Coloured; in general, more heavily coloured.

PIGMY BACKSWIMMERS See Pleidae.

PIGMY MOLE CRICKETS See Tridactylidae.

PIGNORIA, LORENZO (1591–1631) (Rose 1850, *New General Biographical Dictionary* 11: 114–115.) Alt. Laurentius Pignorius.

PILACERORIS Noun. (Pl., Pilacerores.) Coccids: A cuticular extension which lacks a Calyx and has an opening at its free end (MacGillivray).

PILATE, EUGENE (1804–1890) (Brown 1967, Ent. News 78: 57–59.)

PILATE, LOUIS (1816–1852) (Sallé 1852, Bull. Soc. ent. Fr. (2) 10: 1.)

PILE Noun. (Latin, *pila* = pillar, pier; *pilleus* = felt cap. Pl., Piles.) 1. A covering of thick, fine, short, erect Setae which gives the structure a velvet-like or fur-like appearance. Cf. Setose. 2. A quantity or collection of 'things' indiscriminately heaped together or arranged in an orderly manner.

PILEIFORM Adj. (Latin, *pileus, pilleus* = felt cap; *forma* = shape.) Pileus-shaped; with the form of an umbrella-shaped cap. See Pileus. Cf. Agariciform. Rel. Form 2; Shape 2; Outline Shape.

PILEOLATED Adj. (Latin, *pileolus* = little cap.) Descriptive of a structure invested with a small cap or caps.

PILEOLUS Noun. (Latin, *pileolus* = little cap.) A small pileus.

PILEOUS Adj. (Latin, *pilus* = hair; *-osus* = with the property of.) Hairy.

PILEUS Noun. (Latin, *pileus* = cap. Pl., Pilei.) A cap or cap-like structure. *e.g.* Mushroom Pileus. Cf.

Fimbria.

PILI SIMPLICES Coccids: Pygidial Setae (MacGillivray).

PILIFER Noun. (Latin, *pilus* = hair; *ferre* = to carry. Pl., Pilifers.) Lepidoptera: A small sclerite on each side of the Labrum in many Species. The Pilifer resembles a rudimentary Mandible or is lobe-like and setose. Alt. Piliger.

PILIFEROUS Adj. (Latin, *pilus* = hair; *ferre* = to carry; *-osus* = with the property of.) Bearing a vestiture of forming a pile. Alt. Piligerous.

PILIFEROUS TUBERCLES Slightly elevated seta-bearing annular sclerites in larvae (Comstock).

PILL BEETLES See Byrrhidae.

PILLARED EYE Ephemerids. A type of eye placed on a cylindrical stalk or process. Alt. Turbinate eye.

PILLBUGS 1. *Armadillidium* spp. and *Australiodillo bifrons* [Isopoda: Armadillididae]: Body oblong to oval with highly convex body; seven pairs of legs; many roll into tight ball, hence the common name. Pillbugs are typically nocturnal and inhabit damp, moist areas; they often are taken in gardens and sometimes invade homes. Pillbugs feed on roots and tender plant tissue. 2. *Sphaerillos grossus*. Adult ca 15 mm long; body slightly flattened. In eastern Queensland. Syn. Armadillo Bugs. See Common Pillbug; Native Pillbug; Sowbug. Cf. Slaters.

PILLIFORM Adj. (Latin, *pilus* = hair; *forma* = shape.) Seta-shaped. See Seta. Rel. Form 2; Shape 2; Outline Shape.

PILOSE Adj. (Latin, *pilus* = hair; *-osus* = full of.) 1. Pertaining to a surface covered with soft down or short Setae. 2. Pertaining to a surface covered with long, Setae. Alt. Pilous.

PILOSITY Noun. (Latin, *pilus* = hair; English, *-ity* = suffix forming abstract nouns.) 1. The state of being pilose. 2. A covering of long, stout Setae, typically standing above a vestiture of smaller, finer Setae. See Pubescence.

PIMPLY FEVER See Kenya Tick Typhus.

PIN CATERPILLAR MOTH *Dendrolimus spectabilis* Butler [Lepidoptera: Lasiocampidae].

PIN-TAIL BEETLES See Mordellidae.

PINACULUM Noun. (New Latin, from *pinaculum;* Greek, *pinax* = tablet; *-ium* = diminutive form. Pl., Pinacula.) Sclerites (single or paired) which bear Setae on abdominal segments of Lepidoptera larvae. Cf. Chalaza; Scolus; Verruca.

PINARA MOTH *Pinara divisa* (Walker) [Lepidoptera: Lasiocampidae].

PINCER Noun. (French. Pl., Pincers.) Insects, Anal Forceps.

PINCH BEETLES See Lucanidae.

PINCKNEY, JOHN STUART (1901–1940) (Hill 1941, J. Econ. Ent. 34: 131.)

PIN-CUSHION GALL See Bedeguar.

PINCUSHION MILLIPEDES Common name for the order Polyxenida in the Class Diplopoda (millipedes) whose members are minute, soft-bodied and heavily bristled. The Order contains a single Genus, *Polyxenus*, of which there are five Species in North America.

PINDER, JOSEPH EDWIN (1866–1951) (Riley 1952, Proc. ent. Soc. Lond. (C) 16: 87.)

PIN-DOWN® See Lycolure.

PINE ADELGID *Pineus pini* (Macquart) [Hemiptera: Adelgidae].

PINE APHIDS See Adelgidae.

PINE-BARK ADELGID *Pineus strobi* (Hartig) [Hemiptera: Adelgidae].

PINE BARK ANOBIID *Ernobius mollis* (Linnaeus) [Coleoptera: Anobiidae]: In Australia, a pest of untreated exotic pine; sometimes superficial damage to outer sapwood; does not cause structural damage and of only minor economic importance. Adult ca 5 mm long; neonate adult appears golden brown, covered with fine golden Setae, darkening as pale Setae lost; Elytra smooth; Elytra and other parts not strongly scleroitized. Development time one year; adults present only in spring and early summer. Presence of bark essential for development. Larvae burrow long distances in bark and wood. Round flight hole ca 2 mm dima.

PINE BARK WEEVIL *Aesiotes notabilis* Pascoe [Coleoptera: Curculionidae], a pest of hoop pine along coastal Queensland, Australia. Natural distribution corresponds with rainforests in which Species of *Araucaria* and *Agathis* occur; also attacks *Pinus* spp. in plantations. Adult ca 14 mm long, robust, cylindrical, greyish-brown with dark pattern and two spines towards apex of Elytra. Head angled with short, blunt snout. Larva white with yellow-brown head and dark brown Mandibles. Attacks on standing plantation hoop pine trees associated with injuries received during pruning. Eggs usually laid at night on pruned stubs or injured bark. Tunnelling by larvae may result in girdling of stem; injury may allow entry of other borers and pathogens. Neonate larvae emerge from eggs ca eight days, feed on phloem around stub bases; larval development ca 60 days. Pupation in chamber beneath bark. Pupa covered by cocoon of thin wood strips cut by mature larva; pupation ca 14 days; neonate adult may not emerge from chamber for 12 days. Adults feed on bark and foliage of hoop pine branches; may live ca 18 months. Copulation occurs ca 1 day after emergence; oviposition commences 2–7 weeks later. Females lay ca 500 eggs, oviposit for more than one year.

PINE BEETLES See Curculionidae.

PINE BUD-MITE *Trisetacus pini* (Nalepa) [Acari: Nalepellidae].

PINE BUTTERFLY *Neophasia menapia* (Felder & Felder) [Lepidoptera: Pieridae]: A pest of ponderosa pine in western North America. Adults fly in large swarms above forest tree canopy. Eggs emerald-green, laid on pine needles in canopy; larvae feed, drop by silken thread to lower vegetation or trunk and pupate. One generation per year; overwinters in egg stage. Host plants include lodgepole pine, ponderosa pine, western white pine.

PINE CHAFER *Anomala oblivia* Horn [Coleoptera: Scarbaeidae].

PINE-CANDLE MOTH *Exoteleia nepheos* Freeman [Lepidoptera: Gelechiidae].

PINE COLASPIS *Colaspis pini* Barber [Coleoptera: Chrysomelidae].

PINE-CONELET BUG *Platylyus luridus* (Reuter) [Hemiptera: Miridae].

PINE-CONELET LOOPER *Nepytia semiclusaria* (Walker) [Lepidoptera: Geometridae].

PINE ENGRAVER *Ips pini* (Say) [Coleoptera: Scolytidae]: A pest of larch, pine and spruce throughout North America. Adult 2.5–4 mm long, black with Antenna and Tarsi brown; Frons densely and coarsely punctate dorsally and granulate ventrally; male with median Tubercles; Pronotum with fine and coarse punctures; Elytra with coarse strial punctures and four pairs of declivital spines along posterior margin. PE is polygamous; male develops in nuptial chamber in tree and mates with several females; each inseminated female constructs an independent egg-gallery from the nuptial chamber. See Scolytidae; California Five-Spined Ips; Monterey-Pine Engraver; Six-Spined Ips.

PINE FALSE-WEBWORM *Acantholyda erythrocephala* (Linnaeus) [Hymenoptera: Pamphiliidae].

PINE-FLOWER SNOUT BEETLE See Nemonychidae.

PINE GALL-WEEVIL *Podapion gallicola* Riley [Coleoptera: Curculionidae].

PINE JEWEL-BEETLE *Chalcophora japonica* [Coleoptera: Buprestidae]: A pest of pine in the Orient. Adult ca 35 mm long.

PINE LEAF-ADELGID *Pineus pinifoliae* (Fitch) [Hemiptera: Adelgidae]: Overwinter as wax-fringed form; winged adults migrate to alternate host (red and black spruce), form cone-like galls, remain a year. Induces browning and death of new branch tips of white pine in North America.

PINE LOOPER *Bupalus piniarius* Linnaeus [Lepidoptera: Geometridae]

PINE LOOPERS *Chlenias* spp. [Lepidoptera: Geometridae].

PINE MOTH *Dendrolimus pini* Linnaeus [Lepidoptera: Lasiocampidae].

PINE-NEEDLE SCALE *Chionaspis pinifoliae* (Fitch) [Hemiptera: Diaspididae]: A pest of evergreen trees that is endemic and widespread in North America. PNS takes sap from pine and spruce needles. Overwinters in egg stage beneath scale-cover of female. Female deposits 20–30 purple eggs. Crawlers active in mid-spring, settle on needles; mature during late summer. Syn. *Phenacaspis pinifoliae*. See Diaspididae.

PINE NEEDLE SHEATH-MINER *Zelleria haimbachi* Busck [Lepidoptera: Yponomeutidae].

PINE NEEDLE-MINER *Exoteleia pinifoliella* (Chambers) [Lepidoptera: Gelechiidae]: A monovoltine pest of pine in eastern North America. Egg white, inserted in exit hole of abandoned mines. Larvae mine current year's foliage; overwinter as larva; pupation in mine with entrance sealed by silk.

PINE-ROOT COLLAR WEEVIL *Hylobius radicis* Buchanan [Coleoptera: Curculionidae].

PINE ROOT-TIP WEEVIL *Hylobius assimilis* Boheman [Coleoptera: Curculionidae].

PINE ROSETTE-MITE *Trisetacus gemmavitians* Styer [Acari: Nalepellidae].

PINE SPITTLEBUG *Aphrophora parallela* (Say) [Hemiptera: Cercopidae].

PINE-STUMP WEEVIL *Mitrastethus australiae* Lea [Coleoptera: Curculionidae].

PINE TIP-MOTH *Rhyacionia busnelli* Busck [Lepidoptera: Tortricidae].

PINE TORTOISE-SCALE *Toumeyella parvicornis* (Cockerell) [Hemiptera: Coccidae].

PINE TUBE-MOTH *Argyrotaenia pinatubana* (Kearfott) [Lepidoptera: Tortricidae].

PINE TUSSOCK-MOTH *Dasychira pinicola* (Dyar) [Lepidoptera: Lymantriidae].

PINE WEBWORM *Tetralopha robustella* Zeller [Lepidoptera: Pyralidae].

PINE WITCHETYGRUB *Cacodacnus planicollis* (Blackburn) [Coleoptera: Cerambycidae].

PINEAPPLE Noun. (Pl., Pineapples.) *Ananas comosus* (Linnaeus) Merrill [*A. sativus* Schult.] [Bromeliaceae]: Tropical, terrestrial, perennial bromeliad endemic to South America. Varieties fall into five major groups: Spanish, Queen, Pernambuco, Perolera and Cayenne.

PINEAPPLE FALSE SPIDER-MITE *Dolichotetranychus floridanus* (Banks) [Acari: Tenuipalpidae].

PINEAPPLE FLAT MITE *Dolichotetranychus floridanus* (Banks) [Acari: Tenuipalpidae].

PINEAPPLE MEALYBUG *Dysmicoccus brevipes* (Cockerell) [Hemiptera: Pseudococcidae]: Thelytokous, produces live crawlers; three nymphal instars; adult ca 3 mm long, lives ca 90 days, produces ca 230 offspring. Crawlers transmitted between plants by wind, ants, phoresy. A pest of pineapple in Hawaii and Australia. PM vectors 'mealybug wilt' (closterovirus) through feeding on plant fluids.

PINEAPPLE MITE *Phytonemus ananas* (Tryon) [Acari: Tarsonemidae].

PINEAPPLE SAP BEETLE *Urophorus humeralis* (Fabricius) [Coleoptera: Nitidulidae]: A pantropical pest of pineapple and stored dry fruit.

PINEAPPLE SCALE *Diaspis bromeliae* (Kerner) [Hemiptera: Diaspididae]: A pest of pineapple in all regions where the plant is cultivated. PS typically infests bottom leaves of plant. Eggs laid under scale cover, hatch within seven days; egg-to-adult within 60 days. Except at exceptionally high infestations, PS has little effect on crop production; light infestations produce yellow spots on leaves; heavy infestations produce grey coloured foliage.

PINEAPPLE TARSONEMID *Stenotarsonemus ananas* (Tryon) [Acari: Tarsonemidae].

PINEAPPLE WEEVIL *Metamasius ritchiei* Marshall [Coleoptera: Curculionidae].

PINGERON, JEAN CLAUDE DE (1730–1795) (Rose 1850, *New General Biographical Dictionary* 11: 120.)

PINGUIS (Latin, *pinguis* = fat.) Naturally and proportionally plump.

PINK BOLLWORM *Pectinophora gossypiella* (Saunders) [Lepidoptera: Gelechiidae]: A widespread, serious pest of cotton *(Gossypium hirsutum),* which develops on flowers, seeds and seed capsules. Described from collections made in India during 1843, but origin of pest probably Australia. Taken in Egypt ca 1905; subsequently distributed in principal cotton growing areas. Adult moth 8–9 mm long, with wingspan 15–20 mm. Body greyish-brown with dark markings on forewing; wings narrow with wide fringe of Setae and apically pointed. Base of Antenna with 5–6 long stiff Setae. Labial Palpus long and curved. Adult females lay 200–400 eggs on cotton plant near bolls. Eggs ca 0.5 mm long, oval and translucent white when freshly laid, but turn red before larval emergence. Eggs may be laid singly or in batches of 5–10; egg stage 4–5 days. Early instar larvae whitish-yellow with a brown head; fourth instar larva with distinctive pink coloration; overwintering in seed husks and plant litter. Larval stage ca 14–20 days and larvae begin pupation during early spring. Pupae are 7–10 mm long, brown; pupal stage requires ca 10 days. Life-cycle 25–30 days to complete and up to six generations per year. Larvae feed on seeds within cotton boll and on flowers. Mature larvae cut a 2 mm diameter hole in side of boll to emerge and pupate. Effects of seed feeding lead to arrested growth, boll rotting and premature or partial boll opening. Larva feeds on many plants of Family Malvaceae; Primary hosts *Hibiscus* spp. Origin in Australia but only occurs in northern Western Australia and Northern Territory. Cf. Cotton Bollworm; Egyptian Bollworm; Spiny Bollworm; Pink-Spotted Bollworm.

PINK CORNWORM *Pyroderces rileyi* (Walsingham) [Lepidoptera: Cosmopterigidae]. Alt. Pink Scavenger-Caterpillar.

PINK CUTWORM *Agrotis munda* Walker [Lepidoptera: Noctuidae]: A pest of several vegetables in Australia. Adult wingspan 30–40 mm; forewing pale brown to brownish grey with dark mottling; hindwing pale brown with brown veins and apical margin dusky. Caterpillars undergo six larval instars. Typically 2–3 overlapping generations per year. Syn. Brown Cutworm. See Cutworms.

PINK GROUND-PEARL *Eumargarodes laingi* Jakubski [Hemiptera: Margarodidae].

PINK GUM-LERP *Cardiaspina densitexta* Taylor [Hemiptera: Psyllidae].

PINK HIBISCUS MEALYBUG *Maconellicoccus hirsutus* (Green) [Hemiptera: Pseudococcidae]: Endemic to Southeast Asia; recovered in Hawaii 1984; first reported in the Western Hemisphere in 1994 in Grenada and subsequently in 18 Caribbean Islands and Guyana. Reported in western USA in 1999; range expansion continues. Commonly attacks hibiscus and closely related Species; reported attacking over 200 Species of plants, including soursop (*Annona muricata*), sugar apple (*Annona squamosa*), breadfruit (*Artocarpus altilis*), papaya (*Carica papaya*), Acacia spp., Pigeon pea (*Cajanus cajan* = *C. indicus*), grapefruit, pumpkin, squash and cocoa (*Theobroma cacao*). Syn. Hibiscus Mealybug; Pink Mealybug.

PINK MEALYBUG See Grey Sugarcane Mealybug; Pink Hibiscus Mealybug.

PINK SCAVENGER-CATERPILLAR *Pyroderces rileyi* (Walsingham) [Lepidoptera: Cosmopterygidae]: In USA a widespread pest of maize, almonds, walnuts and other fruits in decay. Eggs pearly white, laid individually or in small groups. Larva forms web of frass and silk in space as it feeds. Alt. Pink Cornworm. See Cosmopterygidae.

PINK SUGARCANE-MEALYBUG *Saccharicoccus sacchari* (Cockerell) [Hemiptera: Pseudococcidae].

PINK TEA RUST MITE *Acaphylla theae* (Watt & Mann) [Acari: Eriophyidae].

PINKER, RUDOLF (1847–1934) (Heikertinger 1935, Koleopt. Rdsch. 21: 55–56.)

PINKEYE Noun. An acute and high contagious conjunctivitis of humans and some domestic animals that occurs in Mexico and USA. Pinkeye is caused by bacteria with *Hippelates* spp. (Chloropidae) implicated as vectors. See Yaws.

PINK-SPOTTED BOLLWORM *Pectinophora scutigera* (Holdaway) [Lepidoptera: Gelechiidae]: Endemic to eastern Australia and a very minor, sporadic pest of cotton in Queensland. Closely related and anatomically similar to Pink Bollworm and Species only distinguished through dissection of the genitalia. Adult PSBW 12–15 mm long; body dark grey. Adult females lay eggs singly or in batches of 4–5 on young leaves, stems and at the bases of bolls. Eggs whitish-yellow when first laid; orange-red before larval emergence; eggs 0.5 mm long with longitudinal ridges; egg stage 3–6 days. Larvae usually hatch early in the morning; early instar larvae creamy-white with a dark brown head; mature larvae 15–18 mm long with distinctive pink coloration. Head heavily sclerotised, shiny and dark brown to black. Pupation occurs inside the cotton bolls. Larvae feed within bolls producing webbing that prevents flowers from opening. Entry holes difficult to detect. See Gelechiidae. Cf. Pink Bollworm.

PINK-STRIPED OAKWORM *Anisota virginiensis* (Drury) [Lepidoptera: Saturniidae].

PINK-WAX SCALE *Ceroplastes rubens* Maskell [Hemiptera: Coccidae]: A pest of avocado, citrus, custard apple and mango; ornamental host plants include: *Ficus* spp., holly, ivy, *Pittosporum*,

Schefflera actinophylla, Syzygium spp. and some ferns. PWS possibly endemic to Africa; occurs in southeast Asia, China, islands of the Pacific, Spain, USA (Florida) and Japan. PWS resembles Florida Wax Scale and Hard Wax Scale. Adult female 3–4 mm long, pink, globular, smooth, with two lobes at either end and a depression at centre of cover's top. Parthenogenetic with males rarely produced; female produces 200–900 eggs depending upon nitrogen content of host plant; eggs oviposited beneath scale cover; crawlers prefer to settle near midrib of leaves (both surfaces) or green twigs; 1–2 generations per year depending upon climate and host plant. PWS produces large amounts of honeydew which serve as substrate for sooty mould and other fungi.

PINK-WAX SCALE PARASITE *Aenasoidea varia* Girault; *Anicetus beneficus* Ishii & Yasumatsu [Hymenoptera: Encyrtidae]

PINK-WINGED GRASSHOPPER *Atractomorpha sinensis* Bolivar [Orthoptera: Pyrgomorphidae].

PINNA Noun. (Latin, *pinna* = feather. Pl., Pinnae.) 1. Bird feather; fish fin. Insects: A narrow wing. 2. Oblique ridges extending to median line on posterior Femur of jumping Orthoptera, which somewhat resembles a feather.

PINNATE Adj. (Latin, *pinnatus* = feathered; *-atus* = adjectival suffix.) Feather-like or with markings resembling a feather; with stiff Setae or sclerotized processes occupying opposite sides of a thin shaft. Obsolete: Pinnulate; Pinnulatus. Alt. Penniform.

PINNATIFID Adj. Divided into feathers, as when wings are cleft nearly to base.

PINON CONE-BEETLE *Conophthorus edulis* Hopkins [Coleoptera: Scolytidae].

PINTO, CÉSAR FERREIRA (1896–1964) (Anon. 1964, Studia ent. 7: 486–487.)

PINWORMS Beetles of the Lymexylidae; often grouped with ambrosia beetles due to similar feeding habits. Adult ca 25 mm long, narrow, brown-black with well developed eyes and short Elytra. Larva very long, narrow, cylindrical with globular head and enlarged, dorsally humped Prothorax. Pinworms mainly attack standing trees. Larvae bore into weakened or dead trees and feed on ambrosia fungi which grow on tunnel walls. Adult females carry fungal spores in pouches near apex of Ovipositor; spores deposited in gelatinous matrix with eggs. First instar larvae transport spores into wood on their bodies. Larvae require several years to mature and can tunnel 1–2 metres with tunnels gradually increasing to ca 3 mm diam. Reinfestation by succeeding generations can occur.

PIOCHARD DE LA BRULERIE, CHARLES JACOB (1845–1876) (Simon 1877, Ann. Soc. ent. Fr. (5) 6: 677–688.)

PIOPHILIDAE Plural Noun. Bacon Flies; Cheese Skippers; Skipper Flies. A small Family of schizophorous Diptera assigned to Superfamily Tephritoidea. Adult small, dark blue or black and shiny; head rounded; Antennal Arista bare; Postvertical Bristles well developed; Fronto-orbital bristles weakly developed or absent; Vibrissae well developed; Mesopleural Bristle absent. Larvae saprophagous with most Species living in dead animal matter. Syn. Neottiophilidae. See Cheese Skipper.

PIPE ORGAN MUD DAUBER See Organ Pipe Mud Dauber.

PIPER, CHARLES VANCOUVER (1867–1926) American academic (Washington State University). (Osborne 1937, *Fragments of Entomological History*. 394 pp. Columbus, Ohio.)

PIPERONYL BUTOXIDE A mono-oxygenase inhibitor widely used as a synergist for pyrethrins. See Synergist.

PIPEVINE SWALLOWTAIL *Battus philenor* (Linnaeus) [Lepidoptera: Papilionidae]: A widespread Species in North America. Host plants include pipevine and related Species; PSTs overwinter as Pupae and complete 2–3 generations per year.

PIPUNCULIDAE Plural Noun. Big-Headed Flies; Big-Eyed Flies. Cosmopolitan Family of ca 1,000 Species of cyclorrhaphous Diptera. Pipunculids are closely related to Syrphidae and some taxonomists relate pipunculids to Dolichopodidae and Platypezidae. Adult body size ca 4 mm with strikingly large globular head and enormous compound eyes; wing narrow, Alula reduced, venation like Syrphidae except cell R5 open in wing margin, Spurious Vein absent. Arista dorsal on 3rd Antennal segment, composed of two microscopic basal segments and long whip-like apical segment. Adults hover inconspicuously in low vegetation. Larvae develop as internal parasites of Homoptera (Cicadellidae, Fulgoridae, Cercopidae, Delphacidae). Oviposition unique: female pounces on host, carries it aloft then inserts an egg into Abdomen; oviposition rapid and with little effect on host. After oviposition, host released and drops to ground. Parasite larva free in Abdomen of host feeding on fluids and organs. Emasculation of reproductive organs and genitalia common. Mature larva breaks Integument of host, usually at juncture of Thorax and Abdomen. Pupation occurs in ground. Syn. Dorilaidae.

PIQUETT, PRICE GODMAN (1908–1963) (Walker & Fales 1963, J. Econ. Ent. 56: 907.)

PIRAN® See Dichlorvos.

PIRATE SPIDERS *Australomimetus* spp. [Araneida: Mimetidae].

PIRATE® See Pyrrol.

PIRAZZOLI, ODOARDO (1815–1884) (Mik *et al.*1884, Wien. ent. Ztg. 3: 128.)

PIRIGRAIN® See Pirimiphos-Methyl.

PIRIMIPHOS-ETHYL An organic-phosphate (pyrimidine phosphate) compound {O,O-diethyl O-(2-(diethylamino) 6-methyl-4 pyrimidnyl phosphorothioate} used as a contact insecticide and fumigant against rust mite, wireworms, corn rootworm, carrot fly, sod webworms. Compound applied to fruit trees, bananas, mushroom, vegetables and turf in some countries. Not registered for use in USA. Toxic to bees. Trade names include Fernex®, Primicid® and Solgard®. See Organophosphorus Insecticide.

PIRIMIPHOS-METHYL An organic-phosphate (thiophosphoric acid) compound {O-[2-(diethylamino)-6-methyl-4-pyrimidinyl]O,O-dimethyl phosphorathioate} used as a contact insecticide against stored-grain pests, ants and public health pests. Compound applied to stored corn and sorghum in USA; applied to grapes, rice, cole crops, citrus, olives, and vegetables in other countries. Toxic to bees. Trade names include: Actellic®, Actellifog®, Blex®, Dominator®, Giustiziere®, Nu-Gro®, Silosan®, Sybol®, Pirigrain®, Rotator®. See Organophosphorus Insecticide.

PIRIMOR® See Primicarb.

PIROPLASM Noun. (Latin, *pirum* = pear; Greek, *plasm* = to form. Pl., Piroplasms.) A pear-shaped intracellular stage of protozoan parasites that resides within erythrocytes of vertebrate hosts.

PIROPLASMOSIS Noun. (Latin, *pirum* = pear; Greek, *plasmatos* = formed or moulded, an image; *sis* = a condition or state.) Any disease produced by Piroplasms (protozoan organisms related to those which cause malarial fevers.)

PISAURIDAE Plural Noun. Nursery-web Spiders. Moderate to large-sized spiders which resemble Lycosidae. Ground dwelling hunters, often associated with water; do not build webs. Female carries egg sac with Pedipalps or Chelicerae, constructs small 'tent' in which eggs hatch and spiderlings remain until dispersal.

PISSIS, PIERRE J NOEL AIMÉ (1812–1889) (Papavero 1971, *Essays on the History of Neotropical Dipterology*. 1: 159–160. São Paulo.)

PISSOT, CONSTANT EMILE (1826–1892) (Anon. 1892, Bull. Soc. ent. Fr. 61: cvii.)

PISTACHIO-SEED CHALCID *Megastigmus pistaciae* Walker [Hymenoptera: Torymidae]: A significant pest of pistachio seeds. PSC is endemic to Mediterranean and western Asia; introduced into California during 1967.

PISTAZINUS Adj. (Italian, from *pistacchio*, resembling the colour of pistachio.) Yellowish green, with a slight brownish tinge.

PISTIL® See Ethephon.

PISTILLATE Adj. (Latin, *pistillus* = pestle; *-atus* = adjectival suffix.) Botany: A flower which lacks stamens; sepals and petals present or absent. Cf. Staminate.

PISTOL CASE-BEARER *Coleophora malivorella* Riley [Lepidoptera: Coleophoridae]: A pest of apple, cherry, pear, plum and quince in North America. Eggs are laid on underside of leaves; larva constructs silken case which resembles a pistol; feeds from end of silken case; larva overwinters in silken case. Cf. Cigar Casebearer.

PISULIIDAE Ross 1967. Plural Noun. A small Family of afrotropical caddisflies.

PIT GALL Plant galls characterized by slight arching of leaf blade. Epidermis sometimes splits to form a blister. In their simplest form, PG nearly invisible. Syn. Blister Gall. See Gall. Cf. Bud-and-Rosette Gall; Covering Gall; Filz Gall; Mark Gall; Pouch Gall; Roll-and-Fold Gall.

PIT SCALES See Asterolecaniidae.

PITCH-EATING WEEVIL *Pachylobius picivorus* (Germar) [Coleoptera: Curculionidae].

PITCHER-PLANT MOSQUITO *Wyeomyia smithii* (Coquillett) [Diptera: Culicidae].

PITCH-MASS BORER *Synanthedon pini* (Kellicott) [Lepidoptera: Sesiidae].

PITCH-PINE TIP MOTH *Rhyacionia rigidana* (Fernald) [Lepidoptera: Tortricidae]: A pest of pines in southern USA. Biology resembles Nantucket Pine Tip Moth.

PITCH-TWIG MOTH *Petrova comstockiana* (Fernald) [Lepidoptera: Tortricidae].

PITCHY Adj. Blackish-brown. See Piceous.

PITFALL TRAP An insect collecting device designed to capture insects moving over the ground. PT consists of a dish, can or jar sunk in the earth to the rim or upper margin. A cover may be placed over the open top of container to exclude rain and small vertebrates. Pitfall traps often are baited with various substances and inspected frequently. Syn. Dish Trap; Pit Trap. See Trap.

PITH Noun. (Anglo Saxon, *pitha* = pith.) The central portion of a dicotyledonous stem. Alt. Stelar Parenchyma.

PITHWORMS See Elateridae.

PITTED APPLE BEETLE *Geloptera porosa* Lea [Coleoptera: Chrysomelidae]: A minor pest of *Citrus* in Australia.

PITTIONI, BRUNO (–1952) (R. Z. 1952, Wien ent. Ges. 37: 186.)

PITTIONI, EMANUEL (1879–1955) (Anon. 1955, Wien ent. Ges. 40: 144.)

PITTOSPORUM BEETLE *Lamprolina aeneipennis* (Boisduval) [Coleoptera: Chrysomelidae].

PITTOSPORUM BUG *Pseudapines geminata* (Van Duzee) [Hemiptera: Pentatomidae].

PITTOSPORUM LEAFMINER *Phyliriomyza pittosporphylli* (Hering) [Diptera: Agromyzidae].

PITTOSPORUM LONGICORN *Strongylurus thoracicus* (Pascoe) [Coleoptera: Cerambycidae]: A pest of *Citrus* and *Pittosporum* in Australia. Adult ca 30 mm long, pale brown with a row of white spots on either side of the Thorax. Female lays eggs in cracks on small branches or where other borers have damaged tree; eclosion occurs within 10 days; neonate larvae bore into wood where larvae feed on limbs. Monovoltine.

PLACOID Adj. (Greek, *plako* = flat surface; French, *placa* = shield; *eidos* = form.) Plate-like. See Placoid Sensillum. Rel. Eidos; Form; Shape.

PLACOID SENSILLUM A sense organ formed of a flat, plate-like external membrane covering an enlarged pore canal, the outer surface of which is continuous with the Integument. Placode Sensilla occur in aphids (as thin walled and probably serving in olfaction, in Coleoptera (also an olfactory receptor) Callahan (1975, 1977) contends Placodes are infrared receptors. PS is well represented in apocritous Hymenoptera where they are believed to function in olfaction. Morphogenesis of a Placode Sensillum consists of a plate-like membrane over a pore canal which is elliptical or oval. The Sensillum is usually flush with the Cuticle, but sometimes it is slightly elevated or depressed. Two types of plates are found: One type has thin plates and many pores (in females and males); the other type consists of thick plates and fewer pores (in females only). Syn. Sensillum Placodeum. See Sensillum. Cf. Multiporous Plate Sensillum.

PLAGA Noun. (Latin, *plagatus* = striped, wounded. Pl., Plagae.) A spot, stripe or streak of colour. A longitudinal spot of irregular form.

PLAGATE Adj. (Latin, *plagatus* = striped, wounded; *-atus* = adjectival suffix.) Marked with plagae.

PLAGUE Noun. (Latin, *plaga* = a wound. Pl., Plagues.) A disease, primarily of rodents, caused by the gram-negative bacterium *Yersinia pestis*. Bacterium facultatively anaerobic, non-motile and varies in shape from rod-like to coccobacilliform; bacteria's characteristics change with temperature and optimal growth at 28°C. Plague exists as Sylvatic Zoonosis in colonial burrowing rodents; Bacteriaemia of rodents varies in degree and duration. Plague is vectored to humans by several Species of fleas, primarily the Oriental Rat Flea and to a significantly lesser extent the Human Flea, Dog Flea, Cat Flea and Sticktight Flea. Plague is manifest in five forms: Bubonic, Pneumonic, Pestis Minor, Septicemic and Andean. Plague is endemic to southern Asia and has spread elsewhere in several pandemics with devastating results, most notably 165 AD, 542 AD (First Pandemic, *Y. pestis antiqua*), 14th century and 17th Century (Second Pandemic, *Y. pestis mediaevalis*) and 1892–present (Third Pandemic, *Y. pestis orientalis*.) Syn. Black Death. See: Andean Plague; Bubonic Plague; Pestis Minor; Pneumonic Plague; Septicemic Plague. Rel. Oriental Rat Flea.

PLAGUE LOCUST Any of up to 20 Species of spur-throated grasshoppers (Orthoptera: Acrididae: Cyrtacanthacridinae) whose populations occasionally increase to staggering densities and migrate considerable distances. Historically important in ancient times due to the damage inflicted to crops. *Schistocerca paranensis* (Burmeister) is most likely the plague locust of ancient Egypt. *Locusta migratoria migratorioides* L. and *Schistocerca gregaria* (Forskål) are also known from ancient times. In modern times, the plague of 1874–1877 in North America by the Rocky Mountain grasshopper, *Melanoplus mexicanus mexicanus* (Saussure), caused extensive damage and periodic plagues in Africa and Australia continue to be problematic. See Kirby & Spence for interesting historical accounts of locust plagues.

PLAGUE SOLDIER BEETLE *Chauliognathus lugubris* (Fabricius) [Coleoptera: Cantharidae].

PLAGUE THRIPS *Thrips imaginis* Bagnall [Thysanoptera: Thripidae]: A pest of numerous crops; endemic to Australia and reported in several nearby countries including Papua New Guinea, New Zealand, New Caledonia and Fiji. Female ca 1 mm long, elongate-cylindrical and grey-brown; head yellow and Thorax orange; Ocelli red. Females with four narrow wings with a long fringe of Setae along margins; Tarsi lack claws but end with a small bladder. Male pale yellow. Adult attracted to unopen flower buds. Adult female uses Ovipositor to saw into leaf tissue and deposit ca 200 eggs in slits. Eggs laid in leaves or stems; eggs white, kidney-shaped; egg stage 5–10 days. Eggs laid on flower parts and new leaves; nymphs feed on flowers; pistils and anthers wither; fruit malformed or fails to set. Nymphs resemble adult in appearance and habits. Nymphs lack wings and are paler than adult. Nymphs pass through four instars, two occur in soil and are non-feeding. All life stages occur on plant and overlap; Species undergoes propupal stage. Pupation in soil. Life cycle ca 10 days; multivoltine with about 12 generations per year.

PLAIN PUMPKIN BEETLE *Aulacophora abdominalis* (Fabricius) [Coleoptera: Chrysomelidae].

PLAINS FALSE-WIREWORM *Eleodes opacus* (Say) [Coleoptera: Tenebrionidae].

PLAIT Noun. (Latin, *plicare* = to fold. Pl., Plaits.) Longitudinal folds of the wing which are independent of flexion lines. Plaits are a derived feature of the forewing in several lineages of apocritious Hymenoptera. Cf. Wrapping Flexure, Wing Flexion Line.

PLANCHON, JULES EMILE (1823–1888) (Anon. 1889, Psyche 5: 156.)

PLANE Noun. (Latin, *planus* = flat.) Flat. Level. A surface without elevations, depressions or markings, specifically when no part of a structure is higher than another.

PLANE-OF-SYMMETRY The median plane. See Symmetry.

PLANIDIUM Noun. (Greek, *planos* = wandering; *-idion* = diminutive. Pl., Planidia.) 1. The hypermetamorphic (heterometamorphic), migratory, first-instar larva of some parasitic insects. Morphologically characterized by a legless condition and somewhat flattened body which often displays strongly sclerotized, imbricated integumental sclerites and spine-like locomotory processes. Term most appropriately restricted to Hymenoptera (Eucharitidae, Perilampidae, some Ichneumonidae) and Diptera (Tachinidae). Incorrectly used interchangeably with Triungulin. (Heraty & Darling 1984, Syst. Ent. 9: 309–318.) See Larva. Cf. Triungulin; Triungulinid.

PLANIPENNATE Adj. (Latin, *planus* = flat; *penna* =

feather; *-atus* = adjectival suffix.) Flat-winged.

PLANIPENNIA Noun. A Suborder of Neuroptera including Superfamilies Coniopterygoidea, Hemerobioidea, Ithonoidea, Mantispoidea, Myrmeleontiformia, Osmyleoidea. Wings large and laid flat on body when at repose. Cf. Megaloptera; Rhaphidoidea.

PLANKTON. Noun. (Greek, *planktos* = wandering.) Microscopic plants and animals suspended in the surface layers of open water or passively floating or drifting in these layers. Plankton constitute an important source of food for many aquatic organisms, including insects. Cf. Nekton; Neuston; Seston. Rel. Aeroplankton.

PLANKTONIC Adj. (Greek, *planktos* = wandering; *-ic* = of the nature of.) Passively floating, wandering or drifting in water; pertaining to plankton. Cf. Nektonic.

PLANOFIX® See NAA.

PLANT BUGS See Miridae.

PLANT GROWTH REGULATOR Chemicals produced by plants or synthesized in the laboratory which influence growth and development of plants. PGRs commonly used commercially include: Chlormequat, Chlorpropham, Daminozide, Ethephon, Gibberellic Acid, Indolebutyric Acid, Naa, Tecnazene. Cf. Insect Growth Regulator.

PLANT HOPPERS See Fulgoridae.

PLANT LICE See Aphididae.

PLANT PIN® See Butoxycarboxim.

PLANTA Noun. (Latin, *planta* = sole of foot. Pl., Plantae.) 1. The basal segment of the posterior Tarsus in pollen-gathering Hymenoptera. 2. The ventral surface of the posterior tarsal segments or anal clasping legs of the retractile lobe at the apex of the caterpillar abdominal Proleg. 3. The small chitinized sclerite on the ventral side of the middle of the Articularis at the distal end of the Calcanea (MacGillivray). Alt. Pelma. Syn. Sarothrum. See Pretarsus. Cf. Arolium; Parempodium. Rel. Leg.

PLANTA (REICHENEAU), ADOLPH (1820–1895) (Lorenz 1895, Jber. naturf. Ges. Graubünden 38: 88–102.)

PLANTAR Adj. (Latin, *planta* = sole of foot.) Descriptive of or pertaining to the Planta or ventral surface of the Tarsomeres.

PLANTAR LOBE Hymenoptera: A distal projection from Tarsomeres 1–4 in some Symphyta, Ichenumonoidea and Aculeata. See Plantella.

PLANTAR SURFACE 1. Of the tarsal segments of insects, that surface which is applied to the ground in walking. 2. In general, the lower surface of the Tarsus.

PLANTELLA Noun. A long media projection of the last tarsal segment (MacGillivray).

PLANTENGA, J R (–1958) (Anon. 1958, Ent. Ber. Amst. 18: 125.)

PLANTHOPPERS See Delphacidae.

PLANTIGRADE Adj. (Latin, *planta* = sole of foot; *grada* = step.) Walking on the Planta or ventral surface of the Pretarsus. A term applied to insect Species that walk on the entire ventral surface of the Tarsomeres, not exclusively on the claws.

PLANTULA Noun. (Latin, *plantula* = small sole of foot. Pl., Plantulae.) A lobe of the divided tarsal Pulvillus; a ventral surface or climbing cushion of the insect leg. See Arolium; Pulvillus.

PLAQUES Noun. (French, *placa* = shield.) Some Naucoridae: Small leathery Hemelytra.

PLASMA Noun. (Greek, *plasma* = a thing moulded; form. Pl., Plasmas.) The liquid part of Haemolymph in which ions, cells, organs and inclusions are bathed, suspended or transported. Plasma varies in colour depending upon the pigments and inclusions it contains. Alt. Plasm. See Blood. Cf. Haemolymph. Rel. Circulatory System.

PLASMATIC Adj. (Greek, *plasma* = a thing moulded; *-ic* = consisting of.) Of, pertaining to, or composed of blood plasma.

PLASMATOCYTE Noun. (Greek, *plasma* = a thing moulded; *kytos* = hollow vessel; container. Pl., Plasmatocytes.) A basic anatomical form of insect Haemocyte. Relatively large Haemocytes with irregular outline and basophilic cytoplasm; Plasma Membrane papillate or with irregular processes; nucleus large, round-oval centrally positioned; cytoplasm generally abundant; granular or agranular. Syn. Amoebocyte; Giant Fusiform Cell; Lamellocyte; Leucocyte; Lymphocyte; Micronucleocyte; Nematocyte; Phagocyte; Podocyte; Radiate Cell; Star-shaped Amoebocyte; Vermiform Cell. See Haemocyte. Cf. Adipohemocyte; Coagulocyte; Granulocyte; Nephrocyte; Oenocyte; Prohaemocyte; Spherulocyte.

PLASMID Noun. (Greek, *plasma* = a thing moulded; *-id* > *-is* = belonging to. Pl., Plasmids.) A loop of DNA that occurs in bacteria and yeasts, and which carries non-essential genes and replicates independent of the chromosomes.

PLASMODIIDAE Plural Noun. A Family of parasitic Protozoa assigned to Class Apicomplexa and which includes Species responsible for transmission of human Malaria.

PLASMODIUM Noun. The nominant Genus of Plasmodiidae whose Species infect mammals, birds and reptiles; schizogony occurs in blood and gametocytes develop in mature red blood cells. *Plasmodium* includes four Species which produce Malaria in humans and use Diptera (*e.g. Anopheles*, *Culex*, *Lutzomyia*) as vectors. See Malaria.

PLASMOGAMY Noun. (Greek, *plasma* = form, mould; *gamos* = marriage. Pl., Plasmogamies.) The process of combining the protoplasts. Cf. Karyogamy.

PLASMOLYSIS Noun. (Greek, *plasma* = form; mould; *lysis* = loosening; *sis* = a condition or state.) Contraction of the living cell through loss of water.

PLASON, VICTOR (1843–1904) (Anon. 1904,

Insektenbörse 21: 354.)

PLASTAZOTE A polyethylene, closed-cell foam material.

PLASTER BEETLE *Cartodere constricta* (Gyllenhal) [Coleoptera: Lathridiidae].

PLASTER BEETLES See Lathridiidae.

PLASTERER BEES See Colletinae.

PLASTIC Adj. (Greek, *plassein* = to form or mould; *-ic* = consisting of.) Formative, in an easily moulded condition. In animals, capable of change in characteristics, or in a changeable state.

PLASTICITY Noun. (Greek, *plassein* = to form or mould; English, *-ity* = suffix forming abstract nouns.) The capacity for being formed, moulded or developed.

PLASTOTYPE Noun. (Greek, *plastos* = formed; *typos* = pattern.) A plaster cast of a type-specimen; Plastotype concept used mainly in Palaeontology. See Type. Rel. Nomenclature; Taxonomy.

PLASTRAL Adj. (French, *plastron* = breast plate; Latin, *-alis* = pertaining to.) Descriptive of structure or function associated with a Plastron. See Plastron.

PLASTRON Noun. (French, *plastron* = breast plate. Pl., Plastrons.) Respiration in an aquatic environment or a liquid medium: A mechanism by which a film of air is maintained in place on the body (egg, appendage) by Setae ('hairs') or cuticular modifications. 'Hair Plastrons' are characterized by Setae that are regionally more dense and uniform in distribution than other Setae on the body; Setae forming a Plastron are also of characteristic shapes, lengths and arrangements. The surface film of air is held in place by a series of water-repellent Setae or cuticular modifications that create an environment of hydrophobicity. A Plastron must resist wetting at customary hydrostatic pressure to serve as an efficient respiratory structure; this is achieved by cuticular geometry or Chaetotaxy. Plastrons frequently occur on the body of aquatic insects; Setae of the Plastron are long-and-sparse (a Macroplastron) and short-and-dense (a Microplastron). Plastrons also occur on the eggshell of some Species of parasitic and aquatic insects. Term 'Plastron' was first proposed by Brocher (1912, Ann. Biol. Lacust. 5: 5–26.) for chrysomelid and elmid beetles that respire with a gas film. Syn. Gas Gill; Physical Gill. See Respiratory System. Cf. Respiratory Horn. Rel. Aeropyle.

PLATASPIDAE Plural Noun. Stink Bugs. A small Family of heteropterous Hemiptera assigned to Superfamily Pentatomoidea; Species occur in tropical Africa, Asia and Australia. Head and Pronotum not laminately expanded; Hemelytra longer than Abdomen; Scutellum very large. Syn. Brachyplatidae; Coptosomatidae.

PLATE Noun. (Greek, *platys* = flat. Pl., Plates.) 1. Any broad, flat surface, such as a sclerite. 2. Coccids: A thin projection of the Pygidium (Comstock). By some writers restricted to such as are spine-like in form, also termed Squama. Male Homoptera. Either of a pair of pieces following the last full ventral segment, which are usually preceded by a shorter piece; the valve (Smith).

PLATE ORGAN See Sensillum Placodeum.

PLATE TECTONICS The movement of discrete regions of the earth's crust in relation to one another. Plates may break apart or fuse to form larger plates. PT theory accounts for most volcanoes and earthquakes along curved parts of seafloor and mountain ranges along the margins of continents. See Continental Drift. Rel. Biogeography.

PLATE THIGH BEETLES See Eucinetidae.

PLATEAU, FELIX (1841–1911) (Lambillon 1911, Revue Soc. ent. Namuroise 11: 26–27.)

PLATELET Noun. (Greek, *platys* = flat; *-et* = dim. form.) A small chitinous sclerite embedded in a membrane.

PLATFORM SPIDERS *Corasoides* spp. [Araneida: Agelenidae].

PLATH, OTTO EMIL (1885–1940) (Osborn 1946, *Fragments of Entomological History,* Pt. II. 232 pp. (106) Columbus, Ohio.)

PLATING EFFICIENCY Tissue Culture: The capacity of a cell population to form colonies. PE is expressed as the percentage of cells inoculated into a culture vessel that gives rise to discrete colonies.

PLATO (429–347 BCE) Greek Philosopher and member of the Teleological School.

PLATONOFF, STEPHAN (1917–1944) (Lindberg 1944, Notul. ent. 24: 105–106.)

PLATT, ERNEST E (1874–1966) (Janse 1969, J. Lepid. Soc. 23: 135–136.)

PLATT-BARRETT, J (–1916) (Moore 1917, Entomologist's Rec. J. Var. 29: 43–44.)

PLATYCNEMIDIDAE Plural Noun. A Family of coenagrionoid Odonata. Adults typically small with slender Abdomen; wings wider than related Families; Pterostigma short, rhomboidal; longitudinal veins straight; Cubital and Anal Veins well developed. Male Tibiae frequently enlarged and foliacious. Male superior appendages not forcipate. Larval form variable, Antenna with seven segments, third typically longest; legs long; three Caudal Gills present.

PLATYCRANINAE A Subfamily of Phasmatidae (Phasmatodea) occurring mainly on the islands north of Australia. Only a single Species occurs in Australia, *Megacrania batesii*, on *Pandanus*.

PLATYGASTERIDAE Westwood 1840. Plural Noun. A cosmopolitan Family (ca 300 nominal Species) of apocritous Hymenoptera assigned to Proctotrupoidea. Small to minute, dark coloured body; head without a protuberance or ledge; Antenna with 10 segments, geniculate, distinct or large club; forewing venation absent or reduced to knob; hindwing without an Anal Lobe; Gaster moderately sclerotized. Species endoparasitic, usually univoltine, gregarious; predominantly parasitic on Diptera; most attack gall forming Cecidomyiidae, some mealybugs, whiteflies. Oviposit in egg or early larva, develop in mature host larva; site-specific oviposition (nerve cord, gut, Brain); only one Species polyembryonic, but widely reported; some hymenopteriform; many Species hypermetamorphic. Alt. Platygastridae.

PLATYPEZIDAE Plural Noun. Flat-Footed Flies. A numerically small, geographically widespread Family of acalypterate Diptera assigned to Superfamily Platypezoidea. Body small (< 10 mm long), dark; Frontal Suture absent; Arista terminal; Hind Tarsus with flattened and expanded Tarsomeres; wing with Anal Cell closed and pointed apically. Larvae ovoid, flattened with projections; inhabit fungi. Adults inhabit low vegetation in damp woods; walk or run with jerky motion. Males sometimes swarm above the ground; females enter swarms. Syn. Clythiidae.

PLATYPODINE BORERS Ambrosia beetles with pale to dark brown adult; body small (ca 2–5 mm long), cylindrical and elongate. Larva creamy white, slightly curved with humped Thorax. PB attack many timber Species in forest and mill yard; adult strongly attracted to freshly sawn trees, logging residues, damaged or stressed standing trees. Attack on freshly cut surfaces within minutes of tree felling. Male usually initiates attack, later assisted by female. Female mates, bores tunnel and male removes bore dust. Fungal spores carried into tunnel on body Setae, in Mycangia or in gut; spores germinate, fungal hyphae grow on tunnel walls, provide food for PB adults and larvae. Tunnels extend deep into log, branched or not branched depending on borer Species. Mature larvae usually construct short side tunnels in a herring-bone arrangement and pupate. Neonate adults emerge through parent tunnels. Life cycle varies with Species and climate ca 1–12 months. See Ambrosia Beetle.

PLATYPTERA Noun. Flat-winged and broad-winged. An archaic ordinal term for the Psocoptera, Termitidae, Plecoptera and Mallophaga.

PLATYPUS BEETLE *Platypus subgranosus* Schedl [Coleoptera: Curculionidae].

PLATYPUS TICK *Ixodes ornithorhynchi* Lucas [Acari: Ixodidae].

PLATYSMATIUM Noun. (Greek, *platys* = flat; Latin, *mata* = mat; Latin, *-ium* = diminutive > Greek, *-idion*.) Acarology: the fused Mentum, Labium and Malapophyses as observed in many Anactinotrichida.

PLATYSTICTIDAE Plural Noun. A Family of coenagrionoid Odonata. Adults small to medium sized; Pterostigma small, thick; longitudinal veins straight, crossveins regular; veins R3 and R4 originate near nodal level, Anal Vein short or absent; Abdomen slender. Larva slender with long, thin legs; mask short (quadrate), flat, asetose, apically incised; three Caudal Gills triangular in cross section.

PLATYSTOMATIDAE Plural Noun. Picture-winged Flies. Cosmopolitan Family of acalypterate Diptera consisting of ca 1,000 nominal Species and placed in Superfamily Tephritoidea. Best represented in tropical regions. Costa typically broken beyond Humeral Vein, but not at end of Subcosta; female with vestigial sixth abdominal segment. Adults feed on fresh mammalian excrement; larvae taken in beetle tunnels, on damaged tree trunks, feeding on Orthoptera eggs, in bacterial root nodules. Males of some Species with stalk-eyes, territorial.

PLATYSTOMIDAE Plural Noun. See Anthribidae.

PLAUMANN, FRIEDRICH (1902–1994) Born Eylau, Prussia; died Seara, Nova Teutonia, Brazil. Well known entomological collector of insects in Brazil for many Entomologists. Personal collection, housed in Fritz Plaumann Entomological Museum, consists of 80,000 specimens and 17,000 Species. See 1991, Revta brasil. Entomol 35: 474–478.

PLAVICH'SHCHIKOV, NIKOLAY NIKOLAEVICH (1892–1962) (Kryzhanovsky 1962, Ent. Obozr. 41: 692–695, bibliogr. (Transl. Ent. Rev., Wash. 41: 429–433.))

PLEASING FUNGUS-BEETLES See Erotilidae.

PLECEPHALIC Adj. Insects with a prognathous head orientation, but with the Occipital Foramen in ventral aspect and Alaforamen extending from its ventral to its caudal aspect (MacGillivray).

PLECIODICTYIDAE Rohdendorf. Plural Noun. A monotypic Family of fossil Diptera known from wing fragments of Lower Jurassic Period deposits preserved in Kirghizistan.

PLECIOFUNGIVORIDAE Kovalev 1985. Plural Noun. A Family of ca 60 Species of fossil Diptera known from several Mesozoic deposits. Syn. Fungivoritidae.

PLECOPTERA Burmeister 1839. (Plectoptera.) (Greek, *plekein* = to fold; *pteron* = wing.) (Permian-Recent.) Stoneflies; Plaited Winged Insects. A cosmopolitan Order of net-veined insects subdivided into two Suborders, 14 Families and ca 2,000 nominal Species. Term Plecoptera used by Brauer for Perlidae; Plectoptera by Packard for Ephemeridae; some confusion since, and both have been used *sensu* Brauer. Most Genera found in Holarctic, probably where Order radiated or numerically expanded. Nymphs of North American Species with respiratory gills located on the Thorax and at the base of the legs. Australian Species with respiratory gills located on mouthparts, Cervix, Thorax, Abdomen or within Anus and frequently retained in adult. Adult moderate sized, dorsoventrally compressed; body loosely jointed; typically neuropteroid habitus. Head prognathous, sessile; compound eye well developed; 2–3 Ocelli; Antennae long, slender, many segmented; Mandible usually well developed sometimes vestigial; Maxilla complete, Palpus with five segments;

Labium complete, Palpus with three segments. Prothorax large, mobile; Mesothorax and Metathorax smaller, subequal; wings membranous with numerous veins and crossveins; hindwings longitudinally folded beneath forewings; few Species brachypterous or apterous (usually males); legs with three Tarsomeres. Abdomen with 10 segments, vestiges of segments 11 and 12 sometimes present; spiracles on segments 1–8; Cerci of variable size; genitalia on one of last three segments. Metamorphosis incomplete. Differ from orthopteroids in aquatic nymphal stage, generally soft Integument, similarity of fore and hindwing and conformation of mouthparts. Most Species diurnal as adults; adults taken near water, immatures taken in water. A few Species found in Southern Hemisphere apparently exist in damp terrestrial situations. Plecoptera naiads typically prefer clean, cold, moving freshwater; narrow tolerances to pollution, water temperature and related parameters. Life cycle of most Species one year; some Species live longer; other Species apparently modify cycle in response to changing environmental conditions. Adults short lived; males die soon after copulation; females die soon after egg laying. Evidence suggests females must feed as adults for eggs to develop. Females deposit eggs on surface of water during flight when apex of Abdomen touches surface. Some Species broadcast or deposit eggs beneath surface of water or upon submergent objects. Egg small with an adhesive coating activated by water. Females may lay more than one clutch of eggs; total egg production may reach 1,000. Included Families: Austroperlidae, Capniidae, Chloroperlidae, Diamphipnoidae, Eustheniidae, Gripopterygidae, Leuctridae, Nemouridae, Notonemouridae, Palaeoperlidae, Peltoperlidae, Perlidae, Perlodidae, Perlopsidae, Pteronarcyidae, Scopuridae and Taeniopterygidae. See Paraplecoptera.

PLECTRUM Noun. (Latin, *plectrum* = instrument with which to strike. Pl., Plectra; Plectrums.) 1. A component of a stridulatory device; specifically a small, thin structure that is rubbed against a membrane. 2. Diptera: A strong marginal bristle standing out from the middle of the Costa (Kirby & Spence). See Stridulation. Cf. File; Rasp; Strigil; Pars Stridens. Rel. Sound.

PLEE, AUGUSTE (1787–1825) (Papavero 1971, *Essays on the History of Neotropical Dipterology*. 1: 127–128. São Paulo.)

PLEGMA Noun. (Greek, plaited. Pl., Plegmata.) Scarabaeoid larvae: A single fold pertaining to the plegmatium and proplegmatium (Boving).

PLEGMATIUM Noun. (Greek, *plegma* = plaited; Latin, *-ium* = diminutive > Greek, *-idion*. Pl., Plegmatia.) Scarabaeoid larvae: A lateral paired space with a plicate, somewhat sclerotized surface, bordered by marginal spines with Acanthoparia, with one Plegma inside of each spine; lateral striae of Hayes (Boving).

PLEID WATER BUGS See Pleidae.

PLEIDAE Fieber 1853. Plural Noun. Pleid Water Bugs; Pygmy Backswimmers; Minute Backswimmers. A small Family of heteropterous Hemiptera. Adult small bodied, stongly convex, stout, yellowish to gray in coloration; Ocelli absent; Labium of three segments; forewing membrane lacking veins; hind Tibia and Tarsus lack long fringe of Setae; Abdomen venter lacking median keel. Individuals occur beneath surface of still water of ponds or between plants in protected areas of larger bodies of water. Pleids are predaceous upon small Crustacea; female inserts eggs into plant tissue.

PLEIOMORPHOUS See Pleomorphic.

PLEIOXENOUS Adj. (Greek, *pleion* = more; *xenos* = host, stranger; Latin, *-osus* = with the property of.) A parasitic Species restricted to one Family of hosts. (Jameson 1985, 19: 862.) See Feeding Strategy. Cf. Monoxenous; Polyxenous. Rel. Parasitism.

PLEISIOTYPIC Adj. (Greek, *plesios* = close, near; *typos* = type; *-ic* = characterized by.) 1. Pertaining to a specimen identified as a Species by someone other than the author of that Species. 2. A specimen that is a Homoeotype and Hypotype. See Type; Homoeotype.

PLEISOMORPHY Noun. (Greek, *plesios* = near; *morphe* = form. Pl., Plesiomorphies.) In cladistical theory of classification, a primitive or ancestral character state. Cf. Apomorphy. See Symplesiomorphy.

PLEISTOCENE EPOCH The interval of the Geological Time Scale (1.65 MYBP–10,000 YBP) which is the third Epoch within the Neogene Period. PE was characterized by Charles Lyell (1833). See Geological Time Scale; Neogene Period. Cf. Miocene Epoch; Pliocene Epoch; Recent Epoch. Rel. Fossil.

PLENA Noun. Coccids: The thickened proximal part of the Rostral Furrow against which the end of the Rostralis rests (MacGillivray).

PLENARY POWERS Special powers granted by the International Congress of Zoology to the International Commission on Zoological Nomenclature permitting the suspension of the International Rules of Zoological Nomeclature or decision as to how they shall apply in specific cases. [Bull. Zool. Nomencl. 4: 51–56].

PLEOMETROSIS Noun. (Greek, *pleion* = more; *metre* = mother; *sis* = a condition or state. Pl., Pleometroses.) Social Insects: Foundation of a colony by more than one queen. Cf. Monometrosis.

PLEOMETROTIC Adj. (Greek, *pleion* = more; *metre* = mother; *-ic* = consisting of.) Social Insects: Pertaining to colonies which have more than one functional queen in a nest. Cf. Monogynous.

PLEOMORPHIC Adj. (Greek, *pleion* = more; *morphe* = form; *-ic* = consisting of.) Pertaining to pleomorphism; descriptive of organisms that experience pleomorphism. Syn. Polymorphic;

Pleiomorphous. Cf. Adelomorphic; Monomorphic. Rel. Form; Shape.

PLEOMORPHISM Noun. (Greek, *pleion* = more; *morphe* = form, shape; English, *-ism* = condition. Pl., Pleomorphisms.) 1. Structures or bodies characterized by having many forms. 2. The occurrence of more than one distinct body or structural form during the life cycle of a plant, animal or microorganism. 3. The body of an individual which can assume or display several distinct shapes. Syn. Polymorphism.

PLEON Noun. (Greek, *plein* = to swim. Pl., Pleons.) See Abdomen.

PLEOPOD Noun. (Greek, *plein* = to swim; *pous* = foot. Pl., Pleopods; Pleopoda.) 1. An abdominal leg of larvae. 2. Posterior leg in an adult aquatic insect.

PLERERGATE Formicidae: See Pleurergate; Replete.

PLESIOBIOSIS Noun. (Greek, *plesios* = near; *bios* = life; *sis* = a condition or state.) Social Insects: A close proximity of two or more nests with casual or no contact between the colonies. A primitive form of association approaching symbiosis; neighbourliness. See Biosis; Commensalism; Parasitism; Symbiosis. Cf. Abiosis; Anhydrobiosis; Antibiosis; Archebiosis; Calobiosis; Cleptobiosis; Hamabiosis; Kleptobiosis; Lestobiosis; Parabiosis; Phylacobiosis; Synclerobiosis; Trophobiosis; Xenobiosis. Rel. Accidental Compound Nests; Double Nests.

PLESIOMORPHIC Adj. (Greek, *plesios* = near; *morphe* = form; *-ic* = of the nature of.) 1. Pertaining to structures with similar form. 2. Pertaining to a primitive condition of an anatomical feature or biological attribute. Alt. Plesiomorphous

PLESIOTYPE Noun. (Greek, *plesios* = near; *typos* = pattern. Pl., Plesiotypes.) 1. Nomenclature: The specimen upon which a subsequent or additional description or figure is based (Banks & Caudell). 2. Any specimen identified with a described or named Species by a person other than the describer (Smith). Cf. Hypotype.

PLEURA Noun. (Greek, *pleura* = side. Sl., Pleuron; Pleurum.) 1. Lateral sclerites between the dorsal and sternal parts of the Thorax. 2. In general, the body between the dorsal Tergum and ventral Sternum. Pleuron is the current accepted form, alone or in combination (*e.g.*, Metapleuron; Hypopleuron).

PLEURADEMA Noun. (Greek, *pleura* = side; *dema* = body. Pl., Pleurademae.) A lateral Apodeme of the Thorax (MacGillivray).

PLEURADEMINA Noun. (Greek, *pleura* = side; *dema* = body.) A pit or thickening of the outer surface at the point of invagination of a Pleuradema (MacGillivray).

PLEURAL Adj. (Greek, *pleura* = side; Latin, *-alis* = pertaining to.) Descriptive of structural or process associated with the lateral surfaces of body segments or Pleurites. See Pleurite. Cf. Dorsal; Lateral; Ventral.

PLEURAL ANOJUGALIS See Plica Jugalis.

PLEURAL APOPHYSIS Internal arm of the Pleural Ridge (Snodgrass).

PLEURAL ARCH An internal cuticular ridge (apodeme) of some insects, including fleas. Fleas have lost the ability to fly, but they jump effectively with an estimated accelerative force of ca 140 g. This feat is achieved with the aid of a complex interaction between energy stored in the metathoracic Pleural Arch and the hind leg. (Foreleg and middle leg do not participate in the jump.) Morphologists hypothesize that the Pleural Arch of modern fleas represents a wing-hinge ligament that was modified when the body became laterally compressed and the wing was lost during the course of evolution. The Pleural Arch contains Resilin. When the hindleg is relaxed, the Resilin is uncompressed. Energy is stored by the compression of Resilin in the Pleural Arch through flexure of the the hindleg and Thorax by muscle contraction. Important muscles involved in the process include the depressor of the Trochanter and muscles previously associated with flight in the winged ancestor (Epipleural and Indirect Flight Muscles). By maintaining hind legs in a flexed position, a flea is capable of storing energy for months. The energy stored in the Pleural Arch can be instantaneously released in the jump. Muscles cannot directly hold the Resilin in a compressed condition for long periods, so a series of catch mechanisms have been developed. These include a link plate, thoracic catch, trochanteral hooks and coxo-abdominal catch Working together, these catches hold the thoracic segments firmly together and bond Abdomen to hind Coxa. See Locomotion. Cf. Jump; Hop. Rel. Siphonaptera.

PLEURAL AREAS 1. Hymenoptera: Three spaces on the Metanotum between Lateral and Pleural Carinae. 2. The metameric divisions of the pleural region: First (anterior) is spiracular area; second (middle) is middle pleural area; third (posterior) is angular area. See Pleuron

PLEURAL BASALIS The basal fold of the wing, or line of flexion between the base of the Mediocubital Field and the Axillary Region. The PB forms a prominent convex fold in the flexed wing that extends between the median sclerites from the articulation of Radius with the second Axillary Sclerite to the articulation of the Vannal Veins with the 3rd Axillary Sclerite (Snodgrass).

PLEURAL CARINAE Hymenoptera: Ridges along the exterior (lateral) margin of the Metanotum.

PLEURAL COXAL PROCESS The Condyle at the lower end of the pleural ridge to which the Coxa is articulated (Snodgrass).

PLEURAL FURROW The Pleural Suture when deep and wide (MacGillivray).

PLEURAL JUGALIS The jugal fold of the wing of some insects, or radial line of folding setting off the jugal region from the vannal region. Alt. Axillary Furrow; Plica Anojugalis (Snodgrass).

PLEURAL MUSCLES See Paradorsal Muscles.

PLEURAL PIECES The lateral sclerites of the Thorax. See Pleura.

PLEURAL REGION The part of the body on which legs and wings are attached. Metamerically divided into segmental pleural areas (Snodgrass).

PLEURAL RIDGE The Endopleural Ridge formed by the Pleural Suture, bracing the Pleuron above the leg, or between the coxal articulation and the wing support (Snodgrass). The single Apodeme of the typical wing-bearing segment of most insects, on each side, terminating in the wing-process above and the coxal process below (Imms). (Syn. Entopleuron) (MacGillivray).

PLEURAL SUTURE A thoracic suture in the pleural Integument extending from the base of the wing to the base of the Coxa.

PLEURAL VANNALIS The Vannal Fold of the wing, or radial line of folding usually between the Cubital Field and the first Vannal Vein, but somewhat variable in position, Anal Furrow, Plica Analis (Snodgrass).

PLEURAL WING PROCESS Produced dorsal margin of the Pleuron, at the upper end of the Pleural Ridge, which serves as a fulcrum for the movement of the wing.

PLEURAL WING RECESS Mayflies: A deep, cup-shaped thick-walled cavity in the wing base, opening downward (Needham).

PLEURANOTUM Noun. (Greek, *pleura* = side; *noton* = back. Pl., Pleuranota.) 1. The area along each lateral aspect of the Pronotum (MacGillivray). 2. The area of the Mesoscutum between the Mesepisterna and Mesocoria (MacGillivray).

PLEURELLA Noun. (Greek, *pleura* = side; The 3rd sclerite of the Pleuron.

PLEURERGATE Noun. (Greek, *pleura* = side; *ergon* = work; *-ate*, to become.) A worker ant capable of distending its Gaster with liquid food until the Gaster becomes a spherical sac.

PLEURITE Noun. (Greek, *pleura* = side; *-ites* = constituent. Pl., Pleurites.) 1. Any of the numerous small sclerites that comprise the pleural area of a body segment. Typically, Pleurites are secondarily developed through fragmentation of a segment or part of a segment (Dorsum, Pleuron, Sternum.) 2. The soft, membranous, lateral part of an abdominal segment (Tillyard). See Segment. Cf. Tergite; Sternite.

PLEURITOCAVAE (See Tjeder 1979, Ent. Scand. 10: 109–111.)

PLEUROCOXAL Adj. (Greek, *pleuron* = the side. *coxa* = hip; Latin, *-alis* = pertaining to.) Descriptive of or pertaining to the Pleuron and Coxa together.

PLEURON Noun. (Greek, *pleuron* = side. Pl., Pleura.) 1. 'In the Thorax of pterygote insects the subcoxal sclerotization above, before and behind the Coxa' (Snodgrass). 2. The lateral region of any segment of the insect body, commonly of the thoracic segments; a Pleuropodite. 3. The term Pleuron was used by Audouin (1824) to describe the lateral portion of the body wall. 4. Groundplan: Sclerotized pleural region of Thorax in Apterygota, immature Plecoptera and so-called primitive Pterygota consist of an Anapleurite (dorsal-most), a Katapleurite (Coxopleurite between Coxa and Anapleurite), and a Sternopleurite (Coxosternite beneath Coxa). A Precoxal Suture separates Anapleurite and Katapleurite, but this suture is not well developed or recognizable in many insects. Pleural regions of Thorax in Pterygota (more highly evolved insects) consist of several sclerites and sutures. A Pleural Wing Process is positioned immediately beneath the wing base; PWP forms a fulcrum upon which base of wing rests. Several sclerites (Epipleurites) are positioned in membrane above Pleuron and near wing base. Anterior Epipleurite (Basalare) is located beneath the wing base anteriad of PWP. Posterior sclerite (Subalare) is located beneath the base of wing and posterior PWP. Typically one Basalare and one Subalare, but occasionally a second sclerite of each exists. Muscles are attached to Basalare and Subalare that are important in wing movement. Tegula is a small sclerite or scale-like flap at the base of the forewing of many insects. Pterothoracic Pleuron is the most complicated part of pleural region. Adults, nymphs and active larvae of insects all display extensive sclerotization of pleural area. In adult winged insects, Pteropleuron plays an important role in flight. In all insects, leg functions also influence pleural-wall development. Presumably, sclerites that form Pleuron are derived from the so-called Trochantin (= Subcoxa, Precoxa, Supracoxal Arch). Trochantin is a small sclerite at base of insect leg and is believed to develop into the pleural wall of the Thorax. Trochantin is typical in primitive pterygotes, often fused to Episternum or it is lost in higher orders. Cf. Dorsum; Venter. Rel. Pterothorax.

PLEUROPOD Noun. (Greek, *pleuron* = the side; *podos* = foot. Pl., Pleuropodia.) Evaginations of Ectoderm in the insect embryo.

PLEUROPODITE Noun. (Greek, *pleuron* = the side; *podos* = foot.) The Pleuron *sensu* Torre Bueno.

PLEUROSTERNAL Adj. (Greek, *pleuron* = the side; *sternon* = chest; *-alis* = pertaining to.) Descriptive of structure or function associated with the Pleurosternum.

PLEUROSTERNITE Noun. (Greek, *pleuron* = the side; *sternon* = chest; *-ites* = constituent. Pl., Pleurosternites.) See Laterosternite.

PLEUROSTERNUM Noun. (Greek, *pleuron* = the side; *sternon* = chest.) See Coxosternum.

PLEUROSTICT (Greek, *pleuron* = the side; *stiktos* = pricked.) Having the abdominal spiracles positioned on the dorsal portion of the ventral sclerites, as in lamellicorn beetles.

PLEUROSTOMA Noun. (Greek, *pleuron* = the side; *stoma* = mouth.) The subgenal margin of the cranium bordering the Mandible (Snodgrass).

PLEUROSTOMAL Adj. (Greek, *pleuron* = the side; *stoma* = mouth; *-alis* = pertaining to.) Descriptive of structure or function associated with the Pleurostoma.

PLEUROSTOMAL SUTURE Part of the Subgenal Suture above the Mandible (Snodgrass).

PLEUROTAXIS Noun. (Greek, *pleuron* = the side; *taxis* = arrangement.) The layer of the Cuticle connecting the ventral surface of the wing and the Pleuron (MacGillivray).

PLEUROTERGITE Noun. (Greek, *pleuron* = the side; *tergum* = back; *-ites* = constituent. Pl., Pleurotergites.) 1. Diptera: The Hypopleura; morphologically, lateral division of the Metanotum (Postscutellum), at least in Nematocera (Curran). 2. In general, lateral area of the Postscutellum (Crampton).

PLEUROTROCHANTIN Noun. (Greek, *pleuron* = the side; *trochanter* = runner.) A combination of sclerites in which the true Trochantin has become fused with the Episternum (Crampton).

PLEUROVENTRAL LINE A line of separation between the pleural region and the venter, positioned mesad of the limb bases, but obscured when the latter are fused with the Sterna (Snodgrass).

PLEURUM Noun. (Greek, *pleuron* = the side. Pl., pleura.) An infrequently used form of Pleuron.

PLEXUS Noun. (Latin, *plexus* = interwoven.) A knot; a complicated, network of vessels, nerves, fibres or Tracheae.

PLICA Noun. (Latin, *plicare* = to fold. Pl., Plicae.) 1. General: A fold, convolution or wrinkle in a structure or surface. 2. A longitudinal plait of a wing (Plica Vannalis). 3. The lamellate, infolded thickening of the anterior and posterior margins of the abdominal segments.

PLICAFORM Adj. (Latin, *plicare* = to fold; *forma* = shape.) Folded; resembling a fold; pertaining to structure which is adapted to fold. *e.g.* Orthoptera wings. See Plicate. Rel. Form 2; Shape 2; Outline Shape.

PLICATE Adj. (Latin, *plicare* = to fold; *-atus* = adjectival suffix.) Folded; descriptive of structure with folds. A sculptural pattern impressed with striae to produce the appearance of having been folded or pleated. Alt. Plicatus; Plicaform. Rel. Form 2; Shape 2; Outline Shape.

PLICATION Noun. (Latin, *plicare* = to fold; English, *-tion* = result of an action. Pl., Plications.) 1. A fold or convolution. 2. Orthoptera: The folds on the hindwing.

PLICIPENNA Noun. The Trichoptera *sensu* Latreille.

PLICTRAN® See Cyhexatin.

PLIENINGER, THEODOR WILHELM HEINRICH (1795–1879) (Anon. 1879, Leopoldina 15: 165–167.)

PLINY (Gaius Plinius Caecilius Secundus) (62–113) Also called Pliny the Younger. Roman author and nephew of Pliny the Elder; gave an account of the Elder's life and provided an account of honey bees.

PLINY (Gaius Plinius Secundus) (23–79) Also called Pliny the Elder. Roman author, best known for his *Historia Naturalis* (37 books, published ca 77) which embraces all aspects of natural history, including insects (11th book). Pliny adopted the classification of Aristotle and biology of honey bees.

PLIOCENE EPOCH The interval of the Geological Time Scale (5 MYBP–1.65 MYBP) which is the second Epoch within the Neogene Period. PE was characterized by Charles Lyell (1833). See Geological Time Scale; Neogene Period. Cf. Miocene Epoch; Pleistocene Epoch; Recent Epoch. Rel. Fossil.

PLOT, ROBERT (1640–1696) (Rose 1850, *New General Biographical Dictionary* 11: 162.)

PLOTNIKOV, VASILLIYEVICH (1877–1950?) (Pavlovsky 1950, Ent. Obozr. 31: 297–300, bibliogr.)

PLÖTZ, CARL (1813–1886) (Musgrave 1932, *A Bibliography of Australian Entomology 1775–1930,* viii + 380 pp. (257), Sydney.)

PLUCHE, NOEL ANTOINE (1688–1761) (Wheeler 1931, *Demons of the Dust,* 378 pp. (7–15), London.)

PLUCK® See Ethephon.

PLUM CURCULIO *Conotrachelus nenuphar* (Herbst) [Coleoptera: Curculionidae]: A pest of stone and pome fruits, including plum, peach, pear, apple, cherry and blueberry in eastern North America. PC is a vector of Brown Rot and endemic to North America east of Rocky Mountains north of latitude 28°N. PC completes 1–2 generations per year; overwinters as an adult in litter and causes round feeding punctures in skin of fruit. Female cuts crescent-shaped flap in skin of fruit and oviposits beneath flap; female can lay 500 eggs, typically less than 100; eclosion occurs within 2–12 days. Larva feeds in fruit 14–21 days, emerges and pupates in soil. Infested apple fruit typically falls to ground before PC completes larval development; cherries remain on tree and complete development. PC is the object of quarantine in most countries of world with Methyl Bromide fumigation used as primary disinfestation technique. See Curculionidae. Cf. Apple Curculio; Quince Curculio. Rel. Brown Rot.

PLUM FRUIT MAGGOT *Cydia funebrana* (Treitschke) [Lepidoptera: Tortricidae]: A pest of plums in Europe and Asia.

PLUM GOUGER *Coccotorus scutellaris* (LeConte) [Coleoptera: Curculionidae].

PLUM LEAF MITE *Phyllocptes abaenus* Keifer [Acari: Eriophyidae].

PLUM LEAFHOPPER *Macropsis trimaculata* (Fitch) [Hemiptera: Cicadellidae].

PLUM RUST-MITE *Aculus fockeui* (Nalepa & Trouessart) [Acari: Eriophyidae].

PLUM WEB-SPINNING SAWFLY *Neurotoma inconspicua* (Norton) [Hymenoptera: Pamphiliidae].

PLUM WILT A fungal disease of plums caused by

Lasiodiplodia triflorae and vectored by the Peach-Tree Borer.

PLUMARIIDAE Bischoff 1920 (Andre 1913). Plural Noun. A small Family of apocritous Hymenoptera assigned to Chrysidoidea; placed in Scolioidea by some workers. Rarely collected; known from males taken in Greece, South America and southern Africa. Plumariids adapted to xeric conditions, males taken at night; females apterous with profound changes of Thorax associated with long-term permanent aptery. Male closed cells of hindwing, development of Jugal Lobe, shape of apical Tergum of Metasoma, genitalia and lack of sexual dimorphism in antennal structure suggest plumariids probably an ancient stock of chrysidoids which split early from bethylids.

PLUMATE Adj. (Latin, *pluma* = feather; *-atus* = adjectival suffix.) Feather-like; feathered. Alt. Plumatus. Syn. Plumiform.

PLUMBAGO BLUE *Syntarucus plinius pseudocassius* (Murray) [Lepidoptera: Lycaenidae].

PLUMBATENDONS Noun. Hemiptera: The expanded arms of the Plumblis (MacGillivray).

PLUMBEOUS (Latin, *plumbum* = lead; *-osus* = with the property of.) Lead-coloured, the blue-grey of lead (Kirby & Spence). Alt. Plumbeus.

PLUMBILIS Noun. Hemiptera: A plunger-like cuticular rod projecting from the dorsal end of the Infunda (MacGillivray).

PLUME MOTHS See Pterophoridae.

PLUMERIA BORER *Lagocheirus undatus* (Voet) [Coleoptera: Cerambycidae].

PLUMERIA WHITEFLY *Paraleyrodes perseae* (Quaintance) [Hemiptera: Aleyrodidae].

PLUMILIFORM Adj. (Latin, *pluma* = feather; *forma* = shape.) Plume-shaped; descriptive of structure shaped as a feather or plume. Syn. Plumate; Plumose. See Plume. Cf. Aristiform. Rel. Form 2; Shape 2; Outline Shape.

PLUMMER, CHARLES CARTON (1907–1952) (Torres 1953, J. Econ. Ent. 46: 1126–1127)

PLUMOSE Adj. (Latin, *pluma* = feather; *-osus* = full of.) 1. Feather-like; pertaining to Setae multiply branched or divided. Feathered as a plume. 2. Pertaining to Antennae with long ciliated processes on each side of each segment. Alt. Plumosus. See Cirrate. Cf. Aristate; Cirrose; Pennate. Rel. Form 2; Shape 2; Outline Shape.

PLUMOSE HAIRS 1. Feather-like Setae furnished with thread-like branches, as on the body of bees. 2. Fine, laterally branched Chaetae. 3. Branched or feather-like unicellular processes of the body wall (Snodgrass).

PLUMOSE SCALE *Morganella longispina* (Morgan) [Hemiptera: Diaspididae].

PLUMP Adj. (Middle Low German, *plump* = heavy, clumsy.) Well-rounded. Pertaining to structure or body with full, rounded outlines; enlarged but not obese.

PLUMULE Noun. (Latin, *plumula* = small feather. Pl., Plumules.) Specialized scales of the Androconia of male Lepidoptera.

PLUMULOSE Adj. (Latin, *plumula*, diminutive of *pluma* = feather; *-osus* = full of.) 1. Down-like feather. 2. Pertaining to Setae which branch like feathers.

PLURALIFERA Noun. (Greek, *pleura* = side; *pherein* = to bear, carry.) 1. A pillar-like projection of the Episternum and Epimeron adjacent to the Mesopleural and Metapleural Sutures against which the Pteraliae articulate. 2. Pleural Wing Process; Claicula Alae; Ascending Process; Alar Process (MacGillivray).

PLURIDENTATE Adj. (Latin, *pluri* = many; *dentis* = tooth; *-atus* = adjectival suffix.) Any structure, typically a Mandible, armed with many teeth.

PLURILOBED Adj. (Latin, *pluri* = many; Greek, *lobos* = lobe.) Many-lobed.

PLURISEGMENTAL Adj. (Latin, *pluri* = many; *segmentum* = segment; *-alis* = pertaining to.) Descriptive of structure that contains many segments.

PLURISETOSE Adj. (Latin, *pluri* = many; *seta* = bristle; *-osus* = full of.) Bearing several Setae, as the head in some carabids.

PLURIVALVE Adj. (Latin, *pluri* = many; *valva* = valve.) With several valves or valve-like appendages.

PLUTELLIDAE Guenée 1845. Plural Noun. Diamondback Moths. A Family of ditrysian Lepidoptera assigned to Superfamily Yponomeutoidea. Adult small; wingspan 7–28 mm; head with lamellar scales; Ocelli present; Chaetosemata absent; Antenna shorter than forewing; Proboscis short and lacking scales; Maxillary Palpus present by variable in segmentation and orientation; Labial Palpus variable; fore Tibia with Epiphysis; tibial spur formula 0-2-4. Egg oval in outline shape, flattened and Chorion rough. Eggs typically deposited singly on leaf near larger veins. Adult with Antenna projecting forward and wings tectiform at repose. Larvae feed beneath silk webbing or mine leaves or bore into stems. A few Species of economic importance to agriculture. Common name refers to diamond-shaped pattern of yellow coloration formed when male folds forewings at repose. See Diamondback Moth.

PMP® See Phosmet.

PNEUMOGASTRIC GANGLION The nerve mass supplying nerves for the tracheal and digestive system, the Vagus.

PNEUMONIC PLAGUE A form of plague caused by the bacterium *Yersinia pestis* and vectored by fleas. Disease localized in the lungs, haemorrhaging common and often fatal. PP is transmitted by inhaling the innoculum. See Plague.

PNEUMOPHYSIS Noun. (Greek, *pneuma* = air; *physis* = nature; *sis* = a condition or state. Pl., Pneumophyses.) A paired lateral lobe of the Endophallus in *Apis*.

PNEUSTOCERA Noun. Respiratory siphons or

horns. Erroneously applied to the prolongations of the Ostiolar Peritreme in Berytidae.

PNYSTEGA Noun. (Unknown origin.) Odonata: Applied by Charpentier to a part of Mesonotum (Smith); a scale or sclerite covering mesopleural spiracles (Kirby & Spence).

POCOCK, REGINALD INNES (1863–1947) (Bonnet 1945, *Bibliographia Araneorum* 1: 54)

POCULIFORM Adj. (Latin, *poculum* = cup; *forma* = shape.) Cup-shaped; goblet-shaped. See Cup. Cf. Calyciform. Rel. Form 2; Shape 2; Outline Shape.

POD Noun. (M.E., *pod* = bag. Pl., Pods.) 1. Any containing or protective envelope such as a bag, pouch, or sac. 2. The grouped eggs of some Orthoptera (Acrididae). Verb. Transitive: To gather; Intransitive: To drive or heard into groups (otters, seals).

POD PEA BORER *Etiella zinckenella* (Treitschke) [Lepidoptera: Pyralidae]: A cosmopolitan pest of peas, beans and other legumes. Most common in warm temperate regions including southern Europe and western USA.

PODA VON NEUHAUS, NICOLAUS (1723–1790) (Ambrosi 1889, Bull. Soc. Veneto-Tren. Sci. nat. 4: 150.)

PODEON Noun. (Latin, *podeon* = neck.) Hymenoptera: The Petiole or true second abdominal segment. (Archaic).

PODEX Noun. (Latin, *podex* = rump.) 1. The dorsal (upper) sclerite of the anal opening. 2. Lepidoptera larva: Supra-anal (Supranal) Sclerite.

PODIAL REGION See Plural Region.

PODICAL PLATES The latero-ventral sclerites attached to abdominal segment X of Orthoptera, the two pieces on each side of the vent, which united form the tergite of a rudimentary ring, Anal Valves, Paranal Lobes.

PODITE Noun. (Greek, *pous* = foot; ; *-ites* = constituent.) A clearly defined limb segment, definitely correlated with muscle attachment (Snodgrass). A Podomere. Cf. Antennomere; Arthromere.

PODLER, HAGGAI (1939–1988) Israeli academic (Hebrew University) and specialist in plant protection and field crops.

PODOCYTE See Plasmatocyte.

PODODUNERA Noun. Apterous insects with biting mouth structures.

PODOMERE Noun. (Greek, *pous* = foot; *mere* = part.) See Podite.

PODOSOMA Noun. Acarology: A division of the body bearing the legs.

PODOTHECA Noun. (Greek, *pous* = foot, *theke* = box.) A portion of the Pupa which covers the legs of the subsequent imaginal stage.

PODSUCKING BUG *Melanacanthus scutellaris* (Dallas); *Riptortus serripes* (Fabricius) [Hemiptera: Alydidae]: In Australia, a minor pest of grain legumes.

PODURIDAE Plural Noun. Elongate-bodied Springtails. A small Family of Collembola assigned to Suborder Arthopleona. Body elongate; Prothorax well developed; Furcula long and flattened; Abdomen with six segments. Species occupy diverse habitats including on the surface of water, in leaf litter and on the surface of snow.

POECILOCYTTARES Noun. (Greek, *poikilos* = various; *kyttarous* = honeycomb cell.) Social wasps that build their combs around the branch or other support covered by the envelope. Cf. Stelocyttares, Phragmocyttares.

POECILOGENY Noun. (Greek, *poikilos* = various; *genesis* = descent.) Diptera: A type of larval polymorphism in which some larval forms are paedogenetic and other larval forms develop into winged sexual adults.

POENACK, HANS (–1963) (Anon. 1963, Mitt. dt. ent. Ges. 22: 63.)

POEPPIG, EDUARD FRIEDRICH (1798–1868) (Papavero 1975, *Essays on the History of Neotropical Dipterology*. 2: 300. São Paulo.)

POEY, [Y ALOY] FELIPE (1799–1891) (Anon. 1891, Entomol. mon. Mag. 27: 134.)

POEY, M Author of *Centurie de Lepidopteres de l'Isle de Cuba* (1834).

POGGE, PAUL (–1884) (Kolbe 1884, Berl. ent. Z. 28: 213–214.)

POHL, JOHANN EMMANUEL (1782–1834) (Papavero 1971, *Essays on the History of Neotropical Dipterology.*1: 62–64. São Paulo.)

POICIANA LOOPER *Pericyma cruegeri* (Butler) [Lepidoptera: Noctuidae].

POIKILOTHERMOUS Adj. (Greek, *poikilos* = various; *therme* = heat; Latin, *-osus* = with the property of.) Pertaining to organisms without the ability to internally regulate their body temperature. Without the ability to maintain the body temperature. Body temperature correlated with environmental temperature. Alt. Poikilothermic In the broad sense, cold blooded animals.

POINCIANA LONGICORN *Agrianome spinicollis* (Macleay) [Coleoptera: Cerambycidae].

POINCIANA LOOPER *Pericyma cruegeri* (Butler) [Lepidoptera: Noctuidae].

POINSETTA WHITEFLY See Sweetpotato Whitefly.

POINTED SNAIL *Cochlicella acuta* (Muller) [Sigmurethra: Helicidae].

POINTED-WING FLIES See Lonchopteridae.

POISERS Noun. (Latin, *pensum* = weight.) Plural Noun. The Halteres in Diptera.

POISON Noun. (Old French from Latin, *potio* = a drink.) An agent containing a toxin which, introduced into an organism, chemically produces an injurious or deadly effect.

POISON GLANDS Diptera (Asilidae, Empidae), 1. Heteroptera: Salivary Glands. 2. Larval Tabanidae: head glands which open on the Mandible. 3. Hymenoptera (Aculeata): Abdominal glands connected with the female Sting.

POISON SAC Aculeate Hymenoptera: The sac-like reservoir for poison produced by poison glands. PS discharges into the anterior end of the Sting bulb (Imms).

POISON SETA A hollow Seta through which certain insects discharge an irritating venom from poison gland cells (Snodgrass). See Urticating Seta.

POISSON, RAYMOND ALFRED (1895–1973) (Razet 1974, Bull. ent. Fr. 79: 1–3.)

POKORNY, ALOIS (1826–1887) (Burgenstein 1887, Verh. zool.-bot. Ges. Wien. 37: 673–678.)

POKORNY, EMANUEL (1838–1908) (Mik 1900, Wien. ent. Ztg. 19: 136.)

POKORNY, FRANTISEK (1865–1935) (Novak 1935, Cas. csl. Spol. ent. 32: 145–146, bibliogr.)

POLAK, KAREL (1847–1900) (Kafka 1900, Vesmir 29: 121–122.)

POLAK, R A (1868–1952) (Leefmans 1952, Ent. Ber., Amst. 14: 49–50.)

POLAR BODY Insect egg: One of the daughter nuclei of the primary Oocyte, with very little cytoplasm, expelled from the nuclear material in maturation.

POLAR CAP Protuberances from one or both ends of mayfly eggs. Most polar caps swell when the egg is deposited in water. The process of swelling exposes filaments which hold terminal 'knobs'. Knobbed filaments anchor the egg to vegetation in the water. Other modifications of polar caps as holdfast structures also occur in Ephemeroptera. See Egg. Cf. Aeropyle; Hydropyle; Micropyle. Rel. Attachment Organ; Pedicel.

POLAR CELLS Insect egg: Certain cells of the posterior end from which the primitive germ cells are derived in some insects (Imms).

POLAR PLASM In insect eggs: The cytoplasm at the posterior end of the egg characterized by distinctive small particles that are composed partly of RNA. Alt. Pole Plasm.

POLARITY Noun. (Greek, *polos* = axis; English, *-ity* = suffix forming abstract nouns.) 1. The condition of having poles. 2. In systematic theory, the arrangement of character states from primitive to derived. 3. In cladistical theory, the perceived direction of evolution by a character through identification of primitive (plesiomorphic) and derived (apomorphic) character states. Information used to determine polarity includes the fossil record, development (ontogeny) and outgroup comparison.

POLE Noun. (Greek, *polos* = a pivot. Pl., Poles.) Either end of a primary axis of an egg, body, appendage or structure.

POLE CELL In insect eggs: cells that form at the basal end of the egg, incorporating a part of the polar plasm (see Polar Plasm). The pole cells are distinguished from other cells in the egg by their large size and the polar granules enclosed in their cytoplasm. They are the primordial germ cells that migrate to the gonads and develop into Oocytes or sperm.

POLEFUME® See Metham-Sodium.

POLENZ, GEORG (1880–1965) (Anon. 1965, Mitt. dt. ent. Ges. 24: 79.)

POLINSKI, WLADYSLAW (1885–1930) (Jaczewski 1930, Polskie Pismo Ent. 9: 289–292.)

POLITUS Adj. (Latin, past part. of *polio* = to polish.) Descriptive of structure that is smooth; shiny; polished. See Nitidus.

POLKKINEN, ANTTI ARMAS (ASKO) (1885–1933) (Palmgren 1933, Memo Soc. Fauna Flora Fenn. 9: 208–209, 225–227.)

POLLARD Noun. (Middle English, *polle* = head; *ard* = possessing a quality. Pl., Pollards.) 1. Coarse bran obtained from wheat or finely ground bran mixed with scourings obtained from wheat during milling. Pollard is used as livestock feed. 2. A tree which has been cut back to the trunk to promote the growth of foliage.

POLLEN Noun. (Latin, *pollen* = fine flour. Pl., Pollens.) 1. The microscopic, multinucleate male gametophyte generation of seed plants. Pollen transports the male gametes that are used in sexual reproduction of plants. Pollen features are often Species-specific; fossilized pollen can be identified and used to reconstruct vegetational habitats and provide ecological information. See Reproduction. Cf. Spore. 2. A dusty or pruinose surface covering which is easily rubbed off and used as a descriptive term for some Diptera.

POLLEN BASKET Bees: A vernacular term for an area of the Integument used as a repository for pollen during transport. In most bees, the PB is created by the broadened and somewhat concave distal surface of the posterior Tibia. Apoidea: The Corbicula.

POLLEN BEETLE *Meligethes aeneus* Fabricius [Coleoptera: Nitidulidae]: A major pest of oilseed rape in Europe.

POLLEN BEETLES *Dicranolaius* spp. [Coleoptera: Melyridae].

POLLEN BRUSH Bees: A vernacular term for appendicular structures, typically Setae, which are used to collect pollen from the anthers of flowers. PB is typically on the Basitarsus of the leg and used to transfer Pollen to the Pollen Basket on the Tibia of the opposable leg. See Scopa.

POLLEN COMB See Scopa. Syn. Pectin.

POLLEN PLATE 1. A polished area margined by Setae in the outer face (distal surface) of the Tibia. Syn. Pollen Basket. 2. Genitalia of male Lepidoptera: A finger-like or spiny process arising from the ventral part of the Harpe, usually from the Cuvcullus (Klots).

POLLEN POCKET Bumblebees: A reservoir for pollen inside the nest and beside a cell in some Species. Larvae and adults have access to the contents of the pocket.

POLLEN POT Bumblebees: A pollen storage resevoir in the nest of some Species. Larvae do not have direct access to the pollen.

POLLEN WASPS See Masarinae (Vespidae).

POLLEN-FEEDING BEETLES See Oedemeridae.

POLLENKITT Noun. A lipid and protein rich coating typically found on dandelion pollen (Stanley & Linskens 1974. *Pollen Biology, Biochemistry* *and Management*. Springer Verlag.)

POLLEX Noun. (Latin, *pollex* = thumb. Pl., Pollices.) 1. The apical, dorsal segment of the Abdomen. 2. The thumb or medial digit on the hand of humans.

POLLICATE Adj. (Latin, *pollex* - thumb; *-ate* = to put into action; *-atus* = adjectival suffix.) Produced inwardly into a short, bent spine, as a Tibia. Alt. Pollicatus.

POLLINATION Noun. (Latin, *pollen* = fine flour; English, *-tion* = result of an action.) Angiosperms: The transfer of pollen from the anther to the Stigma before fertilization. Gymnosperms: The transfer of pollen from male cones to female cones before fertilization. Ethodynamic Pollination: An active or dynamic mechanism of pollen transfer. Topocentric Pollination: A passive mechanism of pollen transfer (Galil 1973, 26: 303–311).

POLLINIA Noun. (Late Latin. Sl., Pollinium.) Pollen-masses of flowers.

POLLINIFEROS Adj. (Latin, *pollen* = fine flour; *fero* = to bear.) Formed for collecting pollen; pollen bearing. Alt. Pollinigerous.

POLLINOSE Adj. (Latin, *pollen* = fine flour; *-osus* = full of.) Covered with a loose, mealy, often yellow dust, like the pollen of flowers. Alt. Pollinosus.

POLLMANN, AUGUST (1813–1898) (Anon. 1898, Leopoldina 34: 112.)

POLO® See Diafenthiuron.

POLS Noun. Coccids: Pectinae (MacGillivray).

POLYACETYLENES Plural Noun. A complex of light-activated alkaloid biopesticides known from the Asteraceae, Apiaceae and Araliaceae. Polyacetylenes express some feeding-deterrent and ovicidal activities in target insect pests, but are probably more effective against microorganisms. Representative polyacetylene compounds include Falcarinone and Falcarindiol. Some non-light-activated polyacetylenes are also known. See Biopesticides; Botanical Insecticide; Natural Insecticide; Insecticide; Pesticides. Cf. Furanocumarins; Terthienyl.

POLYANDROUS Adj. (Greek, *polys* = many; *aner* = male; Latin, *-osus* = with the property of.) Pertaining to females which engage in polyandry.

POLYANDRY Noun. (Greek, *poly* = many; *aner* = male.) The condition in which one female copulates with several males. See Polygyny. Cf. Monandry.

POLYBASIC Adj. (Greek, *poly* = many; *basis* = pedestal; *-ic* = characterized by.) Genera originally founded on several Species.

POLYCALIC See Polydomous.

POLYCHLOROTERPENE Noun. A chlorinated hydrocarbon insecticide prepared from camphene, a product derived from conifers.

POLYCHROMATIC Adj. (Greek, *poly* = many; *chromos* = colour; *-ic* = consisting of.) Descriptive of structure with many colours.

POLYCORE TML® See Trimedlure.

POLYCRON® See Profenofos.

POLYCTENIDAE Westwood 1874. Plural Noun. Bat Bugs. A numerically small Family of heteropterous Hemiptera assigned to Superfamily Cimicoidea; sometimes regarded as Cimicidae. Polyctenids are not commonly taken as they live as ectoparasites on bats. Adult lacks compound eyes and Ocelli; wings absent or forewings short and coriaceous; body 3–5 mm long, with short and long bristle-like Setae which often resemble Ctenidia of fleas. Head and Prothorax dorsoventrally flattened; fore leg short, flattened; middle and hind legs long and slender. Species viviparous with female producing nymphs; three nymphal instars.

POLYDOMOUS Adj. (Greek, *polys* = many; *domos* = house; Latin, *-osus* = full of.) Social Insects: pertaining to individual colonies which utilize several nests; applied to ants when one colony has several nests. Syn. Polycalic. Cf. Monodomous.

POLYDOMY Noun. (Greek, *polys* = many; *domos* = house. Pl., Polydomies.) A social condition in which one colony occupies several nests. Cf. Monodomy.

POLYEMBRYONIC Adj. (Greek, *polys* = many; *embryon* = fetus; *-ic* = characterized by.) Pertaining to polyembryony. Cf. Monoembryonic.

POLYEMBRYONY Noun. (Greek, *polys* = many; *embron* = fetus. Pl., Polyembryonies.) A form of reproduction in which several embryos develop from one fertilized egg. Phenomenon occurs in Strepsiptera and parasitic Hymenoptera including Encyrtidae, Platygasteridae and Dryinidae.

POLYESTERS Noun. (Greek, *polys* = many; Latin, *aether* = ether.) A category of plastic embedding media used in microtomy and electron microscopy of bacteria. Not widely used because of difficulty in handling, infiltration and sectioning. Vestopal-W developed by Ryter & Kellenberger (1958, J. Ultrast. Res. 2: 200.) See Histology. Cf. Epoxy Resins; Methacrylates. Rel. Embedding.

POLYETHISM Noun. (Greek, *polys* = many; *ethos* = custom; English, *-ism* = condition. Polyethisms.) Social Insects: A division of labour among members of a colony. See Age Polyethism; Caste Polyethism.

POLYGAMOUS Adj. (Greek, *polys* = many; *gamos* = marriage; Latin, *-osus* = with the property of.) Pertaining to a male which sequesters (isolates) or copulates with many females. Cf. Monogamous.

POLYGAMY Noun. (Greek, *polys* = many; *gamos* = marriage. Pl., Polygamies.) The condition of one male copulating with or inseminating many conspecific females. See Polyandry. Cf. Monogamy.

POLYGENESIS Noun. (Greek, *polys* = many; *genesis* = descent.) The development of a new type at more than one place or more than one time.

POLYGENIC Adj. (Greek, *polys* = many; *genos* = descent; *-ic* = consisting of.) See Polyphyletic.

POLYGONAL Adj. (Greek, *poly* = many; *gonia* = angle; Latin, *-alis* = characterized by.) Descriptive of structure with many angles. Alt. Polygonous.

POLYGONEUTIC Adj. (Greek, *poly* = many; *gone* = generation; *-ic* = consisting of.) Descriptive of insects that produce several broods.

POLYGONEUTISM Noun. (Greek, *poly* = many; *gone* = generation; English, *-ism* = condition. Pl., Polygoneutisms.) The reproductive ability or capacity to produce several broods of progeny during one season.

POLYGYNE FORM Social ant colonies which have more than one inseminated, functional, egg-laying queen. Cf. Monogyne Form.

POLYGYNOUS Adj. (Greek, *polys* = many; *gyne* = woman; Latin, *-osus* = with the property of.) Social Insects: Pertaining to a population with numerous gynes; relating to a female-biased colony of social bees, wasps or ants.

POLYGYNY Noun. (Greek, *polys* = many; *gyne* = woman.) Social Insects: The existence of more than one functional queen in a nest. See Primary Polygyny; Secondary Polygyny; Oligogyny. Cf. Monogyny.

POLYHEDRAL Adj. (Greek, *polys* = many; *hedra* = side, seat; *-alis* = characterized by.) Descriptive of structure with many sides or many angles.

POLYLECTIC Adj. (Greek, *polys* = many; *lekto* = picked; *-ic* = characterized by.) Pertaining to bees which collect several kinds of pollen. Cf. Oligolectic.

POLYMEGALY Noun. (Greek, *polys* = many; *megale* = large. Pl., Polymegalies.) The condition of two or more distinct size-classes of Spermatozoa within a Species. Size can represent tail-length, head-length or overall length of a Spermatozoon. Polymegaly seems widespread among Species of *Drosophila,* and may be attributed to the sex Chromosome complement, temperature or genetic constitution. See Spermatozoa. Cf. Apyrene; Eypyrene; Hyperpyrene; Oligopyrene.

POLYMERASE CHAIN REACTION A laboratory technique for cloning DNA without the use of microorganisms. PCR allows the rapid isolation, selection and amplification of DNA regions from small amounts of tissue. Ideally, the action of an enzyme (polymerase) to produce many copies of a polynucleotide sequence of DNA. Acronym: PCR. See Genetic Engineering. Cf. Microsatellite. Rel. Biotechnology.

POLYMITARCIDAE Plural Noun. A small Family of Ephemeroptera assigned to Superfamily Ephemeroidea. Male with middle and hind legs reduced; all legs of female reduced. Nymphs burrow into mud and clay in large rivers and lakes.

POLYMORPHA Noun. The clavicorn and serricorn Coleoptera. See Coleoptera.

POLYMORPHIC Adj. (Greek, *polys* = many; *morphe* = form; *-ic* = characterized by.) 1. Descriptive of stucture which may occur in several forms. 2. Pertaining to structure which differs by sex, season, locality or age. 3. Undergoing several changes, in this sense applied to insects with a complete Metamorphosis. Alt. Polymorphous. Syn. Pleomorphic. Cf. Dimorphic; Monomorphic. Rel. Form; Shape.

POLYMORPHISM Noun. (Greek, *polys* = many; *morphe* = form; English, *-ism* = condition. Pl., Polymorphisms.) 1. The condition in which one Species displays several anatomical forms or colour variants. Polymorphism has been identified in at least 20 Orders of insects and is probably found in all major insect groups. 2. The occurrence of different body forms or different forms of organs in one individual during different periods of life. 3. Social Insects: The coexistence of two or more functionally different castes within a sex. Knowledge of polymorphism is based almost entirely upon external anatomy of large and conspicuous insects; polymorphism of internal anatomical features remains unknown. Polymorphism is poorly known in microscopic insects. Typically, polymorphism is unmistakable in terms of shape, size and colour. However, polymorphism can involve expression of different forms of organs in one individual during different periods of life. Polymorphism can be expressed within a sex, between sexes and within a developmental stage of a Species. See Sexual Dimorphism. Cf. Dimorphism; Monomorphism. Rel. Form; Shape.

POLYMORPHISM PARSIMONY (Felsenstein 1979, Syst. Zool. 28: 49–62.) See Parsimony.

POLYNEPHRIA Noun. (Greek, *poly* = many; *nephros* = kidney.) Insects with many Malpighian Tubules.

POLYPEPTIDE Noun. (Greek, *polys* = many; *peptein* = to digest. Pl., Polypeptides.) An aggregation of many amino acids linked by peptide bonds.

POLYPHAGA (Triassic Period-Recent.) The largest Suborder of Coleoptera including more than 90% of Taxa. Cervical sclerites present; prothoracic Pleuron fused with Trochantin, concealed and forming Cryptopleuron; hindwing without Oblongum Cell; transverse fold not crossing MP; Metepisternum not meeting middle coxal cavity; hind Coxae mobile, not dividing first Ventrite; Ovarioles telotrophic. See Coleoptera.

POLYPHAGIDAE Plural Noun. Sand Cockroaches. A widespread Family of ca 200 Species of cockroaches assigned to Superfamily Blaberoidea. Body size highly variable (3–30 mm); Pronotum setose; winged and wingless forms; winged males with anal area of hindwing not folded fanlike at repose. SC taken in association with ant nests; some Species desert adapted. Regarded as most primitive Family within Blaberoidea.

POLYPHAGOUS Adj. (Greek, *polys* = many; *phagein* = to eat; Latin, *-osus* = with the property of.) 1. Pertaining to organisms that exhibit polyphagy. Descriptive typically of an herbivorous insect that feeds on many Species of food plants, a predator that feeds on many Species of prey or a parasite that feeds on many Species of host. 2. Pertaining to an omnivorous diet. Syn. Euryphagous. See Feeding Strategy. Cf. Monophagous; Oligophagous; Stenophagous.

POLYPHAGOUS PINHOLE BORER *Platypus australis* Chapuis [Coleoptera: Curculionidae].

POLYPHAGY Noun. (Greek, *polys* = many; *phagein* = to eat.) 1. The lowest level of specialization in food or least limited range of dietary requirements. 2. An unspecialized diet; An unlimited dietary range. Syn. Euryphagy. See Omnivorous. Cf. Monophagy, Oligophagy, Stenophagy.

POLYPHEMUS MOTH *Antheraea polyphemus* (Cramer) [Lepidoptera: Saturniidae].

POLYPHENISM Noun. (Greek, *polys* = many; *phainein* = to appear; English, *-ism* = condition. Pl., Polyphenisms.) The occurrence of several phenotypes in one population which are not controlled genetically.

POLYPHYLETIC Adj. (Greek, *polys* = many; *phylon* = race; *-ic* = of the nature of.) A natural or genetically cohesive lineage of Taxa whose most recent common ancestor has given rise to excluded Taxa of which at least one of the sister groups is only partly included in the group. (Oosterbroek 1987, Syst. Zool. 36: 103–108.) See Sister Group. Cf. Monophyletic; Paraphyletic; Holophyletic.

POLYPHYLY Noun. (Greek, *polys* = many; *phylon* = race. Pl., Polyphylies.) See Polyphyletic; Classification. Cf. Holophyly; Monophyly. Rel. Taxonomy.

POLYPLACIDAE Plural Noun. A small Family of Anoplura whose member Species parasitize lagomorphs (hares, rabbits, pikas). Body with few Setae; head relatively small; compound eyes absent; Mandibles absent, mouthparts forming a beak or rounded anteriorly; middle and hind legs subequal in size; Abdomen with Paratergites.

POLYPLOID Noun. (Greek, *polys* = many, *aploos* = onefold; *eidos* = form. Pl., Polyploids.) An organism whose somatic cells each contain more than two sets of Chromosomes. See Chromosome. Cf. Diploid; Haploid; Tetraploid.

POLYPNEUSTIC Adj. (Greek, *polys* = many; *pneuma* = wind; *-ic* = consisting of.) 1. Pertaining to a system of respiration that is characterized by eight or more pairs of spiracles. PS subdivided into Holopneustic (1 Mesothorax, 1 Metathorax, 8 abdominal), Peripneustic (1 Mesothorax, 8 abdominal, *e.g.* cecidomyids) and Hemipneustic (1 mesothoracic, 7 abdominal, *e.g.* mycetophilids.) 2. Diptera: Containing many respiratory orifices (apertures) in the spiracle of certain larvae. See Respiration. Cf. Apneustic; Oligopneustic; Propneustic.

POLYPNEUSTIC LOBES Diptera: Paired, sclerotized, lobe-like projections from the apical segment of larvae. PL bear the multiple pores of the spiracles (Imms). PL typical of third-instar larvae of Glossinidae.

POLYPOD Noun. (Greek, *polys* = many; *pous* = foot. Pl., polypods.) Any organism with many legs. Specifically, a larval type with a completely segmented Abdomen, and each segment with a pair of rudimentary limbs (*e.g.* caterpillars). Syn. Eruciform larvae of older publications.

POLYPOD LARVA A larva with abdominal limbs and peripneustic tracheal system (Folsom & Wardle).

POLYPOD PHASE Insect embryo: A phase in which the Abdomen has acquired its complete segmentation and full number of appendages and in which the other organs are further advanced (Imms).

POLYPODEIFORM Adj. (Greek, *polys* = many; *pous* = foot; Latin, *forma* = shape.) Hymenoptera: Apocritous endoparasitic, segmented larva with paired, short, flexible projections from thoracic and abdominal segments. Found in Cynipoidea and Proctotrupoidea. Cf. Vesiculate Larva.

POLYPODOUS Adj. (Greek, *polys* = many; *pous* = foot; Latin, *-osus* = with the property of.) Descriptive of an organism with many legs. Term is specifically applied to the Myriapoda, larvae of Lepidoptera and sawflies. Term used in contrast with apodous and hexapodous larvae.

POLYPSOCIDAE See Caeciliidae.

POLYSACCHARIDE Noun. (Greek, *polys* = many; Latin, *saccharum* = sugar.) A carbohydrate, one molecule of which can, by hydrolysis, be split into many molecules of a simple sugar.

POLYSEMA SKIPPER *Proeidosa polysema* (Lower) [Lepidoptera: Hesperiidae].

POLYSPERMY Noun. (Greek, *polys* = many; *sperma* = seed.) The penetration of an egg by many Spermatozoa.

POLYSTOECHOTIDAE Plural Noun. Giant Lacewings. A small Family of Neuroptera assigned to Superfamily Hemerobioidea. Prothorax normal in size; wingspan 40–75 mm; Forewing with recurrent Humeral Vein and Radial Sector. Two Species occur in North America. Larvae are terrestrial predators.

POLYTENE Adj. (Greek, *polys* = many, *tainia* = band.) Giant chromosomes in some somatic cells, such as in the salivary glands of *Drosophila*, which contain many chromonemata arranged in parallel.

POLYTHALAMOUS Adj. (Greek, *polys* = many; *thalmos* = chamber; Latin, *-osus* = with the property of.) Pertaining to a multiple-chambered gall found on plants; typically induced by insects. See Gall. Cf. Monothalamous. Rel. Cynipidae; Cecidomyiidae.

POLYTHETIC Adj. (Greek, *polys* = many; *thetos* = placed; *-ic* = characterized by.) Pertaining to classifications in which members of a group display most but not necessarily all of the characteristics which define that group.

POLYTHORIDAE Plural Noun. A small Neotropical Family of Odonata assigned to the Calopterygoidea Species of which occur in forests. Adults small to moderate-sized; wings variable in most diagnostic characters; many Antenodal Crossveins present; Pterostigma elongated; Abdomen elongate. Larva resembles Epallagidae; Abdomen with lateral pseudopodal appendages.

POLYTRIN® See Cypermethrin.

POLYTROPHIC Adj. (Greek, *polys* = many; *trophe* = nourishment; *-ic* = of the nature of.) 1. Pertaining to Ovarioles. 2. Pertaining to organisms with many trophi or mouthparts.

POLYTROPHIC OVARIOLE An Ovariole which contains a primary Oocyte and Trophocytes (nutritive cells, nurse cells) which are derived from same Oogonium and enclosed by Follicle Cells. The Oocyte and Trophocytes travel the length of the Ovariole surrounded by the Follicle Cells. A derived type of Ovariole which occurs in Phthiraptera, Psocoptera, Dermaptera and most endopterygotes. Syn. Polytrophic Egg Tube. See Meroistic Ovary. Cf. Telotrophic Ovariole. Rel. Panoistic Ovary.

POLYTROPHIC OVARY An Ovary composed of many polytrophic ovarioles.

POLYXENOUS Adj. (Greek, *polys* = many; *xenos* = host; Latin, *-osus* = with the property of.) Pertaining to parasitic Species which have several Species of hosts. See Parasitism. Cf. Monoxenous; Oligoxenous; Pleioxenous.

POLYXENY Noun. (Greek, *polys* = many; *xenos* = host.) A parasite which feeds upon many Species of hosts.

POMACE FLY See Drosophilidae.

POME LOOPER *Chloroclystis testulata* (Guenée) [Lepidoptera: Geometridae].

POMERANTZEV, DMITRIYA VLADIMIROVICH (1869–1952) (Lashchkevich 1953, Ent. Obozr. 33: 380–383, bibliogr.)

POMINEX® See Alphacypermethrin.

POMINI, FRANCESCO PIO (1915–1941) (De Beaux 1943, Annali Mus. civ. Stor. nat. Giacomo Doria 61: 3.)

POMMETROL® See Chlorpropham.

POMPILIDAE Latreille 1804. Plural Noun. Spider Wasps; Tarantula Hawks. Cosmopolitan, numerically large Family with ca 4,000 nominal Species of aculeate Hymenoptera; currently assigned to Vespoidea. Body moderate to large sized; male Antenna with 13 segments; female Antenna with 12 segments; compound eye not notched; Maxillary Palpus with six segments; Labial Palpus with four segments; Pronotum posterior margin overlapping Mesonotum; Mesopleuron with a conspicuous suture or Sulcus. Both sexes macropterous; forewing with 10 closed cells; hindwing with three closed cells and Jugal Lobe; paired Coxae medially juxtaposed; legs long and slender; hind Femur and Tibia spinose; middle and hind Tibiae each with two spurs, hind tibial spur modified into Calcar. Gaster without constrictions; female with six segments male with seven segments. All Species develop as solitary external parasites of spiders. Female pompilid searches for spider, subdues spider with venom from Sting, malaxates spider and transports spider to place of concealment. Many Species construct cells or excavate soil in which larva develops on host. Some Species develop as cleptoparasites of other pompilids with parasitized hosts detected by female who destroys extant egg and lays her own on paralysed spider, or second egg hatches and larva kills the first pompilid and then consumes spider host. A few Species temporarily paralyse spider which recovers but is eventually killed by developing larva. Assigned to the Pompiloidea in some classifications. Syn. Psammocharidae.

POMPILOIDEA Latreille 1804. A Superfamily of aculeate Hymenoptera which includes the Pompilidae and Rhopalosomatidae. Sometimes combined with Vespoidea *sensu stricto*. Adult hindwing without a closed basal cell. See Aculeata; Hymenoptera. Cf. Apoidea; Chrysidoidea; Formicoidea; Sphecoidea; Vespoidea.

POND SKATERS See Gerridae.

PONDERABLE Adj. (Latin, *ponderis* = a weight.) That which may be weighed.

PONDEROSA PINE BARK-BORER *Canonura princeps* (Walker) [Coleoptera: Cerambycidae].

PONDEROSA PINE-CONE BEETLE *Conophthorus ponderosae* Hopkins [Coleoptera: Scolytidae]. Syn. Lodgepole Cone Beetle.

PONE Adj. (Latin, *ponere* = to place.) Behind (the middle).

PONERINAE A Subfamily of ants. Body heavily sclerotized; first gastral segment constricted.

PONEROID COMPLEX A major taxonomic grouping of ants recognized in some classifications. Cf. Myrmecioid Complex.

PONS CEREBRALIS The Protocerebral Bridge.

PONS COXALIS A transverse bridge uniting the Coxopodites (Snodgrass).

PONS GLOMERULUS See Protocerebral Bridge.

PONTA Noun. (Latin, *pontis* = a bridge.) Diptera: A small, transverse sclerite connecting the margin of the Mesoscutum and the Mesoprealifera. (MacGillivray).

PONTICULUS Noun. (Latin, *ponticulus* = small bridge.) The frenulum.

PONTOPPIDAN, ERIC (1698–1764) (Henriksen 1922, Ent. Meddr 15: 72–73.)

POOS, FREDERICK WILLIAM JR (1891–1987) Born Potter, Kansas; died near Washington, DC. American economic Entomologist known for work with many crop pests. President, Entomological Society of Washington and member of many professional societies. First person to use the aspirator for collecting leafhoppers and subsequently called a 'pooter' by English Entomologists. (Wallenmaier 1989, PESW 91: 298–301, photograph, bibliogr.)

POOTER Noun. (Pl., Pooters.) A vernacular term coined by English Entomologists for an aspirating device used to collect small, highly mobile insects. Named in honor of F. W. Poos, an American Entomologist who employed the device to collect Cicadellidae.

POPENHOE, CHARLES HOLCOMB (1884–1933) (Osborn 1937, *Fragments of Entomological History,* 394 pp. (276), Columbus, Ohio.)

POPENHOE, EDWIN ALONZO (1855–1913) (Osborn 1937, *Fragments of Entomological History,* 394 pp. Columbus, Ohio.)

POPLAR Noun. (Latin, *populus* = poplar. Pl., Poplars.) Any tree of the Genus *Populus* (Salicaceae); 30–40 dioecious, soft-wooded Species of Northern Hemisphere, including Aspen, Cottonwood, Tulip-tree, poplar.

POPLAR-AND-WILLOW BORER *Cryptorhynchus lapathi* (Linnaeus) [Coleoptera: Curculionidae]: A widespread pest of alder, birch, poplar and willow in North America and Europe. Adult chews small hole in bark to feed or oviposit; typically feeds on green, smooth barked shoots of current year. Larvae mine bark and wood; six instars. Overwinters in any stage; adults usually emerge in autumn.

POPLAR BORER *Saperda calcarata* Say [Coleoptera: Cerambycidae]: A Species endemic to North America and regarded as a pest of cottonwood, poplars and willows; the most significant pest of poplars in many areas. Larva feeds on large limbs, trunks and felled trees. Adult ca 18–26 mm long; female with long, hornlike Ovipositor that projects from apex of Abdomen. PB overwinters as larva; requires 2–3 years to complete a generation. See Cerambycidae. Cf. Elm Borer; Locust Borer.

POPLAR GALL APHID *Pemphigus bursarius* (Linnaeus) [Hemiptera: Aphididae].

POPLAR LEAF-FOLDING SAWFLY *Phyllocolpa bozemani* (Cooley) [Hymenoptera: Tenthredinidae].

POPLAR PETIOLE-GALL APHID *Pemphigus populitransversus* Riley [Hemiptera: Aphididae].

POPLAR TENT-MAKER *Ichthyura inclusa* (Hübner) [Lepidoptera: Notodontidae].

POPLAR TWIG GALL-APHID *Pemphigus populiramulorum* Riley [Hemiptera: Aphididae].

POPLAR VAGABOND-APHID *Mordvilkoja vagabunda* (Walsh) [Hemiptera: Aphididae].

POPPIUS, BERTIL ROBERT (1876–1906) (Lavender 1906, Luonnon Ystavä 20: 234–236.)

POPPIUS, KARL ALFRED (1846–1920) (Anon. 1921, Notul. ent. 1: 25–26.)

POPULATION Noun. (Latin, *populus* = people; English, *-tion* = result of an action.) 1. A collection of similar entities, persons, zooids or particular things, localized in space and time and possessing unique properties. Cf. Individual. 2. A group of individuals of one Species inhabiting a certain area at a certain time. Cf. Deme.

PORCATE Adj. (Latin, *porca* = ridge between two furrows; *-atus* = adjectival suffix.) Pertaining to a surface with parallel, elevated ridges and furrows, and whose furrows are wider than the ridges. Pertaining to structure formed of ridges or ridged. Alt. Porcated; Porcatus.

PORE Noun. (French, *pore,* Latin, *porus;* Greek, *poros* = channel. Pl., Pores.) 1. An isolated puncture. 2. A minute impression that perforates the surface (Kirby & Spence). 3. Any small, circular opening on a membrane, wall or surface of a structure. Pores typically are an adaptation for absorption, transpiration or the exchange of gas. When used as a connecting form (*e.g.,* Madrepore), pore indicates the opening of a duct. Syn. Foramen; Hole. See Aperture. Cf. Poriform.

PORE CANAL The channel of the Cuticula beneath Setae or other external part of many sense organs (Snodgrass). The name incorrectly given to protoplasmic upgrowth from a underlying epithelial cell (Wardle). Pore Canals penetrate the Endocuticle and Exocuticle, but usually do not penetrate the Epicuticle. Some Pore Canals end abruptly at the interface of the Epicuticle and Exocuticle; other Pore Canals connect with epicuticular Wax Canals; a few Pore Canals penetrate the Epicuticle unchanged and taper before ending beneath the Cuticulin Layer. In some insects the distal end of the Pore Canal connects with microscopic arborizations which penetrate the Epicuticle and terminate on the surface. See Integument. Cf. Basement Membrane; Cement Layer; Cuticulin Layer; Dermal Glands; Epidermal Cells; Epidermis; Endocuticle; Exocuticle; Wax Layer. Rel. Skeleton; Skin.

PORE PLATE SENSILLUM A thickened, elliptical, perforated sclerite found on the Antenna and separated from the antennal Cuticle by an inner, flexible wall and furrow. Typically numerous on the Antenna segments, particularly flagellar segments. Each Pore Plate is innervated by a variable number of sense cells whose Dendrites ramify and occupy an outer receptor lymph cavity adjacent to each perforation. Several accessory cells are associated with each PPS including Thecogen Cells, Trichogen Cells, Tormogen Cells and Envelope Cells. PPS are found in many taxonomic groups of insects, but are especially well represented in apocritous Hymenoptera. Believed to function in olfaction. See Sensillum Placodeum. Syn. Glumes; Multiporous Plate Sensilla; Rhinaria (Antennal); Tyloid (Antennal).

PORIFEROUS (Latin, *porus;* Greek, *poros* = channel; *ferre* = to bear, carry; Latin, *-osus* = with the property of.) Adj. Closely set with deep pittings or punctures.

PORIFORM Adj. (Latin, *porus;* Greek, *poros* = channel; Latin, *forma* = shape.) Pore-shaped; pertaining to a microscopic aperture, opening or hole in a surface. See Pore. Cf. Aperture; Duct. Rel. Form 2; Shape 2; Outline Shape.

PORINA MOTH *Wiseana cervinata.* [Lepidoptera: Hepialidae].

PORKKA, OSMO HANNU (1901–1939) (Anon. 1940, Suomen Hyönt. Aikak 6: 4,6.)

POROUS Adj. (Greek, *poros* = channel; Latin, *-osus* = with the property of.) Pertaining to a surface with pores. Alt. Porose; Porosus. Cf. Cribrate; Punctate.

POROUS AREAS Female ticks: A pair of depressions on the dorsal surface of the Basis Capituli made up of minute open pores (Matheson).

PORRECT Adj. (Latin, *porrectus* = streched out.) Extending forward horizontally; projecting. Cf, Suberect.

PORRITT, GEORGE TAYLOR (1848–1927) (Walker 1927, Proc. Linn. Soc. Lond. 1926–27: 92–93.)

PORTA ATRII An atrial orifice.

PORTA, ANTONIO (1874–1971) (Fiori 1973, Memorie Soc. ent. ital. 52: 72–78, bibliogr.)

PORTER Noun. (Old French, *portier* = porter; Pl., Porters.) A vertebrate animal which carries disease-producing organisms to a host. See Transmission. Cf. Vector.

PORTER, CARLOS EMILIO (1868–1942) (Anon. 1943, Boln Lab. Clin. Luis Razetti 3: 217–218.)

PORTIER, PAUL (1866–1962) (Viette 1962, Bull. Soc. ent. Fr. 67: 8–9.)

PORTLAND, DUCHESS OF See Bentinek, M. C.

PORTSCHINSKY, JOSIF ALILZIEVICH (1848–1916) (Semenov-Tian-Shansky 1916, Revue Russe Ent. 16: 404–406)

PORTULACA Noun. (Latin, *portulaca* = purslane.) A Genus of tropical succulent herbs assigned to Portulacaceae. Flowers with 4–6 petals; partially inferior 1-celled Ovary; fruit is a capsule.

PORTULACA LEAF-MINING WEEVIL *Hypurus bertrandi* Perris [Coleoptera: Curculionidae].

POSITIVE PHOTOTAXIS A behavioural reaction, typically in the form of movement or body orientation, in which an organism responds positively to the stimulus of light. See Phototaxis. Cf. Negative Phototaxis.

POSITIVE TROPISM A behavioural reaction in which the stimulus attracts (Wardle). See Tropism.

POSPELOV, VLADIMIR PETROVICH (1872–1949) (Zverezomb-Zubovskii, Ent. Obozr. 31: 301–314, bibliogr.)

POSSE® See Carbosulfan.

POSSOMPES, BERNARD (1912–1975) (Caussanel 1976, Bull. Soc. ent. Fr. 80: 249–252, bibliogr.)

POSSUM SOFT TICK *Ornithodoros macmillani* Hoogstraal & Kohls [Acari: Argasidae].

POSSUM TICK *Ixodes trichosuri* Roberts [Acari: Ixodidae].

POST, HENRY ELIAS (1906–1965) (Facher 1965, J. Econ. Ent. 58: 1040.)

POSTABDOMEN Noun. (Latin, *post* = after; *Abdomen* = belly.) The more or less modified or slender posterior segments of the Abdomen, including the genital segments, and segments of the female insect Abdomen retracted within a genital sinus (MacGillivray).

POSTACROSTICHALS Diptera: The posterior Acrostichal Bristles behind the Transverse Suture (Comstock).

POSTAL VEIN Hymenoptera: The Costa *sensu* Comstock. See Vein; Wing Venation.

POSTALAR Adj. (Latin, *post* = behind; *ala* = wing.) Behind the wings in position.

POSTALAR BRIDGE A lateral extension of the Postnotum of a wing bearing segment behind the wing base, generally united with the Epimeron; Lateropostnotum (Snodgrass).

POSTALAR CALLOSITIES Rounded processes at the posterior lateral margin of the dorsum, in Diptera.

POSTALAR CALLUS Diptera: A more or less distinct rounded swelling on each side placed between the root of the wing and the Scutellum.

POSTALAR MEMBRANE Diptera: The strip of membrane connecting the Squamae with the Scutellum.

POSTALARE See Postalar Bridge.

POSTALARIA Noun. (Latin, *post* = behind; *ala* = wing. Pl., Postalariae.) The fused adjacent parts of the Postscutellum and the Epimeron forming a connection caudad of the Rotaxis or Axillary Membrane (MacGillivray).

POSTALIFERA Noun. (Latin, *post* = behind; *ala* = wing; *ferens* = bearing.) A long, narrow sclerite on the posterior side of the Pleuralifera, Anterior Subalare, Postparapteron, Epimeral Parapteron, Costale, Postepimeron, Costal Sclerite, Posterior Costal Sclerite (MacGillivray).

POSTANAL FIELD Isoptera: The posterior lobe of the forewing in the Genus *Mastotermes* (Holmgren.)

POSTANAL PLATE Coccids: See Telson (MacGillivray).

POSTANAL SETA Acari: A ventral Idiosomal Seta posterior of the Anus. PS typically resemble dorsal Idiosomal Setae. Syn. Caudal Setae. (Goff *et al.* 1982, J. Med. Ent. 19: 230.)

POSTANNELLUS Noun. (Latin, *post* = behind; *annelus* = ring. Pl., Postannelli.) Hymenoptera: The Antennomere immediately distal of the ring segment(s), usually the second segment of the Flagellum. See Annelus.

POSTANS, ARTHUR THOMAS (1892–1973) (Greaves 1973, Entomologist's Rec. J. Var. 85: 135.)

POSTANTENNAL APPENDAGES The so-called second Antennae in insects. Typically reduced in the embryo and absent in the adult (Snodgrass).

POSTANTENNAL ORGAN Collembola: A cuticular modification, usually present and postioned behind the Antennal Scape. Shape varies from circular depression to multituberculate; presumably olfactory in function. Syn. Organ of Tomosvary.

POSTARTICULAR Adj. (Latin, *post* = behind; *articulus* = joint.) Placed or set posterior of a joint.

POSTARTIS Noun. The swelling on the caudal or ventral aspect of the proximal end of a Mandible articulating in a Postcoila; Condilo Vero, ventral Condyle, Hypocondyle (MacGillivray).

POSTAXIAL SURFACE OF THE LEG The posterior surface of the insect leg when it is extended at a right angle to the body (Snodgrass).

POSTBRACHIAL Adj. (Latin, *post* = behind; Greek, *brachion* = upper arm; Latin, *-alis* = pertaining to.) See Probrachial.

POSTCALCAR Noun. (Latin, *post* = behind; *calcar* = spur. Pl., Postcalcars.) Hymenoptera: The proximal or trailing surface of Sting in the Aculeata or the serrated surface of the Ovipositor in Parasitica and Symphyta. Cf. Precalcar.

POSTCEREBRAL Adj. (Latin, *post* = after; *cerebrum* = brain; *-alis* = characterized by.) Descriptive of structure behind the Cerebrum. Cf. Precerebral.

POSTCEREBRAL GLANDS Salivary glands of the head.

POSTCLYPEUS Noun. (Latin, *post* = after; Clypeus = shield.) 1. General: The posterior or upper part of Clypeus when any line of demarcation exists. 2. Odonata: The upper portion of a transversely divided Clypeus. 3. Psocidae: A peculiar inflated structure behind the Clypeus. Syn. Eupraclypeus; Nasus; Afternose; Paraclypeus; First Clypeus; Clypeus Posterior.

POSTCOILA Noun. (Pl., Postcoilae.) The articulation of the Mandibles to the head, Hypocondyle, ventral Condyle (MacGillivray).

POSTCORNU Noun. (Pl., Postcornua.) Orthoptera: See Lophus. Hymenoptera: A sclerotized caudal projection on larval Symphyta.

POSTCORNUA Noun. Posterior Cornu; the Cornu on the caudal side of the Basipharynx (MacGillivray).

POSTCOSTA Noun. 1. The Subcosta (Comstock). 2. Odonata: The first Anal Vein of Comstock. Trichoptera: The Anal Vein (Smith).

POSTCOSTAL Adj. (Latin, *post* = after; *costa* = rib; *-alis* = characterized by.) Descriptive of structure positioned behind the Costa.

POSTCOSTAL BRIDGE 1. The postcoxal part of the Pleuron, often united with the Sternum behind the Coxa (Snodgrass). The sclerite extending behind the Coxa and connecting the Epimeron with the Furcasternum (Crampton). Postcoxale.

POSTCOSTAL SPACE Odonata: The cell or cells positioned posterior of the Postcosta. The Anal Cell of Comstock.

POSTCOXALE Noun. The Postcoxal Bridge.

POSTCOXALIA Noun. The fused adjacent parts of the Sternellum and the Epimeron posterior to the Coxacoria (MacGillivray).

POSTCRANIUM Noun. (Latin, *post* = after; *cranium* = skull; Latin, *-ium* = diminutive > Greek, *-idion*. Pl., Postcrania.) Diptera: The entire posterior surface of the head. The surface may be flat, concave or convex, depending upon the group of insects. See Head Capsule. Cf. Cervix; Occiput.

POSTCUBITAL CROSSVEINS Postnodal crossveins. See Vein; Wing Venation.

POSTCUBITALS Noun. (Latin, *post* = behind; *cubitus* = the elbow; *-alis* = characterized by.) The postnodal spaces *sensu* Smith.

POSTCUBITUS Noun. (Latin, *post* = behind; *cubitus* = the elbow). 1. The usual sixth vein of the wing, represented by an independent Trachea in most nymphal wings; associated basally with the Cubitus in the adult. 2. The first Anal Vein *sensu* Comstock and Needham in most cases (Snodgrass).

POSTDORSULUM Noun. (Latin, *post* = behind; *dorsulum* = dim. of *dorsum* = the back.) The intermediate sclerite between the Mesophragma and the Postscutellum.

POSTEMBRYONIC Adj. (Latin, *post* = behind; Greek, *embryon* = embryo, foetus; *-ic* = characterized by.) 1. Descriptive of the period of life of an insect immediately following embryonic development within the egg. 2. Pertaining to any stage after the insect has emerged from the egg. See Development.

POSTEPISTOMA Noun. (Latin, *post* = behind; *epi* = upon; *stoma* = mouth. Pl., Postepistomae.) Hymenoptera: The portion of head posteriad of the Clypeus. See Postclypeus.

POSTERIOR Adj. (Latin, *posterior* = latter.) 1. A term of position pertaining to a structure or colour located behind the midline. 2. Toward the rear, caudal or anal end of the insect. 3. Pertaining to the surface of an appendage that is visible from the posterior aspect. See Orientation. Cf. Anterior; Medial; Lateral. Ant. Anterior. Adv. Posteriad.

POSTERIOR ANGLE 1. Hemiptera: Area near base of the Hemelytra; the posterior or anal angle of the wing. 2. Coleoptera: The lateral angle of the Thorax near the base of the Elytra.

POSTERIOR APOPHYSIS Female Lepidoptera: One of a pair of slender, chitinized rods that extend cephalad within the Abdomen from the eighth abdominal segment and provide a surface for muscle attachment (Klots).

POSTERIOR ARCULUS Wings of some insects: That part of the Arculus which is formed by a crossvein (Comstock).

POSTERIOR CALLOSITIES Diptera: The swellings at the posterior corners of the Mesonotum. Present in the Calyptratae and other Families, but typically absent from the Acalyptratae (Curran). Alt. Posterior Calli.

POSTERIOR CELLS Diptera: The first Cell (Williston); the Radial (Comstock); the fifth Cell (Williston); the Cubitus of Comstock (Smith).

POSTERIOR CEPHALIC FORAMEN Odonata: The opening of the head posteriorly through which the cavities of head and Thorax communicate. See Occipital Foramen.

POSTERIOR CLYPEUS See Postclypeus.

POSTERIOR CORNU See Postcornua (MacGillivray).

POSTERIOR CROP The second stomach in Homoptera (Hickernell).

POSTERIOR CROSSVEIN 1. Diptera: The vein or veins closing the discal cell apically (M and M3).

According to Schiner the basal section of Cu of the Comstock-Needham System (Curran). See Vein; Wing Venation.

POSTERIOR DORSOCENTRAL BRISTLES Diptera: A row of transverse bristles behind the transverse suture, Postsutural Dorsocentrals (Comstock).

POSTERIOR EDGE The boundary of the posterior margin of the wing (Say).

POSTERIOR FIELD In Tegmina, the anal field.

POSTERIOR FORAMEN The Foramen Magnum. Syn. Occipital Foramen.

POSTERIOR INTERCALARY Diptera: One of the Anal Veins *sensu* Comstock.

POSTERIOR INTESTINE The terminal section of the Proctodaeum commonly termed the 'Rectum,' but usually divided into an anterior rectal sac and a posterior Rectum proper (Snodgrass).

POSTERIOR LABRAL MUSCLE One of the two pairs of long muscles which move the Labrum, inserted posteriorly, usually on the Epipharyngeal Processes of the Tormae (Snodgrass).

POSTERIOR LAMINA Odonata: See Hamule (Garman).

POSTERIOR LATERAL MARGINS Orthoptera: The cuticular margin that extends from the base of the Pronotum downward to the posterior angle of the sides.

POSTERIOR LATERAL PLATES Coccidae: The caudal sclerites *sensu* MacGillivray.

POSTERIOR LOBE Pronotum in Orthoptera: See Lobe. Diptera: Part of the wing between the axillary incision and the base. The Alar Appendage of Loew.

POSTERIOR MARGIN Inner margin.

POSTERIOR MEDIAN The Media of Comstock (Tillyard).

POSTERIOR MESENTERON RUDIMENT Insect embryo: The posterior group of cells of the ventral Endoderm remnant that regenerates the Mesenteron (Snodgrass).

POSTERIOR NOTAL RIDGE See Notalia (MacGillivray).

POSTERIOR NOTAL WING PROCESS A posterior lobe of the lateral margin of the alinotum supporting the 3rd axillary sclerite of the wing base (Snodgrass).

POSTERIOR ORBITS Diptera: The part of the head immediately behind the eyes (Curran). The Occipital Orbit *sensu* MacGillivray.

POSTERIOR PEREION The Metanotum.

POSTERIOR PHARYNX A pharyngeal chamber of the Stomodaeum behind the Brain, present in Orthoptera, Coleoptera and some other insects (Snodgrass). The modified Stomodaeum.

POSTERIOR PLATE Hymenoptera: See Second Valvifer.

POSTERIOR PLEON A terminal segment of the Abdomen.

POSTERIOR PLEOPODA Lepidoptera larva: The anal clasping legs of caterpillars. See Planta.

POSTERIOR POLE Egg orientation: The end opposite the Pedicel is often causally referred to as the posterior end. However, egg orientation within the Ovary, relative to the orientation of the adult female that carries it, must be included in any discussion of anterior and posterior egg poles. Orientation. Cf. Anterior Pole.

POSTERIOR STIGMATAL TUBERCLE Lepidoptera larva: A tubercle which varies in position on the thoracic and abdominal segments of caterpillars.

POSTERIOR TENTORIAL ARMS The cuticular invaginations arising from the Posterior Tentorial Pits in the lower ends of the Postoccipital Sutures (Snodgrass). The Apodemes placed one on each side of the head, immediately above the articulation of the Cardo of the Maxilla (Imms). Syn. Gular Pit. See Tentorium. Cf. Anterior Tentorial Arms; Corporotentorium; Dorsal Rami. Rel. Endoskeleton.

POSTERIOR TENTORIAL PITS A pair of pits (cuticular invaginations) on the insect head located in the lower end of the Postoccipital Suture. See Tentorium. Cf. Anterior Tentorial Pits. Rel. Endoskeleton.

POSTERIOR TRAPEZOIDAL TUBERCLE Lepidoptera larva: A tubercle on the thoracic and abdominal segments of caterpillars. Positioned subdorsal, posterior and always present.

POSTERIOR TUBEROSITY Most insects: The prominent shoulder-like area at the base of the Anal Veins, sometimes divided into 2–3 tuberosities corresponding to the separate Anal Veins (Comstock).

POSTERIOR VEINS Veins of the insect which separate the Posterior Cells. See Vein; Wing Venation.

POSTERIOR WING Hindwing. The paired wings that originate on the insect's Metathorax. PWs are sometimes smaller or modfied from the size, shape and construction displayed by anterior wings. Syn. Inferior Wings; Secondaries. See Wings. Cf. Haltere; Hindwing. Rel. Anterior Wing; Flight.

POSTERODORSAL Noun. (Latin, *postero-* = comb. form meaning after or latter; *dorsum* = the back; *-alis* = characterized by. Pl., Posterodorsals.) Diptera: A bristle at the meeting of the dorsal and posterior faces of the leg.

POSTEROLATERAL Adj. (Latin, *postero-* = comb. form meaning after or latter; *latus* = a side; *-alis* = characterized by.) Descriptive of structure toward the rear and side.

POSTEROVENTRALS Noun. Latin, *postero-* = comb. form meaning after or latter; *ventral* = below; *-alis* = characterized by. Diptera: Leg bristles at the junction of the ventral and posterior surfaces of the leg.

POSTFRENUM Noun. (Latin, *post* = after; *frenum* = bridle. Pl., Postfrenums. Coleoptera: The part of the Metathorax in which the Postscutellum lies (Kirby & Spence); the Postscutellum of Snodgrass. Alt. Postfroenum.

POSTFRONS Noun. The part of the Vertex between

the median Ocellus and the Frons (MacGillivray).

POSTFRONTAL PHARYNGEAL DILATORS Muscles originating on the postfrontal region of the cranium and inserting on the Pharynx (Snodgrass).

POSTFRONTAL SUTURES Facial sutures present in some insects diverging from the Coronal Suture laterad of the Antennal bases (Snodgrass).

POSTFURCA Noun. (Latin, *post* = after; *furca* = fork. Pl., Postfurcae.) An internal process of the Metasternum to which the muscles of the hind legs are attached.

POSTGENA Noun. (Latin, *post* = after; *gena* = cheek. Pl., Postgenae.) 1. The lateral part and ventral parts of the Occipital Arch (Snodgrass). 2. The portion of the Cranium immediately posteriad of the Gena. 3. Diptera: The part of the Gena behind the Genal Suture (Comstock); the Hypostoma *sensu* MacGillivray. See Head Capsule.

POSTGENACERORES Noun. Coccids: The caudolateral groups of Genacerores (Comstock).

POSTGENAL Adj. (Latin, *post* = after; *gena* = cheek; *-al* = adjectival suffix.) Descriptive of or pertaining to the Postgena. Structure or colour positioned behind the Gena.

POSTGENAL BRIDGE Insect Head: A basic type of sclerotized separation of the Foramen Magnum and Labium. PGB is characterized by medial extension and fusion of the Postgenae, following a union of the Hypostoma. PGB may be interpreted as more derived (apomorphic) or 'advanced' condition of the Hypostoma. PGB appears in adults of Higher Diptera and aculeate Hymenoptera. The Posterior Tentorial Pits retain their placement in the Postoccipital Suture. See Head Capsule. Cf. Gula; Gulamentum; Hypostomal Bridge.

POSTGENITAL SEGMENTS Insect Abdomen: Segments posterior of the genitalia (Snodgrass). Modifications of postgenital sclerites are frequent as adaptations associated with copulation and oviposition. Some modifications include fusion of Tergum, Pleuron and Sternum to form a continuous sclerotized ring; fusion is notable in apterygote and pterygote insects. Fusion of segment 11 occurs in Machilidae (Thysanura); fusion of Terga 9–10 occurs in Acrididae (Orthoptera). Sometimes, fusion is restricted to one sex (*e.g.* male Odonata, Ephemeroptera, Dermaptera, female panorpid Mecoptera). Fusion of Sterna 2–3 occurs in some Coleoptera. Eleventh abdominal segment is last true Somite of insect body. Frequently this segment occurs in embryonic stage of primitive insects when it cannot be located in postemergent stages. Segment 11 forms a conical terminus that bears an Anus at the apex and lateral Cerci. Dorsal surface segment 11 is called an Epiproct; ventrolateral surface is called a Paraproct. Paraprocts are connected ventrally by a longitudinal, medial, membranous area. Primitive groups of living insects (*e.g.* Thysanura, Ephemeroptera) and some fossil groups (Palaeodictyoptera) display a conspicuous, long, median filament that projects from apex of Epiproct, called Appendix Dorsalis (Caudal Style). AD appears annulated and similar in shape to Cerci but function the function of AD not known. The 12th abdominal segment is called a Periproct (Crustacea) or Telson (some embryonic insects) and appears in adult Protura and naiadal Odonata.

POSTGLANDULAR AREA Hymenoptera: An area posterior of the Gradulus of the Tergum or Sternum.

POSTGULA Noun. (Latin, *post* = after; *gula* = throat.) Dermaptera: A sclerite at the extreme base and ventral surface of the head.

POSTHUMERAL BRISTLES Diptera: Usually two bristles, positioned above the Dorsopleural Suture, between Humeral Callus and wing base, and on bottom of presutural depression.

POSTICAL VEIN 1. Diptera: The fifth longitudinal vein *sensu* Meigen. 2. The Media *sensu* Comstock (Smith). See Vein; Wing Venation.

POSTICUS Noun. (Latin, *post* = after; *aceus* = quality.) Posterior.

POSTLABIAL AREA The Postmentum.

POSTLABIUM Noun. (Latin, *post* = after; *labia* = lip; *-ium* = diminutive. Pl., Postlabia.) 1. The lip-like modifications of the Peritreme bounding the posterior side of a spiracle (MacGillivray). 2. The proximal region of the body of the Labium; the Postmentum (Snodgrass).

POSTMANDIBULAR Adj. (Latin, *post* = after; *mandibula* = jaw.) Posterior of or behind the Mandible.

POSTMARGINAL VEIN 1. Hymenoptera (Chalcidoidea): A forewing vein along the anterior margin distad of the Stigmal Vein; Second Abscissa or part of Sc + R in the Comstock-Needham system. See Vein; Wing Venation.

POSTMAXATENDON The tendon attached to the ental surface of the Subgalea (MacGillivray).

POSTMEDIA Noun. (Latin, *post* = after; *medius* = middle.) Ephemeridae: An apparently distinct wing vein positioned between the Media and Cubitus (Comstock).

POSTMEDIANS Diptera: Leg bristles positioned above or behind the middle (Smith).

POSTMENTAL Adj. (Latin, *post* = after; *mentum* = the chin; *-alis* = pertaining to.) Descriptive of structure associated with the Postmentum; structure positioned posterior of or behind the Mentum.

POSTMENTUM Noun. The Postlabium, or basal part of the Labium proximal to the stipital region, or Prementum, when sclerotized, containing either a single Postmental Sclerite, or a Distal Mental Sclerite and a Proximal Submental Sclerite (Submentum, Walker, 1931) (Snodgrass).

POSTMETAMORPHIC Adj. (Latin, *post* = after; Greek, *metamorphosis* = a transformation; *-ic* = characterized by.) Conditions or features which follow Metamorphosis.

POSTNASUS Noun. (Latin, *post* = after; *nasus* = the nose. Pl., Postnasi.) Isoptera: Part of the Face immediately contigous to the Antennae and positioned behind the Nasus. See Nasus.

POSTNODAL COSTAL SPACES Odonata: The cells below costal margin from Nodus to Stigma.

POSTNODAL CROSSVEINS 1. Odonata: Crossveins positioned between the Costa and Radius and between Nodus to Stigma (Garman). 2. Postcubital Crossveins (Smith). See Vein; Wing Venation.

POSTNODAL RADIAL SPACES Odonata: A longitudinal vein positioned between Radial and Medial from Nodus to the outer margin (Comstock); Ultranodal Sector.

POSTNOTAL PLATE A sclerite posterior of each wing-bearing tergal sclerite, which bears the intersegmental attachments of the dorsal muscles (Snodgrass).

POSTNOTUM Noun. (Latin, *post* = after; Greek, *noton* = the back. Pl., Postnota.) 1. An intersegmental sclerite of the thoracic dorsum associated with the Tergum of the preceding segment. The Postnotum bears the Antecosta and usually a pair of internally projecting Phragmatal lobes. Syn. Phragmanotum; Pseudonotum (Snodgrass). 2. Diptera: An area posterior and ventral of the Scutellum; Postnotum is divided into a medial Mediotergite and two Laterotergites. Cf. Laterotergite.

POSTNUCLEAR BODY See Centriole Adjunct.

POSTOCCIPITAL RIDGE The internal aspect of the Postoccipital Suture, often produced into apodemal sclerites (Snodgrass).

POSTOCCIPITAL SUTURE A landmark on the posterior surface of the head and is typically near the Occipital Foramen. The Postoccipital Suture forms a Posterior Submarginal Groove of the head with posterior tentorial pits marking its lower ends on either side of the head. Some morphologists regard this suture as an intersegmental boundary between the first and second Maxillae (= Labium). Internally, the Postoccipital Suture forms the Postoccipital Ridge which serves as an attachment for dorsal prothoracic and cervical muscles of the head. Absence of a Postoccipital Suture in pterygote insects is a derived (apomorphic) condition. See Head Capsule. Cf. Occipital Suture.

POSTOCCIPUT Noun. (Latin, *post* = behind; *occiput* = back of head. Pl., Postocciputs.) The extreme posterior, often U-shaped, sclerite which forms the rim of the Cranium behind the Postoccipital Suture. Generally interpreted as a sclerotic remnant of the Labial Somite in the groundplan head of insects. See Head Capsule. Cf. Occiput; Vertex.

POSTOCELLAR AREA Hymenoptera: The region on the dorsal aspect of the head bounded by the Ocellar Furrow, Vertical Furrows and the caudal margin of the head. See Head Capsule.

POSTOCELLAR BRISTLES Diptera: A pair (or more) of bristles arising just below the Vertex on the Occiput and behind the Ocellar Tubercle, sometimes termed Post-verticals (Curran).

POSTOCELLAR GLANDS Apoidea: A mass of glands positioned near the Ocelli in the drones and queen (Bordas).

POSTOCULAR Adj. (Latin, *post* = behind; *oculus* = eye.) Pertaining to structure or colour posterior of the compound eyes.

POSTOCULAR SPOTS Odonata: Pale spots on the dorsum of the head in Zygoptera, behind and usually laterad of the Ocelli (Garman).

POSTOESOPHAGEAL COMMISSURE Commissure joining the Tritocerebral Lobes and passing immediately behind the Oesophagus.

POSTORAL Adj. (Latin, *post* = behind; *os* = mouth; *-alis* = characterized by.) Descriptive of structure positioned behind the mouth. A term applied to those segments bearing mouth structures.

POSTORBITAL BRISTLES Diptera: See Cilia of the posterior orbit (Comstock).

POSTPARADENSA Noun. Some coccids: The posterior group of paradensae when they are in two groups (MacGillivray).

POSTPARAPTERA Noun. (Latin, *post* = after; Greek, *para* = beside; *pteron* = wing. Sl., Postparapteron.) Epimeral Paraptera (Comstock).

POSTPARAPTERUM Noun. (Latin, *post* = after; Greek, *para* = beside; *pteron* = wing.) The Supraepimeron *sensu* MacGillivray.

POSTPECTUS Noun. (Latin, *post* = after; *pectus* = breast. Pl., Postpectoria.) The ventral surface of the Metathorax. Hymenoptera: The fused Sternum and Pleura.

POSTPEDES Noun. (Latin, *post* = after; *pedes* = feet.) Posterior legs or feet. The 3rd pair of legs in insects, see anal legs.

POSTPEDICEL Noun. (Latin, *post* = after; *pediculus* = small foot.) The 3rd Antennal segment or the segment immediately distal of the Pedicel.

POSTPETIOLE Noun. (Latin, *post* = after; *petiolus* = small foot.) Formicidae: The second segment or posterior segment of the Pedicel when two are present; the 3rd abdominal segment. See Gaster; Petiole.

POSTPHARYNGEAL DILATORS One or more pairs of muscles originating on the Vertex and inserting on the Stomodaeum posterior of the Brain. Alt. Dilatores Postpharyngeales (Snodgrass).

POSTPHARYNX Noun. (Latin, *post* = after; Greek, *pharyngx* = gullet.) The posterior, tubular portion of the Pharynx (MacGillivray).

POSTPHRAGMA Noun. (Latin, *post* = after; Greek, *phragma* = fence. Pl., Postphragmata.) 1. The posterior Phragma or inward projecting partition of any secondary (derived) body segment (Comstock). 2. The Phragma developed in relation with the Postnotum (Imms). 3. The Phragma of the caudal end of the Notum or Postscutellum (MacGillivray).

POSTPHRAGMINA Noun. The transverse external

opening or thickening marking the entrance to the Postphragma (MacGillivray).

POSTPLEURELLA Noun. The fourth sclerite of the Pleuron (Jardine).

POSTPLEURON Noun. Pl., Postpleura.) The Epimeron (MacGillivray).

POSTPRONOTUM Noun. (Latin, *post* = after; Greek, *pro* = before; *notum* = the back. Pl., Postpronota.) The posterior region of the Pronotum (Crampton).

POSTRETINAL Adj. (Latin, *post* = after; *rete* = a net; *-alis* = pertaining to.) Descriptive of structure positioned behind the eye's Retina.

POSTRETINAL FIBRES Nerve fibres arising from the Ommatidia which pass into the ganglionic sclerite of the insect Brain (Packard).

POSTSCUTELLUM Noun. (Latin, *post* = behind; *scutella* = drinking bowl. Pl., Postscutella.) 1. The fourth and posterior sclerite of the dorsum of the thoracic rings (Costal, Subpostdorsum, Metaphragma, Pseudonotum, Postnotum or Acrotergite) in the Metanotum. 2. The Postfroenum (MacGillivray). 3. Diptera: A convex transverse swelling below the Scutellum, actually the upper posteriorly produced section of the Metanotum (Curran). 4. The Pseudonotum.

POSTSCUTUM Noun. (Pl., Postscuta.) Trichoptera: The small sclerite posterior of the mesothoracic Scutellum. Syn. Postscutellum.

POSTSTERNELLUM Noun. (Pl., Poststernella.) The presumed posterior division of the Sternum, positioned behind the Sternellum.

POSTSTERNITE Noun. (Latin, *post* = after; *sternum* = breastbone; *-ites* = constituent. Pl., Poststernites.) The postcostal lip of a definitive sternal sclerite that includes the intersegmental sclerotization following (Snodgrass).

POSTSTETHIDIUM Noun. (Latin, *post* = after; Greek, *stethos* = chest; *idion* = dim. form.) Acarology: The unsclerotized, posterior part of the Prodorsum.

POSTSTIGMATAL CELL Part of the Marginal Cell beyond the Stigma. Bees: the second Radial Cell of Comstock (*teste* Smith).

POSTSTIGMATAL PRIMARY TUBERCLE Lepidoptera larva: A tubercle on the thoracic segment of caterpillars; subprimary, stigmatal, posterior.

POSTSUBTERMINAL Adj. (Latin, *post* = after; *sub-* = below; *terminus* = a boundary; *-alis* = pertaining to.) Lepidoptera: Following the subterminal transverse line.

POSTSUTURAL BRISTLES Diptera: Dorsal bristles behind Transverse Suture.

POSTSUTURAL DORSOCENTRALS Diptera: See Posterior Porsocentrals.

POSTTENTORIUM Noun. (Latin, *post* = behind; *tentorium* = tent. Pl., Posttentoria.) The Posterior Tentorial Arms *teste* Crampton. See Tentorium. Cf. Dorsal Ramus.

POSTTERGA Noun. (Latin, *post* = after; *terga* = (Pl.) backs.) Coleoptera larvae: The posterior Scutes.

POSTTERGITE Noun. (Latin, *post* = behind; *tergum* = back; *-ites* = constituent. Pl., Posttergites.) 1. The narrow postcostal lip of a postnotal thoracic sclerite (Snodgrass). 2. The posterior sclerite of the Eunotum (Crampton).

POSTTRIANGULAR CELLS The Discoidal Areolets.

POSTVERTEX Noun. (Latin, *post* = behind; vertex. Pl., Postvertexes, Postvertices.) The posterior area of the Vertex when divided into two parts by fusion of compound eyes (MacGillivray).

POSTVERTICAL BRISTLES Diptera: Posterior pair of Setae comprising lesser ocellars *sensu* Comstock. The Postocellars *sensu* Curran. Alt. Postverticals.

POSTVERTICAL CEPHALIC BRISTLES Diptera: Setae in the middle of the upper part of Occiput.

POTAMANTHIDAE Plural Noun. A small Family of Ephemeroptera assigned to Superfamily Ephemeroidea. Wing span 7–13 mm; Vein 1A forked near wing margin; hindwing with numerous veins and cross veins; all legs well developed in both sexes. Naiads occur in swift-flowing shallow water.

POTASSIUM Noun. (Modern Latin from potash.) A silvery, soft metal not found in a free state, but important as an element in saltpeter and other common compounds. Symbolised by K.

POTATO APHID *Macrosiphum euphorbiae* (Thomas) [Hemiptera: Aphididae]: A pest of potato in northeastern North America; distrubuted throughout USA. In colder regions, sexually reproducing females lay overwintering eggs typically on rose stems; spring generation migrates to other plants. In warmer regions females are parthenogenetic and produce live young; winter and summer generations on same plant. Host plants numerous and diverse, including field crops, tree crops and flowers. Heavy infestations may result in death of potato plants. Aka: Pink-and-Green Tomato Aphid.

POTATO BUG *Calocoris norvegicus* (Gmelin) [Hemiptera: Miridae].

POTATO FLEA-BEETLE *Epitrix cucumeris* (Harris) [Coleoptera: Chrysomelidae].

POTATO FLEA-BEETLES Members of the Genus *Psyllioides* spp. [Coleoptera: Chrysomelidae].

POTATO LEAFHOPPER. *Empoasca fabae* (Harris) [Hemiptera: Cicadellidae]: A pest of potato in eastern North America; also feeds on many other plants including alfalfa, clover, celery, soybeans and peanuts. PLH produces hopperburn. Adult ca 3 mm long; pale green with white spots on head and Thorax; body wedge-shaped and widest at head; hindlegs saltatorial. Females live up to 60 days and lay 2–3 eggs per day. Egg white, elongate, ca 1 mm long; female's Ovipositor inserts eggs into main veins or petioles on underside of leaves; eclosion occurs within 6–10 days. Neonate nymph wingless, pale; mature nymphs greenish-yellow; five nymphal instars. PLH completes 1–4 generations per year, depending upon climate and host plant. Populations demonstrate migratory tendencies from northern areas toward south; PLH cannot overwinter in northern extremes of distribution or tolerate heat of summer in southern extremes of distribution. Rel. Hopperburn. See Cicadellidae.

POTATO MOTH *Phthorimaea operculella* (Zeller) [Lepidoptera: Gelechiidae].

POTATO MOTH See Potato Tuberworm.

POTATO PSYLLID *Paratrioza cockerelli* (Sulc) [Hemiptera: Psyllidae]: Widespread in North America from British Columbia south to Mexico. Eggs petiolate, orange, attached to the underside of leaves. Five nymphal instars; nymphs transmit psyllid yellows. Migrate from New Mexico and Texas to potato growing areas; attack many solanaceous Species. Syn. Tomato Psyllid.

POTATO STALK-BORER *Trichobaris trinotata* (Say) [Coleoptera: Curculionidae]: A monovoltine pest of eggplant, ground cherry, potato and jimson weed in eastern North America. Overwinters as an adult in old vines. Eggs deposited singly in stem or petiole cavities; eclosion occurs within 7–10 days. Neonate larva bores up and down stem, causing plant wilt. Mature larva packs stem with scrapings from stem; pupation within stem. Adult develops rapidly but delays emergence until following spring.

POTATO STEM-BORER *Hydraecia micacea* (Esper) [Lepidoptera: Noctuidae].

POTATO TUBERMOTH *Scrobipalpopsis solanivora* Povolny [Lepidoptera: Gelechiidae]: A minor pest of potatoes in Central America.

POTATO TUBERWORM *Phthorimaea operculella* (Zeller) [Lepidoptera: Gelechiidae]: Endemic to South America and widespread in USA; a multivoltine pest of potato, tomato, tobacco, eggplant and solanaceous weeds. Adult brownish-grey, wingspan 12–16 mm; hide in foliage or on ground during day and become active at dusk. Preoviposition two days; female lays 50–100 eggs during 14 day period. Eggs pearly then yellowish; laid individually on leaves, stems or eyes of exposed tubers; eclosion occurs ca 5–14 days. Larvae mine stems; feed internally on leaves webbed together with silk; larvae move from leaves to tubers via the stem; four larval instars; mature larva leaves feeding site, spins flimsy cocoon on ground, trash or host plant. PTW undergoes up to 12 generations per year. Syn. Tobacco Splitworm; Potato Moth (Australia).

POTATO WIREWORM *Hapatesus hirtus* Candeze [Coleoptera: Elateridae].

POTATO-SCAB GNAT *Pnyxia scabiei* (Hopkins) [Diptera: Sciaridae].

POTOMAC HORSE FEVER A rickettsia-like disease of horses in USA, Canada and France. Aetiological agent *Ehrlichia risticii* assigned to Tribe Ehrlichieae of Rickettsiaceae. Vectors of disease not known. See Rickettsiaceae. Cf. Canine Ehrlichiosis; Heartwater.

POTRUNCUS Noun. The metathoracic ring (Kirby & Spence).

POTTER WASPS Eumeninae [Hymenoptera: Vespidae].

POTTER, CHARLES (1907–1989) English pesticide scientist. (Needham 1990, Antenna 14 (2): 57–60.)

POTTER, W R R (–1939) (Swezey 1940, Proc. Hawaii ent. Soc. 10: 455.)

POTTINGER, HAROLD LANDSEER (1889–1971) (Anon. 1971, News Bull. ent. Soc. Qld 79: 9.)

POUCH Noun. (Middle English, *pouche* = bag. Pl., Pouches.) 1. A large sac. 2. Trichoptera: A depressed, usually longitudinal area in a wing.

POUCH GALL Plant galls characterized by bulging of the leaf blade. Often induced on young leaves by first stimulation of gall incitor. Subsequent stimulation in same place causes intense arching of leaf tissue with bulge on one side and pouch on other side of leaf. See Gall. Cf. Bud-and-Rosette Gall; Covering Gall; Filz Gall; Mark Gall; Pit Gall; Roll-and-Fold Gall.

POUCHET, FELIX ARCHIM (1800–1872) (Beaurain *et al.* 1877, Bull. Soc. Amis Sci. nat. Rouen (2) 13: 175–229.

POUJADE, GUSTAVE ARTHUR (–1909) (Anon. 1909, Ent. Rdsch. 26: 115.)

POULTON, EDWARD BAGNALL (1856–1943) English academic and specialist in animal coloration. Second Hope Professor of Zoology (Oxford University.) (Carpenter 1944, Nature 153: 15–17.)

POULTRY AIRSAC MITE *Cytodites nudus* (Vizioli) [Acari: Cytoditidae].

POULTRY BODY LOUSE *Menacanthus stramineus* (Nitzch) [Phthiraptera: Menoponidae].

POULTRY BUG *Haematosiphon inodorus* (Gugès) [Hemiptera: Cimicidae].

POULTRY-CHEWING LICE [Phthiraptera: Menoponidae].

POULTRY CYST MITE *Laminosioptes cysticola* (Vizioli) [Acari: Laminosioptidae].

POULTRY FEATHER MITE *Rivoltasia bifurcata* (Rivolta) [Acari: Epidermoptidae].

POULTRY FLUFF LOUSE *Goniocotes gallinae* (De Greer) [Acari: Philopteridae].

POULTRY HEAD LOUSE *Cuclotogaster heterographus* (Nitzsch) [Acari: Philopteridae].

POULTRY-HOUSE MOTH *Niditinea spretella* (Denis & Schiffermüller) [Lepidoptera: Tineidae].

POULTRY RED MITE *Dermanyssus gallinae* (De Greer) [Acari: Dermanyssidae].

POULTRY SHAFT LOUSE *Menopon gallinae* (Linnaeus) [Phthiraptera: Menoponidae].

POULTRY STICKFAST FLEA *Echidnophaga gallinacea* (Westwood) [Siphonaptera: Pulicidae].

POULTRY WING LOUSE *Lipeurus caponis* (Linnaeus) [Phthiraptera: Philopteridae].

POUNCE® See Permethrin.

POUND, C J (1866–1946) Australian founder of Veterinary Entomology and Government Bacteriologist who worked on tick fevers. (Mackerras & Marks 1974, Changing patterns in Entomology. A symposium. iv + 76 pp., p. 7. Brisbane.)

POUPART, FRANCOIS (1661–1709) (Rose 1850, *New General Biographical Dictionary* 11: 215.)

POUTIERS, R (–1970) (Anon. 1970, Bull. Soc. Ent. Fr. 75: 165.)

POWDER-KEG LONGICORN *Gracilia minuta* (Fabricius) [Coleoptera: Cerambycidae].

POWDER-POST BEETLE *Lyctus brunneus* (Stephens) [Coleoptera: Bostrichidae]: A cosmopolitan pest of sapwood of unfinished hardwoods, picture frames, bamboo and baskets; possibly endemic to North America. Larva and adult are pests; larva requires starch for development. Adult 2–7 mm long, elongate, dark brown, shiny; Antennal club with two segments. Female bites wood and lays egg in incision; 1–3 eggs per incision and ca 70 eggs per female during lifetime; incubation 6–15 days. Larva white, scarabaeiform with large spiracles on eighth abdominal segment; larval stage two months to one year, depending upon temperature, humidity and starch in sapwood. Mature larva tunnels toward surface, excavates oval cell for pupation; pupal stage 12–27 days. Adult stage 78–300 days. Reinfestation of timber common. Cf. Small Powder-Post Beetle; Malayan Powder-Post Beetle. See Bostrichidae; Lyctidae.

POWDER-POST BOSTRICHID *Amphicerus cornutus* (Pallas) [Coleoptera: Bostricidae].

POWDER-POST TERMITES See Kalotermitidae.

POWELL, HAROLD (1875–1954) (Buxton 1955, Proc. R. ent. Soc. Lond. (C) 19: 67.)

POWELL, JOHN ARTHUR (1810–1886) (Dunning 1886, Entomologist 19: 193–200, bibliogr.)

PPE See Personal Protection Equipment.

PRADHAN, SHYAN SUNDER LAL (1913–1973) (Kapoor 1975, Entomol. mon. Mag. 110: 125.)

PRADIER, JULES (1807–1858) (Lafont 1858, Bull. Soc. ent. Fr. (3) 6: ccvii-ccviii.)

PRAEAULACIDAE Rasnitsyn 1972. Plural Noun. A Family of parasitic Hymenoptera assigned to the Evanioidea.

PRAEBRACHIAL Adj. (Latin, *prae* = before; *brachialis* = arm; *-alis* = pertaining to.) 1. Descriptive of structure anterior of the Brachium. 2. A longitudinal vein in middle of an ephemerid wing, usually forked; number 6 of some systems. Alt. Prebrachial.

PRAECOSTAL SPUR A false vein in the costal angle at the base of the hindwings. Alt. Precostal spur.

PRAED, C W M See Mackworth-Praed, C. W.

PRAEDORSUM Noun. The anterior part of the dorsum. Alt. Predorsum.

PRAEFURCA Noun. See Prefurca (Curran).

PRAEICHNEUMONIDAE Rasnitsyn 1983. Plural Noun. A small Family of apocritous Hymenoptera assigned to the Ichneumonoidea.

PRAELABRUM Noun. See Prelabium.

PRAEOCULAR Adj. See Preocular.

PRAEPUTIUM Noun. (Latin, *praeputium* = the foreskin.) The external membranous covering of the Penis. Orthoptera: A spherical muscular mass at the base of the Penis. Alt. Preputium.

PRAESCUTELLUM Noun. A sclerite, rarely present, between the Mesoscutum and Mesoscutellum. See Prescutellum.

PRAESCUTUM Noun. The first of the four divisions of the Notum of the thoracic rings. See Prescutum.

PRAESIRICIDAE Rasnitsyn 1968. Plural Noun. A Family of Symphyta (Hymenoptera) assigned to the Pamphilioidea.

PRAESUBTERMINAL Adj. Lepidoptera: Preceding the subterminal transverse line. See Presubterminal.

PRAETARSUS Noun. See Pretarsus.

PRAETERGA Noun. Coleoptera larvae: The anterior thoracic Scutes. See Preterga.

PRAETORNAL See Pretornal.

PRAHWE, KONSTANTIN KONSTANTINOVICH (1868–1910) (Lutshnik 1911, Revue Russe Ent. 11: 309–310.)

PRAIRIE ANT *Formica montana* Emery [Hymenoptera: Formicidae]: A small, chocolate-brown ant assigned to the *F. fusca* group. Typically found on prairies and open-field habitats in North America. PA constructs mounds to two metres in diameter to capture sun's radiant energy and regulate nest temperatures.

PRAIRIE FLEA-BEETLE *Altica canadensis* Gentner [Coleoptera: Chrysomelidae].

PRAIRIE GRAIN-WIREWORM *Ctenicera aeripennis destructor* (Brown) [Coleoptera: Elateridae].

PRAMEX® See Permethrin.

PRASINOUS Adj. (Latin, *prasine* = *verdigris* from Greek *presinos* = leek; Latin, *-osus* = with the property of.) Grass-green; light green tending to yellow. Alt. Prasinus.

PRATINICOLOUS Adj. (Late Latin, *pratincola*, from *pratum* = meadow; *colere* = living in or on; Latin, *-osus* = with the property of.) Descriptive of organisms that visit or live in grassy meadows or bogs. See Habitat. Cf. Agricolous; Caespiticolous; Madicolous; Paludicolous; Ripicolous; Silvicolous. Rel. Ecology.

PRATT, BENJAMIN G (1862–1947) (Anon. 1948, J. Econ. Ent. 41: 840–841.)

PRATT, FREDERICK C (1869–1911) (Hunter 1911, Proc. ent. Soc. Wash. 13: 189–190.)

PRAYING MANTID See European Mantid.

PRAYING MANTIDS Mantodea.

PREABDOMEN Noun. (Pl., Preabdomens.) The unmodified anterior abdominal segments of insects (Snodgrass). Diaspindinae: First four abdominal segments (MacGillivray). Exposed segments of female insect Abdomen when others are retracted within a Genatasinus (MacGillivray).

PREACROSTICHALS Noun. Diptera: Anterior Acrostichal Bristles in front of Transverse Suture (Comstock).

PREADAPTATION Noun. (New Latin.) Any extant anatomical structure, biological feature or behavioural pattern which enhances evolutionary adaptation.

PREALAR Adj. (Latin, *pre* = before; *ala* = wing.) Pertaining to structure, shape or colour anterior of the wing.

PREALAR BRIDGE Extensions from anterior part of Prescutum and Antecosta which serve as anterior supports of Tergum on Pleura. PB heavily sclerotized and forming a rigid support between unsclerotized membrane of Prothorax and Pleuron (Needham). Sclerite extending in front of wing and connecting Notum with Episternum (Crampton).

PREALAR BRISTLE Diptera: Anterior Supraalar Bristle, frequently absent or reduced, used particularly in Muscidae (Curran). A bristle in Anthomyiidae inserted just posterior of Transverse Suture in line with Supraalar Bristles (Comstock).

PREALAR CALLUS Diptera: An inconspicuous projection positioned anterior of wing base on each side of Mesonotum, just posterior of outer end of Transverse Suture (Comstock).

PREALARE See Prealar Bridge.

PREALARIA Noun. (Pl., Prealariae.) The Prealare (MacGillivray).

PREALIFERA Noun. The anterior of the two sclerites, anterior of Pleuralifera; Anterior Basalare; First Parapteron (MacGillivray).

PREANAL Adj. (Latin, *pre* = before; *anus* = anus; *-alis* = pertaining to.) Descriptive of structure above or anterior of the anal opening.

PREANAL AREA Insect wing: Part positioned in front of anal area including all wing except anal area (Comstock).

PREANAL LOBE Hindwing of Hymenoptera: Part of anal area between axillary and preaxillary excisions (Comstock).

PREANAL PLATE or LAMINA See Supraanal Plate.

PREANAL PLATES Coccids: Eleventh pair of dorsal sclerites on seventh abdominal segment anterior of Anal Ring (MacGillivray).

PREANAL REGION See Remigial Region.

PREANTENNA Noun. (Latin, *pre* = before; *antenna* = sail yard. Pl., Preantennae.) Theoretically, a pair of primitive procephalic appendages anterior of the Antennae, possibly represented in *Scolopendra* and *Dixippus* by a pair of embryonic preantennal lobes. Preantennae are absent from all adult arthropods according to Snodgrass.

PREANTENNAL Adj. (Latin, *pre* = before; *antenna* = sail yard; *-alis* = pertaining to.) Descriptive of structure anterior of the Antenna.

PREANTENNAL APPENDAGES Structures in front of the Antennae.

PREAPICAL BRISTLE Diptera: A dorsal, short Seta (bristle) positioned before the apex of the Tibia (Curran).

PREAPICAL Adj. (Latin, *pre* = before; *apex* = tip; *-alis* = pertaining to.) Descriptive of structure or colour positioned anterior of the apex of the structure.

PREARTICULAR Adj. (Latin, *pre* = before; *articulus* = joint.) Placed or set anterior of a joint.

PREARTIS Noun. The Artis articulating in the Precoila. See Epicondyle (MacGillivray).

PREAXIL Adj. (Latin, *pre* = before; *axis* = axis.) Anterior of or placed before the axis.

PREAXIL SURFACE OF THE LEG The anterior surface when the limb is extended at a right angle to the body (Snodgrass).

PREAXILLARY EXCISION Hindwing of Hymenoptera: A second excision of the apex of the first anal fold. Additional to the axillary excision (Comstock).

PREBALANCER Noun. See Prehalter.

PREBASILAIRE Noun. (Latin, *pre* = before; *basis* = pedestal, base.) The Basilare with its anterior raised margin swollen into a thick callosity (Jardine).

PREBASILAR Adj. Before the base.

PRECALCAR Noun. (Latin, *pre* = before; *calcar* = spur. Pl., Precalcars.) Hymenoptera: Distal or leading surface of Sting in Aculeata or serrated surface of Ovipositor in Parasitica and Symphyta. Cf. Postcalcar.

PRECAMBRIAN Noun. (New Latin.) A term which has no formal status in the Geological Time Scale, but which corresponds to the Archean and Proterozoic Eons. See Geological Time Scale. Cf. Archaean Eon; Proterozoic Eon.

PRECEPHALIC Adj. (Latin, *pre* = before; *kephale* = head; *-ic* = of the nature of.) Anterior of the front of the head.

PRECEREBRAL Adj. (Latin, *pre* = before; *cerebrum* = brain; *-alis* = pertaining to.) Descriptive of structure anterior of the Brain. Cf. Postcerebral.

PRECIPITIN Noun. (New Latin, from *praecipitare* > *praeceps* = headlong.) A chemical substance produced in the blood of insects by precipitation through the mixture of different blood-plasmas (Imms).

PRECISION® See Fenoxycarb.

PRECLAVUS Noun. (Latin, *pre* = before; *clava* = club.) The remigial region.

PRECLUDE® See Fenoxycarb.

PRECLUNIAL Adj. (Latin, *pre* = before; *clunes* = rump; *-alis* = pertaining to.) Psocoptera: Pertaining to abdominal segments basad of the Clunium. See Clunium.

PRECLYPEUS Noun. (Latin, *pre* = before; *clypeus* = shield.) The anterior, unpaired sclerite borne by the Postclypeus. Syn. Second Clypeus; Anteclypeus; Infraclypeus; Rhinarium (MacGillivray).

PRECOCENE Noun. (New Latin.) A natural insecticide first isolated from leaves of the Central American composite *Ageratum houstonianum* Miller. Induces antijuvenile hormone effect through precocious development and moulting in immature insects. Mode-of-action through tissue-specific toxic effect on Corpora Allata.

PRECOCIOUS STAGES All the stages of development from the fertilized egg to the Pupa.

PRECOILA Noun. (Pl., Precoilae.) Emargination or Acetabulum on each Clypealia on which each

Mandible articulates; the dorsal Condyle. The Epicondyle (MacGillivray).

PRECOR® See Methoprene.

PRECORNUA Noun. (Latin, *pre* = before; *cornu* = horn.) The Cornua of the cephalic side of the Basipharynx (MacGillivray). Syn. Anterior Cornu.

PRECOSTA Noun. (Latin, *pre* = before; *costa* = rib. Pl., Precostae; Precostas.) 1. A small first vein of the wing in certain fossil insects (Snodgrass). Typically, Precosta is fused with Costal Vein. Precosta represented by a serrated strip in Odonata; in Coleoptera Precosta forms elytral Epipleuron. 2. The well developed marginal process of the Prothorax. The Postscutellum *sensu* Needham.

PRECOSTAL AREA Fulgoroidea: The space between Costal Vein and anterior margin of the wing.

PRECOXA Noun. (Latin, *pre* = before; *coxa* = hip.) Collembola: One of the two joints preceding the Coxa.

PRECOXAL Adj. (Latin, *pre* = before; *coxa* = hip; *-alis* = pertaining to.) Descriptive of or pertaining to the Precoxa; anterior of the Coxa.

PRECOXAL BRIDGE The Precoxale. The Precoxal part of the Pleuron anterior of the Trochantin, usualy continuous with the Episternum, frequently united with the Basisternum, sometimes a distinct sclerite.

PRECOXALE Noun. (Pl., Precoxalia.) The Precoxal Bridge.

PRECOXALIA Noun. The fused adjacent parts of the Sternum and the Episternum anterior of the Coxacoria (MacGillivray).

PREDACEOUS Adj. (Latin, *praeda* = prey; *-aceus* = of or pertaining to.) Pertaining to an animal which typically sustains life by preying upon other organisms. Alt. Predacious; Predaceousness; Predaciousness. See Predator. Cf. Prey; Parasitic.

PREDACEOUS DIVING BEETLES See Dysticidae.

PREDACEOUS MITE See Predatory Mite.

PREDACITY Noun. (Latin, *praedatio*, from *praedari* = to plunder; English, *-ity* = suffix forming abstract nouns. Pl,. Predacities.) A predaceous quality or state of being.

PREDATION Noun. (Latin, *praedatio* > *predatus* > *praedari* = to plunder; English, *-tion* = result of an action. Pl., Predations.) 1. A mode of living in which an animal kills and consumes another animal. See Predator. 2, A type of behaviour associated with the acqusition of animal food. See Carnivory. Cf Herbivory; Parasitism; Saprophagy. Rel. Digestion; Extra-oral Digestion.

PREDATISM Noun. (Latin, *praedatio*, from *praedari* = to plunder; English, *-ism* = condition.) The act or state of being predatory. Alt. Predation. See Symbiosis. Cf. Commensalism; Inquilinism; Parasitism.

PREDATIVE Adj. (Latin, *predaetor* = hunter.) Predatory.

PREDATOR Noun. (Latin, *predaetor* = hunter; *or* = one who does something. Pl., Predators.) A animal that overpowers, kills and consumes other animals (prey). Predators constitute one of three principal categories of Natural Enemies used in applied biological control. See Predation. Cf. Pathogen; Parasite.

PREDATOR® See Chlorpyrifos.

PREDATORY Adj. (Latin, *praedatorius*.) Pertaining to predaceous behaviour.

PREDATORY MITE *Amblyseius fallacis* (Garman); *Amblyseius womersleyi* Schicha; *Phytoseius* spp.; *Typhlodromus occidentalis* Nesbitt; *Typhlodromus pyri* Scheuten [Acari: Phytoseiidae].

PREDATORY SHIELD BUG *Cermatulus nasalis* (Westwood); *Oechalia schellembergii* (Guérin-Méneville) [Hemiptera: Pentatomidae]: In Australia, a predator of cotton pests. Adult 7–12 mm long, elongate-elliptical; body brown with black markings and paler punctations; mouthparts extend to hind Coxa; lateral projections from Pronotum sharply spined; Hemelytron projects beyond apex of Abdomen; membranous part of wing is smoky-brown with dark brown veins. Black eggs laid in 'rafts' usually in multiples of 14, with long, white spines around the rim. Nymphs dark red and brown; lack wings.

PREDOMINANT Adj. (French > Latin, *pre* = before; *dominare* = to rule; *-antem* = adjectival suffix.) Ecology: An organism which predominates over others.

PREDOTA, KARL (1873–1962) (Reisser 1962, Z. wien ent. Ges. 47: 42–44)

PRE-EPISTERNUM Noun. 1. The anterior part of the Episternum marked as a separate sclerite (Imms). 2. A sclerite anterior of the Episternum in some of the more generalized insects (Comstock). 3. The Episternal Laterale *sensu* Crampton.

PRE-ERUCIFORM Proctotrupidae: Before the caterpillar stage, specifically applied to early larval development. See Eruciform. Rel. Form 2; Shape 2; Outline Shape.

PREFEMUR Noun. (Latin, *pre* = before; *femur* = thigh. Pl., Prefemurs.) See Ischiopodite.

PREFORMATION Noun. (Latin, *pre* = before; *formatus* = form, shape. Pl., Preformations.) An hypothesis in biology which asserted that each Ovum of an animal contains a complete, microscopic copy of the parent. Under this concept, development entails only an increase in size. The concept of Preformation is rejected today in favour of the Concept of Epigenesis. See Epigenesis.

PREFRONS Noun. (Latin, *pre* = before; *frons* = forehead.) Psocididae: The Postclypeus; a conspicuous sclerite often presenting an inflated appearance.

PREFURCA Noun. (Latin, *pre* = before; *furca* = fork. Pl., Prefurcae.) Diptera: 'the stem vein in front of a fork, that reaches back to where itself forks from another vein' (Smith). The Petiole of the 2nd and 3rd longitudinal veins; the base of R + , the Radial Sector, Rs (Curran).

PREGENACERORES Noun. The cephalolateral groups of genacerores in coccids (Comstock).

PREGENICULAR Adj. (Latin, *pre* = before; gena = cheek.) Anterior of the knee. Orthoptera: The part of the Femur proximad of the 'knee.'

PREGENICULAR ANNULUS A more or less conspicuously coloured ring on the caudal Femora proximad to the knee in Orthoptera.

PREGENITAL Adj. (Latin, *pre* = before; *genitum* = to begat; *-alis* = pertaining to.) Descriptive of structure anterior of the genital segment.

PREGENITAL SEGMENTS 1. Insect Abdomen: segments anterior of the genitalia, and which contain the viscera. 2. Visceral segments *sensu* Snodgrass.

PREGRADULAR AREA Hymenoptera: An elevated area anteriad of the Gradulus which slides, rests, or articulates with the preceding Tergum or Sternum.

PREGULA Noun. (Latin, *pre* = before; *gula* = throat.) The Submentum 'incorrectly' (MacGillivray). Alt. Gular bar.

PREHALTER Noun. (Latin, *pre* = before; *halter* = weight. Pl., Prehalters.) Diptera: A membranous 'scale' in front of the true Haltere.

PREHENSILE Adj. (Latin, *prehensum* = to lay hold of; *-ilis* = suitable for.) Pertaining to structure adapted for grasping, holding or seizing.

PREHENSION Noun. (Latin, *prehensum* = to lay hold of.) The act of grasping or holding.

PREIMAGINAL Adj. (Latin, *pre* = before; *imago* = image; *-alis* = pertaining to.) Anteceding or preceding the adult stage or Imago of insects.

PREISS, PAUL (1859–1938) (Schultze-Rhonof 1938, Dt. ent. Z. Iris. 52: 47–49.)

PRELABIA Noun. (Latin, *pre* = before; *labium* = lip.) The anterior, lip-like modification of a Peritreme (MacGillivray).

PRELABIUM Noun. (Latin, *pre* = before; *labium* = lip; *-ium* = diminutive > Greek, *-idion*. Pl., Prelabia.) The distal part of the Labium, and comprising the Prementum, Ligula, and Palpi. The Eulabium *sensu* Snodgrass.

PRELABRUM Noun. (Latin, *pre* = before; *labrum* = lip.) Diptera: The Clypeus. Alt. Praelabrum.

PRELARVA Noun. (Latin, *pre* = before; *larva* = a ghost. Pl., Prelarvae) A quiescent, non-feeding stage of some acarines and arachnids. Form highly variable and ranging from a featureless sac to a body with mouthparts and three pairs of legs.

PRE-LINNAEAN NAME A name in Zoology printed before 1758.

PRELL, HEINRICH BERNARD (1888–1962) (Francke-Grosmann 1962, Anz. Schädlingsk. 35: 126–127.)

PRELLER, CARL HEINRICH (1830–1890) (Weidner 1967, Abh. Verh. naturw. Ver. Hamburg Suppl. 9: 117–118.)

PREMANDIBULAR Adj. (Latin, *pre* = before; *mandibulum* = jaw; *-ar* = pertaining to.) Positioned in front of the Mandible, applied to a temporary segment of the embryo, the intercalary segment.

PREMANDIBULAR APPENDAGES See postantennal appendages.

PREMATURATION PERIOD That part of the life-cycle of an insect between emergence from the egg and commencement of sexual maturity (Follsom & Wardle).

PREMAXATENDON Noun. The Maxatendon attached to, or near to, the Entopartartis (MacGillivray).

PREMEDIA Noun. (New Latin.) Ephemeridae: An apparently distinct vein between Radius and Media (Comstock).

PREMENTAL Adj. (Latin, *pre* = before; *mentum* = chin; *-alis* = pertaining to.) Descriptive of structure associated with the Prementum.

PREMENTUM Noun. (Latin, *pre* = before; *mentum* = chin.) The stipital region of the Labium, containing the muscles of the palpi and the Ligular lobes, and giving insertion to the cranial muscles of the Labium (mentum, Walker 1931) (Snodgrass), an appendage of the insect Labium, borne on the Notum (Palpiger of some authors) (Imms), the region of the Labium distal to the Mentum and formed by the Palpigers and labial stipites (Crampton).

PREMGARD® See Resmethrin.

PREMIER® See Imidacloprid.

PREMISE® See Imidacloprid.

PREMORSE Adj. (Latin, *pre* = before; *mordere* = to bite.) Descriptive of structure which terminates in an irregular, truncate apex as if bitten off (Kirby & Spence). Alt. Premorsus.

PREMOULTING Adj. (New Latin > Latin, *pre* = before; *mutare* = to change.) Preceding, or before, moulting.

PRENFISH® See Rotenone.

PRENOTA Noun. The Praescutum. (MacGillivray).

PRENSISETA Noun. (Latin, *prensare* = to seize; *seta* = hair. Pl., Prensisetae.) Short, stout Setae at the ends of flattened, twisted rods on the Surstyli of male tephritid flies (Munro 1947, Mem. Ent. Soc. So. Afr. 1: 57); used to grasp female aculeus for copulation.

PRENSOR Noun. (Latin, *prensare* = to seize; *or* = one who does something.) The genital lateral clasping organ of male Lepidoptera, See Clasper.

PRENTISS, EDWIN CHARLES (1848–1889) (Mann 1880, Psyche, Camb. 3: 128.)

PREOCCUPIED Adj. Taxonomy: A name already in use in another category.

PREOCELLAR BAND Odonata: A dark pigment stripe immediately in front of the Ocelli (Garman).

PREOCELLAR BRISTLES Dipera: A pair of small bristles sometimes found below the Median Ocellus (Comstock).

PREOCULAR Adj. (Latin, *pre* = before; *oculus* = eye.) Before the eyes. Alt. Praeocular.

PREOCULAR ANTENNA An Antenna inserted close to the front of the eyes.

PREORAL Adj. (Latin, *pre* = before; *os* = mouth;

-alis = pertaining to.) Descriptive of structure associated with the mouth.

PREORAL CAVITY See Mouth Cavity.

PRE-ORAL DIGESTION See Extra-Oral Digestion.

PREORAL LOBE See Prostomium.

PREP® See Ethephon.

PREPARADENSA Noun. Some Coccoids: The cephalic portion of the Paradensae when they are in two groups (MacGillivray).

PREPARAPTERON Noun. (Pl., Preparaptera.) See Basalare; Episternal Paraptera (Comstock).

PREPECTUS Noun. (Latin, *pre* = before; *pectus* = breast. Pl., Prepectoria.) 1. An anterior marginal sclerite of the sternopleural areas of a segment, set off by a transverse suture continuous across the Sternum and Episterna (Snodgrass). 2. Hymenoptera: An area along the anterior margin of the Mesepisternum of the Mesothorax which is often separated from the remainder of the sclerite by a suture-like furrow or Sulcus. Cf. Antepectus.

PREPHARYNX Noun. The flaring anterior part of the pharynx (MacGillivray).

PREPHRAGMA Noun. (Latin, *pre* = before; Greek, *phragma* = fence. Pl., Prephragmata.) 1. The anterior Phragma or partition of any segment (Comstock). 2. The Phragma developed in relation with the Notum (Imms). 3. Anterior Phragma in the Pronotum, Prophragma or Proterophragma in the Mesonotum, Mesophragma or Deutophragma in the Metanotum (MacGillivray).

PREPHRAGMINA Noun. The anterior notal ridge (MacGillivray).

PREPLEURON Noun. (Latin, *pre* = before; Greek, *pleura* = rib. Pl., Prepleura.) The first section of the Pleura.

PREPUCE Noun. (Latin, *praeputium* = the foreskin. Pl., Prepuces.) 1. The foreskin of the Penis. 2. The Distiphallus. Alt. Praeputium.

PREPUPA Noun. (Latin, *pre* = before; *pupa* = puppet. Pl., Prepupae; Prepupas.) 1. A quiescent phase (period) between the last larval instar and the pupal stage (Imms). 2. An active but non-feeding stage in the larva of Holometabola; a full-fed larva (Wardle). 3. The last part of the final larval instar in which the larva has completed feeding and defecated.

PREPUTIAL Adj. (Latin, *pre* = before; *praeputium* = the foreskin; *-alis* = pertaining to.) Descriptive of or pertaining to the Prepuce.

PREPUTIAL GLANDS Certain glands associated with the external opening of the Ejaculaty Duct.

PREPUTIAL SAC 1. The Vesica *sensu* Snodgrass; the Endophallus *sensu* Pierce. A specially developed Endophallus. 2. Dermaptera: Hollow sac in Penis lobe carrying sclerotized plate on wall and virga at anterior end. 3. Lepidoptera: Flexible, eversible tube, sometimes long, within the Aedeagus.

PRERECTAL Adj. (Latin, *pre* = before; *rectum* = distal lower intestine; *-alis* = pertaining to.) Descriptive of structure anterior of the Rectum.

PRESCUTAL RIDGE The internal ridge formed by a Prescutal Suture.

PRESCUTAL SUTURE A transverse groove of the Mesonotum or Metanotum behind the Antecostal Suture, setting off a Prescutum from the Scutum, and forming internally a Prescutal Ridge (Snodgrass).

PRESCUTELLAR BRISTLES Diptera: Setae (bristles) in a transverse row in front of the Scutellum.

PRESCUTELLAR CALLUS The Postalar Callus.

PRESCUTELLAR ROW Diptera: A row of conspicuous large Setae in front of the Scutellum consisting of the posteriormost dorsocentrals and the acrostichal row (Comstock).

PRESCUTELLUM Noun. (Latin, *pre* = before; *scutellum* = a small shield. Pl., Prescutella.) The sclerite nearest the head when the upper part of the segment is divided into four parts.

PRESCUTO-SCUTAL Adj. Pertaining to the Prescutum and the Scutum combined.

PRESCUTUM Noun. (Latin, *pre* = before; *scutum* = shield. Pl., Prescuta.) The anterior area of the Mesonotum or Metanotum between the Antecostal Suture and the Prescutal Suture, when the latter is present (Snodgrass). The anterior Scutum of each thoracic segment. In front of the Scutum; the anterior division of the Scutum (Imms). Syn. Praescutum.

PRESERVATION Noun. (Latin, *pre* = before; *servare* = to keep; English, *-tion* = result of an action.) The act of keeping safe from decay. Rel. Cryopreservation.

PRESL, JOHANN SWATOPLUK (1791–1849) (Eiselt 1836, *Geschichte, Systematik und Literatur der Insektenkunde*. 255 pp. (92–101), Leipzig.)

PRESOCIAL Adj. (Latin, *pre* = before; *socialis* = living together; *-alis* = pertaining to.) Social Insects: A condition manifest in organisms that are less than Eusocial; insects which are Parasocial or Subsocial. See Eusocial. Cf. Parasocial; Subsocial.

PRESS Noun. (Latin, *pressum* = to press. Pl., Presses.) A filter or sclerotized, reticulate structure intended to separate particulate matter from fluid or exude filaments from glands. Cf. Silk Press.

PRESSURE PLATE A structure at the base of the Pulvillus which exerts pressure on the sole of the pad.

PRESSURE RECEPTOR Heteroptera: Mechanoreceptors in some aquatic insects, in which the receptors are associated with the Plastron and larger hydrofugal Setae. Pressure receptors among hydrofugal hairs are deflected as the amount of air held by the Plastron changes. Functionally, PRs determine changes in depth via setal deflection in the Hydrofuge. See Sensillum. Cf. Proprioreceptor.

PREST, WILLIAM (1824–1884) (Carrington 1884, Entomologist 17: 119–120.)

PRESTERNAL Adj. (Latin, *pre* = before; *sternon* = chest.) Anterior of the Sternum. Pertaining to the

Presternum.

PRESTERNAL SUTURE A suture produced by an internal submarginal ridge reinforcing the anterior part of the Eusternum. The PS sets off a narrow marginal area of the Sternum, called the Presternum (Snodgrass).

PRESTERNOIDEA Noun. The second antecoxal piece; Episternal Lateral, Precoxale (MacGillivray).

PRESTERNUM Noun. (Latin, *pre* = before; *sternon* = chest. Pl., Presterna.) A narrow, anterior area of the Sternum sometimes set off from the Basisternum by a submarginal suture of the Eusternum (not the Acrosternite) (Snodgrass).

PRESTIGMA Noun. (Latin, *pre* = before; Greek, *stigma* = mark.) Hymenoptera (Braconidae, Bethylidae): An expansion of the venation at the confluence of the Costal/Subcostal and Radial Sector veins basad of the forewing Pterostigma. Syn Prostigma. See Pterostigma.

PRESTOMAL TEETH Some Diptera: Teeth arising from the lateral margin of the discal sclerite of the Labella, positioned between the opening of two Pseudotracheae, used for scraping food before deglutition (Matheson). The rows of teeth posteriorly on the inner wall of the Prestomium (Snodgrass).

PRESTOMIUM Noun. (Latin, *pre* = before; Greek, *stoma* = mouth; Latin, *-ium* = diminutive > Greek, *-idion*. Pl., Prestomia.) Diptera: The cleft between the labellar lobes anterior of the aperture of the food canal (Snodgrass).

PRESTON, THOMAS ARTHUR (–1905) (Anon. 1905, Marlborough Coll. nat. Hist. Soc. 53: 101–104.)

PRESUTURAL BRISTLES Diptera: One or more bristles positioned immediately anterior of the Transverse Suture above the presutural depression (Comstock); lateral bristles anterior of but near the suture (Curran).

PRESUTURAL DEPRESSION Diptera: A depression, usually triangular, at the outer end of the Transverse Suture.

PRETARSUS Noun. (Latin, *prae* = before; Greek, *tarsos* = sole of foot. Pl., Pretarsi.) 1. The distalmost (apical) part of the Tarsus with its appendages (Comstock). 2. The terminal segment (Dactylopodite) of an insect's leg. Structurally, Pretarsus is the most complicated leg part. Primitive apterygotes, Phthiraptera (lice) and mecopteran Bittacidae: Petarsus forms a single median claw that resembles the claw of millipedes and centipedes. Most pterygote insects show a more complicated pattern. Typically, the Pretarsus shows paired lateral claws (Ungues) that are hollow and communicate with Pretarsus lumen. Ungues articulate dorsally with apical part of Tarsus. Ungues are usually smooth, curved and tapered. Ungues of some insects (*e.g.* beetles) may be pectinate; some wasps display a series of tines along the ventral surface. Usually Ungues are symmetrical, but asymmetry is noted in males of some tephritid flies and other insects. Some insects display a Median Unguitractor Plate or a Median Flexor Plate. This forms a ventral, median, sclerotized structure between Ungues at the Pretarsus base. The Pretarsal Depressor Muscle (or its apodeme) attaches to the Unguitractor and depresses the Ungues. Distally the Unguitractor Plate narrows and forms an Empodium in Heteroptera, Plecoptera and Neuroptera. A Planta sometimes is positioned at the distal end of the Median Flexor Plate. Small basal sclerites called Basipulvillus, Auxilia, Distipulvillus occur in various groups of insects. Many insects display a large, fleshy or sclerotized pad between the Ungues; the pad is called the Arolium. Most larval insects display a simple, claw-like Pretarsus or a terminal extension of the Tarsus. Alt. Praetarsus. See Leg. Cf. Coxa; Femur; Tarsus; Tibia; Trochanter; Rel. Arolium; Auxilla; Basipulvillus; Empodium; Euplanta; Parempodium; Planta; Ungues.

PRETENTORIA Plural Noun. (Latin, *pre* = before; *tentorium* = tent. Sl., Pretentorium.) The anterior arms of the Tentorium (MacGillivray, Crampton). See Tentorium.

PRETENTORINA Noun. (Latin, *pre* = before; *tentorium* = tent. Pl., Pretentorinae.) A frontal pit formed by the invagination of the anterior arms of the Tentorium (MacGillivray). See Tentorium.

PRETERGA Plural Noun. (Latin, *pre* = before; *tergum* = the back. Pl., Pretergum.) Coleoptera larvae: The anterior thoracic Scutes. Alt. Praeterga. See Tergum.

PRETERGITE Noun. (Latin, *pre* = before; *tergum* = the back; *-ites* = constituent. Pl., Pretergites.) The anterior marginal sclerite of the Notum (Crampton). See Antecosta; Tergum.

PRETORNAL Adj. (Latin, *pre* = before; *tornare* = turn; *-alis* = pertaining to.) Lepidoptera: Preceding the Tornus. Alt. Praetornal.

PREUDHOMME DE BORRE, FRANCOIS PAUL CHARLES ALFRED (1833–1905) (Lameere 1906, Ann. Soc. ent. Belg. 50: 7–11.)

PREUPSILON Noun. (Latin, *pre* = before; Greek, *upsilon* = the letter U.) Diptera: The stem of the Y in the upsilon or Furca (MacGillivray).

PREVAIL® See Cypermethrin.

PREVERTEX Noun. (Latin, *pre* = before; *vertex* = summit.) The cephalic area of the Vertex when it is divided into two parts by the fusion of the compound eyes (MacGillivray).

PREY Noun. (Latin, *praeda, prehendere* = to control, seize. Pl., Prey.) An animal that is or may be seized, captured or confined by another animal and subsequently eaten or consumed. See Predation. Cf. Host.

PREY-CAPTURE SETA Setae on the Antenna of ground beetle *Loricera pilicornis* specialized for feeding on Collembola. Long Setae on proximal antennal segments form a trap used to capture Collembola. Positions of Setae on antennal segments and their shapes change from proximal to distal along the Antenna. Typical trap Setae occur on four proximal antennal segments, and are characterized by extraordinary length and sockets that have their highest point on the outside of the trap. Resistance to bending trap Setae is about 10 times that of similar Setae not in trap. Trap Setae are bent up to 60° and their resistance increases linearly; other Setae range at 40–50° and resistance increases more sharply. See Seta; Sensilla. Cf. Camouflage Seta; Defence Seta.

PREYSSLER, JOHN DANIEL Author of *Werzeichniss Boehmischer Insecten* (1790). (Eiselt 1836, *Geschichte, Systematik und Literatur der Insektenkunde.* 255 pp (64–65), Leipzig.)

PRICKLY ACACIA SEED BEETLE *Bruchidius sahlbergi* Schilsky [Coleoptera: Bruchidae].

PRICKLY PEAR BUG *Chelinidea tabulata* (Burmeister) [Hemiptera: Coreidae].

PRICKLY PEAR COCHINEAL *Dactylopius confusus* (Cockerell); *Dactylopius opuntiae* (Cockerell) [Hemiptera: Dactylopiidae].

PRICKLY PEAR MOTH-BORER *Tucumania tapiacola* Dyar [Lepidoptera: Pyralidae].

PRICKLY PEAR SCALE *Diaspis echinocacti* (Bouché) [Hemiptera: Diaspididae].

PRICKLY PEAR SPIDER MITE *Tetranychus desertorum* Banks [Acari: Tetranychidae].

PRICKLY PEAR *Opuntia* spp. [Cactaceae]: A complex of Species of cactus native to the New World and adventive elsewhere. PP develops significant rangeland and pasture weed problems in areas such as Australia, India and Sri Lanka. PP controlled with stenophagous Coleoptera, Lepidoptera and Hemiptera. Most important: Common Prickly Pear, Spiny Prickly Pear; less important: Devil's Rope, Drooping Prickly Pear, Tiger Pear, Velvety Tree Pear, White Spine Prickly Pear.

PRIDDEY, T G (1845–1901) (Anon. 1901, Ent. News 12: 192.)

PRIEBISCH, C. H. (Möbius 1943, Dt. ent. Z. Iris 57: 16.)

PRIESNER, HERMANN (1891–1974) (Strassen 1975, Senckenburg. biol. 56: 89–102, bibliogr.)

PRIMACOL® See NAA.

PRIMARIES Noun. Jargon for the anterior wings (forewings). Cf. Secondaries.

PRIMARY AFFERENT NEURONS See Deutocerebrum.

PRIMARY CULTURE A culture of cells taken directly from the tissue of a plant or animal.

PRIMARY CUTICULA See Exocuticula.

PRIMARY EYES Coccoids: The simple eyes of the female which persist throughout all the nymphal stages into the adult.

PRIMARY HOMONYM Taxonomic nomenclature: The earliest published of two or more identical trivial names proposed in combination with the same generic name. Also one of two or more identical generic names or any other higher category. Cf. Secondary Homonym.

PRIMARY INOCULUM Pathology: An inoculum that initiates disease each season. Cf. Secondary Inoculum.

PRIMARY IRIS CELL Compound Eye: Densely pigmented cells that surround the Lens and Cone (Imms). See Primary Pigment Cell.

PRIMARY LOBES Coccids: The lobes *sensu* MacGillivray.

PRIMARY MONOGAMY Social Insects: The presence of only one functional queen in a nest through foundation by one queen. See Monogamy. Cf. Secondary Monogamy.

PRIMARY OCELLI Ocelli of adult insects, nymphs and naiads (Comstock). The dorsal Ocelli typically positioned on the Vertex of the head and between the compound eyes. Variable in number among insects but typically three in number.

PRIMARY PARASITE A protelean parasite which develops on or in a non-parasitic host. Cf. Secondary Parasite.

PRIMARY PHALLIC LOBES Male Reproductive System: PPL produce simple phallic organs (the Penis or Phallomeres) *sensu* Snodgrass. Cf. Periphallic Organs.

PRIMARY PIGMENT CELL 1. Corneagenous pigment cells in the insect compound eye (Snodgrass). 2. Pigment Cells that surround the Cone and proximal portion of the Lens. In Coleoptera, PPC are formed from two Semper Cells. PPC are analogous to the Iris of the vertebrate eye. Semper Cells first secrete the Lens and Cone then subsequently migrate to the periphery of the Cone and surround it. After migration, PPC contain visible pigment granules; pigment usually does not move in response to light. Syn. Corneal Pigment Cell; Corneagenous Pigment Cell; Primary Iris Cell.

PRIMARY POLYGYNY Social Insects. The foundation of a nest by more than one functional queen. Cf. Monogyny. See Secondary Polygyny; Oligogyny.

PRIMARY REPRODUCTIVE Isoptera: The king or queen of the colony; typically one king and queen per colony. Anatomically characterized by compound eyes, heavily sclerotized body, functional wings for dispersal which are shed after colonizing flight. Cf. Secondary Reproductive; Neoteinic.

PRIMARY SCREWWORM *Cochliomyia hominivorax* (Coquerel) [Diptera: Calliphoridae]. See New World Screwworm. Cf. Secondary Screwworm.

PRIMARY SEGMENTATION A fundamental and presumably primitive type of segmentation that occurs in all soft bodied arthropods, larval holometabolous insects and annelid worms. Some morphologists consider PS the embryonic form of segmentation. PS is characterized by longitudinal muscles attached to intersegmental folds or rings around the body wall at the anterior and posterior margin of each Somite. These rings represent intersegmental lines of the body wall and define Somites. Internally, the grooves coincide with lines of attachment of the primary longitudinal muscles. In primary segmentation, the body segments (Somites) correspond with true Somites of the hypothetical ancestral body. Functionally, this intrasegmental, longitudinal musculature permits flexibility and enables the body to move from side-to-side. Contraction (shortening) of the body is achieved by tension of longitudinal muscles. Extension (elongation) of the body is generally achieved by relaxing longitudinal muscles. See Segmentation. Cf. Secondary Segmentation.

PRIMARY SETAE Lepidoptera larva: Setae on cuticular tubercles that are arranged in predictable number and position (Comstock).

PRIMARY SOMATIC HERMAPHRODITE An insect which has the gonad or gonads of one sex only but parts of the secondary sexual apparatus, internal or external, of both sexes. Cf. Sexual Mosiac.

PRIMARY TYPE Taxonomy: The specimens used in the original description of a nominal Species. Included representatives are Syntypes, Holotypes and Neotypes. See Type. Cf. Proterotype; Secondary Type. Rel. Nomenclature; Taxonomy.

PRIMARY VECTOR Medical Entomology: Arthropods which actively transmit a pathogen to humans or other animals. Alt. Important Vector. See Vector. Cf. Secondary Vector.

PRIMARY VEIN A principal vein of the insect wing. See Vein; Wing Venation.

PRIMICARB A carbamate compound {2-dimethylamino-5,6-dimethylpyrimidin-4-yl-dimethylcarbamate} used as a contact insecticide and fumigant for control of aphids. Primicarb applied to fruit trees, vegetables, ornamentals, vegetables and field crops in some countries; expresses some phytotoxicity on ornamentals. Not registered for use in USA. Trade names include: Aficida®, Aphox®, Demo®, Fernos®, Phantom®, Pirimor®, Rapid®, Romicarb®. See Carbamate Insecticide.

PRIMICID® See Pirimiphos-Ethyl.

PRIMITIVE Adj. (Latin, *primitivus* = original.) Not modified or derived. Cf. Derived, Apomorphic.

PRIMITIVE BRISTLETAILS See Lepidotrichidae.

PRIMITIVE CADDISFLIES See Trichoptera; Rhyacophilidae.

PRIMITIVE CHARACTER A morphological feature which is relatively uncomplicated and believed representative of an ancestral form. Syn. Plesiomorphous Character. Cf. Derived Character.

PRIMITIVE CRANE FLIES See Tanyderidae.

PRIMITIVE KATYDIDS See Prophalangopsidae.

PRIMITIVE MOTHS See Micropterigidae.

PRIMITIVE STREAK Embryology: The insectan germ band.

PRIMITIVE WEEVILS See Brentidae.

PRIMITIVELY EUSOCIAL Social Insects: Individuals residing in a eusocial colony in which the çastes are morphologically similar and food exchange is minimal or absent.

PRIMORDIAL Adj. (Latin, *primordium* = beginning; *-alis* = characterized by.) Descriptive of structure that is 'primitive' in development. The first or earliest in point of time of a developing structure or lineage. See Primitive.

PRIMORDIAL GERM CELLS Embryology: Germ cells from which the eggs are developed.

PRIMORDIUM See Anlage.

PRIMULA APHID *Microlophium primulae* (Theobald) [Hemiptera: Aphididae].

PRINACOL® See NAA.

PRINCIPAL COMPONENT ANALYSIS Morphometrics: An analytical method of multivariate path analysis for determining the principle component, also known as the eigenvector. The eigenvector is a linear combination of observed variables which is unaltered, except for scale, under multiplication by some matrix. Acronym PCA.

PRINCIPAL NEURON See Deutocerebrum.

PRINCIPAL SECTOR Odonata: A vein which extends from its point of separation from the Median Sector to the outer margin, at or just below the apex, Media (Comstock).

PRINCIPAL SULCUS Orthoptera: A transverse impression of the Prothorax, at or behind the middle.

PRINCIPAL VEIN Hemiptera: The fused veins of the Tegmen from Sc to Cu (Tillyard). See Vein; Wing Venation.

PRINEX® See Chlorpyrifos-Methyl.

PRINGLE, JOHN (1707–1782) (Anon. 1850, Encyclopaedia Britannica 14th Ed. 18: 498, bibliogr.)

PRINSEPS, JAMES (1800–1840) (Rose 1850, *New General Biographical Dictionary* 11: 232–233.)

PRINTZ, YAKOV IVANOVICH (1891–1966) (Anon., 1966, Phylloxera and measures of control. 91 pp. (3–4, 86–91, bibliogr.) Leningrad.)

PRINZ, JOHANN (1845–1934) (R. Z. 1934, öst. EntVer. 19: 41–42.)

PRIODONT Noun. A form of male Lucanid which has the smallest Mandibles. See Teledont; Mesodont; Amphiodont.

PRIONOGLARIDAE Plural Noun. A small Family of Psocoptera assigned to Suborder Trogiomorpha. Antenna with more than 20 segments; flagellar segments not secondarily annulated; Lacinia missing terminal spines; Forewing Sc curved and joining R1 distally; Tarsi with three segments; Ovipositor Valvulae separated or in contact apically. North American Species cavericolous or associated with palms in southwest.

PRIORITY See Law of Priority.

PRISKE, A R (–1944?) (Hale-Carpenter 1946, Proc. R. ent. Soc. Lond. (C) 10: 54.)

PRISMATIC Adj. (Latin, *prisma* = prism; *-ic* = of the nature of.) 1. Formed like a prism. 2. A play of colours similar to that produced through a prism.

PRITCHARD, ANDREW (1804–1882) (Anon. 1883,

Zool. Anz. 6: 80.)

PRITCHARD, ARTHUR CARL (1915–1965) (Denning 1965, J. Econ. Ent. 58: 807–808)

PRIVET APHID *Myzus ligustri* (Mosley) [Hemiptera: Aphididae].

PRIVET LEAFMINER *Caloptilia cuculipennella* (Hübner) [Lepidoptera: Gracillariidae].

PRIVET MITE *Brevipalpus obovatus* Donnadieu [Acari: Tenuipalpidae].

PRIVET THRIPS *Dendrothrips ornatus* (Jablonowski) [Thysanoptera: Thripidae].

PRIZBRAM'S RULE Body weight doubles for each moult of larva in accord with surface area to volume ratio. In general, probably not a reliable estimate of growth because body components grow at characteristic rates based on individual requirements or constraints (Prizbram & Megusar 1912). Cf. Dyar's Law.

PRIZBRAM'S FACTOR 1.26 or –3/2, the proportional increase in length in insect immature stages between instars, mathematically expressed (Wardle). Cf. Dyar's Law.

PROALA CORIACEA Front wings of a tough substance which bends but does not fold without breaking.

PROALA CRUSTACEA A forewing of a hard, brittle texture, incapable of being bent or folded without injury.

PROBOFOSSA The labial gutter *sensu* MacGillivray.

PROBOSCARIA Diptera: A slender sclerite along each margin of the Proboscella (MacGillivray).

PROBOSCELLA Diptera: The mesial sclerites supporting the Probofossa (MacGillivray).

PROBOSCIDEA An ordinal term for the Coccidae.

PROBOSCIDIAL FOSSA The preoral part of head outside the Os or oral opening. Cf. Buccal Cavity.

PROBOSCIS Noun. (Greek, *proboskis* = trunk. Pl., Proboscises; Proboscides.) 1. General: A trunk-like process of an animal's head. 2. Insect: Any extended or extensible mouth structure. 3. Hemiptera: The Rostrum (Labium) enclosing Maxillary and Mandibular Stylets. 4. Lepidoptera: Tongue-like appendages adapted for nectar feeding. Maxillae are most important in Lepidoptera while the Mandibles and Labium are vestigial or absent. Maxillary Galeae become elongated and form a tubular Proboscis that is coiled at repose. Coiling mechanism has been studied in some detail. An early hypothesis argued that extension of the Proboscis was effected by hydrostatic pressure; coiling was brought about by contraction of short, oblique intrinsic musculature. More recent experimentation suggests that coiling and extension both are brought about by muscle contraction. When musculature seals the proboscidial Haemocoel, a change in cuticular tension in the Galea causes uncoiling. Elasticity of the Galea's cuticular wall is apparently responsible for coiling. To maintain an efficient feeding tube during coiling and uncoiling, Galea is bound dorsally by glandular secretions and ventrally by Setae modified into hooks which medially interdigitate. 5. Hymenoptera: Labium of long-tongued bees. Representatives of at least four Families (Megachilidae, Anthophoridae, Apidae and Feliidae) display mouthparts with Maxillary Galea and Labial Glossa longer than Stipes. Labial Palpi become elongate, flattened and form a sheath. Long-tongued bees are distinguished from short-tongued bees in that the parts are relatively longer. In addition, the Glossa has a terminal Flabellum, an internal sclerotized rod and a long forward-curving subligular process; Axillary Combs also differ in position between the two groups. 6. Adult Diptera: The extensile mouthparts. See Mouthparts. Cf. Labium; Labrum; Mandible; Maxilla.

PROBRACHIAL Noun. Ephemerida: A simple, longitudinal wing vein posteriad of the Prebrachial Vein.

PROCEPHALIC Adj. (Greek, *pro* = before; *kephale* = head; *-ic* = of characterized by.) Relating or belonging to the Procephalon.

PROCEPHALIC ANTENNA Arthropoda: The Antennules.

PROCEPHALIC LOBES In the embryo, part of the anterior overhanging portion of the head.

PROCEPHALON Noun. (Greek, *pro* = before; *kephale* = head. Pl., Procephalons.) The embryonic head segment formed by the coalescence of the first three primitive segments. The embryonic Procephalon shows no clear external signs of segmentation and thus we conclude that this must be the anteriormost segment of the head. See Protocephalon.

PROCEREBRAL Adj. (Greek, *pro* = before; Latin, *cerebrum* = brain; *-alis* = pertaining to.) Descriptive of structure anterior of the Cerebrum. Term pertaining to part of the Brain containing the median Protocerebrum and Optic Ganglia.

PROCEREBRAL LOBES The central portion of the Cerebrum, composed of the fused median lobes and giving rise to the Mushroom Bodies.

PROCEREBRUM Noun. (Greek, *pro* = before; Latin, *cerebrum* = brain.) The anterior part of the brain, formed by the ganglion off the first primary segment, also termed Ocular Lobe, from the part it innervates.

PROCESS Noun. (Middle English, *proces* from Latin, *processus* = to proceed. Pl., Processes.) 1. Anatomy: A projection from the surface, margin or appendage. 2. Any prominent part of the body which projects from the surface but which is not otherwise definable; a bump; a hump; a wart; a tubercle. 3. A progessive change in form, physiology or behaviour from one phase to another phase on the way to completing a developmental cycle.

PROCESS OF LABRUM Apoidea: The Appendicle.

PROCESSIONARY CATERPILLAR *Ochrogaster lunifer* Herrich-Schaffer [Lepidoptera: Thaumetopoeidae].

PROCESSIONARY MOTH *Thaumetopoea pityocampa* [Lepidoptera: Thaumetopoeidae].

PROCESSUS ARTICULARIS Siphonaptera: See Telomere. Hymenoptera: See Dorsal Ramus.

PROCESSUS MEDIANUS Hymenoptera: See Tuxen.

PROCHNOW, OSKAR (1884–1934) (Horn 1934, Arb. morph. taxon. Ent. Berl. 1: 310.)

PROCIDENTIA Noun. (Pl., Procidentiae.) Hymenoptera: The medial production of the seventh abdominal Tergum in some male Symphyta.

PROCLINATE Adj. (Greek, *pro* = before; *klinein* = slope; *-atus* = adjectival suffix.) Directed forward; a term applied to Setae or bristles. Cf. Declinate; Inclinate; Reclinate.

PROCORIA Noun. The Coria anterior of the Prothorax (MacGillivray).

PROCRAMPTONOMYIIDAE Kovalev. Plural Noun. A small Family of fossil Diptera known from two Genera included in Jurassic Period deposits from Siberia and the USA.

PROCRYPTIC COLORATION Combinations of colours or patterns of colour on the body or appendage that serve in protective resemblances and for concealment as a protection against predators. See Coloration. Cf. Advancing Coloration; Alluring Coloration; Anticryptic Coloration; Apetetic Coloration; Aposematic Coloration; Combination Coloration; Cryptic Coloration; Directive Coloration; Disruptive Coloration; Epigamic Coloration; Episematic Coloration; Protective Coloration; Pseudepisematic Coloration; Pseudoaposematic Coloration; Seasonal Coloration; Sematic Coloration. Rel. Crypsis; Mimicry.

PROCTIGER Noun. (Hybrid, Greek, *proktos* = anus; Latin, *gerere* = to bear. Pl., Proctigers.) A small Papilla or conical structure which bears the Anus. Morphologists regard the Proctiger as the reduced tenth abdominal segment.

PROCTODAEAL TROPHALLAXIS See Trophallaxis. Stomodaeal Trophallaxis.

PROCTODAEAL Adj. (Greek, *proktos* = anus; *hodos* = way; *-alis* = pertaining to.) Descriptive of structure associated with the Proctodaeum.

PROCTODAEAL VALVE An anatomical mechanism adapted for closing the entrance into the intestine. PV developed in the pyloric region of the anterior end of the Proctodaeum. Syn. Pyloric Valve *sensu* Snodgrass.

PROCTODAEUM Noun. (Greek, *proktos* = anus; *hodos* = way. Pl., Proctodaea; Proctodaeums.) The hindgut and Malpighian Tubules of an insect. Proctodaeum formed as an invagination of the Epiblast that produces the Anus and intestine; projects as far forward as and including the Malpighian Tubules. Alt. Proctodeum. See Alimentary System. Cf. Mesenteron; Midgut; Stomodaeum.

PROCTOR, WILLIAM (1872–1951) (Alexander 1951, Ent. News 62: 237–241.)

PROCTOTRUPIDAE Latreille 1802. Plural Noun. A numerically small (ca 30 Genera, 300 Species), cosmopolitan Family of Parasitica assigned to the Proctotrupoidea. Characterized by dark, weakly compressed body with non-carainate Gaster, Antenna 13-segmented, non-geniculate, non-clavate; tibial spur formula 1-2-2; Usually macropterrous, rarely micropterous; forewing Subcostal, Costal, and Stigmal Veins well developed; hindwing without closed cell. Typically found in damp habitats where they parasitize Coleoptera larvae. Syn. Serphitidae.

PROCTOTRUPOIDEA Latreille 1802. Hymenoptera. A numerically large, cosmopolitan Superfamily of Parasitica. Fossil record referrable to Cretaceous Period. Adults characterized by reduced wing venation and no closed cells in the hindwing, body typically dark and not metallic, Pronotum extending to the Tegula, lateral surface of Pronotum sometimes sulcate. Included Families: Austroniidae, Diapriidae, Heloridae, Jurapriidae, Mesoserphidae, Monomachidae, Pelecinidae, Peradeniidae, Platygastridae, Proctotrupidae (= Serphitidae), Roproniidae, Scelionidae, Trupochalcididae, Vanhorniidae. Syn. Serphoidea. Cf. Ceraphronoidea; Chalcidoidea; Cynipoidea; Ephialtitoidea; Evanioidea; Ichneumonoidea; Stephanoidea. See Hymenoptera.

PROCUMBENT Adj. (Latin, *pro* = forward; *cumbens* = lying down.) Pertaining to structures or bodies that are trailing, prostrate or lying flat. Cf. Prostrate.

PROCURVED Adj. (Latin, *pro* = forward; *curvus* = curved.) Pertaining to structure (Antenna, Seta) which is basally perpendicular to the body and whose apical part is curved anteriad. Cf. Recurved.

PROCUTICLE Noun. (Latin, *pro* = forward; *cuticula* dim. of *cutis* = skin. Pl., Procuticles.) The area beneath the Epicuticle and represented by chitinous rods embedded in the protein matrix. Procuticle functions in support of epicuticle, protects Epidermis, stores biomechanical energy and stores food. See Cuticle; Integument.

PRODORSUM Noun. (Latin, *pro* = forward; *dorsum* = the back.) Acarology: dorsal surface of the Aspidosoma bordered by the Disjugal Furrow and Abjugal Furrow.

PRODOXIDAE Plural Noun. A small Holarctic Family of monotrysian Lepidoptera assigned to Superfamily Incurvarioidea. Body small; Head tufted; Ocelli and Chaetosemata absent; Antenna smooth; Maxillary Palpus with five segments, Ovipositor present. Larva with thoracic legs present or replaced with Calli; Prolegs absent or vestigial; Crochets typically absent or represented by a transverse row on abdominal segments 3–6. Family includes Yucca Moths symbiotic with Agavaceae and exclusive pollinators of *Yucca* flowers.

PRODROMUS Noun. (Greek, *pro* = before; *dromus* = running, moving.) An introductory or preliminary publication, more frequently used in botanical literature. Alt. Prodromous.

PRODUCED Adj. (Latin, *productum* = to bring forward.) Drawn out, prolonged, extended, shown, disproportionately long. Alt. Producted; Productus.

PRODUCER Noun. (Latin, *producere* = to produce. Pl., Producers.) Ecology: An autotrophic organism which synthesizes or manufacturers organic matter from inorganic matter. Typically, a green plant which photosynthesizes. Cf. Decomposer. Rel. Autotroph.

PRODUCTILE Adj. (Latin, *pro* = forward; *ductilis* from *ducere* = to lead.) Capable of being lengthened out.

PRODUCTION Noun. (Latin, *productum* = to bring forward; English, *-tion* = result of an action. Pl., Productions.) That part of any structure which has been produced or extended; the act of producing or extending.

PROEMINENT Adj. (Latin, *pro* = forward; *eminens* pp of *eminere* = to stand out.) Standing out; an obsolete term used to describe the head when it is horizontal and does not form an angle with the Thorax. Cf. Prognathous. Rel. Orientation.

PROEPIMERON Noun. (Latin, *pro* = forward, + epimeron. Pl., Proepimera.) Odonata: The posteriormost pleural sclerite of the Prothorax (Garman). Alt. Proepimerum. See Epimeron. Cf. Proepisternum.

PROEPISTERNUM Noun. (Latin, *pro* = forward; + episternum.) Odonata: The anteriormost pleural sclerite of the Prothorax (Garman). See Sternum. Cf. Episternum; Proepimeron.

PROFENOFOS An organic-phosphate (phosphoric acid) compound {O-(4-bromo-2-chlorophenyl)O-ethyl S-propyl phosphorothioate} used as a contact insecticide and stomach poison for control of bollworms, leaf-eating insects and mites. Compound applied to cotton, tobacco, soybeans, sugar beets, potatoes and vegetables in many countries. Registered for cotton in USA. Toxic to bees, fishes and wildlife. Trade names include Curacron®, Polycron® and Selecron®. See Organophosphorus Insecticide.

PROFILE Noun. (Italian, *profilare* = to draw in profile. Pl., Profiles.) The outline of a structure or body when viewed in lateral aspect.

PROFIT, JOACHIM (1911–1942) (Kohler 1942, Arb. physiol. angew. Ent. Berl. 9: 135–136.)

PROFOUND Adj. (Old French, *profond* = deep.) Of great depth. Alt. Profundus.

PROFUME® See Methyl Bromide.

PROFURCA Noun. (Latin, *pro* = anterior; *furca* = fork. Pl., Profurcae.) Anterior Furca; Prothoracic Furca. See Furca.

PROGNATHOUS Adj. (Greek, *pro* = anterior; *gnathous* = jaw; Latin, *-osus* = with the property of.) 1. General zoological useage, pertaining to animals with prominent or projecting jaws. 2. Entomological useage, pertaining to one of three primary orientations of the insect head. Characterized by a Foramen Magnum near the Vertex margin, and Mandibles directed anteriad and positioned at the anterior margin of the head. When viewed in lateral aspect, the primary axis of the head is horizontal. Prognathous condition is displayed by some predaceous forms (*e.g.* Carabidae, Bethylidae) and in other insects appears to be a response to living in concealed situations such as between bark and wood or similar confined habitats (*e.g.* Cucujidae). See Head. Cf. Hypognathous; Opisthognatous.

PROGONEATA Noun. (Greek, *pro* = before; *gone* = generation.) A classification of arthropods which groups progoneate Taxa. The Chilopoda, Pauropoda, Symphyla.

PROGONEATE Adj. (Greek, *pro* = before; *gone* = generation; *-atus* = adjectival suffix.) Pertaining to arthropods with the genital opening on an abdominal segment anterior of the apex. Cf. Opisthogoneate.

PROGONIA (Greek, *pro* = anterior; *gonia* = angle.) The anterior angle of the hindwings.

PROGONOCIMICIDAE Plural Noun. A Family of Heteroptera known from late Permian deposits.

PROGREDIENS TYPE Aphididae: Nymphs of a third generation which develop at once into wingless agamic females (Comstock).

PROGREDIENTES Adj. Aphididae: Certain apterous progeny of colonici.

PROGRESSION RULE In biogeography, a principle of the dispersal explanation according to which the direction of movement (dispersal) is indicated by the branching pattern of a cladogram.

PROGRESSIVE PROVISIONING Solitary bees and wasps: The practice of feeding larvae in open cells throughout larval development. Cf. Mass Provisioning.

PROHAEMOCYTE Noun. (Greek, *pro* = before; *haima* = blood; *kytos* = hollow vessel; container. Pl., Prohaemocytes.) A basic anatomical form of insect Haemocyte. Small, round-oval or elliptical cells; Plasma Membrane generally smooth, occasionally vesiculate; Nucleus large, centrally located; with little cytoplasm but contains granules, droplets or vacuoles. Prohaemocytes typically occur in groups. Syn. Formative Cell; Jeune Globule; Jeune Leucocyte; Macroneucleocyte; Micronucleocyte; Nematocyte; Plasmatocyte-like Cell; Prohaemocytoid; Proleucocyte; Proleucocytoid; Young Granulocyte; Young Plasmatocyte. See Haemolymph. Cf. Adipohemocyte; Coagulocyte; Granulocyte; Oenocyte; Plasmatocyte; Spherulocyte.

PROHASKA, KARL (1854–1937) (Ruschnig 1938, Carinthia II 48: 136–138.)

PROHEMEROBIIDAE Handlirsh 1906. Plural Noun. A Family of fossil insects, presumably ancestral to the Hemerobiidae (Neuroptera).

PROHEMOCYTOID See Prohaemocyte.

PROJECTION Noun. (Latin, *projectio*. Pl., Projections.) 1. Something jutting out at a rather sharp angle from a rather flat surface. Syn. Bulge. Cf. Protuberance. 2. The act of being thrown forward; the state of being projected.

PROJECTION INTERNEURON See Deutocerebrum. Syn. Output Neurons; Principal Neurons.

PROKARYOTE Noun. (Greek, *pro* = before; *karyon* = nucleus. Pl., Prokaryotes.) An indeterminant category which includes bacteria and blue-green algae. Cf. Eukaryote.

PROKIL® See Cryolite.

PROKOPOVICH, P I (1775–1850) ((Galton 1971, Survey of a thousand years of beekeeping in Russia. 90 pp. (35–36), London.)

PROLATE Adj. (Latin, *prolatus* = extend, bring forward.) Structure that is stretched, extended or elongated in a direction with an imaginary line connecting the poles. Cf. Oblate. Rel. Form 2; Shape 2; Outline Shape.

PROLATE® See Phosmet.

PROLEG Noun. (Greek, *pro* = before; Middle English, *legge* = leg. Pl., Prolegs.) 1. Any process or appendage that serves the purpose of a leg. 2. Specifically the pliant, non-segmental abdominal legs of caterpillars and some sawfly larvae. Prolegs are not true segmented appendages. Larval Prolegs of terrestrial Lepidoptera and Symphyta are not well developed, but are adapted to grasping the substratum. These structures are considered serially homologous with legs, but Hinton claims they are adaptive structures with no relation to legs. Syn. Abdominal Feet; False Legs. See Leg. Rel. Locomotion.

PROLEUCOCYTE Noun. (Greek, *pro* = before; *leukos* = white; *kytos* = hollow vessel; container. Pl., Proleucocytes.) 1. See Prohaemocyte. 2. An undifferentiated leucocyte in Haemolymph with basophilic cytoplasm. Proleucocytes reproduce by mitotic division and produce other types of Leucocytes. See Haemocyte.

PROLEUCOCYTOID See Prohaemocyte.

PROLIFERATE Verb. To increase in size by a repeated process of budding or cell division.

PROLIFERATION Noun. (Latin, *proles* = offspring; *ferre* = to carry; English, *-tion* = result of an action.) Growing by the rapid production of new cells.

PROLOMA Noun. (Greek, *pro* = before; *loma* = fringe, hem. Pl., Prolomata.) The anterior margin of the hindwings.

PROLONGED Adj. Extended or lengthened beyond ordinary limits.

PROMASTIGOTE Noun. (Greek, *pro-* = before; *mastigote* = protozoa with flagella. Pl., Promastigotes.) An anatomical form manifest at a specific phase in the complex life cycle of some parasitic Protozoa (*Leishmania* and *Trypanosoma*). Promastigote displays Flagellum at anterior end of body with flagellar base and kinetoplast anterior of nucleus. See Leishmania. Trypanosoma. Cf. Amastigote. Opisthomastigote. Epimastigote. Trypomastigote.

PROMECARB A carbamate compound {3-methyl-S-isopropylphenyl methylcarbamate} used as an acaricide, contact insecticide and stomach poison for control of mites, ticks, aphids, corn rootworms, flies, leaf miners, mosquitoes, sawflies and related insects. Applied to fruit trees, citrus and potatoes in some countries. Not registered for use in USA. Toxic to bees. Trade names include Carbamult®, Minacide® and Promecarbe®. See Carbamate Insecticide.

PROMECARBE® See Promecarb.

PROMEROS Noun. (Greek, *pro* = before; *meros* = segment.) The first abdominal segment in Lepidoptera.

PROMET® See Furathiocarb.

PROMETHEA MOTH *Callosamia promethea* (Drury) [Lepidoptera: Saturniidae].

PROMINENT Adj. (Latin, *prominens* from *prominere* to jut out.) Raised, elevated, or produced beyond the level or margin; standing out in relief by colour, shape, size, proportion or other conspicuousness.

PROMINENT MOTHS See Notodontidae.

PROMINENTS Common name for the Family Notodontidae [Lepidoptera].

PROMORPHOLOGY Noun. The compound microscope has allowed morphologists to magnify objects and study cellular detail at about 980 diameters with transmitted light. With this level of magnification available to biologists, morphological research of the 18–19th century shaped a branch of biology which was concerned with structural organic types. Conceptually, this was divided by the German philosopher and biologist Ernst Haeckel (1834–1919) into Promorphology and Tectology. Promorphology considered geometrically the form of an organism and its component parts. See Tectology.

PROMOTER SEQUENCE A regulatory DNA sequence that initiates the expression of a gene. See Molecular Genetics. Rel. Biotechnology.

PROMUSCIDATE Adj. With a Proboscis or extended mouth structure.

PROMUSCIS Noun. An extended mouth structure. A term applied to the long tongue of bees, to the Proboscis in flies, and to the rostrate strucutre in Hemiptera.

PRONE SURFACE The ventral surface of a body. Syn. Superficies.

PRONIN, GEORG (1898–1962) (Leech 1967, J. Lepid. Soc. 21: 74–76.)

PRONOTAL CARINA Orthoptera and certain Heteroptera: The main or median Carina or keel on the Pronotum.

PRONOTAL COMB Siphonaptera: A row of stout spines on the posterior margin of the Pronotum.

PRONOTAL UMBONE A cuticular projection found on the Pronotum of some Cerambycidae.

PRONOTUM Noun. (Greek, *pro* = before; *noton* = back.) The upper or dorsal surface of the first thoracic segment (Prothorax) of an insect's body. See Notum. Cf. Mesonotum; Metanotum. Rel. Tergum.

PRO-NOX FISH® See Rotenone.

PRONTO® See Trichlorfon.

PRONUCLEUS Noun. (Greek, *pro* = before; Latin, *nucleus* = nut, kernel. Pl., Pronuclei.) The nucleus of male Spermatozoa and female ova, the union of which forms the nucleus of a fertilized Ovum.

PRONYMPH Noun. (Greek, *pro* = before, + nymph.) Odonata: Newly hatched nymph, which shows a more or less embryonic appearance, being invested with a shining chitinous sheath, a stage of extremely short duration, from a few seconds (*Anax*) to 2–3 minutes (*Agrion*) (Imms). Some Orthoptera: A stage in certain metabolous insects in which the larval tissues are completely broken down and the imaginal tissues are beginning to build up (Smith).

PROOVIGENIC Adj. (Latin, *pro* = before; *ovum* = egg; Greek, *gen*- to produce; *-ic*, consisting of.) An apparent biological phenomenon in which a female insect transforms into the adult stage with all eggs which will be produced in the reproductive system and fully developed. The phenomenon is displayed by some Families of parasitic Hymenoptera, including Eucharitidae and Mymaridae. (Flanders 1950, Can. Ent. 82: 134–140) Cf. Synovigenic.

PROPAGATIVE TRANSMISSION A form of Biological Transmission in which a pathogen multiplies but does not undergo developmental changes in the body of an arthropod vector. PT typical of bacterial and viral pathogens associated with insect and tick vectors. Cf. Cyclopropagative Transmission; Propagative Transmission.

PROPALTICIDAE Plural Noun. A small Family of polyphagous Coleoptera assigned to Superfamily Cucujoidea. Body broadly oval, flattened, less than 2 mm long and invested with scale-like Setae; head large, prognathous; compound eyes large; Antenna with 11 segments including club of three segments; Pronotum with median Endocarina; Abdomen with five Ventrites; fore coxal cavity open; middle coxal cavity closed; tarsal formula 5-5-5. Adult and larva occur under bark.

PROPAPHOS Noun. An organic-phosphate (phosphoric acid) compound {4-(methylthio) phenyl dipropyl phosphate} used as a stomach poison, systemic insecticide and contact insecticide against rice pests in some countries. Not registered for use in USA. Toxic to fishes. Trade name is Kayaphos®. See Organophosphorus Insecticide.

PROPARAPTERA Noun. The Paraptera of the Prothorax. The term is erroneously applied in this connection (Smith).

PROPARGITE A synthetic, organic hydrocarbon compound {2-(p-tert-butylphenoxy) cylohexly 2-propynl sulphite} used as an Acaricide. Propargite applied to numerous fruit trees, corn, cotton, peanuts, potatoes and sorghum in USA, and additional crops in other countries. Phytotoxic to cotton, citrus, and pears. Trade names include: Acaryl®, Fenpropar®, Comite®, Omite®, Retador®. See Insecticide.

PROPEDES Noun. The anterior legs of an adult insect but also applied to the Prolegs of larvae. (Archaic.)

PROPER, ARGYLE B (1905–1933) (Hungerford 1934, Ann. ent. Soc. Am. 27: 123.)

PROPETAMPHOS An organic-phosphate (phosphoric acid) compound {(E)-O-2-isopropoxy-carbonyl-1-methylvinyl-O-methyl ethylphosphoramidothioate} used as a contact insecticide and stomach poison against public-health pests. Applied in buildings and domiciles for control of cockroaches, lice, fleas, ants, bed bugs, mosquitoes, crickets and houseflies; applied to livestock to control ectoparasites. Toxic to fishes. Trade names include: Blotic®, Detmol®, Ovidip®, Safrotin®, Seraphos®. See Organophosphorus Insecticide.

PROPHALANGOPSIDAE Plural Noun. Hump-Winged Crickets; Primitive Katydids. A small Family of ensiferous Orthoptera, including three Genera, two Species of which occur in northwestern North America. Antennal Torulus positioned halfway between Epistomal Suture and Vertex; male brachypterous; female micropterous; Ovipositor very short. Syn. Haglidae.

PROPHARYNX Noun. The dorsal part of the Prepharynx (MacGillivray).

PROPHRAGMA Noun. The anterior dividing wall of the Mesothorax, which is horny and, at its upper edge, bears the connecting membrane between the Prothorax and Mesothorax.

PROPION® See Proproxur.

PROP-LEG See Proleg.

PROPLEGMATIUM Noun. (Greek, *pro* = before; *plegma* = plaited; *-idion* = dim. Pl., Proplegmatia.) Scarabaeoid larvae: A space with plicate inner surface and usually somewhat in front of a Plegmatium. Syn. Submarginal Striae of Hayes (Boving).

PROPLEURAL BRISTLES Diptera: A bristle or bristles positioned on the Propleura immediately above the front Coxae (Curran). Prothoracic bristles (*sensu* Smith).

PROPLEURON Noun. (Greek, *pro* = before; *pleura* = side. Pl., Propleura.) The lateral portion (sclerite) of an insect's Prothorax. See Thorax. Cf. Pronotum; Prosternum.

PROPNEUSTIC Adj. (Greek, *pro* = before; *pheum* = breath; *-ic* = consisting of.) Pertaining to a system of respiration characterized by larvae in which only the anteriormost pair of spiracles is functional, or Diptera Pupae with a Mesothoracic Spiracle. Cf. Apneustic; Oligopneustic; Polypneustic.

PROPODEON See Propodeum.

PROPODEONIS SCUTUM The Scutum of the Propodeum.

PROPODEUM Noun. (Greek, *pro* = before; *podeon* = neck. Pl., Propodea.) Hymenoptera (Apocrita): The first abdominal segment which has through evolution disassociated from the Abdomen and becomes incorporated into the thoracic region. In Parasitica and Aculeata, the Propodeum is characterized by anterolateral spiracles, a broad attachment to the Metanotum anteriorly and posteriorly separated from the remainder of the Abdomen by a narrow constriction (Petiole). This anatomical reorganization has resulted in problems with terminology. Some specialists use the term Mesosoma when referring to the combined Propodeum and Thorax. The term Metasoma is used when referring to the remaining abdominal segments. Alt. Propodeon. Syn. Latreille's Segment; Epinotum (Formicidae); Median Segment. See Mesosoma. Rel. Thorax.

PROPODITE Noun. (Greek, *pro* = before; *podos* = foot; *-ite* = dim.) 1. The penultimate segment of a generalized limb. 2. The Tarsus *sensu* Snodgrass.

PROPODOSOMA Noun. (Greek, *pro* = before; *podeon* = neck; *soma* = body.) Acarology: The anterior part of the Actinotrichid Podosoma.

PROPOLIS Noun. (Greek, *pro* = for; *polis* = city.) A glue-like or resin-like substance collected from plants and used by bees to serve as a cement in places where wax is not effective in binding a cell.

PROPOXAN® See Proproxur.

PROPOXUR Noun. A carbamate compound {2-(1-methyl ethoxy) phenyl methylcarbamate} used as a contact insecticide and stomach poison. Developed ca 1958 as a cholinesterase inhibitor on insects; purportedly not affecting cholinesterase in mammals and eliminated from mammalian system within 24 hours. Applied as a surface spray which crystallizes to form a residue; used extensively for control of structural pests, cockroaches and other urban pests (earwigs, crickets, ants, fleas, mosquitoes.) Purported beneficial properties as an urban-pest control material include flushing action, fast knockdown and long residual life. Applied to alfalfa, cocoa, corn, cotton, grapes, rice, soybeans, sugarcane and vegetables in some countries. Some phytotoxicity on ornamentals and blossom-thinning of fruit trees. Trade names include: Aprocarb®, Bay-o-cide®, Baygon®, Bripoxur®, Certamate®, Killzone®, Mitoxur®, Propion®, Propoxan®, Proprotox®, Propyon®, Prox®, Sendran®, Suncide®, Tugen®, Unden®. See Carbamate Insecticide.

PROPRIORECEPTION Noun. (Latin, *proprius* = one's own; *capere* = to take; English, *-tion* = result of an action. Pl., Proprioreceptions.) The perception of stimuli originating within the body of an organism.

PROPRIORECEPTOR Noun. (Latin, *proprius* = one's own; *capere* = to take; *or* = one who does something. Pl., Proprioreceptors.) A sensory receptor within the body of an organism which is sensitive and responsive to phasic or tonic internal stimuli. Term proprioreceptor is sometimes applied to Campaniform Sensilla which serve as stretch receptors. Proprioreceptors form 'beds' on the face of the locust. These receptors respond tonically: A stimulus is transmitted as long as a receptor is deflected or stressed. Proprioreceptors enable the insect to orient to wind (hair plates on the locust face), position of the body (cervical receptors of the head) or position of appendages (legs, wings.) Proprioreceptors are stimulated by changes in shape of the body (muscles, tendons, body wall.) The response to stimulation may be phasic or tonic. Phasic receptors involve action potential in response to bending (deflecting) or straightening. Typically, phasic receptors occur on Antennae, Tarsi, and other appendages. Tonic receptors display continuous action potentials for the period of stress or deflection. Examples include facial Setae of the locust which are stimulated by air currents and the Setae positioned between the head and Thorax which function as proprioreceptors informing the insect of its body position. See Sensilla. Cf. Hair Plate; Strech Receptor. Rel. Phasic Receptor; Tonic Receptor.

PROPROTOX® See Proproxur.

PROPULSATORY Adj. (Latin, *propulsare* from *propellere* = propel.) Driving onward or forward, propelling.

PROPUPA Noun. (Latin, *pro* = before; *pupa* = puppet. Pl., Propupae; Propupas.) A stage in the nymphal development of *Aspidiotus* (Diaspididae: Coccoidea) in which wing pads are present and the legs are short and thick. The Propupa precedes the true Pupa (C. V. Riley.) Syn. Semipupa. See Pupa. Rel. Metamorphosis.

PROPYGIDIUM Noun. (Greek, *pro* = before; *pygidion* = buttock; *-idion* = diminutive. Pl., Propygidia.) 1. The Hypopygium. 2. Coleoptera: A dorsal sclerite immediately anterior of the Pygidium.

PROPYON® See Proproxur.

PRORHYACOPHILIDAE Plural Noun. A Family of Trichoptera known from the Triassic Period of Kirgizia.

PROSCUTELLUM Noun. (Latin, *pro* = before; *scutellum* = small shield. Pl., Proscutella.) The Scutellum of the Pronotum.

PROSCUTUM Noun. (Latin, *pro* = before; *scutum* = shield. Pl., Proscuta.) The Scutum of the Pronotum.

PROSOMA Noun. (Greek, *pro* = before; *soma* = body.) 1. The Cephalothorax of arachnids. 2. The head in apocritous Hymenoptera. 3. Acarology: anterior division of the body; one of the chelicerate Tagmata.

PROSOMAL Adj. (Greek, *pro* = before; *soma* = body; *-alis* = pertaining to.) Descriptive of structure associated with the Prosoma.

PROSOPISTOMATOIDEA A Superfamily of Ephemeroptera including Baetiscidae and Prosopistomatidae.

PROSPALTELLINAE Nikolskaya 1966. A Subfamily

of Aphelinidae. Morphological features include: Antenna usually with 7–8 segments (rarely 6-segmented), with 2–3 segments in club; Pronotum divided medially, but Prepectus entire; Scutellum considerably wider than long; Axilla projecting into Parapsides; forewing sparsely setose; Metasoma not modified. Biology: Internal parasites of Diaspididae and Aleyrodidae. Important Genera include *Coccophagoides* (15 Species) and *Encarsia* (= *Aspidiotiphagus*, *Prospaltella*) cosmopolitan, about 160 Species.

PROSTEMMATIC See Anteocular.

PROSTERNAL Adj. (Latin, *pro* = forward; *sternum* > Greek *sternon* = breastbone; *-alis* = pertaining to.) Descriptive of structure associated with the Prosternum.

PROSTERNAL EPIMERA The Epimera of the Prothorax.

PROSTERNAL EPISTERNA The Episterna of the Prothorax.

PROSTERNAL FURROW Some Reduviidae and Phymatidae: A cross-striated longitudinal groove in the Prosternum, by means of which stridulation is caused by rubbing the rugose apex of the Rostrum in it by back-and-forth movements of the head.

PROSTERNAL GROOVES Those which occur laterally in some Coleoptera, *e.g.,* Elateridae, to receive the Antennae.

PROSTERNAL LOBE Some Coleoptera: An anterior prolongation of the Prosternum which more-or-less conceals the mouth from below.

PROSTERNAL PROCESS Aquatic Coleoptera: A structure on the Prosternum.

PROSTERNAL SPINE The curved Mucro in Elateridae, which extends backward into a mesosternal cavity, the cone or tubercle between forelegs in some Orthoptera.

PROSTERNAL SUTURE That suture of the Prothorax which separates the Sternum from the pleural pieces.

PROSTERNELLUM Noun. (Modern Latin.) The Sternellum of the Prothorax.

PROSTERNUM Noun. (Latin, *pro* = forward; *sternum* = breastbone.) The anteriormost sternal sclerite or the sclerite between the forelegs. Cf. Mesosternum; Metasternum.

PROSTETHIUM Noun. (Greek, *pro* = before; *stethos* = breast, *-idion* = dim. Pl., Prostethia.) The Sternum of the Prothorax, Prosternum or Presternum.

PROSTHECA Noun. (Greek, *prostheke* = appendage.) 1. Some Coleoptera adults: A mandibular sclerite along the medial surface between a basal molar and apical tooth. Prostheca is invested with Setae and articulated at Basalis; it resembles the Maxillary Lacinea (Cf. Retinaculum). Some Coleoptera larvae: A lightly sclerotized, flexible, medial, fringed process on the Mandible. 2. Larval nematocerous Diptera: Tuft of Setae near molar surface on upper paralabral surface (Teskey).

PROSTIGMA Noun. (Greek, *pro* = before; *stigma* = mark.) Hymenoptera: An apical enlargement of the forewing Subcostal Vein, separated from the larger Stigma. Cf. Pterostigma.

PROSTOMIAL Adj. (Greek, *pro* = before; *stoma* = mouth; *-alis* = pertaining to.) Descriptive of structure associated with the Prostomium.

PROSTOMIAL GANGLION The Brain or Archicerebrum in annelids.

PROSTOMIAL LOBE One of the paired lateral lobes forming the Protocerebrum.

PROSTOMIUM Noun. (Greek, *pro* = before; *stoma* = mouth; *-idion* = dim. Pl., Prostomia.) The anterior preoral unsegmented part of the trunk of a segmented animal. Acron *sensu* Snodgrass.

PROSTOMUM Noun. Pediculus: See Mouth Cone.

PROSUTURE Noun. The Prothoracic Suture (MacGillivray).

PROTAESTHESIS Noun. (Greek, *protos* = first; *aistheto* = perceptible; *sis* = a condition or state.) A primitive Sensilla or sense-bud (Berlese).

PROTAMPHIBION Noun. A name applied by P. Mayer to the hypothetical common ancestor of the Plecoptera, Ephemeridae and Odonata.

PROTANDROUS Adj. (Greek, *protos* = first; *aner* = male; Latin, *-osus* = with the property of.) Pertaining to males which demonstrate protandry.

PROTANDRY Noun. (Greek, *protos* = first; *aner* = male.) 1. The condition or phenomenon in which the male reproductive organs (gametes) of a hermaphroditic organism mature more rapidly than the female reproductive organs (gametes) and thereby preclude self-fertilization. Protandry occurs in some plants and invertebrates. Cf. Protogyny. 2. The appearance of adult males earlier than adult females of the same Species. The phenomenon is sometimes ascribed to members of a brood or population. A common phenomenon in parasitic Hymenoptera, some aculeate non-social Hymenoptera and some Lepidoptera. Alt. Proterandry; Metagyny.

PROTARSUS Noun. (Greek, *protos* = first; *tarsus* = sole. Pl., Protarsi.) The Tarsus of the anterior leg. Ticks: The segment preceding the Tarsus.

PROTEASE Noun. A digestive enzyme which splits or breaks up proteins.

PROTECTANT FUNGICIDE Any fungicide which provides a protective chemical barrier over the surface of the host and protects it from initial fungal infection. Cf. Eradicant Fungicide. See Fungicide.

PROTECTIVE COLORATION Any pattern or arrangement of colours in an animal that enables it to escape detection by resembling or blending into its surroundings. See Coloration. Cf. Advancing Coloration; Alluring Coloration; Anticryptic Coloration; Apetetic Coloration; Aposematic Coloration; Combination Coloration; Cryptic Coloration; Cuticular Colour; Directive Coloration; Disruptive Coloration; Epigamic Coloration; Episematic Coloration; Procryptic Coloration; Pseudepisematic Coloration; Pseudoaposematic Coloration; Seasonal Coloration; Sematic Coloration. Rel. Crypsis; Mimicry.

PROTECTIVE LAYER Cynipid galls: A sclerified (hardened) tissue that is best developed in the European Subgenus *Cynips* in which the cell-walls are thickened and the cells contain many crystalline materials (Kinsey.)

PROTECTIVE MIMICRY A form of protective coloration, in which one Species has a resemblance to some other Species which is immune from attack by predators. Cf. Crypsis.

PROTEIFORM Adj. Structure or a body capable of assuming many forms or varieties. See Protean. Cf. Amoebiform. Rel. Form 2; Shape 2; Outline Shape.

PROTEIN Noun. (Greek, *proteion* = first. Pl., Proteins.) Any one of a class of complex nitrogenous compounds, such as albumen, which form an essential part of the structure of all animal and vegetable cells. Proteins contain nitrogen and sometimes sulphur in addition to carbon, hydrogen and oxygen, and are characterized by high molecular weight (from 35,000 up to millions).

PROTELEAN PARASITE An insect in which only the immature stage (larva) develops parasitically and eventually kills the host. Attributes of protelean parasites used to distinguish them from other parasitic animals include: (1) larval stage parasitic, adult free living; (2) larva typically kills and consumes one host; (3) body size of parasite and host similar; (4) parasite life cycle relatively simple; (5) parasite shares relatively close taxonomic affinity with host; (6) parasite displays reproductive capacity between so-called true parasites and free-living forms. Syn. Parasitoid. See Parasitism.

PROTELUM Noun. The eleventh segment in insects.

PROTELYTROPTERA Tillyard 1931. An Order of insects extant during the Permian.

PROTEMERIDA Handlirsch 1908. An Order of insects extant during the Carboniferous.

PROTENTOMIDAE Plural Noun. A small Family of Protura assigned to the Acerentomoidea. Divided into Fujientomidae, Hesperentomidae and Protentomidae by authors.

PROTENTOMON Noun. (Greek, *protos* = first; *entomon* = insect.) The hypothetical organism postulated as the archetype or ancestral form of the winged insects (Imms).

PROTEOLYTIC Adj. (Greek, *protein*; *lysis* = to loosen; *-ic* = characterized by.) Protein-splitting, *i.e.,* breaking up proteins into other compounds.

PROTERANDRIC Adj. (Greek, *proteros* = fore; *aner* = male; *-ic* = characterized by.) Having two kinds of males.

PROTERANDRY Noun. (Greek, *proteros* = fore; *aner* = male.) The appearance or emergence of males before females.

PROTERGITE Noun. (Latin, *pro* = in front of; *tergum* = back; *-ites* = constituent. Pl., Protergites.) The Prescutum *sensu* Berlese.

PROTERGUM Noun. (Latin, *pro* = in front of; *tergum* = back. Pl., Proterga.) The anterior Tergum; the Tergum of the Prothorax.

PROTEROPHRAGMA Noun. The prephragma of the Mesothorax.

PROTEROSOMA Noun. (Greek, *proteros* = before, in front of; *soma* = body.) Acarology: A division of the body anteriad of the Sejugal Furrow.

PROTEROTYPES Noun. Taxonomy: Primary types, including all the material upon which the original description is based.

PROTEROZOIC EON The second interval in the earth's Geological Time Scale (2.5 BYBP–570 MYBP). Traditionally viewed as boundary between life (Zoic) and absence of life (Azoic). PE characterized by oragenic activity (Appalachian Mountains), periods of glaciation and sedimentary rocks without fossils that overlay strata invested with diverse invertebrate forms. During PE, Stromatolites were abundant (2.3 BYBP); cellular life forms (eukaryote cells called acritarchs) appear ca 2 BYBP, multicellular algae appear ca 1 BYBP and soft-bodied multicellular invertebrates appear ca 570 MYBP. See Geological Time Scale. Cf. Archaean Eon; Precambrian.

PROTERPIA ORANGE *Eurema proterpia* [Lepidoptera: Pieridae].

PROTHETELY Adj. (Greek, *pro* = before, *thetikos* = to set down.) The possession of two pairs of true external wing pads by a holometabolous insect larva (Folsom & Wardle). This condition occurs in many Families of Coleoptera and some Lepidoptera. Cf. Hysterotely; Metathetely; Paedogenesis.

PROTHIOFOS An organic-phosphate (dithio-phosphate) compound {O-(2,4-dichlorophenyl) O-ethyl-S-propyl phosphorodithioate} used as a broad-spectrum contact insecticide and stomach poison for control of mites, plant-sucking insects (aphids, mealybugs), leaf-eating insects (caterpillars, armyworms, leaf rollers), rootworms, and wireworms. Used as a public-health insecticide for control of mosquitoes and flies. Compound applied to fruit trees (apples, pears, citrus), grapes, sugarcane, sugar beets, tobacco and ornamentals in many countries. Not registered for use in USA. Trade names include: Bideron®, Tokuthion®, Toyodan®, Toyothion®. See Organophosphorus Insecticide.

PROTHOATE An organic-phosphate (dithiophosphate) compound {O,O-diethyl S-(N-isopropylcarbamoylmethyl) phosphorodithioate} used as an acaricide and contact/systemic insecticide for control of mites, plant-sucking insects (to include aphids, leafhoppers, lace bugs), thrips and others. Compound applied to citrus, cotton, fruit trees, grapes, mangos, sugar beets, sugarcane, tobacco, ornamentals and vegetables in some countries. Not registered for use in USA. Toxic to fishes. Trade names include Fac®, Fostion® and Telefos®. See Organophosphorus Insecticide.

PROTHORACIC BRISTLE Diptera: A strong bristle immediately above the front Coxa. See Propleura Bristles.

PROTHORACIC GLANDS Orthoptera: Glands on the sides of the Thorax in certain phasmid Genera.

PROTHORACIC SCUTUM The Scutum of the Prothorax; the Prescutum.

PROTHORACIC SHIELD The cervical shield.

PROTHORACIC SPIRACLES In older reseach papers, the anteriormost pair of thoracic spiracles. Now more commonly termed Mesothoracic Spiracles (Comstock).

PROTHORACOTHECA Noun. The pupal covering of the Prothorax.

PROTHORAX Noun. (Greek, *pro* = before; *thorax* = breast plate.) The first thoracic segment, or portion of the Thorax nearest the head. The Prothorax bears anterior pair of legs but does not bear wings (except some Palaeozoic fossil insects). Size and shape of Prothorax are highly variable. Prothorax is a large segment (Orthoptera, Hemiptera, Coleoptera), or reduced in size to form a narrow, transverse sclerite (Diptera, Hymenoptera). Thoracic sclerites are separated by membrane that may be large and conspicuous in primitive Holometabola (Neuroptera, Coleoptera) or small in highly evolved Holometabola (Diptera, Hymenoptera). Dorsal portion of Prothorax is divided into an anterior Antepronotum and Postpronotum. Antepronotum is well developed in Nematocera; Postpronotum is well developed in higher Diptera. Lateral wall of Prothorax (Propleuron) is not well developed in many insects. The ventral portion of Prothorax is called Prosternum and may be divided into an anterior Presternum and posterior Basisternum. Syn. Corselet (Strauss-Durckheim); Manitruncus (Kirby & Spence). See Thorax; Cf. Mesothorax; Metathorax; Pterothorax.

PROTISTA Noun. A Kingdom of largely unicellular Eukaryotic organisms including Algae, Lower Fungi and Protozoa. Syn. Protoclista.

PROTISTOLOGY Noun. (Greek, *protista* = first, primary; *logos* = discourse. Pl., Protistologies.) The study of Protista.

PROTOARTHROPODAN Noun. (Greek, *proteros* = fore; *arthros* = joint; *podos* = foot; *-an* = belonging to.) The hypothetical primitive ancestral arthropod.

PROTOBIOBIONIDAE Rohdendorf. A small Family of nematocerous Diptera known from two Species taken in Upper Jurassic Period deposits of China and Russia.

PROTOBLATTARIA Brues & Melander 1932. An Order of insects extant during the Upper Carboniferous to the Permian.

PROTOBRANCHIATE Adj. (Greek, *protos* = first in time; *branchia* = gills; *-atus* = adjectival suffix.) Nymphal Odonata: With the respiratory apparatus contained in the Rectum.

PROTOCEPHALIC Adj. (Greek, *protos* = first in time; *kephale* = head; *-ic* = of the nature of.) Descriptive of or pertaining to the insect embryo's head (Protocephalon).

PROTOCEPHALON Noun. (Greek, *protos* = first in time; *kephale* = head. Pl., Protocephalons.) 1. The primary head region of the insect embryo. 2. An early stage in the evolution of the arthropod head, corresponding to the cephalic lobes of the embryo, comprising the Prostomium and first Postoral Somite. 3. The primitive (groundplan) head of insects, and consists of the Labrum, compound eye, Antenna and usually the Postoral Somite of the second Antenna. Protocephalon persists in some Crustacea, but most mandibulate arthropods display a head of more complex structure The Protocephalon is modified in modern insects to include a Procephalon and Gnathocephalon. Alt. Protocephalic Region. See Head.

PROTOCEREBRAL Adj. (Greek, *protos* = first in time; Latin, *cerebrum* = brain; *-alis* = pertaining to.) Descriptive of structure associated with the Protocerebrum. See Cerebrum; Protocerebrum.

PROTOCEREBRAL BRIDGE The mass of the Protocerebrum positioned in the dorsal and posterior part of the Pars Intercerebralis; the Posterior Dorsal Commissure of Thompson (Snodgrass). See Protocerebrum.

PROTOCEREBRAL LOBES The lateral parts of the Protocerebrum. See Protocerebrum.

PROTOCEREBRAL REGION Part of the primitive arthropod Brain which innervates the eyes; the ocular region (Snodgrass). See Protocerebrum.

PROTOCEREBRAL SEGMENT The ocular segment of the insect Brain.

PROTOCEREBRUM Noun. (Greek, *protos* = first in time; Latin, *cerebrum* = brain. Pl., Protocerebra; Protocerebrums.) The primitive anterior cerebral vesicle and first part of arthropod Brain. Anteriormost portion of insect Brain and represents fused ganglia of optical segment. Anatomically, Protocerebrum is most complex part of Brain, and is subdivided into Protocerebral Lobes and Optic Lobes. Protocerebrum is bilobed, positioned above Oesophagus in hypognathous head, and is continuous with ventral portion of optic lobes. Cell bodies are constricted along periphery: Perikaria (nerve cell bodies) are peripheral; nuclei of nerve cells are central. Pars Intercerebralis forms anterior, mediodorsal portion of Protocerebral Lobes. Anterior cells contribute to ocellar neurons which contain Neurosecretory Cells. Protocerebral Bridge connects two hemispheres of Protocerebrum. Central Body forms mass of Neuropile and lies at centre of Protocerebrum. Volume of Central Body is correlated with size of insect. Central Body may be responsible for mediating general arousal or quickness shown by insects. See Brain; Central Nervous System; Nervous System. Cf. Deutocerebrum; Tritocerebrum. Rel. Nerve Cell; Neuropile; Neurosecretory Cells; Pars Intercerebralis; Protocerebral Bridge.

PROTOCORMIC Adj. (Greek, *protos* = first in time; *korm* = tree trunk; *-ic* = characterized by.) Descriptive of or pertaining to the trunk of the insect embryo.

PROTOCORMIC REGION The primary trunk region of the insect embryo.

PROTOCOSTA Noun. (Greek, *protos* = first in time; *costa* = rib, side. Pl., Protocostae.) Lepidoptera: The thickened costal margin of a wing.

PROTOCRANIUM Noun. (Greek, *protos* = first in time; *kranion* = skull; Latin, *-ium* = diminutive > Greek, *-idion*. Pl., Protocrania; Protocraniums.) The posterior part of the Epicranium, sometimes the Occiput.

PROTOCUCUJIDAE Plural Noun. A small Family of polyphgaous Coleoptera assigned to the Cucujoidea. Found in South America and Australia. Adult 3.5–5.5 mm long, elongate, somewhat flattened, with moderate vestiture of stout, decumbent Setae; Antenna with 11 segments including club of three segments; tarsal formula 5-5-5 in females, 5-5-4 in males; Abdomen with five Ventrites.

PROTODERMAPTERA Plural Noun. A Superfamily of the Order Dermaptera. The ear-wings (Burr); equivalent to Labiduroidea *sensu* Tillyard.

PROTODEUTOCEREBRAL Adj. (Greek, *protos* = first in time; *deuteros* = second; *cerebrum* = brain; Latin, *-alis* = pertaining to.) Descriptive of structure associated with the Protocerebrum and Deutocerebrum. See Brain.

PROTODIPTERA Noun. An Order of insects consisting of the Permotanyderidae and Permotipulidae. Regarded as a link between the Mecoptera and Diptera. See Diptera; Mecoptera.

PROTOGONIA Noun. (Greek, *proto* = first; *gonia* = angle.) The apical angle of the forewing.

PROTOGRAPH Noun. (Greek, *proto* = first; *graphein* = to write.) An original description of a Species by a figure or picture made from the original type. See Holotype.

PROTOGYNE Noun. (Greek, *protos* = first in time; *gyne* = woman. Pl., Protogynes.) 1. An individual demonstrating the reproductive phenomenon of Protogyny. 2. The condition or phenomenon in which the female reproductive organs (gametes) of an hermaphroditic organism mature more rapidly than the male reproductive organs (gametes) and thereby preclude self-fertilization. Protogyny is found in some plants and invertebrates. Cf. Protandry.

PROTOHYMENIDAE Plural Noun. An extinct Family of Megasecoptera, the members of which displayed tubular projections from the Abdomen.

PROTOLEPIDOPTERA Noun. Proposed for those forms (Eriocephalidae) in which the Lacinia and the Mandibles are obvious and the spiral tongue is not developed. See Neolepidoptera; Palaeolepidoptera.

PROTOLIGONEURIDAE Rohdendorf 1962. Plural Noun. A Family of Diptera based on fossil material of a monotypic Genus preserved in Lower Jurassic Period deposits from Kirghizistan.

PROTOLOG Noun. (Greek, *protos* = first in time; *logos* = word.) The original description in words.

PROTOLOMA Noun. (Greek, *protos* = first in time; *loma* = fringe, hem.) The anterior margin of the forewings.

PROTOMECOPTERA Noun. A Suborder of Mecoptera including Meropeidae and Notiothaumidae. Cf Eumecoptera.

PROTOMESAL AREOLES Certain Areolets in the wings of the Hymenoptera that are positioned between the Costal Cells and apical margin. Syn. Protomesal Cells.

PROTONEURIDAE Plural Noun. A widespread, predominantly tropical Family of zygopterous Odonata. Adults small to moderate sized, slender with broad head; Pterostigma short; quadrilateral rectangular; longitudinal veins tend to be straight; Anal Vein reduced. Male superior appendages not forcipate. Larvae differ in habitus but tend to be short, slender, with prominent eyes; Antenna with seven segments, third longest; three Caudal Gills long, laemllate, swollen, saclike; gills sometimes constricted at Nodus. Species prefer still or running water in shaded localities or forested situations.

PROTOORTHOPTERA Handlirsch 1908. An Order of insects living during the Carboniferous and Permian.

PROTOPLASM Noun. (Greek, *proto* = first; *plasma* = something formed.) A nitrogenous, glairy compound which forms the soft tissues of animals; of unstable, changing and variable chemical composition, 'the physical basis of life' (Huxley.)

PROTOPLASMIC Adj. (Greek, *protos* = first in time; *plasma* = form, mould; *-ic* = of the nature of.) Descriptive of or pertaining to protoplasm.

PROTOPLAST Noun. (Greek, *protos* = first in time; *-plast* = comb. form meaning a unit of protoplasm. Pl., Protoplasts.) Microbial or plant cells whose cell walls have been removed.

PROTOPLECIIDAE Rohdendorf. Plural Noun. A Family of fossil Diptera known from Mesozoic deposits in Germany, Russia and China.

PROTOPOD LARVA Noun. (Pl., Protopods.) 1. A stage in which the insect larva is characterized by a lack of differentiation of the external and the internal organs (Wardle). 2. First-instar larva of some parasitic Hymenoptera (Proctotrupoidea).

PROTOPOD PHASE Insect embryo: That phase in which the metamerism is incomplete, the Abdomen incomplete and the appendages more or less rudimentary (Imms). See Oligopod.

PROTOPODITE Noun. (Greek, *proto* = first; *podite* = small leg; *-ites* = constituent. Pl., Protopodites.) 1. The first part of the Maxilla. 2. The basal piece of a segmented appendage in the Arthropoda.

PROTOSPECIES Noun. (Greek, *protos* = first in time, + species.) The preexisting stock from which other Species arise (J. C. Chamberlain). In an evolutionary classification, the ancestral form of a lineage from which all subsequent Species are derived. See Prototype.

PROTOTERGITE Noun. (Greek, *proto* = first; Latin, *tergum* = back; *-ites* = constituent.) The foremost dorsal segment of Abdomen.

PROTOTHEORIDAE Plural Noun. A small Family of glossatous Lepidoptera assigned to Superfamily Hepialoidea and sometimes recognized as members of Hepialidae in the broad sense. Small, drab moths; compound eyes large; Ocelli absent; Mandibles vestigial; Proboscis short and extends to distal end of Labial Palpus; Maxillary Palpus minute; Antenna submoniliform; Epiphysis present; tibial spur formula 0-2-4. All Species occur exclusively in South Africa.

PROTOTHORAX Noun. See Prothorax.

PROTOTYPE Noun. (Greek, *protos* = first; *typos* = pattern.) A hypothetical primitive form of development or ancestral lineage from which later forms can be traced. See Type. Cf. Archetype. Rel. Groundplan.

PROTOZOA Noun. (Greek, *proto* = first; *zoon* = animal.) The Phylum of the animal kingdom containing the one-celled animals. One-celled animals in general.

PROTOZOAL Adj. (Greek, *protos* = first in time; *zoon* = animal; Latin, *-alis* = pertaining to.) Pertaining to Protozoa.

PROTRACTED Adj. (Latin, *protractus* > pp of *protatiere*; *pro* = forward; *trahere* = to draw.) Extended. Arcane Protractus.

PROTRACTILE Adj. (Latin, *protractus* > pp of *protatiere*; *pro* = forward; *trahere* = to draw; *-alis* = pertaining to.) Capable of protraction or extension.

PROTRACTOR Noun. (Latin, *protractus* > pp of *protatiere*; *pro* = forward; *trahere* = to draw; *or* = one who does something.) That which extends or lengthens a structure. Cf. Retractor.

PROTRACTOR MUSCLE Descriptive of a muscle that extends or lengthens a structure. Cf. Retractor; Extensor; Flexor.

PROTRUSIBLE Adj. (Latin, *protrudere* > *pro* = forward; *trudere* = to thrust.) Descriptive of something capable of being protruded, or put out. Cf. Eversible.

PROTUBERANCE Noun. (Latin, *protuberans*. Pl., Protuberances.) Any elevation above the surface of which it is a part. Typically, a rather rounded extension from a rather flat surface. Syn. Bulge; Bump; Lump. Cf. Projection.

PROTUBERANT Adj. (Latin, *protuberare*; *-antem* = adjectival suffix.) Descriptive of a conical or mound-like rising above the surface or general level. Cf. Everted.

PROTURA Silvestri 1907. Noun. (Greek, *protos* = first; *oura* = tail.) A Class and Order of entognathous Hexapoda; sometimes considered a primitive Order of Insecta. Cosmopolitan in distribution and consisting of about 500 nominal Species, placed in 4–8 Families. Cryptic lifestyle, inhabiting leaf litter, soil, moss, beneath rocks or under bark; most frequently taken in Berlese samples; may feed on mycorrhizal fungi. Characteristically small bodied (less than 2 mm long) and elongate. Head prognathous with entognathous piercing mouthparts and well developed Maxillary and Labial Palpi; Antenna and compound eye absent; Thorax weakly developed; legs with five segments; Pretarsus with median claw and Empodium. Adult Abdomen with 12 segments; Sterna 1–3 with small eversible Styli; Cerci absent and Gonopore beween segments 11 and 12; Ovaries, paired, sac-like, meroistic; Spermatozoa not motile. Malpighian Tubules present; seven abdominal ganglia; Development anamorphic. Included Families: Acerentomidae, Eosentomidae, Protentomidae and Sinentomidae. See Apterygota. Cf. Diplura.

PROTURAN Adj. A member of the Protura. Descriptive of or pertaining to the Protura.

PROUPSILON Noun. Diptera: The arm of the Y of the upsilon of Furca in or near the furrow separating a Paraglossa and the Mediproboscis; furca-3 (MacGillivray).

PROUT, LOUIS BEETHOVEN (1864–1943) (Breyer 1944, Revta Soc. ent. argent. 2: 72.)

PROVADO® See Imidacloprid.

PROVANCHER, LÉON (1820–1892) (Huard 1894, Naturaliste Can. 21: 38–41, 53–58, 85–88, 101–104, 134–137, 148–152, 182–185.)

PROVENTRICULAR Adj. (Latin, *pro* = before; *ventriculus* = small stomach.) Descriptive of or pertaining to the Proventriculus.

PROVENTRICULAR VALVULE Ptychoptera: A circular fold of the intestinal wall (Packard).

PROVENTRICULUS Noun. (Latin, *pro* = before; *ventriculus* = small stomach.) A typically musculated, complex portion of the Foregut positioned posterior of the Crop and anterior of the circular muscles which provide an anterior constriction for the midgut. A distinct Proventriculus is found in several orders, including Orthoptera, Odonata, Coleoptera, Hymenoptera, Siphonaptera and Mecoptera. The Proventriculus is not sclerotized in Apterygota and most sucking insects (except Siphonaptera). The Proventriculus is often poorly developed or absent from fluid feeding insects and variably developed in immature insects. Longitudinal muscles lie adjacent to the Proventriculus; circular muscles surround the longitudinal muscles; 6–8 well developed longitudinal folds (Plicae) extend the length of the Proventriculus; muscles are found within the folds. Each Plica is divided into three areas from front to back: The Dental Area is toothed, tuberculate or not sclerotized; the Pulvillar Area sometimes displays Microtrichae or Acanthae; the Valvular Area is slender and separated from the Pulvillar Area by an annular groove. Contraction of circular muscles close the Proventriculus and prevent passage of food and enzymes into the midgut. Acanthae project into the lumen of the Proventriculus in adult insects which feed on particulate matter; these Acanthae serve as chitinous teeth or sclerotized plates which masticate food by grinding or sifting. The Proventriculus of fluid feeders has long, branched Setae and the remainder of the Proventricular surface bears Acanthae which serve as triturating (grinding) surfaces. Proventriculus was first described in Mecoptera as 'Faltenmagen' (Ramdor 1811, Abhandl. Uber Verdauungswerk. Ins.), 'Magen' (Loew 1848, Linn. Ent. 3: 345–385), 'le gesier' (Dufour 1841, Mem. Math. Sav. Erang. Acad. Sci Paris 7: 265–647), and 'Haarcylinder' (Brauer 1855, Verh. Zool. Bot. Ges. 5: 701–726.) See Foregut; Alimentary System. Cf. Crop; Stomodeal Valve. Syn. Gizzard.

PROVENTRICULUS ANTERIOR *Blatta*: The more tapering posterior half of the Proventriculus which forms an armature of spiny sclerites (Acanthae).

PROVENTRICULUS POSTERIOR *Blatta*: The more tapering posterior half of the Proventriculus which has a circle of soft cushion-like lobes covered with backward projecting spines.

PROVISION Verb. (Latin, *provisum* from *pro* = before; *visum* = see.) To supply or cache with necessities, as in a hive or larval cell.

PROVISIONAL MANDIBLES Some Coleoptera: Parts of the Mandible found in the Pupa and used for cutting through the cocoon upon emergence of the Imago (Imms).

PROVISIONING Verb. See Mass Provisioning; Progressive Provisioning.

PROWAZAK, STANISLAUS VON (1875–1959) Austrian scientist who died of Epidemic Typhus while studying the disease. (Weidner 1967, Abh. Verh. naturw. Ver. Hamburg Suppl. 9: 342–343.)

PROX® See Proproxur.

PROXACALYPTERON Noun. Diptera: The Squama, Squama Thoracalis, Tegula, Calypteron (MacGillivray).

PROXADENTES Noun. (Sl., Proxadentis.) The dentes, from the mola to the end of the Mandible (MacGillivray).

PROXAGALEA Noun. The proximal segment of the Galea; Basigalea (MacGillivray).

PROXIMAD Adv. (Latin, *proximus* = next; *ad* = toward.) Toward the proximal end of a structure or an appendage. See Orientation. Cf. Anteriad; Apicad; Basad; Caudad; Centrad; Cephalad; Craniad; Dextrad; Dextrocaudad; Dextrocephalad; Distad; Dorsad; Ectad; Entad; Laterad; Mediad; Mesad; Neurad; Orad; Rostrad; Sinistrad; Sinistrocaudad; Sinistrocephalad; Ventrad.

PROXIMAL Adj. (Latin, *proximus* = next; *-alis* = pertaining to.) 1. Pertaining to structure near the base of an appendage. 2. The portion of a segment nearest the body or connection with the body. See Orientation. Cf. Apical; Basal; Distal.

PROXIMAL RACHIS Hymenoptera: The shank of an antennal joint into which the lateral spines or other processes are inserted.

PROXIMAL SENSORY AREA of Hayes. Scarabaeoid larvae: See Haptolachus of Boving.

PROXOL® See Trichlorfon.

PROZONA Noun. Orthoptera and Elateridae: The portion of the Pronotum anterior of the Antecostal Sulcus. Cf. Metazona.

PRÜFFE, JAN (1890–1959) (Gromadska 1960, Prsegl. zool. 4: 159–166, bibliogr.)

PRUINESCENCE Adj. A minute dust or 'bloom' covering certain insects. See Pruinose.

PRUINOSE Adj. (Latin, *pruina* = hoar frost; *-osus* = full of.) Covered with fine dust, as if frosted; with the brightness of a surface somewhat obscured by the appearance of a plum-like bloom, but which cannot be rubbed off (Kirby & Spence). Alt. Pruinosus; Pruinous.

PRUINOSE BEAN WEEVIL *Stator pruininus* (Horn) [Coleoptera: Bruchidae].

PRUINOSE SCARAB *Sericesthis geminata* Boisduval [Coleoptera: Scarabaeidae].

PRUNE LEAFHOPPER *Edwardsiana prunicola* (Edwards) [Hemiptera: Cicadellidae].

PRUNNER, LEONARD DE Author of *Lepidoptera Pedemontana* (1798).

PRUNOSUS (PRUNUS) Adj. The purplish red of plums; plum colour.

PRUSSIC ACID Hydrocyanic acid (HCN).

PRYER, HENRY JAMES STOVIN (1850–1888) (Anon. 1888, Entomol. mon. Mag. 24: 277–278.)

PRYER, WILLIAM BURGESS (1843–1890) (Janson 1899, Entomologist 32: 52.)

PSAMMOBIOTIC Adj. (Greek, *psammos* = sand; *bios* = living; *-ic* = characterized by.) Pertaining to organisms that live in sandy habitats (*e.g.* myrmeleontid larvae). See Biotia.

PSAMMOPHILOUS Adj. (Greek, *psammos* = sand; *philein* = to love; Latin, *-osus* = with the property of.) Sand-loving; living or frequenting sandy places. *e.g.* Psammophilous wasps such as *Bembix*. See Habitat. Cf. Ammophile; Arenicolous; Eremophilous; Rupicolous. Rel. Ecology; Habitat.

PSAMMOPHORE Noun. (Greek, *psammos* = sand; *phorein* = to carry. Pl., Psammophores.) A term proposed by Santschi (1909, Rev. Suisse Zool. 17: 449–458) for groups of Setae on the ventral surface of the head and Mandibles of some desert inhabiting ants and digger wasps. The Setae form baskets that are used to transport sand or dry soil. Sand or dry soil are collected in the basket with the Fore Tibiae and apex of the Gaster. When the basket is full, the Mandibles are 'closed'. Setae on the Mandible's ventral surface serve as a lid for the basket when the Mandibles are closed. Cf. Ammochaetae.

PSAMMOPHYTE Noun. (Greek, *psammos* = sand; *phyton* = plant. Pl., Psammophytes.) Any plant that develops in a sandy or gravelly substrate. See Habitat. Cf. Halophyte.

PSELAPHIDAE Plural Noun. Ant-Loving Beetles. A cosmopolitan Family of Coleoptera including about 5,000 nominal Species; closely related to the Staphylinidae. Adult body to about 3.5 mm long, robust; Antenna clavate with 2–11 segments; Maxillary Palpus usually long; Elytra short, truncate, exposing some abdominal Terga; tarsal formula 3-3-3, rarely 2-2-2; Abdomen inflexible, with six Ventrites. Larva elongate to fusiform with errect Setae; head prognathous; Antenna with 2–3 segments and eversible glands in socket; legs with five segments; Urogomphi absent. Larva and adult predaceous in moist habitats including leaf litter, moss, fungi. Many Species termitophilous or myrmecophilous; trichomes present. Adults attracted to lights in large numbers.

PSELAPHOTHECA Noun. That part of the Pupa which covers the Palpi.

PSEPHENIDAE Plural Noun. A small Family of polyphagous Coleoptera assigned to the Byrrhoidea or Dryopoidea. Body 3–8 mm long, oval, flattened, with short, dense pubescence; head deflexed; Antenna filiform or serrate with 11 segments; tarsal formula 5-5-5, Tarsomeres long; Abdomen with five Ventrites. Larva broadly oval, strongly dorsoventrally flattened, discoid; head prognathous, with numerous Stemmata; Antenna with three segments; legs with five segments, each Pretarsus with one claw; Urogomphi absent; body with several gin traps. Larva aquatic feeding upon algae attached to rocks. Pupation within larval Integument along stream banks. Adults cryptic, taken near water.

PSEUDALIS Noun. Some Orthoptera: The small sclerite laterad of the Tegula, near the Praescutum (MacGillivray).

PSEUDARBELIDAE See Psychidae.

PSEUDAROLIA Noun. (Greek, *pseudo* = false; *arolium* = protection. Sl., Pseudoarolium.) False Arolia or processes of the Tarsus resembling the true Arolia and more-or-less similarly placed, paired structures found beneath the claws in some Hemiptera.

PSEUDARTHROSIS Noun. (Greek, *pseudo* = false; *arthron* = joint; *sis* = a condition or state. Pl., Pseudarthroses.) A false joint or articulation.

PSEUDEPISEMATIC COLORATION (Greek, *pseudo* = false; *epi* = upon; *soma* = body.) Colours displayed in a body that are critical in aggressive mimicry and alluring coloration. See Coloration. Cf. Advancing Coloration; Alluring Coloration; Anticryptic Coloration; Apetetic Coloration; Aposematic Coloration; Combination Coloration; Cryptic Coloration; Directive Coloration; Disruptive Coloration; Epigamic Coloration; Episematic Coloration; Procryptic Coloration; Protective Coloration; Pseudoaposematic Coloration; Seasonal Coloration; Sematic Coloration. Rel. Crypsis; Mimicry.

PSEUDERGATE Noun. (Greek, *pseudo* = false; *ergos* = work; Latin, *-ate* = noun ending meaning agent. Pl., Pseugergates.) A worker-like caste found in lower termites (Mastotermitidae, Kalotermitidae, Termopsidae). Pseudergates are not true workers because the nymphs do not moult into adults that function as a worker caste; instead, nymphs moult into Soldiers, Primary Reproductives (King or Queen) or Neoteinics (Secondary Reproductives); pseudergate nymphs may moult without increasing in size; See Isoptera; Caste. Cf. Soldier. Worker. Rel. Social Insect.

PSEUDIDOUM Noun. See Nymph.

PSEUDIMAGO Noun. See Sub-imago.

PSEUDOAPOSEMATIC COLORATION Combinations of colours or colour patterns that lead to protective mimicry. See Coloration. Cf. Advancing Coloration; Alluring Coloration; Anticryptic Coloration; Apetetic Coloration; Aposematic Coloration; Combination Coloration; Cryptic Coloration; Directive Coloration; Disruptive Coloration; Epigamic Coloration; Episematic Coloration; Procryptic Coloration; Protective Coloration; Pseudepisematic Coloration; Seasonal Coloration; Sematic Coloration. Rel. Crypsis; Mimicry.

PSEUDOBLEPHAROPLAST See Centriole Adjunct.

PSEUDOCAECILIIDAE Plural Noun. A numerically small Family of Psocoptera assigned to the Suborder Psocomorpha. Antenna with 13 segments, without secondary annulations; Tarsi with two segments; wings with long Setae, Pterostigma elongate; Areola Postica present. Species found on fresh foliage and twigs.

PSEUDOCARDIA Noun. The Dorsal Vessel (Kirby & Spence).

PSEUDOCELLI Noun. Sense organs distributed over the body in certain Collembola and Protura.

PSEUDOCELLULA Noun. See Accessory Cell.

PSEUDOCEPHALON Noun. (Greek, *pseudo* = false; *kephale* = head.) The first segment of the body in muscid larvae; generally retracted.

PSEUDOCEPS Noun. (Greek, *pseudo* = false; *kephale* = head.) Hymenoptera: Distal Rachis and Legula.

PSEUDOCERCUS Noun. (Greek, *pseudo* = false; *kerkos* = tail. Pl., Pseudocerci.) The median, segmented, terminal filament on the Abdomen of Thysanura, Archaeognatha and Ephemeroptera. Syn. Appendix Dorsalis; Caudal Style; Media Cercus; Filum Terminale (Folsom & Wardle). See Appendix Dorsalis.

PSEUDOCHRYSALIS See Semipupa.

PSEUDOCOCCIDAE Plural Noun. Mealybugs. The numerically largest Family of coccoid Hemiptera. A cosmopolitan group of sap-sucking phytophages which attacks fruit trees and ornamental trees and plants. Body oval, somewhat flattened, segmented but Tagma not clearly evident; Multilocular Pores present; thoracic spiracles similar in size; abdominal spiracles absent; apex of Abdomen not forming a Pygidium; many Species forming a 'mealy' wax, white, body covering; sexually dimorphic as adults with male small-bodied and winged, female wingless; a few Species parthenogenetic.

Mealybugs retain their legs throughout life and are capable of locomotion; typically sedentary and accumulate in large numbers if left unchecked. Female of some Species lay 600–800 eggs during lifetime; a few Species Ovoviviparous. Mealybugs feed on the roots, stems, leaves and fruit of plants; honeydew accumulates to cause sooty mould; many Species regarded as serious economic pests of ornamental plants and crops through damage to plant, sooty mould and degradation of fruit. See Citrophilous Mealybug; Citrus Mealybug; Grape Mealybug; Long-Tailed Mealybug. Cf. Aleyrodidae; Coccidae; Diaspididae.

PSEUDOCOEL Noun. (Greek, *pseudo* = false; *koilia* = cavity.) A depression or hollow which does not form a tube.

PSEUDOCONE Noun. A soft, gelatinous Cone in the compound eye of some insects, replacing the crystalline Cone of others. Cf. Eucone, Acone.

PSEUDOCONIC EYE Ommatidia of the compound eye in which the cone is an extracellular body formed by a vitreous secretion of Cone Cells. The Pseudocone is soft and clearly distinguishable from the lens or hard and optically continuous with it (Snodgrass). Semper Cells produce a gel or liquid-filled Cone. A Pseudoconic Eye occurs in some Diptera and Odonata. See Compound Eye. Cf. Euconic Eye; Exoconic Eye.

PSEUDOCROP Noun. (Greek, *pseudo* = false, + crop.) Hemiptera: An enlargement of the anterior region of the midgut (Wardle). See Alimentary System.

PSEUDOCUBITUS Noun. (Greek, *pseudo* = false; *cubitus* = elbow.) The highly complex Cubitus (C) of the Chrysopidae (Tillyard).

PSEUDOCULI Noun. Myrientomata: A pair of organs in the head, of undetermined nature (Comstock).

PSEUDO-ELYTRON Noun. (Greek, *pseudo* = false; *elytron* = sheath.) Strepsiptera: Abbreviated, strap-like anterior wings.

PSEUDOGAMY Noun. (Greek, *pseudo* = false; *gamos* = marriage. Pl., Pseudogamies.) See Gynogenesis.

PSEUDOGULA See Hypostomal Bridge.

PSEUDOGYNA Noun. A female insect that reproduces without impregnation. An apterous worker-like ant which combines the size and Gaster of the worker with the thoracic characters of the female.

PSEUDOGYNA FUNDATRIX Aphididae: The immediate issue of a fecundated egg, a stem-mother.

PSEUDOGYNA GEMMANS Aphididae: The wingless descendant of the stem-mother (Fundatrix) or of the winged migrants (Migrantes) which reproduce asexually through a number of generations.

PSEUDOGYNA MIGRANS Aphididae: A winged descendant of the stem-mother (Fundatrix) through which the Species is spread.

PSEUDOGYNA PUPIFERA Aphididae: The last generation of Pseudogynes Gemmantes, which produces the true sexes.

PSEUDOHALTERES Noun. The reduced front wings of the Strepsiptera or Stylops (Comstock). Syn. Pseudoelytra.

PSEUDOHYPOGNATHOUS Adj. (Latin, *-osus* = with the property of.) Having the cephalic part of the head so bent that the insect appears prognathous although actually hypognathous (MacGillivray). See Head. Cf. Hypognathous; Opisthognathous; Prognathous.

PSEUDOLOBES Noun. Species of coccids that transform in a Puparium or do not escape from the last nymphal Exuviae: A few or many lobe like projections in the margin of the Pygidium (MacGillivray).

PSEUDOMEDIA Noun. The highly complex vein M of Chrysopidae (Tillyard).

PSEUDOMYIASIS Noun. (Greek, *pseudo* = false; *myia* = fly; *-iasis* = suffix for names of diseases. Pl., Pseudomyiases.) Accidental ingestion of eggs or larvae of myiasis-producing flies by humans. Most records of Pseudomyiasis involve Sarcophagidae and Muscidae. Term proposed to replace 'Enteric Myiasis' because ingestion is accidental and development does not proceed. (Zumpt 1965). See Myiasis.

PSEUDONEURIUM Noun. (Greek, *pseudo* = false; *neuron* = sinew; Latin, *-ium* = diminutive > Greek, *-idion*. Pl., Pseudoneuria.) A false vein formed by a chitinous thickening of a wing fold.

PSEUDONEUROPTERA Noun. Net-winged insects with incomplete Metamorphosis, including Ephemeridae, Odonata, Plecoptera, Isoptera, Corrodentia and Archiptera.

PSEUDONOTUM Noun. (Greek, *pseudo* = false; *notun* = back.) The Postnotum or Postscutellum of authors (MacGillivray). The Postfroenum *sensu* Kirby & Spence.

PSEUDONYCHIUM Noun. (Greek, *pseudo* = false; *ergos* = work; Latin, *-ium* = diminutive > Greek, *-idion*. Pl., Pseudonychia.) Paronychia, Empodium. Collembolla: The lateral teeth of the ungues.

PSEUDONYM Noun. (Greek, *pseudes* = false; *onyma* = name. Pl., Pseudonyms.) Literally, a false name given in word or writing for authorship or credit for work done, ideas presented or results published. Cf. Name.

PSEUDONYMPH See Semipupa.

PSEUDOPHASMATIDAE Plural Noun. Striped Walkingsticks. A small Family of Phasmida with two Species in USA. Mesothorax less than three times as long as Prothorax; middle and hind Tibiae deeply emarginate apically; Tarsal formula 5-5-5; first abdominal segment as long as thoracic Metanotum. Species occur on gasses in subtropical-tropical areas.

PSEUDOPHLOEINAE Plural Noun. A Subfamily of Coreidae.

PSEUDOPHYLLIDAE Plural Noun. True katydids.

A Family of ensiferous Orthoptera including about 250 Genera and 1,000 Species. Geographically widespread but more Species are found in tropical forests than elsewhere. Few Species found in temperate regions. Head short and rounded, face not slanted or flattened, Antenna longer than body and inserted between compound eyes, Prothoracic Spiracle small but not concealed by the Pronotum, Tibiae lack apical spines and fore Tibia with auditory organs which are covered. The body of pseudophyllids is sometimes quite large, in some Species with a wing span of 20 cm.

PSEUDOPLACENTAL VIVIPARITY A method of viviparous reproduction in which the egg is deficient in yolk, lacks a Chorion, and develops within a maternal genital pouch. Nourishment obtained via a connection between maternal and embryonic tissues. Seen in some Blattodea, Dermaptera, Psocoptera and Hemiptera. See Viviparity. Cf. Adenotrophic Viviparity; Haemocoelic Viviparity.

PSEUDOPOD Noun. (Greek, *pseudes* = false; *pous* = foot. Pl., Pseudopodia.) A soft, flexible, foot-like appendage characteristic of some dipterous larvae. A proleg. See Parapodium.

PSEUDOPOLYGYNY Noun. (Greek, *pseudo* = false; *poly* = many; *gyne* = female.) Social Insects: A condition in which several dealate queens reside with an inseminated, egg-laying queen.

PSEUDOPOMIZIDAE Plural Noun. A small Family of acalypterate Diptera assigned to Superfamily Nerioidea; best represented in Neotropical Realm with Species in New Zealand and a few Species in Australia. Adult minute to moderate sized; proclinate Fronto-orbital Bristles absent; Presutural Bristle present; apical pair of Scutellar Bristles longer than other scutellar bristles.

PSEUDOPOSITOR Noun. (Pl., Pseudopositors.) The slender abdominal tube caudad of the fourth or fifth segment in the female of some insects (MacGillivray).

PSEUDOPTERA Noun. An ordinal name for scale insects (Amoyot 1847).

PSEUDOPUPA Noun. (Pl., Pseudopupae; Pseudopupas.) 1. Thysanoptera: The fourth nymphal instar of Terebrantia or fourth and fifth nymphal instars of Tubulifera during which feeding stops and transformation into the adult form occurs. 2. Coleoptera: A larva in a quiescent coarctate condition preceding one or more larval instars which are followed by the true Pupa (Imms). 3. A Semipupa.

PSEUDOPUPILLA Noun. (Pl., Pseudopupillae.) Odonata: The black spots visible on the Compound Eyes of the living insects.

PSEUDORHINARIA See Scent Plaques.

PSEUDOSCORPIONES Noun. An Order of arthropods assigned to Class Arachnida. Members of the Order are typically small bodied, somewhat compressed with a scorpion-like habitus, large chelate Pedipalps but lacking a Telson. Pseudoscorpions inhabit leaf litter, caves, moist soil and intertidal habitats. Species are predatory or scavengers of small arthropods or algiverous.

PSEUDOSEMATIC COLOURS False warning and signalling colours. See Aposematic Coloration; Mimicry.

PSEUDOSENSORIA See Scent Plaques.

PSEUDOSESSILE Noun. (Greek, *pseudo* = false; Latin, *sessum* = sit.) Hymenoptera: Apocrita in which the Petiole is reduced such that the Propodeum and Gaster are in broad contact. Syn. Subsessile.

PSEUDOSPIRACLE Noun. A false spiracle found in Nepidae and Gryllotalpa.

PSEUDOSTERNITE Noun. (Greek, *pseudo* = false; *sternon* = chest.) See Epiphallus (Snodgrass).

PSEUDOSTIGMATIDAE Plural Noun. A Family of zygopterous Odonata. Adults moderate to large bodied, elongate; Postnodal Crossveins numerous; quadrilateral elongate, sometimes with Crossveins; Cubital and Anal Veins long; Abdomen sometimes exceedingly long. Larvae relatively short; body cylindrical; eyes prominent; Antenna 7-segmented, third and fourth segments longest; legs slender; Tibiae and Tarsi distally holding branched spines.

PSEUDOSUTURAL FOVEAE Diptera: Impressed polished areas on the humeral portion of the Mesonotum, humeral pits, in Tipulidae and other Taxa (Curran).

PSEUDOTETRAMEROUS Adj. (Greek, *pseudes* = false; *tetra-* = comb. form meaning four; *mero* = comb. form meaning part; Latin, *-osus* = with the property of.) Legs displaying four apparent tarsal segments, but five are actually present. Condition seen in some Coleoptera.

PSEUDOTRACHEA Noun. (Pl., Pseudotrachea.) A false Trachea. A structure having the aspect of a Trachea, in the dipterous Labella, the ringed and ridged grooves by means of which they scrape their food.

PSEUDOTRIMEROUS Adj. (Greek, *pseudes* = false; *tri-* = comb. form meaning three; *mero* = comb. form meaning part; Latin, *-osus* = with the property of.) Legs displaying three apparent tarsal segments, but four segments are actually present.

PSEUDOVARY Noun. (Greek, *pseudes* = false, + ovary.) The organ or mass of germ cells of an agamic insect.

PSEUDOVEIN Noun. (Greek, *pseudes* = false, + vein. Pl., Pseudoveins.) Heteroptera: Longitudinal vein-like structures in the wing which are solid. Homology of Pseudoveins and true veins is uncertain.

PSEUDOVITELLUS Noun. Hemiptera: A solid mass of cell-tissue in the Abdomen, in small groups of large rounded conspicuous cells; uncertain function (Imms).

PSEUDOVUM Noun. (Greek, *pseudes* = false; *ovum* = egg. Pl., Pseudova.) 1. The germ-cell produced by agamic Aphididae. 2. An unfertilized egg; 'virgin' egg.

PSILATE Adj. (Greek, *psilos* = smooth, naked; *-atus* = adjectival suffix.) Descriptive of surface sculpture, usually the insect's Integument, that is smooth, not pitted or with submicroscopic pits. See Sculpture; Sculpture Pattern. Cf. Alveolate; Baculate; Clavate; Echinate; Favose; Gemmate; Punctate; Reticulate; Rugulate; Scabrate; Shagreened; Smooth; Striate; Verrucate.

PSILIDAE Plural Noun. Rust Flies. A numerically small, geographically widespread Family of acalypterate Diptera assigned to Superfamily Diopsoidea. Adult small to moderate-sized, slender with long Antennae; Oral Vibrissae absent; Costa broken before end of R1; Mesonotal Bristles above wing; Pleural Bristles absent; Prosternum without Precoxal Bridge. Larvae feed on plant roots or inhabit galls. See Carrot Rust Fly.

PSILOPSOCIDAE Plural Noun. A small Family of Psocoptera assigned to the Suborder Psocomorpha. Antenna with 13 segments, without secondary annulations; forewing and hindwing venation asetose. Nymph with apex of Abdomen heavily sclerotized, Epiprocts and Paraprocts in ventral position. Individuals pale coloured, typically found in dark, damp situations such as cellars and caves.

PSOCATROPIDAE See Psyllipsocidae.

PSOCIDAE Plural Noun. Psocids. The largest Family of Psocoptera in North America; cosmopolitan, assigned to Suborder Psocomorpha. Antenna with 13 segments, without secondary annulations; forewing asetose; Areola Postica joined to M; Tarsi with two segments. Species found on bark and foliage of trees.

PSOCIDS See Psocoptera.

PSOCOMORPHA Noun. (Greek, *psocus* = biting; *morphe* = form.) The largest Suborder of Psocoptera. Antenna with 13 segments, without secondary annulations; Hypopharyngeal Filaments partly separated; Tarsi with 2–3 segments; Pterostigma thickened.

PSOCOPTERA Shipley 1904 (Permian-Recent.) Book Lice; Bark Lice. (Greek, *psocus* = biting; *pteron* = wing.) A cosmopolitan Order of small to moderate sized, soft bodied insects currently containing about 1,800 Species. Apparently, Psocoptera are derived from hemipteroid ancestor and most closely related to Mallophaga-Anoplura. Psocopteran head is large and moveable with Epicranial Suture evident and Clypeus divided into Postclypeus and Anteclypeus. Antenna long, filiform and usually contains 13 segments; Mandibles asymmetrical; Maxillary Palpus with four segments; Labial Palpus with 1–2 segments; Hypopharnyx with two apical Superlinguae; three Ocelli present in winged forms but Ocelli are absent in wingless forms. Prothorax reduced in winged forms; Mesothorax and Metathorax fused in wingless forms; wings membranous with venation reduced; pairs held together when active and at rest; wings held obliquely over body when at repose; two pairs of thoracic spiracles typically present. Legs with Coxae often containing a stridulatory device (Pearman's Organ); Tibiae long, cylindrical with apical spurs; 2–3 Tarsomeres; Pulvillus present, Empodium absent. Abdomen with 10 segments; Epiproct, Paraprocts present; Cerci absent; polytrophic Ovarioles; Spermatophore developed. Paurmetabolous development; viviparity and parthenogenesis reported in some Species; first instar nymph without Ocelli, with two tarsal segments; wing buds develop in second nymphal instar. Included Families: Amphientomidae, Amphipsocidae, Archipsocidae, Asiopsocidae, Caeciliidae, Calopsocidae, Ectopsocidae, Elipsocidae, Epipsocidae, Hemipsocidae, Lachesillidae, Lepidopsocidae, Liposcelidae, Mesopsocidae, Myopsocidae, Pachytroctidae, Peripsocidae, Philotarsidae, Polypsocidae, Prionoglaridae, Pseudocaeciliidae, Psilopsocidae, Psocidae, Psolculidae, Psoquillidae, Psyllipsocidae, Ptiloneuridae, Stenopsocidae, Sphaeropsocidae, Thyrosophoridae, Trichopsocidae, Trogiidae. Syn. Corrodentia.

PSOQUILLIDAE Plural Noun. A numerically small Family of Psocoptera assigned to the Suborder Trogiomorpha. Antenna with more than 20 segments; Tarsi with three segments; Paraprocts with strong posterior spine; forewing well developed, hindwing reduced.

PSORALEN Noun. Any member of a group of linear furanocoumarin compounds found in several plants including celery, parsley and parsnips. Psoralens display a photosensitizing effect on animals; toxic to many kinds of organisms; some carcinogenic to rodents.

PSOTA, FRANK J (–1936) (Mickel 1938, Ann. ent. Soc. Am. 31: 121.)

PSYCHIDAE Herrich-Schaeffer 1845. Plural Noun. Bagworms; Bagworm Moths; Bag Moths; Case Moths. A cosmopolitan Family of ditrysian Lepidoptera assigned to Superfamily Tineoidea; presently containing about 600 Species. Adult wingspan 10–60 mm; head with coarse hair-like or narrow lamelliform scales; Ocelli usually absent; Chaetosemata absent; Proboscis absent; Maxillary Palpus reduced or absent; Labial Palpus of 1–2 segments; wings macropterous, brachypterous or apterous; Epiphysis usually present, long; tibial spur formula 0-2-4, 0-1-1 or 0-0-0. Egg oval with reticulate sculptural pattern on Chorion; eggs laid in masses mixed with scales from Abdomen; egg load of some Species very high, ca 10,000 per female. Eggs often deposited in female's case; neonate larva disperses by ballooning or lowering themselves to vegetation on strand of silk. Larva constructs and resides in case (bag), often incorporating plant material into fabric of bag; anterior aperture for feeding, posterior aperture for ejection of excrement; thoracic legs well developed; Prolegs on abdominal segments 3–6 and anal Prolegs on segment 10; Crochets uniordinal and uniserial. Pupa with dorsal abdominal spines. Adult female of some Species confined to case (bag), eyes and appendages reduced or absent. See Bagworm. Syn. Arrhenophanidae; Pseudarbelidae.

PSYCHODIDAE Plural Noun. Moth Flies; Sand Flies; Owl Midges. A cosmopolitan Family of nematocerous Diptera. Body small; Rs and M each with four branches; Anal Veins reduced. Adults often short-lived, do not feed and found in moist, shaded habitats. Higher classification problematic with two Subfamilies (Psychodinae and Phlebotominae). Psychodinae (moth flies) Antenna long, each Flagellomere with a whorl of Setae; wing with longitudinal veins prominent, crossveins reduced and limited to basal region of wing; wings typically pointed apically, body densely setose and wings held tent-like over Abdomen. Species often achieve large numbers around sewage rendering plants; larvae of some Species develop in bathroom basins and drains. *Psychoda alternata* a cosmopolitan pest in urban homes; larvae implicated in transmission of Pseudomyiasis. Phlebotominae recognized as a Family in some classifications; wings held upward with costal margins forming an angle of 60°; body not as setose as Psychodinae; Species with blood-sucking females; most Species require blood meal before oviposition, some Species autogenous; females serve as vectors in transmission of Leishmaniasis and Sand Fly Fever with *Phlebotomus* and *Lutzomyia* medically important. Males do not take blood and do not transmit disease to humans. Adults of most Species feed at night and seek refuge in protected areas during day. Eggs deposited in small batches, hatch ca 6–17 days; Larvae small, white, setose; feed on decomposing vegetation, dung and sewage; four instars. Life-cycle temperature dependent.

PSYCHODOMORPHA. A Division of nematocerous Diptera including Psychodidae and Tanyderidae.

PSYCHOGENESIS Noun. (Greek, *psyche* = soul, mind; *genesis* = descent; *-sis* = a condition or state.) The origin and development of social and other instincts and habits.

PSYCHOMYIIDAE Plural Noun. Tubemaking Caddisflies; Trumpetnet Caddisflies. A small Family of Trichoptera assigned to Superfamily Hydropsychoidea. Ocelli absent; terminal segment of Maxillary Palpus much longer than penultimate segment; Mesoscutellum with a pair of small wart-like tubercles; Front Tibia lacking preapical spur.

PSYCHOPSIDAE Plural Noun. A rarely collected Family related to lacewings in the Hemerobioidea [Neuroptera]: Representatives occur from the Oriental region, but is best represented in southern Africa and Australia. Adults are cryptic, nocturnal and live for 1–2 months. Eggs are laid singly on vegetation on the apices of long filaments. Larvae are predaceous and occur under bark of eucalyptus. The life-cycle may require two years for completion.

PSYCHROPHILE Noun. (Greek, *psychros* = cold; *philein* = lover. Pl., Psychrophiles.) An organism that lives and reproduces when exposed to relatively low temperatures for prolonged periods of time. Cf. Halophile; Osmophile.

PSYLLIDAE Plural Noun. Jumping Plantlice; Psyllids. A cosmopolitan Family of about 1,250 Species placed in Superfamily Psylloidea. Psyllids superficially resemble aphids or cicadas. Psyllids are monomorphic, reproduce sexually, oviparous and develop through five nymphal instars. Psyllids are phloem feeding, usually stenophagous and found almost exclusively on perennial dicotyledons. Adults inflict little damage; nymphs are often serious pests of cultivated crops and ornamental trees. Psyllid nymphs vector bacterial and viral diseases and salivary injections can induce abnormalities in plant growth. Many forms of galls on a host plant are induced by nymphs. Psyllids secrete honeydew which is predominantly carbohydrate. Adults can disperse but generally not to the extent of aphids.

PSYLLIDS See Psyllidae.

PSYLLIPSOCIDAE Plural Noun. A numerically small Family of Psocoptera assigned to the Suborder Trogiomorpha. Antenna with more than 20 segments; Tarsi with three segments; Paraprocts with strong posterior spine. Pale bodied; long legged; some troglobitic Species; polymorphism in wing development. Syn. Psocatropidae.

PSYLLOIDEA Plural Noun. Jumping Plant Lice; Lerp Insects. A Superfamily of sternorrhychous Hemiptera including Calophyidae, Carsidaridae, Homotomidae, Phacopteronidae, Psyllidae and Triozidae. Adult with 10 antennal segments, three Ocelli, hindcoxa large, hindleg saltatorial, forewing R, M and Cu usually single branched; Clavus present; Proctiger well developed; female usually with Circumanal Wax Gland rings. Eggs Pedicelate, attached to plant tissue via Pedicel or inserted entirely into plant tissue. Nymph dorsoventrally flattened; head and Prothorax fused; non-saltatorial; some Species form Lerps, pit galls or closed woody galls. Presumably the most primitive group of Sternorrhyncha and sister group of Aleyrodidae; more abundant and diverse in Australia than Northern Hemisphere.

PTERALIA Plural Noun. (Greek, *pteron* = wing.) 1. The group of articular sclerites at the wing base, including the Humeral Sclerite (plate) and Axillary Sclerite or Axillary Sclerites (Snodgrass). 2. The Axillary Sclerites *sensu* Imms. See Wing Articulation.

PTERERGATE Noun. (Greek, *pteron* = wing; *ergates* = worker; *-atus* = adjectival suffix. Pl., Pseudergates.) A worker ant or a dinergate with vestigal wings but unmodified Thorax.

PTERIGOSTIUM See Pterygostium.

PTERINES Noun. (Greek, *pteron* = wing. Pl., Pteridines.) Pterines (pteridines) are nitrogen-containing organic compounds important in folic acid, xanthopterin; Pterines are sometimes synthesized from purines. Pterines produce several colours including white (via leucopterin), yellow (via xanthopterin) red (via erythopterin) and fluorescence in ultraviolet (via biopterin). Pterines also occur in primary and secondary pigment cells within the compound eye. The functions of pterines, beyond the visual effect, include serving as cofactors of various enzymes and possibly producing folic acid. See Pigmentary Colour. Cf. Carotinoids; Flavinoids; Melanins; Ommochromes; Quinones. Rel. Structural Colour.

PTERNOTORMA Noun. (Pl., Pternotormae.) Scarabaeoid larvae: A curving stout process at the end of the Laeotorma and sometimes of the Desotorma (Boving).

PTERODICERA Noun. (Greek, *ptero* = wing, *di* = twice; *keras* = horn.) An archane term for insects with wings and two Antennae.

PTEROGOSTIA Noun. (Greek, *pterogostium* = a wing vein.) The wing veins.

PTEROGOSTIC Adj. Descriptive of or referring to the wing structure.

PTEROMALIDAE Plural Noun. Pteromalid Wasps. A cosmopolitan, numerically large (ca 600 Genera, 3,100 Species) Family of apocritous Hymenoptera assigned to Chalcidoidea. Diagnostic features include: Body 1–10 mm long, usually metallic; Antenna with 8–13 segments including 1–3 Anelli; mesoscutal Notauli complete or incomplete; Propodeum often with Carinae or Nucha; wings usually macropterous, Marginal Vein long, Postmarginal, Stigmal Veins usually long; tarsomeral formula usually 5-5-5; Gaster subsessile to strongly petiolate. Pteromalids represent a generalized, polyphyletic group of chalcidoids. Developmental strategies diverse: Solitary or Gregarious, typically ectoparasites. Egg shape, size and number highly variable; reports of up to 700 eggs/female; sometimes spiculate. First instar larva Hymenopteriform, 13 segments; head, Mandibles sometimes large; ectoparasites with open respiratory system; endoparasites with closed respiratory system. Feeding spectrum very diverse: predominantly parasitic of Holometabola, attack concealed host larvae and Pupae in stems, leaf mines, galls, and similar habitats; some larval-pupal parasites; some predators of cecidomyiid larvae; some predators on coccoid, delphacid eggs; some gall formers, also feed on gall tissue.

PTERONARCHIDAE Plural Noun. Giant Stoneflies; Salmonflies. A small Family of Plecoptera including three Genera and 13 Species. Primarily North American, *Pteronarcys* ranging to eastern Asia. Larvae with branched gills on thoracic Sterna and first two abdominal segments; mouthparts holognathous; wings with numerous crossveins in all regions. Larvae herbivorous. Syn. Pteronarcidae.

PTERONARCIDAE See Pteronarchidae.

PTEROPEGA Noun. (Greek, *ptero* = wing; *pegos* = united.) Wing sockets or cavities into which the wings are inserted.

PTEROPHORIDAE Plural Noun. Plume Moths. A Family of ditrysian Lepidoptera assigned to Superfamily Pterophoroidea. Family includes ca 500 Species and is nearly cosmopolitan in distribution; placed within Pyraloidea in some classifications or with Tineodidae and Oxychirotidae. Adult small; wingspan 10–30 mm; head with smooth scales; Ocelli and Chaetosemata absent; Antenna filiform; Proboscis without scales; Maxillary Palpus minute with one segment; Labial Palpus form variable; legs long, slender; fore Tibia with Epiphysis; tibial spur formula 0-2-4; wings narrow and divided into several plume-like regions by clefts. Egg flattened, oval or elliptical in outline shape with Chorion not conspicuously sculptured. Larval head subprognathous, with primary Setae. Pupa rather slender; exposed on leaf surface or in leaf litter. Adult rests with body horizontal and elevated by forelegs and midlegs; hindlegs off substrate and held parallel to Abdomen; wings streched away from body with forewing covering hindwing. First instar larva of some Species mine leaves then feed on the exposed surface; some Species burrow in flower buds, flowers or stems. Cf. Alucitidae.

PTEROPHOROIDEA Noun. A Superfamily of ditrysian Lepidoptera including Pterophoridae.

PTEROPLEURAL BRISTLES Noun. Diptera: Setae inserted on the pteropleura.

PTEROPLEURITE Noun. (Greek, *pteron* = wing; *pleuron* = side of body; *-ites* = constituent. Pl., Pteropleurites.) The upper and lower sections of the Pteropleura (Curran). See Pterothoracic Pleuron.

PTEROPLEURON Noun. (Greek, *pteron* = wing; *pleuron* = side. Pl., Pteropleura.) 1. The upper part of the Epimeron (Anepimeron) of the Mesothorax (Comstock). 2. Diptera: A sclerite below the wing base posterior of the Mesopleural Suture. 3. The posterior lateral sclerite of the Mesothorax *sensu* Lowne. 4. The Episternum of the Mesothorax *sensu* Hammond. 5. The Mesepimeron *sensu* Tillyard. See Pterothoracic Pleuron.

PTEROPLISTIDAE Plural Noun. Feather-winged Crickets. A monogeneric Family of ensiferous Orthoptera that includes about five Species found in the Indomalayan Realm.

PTEROSTIGMA Noun. (Greek, *pteron* = wing; *stigma* = mark. Pl., Pterostigmata.) An enlarged, pigmented area along the costal margin of the wing or at the apex of the Radius. The Pterostigma forms a sinus containing red or reddish-brown pigment and cellular material. Pterostigma is present on both wings of Odonata; present on forewing of Psocoptera, Megaloptera, Mecoptera and Hymenoptera. See: Bathmis;

Wing. Cf. Prostigma; Stigma.

PTEROTHECA Noun. (Greek, *pteron* = wing; *theke* = cover. Pl., Pterothecae.) The part of the pupal case that covers the wings. See Pupa.

PTEROTHORACIC Adj. (Greek, *pteron* = wing; *thorax* = chest.) Descriptive of or pertaining to the Pterothorax.

PTEROTHORACIC PLEURON The lateral aspect of the wing-bearing segments of the Thorax. A conspicuous Pleural Suture extends from the Pleural Wing Process to the wing base in most pterygote insects. PS is a landmark on Pterothorax, but course of this suture is variable: Sometimes vertical, sometimes oblique and straight, sometimes horizontal, and sometimes sinuous. PS develops into a cuticular inflection called Pleural Ridge which serves to reinforce the Pleuron. A Pleural Apophysis (pit) occurs in some insects along course of the Pleural Suture. PS terminates ventrally along dorsal rim of Coxa. This point of termination forms a Coxal Condyle that serves as a point of articulation between Thorax and leg. PS also subdivides Pleuron into an anterior Episternum and posterior Epimeron. In simplest form, the Prealar Bridge meets or connects with dorsal margin of Episternum; Postalar Bridge meets or connects with dorsal margin of Epimeron. Anteroventrally, the Episternum forms a connection with the Sternum in front of Coxa to form a Precoxal Bridge. A Pleurosternal Suture marks boundary between Pleuron and Sternum in this region of Thorax. This suture is lost in some insects and the resultant sclerite is called a Pleurosternite in Hymenoptera such as sawflies. Posteroventrally, Epimeron forms a connection with Sternum behind Coxa called Postcoxal Bridge. The Postcoxal Bridge is less frequently developed and is usually narrower than the Precoxal Bridge. See Pterothorax; Thorax. Cf. Episternum; Epimeron.

PTEROTHORAX Noun. (Greek, *pteron* = wing; *thorax* = chest. Pl., Pterothoraxes. Pterothoraces.) The wing-bearing thoracic segments in winged insects. Typically, Mesothorax and Metathorax in winged insects such as Hymenoptera, Lepidoptera, Thysanoptera. Wing-bearing thoracic segments are subdivided into many sclerites that are separated by sutures. These sutures and sclerites are the evolutionary product of repeated modifications of the Thorax in response to sundry environmental and biomechanical demands placed upon the body. Similar modifications have occurred independently in many groups of insects while some modifications are unique. Generalizations are tenuous given the number of sutures and sclerites, coupled with the number of insect Species involved. Development of Pterothorax varies among winged insects. When both pairs of wings participate more-or-less equally in flight, thoracic segments bearing wings are similar in size. This condition is seen in Odonata, some Lepidoptera and some Neuroptera. When one pair of wings is dominant in flight, the associated thoracic segment is larger and modified for flight while the other thoracic segment is reduced in size. This condition is seen in Hymenoptera when the forewing is large and dominant in flight. The reverse condition is seen in Coleoptera when the hindwing is large and dominant in flight. See Thorax. Cf. Mesothorax; Metathorax. Rel. Wings.

PTEROTHYSANIDAE Plural Noun. A small Family (ca 15 Species) of ditrysian Lepidoptera assigned to Superfamily Calliduloidea or recognized as part of the Callidulidae. Predominantly Oriental in distribution with one Species from Madagascar. Adult relatively large-bodied; nocturnal or diurnal; Ocelli absent; Antenna filiform or pectinate; tibial spur formula or 0-2-2.

PTERYGIUM Noun. (Greek, *pteron* = wing; Latin, *-ium* = diminutive > Greek, *-idion*. Pl., Pterygia.) Lepidoptera: Small wing-lobes at the base of the underwings (Kirby & Spence). Coleoptera: Lateral expansion of the snout (Smith).

PTERYGODES Noun. The Patagia or Tegulae.

PTERYGOGENEA Noun. (Greek, *ptero* = wing, *genesis* = descent.) Winged insects presumed to be derived from ancestors which were winged. See Apterygogenea.

PTERYGOID Adj. (Greek, *pteryx* = wing; *eidos* = form.) Wing-like. Cf. Aliform. Rel. Eidos; Form; Shape.

PTERYGOPHORIDAE Cameron 1878. Plural Noun. A Family of Symphyta (Hymenoptera) assigned to the Tenthredinoidea.

PTERYGOPOLYMORPHISM (Greek, *pteryx* = wing; *poly* = many; *morphos* = shape; English, *-ism* = condition. Pl., Pterygopolymorphisms.) The condition of having several forms of wings in the same Species. Heteroptera: Both in form and length, or in development.

PTERYGOSTIUM Noun. (Greek, *pteryx* = wing; Latin, *-ium* = diminutive > Greek, *-idion*. Pl., Pterygostia.) A wing vein. Alt. Pterygostium.

PTERYGOTA Noun. (Greek, *pteryx* = wing.) A Subclass or Infraclass of the Insecta whose members are primarily winged but sometimes secondarily apterous and display some form of Metamorphosis. Diagnostic features include: Wings restricted to Mesothorax and Metathorax; wings with venation; wings attached high on pleural wall; basal articulation with a complex of sclerites; internal apodeme from Pleural Wing Process to base of Coxa. Cf. Apterygota. See Metamorphosis.

PTERYGOTE Adj. (Greek, *pteryx* = wing.) Wing bearing; a term often applied to insects which bear wings. A member of the Pterygota.

PTHIRIDAE Plural Noun. A small Family of Anoplura whose Species parasitize primates. Body with few Setae; head relatively small; compound eyes absent; Mandibles absent, mouthparts forming a beak or rounded anteriorly; fore legs smaller than middle and hind legs; Abdomen compact with setose lateral tubercles. Family includes Crab Louse.

PTILIIDAE Plural Noun. Minute beetles assigned to the Staphylinoidea. Body flattened, pubescent; Antenna with 10–11 segments, club with 2–3 segments, verticellate whorls of long Setae; Elytra elongate or short and truncate; 1–3 abdominal Terga exposed; marginal fringe of hindwing long; tarsal formula 2-2-2, legs sometimes appearing with one tarsal segment. Larval head hypognathous; Stemmata absent; Antenna with three segments; Pupa obtect. Species common in decaying vegetation, dung, rarely myrmecophilous. Some Species fungivorous.

PTILINAL SUTURE Diptera: The crescentic groove cutting across the Frons above the Antennal bases in flies where the Ptilinum has been withdrawn (Snodgrass).

PTILINUM Noun. (Greek, *ptilon* = feather. Pl., Ptilina.) Schizophorous Diptera: A modification of the Frons that forms a sac-like, inflatable, pulsatile, cuticular organ on the head of teneral flies. Ptilinum is thrust through an arcuate Ptilinal Suture just above the base of the Antennae. Ptilinum is used to aid emergence from the puparium and to burrow through the soil to reach the surface. After emergence, the Ptilinum is retracted by muscles that subsequently atrophy. Ptilinum hardens and becomes invisible on the integument's surface after adult emergence. A crescent-shaped sclerite between the Ptilinial Suture and the Antennal socket is called a Lunule. See Head Capsule; Frons. Cf. Face; Osmeterium. Rel. Hydrostatic Pressure.

PTILODACTYLIDAE Plural Noun. (Greek, *ptilon* = feather; *daktylos* = finger.) Toed-Winged Beetles; Ptilodactylid Beetles. A small Family of polyphagous Coleoptera assigned to Superfamily Dryopoidea. Included Subfamilies: Aploglossinae, Ptilodactylinae. Adult small-bodied (< 9 mm long), pubescent, elongate-oval in outline shape, brownish; head strongly deflexed in profile and not visible from dorsal aspect; male Antenna pectinate; female Antenna serrate; Pronotum with sharp lateral Carina; Scutellum cordiform with notch anterior; tarsal formula 5-5-5, 3rd tarsal segment lobed ventrally, tarsomeres with ventral brushes. Adults inhabit damp, swamp-like habitats; larvae aquatic or inhabit moist, dead wood.

PTILONEURIDAE Plural Noun. A small, predominantly Neotropical Family of Psocoptera assigned to Suborder Psocomorpha. Tarsal formula 3-3-3; forewings always well developed and with two Anal Veins.

PTILOTA Noun. Winged insects.

PTINIDAE Plural Noun. Spider Beetles. A widespread Family of polyphagous Coleoptera. Body small, coarsely pubescent, somewhat globular; head and Prothorax narrower than Elytra; Antenna long, filiform-submoniliform; Scapes close set; Prothorax hood-like but not completely covering head; tarsal formula 5-5-5. Larva scarabaeiform. Typically scavengers of dead animal remains; some pests of stored products; some inquilines of ant nests.

PTUNARRA BROWN BUTTERFLY *Oreixenica ptunarra* Couchman [Lepidoptera: Nymphalidae].

PTYCHOIDY Noun. (Greek, *ptyche* from *ptyssein* = to fold. *oeides* = resembling in form or appearance. Pl., Ptychoides.) Acarology: An articulation between the Prosoma and Opisthosoma that allows the Prosoma to be deflexed and the legs concealed.

PTYCHONOME Noun. (Greek, *ptyche* from *ptyssein* = to fold. Pl., Ptychonomes.) An enclosure formed with leaves by some Lepidoptera (Gracillariiidae). (Hering 1951, *Biology of the Leaf Miners*.) Cf. Tentiform.

PTYCHOPTERIDAE Plural Noun. Phantom Crane Flies. A small Family of nematocerous Diptera assigned to Superfamily Ptychopteroidea. Adult tipulid-like or mosquito-like in appearance. Antenna with more than six, articulated segments; Ocelli absent; Mesonotum with V-shaped suture; legs long; wing with one Anal Vein meeting wing margin; lacking closed Discal Cell. Larvae inhabit marshes and ponds where they feed on decaying plant matter. Syn. Liriopeidae.

PUBES Noun. (Latin, *pubes* = adult.) Short, fine, soft, erect Setae. Alt. Pubescence.

PUBESCENCE Noun. (French.) State of being pubescent.

PUBESCENT Adj. (Latin, *pubescere* = to become mature.) Covered with soft, short, fine, closely set Setae. Cf. Pilosity.

PUBIS Noun. (New Latin.) The lateral region of the Prothorax.

PUBLICATION Noun. (Latin, *publicatio* = publish; English, *-tion* = result of an action. Pl., Publications.) 1. Any writing or text printed in a conventional manner and distributed to a readership. Publications may be non-technical writings for a general readership or scientific writings for a more limited audience. Scientific publications typically follow a prescribed format that includes an Abstract, Introduction, Materials and Methods, Results, Discussion and References. 2. Zoological Nomenclature: The printing of a scientific name with indication that it is new to science and compared with extant nominal Species. Such publication of a name should be denoted by affixing to it 'n. n.,' 'new name,' 'nom. nov.' or in the case of a description, 'n. sp.,' 'n. gen.,' 'new species,' 'new genus,' or 'spec. nov.,' 'gen. nov.,' or 'species nova' or 'genus novum.' Under conditions such as a change of name or replacement in an accepted form of publication, the publication must be printed and offered for sale.

PUCKETT, FELIX S (1885–1941) (Porter 1941, J. Econ. Ent. 34: 592.)

PUECH-DUPONT, RICHARD HENRY (1798–1873) (Anon. 1889, Abeille 26: 273–274, bibliogr.)

PUENGELER, RUDOLF (1857–1927) (Pfaff *et al.* 1934, Festschr. 50 Jahr Bestehan Int. ent. Ver. Frankf. p.6.)

PUG MOTHS See Geometridae.

PUGABORNE, FEDERICO (1856–1920) (Porter 1935, Revta. chil. Hist. nat. 39: 305–312.)

PUGET-SOUND WIREWORM *Ctenicera aeripennis aeripennis* (Kirby) [Coleoptera: Elateridae].

PUGH, CYRIL HENRY WALLACE (1889–1973) (Brindle 1974, Entomol. mon. Mag. 109: 256.)

PUIG Y VALLS, RAFAEL (1845–1920) (Marques de Campo 1924, Mems R. Acad. Cienc. Artes Barcelona 18: 209–218.)

PUJATTI, DOMENICO (1903–1954) (Conci 1955, Memorie Soc. ent. Ital. 33: 138–140, bibliogr.)

PUJOL Y FIOL, DON EMANUEL (1875–1953) (Angenjo 1954, Graellsia 12: 21–28)

PULICID FLEAS See Pulicidae.

PULICIDAE Plural Noun. Pulicid Fleas. A numerically large, cosmopolitan Family of Siphonaptera assigned to Superfamily Pulicoidea. Middle coxa without internal apodeme; hind Tibia without an apical tooth; Sensillum with 8–14 pits on each side; Several Species associated with humans and domestic animals.

PULLEINE, ROBERT HENRY (1869–1934) (B.S.R. *et al.* 1935, Trans. Proc. R. Soc. S. Aust. 59: v–vi, bibliogr.)

PULMONARIUM Noun. (Latin, *pulmo* = lung; *arium* = place of a thing. Pl., Pulmonaria.) Membranous connection between the sclerites of the Terga and pleura of the abdominal rings. The connexivum.

PULMONARY SPACE Noun. See Pulmonarium.

PULS, JACQUES CHARLES (–1889) (Laboulbène 1889, Bull. Soc. ent. Fr. (6)9: lxix.)

PULSATILE Adj. (Latin, *pulsus* = driven.) Having the power of pulsating or moving in a rhythmic manner. Term applied to special organs in the legs, wing bases and Antennae which aid in circulating Haemolymph in these appendages.

PULSATILE ORGAN See Pulsating Membranes.

PULSATING MEMBRANES Small, muscular membranes found in the Thorax, head, and appendages of various insects, the rhythmic contractions of which probably contribute to the circulation of the blood (Snodgrass).

PULSATORY Adj. (Latin, *pulsus* = driven.) Descriptive of or pertaining to the pulse or beat of the Heart or any other chamber of the insect body. Alt. Pulsating; Pulsatile.

PULSE Noun. (Latin, *puls* = porridge of legumes; Greek, *poltos* = porridge. Pl., Pulses.) 1. Edible seed from several leguminous crops (*e.g.* beans, peas). 2. (Latin, *pulsus* = striking, beating. Pl., Pulses.) A periodic recurrent wave or constriction/dilation of a tubular vessel (Heart, Aorta) which results in the conduction of fluid (Haemolymph) through a tubular network in the body.

PULSE BEETLE *Callosobruchus chinensis* Linnaeus [Coleoptera: Bruchidae]. Syn. Oriental Cowpea Bruchid.

PULTENEY, RICHARD (1730–1801) (Rose 1850, *New General Biographical Dictionary* 11: 250.)

PULVERULENT Adj. (Latin, *pulvis* = dust; *ulentus* = that abounds in.) Powdery; dusty; covered with very minute, powder-like scales (Kirby & Spence). Alt. Pulverulentus.

PULVICORIA Noun. The Coria of the distal tarsal segment on each side of the Tubercula produced as a long slender tail-like lobe as long as a claw, along the outer side of the claws, sometimes pectinate (MacGillivray).

PULVILLIFORM Adj. (Latin, *pulvillus* = small cushion; *forma* = shape.) Pulvillus-shaped. See Pulvillus. Rel. Form 2; Shape 2; Outline Shape.

PULVILLUS Noun. (Latin, *pulvillus* = small cushion. Pl., Pulvilli.) 1. Membranous, pad-like structures between the tarsal claws. 2. Cushions of short, stiff Setae or other clothing on underside of tarsal joints rarely fleshy lobes. See Arolium (Smith), 'restricted to the structures in Diptera'.

PULVINARIA SCALE *Pulvinaria cellulosa* Green [Hemiptera: Coccidae]: In Australia, a minor and sporadic pest of citrus. On leaves and twigs PS causes a heavy growth of sooty mould. Alt. Cottony Citrus Scale.

PULVINATE Adj. (Latin, *pulvinus* = a cushion; *-atus* = adjectival suffix.) Moderately convex. Alt. Pulvinatus.

PULVINIS Noun. (Latin, *pulvinus* = a cushion.) The mass of Retineriae covering the ventral side of the Tarsi to form Tullili (MacGillivray).

PULVINULUS Noun. (Latin, *pulvinus* = a cushion.) A soft ball at the end of the Tarsus.

PUMPKIN BEETLE *Aulacophora hilaris* (Boisduval) [Coleoptera: Chrysomelidae]: In Australia, adults defoliate cucurbits; larvae severely damage root system.

PUMPKIN BUG *Megymenum affine* Boisduval [Hemiptera: Dinidoridae]. See Cucurbit Shield Bug.

PUMPKIN CATERPILLAR *Diaphania indica* (Saunders) [Lepidoptera: Pyralidae]: A significant pest of some cultivated curcurbits in India.

PUNCTATE Adj. (Latin, *punctum* = point; *-atus* = adjectival suffix.) 1. Descriptive of surface sculpture, usually the insect's Integument, that is microscopically pitted. See Sculpture; Sculpture Pattern. Cf. Alveolate; Baculate; Clavate; Echinate; Favose; Gemmate; Reticulate; Rugulate; Scabrate; Shagreened; Smooth; Striate; Verrucate. 2. Pertaining to a surface with impressed points, microscopic pits or punctures. Cf. Alveolate.

PUNCTATE-STRIATE Adj. Descriptive of surface sculpture, usually the insect's Integument, that displays rows of punctures which resemble and replace striae.

PUNCTATE SUBSTANCE See Medullary Substance.

PUNCTATION Noun. (Latin, *punctum* = point; English, *-tion* = result of an action. Pl., Punctations.) Being marked with punctures or very small pits or deep depressions. Alt. Punctuation;

Puncturation.

PUNCTIFORM Adj. (Latin, *punctum* = point; *forma* = shape.) Dot-shaped; Point-shaped. Alt. Punctiformis; Punctate. See Punctate; Sculpture. Rel. Form 2; Shape 2; Outline Shape.

PUNCTO-STRIATUS Adj. Punctured in longitudinal straight lines.

PUNCTUATED EQUILIBRIUM A theory of evolution. Cf. Gradualism.

PUNCTULATE Adj. (Latin, *punctum* = paint; *-atus* = adjectival suffix.) With small punctures. Alt. Punctulatus.

PUNCTUM Noun. (Latin, *punctum* = point. Pl., Puncta.) 1. A minute pit or spot, as on the Elytra of Coleoptera. 2. A small area marked off in any way from the surrounding area.

PUNCTURE Noun. (Latin, *punctura* = prick. Pl., Punctures.) A small impression on the hard outer parts of the insect body, like that made by a needle.

PUNCTUREVINE *Tribulus terrestris* Linnaeus [Zygophyllaceae]: A prostrate or decumbent annual herb endemic to Palearctic. Produces woody nutlets with sharp, stout, divergent spines. Behaves as summer annual; seeds germinate during spring, flowers during summer; seeds may remain dormant in soil for several years. Introduced into midwestern USA with livestock; a serious weed in dry southwestern USA, Hawaii and Mexico. Partially controlled by weevils imported from Italy: Seed-infesting *Microlarinus lareynii* (Jacquelin Duval) and crown/stem-infesting *M. lypriformis* (Wollaston).

PUNCTUREVINE SEED-WEEVIL *Microlarinus lareynii* (Jacquelin du Val) [Coleoptera: Curculionidae].

PUNCTUREVINE STEM-WEEVIL *Microlarinus lypriformis* (Wollaston) [Coleoptera: Curculionidae].

PUNGUR, JULIUS (GYUL) (1843–1907) (Anon. 1907, Leopoldina 43: 71.)

PUNICEUS Adj. (Latin.) Bright red with a violet tinge.

PUNKIE See Ceratopogonidae.

PUPA Noun. (Latin, *pupa* = puppet, young girl. Pl., Pupae; Pupas.) 1. A phase of complete Metamorphosis during which larval anatomical features are destroyed and adult features are constructed. 2. Holometabolous insects: The stage between larva and adult. Entomologists agree about physical appearance of the Pupa but do not agree over what the Pupa represents developmentally. The Pupa is a transitional stage of holometabolous development necessary for reconstruction of the larva to adult form. The degree of reconstruction differs among groups. Larval structures are completed within the egg. Muscles are conspicuous structures formed during pupal stage. The fate of larval tissues is not always clear, but larval muscles are broken down (Histolysis) and adult muscles are formed (histogenesis). During this process, the Pupa is relatively immobile. Some larval tissues are converted into corresponding parts of the adult. Some larval tissues are broken down and adult organs are built from undifferentiated embryonic cells called imaginal discs or histoblasts. These are carried in the larva, but are not part of the larval structure. See Metamorphosis. Cf. Chrysalis; Larva; Puparium; Nymph; Naiad. Rel. Egg; Adult.

PUPA ADHERAENA An adherent Pupa; one which hangs perpendicularly, head down.

PUPA ANGULARIS A Pupa with a pyramidal process or nose on the back.

PUPA CONICA A conical Pupa; as opposed to an angular Pupa.

PUPA CONTIGUA A bound Pupa or one which remains upright against a vertical object and is supported by a silk thread across the Thorax.

PUPA CUSTODIATA A guarded Pupa or one in a partly open cocoon.

PUPA DERMATA A Pupa which retains the larval skin and no trace of the position of the future limbs is apparent.

PUPA EXARATA Exarate Pupa; sculptured Pupa; a Pupa in which the limbs of the encased adult lie free but closely attached to the body.

PUPA FOLLICULATA An encased Pupa, a Pupa in a case or a cocoon.

PUPA INCOMPLETA Lepidoptera: A Pupa in which the appendages are often partly free and more than three of the abdominal segments are moveable (Imms).

PUPA LARVATA A masked Pupa or one in which the different parts of the forming adult are traceable as lines on the surface.

PUPA LIBERA Lepidoptera: A Pupa with many free segments (Imms).

PUPA NUDA A naked Pupa; one which lies free of any attachment.

PUPA OBTECTA 1. Lepidoptera: A more 'highly developed' or specialized Pupa which displays smooth and rounded form with only the fourth, fifth and sixth segments free (Imms). 2. A Pupa in which the appendages are fused to the body (Wardle). An Obtect Pupa. Cf. Exarate Pupa.

PUPA SUBTERRANEA Underground Pupa or one which is buried during the transformation.

PUPAL SAC The thin, semi-transparent envelope of the head and Thorax of some mosquitoes (Nuttall & Shipley).

PUPARIATION Noun. (Latin, *pupa* = puppet, young girl; English, *-tion* = result of an action.) Formation of the barrel-shaped Puparium by cyclorrhaphous Diptera (Fraenkel & Bhaskran 1973, AESA 66: 418–422.) The process is characterized by withdrawal of the three anterior-most larval body segments into the body and subsequent shortening of the larval body. Syn. Puparisierung (German); Pupariating (Metcalf & Flint 1939.) See Puparium. Cf. Pupation.

PUPARIUM Noun. (Latin, *pupa* = puppet; *arium* = place of a thing. Pl., Puparia.) 1. Cyclorrhaphous Diptera: The thickened, hardened, barrel-like third larval instar Integument within which the Pupa and adult are formed (Fraenkel & Bhaskaran 1973). 2. The covering of certain coccids. 3. Stylopids: The Integument of the seventh instar larva in which the adult female is enclosed (Comstock). See Pupariation. Cf. Pupa. Rel. Metamorphosis.

PUPATE Intransitive Verb. To become a Pupa.

PUPATION Noun. (Latin, *pupa* = puppet, young girl; English, *-tion* = result of an action.) The act of becoming a Pupa. Cf. Pupariation.

PUPIFEROUS Adj. (Latin, *pupa* = puppet, young girl; *ferre* = to bear; *-osus* = with the property of.) Descriptive of the generation of plant lice which produces sexed individuals.

PUPIFORM Adj. (Latin, *pupa* = puppet; *forma* = shape.) Pupa-shaped; descriptive of structure which resembles a Pupa in outline form or appearance. See Pupa. Rel. Form 2; Shape 2; Outline Shape.

PUPIGENOUS See Pupiparous.

PUPIGEROUS Adj. (Latin, *pupa* = puppet, young girl; *gero* = to bear, carry; *-osus* = with the property of.) Forming a larval Puparium, coarctate, said of dipterous larvae that contract to from an envelope for the enclosed Pupa.

PUPIL Noun. (Latin, *pupilla* = pupil of the eye. Pl., Pupils.) The central spot of an ocellate spot.

PUPILLATE Adj. (Latin, *pupilla* = pupil of the eye; *-atus* = adjectival suffix.) With an eye-like centre, as a spot or mark. Alt. Pupillatus.

PUPIPARA Noun. A series of Diptera, in which the females do not extrude the young until they are ready to pupate.

PUPIPARID Noun. Any pupiparous insect, specifically certain Diptera.

PUPIPAROUS Adj. (Latin, *pupa* = puppet, young girl; *parere* = to bear; *-osus* = with the property of.) Pertaining to insects producing immatures ready to pupate.

PUPIVOROUS Adj. (Latin, *pupa* = puppet, Greek, *vorare* = to eat; Latin, *-osus* = with the property of.) Feeding upon Pupae.

PURCELL, WILLIAM FREDERICK (1866–1919) (Bonnet 1945, *Bibliographia Araneorum* 1: 62.)

PURCHAS, SAMUEL (1515–1626) (Rose 1850, *New General Biographical Dictionary* II: 251.)

PURDEY, WILLIAM (1844–1922) (Entomologist 55: 71–72.)

PURIN Noun. (Latin, *purum* = pure.) A protein decomposition product. A basic compound, which may be regarded as the parent substance from which uric acid and related compounds may be derived by substitution. Alt. Purine.

PURKYNEM, CYRIL (–1963) (Hevrovsky 1963, Cas. csl Spol. ent. 60: 266, bibliogr.)

PURPLE AZURE *Ogyris zosine* Hewitson [Lepidoptera: Lycaenidae].

PURPLE BROOD A disorder of honey bees in which the brood in unsealed chambers become purple and die. PB occurs during early summer in southern USA and results from workers foraging on southern leatherwood (*Cyrilla racemiflora*).

PURPLE BROWN-EYE *Chaetocneme porphyropis* (Meyrick and Lower) [Lepidoptera: Hesperiidae].

PURPLE SCALE *Cornuaspis beckii* (Newman) [Hemiptera: Diaspididae]: A cosmopolitan, pest of avocado, citrus, fig, pecan and many ornamental plants; PS endemic to the Orient. Adult female's cover brown, 3–4 mm long, shaped as a mussel shell. Female oviposits ca 50–100 white eggs in two rows beneath cover; eclosion occurs within two weeks; crawlers settle in sheltered places, often on mature leaves; life cycle completed in 6–8 weeks. PS multivoltine with up to six generations per year, depending upon climate. PS controlled with natural enemies, including Species of coccinellid beetles and parasitic Hymenoptera. Syn. Mussel Scale; *Lepidosaphes beckii* (Newman).

PURPLEBACKED CABBAGEWORM *Evergestis pallidata* (Hufnagel) [Lepidoptera: Pyralidae].

PURPLESPOTTED LILY-APHID *Macrosiphum lilii* (Monell) [Hemiptera: Aphididae].

PURPLESTRIPED SHOOTWORM *Zeiraphera unfortunana* Powell. [Lepidoptera: Tortricidae].

PURPOSIVE Adj. (Middle English, *purpos*.) Having an intended object, effect or result.

PURPURESCENT Adj. (Latin, *purple*.) Becoming purple in shade.

PURPUREUS Adj. (Latin, *purple*.) Purple. Alt. Purpureal; Purpureous.

PUSS CATERPILLAR *Megalopyge opercularis* (J. E. Smith) [Lepidoptera: Megalopygidae].

PUSTULATE Adj. (Latin, *pustule* = blister; *-atus* = adjectival suffix.) Covered with pustules. Alt. Pustulosus; Pustulous.

PUSTULATED HAIRS Mallophaga: Setae arising from unchitinized areas on the Integument.

PUSTULE Noun. (Old French from Latin, *pustule* = blister. Pl., Pustules.) 1. A small blister-like swelling. 2. A coloured point of moderate circumference. 3. A blister-like elevated spot on plant tissue which erupts to release fungal spores. Alt. Pustula.

PUTATIVE Adj. (Latin, *putativus* from *putare* = to reckon, think.) 1. That which is commonly accepted as fact or explanation. 2. Assumed to exist or have existed. 3. Something supposed, presumed or reputed but subject to further examination and analysis. Cf. Theoretical.

PUTATIVE PHYLOGENY A presumed or hypothesized phylogeny. Subject to further examination and analysis.

PUTATIVE SPECIES A presumed or hypothesized Species concept. Subject to further examination and analysis.

PUTNAM SCALE *Diaspidiotus ancylus* (Putnam) [Hemiptera: Diaspididae]: An occasional serious pest of elm and other trees in North America; overwinters as partially developed.

PUTNAM, FREDERICK WARD (1839–) (Weiss 1936, *Pioneer Century of American Entomology*, 320 pp. (249–250), New Brunswick.)

PUTNAM, JOSEPH DUNCAN (1855–1881) (Pratt *et al.* 1883, Proc. Davenport Acad. Sci. 3: 195–248, bibliogr.)

PUTON, JEAN BAPTISTE AUGUSTE (1834–1913) (Sainte-Claire Deville 1913, J. Bull. Soc. ent. Fr. 1913: 173.)

PUTREFACTION Noun. (Latin, *putris* = putrid; *facere* = to make; English, *-tion* = result of an action.) The decomposition of organic matter (proteins) into substance that loses its structural definition and which emanates obnoxious odours. See Decay. Cf. Decompositon; Rot.

PUTZEYS, JULES ANTOINE ADOLPHE HENRI (1809–1882) (Desmarest 1882, Bull. Soc. ent. Fr. (6)2: iii–iv.)

PWAINGYET Noun. Commercial name for cerumen of the East Indian stingless bee *Trigona laeviceps* (Rau).

PYATNITZKII, GEORGII KONSTANTINOVICH (1908–1951) (Stark & Grigorev 1953, Ent. Obozr. 33: 376–379, bibliogr.)

PYCNIDIUM Noun. (New Latin.) Fungi: A microscopic, flask-shaped fruiting body which produces asexual spores. Cf. Perithecium.

PYCNOLEUCOCYTE See Granulocyte.

PYCNOSIS Noun. (Greek, *pyknos* = dense; *-is* = condition.) 1. Cell degeneration to include nuclear condensation and resulting in Chromosomes which stain intensively. 2. Larval Histolysis: Cases in which chromatin becomes distributed in nodules of histolysing tissue (Henneguy).

PYCNOTIC Adj. (Greek, *pyknos* = dense; *-ic* = characterized by.) Descriptive of or pertaining to Pycnosis.

PYDRIN® See Fenvalerate.

PYETT, CLAUDE A (–1903) (Anon. 1903, Entomologist 36: 296.)

PYGAL Adj. (Greek, *pyge* = the rump; *-alis* = pertaining to.) Descriptive of structure associated with the posterior end of the Abdomen, or Pygidium.

PYGIDIAL AREA Some aculeate Hymenoptera: An area on the Pygidium bounded on each side by a Carina, the two Carinae meeting posteriorly on the middle line of the segment (Comstock).

PYGIDIAL FRINGE Coccoidea: The projections and indentations of the lateral margin of the Pygidium (MacGillivray).

PYGIDIAL GLAND 1. A paired gland in male parasitic Hymenoptera (Braconidae, solitary bees) consisting of several tubules connected with a common attachment or resevoir on the Pygidium. Also found in many ants on seventh Tergum, not eighth Tergum (Pygidium) of other insects, and hence not necessarily homologous with them. In ants consists of paired glands which empty into membrane between sixth and seventh Terga with each cell having its own ductile. 2. The structure of the pygidial defence glands of Carabidae (Coleoptera). See Forsyth 1972. Trans. Zool. Soc. London 32: 249–309. Coleoptera: Paired organs secreting pungent and corrosive fluids, which lie near the Anus (Imms).

PYGIDIAL INCISION Some coccids: The deep mesial emargination of the caudal margin of the Pygidium (MacGillivray).

PYGIDIAL MARGIN Coccids: lateris *sensu* MacGillivray.

PYGIDIAL PLATE Hymenoptera (some Aculeata): Portion of the sixth gastral Tergum of females or seventh gastral Tergum of males which is flat and bounded by cuticular ridges, Carinae, Sulci or other notable topographical features.

PYGIDIAL SETAE Coccids: Small Setae on the proximal part of the Pygidium adjacent to the lobes and the Lateres adjacent to the Latadentes (MacGillivray).

PYGIDICRANIDAE Plural Noun. The most primitive Family of Dermaptera, *sensu stricto*. Rare insects, body 10–35 mm long, long Antenna, winged; Species occur in Australia, South Africa.

PYGIDICRANOIDEA Superfamily of Dermaptera containing Diplatyidae and Pygidicranidae.

PYGIDIUM Noun. (Greek, *pygidion* = narrow buttock, rump; *-idion* = diminutive. Pl., Pygidia.) 1. The Tergum of the last segment of the Abdomen, whatever its numerical designation. 2. Coleoptera: The apical abdominal segment exposed by the Elytra. 3. Diaspididae: A strongly sclerotized unsegmented region terminating the Abdomen of the adult female, following the first four abdominal segments; not to be confused with the true Pygidium of other insects (Comstock). See Suranal Plate.

PYGIOPSYLLIDAE Plural Noun. A moderate-sized Family of ceratophylloid Siphonaptera with one Genus in South America and many Genera and ca 50 Species in Asia and Australia; many Species associated with marsupials, some associated with rodents and a few associated with birds.

PYGMY Adj. (Latin, *pygmaeus* from Greek, *pygmaios* from *pygme* = a measurement from elbow to first knuckle.) Descriptive of individuals that are small or diminutive in size or stature.

PYGMY BACKSWIMMERS See Pleidae.

PYGMY GROUSE LOCUSTS See Saltatoria.

PYGMY MOLE CRICKETS See Saltatoria.

PYGMY MOTHS See Nepticuloidea.

PYGOFER Noun. (Greek, *pyge* = rump; Latin, *ferre* = to carry. Pl., Pygofers.) The apical segment of the Abdomen in some Homoptera, especially the lateral margins which appear in the ventral view, hence the term is sometimes used in the plural, Pygofers.

PYGOPHORE Noun. (Greek, *pyge* = rump; *pherein* = to bear. Pl., Pygophores.) The large upper sclerite of the genitalia in Homoptera.

PYGOPOD Noun. (Greek, *pyge* = rump; *pous* = foot. Pl., Pygopods.) 1. Any appendage of the tenth abdominal segment (*sensu* Snodgrass). 2. Paired appendicular processes in Trichoptera, Coleoptera and Lepidoptera. 3. Terminal, eversible fan-like appendages with numerous digitiform tubules in some beetle larvae.

Pygopods may have evolved as modifications of intersegmental membrane. Podia are withdrawn into segment nine and have a common or median stalk. Each Podium has several rows of equally spaced Acanthae that serve as holdfasts. Functionally Pygopods enable a larva to attach to different substrata and move. When a larva walks on a flat substratum, Pygopods are retracted into the body. When the larva walks on the edge of a leaf, Pygopods are everted and used as holdfast on smooth and rough surfaces. Operation of Pygopods is complex; inversion of Podia is achieved by muscle fibres. The musculature is complex and delicate. Eversion is probably achieved by hydrostatic pressure; muscles attaching to glands cannot be located. Hydrostatic pressure must act antagonistically with inversion muscles, and pressure on the abdomen results in eversion of the Podia. Intrinsic and complex musculature associated with inversion suggests a high level of control over individual digits of Podia. Adhesion also is facilitated by surface molecular forces of a liquid film. Pygopods typically are damp. Source of moisture has not been identified, but apparently it originates within the terminal abdominal segments. See Abdominal Appendage.

PYGOSOMAL PLATE Acari: A sclerite on the Pygosoma of some chiggars. PP may be single or paired with Setae. Syn Caudal Plate; Pygidial Plate. (Goff *et al.* 1982, J. Med. Ent. 19: 232.)

PYGOSTYLUS Noun. (Greek, *pyge* = rump; *stylos* = column. Pl., Pygostyli.) Hymenoptera: Appendicular sensory structures on gastral Tergum VIII or IX. Sometimes regarded as homologous with the Cercus. See Cercus.

PYGOTHECA Noun. (Pl., Pygothecae.) The parts containing the genitalia in Homoptera.

PYLORIC Adj. (Greek, *pyloros* = gate keeper; *-ic* = consisting of.) Descriptive of or pertaining to the posterior extremity of the Midgut. The Pylorus.

PYLORIC VALVE A boundary between Midgut and Hindgut. Extrinsic musculature suspends Hindgut in the Abdomen and probably serves to dilate portions of the HG and facilitate excretion. The Pylorus also serves as point-of-contact and holdfast for Malpighian Tubules. MT are diverticula of Hindgut and a true Pylorus is anterior of them. An Analogy of PV with vertebrates is not correct. PV differs among primitive relatives of the Insecta. Collembola have a pyloric ring. See Alimentary System; Hindgut. Cf. Malpighian Tubules. Rel. Excretion. Syn. Pyloric Valvule.

PYLORUS Noun. (Greek, *pyloros* = gate keeper.) The Midgut Valve.

PYMETROZINE A pyridine azomethene compound {4,5-dihydro-6-methyl-4-[(3-pyridinyl-methylene)-amino]-1,2,4-triazin-3(2H)one} used as a contact insecticide and stomach poison against plant-sucking insects (aphids, planthoppers, scale insects, whiteflies.) Compound applied to fruit trees, cotton, ornamentals, rice and vegetables in some countries; some systemic activity has been observed. Experimental registration only for testing in USA. Tradenames include: CGA-215944®, Chess®, Fulfill®, Sterling®.

PYNAMIN-FORTE® See Pyrethrin.

PYNOSECT® See Resmethrin.

PYNOSET® See Permethrin.

PYRACLOFOS An organic-phosphate (thiophosphoric acid) compound {(R,S)-[O-1-(4-chlorophenyl)-pyrazol-4-yl]-O-ethyl-S-n-propyl phosphorothioate} used as an acaricide, contact insecticide and stomach poison against mites, nematodes, thrips, corn borer, budworms, bollworms, armyworms, loopers, cutworms, rice weevil and plant-sucking pests. Compound applied to tea, rice, cotton, corn, potatoes, soybeans and vegetables in some countries. Not registered for use in USA. Toxic to fishes. Trade names include Boltage®, Starlex® and Voltage®. See Organophosphorus Insecticide.

PYRALIDAE Latreille 1809. Plural Noun. Grass Moths; Pyralids; Snout Moths; Wax Moths. A large, cosmopolitan Family of ditrysian Lepidoptera assigned to Superfamily Pyraloidea. One of the largest Families of Lepidoptera with an estimated 30,000 described Species; higher classification problematic with 15–25 Subfamilies recognized. Adult body size highly variable; wingspan 10–95 mm; head with smooth scales; Ocelli typically present, sometimes reduced or absent; small Chaetosemata present in some Subfamilies, but typically absent; Antenna typically filiform but other forms include lamellate and pectinate; Proboscis typically well developed with dense scales near base, sometimes reduced or absent; Maxillary Palpus typically with four scaled segments, but 1–3 segments in some groups; Labial Palpus usually long and porrect; legs long to very long; fore Tibia with Epiphysis; tibial spur formula typically 0-2-4 or rarely 0-2-2; wings variable in size and shape but Pterostigma absent; hindwing broader than forewing and with large anal area. Egg flattened, oval in outline shape typically with longitudinal ribs and transverse ridges, rarely smooth; some Species with eggs flattened and scale like; eggs laid individually or in masses covered with scales from Corethrogyne. Larva typically cylindrical with prognathous head; Larva webs leaves, tunnels, feeds on stored products; larval Nymphulinae aquatic. Pupa typically strongly sclerotized; Antenna long; Maxillary Palpus present (except Epipaschiinae), Pilifers present, meet medially or slightly separated; Abdomen without dorsal spines (except Galleriinae); Cremaster present or absent; Setae at posterior end of Abdomen straight or hooked. Adult resting posture highly variable: some Species with tectiform wings, others with wings flat over Abdomen, some Species with wings spread; leg posture also highly variable. Larvae feed in concealed places, some forming shelters of host plants or food material; a few

Species mine leaves; many bore into buds, flowers, stems while some form galls. Some Species are predaceous on scale insects, the eggs of other Lepidoptera, and ants. Many Species are of considerable economic significance by attacking agricultural crops and stored products.

PYRALOIDEA A Superfamily of ditrysian Lepidoptera, including Hyblaeidae, Lathrotelidae, Oxychirotidae, Pyralidae, Thyrinidae and Tineodidae in some classifications; including Lathrotelidae and Pyralidae or just Pyralidae in other classifications. See Lepidoptera.

PYRAMDRON® See Hydramethylnon.

PYRAMID ANT *Conomyrma insana* (Buckley) [Hymenoptera: Formicidae].

PYRAMIDAL Adj. (Greek, *pyramidos* = a pyramid; *-alis* = characterized by.) Resembling a pyramid in form; pyramid-like; angular conical. Alt. Pyramidate; Pyramiform.

PYRAMIDATE FASCIA An angulate fascia.

PYRANICA® See Fenpyrad.

PYRELLIN EC® A synthetic-pyrethroid combining pyrethrin and rotenone. Used as a foliar insecticide on some crops and ornamentals. Cf. Pyrethrin; Rotenone.

PYRESYN® See Pyrethrin.

PYRETHRIN Noun. (Greek, *pyrethron* = fever-few; Pl., Pyrethrins.) A mixture of esters extracted from pyrethrum flowers (Asteraceae) and used as a botanical insecticide. First reported in Persia ca 1828. Early materials from flower extracts of *Chrysanthemum roseum* Adam and *C. carneum* Steud. More recently, an oil from flowers of *Pyrethrum cinerariaefolium* Vis. (cultivated in Australia, Ecuador and Kenya), which is then combined with other plant materials for use as a botanical insecticide. Pyrethrins registered on most food crops, for application to domestic animals and food-handling areas. Pyrethrins used as space spray to control bed bugs, and insect pests of home, garden industry and stored products. Piperonyl butoxide often used as a synergist. Commercial products display broad-spectrum effectiveness with rapid knockdown; some resistance in cockroaches has been observed. Mode-of-action as a contact insecticide or as a stomach-poison; sometimes used as a fumigant. Characterized as rapid in action, unstable in bright light or high temperatures and ineffective in windy areas; incompatible with Bordeaux mixture, calcium arsenate and alkaline solutions. Minor phytotoxicity observed but highly toxic to cold-blooded vertebrates. Symptoms of human poisoning include: irritation of nose, throat or eyes. Allergic reactions include: numbness of lips and tongue, nausea, vomiting, headache. Skin contact may cause dermatitis. First-aid treatment: Eliminate source of poisoning and seek medical advice. Trade names include: Bug-Buster®, Cinerin®, Cinerolone®, Evergreen Crop Spray®, Excite-R®, Pallethrine®, Pynamin-Forte®, Pyresyn®, Pyrexcel®, Pyrocide®, Sectol®, Sectrol®, Synerol®. See Botanical Insecticide; Natural Insecticide; Insecticide. Cf. Rotenone.

PYRETHROID Noun. (Greek. *pyrethron* = fever-few; *eidos* = form. Pl., Pyrethroids.) An organic synthetic insecticide with a structure based on that of Pyrethrum; a natural botanical insecticide. Any number of a class of synthetic pyrethrum-based insecticides. See Pyrethrum; Synthetic Pyrethroid.

PYREXCEL® See Pyrethrin.

PYRGOMORPHIDAE Plural Noun. Pyrgomorphids. A small Family of caeliferous Orthoptera assigned to Superfamily Pamphagoidea. Pyrgomorphids well represented in Australia but also occur in Africa, Asia and Oceania. Head conical; Prosternum with median process.

PYRGOTA FLY *Pyrgota undata* [Diptera: Pyrgotidae]: A monovoltine fly widespread in eastern North America. Larva endoparasitic on scarabaeid beetles. Adult ca 10–12 mm long; body and legs brown; wing with elaborate patterns; Ocelli typically absent; apex of Abdomen procurved with Ovipositor conical and apically sharp. Female lands on adult beetle and oviposits. See Prygotidae.

PYRGOTIDAE Plural Noun. Cosmopolitan Family of ca 330 Species of Diptera. Assigned to Otitidae in some classifications; placed between Otitidae and Conopidae; closely related to Tephritidae. Recognized by large head; frontal region produced, without bristles; wings long, wide with a well developed Alula; first Posterior Cell open at wing margin; Anal Cell short and closed well before margin of wing; apex pointed; Subcostal Vein often setulose or with one or two bristles; female Ovipositor is large, often heavily sclerotized. Adults nocturnal; taken in light traps. Pyrgotidae larvae are primary internal parasites of adult Scarabaeidae; female oviposits in Abdomen of adult scarabs usually while beetle is in flight. Larvae solitary or gregarious with 20 individuals developing in a host. Pupation within host remains in soil. One generation per year. See Pyrgota Fly.

PYRID® See Fenvalerate.

PYRIDABEN Noun. A pyridazinone compound {2-*tert*-butyl-5-(4-*tert*-butylbenzylthio)-4-chloropyridazin-3(2H)-one} used as a selective acaricide/insecticide on ornamentals, fruit crops, tea, cotton and vegetables in some countries. Trade names include Nexter® and Sanmite®.

PYRIDAFENTHION Noun. An organic-phosphate (thiophosphoric acid) compound {O,O-dimethyl 2,3-dihydro-3-oxo-2-phenyl-6-pyridazinyl phosphorothioate} used as a contact insecticide and stomach poison against plant-feeding mites, plant-sucking insects, bollworms and chewing insects. Compound applied to fruit trees, corn, cotton, grapes, rice, soybeans, sorghum and vegetables in some countries. May be used as public health insecticide. Not registered for use in USA. Trade names include: Ofnack®, Ofnak®, Ofunack®, Oreste®. See Organophosphorus Insecticide.

PYRIFORM Adj. (Latin, *pyrum* = pear; *forma* = shape.) Pear-shaped; descriptive of structure enlarged or globose apicad. See Pear. Cf. Adeniform; Ampulliform; Guttiform; Lageniform; Lobiform; Reniform; Scrotiform. Rel. Form 2; Shape 2; Outline Shape.

PYRIFORM SCALE *Protopulvinaria pyriformis* (Cockerell) [Hemiptera: Coccidae].

PYRINEX® See Chlorpyrifos.

PYRIPROXYFEN Noun. A pyridene compound {4-phenoxyphenyl(RS)-2-(2-pyridyloxy)propyl ester} used as a contact insecticide and stomach poison for control of public health pests. Target pests include fleas, flies, midges, mosquitoes and related insects. Comound applied to breeding sites of pests (stockpiled manure, aquatic sites for mosquitoes and midges) and barns; used in Malaria control programmes in some countries. Pyriproxyfen is not registered for use in USA. Trade names include: Admiral®, Epingle®, Nylar®, Sumilarv®.

PYROCHROIDAE Plural Noun. Fire-coloured Beetles. A small Family of Coleoptera assigned to Superfamily Tenebionoidea. Body 5–20 mm long, dorsoventrally flattened; Pronotum without dents or depressions, narrower than base of Elytra; Elytra rounded basally and widest near midline; tarsal formula 5-5-4. Adults taken on folage and flowers. Larvae occur under bark of dead trees.

PYROCIDE® See Pyrethrin.

PYRRHOCORIDAE Fieber 1860. Plural Noun. Pyrrhocorid Bugs; Cotton Stainers. A moderate-sized Family of heteropterous Hemiptera placed in Superfamily Pyrrhocoroidea including about 40 Genera. Adults rather large and boldly coloured. Most pyrrhocorids feed on plant tissue; some Species are predaceous. Eggs dropped from female but may adhere to plant tissue. Family related to Largidae.

PYRRHOCOROIDEA Southwood 1956. A Superfamily of Hemiptera related to the Correoidea, Pentatomoidea and Lygaeoidea. All characterized by Trichobothria on the Abdomen.

PYRRHOPHYTA Noun. A Division of the algae including the Class Dinophyceae (dinoflagellates), See Dinoflagellate.

PYRROLE (Greek, *pyro* = fire; Latin, *oleum* = oil; Pl., Pyrroles.) A carbonitrile compound {4-bromo-2-(4-chlorophenyl)-1-(ethoxymethyl)-5-(trifluoromethyl) pyrrole-3-carbonitrile} first isolated from fermentation of *Streptomyces*. Pyrrole is used as a contact insecticide, stomach poison and acaricide against mites, plant-sucking insects (aphids, leafhoppers, whiteflies), thrips, Lepidoptera larvae (armyworms, boll worms), Colorado Potato Beetle and related insects; it expresses residual effectiveness and ovicidal activity on some pests. Compound is applied to tree crops, coffee, corn, cotton, grapes, ornamentals, soybeans and vegetables. Phytotoxic to grapes and brassicas; toxic to fish. Pyrrole under experimental registration in USA. Pyrrole does not control Pink Bollworm, Red Mite, Pear Psylla or Rosy Apple Aphid. Trade names include: AC 303 630®, Alert®, Pirate®, Stalker®. See Biopesticide. Cf. Nereistoxin; Nicotine; Pyrethrin; Ryania; Sabadilla; Stemofoline.

PYRROLIZIDINE ALKALOIDS Toxic compounds produced by plants (*e.g.*, Asteraceae, Boraginaceae) and ingested by insects (*e.g.* Lepidoptera, Coleoptera) for defence and production of sex pheromones. (Bernays *et al.* 1977, J. Zool. London 182: 85–87; Boppre 1983, Oecologia 59: 414–416.)

PYTHIDAE Plural Noun. A small Family of polyphagous Coleoptera assigned to Superfamily Tenebrionoidea. Adult body typically dark coloured with brown, black and red; elongate, 3–15 mm long with vestiture of decumbent and erect Setae; eyes prominent but not emarginate; Antenna with 11 segments; frontal coxal cavity open; tarsal formula 5-5-4. Larva elongate, parallel-sided and typcially yellow; Urogomphi forked or with accessory processes. Larvae occur under bark of rotting logs.

QAMLIN® See Permethrin.

Q-FEVER A minor rickettsial disease first noted in Australia and now cosmopolitan in distribution. QF aetiological agent *Coxiella burnetii,* not closely related to other members of Tribe Rickettsieae. In Australia, *C. burnetii* occurs in bandicoots (circulated by ticks *Haemaphysalis humerosa* and *Ixodes holocyclus),* kangaroos (circulated by *Amblyomma triguttatum)* and domestic livestock. Humans infected by inhalation of disease; most common among ranch and slaughterhouse workers. Most cases subclinical or not diagnosed; causes fever, headache and often pneumonia; causes few deaths. See Rickettsiaceae. Cf. Trench Fever.

QFLIES See Queensland Fruit Fly.

QUADRANGLE Noun. (Pl., Quadrangles.) Odonata: A cell in the wing of the Zygoptera bounded by M, Cu, Arculus and a crossvein between M and Cu (similar in position to the triangle of the Anisoptera) (Garman).

QUADRANGULAR Adj. (Latin, *quadrangulus* = four-angled.) A square structure.

QUADRAT Noun. (Latin, *quadratus* = squared. Pl., Quadrats.) The dimensionless sampling unit for biotic study on land; in some studies a square metre, selected as representative of a larger area to be studied (Shelford).

QUADRATE Noun. (Latin, *quadrans* = fourth part; *-atus* = adjectival suffix.) Square or nearly so. Alt. Quadratus.

QUADRATE PLATES Aculeate Hymenoptera: Large sclerites articulating with the triangular sclerite at its dorsal and posterior angle (Imms).

QUADRICAPSULAR Adj. (Latin, *quadrans* = fourth part; *capsula* = small box.) Descriptive of structure which embraces four capsules.

QUADRIDENTATE Adj. (Latin, *quadrans* = fourth part; *dens* = tooth; *-atus* = adjectival suffix.) Four-toothed. A term usually applied to Mandibles. Alt. Quadridentatus.

QUADRIFARIUS Adj. (Latin, *quadrans* = fourth part.) Four-fold or in fours, as rows.

QUADRILATERAL Adj. (Latin, *quadrans* = fourth part; *lateris* = side; *-alis* = characterized by.) 1. Four-sided. 2. Zygopterous Odonata: The Discal Cell of the wing, enclosed by M above, Cu below, the lower part of the Arculus basally and a thickened cross-vein distally.

QUADRIMACULATE Adj. (Latin, *quadrans* = fourth part; *macula* = spot; *-atus* = adjectival suffix.) Pertaining to structure or surface which displays four spots (maculae.) Alt. Quadrimaculatus.

QUADRIPARTITE Adj. (Latin, *quadrans* = fourth part; *partis* = part; *-ites* = constituent.) Descriptive of structure which displays four parts. Alt. Quadripartitus.

QUADRIPINNATE Adj. (Latin, *quadrans* = fourth part; *pinna* = feather; *-atus* = adjectival suffix.) Descriptive of structure, typically an appendage, which displays four feather-like branches or clefts. Alt. Quadripinnatus.

QUADRIVALVATE Adj. (Latin, *quadrans* = fourth part; *valva* = valve; *-atus* = adjectival suffix.) Four-valved. Alt. Quadrivalvular.

QUAIL, AMBROSE (1872?–1905) (Tutt 1905, Entomologist's Rec. J. Var. 17: 304.)

QUAINTANCE, ALTUS LACY (1870–1958) (Mallis 1971, *American Entomologists*. 549 pp. (487–489), New Brunswick.)

QUANT 5® See Isomate-M.

QUANTUM SPECIATION See Founder Model.

QUARANTINE Noun. (Latin, *quarantina* from *quartanta* = forty. Pl., Quarantines.) 1. Medical Entomology: A period of 40 days that was originally imposed upon ships arriving in ports, and whose passengers or crew were suspected of being infected with a contagious disease. The period of 40 days was initially from the period of time necessary to express symptoms of Bubonic plague. After the quarantine period, if no symptoms were expressed the passengers and cargo were then allowed onshore. 2. A regulation which prohibits the movement of people, animals, plants or trade goods which pose a threat to an area, place or country. 3. A place in which people, animals, plants or trade goods are kept while under Quarantine Action. Rel. Quarantine Facility.

QUARANTINE ACTION 1. The exclusion of potential pests which do not occur in an area, the limitation of movement of pests known to occur in an area or the supplement of eradication programmes to eliminate a pest from an area. See Regulatory Control. 2. The treatment of humans, animals, plants and trade goods by various techniques to eliminate the threat posed by organisms under quarantine legislation. QA techniques include fumigation, irradiation and exposure to heat and cold.

QUARANTINE FACILITY 1. Biological Control: A specially designed building, enclosure or containment facility which serves to isolate and permit evaluation of organisms imported from another country or area. QF is typically approved and periodically inspected by a governmental agency.

QUARTAN Noun. Malarial paroxysms recurring at three day intervals.

QUARTAN MALARIA A moderately serious form of human Malaria caused by the protozoan *Plasmodium malariae*. QM cosmopolitan with patchy distrubution and less frequently encountered than other malarias. *Plasmodium malariae* merozoite stage attacks old red blood cells and does not produce Hypnozoites. See Malaria. Cf. Malignant Tertian Malaria.

QUASISOCIAL Adj. (Latin, *quasi* = as if; *sociare* = to associate; *-alis* = characterized by.) Eusocial Bees: Small groups of females of similar age or the same generation which cooperatively construct and provision cells with more than one bee working in a cell. See Communal; Eusocial; Subsocial; Parasocial; Primitively Social; Highly

Social. Cf. Aggregation; Social.

QUATREFAGES DE BREAU, JEAN LOUIS ARMANDE DE (1810–1892) (Anon. 1892, Bull. Soc. zool. Fr. 17: 21–25)

QUAYLE, HENRY JOSEF (1876–1951) American academic (University of California, Riverside) and specialist in Citrus Entomology. (Mallis 1971, *American Entomologists*. 549 pp. (489–490), New Brunswick.)

QUEDENFELDT, FRIEDRICH OTTO GUSTAV (1817–1891) (Anon. 1891, Insektenbörse 8: 23.)

QUEDENFELDT, MAX (1851–1891) (Honrath 1891, Berl. ent. Z. 36: 473–475, bibliogr.)

QUEEN Noun. (Anglo Saxon, *cwen* = woman.) 1. The female member of the reproductive caste in social insects. Cf. Worker Caste. 2. *Danaus gilippus berenice* [Lepidoptera: Danaidae]: A non-migratory butterfly found in eastern United States. Favoured host plant of larval stage includes milkweeds.

QUEEN CONTROL Social Insects: The capacity of a queen to control the reproductive activities of workers and other queens.

QUEEN SUBSTANCE Any chemical the queen uses to attract workers and control their reproductive activity.

QUEENRIGHT Adj. Social Insects: Pertaining to a colony with a functional queen.

QUEENSLAND FRUIT-FLY *Bactrocera tryoni* (Froggatt) [Diptera: Tephritidae]: A polyphagous, important pest of many fruits and vegetables. QFF endemic to eastern Australia and a concern to agriculture and quarantine in tropical and subtropical regions; feeds upon more than 60 native Australian Species of plants and regarded as most serious pest of fruit and vegetables in Australia. Adult 5–7 mm long; head dark brown, compound eye red with metallic reflections; Thorax reddish brown with yellow spots on Prothorax and Mesoscutum laterally, and Scutellum yellow; Abdomen black; wings hyaline with anterior margin dark brown or fuscous; adults strong fliers, capable of travelling long distances during lifetime. Female feeds ca 7 days on microbes, honeydew and protein on surface of fruit and leaves for egg development; Biology resembles Mediterranean Fruit Fly; female punctures fruit, lay eggs, causes damage by oviposition stings. Eggs white, banana-shaped, ca 1 mm long; 10–15 eggs deposited per oviposition session; eclosion occurs within 2–3 days. Neonate maggots burrow into fruit and quickly cause damage partly by feeding and tunnelling and partly as a result of rotting that follows their invasion. Larval development complete ca 10 days; drop to ground and pupate in soil. Multivoltine with 3–6 generations per year depending upon climate and host plant. See Tephritidae. Syn. Qflies. Cf. Mediterranean Fruit Fly; Mexican Fruit Fly; Papaya Fruit Fly.

QUEENSLAND PINE-BEETLE *Calymmaderus incisus* Lea [Coleoptera: Anobiidae]: A destructive borer of house timber (sap wood) and furniture in Australia. Also a pest of hoop and bunya pine timbers; does not usually attack hardwoods, native cypresses, or spotted gum *Eucalyptus maculata*. Adult oval, ca 3 mm long, shining reddish brown; Prothorax narrowed anteriorly and smoothly curved dorsally; antennal club with three segments; body covered with fine Setae and numerous minute punctures; legs tightly folded against body. Egg white, spherical, 0.4 mm diam. Larva soft bodied, curved, wrinkled, white with dark-brown Mandible; mature larva ca 4–5 mm long, body with numerous fine Setae. QPB adult active October to February. Eggs laid in cracks in timber; incubation a few weeks. Larvae burrow long distances in timber and cause most damage. Adults emerge leaving circular 2 mm hole in timber surface. Fine, granular frass in galleries. Life cycle typically 3 years. Susceptible timber, if left untreated, reinvaded until honeycombed and has lost most of its strength.

QUELETOX® See Fenthion.

QUELLE, FERDINAND (1877–1963) (Anon. 1963, Mitt. dt. ent. Ges. 22: 5.)

QUERCI, CLORINDA (1874–1959) (Querci 1959, Entomologist's Rec. J. Var. 2: 215–217.)

QUERCI, ORAZIO (1875–1967) (S.N.A.J. 1968, Entomologist's Rec. J. Var. 80: 87–88.)

QUIBELL, WILLIAM (1877–1959) (Anon. 1960, Entomologist's Rec. J. Var. 72: 79.)

QUICK-PHOS® See AIP.

QUIESCENCE Noun. (Latin, *quiescere* = to become still. Pl., Quiescences.) 1. A condition of interrupted growth due to the direct effect of unfavourable environmental conditions (Shelford 1929, *Laboratory and Field Ecology*). 2. A sudden, short-term, non-cyclical interruption in growth or development of an organism due to one or more environmental factors. See Aestivation; Athermopause; Dormancy; Hibernation. Cf. Diapause; Oligopause.

QUIESCENT Adj. (Latin, *quiescere* = to become still.) 1. Not active. 2. A descriptive term applied to the pupal stage in complete Metamorphosis.

QUIET Adj. Subdued. Archane: Not conspicuous or contrasting in colour or maculation.

QUILIS PEREZ, D. MODESTO (1904–1938) (Sachtleben 1941, Arb. physiol. angew. Ent. Berl. 8: 212.)

QUIMBY, MOSES (1810–1875) (Elwood 1876, Beekeepers' Mag. 4: 145–147.)

QUINAL® See Quinalphos.

QUINALPHOS An organic-phosphate (thiophosphoric acid) compound {O,O-diethyl-O-quinoxalin-2-yl-phosphorothioate} used as a contact insecticide and stomach poison against plant-sucking insects, chewing insects, mites, thrips and other insects. Compound applied to vegetables, citrus, cotton, grapes and ornamentals in many countries. Phytotoxic to some fruit trees. Toxic to bees and fishes to a lesser extent. Not registered for used in USA. Trade names include:

Bayrusil®, Chinalphos®, Ekalux®, Kinalux®, Knave®, Quinatox®, Quinal®, Savall®, Suquin®, Tombel®. See Organophosphorus Insecticide.

QUINATOX® See Quinalphos.

QUINCE Noun. Fruit of the Asian tree *Cydonia oblonga*. Tree low with crooked branches; flowers large, white or yellow. Fruit resembles large, yellow apple but has many seeds in each carpel. Used as dwarf stock for pear; flesh hard, acid, used for marmalade.

QUINCE CURCULIO *Conotrachelus crataegi* Walsh [Coleoptera: Curculionidae]: A pest of hawthorn and quince in eastern USA. Overwinters as larva within soil. Larva causes damage as it bores into fruit. See Curculionidae. Cf. Apple Curculio; Plum Curculio.

QUINCE MOTH *Eusophera bigella* (Zeller) [Lepidoptera: Pyralidae]: A pest of apple, peach and quince around the Mediterranean.

QUINCE TREEHOPPER *Glossonotus crataegi* (Fitch) [Hemiptera: Membracidae].

QUININE Noun. (Spanish, *quina* = cinchona bark.) An alkaloid produced from the bark of *Cinchona* spp. and used for treatment of Malaria.

QUINOMETHIONATE See Chinomethionat.

QUINONES Plural Noun. (Latin, *quinic* = acid; Chemistry, -one.) A group of pigmentary colours that includes Anthraquinones (carminic acid from coccineal) and Aphins. One quinone is a yellow crystalline compound [$CO(CHCH)_2CO$] produced by some tenebrionid beetles. See Pigmentary Colour. Cf. Carotinoids; Flavinoids; Melanins; Ommochromes; Pterines (Pteridines). Rel. Structural Colour.

QUINQUEDENTATE Adj. (Latin, *quinque* = fifth part; *dens* = tooth; *-atus* = adjectival suffix.) 5-toothed.

QUINQUELOCULAR WAX GLAND Pseudococcidae: A Wax Gland composed of only one cell type whose cytoplasm contains a well-developed Endoplasmic Reticulum. QWG are associated with the Stigmatic Furrows, produce hydrophobic wax filaments which extend from the Stigmatic Furrow to the spiracles and protect the passage of air. Cf. Triangular Wax Gland; Tubular Wax Gland; Multilocular Wax Gland.

QUOTIDIAN Noun. Malarial paroxysms recurring every 24 hours. Cf. Tertian; Quertan.

QUOY, JEAN RENÉ CONSTANTIN (1790–1869) (Rochemond 1870, de Ann. Acad. La Rochelle 9: 231–238.)

R Adj. Standard abbreviation for the Radial Vein of the generalized insect wing. Based on the Comstock-Needham nomenclature for insect wing venation.

RAAB, EMMERICH (–1959) (Anon. 1960, Z. wien ent. Ges. 45: 64.)

RABAUD, M ETIENNE (–1956) (Anon. Bull. Soc. ent. Fr. 61: 145.)

RABBANI, MOHAMMAD G (1945–1977) Mosquito cytogeneticist. Born at Faridpur, Bangladesh; died at Manaus, Brazil. Ph. D. University of Illinois 1972. (Anon. 1977, Fla. Ent. 60 (2): 96.)

RABBIT BOT FLY *Cuterebra princeps* [Diptera: Cuterebridae]: A pest of cottontail and jack rabbits in western North America; closely related *C. tenebrosa* attacks pack rats and mice. Adult 1.7–1.8 mm long; body predominantly bluish black; Thorax brownish; venter with white pubescence. Larva develops in 'warbles' under skin of host. See Cuterebridae. Cf. Tropical Warble-Fly.

RABBIT CALICIVIRUS DISEASE Acronym: RCD; also known as Viral Haemorrhagic Disease (VHD) and Rabbit Haemorrhagic Disease (RHD). A newly discovered disease of rabbits with potential use as a biological control agent of introduced or feral rabbit populations. RCD was first observed in China in 1984 and then again in Europe in 1988. RDC appears to be a disease only of rabbits, but possibly crossing from other mammal host Species. RDC is a highly virulent disease in rabbits and appears to be Species specific to the European wild rabbit and its domesticated derivatives. An estimated 30 million rabbits were killed by the disease in Italy. The disease is now endemic in populations of wild rabbits throughout Europe. The Australian government (CSIRO) has been studying the disease for use as a biological control agent against the introduced European rabbit which acts as a pest and has decimated large areas of native grasslands.

RABBIT LOUSE *Haemodipsus ventricosus* (Denny) [Anoplura: Hoplopleuridae].

RABBIT TICK *Haemaphysalis leporispalustris* (Packard) [Acari: Ixodidae].

RABENHORST, GOTTLOB LUDWIG (1806–1881) (Anon. 1881, Leopoldina 17: 102.)

RABON® See Tetrachlorvinphos.

RABOND® See Tetrachlorvinphos.

RACE Noun. (French, *race* = Family. Pl., Races.) An anthropological term used to indicate a population of a Species with constant characters which are not quite specific, usually occurring in a different faunal region from the type and thus geographical; sometimes incorrectly considered synonymous with Subspecies. Cf. Biotype; Variety; Strain.

RACEME Noun. (Latin, *racemus* = a bunch of berries. Pl., Racemes.) A type of inflorescence in which flowers are formed on individual stalks along a main axis or peduncle.

RACEMOSE Adj. (Latin, *racemus* = bunch; *-osus* = full of.) Shaped as or resembling a raceme or bunch of grapes. Alt. Racemous.

RACET® See Acephate.

RACHICERIDAE Plural Noun. See Xylophagidae.

RACHIS Noun. (Greek, diminutive of *rhachion* = a spine. Pl., Rachises.) 1. General: Any axial structure. 2. Hymenoptera: The ridge portion of interdigitating valvifers in the Ovipositor of nonaculeates. Cf. Aulax; Olistheter. 3. A continuation of the leafstalk in some broad-leafed plants. Alt. Rhachis.

RACK, EDMUND (1735–1787) (Rose 1850, *New General Biographical Dictionary* 11: 273.)

RADAR Noun. Acronym for **Ra**dio **D**etecting **a**nd **R**anging. A system used to locate objects by means of radio signals. Signals are emitted from a source in the form of pulses at an ultrahigh frequency. Objects reflect minute signals in response to the pulsed signal. The minute signals are received and analysed at or near the source of the pulsed signals. The range, bearing and other characteristics of the object may be interpreted through the received signal.

RADDATZ, ADOLF (1822–1913) (Bornhöft 1913, Arch. Ver. Freunde Naturg. Mecklenb. 67: [3].)

RADDE, GUSTAV FERDINAND RICHARD (1831–1903) (Adelung 1903, Zool. Zentbl. 10: 829–831.)

RADEMACHER, BERNARD (1901–1973) (Kovacevic 1973, Zast. Bilja 126: 381–382.)

RADEMACHER, PAUL (1843–1906) (Dittrich 1907, Z. Ent. 32: 1.)

RADIAL Adj. (Latin, *radius* = ray; *-alis* = characterized by.) 1. Arranged like rays starting from a common centre. 2. Descriptive of or pertaining to the Radius or Radial Wing-vein.

RADIAL AREA Orthoptera: The space between the Mediastinal (Subcosta) and Radial Veins. See Scapular Area.

RADIAL CELL 1. Any wing cell anteriorly margined by the radius or any of its branches. 2. Hymenoptera: A Marginal Cell.

RADIAL CELLULE An area of wing membrane near the apex, included between the exterior margin and a vein that originates at the Carpus and passes towards the apex. Syn. Marginal Cellule.

RADIAL CROSSVEIN The vein extending between the two principal divisions of Radius, *i.e.* from vein R to vein Rs (Comstock). See Vein; Wing Venation.

RADIAL CUNEATE AREA Some Neuroptera: An expansion of the area of the wing lying between the distal part of vein R and the Media or between branches of vein R (Comstock).

RADIAL PLANATE VEIN Wings of Myrmeleonidae: That which transverses the interradial area; crossing the branches of the Radial Sector in the direction of the wing apex (Needham). See Vein; Wing Venation.

RADIAL SECTOR 1. In general, the lower of the two primary divisions of the Radius (Comstock). 2. Odonata, an indirect branch from the Media, just below and parallel with Media.

RADIAL SPOKE A component of Accessory Doublets in the Sperm Flagellum, first described by Afzelius (1959). RS are short rod-like structures that project perpendicularly from the proximal surface of Subfibre A toward the Central Sheath (which envelopes the Central Tubules). Groups of Spokes demonstrate a spatial periodicity and wind around the Axoneme in helical form. Each Spoke consists of two parts: A Spoke Shaft (composed of six polypeptides) and a Spoke Head (enlargement of the spoke near the Central Sheath). Spokes occur in groups of 2–3, separated by a distance of about 96 nm. Interactions between Radial Spokes and Central Tubules in part regulate flagellar movement. See Axoneme; Microtubule. Cf. Dynein Arm; Nexin. Rel. Sperm Flagellum.

RADIAL SYMMETRY A form of symmetry which represents the correspondence of size, shape, and proportion around the central axis of a body. An imaginary plane divides the body into similar parts. Examples include coelenterates and starfish. See Asymmetry; Symmetry. Cf. Bilateral Symmetry; Pentameral Symmetry; Spherical Symmetry; Zonal Symmetry.

RADIAL VEIN 1. Hemiptera: The first important vein positioned between the Costa and Ulnar. 2. Orthoptera: The Radius of Comstock. 3. Diptera: The second longitudinal vein of Meigen; Radius of Comstock. See Vein; Wing Venation.

RADIALIS Noun. The largest and most important Pteralia of the Rotaxis, formed by the fusion of the Subcosta, the Radius and the Media (MacGillivray).

RADIALLY Adv. In the form or manner of radii or rays from a common centre.

RADIATE Adj. (Latin, *radiare* = to meit rays; *-atus* = adjectival suffix.) Seeming to emit or throw out rays, as a spot, marked with lines proceeding from a common centre. Alt. Radiated; Radiatus.

RADIATE CELL See Plasmatocyte.

RADIATE VEINS The longitudinal veins spreading fan-like in the anal field of the hindwings. Syn. Anal Veins. See Vein; Wing Venation.

RADICAL Adj. (Latin, *radix* = root; *-alis* = characterized by.) The part of a molecule that divides or splits off during chemical reactions, forming new molecules of different substances.

RADICLE Noun. (Latin, *radix* = root.) The basal portion of the Antenna which articulates with the head. Hymenoptera: See Scape. Alt. Radicula See Bulbus. Cf. Antartis.

RADICOLA Noun. (Latin, *radix* = root; *cola* = inhabitant. Pl., Radicolae.) *Phylloxera*: The stage or individual that forms root-gall. See *Phylloxera*.

RADIOMEDIAL CROSSVEIN A vein of the insect wing extending from Radius to Media, when in typical position from vein R4+5 to vein M1+2 (Comstock).

RADISH MAGGOT See Cabbage Maggot.

RADIUS Noun. (Latin, *radius* = staff, rod, ray. Pl., Radii; Radiuses.) 1. Comstock-Needham system of horismology: The 3rd longitudinal vein caudad of the Subcosta. Origin at the wingbase and dividing into not more than five branches. 2. A single subdivision of a digitate wing, as in Pterophoridae (Kirby & Spence). See Vein; Wing Venation. 3. Radial symmetry: An imaginary plane dividing the body into similar parts. See Symmetry.

RADIX Noun. (Latin, *radix* = root. Pl., Radices.) 1. The base or origin of any structure. 2. Base of the wings, and their point of insertion. See Pteropega. 3. Hymenoptera: Basal portion of gonopod.

RADIX FORCIPITIS Hymenoptera: See Gonocoxite.

RADOSZKOWSKI, OCTAVII IVANOVICH BURMEISTER (1820–1895) (Portchensky 1896, Trudy russk. ent. Obshch. 1896: i–vi, bibliogr.)

RADULA See Raster.

RAEBEL, HERMANN (1878–1963) (Pasternak 1963, Z. wien ent. Ges. 48: 158–159.)

RAETZER, AUGUST (1845–1907) (Anon. 1911, Mitt. naturf. Ges. Solothurn 16: 144–148, bibliogr.)

RAFF, JANET WATSON (1885–1973) (Lees 1975, Proc. R. ent. Soc. Lond. (C) 39: 57.)

RAFFRAY, ACHILLE (1844–1923) (Luigioni 1923, Boll. Soc. ent. ital. 55: 153–155.)

RAFINESOUE, CONSTANTIN SAMUEL (1783–1840) (Brendel 1879, Am. Nat. 13: 764–765; Fitzpatrick 1911, Raffinesque: A sketch of his life. 241 pp. bibliogr., Iowa.)

RAFT Noun. (Old Norse, *raptr.*) The massed floating eggs of certain Culicidae.

RAGADAN® See Heptenophos.

RAGNOW, HERMANN (1862–1938) (Anon. 1938, Arb. morph. taxon. Ent. Berl. 5: 296.)

RAGONOT, EMILE LOUIS (1843–1895) (McLachlan 1895, Entomol. mon. Mag. 31: 287; Constant 1896, Ann. Soc. ent. Fr. 65: 1–18, bibliogr.)

RAGUSA, EMILE ENRICO (1849–1924) (Turati 1925, Comment. Soc. ital. Sci. nat. 1925: 8–11.)

RAGWEED BORER *Epiblema strenuana* (Walker) [Lepidoptera: Tortricidae]: In North America, larvae bore into ragweed.

RAGWEED PLANT BUG *Chlamydatus associatus* (Uhler) [Hemiptera: Miridae].

RAGWORT FLEA-BEETLE *Longitarsus jacobaeae* (Waterhouse) [Coleoptera: Chrysomelidae]: A beetle imported into western USA ca 1969 for biological control of tansy ragwort, *Senecio jacobaea* Linnaeus. RFB adults actively feed on foliage during autumn, winter and spring; feed by rasping which causes small circular holes. Larvae bore and feed within roots and leaf petioles during same periods. Adults emerge during early summer; females aestivate and oviposit during late summer. Rel. Cinnabar Moth.

RAILLIET, ALCIDE (1852–1930) (Anon. 1931, J. Parasit. 17: 166.)

RAIMONDI, ANTONIO (1826–1890) (Macedo 1940, Boln. Mus. Hist. nat. Javier Prado 4: 431–443.)

RAINBOW, WILLIAM JOSEPH (1856–1919) (Musgrave 1920, Rec. Aust. Mus. 13: 87–91, bibliogr.)

RAIN BEETLE *Plecoma* spp. [Coleoptera: Scarabaeidae].

RAISIN MOTH *Cadra figulilella* (Gregson) [Lepidoptera: Pyralidae]: A widespread pest of dried fruits and meals. Forewing pale yellow-grey; wingspan 15–20 mm; hindwings whitish. See Pyralidae. Syn. *Ephestia figulilella* Gregson.

RAIT-SMITH, WILLIAM (1875–1958) (Richards 1959, Proc. R. ent. Soc. Lond. (C) 23: 72.)

RAK-5® See Isomate-M.

RAKE Noun. (Anglo Saxon, *raca* = rake. Pl., Rakes.) 1. Any structure consisting of parallel projections. 2. Hymenoptera (Apoidea): Spines along the apex of the hind Tibia used to groom pollen from the contralateral hind leg. Cf Comb.

RALPH KNIFE A long-edged glass knife used in ultramicrotomy and histology (Bennet *et al.* 1976, Stain Technol. 51: 71.) Named in honour of Paul Ralph, who invented the technique of preparing the knife. Cf. Latta & Hartman Knife

RAMACHANDRA RAO, R B Y (1866–1972) (Anon. 1972, Entomologist's Newsl. 2(8).

RAMAKRISHNA AYYAR, TARAKAD VAIDYANATHA (1880–1952) (Mani & Rao 1952, Indian J. Ent. 14: 187–190.)

RAMAL Adj. (Latin, *ramus* = branch; *-alis* = pertaining to.) Pertaining to structure that branches or is branch-like.

RAMBOUSEK, FARANTISEK G (1886–1931) (Stanak 1931, Okhr. Rostlin 11: 145–150, bibliogr.)

RAMBRING, HELGE (1916–1968) (Henriksen 1970, Lepidoptera Kbh. 1: 149–151.)

RAMBUR, JULES PIERRE (1801–1870) (Wallace 1871, Proc. ent. Soc. Lond. 1871: liii; Grasslin 1872, Ann. Soc. ent. Fr. (5)2: 297–312, bibliogr.)

RAMELLUS Noun. Forewings of Ichneumonoidea: The distal stump of the Medial Vein when it is otherwise incomplete (Tillyard).

RAMI OF FIRST VALVULA Hymenoptera: Basal portion of first Valvula.

RAMI OF SECOND VALVULA Hymenoptera: Basal portion of second Valvula.

RAMI VALVULARUM The proximal, often slender, parts of the first and second Valvulae by which the latter are attached to the Valvifers (Snodgrass).

RAMIFICATION Noun. (Latin, *ramus* = branch; English, *-tion* = result of an action.) A branching pattern in several directions.

RAMON Y CAJAL, SANTIAGO (1852–1934) (Porter 1934, Revta. chil. Hist. nat. 38: 249–274.)

RAMOSE Adj. (Latin, *ramus* = branch; *-osus* = full of. Pl., Rami.) Branched, or having long branches. Alt. Ramosus; Ramous.

RAMPART® See Carbofuran; See Phorate; See Terbufos.

RAMSDEN, HILDEBRAND (–1899) (Anon. 1940, Proc. ent. Soc. Lond. 1899: xxxvii.)

RAMUS Noun. (Latin, *ramus* = branch. Pl., Rami.) 1. A branch-like division of any structure or appendage. Collembola: Either one of the free ends of the minute, paired appendages of the 3rd abdominal segment. Lepidoptera: Paired, lateral process of Sternum VIII. Hymenoptera: See Gonocoxite.

RANDAL® See Fenpropathrin.

RANGA RAO, P V (1916–1972) (Dharmaraju 1973, Indian J. Ent. 34: 189)

RANGE CATERPILLAR *Hemileuca oliviae* Cockerell [Lepidoptera: Saturniidae].

RANGE CRANE-FLY *Tipula simplex* Doane [Diptera: Tipulidae]: A Species known from west coast of USA. Adult female 10–12 mm long brachypterous, greyish-brown; male 8–10 mm long, macropterous; legs rather short. Eggs elongate, black, deposited in soil. Larvae active January to March; reside in holes in soil, emerge during night to feed on grass and grain. Pupation occurs in soil. See Tipulidae. Cf. Eastern Crane Fly; European Crane Fly.

RANGGU Common Name for Philippine Neem Tree in Sarawak.

RANGNOW, RUDOLF (1889–1939) (Sachtleben 1940, Arb. morph. taxon. Ent. Berl. 7: 76.)

RANSOM, BRAYTON HOWARD (1879–1925) (Hall *et al.* 1925, Proc. ent. Soc. Wash. 27: 153; Hall *et al.* 1925, Am. J. trop. Med. 5: 389–392.)

RANSOM, EDWARD (–1946) (Anon. 1947, Trans. Suffolk Nat. Soc. 6: ixv.)

RANSOME, ALGERNON LEE (1884–1969) (R.W.W. 1972, Entomologist's Rec. J. Var. 84:56.)

RANTZAU, CARL (1822–1848) (Henriksen 1926, Ent. Meddr 15: 193.)

RAO, SCHRI RAMCHANDRA (1885–1972) (Anon. 1972, Acrida I (4): i–iii)

RAO, V. TIRUMALARAO (–1970) (Hinton 1971, Proc. R. ent. Soc. Lond. (C) 35: 54.)

RAPACIOUS. Adj. (Latin, *rapax* from *repere* = to seize and carry off.) Predatory, capturing and eating prey. See Feeding Strategy.

RAPD PCR Randomly Amplified Polymorphic DNA Polymerase Chain Reaction.

RAPHE Noun. (Greek, *rhaphe* = seam.) Lepidoptera larva: The median sclerotic bar of the dorsal wall of the silk-press of the silk-spinning apparatus.

RAPHIDIIDAE Plural Noun. Snakeflies. A small Family of Rhaphidoidea found in Holarctic and Central America. Adults with Ocelli; front legs not raptorial with Coxa at posterior end of Prothorax; forewing Stigma with cross vein at proximal margin; Hindwing with Cu2 and A partially fused. Life cycle up to 3 years, apparently varies within Species. Some Species apparently gregarious as immatures.

RAPHIDIOPTERA See Raphidoidea.

RAPHIDOIDEA Burmeister 1839. (Greek, *rhaphis* = needle; *eidos* = form.) Snake-flies; Camelneck-flies. (Lower Permian-Recent.) Primitive

Endopterygota; sistergroup Megaloptera. About 200 nominal Species; widespread in Holarctic; not found in southern Hemisphere. Moderate sized; head prognathous, elongate, dorsoventrally compressed; compound eyes bulging laterally; Ocelli present or absent; Mandible with 3–4 teeth; Maxillary Palpus with five segments; Labial Palpus with three segments; Hypopharnyx with sclerotized holdfast; Crop with large diverticulum. Prothorax extremely elongate, fore Coxae articulate at posterior margin; Mesothorax, Metathorax similar in size, shape, each with a pair of spiracles; wings hyaline, subequal in size, tectiform at rest; Pterostigma pigmented; legs cursorial, tarsal formula 5-5-5. Abdomen with 10 visible segments; spiracles on segments 1–8. Nervous system with three thoracic ganglia, eight abdominal ganglia; six Malpighian Tubules. Larva elongate, dorsoventrally compressed; head with several Stemmata; Abdomen with 10 segments. Pupa exarate; decticous. Restricted to forest habitats; larvae in leaf litter, under bark; adults diurnal. Larva and adult predaceous on soft-bodied arthropods; prolonged courtship, male assumes inferior position during copulation; female lays ca 800 eggs during life in batches of ca 100 cemeted to substrate; life cycle typically two years, sometimes one or three years. Included Families: Inocelliidae and Raphidiidae. Syn. Raphidioptera; Rhaphidiodea.

RAPID® See Primicarb.

RAPID PLANT-BUG *Adelphocoris rapidus* (Say) [Hemiptera: Miridae].

RAPISMATIDAE Nevas 1929. Plural Noun. A small, rare Family of large bodied Neuroptera known from mountainous regions of India, Nepal, Burma, Thailand, Malaysia and Borneo. Superficially resembling Ithonidae. Syn. Rhapismatidae.

RAPTORIA Noun. Orthoptera in which the anterior legs are fitted for grasping, *e.g.* Mantidae.

RAPTORIAL Adj. (Latin, *raptor* = robber; *-alis* = characterized by.) Adapted for seizing prey, predaceous; descriptive of legs with opposable spines or elongate protuberances (on Femur and Tibia, Tibia and Tarsus) that are adapted for impaling prey. Examples include the forelegs of Mantidae. Alt. Raptatory; Raptorious. See Leg; Raptor. Cf. Cursorial; Fossorial; Gressorial; Natatorial; Saltatorial; Scansorial. Rel. Locomotion.

RARITAN FORMATION A stratum of Upper Cretaceous Period (Cenomanian) age (ca 94 MYPB) found in New Jersey which bears fossil insects preserved in amber. See Amber.

RASCHKE, JOHANN GOTTFRIED (1763–1815) (Fischer 1817, Magazin Ent. (Germar) 2: 343–344.)

RASMUSSEN, JOHANNES (1869–1937) (Jacobsen 1938, Ent. Meddr. 20: 107.)

RASORIAL Adj. (Latin, *radere* = to scratch, shave; *-alis* = characterized by.) 1. Descriptive of structures or appendages formed or adapted for scratching; term typically applied to leg modifications. 2. Scratching on the ground in search of food.

RASP Noun. (Middle English, *raspe* = rasp. Pl., Rasps.) 1. A type of rough file typically consisting of a series of teeth, denticles, acanthae or spines. Cf. Strigil. 2. A roughened surface adapted for the production of sound by scraping or rubbing against a moveable part such as a membrane or opposable Strigil. See Sound Production. Cf. Plectrum; Strigil.

RASPAIL, XAVIER (1840–1926) (Petit 1927, Bull. Soc. zool. Fr. 52: 33)

RASPBERRY BUD-MOTH *Lampronia rubiella* (Bjerkander) [Lepidoptera: Incurvariidae].

RASPBERRY CANE-BORER *Oberea bimaculata* (Oliver) [Coleoptera: Cerambycidae].

RASPBERRY CANE-MAGGOT *Pegomya rubivora* (Coquillett) [Diptera: Anthomyiidae]: A pest of raspberry, blackberry and rose in USA. Eggs deposited on leaf axils. Larvae girdle tips and bore into new shoots. See Anthomyiidae.

RASPBERRY CROWN-BORER *Pennisetia marginata* (Harris) [Lepidoptera: Sesiidae]: A pest of many Species of berries in Canada and the northern USA. Larva attacks and mines the crowns and lower parts of canes. Syn. *Bembecia marginata.* See Sesiidae. Cf. Strawberry Crown-Moth; Red-Necked Cane Borer.

RASPBERRY FRUITWORM *Byturus unicolor* Say [Coleoptera: Byturidae].

RASPBERRY LEAFROLLER *Olethreutes permundana* (Clemens) [Lepidoptera: Tortricidae].

RASPBERRY SAWFLY *Monophadnoides geniculatus* (Hartig) [Hymenoptera: Tenthredinidae]: A pest of blackberry, raspberry and loganberry in North America. Eggs inserted in leaf tissue. Larva pale green, spiny; feeds on underside of leaves making small holes.

RASPING (Old French, *rasper* > French, *raper* = to scrape.) 1. Scraping or rubbing with a rasp. 2. A grating or scraping sound. Cf. Abrasive.

RASTER Noun. (Greek, a screen. Latin, *raster* > *rastrum* = rake. Pl., Rastri; Rasters.) Scarabaeoid larvae: A complex of specifically arranged bare areas, Setae and spines on ventral surface of last abdominal segment, anterior of Anus; The Raster is divided into Septula, Palidium, Teges, Tegillum and Campus. Radula of Hayes (Boving).

RASTRA Noun. (Latin, *raster, rastrum* = rake.) Row of Setae located on or near the margin of the Lacina or Galea (MacGillivray).

RASTRATE Adj. (Latin, *raster, rastrum* = rake; *atus* = adjectival suffix.) Pertaining to or descriptive of surfaces inscribed with longitudinal scratches. Alt. Rastratus.

RASTROCOCCUS MEALYBUG *Rastrococcus truncatispinus* Williams [Hemiptera: Pseudococcidae]: A pest of citrus, fig, mango, oleander and eucalyptus in eastern Australia. Adult female oval in outline shape, 5–6 mm long, body coloured by longitudinal stripes of grey, red and orange; long filaments of white wax radiate from margin of body with longest filaments at anterior and posterior ends of body. Male considerably smaller than female, fragile, with one pair of functional wings. Female oviposits batches of 30 eggs into egg sac; eclosion occurs within 1–2 weeks; female passes through three moults; male passes through four moults; life cycle requires ca 6 weeks, depending upon season. RMB produces large amounts of honeydew which serve as substrate for sooty mould and other fungi.

RAT FLEAS Any of several Species of *Xenopsylla* [Siphonaptera: Pulicidae], including *X. vexabilis* Jordan (Australia), *X. brasiliensis* (Africa, South America, India) and *X. astia* (southeast Asia), and *X. cheopis* (cosmopolitan.) All Species capable of vectoring Bubonic Plague to humans. See Oriental Rat Flea; Plague.

RAT TYPHUS See Murine Typhus.

RATARDIDAE Plural Noun. A small Family of ditrysian Lepidioptera assigned to Superfamily Bombycoidea.

RATH'S ORGAN See Organ of vom Rath.

RATHJENS, CAR (1887–1966) (Weidner 1967, Abh. Verh. naturw. Ver. Hamburg Suppl. 9: 357.)

RATHKE, JENS (1769–1855) (Henriksen 1925, Ent. Meddr 15: 152–153; Natvig 1944, Norsk ent. Tidsskr. 7: 4–5.)

RATHKE, MARTIN HEINRICH (1793–1860) (Zaddach 1860, Neue Preuss. Provinz. Blätt. (3) 6: 271–312, bibliogr.)

RATHVON, SIMON SNYDER (1812–1891) (Calvert 1930, Ent. News 41: 234–236.)

RATIONAL BEHAVIOR A form of behaviour in which actions are influenced by memory of previous experiences.

RATOON Noun. (Spanish, *retoño*. Pl., Ratoons.) A process of cutting back mature plants to stimulate them to produce a second crop from the same plant.

RATOON SHOOTBORER *Ephysteris promptella* (Staudinger) [Lepidoptera: Gelechiidae]: A pest of sugarcane in Australia, Asia, Africa and the Mediterranean region. Eggs pale green-yellow with iridescent blue, gold and pink; laid on cane stalk. Larvae bore into basal part of stalk or young shoot. Pupation on soil or debris within weak cocoon of silk covered with excrement or debris. See Gelechiidae.

RAT-TAILED MAGGOT Larvae of Syrphidae whose body terminates in a long, flexible, respiratory tube. Syn. Rat-Tailed Larva. See Drone Fly.

RATZEBURG, JULIUS THEODOR CHRISFFAN (1801–1871) Born in Berlin; father was Professor of Botany. Julius awarded M.D. but devoted himself to study of Forest Entomology as member of Forest Academy at Eberswalde. Published several volumes on forest insects (1837, 1839, 1840, 1844); *Die Ichneumonen der Forstinsekten* stands as a comprehensive treatment of ichenumonids. (Wallace 1871, Proc. ent. Soc. Lond. 1871: lii–liii; Danckelmann 1872, Forst. u-Jagdwesen 41: 307–323, bibliogr.)

RAU, PHILIP (1885–1948) American naturalist and student of aculeate wasp behaviour. (Meinert 1948. Lepid. News 2: 62.)

RAUSCHER, HERBERT (1910–1973) (Leib 1973, Anz. Schädlingsk. Pflanzen-Umwelt 46: 89–90.)

RAVION® See Carbaryl.

RAW, FRANK (1919–1967) (Kennedy 1968, Proc. R. ent. Soc. Lond. (C) 32: 59.)

RAY, JOHN (1627–1705) English natural historian and systematist. Educated at Trinity College and subsequently served as Lecturer in Greek (1651), Lecturer in Mathematics (1653), Humanity Reader (1655) and Junior Dean (1658). Ray undertook extensive collecting trips in England, Scotland and Wales; collected in continental Europe with Francis Willughby 1663–1666. Best known for his taxonomic work which anticipates Linnaeus. Ray was author of *Catalogus Plantarum Anglicae* (1670); *Methodus Plantarum Nova* (1682), *Historia Plantarum* (1686–1704), *Methodus Insectorum, seu Insecta in methodum aliqualem digesta* (1705) and the posthumous *Historia Insectorum* (1710). Ray Society founded in London for publication of scientific work (1844). Formerly spelled Wray. (Duméril 1823, *Considérations générale sur la classe des insectes*. 272 pp. (247), Paris; Derham 1844, *Memorials of Ray*. 220 pp., bibliogr.)

RAY, JULES (1815–1883) (Jourdheuille 1883, Ann. Soc. ent. Fr. (6) 3: 565–569.)

RAYMENT, PERCY TARLTON (1882–1969) English born, emigrated to Australia and became amateur Entomologist publishing extensively on bees of Australia.

RAYNOR, GILBERT HENRY (1854–1929) (Burrows 1929, Entomologist's Rec. J. Var. 41: 139–140.)

RAYWARD, ARTHUR LESLIE (1866–1935) (Sheldon 1935, Entomologist 68: 292.)

RAZZAUTI, ALBERTO (1885–1972) (Barsotti 1972, Memorie Soc. ent. ital. 51: 88–90, bibliogr.)

RCD See Rabbit Calicivirus Disease.

REA, GEORGE HAROLD (1880–1964) (Anderson 1958, Pa Beekeeper 33: 18–20.)

REACTION PLUG *Drosophila*: Contents of the Vagina formed after coitus. In intraspecific crosses the plug remains soft, unpigmented and disappears; in interspecific crosses the plug hardens, melanizes and persists, thereby preventing oviposition. Syn. Reaction Mass.

REACTION Noun. (French.) A response to a stimulus.

READ, GRANTLEY DICK (1890–1959) (Uvarov 1960, Proc. R. ent. Soc. Lond. (C) 24: 54.)

REAKIRT, TRYON (1844–) (Essig 1931, *History of Entomology*, 1029 pp. (737), New York.)

REALM Noun. (Middle English, *realme*. Pl., Realms.) Biogeography: A primary marine or terrestrial division of area and consisting of one or more regions. See Biogeography. Cf. Ethiopian Realm; Holarctic Realm; Neotropical Realm; Oriental Realm; Oceanic Realm; Palaearctic

Realm.

RÉAUMUR, RÉNÉ ANTOINE FERCHAULT DE (1683–1757) French chemist, natural historian and perhaps the most notable entomological observer of the 18th century. Elected member of French Academy (1708); developed porcelain; established temperature scale. Published *Memoires pour servir a l'Histoire des Insectes* (6 vols, 1734–1742.) (Dumeril 1823, *Considérations générale sur la classe des insectes.* 272 pp. (247), Paris.)

REBEL, HANS (1861–1940) (Anon. 1931, Ann. naturh. Mus. Wien 45: i–v, bibliogr.)

REBELATE® See Dimethoate.

REBILLARD, PIERRE (1900–1974) (Viette 1976, Bull. Soc. ent. Fr. 80: 293.)

RECENT EPOCH (10,000 YBP–Present.) The interval of the Geological Time Scale (5 MYBP–1.65 MYBP) that represents the fourth Epoch within the Neogene Period. RE was characterized by Charles Lyell (1833). See Geological Time Scale; Neogene Period. Cf. Miocene Epoch; Pliocene Epoch; Pleistocene Epoch. Rel. Fossil.

RECEPTACULA OVORUM The receptacle holding eggs in the female insect.

RECEPTACULUM SEMINIS The Spermatheca.

RECEPTIVE APPARATUS The part of a sense organ primarily responsive to the stimulus transmitted by or through the peripheral parts, formed of the sense cell or cells (Snodgrass).

RECEPTOR Noun. (Latin. Pl., Receptors.) A so-called sense organ, or specialized structure of the Integument responsive to external stimuli (Snodgrass).

RECIPROCAL DENSITY-DEPENDENT MORTALITY A biotic agent whose actions induce change in another organism's population size and through such action causes change in that biotic agent's population size.

RECLAIRE, AUGUST (1881–1949) (Van der Wiel 1949, Ent. Ber., Amst. 12: 409–413, bibliogr.)

RECLINATE Adj. (Latin, *reclinare* from *re-*; *clinare* = lean; *-atus* = adjectival suffix.) Reflexed; directed backward, *e.g.,* the bristles in Diptera. Alt. Reclinatus

RECLIVATE Adj. (Latin, *re-* = back; *clinare* = to lean; *-atus* = adjectival suffix.) Curved into a convex, then into a concave line. Alt. Reclivatus.

RECOGNITION SPECIES CONCEPT A Species concept based upon reproductive mechanisms that facilitate gene exchange, or a field for gene recombination. See Species Concepts.

RECOMBINANT DNA A hybrid protein produced from a foreign DNA sequence that has been introduced into a cell. Alt. Recombinant Product. Rel. Genetic Engineering.

RECONDITE Adj. (Latin, *recondere* = to put up again; *-ites* = inhabitant.) Concealed, *e.g.,* as the Sting within the Abdomen. Alt. Reconditus.

RECRUITMENT Adj. (French, *recruter* from Latin, *re-* = again, *crescere* = to increase; *-ment*, result of an action. Pl., Recruitments.) Social Insects: A form of assembly in which members of a colony are directed to a place where work is required. See Social; Social Insects. Cf. Aggregation; Communal.

RECRUITMENT TRAIL An odour trail made by workers to recruit nestmates to a place with food or where work is needed. Cf. Exploratory Trail.

RECTACUTA Noun. A sclerite in the Coria that connects the head and Mandible (MacGillivray).

RECTAL Adj. (Latin, *rectum* = straight; *-alis* = pertaining to.) Pertaining to the Rectum.

RECTAL CAECUM See Rectal Sac.

RECTAL CAUDA Hemiptera: The terminal tubular process (tail) that terminates the Abdomen of some males (Smith).

RECTAL GILL A type of Tracheal Gill that occurs in the naiad of Anisoptera. RGs occur within the anterior part of the Rectum (Brachial Basket). Anatomically RGs are variable in shape but consist of six longitudinal folds supported by transverse cross-folds. Attachments of the RG are taxonomically important. See Gill. Cf. Caudal Gills; Lateral Abdominal Gill. Rel. Cutaneous Respiration.

RECTAL GLANDS Appendages or thickenings of the Rectum which secrete a lubricating material.

RECTAL MYIASIS A type of Myiasis in which fly larvae invade the intestine via the Anus. Larvae are excrement feeders and complete their development in the Rectum of the host. Common aetiological agents include Drone Fly (Rat-tailed Maggot); False Stable Fly; Latrine Fly; Lesser House Fly. See Myiasis.

RECTAL SAC The enlarged anterior part of the Rectum, sometimes produced into a large rectal Caecum (Snodgrass).

RECTAL TRACHEAL GILLS Lamelliform structures in the Rectum of the nymphs of some Odonata, supplied with Trachea and Tracheoles and serving as respiratory organs.

RECTAL VALVE A circular or lobate fold of the proctodeal wall between the anterior intestine and the Rectum (Snodgrass).

RECTANGULAR Adj. (Latin, *rectus* = right; *angulus* = angle.) In the form of a right angle or rectangle. Alt. Rectangulate; Rectangulatus.

RECTATE Adj. (Latin, *rectus* = straight; *-atus* = adjectival suffix.) Straight.

RECTIGRADE Straight-walking. A term applied to larvae with 16 legs and which walk with a straight body.

RECTILINEAR Adj. (Latin, *rectus* = straight; *linea* = line.) In the form of a straight line.

RECTOTENDON Noun. The tendon to which retractor muscles are attached (MacGillivray).

RECTUM Noun. (Latin, *rectus* = straight.) The posterior part of the Hindgut; term often applied to entire posterior intestine or used as equivalent to the Cloaca. Anatomy of Rectum varies. Collembola: Rectum a flat layer of epithelium cells; Blattaria: rectal pads extend length of Rectum; Orthoptera: A layer of epithelium cells with an additional layer of basal cells embedded in basal surface of rectal pads; Hemiptera: Epithelium cells unspecialized with at least one area which is thickened. Rectal anatomy in Holometabola often more complex. Hymenoptera and some Lepidoptera: rectal pads small, round and numerous. Rectum can contain a layer of columnar epithelium cells and a second layer of smaller cells. Sometimes layers separated by a lumen. Diptera: Columnar epithelium cells arranged in a cone or papilla. Saltwater and freshwater mosquitoes reflect different complexities of rectal epithelium; anterior portion called the Rectal Sac; junction with diverticula frequently forms Rectal Caecae and posterior portion called the Rectum. The Rectum modifies fluid manufactured by Malpighian Tubules and resorbs water, ions, amino acids and other metabolites. Rectum appears to differ from other absorptive epithelia in that it continues to absorb water when Hindgut lumen contains no transportable solute. See Excretion; Hindgut. Cf. Colon; Ileum.

RECTUS Noun. (Latin, *rectus* = straight.) Straight.

RECUMBENT Adj. (Latin, *recumbere* = leaning, lying down.) Lying down; reclining. Applied to Setae which lay on the surface of the body.

RECURRENT Adj. (Latin, *re-* ; *curere* = to run.) Running backward.

RECURRENT NERVE The Median Stomodaeal Nerve extending posteriorly from the Frontal Ganglion (Snodgrass); the Stomogastric Nerve (Wardle).

RECURRENT NERVURE Hymenoptera: Medial Crossvein of Comstock from the point of branching to the junction.

RECURRENT VEIN 1. Many Neuroptera: The Humeral Crossvein when it curves back toward the base of the wing and bears branches (Comstock). 2. Hymenoptera: Medio-cubital Crossvein. See Vein; Wing Venation.

RECURVED Adj. (Latin, *recurvo* = bent.) Bowed or bent downward or backward or outward. Alt. Recurvate; Recurvatus; Recurvus.

RECUSPINE Adj. With points directed posteriad.

RED ADMIRAL *Vanessa atalanta rubria* (Fruhstorfer) [Lepidoptera: Nymphalidae].

RED-AND-BLACK FLAT MITE *Brevipalpus phoenicis* (Geijskes) [Acari: Tenuipalpidae].

RED-AND-BLACK SPIDERS See Nicodamidae.

RED-AND-BLUE BEETLE *Dicranolaius bellulus* (Guérin-Méneville) [Coleoptera: Melyridae]: An Australian Species. Adult 3–6 mm long, elongate and orange-brown with metallic green-blue bands at Elytron base and apex; 3rd antennal segment enlarged in male; body covered in fine Setae. Eggs laid in soil. Larva elongate, slightly flattened and lightly sclerotised with red or pink coloration; larvae live in soil. Pupation occurs in soil. Adults and larvae are predators. Adults occur in flowers feeding on pollen or actively searching for eggs and small caterpillars. Larvae found in soil feeding on larval insects.

RED ASSASSIN BUG *Haematoloecha rubescens* Distant [Hemiptera: Reduviidae].

RED-BACKED CUTWORM *Euxoa ochrogaster* (Guenée) [Lepidoptera: Noctuidae]. See Cutworms.

RED-BACKED OEDEMERID *Eobia bicolor* (Fairmaire) [Coleoptera: Oedomeridae].

RED-BANDED BLISTER BEETLE *Mylabris ligata* Mars. [Coleoptera: Meloidae].

RED-BANDED LEAFROLLER *Argyrotaenia velutinana* (Walker) (Lepidoptera: Tortricidae): A pest of forest trees, orchard trees and fruit-bearing shrubs. RBL sometimes a significant pest of apples when the larval stage skeletonizes underside of leaves near midrib, and binds leaves together with silk. RBL may attack fruit when larvae feed in large numbers. Overwinters as Pupa in leaf litter on ground. Adults appear as trees begin to bud. Eggs deposited in masses (ca 50) on underside of branches; eggs hatch and larvae feed, pupate and produce second generation appears during August. Third generation occurs in warmer climates. Cf. Fruit-Tree Leafroller.

RED-BANDED SHIELD BUG *Piezodorus hyloneri* (Gmelin) [Hemiptera: Pentatomidae]: In Australia, a pest of grasses, beans and lucerne. Adult 9–12 mm long and elongate-oval; green or yellow with a prominent pink or red transverse band across Pronotum. Mouthparts extend past the middle Coxa. No information is available on the egg stage. Nymphs resemble adult but lack wings. Feed on fruit and seeds.

RED-BANDED THRIPS *Selenothrips rubrocinctus* (Giard) [Thysanoptera: Thripidae]. See Cacao Thrips.

RED BERRY-MITE *Acalitus essigi* (Hassan) [Acari: Eriophyidae].

RED BUGS See Pyrrhocoridae.

RED-BUG LEAF-FOLDER *Fascista cercerisella* (Chambers) [Lepidoptera: Gelechiidae].

RED CARPENTER-ANT *Camponotus ferrugineus* (Fabricius) [Hymenoptera: Formicidae]: Endemic to North America; similar in habits and distribution to Black Carpenter Ant; Florida Carpenter Ant. See Formicidae. Cf. Black Carpenter-Ant; Brown Carpenter-Ant; Florida Carpenter-Ant.

RED-CEDAR TIP MOTH *Hypsipyla robusta* (Moore) [Lepidoptera: Pyralidae] an important pest of red cedar (*Toona ciliata* M. Roem.) in Australia. RCTM larvae damage fruits of mature trees and shoots of young trees. Adult life ca 6 days; females mate once and lay ca 450 eggs; eggs often laid on leaf veins near junction of leaflet and leaf rachis; eggs laid individually on young trees or in small clusters on mature trees; 5–6 larval instars; larvae move down trunk to pupate at base of tree or in litter.

RED-CLOVER SEED-WEEVIL *Tychius stephensi* Schnherr [Coleoptera: Curculionidae].

RED COFFEE STEM-BORER *Zeuzera coffeae* Nietner [Lepidoptera: Cossidae].

RED COTTON-BUG. *Dysdercus koenigii*, *Dysdercus cingulatus* [Hemiptera: Pyrrhocoridae].

RED DATE-SCALE *Phoenicococcus marlatti* Cockerell [Hemiptera: Phoenicococcidae].

RED-ELM BARK-WEEVIL *Magdalis armicollis* (Say) [Coleoptera: Curculionidae]: A pest of elm in eastern North America. Larvae tunnel under bark; adults feed on leaves, particularly on upper branches; can transmit Dutch Elm Disease. Cf. Black Elm-Bark Weevil.

RED FLOUR-BEETLE *Tribolium castaneum* (Herbst) [Coleoptera: Tenebrionidae]: Cosmopolitan distribution; endemic to Indo-australian region. RFB an urban pest of milled grain and cereal products; will feed on numerous other foods and leather; often taken in packaged foods. RFB cause flour to acquire objectionable odour and disagreeable taste. RFB probably derived from bark-inhabiting and fungus-feeding ancestor. Larva and adult cannibalistic or facultative predators of egg and pupal stage. Eggs laid directly in food; 7–8 larval instars; mature larvae emerge from food and pupate exposed (not enclosed in cocoon or pupation chamber.) Adults with well developed wings and will fly; typically seen walking on food. Development influenced by type of food, temperature and humidity: Development, biology and anatomy similar to Confused Flour Beetle but RFB adult Antenna with well defined club of three segments. Cf. Confused Flour Beetle.

RED GRASSHOPPER-MITE *Eutrombidium trigonum* (Hermann) [Acari: Trombidiidae].

RED HARVESTER-ANT *Pogonomyrmex barbatus* (F. Smith) [Hymenoptera: Formicidae]: A Species of ant endemic in the southwestern USA and Mexico and the largest representative of *Pogonomyrmex*. Workers monomorphic, 6–12 mm long; bite and sting aggressively. Workers clear vegetation from a circular area that encompasses a few to several feet in diameter; colony entrance at the centre of the circular area but does not display a mound. Food includes grain and seed which is stored in subterranean chambers. Syn. Texas Harvester. See Formicidae. Cf. California Harvester-Ant; Florida Harvester-Ant; Harvester-Ant; Western Harvester-Ant.

RED-HEADED ASH-BORER *Neoclytus acuminatus* (Fabricius) [Coleoptera: Cerambycidae].

RED-HEADED JACK-PINE SAWFLY *Neodiprion rugifrons* Middleton [Hymenoptera: Diprionidae].

RED-HEADED PASTURE COCKCHAFER *Adoryphorus couloni* (Burmeister) [Coleoptera: Scarbaeidae]: A pest of pastures in southeastern Australia. Larvae occupy upper rhizosphere and possess a non-storage type of gut; large quantities of food consumed daily.

RED-HEADED PINE SAWFLY *Neodiprion lecontei* (Fitch) [Hymenoptera: Diprionidae]: A widespread pest of pines in North America. Monovoltine in north; multivoltine in south. Larva strips edges of needles, causing drying; overwinters as mature larva within cocoon. Female monogamous.

RED-HEADED PINE SAWFLY NPV A naturally-occurring, host-specific Nuclear Polyhedrosis Virus used as a biopesticide against larva of Redheaded Pine Sawfly (*Neodiprion lecontei*). Trade names include: Ne NPV®, Biocontrol-1® and Lecontvirus®. See Biopesticide; Nuclear Polyhedrosis Virus. Cf. Beet Armyworm NPV; Douglas-Fir Tussock Moth NPV; European-Pine Sawfly NPV; *Heliothis* NPV; *Mamestra brassicae* NPV; *Spodoptera littoralis* NPV.

RED-HUMPED CATERPILLAR *Schizura concinna* (J. E. Smith) [Lepidoptera: Notodontidae]: A widespread pest of fruit and ornamental trees in North America. Overwinters as mature larva within cocoon on ground; pupation occurs during early summer. Adults active in summer and female oviposits clusters of 50–100 eggs on underside of leaves. Neonate larvae skeletonize leaf with bodies parallel and heads directed toward leaf margin; later instar larvae disperse to other leaves and consume entire leaf. When disturbed, larvae elevate both ends of body and remain attached to leaf via prolegs on middle segments. Cf. Yellow-Necked Caterpillar.

RED IMPORTED FIRE-ANT *Solenopsis invicta* Buren [Hymenoptera: Formicidae]: Native to South America and adventive to the United States where it creates significant social problems in the south. Worker 2–7 mm long; body elongate, tan to dark reddish-brown with an orange band on dorsum of Abdomen. Antennal club of two segments; Pedicel of two segments. Queen initially winged, but wings lost after mated; initially queens lay 75–125 eggs within a brood chamber dug into soil; eggs white, oval, 1 mm long; egg stage 8–10 days. Larva 1–2 mm long, white and lack appendages. Larval stage 6–12 days. Queen rears first brood but workers take over brood care as colony matures. Pupal period 9–16 days. Nest is an above ground mound up to 1 m high. Two sizes occur within worker caste: majors (6.4 mm) and minors (3.2 mm.) Several queens may be present in one colony (polygynous colony) or a single queen may be present (monogynous colony). Fertilized eggs become females, usually worker caste; unfertilized become males. Soldier and worker castes female only. Exists in monogyne and polygyne forms.

RED-LEGGED EARTH MITE *Halotydeus destructor* [Acari: Penthaleidae], accidentally introduced to Western Australia ca 1920, spread to SA and NSW by 1930; now distributed throughout winter-rainfall regions of WA, SA, Vic, Tas and NSW. RLEM an important pest of legume crops in pastures, oats, turnips, wheat, barley, canola and field peas. RLEM attacks reduce ability of plants to fix nitrogen which lowers soil fertility, which in turn causes a loss in yield of grain grown in rotation with pastures. Syn. Earth Flea.

RED-LEGGED FLEA BEETLE *Derocrepis erythropus* (Melsheimer) [Coleoptera: Chrysomelidae].

RED-LEGGED GRASSHOPPER *Melanoplus femurrubrum* (DeGeer) [Orthoptera: Acrididae]: A pest of grains, alfalfa, sorghum and beans in central USA. Female deposits pods each with about 20 eggs; males do not stridulate. See Acrididae.

RED-LEGGED HAM BEETLE *Necrobia rufipes* (DeGeer) [Coleoptera: Cleridae]: A cosmopolitan pest of stored products and carrion; most significant pest of dried meats; occasionally attacks cured meats; a significant pest in copra cargoes where it feeds directly on copra and other insects. Most common in tropical or warm temperate regions. Adult and larva predatory and cannibalistic; adult can enter pupal cocoon to feed on Pupa; sometimes associated with dermestid beetles. Adult 4.0–6.5 mm long, metallic green-blue, legs and antennal bases reddish; Antenna with loosely defined club of three segments. Egg banana shaped, yellowish-white, glued in clusters; incubation 4–15 days. Larva to 10 mm long, purplish, slender and tapering toward head; 3–4 larval instars, larval stage 17–28 days. Pupation away from feeding site within papery cell formed from salivary secretion; pupal stage 11–52 days. Adult stage 120–140 days, longevity increased by predation; life cycle 1–3 months. Syn. Copra Beetle. Cf. Red-Shouldered Ham Beetle.

RED-MARGINED ASSASSIN-BUG *Scadra rufidens* Stål [Hemiptera: Reduviidae].

RED MELON-BEETLE *Aulacophora africana* Weise [Coleoptera: Chrysomelidae]: An Old-World minor pest of curcurbits.

RED MILKWEED-BEETLE *Tetraopes tetrophthalmus* (Forster) [Coleoptera: Cerambycidae].

RED-NECKED CANE-BORER *Agrilus ruficollis* (Fabricius) [Coleoptera: Buprestidae]: A monovoltine pest of blackberry, dewberry and raspberry in eastern North America. Eggs laid on bark of cane near base of leaf. Neonate larva bores into sapwood, around cane causing gall or swelling which may sometimes cause death or breaking of cane at swelling; larva overwinters in pith of cane; completes development and pupates during following spring. See Buprestidae. Cf. Raspberry Crown-Borer.

RED-NECKED PEANUT-WORM *Stegasta bosqueella* (Chambers) [Lepidoptera: Gelechiidae].

RED-OAK BORER *Enaphalodes rufulus* (Haldeman) [Coleoptera: Cerambycidae].

RED ORCHID-SCALE *Furcaspis biformis* (Cockerell) [Hemiptera: Diaspididae].

RED-PINE CONE BEETLE *Conophthorus resinosae* Hopkins [Coleoptera: Scolytidae].

RED-PINE SAWFLY *Neodiprion nanulus nanulus* Schedl [Hymenoptera: Diprionidae].

RED-PINE SCALE *Matsucoccus resinosae* Bean & Godwin [Hemiptera: Margarodidae]. In North America, a serious pest of pines imported into New England. Bivoltine and overwinters as crawler under bark.

RED-PINE SHOOT-MOTH *Dioryctria resinosella* Mutuura [Lepidoptera: Pyralidae].**RED SCALE** See California Red Scale.

RED-SHOULDERED HAM-BEETLE *Necrobia ruficollis* (Fabricius) [Coleoptera: Cleridae]: A widespread, minor pest of stored products including dried and smoked meats, cheese, bones and animal skins; RSHB most common in tropical South America and Africa. Adult and larva may be predaceous; adult feeds on surface of meat while larva bores into meat. Adult 4.0–6.5 mm long, predominantly metallic blackish blue with Thorax, legs and base of Elytra red-brown. Egg banana shaped, yellowish-white; incubation 2–5 days. Larval stage 100–200 days; pupation within papery cell, pupal stage 9–14 days; adult lives to one year; life cycle 1–3 months. Syn. Red-Necked Bacon Beetle.

RED-SHOULDERED LEAF-BEETLE *Monolepta australis* (Jacoby) [Coleoptera: Chrysomelidae]: A minor and sporadic pest of *Citrus* in Australia. RSLB a major pest of avocado and also damages carambola, cashew, cotton, eucalyptus, grapes, lychee, mango and ornamentals. Often found in clusters under leaves. Adult 6–7 mm long, oval and yellow with a red spot in middle of each wing and a red band across wing base. Adult females lay eggs singly or in small batches in bushland soil. Eggs yellow and finely sculptured; eggs laid in pasture soil; egg stage 9–13 days. Larvae small white grubs that live in soil and feed on roots; larval stage ca 34 days. Pupation in soil; pupal stage 5 days. Adult emergence synchronized by heavy rains during spring and summer; beetles migrate in swarms to tree crops where they feed on foliage, flowers and fruit; swarms can severely damage young foliage and cause *Citrus* rind disfigurement within 24 hours. Multivoltine with 2–4 generations per year depending upon climate. Alt. Monolepta Beetle.

RED-SHOULDERED STINK-BUG 1. *Thyanta accerra* McAtee [Hemiptera: Pentatomidae]. 2. *Thyanta pallidovirens* Stål [Hemiptera: Pentatomidae].

RED SUNFLOWER SEED-WEEVIL *Smicronyx fulvus* LeConte [Coleoptera: Curculionidae]: A pest of sunflower in parts of North America. Adults aggregate on developing seed heads, feed on pollen and oviposit in developing achenes. Cf. Gray Sunflower Seed Weevil.

RED-TAILED FLESH FLY *Sarcophaga haemorrhoidalis* Fallén [Diptera: Sarcophagidae]: A widespread urban pest in tropical and warm-temperate regions but not known from Oriental or Australasian regions. RTFF common around human habitation. Adult 10–14 mm long; grey with genitalia of male large and red. Female larviparous; larvae scavenge in carrion and faecal matter; larval stage ca four days; pupal stage ca 10 days.

RED-TAILED SPIDER WASP *Tachypompilus analis* (Fabricius) [Hymenoptera: Pompilidae].

RED-TAILED TACHINA *Winthemia quadripustulata* (Fabricius) [Diptera: Tachinidae].

RED TURNIP-BEETLE *Entomoscelis americana* Brown [Coleoptera: Chrysomelidae].

RED TURPENTINE-BEETLE *Dendroctonus valens* LeConte [Coleoptera: Scolytidae]: A pest of pine, larch, fir and spruce in northern USA and southern Canada. Adult dark red but not black, the largest Species of *Dendroctonus* (ca 4.5–10 mm long); Pronotum densely, irregularly punctate; Elytral Striae indistinct, punctures obscure. Female excavates irregular, longitudinal galleries for eggs; larvae mine between bark and wood. Attack trees in all conditions from stumps and fresh-cut logs to healthy trees. See Scolytidae. Cf. Douglas-Fir Beetle; Eastern Larch Beetle; Mountain Pine Beetle; Southern Pine Beetle; Western Pine Beetle.

RED WAX-SCALE *Ceroplastes rubens* Maskell [Hemiptera: Coccidae]: A polyphagous, widespread pest of economic plants, including *Citrus,* in many regions of the world.

RED WOOD-ANT Any member of the *Formica rufa* group of ants common to the forests of Europe.

REDALE® See Hexaflumuron.

REDBACK SPIDER *Latrodectus hasselti* Thorell [Araneida: Theridiidae]: A venomous spider endemic in Australia and regarded as a serious urban pest. Adult female body ca 15 cm long; predominantly black with red band on dorsum of Abdomen and red hourglass-shaped mark on venter; male considerably smaller with similar markings. Immatures whitish with black spots on Abdomen and white hourglass-shaped spot on venter. Webs typically concealed and constructed on underside of objects; strands of web irregularly spaced and strong. Cf. Black Widow Spider. Brown Widow.

REDI, FRANCESCO (1626–1698) Italian physician, poet, naturalist, early microscopist and experimentalist; best known for his experiment to disprove the concept of spontaneous generation of flies; discovered and described the scabies mite. Author of *Opusculorum pars prior sive Experimenta circa Generationem Insectorum* (3 vols, 1686) and *Esperienze Interno alla Generazione degli Insetti* (1668). Curiously, Redi believed that gall insects occurred spontaneously. (Duméril 1823, *Considérations générale sur la classe des insectes.* 272 pp. (244), Paris.)

REDLICH, HERMANN JULIUS ALBERT (1842–1903) (Hoffman 1903, Ent. Z., Frankf a. M 16: 85.)

REDTENBACHER, LUDWIG (1814–1876) Austrian coleopterist and Director of Royal Vienna Zoological Museum. (Westwood 1876, Proc. ent. Soc. Lond. 1876: xliii–xliv; Katter 1877, Ent. Kal. 2: 69–70; Ganglebauer 1900, Festschr. zool.-bot. Ges. Wien p. 350.)

REDUCE Verb. (Latin, *re-*; *ducere* = to lead.) To lessen or decrease in size, as the parts of an insect. Cf. Enlarge.

REDUCING SUGAR Any sugar which contains the radical CO, which when heated with Fehling's solution (blue) reduces the copper compound in solution so that cuprous oxide (red) is formed. All monosaccharides and a few disaccharides (*e.g.* maltose) are reducing sugars.

REDUCTION Noun. (Latin, *reductus* = reduce; English, *-tion* = result of an action.) 1. Maturation of Ovum and Sperm: The process in which the number of chromosomes in the egg-nucleus is reduced to half the number normal for somatic cells. 2. The act of reducing or retracting; a lessening in size of parts of an insect, as compared with a norm. 3. To combine with or subject to the action of hydrogen; deoxidize. See Oxidation.

REDUCTUS Adj. A zig-zag marking or corrugation.

REDUVIIDAE Latreille 1807. Plural Noun. Assassin Bugs; Ambush Bugs (Phymatinae); Kissing Bugs, Conenoses, Cone-Nose Bugs (Tritominae only); Thread-Legged Bugs. A numerically large (ca 2,500 Species), cosmopolitan Family of Hemiptera, best represented in tropical regions. Body moderate to large sized; head narrow, elongate and neck-like behind eyes; beak slightly curved, three segments with apex fitting into prosternal groove; groove minutely, transversely striate; Abdomen widened with Terga exposed beyond apex of wings. Eggs rather large, barrel-shaped; eggs deposited singly or in small clusters in habitat of adult (on ground, in vegetation, within homes); female lays a few eggs to ca 1,000 during lifetime. Typically five nymphal instars. Predominantly predaceous upon other arthropods; some Species (Triatominae) feed upon vertebrate blood and *Triatoma* spp. transmit Chagas' Disease to humans.

REDUVIOIDEA The Cimicomorpha *sensu* authors. A Superfamily of Hemiptera including Pachynomidae and Reduviidae.

REDWATER DISEASE See Bovine Babesiosis.

REED, EDMUND BAYNES (1837–1917) (Bethune 1917, Can. Ent. 49: 37–39, bibliogr.)

REED, EDWYN CARLOS (1841–1910) (Porter 1911, Revta. chil. Hist. nat. 15: 18–21, bibliogr.)

REED, WALTER C (1851–1902) (Sternberg 1903, Proc. Wash. Acad. Sci. 5: 407–409; McCaw 1904, Pop. Sci. Mon. 65: 262–268; Kelly 1923, *Walter Reed and Yellow Fever.* 355 pp., bibliogr.)

REEKER, ADOLF (1868–1942) (Paul 1942, Stettin. ent. Ztg 103: 155–157.)

REEVE, FREDERICK C (–1970) (Hinton 1971, Proc. R. ent. Soc. Lond. (C) 35: 53.)

REFLECTANCE BASKING Mechanism of thermal regulation in which wings are used as solar reflectors transferring solar radiation onto insect body. (Kingsolver 1985, Oecologia 66: 540–545.)

REFLECTED Adj. (Latin, *reflectere* = to turn back.)

Bent up or back. Alt. Reflex; Reflexed; Reflexus.

REFLECTOR LAYER Cells acting as a background in the light-producing organs of insects, such as Lampyridae, which scatter the incident light and prevent its dispersal internally (Imms).

REFLEX Adj. (Latin, *reflectere* = to turn back.) A reaction following and external stimulus or set of stimuli, acting on and through the nervous and motor mechanism.

REFLEX ARC A sensory nerve path directly connected to a motor path.

REFLEX BLEEDING The ejection of Haemolymph (insect blood) through intersegmental membranes of an appendage or part of the body. Haemolymph may contain distasteful or toxic chemicals. RB generally interpreted as a method of defence employed by some insects (*e.g.* meloid, chrysomelid beetles). In Meloidae, Cantharadin commonly is regarded as a chemical defence substance. RB may involve the loss of considerable amounts of Haemolymph without apparent ill effect. Syn. Autohaemorrhage; Reflexive Bleeding. See Circulatory System. Cf. Haemolymph. Rel. Hydrostatic Skeleton.

REFON® See Ethephon.

REFRACTED Adj. (Latin, *refractus* past part. of *refringere* from *re-* ; *frangere* to break.) Descriptive of that which is bent counter to the natural curvature of the structure, thereby giving the structure a broken appearance. Alt. Refractus.

REFRINGENT Adj. (Latin, *refringens* pres. part of *refringere*. See Refracted.) Refractive or possessing the ability to refract or deflect rays of light. Sometimes used in conjuction with Setae on the body of insects. See Birefringent.

REFUSE Noun. (Old French, *refuser* = refuse. Pl., Refuses.) 1. The valueless or useless part of something. 2. Material that has been discarded as without value. Syn. Debris 2; Garbage; Rubbish; Trash. Cf. Waste.

REGAL MOTH *Citheronia regalis* (Fabricius) [Lepidoptera: Saturniidae]. Syn. Hickory Horned Devil (larva).

REGAN, WILLIAM SWIFT (1884–1959) (Telford 1960, J. Econ. Ent. 53: 700.)

REGEL, EDUARD AUGUST VON (1815–1892) (Ratzeburg 1874, Forstwissenschaftliches Schriftsteller-Lexicon 1: 431–434.)

REGENERATION Noun. (Latin, *re-*; *genesis* = produced; English, *-tion* = result of an action.) The development of new growth in appendages or lost parts by insects. A phenomenon generally occurring in the larval stages wherein growth is incremental and noticeable between instars.

REGENERATIVE CELLS One of three basic types of cell which occur in the midgut epithelium. RC generate the replacement cells of the epithelium. See Alimentary System; Cf. Columnar Cell; Endocrine Cell; Peritrophic Membrane. Rel. Digestion.

REGENERATIVE CRYPTS Pouch-like diverticula or pockets of the stomach wall containing groups of digestive or regenerative cells.

REGENT MC® See Fipronil.

REGIMBART, MAURICE AUGUSTE (1852–1907) (Zaitzev 1907, Revue Russe Ent. 7: 174–175; Zaitzev 1912. Revue Russe Ent. 12: 371–375, bibliogr.)

REGION Noun. (Old French, *regium*. Pl., Regions.) A space or area adjoining a specified point. A part of the body composed of a number of segments, as the head, the Thorax, or the Abdomen. Cf. Tagma.

REGNELL, ANDERS FREDERIK (1807–1884) (Sandahl 1884, Ent. Tidschr. 5: 191–192, 228.)

REGRESSIVE EVOLUTION A controversial concept within the context of evolutionary theory. Regressive evolution theory addresses questions posing reduction in anatomical structure, physiological process and behavioural action as responses to loss or reduction in function over generational time.

REGRESSIVE MOULT The manifestation of supernumerary larval or nymphal moults in the life cycle induced by adverse environmental conditions or nutrition.

REGULAR Adj. (Latin, *regularis*.) 1. Something created or established on the basis of rules, directions, patterns or plans. 2. Something which is normal, correct or standard. Syn. Normal; Typical; Usual. Cf. Irregular.

REGULATION Noun. (Latin, *regulatus* past part. of *regulare* from *regula* = regular.) 1. Biological Control: Any rule which applies to the import, export or movement of living organisms. 2. The control of population density.

REGULATOR® See Fenoxycarb.

REGULATORY CONTROL An approach to pest control which seeks to prevent importation (migration) and establishment of exotic pests in areas where the pests do not occur. Types of regulatory control programmes include Eradication, Containment and Suppression. Cf. Biological Control; Chemical Control; Cultural Control; Integrated Pest Management; Natural Control.

REGURGITATE Verb. (Medieval Latin, *regurgitare* from *re-*; *gurgitare* = thrown back; *-atus* = adjectival suffix.) To voluntarily bring the stomach contents up into the mouth. An action see in social insects during trophallaxis.

REGURGITATION Noun. (Medieval Latin, *regurgitare* from *re-*; *gurgitare* = thrown back; English, *-tion* = result of an action.) The act of bringing up undigested, digested or partly digested food into the mouth voluntarily.

REH, LUDWIG (1867–1940) (Sachtleben 1941, Arb. physiol. angew. Ent. Berl. 8: 68–69; Weidner 1967, Abh. Verh. naturw. Ver. Hamburg Suppl. 9: 306–314.)

REHBINDER, JOHANN V (1757–) (Zimsen 1964, *The Type Material of J. C. Fabricius.* 656 pp. (13), Copenhagen.)

REHN, JAMES ABRAM GARFIELD (1881–1965) American orthopterists and Associate Curator at the Academy of Natural Sciences, Philadelphia. (Bei-Bienko 1965, Ent. Obozr. 44: 714–716, bibliogr. Translation: Ent. Rev., Wash. 44: 417–418; Gurney 1965, J. Econ. Ent. 58: 805–807.)

REIBER, FERDINAND (1849–1892) (Anon. 1892, Leopoldina 28: 163.)

REICHE, LOUIS JEROME (1799–1890) (Bourgeois 1890, Bull. Soc. ent. Fr. (6) 10: lxxxviii–xc; Brisout de Barneville 1890, Ann. Soc. ent. Fr. (6) 10: 559–562)

REICHEL, CHARLES GOTTHOLD (1751–1825) (Weiss 1936, *Pioneer Century of American Entomology,* 320 pp. (62–63), New Brunswick.)

REICHENBACH, H T L Author of *Monographia Pselaphorum*, published in 1816.

REICHENBACH, HEINRICH GOTTLIEB LUDWIG (1793–1879) (Friedrich 1879, Sber. naturw. Ges. Isis Dresden 1879: 97–105; Anon. 1881, Leopoldina 17: 19–22, 34–36, 50–54, bibliogr.)

REICHENSPERGER, AUGUST (1878–1962) (Anon. 1963, Mitt. schweiz. ent. Ges. 36: 143–144.)

REICHERT, ALEXANDER JULIUS (1858–1939) (Emden *et al.* 1929, Z. wiss. InsektBiol. 24: 1–10, bibliogr.)

REICHLIN-MELDEGG, GUSTAV (1880) (Harold 1880, Mitt. münch. ent. Ver. 4: 175.)

REID, PERCY CHARLFF (1858–1923) (J. P. 1923, Entomologist 56: 98.)

REIMARUS, HERMANN SAMUEL (1694–1768) (Rose 1850, *New General Biographical Dictionary* 11: 307–308.)

REIMOSER, EDUARD (1864–1940) (Heikertinger 1940, Koleopt. Rdsch. 26: 92; Pesta 1940, Annln naturh. Mus. Wien 51: 5–7, bibliogr.)

REINDEER WARBLE-FLY See Caribou Warble-Fly.

REINECK, GEORG (1882–1937) (Korschofskay 1938, Ent. Bl. Biol. Syst. Käfer 33: 466–469, bibliogr.)

REINECKE, OTTOMAR (1840–1917) (Anon. 1918, Ent. News 29: 240.)

REINHARDT, JOHANN CHRISTOPHER HAGEMANN (1777–1845) (Henriksen 1923, Ent. Meddr 15: 141–144; Spärk 1933, Vidensk. Meddr dansk. naturh. Foren. 95: 63)

REINHARDT, JOHANNES THEODOR (1816–1882) (Anon. 1882, Zool. Anz. 5: 644; Henriksen 1926, Ent. Meddr 15: 213–214.)

REINICKE, WILLIAM RHODES (1879–1929) (Anon. 1929, Ent. News 40: 134.)

REISS, HUGO (1890–1974) (Harde 1974, Ent. Z., Frankf. a. M. 84: 203–204.)

REISS, R A (1875–1929) (Gradojevic 1931, Acta Soc. ent. jugosl. 5–6: 3–5.)

REISSIG, JACOB (1800–1860) (Ratzeburg 1874, Forstwissenschaftliches Schriftsteller-Lexicon 1: 434–435.)

REITTER, EDMUND (1845–1962) (Wanka 1915, Wien. ent. Ztg. 34: 215–287, bibliogr.; Heikertinger 1920, Wien. Ent. Ztg. 38: 1–20, bibliogr.)

REITTER, RONALD (–1962) (Anon. 1962, Z. wien ent. Ges. 47: 220.)

REJUVENESCENCE Noun. Latin, *re-*; pp of *juvenescere* = to grow young.) A renewal of youth; bringing back to a condition of youth.

RELAPSING FEVER TICK *Ornithodoros turicata* (Dugès) [Acari: Argasidae]: A soft-tick in the USA capable of transmitting relapsing fever.

RELAPSING FEVERS A group of diseases caused by spirochaetes of the Genus *Borrelia* and vectored principally by argasid ticks with two transmitted by ixodid ticks and one by the human louse. Notable diseases transmitted to humans include: Epidemic Relapsing Fever and Lyme Disease.

RELATIONSHIP Noun. (Middle English, *relacion.*) The affinity of organisms determined by genetical relatedness.

RELATIVE Adj. (Latin, *relativus.*) Something not absolute and depending on some other thing or concept as a norm, datum plane, or standard, agreed upon tacitly or by formal adoption. Cf. Absolute.

RELDAN® See Chlorpyrifos-Methyl.

RELEASE® See Gibberellic Acid.

REMARUS, HERMANN SAMUEL (1694–1768) (Weidner 1967, Abh. Verh. naturw. Ver. Hamburg Suppl. 9: 26–32.)

REMIFORM Adj. (Latin, *remus* = oar; *forma* = shape.) Oar-shaped. See Oar. Cf. Ensiform; Fusiform. Rel. Form 2; Shape 2; Outline Shape.

REMIGIAL Adj. (Latin, *remex* = rower; *-alis* = pertaining to.) Descriptive of or pertaining to the Remigium.

REMIGIAL REGION The wing area anterior of the Vannal Fold. RR contains the costal, Subcostal, Radial, Medial, Cubital, and Postcubital Veins. RR chiefly productive of flight movements and is directly affected by the motor muscles of the wing (Snodgrass). Syn. Preanal Region, Preclavus (Snodgrass). See Wing. Cf. Anal Region; Axillary Region; Jugal Region.

REMIGIUM Noun. (Latin, *remex* = rower; Latin, *-ium* = diminutive > Greek, *-idion*. Pl., Remigiums.) The anterior, rigid part of the wing that envelops most of the large wing veins. See Remigial Region.

REMIPED Adj. (Latin, *remi* = oar; *pedis* = foot.) Having oar-shaped or oar-like legs which are used in rowing by aquatic insects.

REMMER, GEORG (–1974) (Gruber 1974, Prakt. SchädlBekämpf. 26: 117.)

REMOTE Adj. (Middle English.) 1. Not near; further than distant. 2. Scarce or scant when referring to pilosity. Alt. Remotus. Cf. Adjacent.

REMPEL, J (1903–1976) (Anon. 1976, Bull. ent. Soc. Can. 8: 10.)

REMY, PAUL A (1895–1962) (Condé 1962, Bull. Soc. ent. Fr. 67: 93–95.)

RENAL CELLS See Nephrocytes.

RENDSCHMIDT, FELIX (1768–1853) (Anon. 1853, Arb. Schles. Ges. Vaterl. Kultur 1853: 185–186.)

RENEGADE® See Alphacypermethrin.

RENGGER, JOHANN RUDOLPH VON BRUGG (1795–1832) (Anon. 1832, Ann. Soc. ent. Fr. 1: 332.)

RENICULUS Adj. (Latin, *renis* = kidney; *culus* = dim. form.) A small kidney-shaped coloured spot.

RENIFORM Adj. (Latin, *renis* = kidney; *forma* = shape.) Bean-shaped; descriptive of structure shaped as a mammalian kidney. See Kidney. Cf. Lobiform; Nephroid; Pyreniform. Rel. Form 2; Shape 2; Outline Shape.

RENIFORM SPOT Some moths: A somewhat kidney-shaped spot at the end of the distal cell (Comstock).

RENK, ALICE V (1908–1960) (O'Neill & Russell 1961, Proc. ent. Soc. Wash. 63: 67–68.)

RENKONEN, OLARI (1907–1959) (Lindroth 1960, Opusc. ent. 25: 153.)

RENNIE, JAMES (1787–1868) (Anon. 1868, Entomol. mon. Mag. 4: 191.)

RENSSELAER, JEREMIAH VAN (1793–1871) (Weiss 1936, *Pioneer Century of American Entomology,* 320 pp. (113–114), New Brunswick.)

REOVIRIDAE Plural Noun. A small Family of arboviruses whose members are 60–80 nm icosahedrons that lack a lipoprotein envelope but have a protein coat. Viruses produce several diseases in humans and livestock, including African Horse Sickness, Bluetongue Disease, Colorado Tick Fever, and Epizootic Haemorrhagic Disease. See Arbovirus. Cf. Bunyaviridae; Flaviviridae; Tongaviridae; Rhabdoviridae.

REPAGULA Noun. (Latin, *repagula* = bolts, limits.) Neuroptera: Rod-like bodies from follicle cells or aborted eggs that are deposited near the place viable eggs are deposited by the same individual. Repagula are known from Ascalaphidae only, and presumably serve as a deterrent of ants and other egg predators. See Egg Defense.

REPAND Adj. (Latin, *repandus* = bent backward.) Wavy, very slightly sinuate, with an uneven sinuous margin. Alt. Repandus.

REPANDULOUS Adj. (Latin, *repandus* = bent backward; Latin, *-osus* = with the property of.) Convexly curved.

REPEL Verb. (Middle English, *repellen.*) Cf. Allure.

REPELLANT Noun. (Latin, *repellere* from *re-*; *pellere* = to drive; *-antem* = an agent of something.) A chemical that causes insects to orient their movements away from a source. See Semiochemical. Cf. Arrestant; Attractant.

REPELLENT Noun. (Latin, *repellere* from *re-*; *pellere* = to drive. Pl., Repellents.) 1. Something which repels. 2. A substance that causes insects to move away from the source. 3. A chemical which causes insects to make oriented movements away from its source.

REPLETE Noun. (Latin, *repletus* = filled up.) Social Insects: A worker ant whose Crop is filled with fluid (honeydew or nectar) to an extent that the gastral segments are distended and the intersegmental membrane stretched. Formicidae: See Pleregate.

REPLETE WORKER Social Insects: A worker filled with fluid and used as a resevoir to regurgitate food upon the demand of nestmates.

REPLICATE Adj. (Latin, *replicare* = to fold back; *-atus* = adjectival suffix.) Refolded, doubled back or down; specifically applied to folded wings as the hindwings of Coleoptera. Alt. Replicatus.

REPLICATILE Adj. (Latin, *replicare* = to fold back.) Capable of being folded back.

REPPEL, PAUL (–1974) (Anon. 1974, Prakt. SchädlBekämpf 26: 132.)

REPPERT, ROY R (1881–1940) (Bilsing 1940, J. Econ. Ent. 33: 707.)

REPRINT Noun. (Pl., Reprints). A copy of a publication produced by a printer and distributed by the author of that publication. See Publication. Cf. Preprint; Offprint; Copy.

REPRODUCTION Noun. (French, *reproduction.* Pl., Reproductions.) The act or process through which organisms replicate themselves. Somatic Cells reproduce via a complex process involving Mitosis; Sex Cells prepare themselves for Syngamy via a process called Meiosis. Reproduction of some animals and plants may be asexual with offspring originating from the cleveage or fission of the parent's body. The dominant form of reproduction in plants and animals is sexual and involves the union a sex cell from each parent to produce an embryo. Cf. Development. Rel. Embryo.

REPRODUCTIVE STAGE Social Insects: A stage in colony development during which males and virgin queens are produced.

REPRODUCTIVE SYSTEM The cells and tissues which collectively constitute the organs that produce, store and conduct gametes (egg and Spermatozoa), attract mates, facilitate copulation and enable oviposition. The RS is characteristically associated with the Abdomen of pterygote insects. Male and female RSs are divided into internal components and external components. Internal fertilization is prevalent among terrestrial organisms and almost universal among Insecta. Some Apterygota (Collembola) engage in external fertilization and presumably this is the ancestral condition to internal fertilization by Pterygota. Internal fertilization is a consequence of living on land and serves to protect Spermatozoa from desiccation. See Aedagus; Ovary; Testis; Ovipositor. Cf. Alimentary System; Circulatory System; Nervous System; Respiratory System.

REPRODUCTIVES Noun. Social Insects: Any member of the colony that is capable of genetically contributing to the production of offspring. Specifically, reproductives include males, egg-laying workers and queens. See Alate. Cf. Caste.

REPUGNATORIAL GLANDS Glands in insects which secrete malodorous or noxious liquids or vapours, as a defence against enemies. Among primitive pterygotes such as the cockroach

Eurycotis floridana abdominal glands produce and store trans-2-henenal and release it through ducts between abdominal Sterna VI and VII. Discharge is achieved through Haemolymph pressure and can cover more than a metre by flicking the Abdomen. RG common and diagnostic in many groups of insects such as Coreidae, Pentatomidae and other Hemiptera. See Defensive Gland. Cf. Tracheal Gland.

REPUGNATORIAL Adj. (Latin, *repugnare* = to resist; *-alis* = characterized by.) Repellent; of such an offensive or defensive nature as to drive away.

RESECTICID® A registered biopesticide derived from *Bacillus thuringiensis* var. *kurstaki*. See *Bacillus thuringiensis*.

RESERVOIR Noun. (Latin, *reservare* = to keep back.) 1. A depression, cavity or sac used for storage of any fluid or secretion. 2. An arthropod which serves as a host for a pathogen although the pathogen may not harm the reservoir. See Zoonosis. Cf. Vector.

RESIDRIN® See Tetramethrin.

RESIDROID® See Permethrin.

RESILIENT Adj. (Latin, *resilire* = to leap back.) Pertaining to structure possessing the power or capacity of returning to or resuming the original position or shape after deformation. See Elastic. Cf. Dissilient.

RESILIN Noun. (Latin, *resilire* = to leap back.) Insect Rubber. A cuticular protein first discovered by the Danish physiologist Weis Fogh (1960). Resilin is a colourless, transparent, gel-like secretion of epidermal cells. When dry, resilin is hard and brittle; when placed in aqueous media, resilin swells and becomes rubbery. Resilin resembles vertebrate protein elastin and is often called 'insect rubber'; only rubbery when swollen because a liquid must be present to solvate peptide chains and break secondary bonds. Resilin confers elastic properties on Integument such that up to 97% of energy stored can be instantly recovered. Resilin is located in wing hinges, thoracic wall, legs and other body regions that store mechanical energy. Resilin is used to construct mechanical springs that are very deformable and endowed with considerable elastic recovery. Resilin functions mechanically to store and release mechanical energy. Sometimes storage can involve very long periods of time. A dragonfly tendon can be stressed under a constant load for months and recover its original shape within milliseconds of load removal. Similarly, Resilin is responsible for providing insects with tremendous acceleration. Energy stored in Resilin for a flea's jump translates into 140 G's of accelerative force and 380 G's in the elaterid beetle's 'click.' This compares with 5 G's of acceleration developed by muscles in the perch (a figure considered high in terms of performance among vertebrates). See Integument; Skeleton. Rel. Hydrostatic Skeleton; Jumping Mechanism.

RESIN Noun. (Latin, *resina* = resin. Pl., Resins.) Vegetable substances that are solid or semisolid, transparent or translucent; mixtures of terpenoids or phenolic compounds soluble in alcohol, ether and similar chemicals but not water. Resins are yellow to brown in colour, melt when heated and are electrically non-conductive. Resins are commonly produced within internal ducts and specialized surface glands of plants or occur as exudates of pine and fir. When combined with essential oils, resins are called oleoresins; when combined with gum, resins are called gum resins; ancient resins are called fossil resins and frequently contain insect inclusions. See Amber; Copal; Gum Arabaic; Gum Damar; Lac; Rosin. Cf. Catechu; Gum; Kino; Latex; Sap.

RESIN BEE Anthophorid bees, particularly members of *Anthophora*, which use resin as cement to construct their nests.

RESIN GNAT A small fly, *Retinodiplopsis resinicola* [Diptera: Chironomidae], the larvae of which induce resin formation at the site where they live.

RESINACEOUS Adj. (Latin, *resinaceus*; *-aceus* = of or pertaining to.) Resiniferous. (Obsolete.)

RESINATE Trans. Verb. (Latin, *resinaceus*; *aceus* = of or pertaining to.) To impregnate with resin.

RESINER Noun. A person who applies resin.

RESINIFY Transitive Verb. (Resinified, Resinifing.) To convert into resin. To treat with resin.

RESINOUS Adj. (Latin, *resinosus* from *resina* = resin; *-osus* = full of.) 1. Pertaining to the properties of resin. 2. Resin-like; with the appearance, texture or aroma of resin. See Resin.

RESISTANCE Noun. (Latin, *resistere* from *re-*; *sistere* to stand.) Pathology: The capacity of a host to prevent or reduce development of a pathogen.

RESISTANCE FACTOR Any condition in plants that protects them from insect attacks, such as structures, chemical substances in the plant or physiological conditions.

RESISTANT HOST Medical Entomology: A vertebrate host which is not affected by a pathogen irrespective of prior exposure. Cf. Amplifying Host; Dead-End Host; Silent Host; Susceptible Host.

RESISTANT VARIETY A variety of plant which is less vulnerable to attack or damage by a pathogen or pest than another variety of the same plant Species. Cf. Susceptible Variety.

RESITOX® See Coumaphos.

RESMETHRIN Noun. A synthetic-pyrethroid compound which acts as a selective, contact insecticide with fast knockdown for control of cockroaches, flies, mosquitoes, midges, fleas and other urban pests. Trade names include: Chrysron®, Crossfire®, Derringer®, Earthfire®, For-Syn®, Pynosect®, Premgard®, Respond®, SBP-1382®, Scourge®, Synthrin®, Vectrin®. See Synthetic Pyrethroid.

RESONATOR Noun. (Latin, *resonatus* = to resound; *or* = one who does something. Pl., Resonators.)

1. An anatomical structure adapted or mechanical device constructed to intensify sound. 2. Insects: general form of a resinator involves a thin, vibrating sclerite or lamella (Imms). See Stridulation. Cf. Plectrum; Tympanum.

RESORB Trans. Verb. (Latin, *resorbens* = to suck in.) To reabsorb.

RESOURCE PARTITIONING Ecology: The phenomenon of sharing resources within a spatially or temporally defined, physically delimited, or otherwise resource -limited niche among individuals of two or more Species co-occurring in that niche. Rel. Species packing; Resource utilization; Interspecific competition.

RESOURCE Noun. (Old French, *resourdre* = to spring forth.) General: A new or a reserve source of supply or support. Ecology: The nutrients and physical space needed for survival. Examples of resources include: Food, water, a place for oviposition, a place for pupation or diapause.

RESOURCE UTILIZATION Ecology: Description and analysis of resource use; typically related to herbivores and specific trophic guilds. See Guild.

RESPIRATION Noun. (Latin, *respiratio* = respire; English, *-tion* = result of an action.) 1. The intracellular process involving the oxidation and the breakdown of carbohydrates into carbon dioxide and water. 2. The process of breathing, oxygenation of tissues and the elimination of gaseous or vaporized waste products from the blood. See Ventilation.

RESPIRATORIA Larvae of some Diptera: Respiratory sclerites. See Polypneustic Lobes.

RESPIRATORY HORN Some Hemiptera, Hymenoptera and Diptera: Chorionic protuberances from the egg surface which function as a Plastron in respiration. Respiratory horns are a morphological solution to the problem of gas exchange involving a Plastron. The respiratory horn forms one to several elongate projections of the eggshell. The Plastron is restricted to the horn area and the remainder of the egg surface is impermeable to water. Water loss is restricted to the cross-sectional area at the base of the horn. Without respiratory horns, the Plastron may cover the entire surface of the egg, or be restricted to a small area of the egg. Eggs with a Plastron distributed over the entire eggshell surface are more likely to desiccate when compared with eggs possessing a respiratory horn. See Respiration. Cf. Plastron.

RESPIRATORY MUSCLES Any of several muscles or muscle groups which are associated with the insect respiratory system. Muscles of terrestrial insects include: Abdominal Expiratory Muscles (Tergo-Sternal Muscles), Abdominal Inspiratory Muscles (Dorsal and Ventral Lateral External Muscles) and Spiracular Muscles (Opener and Closer Muscles.) Muscles of aquatic insects include: Gill Protractor Muscles, Gill Retractor Muscles.

RESPIRATORY SYSTEM The complex association between an anatomical Tracheal System and physiological adaptations which facilitate gas exchange between an organism's body and the environment (atmosphere, water, metazoan host, plant tissue). The basic plan involves the transport of Oxygen and CO_2 through a network of tubes within the body. Typically the Tracheal System tubes are internal, cuticular in origin and ramify or branch into a series tubes which connect with the side of the body via spiracles. Several designs of the RS are manifest among the Insecta. RS design has been influenced by oxygen demand, stage of development, habitat and local edaphic factors. See Cutaneous Respiration; Tracheal Respiration; Tracheal System. Cf. Respiratory Muscles; Spiracle; Trachea; Tracheal Gill; Tracheole. Rel. Metabolism.

RESPIRATORY TRACHEAE See Ventilation Tracheae.

RESPLENDENT SHIELD BEARER *Coptodisca splendoriferella* (Clemens) [Lepidoptera: Heliozelidae].

RESPOND® See Resmethrin.

RESPONSAR® See Cyfluthrin.

RESTRICTED Trans. Verb. (Latin, *restrictus* past part. of *restringere* = to restrain.) Held back; suspended; confined to a limited area.

RESTRICTED-USE PESTICIDE One of two categories of pesticide established by the USA Environmental Protection Agency that may be applied only by applicators certified by the state in which they work. See General-use pesticide.

RESTRICTION ENZYME An enzyme used in genetic engineering to cut through DNA at specific points. See Genetic Engineering. Rel. Biotechnology.

RESTRICTION FRAGMENT LENGTH POLYMORPHISM Fragments of DNA that are of different length and produced by cutting DNA with restriction enzymes. See Genetic Engineering. Cf. Restriction Enzyme.

RESUPINATE Adj. (Latin, *resupinare* = to bend backward; *-atus* = adjectival suffix.) Upside down; horizontally reversed. Alt. Resupinatus.

RESURGENCE Noun. (Latin, *resurgere* = to rise. Pl., Resurgences.) A situation in which a pest population that has been lowered rebounds to achieve densities higher than before an initial suppression occurred.

RETADOR® See Propargite.

RETE Noun. (Latin, *rete* = net.) A net, network or plexus.

RETE MUCOSUM Noun. The Hypodermis. See Epidermis.

RETECIOUS Adj. (Latin, *rete* = net; Latin, *-osus* = with the property of.) Resembling a network.

RETICULATE Adj. (Latin, *reticulatus* = latticed; *-atus* = adjectival suffix.) Descriptive of surface sculpture, usually the insect's Integument, that is covered with net-like (intermeshed) lines; a net-like display of Carinae, incised striate or rugae. Alt. Reticular; Reticulated; Retiiculatus; Reticulose;

Reticulous; Reticulosus. See Sculpture Pattern. Cf. Alveolate; Baculate; Clavate; Echinate; Favose; Gemmate; Psilate; Punctate; Rugulate; Scabrate; Shagreened; Smooth; Striate; Verrucate.

RETICULATE MITE *Lorryia reticulata* (Oudemans) [Acari: Tydeidae].

RETICULATE-WINGED TROGIID *Lepinotus reticulatus* Enderlein [Psocoptera: Trogiidae].

RETICULATED BEETLES See Cupedidae.

RETICULUM Noun. (Latin, *reticulum* = a small net. Pl., Reticula.) 1. A net-like structure or sculptural pattern. 2. Longitudinal and radiating filaments of the insect muscle (Packard).

RETINA Noun. (Latin, *rete* = net.) The light-sensitive part of the insect eye upon which the image is formed. See Compound Eye; Lens; Cone; Rhabdom; Pigment Cells.

RETINACULUM Noun. (Latin, *retinaculum* = tether. Pl., Retinacula.) 1. Collembola: A hook-like holdfast structure on the venter of the 3rd abdominal segment. Retinaculum is adapted to hold the Furca on the fourth abdominal segment and form part of the jumping mechanism of Collembola. See Tenaculum. 2. Coleoptera: A medial, serrated, tooth-like process of the Mandible (Cf. Prostheca). 3. Lepidoptera: A membranous hook, series of hooks or groups of specialized scales along the posterior and ventral part of the forewing into which the Frenulum is fitted. See Hamus. 4. Hymenoptera: Horny scales that move the Sting or prevent its hyperextension from the body.

RETINAL Adj. (Latin, *rete* = net; *-alis* = pertaining to.) Descriptive of or pertaining to the Retina.

RETINAL CELLS The cells composing the Retina of the eye.

RETINAL PIGMENT The pigment layer of the compound eye just above the basilar or fenestrate membrane.

RETINAL PIGMENT CELLS Pigment cells in the retinal region of the eye (Snodgrass). Syn. Secondary Pigment Cells.

RETINERIA Noun. (Pl., Retineriae.) Microscopic seta-like projections on the ventral side of the Tarsi (MacGillivray). See Acanthae.

RETINOPHORA See Retinula.

RETINUE Social Insects: A group of workers which attend the queen.

RETINULA Noun. (Latin, dim. form of *retina*.) 1. A group of two, three, or more, visual cells which surround and secrete a longitudinal optic rod or Rhabdom (Imms). 2. The nerve-end cells of the fibres passing through the periopticon (Packard). 3. The basal part of the Ommatidium, composed of a group of slender pigmented cells (Needham). 4. The sensory element of an Ommatidium. A Retinula is structurally complex and consists of elongate and differentiated nerve cells. Number of Retinula cells ranges from 3–11 and typically 6–8 occur in each Ommatidium (*e.g.* 8–9 occur in hornet *Vespula maculata;* beetles have eight retinular cells; cockroach *Periplaneta americana* has seven retinular cells in each Ommatidium). See Rhabdom; Rhabdomere.

RETINULAR Adj. (Latin, *rete* = net.) Descriptive of or pertaining to the Retinula.

RETORT-SHAPED ORGANS Hemiptera: Oval areas of glandular tissue at the enlarged proximal ends of both pairs of the mouth stylets. The function of the areas has not been established.

RETRACTED Adj. (Latin, *retractus* = withdrawn.) A structure drawn back or into another part. Cf. Prominent.

RETRACTILE Adj. (Latin, *retractus* = withdrawn.) Descriptive of structure that is capable of being advanced forward and retracted backward.

RETRACTOR Noun. (Latin, *retrahere* = to draw back; *or* = one who does something. Pl., Retractors.) Any structure used to withdraw or shorten. Specifically and often applied to a muscle. Cf. Protractor.

RETRACTOR ANGULIS ORIS One of the pair of large muscles arising dorsally on the Frons and inserted in the Hypopharynx (Snodgrass).

RETRACTOR HYPOPHARYNGIS Retractor of Hypopharynx; one of a pair of muscles arising in the Tentorium and inserted lateraly at the base of the Hypopharynx (Snodgrass).

RETRACTOR MUSCLE Any muscle used in drawing rearward or flexing an appendage. Cf. Extensor; Flexor; Protractor Muscle.

RETRACTOR OF THE CLAWS Depressor muscle of the claws.

RETRACTORES ANGULORUM ORIS See Retractors of the mouth angles.

RETRACTORES VENTRICULI The delicate muscles which assist in supporting the Alimentary Canal.

RETRACTORS OF THE MOUTH ANGLES A pair of large muscles arising dorsally on the Frons, inserted on the oral branches of the suspensorial sclerites of the Hypopharynx. Alt. Retractores Angulorum Oris (Snodgrass).

RETROARCUATE Adj. (Latin, *retro* = backward; *arculatus* = curved; *-atus* = adjectival suffix.) Descriptive of structure which is curved backward.

RETROCEREBRAL NERVOUS SYSTEM See Stomatogastric Nervous System.

RETROCESSION Noun. (Latin, *retro* = backward; *cessus*, pp of *cedere* = to go.) Going or moving backward.

RETROGRESSIVE DEVELOPMENT An evolutionary trend that results in simplification of an organism, usually through partial, substantial or complete loss of one or more structures (*e.g.,* loss of wings in many Orders of insects). Syn. Regressive Development.

RETRORSE Adj. (Latin, *retrorsum* = backward.) Backwards; in a backward direction.

RETROTECTUM Noun. (Latin, *retro* = backward; tectum = roof.) Acarology: The collar attached to the proximal part of a leg segment.

RETROVIRUS Noun. (New Latin.) RNA virus that replicates via conversion into a DNA duplex.

RETURNING VEIN Wings of certain polyphagous Coleoptera: The incompletely chitinized Medi (M) of Tillyard. See Vein; Wing Venation.

RETUSE Adj. (Latin, *retusus* = blunted.) Descriptive of structure terminating in an obtuse sinus or broad, shallow notch.

RETZIUS, ANDREAS JOHANN (1742–1821) (Eiselt 1836, *Geschichte, Systematik und Literatur der Insektenkunde.* 255 pp. (46), Lepizig.)

RETZIUS, GUSTAV (1841–1919) (Anon. 1922, K. svenska VetenskAkad Arsb. 1922: 239–243.)

REUSS, ADOLF (1804–1879) (Anon. 1879, Ber. senckenb. naturf: Ges. 1878–1879: 7.)

REUTER, ENZIO RAFAEL (1867–1951) (Soumaleinen 1951, Ann. ent. fenn. 17: 49–51.)

REUTER, ODO MORANNAL (1850–1913) Finnish academic and student of Hemiptera who published more than 500 papers and best known for his extensive and detailed work on the Miridae. (E. B. 1913, Entomol. mon. Mag. 49: 230–231.)

REUTIMEYER, KARL ERNST (1889–1971) (Bros 1971, Mitt. ent. Ges. Basel 21: 126–127, bibliogr.)

REUTTI, CARL (1830–1894) (Hering 1894, Stettin. ent. Ztg 55: 305–307.)

REVEILLET, MARIUS (–1963) (Anon. 1963, Bull. Soc. ent. Fr. 69: 213.)

REVELIERE, EUGENE (1822–1892) (Anon. 1892, Insektenbörse 9: [9].)

REVERDIN, JACQUES LOUIS (1842–1929) (Hemming 1929, Entomologist 62: 93–96, bibliogr.)

REVERSAL Noun. (Middle English, *reversen* = to reverse; Latin, *-alis* = characterized by. Pl., Reversals.) Systematic theory: A form of parallelism in which an acquired character state reverts to an ancestral condition. See Homoplasy; Parallelism.

REVERSE Adj. (Latin, *reversus* = to turn back) 1. Turned toward an unusual or contrary direction, as upside down or inside out. Term applied to wings when they are deflexed, with hindwing margins projecting beyond those of the forewings. 2. A morphological feature or character that has reverted to a primitive state. Syn. Converse; Obverse.

REVERSIBLE COLOUR CHANGE A phenomenon common in Crustacea (*e.g.* lobsters) and some other invertebrate groups, but not common within the Insecta; RCC has been reported in Coleoptera. Scarab genus *Dynastes* has an exocuticle with an upper 3 μ transparent area, a 5 μ yellow spongy area and a black layer beneath. When filled with air, the yellow spongy area is optically heterogenous and yellow is reflected. When filled with water, yellow spongy area becomes optically homogeneous, light is absorbed and black is reflected. During night, relative humidity is higher, Cuticle absorbs moisture and body appears black. During day, relative humidity is lower, yellow colour appears and the animal is less conspicuous. This phenomenon may be an adaptation to avoid predation. Colour change can be rapid, and occur within 30 seconds in dead specimens. See Coloration. Cf. Pigmentary Colour; Structural Colour.

REVERTIBLE THELYTOKY A form of thelytokous parthenogenesis found in some parasitic Hymenoptera. Protobacteria (*Wolbachia* spp.) infect female wasps and maintain all-female progeny production without fertilization of female gametes by male gametes. RT can be 'cured' by treatment of females with diet-induced antibiotics or high temperature. Syn. Microbe-associated Thelytoky. See Parthenogenesis; Thelytoky. Cf. Nonrevertible Thelytoky.

REVISION Noun. (Latin, *revisio* = a seeing again.) A comprehensive taxonomic treatment of a higher taxonomic category such as a Genus, Tribe or Family. Revisions include, but are not restricted to, descriptions of new Taxa, redescriptions of poorly characterized Taxa, clarification of nomenclatural problems, proposal of new synonymy, review of existing synonymy, assessment of previous taxonomic work, statements of geographical distribution, phenology, and summaries of biological information. Cf. Monograph; Synopsis.

REVIVESCENCE Noun. (Latin, *revivere* > *re-*; *vivere* = to live.) Coming back to life; awakening from hibernation or torpor (Say).

REVOLUTE Adj. (Latin, *revolvere* = to roll back.) Spirally rolled backward. Alt. Revolutus.

REVY, D (1900–1954) (Kaszab 1954, Folia ent. hung. 7: 21–28.)

REY, CLAUDIUS (1817–1895) (Anon. 1895, Entomol. mon. Mag. 31: 122–123; Guillebeau 1895, Ann. Soc. ent. Fr. 64: 127–130.)

REY, EUGENE (1866–1949) (Sachtleben 1941, Arb. morph. taxon. ent. Berl. 8: 287.)

REYNOLDS, LAWRENCE R (1878–1922) (Anon. 1922, Science 56: 475.)

RFLP Acronym. See Restriction Fragment Length Polymorphism.

RH-5992® See Tebufenozide.

RH-988® See Triazamate.

RHABDITES Noun. (Greek, *rhabdos* = rod; *-ites* = dim. suffix.) 1. The blade-like elements of the Sting and Ovipositor. 2. Rod-like or blade-like processes projecting from the Epidermis.

RHABDOM Noun. (Greek, *rhabdos* = rod.) 1. Compound eye: The rod-like structure lying in the axis of the Retinula and positioned below the Cone. 2. The collective Rhabdomeres of an Ommatidium. The Rhabdom collects light and propagates electrical signal to the Brain. Gross anatomical study of the Rhabdom shows considerable variation in its cross-sectional form. The Rhabdom is open in the honeybee and closed in the cockroach; sometimes the Rhabdom is twisted and compact. In many Species, Ommatidia are short in the dorsal region of the compound eye and longer elsewhere on the eye. The Rhabdom may be nearly rectangular in dorsal region and differently shaped elsewhere.

RHABDOMERE Noun. (Greek, *rhabdos* = rod; *meros* = part. Pl., Rhabdomeres.) 1. Compound eye: The rod-like distal portion of a Retinula Cell. 2. The Rhabdomere is analogous with the vertebrate Retina. The Rhabdomere forms along one or two longitudinal surfaces of the Retinula Cells. Ultrastructurally, each Rhabdomere consists of microtubules that are hexagonal or round in cross section. Each microtubule is about 500 Å long. Collectively, microtubules are usually positioned at a right angle to primary axis of Retinula Cells. (Exceptions include dolichopodid flies and gerrid bugs whose microtubular borders are parallel to the optical axis.) Microtubules contain rhodopsin at a concentration of ca 17,000 molecules per microlitre. See Retinula; Rhabdom.

RHABDOPODA Noun. (Greek, *rhabdos* = rod; *pous* = foot.) Clasping organs of the ninth abdominal segment of male.

RHABDOVIRIDAE Plural Noun. A small Family of arboviruses which are of relatively minor importance in terms of their impact on humans. Virons bullet-shaped, ca 100–400 nm long and 45–100 nm wide, with lipoprotein envelope and surface projections. Rhabdovirids vectored by blood-feeding nematocerous Diptera. Viruses produce a few diseases including Bovine Ephemeral Fever and Vesicular Stomatitis. See Arbovirus. Cf. Bunyaviridae; Flaviviridae; Tongaviridae; Reoviridae.

RHACHIBEROTHIDAE Tjeder 1959. Plural Noun. A Family of Neuroptera sometimes placed in the Berothidae or Mantispidae. Family restricted to Africa and includes about 10 Species. Biology unknown; adults taken at lights; first instar larva of one Species known.

RHACHIS See Rachis.

RHAGIGASTERINAE Ashmead 1903. Subfamily of aculeate Hymenoptera within the Thynnidae.

RHAGIONIDAE Plural Noun. Snipe Flies. A cosmopolitan, generalized Family of lower brachycerous Diptera, consisting of about 20 Genera and 500 nominal Species. Fossil record known from the Jurassic Period. Related Families include Vermileonidae, Pelecorhynchidae, Tabanidae and Athericidae. Sometimes assigned to Tabanodea; difficult to define based on morphological features. Body moderate-sized, robust, with relatively short Antenna and long legs; sometimes with abdomen apically pointed or resembling wasp or bee. Antenna usually with sharp separation of basal flagellar segment from distal Flagellomeres; Clypeus exposed, strongly convex; Scutellum large, setose; wing with two Submarginal Cells, 4–5 Posterior Cells; Anal Cell closed, open or Anal Vein effaced apically and Stigmal area wing usually more or less pigmented. Adults typically sluggish and found in association with wetlands, meadows and forest habitats; some Species haematophagous as adults. Larvae occur predominantly in decaying vegetation where they feed upon soft-bodied invertebrates. Cf. Vermileonidae.

RHAGOLETIS Loew 1862. Plural Noun. A widespread, economically destructive Genus of true fruit flies [Tephritidae] that includes 65 Species with represenatives in the New World Europe and temperate Asia. Most known Species are stenophagous, attacking the fruits of closely related plant Species. See Tephritidae.

RHAPHIDIODEA Burmeister 1839. (Greek, *rhaphis* = needle; *eidos* = form.) Noun. A synonym of Raphidoidea. Also: Raphidioptera, Rhaphidioptera.

RHAPHIDIOPTERA Noun. (Greek, *rhaphis* = needle; *pteron* = wing.) A synonym of Raphidoidea. Also: Rhaphidiodea, Raphidioptera.

RHAPHIDOPHORIDAE Plural Noun. Camel Crickets. A Family of ensiferous Orthoptera including 30 Genera and about 250 nominal Species, that are assigned within seven Subfamilies. Morphologically Rhaphidophoridae are characterized by the apex of the Metatarsus with one spine at most, and the hindleg is unusually long (except some fossorial North American Species). The Family is widespread and its representative Species can be taken in caves, crevices, and related habitats. The oldest recognizable illustration of an insect (Magdalenian, ca 16,000 BPE) is of *Troglophilus* from a cave in southern France. See Orthoptera.

RHAPHIDOPHOROIDEA A Superfamily of ensiferous Orthoptera whose Taxa are sometimes included within Gryllacridoidea. Rhaphidophorids are apterous, lack auditory and stridulatory devices; Antennae close-set and Tarsi strongly compressed laterally. Superfamily includes Families Macropathidae and Rhaphidophoridae; collectively representing ca 45 Genera and 300 Species. Most Species live in caves, burrows, crevices, or in sand; a few taken on vegetation. Most rhaphidiophorids feed on dead or moribund invertebrates; some Species eat plant material.

RHEGMATOCYTE See Spherulocyte.

RHEOTAXIS Noun. (Greek, *rheo* = current; *taxis* = arrangement.) Orientation or reaction in response to water currents. See Orientation. Cf. Aerotaxis; Anemotaxis; Chemotaxis; Geotaxis; Menotaxis; Notarotaxis; Osmotaxis; Phototaxis; Rotaxis; Scototaxis; Strophotaxis; Telotaxis; Thermotaxis; Thigmotaxis; Tonotaxis; Tropotaxis.

RHEOTROPISM Noun. (Greek, *rheo* = current; *trepein* = to turn; English, *ism* = state.) Orientation to currents of water. See Tropism. Cf. Aeolotropism; Anemotropism; Chemotropism; Electrotropism; Galvanotropism; Geotropism; Heliotropism; Hydrotropism; Phototropism; Stereotropism; Thermotropism; Thigmotropism; Tonotropism. Rel. Taxis.

RHINARIUM Noun. (Greek, *rhis* = nose; *arium* = place of a thing. Pl., Rhinaria.) 1. A nostril piece or portion of the Nasus. 2. Odonata: Lower portion of the Clypeus. See Anteclypeus. 3. Hymenoptera: Elevated, keel-like ridges on the flagellar segments of parasitic Hymenoptera Antennae.

RHINOCEROS BEETLE *Xyloryctes jamaicensis* (Drury) [Coleoptera: Scarabaeidae].

RHINOCEROS BEETLES See Scarabaeidae.

RHINOPHORIDAE Plural Noun. A small Family of schizophorous Diptera assigned to Superfamily Muscoidea. Adults resemble tachinids. Postscutellum weakly developed; Calypters narrow; Frontal Suture present. Two Species occur in USA. Larvae are parasitic on terrestrial isopods.

RHINOTERMITIDAE Plural Noun. A Family of Isoptera. Alates with Ocelli; Fontanelle on head; Antenna with 14–22 segments; Mandible dentition asymmetrical; forewing scale large; wings often recticulate; tarsal formula 4-4-4; Cerci with two segments. Soldiers without eyes; Fontanelle present. Worker caste always present; most Species subterranean, some build mounds or maintain contact with soil via earthen tubes.

RHIPICERIDAE Plural Noun. A cosmopolitan Family of polyphagous Coleoptera with about 160 nominal Species. Placed in Elateroidea, Dascilloidea or form central group in Rhipiceroidea. Body 10–25 mm long, dark with areas of light pubescence; Antenna with 11 to more than 20 segments, flabellate to lamellate in male, serrate to pectinate in female; articulating posternal process absent; tarsal formula 5-5-5; Abdomen with five Ventrites. Larval head prognathous; legs with 3–5 segments, 0–1 tarsal claws; Urogomphi present or absent. Biology poorly known; typically mine rotting wood; one Species ectoparasitic on cicada Pupae.

RHIPIPHORIDAE Plural Noun. A cosmopolitan Family of Coleoptera consisting of ca 300 nominal Species. Classified with Mordelloidea, Meloidea and Cucujoidea; appear related to mordellids and stylopids. Separated from other Coleoptera based on males with pectinate or flabellate Antenna, females with serrate Antenna; both sexes lack apical abdominal spine and preapical ridge on hind Tibia; larviform females distinguished from stylopids by presence of Trochanters on fore and middle legs and with a Prothorax. All Species parasitic and pass at least part of larval life as internal parasites. Biological adaptation almost unique within Coleoptera, providing Strepsiptera considered separate. Rhipiphorinae parasitize larvae of wasps and bees (Macrosiagonini attack wasps, Rhipiphorini attack bees); Rhipidiinae parasitize cockroaches. Rhipiphorids deposit eggs away from body of host. Bee and wasp parasites with triungulin larva phoretically transported to nest of host. Parasite larva penetrates Integument of immature host. Following short feeding period, parasite emerges from host and continues to feed externally. Biology of hyperparasitic groups somewhat more complex. Rhipiphorids which parasitize cockroaches also have triungulin larva which penetrates Integument of host and feeds internally throughout period of development. Most rhipiphorids apparently undergo one generation per year, but cockroach parasites in tropical areas probably go through more than one generation since hosts are active throughout year. Overwintering occurs in various stages, depending on Species of rhipiphorid.

RHIPIPTERA See Strepsiptera.

RHIZOME Noun. (Greek, *rhizoma* = root. Pl., Rhizomes.) A thick horizontal stem of plants, usually underground, which sends roots further into soil and shoots into air. Cf. Stolon.

RHIZOMORPH Noun. (Greek, *rhizoma* = root; *morphe* = form. Pl., Rhizomorphs.) Fungi: A specialized type of fungal mycelium in which strands of hyphae are twisted together and appear rootlike.

RHIZOPHAGIDAE Plural Noun. A cosmopolitan Family consisting of ca 200 nominal Species. Typically less than 2 mm long, resemble colydiids and cucujids. Distinguished by ten segmented Antenna with a two-segmented club, Elytra truncate, apical Terga exposed, front Coxal cavities closed posteriad, first abdominal Sternum (exposed) as long as next two Sterna. Higher classification recognizes 4–6 Subfamilies. Biology poorly studied. Species predaceous on bark beetles, feeding on eggs and larvae of several Species of scolytids. One Species reported as carrion feeder and named 'graveyard beetle' because it supposedly feeds on human cadavers.

RHIZOPON AA® See Indolebutyric Acid.

RHIZOPON-B® See NAA.

RHOADS, SAMUEL NICHOLSON (1862–1952) (P.P.C. & J.A.G.R. 1953, Ent. News 64: 125.)

RHODANASE Noun. An enzyme found in insects which has the capacity to detoxify cyanide (Lang 1933, Biochem. Zeit. 259: 243–256.)

RHODESGRASS SCALE *Antonina graminis* (Maskell) [Hemiptera: Pseudococcidae]: Endemic to Orient; described from grass in Hong Kong during 1897; detected in Texas on rhodesgrass (*Chloris gayana* Kunth) during 1942. Subsequently, recovered from nearly 100 Species of grasses in more than 25 countries. Adult thelytokous, covered with cottony secretion, produces live young (viviparous); crawler cream coloured; second instar sedentary, covered with wax. Controlled in places by a small, flightless, parasitic wasp, *Neodusmetia sangwani* (Rao), collected in India. Alt. Rhodesgrass Mealybug.

RHODESIAN SLEEPING SICKNESS See Sleeping Sickness.

RHODNIUS Stål 1859. [Hemiptera: Reduviidae].

RHODOCIDE® See Ethion.

RHODODENDRON BORER *Synanthedon rhododendri* (Beutenmüller) [Lepidoptera: Sesiidae]: A pest of rhododendron, azaleas and mountain laurel in North America. See Sesiidae. Cf. Currant Borer.

RHODODENDRON GALL MIDGE *Clinodiplosis rhododendri* (Felt) [Diptera: Cecidomyiidae].

RHODODENDRON LACE BUG *Stephanitis rhododendri* Horvath [Hemiptera: Tingidae]: A pest of Rhododendron and Mountain Laurel in North America. Causes white-peppered discoloration on upper surface of leaves. Overwinters in egg stage.

RHODODENDRON WHITEFLY *Dialeurodes chittendeni* Laing [Hemiptera: Aleyrodidae].

RHODOFIX® See NAA.

RHODOPTERA Plural Noun. Apterous insects with sucking mouth structures.

RHOMB Noun. (Greek, *rhombos* = an object that can be turned.) An equal-sided, four-sided figure with opposite sides parallel and with two opposite angles acute and the other two obtuse. Alt. Rhombus.

RHOMBOID Adj. (Greek, *rhomboeides* = rhomboid.) Rhombus-like; descriptive of a structure whose outline shape is a parallelogram in which the angles are oblique and the adjacent sides are unequal. Alt. Rhomboidal. Rel. Eidos; Form; Shape.

RHOPAEA CANEGRUB *Rhopaea magnicornis* (Blackburn) [Coleoptera: Scarabaeidae]: In Australia, larvae feed on roots to damage pineapples, pastures and sugarcane in Queensland. See Canegrub.

RHOPALIDAE Amyot & Serville 1843. Plural Noun. Scentless Plant Bugs; Rhopalid Bugs. A small Family of Hemiptera, closely related to Coreidae. Antenna above midline of head; first antennal segment shorter than head length; Ocelli large and tuberculate; Pronotum with transverse ridge; wing membrane with numerous veins and Corium with several closed cells; thoracic scent gland oriface very small. All Species are phytophagous and most Species occur in weedy fields; a few are arboreal.

RHOPALOCERA Noun. The series of Lepidoptera in which the Antennae are alike in both sexes and form a club at the apex.

RHOPALOMERIDAE See Ropalomeridae.

RHOPALOSOMATIDAE Ashmead 1896. Plural Noun. A small Family of aculeate Hymenoptera assigned to the Pompiloidea.

RHOPHILOUS Adj. (Greek, *rheo* = current; *philos* = loving; Latin, *-osus* = with the property of.) Current loving. Descriptive of insects that live within the currents of moving waters, *e.g.* Odonata. See Habitat. Rel. Rheotropism.

RHOPHITINAE A Subfamily of Halictidae (Apoidea) consisting of about 15 Genera.

RHOPHOTEIRA An ordinal term for fleas (Clairville). See Siphonaptera.

RHUBARB CURCULIO *Lixus concavus* Say [Coleoptera: Curculionidae]: A widespread pest throughout USA. Adults found in stems of rhubarb; larvae found in stems of dock, thistle and sunflower. Overwinters as an adult.

RHUMBLER, LUDWIG (1864–1939) (Escherich 1940, Z. Angew. Ent. 26: 682–683.)

RHYACOPHILIDAE Plural Noun. Primitive Caddisflies. A moderate-sized Family of Trichoptera assigned to Superfamily Rhyacophiloidea; ca 100 Species occur in North America. Adults resemble Glossosomatidae. Ocelli present; Maxillary Palpus with second segment rounded or globose; fore Tibia with preapical spur. Larvae are predatory and do not construct cases.

RHYACOPHILOIDEA A Superfamily of Trichoptera (caddisflies) including Glossosomatidae, Hydrobiosidae, Hydroptilidae and Rhyacophilidae.

RHYNCHITIDAE Plural Noun. Thief Weevils. Tooth-nosed Weevils. A Family of polyphagous Coleoptera assigned to Superfamily Curculionoidea. Adult ca 1.5–6.5 mm long; Beak long, nearly parallel-sided; Mandible flat with teeth along lateral margin; antennal club compact; Tibiae with short, straight apical spur. Inhabit low vegetation. Some Species are economic pests. Larvae feed on buds, fruits and nuts.

RHYNCHOPHORA The section of the Coleoptera in which the head is produced into a Snout, at the end of which the mouth structures are situated; the weevils.

RHYNCHOPHTHIRINA Ferris 1931. Elephant Lice. A small Suborder of Phthiraptera, including the Genus *Haematomyzus* whose Species parasitize African and Asian elephants and the African warthog.

RHYNCHOTA Burmeister. 1. The old Order Hemiptera, in the broad sense. Syn. Rhyngota. 2. Insects in which mouthparts are prolonged into a Beak or Rostrum to protect piercing lancets.

RHYNCHUS Noun. (Greek, *rynchos* = beak.) The Promuscis (Fabricius).

RHYNIE CHERT Devonian fossil bed at Rhynie, Scotland west of Aberdeen. Specimens preserved in Chert (impure flint-like rock), not separated from their matrix. RC well known for fossil plants documenting early vascularized plants; RC first reported by William Mackie (1914). Other significant Devonian deposits include Gilboa (New York, USA) and Alken an der Mosel.

RHYPARDIA BEETLE *Rhypardia* spp. [Coleoptera: Chrysomelidae]: A minor and sporadic pest of *Citrus* in Australia. Adult ca 7 mm long, metallic brown. Eggs laid in pasture soil; larvae feed on grass roots, then pupate in soil. Adult emergence synchronized by heavy rains during spring and summer; beetles migrate in swarms to tree crops where they feed on foliage, flowers and fruit; swarms can severely damage young foliage and cause *Citrus* rind disfigurement within 24 hours. Multivoltine with 2–4 generations per year depending upon climate.

RHYPAROCHROMINAE A large, widespread Subfamily of Lygaeidae consisting of 13 Tribes. Predominantly litter-dwelling in habit. (Sweet 1964, Ent. Amer. 43: 1–124; Ent. Amer. 44: 1–201. Slater 1986, J. N. Y. ent. Soc. 94: 262–280.)

RHYPHIDAE Plural Noun. See Anisopodidae.

RHYSODIDAE Plural Noun. Wrinkled Bark Beetles. A numerically small, widespread Family of adephagous Coleoptera, sometimes placed in Carabidae. Adult body black, shiny, sub-depressed, 5–10 mm long; deep Sulcus on head and Pronotum sometimes lined with Setae and filled with debris. Head strongly constricted posteriorly to form neck; Antenna moniliform with 11 segments. Fore Coxa globose; frontal cavity externally open, internally closed; middle Coxae widely separated, coxal cavity open laterally; transverse Mesosternal Sulcus not present. Legs rather short; tarsal formula 5-5-5. Abdomen with six Ventrites and three connate, fused, basal Ventrites. Adult and larva inhabit decaying logs.

RHYTHM Noun. (Greek, *rhythmos* = measured motion. Pl., Rhythms.) Periodic and predictable changes in physiological process or behavioural action. Rhythms may be phase-set by endogenous (internal) or exogenous factors. Rhythms may be momentary, hourly, daily, monthly, seasonal, annual or generational. See Circadian Rhythm. Diel Periodicity.

RHYTHMIC Adj. (Greek, *rhythmos* = measured motion; *-ic* = of the nature of.) Pertaining to behavioural action or physiological process which is predictable and periodic in manifestation. Alt. Rhythmical.

RIBAGA, COSTANTINO (1870–1949) An assistant of Antonio Berlese and with an interest in mite systematics and Psocoptera.

RIBAULT, H (–1967) Professor at Faculte de Medecine de Toulouse and amateur Entomologist specializing in Homoptera-Heteroptera. Collection deposited in Museum d'Historie Naturelle de Paris. (Wagner 1968, Ent. Z., Frank. a. M. 78: 245–246.)

RIBBANDS, CHARLES RONALD (1914–1967) (Lewis 1967, Nature 214: 1171.)

RIBBE, CARL (1860–1934) (Heller 1934, Dt. Ent. Zeit. Iris 48: 138–143, bibliogr.)

RIBBE, HEINRICH (1832–1898) (Anon. 1898, Societas ent. 12: 165.)

RIBBON CENTIPEDE *Schizoribaubtia aggregatum* [Scolopendromorpha: Scolopendridae]: The Species occurs in Australia within moist leaf litter and under stones; body pale orange with thick Antenna; apparently cannot bite humans. See Chilopoda. Cf. Giant Centipede; House Centipede.

RIBEIRO, ALIPIO MIRANDA (1874–1939) (Anon. 1939, Boln. Biol. (São Paulo) 4: 153–159, bibliogr.)

RICE Noun. (Greek, *oryza* = rice; Latin, *risium* = rice. Pl., Rices.) Currently the most widely grown cereal grain cultivated for human food; grown between 53°N and 40°S latitude with thousands of varieties developed; all cultivars developed from *Oryza sativa* from Asia and *O. glaberrima* from West Africa. Historically, Alexander the Great has been credited with bringing rice from India to the west (ca 344 BC).

RICE BLOOD-WORM *Chironomus tepperi* Skuse [Diptera: Chironomidae], an important pest of rice in Australia. Larvae attack endosperm of seeds and roots of establishing seedlings. Infestations of RBW lead to lower plant densities.

RICE BUG *Leptocorisa oratorius* (Fabricius) [Hemiptera: Alydidae], an important pest of rice grown in Asia.

RICE DELPHACID *Sogatodes orizicola* (Muir) [Hemiptera: Delphacidae].

RICE LEAF-FOLDER *Lerodea eufala* (Edwards) [Lepidoptera: Hesperiidae].

RICE LEAFHOPPER *Nephotettix nigropictus* (Stål) [Hemiptera: Cicadellidae].

RICE LEAF-ROLLER *Gnaphalocrosis medinalis* (Guenée) [Lepidoptera: Pyralidae].

RICE MOTH *Corcyra cephalonica* (Stainton) [Lepidoptera: Pyralidae]: A serious pest of stored products (including rice, dried fruits, nuts and spices); cosmopolitan but common in humid tropical regions, particularly southeast Asia. Direct damage caused by larval feeding; indirect damage caused by contamination with frass and silk webbing. Adult brown; hindwing pale brown; wingspan 15–25 mm. Eggs sticky and laid in produce; eclosion occurs ca four days. Larval body white with brown head capsule and prothoracic shield; seta surrounded by black circle above each spiracle; each spiracle with posterior margin of peritreme thickened; instar number variable, typically seven in male and eight in female. Pupation within produce or on containment. Life cycle ca one month.

RICE ROOT-APHID *Rhopalosiphum rufiabdominalis* (Sasaki) [Hemiptera: Aphididae].

RICE SKIPPER *Parnara guttatus* Bremer & Grey [Lepidoptera: Hesperiidae].

RICE STALK-BORER *Chilo plejadellus* Zincken [Lepidoptera: Pyralidae].

RICE STEM-BORER *Chilo suppressalis* (Walker) [Lepidoptera: Pyralidae]: Native to the Orient and accidentally introduced into Egypt, Spain and Hawaii. A major pest of rice, corn and sorghum. Female nocturnal, larvae gregarious during first three instars, then bore into stems. Pupation within plant stem burrows. Multivoltine with up to four generations per year; can remain in field after harvest. Syn. Asiatic Rice Borer.

RICE STINK-BUG *Oebalus pugnax* (Fabricius) [Hemiptera: Pentatomidae]: A multivoltine pest of rice and grasses in the USA; adult and nymph puncture rice kernel with piercing-sucking mouthparts and secrete saliva which forms a feeding sheath; reduce the yield and quality of rice. Adult 9–12 mm long; body straw-coloured with reddish Antenna; Femora with anterolaterally directed spines. Adults overwinter with spring-summer generations of adults living about 50 days. Egg cylindrical, ca 0.8 mm long, neonate egg green becoming red with development; eggs deposited in two parallel lines on plant; eclosion occurs within 4 days. Five nymphal instars; nymphal stage completed within a month; early instar nymphs predominantly black, abdomen red with black spots; later instars becoming straw-coloured. RSB can undergo up to four generations per year in USA. See Pentatomidae. Rel. Stink Bugs.

RICE WATER-WEEVIL *Lissorhoptrus oryzophilus* Kuschel [Coleoptera: Curculionidae]: A pest of rice, bulrushes, water lily in eastern North America; larva feed on roots of rice plant; reduces plant's ability to absorb nutrients and causes stunting, delays maturity and reduces yield. RWW adult overwinters in matted grass near lakes and ponds; adult migrates to rice fields during spring; adult and larva aquatic. Adult greyish-brown, 2–3 mm long; snout short; feeds on young plants, copulates in water; respires with oxygen in air bubble held beneath Elytra. Egg white, cylindrical, ca 0.5 mm long; eggs deposited in longitudinal slits on leaves; eclosion occurs within 4–9 days. Larval head brown; body white, legless with hook-like projections from tubercles on dorsum of segments 2–7; larval development ca 30 days; mature larva ca 8 mm long. Larva feeds on roots of plant; hook-like spiracles inserted into plant for air. Pupa oval shaped and within watertight cocoon covered with mud; pupation requires ca 5–9 days. RWW with 1–2 generations per year. Syn. Rice-Root Maggot.

RICE WEEVIL *Sitophilus oryzae* (Linnaeus) [Coleoptera: Curculionidae]: A cosmopolitan pest (perhaps endemic to India) of stored grains (rice, wheat) and farinaceous cereal products; prefers whole grain. Direct damage through feeding or contamination of grain with uric acid. Adult 2.5–4.0 mm long, reddish brown with four pale spots on Elytra; Pronotum with shallow, round pits; adults diurnal and actively fly; adult can live 1 year. Female lays ca 150–300 eggs during lifetime; typically chews hold in kernel, oviposits within endosperm; gelatinous plug hardens over oviposition hole; incubation ca 6 days. Four larval instars; mature larva ca 4.0 mm, white with dark head; legless. Larvae cannibalistic and only one adult per grain emerges. Pupation within grain. Life cycle 4–6 weeks; ca 35 days @ 27°C; oviposition in part density and temperature dependent; no oviposition <13 or >35°C. See Curculionidae. Cf. Granary Weevil; Maize Weevil.

RICH, GEORGE BERNARD (1914–1973) (Gregson 1973, Bull. ent. Soc. Can. 5: 40–42.)

RICHARD, LOUIS CLAUDE MARIE (1754–1821) (La Sègue 1845, Musée Botanique de B. Delessert. 588 pp. (474), Paris; Larousse 1876, Grande Dictionaire Universale du 19 siècle 13: 1182.

RICHARDIIDAE Plural Noun. A small-sized Family of schizophorous Diptera assigned to Superfamily Tephritoidea; Family predominantly Neotropical with two Species known from southwestern USA. Adults often taken at fruit-baited traps; wings pictured; Costa spinose; Femora typically incrassate.

RICHARDS, ARTHUR (1885–1973) (Lees 1974, Proc. R. ent. Soc. Lond. (C) 38: 60–61.)

RICHARDS, MAUD J (NÉE NORRIS) (1907–1970) (Anon. 1971, Rep. Anti-Locust Res. Centre 1970–71: 10–11)

RICHARDS, O W (31 December 1901–10 November 1984.) English Entomologist and taxonomic specialist of aculeate Hymenoptera. (Southwood 1987, Biogr. Mem. Fellows R. Soc. 33: 539–571.)

RICHARDSON, A W (1873–1948) (Williams 1949, Proc. R. ent. Soc. Lond. (C) 13: 68.)

RICHARDSON, NELSON MOORE (1855–1925) (Anon. 1925, Entomol. mon. Mag. 61: 207–208; Sheldon 1925, Entomologist 58: 174–175, 200.)

RICHARDSON, WILLIAM ALFRED (1897–1964) (Wigglesworth 1965, Proc. R. ent. Soc. Lond. (C) 29: 54.)

RICHELMANN, GEORG (1851–1924) (Schulze 1924, Dt. Ent. Zeit. Iris 38: 273–278.)

RICHTER, ANDREI ANDREEVICH (1911–1950) (Stakelberg 1952, Ent. Obozr. 32: 341–344, bibliogr.)

RICHTER, PAUL (1841–1891) (Honrath 1891, Berl. ent. Z. 36: 472 473, bibliogr.)

RICINIDAE Plural Noun. A small Family of amblycerous Mallophaga whose Species develop as parasites of hummingbirds in neotropics and passerine birds in Australia. Mandibulate mouthparts; Antenna typically capitate, with four segments, concealed in grooves; Maxillary Palpus with 2–3 segments; Tarsi with paired claws; first abdominal segment fused to Metathorax; spiracles on abdominal segments 2–7.

RICKETTS, HOWARD TAYLOR (1871–1910) American scientist who died of Epidemic Typhus while studying the disease. (Howard 1930, Smithson. misc. Collns. 84: 174.)

RICKETTSIACEAE Plural Noun. A Family of Rickettsiales whose members replicate as parasites or symbionts (endosymbionts) within tissue cells of vertebrates. Tribes within Family include: Ehrlichieae, Wolbachieae and Rickettsieae. Ehrlichieae primarily pathogens of mammals; Wolbachieae are exclusively symbionts of arthropods. Principal Genera of Rickettsieae include *Coxiella, Rickettsia* and *Rochalimaea.* Rickettsiaceae Species serve as aetiological agents for several human diseases, including Epidemic Typhus, Murine Typhus, Q-Fever, Trench Fever and several Tick-Borne Spotted Fevers (Rocky Mountain Spotted Fever, Rickettsial Pox, Scrub Typhus). Rickettsiaceae are replicated and vectored by acarines (mites and ticks) and insects (lice and fleas.) Some acarines maintain Rickettsiaceae via trans-ovarial and trans-stadial transmission, but insects do not. Rel. Endosymbionts.

RICKETTSIAL POX A minor rickettsial disease first reported in humans during the 1940s in the eastern USA and Russia; subsequently reported in South Korea, Equatorial Africa and Italy. Aetiological agent *Rickettsia akari* occurs in house mouse and vectored to humans by the gamasid mite *Liponyssoides sanguineus*. RP transovarially transmitted. Cf. Rocky Mountain Spotted Fever; Scrub Typhus; Tick-Borne Spotted Fever.

RICKETTSIALES (Patronym of H. T. Ricketts.) An Order of prokaryotic gram-negative intracellular parasites or symbionts associated with arthropods, most notably Acari, fleas and lice. Anatomically rod-shaped, coccoid or pleomorphic. Order includes three Families: Anaplasmataceae, Bartonellaceae and Rickettsiaceae. See Anaplasmataceae; Bartonellaceae; Rickettsiaceae. Cf. Arbovirus; Bacteria; Rocky Mountain Spotted Fever; Spirochaetales.

RICKSECKER, LUCIUS EDGAR (1841–1913) (Fall 1913, Ent. News 24: 239–240.)

RICRON® See Dicrotophos.

RIDDELL, JANET (Blair 1942, Proc. R. ent. Soc. Lond. (C) 6: 41.)

RIDDEX® See Dichlorvos.

RIDEOUT, J K (Anon. 1951, Trans. Proc. S. Lond. ent. Hist. Soc. 1950–51: 56.)

RIDGED-WINGED FUNGUS-BEETLE *Thes bergrothi* (Reitter) [Coleoptera: Lathridiidae].

RIDINGS, JAMES (1803–1880) (Cresson 1880, Trans. Am. ent. Soc. 8: xv; Brown 1968, Ent. News 59: 135–137, bibliogr.)

RIDINGS, JAMES H (1842–1908) (Anon. 1908, Ent. News 19: 242.)

RIECKE, HANS (1911–1954) (Sokolowski 1954, Verh. Ver. naturw. Heimatforsch. Hamb. 31: xvii.)

RIEDEL, KARL JULIUS MAX (1862–1937) (Anon. 1938, Arb. morph. taxon. Ent. Berl. 5: 72.)

RIEDEL, MAX PAUL (1870–1941) (Sachtleben 1940, Arb. morph. taxon. Ent. Berl. 7: 27.)

RIEHL, FRIEDRICH (1795–1876) (Dohrn 1876, Stettin. ent. Ztg 37: 189.)

RIEL, PHILIBERT (1862–1943) (Josserland 1944, Bull. mens. Soc. linn. Lyon 13: 33 40, bibliogr.)

RIES, DONALD TIMMERMAN (1904–1968) (Mockford 1968, Ann. ent. Soc. Am. 61: 1630–1631.)

RIES, ERICH (1908–1944) (Sachtleben 1944, Arb. morph. taxon. Ent. Berl. 11: 141.)

RIESEN, AUGUST (1840–1910) (Ziegler 1910, Berl. ent. Z. 55: 264.)

RIFFARTH, HEIRICH H (1860–1908) (Horn 1908, Dt. ent. Z. 1908: 426–427.)

RIFFLE Noun. (Origin obscure. Pl., Riffles.) The surface of a shallow stream which is caused to ripple by current passing over stones and irregularities in the bottom.

RIFFLE BEETLES See Elmidae.

RIFFLE BUGS See Veliidae.

RIFT VALLEY FEVER An arbovirus of the Bunyaviridae known to occur in Africa; acronym RVF. In humans, RVF causes fever with associated aches and pains. Epizootic RVF causes high mortality in young cows and sheep; human epidemics produce a low level of mortality. Vectors of RVF include mosquitoes in several Genera, most often members of the *Culex pipiens* complex. See Arbovirus; Bunyaviridiae. Cf. Oropouche Virus; Sandfly Fever.

RIGID Adj. (Latin, *rigidus* from *rigere* = to be stiff.) Descriptive of structure that is stiff or inflexible. Cf. Flexible.

RIJKSMUSEUM VAN NATURRULIJKE HISTORIE The natural history museum at Leiden, The Netherlands founded by King Willem I in 1820. RMNH serves as the repository of several important collections made in the Dutch East Indies colonies and Japan.

RIKER MOUNT A display containter for relatively large-bodied insects, such as butterflies, larger moths, beetles and dragonflies. Typically a flat cardboard box of variable dimensions, about 3 cm deep, filled with cotton and covered with a pane of glass or plastic set into the cover. Unpinned specimens are placed upside down on glass cover, spread into a display position and held in place with cotton filling the box.

RILEY, CHARLES FREDERICK CURTIS (1872–1933) (Hungerford 1934. Ann. ent. Soc. Am.27: 123.)

RILEY, CHARLES VALENTINE (1843–1895) Dominant figure in late 19th Century Entomology. Born in Chelsea, London, England; Died Washington, D. C. from bicycle accident. Early education at Dieppe, France and Bonn, Germany where he displayed strong artistic talents directed at natural history subjects. Father died; CVR emigrated to American in 1860. Worked in farming near Chicago, Illinois. Began entomological publications during 1863 in 'Prairie Farmer' and during 1864 as staff member of PF. During 1865 served in American Civil War as private in 134th Illinois Volunteer Regiment. Appointed first State Entomologist of Missouri during 1868 through recommendation of B. D. Walsh. Published nine Missouri Reports which are considered classical publications in economic Entomology. Established 'American Entomologist' during 1868 with B. D. Walsh as a rapid outlet for publications; later became 'The American Entomologist and Botanist'; subsequently reverted to original title. Supported Darwinian concept of evolution from its inception. Published influential report on Rocky Mountain Locust in 1877. Instrumental in control of cottony cushion scale in California with imported Vedalia Beetle. Became second USDA Entomologist 1878, replacing Townsend Glover. Resigned 1879; Rehired 1881 by new Commissioner of Agriculture, George B. Loring. Personal collection of CVR (ca 115,000 specimens) became nucleus of U. S. National Collection of insects. Established periodical 'Insect Life' (1888); Founding member 'Association of Official Economic Entomologists' which later became Entomological Society of America. Ultimately resigned as Chief Entomologist, USDA during 1894 in response to appointment of Charles W. Dabney as Assistant Secretary of Agriculture. (Henshaw 1889. Bibliogr. of the more important contributions to American economic Entomology. Pts 2–3. pp. 53–371. Bibliogr. only. Washington; Anon. 1892, Colman's Rural Wld. St. Louis Mon. 12 May; Fletcher 1895, Can. Ent. 27: 273–274; McLachlan 1895, Entomol. mon. Mag. 31: 269–270; Meldola *et al.* 1895, Proc. ent. Soc. Lond. 1895: xxvi–xxx, lxviii–lxix; Goode 1896, Science 3: 217–225.)

RILEY, G (fl 1790) (Lisney 1960, *Bibliogr. of British Lepidoptera 1608–1799*, 315 pp. (246–250), London.)

RILEY, WILLIAM ALBERT (1876–1963) (Hodson 1964, Ann. ent. Soc. Am. 57: 266.)

RIMA Noun. (Latin, *rima* = cleft. Pl., Rimae.) A crack or longitudinal opening with sharp edges.

RIMI® See Chlorpyrifos.

RIMIFORM Adj. (Latin, *rima* = cleft; *forma* = shape.) Structure with a shape resembling narrow fissures. Cf. Sulciform. Rel. Form 2; Shape 2; Outline Shape.

RIMILURE® See Codelure.

RIMILURE-PBW® See Gossyplure.

RIMOSE Adj. (Latin, *rima* = cleft; *-osus* = full of.) 1. With minute, narrow and nearly parallel excavations running into each other. 2. Chinky; resembling the cracked bark of a tree. Alt. Rimosus; Rimous.

RIMSKY-KORSAKOFF, MIKHAIL NIKOLARVICH (1873–1951) (Uvarov 1951, Entomol. mon. Mag. 87: 215; Stakel'berg 1952, Ent. Obozr. 32: 332–340, bibliogr.)

RING GLAND Muscomorphan Diptera larvae: A term coined by Hadorn (1937) for the composite endocrine gland, represented in other insects by the Prothoracic Gland, Corpus Cardiacum and Corpus Allatum. Cf. Weismann Ring.

RING JOINT Insect Antenna: The short, proximal segment or segments of the clavola (Comstock). Hymenoptera: The segment(s) immediately distad of the Pedicel; ring-like in form and smaller than more distal segments of Flagellum. See Annelus.

RING-LEGGED EARWIG *Euborellia annulipes* (Lucas) [Dermaptera: Labiduridae]: Endemic and widespread in North America; common in the southern and southwestern states. Eggs hatch within 14 days; complete development within 80 days; five nymphal moults; feeds upon grains and soil-inhabiting arthropods; occasionally climbs plants to feed upon aphids.

RING VEIN Thysanoptera: The Ambient Vein *sensu* Comstock. See Vein; Wing Venation.

RING WALL Male Lepidoptera: The Penis funnel or Juxta. A chitinized support at the point where the sheath of the Penis joins the body (Imms).

RING-ANT TERMITE *Neotermes insularis* [Isoptera: Kalotermitidae]: An Australian arboreal termite typically found in soft-growth rings of living trees. *Eucalyptus* favoured host; rarely attacks timbers in buildings. Lacks worker caste; nymphs serve as workers but can develop into reproductives or soldiers. A tree and forest pest typically found in upper branches; rarely detected until major damage to tree; contact with soil not required.

RINGDAHL, OSCAR (1885–1966) (Lindroth 1966, Opusc. ent. 31: 118.)

RINGENT Adj. (Latin, *ringi* = to open mouth.) Gaping; pertaining to structure with valves or lips which are separated by a distinct gap. Alt. Ringens.

RIODINIDAE See Lycaenidae.

RION® Malathion.

RIORDON, DEREK F (–1976) (Anon. 1976, Bull. ent. Soc. Can. 8: 10.)

RIPARIAN Adj. (Latin, *ripa* = riverbank.) Descriptive of organisms inhabiting rivers or river banks.

RIPCORD® See Cypermethrin.

RIPICOLOUS Adj. (Latin, *ripa* = riverbank; *colere* = to inhabit.) Pertaining to organisms inhabiting riverbanks. See Habitat; Riparian. Cf. Algicolous; Halophilous; Hydrophilous; Madicolous; Paludicolous; Pratinicolous; Rhophilous; Ripicolous. Rel. Ecology.

RIPPER, WALTER EUGENE (1906–1965) (Pearson 1966, Proc. R. ent. Soc. Lond. (C) 30: 63.)

RIPPLE BUGS See Veliidae.

RIPPON, CLAUDE (–1944) (Cockayne 1945, Proc. R. ent. Soc. Lond. (C)9: 48.)

RIS, FRIEDRICH (1867–1931) (Calvert 1931, Ent. News 42: 181–191; Morton 1931, Entomol. mon. Mag. 67: 65–66.)

RISBEC, JEAN (–1963) (Anon. 1963, Bull. Soc. ent. Fr. 69: 213.)

RISK Noun. The probability that a control method will cause human or environmental harm.

RITCHIE, A. S. (–1870) (Anon. 1870, Can. Ent. 2: 155–156; Anon. 1871, Can. Ent. 3: 177, bibliogr.)

RITCHIE, ARCHIBALD HAMILTON (1887–1936) (Howard 1930, Smithson. misc. Collns. 84: 224, 377, 448; Laing 1930, Entomol. mon. Mag. 72: 120; Horn 1936, Arb. physiol. angew. Ent. Berl. 3: 301.)

RITRON® See Dichlorvos.

RITSCHL, JULIUS (–1899) (Hering 1899, Stettin. ent. Ztg 60: 355–356.)

RITSEMA, CONRAD (1846–1929) (Anon. 1912, Insecta 2: 22.)

RITUALIZATION Noun. (Latin, *ritus* = rite; English, *-tion* = result of an action.) The modification of a behaviour pattern which turns the pattern into a communication signal or improves efficiency of the signal.

RITZEMA BOS, JAN (1850–1928) (Howard 1928, J. Econ. Ent. 21: 636–637.)

RIVER Noun. (Middle English, *rivere*. Pl., Rivers.) A naturally occurring freshwater drainage which is qualitatively larger than a stream. Rivers exhibit aging in the sense of newly formed rivers are often unconfined; mature rivers are confined, often within valleys developed over geological time.

RIVER BLINDNESS See Onchocerciasis.

RIVER SKIMMERS See Macromiidae.

RIVERA, MANUEL JESUS (1875–1910) (Porter 1910, Revta. chil. Hist. Nat. 14: 254–258, bibliogr.; Etcheverry 1987, Revta. chil. Ent. 15: 89–92.)

RIVERS, JAMES JOHN (1824–1913) (Grinnell 1914, Bull. Brooklyn ent. Soc. 9: 72–73.)

RIVINUS, AUGUSTUS QUIRINUS (1652–1723) (Rose 1850, *New General Biographical Dictionary* 11: 356.)

RIVNAY, EZEKIEL (1899–1972) (Harpaz 1973, Israel J. Ent. 8: 183–202, bibliogr.)

RIVOSE Adj. (Latin, *rivus* = stream; *-osus* = full of.) Structure which is marked with sinuate, irregular furrows, not running in a parallel direction. Alt. Rivosus.

RIZAL Y ALONZO, JOSÉ (1861–1896) (Buntug *et al.* 1946, Spec. Suppl. Bull. Bur. Health 22 (6): 1–38.)

RMSF See Rocky Mountain Spotted Fever.

ROACH KIL® See Boric Acid.

ROACHFLIES See Peltoperlidae.

ROACHLIKE STONEFLIES See Peltoperlidae.

ROACH-PRUFE® See Boric Acid.

ROAD BEETLES See Staphylinidae.

ROARK LENNOX E (1909–1963) (Kirkpatrick 1964, J. Econ. Ent. 57: 612–613.)

ROARK, RURIC CREEGAN (1887–1962) (Busbey 1963, Proc. ent. Soc. Wash. 65: 66–71, bibliogr.)

ROBACK, SELWYN (–1988) American museum scientist (Academy of Natural Sciences, Philadelphia) and taxonomic specialist on Chironomidae. (Anon. 1988, Ent. News 99: 198.)

ROBBER FLIES See Asilidae.

ROBBINS, JOHN CUTHBERT (1906–1932) (Anon. 1932, Entomologist's Rec. J. Var. 44: 78.)

ROBBINS, RANDOLPH WILLIAM (1871–1941) (J.A.S. *et al.* 1942, London Nat. 1941: 2–11, bibliogr.)

ROBERT, CHARLES (1802–1837) (Lacordaire 1837, Bull. Soc. ent. Fr. 6: xxxi–xxxiii.)

ROBERT, NICOLAS (1610–1684) (Rose 1850, *New General Biographical Dictionary* 11: 357.)

ROBERTS, ARTHUR WILLIAM PYMER (1879–1955) (Van Emden 1955, Entomol. mon. Mag. 91: 168.)

ROBERTS, FREDERICK HUGH SHERSTON (BOB) (–1972) (Anon. 1972, Circ. ent. Soc. Aust. (N.S.W.) 226: 26–27.)

ROBERTS, WILLIAM REES BREBNER (1881–1941) (Nabours 1942, Science 95: 113–114.)

ROBERTSON, CHARLES (1858–1935) (Parks 1936, Bios 7: 85–96, bibliogr.)

ROBERTSON, JAMES GRANT (1921–1967) (Randall 1968, Can. Ent. 100: 670–672, bibliogr.)

ROBERTSON, ORIS TRIGNE (1905–1969) (Noble & Chapman 1969, J. Econ. Ent. 62: 974.)

ROBERTSON, R BOWEN (1860–1919) (Lucas 1920, Entomologist 53: 96.)

ROBIN, CHARLES PHILIPPE (1821–1885) (Laboulbène 1886, Ann. Soc. ent. Fr. (6) 5: 467–472, bibliogr.)

ROBINEAU-DESVOIDY, ANDRÉ JEAN BAPTISTE (1799–1857) (Bigot 1857, Bull. Soc. ent. Fr. (3) 5: cxxii–cxxxv; Osten Sacken 1893, Berl. ent. Z. 38: 383–386; Osten Sacken 1903, *Record of my life work.* viii + 240 pp. (180–192), Cambridge, Mass.; Sabrosky 1974, Mosquito Syst. 6: 220–221.)

ROBINET, STEPHAN (1796–1869) (Anon. 1870, 0st. Seidenbau-Z. 2: 4.)

ROBININ Noun. (From Jean Robin, a French botanist. Pl., Robinins.) A yellow crystalline glycoside derived from kaempferol and which occurs in the flowers of a locust plant.

ROBINSON, ARTHUR (1865–1948) (Edelsten 1948, Entomologist 81: 288.)

ROBINSON, COLEMAN TOWNSEND (1838–1872) (Grote 1872, Can. Ent. 4: 109–111, bibliogr.; H. S. 1925. Ent. News 36: 309.)

ROBINSON, E W (1835–1877) (Anon. 1877, Entomol. mon. Mag. 14: 118–119.)

ROBINSON, FRANK EDWARD (–1886) (McLachlan 1886, Proc. ent. Soc. Lond. 1886: lxviii.)

ROBINSON, HERBERT CHRISTOPHER (1874–1929) (C. B. K. 1930, J. fed. Malay St. Mus. 16: 1–12, bibliogr.)

ROBINSON, JESSE MATHEWS (1889–1949) (Anon. 1949, J. Econ. Ent. 42: 1000.)

ROBINSON, MARK (1906–1965) (Cartwright 1968, Ent. News 79: 285–286, bibliogr.)

ROBINSON, TANCRED (–1748) (Rose 1850, *New General Biographical Dictionary* 11: 364.)

ROBINSON, WIRT (1864–1929) (Calvert 1929, Ent. News 40: 168.)

ROBLES DISEASE See Onchocerciasis.

ROBOROVSKY, VSEROLOD IVANOVICH (1856–1910) (Semevov-Tian-Shansky 1910, Revue Russe Ent. 10: 247–248.)

ROBSON, JAMES (1733–1806) (Lisney 1960, *Bibliogr. of British Lepidoptera 1608–1799,* 315 pp. (201–203), London.)

ROBSON, JOHN EMMERSON (1833–1907) (Waterhouse 1907, Proc. ent. Soc. Lond. 1907: xcv–xcvi.)

ROBUST Adj. (Latin, *robustus* = oaken, hard.) A subjective term descriptive of bodies, appendages or structures which appear stout or thickened. Ant. Weak.

ROBUST BOT-FLIES See Cuterebridae.

ROBUST CLICK-BEETLES See Cebrionidae.

ROBUST LEAFHOPPER *Penestrangania robusta* (Uhler) [Hemiptera: Cicadellidae].

ROCCI, UBALDO (1885–1943) (Taccani 1946, Natura 37: 68–70.)

ROCHA LIMA, HENRIQUE DA (1879–1956) (Weidner 1967, Abh. Verh. naturw. Ver. Hamburg Suppl. 9: 343.)

ROCHLOR® See Trichlorfon.

ROCK Noun. (Middle English, *rokke*. Pl., Rocks.) A

piece of stony material, excised from a parent geological formation.

ROCK CRAWLERS See Grylloblattoidea.

ROCKWOOD, LAWRENCE PECK (1886–1955) (Wade 1956, Proc. ent. Soc. Wash. 58: 55–57.)

ROCKY MOUNTAIN GRASSHOPPER *Melanoplus sprentus* (Walsh) [Orthoptera: Acrididae].

ROCKY MOUNTAIN SPOTTED FEVER A tick-borne rickettsial disease caused by *Rickettsia rickettsii.* RMSF best known in USA, but also occurs in Mexico, Central America and South America. Incubation period ca seven days; clinical period ca 14–21 days. Symptoms include rash, fever, headache, chills and photophobia; mortality rate noteworthy. Principal vectors in USA include ixodid ticks *Dermacentor andersoni* and *D. variabilis;* vectors elsewhere include *Amblyomma* spp. and *Rhipicephalus sanguineus.* Trans-ovarial and trans-stadial transmission of disease demonstrated; disease passed through feeding but not faeces. Argasid ticks naturally infected but do not appear to transmit RMSF to humans. See Rickettsiaceae. Cf. Rickettsial Pox; Scrub Typhus.

ROCKY MOUNTAIN WOOD-TICK *Dermacentor andersoni* Stiles [Acari: Ixodidae].

ROCQUIGNY-ADANSON, GUILLAUME CHARLES DE (1852–1904) (Abbé Pierie 1904, Revue scient. Bourbon. Cent. Fr. 17: 141–143.)

RODENT CHEWING-LICE See Gyropidae.

RÖDING, PETER FRIEDRICH (1767–1846) (Weidner 1967, Abh. Verh. naturw. Ver. Hamburg Suppl. 9: 54–57.)

ROD-OF-THE-EYE See Rhabdom.

RODRIGUEZ LUNA, JUAN J (1840–1916) (Anon. 1917, Ent. News 28: 335–337; Lameere 1920, Ann. Soc. ent. Belg. 59: 141.)

RODY® See Fenpropathrin.

ROEBER, JOHANNES KARL MAX (1861–1942) (Urban 1943, Stettin. ent. Ztg. 104: 182.)

ROEBUCK, WILLIAM DENISON (1851–1919) (Porritt 1919, Entomol. mon. Mag. 55: 91–92.)

ROEDER, VICTOR VON (–1910) (Anon. 1911, Wien. ent. Ztg. 30: 80.)

ROEDING, GEORGE CHRISTIAN (1868–1928) (G.H.H. 1928, Mon. Bull. Calif. Dep. Agric. 17: 383.)

ROEDINGER, HERMANN (Weidner 1967, Abh. Verh. naturw. Ver. Hamburg Suppl. 9: 254.)

ROELOFS, WILLEM (–1897) (Anon. 1897, Entomol. mon. Mag. 33: 186; Trimen 1897, Proc. ent. Soc. Lond. 1897: lxxiii.)

ROEMER, CARL FERDINAND (1818–1891) (Struckmann 1892, Leopoldina 28: 31–32, 43–46, 63–67.)

ROEMER, JOHANN JACOB (1761–1819) Swiss author of *Genera insectorum Linnei et Fabricii iconibus illustrata* (1789) (Schinz 1819, Meisner's naturw. Anzeiger 2: 89–94, bibliogr.)

ROEMER, JOHN CHARLES Compiler of *Genera Insectorum, Linnaei et Fabricii, Iconiburs Illustrata* (1789). The work is essentially a revision of Sulzer's *Kennziechen der Insecten.*

ROEPKE, WALTER KARL JOHANN (1882–1961) (Anon. 1960, J. Lepid. Soc. 13: 241–242.)

ROESCHKE, HANS (1867–1934) (Anon. 1935, Koleopt. Rdsch. 21: 57.)

ROESEL VON ROSENHOF, AUGUST JOHANN (1705–1759) German miniature painter and engraver. Roesel was not trained as a naturalist but was a careful observer and outstanding 18th century artist. Illustrations distributed bimonthly to subscribers and later formed *Insecten-Belustigung* (1746–1761) (Kleeman 1761, In Roesel & Kleeman. *Insecten-Belustigung* 4: 1–48; Eiselt 1836, *Geschichte, Systematik und Literatur der Insektenkunde.* 255 pp. (440–441), Leipzig; Miall 1912, *Early Naturalists Their Lives and Work (1530–1789),* 396 pp. (293–303), London.)

ROESLERSTAMM, J E FISHER VON See Fischer von Roeslerstamm.

ROESLERSTAMMIIDAE Plural Noun. A numerically small Family of Lepidoptera assigned to Superfamily Tineoidea and known from Palearctic, Oriental and Australian Realms. Adult small; wingspan 10–20 mm; head with elevated, dense scales on Vertex; Frons usually with smooth scales; Ocelli, chaetostemata absent; Compound Eyes sometimes with posterior indentation or divided horizontally with band of scales; Antenna filiform, with Pecten and typically longer than forewing; Proboscis small, without scales; Maxillary Palpus of one segment; Labial Palpus with three segments, long, slender, curved; fore Tibia with Epiphysis; tibial spur formula 0-2-4. Biology poorly known; Palaearctic Species known to deposit eggs at apex of leaf; first and second larval instars mine leaves; subsequent instars live on leaf or in web on underside of leaf. Pupation within folded leaf edge. Syn. Amphitheridae.

ROESSLER, ADOLF (1814–1885) (Pagenstecher 1885, Jb. Nassau Ver. Naturk. 38: 149–152, bibliogr.)

ROGENHOFER, ALOIS FRIEDRICH (1832–1897) (Anon. 1897, Entomol. mon. Mag. 33: 108; Mik 1897, Wien. Ent. Ztg. 16: 44; Anon. 1924, Jber wien. ent. Ver. 30: 17–24, bibliogr.)

ROGER, JULIUS (–1865) (Kraatz 1865, Berl. ent. Z. 9: 1.)

ROGERS ST AUBYN, CANON (1944) (Cockayne 1945, Proc. R. ent. Soc. Lond. (C) 9: 48.)

ROGERS, JAMES SPEED (1891–1955) (Alexander 1957, Ent. News 68: 85–88.)

ROGERS, WILLIAM PRESCOTT (1887–1953) (Learned 1954, Lepid. News 8: 44–45.)

ROGOR® See Dimethoate.

ROGUE Verb. (Unknown origin.) Pathology: To remove a diseased plant or plants with the objective of reducing the spread of disease.

ROHDENDORF, BORIS BORISOVITSCH (1904–1977) Russian Dipterist and Head of the Palaeontological Institute of Academy of Sciences of the USSR. Specialist in Sarcophagidae, Tachinidae and Calliphoridae who described many Species of fossil Diptera.

ROHR, JULIUS PHILIP BENJAMIN VON (1735–1793) (Zimsen 1964, *The Type Material of J. C. Fabricius*. 656 pp. (14–15). Copenhagen; Papavero 1971, *Essays on the History of Neotropical Dipterology*. 1: 20–21. São Paulo.)

RÖHRL, ANTON (1891–1944) (Sachtleben 1944, Arb. morph. taxon. Ent. Berl. 11: 141.)

ROHWER, SIEVERT ALLEN (1888–1951) (Bishopp *et al.* 1951, J. Econ. Ent. 44: 437–439.)

ROJAS, MARCO AURELIO DE (1831–1866) (Sallé 1866, Ann. Soc. ent. Fr. (4) 6: 600.)

RO-KO® See Rotenone.

ROLANDER, DANIEL (1726–1793) (Papavero 1971, *Essays on the History of Neotropical Dipterology*. 1: 8–9. São Paulo.)

ROLANDO, LUIGI (1773–1831) (Bellingeri 1932, Memorie Accad. Sci. Torino 37: 153–193.)

ROLE Noun. (French *role* = role. Pl., Roles.) 1. A function performed by someone or something in a process or situation. 2. Social Insects: A behaviour pattern displayed by some members of a colony such that a division of labour is achieved.

ROLFS, PETER HENRY (1865–1944) (Osborn 1937, *Fragments of Entomological History,* 394 pp. (223–224), Columbus, Ohio; Osborn 1946, *Fragments of Entomological History,* Pt. II. 232 pp. (107–108), Columbus, Ohio.)

ROLL-AND-FOLD GALL Plant galls characterized by several types of rolls and folds in leaves. Leaf roll galls with blade rolled upward from leaf margin toward midrib, and swelling of rolled portion of leaf blade; fold galls with affected tissue swollen and undifferentiated. FRG galls primarily produced by aphids, thrips, psyllids and cecidomyiids. See Gall. Cf. Bud-and-Rosette Gall; Covering Gall; Filz Gall; Mark Gall; Pit Gall; Pouch Gall.

ROLL, HERMANUS FREDERIK (1867–1935) (Olivier 1935, Meded. Dienst Volksgezondh. Ned.-Indië 24: [1–2].)

ROLLASON, WILLIAM ALFRED (1863–1911) (Morice 1911, Proc. ent. Soc. Lond. 1911: cxxii–cxxiii.)

ROLLING WASPS See Tiphiidae.

ROLPH, WILLIAM HENRY (1847–1883) (Kraatz 1884, Dt. ent. Zt. 28: 239.)

ROMAN, PER ABRAHAM (1872–1943) (Lundblad 1944, Ent. Tidschr. 65: 39–45, bibliogr.)

ROMANES, GEORGE JOHN (1848–1894) (Anon. 1894, Entomologist's Rec. J. Var. 5: 176.)

ROMANOVICH, K R See Osten Sacken, C. R.)

RÖMER, FRITZ (1866–1909) (Brauer 1910, Fauna Arctica 5: i–iii)

ROMETA® See Methamidophos.

ROMICARB® See Primicarb.

ROMINE, RAY (1910–1954) (Eff 1955, Lepid. News 9: 22.)

ROMITRAZ® See Amitraz.

RONCHETTI, VITTORIO (1874–1944) (Parisi 1944, Atti Soc. ital. Sci. nat. 83: 257–270, bibliogr.)

RONDANI, CAMILLO (1807–1879) Italian academic (Professor of Zoology, University of Parma), Dipterist and specialist in economic Entomology, describing many Diptera and some parasitic Hymenoptera. (Anon. 1870, Boll. Soc. ent. ital. 2: 297–300, bibliogr.; Meade 1879, Entomol. mon. Mag. 16: 138–139.)

RONDELET, GUILLAUME (1507–1566) (Rose 1850, *New General Biographical Dictionary* 11: 382.)

RONDO® See Permethrin.

RONDON, JACQUES (–1969) (Anon. 1969, Bull. Soc. ent. Fr. 74: 153.)

RONNIGER, HERMAN (–1955) (Reisser 1956, Wien. ent. Ges. 41: 210–212, bibliogr.)

ROOF Noun. (Middle English, *rouf*. Pl., Roofs; Rooves.) 1. A covering structure of any part of the body. 2. Insect anatomy: The upper interior wall of any cavity. Cf. Ceiling; Floor; Wall. Rel. Window. 3. The highest part of a cavity.

ROOS, KARL (1908–1942) (Sachtleben 1943, Arb. physiol. angew. Ent. Berl. 10: 260; Schneider-Orelli 1943, Mitt. schweiz. ent. Ges. 18: 530–531, bibliogr.)

ROOS, KURT L (–1967) (Kennedy 1968, Proc. R. ent. Soc. Lond. (C) 32: 60.)

ROOT, AMOS (1839–1923) (Pellett 1923, Am. Bee J. 63: 292; Pellett 1936, Ohio Farmer 117: 321.)

ROOT CAP The cellular covering of the apex of a plant root.

ROOT EATING BEETLES See Rhizophagidae.

ROOT EATING FLIES See Anthomyiidae.

ROOT, FRANCIS METCALF (1887–1934) (Calvert 1934, Ent. News 45: 285–286, bibliogr.; Hegner 1935, J. Parasit. 21: 67–69, bibliogr.)

ROOT GNAT See Sciaridae.

ROOT, GEORGE A (1890–1942) (Bellis *et al.* 1942, J. Econ. Ent. 35: 953.)

ROOT HAIRS The fine, hair-like outgrowths of the root apex in plants.

ROOT, LYMAN C (1840–1928) (Anon. 1928, Am. Bee J. 68: 497.)

ROOTSTOCK Noun. (Pl., Rootstocks.) The lower portion of a tree, including the root system. The part of a tree to which another variety is grafted.

ROOY, ALEXANDER BENJAMIN VAN MEDENBACH DE (1841–1878) (Kraatz 1878, Dt. ent. Z. 22: 226.)

ROOY, HENRICUS CORNELIUS VAN MEDENBACH DE (1794–1877) (Anon. 1878, Tijschr. Ent. 22: ii.)

ROPALOMERIDAE Plural Noun. A small, predominantly Neotropical Family of brachycerous Diptera assigned to Superfamily Sciomyzoidea. Syn. Rhopalomeridae. Adult moderate sized, typically brownish or grey with eyes bulging; Ocelli present; posterior thoracic spiracle without bristle; Femora incrassate.

ROPRONIIDAE Viereck 1916. Plural Noun. A numerically small, Holarctic and Oriental Family of apocritous Hymenoptera assigned to Superfamily Proctotrupoidea. Adult 7–11 mm long; Antenna with 14 segments but without

Anellus; forewing Stigma present, Medial Cell polygonal; Metasoma strongly compressed laterally. Larvae develop as parasites of sawflies.

ROQUES, XAVIER (1882–1915) (Berland 1920, Ann. Soc. ent. Fr. 89: 430–432.)

RÖRIG, GEORG (1864–1941) (Sachtleben 1941, Arb. physiol. angew. Ent. Berl. 8: 212.)

RORULENT Adj. (Latin, *rorulentus* from *ros* = dew.) Covered like a plum with a bloom that may be rubbed off (Kirby & Spence); pulverulent. Alt. Rorulentus.

ROSA, GABRIELE (1812–1897) (Rosa 1912, Gabriele Rosa. 108 pp. Brescia.)

ROSACEOUS Adj. (Latin, *rosa* = rose; *-aceus* = of or pertaining to.) Resembling a rose in scent, colour or texture. Alt. Rosaceus.

ROSE APHID *Macrosiphum rosae* (Linnaeus) [Hemiptera: Aphididae].

ROSE, ARTHUR J (1859–1945) (Hale-Carpenter 1946, Proc. R. ent. Soc. Lond. (C) 10: 54.)

ROSE CHAFER *Macrodactylus subspinosus* (Fabricius) [Coleoptera: Scarabaeidae]: A monovoltine pest of fruit trees, ornamentals, shrubs and vegetables in eastern USA and Canada. RC overwinters as larva within soil; pupation occurs in May; adults appear in June-July. Adults mate and feed on foliage to skeletonize or defoliate. Eggs deposited in clutches of 5–25 in soil; eclosion occurs within 7–15 days. Larvae feed on roots of host plants. See Scarabaeidae. Cf. Japanese Beetle.

ROSE CURCULIO *Merhynchites bicolor* (Fabricius) [Coleoptera: Curculionidae]: Widespread in North America and Europe. A monovoltine pest of roses and peonies. Eggs laid in buds; larva feeds on seeds; pupation in soil. See Curculionidae. Cf. Plum Curculio.

ROSE-GRAIN APHID *Metopolophium dirhodum*. [Hemiptera: Aphididae].

ROSE LEAF-BEETLE *Nodonota puncticollis* (Say) [Coleoptera: Chrysomelidae].

ROSE LEAFHOPPER *Edwardsiana rosae* (Linnaeus) [Hemiptera: Cicadellidae]: Widespread throughout North America. Bivoltine; eggs of first generation laid in leaf tissue; eggs of second generation laid in bark; overwinters as egg in bark. Nymphs suck plant juices from underside of leaves. Pest of rose, apple, gooseberry, raspberry, sugarbeet and beans.

ROSE MIDGE *Dasineura rhodophaga* (Coquillett) [Diptera: Cecidomyiidae]: A widespread, multivoltine pest of roses, particularly in greenhouses. RM overwinters in cocoon within soil. Female oviposits in buds behind sepals. Larvae feed on new growth inside buds and complete larval stage ca 5–6 days; mature larva drops to ground, spin a cocoon and pupate. Generation completed ca 20 days. See Cecidomyiidae.

ROSE-ROOT GALL-WASP *Diplolepis radicum* (Osten Sacken) [Hymenoptera: Cynipidae].

ROSE SCALE *Aulacaspis rosae* (Bouché) [Hemiptera: Diaspididae].

ROSE SLUG *Endelomyia aethiops* (Fabricius) [Hymenoptera: Tenthredinidae].

ROSE STEM-GIRDLER *Agrilus aurichalceus* Redtenbacher [Coleoptera: Buprestidae].

ROSEATE Adj. (Latin, *rosa* = rose; *-atus* = adjectival suffix.) Rose coloured. Alt. Roseous; Roseus.

ROSELLE, EUGENE JEAN BAPTISTE FERNAND DU (1854–1904) (Hautefeuille 1905, Bull. Soc. linn. N. Fr. 17: 131–l 36)

ROSENBERG, EINAR CARL (1875–1946) (Tuxen 1947, Ent. Meddr. 25: 144–148, bibliogr.)

ROSENBERG, WILLIAM FREDERIK HENRY (1868–1957) (A. G. G. 1951, Entomologist 90: 108.)

ROSENFELD, ARTHUR HINTON (1866–1942) (Wolcott 1943, J. Econ. Ent. 36: 358.)

ROSENHAUER, GUILLAUME DIEUDONNÉ (1811–1838) (Anon. 1887, Abeille (les ent. et leurs écrits) 24: 190–191.)

ROSENHAUER, WILHELM GOTTLOB (1813–1881) (Katter 1881, Ent. Nachr. 7: 231–232; Kraatz 1881, Dt. ent. Z. 25: 342–343.)

ROSENHOF, AUGUST JOHANN ROESEL VON (1705–1759) German miniature painter, natural historian and author of *Insecten Belustigungen* (4 vols. 1746–1761). Illustrations of insects among the best of the 18th century.

ROSENKRANTZ, HANS FREDERICK (1822–1905) (Engelhart 1918, Ent. Meddr. 12: 21–22.)

ROSER, KARL V (1787–1861) German natural historian and Director of Ministry of Foreign Affairs, Wurttemburg. Personal collection became foundation for entomological collection of Staatliches Museum für Naturkunde, Stuttgart.

ROSETTE Noun. (French, from Latin, *rosa;* diminutive of rose.) 1. A short, bunchy type of plant growth. 2. A cluster of leaves that arise in a small circle from a central axis on a stem or branch. 3. Any group of appendages (similar in size and shape) basally attached to a structure.

ROSIN Noun. (Middle English variant of resin. Pl., Rosins.) A hard, brittle, amber coloured to black, tasteless residue from the distillation of volatile oil of turpentine. Used to manufacture soap and varnish, serves as a flux for soldering and for rosining the bow of some stringed musical instruments. See Resin.

RÖSNER, JOSEPH (1860–1920) (Hedwig 1921, Jh. Ver. Schles. Insektenk. 13: 23.)

ROSPIN® See Chlorpropylate.

ROSS RIVER FEVER A non-fatal arboviral disease endemic in Australia and now widespread in Pacific islands. RRF caused by *Alphavirus* of Tongaviridae; enzootic in marsupials, horses and cattle. Mosquito vectors include *Aedes* spp. and *Culex annulirostris*. Disease manifest in headache, fever, arthralgia and rash. See Arbovirus; Tongaviridae. Cf. Chikungunya; O'Nyong-Nyong; Sindbis.

ROSS, RONALD (1857–1932) Pioneer in Medical Entomology; among other accomplishments, Ross first described the complete life cycle of bird malaria. (Bonn 1932, Genesk. Tijdschr. Nederl. Ind. 72: 1330; Watson 1933, Sci. Progr. 27: 377–395; Wenyon 1933, Trans. R. Soc. trop. Med. Hyg. 26: 473–478.)

ROSS, TERRY SPINKS (1903–1928) (Plank 1928, J. Econ. Ent. 21: 440.)

ROSSI, GUSTAV DE (–1903) (Anon. 1903, Wien. ent. Ztg. 22: 36.)

ROSSI, PIETRO (1738?–1804) (Swainson 1840, *Taxidermy, with biogr. of Zoologists*. 392 pp. (p. 311.) London. [Vol. 12, *Cabinet Cyclopedia*. Edited by D. Lardner.]; Baccetti 1962, Frustula ent. 5 (3): 1–30, bibliogr.)

ROSSI, PIETRO (1738–1804) Italian academic, naturalist and author of *Fauna Insecta quae in Provinciis Florentina et Pisana praesertim collegit* (2 vols, 1790) and *Mantissa Insectorum exhibens species nuper in Etruria collectas* (2 vols, 1792–1794.) First Professor of Entomology (University of Pisa, 1801–1804.)

ROSSKOTHEN, PAUL (1894–1973) (Wall 1974, Ent. Bl. Biol. Syst. Käfer 70: 3–4, bibliogr.)

ROSSMÄSSLER, EMIL ADOLF (1806–1867) (Anon. 1867, Zool. Gart. Lpz. 8: 199–200; Ille 1867, Natur 16: 188–190, 193–195, 217–220.)

ROSSOM, AREND JOHAN VAN (1842–1909) (Oudemans 1910, Tijdschr. Ent. 53: 1–7.)

ROSTAGNO, FORTUNATO (1847–1934) (Turati 1934. Boll. Soc. ent. ital. 66: 182.)

ROSTELLUM Noun. (Latin, *rostellum* = small beak. Pl., Rostella.) A small beak; a beak-like process. The tubular mouthparts of sucking lice and Hemiptera. Syn. Rostrum.

ROSTOK, MICHAEL (1821–1893) (Richter 1926, Bautzen. Nachr. 1926: 161–163.)

ROSTRAD Adv. (Latin, *rostrum* = beak; *ad* = toward.) Toward the Rostrum. See Orientation. Cf. Anteriad; Apicad; Basad; Caudad; Centrad; Cephalad; Craniad; Dextrad; Dextrocaudad; Dextrocephalad; Distad; Dorsad; Ectad; Entad; Laterad; Mediad; Mesad; Neurad; Orad; Proximad; Sinistrad; Sinistrocaudad; Sinistrocephalad; Ventrad.

ROSTRAL Adj. (Latin, *rostrum* = beak; *-alis* = pertaining to.) Pertaining to a Rostrum; descriptive of something attached to a Rostrum. Term typically applied to Hemiptera.

ROSTRAL FILAMENTS Coccidae: Four hair-like Trophi which form the tube through which fluid is imbibed.

ROSTRALIS Noun. (New Latin, from Latin, *rostrum* = beak, muzzle. Pl., Rostrales.) Hemiptera: The Mandibles and Maxillae combined to form the 'sucking tube' (MacGillivray). Syn. Beak. See Rostrum.

ROSTRATE Adj. (Latin, *rostrum* = beak; *-atus* = adjectival suffix.) Pertaining to a Rostrum or a long protraction bearing the mouthparts. Alt. Rostratus; Rostriform. See Rostrum.

ROSTRIFORM Adj. (Latin, *rosturm* = beak, muzzle; *forma* = shape.) Beak-shaped; descriptive of structure produced into a snout, muzzle or Rostrum. Alt. Rostrate. See Beak; Rostrum. Rel. Form 2; Shape 2; Outline Shape.

ROSTRULM Noun. (Latin, *rostrulum* = small beak. Pl., Rostrula.) Siphonaptera: The Proboscis.

ROSTRUM Noun. (Latin, *rostrum* = beak. Pl., Rostra.) 1. General: A snout-like prolongation of the head. 2. Hemiptera: The beak; a joined (segmented) sheath formed by the Labium to enclose the maxillary and mandibular Stylets. The Rostrum typically consists of three segments: Two basal segments are short and distal segment is long and slender. A groove may occur on the anterior face into which mandibular and maxillary stylets fit at repose. Lips of the groove oppose, thus closing except basally where there is a triangular opening. Homologies of labial parts are tenuous: First segment may be Prementum with Postmentum incorporated into the head; other segments (when present) may be derived from Labial Palpi. 3. Coleoptera: A tubular extension of the head beneath compound eyes in Rhynchophora. Rel. Rostriform.

ROSTRUP, SOFIE (1857–1940) (Henriksen 1938, Ent. Meddr. 20: 65–66; Bocien 1940, Ent. Meddr. 20: 593–596.)

ROSY APPLE-APHID *Dysaphis plantaginea* (Passerini) [Hemiptera: Aphididae]: A widespread pest of apples and pears in North America. RAA feeding causes leaves to curl and fruit to deform. RAA eggs first green then black, overwintering eggs laid in apple bark fissures, on twigs or within bud axils. Several wingless generations feed on apple; winged generation turns to plantain and several generations develop. Subsequent generation returns to apple; wingless male and oviparous female produce overwintering eggs. See Aphididae.

ROT Noun. (Middle English, *rotien* = to steep flax; Old High German, *rozzen* = to rot. Pl., Rots) 1. The process or action of decay or decomposition. 2. The breakdown or decay of plant tissue through the action of fungi and bacteria. Rel. Decomposition; Putrefaction.

ROTACIDE® See Rotenone.

ROTATE Adj. (Latin, *rota* = wheel; *-atus* = adjectival suffix.) Shaped as a wheel.

ROTATION Noun. (Latin, *rota* = wheel; English, *-tion* = result of an action. Pl., Rotations.) A turning about an axis.

ROTATIVE Adj. (Latin, *rota* = wheel.) Turning entirely around or apparently so. Alt. Rotatory.

ROTATOR Noun. (Latin, *rota* = wheel; *or* = one who does something. Pl., Rotators.) 1. Any part or structure used in or for turning. 2. A muscle that is adapted for rotating the structure to which it is attached. See Muscle. Cf. Abductor; Adductor.

ROTATOR® See Pirimiphos-Methyl.

ROTAXIS Noun. (Latin, *rota* = wheel; *axis* = axle.) 1. The expanded part of the Notacoria which contains the Pteraliae. 2. The Axillary Membrane *sensu* MacGillivray. See Orientation. Cf. Aerotaxis; Anemotaxis; Chemotaxis; Geotaxis; Menotaxis; Notarotaxis; Osmotaxis; Phototaxis; Rheotaxis; Scototaxis; Strophotaxis; Telotaxis; Thermotaxis; Thigmotaxis; Tonotaxis; Tropotaxis.

ROTEFIVE® See Rotenone.

ROTENONE Noun. (Pl., Rotenones.) A botanical insecticide extracted from the roots of *Lonchocarpus utilis* Smith or *L. uruca* Killip & Smith in South America and *Derris elliptica* (Wall.) Benth. or *D. malaccensis* Brain in southeast Asia. Material extracted in South America called Cube Root; material extracted in southeast Asia called Derris Root; also noted in roots of more than 60 Species of legumes. Most material used in USA obtained from Cube Root grown in Peru. Minor phytotoxicity but highly toxic to fishes and pigs; long used as material to eliminate unwanted fish Species before stocking waters with gamefish. Rotenone first applied to crops in British Malaya ca 1848; currently used on numerous vegetable crops, forage crops, fruit trees and ornamental plants in many areas of world. Commercial products effective in control of some plant pests and ectoparasites of domestic animals; may serve as an acaricide or as an insect repellant. Formulated as a dust and effective as a contact or stomach poison. Characteristics include: short shelf-life, sensitive to light; short residual life; slow acting; ineffective if applied in alkaline solution. Acts as respiratory toxin; mode-of-action not specific to insects and mitochondrial respiration adversely affected in many animals; blocks electron transport and prevents oxidation of NADH. Trade names include: Atox®, Barbasco®, Chem-Fish®, Cube Root®, Cuberol®, Cubor®, Derrin®, Derris®, Dri-Kill®, Extrax®, Fish-Tox®, Foliafume®, Haiari®, Mexide®, Nekos®, Nicouline®, Noxfire®, Noxfish®, Prenfish®, Pro-Nox Fish®, Ro-Ko®, Rotefive®, Rotenox®, Rotessenol®, Rotacide®, Timbo®, Tox-R®, Tubatoxin®. See Biopesticide; Botanical Insecticide; Natural Insecticide; Insecticide. Cf. Nereistoxin; Nicotine; Pyrethrin; Pyrroles; Ryania; Sabadilla; Stemofoline. Nereistoxin; Pyrethrin; Nicotine; Ryania; Sabadilla.

ROTENOX® See Rotenone.

ROTESSENOL® See Rotenone.

ROTETRA® See Tetradifon.

ROTH, CARL DAVID EMANUEL (1831–1898) (Trybom 1898, Ent. Tidskr. 19: 187–189.)

ROTH, HENRY LING (1855–1925) (Haddon 1925, Nature 115: 844.)

ROTH, PAUL (–1967) (Anon. 1967, Bull. Soc. ent. Fr. 72: 222.)

ROTHAMSTED LIGHT TRAP A type of light trap with a time-switch control to permit pre-selected automatic sampling. Samples are collected in a detachable killing jar. (C B Williams, Phil. Trans. R. Soc. Lond., 7B, vol. 244, pp 331–378, Nov 1961). The trap is constructed with or without a stand. Standard capacity of collecting jar is 450 ml; overall dimensions of trap: 56 x 56 x 46 cm (22 x 22 x 18 inches); overall height of trap mounted on stand: 1.37 m (4'6'); net weight: 14.54 kg (32 lb). [See: Proc. R. Ent. Soc., London, Ser. A, Gen. Ent. 23: 80–85.]

ROTHENBACH, JOHANN CHRISTIAN (1796–1881) (Jäggi 1881, Mitt. schweiz. ent. Ges. 6: 243–250.)

ROTHNEU, GEORGE ALEXANDER JAMES (1849–1922) (Poulton 1922, Entomol. mon. Mag. 58: 113–114.)

ROTHSCHILD, EDMOND (1915?–1934?) (Anon. 1934, Hadar: Mon. J. Citrus Ind. Palestine 7: 237.)

ROTHSCHILD, LIONEL WALTER (1868–1937) (Imms 1937, Proc. R. ent. Soc. Lond. (C) 2: 62; Jordan 1938, Novit. zool. 41: 1–41, bibliogr.)

ROTHSCHILD, NATHANIEL CHARLES (1877–1923) (Frohawk 1923, Entomologist 56: 284–286.)

ROTOITIDAE Boucek & Noyes 1987. Plural Noun. A monotypic Family of chalcidoid Hymenoptera known only from females of the type Species collected in New Zealand. Adult Antenna with 14 segments (0 ring segments, 6-segmented club); Prepectus absent; tarsal formula 4-4-4; forewing with incipient Basal Vein and Postmarginal Vein exceedingly long; abdominal spiracles on sixth gastral segment. Host unknown.

ROTTENBERG, ARTHUR LÉOPOLD ALBERT MARIE VON (1843–1875) (Kiesenwetter 1875, Dt. Ent. Z. 1875: 439–440, bibliogr.)

ROTULA Noun. (Latin, *rotula* = small wheel. Pl., Rotulae.) A small, round segment sometimes present between the segments of the Antennae or Palpi. See Torquillus.

ROTULIFORM Adj. (Latin, *rotula* = small wheel; *forma* = shape.) Wheel-shaped. See Rotula. Cf. Actiniform; Maniform; Patelliform. Rel. Form 2; Shape 2; Outline Shape.

ROTUND TICK *Ixodes kingi* Bishop [Acari: Ixodidae].

ROTUNDATE Adj. (Latin, *rotundus* = round; *-atus* = adjectival suffix.) Rounded; pertaining to structure which is circular or in the form of a segment of a circle, with rounded angles passing gradually into each other. Alt. Rotundatus.

ROUAST, GEORGES (–1898) (Bouvier 1899, Bull. Soc. ent. Fr. 1899: 152.)

ROUBAL, JAN (Pfeiffer 1950, Cas. csl. Spol. ent. 47: 201–210, bibliogr.)

ROUBAUD, EMIL (–1962) (Anon. 1962, Bull. Soc. ent. Fr. 67: 195.)

ROUGEMENT, PHILIPPE ALBERT DE (–1881) (Katterm 1882, Ent. Nachr. 8: 231; Tribolet 1882, Mitt. schweiz. ent. Ges. 6: 259–261, bibliogr.)

ROUGEMONT, FRÉDERIC (1838–1917) (Anon. 1917, Bull. Soc. Neuchâtel sci. nat. 42: 3–6.)

ROUGET, AUGUSTE (1818?–1886) (Bourgeois 1886, Bull. Soc. ent. Fr. (6) 6: ixxxix.)

ROUGH BOLLWORM *Earias huegeli* Rogenhofer. [Lepidoptera: Noctuidae]: A pest of cotton in

Australia. Adult moth 12–15 mm long with a wingspan of 18–22 mm; Forewings greenish-beige with a dark green wedge in the centre. (Green transverse lines on the forewing of rough bollworm distinguish it from northern rough bollworm). Hindwings are white with greenish-beige markings on the margin. Thorax dark beige and Abdomen white. Antennae long and slender (filiform); Labial palps prominent. Adult females lay small (0.5 mm in diameter), light blue-green eggs on young leaves and stems; eggs with 30 longitudinal ridges with alternate ridges projecting upwards to form a crown. Egg stage ca 3 days. Mature larvae 15–20 mm long; spiny with mottled coloration of brown, orange, yellow and grey. Larval stages ca 14 days. Pupation occurs on the plant; Pupae ca 13 mm long and brown in a greyish silken cocoon of irregular shape. Life-cycle 28–42 days. Larvae enter terminals and tunnel down stems. Larvae enter squares and feeding may cause square to fall. Larvae may also tunnel down the main stem causing destruction of primary growing point at any growth stage including seedlings. Larva attacks plant at any stage of development; seedlings killed; squares and young bolls shed when attacked; severe damage when larva penetrates green bolls; larva may destroy 1–2 locules but also introduce boll rotting organisms which destroy entire boll. Cf. Northern Rough Bollworm. See Noctuidae.

ROUGH HARVESTER-ANT *Pogonomyrmex rugosus* Emery [Hymenoptera: Formicidae].

ROUGH STINK-BUG *Brochymena quadripustulata* (Fabricius) [Hemiptera: Pentatomidae].

ROUGHSKINNED CUTWORM *Athetis mindara* (Barnes & McDunnough) [Lepidoptera: Noctuidae]. See Cutworms.

ROULLARD, FRED PETE (1884–1965) (Simmons 1965, J. Econ. Ent. 58: 1038–1039.)

ROUND FUNGUS BEETLES See Leiodidae.

ROUND SPOT See Orbicular Spot.

ROUNDHEADED BORERS See Cerambycidae. Cf. Buprestidae.

ROUNDHEADED APPLE-TREE BORER *Saperda candida* Fabricius [Coleoptera: Cerambycidae]: A pest of apple and other deciduous trees in North America. RATB overwinters as a larva; larvae of two size classes – smaller larvae derived from eggs laid in the past season, and larger larvae derived from eggs laid a year earlier. Two-year old larvae reside deeper in tree than one-year old larvae. Larva feeds on inner bark and sapwood of tree; typically, burrows made at base of trunk and can be below ground level; tree sometimes girdled by feeding action; feeding burrows extend through heartwood and sapwood, structurally weakening young trees. Pupation occurs duirng spring; pupal period ca 15–30 days. Adults appear June-September; oviposit on bark near base of tree above or below ground. See Cerambycidae. Cf. Flatheaded Apple-Tree Borer; Pacific Flatheaded Borer.

ROUNDHEADED CONE-BORER *Paratimia conicola* Fisher [Coleoptera: Cerambycidae].

ROUNDHEADED FIR-BORER *Tetropius abietis* Fall [Coleoptera: Cerambycidae].

ROUNDHEADED PINE-BEETLE *Dendroctonus adjunctus* Blandford [Coleoptera: Scolytidae].

ROUNDHEADED WOODBORERS See Cerambycidae.

ROUNDUP See Glyphosate.

ROUSSEAU, ERNEST (1872–1920) (Lestage 1921, Bull. Soc. ent. Belg. 3: 35–41, bibliogr.)

ROUSSEAU, JACQUES (1905–1970) (Pomerleau 1971, Naturaliste can. 98: 215–224.)

ROUTLEDGE, GEORGE BELL (1864–1934) (Adkin 1935, Entomologist 68: 24; F.H.D. 1935, NWest Nat. 10: 145–148.)

ROUX, EMILE (1853–1933) (Anon. 1933, Bull. Inst. Pasteur, Paris 31: 1057–1059.)

ROUX, JEAN LOUIS FLORENT POLYDORE (1792–1833) (Barthelemy 1834, Bull. Soc. ent. Fr. 3: xliv–li.)

ROUX, WOUTER KIRSTEIN (1906–1941) (Favre 1941, J. ent. Soc. sth. Afr. 4: 240.)

ROUZET, JEAN HIPPOLYTE (1802–1865) (Desmarest 1866, Ann. Soc. ent. Fr. (4) 6: 135–137.)

ROVE BEETLES See Staphylinidae.

ROWDE, ALFRED OLIVER (1877–1960) (Uvarov 1961, Proc. R. ent. Soc. Lond. (C) 25: 50.)

ROWLAND-BROWN, HENRY (1865–1922) (Bethune-Baker 1922, Entomologist's Rec. J. Var. 34: 119–120; Sheldon 1922, Entomologist 55: 121–123)

ROWLEY, ROBERT RUSSELL (1854–1935) (dos Passos 1951, J. N. Y. ent. Soc. 59: 162.)

ROXBURGH, WILLIAM (1757–1815) (Rose 1850, *New General Biographical Dictionary* 11: 397.)

ROXION® See Dimethoate.

ROYAL CELL 1. Isoptera: A cell in which the queen resides. 2. Hymenoptera (Apidae): A brood cell in which a queen will develop and which is demonstrably larger than brood cells used for workers.

ROYAL JELLY Hymenoptera: A liquid secreted by the Pharyngeal Glands of the honey bee worker-nurses and consumed by larvae in the brood. Composition of RJ depends upon the caste to which adult will belong. Queen-bee larvae receive only RJ.

ROYAL MOTHS See Saturniidae.

ROYAL PAIRS Sexually active males and females of social insects. Alt. Royalties.

ROYAL PALM BUGS See Thaumastocoridae.

ROZEN JEROME G, JR American taxonomic Entomologist (American Museum of Natural History) specializing in the Apoidea.

ROZIER, FRANCIS (1734–1793) (Rose 1850, *New General Biographical Dictionary* 11: 398.)

ROZSYPAL, J (1896–1966) (Povolny 1966, Acta ent. bohemoslovaca 63: 166–169, bibliogr.)

RRF See Ross River Fever.

RSSE Russian Spring-Summer Encephalitis. See

Tick-Borne Encephalitis.

RUBAN® See Bensultap.

RUBER CLEAR Unmixed red.

RUBESCENT Adj. (Latin, *rubescens,* present participle *rubescere* = to grow red, from *rubere* = to be red.) Growing or becoming red; reddening.

RUBIGINOSE Adj. (Latin, *rubigo* = rust.) Descriptive of structure which is rust-like in colour. Alt. Rubiginosus; Rubiginous.

RUBINEOUS Adj. (Medieval Latin, *rubinus,* from Latin, *rubeus* = red; *-osus* = with the property of.) Ruby-like in colour and brilliance. Alt. Rubineus.

RUBIO, JOSÉ MARIA ANDREA (1881–1967) (Agenjo 1967, Graellsia 23: 121–125.)

RUBITOX® See Phosalone.

RUBRICANS Adj. (Latin, *rubric* = red chalk.) Of a bay or grey-black colour.

RÜBSAAMEN, EDWALD HEINRICH (1857–1919) (Musgrave 1932, *A Bibliogr. of Australian Entomology 1775–1930,* viii + 380 pp. (274), Sydney.)

RUBY TAILED WASPS See Chrysididae.

RUBY WASPS See Chrysididae.

RUCKES, HERBERT (1895–1965) (Arnaud 1966, Pan-Pacif. Ent. 42: 156–157)

RUDBECK, OLAÜS (1660–1740) (Fée 1831, Mém. Soc. Sci. agric. Lille 1: 85–86; Rose 1850, *New General Biographical Dictionary* 11: 400–401.)

RUDEL, KURT (1873–1940) (Weidner 1967, Abh. Verh. naturw. Ver. Hamburg Suppls. 9: 264.)

RUDIMENT Noun. (Latin, *rudimentum* = the beginning. Pl., Rudiments.) 1. The beginning of any structure or part before it has developed. 2. Vestigial.

RUDIMENTARY Adj. (Latin, *rudimentum* = the beginning.) 1. Pertaining to something not developed or something in an early stage of development. 2. Something in a primitive stage. Cf. Derived. Ant. Complete; Mature; Perfect.

RUDOLPH, HUGO (–1938) (Nordman 1938, Notul. Ent. 18: 69–70.)

RUDOLPHI, CARL ASMUD (1771–1832) (Lühe 1900, Archs Parasit. 3: 549–577, bibliogr.; Nordenskiöld 1935, *History of Biology,* 625 pp. (352–355) London.)

RUDOW, FERDINAND (1840–1920) (1920, Ent. Z. Frankf. a. M. 34: 57–58.)

RUFAST® See Acrinathrin.

RUFESCENT Adj. (Latin, *rufus* = reddish.) Somewhat reddish. Alt. Rufescens.

RUFFIN, EDMOND (1794–1865) Howard 1930, Smithson. misc. Collns. 84: 12, 147.

RUFOUS SCALE *Selenaspidis articulatus* (Morgan) [Hemiptera: Diaspididae]: An armored scale-insect pest of *Citrus* widely distributed in tropical and subtropical regions. Syn. West Indies Scale; West Indian Red Scale.

RUFOUS Adj. (Latin, *rufus* = reddish; Latin, *-osus* = with the property of.) Pale red.

RUGA Noun. (Latin, *ruga* = wrinkle. Pl., Rugae.) A wrinkle, fold or crenation.

RUGATE Adj. (Latin, *rugare* = to wrinkle; *-atus* = adjectival suffix.) 1. Descriptive of surface sculpture, usually the insect's Integument, that is covered with wrinkles. 2. Pertaining to a wrinkled surface. See Sculpture Pattern. Cf. Alveolate; Baculate; Clavate; Echinate; Favose; Gemmate; Psilate; Punctate; Rugose; Scabrate; Shagreened; Smooth; Striate; Verrucate.

RUGBY® Cadusafos.

RUGGED Adj. Pertaining to surface which is superficially rough or craggy from intermixed Mucros, spines and tubercles.

RUGGLES, ARTHUR GORDON (1875–1947) (Anon. 1948, J. Econ. Ent. 41: 841–842.)

RUGOSE Adj. (Latin, *ruga* = wrinkle; *-osus* = full of.) 1. Wrinkled. Pertaining to a surface covered with wrinkles. 2. Botany: Elevated veinlets with the surrounding surface sunken. See Sculpture Pattern; Rugate. Cf. Favose. Alt. Rugosus; Rugous.

RUGOSISSIMUS Adj. Very rugose or wrinkled

RUGULA Noun. (Latin, diminutive of *ruga* = wrinkle. Pl., Rugulae.) A small wrinkle.

RUGULOSE Adj. (Latin, *rugula* = little wrinkle; *-osus* = full of.) Minutely wrinkled. Alt. Rugulosus.

RÜHL, FRITZ (1836–1893) (Anon. 1893, Ent. News 4: 280.)

RÜHL, MARIE (1868–1930) (Fischer 1930, Societas ent. 45: 29.)

RUHMAN, MAX HERMANN (1880–1943) (Venavle 1944, Proc. ent. Soc. Br. Columb. 41: 35–36.)

RUIZ PEREIRA, HERMANO FLAMINO (1884–1942) (Cortes 1943, Revta. Ent., Rio de J. 14: 324–325; Porter 1943, Revta. chil. Hist. nat. 45: 201.)

RUMBLINE® See Alanycarb.

RUMPHIUS (RUMPH, ROMPF), GEORG EBERHARD (1627–1702) (Rose 1850, *New General Biographical Dictionary* 11: 404–405.)

RUMSEY, FREDERICK WILLIAM (1885–1960) (S. W. 1960, Proc. Trans. S. Lond. ent. nat. Hist. Soc. 1960: xxxv–xxxvii.)

RUMSEY, WILLIAM EARL (1865–1938) (Peiars 1938, J. Econ. Ent. 31: 463; Osborn 1946, *Fragments of Entomological History,* Pt. II, 232 pp. (108) Columbus, Ohio.)

RUMULE Noun. (Unknown origin. Pl., Rumulae.) A teat-like fleshy protuberance on the larval body. (Kirby & Spence); archaic.

RUNCINATE Adj. (Latin, *runcina* = plane; *-atus* = adjectival suffix.) Notched; descriptive of structure cut into several transverse acute segments which point backward. Alt. Runcinatus. See Outline Shape.

RUNNER, GEORGE A (1876–1941) (Porter 1941, J. Econ. Ent. 34: 592; Osborn 1946, *Fragments of Entomological History,* Pt. II, 232 pp. (108) Columbus, Ohio.)

RUPERTSBERGER, MATTHIAS (1843–1931) (Anon. 1932, Jber. Oberöst. Musealver 1932: 84; Anon. 1932, Koleopt. Rdsch. 18: 216.)

RUPICOLOUS Adj. (Latin, *rupes* = rock; *colere* = to

inhabit.) Descriptive of organisms that live among rocks or stones. Alt. Rupestrine. See Habitat. Cf. Arenicolous; Deserticolous; Eremophilous; Ericeticolous; Lapidicolous; Psammophilous; Rel. Ecology.

RUPTOR OVI Egg burster. See Oviruptor.

RUPTURED Adj. (Latin, *ruptura* = to break.) Pertaining to structure broken or separated. See Broken; Fractured.

RURALIDAE See Lycaenidae.

RURSUS Adj. (Unknown origin.) Backwards.

RUSCHENBERGER, WILLIAM SAMUEL WAITHMAN (1807–1895) (Nolan 1895, Proc. Acad. nat. Sci. Philad. 1895: 452–462; Weiss 1936, *Pioneer Century in American Entomology,* 320 pp. (164–166), New Brunswick.)

RUSCHEWEYH, GEORGE (1826–1899) (Dyar 1900, Ent. News 11: 580.)

RÜSCHKAMP, FELIX (1885–1957) (Anon. 1957, Ent. Bl. Biol. Syst. Käfer 53: 130–133, bibliogr.)

RUSCHKE, FRANZ (1882–1942) (Fahringer 1943, Z. Angew. Ent. 30: 151–152, bibliogr.)

RUSH SKELETONWEED GALL-MIDGE *Cystiphora schmidti* [Diptera: Cecidomyiidae].

RUSS, PERCY (Beirne 1944, Irish. nat. J. 8: 208–210.)

RUSSELL, ARCHIBALD GEORGE BLOMEFIELD (1879–1955) (Hall 1956, Proc. R. ent. Soc. Lond. (C) 20: 75.)

RUSSELL, FRANK S (Hale-Carpenter 1946, Proc. R. ent. Soc. Lond. (C) 10: 54.)

RUSSELL, FREDERICK WILLIAM (1844–1915) (Anon. 1916, Psyche 23: 25; Hall 1916, Ent. News 27: 47–48.)

RUSSELL, HARRY MERWIN (1882–1915) (Britton 1915, J. Econ. Ent. 8: 433; Quaintance *et al.* 1916, Proc. ent. Soc. Wash. 18: 3–5.)

RUSSELL, JOHN ANTHONY (–1942) (Anon. 1942, Entomologist 75: 72.)

RUSSELL, SYDNEY GEORGE CASTLE (1866–1955) (Buckhardt 1958, Entomologist's Rec. J. Var. 70: 283–284; Kershaw 1958, Entomologist's Rec. J. Var. 70: 1–4, 37–41, 94, 100, 156–160.)

RUSSET Adj. (Old French, *rousset*, diminutive of *roux* = red.) Reddish brown.

RUSSIAN SPRING-SUMMER ENCEPHALITIS See Tick-Borne Encephalitis.

RUSSIAN WHEAT-APHID *Diuraphis noxia* (Mordvilko) [Hemiptera: Aphididae]: A potential pest of small grain crops in many parts of the world; RWA a particular pest of barley, oats, rye, sorghum and wheat. Toxins injected into plant during feeding that destroy chlorophyll and prevent carbohydrate formation; affected plant appears white and yellow with leaf curling; heavy infestation causes plant death. RWA considered endemic to the southern part of the USSR and spread to Europe, South Africa (1978), Mexico (1980) and the USA (1985.) Adult ca 1.5 mm long, elongate, pale green; Antenna and Cornicle short compared with other aphids. Parthenogenetic female produces ca 2–3 nymphs per day during warm weather; maturation requires ca 7 days. Cf. Greenbug.

RUSSO, GIUSEPPE (1897–1972) (Tremblay 1974, Boll. Lab. ent. agr. Filippo Silvestri 30: i–xiii, bibliogr.)

RUST Noun. (Anglo Saxon, *rust* = redness. Pl., Rusts.) Diseases of grasses and other plants caused by parasitic fungi.

RUST, E W Collected parasites of Black Scale in South Africa for California (1923–1928.)

RUST FLIES See Psilidae.

RUST-RED FLOUR-BEETLE *Tribolium castaneum* (Herbst) [Coleoptera: Tenebrionidae]: A widespread, major pest of stored cereals and other foods in warmer regions and periodic invader of temperate areas; in Australia, a secondary pest of broken and damaged grain, flour and cereals; does not attack intact, sound kernels. Common on farms, at mills, in stores and homes. Adult cannibalistic upon egg and Pupa; Larva and adult predatory upon Lepidoptera and other beetles in stored products. Adult 3.0–4.0 mm long, reddish brown; Antennal club with three segments; Elytra with conspicuous longitudinal grooves (striae); strong flier; lives 9–12 months. Female lays ca 150–600 eggs among foodstuffs during lifetime (ca two months); incubation 2–3 days. Mature larva 6.0 mm long, body white, head dark; two Urogomphi at apex of Abdomen; undergo 7–8 instars; stage ca 13–14 days. Pupal stage 4–5 days; life cycle 4–8 weeks, depending upon diet, temperature and humidity. Cf. Confused Flour Beetle. Dark Flour Beetle. Yellow Mealworm.

RUSTIC BORER *Xylotrechus colonus* (Fabricius) [Coleoptera: Cerambycidae].

RUSTON, ALFRED HAROLD (1856–1929) (Fryer 1930, Entomologist 63: 24.)

RUSTY-BANDED APHID *Dysaphis apiifolia* (Theobald) [Hemiptera: Aphididae].

RUSTY GRAIN-BEETLE *Cryptolestes ferrugineus* (Stephens) [Coleoptera: Cucujidae]: A cosmopolitan, important pest of stored cereal grains, flour, dried fruits and nuts in warm regions. Adults feed on exposed seed; larvae cannot penetrate intact grains, but do attack damaged grain; internal seed feeders. Adult 2.5 mm long, elongate, flat, reddish brown; Antenna long, filiform; head and Thorax rather large; winged but reluctant to fly. Female lays ca 200 eggs in produce. Larva campodeiform with horn-like processes at apex of Abdomen. Pupa enclosed within silken cocoons. Syn. Rust-Red Grain Beetle.

RUSTY PLUM-APHID *Hysteroneura setariae* (Thomas) [Hemiptera: Aphididae]: Endemic to eastern North America. Eggs laid on small twigs; hatch when buds open. Several wingless generations feed on plum or peach, distorting new growth. Winged generation migrates to grasses, feeds and reproduces over summer. Autumn winged generation migrates to plum or peach, mates, lays overwintering eggs.

RUSTY TUSSOCK-MOTH *Orgyia antiqua* (Linnaeus) [Lepidoptera: Lymantriidae].

RUSZKOWSKI, JAN (1889–1961) (Opyrchalowa 1962, Przegl. zool. 6: 211–213.)

RUT Noun. (Latin, *rugire* = to roar.) The period of sexual activity in male animals. Cf. Oestrus.

RUTELLA Acari: Hypostome with Rutella or Corniculi.

RUTENBERG, J G HAGG See Haag-Rutenberg.

RUTHE, JOHANN FRIEDRICH (1788–1859) (Kraatz 1860, Berl. ent. Z. 4: 101–102.)

RUTHERGLEN BUG *Nysius vinitor* Bergroth [Hemiptera: Lygaeidae]: A pest of numerous vegetable, tree and field crops in eastern Australia; major pest of safflower. Adult ca 5 mm long, slender-ovate and greyish-brown with dark markings; males smaller and darker than females; wings nearly transparent; Pronotum and legs with fine Setae. Adult stage to 28 days in summer, but adults overwinter at end of autumn. Overwintering adults lay eggs during spring. Female lays ca 400 eggs during life. Eggs ca 1 mm, cream-coloured, placed singly or in small groups on ground, litter or flower heads; eclosion occurs within a week. Five nymphal instars. Multivoltine; generation time ca four weeks. Adult and nymph feed on stems, foliage or seeds; feeding in large numbers causes extensive plant withering, fruit malformation and plant death. Adults migrate long distances via flight; nymphs move en masse over ground. Cf. Grey Cluster Bug.

RUTILANT Noun. (Latin, *rutilus* = red.) A shining bronze-red colour.

RUTILOUS Adj. (Latin, *rutilus* = red; *-osus* = with the property of.) Descriptive of structure which is shining bronze red in colour.

RUTIN Noun. (Latin, *ruta,* from rue. Pl., Rutins.) A yellow crystalline flavonol glycoside in the leaves of rue and tobacco, flower buds of Japanese Pagoda Tree. Rutin yields quercetin and rutinose on hydrolysis.

RUTTNER, FRANZ (1882–1961) (Schimitschek 1961, Anz. Schädlingsk. 34: 171.)

RUTTY, JOHN (1698–1775) (Rose 1850, *New General Biographical Dictionary* 11: 412.)

RUZICKA, ANTONIN (1883–1943) (Hlouseck & Stehlik 1949, Sborn. pfir. Klubu Trebici 4: 7–11.)

RUZSKII, MICHAIL DIMITRIEVICH (1864–1936) (Berezhkov 1937, Trudy biol. nauchnoissled. Inst. tomsk. gos. Univ. 4: 1–6.)

RVF See Rift Valley Fever.

RYABOV, MIKHAILA ALEKSEEVICH (1890–1962) (Danilevski & Kusnetzov 1963, Ent. Obozr. 42: 473–475, bibliogr. Translation: Ent. Rev., Wash. 42: 258–259.)

RYAN® See Ryania.

RYANIA Noun. An alkaloid extract (dried powder) from the stems and roots of the South American plant *Ryania speciosa* Vahl (Flacourtiaceae). Regarded as a natural insecticide more stable than nicotine and rotenone; at least 11 compounds identified with different insecticidal properties with most common including Ryanodine and dehydroyanodine. A slow-acting stomach poison whose mode-of-action involves the Ca^{2+} release channel in muscle. Registered on apples, citrus, corn, pears and walnuts for control of thrips, codling moth and European corn borer. Trade names include: Ryanodine® and Ryan® See Botanical Insecticide; Natural Insecticide; Insecticide. Cf. Pyrethrin; Nicotine; Rotenone; Sabadilla.

RYANODINE® See Ryania.

RYBINSKI, MICHAEL (1846–1905) (Anon. 1905, Wien. ent. Ztg. 24: 118.)

RYDEN, NILS (1883–1961) (Lindroth 1961, Opusc. ent. 26: 240.)

RYE, BERTRAM GEORGE (1872–1936) (Wolff 1938, Ent. Meddr. 20: 103–105.)

RYE, EDWARD CALDWELL (1832–1885) (Anon. 1885, Entomol. mon. Mag. 21: 238–240; Carrington 1885, Entomologist 18: 79–80.)

RYGGE, JOHAN (1868–1944) (Schöyen 1945, Norsk. ent. Tidsskr. 7: 136–137.)

RYPOPHAGOUS Adj. (Greek, *rhypos* = filth; *phagein* = to eat; Latin, *-osus* = with the property of.) Filth-eating. Alt. Rhypophagus. See Feeding Strategy. Cf. Scatophagous.

RYZ-UP® See Gibberellic Acid.

RZEHAK, EMIL (1856–1934) (Jedlitschka 1935, Mitt. naturw. Ver. Troppau 27: 3–10, bibliogr.)

SAALAS, UUNIO (1882–1969) (Lindroth 1969, Opusc. ent. 34: 172–173.)

SAALMÜLLER, MAX (1832–1890) (Anon. 1892, Dt. Ent. Z. Iris 5: 453–459, bibliogr.)

SABADILLA Noun. (Spanish, *cebadilla, cebada* = barley; Latin, *cibus* = food. Pl., Sabadillas.) An alkaloid biopesticide in the form of dried powder that is extracted from mature seeds of *Schoencaulon officinale* A. Gray (= *Sabadilla officinarum* Brant and *Veratrum sabadilla* Retz). Sabadilla is used as a botanical insecticide against thrips on citrus. Major insecticidal components include veratridine, cervadine and esters of veracevine. Sabadilla probably acts as a neurotoxin that affects sodium-ion channels of excitable membranes. Trade name: Veratrin® See Botanical Insecticide; Natural Insecticide; Insecticide. Cf. Methyllycaconitine; Nereistoxin; Nicotine; Rotenone; Pyrethrin; Pyrroles; Ryania; Stemofoline. Alt. Cebadilla.

SABER® See Lambda Cyhalothrin.

SABINE, EBENEZER (–1906) (Anon. 1906, Insektenbörse 23: 82; Anon. 1906, Entomologist 39: 128.)

SABINE, LLEWELLYN EDMUND (1886–1963) (Worms 1963, Entomologist's Rec. J. Var. 75: 139–140.)

SABULOSE Adj. (Latin, *sabulum* = sand; *-osus* = full of.) Sandy or gritty. A term applied to structures with a sandy texture or surface. Alt. Sabuline; Sabulosus.

SAC Noun. (Latin, *saccus* = sack. Pl., Sacs.) 1. Any small pouch-like, bladder or bladder-like vessel or structure. 2. A pouch-like vessel on or in the body of a plant or animal that often contains fluid. Cf. Bursa. 3. Coccidae: The separate cottony envelope secreted by some Species. See Ovisac.

SAC SPIDER See Clubionidae.

SAC TUBE Mallophaga: A trough composed of a pair of half-tubes, formed through a prolongation of the apex of the Stylet Sac forward into the Buccal Funnel (Imms).

SACBROOD Noun. (Latin, *saccus* = sack; Middle High German, *bruot* = brood. Pl., Sacbroods.) A widespread, pathogenic viral disease of honey bee larvae, first described by White (1917) in America. SB virus infects fat-body of larva; 2-day-old larvae most susceptible; infected larva remains in cell on dorsum with head toward cell cap; fluid accumulates between body and last instar larval Integument; body becomes yellow, then dark brown after death. SB virus common in hives but disease manifest only in genetically susceptible bees; SB spread in hive due to poor management by worker bees in hive, requeening or moving hives. Death occurs during prepupal period after cocoon has been spun and cell capped. SB virus accumulates in Hypopharyngeal Gland of adult worker. Disease does not threaten colony when infected larvae are identified by workers and removed from brood; SB looses virulence in dead and desiccated larvae. Cf. American Brood Disease; European Brood Disease. Rel. Acute Bee-Paralysis Virus; Chronic Bee-Paralysis Virus.

SACCATE Adj. (Latin, *saccus* = sack; *-atus* = adjectival suffix.) 1. Sac-like, Bag-like or Pouch-like in form. 2. Pertaining to structure which is encysted. See Sac. Cf. Bursiform; Pyreniform; Scrotiform.

SACCHAROMYCETES Plural Noun. (Greek, *sakchar* = sugar; *mykes* = fungi.) Organisms related to yeasts and contained within the Mycetome of aphids.

SACCHAROSE See Sucrose.

SACCIFORM Adj. (Latin, *saccus* = sack; *forma* = shape.) Bag-like; shaped as a sack.

SACCIFORM LARVA Hymenoptera: An apocritous, ovoid, featureless larva without segmentation. A SL is characteristic of several Families of parasitic Hymenoptera including the Dryinidae, Mymaridae and Trichogrammatidae. See Larva.

SACCOID GILL A swollen sac-like gill.

SACCULAR Adj. (Latin, *saccus* = sack.) Sac-like; sac-formed. Alt. Sacculated.

SACCULE Noun. (Latin, *sacculus* = little sack. Pl., Saccules.) A small sac or small pouch.

SACCULUS Noun. (Latin, *sacculus* = little sack. Pl., Sacculi.) Genitalia of male Lepidoptera: The sclerotized, ventral part of the Harpe (Klots).

SACCUS Noun. (Latin, *saccus* = sack. Pl., Sacci.) Lepidoptera: A median sclerotized pocket (pouch) in the female genitalia formed by the invaginated sternal region of the ninth abdominal segment. Genitalia of male Lepidoptera: A midventral, anteriorly directed projection of the Vinculum (ninth abdominal Sternum) inside the body that serves as a muscle attachment (Klots).

SACHSE, CARL TRAUGOTT (1815–1863) (Reichenbach 1864, Sber. naturw. Ges. Isis Dresden 1864: 1–6.)

SACHSE, RUDOLPH (–1891) (Kraatz 1891, Dt. ent. Z. 35: 11.)

SACHTLEBEN, HANS (1893–1968) (Herbst 1963, Mitt. dt. ent. Ges. 22: 41–43.)

SACK, PIUS (1865–1946) (Richter & Seitz 1936, Natur Volk. 66: 243.)

SACKEN See OSTEN SACKEN.

SADAO CHENG In Thailand, the common name for 'Philippine Neem Tree.' See Philippine Neem. Cf. Indian Neem; Thai Neem.

SADDLE GALL-MIDGE *Haplodiplosis marginata* [Diptera: Cecidomyiidae].

SADDLE Noun. (Anglo Saxon, *sadol*. Pl., Saddles.) 1. Culicidae: A sclerite on the Anal Siphon of larvae. 2. Any broad, transverse shallow depression on a convex sclerite.

SADDLEBACK CATERPILLAR *Sibine stimulea* (Clemens) [Lepidoptera: Limacodidae].

SADDLED LEAFHOPPER *Colladonus clitellarius* (Say) [Hemiptera: Cicadellidae].

SADDLED PROMINENT *Heterocampa guttivitta* (Walker) [Lepidoptera: Notodontidae].

SADUN, ELVIO HERBERT (1918–1974) (Weinstein 1974, J. Parasit. 60: 897–899.)

SAFIDON® See Phosmet.

SAFRO, VICTOR IRVING (1888–1944) (Leonard 1945, J. Econ. Ent. 38: 727–729.)

SAFROTIN® See Propetamphos.

SAF-T-SIDE® See Petroleum Oils.

SAGA® See Tralomethrin.

SAGARRA Y CASTELLARNAU, IGNACIO DE (Dusmet y Alonso 1919, Boln. Soc. ent. Esp. 2: 181.)

SAGEBRUSH DEFOLIATOR *Aroga websteri* Clarke [Lepidoptera: Gelechiidae].

SAGEDER, FRANZ (1874–1949) (R. 1950, Z. Wien. ent. Ges. 35: 124.)

SAGITTA Noun. (Latin, *sagitta* = arrow. Pl., Sagittae.) 1. Any arrow-like structure. 2. Hymenoptera: See Digitus; Volsella. 3. Arrow-like spots on the wings of some insects.

SAGITTAL Adj. (Latin, *sagittalis* > *sagitta* = arrow; *-alis* = characterized by.) 1. Descriptive of something shaped as an arrowhead. 2. Designating the position of something along the longitudinal plane of a body.

SAGITTAL PLANE The longitudinal, vertical plane that divides a bilaterally symmetrical animal into right and left halves. See Orientation; Section. Cf. Frontal Plane; Parasagittal Planes; Transverse Plane.

SAGITTATE Adj. (Latin, *sagitta* = arrow; *-atus* = adjectival suffix.) Descriptive of structure shaped as an arrow head; more-or-less elongate triangular. Alt. Sagittatus. See Outline Shape. Cf. Semisagittate.

SAGRA, RAMON DE LA (1798–1871) (Urban 1903, Symbolae Antillae 3: 117–118.)

SAHAGUN, BERNARDINO DE (–1590) (Dusmet y Alonso 1919, Boln. Soc. ent. Esp. 2: 79.)

SAHLBERG, C L Finnish academic and author of *Dissertatio Entomolgica Insecta Fennica enumerans* (1817–1823) and *Periculi Entomographici* (1823).

SAHLBERG, CARL REINHOLD (1779–1860) (Toruroth 1861, Acta Soc. Scient. fenn. 6: 1–7.)

SAHLBERG, JOHN REINHOLD (1846–1920) (Böving 1920, Science 52: 216, bibliogr.; Salaas 1920, Ent. Bl. Biol. Syst. Käfer 16: 195–199.)

SAHLBERG, REINHOLD FERDINAND (1811–1874) (Saalas 1958, Acta ent. fenn. 14: 1–255, bibliogr.)

SAILER, MAX (–1909) (Ebner 1909, Ent. Z., Frank. a. M. 23: 136–137.)

SAILER, REECE I (1915–1986) American taxonomic Entomologist specializing in Hemiptera and an administrator with USDA.

SAINT CYR, DOMINIQUE NAPOLÉON (1826–1899), (Anon. 1899, Naturaliste can. 26: 45–47, 59–63; Tretty 1899, Can. Ent. 31: 102.)

SAINT-HILAIRE, AUGUSTINE FRANÇOIS CÉSAR PROVENCAL DE (1779–1853) (Urban 1864, In Martius, C.F.P. Flora Brasiliensis 1: 91–99; Sampaio 1928, Boln. Mus. nac. Rio de J. 4 (4): 1–31.)

SAINT-JOHN'S WORT *Hypericum perforatum* Linnaeus [Hypericaceae]: A Palaearctic perennial plant that has become an introduced rangeland pest in Australia (1880) and northwestern USA (1900). SJW is poisonous to sheep and the plant's chemicals photosensitize white-skin areas of animals. Ecologically, SJW is a pest because it is a competitor of pasture grasses and field crops. SJW is attacked by numerous phytophagous insects and controlled or suppressed with chrysomelid beetles, most notably the leaf-feeding *Chrysolina quadrigemina* (Suffrain) imported from France; beetle larvae attack winter basal-growth of plant. Syn. Goat Weed; Klamath Weed.

SAISO, KAROLY (KARL) (1851–1939) (Szent-Ivany 1941, Folia ent. hung. 6: 41–43.)

SAITO, KOZO (1904–1961) (Hasegawa 1967, Kontyû 35: (Suppl.): 37–38.)

SAKAGUCHI, SOICHIRO (1887–1965) (Hasegawa 1967, Kontyû 35: (Suppl.): 38.)

SAKAI, HISAMA (1896–1948) (Hasegawa 1967, Kontyû 35 (Suppl.): 38, bibliogr.)

SALAVARIAN TRYPANOSOME A section of *Trypanosoma* whose members are transmitted mechanically or undergo cyclical development in dipterous vectors before salivary transmission to a mammalian host. Most ST transmitted by *Glossina* spp. Cf. Stercorarian Trypanosome.

SALDIDAE Amyot & Serville 1843. Plural Noun. Shore Bugs. A small, widespread Family of heteropterous Hemiptera assigned to Superfamily Saldoidea. Adult body small, oval in outline shape, flattened with sombre colour or mottled patterns; Antenna with four segments; Labium elongate with three-segments; Hemelytron with 4–5 closed cells. Species predaceous upon small, sessile or moribund invertebrates, or scavenge upon damp substrate. Saldids are typically littoral and occur along banks of lake, stream, marsh, or salt water; some Species burrow into mud, sand or soil. Eggs are laid in vegetation or substrate; adults often engage in short flights before retreating into vegetation. Copulation side-by-side, not with male-above posture as exhibited in Gerromorpha. Apparently, Saldidae are closely related to Aepophilidae; fossil record to middle Jurassic Period.

SALDOIDEA A Superfamily of Hemiptera, including the Aepophilidae and Saldidae.

SALEBROSE Adj. (Latin, *salebrosus* = rough, uneven.) Rough; rugged. Alt. Salebrous.

SALFI, MARIO (1900–1970) (La Greca 1970, Memorie Soc. ent. ital. 49: 189–194, bibliogr.; La Greca 1970, Boll. Soc. ent. ital. 102: 117.)

SALIBRE® See Hexythiazox.

SALIENT Noun. (Latin, *saliens* = leaping. Pl., Salients.) Structure which projects outward or upward from its surroundings.

SALIS-MARSCHINS, CARL ULYSSES (1762–

1818) (Anon. 1818, Meisner's Naturw. Anz. 1: 72.)

SALITUBA Noun. (Latin, *saliva* = spittle; *tuba* = tube. Pl., Salitubae.) The salivary bulb (MacGillivray).

SALIVA Noun. (Latin, *saliva* = spittle. Pl., Saliviae, Salivas.) 1. A viscous, transluscent or opalescent secretion of variable composition and origin (Salivary Gland, Labial Gland, Crop) that is emitted from the Buccal Cavity of many Species of insects. Saliva contains water, salts, proteins and starch-splitting enzymes. Functions of saliva are diverse: Saliva moistens food, initiates digestion and lubricates food during its passage through the foregut. See Alimentary System; Salivary Gland. 2. A sclerite on each side of the head which supports the anterior half of the Hypopharynx (MacGillivray). Cf. Spittle.

SALIVARIUM Noun. (Latin, *saliva* = spittle; *-arium* = place of a thing. Pl., Salivaria.) The cavity between the base of the Hypopharynx and the two canals between the Maxillary Bristles, through which the salivary secretions are ejected by the Salivary Pump.

SALIVARY Adj. (Latin, *salivarius* = slimy.) Pertaining to saliva; descriptive of glands that produce saliva or the system that transports saliva.

SALIVARY CANAL Hemiptera: A longitudinal groove on the medial surface of the Maxillary Stylets that forms a tube for the passage of salivary secretions from the head to the food object. SC posterior of the Food Canal. Cf. Food Canal.

SALIVARY GLAND 1. Any of several Glands that originate in the head and display different anatomical forms among various insect Species. SG is probably the most commonly encountered head gland found in the Insecta. Terms 'Labial Gland' and 'Salivary Gland' are used interchangeably and sometimes create confusion. 'Labial Gland' suggests position for gland outlet; 'Salivary Gland' suggests role of gland. Flies and some other insects groups produce copious secretions from glands that are suggestive of saliva and in these groups the term Salivary Gland is more common. Because SG is so widespread and diverse in function, anatomy of SG varies considerably among insects. In gross anatomy, SG may resemble clusters of grapes connected by intercalary ducts which in turn connect with Primary Salivary Ducts. This type of SG is sometimes called an Acinar Salivary Gland. In some insects, (*e.g.* higher Diptera) SGs form long tubes that may project into Abdomen. This type of SG is sometimes called a Tubular Salivary Gland Anatomically, in either type, wall of gland consists of a layer of epithelial cells, a layer of Cuticle restricted to duct and selected areas, and a basal membrane. Depending upon insect Species, tracheoles and nerves may be embedded in basal membrane. 2. Cockroaches: Innervated Acinar Glands that contain several types of cells. 3. Hemiptera: Modified Labial Glands that are contained primarily in anterior region of Thorax and associated with gut. SG consists of two Principal SG and two Accessory Glands; duct of each Accessory Gland connects with duct of each Principal SG. A nerve is associated with PSG and AG. Ducts of PSGs unite to form a Common Salivary Duct. CSD discharges into Salivary Pump. SP leads to Salivary Canal that joins with base of Maxillae. 4. Adults of lower Diptera (mosquitoes) display SGs that consist of several tubes or lobes. Lateral tubes consists of transport cells positioned between glandular cells. SGs of some lower Diptera larvae (chironomid midges) are anatomically different. Higher Diptera employ SGs in larval and adult stages. SGs common duct divides in Thorax to form two long, thin tubes that project into Abdomen. Each gland is composed of a layer of epithelial cells. Each gland is differentiated into a short, proximal resorptive area and a long, distal secretory area. Epithelial cells in resorptive area are flattened and lack Canaliculi; mitochondria display normal cristae and secretory granules are not evident; epithelial cells in resorptive area are bound together with pleated Septate Junctions. Secretory area of gland, epidermal cells are characterized by numerous mitochondria, abundant Rough Endoplasmic Reticulum, Golgi Bodies and pale secretory granules. Apical membrane of epithelial cells is strongly convoluted and forms a system of branches called Secretory Canaliculi. Canaliculi open into lumen of Salivary Gland where secretory products accumulate. Epithelial cells in secretory area are bound together with Desmosomes, Pleated Septate Junctions and Gap Junctions. Adult blowflies show paired, elongate, non-innervated tubular glands that connect to Hypopharnyx via a median duct. Median Duct extends posteriad through head and bifurcates in anterior part of Thorax. Anterior portion of paired tubules displays an absorptive region. Paired tubules are adjacent to gut and project posteriad through Thorax and into Abdomen. Tubules in most of Thorax and Abdomen form a secretive region. 5. Lepidoptera larvae display well developed Labial Glands. LG secretions of silkworm *Bombyx mori* used for attachment of fish hooks. Adults of moth *Manduca sexta* display paired, tubular Salivary Glands which secrete invertase through a common salivary duct. Each gland is divided into several distinct regions. See Exocrine Gland; Gland; Saliva. Cf. Mandibular Gland; Labial Gland. Rel. Ingestion; Digestion.

SALIVARY MEATUS A Salivary Duct that connects the Salivary Canal and Salivary Pump.

SALIVARY PUMP 1. Hemiptera: An apparatus provided with stout muscles that operates as a force-pump to impel saliva down the canal and convey it into the opening in tissues made by the Stylets. 2. The chitinous, cup-like structure at the base of the Labial Stylets of piercing Diptera (*e.g.* mosquitoes). See Salivary Syringe.

SALIVARY RECEPTACLE A small cavity above the Salivary Duct opening and between the Labium and Hypopharynx.

SALIVARY SYRINGE 1. Diptera (Mosquitoes): The Salivarium into which the Salivary Duct opens and through which saliva is discharged. 2. The specialized Salivarium (Snodgrass). See Salivary Gland.

SALIVOS Noun. (Latin, *salivosus* = slavering.) The aperture of the Salivary Gland Duct (MacGillivray).

SALIVOUS Adj. (Latin, *saliva* = spittle; *-osus* = possessing the qualities of.) Pertaining to saliva or being composed of saliva.

SALLE, AUGUSTE (1820–1896) (Anon. 1896, Leopoldina 32: 139; Giard 1896, Bull. Soc. ent. Fr. 1896: 213.)

SALMON, DANIEL ELMER (1850–1914) (Anon. 1915, Ent. News 26: 96.)

SALPINGIDAE Plural Noun. Narrow-Waisted Bark Beetles. A numerically small, widespread Family of polyphagous Coleoptera assigned to Superfamily Tenebrionoidea. Adults elongate, flattened and dark coloured; body glabrous or with sparse vestiture of Setae; head prognathous or rostrate, but not constricted behind compound eyes; tarsal formula typically 5-5-4; Elytra somewhat shortened and exposing apical abdominal Terga. Larva elongate, somewhat flattened and parallel-sided; head and apex of Abdomen pigmented, body otherwise pale. Adults typically occur on flowers and foliage, among leaf litter or under rocks; larvae under bark, in dead twigs or vines. Adults and larvae of some Species are predaceous in a diversity of habitats; some Species are phytophagous; a few Species are algal feeders within intertidal habitats.

SALT Noun. (Anglo Saxon, *sealt*, Latin, *sal,* Greek, *Hals.* Pl., Salts.) A colourless or white crystalline solid called sodium chloride (NaCl) abundant in solid and liquid form.

SALT-MARSH CATERPILLAR *Estigmene acrea* (Drury) [Lepidoptera: Arctiidae].

SALT-MARSH CULEX *Culex sitiens* Wiedemann [Diptera: Culicidae]: Head with narrow, curved white scales dorsally, darker laterally; Proboscis dark with narrow pale band near midline; wing with dark scales; Haltere pale. Female feeds on humans during night; also feeds on other mammals and birds; capable of migrating long distances from coastal habitat. Breeds in brackish water, occasionally fresh water. In Australia, not noted as vector of disease but is susceptible to infection by Ross River Fever. See Culicidae.

SALT-MARSH MOSQUITO 1. *Aedes vigilax* (Skuse) [Diptera: Culicidae]: The common name in Australia for one of several Species referred to elsewhere in the world as 'Salt-Marsh Mosquito.' SMM viciously bites day or night, and enters buildings in search of humans. SMM breeds in salt marshes and brackish temporary pools near mangroves. Adult female head with narrow dark scales near margin of eye, pale near occiput; Proboscis pale along basal half; palpi apically white; wings scales dark with pale mottling; apex of Abdomen tapered; Cerci long; Femora and Tibia mottled. Female autogenous; feeds on humans, other mammals and birds. Eggs laid singly in mud, on vegetation or in water; resistant to desiccation and may exist for long periods between heavy rains which fill pools and initiate eclosion. Larval and pupal stages ca 6 days in mid summer. SMM vector of Ross River virus, Murray Valley Encephalitis (lab), and Dog Heartworm in Australia. Serves as mechanical vector of Myxomatosis. 2. *Aedes sollicitians* (Walker) [Diptera: Culicidae] in North America; endemic to eastern North America around salt marshes. Adults strong fliers, can travel 40–50 km; fierce daylight biters of humans and domestic animals. Adult with golden-brown Thorax; abdomen with median longitudinal white line and transverse rings or golden rings; Proboscis and Tibiae with wide white bands; wings speckled with white and brown. See Culicidae. Cf. Black Salt-Marsh Mosquito; California Salt-Marsh Mosquito.

SALTATORIA Plural Noun. (Latin, *saltator* = dancer.) Bush Crickets; Grasshoppers; Katydids; Locusts; Long Horned Grasshoppers; Mole Crickets; Pygmy Grouse Locusts; Pygmy Mole Crickets; Short Horned Grasshoppers; Wingless Camel Crickets. A taxonomic category that includes jumping Orthoptera (grasshoppers, crickets and related Taxa). Cf. Gressoria.

SALTATORIAL Adj. (Latin, *saltare* = to leap; *-alis* = characterized by.) Adapted for leaping; insects with anatomical adaptations and behavioural predisposition to leap, jump or hop. Alt. Saltatory. See Leg. Cf. Cursorial; Fossorial; Gressorial; Natatorial; Raptorial; Scansorial. Rel. Locomotion.

SALTATORY APPENDAGE 1. Any appendage modified for jumping or leaping (*e.g.* the hind leg of a grasshopper). 2. Collembolla: The Furcula.

SALTBUSH PLANTHOPPER *Privesa pronotalis* Distant [Hemiptera: Ricaniidae].

SALTBUSH SCALE See Cottony Saltbush Scale.

SALTER, KEITH ERIC WELLESLEY (1908–1969) (Chadwick 1974, J. ent. Soc. Aust. (N.S.W.) 8: 41–42, bibliogr.)

SALTERO, DON (FL 1790) (Faulkner 1829, *An historical and topographical description of Chelsea…interspersed with biographical anecdotes.* 406 pp. (378–383), Chelsea.)

SALTICIDAE Plural Noun. Jumping Spiders. Small to moderate-sized spiders with large anteromedial eyes, good vision and the capacity to jump. Most Species are active on vegetation during daytime; salticids do not build webs but occupy silken retreat during night. Diverse feeding habits include eating insects, insect eggs and other spiders. Many Species mimic ants, flies and beetles.

SALTMARSH Noun. (Anglo Saxon, *sealt* = salt;

mersc = lake. Pl., Saltmarshes.) Ecology: Coastal saltwater habitats in which angiosperm herbs and shrubs grow and which are periodically inundated by the sea. Most saltmarshes occur in temperate regions, but some saltmarshes may occur on the landward side of mangrove vegetation in some tropical areas. Saltmarsh tends to form along sheltered coast which promotes the accumulation of particulate organic matter along a gently-sloaping plateau. Many insects are habitat-specific to saltmarsh habitats; some insects periodicially invade saltmarsh habitats; some Species are vagrants in saltmarshes. Saltmarsh provides special challenges to insects involving respiration, osmotic and ionic adaptation, and inundation. Species of Diptera are the most frequently collected Order of insects taken in saltmarsh; Coleoptera are also frequently taken in this habitat. See Habitat. Cf. Estuarine; Marsh; Desert. Rel. Biome.

SALTPAN BLUE *Theclinesthes sulpitius* (Miskin) [Lepidoptera: Lycaenidae].

SALT-WATER MOSQUITO *Aedes australis* (Erichson) [Diptera: Culicidae].

SALVAGE, THOMAS (1850–1926) (Anon. 1926, Entomologist 59: 176.)

SALVATION JANE Syn. Patterson's Curse.

SALVERFORM Adj. (Spanish, *salva* = small tray; Latin, *forma* = shape.) Tubular with a spreading limb. Rel. Form 2; Shape 2; Outline Shape.

SALVIN, OSBERT (1835–1898) (McLachlan 1898, Entomol. mon. Mag. 34: 164–165; Salvin 1909, Autobiogr. Jubilee Suppls. Ibis 2: 127–128.)

SALVINIA *Salvinia molesta* Mitchell [Salviniaceae]: A free-floating, mat-forming perennial aquatic fern that is generally regarded as one of the three most important aquatic weeds globally. *Salvinia* is endemic to neotropics (Brazil) and has been introduced into Sri Lanka (1939), Australia (1953), Africa, India, Southeast Asia and New Zealand. Circumstantial evidence suggests the pest has become widely distributed via the aquarium trade. Possibly an accidental horticultural hybrid and sometimes called *S. auriculata.* Sexually sterile pentaploid in which spores do not germinate; reproduction vegetative as a result of fragmentation; each node capable of growing into an independent plant. Plant consists of horizontal rhizomes that form two aerial leaves at each internode; leaves float; lower surface of leaf wettable but upper surface not wettable; plants growing together form dense mats. Natural enemies include grasshopper *Paulimina acuminata* (DeGeer), curculionid *Cyrtobagous salviniae* Calder & Sands, and pyralids *Samea multiplicalis* Guenée and *Nymphula responsalis.* Cf. Alligator Weed; Water Hyacinth; Water Lettuce.

SALVINIA WEEVIL *Cyrtobagous salviniae* Calder & Sands [Coleoptera: Curculionidae]: A highly effective, host-specific biological control agent of *Salvinia.* Adult ca 2 mm long, black with long Rostrum; adult and larva feed on *Salvinia.*

SALZ, HANAN See Bytinski-Salz.

SAMOGGIA, ARRIGO (1904–1939) (Grandi 1939, Boll. Ist. Ent. Univ. Bologna 11: 64–66, bibliogr.)

SAMOUELLE, GEORGE (–1846) English bookseller and member of the staff of the British Museum (Natural History) 1821–1841. (Miller 1973, *That Noble Cabinet.* 400 pp., (231), London. Mackechnie-Jarvis 1976, Proc. Brit. ent. nat. Hist. Soc. 8: 94–95.)

SAMOURAI® See Lambda Cyhalothrin.

SAMPSON, FRANK WINN (1853–1926) (Poulton 1926, Proc. ent. Soc. Lond. 1:77.)

SAN JOSE SCALE *Quadraspidiotus perniciosus* (Comstock) [Hemiptera: Diaspididae]: A widespread, multivoltine pest of fruit trees, shrubs and ornamental trees. If left uncontrolled, SJS can kill trees. Overwinters as second-instar scale (sooty-black stage). Adults appear during spring; sexually dimorphic with male two-winged and insect-like while female wingless and remains under scale-cover. Female produces live crawlers that move to another part of plant or may be transported by wind or phoresey. Crawlers insert mouthpart Stylets into host plant, moult and shed legs, Antennae, and form scale-cover over body. SJS undergoes two nymphal instars and can complete six generations per year. See Diaspididae.

SANBAR® See Permethrin.

SANBORN, CHARLES EMERSON (1877–1944) (Fenton 1944, J. Econ. Ent. 37: 857–858; Stiles 1944, Science 100: 140–141.)

SANBORN, FRANCIS GREGORY (1838–1884) (Anon. 1884, Psyche 4: 175; Dickinson 1884, Proc. Worcester Soc. Antiq. 1884: 1–20.)

SAN-BRUNO ELFIN *Incisalia fotis bayensis* (R. M. Brown) [Lepidoptera: Lycaenidae].

SAND-CHERRY WEEVIL *Coccotorus hirsutus* Bruner [Coleoptera: Curculionidae].

SAND COCKROACHES See Polyphagidae.

SAND-DUNE SNAIL *Theba pisana* (Muller) [Sigmurethra: Helicidae].

SAND FAIRY *Cicadetta arenaria* (Distant) [Hemiptera: Cicadidae].

SAND FLEA *Tunga penetrans* (Linnaeus) [Siphonaptera: Tungidae]: A significant parasite of humans in tropical Africa and Neotropics to southeastern United States. SF is native to South America where it parasitizes pigs and introduced into tropical Africa during 1800s. SF may also attack birds, dogs and cats. Adult SF lacks combs; male and virgin females live as ectoparasites of mammals. Inseminated females burrow under skin of humans and other mammals (usually selecting feet as a site for penetration), feed and becomes the size of a pea. Adult female undergoes transformation process called Neosomy. Eggs develop within Abdomen of gravid female while under skin of host; female oviposits ca 200 eggs, most of which pass out of the body. Larvae typically develop as other fleas;

a few eggs may remain in wound in host's body and larvae may develop in tissue. Body of female expelled from host tissue after oviposition. SF wounds can be point of entry of secondary infections. Syn. Chigoe; Chique; Jigger. See Tungidae. Cf. Sticktight Flea.

SAND FLIES A well defined group of minute, nematocerous Diptera sometimes considered a Family (Phlebotomidae) but more often a Subfamily (Phlebotominae) within Psychodidae. SF are predominantly tropical and subtropical with ca 700 described Species. Fossil record appears in Lower Cretaceous Period. Adult <3.0 mm long, elongate-bodied, brownish coloured; Antenna with 16 segments and lacks sexual dimorphism; palps with five segments; wings held erect above body; fork of R2+3 and R4 near middle of wing. Adults active nocturnally and rest during day in dark, damp habitats. Females feed upon blood with Mandibles; males do not feed on blood and lack Mandibles; both sexes feed upon plant fluids. Males form mating swarms on vegetation or vertebrate hosts visited by females; copulation during or shortly after female takes a blood meal. Most Species are anautogenous; some Species are autogenous. Females oviposit within soil, leaf litter or burrows. SF eggs elongate with rounded poles, one surface flattened and opposite surface convex; Chorion sculptured. Female lays 10–100 eggs. SF larva with 2–4 long caudal Setae and amphineustic respiration; head with chewing mouthparts; larva feeds upon decaying organic matter, including excrement and dead insects; SFs undergo four larval instars. Pupa exarate, stands perpendicular to substrate and displays short prothoracic respiratory horns. Larvae and pupae are terrestrial but susceptible to desiccation and require high humidity. Sand flies are considered medically important as vectors of Leishmania, Sandfly Fever and Bartonelosis. See Psychodidae. Cf. Biting Midge; Ceratopogonidae.

SAND FLY A term often applied to any minute, biting fly. See Diptera. Cf. Midge; Mosquito.

SAND SCORPION *Urodacus novaehollandiae* Peters [Scorpionida: Scorpionidae]: A Species that occurs in Western Australia.

SAND WASP *Bembix pruinosa* Fox [Hymenoptera: Sphecidae]: A wasp widespread and endemic to North America that nests in large aggregations within loose sand. SW eggs are laid in empty cells, and provisioning begins shortly before eclosion; female SW provides numerous paralysed adult flies to larval wasps that are linearly arranged in chambers. See Sphecidae. Cf. Eastern Sand Wasp; Western Sand Wasp.

SAND WASPS Members of Genera *Bembix, Sphex* and *Prionyx.* See Sphecidae.

SAND WIREWORM *Horistonotus uhleri* Horn [Coleoptera: Elateridae].

SANDAHL, OSKAR THEODOR (1829–1894) (Lampa 1894, Ent. Tidskr. 15: 315–323, bibliogr.)

SANDAL-BOX HAWK MOTH *Coenotes eremophilae* (Lucas) [Lepidoptera: Sphingidae].

SANDBERG, GEORG (1842–1891) (Schöyen 1891, Ent. Tidskr. 12: 71–76, bibliogr.)

SANDEMAN, ROBERT GWYNNE CHILDE CRAWSHAY (1899–1952) (Riley 1953, Proc. R. ent. Soc. Lond. (C) 17: 73.)

SANDER, HEINRICH (1754–1782) (Roemer 1785, Neuestes Mag. Liebh. Ent. 2: 81–86.)

SANDERS, GEORGE ETHELBERT (1884–1943) (Davis 1943, J. Econ. Ent. 36: 811–812.)

SANDERS, JAMES G (1881–1957) (Wheeler & Valley 1975, History of Entomology in the Pennsylvania Department of Agriculture. 37 pp. (17–20), Harrisburg.)

SANDERSON, EZRA DWIGHT (1878–1944) (Phillips 1944, J. Econ. Ent. 37: 858–859.)

SANDFLY FEVER Any of eight arboviral diseases of the Bunyaviridae; four diseases occur in New World and four occur in Old World. Diseases typically produce a high fever of short duration in humans; SF is vectored by phlebotomine flies and typically is non-fatal. Five diseases are transmitted trans-ovarially by *Phlebotomus* spp. and recovered from male sandflies. See Arbovirus; Bunyaviridae.

SANDGRINDER *Arenopsaltria fullo* (Walker) [Hemiptera: Cicadidae].

SANDGROPERS See Cylindrachetidae.

SANDHOUSE, GRACE ADELBERT (1896–1940) (Cushman *et al.* 1940, Proc. ent. Soc. Wash. 42: 187–189; Muesebeck 1941, Ann. ent. Soc. Am. 41: 262–263.)

SANDIN, JOHAN EMIL (1852–1923) (Anon. 1952, Opusc. ent. 17: 17–20.)

SANG, JOHN (1828–1887) (Anon. 1887, Entomol. mon. Mag. 23: 261, 278–279; Dimmock 1888, Psyche 5: 36; Robson 1888, Naturalist, Hull 151: 52–54.)

SANGIOVANNI, GIOSUE (1776–1849) (Anon. 1851, Atti Accad. Sci. napoli 6: lxxxiii.)

SANGSTER, JOHN HERBERT (1831–1904) (Anon. 1904, Can. Ent. 36: 72.)

SANGUINARY ANT *Formica sanguinea rubicunda* Emery [Hymenoptera: Formicidae].

SANGUINIVOROUS Adj. (Latin, *sanguis* = blood; *vorare* = to devour; *-osus* = possessing the qualities of.) Pertaining to insects which are blood-eating. See Feeding Strategies. Cf. Haematophagous; Solenophagous.

SANGUINOLENT Adj. (Latin, *sanguis* = blood.) Blood-like in colour or appearance.

SANMITE® See Pyridaben.

SANO, TEIZO (1837–1902) (Hasegawa 1967, Kontyû 35 (Suppl.): 41–42, bibliogr.)

SANOPLANT® See Entomogenous Nematodes.

SANTOCEL C® See Silica Compounds.

SANTOX® See EPN.

SANTSCHI, FELIX (1872–1940) (Donisthorpe 1941, Entomologist's Rec. J. Var. 53: 99–100; Kutter 1941, Mitt. schweiz. ent. Ges. 18: 286–289.)

SANVALERATE® See Fenvalerate.

SANVEX® See Cartap.

SAP Noun. (Latin, *sapare* = to taste of; Old French = *sappe;* Anglo Saxon, *saep.* Pl., Saps.) 1. A fluid that transports water, nutrients and metabolites through the vascular system (xylem and phloem) to tissues of plants. 2. Any fluid regarded as vital to life or health of an organism. Cf. Resin; Gum.

SAP BEETLES See Nitidulidae.

SAP-FEEDING BEETLES See Nitidulidae.

SAP-FLORA BEETLES See Synteliidae.

SAPECRON® See Chlorfenvinphos.

SAPLING BORER *Sahyadrassus malabaricus* (Moore) [Lepidoptera: Hepialidae].

SAPONIFY Verb (Latin, *sapo* = soap.) Intransitive: To undergo saponification. Transitive: The chemical change of a fat into a soap.

SAPONIN Noun. (Latin, *sapo* = soap. Pl., Saponins.) Any of several steroid glucosides found in some plants which produce soapy foaming solutions in water (*e.g.* sarsasparilla). Saponins foam in water and are used as emulsifying agents and detergents.

SAPOR Noun. (Latin, *sapor; or* = a condition. Pl., Sapors.) A property or attribute affecting the sense of taste; savour, flavour.

SAPORIFIC Adj. (Latin, *saporificus; -ic* = of the nature of.) Pertaining to the capacity to produce the sensation of taste.

SAPPHIRINE Adj. (Latin, *sapphiratus* = adorned with sapphire.) Sapphire blue. Alt. Sapphirinus.

SAPROBE Noun. (Greek, *sapros* = rotten; *bios* = life. Pl., Saprobes.) An organism living in an environment rich in organic matter. Alt. Saprobiont. Syn. Sapront. Cf. Heterotroph; Saprogen.

SAPROBIOTIC Adj. (Greek, *sapros* = rotten; *bios* = life; *-ic* = characterized by.) Descriptive of organisms living on, or within decaying matter, or living in organic waste or sewage.

SAPROGEN Noun. (Greek, *sapros* = rotten; *genes* = producing. Pl., Saprogens.) An organism (microbe, fungus) that lives on or within non-living organic matter and which is capable of causing the degradation of that matter. Cf. Decomposer; Heterotroph.

SAPROMYZIDAE See Lauxaniidae.

SAPRONT Noun. (Greek, *sapros* = rotten; *bion* = living. Sapronts.) A saprobe.

SAPROPHAGE Noun. (Greek, *sapros* = rotten; *phagein* = to eat. Pl., Saprophages.) Any organism that feeds on dead and decaying organisms. See Feeding Strategies. Cf. Carnivore; Herbivore; Omnivore; Phytophage. Rel. Ecology.

SAPROPHAGOUS Adj. (Greek, *sapros* = rotten; *phagein* = to eat; Latin, *-osus* = possessing the qualities of.) Descriptive of organisms (plants or animals) that feed upon dead and decaying animals or plant matter. See Cannibalistic; Parasitic; Predaceous. Cf. Carnivorous; Herbivorous; Phytophagous.

SAPROPHAGY Noun. (Greek, *sapros* = rotten; *phagein* = to eat. Pl., Saprophagies.) 1. The act or process of being saprophagous. 2. A type of feeding strategy that requires dead animal tissue or plant matter as a source of nutrition. See Feeding Strategies; Digestion; Extra-oral Digestion. Cf. Cannibalism; Necrophily; Parasitism; Predation. Rel. Carnivory; Herbivory; Phytophagy.

SAPROPHYTE Noun. (Greek, *sapros* = rotten; *phyte, phite* = plant. Pl., Saprophytes.) Any plant or animal subsisting on dead or decaying vegetable matter. Cf. Autophyte, Parasite.

SAPROPHYTIC Adj. (Greek, *sapros* = rotten; *phyton* = plant; *-ic* = of the nature of.) An organism that feeds upon dead organisms or decaying organic matter. Cf. Endophytic; Exophytic. Rel. Necrophilic.

SAPROVORE Noun. (Greek, *sapros* = rotten; Latin, *vorare* = to devour. Pl., Saprovores.) An organism that consumes or feeds upon dead and decomposing organisms. See Feeding Strategies. Cf. Carnivore; Herbivore; Omnivore. Rel. Ecology.

SAPROZOIC Adj. (Greek, *sapros* = rotten; *zoon* = animal; *-ic* = of the nature of.) Pertaining to an animal that feeds on dead or decaying animal matter.

SAPWOOD TIMBERWORM *Hylecoetus lugubris* Say [Coleoptera: Lymexylidae].

SAPYGIDAE Latreille 1802. Plural Noun. A widespread, small Family of aculeate Hymenoptera assigned to the Vespoidea or Scolioidea and which includes about 80 nominal Species. Female Antenna with 12 segments; male Antenna with 13 segments; Mandible with three apical teeth; compound eyes large, medially emarginate; Pronotum with anterior margin truncate, posterior margin extending to Tegula; both sexes macropterous; forewing venation extensive; hindwing with two closed cells, anal and Jugal Lobes; Mesosternum without spines or lamina overlapping base of middle Coxa; middle Coxae juxtaposed or closeset, never widely separated; gastral Terga I and II not separated by constriction; Felt Line absent; Sting sheath barbed along dorsal margin. Sapygids are parasitic within the nests of other aculeate Hymenoptera. The Sting is used to penetrate cell of host; an egg is laid within cell of host near host's egg. The sapygid first-instar larva destroys the host egg with large Mandibles and develops as cleptoparasite in nests on provisions of megachilid bee. Larvae of some sapygids feed on host larva. Subfamilies include Fedtschenkiinae and Sapyginae. See Aculeata.

SARATOGA SPITTLEBUG *Aphrophora saratogensis* (Fitch) [Hemiptera: Cercopidae]: A pest of red pine plantations in the Great Lakes region of North America.

SARCODE Noun. (Greek, *sarkodes* = like flesh. Pl., Sarcodes.) An early term for Protoplasm.

SARCOLEMMA Noun. (Greek, *sarx* = flesh; *lemma* = skin. Pl., Sarcolemmae.) A thin, elastic transparent covering over striated muscular fibres.

SARCOLYSIS Noun. (Greek, *sarx* = flesh; *lysis* = loosing; *sis* = a condition or state. Pl., Sarcolyses.) The breaking down or lysis of muscle tissue.

SARCOLYTE Noun. (Greek, *sarx* = flesh; *lyterios* = loosing. Pl., Sarcolytes.) 1. Haemolymph of Diptera pupae: A muscle fragment in the process of disintegration. 2. A fragment of muscular fibre with a Nucleus (Imms).

SARCOMERE Noun. (Greek, *sarx* = flesh; *meros* = part. Pl., Sarcomeres.) A short length of Myofibril between two Z-discs. See Muscle.

SARCOMERIC Adj. (Greek, *sarx* = flesh; *meros* = part; *-ic* = of the nature of.) Descriptive of or pertaining to a Sarcomere; of the character of Sarcomere.

SARCOPHAGIDAE Plural Noun. (Greek, *sarx* = flesh; *phagein* = to devour.) Flesh Flies. A cosmopolitan Family of about 2,300 Species of muscoid Diptera. Often confused with Muscidae and Tachinidae; sometimes treated as Subfamily of Calliphoridae. Sarcophagids generally are nonmetallic, grey or silver-grey and black with striped Mesonotum and spotted Abdomen; usually strongly setose. Distinguished from other Muscoidea by presence of Hypopleural Bristles, usually with Pteropleural Bristles, without Postscutellum, wing vein M1 angled usually nearer apex of Discal Cell than apex of wing. Saprophagous habits of larvae cause Myiasis. Some Sarcophaginae are parasites of other arthropods; some attack eggs, nymphs and adults of Acridoidea, nymphs and adults of mantids; larvae of bees and social wasps. Predominately larviparous; females deposit first instar larvae; a few Species lay eggs that hatch quickly; some Species lay eggs that require incubation. Some Species larviposit onto apex of quiescent host's Abdomen; others larviposit while host flying; others insert Ovipositor into genital opening of moving but not flying host. Entrance into host through intersegmental membranes, genital or anal openings. Larvae feed primarily on fat body, reproductive tissue and/or muscle tissue. Host usually dies soon after larvae enter body; host may remain alive and reproduce after emergence of mature larvae. Nymphal grasshopper hosts usually do not reach adult stage. Emergent larvae fall to ground and pupate in soil. Species that attack social bees and wasps larviposit near prey egg or in prey brood cell; larvae consume egg then contents of cell. Several Species are internal parasites of isopods and lay membranous eggs near host. Planidiform larvae seek host isopod and enter body through intersegmental membranes separating ventral sclerites; internal parasite larvae feed on host's fat body, weaken female and terminate reproduction.

SARCOPHAGOUS Adj. (Greek, *sarx* = flesh; *phagein* = to devour; Latin, *-osus* = *possessing the* qualities of.) Flesh-eating; feeding on flesh.

SARCOPLASM Noun. (Greek, *sarx* = flesh; *plasma* = mould. Pl., Sarcoplasms.) The undifferentiated protoplasm of the muscle fibre, in which the Sarcostyles are embedded. Alt. Sarcoplasma.

SARCOPLASMA Noun. (Greek, *sarx* = flesh; *plasma* = mould. Pl., Sarcoplasmata.) See Sarcoplasm.

SARCOPTIDAE Plural Noun. Itch Mites. One of two Families in the Suborder Astigmata that cause human dermatitis. Infection as a result of mites burrowing into skin called scabies. Treatment with benzyl benzoate. Additional injury results from scratching due to the severe irritation of the mites in skin.

SARCOSOME Noun. (Greek, *sarx* = flesh; *soma* = body. Pl., Sarcosomes.) Mitochondria of muscle cells.

SARCOSTYLE Noun. (Greek, *sarx,* = flesh; *stylos* = pillar. Pl., Sarcostyles.) Myofibris. One of the highly elastic, longitudinal fibrillae of the insect muscle.

SARCOTOXIN Noun. An antibacterial protein formed by larvae of the sarcophagid fly, *Sarcophaga peregrinia,* following injury or exposure to bacteria.

SARGENT, HOWARD BROUGH (1893?-1975) (Gunn 1976, Proc. R. ent. Soc. Lond. (C) 40: 52.)

SAROLEX® See Diazinon.

SAROTHRIIDAE Plural Noun. See Jacobsoniidae.

SAROTHRUM Noun. (Greek, *saron* = broom; *throna* = flowers. Pl., Sarothrums.) The basal joint of the posterior Tarsus in pollen gathers. A pollen brush. See Metatarsus.

SARRA, RAFFAELE (1861–1938) (Roberti 1973, Memorie Soc. ent. ital. 52: 71–72.)

SARTORIUS, EDUARD (1863–1929) (Hockemeyer 1929, Verh. Ver naturw. Heimatforsch. 21: xxi–xxiii; Weidner 1967, Abh. Verh. naturw Ver. Hamburg Suppl. 9: 192.)

SASAKI, CHIUJIHO (1857–1938) (Ishimori 1938, Kontyû 12: 115–120, bibliogr.; Hasegawa 1967, Kontyû 35 (Suppl. 39, bibliogr.)

SASISAKA, IKUSABURO (–1928) (Hasegawa 1967, Kontyû 35 (Suppl.): 38–39, bibliogr.)

SASKATOON BORER *Saperda candida bipunctata* Hopping [Coleoptera: Cerambycidae].

SASSCER, ERNEST RALPH (1883–1955) (Cory 1955, Proc. ent. Soc. Wash. 57: 309–310.)

SATAKE, SHOICHI (1885–1938) (Hasegawa 1967, Kontyû 35 (Suppl.): 40–41, bibliogr.)

SATIN BLUE *Nesolycaena albosericea* (Miskin) [Lepidoptera: Lycaenidae].

SATIN MOTH *Leucoma salicis* (Linnaeus) [Lepidoptera: Lymantriidae]: A monovoltine pest of shade and ornamental trees; endemic to Palearctic and adventive to North America ca 1920. Overwinters as early-instar larva in webs on bark of host tree; larva with square patch of white Setae on dorsum of each abdominal segment and black-and-grey markings along sides; short Setae arising from tubercles. Eggs deposited in oval clusters on host tree and covered with satin-like accessory gland secretion. See Lymantriidae. Cf. Brown-Tail Moth; Gypsy Moth.

SATISFAR® See Etrimfos.

SATO, SAKAE (1880–1928) (Hasegawa 1967, Kontyû 35 (Suppl.): 41.)

SATOMURA, HIROSHI (1916–1945) (Hasegawa 1967, Kontyû 35 (Suppl.): 41.)

SATTERTHWAIT, ALFRED FELLENBERG (1879–1954) (Fellenberg 1956, Ann. ent. Soc. Am. 49: 301; Wade 1956, Proc. ent. Soc. Wash. 58: 234–235.)

SATTLER, WILHELM (1859–1920) (Bücking 1920, Ent. Bl. Biol. Syst. Käfer 16: 94–96.)

SATUNIN, KONSTANTIN ALEXSYEEVICH (–1915) (Semenov-Tian-Shansky 1915, Rev. Russe Ent. 15: 677–681, bibliogr.)

SATURATE Adj. (Latin, *saturatus* = of full, rich colour; *-atus* = adjectival suffix.) Deep, full; term applied to any colour.

SATURATION Noun. (Latin, *saturatio,* from *satur* = full of food. Pl., Saturations.) 1. The act or process of being saturated. 2. One of three attributes of colour which distinguishes grey of the same brilliance; distinctness of hue. See Coloration. Cf. Brilliance; Hue. 3. Breeding Programmes: The increased resemblance of progeny to the sire with successive generations born to the same parents.

SATURN BUTTERFLIES See Nymphalidae.

SATURNIIDAE Plural Noun. Giant Silkworm Moths; Royal Moths; Emperor Moths. A widespread Family of ditrysian Lepidoptera assigned to Bombycoidea; Family includes ca 1,000 Species and best represented in tropical and subtropical areas. Adult large to very large, stout bodied; wingspan 65–265 mm; head relatively small with piliform scales; Ocelli and Chaetosemata absent; Antenna short, quadripectinate and without scales in most Taxa, bipectinate in *Hemileuca;* Proboscis usually absent; Maxillary Palpus vestigial; Labial Palpus small; Thorax and legs with long piliform scales; fore tibial Epiphysis present; tibial spur formula 0-2-2, 0-2-4, or 0-0-0; forewing often falcate, Retinaculum absent, Areole absent; hindwing with tail in some Species. Egg flattened, broadly oval in outline shape with primary axis horizontal and Micropyle at one end; Chorion smooth; eggs typically laid on host plant usually individually, sometimes in small groups and rarely in large masses. Larva head hypognathous; Setae on Scoli or Chalazae; Setae often branched; abdominal Prolegs on segments 3–6, anal Prolegs very large; abdominal segment 8 with dorsal projection. Pupa lacking Maxillary Palpus; Proboscis short; Cremaster present. Adults of most Species nocturnal and visit lights; some Species diurnal. Syn. Attacidae.

SATYR BUTTERFLIES See Nymphalidae.

SATYRIDAE Plural Noun. Satyr Butterflies. Wood Nymphs. A widespread Family of Lepidoptera. Adults weak fliers, hide near ground when alarmed; wings typically dull coloured with spots on both sides of wings; forelegs atrophied. Larval host plants predominantly grasses; pupation on ground; Chrysalis usually hung by Cremaster.

SATYRS See Satyridae.

SAUBER, CHRISTIAN JOHANNES AMANDUS (1846–1917) (Hasebrook 1917, Int. ent. Z. 11: 41–45; V.B. 1917, Dt. Ent. Z. 1917: 180–181.)

SAUBINET, E (–1897) (Anon. 1897, Leopoldina 33: 56.)

SAUCER BUGS See Naucoridae.

SAUER, H F G Brazilian Entomologist (Instituto Biologico de São Paulo) working with natural enemies of cotton pests.

SAULCY, FELICIEN HENRY CAIGNART DE (–1912) (Kheil 1911, Int. ent. Z. 5: 243–245.)

SAUNDBY, ROBERT (1896–1971) (Butler 1972, Proc. R. ent. Soc. Lond. (C) 36: 62.)

SAUNDER'S CASE MOTH *Oiketicus elongatus* Saunders [Lepidoptera: Psychidae].

SAUNDERS EMBIID *Oligotoma saundersii* (Westwood) [Embiidina: Oligotomidae].

SAUNDERS, CHARLES JAMES (1868–1941) (Blair 1941, Entomol. mon. Mag. 77: 209.)

SAUNDERS, EDWARD (1848–1910) (Anon. 1910, Entomologist's Rec. J. Var. 22: 75–76; Dixey 1910, Proc. ent. Soc. Lond. 1910: lxxxvi; Stebbing 1910, Proc. Linn. Soc. Lond. 1910: 94–98; Waterhouse 1910, Zoologist (4) 14: 77–78.)

SAUNDERS, GEORGE SHARP (1842–1910) (Anon. 1910, Entomol. mon. Mag. 46: 120.)

SAUNDERS, SIDNEY SMITH (1819–1884) (Anon. 1884, Psyche 4: 175; Anon. 1884, Entomol. mon. Mag. 20: 278–279, bibliogr.; Dunning 1884, Proc. ent. Soc. Lond. 1884: xl-xlii, bibliogr.)

SAUNDERS, WILLIAM (1835–1914) (Goding 1894, Rep. ent. Soc. Ont. 25: 1, 120–123; Anon. 1914, Ent. News 25: 480; Bethune 1914, Can. Ent. 46: 333–336.)

SAUNDERS, WILLIAM WILSON (1809–1879) (Anon. 1879, Entomol. mon. Mag. 16: 119–120; Carrington 1879, Entomologist 12: 278–280.)

SAUNT, JOHN WILLIAM (1881–1958) (Benson 1959, Entomol. mon. Mag. 95: 72.)

SAURUCK, FRANZ (–1958) (Anon. 1958, Z. wien. ent. Ges. 43: 136.)

SAUSSURE, HENRI LOUIS FRÉDÉRIC DE (1829–1905) (Anon. 1905, Insektenbörse 22: 61; Bouvier 1905, Bull. Mus. Hist. Nat. Paris 11: 223–225; Burr 1905, Entomologist's Rec. J. Var. 17: 167–170; Bedot 1906, Revue suisse Zool. 14: 1–32, bibliogr.; Essig 1931, *History of Entomology,* 1029 pp., (748–750), New York.)

SAUSSURE, HORACE BENEDICT (1740–1799) (Rose 1850, *New General Biographical Dictionary* 11: 467.)

SAUTER, ANTON ELEUTHERIUS (1800–1881) (Anon. l881, Leopoldina 17: 156; Speiser 1908, Schr. phys.-ökon. Ges. Königsb. 49: 299–301.)

SAUTER, HANS (1871–1948) (Rohlfien 1975, Beitr. Ent. 25: 274–276.)

SAVAGE BEETLES See Carabidae.

SAVALL® See Quinalphos.

SAVEY® See Hexythiazox.

SAVI, PAOLO (1798–1871) (Targioni-Tozzetti 1871, Boll. Soc. ent. ital. 3: 81–82; Anon. 1872. J. Zool., Paris 1: 96–97.)

SAVIGNY, MARIE JULES CÉSAR LELORGNE DE (1777–1851) French naturalist who accompanied Napoleon on the expedition to Egypt (1798). Savigny worked extensively in comparative anatomy of insect mouthparts and published *Memoires sur les animaux sans vertebre. 1. Fascicule: Theorie des organes de la bouche des Crustaces et des Insectes* (1816). Savigny lost his sight due to microscopical work. (Reiche 1851, Bull. Soc. ent. Fr. (2) 9: ci; Westwood 1852, Proc. ent. Soc. Lond. 1852: 136.)

SAVILLE, CHARLES (–1930) (Mitchell 1930, Entomologist 63: 192.)

SAVILLE-KENT, WILLIAM (–1908) (Anon. 1908, Nature 78: 641–642; B.D.J. 1909, Proc. Linn. Soc. Lond. 1908–09: 42.)

SAVIO, AUGUSTE (1882–1935) (Piel 1936, Entomology Phytopath. 4: 44–46.)

SAW Noun. (Anglo Saxon, *sagu*. Pl., Saws.) Hemiptera (Auchenorrhyncha): First and second Valvulae of the Ovipositor. Hymenoptera (Tenthredinoidea): The median pair of flattened, usually serrate, sclerites forming the Ovipositor. Syn. Terebra. See Ovipositor.

SAW GUIDE Hymenoptera: An arcane term for the paired, elongate, external, compressed sclerites forming the Ovipositor sheath. Syn. Gonostylus.

SAWADA, MASATOCHI (1892–1962) (Hasegawa 1967, Kontyû 35 (Suppl.): 42.)

SAWADA, TAKASHIRO (1915–1965) (Hasegawa 1967, Kontyû 35 (Suppl.): 42.)

SAWFLY (Pl., Sawflies.) A common name for Hymenoptera assigned to the Suborder Symphyta. See Symphyta.

SAW-TOOTH GRAIN WEEVIL *Oryzaephilus surinamensis* (Linnaeus) [Coleoptera: Cucujidae]: A cosmopolitan, multivoltine pest of rice, wheat, cereals, packaged food and spices; a significant pest of stored cereal grains and cereal products. STGW cannot complete development on polished rice but can complete development on rice polish. STGW cannot develop without carbohydrates or develop at <17 or >38°C; optimal conditions 30–35°C, 75% RH. Adult 3.0–3.5 mm long, dark brown, body slender, flattened with six saw-toothed projections from lateral margins of Thorax; adult active but does not fly. Female oviposits ca 250–400 eggs during lifetime; eggs laid singly or in small clusters near food; incubation 4–12 days. Larva with brown head, white body, elongate, flattened with six legs; moves actively seeking food; larval stage 14–70 days; typically three larval instars (range 2–4). Pupation beneath food in chamber of salivary cement and food particles; pupal stage 7–27 days. Life cycle 3–6 weeks; adult stage five days to three years; overwinters as adult); 4–6 generations per year. See Cucujidae. Cf. Merchant Grain Beetle.

SAWYER BEETLES See Cerambycidae.

SAXATILE Adj. (Latin, *saxatilis* = found among rocks.) Lithophilous.

SAXESEN, FRIEDRICH WILHELM REISIG (1792–1858) (Ratzeburg 1874, Forstwissenschaftliches Schriftsteller-Lexicon 1: 451–454.)

SAXICOLOUS Adj. (Latin, *saxum* = rock; *colere* = to inhabit.) Pertaining to organisms that inhabit rocky or stony areas. Alt. Saxicoline. Syn. Lithophilous. Cf. Arenicolous; Deserticolous; Eremophilous; Ericeticolous; Lapidicolous; Psammophilous. Rel. Ecology.

SAY STINK-BUG *Chlorochroa uhleri* (Stål) [Hemiptera: Pentatomidae].

SAY, THOMAS (1787–1838) American naturalist and early worker in North American Entomology. Born 27 June 1787; son of Philadelphia apothecary; formal education limited to three years. Associated with William Maclure, William Bartram, Thomas Nuttall, Alexander Wilson and other naturalists of the period. Travelled to Georgia and Florida with Maclure and Titian Peale during 1818; travelled to Colorado as chief Zoologist with Major Stephen Long during the 1819 expedition. Published many articles and *American Entomology* (1824); many illustrations prepared by wife Lucy Say. Thomas Say is generally regarded as 'Father of American Entomology. Died in New Harmony, Indiana. (Anon. 1835, Am. J. Sci. 27: 393–395, bibliogr.; Coates 1835, Biography Acad. nat. Sci. Philad. [31 pp.], bibliogr.; Coates 1835, Waldie's select circ. Libr. 5: 236–239; Coates 1837, Nat. port. Gall. 4 (39): 1–10; Swainson 1840, *Taxidermy; with Biography of Zoologists*. 392 pp. (pp. 317–318, bibliogr.). London. [Volume 12 of *Cabinet Cyclopedia*. Edited by D. Lardner]; Morris 1846, Am. J. Sci. 1: 20–24; Kingsley 1882, Pop. Sci. Mon. 21: 577, 687–691; Schwarz 1887, Proc. ent. Soc. Wash. 1: 81–82; Dall 1888, Proc. biol. Soc. Wash. 4: 98–102; Scudder 1891, Psyche 6: 57–60, 121–124, 137–141, 169–172, 185–187, 297–298, 345–346, 357–358; Webster 1895, Ent. News 6: 1–4, 33–34, 80–81, 101–103; Youmans 1896, *Pioneers of Science in America*. viii + 508 pp. (215–222), New York; Weiss & Ziegler 1931, *T. Say, Early American Naturalist*. 260 pp. Baltimore.)

SAYCE, OCTAVIUS ALBERT (1862–1911) (Anon. 1911, Victorian Nat. 28: 25–26, bibliogr.; B.D.J. 1912, Proc. Linn. Soc. Lond. 1911–12: 63–64.)

SBP-1382® See Resmethrin.

Sc Adj. Standard abbreviation for the Subcostal Vein of the generalized insect wing. Based on the Comstock-Needham nomenclature for insect wing venation.

SCAB Noun. (Middle English, *scabbe;* Latin, *scabies, scaber* = rough. Pl., Scabs.) 1. A crust-like diseased area on the surface of a plant, and caused by various bacterial or fungal agents. 2. A crust-like hardened outer covering of a wound, sore or ulcer during healing.

SCAB MITE *Psoroptes equi* (Raspail) [Acari: Psoroptidae].

SCAB MITES See Psoroptidae.

SCABELLUM Noun. (Latin, *scabellum*. Pl., Scabella.) 1. Diptera: The slightly swollen base of the Petiole of the Haltere which articulates with side of the reduced Metanotum (Tillyard). 2. An ancient muscial instrument consisting of two metal plates connected with a hinge and attached to the foot of the musician.

SCABIES Noun. (Latin, *scabies, scabere* = to scratch.) A contagious disease of animals which is caused by mites which burrow under the skin to feed and reproduce. See Mange.

SCABIES MITES *Sarcoptes scabiei* (DeGeer) [Acari: Sarcoptidae].

SCABRATE Adj. (Latin, *scaber* = rough; *-atus* = adjectival suffix.) Descriptive of surface sculpture, usually the insect's Integument, that is microscopically sculptured. See Sculpture Pattern. Cf. Alveolate; Baculate; Clavate; Echinate; Favose; Gemmate; Psilate; Punctate; Rugose; Shagreened; Smooth; Striate; Verrucate.

SCABRICULOUS Adj. (Latin, *scaber* = rough; *-osus* = *possessing the* qualities of.) Pertaining to surface that is regularly and finely wrinkled. See Sculpture.

SCABROUS Adj. (Latin, *scaber* = rough; *-osus* = *possessing the* qualities of.) Pertaining to a rough surface; irregularly and roughly rugose; invested with rough spines or scales. Alt. Scabrose; Scabrosus.

SCALARIFORM Adj. (Latin, *scala* = ladder; *forma* = shape.) Ladder-shaped; structure with two long parallel tracks connected by numerous, shorter, parallel bars which are perpendicular with the tracks. A term applied to venation when the small veins between two large longitudinal veins are regularly arranged as the rungs of a ladder. Rel. Form 2; Shape 2; Outline Shape.

SCALE Noun. (Anglo Saxon, *sceala* = shell, Latin *scala* = fish scale. Pl., Scales.) 1. General: Any small, flat, cuticular projection from the Integument of an insect. 2. A unicellular outgrowth of the body-wall, wing or appendage of various shapes; a modified seta. 3. Isoptera: The stump of the shed wing of neonates. 4. Coccoids: The protective cover over a scale insect's body, consisting of Exuviae and glandular secretions; the waxy covering of a male lecaniid. See Test. 5. Lepidoptera wing: Flattened, highly modified, ridged, overlapping Setae or Macrotrichia that form wing vestiture; scales often contain pigment and may be attached by a short stalk to the Cuticle. 6. Diptera: The Alula. See Integument. Cf. Seta.

SCALE INSECTS See Coccoidea.

SCALE-EATING CATERPILLAR *Catoblemma dubia* (Butler) [Lepidoptera: Noctuidae].

SCALE-EATING LADYBIRD *Rhyzobius lophanthae* (Blaisdell) [Coleoptera: Coccinellidae].

SCALECIDE® See Petroleum Oils.

SCALIGER, JULIUS CAESAR (1484–1558) (Rose 1850, *New General Biographical Dictionary* 11: 475–476.)

SCALLOP Noun. (Old French, *escalope* = scallop. Pl., Scallops.) 1. One of the valves of the shell of a scallop. 2. One of a series of concentric curves or arcs that form a distinctive ornamental sculpture on a surface.

SCALLOPED Adj. (Old French, *escalope* = scallop.) 1. Descriptive of a wavy or undulous surface, or margin of a surface marked with a sinuous series of rounded hollows, without intervening angles. 2. An object shaped as a scallop shell. See Shape.

SCALPELLUM Noun. (Latin, *scalpellum* = little knife. Pl., Scalpella.) 1. A lancet or a lancet-like structure adapted for piercing. 2. Diptera: Part of the Maxilla.

SCALPRIFORM Adj. (Latin, *scalpellum* = little knife; *forma* = shape.) Chisel-shaped. See Chisel. Rel. Form 2; Shape 2; Outline Shape.

SCALY CRICKETS See Mogoplistidae.

SCALY GRAIN-MITE *Suidasia nesbitti* Hughes [Acari: Acaridae].

SCALY HAIRS Coccids: Pectinae (MacGillivray).

SCALY LEG MITE *Knemidokoptes mutans* (Robin & Lanquetin) [Acari: Knemidokoptidae].

SCANMASK® See Entomogenous Nematodes.

SCANNING ELECTRON MICROSCOPE A microscope that uses a beam of electrons rather than reflected light to examine the surface features of a specimen. The wavelength of the electron beam varies with its acceleration by a potential that is measured in volts. Biological tissues are typically examined with accelerating voltages between 10,000 and 20,000 volts. This results in an electron beam far narrower than that of light, and thus, magnification ranges from 20X to ca. 40,000X. Acronym SEM. Cf. Transmission Electron Microscope.

SCANNING TUNNELING MICROSCOPY An instrument invented by Binnig and Rohrer which is adapted for the inspection of thin films, crystal growth and similar features to the atomic level.

SCANSORIAL Adj. (Latin, *scandere* = to climb; *-alis* = characterized by.) Descriptive of appendages adapted for climbing. Insects: A term applied to legs adapted for climbing or holding mammalian hair. The Tarsus of the scansorial leg is reduced to one or two segments which often work opposite a tibial process. Examples include the Phthiraptera which are ectoparasites of birds and mammals, Mallophaga (chewing, biting lice) and Rhynchophthirina (elephant lice - *Hematomyzinus* which feeds on elephants and wart hogs) and the Anoplura (sucking lice). See Leg. Cf. Cursorial; Fossorial; Gressorial; Natatorial; Raptorial; Scansorial. Rel. Locomotion.

SCAPE Noun. (Latin, *scapus* = stalk, stem, shaft; Greek, *skapos* = stalk. Pl., Scapes.) 1. The basal segment of the Antenna. Typically one of the longest segments of the Antenna; elongate, cylindrical and containing musculature which originates in the head. Articulation of Scape with head via a small sclerotic process called an Antennifer. 2. The three basal antennal segments. Alt. Scapus. See Antenna. 3. Diptera: The Peduncle of the Haltere. 4. The shaft of a feather.

SCAPHIDIIDAE Plural Noun. A small, cosmopolitan Family of Coleoptera. Adult minute to small bodied, convex, fusiform, shining; Antenna slender, loosely clavate; Elytra truncate; legs slender; tarsal formula 5-5-5. Larva fusiform; Antenna with three segments; Urogomphi with two segments. Adult and larva feed on fungus in damp, shady habitats. Related to Staphylinidae and sometimes placed within that Family.

SCAPHIFORM Adj. (Greek, *skaphion* = small boat; Latin, *forma* = shape.) Boat-shaped. Syn. Scaphoid. Rel. Form 2; Shape 2; Outline Shape.

SCAPHIUM Noun. (Greek, *skaphion* = small boat; Latin, *-ium* = diminutive > Greek, *-idion*. Pl., Scaphia.) Male Lepidoptera: A ventral process of the tenth abdominal segment below the Uncus. See Gnathos.

SCAPHOID Adj. (Greek, *skaphion* = small boat.) Boat-shaped. Alt. Scaphiform.

SCAPULA Noun. (Latin, *scapula* = shoulder blade. Pl., Scapulae, Scapulas.) 1. The shoulder blade of mammals. 2. Heteroptera: The inferior lateral face of the Mesonotum. 3. Lepidoptera: The shoulder Tippets, Patagia or Axillae. 4. Hymenoptera: The Parapsides or lateral sclerites of the Mesonotum. Proctotrupidae: The lateral lobes on each side of the Parapsidal Furrow. 5. Acari: The projecting lateral angles of the Scutum (Matheson).

SCAPULAR Adj. (Latin, *scapula* = shoulder blade.) Descriptive of or pertaining to the Scapula.

SCAPULAR AREA Wing: The part nearest the 'shoulder.' Orthoptera: The radial area.

SCAPULAR PIECE 1. The Episternum. 2. The Scapula.

SCAPULAR VEIN Orthoptera: The Radius. See Vein; Wing Venation.

SCAPULARIA Noun. (Latin, *scapula* = shoulder blade. Pl., Scapulariae.) Mesepisternum. See Scapula.

SCARAB BEETLES See Scarabaeidae.

SCARAB FLIES See Pyrgotidae.

SCARABAEIDAE Plural Noun. Chafers; Dung Beetles; Elephant Beetles; Flower Beetles; Goliath Beetles; June Beetles; June Bugs; Lamellicorn Beetles; May Beetles; Rhinoceros Beetles; Scarab Beetles; Unicorn Beetles; White Grubs. A cosmopolitan Family of polyphagous Coleoptera assigned to the Superfamily Scarabaeoidea, and including more than 12,000 Species. Body to 60 mm long, stout or compact, often somber coloured, heavily sclerotized. Head not deflexed; mouthparts visible ventrad; Antenna with 7–10 segments, club lamellate with 3–7 segments; fore Tibia dentate; tarsal formula typically 5-5-5, rarely 0-5-5; apical Sternum of Abdomen exposed. Larva head hypognathous; Stemmata absent or one on each side of head; legs with five segments, two or more tarsal claws; Urogomphi absent. Adults phytophagous feeding upon the foliage and flowers of many Species of plants, or feed as scavengers. Larvae typically are concealed in the soil and feed upon the roots of plants used as food by the adult stage; larvae of some scarab Species feed upon dung or decaying vegetable matter. Root-feeding Species are regarded as pests; dung-feeding Species are regarded as beneficial and in some places rapidly recycle nutrients in excrement of foraging animals and promote pasture productivity.

SCARABAEIFORM Adj. (Latin, *scarabaeus* = a beetle; *forma* = appearance.) Grub-shaped; pertaining to the body of an insect larva which resembles a scarab beetle larva. See Grub. Cf. Larviform. Rel. Form 2; Shape 2; Outline Shape.

SCARABAEIFORMIA Noun. (Latin, *scarabaeus* = a beetle; *forma* = appearance.) A Series of polyphagous Coleoptera including the Scarabaeoidea.

SCARABAEOID Adj. (Latin, *scarabaeus* = a beetle; Greek, *eidos* = form.) Scarab-like in shape or biology. See Scarabaeidae. Cf. Scarabaeiform. Rel. Eidos; Form; Shape.

SCARABAEOIDEA Noun. A numerically large, highly distinctive Superfamily of polyphagous Coleoptera including Acanthoceridae, Geotrupidae, Glaphyridae, Hyboscoridae, Lucanidae, Scarabaeidae, Trogidae. Antennal club lamellate; Prothorax modified for burrowing with large fore Coxa, spinose Tibia with only one spur; hind coxal plate absent; hindwing venation reduced and with spring mechanism for folding; second Ventrite present laterally only; Tergum 8 exposed, forming true Pygidium; four Malpighian Tubules present. Larva grub-like, C-shaped; Antenna well developed, legs well developed; Urogomphi absent. See Coleoptera.

SCARABIDOID (Latin, *scarabaeus* = a beetle; Greek, *eidos* = form.) Scarab-like; a stage of a meloid larval hypermetamorphic development in which the larva resembles a white grub or scarabaeid larva. Rel. Eidos; Hypermetamorphosis; Form; Shape.

SCARIFIED Adj. (Greek, *skariphasthai* to scratch.) Clawed or scratched. A term applied to surfaces with irregular fine grooves.

SCARIOSE Adj. (French, *scarieux* = membranous.) Dry and scaly. Alt. Scarious.

SCARLET OAK SAWFLY *Caliroa quercuscoccineae* (Dyar) [Hymenoptera: Tenthredinidae].

SCATHOPHAGIDAE Plural Noun. Dung Flies. A moderate-sized Family of schizophorous Diptera assigned to Superfamily Muscoidea. Adult with habitus of house fly; wing with vein CuA + 1A;

Scutellum asetose; Sternopleuron with one bristle. Larvae of most Species inhabit dung; some Species are plant feeders; a few are leaf miners and a few are aquatic. Sometimes placed in Anthomyiidae. Syn. Cordyluridae; Scatomyzidae; Scatophagidae; Scopenumatidae.

SCATOMYZIDAE See Scathophagidae.

SCATOPHAGIDAE See Scathophagidae.

SCATOPHAGOUS Adj. (Greek, *skatophagous* = dung eating; Latin, *-osus* = *possessing the* qualities of.) Feeding upon dung or excrement; merdivorous. Syn. Coprophagous. See Feeding Strategies.

SCATOPHILE Noun. (Greek, *skatophagous* = dung eating; *philein* = to love. Pl., Scatophiles.). An organism that is attracted to resides near excrement or animal waste.

SCATOPHILOUS Adj. (Greek, *skatophagous* = dung eating; *philein* = to love; Latin, *-osus* = with the property of.) Excrement-loving; descriptive of organisms that are attracted to excrement or inhabit excrement. See Habitat. Cf. Fimicolous; Muscicolous; Myrmecophilous; Necrophilous; Nitrophilous; Stercoraceous. Rel. Ecology.

SCATOPSIDAE Plural Noun. Minute Black Scavenger Flies; Scavenger Flies. A relatively small Family of nematocerous Diptera assigned to Division Bibionomorpha. Adult small to minute bodied (< 3 mm), brown or black in coloration; Antenna relatively short; veins bold near Costal margin of wing, remaining veins weak. Larvae inhabit decaying vegetation and excrement.

SCATS See Scatophagidae.

SCATTERCARB® See Carbaryl.

SCATTERING COLOUR Structural colour that is produced from the scattering of light from a surface, such as the irregular and diffuse light reflected from ground glass, or the scattering of light within the Integument. Scattering colour produces Tyndall Blue or Structural White. SC can be a thermally important method of heat gain; scattering by surface waxes can significantly reduce heat gain in desert tenebrionid beetles. See Coloration. Cf. Diffraction; Interference; Iridescence. Rel. Pigmentary Colour.

SCAVENGER Noun. (Middle English, *scaveger* = an officer with various duties. Pl., Scavengers.) An animal that feeds upon dead and decaying animal or vegetable matter. See Feeding Strategies. Cf. Herbivore; Carnivore; Saprovore; Predator.

SCAVENGER FLIES See Scatopsidae.

SCAVENGER WATER-BEETLES See Hydrophilidae.

SCELIONIDAE Haliday 1839. Plural Noun. Scelionid Wasps. A cosmopolitan Family of apocritous Hymenoptera assigned to Proctotrupoidea; numerically largest Family of Proctotrupoidea. Adult body 0.5–5 mm long, dark, non-metallic, strongly sclerotized; Antenna geniculate, Torulus near clypeal margin; female Antenna with 11–12 segments, male with 7–10 segments; antennal club well formed, multisegmented; Pronotum extending to Tegula; wing venation reduced to Submarginal and Stigmal Veins; lateral margin of Gaster frequently carinate; Ovipositor concealed at repose. Hypermetamorphic; first-instar larva teleaform. Solitary, internal egg parasites of Hemiptera, Lepidoptera, Mantoidea, Araneae; many Species appear host Family-specific and prefer eggs in aggregations (clumps); Egg marking (semiochemical) reported in several Species. Phoresy widespread in scelionids and probably most common among parasitic Hymenopera. Strongly female-biased sex ratios; male protandrous, combative, with sibling mating common.

SCENOPINIDAE Plural Noun. Window Flies. A cosmopolitan Family of lower brachycerous flies including about 250 described Species in 10 Genera. Adult moderate to small, nondescript; closely related to Bombyliidae and Therevidae by characteristic wing venation: Median vein unbranched and curved forward apically to approach or meet vein R4+5; Anal Cell closed and petiolate; Antenna Style small and enclosed in small subapical pit. Biologically poorly known. Adults visit flowers; some Species taken at windows. Most Species commonly occur in arid to semiarid areas, and frequently near water. Larvae predaceous and associated with wood-boring larvae, termites, woodrat nests, bird nests and carpet beetle larvae. Eggs presumably laid near host. Larvae resemble therevid larvae; pupation probably takes place in substrate in which larva fed. Syn. Omphralidae. See Windowpane Fly.

SCENT Noun. (Old French, *sentir* = to smell. Pl., Scents.) The chemical odour emanated by a plant or animal. Rel. Pheromone; Kairomone; Semiochemical.

SCENT BRUSH A tuft or brush of specialized Setae or scales which facilitate the rapid diffusion of odorous secretions from glands at their bases (Klots). Syn. Scent Tuft.

SCENT GLAND An Exocrine Gland which forms as an invagination of the Integument and which releases volatile chemicals. Typically SGs are invaginations of the Epidermis which are lined with Cuticle and sometimes eversible. Scent Glands are sometimes named by the part of the body on which they originate (Thoracic Scent Gland, Abdominal Scent Gland) or after a person (Brindley's Gland). SG composed of Basement Membrane, Epithelial Cells and Glandular Cuticular Intima. Epithelial Cells form Intersititial (Support) Cells, Secretory Cells, Ducts and Reservoirs. Secretory Cells release acids, alcohols, aldehydes, esters, ketones and ketoaldehydes. SG sometimes occur at the base of Scent Brushes or Hair Pencils. Frequently found in male insects as a secondary sexual character. SG putative functions include defence from predators, defence from microorganisms, alarm pheromones, aggregation pheromones, sex pheromones. See Exocrine Gland. Cf.

Metapleural Gland; Metathoracic Scent Gland. Rel. Defence Gland; Scent Gland System.

SCENT GLAND SYSTEM Heteroptera: The Abdominal Scent Gland (always occurs in the nymph, but rarely in the adult) and Metathoracic Scent Gland (only adult). See Gland. Cf. Abdominal Scent Gland; Metapleural Gland; Metathoracic Scent Gland. Rel. Defence Gland.

SCENT PLAQUE Oviparous aphids: Small, circular organs on the hind Tibia. Syn. Pseudorhinaria; Pseudosensoria.

SCENT PORE Heteroptera: The Ostioles of a Scent Gland.

SCENT SCALE See Androconia.

SCENT TUFT See Scent Brush.

SCENTLESS PLANT BUGS See Rhopalidae.

SCHAEFER, CARL W (1934–) American academic and taxonomic specialist on Hemiptera.

SCHAEFFER, CHARLES FREDERIC AUGUST (1860–1934) (Anon. 1933, American men of science. p. 1977; Davis 1935, Bull. Brooklyn ent. Soc. 30: 32; Osborn 1946, *Fragments of Entomological History,* Pt. II, 232 pp. (109–110) Columbus, Ohio.)

SCHAEFFER, JACOB CHRISTIAN (1718–1790) (Duméril 1823, *Considérations générale sur la classe des insectes.* 272 pp. (253–254), Paris; Eiselt 1836, *Geschichte, Systematik und Literatur der Insektenkunde.* 255 pp. (52), Leipzig.)

SCHAEFFER, JOHN CHRISTIAN (1718–1799) English cleric and author of *Icones Insectorum circa Ratisbonam Indigenorum* (3 vols, 1769). Also published *Elementa Entomologica* (1769, 1777).

SCHÄFFER, CAESAR (1867–1947) (Weidner 1967, Abh. Verh. naturw. Ver. Hamburg Suppl. 9: 220–221.)

SCHAFFNER, JOHN V (1888–1957) (Brown 1958, J. Econ. Ent. 51: 266.)

SCHAJOVSKOY, SERGIO (1902–) (Gentili 1974, Revta. Soc. ent. argent. 34: 202.)

SCHALLER, JOHANN GOTTLOB (1734–1813) (Germar 1815, Magazin Ent. (Germar) I (2): 193.)

SCHALTZ, OLAF (1881–1973) (Tuxen 1974, Ent. Meddr. 24: 168.)

SCHARFF, ROBERT FRANCIS (1858–1933) (Preeger 1934, Ir. nat. J. 5: 153–155.)

SCHATZMAUR (SCHATZMAYR), ARTURO (1880–1950) (Parisi 1951, Atti Soc. ital. Sci. nat.90: 5–12, bibliogr.; Gridelli 1952, Memorie Soc. ent. ital. 30: 145–151, bibliogr.)

SCHAUDINN, FRITZ (1871–1906) (Langeron 1907, Archs Parasit. 11: 388–408, bibliogr.)

SCHAUFUSS, CAMILLO FESTIVUS CHRISTIAN (1862–1944) (Otto 1926, Abh. Ver. Verh. naturg. Grnz. 7: 84–88; Sachtleben 1944, Arb. morph. taxon. Ent. Berl. 11: 57.)

SCHAUFUSS, LUDWIG WILHELM (1833–1890) (Schaufuss 1871, Nunquam Otiosus 1: 24–26, bibliogr.; Anon. 1890, Entomol. mon. Mag. 26: 248; Reitter 1890, Wien. Ent. Ztg. 9: 184.)

SCHAUINSLAND, HUGO (1857–1937) (Anon. 1937, Arb. morph. taxon. Ent. Berl. 4: 241.)

SCHAUM, HERMANN RUDOLPH (1819–1865) (Kiesenwetter 1865, Ann. Soc. ent. Fr. (4) 5: 643–648, bibliogr.)

SCHAUPP, FRANZ G (1840–1904) (Leng 1923, Bull. Brooklyn ent. Soc. 18: 1–12; Anon. 1929, Ann. ent. Soc. Am. 22: 392–394.)

SCHAUS SWALLOWTAIL *Papilio aristodemus ponceanus* Schaus [Lepidoptera: Papilionidae].

SCHAUS, WILLIAM (1858–1942) (Osborn 1937, *Fragments of Entomological History,* 394 pp. (234), Columbus, Ohio; Heinrich *et al.* 1942, Science 96: 244–245; Heinrich *et al.* 1942, Proc. ent. Soc. Wash. 44: 189–195, bibliogr.)

SCHELHAMMER, GUNTHER CHRISTOPHER (1649–1716) (Rose 1850, *New General Biographical Dictionary* 11: 481.)

SCHELLENBERG, J R Swiss painter and engraver. Produced plates, and perhaps text, for *Cimicum in Helveticae aquis et terris degens genus* (1800) and *Genus des Mouches Dipteres* (1803).

SCHENK, ADOLPH (1803–1878) (Anon. 1878, Ent. Nachr. 4: 79; Kraatz 1878, Dt. ent. Z. 22: 225–226.)

SCHENKER, PAUL (–1966) (S. 1967, Mitt. schweiz. ent. Ges. 39: 262.)

SCHENKLING, KARL (–1911) (Soldanski 1911, Dt. Ent. Z. 1911: 730–732.)

SCHENKLING, SIEGMUND (1865–1946) (Sachtleben 1951, Beitr. Ent. 1: 102; Weidner 1967, Abh. Verh. naturw. Ver. Hamburg Suppl. 9: 192.)

SCHEPDAEL, JEAN VAN (1907–1974) (Lessermans 1974, Linneana Belgica 6: 4–10, 31–35, bibliogr.)

SCHERDLIN, PAUL (1872–1935) (Schuler 1936, Miscnea ent. 37: 14–15.)

SCHERNHAMMER, JOSEF (1859–1893) (Anon. 1894, Jber. wien. ent. Ver. 4: 8.)

SCHEUCHZER, JOHANN JACOB (1672–1733) (Rose 1850, *New General Biographical Dictionary* 11: 482; Jaggli 1950, Mitt. schweiz. ent. Ges. 31: 194.)

SCHEUTHLE, WILHELM (–1928) (Lupp 1929, Bienenpflege 51: 8.)

SCHIEFFERDECKER, WILHELM FRIEDRICH (1818–1889) (Stieda 1890, Schr. phys.-ökon. Ges. Königsb. 31: 50–63, bibliogr.)

SCHIFFERMÜLLER, IGNAZ (1727–1809) Austrian Lepidopterist and resident of Linz. Collection of Lepidoptera placed in Naturhistorisches Museum of Vienna 1806 and destroyed by fire 1848. (Eiselt 1836, *Geschichte, Systematik und Literatur der Insektenkunde.* 255 pp. (52–53), Leipzig; Hoffman 1952, Z. wien. ent. Ges. 37: 57–65, 207.)

SCHILDE, JOHANNES GUSTAV (1839–1888) (Moschler 1888, Stettin. ent. Ztg. 49: 315–316.)

SCHILDERS, FRANZ ALFRED (1896–1970) (Petsch 1970, Anz. Schädlingsk. Pflanzenschutz. 43: 188–189.)

SCHILLE, FRYDERYK ERNEST (1850–1931) (Niezabitowski 1931, Polskie Pismo ent. 10: 77–

88, bibliogr.)

SCHILLING, HEINRICH (–1903) (Anon. 1903, Insektenbörse 20: 195.)

SCHILLING, PETER SAMUEL (1773–1852) (Anon. 1852, Arb. schles. Ges. vaterl. Cult. 1852: 17.)

SCHILSKY, JULIUS (1848–1912) (Horn 1912, Ent. Bl. Biol. Syst. Käfer 8: 241–243.)

SCHIMA, KARL (1862–1940) (Kitt 1941, Z. wien. entVer. 26: 34.)

SCHIMKEVITSCH, VLADIMIR MICHAILOVITCH (1858–1923) (Bonnet 1945, *Bibliographia Araneorum* 1: 49.)

SCHINDLER, EMIL (–1880) Stierlin 1882, Mitt. schweiz. ent. Ges. 6: 230–231.

SCHINER, IGNAZ RUDOLPH (1813–1873) (Frauenfeld 1873, Verh. zool.-bot. Ges. Wien 23: 465–468; Brauer 1901, Festschr. Bot.-Zool. Oesterr. 1850–1900 1901: 344–345; Osten Sacken 1903, *Record of my life work in Entomology.* 240 pp (158–164), Cambridge, Mass.)

SCHINZ, HEINRICH RUDOLPH (1777–1861) (Anon. 1863, NeujBl. naturf. Ges. Zurich 65: 1–18; Rudio 1896, Festschr. naturf Ges. Zurich 1746–1896: 81–84.)

SCHIÖDTE, JORGEN MATTHIAS CHRISTIAN (1815–1884) (Dunning 1884, Proc. ent. Soc. Lond. 1884: xxxix–xl. Hanson 1884, Ent. Tijdschr. 5: 101–110, 207–208; Henriksen 1926, Ent. Meddr. 15: 226–241.)

SCHIRACH, ADAM GOTTLOB (–1773) (Buzaires 1866, Apiculteur 10: 110–113.)

SCHISTOCERCA Stål 1873. A Genus in the Subfamily Cyrtacanthacridinae [Orthoptera: Acrididae] that includes many economically important Species whose numbers can increase to devastating proportions and can migrate great distances. Alt. Migratory Locust; Plague Locust.

SCHIZODACTYLIDAE Plural Noun. A small Family of ensiferous Orthoptera assigned to Superfamily Gryllacridoidea, presently consisting of three Genera and seven Species that occur in Asia and South Africa. Individuals live in sandy habitats along river banks.

SCHIZODACTYLOIDEA Splay-footed Crickets. Sometimes considered a Family within the Gryllacridoidea but conspicuous and distinguished by the long hindwing which is coiled into a tight spiral at right angle to the substrate, tarsal segments strongly depressed and broadly expanded, body stout, head large and short with powerful Mandibles, Antenna longer than body, Torulus located beneath compound eye, anterior Tibia without auditory organ, base of Abdomen with minute stridulatory teeth, Cerci long and not annulated, Ovipositor reduced. Consist of three Genera and seven Species which occur in Asia and South Africa. The live in sandy habitats (along river banks).

SCHIZOGONY Noun. (Greek, *schizein* = to cleave; *genesis* = to produce. Pl., Schizogonies.) 1. The phase in the life-cycle of the malarial parasites *Plasmodium* spp. which is passed in the human host. Alt. Schizongony Phase. See Malaria; Cryptozoic Schizogony; Exoerythrocytic Schizogony; Metacryptozoic Schizogony. Cf. merogony. 2. Asexual production of spores.

SCHIZONT Noun. (Greek, *schizein* = to cleave; *onta* = beings. Pl., Schizonts.) The second stage in the development of *Plasmodium*, the protozoan parasite which causes Malaria. The mature trophozoite.

SCHIZOPHORA Noun. A Series of muscomorphan Diptera including the Superfamilies Astioidea, Conopoidea, Drosophiloidea, Heleomyzoidea, Micropezoidea, Muscoidea, Opomyzoidea, Tanypezoidea, Tephritoidea and Sciomyzoidea. See Diptera.

SCHIZOPTERIDAE Reuter 1861. Plural Noun. A small, predominantly tropical, Family of heteropterous Hemiptera assigned to Superfamily Dipsocoromorpha. Body minute (1–2 mm long), first and second antennal segments subequal in length; forewings strongly sclerotized and convex; prothoracic pleural region enlarged ventrally to enclose fore Coxae; medial surface of hind Coxae with roughened pads into which a metasternal spine is positioned as a jumping device; male genitalia asymmetrical. Nymph with one pair of dorsal abdominal scent glands. Schizopterids inhabit forest litter and damp soil; biology poorly known but may be predaceous.

SCHJØT-CHRISTENSEN, KNUD BORGE (1914?–1975) (Gunn 1976, Proc. R. ent. Soc. Lond. (C) 40: 52.)

SCHLECTENDAL, DOEDERICH H R VON (–1916) (Soldanski 1916, Dt. ent. Z. 1916: 364; Taschenberg 1917, Naturw. 86: 321–336, bibliogr.)

SCHLETTERER, AUGUST (1850–1908) (Kohl 1908, Verh. zool.-bot. Ges. Wien 58: 529–553, bibliogr.)

SCHLICK, RASMUS WILLIAM TRAUGOTT (1839–1916) (Kryger 1917, Ent. Meddr. 11: 320–324; Engelhart 1918, Ent. Meddr. 12: 14.)

SCHLIEBEN, LUDWIG HERMAN VON (1832–1903) (Anon. 1903. Insektenbörse 20: 410; Daniel 1904, Münch. koleopt. Z. 2: 95.)

SCHLIER, WALTER (1912–1964) (de Bros 1964, Mitt. ent. Ges. Basel 14: 1.)

SCHLÜBLER, GUSTAV (1787–1834) (Ratzeburg 1874, Forstwissenschaftliches Schriftsteller-Lexicon 1: 466.)

SCHLUGA, JOHANN BAPTIST (fl 1760) (Eiselt 1836, *Geschichte, Systematik und Literatur der Insektenkunde.* 255 pp. (53), Leipzig.)

SCHMARDA, LUDWIG KARL (1819–1908) (Anon. 1908, Zool. Anz. 33: 176.)

SCHMASSMANN-MESMER, WALTER (1890–1971) (Hoffman 1971, Anz. Schänlingsk. Pflanzenschutz 45: 91–92.)

SCHMELTZ, JOHUNNES DIETRICH EDUARD (1839–1909) (Zimmerman 1910, Verh. Ver. naturw. Unterh. Hamb. 14: 208–212, bibliogr.; Weidner 1967, Abh. Verh. naturw. Ver. Hamburg Suppl. 9: 142–148.)

SCHMID, ANTON (1809–1899) (Anon. 1899, Entomol. mon. Mag. 35: 194; Hoffmann 1901, Ber. naturw. Ver. Regensburg 7: 134–138.)

SCHMID-BINDER, WILHELM (1871–1955) (Anon. 1955, Mitt. ent. Ges. Basel 5: 22–23.)

SCHMIDLIN, ANTON (1893–1971) (de Bros 1972, Mitt. ent. Ges. Basel 22: 103–108, bibliogr.)

SCHMIDT, EDUARD OSKAR (1823–1886) (Wildermann 1887, Jb. naturw. 2: 579.)

SCHMIDT, ENRIQUE (HENRY) (1864–1948) (Alexander 1949, Ent. News 60: 261–262.)

SCHMIDT, FERDINAND JOSEPH (–1877) (Kraatz 1878, Dt. ent. Z. 22: 224–225.)

SCHMIDT, FRANZ (1814–1882) (Grumack 1883, Ent. Nachr. 9: 55–56; Staudinger 1883, Stettin. ent. Ztg. 44: 113–114.)

SCHMIDT, HEINRICH JULIUS CARL (1864–1948) (Rohlfien 1975, Beitr. Ent. 25: 276–277.)

SCHMIDT, HERMANN (–1859) (Anon. 1860. Jber. Ges. Freunden Naturw. Gera 3: 51–52.)

SCHMIDT, JOHANN ANDREAS (1652–1726) (Rose 1850, *New General Biographical Dictionary* 11: 486.)

SCHMIDT, KARL (–1916) (Soldanski 1916, Dt. ent. Z. 1916: 227.)

SCHMIDT, WILHELM LUDWIG EWALD (1804–1843) (Dieckhoff 1843, Stettin. ent. Ztg. 4: 194–199.)

SCHMIDT-GOEBEL, HERMANN MAX (1809–1882) (Fauvel 1882, Revue ent. 1: 264; Türckheim 1883, Berl. ent. Z. 27: ii.)

SCHMIEDEKNECHT, OTTO (1847–1936) (Bergmann 1933, Int. ent. Z. 27: 25–33, bibliogr.; Bergmann 1936, Ent. Z., Frankf.: A. M. 49: 536–538.)

SCHMIEDER, RUDOLF GUSTAV (1898–1967) (Anon. 1968, Ent. News 79: 66–70, bibliogr.)

SCHMITT, OTTO (–1964) (Anon. 1964, Z. wien. ent. Ges. 49: 128.

SCHMITT, P JEROME (HIERONYMUS) (1857–1904) (Anon. 1904, Ent. News 15: 225–226.

SCHMITZ, PATER HERMANN (1878–1960) (Colyer 1961, Studia ent. 4: 545–547.)

SCHNABEL, BRUNO (1832–1916) (Nagel 1919, Jh. Ver. schles. Insektenk. 10–12: 25–26.)

SCHNABEL, JOHANN (1838–1912) (Bogdanov 1891, Izv. imp. obshch. Lyub. Estest. Antrop. Etnogr. imp. Mosc. Univ. 77: [175–176], Bibliography (pl. 24).)

SCHNACKENBECK, GUSTAV (1854–1939) (Weidner 1967, Abh. Verh. naturw. Ver. Hamburg Suppl. 9: 297.)

SCHNEIDER, ANTON (1831–1890) (Limpricht 1901, Jber. schles. Ges. Vaterl. Cult. 68: 9–13.)

SCHNEIDER, ERNST (1884–1965) (Schneider 1967, Mitt. ent. Ges. Basel 17: 21–22.)

SCHNEIDER, GUIDO ALEXANDER JOHANN (1866–1948) (Luther 1950, Memo. Soc. Fauna Flora fenn. 25: 196–200.)

SCHNEIDER, HANS (1907–1943) (Escherich 1943, Z. angew Ent. 30: 156–157.)

SCHNEIDER, HANS JACOB SPARRE (1953–1918) (Tullgren 1920, Ent. Tijdskr. 41: 145–146.)

SCHNEIDER, KARL (1876–1952) (Anon. 1952, Mitt. ent. Ges. Basel 2: 87–88.)

SCHNEIDER, LOUIS (1836–1901) (Anon. 1901, Ent. News 12: 256.)

SCHNEIDER, OSKAR (1841–1903) (Anon. 1903, Insektenbörse 20: 313–314.)

SCHNEIDER, WILLIAM GOTTLIEB (1814–1889) (Dittrich 1889, ent. Breslau 14: xxv–xxvii, bibliogr.)

SCHNEIDER-ORELLI, OTTO (1880–1965) (Maksymov 1966, Anz. Schädlingsk 39: 45–46.)

SCHNEIRLA, THEODORE C (Anon. 1968, J. N. Y. ent. Soc. 76: 175.)

SCHNETZLER, JEAN BALTHAZAR (1823–1896) (Dufour 1897, Bull. Soc. vaud. Sci. nat. 33: 1–21, bibliogr.)

SCHNUSE, CARL AUGUST WILHELM (1850–1909) (Papavero 1975, *Essays on the History of Neotropical Dipterology*. 2: 419–423. São Paulo.)

SCHOCH, GUSTAV (1833–1899) (Ris. 1899, Mitt. schweiz. ent. Ges. 10: 211–217, bibliogr.)

SCHOENBORN, HENRY F (1833–1896) (Anon. 1896, Ent. News 7: 256.)

SCHOENE SPIDER MITE *Tetranychus schoenei* McGregor [Acari: Tetranychidae].

SCHOENHERR, CARL JOHANN (1772–1848) (Carlson 1848, Proc. ent. Soc. Lond. 1848: liii–lv; Dohrn 1849, Stettin. ent. Z. 10: 193–199; Mannerheim 1849, Bull. Soc. Nat. Moscou 22: 574–596, bibliogr.)

SCHOENHERR, CHARLES JOHN Swedish Coleopterist and author of *Synonymia Insectorum* (3 vols, 1806–1817), *Curculionidum Disposito methodica* (1826), and *Genera et Species Curculionidum, cum Symonymia hujus familiae*.

SCHOENICHEN, WALTHER (1876–1956) (Anon. 1936. Natur Niederrh. 12: 27–28; Anon. 1956, NachrBl. dt. PflSchutzdienst. Braunschweig 8: 144.)

SCHOLASTICISM Noun. (Greek, *scholastes* = one who lives at ease; English, *-ism* = a doctrine. Pl., Scholasticisms.) A philosophical approach to truth which relies on accepting the views of scholars, and not direct experimentation or personal observation. Scholasticism originated as an attempt to unify theology and philosophy. Scholasticism was popular during the Middle Ages, but subsequently was discounted in the scientific community. Early proponent was Peter Abelard (1079–1142). Rel. Empiricism.

SCHOLECHIASIS Noun. (Greek, *skolex* = a worm; appearing worm eaten; *-iasis* = suffix for names of diseases. Pl., Scholechiases.) The vomiting or evacuation of caterpillars.

SCHOLTEN, LAMBERTUS HENRICUS (–1948) (Lempke 1949, Ent. Ber., Amst. 12: 313–314.)

SCHOLTZ, HEINRICH (1812–1859) (Cohn 1859, Jber. schles. Ges. vaterl. Cult. 37: 34–35; Klette 1860, Verh. siebenbürg nat. Ver. 10: 240–241.)

SCHOLZ, RICHARD (1866–1935) (Anon. 1935. Ent. Bl. Biol. Syst. Käfer 31: 177; Heikertinger 1935, Koleopt. Rdsch. 21: 196.)

SCHOMBERGK, ROBERT HERMANN (1804–1864) (Jardine 1848, Nats Libr. 39: 17–79; Pascoe 1966, Proc. ent. Soc. Lond. 1866: 138.)

SCHOPFER, EDWARD (1858–1928) (Heller 1929, dt. ent. Z. Iris 43: 37–40, bibliogr.; Möbius 1943, Dt. ent. Z. Iris 57: 20–21.)

SCHOTT, ARTHUR CARL VICTOR (1814–1875) (Anon. 1875, Leopoldina 11: 164; Weiss 1946, J. N. Y. ent. Soc. 54: 170–171, bibliogr.)

SCHOTT, FRED M (1887–1946) (Weiss 1946, J. N. Y. ent. Soc. 54: 170–171.)

SCHÖTZE, EDUARD (–1961) (Anon. 1961, Z. wien. ent. Ges. 46: 96.)

SCHOUSBOE, PETER KOFOED ANKER (1760–) (Zimsen 1964, *The Type Material of J. C. Fabricius.* 656 pp. (13), Copenhagen.)

SCHOUTEDEN, HENRI (1881–1972) (Basilewsky 1973, Bull. Soc. ent. Fr. 78: 7–8.)

SCHÖVERS, TIMON CORNELIS (1878–1946) (Van Eydenhoven 1946, Ent. Ber., Amst. 12: 38–39.)

SCHØYEN, THOR HIORTH (1885–1961) (Fjelddalen 1961, Norsk. ent. Tidsskr. 11: 287–291, bibliogr.)

SCHØYEN, WILHELM MARIBO (1844–1918) (Munster 1918, Overs. Vidensk. Möter 1918: 86–88; Munster 1920, Norsk. ent. Tidsskr. 1: 3–7, bibliogr.)

SCHÖYEN, WILHELM MORITZA (1844–1918) (Munster 1928, Norsk. ent. Tidsskr. 1: 3–7, bibliogr.)

SCHRADER, FRANZ (1891–1962) German-born American academic and DaCosta Professor (Columbia University). Cytologist and author of numerous papers on chromosomes, cytological phenomena and sex determination; developed hypothesis on evolutionary origin of haplo-diploid parthenogenesis based on work with coccoids. Author of *Die Geschlechtschromosomen* (1928) and *Mitosis: The Movements of Chromosomes in Cell Division* (1944). Founding editor of journals *Chromosoma* and *Journal of Cell Biology.* Member, U. S. 'National Academy of Sciences,' and 'American Academy of Arts and Sciences.' Cooper 1993, *Biograph. Mem.* 62: 369–380.

SCHRANK, F DE P (1747–) Bavarian academic and naturalist. Author of *Enumeratio Insectorum Austriae indigenorum* (1781).

SCHRANK, FRANZ VON PAULA (1747–1835) (Gistel 1837, Faunus 1: 1, 5–8; Swainson 1840, *Taxidermy: With Biography of Zoologists.* 392 pp. [Vol. 12. of *Cabinet Cyclopedia.* Edited by D. Lardner.].)

SCHREBER, JOHANN CHRISTIAN DANIEL (1739–1810) (Griffin 1939, J. Soc. Biblphy. nat. Hist. 1: 221.)

SCHREIBER, EGID (1836–1913) (Anon. 1914, Schmid & Thesings 1914, Biologen Kalender 1: 365–366.)

SCHREIBERS, CHARLES Director of Imperial Cabinet of Natural History at Vienna. Author of *Coleopters Insects* (volume 6, Linnean Transactions).

SCHREIBERS, KARL FRANZ ANTON VON (1775–1852) (Anon. 1852, Verh. zool.-bot. Ges. Wien 2: 46.)

SCHREITER, CARLOS RUDOLFO (1897–1942) (Descole 1942, Revta. argent. Zoogeogr. 2: 61–62, bibliogr.)

SCHRENK (SCHRENCK), PETER LEOPOLD VON (1826–1894) (Heger 1894, Mitt. anthrop. Ges. Wien 24: 11–12, 13.)

SCHRÖDER, GUSTAV (–1931) (Krüger 1931, Stettin. ent. Ztg. 92: 322–323.)

SCHROEDER VAN DE KOLK, JACOB LUDWIG CONRAD (1797–1862) (Vrolik 1862, Jaarb. K. Akad. Wet. Amst. 1862: 161–191.)

SCHROETER, JOHANN SAMUEL (1735–1808) (Swainson 1840, *Taxidermy with Biography of Zoologists.* 392 pp. (321). London. [Vol. 12. of *Cabinet Cyclopedia*, Edited by D. Lardner.].)

SCHROTTKY, KURT (CARLOS) (1937) (Sachtleben 1938, Arb. morph. taxon. Ent. Berl. 5: 295.)

SCHUBART, OTTO (1900–1962) (Anon. 1964, Revta. bras. ent. 11.)

SCHUBERG, AUGUST (1865–1939) (Sachtleben 1939, Arb. physiol. angew. Ent. Berl. 6: 209.)

SCHUBERT, KARL (1867–1911) (Ross 1911, Ent. Z., Frankf. a. M. 5: 231; Soldanski 1911, Dt. Ent. Z. 1911: 728–730.)

SCHUCKMANN, WALDEMAR HELMUT FRANZ (1883–1939) (Sachtleben 1939, H. Arb. physiol. angew. Ent. Berl. 6: 379.)

SCHUKARD, WILLIAM EDWARD (1803–1868) (Newman 1868, Entomologist 4: 180–182.)

SCHULTES, JOSEPH AUGUST (1773–1831) (Gistl 1832, Faunus 1: 52–53.)

SCHULTHESS (SCHULTHESS-TECHBERG) (SCHULTHESS-SCHINDLER), ANTON (1855–1941) (Schneider-Orelli 1935, Mitt. schweiz. ent. Ges. 16: 301–308, bibliogr.; Nadig 1941, Mitt. schweiz. ent. Ges. 18: 398–399.)

SCHULTZ, ERNST CHRISTOPHS (Weidner 1967, Abh. Verh. naturw. Ver. Hamburg Suppl. 9: 40.)

SCHULTZ, OSCAR OTTO KARL HUGO (1868–1911) (Richter 1911, Int. ent. Z. 5: 183–184.)

SCHULTZE, AUGUST (1837–1907) (Horn 1907, Dt. Ent. Z. 1907: 590; Daniel 1908, Münch. koleopt. Z. 3: 397–399, bibliogr.)

SCHULZ, HANS (1909–1942) (Sachtleben 1942, Arb. physiol. angew. Ent. Berl. 9: 127.)

SCHULZ, HEINRICH E M (1859–1918) (Weidner 1967, Arb. Verh. naturw. Ver. Hamburg Suppl. 9: 193.)

SCHULZ, JOHANN DOMINICUS (Weidner 1967, Arb. Verh. natur. Ver. Hamburg Suppl. 9: 40.)

SCHULZE, CARL MAX WILLY (1881–1940) (Anon. 1941, Ent. Bl. Biol. Syst. Käfer 37: 96; Sachtleben 1941, Arb. morph. taxon. Ent. Berl. 8: 286.)

SCHUMACHER, CHRISTIAN FRIEDERICH (1757–1830) (Henriksen 1923, Ent. Meddr. 15: 110–111.)

SCHUMACHER, HANS (1917–1975) (Anon. 1975,

Prakt. Schädlingsk. 27: 74.)

SCHUMMEL, THEODOR EMIL (1786–1848) (Letzner 1858, 50 jähr. Bestehen schles. Ges. vaterl. Kult. 1858: 16–18.)

SCHÜRHOFF, PAUL NORBET (–1939) (Anon. 1939, Ent. Bl. Biol. Syst. Käfer 35: 128.)

SCHUSTER, ADRIAN (1860–1942) (Anon. 1942, Ent. Bl. Biol. Syst. Käfer 38: 127; Heikertinger 1942, Koleopt. Rdsch. 28: 21.)

SCHUSTER, MORITZ (1823–1894) (Anon. 1894, Ent. News 5: 96.)

SCHUSTER, ROBERT OSCAR (1928–1989) American museum specialist (University of California, Davis) with diverse research interests.

SCHÜTZE, EDUARD (1894–1961) (Anon. 1961, Z. wien. ent. Ges. 46: 96.)

SCHÜTZE, JULIUS (1835–1904) (Geiser 1939, Fld Lab. 7: 46.)

SCHÜTZE, KARL TRAUGOTT (1858–1938) (Starke 1938, Dt. ent. Z. Iris 52: 184–185.)

SCHWAB, ADOLF (1806–1891) (Reitter 1891, Wien. ent. Ztg. 10: 40.)

SCHWANGART, FRIEDRICH (1874–1958) (Haltenroth 1959, Verh. dt. zool. Ges. Zool. Anz. Suppl. 23: 530–531; Petzsch 1964, Anz. Schädlingsk. 37: 60.)

SCHWANVICH (SCHWANWITSCH), BORIS NIKOLAEVICH (1889–1957) (Fedorov & Beklemishev 1958, Zool. Zh. 37: 1422–1425, bibliogr.)

SCHWARDT, HERBERT H (1903–1962) (Isely 1963, J. Kans. ent. Soc. 36: 1–2.)

SCHWARTZ, MARTIN (1880–1947) (Winning 1948, Anz. Schädlingsk. 21: 141–142.)

SCHWARZ, EUGENE AMANDUS (1844–1928) (Horn 1928, Ent. Mitt. 17: 307–310; Howard 1928, Proc. ent. Soc. Wash. 30: 153–183, bibliogr.)

SCHWARZ, HERBERT FERLANDO (1883–1960) (Pallister 1960, J. N. Y. ent. Soc. 68: 187–189.)

SCHWARZ, HERMANN (1876–1940) (Meiners 1941, Ent. News 52: 118–l19.)

SCHWARZ, KARL (1847–1898) (Gerhardt 1898, Z. Ent. 23: 41–42.)

SCHWARZ, OTTO CARL ERNST (1861–1908) (Semenov-Tian-Shansky 1908, Revue Russe Ent. 8: 350; Horn 1909, Dt. Ent. Z. 1909: 170–172.)

SCHWEDER, GOTTHARD (1832–1915) (Kupffer 1915, KorrespBl. NaturfVer. 57: i–xviii; Meder 1924, KorrespBl. NaturfVer. 58: 9–12.)

SCHWING, EDWARD A (1893–1951) (Anon. 1951, J. Econ. Ent. 44: 271–272.)

SCHWINGENSCHUSS, LEO (1878–1954) (Reisser 1956, Z. wien. ent. Ges. 41: 271–272.)

SCIADOCERIDAE Plural Noun. A small Family of acalypterate Diptera assigned to Superfamily Platypezoidea (Phoroidea). A few Species are known from New Zealand, Australia and Patagonia.

SCIAPHILOUS Adj. (Greek, *skiaros* = dark coloured; Latin, *-osus* = possessing the qualities of.) Pertaining to activity pattern or periodicity of an organism. Possibly synonymous with crepuscular.

SCIARIDAE Plural Noun. Dark-Winged Fungus Gnats; Root Gnats. A cosmopolitan Family of nematocerous Diptera, assigned to Superfamily Sciaroidea. Family well represented in the fossil record. Resemble Mycetophilidae but compound eyes medially contiguous near the antennal Torulus. Adults common in moist habitats. Larvae with black head and pale body; occur in fungi and decaying plant material. Some Species are pests in mushroom houses. Syn. Lycoriidae.

SCIENTIFIC NAME The word or words applied to an abstract entity that forms part of the taxonomic hierarchy used in the classification and codification of biological knowledge. The most commonly used names in the hierarchy involve the Species, Genus and Family. Different rules apply to the correct formation of names, depending upon whether the organism is a plant, animal and microorganism. In Zoology, Family names end in the suffix '-idae' and Subfamily names end in '-inae.' Insect scientific names involving Species consist of two parts - the generic name and specific epithet. Generic names are always capitalized; specific epithets are never capitalized Species names are always italicized or underlined, and may or may not be followed by the name of the person(s) who described the Taxon. Author's names are never italicized (*e.g. Bison bison* Linnaeus). The formation of names follows the rules of grammar for Latin. See Description; Systematic Name. Cf. Common Name. Rel. Classification; Nomenclature; Taxonomy.

SCIMITAR® See Lambda Cyhalothrin.

SCIOMYZIDAE Plural Noun. Marsh Flies. Widespread Family of cyclorrhaphous Diptera including about 600 Species. Assigned to Superfamily Sciomyzoidea with Dryomyzidae, Heleomyzidae, Sepsidae, Rhopalomeridae and Coleopidae. Adult 2–12 mm long, pale yellowish-brown, grey-black; distinguished from other acalyptrate Diptera by oral Vibrissae absent; Postvertical Bristles diverging; Costa entire; Subcosta complete; one or more Tibiae with preapical bristles. Adult slow-flying; rest on emergent vegetation along margins of flowing or standing fresh water; often perch with head directed downward. Larvae obligate parasites or predators on freshwater snails, fingernail clams, land snails and slugs. Larvae of some Species feed saprophagously; larvae of some Species change feeding habits from selective to less selective as stadia change. Syn. Tetanoceridae.

SCIOMYZOIDEA Noun. A Superfamily of schizophorous Diptera including Celyphidae, Chamaemyiidae, Dryomyzidae, Helcomyzidae, Lauxaniidae (Sapromyzidae), Ropalomeridae (Rhopalomeridae), Sciomyzidae (Tetanoceratidae, Tetanoceridae) and Sepsidae.

SCION Noun. (French, *scion* = twig, shoot. Pl., Scions.) Botany: A fruiting variety of plant which is budded or grafted onto a rootstock.

SCIRTIDAE Plural Noun. Marsh Beetles; Soft Bodied Plant Beetles. A small Family of polyphagous Coleoptera assigned to the Eucinetoidea. Body ovoid, somewhat flattened, 2–11 mm long; head large, deflexed; Antenna filiform with 11 segments; Prothorax transverse; legs slender; tarsal formula 5-5-5; Abdomen with five Ventrites. Larva elongate, campodeiform, somewhat flattened; head prognathous; Antenna with more than five segments; legs with five segments; Urogomphi present or absent. Adult typically associated with wet habitats; larva aquatic, with retractile anal gills. Syn Cyphonidae, Helodidae.

SCIRTOTHRIPS *Scirtothrips albomaculatus* Bianchi and *Scirtothrips dorsalis* Hood [Thysanoptera: Thripidae]: Pests of *Citrus* in Australia, California, southeast Asia and South Africa. Other host plants include mango and strawberry. Adults yellow, narrow, ca 1.5 mm long. Eggs ca 0.3 mm long, laid in soft tissue, new leaves or fruit near calyx; two larval instars, propupa and pupa. Life cycle ca five weeks; multivoltine with 6–7 generations per year. Cf. *Megalurothrips*.

SCISSURE Noun. (Latin, *scissura, scindere* = to cut. Pl., Scissures.) Acarology: An interruption of the sclerotized Cuticle in the shape of a narrow band.

SCLERITE Noun. (Greek, *skleros* = hard; *-ites* = constituent. Pl., Sclerites.) 1. Any hard portion of the insect Integument separated from similar areas by membrane, apodeme, suture or Sulcus. 2. Hardened areas of an insect's body wall that are consequences of the process of sclerotization. Sclerites (also called 'plates') are variable in size and shape. Sclerites do not define anatomical areas and do not reflect a common plan of segmentation. Sclerites develop in several ways, including as *de novo* hardening of membranous areas of the body wall and as *de novo* separations from larger sclerotized areas of the body. Sclerites receive different names, depending upon the region of the body in which they are located. Tergites are sclerites that form a subdivision of the dorsal part of the body wall (Tergum). Laerotergites are sclerites that form as a subdivision of the lateral portion of the Tergum. Sternites are sclerites that form as a subdivision of the ventral part of the body wall (Sternum), or any of the sclerotic components of the definitive Sternum. Pleurites are sclerites in the pleural region of the body wall that are derived from limb bases. Alt. Sclerome. See Integument. Rel. Segmentation; Suture; Tagma.

SCLERODERM Noun. (Greek, *skleros* = hard; *derma* = skin. Pl., Scleroderms.) A hardened portion of the Integument. See Integument.

SCLEROGIBBIDAE Ashmead 1902. Plural Noun. A widespread, small Family (ca 15 Species) of aculeate Hymenoptera assigned to the Chrysidoidea. Diagnostic features include Antenna with 18–40 segments; Torulus low on face, near Clypeus; strong sexual dimorphism with males winged and females wingless. External parasites of Embioptera.

SCLEROMA Noun. (Greek, *skleroun* = to harden, *skleros* = hard. Pl., Scleromas; Scleromata.) 1. Hardened tissues, especially the Integument. 2. The sclerotized cuticular ring of a body segment in insects. Distinguished from other similar rings by membrane uniting it to others (Snodgrass). Alt. Sclerome. See Sclerite; Segmentation. 3. A tumour-like induration.

SCLEROPTERIDAE Plural Noun. (Greek, *skleros* = hard.) Stiff-winged Crickets. A small Family of ensiferous Orthoptera (3 Genera and about 20 Species) which are restricted to Africa and Asia. Body glabrous or sparsely setose, Frons wide and flat, hind Tibia serrated, wings present in both sexes, and the Ovipositor is 'needle-like'.

SCLEROTIC Adj. (Greek, *skleros* = hard.) Consisting of or pertaining to Sclerotin.

SCLEROTIN Noun. (Greek, *skleros* = hard. Pl., Sclerotins.) A term proposed by Pryor (1945) for any of several kinds of tough proteins that are resistant to chemical and physical degradation. Sclerotins are covalently-bonded through the process of quinone-tanning and form the hard parts of the insect Integument, eggshells, Oothecase and silk structures. Sclerotins are found in other arthropods and vertebrates. See Integument. Cf. Arthropodin; Chitin; Resilin; Vulcanization.

SCLEROTIUM Noun. (Greek, *skleros* = hard; Latin, *-ium* = diminutive > Greek, *-idion*. Pl., Sclerotia.) Fungi: A small, hard, dark structure which survives unfavourable conditions for long periods of time and sends hyphae or spore fruits when conditions are favourable.

SCLEROTIZATION Noun. (Greek, *sklerotes* = hardness; English, *-tion* = result of an action. Pl., Sclerotizations.) The physicochemical process by which sclerites are fomed. See Integument.

SCLEROTIZED Adj. (Greek, *skleros* = hard.) Insect Integument: Hardened in definite areas by deposition or formation of sclerotin or substances other than Chitin.

SCOBINA Noun. (Latin, *scobina* = file. Pl., Scobinae.) Hymenoptera (Symphyta): A rasp on the distal portion of dorsal ramus of Gonapophysis VIII.

SCOBINATE Adj. (Latin, *scobina* = file.) Descriptive of a rasp-like surface; nodulated. See Sculpture Pattern. Cf. Acuductate; Alveolate; Asperites; Baculate; Clavate; Corrugated; Echinate; Favose; Gemmate; Psilate; Punctate; Reticulate; Rugulate; Scabrate; Shagreened; Smooth; Striate; Verrucate; Wrinkle.

SCOBINATION Noun. (Latin, *scobina* = file; English, *-tion* = result of an action. Pl., Scobinations.) A nodule.

SCOLEBYTHIDAE Evans 1963. Plural Noun. A small Family of apocritous Hymenoptera assigned to the Chrysidoidea. Aculeate wasps characterized as 7–10 mm long; macropterous; head hypognathous; Antenna 13 segmented in both sexes; tibial spur formula 1,2,2; Metanotum very thin (strongly transverse); Proepisternum unusually large; forewing with Stigma and closed Marginal Cell; hindwing without closed cells, and with two short veins. Presently including three monotypic Genera, *Scolebythus* from Madagascar, *Clytopsenella* from Brazil, *Ycaploca* from South Africa and Australia. Biology of the Family limited. Probably parasitic on Cerambycidae larvae with development gregarious.

SCOLECIASIS (Greek, *skolekias* = being worm eaten; *-iasis* = suffix for names of diseases.) The invasion of man and animals by Lepidoptera larvae.

SCOLIIDAE Latreille 1802. Plural Noun. Hairy Flower-Wasps. Cosmopolitan, small Family (ca 300 Species) of aculeate Hymenoptera assigned to Superfamily Scolioidea or Vespoidea; predominantly tropical. Large bodied wasps with slight sexual dimorphism; male Antenna with 13 segments; female Antenna with 12 segments; medial margin of compound eye emarginate; Pronotum projecting to Tegula, immovable in relation to Mesonotum; Propodeum with two parallel, dorsal, longitudinal sutures; fore Coxae juxtaposed; middle and hind Coxae separated; Mesosternum and Metasternum each with a laminate process over base of respective Coxae; Wings macropterous in both sexes with surface finely corrugated along distal margin; forewing with 10 closed cells; hindwing with three closed cells, Jugal Lobe; Claval Lobe absent; distal margin of wings striated or corrugated. Propodeum with two dorsal longitudinal sutures. Gastral Sterna I and II separated by deep constriction; apical Sternum of male spinose. Scoliids fossorial, seek Coleoptera larvae in soil. Host larva paralysed by female wasp, Scoliid larva develops as primary, solitary, external parasite of grub in cell. Females sting and paralyse host and lay a single egg upon host larva. Larva develops on body of host.

SCOLIOIDEA Latreille 1802. A Superfamily of aculeata Hymenoptera. Included Families: Bradynobaenidae, Falsiformicidae, Mutillidae, Sapygidae, Scoliidae, Sierolomorphidae and Tiphiidae. Alternative classification places Scolioidea within Vespoidea.

SCOLOPALE Noun. (Greek, *skolos* = stake; *pale* = struggle. Pl., Scolopalia; Scolopales.) A hollow peg-like structure enclosed in a scolophore. A so-called sense rod or minute rod-like capsule enveloping the distal end of the sense cell in certain sense organs. See Scolopal Body; Scolopala; Scolopalae.

SCOLOPENDRA Plural Noun. A Genus of chilopods including some of the largest centipedes. See Centipedes.

SCOLOPHORE Noun. (Greek, *skolos* = stake; *pherein* = to bear. Pl., Scolophores.) A spindle-shaped bundle of Sensilla usually attached to the Integument and regarded as auditory in function. A bipolar nerve cell continuous with a fibre of the Chordotonal Nerve (Imms). Syn. Scolopidium.

SCOLOPIDIUM See Scolophore.

SCOLOPOID Adj. (Greek, *skolos* = stake, *eidos* = form.) Scolop-like. Rel. Eidos; Form; Shape.

SCOLOPOPHORUS ORGAN See Sensillum Scolopophorum.

SCOLOPOPHOROUS SENSILLUM A sense organ in which the sense cells contain 'sense rods' of the colops type (Snodgrass). Syn. Sensillum Scolopophorum.

SCOLOPS Noun. (Greek, *skolops* = anything pointed. Pl., Scolopes.) See Scolopale.

SCOLUS Noun. (Greek, *skolos* = thorn. Pl., Scoli.) Lepidoptera larva: Smooth or rough elevated cuticular protuberances which bear Setae. Cf. Chalaza; Pinaculum; Verruca.

SCOLYTIDAE Plural Noun. Ambrosia Beetles; Bark Beetles; Engravers. Sometimes regarded as a Subfamily of Curculionidae. Adult 1–8 mm long, cylindrical, dark coloured (reddish-brown to black); Antenna geniculate with conspicuous, annulate club. Larval body white, cylindrical, curved and legless, The Family includes several Species that cause significant economic damage to deciduous trees and the most serious pests of coniferous trees. Adults bore into pine tree, oviposit, feed on inner bark or fungi which are cultivated in galleries. Bark Beetles live within bark at the surface of the wood and feed upon phloem tissue; many attack living trees and all attack dying trees. Death of a tree is caused by fungal pathogens which are introduced to the tree by adult beetles and spread by scolytid larvae. Most, probably all, bark beetles transport fungi to tree. Adults and larvae feed on phloem, interrupt flow of nutrients; fungus clogs water transport vessels in sapwood. Each Species of bark beetle has a characteristic gallery pattern for the adult and for the larva. See Douglas-Fir Beetle; Mountain Pine Beetle; Southern Pine Beetle; Spruce Beetle; Western Pine Beetle. Rel. Buprestidae; Cerambycidae.

SCOPA Noun. (Latin, *scopa* = broom. Pl., Scopae.) Hymenoptera: 1. General: Any brush, tuft, mat or pile of Setae found on the body or its appendages which is used to collect pollen. Cf. Corbicula. 2. Apidae: The pollen brush located on the posteriomedial surface of the hind Tibia and Basitarsus. 3. Megachilidae: The mat of Setae on the metasomal venter. 4. Ichneumonidae: A tuft of Setae found on the hind Coxa of females (Heinrich 1961, Can. Ent. 92, Suppl. 15, 1–87). 5. Symphyta: An enlarged, sometimes setose, apicoventral flange on the Gonostylus. 6. Lepidoptera: A fringe of long scales along the posterior margin of abdominal segment VIII.

SCOPATE Adj. (Latin, *scopa* = brush; *-atus* = adjectival suffix.) Descriptive of structure or body covered or furnished with Scopae or brushes.

SCOPENUMATIDAE See Scathophagidae.

SCOPEUMATIDAE Plural Noun. See Anthomyiidae.

SCOPIFEROUS ANTENNA An Antenna which has a thick brush of hair/Setae somewhere on it.

SCOPIFORM Adj. (Latin, *scopa* = brush; *forma* = shape.) Brush-shaped. Cf. Aristiform; Penniform; Plumose. Rel. Form 2; Shape 2; Outline Shape.

SCOPIPED Adj. (Latin, *scopa* = brush.) A Tarsus in which the Pulvilli or Pulvinuli are thickly covered with hair/Setae and appear brush-like.

SCOPOLI, JOHANN ANTON (1723–1788) Italian physician, academic and author of *Entomologia Carniolica* (1763) and *Introductio ad historiam naturalem* (1777). (Duméril 1823, *Considérations générale sur la classe des insectes*. 272 pp. (252–253), Paris; Anon. 1850, Verh. zool.-bot. Ges. Wien. 1: 150; Rose 1850, *New General Biographical Dictionary* 11: 495; Voss 1881, Verh. zool.-bot. Ges. Wien 31: 17–66, bibliogr.; Pavesi 1901, Memorie Soc. ital. Sci. nat. 6: 1–68.)

SCOPULA Noun. (Latin, *scopula* = little broom, small brush. Pl., Scopulas; Scopulae.) 1. A brush. 2. Any small, dense tuft of Setae, frequently bristle-like, stiff, of similar length and shape.

SCOPULIPEDES Noun. (Latin, *scopula* = little broom; *pedes* = feet.) Bees which have pollen-gathering structures on the feet.

SCOPURIDAE Plural Noun. A Family of Plecoptera including one Genus, two Species, known from Japan and Korea. Treated as a Family of Euholognatha or Nemouroidea. Herbivorous but some Species apparently consume animal material at some stage of their lives. See Plecoptera.

SCORIACEOUS Adj. (Greek, *skoria* from *skor* = dung; Latin, *aceus* = of or pertaining to.) Ash-like in colour. See Coloration.

SCORPION-TAILED SPIDER *Arachnura higginsii* (L. Koch) [Araneida: Araneidae].

SCORPIONES Plural Noun. The Order in the Class Arachnida that comprises the Scorpions. True Scorpions.

SCORPIONFLIES See Mecoptera. The common name for any member of the Mecoptera. Syn. Hanging Fly, Hang Fly.

SCORPIONS See Scorpionida.

SCOTCH BROOM *Cytisus scoparius* (Linnaeus) Link.

SCOTCH GREY *Aedes alternans* (Westwood) [Diptera: Culicidae]: A minor pest in Australia, New Guinea and Indonesia. Female is day-biting and attacks humans, other mammals and birds; not known to vector disease. Adult with shaggy appearance due to outstanding scales; Palpus ca 0.6 times as long as Proboscis; apex of Abdomen tapering; Cerci long; legs mottled. Eggs laid singly in mud and can withstand desiccation. Larvae cannibalistic and predaceous on other mosquito larvae; breeds in brackish water or inland flood-filled pools. Larval and pupal stages ca one week during summer. Syn. Hexham Grey (Australia). See Culicidae. Cf. Grey-Striped Mosquito.

SCOTOPHASE Noun. (Greek, *skotos* = darkness; *phasis* = appear. Pl., Scotophases.) The darkness portion of the photoperiod. Under natural conditions, the Scotophase would be some interval of time less than 24 hours (*e.g.* L8:D16; L12:D12; L16:D8). Under experimental conditions the Scotophase may extend beyond 24 hours (*e.g.* L6:D36; L8:D46). See Photoperiod. Cf. Photophase. Rel. Circadian Rhythm; Diel Periodicity.

SCOTOTAXIS Noun. (Greek, *skotos* = darkness; *taxis* = arrangement. Pl., Scototaxes.) A behavioural reaction, typically in the form of movement or body orientation, in which an organism responds positively toward darkness or the absence of light. Alt. Skototaxis. See Orientation. Cf. Aerotaxis; Anemotaxis; Chemotaxis; Geotaxis; Menotaxis; Osmotaxis; Phototaxis; Rheotaxis; Rotaxis; Strophotaxis; Telotaxis; Thermotaxis; Thigmotaxis; Tonotaxis; Tropotaxis.

SCOTT, ALEXANDER WALTER (1800–1883) (Anon. 1890, *In* Scott, Australian Lepidoptera 2(1): 4.)

SCOTT, ERNEST (1887–1964) (J.M.C.H. 1965. Entomologist's Rec. J. Var. 77: 22–23.)

SCOTT, H ELDON (1916–1975) (Anon. 1976, Bull. ent. Soc. Can. 8: 10.)

SCOTT, HUGH (1885–1960) (Benson 1960, Entomol. mon. Mag. 96: 105; Britton 1960, Nature 188: 1070–1071.)

SCOTT, JOHN (1823–1888) (Carrington 1888, Entomologist 21: 288; Sharp 1888, Proc. ent. Soc. Lond. 1888: xlix–l.)

SCOTT, LOUIS M (1886–1944) (Sasscer 1945, J. Econ. Ent. 38: 414.)

SCOURGE® See Resmethrin.

SCOUT Noun. (Middle English, *scouten* = search. Pl., Scouts.) Social Insects: A worker searching outside the nest for food or slaves. See Caste.

SCOUT X-TRA® See Tralomethrin.

SCRAPER Noun. (Anglo Saxon, *scrapian*. Pl., Scrapers.) 1. Any structure or part of a structure modified for or adapted to scraping or rasping. 2. A roughened surface, typically on the wing margin or legs, used to produce sound. 3. Honey bee: A semicircular notch in the fore Basitarsus through which the Antenna passes when being cleaned. 4. The roughened part, file or rasp of a chordotonal apparatus.

SCREW-WORM *Cochliomyia hominivorax* (Coquerel) [Diptera: Calliphoridae] or any of several Species of calliphorid or sarcophagid flies endemic to tropical and subtropical areas and which are obligatory agents of Myiasis and significant pests of warm-blooded animals. SWF particularly destructive to livestock by inflicting damage resulting in loss of physical condition, sterility or death. Most notable Species include New World Screw-Worm and Old World Screw-Worm. See Calliphoridae; New World Screw-Worm; Old World Screw-Worm. Rel. Blow Flies.

SCRIBA, EMIL (1834–1917) (Anon. 1917, Ent. Bl. Biol. Syst. Käfer 13: 238; Reitter 1917, Wien.

ent. Ztg. 36: 228.)

SCRIBA, W (1807–1898) (Kraatz 1898, Dt. ent. Z. 42: 9.)

SCRIBBLYGUM MOTH *Ogmograptis scribula* Meyrick [Lepidoptera: Bucculatricidae].

SCRIPT Adj. (Latin, *scriptus* = written.) Pertaining to marks which resemble letters.

SCROBE Noun. (Latin, *scrobis* = ditch. Pl., Scrobes.) 1. A cuticular impression, Sulcus, depression, invagination or groove formed for the reception, protection and concealment of an appendage. 2. Orthoptera: Pits in which Antennae reside when at repose. 3. Rhynchophora: Grooves at the sides of the Rostrum which receive the antennal Scape, and grooves on the sides of Mandibles. 4. Hymenoptera: Typically circular impressions upon the Frons, in which the Scapes lie at repose. Syn. Scrobal Impressions.

SCROBICULATE Adj. (Latin, diminutive *scrobis* = ditch; *-atus* = adjectival suffix.) Pitted; pertaining to surface with deep circular pits.

SCROFA HAWK MOTH *Hippotion scrofa* Boisduval [Lepidoptera: Sphingidae].

SCROTAL MEMBRANE The envelope covering the testes in some insects.

SCROTIFORM Adj. (Latin, *scrotum* = scrotum; *forma* = shape.) Purse-shaped; pouch-shaped; Scrotum-shaped. See Pouch. Cf. Bursiform; Pyreniform. Rel. Form 2; Shape 2; Outline Shape.

SCROTUM Noun. (Latin, *scrotum* = a pouch. Pl., Scrota; Scrotums.) An external pouch on the trunk of most male mammals that contains the Testes. Syn. Scrotal Membrane. Rel. Scrotiform.

SCRUB TYPHUS A rickettsial disease occurring in Pakistan, India, Russia, China, Korea, Indonesia, northeast Queensland (Australia), some islands of Indian Ocean, south Pacific and Japan. Aetiological agent *Rickettsia tsutsugamushi* occurs in birds and rodents; ST vectored to humans by trombiculid mites (*Leptotrombidium* spp.) that contact infected rodents. Disease is characterized by headache, rash, fever and lymphadenopathy; clinical period for ST is 14–21 days with moderately high mortality (ca 30%) in untreated cases. Cf. Rickettsial Pox; Rocky Mountain Spotted Fever; Tick-Borne Spotted Fever.

SCRUB TYPHUS MITE *Leptotrombidium deliense* (Walsh) [Acari: Trombiculidae] one of the principal vectors of Scrub Typhus in the Orient.

SCUDDER, SAMUEL HUBBARD (1837–1911) American entomologist, and bibliographer; author of more than 120 papers on fossil North American insects. (Dimmock 1878, Dimmock's special bibliographies No. 3: 1–28; Benjamin 1890, Harper's Weekly 34: 925–926; Anon. 1893. National cyclopaedia of American biogr. 3: 99–100; Bethune 1911, Can. Ent. 43: 253–254; Cockerell 1911, Science 34: 339–342, bibliogr.; Jacobson 1911, Rev. Russe Ent. 11: 408; Kingsley *et al.* 1911, Psyche 18: 175–192; Rehn *et al.* 1911, Ent. News 22: 289–292, bibliogr.; Turner 1911, Entomologist's Rec. J. Var. 23: 255–256; Osborn 1937, *Fragments of Entomological History,* 394 pp. (161–162). Columbus, Ohio; Cockerell 1942, Bios 13: 147–151.)

SCULPTURE Noun. (Latin, *sculptura* = sculpture. Pl., Sculptures.) Markings or patterns of impression or elevation on the body surface. See Sculpture Pattern.

SCULPTURE PATTERN Surface features of the insect Integument and eggshell that are sometimes apparent to the unaided eye in large specimens, but which are more often minute and only visible microscopically. SP typically consists of elevations (ridges), pits, grooves and depressions which combine in various ways to form patterns that are often ornate, elaborate or complex and difficult to describe. SP are sometimes diagnostically useful in identification of Species or other Taxa. See Integument. Cf. Acuductate; Alveolate; Asperites; Baculate; Clavate; Echinate; Estriate; Favose; Gemmate; Psilate; Punctate; Reticulate; Rugulate; Scabrate; Shagreened; Smooth; Striate; Verrucate. Rel. Chaetotaxy; Eggshell; Sensilla.

SCULPTURED Adj. (Latin, *sculptura* = sculpture.) Descriptive of structure that is marked superficially with elevations, depressions or both, and arranged in some definite manner.

SCULPTURED PINE-BORER *Chalcophora angulicollis* (LeConte) [Coleoptera: Buprestidae].

SCURF Noun. (Anglo Saxon, *scurf.* Pl., Scurfs.) 1. The scaly covering of some leaves. 2. The dried skin of animals which peels off in scales. Cf. Scute; Slough; Slime.

SCURFY SCALE *Chionaspis furfura* (Fitch) [Hemiptera: Diaspididae]: In North America, a widespread pest of apple, pear, quince, and other deciduous trees. Overwinters in egg stage; eggs reddish-purple laid in clutch of 40 under female; crawlers settle within hours on bark, occasionally fruit. Scale cover greyish-white, pear-shaped in outline. SS resembles Dogwood Scale; Elm Scurfy Scale; San Jose Scale. See Diaspididae.

SCUTA Noun. (Latin, *scutum* = shield. Sl., Scutum.) Chitinous sclerites of the body segments. The sclerites.

SCUTALARIA Noun. 1. The lateral end of the Scutellum extending as a blunt projection into the Notarostaxis. 2. In some insects, the Posterior Notal Wing Process (MacGillivray).

SCUTAREA Noun. (Latin, *scutum* = shield.) Hemiptera: The Scutellum *sensu* MacGillivray.

SCUTATE Adj. (Latin, *scutum* = shield; *-atus* = adjectival suffix.) 1. Descriptive of structure covered with large flat scales. 2. Shield-shaped; scutiform. 3. Descriptive of structure with a Scutum. Alt. Scutatus.

SCUTCHEON Noun. (Latin, *scutum* = shield. Pl., Scutcheons.) 1. Anything shaped as an escutcheon. 2. A Scute; a Scutellum. 3. A term also used by some authors (Walker) for the Pronotum in Homoptera.

SCUTE Noun. (Latin, *scutum* = shield. Pl., Scutes; Scuta.) 1. Chitinous shields or sclerites on the segments of the larval body. 2. A scale of a fish or reptile.

SCUTELA Plural Noun. (Latin, *scutum* = shield.) The Scutellum (Say).

SCUTELLAR ANGLE The angle in the Elytra adjacent to the Scutellum when a wing is expanded.

SCUTELLAR BRIDGE Diptera: A small ridge on each side of the Scutellum connecting with the Scutum and crossing the intervening suture (Comstock).

SCUTELLAR SPACE Mantids: An area between Antennae and Clypeus.

SCUTELLARY Adj. (Latin, *scutellum* = small shield.) Descriptive of or pertaining to the Scutellum.

SCUTELLATE Adj. (Latin, *scutellum* = small shield; *-atus* = adjectival suffix.) Descriptive of an area divided into surfaces resembling small plates. Pertaining to a surface covered with small, overlapping sclerites. Alt. Scutellatus. Cf. Imbricated.

SCUTELLERIDAE Leach 1815. Plural Noun. Jewel Bugs. A Family of pentatomoid Heteroptera. Pronotum without large posterolateral lobes; Hemelytra not longer than apex of Abdomen, not folded between corium and membrane; Mesonotum and Metanotum not visible in lateral aspect; Scutellum large, convex; Tarsi with three segments. Phytophagous.

SCUTELLUM Noun. (Latin, *scutellum* = small shield. Pl., Scutella.) 1. Any small, shield-like sclerite. Cf. Aspid-. 2. Heteroptera: The triangular part of the Mesothorax; generally placed between the bases of the Hemelytra, but in some groups overlapping them. 3. Coleoptera: The triangular sclerite between the Elytra. 4. Hymenoptera: The sclerite posterior of the Transscutal Suture 5. Diptera: A subhemispherical part of the sclerite separated by an impressed line from the Mesonotum. See Thorax. Cf. Scutum.

SCUTIFORM Adj. (Latin, *scutum* = shield; *forma* = shape.) Shield-shaped; plate-shaped. Alt. Scutate. See Shield. Rel. Peltate. Rel. Form 2; Shape 2; Outline Shape.

SCUTOPRESCUTUM Noun. (Latin, *scutum* = shield; *pre-* = before.) Hymenoptera: The anterior sclerite of the Mesonotum (Imms).

SCUTOSCUTELLAR SUTURE The external suture of the U-shaped notal ridge of the Alinotum. The arms divergent posteriorly, dividing the Notum into Scutum and Scutellum (Snodgrass).

SCUTTLE FLY See Phoridae.

SCUTULIS Noun. (Latin, *scutulum* = an oblong shield.) Hemiptera: The shield-shaped part of the Scutellum (MacGillivray).

SCUTUM Noun. (Latin, *scutum* = shield. Pl., Scuta; Scutums.) 1. A shield or shield-like sclerite. 2. The second dorsal sclerite of the Mesothorax and Metathorax (Smith). 3. Hymenoptera: A major portion of the Alinotum anterior of the Transscutal Suture in some Hymenoptera and not identical with that in generalized insects (Snodgrass). 4. Ticks: The middle division of the Notum.

SCYBALUM Noun. (Greek, *skybalon* = offal, dung.) A hardened mass of faeces or excrement.

SCYDMAENIDAE Plural Noun. Stone Beetles. Small to minute bodied beetles assigned to the Staphylinoidea. Adult Antenna with 11 segments; body constricted between Prothorax and base of Elytra. sometimes forming a neck between head and Prothorax. Species occupy decaying vegetation, moss, ant nests. Presumably predaceous.

SCYTHRIDIDAE Plural Noun. Scythridid Moths. A widespread Family of ditrysian Lepidoptera assigned to Superfamily Gelechioidea. Includes ca 400 Species; best represented in Mediterranean and western Asia. Adult small; wingspan 8–20 mm; head smooth scaled; Ocelli sometimes absent; Chaetosemata absent; Antenna shorter than forewing, weakly serrate; Scape usually with Pecten; Proboscis scaled basally; Maxillary Palpus small, with four segments folded at base of Proboscis; Labial Palpus recurved; fore Tibia with Epiphysis; tibial spur formula 0-2-4; forewing lanceolate. Egg upright, elongate-elliptical, flattened on apex and base. Larval head hypognathous; prothoracic and anal shields well sclerotized. Pupation in flimsy cocoon; pupa slightly compressed. Adults of brightly coloured Species diurnal; exhibit thanatotic behaviour when disturbed. Larvae apparently spin webs and some communal; foodplants primarily herbaceous.

SEA MOTHS See Pegasidae.

SEABEACH FLIES See Helcomyzidae.

SEABIRD SOFT TICK *Ornithodoros capensis* Neumann [Acari: Argasidae].

SEABIRD TICK *Amblyomma loculosum* Neumann [Acari: Ixodidae].

SEABRA, ANTERO FREDERICO DE (1874–1953) (Ferreira 1955, Bolm. Soc. Estud. Moçamb. 93: 3–21, bibliogr.)

SEAL-BROWN Adj. A brilliant, deep red-brown; almost castaneous but darker.

SEAM Noun. (Anglo Saxon, *seam;* Greek, *saum.* Pl., Seams.) 1. A crevice, crack or interstice that marks the contact between flattened structures. Syn. Suture. 2. A narrow strip, streak or stripe. 3. A scar left by a cut or wound. 4. A thin layer or stratum of coal or minerals in the earth's crust.

SEAM SQUIRREL See Human Louse.

SEASONAL COLORATION Butterflies: A change or difference in aspect and in colour in two or more succeeding broods in the same Species in the same or in different seasons. See Coloration. Cf. Advancing Coloration; Alluring Coloration; Anticryptic Coloration; Apetetic Coloration; Aposematic Coloration; Combination Coloration; Cryptic Coloration; Directive Coloration; Disruptive Coloration; Epigamic Coloration; Episematic Coloration; Procryptic Coloration; Protective

Coloration; Pseudepisematic Coloration; Pseudoaposematic Coloration; Sematic Coloration. Rel. Crypsis; Mimicry; Morph; Polymorphism.

SEASONAL CYCLE The sequence of events in an insect's life cycle that occurs within a one-year period.

SEASONAL POLYMORPHISM A type of polymorphism expressed by some insects in which several morphs are produced in response to conditions such as photoperiod, crowding and food quality (*e.g.* Green Peach Aphid). Environmental and genetic factors control the endocrine mechanism. See Polymorphism. Cf. Caste Polymorphism; Cyclical Polymorphism; Kentromorphism.

SEAWEED FLIES See Coelopidae.

SEAWEED FLY *Coelopa frigida* (Fabricius) [Diptera: Coelopidae]: A Species endemic to east coastal Canada south to Rhode Island and Europe. Adult ca 2–4 mm long, body predominantly black with reddish Antenna, mouthparts, legs and parts of Abdomen. Larva feed upon kelp. See Coelopidae.

SEBA, ALBERT (1665–1736) Dutch pharmacist. Published Thesaurus in 4 folio volumes (1734, 1765). (Anon. 1742, Acta ephem. Acad. nat. Curios 6: 239–252; Rose 1850, *New General Biographical Dictionary* 11: 502–503.)

SEBACEOUS Adj. (Latin, *sebum* = tallow; *-aceus* = of or pertaining to.) Fatty or oily. Alt. Sebific.

SEBACEOUS GLAND MITE *Demodex brevis* Akbulatova [Acari: Demodicidae].

SEBIFIC DUCT A duct which carries secretions of the Colleterial Gland to the Bursa Copulatrix.

SEBIFIC GLAND See Colleterial Gland.

SEBORRHEA Noun. (Latin, *sebum* = tallow; *rhein* = to follow.) A morbidly increased discharge of sebaceous matter that collects on the skin in scales.

SECOND ANTENNAE The appendages of the Tritocerebral Somite of Crustacea. See Postantennal Appendages (Snodgrass).

SECOND ANTENNAL SEGMENT The third (tritocerebral) segment of head.

SECOND AXILLARY SCLERITE The pivotal sclerite of the wing base resting on the Pleural Wing Process, connected with the base of the Radial Vein (Snodgrass). The sclerite that articulates partly with the First Axillary and as a rule, partly with the base of the Radius (Imms).

SECOND BASAL CELL Diptera: A cell lying immediately behind the first basal, rarely united with it, more often open apically and united with the Discal Cell, but closed in most cases (cell M) (Curran).

SECOND CLYPEUS See Anteclypeus.

SECOND COSTAL CELL Hymenoptera (Packard): The Stigma.

SECOND INNER APICAL VEIN Hymenoptera (Norton): Media of Comstock, to the junction of the Medial Crossvein, also termed in part the Submarginal Nervure (Smith). See Vein; Wing Venation.

SECOND JUGAL VEIN Vena Cardinalis *sensu* Snodgrass. See Vein; Wing Venation.

SECOND LATERAL THORACIC SUTURE Odonata: A suture that extends from the base of the hindwing to the posterior margin of the hind Coxa.

SECOND LONGITUDINAL VEIN Diptera (Williston): Radius of Comstock. See Vein; Wing Venation.

SECOND MAXILLA The Labium.

SECOND MAXILLARY SEGMENT The seventh (labial) segment of the insect head.

SECOND MEDIAN AREA The Areola. See median area of the wing.

SECOND MEDIAN PLATE A variable-sized sclerite at the base of the mediocubital field of the wing. SMP folds convexly along the outer edge of the First Median Sclerite on the Plica Basalis; SMP is often absent, or represented by the united bases of the Medial and Cubital Veins (Snodgrass).

SECOND SPIRACLE The Metathoracic Spiracle, located near the anterior margin of the Metapleuron, between the Mesopleuron and Metapleuron, or in the posterior margin of the Mesopleuron (Snodgrass). See Spiracle.

SECOND SUBMARGINAL NERVURE Hymenoptera: The Radius of Comstock.

SECOND TROCHANTER The second segment of the Telopodite, often not distinct from the base of the Femur. Insects usually fused with the first Trochanter; Prefemur; Ischiopodite (Snodgrass).

SECOND VALVIFERS Ovipositor of pterygotes: The basal pair of lobes or sclerites supporting the base of the Second Valvulae (Snodgrass).

SECOND VALVULAE Ovipositor of pterygotes: The dorsal elongate pair of processes, often united in a single, median dorsal sclerite (Snodgrass).

SECOND VEIN Diptera: The vein (frequently absent), positioned immediately behind the first vein, its base always united with the base of the third vein (Rs and its anterior branch R) (Curran). See Vein; Wing Venation.

SECONDARIES The hindwings; always attached to the Metathorax. See Primaries.

SECONDARY ANAL VEIN Odonata: A longitudinal vein sometimes regarded as the true Anal Vein running from the distal end of the anal crossing towards the base of the wing (Comstock). See Vein; Wing Venation.

SECONDARY COPULATORY APPARATUS Accessory copulatory structures found on the venter of the second and third abdominal segments of Odonata.

SECONDARY CUTICULA See Endocuticula.

SECONDARY INOCULUM Pathology: An inoculum which spreads disease during a season. Cf. Primary Inoculum.

SECONDARY IRIS CELLS Elongated pigment cells that surround the Rhabdom and thus serve to isolate each Ommatidium from its neighbour

(Imms). See Secondary Pigment Cell. Syn. Iris Pigment Cell; Outer Pigment Cells.

SECONDARY MONOGAMY Social Insects: The presence of only one functional queen in a nest through the elimination of multiple founding queens until only one queen remains. See Monogamy. Cf. Primary Monogamy.

SECONDARY PARASITE See Hyperparasite. Cf. Primary Parasite; Tertiary Parasite.

SECONDARY PEST Pest Species that are usually present at low levels and whose population numbers are controlled by the action of natural enemies. Such Species can assume full pest status when the natural enemies are destroyed by a pest management tactic. See Key Pest; Subeconomic Pest.

SECONDARY PEST UPSET The rapid numerical increase to pest status of a noneconomic-pest Species following application of pesticide for the control of another pest. SPU typically occur through elimination of secondary pest's natural enemies by action of the pesticide.

SECONDARY PIGMENT CELLS Pigment cells that surround a Rhabdom and serve to isolate it from light reaching adjacent Rhabdoms. Number of SPCs varies considerably but at least six SPCs are present; Nine SPCs have been reported in Diptera. 18 SPCs have been reported in mecopteran *Panorpa* and 24 SPCs have been reported in mayfly *Baetis*. Typically, 12 SPCs occur in an insect's compound eye. Pigment in SPC can migrate or may be restricted to a portion of the cell. Syn. Iris Pigment Cell; Outer Pigment Cells; Secondary Iris Cell. See Compound Eye. Cf. Basal Pigment Cell; Primary Pigment Cell.

SECONDARY POLYGYNY Social Insects: Existence of more than one functional queen in a nest through subsequent occupancy by at least one queen. See Polygyny; Primary Polygyny; Oligogyny. Cf. Monogyny.

SECONDARY REPRODUCTIVE Isoptera: A king or queen with body not strongly sclerotized and nonalate or with rounded wing buds. Syn. Neoteinic; Supplementary Reproductive. See Neoteinic. Cf. Primary Reproductive.

SECONDARY SCREWWORM *Cochliomyia macellaria* (Fabricius) [Diptera: Calliphoridae]: A New World synanthropic fly of medical and veterinary importance and homologue of Oriental Latrine Fly. A scavenger Species often confused with New World Screwworm (Primary Screwworm); a minor agent of Myiasis and pest of meat in open markets. See Calliphoridae. Cf. Oriental Latrine Fly.

SECONDARY SEGMENTATION 1. Any form of body segmentation that does not strictly conform with the embryonic Metamerism. 2. SS is characteristic of hard-bodied arthropods, including adult and nymphal insects. SS is a derived anatomical feature with two significant aspects. First, SS in adult or nymph does not strictly conform with embryonic Metamerism or Metamerism expressed in soft-bodied forms. Second, SS musculature is intersegmental. The acquisition of secondary segmentation represents an evolutionary step in development of the Arthropoda. The soft-bodied arthropod has primary segmentation and muscles that are intrasegmental. Movement of body parts is relatively simple because the body wall is flexible. However, when the body wall becomes hard, flexibility is restricted to articulation between hardened parts or the extension provided by intersegmental membranes. The arthropod's movement is possible only if soft and flexible membranes are positioned between inflexible (hard) body parts. See Segmentation. Cf. Primary Segmentation.

SECONDARY SETAE Some Lepidoptera larvae: Setae that are scattered and lack a constant or fixed position (Comstock).

SECONDARY SEXUAL CHARACTERS All characters possessed by one sex and not the other sex. SSC are not primarily reproductive in purpose, and typically include such attributes as colour, size, shape, ornamentation, structure or behaviour. Cf. Sexual Dimorphism.

SECONDARY SOMATIC HERMAPHRODITE An insect which has the gonad or gonads of one sex and the secondary sexual apparatus of that sex, but bearing in the same individual some or all of the secondary sexual characters of both sexes (Imms).

SECONDARY TYPE Any type-specimen that is not a primary type. STs include hypotypes, topotypes and homoeotypes. See Type. Cf. Homoeotype; Hypotype; Primary Type; Topotype. Rel. Nomenclature; Taxonomy.

SECONDARY VECTOR Medical Entomology: Arthropods which transmit pathogens to humans or other vertebrate animals but which cannot sustain the pathogen in the vertebrate hosts without the aid of a primary vector. Alt. Minor Vector. See Vector. Cf. Primary Vector.

SECRETION Noun. (Latin, *secretio* = separation: English, *-tion* = result of an action. Pl., Secretions.) 1. Any fluid or volatile substance produced by a gland. 2. A fluid or substance passed through a duct, membrane or aperture. Coccidae: The waxy, fibrous, cottony or silky substances forming the 'scales.' Rel. Accessory Gland.

SECRETIONARY COVERING Diaspine scales: The part of the puparium that covers the Exuviae.

SECRETIONARY SUPPLEMENT Part of a diaspidid scale extending beyond or around the pellicles.

SECRETORY Adj. Concerned in the process of secretion.

SECTAGON II® See Metham-Sodium.

SECTASETAE Noun. Setae distributed over the body of Psylloidea. The shapes of Sectasetae are diverse: Apically pointed, apically truncate or apically oblate. Sectasetae are tubular or flattened (Ferris 1923). In some Psyllidae flattened Sectasetae are arranged in an overlapping manner along the lateral margin of the body. These Setae form an effective seal which protects the ventral surface of the body. (Ferris 1923, Can. Ent. 55: 250–256)

SECTOL® See Pyrethrin.

SECTOR Noun. Longitudinal veins in Odonata which strike the principal veins at an angle, and usually reach the apex or hind margin. They are radial, subnodal, principal, nodal, median, short, and upper and lower of triangle.

SECTORIAL CROSSVEIN Crossvein which extends from the stem R to R4+5 or from R3 to R4. See Vein; Wing Venation.

SECTORIS COCOONIS Tearing or cutting structure used by the Lepidoptera in working out of a cocoon, variously placed.

SECTROL® See Pyrethrin.

SECUND Adj. (Latin, *secundus* = following.) Pointing one way; unilateral.

SECUNDUM NATURAM A Latin phrase meaning 'according to the course of nature.'

SECUNDUM ORDINEM A Latin phrase meaning 'in order.'

SECURIFORM Adj. (Latin, *securis* = ax; hatchet; *forma* = shape.) Ax-shaped; shaped as the blade of a hatchet; triangular-compressed. Cf. Dolabriform. Rel. Form 2; Shape 2; Outline Shape.

SEDANOX® See Chlorfenvinphos.

SEDENTARY Adj. (Latin, *sedere* = to sit.) Not active; settled or remaining in one place. Pertaining to organisms which are sessile.

SEDESATNIKEM, AUGUSTIN HOFER (1910–) (Pfeffer 1970, Acta ent. bohemoslovaca 67: 278–284, bibliogr.)

SEDLACEK, JOSEF (1913–1993) Czech born Australian collector interested in Coleoptera. (Monteith 1993. Myrmecia 29 (4): 5–8).

SEDUCIN Noun. (Latin, *seducere* = to lead apart.) Blattaria: A substance secreted from the tergal glands of male *Nauphoeta cinerea* and necessary to induce copulation by the female. See Tergal Gland.

SEEBER, C ERNST (1833–1895) (Anon. 1895, Ent. News 6: 172, 195–196.)

SEEBOLD, TEODORO (–1915) (Dusmet y Alonso 1919, Boln. Soc. ent. Esp. 2: 180.)

SEED Noun. (Anglo Saxon, *saed* = seed. Pl., Seeds.) 1. The embryonic plant and food reserves surrounded by a protective coat. 2. Semen.

SEED BEETLES See Bruchidae.

SEED BUG Any member of the Genus *Nysius* [Hemiptera: Lygaeidae].

SEED BUGS See Lygaeidae.

SEED CHALCIDS See Eurytomidae.

SEED-CORN BEETLE *Stenolophus lecontei* (Chaudoir) [Coleoptera: Carabidae]: A widespread, sporadic pest of corn in North America.

SEED-CORN MAGGOT *Delia platura* (Meigen) [Diptera: Anthomyiidae]: A widespread fly whose larval stage feeds upon on diverse plant species and decaying organic matter. A pest of beans, beets, corn, cabbage and seed potatoes. Adults 5–7 mm long; resembles housefly; body grey with black stripes on Thorax. Female oviposits in soil rich in organic matter; transmits the bacterium responsible for Stewart's Disease from their excrement to a feeding site. Eggs deposited in decaying plant matter. Larva tunnels into roots; feeds upon seed-corn and plant fails to sprout. SCM overwinters as larva within puparium in soil; completes 2–3 generations per year. Syn. Bean Maggot. See Anthomyiidae.

SEED CRUSHERS Vernacular: Large-headed soldier ants (Wheeler).

SEED-HARVESTING ANT *Pheidole ampla* Forel; *Pheidole anthracina* Forel [Hymenoptera: Formicidae].

SEED LAC The granules of lac from stick lac. See Lac.

SEED MITE *Tyrophagus longior* (Gervais) [Acari: Acaridae].

SEED WEEVILS Members of the Genus *Apion* [Coleoptera: Apionidae]. See Bruchidae. Cf. Sweet Potato Weevils.

SEEDLING BEAN MIDGE *Smittia aterrima* (Meigen) [Diptera: Chironomidae].

SEEDOX® See Bendiocarb.

SEEDOXIN® See Bendiocarb.

SEEVERS, CHARLES H (–1965) (Anon. 1966, Studia ent. 1966: 526.)

SEGALERVA, EDUARDO ZARCO (1908–1957) (Ceballos 1958, Eos, Madr. 34: 1.)

SEGMACORIA Noun. The Coria between segments; the intersegmental membrane *sensu* MacGillivray.

SEGMENT Noun. (Latin, *segmentum*, from *secare* = to cut. Pl., Segments.) 1. Any natural or apparent subdivision of the body. 2. Any subdivision of an arthropod appendage separated from similar structural elements by areas of flexibility or associated with muscle attachments. Syn. Somite; Podite; Embryonic Metamere; Arthromere.

SEGMENTAL APPENDAGES Paired ventrolateral segmental outgrowths of the body wall serving primarily for locomotion (Snodgrass).

SEGMENTAL DORSUM The dorsum of an individual segment (Snodgrass).

SEGMENTAL PLEURAL AREA The pleural area of an individual body segment in an insect (Snodgrass).

SEGMENTAL VENTER The ventral surface of an individual body segment of an insect (Snodgrass).

SEGMENTATED Noun. Composed of rings or segments. See Segment.

SEGMENTATION Noun. (Latin, *segmentum* = piece, *secare* = to cut; English, *-tion* = result of an action. Pl., Segmentations.) 1. Embryological development: A dividing or being divided into segments; the division of an originally single-celled egg into many cells (Blastomeres). See

Blastomere; Cleavage. 2. Jointed structure of the insect body wall which is divided into several rings. Segmentation represents a fundamental morphological characteristic of annelids and arthropods. 3. The process of Metamerism; an early step in body organization that involves the formation of a body composed of 'segments.' First indication of Segmentation in insects occurs in mesodermal layer of the embryo. Mesoderm becomes thickened; layers of thickened Mesoderm are separated by thinner layers with resultant appearance of Segmentation. Subsequently, clefts form in thickened layers of Mesoderm that result in Coelomic Sacs. Segmentation develops only between the mouth and Anus; preoral and postanal segments do not occur among the Insecta. Segmentation is also internal and involves most of the organs (Imms). Some morphologists hypothesize that the groundplan of body segmentation developed during early Palaeozoic. In this hypothesis, the hypothetical ancestral annelid-arthropod body increased in size and became segmented. Some specialists argue that primitive insect body consisted of 18 segments; other specialists propose 20 segments. Primordial segments are called Somites or Metameres. Conjecture holds that Somites are serially repetitive and that each segment is identical in form and function with adjacent segments. This condition is sometimes called Serially Homodynamous and implies that each segment of the body contains intrasegmental musculature. See Integument; Metamerism. Cf. Primary Segmentation; Secondary Segmentation. Rel. Sclerite; Scleroma. 2.

SEGMENTATION CAVITY See Blastocoele.

SEGREGATE Noun. (Latin, *segregatus* pp of *segregare* = to separate. Pl., Segregates.) Something that is set apart and separate. Verb. (Latin, *segregare* = to separate.) To set apart and separate.

SEGREGATED Adj. (Latin, *segregare* = to separate.) Detached or scattered into groups.

SEGREGATION Noun. (Latin, *segregare* = to separate; English, *-tion* = result of an action.) The act or condition of being separated or placed apart.

SEIB, SIMON H M (–1908) (Anon. 1908, Ent. News 19: 396.)

SEIDELIN, HARALD (1878–1932) (Anon. 1932, Trans. R. Soc. Trop. Med. Hyg. 26: 97–98.)

SEIDENKRANZ, HEINZ (1922–1972) (Grüber 1974, Prakt. Schädlingsk. 26: 142.)

SEIDL, WENZEL BENNO (1773–1842) (Anon. 1853, Lotos 3: 188–191.)

SEIDLITZ, GEORG KARL MARIA VON (1840–1917) (Bickhardt 1917, Ent. Bl. Biol. Syst. Käfer 13: 239–248, bibliogr.; Reitter 1917, Wien. ent. Ztg. 36: 228.)

SEIFERT, OTTO (1848–1910) (Anon. 1911, Can. Ent. 43: 16.)

SEIGE® See Hydramethylnon.

SEILER, JAKOB (1847–1923) (Leuthardt 1925, TätBer. naturf. Ges. Baselland 7: 158–162.)

SEISS, COVINGTON FEW (–1915) (Anon. 1915, Ent. News 26: 383–384.)

SEITNER, MORITZ (1862–1936) (Horn 1936, Arb. physiol. angew. Ent. Berl. 3: 301; Schimitschek 1936, Anz. Schädlingsk. 12: 134–136; Schimitschek 1937, Z. angew. Ent. 23: 653–656.)

SEITZ, ADALBERT (1860–1938) (Anon. 1930, Ber. senckenb. naturf Ges. 60: 192; Haimach 1930, Ent. News 41: 206–207; Kleinschmidt *et al.* 1930, Biography. Distributed with Seitz's *Grosschmetterlinge der Erde. Fauna Palaearctica.* 1: Suppl. Bd. I. I I pp.; Draudt *et al.* 1938, Ent. Rdsch. 55: 262–276; Franz 1938, Senckenbergiana 20: 279–286, bibliogr. only.)

SEJUGAL Adj. (Latin, *sex* = six; *jugum* = yoke; *-alis* = characterized by.) Acarology: Pertaining to the furrow that separates the Propodosoma and the Metapodosoma in Actinotrichida.

SEJUNCTUES Adj. (Latin, *sejunctus* = separate.) Separated.

SEKIYA, HIDEO (1911–1950) (Hasegawa 1967, Kontyû 35 (Suppl.): 45 46.)

SELANDER, RICHARD B (1927–) American academic (University of Illinois) and taxonomic specialist on Coleoptera, particularly Meloidae.

SELBY, PRIDEAUX JOHN (1788–1867) (Anon. 1867, Entomologist 3: 276.)

SELECRON® See Profenofos.

SELECTION Noun. (Latin, *selectio.* Pl., Selections.) 1. The act or condition of being chosen from a group of things. 2. Biology: The process which results in some organisms (traits, characteristics) being advantaged (promoted, surviving, propagated) while others are disadvantaged (killed, aborted, eliminated) from a population or group. See Artificial Selection; Natural Selection.

SELF Noun. (Anglo Saxon, *seolf.* Pl., Selves.) The totality of essential physical features, biological characteristics and behavioural attributes that constitute an individual (organism), and which serve to distinguish that individual from other individuals.

SELF GROOMING See Grooming.

SELF, KENNETH (1885–1967) (Worms 1967, Entomologist's Rec. J. Var. 79: 235.)

SELFLESS Adj. Descriptive of behavioural action by an individual without regard for personal interest. Ant. Selfish. Rel. Altruism.

SELLATE Adj. (Latin, *sella* = saddle; *-atus* = adjectival suffix.) Saddle-shaped.

SELLERS, WENDELL FOLSOM (1903–1960) (Shepard 1961, J. Econ. Ent. 54: 613.)

SELLMAN, EINAR (1890–1920) (Lundblad 1920, Ent. Tidskr. 41: 141–143.)

SELLOW (SELLO), FRIEDRICH (1789–1831) (Urban 1864, *In* Martius, C.F.P. *Flora Brasiliensis* 1: 106–111; Papavero 1971, *Essays on the History of Neotropical Dipterology* 1: 56–60, 70–80. São Paulo.)

SELOUS, CUTHBERT FENNESSY (–1946) (Hall 1947, Proc. R. ent. Soc. Lond. (C) 11: 62.)

SELVA Noun. (Latin, *selva* = forest.) Tropical rain forest.

SELYS-LONGCHAMPS, MICHEL EDMOND DE (1813–1900) Kraatz 1900, Dt. Ent. Z. 44: 8–9; Lameere 1900, Ann. Soc. ent. belg. 44: 467–472; Bargagli 1901, Boll. Soc. ent. ital. 33: 36–39; Calvert 1901, Ent. News 12: 32, 33–37; Blasius 1901, R. J. Orn. Lpz. (5) 8: 361–381, bibliogr.; Plateau 1901, Ann. Acad. Belg. 68: 145–157, bibliogr.)

SELZER, AUGUST (1850–1921) (Hasebrock 1921, Int. ent. Z. 15: 49–53; Hasebrock 1923, Sber. Vortr. EntVer. Altona 1921–23: 1–6.)

SEM See Scanning Electron Microscope.

SEMAPHORE Noun. (Greek, *sema* = sign; *pherein* = to carry. Pl., Semaphores.) A form of visual signalling or communication. A term applied to male insects that display one or more (often coloured or adorned) appendages in a ritualistic or stylized manner during courtship.

SEMAPHORONT Noun. (Greek, *sema* = sign; *pherein* = to carry. Pl., Semaphoronts.) 1. Cladistics: A character-bearing state of an organism. See Character 2. 2. Biology: An organism at a particular stage in its life history.

SEMAPORE® See *Nosema locustae*.

SEMATIC COLORATION Warning and signalling colours in insects. See Coloration. Cf. Advancing Coloration; Alluring Coloration; Anticryptic Coloration; Apetetic Coloration; Aposematic Coloration; Combination Coloration; Cryptic Coloration; Directive Coloration; Disruptive Coloration; Epigamic Coloration; Episematic Coloration; Procryptic Coloration; Protective Coloration; Pseudepisematic Coloration; Pseudoaposematic Coloration; Seasonal Coloration. Rel. Crypsis; Mimicry.

SEMATURIDAE Plural Noun. A small Family of ditrysian Lepidoptera of uncertain placement but sometimes placed within Superfamily Geometroidea. Adult lacks abdominal tympana. One Species occurs within the USA.

SEMBLING Verb. (French, *sembler* = gather.) Assembling.

SEMELPAROUS Adj. (Latin, *semel* = once; *parere* = to bear; *-osus* = possessing the qualities of.) Pertaining to adult insects that are short lived (relative to the developmental period for egg to adult) and which deposit all eggs over short period of time. Cf. Iteroparous.

SEMEN Noun. (Latin, *semen* = seed. Pl., Semens.) The fluid secreted within the Seminal Vesicles of the male and surrounding the Spermatozoa. Semen is typically secreted in very low volume. Poorly studied biochemically; Semen probably serves multiple functions in Spermatozoa transport, copulation and Spermatozoa activation. Syn. Seminal Fluid.

SEMENOV-TIAN SHANSKY, ANDREI PETROVICH (1866–1942) (Bogdanov-Katkov 1927, Défense Plantes 4: 167–175; Avinoff 1936, Ann. ent. Soc. Am. 29: 447–560; Hale-Carpenter 1946, Proc. R. ent. Soc. Lond. (C) 10: 53.)

SEMENOV-TIAN-SHANSKY, PETER PETROVICH (1827–1914) (Burr 1914, Entomologist's Rec. J. Var. 26: 127–128.)

SEMEVIN® See Thiodicarb.

SEMI- Latin prefix meaning 'half.'

SEMICIRCULAR Adj. (Latin, *semi* = half; Greek, *kirkos* = a circle.) Pertaining to structure which has an outline shape resembling half of a circle. See Outline Shape.

SEMICOMPLETE Adj. (Latin, semi= half; + complete.) Incomplete Metamorphosis. See Hemimetabolous; Paurometabolous.

SEMICORDATE Adj. (Latin, *semi* = half; *cordis* = heart; *-atus* = adjectival suffix.) Pertaining to structure which is half or partly heart-shaped. See Outline Shape. Cf. Cordate.

SEMICORONATE Adj. (Latin, *semi* = half; *corona* = crown; *-atus* = adjectival suffix.) Pertaining to structure partly surrounded by a margin of spines or hooks and resembling a crown. See Outline Shape. Cf. Coronate.

SEMICORONET Adj. (Latin, *semi* = half; *corona* = crown.) See Semicoronate. See Outline Shape.

SEMICYLINDRICAL Adj. (Latin, *semi* = half; *cylindrus; -alis* = characterized by.) Shaped as half a cylinder. See Outline Shape. Cf. Cylinder.

SEMIGLOBATE Adj. (Latin, *semi* = half; *globus* = globe; *-atus* = adjectival suffix.) Hemispherical, of the form of half a globe. Alt. Semiglobose; Semiglobosus. See Outline Shape. Cf. Globose.

SEMIHYALINE Adj. (Latin, *semi* = half; Greek, *hyalos* = glass.) Hyaline in part only; opaque, not completely transparent. Cf. Hyaline.

SEMILOOPER Noun. (Latin, *semi* = half; + loopers. Pl., Semiloopers.) A caterpillar in which 1–2 pairs of the abdominal legs are absent and movement is restricted to progression only in small loops. Term applied to larvae of some Noctuidae. See Noctuidae. Cf. Looper.

SEMI-LOOPERS See Noctuidae.

SEMILUNAR VALVE The valve guarding the auriculo-ventricular opening of the Heart.

SEMILUNAR Adj. (Latin, *semi* = half; *lunas* = moon.) In the form of half a crescent. See Outline Shape.

SEMINAL CUP Heteroptera: Another name for a circlet of chorionic processes that surround the egg's Operculum.

SEMINAL DUCTS See Vasa Deferentia.

SEMINAL FLUID See Semen.

SEMINAL VESICLE Male reproductive system: An enlarged tube-like or pouch-like portion of the Vas Deferens in which Spermatozoa are densely packed; storage site for Spermatozoa over long periods in some termites and Lepidoptera. Epithelial cells of the Vas Deferens are columnar; epithelial cells of the Seminal Vesicle are cuboidal and secretory. When highly convoluted, the Seminal Vesicle is called an Epididymis. SVs store male seminal fluid and later stages of its development may take place. See Testes. Cf. Vas Deferens; Epididymis. Rel. Gonad.

SEMINIFEROUS Adj. (Latin, *semen* = seed; *ferre* = to bear, carry; *-osus* = possessing the qualities of.) Semen-secreting.

SEMIOCHEMICAL Noun. (Greek, *simeon* = mark, signal; *-alis* = pertaining to. Pl., Semiochemicals.) 1. Any chemical involved in communications among organisms. 2. Naturally produced chemical compounds which influence insect–insect behaviour. Chemicals which mediate interactions between organisms (see Law & Regnier 1971, 40: 533–548). See Alleochemicals. Cf. Allomone; Apneumone; Kairmone; Pheromone; Synomone. (Nordlund *et al.* eds. 1981, *Semichemicals.* John Wiley & Sons).

SEMIPUPA Noun. (Latin, *semi* = half; *pupa* = puppet, young girl. Pl., Semipupae.) The stage of the larva just preceding pupation, more specifically the interpolated stage between the active larva and the true pupa, during Hypermetamorphosis.

SEMISAGITTATE Adj. (Latin, *semi* = half; *sagitta* = arrow; *-atus* = adjectival suffix.) Shaped as the longitudinal half of an arrow head. See Outline Shape. Cf. Sagittate.

SEMISOCIAL Adj. (Latin, *semi* = half; *sociare* = to associate; *-alis* = characterized by.) Pertaining to eusocial bees in which small colonies of females cooperate actively and display a division of labour. All females are of the same generation and each female constructs or provisions one cell during a period of time. See Social. Cf. Highly Social; Primitively Social; Quasisocial; Subsocial. Rel. Aggregation; Communal.

SEMISPECIES Plural Noun. (Latin, *semi* = half; species.) An informal term for populations transitional between Species and Subspecies. Semispecies have no standing in nomenclature.

SEMITROPICAL or GULF STRIP The southern part of the Austro-riparian area which extends from Texas to southern Florida, covers a narrow strip in southern Georgia and probably follows the coastal lowlands into the southern Carolinas.

SEMON, RICHARD WOLFGANG (1859–) (Musgrave 1932, *A Bibliography of Australian Entomology 1775–1930,* viii + 380 pp. (286–287), Sydney.)

SEMPER, CARL GOTTFRIED (1832–1893) (Weidner 1967, Abh. Verh. naturw. Ver. Hamburg Suppl. 9: 159–162.)

SEMPER, JOHANN GEORG (1837–1909) (Timin 1910, Verh. Ver. naturw. Unterh. Hamb. 14: 196–199; Weidner 1967, Abh. Verh. naturw. Ver. Hamburg Suppl. 9: 162–163.)

SEMPER'S RIB A degenerate Trachea present in Lepidoptera alongside the ordinary Trachea within the wing cavity (Imms).

SENAC, HIPPOLYTE (1830–1892) (Lévillé 1894, Ann. Soc. ent. Fr. 63: 449–452.)

SENDRAN® See Proproxur.

SENECIO MOTH *Nyctemera amica* (White) [Lepidoptera: Arctiidae].

SENESCE Verb. (Latin, *senescere* = grow old.) To decline in stature, vigour and capacity following maturity.

SENESCENCE Noun. (Latin, *senescere* = grow old.) The process of aging in an organism with the progressive loss of function or impairment of facilities with time. Rel. Development 2.

SENILITY Noun. (Latin, *senilis* = senile; English, *-ity* = suffix forming abstract nouns. Pl., Senilities.) 1. Old age with the attendant decay of the physical and cognitive faculties of an organism. 2. The act, process or quality of being senile.

SENIOR WHITE, RONALD (–1954) (Buxton 1955, Proc. ent. Soc. Lond. (C) 19: 69.)

SENNA, ANGELO (1866–1952) (Colosi 1952, Monitore zool. ital. 59: 83–84.)

SENSE CELL 1. The Neurocyte of a sensory neuron. 2. The receptive cell of a sense organ, with a proximal nerve-process going to a nerve centre (Snodgrass).

SENSE CONE Scarabaeoid larvae: The Nesium *sensu* Boving. See Sensillum Basiconicum, Sensillum Coeloconium.

SENSE DOME Sensillum Campaniformium.

SENSE HAIR. More-or-less modified seta forming part of a sense organ, at times equivalent to Sensillum Chaeticum.

SENSE ORGAN Any specialized, innervated structure of the body wall receptive to external stimuli; most insect sense organs are innervated Setae. Any structure by which an insect can see, smell, feel or hear.

SENSE PORE A minute pit or depression in the Integument (Cuticula) that externally marks the position of a sense organ (Snodgrass). See Sensillum.

SENSE ROD See Scolopale.

SENSE-BRISTLE See Sensillum Chaeticum.

SENSILLUM Noun. (Latin, *sensus* = sense. Pl., Sensilla.) 1. A simple sense organ or sensory receptor that occurs on various appendages and Tagma of the insect body. Traditional classification of Sensilla was based on the anatomy that was studied imperfectly with optical microscopy. Consequently, many names were provided for Sensilla that were unnecessary. Presently, entomologists are confronted with many different and complex names applied to Sensilla. A perfected anatomical classification of Sensilla requires time and careful systematic study with SEM and TEM. A functional classification of Sensilla is not coincident with an anatomical classification of the receptors. Functionally, Sensilla of various anatomical types respond to different categories of stimuli and have been broadly classed as mechanoreceptors and chemoreceptors. Additional categories that have been recognized more recently include thermoreceptors and adhesive structures. 2. A complex, bilaterally symmetrical sexually-dimorphic structure found on the tenth Tergum of Siphonaptera. The Sensillum is covered with short, tapering spines (Microtichia), circular pits which give rise to longer Setae

(Trichobothria), and a dome-like Cupola. The Sensillum functions as a compound sense organ for contact receptors in sexual behaviour, to align genitalia during copulation, as a mechanoreceptor detecting air movement associated with host detection and as an ultrasonic receptor in flea-to-flea communication. See Chemoreceptor; Mechanoreceptor. Rel. Receptor.

SENSILLUM AMPULLACEUM See Ampulliform Sensillum.

SENSILLUM BASICONICUM See Basiconic Sensillum.

SENSILLUM CAMPANIFORMIUM See Campaniform Sensillum.

SENSILLUM CHAETICUM See Chaeticum Sensillum.

SENSILLUM COELOCONICUM See Coeloconic Sensillum.

SENSILLUM OPTICUM See Compound Eye; Ommatidium.

SENSILLUM PLACODEUM See Placoid Sensillum.

SENSILLUM SCOLOPOPHORUM See Scolopophorous Sensillum.

SENSILLUM SPATULATUM See Spatulate Sensillum.

SENSILLUM SQUAMIFORMIUM See Squamiform Sensillum.

SENSILLUM TRICHODEUM See Trichoid Sensillum.

SENSIM Adj. A Latin term meaning 'gradually.'

SENSITIVITY Noun. (Latin, *sensilis* = sensitive; English, *-ity* = suffix forming abstract nouns. Pl., Sensitivities.) The labile property of protoplasm that makes it responsive to stimuli; sensitivity is highly developed in nerve tissue (Snodgrass).

SENSORIUM Noun. (Latin, *sensus* = sense; Latin, *-ium* = diminutive > Greek, *-idion*. Pl., Sensoria.) Aphididae: A circular opening, covered by a membrane, on the Antenna or legs.

SENSORY Adj. (Latin, *sensus* = sense.) Pertaining to sensation or to the senses; having sensation.

SENSORY BEHAVIOUR Animals: The response to any given stimulus expressed by movements of muscles or glands collectively.

SENSORY CELLS Nerve cells which convey or perceive sensation.

SENSORY-CELLS TYPE I Always bipolar nerve cells positioned either within or just beneath the Epidermis of the body wall, or the epithelium of the ectodermal sense organs (Snodgrass).

SENSORY-CELLS TYPE II Bipolar or multipolar sense cells that are positioned in the inner surface of the body and on the wall of the Alimentary Canal (Snodgrass).

SENSORY CHAETAE Chaetae which contain a nerve process from a sense-cell positioned within the Epidermis, commonly tactile in function. See Chaeta. Cf. Sensory Setae (Wardle).

SENSORY NERVE A nerve that carries sense perceptions.

SENSORY NERVOUS SYSTEM Part of Nervous System which connects with the exterior and transmits external stimuli to the motor system. See Nervous System. Cf. Motor Nervous System.

SENSORY NEURON Nerve cells associated with the sensory receptor part of sense organs. SN display bipolar or multipolar innervation. Most insect sensory axons do not have synapses. Syn. Sensory Neurone. See Nervous System; Nerve Cell; Cf. Afferent Neuron; Interneuron; Motor Neuron.

SENSORY PALPUS Hymenoptera: See Gonostylus.

SENSORY PITTINGS Anatomically, deep pits or punctures through the surface of the Integument which may bear pegs, bristles or Setae, and which may be open or covered by a membrane. Functionally, organs (Sensilla) used to perceive sound or odour.

SENSORY PLATE Siphonaptera: A sclerite on the ninth Tergum which is invested with chemoreceptors and or mechanoreceptors. Syn. Pygidium. See Sensillum.

SENSORY SETAE Setae which receive chemical, tactile, mechanical or temperature-related stimuli and which are always connected with the nervous system.

SENSU LATO Latin phrase meaning 'in the broad sense.' Abbreviated s.l., s. lat., sens. lat.

SENSU STRICTO Latin phrase meaning 'in the strict sense.' Abbreviated s. s., s. str., sens. str.

SENTA SKIPPER *Neohesperilla senta* (Miskin) [Lepidoptera: Hesperiidae].

SENTANG Common name for Philippine Neem Tree in West Malaysia.

SENTHION® See Fenitrothion.

SENTINEL® See Lambda Cyhalothrin.

SENTINEL SPECIES. See Indicator Species.

SENTRICON® See Hexaflumuron.

SEOANE LOPEZ, VICTOR (1834–1900) (Dusmet y Alonso 1919, Boln. Soc. ent. Esp. 2: 188–189.)

SEPARATE Noun. (Pl., Separates.) A printing of a scientific publication which bears the original pagination and plate numbers. Printed at the time of the original work. Cf. Reprint; Offprint; Copy. Alt. Separatum.

SEPP, CHARLES Dutch author of *Beschouwing der Wonderen Gods in de Minstgeachte Schepzeln of Nederlandiche Insecten* (3 vols, 1762).

SEPP, JAN (JOHANN) (1778–1853) (Heyden 1855, Stettin. ent. Ztg. 16: 16–17; Prittwitz 1866, Stettin. ent. Ztg. 27: 276–277.)

SEPSIDAE Plural Noun. Black Scavenger Flies. A numerically small, widespread Family of cyclorrhaphous Diptera assigned to Superfamily Sciomyzoidea. Adult body typically small, sometimes ant-like, usually dark coloured and shining; head spherical; abdomen narrow at base. Adults often move their wings while walking or standing; may occur in large numbers near larval habitat. Larvae inhabit decaying fruit, vegetation and mammal excrement.

SEPTA Noun. (Latin, *septum* = an enclosure. Pl., Septula.) Odonata: The triangular area of the

Mesonotum before the insertion of the primaries. The Axillary Calli.

SEPTASTERNUM Noun. (Latin, *septum* = an enclosure; *sternum* = breast bone.) Coxosternum *sensu* MacGillivray.

SEPTENE® See Carbaryl.

SEPTIC Noun. (Greek, *septikos* from *sepein* = to make putrid.) Causing putrefaction; infective. Cf. Aseptic.

SEPTICAEMIC PLAGUE (Greek, *septikos* = putrefictive.) A form of plague caused by the bacterium *Yersinia pestis* and vectored by fleas. Disease in humans is characterized by high fever and haemorrhages of the nose, mouth and intestine. Death from SP is common. See Plague.

SEPTULA Noun. (Latin, *septum* = an enclosure. Pl., Septulae.) 1. A small septum. 2. Any of the smaller more-or-less elevated ridges of a Phragma which provide a surface for the attachment of muscles. 3. Scarabaeoid larvae: A narrow bare region of the Raster between a single transverse Palidium and the base of the lower anal lip; alternatively, between a pair of oblique Palidia diverging posteriad toward the end of the anal slit, or between a pair of backward diverging, or parallel, or curved Palidia to inside the ends of the anal slit (Boving).

SEPTUM Noun. (Latin, *septum* = an enclosure; partition. Pl., Septa.) A wall, partition or broad internal projection within a cavity or hollow organ. Septa create separations for masses of tissue within an organ (*e.g.* grapefruit) or chambers for organs (*e.g.* septate insect Heart, chambered nautilus shell, corals).

SEPULCIDAE Plural Noun. A fossil Family of Symphyta (Hymenoptera). See Hymenoptera.

SEQUI- Latin prefix meaning 'one-and-one-half.'

SEQUOIA PITCH-MOTH *Synanthedon sequoiae* (Hy. Edwards) [Lepidoptera: Sesiidae].

SERAFINI, GIOVANNI (1782–1850) (Ambrosi 1889, Bull. Soc. veneto trent. Sci. nat. 4: 154–156.)

SERAPHOS® See Propetamphos.

SERDUKOV, GALIN VASILEV (1906–1962) (Pavlovskii & Gutzevich 1966, Ent. Obozr. 45: 917–920, bibliogr.) (Translation: Ent. Rev., Wash. 45: 516–517, bibliogr.).

SERGEANT ANTS Common name for *Myrmecia* spp. [Hymenoptera: Formicidae] in Western Australia. See Bulldog Ants.

SERGENT, ANDRÉ (1903–1937) (Anon. 1937, Archs Inst. Pasteur Algér. 15: 433–437, bibliogr.)

SERGENT, ETIENNE (1878–1948) (Anon. 1948, Archs Inst. Pasteur Algér. 26: 209–241, bibliogr.)

SERGIEV, PETER GRIGOREVICH (1892–1973) (Derbeneva-Ukhova 1974, Parasitologiya 8: 278.)

SERIAL HOMOLOGY The concept of an organ of one body segment being in a state of Homology with another organ on another body segment of the same animal, providing both organs are derived from corresponding parts. Example: Wings serially homologous with Halteres of Diptera; Maxillary and Labial Palpi serially homologous with legs and Cerci. Syn. Homotypy. See Homology. Cf. Special Homology.

SERIAL VEIN A wing vein made up of parts of other veins. See Vein; Wing Venation.

SERIATIM Adv. (Medieval Latin.) Placed in longitudinal rows.

SERIATION Plural Noun. (Latin, *serere* = to bind; to join; *-tion* = suffix. Pl., Seriations.) 1. An arrangement or formation in an orderly sequence. 2. Lines arranged in parallel series, as in the Corixidae in the Heteroptera.

SERICATE Adj. (Latin, *sericus* = silky; *-atus* = adjectival suffix.) 1. Silky or covered with short, thick, silky Setae. 2. Structure giving the sheen of silk. Alt. Sericatus; Sericeous; Sericeus.

SERICIN Noun. (Latin, *sericus* = silky.) The outer covering of the silk thread synthesized by some insect larvae such as the silk moth. Cf. Fibroin.

SERICOSE Noun. (Latin, *sericus* = silk; *-osus* = full of.) Hymenoptera: The slit-like opening of the larval Silk Gland.

SERICTERIUM Noun. (Greek, *serikton* = silk; Latin, *-ium* = diminutive > Greek, *-idion*. Pl., Sericteria.) Lepidoptera larva: The silk-producing gland or glands; the spinning structures.

SERICULTURE Noun. (Latin, *sericus* = silk; *cultura* = civilization.) The production of silk by growing silkworms to the pupal stage and harvesting silk from their cocoons. Technology developed in China ca 2600 BC; silk cloth may have existed in China ca 4000 BC. Technology taken to Korea ca 1200 BC, Japan ca 300 BC, India ca 140 BC and Persia ca 400 AD. Subsequently, technology spread to Europe, Australia and North America but sericulture industry did not establish in these regions. Traditional methods of silkworm culture employ mulberry-tree leaves; artificial diets for silkworm are currently available. Cocoon colours include white, yellow, pink, green, rusty and cinnamon-buff. White cocoons most common in China and Asia; yellow cocoons common in Europe; green cocoons common in India. See. Silkworm.

SERIES (Latin.) Noun. A group of Species, Genera or Families, arranged to show agreement in a common character which is not of sufficient importance to warrant the next higher division.

SERIFIC Adj. (Latin, *sericum* = silk; *facere* = to produce; *-ic* = characterized by.) Pertaining to the production or manufacture of silk.

SERIFIC GLAND A gland which produces a thick, mucous-like secretion which hardens to form silk.

SERINE Noun. (Latin, *serum* = whey.) Beta hydroxyalanine, a non-essential amino acid.

SEROLOGY Noun. (Latin, *serum* = whey; Greek, *logos* = discourse.) The study of serum.

SEROSA Noun. (Latin, *serum* = whey.) The outer membrane that envelops the forming embryo; the amnion and remainder of the egg. See Dorsal Blastoderm.

SEROUS Adj. (Latin, *serum* = whey; *-osus* = *pos-*

sessing the qualities of.) Descriptive of or pertaining to serum; of the nature of serum.

SERPENTINE Adj. (Latin, *serpentinus* = snake-like.) Sinuous; meandering. Winding or turning in seemingly random directions.

SERPENTINE LEAFMINER *Liriomyza trifolii* (Burgess), *Liriomyza brassicae* (Riley) [Diptera: Agromyzidae]: Widespread pests of many ornamental and vegetable crops. See Agromyzidae.

SERPENTINOUS Adj. (Latin, *serpentinus* = snake-like.) A dirty, dark green. Alt. Serpentinus.

SERPHITIDAE Brues 1937. A Family of apocritous Hymenoptera assigned to the Proctotrupoidea.

SERRA Noun. (Latin, *serra* = saw. Pl., Serrae.) A saw or saw-like part. Any saw-like structure.

SERRATE Adj. (Latin, *serra* = saw; *-atus* = adjectival suffix.) 1. Saw-like, with notched edges like the teeth of a saw. 2. Descriptive of structure with conical or tooth-like processes along an apical margin. Alt. Serratus; Serrulous; Serrulosus. See Saw. Cf. Biserrate; Serriform.

SERRATE-DENTATE Structure which is toothed and in which the tooth edges themselves are saw-toothed. Alt. Serratodentatus. Serratodenticulate. Cf. Aculeate-Serrate; Biserrate; Dentate-Serrate; Multiserrate; Subserrate; Uniserrate.

SERRATE PLATES Coccids: The Pectinae *sensu* MacGillivray.

SERRATED DUCTS Coccids: The Pectinae *sensu* MacGillivray.

SERRATION Noun. (Latin, *serra* = saw; English, *-tion* = result of an action. Pl., Serrations.) 1. Any saw-like formation. 2. A tooth, as of a saw; a series of such teeth. Alt. Serrature.

SERRATULATE Adj. (Latin, *serra* = saw; *-atus* = adjectival suffix.) Structure armed with small teeth or serrations.

SERRATURE Noun. (Latin, *serra* = saw.) See Serration.

SERRES, MARCEL DE French academic (Geology) and author of many articles on insect anatomy which were published in Memoires du Museum. Published 'Memoire sur les yeux composes et les yeux lisses des Insects' (1813).

SERRES, OLIVIER DE (1539–1619) (Chavannes 1843, Ann. Soc. Sericole 7: 294–299; Rose 1850, *New General Biographical Dictionary* 12: 3–4; Miall 1912, *Early Naturalists, Their Lives and Work (1530–1789)*, 396 pp. (93–98), London.)

SERRES, PIERRE MARCEL TOUSSAINE (1780–1862) (Gervais 1862, Mém. Acad. Sci. Lett. Montpellier (Sci.) 5: 303–308, bibliogr.)

SERRICORNIA Noun. (Latin, *serra* = saw; *cornu* = horn.) Coleoptera in which the Antennae are serrate or saw-toothed.

SERRIFEROUS Adj. (Latin, *serra* = saw; *ferre* = to carry; *-osus* = *possessing the* qualities of.) Possessing a saw-like structure, appendage or organ, such as the Ovipositor in the female (*e.g.* saw-flies).

SERRIFORM Adj. (Latin, *serra* = saw; *forma* = shape.) Saw-like. Alt. Serrate. See Saw. Rel. Form 2; Shape 2; Outline Shape.

SERRULA Noun. (Latin *serrula* = a small saw. Pl., Serrulae.) 1. Hymenoptera: Aculeata: Serrations on the ventral surface of the Aculeus. 2. Arachnida: A saw-like ridge on the Chelicera.

SERRULATE Adj. (Latin, *serrula* = a small saw; *-atus* = adjectival suffix.) Finely serrated, with minute teeth or notches.

SERRULATE PLATES Coccids: The Pectinae *sensu* MacGillivray.

SERTAN® See European-Pine Sawfly NPV.

SERUM Noun. (Latin, *serum* = whey. Pl., Sera.) Liquid part of the blood which separates on coagulation. A misnomer for the insect blood plasma of Haemolymph.

SERUM ALBUMIN A protein of serum.

SERUM-FREE MEDIUM Tissue Culture: A culture medium in which serum has been replaced by a mixture of chemically defined components.

SERVICE, ROBERT (1854–1911) (C.T.R. 1911, Zoologist 69: 239–240; Gladstone 1911, Ann. Scot. nat. Hist. 1911: 129–132, bibliogr.)

SERVILLE, JEAN GUILLAUME AUDINET See AUDINET-SERVILLE.

SESIIDAE Plural Noun. Clearwings; Clearwing Moths. A cosmopolitan Family of ditrysian Lepidoptera including ca 1,000 Species. Adults are small to moderate in size with wingspan 10–60 mm and head with smooth scales; Ocelli prominent but Chaetosemata weakly developed; Antenna considerably shorter than forewing length and shape variable: pectinate, bipectinate, clavate with apical tuft of Setae or filiform; Proboscis without scales and typically well developed but rudimentary in some Species; Maxillary Palpus small, with 1–3 segments; Labial Palpus recurved; fore Tibia with Epiphysis; tibial spur formula 0-2-4; forewing narrow, elongate and Pterostigma absent. Eggs are flattened, discoid with Chorion weakly sculptured. Larval head subprognathous; larvae bore in stems, twigs, bark or roots and a few Species live in galls; a few Species are predatory on scale insects. Pupation occurs within larval tunnels. Adults are diurnal and fly rapidly; some Species mimic bees and wasps in terms of posture, colour and lack of scales on wings. Some Species are considered economically important. Syn. Aegeriidae. See Lepidoptera. Cf. Apple Bark-Borer; Currant Borer; Peach-Tree Borer; Lilac Borer; Squash Vine-Borer.

SESIOIDEA Noun. A cosmopolitan Superfamily of Lepidoptera including about 1,000 Species assigned to Brachodidae, Choreutidae and Sesiidae. Most Species are active diurnally.

SESQUIALTER Noun. Latin term meaning 'one-and-one-half.'

SESQUIALTEROUS FASCIA A continuous colour band or fascia that extends across both wings. See Fascia.

SESQUIALTEROUS OCELLUS An ocellate spot with a smaller one by it, termed sesquiocellus.

SESQUIOCELLUS Adj. (Latin, *sesqui* = one and one half; *oculus* = eye.) A large ocellate spot including a smaller one.

SESQUITERTIAL Adj. Occupying the fourth part of anything.

SESQUITERTIOUS FASCIA A fascia and a third of a fascia on a wing or wingcover.

SESSILE Adj. (Latin, *sedere* = to sit.) 1. Any body, structure or appendage broadly attached to or resting on a base without a constriction, stalk or pedicel. 2. Pertaining to the Abdomen broadly attached to the Thorax.

SESSILIVENTRES Hymenoptera with a sessile Abdomen (Gaster, Metasoma). A taxonomic category corresponding to Symphyta (Sawflies). Cf. Apocrita; Terebrantia.

SESTON Noun. (Greek, *sesis* = sifting.) All living and non-living bodies (floating or swimming) which occur in water. Cf. Nekton; Neuston; Plankton.

SETA Noun. (Latin, *seta* = bristle. Pl., Setae.) 1. Hollow, often slender, hair-like cuticular projections produced by Epidermal cells of the Integument. A setal complex is composed of three differentiated cells: A Trichogen, a Tormogen and a Sense Cell. The Trichogen is the prominent, cuticular, sensory receptive portion of the device. The Trichogen often forms the long, tapered, hair-like structure which many entomologists erroneously call a 'hair.' Trichogens occur anywhere on the insect body and are sometimes confused with Microtricheae. Microtricheae often are found over the entire surface of the wing but trichodes are only found on wing veins. Trichogens also occur within the insect body. Sensory receptors in the Vagina of the earwig *Labidura riparia* consist of a Trichode Sensillum on the luminal surface near the spermathecal aperture. Stimulation of these receptors may provide tactile information during intromission and the passage of eggs. Setae may also be used in defence (Hastiseta, Nudiseta, Spiciseta), camouflage, adhesion. Syn. Bristle; Hair; Microtrichia. See Adhesive Setae; Camouflage Seta; Chaeta; Defensive Setae. Cf. Acantha; Bristle; Spine; Thorn. Rel. Tormogen; Trichogen. 2. Appendages of the Cuticle through which coccids secrete wax (MacGillivray).

SETACEOUS Adj. (Latin, *seta* = bristle; *-aceus* = of or pertaining to.) Seta-shaped; descriptive of structure or projection which is slender and gradually tapering toward the apex. Alt. Setaceus. See Seta.

SETAL Adj. (Latin, *seta* = bristle; *-alis* = characterized by.) Seta-like. Descriptive of or pertaining to Setae.

SETAL ALVEOLUS A depressed setal socket.

SETAL MEMBRANE The membranous floor of the setal socket, or alveolus, supporting the seta (Snodgrass).

SETAL SENSE ORGANS Setae connected with a sensory nerve cell positioned in or just beneath the Epidermis (Snodgrass).

SETARIOUS Adj. (Latin, *setarius*, from *seta* = bristle; *-osus* = possessing the qualities of.) Aristate. Terminating in a simple, unbranched bristle, as in the Antenna of some Diptera (Say). See Seta. Cf. Aristate.

SETIFEROUS Adj. (Latin, *seta* = bristle; *ferre* = to carry; *-osus* = possessing the qualities of.) Covered with or bearing Setae. Alt. Setigerous; Setose; Setosus. See Seta. Cf. Pile.

SETIFEROUS SENSE-ORGAN A sense-organ bearing or consisting of a sense Seta.

SETIFEROUS TUBERCLE A raised structure bearing a Seta or Setae.

SETIFORM Adj. (Latin, *seta* = bristle; *forma* = shape.) 1. Bristle-shaped; seta-shaped. 2. A term applied to the central lobe of a Ligula when it is very long and thread-like. Alt. Setiformis. See Seta. Cf. Aculeiform; Spiculiform; Spiniform; Spiciform. Rel. Form 2; Shape 2; Outline Shape.

SETIGENOUS Adj. (Latin, *seta* = bristle; *gerere* = to bear; *-osus* = possessing the qualities of.) Pertaining to structure or surface which gives rise to or produces Setae.

SETIGERIS Noun. The tibial comb (MacGillivray).

SETIGEROUS TUBERCLES Diptera: Tubercles occurring on the Scutellum or legs, each bearing a spine or bristle on its top (Curran).

SETIPAROUS Adj. (Latin, *seta* = brisle; *parere* = to produce; *-osus* = possessing the qualities of.) Producing Setae.

SETIREME Noun. (Latin, *seta* = bristle; *remi* = oar. Pl., Setiremes.) The setose, oar-like leg of aquatic insects.

SETOMORPHIDAE See Tineidae.

SETOSE Adj. (Latin, *seta* = bristle; *-osus* = full of.) A surface, structure or appendage covered with Setae or stiff bristles. Alt. Setosus; Setous. See Seta. Cf. Bisetose; Jubate; Setigerous; Spiniferous; Villose.

SETSUKO, NAKATU (1930–1971) (Gressitt & Higa 1971, Pacif. Insects 13: iii–iv.)

SETTELE, LUDWIG (1895–1974) (Wyniger 1975, Mitt. ent. Ges. Basel 25: 29–30.)

SETULA Noun. (Latin, *setula* diminutive of *seta* = bristle. Pl., Setulae.) 1. General: A small, short seta, hair or bristle. 2. Diptera: The small spine at the apex of the Subcosta. Alt. Setule.

SETULE Noun. (Latin, *setula* diminutive of *seta* = bristle. Pl., Setules.) 1. See Setula. 2. Acarina: A small branch or barb on a Seta (Goff *et al.* 1982, J. Med. Ent. 19: 234).

SETULOSE Adj. (Latin, diminutive of *seta* = bristle; *-osus* = full of.) Bearing small Setae; truncated or blunt Setae.

SEU (SIVE) Latin meaning 'either,' 'or.'

SEUBERT, MORITZ, (1818–1878) (Knoblauch 1878, Leopoldina 14: 49, 100–101.)

SEVEN-SPOTTED BEETLE *Coccinella septempunctata brucki* Mulsant [Coleoptera: Coccinellidae].

SEVEN-SPOTTED LADY-BEETLE *Coccinella*

septempunctata Linnaeus [Coleoptera: Coccinellidae].

SEVERIN, GUILLAUME (1862–1938) (Anon. 1938, Arb. morph. taxon. Ent. Berl. 5: 352; Lameere 1938, Bull. Ann. Soc. ent. Belg. 78: 313–314; Porter 1938, Revta. chil. Hist. nat. 42: 350.)

SEVERIN, HARRY CHARLES (1885–1964) (Walstrom 1964, J. Econ. Ent. 57: 1016.)

SEVERIN, HENRY H P American economic Entomologist (University of California).

SEVERINUS, MARCUS AURELIUS (1580–1656) (Rose 1850, *New General Biographical Dictionary* 12: 8.)

SEVILLA, S ISIDORO DE (–674) (Dusmet y Alonso 1919, Boln. Soc. ent. Esp. 2:76.)

SEVIN® See Carbaryl.

SEWAGE SPRINGTAIL *Hypogastrura viatica* (Tullberg) [Collembola: Hypogastruridae].

SEX Noun. (Latin, *sexus*.) 1. The physical difference between male and female. Symbolically represented by the sign of Mars for male, and by the sign of Venus for the female. 2. Workers or undeveloped females in social forms. 3. Intermediates between alate and apterous forms.

SEX PHEROMONE Chemical compounds released from Exocrine Glands which are used to attract conspecific members of the opposite sex. Sex pheromone glands are found on many parts of the body. In the Aphididae, female Oviparae of the aphid *Acyrthosiphon pisum* bear tibial plaques through which sex pheromones are released. Males and virginopara lack Tibial Plaques. Male Osmylidae (Neuroptera) of some Species with Eversible Scent Glands between abdominal segments 8–9. Lepidoptera adults display glands on various parts of the Abdomen, particularly near the apex within intersegmental membrane between the eighth and ninth segments. Many Species of Arctiidae bear Coremata which emerge from pockets concealed between abdominal Sterna 7–8 or within genital valves. The size of Coremata varies considerably: Some are about twice as large as the moth and still capable of eversion. Air enters large Tracheal Sacs connected to Lateral Tracheal Trunks and the base of the tubes. The sac inflates and exposes the Coremata; a small amount of Haemolymph also enters the inflated Coremata. ESG also are found on the head, Thorax, wings and legs of male Lepidoptera. Common names for the structures include Androconial Scales; Coremata; Costal Hairs; Hair Pencils; Scent Fans; Stobbe's Gland. Male phyticines display scent glands at the base of the forewing costal margin: A membranous flap covers specialized scales which are everted during wing fanning. Males of the moth *Aphomia sociella* Linnaeus produce sex pheromones with glands at the base of forewing. The compound has been identified as Mellein (Ochracin), which is typically a fungal metabolite. Males of some saturniid moth Species have hair pencils on the fore Tibia. Volatile glandular secretions are less well documented among the Diptera. In the Tephritidae, sex pheromone glands are suggested in many Species, but unequivocal proof has not been provided in any Species. Sex pheromone glands have been described for *Ceratitis* and *Dacus*. The presumed glands are located on the Rectum near the Anus in the male. *Anastrepha, Dacus, Ceratitis* and *Rhagoletis* males display enlarged salivary glands, but not females. Cf. Aggregation Pheromones; Alarm Pheromone; Trail Pheromone.

SEXTON BEETLE See Silphidae.

SEXUAL DIMORPHISM Polymorphism or differences in size, shape, anatomical features, colour or behaviour between males and females of a Species. SD is probably the most common type of polymorphism. Sometimes polymorphism becomes more complex and with more than one morph in one sex. Sexual Selection may play a role in influencing the expression of Sexual Dimorphism. SD may play a role in predator deception. Coloration dimorphism is common among insects but its function is not clear. Some factors understood for coloration dimorphism in birds may also apply to insects. Some profound sexual dimorphism may reflect differences in biology; some sexual dimorphism is subtle and functional differences may be difficult to measure. Sometimes, sexual dimorphism in coloration will exist, and a portion of a female population will mimic the coloration pattern of conspecific males. This has been described for a few butterflies and Odonata. See Dimorphism. Cf. Monomorphism; Polymorphism; Seasonal Polymorphism. Rel. Androchromotypic; Gynochromatypic.

SEXUAL MOSAIC An individual that expresses a type of Gynandromorphism in which secondary sexual features of both sexes appear in the individual. SM reported in several Orders of insects, including Orthoptera and Hymenoptera. Ant *Phediole dentata* Mayr shows an individual with a head intermediate between a male and worker while external genitalia are predominantly male. This type of gynandromorph may arise from nondisjunction early in mitosis of embryo when cell produces two X chromosomes instead of one, which leads to development of some female tissue in an otherwise male insect. See Gynandromorphism. Cf. Bilateral Gynandromorph; Transverse Gyandromorphism. Rel. Development; Metamorphosis; Mutant.

SEXUALE Noun. (Latin, *sexualis*. Pl., Sexuales.) A stage in the complex life cycle of migratory aphids. Sexuales are characterized by males and oviparous females which are typically apterous. Males are winged or wingless; females differ from apterous, viviparous generations (Fundatrix, Fundatrigena) in displaying a larger body. Cf. Fundatrix; Fundatrigenia; Alienacola; Sexupara; Migrante.

SEXUPARA Noun. (Latin, *sexus* + para. Pl., Sexuparae.) A stage in the complex life cycle of migratory Species. Characaterized by a parthenogenetic, viviparous female which usually develops on secondary host plant Species. Sexupara produce an alate form which migrates to primary host plants which in turn gives rise to the Sexuales. Cf. Fundatrix, Fundatrigenia, Alienacola, Migrante, Sexuale.

SEYCHELLES SCALE *Icerya seychellarum* (Westwood) [Hemiptera: Margarodidae]: A minor pest of *Citrus* in Asia, Australia, South Africa and islands of the Pacific.

SGONINA, KURT (1913–1939) (Sachtleben 1939, Arb. physiol. angew. Ent. Berl. 6: 209.)

SHADFLIES See Ephemeroptera.

SHAFT LOUSE *Menopon gallinae* (Linnaeus) [Mallophaga: Menoponidae]: A cosmopolitan pest of poultry to include chicken, pigeon, turkey, guinea fowl and duck; does not infest young chickens. SL often clings to feather shafts around vent, back and breast. Cf. Brown Chicken-Louse; Chicken Body-Louse; Fluff Louse.

SHAFT OF ANTENNA The portion of the Antenna including all segments distad of the pedicel. Syn. Flagellum. Cf. Funicle.

SHAGREENED Adj. (French, *chagrin*; Venetian, *sagrin*; from Turkish, *sagri* = the back of a horse.) Descriptive of surface sculpture, usually the insect's Integument, that is covered with a closely-set, fine, irregular roughness; like rough-surfaced horse leather (termed shagreen) or shark leather. See Sculpture Patterns. Cf. Acuductate; Alveolate; Baculate; Clavate; Echinate; Favose; Gemmate; Psilate; Punctate; Reticulate; Rugulate; Scabrate; Smooth; Striate; Verrucate.

SHALE Noun. (Greek, *schale* = a skin or bark, a thin layer. Pl., Shales.) A slate-like sedimentary rock formed by the deposition of finely ground clays from the water bearing them, on the bottom of lakes and other bodies of still or slowly moving waters (geology and palaeontology).

SHALLOT APHID *Myzus ascalonicus* Doncaster [Hemiptera: Aphididae].

SHANK Noun. (Middle English, *schanke*, akin to German *schenkel* = thigh.) The Tibia.

SHANKAR RAO, N B (–1965) (Pearson 1967, Proc. R. ent. Soc. Lond. (C) 31: 62.)

SHANNON, RAYMOND CORBETT (1894–1945) (Anon. 1945, Bull. Brooklyn ent. Soc. 40: 7l; Cortes 1945, Agric. tecnica 5: 102; Del Ponte 1947, Revta. Soc. ent. argent. 13: 345–347; McAtee & Wade 1951, Proc. ent. Soc. Wash. 53: 211–222.)

SHANNON, W. P. (1847–1897) (Davis 1931, Proc Indiana Acad. Sci. 41: 54–55.)

SHANNON-MOTH *Asmicridea grisea* (Mosely) [Trichoptera: Hydropsychidae].

SHAPE Noun. (Old English, *sceppan* = to shape, form or create. Pl., Shapes.) 1. A two-dimensional form with characteristic surface area and contour of margin. Common two-dimensional (outline) shapes include: Acuminate (terminating in a sharp point); Aciculate (needle-like); Cordate (heart-shaped); Triangular; Semicircular (half-circle); Round; Oval. See Outline Shape. 2. A three-dimensional form with characteristic volume and contour of surface. See Form; Outline Shape. Common three-dimensional shapes include: Fusiform (spindle-like); Hemispherical (half-sphere); Hypocrateriform (salver-shaped); Limaciform (slug-like); Pyriform (pear-shaped); Globular (spherical); Conical. Ambiguous shape descriptors include: Semicordate (partly heart-shaped). Both definitions apply to an individual, organ, structure, appendage or body but the distinction between shape and form is blurred in most entomological discussions and taxonomic/anatomical descriptions. Cf. Shapeless; Eidos; Form. Rel. Taxonomy; Anatomy; Morphology.

SHAPELESS Adj. (Middle English, *scap, shap* = shape; *less* = without.) 1. Descriptive of a structure (body) without a definite or fixed shape. See Amorphous; Form. 2. Descriptive of a structure that does not conform to the normal (expected) shape.

SHARD Noun. (Anglo Saxon, *sceran* = to shear. Pl., Shards.) 1. A small, broken portion of a brittle substance. 2. A chitinous sheath or Elytron.

SHARD BEETLE Any member of the Genus *Geotrupes* [Coleoptera: Scarabaeidae].

SHARP Adj. (Middle English, *sharpe* = sharp.) Descriptive of structure with a relatively long, thin edge, or structure tapered to a fine (acute) point. Syn. Acute. See Form. Cf. Blunt; Dull.

SHARP, DAVID (1840–1922) (Calvert 1922, Ent. News 33: 318–320; Everts 1922, Tijdschr. Ent. 65: 219–220; Lucas 1922, Entomologist 55: 217–221; Walker 1922, Entomol. mon. Mag. 58: 234–237; Slater 1923, Zool. Record 58: v-vi.)

SHARP, EDWIN P (1869–1936) (Chartres 1936, Entomologist 69: 244.)

SHARP, WILLIAM E (1856–1919) (Fowler 1919, Entomol. mon. Mag. 55: 263.)

SHAW, ALFRED ELAND (1861–1931) Australian MD and amateur Entomologist, publishing on Australian cockroaches. (Kelly 1931, Entomol. mon. Mag. 67: 253; Musgrave 1931, Aust. Mus. Mag. 4: 232; Mackerras & Marks 1974, Changing patterns of Entomology. A symposium. iv + 76 pp. (9), Brisbane.)

SHAW, GEORGE (1751–1813) British naturalist. Author of *Naturalist's Miscellany* (1789), *General Zoology* (14 vols), and *Zoological Lectures*. (Rose 1850, *New General Biographical Dictionary* 12: 19–20; Lisney 1960, *Bibliography of British Lepidoptera* 1608–1799, 315 pp. (227–245), London.)

SHAW, THOMAS (1692–1751) (Rose 1850, *New General Biographical Dictionary* 12: 19.)

SHCHEGHOLEV, VLADIMIR NIKOLAEVICH (1890–1966) (Berim 1967, Ent. Obozr. 46: 483–489, bibliogr. (Translation: Ent. Rev., Wash. 46: 285–289, bibliogr.).)

SHCHETINSKI, ALEXANDER ANTONOVICH (–1907) (Kusnezov 1907, Rev. Russe Ent. 7: 76.)

SHEARWOOD, GEORGE PERRY (–1891) (Adkin 1891, Entomologist 24: 199–200.)

SHEATH Noun. (Anglo Saxon, *sceth* = pod, shell. Pl., Sheaths.) A outer structure which encloses other structures and serves to protect them from damage (*e.g.* neural sheath; Ovipositor sheath).

SHEATH OF PENIS Odonata: A median, hood-like sclerite between the Hàmules, under which the Penis is folded when not in use.

SHEEP-BITING LOUSE *Bovicola ovis* (Schrank) [Mallophaga: Trichodectidae]: A pest of sheep; endemic to North America and adventive to Australia. Eggs attached to hair or wool; postembryonic development complete in 14 days. Syn. Sheep Body Louse. See Trichodectidae. Cf. Cattle-Biting Louse; Horse-Biting Louse; Goat-Biting Louse.

SHEEP BLOW-FLY *Lucilia cuprina* (Wiedemann) [Diptera: Calliphoridae]: A significant pest of sheep, causing Cutaneous Myiasis (blowfly strike) in Australia and South Africa. Females anautogenous and require protein meal before Ovary maturation and mating occur. Female oviposits clusters of 100–300 eggs on carrion or sheep fleece; prefers wool damp with urine or faeces; carrion produces fewer adult flies than live sheep hosts; eclosion occurs within 8–12 hours at sheep body temperature. First instar larva feeds on skin exudates and fleece rot; second and third instar feed on skin causing lesions; feeding produces extensive lesions and sometimes death of sheep. Mature larva drops to ground to pupate in soil; pupal stage ca 6 days. SBF multivoltine with ca eight generations per year. Syn. Australian Sheep Blowfly. See Calliphoridae. Cf. Green Blowfly.

SHEEP BODY-LOUSE See Sheep-Biting Louse.

SHEEP BOT-FLY *Oestrus ovis* Linnaeus [Diptera: Oestridae]: A significant pest of sheep and goats; endemic to Palaearctic and now cosmopolitan through commercial movement of livestock. Adult 12–15 mm long; Head and Thorax dull yellow, Abdomen shiny grey or black with irregular pattern of pale coloration. Frons with black pits between eyes; Scutum and Scutellum brown with yellow Setae and black tubercles. Female produces ca 500 eggs during lifetime that are retained within Uterus. SBF larviparous with ca 50 larvae per oviposition episode; larva deposited near nostril of sheep. Larva creamy white to yellow with dark, wide cross bands on dorsum and spines on venter; larva develops in nasal passage of sheep; mature larva 'sneezed' from body of host. Pupation occurs in soil. Small numbers of larvae cause little harm; heavy infestation may be a serious problem; SBF occasionally infests dogs and humans but larvae do not complete development. Syn. Sheep Nasal Bot-Fly (Australia); Sheep Nostril Fly. See Oestridae; Bot Flies. Cf. Rabbit Bot Fly; Common Cattle Grub; Northern Cattle Grub. Rel. Myiasis.

SHEEP DIP A water-based bath that includes pesticides into which sheep are driven or immersed for the treatment of ectoparasites. See Dip 2.

SHEEP FACE-LOUSE *Linognathus ovillus* (Neumann) [Phthiraptera: Linognathidae].

SHEEP FOLLICLE-MITE *Demodex ovis* Railliet [Acari: Demodecidae].

SHEEP FOOT LOUSE *Linognathus pedalis* (Osborne) [Phthiraptera: Linognathidae].

SHEEP ITCH-MITE *Psorobia ovis* (Womersley) [Acari: Psorergatidae].

SHEEP KED *Melophagus ovinus* (Linnaeus) [Diptera: Hippoboscidae]: A widespread, blood sucking, flightless fly. SK usually of minor importance as an ectoparasite of goats and sheep but important in southern Australia; heavy infestations sometimes cause losses through anaemia and staining of wool. Adult 5–7 mm long, reddish brown; body sac-like, leathery; head small and retracted into Prothorax; Antenna concealed in pit; legs short with Femora incrassate; forewing micropterous; Haltere absent. Larva develops within female; mature within ca 7 days; larva extruded and quickly pupates with secretion covering larva hardening to attach puparium to wool of host. Pupae occur on host throughout year; pupation 3–6 weeks. Female fly mature within 2–4 weeks, live ca 4 months; produce one larva per reproductive cycle (ca 7–8 days) and 10–12 larvae per lifetime. Syn. Sheep Tick. See Hippoboscidae.

SHEEP MYIASIS Cutaneous Myiasis caused in sheep by calliphorid flies, most notably *Lucilia* spp. and *Calliphora* spp. See Sheep Blow-Fly.

SHEEP NASAL BOT-FLY See Sheep Bot-Fly.

SHEEP NOSTRIL-FLY See Sheep Bot-Fly.

SHEEP SCAB-MITE *Psoroptes ovis* (Hering) [Acari: Psoroptidae].

SHEEP STRIKE A significant Myiasis problem of sheep in Australia, South Africa and Great Britain caused by *Calliphora.*

SHEEP SUCKING BODY-LOUSE *Linognathus ovillus* [Anoplura: Linognathidae]: A blood-sucking parasite of sheep in North America. Found in colonies on various parts of the host's body. Eggs attached to hair or wool; eclosion occurs within 18 days; postembryonic development completed within 14 days. See Linognathidae.

SHEEP TICK See Sheep Ked.

SHEERPA® See Cypermethrin.

SHEETS, LOUIS WAYNE (1908–1970) (Keller 1971, J. Econ. Ent. 64: 779.)

SHELDON, WILLIAM GEORGE (1859–1943) (Cockayne 1944, Proc. R. ent. Soc. Lond. (C) 8: 68; Rait-Smith 1944, Entomol. mon. Mag. 80: 43–44.)

SHELFORD, ROBERT WALTER CAMPBELL (1872–1912) (Burr 1912, Entomologist's Rec. J. Var. 24: 205–206; Morice 1912, Proc. ent. Soc. Lond. 1912: clxiv–clxv; Poulton 1912, Zoologist 70: 273–275.)

SHELJOZHKO, LEO (1890–1969) (W.F. 1971, NachrBl. bayer Ent. 20 (5): 81–86, bibliogr.)

SHELKOVNIKOV, ALEXANDER BORISOVICH (1890–1937) (Burr 1937. Entomologist's Rec. J. Var. 49: 39–40.)

SHELL Noun. (Anglo Saxon, *scell* = shell.) The hard, outer covering of an animal or fruit. See Eggshell. Cf. Sheath.

SHELL GLANDS Highly modified Nephridia in Crustacea.

SHELLAC Noun. (French, from *laque en ecailles* = lac in thin plates. Pl., Shellacs.) Commercially cleaned lac made into flakes or sheets. See Lac.

SHELTER TUBE Earthen tubes constructed by termites on building foundations, tree trunks and other exposed habitats. ST serve as avenues for worker and soldier movement to food sources, to protect colony from predators and maintain humidity.

SHEPPARD, EDWARD (1816–1883) (Anon. 1883, Entomol. mon. Mag. 20: 118.)

SHERBORN, CHARLES DAVIS (1861–1942) (Norman 1942, Nature 150: 146–147; Woodward 1942, Proc. Linn. Soc. Lond. 154: 295–296; Norman 1944, *Squire, memories of Charles Davis Sherborn*. 202 pp. bibliogr. London.)

SHERMAN, FRANKLIN (1877–1947) (Bentley 1947, J. Econ. Ent. 40: 610–611.)

SHERMAN, JOHN DEMPSTER (1872–1960) (Uvarov 1961, Proc. R. ent. Soc. Lond. (C) 25: 50.)

SHERMAN, M J (–1962) (Anon. 1962, Bull. Soc. ent. Fr. 67: 45.)

SHESTAKOV, ANDREI VALENTINOVICH (1890–1933) (Gussakovsky 1935, Revue Russe Ent. 25: 324–328, bibliogr.)

SHIBAKAWA, MATANOSUKA (1888–1916) (Hasegawa 1967, Kontyû 35 (Suppl.): 43.)

SHIBAYAMA, NAOKIYO (1855–1914) (Hasegawa 1967, Kontyû 35 (Suppl.): 43.)

SHIBUYA, J (Remington 1946, Ann. ent. Soc. Am. 39: 449.)

SHIELD See Shield Bugs.

SHIELD BEARERS See Heliozelidae.

SHIELD BEETLES See Chrysomelidae.

SHIELD BUGS See Pentatomidae.

SHIELD HUNTSMAN SPIDERS *Neosparassus* spp. [Araneida: Heteropodidae].

SHIELD-BACKED PINE-SEED BUG *Tetyra bipunctata* (Herrich-Schäffer) [Hemiptera: Pentatomidae].

SHIMADA, GURO (1882–1960) (Hasegawa 1967, Kontyû 35 (Suppl.): 43.)

SHIMER, HENRY (1828–1895) (Anon. 1895, Ent. News 6: 240; Anon. 1895, Ent. News 6: 305–306.)

SHIN Noun. (Middle English, *shine*. Pl., Shins.) An archane or colloquial term for the insect Tibia. See Leg. Cf. Tibia.

SHINING FLOWER BEETLES See Phalacridae.

SHINING FUNGUS BEETLES See Staphylinidae.

SHINING SLAVE-MAKER *Polyerges lucidus* (Wheeler) [Hymenoptera: Formicidae]: A slave-making ant. Syn. Shining Amazon.

SHINJI, ORIHEI (1885–1951) (Hasegawa 1967, Kontyû 35 (Suppl.): 44.)

SHINY PASTURE SCARAB *Scitala sericans* Erichson [Coleoptera: Scarabaeidae].

SHINY SPIDER BEETLE *Mezium affine* Boieldieu [Coleoptera: Anobiidae].

SHIP COCKROACH See American Cockroach.

SHIP TIMBER BEETLES See Lymexylidae.

SHIPLEY, ARTHUR EVERETT (1861–1927) (L.O. 1915, Pop. Sci. Mon. 87: 69; Harner 1928, Proc. Linn. Soc. Lond. 1927–28: 130–137.)

SHIPP, JOHN WILLIAM (–1898) (Anon. 1898, Insektenbörse 15: 85; Anon. 1898, Entomologist 31: 100.)

SHIRAIWA, HIDEO (1897–1946) (Hasegawa 1967, Kontyû 35 (Suppl.): 44.)

SHOCH, GUSTAV (1833–1899) (Anon. 1899, Insektenbörse 16: 61, bibliogr.)

SHOEMAKER, ERNEST (Gates-Clarke 1959, Ent. News 70: 220.)

SHOP TYPHUS See Murine Typhus.

SHORE BUGS See Saldidae.

SHORE FLIES See Ephydridae.

SHOREY, HARRY H (1931–1998) American academic (University of California, Riverside.) Specialist in Insect Behaviour and pioneer in control of insect pests with pheromones.

SHORT HEEL FLIES See Sphaeroceridae.

SHORT-FACED SCORPIONFLIES See Panorpodidae.

SHORT SECTOR Odonata: The Media *sensu* Comstock.

SHORT-HORNED GRASSHOPPER Common name for members of Acrididae, a numerically large Family that includes most of the so-called grasshoppers and locusts. They are plant feeders and many Species can be economically important. This Family includes the migratory or plague locusts. See Plague Locust; Migratory Locust; Saltatoria.

SHORT-LEAF PINE-CONE BORER *Eucosma cocana* Kearfott [Lepidoptera: Tortricidae].

SHORT-LEGGED HARVESTMEN [Opiliones: Triaenonychidae].

SHORT-NOSED CATTLE LOUSE *Haematopinus eurysternus* (Nitzsch) [Anoplura: Haematopinidae].

SHORT-TAILED CRICKETS See Gryllidae.

SHORT-TOUNGED BEES Members of the Andrenidae, Colletidae and Halictidae. See Bees. Cf. Long-Tongued Bees.

SHORT-WINGED MOLE CRICKET *Scapteriscus abbreviatus* Scudder [Orthoptera: Gryllotalpidae]: A Species endemic to Brazil, Paraguay and Argentina and adventive to southeastern USA and islands of the Caribbean. Adult colour pattern mottled; Pronotum elongate; Tegmen covering ca 0.3 of Abdomen; hindwing vestigial; fore Tibia with 2 convergent, apical dactyls; hind Femur shorter than Pronotum; males produce courtship song, but do not call. See Gryllotalpidae; Mole Crickets. Cf. Changa Mole Cricket; Imitator Mole Cricket; Tawny Mole Cricket.

SHOT HOLE Pathology: A symptom of disease in which small diseased fragments of leaf fall out and small holes appear on the leaf surface.

SHOTHOLE BORER *Scolytus rugulosus* (Müller) [Coleoptera: Scolytidae]: A multivoltine pest of fruit and ornamental trees, particularly apple cherry, peach and plum; endemic to Europe and adventive to North America. Name derived from the small holes gnawed by adult in bark when entering and leaving tree. Female constructs an egg gallery under bark and parallel with grain; larvae construct parallel burrows which extend across the grain of the wood. Larva feeds on sapwood ca 30 days; beads of gum exude from holes on peach tree. Syn. Fruit-Tree Bark Beetle. See Scolytidae. Cf. Fir Engraver; Hickory Bark Beetle; Peach Bark Beetle; Smaller European-Elm Bark Beetle. Rel. Engravers.

SHOULDER Noun (Middle English, *shulder*. Pl., Shoulders.) 1. A term loosely applied to any obtuse angulation. 2. Generally: The humeral angle of the forewing or Elytron. 3. Lepidoptera: Anterior angles of the Thorax. 4. Heteroptera: Angles of the Prothorax. 5. Orthoptera: Lateral angles of Metazona of the Pronotum.

SHOULDERED BROWN BUTTERFLY *Heteronympha penelope* Waterhouse [Lepidoptera: Nymphalidae]

SHOUMATOFF, N (Brown & Heinemann 1972, *Jamaica and its butterflies*. xv + 478 pp. (20–21), London.)

SHOVEL Noun. (Middle English, *shovele*. Pl., Shovels.) Mayflies: The expanded, flattened leg joints of burrowing nymphs (Klots).

SHROUD Noun (Middle English, *schroud*.) Cf. Cover; Envelope; Mantle.

SHRUB Noun. (Middle English, *shrubbe* > Anglo Saxon, *scrybb*. Pl., Shrubs.) A low-standing woody plant with many stems. See Bush. Cf. Tree. Rel. Plant.

SHUCKARD, WILLIAM EDWARD (1803–1868) (Bates 1868, Proc. ent. Soc. Lond. 1868: lvi-lvii; Newman 1869, Entomologist 4: 180–182.)

SHUGUROV, ALEXANDER MIKAILOVICH (1881–1912) (Burr 1912, Entomologist's Rec. J. Var. 24: 278–279; Laister 1912, Revue Russe Ent. 12: 624–627, bibliogr.)

SHWANVICH (SCHWANWITSCH), BORIS NIKOLAEVICH (1889–1957) (Fedorov & Beklemishev 1958, Zool. Zh. 37: 1422–1425, bibliogr.)

SIACOURT-C® See Chlormequat.

SIALIDAE Plural Noun. Alderflies. A numerically small, widespread Family of Megaloptera (Neuroptera). Adult body ca 10–25 mm long; anocellate; fourth tarsal segment bilobed. Larva aquatic with seven pairs of Abdominal Gills but lack hooked Anal Prolegs; predaceous on aquatic insects and typically reside under rocks in streams.

SIALISTERIUM Noun. (Greek, *sialo* = combining form for saliva; Latin, *teria* = combining form for a place. Pl., Sialisteria.) A Salivary Gland.

SIAMESE GRAIN BEETLE *Lophocateres pusillus* (Klug) [Coleoptera: Lophocateridae]: A pest of stored products in tropical regions. Adult 2.5–3.0 mm long, flattened.

SIB MATING A reproductive strategy in which males copulate with sibling sisters of the same brood. Common in parasitic Hymenoptera and some other insects. Cf. Adelphogamy; Polyandry.

SIBLING SPECIES Species which are morphologically indistinguishable and reproductively incompatible. Regarded by some workers as indicative of recent speciation. Cf. Morphological Species.

SICH, ALFRED (–1943) (Turner 1943, Entomologist's Rec. J. Var. 55: 70; Cockayne 1949, Proc. R. ent. Soc. Lond. (C) 8: 70.)

SICHEL, JULES (1802–1868) (Mulsant 1869, Opusc. ent. 14: 73–102, bibliogr.)

SICHIROLLO, GIUSEPPE (Anon. 1935, Boll. Soc. ent. ital. 67: 49.)

SICUT ANTE (Latin.) As before.

SIDE PIECE Genitalia of male mosquitoes: The main lateral part of the clasping organ or basal part of the clasp.

SIDEBOTHAM, JOSEPH (1823–1885) (Anon. 1885, Entomol. mon. Mag. 22: 46; McLachlan 1885, Proc. ent. Soc. Lond. 1885: xlii.)

SIEBERT, CHRISTIAN (1859–1926) (Rohlfien 1975, Beitr. Ent. 25: 277.)

SIEBKE, JOHAN HEINRICH SPALCKHAWER (1816–1875) (Schneider 1876, Eneum. Insect. Norvegicorum 3: vii–x; Natvig 1944, Norsk. ent. Tidssk. 7: 22–25, bibliogr.)

SIEBOLD, CARL THEODOR ERNST VON (1804–1885) (Anon. 1885, Entomol. mon. Mag. 21: 280; Ehlers 1885, Z. wiss. Zool. 42: i–xxxiv, bibliogr.; McLachlan 1885, Proc. ent. Soc. Lond. 1885: xl–xli.)

SIEBOLD, PHILIPP FRANZ (1796–1866) (Anon. 1866, Zool. Gart. 7: 435; Martius 1867, Sber. bayer. Akad. Wiss. 1: 387–388.)

SIEBOLD'S ORGAN A chordontal organ on the fore Tibia of Locustidae which consists of a series of subintegumental Scolopophores.

SIEGE® See Cypermethrin.

SIENNA Adj. (Italian, *terra de senua* from Sienna.) Brownish orange.

SIEROLOMORPHIDAE Krombein 1951. Plural Noun. A small Family of aculeate Hymenoptera assigned to Superfamily Scolioidea, known from North America, Hawaii and Palaearctic.

SIEVE PLATE A porous cuticular sclerite supported by Trabeculae which covers respiratory Spiracles. Serve to filter air and maintain water balance. See Spiracle; Cf. Atrium. Rel. Tracheal System.

SIEWERS, CHARLES GODFREY (1815–1882) (Drury 1882, Can. Ent. 14: 176.)

SIGLURE® See Trimedlure.

SIGMA Noun. (Greek, *sigma*.) Diptera: The Furca *sensu* MacGillivray.

SIGMOID Adj. (Greek *sigma; eidos* = form.) Shaped as the Greek letter sigma, or English 'S.' Alt. Sigmoidal. Rel. Eidos; Form; Shape.

SIGMOID FUNGUS-BEETLE *Cryptophagus varus* Woodroffe & Coombs [Coleoptera: Cryptophagidae].

SIGMOIDEA The Parapteron, Dens, Humerus, First Axillary of the posterior wings; the Scutellaire or anterior Axillary; in the anterior wings, the grand humeral (MacGillivray).

SIGNAL WORD Noun. A specific designation required on every pesticide label to denote the relative toxicity of the material. Category I pesticides are designated with the signal words of Danger, Poison. Category II pesticides are designated as Warning, Category III and IV are designated as Caution.

SIGNATE Adj. (Latin, *signum* = a sign; *-atus* = adjectival suffix) Marked with signatures or lines resembling letters or characters.

SIGNATURAE Noun. (Pl., Signatures.) Surface markings with a resemblance to letters or characters.

SIGNIPHORIDAE. Plural Noun. (Syn. Thysanidae.) A small (ca 80 Species), cosmopolitan Family of parasitic Hymenoptera assigned to the Chalcidoidea. Adult small to minute size, dorsoventrally compressed; Antenna with 4–7 segments including 1–4 transverse annelli and long, unsegmented club; Parapsidal Sutures absent, but internal weak ridges present; Scutellum strongly transverse ('ribbon-like'); wings macropterous, not brachypterous or apterous; forewing Submarginal and Marginal Veins long, subequal, Remigial Setae sparse or absent, long fringe; tarsal formula 5-5-5; middle Tibia and tibial spur conspicuously spinose; propodeal triangle along anteriomedial margin; Petiole short; Gaster broadly joined to Propodeum. Primary and secondary parasites of whiteflies, scale insects, Diptera puparia and the primary parasites which attack these insects.

SIGNORET, VICTOR ANTOINE (1816–1889) (Distant 1889, Entomol. mon. Mag. 25: 309; Fairmaire 1889, Ann. Soc. ent. Fr. (6) 9: 505–512, bibliogr.)

SIGNUM BURSAE Lepidoptera: One or more heavily sclerotized and often elaborate structures in the wall of the Bursa Copulatrix of the female.

SIGWART, HERMANN FISCHER See FISCHER SIGWART.

SIKORA, FRANZ (–1902) (Daniel 1903, Münch. koleopt. Z. 1: 261–262.)

SIKORA, HILDA (Weidner 1967, Abh. Verh. naturw. Ver. Hamburg Suppl. 9: 344–347.)

SILACEOUS Adj. (Latin, *sil* = yellow ochre; *-aceus* = of or pertaining to.) Of the colour of yellow ochre. Alt. Silaceus.

SILAFLUOFEN Noun. A silane compound {(4-ethoxyphenyl)(3-4(4-fluoro-3-phenoxyphenyl)-propyldimethyl)silane} used as a contact insecticide and stomach poison to control termites, public-health insects and rice pests in some countries. Not registered for use in USA. Trade names include HOE 084498® and Silonen®.

SILANTYEV, ANATOLII ALEXSEEVICH (1868–1918) (Bogdanov-Katkov 1918, Denkschr. Nicil. Versuchsstat 1918: 7–34, bibliogr.)

SILATIN® See Cyhexatin.

SILBERMAN, GUSTAVE Editor and publisher of Revue Entomologique.

SILBERMANN, GUSTAV HENRI RODOLPHE (1801–1877) (Fauvel 1877, Annu. Ent. 5: 127–128; Baruthio 1924, Bull. Soc. Hort. BasRein. 1924: 5–7.)

SILBERNAGEL, A (1892–1944) (Paclt 1945, Cas. csl. Spol. ent. 42: 145–147, bibliogr.)

SILENT HOST Medical Entomology: A vertebrate host which is infected with a pathogen but shows no sign of disease. Cf. Amplifying Host; Resistant Host; Dead-End Host; Susceptible Host.

SILIBATE-PBW® See Gossyplure.

SILICA AEROGEL An inorganic insecticide used against cockroaches and stored product pests. Finely ground silica formulations which act as physical abrasive, absorptive dusts which kill by dehydration. See Inorganic Insecticides.

SILICA COMPOUNDS Various inorganic compounds used as insecticides. Trade-names include: Dri-Die®, Santocel C® and Silica Aerogel® and Silikil®. See Inorganic Insecticide.

SILIKIL® See Silica Compounds.

SILK Noun. (Anglo Saxon, *seoloc* = silk; from Latin, *sericus* = fabric of the *Seres*, Chinese. Pl., Silks.) A tough, lustrous fibre composed of two proteins: Fibroin (70–75%) tough, insoluble, elastic; and Sericin (20–25% amorphous). Fibroin forms the central core of the thread and is surrounded by Sericin. Silk is secreted in liquid form (Bave) which hardens into silk threads when exposed to the air. Functions of silk are diverse. Apterygote males spin a silken thread upon which to collect sperm. Many pterygote insects construct pupal cocoons of silk. Silk can be produced by larvae, naiads or adults. Immature caddisflies construct tubes of silk. Larval Neuroptera and moths construct cocoons for pupation. Weaver ant larvae produce silk used to tie arboreal nests together. Adult signiphorid wasps use silk to sequester parasitized whitefly hosts. See Gland. Cf. Wax.

SILK GLAND Ectodermal Glands modified to produce silk. Anatomically, SGs are called Labial Glands in many Lepidoptera larvae. SGs may consist of Malpighian Tubules (some Coleoptera and Hymenoptera), Dermal Glands (Homoptera), Anal Gland (some Neuroptera larvae), the fore Basitarsus of Embioptera nymphs and adults. SG anatomy has been most intensively studied in the silkworm, *Bombyx mori*. Histological studies show that the posterior silk gland produces silk proteins (Fibroin), the middle silk gland produce gelatinous silk proteins (Sericins) which coat the Fibroin and anterior silk glands are ducts lined with thick Cuticular Intima. See Exocrine Gland; Silk. Cf. Defence Gland; Pheromone Gland; Wax Gland. Rel. Endocrine Gland; Gland Cell.

SILK MOTHS See Bombycidae.

SILK PRESS The highly developed salivary syringe or ejection apparatus in the larvae of Lepidoptera (Snodgrass).

SILKEN FUNGUS-BEETLES See Cryptophagidae.

SILKWORM MOTHS See Bombycidae; See Choreutidae.

SILKWORM *Bombyx mori* (Linnaeus) [Lepidoptera: Bombycidae]: A domesticated moth endemic to China and transported to other countries for sericulture. Commercially reared in Orient for silk threads extracted from cocoon and used as a source of fabric. Egg stage one year (diapause) or 11–14 days (non-diapause). Five larval instars; monovoltine in Europe and Australia; multivoltine in tropical and subtropical countries. Cocoon development requires 2 days; cocoons white, yellow, pink, green or rusty, depending upon biotype. Adult brachypterous and cannot fly. Larval host plants in Moraceae: White Mulberry (*Morus alba* L), Red Mulberry (*Morus rubra* L), Black Mulberry (*M. nigra* L), Paper Mulberry (*Broussonetia papyrifera* (L.) Venten.) and Osage Orange *Maclura pomifera* (Raf.) C. K. Schneider (endemic to North America). See Bombycidae; Sericulture.

SILKY ANT *Formica fusca* Linnaeus [Hymenoptera: Formicidae]: A widespread north-temperate Species; inhabits forest, field, hills and mountains. Workers reddish to black in coloration, 2–4 mm long; construct 'masonry domes' (compact mounds of soil). Food includes honeydew and insects. See Formicidae. Cf. Allegheny Mound-Ant.

SILKY AZURE *Ogyris oroetes* Hewitson [Lepidoptera: Lycaenidae].

SILKY CANE-WEEVIL *Metamasius hemipterus sericerus* (Oliver) [Coleoptera: Curculionidae].

SILKY LACEWINGS See Psychopsidae.

SILKY OAK LEAFMINER *Peraglyphis atimana* (Meyrick) [Lepidoptera: Tortricidae].

SILLIMAM, BENJAMIN (1816–1885) (Anon. 1885, Atheneum No. 2987: 124–125.)

SILONEN® See Silafluofen.

SILOSAN® See Pirimiphos-Methyl.

SILPHIDAE Plural Noun. Carrion Beetles; Burying Beetles; Sexton Beetles. A cosmopolitan Family of about 600 Species of polyphagous Coleoptera; poorly represented in tropics, assigned to Staphylinoidea and closely related to Staphylinidae. Body to 45 mm long, flattened; Antenna short with 11 segments, loosely clavate; segment preceding club concave, cup-like. Fore Coxa prominent; Elytra longitudinally ribbed, nearly reaching apex of Abdomen; tarsal formula 5-5-5, all segments setose ventrally. Abdominal Terga exposed or concealed by Elytra. Species typically associated with carrion. Predatory Species attack snails and muscoid fly larvae; members of *Xylodrepa* arboreal, attack Lepidoptera and Symphyta larvae; a few Species phytophagous.

SILRIFOS® See Chlorpyrifos.

SILT Noun. (Middle English, *sylt*. Pl., Silts.) 1. Unconsolidated sedimentary material whose constitutent elements are less than 0.16 mm in diameter. 2. A sediment of fine mud or earth deposited as by a river.

SILURIAN PERIOD The third Period (439–409 MYBP) of Palaeozoic Era and named for *Silures* (a Celtic tribe) from the part of Wales where SP was first investigated. SP shows an adaptive radiation of lime-secreting organisms in shallow seas, seaweed and reef-forming corals. Marine invertebrate diversification includes coelenterates, echinoderms, bryozoans, brachiopods, molluscs (including ammonoids), arthropods (including eurypterids); aquatic vertebrates diversify in fresh water (ostracoderms, acanthodians and placoderms). First indications of terrestrial plants *(Baragwanathia)* found in Upper Silurian of Victoria, Australia and first vascular land plant *(Cooksonial)*. Oldest terrestrial (or periodically terrestrial) arthropod may be *Necrogrammarus* (originally put in Crustacea) possibly a millipede. See Geological Time Scale; Palaeozoic Era. Cf. Cambrian Period; Devonian Period; Lower Carboniferous Period; Upper Carboniferous Period; Ordovician Period; Permian Period. Rel. Fossil.

SILVANIDAE Plural Noun. See Cucujiidae.

SILVANOL See Lindane.

SILVEIRA CALDEIRA, JOAO DA (1800–1854) (Papavero 1971, *Essays on the History of Neotropical Dipterology*. 1: 144–147. São Paulo.)

SILVER DEWDROP SPIDER *Argyrodes argentiopunctata* Rainbow [Araneida: Theridiidae].

SILVERBIRCH BRANCHCUTTER *Strongylurus cretifer* Hope [Coleoptera: Cerambycidae].

SILVERED SKIPPER *Hesperilla crypsargyra* (Meyrick) [Lepidoptera: Hesperiidae].

SILVERFISH 1. *Acrotelsa collaris* (Fabricius) [Thysanura: Lepismatidae]: A widespread pest of paper, starch, and fabrics in urban habitats in tropical regions. Adult body with scales on dorsum predominantly black, 16–18 mm long; compound eyes present; large Setae with barbs and arranged into combs; Maxillary Palpus with five segments; apterous; abdominal Tergum 10 triangular, fringed with bristle combs; Cerci and Appendix Dorsalis longer than Abdomen. 2. *Ctenolepisma lineata* (Fabricius) [Thysanura: Lepismatidae]: A pest of paper, starch, and fabrics in urban habitats in eastern Australia. Adult dorsal part of body pale to grey with longitudinal band; compound eyes present; Setae with barbs; Maxillary Palpus with five segments; Prosternum lacking tuft of Setae; apterous; abdominal Terga 2–6 each with two pairs of setal combs; Cerci and Appendix Dorsalis longer than Abdomen. 3. *Ctenolepisma longicaudata* Escherich [Thysanura: Lepismatidae]: A common pest in eastern Australia which attacks paper, starch and fabrics in urban habitats and potentially a serious pest in libraries. Adult ca 15 mm long; silver-grey, not banded; compound eyes present; Setae with barbs; Maxillary Palpus with five segments; Labial Palpus with numerous sensory structures; Prosternum lacking tuft of Setae; apterous; abdominal Terga 2–6 each with three pairs of setal combs; Cerci and Appendix Dorsalis longer than Abdomen. Female oviposits in crevices during summer, clutches of 2–20 eggs; ca 50 eggs per year. Eggs ovoid, 1 mm long, first cream coloured then darken; eclosion occurs within 3–7 weeks. Nymph undergoes about 15 moults; development 2–3 years. Adult continues to moult; lives ca four years; can survive ca nine months without food. Syn. *Ctenolepisma urbana* Slabaugh. See Thysanura; Common Silverfish. Cf. Firebrat.

SILVERLEAF WHITEFLY *Bemisia argentifolii* Bellows & Perring [Hemiptera: Aleyrodidae]: A whitefly Species morphologically very similar to sweetpotato whitefly; also called silverleaf whitefly (SLWF) 'type B.' First appeared in USA during 1980s; 1986–1992 SLWF replaced SPWF in southwestern USA. SLWF reproduces faster, reaches higher populations and has wider host range than SPWF. SLWF is a poor vector of lettuce infectious yellows virus (LIYV) but an effective vector for closterviruses and geminiviruses. See Aleyrodidae. Cf. Sweetpotato Whitefly.

SILVER-SPOTTED SKIPPER *Anisyntoides argenteoornatus* (Hewitson) [Lepidoptera: Hesperiidae]. *Epargyreus clarus* (Cramer) [Lepidoptera: Hesperiidae].

SILVER-SPOTTED TIGER MOTH *Lophocampa argentata* (Packard) [Lepidoptera: Arctiidae].

SILVERLOCK, OSCAR C (–1911) (Morice 1911, Proc. ent. Soc. Lond. 191 1: cxxi–cxxii.)

SILVESTRI, FILIPPO (1873–1949) (Porter 1935, Revta. chil. Hist. nat. 39: 225–227, bibliogr.; Anon. 1937, Annuar. pontifica Accad. Sci. 1: 674–704, bibliogr.; Choû, 1949. J. Entomologia Sin. 5: 1–5; Russo 1949, Boll. Lab. ent. agr. Portici 9: iii–xlix, bibliogr.; Carpenter 1950, Proc. Linn. Soc. Lond. 162: 110–113; Wigglesworth 1950, Proc. R. ent. Soc. Lond. (C) 14: 65–66.)

SILVICOLIDAE See Anisopodidae.

SILVICOLOUS (Latin, *silva* = forest; *colere* = to inhabit.) Descriptive of organisms that inhabit forests or woodlands. See Habitat. Cf. Agricolous; Caespiticolous; Madicolous. Rel. Ecology.

SIM, ROBERT J (1881–1955) (Weiss 1955, J. N. Y. ent. Soc. 63: 166–169, bibliogr.)

SIMANTON, FRANK LESLIE (1875–1935) (W.A.S. 1936, J. Econ. Ent. 29: 224.)

SIMES, JAMES APPLEGATE (1874–1951) (Riley 1952, Proc. R. ent. Soc. Lond. (C) 16: 87–88.)

SIMMERMACHER, GEORG (1856–1885) (Anon. 1885, Leopoldina 21: 160; Dimmock 1888, Psyche 5: 36.)

SIMMONDS, HUBERT W (1877–1966) (Pemberton 1966, Proc. Hawaii. ent. Soc. 19: 317.)

SIMON, EUGENE (1848–1924) (Pickard 1924, Bull. Soc. ent. Fr. 1924: 193–194; Anon. 1925. Ent. News 36: 222–224; Berland 1925, Ann. Soc. ent. Fr. 94: 73–100, bibliogr.)

SIMONDETTI, MARIO (–1962) (Anon. 1962, Boll. Soc. ent. ital. 92: 33.)

SIMONOV, NIKOLAI PAVLOVICH (1881–1912) (Derzhavin 1912, Revue Russe Ent. 12: 628.)

SIMPLE Adj. (Latin, *simplex*.) Not complicated; not modified by any condition causing complexity. Ant. Complex.

SIMPLE ANTENNA A capitate Antenna of only one segment.

SIMPLE EYES The Ocelli. See Stemmata. Cf. Compound Eye.

SIMPLE GLAND See Gland; Unicellular Gland.

SIMPLE LATERAL EYES Some adult insects: Single or grouped eyes or true Ocelli (not Ommatidia) placed at the sides of the head (Snodgrass).

SIMPLE OCELLUS An ocellate spot with only the parts termed iris and pupal.

SIMPSON, CHARLES B (1876–1907) (Anon. 1907, Ent. News 18: 112; Horn 1907, Dt. Ent. Z. 1907: 348.)

SIMSON, AUGUSTUS (1836–1919) (Chapman 1919, Proc. Linn. Soc. N. S. W. 44: 18.)

SIMULIIDAE Plural Noun. Black Flies; Buffalo Flies. A cosmopolitan Family of nematocerous Diptera containing ca 1,000 described Species. Adult small bodied (<5 mm long), typically dark coloured; female Mandibles and Maxillae form elongate, serrated, piercing Stylets; male mouthparts reduced; Antenna with 9–12 segments, typically 11 segments; female's compound eyes separated medially; male's compound eyes typically contiguous above antennal toruli; Ocelli absent; Scutum prominent and forming a hump; wing broad, iridescent with veins in anterior area well developed and large anal lobe (Alula). Adult BF males form large, cloud-like swarms near water; mating occurs when females fly into or near swarms; some BF Species may migrate more than 200 km from breeding habitat. Adult male and female daytime feeders, may take nectar; females of most Species require blood meal for egg development. BF frequently pests of vertebrates near water, particularly in northern temperate regions; many *Simulium* Species are pests of domestic livestock or man. Female lays 200–500 eggs on water surface, aquatic vegetation, or along margin of water. Univoltine Species overwinter in egg stage; eggs of multivoltine Species typically hatch within 3–7 days. Larvae aquatic, most abundant in swift-flowing well oxygenated streams; some Species in rivers and a few in temporary ponds; larvae attach via caudal sucker and silken threads to rocks and objects submerged in water. Larvae are filter feeders upon algae, bacteria, protozoa, small Crustacea and decaying organic matter; larval development temperature-dependent, ca 7–12 days at higher water temperatures; most Species with 6–7 larval instars. Pupae aquatic with respiratory filaments on thoracic dorsum; pupae occupy silken cocoon; development requires 2–30 days depending upon temperature and conditions. BF of significant medical concern as vectors of Onchocerciasis to humans. Syn. Melusinidae. Cf. Ceratopogonidae.

SINCIPUT Noun. Coleoptera: Part of the Vertex between the eyes. In general, the Frons or front.

SINDBIS Noun. (Latin, *sinciput* = half a head.) An arboviral disease in Egypt, Uganda, South Africa, Asia and Australia; assigned to *Alphavirus* of Tongiviridae. Disease non-fatal, characterized by low-grade fever and rash. Vectors include *Culex* spp. See Arbovirus. Cf. Chikungunya; O'Nyong-Nyong; Ross River Fever.

SINE TIPO Latin phrase meaning 'without type.' A Species which does not have a name bearer (Holotype, Neotype).

SINGAPORE ANT *Monomorium destructor* (Jerdon) [Hymenoptera: Formicidae]: A tramp Species, probably endemic to India and adventive to tropical Australia. Worker body ca 2–3 mm long; head, Alitrunk, Petiole and Postpetiole uniformly glossy, varying from light yellow to dull brownish-yellow, Gaster dark brown to blackish brown, usually with conspicuous yellowish area mediobasally, but sometimes uniformly dark; head flattened and subquadrate, occipital margin with 2–4 pairs of Setae near midline, pubescence sparse and directed towards midline; Mandible with three strong teeth and a minute basal denticle, usually with distinct longitudinal rugulose or striate sculpture, rarely virtually smooth in smallest workers; eyes very small, with 4–6 Ommatidia in longest row; antennal Scape extending to occipital margin in smallest workers, but shorter in larger individuals; club absent; petiolar node in dorsal view globular to subglobular, not distinctly compressed; propodeal dorsum always finely transversely striolate to rugulose, usually with punctations; first gastral Tergum smooth except setigerous pits. SA an urban pest of hides, skins, book bindings and insect collections but also can attack fabrics, rubber goods and polyethylene. Typically nests in and near buildings and can establish a colony in very small spaces. SA also is an agricultural pest, particularly in developing countries; infests wheat, sorghum, tobacco seeds, sesame and will attack honey bee hives causing bees to abscond. SA can harbour pathogenic bacteria (the plague bacillus) in its gut. See Formicidae.

SINGH, SARDAR MOHAN (1910–1952) (M.V.V. 1953. Indian J. Ent. 15: 77–78.)

SINGH, SHRI R N (1899–1948) (Lal 1948, Indian J. Ent. 10: 289–291.)

SINGHA, PRAKASH CHAND (1924–1955) (Hall 1957, Proc. R. ent. Soc. Lond. (C) 21: 66.)

SINGLE RELEASE In Augmentative Biological Control, a Species (biotype, race, strain, variety) imported and released into an area only once. See Augmentative Biological Control. Cf. Multiple Release.

SINISTRAD Adv. (Latin, *sinister* = left; *ad* = toward.) Toward the left side. See Orientation. Cf. Anteriad; Apicad; Basad; Caudad; Centrad; Cephalad; Craniad; Dextrad; Dextrocaudad; Dextrocephalad; Distad; Dorsad; Ectad; Entad; Laterad; Mediad; Mesad; Neurad; Orad; Proximad; Rostrad; Sinistrocaudad; Sinistrocephalad; Ventrad.

SINISTRAL Adj. (Latin, *sinister* = left; *-alis* = characterized by.) Pertaining to or positioned on the left side of the body. See Orientation. Cf. Dextral.

SINISTRALITY Noun. (Latin, *sinistra* = left handed. Pl., Sinistralities.) 1. The condition or phenomenon of 'handedness' in which the left side of a appendage, body or structure differs from the right side in appearance (size, shape) or functionality (use, disuse). 2. Left handedness; the condition in which the left hand is used in preference over the right hand. See Orientation. Cf. Dextrality.

SINISTROCAUDAD Adv. (Latin, *sinister* = left; *cauda* = tail; *ad* = toward.) Extending obliquely from the left toward the tail. See Orientation. Cf. Anteriad; Apicad; Basad; Caudad; Centrad; Cephalad; Craniad; Dextrad; Dextrocaudad; Dextrocephalad; Distad; Dorsad; Ectad; Entad; Laterad; Mediad; Mesad; Neurad; Orad; Proximad; Rostrad; Sinistrad; Sinistrocephalad; Ventrad.

SINISTROCEPHALAD Adv. (Latin, *sinister* = left; Greek, *kephale* = head; Latin, *ad* = toward.) Extending obliquely from the left toward the head. See Orientation Cf. Anteriad; Apicad; Basad; Caudad; Centrad; Cephalad; Craniad; Dextrad; Dextrocaudad; Dextrocephalad; Distad; Dorsad; Ectad; Entad; Laterad; Mediad; Mesad; Neurad; Orad; Proximad; Rostrad; Sinistrad; Sinistrocaudad; Ventrad.

SINISTRON Noun. (Latin, *sinister* = left.) The left side of the insect body. Cf. Dextron.

SINK Noun. (Anglo Saxon, *sincan*. Pl., Sinks.) A shallow depression in an otherwise flat surface. Syn. Basin. Verb. Taxonomic jargon meaning the identification of synonymy between two Taxa with the suppression of one of the names.

SINORATOX® See Dimethoate.

SINTENIS, FRANZ (1836–1911) (Djakonov 1911, Revue Russe Ent. 11: 310–313, bibliogr.; Lackschewitz 1937, Sber. naturf. Ges. Dorpat 43: xlvii–lv, bibliogr.)

SINTENIS, PAUL ERNEST (1847–) (Papavero 1975, *Essays on the History of Neotropical Dipterology*. 2: 306. São Paulo.)

SINTON, JOHN ALEXANDER (1884–1956)

(Christophers 1956, Biog. Mem. Fellows R. Soc. Lond. 2: 269–290, bibliogr.)

SINUATE Adj. (Latin, *sinuatus*, pp of *sinuare* = to bend.) Descriptive of structures which are long and narrow with a winding or wavy form. Term applied to edges or margins that are wavy. Alt. Sinuated. Sinuatus. See Shape.

SINUATE LADY-BEETLE *Hippodamia sinuata* Mulsant [Coleoptera: Coccindellidae].

SINUATE PEAR-TREE BORER *Agrilus sinuatus* (Oliver) [Coleoptera: Buprestidae]: A pest of pear, cotoneaster, hawthorn and mountain ash in northeastern USA; endemic to Europe and adventive to North America ca 1894. SPTB completes one generation each two years. Overwinters as larva within feeding burrows in tree; exhibits two overwintering-larval sizes: small larvae (ca 12 mm) along inner bark and larger larvae (ca 36 mm) within cells in sapwood of tree. Large larvae pupate in early spring; small larvae resume feeding during spring. Adults emerge during May-June, feed on foliage and oviposit in cracks or fissures in bark. See Buprestidae. Cf. Bronze Birch Borer; Pacific Flathead Borer; Roundheaded Apple-Tree Borer.

SINUATOCONVEX Adj. (Latin, *sinuatus*, + convex.) Sinuate and convex. See Shape.

SINUATOLOBATE Adj. (Latin, *sinuatus*, + lobate.) Sinuate and lobed. See Shape.

SINUATOTRUNCATE Adj. (Latin, *sinuatus*, + truncate.) Truncated with the margin sinuate. See Shape.

SINUATO-UNDULATE Adj. (Latin, *sinuatus*, + undulate.) With linear markings or structures with obtuse sinuses. See Shape.

SINUOUS Adj. (Latin, *sinus* = curve; *-osus* = possessing the qualities of.) Pertaining to structure or outline shape which undulates; curved in and out. See Shape.

SINUS Noun. (Latin, *sinus* = curve. Pl., Sinuses.) A more-or-less profound curvilinear indentation. An excavation as if scooped out, a curved break in a otherwise straight margin. See Shape.

SIPERIN® See Cypermethrin.

SIPHLONURIDAE Plural Noun. A numerically large, geographically widespread Family of baetoid Ephemeroptera. One Species in Australia whose nymphs are algal scrapers. Adults with compound eyes large and often two-parted; resemble Heptageniidae in wing venation. Nymphs found in streams and rivers; some Species predaceous.

SIPHON Noun. (Greek, *siphon* = reed, tube. Pl., Siphons.) 1. A tube-like mouthpart in some insects. 2. The respiratory tube of a culicid larva. 3. Any tubular external process or stucture. Alt. Syphon.

SIPHONAPTERA Latreille, 1825. (Lower Cretaceous Period-Recent). (Greek, *siphon* = tube; *aptera* = wingless.) Fleas. An Order of about 2,400 Species of holometabolous, endopterygote Neoptera, divided into Pulicoidea (2 Families) and Ceratophylloidea (15 Families). Adult body typically less than 6 mm long, laterally compressed; compound eyes absent or replaced by one large Ocellus on either side of head; piercing-sucking mouthparts; Mandibles absent; Maxillary and Labial Palpi well developed; Antennae short, held in fossa. Apterous; Thorax with rearward projecting cuticular spines and bristles; legs spinose, with large Coxae, adapted for jumping. Abdomen 10-segmented with Tergum I reduced and Sternum I absent; Sensillum positioned at apex of Abdomen. Female with panoistic Ovary. Larva 13-segmented; apodous, vermiform with three instars; head capsule sclerotized; eyeless; Antenna 1-segmented; terminal segment with Anal Struts. Pupa adecticous and exarate; pupal wing buds in some Species; cocoon constructed. Adults of all Species external parasites of mammals and birds; larvae live in nests of hosts. Some vector diseases such as plague; some serve as intermediate hosts of tapeworm. Included Families: Ancistropsyllidae, Ceratophyllidae, Chimaeropsyllidae, Coptopsyllidae, Hystrichopsyllidae, Ischnopsyllidae, Leptopsyllidae, Malacopsyllidae, Macropsyllidae, Pulicidae, Pygiopsyllidae, Rhopalopsyllidae, Stephanocircidae, Tungidae, Vermipsyllidae, Xyphiopsyllidae. See Aphaniptera. Syn. Rhophoteira.

SIPHONATA Noun. Hemiptera; specifically, plant lice and leaf hoppers.

SIPHONELLOPSIDAE Plural Noun. See Chloropidae.

SIPHONETS (Greek, *siphon* = reed, tube. Sl., Siphonulus.) See Honey Tubes; Cornicles. Alt. Siphonuli.

SIPHONOPHORA Coccinellidae: A name preoccupied in the Coelenterates.

SIPHUNCLE Noun. (Latin, *siphunculus* = little tube. Pl., Siphuncles; Siphunculi.) A tubular wax-secreting structure on the fifth abdominal segment of Aphididae. Typically paired, occasionally reduced, rarely more than one pair and erroneously believed to secrete honeydew. Cf. Cornicle.

SIPHUNCULATA The sucking lice.

SIPHUNCULATE Adj. (Greek, *siphunculus* = little tube; *-atus* = adjectival suffix.) Having or possessing a siphon or tube.

SIPHUNCULINA *Siphunculina funicola* [Diptera: Chloropidae], a chloropid fly in southeast Asia which annoys domestic animals and humans and serves as mechanical vector for numerous pathogens. Adult attracted to blood from wounds, serous discharge from eyes and wounds and excrement. Plays a role similar to *Hippelates* in New World. See Chloropidae. Cf. Eye Gnats.

SIPHUNCULUS Noun. (Greek, *siphunculus* = little tube. Pl., Siphunculi.) 1. Siphuncle, the instrument of suction in sucking insects. 2. Aphididae: A Cornicle. 3. Acarology: A paired structure that contains the ducts of the salivary glands.

SIRENE® See Gossyplure.

SIREX PARASITE Any member of a complex of parasitic Hymenoptera which attack the sirex wood-inhabiting wasp: *Ibalia leucospoides* Hochenwarth [Hymenoptera: Ibaliidae]; *Megarhyssa nortoni* (Cresson), *Rhyssa persuasoria* (Linnaeus) [Hymenoptera: Ichneumonidae]; *Schlettererius cinctipes* (Cresson) [Hymenoptera: Stephanidae].

SIREX WASP *Sirex noctilio* Fabricius [Hymenoptera: Siricidae]: Endemic to Europe; adventive to Australia and New Zealand as a pest of Monterrey Pine *(Pinus radiata),* other *Pinus* spp., larches and spruce. Female wasp drills oviposition shafts into trunk of tree; toxic accessory gland secretion and fungus (*Amylostereum areolatum*) introduced with wasp egg. Fungus metabolites can kill pine tree; fungi consumed by larval wasps, taken into Mycangia of neonate female wasps. Fungi transmitted to new tree by ovipositing wasp.

SIRICIDAE Plural Noun. Horntails; Woodwasps. A small Family of Symphyta endemic in temperate forests of northern hemisphere; introduced into South America, Australia and New Zealand. Adult 10–55 mm long; Antenna filiform, flattened, with 14–30 segments; head with Hypostomal (Postgental) Bridge; fore Tibia with one subapical spur; middle Tibia without subapical spur; Abdomen cylindrical with Tergum I medially divided, posterior margin of apical Tergum (female) or Sternum (male) projecting into a long spine; Ovipositor projecting well beyond apex of Abdomen; male genitalia orthandrous. Siricinae associated with Pinaceae; Tremicinae associated with angiosperms, particularly Aceraceae, Fagaceae and Ulmaceae. Many Species develop with symbiotic fungi.

SIRICOIDEA A Superfamily of Symphyta (Hymenoptera). Adult head with Postgenal Bridge; Mesonotum typically with transverse suture or Sulcus; Axillae well defined. Larva lack Stemmata; thoracic legs reduced; tarsal claws absent; abdominal Prolegs absent. Larvae associated with wood, symbiotic with fungi; some Species cannot develop without fungal hyphae as food. Adult females possess abdominal Mycangia which connect to base of Ovipositor via ducts. Mycangia contain arthrospores. 'Mucus glands' secretions injected into plant during oviposition. Female larvae possess hypopleural organs between first and second abdominal segments used to culture fungi. Several Families known from Upper Jurassic Period. Extant Families: Siricidae, Orussidae, Xiphydriidae, Syntexidae (Anaxyelidae).

SIRKKA, PAULI WILHELM (1912–1940) (Anon. 1940, Suomen hyönt. Aikak 6: 5, 7.)

SIRRINE, FRANK ATWOOD (1861–1959) (Chapman *et al.* 1960, J. Econ. Ent. 53: 700.)

SISLEY, CLAUDE (1883–1959) (Uvarov 1960, Proc. R. ent. Soc. Lond. (C) 24: 54.)

SISTAN® See Metham-Sodium.

SISTENS TYPE Aphididae: Nymphs of the third generation which remain undeveloped for a time (Comstock).

SISTENTES Noun. (Sl., Sistens.) Aphididae: Apterous progeny of Colonici and similar to them; Alienicolae.

SISTER GROUP In cladistical theory of classification, a Species or higher monophyletic taxon that is hypothesized to be the closest genealogical relative of a given Taxon exclusive of the ancestral Species of both Taxa. (Wiley 1981, *The Theory and Practice of Phylogenetic Systematics*, p. 7). See Monophyletic; Polyphyletic; Paraphyletic; Holophyletic.

SISYRIDAE Plural Noun. Spongillaflies. A numerically small, widespread Family of osmyloid Neuroptera. Antenna small, moniliform; wings not setose; forewing costal cross veins simple; one crossvein between R1 and Rs. Adults occur near water; eggs typically covered in silk and deposited near water; larvae aquatic predators of freshwater sponges.

SIT-AND-WAIT PREDATOR Ecology: A predator whose mode of foraging is such that time or energy is not expended in the search for food that is not used simultaneously for other activities (Schoener 1971).

SITKA-SPRUCE WEEVIL *Pissodes strobi* (Peck) [Coleoptera: Curculionidae]. Syn. Englemann Spruce Weevil; White Pine Weevil.

SITONA EGG-PARASITE *Anaphes diana* (Girault) [Hymenoptera: Mymaridae].

SITONA WEEVIL *Sitona discoideus* Gyllenhal [Coleoptera: Curculionidae].

SITONA WEEVIL PARASITE *Microctonus aethiopoides* Loan [Hymenoptera: Braconidae].

SITOPHORE Noun. (Greek, *sitos* = food; *pherein* = to carry. Pl., Sitophores.) The convex, dorsal surface of the Hypopharnyx, between the Suspensoria, which forms a widening, trough-like channel through which food is conducted to the mouth or Cibarium. See Mouthparts.

SITOWSKI, LUDWIK (1881–1947) (Lenke 1948, Polskie Pismo ent. 18: 5–13, bibliogr.)

SIX LINEBLUE *Nacaduba berenice berenice* (Herrich-Schaffer) [Lepidoptera: Lycaenidae].

SIX-EYED SPIDERS See Dysderidae.

SIX-SPINED IPS *Ips calligraphus* (Germar) [Coleoptera: Scolytidae]: A pest of pines widespread in eastern North America and extending south to Honduras. Adult black, Antenna and legs dark brown; head with medial tubercle; Pronotum with fine to coarse punctures; elytra with soarse strial punctures. SSI attacks green bark of injured trees, stumps and logs. See Scolytidae; Engraver. Cf. California Five-Spined Ips; Monterey-Pine Engraver; Pine Engraver.

SIX-SPOTTED MITE *Eotetranychus sexmaculatus* (Riley) [Acari: Tetranychidae].

SIX-SPOTTED SKIPPER *Hesperilla sexguttata* Herrich-Schaffer [Lepidoptera: Hesperiidae].

SIX-SPOTTED THRIPS *Scolothrips sexmaculatus*

(Pergande) [Thysanoptera: Thripidae]: A predaceous thrips used in biological control of spider mites. Female ca 0.85 mm long, elongate-cylindrical and clear pale yellow with faint dark spots on Thorax; Thorax has six pairs of long Setae. Females with four narrow wings with a long fringe of Setae along margins; forewing with two dark spots or bands; Tarsi lack claws but end with a small bladder; Abdomen has a blunt apex. Males smaller than females. Adult females use Ovipositor to saw into leaf tissue and deposit 4–5 eggs in slit. SST eggs laid in leaves or stems; egg white and kidney-shaped; egg stage 5–10 days. Nymphs resemble adults in appearance and habits. Nymphs lack wings and are paler than adult. Nymphs pass through four instars, two occur in soil and are non-feeding. All life stages occur on plant and overlap. SST is multivoltine with 5–8 generations each year. See Thripidae.

SIX-TOOTHED SPRUCE BARK BEETLE *Pityogenes chalcographus* [Coleoptera: Scolytidae].

SIXTH LONGITUDINAL VEIN Diptera: The first Anal Vein *sensu* Comstock. See Vein; Wing Venation.

SJÖBERG, OSCAR (–1959) (Lindroth 1959, Opusc. ent. 24: 184.)

SJÖSTEDT, BROR JNGVE (1866–1948) (Pasteels 1948, Bull. Ann. Soc. ent. Belg. 84: 221; Malaise 1952, Opusc. ent. 17: 103–104.)

SKALA, HUGO (1875–1953) (Anon. 1952, Z. wien. ent. Ges. 37: 56; Hoffman 1953, Z. wien. ent. Ges. 38: 147–149, bibliogr.)

SKALITZKY, KARL (1841–1914) (Reitter 1914, Wien. ent. Ztg. 33: 210.)

SKARRATT, RONALD JAMES MARTIN (–1965) (Anon. 1965. Nth Gloucester nat. Soc. J. 16: 232–233; Pearson 1966, Proc. R. ent. Soc. Lond. (C) 30: 63.)

SKEETAL®. A registered biopesticide derived from *Bacillus thuringiensis* var. *israelensis*. See *Bacillus thuringiensis*.

SKELETAL Adj. (Greek, *skeletos* = hard; *-alis* = characterized by.) Descriptive of or pertaining to the skeleton.

SKELETAL ASYMMETRY The insect Cuticle is composed of a series of laminar planes. Oblique sections of insect Cuticle reveal parabolic patterns. The cuticle's plane of orientation is always left handed and the Cuticle is asymmetrical with respect to sides of the body. See Asymmetry; Symmetry. Cf. Behavioural Asymmetry; Developmental Asymmetry.

SKELETON Noun. (Greek, *skeletos* = dried, hard. Pl., Skeletons.) Arthropods: The hard chitinous or sclerotized portion of the Integument which forms the external protective covering of the body and which forms an internal surface for the attachment for muscles. See Body. Cf. Integument. Rel. Bone; Cartilage; Chitin.

SKELL, FRITZ (1885–1961) (Forster 1961, NachrBl. bayer. Ent. 10: 57–59, bibliogr.)

SKERTCHLY, SYDNEY BARBER JOSIAH (1850–1926) (Longman 1926, Qd. Nat. 5: 70–71.)

SKIN Noun. (Old Norse, *skinn* > Anglo Saxon, *scinn* = skin. Pl., Skins.) 1. A general term for the external covering of a structure, appendage, body or seed. 2. The tough but flexible integument of an animal. Cf. Hide; Pelt. 3. A membranous film on the surface of some fluids. Syn. Hide; Integument; Pelt; Rind.

SKIN BEETLES See Dermestidae.

SKIN MOTH *Monopsis laevigella* (D. & S.) [Lepidoptera: Tineidae]: An Holarctic pest of animal skins and animal food products.

SKINNER, HENRY (1861–1926) (Calvert 1926, Ent. News 37: 225–249, bibliogr.; Anon. 1927, Entomol. mon. Mag. 63: 17; Williams *et al.* 1927, Ann. ent. Soc. Am. 20: 140; Mallis 1971, *American Entomologists*. 549 pp. (322–332), New Brunswick.)

SKINNER, JAMES WILLIAM (1855–1931) (Davis 1932, Ann. ent. Soc. Am. 25: 251.)

SKIPJACK BEETLES See Elateridae.

SKIPPER *Trapezites sciron* Waterhouse & Lyell [Lepidoptera: Hesperiidae].

SKIPPER® See Thiodicarb.

SKIPPER FLIES See Piophilidae.

SKIPPERS Noun. 1. Dipterous larvae sometimes found in cheese and other food. See Cheese Skipper. 2. Common name for Hesperiidae. See Hesperiidae.

SKOTOTAXIS See Scototaxis.

SKULL Noun. (Middle English, *skulle* = cranium. Pl., Skulls.) The hard part of the head, typically enclosing the Brain. A term not appropriately applied to the insect head. See Head. Cf. Cranium.

SKUSE, FREDERICK A ASKEW (1864–1896) (Fuller 1896, Agric. Gaz. N.S.W. 7: 598; Alexander 1932, Proc. Linn. Soc. N.S.W. 57: 6–7.)

SLADEN, CHARLES ANDREW (1851–1928) (Russell 1928, Entomologist 61: 240; Russell 1929, Entomologist 62: 192.)

SLADEN, FREDERICK WILLIAM LAMBART (1876–1921) (Gibson 1921, Can. Ent. 53: 240.)

SLASH-PINE FLOWER-THRIPS *Gnophothrips fuscus* (Morgan) [Thysanoptera: Phlaeothripidae].

SLASH-PINE SAWFLY *Neodiprion merkeli* Ross [Hymenoptera: Diprionidae].

SLASH-PINE SEEDWORM *Cydia anaranjada* (Miller) [Lepidoptera: Tortricidae].

SLASTSCHEWSK, P (–1935) (Anon. 1935, Arb. morph. taxon. Ent. Berl. 2: 63.)

SLATERS *Porcello scaber* [Isopoda: Porcellionidae]: Crustacea with body oval in outline shape and somewhat flattened dorsoventally; typically grey or slate-coloured; seven pairs of legs. Slaters are restricted to damp habitats; in urban situation typically occur in gardens, compost heaps, under flower pots, in shadehouses and glass houses. Slaters may invade homes in large numbers during periods of heavy rain and soil saturation with water. Typically feed on decaying plant material but will attack seedlings and young plants. Adult female holds eggs in Marsupium on venter of body; eclosion occurs within 60 days; neonates retained within Marsupium for short time. Syn. Woodlice; Sowbugs. Cf. Pillbug. Rel. Centipedes; Millipedes.

SLATY Adj. (From slate.) Coloured with a very dark blackish grey with a reddish tinge.

SLAVERY See Dulosis.

SLE See St Louis Encephalitis.

SLEEPING SICKNESS An often fatal human disease caused by the parasitic protozoan *Trypanosoma brucei* and vectored by tsetse flies *Glossina* spp. [Diptera: Glossinidae]: Disease widespread and often epidemic in Africa (14°N to 29°S). Infected fly bites human for blood; ulcer forms 5–15 days after bite; trypanosomes occur in lymphatic system ca 1–3 weeks and cause splenomegaly and lymphadenopathy; subsequently, trypanosomes invade central nervous system; death occurs within nine months to several years. Two forms of disease: Gambian Sleeping Sickness *(T. brucei gambiense)* and Rhodesian Sleeping Sickness *(T. brucei rhodesiense)*. Anatomical forms of pathogens indistinguishable from *T. brucei brucei* but geographically, ecologically and epidemiologically distinct. *Trypanosoma b. gambiense* in West Africa and northern and western Central Africa; *T. b. rhodesiense* in Central Africa and East Africa. Each Subspecies with different vectorial Species of flies. See Trypanosomiasis. Tsetse Fly. Cf. Nagana.

SLEEPY ORANGE See Coloration. Cf. Nicippe Yellow.

SLEESMAN, JAY PETERSON (1904–1975) (Treece 1976, J. Econ. Ent. 69: 132.)

SLEIGHT, CHARLES EDWIN (1861–1917) (Anon. 1918, Ent. News 29: 280; Davis 1918, J. N. Y. ent. Soc. 26: 47–48.)

SLENDER DUCK-LOUSE *Anaticola crassicornis* (Scopoli) [Mallophaga: Philopteridae].

SLENDER GOOSE-LOUSE *Anaticola anseris* (Linnaeus) [Mallophaga: Philopteridae].

SLENDER GUINEA-LOUSE *Lipeurus numidae* (Denny) [Mallophaga: Philopteridae].

SLENDER GUINEA-PIG LOUSE *Gliricola porcelli* (Schrank) [Mallophaga: Gyropidae].

SLENDER-HORNED FLOUR BEETLE *Gnathocerus maxillosus* (Fabricius) [Coleoptera: Tenebrionidae]: A minor pest of stored products including cereal grains, pumpkins, peanuts; a minor pest of corn in tropical areas. Egg incubation ca four days; 7–9 larval instars, development ca one month; pupal development ca 5–6 days. Cf. Broad-Horned Flour Beetle.

SLENDER MUDNEST BUILDERS *Sceliphron* spp. [Hymenoptera: Sphecidae].

SLENDER PIGEON-LOUSE *Columbicola columbae* (Linnaeus) [Mallophaga: Philopteridae]: A louse endemic to North America and pest of pigeons. SPL ca 2 mm long, undergoes three moults and completed development within 20 days. Cf. Small Pigeon Louse.

SLENDER SAC SPIDERS Members of the Genus *Cheiracanthium* spp. [Araneida: Clubionidae]: Body slender with long, thin legs; fangs and fang-bearing Chelicerae projecting forward tusk-like; spiders found in shrubs and grasses during day and retreat if approached. Cf. Stout Sac Spider. See Clubionidae.

SLENDER SEED-CORN BEETLE *Clivina impresifrons* LeConte [Coleoptera: Carabidae].

SLENDER TURKEY-LOUSE *Oxylipeurus polytrapezius* (Burmeister) [Mallophaga: Philopteridae]: A major pest of turkeys. Cf. Large Turkey-Louse.

SLEVIN, LOUIS STANISLAUS (1879–1945) (Ross 1946, Pan-Pac. ent. 22: 141.)

SLEVOGT, JOHANN BENEDIKT BALDUIN (1847–1911) (Kuznezov 1911, Revue Russe Ent. 11: 162–165, bibliogr.)

SLIFER, ELEANOR H American anatomist (Academy of Natural Sciences, Philadelphia) specializing in the structure of chemoreceptors.

SLIFER'S PATCHES Orthoptera: Slightly depressed areas near the dorsal midline on the integument which lack complex sculpture. SP are paired and arranged on head (antennal crescents), Thorax and abdomen (Fenestrae) of some Orthoptera. Integument in this area lacks true Endocuticle and bears large Pore Canals. SP were originally regarded as temperature sensitive and thermoregulatory in function; patches are more permeable to water than surrounding Cuticle. *Locusta migratoria migratorioides* patches may be important in temperature regulation on the abdomen near Heart. Cf. Thermoregulation.

SLIME Noun. (Middle English, *slim*. Pl., Slimes.) 1. A wet, viscous substance, typically protein, that may be secreted by cells, organs or tissues to cover a body, appendage or structure. Cf. Mucous; Scurf; Scute; Slough. 2. A soft, moist clay, earth or mud.

SLINGERLAND, MARK VERNON (1864–1909) (Anon. 1906, *National Cyclopedia of American Biography* 13: 314–315; Bethune 1909, Can. Ent. 41 170; Comstock 1909, J. Econ. Ent. 2: 195–196.)

SLIT FACED BATS See Nycteridae.

SL-NPV® See *Spodoptera littoralis* NPV.

SLOANE, HANS (1660–1753) (Anon. 1753, Histoire de l'académie des Sciences, Paris. pp. 305–320; Faulkner 1829, an historical and topographical description of Chelsea. interspersed with biographical anecdotes. xii + 406 pp. (338–374), Chelsea; Rose 1850, *New General Biographical Dictionary* 12: 49–50; Anon. 1897, Dictionary of national biogr. 52: 379–380; Brooks 1954, *Sir Hans Sloane the great collector and his circle of friends*. 234 pp. London.)

SLOANE, THOMAS GIBSON (1857–1932) (Carter 1932, Victorian Nat. 49: 191–194; Anderson 1933, Proc. Linn. Soc. N.S.W. 58: iii–iv.)

SLOPER, G ORBY (–1922) (Anon. 1922, Entomologist 55: 288.)

SLOSSON, ANNIE TRUMBULL (1838–1926) (Leng

1918, J. N. Y. ent. Soc. 26: 129–133; Skinner 1919, Ent. News 30: 200; Davis 1926, J. N. Y. ent. Soc. 34: 361–364; Osborn 1937, *Fragments of Entomological History,* 394 pp. (208), Columbus, Ohio.)

SLOUGH Noun. (Middle English, *slogh* = snake skin. Pl., Sloughs.) The shed or detached skin of an animal. Cf. Scurf; Slime. Rel. Cuticle; Integument.

SLUG Noun. (Middle English, *slugge*. Slugs.) 1. General: Any larva that has a slimy, viscid appearance, with the body closely applied to the food plant. Specifically: Larvae of some sawflies and some Coleoptera. Rel. Limaciform; Gastropoda. 2. A terrestrial pulmonate gastropod.

SLUG CATERPILLAR MOTHS See Limacodidae.

SLUGIT® See Metaldehyde.

SM INTERSPACE Submedian interspace.

SMALL ANT-BLUE *Acrodipsas myrmecophila* (Waterhouse & Lyell) [Lepidoptera: Lycaenidae].

SMALL BAMBOO BORERS *Dinoderus* spp. [Coleoptera: Bostrichidae]: A pantropical Genus of minute beetles that bore into bamboo, wood and stored products.

SMALL BOTTLE-CICADA *Chlorocysta vitripennis* (Westwood) [Hemiptera: Cicadidae].

SMALL BROWN MANTIDS *Bolbe* spp. [Mantodea: Mantidae]. See Mantidae.

SMALL BROWN-AZURE *Ogyris otanes* C. & R. Felder [Lepidoptera: Lycaenidae].

SMALL CHESTNUT-WEEVIL *Curculio sayi* (Gyllenhal) [Coleoptera: Curculionidae].

SMALL CITRUS-BUTTERFLY *Eleppone anactus* (W. S. Macleay) [Lepidoptera: Papilionidae]: A minor and frequent pest of citrus endemic to Australia. Caterpillars feed on all cultivated citrus varieties and other Species in Rutaceae. Adult wingspan ca 75 mm; forewing black and grey with white markings; hindwing white with orange, red and blue markings. Eggs yellow, spherical, 2–3 mm diam Caterpillar to 50 mm long, with Osmeterium, brown to black with three rows of orange coloured spots along body and fleshy spines; Five larval instars; all instars feed on young foliage, chewing large areas from the leaf edge. Pupation in upright position on twig and held in place by girdle. Life cycle 2–3 months; multivoltine with at least three generations per year. See Papilionidae. Cf. Large Citrus-Butterfly.

SMALL COPPER *Lucia limbaria* (Swainson) [Lepidoptera: Lycaenidae].

SMALL CROSSVEIN Diptera: The anterior crossvein (r-m) (Curran). See Vein; Wing Venation.

SMALL CYPRESS JEWEL-BEETLE *Diadoxus erythrurus* (White) [Coleoptera: Buprestidae]: In Australia, a pest of native cypress in parts of southern Queensland; attacks introduced *Cupressus* spp. Adult ca 12–18 mm long, torpedo-shaped, body greenish-yellow, thoracic notum brown-black with yellow stripe, Elytra with four irregular yellow spots along side. Larva buprestid form. Attack freshly felled logs, damaged trees; kill damaged, living trees. Eggs laid in bark, neonate larvae tunnel to inner surface, feed in cambial region, pupate in sapwood. Life cycle 1–2 years. Problem in logs and in damaged, standing trees not quickly realized. Adults may emerge after many years from in-service timber. See Buprestidae.

SMALL DARTER *Telicota brachydesma* Lower [Lepidoptera: Hesperiidae].

SMALL DINGY SKIPPER *Hespenlla crypsigramma* (Meyrick & Lower) [Lepidoptera: Hesperiidae].

SMALL DORSAL-RECTI MUSCLES The narrower set of longitudinal dorsal muscles which lie between the great dorsal recti muscles and the dorsoventral muscles of the side of the insect body (Comstock & Kellogg).

SMALL DUNE-SNAIL *Cernuella vestita* (Rambur) [Sigmurethra: Helicidae].

SMALL DUNG FLIES See Sphaeroceridae.

SMALL DUSKY-BLUE BUTTERFLY *Candalides erinus erinus* (Fabricius) [Lepidoptera: Lycaenidae].

SMALL ERMINE MOTHS A Genus (*Yponomeuta*) [Lepidioptera: Yponomeutidae] of 30 Palaearctic Species. See Yponomeutidae.

SMALL-EYED FLOUR BEETLE *Palorus ratzeburgil* (Wissmann) [Coleoptera: Tenebrionidae]: A widespread minor pest of stored grain and processed stored foods Adult 2.5 mm long, reddish brown, flattened; larva 3.5 mm long, whitish. Life cycle 4–6 weeks. See Tenebrionidae. Cf. Confused Flour Beetle; Broadhorned Flour Beetle; Rustred Flour Beetle; Longheaded Flour Beetle.

SMALL FALSE WIREWORM *Gonocephalum misellum* (Blackburn); *Gonocephalum walken* (Champion); *Isopteron punctatissimus* (Pascoe) [Coleoptera: Tenebrionidae].

SMALL FRUIT FLIES See Drosophilidae.

SMALL FRUIT-TREE BORER *Cryptophasa albacosta* Lewin [Lepidoptera: Oecophoridae].

SMALL GRASS YELLOW *Eurema smilax* (Donovan) [Lepidoptera: Pieridae].

SMALL GREEN-BANDED BLUE *Psychonotis caelius taygetus* (C. & R. Felder) [Lepidoptera: Lycaenidae].

SMALL HAIRY-MAGGOT BLOWFLY *Chrysomya varipes* (Macquart) [Diptera: Calliphoridae].

SMALL HEADED FLIES See Acroceridae.

SMALL HIVE BEETLE *Aethina tumida* Murray [Coleoptera: Nitidulidae]: A beetle endemic to sub-Saharan Africa and introduced into the southeastern USA ca 1998. SHB not regarded a pest in Africa; considered a serious pest of honey bee hives in North America. SHB multivoltine with ca 5 generations per year; female oviposits clusters of eggs within hive; egg stage ca 2–3 days; larvae feed on hive products ca 10–14 days; prepupal stage moves outside hive and pupation occurs in soil; pupal stage ca 20 days; adults live ca 180 days. SHB adult and larva feed on pollen, honey, honey bee eggs, larval and pupal stages. Honey seeps from hive frames and ferments, thereby rendering the honey unsuitable. See Nitidulidae.

SMALL INTESTINE The Ileum or Anterior Intestine (Snodgrass). See Hindgut.

SMALL LUCERNE WEEVIL *Atrichonotus taeniatulus* (Berg) [Coleoptera: Curculionidae].

SMALL MAMMAL-SUCKING LICE See Hoplopleuridae.

SMALL MANGO TIPBORER See Mango Tipborer.

SMALL MILKWEED BUG *Lygaeus kalmii* Stål [Hemiptera: Lygaeidae].

SMALL OAKBLUE *Arhopala wildei wildei* Miskin [Lepidoptera: Lycaenidae].

SMALL PASTURE SCARAB *Sericesthis nigra* (Lea) [Coleoptera: Scarabaeidae].

SMALL PIGEON LOUSE *Campanulotes bidentatus compar* (Burmeister) [Mallophaga: Philopteridae]: A pest of pigeon. Adult white, ca 1 mm long with head rounded along anterior margin. See Philopteridae. Cf. Slender Pigeon Louse.

SMALL PLAGUE GRASSHOPPER *Austroicetes cruciata* (Saussure) [Orthoptera: Acrididae].

SMALL POINTED SNAIL *Cochlicella barbara* (Linnaeus) [Sigmurethra: Helicidae].

SMALL POND-SNAIL *Lymnaea peregra* (Muller) [Basommatophora: Lymnaeidae].

SMALL POWDER-POST BEETLE *Lyctus discedens* Blackburn [Coleoptera: Bostrichidae]: A minor pest of hardwoods in Australia. See Bostrichidae. Cf. Powder-Post Beetle; Malayan Powder-Post Beetle.

SMALL PURPLE LINEBLUE *Prosotas dubiosa dubiosa* (Semper) [Lepidoptera: Lycaenidae].

SMALL SOUTHERN-PINE ENGRAVER *Ips avulsus* (Eichhoff) [Coleoptera: Scolytidae].

SMALL STRIPED SLUG *Arion hortensis* Ferussae [Sigmurethra: Arionidae].

SMALLER EUROPEAN-ELM BARK-BEETLE *Scolytus multistriatus* (Marsham) [Coleoptera: Scolytidae]: A monovoltine, significant pest of Elm trees in eastern USA; can complete a partial second generation in some areas. Adult ca 2 mm long, dark reddish-brown, punctate; elytra striated. Adults emerge during spring and feed on crotches of living elm twigs; later bore into wood or recently cut, dead or dying elm trees. Holes resemble those constructed by Shothole Borer. Female constructs egg gallery parallel with grain and larval galleries cut across grain of wood. Overwinters as mature larva in cell at end of tunnel; pupation occurs during following spring; adults appear during summer. Adults transmit Dutch Elm Disease from dead trees to living trees. See Scolytidae. Cf. Douglas-Fir Engraver; Fir Engraver; Shothole Borer. Rel. Engravers.

SMALLER HAWAIIAN CUTWORM *Agrotis dislocata* (Walker) [Lepidoptera: Noctuidae]. See Cutworms.

SMALLER LANTANA BUTTERFLY *Strymon bazochii gundlachianus* (Bates) [Lepidoptera: Lycaenidae].

SMALLER WATER STRIDERS See Veliidae.

SMALLER YELLOW ANT *Acanthomyops claviger* (Roger) [Hymenoptera: Formicidae]: Widespread in North America and most abundant in eastern USA. Worker ca 0.12–0.15 mm long; monomorphic, stingless, smooth, shiny and setose; compound eyes small; Maxillary Palpus short, with three segments; Antenna with 12 segments; pedicel of one segment with long, conical node; Sting absent; mandibular glands secrete citronella when worker disturbed. Nests in soil, logs and under stones. Commonly feeds upon honeydew of subterranean aphids and mealybugs. Nuptial flights during September. See Formicidae. Cf. Larger Yellow Ant.

SMALTINUS Adj. (Italian, *smalto*.) Dull greyish blue.

SMARAGDINE Adj. (Greek, *smaragdos* = emerald.) The brilliant crystalline green of the emerald; emerald colour. Alt. Smaraginus.

SMARDA, JAN (1904–1968) (Pospiscil 1969, Cas. morav. Mus. Brne 54: 223–225.)

SMART, HERBERT DOUGLAS (1880–1945) (H. B. W. 1945. Entomologist 78: 192; Hale-Carpenter 1946, Proc. R. ent. Soc. Lond. (C) 10: 54.)

SMARTWEED BORER *Ostrinia obumbratalis* (Lederer) [Lepidoptera: Pyralidae].

SMASH® See Fenpropathrin.

SMDC® See Metham-Sodium.

SMEARED DAGGER MOTH *Acronicta oblinita* (J. E. Smith) [Lepidoptera: Noctuidae].

SMEATHMAN, HENRY (1750–1787) Published *History of the Termites or White Ants* (Phil. Trans., vol. 71). (Swainson 1840, *Taxidermy; with Biography of Zoologists.* 392 pp. (329–330), London. [Vol. 12 of *Cabinet Cyclopedia.* Edited by D. Lardner]; Griffin 1942, Proc. R. ent. Soc. Lond. (A) 17: 1–9.)

SMEE, C (1896–1947) (Williams 1948, Proc. R. ent. Soc. Lond. (C) 12: 64.)

SMELLIE, WILLIAM (1740–1795) (Rose 1850, *New General Biographical Dictionary* 12: 52.)

SMETANA, OLDRICH (Moucha 1971, Acta ent. bohemoslovaca 68: 430.)

SMETHURST, CHARLES (–1882) (Nelson 1892, Naturalist, Hull 1892: 145–146.)

SMICRIPIDAE Plural Noun. See Nitidulidae.

SMIDT [SCHMIDT] An associate of J. C. Fabricius. (Papavero 1971, *Essays on the History of Neotropical Dipterology.* I: 21–22. São Paulo.)

SMINTHURIDAE Plural Noun. A numerically large, widely distributed Family of symphypleonian Collembola. Characterized by a globose body, Thorax and abdominal segments 1–4 fused. Found in vegetation such as flowers.

SMITE® See Chlorpyrifos-Methyl.

SMITH BLUE *Euphilotes enoptes smithi* (Mattoni) [Lepidoptera: Lycaenidae].

SMITH, ADAM CHARLES (1871–1943) (J.A.S. 1944, Lond. Nat. 1943: 48–49.)

SMITH, ARTHUR (1916–1991) Scientific illustrator (British Museum Natural History 1940–1973) who completed 15,000–20,000 illustrations of insects and other animals. Many illustrations in Hopkins & Rothschild catalogues I-V, 1953–1971. (1992, Ent. mon. Mag. 128: 173).

SMITH, AUDREY Z Librarian of Hope Museum, Oxford University, and author of *A History of the Hope Entomological Collections in the University Museum Oxford* (1986).

SMITH, BENJAMIN HAYES (1841–1918) (Stone 1919, Ent. News 30: 88–90; Hall 1947, Proc. R. ent. Soc. Lond. (C) 11: 62.)

SMITH, BERNARD (–1903) (Anon. 1903, Entomol. mon. Mag. 39: 304.)

SMITH, CHARLES (c. 1715–1762) (Lisney 1960, *Bibliography of British Lepidoptera 1608–1799*, 315 pp., (145–147, bibliogr.), London.)

SMITH, CLAUDE (1922–1949) (Hurd 1950, Pan-Pacif. Ent. 26: 59–60.)

SMITH, CLIVE BRAMWELL (–1945) (J.A.S. 1945, Lond. Nat. 25: 77–78.)

SMITH, DENNIS ALFRED (1913–1972) (H.C.H. 1972, Entomologist's Rec. J. Var. 84: 291; Butler 1973, Proc. R. ent. Soc. Lond (C) 37: 56–57.)

SMITH, EMILY ADELLA (Osborn 1937, *Fragments of Entomological History,* 394 pp. (185), Columbus, Ohio.)

SMITH, ERNEST GEORGE (–1970) (Hinton 1971, Proc. R. ent. Soc. Lond. (C) 35: 53.)

SMITH, FREDERICK (1805–1879) English taxonomist and specialist in Hymenoptera. Purported to have described more Genera of ants than any other Myrmecologist. (Dunning 1879, Proc. ent. Soc. Lond. 1879: lxiv–lxvi; Dunning 1879, Entomologist 12: 88–92.)

SMITH, GEORGE DOLE (1833–1880) (Anon. 1880, Am. Nat. 14: 756; Orne 1880, Proc. Boston. Soc. nat. Hist. 21: 51–53; Mann 1881, Psyche 3: 199.)

SMITH, GEORGE LESLIE (1879–1971) (Ewing & Rainwater 1972, J. Econ. Ent. 65: 936.)

SMITH, HARRY SCOTT (1883–1957) American academic (University of California) and administrator. Developed and directed Department of Biological Control. Responsible for successful control of numerous insect pests of citrus. Credited with promoting the term 'Biological Control.' Developed foundation of modern plant quarantine as implemented in the United States. (Clausen & Flanders 1958, J. Econ. Ent. 51: 267; Mallis 1971, *American. Entomologists,* xvii + 549 pp. (501–503), New Brunswick.)

SMITH, HERBERT DWIGHT (1894–1959) (Haeussler & Parker 1961, J. Econ. Ent. 54: 217.)

SMITH, HERBERT HUNTINGTON (1851–1919) American naturalist and collector. Made extensive collections in tropical America for Biological Centrali Americana, Carnegie Museum, British Museum and other institutions. (Anon. 1919, Ent. News 30: 210; Holland 1919, Ent. News 30: 211–214; Holland 1919, Science 49: 481–483; Osborn 1937, *Fragments of Entomological History,* 394 pp. (230) Columbus, Ohio; Papavero 1975, *Essays on the History of Neotropical Dipterology.* 2: 377–386. São Paulo.)

SMITH, JAMES EDWARD (1759–1828) English Naturalist and Botanist; 1784 purchased Linnean botanical collection. Author of *English Botany* (36 vols, 1792–1807) *Flora Britannica* (1800–1804) and *The Natural History of the rarer Lepidopterous Insects of Georgia, collected from the Drawings and Observations of Mr. John Abbot* (2 vols, 1797). (T. 1829, Mag. nat. Hist. 1: 91–93, bibliogr.; Goode 1897, Rep. Smithson. Instn 2 (2) 418, 429.)

SMITH, JOHN BERNHARD (1858–1912) (Gibson 1912, Can. Ent. 44: 97–99; Grossbeck 1912, Ent. News 23: 193–196; Howard, *et al.* 1912, Proc. ent. Soc. Wash. 14: 111–117; Anon. 1916, National cyclopedia of American biogr. 15: 71–72.)

SMITH, MARION RUSSEL (Smith 1973, Proc. ent. Soc. Wash. 75: 88–95, bibliogr.)

SMITH, R W (1901–1966) (Beirne 1965, Proc. ent. Soc. Ont. 96: 120.)

SMITH, RALPH HENRY (1888–1945) (Essig 1945, J. Econ. Ent. 38: 504–505, 624, 625; Essig 1946, Ann. ent. Soc. Am. 39: 5–6.)

SMITH, RALPH INGRAM (1882–1927) (E.R.S. 1937, J. Econ. Ent. 20: 651–652; Osborn 1937, *Fragments of Entomological History,* 394 pp. (217–218), Columbus, Ohio.)

SMITH, SAMUEL GORDON (1885–1965) (R.C.R.C. 1965. Entomologist's Rec. J. Var. 65: 136–137.)

SMITH, SIDNEY (–1884) (Dimmock 1884, Psyche 4: 266; Anon. 1885, Entomologist 18: 56.)

SMITH, SIDNEY IRVING (1843?–1926) (Verrill 1926, Science 64: 57–58; Osborn 1937, *Fragments of Entomological History,* 394 pp. (217), Columbus, Ohio.)

SMITH, THEOBALD (1859–1934) Pioneer Medical Entomologist who discovered *Babesia bigemina,* causal agent of Texas Cattle Fever, in the host's red blood cells (1889) and subsequently (1893) implicated the cattle tick *Boophilus annulatus* in transmission of the disease. (Howard 1930, Smithson. misc. Collns. 84: 174, 466, 480; Hall 1935, J. Parasit. 21: 231–243.)

SMITH, W P (1891–1963) (Wigglesworth 1964, Proc. R. ent. Soc. Lond. (C) 28: 58.)

SMITH, WALTER REGINALD (–1970) (Butler 1972, Proc. R. ent. Soc. Lond. (C) 36: 62.)

SMOKE FLY *Microsania australis* Collart [Diptera: Platypezidae].

SMOKY BUZZER *Cicadetta waterhousei* (Distant) [Hemiptera: Cicadidae].

SMOKY-BROWN COCKROACH *Periplaneta fuliginosa* (Serville) [Blattodea: Blattidae]: A widespread urban pest; origin uncertain; occurs in southeastern USA and eastern Australia. Adult SBC resembles Australian Cockroach, dark brown to black without pale markings. SBC prefers to inhabit sheds, wall voids, sewers and grease traps; in Australia a pest of glasshouse, garden and plant nursery. Adults macropterous, fly short distances; attracted to light. Adult lives 6–12 months; female lays ca 15 Oothecae during lifetime; SBC oviparous, Oothecae dropped or glued to substrate; Ootheca contains 22–25 eggs; nymphs undergo 9–12 moults; nymphal stage 6–12 months, more rapid when gregarious; life cycle 6–12 months. See Blattidae. Cf. American Cockroach; Australian Cockroach.

SMOLKA, A (–1928) (Anon. 1928, Ent. NachrBl. Troppau 2: 76.)

SMOOTH Adj. (Middle English, *smothe;* Anglo Saxon, *smoth.*) Descriptive of surface sculpture, usually the insect's integument, that is flat, mirror-like or devoid of surface features. Smooth is, perhaps, the groundplan condition of sculpture types or forms. See Sculpture Pattern. Ant. Wrinkled. Alt. Estriate. Cf. Acuductate; Alveolate; Baculate; Clavate; Echinate; Favose; Gemmate; Psilate; Punctate; Reticulate; Rugulate; Scabrate; Shagreened; Striate; Verrucate.

SMOOTH SLATER *Porcellio laevis* Latreill [Isopoda: Porceilioni].

SMOOTH SPIDER-BEETLE *Gibbium psylloides* (Czenpinski) [Coleoptera: Anobiidae]: A pest of stored products, principally Palearctic in Europe and around the Mediterranean. Reported from tomb of King Tutankhamun. SSP often confused with closely related *G. aequinoctiale* Boieldieu; many records published under *G. psylloides* probably refer to *G. aequinoctiale*. Latter Species nearly cosmopolitan and dominant in New World and Asia. Feeds on decaying, moist food, seed, fur, wool carpets; in Australia, feeds on debris from cocoa beans, dried fruits and grains. Adult 1.7–3.2 mm long, body bulbous; shiny dark brown to black; head, Pronotum and Elytra nearly glabrous; head below eyes with narrow, prominent, subparallel ridges that extend to anterior margin of Pronotum; posterior margin of antennal fossa strongly produced laterally with lateral 0.3 sloping toward middle; scale-like Setae near antennal fossa, on Scape and Vertex; legs unusually long; Abdomen with four visible Sterna. Mature larva white, 4 mm long, curled; three pairs of thoracic legs. Syn. Storehouse Beetle. See Anobiidae.

SMOOTH SUCKING-LICE See Linognathidae.

SMUT BEETLE *Phalacrus politus* Melsheimer [Coleoptera: Philacridae].

SMYTH, ELLISON ADGER (1863–1941) (Osborn 1937, *Fragments of Entomological History,* 394 pp. (280) Columbus, Ohio; Calvert 1941, Ent. News 52: 270.)

SNAIL KILLING FLIES See Sciomyzidae.

SNAILCASE CADDIS FLIES See Trichoptera; Heliocopsychidae.

SNAILS Gastropoda. Terrestrial, shell-bearing, pulmonate gastropods.

SNAKE Noun. (Anglo Saxon, *snaca.* Pl., Snakes.) Squamata: Any of numerous limbless reptiles having elongate bodies and which constitute the group Ophidia.

SNAKE MITE *Ophionyssus natricis* (Gervais) [Acari: Macronyssidae]: A mite which is an obligate parasite of snakes and sporadic pest in reptile houses of zoos.

SNAKE TICK *Amblyomma moreliae* (L. Koch) [Acari: Ixodidae].

SNAKEFLIES See Raphidoidea; Inocelliidae.

SNAPPING BEETLES See Elateridae.

SNELGROVE, LOUIS EDWARD (1878–1965) (Pearson 1966, Proc. R. ent. Soc. Lond. (C) 30: 64.)

SNELL, BASIL BASSETT (1905–1956) (Hall 1957, Proc. R. ent. Soc. Lond. (C) 21: 66.)

SNELLEN VAN VOLLENHOVEN, SAMUEL CONSTANT (1816–1880) Dutch Entomologist and Curator of Rijksmuseum van Natuurlijke Historie (1854–1873); president and founding member of Dutch Entomological Society. (Anon. 1880, Ent. Nachr. 6: 147; Anon. 1880, Nature 21: 538–539; Anon. 1880, Am. Nat. 14: 468; Dohrn 1880, Stettin. ent. Ztg. 41: 249; May 1880, Entomologist 13: 117–118.)

SNELLEN, PIETER CORNELIUS TOBIAS (1832–1911) (Fowler 1911, Entomol. mon. Mag. 47: 114; H. 1911, Dt. ent. Z. Iris 25: 125; Wheeler 1911, Entomologist's Rec. J. Var. 23: 232; Piepers 1912, Tijdschr. Ent. 55: 1–8.)

SNIP® See Azamethiphos.

SNIPE FLIES See Rhagionidae.

SNODGRASS, ROBERT EVANS (1875–1962) American government Entomologist (USDA) specializing in Insect Morphology; author of more than 80 scientific papers which totalled about 6000 pages. *Principles of Insect Morphology* was his most influential work. Other significant works important to morphology include his treatment of invertebrate evolution (1938) and textbook of arthropod anatomy (1952). Regrettably, Snodgrass had no formal association with students and did not train university students in his area of study. (Thurman 1959, Smithson. misc. Collns. 137: 1–22, bibliogr.; Ronderos 1962, Revta. Soc. ent. argent. 24: 62; Anon. 1963, Revta. Soc. ent. Uruguay 5: 53–57. Carpenter 1963, Bull. Ann. Soc. r. ent. Belg. 99: 311–315; Denis 1963, Bull. Soc. ent. Fr. 68: 101–104; Schmitt 1963, Ent. News 74: 141–142; Varley 1963, Proc. R. ent. Soc. Lond. (C) 27: 50; Anon. 1963, J. N. Y. ent. Soc. 71: 58; Anon. 1964, Studia ent. 7: 488; Mallis 1971, *American Entomologists.* 549 pp. (165–169), New Brunswick.)

SNOUT Noun. (Middle English, *snute* = nose. Pl., Snouts.) The elongate head in Rhynchophora with mouthparts at the apex. See Rostrum.

SNOUT BEETLES See Curculionidae.

SNOUT BUTTERFLIES See Nymphalidae.

SNOUT MITES Acari: Bdellidae.

SNOUT MOTHS See Pyralidae. Lasiocampidae.

SNOUT WEEVILS See Curculionidae.

SNOW FLIES See Capniidae.

SNOW, FRANCIS HUNTINGTON (1840–1908) American academic and naturalist. Graduate Williams College 1862; Andover Theological Seminary 1866; Professor of Mathematics and

Natural History, University of Kansas 1866; 1886 founded Snow Hall of Natural History; 1890 Chancellor of KU and Professor of Entomology. Discovered fungus of Chinch Bug. (Anon. 1898, Outlook 59: 876; Anon. 1907. National cyclopedia of American biogr. 9: 494–495. Anon. 1908, J. Econ. Ent. 1: 411. Green *et al.* 1908, Grad. Mag. Univ. Kans. 7: 121–146; McClung 1908, Ent. News 19: 447–449; Hunter 1913, Kans. Univ. Sci. Bull. 8: 1–61, bibliogr.)

SNOW POOL MOSQUITOES Northern hemisphere Species of the Genus *Aedes* which appear early during spring in Canada and Alaska; pools created by melting snow. Monovoltine; females oviposit in cold water; eggs and larvae capable of surviving freeze and thaw. Populations reach astronomically high numbers and make life unbearable for animals and humans.

SNOW SCALE See White Louse Scale.

SNOW SCORPIONFLIES See Boreidae.

SNOW, WILLIAM APPLETON (1869–1899) (Mik 1900, Wien. ent. Ztg. 19: 88; Silantjev 1911, Revue Russe Ent. 11: 461–462.)

SNOW, WILLIS EVERETT (1918–1959) (Smith 1959, J. Econ. Ent. 52: 1229.)

SNOWBALL APHID *Neoceruraphis viburnicola* (Gillette) [Hemiptera: Aphididae].

SNOWBERRY FLY *Rhagoletis zephyria* Snow [Diptera: Tephritidae]: A North American Species which develops on *Symphoricarpos albus*.

SNOWY TREE-CRICKET *Oecanthus fultoni* Walker [Orthoptera: Gryllidae]: A monovoltine pest of apples, peach, plum, cherry and prune in North America. Eggs and excrement deposited in pin-size hole in tree bark during fall; hole sealed with secretion, egg hatches in spring. Nymphs feed on leaves, fungi, other insects, fruit. Males stridulate but females cannot hear sound; male provides female with metathoracic gland secretion as nuptial gift.

SNYDER, ARTHUR JOHN (1867–) (dos Passos 1951, J. N. Y. ent. Soc. 59: 175.)

SNYDER, HOWARD AUSTIN (1854–1934) (Anon. 1934, Ent. News 45: 140.)

SNYDER, THOMAS ELLIOTT (1885–1970) (Coaton 1971, J. ent. Soc. sth. Afr. 34: 199–200; Emerson *et al.* 1971, Proc. ent. Soc. Wash. 73: 239–242.)

SOARING FLIGHT A form of passive flight in which the insect engages ascending air currents. SF confined to large insects with low wing loading and seen in migrating butterflies. See Flight. Cf. Passive Flight.

SOBOLOV, ALEKSEII NIKOLAEVICH (1870–1911) (Silantyev 1911, Revue Russe Ent. 11: 461–462.)

SOBOTA, ANTONIN (1898–1974) (Brozik 1974, J. Zpravy Cesk. Spol. ent. 10: 88.)

SOBRERO, ASCANIO (1812–1888) (Cossa 1889, Atti Accad. Sci., Torino 24: 158–163, bibliogr.)

SOCIAL Adj. (Latin, *sociare* = to associate; *-alis* = characterized by.) A term descriptive of organisms living in more-or-less organized communities or aggregations of individuals of the same Species. Several levels of sociality are common among many groups of insects, but social behaviour is highly developed among termites, ants and bees. See Eusocial. Cf. Communal. Rel. Hymenoptera.

SOCIAL BEES See Apoidea.

SOCIAL BUCKET The process by which liquid food is carried between the Mandibles and shared by nestmates during mouth-to-mouth contact.

SOCIAL FACILITATION See Group Effect.

SOCIAL HOMEOSTASIS The maintenance of steady states at the society-level by control of nest climate, population density, behaviour and physiology.

SOCIAL INSECTS Insects that live in organized groups or colonies. SI characterized by: Contact between generations, division of labour among members of the group, progressive provisioning with food or care for the immatures. Higher forms of social development often show differentiated adult castes or forms with specific functions in the group. Social organization most highly developed in Isoptera and Hymenoptera. See Isoptera; Hymentopera.

SOCIAL PARASITE A type of guest within the nests of social insects which feeds upon the food stores of the colony. See Cleptoparasite. Cf. Commensal.

SOCIAL PARASITISM A type of symbiotic relationship in which one Species consumes the food furnished or stored by a second Species. In social parasitism the consumption of food by the guest Species indirectly affects the host Species. See Cleptoparasitism. Cf. Commensalism.

SOCIAL STOMACH Social Insects: The Crop which can be distended to store and regurgitate food to nestmates.

SOCIAL SYMBIOSIS See Social Parasitism.

SOCIAL THRESHOLD See Threshold. Cf. Medical Threshold; Economic Threshold.

SOCIAL WASP *Polistes* spp. [Hymenoptera: Vespidae]. See Vespidae; Sphecidae. Cf. Solitary Wasp.

SOCIETY Noun. (Latin, *societas* = company. Pl., Societies.) A group of genetically related organism (population, Subspecies, Species) which collectively form a community and the members show contact between generations (adult and immature), division of labour and social exchange.

SOCIOBIOLOGY Noun. (Latin, *societas* = company. Pl., Sociobiologies.) The study of the biological basis of social behaviour.

SOCIOGENESIS Noun. (Latin, *societas* = company; *sis* = a condition or state. Pl., Sociogeneses.) Social Insects: The processes and patterns that develop a colony through its life.

SOCIOTOMY See Colony Fission.

SOCKET Noun. (Old French, *soc* = ploughshare. Pl., Sockets.) An opening, hollow or depression that serves to hold something (*e.g.* eye socket). Syn. Acetabulum. See Cotyla; Joint.

Cf. Aditus.

SOCUS Noun. (Pl., Socii.) 1. A lateral appendicular process of the tenth abdominal segment in Trichoptera and Lepidoptera. Believed homologous with the Cercus-like appendage of the tenth segment in lower Hymenoptera by some morphologists (Snodgrass). 2. Male Lepidoptera: A paired, slender prolongation from the caudal margin of the Tegumen near or at the base of the Uncus, extending ventrad and caudad on each side of the Anus often clavate and setose (Klots).

SOD See Turf.

SOD WEBWORM *Herpetogramma licarsisalis* (Walker) [Lepidoptera: Pyralidae]: A pest of turf and lawn in the Orient, Australia and Hawaii. Larvae feed on blades of grass and produce webbing. Larvae feed at night and pupation occurs at night; adults emerge, mate and oviposit at night. Generation time ca 32 days at 25°C; larval development 14 days with five instars; females lay about 250 eggs.

SÖDERMAN, ALEXANDER HENRIK (1876–1933) (Hellén 1934, Notul. ent. 14: 59–61.)

SODIUM Noun. (Latin, from soda.) A silver-white metal found in a free state, but important as an element in salt and other common compounds. Chemical abbreviation, Na.

SODIUM CYANIDE A chemical compound of sodium, nitrogen and carbon ordinarily in white lumps or granular powder, soluble in water and alcohol. SC evolves hydrocyanic acid gas which is highly poisonous and used in insect killing bottles. Chemical abbreviation, NaCN.

SODIUM HEXAFLUOROSILICATE An inorganic insecticide, formulated with a bait, and used in some countries to control ants, cockroaches, earwigs and sowbugs. See Inorganic Insecticide.

SODIUM PUMP A mechanism, typically within the membrane of nerve cells, which actively moves sodium ions out of the cell and thereby maintains a resting action potential across the membrane.

SODOFFSKY, CARL HEINRICH WILHELM (1797–1858) (Anon. 1859, KorrespBl. Naturf Ver. Riga 11: 124–125.)

SOENEN, MARCEL LOUIS (1875–1958) (Anon. 1958. Linneana Belgica 1: 41.)

SOFFNER, JOSEPH (Anon. 1969. Mitt. dt. ent. Ges. 28: 49–50.)

SOFT BODIED PLANT BEETLES. See Dascillidae; Helodidae; Scirtidae.

SOFT BROWN SCALE *Coccus hesperidum* Linnaeus [Hemiptera: Coccidae] is a cosmopolitan pest of citrus, cotton, figs, grape, papaya, passionfruit and ornamental plants; SBS particularly troublesome in tropical and subtropical areas. Adult female brown, oval in outline shape, flat, 3–4 mm long; dorsal longitudinal ridge incomplete. All immature stages with functional legs and shift feeding sites on host plant; crawlers prefer new growth. SBS parthenogenetic but some males are produced; females deposit ca 200 crawlers during lifetime; multivoltine with number of generations depending upon host plant and region; development asynchronous with overlapping generations. SBS produces large amounts of honeydew which attracts ants and serves as substrate for sooty mould and other fungi.

SOFT GREEN-SCALE *Coccus alpinus* De Lotto [Hemiptera: Coccidae]: A pest of coffee, tea and guava in tropical Africa.

SOFT PESTICIDE A chemical used in control of pests which tends to have a short residue life and is relatively specific to the target organism. Soft pesticides, where available, preferred over hard pesticides in IPM strategies. See Chemical Control. Cf. Hard Pesticide.

SOFT SCALES See Coccidae. Cf. Armoured Scales.

SOFT TICKS See Argasidae.

SOFT-WINGED FLOWER-BEETLES See Melyridae.

SOIL Noun. (Middle English, *soilen* > Latin, *solium* = seat > *solum* = base, ground, foundation. Pl., Soils.) 1. The loose, aggregated material forming the surface of the ground in which plants grow. Cf. Humus. 2. Cultivated or tilled ground.

SOIL-BARRIER TREATMENT A pest-control procedure used for the prevention of termite entry into wooden buildings. In some areas procedure involves use of chlorinated hydrocarbon insecticides (Aldrin, Chlordane, Dieldrin, Heptachlor); where CHIs prohibited, replaced with organophosphorus insecticide (Chlorpyrifos). Syn Chemical Soil Treatment.

SOIL GARD® A biopesticide used to control plant pathogens including *Fusarium* spp., *Pythium* spp. and *Sclerotinia* spp. Active ingredient the fungus *Gliocladium virens*. See Biopesticide.

SOIL-PREP® See Metham-Sodium.

SOJIMA, KUMAROKU (1864–) (Hasegawa 1967, Kontyû 35 (Suppl.): 46.)

SOKOLAR, FRANZ (1851–1913) (Anon. 1913. Ent. Z., Frankf. a. M. 27: 163–164, bibliogr.; Heikertinger 1913, Ent. Bl. Biol. Syst. Käfer 9: 265–269, bibliogr.)

SOKOLOV, NIKOLAI NIKOLAEVICH (1866–1936) (Anon. 1936, Arb. physiol. angew. Ent. Berl. 3: 302).

SOKOLOWSKI, K (1888–1960) (Zimmerman 1960, Mitt. dt. ent. Ges. 19: 75.)

SOLACERORIS Noun. Some coccids: The combined Mesogenacerores and Pregenacerores forming a single, large, crescentic, cephalic group (MacGillivray).

SOLANACEOUS TREEHOPPER *Antianthe expansa* (Germar) [Hemiptera: Membracidae].

SOLANDER'S BROWN BUTTERFLY *Heteronympha solandri* Waterhouse [Lepidoptera: Nymphalidae].

SOLANUM FRUIT FLY *Dacus cacuminatus* (Hering) [Diptera: Tephritidae].

SOLAR Adj. (Latin, *sol* = sun.) 1. Pertaining to the sun.

SOLARI, FERDINANDO (1877–1956) (Invrea 1956, Memorie Soc. ent. ital. 34: 1; Invrea 1957, Atti. Accad. naz. ital. Ent. Rc. 5: 25–29.)

SOLDIER Noun. (Old French, from Latin, *solidus* = a piece of money. Pl., Soldiers.) 1. Social Insects: A member of a worker subcaste responsible for colony defence. 2. Isoptera: A termite which is sexually undifferentiated and displays an enlarged head with forcipate Mandible. 3. Formicidae: worker majors in certain ants.

SOLDIER ANT See Army Ant; Dinergate.

SOLDIER BEETLES See Cantharidae.

SOLDIER FLIES See Stratiomyidae.

SOLEAFORM Adj. (Latin, *solea* = sandal; *forma* = shape.) Slipper-shaped; sandal-shaped. See Cf. Calceiform; Panfuriform. Rel. Form 2; Shape 2; Outline Shape.

SOLENARIA Noun. (Greek, *solen* = a pipe; Latin, *-aria* = plural suffix forming groups.) Two subcylindrical air tubes of the suctorial apparatus (Antlia).

SOLENHOFEN BEDS Fossil deposits of Upper Jurassic Period age found in Germany. An estuary-type habitat composed of lithographic stone which includes *Archaeopteryx*, dragonflies, and *Pseudosirex* (Hymenoptera). Comparable aged deposits incude British Lias (Somerset to Gloucestershire), Lias of Dobbertin (Mecklenberg), and Montsech in Spain.

SOLENIDION Noun. (Greek, *solen* = pipe; *-idion* = diminutive. Pl., Solenidia.) Acari: Specialized sensory Setae that are positioned on appendages. Solenidia are typically hollow, slender, cylindrical Setae (parallel sided or tapering apicad) with apex bluntly rounded. Solenidia are called 'bulbapex' when the setal shaft is expanded distally. Solenida are associated with a sensory cell; the shaft is fenestrated and appears striated with light microscopy and scanning electron microscopy. Solenidia are not optically active under cross-polarized light. (Goff *et al.* 1982, J. Med. Ent. 19: 234).

SOLENOPHAGE Noun. (Greek, *solen* = pipe; *phagein* = to eat. Pl., Solenophages.) A blood-sucking insect that feeds from a blood vessel. See Feeding Strategies. Cf. Haematophage.

SOLENOPHAGOUS Adj. (Greek, *solen* = pipe; *phagein* = to eat.) Pertaining to blood-sucking organisms or blood-sucking characteristics. See Feeding Strategies. Cf. Haematophagous; Sanguinivorous.

SOLEUS Noun. (Latin, *solea* = sole of foot. Pl., Solea.) The ventral surface of the insect foot; the underside of the Tarsus, including the Pulvilli.

SOLFAC® See Cyfluthrin.

SOLIFUGAE Plural Noun. An Order of Arachnida with ca 120 Species in North America which occur mainly in arid regions of the West. 20–30 mm long, pale coloured with slight constriction in the middle of body and very large Chelicerae. There are no venom glands. Fourth pair of legs with malleoli on coxa and trochanters. Mostly nocturnal predaceous. Two Families in U.S.: Ammotrechidae and Erembatidae. Alt. Wind Scorpions; Sun Scorpions; Camel Spiders.

SOLGARD® See Pirimiphos-Ethyl.

SOLID Noun. (Latin, *solidus*. Pl., Solids.) 1. A structure or substance which is not hollow. 2. Consolidated. 3. In the taxonomical sense, a usually jointed or segmented structure in which the segments fuse into one mass (*e.g.*, the apical segments of a clavate Antenna).

SOLIER, ANTOINE JOSEPH JEAN (1792–1851) (Mulsant 1852, Opusc. ent. 1: 82–94, bibliogr.)

SOLIMAN, HAMED SELEEM (1894–1970) (Anon. 1971, Bull. Soc. ent. Égypte 54: [1].)

SOLITARY Adj. (Latin, *solitarius* from *solias* = solitude, *solus* = alone.) 1. Being, living or existing alone. In the context of social organisms, individuals of a Species that are not colonial or gregarious. 2. Pertaining to eusocial bees in which each female constructs her nest(s) without regard for the position of other nests constructed by females of the same Species. See Communal; Quasisocial; Semisocial; Subsocial; Primitively Social; Highly Social. Cf. Communal; Social. Noun. (Latin, *solus* = alone. Pl., Solitaries.) Refering to the biology of insect parasitoids: A single parasitoid that develops on a single host. Cf. Gregarious.

SOLITARY ANT See Mutillidae.

SOLITARY BEE Any number of bee Species that represent one extreme of the social spectrum by living singly and not in a hive or nest. See Semisocial. Cf. Communal; Quasisocial; Subsocial; Primitively Social; Highly Social; Aggregation; Communal.

SOLITARY MIDGES See Thaumaleidae.

SOLITARY OAK LEAFMINER *Cameraria hamadryadella* (Clemens) [Lepidoptera: Gracillariidae].

SOLITARY PARASITE A parasite which completes development alone on or in a host. Cf. Gregarious Parasite; Superparasite.

SOLITARY PARASITISM A condition common in parasitic Hymenoptera and some Diptera through which a parasite larva completes development in a one-to-one relationship with its host. If supernumerary parasite eggs or larvae are present on or in the host, then they are eliminated. See Parasitism. Cf. Gregarious Parasitism.

SOLITARY WASP See Mud Wasps; Sand Wasps.

SOLPUGIDA Sun Spiders. See Solifugae.

SOLSKY, SIMON MARTINOVICH (1831–1879) (Anon. 1879, Naturaliste 7: 56; Portschinsky 1896, Trudy russk. ent. Obshch. 30: vii–x, bibliogr.)

SOLTAU, HUGO (1855–1899) (Schwarz 1899, Proc. ent. Soc. Wash. 4: 405; Wickham 1900, Ent. News 11: 450.)

SOLUBLE POWDER Noun. A pesticide formulation consisting of a finely ground solid material that dissolves completely in water or other liquid forming a true solution. Designated as 'SP' on U.S. Environmental Protection Agency label or Material Safety Data Sheet documentation.

SOLUTE Noun. (Latin, *solutus*, past part. of *solvere* = to loosen. Pl., Solutes.) The substance dissolving in a solvent to form a solution.

SOLUTION Noun. (Latin, *solutis* = a loosening; English, *-tion* = result of an action. Pl., Solutions.) A concentrated liquid pesticide formulation that may be used directly or requires diluting. Designated as 'S' on U.S. Environmental Protection Agency label or Material Safety Data Sheet documentation.

SOLVENT Noun. (Latin, *solvere* = to loosen. Pl., Solvents.) 1. Any substance (solid or liquid) in which another dissolves. 2. A liquid in which the active ingredient of a pesticide dissolves. Cf. Active Ingredient; Diluent; Inert Ingredient; Surfactant; Synergist.

SOLVIDAE See Xylomidae.

SOLVIGRAN® See Disulfoton.

SOLVIREX® See Disulfoton.

SOMA Noun. (Greek, *soma* = body. Pl., Somata.) The body of an animal as distinguished from the germ cells.

SOMATIC Adj. (Greek, *soma* = body; *-ic* = consisting of.) Pertaining to the body tissues or body cells, in distinction from germ (reproductive) tissue or cells.

SOMATIC CELLS Cells of the animal's body as distinguished from the germ cells.

SOMATIC LAYER The external layer of the Mesoderm; term pertains to the walls of the Alimentary Canal. Splanchnopleure (Snodgrass).

SOMATIC NERVES A series of paired bundles of nerve fibres emerging from the ganglia of the central nervous system (Wardle).

SOMATIIDAE Plural Noun. A small Family of Neotropical acalypterate Diptera assigned to Schizophora, but placement within the Series not certain.

SOMATOGENESIS Noun. (Greek, *soma* = body; *genesis* = descent; *sis* = act of. Pl., Somatogeneses.) Genetics: The emergence under favourable surroundings of bodily structure out of hereditary sources (Walter).

SOMATOPLASM Noun. (Greek, *soma* = body; *plasma* = something formed. Pl., Somatoplasms.) Genetics: The body-tissues that complete a life cycle and then die.

SOMATOPLEURE Noun. (Greek, *soma* = body; *pleura* = rib. Pl., Somatopleures; Somatopleura.) The somatic layer.

SOMATOTHECA Noun. (Greek, *soma* = body; *theke* = sheath. Pl., Somatothecae.) Part of pupa covering abdominal rings; Gasterotheca.

SOMERSALO, ARNE SAKARI (1891–1941) (Saalas 1944, Suomen hyönt. Aikak. 10: 153–154.)

SOMITE Noun. (Greek, *soma* = body; *-ites* = constituent. Pl., Somites.) A body segment of the adult insect; an Arthromere; in the embryo specifically a Metamere. See Segmentation. Cf. Pleurite; Sternite; Tergite.

SOMMER, CARL OTTO FRIEDRICH (1857–1899) (Dittrich 1910, Z. Ent. 25: 25–26, bibliogr.; Anon. 1943, Dt. Ent. Z. Iris 57: 23.)

SOMMER, MICHAEL CHRISTIAN (–1868) (Henriksen 1925, Ent. Meddr. 15: 165–166; Weidner 1967, Abh. Verh. naturw. Ver. Hamburg Suppl 9: 134–135.)

SONCHUS FLY *Ensina sonchi* (Linnaeus) [Diptera: Tephritidae].

SØNDRUP, HANS PEDER STEFFEN (1870–1954) (Møller 1955, Ent. Meddr. 27: 99–103, bibliogr.)

SONIC Adj. (Latin, *sonare* = to sound; *-ic* = of the nature of.) Pertaining to sound.

SONIFACTION Noun. (Latin, *sonare* = to sound; English, *-tion* = result of an action. Pl., Sonifications.) 1. The production of sound by stridulation. 2. A vibratory process by which insect body parts are cleaned or prepared for Scanning Electron Microscopy. See Stridulation.

SONORAN FAUNAL AREAS See Upper and Lower Sonoran.

SONORAN TENT CATERPILLAR *Malacosoma tigris* (Dyar) [Lepidoptera: Lasiocampidae].

SONORIFIC Adj. (Latin, *sonus* = sound; *facere* = to make.) Sound producing; term applied to stridulating organs.

SOOTHSAYERS See Mantidae.

SOOTY MOULD A fungal bloom on plants caused by *Capnodium* spp., *Fumago* spp. and others. These fungi live in the accumulation of honeydew secreted by aphids, mealybugs and scale insects.

SOPRATHION® Parathion.

SORACI, FRANK A (Creighton 1960, J. N. Y. ent. Soc. 68: 1–2.)

SORAUER, PAUL CARL MORITZ (1839–1916) (Soldanski 1916, Dt. ent. Z. 1916: 87–88.)

SORBA® See Lufenuron.

SORDELLI, FERDINANDO (1837–1916) (De Marchi 1916, Atti Soc. ital. Sci. nat. 55: 1–4; Molon 1916, Risv. Orticolo 2: 1-4.)

SORDID Adj. (Latin, *sordidus* from *sordere* = to be dirty.) Dirty; dull.

SØRENSEN, S JOHES (–1974) (Pedersen 1975, Lepidoptera, Købhvn 9: 258.)

SÖRENSEN, WILLIAM EMIL (1848–1916) (Bonnet 1945, *Bibliographia Araneorum* 1: 43–44; Papavero 1975, *Essays on the History of Neotropical Dipterology*. 2: 367. São Paulo.)

SORENSON, CHARLES J (1884–1969) (Knowlton & Hammond 1970, J. Econ. Ent. 63: 1722.)

SORGHUM-HEAD CATERPILLAR *Cryptoblabes adoceta* Turner [Lepidoptera: Pyralidae]: A minor and sporadic pest of citrus, corn, grape, mango, sorghum and persimmon; SHC endemic to Australia. Adult ca 5 mm long, drab grey-brown. Larvae bore into Navels, Silletas and Joppas (mainly where the fruit touch) causing fruit drop.

SORGHUM MIDGE *Contarinia sorghicola* (Coquillett) [Diptera: Cecidomyiidae]: The major global pest of grain sorghum; only flowering heads attacked. Adult 2 mm long; orange bodied; larvae feed on developing grain, causing it to shrivel.

SORGHUM WEBWORM *Nola sorghiella* Riley [Lepidoptera: Noctuidae].

SORHAGEN, LUDWIG FRIEDRICH (1836–1914) (Anon. 1915, Dt. ent. Z. 1915: 88; Brunn 1915, Int. ent. Z. 13: 5–6; Anon. 1919. Arch. Naturgesch. 85: 209, bibliogr.)

SOROKIN, SERGEY VYACHESLAVOVICH (1906–1970) (Blagoveshtchensky 1972, Ent. Obozr. 51: 692–694, bibliogr.)

SORORIA CANEGRUB *Lepidiota sororia* Moser [Coleoptera: Scarabaeidae] (Australia). See Canegrub.

SORSAKOSKI, ONNI (1867–1936) (Hellén 1936, Notul. ent. 16: 94–95.)

SOUDEK, STÉPAN (1889–1936) (Kratochuil 1936, Cas. csl. Spol. ent. 33: 125–131, bibliogr.

SOUND (Latin, *sonus*. Pl., Sounds.) A mechanical disturbance which is potentially referable to external, localized sources that are transmitted through the air, water or substrate. Many insects utilize sound as a method of communication. Structures used to produce sound are often minute and cryptic; sounds produced are ephemeral, often inaudible to humans and frequently require elaborate instrumentation to record. The duration of sound produced by insects varies among Species. Sound may be continuous (such as a hiss) or produced in discrete time intervals. Unitary sounds are divided into chirps, pulses, syllables and echemes (a group of syllables). The periodicity of sound (Hertz) varies with age, sex, and temperature. In general, a hard body produces more Hertz than soft body and a small body produces more Hertz than a large body. A large insect (butterfly) may produce 20 Hz, a medium sized insect (honey bee) may produce 250 Hz, a small sized insect (mosquito) may produce 280–350 Hz, a minute sized insect (ceratopogonid) may produce 1,000 Hz. Insects use several mechanisms for sound production: Sound is produced by anatomical modifications (vibration of elastic structures, shocks to the substrate, stridulation); sound is produced as a consequence of other activities (gnawing, grooming, locomotion). See Drumming; Stridulation. Rel. Auditory Organ; Seta; Sensilla.

SOURBUSH SEED-FLY *Acinia picturata* (Snow) [Diptera: Tephritidae].

SOURSOB MITE *Aplonobia histricina* (Berlese) [Acari: Tetranychidae] (Australia).

SOUTH AFRICAN EMEX WEEVIL *Perapion antiquum* (Gyllenhal) [Coeloptera: Brentidae] (Australia).

SOUTH AFRICAN TICK-BITE FEVER See Kenya Tick Typhus.

SOUTH COAST CONE-WORM *Dioryctria ebeli* Mutuura & Munroe [Lepidoptera: Pyralidae].

SOUTH, RICHARD (1846–1932) (Anon. 1932, Nature 129: 605; Riley 1932, Entomologist 65: 97–100.)

SOUTHEASTERN BLUEBERRY BEE *Habropoda laboriosa* (Fabricius) [Hymenoptera: Anthophoridae].

SOUTHERN ARMYWORM 1. *Persectania ewingii* (Westwood) [Lepidoptera: Noctuidae]: A serious pest of field crops and pastures in Australia. Resembles Common Armyworm in anatomy, biology and life history. Typically 2–3 overlapping generations per year. See Common Armyworm. 2. *Spodoptera eridania* (Cramer) [Lepidoptera: Noctuidae]: A pest of cotton in USA. SAW egg resembles Fall Armyworm; eclosion requires 2–3 days. Neonate and early-instar larvae purple; older larval instars coffee-coloured with three narrow yellow-orange longitudinal lines dorsomedially and black triangles surrounding white spots between the yellow lines; head reddish-brown to pale yellow; mature larva ca 35 mm long. Larval development requires 14–18 days. Pupation within soil, requires 7–12 days; Pupa resembles Fall Armyworm. Adult grey with pale longitudinally-mottled forewing; hindwing pearly-white and finged; wingspan ca 17–32 mm. Female undergoes two day pre-oviposition period; eggs deposited in clusters on underside of leaves. See Armyworms. Cf. African Armyworm; Beet Armyworm; Fall Armyworm.

SOUTHERN BEET WEBWORM *Herpetogramma bipunctalis* (Fabricius) [Lepidoptera: Pyralidae].

SOUTHERN BUFFALO GNAT *Cnephia pecuarum* (Riley) [Diptera: Simuliidae].

SOUTHERN CABBAGEWORM *Pontia protodice* (Boisduval & LeConte) [Lepidoptera: Pieridae]: A Species widespread in North America with larvae sometimes a pest of crucifers. SCW overwinters as pupa and completes; 2–3 generations per year. Alt. Common White; Checkered White.

SOUTHERN CATTLE TICK *Boophilus microplus* (Canestrini) [Acari: Ixodidae].

SOUTHERN CHINCH BUG *Blissus insularis* Barber [Hemiptera: Lygaeidae]: A Species of Chinch Bug found in southeastern North America as a pest of St. Augustine Grass. See Chinch Bug.

SOUTHERN CORN BILLBUG *Sphenophorus callosus* (Oliver) [Coleoptera: Curculionidae].

SOUTHERN CORN ROOTWORM *Diabrotica undecimpunctata howardi* Barber [Coleoptera: Chrysomelidae]: A bivoltine pest of corn in the southern USA. Adult yellow to green with black spots on Elytra; very similar to Mexican and Southern Corn Rootworm adults. Adult overwinters; female undergoes preoviposition period ca 14 days and will lay 200–1,200 eggs. Egg white, ca 0.1 mm long, oval. Larva resembles Mexican Corn Rootworm but displays two spine-like processes on dorsum of apical abdominal segment. Neonate larva burrows into roots and underground stem. Pupa white, fragile. Syn. Spotted Cucumber Beetle (adult).

SOUTHERN CORNSTALK BORER *Diatraea crambidoides* (Grote) [Lepidoptera: Pyralidae]: A bivoltine pest of corn, sorghum and Johnson grass along the east coast of the United States. SCB causes twisted and stunted stalks, and leaves that are ragged or dangling. Eggs flattened, cream-coloured and laid in overlapping rows (shingle-like) on underside of leaves. Young larvae create webbing, feed on leaf whorls and stalks; do not bore into ears. Overwinter among roots, pupate during spring. Cf. Southwestern Corn Borer.

SOUTHERN COWPEA WEEVIL *Callosobruchus chinensis.* [Coleoptera: Bruchidae]: A widespread pest of legumes; most common in tropical and subtropical regions; not reported from Australia. See Cowpea Weevil.

SOUTHERN DART *Ocybadistes walkeri hypochlorus* Lower [Lepidoptera: Hesperiidae] (Australia).

SOUTHERN DOG FACE *Colias caesonia.* [Lepidoptra: Pieridae].

SOUTHERN FALSE WIREWORM 1. *Gonocephalum macleayi* (Blackburn) [Coleoptera: Tenebrionidae]: An Australian pest of cotton. Adult 8–9 mm long, oval and black; Antenna with loose club of 4 segments; elytra with 9 Striae with large punctures. Adult female lays eggs in soil; eggs 1.25–1.30 mm long, oval, smooth and white. Larva elongate-cylindrical, strongly sclerotised and yellow-brown; head Vertex and Abdomen apically black. Pupation occurs in the soil; pupa elongate, cream and 7.0–8.5 mm long. Eggs laid during summer and autumn; larvae feed on organic matter in soil until they reach full size in spring. False wireworms prefer dry conditions protected by stubble or weeds. 2. *Saragus* spp. [Coleoptera: Tenebrionidae]: An Australian pest of cotton. Adult 10–20 mm long, oval-round and black. Commonly called 'pie dish beetles' because of their shape; Elytra in some Species are strongly striate, but smooth and shiny in others; Antenna serrate with a loose 3-segmented club. Larva hard-bodied, shiny, tan-coloured and ca 30 mm long. Pupation occurs in the soil. Eggs laid in summer and autumn; larvae feed on organic matter in the soil until they reach full size in spring. False wireworms prefer dry conditions protected by stubble or weeds. See Tenebrionidae.

SOUTHERN FIRE ANT *Solenopsis xyloni* McCook [Hymenoptera: Formicidae]: A pest endemic to North America and common in Gulf States. Workers yellowish-red with Gaster darker; polymorphic; body hard, to 6 mm long with dense vestiture of Setae; Antenna with 10 segments including club of two segments; Frontal Carinae widely separated and partially concealing Toruli; Clypeus bicarinate with 2–5 teeth along anterior margin; Mandible with three sharply defined teeth, but not sharply incurved; gastral pedicel with two segments. SFA typically nests in soil in exposed areas and forms small, irregular mounds of loose soil. Workers omnivorous with powerful sting and capable of killing small birds; feeds on honeydew, dead insects and carcasses. See Formicidae. Cf. Imported Fire Ant; Little Fire Ant; Native Fire Ant.

SOUTHERN GARDEN LEAFHOPPER *Empoasca solana* DeLong [Hemiptera: Cicadellidae].

SOUTHERN GREEN STINK-BUG See Green Vegetable Bug.

SOUTHERN HOUSE-MOSQUITO *Culex quinquefasciatus* Say [Diptera: Culicidae] (USA): An abundant pest in southestern USA. SHM acts as vector for Dog Heartworm and St Louis Equine Encephalitis. See Culicidae. Cf. Brown House-Mosquito; Northern House Mosquito.

SOUTHERN LADYBIRD *Cleobora mellyi* Mulsant [Coleoptera: Coccinellidae] (Australia).

SOUTHERN LYCTUS BEETLE *Lyctus planicollis* LeConte [Coleoptera: Lyctidae].

SOUTHERN MASKED CHAFER *Cyclocephala immaculata* (Oliver) [Coleoptera: Scarabaeidae].

SOUTHERN MOLE CRICKET *Scapteriscus acletus* Rhen & Hebard [Orthoptera: Gryllotalpidae]: A Species endemic to South America and adventive to the USA where it displays two colour morphs of the Pronotum ('4-dot' and 'mottled'). Both morphs occur in Uruguay, Paraguay and northern Argentina; the 'mottled' morph also occurs in northern Brazil and Venezuela. See Mole Crickets. Cf. Changa Mole Cricket; Imitator Mole Cricket; Northern Mole Cricket; Short-Winged Mole Cricket; Tawny Mole Cricket.

SOUTHERN ONE-YEAR CANEGRUB *Antitrogus mussoni* (Blackburn) [Coleoptera: Scarabaeidae] (Australia). See Canegrub.

SOUTHERN ORANGE DART *Suniana sunias nola* (Waterhouse) [Lepidoptera: Hesperiidae] (Australia).

SOUTHERN PINE BEETLE *Dendroctonus frontalis* Zimmermann [Coleoptera: Scolytidae]: A pest of pine and spruce in eastern USA. Adult 2–3 mm long; pale brown to black; anterior part of head with pair of tubercles separated by median Sulcus; female's Pronotum with transverse ridge along anterior margin, coarsely and sparsely punctate; Elytral Striae with distinct punctures; intervals with ridges. Syn. Arizona Pine Beetle, *Dendroctonus arizonicus*. Cf. Black Turpentine Beetle.

SOUTHERN PINE CONEWORM *Dioryctria amatella* (Hulst) [Lepidoptera: Pyralidae].

SOUTHERN PINE ROOT-WEEVIL *Hylobius aliradicis* Warner [Coleoptera: Curculionidae].

SOUTHERN PINE SAWYER *Monochamus titillator* (Fabricius) [Coleoptera: Cerambycidae].

SOUTHERN POTATO WIREWORM *Conoderus falli* Lane [Coleoptera: Elateridae].

SOUTHERN RED MITE *Oligonychus ilicis* (McGregor) [Acari: Tetranychidae].

SOUTHERN REPTILE TICK *Aponomma hydrosauri* (Denny) [Acari: Ixodidae] (Australia).

SOUTHERN SALTMARSH MOSQUITO *Aedes camptorhynchus* (Thomson) [Diptera: Culicidae] (Australia).

SOUTHERN SANDY DUNG BEETLE *Euoniticellus pallipes* (Fabricius) [Coleoptera: Scarabaeidae] (Australia).

SOUTHERN SCORPION *Cercophonius* spp. [Scorpionida: Bothriuridae] (Australia).

SOUTHERN WHISTLING MOTH *Hecatesia thyridion* Feisthamel [Lepidoptera: Noctuidae] (Australia).

SOUTHERN WOOD NYMPH *Cercyonis (Minois) pegala pegala* [Lepidoptera: Satyridae]: A Species endemic to eastern North America whose larvae feed on grasses. SWN overwinters as a first-instar larva.

SOUTHWESTERN CORN BORER *Diatraea grandiosella* Dyar [Lepidoptera: Pyralidae]: A pest in midwestern and southwestern states of USA. Cf. Southern Cornstalk Borer.

SOUTHWESTERN HERCULES BEETLE *Dynastes granti* Horn [Coleoptera: Scarabaeidae].

SOUTHWESTERN PINE TIP-MOTH *Rhyacionia neomexicana* (Dyar) [Lepidoptera: Tortricidae].

SOUTHWESTERN SQUASH-VINE BORER *Melittia calabaza* Duckworth & Eichlin [Lepidoptera: Seisiidae].

SOUTHWESTERN TENT CATERPILLAR *Malacosoma incurvum* (Hy. Edwards) [Lepidoptera: Lasiocampidae].

SOWBUGS See Slaters.

SOWDA See Onchocerciasis.

SOWTHISTLE APHID *Nasonovia lactucae* (Linnaeus) [Hemiptera: Aphididae]: A vector of Lettuce Necrotic Yellows Virus Disease. Syn. *Hyperomyzus lactucae* (Linnaeus).

SOYBEAN FLY *Melanagromyza sojae* (Zehntner) [Diptera: Agromyzidae] (Australia).

SOYBEAN GIRDLER *Oberopsis brevis* Swederus [Coleoptera: Lamiidae].

SOYBEAN LOOPER 1. *Pseudoplusia includens* (Walker) [Lepidoptera: Noctuidae]. 2. *Thysanoplusia orichalcea* (Fabricius) [Lepidoptera: Noctuidae] (Australia).

SOYBEAN MOTH *Stomopteryx simplexella* (Walker) [Lepidoptera: Gelechiidae] (Australia).

SOYBEAN NODULE-FLY *Rivellia quadrifasciata* (Macquart) [Diptera: Platysomatidae].

SOYBEAN THRIPS *Scericothrips variabilis* (Beach) [Thysanoptera: Thripidae].

SÖYRINKI, YRJO JOHANNES (1904–1941) (Kangas 1944, Suomen hyönt. Aikak. 10: 152, 156.)

SPADICEOUS Adj. (Latin, *spadix* = palm branch; *-aceus* = of or pertaining to.) Arranged as a spadix.

SPADON Noun. (Latin, *spado*, genitive; Greek, *spadonos* = a eunuch. Pl., Spadones.) The worker or neuter in bees and ants. See Caste.

SPAGNOLINI, ALESSANDRO (1833–1880) (Pavesi 1880, Zool. Anz. 3: 312.)

SPALDING, THOMAS UTTING (1866–1929) (Tanner 1929, Ent. News 40: 343–344.)

SPALLAZANI, LAZZARO (1729–1799) (Rose 1850, *New General Biographical Dictionary* 12: 80–81; Espada 1872, An. Soc. esp. Hist. nat. 1: 163–181; Pavesi 1901, Memorie Soc. ital. Sci. nat. 6 (3): 1–68.)

SPANANDROUS Adj. (Greek, *spanos* = scarce; *aner* = male.) Pertaining to a paucity of males.

SPANANDRY Noun. (Greek, *spanos* = scarce; *aner* = male. Pl., Spanandries.) 1. Scarceness of males. 2. Sex ratios characterized by a strong female bias, or infrequency of males. See Sex Ratio.

SPANGBERG, JACOB (1846–1894) (Sandahl 1894, Ent. Tidsskr. 15: 165–168, bibliogr.)

SPANISH RED SCALE *Chrysomphalus dictyospermi* (Morgan) [Hemiptera: Diaspididae] (Australia).

SPANISHFLY *Lytta vesicatoria* (Linnaeus) [Coleoptera: Meloidae].

SPANNIT® See Chlorpyrifos.

SPANOGAMY Noun. (Greek, *spanos* = scarce; *gamos* = marriage. Pl., Spanogamies.) A progressive decrease in the number of females.

SPÄRCK, RAGNAR (1896–1965) (Tuxen 1965, Ent. Meddr. 34: 175–182, bibliogr.)

SPARRE SCHNEIDER, HANS JACOB See Schneider.

SPARRMANN, ANDERS (1748–1820) (Rose 1850, *New General Biographical Dictionary* 12: 83.)

SPARSE Adj. Scattered, spread irregularly and some distance apart, thin, *e.g.* pile of hairs. Alt. Sparsus; Sparsate.

SPASM Noun. (Greek, *spasmos* = tension. Pl., Spasms.) Involuntary muscular contractions.

SPASSKOGO, SERGHEY ALEXSANDROVICH (1882–1958) (Dobrovskii 1960, Ent. Obozr. 39: 959–962, bibliogr. (Translation: Ent. Rev. Wash., 697–699, bibliogr.).)

SPÄTH, FRANZ (1863–1946) (Sachtleben 1943, Arb. morph. taxon. Ent. Berl. 10: 256; Schedl 1946, Zentbl. Gesamtgeb. Ent. 1: 95.)

SPATHA Noun. (Latin, *spatula*; Greek, *spathe* = a blade. Pl., Spathae.) 1. A sword or sword-shaped structure. 2. A median sclerite in the Genitalia of male aculeate Hymenoptera which covers the bases of the Sagittae (Smith); a dorsal lobe of the Aedeagus in Hymenoptera (Snodgrass). 3. Diptera (Tipulidae): Ejaculator Apodeme.

SPATHAL ROD Hymenoptera: Thickening of the Spatha.

SPATHULATE Adj. (Latin, *spatula* = spoon; *-atus* = adjectival suffix.) Sword-shaped; narrow and flat at the base and enlarged at the apex. See Shape. Cf. Spatulate.

SPATULA Noun. (Latin, *spatula* = spoon. Pl., Spatulae.) 1. A spoon or spoon-shaped unicellular process, structure or projection from the bodywall. 2. Diptera: The 'breast bone' of cecidomyiid larvae.

SPATULATE Adj. (Latin, *spatula* = spoon; *-atus* =

adjectival suffix.) Shaped as a spatula; typically rounded or broad at the apex and tapered or attenuate at the base. See Spatula. Rel. Outline Shape.

SPATULATE SENSILLUM A Sensillum which is fan-shaped with a concave face directed away from the surface from which the Sensillum projects. Reported on the Antenna of some parasitic Hymenoptera (Pompilidae) and Tarsomeres of some Encyrtidae. Syn Sensillum Spatulatum. See Sensillum. Cf. Capitate Seta.

SPAWN, GERALD BISHOP (1907–1963) (Walstrom 1964, J. Econ. Ent. 57: 184–185.)

SPEAR Noun. (Middle English, *spere* = spear. Pl., Spears.) A weapon devised by humans and used for stabbing. Specifically, spears are constructed of metal or wood, long, cylindrical, with a straight shaft used as a handle or part with which to propel the device through the air to stab or impale a victim. The spear is tapered to a sharp point and may be barbed or with tines. Spear is used as a comparative descriptor for some entomological terms that describe shape, form or function. Cf. Acantha; Aculeus; Spine. Rel. Hatchet; Sickle; Sword.

SPEAR-WINGED FLIES See Lonchopteridae.

SPECIAL CREATION A doctrine which maintains that all organisms were created in their present form, and denies the operationality of evolutionary process. Cf. Heredity; Evolution.

SPECIAL HOMOLOGY The concept of Homology developed by Owen (1843) to indicate 'the same organ in different animals under every variety of form and function'. Criteria used to establish Special Homology include (a) interindividual structural correspondence, (b) agreement in relative position and connections of correspondent parts, and (c) agreement of correspondent parts in adult structure, development or both. Special Homology involves interspecific identification of homologous structures, and is used as a guide to genetic relationship. Workers subsequent to Owen have adopted a meaning of Special Homology which emphasizes common ancestry. See Homology. Cf. Serial Homology.

SPECIALIZATION Noun. (Latin, *specialis* = special; English, *-tion* = result of an action. Pl., Specializations.) 1. The restricted adaptation or function of an animal or of a structure.

SPECIALIZED Adj. (Latin, *specialis* = special.) Biology: A comparative term used in contrast with 'generalized' to indicate a relatively recently evolved character or character state. A 'specialized organism' is one in which derived characters predominate. Structure which is derived from the primitive or ancestral condition. That which is adapted to unique or unusual conditions when compared with other structures or related Taxa. See Derived; Primitive. Cf. Generalized. Ant. Unspecialized.

SPECIES Singular and Plural Noun. (Latin, *species* = particular kind.) 1. Species or kind. The primary biological unit, debatably an actual thing or a purely subjective concept. 2. A static moment in the continuum of life; an aggregation of individuals similar in appearance and structure, mating freely and producing young that themselves mate freely and bear fertile offspring resembling each other and their parents, including all varieties and races (Smith). Abbreviated sp. for one Species, and spp. for two or more Species. See Biological Species Concept; Morphological Species; Ecospecies.

SPECIES CONCEPTS Several definitions for Species have been proposed in the biological literature; they have become known as 'Species concepts'. See an individual entry for particular definitions related to each concept: Biological Species Concept, Cohesion Species Concept, Ecological Species Concept, Evolutionary Species Concept, Phylogenetic Species Concept, Recognition Species Concept.

SPECIES INDETERMINATA Taxonomy: An indeterminate Species; abbreviated sp. indet., sp. ind.

SPECIES INQUIRENDA Taxonomy: A Species in question.

SPECIES NOVUM Taxonomy: A Species new to science and described in the text following. (Pl., Species Nova.) Abbreviated 'Spec. Nov.', 'N. Sp.', 'N.S.' or 'Sp. N.'

SPECIES ODOUR Social Insects: The odour on members of a Species which is unique to that Species. See Colony Odour.

SPECIFIC CHARACTER A feature or structure common to all individuals of a Species, by means of which they may be distinguished from all other individuals of other Species; an essential character.

SPECIFIC NAME The single or compound name applied to a Species, which must be only one word, *e.g.* an adjective, which must agree grammatically with the generic name, or a noun in the nominative in apposition with the generic name, or nouns in the genitive form, and all according to the rules of Latin grammar.

SPECIFIC TRIVIAL NAME Taxonomic nomenclature: The second term or epithet of the binomial designation of a Species. See Taxonomy.

SPECIFIC-T® See Codling-Moth Granulosis Virus.

SPECIMEN Noun. (Latin, *specere* = to look; to behold. Pl., Specimens.) 1. A sample. 2. A single example of anything *e.g.*, of an insect, the individual unit of a collection, a single insect of any Species, or a part of an insect properly preserved. Rel. Exemplar; Taxon.

SPECIOGENESIS Noun. (Latin, *species* = particular kind; Greek, *genesis* = being produced. Pl., Speciogeneses.) Species reproduction (J. C. Chamberlain).

SPECKLED COCKROACH *Nauphoeta cinerea* (Oliver) [Blattodea: Blaberidae]: Endemic to Africa and now circumtropical in distribution. Males stridulate during courtship. SC is ovoviviparous; female extrudes Ootheca, rotates Ootheca then retracts it into brood chamber. A minor pest in Australian poultry farms where it infests grain, hay sheds and sometimes bird nests. See Blattodea.

SPECKLED LINEBLUE *Catopyrops florinda halys* (Waterhouse) [Lepidoptera: Lycaenidae] (Australia).

SPECKLED LONGICORN *Paradisterna plumifera* (Pascoe) [Coleoptera: Cerambycidae]: A pest of *Citrus* and Radiata Pine in Australia. Adult ca 12 mm long; grey and speckled with brown spots; Antenna longer than body. Eggs laid on bark of trunk and larger branches; eclosion occurs within 10 days; neonate larvae bore into wood and produce sawdust-like frass which accumulates on trunk. Diseased and older trees seem attractive to adult beetles.

SPECKLED SNOUT-BEETLE *Eremnus cerealis* Marshall [Coleoptera: Curculionidae] (South Africa). Cf. Grey Snout-Beetle.

SPECTACULAR GRASSHOPPER See Leichhardt's Grasshopper.

SPECTRACIDE® See Diazinon.

SPECTRAL VEIN Hymenoptera: An unpigmented vein that is normally invisible by transmitted light but can be seen by light reflected of the wing membrane because its contour is marked by a ridge or furrow on the wing surface (Mason 1986, 88: 2). See Vein; Wing Venation. Cf. Nebulous Veins; Spurious Vein; Trace Vein; Tubular Vein. Rel. Horismology.

SPECTRUM Noun. (Latin, *spectrum* = appearance. Pl., Spectra.) A range of electromagnetic radiation, often colours. The entire range of variables between two opposing extremes. Rel. Range.

SPECULAR Adj. (Latin, *speculum* = mirror.) Mirror-like. Pertaining to a Speculum. See Attribute.

SPECULAR MEMBRANES Male Cicada: The inner or posterior mirror-like membrane of the sound-organ. Syn. The Mirror.

SPECULUM Noun. (Latin, *speculum* = mirror. Pl., Specula.) 1. Birds: A metallic-coloured stripe on the wing. 2. Parasitic Hymenoptera: An oblique, bare stripe on the forewing Cf. Clasura. 3. A transparent area or spot on wings of some adult Lepidoptera or a spot on the neck of some caterpillars. 4. A glassy area at base of Tegmina in male Orthoptera that serves as a sounding board in acoustical communication.

SPEGAZZINI, CARLOS (1858–1926) (Bruch 1926, Revta. Soc. ent. argent. 1: 69–74; Molfino 1929, An. Soc. cient. argent. 108: 7–77, bibliogr.)

SPEMANN, HANS (1869–1941) (Sachtleben 1942, Arb. physiol. angew. Ent. Berl. 9: 55.)

SPENCE, ROBERT HENRY (–1851) (Westwood 1951, Proc. ent. Soc. Lond. 851: 135.)

SPENCE, WILLIAM (1783–1860) English Entomologist and coauthor of *Introduction to Entomology* (4 vols, 1815–1826) with William Kirby. (Eiselt 1836, *Geschichte. Systematik und Literatur der Insektenkunde.* 255 pp. (101–105), Leipzig; Stainton 1861, Proc. ent. Soc. Lond. 1858–61: 92–94; Neave 1933, Centennial history of the Entomological Society of London. xlvi + 224 pp. (127–128), London.)

SPENCE, WILLIAM BLUNDELL (–1900) (Verrall 1900, Proc. ent. Soc. Lond. 1900: xlii–xliii; Griffin 1933, Nature 1933: 178.)

SPENCER, GEORGE JOHNSTON (1880–1966) (Graham 1966, J. ent. Soc. Br. Columb. 63: 42–43.)

SPENGLER, LOREN (1720–1807) (Zimsen 1964, *The Type Material of J. C. Fabricius.* 656 pp. (15), Copenhagen.)

SPERCHEIDAE Plural Noun. See Hydrophilidae.

SPERIMOS® A registered biopesticide derived from *Bacillus thuringiensis* var. *israelensis*. See *Bacillus thuringiensis*.

SPERM Noun. (Greek, *sperma* = seed. Pl., Sperms. Spermatozoa.) Any male gamete. Principal components of sperm include an acrosome, head, nucelus and tail. See Spermatozoon. Rel. Gamete.

SPERM ACTIVATION A process that involves changes in Spermatozoon motility. Insect Spermatozoa may actively move within the female reproductive system, remain quiescent for prolonged periods and become active in the presence of eggs requiring fertilization. This is common in some insects and other arthropods such as spiders.

SPERM ANATOMY Generalized insect Spermatozoa is elongate (filament-like) with two regions: An enlarged head at anterior end and a long Flagellum (tail) projecting from head. Head contains several organelles (Centriole; Centriole Adjunct, Acrosome, Nucleus); Flagellum is usually derived from mitochondral elements. A few insects have double-headed Spermatozoa (Lepismatidae; *Dytiscus marginalis*), but significance of feature is not understood; double-headed Spermatozoa also occur in American marsupials and some gastropods (*Turritella terebra*). Surface of mature Spermatozoon is structurally uniform and consists of a Plasma Membrane surrounded by a coat of varying size and complexity. Outer coat (Glycocalyx) consists of glycoprotein–proteinaceous rods oriented at a right angle to primary axis of Spermatozoon. Plasma Membrane occurs beneath outer covering. PM is trilaminar and asymmetrical but differs in size and physical appearance among insects; PM shows intense phosphatase activity. Head of Spermatozoon typically is elongate or tear-drop shaped, but exceptions occur. Nucleus of Spermatozoon is usually elongate, spindle-shaped anteriad and posteriorly truncate. Mecoptera and Thysanura Nucleus is sometimes helicoidal (coiled) around Flagellum; some Isoptera that lack a Flagellum display Nucleus spheroidal or flattened. Nucleus of mature Spermatozoon contains DNA that stands parallel to primary axis of Spermatozoon. Nucleus also contains proteins that replace histones; changes in nuclear material are noted during Spermatogenesis. Nucleus of mature Spermatozoon does not contain RNA. Some insects only have one Centriole; many animal Spermatozoa have two Centrioles and young spermatids apparently contain two Centrioles. One Centriole serves as Basal Body for the growing Flagellum; other Centriole orients at a right angle to first Centriole. Centrioles disappear during Spermatogenesis, and mature insect Spermatozoa lack a Centriole. Centrioles may not be necessary for initiation of cleavage. See Sperm Flagellum.

SPERM CAPACITATION A process that involves anatomical changes to Spermatozoa inside the female reproductive tract. Capacitation proposed in mammals and may occur in insects, but processes involved between mammals and insects are not identical in origin or function. Mammalian Spermatozoon capacitation involves three processes: Agglutination, activation and acrosome reaction. See Spermatozoa.

SPERM CAPSULE Odonata: Rounded masses of Spermatozoa which have adhered in radiating fashion, somewhat gummy externally (Imms).

SPERM COMPETITION The phenomenon of competition in which one male's sperm establishes precedence over another male's sperm for egg fertilization in a female that has mated more than once. SC typically manifested by the timing of copulation with sperm from the last copulation effectively becomes the first sperm to fertilize eggs as they are oviposited. Cf. Paternity Assurance.

SPERM CYST A cellular envelope (capsule) in the testis of most insects which encloses a group (bundle, packet) of Spermatogonia. Syn. Spermatogonial Cyst.

SPERM DUCT The micropilar canal *sensu* Korschelt (1884, Zool. Anz. 7: 394–398, 420–425); Johannsen & Butt (1941, Embryology of Insects and Myriapods).

SPERM FLAGELLUM A tail-like structure projecting from the head of an insect Spermatozoon. Some insect Spermatozoa lack a Flagellum (*e.g.* some Diptera, Coccoidea and Isoptera); other insects (some Psocoptera, Mallophaga, Anoplura, Thysanoptera and Rhynchota) have two Flagella; primitive termite *Mastotermes darwiniensis* Froggatt has a conical head with about 100 weakly motile Flagella. Typical Spermatozoon Flagellum consists of an Axoneme (Axial Filament) completely enclosed by a Ciliary Membrane; CM serves as a barrier between exterior environment and Axoneme. See Axoneme.

SPERM GUIDE A micropylar device in the form of an external depression in the Chorion adjacent to the Micropyle on Ephemeroptera eggs. (Koss 1968, Ann. Rev. Ent. 61: 696–721.)

SPERM MOVEMENT Directed Spermatozoon movement depends upon Flagellum motion. Flagella provide undulatory motion that consists of symmetrical movements. Straight line motion consists of two symmetrical bends of opposite direction per wavelength, separated by two straight regions of the Flagellum that are of identical length. Circular motion involves a principal bend angle which is greater than the reverse bend angle. Characteristically, wavelength is shorter than the organelle. Flagellum bending is achieved with energy obtained by hydrolysis of ATP in dynein arms. Beat form and beat frequency depend on the temperature, ionic environment, physiological conditions of structure and medium viscosity. See Sperm Flagellum. Cf. Cell Movement.

SPERM TUBE One of the secondary divisions of the testis (Snodgrass).

SPERMALEGE Noun. (Greek, *sperma* = seed; *lego* = to gather. Pl., Spermalegia.) Cimicidae: A structure on the female's Abdomen which receives sperm during traumatic insemination. Typically consists of two components: an integumental part (Ectospermalege) and internal component (Mesospermalege). See Traumatic Insemination. Rel. Reproductive System.

SPERMATHECA Noun. (Greek, *sperma* = seed, *theke* = case. Pl., Spermathecae.) A sac, duct or reservoir within the female insect that receives Spermatozoa during copulation. The Spermatheca is derived from Ectoderm, lined with Cuticle, musculated and attached to the Median Oviduct. The size and shape of the Spermatheca varies from a small, inconspicuous sac to a large, bulbous tube. The generalized spemathecal complex consists of a spermathecal duct opening from the Median Oviduct, a Spermatheca which serves as a sac-like reservoir for the Spermatozoa and a spermathecal gland. The generalized hymenopteran Spermatheca consists of a capsule, gland and duct. Spermatozoa are stored in the capsule; the gland secretes a substance into the ductlets or directly into the capsule. The common duct transmits Spermatozoa into the Median Oviduct. Spermathecae are variable in number and capable of storing Spermatozoa for several years in some Species of insects. Syn. Receptaculum Seminis. Alt. Spermatotheca.

SPERMATHECAL Adj. (Greek, *sperma* = seed, *theke* = case; Latin, *-alis* = characterized by.) Descriptive of or pertaining to the Spermatheca.

SPERMATHECAL GLAND A special gland opening into the duct of the Spermatheca (Imms).

SPERMATID Noun. (Greek, *sperma* = seed. Pl., Spermatids.) The final male reproductive cell, arising by division of the second Spermatocytes; contains the haploid Chromosome number and is converted without further reduction division into a Spermatozoon.

SPERMATOCYST Noun. (Greek, *sperma* = seed; *kystis* = bladder. Pl., Spermatocysts.) The Seminal Sac.

SPERMATOCYTE Noun. (Greek, *sperma* = seed;

kytos = hollow. Pl., Spermatocytes.) A cell which develops from growth and modification of a spermatogonium. The primordial germ cell in the male divides to form two secondary Spermatocytes, each of which develops into two spermatids.

SPERMATOGENESIS Noun. (Greek, *sperma* = seed, *genesis* = origin. Pl., Spermatogeneses.) The process of sperm development through reduction division of chromosomes and transformation of cell content from a Spermatogonium, through primary and secondary Spermatocytes to a Spermatozoon. This process is architecturally more complex than companion process of oogenesis in female. At the end of the developmental transition, Spermatozoon traditionally has been recognized to consist of three regions: Acrosome, nucleus and tail. Syn. Spermiogenesis. Cf. Oogenesis.

SPERMATOGENETIC Adj. (Greek, *sperma* = seed; *genesis* = descent; *-ic* = of the nature of.) Pertaining to sperm formation.

SPERMATOGONIAL CYST See Sperm Cyst.

SPERMATOGONIUM Noun. (Greek, *sperma* = seed; *gonos* = offspring; Latin, *-ium* = diminutive > Greek, *-idion*. Pl., Spermatogonia.) The primordial male germ cell. Cf. Oogonium.

SPERMATOID Adj. (Greek, *sperma* = seed; *eidos* = form.) Sperm-like. Pertaining to structures which resemble Spermatozoa in form or function. See Spermatozoa. Rel. Eidos; Form; Shape.

SPERMATOPHORA See Spermatophore.

SPERMATOPHORE Noun. (Greek, *sperma* = seed; *pherein* = to bear, carry. Pl., Spermatophores.) 1. A capsule (cover, envelope) that surrounds Spermatozoa. 2. A seminal packet composed of seminal fluid mixed with excretions of the Male Accessory Glands (Mesadenia). Spermatophore is typically manufactured by elements of the male repoductive system and transmitted to the female during copulation. Spermatophores are highly variably in size, shape and hardness. Shape can be determined by male or female. Male-determined Spermatophores are produced in the male Ejaculatory Duct before copulation or moulded in the male during copulation. Male-determined Spermatophores constructed in the anterior end of the Ejaculatory Duct before copulation are typically held in place by the female's external genitalia after copulation. The gelatinous matrix and sperm sacs are held outside the female's body. Structurally, this type of male-determined Spermatophore is ornate. The MDS is regarded as more primitive and is found in Blattaria, Orthoptera, Odonata, Neuroptera and chironomid midge *Glyptotendipes paripes*. MDS can also be moulded in the male copulatory organ or within an everted Spermatophore sac of the copulatory organ during copulation. This type is found in Hemiptera (Reduviidae), Diptera (Ceratopogonidae) and Coleoptera (Dytiscidae, Scarabaeidae). Female-determined Spermatophores are formed within the female reproductive system, and their shapes are determined by the shape of the female reproductive system. Male accessory gland secretions encapsulate Spermatozoa and are placed in the Bursa Copulatrix or Vagina. This strategy is seen in all Trichoptera, Species of Lepidoptera that produce Spermatophores, some Coleoptera (Coccinellidae, Meloidae), some Diptera (Simuliidae) and perhaps some Hymenoptera. Outline shapes include (but are not restricted to) stalk-shaped, bottle-shaped and dumb-bell shaped. Spermatophores prevent desiccation of terrestrial arthropod Spermatozoa; Spermatophores also provide nutrients to a female reproductive partner and/or serve as a physical or chemical barrier to subsequent copulation, insemination or fertilization by another male. Not all insect Species produce Spermatophores; some male insects transfer small groups of Spermatozoa to the female as suspensions within a viscous seminal plasma. Spermatophores are regarded as most common and elaborate among lower orders of Pterygota. Other types of sperm-packaging devices include droplet Spermatophores, sperm cysts, sperm bundles and Spermatozeugmata (large aggregates of Spermatozoa). Sperm balls have been described in some Thysanoptera, but these may represent something other than Spermatophores. Alt. Spermatiophore. See Male Accessory Gland. Cf. Droplet Spermatophore.

SPERMATOPHRAGMA Noun. (Greek, *sperma* = seed; *phragma* = fence. Spermatophragmata.) Lepidoptera: A structure beneath the Penis used for the reception of the Spermatophore (Amauris, Acraea) (Chapman 1982, p. 367.)

SPERMATOPHYLAX Noun. (Greek, *sperma* = seed; *phylassein* = to protect. Pl., Spermatophylaxes.) Orthoptera (Tettigoniidae): A sperm-free portion of a Spermatophore which is manufactured by a male and consumed by a female after copulation. (Boldyrev 1915, Hor. Ent. Soc. Ross. 41: 1–245.)

SPERMATOPOSITOR Noun. (Greek, *sperma* = seed; Latin, *ponere* = to place; *or* = one who does something. Pl., Spermatopositors.) Acarology: The male organ for Spermatophore deposition.

SPERMATOTHECA Noun. See Spermatheca.

SPERMATOZOON Noun. (Greek, *sperma* = seed; *zoon* = animal. Pl., Spermatozoa.) 1. A male gamete (cell) consisting of a head (containing a nucleus) with acrosome, and usually a propulsive thread-like tail, which unites with the egg or female ova during the process of fertilization. 2. The mature sperm cell. See Spermatogenesis; Sperm Anatomy.

SPERMIOGENESIS Noun. See Spermatogenesis.

SPERMORA Noun. The mouth of the duct of the Spermatheca (MacGillivray).

SPERMORARIA Noun. The special cuticular area on which the spermora is set (MacGillivray).

SPERMOREA Noun. The mouth of the duct of the Spermatheca (MacGillivray).

SPERONI, ALBERTO P (1893–1944) (Anon. 1945, Revta. Soc. ent. argent. 12: 339.)

SPERRY, JOHN LOVELL (1894–1945) (Comstock 1954, Bull. sth. Calif. Acad. Sci. 53: 58–59; Remington 1954, Lepid. News 8: 30.)

SPESSIVTSEFF, PAUL (1866–1938) (Butubitsch 1938, Z. angew. Ent. 25: 536–537, bibliogr.)

SPEYER, ADOLPH (1812–1892) (Elwes 1893, Proc. ent. Soc. Lond. 1893: lvii–lviii; Speyer 1893, Dt. ent. Z. Iris 6: 37–68.)

SPEYER, ARTHUR JOHANNES (1858–1922) (Weidner 1967, Abh. Verh. naturw. Ver. Hamburg Suppl. 9: 193.)

SPEYER, EDWARD RICHARD (1888–1974) (Lees 1975, Proc. R. ent. Soc. Lond. (C) 39: 57.)

SPEYER, WALTER (1889–1958) (Anon. 1958, Z. wien. ent. Ges. 43: 290; Buhl 1958, Anz. Schädlingsk. 31: 140–141.)

SPHAERIIDAE Minute Bog Beetles. See Microsporidae.

SPHAERITIDAE Plural Noun. A single Species, *Sphaerites politus* Mannerheim, occurs in the Pacific Northwest of U.S. 3.5–5.5 mm long, black with metallic bluish lustre; Antennae not elbowed; Tibiae swollen, lack teeth; last abdominal section exposed beyong elytra. Feed on carrion, manure and rotting fungus. False Clown Beetles.

SPHAEROCERATIDAE Plural Noun. See Sphaeroceridae.

SPHAEROCERIDAE Plural Noun. Dung Flies. Widespread, numerically moderate-sized Family of acalypterate flies. Adults black, brown; recognized by basal segment of hind Tibia short, swollen and longitudinal veins not reaching apex or margin of wing. Occur in areas of standing water near excrement; larvae feed in excrement and refuse.

SPHAEROCOCCINAE A cosmopolitan Subfamily of typically legless mealybugs. Most Species feed on grasses, but Australian Species of *Sphaerococcus* feed on *Casuarina*. Containing about 15 Genera and 40 Species, and apparently most closely related to the Rhizoecinae. (Koteja 1974, Sci. Pap. Agric. Univ. Karkow 89, 162 pp.)

SPHAEROPSOCIDAE Plural Noun. A numerically small, infrequently collected Family of Psocoptera assigned to the Suborder Troctomorpha. Body frequently ornately sculptured; Antenna with fewer than 20 segments and secondary annulations; wings present or absent, forewing elytriform; Tarsi with three segments. Inhabit grasses in Tasmania and Western Australia. See Psocoptera.

SPHECIDAE Latreille 1802. Plural Noun. Beetle Wasps; Burrowing Wasps; Cicada Killers; Digger Wasps; Mud Daubers; Sand Wasps; Square Headed Wasps. A cosmopolitan, large Family (ca 8000 Species) of aculeate Hymenoptera assigned to Superfamily Sphecoidea. Male Antenna with 13 segments, female Antenna with 12 segments; Pronotum not extending to Tegula, but lateral margin forming a lobe which covers Spiracle; both sexes macropterous, forewing with Pterostigma and many closed cells; hindwing with closed cells, jugal lobe and anal lobe; hindleg and Gaster without Scopa or Corbicula; body Setae simple. Adults nectivorious; larvae carnivorous. Prey spectrum broad, but many feed on narrow range of prey Taxa; some feed on prey from broad range of insect Orders. Prey may be one large individual or many smaller individuals, depending upon sphecid Species. Some sphecids capture and provision with many small spiders; some provision nest with large caterpillars; some sphecids take small Lepidoptera adults. Generally, nests provisioned with several smaller prey rather than a small number of larger prey. Female wasp locates prey, stings prey (usually paralysed, sometimes killed), then the prey is flown, carried or dragged to nest. Most sphecids construct nest before searching for prey; some sphecids add prey to nest until filled. One egg laid upon prey, nest sealed and larva develops as it consumes prey. A few sphecids progressively provision nest until larva matures; burrow then sealed and larva pupates. Some females may progressively provision several nests simultaneously. Sphecid nest architecture diverse: Constructed by tunnelling in earth, utilize cavities in rotten wood, plant stems or beetle burrows in timber. Diggers may prefer friable or sandy soil; some excavate tunnels in very hard material. Digging Species often show rows of specialized Setae or spines on forelegs used in excavating tunnel. Typically, tunnel in soil terminates in enlarged brood cell; sometimes apex of tunnel radiates into several cells, or cells may be constructed at intervals off main tunnel. Species nesting in plant cavities may construct linear series of cells; some Species use mud, pith, resin or sawdust to partition cells. Pemphredonines usually prey on aphids and nests in hollow twigs or beetle burrows. Some Species possess specialized Setae on Mandibles used to manipulate cell-closing material. Sphecids in cold, temperate climates usually have only one generation per year. Overwintering usually in prepupal stage. Sphecids frequently nest together or in circumscribed areas; this is thought linked to eusociality or a precursor of it. Aggregative nesting may result from limited availability of nesting sites, or nesting strategies in themselves may be adaptive. Included Subfamilies: Astatinae, Crabroninae, Larrinae, Nyssoninae, Pemphredoninae, Philanthinae, Sphecinae. Cf. Vespoidea.

SPHECOIDEA Latreille 1802. Hymenoptera. A Superfamily of apocritous Hymenoptera. Anatomically similar to bees but differing in not possessing branched Setae on the body. Families include: Ampulicidae, Baissodidae and Sphecidae. Under some classifications only Sphecidae recognized; other classifications included Apoidea.

SPHECOPTERIDAE Plural Noun. A Family of Megasecoptera found in Upper Carboniferous beds of Cemmentry, France.

SPHERICAL Adj. (Greek, *sphaira* = globe; Latin, *-alis* = characterized by.) Descriptive of structure shaped in the form of a sphere. Alt. Sphaericus. See Shape.

SPHERICAL MEALYBUG *Nipaecoccus viridis* (Newstead) [Hemiptera: Pseudococcidae]: A pest of avocado, citrus, fig, grape, mango, papaya and soursop in tropical and subtropical areas of Africa, southeast Asia, Australia and islands of the Pacific. Adult female oval in outline shape, 2.5–4.0 mm long, slightly flattened, body covered with creamy, white wax; body purplish or brown; body fluid (Haemolymph) purple. Male considerably smaller than female, fragile, with non-functional mouthparts, one pair of functional wings. Female oviposits batches of eggs into specially constructed white, striated, hemispherical egg sac; ca 500 purple eggs produced per female during lifetime; crawlers purple; female passes through three moults; male passes through four moults; life cycle requires ca three weeks. SMB produces large amounts of honeydew which serve as substrate for sooty mould and other fungi.

SPHERICAL SYMMETRY A type of symmetry in which an organism or body lacks a definable axis (*e.g.* sponges). See Asymmetry; Symmetry. Cf. Bilateral Symmetry; Pentameral Symmetry; Zonal Symmetry.

SPHERIMOS® See *Bacillus sphaericus*.

SPHEROIDAL Adj. (Greek, *sphaira* = globe; Latin, *-alis* = characterized by.) More or less spherical, sphere-like.

SPHEROIDOCYTE See Adipohaemocyte.

SPHERULATE Adj. (Latin, *sphaerula* = small globe; *-atus* = adjectival suffix.) Provided with one or more rows of minute tubercles or spheres.

SPHERULE Noun. (Latin, *sphaerula* = small globe. Pl., Spherules.) 1. A minute sphere or globule. 2. A refringent granule in leucocyte cytoplasm.

SPHERULES OF GRANULES Pupae of Diptera: Spherical bodies with muscle fragments (Sarcolytes).

SPHERULOCYTE Noun. (Latin, *sphaerula* = small globe; Greek, *kytos* = hollow. Pl., Spherulocytes.) A basic anatomical form of insect Haemocytes. Variable-sized, oval-round Haemocytes; Plasma Membrane with or without micropapillae, filipodia or irregular processes; Nucleus typically small, centrally or eccentrically positioned; Cytoplasm with spherules containing granular, fine-textured, filamentous or flocculent material; Cytoplasm also contains Polyribosomes, Golgi Bodies and Lysosomes. Histochemical analysis shows spherules with lipochrome, tyrosinase, glycoproteins, glycomucoproteins and mucopolysaccharides. Syn. Eleocyte; Eruptive Cell; Hyaline Cell; Rhegmatocyte. See Haemocyte. Cf. Adipohemocyte; Coagulocyte; Granulocyte; Nephrocyte; Oenocyte; Plasmatocyte; Prohaemocyte. Rel. Haemolymph.

SPHINCTER Noun. (Greek, *sphiggein* = to bind tight. Pl., Sphincters.) A muscle which contracts to close or constrict an opening or orifice.

SPHINDIDAE Plural Noun. Dry Fungus Beetles. A small Family of polyphagous Coleoptera assigned to Cucujoidea. Adults are 1–2 mm long, body oval to ovate, convex and typically finely pubescent. Antenna with 10 segments including club of three segments; tarsal formula 5-5-5 in female, 5-5-4 in male; Abdomen with five Ventrites. Larval head prognathous; six Stemmata on each side of head; Antenna with three segments; legs with five segments, two tarsal segments on each Pretarsus; Urogomphi absent. Adults and larvae feed on spores of slime mould. See Cucujoidea.

SPHINGIDAE Plural Noun. Hawk Moths; Sphinx Moths. A predominantly tropical Family of ditrysian Lepidoptera assigned to Superfamily Sphingoidea or placed in Bombycoidea in some classifications. Adults are medium to large bodied, wingspan 36–190 mm and head with short scales and eyes large. Ocelli typically absent but sometimes minute and concealed beneath scales; Chaetosemata absent; Antenna short, variable in form; Proboscis typically large, well developed, coiled at repose, without scales; Maxillary Palpus of one scaled segment; Labial Palpus ascending Frons, medial surface without scales; Thorax without tympanal organ; Epiphysis present; tibial spur formula usually 0-2-4 or 0-2-2; forewing long and narrow; hindwing shorter. Egg flattened, nearly spherical in outline shape with Micropyle at side; Chorion smooth; eggs laid individually on host plant. Larva without conspicuous Setae; abdominal Prolegs on segments 3–6; Anal Prolegs large; segment 8 typically with dorsal horn; feeds diurnally. Pupa fusiform, Maxillary Palpus absent; Labial Palpus concealed; Cremaster conspicuous; pupation within cell in soil or cocoon in debris. Adults fast flying, take nectar while hovering; most crepuscular or nocturnal; some Species diurnal.

SPHINGOIDEA A Superfamily of ditrysian Lepidoptera including the Sphingidae; not recognized in some classifications with Sphingidae placed in Bombycoidea.

SPHINX MOTHS See Sphingidae.

SPHRAGIDAL Adj. (Greek, *sphragis* = seal; *-alis* = characterized by.) Pertaining to an accessory gland fluid secreted by male Lepidoptera which forms a Sphragis.

SPHRAGIS Noun. (Greek, *sphragis* = seal.) Lepidoptera: A male structure which detaches during copulation and remains with the female, forming a seal on the female Bursa Copulatrix.

SPHRAGOPHOR Noun. (Greek, *sphragis* = seal; *pherein* = to bear, carry; *-or* = one who does something.) Hymenoptera: first Valvula.

SPICA Noun. (Latin, *spica* = spike, point. Pl., Spicae.) 1. A spike-like structure. See Shape. 2. A Calcar.

SPICATE Adj. (Latin, *spica* = spike; *-atus* = adjectival suffix.) A surface covered with spikes. Displaying an arrangement of spikes. See Shape.

SPICEBUSH SWALLOWTAIL *Papilio troilus* Linnaeus [Lepidoptera: Papilionidae]: A common Species in eastern North America. Larva lives in folded leaves; overwinters as pupa; 2–3 generations per year. Host plants include Spice bush, sassafras, sweet bay. Alt. Green-clouded Swallowtail; Green-spotted Swallowtail.

SPICIFORM Adj. (Latin, *spica* = spike; *forma* = shape.) Spike-shaped. See Spike. Cf. Acanthiform; Pencilliform; Spiculiform; Spiniform. Rel. Form 2; Shape 2; Outline Shape.

SPICISETA Noun. (Latin, *spica* = spike; *seta* = hair. Pl., Spicisetae.) One of three types of body Setae found on larval Dermestidae. Variable in length and sometimes exceeding the body length; Spicisetae are typically composed of sharply pointed, overlapping scales. A sensory function has been suggested. See Sensillum. Cf. Hastiseta; Nudiseta.

SPICULA Noun. (Latin, *spicula* = small spike. Pl., Spiculae.) See Spiculum.

SPICULE Noun. (Latin, *spicula* = small spike. Pl., Spicules; Spiculi.) 1. Any minute, pointed process or spine. Cf. Bristle; Seta; Spine. 2. Diptera: Extremely short, dark, socketed spines. See Spiculum.

SPICULIFORM Adj. (Latin, *spicula* = small spike; *forma* = shape.) Spicule-shaped; structure shaped as a slender, needle-like process. Alt. Spiculate. See Spicule. Cf. Acanthoid; Aculeiform; Penicilliform; Spiciform; Spiniform. Rel. Form 2; Shape 2; Outline Shape.

SPICULUM Noun. (Latin, *spiculum* = a point or dart. Pl., Spicula.) 1. Any slender, needle-like process. 2. Hymenoptera: The Sting or equivalent of the Ovipositor (archaic); the anterior median apodeme of Sternum IX. Syn. Spicula; Spiculum. Cf. Acantha; Spine.

SPIDERS See Araneae.

SPIDER BEETLE *Gibbum psylloides* (Czenpinski) [Coleoptera: Ptinidae]: A cosmopolitan pest of stored products, wool and leather. SB larva is the injurious stage. Development: Egg stage 13–20 days; larval stage 29–35 days; pupal stage 15–18 days; adults 210–280 days.

SPIDER BEETLES See Ptinidae.

SPIDER LONGICORN See Citrus Longicorn.

SPIDER MITES See Tetranychidae.

SPIDER PARASITE FLIES See Acroceridae.

SPIDER WASPS See Pompilidae.

SPIDERHUNTING SCORPION *Isometroides vescus* (Karsch) [Scorpiones: Buthidae] (Australia).

SPIDER-MITE DESTROYER *Stethorus picipes* Casey [Coleoptera: Coccinellidae].

SPIEGEL, M A (–1961) (Anon. 1961, Bull. Soc. ent. Fr. 66: 127.)

SPIKE Noun. (Middle English, *spike*. Latin, *spica* = head of grain. Pl., Spikes.) 1. An ear of grain. 2. An elongated, indeterminant inflorescence with flowers sessile on the main axis. 3. A very large, metal nail that is often square in cross-section. 4. An elongate, hard projection from the integument or egg. Cf. Spine. Rel. Spiniform.

SPIKED PEA-GALL The gall formed by the cynipid wasp *Diplolepis nervosa* on rose leaves. See Pea Gall.

SPINA Noun. (Latin, *spina* = thorn, spine. Pl., Spinae.) 1. The unpaired, median apodemal process of a Spinasternum (Snodgrass). 2. The median apodeme of the Poststernellum (Imms). See Apodeme; Spinasternum; Postscutellum. Cf. Furca.

SPINACH FLEA-BEETLE *Disonycha xanthomelas* (Dalman) [Coleoptera: Chrysomelidae]: A pest of amaranth, beets and chickweed in eastern USA. SFB resembles Three-spotted Flea Beetle in anatomy, biology and distribution. See Chrysomelidae.

SPINACH LEAFMINER *Pegomya hyoscyami* (Panzer) [Diptera: Anthomyiidae]: A widespread, multivoltine pest in North America, attacking sugar beet, beet, chard, spinach and many Species of weeds; SLM adventive from Europe prior to 1880. Overwinters within puparium in soil; adults appear during April-May. Female oviposits 1–5 small white eggs on underside of host plant leaves. Neonate larva penetrates leaf and feeds between upper and lower surfaces of leaf, causing blotches. SLM completes 3–4 generations per year. See Anthomyiidae.

SPINASTERNUM Noun. (Latin, *spina* = spine; Latin, *sternum* = breast bone. Pl., Spinasterna.) One of the spina-bearing intersegmental sclerites of the thoracic venter, associated or united with the preceding Sternum. A Spinasternum may become a part of the definitive Prosternum or Mesosternum, but not of the Metasternum (Snodgrass).

SPINATE Adj. (Latin, *spina* = thorn; *-atus* = adjectival suffix.) Produced into an acuminate spine.

SPINDLE Noun. (Old English, *spinnan* = to spin. Pl., Spindles.) An elongate, cylindrical structure thickest in the middle and tappering toward each end. Cf. Clavate; Filiform; Fusiform; Subfusiform.

SPINDLE TUBER A viral disease of potatoes which is transmitted by numerous plant-feeding insects.

SPINE Noun. (Latin, *spina* = spine. Pl., Spines.) 1. A stiff, sharp, pointed, tapered process on the surface of a plant or animal. Spines are typically defensive structures. 2. A multicellular, more-or-less thorn-like process or projection of the Cuticula not separated from it by a joint (Comstock). 3. A large seta provided with a calyx or cup by which it is articulated to the Cuticle (MacGillivray). Cf. Acantha; Aculeus; Seta; Spicule; Thorn. Rel. Integument.

SPINE DUCTS Coccoidea: The plates *sensu*

MacGillivray.

SPINED ASSASSIN-BUG 1. *Sinea diadema* (Fabricius) [Hemiptera: Reduviidae]: A predaceous Species endemic to North America. Eggs cylindrical, white, operculate; eggs laid in groups covered with yellowish secretion. Prey include larvae of many economically important insects; all stages of Mexican bean beetle. See Reduviidae.

SPINED CITRUS BUG *Biprorulus bibax* Breddin [Hemiptera: Pentatomidae]: A Species endemic to Australia and which feeds and reproduces on native Rutaceae and citrus; home garden-trees often severely affected. SCB will feed on native plants but does not reproduce on them (e.g. desert lime, *Eremocitrus glauca* and finger lime *Microcitrus australasica*.) Sucking by adults and nymphs causes young fruit to colour prematurely and fall. Adult 15–20 mm long, shield-like with prominent sharp spine projecting from lateral part of thoracic notum. Eggs deposited in rafts of 15–30 over a period of several weeks; female can produce 100–200 eggs per lifetime; neonate egg ca 1 mm diameter, white, later becomes spotted with black and red; first nymphal instar gregarious near eggshells and does not feed; subsequent instars disperse over plant; five nymphal instars; adults can live for 18 months and aggregate during winter. Multivoltine with three overlapping generations per year. See Pentatomidae.

SPINED RAT-LOUSE *Polyplax spinulosa* (Burmeister) [Anoplura: Hoplopleuridae].

SPINED SOLDIER-BUG *Podisus maculiventris* (Say) [Hemiptera: Pentatomidae]: A beneficial predator widespread in North America. Eggs metallic bronze, laid in masses of 20–30; one female can lay 1,000 eggs. First instar nymphs gregarious around eggshells, moult and disperse. Attack sawfly larvae, caterpillars without Setae, leaf-eating beetle larvae. Prey include Colorado potato beetle. Two generations per year. Overwinter as adults. See Pentatomidae.

SPINED STILT-BUG *Jalysus wickhami* Van Duzee [Hemiptera: Berytidae].

SPINELEGGED CITRUS WEEVIL *Maleuterpes spinipes* Blackburn [Coleoptera: Curculionidae]: A minor pest of citrus and other Rutaceae in Australia. Adult ca 3 mm long, brown with greyish markings. Adults do not fly. Larvae feed on roots without apparent damage; adults feed on fruit rind and young leaves. Syn. Dicky Rice Weevil.

SPINETAILED WEEVIL *Desiantha caudata* Pascoe [Coleoptera: Curculionidae] (Australia).

SPINIFEROUS Adj. (Latin, *spina* = spine; *ferre* = to carry; *-osus* = possessing the qualities of.) Pertaining to structure bearing spines. Cf. Setose.

SPINIFEX TERMITE *Nasutitermes triodiae* (Froggatt) [Isoptera: Termitidae] (Australia).

SPINIFORM Adj. (Latin, *spina* = spine; *forma* = shape.) Spine-shaped. See Spine. Cf. Acanthoid; Aculeiform; Penicilliform; Spiculiform. Rel. Form 2; Shape 2; Outline Shape.

SPINNERET Noun. (Anglo Saxon, *spinnan* = to spin. Pl., Spinnerets.) 1. A larval apparatus by means of which silk is spun. 2. In general, any such apparatus in adults or larvae, wherever found. A hollow spine that projects to the opening of Silk Glands. Cf. Filator. 3. Coccidae: A wax pore or Ceroris. 4. Diaspididae: A Genaceroris, (MacGillivray). Syn. Fusulus. See Silk Gland.

SPINNING BRISTLE Embiidae: A long hollow bristle, on the plantar surface of the first and second tarsal segments of the fore leg. Occur in numbers and issue silk threads developed in spinning glands.

SPINNING GLANDS Arachnids; Lepidoptera Larvae; Psocidae: The glands which produce the viscid secretion which forms the silk. See Silk.

SPINOLA, MAXIMILIAN (1780–1857) Genoese aristocrat and naturalist. Author of *Insectorum Liguriae species novae aut rariores* (2 vols, 1806–1808) and Essai d'une nouvelle classification generale des Diplolepaires (IN: Ann. du Mus. d'Hist. Nat.). (Swainson 1840, *Taxidermy; with Biography of Zoologists*. 392 pp. (p.334). London. [Vol. 12 of *Cabinet Cyclopedia*. Edited by D. Lardner]; Saunders 1858, Proc. ent. Soc. Lond. 1856–58: 102; Gestro 1915, Annali Mus. civ. Stor. nat. Giacomo Doria 47: 33–53; Ekis 1975, Boll. Mus. zool. Univ. Torino 1975: 1–80, list of described Cleridae.)

SPINOSE Adj. (Latin, *spinosus,* from *spina* = spine.) Spiny; descriptive of structure or a body covered with spines. Alt. Spined; Spinosus; Spinous. See Spine. Cf. Aculeolate.

SPINOUS-RADIATE Beset with spines in a circle, either concatenate, united at their bases, or setaceous, like bristles. See Shape.

SPINULA Noun. (Latin, *spinula* = a small spine. Pl., Spinulae.) 1. A small spine; a spinous processes at the apex of the Tibia. Also called spines, spurs or heels. 2. A hair-like appendage of the exocuticle (MacGillivray). 3. A hollow cuticular appendage through which coccids secrete wax (MacGillivray). 4. Lepidoptera: Minute projections from the larval integument. Alt. Spinule.

SPINULATE Adj. (Latin, *spinula* = a small spine; *-atus* = adjectival suffix.) Set with little spines or spinules. Alt. Spinulose; Spinulosus.

SPINY ASSASSIN-BUG *Polididus armatissimus* Stål [Hemiptera: Reduviidae].

SPINY BOLLWORM A common name applied to *Earias biplaga* Walker or *E. insulana* (Boisduval) [Lepidoptera: Noctuidae] in Africa. Both Species pests of cotton in Africa and Middle East. *Earias insulana* also a pest of cotton in India. Cf. Cotton Bollworm. Egyptian Bollworm; Pink Bollworm; Spotted Bollworm.

SPINY ELM-CATERPILLAR See Mourningcloak Butterfly.

SPINY LEAF-INSECT *Extatosoma tiaratum* (Macleay) [Phasmatodea: Phasmatidae] (Australia).

SPINY OAKWORM *Anisota stigma* (Fabricius) [Lepidoptera: Saturniidae].

SPINY PRICKLY-PEAR *Opuntia stricta* (Haworth) (Haworth) [Cactaceae]: A Species of cactus native to the New World and adventive elsewhere as a significant rangeland pest. Controlled with *Cactoblastus cactorum* (Bergroth) in Australia. See Prickly Pear, Cactus Moth.

SPINY RAT MITE *Laelaps echidninus* Berlese [Acari: Laelapidae] (Australia).

SPINY SHORE BUGS See Leptopodidae.

SPINY SNOUT MITE *Neomolgus capillatus* (Kramer) [Acari: Bdellidae] (Australia).

SPINY WHITEFLY *Aleurocanthus spiniferus* (Quaintance) [Hemiptera: Aleyrodidae] (Australia).

SPINY-HEADED WORMS See Acanthocephala.

SPIRA Noun. (Greek, *speira* = something wrapped.) Hymenoptera: The coiled Ovipositor of Cynipidae.

SPIRACERORES Coccids: The Cerores on a Canella (MacGillivray).

SPIRACLE Noun. (Latin, *spiraculum* = air hole. Pl., Spiraculae; Spiracles.) A pore, hole or aperture in the integument which serves as an adaptation to permit gas exchange between the body and the environment. Spiracles are typically paired and lateral 'holes' in the plural region, with one pair per segment on the Mesothorax, Metathorax and first eight abdominal segments. Air passes through Spiracles, enters an Atrium, then into Tracheae and subsequently into Tracheoles; CO_2 passes from tissue to Tracheoles, then Tracheae and exits via Spiracles. The Spiracle is often surrounded by a thickened cuticular ring called the Peritreme. Most large-bodied insects display a Spiracle consisting of 1–2 Atrial Valves; the Spiracle is closed by muscle contraction and opens by cuticular elasticity. Spiracles of some Species are invested with Setae, cuticular projections across the aperture or a Sieve Plate; these features serve to prevent the introduction of dirt, debris or unwanted water; aquatic insects may develop peristigmatic glands, hydrofugal Setae or hydrophobic Cuticle surrounding the spiracle as an adaptation to prevent flooding. See Tracheal System. Cf. Atrium; Peritreme; Peristigmatic Gland; Plastron; Sieve Plate; Stigma 1. Rel. Respiratory Muscles; Respiratory System.

SPIRACULA ANTEPECTORIALIA Arcane: Spiracles in the membrane between the Presternum and Mesosternum.

SPIRACULAR Adj. (Latin, *spiraculum* = air hole.) Descriptive of or pertaining to the Spiracles.

SPIRACULAR AREA Hymenoptera: Anteriormost of three areas between the lateral and Pleural Carinae on the Metanotum. Alt: First Pleural Area.

SPIRACULAR ATRIUM The pit-like or tubular chamber forming the exterior part of the Spiracle (Snodgrass).

SPIRACULAR DEPRESSIONS Coccids. See Stigmatic Cleft (MacGillivray).

SPIRACULAR GLANDS Coccids: The Spiracerores (MacGillivray).

SPIRACULAR GROOVES Coccids: See Stigmatic Clefts (MacGillivray).

SPIRACULAR LINE Lepidoptera larva: The lateral line which includes the Spiracles, stigmatal line.

SPIRACULAR MUSCLES Small and few in number, typically 1–2 per segment. One SM is an occulsor; one SM is a dilator.

SPIRACULAR SETA Coccids: The large, spine-like Setae, or spinae in the stigmatic clefts which serve to keep the Canellae open (MacGillivray).

SPIRACULAR SPINE Coccoidea: Spiracular Setae (MacGillivray).

SPIRACULAR SULCUS Hymenoptera: A narrow groove or channel on the Metanotum, extending from Spiracle to apical margin.

SPIRACULAR SYNCHRONIZATION A mechanism involved in Active Ventilation in which the local action of carbon dioxide on the closure muscles of a Spiracle causes the muscle to relax and thereby open the Spiracle. SS seen in some cockroaches, mantids, grasshoppers and moths. See Active Ventilation. Cf. Abdominal Pumping; Autoventilation; Auxiliary Respiratory Mechanisms. Rel. Respiratory Muscles; Tracheal Ventilation.

SPIRACULAR TRACHEA The short, usually unbranched Trachea rising directly from the Spiracle (Snodgrass).

SPIRACULAR VALVE A mechanism for closing the Spiracle with muscle contractions. Closure of the SV seals the respiratory system from unfavourable conditions. See Spiracle.

SPIRACULARIA Noun. Coccids: The prominent Parademe extending from the inner surface of the Peritreme into the body cavity (MacGillivray).

SPIRAL Noun. (Latin, *spira* = coil; *-alis* = characterized by. Pl., Spirals.) A structure or appendage that is rolled like a watch spring, or twisted like a corkscrew. See Shape. Cf. Cercinate; Coil.

SPIRAL FIBRE The spiral thickening or folding of the chitinous lining of a Trachea, which gives to the latter its characteristic microscopic appearance as well as its support and elasticity, taenidium. See Taenidium.

SPIRAL THREAD See Taenidium.

SPIRAL TONGUE The suctorial apparatus of Lepidoptera, so-called because of the manner it is carried when at rest.

SPIRALING WHITEFLY *Aleurodicus dispersus* Russell [Hemiptera: Aleyrodidae]: A pest of numerous fruits and vegetables in Central America, islands of the Pacific, Australia and New Guinea. Adult ca 2 mm long, white; nymphs occur in moderate to high numbers, usually on the underside of leaves and on fruits. Common name from the habit of female laying eggs in a spiral pattern.

SPIRALIS Noun. (Latin, *spira* = coil. Pl., Spirales.) Auxillary cord; ligament; spring-vein (MacGillivray).

SPIREA APHID *Aphis spiraecola* Patch [Hemiptera: Aphididae]: A cosmopolitan pest of apples, citrus and ornamental plants; a significant pest of citrus in Florida and California. Males and oviparious females do not appear on citrus; eggs occur on spirea. Causes stunted terminal growth, leaf curl. Adult green with apex of adbomen black, ca 2 mm long. Adults and nymphs can produce copious amounts of honeydew. Generation can be completed within a week; multivoltine with up to 25 generations per year. Syn. Citrus Aphid.

SPIRIGNATH Noun. (Greek, *speira* = something wrapped; *gnathos* = jaw. Pl., Spirignaths.) Lepidoptera: Spiral or coiled mouthparts. Syn. Spiritrompe.

SPIRITROMPE Noun. See Spirignath.

SPIROCHETE Noun. (Greek, *speira* = something wrapped; *cheta* = hair. Pl., Spirochetes.) A group of highly motile, spiral or undulate bacteria which form a major component of the gut microbiota of termites. Spirochetes exist free in gut fluid of all termites or the cytoplasm of lower termites, and facilitate the propulsion of hindgut Protozoa attached to their surface. Little known about role played in symbiotic relationship within termites; probably not pathenogenic.

SPIROPLASMAS Plural Noun. (Greek, *speira* = coil; *plasma* = form.) Helical or spiral-shaped, motile prokaryotes that occur in plants and insects. Cf. Mycoplasmas.

SPIRULATE Adj. (Latin, *spira* = coil; *-atus* = adjectival suffix.) Descriptive of structure or organisms that are spiral-shaped or coiled.

SPITFIRE GRUBS In Australia, the larval stage of *Perga* spp. [Hymenoptera: Pergidae].

SPITTING SPIDER In Australia, *Scytodes* sp. [Araneida: Scytodidae].

SPITTING SPIDERS See Scytodidae.

SPITTLE BUGS See Cercopidae. A common name derived from the nymphal habit of concealment in a frothy fluid secreted from the Anus.

SPITTLE Noun. (Middle English, *spitte*. Pl., Spittles.) Cf. Cuckoo Spit.

SPITTLEBUGS See Cercopidae.

SPITZ, ROBERT (1859–1954) (R. 1956, Z. wien. ent. Ges. 41: 245–246.)

SPIX, JOHANN BAPTIST VON (1781–1826) (Gistl 1885, Faunus 2 (Suppl.): 7–8, bibliogr.; Papavero 1971, *Essays on the History of Neotropical Dipterology.* 1: 65–69. São Paulo.)

SPLANCHNIC LAYER The inner layer of Mesoderm applied to the walls of the Alimentary Canal (Snodgrass). The Splanchnopleure.

SPLANCHNIC NERVES Nerves arising from the apical abdominal ganglion and passing to the hind intestine and reproductive system (Imms).

SPLANCHNIC Adj. (Greek, *splangchnon* = guts; *-ic* = characterized by.) Pertaining to viscera or structures associated with viscera.

SPLANCHNOCYTE Noun. (Greek, *splangchnon* = guts; *kytos* = hollow. Pl., Splanchnocytes.) The smallest of the Chromophil Leucocytes which take stains more intensely than other Leucocytes (Wardle).

SPLANCHNOPLEURE Noun. (Greek, *splangchnon* = intestine; *pleura* = side. Pl., Splanchnopleures; Splanchnopleura.) See Splanchnic Layer.

SPLEEN Noun. (Greek, *splen* = spleen. Pl., Spleens.) The organ found in vertebrates which functions to produce Lymphocytes and destroy red blood cells.

SPLENDENS Adj. (Latin, *splendens* from *splendare* = to shine.) Shining coloured, with a metallic brightness, reflecting light intensely (Kirby & Spence). Alt. Splendent.

SPLENDID EMERALD WASP *Stilbum splendidum* Fabricius [Hymenoptera: Chrysididae] (Australia).

SPLENDID GHOST MOTHS *Aenetus* spp. [Lepidoptera: Hepialidae] (Australia). See Common Splendid Ghost Moth.

SPLENDID TENEBRIONID *Cyphaleus mastersi* Pascoe [Coleoptera: Tenebrionidae] (Australia).

SPLENIC ORGAN Bilaterally arranged groups of cells located immediately below the pericardial cells on either side of the body, or on the concave surface of the Dorsal Diaphragm. Found in Dermaptera (Forficula), some Orthoptera and Thysanoptera.

SPLITTER Noun. (Middle English, *splitten*. Pl., Splitters.) A taxonomist who describes Taxa (typically Genera and Species) based upon minute character differences which, in the opinion of other taxonomists, seem insufficient to merit generic or specific status. See Taxonomy; Classification. Cf. Lumper. (Taxonomic Jargon).

***SPODOPTERA LITTORALIS* NPV** A naturally-occurring, host-specific Nuclear Polyhedrosis Virus used as a biopesticide against larva of *Spodoptera littoralis*. Trade names include: SL-NPV® and Spodopterin®. See Biopesticide; Nuclear Polyhedrosis Virus. Cf. Beet Armyworm NPV; Douglas-Fir Tussock Moth NPV; European-Pine Sawfly NPV; *Heliothis* NPV; *Mamestra brassicae* NPV; Redheaded Pine Sawfly NPV.

SPODOPTERIN® See *Spodoptera littoralis* NPV.

SPOD-X® A selective biopesticide used for the control of Beet Armyworm larvae attacking ornamental plants, field crops and greenhouse crops. Active ingredient Polyhedral Occlusion Bodies of NPV of *Spodoptera exigua*. See Beet Armyworm NPV. Cf. Gemstar®. Rel. Biopesticide.

SPOILE Noun. (Unknown origin.) The cast skin at moulting.

SPONGE Noun. (Greek, *spongia* = sponge. Pl., Sponges.) 1. The elastic, soft, porous, fibrous mass that constitutes the skeleton of Porifera. Material endowed with the properties of absorbing and holding large quantities of water, becoming soft when wet and retaining its strength. 2. Any porous material. Rel. Spongiform.

SPONGEFLIES See Sisyridae.

SPONGIFORM Adj. (Latin, *spongia* = sponge; *forma* = shape.) Sponge-shaped; descriptive of structure soft and porous. See Sponge. Rel. Form 2; Shape 2; Outline Shape.

SPONGILLAFLIES See Sisyridae.

SPONGING MOUTHPARTS An anatomical type of mouthparts found in Muscomorpha. Some Species (stable flies, tsetse flies) display sponging mouthparts adapted for piercing. Components include a Labrum broad at base and tapering to a blunt point. Posterior surface has a deep median groove that extends into mouth opening and leads into Cibarial Pump. Mandibles are long, flattened and blade-like with sharp tips; Mandibles move in a transverse plane, (*i.e.* they cannot be protracted or retracted). As piercing organs, thrust is made with head and body. Food canal is formed with Madibles closed on Labium. Maxillae form slender pointed blades called Galeae (Laciniae absent). Hypopharynx is a slender stylet containing salivary canal opening at tip, just above base of Hypopharyngeal Duct. Labium is a large, thick appendage suspended by a membranous basal region of Postmentum (Postlabium). Haustellum (Prementum) has a large deep groove (Labial Gutter). Theca is a sclerite forming sides and posterior of Haustellum. Labella are terminal lobes of Proboscis; Labella are soft pads and can be spread into a broad disc with posterior halves united and anterior halves separated. Undersurface of each lobe has a series of transverse grooves or channels (Pseudotrachae) that lead to base of cleft. In their normal positions the Labrum, Mandibles, Hypopharnyx and Galeae lie within Labial Gutter. Maxillary palps usually overlap edges of Labrum. Apex of Labrum when pressed down overreaches end of Labial Gutter. When feeding, Pseudotracheae collect blood exuding from punctures made by Mandibles, Maxillae and Galeae. Blood is taken by labellar channels into base of cleft where it is sucked into Food Canal. Muscoid sponging as in housefly. Proboscis consists of Rostrum, Haustellum and Labella. Rostrum forms a large inverted cone that bears Maxillary Palps. A small sclerite associated with each palpus may be remnant of Stipes; no trace of Mandibles. Haustellum is premental; Labial Gutter forms a deep groove on anterior face of Haustellum. Theca forms a sclerite on posterior side of Haustellum. Hypopharnyx is blade-like, slender, lies within Labial Gutter and contains Salivary Canal. Labrum is attached to distal end of Rostrum and covers Labial Gutter. A groove on inner surface is Food Canal; floor is Hypopharnyx. See Mouthparts.

SPONGIOPLASM Noun. (Greek, *spongia* = sponge; *plasma* = mould. Pl., Spongioplasms) The dark, contractile substance of the fibrillar bundles of insect muscle (Wardle). The reticulum. Alt. Spongioplasma.

SPONGIOSE Adj. (Latin, *spongia* = sponge; *-osus* = full of.) Sponge-like, of a soft elastic tissue resembling a sponge. Alt. Spongy; Spungeous.

SPONGIPHORIDAE Plural Noun. Little Earwigs. A cosmopolitan Family of Dermaptera. Small bodied; winged. North American Species include *Labia minor* (Linnaeus) which is commonly collected. Syn. Labiidae *auctt*.

SPONGIPHOROIDEA A Superfamily classification within the Dermaptera not without controversy (See Günther & Heter 1974, Popham 1985).

SPONGY PARENCHYMA Cynipid galls: The material occupying the central portion of a gall and constituting the major material of all the spongy and more hollow oak-apples of the Genus *Cynips* (Kinsey).

SPOON Noun. (Anglo Saxon, *spon* = chip. Pl., Spoons.) 1. Diptera: small sclerite at the base of the Haltere. 2. Hymenoptera: The Labellum of the honey bee. 3. Lepidoptera: Distal elongation of the Sacculus.

SPOONWINGED LACEWING *Chasmoptera hutti* Westwood [Neuroptera: Nemopteridae] (Australia).

SPOOR FACTOR See Epideictic Pheromone.

SPORANGIUM Noun. (Greek, *spora* = seed; *angeion* = a vessel. Pl., Sporaniga.) Fungi: A structure containing spores produced during the asexual stage of the life cycle.

SPORE Noun. (Greek, *sporos* = seed. Pl., Spores.) 1. A cellular agent of asexual reproduction. 2. A reproductive body of algae, fungi and ferns. Spores are typically microscopic, primitive and unicellular (except phragmospores). Anatomical features highly variable: Pigmented or not pigmented; thin or thick walled; motile or not motile. Spores are subject to fossilization and often distinctive in size, shape and surface feature. See Reproduction. Cf. Pollen. Rel. Seed.

SPOREINE® A registered biopesticide derived from *Bacillus thuringiensis* var. *kurstaki*. See *Bacillus thuringiensis*.

SPORMAN, KARL (1853–1937) (Anon. 1938, Arb. morph. taxon. Ent. Berl. 5: 72; Pafau 1938, Ent. Z., Frankf. a. M. 52: 73–75.)

SPOROBLAST Noun. (Greek, *sporos* = seed; *blastos* = bud. Pl., Sporoblasts.) A stage of spore formation.

SPORONGONY PHASE The phase in the life cycle of *Plasmodium* passed in an insect-host; sexual production of spores.

SPORONGONY Noun. (Greek = *sporos* = seed. Pl., Sporogonies.) The changes that occur within the oocyst of *Plasmodium*, the malarial protozoan which produces the Sporozoites.

SPOROTHECA Noun. (Greek, *sporos* = seed; *theke* = case. Pl., Sporothecae.) A fungal-spore containing structure on the body of some heterostigmatid mites. (Lindquist 1985, Exp. Appl. Acrarol. 1: 73–85).

SPOROZOITE Noun. (Greek, *sporos* = seed; *zoon* = animal; *-ites* = inhabitant. Pl., Sporozoites.) The minute organism produced in the Oocyst of the protozoan malarial parasites *Plasmodium* spp. which is injected into a human by *Anopheles* spp.

mosquitoes. Sporozoites migrate to human liver and develop into latent Hypnozoites or immediately undergo Schizogony. See Malaria; Schizogony.

SPORT Noun. (Middle English; *disport*. Pl., Sports.) An individual within a population perceived as an aberration; a mutation.

SPORULATION Noun. (Greek = *sporos* = seed; English, *-tion* = result of an action. Pl., Sporulations.) Fungi: The production of spores.

SPOTLESS GRASS YELLOW *Eurema laeta sana* (Butler) [Lepidoptera: Pieridae] (Australia).

SPOTTED ALFALFA APHID 1. *Therioaphis maculata* (Buckton) [Hemiptera: Aphididae]: In North America, a widespread pest of alfalfa. Males rare, females typically wingless, parthenogenetic, viviparious; sexual generations in colder climates with overwintering egg state. SAA Feed on underside of leaves, first at base of plant and later on stems. SAA can complete 30 generations per year. 2. *Therioaphis trifolii* Monell [Hemiptera: Aphididae] (Australia).

SPOTTED ALFALFA APHID PARASITE *Trioxys complanatus* Quilis [Hymenoptera: Braconidae] (Australia).

SPOTTED ASPARAGUS BEETLE *Crioceris duodecimpunctata* (Linnaeus) [Coleoptera: Chrysomelidae]: A bivoltine pest of asparagus native to Europe and adventive to North America ca 1880. Biology resembles Asparagus Beetle. Eggs greenish, attached singly to leaves of asparagus; ecolosion ca 7–15 days. Larvae move to berries and feed upon pulp. Cf. Asparagus Beetle.

SPOTTED BEET WEBWORM *Hymenia perspectalis* (Hübner) [Lepidoptera: Pyralidae].

SPOTTED BLISTER-BEETLE *Epicauta maculata* (Say) [Coleoptera: Meloidae].

SPOTTED BOLLWORM A common name applied to *Earias insulana* (Boisd.) and *E. fabia* (Stoll) [Lepidoptera: Noctuidae] in India where both Species are pests of cotton. Cf. Cotton Bollworm; Egyptian Bollworm; Pink Bollworm; Spiny Bollworm; Spotted Pink Bollworm.

SPOTTED BROWN BUTTERFLY *Heteronympha paradelpha* Lower [Lepidoptera: Nymphalidae] (Australia).

SPOTTED COWPEA BRUCHID *Callosobruchus maculatus* (Fabricius) [Coleoptera: Bruchidae]: A widespread pest of pulse crops throughout warmer regions. SCB apparently is endemic to Africa; in North America, a pest of cowpea in the field or beans and peas in storage. Adults are 2–3 mm long, brownish with coloured patches on Elytra and eyes emarginate. Antenna is pectinate in male and serrate in female; hind Femur with parallel ridges on ventral surface, each with an apical spine; Elytra do not reach apex of Abdomen. SCBs are strong fliers but do not feed on stored products; adults are short-lived (ca 12 days). Females lay ca 100 eggs during lifetime; eggs adhere to developing pods in field or upon seeds in storage; incubation 5–6 days. Larva bores into cotyledon and hollows the seed; usually 1–3 larvae per seed; larva scarabaeiform with five instars; ca 20 days for larval development. Pupation within seed with window for emergence; period ca seven days. Life cycle ca 21–36 days; multivoltine (6–7 generations per year). Syn. Cowpea Weevil. Cf. Oriental Cowpea Bruchid.

SPOTTED CUCKOO BEES *Thyreus* spp. [Hymenoptera: Anthophoridae] (Australia).

SPOTTED CUCUMBER BEETLE *Diabrotica undecimpunctata howardi* Barber [Coleoptera: Chrysomelidae]: A.pest of field and garden crops in eastern North America and parts of Mexico. Eggs yellow, oval, laid near base of host plants. Larva feeds on roots and stems for several weeks, constructs cell in soil and pupates. Adult consumes leaves; sometimes migratory over long distances. Syn. Southern Corn Rootworm (larva). Cf. Striped Cucumber Beetle; Western Striped Cucumber Beetle.

SPOTTED CUTWORM *Xestia* spp. [Lepidoptera: Noctuidae]. See Cutworms.

SPOTTED FEVER See Tick-Borne Spotted Fever.

SPOTTED HAIRY FUNGUS BEETLE *Mycetophagus quadriguttatus* P. Müller [Coleoptera: Mycetophagidae].

SPOTTED KATYDID *Epipitytha trigintiduoguttata* (Serville) [Orthoptera: Tettigoniidae]: A pest of *Citrus* in Australia and more common in inland areas. Adult ca 50 mm long; body green and brown with dark brown marks on wings and body. Eggs disc-like, ca 5 mm long, laid in parallel rows on tree trunks or limbs; nymphs feed on young foliage, flowers and young fruit; five instars. Adults easily fly from tree to tree in groves. Monovoltine. Cf. Angular-Winged Katydid; Broad-Winged Katydid; Citrus Katydid; Crested Katydid; Forktailed Bush-Katydid; Inland Katydid; Mottled Katydid; Philippine Katydid.

SPOTTED LEAFHOPPER *Austroagallia torrida* Evans [Hemiptera: Cicadellidae] (Australia).

SPOTTED LOBLOLLY PINE SAWFLY *Neodiprion taedae taedae* Ross [Hymenoptera: Diprionidae].

SPOTTED MEDITERRANEAN COCKROACH *Ectobius pallidus* (Oliver) [Blattodea: Blatellidae].

SPOTTED OLEANDER CATERPILLAR *Empyreuma affinis* Rothschild [Lepidoptera: Arctiidae]: Native of the Caribbean and recorded from Guadeloupe, Martinique, Haiti, Dominican Republic, and Cuba; adventive to the USA (Boca Raton, Florida) during 1978. SOC is limited to the Keys and south Florida. SOC eggs similar in appearance to eggs of oleander caterpillar: spherical, ca 1 mm wide, pearly white and turn yellow before hatching. Larvae are pale orange, hairy caterpillars with tubercles on lateral and dorsal regions of each segment that display tufts of stiff reddish-brown setae; Mesonotum, Metanotum, and eighth abdominal segment display a pair of longer stiff black setae. Six rows of regularly spaced silver-coloured spots are ringed with dark brown to form discontinuous longitudinal bands along the caterpillar's body. Adults are day-flying moths. Male locates a female from several metres via her sex pheromone. Close-range mate location is facilitated by ultrasonic acoustic signals emitted by males and female. Mating occurs during sunrise. Inseminated female searches for a site to lay her eggs, usually the underside of an oleander leaf; eggs laid in a group, as oleander caterpillar, but her progeny feed singly, not gregariously as the oleander caterpillar. Larvae complete six instars within ca 28 days, depending on temperature. Sixth instar larvae leave host plant and search for a suitable site for pupation. SOC pupates alone rather than in a large aggregation as oleander caterpillar.

SPOTTED PINE SAWYER *Monochamus mutator* LeConte [Coleoptera: Cerambycidae].

SPOTTED PINK BOLLWORM *Pectinophora scutigera* (Holdaway) [Lepidoptera: Gelechiidae]: A minor pest of cotton and other Malvaceae. SPB is endemic in eastern Australia and Papua New Guinea and has been reported in Micronesia. SPB biology resembles closely related Pink Bollworm, *P. gossypiella*. Cf. Cotton Bollworm; Egyptian Bollworm; Pink Bollworm; Spiny Bollworm; Spotted Bollworm.

SPOTTED SKIPPER *Hesperilla ornata* (Leach) [Lepidoptera: Hesperiidae] (Australia).

SPOTTED STALK BORER *Chilo partellus* (Swinhoe) [Lepidoptera: Pyralidae]: A pest of sorghum and maize in Asia and Africa. Females do not feed, are short-lived and oviposit soon after emergence. Eggs laid in clusters with total fecundity less than 500 eggs per female.

SPOTTED TENTIFORM LEAFMINER *Phyllonorycter blancardella* (Fabricius) [Lepidoptera: Gracillariidae]: An important pest of commercial apple production in parts of North America. Larva mines leaves; head wedge-shaped; early instars flat, late instars cylindrical; pupates and overwinters in mines. Host plants include hawthorn, plum, quince, wild cherry.

SPOTTED TUSSOCK MOTH *Lophocampa maculata* Harris [Lepidoptera: Arctiidae].

SPOTTED VEGETABLE WEEVIL *Desiantha diversipes* (Pascoe) [Coleoptera: Curculionidae] (Australia).

SPOTTED WILT A viral disease of tomatoes which is transmitted by several Species of thrips.

SPOTTEDGUM PSYLLID *Eucalyptolyma maideni* Froggatt [Hemiptera: Psyllidae] (Australia).

SPOTTED-WINGED ANTLION *Dendroleon obsoletus* (Say) [Neuroptera: Myrmeleontidae].

SPOTTON® See Fenthion.

SPRAGIS See Spermatophragma.

SPRAGUE, FIONA FRANCESCA (1961) (Varley 1962, Proc. R. ent. Soc. Lond. (C) 26: 53.)

SPRAGUE, PHILIP L (1829–1874) (F.G.S. 1875, Can. Ent. 7: 95–96.)

SPRAY-TOX® See Kadethrin.

SPREAD-WINGED DAMSELFLIES See Lestidae.

SPRENGEL, CHRISTIAN KONRAD (1750–1816) German who first noted the role of bees in the pollination of plants and recorded observations in *Das entdeckte Geheimniss der Natur im Bau und in der Befruchtung der Blumen* (1793). (Soldanski 1916, Dt. Ent. Z. 1916: 229–230; Herbst 1966, Mitt. dt. ent. Ges. 25: 31–34; Voderberg 1966, Mitt. dt. ent. Ges. 25: 35–37.)

SPRIGONE® See Tetramethrin.

SPRING BEETLE *Colymbomorpha vittata* Britton [Coleoptera: Scarabaeidae] (Australia).

SPRING CANKERWORM *Paleacrita vernata* (Peck) [Lepidoptera: Geometridae]: A sporadic defoliator of fruit and shade trees in North America. Overwinters as pupa within soil. Adults appear in early sping; female wingless and male winged. Female climbs tree, mates with male, deposits masses of oval, dark-brown eggs under loose bark; eclosion occurs within 30 days. Larva with two pairs of Prolegs; feed ca 30 days on foliage, move to soil and pupate. SC complete one generation per year. Cf. Fall Cankerworm.

SPRING, FREDERIC ANTOINE JOSEPH (1814–1872) (Dewalque *et al.* 1872, Bull. Acad. r. Belg. (2) 33: 93–102; Schwann 1874, Ann. Acad. r. Belg. 40: 251–290, bibliogr.)

SPRING Noun. (Anglo Saxon, *springan*. Pl., Springs.) 1. Any coiled structure. 2. Collembola: The Furcula.

SPRINGER, GIOVANNI (–1905) (Anon. 1905, Boll. Soc. ent. ital. 95: 138.)

SPRINGFLIES See Perloidae.

SPRINGTAILS Minute arthropods 1–3 mm long; colours vary from black, grey, white, yellow, orange, purple, blue, pink and red. Body globular to elongate with undefined segments. Springtails with three pairs of legs; jumping organ (Furcula) is present at the apex of the Abdomen and resembles a 2-pronged fork; ventral tube (Collophore) present on first abdominal segment. Springtails with four segments in Antenna but lack eyes and wings. Adult females lay eggs singly or in batches in soil or leaf litter. Little known about Collembolan eggs. Juveniles and adults similar in appearance but juveniles have fewer abdominal segments. Juveniles undergo 5–6 moults to adulthood; springtails occur mainly in soil and leaf litter and feed on fungal hyphae and spores. They are occasionally seen on young cotton seedlings. See Collembola.

SPRINKLING SEWAGE FILTER FLY See Trickling Filter Fly.

SPRITEX® See Tetramethrin.

SPROUT NIP® See Chlorpropham.

SPRUCE APHID *Elatobium abietinum* (Walker) [Hemiptera: Aphididae] (Australia).

SPRUCE BARK-BEETLE *Ips typographus* [Coleoptera: Scolytidae].

SPRUCE BEETLE *Dendroctonus rufipennis* (Kirby) [Coleoptera: Scolytidae].

SPRUCE BEETLES See Curculionidae.

SPRUCE BUD MIDGE *Dasineura swainei* (Felt)

[Diptera: Cecidomyiidae].

SPRUCE BUD MOTH *Zeiraphera canadensis* Mutuura & Freeman [Lepidoptera: Tortricidae].

SPRUCE BUD SCALE *Physokermes piceae* (Schrank) [Hemiptera: Coccidae].

SPRUCE BUDWORM *Choristoneura fumiferana* (Clemens) [Lepidoptera: Tortricidae]: The most destructive pest of spruce and fir trees in northeastern USA and Canada. Univoltine; overwinters as second instar larva within cocoon on twigs near buds. Eggs green, shingle-like laid on underside of needles. Larvae feed on tree foliage; early instars mine old needles; late instars attack bud, cone or twigs. Needles cut and incorporated into tubular web; pupation occurs within web and complete ca 7–10 days. Adults active July-August. Cf. Jack Pine Budworm.

SPRUCE CONEWORM *Dioryctria reniculelloides* Mutuura & Munroe [Lepidoptera: Pyralidae].

SPRUCE MEALYBUG *Puto sandini* Washburn [Hemiptera: Pseudococcidae].

SPRUCE NEEDLEMINER *Endothenia albolineana* (Kearfott) [Lepidoptera: Tortricidae]: A widespread, monovoltine pest of spruce in North America. Eggs overlapping, laid in groups on spruce needles. Larva bores into needles at base; some needles detached and held to twig by funnel-shaped web. Larvae gregarious within webs; overwinters as larva in needle; resumes feeding during spring; pupation under frass within web during summer.

SPRUCE SEED MOTH 1. *Cydia strobilella* (Linnaeus) [Lepidoptera: Tortricidae]. 2. *Laspeyresia youngana* (Kearfott) [Lepidoptera: Tortricidae].

SPRUCE SPIDER MITE *Oligonychus ununguis* (Jacobi) [Acari: Tetranychidae] (Australia).

SPRY, FRANK PALMER (1858–1922) (Bernard 1922, Victorian Nat. 39: 60–62.)

SPUD NIC® See Chlorpropham.

SPULER, ANTHONY (1889–1932) (Webster 1932, J. Econ. Ent. 25: 939–941; Hungerford 1934, Ann. ent. Soc. Am. 27: 123.)

SPULER, ARNOLD (1869–1937) (W. 1937, Ent. Z. 51: 54.)

SPUMALINE Noun. (Latin, *spuma* = foam, froth, scum; *linea* = line. Pl., Spumalines.) 1. A froth (spume) formed and used by insects to envelope an egg to render it invisible. 2. Trichoptera: An hygroscopic Colleterial Gland secretion of the female which surrounds the eggs. 3. Bombyliidae: Eggs 'dusted' with adherent particles of dirt. 4. Aphididae: *Longistigma caryae* dusts its eggs.

SPUME Noun. (Latin, *spuma* = foam, froth, scum. Pl., Spumes.) A froth, foam or scum formed on the surface of a liquid by the action of boiling, effervescence, or agitation. Spume envelops a life stage of an insect (*e.g.* nymph of spittlebug; egg of bombyliid fly). See Spumaline.

SPUMI- Latin connecting form for the word 'spume'.

SPUR Noun. (Anglo Saxon, *spora* = spur. Pl., Spurs.) 1. Any spine-like appendage of the integument that articulates with the body wall. Spurs frequently occur on leg segments such as the Tibia. 2. Orthoptera: Spine-like cuticular appendages at the apex of the hind Tibia; spurs are larger than spines, articulated and apically curved. See Acantha; Calcar; Spine; Strigil.

SPUR FORMULA A numerical expression of the arrangement of spurs on a structure, *e.g.,* in Trichoptera: 2-3-4, indicates two spurs on fore Tibia, three spurs on middle Tibia and four spurs on hind Tibia. Cf. Palpal Formula; Tarsal Formula.

SPURGE BUGS See Stenocephalidae.

SPURIOUS Adj. (Latin, *spurius* = bastard.) False or accidental. A term applied to any false or adventitious structures, such as the aborted anterior legs in certain diurnal Lepidoptera (Smith), or the vein-like structures of the wing.

SPURIOUS CELLS Diptera: The third Anal Cell of Comstock *teste* Packard.

SPURIOUS OCELLUS An ocellate spot without definite iris or pupil.

SPURIOUS SUTURE An impressed line resembling a suture but which does not cut the surface.

SPURIOUS VEIN Hymenoptera: Some kinds of folds or thickenings in the wing surface that strongly resemble veins and may be readily mistaken for them. When present, SV are constant and are used in classification. See Vein; Wing Venation. Cf. Nebulous Veins; Spectral Veins; Trace Vein; Tubular Vein. Rel. Horismology.

SPUR-LEGGED PHASMATID *Didymuria violescens* (Leach) [Phasmida: Phasmatidae]: An Australian phasmid that may reach plague proportions during some years. Extensive defoliation by SLP may cause death of eucalypts in southeastern Australia.

SPURNY, JOHANN (1805–1940) (Heikertinger 1940, Koleopt. Rdsch. 26: 92.)

SPURRS A kit for embedding biological tissues into an epoxy resin; the moulded resin block is sectioned with an ultra-mictotome to produce sections for examination with a transmission electron miscroscope. Also Spurr's Kit.

SPUR-THROATED LOCUST *Nomadacris guttulosa* (Walker) [Orthoptera: Acrididae]: Endemic and widespread in Australia; primarily found in tropical grasslands of northeastern Australia as a significant pest of sorghum and sunflower. Hoppers do not form bands. Adults overwinter in trees and shrubs. Adult 50–75 mm long (females larger than males), elongate and green to brown; hindwings transparent or tinged with blue; hind legs mauve (red-purple) coloration; conspicuous spur in 'neck' (between forelegs). Antenna short, Pronotum large and hind legs adapted for jumping. Oviposition with onset of summer rains; spined Ovipositor for laying eggs in soil; pods, contain ca 150 eggs, placed in holes in moist soil; egg stage 18–30 days depending on temperature. Nymphs similar to adults but lack wings; nymphal stage 30–60 days. Plagues affect all summer crops. Syn. *Austracris guttulosa* Walker.

Cf. Australian Plague Locust.

SQUAMA Noun. (Latin, *squama* = scale. Pl., Squamae.) 1. A scale or scale-like structure. 2. A broad, flat sclerite attached to an organ, appendage or structure. 3. Odonata: A lateral expansion of the Mentum, which bears the Palpus of the Maxilla and Labium. 4. Coccoidea (Diaspididae): The fimbriated or spine-like marginal processes, other than the lobes and true spines (Syn. 'Plates' of Comstock; 'Scaly Hairs' of Maskell; 'Pectinae' of MacGillivray). 5. Lepidoptera: A scale-like appendage covering the base of the forewing (See Patagium; Tegula). 6. Diptera: A small scale above the Haltere. An Outer Squama of houseflies originates posterior of the third Axillary Sclerite and is probably homologous with the Jugal Lobe of other insects. An Inner Squama is larger and probably originates as a membranous extension of the Scutellum. The Inner Squama protects the Haltere and covers the Outer Squama when the wing is flexed over the body. See Alula; Antisquama; Calypter. 5. Hymenoptera: The scale-like first abdominal segment of ants.

SQUAMA PALPIFERA The third portion of the Maxilla.

SQUAMATE Adj. (Latin, *squama* = scale; *-atus* = adjectival suffix.) 1. Scale-shaped. 2. Pertaining to structure covered with scales. Alt. Squamiform; Squamose; Squamosus; Squamous; Squamulate; Squamulose; Squamulosus; Squamulous. See Shape.

SQUAMIFORM SENSILLUM 1. A modified Trichode Sensillum that forms a sense organ on the wing base and marginal vein of Lepidoptera. The external part of the Squamiform Sensillum is scale-like, elongate or fusiform. It is innervated by one large sense cell. 2. Lepidoptera: A scale-like sense organ on wing base and Marginal Vein. Syn. Sensillum Squamiformium See Sensillum. Cf. Trichode Sensillum.

SQUAMIFORM Adj. (Latin, *squama* = scale; *forma* = shape.) Scale-shaped; descriptive of structure with a scale-like form. Alt. Squammate. See Squama. Cf. Lamelliform; Laminiform; Phylliform. Rel. Shape. See Form 2; Shape 2; Outline Shape.

SQUAMOPYGIDIUM Noun. (Latin, *squama* = scale; *pygidion* = rump; *-ium* = diminutive > Greek, *-idion*. Pl., Squamopygidia.) The anal processes, especially in Elateridae.

SQUAMULA Noun. (Latin, *squama* = a small scale. Pl., Squamulae.) 1. A small corneous scale covering the base of the forewings in some insects. 2. Diptera: The Alula. Alt. Squamule. Syn. Tegula.

SQUAMULA ALARIS Diptera: The Upper Squama (Comstock).

SQUAMULA THORACALIS Diptera: The Lower Squama (Comstock).

SQUAMULATA CANEGRUB *Lepidiota squamulata* Waterhouse [Coleoptera: Scarabaeidae] (Australia). See Canegrub.

SQUARE HEADED WASPS See Sphecidae.

SQUARE-NECKED GRAIN BEETLE *Cathartus quadricollis* (Guérin-Méneville) [Coleoptera: Cucujidae]: A cosmopolitan pest of stored grains and cereals. Development ca 20 days at 28°C, 80% RH.

SQUARE-NOSED FUNGUS BEETLE *Lathridius minutus* (Linnaeus) [Coleoptera: Lathridiidae] (Australia).

SQUARROSE Adj. (Latin, *squarrosus* = scurfy; *-osus* = full of.) Scurfy, ragged, clothed with rough loose scales differing in direction, standing upright, or not parallel to the surface. Alt. Squarrosus; Squarrous.

SQUASH BEETLE *Epilachna borealis* (Fabricius) [Coleoptera: Coccinellidae]: A pest of curcurbits in eastern USA.

SQUASH BUG *Anasa tristis* (DeGeer) [Hemiptera: Coreidae]: A monovoltine pest of vine crops including squash, melons, pumpkin and gourds throughout North America and Central America. SB overwinters as an unmated adult in concealed habitat. Eggs shiny brown, elliptical, laid individually or in clusters on stems or underside of leaves; eclosion occurs within 7–14 days. Neonate nymph head, Thorax and appendages red with abdomen green; older nymphs greyish with with black appendages. Five nymphal instars. Early instar nymphs feed gregariously; later instars disperse. Feeding induces leaves to droop, turn black and become crisp; small plants are killed. See Coreidae. Cf. Stink Bugs.

SQUASH BUGS See Coreidae.

SQUASH-VINE BORER *Melittia curcurbitae* (Harris) [Lepidoptera: Sesiidae]: A pest of curcurbits in eastern North America. SVB overwinters as mature larva or pupa within cocoon in soil. Adults emerge in spring. Eggs 1 mm, oval, flattened and laid on stems and leaf stalks. Neonate larva penetrates leaf or stem, tunnel and cause massive wilting or runners or entire plant. Larvae white with brown head, complete feeding ca 30–45 days, emerge from plant and pupate in soil. See Sesiidae.

SQUEAKER BEETLES See Hygrobiidae.

SRAMEK-HUSEK, RUDOLF (–1957) (Jirovec 1957, Zool. Listy 20: 94.)

SRIVASTAVA, DAYAL SARAN (1908–1972) (Butler 1973, Proc. R. ent. Soc. Lond. (C) 37: 57.)

ST ANDREW'S CROSS SPIDER *Argiope keyserlingii* Karsch [Araneida: Araneidae]: A commonly encountered spider in Australia. Female 12–15 mm long; Abdomen oval dorsum with stripe-and-spot pattern of black, white, yellow and red; venter with two longitudinal yellow stripes and yellow Spinnerets; female constructs large orb-web with cross-shaped zig-zag pattern of white silk (Stabilamentum) and rests with legs stretched in centre of web. Male ca 5 mm long, red-brown without pattern.

ST JOHN'S WORT LEAF BEETLE *Chrysolina hyperici* (Forster); *Chrysolina quadrigemina*

(Suffrian) [Coleoptera: Chrysomelidae] (Australia).

ST JOHN'S WORT MIDGE *Zeuxidiplosis giardi* (Kieffer) [Diptera: Cecidomyiidae] (Australia).

ST JOHN'S WORT ROOTBORER *Agrilus hyperici* (Creutzer) [Coleoptera: Buprestidae] (Australia).

ST JOHN'S WORT STUNT MITE *Aculus hyperici* (Liro) [Acari: Eriophyidae] (Australia).

ST LINE Moths: The subterminal line that crosses the forewings basad of the outer margin.

ST LOUIS ENCEPHALITIS An arboviral disease whose agent belongs to the Family Flaviviridae. An arboviral disease of humans which is endemic to USA; most important encephalitis in USA. SLE vectored by *Culex* Species, most notably *C. tarsalis* and members of *Culex pipiens* complex. Mosquitoes acquire SLE from infected birds transmit viron to humans. Disease often epidemic with numerous fatalities but protection conferred through immunization with vaccine; mortality in humans less than 10%. See Arbovirus; Encephalitis; Flaviviridae. Cf. Japanese Encephalitis; Murray Valley Encephalitis; West Nile Encephalitis.

ST MARK'S FLIES See Bibionidae.

ST SPACE See Subterminal Space.

STABBERS Plural Noun. Lice: Mouthpart Stylets or the sucking apparatus.

STABILE, GIUSEPPE (1827–1869) (Riva 1869, Verh. schweiz. naturf. Ges. 53: 205–209; Sordelli 1869, Atti Soc. ital. Sci. nat.12: 173–179.)

STABILENE® See Chlormequat.

STABILIMENTUM Noun. (Latin, *stabilis* = firm; *mentum* = to project.). A zig-zag shaped ribbon of silk spun across the centre or below the hub of the web of some orb-web weaving spiders. The Stabilimentum differs in size and complexity among Species that construct it.

STABLE Adj. (Latin, *stabilis* = standing firm.) Pertaining to structural form, chemical composition or behavioural condition which is constant, predictable or fixed. 2. Something which is not prone to chemical degradation, biological decay or physical decomposition. Ant. Instable; Labile.

STABLE FLIES See Muscidae.

STABLE FLY *Stomoxys calcitrans* (Linnaeus) [Diptera: Muscidae]: A cosmopolitan pest of domestic animals. Uncommon in domestic habitats except in association with domestic animals; common in dairy and intensive animal production facilities. SF mechanical vector of numerous pathogenic Protozoa, bacteria and viruses, including Surra *(Trypanosoma evansi)* which is usually fatal to horses and mules and also affects other domestic stock. SF sometimes a problem at resort beaches. Adult 5–7 mm long, superficially resembles house fly but larger bodied, more robust, possesses piercing-sucking mouthparts and wings held widespread at rest; both sexes suck blood; female requires blood meal before oviposition. SF prefers sunny, outdoor habitats; strong fliers, but do not disperse long distances. Egg ca 1 mm long, curved on one side, straight and grooved on other side; eggs laid in decaying vegetable material, wet straw and manure in groups of 25–50 with 4–5 oviposition episodes; eclosion occurs within a few days. Larva feeds 1–3 weeks; larva develops in manure-laden straw and similar habitats. Pupation in drier habitat; life cycle 3–5 weeks. Syn. Biting House Fly. See Muscidae. Cf. House Fly.

STACH, JAN (Fudakowski 1958, Polskie Pismo ent. 9–10 (B): 65–68.)

STACKELBERG, ALEKSANDRA ALEXSANDROVICH (1897–1975) (Kryschanovskii 1976, Ent. Obozr. 55: 224–234, bibliogr.)

STADIUM Noun. (Latin, from Greek *stadion* > *stadios* = fixed, stable; Latin, -ium = diminutive > Greek, *-idion*. Pl., Stadia.) 1. The interval of time in development; specifically the time between successive moults during larval development by insects. 2. The interval of time between Ecdyses (Snodgrass 1935; Jones 1978). 3. The duration of an instar (Carlson 1983). See Metamorphosis; Development. Cf. Instar; Stage. Rel. Period; Phase; Interval.

STÄDLER, FRITZ (–1919) (Anon. 1919, Intern. ent. Z. 13: 82.)

STADLER, HANS (1875–1962) (Ade 1955, Nachr. naturw. Mus. Aschaffenb. 47: [1–5].)

STAEGER, RASMUS CARL (1800–1875) (Henriksen 1925, Ent. Meddr. 15: 179–182.)

STAEGER, ROBERT (1867–1962) (Schmidlin 1963, Mitt. ent. Ges. Basel 13: 28–30.)

STAFAST® See NAA.

STAFFORD, ETHELBERT WITTIROE (1886–1963) (Wilson 1965, Ann. ent. Soc. Am. 58: 769.)

STAG BEETLES See Lucanidae.

STAGE Noun. (Old French, *estage* = dwelling, habitation, floor of building. Latin, *stare* = stand. Pl., Stages.) 1. A period or step in a process or action. 2. Any specific period in the development of an insect (*e.g.*, egg stage, caterpillar stage). Hinton (1971, 1974, 1976) synonymized the terms Instar and Stage, but Jones (1978) disagrees and argues that Stage begins at Ecdysis when the Instar emerges from the egg, immature exuvia or pupal case. Jones: Stages occur between Ecdyses, and Instars occur between apolyses. Fink (1983) restricts Stage to mean 'major and minor divisions of an arthropod's life cycle which are not strictly delimited by ecdyses or apolyses.' See Development. Cf. Instar. Rel. Metamorphosis.

STAGHORN FERN BEETLE *Halticorcus platycerii* Lea [Coleoptera: Chrysomelidae] (Australia).

STAINERS See Pyrrhocoridae.

STAINFORTH, THOMAS (–1944) (Walsh 1944, Entomol. mon. Mag. 80: 86.)

STAINTON, HENRY TIBBATS (1822–1892) (Douglas *et al.* 1892, Stettin. ent. Ztg. 53: 323–329, bibliogr.; Bethune 1893, Rep. ent. Soc. Ont. 24: 108–109; Douglas *et al.* 1893, Entomol. mon. Mag. 29: 1–4, bibliogr.)

STÅL, CARL (1833–1878) Swedish Hemipterist and regarded by some specialists as the most accomplished worker on the group. (Bates 1878, Proc. ent. Soc. Lond. 1878: lxvii; Bolivar 1878, An. Soc. esp. Hist. nat. 7: 59–61; Reuter 1878, Entomol. mon. Mag. 15: 72, 94–96; Signoret 1878, Ann. Soc. ent. Fr. (5) 8: 177–186, bibliogr.; Distant 1879, Entomol. mon. Mag. 15: 191–192; Spangberg 1879, Stettin. ent. Ztg. 40: 97–105, bibliogr.)

STALEY, OLIVER JACOB (1869–1894) (Kuze 1895, Can. Ent. 27: 133.)

STALK Noun. (Anglo Saxon, *stel* = stalk. Pl., Stalks.) Alternative term for a Pedicel (Pedicle) or Petiole. Noncommittal with regard to Homology of structures. See Pedicel.

STALK BORER *Papaipema nebris* (Guenée) [Lepidoptera: Noctuidae].

STALKED BODIES See Brain. Cf. Corpora Pedunculata; Gyri Cerebrales.

STALKED EGG Elongate with constricted stalk-like projection from one or both poles of egg. Stalk length variable, sometimes corkscrew shaped, often several times longer than remainder of egg. Type of egg found in some Diptera (Pyrgotidae) and most of major Superfamilies of parasitic Hymenoptera including Chalcidoidea, Chrysidoidea, Cynipoidea, Evanioidea, Ichneumonoidea and Proctotrupoidea. See Egg.

STALKER® See Pyrrol.

STALK-EYED FLIES See Diopsidae.

STALK-EYED FLY *Sphyracephala brevicornis* (Say) [Diptera: Diopsidae]: A Species widespread in USA. Adult black with reddish head; eyes positioned at end of two 'stalks;' Thorax with lateral spines; Tarsi reddish; male with raptorial forelegs. Egg pear-shaped, deposited individually or in small clusters on moss or in mud. Larva cylindrical, smooth and shining white; head telescopes into Prothorax; spiracular stalks at apex of abdomen. Larvae associated with aquatic habitats and feed upon decaying organic material. Overwinter as adults among rocks and crevices.

STAMINATE Adj. (Latin, *stamen* = warp; *-atus* = adjectival suffix.) Botany: A flower which lacks pistils; sepals and petals present or absent. Cf. Pistillate.

STAMM, ROBERT HUTZEN PEDERSEN (1877–1934) (Henriksen 1933, Ent. Meddr. 15: 507–508, bibliogr.)

STAMMER, H. J. (Kirchberg 1964, Mitt. dt. ent. Ges. 23: 182.)

STANCE® See Ethephon.

STANDARD Noun. (Latin, *stare* = to stand; *ard* = possessing a quality. Pl., Standards.) An accepted or agreed object, thing or definition to which all others in its category are referred.

STANDEN, RICHARD S (1835–1917) (Champion 1917, Entomol. mon. Mag. 53: 279; Rowland-Brown 1917, Entomologist 50: 263–264.)

STANDFUSS, GUSTAV (1815–1897) (Schaufuss 1897, Insektenbörse 14: 247; Wocke 1898, Z. Ent. 23: 39–40.)

STANDFUSS, MAXIMILIAN RUDOLPH (1854–1917) (Oberthür 1915, Études de lépidopterologie comparée. 10 (Portraits de lépidoptèristes. 2 Sér.): [6]; Hedicke 1917, Dt. ent. Z. 1917: 325–326; Schellenberg *et al.* 1917, Mitt. ent. Zurich 3: 154–196; Ris 1918, Ver. schweiz. ent. Ges. 100: 136–142, bibliogr.)

STANDISH, BENJAMIN (1783–1866) (Newman 1867, Entomologist 3: 204.)

STANDISH, FRANCIS ORAM (1832–1880) (Carrington 1880, Entomologist 13: 142.)

STANDUP® See Chlormequat.

STAN-GUARD®. A registered biopesticide derived from *Bacillus thuringiensis* var. *kurstaki*. See *Bacillus thuringiensis*.

STANNIUS, FRIEDRICH HERRMANN (1808–1883) (Nordenskiöld 1935, *History of Biology*. 629 pp. (417–418), London.)

STANTON, AMBROSE THOMAS (1875–1938) (Freyer 1938, Proc. R. ent. Soc. Lond. (C) 3: 59–60; Roubaud 1938, Bull. Soc. Path. exot. 31: 179.)

STAPHYLA Noun. (Greek, *staphyle* = bunch of grapes. Pl., Staphylae.) A group of Gongylidia; the enlarged hyphal apices of fungi that live symbiotically with attine ants.

STAPHYLINIDAE Plural Noun. Road Beetles; Rove Beetles; Shining Fungus Beetles. A cosmopolitan Family of polyphagous Coleoptera currently holding more than nine Subfamilies, 400 nominal Genera and 32,000 Species. Body small to moderately large, elongate, parallel sided; Antenna filiform, moniliform or rarely weakly clavate, with 10–11 segments. Elytra short, truncate; hindwings typically membranous, functional, complexly folded beneath Elytra. Abdomen usually flexible; sometimes held arched over back; several Terga exposed. Many Species predaceous, but feeding habits of most Species unknown. Prefer moist habitats, found in nests of social Hymenoptera, decaying vegetation, moss, soil, beneath stones, logs, in rotting logs, associated with flowers, nests of some mammals. Larvae and adults predaceous; some Species termitophilous. Prey include larvae and pupae of Diptera, other Coleoptera, spiders and mites. Myrmecophilous Species range in habits from true inquilines to predators of all stages of host development. Parasitic Species exsanguinate host larva and pupate on, near or within host's remains. Nomenclatural aspects of Family-group name considered by Blackwelder (1952, Bull. U. S. N. M. 200: 1–483) and ICZN (1959, Opinion 546.)

STAPHYLINOIDEA Noun. A numerically large, cosmopolitan Superfamily of polyphagous Coleoptera including Agyrtidae, Anistomidae, Dasyceridae, Hydraenidae, Leiodidae, Leptinidae, Limuloididae, Micropeplidae, Pselaphidae, Ptiliidae, Scydmaenidae, Silphidae and

Staphylinidae. Adult staphylinoids are characterized by fore Coxa strongly projecting; Metasternum typically lacking median suture; Elytra truncate, exposing abdominal Terga; hindwing venation reduced without R-M loop; legs often spinose. See Coleoptera.

STAR JASMINE THRIPS *Thrips orientalis* (Bagnall) [Thysanoptera: Thripidae].

STAR-SHAPED AMOEBOCYTE See Plasmatocyte.

STARCH GEL ELECTROPHORESIS See Electrophoresis.

STARCH Noun. (Anglo Saxon, *stearc* = stiff. Pl., Starches.) A hexosan polysaccharide hydrolysed to maltose by amylase and then glucose by maltase.

STÄRCKE, AUGUST (1880–1954) (Verhoeff 1954, Ent. Ber., Amst. 15: 255–262, bibliogr.)

STARK, JOSEF (1813–1889) (Dohrn 1889, Stettin. ent. Ztg. 50: 320–322.)

STARK, VLADIMIR NIKOLARVICH (1899–1962) (Birevizmam 1959. Ent. Obozr. 38: 702–703; Grigorev & Likventov 1963, Ent. Obozr. 42: 234–241, bibliogr. Translation: Ent. Rev., Wash. 42: 127–131, bibliogr.

STARKE, HERMANN (1870–1954) (Jordan 1955, Natura lusat. 2: 5–7.)

STARKEY, GEORGE (1627 (1628)?-1655) (Wilkinson 1973, Gt. Lakes Ent. 6: 59–64.)

STARLEX® See Pyraclofos.

STARLING MITE *Ornithonyssus bursa* (Berlese) [Acari: Macronyssidae] (Australia).

STARVE Verb. (Old High German, *sterven* = to die.) Starved; Starved; Starving; Starves. 1. To die from a lack of food or nourishment. 2. To kill with hunger or the deprivation of food. 3. To suffer extreme hunger. See Die; Hunger.

STASE Noun. (Etymology obscure. Pl., Stases.) A concept in development. Acarology: Stase is one of the successive forms through which acarines pass during development. Successive stases are identified and differ from each other by discontinuous external features (characters). Stases at the same level of development are identified by homologous external features (characters). Allometric differences or differences in size or shape of a character cannot be used to determine successive Stases. New external characters must be developed to separate Stases. Stase differ from instar in that an instar involves the moulting process and a change of the integument; a Stase is always an Instar but an Instar is not always a Stase (*teste* Grandjean 1970). A maximum of six Stases occur in the Acari: prelarva (deutovum, prolarva), larva (tritovum), protonymph, deuteronymph (hypopus), tritonymph and adult. Cf. Instar. Stasoid.

STASIS Noun. (Greek, *stasis* = standing. Pl., Stases.) 1. The cessation of fluid movement in animals. 2. The retardation of growth. 3. A period of development delimited by moults (Johnston & Wacker 1967). See Stase; Stasoid.

STASOID Noun. (Etymology obscure. Pl., Stasoids.) Acarology: Developmental forms that differ from one another by discontinuous characters but which cannot be homologized with corresponding forms in other Species of the same group. See Homology; Stase; Stasis. Rel. Eidos; Form; Shape.

STATARY PHASE Social Insects: The period during which an army ant colony does not move, the queen oviposits and most of the brood are eggs and pupae. Cf. Nomadic Phase.

STATHER, THOMAS (1812–1878) (Stainforth 1919, Trans. Hull sci ent. Fld nat. Club 4: 281–298.)

STATHMOPODIDAE Plural Noun. See Oecophoridae.

STATIC Adj. (Greek, *statikos* = causing to stand; *-ic* = of the nature of.) Passive, at rest or in equilibrium. Pertaining to a system in equilibrium or at rest.

STATIC ORGAN 1. Any structure which aids in preserving balance. 2. Phylloxera: SO positioned at the antennal base and consisting of a small vesicle enclosing a central body or statolith with nervous connections (Stauffacher). See Statolith.

STATIC SENSE The sense of balance or of maintaining position in the air or otherwise.

STATIONARY PHASE Tissue Culture: The period of population dynamics following the Exponential Phase. During the Stationary Phase, cell populations do not increase. See Growth Curve. Cf. Exponential Phase; Lag Phase.

STATISTICAL METHOD A process involving numerical analysis that is used to determine probabilities of events occuring. SM involves a combination of hypothesis formulation, mathematical computation and comparison of numbers with tabled values. SM has become extensively used in the biological sciences and forms the basis upon which ideas are tested and conclusions (decisions) are made. Cf. Biometry.

STATOCYST Noun. (Greek, *statos* = stationary; *kytos* = hollow. Pl., Statocysts.) A sense organ responsible for maintaining equilibrium and orientation. Statocysts occur on the Thorax of the ant *Dorymyrmex*, positioned above the hind Coxa. Statocysts also occur on the head of the termite *Anoplotermes* and cerci of the cockroach *Arenivaga* where they act as gravity receptors. Anatomically, statocysts are cuticular invagination lined with Setae and containing a few grains of sand. The sand grains are called statoliths Cuticular projections within the chamber of the statocyst maintain the grains of sand relatively immobile, but when they move, they stimulate some receptors and the body is given sensory information important in orientation. Poorly documented in insects but see *Dorymyrmex* (Marcus 1956, Zeitsch. Wiss. Zool. 159: 225–254.) See Sensillum. Cf. Palmen's Organ; Proprioreceptor.

STATODYNAMIC SENSE The sense which regulates the position of an animal, through awareness of its relation in space and surrounding objects.

STATOLITH Noun. (Greek, *statos* = stationary; *lithos* = stone. Pl., Statoliths.) A structure (calcium carbonate, sand grain) found within a Statocyst. Alt. Otolith. See Statocyst. Rel. Sensillum.

STAUDINGER, OTTO (1830–1900) (Elwes 1900, Proc. ent. Soc. Lond. 1900: xxvi–xxxii; S. 1900, Dt. ent. Z. Iris 13: 341–358, bibliogr.)

STAUFFER, JACOB (1809–1880) (Anon. 1880. Am. Nat. 14: 466; Weiss 1936, *Pioneer Century of American Entomology,* 320 pp. (175), New Brunswick.)

STAUNTON, GEORGE LEONARD (1737–1801) (Rose 1850, *New General Biographical Dictionary* 12: 105–106.)

STEAMER See German Cockroach.

STEAMFLY See German Cockroach.

STEARNS, LOUIS AGASSIZ (1892–1960) (MacCreary 1963, J. Econ. Ent. 56: 122.)

STEATOCYTE Noun. (Greek, *spear* = stiff fat; tallow; *kytos* = a hollow.) A form of amoebocyte which detaches itself from the imaginal adipose tissue to destroy the larval fat-cells (Henneguy).

STECK (HOFMANN), THEODOR (1857–1937) (Benson 1937, Entomol. mon. Mag. 73: 46; Schulthess 1937, Mitt. schweiz. ent. Ges. 17: 1–4, bibliogr.; Anon. 1958, Mitt. schweiz. ent. Ges. 31: 109–120.)

STECK, HERMANN (1883–1937) (Anon. 1937. Arb. morph. taxon. Ent. Berl. 4: 242.)

STEDMAN, JOHN GABRIEL (1744–1797) (Rose 1850, *New General Biographical Dictionary* 12: 106.)

STEEL, THOMAS (1858–1925) (Carter 1926, Proc. Linn. Soc. N.S.W. 51: vii.)

STEELBLUE BLOWFLY *Chrysomya saffranea* (Bigot) [Diptera: Calliphoridae] (Australia).

STEELBLUE LADY-BEETLE *Orcus chalybeus* (Boisduval) [Coleoptera: Coccinellidae]: In Australia, a beneficial predator of aphids and other pests. Syn. *Halmus chalybeus* (Boisduval).

STEELBLUE SAWFLY *Perga affinis affinis* Kirby; *Perga dorsalis* Leach [Hymenoptera: Pergidae] (Australia).

STEENBERG, CARL MARINUS (1882–1946) (Degerböl 1946, Vidensk. Meddr dansk. naturh. Foren. 109: v–x; Tuxen 1947, Ent. Meddr. 25: 142–144, bibliogr.)

STEENBURGH, WILLIAM ELGIN VAN (1899–1974) (Glen 1974, Bull. ent. Soc. Can. 6: 84–86.)

STEENSTRUP, JOHANN JAPETUS SMITH (1813–1897) (Henriksen 1925, Ent. Meddr. 15: 153–156; Spärch 1933, Vidensk. Meddr dansk. naturh. Foren. 95: 56–90.)

STEFADONE® See Chlorfenvinphos.

STEFANELLI, PIETRO (1834–1919) (Verity 1919, Boll. Soc. ent. ital. 51: 76–81.)

STEGASIMOUS Adj. (Greek, *stege* = roof; *-osus* = possessing the qualities of.) Acarology: A condition in which the Gnathosoma is retracted and not visible in dorsal view.

STEIN, F (–1931) (Davis 1931, Proc. Indiana Acad. Sci. 41: 55.)

STEIN, JOHANN PHILIP EMIL FRIEDRICH (1814–1882) (Dohrn 1882, Stettin. ent. Ztg. 43: 509–510; Kraatz 1882, Dt. ent. Z. 26: 8.)

STEIN, PAUL (1852–1921) (Kramer 1922, Dt. ent. Z. 1922: 236–241, bibliogr.)

STEIN, SAMUEL FRIEDRICH NATHANIEL VON (1818–1885) (Anon. 1885, Naturaliste 7 (9) 72; Dimmock 1885, Psyche 4: 266.)

STEINBERG, DIMITRIV MAKSIMILIANOVICH (1909–1962) (Bei-Bienko 1963, Ent. Obozr. 42: 468–472, bibliogr. (Translation: Ent. Rev., Wash. 42: 255–257, bibliogr.).

STEINER, CARL (1831–1908) (Vogel 1932, Schr. phys.-ökon. Ces. Königsb. 67: 107.)

STEINER, GOTTHOLD (–1961) (Goffart 1961, Anz. Schädlingsk. 34: 171.)

STEINER, HAROLD METZLER (1909–1973) (Dean *et al.* 1973, J. Econ. Ent. 67: 317.)

***STEINERNEMA* SPP** See Entomogenous Nematodes.

STEINERT, HERMANN (–1898) (Schlopfer 1898, Dt. Ent. Z. Iris 11: 400–401, bibliogr.; Möbius 1943, Dt. Ent. Z. Iris 57: 23.)

STEINHAUS, EDWARD ARTHUR (1914–1969) (Hughes & Marsh 1970, Misc. Publs Centre Pathobiol. 2: 1–54, bibliogr.; Lipa 1970, Polskie Pismo ent. 49: 879–880; Pristarko 1970, Ent. Obozr. 49: 508–510. Translation: Ent. Rev., Wash. 49: 306–307).

STEINHEIL, EDUARD (1830–1878) (Forel 1879, Mitt. münch. ent. Ver. 3: 1–5, bibliogr.; Papavero 1975, *Essays on the History of Neotropical Dipterology.* 2: 305. São Paulo.)

STEJSKAEM, J. V. (1878–1945) (Malac 1947, Ent. Listy 10: 64.)

STELEX Adj. (Greek, *stele* = pillar.) The crown of a tree from which the trunk springs.

STELLATE Adj. (Latin, *stella* = star; *-atus* = adjectival suffix.) Resembling a star; star-shaped; with a star-shaped structure. Alt. Stellated; Stelliform.

STELOCYTTAROUS Adj. (Greek, *stele* = pillar; *kyttarous* = honeycomb cell; Latin, *-osus* = possessing the qualities of.) Social wasps: Pertaining to the comb layers of the nest supported by pillars and not connected with the envelope. See Poecilocyttares; Phragmocyttares.

STEM MOTHER Noun. A wingless, parthenogenetic, viviparous female aphid that arises from an overwintering egg. See Fundatrix.

STEM SAWFLIES See Cephidae.

STEMAPODA Noun. The modified filamentous anal legs of *Cerura* and other notodontid larvae.

STEMERDING, STEVEN (–1949) (Macgillivray 1949, Ent. Ber., Amst. 12: 392.)

STEMMA Noun. (Greek, *stemma* = garland. Pl., Stemmata.) 1. A simple eye or single-lens optical device found on the head of most holometabolous larvae in the region of the head where the compound eye will develop. Stemmata have not been reported in fleas or apocritious Hymenoptera; they are reduced in size or apparently lost in some wood-boring Symphyta larvae, mining Lepidoptera and Brachycera. Stemmata were first described by Malpighi. Variable in number; not homologous with the Ocellus. Called 'lateral Ocellus' by some early 20th century morphologists (Snodgrass). The biconvex lens of the Stemma forms an image on a Rhabdom. However, information collected is not used for image formation. Instead, visual information is used to detect motion. When several Stemmata are clustered together, a mosaic pattern of the environment can be formed. Some larvae use Stemma to detect the plane of polarized light. See Eye; Vision. Cf. Compound Eye; Ocellus. 2. Small tubercles on an Antenna.

STEMMATICUM Noun. (Greek, *stemma* = crown, garland. Pl., Stemmatica.) A conspicuous triangular or rectangular boundary for the Ocelli in many Genera of Mymaridae (Hymenoptera: Chalcidoidea).

STEMOFOLINE Noun. (Pl., Semofolines.) An alkaloid biopesticide compound extracted from the leaves and stems of *Stemona japonica* Miq. (Stemonaceae). Ground roots of stemonaceous plants are used as an agricultural insecticide in China. See Biopesticide. Cf. Nereistoxin; Nicotine; Pyrethrin; Pyrroles; Ryania; Sabadilla.

STEMPEL, WALTER (1869–1938) (Anon. 1938, Arb. morph. taxon. Ent. Berl. 5: 295.)

STENBERG, ISAAC LUDWIG (–1891) (Sandahl 1891, Ent. Tijdschr. 12: 16.)

STENDELL, WALTER (1889–1914) (Anon. 1914, Dt. Ent. Z. 1914: 649; Anon. 1915, Ent. News 26: 240.)

STENOCEPHALIDAE Plural Noun. Spurge Bugs.

STENOCEPHALOUS Adj. (Greek, *stenos* = narrow; *kephale* = head; Latin, *-osus* = possessing the qualities of.) Pertaining to animals with a narrow, elongate head. Alt. Stenocephalus.

STENOGAMOUS Adj. (Greek, *stenos* = narrow; *gamos* = marriage; Latin, *-osus* = possessing the qualities of.) Pertaining to Species which mate in confined areas (*e.g.* mosquitoes).

STENOGASTRIC Adj. (Greek, *stenos* = narrow; *gaster* = stomach; *-ic* = consisting of.) Descriptive of animals with a shortened Abdomen or Gaster.

STENOGASTRINAE Hover Wasps. A behaviourally diverse Subfamily of Vespidae (Hymenoptera: Aculeata), considered intermediate between Eumenidae and more highly evolved Vespidae. See Vespidae.

STENOHALINE Adj. (Greek, *stenos* = narrow; *halinos* = saline.) Pertaining to organisms which cannot adapt to a wide range of salinity. Cf. Euryhaline.

STENOHYGRIC Adj. (Greek, *stenos* = narrow; *hygros* = wet; *-ic* = characterized by.) Pertaining to organisms that cannot adapt to wide ranges of relative humidity. See Ecotype. Cf. Euryhygric.

STENOMICRIDAE See Periscelididae.

STENOMIDAE See Oecophoridae.

STENOPELMATIDAE Plural Noun. Jerusalem Crickets. A Family of ensiferous Orthoptera consisting of 5–6 Genera and 30 Species, most of which occur in Central America. Some Species found in South Africa and southeastern Asia. Stout bodied; enlarged head; reduced compound eye; short Antenna. Tibiae spinose, adapted for fossorial habits; wings typically absent; Cerci not annulated; Ovipositor short. Fossil record dating from Miocene, and apparently a relictual group. Some Species feed on roots and plant parts which contact the soil, but most are nocturnal predators or scavengers.

STENOPELMATOIDEA A Superfamily of ensiferous Orthoptera sometimes placed within the Gryllacridoidea. When separated, includes all winged forms and those with enlarged head, reduced Antenna, and auditory organ at base of fore Tibia. The stridulatory mechanism involves the Femur and Abdomen and the Tarsi are four segmented. The group includes about 125 Genera and 750 Species, widely distributed but poorly represented in the Holarctic.

STENOPHAGOUS Adj. (Greek, *stenos* = narrow; *phagein* = to eat; Latin, *-osus* = possessing the qualities of.) Pertaining to an organism which exhibits Stenophagy. Pertaining typically to an herbivorous organism which feeds on a few Species of food plants, a predator which feeds on a few Species of prey or a parasite which feeds on a few Species of host. By implication, the dietary items are closely related. See Oligophagous. Cf. Euryphagous; Monophagous; Polyphagous. Rel. Oligoxenous.

STENOPHAGY Noun. (Greek, *stenos* = narrow; *phagein* = to eat. Pl., Stenophagies.) 1. A limited range of dietary requirements. 2. A limited range of host plant Species for a herbivore, host Species for a parasite or prey Species for a predator. Syn. Oligophagy. Cf. Euryphagy; Monophagy; Polyphagy.

STENOPSOCIDAE Plural Noun. A Family of Psocoptera assigned to the Suborder Psocomorpha. Antenna with 13 segments, without secondary annulations; forewing with venation between Pterostigma and Rs, and between Areola Postica and M; genitalia reduced. Common on leaves.

STENOPTEROUS Adj. (Greek, *stenos* = narrow; *pteron* = wing; Latin, *-osus* = possessing the qualities of.) Descriptive of abbreviated or narrowed but complete wings. Term often applied to Heteroptera.

STENORHYNCHAN Adj. (Greek, *stenos* = narrow; *rhynchos* = beck.) Narrow beaked or snouted.

STENOTHORAX Noun. (Greek, *stenos* = narrow; + thorax.) An apparent ring-like sclerite between the Prothorax and Mesothorax.

STENOTRITIDAE Plural Noun. A small Family of Apoidea consisting of two Genera. Apparently related to Colletidae. (McGinley 1980. J. Kans. Ent. Soc. 53: 539–552.)

STEP, EDWARD (1855–1931) (Adkin 1931, Entomologist 64: 287–288; Blair 1932, Proc. S. Lond. ent. nat. Hist. Soc. 1931–32: 34–35.)

STÉPAN, VACSLAV J (1873–1941) (Samal 1941, Cas. csl. Spol. ent. 38: 1–2.)

STEPHAN, JULIUS (1877–1954) (Anon. 1937, Ent. Z., Frankf: A. M. 50: 553–554; Anon. 1952. Ent. Z., Frankf. a. M. 61: 185–186; Anon. 1954, Z. wien. ent. Ges. 65: 400.)

STEPHANIDAE Leach 1815. Plural Noun. A Family of parasitic Hymenoptera assigned to Stephanoidea.

STEPHANOCIRCIDAE Plural Noun. Helmet Fleas. A numerically small Family of ceratophylloid Siphonaptera, with representative Species in Australia and South America. Hosts include marsupials and rodents.

STEPHANOIDEA Leach 1814. A Superfamily of parasitic Hymenoptera. Included Families Ceraphronidae, Ichneumonidae, Maimetshidae, Megalyridae, Stephanidae and Trigonalidae. See Ceraphronoidea; Trigonaloidea.

STEPHENS, JAMES ALFRED (1882–1947) (Blair 1947, Entomol. mon. Mag. 83: 135; Williams 1948, Proc. R. ent. Soc. Lond. (C) 12: 64.)

STEPHENS, JAMES FRANCIS (1792–1852) Author of *Systematic Catalogue of British Insects* (1829), *Illustrations of British Entomology*, and *Nomenclature of British Insects.* (Swainson 1840, *Taxidermy; with Biography of Zoologists.* 329 pp. (pp. 336–337, bibliogr.). London. [Vol. 12, *Cabinet Cyclopedia*. Edited by D. Lardner.]; Newman 1850, Zoologist 11: 374–437, 453–464; Anon. 1853, Gardner's Chronicle, 8.1.1853 p.I5; Stainton 1853, *Bibliotheca Stephensiana* (2)-(10); Westwood 1853, Proc. ent. Soc. Lond. 1853: 45–50, bibliogr.; Anon. 1889, Abeille (Les ent. et leurs écrits) 26: 261–265, bibliogr.)

STEPHENS, JOHN WILLIAM WATSON (1865–1946) (Christophers 1947, Obit. Not. Fell. R. Soc. Lond. 5: 525–540, bibliogr.)

STEPHENSEN, KNUTH (1882–1947) (Tuxen 1947, Ent. Meddr. 25: 196–199.)

STERCORACEOUS Adj. (Latin, *stercus* = dung *-aceus* = of or pertaining to.) Pertaining to substance that resembles excrement in texture, colour, shape or odour. Descriptive of material that contains excrement. Syn. Scatophilous. Cf. Fimicolous.

STERCORAIAN TRYPANOSOME A section of *Trypanosoma* whose members are transmitted via contact with faeces of insect vectors to a mammalian host. Cf. Salavarian Trypanosome.

STERCORAL Adj. (Latin, *stercus* = dung.) Relating or pertaining to excrement.

STEREOTROPISM Noun. (Greek, *stereos* = solid; *trope* = turn.) Negative thigmotropism; avoidance of contact. See Tropism. Cf. Aeolotropism; Anemotropism; Chemotropism; Electrotropism; Galvanotropism; Geotropism; Heliotropism; Hydrotropism; Phototropism; Rheotropism; Thermotropism; Thigmotropism; Tonotropism. Rel. Taxis.

STERIGMA Noun. (Latin, from Greek, *sterigma* = a prop. Pl., Sterigmata.) Lepidoptera: A sclerite or sclerites surrounding the female's Ostium Bursa.

STERILE Noun. (Latin, *sterilis* = barren. Pl., Steriles.) Individuals incapable of reproduction. Cf. Fertile. Rel. Neuter.

STERILE INSECT RELEASE METHOD See Sterile Insect Technique.

STERILE INSECT TECHNIQUE A method of insect control in which laboratory-propagated insects are irradiated to the point of sterility and then released into the environment to compete for mates with conspecifics in feral populations. The Sterile Insect Technique (SIT) was first proposed and implemented by Knipling (1955) to successfully eradicate screwworm fly *(Cochliomyia hominivorax)* from southeastern USA. Success of programme prompted subsequent use on numerous Species of Tephritidae and some other insects in various locations throughout the world. SIT involves several activities including Mass Rearing, Sterilization, Transportation, Release and Assessment. Syn. Sterile Insect Release Method.

STERILITY Noun. (Latin, *sterilis* = barren; English, *-ity* = suffix forming abstract nouns. Pl., Sterilities.) Barrenness; a structural or functional inability to reproduce.

STERLER, ALOIS (1787–1831) (Gistl 1832, Faunus 1: 51.)

STERLING® See Pymetrozine.

STERN, RUDOLF (–1958) (Anon. 1958, Z. wien. ent. Ges. 43: 184.)

STERNA See Sternum.

STERNACOILA Noun. (Greek, *sternon* = chest.) A blunt, spine-like projection of the Sternellum adjacent to the Coxa, which serves as a second point of articulation (MacGillivray).

STERNACOSTA Noun. (Greek, *sternon* = chest; *costa* = rib.) The transverse internal ridge of the sternal suture through the bases of the Sternal Apophyses (Snodgrass).

STERNACOSTAL SUTURE The external suture of the Sternacosta, separating Basisternum from Sternellum (Snodgrass). See Suture.

STERNAL Adj. (Greek, *sternon* = chest; Latin, *-alis* = pertaining to.) Descriptive of or pertaining to the Sternum.

STERNAL APODEME Hymenoptera: The anterodorsal apodeme of a Sternum. See Apodeme.

STERNAL APOPHYSES The lateral apodemal arms of Eusternum. In higher insects united on a median base, the whole structure forming the Furca (Snodgrass). See Apophysis.

STERNAL LATERALE Some lower insects: A sclerite located at each side of the Sternum or Presternum (Snodgrass).

STERNAL ORIFICE Perlidae: A peculiar slit of each side of Sternum, extending inward and ending blindly. Syn. Furcal Orifice. See Oriface.

STERNAL SPATULA Diptera larvae: See Anchor

Process; Breastbone; Furca.

STERNANUM Noun. (Greek, *sternon* = chest.) Basisternum; Cephalic part of mesial section of the Sternum (MacGillivray). The point of articulation of Coxa to Sternum (MacGillivray).

STERNAULUS Noun. (Greek, *sternon* = chest; Latin, *-ulous* = characterized by. Pl., Sternauli.) Hymenoptera: The short, sometimes obsolete Sulcus on either side of the Mesosternum.

STERNE, CARUS (–1903) (Anon. 1903, Insektenbörse 20: 306.) (Pseudonym of E. Krause).

STERNECK, JAKOB DOUBLEBSKY VON (1868–1941) (Reisser 1941, Z. Wien. entVer. 26: 265–271, bibliogr.)

STERNELLAR Adj. (Greek, *sternon* = breastbone; *-ellum* = dim. form; *-ar* = adj. suffix.) Pertaining to the Sternellum.

STERNELLUM Noun. (Greek, *sternon* = breastbone; *-ellum* = dim. form.) The second sclerite of the ventral part of each thoracic segment, frequently divided into longitudinal parts which may be widely separated (Smith). An area of the Eusternum posterior of the bases of Sternal Apophyses or Sternacostal Suture (Snodgrass).

STERNEPIMERON Noun. (Greek, *sternon* = breastbone; + epimeron.) The part of the Epimeron on the ventral side of the Epimeral Suture. See Hypoepimeron; Katepimerum; Meropleuron (MacGillivary).

STERNEPISTERNUM Noun. (Greek, *sternon* = breastbone; + episternum.) Part of the Episternum on the ventral side of the Episternal Suture. See Katepisternum; Sternopleurite; Sternopleura (MacGillivray).

STERNITE Noun. (Greek, *sternon* = chest; *-ites* = constituent. Pl., Sternites.) 1. The ventral part of a sclerotized, ring-like body segment which is separated by membrane from sclerotized lateral or dorsal elements of the segment. 2. A sclerotized or hardened subdivision of a Sternum. 3. Any sclerotized component of a definitive Sternum. See Segmentation; Sternum. Cf. Tergite.

STERNOCOXAL Adj. (Greek, *sternon* = chest; Latin, coxa = hip) Descriptive of or pertaining to the Sternum and Coxa together.

STERNOIDEA (Greek, *sternon* = chest.) Two narrow transverse sclerites anterior of the Pretrochantin, in certain generalized insects (MacGillivray).

STERNOPLEURITE Noun. (Greek, *sternon* = chest; *pleura* = side. Pl., Sternopleurites.) 1. The infracoxal sclerotization of a generalized thoracic Pleuron. Generally united with the primary Sternum in definitive eusternal sclerite (Snodgrass). 2. The compound sclerite formed by fusion of the Episternum and Sternum. See Sternopleuron (Imms).

STERNOPLEURON Noun. (Greek, *sternon* = chest; *pleura* = side. Pl., Sternopleura.) 1. The ventral portion of a sclerotized ring or body segment. 2. A subdivision of a sternal sclerite, or any one of the sclerotic components of a definitive Sternum (Pl., Sternopleura). 3. Ventral portion of Episternum (Katepisternum) of the Mesothorax (Comstock). 4. Diptera: Ventral portion of the Pleuron, below the Sternopleural Suture and above the anterior Coxa (Smith). 5. Compound sclerite formed by fusion of Episternum and Sternum (Imms). See Sternopleurite.

STERNOPLURAL SUTURE Diptera: A suture below and nearly parallel with the Dorsopleural Suture, separating Mesopleura and Sternopleura.

STERNOPLURAL Adj. (Greek, *sternon* = chest; *pleura* = side; *-alis* = pertaining to.) 1. Descriptive of or pertaining to the Sternopleura, or Stenopleurite. 2. Diptera: One or several Setae on each Sternopleuron below and posterior of the Sternopleural Suture (Comstock).

STERNORHABDITES Noun. (Greek, *sternon* = chest; *rhabdos* = rod.) Odonata: Style. Blattaria: Valvulae Inferiores. Hymenoptera: Cuticular structures or tubercles in larvae which form the Ovipositor in the adult.

STERNORRHYNCHA Noun. (Greek, *sternon* = chest; *rhyngchos* = snout.) A Suborder of Hemiptera including psyllids, whiteflies, aphids, scale insects, coccoids and mealybugs. Antenna filiform; beak (Rostrum) apparently arises from Sternum between fore Coxae; hindwing lacking Vannus and vannal fold; Tarsi with 1–2 segments. Presumably monophyletic and a sister group of all other Hemiptera.

STERNUM Noun. (Greek, *sternon* = chest; Latin, *sternum* = breast bone. Pl., Sterna.) 1. The entire ventral division of any ring-like body segment. 2. Sternum of the head is not apparent. Sternum of the Abdomen relatively simple and a simple transverse sclerite. Sternum of Thorax complex as ventral surface of insect Thorax between coxal cavities. Sclerotized thoracic Sternum probably evolved simultaneously with pleural region. Groundplan thoracic Sternum consisted of four sclerites: An Intersternite (Spinasternite), two Laterosternites and a Mediosternite. Intersternite develops between prosternal and mesosternal regions; a second Intersternite develops between mesosternal and metasternal regions. Laterosternites develop from Coxosternites along ventral surface of Coxa. Mediosternite fuses with Coxosternites laterally to form a large sclerite (Eusternum). Anterior Intersternite typically remains free but fuses with Prosternum in some Orthoptera. Posterior Intersternite usually fuses with mesothoracic Eusternum. Longitudinal line of union between Mediosternite and each Coxosternite is called the Laterosternal Sulcus (Pleurosternal Suture). Paired Sternal Apophyses (Furcal Pits) occur in Laterosternal Sulcus. A transverse Sternacostal Sulcus extends across Mediosternite. This Sulcus originates with each Furcal Pit and develops mesad; Sulcus provides line-of-contact between Furcal Pits and an internal strengthening for this sternal sclerite. Completion of suclus bissects sternal sclerite into an anterior Basisternite and posterior Furcasternite. In Mesosternum, the posterior Spinasternite fuses with Furcasternite to form a Sternellum. Also, Furcal Pits migrate mesad in some insects and form one common medial apophysis. See Thorax. Cf. Sternite.

STERNUM COLLARE 1. The collar-bone. 2. A prominent narrow part in the ventral surface of the Jugulum.

STERNUM PECTORALE The breast-bone, a prominent Carina on the Pectus (Knoch).

STERTZ, OTTO (1847–1918) (Möbius 1918, Dt. ent. Z. Iris 32: 134; Wolf 1919, Jber. schles. Insekt. Breslau 10–12: 23–25.)

STERZL, OTTO (1901–1969) (Anon. 1969, Z. wien. ent. Ges. 53: 64; Tauber 1969, Ent. NachrBl. 16: 126–127, bibliogr.)

STETHIDIUM Noun. (Greek, *stethos* = the chest; *-idion* = diminutive. Pl., Stethidia.) The entire Thorax including its appendages.

STETHOSOMA Noun. (Greek, *stethos* = breast; *somatos* = body. Pl., Stethosomas; Stethosomata.) Acarology: A division of the body limited by the Circumcapitular and Disjugal Furrows; the Prosoma without the Gnathosoma.

STEUDEL, WILHELM (1829–1903) (Anon. 1903, Insektenbörse 20: 403; Klunzenger 1904, Jh. Ver. vaterl. Naturk. Württ. 60: xxxv–xliii, bibliogr.)

STEVEN, CHRISTIAN VON (1781–1863) Director of Imperial Botanical Garden at Odessa. Author of *Description de quelques Insectes de Caucase et de la Russie Meridionale* (IN: Mem. Imp. Nat. Moscou). (Nordmann 1865, Bull. Soc. Imp. nat. Moscou 38 (1): 101–106.)

STEVENS LEAFHOPPER *Empoasca stevensi* Young [Hemiptera: Cicadellidae].

STEVENS, JOHN SANDERS (–1903) (Anon. 1903, Entomol. mon. Mag. 39: 229.)

STEVENS, NETTIE M (1861–1912) (Calvert 1912, Ent. News 23: 288.)

STEVENS, SAMUEL (1817–1899) (Anon. 1899, Entomol. mon. Mag. 35: 238–239; Anon. 1899, Entomologist 32: 264.)

STEVENSON, ROY HEW RUSSEL (1878–1968) (Hinton 1971, Proc. R. ent. Soc. Lond. (C) 35: 53.)

STEWARD, CHARLES COWLEY (1907–1968) (Haufe 1969, Can. Ent. 101: 186.)

STEWARD® A registered biopesticide derived from *Bacillus thuringiensis* var. *kurstaki*. See *Bacillus thuringiensis*.

STEWART, A M (1862–1948) (Williams 1949, Proc. R. ent. Soc. Lond. (C)13: 68.)

STEWART, MORRIS ALBION (1902–1961) (Furman 1962, Pan-Pacif. Ent. 38: 71–72.)

STEWART'S DISEASE A bacterial disease of corn caused by *Bacterium stewarti* and transmitted by the Corn Flea-Beetle, Corn Rootworm and Seed-Corn Maggot.

STICHA, BOHUMI (Heyrovsky 1950, Cas. csl. Spol. ent. 47: 214–215.)

STICHEL, HANS (1862–1936) (Anon. 1936, Lambillionea 36: 221; Mell 1937, Mitt. dt. ent. Ges. 7: 69–71.)

STICK INSECT See Phasmatoidea; Phasmida.

STICKER Noun. An ingredient added to a pesticide formulation to improve its adherence to a surface, *e.g.* casein.

STICKLAC Noun. The branches or twigs with the dried lac insect on them.

STICKNET, FENNER SATTERTHWAITE (1892–1936) (Boyden 1937, J. Econ. Ent. 30: 220–221; Mickel 1937, Ann. ent. Soc. Am. 30: 183.)

STICKTIGHT FLEA *Echidnophaga gallinacea* (Westwood) [Siphonaptera: Pulicidae]: A Species widespread in North America. Adult dark brown-black; head angulate in profile and divided by a Sulcus; two bristles each anterior and posterior of Antennal Torulus; eyes pigmented; Mandibles tapering apicad and serrated; Egg whitish, oval and dry; falls to ground in nest; eclosion occurs within 3–4 days at 26°C and 85% RH. SF of veterinary importance as a parasite of chickens and turkeys when adults attach to face and wattles in large numbers; also attacks rabbits, rats, dogs, cats, horses and humans. A vector of Epidemic or Murine Typhus. See Pulicidae. Cf. Sand Flea.

STICKTIGHTS See Tungidae.

STICKY TRAP Noun. A substrate coated with a thick, sticky substance to trap flying or crawling insects. Substrates range from paper, plastic strips and glass, to objects such as spheres, cylinders and wire. Most sticky traps are yellow; some are blue or red to attract certain types of insects. Insects trapped by such a method are often destroyed, but many can be preserved by cleaning with a solvent such as toluene, xylene or cellosolve. See Fly Paper; Trap.

STIDHAM, ISAAC FERDINAND (1837–1913) (Skinner 1913, Ent. News 24: 321–322.)

STIDSTON, S T (1881–1963) (Wigglesworth 1964, Proc. R. ent. Soc. Lond. (C) 28: 58.)

STIEBER, ERNST (1829–1906) (A.G. 1906, Insektenbörse 23: 145.)

STIERLIN, WILHELM GUSTAV (1821–1907) (Horn 1907, Dt. ent. Z. 1907: 450–451; Stierlin 1908, Mitt. schweiz. ent. Ges. I1: 267–273, bibliogr.)

STIFF-WINGED CRICKETS See Scleropteridae.

STIGMA METATHORACIS Arcane: Diptera metathoracic Spiracle; SM postioned on both side of the Metanotum, anterior of the Haltere.

STIGMA PLATE Ticks: The sclerite which surrounds a Spiracle (Matheson).

STIGMA Noun. (Greek, *stigma* = mark. Pl., Stigmata.) 1. A Spiracle or respiratory pore of the respiratory system. 2. Hymenoptera: A pigmented spot along the costal margin of a wing, usually at the end of the Radius. (See Anastomosis, Pterostigma.) 1. Diptera: A coloured wing spot near the apex of the Auxiliary Vein. 2. Lepidoptera: The specialized patch of black scales on the forewings of Hesperidae. Syn. Stigmata.

STIGMAL VEIN Hymenoptera (Chalcidoidea): A vein originating along the costal margin of the forewing and which projects at an angle from the Marginal Vein. SV Homology uncertain, but generally considered as the Radial Crossvein, r. Syn. Stigma. See Vein; Wing Venation.

STIGMAPHRONIDAE Kozlov 1975. A Family of parasitic Hymenoptera assigned to the Trigonaloidea.

STIGMATAL LINE Lepidoptera larva: The spiracular line.

STIGMATIC Adj. (Greek, *stigma* = mark; *-ikos* = characteristic of.) Of or pertaining to a Stigma.

STIGMATIC APERTURES Coccids: The lateral aperture of the test of the female, on either side of the Anus into which the stigmatic processes fit (MacGillivray).

STIGMATIC CHAMBER Typically large, coarsely reticulate atrial chamber adjacent to the spiracular aperture of Diptera; taxonomically important in Tephritidae. Function unknown but believed to filter particulate matter. Syn. Felt Chamber.

STIGMATIC CICATRIX or SCAR The remains of an original Spiracle after a moult (Imms).

STIGMATIC PROCESS Coccids: The projections which bear the mesothoracic Spiracles (MacGillivray).

STIGMATIC SPINES Coccids: spiracular Setae (MacGillivray).

STIGMATIFEROUS Adj. (Greek, *stigma* = mark; *pherein* = to carry; Latin, *-osus* = possessing the qualities of.) Bearing Spiracles or Stigmata.

STIGMELLIDAE See Nepticulidae.

STIGMERGY Noun. (Greek, *stigma* = mark; *ergos* = work.) The guidance of work performed by individual colony members by evidence of previous work and not through direct signals from nestmates.

STILES, CHARLES WARDELL (1867–1941) (Cram 1941, Revta. Med. trop. Parasit. Habana 7: 1–2; Usinger 1941, Pan-Pacif. Ent. 17: 84; Wright 1941, J. Parasit. (Cuba) 27: 195–201.)

STILETTO FLIES See Therevidae.

STILL, JOHN NATHANIEL (–1895) (Meldola 1895, Proc. ent. Soc. Lond. 1895: lxxi; South 1895, Entomologist 28: 315.)

STILLER, VIKTOR (1860–1948) (Szekessy 1948, Folia ent. hung. 3: 1–5.)

STILT BUGS See Berytidae.

STILT FLIES See Micropezidae.

STILT LEGGED FLIES See Micropezidae.

STILT PROLEGS Usually long Prolegs which raise the larva when walking.

STIMSON, LAWRENCE GORDON (1930–1964) (Wigglesworth 1965, Proc. R. ent. Soc. Lond. (C) 29: 54.)

STIMUKIL® See Methomyl.

STIMULI Noun. (Latin.) Small, acute spines on some larvae, particularly wood-borers.

STIMULUS Noun. (Latin, *stimulus* = goad. Pl., Stimuli.) Any of the disturbing forces or conditions in an organism, external or internal, which tend to cause a response.

STING Noun. (Anglo Saxon, *stingan* = to sting. Pl., Stings.) Hymenoptera: A sclerotized, tapered, tubular shaft developed as a modification of the female reproductive system of Aculeata. The Sting resembles a hypodermic needle that injects venom into prey, or as a method for defence of self or colony. When not in use, the Sting is retracted into the apex of the Abdomen (Metasoma). Syn. Dart; Gorgeret; Second Valulae. See Aculeus.

STINGER® See Dimethoate.

STINGING ROSE CATERPILLAR *Parasa indetermina* (Boisduval) [Lepidoptera: Limacodidae].

STINGLESS BEES *Trigona* spp. and *Austroplebeia* spp. [Hymenoptera: Apidae]: Several hundred pantropical (predominantly Neotropical), eusocial bees belonging to the Meliponinae (Apidae). Adult black; apex of forewing Marginal Cell open; veins forming Submarginal Cell weak or absent; Sting not functional; hind Tibia with Corbicula, but apical tibial spurs absent. Stingless bees live in large perennial colonies built in tree hollows or logs with plant resins and wax for nest construction construct, some colonies form subterranean nests. SB are important pollinators of cultivated fruits and domesticated crops. Workers collect and store pollen and nectar to feed young; food stored in wax pots (not in combs); workers protect nest by biting. See Apidae. Rel. Honey Bee; Bumble Bee.

STINK BEETLE *Nomius pygmaeus* (Dejean) [Coleoptera: Carabidae].

STINK BUGS See Cydnidae; Pentatomidae; Plataspidae.

STINK GLAND Any glandular structure that secretes the malodorous protective fluids. SG of insects discharge through special openings or other structures. See Gland. Rel. Pentatomidae.

STIPES Noun. (Latin, *stipes* = stalk. Pl., Stipites.) 1. The second segment of the Maxilla in the insect head. Broadly attached to the Cardo basad, bearing the movable palpus laterad, and attached to the Galea and Lacinea distad. Collectively, the Cardo and Stipes are called the coxopodite in the generalized mouthpart. Modified into a piercing structure in some Diptera and into a lever for flexing the Proboscis in others. 2. Either of the pair of forceps in the male genitalia of aculeate Hymenoptera; the Sagittae *sensu* Smith. 3. The stalk of an elevated eye. 4. Distal portion of an Embolus in spiders. See Maxilla. Alt. Stipe.

STIPITAL Adj. (Latin, *stipes* = stalk; *-alis* = pertaining to.) Descriptive of or pertaining to the Stipes and Cardo.

STIPODEMA Some Diptera: The apodeme of the Labrum (MacGillivray).

STIPULA Noun. (Latin, *stipula* = small stalk. Pl.,

Stipulae.) The area attached to the distal end of the Mentum in most insects. See Eulabium; Labiostipites; Labiosternite (MacGillivray).

STIPULARIA Noun. A bar from the projection of the Stipula which articulates with the Subgalea or Stipes (MacGillivray).

STIRPS Noun. (Latin, stock.) 1. A stock or stem. 2. A division used in zoological classification similar to the Superfamily. Archaic.

STIRRUP-PBW® See Gossyplure.

STIRUP M® A synthetic pheromone {multimethylalkenols} used for the attraction and increased searching activity of male spider mites (tetranychids). Compound is applied to fruit trees, nuts, ornamentals, vegetables and vine crops in conjunction with registered acaricides. See Acaricide.

STITT, EDWARD RHODES (1867–1948) (Bishopp 1949, J. Wash. Acad. Sci. 39: 381–382.)

STITZ, H (1868–1947) (Anon. 1938, Arb. morph. taxon. Ent. Berl. 5: 352.)

STOBBE, RUDOLF (1885–1915) (Ramme 1916, Dt. ent. Z. 1916: 371–375.)

STOBBE'S GLAND Lepidoptera: An eversible pheromone gland on the abdominal Sternum 2 of male noctuid and sphingid moths; pheromone released via the Hair Pencil. (Stobbe 1912, Zool. Jahrb. 32: 493–532.) See Coremata; Hair Pencil. Cf. Androconia Scales.

STOCHASTIC MODEL See Model. Cf. Deterministic Model.

STOCK, ALEXANDER (1912–1975) (O'Farrell 1975, News Bull. Aust. ent. Soc. 11: 76–77.)

STOCKADE® See Cypermethrin; Permethrin.

STÖCKEL, KARL (Cleve 1968, Mitt. dt. ent. Ges. 27: 50.)

STOCKHAUSEN, PAUL C (1857–1935), (Calvert 1935, Ent. News 46: 203–204.)

STÖCKLEIN, FRANZ (1879–1956) (Anon. 1957, Ent. Bl. Biol. Syst. Käfer 53: 1–2.)

STOCKMANN, STEN (1902–1975) (Hackman 1975, Notul. ent. 55: 92.)

STODDARD, ALBERT LEE (1889–1970) (Harris 1972, *Butterflies of Georgia*. xvi + 326 pp. (9–14), Norman, Oklahoma.)

STOECKLIN, PETER (1905–1975) (Ueker 1975, Mitt. ent. Ges. Basel 25: 77.)

STOLICZKA, FERDINAND (1838–1874) (Stearns 1874, Proc. Calif. Acad. Sci. 5: 363–364; W.T.B. 1874, Nature 10: 185–186.)

STOLL, CASPAR (–1795) Author of *Representation exactement coloriee d'apres Nature, des Spectres, des Mantes, des Sauterelles, & c* (1787). (Swainson 1840, *Taxidermy with Biography of Zoologists*. 392 pp. (337, bibliogr.). London. [Vol. 12 of *Cabinet Cyclopedia*. Edited by D. Lardner.].)

STOLL, OTTO (1849–1922) (Anon. 1923, Ber. Mus. Ges. Wunthur. 8: 6; Strobl 1923, Verh. schweiz. naturf. Ges. 1923: 55–58.)

STOLON Noun. (Latin, *stolo* = shoot. Pl., Stolons.) A creeping stem or runner which grows horizontally and produces roots and shoots where nodes of the Stolon contact the soil. Cf. Rhizome.

STOLONATE Adj. (Latin, *stolo* = shoot; *-atus* = adjectival suffix.) Resembling a Stolon in growth or form.

STOLTZ, HAMILKAR (1867–1934) (Heikertinger 1934, Koleopt. Rdsch. 20: 244; Anon. 1935, Arb. morph. taxon. Ent. Berl. 2: 63.)

STOMA Noun. (Greek, *stoma* = mouth. Pl., Stomata.) A respiratory or ventilation pore in plant leaves. See Stigma. Cf. Lenticel.

STOMACH Noun. (Greek, *stomachos* = gullet, throat. Pl., Stomachs.) The part of the Alimentary Canal immediately following the Gizzard and feeding the Ileum into which most of the digestive juices are poured. Syn. Chylific Ventricle; Ventriuclus.

STOMACH FLIES See Oestridae.

STOMACH MOUTH A term sometimes given to the proventricular apparatus forming a mouth-like entrance to the Ventriculus (Snodgrass).

STOMACH POISON Noun. An insecticide that kills an insect by being ingested and entering the Haemolymph through the gut. SPs traditionally have been used against biting-chewing insects. See Insecticide. Cf. Contact Poison; Fumigant.

STOMACHIC GANGLION The termination of the Recurrent Nerve in the middle region of the foregut.

STOMACHIC Adj. (Greek, *stomachos* = gullet; *-ic* = characterized by.) Descriptive of or pertaining to the stomach.

STOMATODAEUM See Stomodaeum.

STOMATOGASTRIC NERVOUS SYSTEM Syn. Retrocerebral Nervous System. Cf. Central Nervous System.

STOMATOGASTRIC See Stomogastric.

STOMATOTHECA Noun. (Greek, *stoma* = mouth; *theke* = case.) Part of the pupa covering the mouthparts.

STOMODAEAL NERVOUS SYSTEM The nervous system centring in the ganglia of the Stomodaeum, Stomatogastric, Visceral, or Sympathetic Nervous System (Snodgrass).

STOMODAEAL TROPHALLAXIS Social Insects: The exchange of liquid food mouth-to-mouth. Food is regurgitated from the Crop or released from glands on the head of one individual and passed to the other individual. See Trophallaxis. Cf. Proctodaeal Trophallaxis.

STOMODAEAL VALVE A cylindrical or funnel-shaped invagination of the posterior end of the Stomodaeum which projects into the cardiac part of the Ventriculus. SV acts as a sphincter and is involved in control of passage of food into the midgut. Syn. Cardiac valve.

STOMODAEUM (Greek, *stoma* = mouth; *hodaios* = way.) 1. The anterior region of the Alimentary Canal. 2. The foregut of a insect. 3. The primitive mouth and Oesophagus of the embryo of arthropods; Alt. Stomodeum; Stomatodaeum. See Alimentary System. Cf. Proctodaeum.

STOMOGASTRIC NERVE See Recurrent Nerve.

STOMOGASTRIC NERVOUS SYSTEM See Stomadaeal Nervous System.

STOMOXIN® See Permethrin.

STONE BEETLES See Scydmaenidae.

STONE BROOD A fungal disease of honey bees, most common in temperate northern hemisphere. Larvae most commonly infected but disease can also attack adults. Infected larvae first appear white and fluffy, subsequently they become hard and brown to green-yellow in colour. Death of larvae occurs typically inside capped cells; infected adults become sluggish, flightless and walk away from colony. SB disease usually is transient but colonies can be driven to extinction. Disease caused by *Aspergillus flavus,* or occasionally *A. fumigatus*. Spores occur on the larval integument and may penetrate the body, but more often spores are ingested by larva with food. Cf. Chalk Brood; Calcino. Rel. Pathogen.

STONE FLIES See Plecoptera.

STONE, STEPHEN (1810–1866) (Newman 1866, Entomologist 3: 154–156.)

STONEHAM, HUGH FREDERICK (1889–1966) (Pearson 1967, Proc. R. ent. Soc. Lond. (C) 31: 62.)

STONER, DAYTON (1883–1944) (Snyder 1945, J. Mammal. 26: 111–113, bibliogr. (mammals only).

STOPPER® See Hexythiazox.

STORE BOX Australian jargon: A hinged, wooden box with cork or styrofoam bottom that contains insect specimens preserved on pins. Syn. Schmitt Box.

STORED FOOD MITE See Fodder Mite.

STORED-GRAIN FUNGUS BEETLE *Litargus balteatus* LeConte [Coleoptera: Mycetophagidae] (Australia).

STORED-NUT MOTH *Paralipsa gularis* (Zeller) [Lepidoptera: Pyralidae]: A pest of stored almonds, dried fruits, peanuts, soybeans and walnuts. SNM common in Asia, southern Europe and USA. Adult forewing pale brown with black discal spot; hindwing whitish; wingspan 20–32 mm. Syn *Aphomia gularis* (Zeller).

STOREHOUSE BEETLE See Smooth Spider-Beetle.

STORER, DAVID HUMPHREYS (1804–1891) (Scudder 1893, Proc. Am. Acad. Arts Sci. 27: 388–391; Weiss 1936, *Pioneer Century of American Entomology*, 320 pp. (221–222), New Brunswick.)

STOREY, GILBERT (–1922) (Williams 1922, Entomologist 55: 120.)

STOREY, WILLIAM HENRY (1905-5–1975) (McKechnie Jarvis 1975, Proc. Brit. ent. nat. Hist. Soc. 8: 60. W[orms] 1975, Entomologist's Rec. J. Var. 87: 126–127.)

STORKAN, JAROSLAV (1890–1942) (Komarek 1946, Vest. csl. Spol. zool. 10: 24–26, bibliogr.)

STORM, WILHELM FERDINAND JOHAN (1835–1913) (Soot-Ryan 1942, Tromso Mus. Aarsh. 65(3): 6.)

STOSSICH, MICHELE (1857–1906) (Parona 1906, Boll. Mus. zool. Univ. Genova 3: 1–5, bibliogr.)

STOTT, CHARLES ERNEST (1868–1935) (Anon. 1935, Entomol. mon. Mag. 71: 212–213; Neave 1936, President's address. R. ent. Soc. 1936: 2.)

STOUGHTON-HARRIS, GEOFFREY (1894–1966) (Worms 1966, Entomologist's Rec. J. Var. 78: 269–271.)

STOUT SAC SPIDER Members of the Genus *Clubiona* [Araneae: Clubionidae]: Small-bodied, diurnal spiders that are often taken on building walls and ornamental shrubs. See Clubionidae. Cf. Slender Sac Spiders.

STOUTS Noun. Specifically any member of the Genus *Haematopota*, eyes with brilliant zig-zag bands and speckled wings. See Tabanidae.

STRACENER, CHARLES LYMAN (1878–1941) (Eddy 1942, J. Econ. Ent. 35: 116–117; Osborn 1946, *Fragments of Entomological History,* Pt. II, 232 pp. (114) Columbus, Ohio.)

STRAIGHT SWIFT *Parnara bada sida* (Waterhouse) [Lepidoptera: Hesperiidae] (Australia).

STRAIGHT-SNOUTED WEEVILS See Brentidae.

STRAIN Noun. (Anglo Saxon, *streon* = procreation, stock, race). A traceable lineage of descendants from a common ancestral Species sharing and distinguished by characters or qualities that are often the result of artificial breeding.

STRAINER Noun. (Pl., Strainers.) Mayfly naiads: A row of stiff Setae (bristles) used in straining planktonic organisms. Strainer is positioned on fore Tibiae or mouthparts (Klots).

STRAMINEOUS Adj. (Latin, *stramen* = straw; *-osus* = possessing the qualities of.) Pale yellow; straw coloured. Pale yellow with a very faint tinge of blue (Kirby & Spence). Alt. Stramineus.

STRAND, ANDREAS (1895–1970) (Palm 1970, Ent. Tidsskr. 91: 1–2.)

STRAND, EMBRIK (1876–1947) (Anon. 1937, Festschr. 60 Geburt. E. Strand. 3: 506–553, bibliogr.; Natwig 1944, Norsk. ent. Tidsskr. 7: 58–61, bibliogr.; Anon. 1949, Z. wien. ent. Ges. 34: 44.)

STRANGULATE Adj. (Old French, *estrangler* = to choke; *-atus* = adjectival suffix.) Strongly constricted and contracted, as if by bands or cords, to form a waist.

STRASSEN, OTTO (1869–1961) (E. F. 1959, Ent. Z., Frankf. a. M. 69: 197–198; Merlens 1959, Natur Volk 89: 157–161; Hohorst 1961, Natur Volk 91: 201–204.)

STRATA See Stratum.

STRATHION® Parathion.

STRATIFICATION Noun. (Latin, *stratum* = layer; English, *-tion* = result of an action.) The state of being made up of layers or strata.

STRATIFIED Adv. (Latin, *stratum* = layer.) Arranged or made up in layers.

STRATIOMYIDAE Plural Noun. Soldier Flies. A cosmopolitan Family of brachycerous Diptera including 1,800 Species in 15 Subfamilies. Stratiomyidae, Tabanidae and others with annulated third antennal segments, bridge gap, between Nematocera and Brachycera. Stratiomyids distinct, except close relationship to Xylomyidae. Two Families separated on wing venation. Stratiomyidae first three longitudinal veins crowded upward toward anterior margin of wing; fourth Posterior Cell open, characteristics not in Xylomyidae. Stratiomyid body bare or with soft fine pubescence; third anntennal segment flagelliform, distinctly annulated; wing venation with Praefurca (base of Radial Vein) originating opposite Discal Cell; Discal Cell almost always pentagonal; wing-membrane almost always wrinkled; Costal Vein ends at or before apex of wing; fourth Posterior Cell open at wing margin; Tibiae without apical spurs; Abdomen flattened and rounded or oblong. Predaceous as larvae; a few Species phytophagous or saprophagous. Larvae under bark, in decaying wood, moss, damp earth, water and/or decaying vegetation. Prey of Species found under bark not known. Aquatic Species prey upon other arthropods, small Crustaceans and worms. Several semiaquatic Species live in mud around fringes of bodies of water burrow deeply in mud and survive for long dry periods. Adults common at flowers; sluggish, slow fliers. Eggs usually laid in masses on plants at edge of water or on water. Nonaquatic Species lay eggs singly in masses on nearby vegetation, soil, or substrate in which larvae live. Larvae nearly uniform in appearance, except large size of later instars; well defined head, no legs and tough leathery integument. Larvae hyperextend to extraordinary lengths. Pupation within tough last larval skin. Some aquatic pupae formed in head the last larval instar leaving rest of integument inflated to float pupa on water surface. Aquatic larvae of some Species live in extreme saline or alkaline waters.

STRATIOMYIIDAE See Stratiomyidae.

STRATUM Noun. (Latin, *stratum* = layer. Pl., Strata) 1. A layer of cells, organisms, or objects occupying a vertical division of space. 2. Ecology: groups of Consocies (and animals not so grouped) occupying the recognizable vertical divisions of a uniform area (Folsom & Wardle). Cf. Horizon.

STRAUB, ANTON (1859–1939) (Heyrovsky 1939, Cas. csl. Spol. ent. 36: 75–76.)

STRAUBENZEE, CASIMIR H C VAN (–1943) (Cockayne 1944, Proc. R. ent. Soc. Lond. (C) 8: 70.)

STRAUCH, ALEXANDER (1832–1893) (Anon. 1893, Leopoldina 29: 162.)

STRAUS-DUERCKHEIM, HERCULE EUGENE (1790–1865) Frenchman probably best known for his *Considerations generales sue 'anatomie comparee des animaux articules auxquelles on a joint l'anatomie descriptive du hanneton vulgaire* (1828). Work regarded as first genuine work on comparative insect anatomy. (Anon. 1891, Bull. Soc. sci. Elbeuf 9: 33–34.)

STRAUSS, LEOPOLD (–1940) (Heikertinger 1940, Koleopt. Rdsch. 26: 92.)

STRAVINSKI, KONSTANTY (1892–1966) (Sandner 1966, Polskie Pismo ent. 43–44 (B): i, ii; Tzolukh 1967, Ent. Obozr. 46: 490–492. (Translation: Ent. Rev., Wash. 46: 290–291.)

STRAW ITCH MITE 1. *Pyemotes herfsi* Oudemans; *Pyemotes ventricosus* (Newport) [Acari: Pyemotidae] (Australia). 2. *Pyemotes tritici* (Lagrèze-Fossat & Montané) [Acari: Pyemotidae].

STRAWBERRY APHID *Chaetosiphon fragaefolii* (Cockerell) [Hemiptera: Aphididae]: A pest of strawberries throughout North America and Australia. SA cool-adapted; abundant during spring and autumn. Transmits many viral diseases to strawberries; nymphs and adults infest new shoots and buds in crowns of plants.

STRAWBERRY BUD-WEEVIL *Anthonomus signatus* Say [Coleoptera: Curculionidae].

STRAWBERRY BUG *Euander lacertosus* (Erichson) [Hemiptera: Lygaeidae] (Australia).

STRAWBERRY CROWN-BORER *Tyloderma fragariae* (Riley) [Coleoptera: Curculionidae]: A pest of strawberry and cinquefoil in many parts of USA. Overwinters as an adult among strawberry field trash. Eggs glistening white, elliptical, laid within punctures made in crown and base of leaf stalks during May-August. Larvae feed in tunnels in crowns ca 25–55 days. Pupation inside crown. Adult emerges during autumn, feeds. SCB completes 1–2 generations per year. See Curculionidae. Cf. Strawberry Root-Weevil.

STRAWBERRY CROWN-GIRDLER See Strawberry Root-Weevil.

STRAWBERRY CROWN-MINER *Monochroa fragariae* (Busck) [Lepidoptera: Gelechiidae]: A pest of strawberries in central and northwestern USA; larva tunnels near base of leaves, causes stunting or death of plant; overwinters as mature larva within silken hibernaculum in crown of plant. See Gelechiidae.

STRAWBERRY CROWN-MOTH *Synanthedon bibionipennis* (Boisduval) [Lepidoptera: Sesiidae]: A pest of strawberries and blackberries in North America. SCM eggs laid on underside of leaves. Neonate larva hollows crowns. See Sesiidae. Cf. Currant Moth; Raspberry Crown-Borer.

STRAWBERRY LEAF-ROLLER *Ancylis comptana fragariae* (W. & R.). *Ancylis comptana* (Froelich) [Lepidoptera: Tortricidae]: A widespread pest of blackberry, raspberry and strawberry in North America. SLR eggs laid individually on underside of leaves. Early instar larva feeds on underside of leaves within webbing; later instar larvae feed on upperside of leaves within fold tied together by silk. Pupation within folded leaf; SLR overwinters as mature larva or pupa in folded leaf on ground,

STRAWBERRY ROOT-APHID *Aphis forbesi* Weed [Hemiptera: Aphididae]: A pest of strawberries in eastern North America and California. SRA attacks crowns, stems, new leaf buds, underside of leaves and roots. In north, males and oviparous females mate, lay overwintering eggs; in south and warm areas only females are known and reproduction is continuously parthenogenetic. Cornfield ant tends SRA in some areas.

STRAWBERRY ROOT-WEEVIL *Otiorhynchus ovatus* (Linnaeus) [Coleoptera: Curculionidae]: A pest of apple, grape, raspberry and strawberry throughout North America. SRW overwinters as larva or adult in field trash or crowns of host plant. Adult ca 6–7 mm long, black, flightless (Elytra fused); nocturnal, feed on leaves and berries; male unknown. SRW eggs white, spherical and deposited among roots or crowns of host plants. Larval head brown, body white and legless; larva feeds on roots. Syn. Strawberry Crown-Girdler; *Brachyrhinus ovatus*. See Curculionidae. Cf. Strawberry Crown-Borer.

STRAWBERRY ROOTWORM *Paria fragariae* Wilcox [Coleoptera: Chrysomelidae]: A pest of numerous plants, including rose, raspberry, strawberry, apple peach and grape throughout North America. SRW overwinters as adult in field trash. Eggs deposited in batches of 4–15 eggs on host plant; eclosion occurs within 10–14 days. Neonate larvae penetrate soil to feed on roots; feeding complete within 60 days. Pupation occurs in soil. See Chrysomelidae.

STRAWBERRY ROOTWORMS Any of several Species of chrysomelid beetles whose larvae damage strawberries.

STRAWBERRY SAP-BEETLE *Stelidota geminata* (Say) [Coleoptera: Nitidulidae].

STRAWBERRY SPIDER MITE 1. *Tetranychus lambi* Pritchard & Baker [Acari: Tetranychidae] (Australia). See Banana Spider Mite. 2. *Tetranychus turkestani* Ugarov & Nikolski [Acari: Tetranychidae].

STRAWBERRY THRIPS *Scirtothrips dorsalis* Hood [Thysanoptera: Thripidae] (Australia).

STRAWBERRY WEEVIL *Rhinaria perdix* (Pascoe) [Coleoptera: Curculionidae] (Australia).

STRAWBERRY WHITEFLY *Trialeurodes packardi* (Morrill) [Hemiptera: Aleyrodidae].

STRAWBERRY Noun. (Anglo Saxon, *streawberige; streaw* = hay; *berige* = berry. Pl., Strawberries.) Members of Genus *Fragaria*. Low herbs with trifoliate leaves, cymose white flowers and long slender runners. Fruit an enlarged, pulpy, juicy, edible receptable bearing numerous seed-like achenes; not a berry.

STREBLIDAE Plural Noun. Bat Flies. A numerically small Family of muscoid Diptera, predominantly tropical and subtropical in distribution. Mouthparts piercing; compound eyes small; Ocelli absent; legs and tarsal claws enlarged; wing usually present but sometimes shed after establishing association with host. Species are typically gregarious and larviparous; some Species pupate on floor of roosting site. Adults live as blood-sucking ectoparasites of bats. *Ascodipteron* sp. female pierces the host's skin with her large Proboscis and draws her body into the wound. Next, her legs and wings cast off and her body assumes a flask shape with the posterior end projecting outward. Cf. Nycteribiidae.

STRECKER, (FERDINAND HEINRICH) HERMAN (1836–1902) (Strecker 1878, Butterflies and moths of North America 1: 275–277; Anon. 1902. Ent. News 13: 1–4; Weiss 1936, *Pioneer Century of American Entomology,* 320 pp. (202–203), New Brunswick.)

STRECKFUSS, ADOLF (1823–1895) (Grunack 1893, Ent. Z. 9: 113; Donitz 1895, Berl. ent. Z. 40: 373–374.)

STREITBERG, KARL (1883–1960) (Zimmerman 1960, Mitt. dt. ent. Ges. 19: 89–91.)

STRENZKE, KARL (–1961) (Neumann 1962, Mitt. dt. ent. Ges. 21: 3–4; Lindroth 1963, Opusc. ent. 27: 128.)

STREPSIPTERA Kirby 1813. (Oligocene-Recent). (Greek, *strepsis* = turning; *pteron* = wing.) Twisted-wing insects. A widespread Order of endopterygote Neoptera including nine Families and about 300 Species. Adults non-feeding, lack Tentorium, with strong sexual dimorphism: female of parasitic Species usually eyeless, lack Antennae and legs, head and Thorax fused; habitus larviform, viviparous and developing a brood canal in puparium; female of free-living Species (some Mengeidae) with compound eyes, developed head, Antennae, chewing mouthparts. All males with insect habitus; forewing reduced (elytriform), hindwing enlarged, membranous with radial venation; morphologically unique among insects in lacking Trochanter in adult leg. Immature stages parasitic on Thysanura, Blattodea, Mantodea, Orthoptera, Hemiptera, Diptera and Hymenoptera. Postembryonic development of parasitic Species hypermetamorphic: Triungulin first-instar larva. Sometimes classified among Coleoptera, near Rhipiphoridae. Included Families: Bohartillidae, Callipharixenidae, Corioxenidae, Elenchidae, Halictophagidae, Mengeidae, Mengenillidae, Myrmecolacidae, Stylopidae.

STRETCH, RICHARD HARPER (1837–1923) (Coolidge *et al.* 1920, Ent. News 31: 181–185; Anon. 1926, Pan-Pacif. Ent. 2: 160.)

STRIA Noun. (Latin, *stria* = furrow, streak. Pl., Striae.) 1. General: Any fine, long, impressed line on a surface. 2. Coleoptera: A longitudinal, depressed line or furrow, frequently punctured, extending from the base to the apex of the Elytra. 3. Lepidoptera: A fine, transverse line.

STRIATE Adj. (Latin, *striatus* = grooved; *-atus* = adjectival suffix.) Descriptive of surface sculpture, usually the insect's integument, that is marked with numerous parallel, fine, impressed lines. Lepidoptera: With numerous fine transverse lines. Alt. Striated; Striatus. See Sculpture Pattern. Cf. Acuductate; Alveolate; Baculate;

Clavate; Echinate; Favose; Gemmate; Psilate; Punctate; Reticulate; Rugulate; Scabrate; Shagreened; Striate; Verrucate; Wrinkled.

STRIATE FALSE WIREWORM *Pterohelaeus alternatus* Pascoe [Coleoptera: Tenebrionidae] (Australia).

STRIATED BORDER The inner cytoplasmic layer of the ventricular epithelial cells with fine lines perpendicular to the surface (Snodgrass).

STRIATED HEM The bounding edge of enteric epithelium.

STRIATED MUSCLE See Voluntary Muscle.

STRIATE-PUNCTATE Adj. With loose punctured striae.

STRIATION Noun. (Latin, *stria* = furrow, streak; English, *-tion* = result of an action.) A longitudinal ridge or furrow; the condition of being furnished or marked with striae.

STRICKLAND CODE The first attempt at codification of rules governing zoological nomenclature. The Code was proposed in 1842 by British Association for Advancement of Science as the produce of a committee chaired by H. E. Strickland (pp. 105–121). Committee members drafting the Strickland Code included Charles Darwin and Richard Owen.

STRICKLAND, HUGH EDWIN (1811–1858) English zoologist who published a few papers on fossil insects. (Jardine 1858, *Memoirs of H. E. Strickland*. xxlxv + xvi + 441 pp. London.)

STRICTURE Noun. (Latin, *strictura* = a contraction. Pl., Strictures.) A pressing together, as if tied.

STRIDULATE Adj. (Latin, *stridor* = grating.) Insects: To make a creaking, grating or hissing sound (noise) by rubbing two ridged or roughened surfaces against each other. *e.g.* Crickets stridulate.

STRIDULATING ORGANS Hardened Parts of the insect body that are used in making sounds. Typically, one part is a file-like surface and the opposing one a scraper or rasp (Imms).

STRIDULATION Noun. (Latin, *stridulus* = creaking, squeaky; English, *-tion* = result of an action. Pl., Stridulations.) 1. Friction of rigid parts on modified surfaces. Insects: The sound produced by rubbing a series of hard projections (spines, Acanthae) against a file-like surface. Stridulatory methods of acoustical communication are widespread in the Insecta and probably the most generally used form of sound communication. Anatomy of system consists of a Pars Stridens which forms a rasp (file) composed of tubercles and a Plectrum which forms a scraper (Strigil). 2. The act of stridulating or making creaking sounds. See Sound. Cf. Pars Stridens; Plectrum. Rel. Auditory Organ.

STRIDULATORY Adj. (Latin, *stridentis* = creaking, noisy.) Pertaining to stridulation; capable of stridulation.

STRIGA Noun. (Latin, *striga* = furrow, streak. Pl., Strigae.) A narrow, transverse line or slender streak, either surface or impressed.

STRIGATE Adj. (Latin, *strigatus* = streaked, striped; *-atus* = adjectival suffix.) With Strigae. A term applied to a surface on which the Strigae are impressed as in the Elytra of some beetles, or to an ornamentation composed of fine, short lines.

STRIGIL Noun. (Latin, *strigilis* = comb, from *stringere* = to scrape, graze. Pl., Strigils.) 1. A tibial comb. Hymenoptera, Lepidoptera: A curved, comb-like (pectinate) movable spur on the distal end of the fore Tibia that is adapted for cleaning the Antenna. 2. A scraper or undulating series of parallel ridges and grooves. Cf. Plectrum. 3. Corixidae: A structure on the abdominal dorsum, sometimes shaped as a comb, and opposed by a Seta-fringed concavity at the base of the Metatarsus (Comstock). Alt. Strigile; Strigilis.

STRIGILATION Noun. (Latin, *strigilis* = comb; English, *-tion* = result of an action. Strigilations.) 1. The act of cleaning Antennae. 2. The act of creating sound by rubbing body parts. See Sound.

STRIGILATOR Noun. (Latin, *strigilatus* = furnished with scraper; *or* = one who does something. Pl., Strigilators.) Any member of a group of Synoeketes that licks the surface of ants or termites to obtain food (*e.g.*, the wingless cricket *Myrmecophila*).

STRIGOSE LUNATE VITTA Lygaeidae: A crescent-shaped more-or-less longitudinally placed roughened stripe or area on the Abdomen.

STRIGOSE VENTRAL AREAS Heteroptera: Roughened ventral areas of the fourth and fifth abdominal segments, which produce stridulation by means of the wart-like tubercles bearing a subapical tooth found in the hind Tibiae, in certain Lygaeidae. Similar lunate areas of the Abdomen (of unknown function); the strigose Lunate Vittae.

STRIGOSE Adj. (Latin, *striga* = ridge; *-osus* = full of.) Rough with rigid bristles; hispid.

STRIGULA Noun. (Latin, *striga* = ridge.) A fine, short, transverse mark or line.

STRIGULATED With numerous strigule.

STRIKE® See NAA.

STRINGE, RICHARD (–1924) (Vogel 1924, Ent. Rdsch. 41: 28.)

STRIOLATE Adj. (Latin, *striola* = small channel; *-atus* = adjectival suffix.) Pertaining to structure with finely impressed, parallel lines.

STRIOLE Adj. (Latin, *striola* = small channel.) A rudimentary stria.

STRIOPUNCTATE Adj. (Latin, *striola* = small channel; *punctum* = to prick; *-atus* = adjectival suffix.) Punctate-striate.

STRIP HARVESTING Harvesting of different areas of a single crop at different times to reduce the affects of a pest or enhance the densities and activities of its natural enemies.

STRIPE Noun. (Middle Dutch, *stripe*. Pl., Stripes.) A longitudinal streak of colour different from the ground colour of a structure or body.

STRIPED ALDER SAWFLY *Hemichroa crocea* (Geoffroy) [Hymenoptera: Tenthredinidae].

STRIPED AMBROSIA BEETLE *Trypodendron lineatum* (Oliver) [Coleoptera: Scolytidae].

STRIPED BLISTER-BEETLE *Epicauta vittata* (Fabricius) [Coleoptera: Meloidae].

STRIPED CUCUMBER BEETLE *Acalymma vittatum* (Fabricius) [Coleoptera: Chrysomelidae]: A pest in North America east of the Rocky Mountains and south into Mexico. Adults feed extensively on leaves, stems and blossoms of host plant. Eggs yellow and laid at base of host plants or in soil cracks. Neonate larva penetrates soil and feeds for 15–45 days on roots and subterranean parts of host plant; pupation within soil. Overwinters in unmated adult stage under litter in wooded areas. Larvae feed upon curcurbits; adults feed upon beans, corn, peas and other cultivated plants. SCB transmits bacterial wilt and cucumber mosaic. See Chrysomelidae. Cf. Spotted Cucumber Beetle; Western Striped Cucumber Beetle.

STRIPED CUTWORM *Euxoa tessellata* (Harris) [Lepidoptera: Noctuidae]. See Cutworms.

STRIPED EARWIG *Labidura riparia* (Pallas) [Dermaptera: Labiduridae].

STRIPED EARWIGS See Labiduridae.

STRIPED FIELD SLUG *Lehmannia nyctelia* (Bourguignat) [Sigmurethra: Limacidae] (Australia).

STRIPED FLEA BEETLE *Phyllotreta undulata* (Kutschera) [Coleoptera: Chrysomelidae] (Australia).

STRIPED GARDEN CATERPILLAR *Trichordestra legitima* (Grote) [Lepidoptera: Noctuidae].

STRIPED HORSE FLY *Tabanus lineola* Fabricius [Diptera: Tabanidae]: A pest of mammals that is widespread in North America. Compound eye with three transverse green stripes on purple background; Thorax with conspicuous stripes; abdomen dark grey to brown with median stripe white. See Tabanidae. Cf. Black Horse-Fly. Rel. Deer Fly.

STRIPED LADYBIRD *Micraspis frenata* (Erichson) [Coleoptera: Coccinellidae]: An Australian Species. Adult 3.8–4.9 mm long, round-ovate, with a yellow and black Thorax and yellow Elytra with black markings; black stripe down centre of Elytra and two dark stripes along margins; two dark spots occur in middle of Elytra near wing base; Thorax yellow with dark posterior and 2 dark spots near middle. Adult female lays eggs in batches near larval food. Eggs yellow, spindle-shaped and laid standing on end. Larva elongate with well defined legs; body black with markings and setose; 4 larval instars. Pupal stage 8–9 days. Striped ladybirds predators of aphids in cotton. Adult and larva may feed on pollen when prey is scarce. Adult and larva occur on cotton feeding on aphids, mites and *Helicoverpa* eggs. Cf. Transverse Ladybird; Three-Banded Ladybird.

STRIPED MEALYBUG *Ferrisia virgata* (Cockerell) [Hemiptera: Pseudococcidae] In Australia, a pest of cotton. Female oval, yellow-orange, and covered in white wax. Mealybugs do not look like insects and lack body segmentation. Legs and Antenna only visible with hand lens. Adult feeds by placing mouthparts into plant and sucking sap. Adult sedentary and feeds in one place. Males rarely found. Female lays small round, yellow eggs into wax secreted from body. First instar mealybugs called crawlers and are active stage in life-cycle. Crawlers move from the wax mass to find a place to feed. Once settled, crawlers moult into sedentary nymphs. Nymphs resemble adult in appearance, but smaller. Honeydew produces sticky cotton. Colonies of mealybugs have been collected from stems and leaves of cotton. Infestations are usually near field borders. 2. *Phenacoccus solani* Ferris [Hemiptera: Pseudococcidae] (Australia).

STRIPED RICEBORER *Chilo suppressalis* (Walker) Lepidoptera: Pyralidae] (Australia).

STRIPED STEMBORER *Chilo suppressalis* (Walker) [Lepidoptera: Pyralidae].

STRIPED SWARMING LEAF BEETLE *Rhyparida didyma* (Fabricius) [Coleoptera: Chrysomelidae] (Australia).

STRIPED WALKINGSTICKS See Pseudophasmatidae.

STRIPETAILS See Perloidae.

STRITT, WALTER (Ungerer 1967, Mitt. dt. ent. Ges. 26: 54.)

STROBEL, PELLEGRINO (1828–1895) (Papavero 1975, *Essays on the History of Neotropical Dipterology*. 2: 349. São Paulo.)

STROBL, P GABRIEL (1846–1925) (D.P. 1894, Boll. Soc. malac. ital. 19: 223–224, bibliogr.; Castefrano 1896, Atti Soc. ital. Sci. nat. 36: 40–43; Morge 1974, Beitr. Ent. (Sonderheft) 24: 41–63, bibliogr.)

STROHL, JEAN (1886–1942) (Peyer 1942, Vjschr. naturf Ges. Zurich 87: 533–539; Fisher 1943, Verh. schweiz. naturf Ges. 123: 320–326.)

STRÖM, HANS (1726–1798) (Hagen 1873, Stettin. ent. Ztg. 34: 225–232, bibliogr.; Strand 1917, Arch. Naturgesch. 83 A (6): 27–46, bibliogr.)

STRÖM, VINCENS (1818–1899) (Henriksen 1925, Ent. Meddr. 15: 185–186, bibliogr.)

STROMA Noun. (Greek, *stroma* = bed covering. Pl., Stromata.) 1. A supporting network of connective tissue. 2. The colourless region in which grana are absent in chloroplasts but contain enzymes concerned with carbon dioxide fixation.

STROMATOLITE Noun. (Greek, *stroma* = bed covering; *lithos* = rock. Pl., Stromatolites.) An organically-produced sedimentary structure which consists of alternating layers of organically rich and organically poor sediment. Stromatolites typically form dome-like or hassock-shaped rock deposits principally during the Precambrian and first appear during Archean Eon (ca 3 BYBP). Stromatolites are formed by thread-like blue-green algae (Cyanobacteria) which trap the sediment in organically-poor layers. See Cyano-bacteria.

STROMBERG, CHARLES W (1856–1895) (Anon. 1895, Ent. News 6: 176; Knauss 1895, Can. Ent. 27: 300.)

STRONG, LEE A (1886–1941) American economic Entomologist and head of the USDA Bureau of Entomology.

STRONG, LEE ABRAHAM (1886–1941) (Osborn 1937, Fragments *Fragments of Entomological History,* 394 pp. (282), Columbus, Ohio; Anon. 1941, Science 93: 563; Anon. 1941, J. Econ. Ent. 34: 479–480; Hoyt *et al.* 1941, Proc. ent. Soc. Wash. 43: 156–166, bibliogr.)

STRONGLY Adj. (Anglo Saxon, *strang*.) Markedly or decidedly.

STROPHANDRIA Symphytous Hymenoptera in which the male genitalia has rotated 180° along its primary axis before eclosion. Includes Superfamily Tenthredinoidea and Xyelidae. Cf. Orthandria.

STROPHANDRIOUS COPULATION Hymenoptera. Sawflies in which the genitalia has rotated 180° along its primary axis and the bearer assumes an end-to-end copulatory stance with a female during copulation. Cf. Orthandrous Copulation.

STROPHOTAXIS Noun. (Greek, *strophos* = twisted; *taxis* = arrangement.) The twisting movement behaviour seen in some animals when exposted to the appropriate external stimulus. See Orientation. Cf. Aerotaxis; Anemotaxis; Chemotaxis; Geotaxis; Menotaxis; Osmotaxis; Phototaxis; Rheotaxis; Rotaxis; Scototaxis; Telotaxis; Thermotaxis; Thigmotaxis; Tonotaxis; Tropotaxis.

STROUHAL, HANS (1887–1969) (Schönmann 1969, Ent. Nachr. Dresden 17: 125.)

STRUCTURAL COLOUR Insects: Physical colours produced by surface structure (Setae, scales, sculpture) that reflect or break up light (by refraction or diffraction) into spectral colours or combinations of them. Examples include polarization colours, interference colours, iridescent blues and greens of insect wings, and Tyndall Blues. The production of structural colours can be attributed to three physical phenomena: Interference, diffraction and scattering. See Interference Colour; Iridescence; Scattering Colour. Cf. Pigmentary Colours.

STRUCTURE Noun. (Latin, *structura*, from *structum* = to build, to arrange. Pl., Structures.) 1. The arrangement of constituent elements (parts, cells, tissues, organs) of a body. 2. Any organ, appendage, or part of an insect. Syn. Constitution.

STÜBEL, ALPHONS (Papavero 1975, *Essays on the History of Neotropical Dipterology*. 2: 307–314. Sao Paulo.)

STUBENRAUCH, LUDWIG VON (1865–1940) (Osthelder 1941, Mitt. münch. ent. Ges. 31: 363–364, bibliogr.)

STUDER, SAMUEL (1757–1834) (Anon. 1935. Verh. schweiz. naturf Ges. 1935: 83–89.)

STUDY, EDWARD (1862–1930) (Seitz 1930, Ent. Rdsch. 47: 9.)

STUHLMANN, FRANZ LUDWIG (1863–1928) (Panning 1958, Mitt. hamb. zool. Mus. Inst. St: 22–23.)

STUMPTAILED LIZARD TICK *Amblyomma albolimbatum* Neumann [Acari: Ixodidae] (Australia).

STUPEOUS Adj. (Latin, *stupa* = tow; *-osus* = possessing the qualities of.) Covered with matted filaments or other fibre-like structure. Alt. Stupose.

STUPULOSE Adj. (Latin, *stupa* = tow; *-osus* = full of.) Covered with coarse decumbent Setae. Alt. Stupulosus.

STURANY, RUDOLF (1867–1935) (Adensamer 1936, Annln naturh. Mus. Wien 47: 59–60.)

STURM, JACOB (1771–1848) A German entomological illustrator of the early 19th century who illustrated *Deutschlands Insekten* for Panzer. (Anon. 1849, Stettin. ent. Ztg. 10: 162–167; Eisinger 1919, Int. ent. Z. 13: 105–111, bibliogr.)

STURM, JOHANN HEINRICH CHRISTIAN FRIEDRICH (1805–1862) (Hauck 1862, J. Orn. Leipzig 10: 157–160; Herrich-Schaeffer 1892, KorrespBl. zool.-min. Ver. Regensburg 16: 41–44.)

STUSSINER, JOSEF (1850–1917) (Reitter 1918, Wien. ent. Ztg. 37: 120–122.)

STYLATE Adj. (Latin, *stylus* = pricker; *-atus* = adjectival suffix.) Pertaining to structure provided with a style.

STYLE Noun. (Greek, *stylos* = pillar; Latin, *stylus* = pricker. Pl., Styles.) 1. A short, more-or-less cylindrical appendage. In the plural form (Styles), a term applied to small, usually pointed, exarticulate appendages, most frequently found on the terminal segments of Abdomen (Cerci). See Stylus. 2. Aphididae: The slender tubular process at the Abdomen's apex. 3. Coccidae: A long spine-like appendage at the apex of the male's Abdomen; the genital spike. See Spike. 4. Diptera: The Ovipositor (Loew); the single immovable organ immediately below the forceps. Male Tipulidae: A thickened, jointed Arista at or near the apex of the third antennal segment (Osten-Sacken). See Arista.

STYLE OF THE FLAGELLUM Diptera (Stratiomyidae): The apical antennal segment. See Antenna; Style 1.

STYLES, JOHN HENRY (1923–1973) (Alma & Nuttall 1973, N.Z. Ent. 5: 366–368, bibliogr.)

STYLET Noun. (Latin, *stylus* = pricker. Pl., Stylets.) 1. A small Style. See Stylus. 2. An elongate, slender and relatively rigid process or appendage on the body of an insect. 3. Phthiraptera, some Hemiptera, Diptera: One of the mouthparts modified for piercing. See Mandibular Stylet; Maxillary Stylet. 4. Hemiptera: A median dorsal element in the shaft of the Ovipositor, formed of the fused Second Valvulae (Snodgrass). 5. Hymenoptera: first Valvulae; Aculeata: Fused portions of second Valvulae. Cf. Cultellus.

STYLET® See Petroleum Oils.

STYLET SAC Lice: The lower canal of the head; a ventral diverticulum extending from the Buccal Funnel to the posterior region of the head (Imms).

STYLET SHEATH 1. Hemiptera: The lipoprotein salivary secretion in phytophagous Species that forms a conduit surrounding the mouthparts between plant surface and phloem. 2. Hymenoptera: Third Valvulae in Aculeata. The dorsal part of the Terebra in aculeate Hymenoptera (Imms).

STYLI LINGUALES The supporting rods of the Hypopharynx in diplopods (Packard).

STYLIFORM Adj. (Latin, *stylus* = pricker; *forma* = shape.) Stylus-shaped; descriptive of structure terminating in a long slender point, as the Antenna in some Diptera. See Stylus 1; Stylet. Cf. Aculeiform; Penicilliform; Spiniform; Spiciform. Form 2; Shape 2; Outline Shape.

STYLIFORM APPENDAGE (Ribaga). Psocidae: One of the chisels. See Chisel; Lacinia.

STYLIGER PLATE 1. Ephemeroptera: A broad, well articulated, freely-movable sclerite at whose outer distal angles arise the clasping organs (Forceps, Styli). 2. The Forceps Base *sensu* Needham.

STYLOCONIC SENSILLUM A sense organ consisting of a terminal tooth or peg seated upon a more-or-less conical base (Folsom & Wardle). A cuticular sense organ consisting of one or more pegs of the basiconic type elevated on a Style or Cone (Imms). Lepidoptera: Temperature receptors or host plant odour receptors. Syn. Biarticulate Type of Berlese; Sensillum Styloconicum.

STYLOPIDAE Plural Noun. Twisted-winged Insects. The largest Family of Strepsiptera with about 75 Species in North American and 50 Species in Australia. Males with 4-segmented Tarsi; Antenna 4–6 segmented; female anatomy variable. Parasites of bees and wasps.

STYLOPIZATION Noun. (Latin, *Stylops*; English, *-tion* = result of an action. Pl., Stylopizations.) Strepsiptera: The act or process of parasitizing other insects (Wardle).

STYLOPIZED Adj. (Latin, *Stylops.*) Descriptive of an insect infested by a member of the Stylopidae. Stylopized insects typically are sexually inactive.

STYLOPS Noun. (Latin, *Stylops*.) The large Genus of Stylopidae. Species of *Stylops* develop as endoparasites of bees. See Strepsiptera.

STYLOSE Adj. (Latin, *stylus* = pricker; *-osus* = full of.) Bearing a Style (Curran).

STYLOTRACHEALIS A long tube bearing a Stigma arising from the head case as in the pupae of some Diptera.

STYLUS Noun. (Latin, *stylus* = pricker. Pl., Styli.) 1. A small, pointed, non-articulated process. See Style; Stylet. 2. Any Ovipositor (not a drilling apparatus) such as found in certain primitive insects (*e.g. Machilis*). See Ovipositor. 3. Collembola: The tapering process of the ventral abdominal appendages, freely movable at its base; a small appendage attached to the Coxa of the middle and hind legs. 4. Coccidae: The outer sheath of the male genitalia. 5. Female Odonata: A small rod-shaped projection at the apex of the lateral Gonapophyses of the Ovipositor. Male nymphs: Short acute processes on the ventral surface of the ninth abdominal segment (Garman).

SUAREZ, D FRANCISCO-JAVIER (1926–1985) Spanish Entomologist and taxonomic specialist on Mutillidae (Hymenoptera). (Cobos 1985. Eos 61: 7–11).

SUARS, ERNEST (1857–1909) (Poskin 1909, Revue mens. Soc. ent. namur. 9: 81.)

SUB- (Latin, *sub* = under, below, proximate or nearly.) 1. Prefix used to mean under, slightly less than, or not complete. 2. In taxonomic classification, any group immediately below the status Taxon (*e.g.* Subclass, Subfamily, Subspecies.)

SUBACUTE Adj. (Latin, *sub* = nearly; *acutus* = sharp.) Moderately acute.

SUBADUNCATE Adj. (Latin, *sub* = nearly; *aduncus* = hooked; *-atus* = adjectival suffix.) Somewhat hooked or curved.

SUBALAR Adj. (Latin, *sub* = below; *ala* = wing.) 1. Pertaining to structure positioned below the wings. 2. Pertaining to the Subalare.

SUBALAR MUSCLE A single, usually large, muscle positioned against the epimeral wall of the Pleuron on each side of each wing-bearing segment, attached ventrally on the Meron of the Coxa, a second sometimes is found, arising on the Epimeron and inserted on the posterior end of the Subalare or on a distinct second Subalar Sclerite (Snodgrass).

SUBALAR SCLERITE A sclerite of the pleural region below the wing (Crampton).

SUBALARE Noun. (Latin, *sub* = below; *ala* = wing.) A thoracic sclerite near the base of the wing and posterior of the Pleural Wing Process. Subalare serves as place of insertion for the wing's posterior pleural muscle. Syn. Postparapteron *sensu* Snodgrass. See Pleuron. Cf. Basalare.

SUBANAL LAMINAE The Podical Plates (Heymons).

SUBANAL PLATE Orthoptera: The Subgenital Lamina.

SUBANTENNAL GROOVE Diptera: See Facial Groove (Curran).

SUBANTENNAL SUTURE A prominent suture extending downward from each Antennal Suture to the Subgenal Suture of some insects (*e.g.* cockroaches). SS sometimes termed the Frontogenal Suture (Snodgrass).

SUBAPICAL LOBE Male Culicidae: Inner lobe of the side piece below the apex in genitalia.

SUBAPICAL SETA Some coccids: A group of Setae that appear to form a line continuous with the caudal end of the ventral thickening (MacGillivray).

SUBAPTEROUS Adj. (Latin, *sub* = nearly; Greek, *a* = without; *pteron* = wing; Latin, *-osus* = possessing the qualities of.) Descriptive of insects with rudimentary wings that cannot function in flight. Syn. Brachypterous; Micropterous. Cf. Macropterous; Apterous.

SUBCARDO Noun. (Latin, *sub* = proximate; *cardo* = hinge. Pl., Subcardos.) The proximal sclerite of the Cardo; Paracardo *sensu* MacGillivray.

SUBCIRCULAR SCARS Coccids: See Discaloca (MacGillivray).

SUBCLASS Noun. (Latin, *sub* = below; *classis* = division.) A secondary division of the aggregate termed a Class in biological classification; the Taxon below Class and above Superfamily.

SUBCLAVATE Adj. (Latin, *sub* = nearly; *clavus* = nail; *-atus* = adjectival suffix.) Descriptive of elongate structure thickened toward the apex, but not quite club-shaped. See Clavate.

SUBCLIMAX AREA Ecology: The most stable and permanent type of biotic environment attainable in a relatively smaller area than that of the climax where more or less local environmental conditions prevent the attainment of the full climax conditions (Klots).

SUBCLYPEAL PUMP Some Diptera: The enlarged, more or less bulb-like structure at the anterior entrance of the Oesophagus.

SUBCLYPEAL TUBE Diptera: The Pharynx.

SUBCONTIGUOUS Adj. (Latin, *sub* = nearly; *continguus* from *contigere* = to touch on all sides; *-osus* = possessing the qualities of.) Pertaining to structures which are close, but not quite touching. Rel. Adjacent.

SUBCORDATE Adj. (Latin, *sub* = nearly; *cor* = heart; *-atus* = adjectival suffix.) Descriptive of structure resembling the shape of a heart. Cf. Cordate.

SUBCORIACEOUS Adj. (Latin, *sub* = nearly; *corium* = leather; *-aceus* = of or pertaining to.) Descriptive of structure which is somewhat leather-like in texture or appearance.

SUBCORNEAL Adj. (Latin, *sub* = below; *cornea* = lens; *-alis* = pertaining to.) Lying under or beneath the corneal element of the Ommatidium.

SUBCORTICAL Adj. (Latin, *sub* = under; *corticis* = bark; *-alis* = pertaining to.) Pertaining to something beneath bark, such as insect borings, tunnels or tubes.

SUBCOSTA Noun. (Latin, *sub* = nearly; *costa* = rib.) The longitudinal, generally unbranched, wing vein extending parallel to the Costa and reaching the outer margin before the Costa (Comstock). Hymenoptera: The Radius *sensu* Comstock. Cf. Costa.

SUBCOSTAL AREA Tingitidae: The narrow part or section of the Hemelytra next to the costal area.

SUBCOSTAL CELL Diptera (Schiner): The Marginal Cell of Loew, Radial of Comstock. Hymenoptera: the first Subcostal Cell (Packard), the Radial and First Radial of Comstock. Any one of the cells margined anteriorly by the Subcosta.

SUBCOSTAL CROSSVEIN Odonata: A single crossvein between Subcosta and Radius next to the body or proximad of all other Antenodal Crossveins; SC present in *Progomphus teste* Garman. See Vein; Wing Venation.

SUBCOSTAL FOLD A depression between the Costa and the Radius.

SUBCOSTAL FURROW See Subcostal Fold.

SUBCOSTAL NERVULE Hindwings of Lepidoptera: Media of Comstock, S.C. 1,2,3,4 and 5, Radius of Comstock (Smith).

SUBCOSTAL VEIN 1. Diptera: The first longitudinal vein of Meigen, Radius (R) of Comstock *teste* Schiner. 2. Lepidoptera: Radius of Comstock. See Vein; Wing Venation.

SUBCOXA Noun. (Latin, *sub* = proximate to; *coxa* = hip. Pl., Subcoxae.) The proximal part of the limb base or the Coxa when differentiated or divided from the Coxa. The Subcoxa usually is incorporated into the pleural wall of the body segment (Snodgrass). The coxa is attached to the body wall by a Coxal Corium with one (monocondylic), two (dicondylic) or three points (tricondylic) of articulation. See Coxa. Rel. Leg.

SUBCOXAL Adj. (Latin, *sub* = proximal to; *coxa* = hip.) 1. Descriptive of structure positioned near the Coxa. 2. Descriptive of or pertaining to the Subcoxa. Rel. Coxal.

SUBCOXAL PLEURITES Chilopoda: 'small sclerites of various shapes more or less closely associated with the bases of the Coxae' (Snodgrass).

SUBCRISTATE Adj. (Latin, *sub* = nearly; *crista* = crest; *-atus* = adjectival suffix.) With a moderately elevated ridge or keel on the Pronotum, as in Orthoptera. Rel. Cristate.

SUBCUTANEOUS Adj. (Latin, *sub* = under; *cutis* = skin, New Latin, *cutaneous* = pertaining to skin; *-osus* = possessing the qualities of.) Under or beneath the skin, integument or epidermal cells. Rel. Cutaneous.

SUBCUTICULAR Adj. (Latin, *sub* = under; *cutis* = skin, beneath; *cuticula* = thin skin.) Pertaining to substance beneath the Cuticle. Something under the Cuticle. Rel. Cuticular.

SUBCUTICULAR FAT BODY See Fat Body.

SUBDERMAL MYIASIS See Myiasis.

SUBDISCAL Adj. (Latin, *sub* = nearly; *discus* = disc; *-alis* = pertaining to.) Below or beneath the disc of any surface or structure. See Discal.

SUBDISCAL VEIN Hymenoptera: The wing vein forming the posterior margin of the Third Disocidal Cell. The first branch of the Anterior Cubitus (CuA). See Vein; Wing Venation.

SUBDORSAL Adj. (Latin, *sub* = under, below; *dorsum* = back; *-alis* = pertaining to.) Below the dorsum. See Dorsal.

SUBDORSAL KEELS Coccids: The dorsal plates, (MacGillivray).

SUBDORSAL LINE Lepidoptera larva: A longitudinal line adjacent to the dorsal line and between the DL and lateral line. If an addorsal line exists, the SDL is postioned between it and the lateral line.

SUBDORSAL PLATES Coccids: The dorsal plates

(MacGillivray). See Plate.

SUBDORSAL RIDGE Slug caterpillars: A raised longitudinal line along the subdorsal row of abdominal tubercles.

SUBECONOMIC PEST A pest with a general equilibrium position well below the economic-injury level. The highest pest populations do not reach the economic-injury level (EIL) and thus do not require intervention. See Key Pest; Secondary Pest.

SUBEPIDERMAL Adj. (Latin, *sub* = under; Greek, *epi* = upon; *derma* = skin; Latin, *-alis* = pertaining to.) Lying below or beneath the Epidermis. See Epidermis.

SUBEQUAL Adj. (Latin, *sub* = nearly; *aequalis* = equal; *-alis* = characterized by.) 1. Pertaining to structures, conditions or phenomena which are similar, but not identical. See Equal.

SUBERECT Adj. (Latin, *sub* = nearly; *erigere* = to raise up.) Not quite upright. Descriptive of Setae which stand at less than 90° from the substrate. See Erect. Cf. Porrect.

SUBERODED Adj. (Latin, *sub* = nearly; *erosus* = gnawed off.) Appearing as if somewhat gnawed or indented. See Eroded. Alt. Suberose; Suberosus.

SUBFACIES Adj. (Latin, *sub* = below; *facies* = face.) Pertaining to the lower face; the underside of the head including the Lora and the Jugulum. See Facies.

SUBFALCATE Adj. (Latin, *sub* = nearly; *falcatus* = bent or curved, from *falx* = sickle; *-atus* = adjectival suffix.) Slightly excavated below the apex, in the wing. See Falcate

SUBFAMILY Noun. (Latin, *sub* = under; *familia* = household, from *famulus* = servant.) A division in zoological classification containing a group of closely allied Genera, different from other allied groups, yet not sufficiently so as to make a Family series. An opinionative category named after the type Genus and ending in -inae (Smith). See Family.

SUBFOSSORIAL Adj. (Latin, *sub* = nearly; *fossor* = digger; *-alis* = characterized by.) 1. Descriptive of structure adapted for digging but not greatly modified. See Fossorial. 2. Pertaining to a behaviour less than fossorial.

SUBFRONTAL SHOOT Orthoptera: A dusky band on the hindwing of grasshoppers just posterior of the frontal margin of the wing and extending nearly to the base (Walden).

SUBFRONTAL Adj. (Latin, *sub* = below; *frons* = forehead; *-alis* = pertaining to.) Located near the front; immediately behind the frontal margin. See Frontal.

SUBFULCRUM Noun. (Latin, *sub* = nearly; *fulcrum* = support.) A sclerite between Mentum and Palpiger, rarely present. See Fulcrum.

SUBFUSIFORM Adj. (Latin, *sub* = nearly; *fusus* = spindle; *forma* = shape.) Pertaining to structure nearly spindle-shaped. See Fusiform. Rel. Form 2; Shape 2; Outline Shape.

SUBGALEA Noun. (Latin, *sub* = below; *galea* = helmet.) A Maxillary sclerite or segment, attached to the Stipes; the Parastipes. See Galea.

SUBGENAL Adj. (Latin, *sub* = under, *gena* = cheek; *-alis* = pertaining to.) 1. Pertaining to structure positioned below the Gena. 2. Descriptive of or pertaining to the Subgena. See Gena.

SUBGENAL AREAS The usually narrow lateral marginal areas of the cranium set off by the Subgenal Sutures above the gnatha appendages, including the Pleurostomata and Hypostomata (Snodgrass).

SUBGENAL RIDGE A submarginal structure of the inner surface of the cranium formed by the inward projection of the Subgenal Suture (Snodgrass).

SUBGENAL SUTURE A lateral, submarginal groove or Sulcus of the cranium located just above the bases of the gnathal appendages (Mandible and Maxillae). Continuous anteriad with the Frontoclypeal Suture in the generalized pterygote head. Internally the SGS forms a subgenal ridge which presumably provides structural support for the cranium above the Mandible and Maxillae.

SUBGENERIC NAME See Generic Name.

SUBGENICULATE Adj. (Latin, *sub* = nearly; *geniculum* = little knee; *-atus* = adjectival suffix.) Descriptive of a structure, such as an Antenna, which is nearly geniculate; applied especially to Antennae articulated from a short, thick Scape. See Geniculate.

SUBGENITAL Adj. (Latin, *sub* = below; *gignere* = to beget; *-alis* = pertaining to.) Pertaining to the regions beneath or near the Genitalia.

SUBGENITAL LAMINA A plate-like sclerite or sclerotized process underlying the genital organs in Orthoptera. See Subgenital Plate.

SUBGENITAL PLATE An apical sternal sclerite or sclerotized process which covers and protects the Gonopore. The Hypandrium *sensu* Snodgrass. Syn. Subgenital Shield.

SUBGENITALIS Noun. (Latin, *sub* = below; *gignere* = to beget.) The subgenital sclerite in the eighth abdominal Sternum (MacGillivray).

SUBGENUAL ORGAN See Supratympanal Organ.

SUBGENUS (Latin, *sub* = below; *genus* = race.) A taxonomic subdivision within a Genus, based upon a character not sufficient for generic separation. See Genus.

SUBGLOBOSE Adj. (Latin, *sub* = nearly; *globus* = globe; *-osus* = full of.) Descriptive of structure which is nearly spherical or globular. See Globose.

SUBGLOBULAR Adj. (Latin, *sub* = nearly; *globus* = globe.) Descriptive of structure which is nearly globular or globe-shape. See Globular.

SUBGLOSSA Noun. (Latin, *sub* = under; Greek, *glossa* = tongue. Pl., Subglossae.) Odonata: A sclerite between the two halves of the Mentum (Graber); the true Mentum. See Glossa.

SUBGUSTA Noun. Pl., Subgustae. The membranous after part of the parapharynx (MacGillivray).

SUBHYPODERMAL COLOUR Any colour that is contained in the fat-body and Haemolymph of an insect (Imms).

SUBIMAGINAL Adj. (Latin, *sub* = proximate; *imago* = likeness; *-alis* = pertaining to.) Descriptive of or pertaining to the Subimago. The stage immediately preceding the adult.

SUBIMAGO Noun. (Latin, *sub* = under; *imago* = likeness. Pl., Subimagoes.) The stage in Ephemeroptera between the aquatic naiad and the adult reproductive. The Subimago has functional wings and undergoes Ecdysis to produce the adult stage. Alt. Pseudoimago. See Imago.

SUBINFLUENT Noun. (Latin, *sub* = nearly; *influere* = to flow into.) Ecology: An organism which is inconspicuous and unimportant in some seasons and numerous in other (Shelford). See influent.

SUBINTEGUMENTAL SCOLOPHORE A Scolophore with the nerve ending free within the body-cavity (Comstock).

SUBKLEW, WERNER (–1949) (Schwerdtfeger 1949, Anz. Schädlingsk. 22: 111–112.)

SUBLABRUM Noun. (Latin, *sub* = under; *labrum* = lip.) An alternate term for the Epipharynx.

SUBLATERAL BRISTLES Diptera: Bristles positioned in a line with the intraalars but anterior of the suture. The anterior pair are sometimes included as posthumerals, but the term is deceptive (Curran).

SUBLINGUAL Adj. (Latin, *sub* = under, below; *lingua* = tongue; *-alis* = pertaining to.) Lying beneath the tongue. See Lingua.

SUBLINGUAL GLAND Apoidea: See Ventral Pharyngeal Gland.

SUBMARGIN Adj. (Latin, *sub* = proximate; *margo* = edge.) Part of a surface just within the margin.

SUBMARGINAL AREA Hindwings: A section between the costal margin and the first strong vein.

SUBMARGINAL CELLS 1. Hymenoptera (Norton): Radial Cells of Comstock. 2. Diptera (Williston): Radial of Comstock.

SUBMARGINAL CELLULES The Cubital Cellules (Say).

SUBMARGINAL NERVURE Hymenoptera (Norton): The irregular line of veins extending parallel with the outer margin, composed in part of Media the Medial Crossvein and Cutitus of Comstock (Smith).

SUBMARGINAL SETA Some coccids: A second row of Setae on the ventral aspect a short distance cephalad of the pygidial Setae (MacGillivray).

SUBMARGINAL TUBERCLES Coccids: See Dorsal Tubercles.

SUBMEDIA Noun. (Latin, *sub* = proximate; *medius* = middle.) Submedian Unguiculus: Omoplate, or second Axillary of the articulation of the wings. In the posterior wing, the Diademal or Second Axillary *sensu* MacGillivray.

SUBMEDIAN CELL Hymenoptera (Packard): The First Cell, Cubital plus Cubital of Comstock, the Second Cell, Medial of Comstock, the Third Cell, Second Medial of Comstock.

SUBMEDIAN INTERSPACE Forewings of Lepidoptera: The area between the Median and Submedian Veins (Cubitus and First Anal of Comstock).

SUBMEDIAN VEIN 1. Odonata: The Cubitus. 2. Lepidoptera: The First Anal, which runs from the base of the forewing to the hind angle, near the inner margin. 3. Vein 1 of the numerical series (Smith). See Vein; Wing Venation.

SUBMENTAL Adj. (Latin, *sub* = under; *mentum* = chin; *-alis* = pertaining to.) Pertaining to the Submentum. See Mentum.

SUBMENTAL PEDUNCLE Coleoptera: The prolonged portion of the Gula supporting the Mentum.

SUBMENTUM Noun. (Latin, *sub* = under; *mentum* = chin.) The basal sclerite of the insect Labium. Submentum articulates with and is attached to the head. See Head Capsule. Cf. Labium.

SUBMETALLIC FLEA BEETLE *Nisotra submetallica* Blackburn [Coleoptera: Chrysomelidae] (Australia).

SUBNODAL SECTOR Odonata: The Radial Sector *sensu* Comstock.

SUBNODUS Noun. (Latin, *sub* = proximate; *nodus* = knob.) Odonata: An additional oblique crossvein; the continuation of the Nodus below vein R (Tillyard). See Nodus.

SUBNYMPH Noun. (Latin, *sub* = below; Greek, *nymphe* = chrysalis.) 1. The resting or pupal stage of female Coccidae. 2. A supernumerary stage before the formation of the pupa; pseudopupa.

SUBOCELLATE Adj. (Latin, *sub* = below; *oculus* = eye; *-atus* = adjectival suffix.) Spotted as if with ocellate spots without the pupil.

SUBOCULAR Adj. (Latin, *sub* = below; *oculus* = eye.) Pertaining to structure or colour beneath or below the eyes. See Ocular.

SUBOCULAR SUTURE A suture which extends from the lower angle of each compound eye to the Subgenal Suture above the anterior mandibular articulation (Snodgrass). In some Species the Subocular Suture may extend to the Subgenal Suture; in other Species it may terminate before reaching another landmark. This suture is straight and common in the Hymenoptera where it may provide additional strength for the head. See Head Capsule. Cf. Ocular Suture.

SUBOESOPHAGEAL Adj. (Latin, *sub* = below; Greek, *oisophagos* = gullet; Latin, *-alis* = pertaining to.) Pertaining to structure positioned beneath the Oesophagus.

SUBOESOPHAGEAL COMMISSURE The commissure of the Tritocerebral Ganglion which extends beneath the Foregut and connects with the TG on the opposite side of the body (Snodgrass). See Substomodaeal Commissure.

SUBOESOPHAGEAL GANGLION Insects: The ventral ganglionic nerve mass of the head positioned beneath the Oesophagus. SOG formed by the fusion of the ganglia of the mandibular, maxillary and labial segments. See Ventral Nerve Cord.

SUBORDER Noun. A division of an order higher than a Family, based on a character common to a large series of Species.

SUBOVATE Adj. (Latin, *sub* = proximate; *ova* = egg; *-atus* = adjectival suffix.) Not quite ovate.

SUBPARALLEL Adj. (Latin, *sub* = nearly; Greek, *parallelos* from *para* = beside; *allelon* = of one another.) Nearly parallel.

SUBPECTINATE Adj. (Latin, *sub* = proximate; *pecten* = comb; *-atus* = adjectival suffix.) Somewhat less than fully pectinate.

SUBPEDUNCULATE Adj. (Latin, *sub* = proximate; New Latin, *peduncle* = foot; *-atus* = adjectival suffix.) Not quite pedunculate, *e.g.,* in Coleoptera, when the constriction between the Prothorax and Mesothorax gives the appearance of a narrow waist.

SUBPHARYNGEAL NERVE One of a pair of small nerves given off from the Suboesophageal Tritocerebral Commissure, or from the anterior end of the Suboesophageal Ganglion (Snodgrass).

SUBPRIMARY SETA Lepidoptera larva: Primary Setae found in post first-instar larvae (Comstock).

SUBPUNCTATE Adj. (Latin, *sub* = nearly; *punctum* = to prick; *-atus* = adjectival suffix.) Slightly punctured. Alt. Subpunctatus.

SUBPYRIFORM Adj. (Latin, *sub* = proximate; *pyri* = pear; *forma* = shape.) Descriptive of structure nearly pear-shaped. See Pyriform. Rel. Form 2; Shape 2; Outline Shape.

SUBQUADRANGLE Noun. (Latin, *sub* = nearly; *quad-* = comb. form meaning four; *angulos* = corner.) Zygoptera: The usually quadrangular cubital area of the wing (Comstock).

SUBQUADRATE Adj. (Latin, *sub* = nearly; *quadratus* = squared.) Not quite a square.

SUBRENIFORM SPOT A rounded spot or outline, below and sometimes attached to the reniform spot in *Catocala* and some allied noctuids.

SUBRENIFORM Adj. (Latin, *sub* = proximate; *renis* = kidney; *forma* = shape.) Descriptive of structure nearly kidney-shaped. See Reniform. Rel. Form 2; Shape 2; Outline Shape.

SUBSCAPHIUM Noun. See Gnathos.

SUBSCUTELLA Noun. The ventral surface of the transversely infolded Postscutellum (MacGillivray).

SUBSEGMENT Noun. (Latin, *sub* = nearly; + segment.) Arthropod podite: A secondary division of a segment into two or more non-musculated segments (Snodgrass).

SUBSELLATE Adj. (Latin, *sub* = proximate; *sella* = saddle; *-atus* = adjectival suffix.) Nearly like or approaching the form of a saddle.

SUBSERRATE Adj. (Latin, *sub* = below; *serra* = saw; *-atus* = adjectival suffix.) See Denticulate. Cf. Aculeate-Serrate; Biserrate; Dentate-Serrate; Multiserrate; Uniserrate.

SUBSESSILE Adj. Taxonomy: On casual observation appearing sessile. See Pseudosessile.

SUBSINUATE Adj. (Latin, *sub* = nearly; *sinus* = curve; *-atus* = adjectival suffix.) Slightly sinuate.

SUBSOCIAL Adj. (Latin, *sub* = under; *sociare* = to associate; *-alis* = characterized by.) Eusocial Bees: A Family group in which an adult female protects and progressively feeds her immatures. The mother leaves or dies before the young reach maturity. See Communal; Eusocial; Quasisocial; Subsocial; Primitively Social; Highly Social. Cf. Aggregation; Colonial; Social.

SUBSPECIES Noun. (Latin, *sub* = under; *species* = particular kind.) A geographical or host variation (British Commission on Nomenclature). A part of a Species marked by average differences in characters which intergrade with those of Subspecies occupying different, though usually adjacent, parts of the general range of the Species, along the common boundary of which intergradation is complete (McAtee). The essence of the Subspecies is intergradation, assumed or actual (Ferris).

SUBSPINIFORM Adj. (Latin, *sub* = proximate; *spina* = spine; *forma* = spine.) Descriptive of structure nearly spine-like or spine-shaped. See Spiniform. Rel. Form 2; Shape 2; Outline Shape.

SUBSPIRACULAR LINE Lepidoptera larva: A stripe or line below the spiracles.

SUBSTIGMATAL CELL Part of the Marginal Cell below the Stigma; in bees, First Radial (Comstock).

SUBSTITUTE KING Termites: Complemental male or king (Comstock).

SUBSTITUTE QUEEN Termites: A complemental queen or reproductive female (Comstock).

SUBSTOMODAEAL COMMISSURE Suboesophageal commissure.

SUBSTRIATE Adj. (Latin, *sub* = nearly; *striatus* = grooved; *-atus* = adjectival suffix.) Pertaining to a surface slightly striate. Alt. Substriatus. See Striate.

SUBTEGULA Noun. (Latin, *sub* = below; + tegula. Pl., Subtegulae.) Trichoptera and Lepidoptera: A prealar sclerite associated with the wing base. Cf. Tegula.

SUBTERETE Adj. (Latin, *sub* = nearly; *teres* = round off.) Nearly but not quite cylindrical. Alt. Subteres.

SUBTERMINAL Adj. (Latin, *sub* = below; *terminus* = end; *-alis* = characterized by.) Below the end or not quite attaining the end.

SUBTERMINAL TRANSVERSE SPACE The area between the transverse posterior line and the subterminal line in moths (Smith).

SUBTERRANEAN Adj. (Latin, *sub* = below; *terra* = earth.) Underground, beneath the surface of the soil or ground.

SUBTERRANEAN CLOVER WEEVIL *Listroderes delaiguei* Germain [Coleoptera: Curculionidae] (Australia).

SUBTERRANEAN TERMITE 1. *Heterotermes ferox* (Froggatt) [Isoptera: Rhinotermitidae]: The most commonly encountered Species of *Heterotermes* in southern Queensland, NSW, Victoria, South

Australia and southern Western Australia. Soldier 4–5 mm long; head long, rectangular, parallel sided when viewed from above; Mandible black, elongate, sword-like without teeth; fontanelle small, indistinct and positioned well behind anterior margin of head; Pronotum flat without anterior lobes; Tarsi with four segments; Cerci two segments. The Species does not build mounds and populations are not large; often attacks poles, fences and posts; regarded as a minor structural pest causing superficial damage. Sometimes confused with *Coptotermes* spp. but body more slender. Nests of *H. ferox* often near other Species; may be distinguished because workers and soldiers slow moving and less aggressive in defence of colony when compared with other Species. 2. *Coptotermes acinaciformis* (Froggatt) [Isoptera: Rhinotermitidae]: Occurs in all states of Australia and regarded as the most destructive *Coptotermes* Species in terms of structural damage to wooden buildings. Soldier 5–7 mm long; Mandibles sword-like, without teeth; Labrum not grooved; fontanelle conspicuous and near anterior margin of head; Pronotum flat without anterior lobes; Tarsi with four segments; Cerci with two segments. Typically does not build mounds, except in tropical north; nests in tree stumps and living trees; colony size to a million individuals; attack structures within 50 m radius. 3. *Schedorhinotermes intermedius* (Brauer) [Isoptera: Rhinotermitidae]: A commonly encountered Species throughout Australia with many described Subspecies; regarded as second to *Coptotermes* spp. in terms of economic damage. Soldier dimorphic with distinct forms; major soldier 5–6 mm long with bulbous head; minor soldier 3–4 mm long with smaller head; Mandible with teeth; fontanelle rather small, positioned behind anterior margin of head; Pronotum flat without anterior lobes; Tarsi with four segments; Cerci 2 segments. The Species does not build mounds and nests in tree stumps, root crown of living trees, beneath patios and beneath fireplaces. Colony can include several thousand individuals; vitality of nest reflected in number of major soldiers present.

SUBTERRANEAN TERMITES *Coptotermes* spp., *Heterotermes* spp., *Schedorhinotermes* spp. [Isoptera: Rhinotermitidae] (Australia).

SUBTILE Adj. (French, *subtile*.) Slightly; feebly. Alt. Subtilis.

SUBTRIANGLE Odonata: The cell in the wing behind the triangle (Garman).

SUBTRIANGULAR SPACE Internal triangle. See triangle.

SUBTRIBE Noun. In zoological classification, a division within a Tribe characterized by the possession of important characteristics, secondary to those of a Tribe. See Tribe.

SUBTROPICAL Adj. (Latin, *sub* = below; *tropicus* = belonging to a turn; *-alis* = appropriate to.) Descriptive of distribution or occurrence near (above or below) the tropics.

SUBTROPICAL PINE TIP MOTH *Rhyacionia subtropica* Miller [Lepidoptera: Tortricidae].

SUBTUS Adj. (New Latin.) Beneath, at the under surface.

SUBULATE Adj. (Latin, *subula* = an awl; *-atus* = adjectival suffix.) Awl-shaped; linear at base and attenuate at the apex. Alt. Subulatus; Subuliform.

SUBULICORN ANTENNA An awl-shaped or thin and tapering Antenna.

SUBULICORNIA Odonata and Ephemeridae: considered as one group.

SUBULIPALPIA A Superfamily of Plecoptera including three Families which have mouthparts modified for carnivory. Mandibles and accessory appendages elongate, apically sharp. Most representatives found in Northern Hemisphere; a few Species found south of Equator.

SUBVARIETY Noun. In zoological classification, a refined division of a variety; a term loosely applied according to the judgment of the describer or classifier.

SUBVENTRAL LINE Lepidoptera larva: A stripe or line along the side just above the bases of the feet at the edge between the lateral and ventral lines (Smith).

SUBVENTRAL RIDGE Slug caterpillars: A longitudinal raised line along the ventral series of abdominal tubercles.

SUBVENTRAL SPACE Slug caterpillars: The area on each side, between the lateral ridge and the lower edge of the body, in which Spiracles are located.

SUCCESSION Noun. (Latin, *successio* = a coming into the place of another.) The change in the composition of an ecosystem as organisms compete and respond to enviromental changes.

SUCCINCTI Noun. (Latin, *succingo* = girdle.) Butterfly chrysalids which are held in place by a silken line surrounding the body. See Suspensi.

SUCCINEOUS Adj. (Latin, *sucinum* = amber; *-osus* = possessing the qualities of.) Resembling amber in colour or appearance. See Bark 2. Alt. Succineus.

SUCCIVOROUS See Phytosuccivorous.

SUCKER Noun. (Anglo Saxon, *sucan* from Latin, *sugere* = to suck. Pl., Suckers.) A disc variously placed by which certain subaquatic insects adhere to surfaces.

SUCKFLY *Cyrtopeltis notata* (Distant) [Hemiptera: Miridae].

SUCKING LICE See Anoplura.

SUCKING PUMP Insects which ingest liquid food (sap, blood, etc.): The highly modified Cibarium; the Cibarial Pump, the Pharyngeal Pump. See Cibarium.

SUCKING SPEARS The specialized Mandibles and Maxillae of hemerobiid larva, used for puncturing prey and sucking its juices.

SUCKING STOMACH A thin-walled muscular pouch connected with the end of the Oesophagus; a food reservoir not commonly present except in some Lepidoptera.

SUCRASE Noun. (French, *sucre* = sugar; *-ase* = enzyme.) An enzyme which splits sucrose into glucose and fructose.

SUCROSE Noun. (French, *sucre* = sugar; Latin, *-osus* = full of.) Cane sugar. A complex, crystalline carbohydrate found in many plants, such as sugarcane, beets, etc. Sucrose is a chief source of alcohol by fermentation and an important food for many insects.

SUCTORIA Noun. (Latin, *sugere* = to suck.) An ordinal term proposed for fleas.

SUCTORIAL Adj. (Latin, *sugere* = to suck.) Mouthparts adapted for sucking. See Haustellate.

SUCTORIAL VESICLES Bladder-like structures connected with the Oesophagus in mosquitoes supposed to assist in blood-sucking.

SUEHIRO, AMY (1906–1968) (Bryan 1969, Proc. Hawaii. ent. Soc. 20: 469–470.)

SUFFERT, ERNST (–1907) (Horn 1907, Dt. ent. Z. 1907: 535; Schaufuss 1907, Ent. Wbl. 24: 112.)

SUFFRIAN, CHRISTIAN WILHELM LUDWIG EDUARD (1805–1876) (Dohrn 1877, Stettin. ent. Ztg. 38: 106–117, bibliogr.; Westwood 1877, Proc. ent. Soc. Lond. 1877: xl.)

SUFFULTED PUPIL The pupil of an ocellate spot which shades into another colour.

SUFFUSED Adj. (Latin, *suffusus* past part. of *suffundere* = to overmelt.) Clouded or obscured by a darker colour.

SUFFUSION Noun. (Latin, *suffusus* past part. of *suffundere* = to overmelt.) 1. The act or process of suffusion or being suffused. 2. A clouding or spreading of one colour over another. 2. Spreading body fluids (Haemolymph) into surrounding tissue.

SUFOS® See Monocrotophos.

SUGAR ANT *Camponotus consobrinus* (Erichson) [Hymenoptera: Formicidae] (Australia).

SUGARBAG BEES *Trigona* spp. [Hymenoptera: Apidae] (Australia).

SUGARBEET CROWN-BORER *Hulstia undulatella* (Clemens) [Lepidoptera: Pyralidae].

SUGARBEET ROOT-APHID *Pemphigus populivenae* Fitch [Hemiptera: Aphididae]: A pest of sugarbeets, beets and weeds in western USA. Overwinters in egg stage on Poplar or wingless females on roots of herbaceous plants. Winged females migrate to beets during July. Nymphs and adults feed on the roots of host plant and reduce size and quality of beets. Subsequent generation migrates to Poplar during September-October. Syn. *Pemphigus betae* Doane. See Aphididae.

SUGARBEET ROOT-MAGGOT *Tetanops myopaeformis* (Rörder) [Diptera: Otitidae]: A Species widespread in North America and a serious pest of sugarbeets in parts of the northcentral United States. Adult black bodied with dusky spot on cotal margin of wing. Eggs curved, glossy and laid individually or in small clutches in ground near host plants. Larva feeds on taproot of host plant. Overwinters as larva and pupates during spring. See Otitidae.

SUGARBEET WEEVIL *Bothynoderes punctiventris* Germ. [Coleoptera: Curculionidae].

SUGARBEET WIREWORM *Limonius californicus* (Mannerheim) [Coleoptera: Elateridae].

SUGARCANE AND MAIZE STEMBORER *Bathytricha truncata* (Walker) [Lepidoptera: Noctuidae] (Australia).

SUGARCANE APHID *Melanaphis sacchari* (Zehntner) [Hemiptera: Aphididae] (Australia).

SUGARCANE ARMYWORM *Leucania stenographa* Lower [Lepidoptera: Noctuidae] (Australia). See Armyworms.

SUGARCANE BEETLE *Euetheola humilis rugiceps* (LeConte) [Coleoptera: Scarabaeidae].

SUGARCANE BORER *Diatraea saccharalis* (Fabricius) [Lepidoptera: Pyralidae]: A significant pest of sugarcane in South Africa. In USA, a pest of sugarcane, corn, rice, sorghum along Gulf States. Eggs flattened, oval, laid in clusters on leaves and stems. Young larvae feed on leaves in whorls and leaf stems; older larvae bore into stalk. Summer generation pupates in stalks; autumn generation overwinters as larvae in field residues and pupates during spring. See Pyralidae.

SUGARCANE BUD-MOTH *Neodecadarachis flavistriata* (Walsingham) [Lepidoptera: Tineidae]: A sporadic pest of minor importance on sugarcane and bananas in Australia. Larvae live behind sheathing leaf base on mature stems, cause feeding scars on inside of leaf sheaths, on rind and under bud scales. Pupation occurs under leaf sheaths within coccon covered with debris. Syn. *Opogona glycphaga* (Meyrick).

SUGARCANE BUTT WEEVIL *Leptopius maleficus* Lea [Coleoptera: Curculionidae] (Australia).

SUGARCANE DELPHACID *Perkinsiella saccharicida* Kirkaldy [Hemiptera: Delphacidae].

SUGARCANE FROGHOPPER *Euryaulax carnifex* (Fabricius) [Hemiptera: Cercopidae] (Australia).

SUGARCANE LEAF BEETLE *Rhyparida dimidiata* Baly [Coleoptera: Chrysomelidae] (Australia).

SUGARCANE LEAF-MITE *Oligonychus indicus* (Hirst) [Acari: Tetranychidae].

SUGARCANE LEAF-ROLLER *Hedylepta accepta* (Butler) [Lepidoptera: Pyralidae].

SUGARCANE LOOPER *Mocis frugalis* (Fabricius) [Lepidoptera: Noctuidae] (Australia).

SUGARCANE MEALYBUG See Pink Sugarcane Mealybug.

SUGARCANE MOSAIC A viral disease of sugarcane that is transmitted by several Species of aphids.

SUGARCANE PLANTHOPPER *Perkinsiella saccharicida* Kirkaldy [Hemiptera: Delphacidae]: In Australia, a vector of Fiji Disease virus in sugarcane; also damages cane plants directly by sap sucking, causing yellowing and mottling.

SUGARCANE ROOT APHID *Geoica lucifuga* (Zehntner) [Hemiptera: Aphididae] (Australia).

SUGARCANE SANDGRUB *Dipelicus optatus* (Sharp) [Coleoptera: Scarabaeidae] (Australia).

SUGARCANE SCALE *Aulacaspis madiunensis* (Zehntner) [Hemiptera: Diaspididae] (Australia).

SUGARCANE SHOOT-BORER *Chilo infuscatellus* Snell. [Lepidoptera: Pyralidae].

SUGARCANE SOLDIER-FLY *Inopus rubriceps* (Macquart) [Diptera: Stratiomyidae]: A sporadic pest of sugarcane in Australia. Larvae feed on roots, attack setts and mature cane stools at harvest; cause poor ratooning.

SUGARCANE STALK MITE *Steneotarsonemus bancrofti* (Michael) [Acari: Tarsonemidae] (Australia).

SUGARCANE STALK-BORER *Chilo auricilius* Ddgn. [Lepidoptera: Pyralidae].

SUGARCANE STALK-MITE *Stenotarsonemus bancrofti* (Michael) [Acari: Tarsonemidae].

SUGARCANE THRIPS *Baliothrips minutus* (van Deventer) [Thysanoptera: Thripidae].

SUGARCANE WEEVIL BORER *Rhabdoscelus obscurus* (Boisduval) [Coleoptera: Curculionidae]: In Australia, larvae damage sugarcane by boring into stems.

SUGARCANE WHITEFLY 1. *Aleurolobus barodensis* (Maskell) [Hemiptera: Aleyrodidae]. 2. *Neomaskellia bergii* (Signoret) [Hemiptera: Aleyrodidae] (Australia).

SUGARCANE WIREWORM *Agrypnus variabilis* (Candèse) Coleoptera: Elateridae]: In Australia, an important pest of cane; larvae damage setts and developing shoots.

SUGAR-MAPLE BORER *Glycobius speciosus* (Say) [Coleoptera: Cerambycidae].

SUGAR-PINE CONE BEETLE *Conophthorus lambertianae* Hopkins [Coleoptera: Scolytidae].

SUI GENERIS (Latin.) Of its own kind.

SULC, KAREL (1872–1952) (Vrtis 1943, Ent. Listy 6: 29–32, bibliogr.; Sachtleben 1952, Beitr. Ent. 2: 527–528; Stehik 1953, Cas. morav. Mus. Brne 38: 14–16, bibliogr.)

SULC, KARL (1872–1932) (Obenberger 1932, Cas. csl. Spol. ent. 29: 89–98, bibliogr.)

SULCA (Latin, *sulcus* = furrow.) An incorrect form for the plural of Sulcus.

SULCATE Adj. (Latin, *sulcus* = furrow; *-atus* = adjectival suffix.) Groove-shaped; groove-like. Descriptive of a surface which is deeply furrowed or grooved; with deep grooves. Alt. Sulcated. Syn. Sulciform.

SULCIFORM Adj. (Latin, *sulcus* = furrow; *forma* = shape.) See Sulcate. Cf. Crateriform; Sulciform. Rel. Form 2; Shape 2; Outline Shape.

SULCUS Noun. (Latin, *sulcatus* past participle of *sulcare* = to furrow. Pl., Sulci.) 1. A furrow or groove; a groove-like excavation. Typically, a groove in the insect Integument. 2. Any externally visible line formed by the inflection of Cuticle. Biomechanically, a Sulcus forms a strengthening ridge. Alt. Sulcated. Cf. Apodeme; Line of Weakness; Suture.

SULPHATE Adj. (Latin, *sulphur*, *-atus* = adjectival suffix.) A salt of sulphuric acid.

SULPHUR BUTTERFLIES See Pieridae.

SULPHUR YELLOW The colour of the element sulphur, practically equivalent to flavous.

SULPHUREOUS Adj. (Latin, *sulphur*, *-osus* = possessing the qualities of.) Brimstone or sulphur yellow. Alt. Sulphureus.

SULPHURIC ACID H_2SO_4, a colourless, odourless liquid of oily consistency, very caustic and corrosive. Mixes with water and alcohol with the production of heat; a powerful dehydrator in histology that is very destructive to tissue when strong.

SULPHURS See Pieridae.

SULPHURYL FLUORIDE A colourless and odourless fumigant gas (SO_2F_2) marketed under the trade-name Vikane®. A stable, noncorrosive, non-staining unreactive compound used against structural pests (drywood termites and wood-boring beetles) in place of Methyl Bromide when that compound cannot be used. SF is phytotoxic. See Fumigant. Cf. Methyl Bromide.

SULPROFOS An organic-phosphate (dithio-phosphate) compound {O-ethyl-O-[4-(methylthio) phenyl] S-propyl phosphorodithioate} used as a contact insecticide for control of mites, plant-sucking insects (mealybugs, leaf hoppers, whiteflies), bollworms, budworms and armyworms. Compound applied to cotton in USA, and soybeans, peanuts and vegetables in many countries. Phytotoxic to cole crops, peanuts, peppers and deciduous fruits. Toxic to wildlife and fishes. Trade names include Bolstar®, Helothion® and Merdafos®. See Organophosphorus Insecticide.

SULZER, JOHANN HEINRICH (1735–1813) Swiss physician and author of *Die Kennzeichen der Insecten nach Anleitung des Ritters Carl Linnaeus* (1761) and *Abgekurzte Geschichte der Insecten nach dem Linneischen System* (1776.) (Germar 1815, Magazin Ent. I (2): 193; Eiselt 1836, *Geschichte, Systematik und Literatur der Insektenkunde*. 255 pp. (46), Leipzig.) Alt. John Henry Sulzer.

SUMANONE® See Fenitrothion.

SUBMARGINAL VEIN Hymenoptera: The forewing vein immediately posteriad of the Costal Cell which joins the Marginal Vein in Chalcidoidea and Proctotrupoidea. See Wing Venation.

SUMI-ALPHA® See Esfenvalerate.

SUMIBAC® See Fenvalerate.

SUMICHRAST, ADRIEN LOUIS JEAN FRANÇOIS (1828–1882) (Boucard 1884, Bull. Soc. zool. Fr. 9: 305–312; Papavero 1975, *Essays on the History of Neotropical Dipterology*. 2: 350–355. São Paulo.)

SUMICIDIN® See Fenvalerate.

SUMIDAN® See Esfenvalerate.

SUMIFLEECE® See Fenvalerate.

SUMIFLY® See Fenvalerate.

SUMILARV® See Pyriproxyfen.

SUMIRODY® See Fenpropathrin.

SUMITA, SHIRO (1866) (Hasegawa 1967, Kontyû 35 (Suppl.): 45. (pl. 2, fig. 7).)

SUMITHION® See Fenitrothion.
SUMITHRIN® See Phenothrin.
SUMITICK® See Fenvalerate.
SUMITOK® See Fenvalerate.
SUMITON® See Esfenvalerate.
SUMMER DERMATITIS See *Gasterophilus inermis*.
SUMMER FRUIT TORTRIX *Adoxophyes orana* [Lepidoptera: Tortricidae].
SUMMER FRUIT-TORTRIX GRANULOSIS VIRUS A viral disease of the larval stage of Summer Fruit Tortrix, *Adoxophyes orana* in some countries. Trade names include: Ao-GVS® and Capex® See Biopesticide; Granulosis Virus. Cf. Codling-Moth Granulosis Virus.
SUMMER OIL Noun. A highly refined petroleum oil that can be applied to trees in full foliage without having phytotoxic effects normally associated with dormant oils. See Dormant Oil.
SUMMERS, JOHN NICHOLAS (1884–1941) (Anon. 1941, Newsl. U.S. Dep. Agric. Bur. ent. Pl., Quar. 8 (12): 2; Burgess 1942, Ann. ent. Soc. Am. 35: 127; Burgess 1942, J. Econ. Ent. 35: 294–295.)
SUMNER, HELEN MARGARET (1931–1970) (Hinton 1971, Proc. R. ent. Soc. Lond. (C) 35: 54.)
SUN FLIES See Heleomyzidae.
SUN MOTHS See Castniidae.
SUN SPIDERS See Solifugae.
SUNCIDE® See Proproxur.
SUNDEVALL, CARL JACOB (1801–1875) (Anon. 1875, J. Zool., Paris 4: 61; Bonnet 1945, *Bibliographia Araneorum* 1: 35.)
SUNFLOWER BEETLE *Zygogramma exclamationis* (Fabricius) [Coleoptera: Chrysomelidae].
SUNFLOWER BUD MOTH *Suleima helianthana* (Riley) [Lepidoptera: Tortricidae].
SUNFLOWER HEADCLIPPING WEEVIL *Haplorhynchites aeneus* (Boheman) [Coleoptera: Curculionidae].
SUNFLOWER MAGGOT *Strauzia longipennis* (Wiedemann) [Diptera: Tephritidae]: A pest of sunflower in USA; larva mines the pith.
SUNFLOWER MIDGE *Contarinia schulzi* Gagné [Diptera: Cecidomyiidae].
SUNFLOWER MOTH *Homoeosoma electellum* (Hulst) [Lepidoptera: Pyralidae].
SUNFLOWER ROOT WEEVIL *Baris strenua* (LeConte) [Coleoptera: Curculionidae].
SUNFLOWER SEED MIDGE *Neolasioptera murtfeldtiana* (Felt) [Diptera: Cecidomyiidae].
SUNFLOWER SPITTLEBUG *Clastoptera xanthocephala* Germar [Hemiptera: Cercopidae].
SUNFLOWER STEM WEEVIL *Cylindrocopturus adspersus* (LeConte) [Coleoptera: Curculionidae].
SUNFURAN® See Carbofuran.
SUN-SPRAY® See Petroleum Oils.
SUPER- Latin prefix meaning 'over,' 'above' or 'beyond.'
SUPERANS Adj. Exceeding in size and length.
SUPERB PLANT BUG *Adelphocoris superbus* (Uhler) [Heteroptera: Miridae].
SUPERCILIARY Adj. Pertaining to structure positioned above the eyes.
SUPERCILIUM Noun. (Latin, *super-* = above; *cilium* = the eyelid; Pl., Supercilia.) 1. An arched eyebrow-like line above an eye-spot or Ocellus. 2. A Seta (or Setae) above the upper margin of the compound eye.
SUPERCLASS Noun. Zoological classification: An aggregate of two or more classes which have in common some character of major importance.
SUPERCOLONY Social Insects: A unicolonial population with workers moving freely between nests so the entire population forms one colony.
SUPERCOOL Trans. Verb. To cool below the freezing point without solidification.
SUPERCOOLING See Frost Protection.
SUPERFAMILY Zoological classification: A category below Order and above Family which includes a series of Family groups more closely related to each other than to similar groups within the Order. The Superfamily concept is subjective and opinionative. Superfamily names end in the suffix 'oidea.'
SUPERFICIAL GERM BAND Embryology: One that remains ventral in position and in which Blastokinesis does not take place (Folsom & Wardle).
SUPERFICIES EXTERNA The outer surface of any organ or part.
SUPERFICIES INFERIA The ventral or lower surface of any organ or part.
SUPERFICIES INTERNA. The inner surface of any organ or part.
SUPERFICIES Plural Noun. (Latin, *superficies* = upper face.) The surface.
SUPERGENUS Noun. In zoological classification, an assemblage of several closely related Genera; the structure and composition of a Supergenus is subjective.
SUPERICORNIA Noun. Heteroptera with the Antenna inserted on the upper parts of the sides of the head, *e.g.* Coreidae. See Infericornia.
SUPERIOR (Latin.) A term of position indicating above. Cf. Inferior.
SUPERIOR ANTENNA An Antenna set on the upper part of the head.
SUPERIOR LOBE An upper lobe. Cf. Inferior Lobe.
SUPERIOR ORBIT See Vertical Orbit (MacGillivray).
SUPERIOR WINGS The fore or upper wings. Cf. Inferior Wings.
SUPERIORS Odonata: The dorsal anal appendages (Garman).
SUPERLINGUAE Maxillulae: A pair of dorsolateral lobes attached to the Hypopharynx. A second pair of jaw-like appendages, consisting of lobes arising from the tongue-like Hypopharynx (Wardle). The Paraglossa *sensu* MacGillivray.
SUPERLINGUAL SEGMENT The fifth segment of head.
SUPERNE Adj. (Latin, *supernus* = above.) Above; relating to the upper surface.

SUPERNUMERARY Adj. (Latin, *super-* = above; + numerary.) Additional or added, of cells, veins or other structures.

SUPERNUMERARY CELL Diptera: Additional cells in the wings that are a consequence of extra crossveins. Phenomenon reported in Nemestrinidae, Bombyliidae and related Families (Curran).

SUPERNUMERARY CROSSVEIN Crossveins, other than those normally present (Curran). See Supernumerary Cell.

SUPERNUMERARY SEGMENT. Cecidomyiidae: A segment between the head and the Prothorax.

SUPERORDER Noun. Taxonomy: A group of related Orders or Orders placed together in classification. *e.g.* The Neuroptera *sensu* Linnaeus.

SUPERORGANISM Noun. (Latin, *super-* = above; + organism. Pl., Superorganisms.) Social Insects: Any society which displays features of organization analogous to the physiological properties of an organism. *e.g.* Reproductive Castes (gonads); Worker Castes (somatic tissue); trophallaxis (circulatory system).

SUPERPARASITE Noun. (Latin, *super-* = above; + parasite. Pl., Superparasites.) 1. A hyperparasite in the broad sense. 2. Parasitic Hymenoptera: A supernumerary immature parasite which cannot complete development on a host. Supernumeraries typically are eliminated by larval combat or physiological suppression. Cf. Solitary Parasite.

SUPERPARASITISM Noun. (Latin, *super-* = above; + parasitism.) 1. A condition in which a female parasite oviposits more eggs on or in a host than can hatch and successfully develop to maturity. 2. Oviposition on or in a host previously parasitized by a conspecific female. 3. Parasitism of one host by more larvae than can survive to maturity. See Host Discrimination. Cf. Multiple Parasitism.

SUPERPOSED Adj. (Latin, *super-* = above; *ponere* = to place.) Placed one above the other, as the frontal tufts in some moths.

SUPERPOSITION EYE A form of compound eye typical of insects active in dim light. The eye which permits passage of light through nonpigmented wall of one Ommatidium to the Rhabdom of another Ommatidium. See Compound Eye. Cf. Apposition Eye.

SUPERPOSITION IMAGE An erect visual image produced by overlapping points of light (Imms).

SUPERTRIANGLE Noun. (Latin, *super-* = above; + triangle.) Odonata: The wing cell just anterior of the triangle (Garman); the area from the Arculus to the distal angle of the triangle in Anisoptera.

SUPERTRIBE Taxonomy: A group of related Tribes, classified together on the basis of morphological and/or biological criteria.

SUPINATE Verb. (Latin, *supinaro* = to bend or lay backward; *-atus* = adjectival suffix.) Clockwise rotation of an appendage. Cf. Pronate.

SUPINE SURFACE The upper surface.

SUPINO, FELICE (1871–1946) Student of Giovanni Canestrini who published in taxonomy and embryology of ticks. (Manfredi 1947, Atti Soc. ital. Sci. nat. 86: 101–108, bibliogr.; Ranzi 1947, Rc. Ist. lomb. Sci. lett. 80: 24–29.)

SUPOCADE® See Chlorfenvinphos.

SUPONA® See Chlorfenvinphos.

SUPPLEMENTAL ANAL LOOP Some genera of Aeschininae: A second anal loop (Comstock).

SUPPLEMENTARY RADIUS Odonata: extra longitudinal vein in the wings between M and M and M and Cu (Garman).

SUPPLEMENTARY REPRODUCTIVE See Neotenic.

SUPPLEMENTARY SCALES See Intercalary Plates (MacGillivray).

SUPPLEMENTARY SECTORS Interposed Sectors.

SUPPLEMENTS Adventitious Veins in the wings of certain Odonata (Comstock).

SUPPORTING PLATE Hymenoptera: First Valvifers.

SUPPRESSION Noun. (French from Latin, *suppressio* = a keeping back.) 1. The non-development of a part normally present. 2. A type of regulatory control program in which a target-pest population is decreased within a geographical region. Applied against pests established in the area of concern and for which eradication is not practical or applicable. See Regulatory Control. Cf. Containment, Eradication.

SUPRA-ALAR BRISTLES Diptera: Usually 1–4 bristles above the root of a wing, between the Notopleural and the Postalar Bristles (Comstock).

SUPRA-ALAR CAVITY Diptera: See Supra-alar Groove.

SUPRA-ALAR DEPRESSION Diptera: See Supra-alar Groove.

SUPRA-ALAR GROOVE Diptera: A groove on the Mesothorax immediately above the base of the wing. Syn. Supra-alar Cavity *sensu* Comstock. Hymenoptera: A groove or depression just above the base of the wings (Smith).

SUPRA-ANAL Adj. (Latin, *supra* = above; *anus* = anus.) Located above the Anus; Suranal.

SUPRA-ANAL APPENDAGES Odonata: A pair of appendages arising from the tenth Tergum, well developed in the male, reduced or vestigial in the female (Imms).

SUPRA-ANAL HOOK Male Lepidoptera: A curved hook attached to the sclerite covering the genital cavity; the Uncus.

SUPRA-ANAL MEMBRANE See Supra-anal Pad.

SUPRA-ANAL PAD The reduced Epiproct in certain Orders of insects, beneath the end of the tenth Tergum; Supra-anal Membrane (Snodgrass).

SUPRA-ANAL PLATE A triangular sclerite covering the anal cavity above, present in many insects, sometimes in one sex only, often in both. See Anal Operculum; Telson (Tillyard).

SUPRA-ANALIS (Latin, *supra* = above; *anus* = anus.) The twelfth abdominal segment or the Supra-anal Plate of some workers (MacGillivray).

SUPRA-BRUSTIA Brustia along the mesial margin of the Mandible (MacGillivray).

SUPRA-CEREBRAL GLANDS The pair of Salivary Glands positioned above the Brain in bees.

SUPRACIDE® See Methidathion.

SUPRACIDIN® See Methidathion.

SUPRA-CLYPEAL AREA Hymenoptera: The region of the head between the antennal Toruli, Clypeus and Frontal Crest.

SUPRA-CLYPEAL MARK Hymenoptera: A pale-coloured spot above the Clypeus in bees.

SUPRA-CLYPEUS Noun. Latin, *supra* = above; *clypeus* = shield.) Postclypeus; Nasus. A subdivision of the Clypeus. See Clypeus; Nasus; Postclypeus.

SUPRA-COXAL GLANDS Secretory glands above the Coxae of some mite Species.

SUPRA-COXAL Adj. (Latin, *supra* = above; *coxa* = hip.) Lying over or above the Coxa.

SUPRA-EPIMERON Noun. (Latin, *supra* = above; Greek, *epi* = upon; *meros* = upper thigh.) The upper sclerite of the epimeron, see anepimeron, the infolded lateral part of the epimeron in Coleoptera, postparapterum (MacGillivray).

SUPRA-EPISTERNUM Noun. (Latin, *supra* = above; Greek, *epi* = upon; *sternum* = breastplate.) The Anepisternum or upper sclerite of the Episternum *sensu* Imms. See Anepisternum; Episternum.

SUPRA-NEURAL BRIDGE (Latin, *supra* = above; Greek, *neuron* = nerve.) The fused endosternites of the honey bee.

SUPRA-OESOPHAGEAL Adj. (Latin, *supra* = above; *oisophagos* = gullet.) Positioned above the Oesophagus.

SUPRA-OESOPHAGEAL GANGLION Insect Brain: A nerve-mass in the head and positioned above the Oesophagus.

SUPRA-ORBITAL Adj. (Latin, *supra* = above; *orbis* = eye socket.) Pertaining to a position above the eye.

SUPRA-PEDAL Adj. (Latin, *supra* = above.) Above the pedes or feet.

SUPRA-SCUTELLA The Postscutellum of Hymenopterists (MacGillivray).

SUPRA-SPINAL Adj. (Latin, *supra* = above; *spina* = spine.) Pertaining to a position above the spine or above the Ventral Nerve Cord of insects.

SUPRA-SPINAL CORD A longitudinal cord in Lepidoptera, positioned above the abdominal part of the Ventral Nerve Cord (Packard), sometimes equivalent to the Ventral Heart.

SUPRA-SPINAL VESSEL See Sericardial Sinus.

SUPRA-SPIRACULAR LINE Lepidoptera larva: A line or stripe above the Spiracles.

SUPRA-STIGMATAL LINE Supraspiracular line.

SUPRASTOMODAEAL Adj. (Latin, *supra* = above; Greek, *stoma* = mouth; *hodaios* = way; *-alis* = pertaining to.) Descriptive of structure positioned above the Stomodaeum.

SUPRATENTORIUM Noun. (Latin, *supra* = above; *tentorium* = little tent. Pl., Supratentoria.) Slender and thread-like arms or large arms arising from the lateral margin of a Pretentorium or of the Corpotentorium (MacGillivray).

SUPRATHION® See Methidathion.

SUPRA-TRIANGULAR CROSSVEINS Odonata: Veins crossing the supratriangular space. See Vein; Wing Venation.

SUPRA-TRIANGULAR SPACE Anisoptera: An area just above the triangle that occupies nearly the same position as the quadrilateral of Zygoptera; hypertrigonal space.

SUPRA-TYMPANAL ORGAN A ganglion in the outer space of the insect Tibia positioned a short distance above the Tympanum and below the knee (Comstock).

SUPRE Latin prefix meaning 'above', 'over' or 'beyond'.

SUPRE CITATO Latin phrase meaning 'cited above'. Abbreviated 'supr. cit'.

SUPREAPISTERNUM Noun. See Anepisternum.

SUPREME OILS® See Petroleum Oils.

SUQUIN® See Quinalphos.

SURA-ANAL Adj. See Supra-anal.

SURA-ANAL FORK A structure at the base of the suranal plate of locusts. Syn. Suranal Furcula; Furcula Supraanalis *sensu* Packard.

SURA-ANAL FURCULA See Suranal Fork.

SURA-ANAL PLATE 1. The middle dorsal sclerite attached to the tenth abdominal segment of the male grasshoppers. The sclerite is positioned above the anal opening. 2. Lepidoptera larva: A Supra-anal Tergum.

SURANGIN B See Mammeins.

SURFACE Noun. (Latin, *superficies* = surface. Pl., Surfaces.) 1. The exterior or outside of an object or body. 2. The face or faces of a three-dimensional object.

SURFACE SCULPTURE The superficial appearance of the external surface of a structure. See Sculpture Pattern.

SURFACE SKIMMING A wing-flapping mode of locomotion used by some adult Plecoptera and subadult Ephemeroptera. SS consists of planar movement across water surface via aerodynamic thrust provided by wings; continuous contact with surface of water (through legs) eliminates need for total aerodynamic weight support.

SURFACE, HARVEY ADAMS (1867–1941) (Osborn 1946, *Fragments of Entomological History,* Pt. II. 232 pp. (117). Columbus, Ohio; Wheeler & Valley 1975, History of Entomology in the Pennsylvania Department of Agriculture. 37 pp. (3–19), Harrisburg.)

SURFACTANT Noun. (Latin, *superfices*; *-antem* = an agent of something. Pl., Surfactants.) 1. A substance added to a liquid to affect the surface tension of the liquid. A triglyceride salt. Surfactants affect properties that depend on surface tension, namely dispersability, emulsifiability, wetting and spreading of liquids. 2. Chemicals added to pesticides which affect surface tension of the pesticide. See Emulsifier; Pesticide; Wetting Agent. Cf. Active Ingredient; Diluent; Inert Ingredient; Solvent; Synergist.

SURINAM COCKROACH *Pycnoscelus surinamensis* (Linnaeus) [Blattodea: Blaberidae]: A tropicopolitan Species endemic to eastern Oriental region. Burrows into leaves and debris; an occasional pest in homes. Adult black, 18–24 mm long.

SURRA Noun. (East Indian, Maranthi, *sura* = a wheezing sound.) A disease caused by the flagellated protozoan *Trypanosoma evansi* and transmitted by horseflies to horses, mules and camels in Africa. See Nagana.

SURSTYLUS Noun. (Latin, *sur* = super; *stylus* = stake. Pl., Surstyli.) Paired appendages of the ninth abdominal Tergum (Crampton). Cf. Cercus; Pygostylus.

SURSUM Latin combining form meaning 'upward'.

SURVEILLANCE Noun. (French, *surveiller* = to oversee; *sur* = over; *veiller* = to watch. Latin, *vigilare*. Pl., Surveillances.) 1. To watch, guard of supervise under close observation. 2. Monitoring crops for pest detection and determination of population density and dispersion.

SURVEY Noun. (French, *sur* = over; *voir* = to see; Latin, *videre*. Pl., Surveys.) 1. A detailed inspection or critical examination of an object or area for a specific purpose. 2. A comprehensive but undetailed exposition or essay that considers a large body of information about a concept, topic or period of time. An outline.

SUSCEPT Noun. (Latin, *suscipere* = to undergo; to support. Pl., Suscepts.) 1. A plant or animal that has the capacity to contract a disease. 2. A plant or animal that harbours pathogenic organisms.

SUSCEPTIBLE Adj. (Latin, *susceptibilis, suscipere* = to support, to admit.) 1. Descriptive of an organism that is capable of submitting successfully to an action, process or condition; an hypothesis *susceptible of* proof. 2. Descriptive of an organism of such a constitution as to be vulnerable through weakness. An individual *susceptible to* infection.

SUSCEPTIBLE HOST Medical Entomology: A vertebrate host that is infected with pathogen and becomes a victim of the pathogen. Cf. Amplifying Host; Resistant Host; Silent Host; Dead-End Host.

SUSCEPTIBLE VARIETY A variety of plant that is more vulnerable to attack or damage by a pathogen or pest than another variety of the same plant Species. Cf. Resistant Variety.

SUSCON® See Carbosulfan.

SUSHKIN, PETR PETROVICH (1868–1928) (Menzbir 1928, Ann. Mus. zool. Leningr. 29: 1–12, i–xxxi, bibliogr.)

SUSLIK Noun. (Russian.) A ground squirrel, *Citellus citellus*, in Europe and Asia; serves as a sylvatic resevoir for *Leishmania tropica* and *Pasturella pestis*.

SUSPENDED Adj. (Latin, *suspendere* > *suspensum* > *sus* = for; *pendere* = to hang.) 1. Descriptive of something pendant or held in suspension. 2. Descriptive of something temporarily inactive or inoperative. Alt. Floating; Hanging; Hovering. Cf. Dependent.

SUSPENSI Chrysalids of butterflies which are suspended by the tail only. See Succincti.

SUSPENSION CONCENTRATE A form of insecticide formulation in which the active ingredient is dissolved in a diluent and wetting agent (SC = active ingredient + diluent + wetting agent). Resemble wettable powders but SC supplied as liquid concentrate, contain deionized water, and may contain agents to kill bacteria and antifreeze to reduce mixing problems. See Aerosol; Dust; Emulsifiable Concentrate; Oil Concentrate; Wettable Powder.

SUSPENSION CULTURE Tissue Culture: A culture consisting of cells suspended in a liquid culture medium.

SUSPENSORIUM Noun. (Latin, *suspendere* = to hang up; *-ium* = diminutive > Greek, *-idion*. Pl., Suspensoria.) l. A dorsal strand of cells by which the gonad is attached to the Coelomic wall. 2. Structures or muscles from which others suspend or hang.

SUSPENSORIUM OF HYPOPHARYNX A pair or group of bar-like sclerites in the lateral wall of the adoral surface of the base of the Hypopharynx. Syn. Fulturae (Snodgrass).

SUSPENSORY Adj. (Latin, *suspendere* = to hang up.) 1. Pertaining to a Suspensorium. 2. Any action or process that is incomplete; pending.

SUSPENSORY LIGAMENT The distal union of the Ovariole filaments with one another (Snodgrass). The SL serves to tether the Ovarioles and is often attached to a dorsal region of the body wall or perhaps the Dorsal Diaphragm. See Ovariole. Cf. Terminal Filament. Rel. Reproductive System.

SUSPENSORY MUSCLES Muscles extending from the body wall to the Alimentary Canal; dilator muscles (Snodgrass). See Muscles.

SUSTAIN® See Fenoxycarb.

SUSTENTOR Noun. (Latin, *sustinere* = to sustain; *or* = one who does something. Pl., Sustentors.) Either of two posterior projections (hooked Cremaster) of a butterfly Chrysalis. Alt. Sustentator. See Cremaster.

SUSTER, PETRE (1896–1954) (Jiturau 1954, Revta. Univ. 'Al I. Cuza' Inst. politeh. Iasi 1: 463–466, bibliogr.)

SUSTERY, OLDRICH (1879–1959) (Boucek 1959, Cas. csl. Spol. ent. 56: 375–377, bibliogr.)

SUSVIN® See Monocrotophos.

SUTER-BURGER, RUDOLF (1901–1966) (Anon. 1966, Mitt. ent. Ges. Basel 16: 12.)

SUTHERLAND, J R GORDON (–1961) (Varley 1963, Proc. R. ent. Soc. Lond. (C) 27: 51.)

SUTTON, GEORGE PERCY (1885–1956) (W.B. 1956, Entomologist's Rec. J. Var. 68: 165.)

SUTURAL Adj. (Latin, *sutura* = seam; *-alis* = pertaining to.) Descriptive of or pertaining to a suture or seam.

SUTURAL AREA Tingidae: The area of the

Hemelytra occupying the inner apical parts, narrow in short winged, corresponds to the membrane of the upper wing in the other Heteroptera.

SUTURAL GROOVE A groove formed by a suture.

SUTURE Noun. (Latin, *sutura* = seam. Pl., Sutures.) 1. A seam or seam-like line of contact between two sclerites or hardened body parts that makes those body parts immovably connected. 2. The anatomical feature which serves as a boundary to separate distinct parts of the body wall. 3. The line of contact between the Elytra in Coleoptera, Tegmina in Orthoptera and Hemelytra in Hemiptera. See Segmentation. Cf. Apodeme; Seam; Sulcus. Rel. Sclerite.

SUTURIFORM Adj. (Latin, *sutura* = seam; *forma* = shape.) An impressed line, often sinuous, which marks the fusion of two sclerites. See Suture. Cf. Sulciform. Rel. Form 2; Shape 2; Outline Shape.

SUZUKI, MOTOJIRO (–1942) (Hasegawa 1967, Kontyû 35 (Suppl.): 44–45.)

SVIRIDENKO, PAVEL ALEXANDREVICH (1893–1971) (Dobrovolskii 1964, Ent. Obozr. 43: 482. (Translation: Ent. Rev., Wash. 43: 247); Dobrovolskii 1973, Ent. Obozr. 52: 234–236, bibliogr.)

SWAIN, HUMPHREY DRUMMOND (1902–1959) (Uvarov 1960, Proc. R. ent. Soc. Lond. (C) 24: 54.)

SWAIN, RALPH BROWNLEE (1912–1953) (Becker 1953, J. N. Y. ent. Soc. 61: 185–188; Byers 1954, Ann. ent. Soc. Am. 47: 228–230; Nelson 1954, J. Econ. Ent. 47: 547.)

SWAINE JACK PINE SAWFLY *Neodiprion swainei* Middleton [Hymenoptera: Diprionidae].

SWAINE, JAMES MALCOLM (1879–1955) (Anon. 1955, Can. Ent. 87: 10.)

SWAINSON, WILLIAM (1789–1855) British Ornithologist. Author of *Zoological Illustrations* (3 vols, 1820–1823). (Bell 1857, Proc. Linn. Soc. Lond. 1: xlix–liii, bibliogr.; Carvalho 1918, Revta. Mus. paul. 10: 884–894, bibliogr.)

SWALE, HAROLD (1853–1919) (Andrewes 1919, Entomol. mon. Mag. 55: 140–141; Anon. 1920, Ent. News 31: 30.)

SWALE Noun. (Old Norse, *svalr* = cool. Pl., Swales.) A depression in the land, or a hollow, that is wet and covered with vegetation. Cf. Bog; Fen; Marsh; Swamp. Rel. Community.

SWALLOW BUG *Oeciacus vicarius* Horvath [Hemiptera: Cimicidae].

SWALLOWTAIL BUTTERFLIES See Papilionidae.

SWAMMERDAM, JOHANN JACOB (JAN) (1637–1685) Dutch physician, microscopist and natural historian. Educated for ministry. Regarded as highly skilled in dissections and description of minute anatomical detail; he provided the first report of the honey bee Sting apparatus. His important work laid a foundation for study of Entomology: *Historia Insectorum Generalis* (1669) and '*Biblia Naturae, sive Historia Insectorum in classes certas redacta, & c*', published posthumously (2 vols, 1737–1738), and as '*The Book of Nature*' (1758). (Boerhaave 1758, *In* Swammerdam. *Book of Nature or History of Insects.* pp. i–xvii. London; Eiselt 1836, *Geschichte, Systematik und Literatur der Insektenkunde* 255 pp. (24–26), Leipzig; Jardine 1848, Nats Libr. 28: 17–58; Harting *et al.* 1876, Natur 25: 165–168, 181–182, 202–204; Nordenskiöld 1935, *History of Biology*, 629 pp. (167–171) London; Pöhlmann 1941, *Jan Swammerdam*, 208 pp. Leiden; Schierbeck 1967, *Jan Swammerdam; His Life and Works*, vi + 202 pp. Amsterdam.)

SWAMP Noun. (Dutch, *zwamp* = marsh, English, *sump*. Pl., Swamps.) An ecological habitat or area characterized by wet, spongy land usually covered with water. Cf. Bog; Fen; Marsh; Swale. Rel. Community.

SWAN, DUNCAN CAMPBELL (1907–1960) (Uvarov 1961, Proc. R. ent. Soc. Lond. (C) 25: 50.)

SWARD Noun. See Turf.

SWARM Noun. (Anglo Saxon, *swearm*, Greek, *schwarm*. Pl., Swarms.) 1. A relatively compact group of small, mobile, conspecific individuals viewed collectively, or migrating or moving in a common direction. 2. A group of honey bees moving from a hive to establish another hive. 3. A mass of unicellular organisms. See Organization. Cf. Aggregation; Synhesma.

SWARMING Noun. (Anglo Saxon, *swearm*.) 1. Social Insects: The concerted, organized departure from a hive or nest by workers accompanied by reproductives. Collectively the individuals form the nucleus of a new colony. See Absconding; Budding; Fission. Cf. Nuptial Flight. 2. Solitary Insects: Aggregations of conspecific individuals characterized by periodic flight within a small region.

SWARMING FLIGHT A behavioural category for insect flight which includes individuals of populations relatively dispersed or populations with patchy distribution or many groups with aquatic immatures. Typically males gather in relatively high concentrations in flight and maintain a stationary position. See Flight. Cf. Mass Flight; Migratory Flight; Trivial Flight.

SWARMING LEAF-BEETLE *Rhyparida* spp. [Coleoptera: Chrysomelidae]. Australian Species; adults prefer new leaves and young shoots; cause severe defoliation; wide host range includes avocado, lychee, maize, sugarcane, ornamentals and native trees. See Chrysomelidae.

SWARTZ, OLAF (1760–1818) (Anon. 1818, K. svenska VetenskAkad. Handl. 39: 370–380, bibliogr.)

SWEAT Noun. (Middle English, *swetea*.) The accumulated moisture excreted through pores in the skin of mammals induced through the generation of heat from vigorous exercise of muscles. Rel. Thermoregulation.

SWEAT BEE *Halictus farinosus* Smith [Hymenoptera: Halictidae]: A Species widespread in western North America. See Halictidae.

SWEAT BEES Members of the Halictidae, particularly the Genus *Lasioglossum*. See Halictidae.

SWEBATE® See Temephos.

SWEET, GEORGINA (–1946) (P.C.M. 1946, Victorian Nat. 62: 211–212.)

SWEET, WINFIELD CAREY (1891–1942) (Anon. 1942, Rep. Rockefeller Found Int. hlth Dir. 1942 [I-4]; Cort 1943, J. Parasit. 29: 365–366.)

SWEET-CLOVER APHID *Therioaphis riehmi* (Brüner) [Hemiptera: Aphididae]: In North America, a pest of *Melilotus*. Winged female ovoviviparous, parthenogenetic.

SWEET-CLOVER ROOT-BORER *Walshia miscecolorella* (Chambers) [Lepidoptera: Cosmopterigidae].

SWEETCLOVER WEEVIL *Sitona cylindricollis* Fähraeus [Coleoptera: Curculionidae].

SWEET-FERN LEAF-CASEBEARER *Acrobasis comptoniella* Hulst [Lepidoptera: Pyralidae].

SWEETPOTATO *Manihot esculenta* Cranz [Euphorbiaceae]: A shrub with long, tuberous, edible roots and consumed as a staple in many developing countries. Sweetpotato attacked by many Species of beetles as stored product.

SWEETPOTATO BEETLES See Chrysomelidae.

SWEETPOTATO FLEA-BEETLE *Chaetocnema confinis* Crotch [Coleoptera: Chrysomelidae]: A pest of sweetpotato, corn, grasses, sugarbeets and some weeds in eastern North America. Overwinter as adults in harborage and trash. Attack seedling plants and move to bindweed as oviposition site, lay eggs then die. Larvae feed on roots of bindweed and develop a new generation during July-August which feeds on weeds or small grains and grasses.

SWEETPOTATO HORN-WORM *Agrius cingulata* (Fabricius) [Lepidoptera: Sphingidae].

SWEETPOTATO LEAF-BEETLE *Colasposoma sellatum* Baly [Coleoptera: Chrysomelidae] (Australia). *Typophorus nigritus viridicyaneus* (Crotch) [Coleoptera: Chrysomelidae].

SWEETPOTATO LEAFMINER 1. *Bedellia orchilella* Walsingham [Lepidoptera: Lyonetidae]. 2. *Bedellia somnulentella* (Zeller) [Lepidoptera: Lyonetiidae] (Australia).

SWEETPOTATO LEAFROLLER *Pilocrocis tripunctata* (Fabricius) [Lepidoptera: Pyralidae].

SWEETPOTATO TORTOISE BEETLES *Aspidomorpha* spp. [Coleoptera: Chrysomelidae] (Australia).

SWEETPOTATO VINE-BORER *Omphisa anastomosalis* (Guenée) [Lepidoptera: Pyralidae].

SWEETPOTATO WEEVILS *Cylas formicarius elegantulus* (Summers); *C. puncticollis* Boheman [Coleoptera: Apionidae]: Multivoltine pests of sweetpotato in southeastern USA and tropical regions. Adults ant-like, shiny; eggs minute, white, laid in punctures of roots or vines near ground; larvae feed in stems or roots; pupation within burrow. See Apionidae (Brentidae; Curculionidae). Cf. Seed Weevils.

SWEETPOTATO WHITEFLY *Bemisia tabaci* (Gennadius) [Hemiptera: Aleyrodidae]: A widespread, highly significant economic pest of many field and vegetable crops. Causes damage by feeding, transmission of viral disease and honeydew production of sooty mould. Species resistant to most synthetic pesticides. The recent introduction of Silverleaf Whitefly (*B. tabaci* Biotype B) to Australia is a major concern. This insect has reached major pest status in many cotton-producing countries due to its prolific development rate, huge host range, copious honeydew production, pesticide resistance and transmission of Gemini viruses. B-type *B. tabaci* has been given separate Species status in USA (*Bemisia argentifolii* Bellows & Perring - Silverleaf Whitefly). In Australia, Silverleaf Whitefly has been detected in nurseries, glass-houses and some gardens of many towns servicing cotton districts. Adult white to pale yellow, oval and 1.0–1.5 mm long; wings covered in white waxy powder. Adult stage 8–17 days but can be up to 60 days for some females. Adult female lays 30–300 eggs. Females usually lay 28–80 eggs, but B Biotype can lay twice as many. Eggs 0.2 mm long, oval and laid in batches that form a semicircle on the underside of leaves. Eggs anchored to leaf by a pedicel (stalk) that is inserted into leaf tissue by ovipositing female. Egg stage 5–9 days and is dependent on temperature and humidity. First instar nymph active and called a 'crawler'. Crawler similar to first instar of scale insects. First instar mobile and moves to a feeding site on the lower surface of leaf. Legs lost in following moult. Late instar nymphs sessile, translucent and develop yellow spots on posterior end of Abdomen. Nymphs do not resemble adult and look like scale insects. Nymphal stage 6–12 days. Pupa 0.7 mm long, yellow and opaque. If host leaf is smooth, then pupa lacks Setae on the upper surface. If leaf is hairy, then pupae have 2–8 long Setae on upper surface. Pupal stage ca 6 days. Adult remains on or near pupal case for 10–20 minutes after emergence to spread and dry wings. Life-cycle completed in 14–21 days depending on temperature. See Aleyrodidae. Syn. Cotton Whitefly; Poinsetta Whitefly; Silverleaf Whitefly, Tobacco Whitefly.

SWENK, MYRON HARMON (1883–1941) (Osborn 1937, *Fragments of Entomological History,* 394 pp. (282), Columbus, Ohio; Calvert 1941, Ent. News 52: 240; Tate 1941, J. Econ. Ent. 34: 863–864.)

SWENSON, KNUD GEORGE (1923–1975) (Oman *et al.* 1976, J. Econ. Ent. 69: 130–131.)

SWETT, LOUIS WILLIAM (1880–1930) (Johnson 1930, Psyche 37: 300.)

SWEZEY, OTTO HERMAN (1869–1959) Economic Entomologist (Hawaiian Sugar Planter's Association). (Pemberton 1960, J. Econ. Ent. 53: 932–933; Usinger & Zimmerman 1960, Pan-Pacif.

Ent. 36: 151–153; Mallis 1971, *American Entomologists*. 549 pp. (506–508), New Brunswick.)

SWIFT MOTHS See Hepialidae.

SWIFT SPIDERS *Supunna* spp. [Araneida: Corinnidae] (Australia).

SWIMMERET Noun. (Anglo Saxon, *swimman* > Greek, *schwimmen*. Diminutive of swimmer. Pl., Swimmerets.) 1. Neuroptera: Paired abdominal or thoracic gill-like (plate-like) appendages on aquatic larvae of some Species; swimmerets serve as oar-like organs of locomotion. Cf. Nectopod. 2. Crustacea: One series of paired appendages on the ventral surface of the abdomen that are used to carry eggs or facilitate locomotion

SWIMMING PADDLES Terminal appendages of mosquito pupae. Cf. Trumpet.

SWINHOE, CHARLES (1836–1923) English soldier (Colonel) and amateur Lepidopterist. Most notable work was *Catalogue of the Eastern and Australian Lepidoptera Heterocera in the collection of the Oxford University Museum* (2 vols, 1892, 1900). (Anon 1924, Entomologist's. mon. Mag 60: 19–20; B.D.J. 1924, Proc. Linn. Soc. Lond. 1923–24: 58–59; Riley 1924, Entomologist 57: 23–24; Walker 1924, Nature 113: 21; Takahashi 1935, Q. J. Taiwan Mus. 18: 335–339.)

SWINTON, ARCHBALD HENRY (1845?-1936) Neave 1936, President. Addr. R. ent. Soc. Lond. 1936: 2–3.)

SWINTON, JOHN (1703–1777) (Rose 1850, *New General Biographical Dictionary* 12: 166–167.)

SWIPE® See Methamidophos.

SWORD Noun. (Old Engish, *sweord* = sword. Pl., Swords.) A weapon devised by humans and used for cutting or stabbing. Specifically, swords are constructed of metal, long, with a handle, sharp edges and tapering to a point. Sword is used as a comparative descriptor for numerous entomological terms that describe shape, form or function. Cf. Ensiform; Falcate. Rel. Hatchet; Sickle; Spear.

SWORD-TAILED CRICKETS See Trigonidiidae.

SWORDGRASS BROWN BUTTERFLY *Tisiphone abeona* (Donovan) [Lepidoptera: Nymphalidae] (Australia).

SWYNNERTON, CHARLES FRANCIS MASSEY (1877–1938) (Burkhill 1938, Proc. Linn. Soc. Lond. 1938: 254–256; Fryer 1938, Proc. ent. Soc. Lond. (C) 3: 59.)

SYBOL® See Pirimiphos-Methyl.

SYCAMORE APHID *Drepanosiphum platanoidis* (Schrank) [Hemiptera: Aphididae] (Australia).

SYCAMORE LACE-BUG *Corythucha ciliata* (Say) [Hemiptera: Tingidae]: A pest endemic to North America that overwinters as adult beneath bark on trees. SLB host plants include Ash, Hickory and Sycamore.

SYCAMORE TUSSOCK-MOTH *Halysidota harrisii* Walsh [Lepidoptera: Arctiidae].

SYDNEY AZURE *Ogyris ianthis* Waterhouse [Lepidoptera: Lycaenidae] (Australia).

SYDNEY BROWN TRAPDOOR SPIDER *Misgolas rapax* Karsch [Araneida: Idiopidae] (Australia).

SYDNEY FUNNELWEB SPIDER *Atrax robustus* Cambridge [Araneidae: Hexathelidae] (Australia).

SYDNEY KELP FLY *Chaetocoelopa sydneyensis* (Schiner) [Diptera: Coelopidae] (Australia).

SYDOW, GUSTAV VON (1859–1939) (Weidner 1967, Abh. Verh. naturw. Ver. Hamburg Suppl. 9: 265.)

SYKES, WILLIAM HENRY (1790–1872) (Anon. 1872, The Times, London, 19.6.1892; Anon. 1872, Proc. R. Soc. Lond. 20: xxxiii–xxxiv.)

SYLVA Noun. (Latin, *sylva* = forest.) Forest trees collectively. Alt. Silvan.

SYLVAN Adj. (Latin, *sylva* = forest.) Pertaining to organisms that inhabit forests or woodland areas.

SYLVATICIN See Acetogenins.

SYLVICOLIDAE Plural Noun. See Anisopodidae.

SYMBIOGENESIS Noun. (Greek, *symbionai* = to live with. *genesis* = descent.) The method of origin of social symbiotic relation among ants and other insects.

SYMBIONT Noun. (Greek, *symbionai* = to live with. Pl., Symbionts.) One of the partners in a symbiotic relationship, *e.g.* an insect or other arthropod living in symbiosis with termites, ants or other insects. Alt. Symbiote.

SYMBIOSIS Noun. (Greek, *symbionai* = to live together. Pl., Symbioses.) A protracted, living together (in more-or-less intimate association or dependency) by different Species. The association is not necessarily for mutual benefit. Symbiosis has many variants, the primary of which are Commensalism, Mutualism and Parasitism. See Biosis. Cf. Aggregation; Commensalism; Mutualism; Social Parasitism; Parasitism; Social Symbiosis; Trophic Symbiosis. Rel. Calobiosis; Cleptobiosis; Dulosis; Hamabiosis; Lestobiosis; Parabiosis; Phylacobiosis; Synclerobiosis; Trophobiosis; Xenobiosis.

SYMBIOTE See Symbiont.

SYMBIOTIC Adj. (Greek, *symbionai* = to live with; *-ic* = of the nature of.) Descriptive of organisms living together in a state of Symbiosis. See Symbiosis. Cf. Mutualistic; Parasitic.

SYMES, EDGAR SYMS (1881–1966) (Anon. 1966, Proc. S. Lond. ent. nat. Hist. Soc. 1966: 131–132; Pearson 1967, Proc. R. ent. Soc. Lond. (C) 31: 63.)

SYMES, HAROLD (1866–1969) (A.C.R.R. 1970, Entomologist's Rec. J. Var. 82: 185–186.)

SYMMETRICAL Adj. (Greek, *syn* = with; *metron* = measure; *-alis* = characterized by.) Pertaining to symmetry. Typically viewed as structures uniformly developed in size, shape and superficial features on either side of a bilateral animal. *e.g.* The arms and legs of humans; Antennae of insects. See Symmetry.

SYMMETRY Noun. (Greek, *symmetria* = due proportion. Pl., Symmetries.) 1. The state or quality of divisibility into similar halves. 2. The regular arrangement of organs or parts that are capable of division into similar halves or similar radii. 3. The property in which an object will coincide with its image in a mirror. 4. The correspondence in size, shape, and position of component parts that are on opposite sides of a dividing line or median plane. Several kinds of symmetry have been described: Bilateral Symmetry; Radial Symmetry; Spherical Symmetry; Zonal Symmetry. See Bilateral Symmetry; Radial Symmetry; Spherical Symmetry; Zonal Symmetry. Cf. Asymmetry.

SYMMOMUS SKIPPER *Trapezites symmomus* Hübner [Lepidoptera: Hesperiidae] (Australia).

SYMPATHETIC NERVOUS SYSTEM The nerves and ganglia which innervate the Alimentary Canal, other viscera, Vagus, Visceral Nervous System and Stomodaeal Nervous System.

SYMPATHETIC Adj. (Greek, *syn* = with, *pathos* = feeling.) Pertaining to the nervous system.

SYMPATRIC Adj. (Greek, *syn* = with; *patra* = homeland; *-ic* = characterized by.) 1. Pertaining to organisms (usually populations, Subspecies or Species) with overlapping geographical distributions. 2. Pertaining to organisms that display distributions overlapping in time. Cf. Allopatric. Parapatric.

SYMPHILE Noun. (Greek, *syn* = with; *philein* = to live. Pl., Symphiles.) A symbiont. A solitary insect or other arthropod which lives as a guest in the nest of a social insect Species (ants, bees, termites). Symphiles may be groomed, fed or transported within the nest by workers. See Inquiline; Symbiont. Cf., Synechthran; Synoekete.

SYMPHILY Noun. (Greek, *syn* = with; *philein* = to live. Pl., Symphilies.) The commensalistic relationship between host and guest (Symphile) which results in mutual benefit for both Species. See: Metochy; Synecthry.

SYMPHYLA Noun. (Greek, *syn* = together; *phylon* = race.) A Class of Arthropoda sometimes regarded as myriapods. Soft-bodied, pale coloured, 2–6 mm long, and resembling centipedes; Antennae prominent; adults with 12 pairs of legs and a pair of Spinnerets on the last abdominal segment; genital openings on the last abdominal segment. Symphyla typically occur in damp soil under stones and in leaf litter; scavenge or feed upon rootlets of young plants. Eggs laid in soil; upon eclosion immature has six pairs of legs; body segment, pair of legs and antennal segments added with each successive moult. Genera include *Hanseniella, Scolopendrella*. Cf. Pauropoda.

SYMPHYSIS Noun. (Greek, *symphysis* = a growing together; *sis* = a condition or state. Pl., Symphyses.) A connection of two sclerites by a soft membrane, permitting a motion or flexibility between them.

SYMPHYTA Noun. (Greek, *symphyton* = a medicinal plant, confrey.) Sawflies; Woodwasps. A cosmopolitan Suborder of Hymenoptera. Symphyta constitute a paraphyletic Taxon whose adults are characterized by one or more closed anal cells in the forewing, at least one closed anal cell in the hindwing, metanotum with cenchri and Abdomen broadly attached to the Thorax. Symphyta larvae are caterpillar-like with a well developed head capsule, thoracic legs and Abdomen with 10 segments typically bearing Prolegs on segments 2–8 and 10. Prolegs are without hooks, absent or reduced in some groups. Traditional Superfamilies include Cephoidea, Megalodontoidea, Orussoidea, Siricoidea, Tenthredinoidea and Xyeloidea. Syn. Chalastrogastra; Sessiliventra; Phyllographaga; Xylophaga. See Hymenoptera. Cf. Apocrita.

SYMPLESIOMORPHIC Adj. (Greek, *sym* = together; *pleisio* = original; *morphe* = form; *-ic* = consisting of.) Pertaining to the joint posession of a primitive (ancestral) character state by more than one element of a clade. Cf. Synapomorphic.

SYMPLESIOMORPHY Noun. Greek, *sym* = together; *pleisio* = original; *morphe* = form. Pl., Symplesiomorphies.) In cladistical theory of classification, the possesion of a primitive or ancestral character state by two or more Taxa of a clade. See Plesiomorphy. Cf. Synapomorphy.

SYMPTOM Adj. (Greek, *symptoma* = anything that has befallen one. Pl., Symptoms.) The indication of a disease by the reaction of a host.

SYNANTHROPE Noun. (Greek, *syn* = together; *anthropos* = humans. Pl., Synanthropes.) Insects (or other organisms) which are intimately associated with humans, human habits and human habitations. The extent and complexity of the association between human and synanthrope is sometimes classified as Eusynanthropic and Hemisynanthropic. Adj. Synanthropic.

SYNANTHROPY Noun. (Greek, *syn* = together; *anthropos* = humans. Pl., Synanthropies.) A concept which recognizes the presence of a human-dominated ecological community (human biocenose) and the association of insects (or other organisms) with that community. The association may be developed rapidly through colonization or slowly evolve over longer periods of time through intermittent contact; the association may be partial or complete. Houseflies (Diptera) serve as an excellent example synanthropic organisms. See Eusynanthropy; Hemisynanthropy.

SYNAPOMORPHY Noun. (Greek, *syn* = together; Pl., Synapomorphies.) In cladistical theory of classification, a derived taxonomic character which is shared by 2 or more Taxa forming an inclusive group. Cf. Autapomorphy. See Apomorphy. Plesiomorphy.

SYNAPSE Noun. (Greek, *synapsis* = union.) The central mechanism of intercommunication between terminal fibres of 2 or more neurones (Snodgrass).

SYNAPTERA Noun. (Greek, *syn* = together; *ptera* = wing.) 1. Originally wingless insects without Metamorphosis. 2. The Thysanura.

SYNAPTEROUS Adj. (Greek, *syn* = together; *ptera* = wing; Latin, *-osus* = possessing the qualities of.) Descriptive of or pertaining to the Synaptera.

SYNAPTIC Adj. (Greek, *synaptos* = fastened together; *-ic* = of the nature of.) Descriptive of or pertaining to a synapse.

SYNARTHROSIS Noun. (Greek, *syn* = together; *arthron* = joint; *-osis* = a condition or state. Pl., Synarthroses.) Articulation without motion.

SYNCEPHALON Noun. (Greek, *syn* = together; *kephale* = head. Pl., Syncephalons.) A secondary composite head in insects, composed of the Prostomium and one or more somites following it (Snodgrass).

SYNCEREBRUM Noun. (Greek, *syn* = together; Latin, *cerebrum* = brain.) The compound Brain of insects.

SYNCHRONOUS Adj. (Greek, *syn* = together; *chronos* = time; Latin, *-osus* = possessing the qualities of.) Descriptive of events (biochemical, behavioural, functional) that occur simultaneously.

SYNCHRONOUS GENERATIONS Population dynamics: A circumstance in which all adult females of a population lay eggs at discrete periods and all individuals within that population are at one developmental stage at any point in time. Cf. Asynchronous Generations.

SYNCHRONY Noun. (Greek, *syn* = together; *chronos* = time. Pl., Synchronies.) Synchronism; identity of time.

SYNCLEROBIOSIS Noun. (Greek, *syn* = together; *kleros* = by chance; *bios* = life; *-osis* = a condition or state. Pl., Synclerobioses.) An association between two Species of ants (that usually inhabit independent colonies) for indefinite reasons (Smith). See Biosis; Commensalism; Parasitism; Symbiosis. Cf. Abiosis; Anhydrobiosis; Antibiosis; Archebiosis; Calobiosis; Cleptobiosis; Hamabiosis; Kleptobiosis; Lestobiosis; Parabiosis; Phylacobiosis; Plesiobiosis; Trophobiosis; Xenobiosis.

SYNCYTIAL Adj. (Greek, *syn* = with; *kytos* = hollow; *-alis* = pertaining to.) Descriptive of structure associated with a Syncytium.

SYNCYTIUM Noun. (Greek, *syn* = together; *kytos* = hollow; Latin, *-ium* = diminutive > Greek, *-idion*. Pl., Synctia.) 1. The masses of protoplasm with nuclei scattered through, from which the egg, nutritive and epithelial cells arise in the insect Ovary (Packard). 2. A protoplasmic mass formed by the fusion of several protoplasts without the fusion of the individual nuclei (Daubenmire).

SYNDESIS Noun. (Greek, *syndesai* = to bind together; *sis* = a condition or state. Pl., Syndeses.) 1. A method of articulation where two parts are connected by a membrane which permits of considerable motion between them. 2. The phenomenon by which chromosomes come together during meiosis.

SYNECOLOGY Noun. (Greek, *syn* = together; *oikos* = dwelling; *logos* = discourse. Pl., Synecologies.) Community Ecology: The study of the influence of climatic, chemical and environmental factors upon a population or upon an association of animals or plants (Folsom & Wardle). See Ecology. Cf. Autecology.

SYNECTHRAN Noun. (Greek, *synechthairein* = to join together in hating. Pl., Synecthrans.) Social Insects: A symbiont living in a nest (as scavengers, predators or parasites) which is treated with hostility by members of the host colony. See Symbiont; Inquiline. Cf. Symphile; Synoekete.

SYNECTHRY Noun. (Greek, *syn* = together; *echthros* = hatred. Pl., Synecthries.) 1. A form of commensalism in which organisms living together have mutual dislike. 2. The relation between ants and other Species of insects inhabiting their nests (Symbionts) in spite of efforts by the ants to destroy them. See Symbiosis. Cf. Symphily; Metochy.

SYNEMA See Synnema.

SYNERGISM Noun. (Greek, *synergos* = working together; *-ism* = condition. Pl., Synergisms.) The action or phenomenon in which two or more factors (substances, chemicals) together achieve a greater effect than the sum of the individual effects of the factors

SYNERGIST Noun. (Greek, *synergos* = cooperator. Pl., Synergists.) 1. A substance involved in synergism. 2. A non-toxic chemical substance in a pesticide which increases the potency of the pesticide with which it is mixed. Examples include Piperonyl Butoxide, Sesamex and Sulfoxide. Cf. Active Ingredient; Inert Ingredient; Diluent; Solvent; Surfactant.

SYNEROL® See Pyrethrin.

SYNFUME® See AlP.

SYNGAMY See Conjugation.

SYNHESMA Noun. (Greek, *syn* = with; *hesmos* = a swarm). A coming together by members of a colony. See Organization. Cf. Aggregation; Swarm. Rel. Honey Bee.

SYNISTA Noun. Neuroptera: Arcane term referring to Taxa in which the mouthparts are undeveloped; forming an imperfect tubular structure. Alt. Synistata. See Elinguata.

SYNKLEPTON Noun. (Greek, *syn* = together; *kleptein* = to steal, hide. Pl., Synkleptons.) (Dubois & Gunther 1982, Zool. Jahrb. Syst. 109: 290–305.)

SYNNEMA Noun. (Greek, *syn* = together; *nema* = thread. Pl., Synnemata.) 1. Fungi: Fused, hair-like structures that bear spores. 2. The united stamen filaments of a flower. Alt. Synema.

SYNOECY Noun. (Greek, *synoikia* = community. Pl., Synoecies.) 1. A form of commensalism. 2. Among social insects, the relationship between host and guest in which the host are indifferent to and tolerate the inquiline guest. See Commensalism. Cf. Symphily; Synecthry; Metochy.

SYNOEKETE Noun. (Greek, *syn* = with; *oiketes* = dweller. Pl., Synoeketes.) Social Insects: A symbiont indifferently tolerated as a guest by members of a host colony. See: Inquiline; Symbiont. Cf. Symphile; Synechthran.

SYNOMONE Noun. (Greek, *syn* = with; *hormaein* = to excite.) 1. A chemical substance produced or acquired by an organism, which, when it contacts an individual of another Species, in the natural context, evokes in the receiver a behavioural or physiological response adaptively favourable to both the emitter and receiver. (Nordlund & Lewis 1976, J. Chem. Ecol. 2: 211–220.) 2. A secondary plant chemical which serves to attract entomophagous insects to the plant and subsequently their prey or hosts. (Boethel & Eikenbary (1986: 152). See Semiochemical. Cf. Kairmone.

SYNONYM Noun. (Greek, *syn* = with; *onyma* = name. Pl., Synonyms.) Two or more different names for one and the same thing. In zoological nomenclature, a different scientific name given to a Species or Genus previously named and described. Cf. Homonym. Rel. Taxonomy; Nomenclature.

SYNONYMOUS Adj. Having the character of a synonym or an identity of terminological application.

SYNOPSIS Noun. (Greek, *syn* = together; *opsis* = a view. Pl., Synopses) A taxonomic publication which briefly summarizes current knowledge of a group. New information is not necessarily included in a synopsis. Cf. Monograph; Review; Revision.

SYNOPTIC COLLECTION A group of Taxa which have been authoratively identified and which serves as the basis of comparison for the identification of other specimens.

SYNOVIGENIC Adj. (Greek, *syn* = together; *ovum* = egg; *genesis* = to produce; *-ic* = of the nature of.) Parasitic Hymenoptera: A condition in which females periodically develop eggs during their adult life (Flanders 1950, Canad. Ent. 82: 134–140). See Oogenesis. Cf. Proovigenic. Rel. Ovariole.

SYNPHAGY Noun. (Greek, *syn* = together; *phagein* = to eat.) Two or more Species sharing the same resource. The relationships of such Species.

SYNTELIIDAE Sap Flora Beetles.

SYNTEXIDAE Plural Noun. Syntexid Wasps.

SYNTHEMIDAE Plural Noun. A Family of anisopterous Odonata restricted to Australia, Asia and South Pacific. Adults medium-sized and Abdomen slender. Male hindwing angled and anal margin incurved. Pterostigma variable and numerous antenodal crossveins. Median basal space crossed and discoidal triangles broad. Anal loop short and broad. Male with auricles on second abdominal segment and Tibiae keeled. Anal appendages long and sinuous. Females of some Species with long Ovipositors. Larva resembles Cordulesgasteridae, robust, setose; head broad with frontal ridge; Abdomen fusiform and Antenna similar to Cordulegasteridae; mask deeply concave with renulations and many Setae.

SYNTHETIC Adj. (Greek, *synthetikos; -ic* = of the nature.) 1. That which is artificial. 2. In animal classification, combining in one group or form the structural characters of two or more different groups or forms.

SYNTHETIC INSECTICIDE A category of chemicals produced by man and conferring a high degree of toxicity or deterrence to insects. Many SIs are highly toxic to man and other animals. See Insecticide.

SYNTHETIC PHARMACEUTICAL A drug derived from synthetic chemicals. Cf. Biopharmaceutical.

SYNTHETIC PHEROMONE A synthesized chemical which resembles or mimics the contact or olfactory chemical properties of a natural chemical messenger (pheromone), and which elicits predictable behavioural patterns in the receiver of that synthetic pheromone. Synthetic pheromones are typically used to attract members of a pest Species or to interrupt chemical communication between members of a pest Species. Synthetic Pheromones mimic aggregation pheromones, alarm pheromones and sex pheromones. Synthetic pheromone compounds currently available include APM-Ropes®, Codelure, Disparlure, Farnesene, Gossyplure, Isomate-M, Lycolure, Muscalure, Stirrup M and Trimedlure.

SYNTHETIC PYRETHROIDS A group of synthetic insecticides whose members display low mammalian toxicity and which are used as surface sprays to control household and agricultural pests. Examples include Acrinathrin, Allethrin, Alphacypermethrin, Bifenthrin, Bioresmethrin, Cycloprothrin, Cyfluthrin, Cypermethrin, Deltamethrin, Esfenvalerate, Fenpropathrin, Fenvalerate, Flucythrinaate, Kadethrin, Lambda-Cyhalothrin, Permethrin, Phenothrin, Resmethrin, Tau-Fluvalinate, Tefluthrin, Tetramethrin, Tralomethrin. See Insecticide. Botanical Insecticide; Carbamate; Insect Growth Regulator; Inorganic Insecticide; Organochlorine; Organophosphorus; Pyrethrin. Cf. Pyrethroid.

SYNTHILIPSIS Noun. (Greek, *syntithenai* = put together.) The basal constriction of the Vertex of the head (notocephalon) in the heteropterous Genus *Notonecta*, the nearest approach of the eyes to each other above (Kirkaldy).

SYNTHORAX Noun. (Greek, *syn* = with; thorax = breastplate. Pl., Synthoraxes.) Odonata: The fused Mesothorax and Metathorax.

SYNTHRIN® See Resmethrin.

SYNTONOPTERIDAE Handlirsch 1911. Plural Noun. A small Family of medium sized to large fossil insects known from Upper Carboniferous deposits at Mazon Creek, Illinois. Ordinal placement uncertain.

SYNTYPE Noun. (Greek, *syn* = together; *typos* = pattern. Pl., Syntypes.) Any one of two or more specimens upon which a Species is founded when no holotype has been selected (Banks & Caudell). All the specimens except the Type, upon which a Species is based and described

(English usage) (Jardine). Syn. Cotype. See Type. Cf. Holotype; Lectotype; Paratype. Rel. Nomenclature; Taxonomy.

SYRINGE Noun. (Greek, *syringx* = pipe. Pl., Syringes.) Hemiptera: A chamber into which the salivary ducts open and by means of which the secretion is forced forward between the Setae or lancets.

SYRINGOGASTRIDAE Prado 1969 A small Family of Neotropical schizophorous Diptera assigned to Superfamily Diopsoidea.

SYRPHIDAE Plural Noun. Drone Flies; Hover Flies; Flower Flies. Cosmopolitan Family including about 5,000 Species of cyclorrhaphous Diptera assigned to Superfamily Syrphoidea. Body coloration variable but often with yellow and black markings; some Species mimic wasps or bees. Distinguished primarily on wing venational characteristics: First posterior cell closed before margin of wing, crossveins enclosing first Posterior and Discal Cells parallel to wing margin and near margin; spurious vein ('Vena Spuria,' except in few Genera) passes through first basal cell and first posterior cell. Syrphids show affinity to Stratiomyiidae by spined Scutellum of some Species and poorly developed face in others. Actual relationships of Family not adequately shown; higher classification of Family questioned. Adults swift fliers, some hover; important pollinators of plants. Larvae entomophagous, phytophgaous and scavengers. Most Species phytophagous; many genera composed of aphid-feeding Species and important to biological control. Some Species myrmecophilous. Entomophagous Species all in Syrphinae and predaceous upon Hemiptera, primarily Aphididae and some Thysanoptera.

SYRPHOIDEA A Superfamily of cyclorrhaphous Diptera within the Series Aschiza and including Pipunculidae (Dorilaidae, Dorylaidae) and Syrphidae.

SYSTELLOGNATHA Noun. A Suborder of Plecoptera. Naiads nocturnal, generally carnivorous; eat some detritus and plants, especially during first instar. Adults do not emerge during winter; most adults do not feed. Cf. Euholognatha.

SYSTEM Noun. (Greek, *synistanai,* Latin & Greek, *systema* = to bring together, to combine. Pl., Systems.) 1. An order of arrangement after a distinct plan, or method. 2. The body when considered as a functional unit. 3. Body organs that cooperate to perform one of the vital functions of life. See Respiratory System; Reproductive System; Alimentary System. Cf. Tract.

SYSTEM® See Dimethoate.

SYSTEMATIC NAME In the sense of the Code of Zoological Nomenclature, the designation by which the actual object is known, not the name of our conception or idea of such object.

SYSTEMATICS Noun. (Greek, *systema* = a whole made of several parts.) A subdiscipline of biology which deals with the classification of organisms. Organisms arranged in definite order or arranged according to a system; the order or system are not necessarily based on common descent. See Classification. Cf. Taxonomy.

SYSTEMATIST Noun. (Greek, *systema* = a whole made of several parts. Pl., Systematists.) 1. Zoology: A person who studies Taxa and develops classifications of forms or groups according to biological relationships and/or phenetic affinities. Cf. Taxonomist.

SYSTEMIC INSECTICIDE 1 An insecticide administered to an organism (plant or animal), translocated throughout the organism and poisonous to the insects feeding on that organism. Most SIs are organophosphates. See Organophosphate. 2. An insecticide which enters a plant through leaves, branches or roots and spreads throughout the plant to confer protection from insects.

SYSTOLE Noun. (Greek, *systole* = drawing together. Pl., Systoles.) The regular contaction of the Heart which sends the blood outward. Cf. Diastole.

SYSTOLIC Adj. (Greek, *systole* = drawing together; *-ic* = of the nature of.) Descriptive of or pertaining to the Systole.

SZEDLACZEK, STEFAN (1846–1907) (Aigner 1907, Rovart. Lap. 14: 161–162.)

SZEKESSY, VOLMOS (1907–1970) (Anon. 1971, Folia ent. hung. 24: 5–16, bibliogr.)

SZEPLIGETI, GYÖZÖ VIKTOR (1855–1915) (Csiki 1915, Rovart. Lap. 22: 141–147, bibliogr.; Soldanski 1916, Dt. ent. Z. 1916: 226–227.)

SZIDAT'S RULE An hypothesis which asserts that phylogenetically primitive parasites develop at the expense of primitive hosts. Cf. Eichler's Rule; Emery's Rule; Farenholtz' Rule; Manter's Rules.

SZULCZEWSKI, JERZY WOJCIECH (1879–1969) (Wroblewski 1970, Polskie Pismo ent. 40: 211–216, bibliogr.)

SZUMKOWSKI, WACLAW (1882–1967) (Kennedy 1968, Proc. R. ent. Soc. Lond. (C) 32: 60.)

TAABOR (TAUBER, J), HENRY (1894–1970) (Novak 1971, Acta ent. bohemoslovaca 68: 429–430.)

TABANID Adj. A member of the Tabanidae (Diptera); resembling the Tabanidae.

TABANIDAE Plural Noun. Buffalo Flies; Clegs; Deer Flies; Elephant Flies; Gad Flies; Green Heads; Hippo Flies; Horse Flies; Stouts. A cosmopolitan Family of brachycerous Diptera consisting of ca 3,800 Species assigned to Division Orthorrhapha. Adult large, stout-bodied; head large, much wider than long; compound eyes very large (male holoptic) with brilliant colours in spots, horizontal or zigzag bands; eye colour and pattern disappears after death; palps with two segments; Antennae porrect with first flagellar segment enlarged. Tabanids show affinities with Athericidae and Pelocorhynchidae from which they are separated by coarser antennal annulations, first posterior cell open in wing margin and presence of large upper and lower calypters. Eggs laid by most Species in clusters (200–1,000) upon vegetation overhanging water, or on nearby rocks; eggs deposited in 1–4 layers. Oviposition in some Species only after female has taken blood meal. Eclosion occurs with aid of hatching spines ca 4 days following oviposition. Larvae aquatic or semiaquatic; first instar moults rapidly into second instar; second instar positively phototactic, moves over substrate, does not feed, and moults within 3–6 days; third instar negatively phototactic and burrows into substrate; number of moults variable with as many as 11 instars; larvae feed on animal life associated with habitat. Pupa obtect; pupation typically in drier habitat. Adults of most Species diurnal. Males form swarms typically in morning; virgin females enter swarm, initiate copulation in air and are inseminated on ground. Inseminated females of most Species suck blood from warm-blooded vertebrates; a few attack cold-blooded vertebrates; most Species anautogenous, a few Species autogenous for first clutch of eggs and anautogenous for subsequent clutches; some Genera with non-blood-feeding Species. Males and non-blood-sucking females take nectar from flowers. Tabanids of some medical and veterinary importance. In large numbers they take significant amounts of blood from livestock; serve as mechanical vectors of Anthrax and Anaplasmosis. Tabanids vector the filarial worms *Loa loa* Guyot and bacterial Tularaemia to humans, *Dirofilaria roemeri* to macropod marsupials and *Elaeophora schneideri* to sheep. Also implicated in transmission of sporozoans in turtles and trypanosomes in livestock.

TABANOIDEA A Superfamily of orthorrhaphous Diptera including Acroceridae (Acroceratidae, Cyrtidae, Henopidae, Oncodidae), Athericidae, Coenomyiidae, Nemestrinidae, Pantophthalmidae, Pelecorhynchidae, Rhagionidae (Leptidae), Stratiomyidae (Chiromyzidae, Stratiomyidae), Tabanidae, Vermeleonidae, Xylomyidae (Xylomyiidae) and Xylophagidae (Coenomyiidae, Erinnidae, Rachiceridae).

TABLELAND PASTURE SCARAB *Antitrogus morbillosus* (Blackburn) [Coleoptera: Scarabaeidae] (Australia).

TACHIKAWA, TETSUSABURO Japanese academic (Ehime University) and taxonomic specialist in parasitic Hymenoptera Family Encyrtidae.

TACHINA FLIES See Tachinidae.

TACHINIDAE Plural Noun. Parasite Flies; Tachinid Flies; Tachina Flies. A cosmopolitan Family of muscoid Diptera. The second largest Family of Diptera, exceeded only by Tipulidae in number of Species. Separated from other Muscoidea by presence of Hypopleural and usually Pteropleural Bristles, wing vein M1 nearly always bent forward before apex and well developed Postscutellum. Adults generally covered with stout bristles or Setae; typically drab coloured, but some Species metallic or colourful. Radically different tribal and generic classifications have been proposed for tachinids. Larva robust with thin, transparent Integument and minute spines on each segment. All Species are endoparasites of other insects, particularly sawflies, larval Lepidoptera and adult Coleoptera and Hemiptera. Tachinids are biologically diverse with some Species larviparous. Oviposition site-specificity in some Species; host feeding observed in some Species. Egg types include membranous, macrotype, microtype and pedicellate. Some Species are employed in biological control of agricultural pests.

TACHINIDS See Tachinidae.

TACHINISCIDAE Plural Noun. A small Family of schizophorous Diptera assigned to Superfamily Tephritoidea, know from the Neotropical and Ethiopian Realms.

TACHYGENESIS Noun. (Greek, *tachys* = swift; *genesis* = descent. Pl., Tachygeneses.) The compression in time or reduction of development in immature stages of insects.

TACTILE Adj. (Latin, *tactilis* = that may be touched.) Descriptive of or pertaining to the sense of touch; used for touching.

TACTILE SENSE Touch; perception by physical contact.

TACTILE SENSILLA Sensillum trichodeum; an organ of touch.

TACTOCHEMICAL Adj. (Latin, *tachis* = touching; *-alis* = characterized by.) Descriptive of or pertaining to perception by touch and chemical stimuli.

TAENIDIUM Noun. (Greek, *taenia* = band, ribbon; *-idion* = diminutive. Pl., Taenidia.) 1. Spiral cuticular thickenings of the respiratory system which prevent collapse of the RS during ventilation. 2. Markings such as surface sculpture on the Integument.

TAENIOPTERYGIDAE Plural Noun. Willowflies. A Holarctic Family of Plecoptera including 13 Genera and ca 100 described Species. Adults dark coloured with long Antennae and short Cerci; larvae with respiratory gills at base of legs.

TAFEL-HUMZIKER, KARL (1872–1964) (Wyniger 1964, Mitt. ent. Ges. Basel 14: 29.)

TAFGOR® See Dimethoate.

TAGMA Noun. (Greek *tagma* = corps. Pl., Tagmata.) A group of successive segments (somites) of a metamerically organized animal that form a distinct region of the body (head, Thorax, Abdomen).

TAGMATIZATION Noun. (Greek, *tagma* = corps; English, *-tion* = result of an action. Pl., Tagmatizations.) The active process of Tagmosis. See Tagmosis.

TAGMOSIS Noun. (Greek, *tagma* = corps; *-osis* = act of. Pl., Tagmoses.) 1. The centralization of process. 2. An evolutionary phenomenon involving the organizational process of the insect body in which body somites fuse to form groups of segments that become distinct Tagma (Head, Thorax, Abdomen). See Morphology; Tagma. Cf. Oligomerization. Rel. Groundplan.

TAHITIAN COCONUT WEEVIL *Diocalandra taitensis* (Guérin-Méneville) [Coleoptera: Curculionidae].

TAIGA Noun. (Russian, *taigma* = rocky.) Boreal coniferous forests of Russia, especially in Siberia.

TAIL Noun. (Old English, *taegel* = tail. Pl., Tails.) 1. A slender, rather elongated terminal segment of the Abdomen. See Cauda. Cf. Trunk. 2. The Cauda in plant lice. 3. Some Lepidoptera and Neuroptera: Elongated processes on the hindwings. 4. Something that resembles a tail in shape or position.

TAILED CUPID *Everes lacturnus australis* Couchman [Lepidoptera: Lycaenidae] (Australia).

TAILED EMPEROR BUTTERFLY *Polyura sempronius* (Fabricius) [Lepidoptera: Nymphalidae] (Australia).

TAILED GREEN-BANDED BLUE *Nacaduba cyanea arinia* (Oberthur) [Lepidoptera: Lycaenidae] (Australia).

TAILED SWITCH LOUSE *Haematopinus quadripertusus* Fahrenholz [Phthiraptera: Haematopinidae] (Australia).

TAIMYRIAN AMBER Fossilized resin containing a rich assortment of insects that are referrable to the mid-Cretaceous. Found in the Taimyrian district of northern Russia. See Amber; Fossil. Cf. Baltic Amber; Burmese Amber; Canadian Amber; Chiapas Amber; Dominican Amber; Lebanese Amber.

TAIT, ROBERT (1869–1939) (Anon. 1939, Arb. morph. taxon. Ent. Berl. 6: 349; Crabtree 1939, Entomologist 72: 152.)

TAIT, WILLIAM (–1904) (Anon. 1904, Ann. Scot. nat. Hist. 1904: 137–138.)

TAKACHIHO, NOBUMARO (1865–1950) (Hasegawa 1967, Kontyû 35 (Suppl.): 47, bibliogr.)

TAKAHASHI, HIDEO (1906–1942) (Hasegawa 1967, Kontyû 35 (Suppl.): 49, bibliogr.)

TAKAHASHI, RYOICHI (1898–1963) (Anon. 1963, Mushi 37: 167–190, bibliogr. List of specimens named in his honour; Hasegawa 1967, Kontyû 35 (Suppl.): 49–50, bibliogr.

TAKAHASHI, SUSUMU (1887–1935) (Hasegawa 1967, Kontyû (Suppl.): 35: 48–49, bibliogr.)

TAKANO, TAKZO (1884–1964) (Hasegawa 1967, Kontyû 35 (Suppl.): 47–48, bibliogr. pl. 3, fig. 9.)

TAKASHIMA, HARUO (1907–1962) (Hasegawa 1967, Kontyû 35 (Suppl.): 46–47.)

TAKE-OFF The initial phase of insect flight.

TAKEUCHI, KICHIZO (Issiki 1968, Kontyû 36: 299.)

TAKEUCHI, MOTOFUMI (1868–1923) (Hasegawa 1967, Kontyû 35 (Suppl.): 50.)

TAKIZAWA, MOTAMU (1907–1963) (Hasegawa 1967, Kontyû 35 (Suppl.): 50, bibliogr. pl. 4, fig. 9.)

TAKTIC® See Amitraz.

TALBOT, GEORGE (1882–1952) (Riley 1952, Entomologist 85: 191–192; Remington 1953, Lepid. News 7: 24.)

TALCORD® See Permethrin.

TALLANT, W N (1856–1905) (Anon. 1905, Ent. News 16: 96; Osborn 1937, *Fragments of Entomological History,* 394 pp. (223), Columbus, Ohio.)

TALON® See Chlorpyrifos.

TALSTAR® See Bifenthrin.

TALUNEX® See AIP.

TALUS Noun. (Latin, *talus* = ankle. Pl., Tali.) 1. The ankle. 2. The apex of the Tibia to which the Basitarsus is attached.

TAM® See Methamidophos.

TAMANOX® See Methamidophos.

TAMARGALAY Common name for Philippine Neem Tree in Myanmar.

TAMARIND WEEVIL *Sitophilus linearis* (Herbst) [Coleoptera: Curculionidae].

TAMARIX LEAFHOPPER *Opsius stactogalus* Fieber [Hemiptera: Cicadellidae].

TAMARON® See Methamidophos.

TAME® See Fenpropathrin.

TANAKA, HUSATARO (1862–1950) (Hasegawa 1967, Kontyû 35 (Suppl.): 51, bibliogr.)

TANAKA, TAKAYOSHI (1897–1927) (Hasegawa 1967, Kontyû 35: (Suppl.): 50–51, bibliogr.)

TANAKA, YOSHIO (1838–1916) (Hasegawa 1967, Kontyû 35 (Suppl.): 51–52, bibliogr. (pl. 1, fig. 1.)

TANAOCERIDAE Plural Noun. Long-horned Grasshoppers. A rare Family of nocturnal, wingless caeliferous Orthoptera. Found in southwestern USA.

TANAOSTIGMATIDAE Plural Noun. A small Family (ca 80 Species) of apocritous Hymenoptera assigned to Chalcidoidea. Widespread in distribution; most abundant in tropical areas. Important anatomical features: Antenna with 13 segments including two ring segments and three club segments, male Antenna sometimes ramose; Clypeus typically bilobed; Mandible with three teeth; prepectus enlarged and developed (swollen) anteriad; Parapsidal Sutures sinuate, usually complete; Mesopleuron large, convex, without middle Coxal furrow; middle Coxa bears anterior articulation posterior of Mesopleuron midline; middle tibial spur enlarged; forewing Marginal Vein much longer than wide, typically longer than Submarginal or Stigmal Veins; Stigmal Vein usually curved with Stigma swollen; Linea Calva (Speculum) present, usually open posteriad; squamiform Setae often present on body. All tanaostigmatids apparently gall formers on Fabaceae, Myrtaceae, Rhamnaceae, Polygonaceae and allied Families. Galls typically monothalamous and formed on most plant parts. Ovarian egg encyrtiform, a feature shared with Encyrtidae.

TANDEM CALLING Social Insects: The release of a pheromone by a leader which recruits a nestmate for tandem running.

TANDEM FLIGHT Aerial copulation in caddis flies. Cf. Phoretic Copulation.

TANDEM RUNNING Social Insects: Workers moving with one individual behind another, head to Abdomen tip, during exploration or recruitment.

TANGENTIAL Adj. (Latin, *tangere* = to touch; *-alis* = characterized by.) Touching; set in or meeting at a tangent. A term often descriptively applied to ornamentation and processes.

TANGIUM Noun. (Latin, *tangere* = to touch; *-ium* = diminutive > Greek, *-idion*.) Hymenoptera: See Ventral Ramus.

TANGLE-VEIN FLIES See Nemestrinidae.

TANGLE-WING FLY *Neorhynchocephalus sackenii* [Diptera: Nemestrinidae].

TANGORECEPTOR Noun. (Latin, *tangere* = to touch; *receptor* = a receiver.) An organ of touch having a single sense cell.

TANI, SADAKO (1885–1911) (Hasegawa 1967, Kontyû 35 (Suppl.): 52.)

TANNER, MATHIS CHARLES (1891–1951) (Knowlton 1958, J. Econ. Ent. 51: 117.)

TANNER, TAUNO RUBEN (1905–1940) (Anon. 1940. Suomen hyönt. Aikak. 6: 4–7.)

TANNINS Plural Noun. (Latin, *tannum* = oak bark.) Water soluble phenolic compounds which occur in many vascular plants. Tannins exhibit a strong tendency to absorb oxygen and leave the products strongly coloured, usually brown; abundant in some plant galls. Generally believed to act as a deterrent to foliage feeding by herbivores. (See Martin *et al.*, 1986. J. Chem. Ecol. 11; 485–494.) Cf. Lignin; Saponin.

TANQUARY, MAURICE COLE (1881–1944) (Riley 1944, Science 100: 539–540; Anon. 1945, J. Econ. Ent. 38: 504.)

TANYDERIDAE Plural Noun. Primitive Crane Flies. A small, cosmopolitan Family of nematocerous Diptera assigned to Superfamily Tanyderoidea; presently contains about 50 extant Species and six fossil Species. Immature stages aquatic to semiaquatic in association with soil or decomposing logs in streams. Adult males may congregate in swarms during evening.

TANYPEZIDAE Plural Noun. A small Family of acalypterate Diptera assigned to Superfamily Diopsoidea.

TANYPEZOIDEA A Superfamily of schizophorous Diptera including Diopsidae, Nothybidae, Psilidae and Tanypezidae.

TAPESTRY BEETLE *Attagenus unicolor* [Coleoptera: Dermestidae].

TAPESTRY MOTHS *Trichophaga tapetzella* (Linnaeus) [Lepidoptera: Tineidae]: A cosmopolitan pest of hair, fur and feathers.

TAPETUM Noun. (Latin, *tapete* = carpet. Pl., Tapeta.) 1. A reflecting surface within an eye, formed of pigment or densely massed Tracheae (Snodgrass). In the eye of night-flying insects, a structure which reflects the light that enters the eyes and causes them to shine with a ruby or golden appearance (Imms).

TAPHONOMY Noun. (Greek, *taphos* = the act of burying; *-nomy* = the systematized knowledge of.) The study of the process of fossilization and the sequence of events impacting organisms after death.

TAPINOMA-ODOUR Dolichoderine ants: The peculiar rancid butter odour produced by a secretion from anal glands (Wheeler).

TAPPES, JOSEPH CABRIEL NICOLAS EDOUARD (1815–1885) (Dimmock 1888, Psyche 5: 36; Anon. 1889, Abeille (Les ent. et leurs écrits) 26: 274–275, bibliogr.)

TAPROBANE Noun. (Sanskrit, *tamraparni* = pool covered with red lotus.) Ancient (Greek) and Medieval (Latin) name for Sri Lanka.

TARANTULA HAWK Aculeate wasps of the Pompilidae, particularly New World representatives of the genera *Pepsis* and *Hemipepsis*. See Pompilidae.

TARBAT, JAMES EDWARD (–1937) (Anon. 1937, Arb. morph. taxon. Ent. Berl. 4: 160; St E.W.G. 1937. Entomologist 70: 96.)

TAREDAN® Cadusafos.

TARGIONI-TOZZETTI, ADOLPHO (1823–1902) Italian academic (Professor of Zoology and Comparative Anatomy, University of Florence) Primary contributions through work with coccoids, *Phylloxera* and mites. Influential in founding Agricultural Entomology Station at Florence. (Bargagli *et al.* 1902, Boll. Soc. ent. ital. 34: 113–117, 199–233; Fowler 1902, Proc. ent. Soc. Lond. 1902: lix–lx; Howard 1930, Smithson. misc. Collns. 84: 250–259.)

TARNANI, IVAN KONSTANTINOVICH (1865–1930) (Strawinski 1930, Polskie Pismo ent. 9: 293–295; Troshchanin 1967, Ent. Obozr. 46: 480–482, bibliogr. Translation: Rev. Ent., Wash. 46: 283–284, bibliogr.)

TARNISHED PLANT BUG *Lygus lineolaris* (Palisot de Beauvois) [Hemiptera: Miridae]: A highly polyphagous, widespread insect and pest of numerous agricultural crops throughout North America. More than 300 Species of plants in 55 Families serve as hosts for TPB; important host plants include alfalfa seed, hay, cotton, strawberries, vegetables and numerous weeds. Adult 5–7 mm long, pale brown with small white, yellowish and black spots irregularly distributed over body. TPB eggs elongate, slightly curved, operculate and inserted into plant tissue up to the Operculum; eclosion occurs within 10 days. Neonate nymph ca 1 mm long, pale green; nymphs moult five times; 3–5 generations per year; overwinters as adult or perhaps nymph. See Miridae.

TARO LEAFHOPPER 1. *Tarophagus proserpina* (Kirkaldy) [Hemiptera: Delphacidae] (Australia). Syn. *Tarophagus colocasiae* (Matsumura) [Hemiptera: Delphacidae]. 2. *Tarophagus proserpina taiwanensis* Wilson [Hemiptera: Delphacidae].

TARRIEL, ERNEST (1861–1882) (Anon. 1882, Feuille jeun. Nat. 12: 75.)

TARSAL Adj. (Greek, *tarsos* = sole of the foot; *-alis* = pertaining to.) Descriptive of or pertaining to the Tarsomeres of the insect foot.

TARSAL CLAW The claw or claws at the apex of the Pretarsus. See Tarsus.

TARSAL COMB Corixidae: A row of small, elongated or rounded, chitinous pegs or teeth on the medial surface near the upper margin of the pala or expanded Tarsus of the anterior leg in the males. A structure which forms part of the stridulatory device.

TARSAL FORMULA The sequential number of segments comprising the Tarsus of each insect leg. TF sometimes diagnostically important and provided in taxonomic keys or descriptions. Conventionally given as three numbers (separated by dashes, commas or hyphens) with the foreleg number first, middle leg number second and the hind leg number last. Example: 5-5-4; 3-3-3. Cf. Palpal Formula.

TARSAL LOBES Membranous appendages arising from the ventral surface of the tarsal segments in some Coleoptera.

TARSAL PULVILLI See Euplantulae.

TARSATION Noun. (Greek, *tarsos* = sole of the foot; English, *-tion* = result of an action.) The act by which insects touch a substrate or other organisms with the Tarsi presumably to explore or communicate. Cf. Antennation; Palpation.

TARSOMERE Noun. (Greek, *tarsos* = sole of the foot; *meros* = part. Pl., Tarsomeres.) Any of the subsegments of the Tarsus in the insect foot (Snodgrass). Tarsomeres lack intrinsic musculature. Alt. Tarsite. See Basitarsus; Pretarsus.

TARSONEMIDA Plural Noun. A group of Actinotrichid mites.

TARSULE Noun. (Greek, *tarsos* = sole of the foot. Pl., Tarsuli.) The Pretarsus or distal region of the Tarsus. The tarsule is regarded as homologous with the dactylopodite of a Crustacean limb (Crampton). Alt. Tarsulus. See Unguitractor.

TARSUS Noun. (Greek, *tarsos* = sole of the foot. Pl., Tarsi.) 1. The insect's foot. 2. A jointed appendage basally attached to the Tibia and apically bears claws and other structures. Typically, Tarsus consists of 1–5 'segments' or joints; tarsal segments are not segments in a strict morphological sense because they do not contain intrinsic musculature. Consequently, subdivisions of a Tarsus are sometimes called Tarsomeres. Zoraptera display Tarsi with two segments and most Hymenoptera display five segments. Some immature insects (Trichoptera: Xiphocentronidae) have Tibiae and Tarsi of all legs fused. Most individual insects show a constant number of tarsal segments on all legs. However, many Coleoptera show different numbers of tarsal segments on different pairs of legs. Basal tarsal segment is called the Basitarsus or Metatarsus. Basitarsus frequently is invested with sensory modifications or grooming structures such as the Strigil. Apical segment of Tarsus is called the Pretarsus. Tarsus is functionally diverse: Parts used to collect pollen; ventral surface of tarsomeres in some Orthoptera bear peg-like Euplantulae (tarsal pulvilli). Foreleg Basitarsus of Embioptera distended by glandular cells used for the production of silk. Tarsus also serves as an accessory copulatory structure in some insects. Palar pegs on foreleg Tarsi of male Corixidae enable them to grasp female's Hemelytra. See Leg. Cf. Coxa; Femur; Pretarsus; Tarsomeres; Trochanter. Rel. Tarsal Formula.

TASAR SILKWORM *Antheraea mylitta* Drury [Lepidoptera: Saturniidae].

TASCHENBERG, ERNST LUDWIG (1818–1898) (Anon. 1898, Ent. News 9: 80; Anon. 1898, Nature 57: 300–301; Taschenberg 1918, Leopoldina 54: 13–16; Howard 1930, Smithson. misc. Collns. 84: 262–263.)

TASCHENBERG, ERNST OTTO WILHELM (1854–1922) (Anon. 1922, Ent. News 33: 256.)

TASK® See Dichlorvos.

TASMANIAN BROWN *Argynnina hobartia* (Westwood) [Lepidoptera: Nymphalidae] (Australia).

TASMANIAN CAVE SPIDER *Hickmania troglodytes* (Higgins & Petterd) [Araneida: Austrochilidae] (Australia).

TASMANIAN CUSHION PLANT MOTH *Nemotyla oribates* Nielsen, McQuillan & Common [Lepidoptera: Oecophoridae] (Australia).

TASMANIAN CUTWORM *Dasygaster padockina* (Le Guillou) [Lepidoptera: Noctuidae] (Australia). See Cutworms.

TASMANIAN EUCALYPTUS LEAF BEETLE *Chrysophtharta bimaculata* (Olivier) [Coleoptera: Chrysomelidae] (Australia). TELB is most serious economic pest of production of eucalyptus in Tasmania. Defoliation leads to reduced tree growth and often tree death. See Chrysomelidae.

TASMANIAN FUNNELWEB SPIDER *Hadronyche venenata* (Hickman) [Araneida: Hexathelidae] (Australia).

TASMANIAN HAIRY CICADA *Tettigarcta tomentosa* White [Hemiptera: Tettigarctidae] (Australia).

TASMANIAN PARALYSIS TICK *Ixodes cornuatus* Roberts [Acari: Ixodidae] (Australia).

TASMANICA SKIPPER *Pasma tasmanica* (Miskin) [Lepidoptera: Hesperiidae] (Australia).

TASSART, OLIVE FLORENCE (–1953) (I.R.B. 1953, Entomol. mon. Mag. 89: 89.)

TASTE CUPS Specialized pits or cuticular depressions (with or without a peg or seta) connected with ganglionated nerve cells. TC occur on the mouth appendages and function in the sense of taste.

TASTER Noun. (Old Fench, *tast.*) A Palpus or contact chemoreceptor.

TATCHELL, LEONARD (1877–1963) (Worms 1963, Entomologist's Rec. J. Var. 75: 140–141; Wigglesworth 1964, Proc. R. ent. Soc. Lond. (C) 28: 58.)

TATERPIX® See Chlorpropham.

TAUARES, JOACHIM DA SILVA (1866–1931) (Dusmet y Alonso 1919, Boln. Soc. ent. Esp. 2:193–194; Navas 1931, Atti Accad. Lincei 84: 652–653; Navas 1931, Boln. Soc. ent. Esp. 14: 98–99; Houard 1932, Marcellia 27: 107–119.)

TAUBER, J See Taabor, H.

TAU-FLUVALINATE Noun. A synthetic-pyrethroid insecticidal compound with contact activity and used as a stomach poison applied to ornamental plants and turf in many countries. Minor phytotoxicity. Trade names include: Aspitan®, Klartan®, Mavrik®, Yardex®. See Synthetic Pyrethroids.

TAUTONOMY Noun. (Greek, *tauto* = the same; *onoma* = name.) Double-naming. A term applied to one nominal Species with the same specific epithet and generic name. *e.g. Bison bison.*

TAWNY Adj. (Old French, *tanné.*) Brownish yellow; resembling colour of a tanned hide.

TAWNY MOLE CRICKET *Scapteriscus vicinus* Scudder [Orthoptera: Gryllotalpidae]: A Species endemic to Argentina, Uruguay, Paraguay, southern Brazil; adventive to southeastern USA where it may be a minor pest. TMC body robust with short, ovoid Pronotum; fore tibial Dactyls nearly touching; Ocelli small, circular in outline shape, widely separated. See Gryllotalpidae; Mole Crickets. Cf. Changa Mole Cricket; Imitator Mole Cricket; Short-Winged Mole Cricket.

TAXA Noun. See Taxon.

TAXIS Noun. (Greek, *taxis* = arrangement. Pl., Taxes.) A reflex movement exhibited by an organism in response to an external stimulus (*e.g.* light, temperature, air current). See Orientation. Cf. Kinesis; Tropism. Rel. Aerotaxis; Anemotaxis; Chemotaxis; Geotaxis; Menotaxis; Osmotaxis; Phototaxis; Rheotaxis; Rotaxis; Scototaxis; Strophotaxis; Telotaxis; Thermotaxis; Thigmotaxis; Tonotaxis; Tropotaxis.

TAXON Noun. (Greek, *taxis* = arrangement. Pl., Taxa.) Any definite entity of classification formally recognized by taxonomists. Taxa include the Species, Genus, Family, Order, Class, Phylum, Kingdom and interpolated categories. See Classification.

TAXONOMIC Adj. (Greek, *taxis* = arrangement; *nomos* = law; *-ic* = consisting of.) Pertaining to systematics or classification. Alt. Taxonomical.

TAXONOMIC SPECIES Forms deemed discrete because of a consensus of structural (morphological) characters (Kinsey).

TAXONOMIST Noun. (Greek, *taxis* = arrangement; *nomos* = law. Pl., Taxonomists.) A person who works in the identification and description of organisms or groups of organisms. Cf. Systematist.

TAXONOMY Noun. (Greek, *taxis* = arrangement; *nomos* = law. Pl., Taxonomies.) 1. The practice of identifying, describing and naming organisms based on the rules of zoological nomenclature. See Classification. Cf. Systematics.

TAYLOR, CHARLOTTE DE BERNIER (1806–1867) (Anon. 1899, National Cyclopedia of American Biography. 2: 165; Weiss 1936, *Pioneer Century of American Entomology,* 320 pp. (186), New Brunswick.)

TAYLOR, FRANK HENRY (1886–1945) (Browne 1946, Proc. Linn. Soc. N.S.W. 71: ii; Hale-Carpenter 1947, Proc. R. ent. Soc. Lond. (C) 11: 62.)

TAYLOR, GEORGE WILLIAM (1851–1912) (Anon. 1904, Can. Ent. 36: 1–2; Hanham *et al.* 1912, Proc. ent. Soc. Br. Columb. 2: 1–4; Hatch 1949, *Century of Entomology in the Pacific Northwest.* viii + 43 pp., 6–7.)

TAYLOR, JOHN (–1920) (Anon. 1920. Entomologist's Rec. J. Var 32: 194.)

TAYLOR, JOHN KIDSON (1839–1922) (Standon 1922, Lancs. Chesh. Nat. 45: 34–36; Anon. 1923, Entomol. mon. Mag. 59: 21.)

TAYLOR, JOHN SNEYD (1900–1973) (G.A.H. 1974, J. ent. Soc. sth. Afr. 37: 199; J[acobs] 1974, Entomologist's Rec. J. Var. 86: 29–30.)

TAYLOR, THOMAS HUGH COLEBROOK (1901–1972) (Adamovic 1972, Acta ent. jugosl. 7: 93–94; Haskell 1972, Nature 237: 415.)

TBE See Tick-Borne Encephalitis.

TCNB® See Tecnazene.

TEA RED SPIDER MITE *Oligonychus coffeae* (Nietner); *Tetranychus kanzawai* Kishida [Acari: Tetranychidae] (Australia).

TEA SCALE *Fiorinia theae* Green [Hemiptera: Diaspididae].

TEAK SKELETONIZER *Eutectona machaeralis* Walker [Lepidoptera: Pyralidae].

TEAR DROP SPIDER *Argiope protensa* L. Koch [Araneida: Araneidae] (Australia).

TEAR GAS® See Chloropicrin.

TEATREE ITCH MITE *Eutrombicula samboni* (Womersley) [Acari: Trombiculidae] (Australia).

TEATREE MOTH *Catamola marmorea* (Warren) [Lepidoptera: Pyralidae] (Australia).

TEATREE WEB MOTH *Catamola thyrisalis* Walker [Lepidoptera: Pyralidae] (Australia).

TEBUFENOZIDE A benzoyl hydrazine compound {N-*tert*-butyl-*N*-(4-ethylbenzoyl)-3,5-dimethylbenzohydrazide} used as an Insect Growth Regulator, contact insecticide and stomach poison for the control of Lepidoptera. Applied to fruit tree crops, forests grapes, rice and vegetables in some countries. Tebufenozide is not effective on adults and is most effective when applied during egg stage. Apparently, compound is selective on Lepidoptera and does not affect non-target insects. Not registered for use in USA. Trade names include: Confirm®, Mimic® and RH-5992®. See Insect Growth Regulator.

TEBUFENPYRAD See Fenpyrad.

TEBUPIRIMFOS An organic-phosphate (thiophosphoric acid) compound {O,(2-tert-butylpyrimidin-5-yl)-O-ethyl O-isopropyl) phosphorothioate} used as a contact soil insecticide against rootworms, wireworms, grubs and other soil-dwelling pests. Toxic to fishes. Experimental registration in USA. Trade name is Bay Mat 7484®. See Organophosphorus Insecticide.

TECHNICAL-GRADE MATERIAL The relatively pure form of a pesticidal compound that comprises the active ingredient in the final pesticide mixture.

TECKNAR®. A registered biopesticide derived from *Bacillus thuringiensis* var. *israelensis*. See *Bacillus thuringiensis*.

TECNAZENE A chlorinated hydrocarbon {1,2,4,5-tetrachloro-3-nitrobenzene} used as a fungicide and Plant Growth Regulator. No longer registered for use in USA. Trade names include: Arena®, Bygran®, TCNB®, Easytec®, Folosan®, Fumite®, Fusarex®, Hytec®, Nebulin®, Trim®, Turbostore®. See Organochlorine Insecticide. Cf. Insect Growth Regulator.

TECTATE Adj. (Latin, *tectum* = roof; *-atus* = adjectival suffix.) Covered; concealed.

TECTIFORM Adj. (Latin, *tectum* = roof; *forma* = shape.) Roof-shaped; descriptive of structure that is sloping from a medial ridge. Descriptive of wings held at repose in insects such as forewings of cicadas and psocids. See Roof. Rel. Form 2; Shape 2; Outline Shape.

TECTOLOGY Noun. The compound microscope has allowed morphologists to magnify objects and study cellular detail at about 980 diameters with transmitted light. With this level of magnification available to biologists, morphological research of the 18th–19th century shaped a branch of biology which was concerned with structural organic types. Conceptually, this was divided by the German philosopher and biologist Ernst Haeckel (1834–1919) into Promorphology and Tectology. Promorphology considered geometrically the form of an organism and its component parts; Tectology divided an organism into morphons of hierarchical order. Morphons included the categories called Plastids (elementary structural units), Organs (Heart, eye, *etc*.), Persons (functional individuals) and Corms (stocks or colonies.) This view of morphology was soon forgotten by biologists and today Haeckel is probably better known for his Theory of Recapitulation. See Promorphology.

TEDION® See Tetradifon.

TEFLUBENZURON A benzoylurea compound {1-(3,5-dichloro-2,3-difluorophenyl)-3-(2,6-difluorobenzoyl)-urea} used as an Insect Growth Regulator against immatures of many insect Species. Primary targets include: Alfalfa Weevil, Boll Weevil, Codling Moth, Colorado Potato Beetle, Diamondback Moth, European Corn Borer, Fall Armyworm, Gypsy Moth. Compound applied to stone and pome fruits, forestry, cabbage, cotton, grapes, ornamentals, potatoes, sorghum, tobacco and vegetables in some countries. Teflubenzuron is not registered for use in USA. Mode-of-action as Chitin Synthesis Inhibitor; non-systemic action with good residual activity; treated females lay sterile eggs. Trade names include: Dart®, Diaract®, Nemolt®, Nomolt®. See Insect Growth Regulator.

TEFLUTHRIN A synthetic pyrethroid compound used as a soil insecticide in some regions of world. Toxic to fishes. Trade names include: Force®, Forza®, Komet®. See Synthetic Pyrethroids.

TEGES Noun. (Pl., Tegites.) Scarabaeoid larvae: A continuous, dense or sparse, patch of hooked or straight, larger or minute, outward pointing or erect Setae, occupying the posterior part, or almost the whole, of the tenth abdominal venter when the palidium is absent. Alternatively, single and transverse, or paired, longitudinal and short, occasionally divided toward the head into two parts with a median intect field, the campus, between, a component of the raster (Boving).

TEGILLUM Noun. (Pl., Tegilla.) Scarabaeoid larvae: A paired patch on each side of the venter of the tenth abdominal segment, consisting of hooked or straight outwardly pointing or erect Setae on each side of a paired and well developed set of palidia, a component of the raster (Boving).

TEGMEN Noun. (Latin, *tegmen* = covering. Pl., Tegmina.) 1. A covering. 2. The hardened, leathery or horny forewing in Orthoptera, Blattaria and some Hemiptera. Term sometimes employed also in Heteroptera for Hemelytra. See Wing; Wing Modification. Cf. Elytron; Hemelytron. 3. Ring-like or tube-like genitalia in Coleoptera; the Phallobase *sensu* Sharp & Muir.

TEGULA Noun. (Latin, *tegula* = tile. Pl., Tegulae.) The anteriormost independent sclerite associated with the wing base. The Tegula is typically scale-like, articulates with the humeral sclerite and protects the wing base from physical damage. A Tegula is not found in Coleoptera and is absent from the Metathorax in most Orders. A Tegula occurs on the Mesothorax and Metathorax of Plecoptera and Dermaptera. The mesothoracic Tegula is well developed in Hymenoptera, Lepidoptera and Diptera. The Tegula is incorrectly called the Alula in Diptera (Comstock), Parapteron of some authors, Cingulum in Orthoptera and Patagium in Lepidoptera. Syn. Hypopterum. See Epaulette;

Parapteron; Patagium; Pterygoda; Squamula; Tegular Plate. Rel. Wing Articulation.

TEGULAR Adj. (Latin, *tegula* = tile.) Pertaining to the Tegula.

TEGULAR ARMS The internal structures that support the tegular sclerites (Comstock).

TEGULAR PLATES Lepidoptera: Structures of the notum that bear the forewing Tegulae (Comstock).

TEGUMEN Noun. Lepidoptera: The Tergum (Snodgrass) in male genitalia. A structure shaped as a hood or inverted trough, positioned dorsad of the Anus; the Uncus articulates with its caudal margin, derived from the ninth abdominal Tergum (Klots).

TEGUMENT Noun. (Latin, *tegumentum* = covering. Pl., Teguments.) A covering surface of skin. See Integument.

TEGUMENTARY Adj. (Latin, *tegumentum* = covering.) Descriptive of or pertaining to the Integument or covering the Integument.

TEGUMENTARY NERVES A pair of slender nerve-strands arising from the dorsal lobe of the Deutocerebrum and passing to the Vertex (Imms).

TEICH, KARL AUGUST (1838–1908) (Schweder 1909, KorespBl. naturfVer. Riga 52: 1–2.)

TEIJERO, JOSÉ ARIAS (1800–1867) (Dusmet y Alonso 1919, Boln. Soc. ent. Esp. 2: 82.)

TEKKAM® See NAA.

TEKNAR® A biopesticide used for the control of mosquito and blackfly larvae. Active ingredient *Bacillus thuringiensis* var. *israeliensis*. See *Bacillus thuringiensis*. Cf. Agree®; Javelin®. Rel. Biopesticide.

TEKNONYMY Noun. (Greek, *teknon* = child; *onyma* = name. Pl., Teknonymies.) The naming of the parent from the child.

TELA Noun. (Latin, *tela* = web.) Any web-like tissue.

TELARIAN Adj. (Latin, *tela* = web.) Web-spinning.

TELEA Noun. (Latin, *tela* = web.) A web-like tissue.

TELEAFORM LARVA Hymenoptera: In hypermetamorphic forms (*e.g.,* Scelionidae: Proctotrupoidea) an unsegmented, weakly cephalized larva with prominent protuberances or curved hooks at the cephalic extremity, and the body posteriorly prolonged into a caudal process, and having one or more girdles or rings of Setae around the Abdomen. See Hypermetamorphosis; Planidium. Cf. Mymariform Larva; Polypodeiform Larva.

TELEFOS® See Prothoate.

TELEGEUSIDAE Plural Noun. Long Lipped Beetles. A Family of Coleoptera consisting of three Species found in North America. Superficially resemble small staphylinids in that body is somewhat elongate with truncate, short Elytra. Telegeusids are about 5 mm long, Maxillary and Labial Palpi have apical segments distinctively enlarged, tarsal formula 5-5-5, and eyes large and nearly contiguous. Biology poorly known; adults presumably predatory and live under bark. Syn. Telegusidae.

TELEMORPH Noun. (Greek, *telos* = an end part; *morphos* = form. Pl., Telemorphs.) The sexual form or perfect form in the life-cycle of a fungus in which sexual spores (ascospores) are formed after sexual reproduction.

TELENGA, NIKOLAYA ABRAMOVICH (1905–1966) (Dyadechko & Sikura 1967, Ent. Obozr. 46: 916–921, bibliogr. (Translation: Ent. Rev., Wash. 46: 550–553, bibliogr.)

TELEODONT Noun. (Greek, *telos* = an end part; *odontos* = tooth. Pl., Teledonts.) A morph of male lucanid bearing the largest Mandibles. See Amphiodont; Mesodont; Priodont.

TELEOLOGICAL Adj. (Greek, *telos* = end, purpose; *logos* = discourse.) Purposive or purposeful. Pertaining to Teleolgy.

TELEOLOGY Noun. (Greek, *telos* = end, purpose; *logos* = discourse. Pl., Teleologies.) The principle of purposiveness; the science of final causes. The mode of explanation or doctrine that asserts the phenomena of life and the development of life are due to a conscious motive or purpose of some intelligence in the world or beyond the world. Teleology developed by Aristotle and Plato. Opposing view, a mechanical explanation of life and its processes, developed by Democritus and the Atomistic school of philosophy.

TELEPHONE POLE BEETLES See Micromalthidae.

TELESCOPE Noun. (Greek, *tele* = far off; *skopein* = to look at. Pl., Telescopes.) 1. Cylindrical body parts or segments which overlap when retracted and do not overlap when extended. Extension movtivated by hydrostatic pressure or mucles of secondary segmentation. 2. Stages in the development of an animal which are merged or compressed.

TELESCOPIC Adj. (Greek, *tele* = far off; *skopein* = to look at; *-ic* = of the nature of.) Arranged so that one portion of an organ or process may be drawn into another, resembling the joints of a telescope.

TELESIS Noun. (New Latin, completion; *sis* = a condition or state. Pl., Teleses.) Progress planned and accomplished by consciously directed effort. Cf. Chaos.

TELFORD, PAUL E (1922–1950) (Knowlton 1950, J. Econ. Ent. 43: 405.)

TELI CORCULUM The Corcula of the Telum or thirteenth segment of an insect.

TELLE, HANS JOACHIM (–1971) (Weicher 1972. Anz. Schädlingsk. Pflanzenschutz. 45: 62.)

TELOCENTRIC Adj. (Greek, *telos* = end; *kentron* = centre; *-ic* = characterized by.) Pertaining to Chromosomes with a Centromere positioned on an end. Cf. Acentric; Metacentric.

TELOCOPRID Dung beetle. One of four groups of dung beetles that attack an individual dung pat. Telocoprid beetles pack dung into spherical masses, brood balls, that are rolled away from the originating dung pat for burial. The dung pat is eventually fragmented and dispersed laterally and vertically by groups of dung beetles.

TELOFILUM Noun. (Greek, *telos* = end; Latin, *filum* = thread. Pl., Telofila.) Ephemeroptera: An arcane term for a short, median caudal filament (Appendix Dorsalis) (Needham).

TELOGANODIDAE Plural Noun. A numerically small Family of Ephemeroptera known from tropical Africa, the Orient and southeastern Queensland and New South Wales, Australia; regarded as a Subfamily of Ephemerellidae in earlier classifications.

TELOGENESIS Noun. (Greek, *telos* = end; *genesis* = descent. Pl., Telogeneses.). See Acrogenesis.

TELOLECITHAL Adj. (Greek, *telos* = end; *lekithos* = yolk; Latin, *-alis* = characterized by.) Eggs with considerable yolk which has accumulated at one pole. Cf. Alecithal; Centrolecithal.

TELOPHRAGMA Noun. (Greek, *telos* = end; *phragma* = to enclose.) The transverse membrane separating Sarcomeres one from another (Folsom & Wardle), Krause's Membrane.

TELOPODITE Noun. (Greek, *telos* = end; *podos* = foot; *-ites* = constituent. Pl., Telopodites.) The primary 6-segmented shaft of the insect limb distal of the coxopodite. The basal segment of the Telopodite called the first Trochanter. The Basipodite (Snodgrass).

TELOTARSUS Noun. (Greek, *telos* = end; *tarsus* = plantar of the foot. Pl., Telotarsi.) The distal of the two principal subsegments of the Tarsus in Arachnida and Chilopoda (Snodgrass).

TELOTAXIS Noun. (Greek, *telos* = end; *taxis* = arrangement. Pl., Telotaxes.) Orientation as if with a purposive end in view. See Orientation. Cf. Aerotaxis; Anemotaxis; Chemotaxis; Geotaxis; Menotaxis; Osmotaxis; Phototaxis; Rheotaxis; Rotaxis; Scototaxis; Strophotaxis; Thermotaxis; Thigmotaxis; Tonotaxis; Tropotaxis.

TELOTROPHIC Adj. (Greek, *telos* = end; *trephein* = feeding; *-ic* = consisting of.) Pertaining to meroistic-type Ovarioles which maintain Trophocytes (nurse cells) within the apical chamber of the Ovariole. All Heteroptera display a telotrophic Ovariole. Coleoptera apparently have a mixture of Polytrophic Ovarioles and Telotrophic Ovarioles. A few Neuroptera, such as Species of *Sialis,* may have telotrophic Ovarioles. Syn. Acrotrophic Ovariole. See Ovariole. Cf. Polytrophic. Rel. Panoisitic Ovary.

TELSON Noun. (Greek, *telson* = boundary. Pl., Telsons.) 1. The primitive terminal body segment in arthropods which contains the Anus. Telson corresponds to the periproct of anellids and is probably not a true body somite (Snodgrass). 2. The terminal region of the Abdomen, found in the embryos of many insects, but rarely in the adult (Imms). 3. The terminal part of the insect Abdomen which bears the Anus (Wardle). 4. Coccids: The single sclerite representing the continuous lateral pilacerores of the meson of the eighth segment; the Postanal Plate *sensu* MacGillivray.

TELUM Noun. (Greek, *telos* = end.) A last abdominal segment of insects.

TEMEGUARD® See Temephos.

TEMEPHOS An organic-phosphate (dithiophosphate) compound {O,O',-(thiodi-4-1-phenylene)-O,O,O,O-tetramethyl phosphorodithioate} used as a selective insecticide for control of aquatic nematocerous Diptera larvae (including mosquitoes, midges, sandflies and punkies). Used as public-health insecticide in many countries. Compound applied to surface of pest breeding sites (ponds, lakes, marshes, swamps, estuaries). Temephos is regarded as a non-phytotoxic material with a long residual life. Toxic to bees, birds, fishes and Crustacea. Trade names include: Abat®, Abate®, Abathion®, Difenthos®, Ecopro®, Lypor®, Nimitex®, Swebate®, Temeguard®, Tiempo®. See Organophosphorus Insecticide.

TEMIK® See Aldicarb.

TEMMINCK, C J (1778–1858) First director of the Rijksmuseum van Natuurlijke Historie.

TEMPERATE POULTRY MITE *Ornithonyssus sylviarum* (Canestrini & Fanzago) [Acari: Macronyssidae] (Australia).

TEMPLE Noun. (French, *tempe*. Pl., Temples.) The posterior part of the gena behind, before or beneath the eye.

TEMPO® See Cyfluthrin.

TEMPORA Plural Noun. (Latin, *templum* = circumscribed space.) The temples; the posterior part of the sides of the head above and behind the eyes.

TEMPORAL MARGINS Mallophaga: postero-lateral part of the head.

TEMPORAL POLYETHISM See Age Polyethism.

TEMPORARY SOCIAL PARASITISM Social Insects: A phenomenon in which a queen enters an alien nest, kills the resident queen and takes over reproductive control of the nest. Resident workers die naturally and the nest becomes dominated by workers of the parasite queen. See Parasitism.

TEN-LINED JUNE BEETLE *Polyphylla decemlineata* (Say) [Coleoptera: Scarabaeidae].

TEN-SPOTTED LADY-BEETLE *Coelophora pupillata* (Swartz) [Coleoptera: Coccinellidae].

TENACULUM Noun. (Latin, *tenax* = holding. Pl., Tenacula.) 1. Collembola: A ventromedial cuticular projection on the third abdominal segment which bifurcates apically to form a holdfast for the Furcula. See Retinaculum; Furcula. 2. Hymenoptera: The Genostylus. Alt. Tenacle. See Catch.

TENANT HAIR See Tenent Hair.

TENDINOUS Adj. (Latin, *tendere* = to strech; Latin, *-osus* = possessing the qualities of.) Descriptive of or pertaining to a tendon. Descriptive of structure with the physical characteristics of a tendon.

TENDIPEDIDAE Plural Noun. See Chironomidae.

TENDO Noun. (Latin, *tendo* = tendon.) The anal area

of the hindwings when it forms a groove for the Abdomen; also called frenum and frenulum. Trichoptera: A small, elliptical space at the base of the hindwings near the base of the Anal Veins and behind the Trochlea.

TENDON Noun. (Latin, *tendo* = tendon, from *tendere* = to strech. Pl., Tendons.) 1. A slender, sclerite (band, strap- or cup-shaped structure) to which muscles are attached for moving appendages or apodemes. 2. A strong bristle or bristles in the wings of Lepidoptera.

TENDON HAIR Specialized sticky Setae adapted for clinging or clasping.

TENEBRIONIDAE Plural Noun. Comb Clawed Beetles; Darkling Beetles (adult); False Wireworm; False Ladybird Beetles; Lively Ant Guest Beetles; Long Jointed Beetles; Meal Worm Beetles; Mealworm (larva). A numerically large (ca 15,000 Species), cosmopolitan Family of polyphagous Coleoptera assigned to Superfamily Tenebrionoidea. Adult 1.5–36 mm long, typically dark coloured, often heavily sclerotized; Antenna often glabrous, typically with 11 segments and a club, rarely with 3, 8 or 10 segments; Mandible short, blunt often apically notched, sometimes concealed beneath Labrum. Elytra often fused and Species flightless; tarsal formula 5-5-4, rarely 4-4-4. Abdomen with five Ventrites, often with quinone-derived defensive secretions. Larva elongate, cylindrical; 0–5 Stemmata; Antenna with 1–3 segments; legs with five segments, Pretarsi each with two claws; Urogomphi present or absent. Primarily detritus feeders, some Species feed on living plants and a few Species predaceous; many Species highly adapted to xeric habitats; ca 100 Species reported as pests of stored products. See Dark Mealworm; Lesser Mealworm; Yellow Mealworm.

TENEBRIONOIDEA A Superfamily of polyphagous Coleoptera. Tarsal formula typically 5-5-4 in both sexes, rarely 4-4-4; four or fewer veins behind MP in hindwing. Included Families: Aderidae, Anthicidae, Archeocrypticidae, Chalcodryidae, Ciidae, Colydiidae, Melandryidae, Meloidae, Monommidae, Mordellidae, Mycteridae, Mycetophagidae, Oedemeridae, Prostomidae, Pythidae, Rhipiphoridae, Salpingidae, Scraptiidae, Tenebrionidae, Zopheridae.

TENERAL Adj. (Latin, *tener* = delicate, tender.) Pertaining to the adult soon after emergence from the nymphal or pupal stage when the Integument is not hardened or its definitive colour has not been completed. See Metamorphosis. Cf. Callow; Pharate. Rel. Pharate.

TENGSTRÖM, JOHAN MARTIN JAKOB VON (1821–1890) (Anon. 1891. Entomol. mon. Mag. 27: 111–112; Sahlberg 1891, Ent. Tidskr. 12: 177–190, bibliogr.)

TENNSTEDT, AUGUST (–1896) (Fauvel 1876, Annu. ent. 4: 129.)

TENSON, ALFONSO DAMPF (1884–1948) (Alexander 1948, Ent. News 59: 89–91.)

TENSOR Adj. (Latin, *tendere* = to strech; *or* = one who does something.) Pertaining to a muscle which stretches a membrane or parts of the body.

TENT Noun. (Latin, *tendere* = to strech. Pl., Tents.) 1. A demountable habitation or dwelling made of fibre. 2. The web of a tent caterpillar. Rel. Tentiform.

TENT CATERPILLARS See Lasiocampidae.

TENT SPIDERS *Cyrtophora* spp. [Araneida: Araneidae] (Australia).

TENTACLE Noun. (Latin, *tentaculum* = feeler. Pl., Tentaculi; Tentacles.) A flexible organ of touch, sometimes retractile, as in the larvae of Lepidoptera. Alt. Tentacule.

TENTACULAR Adj. (Latin, *tentaculum* = feeler.) Descriptive of or pertaining to organs of touch or tentacles.

TENTACULAR ORGAN A cylindrical, eversible organ found on the dorsum of the eighth abdominal segment of many lycaenid caterpillars. The organ is rapidly everted when the larva is disturbed. Cf. Newcomer's Organ; Perforated Cupola Organ.

TENTACULATE Adj. (Latin, *tentaculum* = feeler; *-atus* = adjectival suffix.) Structure adorned with pliant, elongate growths or tactile processes; sometimes applied to a margin or fringe.

TENTACULIFEROUS Adj. (Latin, *tentaculum* = feeler; *fere* = bear, carry; *-osus* = possessing the qualities of.) Descriptive of structure with tentacles.

TENTHREDINID Adj. Descriptive of or pertaining to the hymenopterous Family Tenthredinidae. The sawflies.

TENTHREDINIDAE Latreille 1802. Plural Noun. Tenthredinid Sawflies; True Sawflies. The largest Family of Symphyta with more than 5,000 described Species. Tenthredinids are cosmopolitan in distribution but best represented in north temperate regions. Adult body 2–15 mm long; Antenna with 7–15 segments and typically with nine segments; Antenna usually setaceous, filiform or weakly clavate; head without Hypostomal Bridge; Pronotum short with posterior margin curved; scutellum with transverse Sulcus separating post-Tergum posteriorly; fore Tibia with two spurs, medial spur bifid; middle Tibia without subapical spurs. Ovipositor usually weakly projecting beyond apex of Abdomen; male genitalia strophandrous. Larvae with Stemmata; Antenna with 1–5 segments; typically with 5-segmented thoracic legs; abdominal Prolegs on segments 2–7. Allantinae feed externally on many dicotyledons. Blennocampinae feed externally upon leaves of angiosperms and some monocotyledons. Doleriinae feed on Cyperaceae, Equisitaceae, Gramineae and Juncaceae. Heterarthrinae feed primarily upon Betulaceae, Rosaceae and Salicaceae. Nematinae feed on many different Families of host plants. Selandriinae feed primarily on ferns, some Species feed on Cyperaceae, Gramineae, and

Juncaceae. Tenthredininae feed on 28 Families of vascular plants. See Symphyta.

TENTHREDINOIDEA Latreille 1802. Hymenoptera: A Superfamily of Symphyta characterized by Mesoscutellum and Scutum continuous laterad, auxillae not defined anteriad, and subantennal grooves absent. Included Families: Argidae, Blasticotomidae, Cimbicidae, Diprionidae, Electrotomidae, Pergidae, Pterygophoridae, Tenthredinidae, Xyelotomidae.

TENTIFORM Adj. (Latin, *tente* = streched out; *forma* = shape.) Tent-shaped. Term descriptive of the nest or habitation constructed by some insects such as leafminers (Gracillariidae) or caterpillars that spin silken retreats. See Tent. Rel. Form 2; Shape 2; Outline Shape. Cf. Ptychonome.

TENTORIAL ARMS See Anterior and Posterior Tentorial Arms.

TENTORIAL BRIDGE The Posterior Tentorial Arms that form a continuous, transverse sclerotized bar across the back of the head beneath the Occipital Foramen (Snodgrass). See Tentorium.

TENTORIAL FOVEAE Hymenoptera: The pit-like openings positioned between the antennal sockets and the dorsal margin of the Clypeus. Syn. Anterior Tentorial Pits.

TENTORIAL MACULA One of the dark spots where the dorsal arms of the Tentorium unite with the epicranial wall near the Antennae (Snodgrass).

TENTORIAL PITS Two pairs of external depressions in the cranial wall at the anterior and posterior end of the tentorial arms. Anterior Tentorial Pits are positioned in the Subgenal Suture or (usually) in the Epistomal Suture; Posterior Tentorial Pits are positioned in the lower ends of the Postoccipital Suture (Snodgrass). See Tentorium.

TENTORIUM Noun. (Latin, *tentorium* = tent. Pl., Tentoria.) 1. The internal skeleton of the insect head. The Tentorium is marked externally by Anterior Tentorial Pits in the Epistomal Suture (Frontoclypeal Suture) and Posterior Tentorial Pits in the Postoccipital Suture or Subgenal Suture. The internal skeleton consists of 2–3 pairs of apodemes which are derived from inflections of the Tentorial Pits and which coalesce internally. Elements include the Epistomal Ridge, Anterior Tentorial Arms, Dorsal Tentorial Arms and Tentorial Bridge (Corportentorium). An additional pair of tentorial arms occurs behind the posterior pair in Collembola, Psocoptera and Thysanura. Elements of the Tentorium provide rigidity and strength to the head, support the Brain and foregut and provide attachment surface for some cephalic muscles. Several muscles attach to the Tentorium. In the lower Orders, ventral adductors of the mandible, maxilla and labium have their origin on the Tentorium. The dilator of the Stomodaeum, retractor of the Hypopharynx and antennal muscles also attach to the Tentorium. 2. Diptera: One of two hollow, cylindrical struts which pass from the ventral border of the Occipital Foramen to the cheeks (Smith).

TENUIS Adj. (Latin, *tenuis* = slight.) Thin, slender, long and drawn out.

TENURE® See Chlorpyrifos.

TENZEL (TENTZEL), WILHELM ERNST (1659–1707) (Rose 1850, *New General Biographical Dictionary* 12: 201–202.)

TEPAL Noun. (Fr. *tepale*, from *petale*; Latin, *-alis* = characterized by. Pl., Tepals.) Perianth not differentiated into petal or sepal.

TEPHRITIDAE Plural Noun. Fruit Flies; Gall Flies; Peacock Flies; Picture-Winged Flies. A cosmopolitan Family of muscomorphan Diptera containing ca 500 Genera and 4,500 nominal Species; Dacinae include ca 800 Species in Asia and Pacific. Tephritidae wings maculated, spotted or with ornate coloration pattern; Sc apically bends forward toward Costa almost perpendicular; abdominal segment VI well developed; Ovipositor typically long and used to pierce host plant tissue. Adults emerge from pupation site (typically soil), rest on flowers, fruit or vegetation and flex their wings. Female mates, oviposits in host plant tissue; eggs hatch within several days; larvae feed on fresh fruit, form galls, mine leaves; damage due to female oviposition, larval feeding and associated bacterial rots. Family considered most important global insect pests at risk for export fruits; economically damaging Species include: Apple Maggot, Mediterranean Fruit Fly, Melon Fly, Mexican Fruit Fly, Oriental Fruit Fly, Papaya Fruit Fly. Tephritinae typically oviposit in flower heads, mostly of Asteraceae. Syn. Trypetidae; Trypaneidae; Trupaneidae. See Diptera. Rel. Quarantine.

TEPHRITOIDEA A Superfamily of schizophorous Diptera including Otitidae (Cephaliidae, Ceroxydidae, Doryceridae, Ortalidae, Ortalididae, Ulidiidae), Platystomidae, Pyrgotidae, Richardiidae, Tachiniscidae and Tephrididae.

TEPPER, JOHANN GOTTLIEB OTTO (1841–1922) (Anon. 1923, Aust. Nat. 4 (2): 113; Howard 1930, Smithson. misc. Collns. 84: 393–394.)

TEPPER'S SKIPPER *Motasingha trimaculata* (Tepper) [Lepidoptera: Hesperiidae] (Australia).

TERANISHI, NOBORU (1896–1938) (Hasegawa 1967, Kontyû 35 (Suppl.): 53–54, bibliogr.)

TERATEMBIIDAE Plural Noun. A small Family of Embiidina (Embioptera) characterized by left Cercus with two segments and male with vein R4+5 forked. Fewer that five Species in USA. Syn. Oligembiidae.

TERATOCYTE Noun. (Greek, *teras* = monster; *kytos* = hollow. Pl., Teratocytes.) Type of cell in the parasitized egg or body of hosts attacked by some parasitic Hymenoptera. Teratocytes are derived from disassociated cells of extra-embryonic tissue (serosa) in Braconidae, Scelionidae, Trichogrammatidae. Teratocytes do not divide but increase significantly in size; number of teratocytes in host varies from 12–800, and may be Species specific. Teratocytes appear to function in secretion, nutrition and immunology. Syn. Giant Cells; Trophamnion Cells; Trophic Cells; Trophoserosal Cells.

TERATOGEN Noun. (Greek, *teras* = monster; *genos* = birth. Pl., Teratogens.) Any material that causes birth defects.

TERATOLOGY Noun. (Greek, *teras* = monster; *logos* = discourse. Pl., Teratologies.) The study of developmental abberations or structural monstrosities. Several physical factors can produce malformations including radiation, accidents in moulting and genetic malformation See Development. Cf. Gynandromorphy; Hysterotely; Mesomely; Ternaer Schistomely.

TERATOMYZIDAE Plural Noun. A small, widespread Family of schizophorous Diptera assigned to Superfamily Asteioidea. Species occur in South America, Orient and Australia. Adult with Postvertical Bristles small; Vibrissae present; Antenna porrect; Prosternum small without Precoxal Bridge. Adults inhabit cool forests and rest on ferns, presumably feeding on the surface flora.

TERBUFOS An organic-phosphate (dithiophosphate) compound {S[[(1,1-dimethylethyl) thio]-methyl] O,O-diethylphosphorodithioate} used as a soil insecticide against rootworms, wireworms, scarab larvae, corn-seed maggots and corn-seed beetles. Compound applied to bananas, corn, sorghum, sugarbeets, and vegetables. Toxic to fishes and wildlife. Trade names include: Geomet®, Geophos®, Granutox®, Rampart®, Terrathion®, Thimet®, Timet® and Volphor®. See Organophosphorus Insecticide.

TERCYL® See Carbaryl.

TEREBELLA Noun. (Latin, *terebra* = borer. Pl., Terebellae.) A saw-like Ovipositor.

TEREBRA Noun. (Latin, *terebra* = borer. Pl., Terebrae.) 1. A borer or piercer. 2. An Ovipositor fitted for boring or cutting, as in sawflies. 3. A mandibular sclerite articulated to the basalis, forming the point of the structure and equivalent to the Galea of the Maxilla. 4. Hymenoptera the Sting or the piercing Ovipositor.

TEREBRANT Noun. (Latin, *terebra* = borer; *-antem* = an agent of something. Pl., Terebrants.) With an Ovipositor adapted for piercing or boring.

TEREBRANTIA Hymenoptera: Symphyta with sessile Abdomen and valved Ovipositors adapted for boring. Thysanoptera: A Suborder of Thrips in which the Ovipositor is borer-like. Cf. Tubulifera.

TERETE Adj. (Latin, *teres* = rounded off, smooth.) Tapering and cylindrical in cross section.

TERGAL GLAND Blattaria: A gland on Tergum 7 and/or Tergum 8 of male cockroaches. TG vary in size, shape and features; ornately sculpture in some Species to densely setose areas in other Species. TG provide secretions/pheromones upon which the female feeds during courtship and copulation. Pheromone enables female to correctly position the female above the dorsum of the male during intromission. Ingestion of pheromone by some Species necessary for copulation. See Accessory Gland. Cf. Seducin.

TERGAL Adj. (Latin, *tergum* = back; *-alis* = pertaining to.) Descriptive of structure or process associated with the upper (dorsal) surface of a Tergum or the insect body. See Tegum. Cf. Dorsal; Sternal.

TERGAL SUTURE The Y-shaped dorsal suture on the head of many insect larvae.

TERGIFEROUS Adj. (Latin, *tergum* = back. Greek, *ferre* = to bear, carry; Latin, *-osus* = possessing the qualities of.) Bearing on the back.

TERGITE Noun. (Latin, *tergum* = back; *-ites* = constituent. Pl., Tergites.) 1. A dorsal sclerite which is part of a body segment or sclerotized ring of a generalized body segment. 2. A sclerotized subdivision of a body Tergum. See Segmentation. Cf. Tergum; Laterotergite; Laterosternite; Pleurite. Ant. Sternite.

TERGONTA Noun. A prominent projection of the caudo-lateral angle of the metapostscutellum which articulates against or is fused with a Unaterga (MacGillivray).

TERGOPLEURAL Adj. (Latin, *tergum* = back; Greek, *pleuron* = rib, side.) Descriptive of or pertaining to the upper and lateral portion of a segment.

TERGORHABDITES Noun. (Latin, *tergum* = back; Greek, *rhabdom* = rod.) The lower pair of corneous appendages forming the Ovipositor in grasshoppers. Sclerites on the inner dorsal surface of the abdominal wall.

TERGOSTERNAL MUSCLES The wing-muscles positioned to the sides of the median dorsal muscles in the anterior part of the segment. TM are attached dorsally on the anterolateral areas of the Tergum and ventrally on the Basisternum anterior of the Coxa (Snodgrass).

TERGOSTERNAL Adj. (Latin, *tergum* = back; Greek, *sternum* = chest; Latin, *-alis* = pertaining to.) Descriptive of structure or process associated with the Tergum and Sternum together.

TERGUM Noun. (Latin, *tergum* = back. Pl., Terga.) 1. The upper or dorsal surface of any body segment of an insect, whether it consists of one or several sclerites. 2. The dorsum of the head, Thorax or abdomen. Thoracic Terga are more complex that Terga of head or Thorax. Tergum of Apterygota and many nymphal pterygotes insects are similar to the Terga of the Abdomen with typical secondary segmentation. Tergum of Pterygota reveals Mesothorax and Metathorax are modified for involvement with flight. Principal features of thoracic Tergum include a transverse suture along the anterior margin of the sclerite. Externally, this forms an Antecostal Suture; internally this suture develops into a submarginal Antecosta The Antecosta of wing-bearing segments often becomes considerably enlarged to provide an area for attachment of dorsolongitudinal flight muscles. When so developed, the apodemal projection is called a Phragma Typically a Phragma exists on Mesothorax, Metathorax and first abdominal segment. Each Phragma may be bilobed, probably to accommodate the dorsolongitudinal aorta (dorsal vessel) which is positioned medially. See Segmentation. Cf. Pleuron; Tergite; Sternum. Rel. Tagma.

TERMEN Noun. (Latin, *termen* = boundary, end. Pl., Termens.) The outer margin of a wing, between the apex and the posterior or anal angle.

TERMEX® See Chlordane.

TERMEYER, RAIMONDO MARIA DE (1740?–) (Dusmet y Alonso 1919, Boln. Soc. ent. Esp. 2: 78.)

TERMIDAN® See Chlordane.

TERMINAL Adj. (Latin, *terminus* = end; *-alis* = pertaining to.) Pertaining to the apex or distal end of a structure or appendage. Opposed to 'basal.' Syn. Apical. See Orientation.

TERMINAL ANASTOMOSIS. Insect Brain: A complex layer of postretinal fibres of the optic lobe.

TERMINAL ARBORIZATIONS Nerve Cells: Fine branching fibrils of the axon and the collateral ends.

TERMINAL FILAMENT A thread-like part of the suspensorial apparatus of the Ovary (Snodgrass); the cellular end thread of an Ovariole (Snodgrass). A slender thread-like apical prolongation of the peritoneal layer of an Ovariole (Imms). Syn. Filum Terminale.

TERMINAL LINE Lepidoptera: A line along the outer margin of the wings.

TERMINAL SPACE The area between the subterminal transverse line (s.t. line) and the terminal line in certain Lepidoptera (Smith).

TERMINALIA Plural Noun. (Latin, *terminus* = end.) 1. General: The terminal segments of parts or structures taken together. 2. Insect wing: The metapterygium, deltoid, mesoptera, onguiculus or third Axillary, in the hindwing, the fourth Axillary (MacGillivray). 3. Terminal abdominal segments (and their parts) modified to form the genital segments of the sex (Crampton).

TERMINATOR® See Diazinon.

TERMITAPHIDIDAE Plural Noun. A small, circumtropical Family of heteropterous Hemiptera assigned to Superfamily Aradoidea. Body flattened, eyeless, mouthpart Stylets long and coiled; apterous; dorsal abdominal scent glands absent. Associated with termite nests and presumably feed on fungal mycelia within termite nest.

TERMITARIUM Noun. (Latin, *termes* = woodworm; *arium* = place of a thing. Pl., Termitaria.) A natural or artificial nest or colony of termites.

TERMITES See Isoptera.

TERMITIDAE Plural Noun. Soldierless Termites; Nasutiform Termites. The largest Family of Isoptera. Cosmopolitan and most highly evolved. Alates with Ocelli; Antenna with 13–18 segments; fontanelle usually present. Tarsal formula 4-4-4; wings not reticulate; anterior wing scales short, not reaching hindwing scales; hindwings without anal lobe. Most Species subterranean, some build mounds, a few are arboreal. Most Species feed on wood or grass; a few Species cultivate fungi. See Isoptera.

TERMITOLOGY Noun. (Latin, *termes* = woodworm; Greek, *logos* = discourse.) The study of termites.

TERMITOMYCES A Genus of basidiomycete fungi cultivated within nest of higher termites (Macrotermintinae). Termites deposit faeces of finely comminuted, partially digested plant material infiltrated with *Termitomyces*. Fungal mycelia partially digest plant material and form convoluted greyish-brown combs; fungus develops round, white nodules (mycotetes) composed of masses of asexual spores (conidia) on comb; termites feed upon older parts of comb and nodules. *Termitomyces* may partially digest plant polysaccharides and lignin.

TERMITOPHILE Noun. (Latin, *termes* = woodworm; Greek, *philein* = to love. Pl., Termitophiles.) An inquiline of termite nests. *e.g.* Some Staphylinidae (Termitonannini). See Inquiline; Symbiosis.

TERMITOPHILOUS Adj. (Latin, *termes* = woodworm; Greek, *philein* = to love; Latin, *-osus* = possessing the qualities of.) Termite loving. A term descriptive of some insects and other organisms that act as guests of termites and habitually live in a termite colony with and among the termites. See Inquiline. Cf. Entomophilous; Myrmecophilous. Rel. Commensalism; Mutualism; Parasitism.

TERMOPSIDAE Plural Noun. Dampwood Termites. A small Family of termites that includes fewer than 25 Species in three Subfamilies. Termopsids infest rotten wood and are considered among the most primitive Isoptera. Several Species are reported to include reproductive soldiers. See Caste.

TERRA INCOGNITA Latin phrase meaning 'an unknown country.'

TERRAPIN SCALE *Mesolecanium nigrofasciatum* (Pergande) [Hemiptera: Coccidae]: A univoltine soft-scale pest of fruit, shade and ornamental trees in eastern and southern USA. TS overwinters as an inseminated female beneath the scale cover; females resume feeding during spring; live crawlers produced during May-June. Crawlers accumulate beneath mother's scale cover and disperse after her death. Initial feeding on leaves and subsequently move to wood. Sooty mould assocated with honeydew secreted by TS. Syn. *Lecanium nigrofasciatum*. See Coccidae.

TERRATHION® See Phorate; Terbufos.

TERRESTRIAL Adj. (Latin, *terra* = earth; *-alis* = characterized by.) Living on or in the land. Cf. Aquatic.

TERRITORY Social Insects: An area of variable size occupied and defended by a colony using behavioural aggression and/or chemical repulsion.

TERRIX® See Entomogenous Nematodes.

TERRY, FRANK WRAY (1877–1911) (Morice 1911, Proc. ent. Soc. Lond. 1911: cxxvi; Bueno 1912, Ent. News 23: 47–48.)

TERTHIENYL Noun. A light-activated alkaloid biopesticide extracted from the roots of African marigold *Tagetes* and several Species of Compositae. Active component may affect mosquito larvae and nematodes. See Biopesticide; Botanical Insecticide; Natural Insecticide; Insecticide; Pesticides. Cf. Furanocumarins; Polyacetylenes.

TERTIARY PARASITE A parasite which develops hyperparasitically upon another hyperparasite. See Hyperparasite; Secondary Parasite.

TERTIARY PARASITISM A condition of parasitism in which a hyperparasitic individual attacks another hyperparasite. Conceptually divided into Interspecific Tertiary Hyperparasitism (Allohyperparasitism) and Intraspecific Tertiary Hyperparasitism (Autohyperparasitism). See Hyperparasitism.

TERTRAPYRROLES Plural Noun. Pigmentary colours including porphyrins and bilins found within the insect body. See Pigmentary Colour.

TERTRE, JEAN BAPTISTE DU (1611–1687) (Rose 1850, *New General Biographical Dictionary* 12: 204.)

TERTTI, MARTTI (1898–1940) (Saalas 1940, Erip. huonnon Ystävä 1940: 1–12.)

TERZI, AMEDEO JOHN ENGEL (1872–1956) (Mattingly 1976, Mosquito Syst. 8: 114–120.) Scientific illustrator at the British Museum of Natural History.

TESAR, ZDENEK (1907–1967) (Balthasar 1967, Acta ent. bohemoslovaca 64: 322–323, bibliogr.)

TESSARATOMIDAE Plural Noun. A small, tropical Family of heteropterous Hemiptera assigned to Superfamily Pentatomoidea; sometimes combined with Pentatomidae. Ocelli widely separated; Antenna typically with five segments sometimes with four segments; Hemelytron without reticulate venation. See Bronze Orange Bug.

TESSELATE Adj. (Latin, *tessella* = small stone cube; *-atus* = adjectival suffix.) Checkered, more-or-less like a chessboard. Alt. Tessellated; Tessellatus.

TESSELLATED PHASMATID *Ctenomorphodes tessulatus* (Gray) [Phasmida: Phasmatidae]: A minor pest of eucalypts in Australia.

TESSELLATED SCALE *Eucalymnatus tessellatus* (Signoret) [Hemiptera: Coccidae] (Australia).

TESSERAE Plural Noun. (Latin, *tessera* = square block.) Some coccids: Adjacent polygonal areas formed by connecting Cellulae (MacGillivray).

TESSIEN, JOHANN HEINRICH CONRAD (1814–1868) (Weidner 1967, Abr. Verh. naturw. Verh. Hamburg Suppl. 9: 1123; Butler 1973, Proc. R. ent. Soc. Lond. (C) 37: 57.)

TEST Noun. (Latin, *testa* = shell.) Coccidae: Scale cover of scale insects. A glandular secretion of wax, proteins and polyphenols; varying in consistency, texture and appearance. See Scale Cover.

TESTA Noun. (Latin, *testa* = shell; *aceus* = of or pertaining to.) A hard covering, coat or Integument.

TESTACEOUS Adj. (Greek, *testa* = burnt piece of clay; *-aceous* = like.) A dull brick red or brownish red colour.

TESTE (Latin.) According to (verbal, not written). Cf. In Litt.

TESTES Plural Noun. (Latin, *testis* = testicle. Sl., Testis.) The functional homologue of the female's Ovary. Testes are positioned above or below the gut, depending upon the insect, and are in the same relative position as the Ovary. Male insects typically have two Testes. Lepidoptera usually possess a single large testis. A Peritoneal Sheath may surround each Testis or both Testes. Testes produces Spermatozoa within tubular Follicles. See Follicle. Cf. Ovary. Rel. Gonad.

TESTICULAR Adj. (Latin, *testis* = testicle.) Descriptive of or pertaining to the Testes.

TESTICULAR FOLLICLES See Follicle.

TESTICULAR TUBE A Testicular Follicle.

TESTICULATE Adj. (Latin, *testis* = testicle; *-atus* = adjectival suffix.) Descriptive of or pertaining to the Testes. Shaped as a Testicle.

TESTOUT, HENRI (1884–1950) (Schaefer *et al.* 1951, Bull. mens. Soc. linn. Lyon 20: 73–78, bibliogr.)

TESTUDINARIOUS Adj. (Latin, *testudo* = tortoise; *-osus* = possessing the qualities of.) Red, black and yellow; resembling a tortoise-shell in coloration. Alt. Testudinarius.

TESTUDINATE Adj. (Latin, *testudo* = tortoise; *-atus* = adjectival suffix.) Resembling the shell of a tortoise in form; a hard and protective covering as a tortoise shell. Alt. Testudinatus.

TETANOCERATIDAE Plural Noun. See Sciomyzidae.

TETANOCERIDAE Plural Noun. See Sciomyzidae.

TETHINIDAE Plural Noun. A small, widespread Family of schizophorous Diptera assigned to Superfamily Ephydroidea (Chloropoidea). Adult small bodied, Postvertical Bristles typically convergent; Postocellar Bristles divergent; Vibrissae present. Species typically found along sea coast in association with seaweed or grasses, salt marshes or alkaline inland habitats.

TETRACAMPIDAE Förster 1856. Plural Noun. An Old World Family of parasitic Hymenoptera (Chalcidoidea) with about 15 Genera and 40 nominal Species. Probably related to Pteromalidae and Eulophidae. One Species introduced into the Nearctic. Morphologically characterized by Antenna with 10–12 segments including a small ring segment (Anellus); Pronotum large and campanulate, about as long as Mesoscutum; Parapsidal Sutures complete; Propodeum frequently with medial Setae; forewing Marginal Vein long, Stigmal and Post Marginal Veins relatively long; foretibial spur weakly developed and straight; males usually with four Tarsomeres; females with five Tarsomeres. Biology poorly known; primary parasites of Coleoptera and Hymenoptera eggs; parasites of Diptera larvae mining leaves and twigs. Included Subfamilies: Mongolocampinae, Platynocheilinae and Tetracampinae.

TETRACHAETAE Adj. (Greek, *tetra* = four; *chaite* = hair.) Diptera in which the mouth structures consist of four longitudinal blades or piercing structures.

TETRACHLOROMETHANE® See Carbon Tetrachloride.

TETRACHLORVINPHOS An organic-phosphate (phosphoric acid) compound {(S)-2-chloro-1-(2,4,5-trichlorophenyl) vinyl dimethylphosphate} used as an acaricide, contact insecticide and stomach poison for control of bollworms, armyworms, borers, leaf-eating insects and public health pests. Compound applied to numerous crops, vegetables, ornamentals and livestock in many countries. Registered as livestock and poultry insecticide in USA. Toxic to bees and fishes. Trade names include: Appex®, Debantic®, Gardcide®, Gardona®, Rabon®, Rabond®. See Organophosphorus Insecticide.

TETRADACTYLE Adj. (Greek, *tetra* = four; *daktylos* = digit, finger.) With four fingers or finger-like processes.

TETRADIFON A chlorinated hydrocarbon {4-chlorophyeny 2,4,5-trichlorophenyl sulphone} used as an acaricide in some countries. Not registered in USA. Effective on immature stages only; some phytotoxicity. Trade names include: Childion®, Mixan®, Murfite®, Rotetra®, Tedion®. See Organochlorine Insecticide.

TETRAGNATHIDAE Plural Noun. Long Jawed Spiders; Long-Jawed Orb Weavers. Spiders characterized by long, conspicuous Chelicera, particuarly in the male.

TETRAGONAL Adj. (Greek, *tetras* = four; *gonia* = angle; *-alis* = characterized by.) Descriptive of structure with four sides or angles; quadrangular. Alt. Tetragonum.

TETRAMERA Plural Noun. (Greek, *tetras* = four; *meros* = part.) Coleoptera with four-segmented Tarsi.

TETRAMEROUS Adj. (Greek, *tetras* = four; *meros* = part.) Pertaining to four-segmented Tarsi.

TETRAMETHRIN Noun. A synthetic-pyrethroid compound used as contact insecticide with fast knockdown for control of flies, mosquitoes, midges, wasps and other urban pests. Toxic to fishes. Trade names include: Butamin®, Doom®, Duracide®, Ecothrin®, Neo-Pynamin®, Phinco-Tzz®, Phthalthrin®, Residrin®, Sprigone®, Spritex®. See Synthetic Pyrethroid.

TETRAPODA Noun. (Greek, *tetras* = four; *pous* = foot.) Butterflies in which the anterior legs are atrophied and the insect appears to possess four legs.

TETRAPTERA Noun. (Greek, *tetras* = four; *pteron* = wing.) An arcane term proposed for all insects with four naked, membranous, reticulated wings.

TETRATEMBIIDAE Plural Noun. A small Family of Embioptera.

TETRATOMIDAE Plural Noun. See Melandryidae.

TETRIGIDAE Plural Noun. Pigmy Grasshoppers; Grouse Locust. A Family of caeliferous Orthoptera characterized by the Pronotum extending posteriad over Abdomen.

TETRIGOIDEA A Superfamily of caeliferous Orthoptera including Batrachideidae, Tetrigidae.

TETTIGOMETRIDAE. Plural Noun. A small Family of auchenorrhynchous Hemiptera assigned to Superfamily Fulgoroidea.

TETTIGONIIDAE Plural Noun. Long-horned Grasshoppers; Katydids. A Family of ensiferous Orthoptera which includes about 80 Genera and 375 Species. Widespread or cosmopolitan in distribution. Head short with Vertex forming a short Rostrum, Antenna longer than body in most Species and inserted between compound eyes, scrobes weakly margined, prothoracic Spiracle large, elongate and frequently concealed by the Pronotum; Prosternum with a pair of spines, fore Tibia usually with apicodorsal spine and basal auditory organ concealed, first and second tarsal segments laterally grooved. Ecologically, most Species are ground inhabitants except Tettigoniini and Onconotini which live on shrubs and bushes. Most Species oviposit in the soil even if they are not ground dwellers. Some Saginae are parthenogenetic. Most Species are carnivorous but some will feed on plant material and achieve pest status. Cf. Mormon cricket.

TETTIGONIOIDEA Katydids, Long-horned grasshoppers. A Superfamily of ensiferous Orthoptera, including Acridozenidae, Bradyporidae, Conocephalidae, Meconematidae, Mecopodidae, Phaneropteridae, Phasmodidae, Phyllophoridae, Prophalangopsidae, Pseudophyllidae, Tettigoniidae, Tympanophoridae. Recognized by four-segmented Tarsi, body coloration typically green, but dull browns and grey are common. Antenna usually long, Prothoracic Spiracle and associated Tracheae modified into an auditory system in some Species, and fore Tibia usually with an auditory organ. Hind Femur not modified for stridulation. Tegmina (forewings) are held over the Abdomen in a vault-like manner. Cerci not long or particularly flexible; Ovipositor usually long and bears six well-developed valves.

TEUNISSEN, HERMANUS GERARDUS MARIA (1914–1992) (Van Achterberg 1992. Zool. Meded. 66 (16–40): 309–311.)

TEXAS CITRUS MITE *Eutetranychus banksi* (McGregor) [Acari: Tetranychidae].

TEXAS HARVESTER See Red Harvester-Ant.

TEXAS LEAF-CUTTING ANT *Atta texana* (Buckley) [Hymenoptera: Formicidae]: Endemic to USA (TX, LA). Workers highly polymorphic; hard-bodied, pale brown, 10–12 mm long, with numerous spines on head and Thorax; Antenna with 11 segments, without distinct club; legs long; gastral

Pedicel with two segments. Constructs large subterranean nests (3–7 metres below ground); chambers to one metre filled with fungal gardens. TLCA workers take leaves from plants to nest, malaxate them for substrate of fungus gardens; another caste maintains fungal garden; soldier caste defends colony. Queen carries fungal pellet on nuptial flight to establish new garden. Populations can become very large and workers can become pests of garden and field crops; can invade houses. See Formicidae. Cf. Larger Yellow-Ant; Little Black Ant; Odorous House Ant; Pavement Ant; Pharaoh Ant; Thief Ant.

TEXTURE Noun. (Latin, *textura* = web > *textus,* past participle of *texere* = to weave. Pl., Textures.) 1. The basic (often complex) structure, pattern, scheme or composition of an object. 2. The size and organization of the microscopical parts of a structure or surface. The essence of an anatomical surface. 3. The essential part of something (substance, nature). 4. Something composed of closely interwoven elements (*e.g.* the threads of a fabric; a strand of silk). See Structure. Cf. Smooth; Granulose; Polished; Punctate. Rel. Sculpture.

TEYROVKSY, VLADIMIR (–1963) (Novak 1963, Cas. csl. Spol. ent. 60: 267–269, bibliogr.)

THAI NEEM TREE See Sadao Cheng. Cf. Indian Neem Tree; Philippine Neem Tree. Rel. Azadirachtin.

THALES (625–547 BC) Greek philosopher living in Ionian city of Miletus and teacher of Anaximander. Thales believed the elemental substance was water and attempted to explain natural phenomena.

THAMNOPHILOUS Adj. (Greek, *thamnos* = bush; *philein* = to love. Latin, *–osus* = with the property of.) Descriptive of organisms that inhabit or live within dense shrubbery or wooded thickets. See Habitat. Cf. Agricolous; Caespiticolous; Dendrophilous; Lignicolous; Silvicolous. Rel. Ecology.

THANATOSIS Noun. (Greek, *thanatos* = death; *-osis* = a condition or state. Pl., Thanatoses.) Death-feigning. A behaviour common among many groups of insects. Chrysidid wasps roll into a ball and drop to the ground when threatened. Zygaenid moths may be physically damaged by potential predators, and apparently feign death to avoid harm. In the process, the moths release hydrocyanic acid from muscles and Haemolymph. (Haemolymph is released in reflexive bleeding areas; cyanide is derived from the hydrolysis of cyanoglucosides, linamarin and lotoaustralin.) Thanatosis varies in complexity and may sometimes be characterized as 'Mild Thanatosis' and 'Severe Thanatosis'. Alt. Letisimulate.

THAUMALEIDAE Plural Noun. Solitary Midges. A numerically small, geographically widespread Family of nematocerous Diptera assigned to Superfamily Chironomoidea in some classifications. Adults small bodied and occur near streams in forested habitats. Larvae aquatic. Syn. Orphnephilidae.

THAUMASTOCORIDAE Kirkaldy 1907. Plural Noun. Royal Palm Bugs; Thaumastocorids. A small Family of heteropterous Hemiptera assigned to the Superfamily Thaumastocoroidea. Family with predominantly Gondwanan distribution. Adult body elongate, dorsoventrally flattened with protuberant compound eyes and juga parallel sided; labium of four segments; Tarsi with two segments. All Species are phytophagous.

THAUMASTOCOROIDEA A Superfamily of Hemiptera containing Thaumastocoridae.

THAUMASTOTHERIIDAE Kirkaldy 1907. See Kormilev 1955. Una curiosa familia de Hemipteres nueva para la fauna Argentina. Thaumastotheriidae (Kirkaldy), 1907. Rev. Soc. Ent. Argent. 18: 5–10.

THAUMETOPOEIDAE Kiriakoff 1970. Plural Noun. A small Family of Ditrysian Lepidoptera assigned to Superfamily Noctuoidea with Species in the Palaearctic, Ethiopian and Oriental Realms; placed in Noctuidae in some classifications. Adult size variable; wingspan 20–80 mm; body stout with dense vestiture of long, piliform Setae; Ocelli vestigial or absent; Chaetosemata absent; Antenna typically bipectinate to apex, occasionally lamellate in male or filiform in female; Proboscis typically vestigial or absent; Maxillary Palpus small with 2–4 segments or absent; Labial Palpus vestigial; Metathorax with pair of tympanal organs; foretibial Epiphysis present in male, absent in female; tibial spur formula 0-2-2 or 0-2-4; apex of female Abdomen with deciduous piliform scales used to cover or mix with egg mass. Egg usually globular with flattened base; laid in masses on host plant. Larval head hypognathous; body with sense vestiture of secondary Setae, many arranged on verrucae. Pupa heavily sclerotized; cremaster bent, with apical divergent slender spines. Adult typically nocturnal or crepuscular, some diurnal; rests with wings tectiform over Abdomen; Larvae gregarious and often processionary; some Species live in communal nests through larval life; larval Setae often urticarious.

THAXTER, ROWLAND (1858–1932) (Setchell 1933, Proc. Am. Acad. Arts. Sci. 68: 678–682; Clenton 1937, Biogr. Mem. nat. Acad. Sci. 17 (3): 55–68, bibliogr.)

THECA Noun. (Greek, *theke* = case. Pl., Thecae.) A case or covering. A term specifically applied to: 1. The fleshy covering of the fly-mouth; 2. the cases of Trichoptera larvae; 3. the lower sclerites of the male genitalia in Hemiptera; 4. the outer covering of the pupa; 5. the phallic tube (Snodgrass); 6. the crumena (MacGillivray).

THECAL Adj. (Greek, *theke* = case; *-alis* = pertaining to.) Descriptive of or pertaining to a Theca.

THEDENIUS, KNUT FREDERIK (1814–1894) (Sandahl 1894, Ent. Tijdskr. 15: 191–199,

bibliogr.)

THEILERIASIS See Theileriosis.

THEILERIIDAE Plural Noun. A Family of parasitic Protozoa assigned to the Subclass Piroplasmasina and includes the Genus *Theileria.* Theileriidae lack pigment, exhibit a Piroplasm, undergo Merogony within erythrocytes or lymphocytes of vertebrate host, and Gametogony and Sporogony within ticks. See Theileriosis.

THEILERIOSIS Noun. (Theileria; Greek, *-osis* = condition. Pl., Theilerioses.) A term which includes several clinical diseases caused by infection with parasitic Protozoa assigned to *Theileria. Theileria* contains several Species which have been associated with ruminants; one Species *(T. parva)* causes high mortality in European cattle *(Bos taurus*). Other Species of *Theileria* widespread, attack cattle and sheep but are not as virulent as *T. parva.* Vectors of theileriosis include ixodid ticks (*Haemaphysalis* spp., *Rhipicephalus* spp. and *Hyalomma* spp.) Life history of disease agent similar to *Babesia.* Transmission of theileriosis is trans-stadial but not transovarial. Alt. Theileriasis. See Theileriidae. Cf. Babesiosis.

THELAZIASIS Noun. (*Thelazia*; Greek, *-asis* = disease.) A cosmopolitan disease of mammals caused by eyeworms *(Thelazia* spp.) which occur in the conjunctival sac of the host. Worm larvae ingested by flies, such as *Drosophila* and *Musca.* Larvae penetrate midgut and enter the Haemocoel of fly; in male fly the larvae migrate to testis and encapsulate while in female fly the larvae migrate to Haemocoel wall and encapsulate. Larvae develop to infective stage in fly then migrate to Proboscis where they are transmitted to eye of mammal host.

THELYTOKOUS Adj. (Greek, *thelys* = female; *tokos* = offspring; Latin, *-osus* = possessing the qualities of.) Pertaining to female offspring produced without fertilization of their egg stage. Alt. Thelyotokous.

THELYTOKY Noun. (Greek, *thelys* = female; *tokos* = offspring.) A form of parthenogenetic reproduction which does not involve genetic fertilization of the female gamete and in which all progeny are female. Forms of thelytoky include: revertible (microbe-associated) thelytoky and nonrevertible thelytoky. Common in parasitic Hymenoptera. Alt. Thelyotoky. Cf Arrhenotoky; Deuterotoky; Heterogony. See Revertible Thelytoky; Nonrevertible Thelytoky.

THEOBALD, FREDERIC VINCENT (1868–1930) (Howard 1915, Pop. Sci. Mon. 87: 70; Edwards 1930, Entomologist 63: 95–96; Duffield 1931, Books, reports, pamphlets by F. V. Theobald. 16 pp. bibliogr. Wye, Kent.)

THEOBALD, NICOLAS (1903–1981) French specialist in fossil organisms of central and south-central France.

THEOPHRASTUS, ERESIUS (372–288 BC) Greek philosopher and member of the Teleological School. Theophrastus wrote extensively on botany and made observations concerning insects as pests of plants. See Teleology. (Rose 1850, *New General Biographical Dictionary* 12: 216–217.)

THEORETICAL Adj. (Greek, *theoretikos* = behold, consider, contemplate; Latin, *-alis* = pertaining to.) 1. Concerning abstract knowledge in contrast to applied knowledge. 2. Ideas or facts that are related to theory and not practical application of the facts or ideas. Alt. Theoretic. Cf. Heurisitic; Empirical; Practical; Putative.

THEORY Noun. (Greek, *theoria* = contemplation, consideration. Pl., Theories.) 1. A systematic analysis of facts or observations, through which an hypothesis is rigorously tested and validated, modified or rejected. 2. A policy, procedure or plan of action which is accepted without question. 3. The hypothetical, conceptual and pragmatic elements of a discipline which form the basis for inquiry in that discipline. Theory formation involves observation, hypothesis formulation, hypothesis testing and the application of hypothetico-deductive reasonsing. Cf. Hypothesis.

THEORY OF PROBABILITIES A mathematical theory acted on implicitly by descriptive entomologists, whereby they assume that no two entities will simultaneously have the same combination of characters as those of a given Species; in mathematics, the theory of chance.

THEREVIDAE Plural Noun. Stiletto Flies. A medium-sized, cosmopolitan Family of orthorrhaphous Diptera, especially rich in Genera and Species that occur in semi-arid regions. Adults small to medium sized flies resembling asilids, rhagionids, apiocerids; some subtropical Species mimic wasps; elongate pilose bodies often with shining pruinose areas; body Setae usually prominent; wing venation fairly constant, vein R4 elongate and sinuous; Discal Cell elongate with three veins proceeding from apex; Anal Cell always closed; Empodia not developed, pulvilliform. Biologically comparatively poorly known. Adults not highly active but visit flowers, take nectar, imbibe water. Larva elongate, slender, fossorial and free living in soil, decaying vegetation or rotting wood; voracious feeders on soil-inhabiting arthropods and earthworms; prefer Coleoptera larvae, especially Elateridae, Scarabaeidae and Tenebrionidae; larvae cannibalistic under some circumstances. Adult female deposits about 50 eggs in soil; five larval instars, last instar pupates or enters diapause; diapause may last two years; most therevids univoltine.

THERIDIIDAE Plural Noun. Cobweb Spiders; Comb-footed Spiders. Construct tangled, irregular web. Species include Black Widow and Redback.

THERMAL CONSTANT The number of degree days required by an insect to complete a developmental event.

THERMAN SENSE Perception of temperature.

THERMOBIOLOGY Noun. (Greek, *therme* = heat, *bios* = life; *logos* = discourse. Pl., Thermobiologies.) The study of thermal heat in biological systems.

THERMOCLINE Noun. (Greek, *therme* = heat; *klinein* = to swerve. Pl., Thermoclines.) A layer of water whose temperature is rapidly changing, typically during seasonal adjustment, and which is positioned between thermally stable layers.

THERMOMETABOLISM Noun. (Greek, *therme* = heat; plus metaboly; English, *-ism* = state or condition. Pl., Thermometabolisms.) The dependence of metabolic activity on temperature.

THERMOPHILE Noun. (Greek, *therme* = heat; *philaein* = to love. Pl., Thermophiles.) An organism which lives and reproduces at relatively high temperatures. Cf. Halophile; Mesophile; Osmophile; Psychrophile. Rel. Ecology.

THERMOPHILIC Adj. (Greek, *therme* = heat; *philaein* = to love; *-ic* = of the nature of.) Descriptive of organisms or tissues that are tolerant of high temperatures. Cf. Cryostatic; Thermophilic

THERMORECEPTOR Noun. (Greek, *therme* = heat; Latin, *recipere* = to receive; *or* = one who does something. Pl., Thermoreceptors.) A sensory receptor which is sensitive to changes or values of temperature. Thermoreceptors have been discovered in several groups of insects. The cockroach *Periplaneta americana* has temperature receptors on the Antenna. The hemipteran *Rhodnius prolixus* Stål has apparent thermo-/hygrosensilla on the Antenna of the adult. The carabid beetle *Agonum (Sericoda) quadripunctatum* (DeGeer) is pyrophilous and occurs only at sites where forest or peat has been burned; adults are attracted to fires. Thermal receptors occur on Antenna of cave beetle *Speophyes lucidulus* Delar. Three sensory cells are associated with each black seta: a cold receptor, a heat receptor and a receptor whose function has not been determined. Host-finding is mediated via heat perception in the braconid wasp *Coeloides brunneri*. See Sensillum. Cf. Anemoreceptor; Chemoreceptor; Hygroreceptor; Mechanoreceptor; Osmoreceptor; Photoreceptor; Thigmoreceptor.

THERMOREGULATION Noun. (Greek, *therme* = heat; *regulare* = to regular; English, *-ion* = result of an action. Pl., Thermoregulations.) The process or mechanism by which body temperature is regulated in response to ambient environmental temperatures. The body temperature exists in a range slightly above freezing (ca 1–2°C) and a Critical Thermal Maximum (ca 40–45°C) at which coma occurs and death will eventuate without temperature correction. Thermoregulation by the insect is achieved through various ectothermic and endothermic processes. See Circulatory System; Haemolymph. Cf. Basking; Critical Thermal Maximum; Evaporative Cooling; Frost Protection. Rel. Physiology; Slifer's Patches.

THERMOSTAT Noun. (Greek, *therme* = heat; *statos* = standing. Pl., Thermostats.). Cf. Cryostat.

THERMOTAXIS. Noun. (Greek, *therme* = heat; *taxis* = arrangement. Pl., Thermotaxes.) Locomotor reaction to temperature stimulus. See Orientation. Cf. Aerotaxis; Anemotaxis; Chemotaxis; Geotaxis; Menotaxis; Osmotaxis; Phototaxis; Rheotaxis; Rotaxis; Scototaxis; Strophotaxis; Telotaxis; Thigmotaxis; Tonotaxis; Tropotaxis.

THERMOTROPISM Noun. (Greek, *therme* = heat; *trope* = turn; English, *-ism* = state or condition. Pl., Thermotropisms.) A turning response by a plant in reaction to temperature changes. See Tropism. Cf. Aeolotropism; Anemotropism; Chemotropism; Electrotropism; Galvanotropism; Geotropism; Heliotropism; Hydrotropism; Phototropism; Rheotropism; Stereotropism; Thigmotropism; Tonotropism. Rel. Taxis.

THERY, ANDRÉ (1864–1947) (Boudy 1947, Bull. Soc. Sci. nat. Maroc 25–27: 105–107, bibliogr.)

THEVENET, JULES (1826–1875) (Tappes 1875, Ann. Soc. ent. Fr. (5) 5: 253–256; Fauvel 1876, Annu. Ent. 4: 127–128.)

THIBAULT, J (–1839) (Anon. 1939, Bull. Soc. ent. Fr. 44: 202.)

THICK HEADED FLIES See Conopidae.

THICKENED LATERAL MARGIN Coccids: Lateris *sensu* MacGillivray.

THICKSET SCOLYTID BORER *Xyleborus solidus* Eichhoff [Coleoptera: Scolytidae] (Australia).

THIEF ANT *Solenopsis molesta* (Say) [Hymenoptera: Formicidae]: A widespread pest endemic to North America. Workers monomorphic; Antenna with 10 segments, including club of two segments. TA typically nests in soil or wood, occasionally invades houses. TA is omnivorous and in homes will take meats, milk products, sugar and grease; takes food and brood of other ant Species. In the field, TA attacks corn, millet and sorghum seeds, causing them not to germinate. Common name ascribed to habit of workers preying upon the larvae and pupae of other ant Species. See Formicidae. Cf. Argentine Ant; Black Carpenter-Ant; Cornfield Ant; Fire Ant; Larger Yellow-Ant; Little Black Ant; Odorous House Ant; Pavement Ant; Pharaoh Ant.

THIEF WEEVILS See Rhynchitidae.

THIELE, CARL (–1946) (Warnecke 1949, Verh. Ver. naturw. Heimatforsch. 30: xv–xvii.)

THIELE, HERMANN (1841–1918) (Stichel 1919, Int. ent. Z. 13: 81–82.)

THIEME, OTTO (1837–1907) (Horn 1907, Dt. ent. Z. 1907: 534; Röhmer 1907, Ent. Wbl. 24: 143; Ziegler 1907, Berl. ent. Z. 52: 114–116, bibliogr.)

THIENEMANN, AUGUST (1882–1960) (Brundin 1960, Ent. Tidsskr. 81: 144–146; Lindroth 1960, Opusc. ent. 25: 246.)

THIERSCH, ERNST LUDWIG (1786–1869) (Ratzeburg 1874, Forstwissenschaftliches Schriftsteller-Lexicon 1: 483–484.)

THIERY, NICOLAS JOSEPH THOMAS DE

MENONVILLE (1739–1780) (Urban 1898, Symbollae Antillanae 3: 136.)

THIFOR® See Endosulfan.

THIGH See Femur.

THIGMOKINESIS Noun. (Greek, *thigema* = touch; *kinesis* = movement; *sis* = a condition or state.) Movement or inhibition of movement in response to a contact stimulus. Alt. Stereokinesis. See Kinesis. Cf. Allokinesis; Blastokinesis; Diakinesis; Klinokinesis; Ookinesis; Orthokinesis. Rel. Taxis; Tropism.

THIGMOTACTIC Adj. (Greek, *thigema* = touch; *taxis* = arrangement; *-ic* = of the nature of.) Contact-loving or pertaining to Species that tend to live in close proximity or in contact with a surface or in a crevice.

THIGMOTAXIS Noun. (Greek, *thigema* = touch; *taxis* = arrangement.) A behavioural reaction in response to physical contact. See Orientation. Cf. Aerotaxis; Anemotaxis; Chemotaxis; Geotaxis; Menotaxis; Osmotaxis; Phototaxis; Rheotaxis; Rotaxis; Scototaxis; Strophotaxis; Telotaxis; Thermotaxis; Tonotaxis; Tropotaxis.

THIGMOTROPISM Noun. (Greek, *thigema* = touch; *trope* = turn; English, *-ism* = state.) Reaction to contact or touch, typically manifest in plants which spiral onto an object. See Tropism. Cf. Aeolotropism; Anemotropism; Chemotropism; Electrotropism; Galvanotropism; Geotropism; Heliotropism; Hydrotropism; Phototropism; Rheotropism; Stereotropism; Thermotropism; Tonotropism. Rel. Taxis.

THIJSSE, JAC P (1865–1945) (MacGillivray 1945, D. Ent. Ber., Amst. 11: 260–262.)

THIMET® See Phorate; Terbufos.

THIMUL® See Endosulfan.

THIN STRAWBERRY WEEVIL *Rhadinosomus lacordairei* Pascoe [Coleoptera: Curculionidae] (Australia).

THIN-N-STOP-DROP® See NAA.

THINSEC® See Carbaryl.

THIOBEL® See Cartap.

THIOCYANATE Noun. An organic insecticide that contains the Sulphur-Carbon-Nitrogen group. Has a creosote-like odour and interferes with insect cellular respiration and metabolism.

THIODAN® See Endosulfan.

THIODICARB Noun. An oxime carbamate compound {dimethyl N,N'(thiobis((methylimino) carbonoyloxy)) bisethanimidothioate} used as a contact insecticide and stomach poison. Applied to corn, cole crops, lettuce, spinach, celery, cotton, and soybeans. Effective against numerous coleopterous, dipterous and lepidopterous pests. Trade names include: Genesis®, Larvin®, Magnum®, Nirval®, Semevin®, Skipper®. See Carbamate Insecticide.

THIOFANOX An oxime compound {3-3-dimethyl-1-(methylthio)-2-butanone O-[(methylamino)carbonyl] oxime} used as a systemic contact insecticide and stomach poison against plant-sucking insects and thrips in some countries. Not registered for use in USA. Trade names include Beneluz® and Decamox®.

THIOMETON An organic-phosphate (dithlophosphate) compound {S-2-ethylthioethyl O,O-dimethyl phosphorodithioate} used as a systemic insecticide and acaricide against mites, aphids, psyllids, thrips, sawflies and plant-sucking insects in some countries. Compound not registered for use in USA. Phytotoxic to ornamentals including cyclamen, ferns and roses in glasshouse. Trade names include: Aphicide®, Dithiomethon®, Ebicid®, Ekatin®, Thiotox®, and Tombel®. See Organophosphorus Insecticide.

THIONEX® See Endosulfan.

THIOPHOS® See Parathion.

THIOSULFAN® See Endosulfan.

THIOTOX® See Thiometon.

THIOXAMYL® See Oxamyl.

THIRD ANAL VEIN The last wing vein, when there are three Anal Veins present. See Vein; Wing Venation.

THIRD AXILLARY SCLERITE The flexor sclerite of the wing base which articulates with the posterior notal process, the sclerite on which the flexor muscle is inserted (Snodgrass). See Wing Articulation.

THIRD LONGITUDINAL VEIN Diptera (Williston): Radius of Comstock (Smith). See Vein; Wing Venation.

THIRD POSTERIOR CELL Diptera: 2nd Medial of Comstock (Smith).

THIRD SUBMARGINAL CROSS-NERVURE Hymenoptera (Norton). Radius of Comstock.

THIRD VALVULAE The accessory lobes of the Ovipositor in pterygote insects, borne on the posterior end of the Second Valvifers (Snodgrass).

THIRTEEN-SPOTTED LADY-BEETLE *Hippodamia tredecimpuntata tibialis* (Say) [Coleoptera: Coccinellidae].

THISTLE APHID 1. *Brachycaudus cardui* (Linnaeus) [Hemiptera: Aphididae]. 2. *Capitophorus elaeagni* (del Guercio) [Hemiptera: Aphididae] (Australia).

THISTLE ROOT APHID *Dysaphis lappae* (Koch) [Hemiptera: Aphididae] (Australia).

THISTLE-DOWN MUTILLID *Dasymutilla gloriosa* [Hymenoptera: Mutillidae]: A North American Species which occurs near creosote, and the female resembles creosote fruit. Female apterous and white; male macropterous and red. Female tumbling movement resembles creosote fruit blowing in the wind.

THOMANN, HANS (1874–1959) (Reiss 1955, Z. wien. ent. Ges. 66: 212–214, bibliogr.; Reiss 1956, Mitt. schweiz. ent. Ges. 29: 324–325, bibliogr.)

THOMAS, CHARLES AUBREY (1895–1962) (Snetsinger 1962, J. Econ. Ent. 55: 576.)

THOMAS, CYRUS (1825–1910) (Anon 1906, National Cyclopedia of American Biography 13: 528–529; Forbes 1910, J. Econ. Ent. 3: 383–384; Weiss 1936, *Pioneer Century of American Entomology.* 320 pp. (220–221), New Brunswick.)

THOMAS, FRIEDRICH AUGUST WILHELM (1840–1918) (Hedicke 1919, Dt. ent. Z. 1919: 233; Hübenthal 1919, Ent. Bl. Biol. Syst. Käfer 15: 87–88.)

THOMAS, JOSEPH (1820–1908) (Anon. 1908, Ent. News 19: 142.)

THOMAS, LANCASTER (1838–1910) (Anon. 1910, Ent. News 21: 290.)

THOMAS, WILLIAM ANDREW (1883–1951) (Stahl & Reid 1951, J. Econ. Ent. 44: 21–22.)

THOMISIDAE Plural Noun. Crab Spiders. Moderate-sized spiders with crab-like appearance. CS do not construct webs; wait in ambush for insect visitors of flowers.

THOMPSON, CAROLINE BURLING (1869–1921) (Calvert 1922, Ent. News 33: 62–63.)

THOMPSON, D'ARCY WENTWORTH 1860–1948 British academic (Professor of Biology, University College, Dundee 1884–1917). Classical scholar with exceedingly broad interests, but perhaps best known for his work *On Growth and Form* (1917).

THOMPSON, EDWARD HENRY (1850–) (Johns 1908, *Notable Australians* (303), Sydney.)

THOMPSON, J ASHBURTON (1848–1915) (Taylor 1943, Proc. Linn. Soc. N.S.W. 68 (1–2): viii.)

THOMPSON, JOHN ANTHONY (1907–1956) (Williams 1956, Entomologist 89: 132.)

THOMPSON, MILLETT TAYLOR (1875–1907) (Felt 1915, In: Thompson *Illustrated Catalogue of American Galls*, 116 pp. (3–4), New York.)

THOMPSON, WILLIAM (–1892) (Anon. 1892, Entomologist 25: 328.)

THOMPSON, WILLIAM (1805–1852) (Westwood 1853, Proc. ent. Soc. Lond. 1853: 50–51.)

THOMPSON, WILLIAM (1824–1899) (Bengtsson 1900, Ent. Tijdskr. 21: 1–16, bibliogr.)

THOMPSON, WILLIAM LOUDEN (1896–1974) (Anon. 1975, Fla. Ent. 58: 216.)

THOMPSON, WILLIAM ROBIN (1887–1972) Advocate of the importance of abiotic factors in explaining population dynamics. Cf. Nicholson. (Downes 1972, Bull. ent. Soc. Can. 4: 38–47, bibliogr.; Thorpe 1973, Biogr. Mem. Fellows R. Soc. Lond. 19: 655–678, bibliogr.)

THOMSEN, MATH (1876–) (Tuxen 1968, Ent. Meddr. 36: 98 only.)

THOMSON, CARL GUSTAF (1824–1899) (Bengtsson 1900, Ent. Tidskr. 21: 1–16, bibliogr.)

THOMSON, CHARLES WYVILLE (1882) (Anon. 1882, Entomol. mon. Mag. 18: 264.)

THOMSON, GEORGE MALCOLM (1848–1935) (Anon. 1936, Arb. morph. taxon. Ent. Berl. 3: 64.)

THOMSON, JAMES (–1897) (Bouvier 1898, Bull. Soc. ent. Fr. 1898: 5; Kraatz 1898., Dt. ent. Z. 42: 9.)

THOMSON, JOHN GORDON (–1937) (Anon. 1937, Nature 140: 495.)

THON, KARL (1879–1906) (Korschelt *et al.* 1907, Verh. dt. zool. Ges. 17: 20, bibliogr.)

THOR, SIGMUND (1856–1937) (Lundblad 1938, Ent. Tidskr. 59: 107–111, bibliogr.)

THORACIC Adj. (Greek, *thorax* = chest; *-ic* = of the nature of.) Pertaining to the Thorax; descriptive of structure attached to the Thorax (*e.g* the wing is a thoracic appendage.)

THORACIC DORSAL BRISTLES Diptera: The specialized bristles on the dorsum of the Thorax.

THORACIC FEET The jointed (segmented) legs on the thoracic segments of the larvae, as distinguished from abdominal legs (Prolegs).

THORACIC GANGLIA Ganglia of Ventral Nerve Cord positioned in thoracic region of body. Most insects contain three TG with one positioned in each thoracic segment. Each Ganglion has 5–6 large motor neurons which connect with muscles to thoracic appendages. Mesothoracic and Metathoracic Ganglia service musculature associated with wings. See Central Nervous System. Cf. Abdominal Gangia; Brain.

THORACIC PLEURAL BRISTLES Diptera: The specialized bristles positioned on the pleural region of the Thorax.

THORACIC REGION The second of the three regions into which the embryonic trunk segments become segregated. The TR is the future locomotor centre of the insect with development of appendages as locomotory organs (Snodgrass). Rel. Head; Abdomen.

THORACIC SALIVARY GLANDS Bees: Glands in the Thorax, corresponding to the Salivary Glands.

THORACICO-ABDOMINAL SEGMENT Hymenoptera: First segment of the Abdomen when united with Thorax to form part of it. The Propodeum. See Propodeum.

THORACOTHECA See Cytotheca.

THORAX Noun. (Greek, *thorax* = chest, breastplate. Pl., Thoraces, Thoraxes.) The second or middle Tagma of the insect body. Thorax evolved early in insect history; fossils show Thorax well developed and differentiated from head and abdomen in most Palaeozoic insects. Three body Tagma probably developed during Devonian. Thorax typically composed of three segments: Prothorax, Mesothorax and Metathorax. Each of these segments is composed of three components: a dorsal sclerite, a lateral or pleural sclerite and a ventral sclerite. Dorsal sclerite is called the Notum. By convention, notal elements of Thorax are called Pronotum, Mesonotum and Metanotum. Notum can be divided into a Prescutum, Scutum and Scutellum. Pleural sclerites are called Propleuron, Mesopleuron and Metapleuron. Ventral sclerites are called Prosternum, Mesosternum and Metasternum. Mesothorax and Metathorax collectively are called the Pterothorax. Thorax bears appendages of locomotion (legs and wings). When Prothorax is free (*e.g.* Coleoptera, Orthoptera, Hemiptera), term 'Thorax' commonly used for that segment only. In Odonata (where Prothorax is small and not fused with larger Mesothorax and Metathorax) term 'Thorax' commonly used for these segments, excluding Prothorax. Primary role of Thorax has been locomotion because primary modifications have been for locomotion (first walking, then flight). Locomotion probably developed before other adaptations (such as Metamorphosis). Diverse independent and interdependent mechanisms for locomotion have evolved throughout Insecta, including walking, jumping and flying. Walking involves legs; jumping also involves legs and most insects are capable of jumping to some extent. See Tagma. Cf. Abdomen; Head. Rel. Leg; Wing.

THORELL, TORD TAMERLAU TEODOR (1830–1901) (Bonnet 1945, *Bibliographia Araneorum* 1: 39.)

THORICTIDAE See Dermestidae.

THORN, LEONARD B (1891–1924) (Anon. 1924, Victorian Nat. 41: 124–125.)

THORNEWELL, CHARLES FRANCIS (1840–1929) (Walker 1939, Entomol. mon. Mag. 65: 263–264.)

THORNLEY, ALFRED (1855–1947) (Scott 1947, Entomol. mon. Mag. 83: 109–110.)

THORNTHWAITE, WILLIAM HENRY E (1850–1908) (Anon. 1908, Entomologist's Rec. J. Var. 20: 190–191.)

THORNTON, JOSEPH NORMAN (1892–1956) (Hall 1957, Proc. R. ent. Soc. Lond. (C) 21: 66.)

THORNY-HEADED WORM See Acanthocephala.

THORSTENSEN, T D (1940) (Knaben 1941, Norsk. ent. Tidsskr. 6: 50.)

THOUSAND-LEGGED WORMS See Diplopoda.

THREAD BUG *Empicoris rubromaculatus* (Blackburn) [Hemiptera: Reduviidae].

THREAD PLATE An epithelial plate of the embryo from which terminal threads of the Ovarioles originate.

THREAD PRESS Lepidoptera larvae: Posterior part of silk-spinning apparatus, positioned within the Spinneret.

THREAD-LEGGED BUGS See Reduviidae.

THREE-BANDED LADYBIRD *Harmonia octomaculata* (Fabricius) [Coleoptera: Coccinellidae]: An Australian Species. Adult 4.6–7.5 mm long, round-ovate, with red-yellow Thorax and Elytra; black markings (median longitudinal stripe on the Elytra, two spots at wing base, four spots in middle of Elytra and two spots at wing tip) give appearance of three bands; Thorax red-yellow with a large dark spot in the middle. Adult females lay eggs in batches near larval food. Eggs yellow, spindle-shaped and laid standing on end. Larva elongate, with well defined legs; body black with markings and fleshy spines; four larval instars. Pupal stage 8–9 days. Three-banded ladybirds are predators of aphids in cotton. Adults and larvae occur in cotton feeding on aphids, mites and *Helicoverpa* eggs. Cf. Striped Ladybird; Transverse Ladybird.

THREE-BANDED LEAFHOPPER *Erythroneura tricincta* Fitch [Hemiptera: Cicadellidae].

THREE-CORNERED ALFALFA HOPPER *Spissistilus festinus* (Say) [Hemiptera: Membracidae]: A multivoltine, polyphagous pest of soybeans in North America. Adult and nymph take plant fluid from main stem near soil level; seedlings can become girdled and die. Overwinters as an adult in plant trash or pine needles; first generation appears during March and completes development on alternative host plant such as clover or vetch; adults move to soybeans during May-June. Adult bright green to brown, 6–7 mm long, triangular in outline shape when viewed from dorsal aspect. Female oviposits eggs individually or in small clusters just below soil level at main stem. Egg ca 1 mm long, white; eclosion is climate-dependent and requires ca 7–17 days. Neonate nymph nearly transparent; subsequent instars cream-coloured, buff-coloured and finally pale green; undergoes five instars with nymphal stage requires 12–16 days. TCLH completes two generations per year in soybeans.

THREE-LINED LEAFROLLER *Pandemis limitata* (Robinson) [Lepidoptera: Tortricidae].

THREE-LINED POTATO BEETLE *Lema trivittata* Say [Coleoptera: Chrysomelidae] (Australia).

THREE-SPOTTED FLEA BEETLE *Disonycha triangularis* (Say) [Coleoptera: Chrysomelidae]: A pest of beets, chickweed and spinach in USA.

THREE-STRIPED BLISTER-BEETLE *Epicauta lemniscata* (Fabricius) (Coleoptera: Meloidae].

THREE-STRIPED LADY-BEETLE *Brumoides suturalis* (Fabricius) [Coleoptera: Coccinellidae].

THRESHOLD Noun. See Medical Threshold; Economic Threshold; Social Threshold.

THRIPIDAE Plural Noun. Common Thrips. A cosmopolitan Family of Thysanoptera and the largest Family of Suborder Terebrantia. Body dorsoventrally compressed; Tentorium reduced; Antenna with 6–9 segments; forewing with one cross vein; abdominal segment 10 divided ventrally in female. Thripidae include many leaf-feeding Species of economic importance. See Cotton-Seedling Thrips; Banded Greenhouse Thrips; Citrus Rust-Thrips; Gladiolus Thrips; Greenhouse Thrips; Pear Thrips.

THRIPS Singular & Plural Noun. (Greek, *thrips* = woodworm.) The common name for Thysanoptera. See Thysanoptera.

THRIPSTIK® See Deltamethrin.

THROAT BOT-FLY *Gasterophilus nasalis* (Linnaeus) [Diptera: Oestridae]: A pest of horses which under heavy infestations causes digestive disorders. TBF second-most common aetiological agent of Enteric Myiasis in horses in many parts of world. Adult Thorax red; wing hyaline; abdomen with band of black Setae around middle. Life history similar to Horse Bot-Fly except female lays ca 300–500 yellow eggs during lifetime; eggs attached to hair under jaw or on throat of host; eclosion occurs within 4–5 days without stimulus from host. First-instar larva moves into mouth of host; larval development in anterior part or pyloric region of stomach; mature larva passed through Anus onto ground to pupate in soil. See

Oestridae. Cf. Horse Bot-Fly; Nose Bot-Fly. Rel. Gasterophilus Enteric Myiasis.

THROMBOCYTOID Noun. (Greek, *thrombos* = clot of blood; *kytos* = hollow vessel; container; *eidos* = form. Pl., Throbocytoids.) See Coagulocyte. Rel. Eidos; Form; Shape.

THROSCIDAE Plural Noun. False Metallic Woodboring Beetles. A numerically small group of oblong-oval, brownish to black beetles, 5 mm long or less. Similar to elaterids and some can 'click' or 'jump' even with Prothorax and Mesothorax fused. Adults found mostly on vegetation, little known of biology.

THROTTLE® See Carbofuran.

THRUST Noun. (Middle English, *thristen*.) Cf. Drag; Lift. Rel. Airfoil.

THUMB Noun. (Anglo Saxon, *thuma*. Pl., Thumbs.) 1. Digitate organ or structure: The thick outer finger. 2. Any thick, blunt branch arising from a structure. See Pollex.

THUMB-CLAW COMPLEX OF PALP Acarology: A compound structure that is adapted for grasping and comprised of a tibial Seta and a palpal Tarsus.

THUNBERG, CARL PETER (1743–1828) Swedish academic, traveller and author of *Characteres Generum Insectorum* (reprinted by Meyer 1791). Thunberg was a pupil of Linnaeus and succeeded him at Uppsala. (Duméril 1823, *Considérations générale sur la classe des insectes*. 272 pp. (258), Paris. Anon. 1829, K. svenska VetenskAkad. Handl. 1829: 242–267, bibliogr.; Fée 1832, Mém. Soc. Sci. agric. Lille 1: 196–219; Hagen 1857, Stettin. ent. Ztg. 18: 5–12, 200–204.)

THUNBERG, CHARLES PETER See Thunberg, Carl Peter.

THURAU, FRIEDRICH (1843–1913) (Anon. 1913, Ent. Z., Frank. a. M. 1913: 26.)

THURBERIA WEEVIL *Anthonomus grandis thurberiae* Pierce [Coleoptera: Curculionidae].

THURICIDE® Noun. A registered biopesticide derived from *Bacillus thuringiensis* var. *kurstaki*. See *Bacillus thuringiensis*.

THURINGIENSIN® See *Bacillus thuringiensis*.

THURMAN, DEED CLEVELAND (1921–1953) (Trembley 1953, J. Econ. Ent. 46: 1126.)

THURMAN-SWARTZWELDER, ERNESTINE (1920–1987) American academic (Louisiana State University Medical Center) and USPHS researcher specializing in Culicidae. (Johnson 1987, Proc. ent. Soc. Wash. 89 (4): 848–849.)

THURNALL, ALFRED (1858–1929) (Sheldon 1929, Entomologist 62: 192.)

THWAITES, GEORGE HENRY KENDRICK (1811–1882) (Anon. 1882, Entomol. mon. Mag. 19: 142–143.)

THYATIRIDAE Plural Noun. A small Family of ditrysian Lepidoptera assigned to Superfamily Geometroidea. Thyatridids sometimes placed in Drepanidae. Adult Forewing Cubitus with three branches; Hindwing Cubitus with four branches; Hindwing Sc + R1 and Rs parallel along anterior margin of Discal Cell; Thoracic typanal organs absent; abdominal tympanal organs present. Larva of North American Species feed on trees and shrubs. Syn. Cymatophoridae.

THYLACIUM Noun. (Greek, *thylakos* = pouch; Latin, *-ium* = diminutive > Greek, *-idion*. Pl., Thylacia.) An external gall-like cyst on the Abdomen of a host insect. Often seen on auchenorrhynchous Hemiptera and typically containing the parasitic larva of Dryinidae.

THYLODRIIDAE See Dermestidae.

THYNNIDAE Shuckard 1840. Plural Noun. A group of aculeate Hymenoptera sometimes placed as a Subfamily of the Tiphiidae. Member of the Scolioidea.

THYREOCORIDAE Amyot & Serville 1843. Plural Noun. Negro Bugs; Thyreocorids. A small Family of heteropterous Hemiptera assigned to Superfamily Pentatomoidea. Body black, oval, often shining; Scutellum large and covering entire Abdomen; Tibiae spinose but fore Tibia not flattened. Most Species occur on weedy plants (*e.g. Nicotiana*) in open-field habitats.

THYREOPHAGUS FLOUR MITE *Thyreophagus entomophagus* (Laboulbene) [Acari: Acaridae] (Australia).

THYRETIDAE Plural Noun. A small Family of ditrysian Lepidoptera assigned to Superfamily Noctuoidea.

THYRIDIAL CELL Trichoptera: Cell formed by first fork of Median Vein; the cell behind the Thyridium.

THYRIDIATE Adj. (Greek, *thyra* = door; *-atus* = adjectival suffix.) Pertaining to structure broken to permit folding or bending, *e.g.,* thyridiate wing vein. Alt. Thyridiatus.

THYRIDIDAE Plural Noun. A Family of ditrysian Lepidoptera assigned to Superfamily Thyridoidea or Pyraloidea. Thyridids are widespread, common in tropical areas and the Oriental Region; about 600 Species are recognized. Adult small to medium sized and stout bodied; wingspan 15–70 mm; head scales smooth; Ocelli usually present; Chaetosemata absent; Antenna form variable; Proboscis nearly always present and without scales; Maxillary Palpus very small, 1–2 segments; Labial Palpus typically of three segments, form variable; fore Tibia with Epiphysis; tibial spur formula variable; males of some Species with hair pencil at base of hind Tibia; forewing broadly triangular, without Pterostigma; hindwing rounded, broader than forewing; Abdomen lacks tympanal organ. Egg upright, flat at base and rounded apically; Chorion with ribs or reticulate sculpture; eggs laid individually. Larval head subprognathous; secondary Setae absent; body densely spinulose except where sclerotized. Pupa strongly sclerotized; Maxillary Palpus absent; Cremaster small; pupation occurs within silken cocoon in leaf litter. Adults nocturnal; rest with front part of body elevated by extended forelegs and midlegs, wings stretched away from body with apices curved upward or downward, depending upon Species. Larvae of some Species feed in shelters, some Species form galls, other Species bore into stems.

THYRIDIUM Noun. (Greek, *thyris* = window; *-idion* = diminutive. Pl., Thyridia.) 1. Neuroptera: Small, pale or almost transparent spots near the anastomosis of wing disc. 2. Trichoptera: Apical margin of gastrocoeli, often alone visible; a hyaline spot on second fork of median vein. 3. Hymenoptera: Scar-like area on the lateral portion of second gastral Tergum (first post-petiolar) segment of Ichneumonidae (See Gastrocoelus). Occasionally found on third Tergum; a Thyridium occurs in the Recurrent Veins bordering the Cubital Cell of some Hymenoptera.

THYROID Noun. (Greek, *thyreos* = shield shaped as a door; *eidos* = form. Pl., Thyroids.) Diptera: A prominent sclerite on the posterior wall of the Haustellum (Snodgrass).

THYRRHUS SKIPPER *Toxidia thyrrhus* Mabille [Lepidoptera: Hesperiidae] (Australia).

THYRSOPHORIDAE Plural Noun. A small Family of Psocoptera.

THYRSUS Noun. (Greek, *thrysos* = wand.) A cluster.

THYSANOPTERA Haliday 1836. (Permian–Recent.) (Greek, *thysanos* = fringe; *pteron* = wing.) Fringe-winged insects; thrips. A cosmopolitan Order of exopterygote Neoptera consisting of about 4,500 nominal Species. Closely related to Hemiptera and probably share common psocopteroid ancestor with Hemiptera. Body 0.5–12 mm long, slender. Head hypognathous or opisthognathous; Ocelli present in winged adult only; right Mandible absent; left Mandible forms Stylet adapted for rasping; Labrum and Labium forming a cone; Maxillae form Stylets adapted for piercing; Maxillary Palpus with 2–8 segments; Labial Palpus with 1–4 segments. Pronotum conspicuous; wing polymorphism (Apterae, Macropterae, Hemimacropterae), wings narrow with reduced venation and long marginal fringe; wings held over Abdomen at repose, parallel or overlapping, but not folded; Tarsi with 1–2 segments; Pretarsus forming an eversible bladder (Arolium) for adhesion; mesothoracic and metathoracic Spiracles present; Suboesophageal and Prothoracic Ganglia fused. Abdomen with 11 segments, apical segment usually concealed; Spiracles on segments 1–8; Ovipositor present or absent; Cerci absent; abdominal ganglia in segment 1. Metamorphosis complex and intermediate between Paurometabolous and Holometabolous development. The nymphal (larvae) instars 1–2 are active and feeding, wingless or wing buds internal; instar three (prepupa) and instar four (pupa) quiescent and nonfeeding, wing buds external; pupation near feeding site or in soil. Arrhenotokous parthenogenesis common; viviparity recognized in some Species. Sociality in some Species. Evolved from fungivorous or detritus-feeding forms; modern forms predominantly phytophagous, some Species fungivorous; a few Species predaceous. Eggs relatively large, often operculate and inserted into plant tissue, under bark or in crevices. Thysanoptera are typically multivoltine. Included Families: Adiheterothripidae, Aeolo-thripidae, Fauriellidae, Heterothripidae, Merothripidae, Phaleothripidae, Thripidae and Uzelothripidae. See Terebrantia. Tubulifera. Rel. Metamorphosis.

THYSANURA Latreille 1796. (Oliogocene–Recent.) Bristletails; Firebrats; Silverfishes. (Greek; *thysanos* = fringe, *oura* = tail. 1. A name applied to older grouping which included Archaeognatha or Microcoryphia, in part. 2. An Order of small to moderate sized, fusiform, dorsoventrally compressed insects. Body usually covered with scales; Mandible dicondylic with two points of articulation; compound eye small or absent, eyes never medially contiguous; Ocelli sometimes present; Antenna elongate, filiform; Maxillary Palpus with five segments. Thoracic segments similar; Coxal Styli absent. Abdomen with 11 segments, some Sterna with Styli; Appendix Dorsalis and Cerci present, similar in size. Simple Metamorphosis. Included Families: Lepidothrichidae, Lepismatidae, Maindroniidae, Nicoletiidae. See Apterygota. Cf. Archaeognatha.

THYSANURA ENTOTROPHICA The Diplura; an arcane taxonomic grouping which is characterized by Mandibles and Maxillae retracted into a pouch above the Labium.

THYSANURIFORM Adj. (Greek, *thysanos* = fringe; Latin, *forma* = shape.) Resembling a thysanuran. Term used to describe the form of some larvae. See Campodeiform.

TIARATE Adj. (Unknown origin; *-atus* = adjectival suffix.) Turbanate or tiara-like. Alt. Tiaratus.

TIBIA Noun. (Latin, *tibia* = shin. Pl., Tibiae.) The fourth segment of insect leg. Tibia is typically long, slender, articulated proximally with Femur and distally with Basitarsus. Basal portion called the Moula which forms the convex, proximal end of Tibia. A femoro-tibial joint develops between Femur apex and Tibia base. Bees collect pollen with a Corbicula (pollen basket) on hind Tibia. Distal part of Tibia usually modified for grooming. Many insects possess articulate spurs or spines at apex of Tibia; Many insects possess articulate spurs or spines at the apex of the Tibia (*e.g.* Calcar, Epiphyses). Syn. Shin. See Leg. Cf. Coxa; Femur; Pretarsus; Trochanter. Rel. Scent Plaques (Pseudosensoria; Pseudorhinaria).

TIBIA FLEXIS The femoro-tibial flexure (MacGillivray). Rel. Coxal Corium.

TIBIAL Adj. (Latin, *tibia* = shin; *-alis* = pertaining to.) Descriptive of or pertaining to the Tibia.

TIBIAL EPIPHYSIS A moveable process attached near the base of the medial surface of the fore

Tibia in many Lepidoptera. See Epiphysis.

TIBIAL EXTENSION The basic mechanism by which most insects jump, typically involving the middle Tibia and seen in Orthoptera, Hemiptera, Psocoptera, Siphonaptera, Mecoptera (*Boreus*), Diptera (Chloropodae, Tipulidae.) See Jumping Mechanism.

TIBIAL MEMBRANE Male Cicada: The drum-like vibratory membrane that produces the sound. See Mirror; Tymbal; Stirrup.

TIBIAL SPURS The spur or spurs frequently borne near to or at the distal end of the Tibia.

TIBIAL THUMB *Pediculus* and related lice: An extension of medial, distal end of Tibia; apposed to the claw for grasping hairs of mammal hosts.

TIBIAROLIUM Noun. (Latin, *tibia* = shin; Greek, *arole* = protection; Latin, *-ium* = diminutive > Greek, *-idion*. Pl., Tibiarolia.) A transparent pad of the distal ventral end of the Tibia; resembling an Arolium closely covered with Retiniariae (MacGillivray).

TIBIOFEMORAL Adj. (Latin, *tibia* = shin; *femur* = thigh; *-alis* = pertaining to.) Descriptive of, pertaining to, or connecting the Tibia and Femur.

TIBIOTARSAL ORGAN Collembola: A sac-like swelling and an enlarged seta on the medial surface of the hind Tibio-tarsus in Sminthuroides.

TIBIOTARSAL SEGMENT The fused Tibia and Tarsus in some insects.

TICK FLIES See Hippoboscidae.

TICK-BORNE ENCEPHALITIS A tick-borne arboviral disease whose agent belongs to the Family Flaviviridae. TBE occurs in north Palearctic and manifest in two forms: Central European Subtype (CEE, with *Ixodes ricinus* as vector) and Far Eastern Subtype (with *I. persulcatus* as vector.) Far Eastern Subtype also called Russian Spring-Summer Encephalitis (RSSE) and more severe disease with mortality ca 20–30%. Both forms can be trans-ovarially or trans-stadially transmitted; epizootics in mammals and birds associated with forests. See Arbovirus. Flaviviridae. Cf. Kyasanur Forest Disease; Louping Ill; Omsk Haemorrhagic Fever.

TICK-BORNE SPOTTED FEVER Rickettsial disease caused by any of five aetiological agents [*Rickettsia australis, R. conorii, R. rickettsii, R. sibirica* and *R. japonica*] which can be transmitted trans-ovarially or trans-stadially by ixodid ticks to humans. See Rickettsiaceae. Rickettsial Pox. Rocky Mountain Spotted Fever.

TICKER *Pauropsalta mneme* (Walker) [Hemiptera: Cicadidae] (Australia).

TICKLE, J D (–1959) (Varley 1962, Proc. R. ent. Soc. Lond. (C) 26: 54.)

TICKS See Acari.

TICK-TOCK *Cicadetta quadricincta* (Walker) [Hemiptera: Cicadidae] (Australia).

TIEF, WILHELM (1846–1896) (Kittner *et al.* 1896, Carinthia 11 86: 137–144.)

TIEFFENBACH, H (1820–1892) (Kraatz 1892, Dt. Ent. Zeit. 36: 12–13.)

TIEGS, OSCAR WERNER (1897–1956) (Pantin 1957, Biogr. Mem. Fellows R. Soc. Lond. 3: 247–255. bibliogr.)

TIEMPO® See Abate; Temephos.

TIGER BEETLE Any of several Species of *Cicindela* found in North America.

TIGER BEETLES See Carabidae; Cicindelidae.

TIGER LONGICORN *Aridaeus thoracicus* (Donovan) [Coleoptera: Cerambycidae] (Australia).

TIGER MOTHS See Arctiidae.

TIGER PEAR *Opuntia aurantiaca* Lindley [Cactaceae]: A Species of cactus native to the New World and adventive elsewhere as a minor rangeland pest. In Australia limited control with *Cactoblastus cactorum* (Bergroth) and *Dactylopius* sp. See Prickly Pear.

TIGER PEAR COCHINEAL *Dactylopius austrinus* De Lotto [Hemiptera: Dactylopiidae] (Australia).

TIGER PRINCE *Macrotristria godingi* Distant [Hemiptera: Cicadidae] (Australia).

TIGER SWALLOWTAIL *Papilio glaucus* Linnaeus [Lepidoptera: Papilionidae]: A common butterfly of eastern North America. Host plants include ash, birch, poplar, wild cherry. TST overwinters as a pupa. Adults sometimes taken at decaying animal matter or smoke.

TIGUVON® See Fenthion.

TILDEN, JAMES WILSON (1904–1988) American academic (San Jose State University) and taxonomic specialist of Lepidoptera. (Arnaud 1992, Pan-Pac. Ent. 68 (1): 27–37.)

TILEHORNED PRIONUS *Prionus imbricornis* (Linnaeus) [Coleoptera: Cerambycidae].

TILLER Noun. (Anglo Saxon, *telgor* = small branch.) Wheat: A sucker or shoot from the root or near the main stalk.

TILLYARD, ROBIN JOHN (1881–1937) Englishman by birth, educated at Cambridge University and employed in Australia and later New Zealand. Wrote extensively on fossil insects and produced several treatises on extant insects including *The Biology of Dragonflies* and *The Insects of Australia and New Zealand.* (Anon. 1937, Ent. News 48: 42; Hale-Carpenter 1937, Proc. Linn. Soc. Lond. 1936–37: 212–218; Imms 1937, Proc. R. ent. Soc. Lond. (C)I:56; Hull 1937, Aust. Zool. 8: 343; Imms 1938, Obit. Not. Fell. R. Soc. 2: 339–345.)

TILLYARD'S SKIPPER *Anisynta tillyardi* Waterhouse and Lyell [Lepidoptera: Hesperiidae] (Australia).

TILTSCHER, PAUL (1891–1917) (Esiki 1917, Rovart. Lap. 24: 179–180, 196.)

TIMBAL Noun. (French, *timbale* = kettle drum. Pl., Timbals.) Cicadidae: The membranous area in the lateral cavity on the lateral wall of the partition separating the two cavities of the chordotonal organ (Comstock). The shell-like drum in cicadas, at the base of the Abdomen used to produce sound. Syn: Tympanum. Alt. Tymbal.

TIMBER BEETLES See Curculionidae; Lymexylidae.

TIMBERLAKE, PHILIP H (1884–1981) American

taxonomic Entomologist specializing in parasitic Hymenoptera early in his career and bees for more than 50 years. Employed by USDA (Melrose Highlands Laboratory, Massachusetts; Salt Lake City, Utah), Hawaiian Sugar Planters Association (Honolulu) and University of California (Riverside).

TIMBO® See Rotenone.

TIM-BOR® See Inorganic Borate.

TIMEMIDAE Plural Noun. Timema Walkingsticks. A small Family of Phasmatoidea, the Species of which resemble earwigs. Adult 15–30 mm long; apterous; legs short; tarsal formula 3-3-3. North American Species restricted to southwest where they feed on deciduous tree foliage.

TIMET® See Phorate; Terbufos.

TIMM, RUDOLF (1859–1936) (Weidner 1967, Abh. Verh. naturw. Ver. Hamburg Suppl. 9: 193.)

TIMM, WULF WILHELM (1850–1924) (Anon. 1924, Sber. vorträge Ent. Ver. Hamburg 1924: 56–57, bibliogr.)

TIMMS, CARTWRIGHT (1901–1970) (Smith 1970, Entomologist's Rec. J. Var. 82: 154–155.)

TIMON-DAVID, JEAN (1902–1968) (Léonide 1970, Annls. Fac. Sci. Marseille 43: 5–19, bibliogr.)

TIMYRIDAE See Lecithoceridae.

TINCTORIAL Adj. (Latin, *tinctorius* = dyeing; *-alis* = pertaining to.) Pertaining to the production of dyes.

TINCTORIAL PATTERN Colour pattern.

TINDALE'S GRASSGRUB *Oncopera tindalei* Common [Lepidoptera: Hepialidae] (Australia).

TINEIDAE Latreille 1810. Plural Noun. Clothes Moths. A cosmpolitan Family of ditrysian Lepidoptera assigned to Superfamily Tineoidea and containing ca 2,000 Species. Adult body small to moderate size; wingspan 5–35 mm; head typically appearing coarse with piliform scales, some Species with overlapping lamellar scales; Chaetosemata, Ocelli absent; Antenna filiform, Scape typically with Pecten; Proboscis short and without scales or absent; rudimentary Mandible present in a few Species; Maxillary Palpus typically with five Segments, folded; Labial Palpus with three Segments, porrect or pendulous; Epiphysis usually present; tibial spur formula 0-2-4; hind Tibia never with smooth scales; wings typically narrow with long marginal fringe. Adult does not feed. Egg oval with reticulate pattern of sculpture. Larva cylindrical; head hypognathous; prothoracic shield present; sometimes bears case; typically feeds on fungi, dried animal or vegetable matter; several Species regarded as significant pests of grain and animal fibres. Pupa typically with two transverse rows of spines on Terga; segments 4–6 mobile in female, 4–7 mobile in male. Adult nocturnal, rests with body flat, Antenna curved rearward, wings held tent-like over Abdomen; prefer to run when disturbed. Syn. Acrolophidae (sometimes); Amydriidae; Hieroxestidae; Oinophilidae; Setomorphidae; Tinaeidae. See Casemaking Clothes Moth.

TINEODIDAE Plural Noun. A small Family of ditrysian Lepidoptera containing ca 15 Species and assigned to Superfamily Alucitoidea. Nearly all Species from Australia with a few sound in Asia. Adult small; wingspan 10–35 mm; head with smooth scales; Ocelli present in some Species; Chaetosemata absent; Antenna 0.6 to slightly longer than forewing; Proboscis well developed, unscaled; Maxillary Palpus with four segments; Labial Palpus porrect; legs long; fore Tibia with Epiphysis; tibial spur formula usually 0-2-4. Larval head subprognathous. Pupa slender, well sclerotized. Adult rests with wings spread and body elevated.

TINEOIDEA A Superfamily of ditrysian Lepidoptera including Arrhenophanidae, Eriocottidae, Gracillariidae, Lyonetidae, Phyllocnestidae, Pseudarbelidae, Psychidae and Tineidae. Adult body small to moderate size; Ocelli present or absent; Chaetosemata absent; Proboscis reduced, without scales; Maxillary Palpus of 1–5 segments, not folded over base of Proboscis; Epiphysis typically present; tibial spur formula 0-2-4. Larvae feed in concealed habitats, portable cases, tunnels, and leaf mines. Pupa with dorsal abdominal spines; male Abdomen mobile; pupation within larval shelter or silken cocoon. See Lepidioptera.

TINGIDAE Plural Noun. Lace Bugs. A small Family of heteropterous Hemiptera. Adult small, oval or rectangular in outline; dorsum with elaborate reticulate pattern of elevated ridges and sunken membranous oval areas; Ocelli absent; Antenna with four segments; Tarsi with two segments; nymphs spiny, do not resemble adults. Nymphs, adults found on underside of leaves, consume plant juices. Eggs laid in groups inserted within underside of leaf near midrib, sometimes covered with accessory gland secretion which hardens, becoming cone-like.

TINGOIDEA A Superfamily of Hemiptera including Joppeicidae, Tingidae and Vianaididae. See Hemiptera.

TINSEAU, ROBERT DE (–1882) (Percrin 1882, Revue Ent. 1: 95–96.)

TINY GRASS-BLUE *Zizula hylax attenuata* (Lucas) [Lepidoptera: Lycaenidae] (Australia).

TIONEL® See Endosulfan.

TIP Adj. (Middle English, *tip*, *tippe*.) The extremity of a structure; the part farthest from the base of a structure or appendage. Syn. Apex. Cf. Base.

TIP DWARF MITE *Calepiterimerus thujae* (Garman) [Acari: Eriophyidae].

TIPHIIDAE Leach 1815. Plural Noun. Rolling Wasps; Tiphiid Wasps. A cosmopolitan Family of aculeate Hymenoptera consisting of about 1,500 nominal Species placed within about 90 Genera and currently assigned to Superfamily Vespoidea; tiphiids placed within Scolioidea in some classifications. Adult moderate to large bodied (ca 30 mm long); male Antenna with 13 segments, female Antenna with 12 segments;

Pronotum, Mesonotum separated by suture, not fused; sexual dimorphism variable (slight to extreme), males macropterous, females, macropterous or apterous; macropterous forewing wing with extensive venation; hindwing with closed cells, claval and anal lobes; middle, hind legs spinose; middle Coxae conspicuously separated; Mesepisternum with spines or lamina near middle Coxa; first and second gastral Terga separated by constriction; male Gaster with apical spines. Larvae develop as parasites of Coleoptera, usually Scarabaeidae, Tenebrionidae, Cicindelidae and sometimes Curculionidae; one Genus, *Diamma,* attacks mole crickets. Female displays complex fossorial behaviour: Searches for grub in soil; host is paralysed with sting and malaxated with mandibles; oviposition on host in host's cell or burrow. Males visit flowers. Fossil record (Baltic Amber, Florissant, USSR). Tiphiidae Subfamilies: Anthoboscinae, Brachycistidinae, Diamminae, Methocinae, Myzininae, Thynninae, Tiphiinae.

TIPOFF® See NAA.

TIP-OFF® See NAA.

TIPULID Adj. Descriptive of or pertaining to the dipterous Family Tipulidae.

TIPULIDAE Plural Noun. Crane Flies, Daddy Long Legs. A cosmopolitan Family of nematocerous Diptera consisting of about 4,000 nominal Species, most described by C. P. Alexander. Adult body elongate; Ocelli absent; 'V' shaped mesonotal suture strong; wing with two Anal Veins; legs long, slender. Adults inhabit moist, shaded habitats; larvae live in water, wet soil or decomposing vegetation. Classification of Tipulidae remains disputed; some taxonomists remove Cylindrotomatinae and Limoniinae and recognize them as separate Families. Fossil record extensive with more than eight Genera and 100 Species known from numerous deposits in many regions. See Cylindrotomatidae; Limoniidae.

TIPULODICTYIDAE Rohdendorf 1962. A Family of Diptera known only from the type Species preserved in Lower Jurassic Period deposits at Kirghizistan

TIPULOMORPHA A Division of nematocerous Diptera including Ptychopteridae (Liriopeidae), Tipulidae, Trichoceridae; (Petauristidae, Trichoceratidae).

TIPULOPLECIIDAE Rohdendorf. Plural Noun. A Family of fossil Diptera known only from an impression of one specimen from Upper Jurassic Period deposits of Kazakhstan.

TIRADE® See Fenvalerate.

TIRELLI, ADECHI (1868–1945) (Tirelli 1972, Memorie Soc. ent. ital. 51: 156–160.)

TISCHBEIN, PETER FRIEDRICH LUDWIG (1813–1883) (Ratzeburg 1874, Forstwissenschaftliches Schriftsteller-Lexicon 1: 484–486.)

TISCHER, CARL FRIEDRICH AUGUST (1777–1849) (Anon. 1850, Stettin. ent. Ztg. 11: 32.)

TISCHERIIDAE Plural Noun. Tischeriid Moths. A small Family of monotrysian Lepidoptera assigned to Superfamily Nepticuloidea. Adult maxillary Palpus reduced or absent; costal margin of forewing arched and apex of wing pointed; hindwing long, narrow with reduced venation. Larvae form blotch mines on Oak tree leaves or leaves of fruit trees and bushes. See Apple-Leaf Trumpet Miner.

TISCHLER, THEODOR (1864–1919) (Hedwig 1919, Jh. Ver. schles. Insektenk. 10–12: 26.)

TISSUE Noun. (Middle English, *tissu* > French, *tissu* = woven > Latin, *texere.* Pl., Tissues.) An aggregation of cells of similar structure and function that collectively form specific parts of a plant or animal (*e.g.* connective tissue). See Organization. Cf. Cell; Gland; Organ; Parenchyma.

TISSUE CULTURE An *in vitro* technique by which cells are maintained or grown. Tissue Culture preserves differentiated cells and maintains the intercellular arrangements and cell architecture. See Genetic Engineering. Rel. Biotechnology.

TISSUE WATER Water produced in the cells by chemical reaction. See Metabolic Water.

TITAN® See Chlormequat.

TITANIIDAE See Chloropidae.

TITILLATOR Noun. (Latin, *titillare* = to tickle; *-or* = one who does something. Pl., Titillators.) Orthoptera: A small process at the apex of the Aedeagus.

TITSCHACK, ERICH HANS WOLDENMAR (1892–) (Cleve I962, Mitt. dt. ent. Ges. 21: 33–35.)

TITZE, ALBERT (–1899) (Wocke 1900, Z. Ent. 25: 24–25.)

TM BIOCONTROL-1® See Douglas-Fir Tussock Moth NPV.

TMV Tobacco Mosiac Virus.

TOAD BUG 1. *Gelastocoris oculatus* (Fabricius) [Hemiptera: Gelastocoridae]: A stream-bank inhabiting Species that is widespread in North America. TB eggs are laid in sand, mud and under stones away from the water's edge. TB nymphs and adults are toad-like in appearance with hopping habits and cryptic coloration to resemble pebbles and mud. TBs are predaceous upon insects along stream margin. TB overwinter as adults. 2. *Nerthra grandis* (Montandon) [Hemiptera: Gelastocoridae] (Australia). See Gelastocoridae.

TOAD BUGS See Gelastocoridae.

TOBA, GENZO (1872–1946) (Hasegawa 1967, Kontyû 35 (Suppl.): 55, bibliogr.)

TOBACCO APHID *Myzus nicotianae* Blackman [Hemiptera: Aphididae].

TOBACCO BEETLE See Cigarette Beetle.

TOBACCO BUDWORM 1. *Helicoverpa virescens* (Fabricius) [Lepidoptera: Noctuidae]: A multivoltine, polyphagous and significant pest of cotton in the southeastern USA. TBW has developed resistance to many synthetic pesticides including carbamates, cyclodienes, organophosphates and pyrethroids. TBW undergoes five instars. First generation typically feeds on legumes and tobacco; second generation prefers corn and subsequent generations prefer cotton. Adult ca 18 mm long with 35 mm wingspan; forewing with three pale transverse marks each with black margin and wing margin pale; hindwing pale with dark transverse band; adults are active nocturnally. TBW overwinters as a Pupa in soil. Female can lay ca 1,000–3,000 eggs during lifetime. Eggs are spherical with radiating ridges, ca 0.5 mm and white when laid and becoming grey or brown with development. Eggs are laid individually on plants during night; eclosion occurs within 2–4 days. Neonate larvae feed on apical meristem and terminal buds of cotton; later instars feed on cotton squares, flowers and bolls. Mature larvae are ca 40 mm long and resemble Corn Earworm in coloration except with vestiture of tuberculate microspines. Pupa as Corn Earworm except more reddish. Cf. Cotton Bollworm. 2. *Helicoverpa armigera* (Hübner) [Lepidoptera: Noctuidae] (Australia). See Noctuidae.

TOBACCO CATERPILLAR *Spodoptera litura* (Fabricius) [Lepidoptera: Noctuidae]. Syn. Cluster Caterpillar; Cotton Leafworm; Tobacco Cutworm.

TOBACCO CUTWORM *Spodoptera litura* (Fabricius) [Lepidoptera: Noctuidae]. Syn. Cluster Caterpillar; Cotton Leafworm; Tobacco Caterpillar. See Cutworms.

TOBACCO FLEA-BEETLE *Epitrix hirtipennis* (Melsheimer) [Coleoptera: Chrysomelidae]: A serious pest of tobacco in some regions of USA but also attacks eggplant, pepper, tomato and potato. Eggs are deposited on soil surface beneath host plant; eclosion occurs within 7 days and neonate larvae burrow into soil to feed on rootlets; larval development complete within 15 days. Adults chew small holes in leaves and also attack young plants in seedbeds; larvae typically feed on roots of tobacco plant and occasionally tunnel in stems. TFB multivoltine with 3–4 generations per year.

TOBACCO HORNWORM *Manduca sexta* (Linnaeus) [Lepidoptera: Sphingidae]: A widespread and significant pest of many plants, including tomato, potato, tobacco, pepper, eggplant. Eggs spherical, greenish and deposited individually on underside of leaves; eclosion occurs within 7 days. Larva feeds externally on foliage; undergoes five moults; mature larva large-bodied (10–12 cm), green with seven oblique white stripes on side of body and a red 'horn' at the apex of the abdomen. Pupation occurs in soil. TH often bivoltine. Syn. *Protoparce sexta* (Johanssen). See Hawk Moth. Cf. Tomato Hornworm.

TOBACCO LEAFMINER *Phthorimaea operculella* (Zeller) [Lepidoptera: Gelechiidae] (Australia).

TOBACCO MOTH *Ephestia elutella* (Hübner) [Lepidoptera: Pyralidae]: A widepread pest of flour, stored grains, dried vegetables and cured tobacco. Female lays 150–200 eggs; egg elliptical, granular and greyish-brown; eclosion occurs within 4–7 days; larva whitish with yellow or brown spots; feeding requires 40–45 days; pupation occurs in crevices or concealed places and requires 7–19 days; adults live three weeks at 30°C, 70% RH. Life cycle 42–49 days at 25°C. 70% RH; 77–84 days at 20°C. Diapause at lower temperatures; diapause larvae survive 450 days. Life history resembling Mediterranean Flour Moth. See Pyralidae. Cf. Almond Moth; Raisin Moth.

TOBACCO SPLITWORM See Potato Tuberworm.

TOBACCO STEMBORER *Scrobipalpa aptotella*. (Lower) [Lepidoptera: Gelechiidae] (Australia).

TOBACCO THRIPS 1. *Frankliniella fusca* (Hinds) [Thysanoptera: Thripidae]: A pest of seedling cotton, peanuts, tobacco and flowers in eastern and southern North America. Syn. Onion Thrips; Potato Thrips. See Onion Thrips. Cf. Flower Thrips; Western Flower Thrips. 2. *Anaphothrips cecili* Girault [Thysanoptera: Thripidae] (Australia).

TOBACCO WHITEFLY *Bemisia tabaci* (Gennadius) [Hemiptera: Aleyrodidae] (Australia). See Sweetpotato Whitefly.

TOBACCO WIREWORM *Conoderus vespertinus* (Fabricius) [Coleoptera: Elateridae]: A pest of beans, cotton, corn, tobacco, potatoes and truck crops in eastern and southern USA.

TOCOSPERMIA Noun. (Greek, *toketos* = bearing offspring; *sperma* = seed.) Acarology: Sperm transfer in which the male Chelicera transports sperm directly to the female's Vagina.

TOCOSPERMY Noun. (Greek, *tokos* > *tiktein* = to beget, to bear; *sperma* = seed. Pl., Tocospermies.) The phenomenon, act or process of introducing sperm into a female without copulation.

TODD, DOUGLAS HAIG (1917–1969) (Esson 1970, N.Z. Ent. 4: 94–96; McGregor 1970, J. Econ. Ent. 63: 346.)

TODD, EDWARD L American government taxonomist (Systematic Entomology Lab, U.S. National Museum) specializing in Noctuidae and Gelastocoridae.

TODD, FRANK EDWARD (1895–1969) (McGregor 1970, J. Econ. Ent. 63: 346.)

TODD, ROBERT GEOFFREY (1886–1967) (Worms 1967, Entomologist's Rec. J. Var. 79: 208.)

TOE BITERS See Belostomatidae.

TOE-BITER *Laccotrephes tristis* (Stål) [Hemiptera: Nepidae] (Australia).

TOED-WINGED BEETLES See Ptilodactylidae.

TOJYO, MISAO (1884–1966) (Hasegawa 1967, Kontyû 35 (Suppl.) 55.)

TOKUTHION® Prothiofos.

TOLERANCE Noun. (Latin, *tolerare* = to endure. Pl., Tolerances.) 1. The ability of a host plant to withstand injury due to pest activity. 2. The ability to endure the effects of pesticides or poisons. 3.

The permissible deviation from a standard.

TÖLG, FRANZ (1877–1917) (Heikertinger 1917, Wien. ent. Ztg. 36: 117–120, bibliogr.)

TOLL, SERGIUSZ (GRAF VON) (1893–1961) (Reisser 1962, Z. wien. ent. Ges. 47: 79–84, bibliogr.)

TOLUNAY, MITHAT ALI (–1962) (Alkan 1962, Z. angew. Ent. 50: 356–357.)

TOMASSETTI, MARIO (–1959) (Anon. 1959, Boll. Soc. ent. ital. 89: 97.)

TOMASZEWSKI, WALTER (1903–1955) (Hase 1955, Anz. Schädlingsk. 28: 62.)

TOMATHREL® See Ethephon.

TOMATO APHID *Macrosiphum euphorbiae* (Thomas) [Hemiptera: Aphididae] (Australia).

TOMATO BUG *Cyrtopeltis modesta* (Distant) [Hemiptera: Miridae].

TOMATO ERINEUM-MITE *Eriophyes lycopersici* (Wolffenstein) [Acari: Eriophyidae] (Australia).

TOMATO FRUITWORM *Helicoverpa zea* (Boddie) [Lepidoptera: Noctuidae]: A significant pest of numerous field crops in many regions of the world. Syn. Bollworm; Corn Earworm; *Heliothis zea* (Boddie).

TOMATO GRUB *Helicoverpa armigera* (Hübner) [Lepidoptera: Noctuidae] (Australia). See Cotton Bollworm.

TOMATO HORNWORM *Manduca quinquemaculata* (Haworth) [Lepidoptera: Sphingidae]: A sporadic pest of tomatoes in the USA. Larva greenish, large bodied with eight 'L' or 'V' shaped stripes on side of body and a black 'horn' at apex of abdomen. Cf. Tobacco Hornworm.

TOMATO LEAFHOPPER *Austroasca viridigrisea* (Paoli) [Hemiptera: Cicadellidae] (Australia).

TOMATO LEAFMINER 1. *Liriomyza bryoniae* Kaltenbach [Diptera: Agromyzidae]: A sporadic pest of tomatoes, cucumbers and melons in glasshouse cultivation in northwestern Europe. Larvae polyphagous, feed on most plant parts except roots. Host plants include tomato (*Lycopersicon esculentum* Miller), potato (*Solanum tuberosum* Linnaeus), eggplant (*Solanum melongea* Linnaeus) and numerous other plants. 2. *Scrobipalpula absoluta* (Meyrick) [Lepidoptera: Gelechiidae].

TOMATO LOOPER *Plusia chalcites* [Lepidoptera: Noctuidae]

TOMATO MIRID *Engytatus nicotianae* (Koningsberger); *Nesidiocoris tenuis* (Reuter) [Hemiptera: Miridae] (Australia). A pest of tobacco, tomatoes and other solanaceous plants; feeding activity results in fruit blemish and dieback.

TOMATO MOTH *Lacanobia oleracea* [Lepidoptera: Noctuidae]

TOMATO PINWORM *Keiferia lycopersicella* (Walsingham) [Lepidoptera: Gelechiidae]: A widespread pest of tomatoes in South America, Central America and southern USA; endemic to Mexico. Early instar larvae are leafminers; later instar larvae are leafrollers; pest status achieved by larva entering fruit below calyx.

TOMATO PSYLLID *Paratrioza cockerelli* (Sulc) [Hemiptera: Psyllidae].

TOMATO-RUSSET MITE *Aculops lycopersici* (Massee) [Acari: Eriophyidae].

TOMATO STEMBORER *Symmetrischema tangolias* (Geyen) [Lepidoptera: Gelechiidae] (Australia).

TOMATO THRIPS *Frankliniella schultzei* (Trybom) [Thysanoptera: Thripidae] (Australia). In Australia, a sporadic pest of early season cotton. Adult 1.0–1.2 mm long, pale yellow to dark brown. Females with four wings, narrow with a long fringe of Setae along margin; Tarsi lack claws but end with a small bladder. Adult female uses Ovipositor to saw into leaf tissue and deposit eggs in slit. Eggs laid in leaves or stems; eggs white and kidney-shaped; egg stage 5–10 days. Nymphs resemble adults in appearance and habits. Nymphs lack wings and are paler than adults. Nymphs pass through four instars, two occur in soil and are non-feeding. Pupation occurs in soil. All generations occur on plant and overlap. Multivoltine with 5–8 generations per year; life-cycle is completed in 15–21 days. Adults puncture leaves and stems to feed on exuding sap. Tomato thrips is polyphagous and can spread Tomato Spotted Wilt Virus. On cotton, damage is similar to Onion Thrips and produces malformation of the leaf, flower, flower bud and young seedling. Syn. Cotton Bud Thrips; Common Blossom Thrips. Cf. Onion Thrips.

TOMBEL® See Quinalphos; Thiometon.

TOMENTOSE Adj. (Latin, *tomentum* = stuffing; *-osus* = full of.) Covered with a tomentum. Alt. Tomentosus.

TOMENTUM Noun. (Latin, *tomentum* = stuffing. Pl., Tomenta.) 1. A form of pubescence composed of matted, woolly Setae. 2. Diptera: A covering of short, flattened, more-or-less recumbent, scale-like Setae which merges gradually into dust or pollen.

TOMLIN, J R le B (1863–1954) (Buxton 1955, Proc. R. ent. Soc. Lond. (C) 19: 69–70.)

TÖMÖSVARY, ÖDÖN (1852–1884) (Anon. 1884, Zool. Anz. 7: 688; Mik *et al.* 1884, Wien. ent. Ztg. 3: 288; Otto 1885, Rovart. Lap. 2 (1): 1–14, bibliogr.)

TOMSIK, BOLESLAVEM (1907–1970) (Povolny & Vesely 1971, Acta ent. bohemoslovaca 68: 285–287. bibliogr.)

TONGAVIRIDAE Plural Noun. A Family of viruses which contains nearly 30 arboviruses in the Genus *Alphavirus*. Virons ca 50–70 nm diameter, spherical with lipoprotein envelope. Most *Alphavirus* virons vectored by mosquitoes. Viruses produce several diseases in humans and livestock, including Chikungunha, Eastern Equine Encephalitis, Venezuelan Equine Encephalitis, Western Equine Encephalitis, O'Nyong-Nyong, Ross River Fever and Sindbis. See Arbovirus. Cf. Bunyaviridae; Flaviviridae; Reoviridae; Rhabdoviridae.

TONGE, ALFRED ERNEST (1869–1939) (Blair 1939, Entomol. mon. Mag. 75: 203.)

TONGUE Noun. (Anglo Saxon, *tunge*. Pl., Tongues.) 1. An elongate, sensory structure attached at one end to the floor of the mouth of vertebrates and which contains intrinsic musculature. 2. An indefinite term, applied usually to the coiled mouthparts of Lepidoptera. 3. The lapping organ of flies. 4. The Ligula of bees and wasps. 5. The Hypopharynx of some insects.

TONGUE WORMS See Pentastomidae.

TONIC Adj. (Latin, *tonos* = tension; *-ic* = characterized by.) Pertaining to the tone or strength of the system or of a structure or organ.

TONNOIR, ANDRÉ LEON (1885–1940) (Edwards 1940, Entomol. mon. Mag. 76: 118–120; Holmes 1940, Proc. Linn. Soc. N.S.W. 65: ii; Nicholson 1940, Nature 145: 453–454.)

TONOFIBRILLAE Plural Noun. (Greek, *tonos* = tension; Latin, *fibrilla* = small fibre.) Cuticular or integumentally derived support non-contractile fibrils which connect muscle fibres with the inner surface of the Integument. Cf. Myofibrillae.

TONOTAXIS Noun. (Greek, *tonos* = tension; *taxis* = arrangement. Pl., Tonotaxes.) A tactical response to a change in density of the medium surrounding an organism. See Orientation. Cf. Aerotaxis; Anemotaxis; Chemotaxis; Geotaxis; Menotaxis; Osmotaxis; Phototaxis; Rheotaxis; Rotaxis; Scototaxis; Strophotaxis; Telotaxis; Thermotaxis; Thigmotaxis; Tropotaxis.

TONOTROPISM Noun. (Greek, *tonos* = tension; *trope* = turn; English, *-ism* = condition. Pl., Tonotropisms.) A behavioural response to sound. See Tropism. Cf. Aeolotropism; Anemotropism; Chemotropism; Electrotropism; Galvanotropism; Geotropism; Heliotropism; Hydrotropism; Phototropism; Rheotropism; Stereotropism; Thermotropism; Thigmotropism. Rel. Taxis.

TONSIL Noun. (Latin, *tonsilla* = tonsil.) Any of several masses of lymphoid tissue.

TOOTH Noun. (Anglo Saxon, *toth*. Pl., Teeth.) 1. An acute angulation. 2. Any short, pointed process from an appendage or margin or a structure.

TOOTH NECKED FUNGUS BEETLES See Derodontidae.

TOOTHED FLEA BEETLE *Chaetocnema denticulata* (Illiger) [Coleoptera: Chrysomelidae].

TOOTH-NOSED WEEVILS See Rhynchitidae.

TOPAZINE Adj. Topaz-coloured; a light crystalline golden yellow. Alt. Topazinus.

TOPLINE® See Amitraz.

TOPOCLINE Noun. (Greek, *topos* = place; *klinein* = to slant. Pl., Topoclines.) A geographical variation attributable to climate or topography.

TOPODEME Noun. (Greek, *topos* = place; *demos* = people. Pl., Topodemes.) A deme found in a given geographical region or area.

TOPOGRAPHY Noun. (Greek, *topos* = place; *graphein* = to write. Pl., Topographies.) The practice of graphic representation in exact and minute detail any aspects of the essential physical features and surface contours of a body, structure or appendage. Cf. Eggshell Topography; Sculpture Pattern. Rel. Eggshell; Integument.

TOPOMORPH Noun. (Greek, *topos* = place; *morphe* = shape. Pl., Topomorphs.) A geographic form, variety or Subspecies of a widely distributed Species developed by local environmental conditions.

TOPOMORPHIC Adj. (Greek, *topos* = place; *morphe* = shape; *-ic* = of the nature of.) Of, pertaining to, or resembling a topomorph.

TOPONYM Noun. (Greek, *topos* = place; *onyma* = name. Pl., Toponyms.) The scientific name of an organism based in part on the presumed home or geographical origin of that organism. Rel. Taxonomy; Nomenclature.

TOPOTYPE Noun. (Greek, *topos* = place; *typos* = pattern. Pl., Topotypes.) Any specimen of a nominal Species that was collected at the type-locality of that nominal Species. See Type. Cf. Holotype; Lectotype; Neotype. Rel. Nomenclature; Taxonomy.

TOPPEL® See Cypermethrin.

TOPPLE® See Cypermethrin.

TOPSSELL, EDWARD (1557–1625) (Lisney 1960, *A Bibliography of British Lepidoptera 1608–1799,* 315 pp., London.)

TORMA Noun. (Greek, *tormos* = socket. Pl., Tormae.) 1. Scarabaeoid larvae: A dark sclerome at each end of the clypeo-labral suture, extending transversely toward the middle line of the Epipharynx, varying in size and shape according to the Species, two asymmetrical tormae present, the dexiotorma to the right and the laeotorma to the left (Hayes' torma, without special name for the right or the left torma) (Boving). 2. Diptera: A sclerotic process at the base of the epipharyngeal wall of the Labrum. 3. One of a pair of small sclerites positioned in the lateral angles between the Labrum and Clypeus, which usually extend into the epipharyngeal surface of the Clypeus (Snodgrass). 4. The lateral chitinized pieces borne on the pharyngeal surface of the labrum in the region of the suture (Imms).

TORMOGEN Noun. (Greek, *tormos* = socket; *genes* = producing. Pl., Tormogens.) The Epidermal Cell which produces a cuticular depression that contains a Seta. The Tormogen is sometimes a membranous, flexible socket. Typically the Tormogen surrounds the base of the Trichogen and anchors it in place. A sense cell is associated with the Trichogen may be comprised of one or more glial cells which ensheathe the axon of the the neuron that innervates the receptive Trichogen. Cells are penetrated by dendrites which develop from axons derived from mitosis of one Epithelium Cell. Syn. Accessory Cell. See Trichogen. Cf. Seta. Rel. Sensillum.

TORNADE® See Permethrin.

TORNADO® See Acephate.

TORNAL Adj. (Latin, *tornare* = to turn; *-alis* = pertaining to.) Pertaining to the Tornus. See Tornus.

TORNATE Adj. (Latin, *tornare* = to turn.) Pertaining to an elongate, taperering structure which is apically blunt or rounded.

TORNIER, GUSTAV (–1938) (Anon. 1938. Ent. Z., Frankf. a. M. 52: 93.)

TORNUS Noun. (Latin, *tornare* = to turn.) Lepidoptera: The junction of the Tegmen and dorsum of the wing, hind or anal angle.

TOROSE Adj. (Latin, *torus* = swelling; *-osus* = full of.) Swelling in knobs, knots or protuberances; knobby. Alt. Torosus; Torous.

TORPEDO See Permethrin.

TORPID Adj. (Latin, *terpere* = to be stiff.) Dormant from cold or other natural conditions which unfavourably affect the organism. See Aestivation; Hibernation.

TORQUE® See Fenbutatin-Oxide.

TORQUEATE Adj. (Latin, *torques* = a natural collar.) With a ring collar.

TORQUILLUS Adj. (Latin, *torguere* = to twist.) Rotula.

TORRE E TASSO, ALESSANDRO CARLO DELLA (1881–1937) Anon. 1937, Pubbl. Mus. ent. Pietro Rossi Duino 2: v–xi; Koch 1937, Ent. Z., Frankf. a. M. 51: 197–200.)

TORRE-BUENO, JOSE ROLLIN DE LA (1871–1948) American businessman and amateur Entomologist specializing in aquatic Hemiptera. Author of an English-language *Glossary of Entomology* and long-time editor of the Bulletin of the Brooklyn Entomological Society. (Olsen 1948, Bull. Brooklyn ent. Soc. 43: 135–137; Sherman 1948, Bull. Brooklyn ent. Soc. 43: 154–156.)

TORRES SALA, JUAN (–1974) (Anon. 1974, Shilap 2: 273.)

TORRES, BELINDO ADOLFO (1917–1965) (De Santis 1965, Revta. Soc. ent. argent. 28: 95–96.)

TORRIDINCOLIDAE Plural Noun. A small Family of myxophagous Coleoptera assigned to Superfamily Microsporoidea.

TORSALO See Tropical Warble-Fly.

TORSION Adj. (Latin, *torquere* = to twist.) A twisting or spiral bending.

TORSIONAL SPRING See Taenidium.

TORTILIS Adj. (French, *tortiller*.) Twisted.

TORTOISE BEETLES See Chrysomelidae.

TORTOISE SCALE See Brown Soft Scale.

TORTRICIDAE Latreille 1802. Plural Noun. Leaf Rollers; Leaf Tiers; Leaf Roller Moths. A cosmopolitan, economically important Family (ca 4,000 Species) of ditrysian Lepidoptera assigned to Tortricoidea; primitive and considered sister-group of Cossidae. Adult small; wingspan 8–35 mm; head with short, coarse scales; Frons typically with long scales directed forward; Ocelli usually present; Chaetosemata present; Antenna typically filiform, about half as long as forewing; Scape typically lacks Pecten; Proboscis without scales; Epiphysis present; tibial spur formula 0-2-4. Eggs flat, scale-like, with reticulate sculptural pattern, laid singly or masses up to 100; eggs sometimes covered with scales from Corethrogyne or surrounded by upright scales. Larval head subprognathous; prothoracic and anal shields present; thoracic and abdominal Prolegs present. Pupal head often with spines and Abdomen with two transverse rows of spines on Terga 3–7; pupation occurs within silk-lined larval habitat. Larva typically feed in concealed places, ties leaves or shoots; early instars of some Species tunnel stems, fruits or galls. Adult holds wings tectiform over body. See Tortricoidea.

TORTRICOIDEA A Superfamily of ditrysian Lepidoptera including Phaloniidae and Tortricidae. See Lepidoptera.

TORTULOSE Adj. (Latin, *torulus* = small swelling; *-osus* = full of.) 1. With small swellings. 2. Hump-backed, somewhat tumid. 2. Superficially with a few large elevations; when applied to segments or joints then beaded, moniliform, as in Antennae (Smith). Alt. Tortulosus; Tortulous.

TORTUOSE Adj. (Latin, *torquere* = to twist; *-osus* = full of.) Irregularly curved and bent; snake-like. Alt. Tortuosus; Tortuous.

TORULA Noun. (Latin, *torulus* = small swelling.) A small Torulus.

TORULOSE ANTENNA An Antenna in which the segments have swellings or knob-shaped joints.

TORULOSE Adj. (Latin, *torulus* = small swelling; *-osus* = full of.) Knobby; with knob-like swellings. Alt. Torulosus; Torulous.

TORULUS Noun. (Latin, *torulus* = small swelling. Pl., Toruli.) 1. The antennifer. 2. The basal socket joint of the Antenna upon which the organ is articulated for movement in all directions.

TORUS® See Fenoxycarb.

TORYMID WASPS See Torymidae.

TORYMIDAE Plural Noun. Torymid Wasps. A cosmopolitan, moderately large (ca 100 Genera, 1,500 Species) Family of apocritious Hymenoptera assigned to the Chalcidoidea. Diagnositic features include body moderate sized (1–8 mm long), typically metallic blue-green; Antenna usually with 13 segments including 1–3 Anelli; Mesoscutum with complete Notauli (Parapsidal Sutures); macropterous, Marginal Vein long, sometimes stigmated, Postmarginal and Stimal Veins short or long; hind Coxa several times larger than fore Coxa; Tarsomere formula 5-5-5; Gaster sometimes petiolate; Ovipositor several times longer than Metasoma; Higher classification controversial, but most schema point toward diverse biology: Phytophagy to Hyperparasitism; most larvae feed externally as parasites. Egg shape variable, typically saussage-like; Larva hymenopteriform; Parasitic Species often have spines; Phytophagous Species spineless. Developmental strategies include: phytophagous gall former; phytophagous feeding on seed endosperm; primary ectoparasite of gall formers; ectoparasite of gall former, then phytophagous; ectoparasite of aculeate Hymenoptera and Coleoptera.

TOSI, ALESSANDRO (1895–1949) (Anon. 1949, Boll. Soc. ent. ital. 80: 18.)

TOSOUINET, PIERRE JULES (1824–1902) (Fowler 1902, Proc. ent. Soc. Lond. 1902: lix; Severin 1903, Mém. Soc. ent. Belg. 10: 1–12. bibliogr.)

TOTACORIA Noun. The combined Sternacoria and Trocacoria of each side (MacGillivray).

TOTAGLOSSA Noun. The fused Glossae and Paraglossae without line of division (MacGillivray).

TOTALENE® See Trichlorfon.

TOTASUTURE Noun. The suture separating the Sternum and Episternum (MacGillivray).

TOTHILL, JOHN (–1969) (Anon. 1969, The Times, London. 18.7.1969.)

TOTIDEM Adj. (Latin, *toti* combining form meaning whole.) In all parts; entirely.

TOTIPOTENT Social Insects: Capable of performing all tasks; pertaining to founding queens of some Species.

TOUMANOFF, CONSTANTIN (1903–1967) (Anon. 1968, J. Invert. Path. 11: 149–151.)

TOURRETTE, MARC ANTOINE LOUIS CLARET DE LA (1729–1793) (Rose 1850, *New General Biographical Dictionary* 12: 270.)

TOWBUG See Cigarette Beetle.

TOWNES, HENRY KEITH JR (1913–1990) American USDA taxonomist, academic (North Carolina State University) and ultimately an independent researcher involved with systematics of Hymenoptera, particularly Ichneumonoidea. Founder of American Entomological Institute; Publisher of Memoirs and Contributions to American Entomological Institute. Author of more than 135 publications, including many book-length revisions of major groups of Hymenoptera. (Wahl 1992, PESW 94 (2): 289.)

TOWNSEND, CHARLES HENRY TYLER (1863–1944) (Borgmeier 1944, Revta. Ent., Rio de J. 15: 236–239; Arnaud 1958, Microentomology 23: 1–63, bibliogr., list of described genera etc.)

TOXIC Adj. (Greek, *toxikon* = poison; *-ic* = of the nature of.) Poisonous.

TOXICOLOGY Noun. (Greek, *toxikon* = poison; *logos* = discourse. Pl., Toxicologies.) The study of poisons, their mode-of-action and effects on organisms which receive them.

TOXIN Noun. (Greek, *toxikon* = poison. Pl., Toxins.) Any poison derived from plant or animal or synthesized chemical which has an adverse effect on an organism. Rel. Haemotoxin; Neurotoxin.

TOXOGNATH Noun. (Greek, *toxikon* = poison; *gnathos* = jaw. Pl., Toxognaths.) The anteriormost leg of centipedes which has become modified from a walking appendage to a poison-bearing claw. The toxocognath is analogous (but not homologous) with the fang of Chelicerata.

TOXOPEUS, LAMBERTUS JOHANNES (1894–1951) (Roepke 1951, Ent. Ber., Amst. 13: 289; Anon. 1952, Indones. J. nat. Sci. 108: 1–5.)

TOX-R® See Rotenone.

TOYAMA, KAMETARO (1867–1918) (Hasegawa 1967, Kontyû 35 (Suppl.): 55–56, bibliogr.)

TOYODAN® Prothiofos.

TOYOTHION® Prothiofos.

TOZZETTI, A T See Targioni-Tozzetti.

TRABECULA Plural Noun. (Latin, *trabecula* = little beam. Pl., Trabeculae.) 1. Rounded, lobular masses of the protocerebrum, from which arise stalks bearing the mushroom bodies (Smith). 2. A paired, movable appendage anterior of the Antennae in certain bird-lice.

TRABECULATE Adj. (Latin, *trabecula* = little beam; *-atus* = adjectival suffix.) Pertaining to structure with trabecula or a cross-barred framework. Alt. Trabecular.

TRACE FOSSIL A burrow, track or trail of an organism left as an impression in sedimentary rock. TFs are indications of life that can be used to reconstruct a habitat, mode or existence or method of feeding. In some instances, TF may be the only indication of life (*e.g.* Proterozoic Eon ca 600 MYBP). See Fossil. Rel. Index Fossil.

TRACE VEIN Any reduced, evanescent, atrophied or relictual wing veins. See Vein; Wing Venation. Cf. Nebulous Veins; Spectral Veins; Spurious Vein; Tubular Vein. Rel. Horismology.

TRACHEA Noun. (Latin, *trachia* = windpipe. Pl., Tracheae.) An elongate, tubular, spirally ringed, elastic element of the Tracheal System. Typically, the Trachea is derived from the invagination of Ectoderm; tracheal diameter ranges from ca 1 μm to 1 mm, depending upon body size and tracheal position. In cross-section the tracheal wall consists of an epithelial layer 1–3 cells thick; the lumen is lined with an epicuticular intima; the spiral ring within the tracheal wall is called a Taenidium. Dilations in the Trachea are called Tracheal Air Sacs. Branches of Tracheae form as shorter, more narrow tubules called Tracheoles. See Tracheal System. Cf. Atrium; Taenidium; Tracheal Air Sac; Tracheole. Rel. Respiratory System.

TRACHEA CEPHALICA DORSALIS The Dorsal Cephalic Trachea.

TRACHEA CEPHALICA VENTRALIS The Ventral Cephalic Trachea.

TRACHEAL AIR SAC Dilations or conspicuous enlargements of Tracheae; TAS are best developed in larger flying insects, especially cyclorrhaphous Diptera and bees. TAS is delicate with Taenidia absent or poorly developed. TAS provide increased air supply for strong-flying insects, provide area for growth of organs in immatures and permit change in organ size with a change in body form. In some noctuid moths tympanal organs are backed by TAS which allow vibrations with a minimum of damping. See Tracheal System. Cf. Ventilation. Rel. Respiratory System.

TRACHEAL CAPILLARY A Tracheole.

TRACHEAL COMMISSURES Transverse tracheal trunks continuous from one side of the body to the other (Snodgrass). See Tracheal System.

TRACHEAL GILL Flattened or hair-like processes on the body of aquatic immature insects through which oxygen is diffused from water. TG typically displays a thin Cuticle and contains many Tracheoles but does not contain structures with high metabolic rate. Tracheoles of TG often produce ventilatory current which reduces the Boundary Layer. See Gill. Cf. Lateral Abdominal Gill. Rel. Cutaneous Respiration.

TRACHEAL GILL THEORY More correctly termed Tracheal Gill Hypothesis. An explanation for the origin of the insect wings, which derives wings from thoracic tracheal gills of aquatic insects. The hypothesis asserts that tracheal gills have lost their original function and have become adapted for the purposes of fight in the migration of the insects to land (Gegenbaur). See Paranotal Lobe Hypothesis.

TRACHEAL GLAND Orthopteran *Romalea* (Acrididae): A Defensive Gland associated with the metathoracic spiracle (which is not used for respiration). The epithelium of the Trachea produces a secretion which is stored within the Trachea. When discharging the defensive chemical, all spiracles except the metathoracic are closed and muscular contractions increase hydrostatic pressure. The pressure compresses the air sacs joining the glandular Trachea and expel the secretion through the spiracle as a foam. See Gland. Cf. Defensive Gland.

TRACHEAL MITE *Acarapis woodi* (Rennie) [Acari: Tarsonemidae]: The most frequently encountered and widely distributed of mite Species that infest honey bees. Primary host *Apis mellifera,* but also reported on *A. cerana* and *A. dorsata.* TM first detected in Europe, subsequently throughout New World, and later in Africa and Asia; TM not reported from Australia or New Zealand. TM typically infest the Tracheae associated with the Prothoracic Spiracles in adult honey bees; occasionally, TM infests tracheal air sacs in head and abdomen. Mites pierce the tracheal Intima and epithelial cells to feed on bee's Haemolymph. TM affects respiration by blocking Tracheae, shortens worker lifespan, reduces queen productivity and drives colonies to extinction during cold weather. Female and male mites invade Trachea of worker bees typically less than 10 days old; female oviposits 5–7 eggs; eclosion occurs within 3–4 days; mites may leave spiracle, climb upon Setae on bee's Thorax, transfer to another worker bee and then enter spiracle of new host. TM association with honey bee first reported in Isle of Wight ca 1904; subsequently spread throughout Europe; during 1960s moved into India and transferred to Asian Bee, *A. cerana*. Other Species of TM include *A. externus* and *A. dorsalis.* Syn. *Tarsonemus woodi.* See Honey Bee Mite. Cf. *Acarapis dorsalis*; *Acarapis externus*; Varroa Mite.

TRACHEAL ORIFICE The primary opening at the point of formation of a Trachea, whether exposed externally or concealed in a secondary atrial depression of the body wall (Snodgrass). See Spiracle.

TRACHEAL SYSTEM The anatomical part of the Respiratory System composed of the Spiracles, Tracheae and Tracheoles. The TS is bilateral and arranged on a segmental plan. The groundplan suggests one pair of Spiracles per body segment, with all Spiracles positioned laterally on each body segment. In extant insects, all Spiracles have been lost from the head and most insects have a reduced number of Spiracles on the Thorax. The Abdomen displays up to eight pairs of Spiracles. In most insects the Trachea project inward from the Spiracle as a trifurcation and consecutive spiracles are connected by Longitudinal Trunks. Typically, insects display a Dorsal, Lateral and Ventral Longitudinal Trunk which extends from the apical region of the Abdomen into the head. The Tracheal Systems of the left and right sides of the body are mirror images; the left and right systems are connected medially by smaller tubes called Transverse Commissures. TS are sometimes classified as Closed and Open Tracheal Systems. See Respiratory System. Cf. Spiracles; Tracheae; Tracheoles. Rel. Closed Tracheal Systems; Open Tracheal Systems.

TRACHEAL VENTILATION A type of gas exchange which involves an Open Tracheal System consisting of Tracheae and Spiracles. See Ventilation. Cf. Active Ventilation; Autoventilation; Auxiliary Pumping Ventilation; Passive Suction Ventilation. Rel. Respiration; Respiratory Muscles; Respiratory System.

TRACHEARY Adj. (Latin, *trachia* = windpipe.) Relating to or composed of Tracheae.

TRACHEATE Adj. (Latin, *trachia* = windpipe; *-atus* = adjectival suffix.) Supplied with Tracheae. A general term applied to all arthropods that breathe by means of spiracular openings into a system of tubular structures that extend into all parts of the body.

TRACHEATION Noun. (Latin, *trachia* = windpipe; English, *-tion* = result of an action. Pl., Tracheations.) The arrangement or system of distribution of Tracheae.

TRACHEOBLAST Noun. (Latin, *trachia* = windpipe; *blastos* = bud. Pl., Tracheoblasts.) Epidermal cells of the respiratory system which form Tracheoles.

TRACHEOLE Noun. (Latin, *trachia* = windpipe. Pl., Tracheoles.) Smaller tubules of the Tracheal System (ca 0.1–0.5 μm diam, 50 μm long), but not a sharply differentiated morphological region. Tracheoles contain Taenidia but lack a wax layer; they maintain their cuticular lining after moulting and are permeable to water. Initially, Tracheoles are not connected to Tracheae; form connection at Ecdysis. Tracheoles are closely associated with metabolically active tissues including muscles, salivary glands, ganglia of the nervous system, digestive system, Malpighian Tubules and reproductive organs. Terminations of Tracheoles are open-ended, close-ended or anastomose when united to form capillary reticulum. See Tracheal System. Cf. Trachea. Rel. Respiratory System.

TRACKER® See Tralomethrin.

TRACT Noun. (Latin, *tractus* = region. Pl., Tracts.) 1. An area or region of undefined size but considered in its entirety. *e.g.* A tract of land. 2. A group of organs that act together to serve a specific function. *e.g.* Reproductive Tract; Digestive Tract. Cf. System.

TRACTICUM Noun. Hymenoptera: Ventral Ramus.

TRADE NAME One of three names for a pesticide. The trade name is given to a pesticide by the manufacturer or formulator. Trade names bear the U.S. registration mark for the duration of the patent. Alt. Brand Name; Proprietary Name. See Approved Common Name; Chemical Name.

TRAGÅRDH, IVOR (1878–1951) (Eyndhoven 1951, Ent. Ber., Amst. 13: 335–336; Turk 1951, Entomol. mon. Mag. 87: 240; Ahlburg 1952, Opusc. ent. 17: 100–102; Tullgren 1952, Ent. Tidskr. 73: 1–3, bibliogr.)

TRÄGÅRDH'S ORGAN A long, conical protuberance on the lateral aspect of a Chelicera.

TRAIL, JAMES WILLIAM HELENUS (1851–1919) (Anon. 1919, Nature 104: 76; Ritchie 1920, Scot. Nat. 1920: 1–5.)

TRAIL PHEROMONE A chemical deposited on the substrate by one member of a Species and followed by another member of the same Species.

TRAJECTORY Noun. (Latin, *trajectus.*) The path in the air of a moving object.

TRALATE® See Tralomethrin.

TRALEX® See Tralomethrin.

TRALLES, BALTHASAR LUDWIG (1708–1797) (Rose 1850, *New General Biographical Dictionary* 12: 274.)

TRALOMETHRIN Noun. A synthetic-pyrethroid compound used as a contact and stomach poison; applied to ornamentals, fruits, vegetables and cotton; also used as public-health insecticide. Trade names include: D-End®, Saga®, Scout X-tra®, Tracker®, Tralate®, Tralex®, Velstar®. See Synthetic Pyrethroids.

TRAMOSERICEOUS Adj. (Latin, *tramo* from *trans* = across; *seri* = silk; *-aceus* = of or pertaining to.) Satiny. Alt. Tramosericeus.

TRAMPOLINE SPIDERS *Corasoides* spp. [Araneida: Agelenidae] (Australia).

TRANSAMINASE Noun. (Latin, *trans* = across; Greek, *ammoniakon* = resinous gum.) A group of enzymes which catalyse transamination.

TRANSAMINATION Noun. (Latin, *trans* = across; Greek, *ammoniakon* = resinous gum; English, *-tion* = result of an action.) The transfer of an amino group from one molecule to another molecule.

TRANSDUCTION Noun. (Latin, *transducere* = to transfer; English, *-tion* = result of an action. Pl., Transductions.) 1. The process of moving from one state to another. 2. The use of a virus (bacteriophage) to transfer genes from one bacterium to another bacterium. See Genetic Engineering. Rel. Biotechnology.

TRANSECT Noun. (Latin, *trans* = across; *secare* = to cut. Pl., Transects.) A narrow area, of varying width, within which quantitative samples of physical and biological data are collected.

TRANSECTION Noun. (Latin, *trans* = across; *sectio* = to cut; English, *-tion* = result of an action.) A cut or section made at a right angle to the longitudinal axis of a body. Syn. Transverse Section. Cf. Saggital Section; Frontal Section.

TRANSFECTION Noun. (Latin, *trans* = across; *inficere* = taint.) The transfer of foreign (new) genetic material into animal cells. Cf. Transformation.

TRANSFORMATION Noun. (Latin, *transformare* = to change shape. Pl., Transformations.) The transfer of genetic material into Bacteria. Cf. Transfection.

TRANSFORMATION SERIES A hypothetical series of events which depict changes in biological conditions, behavioural actions or anatomical features in a lineage of organisms through time and evolution. A TS only considers structures or organisms that are genetically related. Relatedness is determined by evidence collected from morphology, embryology, palaeontology and molecular data. Determining common descent has been a pivotal problem in biology. Richard Owen (1804–1892) proposed the term 'Homology' to express a fundamental similarity in structure among organisms. See Evolution; Homology; Organism. Cf. Morphological Transformation. Rel. Groundplan.

TRANSFRONTAL BRISTLES Diptera: The lower frontal bristles which are often directed across the Frontal Vitta (Comstock).

TRANSITION ZONE The transcontinental belt in which the austral and boreal elements overlap, divided into a humid or Alleghanian area, a western arid area, and a Pacific Coast humid area.

TRANSLUCENT Adj. (Latin, *trans* = across; *lucere* = to shine.) Pertaining to structure which is semitransparent; structure which permits the passage of light but diffusing it such that objects cannot be resolved. Cf. Transparent.

TRANSLUCID Adj. (Latin, *translucidius* = to shine.) Translucent.

TRANSMISSION Noun. (Pl., Transmissions.) 1. The act or process of transmitting. 2. Medical entomology: The passage of a parasite from an intermediate host (vector) to a definitive host. Transmission may be Mechanical or Biological; Biological Transmission may be Propagative, Cyclopropagative or Cyclodevelopmental. See Biological Vector; Mechanical Vector. Cf. Horizontal Transmission; Vertical Transmission; Propagative Transmission; Cyclopropagative Transmission; Cyclodevelopmental Transmission.

TRANSMISSION ELECTRON MICROSCOPE A microscope that uses a transmitted beam of electons rather than transmitted light to examine thin-sections of biological tissues. The wavelength of the electron beam varies with its acceleration by a potential that is measured in volts. Biological tissues are examined with accelerating voltages up to 60,000 volts. Cf. Scanning Electron Microscope. See Light Microscope.

TRANSMONTANE Adj. (Latin, *trans* = other side; *mons* = mountain.) Pertaining to the other side of the mountains. Cf. Cismontane.

TRANSOVARIAL TRANSMISSION The capacity of a disease vector (acarine) to retain a disease agent (arbovirus) between generations and effectively transmit the disease from the adult female stage to her eggs. Cf. Trans-stadial Transmission.

TRANSPARENT SCALE *Aspidiotus destructor* Signoret [Hemiptera: Diaspididae] (Australia).

TRANSPARENT-WINGED PLANT BUG *Hyalopeplus pellucidus* (Stål) [Hemiptera: Miridae].

TRANSPIRATION Noun. (Latin, *trans* = across; *spirare* = to breathe; English, *-tion* = result of an action.) The act or process of exhaling or passing off liquids as vapour. In plants, the passing of water vapour through the stomata.

TRANSPOSABLE ELEMENT One of a large group of mobile genetic elements. Unlike viruses, TEs are constrained to move from place to place within the chromosomes of a single cell and the cell's progeny. Movement within the cell takes place by a site-specific recombination system that specifically recognizes its own DNA.

TRANSSCUTAL SUTURE Hymenoptera: A transverse thoracic suture which separates the posterior part of the Mesoscutum from the anterior part of the Scutellum. A landmark in many groups. Syn. Scutoscutellar Suture.

TRANSSCUTELLAR SUTURE Higher Diptera: A suture which cuts through the anterior part of the Scutellum between lateral extremities of the Scutoscutellar Suture (Snodgrass).

TRANS-STADIAL TRANSMISSION The capacity of a disease vector (acarine) to retain a disease agent (arbovirus) between life stages and effectively transmit the disease from the larva to nymph or nymph to adult. Cf. Transovarial Transmission.

TRANSTARSUS Noun. (Latin, *trans* = across; *tarsus* = sole of foot.) Collembola: The small apical Pretarsus.

TRANSTILLA Noun. Male Lepidoptera: A process arising from the dorsobasal angle of the Harpe; in some cases it appears to serve for the articulation of the Harpe with the Vinculus or Tegumen, sometimes the two Transtillae extend mesad and their ends fuse together to form a transverse band. The transtilla may be more-or-less finger-like (ampulla) (Klots).

TRANSVERSA Noun. Heteroptera: The projection of the Pronotum over the Mesonotum (MacGillivray).

TRANSVERSE Adj. (Latin, *transversus* = across.) Pertaining to structures which are wider than long; running across or cutting the longitudinal axis at right angles. Alt. Transversus. Ant. Longitudinal.

TRANSVERSE ANTERIOR LINE Lepidoptera: The line that crosses the forewings of certain moths one-third or less from the base; the antemedial line (Smith). Abbreviated, t.a. line.

TRANSVERSE BASAL TRACHEA Generalized insects: The Trachea connecting the costo-radial and the cubito-anal groups of Tracheae (Comstock).

TRANSVERSE CORD Some Plecoptera: A nearly continuous series of crossveins extending across each wing just beyond the middle (Comstock).

TRANSVERSE COSTAL VEIN Hymenoptera: A short second branch of the Subcosta which fuses with the Radius. See Vein; Wing Venation.

TRANSVERSE CUBITAL VEIN Hymenoptera: A section of the Radial Sector or a Radio-Medial Crossvein. See Vein; Wing Venation.

TRANSVERSE GYNANDROMORPH An individual that expresses a type of gynandromorphism in which one half (anterior or posterior) of an individual shows female features and the other half shows male features. TG have been reported among social insects, including ants. See Gynandromorphism. Cf. Bilateral Gynandromorph; Sexual Mosaic. Rel. Development; Metamorphosis; Mutant.

TRANSVERSE IMPRESSION Diptera: See Cheek Grooves (Comstock).

TRANSVERSE INCISION A transverse Sulcus.

TRANSVERSE LADYBEETLE 1. *Coccinella transversoguttata richardsoni* Brown [Coleoptera: Coccinellidae]. 2. *Coccinella transversalis* Fabricius [Coleoptera: Coccinellidae]: An Australian Species. Adult 3.8–6.7 mm long, round-ovate; black head and orange or orange-red Elytra with black markings; markings on transverse ladybird are a black stripe down centre of Elytra having large spots at wing base and apex; two large spots and two large splotches occur in middle of Elytra also at wing base and apex; life-cycle from egg to adult 38–45 days. Adult female lays eggs in batches near larval food. Eggs yellow, spindle-shaped and laid standing on end; egg stage 8–10 days. Larva elongate, with well defined legs; body black with red markings and setose; four larval instars; larval stage 19–23 days. Pupal stage 8–9 days. Transverse ladybirds are common predators of aphids. A single larva can consume ca 525 aphids. Adult and larva occur on cotton feeding on aphids, mites and *Helicoverpa* eggs. Cf. Striped Ladybird.

TRANSVERSE MEDIAN VEIN Hymenoptera: A crossvein between the Discoidal Vein and the Anal Vein. See Vein; Wing Venation.

TRANSVERSE MUSCLES Muscles which lie internal to the longitudinals, including dorsal transverse and ventral transverse muscles (Snodgrass).

TRANSVERSE NERVES The respiratory nerves.

TRANSVERSE PILACERORES Coccids: The band of Pilacerores in the ovisac which secrete the transverse plate (MacGillivray).

TRANSVERSE PLANE Any one of several mutually parallel planes (cross sections) that pass through the body at a right angle to the long axis of the body and at a right angle to the sagittal plane. Cf Parasagittal Plane; Saggital Plane.

TRANSVERSE PLATE Coccids: The transverse, cephalic portion of the ovisac (MacGillivray).

TRANSVERSE POSTERIOR LINE A line crossing the forewings of certain Lepidoptera, two-thirds or more from the base. The postmedial line (Smith). Abbreviated: T.p. line.

TRANSVERSE RADIAL VEIN Hymenoptera: The transverse Marginal Vein *sensu* Borror *et al.* See Vein; Wing Venation.

TRANSVERSE SULCI Orthoptera: The transverse grooves of the Pronotum. See Sulcus.

TRANSVERSE SUTURE Diptera: The suture between the Prescutum and the Scutum of the Mesothorax (Comstock). A division across the body; more particularly across the dorsum of the Thorax.

TRAP Noun. (Middle English, *trappe*; Anglo Saxon, *treppe*. Pl., Traps.) 1. A device or machine designed or adapted for the collection of insects. Cf. Assateague Insect Trap; Flight Intercept Trap; Light Trap; Lumsden Suction Trap; Malaise Trap; McPhail Trap; New Jersey Trap; Pit Trap; Pitfall Trap; Rothamsted Light Trap; Sticky Trap; Wilkinson Trap.

TRAP CROP Noun. A small area of a crop used to divert pests from a larger more susceptible area of the same or different crop.

TRAPDOOR SPIDERS See Ctenizidae; Idiopidae; Nemesiidae.

TRAPEZIFORM Adj. (Greek, *trapezion* = small table; Latin, *forma* = shape.) Trapezium-shaped. Descriptive of structure with the outline shape of a quadrilateral with no two sides parallel. Cf. Trapezoid. Rel. Form 2; Shape 2; Outline Shape.

TRAPEZIUM Noun. (Greek, *trapezion* = small table; *-idion*. Pl., Trapezia; Trapeziums.) A four-sided figure in which no two sides are parallel.

TRAPEZOID Adj. (Greek, *trapezion* = small table; *eidos* = form.) Trapeze-like; descriptive of outline shape of four sides, with two sides parallel and two are not parallel. Alt. Trapezoidal. Cf. Trapeziform. Rel. Eidos; Form; Shape.

TRAUB, ROBERT S (1916–1996) American medical Entomologist (U.S. Army, Univ. of Maryland, Smithsonian Institution.) Author of ca 200 papers and specialist in Siphonaptera. (Lewis, 1996, Flea News no 53, 618–625.)

TRAUB Cf. Organ.

TRAUMATIC INSEMINATION Hemiptera: An aberrant form of 'copulation' in which the males of some Nabidae and most Cimicidae penetrate the female's body wall to deposit sperm in a special pouch (Spermalege) or females internal reproductive system. See Spermalege.

TRAUMATIC MYIASIS A type of facultative Myiasis in which fly larvae that typically feed on carrion will feed and develop in wounds and sores. Flies frequently involved in this activity include several Species of *Calliphora, Chrysomyia, Cochliomyia, Lucilia,* and *Phaenicia*. See Myiasis.

TRAUTMANN, WOLDEMAR (1880–1929) (Hepp 1930, Ent. Z., Frankf: A. M. 43: 267.)

TRAVER, JAY R (1894–1974) (Alexander 1975, Eatonia 20: 2–5, bibliogr.)

TREAT, MARY LUA ADELIA ALLEN (1835–) (Howard 1930, Smithson. misc. Collns. 84: 22–23.)

TREBON® See Etofenprox.

TRECUT® See NAA.

TREDL, RUDOLF (1871–1921) (Kleine 1922, Ent. Bl. Biol. Syst. Käfer 18: 65.)

TREE CRICKET *Paragryllacris combusta* Germar [Orthoptera: Gryllacrididae] (Australia).

TREE CRICKETS See Gryllacrididae.

TREE FUNNELWEB SPIDERS *Hadronyche cerberea* L. Koch [Araneida: Hexathelidae] (Australia).

TREE LUCERNE MOTH *Uresiphita ornithopteralis* (Guenée) [Lepidoptera: Pyralidae] (Australia).

TREE Noun. (Middle English, *tree* > Anglo Saxon, *Treo* > Greek, *drys* > a tree. Pl., Trees.) A woody perennial plant that has a single stem (trunk), assumes considerable height at maturity (ca 3+ metres), and displays branches with leaves on the crown. The distinction between tree, shrub and bush is not complete. See Plant. Cf. Bush; Grass; Shrub.

TREEHOPPERS See Eurymelidae; Membracidae.

TREE-PEAR BEETLE *Archlagocheirus funestus* (Thomson) [Coleoptera: Cerambycidae] (Australia).

TREFFRY, JOSEPH EVELEIGH (–1900) (T.W.F. 1900, Rep. ent. Soc. Ont. 31: 105.)

TREHERNE, REGINALD CHARLES (1886–1924) (Gibson *et al.* 1924, Can. Ent. 56: 151–153, bibliogr.)

TRE-HOLD® See NAA.

TREITSCHKE, FREDERICK A German naturalist who continued the publication of Ochsenheimer's *Lepidoptera of Europe*.

TREMBLEY, ABRAHAM (1710–1784) (Rose 1850, *New General Biographical Dictionary* 12: 276; Ratzeburg 1874, Forstwissenschaftliches Schriftsteller-Lexicon 1: 60 (footnote); Miall 1912, *Early naturalists, their lives and work*. (1530–1789.) 369 pp. (279–284), London; Nordenskiöld 1935, *History of Biology*. 625 pp. (232–233) London.)

TREMOLERAS, JUAN (1870–1934) (Anon. 1934, Revta. Soc. ent. argent. 6: 284–286, bibliogr.; Cordero 1935, Physis, B. Aires 11: 541–544,

bibliogr.)

TREMOR Noun. (Latin, *tremere* = to tremble; *or* = one who does something. Pl., Tremors.) Trembling or shaking, typically induced by fixed behavioural pattern, pesticide poisoning or disease.

TREMULOUS Adj. (Latin, *tremulus*, from *tremere* = to tremble; *-osus* = possessing the qualities of.) 1. Pertaining to an organism affected with shaking or tremors. 2. Pertaining to an organism exceedingly sensitive or easily shaken.

TRENCH FEVER A minor rickettsial disease first reported in humans during WWI when it reached epidemic proportions and affected more than a million soldiers; subsequently reported in Russia during WWII. Aetiological agent *Rochalimaea quintana* and vectored among humans by the body louse *Pediculus humanus*. TF acquired by louse who feeds on infected human; rickettsia replicates along cuticular margin of midgut epithelial cells; disease does not affect louse and remains in louse for its lifetime; no trans-ovarial transmission. TF passed with faeces after ca 10 days in louse; human acquires disease by scratching faeces into wound; possibly acquired through inhalation of infected louse faeces. See Rickettsiaceae. Cf. Epidemic Typhus; Murine Typhus.

TRENER, GIOVANNI BATTISTA (1877–1954) (Dalla Fior 1954, Studi trentini Sci nat. 31: 3–8; Ferrari 1954, Natura alpina 5: 1–6.)

TRENTEPOHL, JOHANN JACOB (1800–1830) (Henriksen 1926, Ent. Meddr. 15: 201, bibliogr.)

TREPARIDIIDAE Plural Noun. See Micropezidae.

TRESSE-BEALE, S C (–1885) (Anon. 1886, Entomol. mon. Mag. 22: 216.)

TRESSENS, FERNAND (1891–1975) (Péricart 1976, Bull. ent. Soc. Fr. 8: 298; Villiers 1976, Entomologiste 32: 95.)

TREUGE, EMIL (1836–1876) (Anon. 1876, Jber. westf. ProvVer. Wiss. Kunst. 5: 35–37.)

TREVIRANUS, GOTTFRIED REINHOLD (1776–1837) (Focke 1879, Abh. naturw. Ver. Bremen 6: 11–48.)

TREVIRANUS, LUDOLPH CHRISTIAN (1779–1864) (Anon. 1865, Sber. bayer. Akad. Wiss. 1865: 264–287, bibliogr.; Ratzeburg 1874, Forstwissenschaftliches Schriftsteller-Lexicon 1: 494–495.)

TREVISAN DI SAINT-LEOW, VITTORE BENEDETTO ANTONIO CONTE (1818–1897) (Anon. 1883, Atti Accad. fisio-med. Milano (4) 1: 174.)

TREW, CHRISTOPH JACOB (1695–1769) (Rumpell 1770, Nova Acta Acad. Caesar. Leop. Carol. Nat. Cur. 4 (Suppl.): 315–332; Rose 1850, *New General Biographical Dictionary* 12: 278–279.)

TRIACT® A broad-spectrum fungicide, miticide, insecticide applied to ornamental, landscape and nursery plants. Active ingredient 90% Clarified Hydrophobic Extract of Neem Oil. Toxic to bees; may be hazardous to fish and aquatic invertebrates. See Azadirachtin.

TRIAD Noun. (Greek, *trias* = three.) A union or group of three. Morphology: An arrangement in threes of the veins of the insect wing.

TRIANGLE Noun. (Latin, *triangulus* = triangular. Pl., Triangles.) 1. A planar polygon with three sides and forming three angles which total 180°. 2. Odonata: A triangular cell in the wing base formed by the Cutitus and two converging thickened crossveins between M and C (Garman); discoidal triangle; cardinal cell.

TRIANGULAR PLATES Aculeate Hymenoptera: The second pair of plate-like sclerites associated with the Sting; the reduced Sternum of the eighth abdominal segment (Imms).

TRIANGULAR SPIDERS *Arkys* spp. [Araneida: Mimetidae] (Australia).

TRIANGULAR WAX GLAND The most common type of wax gland on some mealybugs. TWG is positioned on all segments and aspects of the body. TWG is pear-shaped and consists of five cells suspended from a chitinized constriction which terminates in a integumental papilla. Constricted portion of gland is composed of one cell (neck cell or terminal cell). Aperture of TWG displays four openings, three comparatively large elongate which form a triangle in outline; a minute, circular fourth opening is surrounded by three larger openings. Bulk of gland is positioned beneath epidermal layer of Integument and consists of a large, central cell which is surrounded by three peripheral cells. Central cell contains one large, basal nucleus and two smaller nuclei positioned near Epidermis. A clear space (the reservoir) is above the large, basal nucleus. A system of branched tubules penetrates the reservoir. Tubules continue toward epidermal layer and reduce in number until they form one principal Efferent Duct that opens on body surface. Three peripheral cells are each uninucelate and volumetrically smaller than the central cell. Each peripheral cell contains vacuoles associated with the nucleus. In turn, vacuoles are associated with Efferent Ducts which continue to the aperture where they each form an opening. TWG produces small, curled 'flakes' of wax when distributed over body or they form long filaments of wax pushed along by long Setae of the cerarian complex. Cf. Multilocular Wax Gland; Quinquelocular Wax Gland; Tubular Wax Gland.

TRIANGULATE Adj. (Latin, *triangularis* = triangular; *-atus* = adjectival suffix.) Pertaining to form which displays three angles.

TRIANGULIN LARVA Noun. A hypermetamorphic first-instar of Strepsiptera and some Coleoptera. TL is characterized by a hardened exoskeleton, clasp-like claws, and a body form adapted for obtaining food or host by way of phoresy. See Planidium.

TRIAOTION® See Azinphos-Ethyl.

TRIARTICULAR Adj. Composed of three joints, segments or articles. Alt. Triarticlate;

Triarticulatus.

TRIASSIC PERIOD The first geological Period (245–208 MYBP) of the Mesozoic Era. Originally named Trias by Friedrich August von Alberti (1834) for the natural stratigraphic divisions in southern Germany. TP characterized by the formation of the Tethys Sea (across southern Europe to East Indies). Principal geological deposits in the southern Hemisphere include: Mendoza (Argentina), Queensland (Australia), and the Upper Triassic Molteno Formation, Karoo Basin (South Africa). First true mammals appeared during TP. Seed ferns became less common during TP while ferns became dominant terrestrial plants. Trees growing above ferns include conifers, cycadeoids, conifers and ginkgos. Land fauna of invertebrates varied. Insect-bearing deposits include the Upper Triassic Stormberg Series (South Africa, with Archaeogrylloidea), Mendoza, Argentina (prophalangopsid) and the first Hymenoptera (Xyelidae) in Australia. Mass extinction during end of TP eliminated 20% of animal Families, including conodont and placodont reptiles. See Geological Time Scale; Mesozoic Era. Cf. Jurassic Period; Cretaceous Period. Rel. Fossil.

TRIATIX® See Amitraz.

TRIAZAMATE Noun. A triazole compound {ethyl (3-tert-butyl-1-dimethylcarbamoyl-1H-1,2,4-triazol-5-ylthio) acetate} used as a systemic aphicide. Triazamate is selective on aphids. Compound is applied to fruit crops, cotton, ornamentals, potatoes, sugarbeets, and tobacco in some countries. Triazamate shows good systemic activity within plant, moving upward and downward; fast-acting with residual activity. Compound with experimental registration in USA. Triazamate is phytotoxic to alfalfa when applied at high rates and toxic to fishes. Trade names include Aphistar®, Aztec® and RH-988®.

TRIAZID® See Amitraz.

TRIAZOPHOS Noun. An organic-phosphate (thiophosphoric acid) compound {O,O-diethyl O-(1-phenyl-IH-1,2,4-triazol-3yl) thiophosphate} used as an acaricide, contact insecticide and stomach poison against thrips, plant-sucking insects, leaf-eating insects and nematodes in many countries. Not registered for use in USA. Toxic to fishes. Trade names is Hostathion®. See Organophosphorus Insecticide.

TRIBACTUR® A registered biopesticide derived from *Bacillus thuringiensis* var. *kurstaki*. See *Bacillus thuringiensis*.

TRIBE Noun. (Latin, *tribus* = tribe.) A category in zoological classification below a Subfamily, above the Supergenus and identified by the suffix -ini.

TRI-BROME® See Methyl Bromide.

TRIBUTE® See Fenvalerate.

TRICARINATE Adj. (Latin, *tri-* = three; *carine* = keel; *-atus* = adjectival suffix.) A structure or surface with three Carinae or keels.

TRICARNAM® See Carbaryl.

TRICERORES *Pseudococcus*: The triangular wax pores (MacGillivray).

TRICHIA *-trichia,* combining form (New Latin, from Greek, *thrix, trichos* = hair; *ia* = a condition.) A hair. See Seta; Microtrichia.

TRI-CHLOR® See Chloropicrin.

TRICHLORFON Noun. An organic-phosphate (phosphoric acid) compound {Dimethyl (2,2,2-trichloro-1-hydroxyethyl) phosphonate} used as a contact insecticide and stomach poison. Compound applied to numerous agricultural crops, ornamentals and livestock as a public health insecticide for control of mites, plant-sucking and leaf-chewing insects, flies, cockroaches, bots, pinworms, screwworms, lice, ticks and fleas. Trade names include: Anthon®, Cekufon®, Chlorofos®, Denkaphon®, Dep®, Dipterex®, Dylox®, Masoten®, Neguvon®, Pronto®, Proxol®, Rochlor®, Totalene®, Trichlorphon®, Trinex®, Tugon®. See Organophosphorus Insecticide.

TRICHLORPHON Noun. An organophosphorus insecticide applied as surface spray or bait for control of cockroaches, silverfish and adult and larval flies. Marketed under tradename Abate 100. See Organophosphorus Insecticide.

TRICHLORPHON® See Trichlorfon.

TRICHOBOTHRIUM Noun. (Greek, *thrix* = hair; *bothros* = pit; Latin, *-ium* = diminutive > Greek, *-idion*. Pl., Trichobothria.) 1. Long slender bristles located in centre of depressed ring in Cuticle. 2. Heteroptera: Seta-bearing spots on the venter of the Abdomen in many Species. See Bristle; Seta. Cf. Ctenoidiobothrium.

TRICHOCERATIDAE Plural Noun. See Trichoceridae.

TRICHOCERIDAE Plural Noun. Winter Crane Flies; Winter Gnats. A numerically small, widespread Family of nematocerous Diptera assigned to Superfamily Trichoceroidea. Adult small-bodied, delicate with long legs and resembling Tipulidae; Ocelli present. Adults crepuscular or nocturnal and occupy damp, shaded habitats; sometimes taken in swarms. Larva eucephalic, taken in rotting vegetation. Syn. Petauristidae.

TRICHODE SENSILLUM The most generalized kind of sensillum (sense organ) whose external part forms a tapering, hair-like Seta articulated at the base set, below the general surface of the Cuticle. TS may be obliquely ridged. Trichodes are abundant on the Antennae of Lepidoptera and are believed to function as pheromone receptors. Syn. Sensillum Trichodeum. See Sensilla.

TRICHODECTIDAE Plural Noun. Mammal Chewing Lice. A Family of ischnocerous Mallophaga whose members attack mammals. Head relatively large; Antenna not concealed, with three segments; Mandibles vertical; Maxillary Palpus absent; Mesothorax and Metathorax fused; Tarsi with one claw.

TRICHODES Noun. The setal tufts of highly

coadapted ant inquilines. Presumably secretory in nature, Trichodes are malaxated by worker ants with the result of a diffuse aromatic secretion. Syn. Trichomes.

TRICHOGEN Noun. (Greek, *thrix* = hair; *genes* = producing. Pl., Trichogens.) An epidermal cell that generates a seta (Snodgrass). Cf. Tormogen.

TRICHOGENOUS CELL See Trichogen.

TRICHOGRAMMATIDAE Plural Noun. Trichogrammatid Wasps. A moderate sized (ca 75 Genera, 600 nominal Species), cosmopolitan Family of parasitic Hymenoptera assigned to Chalcidoidea. Body minute, weakly sclerotized, without metallic coloration or ornate, bold microsculpture; Antenna with 5–9 segments including funicle with two or fewer ring-like segments and club with 1–5 segments; wings usually macropterous, rarely brachypterous; forewing Stigmal Vein sometimes enlarged, Postmarginal Vein absent, often with distinct setal tracts; Tarsomere formula 3-3-3; foreleg without Strigil, Gaster not petiolate, broadly joined to Propodeum. Biology: Primary, solitary or gregarious endoparasites of egg stage. Hypermetamorphic larval development with first instar sacciform or mymariform, and last instar segmented, robust, without spines. Trichogrammatids display a broad host spectrum to include principal Orders of Holometabola, Hemiptera and Thysanoptera. Extensively used in biological control programmes. Unusual biological features include phoresy on Tetigoniidae and Nymphalidae; *Prestwichia* and *Hydrophilita* parasitize submerged eggs, and representative Taxa are among smallest insects (*Megaphragma* ca 0.20 mm long.)

TRICHOID Adj. (Greek, *trichoeides, trich* = hair, *eidos* = form.) Hair-like; capillary. Cf. Acanthoid; Acicular; Acuminate. Rel. Eidos; Form; Shape.

TRICHOIDE Noun. (Greek, *thrix* = hair, *eidos* = form. Pl., Trichodes.) Formed like a trichia or hair; trichia-like.

TRICHOME Noun. (Greek, *trichoma* = a growth of hair. Pl., Trichomes.) 1. Glandular hairs found on some plants which provide exudates which can repel insects. See Trichodes. 2. Tufts of golden yellow Setae found on bodies of many myrmecophilous and termitophilous insects. Glandular cells associated with trichomes and serve to disperse nest odours or provide appeasement substances for workers who might damage myrmecophiles.

TRICHOPHORE Noun. (Greek, *thrix* = hair; *pherein* = to bear. Pl., Trichopores.) A cylindrical, internal cavity of the cuticula beneath the base of a seta which contains the distal parts of cells associated with the seta (Snodgrass).

TRICHOPSOCIDAE Plural Noun. A numerically small Family of Psocoptera resembling Pseudocaeciliidae and assigned to Suborder Psocomorpha. Antenna with 13 segments, without secondary annulations; Tarsi with two segments; wings with normal Pterostigma and Aroela Postica. Species inhabit fresh foliage and twigs.

TRICHOPTERA Kirby 1826. (?Permian, Triassic Period–Recent.) (Greek, *thrix* = hair; *pteron* = wing) Caddisflies; Caddis-flies; Microcaddis Flies; Northern Caddis Flies; Primitive Caddis Flies; Snailcase Caddis Flies. A cosmopolitan Order of endopterygote neopterous insects containing ca 20 Families and 7,000 Species. Higher classification is debated at Suborder and Superfamily level. Adult anatomy suggests close relationship with Lepidoptera. Adult soft-bodied, <40 mm long, moth-like with dense vestiture of Setae on body and appendages; head not fused to Thorax; compound eyes large; Ocelli typically absent; Antennae long, slender; Mandibles vestigial; Maxilla and Galea reduced; Maxillary and Labial Palps long, flexible, multisegmented. Thorax well developed; two pairs of membranous wings, subequal in size, held roof-like over Abdomen at rest; setose forewing with many longitudinal veins and cells; Pterostigma usually absent; Nygma usually present; Anal Veins looped, only one Anal Vein extending to posterior margin of wing; hindwing broad, usually folded lengthwise; hamulate or jugate wing coupling. Internal anatomy with six Malpighian Tubules; three thoracic ganglia; seven abdominal ganglia; collaterial glands paired; testes ovoid sacs; numerous polytrophic Ovarioles. Adult mouthparts adapted for feeding on liquids; larval mouthparts adapted for chewing. Metamorphosis complete; pupa decticous, exarate. Adults weak fliers, typically crepuscular; during day rest on vegetation or other objects near ponds and streams; often attracted to lights, particularly near water; copulation in flight or on ground. Larvae are campodeiform or eruciform, and also construct a case for protection; cases constructed of silk, pebbles, sand grains, twigs or other materials in ponds and streams. Larvae aquatic, typically in freshwater; resemble Lepidoptera larvae but lack Prolegs on segments 1–8, display pair of hook-like appendages at apex of Abdomen; Abdominal hooks used to drag case as larva feeds. When larva pupates, case attached to rock or other fixed object in water. Pupa exarate; Mandibles strong. Case-making larvae primarily detritus or plant feeders; some larvae do not make cases but spin silken webs to capture food drifting in water. A few Species free living and predaceous. Included Families: Anomalopsychidae, Antipodoeciidae, Arctopsychidae, Atriplectididae, Barbarochthonidae, Beraeidae, Brachycentridae, Calamoceratidae, Calocidae, Chathamiidae, Conoesucidae, Dipseudopsidae, Ecnomidae, Glossosomatidae, Goeridae, Helicophidae, Helicopsychidae, Hydrobiosidae, Hydropsychidae, Hydroptilidae, Hydrosalpingidae, Kokiriidae, Lepidostomatidae, Leptoceridae, Limnephilidae, Limnocentropodidae, Molannidae, Necrotauliidae

(Triassic Period fossil), Odontoceridae, Oeconesidae, Petrothrincidae, Philopotamidae (Triassic Period fossil), Philorheithridae, Phryganeidae, Phryganopsychidae, Plectrotarsidae, Polycentropodidae, Prorhyacophilidae (Triassic Period fossil), Psychomyiidae, Rhyacophilidae, Sericostomatidae, Stenopsychidae, Tasimiidae, Uenoidae, Xiphocentronidae.

TRICHORYTHIDAE Plural Noun. A small, widespread Family of Ephemeroptera whose members lack hindwings but possess a median caudal filament (Appendix Dorsalis). Naiads inhabit streams and rivers.

TRICHOSCELIDIDAE Plural Noun. A small Family of acalypterate Diptera assigned to Superfamily Sphaeroceroidae. Syn. Trixoscelididae.

TRICHOSOR Noun. (Greek, *thrix* = hair; *or* = a condition.) Neuroptera: A setose thickening of the wing membrane along the margin. Rel. Marginal Fringe Seta; Wing Membrane.

TRICHOSTICHAL BRISTLES Diptera: Metapleural Bristles *sensu* Comstock.

TRICHOTOMOUS Adj. (Greek, *tricha* = threefold; *tome* = cut; Latin, *-osus* = possessing the qualities of.) Pertaining to a trichotomy; an object, structure, concept or process divided into three parts. Cf. Entire; Dichotomous.

TRICHOTOMY Noun. (Greek, *tricha* = threefold; *tome* = cut. Pl., Trichotomies.) The condition or state of something divided into three parts. Cf. Dichotomy.

TRICHROISM Noun. (Greek, *tria* = three; *chros* = colour; English, *-ism* = condition.) A condition of any given part exhibiting three different colours in different individuals of the same Species, *e.g.* in Lepidoptera, the hindwings of certain Heliconidae.

TRICKLING FILTER FLY *Psychoda alternata* [Diptera: Psychodidae]: A cosmopolitan pest commonly associated with sink and bath drains. Larvae build to high populations feeding in gelatinous material associated with sewage-effluent trickling filters. Males emerge before females. Adult does not suck vertebrate blood; body densely setose; Antenna long; wings held tectiform over Abdomen. Eclosion occurs within 24 hours following oviposition. Larva aquatic with four instars; larvae apparently not cannibalistic. TFF regarded as pest through Pseudomyiasis and implicated in mechanical transmission of parasitic nematodes to livestock; also plays important role in degradation of sewage effluent. Life cycle ca 8–27 days. Syn. Sprinkling Sewage Filter Fly. See Psychodidae; Moth Flies.

TRICONDYLIC ARTICULATION A joint with three points of articulation between adjacent segments. Tricondylic articulations may be limited to the Coxa articulation with the pleural wall and provide three points of contact between the leg and the Thorax. Syn. Tricondylic Joint *sensu* Snodgrass. See Leg Articulation. Cf. Dicondylic Articulation; Monocondylic Articulation.

TRICTENOTOMIDAE Plural Noun. A small Family of Coleoptera assigned to Superfamily Tenebrionoidea.

TRICUSPID Adj. (Latin, *tres* = three; *cuspis* = point.) Divided into three cusps or points. Three-toed or three-clawed. See Tridactyle; Tridactylous; Tridactylus. Alt. Tricuspidate; Tricuspidatus.

TRIDACTYLIDAE Plural Noun. Pigmy Mole Crickets. A small Family of caeliferous Orthoptera assigned to Superfamily Tridactyloidea. Adult small, smooth and shiny bodied; Pronotum overlapping Mesonotum; forewing tegminized, with reduced venation, not extending to apex of Abdomen; fore and middle Tarsi of two segments; hind Tarsus absent or of one segment; hind Femur enlarged; hindlegs saltatorial; tympanal organs absent; four stylus-like appendages at apex of Abdomen. PMC burrow in moist sand or soil near stream or lake.

TRIDACTYLOIDEA. A Superfamily of caeliferous Orthoptera that includes the Families Cylindrachetidae, Ripipterygidae, Tridactylidae, Trigonopterygoidea, Borneacrididae and Trigonopterygidae.

TRIDENTATE Adj. (Latin, *tridens* = three-pronged; *-atus* = adjectival suffix.) Pertaining to structures with three tooth-like projections; Three-toothed. A tooth in the mayfly Mandible with three long points (Needham).

TRIEWALD, MARTIN (1691–1747) (Rose 1850, *New General Biographical Dictionary* 12: 79.)

TRIFASCIATE Adj. (Latin, *tri-* = three; *fascia* = band, sash; *-atus* = adjectival suffix.) With three fascia or bands of colour. Alt. Trifasciatus.

TRIFID Adj. (Latin, *trifidus* = three-forked.) Pertaining to structure cleft in three parts.

TRIFLURMURON Noun. A Chitin synthesis inhibitor used against mosquito larvae. Less effective against pupae; ovicidal at high concentrations. Temperature dependent; effective when ingested; some residual activity on plants. Developed by Bayer Laboratories (1975). See Insect Growth Regulator.

TRIFURCATE Adj. (Latin, *trifurcatus* = three-forked; *-atus* = adjectival suffix.) Pertaining to structure with three forks or branches. Alt. Trifurcatus.

TRIGAMMA Noun. (Greek, *tria* = three; *gamma* = third letter of Greek alphabet; ref. Y-shaped or pronged. Pl., Trigammae.) Lepidoptera wing: A three-pronged fork, the prongs of which are vein M , C and Cu (Imms).

TRIGARD® See Cyromazine.

TRIGONAL Adj. (Greek, *trigonos* = triangular; Latin, *-alis* = characterized by.) Triangular in appearance; an area bounded by a triangle.

TRIGONALIDAE Cresson 1867. Plural Noun. Trigonalids. A cosmopolitan Family of apocritious Hymenoptera consisting of about 70 nominal Species. Placed in various Superfamilies including Stephanoidea and regarded as an independent lineage under some classifications. Body medium sized; Antenna with 26–27 segments;

Toruli on a facial elevation; Mandibles large with asymmetrical dentition in some Species; forewing with 10 closed cells; hindwing with two closed cells; Tarsomeres with plantar lobes; second gastral Tergum and Sternum largest; Spiracles on Tergum VII; Ovipositor concealed and projecting from apex of Gaster. All Species parasitic on larval Hymenoptera and Tachinidae. Egg deposited on vegetation, frequently in large numbers (ca several thousand); eggs hatch when consumed by larval Symphyta or Lepidoptera. Parasite larva penetrates Haemocoel from gut. Most Species complete development when herbivorous larva is attacked by a parasitic hymenopteran or tachinid.

TRIGONALIDS See Trigonalidae.

TRIGONALOIDEA Hymenoptera: A Superfamily of Apocrita characterized by adults with numerous closed cells in the forewing and hindwing; Pronotum without a median dorsal surface and not reaching above the Tegula, pronotal spiracuar lobe projecting to the Tegula; hind Tibia not modified for grooming; antennal Sulcus absent. Included Families are Trigonalidae and Ichneumonidae. See Stephanoidea.

TRIGONATE Adj. (Greek, *trogonos* = triangular; *-atus* = adjectival suffix.) Three-angled. Alt. Trigonatus; Trigonous.

TRIGONEUTISM Noun. (Greek, *tria* = three; *gone* = offspring; English, *-ism* = condition.) The occurrence of three broods in one season. Cf. Bivoltinism; Multivoltinism.

TRIGONIDIIDAE Plural Noun. (Greek, *trigon* = angular.) Sword-tailed Crickets. A Family of ensiferous Orthoptera containing about 25 Genera and 275 Species, included within two Subfamilies (Trigonidinae and Phylloscirtinae.) Trigonidines are cosmopolitan in distribution, and phylloscirtines appear restricted to America. Head wider than anterior margin of Pronotum, eyes conspicuous and legs moderately long and slender. Male Tegmina resemble beetle Elytra. Trigonidiidae are found in humid habitats and some Species are known to skate on the surface of water. Eggs are deposited in plant tissue.

TRIGONULUM Noun. (Greek, *trigonon* = triangle.) Odonata: The triangle.

TRILATERAL Adj. (*tres* = three; *latus* = side.) Descriptive of structure with three sides.

TRILINEATE Adj. (*tres* = three; *linea* = line.) Three-lined. Alt. Trilineatus.

TRILOBATE Adj. (Greek, *tria* = three; *lobos* = lobe; *-atus* = adjectival suffix.) Having three lobes. Alt. Trilobatus; Trilobe; Trilobed.

TRILOBITE SCALE *Pseudaonidia trilobitiformis* Green [Hemiptera: Diaspididae]: A minor polyphagous pests of citrus, tea, guava and mango in tropical and subtropical areas.

TRILOGY® A broad-spectrum fungicide/miticide which can be applied to citrus, stone, pome and tropical fruits, curcurbits and most kinds of vegetables. Not for application to nurseries, turf or landscape plants. Active ingredient Clarified Hydrophobic Extract of Neem Oil 70%. Marketed in USA by Thermo Trilogy Corporation. See Azadirachtin.

TRIM® See Tecnazene.

TRIMATON® See Metham-Sodium.

TRIMBLE, ISAAC PIM (1802–1889) (Johnson 1901, Proc. ent. Soc. Wash. 4: 230–233; Weiss 1918, Ent. News 29: 29–32, bibliogr.; Weiss 1936, *Pioneer Century of American Entomology,* 320 pp. (257–259), New Brunswick.)

TRIMEDLURE A synthetic pheromone {test-butyl 4 (or 5)-chloro-2 methyl cyclohexycarboxylate} used as an attractant (trap) for Mediterranean Fruit Fly. Trade names include Capilure®, Medlure®, Polycore TML® and Siglure. See Synthetic Pheromone. Rel. Parapheromone.

TRIMEN, ROLAND (1840–1916) (Anon. 1916, Entomol. mon. Mag. 52: 209–210; Bethune-Baker 1916, Entomologist's Rec. J Var. 28: 231–236; Poulton 1916, Nature 79: 485–486; Poulton 1917, Proc. Linn. Soc. Lond. 129: 76–78.)

TRIMENOPONIDAE Plural Noun. A small Family of lice known from South America and assigned to Suborder Amblycera. Known Species parasitic on marsupials.

TRIMERA Plural Noun. 1. Series of Coleoptera whose members display only three Tarsomeres. 2. A Superfamily of Psocoptera whose members have three-segmented Tarsi.

TRIMEROUS Adj. (Greek, *tria* = three; *meros* = part; Latin, *-osus* = possessing the qualities of.) 1. Pertaining to structure composed of three parts or pieces. 2. Legs with three-segmented Tarsi.

TRIMETHION® See Dimethoate.

TRIMORPHISM Noun. (Greek, *tria* = three; *morphos* = form; English, *-ism* = condition. Pl., Trimorphisms.) The phenomenon of having three forms in colour or structure in one Species, or, in ants and termites, in one worker caste.

TRINEX® See Trichlorfon.

TRINOMIAL Adj. (Latin, *tres* = three; *nomen* = name; *-alis* = appropriate to.) Zoological nomenclature: Descriptive of organisms whose scientific names consist of three parts, *e.g.,* a varietal name, generic name and specific name. See Nomenclature; Taxonomy. Cf. Binomial; Uninomial.

TRIORDINAL Adj. (Latin, *tri* = three; *ordo* = order; *-alis* = pertaining to.) Lepidoptera: Pertaining to a larval proleg with Crochets that display three alternating sizes. Cf. Multiordinal; Uniordinal.

TRIOZIDAE Plural Noun. A small, widespread Family of sternorrhynchous Hemiptera assigned to Superfamily Psylloidea. Forewing Vena Spuria, Pterostigma and costal break absent; veins R, M, CuA originating near same place on wingbase. Trioxids form galls on the leaves of diverse host plants.

TRIPARTITE Adj. (Latin, *tres* = three; *partitus* = separated; *-ites* = constituent.) In three parts.

TRIPECTINATE Adj. (Latin, *tri* = three; *pecten* =

comb.) Having three branches or processes to each joint of an Antenna.

TRIPLOBLASTIC Adj. (Greek, *tria* = three; *blastos* = bud; *-ic* = consisting of.) Embryology: Consisting of three germ layers (Ectoderm, Mesoderm and Endoderm) from which complex organ-systems are developed. The primitive embryonic cellular pattern (Klots). Cf. Diploblastic; Holoblastic.

TRIPUPILLATE Adj. (Latin, *tri* = three; *pupilla* = pupil of the eye; *-atus* = adjectival suffix.) Possessing three pupils in the eye; applied to ocellar spots having three inner spots or pupils.

TRIQUETRAL Adj. (Latin, *triquetrus* = having three sides; *-alis* = characterized by.) Triangular in section, with three flat sides. Alt. Triquetrous; Triquetrum.

TRIRADIATE Adj. (Latin, *tri* = three; *radius* = ray; *-atus* = adjectival suffix.) Three-rayed.

TRIRAN® See Cyhexatin.

TRIREGIONAL Adj. (Latin, *tres* = three; *regio* = territory; *-alis* = characterized by.) Descriptive of structure that is divided into three distinct parts or regions.

TRISTAN, JOSÉ FIDEL (1874–1932) (Calvert 1932, Ent. News 43: 197–200.)

TRITOCEREBRAL Adj. (Greek, *tritos* = third; Latin, *cerebrum* = brain; *-alis* = pertaining to.) Pertaining to the Tritocerebrum.

TRITOCEREBRAL COMMISURE See Tritocerebrum.

TRITOCEREBRAL SEGMENT The second part of the insect Brain.

TRITOCEREBRUM Noun. (Greek, *tritos* = third; Latin, *cerebrum* = brain.) Posterior division of insect Brain, formed by ganglia of third (intercalary) segment of head; represented by a small pair of lobes which presumably represent third segment of hypothetical head. Paired lobes of Tritocerebrum are positioned beneath Deutocerebrum and near Aorta. Tritocerebrum sends sensory and motor neurons to Frontal Ganglion via Frontal Ganglion Connectives. Tritocerebrum sends sensory and motor neurons to Labrum via Labral Nerve. Other components include Tritocerebral Commissure which loops under Pharynx and attaches to Tritocerebral Lobe on opposite side of Brain. Circumoesophageal Connectives project ventrad and posteriad on either side of Pharynx and connect to Subesophageal Ganglion. Syn. Labrofrontal Lobe (Smith). See Brain; Central Nervous System. Cf. Deutocerebrum; Protocererbrum.

TRITONYMPH Noun. The third nymphal stage in mites. Cf. Protonymph; Deutonymph; Adult.

TRITURATE Noun. (Latin, *trituratus*; *-atus* = adjectival suffix.) A triturated substance.

TRITURATE Trans. Verb (Latin, *triturare* = to thresh grain; *-atus* = adjectival suffix.) To rub, thresh, grind or pulverize. Triturated, Triturating.

TRITURATING BASKET 1. Adapted for grinding or crushing. 2. A cuticular network in the foregut of bees adapted for filtering particulate material from nectar and other liquid nutrients. See Bolus.

TRITURATION Noun. (Latin, *trituratio;* English, *-tion* = result of an action. Pl., Triturations.) The act of triturating or state of being triturated.

TRIUMPH® See Isazofos.

TRIUNDULATE Adj. (Latin, *tres* = three; *undulatus* = wave-like; *-atus* = adjectival suffix.) Descritpive of structure or process with three waves or undulations.

TRIUNGULID Noun. (Latin, *tres* = three; *ungula* = claw. Pl., Triungulids.) The active campodeiform larva of Strepsiptera or *Stylops*. Cf. Triungulin. See Heteromorphosis; Hypermetamorphosis. Cf. Triungulin; Triungulinid.

TRIUNGULIN Noun. (Latin, *tres* = three; *ungula* = claw. No Plural.) The free-living, hypermetamorphic first-instar larva of some Coleoptera (all Rhipiphoridae, Drilidae, Meloidae; some Colydiidae, Carabidae, Staphylindae, and at least one Eucnemidae). Triungulin is transported by adult insects to a suitable host stage of a host Species. See Heteromorphosis; Hypermetamorphosis. Cf. Planidium; Triungulid; Triungulinid. Rel. Phoresy.

TRIUNGULINID Noun. (Latin, *tres* = three; *ungula* = claw. Pl., Triungulinids.) The free-living, hypermetamorphic first-instar larva of Strepsiptera. The larva resembles the triungulin of Coleoptera, but lacks a Trochanter in all legs, well formed Antennae and Mandibles. Cf. Triungulin.

TRIVIAL Adj. (Latin, *trivialus* = within the crossroads, *i.e.* everywhere > *trivium* = crossroads; *-alis* = characterized by.) Trifling. Applied to the name of an organism, specific as opposed to generic, or popular as opposed to technical.

TRIVIAL FLIGHT A behavioural category for insect flight which includes individuals moving relatively short distances for feeding, patrolling territory, acquiring sexual partners and seeking ovipositional sites. See Flight. Cf. Mass Flight; Migratory Flight; Swarming Flight.

TRIVIAL MOVEMENT Displacement of insects within or near the breeding habitat. Non-migratory movement. See Migration.

TRIVITTATE Adj. (Latin, *tres* = three; *vitta* = band; *-atus* = adjectival suffix.) With three stripes or vitta. Alt. Trivittatus.

TRIVOLTINE Adj. (Latin, *tri* = three; Italian, *volta* = time.) Having three generations in a year or season. *e.g.* Silkworms (*Bombyx mori*).

TRIXAGIDAE Plural Noun. A cosmopolitan Family of Coleoptera including about 200 nominal Species currently and assigned to Superfamily Elateroidea. Trixagids are similar in habitus to elaterids and eucmenids. Adult distinguished by parallel-sided hind femoral plates, prosternal process broad and flat, Prothorax and Mesothorax firmly united, and body sombre coloured. Biologically rather poorly studied. Adults taken at flowers; larvae found in wood. Immatures suspected of carnivory, based on dentate outer margin of Mandible. Syn: Throscidae.

TRIXOSCELIDAE See Heleomyxidae.

TROCACOILA Noun. The mesial end of the Trochanter modified into a Condyle (MacGillivray).

TROCASUTURE Noun. A suture of the Trochanter (MacGillivray). Alt. Trochasuture.

TROCHACORIA Noun. The Coria of the Trochanter in the insect leg (MacGillivray).

TROCHALOPODA Heteroptera in which the hind Coxa is nearly globose and the articulation is a ball and socket joint. See Pagiopoda.

TROCHANTELLUS Noun. (Greek, *trochanter* = runner.) Hymenoptera: The constricted portion of the base of the Femur which superficially appears to represent a second Trochanter. Found in Symphyta and some Parasitica.

TROCHANTER Noun. (Greek, *trochanter* = runner. Pl., Trochanters.) The second and typically smallest segment of the insect leg. Positioned between the Coxa and Femur; fused with the Femur in some Species and apparently two-segmented in other Species. Trochanter is the basal segment of the telopodite and typically inconspicuous, rigid and dicondylic. Trochanter on all legs of most insects is composed of one segment. Exceptions occur: Odonata have two Trochanters; Strepsiptera adults lack a Trochanter. Parasitic Hymenoptera show a Trochantellus or apparent second Trochanter distad of a true Trochanter. Very important to insect and provides leg with considerable flexibility and rotation via coxo-trochanteral joint. This 'joint' probably serves as a rotational device because articulation points are rotated 90° from those on the Coxa. Trochantellus and second Trochanter probably provide additional flexure. Trochanter is frequently invested with Campaniform Sensilla that form a complex of stretch receptors which inform insect where the distal portion of the leg is positioned with respect to remainder of body. See Leg. Cf Coxa; Femur; Tibia; Pretarsus

TROCHANTERAL Adj. (Greek, *trochanter* = runner; *-alis* = pertaining to.) Descriptive of structure associated with the Trochanter.

TROCHANTERELLUS See Apophysis.

TROCHANTEROFEMORAL Adj. (Greek, *trochanter* = runner; *femur* = thigh; *-alis* = pertaining to.) Descriptive of structure or process involving the Trochanter and Femur.

TROCHANTIN Noun. (Greek, *trochanter* = runner. Pl., Trochantins.) 1. A small sclerite at the base of the insect leg of some insects. In some hypotheses, the Trochantin is believed to develop into the pleural wall of the Thorax. 2. Coleoptera: A sclerite often present on the outer side of, and sometimes moveable on, the Coxa; also a small sclerite connecting the Coxa with the Sternum in Dytiscidae. 3. Neuroptera and Trichoptera: The posterior separated part of the Coxa. 4. Orthoptera: A narrow longitudinal sclerite between Mandible and Gena. 5. Any small, intercalated sclerite of an appendage. Alt. Trochantine.

TROCHANTIN OF THE MANDIBLE Orthoptera: A small sclerite adjacent to the Mandible and separated by a suture from a Gena (Imms). Cf. Mandibularia.

TROCHANTINAL Adj. (Greek, *trochanter* = runner; *-alis* = pertaining to.) Descriptive of or pertaining to the Trochantin.

TROCHANTINOPLEURA. See Coxopleurite.

TROCHELLA Noun. (Pl., Trochellae.) A small, chitinized sclerite on the dorsal part of the Coria of the Trochanter (MacGillivray).

TROCHIFORM Adj. (Latin, *trochus* = Genus of marine mollusc; *forma* = shape.) Shell-shaped; descriptive of structure resemling the shell of *Trochus, i.e.* a conical operculate shell flattened at the base and an oblique aperture. Cf. Chonchiform; Conriform; Heliciform; Mytiliform; Pectiniform. Rel. Form 2; Shape 2; Outline Shape.

TROCHIFORMIS Adj. Cylindroconic.

TROCHLEA Noun. (Greek, *trochilia* = pulley. Pl., Trochleae.) The thickened base of the hindwings in Cicada. Trichoptera: A small elliptical space at base of hindwing behind origin of median vein.

TROCHLEARIS Adj. (Greek, *trochilia* = pulley.) Pulley-shaped, like a cylinder contracted medially.

TROCHLIA Noun. (Greek, *trochos* = wheel. Pl., Trochliae.) Any distal segment of a leg enlarged into a plate or flange (MacGillivray).

TROCHOID Adj. (Greek, *trochoeides* = round as a wheel; *eidos* = form.) Wheel-like; circular; descriptive of joints capable of rotation. Cf. Cycloid; Deltoid. Rel. Eidos; Form; Shape.

TROCHOPHORE Noun. (Greek, *trochos* = wheel; *phora* = bearing. Pl., Trichophores.) The free swimming, pelagic larval stage of some annelids and molluscs. So named because the stage bears a preoral whorl of Setae.

TROCHUS Noun. (Greek, *trochos* = wheel. Pl., Trochi; Trochuses.) 1. The inner, anterior, coarse ciliary zone of the rotifer's disc. 2. Any small segment intercalated between the normal segments in an articulated structure or part.

TROCTOMORPHA A Suborder of the Psocoptera that contains four Families, Amphientomidae, Pachytroctidae, Liposcelidae, and Sphaeropsocidae. Leposcelidae contains the traditional 'booklouse.'

TROGIDAE Plural Noun. A widespread Family of polyphagous Coleoptera assigned to Scarabaeoidea. Body to 30 mm long, stout, Integument heavily sclerotized, embossed or tuberculate. Head deflexed, mostly concealed when viewed from above; Antenna short, lamellate, with 10 segments including three segments in club; fore Tibia weakly dentate; fore Tarsi short; tarsal formula 5-5-5; Abdomen with five Ventrites; apex of Abdomen covered by Elytra. Larva scarabaeiform; head hypognathous, darkly pigmented; at most one stemma on side of head;

Antenna with three segments; three pairs of well developed legs present each with five segments and two tarsal claws; Urogomphi absent. Adults slow-moving; winged adults attracted to lights; stridulate with *plectrum* on penultimate abdominal segment against *pars stridens* on internal surface of Elytra. Typically found beneath dried animal carcass; larvae construct burrows perpendicular to carcass. Adults and larvae feed upon dried carrion or sometimes dried excrement.

TROGIIDAE Plural Noun. Trogiids. A cosmopolitan Family of Psocoptera assigned to the Suborder Trogiomorpha. Antenna with more than 20 segments; Ocelli absent; forewing not well developed, hindwing absent; Tarsi with three segments; paraprocts with strong posterior spine. Some domicillary Species.

TROGIOMORPHA Plural Noun. A Suborder of Psocoptera containing the Lepidopsocidae, Psyllipsocidae, Trogiidae.

TROGOSSITIDAE Plural Noun. Bark Gnawing Beetles; Trogositid Beetles. A widely distributed Family of Coleoptera with about 600 Species. Higher classification unstable and disputed; members of Cleroidea. Family name spelled Trogositidae or considered synonymous with Lophocateridae, Ostomidae, Peltidae, Temnochilidae. Adult 1–18 mm long; somewhat flattened, heavily sclerotized; lack conspicuous Setae over body, (sometimes with scales); Antenna short, with 7–11 segments, moniliform, clavate. Fore Coxa transverse, not projecting; tarsal formula 5-5-5, first Tarsomeres minute and easily overlooked; Tarsomeres not lobed; Pretarsus with Empodium; Elytra with parallel sides; Abdomen with five Ventrites. Larva campodeiform; head prognathous, 2–5 Stemmata on each side; Antenna with three segments; legs with five segments, each Pretarsus with one tarsal claw; Urogomphi present. Adults slow moving; biology poorly known; *Nemosoma* and related genera found beneath bark in galleries, predaceous on scolytids. *Tenebroides mauritanicus* (L.) a pest of stored products and predaceous on other pests of stored products.

TROJAN-10® See *Nosema locustae*.

TRONICEK, EDVARD (Paclt 1968, Acta Mus. Regenaehr. Sci. nat. 9: 195–198.)

TROOP, JAMES (1853–1941) (Davis 1941, J. Econ. Ent. 34: 865–866.)

TROOST, GERARD (1776–1850) (Youmans 1869, *Pioneers of Science in America.* viii + 508 pp. (119–127), New York.)

TROPHALLACTIC APPEASEMENT Social Insects: The use of liquid food (trophallaxis) to appease potentially hostile workers.

TROPHALLACTIC GLANDS Glands which produce the exudate or secretion given by larvae to the adults after being fed.

TROPHALLAXIS Noun. (Greek, *trophe* = nourishment; *allaxis* = interchange, barter. Pl., Trophallaxes.) 1. The exchange of nourishment between insects of the same Species or different Species. The term was first used by Wheeler (1918. Proc. Amer. Phil. Soc. 57: 293–343). 2. Social Hymenoptera: A mutual exchange of food between adults and their larvae. Adult provides proteins to larva; larva provides carbohydrates to adult. Regurgitation of a droplet of liquid by a larva and taken by an adult worker. Posssible significance: Method of oral excretion; method for temperature and humidity control; method of food distribution. Data on vespids suggest larval salivary secretions essential for colony survival because secretions are only source of compounds necessary for nitrogen metabolism and egg production. Larval salivary secretions are only source of nutrition for queen who will die if deprived of secretion. See Proctodaeal Trophallaxis; Stomodaeal Trophallaxis. Rel. Nuptial Gift.

TROPHAMNION Noun. (Greek, *trophe* = nourishment; *amnion* = fetal membrane. Pl., Trophamnions.) An envelope surrounding the egg of polyembryonic parasitic Hymenoptera which provides nourishment to the developing embryos contained within the envelope. Trophamnion Cell. See Teratocyte.

TROPHI Plural Noun. (Greek, *trophe* = nourishment.) 1. The mouthparts of arthropods. 2. An obsolete term for the insect mouthparts. Collectively, the Mandibles, Maxillae and Labium. See Buccal Appendages.

TROPHIC Adj. (Greek, *trophe* = nourishment; *-ic* = characterized by.) 1. Descriptive of or pertaining to the trophi. 2. Descriptive of or pertaining to foods or eating.

TROPHIC CELLS See Teratocyte.

TROPHIC EGG Social Insects: An inviable, degenerate egg which is fed to colony members.

TROPHIC GUILD See Guild.

TROPHIC LEVEL The level of a group of organisms in the food chain or process of energy transfer within an ecosystem. TL typically identified by the method of obtaining food: Primary producer (green plant), primary consumer (herbivore), secondary consumer (predator) and tertiary consumer (parasite).

TROPHIC PARASITISM Social Insects: The intrusion of one Species into the social system of another Species for the acquisition of food.

TROPHIC PLASTICITY In insects, the ability to ingest, digest and metabolize novel foods.

TROPHIC SYMBIOSIS Social Insects: A form of symbiosis between ants and other insects (sought and attended by ants) for the food or secretions ants derive from these insects. Example: Ants and aphids. See Biosis; Commensalism; Parasitism; Symbiosis. Cf. Myrmecophily.

TROPHIFER Noun. (Greek, *trophe* = nourishment; Latin, *ferre* = to carry. Pl., Trophifers.) The genal region of the Postgena (MacGillivray).

TROPHOBIONT (Greek, *trophe* = nourishment; *bion* = living. Pl., Trophobionts.) Noun. An insect which provides food (honeydew) to ants.

TROPHOBIOSIS Noun. (Greek, *trophe* = nourishment; *bios* = living; *-osis* = state, condition. Pl., Trophobioses.) Trophic symbiosis. Social Insects: A relationship in which social insects obtain food from non-social insects and in return protect the non-social form (trophobiont). *e.g.* Aphids provide honeydew to ants and are protected from predators and parasites by worker ants. See Biosis; Commensalism; Parasitism; Symbiosis. Cf. Abiosis; Anhydrobiosis; Antibiosis; Archebiosis; Calobiosis; Cleptobiosis; Hamabiosis; Kleptobiosis; Lestobiosis; Parabiosis; Phylacobiosis; Plesiobiosis; Synclerobiosis; Xenobiosis.

TROPHOCYTE Noun. (Greek = *trophos* = one who feeds. *trophon* = food, that which feeds; *kytos* = a hollow vessel; container, cell. Pl., Trophocytes.) Principal cell type comprising insect fat body. Pleiomorphic form; enlarged by vacuoles and distension with glycogen and nutrients. Nuclei of Trophocytes change shape with change in cell size; cell boundaries become less apparent as Trophocyte increases in size. Important role in intermediary metabolism; concerned with production of trehalose (primary circulatory carbohydrate). Site for storage of lipids (as droplets of triglycerides) and carbohydrate (as granules of glycogen). See Fat Body.

TROPHOGENESIS Noun. (Greek, *trophe* = nourishment; *genesis* = be produced; *-sis* = a condition or state. Pl., Trophogeneses.) Social Insects: The origin of different caste traits due to differential feeding of the immatures. Cf. Blastogenesis.

TROPHOSEROSAL CELLS See Teratocyte.

TROPHOTHYLAX Noun. (Greek, *trophe* = nourishment; *thylax* = sack.) Hymenoptera: An invagination in the first abdominal segment of larval ants assigned to the Subfamily Pseudomyrminae.

TROPHOZOITE Noun. (Greek, *trophe* = nourishment; *zoon* = animal; *-ites* = inhabitant. Pl., Trophozoites.) The first stage of the asexual development of *Plasmodium*, the protozoan parasite which causes Malaria.

TROPIC Adj. (Greek, *tropikos* = turn; *-ic* = of the nature of.) 1. Pertaining to or of the nature of a tropism. 2. A movement in response to a unidirectionál stimulus.

TROPIC BEHAVIOUR A form of behaviour which is composed predominantly of simple reflex actions (Wardle).

TROPICAL BED BUG *Cimex hemipterus* (Fabricius) [Hemiptera: Cimicidae]: A tropicopolitan Species of cimicid bug parasitic on humans, chickens and rarely bats. Anatomically similar to 'Bed Bug' (*C. lectularius*) and similar in biology; generally allopatric from Bed Bug, but sympatric in South Africa. Male *C. hemipterus* and female *C. lectularius* copulate but few eggs produced and most are inviable. Reciprocal cross-mating produces many eggs which are abnormal. See Cimicidae. Cf. Bed Bug.

TROPICAL CASE-BEARING CLOTHES MOTH *Tinea translucens* Meyrick [Lepidoptera: Tineidae]. A cosmopolitan pest of products made of hair, feathers and wool; pest is more common in tropics. Cf. Casemaking Clothes Moth; Large Pale Clothes-Moth; Webbing Clothes Moth.

TROPICAL FOWL-MITE *Ornithonyssus bursa* (Berlese) [Acari: Macronyssidae].

TROPICAL GREY CHAFF SCALE *Parlatoria cinerea* Doane & Hadden. [Hemiptera: Diaspididae]. An occasional pest of citrus which has frequently been confused with the Chaff Scale.

TROPIÇAL HORSE-TICK *Anocentor nitens* (Neumann) [Acari: Ixodidae].

TROPICAL LEGUME LEAF BEETLE *Pagria signata* (Motschulsky) [Coleoptera: Chrysomelidae] (Australia).

TROPICAL LOG BEETLES See Discolomidae.

TROPICAL RAT FLEA See Oriental Rat Flea.

TROPICAL RAT LOUSE *Hoplopleura pacifica* Ewing [Anoplura: Hoplopleuridae].

TROPICAL RAT MITE *Ornithonyssus bacoti* (Hirst) [Acari: Macronyssidae].

TROPICAL REGION See Neotropical Region.

TROPICAL REPTILE TICK *Aponomma fimbriatum* (C.L. Koch) [Acari: Ixodidae] (Australia).

TROPICAL SCRUB ITCH MITE *Eutrombicula hirsti* (Sambon) [Acari: Trombiculidae] (Australia).

TROPICAL SOD-WEBWORM *Herpetogramma phaeopteralis* Guenée [Lepidoptera: Pyralidae].

TROPICAL SPLENOMAGALY See Kala-azar. Syn. Black Disease; Dumdum Fever

TROPICAL WARBLE-FLY *Dermatobia hominis* (Linnaeus, Jr.) [Diptera: Cuterebridae]: An economically highly-significant pest of many domestic livestock and aetiological agent of Cutaneous Myiasis in humans; endemic to Neotropical Realm. Adult head and legs yellow; does not feed, lives 1–9 days; female mates within day of emergence, produces 800–1,000 eggs during lifetime; eggs attached to another insect, typically mosquito or anthomyiid fly; incubation 4–9 days; eclosion is stimulated by body temperature of warm-blooded host. Neonate larva penetrates skin of host and feeds on exudates, not blood; development within cyst-like swelling; larval body bears spines which serve as holdfast structures; first-instar larva subcylindrical; second-instar pyriform; third-instar elongate; respiration through opening created by penetration of skin, and odour frequently emanated through wound which serves to attract other Myiasis agents; larval development requires 4–18 weeks. Prepupa emerges from wound, drops to ground and pupation occurs in soil; pupal stage 4–11 weeks. Syn. Torsalo; Human Bot Fly. See Cuterebridae. Cf. Rabbit Bot Fly.

TROPICAL WAREHOUSE-MOTH See Almond

Moth.

TROPICOPOLITAN A term used to describe organisms which occur in all tropical regions. Cf. Pantropical.

TROPILAELAPS CLAREAE A laelapid mite that develops as an external parasite of several Species of honey bees (*Apis cerana, A. dorsata, A. florea, A. laboriosa, A. mellifera*) in Asia. Mites feed on the Haemolymph of bee larvae, especially drones and to a lesser extent, worker larvae. The mites move actively on the comb and may steal food from the mouths of adult bees but cannot pierce the Integument to take Haemolymph. Cf. Tracheal Mite; Varroa Mite.

TROPISM Noun. (Greek, *trope* = to turn; English, *-ism* = condition. Pl., Tropisms.) 1. A forced or involuntary movement in an animal or plant that is caused by outside sources of energy (Loeb). 2. The response evoked in animals by a chemical or physical stimulus. See Orientation. Cf. Kinesis; Taxis. Rel. Aeolotropism; Anemotropism; Chemotropism; Electrotropism; Galvanotropism; Geotropism; Heliotropism; Hydrotropism; Phototropism; Rheotropism; Stereotropism; Thermotropism; Thigmotropism; Tonotropism.

TROPOTAXIS Noun. (Greek, *tropos* = turn; *taxis* = arrangement. Pl., Tropotaxes.) A taxis in which an organism orients to a source of stimulation by simultaneous comparison of the amount of stimulation to symmetrically arranged sensory receptors on either side of the body. See Orientation. Cf. Aerotaxis; Anemotaxis; Chemotaxis; Geotaxis; Menotaxis; Osmotaxis; Phototaxis; Rheotaxis; Rotaxis; Scototaxis; Strophotaxis; Telotaxis; Thermotaxis; Thigmotaxis; Tonotaxis.

TROSCHEL, FRANZ HERMAN (1810–1882) (Dechen 1883, Verh. naturh. Ver. preuss. Rhein. 40: 35–54, bibliogr.)

TROST, ALOIS (1850–1909) (Anon. 1910, Ent. Rdsch. 27: 10.)

TROTTER, ALESSANDRO (1874–1967) Italian academic (Professor of Plant Pathology, University of Naples.) Influenced by Caro Massalongo and published research interest in eriophyid galls and their affect on plants. (Stefano 1968, Marcellia 35: 3–44, bibliogr.)

TROUESSART, EDOUARD LOUIS (1842–1927) (Bourdelle 1928, Archs Mus. Hist. nat., Paris (6) 3: 1–18, bibliogr.)

TROUGHT, TREVOR (1891–1970) (Hinton 1971, Proc. R. ent. Soc. Lond. (C) 35: 54.)

TROUNCE® See Dimethoate.

TROUT STREAM-BEETLES See Amphizoidae.

TRUCIDOR® See Vamidothion.

TRUE BUGS See Hemiptera.

TRUE FLIES See Diptera.

TRUE KATYDIDS See Pseudophyllidae.

TRUE LICE See Anoplura.

TRUE SAWFLIES See Tenthredinidae.

TRUE SCORPIONS See Scorpiones.

TRUE SPIDERS See Araneae.

TRUE WASPS See Vespoidea.

TRUE WATER BEETLES See Dytiscidae.

TRUE WEEVILS See Curculionidae.

TRUE WIREWORM *Agrypnus variabilis* (Candeze) [Coleoptera: Elateridae]: An Australian Species sometimes taken in cotton. Adults commonly called 'click beetles.' Adult elongate, 7–12 mm long and dark brown; Prothorax with two spines on apex; when beetle is upside down it engages the spines to Elytra and flips itself into air, thus righting itself and making a clicking sound. Adult female lays eggs singly or in batches within soil. Eggs opaque, pearly-white, cylindrical and 0.5 mm long. Egg stage 14–30 days. Larval stage 1–5 years with many generations overlapping. Larva elongate to cylindrical, 7–20 mm long and vary from yellow to brown; larval Integument very hard. Head and last abdominal segment reddish-brown; Urogomphi present; head flattened. Pupal stage occurs in soil, with adults emerging to mate and oviposit. Larvae feed on organic material in soil and reach full size in spring. Adults emerge in November-December. True wireworms prefer wet soil for egg laying (*e.g.* irrigated summer crops).

TRUENO® See Hexaflumuron.

TRULLAE Coccids: The lobes, (MacGillivray).

TRUMAN, PHILETUS CLARK (1841–1901) (Anon. 1901, Ent. News 12: 327–328.)

TRUMPET Noun. (Middle English, *trumpete*.) The respiratory horn or tube of the mosquito pupa.

TRUMPET-NET CADDISFLIES See Psychomyiidae.

TRUNCATE Adj. (Latin, *truncatus* = cut off.) Pertaining to structures which end abruptly as if cut at a right angle to the longitudinal axis.

TRUNCATE WING Insects: one shortened straight across, as in *Drosophila* laboratory mutants.

TRUNCATION Noun. (Latin, *truncatus* = cut off; English, *-tion* = result of an action. Pl., Truncations.) A transverse abrupt termination of a structure or appendage. Alt. Truncature.

TRUNCUS Noun. (Latin, *truncus* = a tree's trunk. Pl., Trunces.) The insect trunk or Thorax.

TRUNK Noun. (Latin, *truncus* = a trunk. Pl., Trunks.) 1. The Thorax or middle portion of the body. Cf. Cauda; Tail. 2. The central part of anything.

TRUPOCHALCIDIDAE Kozlov 1975. A Family of apocritous Hymenoptera assigned to the Proctotrupoidea, sometimes placed in the Austrioniidae.

TRUSS CELL Wings of Myrmeleontidae: A long cell immediately following the point of fusion of the Subcostal and Radial Veins (Wardle).

TRYBOM, FILIP (1850–1913) (Aurivillius 1914, Ent. Tidskr. 35: 81–86, bibliogr.)

TRYON, HENRY (1856–1943) First professional Entomologist in Queensland, Australia: Assistant Curator, Queensland Museum (1883), Government Entomologist (1894, Vegetable Pathologist (1901). (Anon. 1929, Qd. Agric. J. 32: 176–183, bibliogr.)

TRYPANOSOMA (Greek, *trypanon* = borer; *soma* = body. Pl., Trypanosomae; Trypanosomas.) A

Genus of parasitic Protozoa which occurs in the blood of many vertebrates including fish and mammals. *Trypanosoma* spp. utilize flies, bugs, lice and fleas as vectors and undergo a complex life cycle which includes several phases (amastigote, promastigote, opisthomastigote, epimastigote and trypomastigote.) Genus divided into two Sections: Salivaria and Stercoraria.

TRYPANOSOMIASIS Noun. (Greek, *trypanon* = to bore; *soma* = body; *myia* = fly; *sis* = a disease. Pl., Trypanosomiases.) An assemblage of diseases of humans and domesticated animals which are induced by parasitic Protozoa assigned to the Order Kinetoplastida. Cf. Leishmaniases. See Human Trypanosomiasis.

TRYPOMASTIGOTE Noun. (Greek, *trypan* = borer; mastigote = flagellate protozoa. Pl., Trypomastigotes.) An anatomical form manifest at a specific phase in the complex life cycle of some parasitic Protozoa (*Leishmania* and *Trypanosoma*). Trypomastigote displays a flagellar base posterior of nucleus from which Flagellum emerges laterally and forms a long, undulating membrane. See Leishmania; Trypanosoma. Cf. Amastigote; Epimastigote; Opisthomastigote; Promastigote.

TRYPSIN Noun. (Greek, *tryein* = to rub down; *pepsis* = digestion. Pl., Trypsins.) A proteolytic enzyme of the pancreatic fluid which splits or breaks proteins, proteoses and peptones. Tissue Culture: Trypsin is used to remove anchorage-dependent cells from their substrate.

TRYPTASE Noun. (Greek, *tryein* = to rub down; *pepsis* = digestion. Pl., Tryptases.) A proteolytic digestive enzyme.

TRYPTIC Adj. (Greek, *tryein* = to rub down; *pepsis* = digestion; *-ic* = of the nature of.) Acting like trypsin; the proteolytic ferment of the pancreatic fluid.

TRYPTOPHANE Noun. (Greek, *tryein* = to rub down; *pepsis* = digestion; *phainein* = to appear. Pl., Tryptophanes.) An end-product of the proteolytic action of the trypsin in the gastric fluid of insects.

TSCHITSCHERINE, TICHON SERGEIEVICH (1869–1904) (Semenov 1904, Revue Russe Ent. 4: 69–76; Semenov 1908, Trudy russk. ent. Obshch. 38 (4): 1–45, bibliogr., list of insects described.)

TSCHUMI, LOUIS (1879–1949) (Anon. 1949, Anz. Schädlingsk. 22: 64.)

TSETSE Singular and Plural Noun. (Sechuana, *Tsetse* = fly destructive of cattle.) Broadly, members of muscoid fly Genus *Glossina* (ca 25 Species) [Diptera: Glossinidae] found in tropical and subtropical Africa and adjacent islands. Three Species groups or Subgenera *(Austenina, Glossina, Nemorhina)* differ in ecology and role as disease vectors. Most notable, the *morsitans* group includes several Species which occupy woodland-savanna habitat and serve as main vectors of *Trypanosoma brucei rhodesiense,* the agent which causes African Sleeping-Sickness (trypanosomiasis) and Nagana in cattle. *Glossina morsitans morsitans* Westwood best studied. Adult 7–14 mm long, brown with wings crossed at rest and projecting beyond apex of Abdomen; wing venation with Discal Cell cleaver-shaped; Palpi nearly as long as Proboscis; Proboscis directed anteriad; antennal Arista with bilaterally branched. Male and female feed on vertebrate blood during day; female larviparous and only produces one larva during reproductive interval (8–25 days); female produces 8–10 larvae (20 maximum) during lifetime. Instars 1, 2 and part of three feed on intrauterine glandular secretion (milk gland); female requires blood meal at regular intervals to complete development of larva; mature larva released from female body and burrows into soil to pupate; fourth larval instar occurs within puparium. Pupal stage ca 2–4 weeks. See Glossinidae; Naganal Sleeping Sickness.

TSETSE FLIES See Glossinidae.

TSETSE TARGETS Odour-baited targets impregnated with deltamethrin placed in a band that serves as an invasion barrier against dense populations of *Glossina morsitans morsitans* Westwood and *G. pallidipes* Austen (Diptera: Glossinidae).

TSETSE TRAPS A biconical trap made of blue cloth that attracts adult tsetse flies. Flies tend to move upward enter holes in the bottom of the lower cone and are trapped. A killing agent usually is placed the apex of the upper cone. The effectiveness of the traps can be improved with olfactory attractants such as acetone, carbon dioxide gas or the urine of ungulates. See Biconical Traps.

TSHEKARDOCOLEIDAE Plural Noun. A small Palaeozoic Family of Coleoptera or coleopteroid Taxa known from Permian deposits in Russia.

TSHEKARDOPERLIDAE Plural Noun. A small Palaeozoic Family of Plecoptera known from Permian deposits in Russia.

TSUCHIDA (1871–1945) (Hasegawa 1967, Kontyû 35 (Suppl.): 53, bibliogr.)

TSUCHIDA, TOSHIZO (–1928) (Hasegawa 1967, Kontyû 35 (Suppl.): 53.)

TSUMACIDE® See Metolcarb.

TSUNEKI, KATSUJI (1908–1994) Japanese academic (Fukui University) and specialist in biology, behaviour and taxonomy of aculeate Hymenoptera.

TUART BUD WEEVIL *Cryptoplus tibialis* (Lea) [Coleoptera: Curculionidae] (Australia).

TUART LONGICORN *Phoracantha impavida* Newman [Coleoptera: Cerambycidae] (Australia).

TUBANGUI, MARCOS A (1893–1949) (Pesigan 1949, Philipp. J. Sci. 78: 367–372, bibliogr.)

TUBATOXIN® See Rotenone.

TUBE Noun. (Latin, *tubus* = pipe. Pl., Tubes.) 1. A slender, hollow, cylindrical body. 2. The anal siphon or respiratory tube of mosquito larvae.

TUBE SPIDER *Misgolas robertsi* (Main & Mascord)

[Araneida: Idiopidae] (Australia).

TUBE THRIPS See Phlaeothripidae (Australia).

TUBEMAKING CADDISFLIES See Psychomyiidae.

TUBER FLEA-BEETLE *Epitrix tuberis* Gentner [Coleoptera: Chrysomelidae].

TUBER MEALYBUG *Pseudococcus affinis* (Maskell) [Hemiptera: Pseudococcidae] (Australia).

TUBERCLE Noun. (Latin, *tuberculum* = a small hump. Pl., Tuberculi, Tubercles.) 1. A small, rounded protuberance. 2. Hymenoptera: A projection in adult Sphecoidea: Rounded lobes of the dorsolateral margin of the Pronotum. 3. Lepidoptera larva: Wart-like cuticular body structures that sometimes bear Setae and are diagnostically useful. Syn: Verruca. Alt. Tubercule; Tuberculum; Tuberculus. See Process 1,2. Cf. Acantha; Spike; Spine. Rel. Integument.

TUBERCULA Noun. (Latin, *tuberculum* = a small hump. Pl., Tuberculae.) 1. Hymenoptera: An elevated triangular process at the anterior angle of the Thorax. 2. The infolded and/or thickened dorsal side of the distal segment of the Tarsus bearing projections or emarginations against which the claws and Orbicula articulate (MacGillivray).

TUBERCULAR Adj. (Latin, *tubercularis* = tubercle.) Descriptive of or relating to a tubercle. Syn. Tuberculate.

TUBERCULATE PITS Tipulidae: Paired, spiny spots at or near the anterior margin of the Mesonotum, one on either side of the median line (Curran).

TUBERCULATE Adj. (Latin, *tuberculum* = a small hump; *-atus* = adjectival suffix.) Covered or furnished with tubercles; formed like a tubercle. Alt. Tuberculose; Tuberculous.

TUBERCULIFORM Adj. (Latin, *tuberculum* = a small hump; *forma* = shape.) Tubercle-shaped; Wart-shaped. Descriptive of structure that forms a short, blunt process. Alt. Tubercular; Tuberculate. See Tubercle. Cf. Bump; Hump. Rel. Form 2; Shape 2; Outline Shape.

TUBERIFEROUS Adj. (Latin, *tuberculum* = a small hump; *ferre* = to bear; *-osus* = possessing the qualities of.) Pertaining to a surface bearing tubercles.

TUBULAR BLACK THRIPS *Haplothrips victoriensis* Bagnall [Thysanoptera: Phlaeothripidae] (Australia).

TUBULAR DERMAL GLAND See Tubular Wax Gland.

TUBULAR GLANDS See Tubular Wax Gland.

TUBULAR SPINNERETS Coccids: See Dorsal Pores (MacGillivray).

TUBULAR VEIN Hymenoptera: A rigid, elongate, tube-like wing vein with sharply defined edges; TV is usually yellow, brown or black, but sometimes milky or clear (Mason 1986, Proc. ent. Soc. Wash. 88: 2.) TV can contain Haemolymph, Tracheae or nerves. See Vein; Wing Venation. Cf. Nebulous Veins; Spectral Veins; Spurious Vein; Trace Vein. Rel. Horismology.

TUBULAR WAX GLAND Pseudococcidae: An abundant flask-shaped gland that consists of 12 cells. TWG has a relatively large, circular aperture positioned at the summit of a papilla. TWG neck is chitinous and consists of one cell; The bulk of the gland is suspended beneath the Epidermis with 10 peripheral cells surrounding one larger central cell. The anatomy of the central and peripheral cells resembles that of the Triangular Wax Gland. TWG extrudes large cylinders of wax. Syn. Tubular Dermal Gland (Coccidae); Tubular Gland (Asterolecaniinae). See Gland. Cf. Triangular Wax Gland; Multilocular Wax Gland; Quinquelocular Wax Gland.

TUBULARIS Noun. (Latin, *tubulus* = small tube.) The retracted sucking tube of Anoplura *sensu* MacGillivray.

TUBULI Noun. Coccids: See Dorsal Pores (MacGillivray).

TUBULIFERA Noun. (Latin, *tubulus* = small tube. Pl., Tubuliferae.) 1. An obsolete taxonomic division of Hymenoptera in which members display terminal abdominal segments that are retracted but may be extended to display a tube-like appearance. 2. Thysanoptera: A Suborder in which Species lack an Ovipositor and the terminal abdominal segments are tubular. See Thysanoptera. Cf. Terebrania.

TUBULOSE Adj. (Latin, *tubulus* = small tube; *-osus* = full of.) Descriptive of something formed as a tube; tube-like; fistulous. Alt. Tubulosus; Tubulous.

TUBULUS Noun. (Latin, *tubulus* = small tube.) Diptera: The slender, flexible, abdominal segments that form an Ovipositor.

TUBUS Noun. (Latin, *tubus* = tube.) 1. The corneous base of a Ligula; the sheath of the tongue.

TUCCIMEI, GIUSEPPE (1851–1915) (Meli 1916, Boll. Soc. geol. ital. 35: lxxxix–xcviii, bibliogr.)

TUDEER, ALF ENUL (1885–1968) (Leppo 1971, Sber. finn. Akad. Wiss. 1969: 87–92.)

TUELY, N. C. (1878) (Bates 1878, Proc. ent. Soc. Lond. 1878: lxvi.)

TUFT Noun. (Middle English, from Old French, *tufe*, Greek, *zopf* = wisp of hair. Pl., Tufts.) A small number of closely arranged, elongate, flexible structures (Setae, cuticular evaginations, appendages) whose bases are juxtaposed but whose apices are free and widely spaced.

TUFTED APPLE-BUDMOTH. *Platynota idaeusalis* (Walker) [Lepidoptera: Tortricidae].

TUGEN® See Proproxur.

TUGON® See Trichlorfon.

TUGWELL, WILLIAM HENRY (1831–1895) (Meldola 1895, Proc. ent. Soc. Lond. 1895: lxxii.)

TUINEN, KLAAS BISSCHOP VAN (1840–1905) (Van Rossum 1906, Tijschr. Ent. 49: 234–236.)

TULARAEMIA Noun. A disease of humans and other warm-blooded animals including birds, sheep, horses, cattle and pigs. Tularaemia is most severe in sheep with ca 40% mortality; human tularaemia with mortality less than 8% in untreated cases. Disease occurs in northern hemisphere between latitudes 30° and 71°, but not known in Spain, Portugal or Great Britain. Etiological agent is a gram-negative, obligatively aerobic bacterium: *Francisella tularensis tularensis* in Nearctic and *F. t. palaearctica* in Holarctic. More than 100 Species of animals naturally are infected with this Species of bacterium. Tularaemla is transmitted by blood-sucking arthropods (especially ticks and tabanid flies), contact (penetration of skin), inhalation or via food or water; *F. tularensis* persists in ixodid ticks and transovarial transmission has been demonstrated. Alt. Tularemia. Syn. Deer Fly Fever; Pahvant Valley Plague.

TULE BEETLE *Tanystoma maculicolle* (Dejean) [Coleoptera: Carabidae].

TULIP-BULB APHID *Dysaphis tulipae* (Boyer de Fonscolombe) [Hemiptera: Aphididae].

TULIP-TREE APHID *Illinoia liriodendri* (Monell) [Hemiptera: Aphididae]: A significant pest of street trees in the western USA. Syn. *Macrosiphum liriodendri* (Monell).

TULIP-TREE SCALE *Toumeyella liriodendri* (Gmelin) [Hemiptera: Coccidae]: In North America, a pest of Tulip Tree that overwinters as a first instar nymph. TTS honeydew produces sooty mould; large populations of TS can kill trees.

TULLBERG, TYCHO FREDRIK HUGO (1842–1920) (Wallgren 1921, Ent. Tidskr. 42: 60–64.)

TULLGREN FUNNEL A collection device which is used to separate or extract small arthropods from soil and leaf litter samples. TF is based upon the basic design of the original Berlese Funnel, except the water jacket was eliminated and a light bulb was placed above the funnel. Most modern funnels currently used for processing soil samples are modified from Tullgren's apparatus. (Tullgren 1917. Z. angew. Ent. 4: 149–150.) See Berlese Funnel.

TULLILUS Noun. (Pl., Tullili.) The Pulvillus *sensu* MacGillivray.

TUMBLING FLOWER-BEETLES See Mordelidae.

TUMBU FLY *Cordylobia anthropophaga* [Diptera: Calliphoridae]: In Africa, a Myiasis producing fly which attacks numerous Species of domestic and wild vertebrate hosts. Female deposits batches of 100–300 eggs in dry sand contaminated with excrement or urine; eggs hatch and larva burrows into skin of host and creates boil-like swelling.

TUMESCENCE Adj. (Latin, *tumidus* = swollen.) A swelling or tumid enlargement; an inflated or swollen area.

TUMESCENT Adj. (Latin, *tumidus* = swollen.) Somewhat swollen.

TUMID Adj. (Latin, *tumidus* = swollen.) Pertaining to bodies or structures which are swollen or enlarged.

TUMID SPIDER-MITE *Tetranychus tumidus* Banks [Acari: Tetranychidae].

TÜMPEL, RUDOLF (1863–1938) (Schmidt 1957, Ent. Z. Frankf. a. M. 67: 202–208, 209–215.)

TUMULUS Noun. (Latin, *tumere* = to swell. Pl., Tumuli.) 1. A dome of lava. 2. An artificial mound of earth, such as the earth collected at the entrance of a ground-nesting ant or bee. 3. A mound of soil over a grave. Cf. Cumulus.

TUNGIDAE Plural Noun. Chigoe Fleas; Jiggers; Sand Fleas; Sticktights. A numerically small, tropical Family of Siphonaptera assigned to Superfamily Pulicoidea. Adult Thorax small; legs short; middle Coxa without an internal ridge (Apodeme); hind Coxa with apex tooth-like; hind Tibia without an apical tooth; Sensillum with ca eight pits on each side. See Sand Flea; Sticktight Flea.

TUNICA EXTERNA Diptera: The outer and the inner (tunica interna) envelope surrounding the single sac-like testis (Snodgrass).

TUNICA INTIMA 1. The inner layer of the Silk Glands. 2. An inner lining or membrane.

TUNICA PROPRIA 1. A noncellular envelope of the Ovariole which contains mucopolysaccharides or mucoproteins. The TP surrounds the Ovariole and terminal filament; the TP thickens and stretches as the Oocyte develops. The thickness of the TP is contributed by the underlying cells of the Ovariole wall. The TP provides support, maintains serial arrangement of follicle cells, and probably acts as a dialytic membrane in egg movement because elastic and muscle fibres are absent from the Ovariole. 2. The outer layer of the Silk Glands. 3. A covering or investing membrane.

TUNICATE Adj. (Latin, *tunica* = coating; *-atus* = adjectival suffix.) Composed of concentric layers, enveloping one another. A term applied to Antennae when each successive segment is concealed within the preceding funnel-shaped segment. Alt. Tunicatus.

TUPELO LEAFMINER *Antispila nysaefoliella* Clemens [Lepidoptera: Heliozelidae].

TURATI, EMILIO (1858–1938) (Mariani 1939, Folia zool. hydrobiol. 9: 314–316, bibliogr.)

TURATI, GIANFRANCO (1861–1905?) (H.T. 1939. Mitt. schweiz. ent. Ges. 1939: 472–473.)

TURBERCLE-BEARING LOUSE *Solenopotes capillatus* Enderlein [Phthiraptera: Linognathidae] (Australia).

TURBINATE Adj. (Latin, *turbo* = whirl; *-atus* = adjectival suffix.) Top-shaped; nearly conical. Turbinate differs from pyriform in being shorter and more suddenly attenuated at the base. Alt. Turbinatus.

TURBINATE EYE A pillared eye.

TURBOSTORE® See Tecnazene.

TURCAM® See Bendiocarb.

TURCKHEIM, HANS VON (1814–1892) (Honrath 1892, Berl. ent. Z. 37: 511–512.)

TÜRCKHEIM, HANS VON FREIHERR (1814–1892) (Horvath 1892, Berl. Ent. Z. 37: 511–512.)

TUREX® A registered biopesticide derived from

Bacillus thuringiensis var. *aizawai* strain GC-91. See *Bacillus thuringiensis*.

TURF Noun. (Old Norse, *torf*. Pl., Turfs, Turves.) A layer of soil containing the matted roots of grass plants and compact blades. Syn. Sod; Sward.

TURF PLANTHOPPER *Toya dryope* (Kirkaldy) [Hemiptera: Delphacidae] (Australia).

TURGID Adj. (Latin, *turgidus* > *turgere* = swollen.) 1. Descriptive of structure expanded through the action of internal force or fluid pressure. 2. A natural or unnatural state of distension. Ant. Flaccid. Cf. Turgor.

TURGOR Adj. (Latin, *turgescere* = to swell; *-or* = one who does something.) The natural distension or swelling in plant tissue due to internal pressure. Turgor is responsible for many kinds of plant growth or movement. Alt. Turgescence. Ant. Wilt.

TURKESTAN COCKROACH *Blatta (Shelfordella) lateralis* (Walker) [Blattaria: Blattidae].

TURKEY CHIGGAR *Neoschoengastia americana* (Hirst) [Acari: Trombiculidae].

TURKEY GNAT *Simulium meridionale* (Riley) [Diptera: Simuliidae].

TURNER, ALFRED JEFFERIS (1861–1947) Australian MD and prolific collector of Lepidoptera. Published 121 papers (1894–1947) and described more than 3,500 new Species, 450 new Genera and four Families of Lepidoptera. (Remington 1948, Lepid. News 2: 48, 82; Mackerras 1949, Proc. R. Soc. Qd 60: 69–70, 73–87, bibliogr.)

TURNER, ARTHUR HENRY (1895–1962) (Varley 1963, Proc. R. ent. Soc. Lond. (C) 27: 51.)

TURNER, CHARLES (1809–1868) (Smith 1868, Entomologist 4: 107–108; Mackechnie-Jarvis 1976, Proc. Brit. ent. nat. Hist. Soc. 8: 105.)

TURNER, CHARLES HENRY (1867–1923) (Pohlman 1923, Trans. Acad. Sci. St. Louis 29 (9): 1–54, bibliogr.)

TURNER, HENRY JEROME (1856–1951) (Donisthorpe 1951, Entomol. mon. Mag. 87: 94–95; Anon. 1952, Proc. Trans. S. Lond. ent. nat. Hist. Soc. 1951–52: xiii.)

TURNER, JAMES ASPINALL (1797–1867) (Anon. 1968, Entomol. mon. Mag. 4: 141; Weiss 1926, Can. Ent. 58: 287–289.)

TURNER, JOHN PATILLO (1902–1940) (Anon. 1940, Ent. News 52: 30.)

TURNER, ROWLAND EDWARDS (1863–1945) English taxonomist (BMNH) specializing in aculeate Hymenoptera, particularly Thynnidae. (Benson 1946, Entomol. mon. Mag. 82: 47; Hall 1947, Proc. R. ent. Soc. Lond. (C) 11: 62.)

TURNER, WILLIAM FRANKLIN (1887–1965) (Alden & Rue 1966, J. Georgia ent. Soc. 1: 22–23; Bruer 1966, J. Econ. Ent. 59: 1551.)

TURNIP APHID *Lipaphis erysimi* (Kaltenbach) [Hemiptera: Aphididae]: In North America, a widespread pest of turnip, lettuce, cabbage, collards, kale, mustard, raddish and rutabaga. TA can complete 46 generations per year in southern states. Syn. *Hyadaphis* (*Rhopalosiphum*) *pseudobrassicae*. Alt. False Cabbage Aphid. See Aphididae.

TURNIP GALL WEEVIL *Ceutorhynchus pleurostigma* (Marshall) [Coleoptera: Curculionidae]: A pest of turnips and Cruciferae.

TURNIP MAGGOT *Delia floralis* (Fallén) [Diptera: Anthomyiidae].

TURNIP MOTH *Agrotis segetum* [Lepidoptera: Noctuidae].

TURNIP SAWFLY *Athalia lugens infumata* (Marlatt) [Hymenoptera: Tenthredinidae].

TURNIPSEED, GEORGE FRANKLIN (1906–1965) (Rabb *et al.* 1966, J. Econ. Ent. 59: 771.)

TURPENTINE BORER *Buprestis apricans* Herbst [Coleoptera: Buprestidae].

TURPLEX® See Azadirachtin.

TURRETED Adj. (Middle English, *turet* = tower.) Descriptive of a head which is produced anteriorly and in a triangular point above.

TURRITUS Adj. (Latin, *turris* = tower.) Towering; rising cone-like from a flattened surface.

TUSSOCK MOTHS See Lymantriidae.

TUTT, JAMES WILLIAM (1858–1911) (Chapman 1911, Zoologist 69: 75–76; Dixey 1911, Proc. ent. Soc. Lond. 1910–1911: lxxxvii–lxxxviii.)

TUXEN, S L Danish Entomologist and Keeper, Entomological Department, Zoological Museum of Copenhagen. Editor of two editions of a Glossary of Insect Genitalia.

TWENTYEIGHT-SPOTTED POTATO LADYBIRD 1. *Epilachna vigintioctopunctata* (Fabricius) [Coleoptera: Coccinellidae]: In Australia, adults and larvae defoliate pumpkin, potatoes, tomatoes, melons and cucumbers. See Coccinellidae. 2. *Epilachna vigintisexpunctata vigintisexpunctata* (Boisduval) [Coleoptera: Coccinellidae] (Australia).

TWENTYSIX-SPOTTED POTATO LADYBIRD *Epilachna vigintioctopunctata pardalis* (Boisduval) [Coleoptera: Coccinellidae] (Australia).

TWICE-STABBED LADY-BEETLE *Chilocorus stigma* (Say) [Coleoptera: Coccinellidae].

TWIG GIRDLER *Oncideres cingulata* (Say) [Coleoptera: Cerambycidae].

TWIG LOOPER *Ectropis excursaria* (Guenée) [Lepidoptera: Geometridae] (Australia).

TWIG PRUNER *Elaphidionoides villosus* (Fabricius) [Coleoptera: Cerambycidae].

TWILIGHT Noun. (Anglo Saxon, *twi* = two; *leoht* = light. Pl., Twilights.) A light of weaker intensity than that experienced by an organism at sunrise or sunset. Generally, twilight represents the light in the sky between full night and sunrise or sunset and full night. Specifically, three types of twilight are recognized. Civil Twilight is the time required for the upper limb of the sun to traverse an arc from the horizon to a point 6° below the horizon. Nautical Twilight is the time required for the sun's upper limb to travel from the horizon to a point 12° below the horizon. Astronomical Twilight involves an angle of 18° between the sun and horizon. Rel. Diel Periodicity; Photoperiod.

TWIN OCELLUS Two ocellate spots joined together.

TWIN-SPOTTED BUDWORM *Hedya chionosema* (Zeller) [Lepidoptera: Tortricidae].

TWISTED-WINGED INSECTS See Stylopidae.

TWISTER® See Carbaryl.

TWO PRONGED BRISTLETAILS See Diplura.

TWO-BANDED FUNGUS BEETLE *Alphitophagus bifasciatus* (Say) [Coleoptera: Tenebrionidae].

TWO-BANDED JAPANESE WEEVIL *Callirhopalus bifasciatus* (Roelofs) [Coleoptera: Curculionidae].

TWO-BRAND CROW BUTTERFLY *Euploea sylvester sylvester* (Fabricius) [Lepidoptera: Nymphalidae] (Australia).

TWO-LINED CHESTNUT BORER *Agrilus bilineatus* (Weber) [Coleoptera: Buprestidae]: A monovoltine pest of beech, chestnut and oak in eastern North America. Eggs laid in small clutches in bark fissures. Larva bores into cambium; overwinters as mature larva in sapwood cell.

TWO-LINED SPITTLEBUG *Prosapia bicincta* (Say) [Hemiptera: Cercopidae].

TWO-MARKED TREEHOPPER *Enchenopa binotata* (Say) [Hemiptera: Membracidae].

TWO-SPINED AUGER BEETLE *Xylobosca bispinosa* (Macleay) [Coleoptera: Bostrichidae] (Australia).

TWO-SPINED SPIDER *Poecilopachys australasia* (Griffith & Pidgeon) [Araneida: Araneidae] (Australia).

TWO-SPOTTED CRICKET *Gryllus bimaculatus* DeGeer [Orthoptera: Gryllidae].

TWO-SPOTTED LADY-BEETLE *Adalia bipunctata* (Linnaeus) [Coleoptera: Coccinellidae].

TWO-SPOTTED LADYBIRD *Adalia bipunctata* (Linnaeus) [Coleoptera: Coccinellidae] (Australia).

TWO-SPOTTED MITE See Two-Spotted Spider Mite.

TWO-SPOTTED SPIDER MITE *Tetranychus urticae* Koch [Acari: Tetranychidae]: A widespread, polyphagous pest of numerous agricultural crops including cotton, peanuts and strawberries. In deciduous trees, TSSM causes stippling on underside of leaves. Overwinters as fertilized female. Adult 0.3–0.5 mm long; body oval (slender to diamond shaped in male), pale yellow to green with 2 dark spots on sides of Abdomen. During autumn, adults overwinter and become bright orange to red. This stage is non-feeding and occurs in ground litter or on tree bark. Mouthparts for piercing and sucking plant material; Chelicerae for piercing and lacerating plant tissue; Stylets inserted into plant to suck out macerated material. Eyes small, limited to simple Ocelli. Abdomen globular with four pairs of legs, covered in long Setae. Adults live 9 weeks; female lays ca 70 eggs during life-time. Eggs spherical, pale yellow to translucent. In spring eggs are laid singly on the underside of leaves. Egg stage 3–10 days. First instar is larval stage and has three pairs of legs. The two following instars are nymphal stages (Protonymph and Deutonmyph) and have four pairs of legs. Larvae and nymphs resemble the adult stage. All stages feed on epidermal and mesophyll cells of the leaf. Damage to these cells leads to water loss and reduced photosynthesis. The life-cycle requires 7–21 days. Syn. Two-Spotted Mite. See Tetranychidae.

TWO-SPOTTED STINK BUG *Perillus bioculatus* (Fabricius) [Hemiptera: Pentatomidae]: A cosmopolitan Species widespread in North America. Female can lay 250 eggs near prey. Early instar nymphs feed on eggs; older nymphs and adults attack caterpillars, beetle larvae. See Pentatomidae

TWO-STRIPED GRASSHOPPER *Melanoplus bivitattus* (Say) [Orthoptera: Acrididae]: A pest of alfalfa, grasses, corn, barley, vegetables and deciduous trees. TSG occurs throughout North America. Female lays 12 pods each with about 100 eggs. See Acrididae.

TWO-STRIPED WALKINGSTICK *Anisomorpha buprestoides* (Stoll) [Phasmatodea: Pseudophasmatidae].

TWO-TAILED SPIDER *Tamopsis* spp. [Araneida: Hersiliidae] (Australia).

TWO-TONED CATERPILLAR PARASITE *Heteropelma scaposum* (Morley) [Hymenoptera: Ichneumonidae]: An Australian Species. Adult wingspan 6–15 mm; body elongate, black with yellow markings. Head black with yellow mouthparts; Antenna longer than head and Thorax; Fore and middle legs yellow, hind leg reddish-brown; Abdomen reddish-brown, thin and elongate; two long, thin abdominal segments produce a 'waist'. Female's Ovipositor long and distinct. Eggs laid within host. Larvae live within host feeding on organs. Pupation occurs within host. Adults do not emerge until after host has pupated. See Ichneumonidae.

TWO-TOOTH LONGICORN *Ambeodontus tristis* (Fabricius) [Coleoptera: Cerambycidae] (Australia).

TYCAP® See Fonofos.

TYCHOPARTHENOGENESIS Noun. (Greek, *tyche* = change; *parthenos* = virgin; *genesis* = descent. Pl., Tychoparthenogeneses.) 'Occasional' parthenogensis. Soumalainen *et al.* 1976. Evol. Biol. 9: 209.

TYCHSEN, OLAF (OLAUS) GERHARD (1734–1815) (Rose 1850, *New General Biographical Dictionary* 12: 297.)

TYKAC, JAROSLAV (1866–1959) (Anon. 1960, Cas. csl. Spol. ent. 57: 409–411, bibliogr.)

TYL, HEINRICH (–1913) (Anon. 1918, Wien. ent. Ztg. 37: 178.)

TYLIDIDAE Plural Noun. See Micropezidae.

TYLOID Noun. (Greek, *tylos* = knot on a club, callus; *eidos* = form. Pl., Tyloids.) Hymenoptera: An elliptical or oval cuticular elevated area on the Flagellum of many Ichneumonidae (Gelinae,

Ichneumoninae, Diplazontinae, some Microleptinae) and Trigonalidae. Typically numerous, sometimes diagnostic and presumably sensory in function. See Glume; Multiporous Plate Sensillum; Rhinaria. Rel. Eidos; Form; Shape.

TYLUS Noun. (Latin, *tylus* from Greek, *tylos* = knob, knot. Pl., Tyli.) Heteroptera: The distal part of the Clypeus or anteclypeal region, margined by deep clefts that separate it from the lateral lobes of the head called the juga. Scarabaeoid larvae: A sclerome covering, completely or partly, the fused epizygal, coryphal and haplomeral elements, produced toward the pedium as a single obtuse point or a few rounded lobes (Boving).

TYMPANAL Adj. (Greek, *tympanon* = drum; Latin, *-alis* = pertaining to.) Descriptive of structure or function associated with a Tympanum or stretched membrane. Alt. Tympanic. See Tympanal Organ.

TYMPANAL AIR-CHAMBER A space inside the Tympanum into which air is admitted by a Spiracle near its margin (Comstock).

TYMPANAL FRAME Lepidoptera: A supporting framework on dorsal, posterior and ventral sides of the tympanal membrane of the thoracic tympana, morphological part postnotal, part epimeral (Richards).

TYMPANAL ORGAN A structurally complex scolophorous device found in Hemiptera, Lepidoptera and Neuroptera. TOs are located on wing, Abdomen and Thorax where they function as auditory organs. TO elements include: Sensory cells adjacent or attached to a thin, flexible, chitinous membrane (Tympanum), a chitinous frame surrounding the Tympanum, and a Trachea or air sac adjacent to the inner surface of the Tympanum. See Auditory Organ. Cf Chordotonal Organ; Johnston's Organ; Subgenual Organ.

TYMPANAL POCKETS Lepidoptera: Pockets in the tympanal frame, typically four in number (never more) and designated by numbers I-IV from antero-dorsal to postero-ventral (Richards.)

TYMPANIC SPIRACLE Diptera: The thoracic Spiracle at the base of the wing.

TYMPANOPHORIDAE. Plural Noun. A very small group of ensiferous Orthoptera (three Genera, 12 Species) found in Malaya, Australia and Fiji. Sometimes placed in Conocephalidae. Small body, wide head, globular eyes, scrobes weakly margined, Pronotum saddle-shaped, Prosternum spineless and fore Tibia with rows of short, strong ventral spines. Wing development variable. Male Tegmina typically with well developed stridulatory organ. Wings sexually dimorphic. Species are presumably carnivorous.

TYMPANULE Noun. (Greek, *tympanon* = drum; *-ules* = diminutive form. Pl., Tympanules.) 1. Small openings covered by a membrane. 2. A small Tympanum. 3. A structure with otoliths and serving as ears.

TYMPANUM Noun. (Greek, *tympanon* = drum; Pl., Tympana, Tympanums.) 1. Any membrane stretched like the head of a drum. Specifically applied to the membrane covering the auditory organs. 2. The membrane of a sound-producing organ which serves as a resonator. See Auditory Organ.

TYNDALL BLUE A colour produced by the scattering of light particles that have a diameter about the same as the wavelength of light being transmitted. Shorter wavelengths are scattered; longer wavelengths are transmitted. Blues are scattered; yellow and reds are transmitted. Particles less than 0.7 μm reflect blue-violet light; particles ca 0.7 μm reflect blue light; particles more than 0.7 μm reflect white light. The Tyndall blues are seen in waxy secretions of dragonflies and lasiocampid larvae. See Pigmentary Colour; Structural Colour. Rel. Reversible Colour Change.

TYPE Noun. (Greek, *typos* = pattern; Latin, *typus* = pattern. Pl., Types.) 1. Something that serves as the symbolic representation of a thing or concept. 2. Taxonomy: The single specimen, or member of a series of specimens, that is used as the basis for description of a Species that is new to science. The name-bearer of a Species. 3. The Species upon which a Genus is founded, or which is selected as the type of a Genus; a Holotype. See: Allelotype; Allotype; Androtype; Apotype; Autotype; Biotype; Chirotype; Cotype; Ecotype; Genoholotype; Genolectotype; Genosyntype; Genotype; Gynetype; Haptotype; Holotype; Homoeotype; Hypoparatype; Hypotype; Icotype; Ideotype; Karyotype; Lectotype; Metatype; Morphotype; Neotype; Nepionotype; Orthotype; Paralectotype; Paratype; Phenotype; Plastotype; Prototype; Syntype; Topotype. 4. The anatomical, behavioural, physiological or ecological characteristics by which an organism may be recognized.

TYPE BY ABSOLUTE TAUTONOMY See Type by Original Description.

TYPE BY ELIMINATON Taxonomy: A nomenclatural type arrived at when some of the original Species of a Genus have been transferred to other Genera. The type of the Genus as selected from among the original Species remaining in the Genus (International Code).

TYPE BY ORIGINAL DESIGNATION Taxonomy: A single Species designated as Type in the original publication of a Genus, when in the original publication of a Genus, Typicus or Typus is used for any of the Species, the single Species in a proposed new Genus (monotypical Genus), when in a Genus containing a number of Species, one original Species has the generic name as its specific or subspecific name, whether a valid name or a synonym (type by absolute tautonomy) (International Code).

TYPE BY VIRTUAL TAUTONOMY Taxonomy: An original Species of a Genus which has a specific or subspecific name, either as a valid name or a synonym, which is virtually the same as the generic name, or of the same origin or meaning (International Code).

TYPE GENUS Taxonomy: The particular Genus upon which the Family is founded; the Type Genus is not necessarily the oldest Genus in the Family. In instance where the Family takes its name from the Type Genus, the Family name follows any nomenclatural changes.

TYPE SPECIES A Species which is the Type of a Genus.

TYPEWRITER *Pauropsalta extrema* (Distant) [Hemiptera: Cicadidae] (Australia).

TYPHOID FEVER An infectious disease caused by *Bacillus typhosus* and ingested with food or drinking water. TF causes high temperature, englargment of spleen, inflammation of the intestine and other symptoms. May be mechanically transmitted by insects such as cockroaches and flies.

TYPHUS See Epidemic Typhus. Murine Typhus.

TYPICAL Adj. (Greek, *typos* = pattern; Latin, *-alis* = characterized by.) 1. Descriptive of the normal or usual form of a Species; a specimen or part which agrees with the Type form. 2. The most frequently encountered condition (state) for a feature (colour, size, shape, number) that exhibits its variation within a population, Subspecies or Species. See Character; Character State. Cf. Atypical.

TYPOLOGY Noun. (Greek, *typos* = pattern; *logos* = discourse. Pl., Typologies.) A concept in which variation is disregarded and members of a Taxon are regarded as replicas of the 'type.' See Essentialism.

TYRAMINE Noun. (Greek, *tyros* = cheese; *ammoniakon* = resinous gum.) An insect-specific neuromodulator that is derived from dopamine and is an immediate precursor of octopamine.

TYROSINASE Noun. (Greek, *tyros* = cheese.) A salt of tyrosine and only amino acid containing the hydroxy-phenyl group.

TYROSINE Noun. (Greek, *tyros* = cheese.) An end-product of the proteolytic action of the trypsin in the gastric fluid of insects. An amino-acid derived from food and reacting with the dioxyphenylalaline in insect blood to produce melanin (Wardle). Alt. Tyrosin.

TYTLER, HARRY C (–1939) (Talbot 1939, J. Bombay nat. Hist. Soc. 41: 409.)

TYZENHAUS, CONSTANTIN (1786–1853) (Adamowicz 1953, Bull. Soc. nat. Moscou 26 (2): 517–529. bibliogr.)

U Symbolically, the chemical notation for Uranium.

UCHIYAM, SHIGETARO (1885–1967) (Hasegawa 1967, Kontyû 35 (Suppl.): 15–16.)

UDEKEM GÉRARD MARIE GHISLAIND (1824–1864) (Lambotte 1865, Bull. Séanc Soc. r. malac. Belg. 1: xxiii–lxxiv. Quetelet 1865, Annals Acad. Belg. 31: 127–130.)

UE, TENJE (1892–1965) (Hasegawa 1967, Kontyû 35 (Suppl.): 15.)

UEHARA, MAGOICHI (1854–1907) (Hasegawa 1967, Kontyû 35 (Suppl.): 15.)

UEKULL, JAKOB JOHANN BARON VON (–1934) (Brock 1934, Arch. Gesch. med. Naturw. 27: 193–212, bibliogr.)

UGLYNEST CATERPILLAR *Archips cerasivorana* (Fitch) [Lepidoptera: Tortricidae].

UHAGON, Y VEDIA, SERAFIND DE (1845–1904) (Martinez de la Escalera 1904, Boln. R. Soc. esp. Hist. nat. 4: 287–291, bibliogr.)

UHLER, PHILIP REESE (1835–1913) American amateur taxonomist specializing in Hemiptera and Provost, Peabody Institution. Regarded as founder of American school of Hemiptera. (Howard 1913, Ent. News 24: 433–443; Champion 1914, Entomol. mon. Mag. 50: 43; Schwartz *et al.* 1914, Proc. ent. Soc. Wash. 16: 1–7, bibliogr.)

UHLMER, G (1877–1963) (Wigglesworth 1964, Proc. R. ent. Soc. Lond. (C) 28: 57.)

UHMANN, ERICH (–1968) (Anon. 1966, Mitt. dt. ent. Ges. 27: 49–5l.)

UHRYR, FERDINAND (–1909) (Losy 1910, Rovart. Lap. 17: 145–147, 160, bibliogr.)

UJHELYI, JOSEPH (1822–1935) (W.F.S. 1936, J. Econ. Ent. 29: 226–227.)

ULBRICH, EDE (EDUARD) (1854–1917) (Kertesz 1917, Rovart. Lap. 24: 97–100, 131, bibliogr.)

ULE, OTTO EDUARD VICENZ (1820–1876) (Anon. 1876, Zool. Gart. Frankf. 17: 344; K.M. 1876, Natur 25: 405–406, 416–417, 431–432, 442–443.)

ULIDIIDAE See Otitidae.

ULIGINOUS Adj. (Latin, *uliginosus* = oozy; *-osus* = possessing the qualities of.) Muddy; pertaining to mud. Alt. Uliginose.

ULKE, HENRY (1821–1910) An American portrait painter and amateur Coleopterist. (Anon. 1910, Ent. News 21: 99–100; Banks *et al.* 1910, Proc. ent. Soc. Wash. 12: 105–111, bibliogr.; Hopkins *et al.* 1912, Ann. ent. Soc. Am. 5: 74.)

ULLMANN, AXEL CONRAD (1840–) (Munster 1923, Norsk. Ent. Tidsskr. 1: 209–211.)

ULLRICH, GÜNTER (1919–1945) (Weidner 1967, Abh. Verh. naturw. Ver. Hamburg Suppl. 9: 265.)

ULLYETT, GERALD CUMMINGS (1901–1951) (J.C.F. 1951, J. ent. Soc. sth. Afr. 14: 204–205.)

ULMER, GEORG (1877–1963) (Illies 1962, Mitt. dt. ent. Ges. 21: 1–2; Kimmins 1963, Revue Ges. Hydrobiol. 48: 523–524.)

ULNA Noun. (Latin, *ulna* = elbow.) A long bone in the forearm of vertebrates but used in an entomological context to refer to venation or region of the wing.

ULNAR AREA Orthoptera: The median area.

ULNAR VEIN 1. Hemiptera: A wing vein between The Radial Vein and Claval Suture. 2. Orthoptera: The Cubitus. See Vein; Wing Venation.

ULOBORIDAE Plural Noun. Humped Spiders. A widespread Family of spiders which lack poison glands. Nest orb-like with egg sacs of *Uloborus* papery and stellate-shaped.

ULONA Noun. (Greek, *oula* = mouth gums.) The thick, fleshy mouthparts of Orthoptera.

ULONATA Noun. (Greek, *oula* = mouth gums.) A Fabrician term for Orthoptera, based on the character or the mouth structures.

ULTRA LOW VOLUME Pesticide application by spraying at 0.6 to 4.7 litres per hectare or sprays applied in an undiluted, low-volume formulation.

ULTRACIDE® See Methidathion.

ULTRAFILTRATION Noun. (Latin, *ultra* = beyond; French, *filtrer* = to strain. Pl., Ultrafiltrations.) Tissue Culture: Filtration under pressure which is used to concentrate culture medium as a preliminary process before product extraction.

ULTRAMARINE Adj. (Latin, *ultra* = beyond; *marinus* > *mare* = sea.) An intense deep blue.

ULTRANODAL SECTOR Odonata: a vein which runs parallel with and between Media, or the Principal and Nodal Sectors; the Postnodal Sector (Smith).

ULTRASONIC Adj. (Latin, *ultra* = beyond; *sonus* = sound; *-ic* = of the nature of.) Pertaining to high-frequency sounds, typically inaudible to human hearing and often produced by insects. Rel. Stridulation.

ULTRASTRUCTURE Noun. (Latin, *ultra* = beyond; *structura* = structure. Pl., Ultrastructures.) The microscopic aspects of anatomical parts, such as cells, surface features of Integument or Setae.

ULTRAVIOLET Noun. (Latin, *ultra* = beyond; *viola* = violet. Pl., Ultraviolets.) Rays of electromagnetic radiation below the colour violet (ca 300 nm).

ULYSSES BUTTERFLY *Princeps ulysses joesa* (Butler) [Lepidoptera: Papilionidae] (Australia).

UMBER Adj. (French, *terre d'ombre* earth from Umbria.) Dark brown with yellow.

UMBETHION® See Coumaphos.

UMBILICATE Adj. (Latin, *umbilicus* = navel; *-atus* = adjectival suffix.) Naval-shaped; resembling a navel.

UMBILICUS Noun. (Latin, *umbilicus* = navel.) A navel or navel-like depression.

UMBO Noun. (Latin, *umbo* = shield boss. Pl., Umboes.) A boss, elevated process or knob. In the plural, two movable spines on the sides of the Prothorax in some Coleoptera (Smith).

UMBONATE Adj. (Latin, *umbo* = shield boss; *-atus* = adjectival suffix.) Bossed; with an elevated knob in the centre. Alt. Umbonatus.

UMBONE Noun. (Latin, *umbo* = shield boss. Pl., Umbones.) An embossed, elevated knob positioned on humeral angle of Elytra.

UMBRA Adj. (Latin, *umbra* = shade.) A shadow; a slight shade of colour on a paler ground, not easily distinguishable.

UMBRACULATE (Latin, *umbraculum* = sunshade; *-atus* = adjectival suffix.) Adj. Pertaining to umbrella like structures. Bearing or carrying an umbrella-like structure, generally over the head (Kirby & Spence).

UMBRELLA ORGAN Sensillum Campaniformium.

UMECRON® See Phosphamidon.

UMEMURA, JINTARO (Hasegawa 1967, Kontyû 35 (Suppl.): 16–17, bibliogr.)

UMEYA, YOSHICHIRO (1890–1962) (Hasegawa 1967, Kontyû 35 (Suppl.): 17, bibliogr.)

UNACORIA Noun. The Coria or membrane of the first abdominal segment (MacGillivray).

UNAPECTINAE Noun. Coccids: pectinae similar in size and form to the latapectinae with the teeth limited to one side of the shaft (MacGillivray).

UNARMED Adj. Without spurs, spines or armatures.

UNARTICULATE Adj. Not jointed or segmented.

UNASPIRACLE Noun. A Spiracle of the first abdominal segment of an insect (MacGillivray).

UNASUTURE Noun. The suture of the first abdominal segment (MacGillivray).

UNCIFORM Adj. (Latin, *uncus* = hook; *forma* = shape.) Hook-shaped. Alt. Uncinate. See Uncus. Cf. Barbed; Hamiform. Rel. Form 2; Shape 2; Outline Shape.

UNCINATE Adj. (Latin, *uncinus* = hook.) Hooked; bent at the apex; hook-like in shape or action, *e.g.*, an uncinate spine. Alt. Unciform. See Uncus. Cf. Aduncate; Ankistroid; Ankyroid; Hamate. Rel. Form 2; Shape 2; Outline Shape.

UNCONVOLUTED Adj. Not convoluted.

UNCUS Noun. (Latin, *uncus* = hook. Pl., Unci.) Lepidoptera, Diptera and some other groups: The curved hook directed downward from a triangular dorsal sclerite in the male, shielding the Penis. The genital hamule. The hooked processes forming the border of the anal opening. Male Lepidoptera. The most posterior, middorsal, unpaired structure, which articulates basally with the tegumen, probably derived from the tenth abdominal somite; lies dorsal and often caudal of the Anus (Klots).

UNDATE Adj. (Latin, *undare* = to rise in waves; *-atus* = adjectival suffix.) Wavy or waved.

UNDATERGUM Noun. The telson (MacGillivray).

UNDEN® See Proproxur.

UNDERGROUND GRASSGRUB (SA) 1. *Oncopera fasciculata* (Walker) [Lepidoptera: Hepialidae] (Australia). 2. *Oncopera rufobrunnea* Tindale [Lepidoptera: Hepialidae] (Australia).

UNDERHILL, GROVER WILLIAM (1880–1970) (Woodside 1971, J. Econ. Ent. 64: 779.)

UNDERWINGS See Noctuidae.

UNDOPTERIGIDAE Kozlov 1988. Plural Noun. A Family of eriocraniid moths.

UNDOSE Adj. (Latin, *undosus* = billowy; *-osus* = full of.) With undulating, broadish, nearly parallel depressions running more-or-less into each other; wavy, resembling ripple-marks on a sandy beach.

UNDULATED Adj. (Latin, *undulatus* = risen like waves.) Wavy, obtusely waved in segments of circles.

UNEQUAL DIGITULE Coccidae: A digitule or seta on the claws (MacGillivray).

UNEQUAL Adj. Descriptive of two or more physical structures, biological characteristics or behavioural attributes that are dissimilar.

UNGER, FRANZ (1800–1870) (Anon. 1870. J. Bot., Lond. 8: 192–203; Hauer 1870, Verh. geol. Reichsanst. (StAnst.-Landesanst.), Wien 1870: 57–58; Leitgeb 1870, Mitt. naturw. Ver. Steierm. 2: 270–294, cxlvii-cliv, bibliogr.)

UNGUICULATE Adj. (Latin, *unguiculus* = little nail; *-atus* = adjectival suffix.) Furnished with claws or nails.

UNGUICULUS Noun. (Latin, *unguiculus* = little nail. Pl., Unguiculi.) 1. A small terminal claw or nail-like process. 2. Collembola: The small tarsal claw.

UNGUIFER Noun. (Latin, *ungula* = hoof.) The medial dorsal process or sclerite on the end of the Tarsus to which the pretarsal claws are articulated (Snodgrass).

UNGUIFLEXOR Noun. (Latin, *ungula* = hoof; *flexum* = to bend; *or* = one who does something. Pl., Unguiflexors.) The muscle moving or extending the Ungues in the claws of an insect.

UNGUIFORM Adj. (Latin, *ungula* = hoof; *forma* = form.) Shaped as a hoof.

UNGUIS Noun. (Latin, *unguis* = claw. Pl., Ungues.) 1. A claw at the apex of the Pretarsus. See Pretarsus 2. Aphididae: A short process on the sixth antennal segment. 3. Collembola: The large tarsal claw. 4. The claw-like structures of the Maxilla (Kirby & Spence).

UNGUITRACTOR PLATE The ventral sclerite of the Pretarsus from which arises the tendon-like apodeme of the retractor muscle of the claws (Snodgrass).

UNGUITRACTOR Noun. (Latin, *unguis* = claw; *tractus* = pull; *or* = one who does something. Pl., Unguitractors.) The muscle (tendon) which pulls (flexes) the Ungues (claws) toward the insect's body. The Unguitractor extends from the Unguitractor Plate to the tibial muscles. Syn. Unguitractor Apodeme; Unguitractor Tendon. See Leg. Rel. Protractor; Retractor.

UNGULA Noun. (Latin, *ungula* = hoof.) A claw, talon or hoof. The terminal segment of the Tarsus. See Leg.

UNGULATE Adj. (Latin, *unguiculus* = little nail; *-atus* = adjectival suffix.) 1. Pertaining to a structure or appendage bearing claws.

UNGULIFORM Adj. (Latin, *ungula* = hoof; *forma* = shape.) Hoof-shaped; descriptive of structure shaped as a horse's hoof. See Hoof. Cf. Calceiform; Panduriform; Soleaiform. Rel. Form 2; Shape 2; Outline Shape.

UNIAXIAL Adj. (Latin, *unus* = one; *axis* = axis; *-alis*

= characterized by.) Descriptive of structure with one axis. Term often applied to hinged structures with movement restricted to one plane of motion. Alt. Monaxial. See Axis; Orientation.

UNICAPSULAR Adj. (Latin, *unus* = one; *capsula* = small capsule.) Descriptive of structure with only one capsule. Cf. Multicapsular.

UNICELLULAR Adj. (Latin, *unus* = one; *cellula* = cell.) Pertaining to structure consisting of one cell only. Alt. Monocellular. Cf. Multicellular.

UNICELLULAR GLAND A gland consisting of one cell. Typically Exocrine Unicellular Gland cells are larger than the surrounding epidermal cells and discharge their secretion through the Integument. Syn. Simple Gland. See Exocrine Gland; Gland. Cf. Multicellular Gland.

UNICILIATE Adj. (Latin, *unus* = one; *cilium* = eyelash; *-atus* = adjectival suffix.) Descriptive of Protozoa or cells with one Cilium or Flagellum.

UNICOLONIAL Adj. (Latin, *unus* = one; *colonia* = farm; *-alis* = characterized by.) Social Insects: Descriptive of a population that lacks behavourial boundaries between colonies. Cf. Multicolonial.

UNICOLOUR Adj. (Latin, *unicolor* = one colour; *or* = a condition.) Descriptive of organisms, structures or appendages composed of one colour. Alt. Monochromatic; Unicolorous; Unicolorate. Cf. Discolour; Multicolour.

UNICORN Adj. (Latin, *unus* = one; *cornu* = horn.) Descriptive of organisms or structure with a single horn-like process.

UNICORN BEETLES See Scarabaeidae.

UNICORN CATERPILLAR *Schizura unicornis* (J. E. Smith) [Lepidoptera: Notodontidae].

UNICORNOUS Adj. (Latin *unus* = one; *cornu* = horn; *-osus* = possessing the qualities of.) Descriptive of an organism or structure with only one horn.

UNICOSTATE (Latin, *unus* = one; *costa* = rib; *-atus* = adjectival suffix.) Descriptive of leaves with a single, prominent midrib.

UNICUSPID Adj. (Latin, *unus* = one; *cuspis* = point.) Descriptive of structure tapering to a point; often applied to teeth. Cf. Bicuspid.

UNIDACTYL Adj. (Latin, *unus* = one; Greek, *daktylos* = finger.) Descriptive of structure with one finger-like process. Alt. Unidactylous. Rel. Unidentate.

UNIDENTATE Adj. (Latin *unus* = one; *dens* = tooth; *-atus* = adjectival suffix.) Descriptive of Mandibles displaying one tooth only. Rel. Unidactyl.

UNIFACIAL Adj. (Latin, *unus* = one; *facies* = face; *-alis* = characterized by.) 1. Descriptive of structure with one 'face' or primary surface. 2. Descriptive of structure with two 'faces' which are similar in composition, design and features.

UNIFOLLICULAR Adj. (Latin *unus* = one; *folliculus* = small sac.) Consisting of one follicle.

UNIFORATE Adj. (Latin, *unus* = one; *foratus* = pierced; *-atus* = adjectival suffix.) Descriptive of structure with one aperture or opening.

UNIFOSZ® See Dichlorvos.

UNIFUME® See Metham-Sodium.

UNILABIATE Adj. (Latin *unus* = one; *labium* = lip; *-atus* = adjectival suffix.) Descriptive of structure with one 'lip' or Labium; Labium a continuous structure and not subdivided.

UNILAMINATE Adj. (Latin, *unus* = one; *lamina* = layer; *-atus* = adjectival suffix.) Histology: Pertaining to ultrastructure composed of one layer only. See Histology. Cf. Multilaminate. Rel. Organization.

UNILATERAL Adj. (Latin *unus* = one; *latus* = side; *-alis* = characterized by.) Descriptive of structure on one side of a bilaterally symmetrical organism, or features on one side of a bilateral organization. See Symmetry. Cf. Bilateral. Rel. Anatomy.

UNILOCULAR Adj. (Latin *unus* = one; *loculus* = compartment.) With one cell or cavity.

UNIMUCRONATE Adj. (Latin, *unus* = one; *mucro* = sharp point; *-atus* = adjectival suffix.) Descriptive of structure with one sharp point.

UNINOMINAL Adj. (Latin *unus* = one; *nomen* = name; *-alis* = characterized by.) Descriptive of organisms with one name or a single nomenclatural epithet. One-named; monomial (Ferris). See Nomenclature. Cf. Binomial; Polynomial. Alt. Mononomial.

UNINUCLEAR Adj. (Latin, *unus* = one; *nucleus* = nucleus.) Descriptive of cells containin one nucleus. See Chromosome. Cf. Multinucleate. Rel. Cell. Alt. Mononuclear.

UNION JACK BUTTERFLY *Delias mysis mysis* (Fabricius) [Lepidoptera: Pieridae] (Australia).

UNIORDINAL CROCHETS Lepidoptera larva: Proleg Crochets of uniform in length in a uniserial circle. See Crochet. Cf. Biordinal Crochet; Multiordinal Crochet.

UNIPAROUS Adj. (Latin, *unus* = one; *parere* = to beget; *-osus* = possessing the qualities of.) 1. Descriptive of organisms producing one offspring at birth. 2. Parthenogenetic organisms which produce only one sex. Cf. Multiparous. 3. Botany: A cymose inflorescence with one axis at each branch.

UNIPENNATE Adj. (Latin, *unus* = one; *penna* = feather; *-atus* = adjectival suffix.) Descriptive of muscle whose tendon of insertion extends along one side of a bone (vertebrate) or apodeme/sclerite (invertebrate). See Muscle. Cf. Bipennate. Rel. Musculature. Alt. Monopennate.

UNIPLICATE Adj. (Latin *unus* = one; *plicare* = to fold; *-atus* = adjectival suffix.) Descriptive of structure displaying a single fold or line of folding. See Plicate. Cf. Multiplicate. Alt. Monoplicate.

UNIPOLAR Adj. (Latin *unus* = one; Greek, *polos* = pole.) Pertaining to cells or structures with only one pole. See: Symmetry. Cf. Bipolar; Multipolar. Rel. Organization. Alt. Monopolar.

UNIPOLAR CELL A nerve cell with only one neuron proceeding from the cell body. See Neuron; Nerve Cell; Nervous System. Cf. Multipolar Cell; Unipolar Cell. Rel. Muscle.

UNIQUE Adj. (French, from Latin, *unicus* = single.)

One only, unlike any other. In general, applied to an only specimen of a Species or Group.

UNIQUE-HEADED BUGS See Enicocephalidae.

UNIRAMOUS Adj. (Latin, *unus* = one; *ramus* = branch; Latin, *-osus* = possessing the qualities of.) Descriptive of elongate appendages or structures which are composed of only one branch. Cf. Biramous; Multiramous. Alt. Monoramous.

UNISEGMENTAL Adj. (Latin, *unus* = one; *segmentum* = a slice, a zone; *-alis* = characterized by.) Descriptive of a body, body region, appendage or structure composed of one segment (*e.g.* maxillary palpus of one segment). See Segmentation. Cf. Multisegmental. Alt. Monosegmental.

UNISEPTATE Adj. (Latin, *unus* = one; *septum* = hedge; *-atus* = adjectival suffix.) Descriptive of structure which has one septum or internal partition with a cell, cavity or coelome. See Septum. Cf. Multiseptate. Alt. Monoseptate.

UNISERIAL Adj. (Latin, *unus* = one; *series* = rank; *-alis* = pertaining to.) Descriptive of structures which are arranged in a row or series. Term is sometimes applied to rows of Setae on appendages or tubercles on sclerites. Alt. Monoserial. Cf. Multiserial.

UNISERIAL CIRCLE Lepidoptera larva: Prolegs in which the outer circle of smaller plantar crochets is absent (Imms).

UNISERRATE Adj. (Latin, *unus* = one; *serra* = saw; *-atus* = adjectival suffix.) Descriptive of structure with one row of serrations along a margin. Cf. Aculeate-Serrate; Biserrate; Dentate-Serrate; Multiserrate; Subserrate. Alt. Monoserrate.

UNISERRULATE Adj. (Latin, *unus* = one; *serra* = saw; *-atus* = adjectival suffix.) Descriptive of structure with one row of small serrations along a margin. Alt. Monoserrulate.

UNISETOSE Adj. (Latin *unus* = one; *seta* = bristle; *-osus* = full of.) With only one bristle or seta.

UNISEXUAL Adj. (Latin, *unus* = one; *sexus* = sex; *-alis* = characterized by.) 1. Descriptive of a condition, process or phenomenon limited to one sex only and without regard to gender. 2. Parthenogenetic Species in which only females are extant, or female-only generations in Species which cyclically produce males. This biological feature is distributed widely within the Insecta, but is best studied in Hymenoptera Families Aphididae and Cynipidae. See Parthenogenesis. Cf. Bisexual. Rel. Deuterotoky; Heterogony; Thelytoky.

UNITOX® See Chlorfenvinphos.

UNIVOLTINE Adj. (Latin, *unus* = one; Italian, *volta* = time.) Pertaining to Species with one generation per year or season. Term sometimes restricted to a cohort of eggs which undergo winter dormancy or diapause. See Voltinism. Cf. Multivoltine. Alt. Monovoltine.

UNSCLEROTIZED Adj. (Middle English, *un* = not; Greek, *skleros* = hard.) Descriptive of Integument or Cuticle that lacks sclerotin. In general, Integument which is not hardened by the deposition of sclerotin.

UNSPECIALIZED Adj. (Middle English, *un* = not; Latin, *specialis* = special.) Generalized; not restricted or adapted to particular functions or conditions.

UNWIN, WILLIAM CHARLES (1811–1887) (Anon. 1887, Entomol. mon. Mag. 24: 47.)

UPPER AUSTRAL ZONE The transcontinental climate belt below the Transition Zone divided into an eastern humid or Carolinian area (the Lower Austral), and a western arid or Upper Sonoran area, which pass insensibly into each other near the 100th meridian. See Carolinian and Upper Sonoran.

UPPER CARBONIFEROUS PERIOD The sixth Period (323–290 MYBP) of the Palaeozoic Era and named for the continued deposition of coal initiated during LCP. Half of the world's workable coal is found in this stratum. Most coal formations are located in the northern hemisphere because climatic conditions favoured development. Terrestrial plants during UCP include evergreens; no deciduous trees or true flowering plants were extant during UCP. Primary coal-forming plants include the lycopod *Lepidodendron* (ca 100' tall with relatively slender trunks, forked roots and cones borne at end of leafy shoots). (Club mosses are modern representatives of lycopods). Horse-tail trees *(Calamites)* grew in bamboo-like thickets 40' tall and 1' in diameter with leaves in whorls and hollow jointed stems. Seed ferns (pteridiosperms) also diversify with *Glossopteris* and *Neuropteris* common. Spore plants (sphenoids) were more common on higher ground and conifers *(Walchia)* appear. Insects display continued refinement of the wing through Ephemerida, Meganisoptera (Upper Carboniferous–Lower Triassic Period), Protemerida (Carboniferous), 'Protorthoptera' (Paoliidae, Cacurgidae), Palaeodictyoptera Protoblattaria (Upper Carboniferous–Permian), Blattaria and Homoptera. Spiders lacked Spinnerets and did not spin webs. Syn. Late Carboniferous; Pennsylvanian. See Geological Time Scale; Palaeozoic. Cf. Cambrian Period; Devonian Period; Lower Carboniferous Period; Ordovician Period; Permian Period; Silurian Period.

UPPER FIELD In Tegmina, the anal field.

UPPER MARGIN In Tegmina (Thomas); corresponds to the posterior or anal margin of most authors.

UPPER MEDIAN AREA See Areola.

UPPER RADIAL Lepidoptera: The Media *sensu* Comstock; vein five of the numerical series; the 'Independent Vein' *sensu* Smith.

UPPER SECTOR OF TRIANGLE. Odonata: The Cubitus *sensu* Comstock.

UPPER SONORAN FAUNAL AREA The arid part of the upper austral west of 100th meridian; covers most of plains in eastern Montana and Wyoming, southwestern South Dakota, western Nebraska, Kansas, Oklahoma and Texas, and eastern Colorado and New Mexico; covers the plains of the Columbia, Malheur and Harney in Oregon and Washington; in California, encircles the Sacramento and San Joaquin Valleys and forms a narrow belt around the Colorado and Mohave deserts, in Utah it covers Salt Lake and Sevier deserts, in Idaho the Snake plains, in Nevada and Arizona irregular areas of suitable elevation.

UPSIDE-DOWN FLIES See Neurochaetidae.

UPSILON 1. A frequently Y-shaped sclerite which supports a Paraglossa in specialized Diptera. 2. A Furca *sensu* MacGillivray.

URANIDIN Adj. (Greek, *ouranos* = heaven.) A yellow colouring matter in some Coleoptera and Lepidoptera.

URANIIDAE Plural Noun. A small Family of ditrysian Lepidoptera assigned to Superfamily Uranioidea; uraniids widespread, predominantly tropical and contain about 650 Species. Adult size highly variable; wingspan 15–160 mm; body often brightly coloured, day-flying; head small with smooth scales; Ocelli absent, Chaetosemata present; Proboscis without scales; Maxillary Palpus minute, with one segment; Labial Palpus small, upturned; Thorax with smooth scales; fore Tibia with Epiphysis; tibial spur formula 0-2-3 or 0-2-4; Abdomen with paired tympanal organs at base of Abdomen in female and between segments 2 and 3 in male. Egg shape variable. Larva cylindrical in cross-section; abdominal Prolegs on segments 3–6 and 10. Pupa in cocoon.

URATE Adj. (Greek, *ouron* = urine; *-atus* = adjectival suffix.) A salt of uric acid. Term applied to excretory cells of the fat body found in insects which lack Malpighian Tubules.

URATE CELLS Cells which sequester uric acid as crystalline urate. UC have been discovered in Collembola, some Blattaria and larval Hymenoptera (Apocrita).

URBAHN, ERNST (Alberti 1963, Mitt. dt. ent. Ges. 22: 23; Alberti 1964, Mitt. dt. ent. Ges. 23: 21–23.)

URBAN Adj. (Latin, *urbanus* = belonging to the city.) Pertaining to a city or town; characteristic of a city of town. Noun. One living in a city.

URBAN, CARL (1865–1941) (Botchert 1941, Ent. Bl. Biol. Syst. Käfer 37: 170–173, bibliogr.)

URBAN ECOSYSTEM An ecosystem constructed by humans that consists of physical structures (buildings), modes of transportation and their facilities, plants and animals. UE is often considered as developed in units called villages, towns and cities. See Ecology. Cf. Agroecosystem.

URBAN ENTOMOLOGY The subdiscipline of entomology concerned with the study of insects affecting humans who live in communities. Concerned with study and control of pests of houses, hospitals, municipal buildings, buildings of enterprise and insects affecting social infrastructure. See Entomology. Cf. Agricultural Entomology; Applied Entomology.

URBAN, EMANUEL (1822–1901) (Anon. 1901, Wien. ent. Ztg. 20: 120.)

URCEOLATE Adj. (Greek, *urceolus* = little pitcher; *-atus* = adjectival suffix.) Pitcher-shaped; descriptive of structure that is dilated or swells in the middle as a pitcher.

UREA Noun. (Greek, *ouron* = urine.) Carbamide. An important physiological compound in urine of mammals, carnivorous birds and reptiles. Urea crystallizes in colourless needles, readily soluble in water and has basic properties; forms some colours in insects.

URETER Noun. (Greek, *oureter* = ureter. Pl., Ureters.) A common Excretory Duct for paired Malpighian Tubules. See Excretion. Cf. Malpighian Tubules.

URIC ACID The primary nitrogenous waste product of some vertebrates (birds, reptiles) and insects. UA excreted from insects via Malpighian Tubules. UA acts as yellow or red pigment in some insects by deposition in Cuticle and cuticular structures.

URICH, FREDERICK WILLIAM (1870–1937) (Busck 1937, Proc. ent. Soc. Wash. 39: 192–193.)

URINARY VESSELS The Malpighian Tubules. Term also applied to anal glands by earlier authors.

URINOGENITAL MYIASIS A type of Myiasis in which fly larvae invade the urinary bladder or urinary tract. Larval infestation causes pain, bleeding and a desire to urinate. Common aetiological agents include Lesser House Fly, Latrine Fly, House Fly and False Stable Fly. See Myiasis.

URITE Noun. (Greek, *oura* = tail; *-ites* = constituent. Pl., Urites.) An abdominal segment, specifically, its ventral portion.

UROCYTE Noun. (Greek, *ouron* = urine; *kytos* = hollow. Pl., Urocytes.) A type of cell in Fat Body which stores uric acid. Cytoplasm typically with little endoplasmic reticulum, few mitochondria and large vacuoles. Occur in Collembola, Thysanura, Blattaria. See Fat Body.

UROGOMPHUS Noun. (Greek, *oura* = tail; *gomphos* = nail. Pl., Urogomphi.) Coleoptera larvae: Fixed or mobile cuticular processes on the apical abdominal segment of some Species (Boving & Craighead). Resemble but not necessarily homologous with Styli, Cerci, Pseudocerci or Corniculi in other taxonomic groups. Cf. Cercus, Pygostylus.

UROMERE Noun. (Greek, *oura* = tail; *meros* = part. Pl., Uromeres.) Any abdominal foot-like appendages of arthropods. Syn. Urite.

UROMORPHIC Adj. (Greek, *oura* = tail; *morphe* = shape; *-ic* = of the nature of.) Tail-like in shape or appearance. Syn. Uromorphous.

UROPORE Noun. (Greek, *ouron* = urine; *poros* = passage. Pl., Uropores.) Acarology: An external opening of the excretory duct where the gut ends blindly. See Anus, 3.

UROPYGI Noun. (Greek, *oura* = tail; *pyge* = rump.)

Whip Scorpions. An Order of Arachnida of mainly tropical Species. Body elongate, dorsoventrally flattened; tail as long as body, slender, segmented. Pedipalps large, powerful; total length 50 mm or more. They are stingless, 1st legs slender, scansorial; remaining three pairs of legs used for walking. Emit or spray defensive substance from caudal glands, has vinegar-like odour. Nocturnal and predaceous. Eggs carried in sac under opisthosoma; newly emerged young tergiferous for a time. See Vinegaroon.

UROPYGIUM Noun. (Greek, *oura* = tail; *pyge* = rump; Latin, *-ium* = diminutive > Greek, *-idion*. Pl., Uropygidia.) The Ovipositor when it is a small extension of the abdominal segments.

UROSOME Noun. (Greek, *oura* = tail; *soma* = body. Pl., Urosomes.) The Abdomen of some arthropods.

UROSTERNITE Noun. (Greek, *oura* = tail; *sternon* = chest; *-ites* = constituent. Pl., Urosternites.) The sternal or ventral sclerite of a Uromere. Cf. Urotergite.

UROSTYLIDAE Plural Noun. A small, tropical Family of heteropterous Hemiptera assigned to Superfamily Pentatomoidea. Body delicate, elongate, greenish and coreid-like in habitus; Scutellum triangular, not covering Abdomen; Tibiae lacking strong spines.

UROTERGITE Noun. (Greek, *oura* = tail; Latin, *tergum* = back; *-ites* = constituent. Pl., Urotergites.) The tergal or dorsal sclerite of a Uromere. Cf. Urosternite.

URQUHART, ARTHUR T (–1916) (Bonnet 1945, *Bibliographia Araneorum* 1: 43.)

URTICANT Adj. (Latin, *urtica* = nettle; *-antem* = adjectival suffix.) Stinging or irritating.

URTICARIA Noun. (Latin, *urtica* = nettle. Pl., Uritcariae.) 1. Setae that are specifically adapted to sting, irritate or disrupt another organism. 2. An itchy, sometimes purulent, skin rash caused by the Setae or scales of insects. Cf. Caltrops Spines.

URTICARIAL Adj. (Latin, *urtica* = nettle; *-alis* = characterized by.) Descriptive of Setae specially modified to sting, irritate or disrupt.

URTICATING ANTHELID *Anthela nicothoe* (Boisduval) [Lepidoptera: Anthelidae] (Australia).

URTICATING HAIRS In some caterpillars and adult insects, Setae (Chaetae) connected with Dermal Glands, through which the venom passes. Barbed Setae which cause discomfort presumably induced by mechanical irritation. Cf. Aerophore.

URTICATING Adj. Nettling, causing a stinging or burning sensation of the skin.

URTICATION Noun. (Latin, *urtica* = nettle; English, *-tion* = result of an action. Pl., Urtications.) The rash produced from poisonous Setae or secretions by some insects.

USDA The acronym for United States Department of Agriculture. Cf. ARS; APHIS.

USINGER, ROBERT LESLIE (1912–1968) American academic (University of California, Berkeley) and student of Heteroptera, particularly Aradidae and Cimicidae. (Aguilar 1968, Riv. Peruana Ent. 11: 1–2; China 1968, Entomol. mon. Mag. 104: 287–288; Linsley & Ashlock 1969, Pan-Pacif. Ent. 45: 167–203, bibliogr.)

USTINOV, ALEXSANDR ALEXKSANDRIOVICH (1894–1964) (Medvedev & Kiryanova 1966, Ent. Obozr. 45: 921–924, bibliogr. Translation: Ent. Rev. Wash. 45: 518–520, bibliogr.)

USTULATE Adj. (Latin, *ustilare* = to scorch; *-atus* = adjectival suffix.) Marked with brown and giving or having scorched appearance. Alt. Ustulatus.

USURPATION Noun. (Latin, *usurpare* = to make use of; English, *-tion* = result of an action. Pl., Usurpations.) Displacement of a founder queen by another queen in social Hymenoptera.

UTA See American Mucocutaneous Leishmaniasis. Syn. Espundia.

UTERINE Adj. (Latin, *uterus* = womb.) Descriptive of or pertaining to the Uterus.

UTERUS MASCULINUS Symphyla: A pouch or sac into which the ejaculatory duct opens.

UTERUS Noun. (Latin, *uterus* = womb. Pl., Uteruses.) Insects: The enlarged portion of the Median Oviduct in viviparous insects. Sometimes enlarged portion of the Bursa Copulatrix at the junction of the Oviducts. See Calyx.

UTILITARIAN Adj. (Latin, *utilitas* = useful.) For useful ends or purposes.

UTRICLE Noun. (Latin, *utriculus* = little bag. Pl., Utricli.) A small sac or bladder; a cell. Alt. Utriculus.

UTRICULAR Adj. (Latin, *utriculus* = little bag.) Furnished with Utricles. Resembing a Utricule. Alt. Utriculate; Utriculoid; Utriculatus. See Utriculus. Cf. Utriculiform; Utriculoid.

UTRICULI BREVIORES Small vesicular sacs connected with the seminal vesicles in crickets and some other insects.

UTRICULI MAJORES Large vesicular sacs or tubular structures connected with the seminal vesicles in crickets and some other insects.

UTRICULIFORM Adj. (Latin, *utriculus* = little bag; *forma* = shape.) Shaped as a small bag or pouch. Alt. Utriculoid. Cf. Utricular; Utriculoid.

UTRICULOID Adj. (Latin, *utriculus* = little bag; Greek, *eidos* = form.) Descriptive of structure resembling a small bag, pouch or bladder. Alt. Utriculiform. See Utriculus. Cf. Bursiform; Scrotiform. Rel. Form 2; Shape 2; Outline Shape.

UTRICULUS Noun. (Latin, *utriculus* = little bag; small uterus. Pl., Utriculi.) A small pouch or bag. See Uterus.

UTRIMQUE Adj. Similarly placed on both sides.

UVARICIN See Acetogenins.

UVAROV, BORIS PETROVITCH (1888–1970) Russian-born Orthopterist specializing in locust and grasshopper biology and control. Uvarov coined the term 'Acridology.' (Bei-Bienko 1970, Ent. Obozr. 49: 915–922. (Translation: Ent. Rev., Wash. 49: 559–562.); Haskell 1970, In memoriam. 163 pp., bibliogr. Anti-Locust Research Centre [later Centre for Overseas Pest Research]. London; Waloff & Popov 1990, Ann. Rev. Ent. 35: 1–24)

UVEX® See Gibberellic Acid.

UYTTENBOOGAART, DANIEL LOUIS (1872–1947) (MacGillavry 1946, Tijdschr. Ent. 89: 1–9, bibliogr.)

UZEL, JINDRICH (1868–1946) (Kermak 1947, Cas. csl. Spol. ent. 11: 13–17, bibliogr.)

UZELLE, THÉOPHILE BRUAND DE (1808–1861) (Millière 1861, Annls. Soc. ent. Fr. (4) 1: 651–656, bibliogr.)

UZELOTHRIPIDAE Plural Noun. A numerically small Family of Thysanoptera assigned to Suborder Terebrantia.

UZI FLY *Tricholyga bombycis* Beck; *Exorista sorbillans* (Wiedemann) [Diptera: Tachinidae].

VACHAL, JOSEPH (1838–1911) (Janet 1911, Bull. Soc. ent. Fr. 16: 4l; Kuhnt 1911, Dt. ent. Z. 1911: 353–354; Buysson 1913, Annls. Soc. ent. Fr. 82: 778–784, bibliogr.)

VACUOLATE Adj. (Latin, *vacuus* = empty; *-atus* = adjectival suffix.) Structure with vacuoles or small cavities that are empty, contain fluid or solid substance.

VACUOLE Noun. (Latin, *vaccus* = empty. Pl., Vacuoles.) A space within the Cytoplasm of a cell which contains crystalline material, a droplet of fluid, or other material.

VACUOLIZATION Noun. (Latin, *vacuus* = empty; English, *-tion* = result of an action.) The process of forming vacuoles.

VADON, JEAN (1904–1970) (Paulian 1971, Bull. Soc. ent. Fr. 76: 171–174, bibliogr.)

VAGABOND CRAMBUS *Agriphila vulgivagella* (Clemens) [Lepidoptera: Pyralidae].

VAGILE Adj. (Latin, *vagus* = wandering.) Able to move or migrate.

VAGINA Noun. (Latin, *vagina* = sheath. Pl., Vaginas.) 1. A sheath covering or receiving an organ or structure. 2. The tubular structure formed by the union of the Oviducts in the female reproductive system. Opening externally to admit the passage of the egg to the Ovipositor. The Vagina receives the Penis of the male in copulation. Sometimes called oviduct; 'every part, the office of which is to cover, protect or defend the tongue'; 'the bivalve coriaceous sheath or cover of the spicula.'

VAGINAL Adj. (Latin, *vagina* = sheath; *-alis* = pertaining to.) Resembling a sheath; pertaining to the Vagina.

VAGINAL AREOLES Coccids: See Discaloca (MacGillivray).

VAGINATA (Latin, *vagina* = sheath.) Sheathed. Obsolete Ordinal term for Coleoptera.

VAGINATE Adj. (Latin, *vagina* = sheath; *-atus* = adjectival suffix.) Enclosed in a bivalved sheath. Alt. Vaginatus.

VAGINIPENNATE Adj. (Latin, *vagina* = sheath; *penna* = wing; *-atus* = adjectival suffix.) Sheath-winged; descriptive of wings that are covered with a hard sheath. Alt. Vaginipennous.

VAGINULA Noun. (Latin, *vaginula* diminutive of *vagina* = sheath. Pl., Vaginulae.) Hymenoptera: The sheath or covering of the Terebra *sensu* Kirby & Spence. The Gonostylus.

VAGRANT GRASSHOPPER *Schistocerca nitens nitens* (Thunberg) [Orthoptera: Acrididae].

VAGRANT SPIDERS See Clubionidae.

VAGUS Noun. (Latin, *vagus* = wandering.) The visceral, sympathetic, or stomogastric nervous system of insects.

VAGUS NERVE The system of nerves and ganglia cephalad of the Supraoesophageal Ganglion (Comstock & Kellogg).

VAHERI, ERKKI BENJAMIN (1907–1940) (Anon. 1940, Suomen hyönt. Aikak. 6: 5–7.)

VAHL, MARTIN (1749–1805) (Zimsen 1964, *The Type Material of J. C. Fabricius*. 656 pp. (13), Copenhagen.)

VAHLKAMPFIA MELLIFICAE See Malpighamoeba mellificae.

VAILLANT, MARÉSCHAL (–1872) (Anon. 1872, Petites Nouv. Ent. 4: 216.)

VALASTEGUI, SEGUNDO E (1910–1972) (Arnaud l973, Pan-Pacif. Ent. 49: 87–88.)

VALCK, FRANS TITUS LUCUSSEN (–1939) (De Beaux 1940, Annali Mus. civ. Stor. nat. Giacomo Doria 60: (3).)

VALEFOS® See Azinphos-Methyl.

VALENCIENNES, ACHILLE (1794–1865) (J. v. d. H. 1866, Ned. Tijdschr. Dierk. 3: 71–72; Vibraye 1868, Mém. Agric. Econ. Rurale 1867: 45–59.)

VALGATE Adj. (Latin, *valgus* = turned abnormally inward or outward; *-atus* = adjectival suffix.) Enlarged at the bottom; club-footed. Alt. Valgatus.

VALID Adj. (Latin, *validus* = strong.) 1. Founded on true or factual information; capable of being justified. 2. In zoological nomenclature, the properly proposed and accepted name of a Species or other taxonomic category.

VALID NAME The legally recognized scientific epithet of a Subspecies, Species or Genus; the scientific name under which the entity was first designated in accordance with the rules specified under the International Code of Zoological Nomenclature. See Scientific Name; Common Name. Cf. Invalid Name. Rel. Nomenclature; Synonym.

VALLE, KAARLO JOHANNES (1887–1956) (Anon. 1956, Sber. finn. Akad. Wiss. 1956: 71–82; Lindroth 1957, Opusc. ent. 22: 125.)

VALLINS, FREDERICK THOMAS (1900–1972) Gardner 1974, Proc. Brit. ent. nat. Hist. Soc. 6: 124–125.

VALLISNIERI, ANTONIO (1661–1730) Italian aristocrat and naturalist. Author of *Dialoghi sopra la curiosa origine, sviluppi e costumi di varii Insetti* (1700). Provided detailed studies of horse bot fly, sheep nasal bot fly and cattle warble fly. (Camerano 1905, Memorie Accad. Sci. 55: 69–112; Howard 1930, Smithson. misc. Collns. 84: 202, 498; Stroppiana 1958, Riv. Parassit. 19: 1–5; Savtelli 1961, Physis 3: 269–308.)

VALVA Noun. (Latin, *valva* = valve. Pl., Valvae.) Harpagones or two lateral sclerites which cover the Ovipositor when not in use.

VALVATE Adj. (Latin, *valva* = fold; *-atus* = adjectival suffix.) Pertaining to structures hinged at the margin only; structure possessing or resembling a valve. A condition in which parts meet edge-to-edge, but do not overlap. Cf. Imbricate.

VALVE Noun. (Latin, *valva* = fold. Pl., Valves.) 1. A lid, cover or similar structure, opposed to an aperture or oriface, which opens in one direction thereby permitting directional flow, and closes in the opposite direction thereby preventing flow in the another direction. 2. Hemiptera: A small triangular sclerite posteriad of the last full ventral segment, at the base of the plates in male

Cicadellidae and allies (Smith). 3. Orthoptera: One of the corneous pieces of the Ovipositor; a Corniculus. 4. Lepidoptera: The Harpes. 5. Hymenoptera: The expanded plate-like Galea of the Maxilla; a Valvula *sensu* Smith.

VALVE OF OPERCULUM Coccids: See Operculum (Green).

VALVES OF THE HEART Internal valve-like lobes of the Heart walls between the chambers, said to be present in certain dipterous larvae (Snodgrass).

VALVIFER Noun. (Latin, *valva* = fold; *ferre* = to bear. Pl., Valvifers.) The basal sclerites of the Ovipositor, probably derived from the Coxopodites of the Gonopods, carrying the Valvulae including First Valvifers of the eighth abdominal segment, and Second Valvifers of the ninth segment (Snodgrass).

VALVULA Noun. (Latin, diminutive of *valva*. Pl., Valvulae.) 1. A small valve. 2. Any small, valve-like process. 3. A coriaceous structure covering sucking mouthparts. 4. Diptera: The Vagina. 5. Hymenoptera: a podical plate (Burmeister). The expanded plate-like Galea of the Maxilla. A branch of the male genital forceps (Smith). 6. Male Lepidoptera: The central part of the Harpe (Klots).

VALVULAR Adj. (Latin, diminutive of *valva*.) Valve-like; with the functions of a valve. Pertaining to a valve.

VALVULAR PROCESS Odonata: A slender, unjointed process at the apex of each genital valve.

VAMIDOTHION Noun. An organic-phosphate (thiophosphoric acid) compound {O,O-dimethyl S-(1-methylcarbamoylethylthioethyl) phosphorothioate} used as a systemic insecticide and acaricide against mites, plant-sucking insects, thrips and other pests. Compound applied to ornamentals, fruit crops, cotton, field crops, rice, tobacco and vegetables in some countries. Not registered for use in USA. Trade names include Kilval®, Kilvar® and Trucidor®. See Organophosphorus Insecticide.

VAN ALDENBURG BENTINCK, GODARD ADRIAAN (HENRI JULES GRAAF) (1887–1968) (Barendrecht 1968, Ent. Ber., Amst. 28: 201–202.)

VAN ASPEREN, KLAAS (1921–1964) (Perry & O'Brien 1965, J. Econ. Ent. 58: 807.)

VAN DER HOEVEN, JAN (1801–1868) (Hartig 1868, Jaarb. K. Akad. Wet. Amst. 1868: 1–34, bibliogr.)

VAN DER WEELE, HERMAN WILLEM (1878–1910) (Everts 1911, Tijdschr. Ent. 54: 1–5, bibliogr.)

VAN DER WIEL, PIETER (–1962) (Anon. 1962, Ent. Ber., Amst. 22: 133–148, bibliogr.)

VAN DER WULP, FREDERICK MAURITS (1818–1899) (Mik 1899, J. Wien. ent. Ztg. 18: 300; Verrall 1899, Proc. ent. Soc. Lond. 1899: xxxvii–xxxviii.)

VAN DOESBURG, P H (1892–1971) (Van Doesburg 1971, Ent. Ber., Amst. 31: 207–214, bibliogr., list of described Species and type-localities.)

VAN DUZEE, EDWARD PAYSON (1861–1940) Professional librarian in New York and later Curator of Insects at California Academy of Sciences. Student of Hemiptera and published a catalogue of North American Hemiptera. (Gunder 1929, Ent. News 40: 103–104; MacFarland 1940, Science 92: 99; Usinger 1940, Pan-Pacif. Ent. 16: 123; Essig 1941, Pan-Pacif. Ent. 16: 145–177, bibliogr.; Essig 1941, Ann. ent. Soc. Am. 34: 263; Torre-Bueno 1941, Bull. Brooklyn ent. Soc. 36: 80–81.)

VAN DUZEE, MILLARD CARR (1860–1934) (Anon. 1934, Ent. News 45: 202; Van Duzee 1934, Pan-Pacif. Ent. 10: 90–96; Hungerford 1935, Ann. ent. Soc. Am. 28: 179–180.)

VAN DUZEE TREEHOPPER *Vanduzea segmentata* (Fowler) [Hemiptera: Membracidae].

VAN DYKE, EDWIN COOPER (1869–1952) (Essig 1953, Pan-Pacif. Ent. 29: 73–108, bibliogr.)

VAN ELDIK, HENRI CHARLES LOUIS (1890–1953) (Hardonk 1953, Ent. Ber., Amst. 14: 389–390.)

VAN OORT, EDUARD DANIEL (1876–1933) (Bayer 1933, Zoöl Meded., Leiden 16: 263–265.)

VAN ROSSEM, GERARD (1919–1990) (Van Achterberg 1992. Zool. Meded. 66 (16–40): 303–308.)

VAN SON, GEORGES (1898–1967) (Janse 1968, J. ent. Soc. sth. Afr. 31: 237–239.)

VAN VOORST, JOHN (1804–1898) (McLachlan 1898, Entomol. mon. Mag. 34: 214–215.)

VAN ZWALUWENBURG, REYER H (1891–1970) (Beardsley 1971, Proc. Hawaii. ent. Soc. 21: 129–131; Simmons 1971, J. Econ. Ent. 64: 994–1005.)

VANDA THRIPS *Dichromothirps corbetti* (Priesner) [Thysanoptera: Thripidae].

VANDELLI, DOMENICO (1735–1816) (Rose 1850, *New General Biographical Dictionary*. 12: 230.)

VANELLA Noun. (Pl., Vanellae.) Unseparated modifications of the proximal parts of the wings. Named after the veins of which they are projections (MacGillivray).

VANHESPEN, EMILE (1902–1964) (Schepdael 1965, Linneana Belgica 3: 50–51.)

VANHORNIDAE Crawford 1909. Plural Noun. A small Family (two Genera, two Species) of apocritous Hymenoptera assigned to the Proctotrupoidea. Sometimes placed in Proctotrupidae or Heloridae. Found in Sweden, east USA and Chile. Antenna 13-segmented; forewing Pterostigma large, venation extensive; Notaulices deep, nearly complete; transverse, foveate scutellar Sulcus; Propodeum with median longitudinal Carina. Internal larval parasites of Eucnemidae.

VANNAL FOLD See Plica Vannalis.

VANNAL REGION The wing area containing the Vannal Veins, or veins directly associated with the third Axillary Sclerite. When large, VR usually separated from the Remigium by the Plica

Vannalis, often forming an expanded fan-like area of the wing (Snodgrass). Syn. Anal Region. See Wing. Cf. Axillary Region; Jugal Region; Remigium.

VANNAL VEIN 1. Any of several veins associated at their bases with the third Axillary Sclerite, and occupying the Vannal Region of the wing. 2. The 'anal' veins except the first, or Postcubitus (Snodgrass). See Vein; Wing Venation.

VANNUS Noun. (Latin, *vannus* = fan.) A large fan-like expansion of the posterior part of the insect wing (Snodgrass). See Vannal Region.

VAPAM® See Metham-Sodium.

VAPCOZIN® See Amitraz.

VAPONA® See Dichlorvos.

VAPONITE® See Dichlorvos.

VAPOROOTER® See Metham-Sodium.

VAPPULA, NIILO A (1897–1971) (Kanervo 1972, Suomen hyönt. Aikak. 38: 111.)

VARIABLE CUTWORM *Agrotis porphyricollis* Guenée [Lepidoptera: Noctuidae] (Australia). See Cutworms.

VARIABLE LADYBIRD *Coelophora inaequalis* (Fabricius) [Coleoptera: Coccinellidae]: An Australian Species. Adult 4.1–6.0 mm long, round-ovate, with yellow-orange Thorax and Elytra and black markings; markings on Variable Ladybird range from: (i) completely black Elytra and two dark spots on Thorax, (ii) two central spots and seven marginal spots on Elytra and three spots on Thorax, (iii) a central stripe and four longitudinal spots in centre of Elytra and four spots on Thorax, and (iv) a central stripe and eight spots on Elytra and a black Thorax. Adult females lay eggs in batches near larval food. Eggs yellow, spindle-shaped and laid standing on end. Larva elongate, with well defined legs; body black with markings and fleshy spines; four larval instars. Pupal stage 8–9 days. Adults and larvae feed on aphids and *Helicoverpa* eggs. Cf. Common Spotted-Ladybird; Minute Two-Spotted Ladybird; Mite-Eating Ladybird; Striped Ladybird; Three-Banded Ladybird; Three-Spotted Ladybird.

VARIABLE OAKLEAF CATERPILLAR *Heterocampa manteo* (Doubleday) [Lepidoptera: Notodontidae].

VARIABLE SHIELD BUG *Kapunda troughtoni* (Distant) [Hemiptera: Pentatomidae] (Australia).

VARIATION Noun. (Latin, *variatio, variare* = to change; English, *-tion* = result of an action. Pl., Variations.) 1. The act, process or condition or being varied. 2. Genetics: Alternative phenotypic expressions of colour, shape or form compared with a 'normal' phenotype. 3. Taxonomy: The sum of the departures from a mean type of any Species; continuous when there is no break between the extremes, or discontinuous when there are gaps without intermediate forms. Variation can be expressed within a population or among populations within the geographical range of a Species.

VARICOSE Adj. (Latin, *varicosus* from *varix* = dilatation of a vein; *-osus* = full of.) 1. Irregularly swollen. A term descriptive of veins or vessels with irregular abnormal enlargements. 2. Botany: Abnormally enlarged veins on a leaf. Alt. Varicosus.

VARIED CARPET BEETLE *Anthrenus verbasci* (Linnaeus) [Coleoptera: Dermestidae]: A cosmopolitan pest of museum collections, bristles, wool, feathers, fur, hair, horns, hides and stored products. Adults visit flowers; larvae act as pests. Adult 2–5 mm long, oval in outline shape; body mottled yellow, white and black dorsally; dorsum covered with flattened overlapping Setae giving a scaly appearance; no cleft at apex of Abdomen where Elytra meet. Larva reddish to red brown with bands of stiff bristles and stiff Setae at posterior end; body widest at posterior end. Larvae often very destructive. Natural food includes fresh and dry carcasses, animal wastes and debris in bird nests. Eggs laid on food in dark, undisturbed area. Larvae feed to nine months. Adults emerge within 3 weeks and live several weeks. Active all year. Egg stage 17–18 days; larval stage 222–322 days; pupal stage 10–13 days; adults 251–347 days. Syn. Variegated Carpet Beetle. Cf. Australian Carpet Beetle; Black Carpet Beetle; Common Carpet Beetle; Furniture Carpet Beetle.

VARIEGATED Adj. (Latin, *variegare* = to make various.) 1. Varied in colour, of several colours in indefinite pattern. 2. Descriptive of structure with irregular colour patterns or marks.

VARIEGATED CARPET BEETLE See Varied Carpet Beetle.

VARIEGATED CUTWORM *Peridroma saucia* (Hübner) [Lepidoptera: Noctuidae]: A cosmopolitan, multivoltine pest of garden crops, some ornamental (a major pest of chrysanthemums in Hawaii) and fruit trees. Eggs deposited in groups of 50–60 on branches, stems and leaves. Larva smooth bodied with yellow spot on middorsal line and a dark 'W' on eighth abdominal dorsum. VC overwinters as pupa and may undergo 3–4 generations per year, depending upon climate and host plant. See Noctuidae. Cf. Cutworms.

VARIEGATED GRAPE LEAFHOPPER *Erythroneura variabilis* Beamer [Hemiptera: Cicadellidae]: An incidental, sometimes serious, pest of grapes in western North America. Damage induced by defoliation of the infested plant. See Cicadellidae.

VARIEGATED MUD LOVING BEETLES See Heteroceridae.

VARIEGATED RAGWEED BEETLE *Zygogramma bicolorata* Pallister [Coleoptera: Chrysomelidae] (Australia).

VARIEGATION Noun. (Latin, *variegare* = to make various; English, *-tion* = result of an action. Pl., Variegations.) Variation in pigmentation.

VARIETA Latin meaning 'variety.' Abbreviated v., var.

VARIETY Noun. (Latin, *varietas* = variety. Pl., Varieties.) A taxonomically ambiguous term 'used to designate variants in size, structure and colour and varying ranks of all these' (McAtee), or indicate a form for which the taxonomic status is uncertain. Varieties have no standing in zoological nomenclature.

VARIKILL® See Fenoxycarb.

VARIN, GILBERT (1899–1969) (De Bros 1969, Mitt. ent. Ges. Basel 19: 119–120.)

VARIN, M G (–1969) (Anon. 1969, Bull. Soc. ent. Fr. 74: 7.)

VARIOLA Noun. (Latin *variola* = small pox.) A small pit or pock-like mark. Alt. Variole. See Punctation.

VARIOLATE Adj. (Latin, *variola* = smallpox.) Pitted as by small pox, full of varioles. Alt. Variolose; Variolosus; Variolous.

VARIORUM NOTAE (Latin.) The notes of various authors.

VARLEY, GEORGE COPLEY (1910–1983) English academic and Hope Professor of Zoology (Oxford University). Specialist in Insect Ecology with particular reference to population dynamics and the interactions between insects and their parasites.

VARROA MITE *Varroa jacobsoni* (Oedemans) [Acari: Varroidae]: A serious pest of honey bees that was first reported in brood cells of *Apis cerana* drone larvae on Java. VM has become widespread; first observed attacking European Honey Bee in Europe ca 1970 and USA during 1987. Within USA, VM is widely distributed by colony transport for pollination and sale of packaged bees and queens. VM female dark reddish brown, ca 2 mm diam; reproduces on drone pupae of *A. cerana* and worker pupae of *A. mellifera;* mites occur within sealed brood cells; female lays 1–12 eggs in cell; first egg is male; nymphs pierce Integument of pupae and feed on Haemolymph; male mites mate and die within sealed cell. VM reduces longevity of adult bees and causes colony extinction within 3–4 years if left untreated. VM transmits Black Queen-Cell Virus, Sacbrood, Deformed-Wing Virus and activates Acute Paralysis Virus. A complex of closely related Species of mites attack *Apis cerana, A. florea* and *A. mellifera*. Cf. Tracheal Mite.

VAS Noun. (Latin, *vas* = canal, vessel. Pl., Vasa.) Any short or narrow vessel; a duct, canal or blind tube.

VAS DEFERENS (Pl., Vasa Deferentia), A portion of the male's internal reproductive system that consists of one of the paired canals (tubes). VD acts as a conduit for Spermatozoa from the Testes. Distal ends of paired VD unite to form an Ejaculatory Duct. Ultrastructure reveals that VD is tubular and consists of an epithelial cell layer and Basement Membrane; VD is surrounded by circular muscles. A Seminal Vesicle forms an enlargement or dilation of VD near its junction with the Ejaculatory Duct. VD is homologous with the female's Lateral Oviduct. See Testes. Cf. Seminal Vesicle; Epididymis. Rel. Gonad.

VAS EFFERENS (Pl., Vasa Efferentia.) A slender tube which connects each Follicle of a Testis with the Vas Deferens (Imms). Regarded by Snodgrass as homologous with the Pedicel of the Ovariole. Some insects (Lepidoptera) lack a Vas Efferens. See Testes. Cf. Follice; Vas Defferens. Rel. Gonad.

VASA MUCOSA The Malpighian Tubules.

VASCO, AMEDEO (–1897) (Anon. I897, Riv. ital Sci. nat. 17: 71.)

VASCULAR Adj. (Latin, *vasculum* = small vessel.) 1. Relating to the blood-vessels or circulatory system. 2. Pertaining to the conductive tissue of plants.

VASCULUM Noun. (Latin, *vas* = vessel, *vasculum* = small vessel. Pl., Vascula.) An elongate cylindrical or rectangular box with a length-wise top that is used to collect plant specimens.

VASE Noun. (Latin, *vas* = vessel. Pl., Vases.) A decorative vessel that is closed and flat at the bottom, deeper than wide with rounded sides and sometimes constricted toward the opening. Cf. Adeniform; Ampulliform; Vasiform.

VASIFORM Adj. (Latin, *vas* = vessel; *forma* = shape.) Vase-shaped; with the form of a hollow tube and duct. See Vase. Cf. Adeniform; Ampulliform. Rel. Form 2; Shape 2; Outline Shape.

VASIFORM ORIFICE Aleurodidae: An ovate, triangular or semicircular opening on the dorsum of the last abdominal segment.

VASILEV, EVGENII MICHAELOVICH (1856–1922) (Boghdanov 1922, Izv. Otd. Prikl. Ent. 2: 41–48, bibliogr.)

VASIL'YEV, IVAN VASIL'YEVICH (1851–) (Anon. 1961, Ent. Obozr. 40: 934–935. (Translation: Ent. Rev., Wash. 40: 536).)

VASSEUR, RENÉ (1910–1953) (F.C. 1953, Revue zool. agric. appl. 52: 73–75, bibliogr.)

VAUGHAN, HOWARD W J (1846–1892) (C.A.B. 1892, Entomologist 25: 300; Godman 1892, Proc. ent. Soc. Lond. 1892: iv.)

VAULOGER DE BEAUPRE, MARCEL (1862–1904) (A.C. 1904, Revue Russe Ent. 4: 253; Léveille 1904, Bull. Soc. ent. Fr. 1904: 125.)

VAULT® A registered biopesticide derived from *Bacillus thuringiensis* var. *kurstaki*. See *Bacillus thuringiensis.*

VAZQUEZ FIGUEROA Y CANALES, AURELIO (–1910) (Dusmet y Alonso 1919, Boln. Soc. ent. Esp. 2: 180.)

VECCHI, ANITA (1893–1953) (Grandi 1952, Atti Accad. naz. ital. Ent. Rc. 1: 21–23.)

VECHT, JACOBUS VAN DER (1906–1992) Dutch entomologist specializing in taxonomy of aculeate Hymenoptera.

VÉCSEY, STEFAN (ISTRAN) (1863–1910) (Schmidt 1911, Rovart. Lap. 18: 97–99, 112.)

VECTOBAC® A registered biopesticide derived from *Bacillus thuringiensis* var. *israelensis*. See *Bacillus thuringiensis.*

VECTOCIDE® A registered biopesticide derived from *Bacillus thuringiensis* var. *israelensis*. See *Bacillus thuringiensis.*

VECTOR Noun. (Latin, *vector* = bearer; *or* = one who does something. Pl., Vectors.) 1. Medical Entomology: An arthropod which carries disease-producing organisms to a vertebrate host. Vectors classified as Primary and Secondary; modes of disease transmission classified as Mechanical and Biological. 2. Any organism which transports or transmits a parasite to a host. See Transmission. Cf. Porter.

VECTOR® See Entomogenous Nematodes.

VECTOR COMPETENCE Vectoring ability of a disease carrying arthropod. Several factors are required for an arthropod to be considered competent: (1) Regular laboratory isolation of the pathogen from the host; (2) distribution of the vector overlaps with a diseased population; (3) host-seeking and blood-feeding coincide with disease incidence; (4) disease incidence varies with host choice; and (5) demonstration of the capacity for disease vectoring in the laboratory.

VECTOR POTENTIAL The ability of an arthropod to biologically vector a pathogen from one host to another.

VECTRIN® See Resmethrin.

VECTROLEX® See *Bacillus sphaericus*.

VECTRON® See Etofenprox.

VEDALIA BEETLE *Rodolia cardinalis* (Mulsant) [Coleoptera: Coccinellidae]: A predaceous coccinellid beetle native to Australia and important in the biological control of cottony cushion scale, *Iceria purchasi* Maskell in California and other areas where the CCS became a pest of citrus. See Albert Koebele.

VEDALIA LADYBIRD *Rodolia cardinalis* (Mulsant) [Coleoptera: Coccinellidae] (Australia).

VEE See Venezuelan Equine Encephalitis.

VEGETABLE BEETLE *Gonocephalum elderi* (Blackburn) [Coleoptera: Tenebrionidae] (Australia).

VEGETABLE LEAFHOPPER *Austroasca viridigrisea* (Paoli) [Hemiptera: Cicadellidae]: An Australian Species sometimes locally common on vegetables and cotton. Adult ca 3 mm long, pale green to brown and wedge-shaped; adults will hop and fly when disturbed. Eggs 0.7–0.9 mm long, curved and greenish-yellow; eggs embedded singly in midrib or Petiole of a young stem. Eggs stage 4–11 days. Nymphs similar to adults but paler and lack wings. Nymphs 0.5–2.0 mm long, pale green and wedge-shaped. Nymphal stage lasts 7–21 days depending on temperature and food. Leafhoppers insert mouthparts into underside of leaves for feeding. Feeding causes yellowing on underside and red-dish-brown coloration on upperside of leaf. Leaves puckered and curl downwards.

VEGETABLE LEAFMINER *Liriomyza sativae* Blanchard [Diptera: Agromyzidae].

VEGETABLE SPIDER MITE *Tetranychus neocaledonicus* (Andre) [Acari: Tetranychidae] (Australia).

VEGETABLE WEEVIL *Listroderes difficilis* Germar [Coleoptera: Curculionidae]. Endemic to South America. A pest of vegetables in southwestern USA. Parthenogenetic; males unknown. Eggs laid on plant during September. Larva feeds on leaves and roots; pupation within earthen cell. Adults thanototic, live two years, cause most damage. May be important in seed beds.

VEGETATIVE FUNCTIONS In the living organism, all those functions which together work to maintain its life, *e.g.*, respiration and digestion in their most ample sense (Snodgrass).

VEGETATIVE PROPAGATION The use of cuttings, bulbs, tubers or corms to grow new plants.

VEIN Noun. (Latin, *vena* = vein. Pl., Veins.) Any chitinous rod-like or hollow tube-like structure supporting and stiffening the insect wings, especially those extending longitudinally from the base of the wing to the outer margin. Veins are generally classified as longitudinal veins (which extend along the primary axis of the wing) and cross veins. Syn. Wing Nerves; Wing Nervures; Wing Nervules. See Cross Vein; Longitudinal Vein. Cf. Adventitious Vein; Nebulous Vein; Spectral Vein; Tubular Vein.

VEINLET Noun. (Latin, *vena* = vein. Pl., Veinlets.) Any relatively small vein in the insect wing. Orthoptera: minute, transverse ribs or ridges between longitudinal veins.

VEITCH, ROBERT (1890–1972) (Weddell 1972, News Bull. ent. Soc. Qd 90: 21; Weddell 1973, J. Aust. Inst. agric. Sci. 39: 66.)

VELEZ DE ARCINIEGA, FRANCISCO (Dusmet y Alonso 1919, Boln. Soc. ent. Esp. 2: 77.)

VELIIDAE Amyot & Serville 1843. Plural Noun. Broad-Shouldered Water Striders; Riffle Bugs; Ripple Bugs; Smaller Water Striders; Water Crickets. A cosmopolitan Family of heteropterous Hemiptera assigned to Superfamily Gerroidea. Adult body size variable (ca 1.5–6.0 mm long); head with impressed line on dorsal surface; Pronotum and Mesonotum fused; wings typically absent; legs thin, not exceptionally long; fore Coxa relatively near middle Coxa; hind Femur wider than middle Femur; hind Tibia projecting beyond apex of Abdomen; tarsal claws preapical on last Tarsomere of each leg; abdominal scent gland orifice absent. Most Species of veliids are associated with fresh water; a few Species occur on salt water; a few Species are associated with bromeliads. Nymphs and adults 'skate' on the surface of water. Cf. Mesoveliidae.

VELLAY, IMRE (EMERICH) (1850–1898) (Jablonowski 1898, Rovart. Lap. 5: 24, 180–186.)

VELOCIPEDIDAE Plural Noun. A small Family of heteropterous Hemiptera assigned to Superfamily Cimicoidea. (See Zool. Meded. 1980, 55: 298–299.)

VELSTAR® See Tralomethrin.

VELUM Noun. (Latin, *velum* = a veil. Pl., Vela.) 1. A membrane. 2. A membranous appendage of the spurs at the apex of the fore Tibia. 3. Neuroptera: A pair of processes projecting from the dorsal surface of the Spermatheca. 4. Siphonaptera: An arched enlargement of dorsal wall of Aedeagus. 5. Lepidoptera: sclerotized structure of Sternum VIII. 6. Hymenoptera: a broad process along the apicomedial surace of the fore Tibia in bees.

VELUM PENIS The thin, membranous covering of the male intromittent organ; also applied to other covering or shield-like structures of the Penis.

VELUTINOUS Adj. (Italian, *velluto* = velvet; Latin, *-osus* = possessing the qualities of.) Descriptive of structure that is velvet-like in texture or appearance; structure covered with dense, soft, short Setae.

VELVET ANTS See Mutillidae.

VELVET WATER BUGS See Hebridae.

VELVET WORMS See Onychophora.

VELVETBEAN CATERPILLAR *Anticarsia gemmatalis* (Hübner) [Lepidoptera: Noctuidae]: A pest of peanuts and soybeans in the southeastern USA. Overwinters in south Florida and migrates into Alabama and Georgia. Egg ca 2.5 mm in diameter. Eggs laid on underside of leaves individually or groups of 2–3; eclosion occurs within 3 days. Larva with five pairs of prolegs, body yellow-green to brown with longitudinal stripes that are yellow-white; mature larva ca 40 mm long. Neonate larva feeds on Epidermis and mesophyll of upper side of leaves; mature larva consumes leaf surface and leaf midribs and major veins. VBC with six larval instars; larval development ca 30 days. Pupation requires 7–12 days; occurs in soil. Adult yellow-brown to dark grey; forewing with a row of pale spots and brownish diagonal stripes.

VELVETY SHORE BUGS See Ochteridae.

VELVETY TREE PEAR *Opuntia tomentosa* Salm.-Dyck. [Cactaceae]: A Species of cactus native to the new world and adventive elsewhere as a minor rangeland pest. Controlled with phytophagous insects. See Prickly Pear.

VENA Noun. (Latin, *vena* = vein. Pl., Venae.) A vein. See Vein.

VENA ARCUATA The first jugal vein (Snodgrass).

VENA CARDINALIS The second jugal vein, usually appearing as a basal branch of the Vena Arcuata (Snodgrass).

VENA DIVIDENS Some Orthoptera: A secondary vein positioned in the fold between the Remigium and Vannus (Snodgrass). The concave vein Cu when it lies in the furrow dividing a definite clavus from the rest of the wings as in cockroaches (Tillyard). The longitudinal vein of the hindwings which marks the beginning of the anal area; anal of Comstock (Smith).

VENA MEDIA The Media.

VENA PLICATE Wings of Dermaptera: The vein around which the folding occurs.

VENA SPURIA Syrphidae: A vein-like thickening of the wing membrane between vein R and M (Wardle). Syn. Spurious Vein.

VENABLES, EDMUND PETER (1881–1966) (Ross 1966, J. ent. Soc. Br. Columb. 63: 45.)

VENATION Noun. (Latin, *vena* = vein; English, *-tion* = result of an action.) The complete system of veins of an insect wing. Syn. Neuration.

VENDEX® See Fenbutatin-Oxide.

VENET, HENRI (–1958) (Anon. 1958, Bull. Soc. ent. Fr. 63: 57.)

VENEZUELAN EQUINE ENCEPHALITIS A tropical arboviral disease of horses and rarely humans; low mortality in horses and humans; endozootics reported in rodents. VEE localized in tropical Americas. Virus assigned to Genus *Aphavirus* of Tongaviridae and vectored by mosquitoes (*Aedes* spp., *Culex* spp., *Psorophora* spp.) See Arbovirus; Tongaviridae. Cf. Eastern Equine Encephalitis; Western Equine Encephalitis.

VENOM Noun. (Latin, *venenum* = poison. Pl., Venoms.) 1. Toxic or poisonous compounds manufactured by glandular cells and secreted or injected into another organism. 2. Hemiptera: Salivary Gland secretions of some predatory Heteroptera which are used to subdue prey. Secretions also classified as defensive materials. Venoms include alkanes, alkanals, dicarboxylates, steroids and aromatic compounds. (Weatherston & Percey 1978, in *Arthropod Venoms*.) 3. Hymenoptera: Toxic fluid, consisting of biologically active peptides, secreted from accessory glands and injected into prey or hosts by the Sting in Aculeata or Ovipositor in Parasitica. Venoms induce death or paralysis, cause pain, and edematious reactions and swelling of skin. See Waspkinins; Mastoparan. 4. Coleoptera: The term venom has been been applied to several Species which employ defensive secretions. 5. Lepidoptera: Several Species secrete material in Urticating Setae.

VENOMOUS Adj. (Latin, *veneum* = poison; *-osus* = possessing the qualities of.) 1. Descriptive of organisms with venom glands. 2. Poisonous organisms.

VENOSE Adj. (Latin, *vena* = vein; *-osus* = full of.) Appendages or structures furnished with veins or vein-like markings. Of or pertaining to veins. Alt. Venosus; Venous.

VENPLICA Noun. (Pl., Venplicae.) The name applied to a Sternal Plica (MacGillivray).

VENT Noun. (Latin, *findere* = to cleave. Pl., Vents.) The Anus.

VENTER Noun. (Latin, *venter* = belly. Pl., Venters.) 1. The ventral surface of a segment. 2. The Abdomen. 3. The ventral surface of the Abdomen. See Segmentation. Cf. Dorsum; Pleuron.

VENTILATION Noun. (Latin, *ventilus* = breeze > *ventilare* = to brandish in the air.) 1. The circulation or movement of air. 2. The physical act or process of being ventillated. 3. The active process by which the respiratory medium (air) is moved within the insect body. The mechanisms of Ventilation vary, depending upon Open and Closed Tracheal Systems. OTS may involve Active Ventilation, Autoventilation or Passive Suction Ventilation. The direction of gas flow within the Tracheal System is typically anterior to posterior with air (oxygen) entering through spiracles on the Thorax and anterior part of the Abdomen and leaving through spiracles on the posterior part of the Abdomen. See Respiration. Cf. Active Ventilation; Autoventilation; Passive Suction Ventilation. Rel. Respiratory Muscles; Tracheal System; Air Sac.

VENTILATION TRACHEAE Tracheae with collapsible walls, responding to varying surrounding pressure (Snodgrass).

VENTOSE Adj. (Latin, *ventrosus* = windy; *-osus* = full of.) Inflated.

VENTRAD Adv. (Latin, *venter* = belly; *ad* = to.) Toward the venter; in the direction of the venter. See Orientation. Cf. Anteriad; Apicad; Basad; Caudad; Centrad; Cephalad; Craniad; Dextrad; Dextrocaudad; Dextrocephalad; Distad; Dorsad; Ectad; Entad; Laterad; Mediad; Mesad; Neurad; Orad; Proximad; Rostrad; Sinistrad; Sinistrocaudad; Sinistrocephalad.

VENTRAL Adj. (Latin, *venter* = belly; *-alis* = pertaining to.) Descriptive of the ventral surface of Abdomen. See Orientation. Cf. Dorsal; Lateral.

VENTRAL APPENDIX Hymenoptera: The proximal part of the ventral ramus of the Gonapophysis.

VENTRAL BRIDGE OF PENIS VALVES Hymenoptera: A ventral sclerotic bridge between Penis valve bases.

VENTRAL CHAIN The Ventral Nerve Cord.

VENTRAL CHITINOUS PROCESS Coccids: See Ventral Thickening (MacGillivray).

VENTRAL COMB Trichoptera: A transverse row of fine teeth on the venter.

VENTRAL DIAPHRAGM A Diaphragm of varying thickness which extends horizontally across the abdominal cavity just above the ganglia of the ventral nerve cord. VD can extend from head to posterior end of abdomen, and usually is attached to the body wall at two points on each segment. VD is comparable to the Dorsal Diaphragm in that it forms a delicate membrane containing muscle strands. Undulating, rhythmic movement of the ventral diaphragm helps move blood into and out of the ventral sinus and promotes circulation. VD is not found in some Thysanoptera or Orthoptera and related groups; VD is innervated in cockroaches and bees. Syn. Ventral Heart. See Circulatory System. Cf. Dorsal Diaphragm.

VENTRAL FACE OF THE LEG Diptera: The lower aspect of the leg when it is laterally extended. See Orientation.

VENTRAL GLANDS Diaspine coccids: Genacerores (MacGillivray).

VENTRAL GONOCOXAL BRIDGE Hymenoptera: basal sclerotic bridge between the Gonocoxites.

VENTRAL GROOVE Pentatomidae: A median lengthwise groove on the Abdomen. Collembola: a cuticular channel passing down the midline of the body.

VENTRAL GROUPED GLANDS Diaspine coccids: Genacerores (MacGillivray).

VENTRAL HEART See Ventral Diaphragm.

VENTRAL MUSCLES Insects: Fibres typically longitudinal and attached on the intersegmental fold or on the Antecostae of successive Sterna (Snodgrass).

VENTRAL NERVE CORD The chain of connected ventral ganglia, beginning with Tritocerebrum of Brain. In Entomology, term VNC is usually applied to thoracic and abdominal ganglia only (Snodgrass). Suboesophageal Ganglion is anteriormost ganglion of Ventral Nerve Cord; SG represents fusion of Mandibular, Maxillary and Labial Ganglia. SG is connected with Tritocerebrum via Circumoesophageal Ganglion and sends motor and sensory neurons to Maxilla, Labium and Salivary Glands. SG continues rearward to Prothoracic Ganglion as a pair of Nerve Cord Connectives. See Central Nervous System. Cf. Abdominal Ganglia; Brain; Thoracic Ganglia. Rel. Circumoesophageal Ganglion; Suboesophageal Ganglion.

VENTRAL PHARYNGEAL GLAND Bees: A transverse row of cells opening into the floor of the Pharynx between the ducts of the lateral Pharyngeal Glands (Snodgrass).

VENTRAL PILACERORES Coccids: The continuous band of Pilacerores located on the lateral portions of the ventral aspect of the Abdomen (MacGillivray).

VENTRAL PLATE A layer of columnar cells of the blastoderm on the ventral side of the egg, which later becomes the germ band (Imms).

VENTRAL PLATES Coccids: The plates of the ovisac excreted by the continuous band of pilacerores of the ovisac (MacGillivray).

VENTRAL PTYCHE Hymenoptera. Longitudinal mesial fold in the corium of each Gonapophysis (E. L. Smith). Ventral limit of egg canal. Syn. Crepidium; Examinium.

VENTRAL SCALE Diaspinae: The under part of the puparium, interposed between the insect and the plant.

VENTRAL SCARS Coccids: See Discaloca (MacGillivray).

VENTRAL SINUS The space of the definitive body cavity below the ventral diaphragm, containing the nerve cord (Snodgrass). Syn. Perineural Sinus.

VENTRAL SPINE Pentatomidae: A spine-like projection anterior from the first or second ventral segment; directed toward the head and sometimes positioned between the Coxae.

VENTRAL STYLET Lice: The lower paired Stylet, basally attached to the Stylet Sac (Imms).

VENTRAL SUTURES The incised lines between the rings of the Abdomen.

VENTRAL THICKENING Coccids: A distinct longitudinal thickening on the lateral part of the ventral aspect of each Operculum (MacGillivray).

VENTRAL TRACHEA The ventral segmental Trachea originating at a Spiracle (Snodgrass).

VENTRAL TRACHEAL COMMISSURE A commissure that crosses below the Ventral Nerve Cord (Snodgrass).

VENTRAL TRACHEAL TRUNK A longitudinal ventral tracheal trunk uniting the series of ventral Tracheae (Snodgrass).

VENTRAL TUBE Collembola: A tube or tubercle proceeding from the ventral side of the first abdominal segments (MacGillivray).

VENTRICLE Noun. (Latin, *ventriculus* diminutive of *venter* = belly.) A chamber of the insect Heart.

VENTRICOSE Adj. (Latin, *venter* = belly; *-osus* = full of.) Pertaining to structure which is distended or inflated in the middle.

VENTRICULAR Adj. (Latin, *ventriculus* ventricle.) Descriptive of or pertaining to a ventricle.

VENTRICULAR GANGLION The stomach ganglion.

VENTRICULAR VALVE A small, internal circular fold or ring of long cells projecting from the posterior margin of the mesenteric epithelium and acting as an occlusor mechanism (Snodgrass). A mechanism of the insect Heart which prevents the backward flow of blood into the Heart itself (Imms).

VENTRICULUS Noun. (Latin, *ventriculus* ventricle.) The true stomach of an insect. See Chylific Ventricle; Midintestine; Midgut.

VENTRILOQUIAL Adj. (Latin, *ventriloquus* = one who speaks from the belly; *-alis* = pertaining to.) Descriptive of insects, such as Orthoptera, with the ability to project sound as if it originates in a place other than the source.

VENTRIMESON Noun. (Latin, *venter* = belly; *meson* = middle.) The imaginary midline along the ventral surface of the body.

VENTRITE Noun. (Latin, *venter* = belly; *-ites* = constituent. Pl., Ventrites.) A ventral, sclerotized part of a body segment. The ventral aspect of a body-ring of an insect (Ferris). See Segmentation. Cf. Sternite. Rel. Sclerotization.

VENTROCEPHALAD Adv. (Latin, *venter* = belly; Greek, *kephale* = head; Latin, *ad* = toward.) Toward the lower side and anteriorly.

VENTRODORSAD Adv. (Latin, *venter* = belly; *dorsum* = back; *ad* = toward.) Extending from belly to back or from the lower to upper surface of a body whose primary orientation is horizontal.

VENTROLATERAL Adj. (Latin, *venter* = belly; *latus* = side; *-alis* = pertaining to.) Toward the venter on the side. Pertaining to the venter and the side of a body.

VENTROVALVULAE Plural Noun. (Latin, *venter* = belly; *valva* = fold.) The ventral valves of the Ovipositor (Crampton).

VENULES Plural Noun. (Latin, *vena* = vein; *-ules* = dim. form.) Branches of the main veins of the insect wing.

VENUS, CARL EDUARD (1816–1889) (Staudinger 1889, Dt. ent. Z. Iris 2: 278–279.)

VERATRIN® See Sabadilla.

VERBENA BUD MOTH *Endothenia hebesana* (Walker) [Lepidoptera: Tortricidae].

VERDIGRIS Noun. (Old French, *vert de Grice* = green of Greece.) 1. A blue-green deposit formed on copper, brass or bronze surfaces. The blue component of verdigris consists of $Cu(C_2H_3O_2)_2.CuO$; the green component of verdigris consists of $2Cu(C_2H_3O_2)_2.CuO$. 2. Greenish colour. 3. In museum collections, the material which accumulates on the body of an insect specimen around the insect pin which holds the specimen. See Green. Cf. Aeneous.

VERHEIJ, CORNELIUS JOHANNES (1917–1966) (Lempke 1966, Ent. Ber., Amst. 26: 213.)

VERHOEFF, KARL WILHELM (1867–1944) (Burr 1946, Entomologist's Rec. J. Var. 58: 157.)

VERITY, ROGER (1883–1959) (Beer 1959, Memorie Soc. ent. ital. 38: 137–148, bibliogr.; Baccetti 1963, Atti Accad. Naz. ital. ent. Rc. 11: 17–33, bibliogr.)

VERLAINE, LOUIS (1889–1939) (Anon. 1939, Bull. Annls. Soc. ent. Belg. 79: 231.)

VERMIAN Worm-like.

VERMICULAR Adj. (Latin, *vermes* = worm.) Worm-like; with tortuous markings resembling the tracks of a worm. Alt. Vermicuate; Vermicultus.

VERMICULE (Latin, diminutive of *vermes* = worm. Pl., Vermcules.) A small worm or grub.

VERMICYTE Noun. (Latin, *vermes* = worm; Greek, *kytos* = hollow. Pl., Vermicytes.) A form of Haematocyte or a variant of a Plasmatocyte. An elongate cell with agranular or slightly granular Cytoplasm.

VERMIFORM Adj. (Latin, *vermes* = worm; *forma* = shape.) Worm-shaped. An elongate, cylindrical body with more-or-less parallel sides and tapered or rounded ends. Descriptive of flea larvae. See Worm. Cf. Eruciform; Flagelliform; Filiform; Fusiform. Rel. Form 2; Shape 2; Outline Shape.

VERMIFORM CELL See Plasmatocyte.

VERMILEONIDAE Plural Noun. Wormlions. A small, widespread Family of brachycerous Diptera consisting of three Genera and about 30 Species. Vermileonidae sometimes considered part of Rhagionidae; now a distinct Family. Adults rounded head; weak anal angle in wing; apical narrowing of fourth posterior and anal cells; slender petiolate Abdomen. Immature habits parallel Myrmeleontidae (ant lions). Adult female deposits eggs beneath surface of sand or dust; larvae form pits. Vermiform larva constructs pit by throwing sand or dust particles from pit with its head; larva lies in bottom of pit awaiting prey (commonly ants). Prey seized and dragged under sand or dust, soft parts consumed, remains thrown from pit. Maturation of larva requires about a year; pupation in soil. Adults feed on nectar. Cf. Rhagionidae.

VERMIPSYLLIDAE Plural Noun. A small Family of Siphonaptera assigned to Superfamily Ceratophylloidea.

VERNAL Adj. (Latin, *vernalis* = spring; *-alis* = belonging to.) Descriptive of biological process or behavioural phenomenon that occurs during spring.

VERNANTIA Trans. Verb. (Latin, *vernalis* = spring.) The moulting or shedding of the skin.

VERRALL, GEORGE HENRY (1848–1911) (Jacobson 1911, Revue Russe Ent. 11:462; Porritt 1911, Entomol. mon. Mag. 47: 262–264; Neave 1933, Centennial history of the Entomological Society of London 1832–1933, xliv + 224 pp. (151–152), London.)

VERRICULATE Adj. (Latin, *verruca* = wart; *atus* = adjectival suffix.) With thick-set tufts of parallel Setae. Alt. Verriculatus.

VERRICULE Adj. (Latin, *verruca* = wart.) A dense tuft of upright Setae. Alt. Verriculus.

VERRILL, ADDISON EMORY (1839–1927) (Osborn 1937, *Fragments of Entomological History,* 394 pp. (172–173), Columbus, Ohio.)

VERRUCA Noun. (Latin, *verruca* = wart. Pl., Verrucae.) Lepidoptera larva: A group of radiating Setae (simple or plumose) on an elevated sclerite. Cf. Chalaza; Pinaculum; Scolus. 2. A wart or wart-like prominences.

VERRUCATE Adj. (Latin, *verruca* = wart; *atus* = adjectival suffix.) Descriptive of surface sculpture, usually the insect's Integument, that is wart-like or covered with rounded protuberances See Sculpture Pattern. Cf. Alveolate; Baculate; Clavate; Echinate; Favose; Gemmate; Psilate; Punctate; Reticulate; Rugulate; Scabrate; Shagreened; Smooth; Striate.

VERSATILE Adj. (Latin, *versatilis* = turning.) Pertaining to structure moving freely in every direction.

VERSICOLOURED Adj. (Latin, *versicolor* = changing colour; *or* = a condition.) Descriptive of structure with several colours; pertaining to changing colours. Alt. Versicolorate; Versicoloratus; Versicolorous.

VERSLUYS, JAN (1873–1939) (Grobben 1940, Alm öst. Akad. Wiss. Wien 1939: 206–208.)

VERSONIAN CELL See Apical Cell (Snodgrass). Syn. Verson's Cell.

VERSONIAN GLANDS The Epidermal Glands of lepidopterous larvae to which the formation of a moulting liquid is ascribed (Snodgrass).

VERTALEC A registered trade name for *Cephalosporium lecanii*, an aphid-specific fungus which is used commercially in Europe to control aphids.

VERTALEX® See *Verticillium lecanii.*

VERTEBRA Noun. (Latin, *vertebra* = turning joint).

VERTEBRAL Adj. (Latin, *vertebra* = turning joint; *-alis* = pertaining to.) Descriptive of vertebra. A term generally used to indicate spots, lines or a different shading of colour, *etc.*, immediately over the part corresponding with the vertebral column in a vertebrate animal (Say).

VERTEBRATA Noun. (Latin, *vertebra* = turning point.) The large category of animals with backbones. Cf. Invertebrata.

VERTEX Noun. (Latin, *vertex* = top. Pl., Vertexes; Vertices.) 1. The top of the head between the eyes, front and occiput for insects with an hypognathous or opisthognathous head. This definition does not apply to prognathous heads because the primary axis of the head has rotated 90° to become parallel to the primary axis of the body. Ocelli are usually in the Vertex. 2. Bees: Part of the head adjacent to and occupied by the Ocelli. 3. Notonectids: 'the imaginary anterior margin of the notocephalon' (Kirkaldy). See Head Capsule. Cf. Crown; Face; Frons; Gena.

VERTEXAL Adj. (Latin, *vertex* = top; *-alis* = pertaining to.) Descriptive of structure occurring on or near the Vertex; directed toward the Vertex.

VERTHION® See Fenitrothion.

VERTICAL BRISTLES Diptera: Two pairs of bristles, an inner and an outer pair, more-or-less behind the upper and inner corners of the eyes, vertical cephalic bristles (Comstock).

VERTICAL CEPHALIC BRISTLES Diptera: See Vertical Bristles.

VERTICAL FURROWS Hymenoptera: The portions of the antennal furrows positioned on the dorsal aspect of the head, extending from near the lateral Ocelli to the caudal aspect of the head, rarely wanting and usually more distinctly marked than the other portions of the antennal furrows.

VERTICAL MARGIN Diptera: The limit between front and occiput.

VERTICAL ORBIT The margin adjacent to the dorsal aspect of a compound eye (MacGillivray).

VERTICAL RESISTANCE Host plant resistance to one or a narrow group of pest genotypes.

VERTICAL TRANSMISSION Medical Entomology: Passage of a pathogen (virus, bacteria, protozoa, *etc.*) between members of an arthropod-host Species (vector) from parent to offspring (Transovarial Transmission), or between stages/instars of one generation (Trans-stadial Transmission). Vertical transmission occurs without movement through a second host Species (arthropod to vertebrate). See Transmission; Transovarial Transmission; Trans-stadial Transmission; Vector. Cf. Circuitous Transmission; Horizontal Transmission.

VERTICAL TRIANGLE Diptera: See Ocellar Triangle (Comstock).

VERTICIL Noun. (Latin, *verticillus* = little whorl. Pl., Verticils.) A whorl of long, fine Setae arranged symmetrically on antennal segments of some insects.

VERTICILLATE Adj. (Latin, *verticillus* = little whorl; *atus* = adjectival suffix.) Whorled; set in whorls. Descriptive of setal arrangements. Alt. Verticillatus.

VERTICILLATE ANTENNA An Antenna in which the joints or segments have rings of long fine Setae, as in Cecidomyiidae.

VERTICILLIN® See *Verticillium lecanii.*

VERTICILLIUM LECANII A fungal pathogen whose spores are used as a biopesticide in the control of aphids and whiteflies in some countries. Not registered in USA. Applied in glasshouses for control of pests on ornamental plants and vegetables. Regarded as non-phytotoxic and not toxic to warm-blooded animals; relatively slow-acting. Trade names include: Mycotal®, Microgermin-F®, Microgermin-G®, Vertalex®, Verticillin®. See Biopesticide. Cf. *Beauveria brongniartii*; *Metarhizium anisopliae* ESF 1®; *Nosema locustae*.

VERTIMEC® See Avermectin.

VESEY-FITZGERALD, LESLIE DESMOND FOSTER (1910–1974) (Lees 1975, Proc. R. ent. Soc. Lond. (C) 39: 57.)

VESICA Noun. (Latin, *vesica* = bladder. Pl., Vesicae.) Lepidoptera: The Penis, or terminal part of the Aedeagus. Vesica is membranous and eversible; typically held within the tubular part of the Aedeagus but everted and inflated during copulation. Preputial Membrane *sensu* Snodgrass. Syn. Penis.

VESICANT Noun. (Latin, *vesica* = blister; *-antem* = an agent of something. Pl., Vesicants.) Any substance which causes blistering in humans. Alt. Vesicating; Vesicatory.

VESICLE Noun. (Latin, *vesicula* = small bladder. Pl., Vesicles.) 1. A small sac, bladder or cyst. 2. An extensible organ producing odours or secretions, as in some beetles and caterpillars.

VESICLE OF PENIS Odonata: A sac with chitinous walls, attached to the Sternum behind the Penis.

VESICULA SEMINALIS In the male reproductive system, an enlarged sac in the Vas Deferens, in which Spermatozoa are stored. Syn. Seminal Vesicle.

VESICULAR STOMATITIS A virulent disease of livestock in the New World caused by any of several arboviruses assigned to the Rhabdoviridae. Transmission to humans by phlebotomine sand flies (*Lutzomyia* spp.) implicated during epizootics. Acronym VS.

VESICULAR Adj. (Latin, *vesicula* = small bladder.) Pertaining to or consisting of vesicles or small sacs or bladders. Alt. Vesiculous.

VESICULATE Adj. (Latin, *vesicula* = small bladder; *atus* = adjectival suffix.) 1. Bladder-like or with a swollen appearance. 2. Hymenoptera: Apocritous larva with the proctodaeum everted. A condition noted in some endoparasitic larval Braconidae and some Ichneumonidae larvae displaying short, caudal appendages with vesicles at the bases. Cf. Caudate Larva.

VESPIDAE Plural Noun. Hornets; Potter Wasps; Paper Wasps; Yellow Jackets. Cosmopolitan, moderate sized Family (ca 800 Species) of aculeate Hymenoptera assigned to Superfamily Vespoidea. Female Antenna with 12 segments; male Antenna with 13 segments; medial margin of compound eye emarginate (notched); Mandibles not elongate but meeting apicad; Pronotum extending to Tegula; middle Coxae juxtaposed; middle Tibia with two spurs; tarsal claws simple; wings folded longitudinally at repose; forewing Discoidal Cell elongate; hindwing with closed cells, with or without Jugal Lobe; gastral Tergum I and Sternum I partly fused, Sterna I and II separated by a constriction. Eusocial wasps which construct nests of various plant materials (carton) in places of protection or concealment. Queens and males seasonally produced and develop in special cells. Nest not long-lived, composed of many cells and established by one or more inseminated queens (foundress). Oviposition largely confined to queen with subordinate queens or first brood female workers enlarging nest and provisioning cells; one egg per cell. Adult carnivorous, frugiverous, melliferous; larva carnivorous. Subfamilies include: Masarinae, Polistinae, Vespinae; each Subfamily considered a Family in some classifications. Masarinae solitary and provision cells with pollen and nectar; Polistinae (paperwasps) nest typically consists of a single comb not enclosed within a paper carton. Prey for Polistinae usually caterpillars; sting inflicts pain and potentially lethal.

VESPIFORM THRIPS *Franklinothrips vespiformis* (D. L. Crawford) [Thysanoptera: Aeolothripidae].

VESPOIDEA Laicharting 1781. True Wasps. A Superfamily of aculeate Hymenoptera. Medium to large sized wasps; characters vary with interpretation of composition. Vespoidea classifications: *Sensu lato* Bradynobaenidae, Eumenidae, Formicidae, Mutillidae, Pompillidae, Rhopalosomatidae, Sapygidae, Scoliidae, Sierolomorphidae, Tiphiidae, Vespidae; *Sensu Stricto* Eumenidae, Masaridae, Vespidae. See Aculeata; Hymenoptera. Cf. Apoidea; Chrysidoidea; Formicoidea; Sphecoidea.

VESTIBULE Noun. (Latin, *vestibulum* = porch. Pl., Vestibules.) The space surrounding the Ovipositor formed by the projecting margins of the surrounding abdominal segments. See Vestibulum.

VESTIBULUM Noun. (Latin, *vestibulum* = porch. Pl., Vestibula.) An external genital cavity formed above the seventh abdominal Sternum when the latter extends beyond the eighth (Snodgrass).

VESTIGE Noun. (Latin, *vestigum* = trace. Pl., Vestiges.) A small, reduced or degenerate organ, part or structure. Vestige identifed by comparison with similar organ, part or structure in other Taxa or organisms. Vestige presumably represents the remains of a previously functional organ, part or structure in the process of disappearing. Vestige noncommittal with regard to Homology. See Homology. Cf. Rudimentary

VESTIGIAL Adj. (Latin, *vestigum* = trace; *-alis* = characterized by.) Descriptive of structure or process that is a vestige; small or degenerate.

VESTITURE Noun. (Latin, *vestitus* = garment.) A surface covering. Insects: Typically a surface covering of scales or Setae on the body, wing or other appendage. Alt. Vesture.

VETARON® See Methamidophos.

VETCH BRUCHID *Bruchus brachialis* Fåhraeus

[Coleoptera: Bruchidae].

VETERINARY ENTOMOLOGY The subdiscipline of Entomology concerned with the study or control of insects affecting animals other than humans. See Entomology. Cf. Agricultural Entomology; Applied Entomology; Medical Entomology.

VETH, HUIBERT JOHANNES (1846–1917) (Everts 1917, Tijdschr. Ent. 60: 271–274.)

VETRAZINE® See Cyromazine.

VEXANS MOSQUITO *Aedes vexans* (Meigen) [Diptera: Culicidae].

VEXILLATE Adj. (Latin, *vexillum* = standard; *atus* = adjectival suffix.) Pertaining to or bearing a Vexillum.

VEXILLUM Noun. (Latin, *vexillum* = standard. Pl., Vexilla.) 1. A banner or standard. 2. Siphonaptera: A lateral, apicodorsal on Sternum VIII. 3. Hymenoptera: An expansion of the apex of Tarsi in some fossorial groups.

VEXTER® See Chlorpyrifos.

VIABILITY (French, *viable* = likely to live.) Tissue Culture: A measure of the proportion of living cells in a population.

VIALE, EMILIO (1921–1959) (Pierce 1960, J. Econ. Ent. 53: 485.)

VIALLANES, HENRI (1856–1893) (Bouvier 1893, Annls. Sci. nat. (Zool.) (7) 15: 353–397, bibliogr.)

VIBERT, LEON (1863–1914) (Alluaud 1914, Bull. Soc. ent. Fr. 1914: 435.)

VIBRANT Adj. (Latin, *vibranus* = vibrating; *-antem* = adjectival suffix.) Having a rapid oscillatory motion.

VIBRATILE Adj. (Latin, *vibrare* = to quiver.) Pertaining to structure involved in vibratory motion.

VIBRISSA Noun. (Latin, *vibrissa* = nasal hair. Pl., Vibrissae.) Curved bristles or Setae in some Diptera, positioned between the Mystax and the Antennae. Whiskers; in some older writings the Mystax proper.

VIBRISSAL ANGLES. Diptera: The more or less rounded angles formed by the facial ridges just above the oral margin (Curran).

VIBRISSAL RIDGES Diptera: Two ridges, one on each side, inside the arms of the Frontal Suture. Syn. Facilia; Facial Ridges (Comstock).

VIBURNUM APHID *Aphis viburniphila* Patch [Hemiptera: Aphididae].

VIBURNUM LEAF BEETLE *Pyrrhalta humeralis* (Chen) [Coleoptera: Chrysomelidae].

VIBURNUM WHITEFLY *Aleurotrachelus jelinekii* (Frauenfeld) [Hemiptera: Aleyrodidae].

VICARIANCE BIOGEOGRAPHY In biogeography analysis, an explanation for the current distribution of extant Taxa as a consequence of ancient distributions which have been distrupted due to response to changes in the environment or geology. See Panbiogeography. Cf. Dispersal Biogeography; Cladistic Vicariance Biogeography.

VICEROY *Basilarchia archippus* (Cramer) [Lepidoptera: Nymphalidae].

VICKERY, ROY ALBION (1884–1938) (Snapp 1938, J. Econ. Ent. 31: 637.)

VICTENON® See Bensultap.

VICTOR, T See Motschulsky, T. V.

VICTOR® See Isazofos.

VICTORIAN PREDATOR MITE *Amblyseius victoriensis* (Womersley) [Acari: Phytoseiidae] (Australia).

VICTORY® A registered biopesticide derived from *Bacillus thuringiensis* var. *kurstaki*. See *Bacillus thuringiensis*.

VIDA, MARCUS HIERONYMUS (MARCO GIVOLAMO) (1490–1566) (Rose 1850, *New General Biographical Dictionary* 12: 363–364.)

VIDGHAL' IGHNATII MARTINOVICH (1835–1903) (Shugarov 1903, Revue Russe Ent. 3: 421; Ghol'd 1966, Ent. Obozr. 45: 457–460, bibliogr. Translation: Ent. Rev., Wash. 45: 249–251, bibliogr.)

VIEHMEYER, HUGO (1868–1921) (Heller 1921, Isis Budissino. 1921: xviii–xx, bibliogr.)

VIEILLARD, M R (–1960) (Anon. 1960, Bull. Soc. ent. Fr. 65: 255.)

VIERECK, HENRY LORENZ (1881–1931) (Davis 1932, Ann. ent. Soc. Am. 25: 251; Rehn 1932, Ent. News 43: 141–148; Wheeler & Valley 1975, History of Entomology in the Pennsylvania Department of Agriculture. 37 pp. (15), Harrisburg.)

VIERICK, H L (1881–) American specializing in Hymenoptera.

VIERTL, ADALBERT (1831–1900) (Aigner 1900, Rovart. Lap. 7: 11, 112–113.)

VIGANO, ANTONIO (1926–1972) (Anon. 1972, Boll. Soc. ent. ital. 104: 133.)

VIGELIUS, LUDWIG CHRISTIAN (1797–1857) (Thomä 1857, Jb. Ver. Naturk. Wiesbaden 12: 424–437.)

VIGILANTE® See Diflubenzuron.

VIGORS, NICHOLAS AYLWARD (1787–1840) (Wilmot 1841, Proc. Linn. Soc. Lond. 1: 106–107; Rose 1850, *New General Biographical Dictionary* 12: 366.)

VIKANE® See Sulfuryl Fluoride.

VIKTOROV, GEORGII ALEKSANDROVICH (1925–1974) (Mazokhin-Porschnyakov 1975, Ent. Obozr. 54: 237–242, bibliogr.)

VILLA, ANTONIO (–1885) (Stoppani 1885, Atti Soc. ital. sci. nat. Milano 28: 138–141; Dimmock 1888, Psyche 5: 36.)

VILLENEUVE DE JANTI, JOSEPH (1868–1944) (Mesnil 1946, Bull. Soc. sci. nat. Méd. Seine-et-Oise (3) 8: 25–30, partial bibliogr.)

VILLERS, CHARLES DE French naturalist and author of C. *Linnaei Entomologia* (4 vols, 1789).

VILLERS, CHARLES JOSEPH DE (1724–1810) (Eiselt 1836, *Geschichte, Systematik und Literatur der Insektenkunde,* 255 pp. (53), Leipzig; Mulsant 1840, Annls. Soc. agric. Lyon 3: 243–253, bibliogr.)

VILLIERS, FRANCOIS DE (1790–1847) (Guénée 1847, Annls. Soc. ent. Fr. (2) 5: 619–624.)

VILLIFORM Adj. (Latin, *villus* = shaggy hair; *forma* = shape.) Villus-shaped; structure with the appearance of velvet or a pile of velvet; pertaining to a surface with soft, fine hair. See Villus. Cf. Pile. Rel. Form 2; Shape 2; Outline Shape.

VILLOSE Adj. (Latin, *villus* = shaggy hair; *-osus* = full of.) 1. Shaggy; covered with long Setae which give a woolly appearance. 2. Covered with soft, flexible thick-set Setae (Kirby & Spence). Alt. Villosate; Villosus; Villous. Cf. Setose.

VILLUS Noun. (Latin, *villus* = tuft of hair; shaggy hair. Pl., Villi.) 1. Minute projections or finger-like process which give the appearance of velvet when closely spaced. 2. A fine, small seta. 3. A short, hair-like or papillate process on the surface of certain absorbent structures and sensory organs.

VIMMER, ANTON (1864–1941) (Maron 1946, Vest. csl. zool. Spol. 10: 11–19, bibliogr.)

VINAL, STUART CUNNINGHAM (1895–1918) (Anon. 1918, J. Econ. Ent. 11: 437, bibliogr.)

VINBLASTINE Noun. A plant-derived alkaloid compound used in chemotherapy for some forms of cancer. The compound has an affinity for actin protein molecules and experimentally alters microtubule structure.

VINCULATE Adj. (Latin, *vinculatus* = chained; *atus* = adjectival suffix.) Cord-like; chain-like.

VINCULUM Noun. (Latin, *vinculum* = a bond, a cord. Pl., Vincula.) Lepidoptera: The coxosternal sclerite of the ninth abdominal segment (Snodgrass). Genitalia of male Lepidoptera: a U-shaped sclerite derived from the ninth abdominal Sternum, its arms articulating dorsally with the tegumen and the Harpes with its caudal margin (Klots). Syn. Subgenital Plate.

VINE HAWK MOTH *Theretra oldenlandiae* (Fabricius) [Lepidoptera: Sphingidae] (Australia).

VINE SCALE *Eulecanium persicae* (Fabricius) [Coccidae: Homoptera]: A minor pest of grape and plum in Australia.

VINE WEEVIL *Orthorhinus klugi* Boheman [Coleoptera: Curculionidae] (Australia).

VINE, ARTHUR CHARLES (1844–1917) (Adkin 1917, Entomologist 50: 240.)

VINEGAR FLIES Any of several Species of Drosophilidae. Annoying in large numbers or a threat to human health as mechanical vectors of enteric diseases. An urban pest around produce and food-handling facilities, canneries, breweries and wineries. Adult 2.5–4.0 mm long, brownish yellow to dark brown in colour; wings folded at repose. Adults typically crepuscular, attracted to fruit, vegetables and fermenting material, lay 25–35 eggs per day; larvae feed in moist, overripe fruit and decaying plant material; pupation in dry habitat. Life cycle 8–12 days. See Drosophilidae.

VINEGAROON *Mastigoproctus giganteus* (Lucas) [Arachnida: Pedipalpida (Uropgyi)]: A large bodied whip scorpion occurring in the southern and southwestern USA. Nocturnal and predaceous upon arthropods; defend themselves via forceful ejection of acetic acid and caprylic acid from the 'tail.'

VINEYARD SNAIL *Cernuella virgata* (Da Costa) [Sigmurethra: Helicidae] (Australia).

VINOGRADOV, STEPHANOVIC (1891–1959) (Stroganova 1959, Zool. Listy 8: 289–290.)

VINOGRADOV-NIKITIN, PAVEL ZACHAROVITSCH (1869–1938) (Anon. 1939. Arb. morph. taxon. Ent. Berl. 6: 69.)

VINOKUROV, GRIGORIY MAKOROVICH (1886–1956) (Shumakova 1958, Ent. Obozr. 37: 947–949, bibliogr.)

VINOUS Adj. (Latin, *vinum* = wine; *-osus* = possessing the qualities of.) Wine-coloured; a deep, transparent red-brown.

VINSON, LUCIEN JEAN (1906–1966) (Mamet 1967, Bull. Maurit. Inst. 4: 375–391, bibliogr.)

VINYLPHATE® See Chlorfenvinphos.

VINZANT, JOHN PAUL (1907–1962) (Powell 1963, J. Econ. Ent. 56: 424.)

VIOLACEOUS Adj. (Latin, *viola* = violet; *-aceus* = of or pertaining to.) Violet-colour. Alt. Violaceus.

VIOLET Adj. (Latin, *viola* = violet.) A purplish-red, the high vibration colour of the solar spectrum.

VIOLET APHID *Micromyzus violae* (Pergande) [Hemiptera: Aphididae].

VIOLET SAWFLY *Ametastegia pallipes* (Spinola) [Hymenoptera: Tenthridinidae].

VIRGA Noun. (Latin, *virga* = rod.) A sclerotized terminal rod usually arising from the wall of the endophallus or Ejaculatory Duct. Found in male Dermaptera.

VIRGILIUS MARO, PUBLIUS (70BC-19BC) (Rose 1850, *New General Biographical Dictionary* 12: 374–375; Pellett 1929, Am. Bee J. 69: 238–240.)

VIRGIN Noun. (Latin, *virgo* = young woman. Pl., Virgins.) 1. An individual who has not engaged in sexual intercourse. 2. A female who has not copulated. 3. Habitat that has not been disturbed by human activities. Rel. Ecclogy; Reproduction.

VIRGINIA CREEPER LEAFHOPPER *Erythroneura ziczac* Walsh [Hemiptera: Cicadellidae].

VIRGINIA CREEPER SPHINX *Darapsa myron* (Cramer) [Lepidoptera: Sphingidae].

VIRGINIA PINE SAWFLY *Neodiprion pratti pratti* (Dyar) [Hymenoptera: Diprionidae].

VIRGULA Noun. (Latin, diminutive of *virga* = rod.) 1. A small rod. 2. An oblique mark (symbol) used in printing.

VIRIDIS Adj. (Latin, *viridis* = green.) Green as verdigris.

VIRIESCENT Adj. (Latin, *viridis* = green.) A greenish colour or becoming green. Alt. Viridescent.

VIRIN KS® See *Mamestra brassicae* NPV.

VIROID Noun. (Latin, *virus* = poison; Greek, *eidos* = type. Pl., Viroids.) A minute virus-like organism which lacks the protein coat of a virus. Smaller than smallest virus; nucleic acid circular. Viroids cause diseases including Avocado Sun Blotch and Citrus Exocortis. See Virus. Cf. Bacteroid; Parasitoid. Rel. Eidos; Form; Shape.

VIROLOGY Noun. (Latin, *virus* = poison > Greek, *ios; logos* = discourse. Pl., Virologies.) The study of viruses. See Pathogen. Cf. Bacteriology. Rel.

Entomology; Parasitology.

VIROX® See European-Pine Sawfly NPV.

VIRTUSS® See Douglas-Fir Tussock Moth NPV.

VIRUS Noun. (Latin, *virus* = poison. Pl., Viruses.) An intracellular, obligate parasite that is visible with electron microscopy but not visible with light microscopy. A virus consists of infectious nucleic acid that is surrounded by a protective protein coat. Viruses are pathogenic in plants and animals, and multiply naturally within living cells. Some viruses are transmitted from a diseased organism to an unaffected organism by organisms called vectors, such as plant sap-sucking insects (aphids, leafhoppers) or vertebrate blood-feeding insects (mosquitoes). See Pathogen. Cf. Bacterium; Rickettsia.

VIS FORMATRIX The creative or formative force (Smith).

VISCERA Plural Noun. (Latin, *viscera* = bowels; Sl., Viscus.) The internal organs of the body. Syn. Intestines; Guts.

VISCERAL Adj. (Latin, *viscera* = bowels; *-alis* = pertaining to.) Descriptive of structure or process associated with the Viscera.

VISCERAL FAT BODY See Fat Body. Cf. Peripheral Fat Body.

VISCERAL LEISHMANIASIS A widespread form of Human Leishmaniasis which causes weight loss, hepatosplenomegaly and anaemia. VL can cause death in untreated cases. VL caused by protozoan parasites *(Leishmania)* and vectored by blood-sucking phlebotomine sandflies. Protozoans undergo a flagellate stage in the gut of sandflies. Disease anthroponotic in India and Africa; zoonotic in Palaearctic and Neotropical regions. Syn. Kala-azar. See Leishmaniasis. Cf. Cutaneous Leishmaniasis; Mucocutaneous Leishmaniasis.

VISCERAL NERVOUS SYSTEM See Stomodaeal Nervous System.

VISCERAL SEGMENTS See Pregenital Segments.

VISCERAL SINUS The large central cavity between the dorsal and ventral sinuses, which contains the principal internal organs.

VISCERAL TRACHEA The median segmental Trachea originating at a Spiracle, branching to the Alimentary Canal, the fat tissue, and the reproductive organs (Snodgrass).

VISCERAL TRACHEAL TRUNK A longitudinal tracheal trunk closely associated with the walls of the Alimentary Canal (Snodgrass).

VISCID Adj. (Latin, *viscum* = mistletoe.) Sticky. Pertaining to structure covered with a shiny, resinous or greasy substance. Jelly-like. Alt. Gelatinous.

VISCOUS Adj. (Latin, *viscum* = mistletoe; *-osus* = possessing the qualities of.) Descriptive of substance that is thick, sticky or semi-fluid.

VISION Noun. (Middle English, *visioun* > Latin, *visio* > *visum* = to see. Pl., Visions.) The sense by which colour and light in the environment are interpreted. Rel. Compound Eye; Ocellus.

VISITING ANT See Army Ant.

VISUAL Adj. (Middle English, *visioun*; Latin, *visio*, *visum* = to see; *-alis* = pertaining to.) Descriptive of or pertaining to vision (sight).

VISUAL CELL Insect eye: A nerve cell in direct connection with the termination of a fibre of the ocellar nerve, which grouped with others forms the Retina (Imms).

VISUAL ORGAN The lens of the eye, the crystalline humour of the eye (J. E. V. Boas).

VISUAL SENSE Sight, perception of light waves through appropriate organs.

VITAL® See Acephate.

VITALE, FRANCESCO (1861–1953) (Conci 1953, Memorie Soc. ent. ital. 32: 60–61.)

VITAMIN Noun. (Latin, *vita* = life; *ammoniacum* = resinous gum. Pl., Vitamins.) Any of several organic substances found in food which are necessary in trace-level quantities for body-growth and development. Typically of unknown composition, but probably nitrogenous.

VITELLARIUM Adj. (Latin, *vitellus* = yolk; *arium* = place of a thing.) The part of the Ovariole that contains the developing eggs; the Zone of Growth. Vitellarium forms a large part of the Ovariole in which the hormone-dependent process of Vitellogenesis occurs. Blood proteins are transferred directly to the developing yolk in the Oocytes. Several blood proteins are synthesized by insects, the site of any one not being definite. Vitellogenic growth of the Oocyte involves sequestering the yolk precursor Vitellogenin from the Haemolymph. The sequestering process is called Micropinocytosis. Ultrastructurally, the process involves the appearance of specialized vesicles in the cortical Ooplasm and coated pits with the development of wide extracellular spaces in the follicular Epithelium and reduction of Gap Junctions. See Ovary; Ovariole. Cf. Germarium; Pedicel; Terminal Filament. Rel. Micropinocytosis; Vitellogenesis.

VITELLIN Noun. (Latin, *vitellus* = yolk. Pl., Vitellins.) The major nutritive yolk proteins; phosphoroprotein of egg yolk. Vitellinis are female-specific Vitelloginins that have been transported into the Oocyte and stored in a crystalline form. See Vitellogenesis. Cf. Vitellogenin; Yolk Protein. Rel. Oogenesis.

VITELLINE Adj. (Latin, *vitellus* = yolk.) Pertaining to yolk or yolk producing tissue. Adj. Vitellinus.

VITELLINE BODIES See Vitelline Membrane.

VITELLINE MEMBRANE The delicate homogeneous tissue surrounding the yolk of an insect egg which forms the innermost layer of the eggshell. The Vitelline Membrane surrounds the Oocyte and is a product of the follicular-cell Epithelium. The VM is composed of protein, lipids and carbohydrates. The VM is formed within the Follicle Cell by Rough Endoplasmic Reticulum, condensed by Dictyosomes and stored within secretory granules. Developmental precursors of the VM sometimes are called Vitelline Bodies

During development of the eggshell, Vitelline Bodies migrate over the Plasma Membrane and fuse. Ultrastructurally, the VM has been studied in a few insects. For instance, the VM of the gyrinid beetle *Dineutes horni* has an electron-dense upper surface and spherical, vesicle-like structures dispersed throughout the membrane. The VM is 300–500 nm thick and granular but without clearly defined substructure in *Drosophila melanogaster*. In Hymenoptera, the parasitic wasp *Nasonia vitripennis* VM contains protein and carbohydrate. The function of the VM is not clearly established for any Species; presumably, the VM provides elasticity to the eggshell which is important during oviposition and periods of mechanical stress. See Egg Shell. Cf. Chorion; Wax Layer.

VITELLOGENESIS Noun. (Latin, *vitellus* = yolk; Greek, *genesis* = descent. Pl., Vitellogeneses.) The physiological process by which yolk is accumulated around a Primary Oocyte for subsequent development of the embryo. Vitellogenesis is an heterosynthetic process in which precursors of yolk proteins are produced by Fat Body. Autosynthesis of yolk proteins has been suggested, but not confirmed, in a few Species of Apterygota (Collembola, Protura).

VITELLOGENIN Noun. (Latin, *vitellus* = yolk; *gignere* = to produce. Pl., Vitellogenins.) A major category of female-specific, yolk-transport proteins. Vitellogenins are large, oligomeric glycolipophosphoproteins synthesized in Fat Body, transported into the Ovariole and selectively sequested by developing Oocytes. See Embryo. Cf. Lipophorins; Vitellins; Yolk Proteins.

VITELLOGENOUS Adj. (Latin, *vitellus* = yolk; *gignere* = to produce; *-osus* = possessing the qualities of.) Pertaining to certain cells in the Ovaries.

VITELLOPHAGE Noun. (Latin, *vitellus* = yolk; Greek, *phagein* = to eat. Pl., Vitellophages.) Embryology: An extraembryonic cell type that serves as an intermediary between the embryo and its yolk. The Vitellophage invades and actively digests yolk. Endoderm cell proliferates into the yolk of the insect egg, which partly digests it. Syn. Vitellophag.

VITELLUS Adj. (Latin, *vitellus* = yolk.) The yolk of the insect egg.

VITEX® See Dimethoate.

VITRELLA Adj. (Latin, *vitrum* = glass.) The vitreous body of the insect eye (Packard).

VITREOUS Adj. (Latin, *vitreus* = glassy; *-osus* = possessing the qualities of.) Descriptive of structures which appear glass-like or transparent. Alt. Vitreus.

VITREOUS BODY 1. The elongated and grouped cells of the corneagen layer which supplement the Lens of the eye in its function (Imms). 2. The Crystalline Cone *sensu* Snodgrass. See Crystalline Cone.

VITREOUS LAYER See Lentige layer.

VITTA Noun. (Latin, *vitta* = band. Pl., Vittae.) A broad, longitudinal stripe (Kirby & Spence).

VITTA FRONTALIS Frontal stripe.

VITTATE Adj. (Latin, *vitta* = band; *-atus* = characterized by.) Striped. Alt. Vittatus.

VITZTHUM VON ECHSTAEDT, HERMANN LUDWIG WILHELM (1876–1942) (Hase 1942, Parasitenk. 12: 501–506.)

VIVIAN, HENRY WYNDHAM (1868–1901) (Richardson 1902, Entomologist 's mon. Mag. 38: 13.)

VIVIPARITY Noun. (Latin, *vivus* = living; *parere* = to beget.) A method of reproduction in which larvae or nymphs are extruded from the female's body. Viviparity occurs in many groups of insects and varies in complexity. Offspring obtain nutrients directly from 'milk glands' within the mother. Egg contains little or no yolk, lacks a Chorion, develops within Ovariole and Follicle Cells supply some nutrients. Similar to Ovoviviparity in that there is a reduction in number of offspring produced per female. See: Adenotrophic Viviparity; Haemocoelic Viviparity; Pseudoplacental Viviparity. Cf. Ovoviviparity; Oviparity; Parthenogenesis; Paedogenesis; Polyembryony.

VIVIPAROUS Adj. (Latin, *vivus* = living; *parere* = to beget; *-osus* = possessing the qualities of.) Pertaining to organisms which bear living young; opposed to organisms which lay eggs. See Reproduction. Cf. Oviparous; Ovoviparous.

VLACH, VILEM (1875–1959) (Anon. 1960, Cas. csl. Spol. Ent. 57: 93, bibliogr.)

VLACOVICH, GIAMAOLO PADOUE (1825–1899) (Anon. 1899, Arch. ital. Biol. 31: 485–488, bibliogr.)

VOCAL CORDS Diptera: Organs on thoracic spiracles which produce a humming sound.

VODOZ, GEORGES (1873–1903) (Ste Claire Deville 1903, Riv. coleott. ital. 1: 240–242, bibliogr.)

VOET, J E Author of *Catalogue raisonne ou systematique du genre des Insectes qu'on appelle Coleoptrees* (1766).

VOGEL, F (1860–) (Anon. 1915, Ent. News 26: 240.)

VOGEL, HERMANN KARL (1842–1907) (Horn 1907, Dt. ent. Z. 1907: 591.)

VOGEL, RICHARD (1881–1955) (Pflugfelder. 1955, Anz. Schädlingsk. 28: 28.)

VOGELSANG, ENRIQUE GUILLERMO (–1969) (Anon. 1969, Mitt. dt. ent. Ges. 28: 50.)

VOGT, ALFRED (1879–1943) (Fisher 1944, Vjschr. naturf. Ges. Zürich 89: 66–67.)

VOGT, CARL CHRISTOPH (1817–1895) (Taschenburg 1902, Leopoldina 56: 10–12, 18–24, 51–54, 57–62, 73–74.)

VOGT, CHARLES J (1885–1964) (Willing 1964, J. Lepid. Soc. 18: 239–240.)

VOGT, OSKAR (–1959) (Kruseman 1959, Ent. Ber., Amst. 19: 170.)

VOID Trans. Verb. (Old French, *voider*.) To excrete waste products.

VOIGT, FRIEDRICH SIEGMUND (1781–1850) (Ratzeburg 1874, Forstwissenschaftliches Schriftsteller-Lexicon 1: 86–87 (footnote).)

VOIGT, WALTER (1856–1928) (Anon. 1940, Natur

Niederheim 16: 22–28.)

VOIGTS, HANS (–1905) (Anon. 1905, Insektenbörse 22: 62.)

VOLANT Adj. (Latin, *volatus* = a flight.) Flying or capable of flight.

VOLATILE Adj. Readily vaporizable.

VOLCK OILS® See Petroleum Oils.

VOLCK, WILLIAM HUNTER (1879–1943) (Essig 1943, J. Econ. Ent. 36: 484–486.)

VÖLKEL, HERMANN (1888–1955) (Hase 1955, Anz. Schädlingsk. 28: 75.)

VOLKONSKY, MICHEL (1907–1942) (Anon. 1942, Archs Inst. Pasteur, Alger 20: 292.)

VOLLENHOVEN S C SNELLEN VAN See Snellen van Vollenhoven, S.C.

VOLLMER, MAX (–1965) (Anon. 1965, Mitt. dt. ent. Ges. 24: 79.)

VOLLRATH, GEORG (1895–1975) (Rössler 1976, Atalanta, München 7: 1–2, bibliogr.)

VOLPHOR® See Phorate; Terbufos.

VOLSELLA Noun. (Latin, *volsella* = forceps. Pl., Volsellae.) Hymenoptera: Either of the median pair of genital appendages; Parameres of the Phallobase.

VOLSELLAR BRIDGE Hymenoptera: A sclerotized bridge between the Volsellae.

VOLTAGE® See Pyraclofos.

VOLTAIRE, FRANCOIS MARIE AROUET (1694–1778) (Rose 1850, *New General Biographical Dictionary* 12: 382–384.)

VOLTINISM Noun. (Italian, *volta* = time; English, *-ism* = condition.) Polymorphism of diapause. See Bivoltine; Multivoltine; Univoltine.

VOLXEM, CAMILLE VAN (1848–1875) (Putzeys 1875, C.r. Soc. ent. Belg. 18: ciii–cvi.)

VOM RATH'S ORGAN See Organ of vom Rath.

VOMIT DROP A drop of regurgitated substance that appears at the end of the Proboscis of flies (Matheson).

VORBRINGER, GUSTAV (1846–1910) (Dampf 1910, Dt. ent. NatlBl. 1: 72; Kuhnt 1910, Dt. ent. Z. 1910: 716–717, bibliogr.)

VORBRODT, KARL (1864–1932) (Anon. 1932, Insektenbörse 49: 153.)

VORHIES, CHARLES TAYLOR (1879–1949) (Wehrle 1950, J. Econ. Ent. 43: 573.)

VORIS, RALPH (1902–1940) (Anon. 1940, Ent. News 51: 210; Kinsey 1941, Ann. ent. Soc. Am. 34: 263–264.)

VORMANN, BERNARD (1843–1902) (Landois 1902, Jber. westf: ProvVer Wiss. 30: 35–37.)

VORTEX FEEDING A feeding strategy adapted by aquatic and terrestrial insects. Some Ephemeroptera naiads vortex-feed from pits in the sand as an unusual method of suspension feeding for stream invertebrates. Suspension feeding relies upon hydrodynamic properties of vortices and pits in the sediment to facilitate capture of fine particles from flowing water. Modified Setae on the forelegs function as aerosol filters and initiate formation of a vortex that spirals within a specially constructed pit. The vortex concentrates seston (= all bodies floating or swimming in water) and transports it to a filtering device located below the surface of the sediments. See Feeding Strategy.

VORTIS SAMPLER A light-weight, portable, suction sampling-system designed to extract insects from debris. Improved features over previous designs include: The insects are deposited directly into a detachable transparent collecting vessel; no nets, bags or filters; a wide range of sizes and shapes of collecting vessels may be used; insects are unharmed by collecting; collected material does not impede air flow. General Specifications: Air flow: 10.5 m^3/min. (370 cfm); weight including engine: 7.8 kg (17.2 lb); overall height: 930 mm (36 inches); collecting area: 0.2 m^2 (2.2 ft^2).

VOSLER, EVERETT JAY (1890–1919) Entomologist and an early American foreign explorer who obtained *Metaphycus lounsburyi* (Howard) from Australia which was imported into California for biological control of black scale *Saissetia oleae*. Died during influenza epidemic (Smith 1918, J. Econ. Ent. 11: 485–486; Smith 1918, Mon. Bull. Calif. Common Hort. 7.)

VOSNESENSKY, ILYA GARRILOVICH (1816–1871) (Essig 1931, *History of Entomology*. 1029 pp. (777–789), New York.)

VOSSELER, JULIUS (1861–1933) (Lindner 1934, Jh. Ver. vaterl. Naturk. Württ. 90: xl–xlv, bibliogr.)

VOUAUX, L (–1914) (Anon. 1915, Miscnea ent. 22(11): 60; Berland 1920, Ann. Soc. ent. Fr. 89: 432–434.)

VPM® See Metham-Sodium.

VROLIK, GERARDUS (1775–1859) (Van der Hoeven 1859, Jaarb. K. ned. Akad. 1859: 116–134, bibliogr.)

VRYDAGH, M J M (–1962) (Anon. 1962, Bull. Soc. ent. Fr. 67: 93.)

VRZAL, ANTON (1892–1935) (Roubal 1935, Cas. csl. Spol. ent. 32: 146.)

V-SHAPED NOTAL RIDGE The endoskeletal ridge of the Mesonotum or Metanotum, its arms divergent posteriorly and marked externally by the Scutoscutellar Suture (Snodgrass).

VS See Vesicular Stomatitis.

VUILLET, ANDRÉ (1883–1914) (Anon. 1915, Science 41: 91; Marchal 1915, Annls. Epiphyt. 6: 1–4, bibliogr.; Berland 1920, Annls. Soc. ent. Fr. 89: 434–436.)

VULGAR Adj. (Latin, *vulgaris* = common, usual.) Common; not conspicuous; obscure in appearance and abundant in number (Smith).

VULTUS Noun. (Etymology obscure.) Face; part of head below the Front and between the compound eyes.

VULVA Noun. (Latin, *vulva* = womb, covering, Integument. Pl., Vulvae.) 1. Vertebrates: The external parts of female genital organs. 2. Insects: The orifice of the female's Vagina. 3. Ticks: The female genitalia *sensu* Matheson.

VULVAR LAMINA Odonata: A spine on the venter of the Abdomen immediately anterior of the Ovipositor (Garman).

VYDATE® See Oxamyl.

WACHANRU, MARIE ROSE GAUDEMARD (1821–1853) (Mulsant 1853, Opusc. ent. 2: 145–154.)
WACHSMANN, FRANZ (FERENCZ) (1837–1911) (Cziko 1911, Rovart, Lap. 18: 81–84.)
WACHTL, FRIEDRICH A (1840–1913) (Bohmerle 1913, Verh. forst. Mähren Schlesien 1913: 135–141; Reitter 1913, Ent. Bl. Biol. Syst. Käfer 9: 201–203, bibliogr.; Reitter 1913, Wien. ent. Ztg. 32: 187–189 bibliogr.)
WADE, JOSEPH SANFORD (1880–1961) (Leonard & Larrimer 1961, Proc. ent. Soc. Wash. 63: 219–222.)
WADLEY, FRANCIS MARION (1892–1969) (App 1970, Proc. ent. Soc. Wash. 72: 270–271.)
WAFA, ABDEL-KHALEK KHALIL (1912–1970) (Anon. 1971, Bull Soc. ent. Egypte 54: [iii]; Butler 1973, Proc. R. ent. Soc. Lond. (C) 37: 57.)
WAGA, ANTON (1799–1890) (Larousse 1876, Grande dictionnaire universal du XIX siècle 15: 1243; Anon. 1890, Nature 43: 131; Mabille 1890, Bull. Soc. ent. Fr. (6) 10: cxi.)
WAGENER, GUIDO RICHARD (1822–1896) (Anon. l896, Leopoldina 32: 59; Gurlt 1897, Arch. path. anat. 148: 180.)
WAGNER PARSIMONY (Kluge & Farris 1969, Syst. Zool. 18: 1–32.) See Parsimony.
WAGNER, ANDREAS CHRISTIAN WILHELM (1866–1942) (Hering 1942, Mitt. dt. ent. Ges. 11: 31; Kröber 1947, Verh. Ver. naturw. Heimatforsch. 29: x-xiv, bibliogr.)
WAGNER, ARNO (–1919) (Astfäller 1920, Int. ent. Z. 14: 129–130.)
WAGNER, FRITZ (1873–1938) (Anon. 1938, Arb. morph. taxon. Ent. Berl. 5: 296; Lindner 1938, Konowia 17: 1–4.)
WAGNER, HANS (1884–1949) (Skoraszewsky 1950, Ent. Bl. Biol. Syst. Käfer 45–46: 155–159.)
WAGNER, NICOLI PETROWITSCH (1829–1907) (Nasonov 1907, Izv. Akad. Nauk. SSSR. (6) 1: 203–204.)
WAGNER, RUDOLPH (1805–1864) (Anon. 1865, Sber. bayer. Akad. Wiss. 1865: 287–294; Wagner 1864, Nachr. Ges. Wiss. Göttingen 1864: 450–451.)
WAGNER, SAMUEL (1798–) (Pellett 1929, Am. Bee J. 69: 504–505.)
WAGNER, WLADIMIR ALEXANDROVITCH (1849–1934) (Bonnet 1945, *Bibliographia Araneorum* 1: 49, bibliogr. only.)
WAGNER'S ORGAN Siphonaptera: Paired, sac-like structures lined with minute, inward directed spines, laterad of the base of Sternum VIII in males in most genera of ceratophyllids. Syn. X-glands (Wagner 1932).
WAHL, BRUNO (1876–) (Anon. 1952, Anz. Schädlingsk. 25: 14.)
WAHLGREN, EINAR (1874–1962) (Lindroth 1937, Opusc. ent. 28: 161.)
WAHNES, CAR (1835–1910) (Kuhnt 1910, Dt. ent. Z. 1910: 330.)
WAHNSCHAFFE, MAX (1823–1884) (Kraatz 1884, Dt. ent. Z. 28: 439–440.)
WAILES, GEORGE (1802–1882) (Stainton 1883, Entomol. mon. Mag. 19: 211–212.)
WAINWRIGHT, COLBRAN JOSEPH (1867–1949) (Fraser 1949, Entomol. mon. Mag. 85: 49–50.)
WAIST Noun. (Middle English, *wast* = growth.) Ants: The Petiole and Postpetiole.
WAITE, MERTON BANWAY (–1945) (Anon. 1945. Washington Post 7.1.1945. p.12.)
WAKEFIELD, CHARLES MARCUS (–1902) (Fowler 1902, Proc. ent. Soc. Lond. 1902: lviii.)
WAKELAND, CLAUDE (1888–1960) (Shockley 1963, J. Econ. Ent. 56: 423.)
WALCKENAER, CHARLES ATHANASE (1771–1852) French author of *Faune Parisienne and Memoires pour servir a l'Historie Naturelle des Abeilles solitaires* (1817). (Westwood 1853, Proc. ent. Soc. Lond. (2) 2: 51–53, bibliogr.; Anon. 1889, Abeille (Les ent. et leurs écrits) 26: 280–281; Bonnet 1945, *Bibliographia Araneorum* 1: 31–32.)
WALCOTT, W H L (1790–1869) (Newman 1869, Entomologist 4: 294.)
WALDEN, BENJAMIN HOVEY (1879–1946) (Friend 1946, J. Econ. Ent. 39: 423.)
WALDHEIM, G F See Fischer von Waldheim.
WALDHEIM, GOTTHELF FISCHER DE See Fischer von Waldheim.
WALES, ROBERT WEBSTER (1889–1948) (Gerry 1949, Univ. Mass. Fernald Clb. Yb. 18: 37–38.)
WALKDEN, HERBERT HALDEN (1893–1967) (White & Wilbur 1968, J. Econ. Ent. 61: 343.
WALKER, EDMUND MURTON (1877–1969) (Wiggins 1966, Centennial of Entomology in Canada. 94 pp. (14–42, bibliogr.), Toronto.)
WALKER, FRANCIS (1809–1874) English taxonomist (BMNH) and prolific describer of Species and variants. Walker is credited with more than 20,000 scientific names associated with his descriptions. (Anon. 1874, Can. Ent. 6: 220; Carrington 1874, Entomol. mon. Mag. 11: 140–141; Newman 1874, Can. Ent. 6: 225–259; Saunders 1874, Proc. ent. Soc. Lond. 1874: xxxvi–xxxvii; Newman 1907, In: Distant, *Insecta Transvaaliensia* iv + 299 pp. (197), London.)
WALKER, FRANCIS AUGUSTUS (1841–1905) (B.D.J. 1904, Proc. Linn. Soc. Lond. 1904: 55–56; Anon. 1905, Entomol. mon. Mag. 41: 97.)
WALKER, FRED WINTER (1892–1934?) (Anon. 1934, Fla Ent. 18: 28–29.)
WALKER, JAMES JOHN (1851–1939) English sailor (Commander) and amateur coleopterist. Editor-in-Chief, Entomologists Monthly Magazine (1927–1939). (Blair 1939, Entomologist 72: 48; Hale-Carpenter 1939, Proc. Linn. Soc. Lond. 1939: 260–262.)
WALKINGSTICK *Diapheromera femorata* (Say) [Phasmatodea: Heteronemiidae].
WALL Noun. (Anglo Saxon, *weall* = rampart, from Latin, *vallum* a wall. Pl., Walls.) 1. Insect Anatomy: The retaining sides of an organ or structure. Cf. Ceiling; Floor; Roof. Rel. Window. 2. The boundary of a cell or cavity.

WALL, ROBERT EMERSON (1903–1933) (Hoffman 1934, Lingnan Sci. J. 13: 333–334, bibliogr.)

WALL SPIDER *Oecobius annulipes* Lucas [Araneida: Oecobiidae] (Australia).

WALLABY FLIES See Hippoboscidae.

WALLABY TICK *Haemaphysalis bancrofti* Nuttall & Warburton [Acari: Ixodidae] (Australia).

WALLACE, ALEXANDER (1829–1899) (Anon. 1899, Entomol. mon. Mag. 35: 275–276.)

WALLACE, ALFRED RUSSEL (1823–1913) English collector and natural historian. Collected, observed and described natural hisotry of Amazon rain forests with Henry Bates (1848–1950); subsequently they travelled separately and Wallace studied the Rio Negro basin (1850–1852); Wallace published *A Narrative of Travels on the Amazon and Rio Negro* (1853). Resident of Malay Archipelago (1854–1862) which further developed his ideas about natural selection. Wallace's work strongly influenced Darwin's publication of *The Origin of Species*. Wallace used evolutionary theory to explain biogeography and endemism; Wallace's Line (between Lombok and Bali) marks the separation of Asian and Australian fauna. Wallace argued that through natural selection, dominant Species of plants arise in small centres of origin, spread and diversify. Wallace was a prolific writer on many topics beside natural history and the recipient of numerous medals and awards. (Bethune-Baker 1913, Proc. ent. Soc. Lond. 1913: cxlii; Poulton 1913, Nature 92: 347–349; Poulton 1913, Zoologist 71: 468–471; Walker 1913, Entomol. mon. Mag. 49: 276–277; Anon. 1914, Ent. News 25: 34–37; Turner 1914, Entomologist's Rec. J. Var. 26: 27–28.)

WALLACE, CHARLES RONALD (1909–1971) (Chadwick 1972, J. ent. Soc. Aust. (N.S.W.) 8: 43–46, bibliogr.)

WALLACE, MURRAY McADAM HAY (1922–1994) Australian economic entomologist (CSIRO) published extensively on mites, and insects and mites associated with dung.

WALLENGREN, HANS DANIEL JOHAN (1823–1894) (Bergroth 1874, Wien. ent. Z. 13: 295; Aurivillius 1895, Ent. Tidskr. 16: 97–110, bibliogr.; Kraatz 1895, Dt. ent. Z. 39: 279; McLachlan 1895, Entomol. mon. Mag. 31: 53–54.)

WALLENTIN, IGNAZ (–1934) (W. Z. 1934, öst. EntVer 19: 62–63.)

WALLER, NATHAN ERNEST (–1958) (Harper 1959, Entomologist 92: 44.)

WALLER, THOMAS NAUNTON (1863–1942) (Anon 1942, Trans. Suffolk nat. Soc. 5: xxviii–xxix.)

WALLIS, JOHN (1714–1793) (Lisney 1960, *Bibliography of British Lepidoptera. 1608–1799*, 315 pp. (197–198), London.)

WALLIS, JOHN BRAITHWAITE (1877–1961) (Bird 1961, Proc. ent. Soc. Manitoba 17: 5–6.)

WALLUM CICADA *Cicadetta stradbrokensis* (Distant) [Hemiptera: Cicadidae] (Australia).

WALMSLEY, THOMAS (1781–1806) (Weiss 1936, *Pioneer Century of American Entomology*, 320 pp. (63), New Brunswick.)

WALNUT APHID *Chromaphis juglandicola* (Kaltenback) [Hemiptera: Aphididae]: In North America, a significant pest of walnut along west coast. Females typically ovoviviparous, parthenogenetic, sometimes winged, throughout summer. Wingless females of autumn lay eggs.

WALNUT BLISTER MITE *Eriophyes erineus* (Nalepa); *Eriophyes tristriatus* (Nalepa) [Acari: Eriophyidae] (Australia).

WALNUT CATERPILLAR *Datana intergerrima* Grote & Robinson [Lepidoptera: Notodontidae]: A pest of walnut, apple, peach, pecan and ornamental trees in the eastern USA. WC univoltine or bivoltine; overwinters as Pupa within soil near host trees. Adults appear during June-July; female oviposits eggs in clusters of 200–300 on underside of leaves of host plant; eclosion occurs within 10–14 days. Larvae feed gregariously and can defoliate branches; larvae move synchronously down tree to moult on trunk, then move up tree to resume feed. See Notodontidae.

WALNUT HUSK-FLY *Rhagoletis completa* Cresson [Diptera: Tephritidae]: A monovoltine pest of walnuts and peach, widespread in western USA. Adult 5–8 mm long; body brown with yellowish markings; eyes blue; wings with black markings. Eggs white-pearlescent, reticulate, deposited in stem region of husk. Larva yellow with black mouth hooks; leaves husk to pupate in soil. See Tephritidae.

WALNUT PINHOLE BORER *Diapus pusillimus* Chapius [Coleoptera: Curculionidae]: In Australia, a small (2 mm long), elongate beetle which attacks damaged or unhealthy standing trees and logs of many valuable rainforest cabinet woods. See Curculionidae.

WALNUT SCALE *Quadraspidiotus juglansregiae* (Comstock) [Hemiptera: Diaspididae]. In western North America, a pest of apple, apricot, cherry, peach, pear, plum and walnut. Resembles San Jose Scale. See Diaspididae.

WALNUT SHOOT-MOTH *Acrobasis demotella* Grote [Lepidoptera: Pyralidae].

WALNUT SPHINX *Laothoe juglandis* (J. E. Smith) [Lepidoptera: Sphingidae].

WALSCH, JOSEF (1866–1934) (Hornstein 1935, öst. EntVer 20: 29.)

WALSH, BENJAMIN DANN (1808–1870) Born in Frome, Worchestershire, England; educated at Trinity College (M.A. 1830); classmate of Charles Darwin. Emigrated to America 1838 and purchased 300-acre farm near Cambridge, Henry County, Illinois where he farmed until 1851. Operated lumber business 1851–1858 at Rock Island, Illinois. First state entomologist of Illinois; active in Entomology late in life and therefore not a prolific writer. Strong influence on C. V. Riley. (Anon. 1870, Can. Ent. 1: 42–43; Anon. 1870, Entomol. mon. Mag. 6: 218–219; Anon. 1870, Naturaliste Can. 2: 94; Hagen 1870,

Stettin. ent. Ztg. 31: 354–356; Goding 1887, Trans. Ill. Hort. Soc. 21: 152; Henshaw 1889, Bibliography of the more important contributions to American Economic Entomology. 95 pp. Washington. Bibliography only; Anon. 1894, Ent. News 5: 269–270; Osten Sacken 1903, *Record of my Life Work in Entomology,* 240 pp. (38–39), Cambridge, Mass.)

WALSH, GEORGE BECKWORTH (1880–1957) (Anon. 1957, Scarborough Evening News 25.10.1957.)

WALSINGHAM, THOMAS DE GREY (1843–1919) (Busck 1920, Proc. ent. Soc. Wash. 22: 41–43; Durrant 1920, Entomol. mon. Mag. 56: 17, 25–28; Rowland-Brown 1920, Entomologist 53: 23–24; Essig 1931, *History of Entomology.* 1029 pp. (pp. 791–792, bibliogr.), New York.)

WALTER-GROSSMANN, CHARLES (1884–1946) (Lundblad 1948, Ent. Tidschr. 69: 69–71, bibliogr.)

WALTHER, HANS (1868–1933) (Möbius 1933, Dt. ent. Z. Iris 47: 191–194, bibliogr.)

WALTON, JOHN (1784–1862) (Smith 1862, Proc. ent. Soc. Lond. 1862: 125–127.)

WALTON, LEE BARKER (1871–1937) (Anon. 1937, Arb. morph. taxon. Ent. Berl. 4: 241; Osborn 1938, Science 88: 49.)

WALTON, WILLIAM RANDOLPH (1873–1952) (Anon. 1953, J. Econ. Ent. 46: 532–533. Wade *et al.* 1953, Proc. ent. Soc. Wash. 55: 103–108.)

WANDERER BUTTERFLY *Danaus plexipus plexipus* (Linnaeus) [Lepidoptera: Nymphalidae] (Australia).

WANGERIN, WALTHER (–1938) (Anon. 1938, Ent. Z., Frankf. a. M. 52: 93.)

WANKA VON LENZENHEIM, THEODOR (1871–1932) (Hetschenko 1932, Wien. ent. Ztg. 49: 186–18, bibliogr.; Hetschenko 1932, Ent. NachrBl. Troppau 6: 90.)

WANKOWITZ, JEAN DE (1835–1885) (Anon. 1885, Bull. Soc. ent. Fr. 1885: clxiii; Anon. 1885, Naturaliste 3: 144; McLachlan 1885, Proc. ent. Soc. Lond. 1885: xliii.)

WAR Noun. (Anglo Saxon, *werre.*) Social Insects: Overt aggression between groups of workers of different colonies.

WAR FEVER See Epidemic Typhus.

WARBEX® See Famphur.

WARBLE FLIES Members of the Genus *Hypoderma* whose larvae attack cattle and deer. See Oestridae; Hypodermatidae.

WARBUTON, CECIL (1854–1958) (Richards 1959, Proc. R. ent. Soc. Lond. (C) 23: 74; Scott 1959, Entomol. mon. Mag. 95: 16–17.)

WARD, CHRISTOPHER (1837–1900) (Anon. 1900, Entomol. mon. Mag. 36: 213.)

WARD, HENRY BALDWIN (1865–1945) (Anon. 1946, Am. Nat. 80: 25–26.)

WARD, IVOR JESMOND (–1947) (Buckell 1947, Can. Ent. 79: 39; Buckell 1947, Proc. ent. Soc. Br. Columb. 44: 35.)

WARD, J. J. (–1946) (Williams 1948, Proc. R. ent. Soc. Lond. (C) 12: 65.)

WARDLE, R A (1890–1974) (McLeod & Welch 1974, Bull. ent. Soc. Can. 6: 52–53.)

WARDROBE BEETLE *Attagenus fasciatus* (Thunberg) [Coleoptera: Dermestidae].

WAREHOUSE BEETLE *Trogoderma variabile* Ballion [Coleoptera: Dermestidae]: A widespread, significant pest of stored food, including whole wheat flour, processed grains and whole kernels of barley and wheat. Adult 1.5–4.0 mm long, oval, dark brown with pale brown markings; larva to 10 mm long, cream coloured with pale-brown markings. Life cycle 1.5–6 months. Eggs laid singly in crushed grain or concealed in crevices of whole kernels; incubation ca seven days. Typically six larval instars; diapause and non-diapause larvae determined by complex of conditions; non-diapause larval development ca 25 days; larva bristly with urticating Setae which cause dermatitis. Pupation near surface of food; pupa remains in last larval Integument; stage ca four days; teneral adult within larval Integument. Neonate adults mate; adults live 9–80 days.

WAREHOUSE MOTH See Tobacco Moth.

WAREHOUSE PIRATE-BUG *Xylocoris flavipes* (Reuter) [Hemiptera: Anthocoridae].

WARION, GUSTAVE (1839–1871) (Bellevoye 1871, Bull. Soc. hist. nat. Metz (2) 14: 197–199; Fallou 1871, Bull. Soc. ent. Fr. (5) 1: xiv–xv.)

WARLOE, HANS (1852–1939) (Natvig 1940, Norsk. ent. Tidsskr. 5: 190–192, bibliogr.; Hellén 1943, Notul. ent. 23: 59.)

WARNE, NORMAN DALZIEL (1868–1905) (Anon. 1905, Entomologist 38: 288.)

WARNECKE, GEORG (1884–1962) (Albert 1962, Mitt. dt. ent. Ges. 21: 18; Lattin 1962, Mitt. hamb. zool. Mus. Inst. 60: 325–328.)

WARNER, WILLIAM W (1929–1962) (Rogers 1963, Fla. Ent. 46: 5.)

WARNIER, ADOLPHE (–1914) (Anon. 1915, Miscnea ent. 22: 50.)

WARNING COLORATION Conspicuous colours or patterns of colour which are frequently associated with qualities which render their possessor unpalatable, offensive or dangerous to predators. See Coloration. Cf. Aposomatic Coloration; Crypsis; Protective Coloration.

WARREN, WILLIAM (1839–1914) (Jordan 1915, Novit. zool. 22: 160–166; Prout 1915, Entomologist's Rec. J. Var. 26: 258–260.)

WART Noun. (Anglo Saxon, *wearte.* Pl., warts.) 1. A spongy excrescence from tissue or Cuticle, more-or-less cylindrical, with apex nearly truncated. 2. The enlarged, common base of a group of Setae. 3. Trichoptera: A pitted elevation.

WARTY GRAIN-MITE *Aeroglyphus robustus* (Banks) [Acari: Glycyphagidae].

WASASTJERNA BJÖRN RUDOLF (1860–1928) (Forsius 1928, Notul. ent. 8: 65–67.)

WASHBURN, FREDERICK LEONARD (1860–1927) (Anon. 1927, J. Econ. Ent. 20: 849–850.)

WASMANN, ERICH (1859–1931) (Borgmeier 1931,

Revta. Ent., Rio de J. I: 107; Donisthorpe 1931, Entomologist's Rec. J. Var. 43: 76; Gremelli 1931, Atti Accad. Lincei 84: 489–508; Schmitz 1932, Tijdschr. Ent. 75: 1–57, bibliogr.)

WASP Noun. (Pl., Wasps.) Correctly: A nest of wasps. See Vespoidea. Cf. Ant; Bee.

WASP BEETLES See Cerambycidae.

WASP FLIES See Conopidae.

WASP MOTHS See Ctenuchidae; Hypodermatidae.

WASPKININS Plural Noun. Peptide components of venom in many Species of Vespidae; compounds analogous to bradykinins and cause blood pressure to lower and smooth muscles to contract. Waspkinin may be molecularly different from vespulakinins.

WASTE Noun. (Middle English, *waste* from Latin, *vastus* = unoccupied. Pl., Wastes.) 1. A gradual loss of substance through use, wear or decay. See Debris. 2. The physical loss of substance through the breaking down of tissue. See Atrophy; Debris. 3. Excrement or the final product of digestion. 4. Material that may be discarded as unwanted but which may have value or purpose in another context. See Refuse.

WATARI, SHOKAN MUSAMI (1897–1953) (Anon. 1953, Lepid. News 7:166; Hasegawa 1967, Kontyû 35 (Suppl.): 95–96.)

WATASE, SHOSUBURO (1862–1929) (Lillie *et al.* 1930, Science 71: 577–578; Hasegawa 1967, Kontyû 35 (Suppl.): 95.)

WATER BEETLES See Dytiscidae; Hydrophilidae.

WATER BOATMEN See Corixidae.

WATER CRICKETS See Veliidae.

WATER HYACINTH *Eichhornia crassipes* (Martius) Solms-Laubach [Pontenderiaceae]: Arguably the world's most significant aquatic weed. Endemic to South America; widespread in tropical and subtropical areas; introduced into Australia 1894; moved to many places as ornamental aquarium plant. WH free-floating with purple, erect flowers emerging from large, spongy leaf-stalks and feathery, trailing roots. WH causes problems by restricting water flow, impedes navigation, reduces fish populations and vectors disease. Natural enemies in USA and Australia include weevils *Neochaetina eichhorniae* Warner, *N. bruchi,* moth *Sameodes albigutattus* and mite *Orthogalumna terebrantis*. Cf. Alligator Weed; Salvinia; Water Lettuce.

WATER LETTUCE *Pistia stratiotes* Linnaeus [Araceae]: A floating weed common in tropical and subtropical regions; interdicted into Australia (1946). An aquatic weed and habitat for larva and pupa of *Mansonia* spp. which vector some diseases of humans. Cf. Alligator Weed; Salvinia; Water Hyacinth.

WATER LOVERS *Hydrophilus* spp. [Coleoptera: Hydrophilidae] (Australia).

WATER MEASURERS See Hydrometridae.

WATER PENNIES See Psephenidae.

WATER PENNY BEETLES See Psephenidae.

WATER SCAVENGER BEETLES See Hydrophilidae; Hydrophiloidea.

WATER SCORPION *Laccotrephes tristis* (Stål) [Hemiptera: Nepidae] (Australia).

WATER SCORPIONS See Nepidae.

WATER SPIDERS See Gerridae.

WATER STRIDERS See Gerridae; Mesoveliidae.

WATER TREADERS See Hydrometridae; Mesoveliidae.

WATERBOATMEN See Notonectidae.

WATERCRESS LEAF-BEETLE *Phaedon viridus* (Melsheimer) [Coleoptera: Chrysomelidae].

WATERCRESS SHARPSHOOTER *Draeculacephala mollipes* (Say) [Hemiptera: Cicadellidae].

WATERHOUSE, CHARLES OWEN (1843–1917) (Champion 1917, Entomol. mon. Mag. 53: 67–68; Distant 1917, Entomologist 50: 71–72.)

WATERHOUSE, EDWARD A (1851–1916) (Distant 1916, Entomologist 49: 71–72.)

WATERHOUSE, FREDERICK GEORGE (1815–1879) (Anon. 1889, Entomol. mon. Mag. 35: 16.)

WATERHOUSE, FREDERICK HERSCHEL (1845–1919) (Anon. 1920, Ent. News 31:149.)

WATERHOUSE, GEORGE ROBERT (1810–1888) (Anon. 1888, Entomol. mon. Mag. 24: 233–234; Harting 1888, Zoologist (3) 12: 99–100.)

WATERHOUSE, GUSTAVUS ATHOL (1877–1950) (Anon. 1949, Proc. R. Soc. Qd. 60: 72; Walkom 1950, Nature 166: 335–336; A.B.W. & A.J.N. 1954, Proc. Linn. Soc. N.S.W. 78: 269–275, bibliogr.)

WATER HYACINTH MOTH *Sameodes albiguttus* (Warren) [Lepidoptera: Pyralidae] (Australia).

WATER HYACINTH WEEVIL *Neochaetina eichhorniae* Warner [Coleoptera: Curculionidae] (Australia).

WATER LILY APHID *Rhopalosiphum nymphaeae* (Linnaeus) [Hemiptera: Aphididae].

WATER LILY LEAF-BEETLE *Galerucella nymphaeae* (Linnaeus) [Coleoptera: Chrysomelidae].

WATER LILY LEAF-CUTTER *Synclita obliteralis* (Walker) [Lepidoptera: Pyralidae].

WATERS, EDWIN GEORGE ROSS (1890–1930) English academic (Professor of Romance Languages) and amateur lepidopterist. (Blair 1930, Entomol. mon. Mag. 66: 91; Walker 1930, Entomol. mon. Mag. 66: 102–104.)

WATERSTON, JAMES (1879–1930) English taxonomist specializing in parasitic Hymenoptera, particularly the Chalciodidea. (Grimshaw 1930, Scott. Nat. 183: 65–67; Laing 1930, Entomol. mon. Mag. 66: 141–142; Riley 1930, Entomologist 63: 143–144; Jordan 1931, Proc. ent. Soc. Lond. 5: 129–130.)

WATERSTON'S ORGAN Hymenoptera: a glandular structure on the sixth abdominal Tergum of ceraphronids. (Ogloblin 1944, Proc. ent. Soc. Wash. 46: 155–158.) Cf. Hagen's Gland.

WATERTON, CHARLES (1782–1865) (Anon. 1865, Ibis 1: 364; Anon. 1866, Naturalist, Hull 2: 53; Carvalho 1918, Revta. Mus. paul. 10: 895–903.)

WATKINS, CHARLES JAMES (1846–1906) (Anon.

1907, Entomologist's Rec. J. Var. 19: 194–195; Anon. 1907, Entomologist 40: 168; Horn 1907, Dt. ent. Z. 1907: 535; Waterhouse 1907, Proc. ent. Soc. Lond. 1907: xcvii.)

WATKINS, WILLIAM (1849–1900) (Anon. 1900, Entomologist 33: 208.)

WATSON, EDWARD YERBURY (–1897) (Trimen 1897, Proc. ent. Soc. Lond. 1897: lxxi–lxxii.)

WATSON, ERIC B (1898–1975) (Pebble 1975, Bull. ent. Soc. Can. 7: 96.)

WATSON, FRANK EDWARD (1877–1947) (dos Passos 1958, J. N. Y. ent. Soc. 66: 1–6, bibliogr.)

WATSON, JOHN ANTHONY LINTHORNE (1935–1993) Australian research entomologist (CSIRO) specializing in biology, and taxonomy of Thysanura, Isoptera and Odonata.

WATSON, JOHN HENRY (1866–1951) (Riley 1953, Proc. R. ent. Soc. Lond. (C)17:74.)

WATSON, JOSEPH RALPH (1874–1946) (Tissot 1946, Fla. Ent. 28: 57–59; Tissot 1946, Ann. ent. Soc. Am. 39: 345–346.)

WATSON, NORA ISABEL (1927–1964) (Wigglesworth 1964, Proc. R. ent. Soc. Lond. (C)29: 54.)

WATSON, WILLIAM (1715–1787) (Rose 1850, *New General Biographical Dictionary* 12: 428–429.)

WATT, MORRIS NETTERVILLE (1892–1973) (Ordish 1973, N. Z. ent. 5: 368; Lees 1974, Proc. R. ent. Soc. Lond. (C) 38: 61.)

WATTENWYL, C BRUNNER See Brunner von Wattenwyl, C.

WATTLE APPLE-GALL WASP *Trichilogaster acaciaelongifoliae* (Froggatt) [Hymenoptera: Pteromalidae] (Australia).

WATTLE CICADA *Cicadetta oldfieldi* (Distant) [Hemiptera: Cicadidae] (Australia).

WATTLE GOAT MOTH *Xyleutes eucalypti* (Herrich-Schaffer) [Lepidoptera: Cossidae] (Australia).

WATTLE LEAFMINER *Acrocercops plebeia* (Turner) [Lepidoptera: Gracillariidae] (Australia).

WATTLE MEALYBUG *Melanococcus albizziae* (Maskell) [Hemiptera: Pseudococcidae] (Australia).

WATTLE ROOT LONGICORN *Eurynassa australis* (Boisduval) [Coleoptera: Cerambycidae] (Australia).

WATTLE TICK SCALE *Cryptes baccatus* (Maskell) [Hemiptera: Coccidae] (Australia).

WATTS, CARL N (1921–1963) (Kerbey 1963, J. Econ. Ent. 56: 546.)

WAUTIER, ABBÉ (1883–1910) (Poskin *et al.* 1910, Revue mens. Soc. ent. namur. 10: 57–58.)

WAVE MOTHS See Geometridae.

WAX Noun. (Anglo Saxon, *weax* = wax. Pl., Waxes.) Esters of long-chain alcohols higher than glycerol and fatty acids; waxes are insoluble in water and difficult to hydrolyse. Waxes are secreted by glands in many parts of the body. Early entomologists believed that insects ingested waxes from plants and then secreted the waxes as waste product. However many insects synthesize the hydrocarbon components of their waxes and do not incorporate these elements from the plant material which they ingest. Architecture of wax filaments is highly variable, and presumably function-specific. The form of the wax may be determined by mechanical constraints. Pseudococcidae produce several types of wax, including: long filaments that are longitudinally ribbed; short, curled filaments which are hollow on one side and resemble a gutter; flat microfilaments that are edged on one side and intertwined in a spiral; and amorphous wax. The functions of insect waxes are diverse. Honey bee workers manufacture and secrete wax to construct their nests. Some aphids cover their eggs with wax, presumably to protect them from desiccation or to facilitate respiration. The wax layer of the Integument maintains water balance. Wax on body may prevent predation in several ways Some coccinellid larvae, scale insects and woolly aphids reflect ultraviolet in strong light. The colour of the wax cover is seen as 'insect white' and may serve as a cryptic coloration or warning coloration for predators. Waxes also protect body structures such as the mesothoracic sense organs of the bark beetle *Melanophila*. These organs are infrared receptors which detect recent fires. See Wax Gland.

WAX CUTTER See Wax Pincers.

WAX GLAND Any Exocrine Gland which secretes a wax product in the form of a scale, string or powder. Wax Glands have been identified on all Tagmata and appendages of the insect body and given different names. Coccidae: Circumgenital and Parastigmatic Glands; Diaspididae: Genacerores (MacGillivray); Hymenoptera (Apoidea): Abdominal Glands that produce beeswax. Neuroptera (Coniopterygidae): glands on head, Thorax and Abdomen which produce a meal-like wax. Generalizations regarding Wax Glands are difficult because they occur in many paurometabolous and holometabolous groups of insects and the glands have not been studied in anatomical detail or systematically compared. Minute canals (ca 60–130 Å diam) extend between epidermal cells and the epicuticle in the larval hesperiid *Calpodes ethlius* Stoll. Similar structures have been reported in other Lepidoptera, the honey bee *Apis mellifera* Linnaeus, the beetle *Tenebrio molitor* Linnaeus and the coccid *Ceroplastes pseudoceriferus* Green. Based on current studies, the anatomy of Wax Glands appears relatively simple. Wax Glands are composed of modified epidermal cells. Smooth Endoplasmic Reticulum within these modified epidermal cells generally is believed responsible for lipid and wax production. Oenocytes clustered beneath epidermal cells have been suggested as sites of wax synthesis or synthesis of wax precursors. See Gland. Cf. Triangular Wax Gland; Tubular Wax Gland; Multilocular Wax Gland; Quinquelocular Wax Gland.

WAX LAYER 1. Integument: A layer of long chain

hydrocarbons, fatty acids and alcohols emitted from pore canals. The Wax Layer is about 0.25 µ thick, and functions in water balance (waterproofing Cuticle and preventing desiccation). Waxes used by insects to maintain water balance vary considerably. The physical characteristics of the waxes vary from a soft grease in cockroaches to hard waxes in some Lepidoptera pupae. The influence of temperature on the wax layer can be dramatic. Once a critical temperature has been surpassed (ranging from 31°C in the cockroach genus *Blatella* to 58°C in pupae of the butterfly genus *Pieris*), the amount of water lost through transpiration increases sharply. See Cuticle; Integument. Cf. Cement Layer; Cuticulin Layer. 2. Egg: A layer of the insect eggshell which surrounds the Vitelline Membrane and is formed when the Vitelline Membrane is completed. Questions exist as to whether this is a distinct layer or a series of lipids impregnated into the Vitelline Membrane. Lipids are produced by follicle cells in the Ovariole which are deposited as plaques. Freeze fracture reveals a smooth-surfaced material arranged longitudinally in multiple layers. Smoothness suggests the material is not protein; longitudinal fractures of the layers suggest hydrophobicity. Conventional opinion holds that the wax layer is responsible for waterproofing the egg, and thereby resisting desiccation and drowning. See Eggshell. Cf. Chorion; Vitelline Membrane.

WAX MIRRORS Honey bees: Paired, oval shaped, smooth areas of Integument on gastral Sterna through which beeswax is secreted. Wax manufactured by glandular epithelium positioned dorsad of the Wax Mirror. Paired wax glands occur on the ventrum of gastral segments 4–7 of worker honey bees. The glands are best developed in young workers who remain in the nest and construct wax combs; the wax glands degenerate in older workers which forage and do not participate in nest maintenance. Metabolically active cells are tall and columnar; they are filled with mitochondria and Rough Endoplasmic Reticulum, but Smooth Endoplasmic Reticulum is absent. Pore canals from the gland cells penetrate the Integument. The ultrastructure of the wax glands has been described but it is not clear whether the pore canals are composed of microfilaments of microtubules. Adjacent epithelium cells of the wax gland are bound together with Macula Adherens (Desmosomes) and Zonula Occludens (Tight Junctions) in the Plasma Membranes. See Beeswax. Cf. Wax Plate.

WAX MOTH *Galleria mellonella* (Linnaeus) [Lepidoptera: Pyralidae]: A widespread pest of honey bees. Eggs laid in hive during night when bees are inactive; larvae feed on wax of comb at night; pupate in tough cocoon on side of hive; larvae capable of destroying hive, but not harming bees.

WAX MOTHS See Pyralidae.

WAX PICK Hymenoptera: A spur on the middle leg of honey bees which is used to groom pollen and wax from the ventral surface of the body.

WAX PINCER A pincer-like structure of the hind leg of the honey bee; regarded by early workers as used in transfer of wax scales (Folsom & Wardle). Syn. Wax Cutter; Wax Shear.

WAX PLATES. Whiteflies: Paired areas on ventrolateral abdominal surface through which wax particles are extruded. WP on segments 2–3 of females, and segments 2–5 of males. Each WP consists of numerous Microtrichia arranged in compact rows. Microtrichia serve as templates for sculpting wax filaments. Secretory vesicles beneath WP manufacture wax; canals conduct wax to plate surface. Each WP contains numerous pore canals. Wax extruded through canal in continuous ribbons until broken by hind Tibia when passed over WP during grooming and wax distribution. See Wax. Cf. Wax Mirror. Rel. Gland.

WAX PORE Coccids: a cuticular pore through which wax is secreted. The cerores of MacGillivray.

WAX SCALE A scale secreted in the wax pocket or gland of a worker bee.

WAXEN Adj. Wax-like; waxy. Of the colour or appearance of wax.

WEAKLY Adv. (Middle English, *weik*.) Feebly; not strong.

WEAVER ANT *Oecophylla smaragdina* (Fabricius) [Hymenoptera: Formicidae] (Australia). Ants (*Oecophylla* spp.) which use larval silk to construct nests. See Citrus Ant.

WEB Noun. (Anglo Saxon, *webbe*. Pl., webs.) A thread-like device for capturing prey.

WEB SPINNERS See Embioptera; Oligotomidae; Yponomeutoidea.

WEB SPINNING See Telarian.

WEBB, HARRY E (1885–1970) (Hinton 1971, Proc. R. ent. Soc. Lond. (C) 35: 54.)

WEBB, JESSE LEE (1878–1942) (Bishopp 1942, Proc. ent. Soc. Wash. 44: 31–33.)

WEBB, SIDNEY (1837–1919) (Rowland-Brown 1919, Entomologist 52: 119–120; Turner 1919, Entomologist's Rec. J. Var. 31: 100.)

WEBB, THOMAS HOPKINS (1801–1866) (Geiser 1937, *Naturalists of the Frontier*. 341 pp. (335), Texas.)

WEBBER, RAY TRASK (1884–1948) (Grossman *et al.* 1948, J. Econ. Ent. 41: 340.)

WEBBING CLOTHES-MOTH *Tineola bisselliella* (Hummel) [Lepidoptera: Tineidae]: A cosmopolitan pest endemic in Africa. WCM a common and serious problem in domestic habitats; also found in bird nests and carrion. WCM attacks wool and natural bristles; in clothing fabrics of mixed fibres, bites but does not feed on cotton or synthetic fibres; larva cannot digest cotton or linen. Adult ca 8–12 mm long, poor fliers, not attracted to light; female often seen in folds of fabrics; female lays ca 40–100 eggs during 3–14 days; eggs oval, sticky, laid individually on food; incubation 4–10 days. Larva white with brown head capsule; spins silk webbing and forms feeding tube; larva prefers fabric stained with fruit juices, or containing human sweat or urine for B-complex vitamins contained therein; mature larva ca 10 mm long; larval stage 30–35 days. Pupation in enlarged tube; pupal stage 8–28 days. Generation-time temperature dependent, three months to a year. Biology and anatomy resemble Casemaking Clothes-Moth. Cf. Casemaking Clothes Moth; Large Pale Clothes-Moth; Tropical Case-Bearing Clothes Moth.

WEBBING CONEWORM *Dioryctria disclusa* Heinrich [Lepidoptera: Pyralidae].

WEBER, EDUARD (1811–1871) (Anon. 1872, Jber. mannheim Ver. Naturk. 38: 90–93, bibliogr.)

WEBER, FRIEDRICH (1781–1823) (Zimsen 1964, *The Type Material of J. C. Fabricius.* 656 pp. (15), Copenhagen.)

WEBER, HERMANN (1899–1956) (Alam 1956, Indian J. Ent. 18: 311; Roonwal 1956, Indian J. Ent. 18: 309–310; Lindroth 1957, Opusc. ent. 22: 50.)

WEBER, LOUIS (1886–1965) (de Bros 1966, Mitt. ent. Ges. Basel 16: 89–90.)

WEBER, LUDWIG (1855–1922) (Müller 1922, Ent. Bl. Biol. Syst. Käfer 18: 113–114.)

WEBER, MARCEL (–1957) (Anon. 1957, Bull. Soc. ent. Fr. 62: 221.)

WEBER, PAUL (1881–1968) (de Bros 1968, Mitt. ent. Ges. Basel 18: 148; Anon. 1969, Z. wien. ent. Ges. 53: 64.)

WEB-SPINNING SAWFLIES See Pamphiliidae.

WEB-SPINNING TREE CRICKETS See Gryllacrididae.

WEBSTER FRANCIS MARION (1849–1916) (Forbes 1916, J. Econ. Ent. 9: 239–241; Hewitt 1916, Can. Ent. 48: 73–74; Howard 1916, Proc. ent. Soc. Wash. 18: 78–83.)

WEBSTER, ROBERT LORENZO (1885–1966) (Johnson 1966, Ann. ent. Soc. Am. 59: 1030.)

WEBWORMS Species of Pyralidae that attack several Species of economically important crops, particularly grasses to include corn, field grasses and pasture grasses. See Ailanthus Webworm; Alfalfa Webworm; Beet Webworm; Bluegrass Webworm; Buffalo-Grass Webworm; Cabbage Webworm; Christmas-Berry Webworm; Corn Root Webworm; Couchgrass Webworm; Sod Webworm; Sorghum Webworm; Southern Beet Webworm; Spotted Beet Webworm; Tropical Sod-Webworm.

WEDELL-WEDELLSBORG, AUGUST FREDERICK (1844–1923) (Henriksen 1925, Ent. Meddr. 14: 453–454.)

WEDGE SKIPPER *Anisynta sphenosema* (Meyrick & Lower) [Lepidoptera: Hesperiidae] (Australia).

WEDGE-SHAPED BEETLES See Rhipiphoridae.

WEDGE-SHAPED PLATES The mesial plates (MacGillivray).

WEE See Western Equine Encephalitis.

WEED Noun. (Middle English, *wede* = weed.) A plant growing in a place where it is not wanted.

WEED WEB MOTH *Achyra affinitalis* (Lederer) [Lepidoptera: Pyralidae] (Australia).

WEED WEEVIL *Lixus mastersi* Pascoe [Coleoptera: Curculionidae] (Australia).

WEED, CLARENCE MOORES (1864–1947) (Anon. 1947, Science 106: 125; Bradley 1960, Trans. Am. ent. Soc. 85: 291.)

WEEKS, ANDREW GRAY (1861–1931) (Anon. 1932, Ent. News 43: 28; Davis 1932, Ann. ent. Soc. Am. 25: 251.)

WEELE, H W VAN DER See van der Weele, H. W.

WEEVIL Noun. (Anglo Saxon, *wifel.* Pl., Weevils.) Any member of the polyphagous Coleoptera Superfamily Curculionoidea. Adult with an elongate, downward curved head forming a 'snout' (Rostrum) with mouthparts at the distal end. See Curculionoidea.

WEEVILS See Curculionidae.

WEGNER, ALFRED (–1915) German meterologist and author of *The Origin of Continents and Oceans* (1915) which developed hypothesis of Contentinal Drift and proposed term 'Pangaea' for first large continental landmass (supercontinent). See Gondwanaland; Laurasia.

WEHNCKE, ERNST (1835–1883) (Regembart 1883, Bull. Soc. ent. Fr. (6) 3: cxxx; Anon. 1884, Dt. ent. Z. 28: 240; Anon. 1884, Leopoldina 20: 166; Anon. 1884, Zool. Anz. 7: 408; Kolbe 1884, Berl. ent. Z. 28: 213–214.)

WEHRLE, LAWRENCE PAUL (1887–1950) (Wehrle 1950, J. Econ. Ent. 43: 963.)

WEHRLI, EUGENE (1871–1958) (Lempke 1958, Ent. Ber., Amst. 18: 141; Reisser 1958, Z. wien. ent. Ges. 43: 271–272.)

WEIDEMEYER, JOHAN WILLIAM (1819–) (dos Passos 1951, J. N. Y. ent. Soc. 59: 139.)

WEIDENBACH, LUDWIG VON (–1830) (Gistl 1832, Faunus I (1): 52.)

WEIGEL, CHARLES ADOLPH (1887–1965) (Nelson & Smith 1966, J. Econ. Ent. 59: 1549–1550.)

WEIGEL, JOHANN ADAM VALENTIN (1740–1806) (Letzner 1858, Ent., Breslau 12: 13–15.)

WEINMAN, CARL J. (1915–1952) (Mills 1953, J. Econ. Ent. 46: 188.)

WEIR, JOHN JENNER (1822–1894) (Anon. 1894. British Nat. 31: 134–136; Anon. 1894, Leopoldina 30: 108; Anon. 1894, Zool. Anz. 17: 136; Anon. 1894, Proc. Linn. Soc. Lond. 1892–94: 37–38; Anon. 1894, Entomologist 27: 157–159; Tutt 1894, Entomologist's Rec. J. Var. 5: 103–105.)

WEIS, ALBRECHT (1839–1914) (Heyden 1914, Ent. Bl. Biol. Syst. Käfer 10: 128; Schnaudigel 1914, Ber. senckenb. naturf. Ges. 45: 99–109.)

WEIS, HOLGER (1854–1933) (Wolff 1934, Ent. Meddr. 18: 494–496.)

WEISE, ERNST (1903–1973) (Papperitz 1974, Ent. Bl. Biol. Syst. Käfer Suppl. 70: 1–2.)

WEISE, JULIUS (1844–1925) (Korschefsky 1928, Ent. Bl. Biol. Syst. Käfer 24: 175–186, bibliogr.)

WEISE, PAUL (–1902) (Kraatz 1903, Dt. ent. Z. 47: 174.)

WEISER, VLADIMIR (1878–1926) (Rambousek

1926, Cas. csl. Spol. ent. 23: 78.)

WEIS-FOGH, TORKEL (1922–1975) (Gunn 1976, Proc. R. ent. Soc. Lond. (C) 40: 52–53.)

WEISMANN, FRIEDRICH LEOPOLD AUGUST (1834–1914) (Anon. 1915, Ent. Bl. Biol. Syst. Käfer 11: 191; Anon. 1915, Wien. ent. Ztg. 34: 68; Calvert 1915, Ent. News 26: 44–47; Parker 1915, Proc. Linn. Soc. Lond. 127: 33–37; Poulton 1917, Proc. R. Soc. Lond. (B) 89: xxvii–xxxiv; dos Passos 1951, J. N.Y. ent. Soc 59:152.)

WEISMANN, ROBERT (1899–1972) (Wyniger 1972, Mitt. ent. Ges. Basel 22: 69–70.)

WEISS, HARRY B (1881–1972) (Wheeler & Valley 1975, History of the Entomology in the Pennsylvania Department of Agriculture. 37 pp. (16), Harrisburg.)

WEISSMANTEL, VILMOS (WILHELM) (1837–1901) (Aigner-Abofi 1902, Rovart. Lap. 9: 29–32.)

WEITENWEBER, WILHELM RUDOLPH (1804–1870) (Möbius 1943, Dt. ent. Z. Iris 57: 26.)

WEITH, J R (1847–1902) (Anon. 1902, Can. Ent. 34: 278; Anon. 1902, Ent. News 13: 298; Needham 1903, Can. Ent. 35: 36–37; Davis 1931, Proc. Indiana Acad Sci. 41: 54.)

WELD, ISAAC (1774–1856) (Weiss 1936, *Pioneer Century of American Entomology,* 320 pp. (49), New Brunswick.)

WELD, LEWIS HART (1875–1964) (Burks 1965, Ann. ent. Soc. Am 58: 133–134, bibliogr.; Burks *et al.* 1965, Proc. ent. Soc Wash. 68: 121–125, bibliogr.; Mallis 1971, *American Entomologists.* 549 pp. (371), New Brunswick.)

WELITSCHKOWSKY, VLADIMIR ALEXEVICH (1857–1927) (Konakov 1937, Acta Univ voroneg. 9 (Zool.): 174–182, bibliogr.)

WELLCOME, HENRY SOLOMON (1853–1936) (Anon. 1936, Am. J. Pharmacy 108: 313–319; Anon. 1936, Clinical Med. Surg. 43: 472; Defosses 1936, La Presse Med. 44: 1363–1364; Wenyon 1938, Obit. Not. Fell. R. Soc. 2: 229–238.)

WELLER, JEREMY DAVID (1932–1965) (Pearson 1966, Proc. R. ent. Soc. Lond. (C) 30: 64.)

WELLES, CHARLES SALTER (1847–1914) (Calvert 1914, Ent. News 25: 192.)

WELLMAN, JOHN RICHARD (1832–1894) (Adkin 1894, Entomologist 27: 36; Anon. 1895. Entomol. mon. Mag. 31: 31.)

WELLS, M M (–1930) (Davis 1931, Ann. ent. Soc. Am. 24: 188.)

WELSH, I H (–1955) (Hall 1956, Proc. R. ent. Soc. Lond. (C) 20: 76.)

WELTI, ARTHUR (–1970) (Lees 1974, Proc. R. ent. Soc. Lond. (C) 38: 61.)

WELTI, ILSE MARIE THECLA (1896–1955) (Hall 1956, Proc. R. ent. Soc. Lond. (C) 20: 76.)

WELTNER, CARL WILHELM HERMANN (1854–1917) (Collin 1919, Mitt. zool. Mus. Berl. 9: 61–70, bibliogr.; Thienemann 1919, Arch. Hydrobiol. 12: 698–702.)

WENCK, ERNST (1906–1970) (Christen 1971, Mitt. ent. Ges. Basel 21: 23–24.)

WENCKER, JOSEPH ANTOINE (1824–1873) (Newman 1873, Entomologist 6: 368; Fauvel 1874, Annu. ent. 2: 121.)

WENDELER, HANS (1886–1967) (Korge 1966, Mitt. dt. ent. Ges. 25: 17–18.)

WENDT, ALBERT (1887–1968) (Müller 1968, Ent. Berichte. Berl. 1968: 137.)

WENE, GEORGE PETER (1913–1971) (Carruth 1971, J. Econ. Ent. 64: 1342.)

WENIGEM, KERLEM (1906–1970) (Janda 1971, Acta ent. bohemoslovaca 68: 284–285, bibliogr.)

WENNINGER, FRANCIS (1888–1940) (Anon. 1941, Proc. Indiana Acad. Sci. 50: 12–13.)

WENZEL, HENRY W. (1857–1925) (Haimbach 1926, Ent. News 37: 29–31.)

WERMELIN, JOHAN HENRIK (1850–1910) (Meves 1911, Ent. Tidskr. 37: 107–108.)

WERNEBURG, ADOLF (–1886) (Kraatz 1886, Dt. ent. Z. 30: 254; McLachlan 1886, Proc. ent. Soc. Lond. 1886: lxx.)

WERNER, FRANZ (1867–1939) (Antonius 1939, Zool. Gart. Lpz. 11: 186–188; Mahandra 1939, Nature 143: 711.)

WERNER, JON (1884–1937) (Anon. 1938, Arb. morph. taxon. Ent. Berl. 5: 186; L.R.N. 1938, Norsk. ent. Tidsskr. 5: 40–41, bibliogr.)

WESCHE, WALTER FRANCIS FREDERICK (1857–1910) (Anon. 1910, Queckett. microsc. Club 11: 145–146.)

WESENBERG-LUND, CARL JØRGEN (1867–1955) (Berg 1956, Hydrobiologia 8: 191–192; Lemche 1956, Vidensk. Meddr dansk. naturh. Foren. 18: vii–xi, bibliogr.)

WESMAEL, CONSTANTIN (1798–1872) (Sauver 1872, Ann. Soc. ent. Belg. 15: 213–233; Westwood 1872, Proc. ent. Soc. Lond. 1872: li; Selys Longchamps 1874, Annu. Acad. r. Belg. 40: 229–250, bibliogr.)

WESSEL, A B (–1940) (Natvig 1942, Norsk. ent. Tidsskr. 6: 123.)

WEST INDIAN CANE WEEVIL *Metamasius hemipterus hemipterus* (Linnaeus) [Coleoptera: Curculionidae].

WEST INDIAN DRYWOOD TERMITE *Cryptotermes brevis* (Walker) [Isoptera: Kalotermitidae]: An urban pest endemic to tropical America and adventive elsewhere. Soldier 4–6 mm long; head phragmotic (sloping and rough) and constricted behind frontal lobes, Fontanelle absent; Mandible very short and broad; Tarsi with four segments, Cerci of two segments. WIDT Regarded as world's most serious termite pest of homes and often undetected until timber collapse. Colony not a single nest but several small nests in sapwood and heartwood; colony easily transported with high potential of small colonies to infest small articles of wood; does not require contact with substrate; exists on moisture content of wood; damage difficult to detect until collapse. Syn. West Indian Powderpost Termite. Cf. Formosan Subterranean Termite.

WEST INDIAN FLATID *Anormenis antillarum* (Klrkaldy) [Hemiptera: Flatidae].

WEST INDIAN FRUIT-FLY *Anastrepha obliqua* (Macquart) [Diptera: Tephritidae].

WEST INDIAN SWEETPOTATO WEEVIL *Euscepes postfaciatus* (Fairmaire) [Coleoptera: Curculionidae].

WEST NILE FEVER An arboviral disease (Family Flaviviridae), endemic to southern Europe, Middle East, India and Africa. WN a childhood disease, often fatal and vectored by several *Culex* Species. *Culex univittatus* serves as principal vector in Middle East and South Africa. WN incubation period 3–6 days in humans, followed by obvious disease for 3–6 days. Mosquitoes acquire WN from infected birds during nesting period and transmit disease to children. See Arbovirus; Flaviviridae. Cf. Dengue Haemorrhagic Fever.

WEST, AUGUST (1874–1949) (Hansen 1949, Ent. Meddr. 25: 333–338, bibliogr.)

WEST, JAMES ALEXANDER (1877–1910) (Forbes 1910, J. Econ. Ent. 3: 384–385.)

WEST, RICHARD MILBOURNE (1867–1936) (Imms 1937, Proc. R. ent Soc. Lond. (C) 1: 55.)

WEST, THOMAS (–1879) (Sharp 1879, Entomologist 12: 208.)

WEST, WILLIAM (1836–1920) (Adkin 1920, Entomologist 53: 215–216.)

WEST, WILLIAM RAY (1920–1944) (Lyle 1944, J. Econ. Ent. 37: 329.)

WESTCOTT, OLIVER SPINK (1834–1919) (Anon. 1920, Ent. News 31: 119–120.)

WESTERMANN, BERNT WILHELM (1781–1868) (Dohrn 1868, Stettin. ent. Ztg. 29: 215–218; Henriksen 1925, Ent. Meddr 15: 161–164.)

WESTERN AUSTRALIAN BROWN BLOWFLY *Calliphora varifrons* Malloch [Diptera: Calliphoridae] (Australia).

WESTERN AVOCADO-LEAFROLLER *Amorbia cuneana* (Walshingham) [Lepidoptera: Tortricidae]: An indigenous and sporadic pest of avocado and citrus in California.

WESTERN BALSAM BARK-BEETLE *Dryocoetes confusus* Swaine [Coleoptera: Scolytidae].

WESTERN BEAN-CUTWORM *Loxagrotis albicosta* (Smith) [Lepidoptera: Noctuidae]. See Cutworms.

WESTERN BENTWINGED CICADA *Froggatoides pallida* (Ashton) [Hemiptera: Cicadidae] (Australia).

WESTERN BIGEYED-BUG *Geocoris pallens* Stål [Hemiptera: Lygaeidae].

WESTERN BLACK FLEA-BEETLE *Phyllotreta pusilla* Horn [Coleoptera: Chrysomelidae].

WESTERN BLACK-HEADED BUDWORM *Acleris gloverana* (Walsingham) [Lepidoptera: Tortricidae].

WESTERN BLACK-LEGGED TICK *Ixodes pacificus* [Acari: Ixodidae]. An ixodid implicated in the transmission of Lyme Disease in California.

WESTERN BLOOD-SUCKING CONENOSE *Triatoma protracta* (Uhler) [Hemiptera: Reduviidae].

WESTERN BOX-ELDER BUG *Boisea rubrolineata* Barker [Hemiptera: Rhopalidae].

WESTERN BROWN BUTTERFLY *Heteronympha merope* (Fabricius) [Lepidoptera: Nymphalidae] (Australia).

WESTERN CEDAR BARK-BEETLE *Phloeosinus punctatus* LeConte [Coleoptera: Scolytidae].

WESTERN CHERRY FRUIT FLY *Rhagoletis indiferrens* Curran [Diptera: Tephritidae]: A pest of bitter cherry and endemic to western USA. See Tephritidae. Cf. Cherry Fruit Fly; Black Cherry Fruit Fly.

WESTERN CHICKEN-FLEA *Ceratophyllus niger* (Fox) [Siphonaptera: Ceratophyllidae]: A pest of rats, humans and birds in western North America. Adult with compound eye relatively large and egg-shaped; Pronotal Comb with 26–28 spines. Only occurs on host when feeding. See Ceratophyllidae. Cf. European Chicken Flea.

WESTERN CHINCH-BUG *Blissus occiduus* Barber [Hemiptera: Lygaeidae]: A Species of Chinch Bug found in western North America. See Chinch Bug.

WESTERN CORN ROOT-WORM *Diabrotica virgifera virgifera* LeConte [Coleoptera: Chrysomelidae]: A serious pest of corn in the midwestern United States. Eggs overwinter in soil and hatch in spring. Close relative of Mexican Corn Rootworm.

WESTERN DAMSEL-BUG *Nabis alternatus* Parshley [Hemiptera: Nabidae].

WESTERN DRYWOOD-TERMITE *Incisitermes minor* (Hagen) [Isoptera: Kalotermitidae].

WESTERN EQUINE ENCEPHALITIS A minor arboviral disease of birds and humans in North America. Virus assigned to Genus *Alphavirus* of Tongaviridae. Disease vectored by mosquitoes (*Culex* spp.); WEE not common in humans. See Arbovirus; Tongaviridae. Cf. Eastern Equine Encephalitis; Venezuelan Equine Encephalitis.

WESTERN FIELD-WIREWORM *Limonius infuscatus* Motschulsky [Coleoptera: Elateridae].

WESTERN FLAT *Exometoeca nycteris* Meyrick [Lepidoptera: Hesperiidae] (Australia).

WESTERN FLOWER-THRIPS *Frankliniella occidentalis* (Pergande) [Thysanoptera: Thripidae]: Attacks seedling cotton, onions, lettuce, potatoes, alfalfa in North America; an occasional pest of grapes and predator of spider mite eggs. Cf. Tobacco Thrips; Flower Thrips.

WESTERN GOLDENHAIRED BLOWFLY *Calliphora albifrontalis* Malloch [Diptera: Calliphoridae] (Australia).

WESTERN GRAPE-LEAF SKELETONIZER *Harrisina brillians* Barnes & McDunnough [Lepidoptera: Zygaenidae]: A moth whose larvae defoliate grape plants; considered a significant pest of commercial vineyards.

WESTERN GRAPE-LEAFHOPPER *Erythroneura elegantula* Osborn [Hemiptera: Cicadellidae].

WESTERN GRAPE-ROOTWORM *Bromius obscurus* (Linnaeus) [Coleoptera: Chrysomelidae]: A pest of grape in western USA. Syn. *Adoxus obscurus* (Linnaeus). See Chrysomelidae. Cf. Grape Rootworm.

WESTERN GRASSDART *Taractrocera papyria agraulia* (Hewitson) [Lepidoptera: Hesperiidae] (Australia).

WESTERN HARVESTER-ANT *Pogonomyrmex occidentalis* (Cresson) [Hymenoptera: Formicidae]: A Species in southwestern USA at higher altitudes. Workers ca 4–5 mm long, resemble Red Harvester-Ant in habitus and habits; constructs a mound at the entrance of nest. See Formicidae. Cf. California Harvester-Ant; Florida Harvester-Ant; Red Harvester Ant; Harvester-Ant.

WESTERN HORN LERP *Creiis periculosa* (Olliff) [Hemiptera: Psyllidae] (Australia).

WESTERN INDIAN DRYWOOD TERMITE *Cryptotermes brevis* (Walker) [Isoptera: Kalotermitidae] (Australia).

WESTERN JEWEL *Hypochrysops halyaetus* Hewitson [Lepidoptera: Lycaenidae] (Australia).

WESTERN LAWN-MOTH *Tehama bonifatella* (Hulst) [Lepidoptera: Pyralidae].

WESTERN LILY-APHID *Macrosiphum scoliopi* Essig [Hemiptera: Aphididae].

WESTERN OAK-LOOPER *Lambdina fiscellaria somniaria* (Hulst) [Lepidoptera: Geometridae].

WESTERN PINE SHOOT-BORER *Eucosma sonomana* Kearfott [Lepidoptera: Tortricidae].

WESTERN PINE-BEETLE *Dendroctonus brevicomis* LeConte [Coleoptera: Scolytidae]: A pest of Ponderosa Pine and Coulter Pine in western North America. WPB completes 1–4 generations per year, depending upon climate. Adult 2–4 mm long, pale brown to black in coloration; female's pronotum with transverse ridge along anterior margin; male's Pronotum with corresponding transverse Sulcus; Elytra microscopically punctate and faintly striate. WPB constructs winding egg galleries between inner bark and sapwood. Syn. *Dendroctonus barberi*. Cf. Southern Pine Beetle; Eastern Larch Beetle; Red Turpentine Beetle.

WESTERN PLANT-BUG *Rhinacloa forticornis* Reuter [Hemiptera: Miridae].

WESTERN POTATO FLEA-BEETLE *Epitrix subcrinita* LeConte [Coleoptera: Chrysomelidae].

WESTERN POTATO-LEAFHOPPER *Empoasca abrupta* DeLong [Hemiptera: Cicadellidae].

WESTERN PREDATORY-MITE *Galandromus occidentalis* (Nesbitt) [Acari: Phytoseiidae].

WESTERN SAND WASP *Bembix comata* Parker [Hymenoptera: Sphecidae]: A wasp endemic to western North America, particularly on coastal sand dunes. Adult stout, ca 12–15 mm long, predominantly black with relatively long, white pubescence and pale transverse, interrupted stripes on Gaster; wings hyaline with pale brown venation; Tarsi pale yellow to greenish. Female constructs nest in sand ca 20–25 cm long with single terminal brood chamber. Single fly placed in chamber with egg attached; fly serves as support for developing larva; female sleeps in antechamber and progressively provisions larva with flies. See Sphecidae. Cf. Eastern Sand Wasp; Sand Wasp.

WESTERN SPOTTED CUCUMBER-BEETLE *Diabrotica undecimpunctata undecimpunctata* Mannerheim [Coleoptera: Chrysomelidae]. See Spotted Cucumber Beetle.

WESTERN SPRUCE-BUDWORM *Choristoneura occidentalis* Freeman [Lepidoptera: Tortricidae].

WESTERN STRIPED CUCUMBER-BEETLE *Acalymma trivittatum* (Mannerheim) [Coleoptera: Chrysomelidae]: Resembles Striped Cucumber Beetle in biology and anatomy; found in western USA. See Striped Cucumber Beetle.

WESTERN STRIPED FLEA-BEETLE *Phyllotreta ramosa* (Crotch) [Coleoptera: Chrysomelidae].

WESTERN SUBTERRANEAN-TERMITE *Reticulitermes hesperus* Banks [Isoptera: Rhinotermitidae]: A Species endemic to the Pacific coast of North America. Swarming initiated during third or fourth year of colony development. Pest of grapes; workers invade dead heartwood of vine and weaken plant. Cf. Eastern Subterranean Termite.

WESTERN TENT-CATERPILLAR *Malacosoma californicum pluviale* Dyar, *Malacosoma californicum* (Packard) [Lepidoptera: Lasiocampidae].

WESTERN THATCHING-ANT *Formica obscuripes* Forel [Hymenoptera: Formicidae]: Endemic to North America. WTA feeds upon insects and workers construct large mounds of twigs and grass. See Formicidae. Cf. Allegheny Mound-Ant; Silky Ant.

WESTERN TREEHOLE-MOSQUITO *Aedes sierrensis* (Ludlow) [Diptera: Culicidae].

WESTERN TUSSOCK-MOTH *Orgyia vetusta* (Boisduval) [Lepidoptera: Lymantriidae].

WESTERN WHEAT-APHID *Diuraphis (Holcaphis) tritici* (Gillette) [Hemiptera: Aphididae].

WESTERN W-MARKED CUTWORM *Spaelotis havilae* (Grote) [Lepidoptera: Noctuidae]. See Cutworms.

WESTERN XENICA *Geitoneura minyas* (Waterhouse & Lyell) [Lepidoptera: Nymphalidae] (Australia).

WESTERN YELLOWJACKET *Vespula pensylvanica* (Saussure) [Hymenoptera: Vespidae]: A significant urban pest in western North America and Hawaii. Workers usually predatory but will scavenge human food; a significant nuisance in campgrounds, outdoor public exhibitions and recreational areas; nests typically subterranean, occasionally in wall voids and attics. See Yellowjacket. Cf. Aerial Yellowjacket; Common Yellowjacket; German Yellowjacket.

WESTERN-CEDAR BORER *Trachykele blondeli* Marseul [Coleoptera: Buprestidae].

WESTERN-HEMLOCK LOOPER *Lambdina fiscellaria lugubrosa* (Hulst) [Lepidoptera: Geometridae].

WESTERN-POPLAR CLEARWING *Paranthrene robiniae* (Hy. Edwards) [Lepidoptera: Sesiidae].

WESTERN YELLOWSTRIPED ARMYWORM *Spodoptera praefica* (Grote) [Lepidoptera: Noctuidae]. See Armyworms.

WESTHOFF, FRIEDRICH (FRITZ) (1857–1896) (Anon. 1897, Zool. Anz. 20: 16. Anon. 1897, Ent. News 8: 72.)

WESTON, WALTER PHILIP (1853–1881) (Carrington 1881, Entomologist 14: 72–96; Anon. 1882, Zool. Jber. Neapel 1881: 6.)

WESTRING, NIKLAS (1797–1882) (Sandahl 1882, Ent. Tidskr. 3: 9–12, 99.)

WESTWOOD, JOHN OBADIAH (1805–1893) English academic and general entomologist. Taxonomic expert in many insect groups and accomplished natural history artist. A founding member of the Entomological Society of London (1833), and served as president for three terms. JOW wrote *An Introduction to the Modern Classification of Insects* (2 vols, 1838–1840), for which he was later awarded Gold Medal by the Royal Society (1855). Westwood wrote on many subjects and was highly influential, but he was not a proponent of Natural Selection. JOW served as first Hope Professor of Zoology (Oxford University). (Anon. 1893, Insektenbörse 10: 1; Anon. 1893, Insect Life 5: 285–286; Anon. 1893, Can. Ent. 25: 261–262; Anon. 1893, London News 102: 38; Bethune 1893, Rep. ent. Soc. Ont. 24: 107–108; Distant 1893, Entomologist 26: 25–26; Cambridge 1893, Entomologist 26: 74–75; McLachlan 1893, Entomol. mon. Mag. 29: 49–51; Wandolleck 1893, Berl. ent. Z. 38: 392–396; Neave 1933, *Centennial history of the entomological Society of London. 1832–1933.* xliv + 224 pp. (131–132), London.)

WETAS Noun. (Maori, *wetas.* Pl., Wetas.) Large, wingless stenopelmatine Orthoptera found in New Zealand. Wetas occur in caves, burrows and rotted logs. They have evolved to fill the niche associated with mice and other gnawing-type small mammals as New Zealand has no endemic terrestrial mammalian fauna.

WETTABLE POWDER A form of insecticide formulation in which the active ingredient is finely divided and mixed with an appropriate solid diluent (talc, diatomaceous earth) and a wetting agent (detergent-like substance which permits the powder to mix with water to form a suspension.) WP = active ingredient + diluent + wetting agent. Diluent may be coated or impregnated with the active ingredient. Syn. Water-Dispersable Powders. Designated as 'WP' or 'W' on U.S. Environmental Protection Agency label or Material Safety Data Sheet documentation. Rel. Aerosol; Dust; Emulsifiable Concentrate; Microencapsulated Concentrate; Oil Concentrate; Suspension Concentrate.

WETTING AGENT A type of surfactant added to pesticides which affect the mixing of solid pesticide particles in water. WA cause pesticides to disperse uniformly in water. See Pesticide; Emulsifier.

WEYENBERGH, HENDRIK (1842–1885) (Wulp 1885. Wien. ent. Ztg. 4: 225–227; Dimmock 1888. Psyche 5: 36.)

WEYMER, GUSTAV (–1914) (Kuhnt 1914, Dt. Ent. Z. 1914: 356.)

WEYRAUCH, WOLFGANG K (1907–1970) (Lamas 1970, Boln. Soc. ent. Peru 5: 33–34.)

WHARF BORER *Nacerdes melanura* (Linnaeus) [Coleoptera: Oedemeridae].

WHEAT APHID *Rhopalosiphum padi* (Linnaeus) [Hemiptera: Aphididae] (Australia).

WHEAT CURL MITE *Eriophyes tulipae* Keifer [Acari: Eriophyidae].

WHEAT HEAD ARMYWORM *Faronta diffusa* (Walker) [Lepidoptera: Noctuidae]. See Armyworms.

WHEAT JOINTWORM *Tetramesa tritici* (Fitch) [Hymenoptera: Eurytomidae]: A significant, monovoltine pest of wheat in eastern North America. Eggs inserted into wheat stems above nodes. Larva forms cell in wall of stem and feeds on sap. Several larvae form hard galls above second or third node, causing stem to break; larva remains in stubble after harvest; overwinters as pupa within larval cell. Cf. Hessian Fly; Wheat-Straw Worm.

WHEAT MIDGE *Sitodiplosis mosellana* (Géhin) [Diptera: Cecidomyiidae].

WHEAT WIREWORM *Agriotes mancus* (Say) [Coleoptera: Elateridae]: A pest of grains and potatoes in central and eastern USA.

WHEAT-POLLARD ITCH A hypersensitivity in humans to stored products caused by the mite *Suidasia nesbitti* Hughes. Syn. Grocers' Itch.

WHEAT-ROOT SCARAB *Sericesthis consanguinea* (Blackburn) [Coleoptera: Scarabaeidae] (Australia).

WHEAT-STEM BORER *Cephus pygmaeus* Linnaeus [Hymenoptera: Cephidae]. See European Wheat-Stem Sawfly.

WHEAT-STEM MAGGOT *Meromyza americana* Fitch [Diptera: Chloropidae]: In North America, a pest of barley, oats, rye and wheat. Attacks wheat during autumn or summer; winter damage resembles damage of Hessian Fly; injured stem is partially severed, head becomes white and then dies. Adults emerge during midsummer. Female oviposits on stem or leaves of host plant. Neonate larva crawls behind leaf sheath to feed on soft part of stem, and then tunnels into stem; Larva pale green; feeds inside lower part of stem, or crown of plant. Summer generation completes development; autumn generation develops on winter wheat; overwinters as larva inside lower part of stem; pupation occurs inside green puparium during spring. See Hessian Fly; Wheat-Stem Sawfly.

WHEAT-STEM SAWFLY *Cephus cinctus* Norton [Hymenoptera: Cephidae]: A pest of wheat, barley, rye and other grasses in eastern North

America. Typically overwinters in pupal cocoon and pupation occurs during following spring. Adults appear during June; eggs inserted into plant stem just above joints; female lays 1–25 eggs during one ovipositional session. Larva slender, S-shaped; feeds within stem by boring through nodes; mature larva resides in base of plant and forms a V-shaped groove on inner side of stem. Syn. Western Grass-Stem Sawfly. Cf. Hessian Fly; European Wheat Stem Sawfly; Black Grain-Stem Sawfly.

WHEAT-STRAW WORM *Tetramesa grandis* (Riley) [Hymenoptera: Eurytomidae]: A pest of wheat, endemic to North America and with alternate generations. Overwinters as larva or pupa in stubble or straw; adults emerge during March-April. First generation in spring; adult minute, apterous, lays eggs in short stems; larvae cause stunting of wheat plants or destroy developing heads. Second-generation adult larger bodied, macropterous and causes less damage. Syn. *Harmolita grandis* (Riley). Cf. European Wheat Sawfly; Hessian Fly; Wheat Jointworm.

WHEATFEED Fine bran with a small quantity of endosperm.

WHEEL BUG *Arilus cristatus* (Linnaeus) [Hemiptera: Reduviidae]: A predaceous Species found in eastern and southern North America. Eggs laid in clusters; nymphs and adults exsanguinate prey; can inflict painful bite to humans.

WHEELER, GEORGE CARLOS (1897–1991) American specialist in morphology and systematics of ant larvae.

WHEELER, GEORGE DOMRILE CHETWYND (1858–1947) (Turner 1948, Entomologist's Rec. J. Var. 60: 55–56.)

WHEELER, LIONEL RICHMOND (1888–1949) (Wheeler 1949, Proc. Linn. Soc. Lond. 161: 253–254.)

WHEELER, WILLIAM MORTON (1865–1937) American academic (Harvard University) known principally for his work on social insects and systematics of ants. (Essig 1931, *History of Entomology*. 1029 pp, (793–796), bibliogr.), New York; Brues 1937, Psyche 44: 61–96, bibliogr.; Melander *et al.* 1937, Ann. ent. Soc. Am. 30: 433–437; Evans & Evans 1970, *William Morton Wheeler, Biologist.* 363 pp. Cambridge, Mass.)

WHEELWRIGHT, HORACE WILLIAM (1815–1865) (Froggatt 1924, Aust. Nat. 11: 186–187.)

WHELAN, DON BION (1887–1969) (Hill 1969, J. Econ. Ent. 62: 1525.)

WHERRYMEN See Gerridae.

WHIFFIN, WALTER HARRY (–1887) (Riley 1924, Proc. S. Lond. ent. nat. Hist. Soc. 1924: 75.)

WHIP SCORPIONS See Uropygi.

WHIP SPIDER *Ariamnes* spp. [Araneida: Therididae] (Australia).

WHIPLASH ROVE BEETLE *Paederus australis* Guérin-Méneville; *Paederus cruenticollis* Germar [Coleoptera: Staphylinidae] (Australia).

WHIPSCORPIONS See Palpigradi.

WHIRLIGIG BEETLES See Gyrinidae.

WHIRLIGIG MITES See Anystidae.

WHISTLING SPIDERS *Selenocosmia* spp. [Araneida: Theraphosidae] (Australia).

WHITE, ADAM (1817–1879) (Dunning 1879, Proc. ent. Soc. Lond. 1879: lxiv; McLachlan 1879, Entomol. mon. Mag.15: 210–211.)

WHITE ALBATROSS *Appias albina albina* (Boisduval) [Lepidoptera: Pieridae] (Australia).

WHITE ANTS Termites. See Isoptera.

WHITE APPLE-LEAFHOPPER *Typhlocyba pomaris* McAtee [Hemiptera: Cicadellidae].

WHITE ARSENIC See Arsenious Trioxide.

WHITE, ARTHUR (1871–1919) (Anon. 1918. Pap. Proc. R. Soc. Tasm. 128: bibliogr.; Musgrave 1932, *A Bibliography of Australian Entomology 1775–1930,* viii + 380 pp. (353, bibliogr.), Sydney.)

WHITE-BACKED GARDEN SPIDER *Argiope trifasciata* (Forskal) [Araneae: Araneidae].

WHITE-BACKED PLANTHOPPER *Sogatella furcifera* (Horvath) [Hemiptera: Cicadellidae).

WHITE-BANDED CLERID *Paratillus carus* (Newman) [Coleoptera: Cleridae] (Australia).

WHITE-BANDED ELM LEAFHOPPER *Scaphoideus luteolus* Van Duzee [Hemiptera: Cicadellidae].

WHITE BRADYBAENA SNAIL *Bradybaena similaris* (Ferussac) [Sigmurethra: Bradybaenidae] (Australia).

WHITE-BRAND SKIPPER *Toxidia rietmanni* (Semper) [Lepidoptera: Hesperiidae] (Australia).

WHITE BUTTERFLIES See Pieridae.

WHITE CEDAR MOTH *Leptocneria reducta* (Walker) [Lepidoptera: Lymantriidae] (Australia).

WHITE, CHARLES ABIATHAR (1826–) (Marcou 1885, Bull. U.S. natn. Mus. 30: 113–181; Stanton 1898, Proc. U.S. natn Mus. 20: 627–642, bibliogr.)

WHITE CLOVER-HEAD WEEVIL *Apion flavipes* (Payk.) [Coleoptera: Curculiondae].

WHITE CUTWORM *Euxoa scandens* (Riley) [Lepidoptera: Noctuidae]. See Cutworms.

WHITE-CROSSED SEED BUG *Neacoryphus bicrucis* (Say) [Hemiptera: Lygaeidae].

WHITE CYPRESS LONGICORN *Uracanthus pallens* Hope [Coleoptera: Cerambycidae] (Australia).

WHITE DRUMMER *Arunta perulata* (Guérin-Méneville) [Homoptera; Cicadidae] (Australia).

WHITE, E BARTON (–1954) (Buxton 1955, Proc. R. ent. Soc. Lond. (C) 19: 68.)

WHITE-FACED HORNET *Dolichovespula maculata* [Hymenoptera: Vespidae].

WHITE-FIR NEEDLEMINER *Epinotia meritana* Heinrich [Lepidoptera: Tortricidae].

WHITEFLIES See Aleyrodidae; Aleyrodoidea.

WHITE-FOOTED HOUSE ANT *Technomyrmex albipes* (F. Smith) [Hymenoptera: Formicidae]: A widespread pest, probably endemic to Old World and adventive to New Zealand and Australia as a coastal pest; Workers often found on wooden structures, tree-trunks; typically nests around buildings, in wall voids, near kitchens. Workers move rather slowly and are omnivorous with preference for sweets. WFHA an occasional pest of electrical switches. Workers 2.5–3 mm long; body shiny black with pale Tarsi; Pedicel of one segment, without Node. Viewed dorsally, the worker Gaster is distinctly heart-shaped, and this Species may be mistaken for the myrmicine Genus *Crematogaster*. The latter, however has a two-segmented waist. When alarmed, worker raises the Gaster toward the source of disturbance in a manner similar to *Crematogaster*. Syn. Black Household Ant (Victoria, Australia).

WHITE, FRANCIS BUCHANAN (1842–1894) (Coates 1895, Trans. Proc. Perthsh. Soc. nat. Sci. 2: lv–lxvi, bibliogr.; Trail 1895, Ann. Scott. nat. Hist. 1895: 73–91, bibliogr.)

WHITE-FRINGED BEETLES *Graphognathus* spp. [Coleoptera: Curculionidae].

WHITE-FRINGED SWIFT *Sabera fuliginosa fuliginosa* (Miskin) [Lepidoptera: Hesperiidae] (Australia).

WHITE-FRINGED WEEVIL *Graphognathus leucoloma* (Boheman) [Coleoptera: Curculionidae]: A pest of sugarcane in Australia; accidentally introduced from South America. Adults feed on foliage, cause non-economic damage to leaves; most damage caused by larvae feeding on roots of germinating setts and ratoons, causing plant death or reduced vigour. Wide host range including lucerne, peanuts, legumes, corn, tobacco and vegetables.

WHITE, GERSHOM FRANKLIN (1873–1937) (Bishopp 1937, Proc. ent. Soc. Wash. 39: 184–188; Dove 1937, J. Parasit. 23: 579–580.)

WHITE, GILBERT (1720–1793) (Rose 1850, *New General Biographical Dictionary* 12: 473; Newton 1877, Notes and Queries (5) 7: 241–243, 264–265, bibliogr.)

WHITE GRASSDART *Taractrocera papyria papyria* (Boisduval) [Lepidoptera: Hesperiidae] (Australia).

WHITE GROUND PEARL *Promargarodes australis* Jakubski [Hemiptera: Margarodidae] (Australia).

WHITE GRUB See Scarabaeidae.

WHITE GRUBS Larvae of the scarab beetle Genera *Lachnosterna* or *Phyllophaga* which are pests of wheat, strawberries and other crops or ornamental plants. Larvae damage the roots of plants during the autumn or cause more serious damage during the spring. Heavily infested plants are killed. See Scarabaeidae. Syn. Grubworms. Rel. June Beetle.

WHITE, HENRY G (1850–1899) (Newcomb 1899, Ent. News 10: 110.)

WHITE ITALIAN SNAIL *Theba pisana* (Müller) [Sigmurethra: Helicidae] (Australia).

WHITE, JOHN (1550–1606) (Hulton & Quinn 1964, The American drawings of John White. Vol. 1. 179 pp. (12–14), London.)

WHITE KNEED SPIDER *Zosis geniculatus* (Oliver) [Araneida: Uloboridae] (Australia).

WHITE LACE LERP *Cardiaspina albitextura* Taylor [Hemiptera: Psyllidae] (Australia).

WHITE LEAFHOPPER *Cofana spectra* (Distant) [Hemiptera: Cicadellidae] (Australia).

WHITE-LINE DART MOTH *Scotia segetum* Schiffermuller [Lepidoptera: Noctuidae].

WHITE LINEBLUE *Nacaduba kurava parma* Waterhouse & Lyell [Lepidoptera: Lycaenidae] (Australia).

WHITE-LINED HAWK MOTH *Hyles lineata* (Fabricius) [Lepidoptera: Sphingidae] (Australia).

WHITE-LINED SPHINX *Hyles lineata* (Fabricius) [Lepidoptera: Sphingidae]. See White-lined Hawk Moth.

WHITE LOUSE SCALE *Unaspis citri* (Comstock) [Hemiptera: Diaspididae]: Endemic to southeast Asia and now a pest of *Citrus* in Australia and USA (Florida). Named from the appearance of the male scale cover: white, narrowly rectangular ca 1.0 mm long with three longitudinal ridges; female adult dark brown, ca 2 mm long, mussel-shaped with longitudinal ridge extending length of body. Female oviposits ca 150 eggs during 2–3 months; eclosion occurs immediately following oviposition; crawlers disperse over host plant but prefer trunks and mature wood of limbs. Multivoltine with up to six generations per year, depending upon climate. WLS apparently restricted to *Citrus* as a host plant. Syn. Citrus Snow Scale.

WHITE LOUSE-SCALE PREDATOR *Chilocorus circumdatus* Gyllenhal [Coleoptera: Coccinellidae] (Australia).

WHITE-MARGINED COCKROACH *Melanozosteria soror* (Brunner) [Blattaria: Blattidae].

WHITE-MARKED FLEAHOPPER *Spanagonicus albofasciatus* (Reuter) [Hemiptera: Miridae].

WHITE-MARKED SPIDER BEETLE *Ptinus fur* (Linnaeus) [Coleoptera: Anobiidae]: A widespread pest of stored products. Adult 2–4 mm long; Prothorax with two patches of white scales and white scales on the Elytra. Cf. Australian Spider Beetle.

WHITE-MARKED TREEHOPPER *Tricentrus albomaculatus* Distant [Hemiptera: Membracidae].

WHITE-MARKED TUSSOCK MOTH *Orgyia leucostigma* (J. E. Smith) [Lepidoptera: Lymantriidae]: A pest of ornamental trees in North America; completes 2–3 generations per year, depending upon climate. Adults dimorphic with male macropterous and female brachypterous. Overwinters in egg stage; eggs deposited in masses ca 50–100 on branches, trunk or other objects; eggs covered with accessory gland material which hardens to form protective cover; eclosion during spring. Larvae skeletonize leaves. See Lymantriidae. Cf. Gypsy Moth.

WHITE MILLIPEDE *Blaniulus guttulatus* Bosc [Julida: Blaniulidae] (Australia).

WHITE NYMPH *Mynes geoffroyi guerini* Wallace [Lepidoptera: Nymphalidae] (Australia).

WHITE-OAK BORER *Goes tigrinus* (DeGeer) [Coleoptera: Cerambycidae].

WHITE PALM SCALE *Phenacaspis eugeniae* (Maskell) [Hemiptera: Diaspididae] (Australia).

WHITE PEACH-SCALE *Pseudaulacaspis pentagona* (Targioni-Tozzetti) [Hemiptera: Diaspididae]: A serious pest of woody ornamentals and fruit trees in many parts of the world. In North America, a pest of stone fruits. Males white; females brownish-white. Endemic in China; distributed throughout Asia, Australia, Africa, Americas, western Europe, Caribbean and Pacific islands.

WHITE-PINE APHID *Cinara strobi* (Fitch) [Hemiptera: Aphididae].

WHITE PINE-CONE BEETLE *Conophthorus coniperda* (Schwarz) [Coleoptera: Scolytidae].

WHITE PINE-CONE BORER *Eucosma tocullionana* Heinrich [Lepidoptera: Tortricidae].

WHITE-PINE SAWFLY *Neodiprion pinetum* (Norton) [Hymenoptera: Diprionidae].

WHITE-PINE WEEVIL *Pissodes strobi* (Peck) [Coleoptera: Curculionidae]: A pest of pines and Norway spruce in eastern North America. Eggs laid in small pits made in leader bark; adults and larvae feed on inner bark, killing the leader. Overwinter as adults. Syn: Englemann Spruce Weevil; Sitka Spruce Weevil.

WHITE PRUNICOLA-SCALE *Pseudaulacaspis prunicola* (Maskell) [Hemiptera: Diaspididae]: A widespread, polyphagous pest of moderate significance.

WHITE RICE STEMBORER *Scirpophaga innotata* (Walker) [Lepidoptera: Pyralidae] (Australia).

WHITE-SHOULDERED HOUSE MOTH *Endrosis sarcitrella* (Linnaeus) [Lepidoptera: Oecophoridae]: A cosmopolitan, multivoltine pest of wool, feathers and grains. Male ca 6 mm long (wingspan 15 mm); female ca 10 mm long (wingspan 17–20 mm); head, Prothorax and base of wing white. Larva white.

WHITE-SPINE PRICKLY PEAR *Opuntia streptacantha* Lam. [Cactaceae]: A Species of cactus native to the New World and adventive elsewhere as a minor rangeland pest. Controlled with phytophagous insects. See Prickly Pear.

WHITE-SPOTTED SAWYER *Monochamus scutellatus* (Say) [Coleoptera: Cerambycidae].

WHITE SPRINGTAIL *Folsomia candida* (Willem) [Collembola: Isotomidae] (Australia).

WHITE-STEMMED GUM MOTH *Chelepteryx collesi* Gray [Lepidoptera: Anthelidae] (Australia).

WHITE-STRIPED WEEVIL *Perperus lateralis* Boisduval [Coleoptera: Curculionidae] (Australia).

WHITE SUGARCANE-SCALE *Aulacaspis tegalensis* (Zehn.)

WHITE-TAILED SPIDER *Lampona cylindrata* (L. Koch) [Araneida: Lamponidae] (Australia).

WHITE-TAILED SPIDERS See Drassidae.

WHITE-VEINED SKIPPER *Anisynta albovenata* Waterhouse [Lepidoptera: Hesperiidae] (Australia).

WHITE WAX PARASITE *Anicetus communis* (Annecke); *Paraceraptrocerus nyasicus* (Compere) [Hymenoptera: Encyrtidae].

WHITE WAX SCALE *Ceroplastes destructor* Newstead [Hemiptera: Coccidae]: A pest of citrus, mango, pear, persimmon, quince and numerous subtropical ornamental plants. Endemic to South Africa and adventive to Australia. Adult female white, globular, ca 6 mm diameter with relatively soft wax cover; resembles Chinese Wax Scale, Pink Wax Scale and Florida Wax Scale. Parthenogenetic; female lays ca 3,000 yellow eggs beneath cover; crawlers prefer to settle along midrib of host plant leaves; crawlers remain on citrus leaves 4–5 weeks before moving to permanent location on twigs (adult WWS typically occur on citrus twigs less than two years old - scale behaviour on other host plants may be different); wax cover developed after permanent settling; two nymphal instars; 1–2 generations per year depending upon climate and host plant. WWS produces honeydew which serves as substrate for sooty mould and other fungi.

WHITE, WILLIAM FARREN (1834–1899) (Anon.1899, Leopoldina 35: 184; Anon. 1899, Entomol. mon. Mag. 35: 216.)

WHITE, WILLIAM HENRY (1892–1951) (Hoyt & Caffrey 1951, J. Econ. Ent. 44: 435–436.)

WHITECAP Noun. (Pl., Whitecaps.) The flocculent, white, waxy covering produced by the crawler stage of some diaspidid scale insects.

WHITEHOUSE, FRANCIS CECIL (1879–1959) (Spencer 1963, Proc. Ent. Soc. Br. Columb. 60: 52.)

WHITEHOUSE, HAROLD BECKWITH (1883–1943) (Cockayne 1943, Entomologist's Rec. J. Var. 55: 94.)

WHITELEGGE, THOMAS (1850–1927) (McNeill 1929, Rec. Aust. Mus. 17: 265–277, bibliogr.)

WHITELY, HENRY (1844–1892) (Godman 1892, Proc. ent. Soc. Lond. 1892: liv.)

WHITEMOTH *Asmicridea edwardsi* (McLachlan) [Tricoptera: Hydropsychidae] (Australia).

WHITES See Pieridae.

WHITLOWS Paronycia.

WHITMAN, ALLEN (1836–1881) (Anon. 1881, Boston Advertiser 138 (115): 2.)

WHITMARSH, RAYMON DEAN (1885–1946) (Anon. 1947, Yrbk. Mass. Sta. Coll. Fernald Club 16: 25.)

WHITNEY, CHARLES PLINY (1838–1928) (Bequart 1933, Occ. Pap. Boston Soc. nat. Hist. 8: 81–88.)

WHITNEY, LEHR ARTIMAN (1890–1945) (Fullaway 1946, Proc. Hawaii. ent. Soc. 12: 492a-492b.)

WHITTINGHAM, WALTER GODFREY (1861–1911) (Metcalfe 1941, Entomologist 74: 191–192.)

WHITTLE, FRANCIS GEORGE (1854–1921) (Sheldon 1921, Entomologist 54: 302–303.)

WHORL Noun. (Anglo Saxon, *hweorfan* = to turn. Pl., Whorls.) 1. A ring or circle of Setae surrounding a body or appendage segment 2. A circular arrangement of structures which radiate outward from an imaginary centre as the spokes of a wheel.

WHORL MAGGOT *Hydrellia philippina* (Ferin) [Diptera: Ephydridae].

WHYMPER, EDWARD (1840–1911) (Soldanski 1912, Dt. ent. Z. 1912: 99.)

WICHMANN, HEINRICH E. (1889–1967) (Freude 1968, NachrBl. bayer. Ent. 17: 14–15.)

WICKHAM, ARCHDALE PALMER (1855–1935) (Hayward 1935, Entomologist 68: 291–292.)

WICKHAM, HENRY FREDERICK (1866–1933) (Buchanan 1934, Proc. ent. Soc. Wash. 36: 60–64.)

WIEDEMANN, CHRISTIAN RUDOLPH WILHELM (1770–1840) German academic (Professor of Natural History at Kiel) and author of *Diptera Exotica* (1821) and *Analecta Entomologica* (1824). (Papavero 1971, *Essays on the History of Neotropical Dipterology.* I: 111–113. São Paulo.)

WIEL, P VAN DER (–1962) Kruseman 1962, Ent. Ber., Amst. 22: 145–148, bibliogr.)

WIEPKIN, CARL FRIEDRICH (1815–1897) (Anon. 1897, Ent. News 8: 120.)

WIERZEJSKI, ANTONI (1843–1916) (Fedorowcz 1970, Memorab. zool. 21: 3–81.)

WIESMANN, ROBERT (1899–1972) (Büttiker 1972, Mitt. schweiz. ent. Ges. 45: 219–220.)

WIGGLESWORTH, VINCENT B (1900–1994) English academic and professor of biology at Cambridge University.

WIGHT, ROBERT ALLEN (1823–1896) (Howard 1897, Ent. News 8: 156–159.)

WIGHTMAN, ARCHIBALD JOHN CREWE (1884–1971) (Butler 1972, Proc. R. ent. Soc. Lond. (C) 36: 62; Haggett 1972, Proc. Brit. ent. nat. Hist. Soc. 5: 75–80.)

WIJVERKENS, FELIX (1900–1970) (Schepdael 1970, Linneana Belgica 4: 138–141.)

WILD SERVICE APHID *Dysaphis aucupariae* (Buckton) [Hemiptera: Aphididae] (Australia).

WILDER, GERRIT PARMELE (1863–1935) (Swezey 1936, Proc. Hawaii. ent. Soc. 9: 189–190.)

WILDING, RICHARD (1858–1950) (Harwood 1950, Entomol. mon. Mag. 84: 95.)

WILEMAN, ALFRED ERNEST (1860–1929) (Riley 1929, Entomologist 62: 215–216; West 1930, Zephyrus 2: 140–144, bibliogr.)

WILEY, ALEXANDER JOHN (–1959) (Uvarov 1960, Proc. R. ent. Soc. Lond. (C) 24: 54.)

WILKE, OTTO AMANDUS (1830–1898) (Anon. 1899, Z. Ent. Breslau 24: xxiii–xxiv.)

WILKES Published 120 copperplates of English Moths and Butterflies. (1773).

WILKES, BENJAMIN (–1749) (Lisney 1960, *Bibliography of British Lepidoptera 1608–1799,* 315 pp., (121–126, bibliogr.), London; Whalley 1972, J. Soc. Biblphy. nat. Hist. 6: 127.)

WILKIN, SIMON (1790–1862) (Anon. 1863, Proc. Linn. Soc. Lond. 1863: xlvi-xlix.)

WILKINSON, SAMUEL JAMES (1816–1903) (Anon. 1903, Entomologist's Rec. J. Var. 15: 275.)

WILKINSON, THOMAS (1818–1876) (Anon. 1876, Entomol. mon. Mag. 12: 279.)

WILL, HOMER CHRISTIAN (1898–1966) American academic (Juniata College, Huntingdon, Pa), and specialist on Symphyta. (Bull. ent. Soc. Am. 32 (2): 126.)

WILLARD, HAROLD FRANCIS (1884–1939) (Mason 1939, J. Econ. Ent. 32: 893; Pemberton *et al.* 1940, Proc. Hawaii. ent. Soc. 10: 447–449, bibliogr.)

WILLE, JOHANNES (–1959) (Anon. 1960, Boln. Soc. ent. agric. Peru 2: i, 1–2.)

WILLEMSE, CORNELIS JOSEPH MARIA (1888–1962) (Anon. 1962, Bull. Soc. ent. Fr. 67: 93; Boven 1966, Publtiës natuurh. Genoot. Limburg 16: 1–73, bibliogr.)

WILLET, JOSEPH EDGERTON (1826–1897) (dos Passos 1951, J. N. Y. ent. Soc. 59: 166.)

WILLEY, ARTHUR (1874–1942) (Cockayne 1944, Proc. R. ent. Soc. Lond. (C) 8: 70.)

WILLIAMS, BENJAMIN SAMUEL (1891–1941) (Harwood 1941, Entomol. mon. Mag. 77: 93.)

WILLIAMS, FRANCIS XAVIER (1882–1967) American natural historian, biological control worker for Hawaiian Sugar Planter's Association and specialist in the biology of bees and wasps. (Pemberton 1968, Proc. Hawaii. ent. Soc. 20: 247–255. bibliogr.)

WILLIAMS, JAMES TRIMMER (1894) (Adkin 1894, Entomologist 27: 228.)

WILLIAMS, JOHN BICKERTON (1848–1916) (Anon. 1916, Can. Ent. 48: 248.)

WILLIAMS, JOSEPH LEROY (1906–1965) (Pearson 1966, Proc. R. ent. Soc. Lond. (C) 30: 64; Schmeider 1968, Ent. News 79: 261–264, bibliogr.)

WILLIAMS, L (–1947) (Williams 1948, Proc. R. ent. Soc. Lond. (C) 12: 65.)

WILLIAMS, ROSWELL CARTER (1869–1946) (Bell 1946, Ent. News 57: 167–171; Hayward 1947, Revta. Soc. ent. argent. 13: 344–345.)

WILLIAMSON, EDWARD BRUCE (1877–1933) (Davis 1932, Proc. Indiana Acad. Sci. 41: 51; Davis 1934, Proc. Indiana Acad. Sci. 43: 23–25; Hungerford 1934, Ann. ent. Soc. Am. 27: 124; Calvert 1935, Ent. News 46: 1–13.)

WILLING, THOMAS NATHANIEL (1858–1920) Canadian naturalist and academic (University of Saskatchewan). (Riegert 1983, Bull. Ent. Soc. Canada 15 (4): 127–130.)

WILLIS, WARREN JENNISON (–1941) (Mickel 1942, Ann. ent. Soc. Am. 35: 128.)

WILLISTON, SAMUEL WENDELL (1852–1918) (Aldrich 1918, Ent. News 29: 322–327; Woodward 1919, Q. J. geol. Soc. Lond. 75: lv–lvi. Lull 1924, Mem. Acad. nat. Sci. 17: 1 15–141, bibliogr.)

WILLKOMM, HEINRICH MORITZ (1821–1895) (Möbius 1943, Dt. ent. Z. Iris 57: 26–27.)

WILLOCKS, FRANK C (1883–1955) (Anon. 1956, Bull. Soc. ent. Egypte 50: ii–iii, bibliogr.)

WILLOUGHBY-ELLIS, HERBERT (1869–1943) (Blair 1943, Entomol. mon. Mag. 79: 280.)

WILLOW BEAKED-GALL MIDGE *Rabdophaga rigidae* (Osten Sacken) [Diptera: Cecidomyiidae].

WILLOW FLEA WEEVIL *Rhynchaenus rufipes* (LeConte) [Coleoptera: Curclionidae].

WILLOW REDGALL SAWFLY *Pontania promixa* (Lepeletier) [Hymenoptera: Tenthredinidae].

WILLOW SAWFLY *Nematus ventralis* Say [Hymenoptera: Tenthredinidae].

WILLOW SHOOT SAWFLY *Janus abbreviatus* (Say) [Hymenoptera: Cephidae].

WILLRUTH, HERMANN (–1974) (Grüber 1974, Prakt. Schädlingsk. 26: 42.)

WILLUGHBY, FRANCIS (1635–1672) (Rose 1850, *New General Biographical Dictionary* 12: 506; Miall 1912, *Early Naturalists, Their Lives and Work.* 396 pp. (99–130), London.)

WILSE, JACON NICOLI (1736–1801) (Strand 1919, Arch. naturgesch. (A) 83: 154–156.)

WILSON (NÉE DALY), EILEEN VICTORIA (1900–1972) (McFarland 1974, Aust. ent. Mag. 2: 40–42.)

WILSON SPHINX *Hyles wilsoni* (Rothschild) [Lepidoptera: Sphingidae].

WILSON, CECIL CLINE (1896–1948) (Noble *et al.* 1948, J. Econ. Ent. 41: 525–526.)

WILSON, CHARLES BRANCH (1861–1941) (Calvert 1941, Ent. News 52: 269–270.)

WILSON, EDWARD O American academic (Harvard University) and specialist on ants, sociobiology and biodiversity.

WILSON, EDWIN (–1915) (Scott 1915, Entomol. mon. Mag. 51: 223.)

WILSON, GEORGE F See Fox Wilson, G.

WILSON, GEORGE RINGO (1885–1938) (Armitage 1938, J. Econ. Ent. 31: 549–550.)

WILSON, HARLEY FROST (1883–1959) (Allen 1959, J. Econ. Ent. 52: 788–789.)

WILSON, JAMES OTTO (1898–1972) (McFarland 1974, Aust. ent. Mag. 2: 40–42, bibliogr.)

WILSON, JOHN W (1902–1974) (Kuitert 1974, J. Econ. Ent. 67: 807.)

WILSON, OWEN S (–1890) (Anon. 1890, Entomol. mon. Mag. 26: 305)

WILSON, THOMAS (1836–1887) (Anon. 1887, Leopoldina 23: 111; Walker 1887, Entomologist 20: 168.)

WILSON, TOM (1856–1919) (Treherne 1917, Proc. ent. Soc. Br. Columb. 10: 30–31.)

WILSON, W A (1890–1968) (A.D.H. 1969, Proc. Somerset. archaeol. nat. Hist. Soc. 113: 112–113.)

WILT Verb. (Middle English, *welken* = to wither.) Loss of rigidity and drooping of a plant due to inadequate or excessive water.

WILT, CHARLES (1821–1886) (Horn 1886, Trans. Am. ent. Soc. 13: vi, xxi.)

WIMMEL, THEODOR FRIEDRICH (1860–1952) (Weidner 1967, Abh. Verh. naturw. Ver. Hamburg Suppl. 9: 193.)

WIMMERS, CARL ERNST (1873–1932) (Pfaff 1932, Ent. Z., Frankf. a. M. 46: 85.)

WINBLAD, GUSTAF (1893–1968) (Lindroth 1969, Opusc. ent. 34: 10.)

WINDOW FLIES See Scenopinidae; Anisopodidae.

WINDOW FLY *Scenopinus fenestralis* (Linnaeus) [Diptera: Scenopinidae]: A widespread pest distributed through commerce and frequently encountered on the windows of produce shops, mills and warehouses. Adult 3–6 mm long, dark bodied with red legs; Antenna brownish black; Thorax with short, recumbent pile. Larva thread-like, inhabits stored grain, flour and grain debris where it is predaceous upon larvae and pupae of Coleoptera, Lepidoptera and other stored product pests. Syn. Windowpane Fly. See Scenopinidae.

WINDOW MIDGES See Anisopodidae.

WINDOWPANE FLY See Window Fly.

WINDRATH-FREY, WALTER (1947) (Culatti 1948, Mitt. schweiz. ent. Ges. 21: 297.)

WINE FLIES See Drosophilidae.

WING Noun. (Middle English, *winge* > Old Norse, *vaengr*. Pl., Wings.) Insect wings are paired, often membranous and reticulated cuticular expansion of dorsolateral portion of Mesothorax and Metathorax. Wings are appendages derived from Integument and consist of a dorsal and ventral lamina. Forewings (primaries) are attached to Mesothorax; hindwings (secondaries) are attached to Metathorax. Wings are specifically adapted as organs of flight in insects, or modified to protect the pair of wings involved in flapping flight. Wings first appear in fossils during the Carboniferous, and were present in most Species collected from that Period. Insect groups (*e.g.* Families) that do not bear wings show many modifications to the Thorax. Environmental factors can promote or maintain flightlessness (*e.g.* island dwelling insect; cave-dwelling insect). Parasitism is a lifestyle in which wings may be a liability, particularly when living on a vertebrate host. Wingless parasitic insects comprise entire Orders (*e.g.* Mallophaga, Anoplura and Siphonaptera). See Apterygota; Pterygota; Wing Membrane. Cf. Forewing; Hindwing. Rel. Elytron; Haltere; Paranotal Lobe; Tegmina.

WING ARTICULATION Points of articulation with the Thorax that are necessary for wing movement. Thoracic components of these articulations include the Prealar Bridge, Anterior Notal Wing Process, Medial Notal Process and Posterior Notal Wing Process. See Wing. Rel. Humeral Plate; First Axillary Sclerite; Second Axillary Sclerite; Third Axillary; Sclerite.

WING BASE The proximal part of the Wing between the bases of the veins and the body. The WB contains the Humeral and Axillary Sclerites (Snodgrass). The part of a wing nearest the body. Cf. Articulatory Epideme.

WING BONES Nerves of the insect wings.

WING BUD The larval histoblast from which the wings of the adult develop.

WING CASE Coleoptera: The Elytron. Hemimetabola: Wing pad (Tillyard).

WING CELLS Areas of the wing inclosed by veins.

WING-COUPLING Any of several physical mechanisms that attach forewing to hindwing. Uncoupled wings are typical of insects with slow movement and erratic flight. Coupling mechanisms improve aerodynamic efficiency and increase flight speed. A Claustrum unites forewing and hindwing during flight. Types of wing coupling mechanisms include hooks of one wing attached to a thickened margin of the other wing. A Hamulus consists of curved spines along the costal margin of the hindwing that engage a Retinaculum of the forewing (Hymenoptera). Jugate Coupling occurs in Trichoptera and primitive moths. Amplexiform Jugate Coupling involves a forewing overlapping a developed Humeral Lobe of the hindwing. More advanced Jugate Coupling involves a slit in the forewing Jugum and either side of the slit covers the dorsal and ventral surfaces of the hindwing's costal margin. Primitive Frenate Coupling occurs in some Mecoptera. Frenular bristles of the hindwing's Humerus overlap the forewing Jugal Lobe to keep the hindwing in phase with the forewing. Frenate Coupling is common in Lepidoptera. Trichoptera possess two coupling mechanisms. Some Families (Hydrobiosidae, Philopotamidae) show a forewing's Jugum engaging Michrotrichia along the hindwing's Costa. In more powerful flying Trichoptera, Hamuli along the Costal margin of the hindwing engage a ventral longitudinal ridge along the posterior margin of forewing's Anal Veins. Some insects lock their wings together while on the ground, when burrowing through detritus or moving into compact spaces. External forces applied when squeezing into 'tight places' may cause wings to move or separate. Some Coleoptera (*e.g.* ground beetles) have developed mechanisms for locking the wings together or to the body while moving on the ground. Some Mecoptera display a wing-body locking mechanism that involves the forewing Jugal Lobe and Metascutellum. The forewing Jugal Lobe is thicker and more sclerotized than the wing membrane. The JLs ventral surface is covered with rows of Carinae that are angled posteriorly. These Carinae contact anteriorly projecting Carinae on the Mesoscutellum when wings are folded over the body. The result is a locking mechanism that prevents wings from moving away from the body when the insect moves through confined spaces. See Claustrum; Frenulum 1; Hamulus 5; Retinaculum.

WING COVERS 1. Parts of the chitinous Cuticle of larvae, nymphs or pupae that cover the rudiments of the wings of the adult. 2. Forewings of an adult when they are thicker than the hindwings and cover them when at rest. See Elytra; Tegmina; Hemelytra.

WING ENANTION A dipterous wing display: The scissor-like, cyclical movement of both wings parallel to the substrate above the Abdomen (Headrick & Goeden 1995). Cf. Wing Hamation; Wing Supination.

WING FLEXION LINE An imaginary line upon which the insect wing is folded at repose. Wing folding on flexion lines is common in the Insecta. (See Wootton 1979, Syst. Ent. 4: 81–93; 1981, J. Zool. 193: 447–468.) Cf. Plait; Wrapping Flexure.

WING HAMATION A wing display of some Diptera: The coordinated unidirectional, synchronous movement of both wings, in the same plane, above the Abdomen and parallel to the substrate (Headrick & Goeden 1995). Cf. Wing Enation; Wing Supination.

WING HEART Pulsatile Organs are formed from specialized skeletal muscles which contract rhythmically. WH have been found in Blattaria, Neuroptera, Coleoptera and Hymenoptera. See Accessory Circulatory Organ; Cercal Heart; Leg Heart.

WING INFLATION The biomechanical process by which wings are expanded after emergence from the pupal case or final nymphal instar. Haemolymph and hydrostatic pressure are involved in the process of inflation by some insects. See Accessory Circulatory Organ; Hydrostatic Skeleton.

WING LOUSE *Liperus caponis* (Linnaeus) [Mallophaga: Philopteridae]: A pest of chickens and other fowl. WL inhabits large wing feathers and feeds on hooklets. Cf. Brown Chicken-Louse; Chicken Head-Louse; Chicken Body-Louse; Fluff Louse; Shaft Louse.

WING MEMBRANE The thin, flexible, laminate, often translucent structural component of the insect wing. The membrane is ca 0.5 μ thin in some insects; beetle elytra may be more than 1 mm thick. In a very thin condition, the wing membrane may consist only of epicuticle. In thicker conditions, the membrane also contains exocuticle; Endocuticle may be absent from some areas of the forewing of some insects. Wing membrane laminae are a product of epidermal cells; epidermal cells become elongated and basement membranes of the dorsal and ventral surfaces fuse to form a middle membrane. Channels develop in areas where basement membranes do not fuse. These channels form a complex network of nerve cells, Haemolymph sinuses and tubular vessels (tracheoles). During subsequent development of the wing bud, epidermal cells near channels form a thick cuticular layer that ultimately becomes the vessel walls. When wing development is completed, the epidermal cells have disappeared, leaving the bilaminate cuticular wing structure. The wing surface generally has Setae, scales, or sculpturing. Setae are found in nearly all groups of insects and are the most commonly encountered structure on the insect wing. Microchaetae (Microtrichae, Aculeae) are short, non-articulated Setae on forewings of Diptera (*e.g.* mosquitoes). Macrochaetae are long, non-articulated Setae on the forewing of Trichoptera. A marginal fringe often occurs along the edge of wings. Some insects (*e.g.* Thysanoptera) display narrow wings with a long fringe; Hymenoptera possess a marginal fringe that is inconspicuous. The fringe reaches elaborate expression among Mymaridae (parasitic wasps) and Ptinidae (beetles) whose Species have reduced the wing blade to a narrow strip and the fringe is several times longer than the wing's width. Superficially, the wing resembles the wings of thrips. Coleoptera such as the Ptiliidae also have a conspicuous marginal fringe. Sculpture is common on insect wings, especially Species that have undergone significant modification not associated with flight (*e.g.* Isoptera). Tegmina of Orthoptera are often elaborately sculptured. Similarly, Elytra of beetles are often scluptured in ways that are diagnostically important. See Veins; Wing.

WING MODIFICATION Wings have become adapted for purposes other than flight. Many insects do not use the forewing as an active participant in flight. Instead, this wing has become structurally modified to protect the hindwing (*e.g.* Orthoptera, Hemiptera, Coleoptera, Dermaptera). These modifications give the wing a distinctive appearance and hence different names have been used. Orthopteran forewing is called the Tegmen; Hemipteran forewing is called the Hemelytron; Coleopteran forewing is called the Elytron. Forewing of these groups is relatively rigid or inflexible, held forward at a fixed angle during flight and over the abdomen at repose. Some insects have profoundly modified hindwings (*e.g.* Haltere of Diptera). See Wing. Cf. Elytron; Haltere; Hemelytron; Tegmen.

WING MUSCLES OF THE HEART See Alary Muscles (Snodgrass).

WING PADS Hemimetabola: The encased undeveloped wings of nymphs. WP appear behind the Thorax as two lateral, flattened structures.

WING REGIONS The principal areas of the wings including Axillary, Remigial, Vannal, Jugal and Anal. WR differentiated in the wing-flexing insects and often separated by distinct lines of folding (Snodgrass). See Anal Region; Axillary Region; Jugal Region; Remigium; Vannal Region.

WING SCALE 1. Hymenoptera: The Tegula. 2. Scales are common on the wings of Lepidoptera, some Coleoptera and Thysanoptera. Scales are anatomically complex structures that develop from modified Setae.

WING SHEATH A wing pad.

WING SIZE See Apterous; Brachypterous; Macropterous.

WING SUPINATION A wing display by Diptera: The clockwise rotation of the wing during extension. Wing supination is either synchronous (each wing is extended simultaneously) or asynchronous (each wing is extended in series) (Headrick & Goeden 1995). See Wing. Cf. Wing Enation; Wing Supination.

WING TEETH Collembola: The large tarsal teeth in some Species.

WING TRACHEAE The longitudinal, air-conducting vessels or tubes of the wings (Wardle). See Vein.

WING VEIN See Vein.

WING VENATION 1. The pattern of tubular vessels positioned between the two membranous layers of the wing in the adult insect (Wardle). 2. Veins are the most common element of wing design and more has been written about wing venation than any other aspect of wing anatomy. Naming of veins has been based on putative Homology among Taxa of insects... a foundation which is problematical. One source of argument comes from the problem of defining a vein. Unfortunately, an unambiguous definition of a vein is difficult to provide because ultrastruture of veins shows considerable variation. Homologizing veins among groups is also difficult owing to different load requirements of flight. Also, wings have become modified to serve functions in addition to flight, which has caused confusion and masked Homologies. Wing veins are always cuticular in construction and sometimes tubular in design. Chitin microfibrils of exocuticle in the veins are helicoidal; Chitin microfibrils of Endocuticle alternate in layers of helicoidal and unidirectional patterns. This type of information provides insight into vein strength. Tubular veins can contain Haemolymph, Tracheae or nerves. Veins are classified as Longitudinal Veins or Cross Veins.

WINGED WALKINGSTICKS See Phasmatidae.

WINGELMÜLLER, ALOIS (1848–1920) (Holdhaus 1920, Verh. zool.-bot. Ges. Wien 70: 107–108.)

WINGLESS Adj. Descriptive of an organism that lacks wings. See Apterous. Cf. Brachypterous; Macropterous.

WINGLESS CAMEL CRICKETS See Saltatoria.

WINGLESS COCKROACH *Calolampra elegans* Roth & Princis; *Calolampra solida* Roth & Princis [Blattodea: Blaberidae] (Australia).

WINGLESS GRASSHOPPER *Phaulacridium vittatum* (Sjostedt) [Orthoptera: Acrididae]: A pest endemic in Australia and a serious problem to improved pastures in higher rainfall regions of NSW, Vic, SA, Tas and WA. Univoltine with cyclical outbreaks ca 4–5 years depending on region and seasonal weather conditions. Pasture damage caused by feeding on clovers and other broad-leaved plants during dry summer weather; causes substantial reduction in green feed available for sheep and cattle.

WINGLESS LONG-HORNED GRASSHOPPERS See Gryllacrididae.

WINGLESS PRICKLY PEAR LONGICORN *Moneilema ulkei* Horn [Coleoptera: Cerambycidae] (Australia).

WINGLESS SOLDIER FLY *Boreoides subulatus* Hardy [Diptera: Stratiomyidae] (Australia).

WINGLETS Noun. 1. Small, concave-convex scales that are generally fringed at apex. 2. Dytiscidae: Rudimentary wings under the base of the Elytra.

WINGS OF THE HEART The series of diagonal and other muscular fibres above the diaphragm in the pericardial cavity. See Pericardial Diaphragm.

WINKELMANN, AUGUST (1899–1964) (Heddergott 1964, Anz. Schädlingsk. 37: 122.)

WINKLER, ALBERT (–1945) (Jarvis 1946, Entomol. mon. Mag. 82: 160.)

WINN, ALBERT F. (–1935) (Moore 1936, Can. Ent. 67: 255; Moore 1948, Can. Ent. 80: 22–23.)

WINNERTZ, JOHANNES (1800–1896) (Osten Sacken 1903, *Record of my Life Work in Entomology,* 240 pp. (44–47), Cambridge, Mass.)

WINN-SAMPSON See SAMPSON.

WINSLOW, JAMES BENIGNUS (1669–1760) (Rose 1850, *New General Biographical Dictionary* 12: 514.)

WINSPIT, WILLIAM JEREMIAH (–1966) (R. W. W. 1966, Entomologist's Rec. J. Var. 78: 271–272.)

WINTER CORBIE *Oncopera rufobrunnea* Tindale [Lepidoptera: Hepialidae] (Australia, Tasmania).

WINTER CRANE FLIES See Trichoceridae.

WINTER FORM PEAR PSYLLA *Psylla pyricola* Förster [Hemiptera: Psyllidae].

WINTER GNATS See Trichoceridae.

WINTER GRAIN-MITE *Penthaleus major* (Dugès) [Acari: Eupodidae].

WINTER MOTH *Operophtera brumata* (Linnaeus) [Lepidoptera: Geometridae]: A polyphagous moth whose larval stage is considered a serious defoliator of apple, oak, elm, maple and Sitka spruce. WM endemic to Palearctic and accidentally introduced into North America during 1949. WM regarded as a threat to orchards in Oregon, shade trees in British Columbia and hardwood forests in Nova Scotia. Females are flightless and emerge during autumn or early winter. WM overwinters in egg stage; mature larvae pupate in soil. Regarded as successfully controlled in Canada with biocontrol program involving a high-density parasite [*Cyzenis albicans* Fallen], a low density parasite [*Agrypon flaveolatum* (Gravenhorst)] and the action of endemic predators at low population density of WM.

WINTER STEM-WEEVIL *Ceutorhynchus picitarsis* Gyllenhal [Coleoptera: Curculionidae].

WINTER STONEFLIES See Capniidae.

WINTER TICK *Dermacentor albipictus* (Packard) [Acari: Ixodidae].

WINTER, GERTRUD (–1947) (Anon. 1948, Anz. Schädlingsk. 21: 15.)

WINTERS (WINTERSTEINER) FREDERICK EDWARD (1885–1946) (Chamberlain 1947, Bull. Brooklyn ent. Soc. 42: 72–74, bibliogr.)

WINTHEM, WILHELM VON (1800–1848) (Weidner 1967, Abh. Verh. naturw. Ver. Hamburg Suppl. 9: 101–102.)

WIPEOUT® See Hydramethylnon.

WIREN, EINAR (1889–1972) (Lindroth 1972, Entomologen 1: 27.)

WIREWORMS See Elateridae. Cf. False Wireworms.

WIRTH, WILLIS W (–1994) American taxonomic entomologist (USDA) and specialist in nematocerous Diptera (Ceratopogonidae).

WIRTHUMER, JOHANN (1886–1961) (Mayer 1975, Beitr. Landesk. oberösterreich Naturwiss. Reich 2 (2): 1–2.)

WISECUP, CLELL BURNS (1902–1959) (Lindquist & Reed 1959, J. Econ. Ent. 52: 789–790.)

WISKOTT, MAX (1840–1911) (Dittrich 1911, Jber. Ver. Schles. Insektenk. 1: xxvi–xxix.)

WISLIZENUS FREDERICK ADOLPHUS (1810–1889) (Weiss 1936, *Pioneer Century of American Entomology,* 320 pp. (231–232), New Brunswick.)

WISSELINGH, TACO HAJO VAN (1890–1968) (Barendrecht 1968, Ent. Ber., Amst. 29: 41–42.)

WISTRÖM, JOHAN ALFRED (1830–1896) (Aurivillius 1896, Ent. Tidskr. 17: 293–297.)

WITCH HAZEL LEAF-GALL APHID *Hormaphis hamamelidis* [Hemiptera: Aphididae].

WITHYCOMBE, CYRIL LUCKES (1898–1926) (Anon. 1926, Entomologist's Rec. J. Var. 37: 178; Poulton 1926, Proc. ent. Soc. Lond. 1: 77–78.)

WITJUTI GRUB *Xyleutes* sp. from witjuti tree *(Acacia kempeana* F. Muell.) [Lepidoptera: Cossidae] (Australia).

WITLACZIL, EMANUEL (–1926) (Anon. 1926, Wien. ent. Ztg. 43: 194.)

WITTFELD, ANNIE M (1865–1888) (dos Passos 1951, J. N. Y ent. Soc. 59: 162.)

WITTPEN, JAN HENDRICK ERNST (1878–1956) (Wiel 1956, Ent. Ber., Amst. 16: 73–75.)

WITTROCK, VIET BREECHER (1839–1914) (Palmen 1915, Memo Soc. Fauna Flora fenn. 41: 104.)

WITTSTADT, HEINRICH (1888–1974) (Riesch 1974, Ent. Z., Frankf. a. M. 84: 270; Garthe 1975, Atalanta, Munich 6: 61–62.)

W-MARKED CUTWORM *Spaelotis clandestina* (Harris) [Lepidoptera: Noctuidae]. See Cutworms.

WN See West Nile Fever.

WOCKE, GEORG (1849–1907) (Anon. 1907, Ent. Z., Frankf. a. M. 21: 182.)

WOCKE, MAXIMILIAN FERDINAND (1820–1906) (Standuff 1906, M. Dt. ent. Z. Iris 19: 145–157.)

WOGHLUM, RUSSEL SAGE (1882–1968) (Lewis & Lafollette 1968, J. Econ. Ent. 61: 1476.)

WOLBACHIA A Genus of endocellular protobacteria that occurs within arthropods, including insects, mites and isopods. *Wolbachia* Species manipulate the reproductive biology of their hosts. Signs of manipulation include Cytoplasmic Incompatibility in all hosts, parthenogenesis in haplo-diploid hosts, and the expression of feminine characteristics in isopods. See Cf. Rickettsia.

WOLCOTT, GEORGE N American economic entomologist, most noted for 'The Insects of Puerto

Rico' (1948–1952, J. Agric. Univ. Puerto Rico 32: 1–975).

WOLCOTT, ROBERT HENRY (1868–1934) (Anon. 1934, Ent. News 45: 112; Hungerford 1935, Ann. ent. Soc Am. 28: 180.)

WOLDSTEDT, FREDERIK WILHELM (1847–1884) (Hellen 1950, Notul. ent. 30: 26–31.)

WOLF SPIDER See Lycosidae.

WOLFAHRTIA TRAUMATIC MYIASIS A form of Myiasis in humans and domestic animals occurring in warm parts of the Palaearctic Realm and caused by sarcophagid larvae of *Wolfahartia magnifica*. Fly larviposits into wounds, mucous membranes, eyes, ears or genital openings. Larval development requires 5–7 days. See Myiasis.

WOLFF, JOHN FREDERICK German physician and author of *Icones Cimicum descriptionibus illustratae* (1804).

WOLFF, MAX (1879–1963) (Petzsch 1964, Anz. Schädlingsk. 37: 11.)

WOLFF, NIELS, L (1900) (Tuxen 1968, Ent. Meddr. 36: 95 only.)

WOLFSCHLÄGER, ROMAN (-1950) (Anon. 1958, Z. wien. ent. Ges. 43: 32; Klimesch 1958, Z. wien. ent. Ges. 43: 82–84.)

WOLLASTON, EDITH (–1911) (Bethune-Baker 1911, Entomologist's Rec. J. Var. 23: 325.)

WOLLASTON, THOMAS VERNON (1822–1878) (Kraatz 1878, Dt. ent. Z. 22: 228–229; Westwood 1878, Proc. ent. Soc. Lond. 1877: xxxviii.)

WOLLENWEBER, H W (–1949) (Anon. 1949, Anz. Schädlingsk. 22: 78.)

WOLLEY-DOD, FREDERIC HOVA (1919) (F.C.W. 1919. Can. Ent. 51: 239–240.)

WOMBAT TICK *Aponomma auruginans* Schulze [Acari: Ixodidae] (Australia).

WOMERSLEY, HERBERT (1889–1962) (Varley 1963, Proc. R. ent. Soc. Lond. (C) 27: 51; Southoot 1964, Rec. S. Aust. Mus. 14: 603–632, bibliogr.)

WONDER BROWN BUTTERFLY *Heteronympha mirifica* (Butler) [Lepidoptera: Nymphalidae] (Australia).

WONFOR, THOMAS W (1828–1878) (Goss 1878, Entomologist 11: 259–260.)

WOOD Noun. (Old English, *wudu* = wood. Pl., Woods.) The hard, fibrous substance which comprises the bulk of trees and shrubs; wood occurs beneath the bark and consists of the xylem elements. Wood is arbitrarily divided into hardwoods and softwoods. See Cellulose; Lignin; Tannin.

WOOD COCKROACHES See Blattellidae.

WOOD CRICKETS See Gryllacrididae.

WOOD GNATS See Anisopodidae.

WOOD MOTHS See Cossidae.

WOOD NYMPH BUTTERFLIES See Nymphalidae.

WOOD NYMPHS See Satyridae.

WOOD SCORPION *Cercophonius* spp. [Scorpionida: Bothriuridae]. (Australia).

WOOD WASPS See Siricidae; Xiphydriidae.

WOOD WHITE BUTTERFLIES *Delias aganippe* (Donovan) [Lepidoptera: Pieridae] (Australia).

WOOD, GORDON F (1933–1970) (Butler 1972, Proc. R. ent. Soc. Lond. (C) 36: 62.)

WOOD, H WORSLEY (1878–1943) (E.A.C. 1943, Entomologist's Rec. J. Var. 55: 78.)

WOOD, HERBERT POLAND (1883–1925) (Bishop 1926, J. Econ. Ent. 19: 574.)

WOOD, HORATIO CHARLES (1841–1919) (Gunthrop 1920, Can. Ent. 52: 112–114, bibliogr.)

WOOD, JOHN GEORGE (1827–1889) (Anon. 1889, Entomol. mon. Mag. 25: 262.)

WOOD, JOHN HENRY (1841–1914) (Anon. 1914, Entomologist's Rec. J. Var. 26: 256–258, bibliogr.)

WOOD, LEONARD (1874–1941) (Watkins 1942, Entomologist 75: 24.)

WOOD, SAMUEL T (1860–1917) (Anon. 1918, Can. Ent. 50: 34–35.)

WOOD, THEODORE (1862–1923) (Anon. 1924, Entomol. mon. Mag. 60: 20; Anon. 1924, Nature 115: 21.)

WOOD, W English author of *Illustrations of the Linnean Genera of Insects* (2 vols, 1821) and *Index Entomologicus*, or a complete illustrated *Catalogue of the Lepidopterous Insects of Great Britain*.

WOOD, WILLIAM (fl 1629–1671) (Weiss 1936, *Pioneer Century of American Entomology,* 320 pp. (123), New Brunswick.)

WOODBRIDGE F C (1865–1950) (Wigglesworth 1951, Proc. R. ent. Soc. Lond. (C) 15: 76.)

WOODCOCK, GEORGE S (1886–1954) (Hall 1956, Proc. R. ent. Soc. Lond. (C) 20: 76.)

WOODFORDE, FRANCIS CARDEW (1846–1928) English educator (Principal, Market Drayton Grammar School) and amateur lepidopterist. (Walker 1928, Entomol. mon. Mag. 64: 237–238.)

WOODHILL, ANTHONY REEVE (1900–1965) (Salter 1965, J. ent. Soc. Aust. (N.S.W.) 2: 52–56, bibliogr.)

WOODLAND EARWIG *Chelidurella acanthopygia* Gene [Dermaptera: Forficulidae].

WOODLICE Members of the Crustacean Suborder Oniscidea which have invaded terrestrial habitats. Cf. Slaters.

WOODLOUSE FLIES See Rhinophoridae.

WOOD-MASON, JAMES (1846–1893) (Anon. 1893, Entomol. mon. Mag. 29: 145–146; Elwes 1893, Proc. ent. Soc. Lond. 1893: lvi.)

WOODROFFE, GERALD ERNEST (1923–1975) (Coombs 1976, Entomol. mon. Mag. 111: 97–98.)

WOODROSE BUG *Graptostethus manillensis* (Stål) [Hemiptera: Lygaeidae].

WOODRUFF, LAURENCE CLARK (1902–1986) American academic (University of Kansas) and administrator. Publications deal with Blattaria. Editor, Journal Kansas Entomological Society (1971–1982). (Byers 1987, JKES 60: 1–3).

WOODRUFF, LEWIS BARTHOLOMEW (1868–1925) (Davis 1926, J. N. Y. ent. Soc. 34: 23–25.)

WOODS WEEVIL *Nemocestes incomptus* (Horn) [Coleoptera: Curculionidae].

WOODTHORPE, EDWARD (–1899) (J.E.M. 1899, Naturalist, Hull 1899: 288.)

WOODWARD, GEORGE CALTHORPE (1882–1955) (Hall 1956, Proc. R. ent. Soc. Lond. (C) 20: 76.)

WOODWARD, SAMUEL PICKWORTH (1821–1865) (Anon. 1866, Proc. Linn. Soc. Lond. 1865–66: lxxxvi–lxxxvii.)

WOODWORTH, CHARLES WILLIAM (1865–1940) (Essig 1940, Science 92: 570–572; Essig 1941, J. Econ. Ent. 34: 128–129.)

WOOL MOTHS See Tineidae.

WOOLEY, PAUL H (1925–1968) (Anon. 1968, Newsl. Michigan ent. Soc. 13(4): 4.)

WOOLLATT, LEIGHTON H (1905–1974) (Turner 1976, Entomol. mon. Mag. 111: 63–64.)

WOOLLETT, GEOFFREY F C (1884–1965) (Pearson 1965, Proc. R. ent. Soc. Lond. (C) 30: 64.)

WOOLLY ALDER-APHID *Paraprociphilus tessellatus* (Fitch) [Hemiptera: Aphididae].

WOOLLY APHID *Eriosoma lanigerum* (Hausmann) [Hemiptera: Aphididae] (Australia). See Woolly Apple Aphid.

WOOLLY APHID PARASITE *Aphelinus mali* (Haldeman) [Hymenoptera: Aphelinidae]: In Australia, an effective parasite of the woolly aphid (*Eriosoma lanigerum*).

WOOLLY APPLE-APHID *Eriosoma lanigerum* (Hausmann) [Hemiptera: Aphididae]: A cosmopolitan pest whose nymph and adult suck fluid from roots and bark of branches; feeding induces roots to develop nodules, twigs swell and knot. WAA is a vector of Perennial Canker disease. WAA overwinters as an egg or nymph in cold climates; overwinters as an adult in warmer climates. In North America, eggs typically deposited in bark of elm during fall; hatch during spring, feed for two generations, develop winged adults which migrate to apple, hawthorn, mountain ash or other host plants. Some aphids move to roots and feed; wingless males and females appear during autumn, mate; female lays one egg on bark. Some resistant rootstocks have been developed for apples. Syn. Woolly Aphid (Australia). See Perennial Canker.

WOOLLY BEAR See Furniture Carpet Beetle; Arctiidae.

WOOLLY BEAR CATERPILLAR *Spilosoma glatignyi* (Le Guillou) [Lepidoptera: Arctiidae] (Australia).

WOOLLY ELM-APHID *Eriosoma americanum* (Riley) [Hemiptera: Aphididae].

WOOLLY PEAR-APHID *Eriosoma pyricola* Baker & Davidson [Hemiptera: Aphididae].

WOOLLY PINE APHID (Qld) *Pineus pini* (Macquart) [Hemiptera: Adelgidae] (Australia).

WOOLLY WHITEFLY *Aleurothrixus floccosus* (Maskell) [Hemiptera: Aleyrodidae]: A New World pest of *Citrus* which has become widespread; origin uncertain. Common in Caribbean, Central and South America. Controlled in some areas with parasitic wasps. Cf. Citrus Blackfly.

WORKER Noun. (Middle English, *werk*, *weore* = work. Pl., Workers.) 1. The neuter or sterile individuals within the colony of a social Species. 2. Isoptera: The sexually undeveloped forms, except the so-called soldiers. 3. Hymenoptera: Individuals anatomically female which lack the capacity to reproduce or lay unfertilized eggs which produce only males. Workers are responsible for nest building, brood care and colony defence. See Social Insect; Division of Labour; Eusociality.

WORM BUSTER® A registered biopesticide derived from *Bacillus thuringiensis* var. *kurstaki*. See *Bacillus thuringiensis*.

WORM, OLAUS (1588–1654) (Rose 1850, *New General Biographical Dictionary* 12: 430; Henriksen 1921, Ent. Meddr. 15: 17–19.)

WORM-HANSEN, JOHAN GEORG (1875–1961) (Tuxen 1961, Ent. Meddr. 31: 153–156, bibliogr.)

WORMLION Noun. Larval stage of *Vermileo* which construct conical pits in sand. See Vermileonidae.

WORMSBACHER, HENRY (–1934) (Anon. 1934, Ent. News 45: 202.)

WORTHLEY, HARLAN NOYES (1895–1967) (Campbell 1967, J. Econ. Ent. 60: 1769–1770.)

WORTHLEY, LEON HOWARD (1877–1937) (Anon. 1937, Can. Ent. 69: 276; Walton 1938, J. Econ. Ent. 31: 132–135.)

WOTTON, EDWARD (1492–1555) (Eiselt 1836, *Geschichte, Systematik und Literatur der Insektenkunde*. 255 pp. (15–16), Leipzig; Rose 1950, *New General Biographical Dictionary* 12: 531.)

WOUNDED TREE BEETLES. See Nosodendridae.

WOYTKOWSKI, FELIX (1892–1936) (Anon. 1970, Boln. Soc. ent. Peru 5: 32; Wielopolska 1974, *Peru, my unpromised land*. [In Polish]. 304 pp. Warsaw.)

WRANGELL, FERDINAND PETROVICH (1796–1870) (Essig 1931, *History of Entomology*, 1029 pp. (802), New York.)

WRAPPING FLEXURE A form of wing bending in which the apical portion of the wing, in the flexed position, is notched with the region on either side of the notch wrapped downward over the dorsum of the metasoma. Flexure is facilitated by two other convex folds, one at the costal margin and one at the vannal margin. The phenomenon appears unique to the Eucoilidae (Hymenoptera). Danforth & Michener 1988, 81: 345.

WRATISLAW, ALBERT HENRY (1822–1892) (Anon. 1892, Entomologist 25: 328.)

WRIGHT, ALBERT ALLEN (1846–1905) (Anon. 1905, Ent. News 16: 160; Wilder 1906, Bull. geol. Soc. Am. 17: 687–690, bibliogr.)

WRIGHT, ALBERT EDWARD (1872–1950) (Anon. 1950, Rep. Raven ent. nat. Hist. Soc. 1950: 43; Wigglesworth 1951, Proc. R. ent. Soc. Lond. (C) 15: 76.)

WRIGHT, CHANCEY (1830–1875) (Anon. 1876,

Proc. Am. Acad. Arts Sci. 11: 350.)

WRIGHT, F R ELLISTON See Elliston-Wright, F. R.

WRIGHT, JOHN CASSIMIR (1812–1924) (Engelhardt 1926, Bull. Brooklyn ent. Soc. 21: 128.)

WRIGHT, JOHN McMASTER (1917–1961) (Apple *et al.* 1962, J. Econ. Ent. 55: 151.)

WRIGHT, WILLIAM GREENWOOD (1830?-1912) (Coolidge 1912, Ent. News 22: 11–13; Grinnell 1913, Ent. News 24: 91–92.)

WRIGHT, WILLIAM SHERMAN (1866–1933) (Essig 1934, Ent. News 45: 27–28, bibliogr.)

WRINKLE Noun. (Anglo Saxon, *wrincle, wrencan* = to twist. Pl., Wrinkles.) 1. A ridge, furrow or prominence formed on an otherwise smooth surface. 2. A twisting, winding or sinuous formation. 3. A ripple on the surface of a liquid. Syn. Corrugated. See Surface Sculpture.

WRINKLED BARK BEETLES See Rhysodidae.

WRINKLED DUNE SNAIL *Candidula intersecta* (Poiret) [Sigmurethra: Helicidae] (Australia).

WRINKLED SUCKING LICE See Haematopinidae.

WRITHLED Adj. An arcane term for wrinkled. See Wrinkle.

WROUGHTON, ROBERT CHARLES (1849–1921) (Anon. 1921, Entomol. mon. Mag. 57: 161.)

WULFEN, FRANZ XAVER VON (1728–1805) (Nardo 1868, Comment. Fauna Flora Veneto 1: 201–210.)

WULFF, KURT (1881–1939) (West 1940, Ent. Meddr. 20: 587–589.)

WULLSCHLEGEL, ARNOLD (1849–1912) (Wheeler 1912, Entomologist's Rec. J. Var. 24: 317.)

WULLSCHLEGEL, JAKOB (1818–1905) (Thut 1909, Mitt. aargau. naturJ: Ges. 11: 114–117, bibliogr.)

WULP, F M VAN DER See van der Wulp, F. M.

WÜNN, HERMAN (–1954) (Reyne 1954, Ent. Ber., Amst. 15: 180.)

WUTZDORFF, HERLNAN (1843–1909) (Horn 1910, Dt. Ent. Z. 1910: 113.)

WYATT (KWIAT), ALEX K (1878–1971) (Irwin 1964, J. Lepid. Soc. 28: 368–371, bibliogr.)

WYATT, COLIN W (–1974) (Bustillo 1975, Shilap 12: 240.)

WYGODZINSKY, PETR WOLFGANG (1916–1987) German born and Swiss educated general entomologist with diverse interests in Apterygota, Heteroptera, Diptera and Coleoptera. Employed in Brazil 1941–1947, Argentina 1948–1962, American Museum of Natural History from 1962. Author of more than 250 scientific papers. (Schuh & Herman 1988, J. N. Y. ent. Soc. 96 (2): 227–244).

WYMAN, JEFFRIES (1814–1874) (Gray 1874, Am. J. Sci. (3) 8: 323–324; Anon. 1875, Proc. Linn. Soc. Lond. 1874–75: lxx–lxxv.)

WYTSMAN, PHILOGENE AUGUSTE GALILEE (1866–1925) (Hedicke 1925, Dt. Ent. Z. 1925: 250.)

X CHROMOSOME A single unpaired sex Chromosome.

X GLAND See Wagner's Organ.

XAMBEU, PIERRE (1837–1917) (Desbordes 1917, Bull. Soc. ent. Fr. 1917: 189.)

XANTHIC Adj. (Greek, *xanthos* = yellow; *-ic* = of the nature of.) Yellowish.

XANTHIN Noun. (Greek, *xanthos* = yellow. Pl., Xanthins.) A yellow carotinoid pigment that occurs in flowers. See Pigmentary Colour. Cf. Carotin.

XANTHOCHROIC Adj. (Greek, *xanthos* = yellow; *chros* = skin colour; *-ic* = of the nature of.) Pertaining to a yellow Integument.

XANTHOCHROISM Noun. (Greek, *xanthos* = yellow; *chros* = skin colour; English, *-ism* = condition.) Hymenoptera: Male Sphecidae that are normally black and yellow bodied, but which are occasionally yellow.

XANTHOMERA SKIPPER *Neohesperilla xanthomera* (Meyrick & Lower) [Lepidoptera: Hesperiidae] (Australia).

XANTHOMMATIN Noun. (Greek, *xanthos* = yellow; *omma* = eye.) A brown eye colour in *Drosophila* consisting of ommachorome pigment formed from the condensation of 2 hydroykynure 9 molecules.

XANTHOPHYLL Noun. (Greek, *xanthos* = yellow; *phyllon* = leaf. Pl., Xanthophylls.) A member of a group of yellow or brown carotinoid pigments formed from the oxygenation of carotenes. Widely distributed in plants and consumed by insects, and one of the colouring substances in insect blood.

XANTHOPISIN Noun. (Greek, *xanthos* = yellow; *opsis* = sight.) A colouring substance impregnating the retinular elements in night eyes of insects (Imms).

XANTHOS Adj. (Greek) Yellow. Alt. Xanthous.

XANTUS DE VESEY, LOUIS JOHN (1825–1894) (Essig 1931, *History of Entomology,* 1029 pp. (804–808), New York.)

XENASTEIIDAE Hardy 1980. Plural Noun. A Family of acalyptrate Diptera consisting of less than 10 Species found on islands in Indian and Pacific Oceans and coastal Australia. Body minute; wing venation reduced without crossveins; CuA + 1A absent. Biology and habits unknown. Syn. Tunisimyiidae.

XENOBIOSIS Noun. (Greek, *xenos* = guest, stranger; *bios* = living; *-osis* = condition. Pl., Xenobioses.) A commensalistic relationship in which one ant Species lives as a guest in the nest of another ant Species in mutual toleration and mingling freely with the host Species, but each Species maintaining its own brood. See Biosis; Commensalism; Inquilinism. Cf. Abiosis; Anhydrobiosis; Antibiosis; Archebiosis; Calobiosis; Cleptobiosis; Hamabiosis; Kleptobiosis; Lestobiosis; Parabiosis; Phylacobiosis; Plesiobiosis; Synclerobiosis; Trophobiosis.

XENOBIOTIC Adj. (Greek, *xenos* = guest, stranger; *biosis* = living.) Pertaining to Xenobiosis.

XENOMONE Noun. (Greek, *xenos* = guest; *hormaein* = to excite. Pl., Xenomones.) (Chernin 1970, Bioscience 20: 845.). A chemical mediating interspecific reactions.

XENOSOME Noun. (Greek, *xeno* = guest; *soma* = body. Pl., Xenosomes.) All DNA-containing, membrane-bounded bodies or organelles found within eukaryotic cells. (Corliss 1985, J. Protozool. 32: 373–376.) Cf. Soldo & Godoy 1973, Prog. Protozool. 4th Inter. Congr. Protozool., p 390.

XENTARI® A registered biopesticide derived from *Bacillus thuringiensis* var. *aizawai* serotype H-7. See *Bacillus thuringiensis.*

XERIC Adj. (Greek, *xeros* = dry; *-ic* = characterized by.) Pertaining to dry conditions or an absence of moisture. Cf. Hygric; Mesic. Rel. Habitat.

XEROMELISSINAE A Subfamily of Colletidae (Apoidea) consisting of about five Genera.

XEROPHILE Noun. (Greek, *xeros* = dry; *philein* = to love. Pl., Xerophiles.) A terrestrial organism which lives and reproduces under relatively dry conditions. See Xerotolerant. Cf. Halophile; Mesophile; Osmophile; Thermophile; Psychrophile. Rel. Ecology.

XEROPHILOUS Adj. (Greek, *xeros* = dry; *philein* = to love; Latin, *-osus* = possessing the qualities of.) Pertaining to organisms that live in arid or dry places. Alt. Xerophytic. See Habitat. Cf. Arenicolous; Deserticolous; Eremophilous; Ericeticolous; Lapidicolous; Psammophilous; Saxicolous. Rel. Ecology.

XEROPHYTE Noun. (Greek, *xeros* = dry; *phyton* = a plant. Pl., Xerophytes.) A plant that is structurally and physiologically adapted to live in arid or dry places.

XEROTOLERANT Noun. (Greek, *xeros* = dry; Latin, *tolerare* = to bear; *-antem* = an agent of something.) The capacity of an organism to survive when exposed to dry conditions. Cf. Xerophile.

XIPHIPHORA SKIPPER *Neohesperilla xiphiphora* (Lower) [Lepidoptera: Hesperiidae] (Australia).

XIPHOSURAN Plural Noun. (Greek, *xipho* = sword; *sura* = a tail.) An order of primitive Arachnida made up of king crabs.

XIPHYDRIIDAE Leach 1815. Plural Noun. Wood Wasps. A small Family of Symphyta (Hymenoptera) assigned to the Superfamily Siricoidea. Wasps typically uncommon but widespread except in Africa. Body more than 14 mm long; Antenna setaceous with 13–19 segments; Hypostomal Bridge present; cervical sclerites elongate; fore Tibia with one spur or two spurs of unequal length; middle Tibia without subapical spur; Abdomen cylindrical with Tergum medially divided; apical Tergum not elongated into spine-like process; Ovipositor projecting well beyond apex of Abdomen; male genitalia orthandrous.

XMC A carbamate compound {3,5-Xylyl-N-methylcarbamate} used as a systemic insecticide for control of leafhoppers, planthoppers, snails and slugs. Not registered for use in USA. Trade names include Cosban® and Macbal®. See Carbamate Insecticide.

XTARGRO® See Ethephon.

XYELIDAE Newman 1834. Plural Noun. Xyelid Sawflies. A Holarctic Family of Symphyta. Body usually less than 5 mm long; third antennal segment longest and formed by fusion of several segments; at least nine distal segments considerably smaller; Hypostomal Bridge absent; Pronotum long with posterior margin nearly transverse; forewing Rs branched near apex; middle and posterior Tibiae with subapical spurs; Ovipositor projecting beyond apex of Gaster. Male genitalia strophandrous or orthandrous. Xyelini associated with staminate pine cones; Xyelectiini and Pleroneurini live in developing buds or shoots of pines; Macroxyelini live on leaves of Ulmaceae or Juglandaceae.

XYELOIDEA Newman 1834. A Superfamily of Symphyta (Hymenoptera) which includes Xyelidae. Generally regarded as most primitive of Hymenoptera with fossils dating from Upper Jurassic Period.

XYELOTOMIDAE Rasnitsyn 1968. Plural Noun. A fossil Family of Symphyta (Hymenoptera) assigned to the Tenthredinoidea.

XYELYDIDAE Rasnitsyn 1968. Plural Noun. A fossil Family of Symphyta (Hymenoptera) assigned to the Superfamily Pamphilioidea.

XYLAN Noun. (Greek, *xylon* = wood; *-an* = adjectival suffix meaning characteristic of.) Wood-gum.

XYLANASE Noun. (Greek, *xylon* = wood; *-ase* = chemical suffix for an enzyme.) An enzyme in wood-feeding Coleoptera which has the property of hydrolysing xylan.

XYLEM Noun. (Greek, *xylon* = wood.) A complex tissue found in higher plants which provides structural support and forms Tracheae that conduct water, amino acids and minerals from roots up the stem. Used as a source of nutrition by some fluid-feeding insects (Cicadellidae, Cercopidae, Coccidae, Cicadidae). Cf. Phloem.

XYLENE (XYLOL) Noun. (Greek, *xylon* = wood; *-ene* = chemical suffix.) A colourless, transparent, volatile liquid which is insoluble in water and very soluble in alcohol. Used in clearing tissues for microscopic mounts and as a solvent for Canada Balsam.

XYLOCOPINAE A Subfamily of Anthophoridae consisting of about 20 Genera. See Apoidea. Cf. Carpenter Bees. Rel. Bumble Bee; Honey Bee.

XYLOID Adj. (Greek, *xylon* = wood; *eidos* = shape.) Wood-like; resembling wood in texture or physical characteristics. See Wood. Rel. Eidos; Form; Shape.

XYLOMYCETOPHAGOUS Adj. (Greek, *xylon* = wood; *mykes* = fungus; *phagein* = to devour; Latin, *-osus* = possessing the qualities of.) Pertaining to insects which feed on wood and fungi (after Francke-Grossmann 1963, Ann. Rev. Ent. 8: 421.) See Feeding Strategy.

XYLOMYIDAE Plural Noun. A numerically small, geographically widespread Family of brachycerous Diptera assigned to Superfamily Stratiomyoidea. Adult wasp-like, slender, 5–15 mm long; blackish coloured with bright markings; forewing cell M3 closed; hind Femur enlarged. Adults inhabit wooded areas; larvae occur under bark and are presumed predators or scavengers. Syn. Solvidae; Xylomyiidae.

XYLOMYIIDAE See Xylomyidae.

XYLOPHAGA (Greek, *xylon* = wood; *phagein* = to eat.) Wood-eaters, a term applied in several orders.

XYLOPHAGIDAE Plural Noun. (Greek, *xylon* = wood; *phagein* = to eat.) A small Family of flies principally found in Western Hemisphere. Formerly placed in Rhagionidae; Genera have been placed in Coenomyiidae, Xylophagidae, Erinnidae. Species rhagionid-like, slender, wasp-like flies. Body not conspicuously setose, lacks bristles. Rhagionids and xylophagids separated by Clypeus recessed in deep facial groove and Antenna has at least eight segments, tapering apically and terminal segments never form a slender style or Arista. Little known of biology. Adults usually found in wooded or forested areas, particularly near water. Larvae live under bark and in decaying wood; some larvae recovered from soil rich in decaying vegetable matter. Larvae presumed predaceous upon larvae of wood boring beetles. Syn. Erinnidae.

XYLOPHAGOUS Adj. (Greek, *xylon* = wood; *phagein* = to eat; Latin, *-osus* = possessing the qualities of.) Wood-eating. Descriptive of organisms that feed in or upon woody tissue. See: Feeding Strategies. Cf. Cannibalistic; Carnivorous; Fungivorous; Parasitic; Phytophagous; Predaceous; Saprophagous. Rel. Ecology.

XYLOPHILOUS Adj. (Greek, *xylon* = wood; *philen* = to love; Latin, *-osus* = possessing the qualities of.) Wood-loving. Pertaining to organisms associated with wood. See Habitat. Cf. Agricolous; Caespiticolous; Dendrophilous; Lignicolous; Nemoricolous; Silvicolous. Rel. Ecology.

XYLORYCTIDAE Meyrick 1890. Plural Noun.

XYLYLCARB® See MPMC.

XYPHIOPSYLLIDAE Plural Noun. A small Family of Siphonaptera assigned to Superfamily Ceratophylloidea.

XYPHUS Noun. A spinous or triangular process of the Mesosternum in many Hemiptera, and some other insects.

XYRONOTOIDEA Noun. A Superfamily of caeliferous Orthoptera including Pneumoridae, Tanaoceridae and Xyronotidae.

Y Symbolically in medieval Roman, 150. Symbolically in Chemistry, Yttrium.

YAGI, NOBUMASA (1894–1967) (Hasegawa 1967, Kontyû 35 (Suppl.): 87–88.)

YAGO, MASATOSHI (1887–1963) (Hasegawa 1967, Kontyû 35 (Suppl.): 88.)

YAKOBSON, G G See Jacobson, G. G.

YAKOVLEV, ALEXANDER IVANOVICH (1863–1909) (Anon. 1910, Revue Russe Ent. 10: lxixxiv, bibliogr.)

YALTOX® See Carbofuran.

YAMADA, SHINICHIRO (1883–1937) (Hasegawa 1967, Kontyû 35 (Suppl.): 91.)

YAMADA, TANEZBURO (1895–1927) (Hasegawa 1967, Kontyû 35 (Suppl.): 91.)

YAMAGUCHI, SUTEO (1908–1939) (Anon. 1939, Kontyû 13: 138; Hasegawa 1967, Kontyû 35 (Suppl.): 90–91.)

YAMAKAWA, SHIZUKA (1886–1966) (Hasegawa 1967, Kontyû 35 (Suppl.): 90.)

YAMAMAI Noun. (Japanese). *Antheraea jamamai* [Lepidoptera: Saturniidae], a large silkworm which feeds on oak trees in Japan.

YAMAMURA, SHOZABURO (1894–1915) (Hasegawa 1967, Kontyû 35 (Suppl.): 92–93.)

YAMANISHI, SEIHEI (1899–1966) (Hasegawa 1967, Kontyû 35 (Suppl.): 92.)

YAMAUCHI, JINTARO (1886–1945) (Hasegawa 1967, Kontyû 35 (Suppl.): 92.)

YANO, NOBUYOSHI (1859–1928) (Hasegawa 1967, Kontyû 35 (Suppl.): 89–90, bibliogr. (pl. 2, fig.1.)

YANONE SCALE *Unaspis yanonensis* (Kuwana) [Hemiptera: Diaspididae]: An armored scale insect native to China; YS considered a serious pest of citrus in Japan and southern France. Syn. Arrowhead Scale.

YARDEX® See Tau-Fluvalinate.

YAROSHEVSKII, VASILII ALEKSYEEVICH (1841–1901) (Scheyrev 1901, Revue Russe Ent. 1: 70–71, bibliogr.)

YARRELL, WILLIAM (1784–1856) (Newman 1856, Zoologist 14: 5257–5258; Saunders 1856, Zoologist 14: 5304.)

YARROW, IAN HARLEY HAYNES (Day 1990, Ent. Mon. Mag. 126: 253–257.)

YASHIRO, HIROTAKA (1896–1961) (Hasegawa 1967, Kontyû 35 (Suppl.): 89.)

YATES, JAMES (1789–1871) (W.C. 1871. Geol. Mag. 8: 480.)

YATES, WILLARD WILSON (1888–1960) (Gjullin & Lindquist 1960, J. Econ. Ent. 53: 1143–1144.)

YAVA SKIN See Elephantiasis.

YAWATA, HIDEO (1921–1945) (Hasegawa 1967, Kontyû 35 (Suppl.): 93.)

YAWS (Prob. West Indian.) A contagious, pantropical, rural disease of humans characterized by skin ulcers on face, feet and hands. Yaws caused by the spirochete *Treponema pertenue* and sometimes probably mechanically transmitted by chloropid flies (*Siphunculina* in Old World; *Hippelates* in New World) or muscid flies *(Musca domestica* and *M. sorbens)*. Syn. Bouhas; Frambesia; Pian. Cf. Pinkeye.

YAZAKI, IHACHI (1868–1931) (Hasegawa 1967, Kontyû 35 (Suppl.): 88–89.)

YAZAWA, YONEZABURO (1868–1942) (Hasegawa 1967, Kontyû 35 (Suppl.): 89.)

YEAST Noun. (Anglo Saxon, *gist.* Pl., Yeasts.) Any member of the Saccharomycetaceae; microscopic fungi whose cells contain two soluble enzymes that act upon sugar to produce alcohol and carbon dioxide.

YELLOW-AND-BLACK POTTER WASP *Delta campaniformis campaniformis* (Fabricius) [Hymenoptera: Vespidae].

YELLOW AUGER-BEETLE *Xylotillus lindi* (Blackburn) [Coleoptera: Bostrichidae] (Australia).

YELLOW-BANDED DART *Ocybadistes walkeri sothis* Waterhouse [Lepidoptera: Hesperiidae] (Australia).

YELLOW BUMBLE BEE *Bombus fervidus* (Fabricius) [Hymenoptera: Apidae]: A bee widespread on prairies in North America. Adult predominantly black; most of Thorax and dorsum of gastral Terga 1–4 yellow; body with dense vistiture of coarse, moderately long Setae. Queens 15–22 mm long; appear during May; nest in dry grass on ground. See Bumble Bee. Cf. American Bumble Bee; Black-Faced Bumble Bee.

YELLOW CLOVER-APHID *Therioaphis trifolii* (Monell) [Hemiptera: Aphididae]: In North America, a pest of red clover *(Trifolium)*. YCA is closely related to Spotted Alfalfa Aphid.

YELLOW DUNG-FLY *Scatophaga stercoraria* [Diptera: Anthomyiidae].

YELLOW DWARF A viral disease of onions which is transmitted by many Species of aphids.

YELLOW-FACED BEES Members of Subfamily Hylaeinae; body small, sparsely setose; dark coloured and face usually with yellow markings; forewing with two Submarginal Cells; hindleg Corbicula absent, pollen and nectar carried in Crop. Some Species nest in ground; some Species nest in natural cavities or crevices; some Species nest in plant stems. See Colletidae.

YELLOW-FACED BUMBLE BEE *Bombus vosnesenskii* Radoszkowski [Hymenoptera: Apidae]: A Species widespread in western North America. Adult predominantly black with yellow on face and dorsum of head, Pronotum, Mesonotum and fourth gastral segment; body with dense vestiture of short, fine Setae. See Bumble Bee. Cf. Yellow Bumble Bee; Black-Faced Bumble Bee.

YELLOW-FACED LEAFHOPPER *Scaphytopius loricatus* (Van Duzee) [Hemiptera: Cicadellidae].

YELLOW FEVER An arbovirus of the Flaviviridae thought endemic to tropical West Africa and transported to Central and South America through slave trade; YF not present in Orient. Monkeys are principal vertebrate hosts in New

World; monkeys not affected in Africa. More than 12 *Aedes* Species transmit YF in Africa; several *Haemogogus* Species transmit YF in South America. Principal vector to humans *Aedes aegypti*. YF has high mortality rate among humans; death often within one week of onset of symptoms. Mosquitoes acquire YF when they feed on infected humans early in the course of the disease. Virus replicates in midgut epithelium and salivary glands of mosquito which retains virus for life. See Arbovirus; Flaviviridae. Cf. Dengue Fever.

YELLOW FEVER MOSQUITO *Aedes aegypti* (Linnaeus) [Diptera: Culicidae]: A widespread, sporadically distributed, serious pest in tropical and subtropical areas; YFM exists between 40°N and 40°S latitude but does not do well in hot dry climates. Probably endemic to tropical Africa and transported via slave trade and commerce. YFM highly adapted to urban environments and can complete life cycle without leaving house of human. Adults with distinctive yellow-white bands on legs and spots on body. Adult female bites during day or night and enters buildings. Eggs typically laid singly on standing water; after some embryonic development egg can resist desiccation and enter diapause for more than one year; female fed on human blood can lay ca 140 eggs; hatch within four days. Larva lives in water held in cans, bottles, potted plants, car tyres, gutters and drains. Larval stage temperature dependent, ca nine days; pupal stage 1–5 days. Male orients to female wingbeat and identity confirmed by contact pheromone; female feeds on blood but male does not take blood. Female transmits Yellow Fever, Chikungunya and Dengue to man; transmits the nematode *Dirofilaria immitis* (heartworm) to dogs, cats and wild carnivores. Syn. Dengue Mosquito. Cf. Asian Tiger Mosquito.

YELLOW FLOWER-WASP *Campsomeris tasmaniensis* Saussure [Hymenoptera: Scoliidae] (Australia).

YELLOW GARDEN-SPIDER *Argiope aurantia* Lucas [Araneae: Araneidae].

YELLOW-HEADED BORER *Dirphya nigricornis* Olivier [Coleoptera: Cerambycidae].

YELLOW-HEADED COCKCHAFER (SA) *Sericesthis harti* (Sharp) [Coleoptera: Scarabaeidae] (Australia).

YELLOW-HEADED CUTWORM *Apamea amputatrix* (Fitch) [Lepidoptera: Noctuidae]. See Cutworms.

YELLOW-HEADED FIREWORM *Acleris minuta* (Robinson) [Lepidoptera: Tortricidae].

YELLOW-HEADED SPRUCE SAWFLY *Pikonema alaskensis* (Rohwer) [Hymenoptera: Tenthredinidae]: A significant pest of white spruce in North America. Univoltine, overwinter as prepupae; female oviposits in needles and expanding shoots during spring; larva defoliates tree, spins cocoon in soil.

YELLOW-HORNED CLERID *Trogodendron fasciculatum* (Schreibers) [Coleoptera: Cleridae] (Australia).

YELLOW LEAFHOPPER *Zygina zealandica* (Myers) [Hemiptera: Cicadellidae] (Australia).

YELLOW LONGICORN *Phoracantha recurva* Newman [Coleoptera: Cerambycidae]: Endemic to Australia; larvae attack *Eucalyptus*.

YELLOW-MARGINED LEAF BEETLE *Microtheca ochroloma* Stål [Coleoptera: Chrysomelidae].

YELLOW MEALWORM *Tenebrio molitor* (Linnaeus) [Coleoptera: Tenebrionidae]: A widespread, conspicuous but minor pest of cereal products, grain debris and material of animal origin; damage through feeding and odour imparted by high populations. YM apparently endemic to Europe and restricted to temperate regions (North America, Europe, northern Asia, Australia). YM a scavenger around damp and decaying stored products in Australia; found in chicken houses among grain, feathers and droppings. Adult 12–15 mm long, dark brown to black; somewhat flattened; Antenna compact; lives 2–3 months. Female lays ca 500 eggs during lifetime, singly or in small groups within food; eggs sticky and become covered with debris; incubation 10–12 days. Larva undergoes 9–20 instars over 4–18 months; mature larva ca 25 mm long, cylindrical, shiny yellow with brown bands. Pupation within larval food; stage 10–20 days; pupa white and naked. Life cycle 9–12 months. Syn. Yellow Mealworm Beetle. See Tenebrionidae. Cf. Dark Mealworm; Lesser Mealworm.

YELLOW MIGRANT *Catopsilia gorgophone gorgophone* (Boisduval) [Lepidoptera: Pieridae] (Australia).

YELLOW MIRID *Campylomma liebknechti* (Girault) [Hemiptera: Miridae]: Adult 1–2 mm long, oval and yellow to pale mottled-green; second antennal segment longest; two dark spots occur on hind margin of forewing; dark spines on legs. Female lays white, oval eggs in terminal buds of cotton; egg stage 4–6 days. Nymphs similar to adults but lack wing buds. Nymphs yellow with red eyes; nymph stage ca 28 days. YM feed on flower buds when they are extremely small and cause shedding. YM predatory on *Helicoverpa* larvae and mites. Generally found on the top 10 cm of terminals. Syn. Apple Dimpling Bug. Cf. Brown Mirid; Green Mirid.

YELLOW MONDAY *Cyclochila australasiae* (Donovan) [Hemiptera: Cicadidae] (Australia).

YELLOW-NECKED CATERPILLAR *Datana ministra* (Drury) [Lepidoptera: Notodontidae]: A widespread pest of fruit and ornamental trees in North America. Overwinters as pupa within soil. Adults active in summer and female oviposits clusters of 50–100 eggs on underside of leaves. Neonate larvae skeletonize leaf with bodies parallel and heads directed toward leaf margin; later instar larvae disperse to other leaves and consume entire leaf. When disturbed, larvae elevate both ends of body and remain attached to leaf via prolegs on middle segments. Cf. Red-Humped Caterpillar.

YELLOW NIGHTSTALKING SAC SPIDER *Cheiracanthium mordax* L. Koch [Araneida: Clubionidae] (Australia).

YELLOW ORCHARD-APHID *Sitobion luteum* (Buckton) [Hemiptera: Aphididae] (Australia).

YELLOW PALMDART *Cephrenes trichopepla* (Lower) [Lepidoptera: Hesperiidae] (Australia).

YELLOW PAPER-WASP *Polistes dominulus* (Christ) [Hymenoptera: Vespidae]: A social wasp adventive in Western Australia (ca 1977) where it is regarded as an urban pest. Adult workers ca 15 mm long; Antenna orange-brown apically with black Scape and basal flagellar segments; body black with yellow markings; in WA more common than Common Paper-Wasp and workers fly rather slowly when patrolling an area; hindlegs hang conspicuously suspended when in flight. Nest typically concealed and often under roof eaves or tiles or behind wall cladding; nest used for one year, sometimes used through warm winter (subtropical areas); old nests may be reactivated or new nests constructed near previous season's nest. See Vespidae. Cf. Common Paper-Wasp; European Wasp; Golden Paper-Wasp.

YELLOW PEACH MOTH *Conogethes punctiferalis* (Guenée) [Lepidoptera: Pyralidae]: A pest of numerous fruit and nut crops, including custard apples, corn, mango, papaya, peach and sorghum. Endemic to southeast Asia; In Australia, a minor and sporadic pest of citrus. Adult moderate-sized (ca 12–15 mm long), yellow with numerous black spots on wings and body. Eggs oval, ca 1 mm long and laid individually on fruit or flower. Larvae pink with brown rectangular spots on dorsum of body; neonate larva bores into fruit; site of attack where two fruits touch or at stem end. Infested fruit recognized by webbed frass around entry hole; damaged fruit unmarketable. Feed ca 3 weeks, emerges to pupate on surface of fruit in webbed frass. Life cycle 3–5 weeks. Multivoltine with 5–6 generations per year.

YELLOW PEACH-MOTH PARASITE *Argyrophylax proclinata* Crosskey [Diptera: Tachinidae]: An Australia Species. Adult 5–11 mm long, round-elongate, pale brown with silver, grey and black markings; head with silver Setae and a dark median stripe; Thorax pale; legs dark with pale Tibiae; wings brownish; Abdomen brown with dark median stripe and bronze Setae. Eggs laid on larval host. Eggs are white, oval and 1–5 mm long. Larva typically maggot-like and creamy-white with pigmented head and spines. Larva burrows into host and feed on internal organs. Pupation occurs within host. Tachinids are parasites of lepidopteran caterpillars. In Australia, parasitizes yellow peach moth larvae *(Conogethes punctiferalis)*.

YELLOW PECAN-APHID *Monelliopsis pecanis* Bissell [Hemiptera: Aphididae].

YELLOW RICE-BORER *Tryporyza incertulas* (Walker) [Lepidoptera: Pyralidae] (Alt. *Scirpophaga incertulas*). An oriental moth and significant pest of rice. Eggs deposited on leaf tip; larvae bore into stems, feed and migrate via flotation to new host-plant stem. Pupation in stem; to six generations per year.

YELLOW ROSE-APHID *Acyrthosiphon (Rhodobium) porosum* (Sanderson) [Hemiptera: Aphididae].

YELLOW SAGE See Lantana.

YELLOW SAND SCORPION *Urodacus armatus* Pocock [Scorpionida: Scorpionidae] (Australia).

YELLOW SCALE *Aonidiella citrina* (Coquillett) [Hemiptera: Diaspididae]: A widespread pest of citrus and ornamental plants; native to Orient and widely distributed in *Citrus* growing regions of world. Resembles California Red Scale. Adult female scale cover yellowish, circular, flattened, ca 2.0–2.5 mm diameter; male cover elongate, somewhat pale; male undergoes prepupal and pupal stage with winged adult; female moults from second instar nymph directly into wingless, sedentary adult beneath scale cover of earlier stages. YS prefers lower parts of tree on underside of leaves not exposed to strong sunlight; rarely attacks fruit or wood. Adult female viviparous and produces ca 100–150 crawlers during 6–8 weeks; crawlers disperse, insert mouthparts into host plant and produce white, flocculent, waxy 'whitecap.' YS multivoltine with up to four generations per year depending upon climate. Parasitic Hymenoptera control YS in some regions. See California Red Scale.

YELLOW-SHOULDERED DUNG BEETLE *Liatongus militaris* (Castelnau) [Coleoptera: Scarabaeidae] (Australia).

YELLOW-SHOULDERED LADY-BEETLE *Scymnodes lividigaster* (Mulsant) [Coleoptera: Coccinellidae] (Australia).

YELLOW SOLDIER FLY *Inopus flavus* (James) [Diptera: Stratiomyidae] (Australia).

YELLOW SPIDER-MITE *Eotetranychus carpini borealis* (Ewing) [Acari: Tetranychidae].

YELLOW SPOT A viral disease of pineapple which is transmitted by thrips.

YELLOW-SPOT BLUE BUTTERFLY *Candalides xanthospilos* (Hübner) [Lepidoptera: Lycaenidae] (Australia).

YELLOW-SPOT JEWEL *Hypochrysops byzos* (Boisduval) [Lepidoptera: Lycaenidae] (Australia).

YELLOW STARTHISTLE *Centaurea solstitialis* Linnaeus (Asteraceae), a Eurasian annual or biennial weed which occupies about three million hectares in the western USA. Regarded as beneficial to some beekeepers; regarded as a rangeland weed by graziers.

YELLOW-STRIPED ARMYWORM *Spodoptera ornithogalli* (Guenée) [Lepidoptera: Noctuidae]: A multivoltine, diurnal pest of cotton in the USA;

also feeds on other crops. Female oviposits eggs in clusters on foliage and covers them with scales. Larva with a pair of triangular black spots on dorsum of most body segments and a bright orange stripe laterad of the spots. Overwinters as Pupa. See Armyworms.

YELLOW SUGARCANE APHID *Sipha flava* (Forbes) [Hemiptera: Aphididae].

YELLOW SUGARCANE CICADA *Parnkalla muelleri* (Distant) [Hemiptera: Cicadidae] (Australia).

YELLOW SWIFT *Borbo impar lavinia* (Waterhouse) [Lepidoptera: Hesperiidae] (Australia).

YELLOW-WINGED LOCUST *Gastrimargus musicus* (Fabricius) [Orthoptera: Acrididae]: An important pest of pastures and pasture seed crops in Australia; swarms only in northern Australia.

YELLOW WOOLLYBEAR *Spilosoma virginica* (Fabricius) [Lepidoptera: Arctiidae].

YELLOWBELLY *Psaltoda harrisii* (Leach) [Hemiptera: Cicadidae] (Australia).

YELLOWBOX LERP *Lasiopsylla rotundipennis* Froggatt [Hemiptera: Psyllidae] (Australia).

YELLOWISH SKIPPER *Hesperilla flavescens* Waterhouse [Lepidoptera: Hesperiidae] (Australia).

YELLOWJACKET Noun. Aculeate wasps of the Genera *Vespula* and *Dolichovespula* with ca 35 Species in Holarctic. Typically 10–15 mm long with yellow and black Thorax and Gaster; nest in hollow logs and stumps near the ground; colony size from a few individuals to several hundred thousand workers. Alt. Yellow Jacket; Yellow-Jacket. See Aerial Yellowjacket; Common Yellowjacket; German Yellowjacket; Western Yellowjacket. Cf. Hornet.

YELLOWS See Pieridae.

YEPEZ, FRANCISCO JOSE FERNANDEZ (1923–1986) Venezuelan academic (Universidad Central de Venezuela), administrator and insect collector. (Lamas 1986, Rev. Peru. Ent. 29: 141–142.).

YERBURY, JOHN WILLIAM (1847–1927) (Collin 1927, Proc. ent. Soc. Lond. 2: 102–103.)

YERSIN, ALEXANDER (1829–1863) (Forel 1864, Bull. Soc. Vaud. Sci. nat. 8: 228–234; Saussure 1866, Mitt. schweiz. ent. Ges. 2: 75–106, bibliogr.)

YETTER, WILLIAM P (1894–1950) (List & Allen 1950, J. Econ. Ent. 43: 962.)

YF See Yellow Fever.

YLÖNEN, PEKKA (1878–1929) (Forsius 1928, Notul. ent. 8: 64.)

YOKOYAMA, KIRIO (1896–1932) (Hasegawa 1967, Kontyû 35 (Suppl.): 94.)

YOLK Noun. (Anglo Saxon, *geoleca*, fr *geolu* = yellow. Pl., Yolks.) The nutritive matter of an egg as distinguished from the living, formative material (embryo). Yolk is supplied to ovarian follicles by autosynthesis or heterosynthesis. Yolk granules are relatively uniformly distributed within the centrolecithal insect egg. Syn. Vitellus. See Embryogenesis. Rel. Vitellophage.

YOLK CELLS Cleavage cells remaining in the yolk and taking no part in the blastoderm formation (Snodgrass).

YOLK CLEAVAGE The division of the yolk into masses containing from one to several cleavage nuclei (Snodgrass).

YOLK PROTEINS A major category of yolk-storage proteins found in the Oocyte of some insects (Diptera). Cf. Vitellogenins.

YOSEMITE BARK WEEVIL *Pissodes schwarzi* Hopkins [Coleoptera: Curculionidae].

YOSHIDA, SHOSHICHIRO (1851–1904) (Hasegawa 1967, Kontyû 35 (Suppl.): 94–95.)

YOTHERS, WILLIAM WALTER (1879–1971) (Porter & Cooper 1972, J. Econ. Ent. 65: 627.)

YOUNG GRANULOCYTE See Prohaemocyte.

YOUNG PLASMATOCYTE See Prohaemocyte.

YOUNG, DOUGLAS BARZILLAR (1860–1925) (Felt 1926, J. Econ. Ent. 19: 419.)

YOUNG, JOHN N (–1898) (Anon. 1898, Entomologist 31: 100.)

YOUNG, LAWRENCE H (–1907) (Anon. 1907, J. Bombay nat. Hist. Soc. 18: 184–185.)

YOUNG, MORRIS (1821–1897) (Dunsmore 1898, Ann. Scot. nat. Hist. 1898: 1–6.)

YPONOMEUTIDAE Plural Noun. Ermine Moths. A small Family of ditrysian Lepidoptera assigned to Superfamily Yponomeutoidea. Adults are small bodied with wingspan 5–35 mm and head typically with smooth, lamellar scales; Ocelli absent and Chaetosemata typically absent; Antenna shorter than forewing and Pecten sometimes present; Proboscis usually present and always without scales; Maxillary Palpus small with 1–3 segments; Labial Palpus curved upward; fore Tibia with Epiphysis; tibial spur formula usually 0-2-4. Eggs are somewhat flattened, oval-elliptical in outline shape and Chorion sculpture reticulate. Eggs typically are deposited overlapping in masses on twig of host plant. Larvae feed externally on leaves of host plant under webbing; sometimes communal feeding occurs under more extensive webbing. See See Yponomeutoidea.

YPONOMEUTOIDEA Webspinners. A Superfamily of ditrysian Lepidoptera including Aegeriidae, Argyresthiidae, Douglassidae, Epermeniidae, Glyphipterigidae, Heliodinidae, Lyonetidae and Yponomeutidae. See Lepidoptera.

YUASA, HIROHARU (1900–1953) (Gressitt 1954, Ann. ent. Soc. Am. 47: 1228; Roberts 1954, J. Econ. Ent. 46: 1127–1128; Hasegawa 1967, Kontyû 35 (Suppl.): 93–94.)

YUCCA HORNWORM *Erinnyis ello* (Linnaeus) [Lepidoptera: Sphingidae].

YUCCA MOTH *Tegeticula yuccasella* (Riley) [Lepidoptera: Incurvariidae]: Pollinator of spanish bayonet (*Yucca aloifolia* Linnaeus, *Y. baccata* Torrey) in southwestern USA and Mexico. Adult moth with modified mouthparts for collecting pollen from one plant and transferring to stamen of another plant. Eggs inserted into Ovary while pollen transferring down stigmatic tube. Larva feeds on developing seeds, but does not destroy all seeds.

YUCCA PLANT BUG *Halticotoma valida* Townsend [Heteroptera: Miridae]: A pest of ornamental *Yucca* in North America. Nymphs scarlet coloured, cause stippling and yellowing of leaves.

YUMA SPIDER MITE *Eotetranychus yumensis* (McGregory) [Acari: Tetranychidae].

ZABRISKIE, JEREMIAH LOTT (1835–1910) (Anon. 1910, Can. Ent. 42: 168; Anon. 1910, J. N.Y. ent. Soc. 18: 127.)

ZACATILLA Noun. (Spanish.) The highest quality of cochineal.

ZACHARIAS, OTTO (1846–1916) (Soldanski 1916, Dt. Ent. Z. 1916: 605; Thienemann 1917, Arch. Hydrobiol. 11(4): i–xxiv, bibliogr.)

ZACHER, FRIEDRICH (1884–1961) (Plate 1959, Mitt. dt. ent. Ges. 18: 40–41; Anon. 1961, Mitt. dt. ent. Ges. 20: 67–68; Pax 1962, Verh. dt. zool. Ges. 26: 720–723.)

ZADDACH, ERNST GUSTAV (1817–1881) (Anon. 1881, Psyche 3: 259; Albrecht 1881, Schr. phys.-ökon Ges. Königsb. 22: 119–128.)

ZAITZEV, FILIPP ADAMOVICH (1872–1957) (Arnoldi 1958, Ent. Obozr. 37: 940–946, bibliogr.)

ZAKHVATKINA, ALEXKSIYA ALEKSEEVICH (1906–1951) (Pavlovskii 1951, Ent. Obozr. 31: 629–633, bibliogr.)

ZALESKY, MILOS (–1944) (Stepanek 1946, Vest. csl. zool. Spol. 10: 39–40.)

ZANDER, ENOCH (1873–1957) (Frickhunger 1933, Z. angew. Ent. 20: 326–327; Zander 1948, Bienen Ztg. 62: 83–88, bibliogr.; Escherich 1949, Z. angew. Ent. 31: 175–179; Zwölfer 1957, Z. angew. Ent. 40: 582–583.)

ZANG, RICHARD (1884–1906) (Horn 1906, Dt. ent. Z. 1906: 12; Daniel 1908, Münch. koleopt. Z. 3: 399.)

ZANGOBIIDAE Evenhuis 1994. A replacement name for the preoccupied Palaeolimnobidae Zhang *et al.* (*non* Palaeolimnobiidae Bode). The Family consists of two monotypic Genera of fossil Diptera taken from Upper Jurassic Period deposits in China. Zangobiidae are possibly related to Tipulidae.

ZANON, VITO (1875–1949) (Brunelli 1949, Boll. Pesca Piscic. Idrobiol. 25: 217–220, bibliogr.)

ZANTHOPSIN Noun. (Greek, *xanthos* = yellow; *opsis* = sight. Pl., Zanthopsins.) A pigment found in the eyes of night-flying insects, impregnating the retinular elements (Imms).

ZAPATER Y MACONELL, PETER BERNARDO (1816–1907) (Anon. 1908, Entomologist's Rec. J. Var. 20: 69; Horn 1908, Dt. ent. Z. 1908: 294.)

ZAPOTOCY, ANTONIN (1884–1957) (Anon. 1957, Zool. Listy 20: 292 only.)

ZAPROCHILIDAE Plural Noun. A small Family of Ensifera found in Australia. Representatives are anthophilous, feeding on pollen and nectar. Syn. Phasmodidae.

ZAUREL, JAN (1879–1946) (Hoffer 1939, Vest. csl. Zool. Spol. 36: 14, bibliogr.)

ZAVADIL, VILÉM (1876–1953) (Sustera 1953, Roc. csl. Spol. ent. 50: 242–243.)

ZAVADSKY, KAREE (Teyrovsky 1949, Ent. Listy 12: 158.)

ZAVATTARI, EDOARDO (1883–1972) (Anon. 1972, Boll. Soc. ent. ital. 104: 22.)

ZAVREL, JAN (1879–1946) (Hrabé 1947, Vest. csl. zool. Spol. 11 18–14, bibliogr.)

ZAWADZKI, ALEXANDER (1798–1868) (Frey 1869, Verh. naturf ver. Brünn. 7: 22–25.)

Z-DISC Noun. (German, *zwischenscheibe* = intermediate.) A membrane or disc which separates Sarcomeres of the muscle fibre. See Muscle Cell. Cf. A Disc; I Disc.

ZEBRA Noun. (Abyssinian, *zebra* = striped.) An Afrotropical grazing mammal realated to and resembling the horse. Body with dark stripes on a whitish background. Cf. Disruptive Coloration.

ZEBRA BLUE *Syntarucus plinius pseudocassius* (Murray) [Lepidoptera: Lycaenidae] (Australia).

ZEBRA CATERPILLAR *Melanchra picata* (Harris) [Lepidoptera: Noctuidae]: A North American moth.

ZEBRA SWALLOWTAIL *Iphiclides* spp. *Graphium (Papillo) marcellus.* [Lepidoptera: Papillionidae]. Native to North America.

ZEBRINE Adj. (Abyssinian, *zebra* = striped.) Striped as a zebra.

ZEBUB Noun. (Arabic.) A fly of veterinary importance.

ZECK, E H (1891–1963) (Anon. 1964, J. ent. Soc. Aust. (N.S.W.) 1: 37–44, bibliogr.)

ZEHNTNER, LEO (1897–1961) (Anon. 1962, Verh. schweiz. naturf. Ges. 141: 270–272; Ryen 1961, Ent. Ber., Amst. 21: 95.)

ZELANIAN Adj. Zoogeography pertaining to New Zealand.

ZELDOX® See Hexythiazox.

ZELEBOR, JOHAN (Papavero 1975, *Essays on the History of Neotropical Dipterology.* 2: 286–288. São Paulo.)

ZELLER, PHILIPP CHRISTOPH (1808–1883) (Strecker 1878, Butterflies and moths of North America 1: 279–283, bibliogr.; A. R.G. 1883, Can. Ent. 15: 176–177; Carrington 1883, Entomologist 16: 120; Dohrn *et al.* 1883, Stettin. ent. Ztg. 44: 406–418, bibliogr.)

ZEMENA, JOSEFA (1859–1925) (Rambousek 1925, Vest. csl. Spol. Zool. 22: 4–5.)

ZEMPEL, E BAUDYS (1886–1968) (Skubrava 1968, Acta ent. bohemoslovaka 65: 399.)

ZEMREL, HERMANN WEBER (1956) (Skuhravy 1957, Cas. csl. Spol. ent. 54: 297.)

ZENKER, CHRISTIAN DANIEL (1756–1819) (Germar 1821, Magazin Ent. (Germar) 4: 443.)

ZEPHYR® See Avermectin.

ZERKOWITZ, ALBERT (–1963) (Anon. 1963, Bull. Soc. ent. Fr. 69: 213.)

ZERNY, HANS (1887–1945) (Reisser 1946, Z. wien ent. Ges. 30: 49–51; Pittioni 1948, Annln naturh. Mus. Wien 56: 558–563, bibliogr.)

ZERTELL® See Chlorpyrifos-Methyl.

ZETA-CYPERMETHRIN A synthetic-pyrethroid compound used as a contact and stomach poison against Lepidoptera larvae, thrips and some beetles. Toxic to bees and fishes; not phytotoxic. Trade name Fury®. See Synthetic Pyrethroid.

ZETEK, JAMES (1886–1959) (Snyder *et al.* 1959, J. Econ. Ent. 52: 1230–1232, bibliogr.)

ZETTERSTED, JOHN WILLIAM Swedish naturalist and author of *Orthoptera Suecica* (1821) and *Fauna Lapponica* (1828).

ZETTERSTEDT, JOHANN WILHELM (1785–1874) (Dohrn 1875, Stettin. ent. Ztg. 36: 192–193; Lichtenstein 1875, Ann. Soc. ent. Fr. (5) 5: 9–10.)

ZEUGL- (Greek, *zeugle* = loop.) Strap of a yoke.

ZEUGLOPTERA (Greek, *zeugle* = loop; *pteron* = wing.) Suborder of primitive Lepidoptera, characterized by diurnal activity, small size, metallic coloration, and functional Mandibles. Included Family Micropterygidae. Cf. Dachnonypha; Monotrysia; Ditrysia.

ZEUNER, FREDERICK EBERHARD (1905–1963) (Wigglesworth 1964, Proc. R. ent. Soc. Lond. (C) 28: 58.)

ZEYHER, CARL LUDWIG PHILIPP (Weidner 1967, Abh. Verh. naturw. Ver. Hamburg Suppl. 9: 136.)

ZIDIL® See Chlorpyrifos.

ZIEGLER, DANIEL (1804–1876) (Morris 1885, Can. Ent. 17: 132–133; Hagen 1886, Rep. ent. Soc. Ont. 16: 22.)

ZIEMANN, HANS (–1939) (Sachtleben 1940, Arb. physiol. angew. Ent. Berl. 7: 80.)

ZIG-ZAG RICE LEAFHOPPER *Recilia dorsalis* Motschulsky [Hemiptera: Deltocephalidae].

ZIKAN, JOSE FRANCISCO (1881–1949) (Borgmeier 1949, Revta. Ent., Rio de J. 20: 647–652, bibliogr.; Araujo 1954, Revta. bras. Ent. 1: 119–128, bibliogr.)

ZILAHI-SEBESS, GEZA (1905–1962) (Ferenc 1962, Folia ent. hung. 15: 265–270, bibliogr.)

ZIMECTRIN® See Avermectin.

ZIMIN, LEONID SERGHEEVICH (1902–1970) (Bei-Bienko 1970, Ent. Obozr. 49: 923–927, bibliogr. Translation: Ent. Rev. Wash. 49: 563–565.)

ZIMMERMAN, PHILIPP WILHELM ALBRECHT (1860–1931) (Braun 1931, Afrika Nachr. 12: 127–128; Correns & Morstatt 1931, Ber. dt. bot. Ges. 49: 220–243, bibliogr.)

ZIMMERMAN PINE MOTH *Dioryctria zimmermani* (Grote) [Lepidoptera: Pyralidae].

ZIMMERMAN, ALOIS (1871–1929) (Daniel 1930, Koleopt. Rdsch. 15: 240–243, bibliogr.)

ZIMMERMAN, CHARLES CHRISTOPH ANDREW (1800–1867) (Hagen 1889, Rep. ent. Soc. Ont. 20: 101–103; Dow 1913, Bull. Brooklyn. ent. Soc. 8: 110–114.)

ZIMMERMAN, FRIEDRICH (–1961) (Kirchberg 1960, Mitt. dt. ent. Ges. 19: 57–60.)

ZIMMERMAN, JOHANN CARL HEINRICH (1845–1930) (Zimmerman 1930, Verh. Ver. naturw. Heimatforsch. 22: xv–xvii.)

ZINK, CARL W (1860–1932) (Weidner 1967., Abh. Verh. naturw. Ver. Hamburg Suppl. 9: 266.)

ZINNA, G Italian entomologist interested in morphology and systematics of parasitic Hymenoptera.

ZINSER, HANS (1878–1940) (Mueller 1940, J. Bacteriol. 40: 747–753; Strong 1940, Science 92: 276–279.)

ZIPCIDE® See Coumaphos.

ZITHIOL® See Malathion.

ZNAMENSKIIO, ALEXSANDRA VASIL'EVICH (1891–1942) (Schtakelberg 1968, Ent. Obozr. 47: 244–247, bibliogr. Translation: Ent. Rev., Wash. 47: 143–145, bibliogr.)

ZNOJKO, DIMITRI VASILEVICH (1903–1933) (Semenov-Tian-Shansky 1934, Trudy Zashch. Rast. (1)8: 73–78, bibliogr.)

ZODARIIDAE Plural Noun. Small Family of spiders. Species live under stones and among leaf litter; zodariids apparently eat ants.

ZODIOPHILOUS See Zoophilous.

ZOHEIRY, MOHAMED SOLIMAN EL (1895–1973) (Anon. 1974, Bull. Soc. ent. Egypte 57: [1–2].)

ZOLLIKOFER, CASPAR TOBIAS (1774–1843) (Jaggli 1950, Mitt. schweiz. ent. Ges. 31: 201–202.)

ZOLONE® See Phosalone.

ZOLOTAREVSKY, BORIS NIKOLAYEVICH (1892?-1964) (Bei-Bienko 1965, Ent. Obozr. 44: 221–222. Translation: Ent. Rev., Wash. 44: 120–121.)

ZONA Noun. (Latin, *zona* = girdle. Pl., Zonae.) A belt, zone or area of distribution.

ZONAL SYMMETRY A type of symmetry involving metamerism and serial segmentation. See Asymmetry; Symmetry. Cf. Bilateral Symmetry; Pentameral Symmetry; Spherical Symmetry.

ZONATE Adj. (Latin, *zona* = girdle.) Descriptive of structure or surface marked with conspicuous ring-like regions.

ZONE Noun. (Latin, *zona* > Greek, *zoma* = girdle. Pl., Zones.) 1. An area or region of a body that is differentiated from surrounding area by some demonstrable physical characteristic or feature. 2. Any band or stripe that encircles a plant or animal's body. 3. Geology: A portion of the stratigraphic record that is characterized by upper and lower boundaries, the area between which contains distinctive index fossils. Zones may be defined on the basis of one or more index fossils. See Fossil. 4. Any of the five regions of the globe, based on temperature and latitude (*e.g.* North Temperate Zone; South Frigid Zone). See Biogeography.

ZONE OF GROWTH Part of the sperm tube beyond the Germarium in which the Spermatogonia enter a stage of multiplication and are usually encysted. The Vitellarium *sensu* Snodgrass.

ZONE OF TRANSFORMATION The part of the sperm tube in which the Spermatocytes develop into Spermatids (Snodgrass). See Follicle Cell.

ZONITE Noun. (Latin, *zona* = zone; Greek, *zoma* = girdle; *-ites* = constituent. Pl., Zonites.) 1. A body segment of Diplopoda. 2. An arthromere or somite. Alt. Zoonite; Zoonule.

ZONOID Adj. (Greek, *zoma* = girdle; Latin, *zona* = girdle; *eidos* = form.) Zone-like; pertaining to something which resembles a zone. See Zone. Rel. Eidos; Form; Shape.

ZOOGEOGRAPHICAL Adj. (Greek, *zoon* = animal; *ge* = earth; *graphein* = to write; Latin, *-alis* = belonging to.) Descriptive of things pertaining to zoogeography or the territorial (spatial) distribution of animals. Cf. Phytogeographical.

ZOOGEOGRAPHY Noun. (Greek, *zoon* = animal; *ge* = earth; *graphein* = to write. Pl., Zoogeographies.) The study of geographical distribution and distributional patterns and association or groupings of animals in space, realm or region. See Realm. Cf. Biogeography; Phytogeography.

ZOOLOGY Noun. (Greek, *zoon* = animal; *logos* = discourse.) The branch of biology that involves the study of animals, including insects. See Biology. Cf. Entomology; Zoology. Rel. Anatomy; Behavior; Biochemistry; Ecology; Genetics; Morphology; Physiology.

ZOOMETER Noun. (Greek, *zoon* = animal; *metron* = measure. Pl., Zoometers.) An animal biometer which indicates conditions through abundance or number of individuals and other factors, as survivial, size, *etc.* (Shelford).

ZOONITE Noun. (Greek, *zoon* = animal; *-ites* = constituent. Pl., Zoonites.) The body segment of an arthropod. See Zonite.

ZOONOSIS Noun. (Greek, *zoon* = animal; *-nosos* = disease. Pl., Zoonoses.) Medical Entomology: Any pathogen-induced disease of wild or domestic animals which is periodically transmitted to humans. Syn. Anthropozoonosis. See Euzoonosis; Parazoonosis. Cf. Anthroponosis.

ZOOPHAGOUS Adj. (Greek, *zoon* = animal; *phagein* = to eat; Latin, *-osus* = possessing the qualities of.) Animal-eating; a term applied to insects which feed on animals or animal products. See Feeding Strategy. Cf. Entomophagous; Herbivorous.

ZOOPHILOUS Adj. (Greek, *zoon* = animal; *philein* = to love; Latin, *-osus* = possessing the qualities of.) Animal-loving; a term applied to insects that feed on animals other than humans. See Habitat. Cf. Anthropophagous; Entomophytous; Ornithophilous. Rel. Ecology.

ZOOPHILIC Adj. (Greek, *zoon* = animal; *philein* = to love.) 1. Pertaining to Zoophily. 2. Descriptive of plants that are typically adapted for pollination by animals other than insects. Alt. Zodiophilous.

ZOOPHILY (Greek, *zoon* = animal; *philaein* = to love.) Medical Entomology: Pertaining to Species of organisms, especially insects and ticks, which bite or feed upon humans. Cf. Anthropophily; Ornithophily.

ZOOPHOBIC Adj. (Greek, *zoon* = animal; *phobos* = fear; *-ic* = of the nature of.) Organisms that avoid other animals; organisms that are avoided by animals.

ZOOPROPHYLAXIS Noun. (Greek, *zoon* = animal; *prophylaktikos* = take precautions against. Pl., Zooprophylaxes.) Guarding from the contraction or transmission of disease among animals.

ZOOSPORE Noun. (Greek, *zoon* = animal; *sporos* = seed. Pl., Zoospores.) Fungal spores with Flagella and capable of movement in water.

ZOOSUCCIVOROUS Adj. (Greek, *zoon* = animal; Latin, *succus* = sap; *vorare* = to devour; *-osus* = possessing the qualities of.) Animal-fluid feeders or insects which imbibe blood or other body-fluids of their hosts.

ZOOTAXY Noun. (Greek, *zoon* = animal; *taxis* = arrangement.) The classification of animals. Cf. Phytotaxy.

ZOOTOMY Noun. (Greek, *zoon* = animal; *temnein* = to cut.) The anatomy or dissection of animals other than man.

ZOPHERIDAE Plural Noun. A small Family of adephagous Coleoptera sometimes included in the Tenebrionidae. Antenna short, weakly clavate; tarsal formula 5-5-4 or 4-4-4. Adults feed upon fungus.

ZORAPTERA Silvestri 1913. (Greek, *zoros* = pure; *pteron* = wing.) The smallest Order of insects, consisting of one Family and about 30 nominal Species. Geographically widespread. One fossil Species in Dominican Amber (lower Miocene–upper Eocene). Adults minute to small bodied. Head hypopgnathous; Epicranial Suture; mouthparts mandibulate; adult Antenna moniliform with nine segments; nymph with eight antennal segments; Maxillary Palpus with five segments; Labial Palpus with three segments. Prothorax well developed; Mesothorax and Metathorax not differentiated in primitively wingless forms; wings present or absent, when present membranous with reduced venation; wings shed at base; wingless forms without compound eyes or Ocelli; winged forms with compound eyes and Ocelli. Coxae large; Tarsi with two segments. Abdomen with 11 segments; Gonopore behind Sternum VII; Ovipositor vestigal or absent; Cerci present, not segmented; panoistic Ovarioles. Thorax with three ganglia; Abdomen with two ganglia. Development gregarious in rotting wood, sawdust, under bark or in assocation with termites. Apparently fungivorous and/or necrophilous. Included Family: Zorotypidae.

ZORAPTERANS See Zorotypidae.

ZOROTYPIDAE Silvestri 1913. Plural Noun. Zorapterans. The nominant Family of Zoraptera, whose members are predominantly tropical.

ZOUFAL, VLADIMIR (1856–1932) (Rambousek 1931, Cas. csl. Spol. ent. 28: 45–46; Heikertinger 1933, Koleopt. Rdsch 19: 150; Maran 1933, Cas. csl. Spol. ent. 30: 5–8.)

ZSIRKO, GIZELLA (1919–1970) (Anon. 1971, Folia ent. hung. 24: 15–16.)

ZUKOWSKY, BERNHARD (1886–1949) (Koch 1949, Ent. Z., Frank. a. M. 59: 65; Weidner 1967, Abh. Verh. naturw. Ver. Hamburg Suppl. 9: 290–291.)

ZÜRN, FREDERIC ANTOINE (1835–1900) (Blanchard 1900, R. Arch. parasit. 3: 644–645.)

ZVEREZOM-ZUBOVSKIY, YEVGENIY VASIL'YEVICH (1890–1967) (Kryshtal 1968, Ent. Obozr. 47: 235–243, bibliogr. Translation: Ent. Rev., Wash. 47: 137–142.)

ZWÖLFER, WILHELM (Wellenstein 1967, Z.

angew. Zool. 54: 303–304.)

ZYGAENIDAE Latreille 1809. Plural Noun. Burnet Moths; Leaf Skeletonizer Moths. A Family of 400 Species of ditrysian Lepidoptera assigned to Superfamily Zygaenoidea. Adult size variable; wingspan 10–110 mm; usually with bright or metallic coloration; head with smooth scales; Ocelli and Chaetosemata present; Antenna clubbed in male and female or bipectinate in male and female form variable; Proboscis with scales; Maxillary Palpus small with 1–2 segments; Labial Palpus short; Epiphysis present or absent; tibial spur formula 0-2-4 or 0-2-2 or absent; forewing broadly, narrowly triangular and apically rounded. Eggs flattened with Chorion weakly sculptured and easily impressed; eggs laid in clusters or near or on host plant. Head retractile; prothoracic shield well developed; Antenna long; primary Setae on head; body spinulose often with secondary Setae on Verrucae. Pupa stout, somewhat flattened. Many Species are day fliers. See Zygaenoidea.

ZYGAENOIDEA Noun. A primitive Superfamily of ditrysian Lepidoptera including Chrysopolomidae, Cyclotornidae, Dalceridae, Epipyropidae, Heterogynidae, Limacodidae, Megalopygidae, Sombrachyidae and Zygaenidae. See Lepidoptera.

ZYGENTOMA See Thysanura.

ZYGOMATIC ADDUCTORS Two sets of muscles positioned between the paired Mandibles of insects (Snodgrass).

ZYGOMORPHIC Adj. (Greek, *zygon* = a yolk; *morphos* = form; *-ic* = of the nature of.) Descriptive of structures or organisms that are Bilaterally Symmetrical. See Symmetry. Alt. Zygomorphous. Cf. Actinomorphic.

ZYGOMORPHY Noun. (Greek, *zygon* = a yolk; *morphos* = form. Pl., Zygomorphies.) The act or state of being Bilateral Symmetrical. Alt. Zygomorphism. See Symmetry. Cf. Actinomorphy.

ZYGOPTERA Noun. Damselflies. Odonata with forewing and hindwing similar in shape, venation comprising a quadrilateral, but not a triangle; wings held over Abdomen at repose; nymphs with caudal tracheal gills. See Odonata. Cf. Anisoptera, Anisozygoptera.

ZYGOPTEROID Adj. (Greek, *zygon* = a yolk; *pteron* = wing; *eidos* = form.) 1. Dragonfly-like in form, biology or behaviour. 2. Equally-winged; a term descriptive of zygopterous dragonflies. See Zygoptera. Rel. Eidos; Form; Shape.

ZYGOPTEROUS Adj. (Greek, *zygon* = a yolk; *pteron* = wing; Latin, *-osus* = possessing the qualities of.) Descriptive of or pertaining to the Zygoptera (damselflies).

ZYGOTE Noun. (Greek, *zygotos* = yolked. Pl., Zygotes.) The cell formed from the union of male and female gametes. The fertilized egg.

ZYGUM Noun. (Pl., Zyga.) Scarabaeoid larvae: A sclerome pertaining to the region Haptomerum and forming its anterior margin. When typically developed, appearing as a convex cross bar anterior of the Sensilla and heli, but often enlarged and carrying these structures (Boving).

ZYMASE Noun. (French > Greek, *zyme* = a leaven.) An enzyme that decomposes sugar into alcohol and carbon dioxide.

ZYMOGEN Noun. (French, *zymogone*. Pl., Zymogens.) An inert substance in the living cell from which an enzyme is produced (Wardle).

ZYMOGRAM Noun. (Greek, *zyme* = a leaven; -*gram* = something written or drawn.) An electrophoretic banding pattern of an isozyme. A Zymogram is used to identify the origin of a cell line.

Appendix 1: Journal Titles

Journal titles appear in their abbreviated form in the text of the dictionary. The full journal title for most is provided as a reference aid. This list is not complete. The dictionary entries were compiled over a 30-year period and consistent abbreviation conventions were not followed by the authors, thus compiling the list of full journal titles was difficult. Further, many of the older journal titles were not traceable using current citation references. For accuracy, we do not include full journal titles for those entries that did not clearly match the abbreviation in standard citation references. The journal titles listed represent 75% of the journal title abbreviations that appear in the text.

Abhandlungen Herausgegeben vom Naturwissenshcaftlichen Verein zu Bremen
Abhandlungen und Berichte aus dem Staatlichen Museum fuer Tierkunde in Dresden
Abhandlungen und Berichte des Vereins fuer Naturkunde zu Cassel
Abhandlungen und Berichte des Zoologischen und Anthroplogish-Ethnographischen
    Museums zu Dresden
Abhandlungen und Verhandlungen des Naturwissenschaftlichen Vereins in Hamburg
Abhandlungen und Verhandlungen. Naturwissenschaftlicher Verein in Hamburg
Acta Entomologica Bohemoslovaca
Acta Entomologica Chilena
Acta Entomologica Jugoslavica
Acta Entomologica Lithuanica
Acta Entomologica. Musei Nationalis (Prague)
Acta Societatis Scientiarum Fennicae
Acta Universitatis Voronegiensis
Acta Zoologica
Advances in Insect Physiology
Agricultura Tecnica
Agricultural Gazette New South Wales
Allgemeine Deutsche Naturhistorische Zeitung
Allgemeine Forst- und Jagdzeitung
American Bee Journal
American Journal of Botany
American Journal of Pharmacy
American Journal of Science
American Journal Tropical Medicine and Hygiene
American Men of Science
American Midland Naturalist
American Naturalist
American Scientist
Anais. Faculdade de Ciencias. Universidade do Porto
Anais. Instituto de Higienee Medicina Tropical (Lisbon)
Anais. Instituto de Medicina Tropical (Lisbon)
Anales de la Sociedad Espanola de Historia Natural
Anales. Instituto de Biologia. Universidad Nacional Autonoma de Mexico
Anales. Sociedad Cientifica Argentina
Annalen des Naturhistorischen Museums in Wien
Annales de la Societe d'Agriculture, Sciences et industrie de Lyon
Annales de la Societe Malacologique de Belgique
Annales de Parasitologie Humaine et Comparee
Annales de Zoologie-Ecologie Animale
Annales des Epiphyties [Paris]

Annales des Sciences Politiques
Annales. Entomological Society of Quebec
Annales. Historico-Naturales. Musei Nationalis Hungarici
Annales. Institut Pasteur (Paris)
Annales. Museum d'Historie Naturelle
Annales. Societe Entomologique de France
Annales. Societe Entomologique du Quebec
Annales. Societe Linneene de Lyon
Annali del Museo Libico di Storia Naturale
Annali. Museo Civico di Storia Naturali Genova
Annali. Museo Civico di Storia Naturali "Giacomo Doria"
Annals and Magazine of Natural History
Annals of Applied Biology
Annals of the Entomological Society of America
Annals of Tropical Medicine and Parasitology
Annuaire. Societe Entomologique de France
Annual Report and Transactions. Manchester Entomological Society
Annual Review of Entomology
Annual Review of Systematics and Ecology
Antenna
Anzeiger fuer Schaedlingskunde
Anzeiger fuer Schaedlingskunde und Pflanzen- und Umweltshutz
Anzeiger fuer Schaedlingskunde und Pflanzenschutz
Apiculteur
Apiculture
Apidologie
Archiv fuer Anthropologie und Voelkerforschung
Archiv fur Hydrobiologie
Archives de Parasitologie
Archives de Zoologie Experimentale et Generale
Archives. Institut Pasteur d'Algerie
Archives. Museum National d'Historie Naturelle (Paris)
Atalanta (München)
Atheneum
Atti della Accademia Roveretana degli Agiati
Atti. Accademia della Scienze di Torino
Atti. Accademia Fisio-Medico-Statisticadi Milano
Atti. Accademia Nazioinale dei Lincei
Atti. Accademia Nazionale Italiana de Entomologia, Rendiconti
Atti. Accademia Nazionale Lincei Classe di Scienze Fisiche, Matematiche, Naturali
Atti. Instituto Botanico Universita Laboratorio. Crittogamico. Pavia
Atti. Instituto Veneto di Scienze, Lettere, de Arti
Atti. Memorie. Accademia di Agricoltura, Scienze, e Lettere (Verona)
Atti. Reale Accademia Nazionali dei Lincei
Atti. Societa dei Naturalisti e Matematici di Modena
Atti. Societa Toscana di Scienze Naturali Residente in Piza
Australian Encyclopaedia
Australian Entomological Magazine
Australian Entomological Society News Bulletin
Australian Journal of Zoology
Australian Naturalist
Bee World
Beitrage Fauna Thuerungen
Beitrage zur Entomologie
Beitrage zur Landeskunde von Oberösterreich Naturwissenshaftliche Reich II

Beitrage zur Naturkunde Niedersachsens
Beitrage zur Naturkundlichen Forschung in Südwestdeutschland
Bericht der Senckenbergeschen Naturforschenden Gesellschaft (Frankfurt)
Bericht des Naturwissenschaftlichen Vereines zu Regensburg
Berichte des Vereins fuer Naturkunde zu Cassel
Berichte. Deutsche Botanische Gesellschaft
Berliner Entomologische Zeitschrift
Bienenvater Wien
Biennial Report Montana State Board of Entomology
Biographical Memoirs. Fellows of the Royal Society
Biographical Memoirs. National Academy of Science
Biologia
Biological Reviews. Cambridge Philosophical Society
Biometrics
BioScience
Bitidningen
Boletim do Museu Paraense Emilio Goeldi
Boletim Museu Nacional (Rio de Janeiro)
Boletim. Sociedad de Estudos de Moçambique
Boletin de Historia Naturale Sociedad "Filipe Poey"
Boletin de Patologia Vegetal y Entomologia Agricola
Boletin. Real Sociedad Espanola de Historia Natural
Boletin. Servicio de Plagas Forestales
Boletin Sociedad Venezolana de Ciencias Naturales
Bollettino dei Musei di Zoologia e Anatomia Comparate della Reale Universita di Genova
Bollettino dei Musei e Laboratorii de Zoologia e Anatomia Comparate della Reale Universita di Genova
Bollettino del Laboratorio di Zool e Bachicoltura del Reale Instituto Superiore Agrario di Milano
Bollettino del Laboratorio di Zoologia Generale e Agraia della Reale Scuola Superiore d'Agricultura in Portici
Bollettino dell'Instituto di Entomologia della Reale Universita degli Studia di Bologna
Bollettino dell'Instituto Zoologico della Reale Universita di Roma
Bollettino della Societa di Naturalisti in Napoli
Bollettino della Societa Veneziana di Storia Naturale e del Museo Civico di Storia Naturale
Bollettino di Pesca, Piscicoltura, e Idrobiologia
Bollettino. Associazione di Romana Entomologia
Bollettino. Laboratorio di Entomologia Agraria "Filippo Silvestri" di Portici
Bollettino Museo di Zoologia. [Anat comp R] Universita di Torino
Bollettino. Museo Zoologia Universita di Torino
Bollettino. Societa Entomologia Italiana
Bollettino. Societa Geografica Italiana
Bollettino. Societa Sarda di Scienze Naturali
Boston Advertiser
Brasil-Medico
British Arachnology Society Newsletter
British Medical Journal
Broteria
Bulletin Amateur Entomologists Society
Bulletin Brooklyn Entomological Society
Bulletin de l'Institut Royale des Sciences Naturelles de Belgique
Bulletin de la Societe d'Etude des Sciences Naturelles de Reims
Bulletin de la Societe de l'Histoire Naturelle de Rheims
Bulletin de la Societe Imperiale des Naturalistes de Moscou
Bulletin de la Societe Linneene du Nord de la France

Bulletin de Museum d'Histoire Naturelle (Paris)
Bulletin des Recherches Agronomiques de Gembloux
Bulletin et Annales. Societe Entomologique de Belgique
Bulletin et Annales. Societe Royale d'Entomologique de Belgique
Bulletin Mensuel Societe Linneene de Lyon
Bulletin of the Entomological Society of America
Bulletin of the Entomological Society of Canada
Bulletin of the Hill Museum: A Magazine of Lepidopterology
Bulletin of Zoological Nomenclature
Bulletin Scientifique de la France et de la Belgique
Bulletin Societe de la Entomologique d'Egypte
Bulletin Societe de la Entomologique de France
Bulletin Societe des Naturalistes de Moscou
Bulletin. Allyn Museum
Bulletin. American Museum of Natural History
Bulletin. Association Minnesota Entomologists
Bulletin. Boston Society of Natural History
Bulletin. British Arachnological Society
Bulletin. British Museum (Natural History) Entomology. Supplement
Bulletin. British Museum (Natural History) Historical Series
Bulletin. British Museum (Natural History) Entomology
Bulletin. California Department of Agriculture
Bulletin. Cercle Zoologique Congolais
Bulletin. Institut d'Egypte
Bulletin. Institucio Catalana d'Historia Natural
Bulletin. Musee d'Histoire Naturelle de Marseille
Bulletin. Musee Royale d'Histoire Naturelle de la Belgique
Bulletin. Philosophical Society of Washington
Bulletin. Research Council of Israel Section B. Zoologie
Bulletin. Societe d'Histoire Naturelle de l'Afrique du Nord
Bulletin. Societe d'Histoire Naturelle de Metz
Bulletin. Societe d'Histoire Naturelle de Toulouse
Bulletin. Societe de Pathologie Exotique
Bulletin. Societe des Lepidopteristes Genève
Bulletin. Soceite des Sciences Naturelles du Maroc
Bulletin. Societe Entomologique de France
Bulletin. Societe Entomologique de Mulhouse
Bulletin. Societe Imperiale des Naturalistes de Moscou
Bulletin. Societe Linneene de Normandie
Bulletin. Societe Sciences Naturelle de l'Ouest de la France
Bulletin. Societe Sciences Naturelle et Physique du Maroc
Bulletin. Societe Vaudoise des Sciences Naturelles
Bulletin. Societe Zoologique de France
Bulletin. Societes Sciences Naturelles
Bulletin. Southern California Academy of Sciences
Bulletin. United States National Museum
Bulletins. Acadamie Royale de Belgique
Cahiers des Naturalistes
Canadian Entomologist
Canadian Journal of Zoology
Carinthia
Carinthia 2 Sonderheft
Casopis Ceske Spolecnosti Entomologicke
Casopis Moravskeho Musea (Brne)
Cassinia. A Journal of Ornithology of Eastern Pennsylvania, Southern New Jersey, and

Delaware
Catalogue of the Coleoptera of America, North of Mexico, Supplement
Ceskoslovenska Mykologia
Ciencia
Circular. Entomological Society of Australia (New South Wales)
Clinical Medicine and Surgery
Coleopterist's Bulletin
Commonwealth Science and Industrial Research Organization. Division of Entomology. Report
Comptes Rendus. Academie des Sciences
Comptes Rendus. Societe de Entomologique de Belgique
Curso de Entomologia
Deutsche Entomologische Zeitschrift
Die Bienenpflege
Die Milben als Parsiten der Wirbellosen, ins Besondere der Arthropoden
Die Neue Buecherei
Dimmock's Special Bibliographies
Dresden
Eatonia
Emu, The
Entomologen
Entomologica Americana
Entomologica Germanica
Entomological News
Entomological Review
Entomological Review (English Translation of Entomologicheskoye Obozreniye)
Entomologicheskoe Obozrenie
Entomologicke Problemy
Entomologische Abhandlungen (Dresden)
Entomologische Arbeiten. Museum Georg Frey (Tutzing-bei Munchen)
Entomologische Berichten (Amsterdam)
Entomologische Berichten (Berlin)
Entomologische Blaetter fuer Biologie und Systematik der Käfer
Entomologische Mitteilungen. Zoologischen Museum (Hamburg)
Entomologische Mitteilungen. Zoologischen Staatsinstitut und Zoologischen Museum (Hamburg)
Entomologische Nachrichten
Entomologische Nachrichten und Berichte
Entomologische Zeitschrift
Entomologische Zeitung (Wein)
Entomologisk Tidsskrift
Entomologiske Meddelelser
Entomologist, The
Entomologist's Monthly Magazine
Entomologist's Record and Journal of Variation
Entomologist's Gazette
Entomology and Phytopathology
Entomophaga
Eos. Revista Espanola de Entomologia
Ergebnisse Der Biologie
Etudes de Lepidopterologie Comparee
Etudes Entomologiques
Etudes sur les Coleopteres Lucanides du globe
Evolution
Exchange

Fabreries
Fauna Ver Luxemburg
Faunus. Zeitschrift fuer Zoologie und Vergleichende Anatomie
Fernald Club Yearbook
Finskt Tidsskrift
Flea News
Flora og Fauna
Florida Entomologist
Florida Historical Quarterly
Florida Naturalist
Folia Entomologica Hungarica
Folia Zoologica
Forst und Jagd
Forst- und Jagd-Wissenschaft
Forstliche Blätter: Wochentliche Rundschau auf dem Gabiete der Forstwritschaft un Forstuirtschaft
Forstwissenschaftliches Schriftsteller Lexicon
Forstwissenschaftliches Zentralblatt
Gartenzeitschrift Illustrierte Flora
Genetics
Gentleman's Magazine
Geological Magazine
Georgia Entomological Society. Journal
Giornale Botanico Italiano
Giornale dell'I. Reale Instituto Lombardo di Scienze, Lettere ed Arti
Giornale Storico della Lunigiana
Giornale. Accademica di Medicina di Torino
Glasgow Naturalist
Glasnik Zemaljskog Muzeja Bosne I Hercegovine u Sarajevu. Prirodne Nauke
Graduate Magazine of the University Kansas, The
Graellsia
Great Basin Naturalist
Great Lakes Entomologist
Haiso Senri
Hamilton National Zeitung
Harper's Monthly Magazine
Harper's Weekly
Herpetologica
History. Berwickshire Naturalists' Club
Horae Societatis Entomologicae Rossicae, Variis Sermonibus in Rossia Usitatis Editae
Hydrobiologia
Ibis
Il Naturalista siciliano: Organo della Società Siciliana di Scienze Naturali
Illinois Teacher
Indian Forester
Indian Journal of Entomology
Indonesian Journal of Crop Science
Insect Life
Insect World Digest
Insecta Matsumurana
Insectes Sociaux
Insektenbörse
International Journal of Entomology
International Journal of Insect Morphology and Embryology
Iris

Irish Naturalists' Journal
Isis Budissina
Israel Journal of Entomology
Israel Journal of Zoology
Izvestiya Akademii Nauk SSSR
Izvestiya Akademii Nauk Tadzhikskoi SSSR
Jaarboek. Van de Koninklijke Akademie van Wetenschappen, Gevestigd te Amsterdam
Jahrbuch der schweizerischen Wald- und Holzwirtschaft. Annuaire suisse de l'économie forestière et de l'industrie du bois
Jahrbuecher. Nassauischer Verein fuer Naturkunde
Jahresberichte der Gesellschaft von Freunden der Naturwissenschaften in Gera (Reuss)
Jahresberichte des Obergymnasiums der Benediktiner zu Kremsmünster, Schuljahr
Jahresberichte des Vereins fuer Naturwissenschaft zu Braunschweig
Jahresberichter der Schlesichen Gesellschaft fuer Vaterlaendische Kultur
Japanese Journal of Genetics
Johns Hopkins Hospital Bulletin
Journal East Africa Natural History Society Uganda
Journal fuer Ornithologie
Journal of Bacteriology
Journal of Biophysical and Biochemical Cytology
Journal of Botany British and Foreign (London)
Journal of Chemical Ecology
Journal of Conchology
Journal of Economic Entomology
Journal of Infectious Diseases
Journal of Invertebrate Pathology
Journal of Mammalogy
Journal of Medical Entomology
Journal of Microscopy
Journal of Modern Literature
Journal of Molecular Evolution
Journal of Natural History
Journal of Parasitology
Journal of Protozoology
Journal of Stored Products Research
Journal of the Kansas Entomological Society
Journal of Zoology Paris
Journal. Bombay Natural History Society
Journal. Entomological Society of Australia (New South Wales)
Journal. Entomological Society Southern Africa
Journal. Georgia Entomological Society
Journal. Lepidopterists Society
Journal. New York Entomological Society
Journal. Presbyterian Historical Society
Journal. Royal Microscopical Society
Journal. Siam Society
Journal. Society for the Bibliography of Natural History
Journal. Washington Academy of Science
Kansas University Science Bulletin
Koelopterologische Zeitschrift
Konowia
Kontyu
Korrespondenzblatt des Naturforschenden Vereins zu Riga
Korrespondenz-Blatt des Zoologische-Mineralogischen Vereins in Regensburg
Kungliga Svenska Vetenskapsakademiens Handlinger

L'Entomologiste
La Presse Medicale
Lambillionea
Leopoldina
Limnological Society of America Special Publications
Lingnan Science Journal
Linneana Belgica
London News
Lotos
Lotus
Lounais-Hameen Luonto
Luonnon Ystävä
Mainzer Kalender
Manitoba Entomologist
Marcellia
Memoires d'Agriculture, d'Economie Rurale et Domestique
Memoires de l'Institut Francais d'Afrique Noire
Memoires de la Societe Imperiale des Sciences, de l'Agriculture et des Arts, de Lille
Memoires. Academie de Vaucluse
Memoires. Academie des Sciences et Lettres Montpellier
Memoires. Academie des Sciences Inscriptions et Belles-Lettres de Toulouse
Memoires. Academie des Sciences. Institut de France
Memoires. Societe Linneene de Normandie
Memoires. Societe Royale Entomologique de Belgique
Memoirs. American Entomological Institute
Memoirs. Cornell University Agricultural Experiment Station
Memoirs. National Museum of Victoria
Memoirs. Queensland Museum.
Memorias e Estudos. Museu Zoologico. Universidade de Coimbra
Memorias. Instituto Oswaldo Cruz
Memorias. Sociedad Cubana de Historia Natural "Filipe Poey"
Memoriaux Societe Royale d'Emulation d'Abbeville
Memorie. Accademia della Scienze di Torino
Memorie. Reale Accademia di Scienze, Lettere, ed Arti (Modena)
Memorie. Societa Entomologica Italiana
Memorie. Societa Italiana di Scienze Naturale e Museo Civico di Storia Naturale di Milano
Memorie. Societa Toscana di Scienze Naturali Residenti in Pisa
Michigan Entomologist
Microentomology
Mitteilungen der Aargauischen Naturforschenden Gesellschaft
Mitteilungen der Thurgauischen Naturforschenden Gesellschaft
Mitteilungen Deutsche Entomololgische Gesellschaft
Mitteilungen Entomologische Gesellschaft (Basel)
Mitteilungen Hamburgisches Zoologische Museum
Mitteilungen Münchener Entomologische Gesellschaft
Mitteilungen Naturhistorischen Gesellschaft Colmar
Mitteilungen Naturwissenschaftlichen Vereins fuer Steiermark
Mitteilungen Naturwissenschaftliches Gesellschaft Winterthür
Mitteilungen Naturwissenschaftliches Museum der Stadt Aschaffenburg
Mitteilungen Schweizerische Entomologische Gesellschaft
Mitteilungen. Zoologisches Museum in Berlin
Monitore Zoologico Italiano
Monthly Bulletin California State Commission of Horticulture
Monthly Bulletin Department of Agriculture (California)
Monthly Bulletin of the State Plant Board of Florida, The

Monti e Boschi
Mosquito News
Mosquito Systematics
Museums Journal
Mushi
Myia
Myrmecia
Nachrichten. Gesellschaft Akademie der Wissenschaften Goettingen
Nachrichten. Naturwissenschaftliches Museum der Stadt (Aschaffenburg)
Nachrichtenblatt der Bayerischen Entomologen
Nachrichtenblatt. Deutschen Pflanzenschutzdienst (Braunschweig)
Natur und Museum (Frankfurt)
Natur und Volk: Bericht der Senckenbergischen Naturforschenden Gesellschaft
Natura Alpina
Natura (Milano)
Natural History Museum. Bulletin. Entomology
Natural Science London
Naturalist
Naturalist, The
Naturaliste
Naturaliste Canadien
Nature
Naturvissenschaften
Neuestes Magazin fuer die Liebhaber der Entomologie
Neujarsblatt Herausgegeben von der Naturforschenden Gesellaschaft in Zurich
New England Journal of Medicine
New Scientist
New York Times
New Yorker Magazine
News Bulletin. Australian Entomological Society
News Bulletin. Entomological Society of Queensland
News. Lepidopterists' Society
Newsletter. Michigan Entomological Society
Norsk Entomologisk Tidsskrift
Northwest Naturalist
Notes and Queries
Notulae Entomologicae
Nouvelle Revue d'Entomologie
Nova Acta Regiae Societatis Scientarum Uppsaliensis
Novitates Zoologicae
Nunquam Otiosus
Obituaries and Notes. Fellows of the Royal Society London
Occasional Papers Boston Society of Natural History
Odonatologica
Oecologia
Oesterreichische Forstzeitung
Ohio Journal of Science
Opuscula Entomologica
Outdoor Indiana
Outlook
Pacific Insects
Pacific Rural Press
Pan-Pacific Entomologist
Papers and Proceedings Royal Society of Tasmania
Parasitologia

Parasitology
Passenger pigeon
Peking Natural History Bulletin
Pfaelzer Heimat
Philippine Journal of Science
Physiological Plant Pathology
Phytopathology
Plant Protection (Leningrad)
Polski Pismo Entomologiczne
Popular Science Monthly
Praktische Schaedlingsbekampfer
President's Address Royal Entomological Society London
Proceedings and Transactions. British Entomology and Natural History Society
Proceedings. Academy of Natural Sciences of Philadelphia
Proceedings. American Academy of Arts and Sciences
Proceedings. Bath Natural History and Antiquarian Field Club
Proceedings. Biological Society of Washington
Proceedings. Birmingham Natural History and Microscopical Society
Proceedings. Bristol Naturalist's Society
Proceedings. California Academy of Sciences
Proceedings. Entomological Society London
Proceedings. Entomological Society of British Columbia
Proceedings. Entomological Society of Manitoba
Proceedings. Entomological Society of Ontario
Proceedings. Entomological Society of Washington
Proceedings. Hawaii Entomological Society
Proceedings. Indiana Academy of Science
Proceedings. Iowa Academy of Science
Proceedings. Linnean Society of London
Proceedings. Linnean Society of New South Wales
Proceedings. National Academy of Science
Proceedings. New Jersey Historical Society
Proceedings. New Zealand Institute of Agricultural Sciences
Proceedings Pacific Coast Entomological Society
Proceedings. Royal Entomological Society of London C
Proceedings. Royal Society of New Zealand
Proceedings. Royal Society of Tasmania
Proceedings. Royal Society Queensland
Proceedings. Society for Experimental Biology and Medicine
Proceedings. Somerset Archaeology and Natural History Society
Proceedings. United States National Museum
Proceedings. Washington State Entomological Society
Psyche
Publicacoes Culturais. Companhia de Diamantes de Angola
Publicaties. Natuurhistorisch Genootschap in Limburg
Quaestiones Entomologicae
Quarterly Journal of Microscopical Science
Quarterly Journal. Geological Society of London
Quarterly Journal. Taiwan Museum (Taipei)
Quarterly Review of Biology
Queensland Agricultural Journal
Queensland Naturalist
Records. Australian Museum
Records. Indian Museum
Records of the Geological Survey of India

Records. South Australian Museum
Redia
Rendiconti. Istituto Lombardo di Scienze e Lettere
Rendiconto delle Sessioni dell'Accademia delle Scienze dell'Istituto di Bologna
Report and Transactions - Guernsey Society of Natural Science and Local Research
Report. Entomological Society of Ontario
Revista Universidad de Oriente (Santiago de Cuba, Cuba)
Revista Brasileira de Biologia
Revista Brasileira de Entomologia
Revista Chilena de Entomologia
Revista Chilena de Historia Natural
Revista de Biologia Lisbon
Revista de Ciencias Veterinarias
Revista de Entomologia (Rio de Janiero)
Revista de la Sociedad Uruguaya de Entomología
Revista Ecuatoriana de Entomologia y Parasitologia
Revista Iberica de Parasitologia
Revista Museo du Paulista (Sao Paulo)
Revista. Museo de La Plata
Revista Muzeelor
Revista Peruana de Entomologia
Revista. Sociedad Entomologica Argentina
Revista. Sociedad Mexicana de Historica Natural
Revista Universitati "Al. I. Cuza" si a Institutului Politehnic Iasi
Revue d'Ecologie et de Biologie du Sol
Revue de Zoologie Agricole et Appliquee
Revue de Zoologie et de Botanique Africaines
Revue des Sciences Naturelles d'Auvergne
Revue Entomologique
Revue Generale d'Agriculture Gironde
Revue Suisse de Zoologie
Rivista de Biologia
Rivista di Parassitologia
Rivista Malariologia
Rovartani Lapok
Royal Archives of Parasitology
Royal Entomological Society London. President's Address
Samab
Sbornik Prirodovedeckeho Klubu v Trebici
Schriften der Naturforschenden Gesellschaft zu Danzig
Schriften der Physikalisch-Oekonomischen Gesellaschaft zu Koeningsberg
Schweizer Archiv fuer Tierheilkunde
Science
Scientia Genetica
Scientific American
Scientific Monthly
Scottish Naturalist
SHILAP: Revista de Lepidopterologia
Sitzungsberichte der Naturforscher-Gesellschaft zu Dorpat
Sitzungs-Berichte der Naturwissenschaftlichen Gesellschaft Isis zu Dresden
Sitzungsberichte der Physiklakisch-Medicinischen Societaet zu Erlangen
Sitzungsberichte Gesellschaft Naturforschender Freunde zu Berlin
Sitzungsberichte und Mitteilungen der Braunschweigischen Wissenschaftlichen Gesellschaft
Sitzungsberichte, Herausgegeben vom Naturhistorischen Verein der Preussischen Rheinland and Westfalens

Sitzungsberichte. Bayerische Akademie der Wissenschaften zu Muenchen
Smithersia
Smithsonian Miscellaneous Collections
Societas Entomologica
Societas Pro Fauna et Flora Fennica
Society for the Bibliography of Natural History. Journal
South African Journal of Natural History
Southwest Review
Sprawozdanie Komisji Fizjograficznej
Stettiner Entomologische Zeitung
Studi Trentini di Scienze Natural
Studia Entomologica
Studies on the Neotropical Fauna
Suomen Hyonterstieteellinen Aikakauskirja
Systematic Zoology
Tätigkeitsbericht der Zoologischen Station Neapel
Tätigkeitsberichte der Natursforschenden Gesellschaft Baselland
Tijdschrift voor Entomologie
Times, The (London)
Tombo
Transactions and Proceedings. Perthshire Society of Natural Science
Transactions and Proceedings. Royal Society New Zealand
Transactions and Proceedings. South London Entomology and Natural History Society
Transactions and Reports. Manchester Microscopical Society
Transactions of the Natural History Society of Northumbria
Transactions of the Norfolk and Norwich Naturalists' Society
Transactions Royal Society Tropical Medicine and Hygiene
Transactions South African Philosophical Society
Transactions. Academy of Science of St. Louis
Transactions. American Microscopical Society
Transactions. American Philosophical Society
Transactions. Hull Scientific and Field Naturalists' Club
Transactions. Illinois Academy of Science
Transactions. Illinois State Horticultural Society
Transactions. Kansas Academy of Science
Transactions. Natural History Society of Northumberland, Durham, and Newcastle-upon-Tyne
Transactions. New York Agricultural Society
Transactions. Royal Entomological Society of London
Transactions. Royal Society of Edinburgh
Transactions. Society for British Entomology
Transactions. Suffolk Naturalists' Society
Travaux. Laboratorie d'Hydrobiologie et de Pisciculture. Universite de Grenoble
Travaux. Museum d'Historie Naturelle "Grigore Antipa"
Treubia
Tromso Museums Arsheretning
Tropical Agriculture
Tropical Agriculture (Trinidad)
Trudy Russkogo Entomolgischeeskogo Obshchestva
University of Massachusetts Fernald Club Yearbook
Vasculum (Substitute)
Verhandlungen. Deutsche Zoologische Gesellschaft
Verhandlungen der Geologischen Bundesanstalt (Wien) [alt. K.K. Geologische Reichsanstalt (Austria)]
Verhandlungen. Naturforschen der Verein in Brünn

Verhhandlugen Schweizerische Naturforschende Gesellschaft
Verhandlungen des Vereins fuer Naturwissenschaftliche Unterhaltung zu Hamburg
Verhandlungen Ver Naturwissenschaftenliche Heimatforschung zu Hamburg
Verhandlungen Zoologisch-Botanische Gesellschaft in Wien
Vestnik Ceskoslovenske Spolecnosti Zoologicke
Victorian Naturalist
Videnskabelige Meddelelser fra Dansk Naturhistorisk Forening
Vierteljahrsschrift Naturforschende Gesellschaft (Zuerich)
Virshows Archiv fuer Patologische Anatomie und Physiologie
Waldie's Select Circulating Library
Ward's Bulletin
Washington Post
West Indian Bulletin
Wiener Entomologische Gesellschaft. Zeitschrift
Wiener Entomologische Monatsschrift
Wiener Entomologische Zeitung
Year-book of the Royal Society of Edinburgh
Zashchita Rastenii (Kiev)
Zastita Bilja
Zeitschrift der Wiener Entomologischen Gesellschaft
Zeitschrift fuer Angewandte Entomologia
Zeitschrift fuer Forst- und Jagdwesen
Zeitschrift fuer Gesellschaft der Naturwissenschaften, Tecnik, und Medizin
Zeitschrift fuer Wissenschaftliche und fuer Mikroskopische Tecnik
Zeitschrift fuer Wissenschaftliche Zoologie
Zeitschrift für Systematische Hymenopterologie und Dipterologie
Zeitschrift. Oesterreichischer Entomologe-Verein
Zeitschrift. Wiener Entomologe-Verein
Zeitschrift. Wiener Entomologische Gesellschaft
Zentralblatt fuer das Gesamte Forstwesen
Zoological Journal. Linnean Society
Zoologicheskii Zhurnal
Zoologicke Listy
Zoologische Beitraege
Zoologische Bijdragen
Zoologische Gaerten
Zoologische Gaerten (Frankfurt)
Zoologische Gaerten (Leipzig)
Zoologische Jahrbücher. Abteilung für Systematik, Oekologie und Geographie der Tiere
Zoologische Jarbuecher
Zoologische Mededelingen (Leiden)
Zoologischer Anzeiger
Zoologist, The
Zpravy Ceskoslovenske Spolecnosti

Appendix 2: Common Names References

Author: Miller, Jacqueline Y.
Title: The Common names of North American butterflies
Publisher: Washington: Smithsonian Institution Press, 1992. ix, 177 p.: map; 23 cm
State/Country: District of Columbia
Date: 1992
Language: English
Standard No: ISBN: 1560981229

Author: Guyton, Faye E.
Title: Insects of economic importance in Alabama listed by common names
Publisher: [S.l.: s.n., 1965?] [9] leaves; 36 cm
Date: 1965
Language: English
NAL Call No: Film 1746

Accession No: CAT10842533
Author: Bosik, Joseph John; Entomological Society of America. Committee on Common Names of Insects
Title: Common names of insects & related organisms, 1997
Publisher: [Lanham, Md.]: The Society, c1997. 232 p.: ill.; 26 x 19 cm
State/Country: Maryland
Date: 1997
Language: English
Standard No: ISBN: 0938522647

Accession No: CAT10838196
Author: Arroyo Varela, Manuel
Title: Nombres vulgares de insectos de interes agricola: espanol/aleman/frances/ingles/italiano. Common names of insects of agricultural interest
Publisher: Madrid: Ministerio de Agricultura, Pesca y Alimentacion, Secretaria General Tecnica, [1995] 155 p.; 24 cm
State/Country: Spain
Date: 1995
Language: Spanish, English, French, German, Italian
Standard No: ISBN: 8449101417

Accession No: IND20502567
Author: Nystrom, K.L. Britnell, W.E.
Title: Insects and mites associated with Ontario forests: classification, common names, main hosts, and importance
Source: Information report / 1994. (O-X-439) 136 p.
Publisher: [Sault Ste. Marie, Ont.]: The Centre, 1985-
State/Country: Ontario
Source Author: Information report (Great Lakes Forestry Centre)
Date: 1994
Language: English
Standard No: ISSN: 0832-7122

Accession No: CAT91952087
Author: Seymour, Paul R. (Paul Roy); Great Britain. Ministry of Agriculture, Fisheries and Food
Title: Invertebrates of economic importance in Britain: common and scientific names 4th ed
Publisher: London: H.M.S.O., 1989. viii, 146 p.; 25 cm
State/Country: England
Date: 1989
Language: English
Standard No: ISBN: 0112428290

Accession No: CAT87886722
Author: Greiff, Margaret; Commonwealth Institute of Entomology
Title: Spanish–English–Spanish lexicon of entomological and related terms: with indexes of Spanish common names of arthropods and their Latin and English equivalents
Publisher: Slough, U.K.: Commonwealth Agricultural Bureaux, c1985. 102, 115, 43 p.; 25 cm
State/Country: England
Date: 1985
Language: English, Spanish
Standard No: ISBN: 0851985602

Accession No: CAT86864144
Author: Yusha, Alex Martignoni, Mauro E., 1926- Iwai, Paul J.; Pacific Northwest Forest and Range Experiment Station (Portland, Or.)
Title: An English translation of Russian common names of agricultural and forest insects and mites
Publisher: Portland, Or.: Springfield, Va.: U.S. Dept. of Commerce, National Technical Information Service [distributor, U.S. Dept. of Agriculture, Forest Service, Pacific Northwest Forest and Range Experiment Station; 1985] 84 p.; 22 x 29 cm
State/Country: Oregon
Date: 1985
Language: English
Series: General technical report PNW; 183

Accession No: CAT85828325
Author: Varshney, R. K.
Title: Index Rhopalocera Indica, part-II: common names of butterflies from India and neighbouring countries
Publisher: Calcutta, India: Zoological Survey of India, 1983. 49 p.; 25 cm
State/Country: India
Date: 1983
Language: English
Series: Records of the Zoological Survey of India. Miscellaneous publication. Occasional paper; no. 47
Other Titles: Title Traced Differently: Common names of butterflies from India and neighbouring countries

Accession No: IND82119341
Author: Ayquipd A., G. E. Cueva C., M. A.
Title: Nombres cientificos y comunes de los insectos que atacan a la cana de azucar en el Peru. Scientific and common names of insects attacking sugarcane in Peru
Source: Revista peruana de entomologia. 1979 (pub. 1980). v. 22 (1) p. 95-97
Publisher: Lima, Sociedad Entomologica del Peru
Date: 1980 1979
Language: Spanish (Summaries or abstracts in English)

Pub. Type: Article
Pub. Agency: Non-U.S. Imprint, not FAO
Subfile/Locat.: IND
Standard No: ISSN: 0080-2425
Identifiers: Peru
Subj. Category: F821 Pests of Plants, Insects

Accession No: IND81113258
Author: Ramsay, G. W.
Title: Common and scientific names of New Zealand mites of agricultural, horticultural, medical, stored products, or veterinary importance, or of general interest
Source: DSIR information series – New Zealand, Dept. of Scientific and Industrial Research, 1980, (139) 32 p.
Publisher: Wellington, The Department
Date: 1980
Language: English
Pub. Type: Article
Standard No: ISSN: 0077-9636

Accession No: IND80103824
Author: Greene, A. Caron, D. M.
Title: Entomological etymology: the common names of social wasps
Source: Bulletin of the Entomological Society of America. June 1980. v. 26 (2) p. 126-130
Publisher: College Park, Md., The Society
Date: 1980 06
Language: English
Pub. Type: Article
Standard No: ISSN: 0013-8754

Accession No: CAT80731457
Author: Hsin, Chiai-lu. Hsia, Sung-yun
Title: Ying han k'un ch'ung su ming tz'u hui = Glossary of common names of insects in English and Chinese
Publisher: Ch'ang-sha: Hu-nan jen min ch'u pan she, 1978. [v], 441, [1] p
State/Country: Mainland China
Date: 1978
Language: English, Latin, Chinese
Pub. Type: Book
Pub. Agency: Non-U.S. Imprint, not FAO
Subfile/Locat.: CAT
Other Titles: Title Traced Differently: Glossary of common names of insects in English and Chinese
Descriptors: Insects – Nomenclature
Insects – Dictionaries – English
English language – Dictionaries – Chinese
Chinese language – Dictionaries – Latin
Latin language – Dictionaries – Chinese
Subj. Category: L700 Animal Taxonomy and Geography
L001 Entomology Related

Accession No: IND79060255
Author: Wentzel, G.
Title: Vulgarnamen von Vorrats-und Materialschadlingen in verschiedenen Sprachen
Common names of stored product and wood-boring insects in different languages
Source: Anzeiger fur Schadlingskunde, Pflanzen- und Umweltschutz vereinigt mit

Schadlingsbekampfung Dec 1978. v. 51 (12) p. 184-188. ill
Publisher: Berlin, Parey
Date: 1978 12
Language: German (Summaries or abstracts in English)
Pub. Type: Article
Subfile/Locat.: IND
Standard No: ISSN: 0340-7330

Accession No: 78-9061320
Author: Ferro, D. N. Lowe, A. D. Ordish, R. G. Somerfield, K. G. Watt, J. C
Title: Standard Names for Common Insects of New Zealand
Source: Bull Entomol Soc N Z 4, 42 p. 1977
Date: 1977
Language: English
Pub. Type: Article
Descriptors: New Zealand

Accession No: 72-9032058
Author: Libermann, J.
Title: Sobre Algunos Nombres Vulgares de Ortopteros Argentinos. About Some Common Names of Argentine Orthoptera
Source: Idia (Inform de Invest Agr) 283: 78-80 July 1971
Date: 1971
Language: Spanish
Pub. Type: Article

Accession No: 71-9015962
Author: Schmidt, G.
Title: Die Deutschen Namen Wichtiger Arthropoden. Common German Names of Important Arthropods
Source: Berlin Biol Bundesanst Land-Forstwirtsch Mitt 137, 222 p. Aug 1970
Date: 1970
Language: German
Pub. Type: Article

Accession No: 38578071
Author: Japan. Nogyo Kairyokyoku. Kenkyubu
Title: Common names of major diseases and insect pests of crops in Japan
Place: [n.p.]
Publisher: Research Division, Agr. Imp. Bureau, Minist. of Agr. & Forest
Year: 1951
Pub. Type: Book
Format: 205 p. 10 x 15 1/2 cm
Notes: In Japanese and English with Latin nomenclature. Includes indexes
Subject: Plant diseases – Japan

Accession No: 30218871
Author: Great Britain. Ministry of Agriculture, Fisheries and Food
Title: Common names of British insect and other pests
Publisher: H.M.S.O.
Year: 1957
Pub. Type: Book
Format: 49 p.; 25 cm
Series: Technical bulletins; no. 6

Accession No: 26902245
Author: Association of Applied Biologists
Title: Common names of British insect and other pests, part two
Place: [s.l.]
Year: 1952
Pub. Type: Book
Format: 40 p. 8vo

Accession No: 40716676
Author: Paulian, Renaud
Title: Les insectes de Tahiti
Place: Paris
Publisher: Société nouvelle des Editions Boubée
Year: 1998
Pub. Type: Book
Format: 331 p.: ill. (some col.); 24 cm
Notes: Includes bibliographical references (p. [295]–305) and indexes
ISBN: 2850040908
Subject: Insects – French Polynesia – Tahiti. Entomology
Vendor: Jean Touzot Libraire Editeur
Control number: JTL00013818

Accession No: 35355965
Title: Lexicon of insects = Lexique des insectes: English, French, Arabic
Place: Rabat
Publisher: League of Arab States, Education, Culture, and Sciences Organization, The Permanent Bureau of Arabisation
Year: 1972
Pub. Type: Book
Format: 11 p.; 27 cm
Notes: Cover title
Subject: Arabic language – Dictionaries. Entomology – Dictionaries – Polyglot. Dictionaries, Polyglot
AltTitle: Mu‘jam al-hasharat: Injilizi, Faransi, ‘Arabi Lexique des insectes
Other: Maktab al-Da'im li-Tansiq al- Ta‘rib fi al-‘Alam al-‘Arabi. League of Arab States

Accession No: 25444756
Title: Noms français d'insectes au Canada, avec noms latins et anglais correspondants = French names of insects in Canada, with corresponding Latin and English names
Edition: 4e éd
Place: [Québec]
Publisher: Ministère de l'agriculture
Year: 1975
Pub. Type: Book
Format: 214p.; 23cm
Notes: "Publication QA38-R4-30." Première éd. en 1947
ISBN: 0775424293

Accession No: 24991460
Author: Debrie, René
Title: Recherches sur les noms de plantes et les noms d'insects dans le parlers de la région d'Amiens
Place: Amiens
Publisher: C.R.D.P., 33, rue des Minimes
Year: 1969

Pub. Type: Book
Format: 83 p. 27 cm
Series: Lingua picardica, 5 Annales du Centre régional de documentation pédagogique Annales du Centre régional de documentation pédagogique d'Amiens
Notes: At head of cover title: Ministère de l'éducation nationale. Institut pédagogique national. Academie d'Amiens. Bibliography: p. [5]
Subject: French language – Dialects – France – Amiens. French language Plant names, Popular. Insects

Accession No: 18713895
Author: Remillet, Michel
Title: Catalogue des insectes ravageurs des cultures en Guyane française
Place: Paris
Publisher: Editions de l'ORSTOM
Year: 1988
Pub. Type: Book
Format: 235 p., [26] p. of plates: ill.; 27 cm
Series: Collection Etudes et thèses, 0767–2888
Notes: Includes bibliographical references and indexes
ISBN: 2709908913
Subject: Insect pests – French Guiana. Insects – Host plants – French Guiana

Accession No: 17497353
Author: Delplanque, A.
Title: Insectes de la Guadeloupe et iles avoisinantes
Edition: Ed. 1975 corr
Place: [Pointe-à-Pitre]
Publisher: CNDP, CDDP Guadeloupe
Year: 1985
Pub. Type: Book
Format: 45 p., [4] p. of col. plates: ill.; 30 cm
Notes: Bibliography: p. 43-45
ISBN: 2903649324
Subject: Insects – Guadeloupe. West Indies, French

Accession No: 10458916
Author: Larsen, Torben B. (Torben Bjørn), 1944-
Title: The butterflies of the Yemen Arab Republic
Place: København
Publisher: Det Kongelige Danske Videnskabernes Selskab
Year: 1982
Pub. Type: Book
Format: 87 p.: ill. (some col.); 27 cm
Series: Biologiske skrifter, 23:3 0366-3612;
Notes: Includes index. Bibliography: p. 76
ISBN: 8773041262 (pbk)
Subject: Butterflies – Yemen. Insects – Yemen
Other: Rydon, A. H. B. (Arthur H. Bruce) Review of species in the Charaxes viola-Group from Arabia and East Africa. 1982

Accession No: 10309858
Author: Al-Humiari, Amin Abdulla
Title: A first guide to the agricultural insect pests of the Yemen Arab Republic and their management
Year: 1982

Pub. Type: Book
Format: x, 177 leaves: ill. (some col.); 28 cm
Notes: Thesis (M.S. – Entomology) – University of Arizona, 1982. Bibliography: leaves 170-177. Microfiche. Ann Arbor, Mich.: University Microfilms International, 1983. 2 microfiches; 11 x 15 cm
Subject: Agricultural pests – Yemen. Pest control, Integrated – Yemen. Insect pests – Control – Yemen. Insects, Injurious and beneficial – Yemen

Accession No: 17731445
Author: Walker, D. H. (Don H)
Title: Insects of eastern Arabia
Publisher: Macmillan
Year: 1987
Pub. Type: Book
Format: xvi, 175 p.
ISBN: 0333432142 (pbk)
Subject: Insects – Arabian Peninsula – Identification. Eastern Arabia – Insects – Field guides
Other: Pittaway, A. R

Accession No: 17016475
Author: Le Pelley, R. H.
Title: Agricultural insects of East Africa; a list of East African plant feeding insects and mites ... mainly covering the period 1908 to 1956
Place: Nairobi, Kenya
Publisher: East Africa High Commission
Year: 1959
Pub. Type: Book
Format: 307 p. 23 cm
Notes: Includes bibliography
Subject: Insect pests – Africa, East

Accession No: 8791169
Author: Boorman, John
Title: West African insects
Place: Harlow, Essex, UK
Publisher: Longman
Year: 1981
Pub. Type: Book
Format: viii, 88 p.: ill. (some col.); 20 cm
Series: West African nature handbooks
Notes: Includes index. Bibliography: p. 82–83
ISBN: 0582606268 (pbk.)
Subject: Insects – Africa, West

Accession No: 9247333
Author: Annecke, D. P. (David P.)
Title: Insects and mites of cultivated plants in South Africa
Place: Durban; Woburn, Mass
Publisher: Butterworths
Year: 1982
Pub. Type: Book
Format: xiv, 383 p.: ill.; 25 cm
Notes: Includes index. Bibliography: p. 336–339
ISBN: 0409083984

Accession No: 5190542
Author: Laux, Wolfrudolf
Title: Russische Namen von Arthropoden pflanzenschutzlicher Bedeutung
Place: Berlin
Publisher: Biologische Bundesanstalt für Land- und Forstwirtschaft Berlin-Dahlem: Kommissionsverlag P. Parey
Year: 1979
Pub. Type: Book
Format: 86 p.; 24 cm
Series: Mitteilungen aus der Biologischen Bundesanstalt für Land- und Forstwirtschaft Berlin- Dahlem, Heft 188 0067-5849;
Notes: Summary in English
ISBN: 3489188004
Subject: Insect pests – Nomenclature (Popular) – Dictionaries – Russian. Plant mites – Nomenclature (Popular) – Dictionaries – Russian. Insects – Nomenclature (Popular) – Dictionaries – Russian. Mites – Nomenclature (Popular) – Dictionaries – Russian. Russian language – Dictionaries – Latin
Other: Schmidt, Günther, 1909-

Accession No: 23150478
Title: Check list of insect pests of food crops in Indonesia
Place: Bangkok, Thailand
Publisher: Regional Office for Asia and the Pacific, Food and Agriculture Organization of the United Nations
Year: 1987
Pub. Type: Book
Format: [iii], 12 p.; 27 cm
Series: Technical document, no. 136/1987 1014-3351; Technical document (FAO Regional Office for Asia and the Pacific. Asia and Pacific Plant Protection Commission); no. 136/1987
Notes: Includes bibliographical references (p. 12)
Subject: Insect pests – Indonesia. Food crops – Diseases and pests – Indonesia
Other: FAO Regional Office for Asia and the Pacific

Accession No: 20991626
Title: Insect pests of economic significance affecting major crops of countries in Asia and the Pacific region
Place: Bangkok, Thailand
Publisher: Asia and Pacific Plant Protection Commission, Regional Office for Asia and the Pacific, Food and Agriculture Organization of the United Nations
Year: 1987
Pub. Type: Book
Format: 56 p.; 27 cm
Series: Technical document; no. 135/1987 Technical document (Fao Regional Office for Asia and the Pacific. Asia and Pacific Plant Protection Commission); no. 135
Subject: Insect pests – Asia. Insect pests – Australasia. Plant parasites – Asia. Plant parasites – Australasia
Other: FAO Regional Office for Asia and the Pacific. Asia and Pacific Plant Protection Commission

Accession No: 6379134
Author: Food and Agriculture Organization of the United Nations. Plant Protection Committee for the South East Asia and Pacific Region
Title: A partial list of insects in Cambodia
Place: Bangkok

Publisher: Food and Agriculture Organization of the United Nations, Regional Office for Asia and the Far East
Year: 1975
Pub. Type: Book
Format: 13 p.; 29 cm
Series: Technical document - FAO Plant Protection Committee for the South East Asia and Pacific Region; no. 98 Technical document (Food and Agriculture Organization of the United Nations. Plant Protection Committee for the South East Asia and Pacific Region); no. 98
Notes: Cover title
Subject: Insect pests – Cambodia. Insects – Host plants – Cambodia. Host plants – Cambodia

Accession No: 4697341
Author: Stapley, J. H.
Title: Check list of insect pests in the British Solomon Islands
Place: Bangkok
Publisher: FAO, Regional Office for Asia and the Far East
Year: 1976
Pub. Type: Book
Format: 6 p.
Series: Food and Agriculture Organization of the United Nations. Plant Protection Committee for the South East Asia and Pacific Region. Technical document; no. 102